Concise Encyclopedia Biochemistry and Molecular Biology

Concise Encyclopedia Biochemistry and Molecular Biology

Third Edition

Revised and expanded by
Thomas A. Scott and E. Ian Mercer

Walter de Gruyter Berlin · New York 1997

Title of the original, German language edition
ABC Biochemie
Edited by
Hans-Dieter Jakubke, Leipzig, Germany
Hans Jeschkeit (†), Halle, Germany
VEB F. A. Brockhaus Verlag, Leipzig 1976; 1981

Translated into English, revised and expanded by
Thomas A. Scott, Ph. D.
10, Lidgett Walk
Leeds, LS8 1NW
Great Britain

E. Ian Mercer, Ph. D.
Institute of Biological Sciences
Sir George Stapledon Building
The University of Wales
Aberystwyth, Ceredigion SY23 3DD
Great Britain

Printed on acid-free paper, which falls within the guidelines of the ANSI to ensure permanence and durability.

Libary of Congress Cataloging in Publication Data

Brockhaus ABC Biochemie. English.
Concise encyclopedia biochemistry and molecular biology. – English language ed., 3rd ed. / [translated into English], revised and expanded by Thomas A. Scott and E. Ian Mercer.
Rev. ed. of: Concise encyclopedia biochemistry. English language ed., 2nd ed. 1988.
ISBN 3-11-014535-9 (hardcover : alk. paper)
1. Biochemistry–Dictionaries. I. Scott, Thomas, 1935– .
II. Mercer, E. I. (Eric Ian) III. Concise encyclopedia biochemistry.
IV. Title.
QD415.A25B713 1997
572'.03–dc21

Die Deutsche Bibliothek – Cataloging-in-Publication Data

Concise encyclopedia biochemistry and molecular biology / rev. and expanded by Thomas A. Scott and E. Ian Mercer. – 3. ed. – Berlin ; New York : de Gruyter, 1997
Einheitssacht.: ABC Biochemie <engl.>
Früher u. d. T.: Concise encyclopedia biochemistry
ISBN 3-11-014535-9
NE: Scott, Thomas [Bearb.]; EST

Preface to the third edition

Once again we have attempted to keep pace with published work in Biochemistry, Biophysics, Molecular Biology and Cell Biology. Existing entries have been updated, and some (e.g. *Photosynthesis, Steroids, Immunoassay*) have been rewritten. Material has also been reorganized, thus, *Phospholipids* is now subsumed into *Membrane lipids; Nucleic acid sequencing,* greatly revised and expanded, now forms a separate entry. There are many entirely new entries, like *Apoptosis, CD-spectroscopy, Cell adhesion molecules, Coated vesicle, Databases, Defective viruses, DNA binding proteins, DNA fingerprinting, DNA methylation, DNA repair, Erythrocyte membrane, ESR-spectroscopy, Human genome project, Molecular chaperones, Multipartite viruses, NMR-spectroscopy, Prion, Proteasome, Protein engineering, Protein folding, Recombination, Viroids, X-ray spectroscopy,* and others.

At the same time, advances in our knowledge of metabolic pathways and natural products have not been neglected. Thus, *Hopanoids* are described, *Nitric oxide* has been accorded a separate entry, and the five carbon pathway of 5-aminolevulinate synthesis is presented under Porphyrins. *Chemiosmosis* has been rewritten, and the classification and study of different *ATPases* is treated more thoroughly. Plant biochemists should note that all aspects of *Photosynthesis* in higher and lower plants and fungi have been reorganized and rewritten, and that *plant hormones* (in particular, *ethylene* and *auxins*) and *phytochromes* have been brought up to date.

The EC numbers of enzymes are from the Recommendations (1992) of the Nomenclature Committee of the International Union of Biochemistry and Molecular Biology ("Enzyme Nomenclature", Academic Press, 1992).

We are indebted to colleagues who suggested corrections and additions. If the book has any shortcomings, however, we alone are responsible. Thanks are due to Mrs. Ingeborg Klak, who oversaw the production of the galley and page proofs, and to Dr. Mario Noyer-Weidner, Natural Sciences Section of Walter de Gruyter, for his guidance and encouragement.

September 1996

Thomas A. Scott
Leeds, Yorkshire, England, UK

E. Ian Mercer
Aberystwyth, Ceredigion, Wales, UK

Preface to the second edition

In the preparation of the first edition, much academic energy and real time were consumed by the translation exercise. For the second edition we have been able to devote ourselves exclusively to collecting and classifying new material, and to revising old material. It is a measure of the pace of development in biochemistry that genetic engineering and the cloning of DNA, which are well represented in this second edition, received scant attention in the first edition. Entries on proteins listed physical properties, purification methods, structure and function. Now it is always necessary to ask whether a protein has been studied by recombinant DNA technology and, if so, to what purpose. In fact, the primary structures of many recently studied proteins have not been determined by direct analysis, but by prediction from the nucleotide sequence of the cloned gene. Here lies a cautionary tale: the entry on Lectins describes a newly discovered type of posttranslational protein modification, resulting in a primary structure that could not have been predicted from the nucleotide sequence of the gene!

Our policy has been to include all areas of biochemistry. Most of the new material can be classified as metabolism, metabolic regulation, molecular biology, enzymology, nonenzymatic protein function, or natural products; moreover we have attempted to give fair (but obviously not equal) coverage to animal, medical, microbiological and plant biochemistry.

The EC numbers of enzymes are from the Recommendations (1984) of the Nomenclature Committee of the International Union of Biochemistry on the Nomenclature and Classification of Enzyme-Catalysed Reactions ("Enzyme Nomenclature", Academic Press, 1984).

Thanks are due to colleagues who suggested new entries, provided information from their own areas of expertise, and criticized the manuscript. Comments from readers in different countries have been most useful, and we hope that this second edition will generate a similarly large response.

Mrs. Ingrid Ullrich of de Gruyter publishers oversaw the production of all the galley and page proofs, and the graphic reproduction of a very large number of new diagrams. The time table for proof reading and circulation of the manuscript between USA, Berlin and England was devised by Dr. Rudolf Weber of de Gruyter publishers, and we are grateful for his guidance and support.

March 1988

Mary Eagleson
Sandy Hook, Connecticut, USA

Thomas Scott
Leeds, Yorkshire, England

Preface to the first edition

The "Brockhaus ABC Biochemie" was first published in 1976 in Leipzig, the second edition followed in 1981. When we undertook to translate this book, based on the second German edition, it was clear that our work would also involve considerable updating of existing entries and the introduction of new material. Such a task can, of course, never be complete. It is a rare and fortunate author or editor in the life sciences, and particularly in biochemistry, whose material is still completely up to date at the time of publication; progress in this field is so rapid and shows no sign of abating. Therein, however, lies the excitement and challenge of this venture. Already we have started collecting, classifying and revising in preparation for a subsequent edition.

We have departed from the style of the German edition by quoting a few literature references. These have been included with some of the new material, and we hope they will be useful to readers who want more information that can be fitted into a work of this sort. Where possible, we have also given each enzyme its EC (Enzyme Commission) Number, according to the Recommendations (1978) of the Nomenclature Committee of the International Union of Biochemistry (published in "Enzyme Nomenclature" Academic Press, 1979).

We apologize to any biochemist whose pet compound, mechanism or pathway has been overlooked, and we should be grateful to receive suggestions for new entries. It is also recognized that a reference work should reach into the past, defining terms no longer used, but encountered when using the older literature. In this respect, suggestions from our more "senior" readers would be most welcome.

Finally, thanks are due to Dr. Rudolf Weber of de Gruyter Publishers for his guidance and encouragement in the preparation of the manuscript and the production of this book.

January 1983

Mary Brewer
Menlo Park, California, USA

Thomas Scott
Leeds, Yorkshire, England

Using this book

Cross referencing is indicated by the word "see", and the subject of the cross reference starts with a high case letter, e.g. ... in the Posttranslational modification of proteins (see), or ... see Enzyme induction. Numbers, Greek letters and configurational letters at the beginning of names are ignored in the allocation of alphabetical order, e.g. β-Galactosidase is listed under G; L-Histidine under H, *N*-2-Hydroxyethylpiperazine under H. The main entry title is printed in bold type, followed by synonyms in bold italics. The remaining text uses only two further types, normal and italics.

Abbreviations: (The standard biochemical abbreviations, e.g. ATP, NAD, etc. are found as entries in the appropriate alphabetical positions).

abb.	abbreviation
[α]	specific optical rotation
b.p.	boiling point
c	concentration
°C	degrees Celsius
(d.)	with decomposition
ρ	density
E. coli	*Escherichia coli*
pI	isoelectric point
M	molar
m.p.	melting point
M_r	relative molecular mass
n	refractive index
syn.	synonym

A

A: 1. a nucleotide residue (in a nucleic acid) in which the base is adenine. 2. Abb. for adenosine (e.g. ATP is acronym of adenosine triphosphate). 3. Abb. for adenine. 4. Abb. of absorbance.

Å: see Angstrom unit.

Abiogenesis: the development of living organisms from inorganic and organic materials, and not by the reproduction of other living organisms. A. is the subject of several scientific theories. By providing a mechanism for the generation of the simplest life forms from nonliving organic compounds, A. possibly explains the origin of life.

According to modern concepts, four thousand million (4×10^9) years ago the composition of the primitive atmosphere of the earth was very different from that of the present atmosphere. In particular, it was a reducing atmosphere, lacking oxygen (which did not begin to appear until about 3×10^9 years ago) and containing much nitrogen. Under the highly energetic conditions of the time (high temperatures, UV irradiation, electric discharge), chemical reactions occurred that eventually led to the generation of the precursors of living organisms. Oparin named this process abiogenic or prebiotic organic evolution. Thus, in the primitive atmosphere, polymeric hydrocarbon derivatives were formed by the action of UV-light on water, methane, ammonia, hydrogen sulfide and carbon monoxide. Hydrogen cyanide was also important in the synthesis of biomolecules. Under simulated primitive earth conditions, three immediate products of HCN (cyanoacetylene, nitriles, cyanamide) can interact with aldehydes, ammonia and water to form various organic compounds, notably amino acids, pyrimidines, purines and porphyrins. In model experiments, Miller synthesized 14 of the 20 proteogenic amino acids by passing an electric discharge through a simulated primitive atmosphere. Von Oró et al. demonstrated the formation of adenine and guanine by the heat polymerization of ammonium cyanide in aqueous solution. Furthermore, formaldehyde is produced readily in simulated primitive atmosphere experiments, and it yields several sugars when heated with calcium carbonate, thus opening the way for the production of nucleosides and nucleotides.

It is assumed that polypeptides were formed by self condensation of abiotically produced amino acids, and polynucleotides by self condensation of abiotically produced nucleotides. Several mechanisms are possible for the promotion of such condensations. In particular, it is known that polyphosphates or polyphosphate esters act as dehydrating agents for the formation of peptides from amino acids when heated or irradiated with UV-light. Such polyphosphates, which are strong candidate forerunners of ATP, could have arisen from phosphate minerals by the action of cyanoacetylene or cyanoguanidine, or under the influence of high temperatures (e.g. near to volcanoes). For the development of a true life form, capable of further development and evolution, with the indispensable coupling of nucleic acids and protein synthesis, it was necessary for prebiotic material to become separated into discrete units. It is therefore thought that the formation of internally organized cells was the next significant step in the evolution of life forms. In this respect, special importance is attached to the formation of precellular structures, such as Oparin's coacervates and Fox's microspheres. It is highly probable that prebiotic membranes were formed by the aggregation of prebiotic lipids.

The direct route from abiotic high molecular mass compounds of the primitive earth to the first true living organisms is, however, still largely unexplained, since it is very difficult to explore experimentally. Essentially, there are two hypotheses for the origin of protobionts, i.e. the metabolic or protein hypothesis and the gene or nucleic acid hypothesis. In the protein hypothesis, primacy is given to the role of proteins in prebiotic evolution. Thus, membrane-bounded, replicative protocells containing catalytically active proteinoids subsequently acquire a coding system and become primitive living cells. In the gene hypothesis, primacy is given to nucleic acids. Originally proposed by H. Muller in 1929, this hypothesis states that life had its origin in the prebiotic formation of genes, which encoded the potential for metabolism and self-replication. The necessary amino acids were then acquired by random encounter. Further elaboration of this hypothesis became possible when the chemical structure of genetic material was established by Watson and Crick and others.

Abrin: see Ricin.

Abscisic acid, *ABA, abscisin, dormin:* (*S*)-(+)-5-(1′-hydroxy-4′-oxo-2′,6′,6′-trimethyl-2-cyclohexen-1-yl)-3-methyl-*cis,trans*-2,4-pentadienoic acid, a sesquiterpene plant hormone of ubiquitous occurrence in higher plants. It is also present in the *Bryophyta,* but was previously reported to be absent from liverworts, where lunularic acid (a dihydro-stilbene) was thought to perform the same function. ABA, in concentrations similar to those in higher plants, has now been unequivocally identified in the gametophyte of the liverwort, *Marchantia polymorpha,* where it probably also functions as a hormone [Xiaoyue Li et al. *Phytochemistry* **37** (1994) 625–637].

Two stereoisomeric forms are possible, depending on the *cis* or *trans* orientation of the $\Delta^{2,3}$ double bond; the *cis* isomer predominates in all plants; the *trans* isomer is occasionally found in small quantities, but does not appear to be biologically active. Only the (+)-form occurs naturally, and its concentration depends on the plant organ and its stage of development (average about 100 µg/kg fresh weight); relatively large quantities are present in fruits, dormant seeds, buds and wilting leaves.

ABA antagonizes the action of auxin, gibberellins and cytokinins, and in concert with these other phytohormones it is involved in the regulation of important growth and developmental processes, such as seed and bud dormancy, stomatal transpiration, flowering, germination and aging. ABA functions in the adaptation of plants to environmental change. Thus, it occurs in the xylem sap where it serves as a chemical signal between root and shoot [A. Bano et al. *Aust. J. Plant Physiol.* **20** (1993) 109–115; A. Bano et al. *Phytochemistry* **37** (1994) 345–347]. Water stress (water shortage or soil flooding) is first perceived by the roots; the resulting transport of ABA to the shoot induces stomatal closure and sometimes also the synthesis of proteins that possibly protect tissues from the effects of desiccation. As seeds develop in the plant, their premature germination is prevented by the presence of ABA, which also appears to induce the synthesis of certain seed proteins. Induction by gibberellic acid of hydrolytic enzymes in the aleurone layer of cereal grains is restrained by the presence of ABA. ABA is also involved in the response of plant tissues to physical damage, possibly in concert with Jasmonic acid (see). Stomatal closure by ABA is due mainly to an efflux of K^+. ABA may directly affect K^+ channels, or the loss of K^+ may be caused by an increase in the Ca^{2+} concentration in the guard cell cytoplasm. Many genes encoding stress proteins (proteins induced by desiccation, low temperature, wounding, etc.), storage proteins and proteins conferring desiccation tolerance display an increased rate of expression in the presence of ABA. Promoter regions of several ABA-induced genes contain ABA response elements, e.g. the consensus sequence, CACGTG, has been found in wheat *Em* (early methionine labeled), rice *Rab* (responsive to ABA) and cotton *LEA* (late embryo ABA). DNA binding proteins of the Leucine zipper (see) type bind to the consensus sequence and promote expression [A. M. Heatherington & R. S. Quatrano *New Phytol.* **119** (1991) 9–32]. The molecular basis of ABA action is unknown, and no receptor has been demonstrated.

ABA is synthesized primarily in leaf chloroplasts and in the root. It is a true terpenoid, being derived

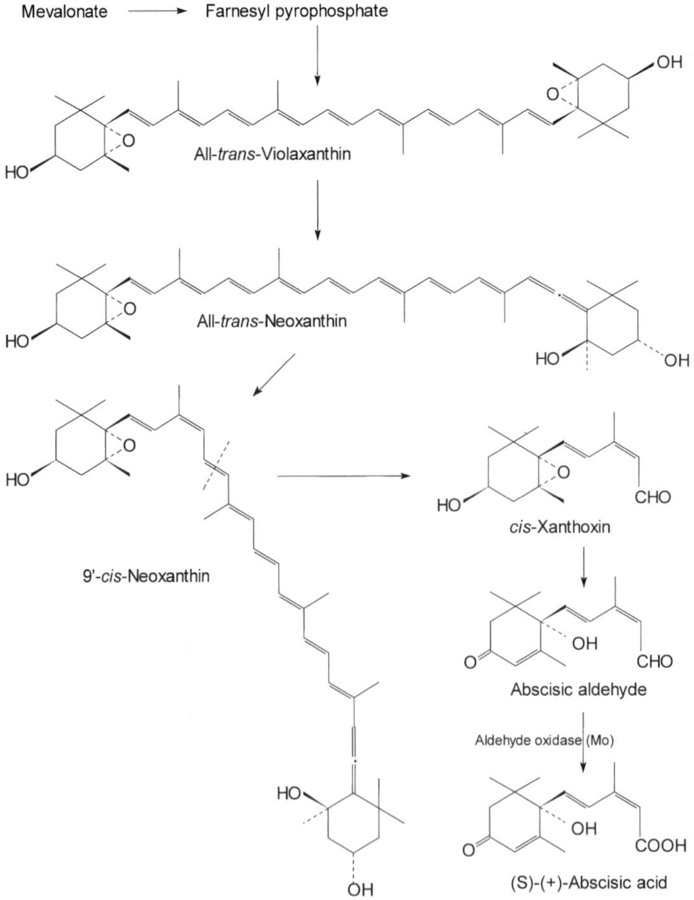

Mevalonate → Farnesyl pyrophosphate

All-*trans*-Violaxanthin

All-*trans*-Neoxanthin

9'-*cis*-Neoxanthin

cis-Xanthoxin

Abscisic aldehyde

Aldehyde oxidase (Mo)

(S)-(+)-Abscisic acid

Biosynthetic pathway of abscisic acid. All-*trans*-neoxanthin is assumed to be an intermediate in the demonstrated conversion of all-*trans*-violaxanthin into 9′-*cis*-neoxanthin.

from mevalonic acid via isopentenyl pyrophosphate. However, it is not derived directly via a C_{15} (sesquiterpenoid) route, but indirectly via a C_{40} (apo-carotenoid) by cleavage. The rate-limiting step of biosynthesis appears to be the conversion of all-*trans*-violaxanthin into 9′-*cis*-neoxanthin, and this conversion is promoted by water deficiency. The pathway (Fig.) concludes with the oxidation of ABA-aldehyde to ABA by a Mo-containing aldehyde oxidase [A. D. Parry & R. Horgan *Physiol. Pl.* **82** (1991) 320–326]. Direct synthesis of ABA via mevalonic acid and geranyl and farnesyl pyrophosphates has been reported in the fungi, *Cercospora rosicola* and *C. cruenta.* In higher plants ABA is accompanied by its glucose esters and its *O*-glucoside, which are thought to be transport and storage forms. It is metabolized and inactivated by 8′-oxidation to phaseic acid.

Absolute oils: see Essential oils.

Absorbance, *extinction, optical density:* a measure of the quantity of light absorbed by a solution. It is equal to log I_0/I, where I_0 is the intensity of the incident light, and I is the intensity of the transmitted light.

Absorptivity, *absorbance index, absorption coefficient:* the proportionality constant ε, in Beer's law for light absorption: $A = \varepsilon lc$, where A is absorbance, l the length of the light path, and c the concentration. If concentration is expressed on a molar basis, ε becomes the *molar absorptivity, molar absorption coefficient* or *molar extinction coefficient,* i. e. $\varepsilon = A/lc$, where l is the length of the light path in cm, and c is the molar concentration.

Abzyme: see Catalytic antibody.

Acatalasia: see Inborn errors of metabolism.

Acetaldehyde, *ethanal:* CH_3-CHO, an important intermediate in the degradation of carbohydrates. In its activated form (see Thiamin pyrophosphate), it is involved in a number of reactions (see Alcoholic fermentation). Two molecules of A. can undergo acyloin condensation to form Acetoin (see).

3′-Acetamido-3′-deoxyadenosine: see 3′-Amino-3′-deoxyadenosine.

Acetate kinase, *acetokinase* (EC 2.7.2.1): see Acetylphosphate, Phosphoroclastic pyruvate cleavage.

Acetic acid, *ethanoic acid:* CH_3-COOH, a common monocarboxylic acid which occurs in the free form as the end product of fermentation and oxidation reactions in some organisms. Acetate is formed metabolically by dehydrogenation of acetaldehyde, catalysed either by aldehyde oxidase (EC 1.2.3.1) or $NAD(P)^+$-dependent aldehyde dehydrogenase (EC 1.2.1.3). The activated form of A. a., Acetyl-coenzyme A (see), is a key compound of intermediary metabolism.

Acetogenins: see Polyketides.

Acetoin, *3-hydroxy-2-butanone, acetyl methyl carbinol:* CH_3-CO-CHOH-CH_3, a reduction product of diacetyl which arises under certain conditions as a side product of the pyruvate decarboxylase (EC 4.1.1.1) reaction. A. is also formed by decarboxylation of acetolactate by acetolactate decarboxylase (EC 4.1.1.5). It is oxidized in a reversible reaction to diacetyl by acetoin dehydrogenase (EC 1.1.1.5), and in some microorganisms it is converted to 2,3-butanediol by D(–)-butanediol dehydrogenase (EC 1.1.1.4).

Acetylcarnitine: see Carnitine.

Acetylcholine: a vertebrate and invertebrate Neurotransmitter (see) in neuromuscular synapses, and in synapses of the following nerves: all motor nerves to skeletal muscle; all preganglionic nerves, including the nerve supply to the adrenal medulla; all postganglionic parasympathetic nerves; postganglionic sympathetic nerves to sweat glands; and some postganglionic sympathetic nerves to blood vessels in skeletal muscle. After release from the nerve terminal at the synapse, A. binds to, and triggers a response by, receptors in the postsynaptic neuron; it then leaves the receptor and is rapidly degraded by Acetylcholinesterase (see). Nerves that employ acetylcholine for their chemical transmission are called *cholinergic nerves.*

Depending on its concentration, A. exerts two different physiological effects. Injection of small amounts of A. produces the same response as the injection of Muscarin (see), i. e. a fall in blood pressure (due to vasodilation), slowing of the heart beat, increased contraction of smooth muscle in many organs, and copious secretion from exocrine glands; this muscarinic effect of muscarin or A. is abolished by atropine. After administration of atropine, larger amounts of A. cause a rise in blood pressure, similar to that caused by nicotine. Nicotinic cholinergic synapses are found in vertebrate neuromuscular junctions, certain ganglia, central synapses, and the electroplax of *Torpedo.* Muscarinic cholinergic synapses operate in smooth muscle, cardiac muscle, ganglia, and many central brain regions. In the brain and central nervous system, muscarinic synapses outnumber nicotinic synapses by 10–100 fold. In ganglia, nicotinic cholinergic receptors are blocked by tetraethylammonium, and in neuromuscular synapses they are blocked by Curare (see), and irreversibly occupied by the snake venom constituent, α-bungarotoxin. Muscarinic cholinergic receptors of the postganglionic parasympathetic systems are blocked by atropine and scopolamine, which are therefore parasympatholytic agents. Other

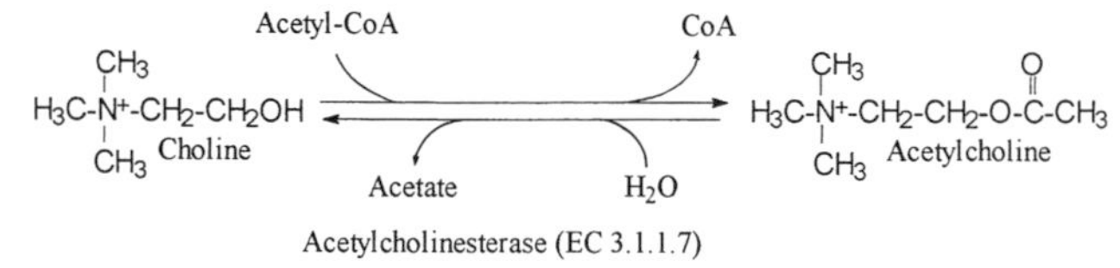

Fig. 1. *Acetylcholine.* Synthesis (in the nerve terminal) and degradation (in the synapse) of acetylcholine.

substances inhibit the activity of acetylcholinesterase, thereby causing nerve paralysis; for example, physostigmine is a reversible inhibitor, whereas certain organic phosphates, which are used as insecticides, are irreversible inhibitors. Other acetylcholinesterase inhibitors, such as the organic fluorophosphates, are among the most potent chemical warfare agents.

Nicotinic and muscarinic effects are mediated by *nicotinic* and *muscarinic receptors*, respectively. These receptors are the products of two distinct gene superfamilies, and their only common property is that they are activated by A. The slower muscarinic receptor response operates via G-protein-coupled receptors (see G-proteins). Depending on the receptor subtype, the subsequent effector mechanism involves inhibition of adenylate cyclase, formation of inositol 1,4,5-*tris*phosphate from phosphoinositides by a specific phospholipase C, or modulation (opening) of certain K^+ channels [D. Brown *Nature* **319** (1986) 358–359].

Muscarinic receptors display a high degree of heterogeneity, but in all forms the primary sequence shows similarities with those of the β_2 adrenoreceptor and rhodopsin, suggesting variations on a common structural theme: 1) there are seven transmembrane segments; 2) the *N*-terminal region lacks a signal sequence and contains two *N*-glycosylation sites; 3) the *C*-terminal region contains several threonine and serine residues (phosphorylation sites); and 4) all are cell membrane proteins interacting with G-proteins.

Orthodox cloning strategies have been employed, involving: 1) purification of receptor from cerebral cortex by affinity chromatography; 2) amino acid sequencing of purified receptor; 3) construction of oligonucleotide probes on the basis of partial amino acid sequences; 4) the use of these probes to screen cDNA libraries. In addition, an oligonucleotide probe has been constructed for a region of sequence homology between the 2nd transmembrane domain of rat m1 receptor and hamster β2 adrenergic receptor, and used to screen rat cortex cDNA libraries. Five pharmacologically distinct muscarinic receptor subtypes (m1-m5) have been cloned and sequenced from rat and porcine cerebral cortex. Using the sequence homology probe, other hybridizing bands are evident in digests of rat and human DNA, indicating the possible existence of further receptor subtypes [T. Kubo et al. *Nature* **323** (1986) 411–426; E. C. Hulme et al. *Annu. Rev. Pharmacol. Toxicol.* **30** (1990) 633–673]. m1, m3 and m5 are coupled to inositol 1,4,5-*tris*phosphate production, m2 and m4 to adenylate cyclase inhibition and modulation of K^+ channels. All are activated by muscarine and oxotremorine, and blocked by atropine. They also show considerable sequence homology, and the sequence of each can be interpreted in terms of seven transmembrane domains containing the acetylcholine binding site.

Occupancy of the nicotinic receptor (by A.) triggers a rapid response (1–2 ms) by direct activation of cation-selective ion channels, thereby causing depolarization of the postsynaptic membrane. The nicotinic cholinergic receptor has been isolated from the electroplax of *Torpedo californica* (electric ray) and *Electrophorus electricus* (electric eel) and from vertebrate muscle. From all three tissues, it is a single membrane protein, M_r 250,000, consisting of 4 glycoprotein subunits: M_r 40,000 (50,116) (α), 50,000 (53,681) (β), 60,000 (56,279) (γ) and 65,000 (57,565) (δ) (the first value is from SDS; the value in brackets is the exact M_r based on amino acid composition) in the ratio 2:1:1:1, with an average 40% amino-acid sequence identity between all 4 chains [B. M. Conti-Troconi et al. *Science* **218** (1982) 1227–1229]. DNA for all 4 subunits has been cloned and sequenced [M. Noda et al. *Nature* **301** (1983) 251–254]. The covalent affinity probes, [^3H]-bromoacetylcholine and 4-(*N*-maleimideo)-^3H benzyl trimethylammonium iodide, label the α-subunit by reacting with cysteine residues 192 and 193. Thus, each of the two α-subunits carries an acetylcholine binding site, and there are 2 sites per oligomeric receptor. When incorporated into liposomes or planar lipid bilayers, the nicotinic receptor permits a flux of $^{22}Na^+$, which is promoted

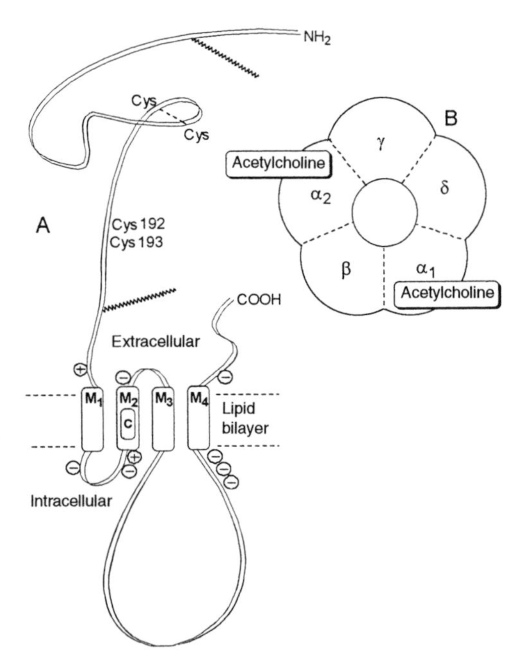

Fig. 2. *Acetylcholine.*
A. Schematic representation of the α-subunit of the nicotinic acetylcholine receptor from *Torpedo.* M_1, M_2, M_3 and M_4 represent transmembrane helical regions, each consisting of at least 20 hydrophobic amino acids, corresponding to about $5^1/_2$ spirals of an α-helix. ~~~~ represents carbohydrate residues. C is a region of M_2 in subunits β and δ, which irreversibly binds the channel blocker, chlorpromazine (Ser_{254} and Leu_{257} of the β-subunit; Ser_{262} of the δ- subunit). The Cys-Cys loop in the extensive *N*-terminal extracellular region is characteristic of all ligand-gated ion channels. The adjacent Cys residues (-Cys_{192}-Cys_{193}-) are unique to all α-subunits.
B. Arrangement of subunits of the nicotinic acetylcholine receptor, viewed perpendicular to the membrane surface. The binding sites for acetylcholine and α-toxin are situated on the α-subunits.

by acetylcholine and blocked by α-bungarotoxin. Nicotinic cholinergic receptor is therefore one of the group of receptors known as *gated ion channels* (others include receptors for gamma-aminobutyric acid, glycine and 5-hydroxytryptamine).

Nicotinic receptors of the vertebrate neuromuscular junction and the electric organs of fishes represent a single subtype, and are known as *peripheral receptors.* Nicotinic receptors are also found in autonomic ganglia and in the central nervous system, where they are collectively known as *neuronal receptors.* The latter are also gated cation channels, but structurally they are a heterogeneous group, and less well characterized than peripheral receptors.

All 4 subunits of the *Torpedo* and other peripheral nicotinic receptors show a high degree of amino acid sequence identity, with particularly close homologies between the α and β and between the γ and δ polypeptides, suggesting a single ancestral gene with evolutionary branching into the αβ- and γδ-type chains. Hydrophobicity analysis predicts the existence of 4 transmembrane α-helical segments (M1-M4) (Fig. 2).

Table. *Drugs which affect cholinergic systems.*
*: action mainly at peripheral ganglia.
Formulae not shown in this table may be found under separate entries.

Muscarinic agonists: acetylcholine, muscarin, carbachol, methacholine, bethanechol, pilocarpine, arecoline, oxotremorine.

Muscarinic antagonists: atropine, scopolamine, benztropine (also blocks dopamine uptake), quinuclidinyl-bromide, pirenzipine.

Nicotinic agonists: acetylcholine, nicotine*, carbachol, arecoline, suberyldicholine, tetramethylammonium*, phenyltrimethylammonium*, dimethylphenylpiperazine*.

Nicotinic antagonists: D-tubocurarine, succinylcholine (depolarizing, desensitizing), gallamine, pempidine*, mecamylamine*, hexamethonium*, pentolinium*, pancuronium, α-bungarotoxin.

Inhibitor of acetylcholine synthesis: 4-naphthylvinylpyridine.

Pump inhibitors (prevent entry of choline into nerve cell, leading to failure to synthesize acetylcholine): triethylcholine, hemicholinium.

Cholinesterase inhibitors (used to eliminate cholinesterase in the histochemical detection of acetylcholinesterase): diisopropylphosphofluoridate, neostigmine, physostigmine, edrophonium.

Release inhibitor: *Botulinum* toxin.

Specific binding agents: α-bungarotoxin, propylbenzilylcholine mustard, quinuclidinyl benzilate.

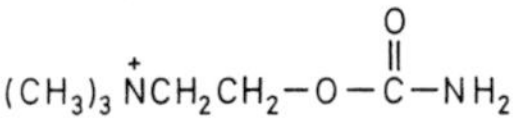

$(CH_3)_3\overset{+}{N}CH_2CH_2-O-\overset{O}{\overset{\|}{C}}-NH_2$

Carbamylcholine (carbachol) (first stimulates skeletal muscle, then blocks neuromuscular transmission)

$(C_2H_5)_4\overset{+}{N}$

Tetraethylammonium

$(CH_3)_3\overset{+}{N}(CH_2)_5\overset{+}{N}(CH_3)_3$

Pentamethonium

$(CH_3)_3\overset{+}{N}(CH_2)_6\overset{+}{N}(CH_3)_3$

Hexamethonium

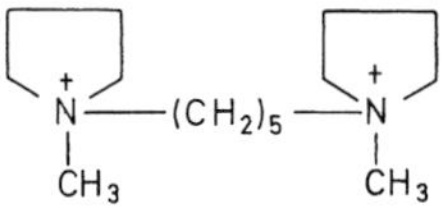

Pentolinium

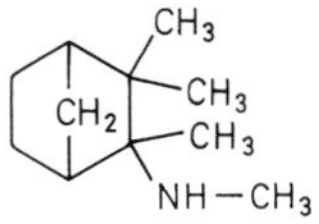

Mecamylamine

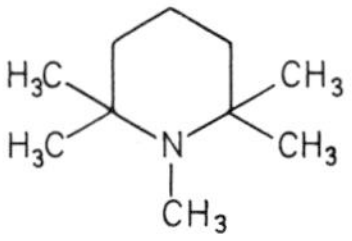

Pempidine

Ganglion blockers

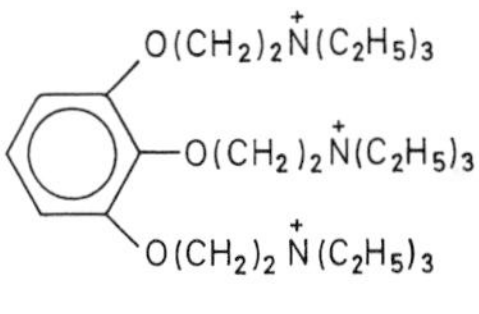

Gallamine (blocks neuromuscular transmission without prior stimulation)

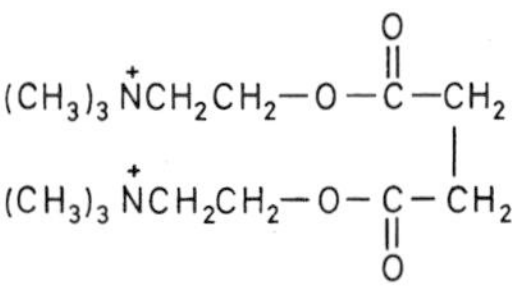

Succinylcholine (suxamethonium) (first stimulates skeletal muscle, then blocks neuromuscular transmission)

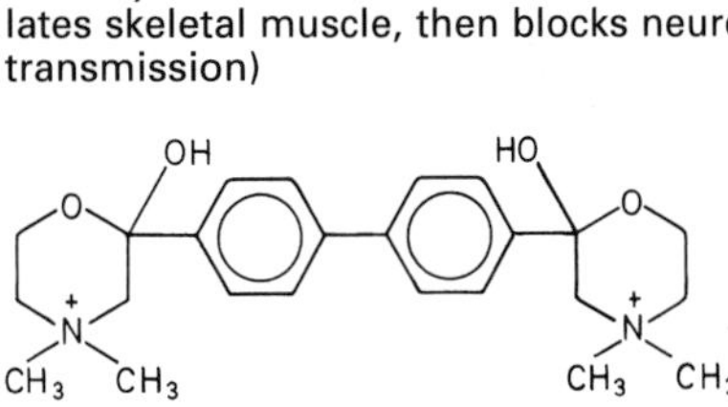

Hemicholinium (prevents entry of choline into nerve cell)

$(C_2H_5)_3\overset{+}{N}CH_2CH_2OH$

Triethylcholine (prevents entry of choline into nerve cell)

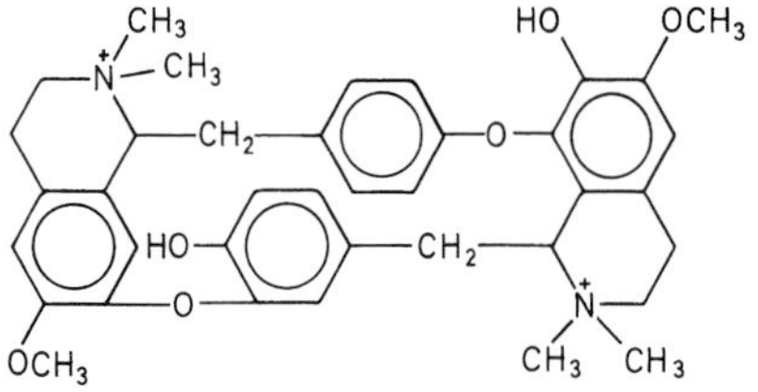

Tubocurarine (blocks neuromuscular transmission without prior stimulation)

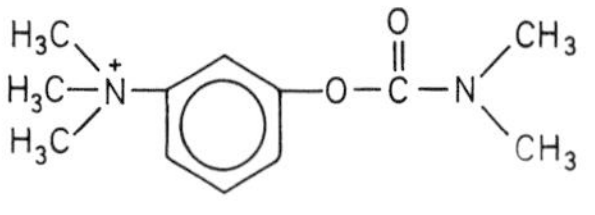

Neostigmine

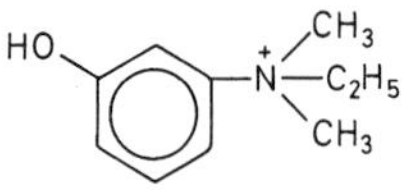

Edrophonium.

The hydrophilic *N*-terminus of each subunit contains consensus sequences for *N*-glycosylation, consists of at least 200 residues, is predicted to be extracellular, and contains a Cys-Cys loop domain known to be present in other gated ion channels (Cys-X-X-X-X-X-X-hydrophobic-Pro-hydrophobic-Asp-X-X-X-Cys). X-ray analysis is not possible because the proteins have not been crystallized, but a structural model has been developed with the aid of electron microscopy of negatively stained preparations and electron image analysis of rapidy frozen receptor proteins. Thus, perpendicular to the plane of the membrane, the receptor is an 80Å diameter rosette consisting of 5 peaks of electron density around a central pore (Fig. 2). This rosette represents the end of a cylinder which spans the membrane, protruding mostly on the extracellular surface, i.e. into the synaptic cleft. [E. S. Deneris et al. *Trends Pharmacol. Sci.* **12** (1991) 34–40]

The autoimmune disease, myasthenia gravis, is due to the presence of circulating antibodies to the peripheral nicotinic receptor. Binding of these antibodies to the receptor results in increased receptor degradation and a decreased efficiency of neuromuscular transmission. The condition can be partly relieved by administration of acetylcholinesterase inhibitors. [J. Newsom-Davis et al. in *Clinical Aspects of Immunology,* P. J. Lachmann et al. (eds), Blackwell Scientific publications, Oxford, 1993: pp. 2091–2113]

Histochemical localization of acetylcholinesterase serves to identify cholinergic synapses. It is based on the technique of Koelle and Friedenwald (G. B. Koelle *Handb. Exp. Pharmakol.* **15** (1963) 187–298]. The substrate used is acetyl- or butylthiocholine, and the product, thiocholine, is visualized by precipitation with lead or copper salts. A more specific marker for cholinergic neurons (acetylcholinesterase is also present in dopaminergic cells of the substantia nigra) is choline acetylase (EC 2.3.1.6).

A. is a phylogenetically ancient hormone which also appears in protists. It may be an evolutionary precursor of the neurohormones. Traces of A. also occur in plants, e.g. in the hairs of the stinging nettle. Related compounds, e.g. murexin in the glands of certain gastropods, are probably venoms. [D. S. McGehee & L. W. Role 'Physiological Diversity of Nicotinic Acetylchline Receptors Expressed by Vertebrate Neurons' *Annu. Rev. Physiol.* **57** (1995) 521–546; D. R. Groebe & S. N. Abramson 'Lophotoxin is a Slow Binding Irreversible Inhibitor of Nicotinic Acetylcholine Receptors' *J. Biol. Chem.* **270** (1995) 281–286; C. Czajkowski & A. Karlin 'Structure of the Nicotinic Receptor Acetylcholine-binding Site' *J. Biol. Chem.* **270** (1995) 3160–3164]

Acetylcholinesterase (EC 3.1.1.7): an enzyme which catalyses the hydrolysis of acetylcholine into choline and acetate. Due to the high turnover number of A. ($0.5–3.0 \times 10^6$ molecules substrate/molecule enzyme/min), the acetylcholine released at a synapse is hydrolysed within 0.1 ms. This enzyme is found in the central nervous system, particularly in postsynaptic membranes of striated muscles, parasympathetic ganglia, erythrocytes and electric organs of fish. Crystalline A. (M_r 330,000) has been isolated from the electric organ of the electric eel *(Electrophorus electricus).* It consists of 4 identical, enzymatically inactive subunits of M_r 82,500; half molecules consisting of 2 covalently bound subunits (M_r 165,00) are enzymatically active. Proteolytic attack on the subunits produces two fragmants of M_r 60,000 and 22,500.

The active center of A. has two parts, the anionic binding site for the quaternary nitrogen (which is responsible for the alcohol specificity) and the esterase center (where a catalytic serine and histidine lyse the ester bond). The enzyme is inactivated by blockage of either the serine hydroxyl (by organic phosphate esters, such as diisopropylfluorophosphate or diethyl *p*-nitrophenylphosphate), or the anionic center by trimethylammonium derivatives. If the enzyme has been blocked by organophosphates, it can be reactivated by pralidoxime salts, which are therefore used as antidotes to organophosphate poisoning.

A. is sometimes called "true cholinesterase" in contrast to relatively unspecific acylcholinesterases (see Cholinesterase). [D. K. Getman et al. 'Transcriptional Factor Repression and Activation of the Human Acetylcholinesterase Gene' *J. Biol. Chem.* **270** (1995) 23511–23519]

Acetyl-coenzyme A, *acetyl-CoA, active acetate:* $CH_3CO \sim SCoA$ (M_r 809.6, λ_{max} 260 nm), a derivative of acetic acid in which the acetyl residue is bound by a high energy bond to the free SH-group of coenzyme A. By virtue of the high transfer potential of its acetyl group, acetyl-CoA provides the C_2 fragment for numerous syntheses. The free energy of the thioester bond (34.3 kJ = 8.2 kcal/mol), however, has no significance as a form of energy storage. In transfer reactions involving acetyl-CoA, either the carboxyl group (electrophilic reaction) or the methyl group (nucleophilic reaction) can react.

The most important pathways for the synthesis of acetyl-CoA (Table) are 1) oxidative decarboxylation of pyruvate, 2) degradation of fatty acids and 3) degradation of certain amino acids. Formation of acetyl-CoA involves either 1) transfer of an acetyl resi-

Reactions in which acetyl-coenzyme A is synthesized

Enzyme	Reaction	Occurrence/significance
Acetyl-CoA-synthetase (EC 6.2.1.1)	$CH_3COO^- + ATP + CoA \rightleftharpoons CH_3CO–CoA + AMP + PP_i$	Yeasts, animals, higher plants
Acyl-CoA synthetase (GDP-forming) (EC 6.2.1.10)	$CH_3COO^- + GTP + CoA \rightleftharpoons CH_3CO–CoA + GDP + P_i$	Liver
Acetate kinase (EC 2.7.2.1)	$CH_3COO^- + ATP \rightleftharpoons CH_3CO–O–PO_3H_2 + ADP$	Microorganisms
Phosphate acetyltransferase (EC 2.3.1.8)	$CH_3CO–O–PO_3H_2 + CoA \rightleftharpoons CH_3CO–CoA + P_i$	Microorganisms
ATP citrate *(pro-3S)*-lyase (EC 4.1.3.8)	Citrate + ATP + CoA ⇌ CH_3CO–CoA + oxaloacetate + ADP + P_i	Outside mitochondria
Pyruvate dehydrogenase complex (EC 1.2.4.1; 2.3.1.12; 1.6.4.3)	Pyruvate + NAD^+ + CoA ⇌ $CH_3CO–CoA + CO_2 + NADH + H^+$	Mitochondrial particles; involves TPP, $LipS_2$
Acetyl-CoA transacetylase (EC 2.3. 1.9)	Acetoacetyl-CoA + CoA ⇌ $2CH_3CO$–CoA	Fatty acid degradation

due from a suitable donor such as pyruvate, and simultaneous reduction of NAD^+, or 2) activation of free acetate in a one- or two-step process, which requires ATP and free coenzyme A.

Acetyl-CoA is the hub of carbohydrate metabolism and has a central position in overall metabolism. Products of carbohydrate, fat and protein metabolism are channeled via acetyl-CoA into oxidative degradation in the tricarboxylic acid cycle. The acetyl residue is used in the synthesis of esters and amides (e.g. acetylcholine, *N*-acetylglucosamine, *N*-acetylglutamate). Acetyl-CoA is also the starting point for isoprenoid synthesis via mevalonic acid and for fatty acid synthesis.

***N*-Acetylglutamic acid, N-*acetylglutamate*, Ac-Glu:** HOOC–CH($NHCOCH_3$)–CH_2–CH_2–COOH, the acetylated form of glutamic acid. It is the allosteric activator of carbamoyl phosphate synthetase (ammonia) (EC 6.3.4.16). See Carbamoyl phosphate.

Acetyl methyl carbinol: see Acetoin.

Acetylphosphate: CH_3–COOPO$(OH)_2$, a high energy acyl phosphate. It is the product of acetate activation in some organisms: Acetate + ATP ⇌ A.p. + ADP, a reaction catalysed by acetate kinase (EC 2.7.2.1). The back reaction is sometimes exploited for ATP synthesis, e.g. in the phosphoroclastic fission of pyruvate.

Acidic α_1-glycoprotein: see Orosomucoid.

Acid lipase deficiency: see Inborn errors of metabolism.

Acidosis: a decrease in the pH of body fluids. It is corrected by excretion of acid via the lungs and kidneys. There are 2 types of A.:

a) *Metabolic A.* is caused by a decrease in the bicarbonate fraction, with little or no change in the carbonic acid fraction. There are several possible causes. 1) Severe diarrhea results in the loss of gastrointestinal secretions containing high concentrations of bicarbonate (the resulting acidosis is a contributory factor to infant death in the developing world). 2) Acetazolamide (Diamox), a drug used to promote diuresis, inhibits carbonic anhydrase in the brush border of the proximal tubule epithelium. Reabsorption of bicarbonate is therefore retarded, leading to acidosis. 3) Severe renal disease, which may impair the ability of the kidney to remove acids formed in the normal course of metabolism. 4) Vomiting usually leads to the loss of bicarbonate from the upper intestine, as well as the acidic stomach contents. Since the loss of alkali exceeds the loss of acid, the net result is acidosis. 5) Diabetes mellitus results in the excessive formation of acetoacetic acid, which accumulates and causes acidosis in the extracellular fluids; as much as 500–1,000 mmoles of acid may be excreted per day.

b) *Respiratory acidosis* is caused by an increase in carbonic acid in relation to bicarbonate. It may occur when alveolar ventilation is impaired, e.g. in pneumonia and asthma, and it can be caused by depression of the respiratory center, e.g. by morphine poisoning.

Acid phosphatase deficiency: see Inborn errors of metabolism.

Acid plants, *ammonium plants:* plants which accumulate organic acids in their leaf cells. Within the cells, these acids are neutralized by ammonium ions.

Aconitate hydratase, *aconitase* (EC 4.2.1.3): a hydratase which catalyses one stage of the tricarboxylic acid cycle, the reversible interconversion of citrate and isocitrate. The reaction proceeds via the enzyme-bound intermediate, *cis*-aconitate. At equilibrium, the relative abundances are 90 % citrate, 4 % *cis*-aconitate, 6 % isocitrate. Thus citrate is favored at equilibrium, but in respiring tissues the reaction proceeds from citrate to isocitrate, as isocitrate is oxidized by isocitrate dehydrogenase. The enzyme requires Fe(II) and requires a thiol such as cysteine or reduced glutathione. The Fe(II) ion forms a stable chelate with citric acid. X-ray analysis of Fe(II) complexes of tricarboxylic acids suggested the "ferrous wheel" hypothesis of aconitase action. According to this mechanism, 3 points on the *cis*-aconitate molecule are bound at separate sites on the enzyme surface; in addition the molecule is also complexed with the Fe(II) atom at the active center. Stereospecific *trans* addition of water to *cis*-aconitate to form either citrate or isocitrate is achieved by rotation of the ferrous wheel, which can add OH to either side of the molecule. Aconitase is inhibited by fluorocitrate. Two isoenzymes are present in animal tissues, one in the cytosol and one in the mitochondria. [J. P. Glusker in P. D. Boyer (ed) *The Enzymes,* **5,** 434, Academic Press, 1971]

Aconitic acid: an unsaturated tricarboxylic acid, usually occurring in the *cis* form (m.p. 130 °C), but sometimes in the *trans* (m.p. 194–195 °C). Free A.a. was first discovered in aconite, *Aconitum napellus.* The anionic form of *cis*-A.a. (propene-*cis*-1,2,3-trioic acid) is important as an intermediate in the isomerization of citrate to isocitrate in the Tricarboxylic acid cycle (see).

Aconitine: an extremely poisonous ester and alkaloid (see Terpene alkaloids) from the roots of aconite *(Aconitum napellus)* and other *Aconitum* and *Delphinium* species. Between 1 and 2 mg A. causes death in adult humans by paralysing the heart and respiration. In spite of useful physiological properties, A. is rarely used clinically, due to its toxicity. A. is sometimes used internally as a tincture for rheumatism and neuralgias, and externally as a pain-killing salve. In antiquity, aconite preparations were used as arrow poisons by the Greeks and (East) Indians. Its hydrolysis products are only slightly toxic.

Aconitum alkaloids: a group of terpene alkaloids, some of them very poisonous, from various aconite *(Aconitum)* species. The best known representative is Aconitine (see).

ACP: acronym of Acyl carrier protein (see).

Acrasin: an attractant secreted by aggregation centers of slime molds, which stimulates single cells to aggregate and form fruiting bodies. The A. of *Dictostelium* is cyclic AMP (see), which attests to the antiquity of this substance as a hormone. The A. of *Polysphondium violaceum* is a dipeptide called "glorin":

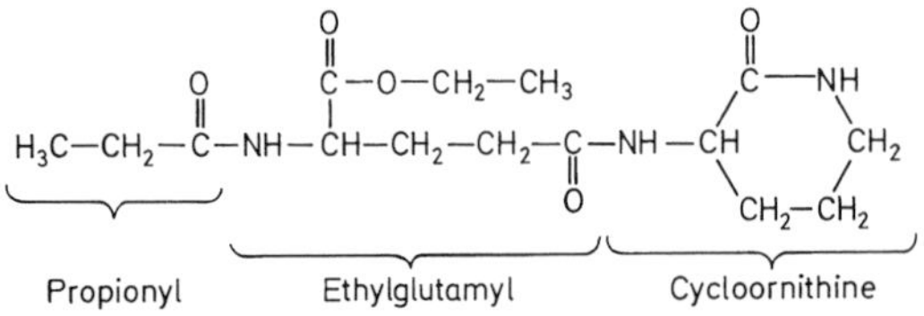

Aggregation attractant ("glorin") of Polysphondium violaceum

ACTH: acronym of adrenocorticotropic hormone. See Corticotropin.

Actin: see Actins.

Actinomycins: a large group of peptide lactone antibiotics produced by various strains of *Streptomyces.* These highly toxic red compounds contain the chromophore, 2-amino-4,6-dimethyl-3-ketophenoxazine-1,9-dioic acid (actinocin), which is linked to two 5-membered peptide lactones by the amino groups of two threonine residues. The various A. differ only in the amino acid sequence of the lactone rings. In vivo, A. inhibit DNA-dependent RNA synthesis at the level of transcription by interacting with the DNA. The effective inhibitory concentration increases as the guanine content of the DNA decreases. A. are pharmacologically important due to their bacteriostatic and cytostatic effect. The spatial structure of actinomycin D has been elucidated by NMR, and the specificity of its interaction with deoxyguanosine demonstrated by X-ray analysis. Actinomycin D (Fig.), one of the commonest A., is used as a cytostatic, e.g. in the treatment of Hodgkin's disease.

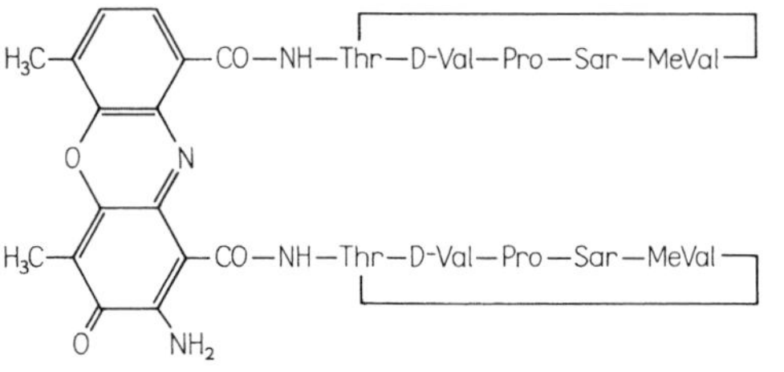

Actinomycin D

Actins: contractile proteins found in many cell types. Actin is an essential component of the contractile complex of Muscle proteins (see). Microvilli, microspikes (filopodia) and stereocilia (hair cells in the cochlea of the ear and related organs) consist of actin associated with other proteins. Monomeric actin (G-actin) has M_r 41,720. Microfilaments in the cell cytoplasm consist of filaments of polymerized actin (F-actin) (see Cytoskeleton). A consensus model of F-actin shows a helical filament with a diameter of 90–100 Å, in which the long axes of the monomers are nearly perpendicular to the filament axis. The positions of

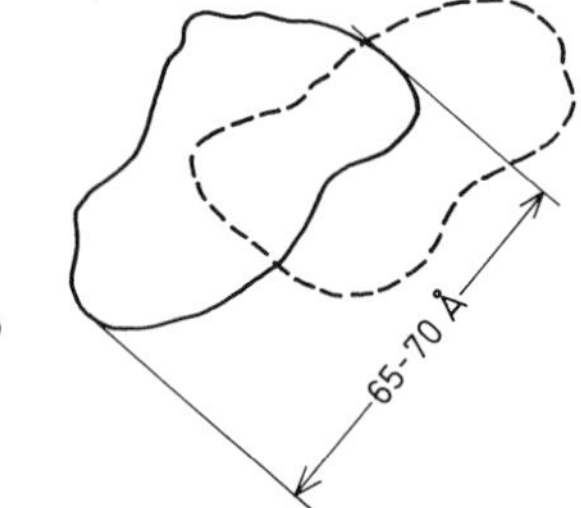

Polymerized actin
a. Idealized model showing arrangement of actin monomers, represented here as two fused spheres. [from D. J. DeRosier The Cytoskeleton. vol 5 of *Cell and Muscle Motility* (Plenum Press, New York, 1985) pp. 139–169 with permission]
b. Cross sections through two successive monomers. The solid and dashed outlines represent sections in two different planes, about 27 Å apart and rotated through 167° around the helix axis. [from E. H. Egelman *J. Musc. Res. Cell Motil. 6* (1985) 129–151, with permission].

the monomers within the filament are flexible, so that binding of proteins (e.g. tropomyosin) may impose a periodic but nonhelical structure; the repeat distance is 7 monomers. The structure of actin has been highly conserved in the course of evolution, possibly due to the large number of proteins with which it interacts specifically.

Each actin monomer binds one molecule of ATP. When polymerization occurs, this ATP is hydrolysed and the resulting ADP remains bound to the actin. However, the hydrolysis is not coupled to the polymerization itself; it occurs about 10 seconds after the monomer has been added to the polymer. Growth of a filament produces an "ATP cap" or terminal region in which the monomers still carry bound ATP. These monomers dissociate more slowly than ADP-bound monomers, so that the ATP cap promotes further growth of the polymer. Conversely, a shrinking polymer, from which monomers dissociate more rapidly than they are added, has a region of ADP-bound monomers at its ends, and this region increases the rate of dissociation.

Addition and release of monomers can occur at either the "pointed" or the "barbed" end of F-actin, but the processes of addition and release are about 10 times faster at the barbed ends. This led to the earlier inference that G-actin was added only to the barbed ends and released from the pointed ends in a process of "treadmilling" [A. Wegner *Nature* **313** (1985) 97–98].

G-Actin displays the interesting property of forming a 1:1 complex with DNA polymerase I (DNase I), which together with one Ca^{2+} ion and one molecule of ATP or ADP, can be crystallized. This enabled the X-ray analysis of G-actin, which is otherwise impossible to crystallize on account of its marked tendency to polymerize. It is not thought that binding to actin has any significance in vivo. X-ray crystallographic studies by W. Kabsch and K. Holmes (Max-Planck-Institute for Medical Research, Germany) reveal that G-Actin consists of 2 domains (known as the "large" and "small" domains, although the former is only slightly larger), each divided into 2 subdomains. Two of the subdomains (one in each domain) display similar structure, i.e. a 5-stranded β-sheet consisting of a β-hairpin motif followed by a right-handed βαβ motif; these subdomains may therefore have arisen by gene duplication, but their primary sequences are not significantly similar. Models of F-actin generated from X-ray crystallographic data of the G-actin: DNase I: ADP: Ca^{2+}-complex agree with the X-ray fiber diagram of oriented F-actin gels, with regard to filament polarity, monomer orientation and 3-dimensional location of the *C*-terminus of the monomer; all 4 subdomains of the monomer are in contact with neighboring monomers. [T. D. Pollard et al. 'Structure of Actin-binding proteins: Insights About Function at Atomic Resolution' *Annu. Rev. Cell Biol.* **10** (1994) 207–249]

Activated amino acids: see Aminoacyl adenylate.

Activated fatty acids: fatty acyl coenzyme A thioesters which, as high energy compounds, have a high potential for group transfer. They are formed during fatty acid biosynthesis, or by the activation of free fatty acids. Acyl CoA synthetases catalyse formation of the CoA derivatives according to the reaction: $CH_3(CH_2)_nCOO^- + ATP + HS\text{–}CoA \rightleftharpoons CH_3(CH_2)_nCO \sim SCoA + AMP + PP_i$. The reaction involves acyladenylate as an intermediate, which is cleaved by coenzyme A to form acyl-CoA and AMP. Several such enzymes are known, and they are named according to the length of carbon chain that shows optimal activity, e.g. acetyl-CoA synthetase, which converts C_2 and C_3 fatty acids, octanoyl-CoA synthetase which converts fatty acids with chain lengths in the range C_4 to C_{12}, and dodecanoyl-CoA synthetase (C_{10} to C_{18}). Mitochondria also contain an acyl-CoA synthetase that cleaves GTP to GDP + P_i. Acyl CoA derivatives of short chain fatty acids can also be formed in a transfer reaction involving succinyl-CoA, catalysed by thiophorases: Succinyl ~ SCoA + R–COOH ⇌ succinic acid + R–CO ~ SCoA. Activated fatty acids are in equilibrium with acylcarnitine in the organism. They are the starting point for fatty acid degradation.

Activated succinate: see Tricarboxylic acid cycle, Succinate-glycine cycle, Fatty acid degradation.

Activation hormone: see Insect hormones.

Activator protein: see Calmodulin.

Active acetaldehyde: see Thiamin pyrophosphate.

Active acetate: see Acetyl-coenzme A.

Active aldehyde: see Thiamin pyrophosphate.

Active carbon dioxide: see Biotin enzymes.

Active center: that part of an enzyme or other protein which binds the specific substrate and converts it to product or otherwise interacts with it. The A.c. thus consists of the actual catalytic center, which is relatively unspecific, and the substrate-binding site, which is responsible for the specificity of the enzyme. Usually only a few amino acid residues interact directly with the substrate in the A.c.; the rest of the protein molecule serves to hold these few in the proper configuration. The amino acids involved in catalysis may lie at a considerable distance from each other in the absence of substrate; they are brought into play by conformational changes induced by substrate binding (see Cooperativity model, Chymotrypsin, Serine proteases).

Amino acid residues of the A.c. are identified by specific labeling with coenzyme, or by reaction with inhibitors or reagents specific for particular side chains. Some widely used irreversible inhibitors of the catalytic center of serine proteases are tosyllysine chloromethyl ketone (TLCK), which reacts with histidine, and diisopropylfluorophosphate (DFP) and phenylmethane sulfonyl fluoride (PMSF), which forms esters with serine residues.

Active formaldehyde: see Active one-carbon units, Thiamin pyrophosphate.

Active formate: see Active one-carbon units.

Active glucose: see Nucleoside diphosphate sugars.

Active glycolaldehyde: see Thiamin pyrophosphate.

Active methionine: see *S*-Adenosyl-L-methionine.

Active one-carbon units, *C_1 units:* molecular groupings containing a single carbon atom, which are activated by binding to tetrahydrofolic acid, or less commonly to thiamin pyrophosphate. C_1 units are bound to N5 and/or N10 of tetrahydrofolic acid (THF) (Fig. 1).

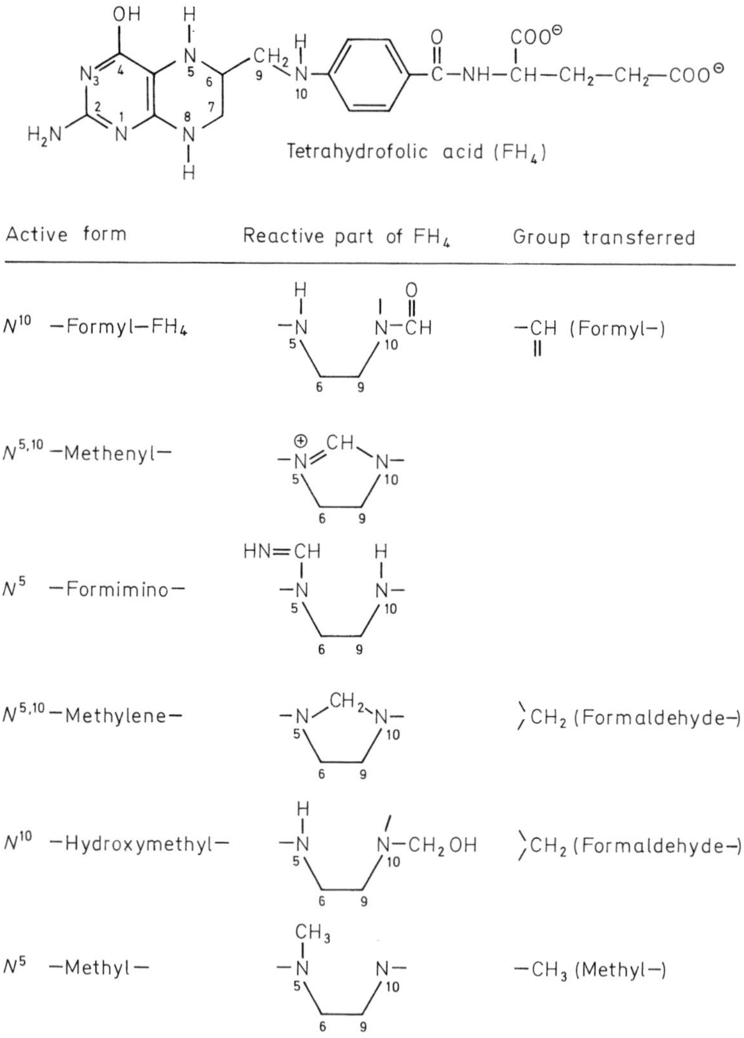

Tetrahydrofolic acid (FH_4)

Active form	Reactive part of FH_4	Group transferred
N^{10}–Formyl–FH_4		–CH (Formyl–) ‖ O
$N^{5,10}$–Methenyl–		
N^5–Formimino–		
$N^{5,10}$–Methylene–		>CH_2 (Formaldehyde–)
N^{10}–Hydroxymethyl–		>CH_2 (Formaldehyde–)
N^5–Methyl–		–CH_3 (Methyl–)

Fig. 1. *Active one-carbon units.* Structure of tetrahydrofolic acid and activated one-carbon units.

The main source of C_1 units is the hydroxymethyl group of serine, which is transferred to THF by serine hydroxymethyltransferase (EC 2.1.2.1), forming N^{10}-hydroxymethyl-THF (activated formaldehyde). Production of C_1 units during histidine catabolism and in the anaerobic degradation of purines is of particular importance. C_1 units are incorporated during purine biosynthesis, and they provide the 5-methyl group of thymine. C_1 units are interconverted while attached to THF (Fig. 2). For other metabolic sources and uses of C_1 units, see legend to Fig. 2.

Active pyruvate: see Thiamin pyrophosphate.

Active sulfate: see Phosphoadenosine phosphosulfate.

Active transport: a process in which solute molecules or ions move across a biomembrane against a concentration gradient. Since thermodynamic work is involved, A.t. must be coupled to an exergonic reaction. In primary A.t. the coupling is direct, e.g. transport of Na^+ and K^+ ions across a cell membrane by the Na^+/K^+-ATPase system requires the simultaneous hydrolysis of ATP. Secondary A.t. utilizes the energy of an electrochemical gradient of a second solute to transport the first. In the secondary A.t. known as *cotransport,* transport of one solute drives the other, e.g. the Na^+-dependent transport of certain sugars and amino acids in animal cells: the intracellular concentration of Na^+ is maintained at a level far below the intercellular concentration by the Na^+/K^+ pump. A specific transport protein (carrier) binds both glucose and Na^+ outside the cell and releases them on the inside; this process is energetically favorable because the Na^+ is moving from a region of higher concentration to one of lower concentration. In

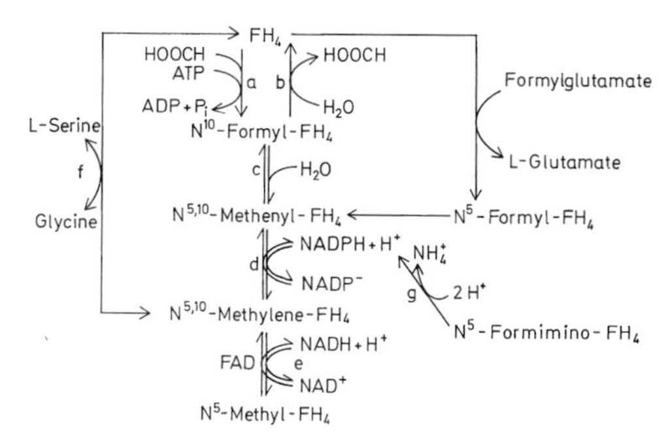

Fig. 2. *Interconversion of active one-carbon units*
a. Formyl-FH_4 synthetase (EC 6.3.4.3)
b. Formyl-FH_4 deformylase (EC 3.5.1.10)
c. Methenyl-FH_4 cyclohydrolase (EC 3.5.4.9)
d. Methylene-FH_4 dehydrogenase ($NADP^+$) (EC 1.5.1.5)
e. 5,10-Methylene-FH_4 reductase (FADH2) (EC 1.7.99.5)
f. Serine hydroxymethyltransferase (EC 2.1.2.1)
g. Formimino-FH_4 cyclodeaminase (EC 4.3.1.4)
Associated systems and reactions:
1. N^5-Formimino-FH_4 is formed from FH_4 and formiminoglycine (from bacterial fermentation of purines) and from FH_4 and formiminoglutamate (see Histidine).
2. In *Clostridium,* reversal of reaction *a* serves to generate ATP.
3. $N^{5,10}$-Methylene-FH_4 acts as a reducing agent as well as a source of one- carbon units in the synthesis of thymidylic acid (see Pyrimidine biosynthesis).
4. The Glycine cleavage system (see) also converts FH_4 to $N^{5,10}$-methylene- FH4.
5. N^5-Methyl-FH_4 serves as a source of methyl groups for conversion of L-homocysteine to L-methionine (5-methyltetrahydrofolate-homocysteine methyltransferase, EC 2.1.1.13) (see L-Methionine).
6. See Purine biosynthesis for action of the enzymes EC 2.1.2.2 and 2.1.2.3.
7. N^5-Methyl-FH_4 is a substrate for methane formation in methanogenic bacteria.
8. See One-carbon cycle for other important reactions
THF and FH_4 are both commonly used abbreviations of tetrahydrofolic acid.

other cases, the membrane potential generated by electron flow along the respiratory chain drives the active transport of sugars or amino acids.

A third type of A.t. is called *group translocation,* because the solute is changed in the course of transport, e.g. the *phosphotransferase* of some bacteria, in which sugars are phosphorylated during transport. An interesting feature of this system is that phospho*enol*pyruvate rather than ATP is the phosphate donor.

A.t. processes are highly specific, and they are saturable. This implies that transport is mediated by enzyme-like proteins or *carriers.* The term *carrier* is also used in the description of Facilitated diffusion (see).

Bacterial transport systems, called *permeases,* have been extensively studied by genetic and other means. The protein products of several permease genes have been isolated, e.g. the product of the lactose permease (*y*) gene.

Activin: see Inhibin.

Actomyosin: see Muscle proteins.

Acylcarnitine: see Carnitine.

Acyl carrier protein, *ACP:* a small, acidic, heat-stable globular protein which is part of the fatty acid synthesizing complex in *E. coli* and other bacteria, yeast and plants. It carries the fatty acid chain during biosynthesis of the latter. The ACP from *E. coli* contains 77 amino acids (M_r 8,847), and the primary structure is known. Sulfur-containing amino acids are absent, but a molecule of phosphopantetheine (which possesses -SH) is linked to the protein via a phosphate ester to the hydroxyl of serine 36. All acyl residues formed during fatty acid biosynthesis are bound as thioesters to the SH of this prosthetic group. The M_r of isolated ACP lie between 8,600 *(Clostridium butyricum)* and 16,000 (yeast).

Synthetic apo-ACP protein, a polypeptide representing amino acids 2–74 of the *E. coli* protein, functions as substrate for the holo-acyl-carrier protein synthase (EC 2.7.8.7); the product has the same biological activity as natural holo-ACP.

Acylglycerols, *glycerides:* esters of fatty acids with glycerol. Mono- and diacylglycerols usually occur only as metabolic intermediates. Mixtures of triacylglycerols are neutral fats (see Fats). The IUPAC-IUB Commission on Biochemical Nomenclature discourages the use of the terms *mono-, di-* and *triglycerides* in favor of *mono-, di- and triacylglycerols,* respectively.

In the intestine, triacylglycerols are hydrolysed to monoacylglycerols, which are re-esterified to triacylglycerols in the intestinal mucosa (Fig. 1). In other tissues (notably liver and adipose tissue), triacylglycerols are synthesized from glycerol 3-phosphate and fatty acyl-CoA (Fig. 2). In adipose tissue, the rates of breakdown and synthesis of triacylglycerols are under hormonal control, resulting in fat storage and/or release of fatty acids, depending on nutritional state, exercise and stress (Fig. 3). Plasma Lipoproteins (see) are responsible for transport and deposition of triacylglycerols in the body.

Role of triacylglycerol synthesis and degradation in adipose tissue (see Fig. 3): Glycerol formed during triacylglycerol degradation cannot be reutilized because adipose tissue does not contain glycerol kinase (EC 2.7.1.30). Synthesis of triacylglycerols therefore depends on a continuous supply of glucose for the production of glycerol 3-phosphate. Conversion of triacylglycerol to diacylglycerol is relatively slow, and is the the rate limiting-step of triacylglycerol degradation. The lipase catalysing this reaction is activated by phosphorylation, a process indirectly under hormonal control via a cAMP-dependent protein kinase. In addition, glucocorticoids stimulate the lipase independently of cAMP, and this stimulation is prevented by insulin.

In a state of caloric excess, high insulin levels promote glucose uptake and prevent activation of mobilizing lipase. Glycerol 3-phosphate is therefore abundant, the rate of triacylglycerol synthesis is high, ex-

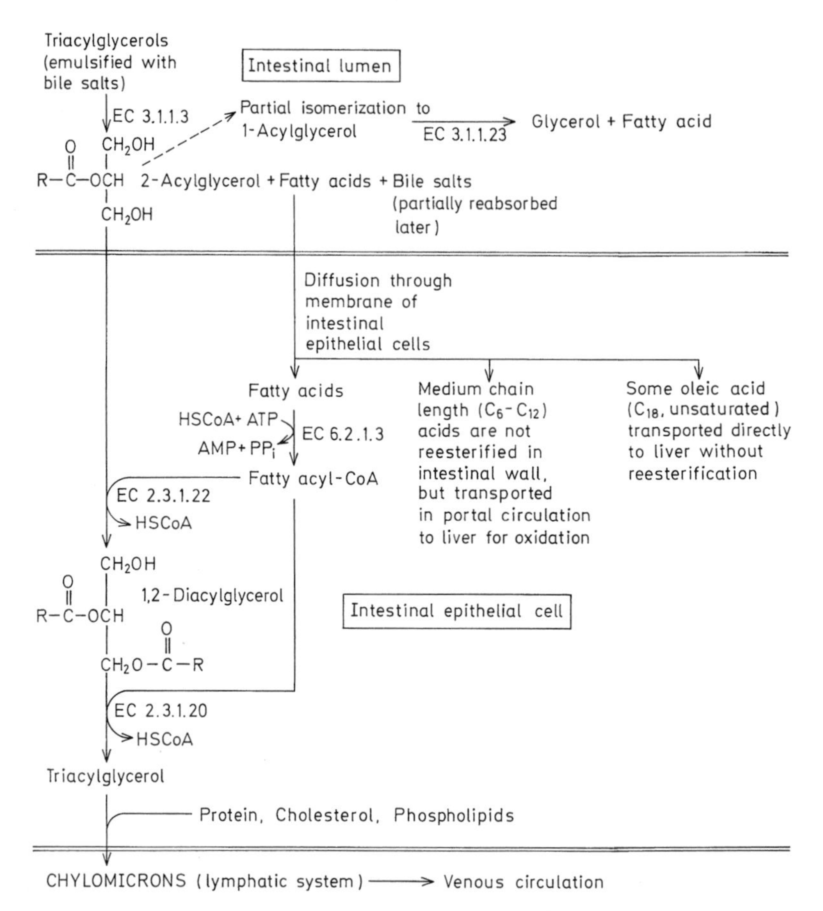

Fig. 1. *Acylglycerols.* Degradation and resynthesis of triacylglycerols in the duodenum. EC 2.3.1.20: Diacylglycerol acyltransferase. EC 2.3.1.22: Acylglycerol palmitoyltransferase. EC 3.1.1.3: Pancreatic triacylglycerol lipase. EC 3.1.1.23: Acylglycerol lipase. EC 6.2.1.3: Long-chain-fatty-acid-CoA ligase.

port of free fatty acids is minimal, and the quantity of stored triacylglycerols shows a net increase. High insulin levels cause an increase in the activity of lipoprotein lipase; this is a tissue-specific effect, and the lipoprotein lipase activity of other tissues does not increase in response to insulin.

In short-term starvation, low insulin levels lead to a decreased glucose uptake, with a consequent decrease in the supply of glycerol 3-phosphate. Re-esterification is retarded, and fatty acids are exported. Activation of mobilizing lipase is not significant in short-term starvation.

Long-term starvation, exercise or stress leads to increased activity of mobilizing lipase. In stress and exercise, the catecholamines (adrenalin and noradrenalin) are mainly responsible for the observed increase of mobilizing lipase activity, by their stimulation of adenylate kinase. Insulin reverses this activation by catecholamines. In long-term starvation, lack of insulin and excess growth hormone cause increased cAMP synthesis, leading to stimulation of mobilizing lipase. Re-esterification is also retarded (lack of insulin prevents glucose uptake), and free fatty acids are exported.

Acylmercaptan: see Thioester.

Adair-Koshland-Nemethy-Filmer model: see Cooperativity model.

Adaptive enzyme: obsolete term for inducible enzyme. See Enzyme induction.

Addison's disease: see Adrenal corticosteroids.

Adenine: 6-aminopurine (Fig.), one of the common nucleic acid bases. An A. residue is also present in the structure of the adenosine phosphates and other physiologically active substances, including Ni-

Amino form   Imino form

Tautomeric forms of adenine

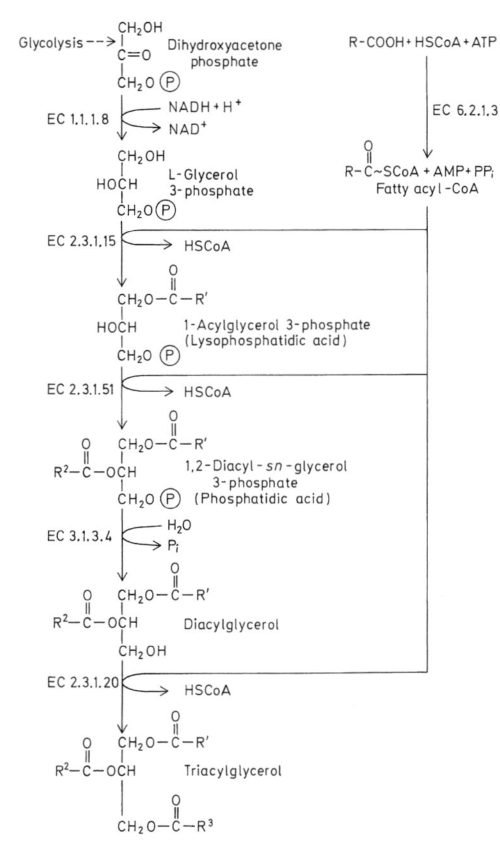

Fig. 2. *Acylglycerols.* Biosynthesis of triacylglycerols in liver and adipose tissue cells. EC 1.1.1.8: Glycerol-3-phosphate dehydrogenase (NAD^+). EC 2.3.1.15: Glycerol-3-phosphate acyltransferase. EC 2.3.1.20: Diacylglycerol acyltransferase. EC 2.3.1.51: 1-Acylglycerol-3-phosphate acyltransferase. EC 3.1.3.4: Phosphatidate phosphatase. EC 6.2.1.3: Long-chain-fatty acid-CoA ligase.

cotinamide-adenine-dinucleotide (see), Flavin-adenine-dinucleotide (see) and various Nucleoside antibiotics (see). A. occurs in free form in various plants, especially yeasts. It is biosynthesized de novo via adenosine monophosphate, or is formed by degradation of nucleic acids. Adenine deaminase (EC 3.5.4.2) removes the 6-amino group to yield hypoxanthine.

Adenine arabinoside: see Arabinosides.

Adenine deaminase, *adenase* (EC 3.5.4.2): see Purine degradation.

Adenine xyloside: see Xylosylnucleosides.

Adenosine: 9-β-D-ribofuranosyladenine. Phosphorylated derivatives of adenosine are metabolically important. See Adenosine phosphates, Nucleosides.

Adenosine deaminase (EC 3.5.4.4): an enzyme, M_r 217,000 (2 subunits M_r 103,000 each) which deaminates adenosine to inosine. It is present in taka-diastase preparations from *Aspergillus oryzae,* and is sometimes confused with Taka amylase (see).

Adenosine 3′-phosphate 5′-phosphosulfate: see Phosphoadenosine-phosphosulfate.

Adenosine phosphates, *adenine ribonucleotides:* components of nucleic acids, and the major form in which chemical free energy is stored and transferred. They also serve as metabolic regulators, e.g. in glycolysis and the tricarboxylic acid cycle. The phosphate ester is carried on C5 of the ribose.

1. *Adenosine 5′*-monophosphate (AMP) is synthesized de novo from inosinic acid (see Purine biosynthesis) and also arises in reactions in which pyrophosphate and AMP are formed from adenosine triphosphate (e.g. in the synthesis of aminoacyl-tRNA).

2. *Adenosine 5′*-diphosphate (ADP) is formed either by adding a second phosphate to AMP (see Adenylate kinase), or by removal of a phosphate from ATP; the latter conversion may be catalysed by adenosine triphosphatases (EC 3.6.1.3), or by kinases which transfer the phosphate to another organic molecule. The energy stored in the anhydride bond of ADP is made available by the reaction $2ADP \rightleftharpoons ATP + AMP$, catalysed by adenylate kinase. ADP is the phosphate acceptor (i.e. it is converted into ATP) in Substrate-level phosphorylation (see), Oxidative phosphorylation (see) and Photophosphorylation (see).

3. *Adenosine 5′*-triphosphate (ATP) is the "energy currency" of every living cell (see Energy-rich phosphates).

Biosynthesis of ATP. ATP is the immediate product of all cellular processes leading to the chemical storage of energy. It is biosynthesized by phosphorylation of ADP in the course of Substrate phosphorylation (see), Oxidative phosphorylation (see) and noncyclic Photophosphorylation (see) in plants. Energy in the form of a third phosphate may also be transferred to ADP from other high-energy phosphates, such as creatine phosphate (see Creatine) or other nucleoside triphosphates, or in the adenylate kinase reaction.

Table 1. *Standard free energy of hydrolysis of ATP in kJ/mol (kcal/mol)*

Removal of orthophosphate:	
$ATP \rightarrow ADP + P_i$	29.4 (7.0)
Removal of pyrophosphate:	
$ATP \rightarrow AMP + PP_i$	36.12 (8.6)
$PP_i \rightarrow P_i + P_i$	28.14 (6.7)

Cleavage of ATP. ATP has a high potential for group transfer (Fig. 1, Table 1):

a) Transfer of orthophosphate to alcoholic hydroxyl groups, acid groups or amide groups, with release of ADP. The enzymes that catalyse these reactions are called kinases; in some cases (e.g. creatine kinase) they also catalyse the synthesis of ATP from ADP.

b) Transfer of the pyrophosphate residue and release of AMP, e.g. in the synthesis of 5-phosphoribosyl-1-pyrophosphate from ribose 5-phosphate in the course of purine biosynthesis.

c) Transfer of the AMP residue and release of pyrophosphate. The receiving group thereby acquires

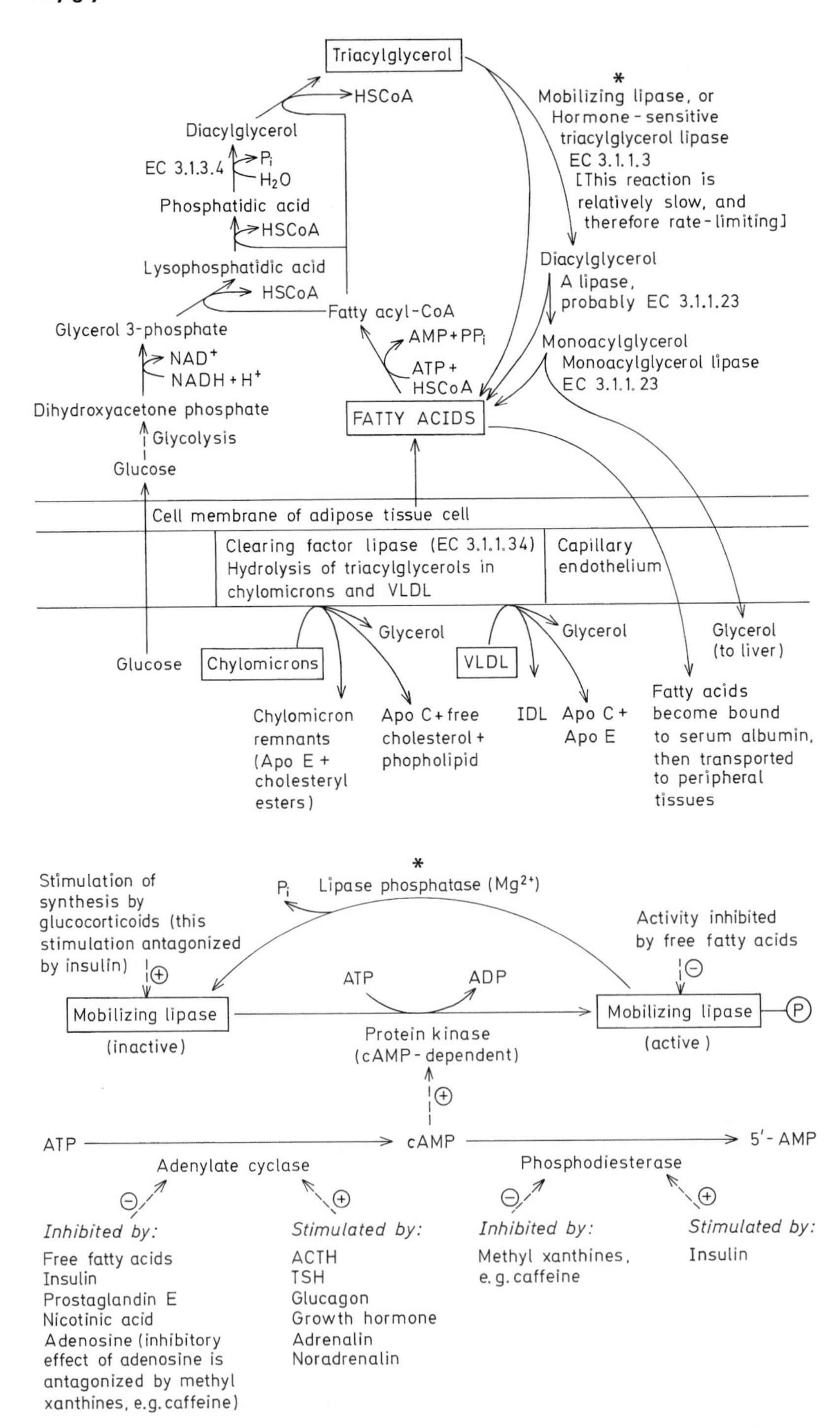

Fig. 3. *Acylglycerols.* Triacylglycerol synthesis and degradation in adipose tissue.

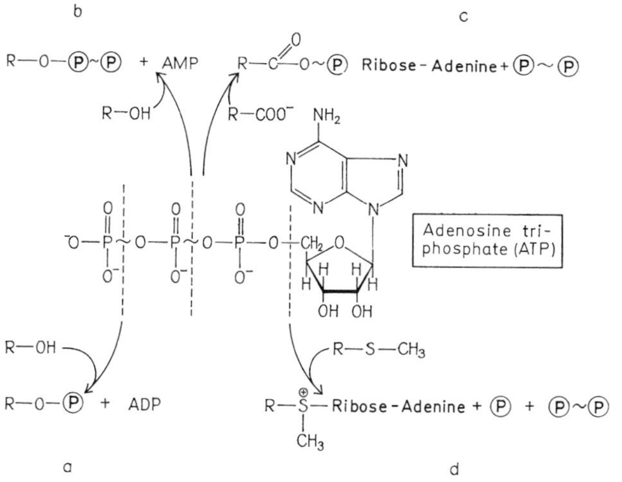

Fig. 1. *Adenosine phosphates.* Different cleavage sites of adenosine 5′-triphosphate (ATP).

a higher group-transfer potential, e.g. in the activation of fatty acids and amino acids. The released pyrophosphate is hydrolysed by inorganic pyrophosphatase (EC 3.6.1.1), thereby making the transfer reaction essentially irreversible.

d) Transfer of the adenosyl residue and release of both orthophosphate and pyrophosphate, e.g. in the synthesis of *S*-adenosyl-L-methionine.

Uses of ATP. The chemical energy stored in ATP is used e.g. in the synthesis of macromolecules from monomeric precursors, and in the activation of various compounds. Often an endergonic reaction is driven by enzymatic coupling to the hydrolysis of ATP. Many catabolic pathways, including glycolysis, require an investment of ATP which is later resynthesized. Essentially all anabolic pathways require ATP, either directly or indirectly.

ATP provides energy for contraction of muscle and the motion of cilia and flagella. In some organisms it provides energy for Bioluminescence (see), which has been exploited in a very sensitive assay for ATP. Electric fish generate current by hydrolysing ATP. Active transport of many substances across membranes depends on a source of ATP. ATP is also released together with acetylcholine at synapses in the peripheral nervous systems of vertebrates. It may be a modulator of nervous transmission, as it inhibits the release of acetylcholine [E.M. Silinsky & B.L. Ginsborg *Nature* **305** (1983) 327–328].

Other nucleoside triphosphates, which are energetically equivalent to ATP, are important in some metabolic reactions: cytidine triphosphate in phospholipid biosynthesis, guanosine triphosphate in protein biosynthesis and oxidative decarboxylation of 2-oxoacids (see Tricarboxylic acid cycle), inosine triphosphate in certain carboxylations, uridine triphosphate in polysaccharide biosythesis.

In living organisms, the adenosine phosphates are in equilibrium and are regarded collectively as the adenylic acid system. The physiological concentrations of ADP and ATP are around 10^{-3} mol/l. The ratio of the forms is called the *energy charge* (EC), and is given by the equation: $EC = ([ATP] + 0.5[ADP])/([ATP] + [ADP] + [AMP])$. The square brackets indicate molar concentrations. If the total complement of A. is in the form of ATP, the energy charge is 1; otherwise it is smaller than 1.

4. *Cyclic adenosine* 3′,5′-monophosphate (3′,5′-AMP, cyclo-AMP, cAMP) is generated from ATP (Fig. 2) by adenylate cyclase (EC 4.6.1.1), and is converted into AMP by 3′:5′-cyclic-nucleotide phosphodiesterase (EC 3.1.4.17), which is specific for cyclic nucleotides. The activities of these two enzymes determine the intracellular level of cAMP. Physiological amounts of various substances, e.g. pyridoxal phosphate in *E. coli,* can reduce the activity of adenylate cyclase. Mammalian phosphodiesterase is inhibited by nucleoside triphosphates, pyrophosphate, citrate and methylated xanthines (especially theophylline) and stimulated by nicotinic acid. In many cells, adenylate cyclase is located just inside the plasma membrane. Receptors for hormones and other chemical signals are located on the outside of the plasma membrane of their target cells. In many systems, binding of a hormone or other activator to its receptor leads to activation of the adenylate cyclase and an increase in the cAMP concentration. The cAMP often serves as an effector molecule which increases the activity of a protein kinase or other enzyme, which in turn regulates some other cellular process by Covalent modification of enzymes (see). Thus, cAMP serves as a "second messenger" for a variety of hormones. In addition, it affects the production and release of hormones, e.g. acetylcholine, glucagon, insulin, melanotropin, parathyrin, vasopressin and corticotropin. It also affects equilibria among various metabolic pathways, e.g. glycogen breakdown and synthesis. In many cases, the physiological effects of cAMP are only seen in the presence of calcium ions. Exogenous "artificial" control of the intracellular cAMP concentration is medically important. Substances which raise this level have been successfully used therapeutically, e.g. psoriasis is treated with papaverine which inhibits cyclic nucleotide phosphodiesterase, and with the tis-

Table 2. *Some examples of the occurrence and effects of cyclic adenosine 3′,5′-monophosphate (after Hardeland)*

Organisms	Effect
1) Protozoa	
Paramecium	Activation of protein kinase
2) Bacteria	
Escherichia coli	Release of glucose inhibition of enzyme induction (see Catabolite repression). Initiation of mRNA synthesis mediated by a specific cAMP receptor protein (catabolite gene activator protein). Inhibition of the degradation of mRNA bound to ribosomes. Stimulation of the synthesis of many enzymes
Serratia marcescens *Salmonella typhimurium* *Proteus inconstans* *Aerobacter aerogenes* *Brevibacterium liquefaciens*	Release of catabolite repression and stimulation of the synthesis of β-galactosidase. (*Brevibacterium liquefaciens* excretes cAMP into the medium).
Photobacterium fischerei	Release of catabolite repression and production of bioluminescence
3) Fungi	
Slime molds	
Dictyostelium discoideum	Extracellular signal transmitter. Cell aggregation in response to chemotactic stimuli.
Polysphondylium pallidium	Does not respond to cAMP, has a different acrasin.
Yeasts *Saccharomyces cereviseae*	Affects the oscillation and redox equilibrium of glycolysis. Affects sporulation.
4) Invertebrates	
e.g. annelids *(Golgingia, Nereis)*, starfish.	Activation of protein kinases.
Liver fluke *(Fasciola)*, blowfly *(Calliphora)*.	Transmission of the effect of serotonin.
5) Vertebrates	
Frog, toad, turkey, pigeon, rat, mouse, guinea pig, rabbit, rabbit, human.	Second messenger in the transmission of hormone stimuli.
6) Higher plants	Enzyme stimulation during germination.
Barley (endosperm)	Stimulation of amylase synthesis.
Peas, lettuce	Cell extension growth, especially in dwarf varieties of *Pisum sativum*.
Weeds	Effects on germination.

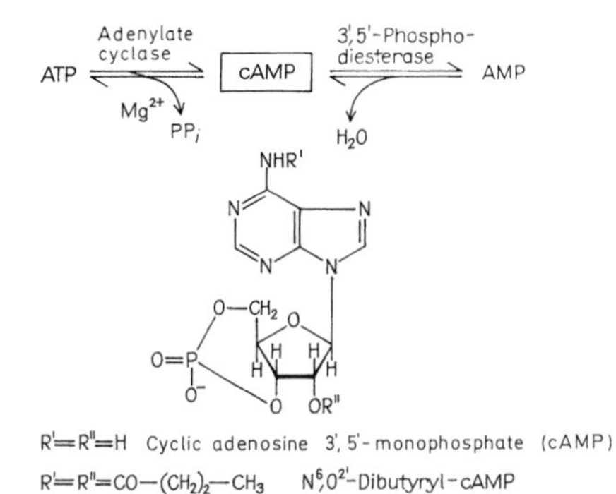

Fig. 2. *Adenosine phosphates.* Synthesis of cyclic adenosine 3′,5′- monophosphate (cAMP), and the structure of cyclic N^6,$O^{2'}$-dibutyryladenosine 3′,5′-monophosphate.

sue hormone, dopamine, which stimulates cAMP formation in the epidermis. cAMP also inhibits growth of certain tumors.

Because it is very polar, cAMP hardly penetrates the cell membrane. Synthetic derivatives of cAMP with organic acid substituents are more lipophilic, and therefore display greater permeation. Most commonly used of these is N^6,$O^{2'}$-dibutyryladenosine 3′,5′-monophosphate (DBcAMP). A number of other 3′,5′-mononucleotides with special functions (e.g. 3′,5′-GMP) occur naturally.

Adenosine triphosphatase, ***ATPase*** (EC 3.6.1.3): an enzyme catalysing the cleavage of ATP to ADP + P_i. Physiologically, this is generally coupled with another process, which is endergonic and driven by the free energy of ATP cleavage. Two of the best studied examples are the actomyosin complex (see Muscle proteins) in which ATP cleavage provides energy for contraction, and the Na^+/K^+ pump in cell membranes in which ATP cleavage provides the energy for ion transport against a concentration gradient (see Transport). In intact mitochondria, the ATPase of the inner mitochondrial membrane functions as an ATP synthase; see Chemiosmotic hypothesis, F-type ATPases, Ion-motive ATPases, P-type ATPases, V-type ATPases.

Adenosine triphosphate: see Adenosine phosphates.

***S*-Adenosyl-L-homocysteine:** see *S*-Adenosyl-L-methionine.

S*-Adenosyl-L-methionine, *S-(5′-deoxyadenosine-5′)-methionine, active methionine, active methyl, SAM: a reactive sulfonium compound (Fig.) (M_r of free cation 398.4), and the most important methylating agent in cellular metabolism (see Transmethylation).

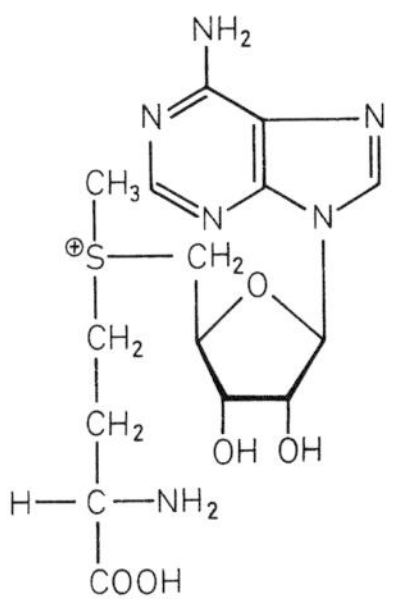

S-Adenosyl-L-methionine

The natural form is the L-(+)-isomer. Due to the asymmetry of the sulfonium group, there are 4 stereoisomers. SAM is unstable at room temperature, both as a solid and in aqueous solution. It is synthesized in the cell by transfer of the adenosine residue of ATP to L-methionine: Met + ATP $\rightarrow$ SAM + PP_i + P_i. When SAM donates a methyl group it is converted into *S*-adenosyl-L-homocysteine, which is then cleaved to adenosine and L-homocysteine. The latter is remethylated to L-Methionine (see). [F. Takusagawa et al. 'Crystal Structure of *S*-Adenosylmethionine Synthetase' *J. Biol. Chem.* **271** (1996) 136–147]

Adenylate cyclase (EC 4.6.1.1): see Adenosine phosphates.

Adenylate kinase, *myokinase* (EC 2.7.4.3): A trimeric enzyme (M_r 68,000, subunit M_r 23,000) found in the mitochondria of muscles and other tissues. It is resistant to heat and acid. A. k. catalyses the reaction 2ADP $\rightleftharpoons$ ATP + AMP. At equilibrium, the concentrations of the 3 adenosine phosphates are nearly equal. In many energy-requiring reactions coupled with ATP, the ATP is converted into pyrophosphate and AMP (see Adenosine phosphates). A.k is important because it catalyses the first stage (AMP to ADP) of the reconversion of this AMP into ATP.

Adenylic acid: see Adenosine phosphates.

Adenylosuccinate, *N-succinyladenylate, sAMP, 5-aminoimidazole-4-N-succinocarboxamide ribonucleotide:* an intermediate in purine biosynthesis, M_r 463.3.

Adenylylsulfate reductases: enzymes of sulfur metabolism which reduce either phosphoadenylylsulfate (APS reductase) or adenylylsulfate. Adenylylsulfate reductase (EC 1.8.99.2) is identical with one component of the sulfate reductase in sulfate assimilation, since adenylylsulfate is the donor of the sulfate group. Properties of some of these reductases are shown in the Table. In every case, the reductase is a complex of 3 components, an adenylylsulfate transferase (see Sulfate assimilation, Fig. 1), a low-molecular-mass carrier, and the actual adenylylsulfate reductase. Phosphoadenylylsulfate reductase from *Saccharomyces cerevisiae* requires NADPH, and has been partly purified.

Adermine: vitamin B_6. See Vitamins.

ADH: acronym of antidiuretic hormone. See Vasopressin.

Adipokinetic hormone: see Lipotropin.

Adiuretin: see Vasopressin.

Adjuvant: a mixture of oils, emulsifiers, killed bacteria and other components, which serves to intensify unspecifically the immune response. The A., which is not (supposed to be) itself antigenic, is injected several times intramuscularly or subcutaneously into an animal to produce the maximal yield of antibodies. Freund's incomplete A., an emulsion of paraffin oils, protects the antigen from rapid degradation. Freund's complete A. contains in addition killed mycobacteria or tuberculosis bacteria. Both are used widely in experimental immunology. In vaccines, the A. is usually aluminium hydroxide or calcium phosphate gel.

Ado: abb. for Adenosine.

ADP: acronym of adenosine 5′-diphosphate. See Adenosine phosphates.

Properties of adenylylsulfate reductases from various organisms

Organism	pH optimum	M_r	Comments
Desulfovibrio[1]	7.4	220,000	Contains 1 molecule FAD and 6–8 atoms nonheme iron
Thiobacillus thioparus[1]	7.4	170,000	Contains 1 molecule FAD and 8–10 atoms nonheme iron
Thiocapsa roseopersicina[1]	8.0	180,000	Contains 1 molecule FAD, 4 atoms nonheme iron and 2 atoms heme iron
Chlorella pyrenoides[2]		330,000	Partly purified enzyme

[1] With $Fe(CN)_6^{3-}$. [2] With a thiol as electron donor; the enzyme from *Chlorella* is active with phosphoadenylylsulfate only in the presence of 3′-nucleotidase.

ADP-ribosylation of proteins: attachment of monomeric or polymeric ADP-ribosyl groups to a protein by transfer from NAD^+:

Adenine Nicotinamide$^+$
| OH O^- |
(ribose-O–P–O–P–O-ribose)$_n$ + Protein ⟶
‖ ‖
O O

Adenine
| OH O^-
Protein-(ribose-O–P–O–P–O-ribose)$_n$
‖ ‖ + Nicotinamide + H^+
O O

where n can vary from 1 to 50. Poly ADP-ribosyl groups represent a novel homopolymer of repeating ADP-ribose groups linked 1′ → 2′ between respective ribose moieties:

↓2′ OH OH
Adenine-ribose-O–P–O–P–O-ribose
‖ ‖ |1′
O O ↓2′ OH OH
Adenine-ribose-O–P–O–P–O-ribose
‖ ‖ |1′
O O ↓2′

The free energy of hydrolysis of the β-*N*-glycosidic linkage of NAD^+ is –34.4 kJ (–8.2 kcal)/mol at pH 7 and 25 °C; it is therefore a high energy bond and NAD^+ can act as an ADP-ribosyl transferring agent. Transfer of one ADP-ribosyl group (n = 1 in the above equation) is catalysed by ADP ribosyl transferase. Formation and concomitant transfer of poly (ADP-ribose) to an acceptor is catalysed by poly (ADP-ribose) synthetase (n is greater than 1 in the above equation).

The A domain of diphtheria toxin, (produced by strains of *Corynebacterium diphtheriae* carrying β phage) inhibits eukaryotic protein synthesis by catalysing ADP-ribosylation of elongation factor 2. *Pseudomonas* toxin catalyses a similar reaction. T4 phage catalyses the monomeric ADP-ribosylation of RNA polymerase and other proteins in *E. coli.* Cholera toxin and related toxins of enterobacter catalyse ADP-ribosylation of an arginine residue in the α-subunit of the GTP-binding protein G_S; the GTPase activity of the G_S is thereby blocked, so that the G_S is permanently active in the stimulation of adenylate cyclase; this results in turn in an increase of cAMP, leading to disruption of ion flow into and out of the cell. Pertussis toxin (from *Bordatella pertussis,* the causative organism of whooping cough) catalyses ADP-ribosylation of a cysteine residue of G_i (a GTP-binding protein), which then no longer functions as a natural inhibitor of adenylate cyclase. Cholera and pertussis toxins catalyse ADP-ribosylation of most G-proteins.

Poly ADP-ribose groups are found in eukaryotic chromosomal proteins, mitochondrial proteins and histones. The biological function of the ADP-ribosylation of proteins in eukaryotic cells is not known, but the occurrence of poly ADP-ribosyl groups in nuclear proteins, particularly in association with chromatin, suggests a regulatory role in nuclear function. [O. Hayaishi & K. Ueda *Annu. Rev. Biochem.* **46** (1977) 95–116; M. R. Purnell et al. *Biochemical Society Transactions* **8** (1980) 215–227; K. Ueda & O. Hayaishi 'ADP-ribosylation' *Annu. Rev. Biochem.* **54** (1985) 73–100; J. Moss & M. Vaughan 'Structure and Function of ARF Proteins: Activation of Cholera Toxin and Critical Components of Intracellular Vesicular Transport Processes' *J. Biol. Chem.* **270** (1995) 12327–12330; W. Mosgoeller et al. 'Nuclear architecture and ultrastructural distribution of poly(ADP-ribosyl)transferase, a multifunctional enzyme' *Journal of Cell Science* **109** (1996) 409–418]

Adrenal corticosteroids, ***adrenocorticoids, corticosteroids, corticoids, cortins:*** an important group of steroid hormones, formed in the adrenal cortex in response to adrenocorticotropic hormone [ACTH, Corticotropin (see)]. A. c. are structurally related to pregnane (see Steroids); they contain a carboxyl group with a neighboring α,β-unsaturated bond in ring A, a ketol side chain in position 17, and other oxygen functions, particularly in positions 11, 17, 21 and 18.

More than 30 steroids have been found in the adrenal cortex; the following show marked A. c. activity: cortisol, cortisone, cortexolone, 11-dehydrocorticosterone, corticosterone and aldosterone (see separate entries). The most abundant A. c. are cortisol, corticosterone and aldosterone, which are secreted daily into the blood in quantities of 15, 3 and 0.3 mg, respectively.

A.c. production is increased in physical and/or psychological stress. A. c. deficiency, e. g. due to pathological changes in the adrenal glands, results in Addison's disease, a condition characterized by tiredness, emaciation, decrease of blood sugar, and dark pigmentation of those areas of the skin exposed to light. Adrenalectomy of experimental animals leads to rapid death unless exogenous A. c. are given. A. c. control mineral metabolism by causing retention of Na^+, Cl^- and water, with a simultaneous K^+ diuresis (mineralocorticoidal or mineralotropic action). They also regulate carbohydrate metabolism, in particular glycogen synthesis in the liver (glucocorticoidal or glucotropic action). Depending on the predominant activity, A. c. are classified as *mineralocorticoids* (aldosterone, cortexone, cortexolone) or *glucocorticoids* (cortisol, cortisone, corticosterone, 11-dehydrocorticosterone).

Adrenal insufficiency, previously treated with extracts of adrenal cortex, is now treated by the administration of pure A. c. High doses reveal other pharmacological properties, especially anti-inflammatory and antiallergic effects. This discovery led to the development of highly active synthetic derivatives, which are now used to treat rheumatism, asthma, allergies, eczema, etc., e. g. prednisone (see Prednisolone), Dexamethasone (see) and Triamcinolone (see).

Adrenal gland, ***suprarenal gland, glandula suprarenalis:*** a well vascularized, vertebrate endocrine gland, weighing about 15 g in the adult human. There are 2 A. g., one just above each kidney. The A. g. consists of 2 developmentally and functionally distinct parts: the mesodermal adrenal cortex (AC) and the

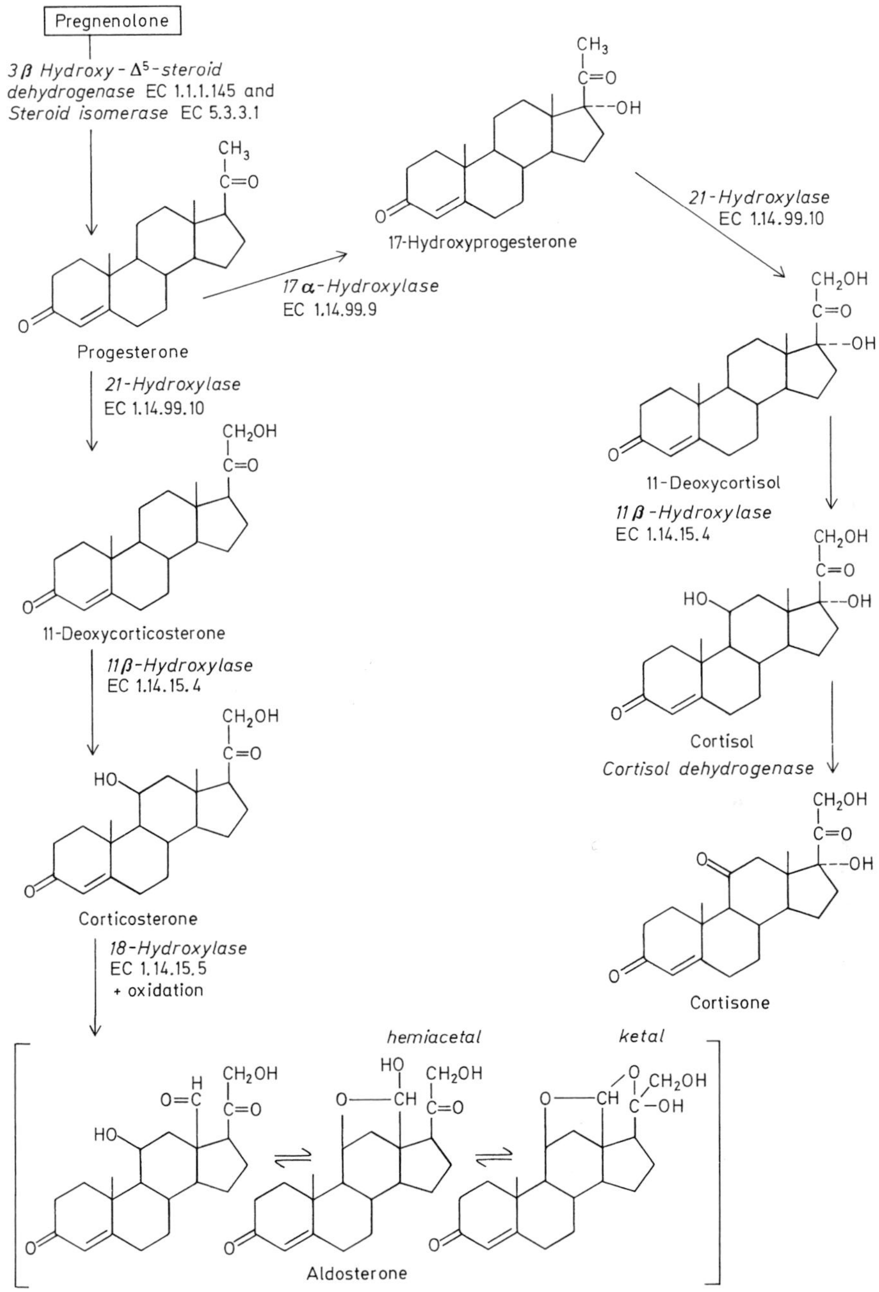

Pregnenolone

3β Hydroxy-Δ⁵-steroid dehydrogenase EC 1.1.1.145 and *Steroid isomerase* EC 5.3.3.1

Progesterone

17α-Hydroxylase EC 1.14.99.9

17-Hydroxyprogesterone

21-Hydroxylase EC 1.14.99.10

11-Deoxycortisol

11β-Hydroxylase EC 1.14.15.4

Cortisol

Cortisol dehydrogenase

Cortisone

21-Hydroxylase EC 1.14.99.10

11-Deoxycorticosterone

11β-Hydroxylase EC 1.14.15.4

Corticosterone

18-Hydroxylase EC 1.14.15.5 + oxidation

hemiacetal *ketal*

Aldosterone

Biosynthesis of adrenal corticosteroids. The primary precursor is acetyl-CoA and biosynthesis proceeds via cholesterol (see Terpenes, Steroids). The adrenal cortex also utilizes cholesterol which it receives as cholesterol esters from extra- adrenal sources (see Lipoproteins). The adrenal cortex is differentiated into 3 concentric layers. The outermost layer (zona glomerulosa) is primarily responsible for synthesis and secretion of aldosterone; it contains the 18-hydroxylase and lacks the 17α-hydroxylase. The intermediate layer (zona fasciculata) and innermost layer (zona reticularis) are responsible for the synthesis of glucocorticoids (mainly cortisol) and adrenal androgens; they possess the 17α-hydroxylase and lack the 18-hydroxylase.

ectodermal adrenal medulla (AM). The AC, which contains 3 histologically distinct zones, produces and exports glucocorticoids and mineralocorticoids (see Adrenal corticosteroids) in response to the pituitary hormones, corticotropin and renin/angiotensin II, respectively. The AC also produces sex steroids (see Androgens). The AM *(paraganglion suprarenale)* is the largest (but not the only) ganglion of the sympathetic nervous system; it produces Adrenalin (see) and Noradrenalin (see) and it is a model example of the close association of the sympathetic nervous system with an endocrine system. The secretory cells are richly innervated by cholinergic, preganglionic, sympathetic nerve fibers. No nerve supply to the AC has been demonstrated.

Adrenal hyperplasia: see Inborn errors of metabolism.

Adrenalin, ***epinephrin, 4-[1-hydroxy-2-(methylamino)ethyl]-1,2-benzenediol:*** a catecholamine hormone, M_r 183.2, synthesized in the adrenal medulla from L-tyrosine (via dopa, dopamine and noradrenalin). It is stored in the chromaffin granules and released into the circulation in response to stimulation by the splanchnic nerve. It is also an adrenergic neurotransmitter, synthesized in, and released from, neurons of the sympathetic nervous system. A. promotes glycogenolysis by activating (via the adenylate cyclase system) liver and muscle phosphorylases (EC 2.4.1.1). It also activates the lipase of adipose tissue. It therefore causes elevated blood concentrations of glucose, lactate and free fatty acids, whose subsequent respiration leads to an increased oxygen consumption by the body.

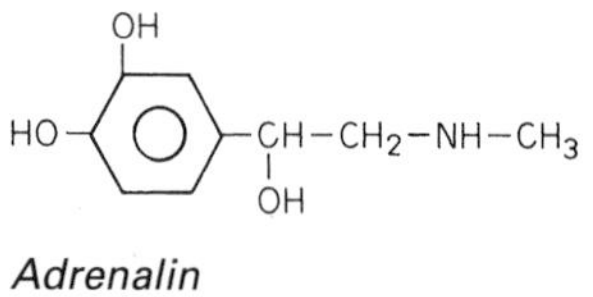

OH
HO–⟨benzene ring⟩–CH–CH$_2$–NH–CH$_3$
OH

Adrenalin

A. is degraded by *O*-methylation and by oxidative deamination by a monoamine oxidase. Its main urinary excretory product is 3-methoxy-4-hydroxymandelic acid (vanillinmandelic acid).

Analogs of A. are used to control blood pressure, counteract depression, stimulate the appetite and relieve asthma.

Adrenocorticotropin, ***adrenocorticotropic hormone:*** see Corticotropin.

Adrenosterone, ***adrost-4-ene-3,11,17-trione:*** a steroid, M_r 300.9, structurally related to androstane. A. is synthesized in the adrenal cortex. In view of its weak androgenic effect, it is considered to be one of the male gonadal hormones (see Androgens).

Adsorption chromatography: see Chromatography.

Aequorin: a photoprotein from the jellyfish *Aequorea.* It consists of an apoprotein (M_r 21,000) linked covalently to a hydrophobic prosthetic group, coelenterazine. Binding of Ca^{2+} to A. causes an irreversible reaction with production of light in the visible range. The fractional rate of A. consumption (and therefore light production) is proportional to the Ca^{2+} concentration in the physiological range. A. can therefore be used as a Ca^{2+} indicator, and has been used for this purpose since the early 1960s. [M. Brini et al. 'Transfected Aequorin in the Measurement of Cytosolic Ca^{2+} Concentrations' *J. Biol. Chem.* **270** (1995) 9896–9903]

Affinity chromatography: see Proteins.

Aflatoxins: mycotoxins produced by *Aspergillus flavus, A. parasiticus* and *A. oryzae,* as well as some *Penicillium* strains. A. have been found in a variety of foodstuffs, especially in damp, tropical environments, which favor growth of the producer microorganisms. Many liver cancers in the tropics are attributable to ingestion of A., and A. poisoning in malnourished children is thought to be the cause of Kwashiorkor (see). Unmodified A. are relatively nontoxic *per se,* but they are converted into potent toxins and carcinogens by monofunctional oxygenases in the liver (Fig.) (see Cytochrome P450). LD_{50} values for ducklings (μg per 50 g body weight) are: A.B_1 18.2; A.B_2 84.8; A.G_1 39.2; A.G_2 172.5; A.M_1 16.6; A.M_2 62. The mechanism of toxicity (Fig.) requires epoxide formation by addition of oxygen at a double bond ($\Delta^{9,10}$ in A.G_1; $\Delta^{8,9}$ in A.B_1 and A.M_2). This double bond is absent from the A_2 compounds, which are probably oxidized to the A_1 type in the body, or are bound to DNA by a different mechanism. [R. Langenbach et al. *Nature* **276** (1978) 277–280]

Afrormosin: 7-hydroxy-6,4′-dimethoxyisoflavone. See Isoflavone.

AGA: acronym of *N*-acetylglutamate.

Agar-agar: a polysaccharide plant mucilage from various red algae. It consists of about 70 % agarose and 30 % agaropectin. Agarose is a linear polymer of alternating D-galactose and 3,6-anhydrogalactose. Agaropectin consists of β-1,3 glycosidically linked D-galactose units. Position 6 of some of the galactose units is esterified with sulfate. A. is obtained by hot water extraction of bleached algae, which may contain up to 40 % A. It is used as a gelling agent in the pharmaceutical and food industries, and in the preparation of solid media for microorganisms and tissue cultures.

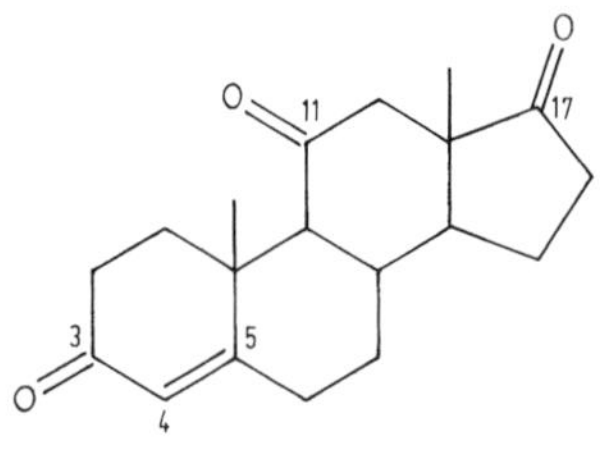

O
O 11 17
3 5
O 4

Adrenosterone

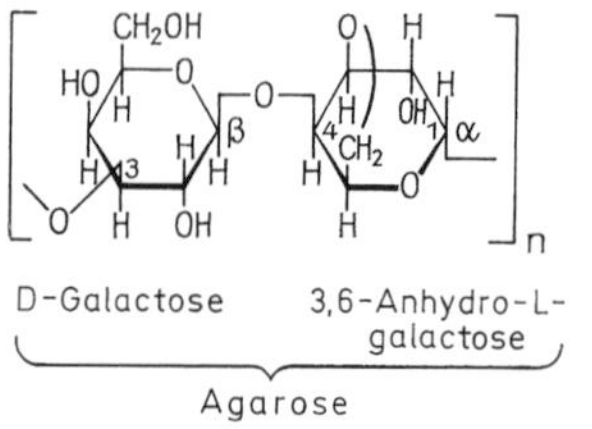

CH$_2$OH O H
HO O H
H O β OH α
H H 4 CH$_2$ 1
H 3 H H O
O H OH H n

D-Galactose 3,6-Anhydro-L-galactose

Agarose

Agarose

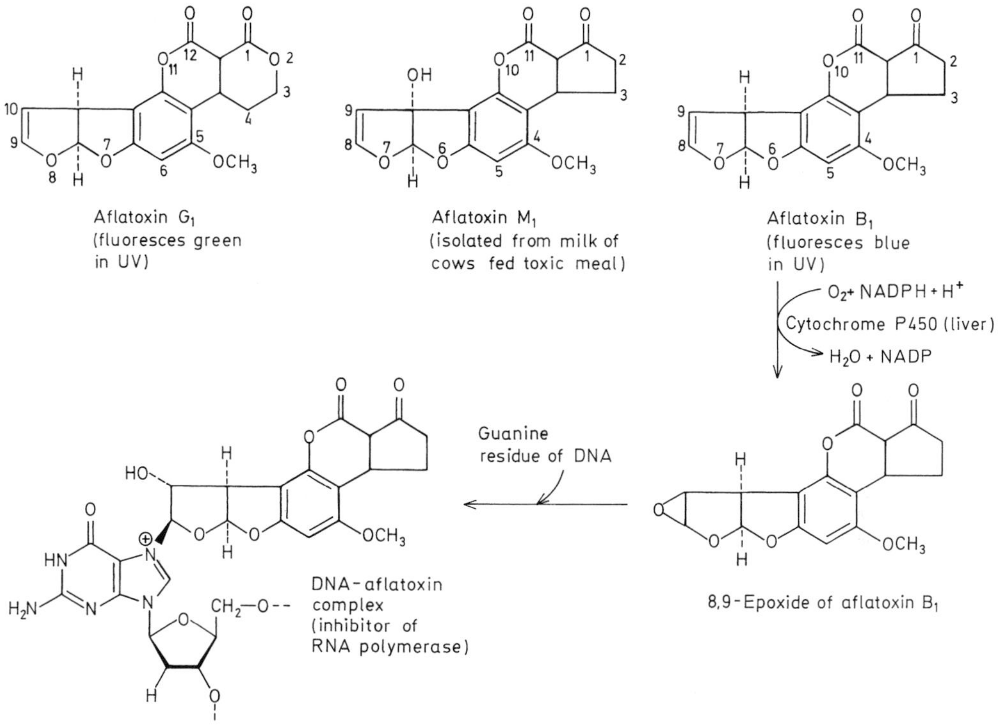

Aflatoxin G_1 (fluoresces green in UV)

Aflatoxin M_1 (isolated from milk of cows fed toxic meal)

Aflatoxin B_1 (fluoresces blue in UV)

O_2 + NADPH + H^+ / Cytochrome P450 (liver) / H_2O + NADP

8,9-Epoxide of aflatoxin B_1

Guanine residue of DNA

DNA-aflatoxin complex (inhibitor of RNA polymerase)

Aflatoxins and their conversion to carcinogenic and toxic derivatives. Aflatoxins B_2, M_2 and G_2 do not possess a double bond at position 9,8 (B_2, M_2) or 9,10 (G_2).

Agathisflavone: see Biflavonoids.

Agglutination: the clumping that occurs when antibodies bind to particles such as bacteria, viruses or erythrocytes. The antibodies must be at least bivalent, in order to bind the antigen-carrying particles together. A. (lower detection limit 0.01 μg/ml serum) is much more sensitive than precipitation (lower detection limit ~ 10 μg/ml serum) because the antigen-antibody reaction occurs on the surface of larger particles.

Passive hemagglutination has a lower detection limit of 3–6 ng antibodies per ml serum. In this technique, soluble antigens are bound to the surface of erythrocytes, which are agglutinated when the antigen-antibody reaction occurs.

Agglutinins: see Lectins.

Aglycon, ***aglycone, genin:*** the noncarbohydrate part of a glycoside. A. are released by hydrolysis (e.g. by acid or enzymes) of the *C*-, *N*- or *S*-glycosidic linkage. See Glycosides, Glucosinolate.

Agmatine, ***4-(aminobutyl)guanidine, 1-amino-4-guanido-butane:*** $H_2N\text{-}C(=NH)\text{-}NH\text{-}(CH_2)_4\text{-}NH_2$, M_r 130.19, a guanidine derivative formed by the amidination of putrescine, or by the decarboxylation of L-arginine. A. has been isolated, e.g. from pollen of *Ambrosia artemisifolia (Compositae),* ergot, sponges, herring sperm and octopus muscle. It is an intermediate in the biosynthesis of Arcain (see).

Agnosterol, ***5α-lanosta-7,9(11),24-trien-3β-ol:*** a tetracyclic triterpene alcohol, M_r 424.7, structurally related to 5α-lanostane (see Lanosterol). A. is a zoosterol (see Sterols) present in the sebaceous oil of sheep's wool.

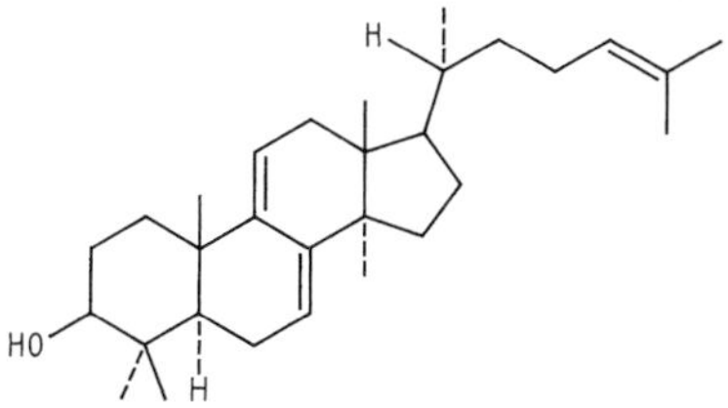

Agnosterol

α_1AGp: see Orosomucoid.

AICAR: acronym of 5(4)-aminoimidazole-4(5)-carboxamide ribotide. See Purine biosynthesis.

AIR: acronym of 5-aminoimidazole ribotide. See Purine biosynthesis.

Ajmaline: a Rauwolfia alkaloid. A. is used clinically to normalize heart rhythm. In high doses it has the tranquilizing effect of Rauwolfia alkaloids.

Alanine, ***aminopropionic acid,*** M_r 89.1.

1. ***L-α-alanine, 2-aminopropionic acid, Ala:*** $CH_3\text{-}CH(NH_2)\text{-}COOH$, a glucogenic, proteogenic amino acid. Ala is one of the main components of silk fibroin. Free Ala, together with glycine, occurs in relatively high concentrations in human blood plasma. It is produced from pyruvate by transamination, and in

some microorganisms, e.g. bacilli, by reductive amination catalysed by alanine dehydrogenase (EC 1.4.1.1). The same enzyme has been reported to be a protomer of the oligomeric glutamate dehydrogenase (EC 1.4.1.2). Ala is degraded to pyruvate and ammonia by alanine oxidase (see Flavin enzymes), or it is converted into pyruvate by transamination.

2. ***β-Alanine, 3-aminopropionic aid:*** $H_2N\text{-}CH_2\text{-}CH_2\text{-}COOH$, a nonproteogenic amino acid. It occurs in the free form, e.g. in human brain, and it is a component of the dipeptides, carnosine and anserine, and of coenzyme A. It is not usually formed by decarboxylation of L-aspartate, but rather in the course of reductive pyrimidine degradation. It can be further metabolized to acetate by deamination, decarboxylation and oxidation.

Alar 85: see Succinic acid 2,2-dimethylamide.

Albizzin, ***2-amino-3-ureidopropionic acid:*** $H_2N\text{-}CO\text{-}NH\text{-}CH_2CH(NH_2)\text{-}COOH$, a nonproteogenic amino acid occurring primarily in species of the genus *Albizzia.* It is presumably formed from carbamoylphosphate and 2,3-diaminopropionic acid by transcarbamylation. It is an antagonist of glutamine.

Albomycin: an antibiotic synthesized by *Actinomyces subtropicus.* A. is a cyclic polypeptide containing a pyrimidine base (cytosine) and 4.16 % Fe in the form of a hydroxamate-Fe(III) complex (Fig.). It is one of the sideromycins (similar to, and possibly identical to, grisein), and it interferes with iron metabolism as an antimetabolite of the sideramines. A. is effective against both Gram-positive and Gram-negative bacteria, and inhibits the aerobic metabolism of *Staphylococcus aureus* and *E. coli.*

Albumins: a group of simple proteins, found in the body fluids and tissues of animals and in some seed plants. In contrast to the globulins, A. have a low molecular mass, are water-soluble and easily crystallizable, and contain an excess of acidic amino acids. High concentrations of neutral salts must be used for the 'salting out' of A. They are rich in glutamate and aspartate (20–25 %) and leucine and isoleucine (up to 16 %) but contain little glycine (1 %). Important examples are serum albumin, α-lactalbumin (a milk protein) and ovalbumin from animals, and the poisonous ricin (from *Ricinus* seeds), leucosin (from seeds of wheat, rye and barley) and legumelin (from legumes).

Serum albumin (plasma albumin) accounts for 55–62 % of serum protein, and is one of the few nonglycosylated proteins in blood plasma or serum. Due to its relatively low M_r (67,500) and high net charge (pI 4.9), serum albumin has a high binding capacity for water, Ca^{2+}, Na^+, K^+, fatty acids, bilirubin, hormones and drugs (when using serum albumin in defined media, it should be remembered that this protein is generally 'sticky'). Its main function is regulation of the colloidal osmotic pressure of the blood. Bovine and human serum albumins contain 16 % nitrogen and are easily purified and crystallized; they are therefore used as standard proteins for calibration.

Human serum albumin consists of a single polypeptide chain of 584 amino acids, which is stabilized by 17 disulfide bridges. Ovalbumin contains one oligosaccharide chain, coupled via an aspartate residue, as well as a phosphorylated serine residue.

Alcaptonuria, ***alkaptonuria:*** see L-Phenylalanine.

Alcohol dehydrogenase, ***ADH*** (EC 1.1.1.1): a zinc-containing oxidoreductase, which in the presence of NAD^+ reversibly oxidizes primary and secondary alcohols to the corresponding aldehydes and ketones. ADH occurs in bacteria, yeasts, plants and the liver and retina of animals. Yeast ADH, distinguished by its high affinity for ethanol, catalyses the last reaction in alcoholic fermentation. Oxidation by liver ADH makes an important contribution to the clearance of blood ethanol. ADH in the retina converts retinal (vitamin A aldehyde) to retinol (see Vitamins). ADH from animal organs and yeast has low substrate specificity; it dehydrogenates both short (C_2 to C_6) and long-chain alcohols. Yeast ADH (M_r 145,000) consists of 4 catalytically active, zinc-containing subunits (M_r 35,000) with 4 NAD^+ or NADH binding sites per tetrameric molecule. The dimeric horse liver enzyme (M_r 80,000) contains 2 zinc atoms (one is essential for catalysis) and one coenzyme binding site per subunit (M_r 40,000, 374 amino acids, sequence known: Cys 46 is the site of binding and catalysis). In the dehydrogenation process, a ternary complex of ADH, NAD^+ and ethanol is formed, in which both the coenzyme and the substrate are bound to the reactive SH of Cys 46 via a Zn atom. Due to the existence of 2 very similar polypeptide chains, E and S, there are 3 types of liver ADH: isoenzyme EE (preferentially

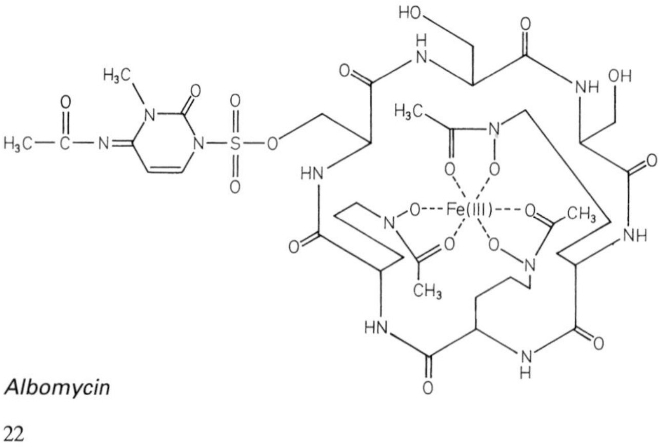

Albomycin

dehydrogenates ethanol), isoenzyme SS (active with sterols), and the hybrid ES. The smallest known ADH is that of *Drosophila melanogaster* (M_r 60,000, 8 subunits of M_r 7,400). [N Y. Kedishvili et al. 'Expression and Kinetic Characterization of Recombinant Human Stomach Alcohol Dehydrogenase' *J. Biol. Chem.* **270** (1995) 3625–3630]

Alcoholic fermentation: the anaerobic formation of ethanol and carbon dioxide from glucose. Two molecules of ATP are produced per molecule of glucose fermented.

A.f. is largely performed by yeasts and other microorganisms, but it can also occur in the tissues of higher plants, e.g. carrots and maize roots. Animals lack pyruvate decarboxylase (see below) and are therefore unable to perform A.f. The starting point for A.f. is glucose 6-phosphate, which is converted by reactions of Glycolysis (see) to pyruvate. Pyruvate is decarboxylated by pyruvate decarboxylase (EC 4.1.1.1) to acetaldehyde, which is then reduced to ethanol by alcohol dehydrogenase (EC 1.1.1.1) (Fig.). Balance: $C_6H_{12}O_6 + 2P_i + 2ADP \rightleftharpoons 2CH_3CH_2OH + 2CO_2 + 2ATP + 2H_2O$. A.f. has long been exploited by man and is now carried out on an industrial scale. The most important substrates are the monosaccharides D-glucose, D-fructose, D-mannose and sometimes D-galactose. In some cases, the disaccharides sucrose and maltose and the polysaccharide starch can serve as substrates. Formation of fusel oils is a side reaction of A.f.

Historical. The simple equation for A.f., 1 glucose $\rightarrow 2CO_2 + 2$ ethanol, was established in 1815 by Gay-Lussac. In 1857 Pasteur proposed that A.f. could only be carried out by living organisms (vitalistic theory of fermentation). This was disproved in 1897 by Buchner, who established that a cell-free filtrate of disrupted yeast cells was capable of A.f. This discovery marks the beginning of modern enzymology. The enzyme system responsible for A.f., which was originally thought to be one enzyme, was called zymase. In 1905 the role of phosphate in A.f. was described by Harden and Young. In 1912 Neuberg proposed the first fermentation scheme, which was revised in 1933 by Embden and Meyerhof.

Alcohols: hydrocarbon derivatives carrying one or more hydroxyl (-OH) groups. The ending "-ol" in a systematic or trivial name of an organic compound indicates that it is an alcohol. A "diol" has 2 OH-groups, a "triol" has 3, and so on. A. can be primary (RCH_2OH), secondary (R^1R^2CHOH) or tertiary ($R^1R^2R^3COH$). Esterified A. are important components of essential oils, fats and waxes. A number of lower A., e.g. ethanol, are formed by fermentation from carbohydrates (themselves polyalcohols) and proteins.

Aldehyde oxidase (EC 1.2.3.1): see Molybdenum enzymes.

Aldoketomutase: see Lactoyl-glutathione lyase.

Aldolase: see Fructose-*bis*phosphate aldolase (EC 4.1.2.13).

Aldonic acids: monocarboxylic acids derived from aldose sugars by oxidation of the aldehyde group. The names of these acids are formed by replacing the "-ose" of the parent aldose with "-onic acid". A.a. may form a 1,4-lactone ring (γ-lactones) or the more stable 1,5-lactone (δ-lactones). Some important A.a. are L-arabonic acid, xylonic acid, D-gluconic acid, D-mannonic acid and galactonic acid.

Aldoses: polyhydroxyaldehydes, one of the two main subdivisions of the monosaccharides (the other is the ketoses; see Carbohydrates). A. are characterized by a terminal aldehyde group -CHO, which is always given the number 1 in systematic nomenclature. The A. are formally derived from their simplest representative, glyceraldehyde, by chain extension. According to the number of carbon atoms in their chain, they are classified as trioses, tetroses, etc. The pentoses and hexoses are particularly important metabolically.

Aldosterone, ***11β,21-dihydroxy-3,20-dioxopregn-4-en-18-al-18 → 11 hemiacetal:*** a highly active mineralocorticoid from the adrenal cortex. Unlike other hormones of the adrenal cortex, A. has a carbonyl group on C18, which forms a hemiacetal with the 11β-hydroxyl group. It is the most important mineralocorticoid, regulating NaCl resorption and K^+ excretion. It also has some glucocorticoid activity. For structure and biosynthesis, see Adrenal corticosteroids.

Alcoholic fermentation. Formation of ethanol from pyruvate.

Alginic acid: a polyuronic acid, M_r ~ 120,000, composed of varying proportions of β-1,4-linked D-mannuronic acid and L-guluronic acids. It is extracted with NaOH from brown algal seaweeds, where it replaces the pectin of higher plants. It can absorb up to 300 times its weight of water. Because it is easily digested, A. a. is widely used in the food industry, in surgery as resorbable sutures, and in the pharmaceutical and cosmetic industries.

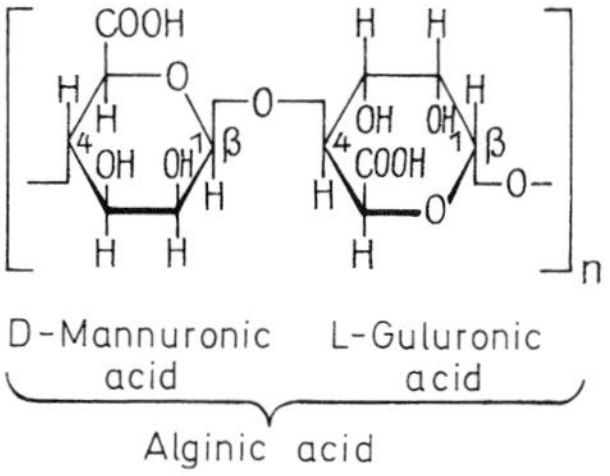

D-Mannuronic acid L-Guluronic acid

Alginic acid

Alginic acid

Alizarin, *1,2-dihydroxyanthraquinone:* a red dye from the root of the madder plant *(Rubia tinctorum* L.) and other *Rubiaceae,* where it is combined with 2 molecules of glucose as ruberythric acid. A. and several of its derivatives are widely used as A. dyes. It has been made synthetically since 1871, and production from madder has died out.

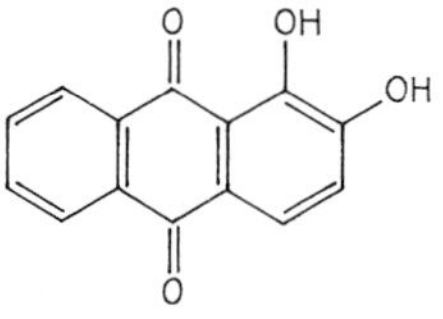

Alizarin

Alkaloids: basic natural products occurring primarily in plants. They contain one or more heterocyclic nitrogen atoms and are generally found as salts of organic acids. Several thousand A. are known. They usually have trivial names derived from their plants of origin.

Classification. It is difficult to define A. to the exclusion of other nitrogen-containing plant products. Classification according to occurrence and function excludes animal and microbial A. Some A. (e. g. nicotine) occur so widely that classification by botanical origin is not possible. Chemical classification excludes the Colchicum A. because they lack a heterocyclic nitrogen. More recently, A. have been classified according to their biogenesis, as protoalkaloids (see Biogenic amines), pseudoalkaloids and A. in the strict (narrow) sense. Protoalkaloids include e. g. decarboxylation products of amino acids, while pseudoalkaloids include compounds that are structurally related to other classes of natural products, such as terpenes. A. in the strict sense can be subdivided into derivatives of ornithine, lysine, phenylalanine, tryptophan and anthranilic acid. The classification shown in the Table (see) takes into consideration both structural and biogenetic features.

Occurrence. Probably 10–20 % of all higher plants contain A., and they are particularly numerous and abundant in certain dicotyledonous families. Closely related families often produce similar A. Usually several structurally related A. are present as hydrophilic salts dissolved in the vacuolar sap. The predominant members of such mixtures are known as primary A., while those present in lower concentrations are secondary A. All parts of the plant may contain A., but they are particularly abundant in seeds, bark and roots. Heterocyclic compounds similar to A. are found in many microorganisms and a few animals (e. g. salamander A.).

Biosynthesis. A. are the end products of secondary metabolism, and are not subject to significant degradation. They accumulate because the plant has no excretory organs. Most A. are derivatives of amino acids (Table), which provide the heterocyclic nitrogen. In addition, acetate, mevalonate and one-carbon units may be involved in the synthesis. The connection between amino acid metabolism and A. synthesis was discovered around the turn of the 20th century, when attempts were made to synthesize A. under physiological conditions, i. e. with native reactants, without high pressure or temperature, and at neutral pH. Later, the biosynthesis of A. in vivo was investigated by isotopic tracing, by administering amino acids labeled with ^{13}C, ^{14}C, ^{15}N or ^{3}H. It was found that plants generate a large number of structures from only a few components, using only a few mechanisms of cyclization: *N*-heterocyclic rings are made by Mannich condensation, or by the formation of amides or Schiff's bases. Secondary cyclizations, i. e. ring closures not involving nitrogen, are the result of oxidative coupling (phenol oxidation).

Synthesis. The first synthesis of an A. was reported in 1886 by Ladenburg, who generated coniine from α-picoline. On the hypothesis that A. are synthesized in the plant as amino acid derivatives, Robinson and Schöpf developed corresponding synthetic pathways and tested them under physiological conditions (see Tropinone). These studies also inspired laboratory syntheses, and some compounds were synthesized for the first time by routes similar to those occurring in the plant. Most A. for medical use are prepared from plant sources.

Biological and economic importance. There is no general explanation for the biological significance of A. Protection against consumption of the plant by animals has been proven in a few cases, but the fact that most insects are limited to one or a few food plants may actually be due to the alkaloid content of these particular species. Many A. have a strong and very specific effect on certain centers of the nervous system (e. g. opiates, see Endorphins). Therefore A. are widely used clinically, as combinations of pure compounds, as extracts of total alkaloids, or as synthetic analogs. However, their use is often accompanied by side effects, mainly toxicity and narcosis.

Historical. Most of the A. plants were known in early folk medicine for their toxicity or useful pharmacological properties. Morphine was first isolated from poppies as the "sleep-inducing principle" by F. W. Sertuerner in 1806. The term A. was coined in 1819 by C. F. W. Meissner.

The main classes of alkaloids and their biogenetic precursors

Class of compound	Structural type (main precursor emphasized)	Biogenetic precursors
Pyrrolidine	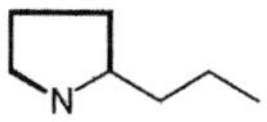	Ornithine and acetate
Pyrrolizidine	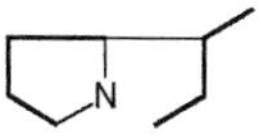	Ornithine
Tropane		Ornithine and acetate
Piperidine *(Conium)*	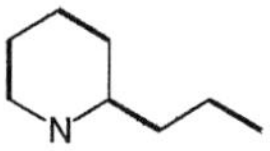	Acetate
Piperidine *(Punica, Sedum, Lobelia)*	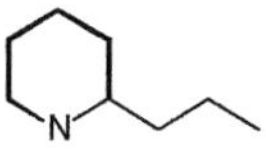	Lysine, acetate or phenylalanine
Quinolizidine	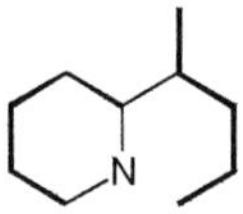	Lysine
Isoquinoline	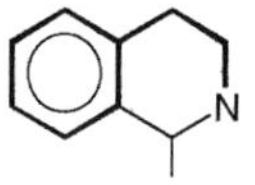	Phenylalanine or tyrosine
Indole	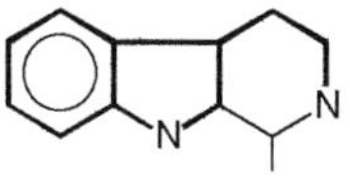	Tryptophan
Quinoline *(Rutaceae)*	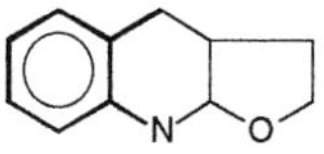	Anthranilic acid
Terpene	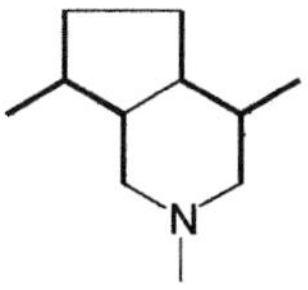	Mevalonic acid

Alkalosis: an increase in the pH of body fluids. It is corrected by excretion of bicarbonate and retention of acid. There 2 types:

a) *Metabolic A.* is caused by an increase in the bicarbonate fraction, with little or no change in the carbonic acid fraction. There are several possible causes: 1) Ingestion of alkaline drugs, e.g. $NaHCO_3$ for treatment of peptic ulcer. 2) Excessive vomiting of stomach contents only (and not the alkaline contents of the intestine) results in loss of HCl secreted by the gastric mucosa; HCO_3^- replaces the lost Cl^-, leading to A., also called hypochloremic A. Vomiting of stomach contents only is a feature of pyloric obstruction in newborns, caused by hypertrophy of the pyloric sphincter muscle. 3) Excess secretion of aldosterone by the adrenal glands promotes increased reabsorption of Na^+ from the distal tubules of the nephrons. Since this process is coupled with increased H^+ secretion, the result is A.

b) *Respiratory A.* is caused by a decrease in the carbonic acid fraction, with little or no change in the bicarbonate. It can result from hyperventilation, which may occur in central nervous system diseases affecting the nervous control of breathing, and in early stages of salicylate poisoning. It can also be promoted by voluntary hyperventilation.

Alkannin: see Naphthoquinones (Table).

Alkylating agents: chemical compounds which can donate alkyl groups, usually methyl or ethyl. Monofunctional A. a., like dimethylsulfate or ethylmethane sulfonate, transfer a single functional group, whereas bifuncional A. a., like mustard gas, nitrogen mustard gas or cyclophosphamide, can react with several molecules or parts of a macromolecule, thus cross-linking them. A. a. are frequently carcinogenic and mutagenic, but some are used in the chemotherapy of cancer (see Mitomycin C). Activated one-carbon units (see) are biological A. a. Chemical A. a. have been widely used in the laboratory synthesis of macromolecules, to protect reactive groups, and in the elucidation of Active centers (see) of enzymes and receptor molecules.

Allantoic acid, ***diureidoacetate:*** a degradation product (M_r 176.1) of allantoin. See Purine degradation.

Allantoin, ***5-ureidohydantoin, glyoxyldiureide:*** an intermediate (M_r 158.1) in aerobic purine degradation. A. was discovered in 1799 in the allantoic fluid of the cow. It is the end product of purine metabolism in most mammals and some reptiles, and is excreted in their urine. A. is also found widely in plants. In some plants (e. g. members of the *Boraginaceae),* known as ureide plants, A. is particularly abundant in the soluble nitrogen pool. In *Arthrobacter allantoicus* and *Streptococcus allantoicus,* A. can serve as C, N and energy source under anaerobic conditions. A. is degraded anaerobically by conversion to allantoic acid (by allantoinase, EC 3.5.2.5), which in turn is converted to ureidoglycine, NH_3 and CO_2 (by allantoate deiminase, EC 3.5.3.9). See Purine degradation.

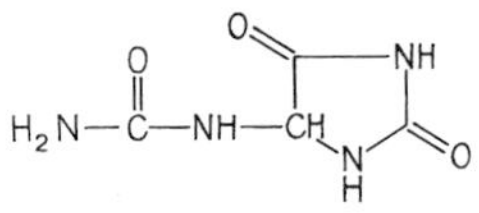

Allantoin

Allantoinase (EC 3.5.2.5): see Purine degradation.

Allelochemicals: see Pheromones.

Allen-Doisy test: see Estrogens.

Allergy: a hypersensitivity of the immune apparatus, a pathological immune reaction induced either by antibodies (immediate hypersensitivity) or by lymphoid cells (delayed type A.). Unlike the delayed type, immediate hypersensitivity can be passively transmitted in the serum. Symptoms of immediate hypersensitivity begin shortly after contact and decay rapidly, but delayed type symptoms do not attain a maximum for 24–48 hours, then decline slowly over days or weeks. Examples of immediate type A. are anaphylaxis, the Arthus reaction and serum sickness. The best known A., anaphylaxia, can occur as a local (cutaneous) reaction (e. g. a rash with blisters) or as a systemic reaction (anaphylactic shock). Asthma, hay fever and nettle rashes are also examples of local anaphylactic reactions which are induced by reagins (see Immunoglobulins; IgE). Only primates can be sensitized by injection with human reagins. An example of delayed type A. is the tuberculin reaction, which is based on a cellular immune response.

Allocholane: an obsolete term for 5α-cholane. See Steroids.

Allodeoxycholic acid, ***3α,12α-dihydroxy-5α-cholan-24-oic acid:*** a bile acid, M_r 392.6. It differs from most other Bile acids (see), in that rings A and B show *trans* coupling. It is present in rabbit bile and feces.

Allogibberellic acid: see Gibberellins.

Allomones: see Pheromones.

Allophanate hydrolase: see Urea amidolyase.

Allophanic acid: see Urea amidolyase.

Allopregnane: obsolete term for 5α-pregnane. See Steroids.

All-or-nothing model: see Cooperativity model.

Allosteric enzymes: see Cooperative oligomeric enzymes.

Allostery: changes in the conformation of proteins with quaternary structure in response to the binding of certain low molecular mass ligands. A. is important in the regulation of enzyme activity (see Effectors) and in the uptake of oxygen by hemoglobin.

Alloxan: a compound discovered by Liebig in mucus secreted during dysentery. It is used to produce experimental diabetes in animals, since it preferentially attacks pancreatic β-cells.

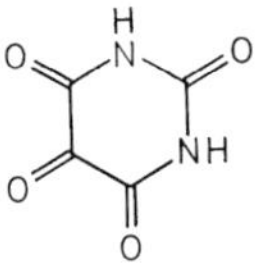

Alloxan

Alnulin: see Taraxerol.

ALS: the acronym of antilymphocyte serum. The latter has largely been replaced by monoclonal antibodies specific for cell-surface determinants of particular types of lymphocytes, such as T-cells, B-cells or macrophages.

Alternaria alternata toxins: toxins produced by the fungus, *Alternaria alternata,* which causes a num-

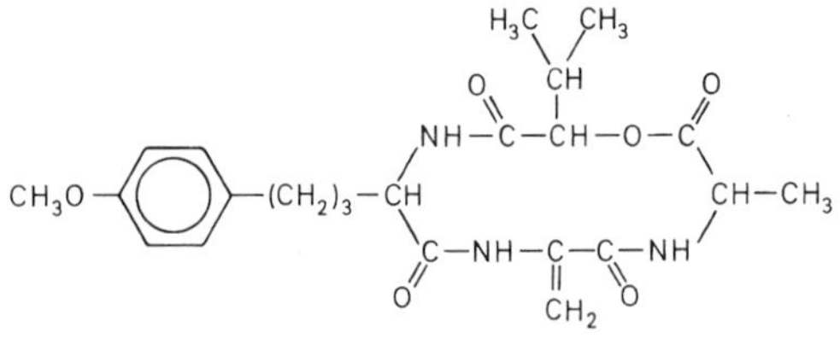

Fig. 1. *Alternariolide* [S. Lee *Tetrahedron Lett.* No. *11* (1979) 843–846].

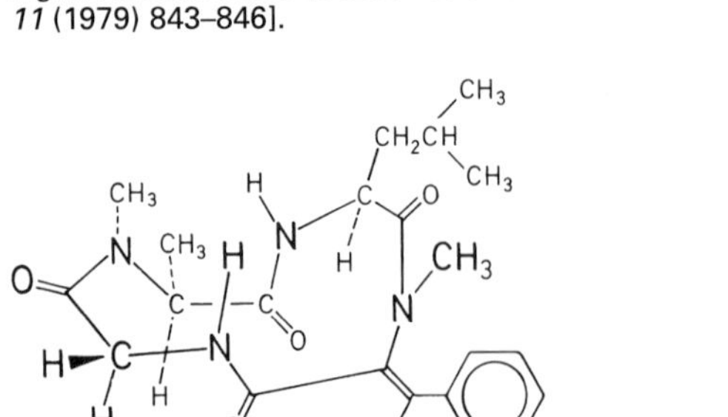

Fig. 2. *Tentoxin* [D. H. Rich J. Am. Chem. Soc. *100* (1978) 2212–2218].

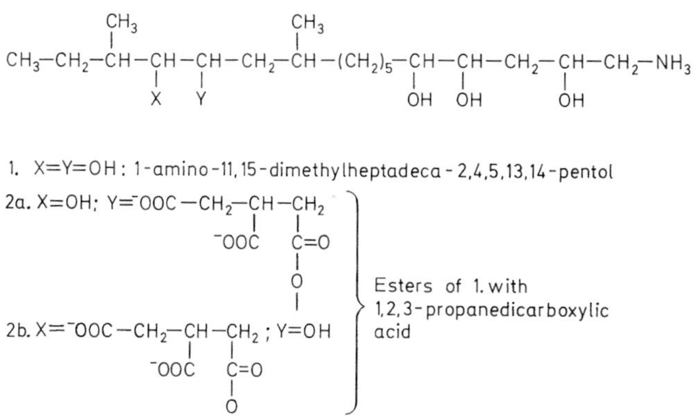

$$CH_3-CH_2-\underset{X}{CH}(CH_3)-\underset{Y}{CH}-CH-CH_2-CH(CH_3)-(CH_2)_5-\underset{OH}{CH}-\underset{OH}{CH}-CH_2-\underset{OH}{CH}-CH_2-NH_3$$

1. X=Y=OH: 1-amino-11,15-dimethylheptadeca-2,4,5,13,14-pentol

2a. X=OH; Y=$^-OOC-CH_2-CH(COO^-)-CH_2-C(=O)-O-$

2b. X=$^-OOC-CH_2-CH(COO^-)-CH_2-C(=O)-O-$; Y=OH

Esters of 1. with 1,2,3-propanedicarboxylic acid

Fig. 3. *Toxins of Alternaria alternata F. sp. Lycopersici* (A. T. Bottini et al. *Tetrahedron Lett.* No. *22* (1981) 2723–2726].

ber of plant diseases. There are several different pathotypes, each infecting a different species and producing different host-specific phytotoxins, several of which have been characterized.

The pathotype attacking apples produces a toxin consisting of at least 6 components, of which 2 have been identified: the depsipeptide ***alternariolide*** (Fig. 1) and its demethoxy derivative. In susceptible apple trees, nM concentrations of alternariolide cause interveinal chlorosis, whereas about 1 μM is required in resistant varieties. The toxin causes immediate release of electrolytes, suggesting the plasmalemma as the toxin-sensitive site.

Various pathotypes also produce a cyclic tetrapeptide known as ***tentoxin*** (Fig. 2) which induces chlorosis in many plants (e.g. lettuce, potato, cucumber, spinach), but not in *Nicotiana,* tomato, cabbage or radish. It binds to chloroplast coupling factor 1 (one toxin-binding site per αβ subunit complex), inhibiting photophosphorylation and the Ca^{2+}-dependent ATPase. Species specificity is due to different binding affinities for coupling factor 1 ($1.3–20 \times 10^{-7}$M for 50 % inhibition in sensitive species, and 20-fold higher for insensitive species).

The pathotype responsible for tomato stem canker produces 3 related toxins (Fig. 3). Aspartate and certain products of aspartate metabolism (e.g. orotic acid) protect tomato plants against the toxins, and it is thought that toxicity may be due to inhibition of aspartate transcarbamylase. Susceptibility to the disease is controlled by a single genetic locus with 2 alleles.

Amanitin: see Amatoxins.

Amarinthin: a red betacyanin dye. It shares the same aglycon with Betanin (see), but in place of glucose it carries a disaccharide: 5-*O*-(β-D-glucopyranosyluronic acid)-5-*O*-β-D-glucopyranoside. It is found in *Amaranthus* species, e.g. the foxtail, *Celosia argentea.*

Amaryllidaceae alkaloids: a group of alkaloids of complicated structure, found only in the plant family, *Amaryllidaceae.* They can be classified as phenylisoquinoline alkaloids (see Isoquinoline alkaloids), because their biosynthesis (Fig.) is similar to that of the isoquinoline alkaloids, beginning with phenylethylamine or tyramine and a carbonyl compound. The final structures are generated by secondary ring cleavage and cyclization. The last step of the biosynthesis is catalysed by a phenol oxidase. The main alkaloid, galanthamine, is isolated from Caucasian snowdrops, *Galanthus woronowii,* and used therapeutically as an inhibitor of acetylcholinesterase.

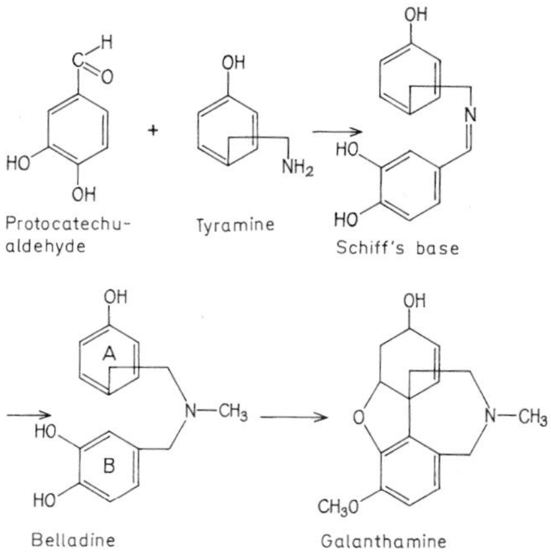

Protocatechualdehyde + Tyramine → Schiff's base → Belladine → Galanthamine

Biosynthesis of belladine and galanthamine alkaloids

Amatoxins: a group of bicyclic octapeptides (Fig.) which, together with the phallatoxins, are the most important poisons in the death cap fungus, *Amanita phalloides.*

These poisons inhibit the nucleoplasmic RNA polymerase II (EC 2.7.7.6) of eukaryotic cells, leading to necrosis of liver and kidney cells. The toxic effects of A. and phallatoxins can be inhibited by simultaneous application of antamanide. More than 90 % of fatal mushroom poisonings are due to consumption of *Amanita phalloides* and related species. [D. R. Chafin et al. 'Action of α-Amanitin during Pyrophorolysis and Elongation by RNA Polymerase II' *J. Biol. Chem.* **270** (1995) 19114–19119]

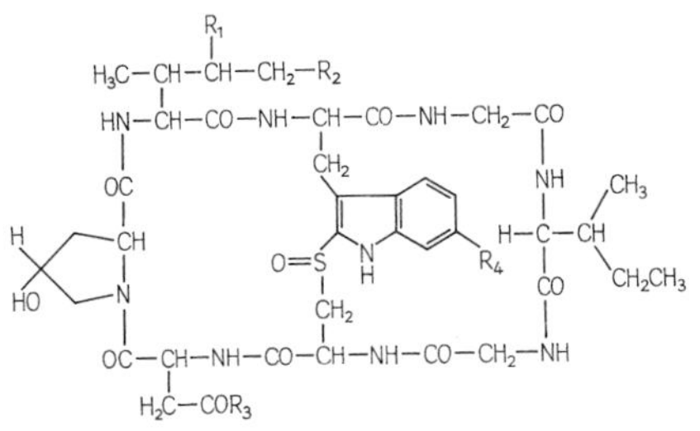

	R_1	R_2	R_3	R_4
α-Amanitin	OH	OH	NH_2	OH
β-Amanitin	OH	OH	OH	OH
γ-Amanitin	OH	H	NH_2	OH
Amanin	OH	OH	OH	H
Amanullin	H	H	NH_2	OH

Amatoxins

Ambanol: the only known naturally occurring isoflavonol (Fig.). It occurs in the roots of *Neorautanenia amboensis.* [M. E. Oberholzer et al. *Tetrahedron Lett.* (1977) 1165–1168]

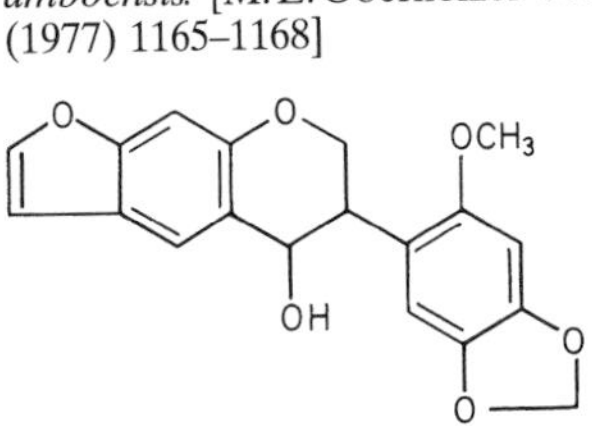

Ambanol

Amber codon, *nonsense codon:* the sequence UAG in mRNA. It does not code for any of the 20 proteogenic amino acids, and it results in the termination of protein synthesis (premature termination when UAG is produced by mutation of a sense codon). Potential precursors for the production of UAG by mutation are UCG (serine), UAU and UAC (tyrosine) and CAG (glutamine).

Amber mutants: mutant bacteria in which the mRNA contains the codon UAG as a result of a point mutation (see Amber codon). The mutation is not necessarily lethal, because a compensatory suppressor mutation in a tRNA may enable the protein synthesizing system to recognize the amber codon as a sense codon. The term *amber* was arbitrarily chosen by the discoverer of A. m.

Amentoflavone: see Biflavonoids.

Amicetins: pyrimidine antibiotics (see Nucleoside antibiotics) synthesized by various *Streptomyces* species. Amicetin A (amicetin, allomycin, sacromycin; M_r 618.7) from *S. fasciculatus, S. plicatus* and *S. vinaceus-drappus* is primarily bacteriostatic, especially against Gram-positive bacteria. 0.5 μg Amicetin A/ml inhibits growth of *Mycobacterium tuberculosis.* Amicetin B (plicacetin) (M_r 517.6) from *S. plicatus* is identical with amicetin A minus the α-methylserine moiety, and it is presumably the precursor of amicetin A. Its activity spectrum is the same as that of amicetin A, but its effect is weaker. *Cytimidine* (consisting of cytosine, 4-aminobenzoic acid and 2-methyl-D-serine; M_r 331.3) is a degradation product of the amicetins. See Bamicetin. [T. H. Haskell et al. *J. Amer. Chem. Soc.* **80** (1958) 743–747 & 747–751; S. Hanessian *Tetrahedron Lett.* (1964) 2451–2460]

Amicetose: 2,3,6-trideoxy-D-erythro-hexopyranose, the sugar moiety located between amosamine and cytosine in the structures of Amicetin (see), bamicetin and plicacetin.

Amidinotransferases, *transamidinases* (EC 2.1.4): a group of enzymes catalysing transamidination, e. g. the biosynthesis of guanidoacetate (the precursor of creatine) by transfer of the amidine group of arginine to glycine. Transfer of the intact amidine group of arginine has been demonstrated by double labeling with C^{14} and N^{15}. An A. from *Streptomyces griseus* and *S. baikiniensis* is involved in the biosynthesis of streptidine.

Amiloride: a drug which inhibits the influx of Na^+ into cells. It was discovered as a natriuretic agent which increases Na^+ excretion but does not affect K^+ excretion [Baer et al. *J. Pharmacol. Exp. Ther.* **157** (1967) 472]. Animal cells have 2 systems for Na^+ transport. The Na^+/K^+ ATPase (inhibited by ouabain) exports Na^+ and imports K^+, each against its concentration gradient. In addition, there is a pump which imports Na^+ and exports H^+, and this is inhibited by A., hence the designation "amiloride-sensitive Na^+ pump". A. has subsequently been found to be relatively unspecific, e. g. it also inhibits protein synthesis. Its derivative, dimethylamiloride, is a more specific

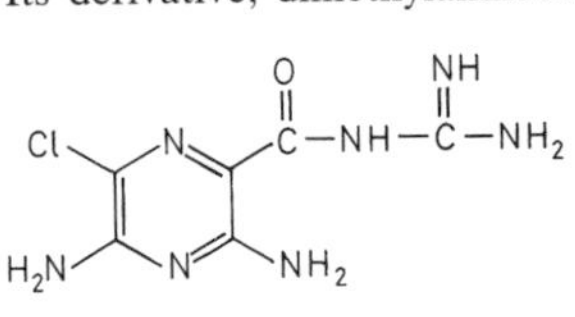

Amiloride

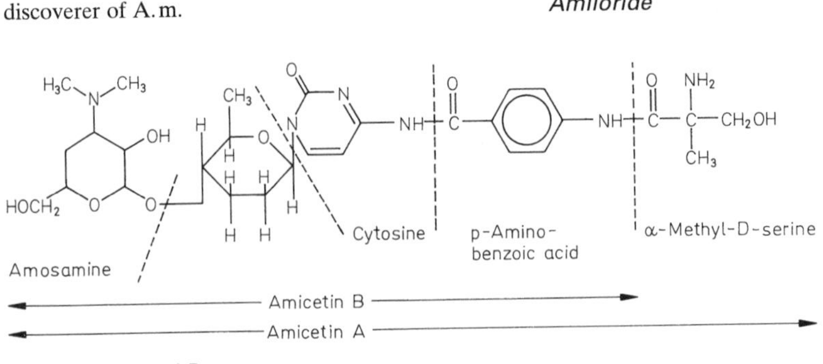

Amosamine — Cytosine — p-Amino-benzoic acid — α-Methyl-D-serine

Amicetin B

Amicetin A

Amicetins A and B

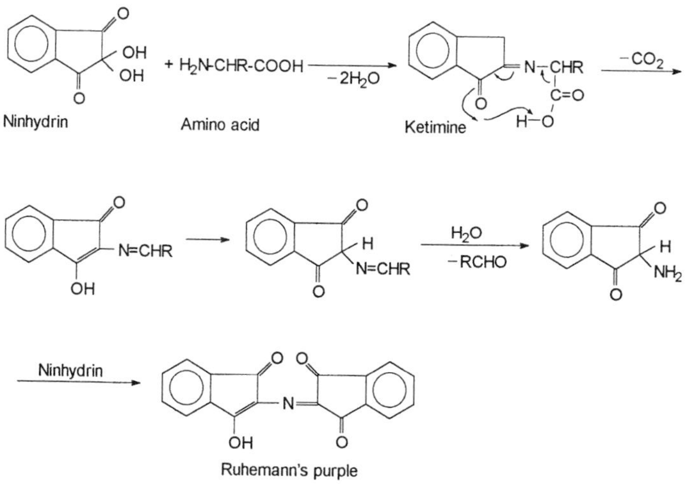

Ninhydrin + H_2N-CHR-COOH —($-2H_2O$)→ Ketimine —($-CO_2$)→ ... → (H₂O, −RCHO) → ... —Ninhydrin→ Ruhemann's purple

The ninhydrin reaction

inhibitor of Na^+ influx. [J.B.Smith & E.Rozengurt *Proc. Natl. Acad. Sci. USA* **75** (1978) 5560–5564; E.Rozengurt *Adv. Enz. Regul.* **19** (1981) 61–85]

Amination: introduction of an amino group ($-NH_2$) into an organic compound. This may be accomplished by reductive A. or by Transamination (see). Reductive A. requires a reduced pyridine nucleotide as a reducing agent, e.g. the most common reductive A. of 2-oxoglutarate by L-glutamate dehydrogenase (EC 1.4.1.4) requires NADPH.

Amine precursor uptake decarboxylase system, *APUD system:* the nervous system and part of the gastrointestinal tract derived from the neuroectodermal cells of the primitive neural crest. Gut and brain produce several identical peptides, e.g. bombesin, cholecystokinin, neurotensin, substance P, enkephalin, gastrin, vasoactive intestinal peptide, somatostatin, etc., reflecting the common embryological origin of the two tissues. These peptides function specifically in both gut and nervous system, but their functions are generally distinct, because they are largely excluded by the blood-brain barrier.

APUD cells are descended from an ancestral neuron; although the axon has been lost, innervation is still possible via synaptic terminals, or nervous control may be indirect. Six modes of APUD cell expression are recognized: neurocrine (into neurons), neuroendocrine (via axons), endocrine (into the blood stream), paracrine (into the intercellular space), epicrine (into somatic cells), and exocrine (to the exterior).

The APUD concept of a diffuse neuroendocrine system has been defined as follows by A.G.E.Pearse [in *Centrally Acting Peptides,* J.Hughes (ed.), MacMillan, 1978]: "The cells of the APUD series, producing peptides active as hormones or as neurotransmitters, are all derived from neuroendocrine-programed cells originating from the ectoblast. They constitute a third (endocrine or neuroendocrine) division of the nervous system whose cells act as third line effectors to support, modulate or amplify the action of neurons in the somatic and autonomic divisions, and possibly as tropins to both neuronal and non-neuronal cells".

Aminoacetic acid: see Glycine.

Amino acid activating enzymes: see Aminoacyl-tRNA synthetases.

Amino acid analyser: see Proteins.

Amino acid oxidases: see Flavin enzymes.

Amino acid reagents: reagents for the colorimetric identification and quantitation of amino acids. One of the most important is ninhydrin (2,2-dihydroxy-1H-indene-1,3(2H)-dione), which reacts with amino acids to form a blue-violet dye called Ruhemann's purple (absorbance maximum 570 nm). With the imino acid, proline, ninhydrin forms a yellow product, absorbance maximum 440 nm.

In the fluorescamine technique, amino acids are converted by reaction with 4-phenyl[furan-2H(3H)-1′-phthalane]-3,3′-dione (fluorescamine) into strongly fluorescent compounds which can be detected in nanomole quantities at 336 nm. The reagent itself is not fluorescent, and in contrast to ninhydrin, it is not sensitive to ammonia.

Other highly sensitive reagents are 2,4,6-trinitrobenzosulfonic acid, 1,2-naphthoquinone-4-sulfonic acid (Folin's reagent), and 4,4′-tetramethyldiaminodiphenylmethane (TDM). Intensely fluorescing amino acid derivatives are formed by reaction with *o*-phthalaldehyde in the presence of reducing agents; with pyridoxal and zinc ions; and with dansyl chloride (5-dimethylaminonaphthalene sulfonyl chloride).

Amino acids: organic acids that also possess at least one, and usually not more than two, amino groups. A.a. are classified as α-, β-, γ-A.a., etc, according to the position of the carbon that carries the $-NH_2$ (Fig.1).

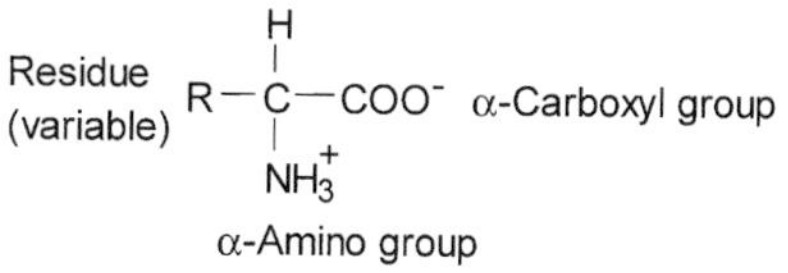

Residue (variable) R—C(H)(NH₃⁺)—COO⁻ α-Carboxyl group; α-Amino group

Fig. 1. *Amino acids.* Structure of an α-amino acid

Proteins and peptides are polymers of α-A. a., which are also occur in the free form in all living cells and body fluids. The twenty A. a. encoded in messenger RNA (see protein biosynthesis, Genetic code) occur in proteins and are known as proteogenic or proteinogenic A. a. (Table 1). The occurrence of nonproteogenic A. a. in proteins is due to Post-translational modification of proteins (see). Further information on individual A. a. can be found in the appropriate entries.

Amino acids formed by post-translational modification, and present only in special proteins

Amino acid	Occurrence
δ-Hydroxy-L-lysine	Fish collagen
L-3,5-Dibromotyrosine	Skeleton of *Primnoa lepadifera* (coral)
L-3,5-Diiodotyrosine	Skeleton of *Gorgonia cavolinii* (coral)
L-3,5,3′-Triiodothyronine	Thyroglobulin (tissue protein in thyroid gland)
L-Thyroxin	Thyroglobulin
Hydroxy-L-proline	Collagen, gelatins

A. a. are classifed as acidic or basic, depending on their isoelectric points. Alternatively, according to the nature of their side chains, they are divided into 4 groups designated by Roman numerals I-IV (see Table 1):

I. A. a. with neutral, hydrophobic (nonpolar) side chains.

II. A. a. with neutral and hydrophilic (polar) side chains.

III. A. a. with acid and hydrophilic (polar) side chains.

IV. A. a. with basic and hydrophilic (polar) side chains.

In addition to this chemical classification, A. a. can be divided according to their degradation into *glucogenic* and *ketogenic* A. a. Glucogenic A. a. are degraded to C_4-dicarboxylic acids or pyruvate, which are intermediates in the tricarboxyic acid cycle. This cycle provides oxaloacetate for Gluconeogenesis (see), so that the carbon chains from this group of A. a. can be incorporated into glucose. Ketogenic A. a., on the other hand, are degraded to ketones, in particular acetoacetic acid. Finally, A. a. can be classified as *essential (indispensable)* and *nonessential (dispensable)*, depending on whether the organism in question is able to synthesize them in amounts adequate for its needs. Essential A. a. must be supplied by the diet, and an inadequate supply leads to negative Nitrogen balance (see). Half-essential A. a. can be synthesized, but in insufficient quantity for all physiological requirements. An inadequate dietary supply of a single essential A. a. inhibits protein synthesis because the ribosome-mRNA-nascent polypeptide complex must suspend its operation at the point where the missing A. a. should be incorporated. The other A. a. then accumulate and are shunted into degradative metabolic pathways; hence the loss of nitrogen. Human requirements for essential A. a. are listed in Table 2.

Table 1. *Proteogenic amino acids.* The three-letter abbreviations are universally accepted and are routinely used for depicting protein and peptide sequences (see Peptides). One-letter notations (recommended by the IUPAC-IUB Commission on Biochemical Nomenclature *[Eur. J. Biochem.* *5* (1968) 151–153] should not be used for the publication of sequences, but are intended to facilitate storage of sequence information and sequence comparison by computer.

Amino acid	3-letter abb.	1-letter abb.	Class
L-Alanine	Ala	A	I
L-Arginine	Arg	R	IV
L-Asparagine	Asn*	N*	II
L-Aspartic acid	Asp*	D*	III
L-Cysteine	Cys	C	II
L-Glutamic acid	Glu	E	III
L-Glutamine	Gln	Q	II
Glycine	Gly	G	I
L-Histidine	His	H	IV
L-Isoleucine	Ile	I	I
L-Leucine	Leu	L	I
L-Lysine	Lys	K	IV
L-Methionine	Met	M	I
L-Phenylalanine	Phe	F	I
L-Proline	Pro	P	I
L-Serine	Ser	S	II
L-Threonine	Thr	T	II
L-Tryptophan	Trp	W	I
L-Tyrosine	Tyr	Y	II
L-Valine	Val	V	I
Unknown or other	Xaa	X	

* When it is not known whether the A. a. in the original protein is Asn or Asp, the abbreviation Asx or B is used; Glx or Z indicates Glu, Gln, Gla (L-4-carboxyglutamic acid) or Glp (pyroglutamic acid). These ambiguities arise because chemical hydrolysis of peptide bonds also hydrolyses Asn, Gln, Gla and Glp to the corresponding acid.

Except for glycine, all A. a. contain at least one asymmetric C-atom. With few exceptions, naturally occurring A. a. have the L-configuration. D-A. a. occur in the cell walls, capsules and culture media of some microorganisms, and in many antibiotics. A. a. lacking an α-amino group (β-, γ-, δ-A. a., e. g. β-alanine) occur as the free acids or as components of natural products, but not in proteins.

Properties. A. a. are amphoteric, since they possess both NH_2 and COOH groups, and their solutions are ampholytes. In the solid state and in aqueous solution between pH 4 and 9, they are zwitterions: H_3^+N-CHR–COO^- (Fig. 1.). With a few exceptions, they are highly soluble in water, ammonia and other polar solvents, but barely so in nonpolar and less polar solvents such as ethanol, methanol and acetone. A. a. with hydrophilic side chains are more soluble in water. The water solublility of an A. a. is lowest at its isoelectric point, because the predominant zwitterionic form decreases the hydrophilicity of the amino and carboxyl groups.

Table 2. *Minimal requirements of human beings for essential amino acids (mg/kg/day)*

	Ile	Leu	Lys	Phe (Tyr)	Met	Cys	Thr	Trp	Val
Child	90	120	90	90[1]	85		60	30	85
Man	10.4	9.9	8.8	4.3[2] 13.3[3]	1.5 13.2*	11.6	6.5	2.9	8.8
Woman	5.2	7.1	3.3	3.1[4]	4.7	0.5	3.5	2.1	9.2
WHO norms	3.0	3.4	3.0	2.0[5]	1.6	1.4	2.0	1.0	3.0

[1] In the presence of L-tyrosine. [2] Tyr 15.9 mg/kg: Tyr/Trp = 5.5. [3] In the absence of L-tyrosine. [4] Tyr 15.6 mg/kg: Tyr/Trp = 7.4. [5] Tyr 5.0 mg/kg: Tyr/Trp = 2.0. * 80–90 % of the methionine requirement can be covered by cysteine. WHO = World Health Organization.

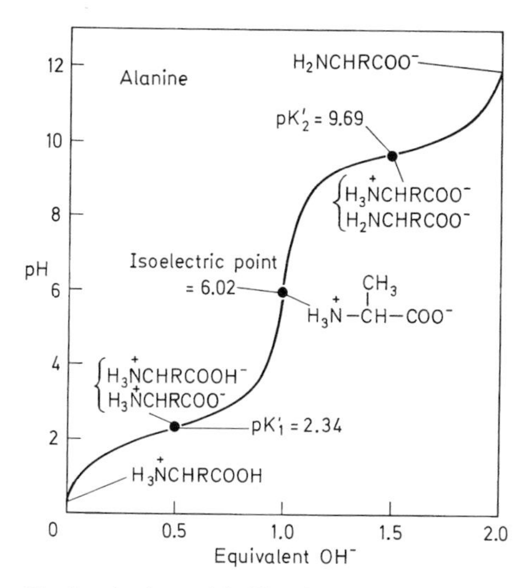

Fig. 2a. *Amino acids.* Titration curve of alanine

Dissociation of A. a. depends strongly on the pH value of the solution; the zwitterionic form is present only between pH 4 and 9. In the more acidic range, A. a. exist as cations (H_3^+N–CHR–COOH) and in more alkaline solutions as anions (H_2N–CHR–COO^-). Titration curves of A. a. therefore show two buffering zones, and are also affected by the dissociation behavior of the side chains, especially the acidic and basic ones (Fig. 2). The acid-base behavior of A. a. is a model for that of peptides and proteins, and it forms the basis for their separation by electrophoresis and ion-exchange chromatography. The UV-absorption of A. a. with chromophoric side-chain functions (tryptophan, tyrosine, phenylalanine) enables their quantitation, both as the free A. a. and in proteins and peptides.

Proteogenic A. a. can be grouped into families according to the biosynthetic sources of their carbon skeletons:

1. The serine family includes A. a. derived from triose phosphate: serine, glycine, cysteine and cystine;

2. The ketoglutarate or oxoglutarate family consists of A. a. whose carbon skeletons are derived from oxoglutarate supplied by the tricarboxylic acid cycle: glu-

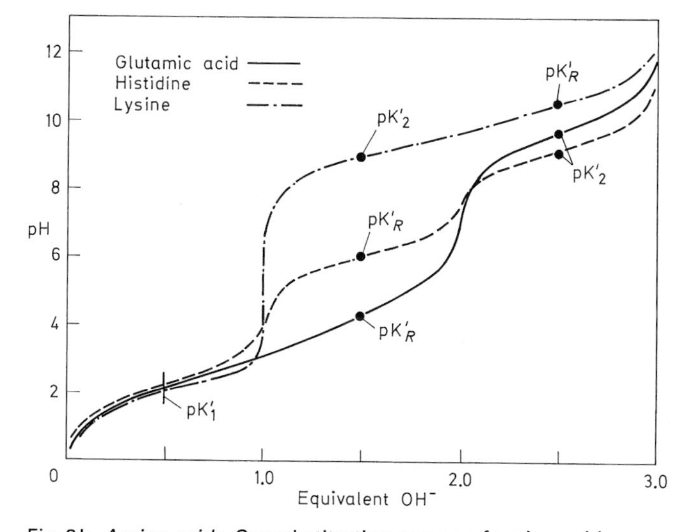

Fig. 2b. *Amino acids.* Sample titration curves of amino acids

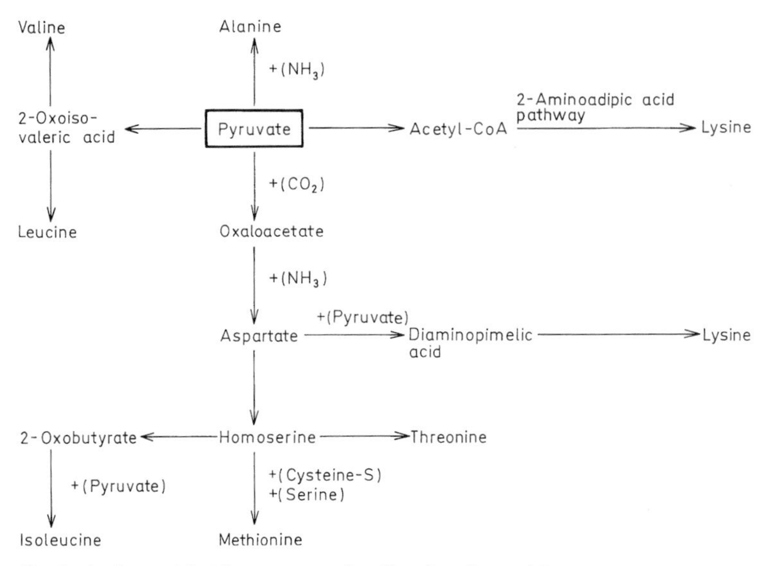

Fig. 3. *Amino acids.* The pyruvate family of amino acids

tamate, glutamine, ornithine, citrulline, arginine (see Urea cycle), proline and hydroxyproline;

3. The pyruvate family is derived from pyruvate and oxaloacetate (Fig. 3);

4. The pentose family includes histidine and the 3 aromatic A. a. (see Aromatic biosynthesis), phenylalanine, tyrosine and tryptophan.

Microbial synthesis of A. a. is exploited industrially by growing special *production strains* of microbes (usually bacteria) on synthetic medium. In such strains, metabolic control is impaired by mutation, resulting in massive overproduction of a particular A. a. The excreted A. a. is harvested from the medium.

The A. a. pool (or pools in the case of compartmentation) of a cell is the totality of free A. a. that are available for metabolism at any one time. This pool is supplied by nutritional sources, proteolysis and de novo synthesis. A. a. of a cell's A. a. pool are used for the synthesis of protein, or they are degraded, or they may serve as precursors of special metabolites like hormones. The major metabolic pathways are 1. transamination to 2-oxoacids, 2. decarboxylation, 3. transformation of the side chain, 4. oxidative deamination to 2-oxoacids (Table 3).

Nonproteogenic A. a. (nonprotein A. a.) are not usually incorporated into proteins. They include A. a. which are intermediates in the biosynthesis of proteogenic A. a., for example δ-aminoadipic acid, diaminopimelic acid and cystathionine. About 250 nonprotein A. a. are known, most of them occurring in plants and limited in each case to certain taxonomic groups. The majority can be grouped according to their biosyntheses into the 4 groups of biogenetically related A. a. In exceptional cases, A. a. which are not normally considered to be proteogenic are incorporated during translation, e. g. L-citrulline in the protein of hair follicles, and δ-aminoadipic acid in maize protein. The occurrence of rare natural A. a. can be used in chemotaxonomy. The terms *rare* or *unusual* refer only to their sporadic occurrence and their structural differences from proteogenic A. a. Many rare, naturally occurring A. a. are structurally related to proteogenic A. a. Thus, more than 20 nonprotein A. a. are known which differ from alanine by substitution of one hydrogen of the methyl group. Nonprotein A. a. may also act as A. a. antagonists, e. g. azetidine-2-carboxylic acid, a toxic constituent of lily of the valley, is a structural analog of proline, in which the ring is contracted by one C-atom. In lily of the vally, uncontrolled incorporation of azetidine-2-carboxylic acid into the plant's own protein is avoided by the highly specific synthesis of prolyl-tRNA, but in other organisms azetidine-2-carboxylic acid is incorporated in place of proline, leading to marked alterations in the tertiary structure and biological activity of proteins. Nonprotein A. a. are particularly common in certain plant families, e. g. the *Mimosaceae* contain 2-diaminopropionic acid and its derivatives, thioether derivatives of L-cysteine, and derivatives of lysine and glutamic acid. Some nonprotein A. a. are biologically active, e. g. Lathyrogenic A. a. (see) and indospicine (L-2-amino-6-amidinocaproic acid) from *Indigofera spicata,* which is a liver toxin and causes growth deformities.

Aminoacyladenylate, ***activated amino acid:*** the product of the first enzymatic reaction of protein biosynthesis. It consists of an amino acid linked by an acid anhydride bond to the phosphate of AMP. In the cell, these compounds are always associated with aminoacyl-tRNA synthetases, which also catalyse a further reaction, the transfer of the aminoacyl residue of the A. to a specific tRNA. AMP is released in this second step. See Aminoacyl-tRNA synthetases.

Aminoacyl-tRNA: a transfer RNA charged with a specific amino acid; the transport form in which the amino acid is brought to the specific acceptor site on the ribosome. The carboxyl group of the amino acid

Table 4. *Metabolic reactions of amino acids*

Type of reaction	Equation
Transamination	$R\underset{NH_2}{\underset{\mid}{C}}HOOH + R'\underset{O}{\underset{\Vert}{C}}COOH \rightleftharpoons R\underset{O}{\underset{\Vert}{C}}COOH + R'\underset{NH_2}{\underset{\mid}{C}}HCOOH$
Decarboxylation	$R\underset{NH_2}{\underset{\mid}{C}}HCOOH \rightarrow RCH_2NH_2 + CO_2$
Amination	$R\underset{O}{\underset{\Vert}{C}}COOH + NH_3 + NAD(P)H_2 \rightarrow R\underset{NH_2}{\underset{\mid}{C}}HCOOH + NAD(P) + H_2O$
Deamination	$R\underset{NH_2}{\underset{\mid}{C}}HOOH \xrightarrow{-2[H]} R\underset{NH}{\underset{\Vert}{C}}COOH \xrightarrow{+H_2O} R\underset{O}{\underset{\Vert}{C}}COOH + NH_3$
Modification of side chain:	
Hydroxyl group	$R{-}OH \xrightarrow{ATP} R{-}O \sim PO_3H_2$ (phosphorylation)
α-Amino group	$R{-}NH_2 \rightarrow R{-}NH{-}COCH_3$ (acetylation)
α-Carboxyl group	$R{-}COOH \xrightarrow[NH_3]{ATP} R{-}CONH_2$ (amide synthesis)
Peptide formation	$R\underset{NH_2}{\underset{\mid}{C}}HCOOH + R'\underset{NH_2}{\underset{\mid}{C}}HCOOH \xrightarrow{-H_2O} RCHCO{-}NH\underset{R'}{\underset{\mid}{C}}HCOOH$
Amino acid activation (Protein biosynthesis 1)	$R\underset{NH_2}{\underset{\mid}{C}}H\overset{O}{\overset{\Vert}{C}}{-}OH + AMP \sim P \sim P \xrightarrow{Enz.} Enz.AMP \sim \overset{O}{\overset{\Vert}{C}}\underset{NH_2}{\underset{\mid}{C}}HR + PP_i$
Amino acid transfer (Protein biosynthesis 2) (Synthesis of aminoacyl-tRNA)	$Enz.AMP \sim \overset{O}{\overset{\Vert}{C}}\underset{NH_2}{\underset{\mid}{C}}HR \sim tRNA{-}OH \rightarrow AMP + Enz. + tRNA{-}O \sim \overset{O}{\overset{\Vert}{C}}\underset{NH_2}{\underset{\mid}{C}}HR$
Gramicidin S synthesis	$Enz.AMP \sim \overset{O}{\overset{\Vert}{C}}\underset{NH_2}{\underset{\mid}{C}}HR + E^{-SH} \rightarrow AMP + Enz. + E^{-S} \sim CO\underset{NH_2}{\underset{\mid}{C}}HR$

Enz. = Enzyme; E = Protein II of gramicidin synthetase.

is esterified to either the 2′ or the 3′ OH group of the ribose of the terminal adenosine of the tRNA. The free energy of hydrolysis of this ester bond is 29.0 kJ/mol. See High-energy bonds, Aminoacyl-tRNA synthetases.

Aminoacyl-tRNA synthetases, ***amino-acid activating enzymes*** (EC 6.1.1): a group of enzymes that activate amino acids and transfer them to specific tRNA molecules as the first step in protein biosynthesis. The process consists of 2 steps, illustrated here with leucine:

1) Leu + ATP + leucyl-tRNA synthetase → [Leu-AMP-enzyme] + PP_i

2) [Leu-AMP-enzyme] + $tRNA^{Leu}$ → leucyl-$tRNA^{Leu}$ + AMP + enzyme.

A. are highly specific with respect to the amino acid they activate, and they also recognize the tRNA with great precision. Their accuracy in loading the correct amino acid depends on an editing process. The active site of the enzyme, which catalyses formation of the ester bond between the amino acid and the tRNA, is unable to distinguish completely between the correct amino acid and a smaller homolog, i.e. valyl-tRNA synthetase may attach Ala or Thr to $tRNA^{Val}$. This is compensated by the existence of a second site on the synthetase which has esterase activity and is much more active with the wrong amino acid than the correct one.

A. may consist of either one polypeptide chain or of 2 or 4 homologous or heterologous subunits. Eukaryotic cells contain more than 20 different A., because mitochondria and plastids have their own amino-acid-specific A., which differ in their specificity toward homologous tRNA from those of the cytoplasm. Some A. are able to load several amino-acid-specific tRNAs, e.g. leucyl-tRNA synthetase of *E. coli*, which can load 5 different species of $tRNA_{E.coli}^{Leu}$.

2-Aminoadipic acid, ***Aad:*** $HOOC{-}(CH_2)_3{-}CH(NH_2){-}COOH$ (M_r 161.1), an amino acid which is proteogenic only in maize. Aad is an intermediate in the biosynthesis of L-lysine by the Aad pathway. In boiling water, the free acid cyclizes to piperidone carboxylic acid.

2-Aminoadipic acid pathway: see L-Lysine.

4-Aminobutyrate pathway

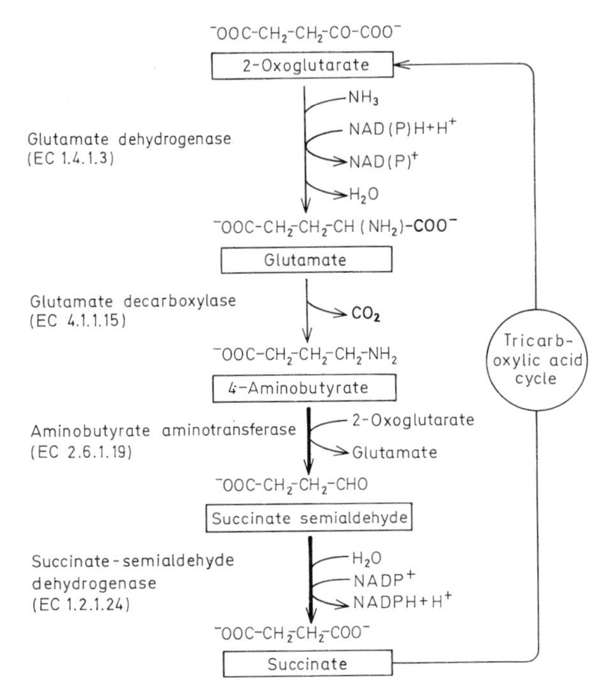

4-Aminobutyrate pathway

4-Aminobutyrate pathway, ***γ-aminobutyrate pathway:*** see 4-Aminobutyric acid.

4-Aminobutyric acid, ***4-Abu, γ-aminobutyric acid, GABA:*** $H_2N-(CH_2)_3-COOH$ (M_r 103.1), a nonproteogenic amino acid. Formation of GABA from L-glutamate by the action of glutamate decarboxylase (EC 4.1.1.15) has been demonstrated in brain, various microorganisms (e.g. *Clostridium welchii, E. coli),* higher plants (e.g. spinach, barley) and other animal tissues (liver and muscle). It can also be formed from 4-guanidobutyric acid (see Guanidine derivatives) in higher fungi *(Basidiomycetes)* and *Streptomycetes.* Degradation of GABA proceeds by transamination to succinic semialdehyde and subsequent oxidation to succinic acid, which is oxidized in the tricarboxylic acid cycle. Synthesis of GABA is particularly important in the brain, where it functions as an inhibitory neurotransmitter. On account of its neural activity, GABA is used to treat epilepsy, cerebral hemorrhage, etc. The 4-aminobutyrate pathway (Fig.) represents a bypass of the oxidative decarboxylation of 2-oxoglutarate in the tricarboxylic acid cycle. Only some brain cells make GABA, and only about 25 % of 2-oxoglutarate produced in these cells is converted to GABA. The 4-aminobutyrate pathway accounts for less than 10 % of the total oxidative metabolism of the brain.

Aminocarboxylic acids: see Amino acids.

Aminocitric acid:

$$HOOC-CH(NH_2)-\underset{\displaystyle COOH}{\underset{|}{C}}(OH)-CH_2-COOH,$$

an acidic amino acid identified in acid hydrolysates of ribonucleoproteins from calf thymus, bovine and human spleen, *E. coli* and *Salmonella typhi.* It elutes before cysteic acid from the amino acid analyser, and gives a characteristic yellow color with ninhydrin. [G. Wilhelm & K. D. Kupka *FEBS Letters* **123** (1981) 141–144]

3′-Amino-3′-deoxyadenosine: a purine antibiotic synthesized by *Cordyceps militaris* and *Helminthosporium* species (see Nucleoside antibiotics). It has antitumor activity. The acetylated derivative, 3′-acetamido-3′-deoxyadenosine, has also been isolated from *Helminthosporium* species.

2-Amino-2-deoxy-D-galactose: see Galactosamine.

Aminoethanol: see Ethanolamine.

L-α-Aminoglutaric acid: see L-Glutamic acid.

5(4)-Aminoimidazole-4(5)-carboxamide ribonucleotide, ***AICAR:*** 5′-phosphoribosyl-5-amino-4-imidazolecarboxamide. See Purine biosynthesis, Fig. 1.

5(4)-Aminoimidazole-4(5)-carboxyribonucleotide: 5′-phosphoribosyl-5-amino-4-imidazolecarboxylate. See Purine biosynthesis, Fig. 1.

5-Aminoimidazole ribonucleotide, ***AIR:*** 5′-phosphoribosyl-5-aminoimidazole, an intermediate in Purine biosynthesis (see) and in the biosynthesis of thiamin. In certain microbial mutants, blocked in purine biosynthesis, AIR polymerizes to a red pigment.

5-Aminoimidazole-4-N-succinocarboxamide ribonucleotide: 5′-phosphoribosyl-4-(*N*-succinocarboxamide)-5-aminoimidazole. See Purine biosynthesis, Fig. 1.

Aminoisobutyric acid: 2-methyl-β-alanine, $H_2N-CH_2-CH(CH_3)-COOH$, a product of the reductive degradation of thymine (see Pyrimidine degradation). The α-form of A.a. (2-methyl-α-alanine, H_2N-$C(CH_3)_2$-COOH) does not occur naturally, and is metabolized only to a negligible extent.

5-Aminolevulinic acid, ***δ-aminolevulinic acid, ALA:*** an intermediate in porphyrin biosynthesis, and part of the Shemin cycle (see Succinate-glycine cycle). ALA is biosynthesized by at least two distinct pathways, which are described in Porphyrins (see).

Aminopeptidases: a group of exopeptidases with relatively wide and often overlapping substrate specificities, which cleave peptide bonds near the *N*-terminus of polypeptides. They can be subdivided into A. which hydrolyse the first peptide bond (aminoacyl-peptide hydrolases and iminoacyl-peptide hydrolases) and those which remove dipeptides from polypeptide chains (dipeptidyl-peptide hydrolases). A. are involved in protein maturation, terminal protein degradation, regulation of the levels of peptide hormones, and protein digestion in the intestine.

A. are generally zinc-metalloenzymes, containing a highly conserved Zn-binding motif (-His-Glu-X-X-His-18aa-Glu-), which is crucial for enzyme activity, and which is present in other Zn metallopeptidases, e.g. thermolysin. This motif is not conserved in all A., e.g. leucine aminopeptidase uses different Zn-coordinating ligands. Several different gene loci are responsible for the biosynthesis of A. that differ in immunological properties, substrate specificity, pH optimum, activators, etc. (Table). In addition, post-trans-

Some aminopeptidases (AA = amino acid that is released; X = amino acid that becomes new *N*-terminus)

Enzyme	Substrate specificities	Main source	Remarks
Leucine aminopeptidase EC 3.4.11.1	Leu–X– (AA–X–)	Bovine and pig lens and kidney	Activated by Mg^{2+}, Mn^{2+}; basic pH optimum; chromogenic substrates not hydrolysed; inhib. by 1,10 phenanthroline, actinonin, bestatin; 6 subunits of M_r 53 000.
Alanine aminopeptidase EC 3.4.11.2	Ala–X– (AA–X–)	Rat kidney	Some activation by Co^+; several human isoforms due to post-translational modification; inhib. by 1,10 phenanthroline, actinonin, amastatin, bestatin; 2 subunits of M_r 110 000.
Cystyl aminopeptidase, Oxytocinase EC 3.4.11.3	Leu–X– (Cys–X–, AA–X–)	Amniotic fluid, pregnancy serum	Not inhibited by bestatin or amastatin; heat labile.
Aminopeptidase A, Angiotensinase, Angiotensin converting enzyme EC 3.4.11.7	Asp–X– Glu–X–	Human kidney	Activated by Ca^{2+}; cleavage of Glu-substrates also activated by Ba^{2+}; inhib. by 1,10 phenanthroline; 2 subunits of M_r 160 000; best known substrate angiotensin II.
Aminopeptidase B EC 3.4.11.6	Lys-X– Arg–X–	Rat liver	Activated by Cl^- and Br^-; inhib. by 1,10 phenanthroline, EDTA, arphamenines A and B, bestatin; unstable; monomer M_r 95 000.
Aminopeptidase W EC 3.4.11.16	X–Trp–	Pig kidney and intestine	Inhib. by amastatin and bestatin; 2 subunits of M_r 130 000.
Aminopeptidase P EC 3.4.11.9	X–Pro–	Pig kidney	Activated by Mn^{2+}; inhib. by EDTA, enalaprilat, 1,10 phenanthroline; 2 (?) subunits M_r 91 000; natural substrates possibly bradykinin and substance P.
Leukotriene A_4 hydrolase EC 3.3.2.6	Ala–X–	Human and mouse leukocytes	Inhib. by 1,10 phenanthroline; monomer, M_r 68–70 000; cytosolic; ubiquitous in mammalian tissues; converts leukotriene A_4 to B_4. 1 mol Zn/mol protein.
Methionine aminopeptidase EC 3.4.11.?	Met–X–	Yeast	Monomer, M_r 43 000.
		E. coli	Monomer, M_r 29 000.
Tripeptidase EC 3.4.11.4	Leu–(Gly–Gly) Gly–(Gly–Gly)	Human and animal liver cytosol	Not inhibited by amastatin; inhib by bestatin; monomer, M_r 50 000–70 000; only attacks tripeptides.

lational modifications (described in particular for human alanine aminopeptidase) are responsible for the existence of multiple forms. Classification is generally based on substrate specificity and behavior toward certain inhibitors. An attempted classification based on amino acid sequences reveals 84 different families with similar structures and catalytic mechanisms, but which show no correspondence with the classification based on substrate specificity [N. D. Rawlings & A. J. Barret *Biochem J.* **290** (1993) 205–218].

Leucine aminopeptidase preferentially hydrolyses peptide bonds adjacent to an *N*-terminal residue that carries a large hydrophilic side chain, in particular a leucyl residue. Commonly used synthetic substrates are leucinamide, leucine 4-nitroanilide and leucine hydrazide. This cytosolic zinc-metalloenzyme has been identified in virtually all animal tissues, and most studies have been performed on the enzyme (M_r 324,000) from bovine lens, which has been crystallized. It consists of 6 identical subunits (M_r 54,000; 487 amino acids; 2 Zn^{2+} per subunit), and its catalytically active site is in the *C*-terminal domain. The enzyme is present in many other cells and tissues, e. g. lung, stomach, kidney intestine, serum and leukocytes. In clinical chemistry, this enzyme is a marker for hepatic cell lysis, and may even be a more sensitive marker for acute hepatitis than the aminotransferases.

Alanine aminopeptidase (arylamidase, aminopeptidase M, aminopeptidase N) preferentially hydrolyses natural and synthetic substrates with an *N*-terminal alanine residue. Other amino acids, especially leucine, may also be removed, but *N*-terminal prolyl residues are not attacked. Biological substrates include Met-Lys-bradykinin, Lys-bradykinin, (Met^5)-enkephalin and (Leu^5)-enkephalin. In enzyme assays, the most frequently used substrates are 4-nitroanilides and β-naphthylamides of alanine and leucine. The enzyme is present in virtually all tissues studied, and displays particularly high specific activity in brush border membranes of kidney proximal tubules, intestine and bile canaliculi. Native liver alanine aminopeptidase of human and rat is an integral membrane-bound ectoenzyme anchored to the outer side of the canalicular plasma membrane by a hydrophobic *N*-terminal sequence. The rat and human enzymes (M_r 110,000) display 77 % homology and consist of 967 amino acids. The enzyme has been cloned from several species, including *E. coli* and human intestine. All the human isoforms of alanine aminopeptidase arise from a single gene product, and the observed heterogeneity is due to three types of post-translational modification: glycation, limited proteolysis and aggregation with other molecules.

Aminopeptidase P is present in the microvillar membranes of pig and human kidney, and membrane-bound forms are also present in rat intestine and lung, bovine lung and guinea pig kidney. Soluble forms occur in rat serum and brain, human platelets and guinea pig serum, and it has also been characterized from human leukocytes. It is one of a group of cell-surface proteins anchored in the lipid bilayer by glycolyl-phosphatidylinositol (see Membrane lipids).

Cystyl aminopeptidase (oxytocinase) is localized mainly in placental lysosomes, and these are generally thought to be the source of the enzyme activity observed in pregnancy serum and amniotic fluid. The physiological substrate appears to be oxytocin, and the enzyme also displays angiotensinase activity. During normal pregnancy, serum levels of the enzyme increase until shortly before onset of labor, whereas the activity in amniotic fluid decreases. High levels of the enzyme are present in pregnancy sera in cases of pre-eclampsia, with low levels in the late stages of severe pre-eclampsia. [G.-J. Sanderink et al. *J. Clin. Chem. Clin. Biochem.* **26** (1988) 795–807; D. Hendricks et al. *Clin. Chim. Acta* **196** (1991) 87–96; I. Rusu & A. Yaron *Eur. J. Biochem.* **210** (1992) 93–100; J. Wang & M. D. Cooper in Zinc Metalloproteases in Health and Disease (edit. N. M. Hooper), Ellis Horwood, Chichester, 1995; T. Rogi et al. 'cDNA Cloning of Human Placental Leucine Aminopeptidase/Oxytocinase' *J. Biol. Chem.* **271** (1996) 56–61]

Aminopropionic acid: see Alanine.

Aminopterin: 4-amino-4-deoxyfolic acid (see Vitamins, folic acid), a cytostatic agent used in the management of some cancers. It inhibits the enzyme dihydrofolate reductase, which reduces the folate coenzymes required for Purine biosynthesis (see) and thymine production (see Pyrimidine biosynthesis), and thus prevents DNA synthesis. However, it is toxic to nondividing cells as well, and cannot be tolerated indefinitely. ***Methotrexate*** (amethopterin) has similar activity.

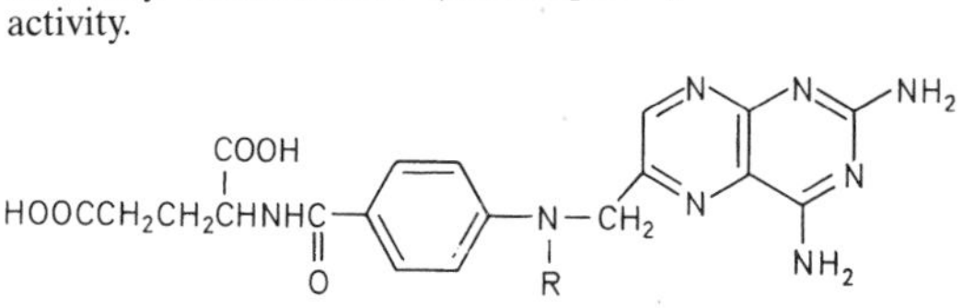

Aminopterin (R = H) and *Methotrexate* (R = CH_3).

Amino sugars: monosaccharides in which a hydroxyl group is replaced by an amino group. The amino group is often acetylated. 2-Amino-2-deoxyaldoses (e. g. D-galactosamine, D-glucosamine, D-mannosamine, neuraminic acid, muramic acid) are particularly important as components of bacterial cell walls, of some antibiotics, blood group substances, milk oligosaccharides and high molecular mass natural products such as chitin.

In the biosynthesis of A. s., the amino group is supplied by transamination from glutamine. Fructose 6-phosphate is aminated to D-glucosamine 6-phosphate by a hexose-phosphate transaminase. Glucosamine phosphate can then be converted by a transacetylase into the *N*-acetyl derivative. The latter is isomerized to the 1-phosphate, then activated by reaction with UTP, to form UDP-*N*-acetylglucosamine, which can be isomerized to UDP-*N*-acetylgalactosamine. See Muramic acid, Neuraminic acid.

Ammonia, *NH_3*: a colorless gas with a sharp, characteristic smell; condensation temperature at normal pressure and about −40 °C. It is very soluble in cold water, but completely driven off by boiling. The aqueous solution is weakly basic, because NH_3 takes up a proton to form an ammonium ion: $NH_3 + H_2O \rightleftharpoons NH_4^+ + OH^-$. The reaction equilibrium lies far to the left, so that NH_3 can be displaced from ammonium compounds by bases. The toxicity of NH_3 is related to the high permeation rate of the nonprotonated form and its tendency to become protonated.

Occurrence: NH_3 is the end product of the degradation of nitrogen-containing organic matter. It therefore occurs as ammonium salts in the soil. In animals NH_3 is eliminated by detoxification reactions (see Ammonia detoxification), while in plants and bacteria it is assimilated for the synthesis of nitrogenous compounds. The concentration of ammonium ions in body fluids and all living cells is therefore relatively low.

Metabolism. NH_3 is the product of nitrate reduction, biological nitrogen fixation, deamination of amino acids, and various catabolic pathways, e.g. oxidative purine degradation and reductive pyrimidine degradation. In this sense, NH_3 is the nitrogenous, inorganic end product of protein and nucleic acid degradation. Its nitrogen is incorporated into the pool of organic nitrogen by various reactions of primary nitrogen assimilation, then further distributed by reactions that transfer nitrogen-containing groups (see Group transfer), e.g. ammonium ions are incorporated in the biosynthesis of carbamoyl phosphate and glutamine, which in turn serve as nitrogen sources in the biosynthesis of many different nitrogenous compounds of the cell. Metabolism of inorganic nitrogen is particularly well developed in plants. Many microorganisms can grow at the expense of ammonium salts as their only source of nitrogen. Higher green plants have only a limited ability to take up ammonium ions from the soil; they depend largely on nitrate which is formed in the soil by nitrification (microbial oxidation of ammonium). After entering the plant cell cytoplasm, nitrate is reduced to ammonium (see Nitrate reduction), which is then assimilated.

Ammonia assimilation: utilization of ammonia in the net synthesis of the nitrogen-containing groups of nitrogenous cell constituents, e.g. amino acids, amides, carbamoyl and guanido compounds. Incorporation of ammonia into the amide group of glutamine, catalysed by glutamine synthetase (EC 6.3.1.2) is of central importance: L-glutamate + NH_3 + ATP $\rightleftharpoons$ L-glutamine + ADP + P_i. The amide nitrogen of L-glutamine is then used in various syntheses:

1. L-Glutamine + α-ketoglutarate + $2H^+$ + $2e^-$ $\rightleftharpoons$ 2L-glutamate (glutamate synthase). The glutamate takes part in the synthesis of other amino acids by transamination: thus a series of coupled reactions results in net assimilation of ammonia into the amino groups of amino acids (Fig.).

Reducing power for bacterial glutamate synthase (EC 1.4.1.13) is provided by NADPH, whereas chloroplast glutamate synthase (EC 1.4.7.1) utilizes reduced ferredoxin. Glutamine synthetase and glutamate synthase occur in plant chloroplasts, where ATP and reduced ferredoxin are supplied directly by the light reaction of photosynthesis. Animals lack glutamate synthase, and they cannot achieve the net synthesis of amino groups from ammonia.

2. L-Glutamine + HCO_3^- + 2ATP + H_2O $\rightleftharpoons$ carbamoyl phosphate + L-glutamate + 2ADP + P_i (see Carbamoyl phosphate). *N*-Acetylglutamate is an essential positive allosteric effector of this enzyme (carbamoyl phosphate synthetase, EC 6.3.5.5). In eukaryotes, the enzyme is located in the cytoplasm. The carbamoyl phosphate provides C2 and N3 in pyrimidine biosynthesis, and it contributes to the synthesis of the guanido group of arginine in plants and bacteria.

3. The amide group of glutamine is used in purine biosynthesis, where it provides N3 and N9 of the purine ring, and the 2-NH_2 group of guanine.

4. In several syntheses, nitrogen is derived directly from the amide group of L-glutamine, e.g. in histidine synthesis; conversion of chorismate into anthranilate (see Tryptophan synthesis); synthesis of amino sugars; amination of UTP to CTP.

5. In some organisms, the amide group of glutamine is transferred to aspartate by the action of asparagine synthetase (glutamine hydrolysing) (EC 6.3.5.4): L-glutamine + L-aspartate + ATP $\rightleftharpoons$ L-glutamate + L-asparagine + AMP + PP_i.

In mammalian liver mitochondria, ammonia is converted directly into carbamoyl phosphate: NH_3 + HCO_3^- + 2ATP $\rightleftharpoons$ carbamoyl phosphate + 2ADP + P_i. The reaction is catalysed by carbamoyl phosphate synthetase (ammonia) (EC 6.3.4.16), and *N*-acetylglutamate is an essential positive allosteric effector. It introduces ammonia into the urea cycle, and it may or may not result in the net synthesis of arginine, depending on the ability of the animal to synthesize ornithine. Much of this ammonia nitrogen may therefore be excreted as urea, without contributing to the net synthesis of guanido groups.

In plants, molds and bacteria, A.a. into the amino group of glutamate is also possible by NADPH-dependent glutamate dehydrogenase (EC 1.4.1.3); the enzyme is most effective when ammonium salts are available directly from the environment in relatively

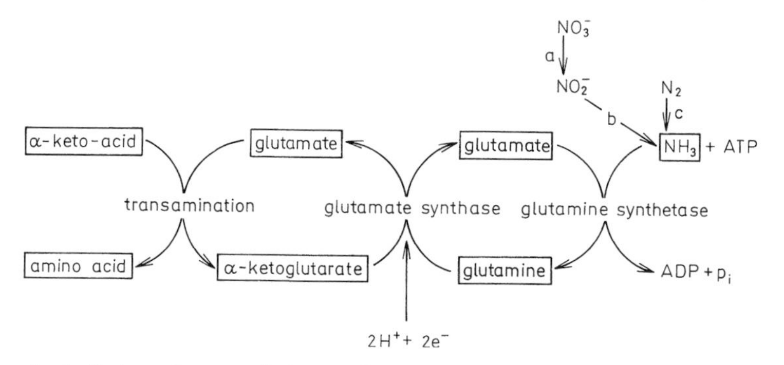

Assimilation of ammonia into amino groups of amino acids. a, Nitrate reductase. b, Nitrite reductase. c, Nitrogenase.

high concentration. In many organisms its activity is lower than that of glutamine synthetase, which also has the lower K_m for ammonia. Thus the glutamine synthetase-glutamate synthase system is the more effective and the more important, especially when the ammonia is derived from nitrate reduction or nitrogen fixation.

In some microorganisms, ammonia assimilation may also occur by alanine synthesis, catalysed by alanine dehydrogenase (EC 1.4.1.1).

Asparagine is usually synthesized by the action of asparagine synthetase (ADP-forming) (EC 6.3.1.4): L-aspartate + NH_3 + ATP $\rightleftharpoons$ L-asparagine + ADP + P_i. Asparagine, like its homolog, glutamine, is important in the storage and transport of amino nitrogen, as well as being a proteogenic amino acid. Metabolically, it is less versatile than glutamine, and its amide nitrogen does not appear to be transferred directly in biosyntheses.

Ammonia detoxification: detoxification of the toxic, nondissociated compound by formation of ammonium salts and nitrogenous excretion compounds. Ammonia produced by catabolism in animals is either excreted directly, or it is converted into other nitrogenous compounds for excretion; it is not reassimilated. Excretion of ammonia as the major end product of nitrogen metabolism (ammonotelism) is limited to a few aquatic organisms. Most animals rid themselves of catabolic ammonia by synthesizing nitrogenous excretion products, such as urea, uric acid, guanine (spiders) or allantoin *(Diptera)*. Animals that excrete urea are called ureoteles, and those excreting uric acid, uricoteles. The main nitrogenous excretory product can change in the course of ontogenesis: the frog larva (tadpole) is ammonotelic, but the adult frog is ureotelic. In this case, ureotelism develops during metamorphosis; enzymes of the urea cycle are induced, having been repressed during the larval stage. The type of nitrogen excretion also depends on the ecological conditions, but the nature of the main nitrogenous excretory product is largely phylum-specific, and in some cases it is relatively specific for smaller taxa (Table 1).

Ureoteles and uricoteles use pre-existing, very ancient biochemical mechanisms for nitrogen excretion: urea synthesis by the urea cycle, which first evolved for the synthesis and degradation of arginine; and uric acid formation, which represents oxidative purine degradation adapted for the purpose of nitrogen excretion, i.e. the existing reaction sequences of purine synthesis and oxidative purine degradation have been combined to form an A.d. system. The particular form of nitrogen excretion in organisms using the purine pathway depends on the last step of purine degradation possessed by the organism. In uricoteles it is uric acid; in other animals, oxidative purine degradation may result in other end products, because the reaction chain has been shortened during evolution by the loss of certain enzymes (Table 2).

Table 2. *End products of the degradation of purines (from nucleic acids) in animals*

End product	Enzyme presumed to have been lost	Animal groups
Urea		Fish, amphibians, mussels
Allantoin	Allantoinase, etc.	Mammals (except primates), snails, dipterans
Uric acid	Uricase, etc.	Primates, birds, reptiles, insects (except dipterans)
Guanine	Guanase, etc.	Spiders
Adenine	Adenase, etc.	Flatworms, annelids

There is a striking correlation between the form of nitrogen excretion and the pathway used by an animal to deaminate amino acids arising from proteolysis: in ureoteles amino acids are transaminated with 2-oxoglutarate to form glutamate, which in turn is attacked by glutamate dehydrogenase, whereas in uricoteles amino acids are attacked by amino acid oxidases.

Plants generally lack the excretory mechanisms that are present in animals. Nitrogen is usually a growth-limiting factor for plants, which convert ammonia into reutilizable nitrogenous compounds. These reserve compounds (which also serve for nitro-

Table 1. *Forms of ammonia detoxification and nitrogen excretion*

Nitrogen excretion	Excretion type	Synthesis pathway	Occurrence
Ammonia	Ammonotelia	Deamination of amino acids	Octopi, marine mussels, crustaceans, etc.
Trimethylamine oxide			Marine teleosts
Urea	Ureotelia	Urea cycle	Amphibians, mammals, marine elasmobranchs
Urea	Ureotelia	Purine cycle	Lungfish
Uric acid	Uricotelia	Purine synthesis and degradation	Terrestrial reptiles, birds, insects (except *Diptera)*
Guanine		Purine synthesis and degradation	Spiders

gen transport) accumulate in the plant's storage organs. They can be metabolized to release ammonia, which is then reassimilated. Nitrogen storage compounds of some plants are shown in Table 3. They are generally analogous to animal excretory compounds; thus allantoin and allantoic acid (plants) belong to the same metabolic pathway as uric acid (animals); citrulline, arginine and canavanine (plants) are metabolically related to urea (animals); urea itself is stored by bovists and puff-balls.

Table 3. *Nitrogen storage compounds of some plants*

N-storage compound	Occurrence in plants
Urea	Bovists and puff-balls *(Gasteromycetales)*
L-Arginine	Apples *(Malaceae), Saxifragaceae,* etc.
L-Citrulline	Birches *(Betulaceae),* Walnuts *(Juglandaceae)*
Allantoin	Borage *(Boraginaceae)*
Allantoic acid	Maples *(Aceraceae)*
L-Glutamine	Many plants
L-Canavanine	Legumes

Ammonia fixation: see Ammonia assimilation.

Ammonium plants: see Acid plants.

Ammonoteles: see Ammonia detoxification.

AMO 1618, ***(2-isopropyl-5-methyl-4-trimethylammonium chloride) phenyl-1-piperidinocarboxylate:*** a synthetic plant growth retardant. Because it inhibits growth of the shoot, it is used on nursery plants to make them bushy. It is a gibberellin antagonist, inhibiting the biosynthesis of gibberellin A_3.

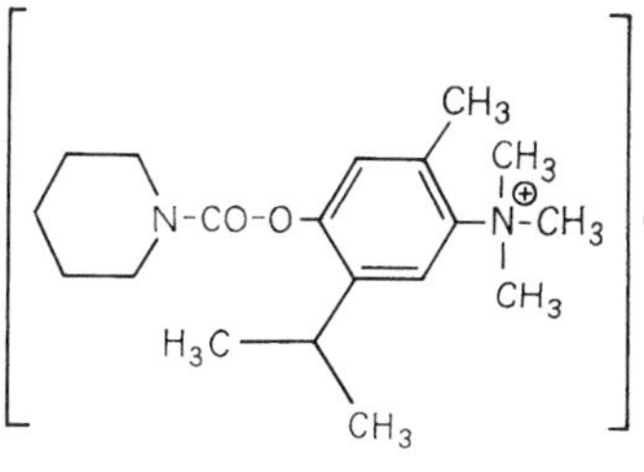

AMO 1618

AMP: acronym of adenosine 5′-monophosphate. See Adenosine phosphates.

3′,5′-AMP: acronym of cyclic adenosine-3′,5′-monophosphate. See Adenosine phosphates.

Amphetamine: see Antidepressants.

Amphibian toxins: a group of chemically very heterogeneous toxins (biogenic amines, peptides, alkaloids, steroids) produced by toads, salamanders, frogs (see Batrachotoxin) and newts (see Tarichatoxin). Pharmacologically, they include heart, muscle and nerve toxins, sympathicomimetic agents, cholinomimetic agents, local anesthetics and hallucinogens. A.t. protect against predators and against bacteria and fungi that might attack the animal's skin.

Amphibolic pathways: metabolic pathways serving both degradation and synthesis. See Metabolic cycle.

Amygdalin: see Cyanogenic glycosides (Table).

Amylases, ***diastases*** (EC 3.2.1.1, 3.2.1.2 and 3.2.1.3): a widely occurring group of hydrolases, which cleave the α-1,4 glycosidic bonds in oligosaccharides (trisaccharides and upwards) and polysaccharides like starch, glycogen and dextrins. α-A. is an endoamylase, whereas β-A. (saccharogenic A.) and γ-A. (glucoamylase) are exoamylases. The action of α-A. produces first dextrins, which are secondarily cleaved to maltose (87 %), glucose (10 %) and branched oligosaccharides (3 %). β-A. and γ-A. attack the substrate at its nonreducing end; β-A. produces maltose units (in β configuration after inversion), while γ-A. produces glucose units (even from α-1,6 glycosidic bonds, if they are adjacent to a 1,4 bond). The A. differ in occurrence, structure and action mechanism. α-A. and γ-A. occur both in animals (α-A. in salivary glands and pancreas of omnivores; γ-A. in liver) and plants, whereas β-A. are found only in plant seeds. Mammalian α-A. are dependent on chloride. Animal and plant α-A. contain calcium. The α-A. of *Bacillus subtilis* contains zinc. Plant β-A. are formed as insoluble zymogens during seed maturation.

α-Amylase test: see Gibberellins.

Amylo-1,6-glucosidase, ***debranching enzyme*** (EC 3.2.1.33): an endoglucosidase which cleaves the 1→6 glycosidic bonds at the branching points of glycogen and amylopectin. In mammals and yeast it is associated with a glycosyl transferase which removes all the glucose residues up to the 1→6 bond. The muscle complex (M_r 237,000) consists of 2 subunits of M_r 130,000, while the yeast complex (M_r 210,000) has 3 subunits of M_r 120,000, 85,000 and 70,000.

Amyloid protein: a pathological, fibrillar, low molecular mass protein. In amyloidosis it is deposited, together with glycoproteins and proteoglycans, primarily in the spleen, liver and kidneys. There are 2 types of amyloid protein, differing in amino acid composition, one occurring in chronic amyloidosis (amyloid protein A), the other in paramyloidosis (amyloid protein IV). Amyloid protein A consists of a single chain containing 77 amino acid residues and no disulfide bridges. It is not structurally related to the immunoglobulins. The primary structure of ape amyloid protein A (M_r 8,500) is known. Amyloid protein IV is an amyloid-like protein containing 40–45 amino acid residues (M_r 6,000), which is both structurally and immunologically very similar to the kappa and lambda chains of the immunoglobulins.

There are morphologial similarities between amyloid protein and infectious protein in scrapie and Creutzfeld-Jakob disease, and protein deposits in Alzheimer's disease, all of which are associated with progressive and irreversible degeneration of nervous function, leading to death. [P.A.Merz et al. *Nature* **306** (1983) 474–476; H.Diringer et al. *ibid.* 476–478] In Alzheimer's disease there is invariably a progressive accumulation of filamentous aggregates of amyloid β-protein (40–42 residues) in the limbic and cerebral cortex, and these aggregates eventually become intimately associated with (surrounded by) dystrophic neurites and glial cells [D.J.Selkoe *Annu. Rev. Cell Biol.* **10** (1994) 373–403]. Amyloid β-protein is a hydrophobic proteolytic fragment of a ubiquitous integral membrane polypeptide known as amyloid β-protein precursor. Missense mutations have been found in or flanking the amyloid β-protein coding region of the amyloid β-protein precursor gene on chromosome 21 of patients with familial (autosomal dominant) Alzheimer's disease [J.Hardy *Nature Genet.* **1** (1992) 233–234]. In addition, transgenic mice overexpressing a familial Alzheimer's disease-linked mutant amyloid β-protein precursor display dense, filamentous, cerebral deposits of aggregated amyloid β-protein. Processing of amyloid β-protein precursor therefore appears to play a key role in the pathogenesis of Alzheimer's disease [G.G.Glenner & C.W.Wong *Biochem. Biophys. Res. Commun.* **120** (1984) 885–890]. Since amyloid β-protein also occurs as a normal soluble product of cellular metabolism and is present in cerebrospinal fluid and plasma, the question arises as to how and why it forms insoluble aggregates in Alzheimer's disease. Stable oligomers (M_r 6,000, 8,000 and 12,000; precipitated with amyloid β-protein-specific antibodies) of amyloid β-protein have been demonstrated in the conditioned media of cells transfected with amyloid β-protein precursor. Such oligomers may be related to the filamentous deposits of Alzheimer's disease [D.J.Selkoe 'Amyloid β-Protein and Alzheimer's Disease' *Annu. Rev. Cell Biol.* **10** (1994) 373–403; S.A.Gravina et al. 'Amyloid β Protein (Aβ) in Alzheimer's Disease Brain' *J. Biol. Chem.* **270** (1995) 7013–7016; M.B.Podlisny et al. *J.Biol. Chem.* **270** (1995) 9564–9570].

Amylopectin: a component of starch (the other is amylose). A. is a branched, water-insoluble polysaccharide (M_r 500,000–1,000,000) consisting of a main chain of α-1,4-linked D-glucose units with side chains (15–25 D-glucose units) attached α-1,6 to every 8th or 9th glucose. A. forms violet to red-violet inclusion compounds with iodine. It swells in water, and upon heating it forms a paste.

β-Amylase (EC 3.2.1.2) degrades A. to limit dextrins, whereas α-amylase (EC 3.2.1.1) degrades it to about 70% maltose, 10% isomaltose and 20% glucose. Hydrolysis with dilute acid yields D-glucose.

Amylopectinosis: see Glycogen storage disease.

Amylose: a component of starch (the other is amylopectin). A. is an unbranched, water-soluble polysaccharide (M_r 50,000–200,000) consisting of 300–1,000 α-1,4-linked D-glucopyranoside residues. The disaccharide unit of A. is maltose. In the hydrated state, the polysaccharide chain is held in a left-handed spiral arrangement by hydrogen bonds, with 6 monosaccharide units per turn of the helix. The central channel of the helix (diam. 0.6 nm) can incorporate appropriately sized molecules, e.g. iodine. A. forms characteristic blue inclusion compounds with iodine, in which the iodine atoms form linear chains with an I-I separation of 0.31 nm (separation distance in the I_2 molecule is 0.27 nm). Degradation of A. by α-amylase (EC 3.2.1.1) produces about 90% maltose and 10% glucose.

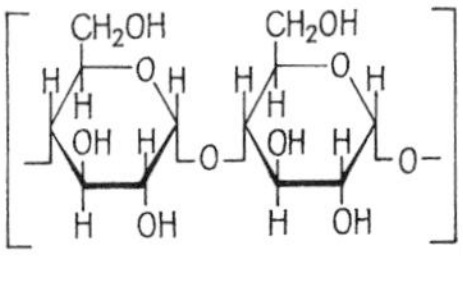

Amylose

β-Amyrenol: see Amyrin.

Amyrin: a pentacyclic triterpene alcohol with one double bond. α-A. (a formal derivative of the hydrocarbon, 5-α-ursane) is found free, esterified and as the aglycon of triterpene saponins of many plants, and has been isolated from many latexes, e.g. latex of the dandelion, *Taraxacum officinale.* β-A. (β-amyrenol, α-viscol) (a formal derivative of the hydrocarbon, 5-α-oleane) is present in mistletoe leaves, grapeseed oil, latex of *Taraxacum officinale;* it is found free, esterified and as the aglycon of triterpene saponins in many other plants, especially in species which yield caoutchouc and gutta percha.

Anabasine: a Nicotiana alkaloid, M_r 162.2. It is isolated in the L-form, primarily from *Nicotiana glauca* and the asiatic *Anabasis aphylla,* being the most abundant alkaloid in both of these species. In other tobacco species it is only a secondary alkaloid. Its physiological effects are similar to those of nicotine, and like nicotine, it is used as an insecticide. For biosynthesis, see Nicotiana alkaloids.

Anabolic steroids: a group of synthetic steroids which stimulate production of body protein. Male gonadal hormones have this effect, as first demonstrated in 1935 for testosterone. By structural modification of natural androgens, it is possible to separate the anabolic from the androgenic effects, leading to

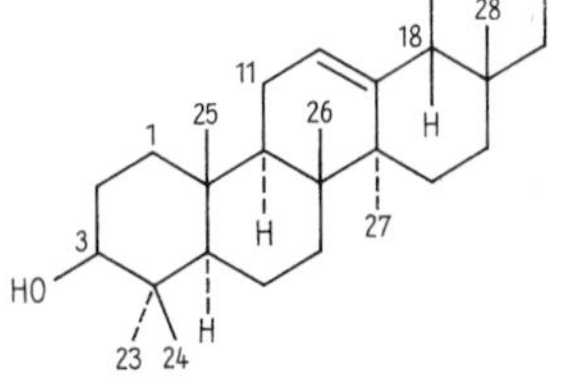

α-Amyrin

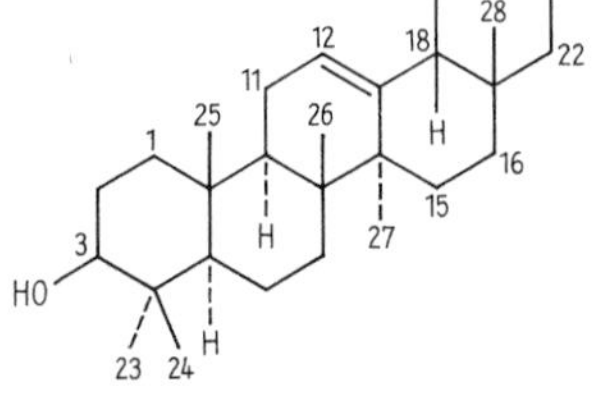

β-Amyrin

therapeutically valuable A.s., e.g. methenolone, 4-chlorotestosterone and androstanozole. A.s. are used in conditions of increased protein turnover or degradation, e.g. after extensive surgical operations, in osteoporosis, burns, nutritional and growth impairments and rickets. A.s. activity is measured in the Herschberg test, in which castrated infant male rats are treated with the substance under investigation; after a certain time the ratio of the increase in mass of the musculus levator ani (anabolic effect) to that of the seminal vesicles (androgenic activity) is measured.

Anabolism: the sum of synthetic metabolic reactions. See Metabolic cycle, Metabolism.

Anaplerotic sequence: see Metabolic cycle.

Anchorin: see Ankyrin.

Anderson's disease: see Glycogen storage disease.

Androcymbin: see Colchicum alkaloids.

Androgens: a group of male gonadal hormones, including testosterone, androsterone and androstenolone, which are formed in the intermediary cells of the testes tissue, and a number of less active A. produced in the adrenal cortex, e.g. androstenedione and adrenosterone (Fig.). Since A. are precursors of Estrogens (see), they are also synthesized in the ovaries and fetoplacental unit. At present more than 30 naturally occurring A. are known, all structurally related to the parent hydrocarbon, androstane (see Steroids).

A. are found in semen, blood and urine. In urine, A. are present partly as conjugates with glucuronic acid, sulfuric acid or protein. The biological function of A. is to induce development of secondary male sex characteristics; they are also required for sperm maturation and the activity of accessory glands of the genital tract. In addition to these sex-specific effects, A. stimulate anabolic processes, involving a net increase in body protein, and they increase nitrogen retention (see Anabolic steroids).

Loss of secondary sex characteristics by castration can be counteracted by administration of A. Thus, administration of A. to a capon causes its degenerated comb to grow. This formed the basis of the capon comb biological assay for A., in which one capon unit was the amount of A. producing an increase in comb surface of 20 % (e.g. 15 mg testosterone). Immunoassays (see) are now available for the assay of individual A.

Highly effective oral A. can be obtained by structural modification, e.g. methyltestosterone and mesterolone. A. are used therapeutically, in particular to correct for deficiency symptoms following castration, hypogenitalism, impotency due to lack of hormones, climacterium, mammary carcinoma and peripheral circulatory impairment. Some synthetic testosterone analogs with a 1,2-cyclopropane ring have antiandrogenic effects. Inappropriate administration of anabolic steroids (e.g. to athletes) can lead to impotence or sterility.

Androisoxazole: see Androstanazole.

Androstanazole: a synthetic and highly active anabolic steroid. It is used, e.g. in the therapy of inflammation and tumors, and in convalescence. The similarly used androisoxazole differs from A. in having an oxygen atom in place of the NH group.

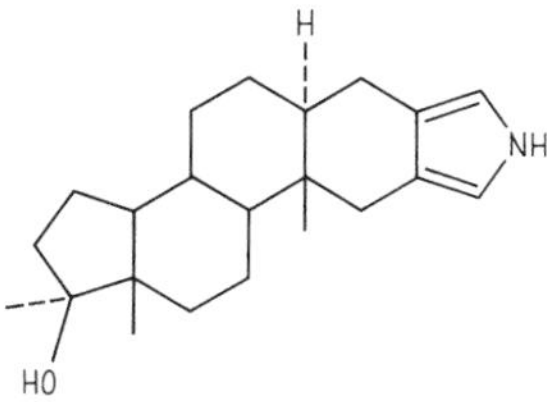

H
NH
HO

Androstanazole

Androstane: see Steroids.

Androstenedione, ***androst-4-ene-3,17-dione:*** a weak androgen, and an intermediate in testosterone biosynthesis (see Androgens).

Androstenolone, ***dehydroepiandrosterone, 3β-hydroxyandrost-5-ene-17-dione:*** an androgen less potent than testosterone, but with similar physiological effects. It is an intermediate in the biosynthesis of testosterone in the adrenal cortex. For structure and biosynthesis, see Androgens.

Androsterone, ***3α-hydroxy-5α-androstan-17-one:*** an androgen, M_r 290.5, formed in the interstitial cells of the testes. Its androgenic activity is 7 times weaker than that of testosterone. It is biosynthesized from progesterone via 17α-hydroxyprogesterone and androstenedione. A. was first isolated by Butenandt in 1931 from human urine (15 mg A. from 15,000 l urine).

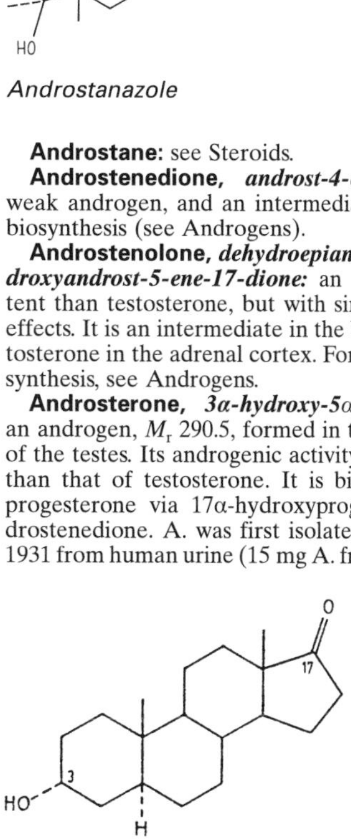

O
17
3
HO
H

Androsterone

Aneurin: see Vitamins (vitamin B_1).

Aneurin pyrophosphate: see Thiamin pyrophosphate.

Angiokeratoma corporis diffusum: see Inborn errors of metabolism.

Angiotensin, ***angiotonin, hypertensin:*** a tissue peptide hormone affecting blood pressure. The kidney protease renin (EC 3.4.99.19) releases the decapeptide A.I (Asp-Arg-Val-Tyr-Ile-His-Pro-Phe-His-Leu) from a plasma protein of the α_2-globulin fraction. Active A.II is then produced by enzymatic cleavage of the His-Leu bond. A.II strongly increases blood pressure, much more effectively than noradrenalin. It also stimulates aldosterone production in the adrenal cortex. A. is inactivated by an angiotensinase (EC 3.4.99.3) in the blood.

Angiotensinase: see Aminopeptidases.

Angiotensin converting enzyme: see Aminopeptidases.

Angiotonin: see Angiotensin.

Angolamycin: see Macrolide antibiotics.

Angstrom unit, ***Å:*** unit of length widely used for light wavelengths and atomic dimensions. 1 Å = 10^{-10} m.

Androgens

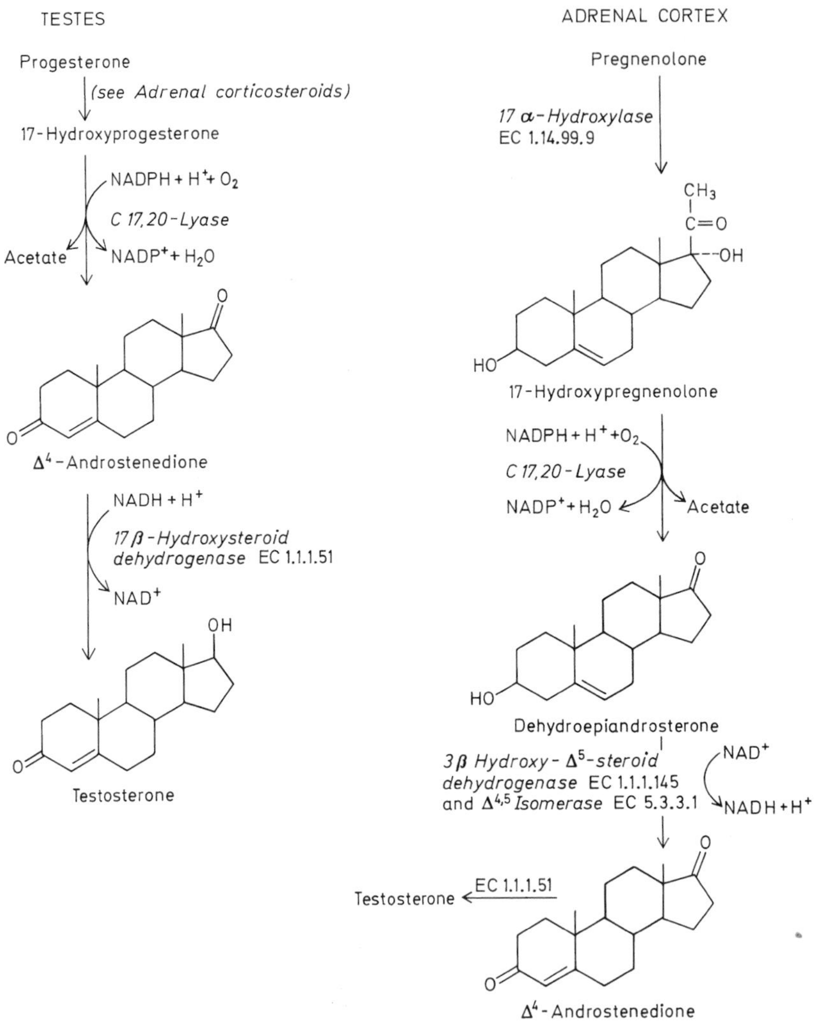

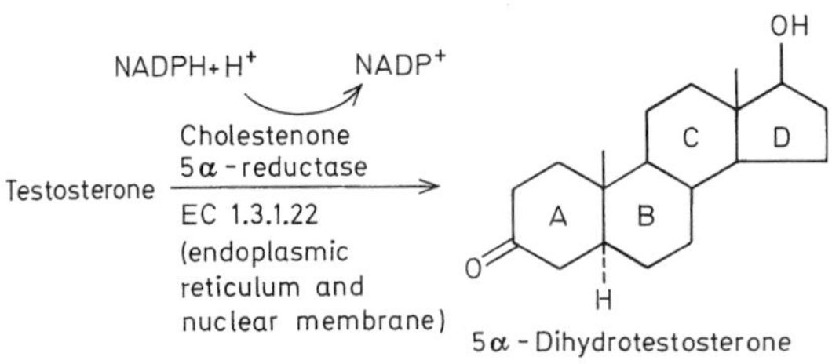

TESTES

Progesterone

(see *Adrenal corticosteroids*)

17-Hydroxyprogesterone

$NADPH + H^+ + O_2$

C 17,20-Lyase

Acetate

$NADP^+ + H_2O$

Δ^4-Androstenedione

$NADH + H^+$

17β-Hydroxysteroid dehydrogenase EC 1.1.1.51

NAD^+

Testosterone

ADRENAL CORTEX

Pregnenolone

17 α-Hydroxylase
EC 1.14.99.9

17-Hydroxypregnenolone

$NADPH + H^+ + O_2$

C 17,20-Lyase

$NADP^+ + H_2O$

Acetate

Dehydroepiandrosterone

3β Hydroxy-Δ^5-steroid dehydrogenase EC 1.1.1.145
and $\Delta^{4,5}$ *Isomerase* EC 5.3.3.1

NAD^+

$NADH + H^+$

Testosterone ← EC 1.1.1.51

Δ^4-Androstenedione

$NADPH + H^+$ → $NADP^+$

Testosterone → Cholestenone 5α-reductase EC 1.3.1.22 (endoplasmic reticulum and nuclear membrane)

5α-Dihydrotestosterone

Biosynthesis of testosterone. The primary precursor is acetyl-CoA, and biosynthesis proceeds via cholesterol (see Terpenes, Steroids). Testes and adrenal cortex also utilize cholesterol which they receive as cholesterol esters from plasma lipoproteins (see Lipoproteins). A second androgen, more potent than testosterone, is 5α-dihydrotestosterone; this is formed from circulating testosterone at peripheral sites, and not by the Leydig cells of the testes. Ring A of 5α-testosterone cannot be aromatized, so this hormone cannot serve as an estrogen precursor.

Angustmycins: purine antibiotics (see Nucleoside antibiotics) synthesized by various species of *Streptomyces.* Angustmycin A (decoyinin; 9-β-D-5,6-didehydropsicofuranosyladenine; M_r 279.2), from *Streptomyces hygroscopius* var. *angustmyceticus,* is specific against mycobacteria, and it does not affect Gram-positive or Gram-negative bacteria or fungi. It acts by inhibiting formation of 5-phosphoribosyl-1-pyrophosphate in purine biosynthesis.

Streptomyces hygroscopicus also synthesizes the structurally related angustmycin C (9-β-D-psicofuranosyladenine, M_r 297.3). Angustmycin C is identical with psicofuranine from *Streptomyces hygroscopicus* var. *decoyicus.* It specifically inhibits XMP aminase in purine biosynthesis and has antibacterial and antitumor activity.

Anhaline: see Hordenine.

Anhalonidine: see Anhalonium alkaloids.

Anhalonium alkaloids: isoquinoline alkaloids found primarily in cacti. They all have a tetrahydroisoquinoline skeleton carrying several phenolic hydroxyl groups (which may be etherified) on Ring B.

Both the biosynthesis and chemical synthesis are based on a Mannich condensation of a β-phenylethylamine derivative with a carbonyl component (Fig.). Depending on the latter, the substituent at C1 can be H-, CH_3- or an isoprenoid residue. Several molecules of A.a. can be linked to form oligomers by phenol oxidation, as in pilocereine. As secondary alkaloids, cacti also produce derivatives of β-phenylethylamine, e.g. hordenine and mescaline.

A.a. are weak narcotics and paralysing agents. Pellotine has a cramp-relaxing effect similar to that of acetylchoine. The most toxic A.a. is lophophorin. The heterocyclic A.a. are not directly responsible for the intoxication following ingestion of cactus preparations.

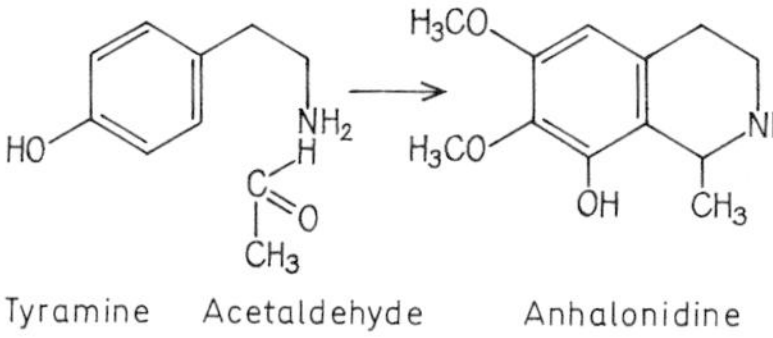

Tyramine Acetaldehyde Anhalonidine

Biosynthesis of the anhalonium base, anhalonidine, by a Mannich condensation

Animal protein factor: see Vitamins (vitamin B_{12}).

Ankyrin, ***anchorin, syndein:*** a membrane protein found in erythrocytes and brain, which binds Spectrin (see) to the membrane. Spectrin-binding activity was first isolated as a proteolytic fragment of M_r 72,000; the M_r of the entire protein is 215,000. There are about 10^5 copies of A. per erythrocyte ghost, i.e. the number required to bind all the spectrin dimers in a 1:1 complex. In the erythrocyte, A. links spectrin to the cytoplasmic part of Band 3 (a transmembrane protein), to the anion-transport protein, and to microtubules. A. is also able to bind unpolymerized tubulin. [V. Bennett, *Annu. Rev. Biochem.* **54** (1985) 273–304; V. Bennet 'Ankyrins: adaptors between diverse plasma membrane proteins and the cytoplasm' *J. Biol. Chem.* **267** (1992) 8703–8706; L. L. Peters & S. E. Lux 'Ankyrins: Structure and function in normal cells and hereditary spherocytes' *Semin. Hematol.* **30** (1993) 85–118]

Annealing of nucleic acids: see Hybridization, Melting point.

Anomers: see Carbohydrates.

Anserine: see Peptides.

Antagonists: see Inhibitors.

Antamanide: a cyclic decapeptide (all the amino acid residues have the L-configuration) isolated from the death cap fungus, *Amanita phalloides.* In experimental animals it acts as an antidote to phallatoxins and amatoxins, providing it is administered no later than the toxins. Like valinomycin, it binds alkali metal ions, but in contrast to valinomycin it shows a strong preference for sodium over potassium. It readily forms stable complexes with Na^+ or Ca^{2+} ions. On account of its sodium selectivity and its highly lipophilic nature, it also has the properties of a sodium ionophore. The conformation of the free peptide and its Na^+ complex have been determined by X-ray diffraction. Replacement of any Pro residue or of Phe residues 9 or 10 abolishes the antidotal activity. Replacement of other amino acid residues has less effect. Since the imino nitrogen and carboxyl of proline are part of a relatively rigid structure (Pro does not permit α-helix formation in protein and is known as a "helix breaker"), replacement of Pro by any other amino acid will alter the conformation of the molecule. It is suggested that Phe 9 and Phe 10 in correct alignment occupy a target receptor on the liver cell membrane, and thereby prevent entry of phalloidin into the cell. See Silybin. [I. L. Karle et al. *Proc. Natl. Acad. Sci. USA* **76** (1979) 1532–1536; H. L. Lotter, *Zeitschrift für Naturforsch.* **39 c** (1984) 535–542]

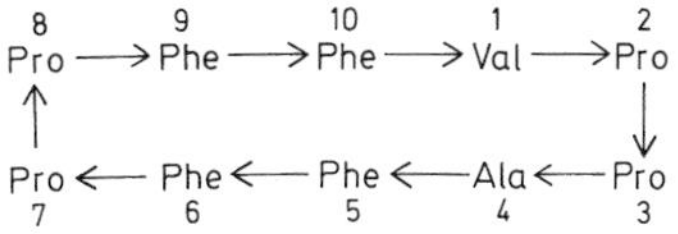

8 Pro → 9 Phe → 10 Phe → 1 Val → 2 Pro
↑ ↓
7 Pro ← 6 Phe ← 5 Phe ← 4 Ala ← 3 Pro

Antamanide

Antheraxanthin: see Zeaxanthin.

Antheridogen: a phytohormone, M_r 330, which is chemically a diterpenoid. A. is derived biogenetically from from the gibberellins. It was isolated from the fern *Aneimia phyllitidis,* in which it stimulates antheridia formation, even at a dilution of 10 µg/l. Other A. have been found in ferns of the families *Schizaeaceae* and *Polypodiaceae.*

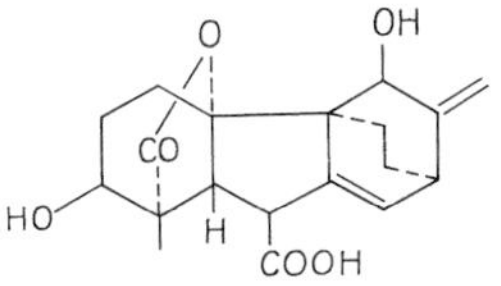

Antheridiogen

Antheridiol: a steroid plant hormone, and the first plant sex hormone to be discovered. It is structurally related to the hydrocarbon stigmastane (see Steroids), and it contains a lactone group. It is secreted by, and was isolated from, the female mycelium of the aquatic fungus *Achlya bisexualis.* Even at high dilution, it stimulates formation of hyphae in male parts of the plant.

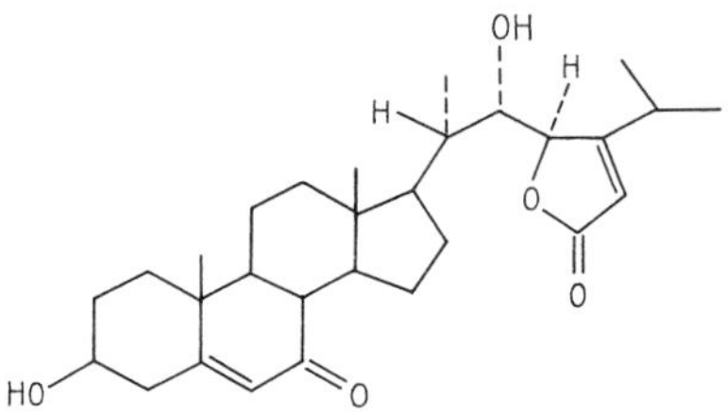

Antheridiol

α-Anthesterol: see Taraxasterol.

Anthocyanidins: see Anthocyanins.

Anthocyanins: widely occurring flavonoid plant pigments responsible for the red, violet, blue or black colors of blossoms, leaves and fruits of higher plants. The variety of colors and patterns is due to the occurrence of A. either alone or in combination with other pigments, usually other flavonoids. A. are the water-soluble glycosides of hydroxylated 2-phenylbenzopyrylium salts; the carbohydrate can be removed by treatment with acid or enzymes, leaving the water-insoluble, unstable aglycon (anthocyanidin).

The basic structure of the A. is the C_{15} flavylium cation. Ring B and the C-atoms 2, 3 and 4 of this structure are biosynthesized, as are other flavonoids, from a C_6-C_3 unit. Ring A is derived from a C_6 unit formed from 3 molecules of acetate (see Stilbenes). The last genetically defined step in A. biosynthesis appears to be conjugation with glutathione as a prelude to transfer of the A. into the vacuole; the gene controlling this reaction has been cloned from maize (gene *Bronze 2, Bz2*), and the deduced primary sequence of the enzyme shows similarity with that of glutathione *S*-transferase [K. A. Marrs et al. *Nature* **375** (1995) 397–400].

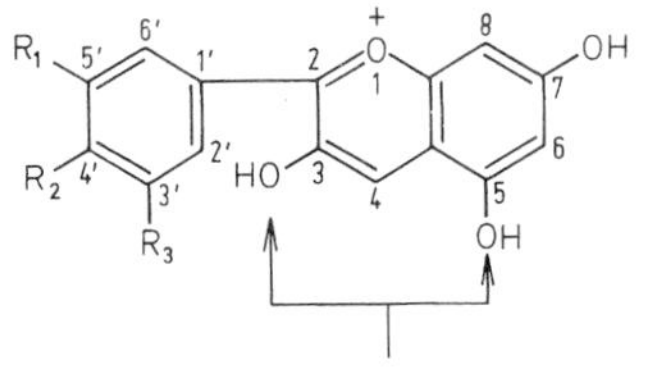

Glycosylation at one or both these sites with glucose, galactose, rhamnose or arabinose, or various oligosaccharides gives the water-soluble anthocyanins.

Anthocyanins
R_1 = H, R_2 = OH, R_3 = OH: Cyanidin (blue) (cornflowers)
R_1 = H, R_2 = OH, R_3 = H: Pelargonidin (red) (geraniums)
R_1 = H, R_2 = OH, R_3 = OCH_3: Peonidin (red) (peonies)
R_1 = OH, R_2 = OH, R_3 = OCH_3: Petunidin (red) (petunias)
R_1 = OCH_3, R_2 = OH, R_3 = OCH_3: Malvidin (blue) (mallow)
R_1 = OH, R_2 = OH, R_3 = OH: Delphinidin (blue) (delphiniums)

Anthocyanidins differ in the number and positions of hydroxyl groups, which may be esterified (e.g. with *p*-cumaric or ferulic acid) or replaced by methoxy groups. Most anthocyanidins are hydroxylated at positions 3, 5 and 7. Those lacking the 3-hydroxyl group, e.g. 3-deoxypelargonidin, 3-deoxycyanidin, 3-deoxydelphinidin, are less common. The 3 basic types, pelargonidin (4′-hydroxy), cyanidin (3′,4′-dihydroxy) and delphinidin (3′,4′,5′-trihydroxy) are formed by addition of hydroxyl groups to ring B; glycosides of these compounds are the most abundant and most widely distributed A.

The carbohydrate moiety of the A. is usually linked β-glycosidically to C3, more rarely to C5; in diglycosides it is linked to both. The commonest carbohydrate residues are glucose, galactose and rhamnose. Xylose and arabinose are less common. Di- and trisaccharide residues are rare, and the resulting A. are called biosides and triosides, respectively. More than 20 types of A. are known, differing in the number, type and position of their carbohydrate residues. Well over 100 natural A. have been isolated and structurally characterized.

A. are amphoteric. Their salts with acids are red, at neutral pH they form colorless pseudo-bases, and in the alkaline range they form unstable, violet anhydrobases. Their absorption maxima lie in the range 475–550 nm and around 275 nm.

Anthraquinones: yellow, orange, red, red-brown or violet derivatives of anthraquinone (9,10-anthracenedione), the largest group of naturally occurring quinones. With a few exceptions, e.g. 2-methylanthraquinone, all natural A. are hydroxylated. Other common substituents are methyl, hydroxymethyl, methoxy, formyl, carboxyl, benzyl and long-chain alkyl groups. Some A., are dimeric.

Of the more than 170 known natural A., which occur either in the free form or as glycosides, more than half are found in lower fungi, particularly in *Penicillium* and *Aspergillus* species, and in lichens. Others are found in higher plants, and in isolated instances in insects. The *Rubiaceae, Rhamnaceae, Leguminosae, Polygonaceae, Bignoniaceae, Verbeniaceae* and *Liliaceae* are particularly rich in A. The best known insect A. are carminic acid, kermesic acid and laccaic acid. The commonest A. of higher plants are emodin, alizarin, rhein, purpurin and morindon. Helminthosporin, skyrin and julichrom are found in fungi. Particularly large quantities of A. are present in madder root *(Rubia tinctorum),* the bark of *Coposma australis* and the mycelium of *Helminthosporin gramineum.*

Due to the laxative properties of A., a number of drugs, such as *Radix Rhei, Cortex Frangulae, Fructus* and *Folia Sennae* are used pharmaceutically. A few A., such as carminic acid and alizarin, are still used to some extent as dyes. [J. Schripsema & D. Dagnino 'Anthraquinones: Elucidation of the substitution pattern of 9,10-Anthraquinones through chemical shifts of peri-hydroxyl protons' *Phytochemistry* **42** (1996) 177–184]

Anthraquinone

Antiandrogens: a group of chemical compounds which reversibly inhibit the effect of male sex hormones by competing with them for their receptors. Several testosterone and gestagen analogs with 1α,2α-methylene groups, e.g. cyproterone acetate, are particularly effective.

Antianemia factor, *vitamin B_{12}*: see Vitamins.

Antiauxins: see Auxin antagonists.

Anti-beri-beri-factor, *vitamin B_1*: see Vitamins.

Antibiotics (Greek *antibios* against life): substances produced by microorganisms (both bacteria and fungi), which kill or inhibit the growth of other microorganisms. In contrast to general cell poisons, A. are selective. The genus *Bacillus* yields a number of therapeutically important A. (bacitracin, gramicidin, polymyxin, tyrocidin), as do *Streptomyces* and *Actinomyces* species (streptomycin, tetracycline, actinomycin, chloramphenicol, macrolides, neomycin) and molds of the genera *Penicillium* and *Aspergillus* (penicillin, griseofulvin, xanthocillin, helvolinic acid). Some of these are peptides (see Peptide antibiotics).

A. are chemically heterogeneous. Important components include amino acids, often nonproteogenic D-amino acids (e.g. in gramicidin), acetate/malonate units (griseofulvin, tetracycline), sugars and sugar derivatives (streptomycin), tetracyclic triterpenes (fusidinic acid, helvolinic acid, cephalosporin P_1). There are also many Nucleoside A. (see). A. act by various mechanisms. Many interfere with protein biosynthesis, but others, e.g. penicillin, inhibit cell wall synthesis in bacteria.

Most A. are produced industrially from microbial cultures (see Industrial microbiology); highly productive strains of microorganisms are used for this purpose, and such strains are maintained in culture collections. Chemical synthesis of A. is also used to a small extent. As far as possible, nutrient media for industrial microbial A. production are prepared from readily available, cheap raw materials, e.g. various sugars, starch, soybean millings, corn steep liquor. The yield of penicillin G can be increased by adding phenylacetic acid to the medium as a precursor. As is typical for the production of secondary metabolites, A. production becomes significant only toward the end of the logarithmic growth phase. Therefore the medium is designed so that cell division is limited by the lack of a nutrient before the sugar has been consumed. Under these conditions, production of A. begins and continues until the energy supply is exhausted. The productivity of the microbial strain, management of the fermentation and recovery of the products are equally important in the commercial production of A.

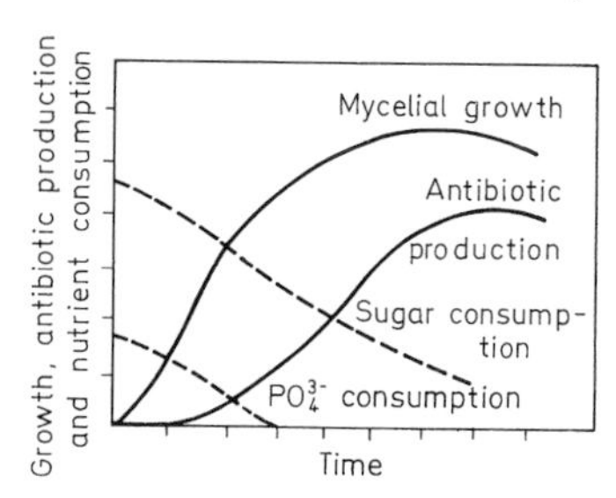

Kinetics of growth and antibiotic production

Antibodies: see Immunoglobulins.

Anticodon: three sequential nucleotide residues in one loop of a transfer RNA molecule, which recognize (i.e. form H-bonded base pairs with) the codon nucleotides in messenger RNA. This anticodon loop thus insures that the correct amino acid is incorporated into the polypeptide sequence. See Wobble hypothesis, Transfer RNA.

Anticytokinins: agents that partially negate the physiological effects of cytokinins, e.g. the synthetic kinetin inhibitor, 6-methylpurine.

Antidepressants: drugs with stimulatory and antidepressant action. They decrease fatigue, reduce appetite and decrease sleeping time. These effects are the result of reduced autonomic activity, especially of cholinergic systems, and of central sympathetic activation. A. are inhibitors of monoamine oxidase (EC 1.4.3.4), and they are all amines, presumably acting as analogs of natural monoamine oxidase substrates; inhibition of monoamine oxidase delays or prevents destruction of the natural catecholamines (adrenalin, noradrenalin, dopamine) which therefore persist at elevated concentrations and cause stimulation. Examples of A. are amphetamine (β-phenylisopropylamine), ephedrine, harmine, etc. The first A. to be discovered was iproniazid (*N'*-isonicotinyl-*N'*-isopropylhydrazide); during testing of this compound as an antitubercular drug, it was found that patients became elated. A. are used for treatment of mental depression. They have also been used for doping race horses and athletes. A. can be addictive, especially amphetamine.

Antidermatitis factor: see Vitamins.

Antidiuretic hormone: see Vasopressin.

Antidiuretin: see Vasopressin.

Antienzymes: polypeptides or proteins which act as enzyme inhibitors, including antibodies against antigenic proteins or coenzymes. A. include many animal and plant protease inhibitors (see Toxic proteins), such as soybean trypsin inhibitor and serum antitrypsin, which form tight complexes with the corresponding proteases. Specific antibodies are used extensively in the purification and characterization of enzymes.

Antigen-antibody reaction, *AAR*: together with phagocytosis, the most important protective mechanism of the animal organism against invading foreign material. The AAR is the specific formation of an insoluble antigen-antibody complex. Soluble antigens are precipitated, and cells carrying antigen on their surfaces are agglutinated. Since the specificity and sensitivity (in the picogram range) of the AAR is very great, even in vitro, it is used in diagnosis. The bonds between antigen and antibody are almost exclusively noncovalent, i.e. ionic, hydrophobic or hydrogen bonding between the amino, carboxyl, hydroxyl and aliphatic groups is involved. See Immunoglobulins.

Antigens: substances that induce an immune response. They are normally foreign to the body, and may be natural or synthetic macromolecules, particularly proteins and polysaccharides (M_r < 2,000), or surface structures on foreign particles. An A. consists of a high molecular mass carrier and, usually, several low molecular mass groups which are responsible for the specificity of the immune response and reaction of the A. with the corresponding immunoglobulins. These small groups, called antigenic determinants, or *haptens,* lie on the surface of the molecule and determine the valence of the A. Nearly all A. are polyvalent and therefore induce more than one species of antibody. The largest determinant groups occur in protein A., and may comprise up to 30 amino acid residues. Simple polysaccharide determinants involve 2 to 7 sugar residues. Much structural work on immunoglobulins uses antibodies to artificial haptens coupled to various proteins; thus, compounds such as phosphorylcholine, *p*-azophenylarsonate, dinitrophenol and *m*-aminobenzene sulfonate are not immunogenic on their own, but they react with preformed antibodies induced by injection of the artificial hapten linked to a carrier (e.g. dintrophenyl derivatives of bovine serum albumin injected into a rabbit). Antigenic determinants on proteins can be either conformational or sequential; the former are destroyed by protein denaturation.

Antigibberellins: see Gibberellin antagonists.

Anti-gray-hair-factor: a member of the vitamin B_2 complex. See Vitamins.

Antihemophilic factor, ***factor VIII:*** an oligomeric β_2-glycoprotein [M_r 1.12×10^6 (human), M_r 1.2×10^6 (bovine); 6 % carbohydrate] consistng of covalently bound subunits. It activates factor X in the process of blood clotting, and is thereby completely consumed. A.f. is stabilized by Ca^{2+}, but is inactivated when blood is stored. DNA clones encoding the complete 2351 amino acid sequence of human A.f. subunits have been isolated. The recombinant protein (M_r 267,039) decreases the clotting time of hemophilic plasma, and it is biochemically and immunologically similar to serum-derived A.f. It has structural similarities with factor V and ceruloplasmin. [J. Gitschier et al. *Nature* **312** (1984) 326–330; W. I. Wood et al. *ibid.* 330–337; G. A. Vehar et al. *ibid.* 337–342; J. J. Toole et al. *ibid.* 342–347]

Antihemorrhagic vitamin: see Vitamins.

Antilymphocyte globulin: see Antilymphocyte serum.

Antilymphocyte serum, ***ALS:*** an immune serum with an opsonizing (see Opsonization) and cytotoxic effect on lymphocytes. It was used for immune suppression before Monoclonal antibodies (see) to specific classes of lymphocytes became widely available. The antibodies were raised in other species of animals by injection of human lymphocytes from blood, thymus, spleen or the thoracic duct. To avoid undesired side effects, the A.s. was purified and administered primarily as antilymphocyte globulin (ALG). Treatment with ALG affects mainly cellular immunity, particularly that provided by long-lived circulating lymphocytes.

Antimetabolite: a compound so similar in structure to a metabolite that it occupies the same specific enzyme binding sites. This results either in inhibition of the enzyme or substitution of the A. for the metabolite in the reaction, possibly leading to incorporation of the A. into cell components. Both effects lead to metabolic disturbances or inhibition of cell division. Many A. are used therapeutically, e.g. analogs of purines and pyrimidines and folic acid antagonists are used as carcinostatic agents.

Antineuritic vitamin, ***vitamin B_1:*** see Vitamins.

Antipain: see Inhibitor peptides.

Antiparallel arrangement: see Strand polarity.

Antipurines: see Purine analogs.

Antipyrimidines: see Pyrimidine analogs.

Antipyrine, ***1,2-dihydro-1,5-dimethyl-2-phenyl-3H-pyrazol-3-one:*** a weak base, pK_a 1.4. After oral administration, A. is rapidly absorbed and becomes distributed throughout the total body water within 2 h. It was first synthesized in 1884, used as an antipyretic and later as an analgesic, then went out of favor in the 1930s as new analgesics became available.

Antirachitis vitamin, ***vitamin D:*** see Vitamins.

Antiscorbutic vitamin, ***vitamin C:*** see Vitamins.

Antiserum, ***immune serum:*** the serum of an animal (or human) that has been immunized against an antigen. An A. can be monovalent or polyvalent, i.e. contain antibodies specific for one or several antigenic determinant(s), depending on whether the animal was immunized with a purified antigen or a mixture of antigens. In addition to the antibodies, the A. contains all other serum proteins. These can be removed by ion exchange or immune adsorption chromatography.

Antisterility factor, ***vitamin E:*** see Vitamins.

Antitumor proteins: protein antibiotics which inhibit tumor growth. A.p. have been isolated from culture filtrates of various *Streptomyces* strains. The most thoroughly examined A.p., neocarcinostatin, is an acidic, single-chain protein (M_r 10,700; primary structure known; 109 amino acid residues; His and Met absent) with typical antibiotic activity against Gram-positive bacteria, such as *Sarcina lutea* and *Staphylococcus aureus.* It is also highly effective against experimental tumors and seems particularly well suited for treatment of tumors of the rectum, stomach, gall bladder and penis. The main effect of A.p. is inhibition of mitosis through inhibition of DNA synthesis and accelerated degradation of existing DNA.

Knowledge of the primary structure of the A.p. has enabled the chemical synthesis of A.p. that are highly effective against tumors. For example, reaction of neocarcinostatin with fluorescein isothiocyanate greatly reduces its toxicity without affecting its high antitumor activity. The phytotoxins, abrin and ricin, must also be classified as A.p., because they inhibit growth of certain tumor cells by inhibiting protein synthesis.

Antivitamins: antimetabolites of vitamins. They inhibit growth of vitamin-dependent microorganisms and cause symptoms of vitamin deficiency in animals.

Apigenin: see Flavones (Table).

Apiin: see Flavones (Table).

D-Apiose: a branched monosaccharide, M_r 150.13, $[\alpha]_D^{19}$ +9° (pure syrup). D-A. is found in various glycosides and as a component of various polysaccharides. It is biosynthesized from D-glucuronic acid.

Apoenzyme: see Coenzyme.

Apoprotein: the protein component of a conjugated protein.

Apoptosis, ***programmed cell death:*** cell death that is regulated by the cell itself, i.e. it is genetically programmed. There is now sufficient evidence to indicate that the molecular basis of A. has been highly conserved during evolution. It is a form of regulated cell suicide that is necessary for orderly embryogenesis and metamorphosis, tissue homeostasis and the function of the immune system in metazoan organisms. At the cell microscopic level, A. involves loss of cell junctions and microvilli, chromatin condensation, DNA fragmentation, cytoplasmic contraction, dense compaction of mitochondria and ribosomes; and membrane blebbing, in which the endoplasmic reticulum fuses with the cell plasma membrane, and the cell separates into several membrane-bounded vesicles, known as apoptotic bodies. The latter are usually phagocytosed by adjacent cells. A variety of factors that activate A. have been identified. It can be induced by cytotoxic agents, but natural, genetically programmed A. appears to depend on the initiation of a signaling pathway by binding of ligands to specific receptors on the cell surface. Evidence is accumulating for a primary role of cytoplasmic proteases in A., notwithstanding the fact that DNA cleavage patterns can be used for the in vitro monitoring of A. The complement of proteases responsible for the onset of A. has been dubbed the 'executioner', but there is no evidence that the actions of these proteases are coordinated within a single unit or complex.

An early indication of protease involvement in A. was provided by the identification of a pore-forming protein, perforin, and a series of serine proteases in the cytoplasmic granules of cytotoxic T-lymphocytes and natural killer cells (both kill by binding to, and inducing A. in target cells). One of these serine proteases, known as granzyme B or fragmentin-2, has the unusual property of cleaving after Asp residues. Exposure of target cells to a combination of perforin and granzyme B induces A.

Apoptotic cell death is also initiated (by an as yet unelucidated mechanism) when the cell surface receptors, Fas (also known as Apo-1) (M_r ~ 45,000) and type 1 tumor necrosis factor receptor (TNF-1; M_r ~ 55,000), bind their respective ligands or cross-linking antibodies. The natural ligand of Fas is a glycosylated type II transmembrane protein, inferring that A. is initiated by cell-cell contact, although shed ligand may also be effective in stimulating A. Cytotoxic T cells also induce A. in target cells by binding to Fas. Fas and TNF-1 display extensive sequence similarity in both their cysteine-rich extracellular domains and their intracellular 'death domains' of some 80 amino acid residues. The gene for Fas is defective in mice that have mutations at the *lpr* (lymphoproliferation) locus; depending on the site of mutation, receptor expression may be prevented or reduced, or the receptor may be nonfunctional. Fas-induced A. does not occur in homozygous mutant mice, which suffer from a complicated immunological disorder involving defective B- and T-lymphoid compartments.

Interleukin-1β converting enzyme (ICE, which converts the M_r 33,000 protease form of interleukin-1β to the active M_r 17,500 form by cleaving after Asp residues) and other members of the ICE-like family of proteases appear to play a significant role in A. For example, expression of CrmA protein (a protease inhibitor that inhibits ICE) potently blocks the A. induced by stimulation of Fas or TNF-1, implying that ICE-type proteases are involved in these suicide pathways. Proteins identified as early targets or 'death substrates' associated with the onset of A. are poly(ADP-ribose) polymerase, lamin B1, α-fodrin, topoisomerase I, β-actin, and the M_r 70,000 component of the U1 small ribonucleoprotein (U1-70kD).

A number of genes controlling different aspects of A. in the nematode *Caenorhabditis elegans* have been identified, and two of these, *ced-3* and *ced-4,* are required for A. during development. The gene product of *ced-3* displays significant homology with ICE.

[R.E. Ellis et al. *Annu. Rev. Cell Biol.* **7** (1991) 663–698; S. Cory *Nature* **367** (1994) 317–318; S.J. Martin & D.R. Green *Cell* **82** (1995) 349–352; M. Tewari et al. *J. Biol. Chem.* **270** (1995) 18738–18741; C.D. Gregory (ed.) *Apoptosis and the Immune Response Wiley-Liss.,* New York, 1995]

Aporepressor: see Enzyme repression.

APP: acronym of Aneurin pyrophosphate. See Thiamin pyrophosphate.

APS: acronym of Adenosine 5′-phosphosulfate. See Sulfate assimilation.

APS reductase: see Adenylsulfate reductase.

APUD system: see Amine precursor uptake decarboxylase system.

Apurine acids: polynucleotides which have been subjected to brief treatment with mild acid, which removes the purines and leaves the phosphate, pentose and pyrimidines.

Apyrimidine acids: nucleic acids from which the pyrimidines have removed by chemical treatment, e.g. by exposure to hydrazine.

Arabans: high molecular mass, branched polysaccharides composed of L-arabinose linked 1,5 and 1,3 in furanose form (see carbohydrates). A. are found widely as components of plant hemicelluloses.

1-β-D-Arabinofuranosyl derivatives of adenine, cytosine, thymine and uracil: see Arabinosides.

Arabinose: a pentose, M_r 150.13, occurring naturally in both the D- and L-forms.

L-A., β-form, m.p. 160 °C, $[\alpha]_D^{19}$ +190 ° → +105 ° (water) is a component of hemicelluloses, e.g. the araban of cherry gum, and is found in plant mucilages, glycosides and saponins.

D-A., β-form, m.p. 160 °C, $[\alpha]_D^{19}$ −175 ° → −105.5 ° (water) has been isolated from certain bacteria and is found in some glycosides.

```
    CHO              CHO
     |                |
HO—C—H            H—C—OH
     |                |
 H—C—OH          HO—C—H
     |                |
 H—C—OH          HO—C—H
     |                |
    CH2OH            CH2OH
```

D-Arabinose L-Arabinose

Arabinosides, ***arabinonucleosides:*** structural analogs of the ribonucleotides, in which the ribofuranose is replaced by arabinofuranose. 1-β-D-Arabinofuranosylcytosine (cytosine arabinoside) inhibits the reduction of cytidine diphosphate to the corresponding deoxynucleoside in the course of DNA synthesis. It is rapidly deaminated to the corresponding uracil derivative. 1-β-D-Arabinofuranosyluracil (spongouridine) and 1-β-D-arabinofuranosylthymine (spongothymidine) occur as natural pyrimidine analogs in various sponges. A. are therefore often called spongonucleosides.

A. may also contain purine bases, e.g. 1-β-D- and 9-β-D-arabinofuranosyladenine (Ara-A) and 9-[(2-hydroxyethoxy)methyl]guanine (Acyclovir). All of these are phosphorylated by a viral thymine kinase to triphosphate esters which inhibit viral DNA polymerase. They are thus interesting as antiviral agents, but unfortunately they are rapidly deaminated by adenine deaminase. Esterification of the sugar hydroxyls enables these compounds to penetrate the cell membrane more easily, which enhances their efficacy. Other synthetic A., e.g. 2′-fluoro-5-iodoarabinosylcytosine, have been investigated as potential antiviral agents. [ACS Highlights: *Science* **220** (1983) 292–293]

Structures of known pyrimidine arabinosides

Arabinoside	R_5	R_6
1-β-D-Arabinofuranosylcytosine	NH_2	H
1-β-D-Arabinofuranosyluracil	H	OH
1-β-D-Arabinofuranosylthymine	OH	CH_3

Arachidic acid, ***icosanoic acid:*** CH_3-$(CH_2)_{18}$-COOH, a fatty acid, M_r 312.5, m.p. 75.3 °C. It occurs widely as a component of glycerides (acylglycerols), but is usually present only in low concentrations. In sunflower oil, soybean oil, milk fat and peanut oil, A.a. may represent up to 3 % of the esterified fatty acids.

Arachidonic acid, ***all-cis-5,8,11,14-eicosatetraenoic acid:*** the biosynthetic precursor of several groups of regulatory substances: Prostaglandins (see), Thromboxanes (see), Leukotrienes (see), and oxidized eicostrienoic and eicosatetraenoic acids. A.a. is derived from the diet, and from linoleic acid by desaturation and chain elongation. It is incorporated into membrane phospholipids, especially phosphatidylinositols (see Inositol phosphates); when released from these by phospholipase A_2, it is available for oxidation by several enzymes. Phosphatidylinositols are cleaved in response to the binding of hormones by their specific receptors (see Second messengers). The resulting A.a. and its metabolites then amplify hormonal signals and "rebroadcast" them into the surrounding tissue.

A.a. is attacked by 2 types of enzymes: lipoxygenase and cyclo-oxygenase. Lipoxygenases, specific for positions 15 or 12, produce 15- or 12-hydroperoxyeicosatetraenoic acids (HPETE), which may be further metabolized to hydroxyeicosatetraenoic acids (HETE). These substances are pharmacologically active and may inhibit production of leukotrienes. Introduction of a hydroperoxy group at position 5, by the action of a third lipoxygenase, leads to 5-HPETE, the precursor of the Leukotrienes (see). The cyclooxygenase forms endoperoxides, which after addition of oxygen at position 15, become Prostaglandins (see). Finally, the endoperoxide, PGH_2, is the precursor of the Thromboxanes (see). [S. Moncada & J. R. Vane, *Pharmacol. Rev.* **30** (1979) 293–331]

Arachin: a groundnut *(Arachis hypogaea)* protein (M_r 345,000) composed of 6 subunits. Each subunit is composed of 2 equal-sized, covalently bound polypeptide chains. A. is very similar to edestin.

Arcain, ***1,4-diguanidobutane:*** H_2N-C(=NH)-NH-$(CH_2)_4$-NH-C(=NH)-NH_2, a strongly basic guanidine derivative first isoated from a mussel *(Arca noae)* but also found in higher fungi, e.g. *Panus tigrinus.* A. is biosynthesized by amidination of putrescine to form agmatine, followed by amidination of agmatine to form A.; in both stages, the amidine group is derived from L-arginine (see Transamidination). A. decreases the blood sugar level of mammals.

Areca alkaloids: pyridine alkaloids in which the pyridine ring is partially hydrated (Fig.). The A.a. are obtained from betel nuts (seeds of the betel palm, *Areca catechu).* In the plant, the A.a. are bound to tannins. The main alkaloid is arecoline. A.a. are probably derived from nicotinic acid or its precursors. The significance of A.a. in human and veterinary medicine has declined greatly, but betal is still widely used as a mild stimulant. Betel nuts are chewed with lime (to release the alkaloids) and leaves of the betel pepper *(Piper betle).* The practice, at least 2,000 years old, is common in East Africa, India and Oceana. The number of betel nut users, recognizable by their red-stained teeth, is estimated at about 200 million.

	R_1 = H	R_1 = CH_3
R_2 = H	Guvacine	Arecaidine
R_2 = CH_3	Guavacoline	Arecoline

Structures of Areca alkaloids

Arecaidine: see Arecoline.

Arecoline, ***1,2,5,6-tetrahydro-1-methyl-3-pyridinecarboxylic acid methyl ester:*** the main areca alkaloid (M_r 155.19, b.p. 209 °C) and the methyl ester of the alkaloid arecaidine (M_r 141.9, m.p. 232 °C). A. is responsible for the physiological effect of betel nuts. It acts as a parasympatheticomimetic agent, but due to its high toxicity is now used only in veterinary medicine.

Arginase (EC 3.5.3.1): a highly active and specific liver enzyme which catalyses a reaction in the urea cycle (L-arginine + $H_2O \rightarrow$ L-ornithine + urea) in ureotelic animals. In terrestrial uroteles, such as mammals, frogs and swamp turtles, A. is found practically only in the liver, with traces in the pancreas, mammary glands, testes and kidneys. In ureotelic elasmobranch fish (sharks and rays) the enzyme is not restricted to the liver. In these fish it serves to generate a high blood urea concentration (2–2.5 %) which is needed for the maintenance of osmotic pressure.

Liver arginase is present in larger amounts when the diet is high in protein, and in subnormal amounts in patients with liver carcinoma. It is a tetrameric molecule (M_r 118,000) binding 4 Mn^{2+} ions. Removal of the metal ions, e.g. by EDTA, causes the enzyme to dissociate into its 4 inactive subunits (M_r per subunit 30,000); this process is reversed by addition of Mn^{2+}. A. is stable for months at pH 7.0 and +4 °C. Below pH 6 it is increasingly inactivated, as it reversibly dissociates into dimers at pH 4 and monomers at pH 2. The enzyme is highly specific, hydrolysing only canavanine and L-arginine, but not D-arginine or other guanido compounds. Its pH optimum is dependent on metal ions (about pH 10 in the presence of Mn^{2+}). It is inhibited by L-ornithine and L-lysine. Uricotelic vertebrates (birds, reptiles) also have an L-arginine-specific A., but its activity is very low, and it differs from the ureotelic liver A. in K_m (50–100 times that of rat liver A.), M_r (276,000), susceptibility to inhibition by ornithine, and immunological behavior.

L-Arginine, ***Arg, 2-amino-5-guanidovaleric acid:*** the most basic proteogenic amino acid, M_r 174.2, m.p. 238 °C (d.), $[\alpha]_D^{20}$ + 26.9 (c = 1.65 in 6 M HCl). Arg is unstable in hot alkaline solution, and forms nearly insoluble nitrates, picrates and picrolonates, and a particularly insoluble salt with flavianic acid. The Sakaguchi reaction can be used for the detection and quantitative determination of Arg: 1 ml 5 % sodium hydroxide solution is added to 3 ml of the solution under test, followed by 2 drops of 1 % ethanolic α-naphthol; a single drop of 10 % sodium hypochlorite is then added and the solution shaken; development of a bright red color indicates the presence of Arg. Particularly large amounts of Arg are present in protamines and histones. In many plants, Arg serves for nitrogen transport and storage, high concentrations of free Arg being found in many plants, including red algae, *Cucurbitaceae* and conifers, and in particularly high concentrations in seedlings and storage organs.

Arg is a glucogenic amino acid, and it is semi-essential for humans, rats and chicks, i.e. it is not required by the adults of these species, but the young animals are unable to synthesize it rapidly enough to satisfy their total needs for growth and development. Arg is synthesized in the Urea cycle (see) where it serves as an important intermediate. It is attacked by various enzymes, depending on the species: 1. arginase (EC 3.5.3.1) hydrolyses Arg that is released proteolytically or produced in the urea cycle; 2. transamidinases (EC 2.1.4) transfer the amidine group to various amino acids and amines, forming guanidine derivatives, an important reaction in the synthesis of phosphagens; 3. arginine deiminase (EC 3.5.3.6) hydrolyses Arg to L-citrulline and NH_3; 4. L-amino-acid oxidase (EC 1.4.3.2) deaminates Arg to 2-oxo-5-guanidovaleric acid; 5. arginine decarboxylase (EC 4.1.1.19) decarboxylates Arg to agmatine; 6. arginine 2-monooxygenase (EC 1.13. 12.1) decarboxylates and oxidizes Arg to 4-guanidobutyramide.

Arg increases spermatogenesis. Various natural products contain the amino acid, e.g. octopine and the phosphagen, arginine phosphate. Arg was first isolated from lupin seedlings in 1886 by Schulze and Steiger.

Argininemia: see Inborn errors of metabolism.

Arginine phosphate: see Phosphagens.

Arginine-urea cycle: see Urea cycle.

Argininosuccinic aciduria: see Inborn errors of metabolism.

Aristolochic acids: a group of related aromatic nitro compounds from *Aristolochia* spp., the most abundant being aristolochic acid I. The A.a. are biosynthesized from isoquinoline alkaloids of the norlaudanosine type by oxidation of the nitrogen-containing ring (Fig.). Aristolochia drugs are among the oldest known, but due to their toxicity, their clinical use has declined markedly. Several insects feeding on *Aristolochia clamatis* or *A. rotundo* accumulate aristolochic acid I, which serves as a defense against predation; particularly investigated is the butterfly, *Pachlioptera aristolochiae,* which retains aristolochic acid I consumed by its caterpillar [J. von Euw et al. *Nature* **214** (1967) 35–39]

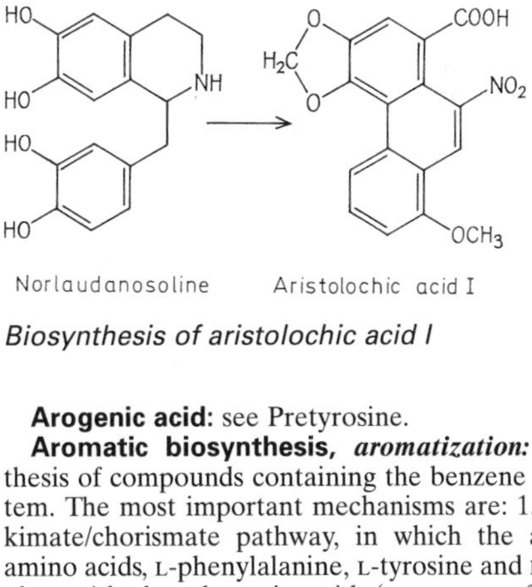

Norlaudanosoline → Aristolochic acid I

Biosynthesis of aristolochic acid I

Arogenic acid: see Pretyrosine.

Aromatic biosynthesis, ***aromatization:*** biosynthesis of compounds containing the benzene ring system. The most important mechanisms are: 1. the shikimate/chorismate pathway, in which the aromatic amino acids, L-phenylalanine, L-tyrosine and L-tryptophan, 4-hydroxybenzoic acid (precursor of ubiquinone), 4-aminobenzoic acid (precursor of folic acid) and the phenylpropanes, including components of lignin, cinnamic acid derivatives and flavonoids are synthesized; and 2. the polyketide pathway (see Polyketides) in which acetate molecules are condensed and aromatic compounds (e.g. 6-methylsalicylic acid) are synthesized via poly-β-keto acids. Biosynthesis of flavonoids (e.g. anthocyanidins) can occur by either pathway.

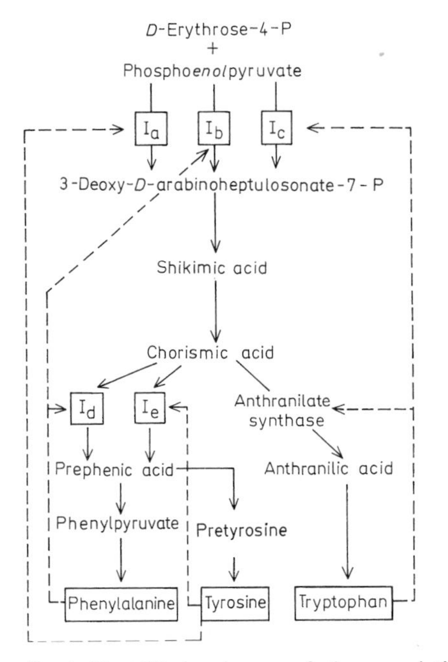

Fig. 1. *Simplified scheme of the regulation of aromatic biosynthesis.* I_a–I_c = isoenzymes of phospho-2-keto-3-deoxyheptonate aldolase (EC 4.1.2.15). I_d and I_c = isoenzymes of chorismate mutase (EC 5.4.99.5).

Shikimate and chorismate are common precursors of all aromatics synthesized by the shikimate/chorismate pathway (Fig. 2). Figure 3 shows the sequence of conversions up to chorismate, which is the branching point in the biosynthesis of different aromatics.

Aromatic biosynthesis is regulated by feedback mechanisms. The first step in the biosynthesis of the 3 aromatic amino acids is catalysed by phospho-2-keto-3-deoxy-heptonate aldolase (EC 4.1.2.15). For each aromatic amino acid there is a separate isoenzyme which is subject both to end product inhibition and to repression of its synthesis by the corresponding amino acid (Fig. 3). Regulatory isoenzymes are also involved in the transformation of chorismate into prephenate.

Arrow poisons: natural toxins used for coating arrows, spears or blow pipe darts. In ancient Greece, extracts containing aconitine were used. Ouabain (see) and similar cardiac glycosides were (are) used in Africa. The jungle Indians of South America use various kinds of Curare alkaloids (see), and tribes in Columbia use Batrachotoxin (see) from the Columbian arrow poison frog.

Arylamidases: a ubiquitous group of Aminopeptidases (see) which cleave synthetic arylamides, e.g. alanine β-naphthylamide. These unnatural substrates are therefore used for the assay of A., in particular in serum and urine where the occurrence of A. is indicative of certain liver diseases. The majority of A. are bound to particles or membranes, and are therefore thought to be involved in resorptive and secretory protein transport. It is also generally assumed that A. are involved in the degradation of peptide hormones and in the final phase of intracellular protein degradation. They can be classified according to their specificity: A.A hydrolyse Asp or Glu arylamides, A.B hydrolyse Lys or Arg arylamides, while A.N hydrolyse Ala or Leu arylamides. High activities of A. are found in the brush borders of the intestinal mucosa and kidney tubules, and in hepatocyte plasma membranes.

2-Arylbenzofurans: a biosynthetically heterogeneous group of natural compounds containing the ring system shown (Fig.). Thus, egonol (5-(3-hydroxypropyl)-7-methoxy-3′:4′-methylenedioxy-2-arylbenzofuran), a constituent of the unsaponifiable fraction of the seed oil of *Styrax japonicum,* may be derived by loss of a carbon atom from a *bis*arylpropanoid, i.e. of lignan or neolignan origin. 6,3′5′-Trihydroxy-2-aryl-

2-Arylbenzofuran ring system

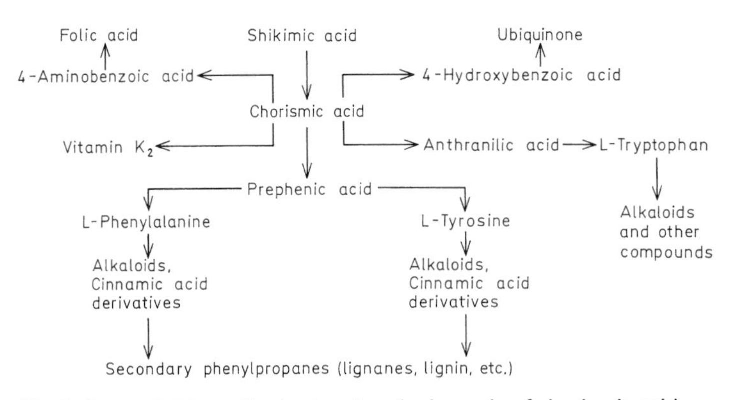

Fig. 2. *Aromatic biosynthesis showing the key role of chorismic acid*

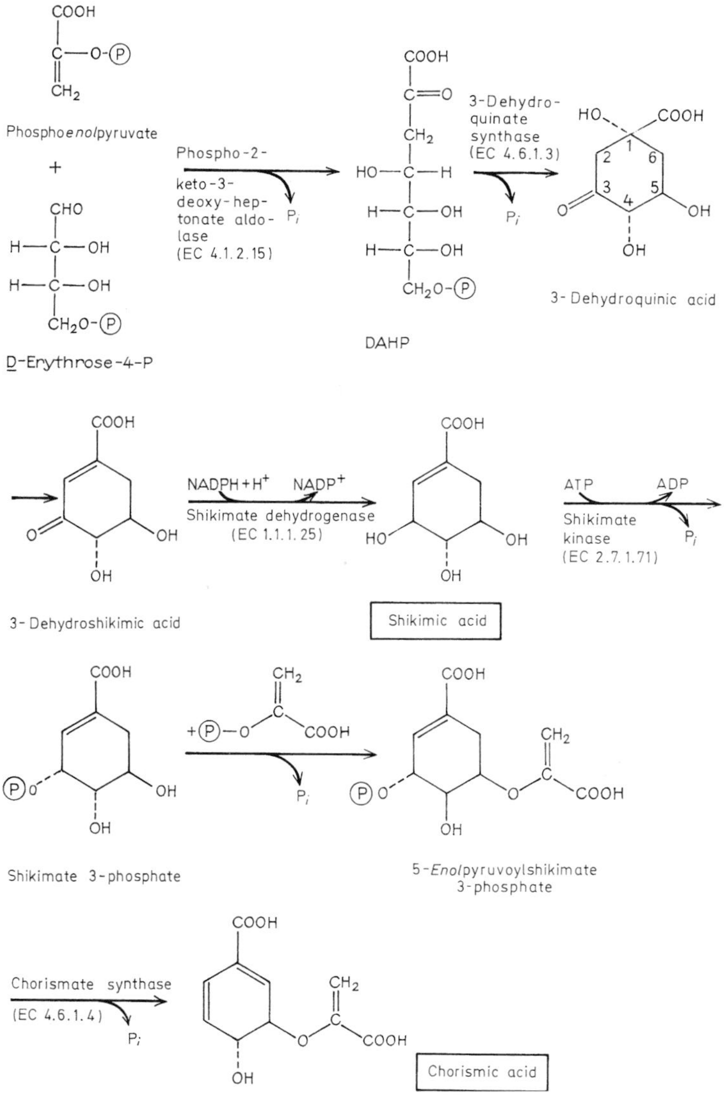

Fig. 3. *Biosynthesis of chorismic acid.* DAHP = 3-deoxy-D-arabinoheptulosonic acid 7-phosphate.

benzofuran occurs with the stilbene, oxyresveratrol, and is probably derived from it by oxidative cyclization. Other 2-arylbenzofurans are found in members of the *Leguminoseae,* and several are phytoalexins. They are accompanied by isoflavonoids with similar substitution patterns, and therefore appear to be derived from isoflavonoids by loss of a carbon atom, e.g. vignafuran (6,2′-dimethoxy-4′-hydroxy-2-arylbenzofuran) from *Lablab niger* and *Vigna unguiculata.* [P.M. Dewick in J.B. Harborne & T.J. Mabry (eds), *The Flavonoids: Advances in Research,* Chapman and Hall, 1982]

3-Aryl-4-hydroxycoumarins: a closely related group of naturally occurring Isoflavonoids (see), e.g. ***scandenin*** (Fig.) from *Derris scandens.* [C.P. Falshaw et al. *J. Chem Soc.* (C) (1969) 374–382]

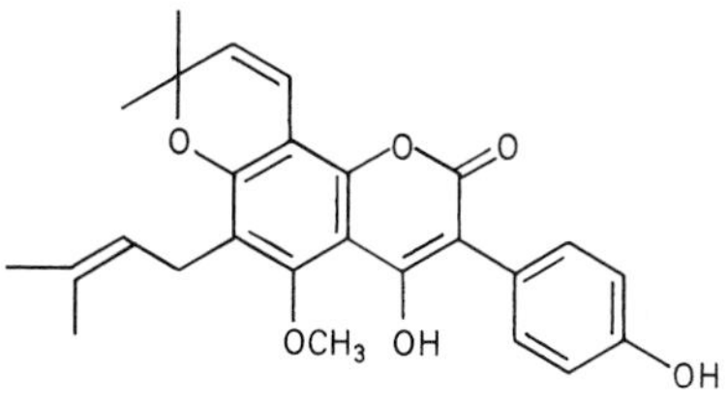

Scandenin

Arylsulfatases: a well studied group of sulfatases that hydrolyse the O-S bond of aromatic sulfate esters: $R\text{-}O\text{-}SO_3^- + H_2O \rightarrow R\text{-}OH + H^+ + SO_4^{2-}$, where R-OH is a phenol. Typical substrates are 2-hydroxy-5-nitrophenylsulfate (nitrocatechol sulfate) and 2-hydroxy-4-nitrophenylsulfate. A. can be classified according to whether they are inhibited (type I) or not inhibited (type II) by sulfate. Most microbial A. are type I. The only other member of this group is the vertebrate microsomal A. (A.C), which is usually contaminated by other sulfatases. The A. of *Aspergillus aerogenes* (M_r 40,700; high substrate specificity) and *Aspergillus oryzae* (M_r 90,000; low substrate specificity) have been extensively studied. Type II A. are found primarily in animal lysosomes, and the A. of beef liver has been highly purified; it consists of the abundant A.A (M_r 107,000; 4 subunits, each of M_r 27,000; pI = 3.4) and the minor form A.B (M_r 45,000; single chain (?); pI = 8.3).

Ascorbate oxidase (EC 1.10.3. 3): see Vitamins.

Ascorbate shuttle: a shuttle system in which cytosolic ascorbate provides reducing equivalents for regeneration of ascorbate in the matrix of adrenal chromaffin granules. Ascorbate in chromaffin granules is oxidized during noradrenalin biosynthesis by intragranular Dopamine β-hydroxylase (see). Electrons are transported across the granule membrane by membrane-bound cytochrome b_{561}. The process is driven by the proton motive force generated by a membrane ATPase which transports protons from the cytosol to the matrix of the granule. Cytosolic ascorbate is regenerated by mitochondrial semidehydroascorbate reductase. Since the interconversion of ascorbate and its semidehydro free radical involves a proton, the pH gradient across the granule membrane ($\Delta pH = 1.5$ units) cooperates with the transmembrane potential ($\Delta\psi = +0.06V$) to shift the E_h of the ascorbate/semidehydroascorbate couple inside the granule ($E_{h(inside)} = E_{h(outside)} + \Delta p$) from + 0.07V (cytosol) to + 0.22V, thereby favoring flow of electrons into the granule (Fig.). [M.F. Beers et al. *J. Biol. Chem.* **261** (1986) 2529–2535; L.M. Wakefield et al. *ibid.* 9739–9745 and 9746–9752]

Ascorbic acid, *vitamin C:* see Vitamins.

Asn, *Asp-NH$_2$:* abb. for L-asparagine.

L-Asparaginase, *L-asparagine amidohydrolase* (EC 3.5.1.1): a widely occurring enzyme which catalyses the hydrolysis of L-asparagine to L-aspartate and ammonia. Its earlier promise as an agent for the therapy of acute lymphoblastic leukemia and lymphosarcomas has not been realized. Its proposed use in cancer treatment was based on the observation that while certain tumor cells possess asparagine synthetase, they are unable to synthesize glycine from serine, so they cannot synthesize asparagine from glycine via glyoxylic acid and oxaloacetate. Such cells find it more difficult to compensate loss of asparagine caused by the action of L-A. Many normal tissues are also sensitive to L-A.; its toxic effects include impairment of the synthesis of secreted proteins, e.g. insulin, clotting factors, albumin, parathyroid hormone.

Bacterial L-A. have been highly purified from *E. coli, Actinetobacter* and *Erwinia carotovora.* All have M_r in the range 135,000–138,000 and are composed of 4 identical subunits (M_r 33,000).

L-Asparagine, *Asn, Asp-NH$_2$:* the β-half-amide of L-aspartic acid: $H_2N\text{-}OC\text{-}CH_2\text{-}CH(NH_2)\text{-}COOH$, M_r 132.1, m.p. (hydrate) 236°C (d.). Both Asn and aspartic acid have a ubiquitous occurrence, both as free amino acids and as protein components. Asn plays a role in the metabolic control of cell functions in nerve and brain tissue. In many higher plants, Asn and glutamine serve as soluble nitrogen reserves. Asn is biosynthesized from L-aspartic acid and ammonia by an asparagine synthetase (EC 6.3.1.1 or 6.3.1.4). Biodegradation of Asn usually starts with cleavage of the amide bond by asparaginase (EC 3.5.1.1).

Aspartate ammonia lyase, *aspartase* (EC 4.3.1.1): an enzyme found in bacteria, higher plants and a few lower animals, which catalyses the reversible interconversion of aspartate and fumarate plus ammonia:

$$^-OOC\text{–}CH_2\text{–}CH(NH_3^+)\text{–}COO^- \rightleftharpoons \; ^-OOC\text{–}CH=CH\text{–}COO^- + NH_4^+$$

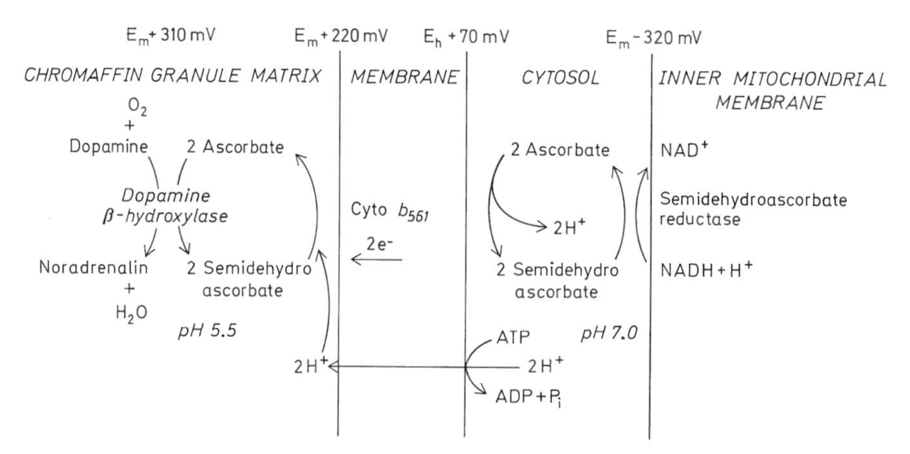

Ascorbate shuttle. E_m, midpoint potential. E_h, standard half-cell potential. Chromaffin granules contain 40 nmol ascorbate/mg protein, representing a concentration of 20 mM if all is in solution.

At higher temperatures, the deamination reaction occurs spontaneously. This pathway, which exists only for aspartate and for no other proteogenic amino acid, is lacking in vertebrates. Bacterial A. (M_r 180,000) is composed of 4 equal-sized subunits (M_r 45,000).

L-Aspartic acid, *Asp:* 2-aminosuccinic acid, HOOC-CH_2-CH(NH_2)-COOH, a proteogenic, acidic amino acid which is not essential for mammals. M_r 133.1, m.p. 269–271 °C, $[\alpha]_D^{25}$ + 5.05 (c = 2, water). Asp and oxaloacetate are interconvertible by transamination. Asparate ammonia lyase (see) is more important for the deamination of Asp than for its synthesis. Asp is important in the urea cycle and in purine and pyrimidine biosynthesis, the latter by way of orotate. In the synthesis of urea and purines, Asp donates its α-amino group in a transamination which is dependent on ATP, but does not require pyridoxal phosphate. N1 of the purine ring system and the 6-amino group of adenine are derived from the amino group of Asp.

Aspartylglycosamine: see Inborn errors of metabolism.

Aspartyl glycosaminuria: see Inborn errors of metabolism.

Aspergillic acid: an antibiotic, M_r 224.3, produced by *Aspergillus flavus.* See Hydroxamic acid.

Aspiculamycin: see Gougerotin.

Assimilate: in the wide sense, a product of any assimilatory process. In the narrow sense, a stable end product of Photosynthesis (see).

Assimilation: the incorporation of nutrients into an organism. See Carbon dioxide assimilation.

Assimilatory power: see Photosynthesis.

Assimilatory sulfate reduction: see Sulfate assimilation.

Astaxanthine: 3,3′-dihydroxy-β,β-carotene-4,4′-dione, M_r 596.82, m.p. 216 °C. It occurs widely as a red animal pigment, especially in crustaceans, echinoderms and tunicates. It is also present in the feathers and skin of flamingos and other birds, but is not found in many plants. Naturally occurring free and esterified (e.g. the dipalmitate) A. are red in color, whereas native A.-containing chromoproteins may be blue, green or brown. The dark, blue-black pigment in lobster shell is a complex of A. with protein, which upon denaturation (e.g. by cooking) releases red A. and colorless protein.

Asteromycin: see Gougerotin.

Asterosaponin A: a steroid saponin (see Saponins). The aglycon is 3β,6α-dihydroxypregn-9(11)-en-20-one, which contains a pregnane ring system (see Steroids), and which is linked to 2 molecules each of 6-deoxy-D-glucose and 6-deoxy-D-galactose, and one molecule of sulfate. A. was first isolated from the starfish *Asterias amurensis.*

AT-content: see GC-content.

Atmungsferment: cytochrome oxidase.

Atomic mass unit: see Dalton.

ATP: acronym of adenosine 5′-triphosphate (see Adenosine phosphates).

ATPase: see Adenosine triphosphatase, F-type ATPases, Ion-motive ATPases, P-type ATPases, V-type ATPases.

ATP citrate (pro-3*S*)-lyase, *citrate cleavage enzyme* (EC 4.1.3.8): a cytosolic enzyme converting citrate to acetyl-CoA and oxaloacetate and simultaneously cleaving one ATP to ADP: Citrate + ATP + CoA → Acetyl-CoA + oxaloacetate + ADP + P_i.

ATP-imidazole cycle: see L-Histidine.

ATP:urea amidolyase: see Urea amidolyase.

Atractyloside: a glucoside from the Mediterranean thistle *Atractylis gummitera.* It is a competitive inhibitor of adenine nucleotide binding and transport across the inner mitochondrial membrane. The closely related carboxyatractylate binds with higher affinity (K_d 10^{-8} M) and is not displaced by adenine nucleotides.

Atrial natriuretic factor, *atrionatriuretic factor, ANF, atriopeptin:* a polypeptide hormone produced by atrial heart muscle. Immunocytochemical studies show that pro-ANF is stored within specific granules of atrial cardiocytes. Immunocytochemistry and radioimmunoassay indicate the existence of ANF in the central nervous system and kidney, but in mammals the quantity of ANF in the atria is orders of magnitude higher than in extracardiac tissues. Atria also contain the greatest amount of ANF precursor mRNA. In nonmammals, ANF may also be synthesized in nonventricular muscle, and its production in tissues other than the heart may be quantitatively different from that in mammals.

Cloning and sequence analysis of ANF precursor cDNA show that ANF is synthesized in a prepro form containing 152 (rat) or 151 (human) amino acids. A disulfide-looped sequence of 17 amino acids with various *C*- and *N*-terminal extensions is necessary for activity. Pro-ANF is present in the atrial granules of rat and human perinuclear cardiocytes as a 126-residue peptide, also called atriopeptigen, cardionatrin IV or γ-atrial natriuretic peptide (γ-ANP), which is derived from prepro-ANF by processing. Human circulating ANF is a 28-residue peptide called cardionatrin I or α-atrial natriuretic peptide (α-ANP). A previously described β-ANP is probably an artifact formed by proteolysis during isolation. Studies with cultured cardiocytes suggest that, in the intact animal, pro-ANF is secreted from the atria, then cleaved to α-ANP by a blood protease.

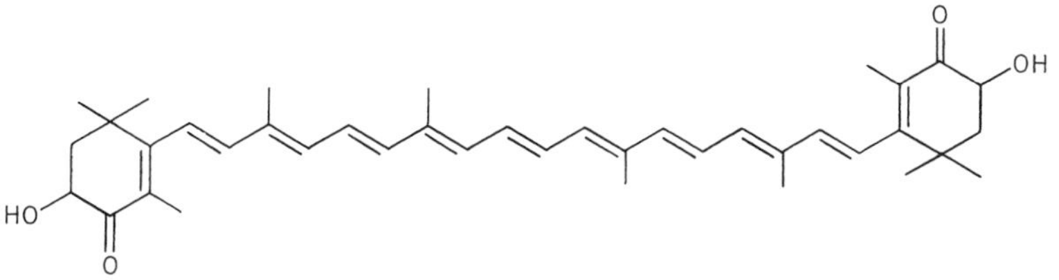

HO O O OH

Astaxanthine

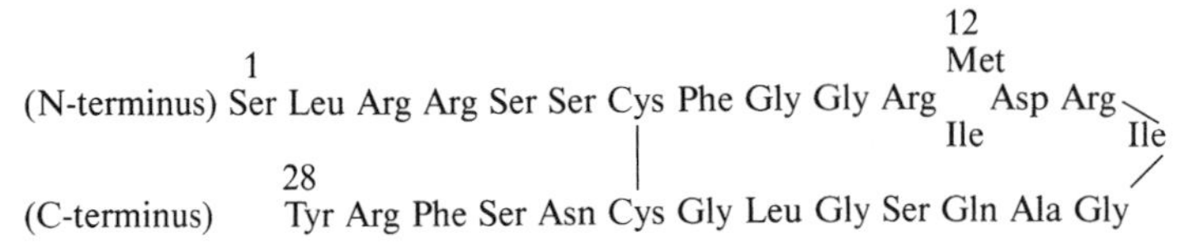

Amino acid sequence of the circulating form of atrial natriuretic factor, also called cardionatrin I or α-atrial natriuretic peptide (α-ANP). Rat α-ANP contains Ile at position 12, while the human form contains Met in this position.

Specific receptors for ANF have been demonstrated in kidney, blood vessels and adrenal cortex. The receptor binding site appears to be the extracellular domain of a transmembrane protein guanylate cyclase. Binding of ANF stimulates intracellular production of cGMP, which in turn activates protein kinase G. In vitro studies show that ANF release is stimulated by adrenalin, arginine vasopressin, acetylcholine and atrial distension. ANF is a potent diuretic (natriuretic) and hypotensive agent. It acts on the adrenal cortex to decrease aldosterone release, and on the kidney glomeruli and papilla to increase the glomerular filtration rate, renal blood flow, urine volume and sodium excretion; it also decreases plasma renin secretion. Decreased vascular volume serves as a negative feedback signal that decreases the concentration of circulating ANF. Circulating plasma levels are 25–100 pg/ml in humans and 100–1,000 pg/ml in rats. ANF probably functions in the short- and long-term control of water and electrolyte balance. [A.J. deBold, *Science* **230** (1985) 767–770 (rat ANF); K.Kangawa et al. *Nature* **313** (1985) 397–400 (human ANF); K.D.Bloch et al. *Science* **230** (1985) 1168–1171; T.G.Flynn & P.L.Davis, *Biochem J.* **232** (1985) 313–321; P.Needleman & J.E.Greenwald, *N. Engl. J. Med* **314** (1986) 826; B.Liu et al. *Biochemistry* **28** (1989) 5599]

Atrial peptide: see Atrial natriuretic factor.

Atrionatriuretic factor: see Atrial natriuretic factor.

Atriopeptin: see Atrial natriuretic factor.

Atromentin: see Benzoquinones.

Atropine, *DL-hyoscyamine:* the ester of DL-tropic acid with tropine. It is a racemate (M_r 289.48, m.p. 115–116 °C) formed during alkaline treatment of L(–)-hyoscyamine. Racemization occurs so readily that most A. from plant material is probably derived from L-hyoscyamine during the isolation procedure. It specifically inhibits those cholinergic neurons which are activated by muscarine. L-Hyoscyamine [m.p. 108–111 °C, $[\alpha]_D$ –21 ° (ethanol)] is the tropine ester of L-tropic acid. The pupil-dilating effect is due only to the L-form, but only racemic A. is used medicinally. It is also used to inhibit salivation and sweating, and to relax cramps in the gastrointestinal tract and bronchi (in asthma). A. is highly toxic. For formula, occurrence and biosynthesis, see Tropane alkaloids.

Atroscine: see Scopolamine.

Attenuation: a regulatory mechanism employed by the bacterial cell. Whereas enzyme repression al-

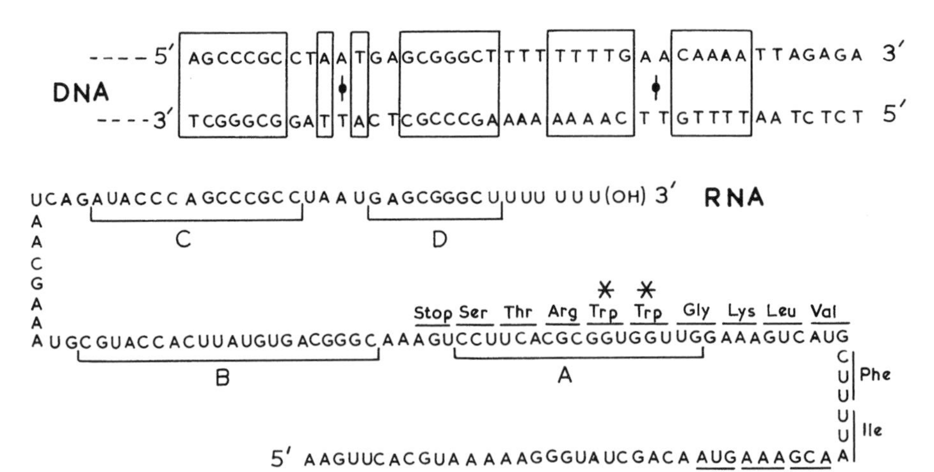

Fig. 1. *Attenuator site in the leader of the tryptophan synthesis operon of E. coli, and the complete sequence of the terminated leader RNA.* Two regions in the *trp* attenuator DNA have a 2-fold axis of symmetry. A, B, C, D in the RNA correspond to similarly labeled regions in Fig. 2. Base pairing with resulting stem and loop formation is possible between A and B (free energy of formation –46.9 kJ/mol), B and C (–49 kJ/mol), C and D (–83.7 kJ/mol). Codons for the leader peptide are also shown and the two strategic Trp codons are indicated by asterisks.

lows the cell to respond to extreme concentrations of metabolites, A. probably represents a means of fine tuning to relatively mild fluctuations in the concentrations of metabolites. A. has been shown to operate in the synthesis of tryptophan, phenylalanine, histidine, leucine, threonoine, isoleucine, leucine and valine in *E. coli,* and in the synthesis of histidine, leucine and tryptophan in *Salmonella typhimurium,* but A. sites are probably present in all amino acid synthesis operons in all bacteria.

The first structural gene of the operon is separated from the promoter-operator by a length of DNA called the leader. Transcription of the operon proceeds via this leader and into the structural genes. For example, in the case of the *trp* operon of *E. coli,* a single continuous 7,000 nucleotide *trp* mRNA transcript is formed, which includes leader RNA (162 nucleotides) at the 5′ end, followed by mRNA sequences for all the enzymes in the biosynthetic pathway. This transcript is formed only when tryptophan is re-

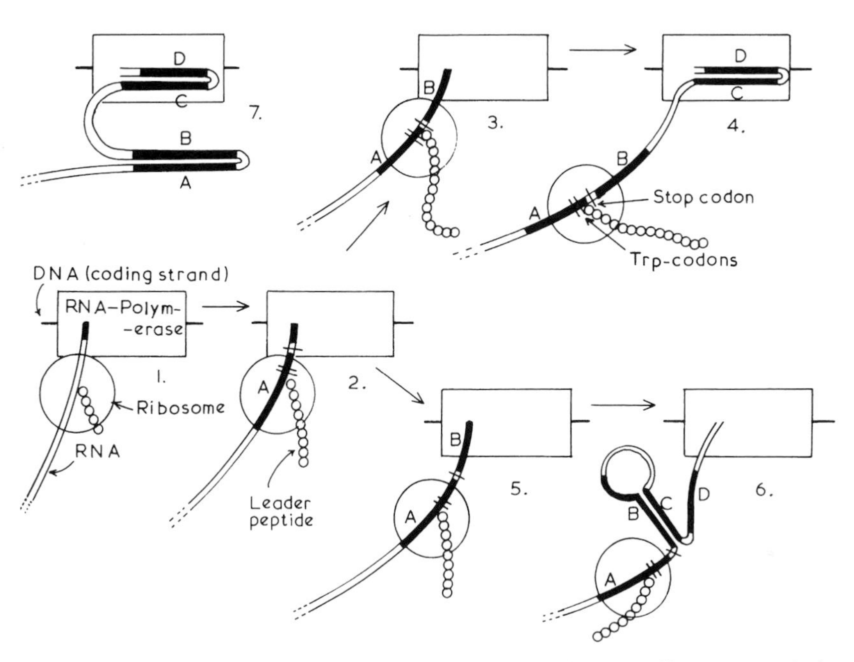

Fig. 2. *Diagrammatic representation of attenuation during coordinate transcription and translation, based on attenuation of the tryptophan synthesis operon in E. coli.*
1. RNA-polymerase has left the operater-promotor site, and has started to transcribe the leader region as it proceeds in the direction of the first structural gene of the operon (component I of anthranilate synthetase). The start codon for the leader peptide has appeared, a ribosome has become attached, and translation has started.
2. As more RNA is synthesized, translation keeps pace with transcription. Region A has been synthesized and B is just beginning. The ribosome is about to translate the Trp codons of the leader peptide message.
3. Tryptophan-aminoacyl-tRNA is plentiful, so translation continues and keeps pace with transcription. The ribosome has just reached the stop codon, and partly covers region B which is now fully synthesized.
4. The ribosome can proceed no further, but transcription continues. When region C first appears, it cannot base-pair with B, due to the presence of the ribosome. As D appears, it base-pairs with C. The resulting secondary structure (CD loop and stem) acts as a signal for the termination of transcription. The RNA-polymerase and its transcript leave the DNA.
5. Tryptophan-aminoacyl-tRNA is scarce, so the ribosome pauses at the Trp codons. As B is synthesized it is not partly covered by the ribosome. Translation is not keeping pace with transcription.
6. As region C emerges from the RNA-polymerase, it is able to base-pair with B, forming a stem-loop structure, known as a preemptor (it preempts the formation of CD). As D is formed, C is not available for base-pairing, so the transcription termination signal (CD loop and stem) cannot be formed. Transcription continues into the first structural gene, and the entire operon is transcribed.
7. If other aminoacyl-tRNA species are scarce, the ribosome will not even reach A by the time D has appeared. The thermodynamically favored structure then consists of the stem and loop structures AB and CD. Transcription is terminated by CD. Thus, a deficiency of other amino acids can attenuate the synthesis of tryptophan.

latively scarce. When the tryptophan level is high, transcription proceeds only part of the way through the leader and is then terminated, producing a small transcript of 140 nucleotides. The controlling factor is not tryptophan itself, but tryptophan-aminoacyl-tRNA, the level of which reflects the availability of tryptophan. Part of the leader RNA is translated into a peptide of 14 amino acid residues, two of which are tryptophan residues. Codons for these two Trp residues occupy strategic positions on the leader RNA (see Figs. 1 and 2). Analysis of leader RNA reveals more than one possible secondary structure. Which secondary structure actually occurs is determined by the progress of ribosomes along the RNA. If tryptophan-aminoacyl-tRNA is plentiful (it is assumed that all the other aminoacyl-tRNA species needed are also plentiful), translation proceeds smoothly and the ribosome reaches the stop codon. At this point, the ribosome physically overlaps two regions of the RNA that are potentially capable of taking part in secondary structure formation (regions A and B in Figs. 1 and 2). The only possible secondary structure is therefore the stem and loop formed by regions C and D, which acts as a transcription termination signal. If tryptophan-aminoacyl-tRNA is scarce, the ribosome halts earlier at the *trp* codons, only region A is covered, and a noninhibitory secondary RNA structure is formed; transcription then proceeds into the structural genes. The mechanism is better appreciated by considering the coordinate progress of transcription and translation as shown in Fig. 2. It can be seen that A. represents a graded response to slight changes in the level of the control amino acid. This is not an "all or nothing" or threshold type of control, unlike repression or feedback inhibition which regulate by an on-off switch mechanism. It is unlikely that the ribosome actually stops at the *trp* codon; it may slow down or pause, depending on the relative supply of aminoacyl-tRNA of the control amino acid. In view of the speed of transcription and coordinate translation (45 nucleotide residues incorporated per second; about 4 min for transcription of the entire *trp* operon; about 3 min for complete degradation of the *trp* operon transcript) a mere slowing of the ribosome will be sufficient to favor the noninhibitory structure of the leader RNA. The two secondary RNA structures must therefore be considered as two thermodynamic extremes that are, to a certain extent, in equilibrium.

In all other A. systems studied so far, the mechanism is analogous to that described here for tryptophan synthesis in *E. coli.,* i.e. codons for the control amino acid are located at strategic positions in the leader RNA. As many as seven strategic codons may be present (e.g. *Phe* and *His* A. in *Salmonella).* [D. L. Oxender et al. *Proc. Natl. Acad. Sci. USA* **76** (1979) 5524–5528; R. M. Gemmill et al. *Proc. Natl. Acad. Sci. USA, ibid.* 4941–4945; H. M. Johnston et al. *Proc. Natl. Acad. Sci. USA* **77** (1980) 508–512; C. Yanofsky & R. Kolter, *Annu. Rev. Genet.* **16** (1982) 113–134]

Aucubin: see Iridoids.

Auriculoside: 7,5′-dihydroxy-4′-methoxy-3′-*O*-β-D-glucosylflavan. See Flavan.

Aurones, *2-benzylidene-3-cumarones:* yellow flavonoid plant pigments. A. are very common in *Compositae, Leguminoseae, Hepaticae* and *Anacardiaceae,* and have been found sporadically in a few other plant families. Individual A. differ with respect to the substituents in rings A and B; common substituent groups are hydroxy, methoxy, methylenedioxy and furano. Examples are sulfuretin (6,3′,4′-trihydroxyaurone, a petal pigment in many members of the *Compositae)* and furano (2″,3″,6,7):3,4-methylene-dioxyaurone from *Derris obtusa.* A. frequently occur as glycosides. [B. A. Bohm in *The Favonoids: Advances in Research,* pp. 336–337 (J. B. Harborne & T. J. Mabry, eds.), Chapman and Hall, 1982]

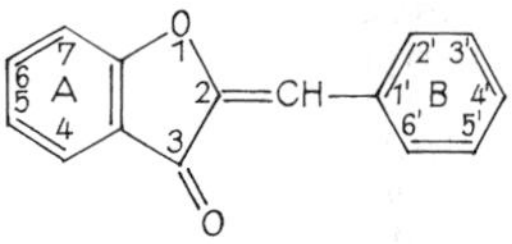

Aurone ring system

Autoimmune diseases: conditions resulting from a lack of immune tolerance for the organism's own components. They may arise by 1. cross reactions between body substances and foreign materials, 2. lack of immune tolerance for body substances which are normally not exposed to the immune apparatus, 3. structural changes in body substances, or 4. reactivation of certain cells, leading to a breakdown of tolerance. A. d. include neurological conditions like multiple sclerosis, bacterial eye inflammation (ophthalmia sympathica), myasthenia gravis, erythematodes (a skin disease), chronic rheumatism, chronic liver inflammation and certain forms of chronic kidney inflammation.

Autolysis: self digestion of dying cells, which occurs when cathepsins (see Proteolysis) are released from lysosomes and attack cytoplasmic proteins. In the living cell these proteases are sequestered inside the lysosomes, but at death the lysosomal membranes break down. See Intracellular digestion.

Autonomously replicating sequence, *ARS:* a DNA sequence that confers the ability to undergo high frequency autonomous transformation. Such sequences were first identified in the yeast, *Saccharomyces cerevisiae,* in which they represent chromosomal origins of replication, and they have since been derived from a variety of eukaryotes including humans. Two important sequences are important for ARS function: domain A [5′ (A/T)TTTAT(A/G) TTT(A/T) 3′] and domain B [5′ CTTTTAGC(A/T) (A/T)(A/T) 3′]. Domain B is located 50–100 bp 3′ to domain A. See Yeast artificial chromosome.

Autophagy: see Intracellular digestion, Proteolysis.

Autotrophy: The ability to synthesize all organic components from simple inorganic compounds like carbon dioxide, ammonia, nitrate and sulfate. The term originally applied only to *carbon autotrophy,* the ability of green plants to assimilate CO_2 from the air (see Carbon dioxide assimilation). *Nitrogen autotrophy* refers to nitrate assimilation (see Nitrate reduction) or to Nitrogen fixation (see), both being processes in which ammonia is formed and assimilated (see Ammonia assimilation). Sulfur autotrophy depends on the ability of plants and microorganisms to reductively assimilate sulfate (see Sulfate assimilation). The converse of A. is heterotrophy.

Auxin antagonists: inhibitors of the effects of auxins on growth and development. The action of A.a. can be at least partly reversed by auxins. The term is used independently of the mechanism of inhibition; competitive inhibitors are called antiauxins. The A.a. include many structurally different kinds of compounds, including some synthetic auxins, e.g. phenylacetic acid and phenylbutyric acid.

Auxin conjugates: see Auxins.

Auxins: a group of plant hormones which are among the most important regulators of plant growth. Natural A. are indole derivatives, previously thought to be biosynthesized from L-tryptophan. However, it may be necessary to modify this view, since a tryptophan-independent route of IAA synthesis has been demonstrated in maize and *Arabidopsis*, in which indole-3-acetonitrile is converted into indole-3-acetic acid by the action of nitrilase [J. Normanly et al. *Proc. Natl. Acad. Sci. USA* **90** (1993) 10355–10359]. The most important A. is indole-3-acetic acid (IAA, heteroauxin) (Fig.), which is taken as the standard for comparison of the activities of other growth stimulants. IAA is synthesized in all higher plants (e.g. pineapple plants contain 6 mg/kg fresh weight); it has also been found in many lower plants, as well as occurring as a bacterial metabolite. The Oat coleoptile test (see) has been used extensively for the quantitative determination of A., but it is now possible to use sensitive physical methods such as high performance liquid chromatography (or gas chromatography) combined with mass spectrometry. Other natural A. include: 2-(-indolyl)ethanol (tryptophol); 2-(3-indolyl)-acetaldehyde; 2-(3-indolyl)acetonitrile; 2-(3-indolyl)-acetamide; 3-(3-indolyl)propionic acid; 4-(3-indolyl)-butyric acid; 3-(3-indolyl)succinic acid; 2-(3-indolyl)-glycolic acid; 2-(3-indolyl)acetyl-β-D-glucose; [2-(3-indolyl)acetyl]aspartic acid; 2-(3-indolyl)acetyl-meso-inositol; 2-(3-indolyl)acetyl-mesoinositol galactoside; 2-(3-indolyl)acetyl-mesoinositol arabinoside. Those containing sugars or amino acids are called auxin conjugates and are probably storage and transport forms.

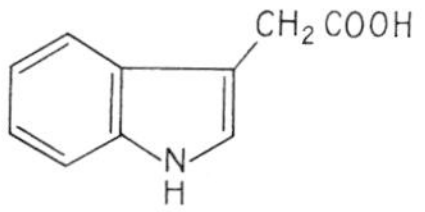

2-(3-Indolyl)acetic acid

Auxins modulate root and shoot formation by stimulating extension growth and cell division in the cambium. Cell wall acidification in response to auxins (probably due to stimulation of the plasma membrane H^+ATPase) at least partly accounts for the loosening of cell walls and stimulation of elongation growth. Auxins also cause membrane hyperpolarization.

Several auxin-binding proteins (ABPs, M_r 20,000–22,000) have been identified. These small proteins are characterized by the presence of an auxin-binding motif, a glycosylation site, and the carboxy-terminal sequence –Lys–Asp–Glu–Leu–COOH. At least 5 ABPs are present in *Zea mays*. In the *Baculovirus* expression system, *Zea* ABP1 (*Zm*ABP1) is targeted to the endoplasmic reticulum of the insect cells used for expression [H. MacDonald et al. *Plant Physiol.* **105** (1994) 1049–1057]. When *Zm*ABP1 is translated in vitro, it is translocated to microsomes derived from the endoplasmic reticulum, where it is processed and glycosylated. Parallel studies on the in vitro translation of *Zm*ABP4 showed that only one translation product is translocated to the microsomes (the *Zm*ABP4 transcript has a different signal sequence from that of *Zm*ABP1, and it encodes two translation products) [N. Campos et al. *Plant Cell Physiol.* **35** (1994) 153–161]. Packaging of ABP into the endoplasmic reticulum is at present difficult to reconcile with the clear evidence that auxin signal is transmitted following binding at the plasma membrane. Since ABPs do not possess a hydrophobic domain long enough to span a membrane, it has been suggested that a transmembrane docking protein is involved in transmission of the ABP signal into the cell. The *AUX1* gene (mutation causes agravitropism and resistance to auxin) encodes a protein with a large hydrophilic *N*-terminus and 7 predicted transmembrane domains. This is strikingly similar to the gonadotropin receptors, whose ligands are glycoproteins of M_r 23,000–30,000. The auxin signaling pathway within the plant cell is unclear, but a G protein is very probably involved.

Binding of A. to *Zm*ABP1 has been investigated by photoaffinity labeling of the mature protein (163 amino acid residues), using the auxin homolog 5-[7-H-3]azidoindole-3-acetic acid. Analysis of tryptic fragments from the labeled protein showed that Asp 134 is the labeled residue, while Trp 136 probably also contributes by hydrophobic binding of the auxin aromatic ring.

There is conclusive evidence that at least the ABP(s) involved in hyperpolarization of protoplasts is localized at the intercellular surface of the plasma membrane, i.e. it functions extracellularly. Antibodies against *Zm*ABP1 block auxin-induced activation of the H^+ATPase, but have no effect on its activation by fusicocci [H. Barbier-Brygoo et al. *Plant J.* **1** (1991) 83–93; A. Ruck et al. *Plant J.* **4** (1993) 41–46].

Certain genes are transcribed rapidly in response to auxin. In particular, these include the small auxin-upregulated RNA (SAUR) genes of unknown function. The promoter for soybean *SAUR15A* gene contains 2 adjacent, auxin-responsive sequences, TGTCTC and GGTCCCAT [Y. Li et al. *Plant Physiol.* **106** (1994) 37–43]. Other auxin-inducible genes have been shown to contain auxin-responsive elements either within the reading frame or within untranscribed regions such as the promoter. It seems likely that some proteins, whose genes contain auxin-responsive elements, are activators or repressors of genes that mediate auxin responses, e.g. proteins PSIAA6 and PSIAA4 of peas; both of these proteins are transcribed extremely rapidly following auxin induction, both show very high turnover (half-lives 8 and 6 min, respectively), and both carry putative nuclear location signals and a βαα DNA-binding structural motif similar to that in the Arc family of prokaryotic repressors [S. Abel et al. *Proc. Natl. Acad. Sci. USA* **91** (1994) 326–330]. Also, protein-binding regions have been identified (using DNAase I and gel mobility shift assays) in the promoter region of the soybean auxin-responsive gene, *GmAux28* [R.T. Nagao *Plant Mol. Biol.* **21** (1993) 1147–1162].

[Reviews: K. Palme *J. Plant Growth Reg.* **12** (1993) 171–178; S. Brunn et al. *ACS Symp. Series* **557** (1994) 202–211; M. Blatt & G. Thiel *Annu. Rev. Plant Physiol. Plant Mol. Biol.* **44** (1993) 543–567; L. Hobbie & M. Estelle *Plant Cell Envir.* **17** (1994) 525–540; Y. Takahashi et al. *Plant Res.* **106** (1993) 367–367; C. Garbers & C. Simmons *Trends in Cell Biology* **4** (1994) 245–250; P. A. Millner *Current Opinion in Cell Biology* **7** (1995) 224–231]

Auxochromes: see Pigment colors.

Auxotrophic mutants, *deficiency mutants:* mutants with nutritional requirements not present in the parental or wild-type organism. The substances which cannot be synthesized by the mutant must be supplied as vitamins or substrates in the growth medium. A. m. have an important role in the elucidation of biosynthetic pathways (see Mutant technique). They may be monoauxotrophic, requiring only one substance as a growth supplement, or polyauxotrophic. In the latter case, the defects may be polygenic, or the loss of a single gene may affect the synthesis of several products. Single gene polyauxotrophy occurs (a) when the block occurs before the branching point of a branched synthesis chain, or (b) an enzyme required in two or more parallel synthetic pathways is defective. The first case (a) is exemplified by methionine-isoleucine double auxotrophs, which require supplementation of a minimal medium with methionine + isoleucine (or threonine, which can be converted into isoleucine). The metabolic block occurs before the formation of homoserine, the branching point in this biosynthetic pathway:

Precursors → L-homoserine → L-threonine
↓ ↓
L-methionine L-isoleucine

An example of the second type (b) is the mutation leading to loss of the keto-acid reductoisomerase (EC 1.1.1.86) in the parallel pathways leading to valine and isoleucine. This enzyme catalyses an isomerization coupled to a reduction, producing 2,3-dihydroxyisovalerate in the valine pathway, or 2,3-dihydroxy-3-methylvalerate in the leucine pathway. A single mutation of the gene for this enzyme produces val^-/ile^- doubly auxotrophic organisms.

(1) (2) (2) (3) (4)
Pyruvate → I_1 → I_2 → I_3 → I_4 → L-valine
⇑ ⇑
Reductoisomerase
⇓ ⇓
Keto- → I_a → I_b → I_c → I_d → L-isoleucine
butyrate (1) (2) (2) (3) (4)

Here, (2) is the reductoisomerase, I_3 is 2,3-dihydroxyisovalerate, and I_c is 2,3-dihydroxy-3-methylvalerate.

Polyauxotrophic nutritional requirements cannot be induced by the mutation of a single gene and loss of a single protein if the metabolic step in question is catalysed by isoenzymes encoded in different genes.

Auxotrophy: the condition of requiring nutritional supplements for growth (see growth factors). A. can arise through mutation (see Auxotrophic mutants). The converse of A. is Prototrophy (see).

Avena coleoptile test: see Oat coleoptile test.

Avenasterol, *28-isofucosterol:* an isomer of Fucosterol (see) found in green marine algae (see Sterols).

Avermectins: a class of macrocyclic lactones produced by the actinomycete *Streptomyces avermitilis.* A. are used in low dosage against nematode and arthropod parasites of cattle, horses, sheep, dogs and pigs. They are also potentially useful against human parasites. A. appear to interfere with the 4-aminobutyric acid (GABA) (see) receptors in synapses that employ GABA as a neurotransmitter. In mammals, such neurons occur only in the central nervous system, where they are protected by the blood-brain barrier. [W. C. Campbell et al. *Science* **221** (1983) 823–827]

Avidin: a basic glycoprotein in the egg whites of many birds and amphibians. The primary structure of chicken A. is known: M_r 66,000, pI 10, 10.5 % threonine. There are 4 identical subunits of M_r 14332 (without the carbohydrate) (128 amino acids). A. forms a stoichiometric, noncovalent complex with 4 molecules of the vitamin, biotin; the complex is not attacked by proteolytic enzymes and therefore not resorbed. Feeding of A. or raw egg white can therefore result in experimental biotin deficiency (see Vitamins, biotin). A. is denatured and thus inactivated by heating. Like the unrelated lysozyme and conalbumin (see Siderophilins), A. protects the egg white against bacterial invasion.

Avitaminosis: see Vitamins.

Axerophthol: see Vitamins (vitamin A).

5-Azacytidine, *1-β-D-ribofuranosyl-5-azacytosine:* a pyrimidine antibiotic (see Nucleoside antibiotics) synthesized by *Streptoverticillius lakadamus* var. *lakadamus.* It is effective against Gram-negative bacteria.

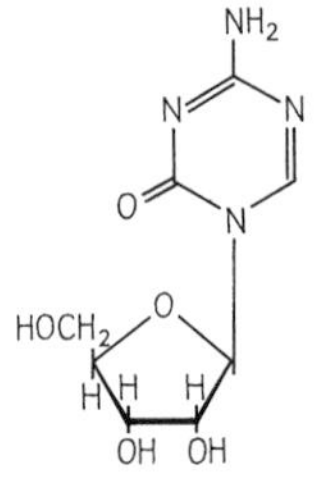

NH_2, N, N, O, N, $HOCH_2$, O, H, H, H, OH, OH

5-Azacytidine

8-Azaguanine, *pathocidin, 5-amino-7-hydroxy-1-v-triazolo [d] pyrimidine:* a purine antagonist first synthesized in 1945 [Roblin et al. *J. Am. Chem. Soc.* **67** 290–294]: M_r 152,04; decomposition without melting above 300 °C. It was shown later to be identical with pathocidin from *Streptomyces spectabalis.* 8-A. affects many different enzymes of purine metabolism and synthesis. In particular it inhibits translation and causes errors of translation, due to its incorporation into mRNA. It is converted by the cell into the corresponding nucleoside 5′-triphosphate, which feedback inhibits the first enzyme of purine biosynthesis, 5-phosphoribosylpyrophosphate amidotransferase (EC 2.4.2.14). 8-A. is inhibitory and toxic to a wide variety of living systems, including bacteria, protozoa and higher animals. It inhibits the growth of several murine adenocarcinomas, but has found only limited application in the treatment of human cancers.

A-Azahomosteroids: see Salamander alkaloids.

L-Azaserine, ***O-diazoacetyl-L-serine:*** $N^- = N^+ = CH-CO-OCH_2-CH(NH_2)-COOH$, a glutamine analog, which inhibits transfer of the amide group (see Transamidation) from glutamine to formylglycinamide ribotide. It is mutagenic and has antitumorigenic activity. L. A. is synthesized by *Strepomyces* spp. and is effective against Clostridia, *Mycobacterium tuberculosis* and Rickettsias.

Azofer: see Nitrogenase.

Azofermo: see Nitrogenase.

Azomycin, ***2-nitroimidazole:*** an antibiotic synthesized by *Nocardia mesenterica* and *Streptomyces eurocidicus;* m.p. 281–283 °C. Both Gram-negative and Gram-positive bacteria are highly sensitive to the compound; 0.5 μg/ml inhibits growth of *Trichomonas vaginalis;* 3 μg/ml inhibits growth of *Salmonella paratyphi.*

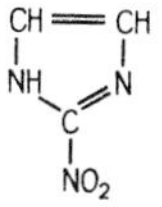

CH = CH, NH, N, C, NO_2

Azomycin

A-Z solution: see Nutrient medium, Table 3.

Azulene, ***cyclopentacycloheptene:*** the parent compound of a group of blue to violet, nonbenzoid aromatic compounds. The fused 5- and 7-membered rings are stabilized by a π-electron sextet.

Originally the term was used for the blue, high boiling fraction of camomile oil. A. are artifacts, produced from colorless sesquiterpenes, the proazulenes (Fig.). The compounds have antiinflammatory properties. Prepared camomile oil contains up to 15 % chamazulene (R = H). Guaiazulene ($R = CH_3$) is present in geranium oil.

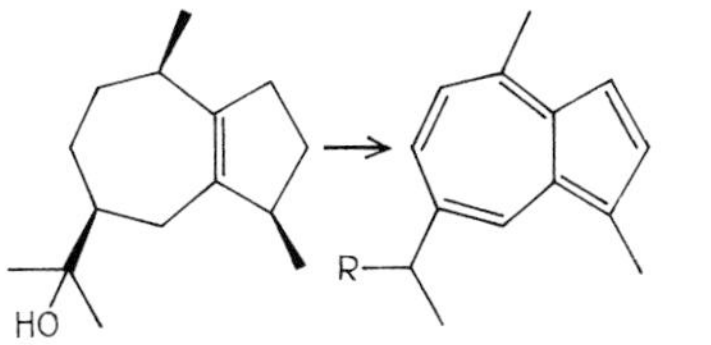

HO, R–

Guaiol (Proazulene) → Guaiazulene, $R = CH_3$

Formation of azulenes

Azurin: a family of blue, copper-containing proteins from *Pseudomonas, Alcaligenes* and *Bordetella* species. A. from *Pseudomonas fluorescens* contains 128 amino acid residues of known sequence, and one intrachain disulfide bond. A. from all sources has M_r 14,000–16,000 and homologous primary and tertiary structure, but different species of A. have different redox potentials.

A. contains 1 atom of Cu^{2+} per molecule, bound in a trigonal pyrimidal array, involving a distorted N_2SS donor set of Cys (S), Met (S) and 2 His (N). With respect to the size of the protein molecule and the nature of the Cu binding, A. is similar to plastocyanin (green plants) and stellacyanin (latex of Chinese laquer tree). Together with cytochrome c_{551}, A. is belived to function in a terminal respiratory network, using either O_2 or nitrate as electron acceptor. On account of its blue color (λ_{max} 625–630 nm; ε about 7,000) and its pI (pH 5.65), A. from *Pseudomonas aeruginosa* is a useful standardization marker in isoelectric focusing. [P. Rosen et al. *Eur. J. Biochem.* **120** (1981) 339–344; P. Frank et al. *J. Biol. Chem.* **260** (1985) 5518–5525]

B

B9: see Succinic acid 2,2-dimethylamide.

B 995: see Succinic acid 2,2-dimethylhydrazide.

Bacimethrin, ***4-amino-2-methoxy-5-pyrimidinemethanol:*** an antibiotic synthesized by *Bacillus megatherium.* It is antagonized by thiamin and pyridoxine, and is active against some yeasts and bacteria.

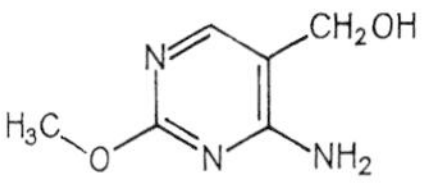

Bacimethrin

Bacitracins: branched, cyclic peptides produced by various strains of *Bacillus licheniformis.* They are effective against Gram-positive bacteria, in which they interfere with Murein (see) synthesis. The most important of the group is B.A, which contains an unusual thiazoline structure synthesized from the *N*-terminal lysine and the neighboring cysteine. B.F is a rearrangement product of B.A., in which the amino group of the heteroproduct is oxidatively removed and the thiazoline ring system is dehydrogenated.

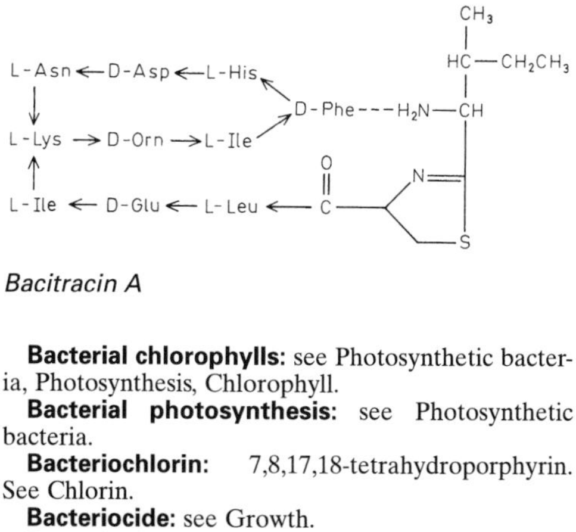

Bacitracin A

Bacterial chlorophylls: see Photosynthetic bacteria, Photosynthesis, Chlorophyll.

Bacterial photosynthesis: see Photosynthetic bacteria.

Bacteriochlorin: 7,8,17,18-tetrahydroporphyrin. See Chlorin.

Bacteriocide: see Growth.

Bacteriophage: see Phages.

Bacteriorhodopsin: a retinaldehyde-containing purple membrane protein (M_r 26,000; 248 amino acid residues) first discovered in the halophilic bacterium *Halobacterium halobium.* The primary structure of B. is not homologous with that of vertebrate rhodopsin (M_r 40,000), but the tertiary structures of the two proteins are similar. B. consists of 7 α-helical regions which lie in the membrane. These are connected by nonhelical regions which protrude into the cytoplasm and extracellular space. The membrane portions of the molecules are elongated ovoids with their long axes roughly perpendicular to the plane of the membrane.

The retinaldehyde is linked to the protein via a Schiff's base with lysine 216. When activated by light, the protein acts as a proton pump. Mutants lacking B. are still able to extrude Na^+, and to generate ATP in the light with the aid of another retinaldehyde-containing protein, ***haloopsin.*** [M. A. Keniry et al. *Nature* **307** (1984) 383–386]

Bacteriostatic agents: see Growth.

Bacteroids: symbiotic, nitrogen-fixing, intracellular forms of *Rhizobium* spp. in root nodules of leguminous plants (see Rhizobia). A root nodule of soybean contains several thousand cells. Each cell is tightly packed with membrane vesicles *(syn.* membrane envelope); the vesicle membrane consists of a double layer of phospholipid derived from the plasmalemma of the host cell, and each vesicle contains about 5 B. bathed in a solution of Leghemoglobin (see). B. will not redifferentiate into free-living Rhizobia; cultures of *Rhizobium* obtained from nodules are derived from unchanged bacterial cells remaining in the infection threads of the host tissue.

L-Baikiain: see Pipecolic acid.

Balata: a rubber-like polyterpene of low molecular mass from the latex of certain tropical trees, in particular from *Mimusops balata.* The double bonds of B. have *trans* configuration (see Polyterpenes, Fig.), and it is therefore not very elastic (in contrast to rubber). It softens when heated.

Balbiani rings: see Giant chromosomes.

Baldrianal: see Valtratum.

Balsams, ***oleoresins:*** solutions of resins in volatile oils. B. are produced either as normal plant constituents, or in response to pathological conditions or injury. Commercially, the most important B. is turpentine, produced (1–2 kg/tree/year) by conifers in reponse to bark injury. Steam distillation of the crude B. yields turpentine oils; the residue is colophoy (rosin). Other B., usually named after the country of origin (e.g. Peru B., Canada B.), are used in perfumes and pharmaceuticals.

Bamicetin: a pyrimidine antibiotic (see Nucleoside antibiotics) synthesized by *Streptomyces plicatus.* It is effective in low concentrations against *Mycobacterium tuberculosis.* Structurally, it is very similar to Amicetin (see), differing merely in the structure of the amosamine moiety which possesses a monomethylamino group in place of the dimethylamino group.

Base pairing: specific hydrogen bonding between adenine and uracil, adenine and thymine, and cytosine and guanine in a double stranded molecule of RNA or DNA, or in the interaction between DNA and RNA. B.p. is the basis of the formation of the double helix of DNA from two complementary single strands (see Deoxyribonucleic acid).

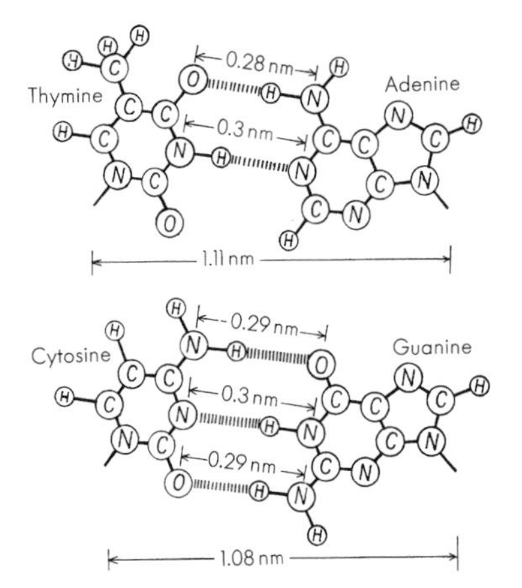

Hydrogen bonding between complementary bases in DNA

Basic proteins: a group of small proteins rich in arginine and lysine, and found in the cell nucleus (see Histones) and in sperm (see Protamines). They form complexes with nucleic acids.

Batatasins: see Orchinol.

Batch culture: see Fermentation technology.

Bathochrome: see Pigment colors.

Batrachotoxin: a neurotoxin from the skin of the Columbian arrow poison frog, *Phyllobates*. It causes a selective and irreversible increase in the permeability of nerve membranes to Na^+. LD_{50} in mice 2 µg/kg i.v. In order to be effective, B. must be injected or gain entry via damaged tissue. It is nontoxic when ingested, providing there are no scratches or ulcers in the gastrointestinal tract. Batrachotoxin A from the same source in less toxic (LD_{50} in mice 1 mg/kg i.v.).

Batrachotoxin.
In batrachotoxin, R = -O-C(=O)- (2,4-dimethylpyrrol-3-yl)

In batrachotoxin A, R = OH

Bavachin: 7,4′-dihydroxy-6-prenylflavanone. See Flavanone.

Bay-region theory of carcingonesis: a hypothesis that polycyclic hydrocarbons possessing a bay region are more potent procarcinogens than those lacking a

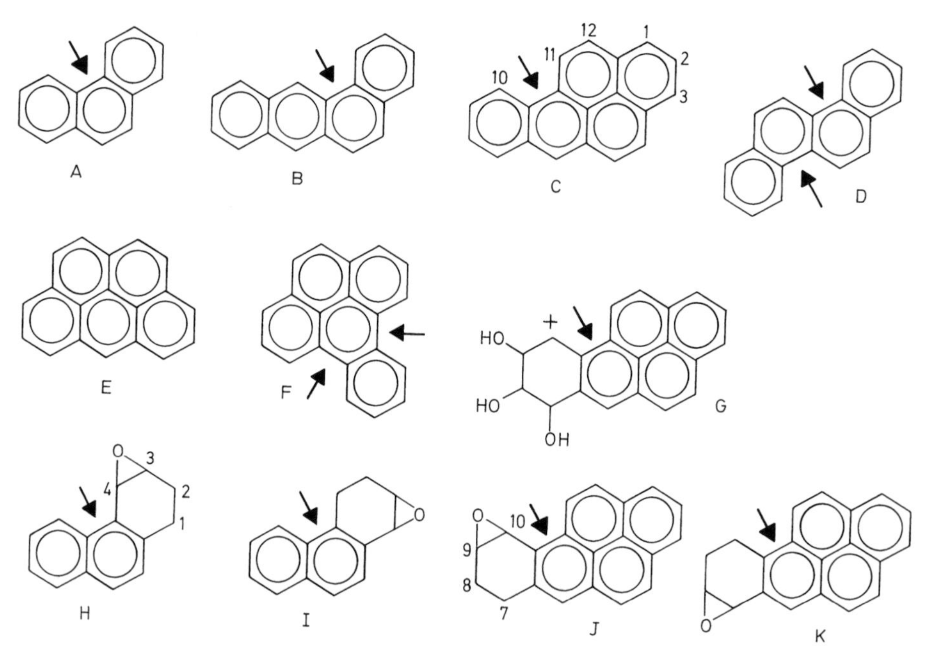

Polycyclic hydrocarbons and their derivatives. Bay regions are indicated by arrows.
A Phenanthrene. *B* Benzanthracene. *C* Benz[a]pyrene. *D* Chrysene. *E* Benz[c,d]pyrene. *F* Benz[e]pyrene. *G* Carbonium ion derived from benz[a]pyrene-7,8-diol-9,10-epoxide. *H* 1,2,3,4-Tetrahydrophenanthrene-3,4- epoxide. *I* 1,2,3,4-Tetrahydrophenanthrene-1,2-epoxide. *J* 7,8,9,10- Tetrahydrobenz[a]pyrene-9,10-epoxide. *K* 7,8,9,10-Tetrahydrobenz[a]pyrene- 7,8-epoxide.

bay region, e.g. phenanthrene is more carcinogenic than anthracene, and benz[a]pyrene is more potent than benz[c,d]pyrene. The actual carcinogens are epoxides formed from polycyclic hydrocarbons by the action of Cytochrome P450 (see), and in the most powerfully carcinogenic of these the epoxide group is situated on the rim of a bay region. Thus, the ultimate carcinogen formed metabolically from benz[a]-pyrene is the 7,8-diol-9,10-epoxide. Carcinogenesis is due to alkylation of DNA by a carbonium ion derived from the epoxide. Perturbational molecular orbital calculations suggest that formation of benzylic carbonium ions is facilitated if the carbonium ion is part of a bay region of a polycyclic hydrocarbon. In further support of the theory, synthetically prepared 1,2,3,4-tetrahydrophenanthrene-3,4-epoxide (epoxide group on the rim of the bay) is more carcinogenic than 1,2,3,4-tetrahydrophenanthrene-1,2-epoxide (epoxide group distal to the bay); similarly, 7,8,9,10-tetrahydrobenz[a]pyrene-9,10-epoxide is more carcinogenic than 7,8,9,10-tetrahydrobenz[a]pyrene-7,8-epoxide. Exceptions include chrysene and benz[e]pyrene, which possess bays but are only weakly carcinogenic; these compounds are readily attacked in regions distal to the bays (regions of relatively high electron density and double bond character, known as "K" regions), forming diols which are rapidly conjugated with sulfate or glucuronate, then excreted. See also Aflatoxins. [M. C. Macleod et al. *Cancer Res.* **39** (1979) 3463–3470]

B-cell growth factor, *BCGF:* see Lymphokines.

B_{12}-coenzyme: see 5′-Deoxyadenosylcobalamine.

Bdellins: a group of protease inhibitors from leeches. Particularly high activities of B. are found in the region of the outer sex organs and in the salivary glands of the leech, *Hirudo medicinalis.* B. inhibit trypsin and plasmin, and they show strong inhibition of the trypsin-like protease, acrosin, present in the acrosomes of spermatozoa.

Bee toxin: a defense secretion produced in an abdominal gland of queen and worker bees *(Apis mellifica* L.) and delivered by the sting. It contains 3 types of active principle: 1. biogenic amines, including histamine, which cause pain and dilate blood vessels, allowing wider distribution; 2. biologically active peptides such as Mellitin (see) and apamin; and 3. enzymes like hyaluronidase and phospholipase A.

Belladonna alkaloids: see Tropane alkaloids.

Bence-Jones proteins: proteins excreted in the urine of patients with multiple myeloma, a malignancy of antibody-producing cells. The proteins were first noticed because they precipitate on heating to 50 °C, but redissolve on further heating to 100 °C. Each B-J.p. consists of a dimer of immunoglobulin L chains (held together by disulfide bridges), and it is synthesized by a clone of identical cells. Each patient therefore produces only one polypeptide in sufficient quantity for sequence determination. B-J.p. have been valuable in the study of the structure of Immunoglobulins (see). Each light (L) chain, $M_r \sim 22{,}500$, consists of 214 amino acids. Methionine is absent, and there are no regions of α-helix.

Benzoquinones: compounds derived from *p*-benzoquinone.

p-Benzoquinone and its mono-, di- and trimethyl, ethyl, methoxy and 2-methoxy-3-methyl derivatives

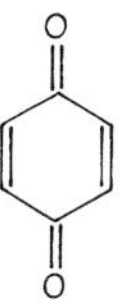

p-Benzoquinone

are found in defense secretions of certain arthropods. More than 90 different B. have been isolated from higher plants, fungi and molds. These include the yellow skin irritant, primin, from *Primula obconica,* the yellow perezone from various Mexican *Perezia species,* the orange embelin (which has antihelminthic and antibiotic activity) and the orange rapanon. The latter two B. occur in many members of the *Myrsinaceae.* Fungal B. include the maroon fumigatin, which is excreted into the culture media of *Aspergillus fumigatus* and *Penicillium spinulosum,* the bronze-colored polyporinic acid from the parasitic *Polyporus nidulans,* the bronze-colored atromentin from *Paxillus atromentosus,* and the red spinulosin from *Penicillium* spp. and *Aspergillus fumigatus.* The B. ring is also present in Ubiquinone (see) and Plastoquinone (see).

***N*-Benzyladenine:** see 6-Benzylaminopurine.

6-Benzylaminopurine, *N-benzyladenine:* a frequently used synthetic cytokinin. In B. the furfuryl residue of natural kinetin is replaced by a benzyl group. In the plant, B. is metabolized to 6-benzylamino-7-glucofuranosylpurine, which lacks cytokinin activity. 6-(2-Hydroxybenzylamino)purine riboside has been isolated as a natural cytokinin from poplar.

6-Benzylaminopurine

6-Benzylamino-7-glucofuranosylpurine

Benzylisoquinoline alkaloids: a group of alkaloids found mainly in members of the poppy family (*Papaveraceae*). The benzyl substituent on C1 of the isoquinoline radical can enter various secondary cyclizations through phenol oxidation. The ring structures of the medically important Papaveraceae alkaloids (see), Erythrina alkaloids (see) and the *bis*benzylisoquinoline Curare alkaloids (see) arise in this way. Biosynthetically, papaverine is derived from 2 molecules of dopa, one being converted to dopamine, the other to 3,4-dihydroxyphenylacetaldehyde. Mannich condensation of these two leads first to norlaudanosine, which is dehydrated to papaverine (Fig. 1).

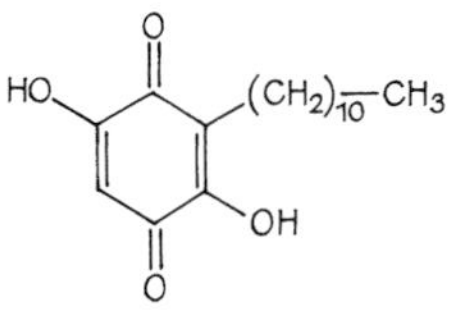

Embelin

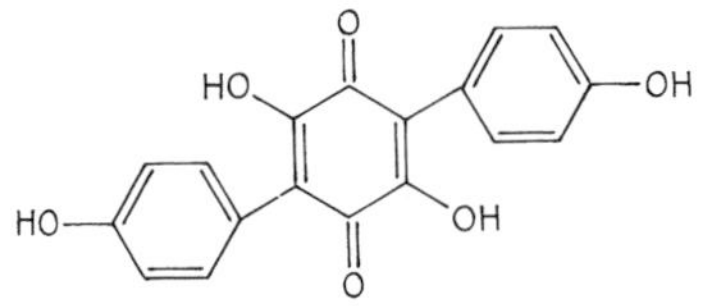

Atromentin

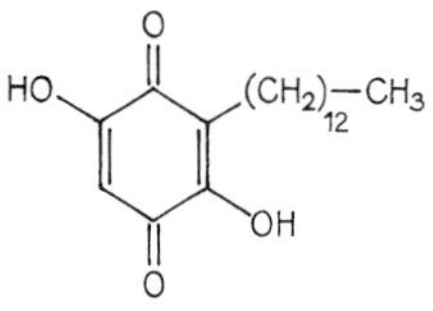

Rapanone

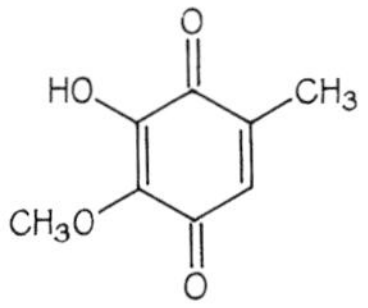

Fumigatin

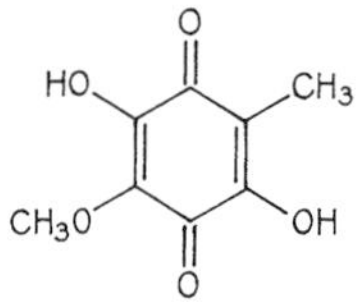

Spinulosin

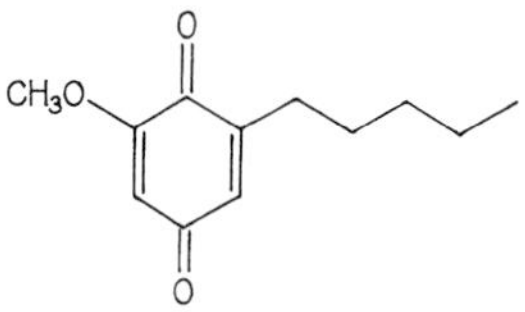

Primin

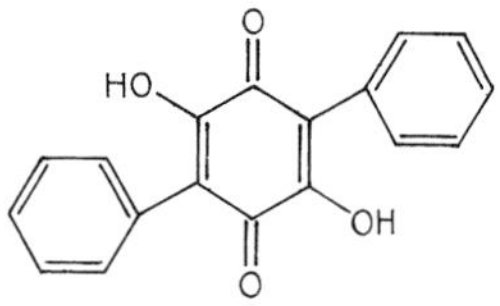

Polyporic acid

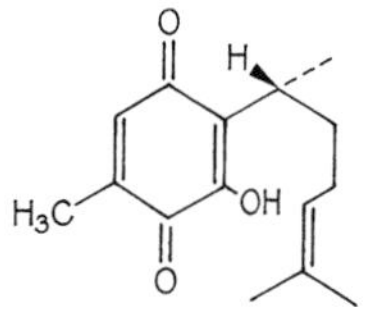

Perezone

Some naturally occurring benzoquinones

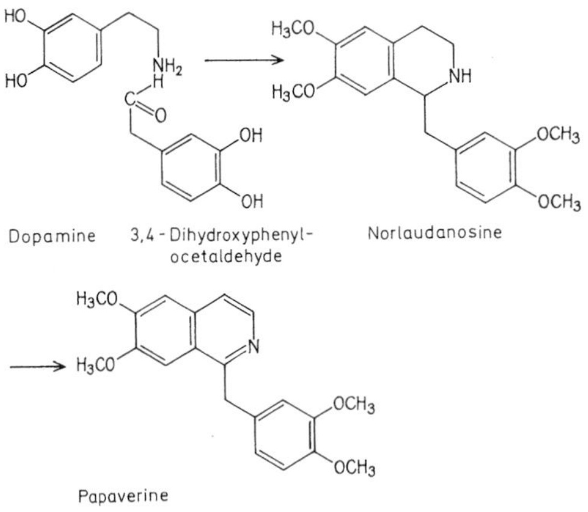

Dopamine 3,4-Dihydroxyphenylacetaldehyde Norlaudanosine

Papaverine

Fig. 1. *Benzylisoquinoline alkaloids.* Biosynthesis of papaverine.

Tetrahydroisoquinoline bases are also precursors of those alkaloids containing the morphine ring system (e.g. thebaine, codeine, morphine). The precursor, reticuline, is formed via a biradical generated by phenol oxidases. It is converted via the tetracyclic alkaloid I into thebaine, from which codeine and morphine are derived by hydrolysis of the methoxy group(s) (Fig. 2).

*Bis*benzylisoquinoline derivatives, including the curare alkaloids, are formed by oxidative coupling of two benzylisoquinoline molecules.

Beri-beri: deficiency disease resulting from lack of vitamin B_1. See Vitamins.

Betacyanins: members of the betalain group of plant pigments with absorption maxima between 534 and 552 nm. They are derived from betanidin or isobetanidin, and differ from one another with respect to their glycosylation. Examples are Betanin (see) and Amaranthin (see).

Betaines: widely occurring biogenic amines. The simplest is betaine (glycine betaine), $(CH_3)_3N^+$–CH_2COO^- [M_r 117.2, m.p. of hydrochloride 227–228 °C (d.)]. B. are synthesized in the ethanolamine cycle, and are metabolically related to mono- and dimethylglycine. They can serve as methyl donors in methylation. The betaine structure, which characterizes all betaines, is the peralkylated zwitterionic form, R_3N^+–CHR'–COO^-.

Betalains: a group of nitrogenous plant pigments found almost exclusively in the order Centrospermae.

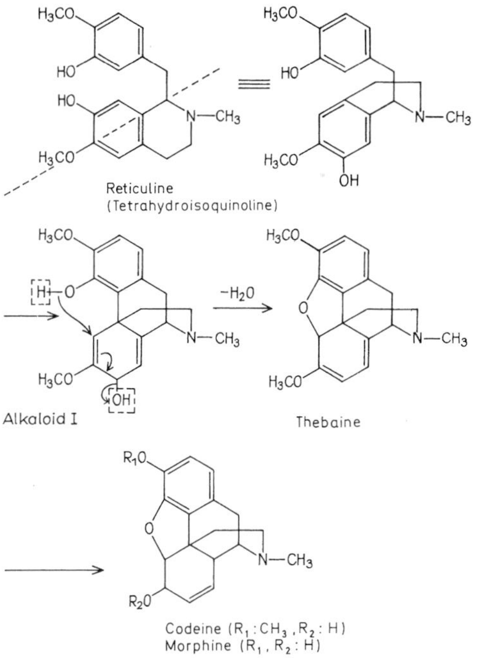

Fig. 2. *Benzylisoquinoline alkaloids.* Biosynthesis of the morphine skeleton.

They do not occur in the same plant with anthocyanins, and are therefore taxonomically useful. The red pigment in the cap of the fly agaric fungus is also a B. derivative. All B. are derivatives of Betalaminic acid (see), e.g. Betanin (see), Indicaxanthin (see) and Muscaaurin (see). Red B. are called Betacyanins (see). Yellow B. are called Betaxanthins (see).

Betalaminic acid: a cyclic dicarboxylic acid aldehyde, which contributes to the structure of all Betalains (see). It is biosynthesized from tyrosine via dopa. Linkage to various amino acids, such as cyclodopa or proline, leads to the Betacyanins (see) and Betaxanthins (see).

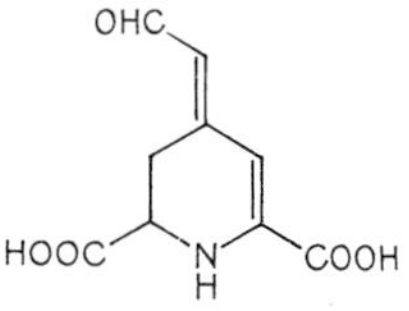

Betalaminic acid

Betamethasone, ***9-fluoro-11β,17,21-trihydroxy-16β-methylpregna-1,4-diene-3,20-dione:*** a synthetic pregnane derivative (see Steroids) with high antiinflammatory activity, which is used, e.g. against arthritis. It differs from Dexamethasone (see) only in the configuration of the 16 methyl group.

Betanidin: see Betanin.

Betanin: the red pigment in beetroot *(Beta vulgaris).* It is a highly water-soluble Betacyanin (see), and its aglycon is betanidin. B. is a zwitterion; it exists as a violet cation below pH 2, and the inflexion point to the red form is at pH 4. See Betalains.

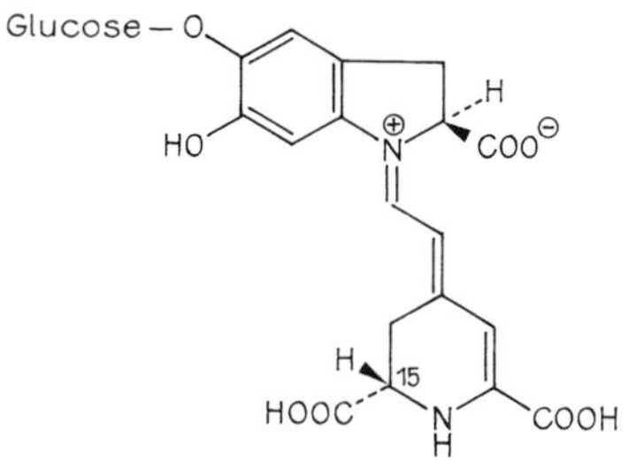

Betanin

Beta structure: see Proteins.

Betaxanthins: yellow Betalain (see) plant pigments with absorption maxima between 474 and 486. Examples are Indicaxanthin (see), Miraxanthins (see) and Vulgaxanthins (see).

Betel: see Areca alkaloids.

Betonicine: see Pyrrolidine alkaloids.

Betulaprenols: see Polyprenols.

Betulenols: optically active, isomeric, bicyclic sesquiterpene alcohols from birch bud oil. They are derived from caryophyllene, and their structures are species-specific for *Betula alba* and *Betula lenta.*

Betulin: a pentacyclic triterpene diol, which differs from Lupeol (see) in possessing a second hydroxyl group at position 28. It is found in birch and hazel bark, in rose hips, and in the cactus *Lemaireocereus griseus.*

Betulinic acid: a pentacyclic triterpene carboxylic acid found in many plants, e.g. *Gratiola officinalis, Melaleuca* spp, various cacti, bark of *Platanus* spp. and bark of the pomegranate *(Punica granatum).*

Biapigenin: see Biflavonoids.

Bicuculline: an alkaloid from *Dicentra cucullaria, Adlumia fungosa* and several species of *Corydalis.* B. is a neurotoxin and a convulsant, which acts as a specific antagonist of the neurotransmitter, 4-Aminobutyric acid (see).

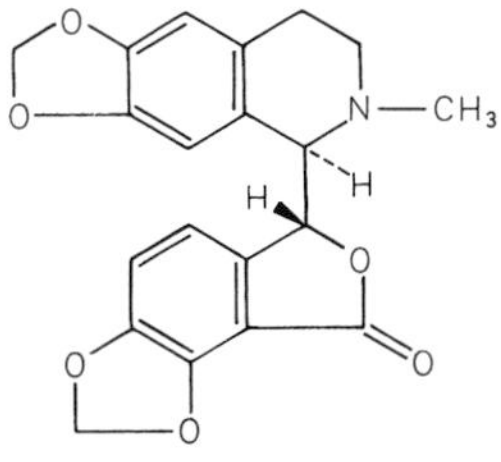

Bicuculline

Biflavonoids, ***biflavenyls:*** dimers of flavonoid units. The interflavonoid linkage may be a carbon-carbon bond or an ether bond (other types of linkage in naturally occurring B. probably remain to be discovered). Ten types of B. have been reported, classified according to the sites and type of linkage:

5′,4‴ (ether), e.g. ochnaflavone (Fig.) (apigenin dimer)
4′,6″ (ether), e.g. hinekiflavone (apigenin dimer)
5′,6″ (C-C), e.g. robustaflavone (apigenin dimer)
5′,8″ (C-C), e.g. amentoflavone (apigenin dimer)
8,6″ (C-C), e.g. agathisflavone (apigenin dimer)
8,8″ (C-C), e.g. cupressuflavone (apigenin dimer)
3,3″ (C-C), e.g. biapigenin (apigenin dimer)
3,3‴ (C-C), e.g. taiwaniaflavone (apigenin dimer)
6,6″ (C-C), e.g. succedaneaflavone (naringenin dimer)
3,8″ (C-C), e.g. xanthochymusside (Fig.)

Formation of B. can be explained by oxidative coupling (pairing of radicals) of two chalcone units, with subsequent cyclization of the C_3 chains, or by electrophilic attack of one radial on the phloroglucinol nucleus of a chalcone or flavone.

Most B. are dimers of apigenin, and the precursor chalcone is generally thought to be the similarly substituted naringenin chalcone. However, different ring substitution patterns do occur [e.g. xanthochymusside (Fig.)]. Most B. are biflavones, i.e. each flavonoid moiety possesses a C2,3 double bond, but flavanone-flavones [e.g. C-C-linked volkensiflavone, a dimer of naringenin(3) and apigenin(8″)] and biflavanones [e.g. xanthochymusside (Fig.) and succedaneaflavone (Fig.)] are also found. Glycosides of B. are rare, e.g. xanthochymusside.

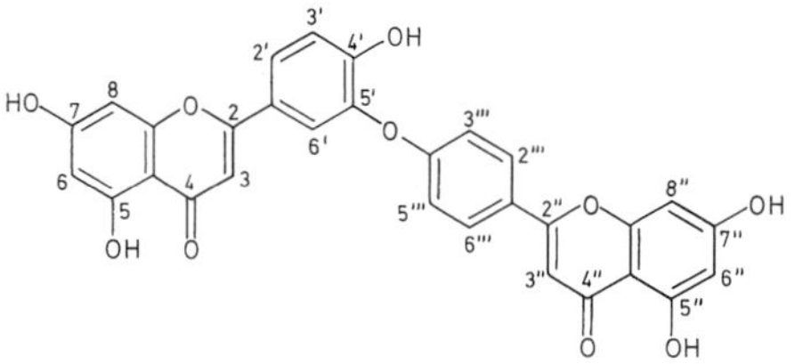

Ochnaflavone (a biflavonoid from *Ochna squarrosa*)

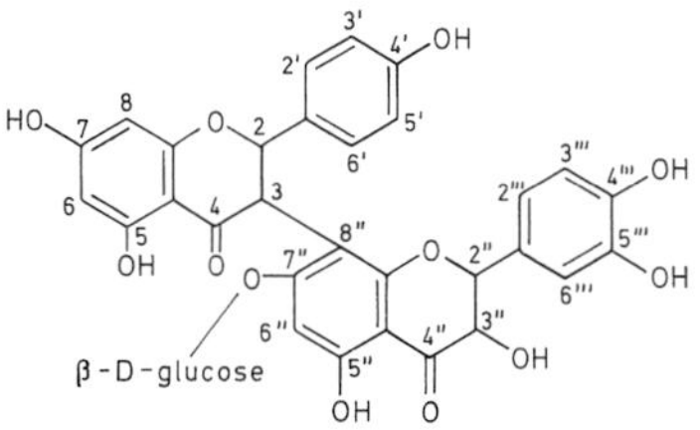

Xanthochymusside (a biflavonoid from *Garcinia*)

The ability to form B. appears to be an early evolutionary development in vascular plants, which has been lost in most angiosperms and some gymnosperms. B. have been isolated from 4 of the 5 orders of gymnosperms, from every family of the Coniferales, except the *Pinaceae,* and from 11 angiosperm families. They are absent from the Gnetales, which supports the exclusion of this order from the gymnosperms and its classification as a separate group, the chlamydosperms. [Geiger & Quinn in *The Flavonoids: Advances in Research* (ed. J.B.Harborne & T.J.Mabry), Chapman and Hall, 1982, pp.525–534]

Bile acids: components of bile which serve as emulsifying agents. They are steroid carboxylic acids linked to taurine or glycine by a peptide bond. According to the constituent amino acid, they are called *glycocholic* or *taurocholic acids.* The former predominate in human and bovine bile, the latter in canine bile. B.a. are characteristic of mammals; in lower vertebrates the same function is performed by Bile alcohols (see). In mammals, the pattern of B.a. is often specific for the species.

Salts of B.a. reduce surface tension and emulsify fats, so that they can be enzymatically degraded and absorbed in the intestine. Lipases are also activated by B.a. Humans produce 20–30 g B.a. per day, and 90 % of this is resorbed in the intestine and returned to the liver in the enterohepatic circulation. One liter bile contains about 30 g B.a.

B.a. are biosynthesized from cholesterol by 7α-hydroxylation, reduction of the double bond at position 5, and epimerization at position 3. The C_{27} side chain is shortened by β-oxidation. Free B.a. can be prepared by alkaline hydrolysis of animal bile. They are used as starting materials for the partial laboratory synthesis of important steroid hormones. See Cholic acid, Deoxycholic acid, Lithocholic acid.

Bile alcohols: polyhydroxylated steroids structurally related to cholestane. They occur as sulfate esters in the bile of lower vertebrates, e.g. *scymmol* (3α,7α,12α,24,25,27-hexahydroxy-5β-cholestane) from shark bile.

Bile pigments: linear tetrapyrroles formed by degradation of porphyrins, especially heme. The α-methine bridge between rings A and B of protoheme is oxidatively cleaved by microsomal hydroxylases [microsomal heme oxygenase (decyclizing), EC 1.14.99.3, which exhibits an absolute requirement for NADPH-ferrihemoprotein reductase, EC 1.6.2.4] to form carbon monoxide and *biliverdin IX.* The iron is reduced to Fe(II) and returned to the body's iron pool, where it is reused or stored in hemosiderin and ferritin. Biliverdin is reduced to *bilirubin* and transported to the liver as a complex with serum albumin.

Bilirubin arises mainly from the hemoglobin of aged erythrocytes in the reticuloendothelial system (spleen), bone marrow and liver. In pernicious anemia, some bilirubin is formed by degradation of myoglobin and cytochromes. It occurs in large quantities in serum and tissues in hemolytic jaundice, and is also found in urine and feces of infants. In free form (i.e. not bound to serum albumin), it is highly toxic, especially in newborns, in whom it can cross the blood-brain barrier more easily than in adults. It acts as an uncoupler of respiration. In the liver, bilirubin is released from serum albumin and concentrated in hepatocytes. It is then conjugated with two glucuronic acid residues (Fig.2) (catalysed by glucuronosyltransferases EC 2.4.1.76 and 2.4.1.77, located mainly in the smooth endoplasmic reticulum). Conjugated bilirubin is excreted in bile, probably bound to lecithin or bile proteins. In the large intestine, most of the conjugate is hydrolysed, and the resulting free bilirubin reduced by intestinal bacteria to *d*-urobilinogen, stercobilinogen and *meso*-bilirubinogen. These colorless compounds are oxidized by oxygen to stercobilin and urobilin, which are responsible for the brown color of feces.

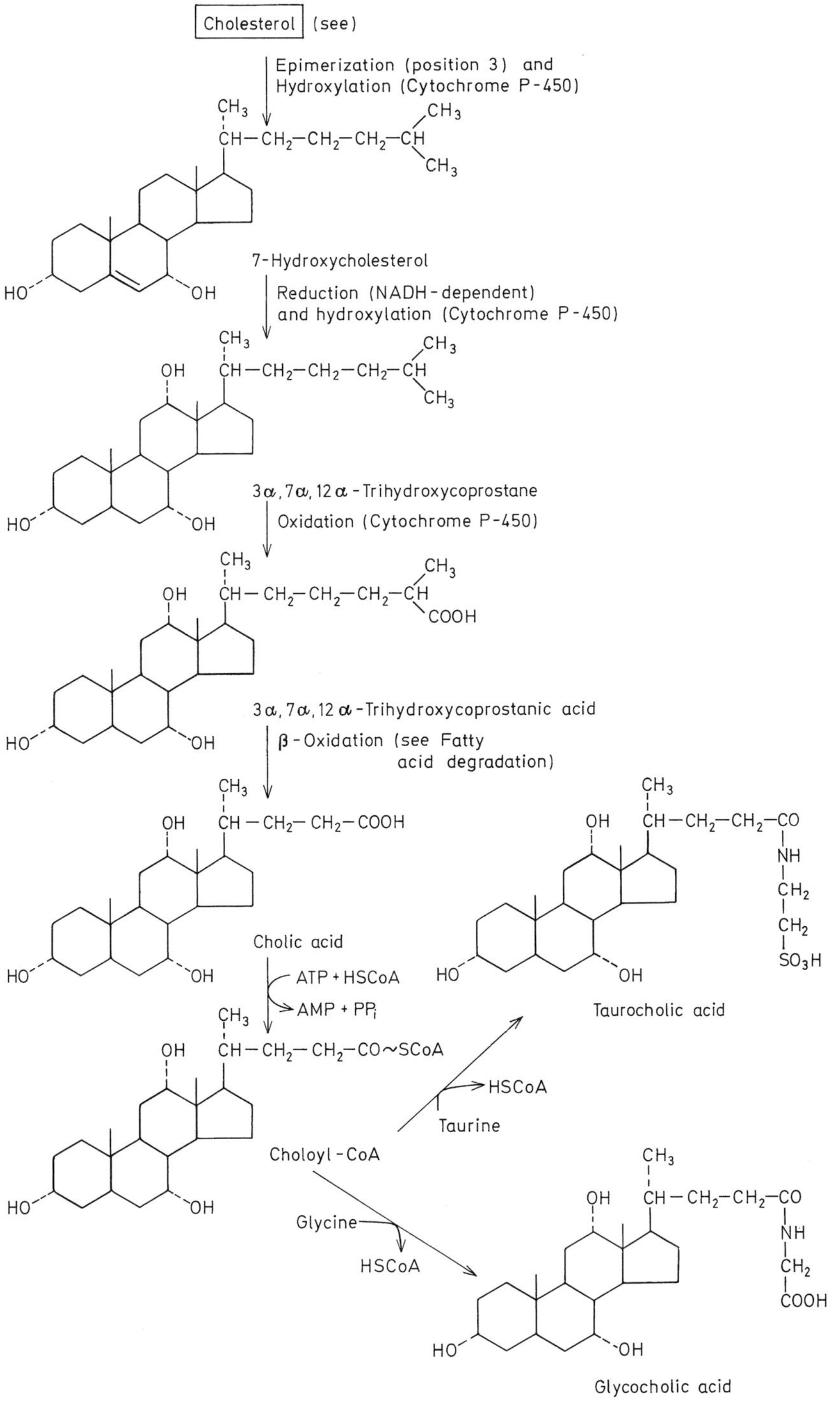

Cholesterol (see)

Epimerization (position 3) and Hydroxylation (Cytochrome P-450)

CH_3 CH–CH_2–CH_2–CH_2–CH(CH_3)CH_3

HO OH

7-Hydroxycholesterol

Reduction (NADH-dependent) and hydroxylation (Cytochrome P-450)

OH CH_3 CH–CH_2–CH_2–CH_2–CH(CH_3)CH_3

HO OH

3α, 7α, 12α-Trihydroxycoprostane

Oxidation (Cytochrome P-450)

OH CH_3 CH–CH_2–CH_2–CH_2–CH(CH_3)COOH

HO OH

3α, 7α, 12α-Trihydroxycoprostanic acid

β-Oxidation (see Fatty acid degradation)

OH CH_3 CH–CH_2–CH_2–COOH

HO OH

Cholic acid

ATP + HSCoA → AMP + PP_i

OH CH_3 CH–CH_2–CH_2–CO~SCoA

HO OH

Choloyl-CoA

Taurine → HSCoA

OH CH_3 CH–CH_2–CH_2–CO–NH–CH_2–CH_2–SO_3H

HO OH

Taurocholic acid

Glycine → HSCoA

OH CH_3 CH–CH_2–CH_2–CO–NH–CH_2–COOH

HO OH

Glycocholic acid

Biosynthesis of bile acids

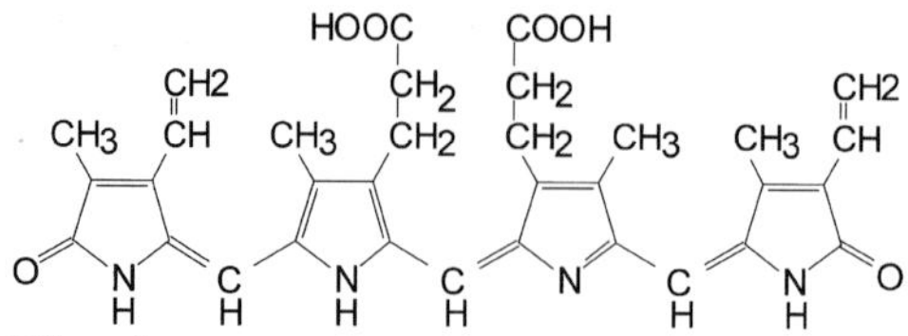

Biliverdin

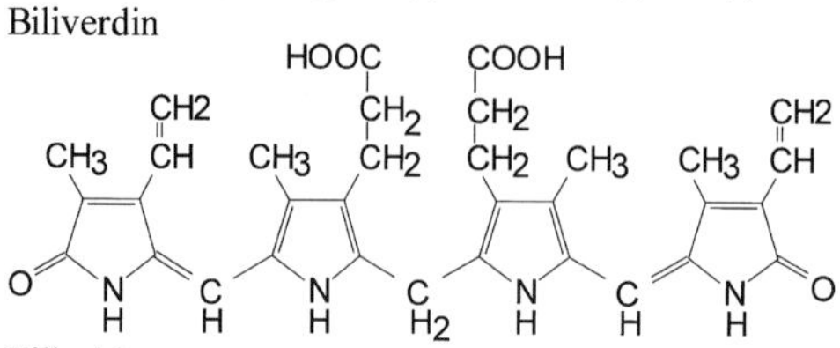

Bilirubin

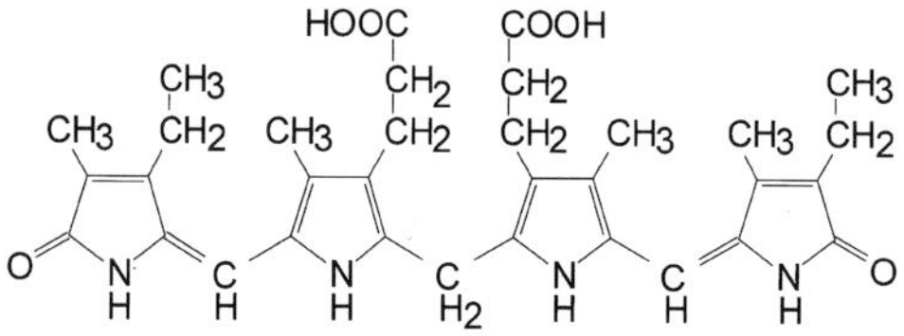

Mesobilirubin

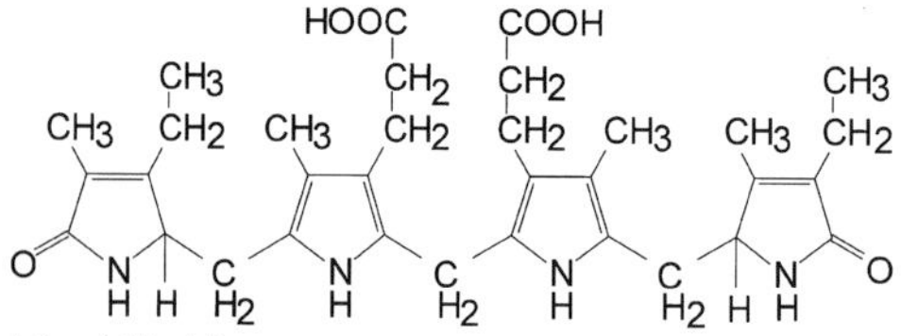

Mesobilirubinogen

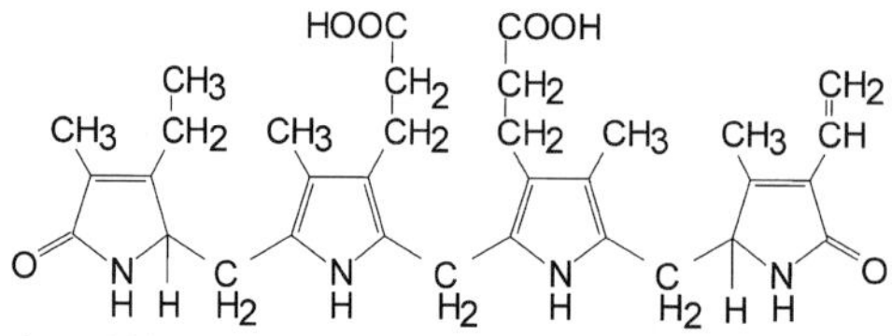

d-Urobilinogen

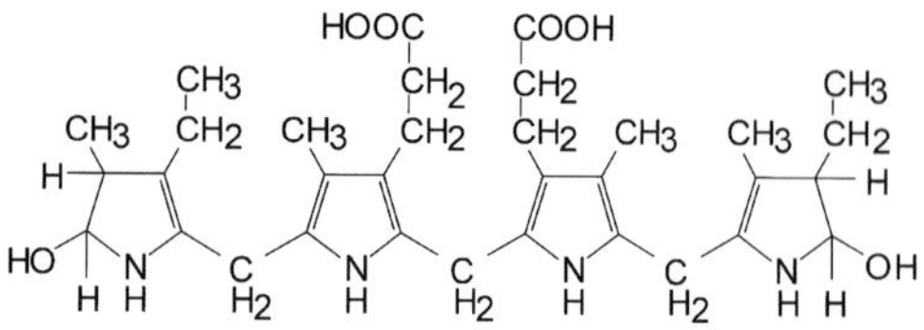

Stercobilinogen

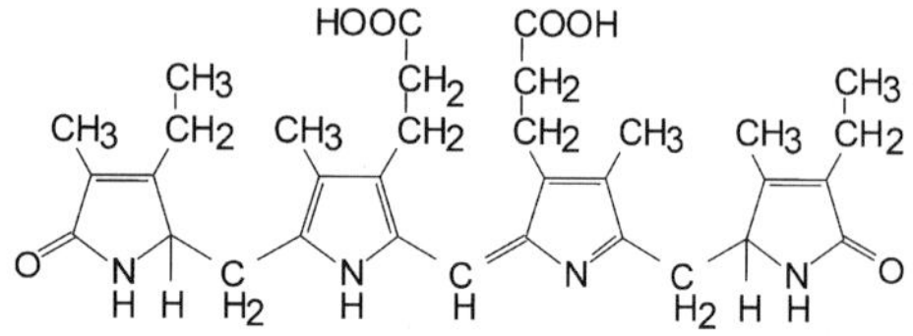

i-Urobilin

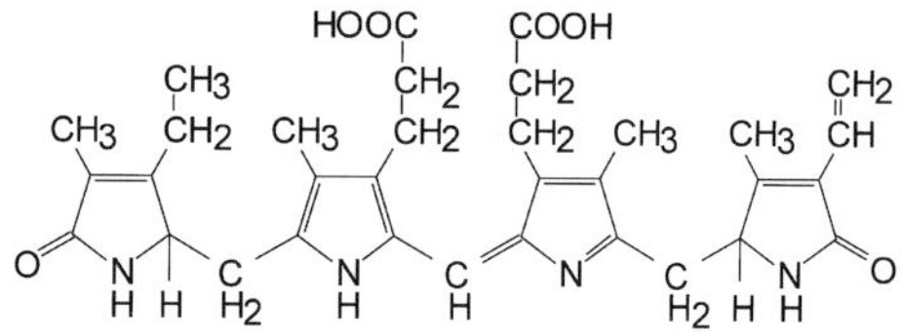

d-Urobilin

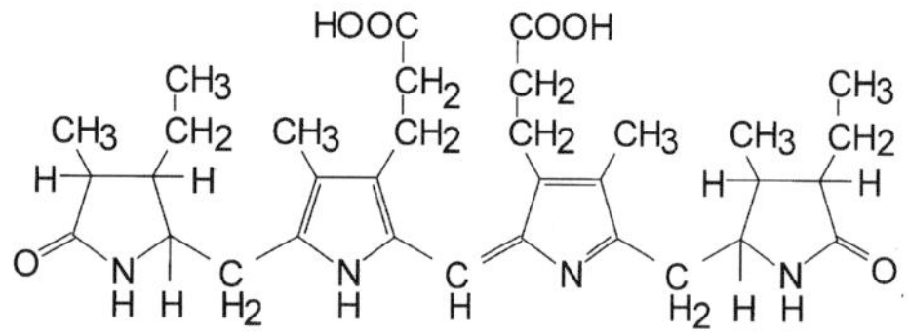

Stercobilin

Fig. 1. *Bile pigments and related compounds*

Biliverdin is present in the bile of some animals, in the placenta of some mammals (uteroverdin) and in egg shells of many birds (oocyan). It also occurs in the meconium of fetuses and neonates, and post mortem bile.

Reduction of the two vinyl substituents of bilirubin to ethyl produces *meso*-bilirubin. Further reduction of methine bridges produces *meso*-bilirubinogen, which is present in normal bile, urine and feces, and is increased in pathological liver conditions. Traces of urobilinogen are absorbed into the portal blood, returned to the liver and excreted in the bile. Part of this urobilinogen, however, gets into the systemic blood circulation and eventually appears in the urine (up to 4 mg/24 h). In obstructive jaundice there is practically no urobilinogen in the urine and feces. In hemolytic states, fecal and urinary urobilinogen are increased. Stercobilinogen is the main excretory product of hemoglobin in most vertebrates, and its oxidation product, stercobilin (*l*-urobilin) is a constituent of normal urine and feces. *i*-Urobilin (urobilin IXα) (the oxidation product of *meso*-bilirubinogen) and *d*-Urobilin (the oxidation product of *d*-urobilinogen) are present in normal urine and feces.

Linear tetrapyrroles can form metal complexes, in which the metal ion is bound to all four nitrogen atoms in a nearly planar ring structure. This is possible because the pyrrole ring can exist in a lactim or lactam form. The ring formed by the metal complex is presumably stabilized by hydrogen bonds, e.g. verdoheme, biliverdin heme.

Biliproteins: see Phycobilisome.

Bilirubin: see Bile pigments.

Biliverdin: see Bile pigments.

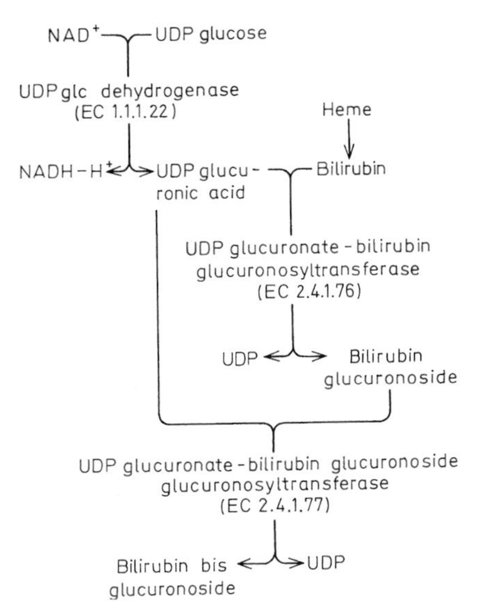

Fig. 2. *Bile pigments.* Biosynthesis of bilirubin *bis*glucuronoside

Biochanin A: see Isoflavone.

Biochemical oxygen demand, *BOD:* the rate at which the oxygen dissolved in water is consumed in the oxidation of organic compounds by microorganisms in the water.

Biochromes: see Natural pigments.

Biocytin: see Vitamins (biotin).

Bioelectronics: the incorporation of biological materials into electronic devices, e.g. Enzfets (see Field effect transitors). See Biosensor. The term is also used in a general sense for the application of electronics to the investigation of biological processes.

Bioelements: those chemical elements required by living organisms. The elemental composition of organisms is considerably different from that of the earth's crust (Table 1); there are more than 90 crust elements, but only 40 of these are represented in living material. The 6 elements C, O, H, N, S and P account for about 90 % of living material. These 6 main elements are present as constituents of biomolecules, in inorganic matrix substances, and in water. Minerals (see) are rarely present in large amounts. The above 6 B. plus Ca, K, Na, Cl, Mg and Fe make up 99.9 % of the biomass. The remaining elements occur mainly as Trace elements (see), which are needed only in catalytic quantities. While the light metals are usually present as mobile cations (see Minerals), the heavy metals are generally fixed as stable components of organic complexes. Table 2 shows the elemental composition of the human body.

Carbon (C) forms the skeleton of all organic molecules. All C in the biomass is ultimately derived from CO_2 fixed by photosynthesis, so that all biomolecules are ultimately biosynthesized from carbohydrates.

Oxygen (O) is a component of nearly all biomolecules, providing a reactive site for metabolic transformations of acids, aldehydes, ketones, alcohols and ethers. Thus, hydrocarbons are not generally biodegradable, unless they can first be converted into a compound carrying an oxygen function, e.g. by hydroxylation. Oxygen is also part of the hydroxyapatite of bones and, very importantly, of water. Atmospheric oxygen arose from the photolytic activity of plants, thereby transforming the reducing atmosphere of the primitive earth to an oxidizing one.

Table 1. *Relative abundance of chemical elements in the earth's* crust and in the human body (adapted from Rapoport).

Element	Earth's crust (%)	Human body (%)	Enrichment (-fold)
Oxygen	50	63	–
Silicon	28	0	–
Aluminium	9	0	–
Iron	5	0.004	–
Calcium	3.6	1.5	–
Potassium	2.6	0.25	–
Magnesium	2.1	0.04	–
Hydrogen	0.9	10	10
Carbon	0.09	20	200
Phosphorus	0.08	1	10
Sulfur	0.05	0.2	4
Nitrogen	0.03	3	100

Hydrogen (H) is present in all biomolecules, linked to C, N, O and S. Removal of H is equivalent to oxidation; when the H is combined with O_2 by operation of the electron transport chain of respiring cells, ATP is generated. In most biological reactions, H participates as the ion $H^+ + e^-$; the coenzymes NADH and NADPH are carriers of $H^+ + 2e^-$ (equivalent to a hydride ion). Reactions involving H_2 are rare (see Hydrogen metabolism).

Table 2. *Elemental composition of the human body, based on dry mass.*

Element	Percent	Element	Percent
Carbon	50	Potassium	1.0
Oxygen	20	Sodium	0.4
Hydrogen	10	Chlorine	0.4
Nitrogen	8.5	Magnesium	0.1
Sulfur	0.8	Iron	0.01
Phosphorus	2.5	Manganese	0.001
Calcium	4.0	Iodine	0.00005

Nitrogen (N) is a component of many biomolecules, especially of proteins and nucleic acids. Molecular N is reduced by certain free-living and symbiotic microorganisms (see Nitrogen fixation) to ammonia, which is the first and final product of nitrogen metabolism.

Sulfur (S) is present in 2 amino acids, cysteine and methionine, and in certain coenzymes. It is assimilated by plants in the form of sulfate (see Sulfate assimilation); animals must obtain organically bound S from plants.

Phosphorus (P) is present in both inorganic and organic compounds as phosphate. Nucleotide residues in nucleic acids are linked by phosphate bonds. Energy is transferred from one molecule (usually ATP) to another in the form of a high-energy phosphate bond. Many coenzymes also contain phosphate.

Biogenic amines, *biological amines:* biologically and pharmacologically important naturally occurring amines, occurring widely in plants and animals. They can be divided into: 1. derivatives of *ethanolamine,* e.g. choline, acetylcholine, muscarine; 2. *polymethylene diamines,* e.g. putrescine, cadaverine; 3. *polyamines,* e.g. spermine; 4. *imidazolylalkylamines,* e.g. histamine; 5. *phenylalkylamines,* e.g. mescaline, tyramine, hordenine; 6. *catecholamines,* e.g. adrenalin, noradrenalin and dopamine; 7. *indolylalkylamines,* e.g. tryptamine, serotonin; and 8. *betaines,* e.g. carnitine.

Many B.a. present in plants, but lacking a heterocyclic ring, are also called protoalkaloids. When such protoalkaloids occur in the same genus or family as true alkaloids to which they are biogenetically related (e.g. hordenine, candicine and mescaline in *Cactaceae* are related biogenetically to the isoquinoline alkaloids), they are then usually also classified as alkaloids. Some B.a. are hormones (e.g. adrenalin), neurotransmitters (e.g. acetylcholine, noradrenalin), or components of coenzymes (e.g. cysteamine and β-alanine are precursors of coenzyme A), phospholipids (e.g. ethanolamine), vitamins (e.g. propanolamine is a precursor of B_{12}), or ribosomes (e.g. cadaverine and putrescine). For biosynthesis, see separate entries for each B.a.

Bioinformatics: see Databases.

Bioluminescence: emission of visible light by an organism. Light is emitted as a result of a redox reaction catalysed by Luciferase (see). The energy from this reaction excites electrons of an oxidation product of luciferin to a higher electronic state. Light is emitted as the oxyluciferin returns to the ground state (Fig.)

In warm regions B. is fairly common. especially among marine animals. Of the vertebrates, only a few fish are luminescent, as are a few fungi and bacteria. B. can be generated either extracellularly, in which case the luciferin and luciferase are secreted by glands, or intracellularly. In the latter case, the components react in special cells. Light emitted by marine animals may also be generated by symbiotic bacteria, which are usually harbored in small pockets on the body surface.

Luciferins are species-specific. They vary widely in structure, and they may be peroxidases, monooxygenases or dioxygenases, showing various reaction mechanisms and cofactor requirements (Table). The system of the firefly, *Photinus pyralis,* requires O_2, ATP and Mg^{2+}, in addition to luciferin and luciferase (Fig.). The light is generated by an intermediate of the process (probably a peroxide) with a quantum yield of 1 (i.e. 1 photon per molecule of luciferin). Bacterial luciferase catalyses a 2-stage reaction: 1. E-$FMNH_2 + O_2 \rightarrow$ E-FMNH–OOH; 2. E-FMNH–OOH + RCHO $\rightarrow$ E + FMN + RCOOH + $h\nu$, where E represents the enzyme, RCHO is a long chain aldehyde, and E-FMNH–OOH is a 4-hydroperoxyflavin [M. Kurfürst et al. *Eur. J. Biochem.* **123** (1982) 355–361]

Bioluminescent systems

Reaction characteristics	Organism	λmax (nm)
Requires NADH and FMNH	*Photobacterium*	470–505
Requires ATP	*Photinus* (firefly)	552–582
	Renilla reniformis (sea pansy, a coelenterate)	509
No cofactors	*Cypridina* (an ostracod crustacean)	460
	Latia (limpet)	535
Photoprotein	*Aequorea*	469

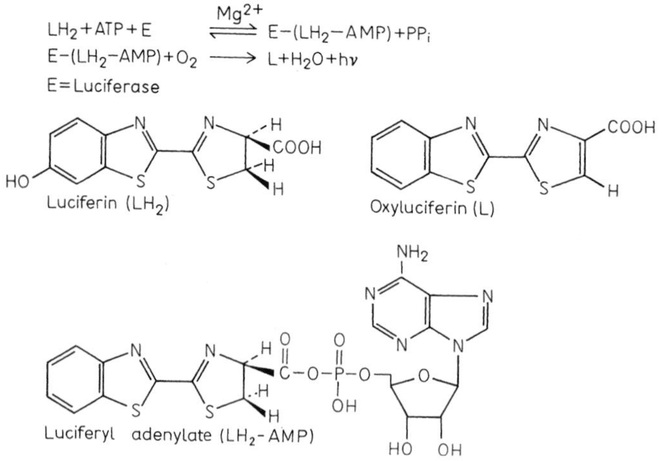

$$LH_2 + ATP + E \underset{}{\overset{Mg^{2+}}{\rightleftharpoons}} E\text{-}(LH_2\text{-}AMP) + PP_i$$

$$E\text{-}(LH_2\text{-}AMP) + O_2 \longrightarrow L + H_2O + h\nu$$

E = Luciferase

Luciferin (LH_2)

Oxyluciferin (L)

Luciferyl adenylate (LH_2-AMP)

Mechanism of light emission by Photinus luciferin

In some cases, B. serves as a protection from predators or as a bait for prey. Many species use B. for communication. Information is encoded in the spectrum, frequency or rhythm of flashing, or the pattern of the light organs on the body. For example, each species of *Photinus* uses a different flashing frequency to attract mates, and some predatory species attract prey by mimicking their flashing pattern.

The B. of *Photinus* is used as the basis of an extremely sensitive assay for ATP in biological samples. When mixed with a sample containing ATP, a luciferin-luciferase system (or powdered firefly abdomens) generates an amount of light proportional to the amount of ATP, and this can be measured in a spectrophotometer. The assay is sensitive in the nanogram range, and can therefore even be used to count bacteria on the basis of their ATP content. Similarly, preparations from luminescent bacteria are used for NADH assays, and the photoprotein from the jellyfish, *Aequorea,* for calcium or strontium determination. See Aequorin.

Horseradish peroxidase can act as a luciferase in the presenceof H_2O_2, a cyclic hydrazide such as luminol and synthetic firefly luciferin (commercially available). By coupling this enzyme to proteins, in particular antibodies, a number of immunoassays have been designed, based on the measurement of emitted light. [T. P. Whitehead et al. *Nature* **305** (1983) 158–159]

Biomacromolecule: see Biopolymer.

Biomass: a term used in ecology, meaning the amount of organic substance in the form of living organisms in a given area.

Biomembrane: a structure containing lipids, glycolipids, proteins and glycoproteins, bounding the cell (cell membrane) or subdividing it into compartments. It is a sheet-like structure, 60–100 Å thick. The lipids have hydrophilic "head" groups (indicated by black spheres in the Fig.) and hydrophobic "tail" regions. In bulk aqueous solution, they spontaneosly form bilayers, in which the molecules are lined up side by side and tail to tail, with the heads pointing toward the water phase on each side of the bilayer, but excluding water from the tail regions. This structure, when stained for electron microscopy with osmium tetroxide or uranyl acetate, is seen as two dark lines separated by an unstained gap. This image was pre-

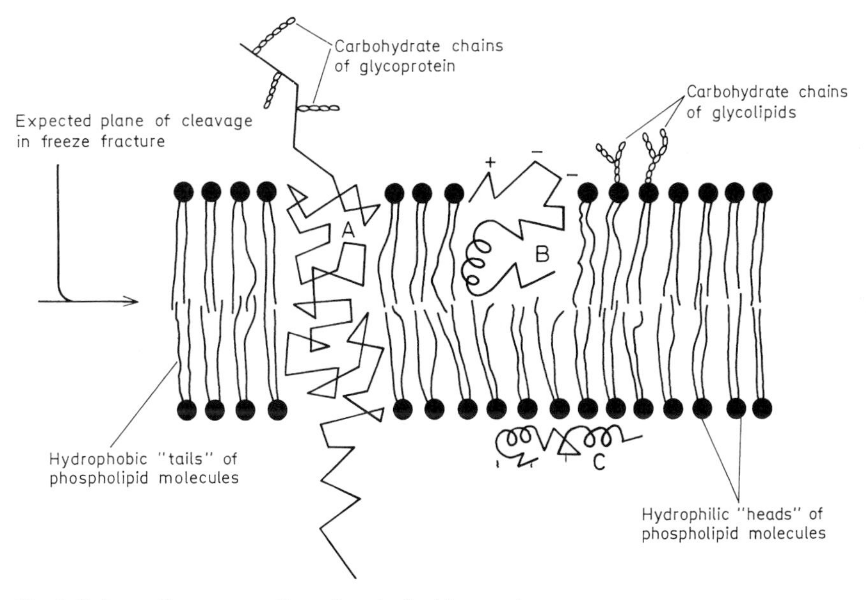

Fig. 1. *Schematic cross section of part of a biomembrane*
A represents an intrinsic (integral) protein that completely spans the membrane; in the example shown the protein chain extends beyond the membrane surface on both sides; it is a glycoprotein, and carbohydrate residues are present only on that segment of the protein protruding from the outer membrane surface. The illustrated model is roughly equivalent to glycophorin, an erythrocyte membrane protein; the protein chain protruding from the inside surface of the erythrocyte membrane is thought to be associated with an extrinsic protein called spectrin.
B represents an intrinsic (integral) membrane protein partly buried and partly exposed on the membrane surface.
C is representative of many extrinsic (peripheral) proteins that are more or less firmly associated with membranes, but appear not to be integrated into the phospholipid layer, e. g. spectrin of the erythrocyte membrane. Extrinsic proteins are usually associated with intrinsic proteins, rather than simply adhering to the hydrophilic heads of the phospholipid molecules as shown here.
Cytochrome *c* may represent an association intermediate between B and C; it is an important functional constituent of the inner mitochondrial membrane, but extremely easily removed.

viously referred to as the *unit membrane,* but the term is now obsolete. Mitochondria and plastids are surrounded by 2 membranes, and nuclei by a membrane that effectively doubles back on itself. The cytoplasm of eukaryotic cells is characterized by extensive membranous structures, e.g. the Endoplasmic reticulum (see), Golgi apparatus (see) and vacuoles. Prokaryotes, in contrast, have no internal membranes, although in some the cell membrane is extensively invaginated.

The major classes of lipids are phospholipids, glycolipids, cholesterol and cholesterol esters (see Membrane lipids). There are also many minor components, and the exact composition depends on the species and type of cell. The type and quantity of proteins depend on the function of the membrane, e.g. myelin membranes contain very little protein (18 %), whereas the inner mitochondrial membrane contains about 75 % protein. Membrane proteins have a variety of functions, e.g. as mediators of both active and passive transport of lipid-insoluble substances across the membrane, as receptors for hormones and other informational molecules, and as enzymes. In certain cases they may also have a structural role.

The currently accepted structure of B. is the *fluid mosaic* model. Lipid molecules and membrane proteins are free to diffuse laterally and to spin within the bilayer in which they are located. However, a flip-flop motion from the inner to the outer surface, or vice versa, is energetically unfavorable, because it would require movement of hydrophilic substituents through the hydrophobic phase. Hence this type of motion is almost never displayed by proteins, and it occurs much less readily than translational motion in the case of lipids. Since there is little movement of material between the inner and outer layers of the bilayer, the two faces of the B. can have different compositions. For membrane proteins, this asymmetry is absolute, and, at least in the plasma membrane, different proportions of lipid classes exist in the two monolayers. Attached carbohydrate residues appear to be located only on the noncytosolic surface. Carbohydrate groups extending from the B. participate in cell recognition, cell adhesion, possibly in intercellular communication, and they also contribute to the distinct immunological character of the cell.

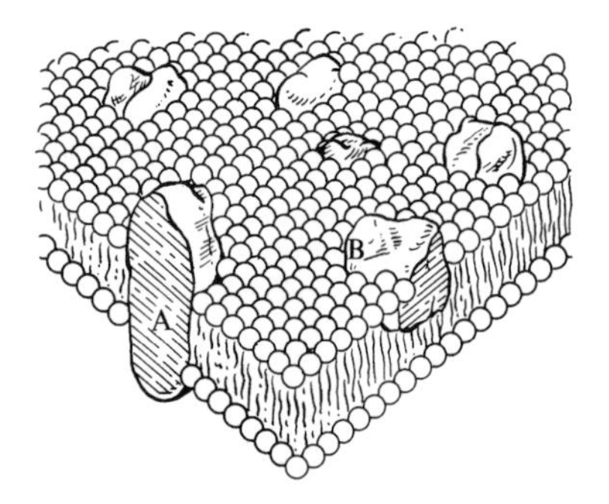

Fig. 2. *Biomembrane.* Three dimensional artistic impression of a phospholipid bilayer with intrinsic proteins. A and B as in Fig. 1.

Integral or *intrinsic* proteins of the B. cannot be removed except by dissolution of the B. with organic solvents or detergents; these proteins are responsible for the bumps observed in freeze-fractured B. preparations under the electron microscope. *Peripheral* or *extrinsic* proteins can be stripped from the B. by altered ionic conditions or extremes of pH.

See Membrane lipids. See Erythrocyte membrane.

Biopolymer, *biomacromolecule:* a biological molecule which can be regarded as a chain (polymer) of identical or similar subunits (monomers). *Periodic* B. are formed from identical subunits, e.g. cellulose which is a polymer of glucose. *Aperiodic* B. are composed of similar but nonidentical monomers. The following classes of B. are listed under separate entries: Proteins, Nucleic acids, Carbohydrates, Lignins, Polyprenols and Polyterpenes. The monomer sequences of nucleic acids and proteins are determined by nucleic acid templates; living organisms have elaborate systems for synthesizing these B. with very low rates of error. The overall composition of polysaccharides and lignins is determined by the enzymes that catalyse their synthesis, but their size and exact monomer sequence may be determined by factors such as the juxtaposition of other structural molecules and the growth rate of the organism or cell (e.g. cell wall polysaccharides and lignin), or the rate of degradation and availability of precursor monomers (e.g. storage polysaccharides). Blood group antigens (see) and presumably other cell-surface recognition oligo- and polysaccharides, however, have precise sizes and monomer sequences. Examples of mixed B. are Glycoproteins (see) in which carbohydrate monomers or oligomers are linked to amino acid side chains of polypeptides, and Glycolipids (see Membrane lipids) in which carbohydrates are linked to a lipid.

Biopterin: see Pteridines.

Biosensor: a biological component (e.g. enzyme, antibody) in combination with a transducing system. Interaction of an analyte with the biological component generates an electrical signal, which is amplified then recorded, or displayed. In most B. the biological component is immobilized on the surface of the transducer. The signal from the biological material to the transducer may be a change in charge, potential, heat, light, mass or other quantity. An amperometric B. produces an electric current on application of a potential difference between two electrodes, whereas a potentiometric B. produces a potential gradient. In both cases the electrical effect is due to a biological reaction. A typical amperometric device is a glucose sensor using glucose oxidase, whereas urease is used in a typical potentiometric sensor for urea. The term B. is sometimes used in the broader sense of sensors that detect biological changes or materials, but do not incorporate biological material in their sensing mechanism, e.g. oxygen electrodes. See Field effect transistor, Immunosensor, Oxygen electrode, Piezoelectric sensors.

Bios II: see Vitamins (biotin).

Biosynthesis, *biogenesis:* synthesis by the metabolic reactions of a living organism. B. is distinct from the formation of compounds that appeared during prebiotic evolution by geological processes, or later in the laboratory under simulated primitive earth conditions. It must also be distinguished from the laboratory synthesis of compounds found in living organisms.

B. can occur either in vivo (in a living organism or in a cultivated part of it, e.g. a perfused organ), or in vitro (in cell homogenates, cell extracts and enzyme preparations). Although the mechanism of B. may be the same in vivo and in vitro, the conditions are usually quite different (see Methods of biochemistry).

The sum of all B. in an organism is referred to as anabolism.

Biotechnology: the application of living organisms or systems derived from them in manufacturing industry. The subject has ancient origins, embracing agriculture, cheese production, fermentation of alcoholic beverages, leather tanning, etc. However, B. is a modern term, referring to particular types of exploitation of biological processes, and the scope of the subject has been greatly expanded by developments in molecular biology. Important areas of B. are: 1. Genetic engineering or Recombinant DNA technology (see). 2. Process engineering (e.g. production of antibiotics by fermentation; microbiological conversion of simple chemical source materials such as methanol and ammonia into animal feed; production of fuel alcohol by fermentation). 3. Concentration of minerals, or "microbiological mining". 4. Biosensors (see). 5. Biocatalysts (see Immobilized enzymes). 6. Waste treatment and biodegradation. 7. Production of monoclonal antibodies for medical research. 8. Study of the relationship between protein structure and function, and the nature of molecular interactions, leading to designed (nonbiological) synthesis of model catalysts and new drugs.

Biotest, *bioassay:* an assay for a biological or nonbiological substance, in which the response of a living organism, organ or tissue (sometimes in tissue culture) to the substance under investigation is measured. B. are used for hormones, vitamins, toxins, antibiotics, etc.

Biotin: see Vitamins.

Biotin carboxylase: see Biotin enzymes.

Biotin enzymes: carboxylases that use biotin as a cofactor. The biotin is bound via an amide bond to the ε-amino group of a specific lysine residue in the enzyme protein, i.e. B.e. contain a biotinyllysyl residue. Free (+)-ε-*N*-biotinyl-L-lysine actually occurs in yeast extract, and is known as biocytin. During catalysis, N-atom 1′ of the biotin residue is carboxylated in an ATP-dependent reaction: ATP + HCO_3^- + biotinyl-enzyme (I) → ADP + P_i + carboxybiotinyl-enzyme (II). The carboxyl group is then transferred from (II) to the carboxylase substrate: (II) + substrate → (I) + carboxylated substrate.

Carboxylation reactions occur in the biosynthesis of fatty acids, in the degradation of leucine and isoleucine, and in the degradation of fatty acids with uneven numbers of C atoms. Acetyl-CoA carboxylase (EC 6.4.1.2) catalyses the following reaction in the biosynthesis of fatty acids: Acetyl-CoA + HCO_3^- + H^+ + ATP → malonyl-CoA + ADP + P_i. The monomer of this enzyme is composed of 4 different subunits. One of these, *biotin carboxylase* (III), catalyses the carboxylation of the biotin residue. This biotin residue is covalently bound to the second subunit, which is known as the *biotin-carboxyl carrier protein* (biotin-CCP): Biotin-CCP + HCO_3^- + H^+ + ATP → carboxybiotin-CCP + ADP + P_i. In the second step of the reaction, the third subunit, the *carboxyl transferase* (IV) catalyses the transfer of the carboxyl group to acetyl-CoA: Carboxybiotin-CCP + acetyl-CoA → biotin-CCP + malonyl-CoA. In the biotin-carboxyl carrier protein the biotin residue acts as a swinging arm, transferring the hydrogen carbonate from the biotin carboxylase to the acetyl-CoA, which is bound to the active center of the carboxyl transferase.

Another example of a carboxylation reaction is the formation of oxaloacetate from pyruvate. Pyruvate carboxylase (EC 6.4.1.1) consists of 4 subunits, each covalently bound to one molecule of biotin and containing one Mg^{2+} ion: 1. Biotinyl-enzyme + ATP + CO_2 + H_2O → carboxybiotinyl-enzyme + ADP + P_i; 2. Carboxybiotinyl-enzyme + pyruvate → biotinyl-enzyme + oxaloacetate. In the degradation of fatty acids with odd numbers of C atoms, carboxylation of propionyl-CoA to methylmalonyl-CoA is also catalysed by B.: Carboxybiotinyl-enzyme + CH_3–CH_2–CO ~ SCoA → biotinyl-enzyme + CH_3–CH(COOH)–CO ~ SCoA. The same reaction occurs in the degradation of isoleucine, leucine and valine.

Bisabolane: see Sesquiterpenes (Fig.).

***Bis*benzylisoquinoline alkaloids:** isoquinoline alkaloids containing 2 fused benzylisoquinoline nuclei joined by one, two or three diaryl ether linkages. The ether linkages arise by phenol oxidation. B.a. occur in plants of the *Menispermaceae* and related families, which include the genera *Berberis, Magnolia, Daphnandra* and *Strychnos.*

2,3-*Bis*phosphoglycerate, *glycerate 2,3-bisphosphate, 2,3-diphosphoglycerate:* see Glycolysis, Hemoglobin, Rapoport-Luerbing shuttle.

Bitter peptides: bitter-tasting peptides that may spoil the palatability of some foods, e.g. B.p. sometimes arise during ripening of certain types of cheese, and they have also been found in fermented soybean products. B.p. have also been isolated from controlled enzymatic digests of pure proteins, e.g. chymotryptic hydrolysis of casein. Bitterness is thought to be related to the average hydrophobicity of the peptides, so that peptides with a high content of Val, Leu, Phe and Tyr are likely to be bitter. B.p. from peptic hydrolysates of soybean protein are Leu-Phe, Leu-Lys, Phe-Ile-Leu-Glu-Gly-Val, Arg-Leu-Leu, and Arg-Leu. In contrast, various markedly hydrophilic peptides from spinach are not bitter, e.g. Glu-Gly, Asp(Glu,Gly,Ser_2), Ala(Glu_2,Gly, Ser).

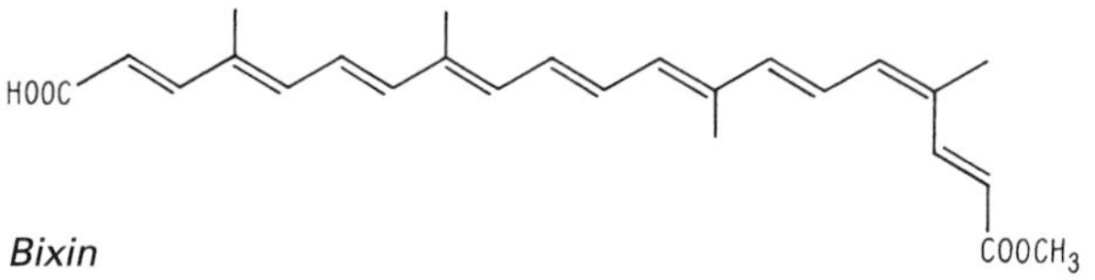

Bixin

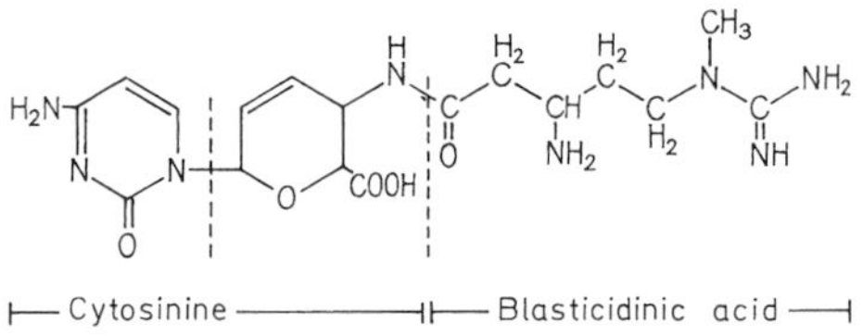

Bitter principles: bitter-tasting substances, found particularly in members of the *Compositae, Gentianaceae* and *Labiatae,* which produce a reflexive increase in the secretion of saliva and digestive juices. Extracts from such plants are used as bitter spices to increase appetite and promote digestion, and they are used in the preparation of stomachic bitters. Chemically, B.p. are very diverse. Many are terpene lactones, e.g. gentiopicroside from gentian root. Bitter-tasting substances with other physiological effects, e.g. quinine alkaloids and cucurbitacins, are not classified as B.p.

Bixin: the monomethyl ester of the C_{24} dicarboxylic acid, norbixin. Norbixin has 4 methyl side branches, 2 terminal carboxyl groups and 9 conjugated double bonds. The C_{20} chain corresponds to the middle part of β-carotene. In naturally occurring B. (m.p. 198 °C) the Δ^{16} double bond is *cis,* but it easily isomerizes to form the more stable all-*trans* B. (m.p. 217 °C). B. is a diapocarotenoid formed by oxidation of a C_{40} carotenoid. It is a yellow to red-orange pigment found in the seeds of *Bixa orellana,* from which it is extracted for use as a food colorant.

Blasticidins: pyrimidine antibiotics synthesized by *Streptomyces griseochromogenes* (see Nucleoside antibiotics). They inhibit the growth of fungi, e.g. the rice fungus, *Piricularia oryzae,* and a few bacteria. The antibiotic effect is due to suppression of polypeptide chain elongation during protein biosynthesis.

Cytosinine — Blasticidinic acid

Blasticidin S

Bleomycins: glycopeptide antibiotics produced by *Streptomyces verticillus* and used widely for treatment of squamous cell carcinomas, lymphomas and testicular cancer. About 200 different B. are known, differing mostly in the nature of the *C*-terminal substituent, which can be varied by adding different amines to the bacterial culture medium. The clinically used preparation is a mixture of 11 B., known by the trade name "Blenoxane"; the chief constituents are B.A_2 (60–70 %) and B.B_2 (20–25 %) (Fig.).

B. cause strand breakage in double-stranded DNA in solution, and in DNA in intact cells. All types of DNA (mammalian, viral, bacterial and synthetic polymers) are attacked. B. also bind to RNA, but do not cause degradation. In RNA-DNA hybrids, B. degrade the DNA and not the RNA. This specificity is determined by the presence of 2-deoxyribose. DNA degradation by B. requires molecular oxygen, and it is prevented by metal chelating agents, e.g. EDTA,

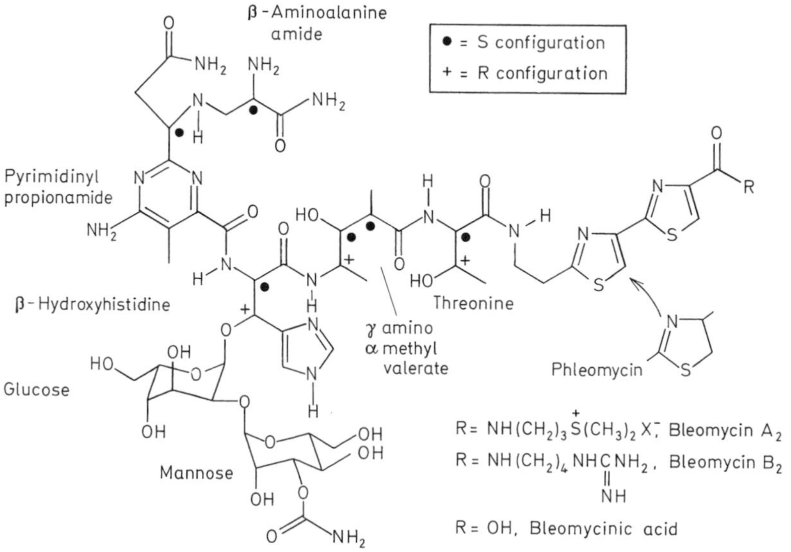

β-Aminoalanine amide; Pyrimidinyl propionamide; β-Hydroxyhistidine; Glucose; Mannose; γ amino α methyl valerate; Threonine; Phleomycin

• = S configuration
+ = R configuration

R = $NH(CH_2)_3\overset{+}{S}(CH_3)_2X^-$, Bleomycin A_2
R = $NH(CH_2)_4NHC(=NH)NH_2$, Bleomycin B_2
R = OH, Bleomycinic acid

Bleomycins

or the iron-specific reagent, deferoxamine. The biologically active form of B. is a complex with Fe(II). Stopped flow spectroscopy shows 2 distinct kinetic events after the Fe(II)-B. complex is exposed to O_2: 1. formation of an unstable intermediate complex, Fe(II)-B.-O_2; 2. decomposition of this complex to an ESR-active species, known as "activated bleomycin", formulated as Fe(III)-B.(O_2H) or Fe(III)-B.(O_2^{2-}), which attacks DNA.

Phleomycins lack the double bond at C4′ of the bithiazole moiety, are otherwise identical to B., but are too toxic for use as antibiotics. Other related antibiotics are zorbamycin, zorbonomycin, victomycin and platomycin [J. C. Dabrowiak in *Advances In Inorganic Biochemistry* vol. 4, Elsevier Biomedical, 1982, pp. 69–113; S. A. Kane & S. M. Hecht 'Polynucleotide Recognition and Degradation by Bleomycin' *Prog. Nucl. Acid Res. Mol. Biol.* **49** (1994) 313–352].

Blitz blot: see Southern blot.

Blood-brain barrier: the barrier that must be crossed when water and solutes are exchanged between blood and extracellular fluid of the central nervous system. Many substances, when injected intravenously, become distributed throughout the various tissues and organs of the body, except for the central nervous system, e. g. acidic dyes, bile acids, tetanus toxin, sodium ferricyanide, trypan blue.

The permeability properties of cerebral capillaries are quite different from those of capillaries in other organs and tissues. Water-filled channels, which can be demonstrated in the walls of most non-nervous capillaries, are absent from cerebral capillaries. A hydrostatically promoted transcapillary flow of fluid into tissue on the arterial side, with the reverse process on the venous side (i. e. Starling-type transcapillary flow) has never been demonstrated in the brain. The central nervous system does not possess lymphatics. Cerebral capillary endothelial cells have no pinocytotic activity. Furthermore, the central nervous system and various compartments of cerebrospinal fluid are also excluded from the extracellular fluid of the rest of the body by the tight choroid epithelium and the tight layer of arachnoid.

The endothelial cells possess mitochondria which probably provide the necessary energy for facilitated diffusion of sugars and other transported metabolites, e. g. amino acids. In the absence of a specific transport mechanism, ions and low M_r polar nonelectrolytes are also unable to cross the endothelial cells of cerebral capillaries.

The lipid membrane of cerebral capillary endothelial cells does not present a barrier to lipid-soluble materials, including O_2 and CO_2. The ability of drugs and other xenobiotics to cross the B. b. b. is strongly correlated with their lipid solubility.

All vertebrates and insects possess a B. b. b. It is present in the embryonic brain, and it can even be demonstrated a few hours after brain death. The B. b. b. clearly serves a protective function, preventing exposure of the central nervous system to a wide variety of electrolytes and macromolecules which circulate in the blood in health and disease. However, its most important function may be the maintenance of a constant environment in the central nervous system; homeostasis of K^+, Ca^{2+}, Mg^{2+} and H^+ in cerebrospinal fluid and cerebral interstitial fluid is probably necessary for the correct function of the central nervous system. Other solutes, e. g. vitamins, may also be subject to homeostasis in the brain and cerebrospinal fluid. In addition, the B. b. b. prevents loss of active compounds, e. g. neurotransmitters, from the brain. [L. Bakay *The Blood Brain Barrier,* Charles C. Thomas, 1956; M. Bradbury *The Concept of the Blood-Brain Barrier,* John Wiley, 1979]

Blood coagulation, ***blood clotting:*** a process in which blood forms a gel (the clot) sufficiently dense to prevent bleeding from a wound. In most invertebrates the process only involves clumping of agglutinized blood cells, but in vertebrates and a few crustaceans the clot is formed from Fibrin (see) as well as trapped blood cells. Fibrin is an insoluble polymer derived from Fibrinogen (see), a large but soluble plasma protein. Fibrinogen is converted to fibrin by the action of a highly specific serine protease called Thrombin (see) and Factor XIII_a. The latter catalyses transamidation between glutamine and lysine residues, thereby forming links between adjacent polypeptide chains in the precipitated fibrin.

Release of thrombin from its inactive zymogen, Prothrombin (see), is the penultimate step in a series of reactions, each of which releases an active serine protease from an inactive precursor in the blood. It is an example of cascade regulation, in which each activated protease activates the next precursor down the line, increasing amounts of material being involved at each step (i. e. there is amplification). The clotting factors are listed in the Table. The assigned Roman numerals are historical, and the activated form is indicated by the subscript "a" (e. g. XII_a).

Traditionally, 2 systems of B. c. are recognized, the intrinsic and extrinsic systems. Blood plasma clots slowly (in several minutes) in the presence of a foreign surface like kaolin or glass. [For this reason, blood for clinical purposes is withdrawn into vessels

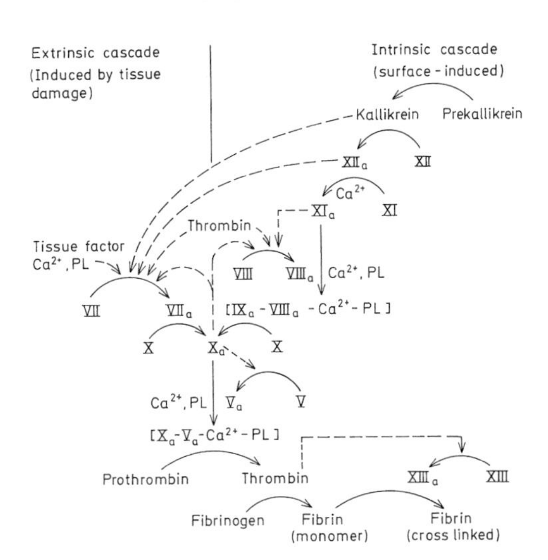

The coagulation cascade. PL = phospholipid or cell membrane. Solid arrows indicate reactions; dotted arrows indicate catalysis. Adapted from L. Lorand *Methods Enzymol. 45B* (1976) 31–37.

Factors in the coagulation cascade

No.	Name	Properties and functions
I	Fibrinogen	M_r 340000, composed of 6 chains: $(A\alpha)_2(B\beta)_2\gamma_2$. Converted to fibrin by removal or 2A and 2B peptides. Fibrin is dissolved by plasmin, a protease related to coagulation factors and released from plasminogen by its own activation cascade.
II	Prothrombin (see) II_a is Thrombin (see)	M_r 72500, 582 amino acids with 12 disulfide bridges and 10 Gla* residues. A glycoprotein. Converted to thrombin by X; thrombin is inhibited by antithrombin III, α_2-macroglobulin, α_1-antitrypsin and hirudin.
III	Tissue factor	A specific protein + phospholipids
IV	Calcium ions	Mediate binding of IX, X, VII and prothrombin to acidic phospholipids of cell membranes, where activation occurs. Stabilize V, fibrinogen and possibly other proteins involved in activation and subunit dissociation of XIII.
V	Proaccelerin V_a is accelerin	Labile glycoprotein, M_r 350000. V_a mediates binding of X and prothrombin to platelets, where X is activated and prothrombin is converted to thrombin.
VII	Proconvertin	M_r 45–54000. Contains Gla*. Part of extrinsic cascade. Released by tissue trauma.
VIII	Antihemophilic factor (see)	Several forms, M_r 10^5 to 2×10^6. Controversy over which is active in coagulation. VIII is an accessory in activation of X by IX_a. Its absence causes hemophilia A.
IX	Christmas factor	M_r 55400 (bovine), 57000 (human). Single chain. IX_a contains heavy chain homologous to serine proteases and light chain with Gla*. Its absence causes hemophilia B.
X	Stuart factor	M_r 54500 (bovine), 59000 (human). 12 Gla* residues present. Activated by $IX_a + VIII_a + Ca^{2+}$, or by VII_a + tissue factor + Ca^{2+}, or by platelet membrane protease.
XI	Plasma thromboplastin antecedent	M_r 124000 (bovine), 160000 (human). Glycoprotein composed of 2 similar or identical polypeptides joined by disulfide bond(s). Activates IX. IXa is inhibited by antithrombin III, trypsin inhibitors, α_1-trypsin inhibitor and C1 inhibitor.
XII	Hageman factor	The first factor in the intrinsic pathway. M_r 74000 (bovine), 76000 (human). Single chain glycoprotein. Activated by plasmin, kallikrein and XII_a. Inhibited by antithrombin III (inhibition accelerated by heparin), C1 esterase inhibitor and lima bean trypsin inhibitor. Activation of XII initiated by contact with abnormal surfaces.
XIII	Fibrin-stabilizing factor (Laki-Lorand factor)	M_r 320000 (tetrameric plasma form), 160000 (dimeric platelet form). Both forms contain two *a* subunits, which are cleaved on activation. Plasma form also contain two *b* subunits. XIII_a is the transpeptidase responsible for cross-linking fibrin fibers.
	Prekallikrein	Activated to kallikrein, a serine protease which activates XII.
	HMW (high molecular weight) kininogen, contact activation cofactor, Fitzgerald factor, Flaujeac factor.	Activated to a kinin involved in activation of XII, at least in vitro.

* Gla = 4-carboxyglutamic acid

coated with Heparin (see) or a Ca^{2+}-binding agent like citrate or EDTA.] The factors involved are all intrinsic to the plasma, hence the name. The physiological significance of the activation of the intrinsic system by foreign surfaces is not known. The proteins involved are XII, XI, high molecular weight kininogen and prekallikrein (Fig.).

The extrinsic system, activated by a specific proteinaceous tissue factor and phospholipids, produces a clot in seconds rather than minutes. VII is the only plasma factor belonging exclusively to the extrinsic system (Fig.). However, it may be activated by XII_a or kallikrein, both members of the intrinsic pathway. In addition, low concentrations of prostacyclin or prostaglandin E_1 (too low to increase the level of cyclic AMP) stimulate the cysteine protease of purified platelets so that it activates factor X; thus the distinction between intrinsic and extrinsic pathways may be artificial [A.K. Dutta-Roy et al. *Science* **231** (1986) 385–388].

Under physiological conditions, conversion of prothrombin to thrombin by X appears to require binding to a cell membrane. V is the platelet cell membrane receptor for X, and it increases the proteolytic activity of the latter.

The extent of B.c. is limited in vivo by activation of vitamin K-dependent protein C. This is a serine protease consisting of 2 polypeptide chains, M_r 21,000 and 35,000, linked by disulfide bond(s). It has a high degree of sequence homology to the Gla-containing clotting factors, and like them, exists in the plasma as a zymogen. It is activated by thrombin bound to an endothelial cell-surface receptor. Activated protein C inactivates V_a and VIII_a by limited proteolysis; these two factors are required for efficient conversion of X to X_a. Thus protein C, which is identical to autoprothrombin IIA, serves as a negative feedback control element for B.c. [T. Drakenberg et al. *Proc. Natl. Acad. Sci. USA* **80** (1983) 1902–1806; L.M. Jackson & Y. Nemerson *Ann. Rev. Biochem.* **49** (1980) 765–811; L. Lorand *Methods Enzymol.* **45B** (1976) 31–37].

The B.c. factors were discovered by analysis of hemophilias, the most common of which is caused by lack of VIII. The other factors were discovered by complementation in mixtures of non-clotting plasmas. Prekallikrein, prothrombin and factors XII, XI, X, IX and VII all give rise to Serine proteases (see) when activated by cleavage of specific peptide bonds. Prothrombin and factors VII, IX, X and XI have homologous *N*-terminal regions containing multiple Gla residues. Gla residues arise post-translationally by carboxylation of Glu residues, a process catalysed by a carboxylase that uses vitamin K as a cofactor (hemorrhage is the major symptom of vitamin K deficiency). The Gla residues are essential for the binding of Ca^{2+}, and for the Ca^{2+}-mediated binding of these proteins to phospholipids (e.g. platelet membranes). [C.T. Esmon 'Cell Mediated Events that Control Blood Coagulation and Vascular Injury' *Annu. Rev. Cell Biol.* **9** (1993) 1–26]

Blood group antigens: specific oligosaccharide moieties of glycoproteins in the membranes of blood cells, which are recognized as antigens by the immune systems of other individuals or organisms. The antigens are attached to the protein, glycophorin, in erythrocytes, and to both proteins and lipids in other parts of the body. In humans, 5 antigen systems have been identified: the ABO, MN, P, rhesus and Lutheran systems. Only the ABO and rhesus systems affect blood transfusion between humans. The other systems were identified using animal antibodies against human blood.

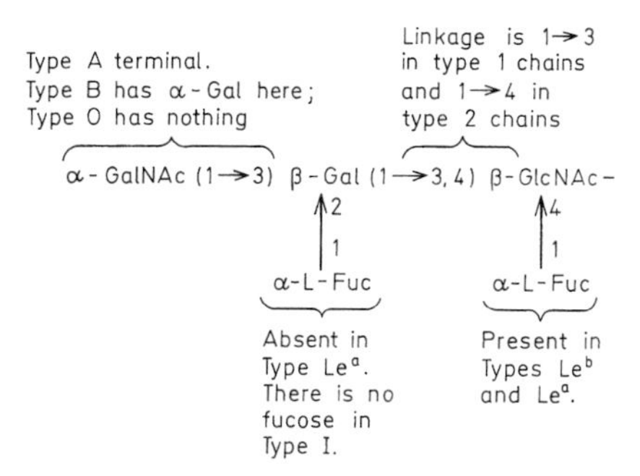

Fig. 1. *Blood group antigens.* Ends of oligosaccharide chains in individuals with different blood groups.

Structural differences between the ABO oligosaccharides are shown in Fig. 1. The genetic basis for these groups is the existence of 3 alleles of a gene coding for the synthesis of a glycosyltransferase. In A-type individuals, the enzyme transfers *N*-acetylgalactosamine onto the terminal positions of the oligosaccharide chains, whereas in B-type individuals, the enzyme is specific for galactose. The *O* gene appears to produce an inactive enzyme. Another gene (the *H* gene) codes for a fucosyltransferase, which places L-fucose on the oligosaccharide. When the *H* gene is inactive, the individual has a rare type I blood, unless an active *Le* gene is present (Le stands for Lewis factor). The *Le* gene encodes an enzyme that adds fucose to the *N*-acetylglucosamine. Individuals lacking an active *H* gene but possessing and active *Le* gene have type Le^a blood. If both *H* and *Le* genes are active, the individual has type Le^b blood.

About 80 % of the population has an active Se (secretion) gene, so that they secrete glycoproteins bearing the blood group substance into saliva and other body fluids. The structure of the carbohydrate portion of such a glycoprotein is shown in Fig. 2.

Blood sugar: see D-glucose.

Blue-green bacteria, ***cyanobacteria, blue-green algae, Cyanophyta, Cyanophyceae, Schizophyceae, Mixophyceae:*** photosynthetic prokaryotic organisms using H_2O as hydrogen donor. Many are also able to fix atmospheric nitrogen. Since they are prokaryotes, they do not possess a nuclear membrane, and mitochondria and chloroplasts are also absent. They were originally classified as plants because they are capable of photosynthesis, but they are phylogenetically unre-

```
                         α-Fuc                                                            Protein
      (1→2)      (1→3)    ↓(1→4)                                                              }
α-Fuc ——→ β-Gal ——→ β-GlcNAc                                                                  }
                    (1→3)↘      (1→3)          (1→3)        (1→3)                              }
                          β-Gal ——→ β-GlcNAc ——→ β-Gal ——→ GalNAc — O — CH2 — Serine
      (1→2)      (1→4)  (1→6)↗    ↑(1→4)                          ↑(1→6)              or
α-Fuc ——→ β-Gal ——→ β-GlcNAc  β-GlcNAc                          β-GlcNAc           Threonine
                     ↑(1→3)    ↑(1→3)                             ↑(1→3)              }
                    α-Fuc      β-Gal                              β-Gal             Protein
```

Fig. 2. *Blood group antigens.* Structure of the carbohydrate component of the Le^b glycoprotein of human blood. GlcNAc = *N*-acetyl-D-glucosaminyl; GalNAc = *N*-acetyl-D-galactosaminyl; Gal = D-galactosyl; Fuc = L-fucosyl.

lated to the green algae, which are eukaryotes. There are many similarities between B. g. b. and chloroplasts, and the endosymbiotic theory of evolution proposes that chloroplasts are descended from symbiotic B. g. b.

In B. g. b. photosynthesis takes place in the thylakoids or thylakoid stacks, which are derived from the cell membrane. The photosynthetic pigments are chlorophyll *a*, carotenoids and biliproteins (phycocyanin and phycoerythrin). The cytoplasm contains prokaryotic-type ribosomes. Typical of most B. g. b. are cyanophycin granules; these are colorless and spherical or polyhedral, and visible under the light microscope. Cyanophycin granules contain storage material consisting of a copolymer of arginine and aspartic acid. The storage carbohydrate of B. g. b. is a polymer of glucose with a degree of branching intermediate between those of glycogen and amylopectin. Cytochemically it resembles glycogen: it stains brown with iodine and occurs as discrete granules (diam. 25–30 nm) located between the thylakoids. Polyphosphate granules *(syn.* metachromatin, volutin, metachromatic granules) are also present in the cytoplasm; these are spherical and vary in size from sub-light microscopic to several microns in diameter; they are concretions of the potassium salts of high M_r linear polyphosphates.

The inner cell wall, like that of bacteria, consists of murein anchored in the cell membrane. It is attacked by lysozyme. Outside the murein layer is a plasmatic layer, and beyond that there may be a slime capsule.

There are 2 major classes: 1. *Chroococcales,* in which the cells are solitary (e. g. *Anacystis)* or colonial and held together by mucoid envelopes, and 2. *Hormogonales,* which grow in filaments (trichomas), often enclosed in a sheath. Cells of hormogonal B. g. b. communicate with each other and form a physiological unit. In trichomas there is a certain degree of specialization; heterocysts, characterized by thick, highly refractory cell walls, are the site of nitrogen fixation.

BOD: acronym of Biochemical oxygen demand (see).

Bohr effect: see Hemoglobin.

Bombesin: a peptide isolated from frog skin. Amino acid sequence: Pyr-Glu-Arg-Leu-Gly-Asn-Glu-Trp-Ala-Val-Gly-His-Leu-Met-NH_2. B. stimulates release of gastric, pancreatic and adenohypophyseal hormones in mammals, and causes contraction of smooth muscles in the gastric and urinary tracts, and in the uterus. When injected peripherally, B. inhibits the secretion of growth hormone. The explanation of this action is the homology of B. with a number of peptides found in mammalian tissues, including the hypothalamus, e. g. the sequence of 10 *C*-terminal amino acids of B. is 90-% identical to the 10 *C*-terminal amino acids of porcine gastrin-releasing hormone. [W. A. Murphy et al. *Endocrinology* **117** (1985) 1179–1183]

Bombykol, *10-trans-12-cis-hexadecadienol-(1):* a pheromone exuded by female silk moths *(Bombyx mori)* to attract males. B. is an oil, n_D^{20} 1.4835. The first structural determination (by Adolf Butenendt) was performed on 15 mg B. isolated from the abdominal glands of 500,000 female moths. The configuration was established by comparison of the biological activities of synthetic compounds with the natural product.

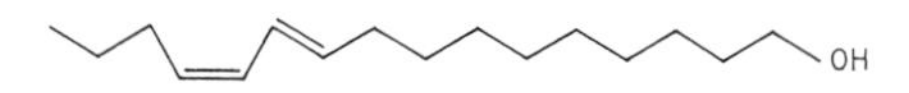

Bombykol

Bongkrekic acid: 3-carboxymethyl-17-methoxy-6,18,21-trimethyldocosa-2,4,8,12,14,18,20-heptaene-dioic acid, M_r 486.61. B. a. is one of 2 toxic antibiotics produced by *Pseudomonas cocovenenans* in spoiled bongkrek (a coconut product consumed in Indonesia). It is an inhibitor of adenine nucleoside translocation, and it affects carbohydrate metabolism.

Bornane: see Monoterpenes, Fig.

Boron, *B:* a nonmetallic element, and an essential micronutrient for plants. It is taken up by the roots as the borate anion, and this uptake is inhibited by the presence of excess calcium. Within the plant, there is also an interesting and incompletely understood interaction of calcium and borate. Both B and calcium are required for meristem growth and for normal root development. The B content of plants is generally in the range 5–60 mg/kg dry mass. The physiological activity of B differs fundamentally from that of the other micronutrients, but it bears a certain similarity to that of phosphorus. B is an indispensable structural component of plants, and fine, differentiated cell wall structure cannot develop without the prior incorporation of B. Carbohydrate metabolism also displays a striking dependence on the supply of B, which promotes sugar transport and assimilatory processes. The influence of B on the formation of differentiated structures and on the distribution of carbohydrates is undoubtedly associated with its other known requirements for e. g. pollen germination, pollen tube growth, flowering and fruit setting, and its influence on the water economy of the plant.

B deficiency results in various symptoms, such as loss of color via gray-green to yellow, followed finally by leaf drop. The apical meristem dies and the root tips become necrotic. B deficiency symptoms are exacerbated by drought. The most familiar B deficiency disease is heart rot and dry rot of beet, in which the heart leaves become brown and die, and the beet putrifies.

Botulin: see Toxic proteins.

Bowman-Birk inhibitor: see Soybean trypsin inhibitor.

Bradykinin, *kallidin I, kinin 9:* Arg-Pro-Pro-Gly-Phe-Ser-Pro-Phe-Arg, one of a group of plasma hormones called kinins. Like the other kinins, it is produced from a plasma precursor by the action of Kallikrein (see), Trypsin (see) or Plasmin (see). It causes dilation of blood vessels (and thus a decrease of blood pressure), causes the smooth muscles of the bronchia, intestines and uterus to contract, and is a potent pain-producing agent. Lysylbradykinin has similar activity.

Brassicasterol, *ergosta-5,22-dien-3β-ol:* a plant sterol (see Sterols), M_r 398.69, $[\alpha]_D$ −64°, m.p. 148 °C, first isolated form rapeseed *(Brassica campestris)* oil.

Brassinosteroids: a class of growth promoting steroids widely distributed in higher plants (pollen & seeds, 1–1,000 ng/kg; shoots, 100 ng/kg; fruit & leaves, 1–10 ng/kg) and which have also been found in lower plants, including cyanobacteria. The first B., named brassinolide (2α,3α,22(*R*),23(*R*)-tetrahydroxy-24(*S*)-

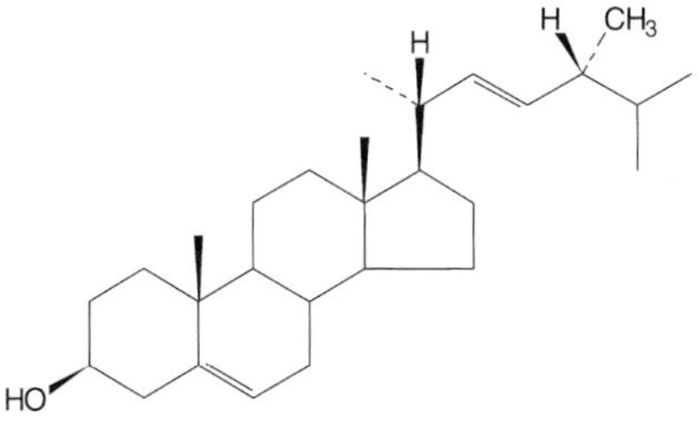

Brassicasterol

methyl-B-homo-7-oxa-5α-cholestan-6-one) was isolated from the pollen of rape (*Brassica napa*) in 1979 [M.D. Grove et al. *Nature* **281** (1979) 216–217]. To date about 30 naturally occurring B. are known, and a number of analogs have been synthesized. By subjecting these compounds to the 'bean second internode' bioassay, which differentiates 'brassin activity' from that of other plant hormones, it has been shown that the structural features required for their unique

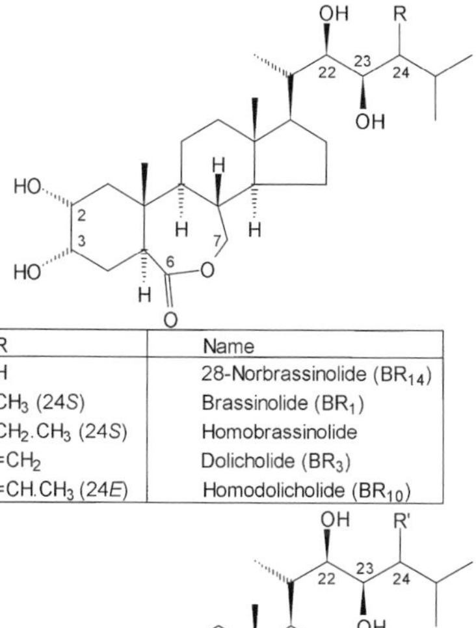

R	Name
H	28-Norbrassinolide (BR_{14})
CH_3 (24*S*)	Brassinolide (BR_1)
$CH_2.CH_3$ (24*S*)	Homobrassinolide
$=CH_2$	Dolicholide (BR_3)
$=CH.CH_3$ (24*E*)	Homodolicholide (BR_{10})

R	R'	Name
H	CH_3 (24*S*)	6-Deoxobrassinosterone (BR_5)
H	$=CH_2$	6-Deoxodolichosterone (BR_6)
H	$=CH.CH_3$ (24*E*)	6-Deoxohomodolichosterone (BR_{13})
=O	H	28-Norbrassinosterone (BR_{15})
=O	CH_3 (24*S*)	Brassinosterone (BR_2)
=O	CH_3 (24*R*)	24-Epibrassinosterone (BR_9)
=O	$CH_2.CH_3$ (24*S*)	Homobrassinosterone (BR_{12})
=O	$=CH_2$	Dolichosterone (BR_4)
=O	$=CH.CH_3$ (24*E*)	Homodolichosterone (BR_{11})

Brassinosteroids

biological activity are: (i) a *cis*-vicinal glycol moiety at C2/C3, (ii) a *trans* junction between rings A and B, (iii) an oxygen function at C6 in the form of a ketone or a lactone, (iv) a vicinal glycol moiety at C22/C23; for maximum activity the configuration at both these carbon atoms should be *R* (i.e. both hydroxyls should have the α-orientation, using the modified Fieser notation), and (v) a short alkyl group on C24 giving the 24*S* configuration (i.e. having the α-orientation); activity follows the order Me > Et > H. The bean second internode bioassay involves the painting of the test compound, dispersed in lanolin, onto the second ~ 2 mm long) internode of 6-day old *Phaseolus vulgaris* seedlings; B. activity is indicated by an elongation, curvature, swelling and splitting of the internode which is apparent after 4 days. Although B. have been shown to promote growth by a combination of cell division and enlargement in young growing tissues, particularly meristems, and to be translocated in plants, their physiological role is not clear. However, their growth promotional activity when applied to plants at low concentrations (0.001–0.1mg/l) has led to their being tested for use in agriculture. Little is known of their biosynthesis although it is assumed that that they are derived from common plant sterols by hydroxylations and, in the case of the ring-B lactones, a Baeyer-Villiger type oxidation of a 6-oxo steroid. [N.B. Mandava *Annu. Rev. Pl. Physiol. Pl. Mol. Biol.* **39** (1988) 23–52; H.G. Cutler et al. (Eds.) (1991) *Brassinosteroids: chemistry, bioactivity and applications* ACS Symposium Series No. 474, American Chemical Society, Washington, DC]

Brevifolin carboxylic acid: see Tannins.

Brimacombe fragments: see Ribosomal proteins.

Bromelain (EC 3.4.22.4): a thiol enzyme from the stems and fruit of the pineapple plant. The stem enzyme is a basic glycoprotein (M_r 33,000, pI 9.55) structurally and catalytically similar to papain. B. is activated by mercaptoethanol and other SH compounds, and it is irreversibly inhibited by reagents that block SH groups. It is an endopeptidase, and it is used in protein chemistry to hydrolyse polypeptide chains into large fragments.

Broussin: 7-hydroxy-4′-methoxyflavan. See Flavan.

Brucine: 2,3-dimethoxystrychnine, M_r 394.47, m.p. 105 °C (tetrahydrate), 178 °C (anhydrous). It is used in preparative chemistry to separate racemic acids into optical isomers. B. has one tenth of the toxicity of strychnine, from which it is derived. The main physiological effect is paralysis of smooth musculature. For formula and biosynthesis, see Strychnos alkaloids.

Bryokinin: see N^6-(γ,γ-dimethylallyl)adenosine.

Bufadienolides: see Toad poisons, Cardiac glycosides.

Buffer: a system that resists and minimizes changes of pH caused by addition or loss of acid and base. Physiologically, buffers stabilize the pH of cells and body fluids during metabolic under- or overproduction of acids and bases, or during environmental changes of hydrogen ion concentration. For laboratory purposes, buffers are used to maintain constant and favorable pH values for enzymatic reactions, to prevent protein denaturation, and to provide appropriate pH conditions for culture of microorganisms and tissues. Buffered solutions are also important as chromatography eluants and electrophoresis media, and for the performance of some chemical reactions, such as the Edman degradation of proteins.

Buffer action is illustrated (Fig.) by the titration of a weak acid (NaH_2PO_4) with a strong base (NaOH): $OH^- + H_2PO_4^- \rightleftharpoons HPO_4^{2-} + H_2O$. As NaOH is added, the ratio $[HPO_4^{2-}]/[H_2PO_4^-]$ (square brackets indicate concentrations) increases, starting at almost zero and becoming progressively larger until it is approximately infinity at the end point of the titration. The resulting titration curve is a graphic representation of the Henderson-Hasselbach equation: pH = pK + log ([salt]/[acid]), or pH = pK + log ([conjugate base]/[acid]).

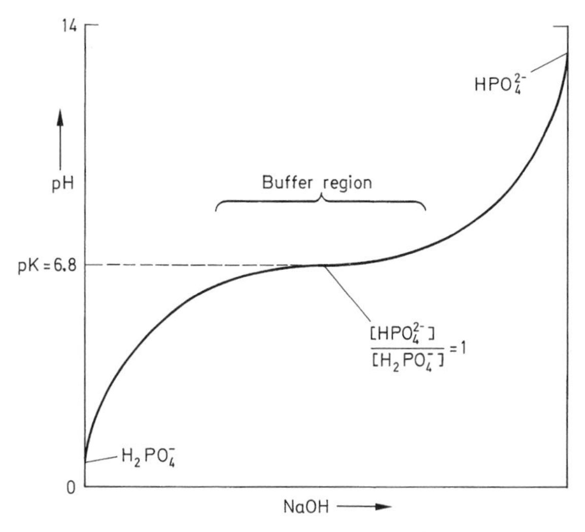

Fig. 1. *Buffer.* Titration curve of $H_2PO_4^-$

Therefore, in the present case, pH = pK + log ($[HPO_4^{2-}]/[H_2PO_4^-]$). When $[HPO_4^{2-}]/[H_2PO_4^-]$ is unity, i.e. one half of the $H_2PO_4^-$ has been titrated, then pH = pK. The intermediate plateau of the titration curve, where pH is relatively unaffected by addition of acid or base, is known as the buffer region, and is defined by the pK value (see also titration curves in the entry Amino acids). As a rule, the useful buffer region is the pK′ value ±1 pH unit. (pK is an absolute dissociation constant corrected to zero ionic strength, whereas pK′ is the apparent or concentration dissociation constant determined at known ionic strength and, where appropriate, in the presence of other solutes).

The degree of dilution and the presence of other ions may affect pH, e.g. if NaCl is added to phosphate buffer, the pH falls because Na^+ tends to associate with the HPO_4^{2-}, and Cl^- with H^+, thereby increasing the dissociation of $H_2PO_4^-$. In very dilute solution, this effect may be unrecognizable because the ions are too far apart for interaction of their electric fields. Thus the pK of $H_2PO_4^-$ in infinitely dilute solution is 7.2, whereas in blood the value is 6.8. This phosphate system is one of the most popular and widely used biochemical buffers, but it suffers from the disadvantages that it has poor buffering capacity above pH 7.5, it tends to precipitate polyvalent cations, and phosphate is a metabolite in some systems.

An ideal buffer should have a pK between 6 and 8, the pH range for most biological reactions. It should also be very soluble in water, and should not cross biological membranes. If the buffer is used in a spectrophotometric assay, it should not absorb at the measurement wavelength. Minimal effects of temperature, concentration and ionic composition of the medium are also desirable. Since no buffer system satisfies all these requirements, more than one buffer should be tested for use with biochemical systems. Buffers such as borate, imidazole, veronal (5,5-diethylbarbiturate), maleate and others are suitable for specific biological systems, but all have potential disadvantages. For example, borate forms complexes with *cis*-diols, which includes many sugars and respiratory metabolites. On the other hand, this property has been exploited in the chromatographic separation and analysis of sugars.

All buffers show a temperature effect, so that the pH recorded at room temperature may be significantly different from that at the working temperature. For example, tris(hydroxymethyl)aminomethane (TRIS) has a high temperature/pH gradient, so that 0.05 M TRIS, pH 7.05 (adjusted with HCl) at 37 °C has pH 7.20 at 23 °C. Since the logarithmic pH scale covers a very wide range of hydrogen ion concentration, apparently small changes of pH should not be dismissed as unimportant; e.g. at pH 7.05 and pH 7.20, the respective hydrogen ion concentrations are $[H^+] = 8.912 \times 10^{-8}$ M and 6.309×10^{-8} M, a difference of 41 %.

Buffer capacity. Whereas pH depends on pK′ and the ratio [proton acceptor]/[proton donor], buffer capacity depends on the quantity of buffer components present. Buffer capacity is defined as the minimal quantity of acid or base that must be added or removed in order to cause a significant pH change. For example, two solutions of 4-(2-hydroxyethyl)-piperazine-1-ethanesulfonic acid (HEPES) of pH 7.05 and of equal volumes may contain different quantities of zwitterion and base.

pH (7.05) = pK (7.55) + log [base]/[zwitterion].

Solution 1 (0.05 M): [base]/[zwitterion] = 0.038/0.012.

Solution 2 (0.10 M): [base]/[zwitterion] = 0.076/0.024.

Solution 2 has the same pH but twice the buffering capacity of solution 1, i.e. solution 2 is able to absorb twice as many protons as solution 1 before a significant pH change occurs. In medical physiology, buffer

$$HOH_2C{-}CH_2{-}\overset{+}{N}H(C_4H_8)N{-}CH_2{-}CH_2{-}SO_3^- \underset{}{\overset{OH^-}{\rightleftharpoons}} HOH_2C{-}CH_2{-}N(C_4H_8)N{-}CH_2{-}CH_2{-}SO_3^- + H_2O$$

Zwitterion of HEPES — *base of HEPES*

capacity is represented by the amount of acid or base that body fluids can accommodate before the pH becomes dangerously high or low.

Buffers of body fluids. Acids and bases, produced by metabolism, pass into body fluids and are ultimately excreted. Body fluids are temporarily protected (more commonly against acidity, because normal metabolism results in a net production of acid) by their buffer systems. The most important body fluid buffers are: 1. plasma [HCO_3^-/CO_2; protein ($nH^+ - 1H^+$)/protein(nH^+); $HPO_4^{2-}/H_2PO_4^-$]; 2. erythrocytes [HCO_3^-/CO_2; deoxyhemoglobin/deoxyhemoglobin · H^+; oxyhemoglobin/oxyhemoglobin · H^+; $HPO_4^{2-}/H_2PO_4^-$]; 3. interstitial fluid (similar to plasma but with lower protein concentration; 4. intracellular fluid (excluding erythrocytes) which contains much higher concentrations of phosphate and protein than plasma or interstitial fluid.

In blood, the main buffer system is bicarbonate at a concentration of [HCO_3^-] = 0.02–0.03 M (20–30 mEq/l). Hemoglobin provides a further 10 mEq/l buffer capacity, and phosphate makes a small contribution of 1.5 mEq/l. The 5 liters of blood in an average adult human are thus able to absorb about 0.15 mole H^+ before the pH becomes dangerously low. The major buffers of the body are, however, present in other tissues. The total musculature of the body, for example, can neutralize about 5 times as much acid as the blood, and the blood HCO_3^-/CO_2 system represents only about a tenth of the total buffer capacity of the body. Since all the buffer systems of the body are able to interact and buffer each other, all changes in the acid/base balance of the body are reflected in the blood. This mutual buffering by the shift of H^+ from one body system to another is known as the isohydric principle.

Bicarbonate acts not only as a blood buffer system, but also represents the main form in which CO_2 is transported from respiring tissues to the lungs for expiration. Some CO_2 is transported as carbamino groups of proteins: Protein-NH_2 + CO_2 ⇌ Protein-NH–COOH ⇌ Protein-NH–COO^- + H^+, and about 80 % of CO_2 is transported as bicarbonate. The different forms of blood CO_2 are in equilibrium:

In order to achieve high concentrations of HCO_3^- from dissolved CO_2, the resulting protons must be removed, i.e. accommodated by a buffer system. The principal buffers serving this function are plasma proteins (accounting for about 10 % of the protons), erythrocyte phosphate (about 20 %) and erythrocyte hemoglobin (60–70 %). For the role of hemoglobin, see Bohr effect, Hemoglobin.

At any time, the extracellular fluids contain about 1.2 mmol/l CO_2 (in its various forms) in transit between respiring tissues and lungs.

Proteins as buffers. Many of the acidic and basic amino acid side chains on the surface of proteins possess pK′ values that permit them to contribute to physiological buffer systems (Fig. 2).

Control of pH by excretion

a) *via the lungs.* The respiratory center in the medulla oblongata responds to increased [H^+] by increasing alveolar ventilation, and vice versa. Increased alveolar ventilation increases the rate of removal of CO_2 from the blood, thereby increasing blood pH. This regulatory feedback for the control of [H^+] is 50–75 % efficient, with a feedback gain of 1–3, i.e. if the blood pH is suddenly decreased from 7.4 to 7.0, the resulting increase in alveolar ventilation adjusts the pH to 7.2–7.3 in 3–12 minutes. As a rough approximation, a decrease in the pH of extracellular fluids of about 0.23 causes a twofold increase in the rate of alveolar ventilation. The alveolar ventilation rate can be varied from about zero to fifteen times normal.

b) *via the kidneys.* The inorganic composition of glomerular filtrate is almost the same as that of plasma, but it has only a very low protein concentration (0.03 %). In transit through the nephron, water and various solutes are reabsorbed and returned to the blood. The mechanism for reabsorption of bicarbonate also plays a role in the loss or retention of H^+. The epithelial cells of most of the nephron (proximal tubule, thick segment of the ascending loop of Henle, distal tubule, collecting tubule and collecting duct) all secrete H^+ into the tubular fluid in exchange for the uptake of Na^+. The bicarbonate ions in the tubular fluid (from the plasma via glomerular filtrate) are

$$\begin{array}{c} \text{Gaseous } CO_2 \\ \Updownarrow \\ \text{Carbamino groups} \rightleftharpoons \text{dissolved } CO_2 + H_2O \rightleftharpoons H_2CO_3 \rightleftharpoons H^+ + HCO_3^- \\ \Updownarrow \\ CO_2 \text{ from respiration} \end{array}$$

Cysteine pK′: 9 – 11; Aspartate or Glutamate 3 – 5; Tyrosine 10 – 10·5; Arginine 12 – 13; Lysine 7·5 – 10·5; Histidine 5·5 – 7

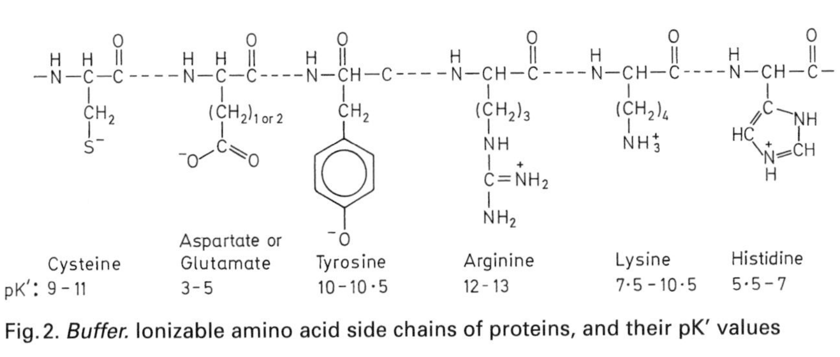

Fig. 2. *Buffer.* Ionizable amino acid side chains of proteins, and their pK′ values

therefore in the presence of increasing concentrations of H^+, which favors formation of H_2CO_3. Carbonic anhydrase in the brush border of the epithelial cells in the proximal (not the distal) tubules accelerates formation of CO_2 and H_2O from the H_2CO_3 in the tubular fluid. The CO_2 (which diffuses readily across all biological membranes) diffuses into the tubular cells, where it forms H_2CO_3 (a reaction catalysed by intracellular carbonic anhydrase); this H_2CO_3 dissociates into H^+ and HCO_3^-.

Na^+ taken up from the tubule is removed from the epithelial cell into the extracellular fluid by active transport; charge neutrality is maintained because the export of Na^+ is accompanied by export of HCO_3^-. The net effect is that for every H^+ formed and secreted into the tubule, a bicarbonate ion diffuses into the extracellular fluid in combination with Na^+ absorbed from the tubule. In effect, bicarbonate is reabsorbed from the glomerular filtrate.

In the collecting ducts, the pH of the tubular fluid can fall as low as 4.5, i.e. $[H^+]$ is 900 times greater in the tubular fluid than in the extracellular fluid. The proximal tubule secretes about 84 % of all H^+, but it can only maintain a 3–4 fold gradient, i.e. a tubular fluid of pH 6.9 compared with pH 7.4 for the extracellular fluid. Distal tubules occupy an intermediate position, being able to decrease the pH of tubular fluid to 6.0–6.5. In Acidosis (see), the ratio $[HCO_3^-]/[CO_2]$ decreases in the extracellular fluid and in the glomerular filtrate. The H^+ secretion rate therefore exceeds the glomerular filtration rate of HCO_3^-, with the following consequences: a) there is a net increase in the quantity of HCO_3^- and Na^+ in the extracellular fluid, which corrects the acidosis by increasing pH; and b) excess H^+ associates with other buffer anions (mainly HPO_4^{2-}) and with ammonia, and is excreted in the urine. HPO_4^{2-} and $H_2PO_4^-$ are poorly resorbed by the tubule, so that they become concentrated in the tubular fluid, forming a strong buffer system, pK 6.8. In the tubule, H^+ reacts with anions other than HCO_3^-, thereby making a contribution to urinary titratable acidity which is equivalent to the amount of alkali required to return the urine to pH 7.4, i.e. the pH of the glomerular filtrate.

All tubule epithelial cells and those of the thin segment of the loop of Henle continually produce ammonia, 60 % arising from the action of glutaminase on L-glutamine, and 40 % from other amino acids (e.g. by the action of glutamate dehydrogenase on L-glutamate). This ammonia associates with H^+, forming NH_4^+, which is excreted with Cl^- and other anions. Local acidosis of tubular cells promotes ammonia production by inducing the production of glutaminase.

In Alkalosis (see), the ratio $[HCO_3^-]/[CO_2]$ increases in the extracellular fluid, so that the HCO_3^- of the glomerular filtrate is in excess of the H^+ secreted from the tubule epithelium. The excess HCO_3^- passes into the urine together with Na^+ and other cations. Thus part of the bicarbonate complement of the blood is removed, and the pH decreases, correcting the alkalosis. [G. Gomori *Methods in Enzymology* **1** (1955) 138–146; N.E. Good et al. *Biochemistry* **5** (1966) 467–477; N.E. Good & S. Izawa *Methods in Enzymology* **24B** (1972) 53–68; W.J. Fergusun et al. *Anal Biochem.* **104** (1980) 300–310]

Bufogenins: see Cardiac glycosides.

Bufotenine, ***5-hydroxy-N-dimethyltryptamine:*** a Toad toxin (see).

Bufotoxin: the main toxin in the venom of the European toad, *Bufo vulgaris.* The minimal lethal dose in cats is 390 µg/kg. See Toad toxins.

Buoyant density: see Density gradient centrifugation.

2,3-Butanediol: see Fermentation products.

2,3-Butylene glycol: see Fermentation products.

n-Butyric acid, ***butanoic acid:*** $CH_3–(CH_2)_2–COOH$, a fatty acid, M_r 88.1, m.p. –5 °C. B.a. accounts for 3–5 % of the fatty acids esterified to glycerol in butterfat. When butter becomes rancid, it is the free B.a. produced by hydrolysis which is responsible for the unpleasant odor. B.a. occurs free in many plants and fungi, and in traces in sweat. Esters of B.a. are present in many essential oils.

Buxus alkaloids: steroid alkaloids characteristic of plants in the boxwood genus *(Buxus).* They are structurally related to pregnane (see Steroids), with additional methyl groups on C4 and C14, amino or methylated amino functions on C3 and C20, and usually a 16α-hydroxyl group and a 9,10-cyclopropane ring. B.a. are biosynthesized via cycloartenol or a similar triterpenoid precursor.

Cyclobuxamide H

C

C: 1. a nucleotide residue (in a nucleic acid) in which the base is cytosine. 2. Abb. for cytidine (e.g. CTP is acronym of cytidine triphosphate). 3. Abb. for cytosine.

Cachectin: see Tumor necrosis factors.

C_4-acid cycle: see Hatch-Slack-Kortschack cycle.

Cactus alkaloids: see Anhalonium alkaloids.

Cadalene precursor: see β-Cadinene.

Cadaverine, *1,5-diaminopentane:* a poisonous biogenic amine produced enzymatically by decarboxylation of lysine. C. is a precursor of certain alkaloids, and it is one of the compounds responsible for the odor of decaying meat and fecal matter. It is the preferred substrate of the amine oxidase EC 1.4.3.6.

Cadherins: see Cell adhesion molecules.

Cadinane: see Sesquiterpenes, Fig.

β-Cadinene: an optically active sesquiterpene found in the essential oils of juniper, cedar, cade and cubebs. Together with its isomers and their hydroxyl derivatives (cadinols), C. is representative of the cardinanes, which are the best known and most widespread sesquiterpenes. Cadinanes are also known as *cadalene precursors* because they can be dehydrogenated to cadalene (4-isopropyl-1,6-dimethylnaphthalene). For formula and biosynthesis, see Sesquiterpenes.

Cadinols: see β-Cadinene.

Caffeine, *1,3,7-trimethylxanthine:* a purine derivative (see Methylated xanthines) found in coffee beans and leaves, tea leaves, cocoa beans and cola nuts. It is usually produced from tea leaves (1.5–3.5 % caffeine content) and as a byproduct from the production of decaffeinated coffee. Due to its stimulatory effects on the central nervous system, C. and C.-containing beverages are used to stimulate the heart and circulation. Its effects are mainly due to inhibition of the phosphodiesterase that degrades cyclic AMP to AMP in adrenalin-producing cells, thereby prolonging adrenalin action.

Calciferol: vitamin D. See Vitamins.

Calcitonin, *thyreocalcitonin:* a polypeptide hormone, M_r (human) 3,420, containing 32 amino acid residues. It is formed in the parafollicular cells of the mammalian thyroid, and in the ultimobranchial gland of nonmammalian species (both glands are derived from the 5th gill pocket of the embryonic gut). C. causes a rapid but short-lived decrease in blood calcium and phosphate, by promoting incorporation of these ions into bones. C. is released in response to a rising Ca^{2+} level, and it is antagonized by Parathormone (see). It is determined by radioimmunological methods.

Calcium, *Ca:* an alkaline earth element which, as its divalent cation, is ubiquitous in nature. As $CaPO_4$ and $CaCO_3$, it provides rigidity and hardness to shells and bones. As a readily chelated ion, Ca^{2+} provides structural stability to proteins and lipids in cell membranes, cytoplasm, organelles and chromosomes. It is a necessary cofactor for a number of extracellular enzymes, including prokaryotic and eukaryotic digestive enzymes, factors II, VII, IX and X in Blood coagulation (see), and for complement activation by antigen-antibody complexes. Ca^{2+} binds to tubulin with high affinity, and is required for entry of cells into the S-phase of the Cell cycle (see). Contractility, secretion, chemotaxis and aggregation of cells are regulated by Ca^{2+} and arachidonic acid metabolites. Ca^{2+} plays an essential role in the propagation of nerve impulses and in muscle contraction.

Ca^{2+} serves as second messenger (see Hormones) in animal cells. When the calcium-system receptor is activated by a molecular stimulus, 4,5-diphospho-phosphatidylinositol (4,5-PIn) is hydrolysed to inositol 1,4,5-*tris*phosphate and diacylglycerol. Diacylglycerol is also a second messenger, and promotes phosphorylation of several proteins by activating protein kinase C (see). At the same time Ca^{2+} is either allowed to enter the cell from the outside or it is mobilized from reserves within the cell. The intracellular Ca^{2+} concentration rises, briefly, to about 1.0 μmolar. At this concentration Ca^{2+} activates some proteins directly, and others indirectly after binding to Calmodulin (see). Some calmodulin-activated proteins are kinases which phosphorylate a different group of proteins from those phosphorylated by protein kinase C. Protein kinase C is itself activated by high concentrations of Ca^{2+}. If it is first sensitized by diacylglycerol, however, it may be activated by Ca^{2+} concentrations only slightly higher than those in the unstimulated cell. Thus, once a cell has been activated by a brief surge of Ca^{2+}, the activation may persist at much lower Ca^{2+} concentrations, because protein kinase C has been activated.

In many cases a given stimulus will activate both cAMP and Ca^{2+}/diacylglycerol as second messengers. H. Rasmussen has proposed the term "synarchic" to describe the interactions which then occur. In some cases, artificial stimulation of just one of the two systems (e.g. stimulation of Ca^{2+} influx by an ionophore, or injection of cAMP) leads to phosphorylation of the same proteins. Often, a Ca^{2+} influx leads to brief stimulation, whereas injection of cAMP produces a slow but long-term stimulation; normal and simultaneous stimulation of both systems produces a long-term, rapid response. The presence of two mutually regulatory second messenger systems allows for greater plasticity in the response of the cell to the primary stimulus.

Plants also contain calmodulin, and they employ Ca^{2+} as a second messenger. (cAMP is not a second messenger in plants.) Both gravitropic and phototropic responses in plants appear to depend on Ca^{2+}.

Excess Ca^{2+} is toxic to cells, which therefore have efficient mechanisms for removing it, e.g. a calcium pump (activated by high intracellular Ca^{2+} concentrations), which removes Ca^{2+} to the cell exterior. Under normal conditions, Ca^{2+} may be exchanged for Na^+ leaking into the cell. Another means of handling large amounts of Ca^{2+} is sequestration within mitochondria. The normal intracellular concentration of Ca^{2+} is 0.1 μmolar; the concentration of free Ca^{2+} in the blood is about 1.5 μmolar. [A. K. Campbell, *Intracellular Calcium, Its Universal Role as Regulator,* Wiley & Sons, 1983; D. Marmé (ed.) *Calcium and Cell Physiology,* Springer, 1985; W. Y. Cheung (ed.) *Calcium and Cell Function, Vol IV,* Academic Press, 1983]

Calcium-dependent regulator protein: see Calmodulin.

C-alkaloids: see Curare alkaloids.

Callistephin: see Pelargonidin.

Calmodulin: a calcium-binding protein which mediates the functions of Ca^{2+} in eukaryotes. It consists of 148 amino acid residues; its primary sequence has been highly conserved throughout eukaryote evolution, and is about 70 % homologous with troponin C, the Ca^{2+}-binding unit in the muscle contractile apparatus. C. has 4 binding sites for Ca^{2+}. Binding of Ca^{2+} causes C. to change conformation, exposing a hydrophobic site which interacts with the regulated protein.

Although C. is apparently ubiquitous, it mediates different cellular responses to Ca^{2+} in different tissues. This specificity is due to the presence of different C.-activated proteins in different tissues. Some proteins activated by interaction with the C.-Ca^{2+} complex are cyclic nucleotide phosphodiesterase, adenylate cyclase and guanylate cyclase. These enzymes are active in the metabolism of cAMP and cGMP, which, like Ca^{2+}, serve as second messengers for hormones (see Calcium, Adenosine phosphates). Their regulation by C. allows for interaction of the two messenger systems. Other proteins activated by C.-Ca^{2+} are muscle phosphorylase kinase and myosin light chain kinase, glycogen synthase kinase, and a variety of other regulatory protein kinases in different tissues. C. also binds to nonenzyme proteins such as histones, myelin basic protein, spectrin, tubulin and dynein ATPase. In mitotic cells, most of the C. is localized on the mitotic spindle (which consists largely of tubulin). Since C. can substitute for other Ca^{2+}-binding proteins in vitro, it is possible that some of the proteins activated by C. in vitro are activated by other Ca^{2+}-binding proteins in vivo. [L. J. Van Eldik & D. M. Watterson in *Calcium and Cell Physiology* (ed. Marmé) Springer, 1985, pp. 105–126; R. C. Brady et al. *ibid.* pp. 140–147; D. Marmé & P. Dieter in *Calcium and Cell Function, Vol IV* (ed. W. Y. Cheung), Academic Press, 1983, pp. 246–311]

Calvin cycle, ***reductive pentose phosphate cycle, photosynthetic carbon reduction cycle:*** a series of 13 enzyme-catalysed reactions, occurring in the chloroplast stroma in plants or the cytoplasm in photosynthetic bacteria, which are organized into a cycle, the purpose of which is to convert CO_2 into carbohydrate using the reduced pyridine nucleotide (NADPH in plants, NADH in photosynthetic bacteria) and ATP generated in the light phase of photosynthesis (see Photosynthesis). The cycle was discovered by Melvin Calvin, research that earned him the Nobel Prize for chemistry in 1961. The cycle operates in all photosynthetic organisms except the green sulfur bacteria *(Chlorobiaceae),* which fix CO_2 via the Reductive citrate cycle (see). Although the atmosphere is the ultimate source of the CO_2 fixed in the Calvin cycle, it is the immediate source only in C_3-organisms (see); in C_4-plants (see Hatch-Slack-Kortschak cycle) and CAM-plants (see Crassulacean acid metabolism) atmospheric CO_2 is initially incorporated into oxaloacetic acid (OAA) by reaction (as HCO_3^-) with phospho*enol*pyruvate; it is subsequently regenerated from OAA (or from malic acid derived from OAA), then fixed in the Calvin cycle.

The cycle can be considered to be composed of two phases, namely: (i) a *CO_2 fixing, carbohydrate forming phase,* in which CO_2 combines with a CO_2-acceptor molecule to form a C_3-carboxylic acid which is then converted into the triose phosphates, 3-phosphoglyceraldehyde (3-PGAld) and dihydroxyacetone phosphate (DiHOAcP) by a reaction sequence that uses NAD(P)H and ATP (A-D, Fig. 1), and (ii) a *regenerative phase,* in which the triose phosphate molecules that remain after the removal for general biosynthetic purposes of one molecule of DiHOAcP per three molecules of CO_2 fixed, undergo a series of reactions (E-M, Fig. 1) whose purpose is to regenerate the number of CO_2-acceptor molecules required by the stoichiometry of the cycle, i. e. 3 for each molecule of DiHOAcP removed for general biosynthesis. In plants the latter is physically removed from the cycle by exporting it from the chloroplast to the cytosol where it is used to form sucrose. This sucrose is then exported from the cell and transported via the phloem to those parts of the plant that have need of it, e. g. meristems, roots, developing seeds and tubers; however such physical separation is not possible in photosynthetic bacteria because they do not have chloroplasts or any other organellar structure.

Phase (i) begins with the carboxylation reaction (A, Fig.), which is catalysed by ribulose-1,5-*bis*phosphate carboxylase (see). The CO_2-acceptor is D-ribulose-1,5-*bis*phosphate (Ru-1,5-bP) which is believed to react, as its enediol, with CO_2 to form the 6C intermediate 2-carboxy-3-keto-D-arabinitol-1,5-*bis*phosphate which is then hydrolytically cleaved to yield two molecules of 3-phosphoglyceric acid (3-PGA) (see Ribulose-1,5-*bis*phosphate carboxylase for the reaction mechanism). If one molecule of DiHOAcP is regarded as the product of one turn of the cycle the stoichiometry requires that 3 molecules of CO_2 react with 3 molecules of Ru-1,5-bP to generate 6 molecules of 3PGA; note that the rest of this article assumes such a stoichiometry, as does Fig. 1, where the number of molecules participating in a given reaction is indicated by a numeral adjacent to the reaction arrow. All the 3PGA molecules are then phosphorylated by phosphoglycerate kinase using ATP, generated in the light phase of photosynthesis, to give an equivalent number of 1,3-*bis*phosphoglyceric acid (1,3-bPGA; more correctly called 3-phosphoglyceroyl phosphate) molecules (B, Fig.); this reaction activates the carboxyl group of 3PGA by converting it to a $-CO(OPO_3^{2-})$ group allowing it to be reduced to a –CHO group in the next reaction. In this reaction all the 1,3-bPGA molecules are reduced by 3-phosphoglyceraldehyde dehydrogenase (using NAD(P)H,

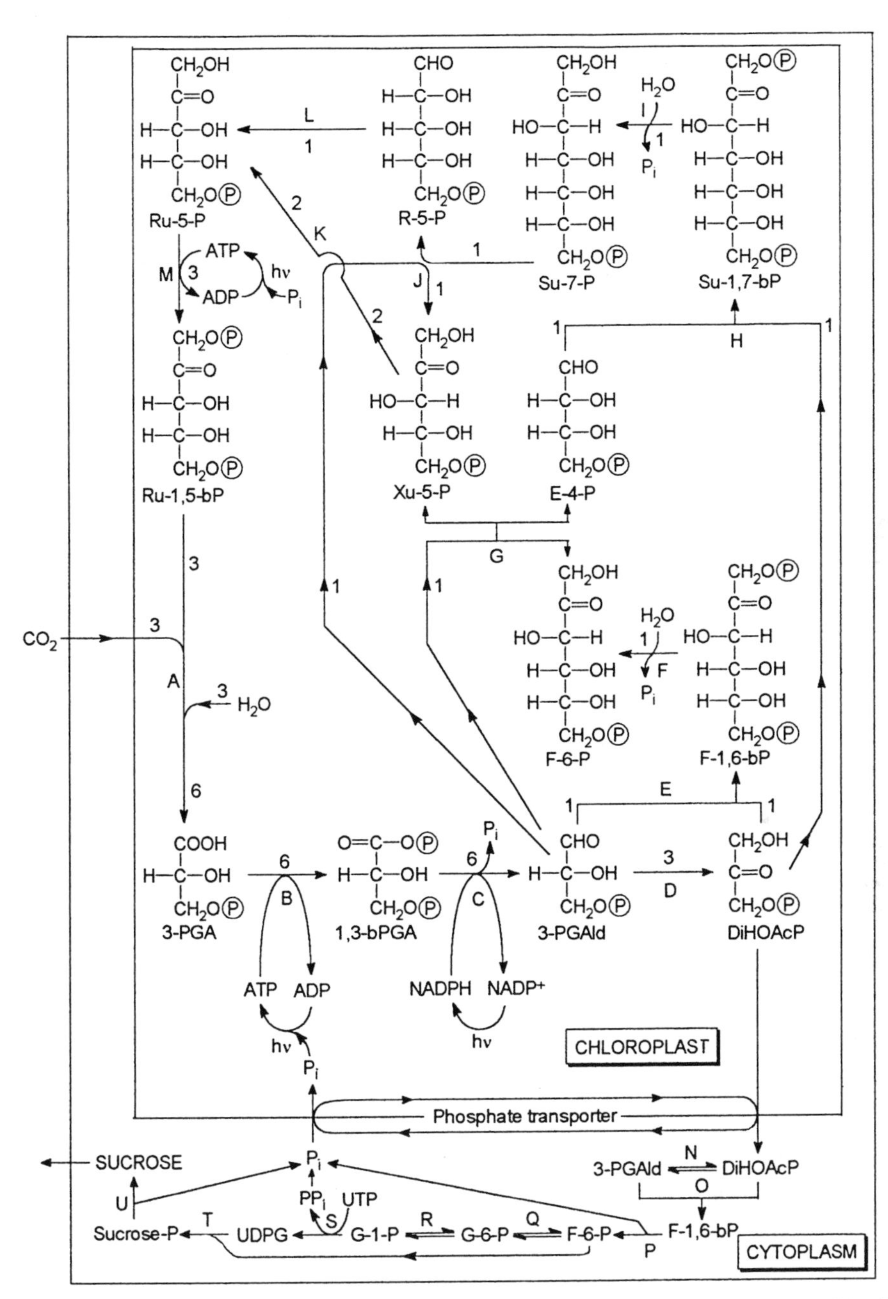

The chloroplastidic Calvin reductive pentose phosphate cycle and the related cytoplasmic synthesis of sucrose. The stoichiometry of the cycle is indicated by the numbers adjacent to the reaction arrows; A = Ribulose *bis*phosphate carboxylase, EC 4.1.1.39; B = phosphoglycerate kinase, EC 2.7.2.3; C = 3-phosphoglyceraldehyde dehydrogenase, EC 1.2.1.13; D & N = triosephosphate isomerase, EC 5.3.1.1; E, H & O = fructose *bis*phosphate aldolase, EC 4.1.2.13; F & P = fructose *bis*phosphatase, EC 3.1.3.11; G & J = transketolase, EC 2.2.1.1; I = sedoheptulose *bis*phosphatase, EC 3.1.3.37; K = ribulose phosphate 3-epimerase, EC 5.1.3.1; L = ribose phosphate ketol- isomerase, EC 5.1.3.6; M = phosphoribulokinase, EC 2.7.1.19; Q = glucose phosphate ketol-isomerase, EC 5.3.1.9; R = phosphoglucomutase, EC 2.7.5.1; S = glucose phosphate uridylyltransferase, EC 2.7.7.9; T = sucrose phosphate synthase, EC 2.1.4.14; U = sucrose phosphatase, EC 3.1.3.24.

which is also generated in the light phase of photosynthesis) to give an equivalent number (6) of 3-PGAld molecules (C, Fig.); 3 of these 3-PGAld molecules undergo an aldose-to-ketose isomerization, catalysed by triose phosphate isomerase, to yield 3 molecules of DiHOAcP (D, Fig.). Thus at the end of phase (i) of the cycle 3 molecules each of 3PGAld and DiHOAcP have been formed. Of these one molecule of DiHOAcP is removed from the cycle as the photosynthetic product (representing the fixation of 3 molecules of CO_2), thereby leaving 3 molecules of 3PGAld and 2 molecules of DiHOAcP with which to start phase (ii) of the cycle. Thus in phase (ii) the carbon skeletons of 5 molecules of triose phosphate, a total of 15 carbon atoms, must be reorganized in such a way that 3 molecules of pentose phosphate, in the form of Ru-1,5-bP, are created; these are the CO_2-acceptor molecules for the next turn of the cycle.

Phase (ii) of the cycle begins with the fructose *bis*phosphate aldolase-catalysed condensation of 1 molecule of 3PGAld with one molecule of DiHOAcP to form one molecule of D-fructose-1,6-*bis*phosphate (F-1,6-bP) (E, Fig.). This is followed by the fructose *bis*phosphatase-catalysed hydrolytic removal of the C1 phosphate group of F-1,6-bP to yield D-fructose-6-phosphate (F-6-P) (F, Fig.). The latter then reacts with one of the two remaining molecules of 3PGAld to form one molecule each of D-xylulose-5-phosphate (Xu-5-P) and D-erythrose-4-phosphate (E-4-P) (G, Fig.). This reaction is catalysed by transketolase which, in the presence of thiamine pyrophosphate, transfers the ketol group ($CH_2OH.C{=}O$) at the C1 & C2 positions of F-6-P to 3PGAld in such a way that it becomes the ketol group of Xu-5-P; carbon atoms 3, 4, 5 & 6 of the F-6-P become E-4-P. The E-4-P then condenses with the one remaining molecule of DiHOAcP under the catalytic influence of fructose *bis*phosphate aldolase, to form one molecule of D-sedoheptulose-1,7-*bis*phosphate (Su-1,7-bP) (H, Fig.). This is followed by the fructose *bis*phosphatase-catalysed hydrolytic removal of the C1 phosphate group of Su-1,7-bP to yield D-sedoheptulose-7-phosphate (Su-7-P) (I, Fig.). The latter then reacts with the one remaining molecule of 3PGAld to form one molecule each of D-xylulose-5-phosphate (Xu-5-P) and D-ribose-5-phosphate (R-5-P) (J, Fig.). This reaction is also catalysed by transketolase and is mechanistically identical to reaction G (Fig.). At this point in phase (ii) there are 2 molecules of Xu-5-P, formed by the two transketolase-catalysed reactions, (G & J, Fig.) and one molecule of R-5-P. All three of these pentose phosphates are now isomerized to D-ribulose-5-phosphate (Ru-5-P). The ribose phosphate ketol-isomerase-catalysed isomerization of R-5-P involves an aldose-to-ketose conversion (L, Fig.), while the ribulose phosphate epimerase-catalysed isomerization of Xu-5-P involves the reversal of the orientation of the H and OH groups at C3 (3*S* to 3*R*) (K, Fig.). The cycle is completed by the phosphoribulokinase-catalysed phosphorylation of the 3 molecules of Ru-5-P, using ATP generated in the light phase of photosynthesis and forming the 3 molecules of Ru-1,5-bP required to start the next cycle.

Several enzymes of the Calvin cycle, namely ribulose-1,5-*bis*phosphate carboxylase, fructose *bis*phosphatase, 3-phosphoglyceraldehyde dehydrogenase and phosphoribulokinase, have been shown to be activated by light [B. B. Buchanan *Annu. Rev. Plant Physiol.* **32** (1981) 349–383].

The main evidence for the Calvin cycle came from: (i) studies of time-dependent variations in the labeling of intermediates when $^{14}CO_2$ was administered to an organism carrying out photosynthesis under steady state conditions; all the intermediates were labeled after a 30 second exposure but only 3PGA was labeled when the exposure was reduced to 5 seconds, proving that 3PGA was the product of the carboxylation reaction , (ii) the labeling pattern of the carbon chain of the intermediates formed after a short period of photosynthesis in the presence of $^{14}CO_2$, which was that expected from the cycle as formulated in the Fig., (iii) the demonstration that the steady state level of 3PGA rose and that of Ru-1,5-bP fell to zero when a photosynthesizing organism was subjected to a period of darkness (when no NAD(P)H or ATP could be formed), but returned to normal when the light was restored; moreover this pattern of 3PGA and Ru-1,5-bP levels was repeated indefinitely during successive light-dark cycles, thereby proving that these two compounds were related in a cyclic, rather than a linear, sequence of reactions, and (iv) the demonstration that all the enzymes required by the cycle shown in the Fig. are present in chloroplasts of plants and in the cytoplasm of photosynthetic bacteria that operate the Calvin cycle. [W. Martin et al. 'Microsequencing and cDNA cloning of the Calvin cycle/oxidative pentose phosphate pathway enzyme ribose-5-phosphate isomerase (EC 5.3.1.6) from spinach chloroplasts' *Plant Molecular Biology* **30** (1996) 795–805]

Calvin plants: see C_3 plants.

Calycosin: see Pterocarpans.

CAM: acronym of Cell adhesion molecule (see) or Crassulacean acid metabolism (see).

cAMP: acronym of cyclic adenosine 3′,5′-monophosphate. See Adenosine phosphates.

Campesterol, ***campest-5-en-3β-ol:*** a plant sterol (see Sterols), M_r 400.68, m.p. 158 °C, $[\alpha]_D^{23}$ −33 ° (22.5 mg in 5 ml $CHCl_3$). C. is found in the oils of rapeseed *(Brassica campestris),* soybean and wheat germ, and in some mollusks.

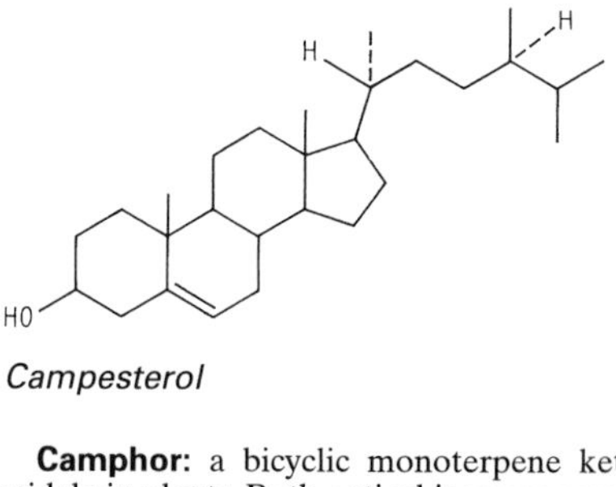

Campesterol

Camphor: a bicyclic monoterpene ketone found widely in plants. Both optical isomers occur naturally: (+)-C. (Japan C.), m.p. 180 °C, b.p. 204 °C, $[\alpha]_D^{20}$ +43.8 ° (c = 7.5 in abs. ethanol), and (−)-C. (Matricaria C.), m.p. 178.6 °C, b.p. 204 °C, $[\alpha]_D^{23}$ −44.2 ° (ethanol). C. is obtained commercially from camphor trees *(Cinnamomum camphora)* native to coastal areas of East Asia. Partial synthesis from pinene is also commercially important, the product being a racemic mixture used mainly in plastics. Natural C. is used pharmaceutically in salves. For formula and biosynthesis, see Monoterpenes.

cAMP receptor protein: see Adenosine phosphates, Table 2.

Camptothecin: 4-ethyl-4-hydroxy-1H-pyrano-[3′,4′:6,7]indolizinol[1,2-b]quinoline-3,14(4H,12H)-dione, the main alkaloid from the wood and bark of the Chinese tree, *Camptotheca acuminata* Decsne *(Nyssaceae);* it is also present in *Mappia foetida* and *Ervatamia heyneana.* M_r 348, m.p. 264–267 °C (d.), $[\alpha]_D^{25}$ +31.3 ($CHCl_3/CH_3OH$, 8:2). C. is one of the most active natural products against leukemia and malignant tumors; it is used as an anticancer agent particularly in the People's Republic of China.

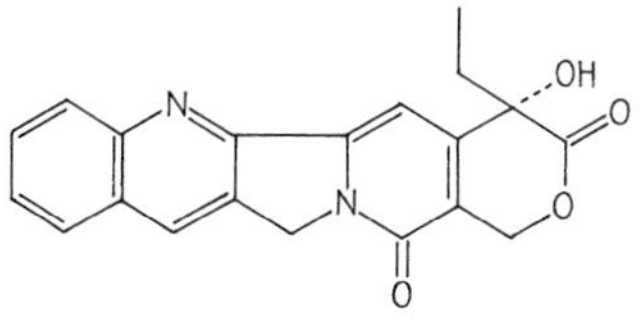

Camptothecin

L-Canaline: L-2-amino-4-(aminooxy)butyric acid, M_r 134.12, m.p. 213 °C (d.), $[\alpha]_D^{22}$ (1.4 % H_2O) = +7.7 ± 0.2°. It is a hydrolysis product of L-Canavanine (see) and is found in all canavanine-containing legumes that have arginase. It is a potent inhibitor of pyridoxal phosphate-containing enzymes by reacting irreversibly with the vitamin B_6 moiety to form a stable, covalently bound oxime. [G. A. Rosenthal et al. *J. Biol. Chem.* **265** (1990) 863–873]

L-Canavanine: L-2-amino-4-(guanidooxy)butyric acid, M_r 176.12, m.p. 172 ± 0.5 °C (d.), $[\alpha]_D^{20}$ +7.9° (c = 3.2), a structural analog of Arginine (see) found only in certain legumes. The presence or absence of L-C. is a useful trait in the Chemical taxonomy (see) of the legumes. It is a nonproteogenic amino acid, and in fact is poisonous to most animals and to L-C.-free plants grafted onto L-C.-containing plants. In monkeys, L-C. induces hematological and serological abnormalities characteristic of systemic lupous erythematosus. The toxicity of L-C. is due in part to the fact that it is incorporated into protein in place of arginine. It is much less basic than arginine, and under physiological conditions it has a less positive charge. Thus it can be expected to disrupt the tertiary and/or quaternary structures of proteins; and in those enzymes in which the arginine side chain is part of, or supports, the active site, L-C. can produce a less active or an inactive enzyme.

In those plants which produce it, L-C. may be a major nitrogen storage compound in the seeds, e. g. it accounts for 55 % of the nitrogen in seeds of *Dioclea megacarpa,* a tropical legume. Production of poisonous L-C. is apparently an adaptive advantage for *D. megacarpa,* as this plant has only one insect predator, the beetle *Caryedes brasiliensis,* which is totally dependent on it. In L-C.-containing plants and in *C. brasiliensis* poisoning is avoided by possession of arginyl-tRNA synthetases which distinguish between arginine and L-C., and therefore do not incorporate L-C. into protein. Degradation of L-C. is similar in germinating seeds and in the beetle, and both utilize the nitrogen of L-C. (Fig.). [G. A. Rosenthal, *Scientific American* **249** (1983) 164–171]

Cancer: a malignant growth. C. is distinct from a *tumor,* which strictly speaking is any lump or swelling, whereas a *neoplasm* results from uncontrolled growth of cells. Some authors consider a neoplasm benign if it does not metastasize, or shed cells capable of colonizing other parts of the body, and reserve the term "malignant" for a metastasizing growth. There are 3 major types of malignant neoplasm: a) *carcinomas,* which arise from epithelial tissue, b) *sarcomas,* which arise from connective tissue, muscle, cartilage, fat or bone, and c) *leukemias* and *lymphomas,* which arise from hematopoietic stuctures. The most common type is the carcinoma.

C. research includes the following areas: 1. etiology, including epidemiology and prevention; 2. transformation of cells in culture; 3. genetics and phenotypes of C. cells; 4. carcinogens, metabolic activation of carcinogens and tumor promotion; 5. treatment of C. The

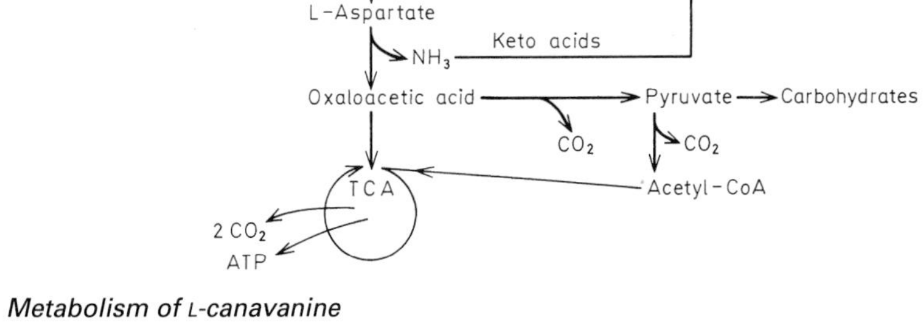

Metabolism of L-canavanine

study of animal C. viruses and their transformation of cells in tissue culture led to the discovery of Oncogenes (see).

It has long been recognized, from both epidemiology and laboratory experiments, that C. can be caused by viruses, certain chemical substances called *carcinogens,* and by ionizing radiation. The effective viruses belong to the group known as *retroviruses,* in which the genetic material is RNA; this is transcribed into DNA which may be inserted into the genome of the host cell. Such insertions may disrupt the control of gene expression, in some cases leading to uncontrolled growth. In other cases, the virus may insert a C.-causing gene (see Oncogene) acquired from a previous host. Similarly, ionizing radiation is thought to induce C. by causing mutations that derepress growth-promoting genes. Carcinogen action is often subtle. Some of these compounds are Mutagens (see), which act randomly on the genome, occasionally altering genes responsible for maintenance of growth arrest. Other compounds, called *tumor promoters* or *cocarcinogens,* evidently do not interact directly with chromosomes, but make them more susceptible to the effects of carcinogens. Such compounds, e.g. Phorbol esters (see), may interact with membrane systems whose physiological function is to trigger growth and division in the presence of appropriate signals, such as hormones.

Transformation of a cell or clone into a C. is rarely, if ever, a single-step event. Cell growth and division are controlled by several to many genes, and a change in any single gene is insufficient to induce uncontrolled cell division. Correspondingly, there may be clones of cells in a normal body which are "precancerous", i.e. they may become malignant if exposed to further injury or mutation. There is evidence that precancerous (benign) tumors are polyclonal, and that the most aggressively dividing of these clones multiply at the expense of the others.

Candicine, N,N,N-*trimethyltyramine:* a biogenic amine found especially in grasses and cacti.

Cane sugar: see Sucrose.

Cantharidin: a toxic agent, M_r 196.21, m.p. 218°C, produced by the beetle *Cantharis vesicatoria* (Spanish fly, blistering beetle), native to southern and central Europe. C. is biosynthesized from mevalonic acid. Its use as an aphrodisiac has caused fatalities.

Cantharidin

Caoutchouc: elastic, high molecular mass polyterpenes, which can be converted into rubber by vulcanization. Natural C. is a mixture of polyisoprenoids with varying molecular masses, ranging from 300,000 to 700,000. According to X-ray and IR data, the double bonds are *cis* oriented, whereas in the C.-like polyterpenes, gutta and balata, they are *trans.* Hundreds of species of plants contain C. in their latex, but it can only be obtained on a large scale from a few members of the *Euphorbiaceae, Apocyanaceae, Asclepiadaceae* and *Moraceae.* The rubber tree, *Hevea brasiliensis,* native to the Amazon region, but also grown in India and Indonesia, is quantitatively the most important world source of C. *Parthenium argentatum* of Mexico and California is the source of guayule C., which is similar to that from *Hevea,* but has a lower molecular mass. C. is harvested by cutting the bark of the trees without injuring the cambium. Latex flows from the cut latex canals, yielding 40–80 ml/tree/harvest. The latex contains 25–35 % C., 60–75 % water, 2 % protein, 2 % resin, 1.5 % sugar and 1 % ash. Coagulation is prevented by the protein. The latter is precipitated with dilute acid or sodium fluorosilicate, and the coagulated C. pressed into sheets to remove the water. The resulting lightweight, very elastic and flexible material is processed to rubber by vulcanization. After 1935, kok-saghys C. was produced in the USSR, and now in the states of the former USSR, from roots of the dandelion species *Taraxacum kok-saghys.*

Biosynthesis. The initiating species in C. biosynthesis in *Hevea brasiliensis* is a derivative of *trans-trans*-farnesyl diphosphate (i.e. a C_{15} allylic diphosphate), which is presumably modified at the methyl carbon of the dimethylallyl group. [Y. Tanaka et al. 'Initiation of Rubber Biosynthesis in *Hevea brasiliensis:* Characterization of Initiating Species by Structural Analysis' *Phytochemistry* **41** (1996) 1501–1505]

CAP: acronym of Catabolite-gene activator protein (see).

Capnine: 2-amino-3-hydroxy-15-methylhexadecane-1-sulfonic acid, or 1-deoxy-15-methylhexadecasphinganine-1-sulfonic acid, the main sulfonolipid from the bacterial genera, *Cytophaga, Capnocytophaga, Sporocytophaga* and *Flexibacter* (all notable for their gliding motility). Other sulfonolipids in these genera are probably *N*-acylated derivatives of C.

$$(H_3C)_2CH-(CH_2)_{11}-CHOH-CH(NH_3^+)-CH_2-SO_3^-$$

Capnine

Sulfonolipids constitute up to 20 % of the cellular lipids of these microorganisms, and are major components of the cell envelope. C. is a structural analog of sphinganine; similarly, *N*-acylcapnines are analogs of ceramides. The only other known source of a similar sulfonolipid is the diatom, *Nitzschia alba* (a eukaryote), which contains *N*-acylated 2-amino-3-hydroxy-4-trans-octadecene-1-sulfonic acid; the nonacylated moiety is an analog of sphingosine. [W. Godchaux III & E. R. Leadbetter *J. Biol. Chem.* **259** (1984) 2982–2990]

Capon-comb unit: see Androgens.

Capon test: see Androgens.

n-Caproic acid, *decanoic acid:* $CH_3-(CH_2)_8-COOH$, a fatty acid, M_r 172.3, m.p. 31.3°C, b.p. 268°C. It occurs in milk fat (2 %), coconut oil <1 %) and various other seed and essential oils.

n-Capronic acid, *hexanoic acid:* $CH_3-(CH_2)_4-COOH$, a fatty aid, M_r 116.16, m.p. –1.5°C, b.p.

207 °C. It is found in milk fat (2 %) and in small amounts in plant fats, e.g. coconut oil and other palm oils, as well as in some essential oils.

n-Caprylic acid, ***octanoic acid:*** CH_3–$(CH_2)_6$–COOH, a fatty acid with antifungal properties, M_r 144.21, m.p. 16.5 °C, b.p. 237 °C. It is found in various animal and plant glycerides, e.g. 1–2 % in milk fat, 6–8 % in coconut oil.

Capsaicin: a pungent principle in the fruits of some peppers *(Capsicum),* M_r 305.42, m.p. 64–65 °C. The aromatic part is biosynthesized from phenylalanine. It is occasionally used as a counter-irritant.

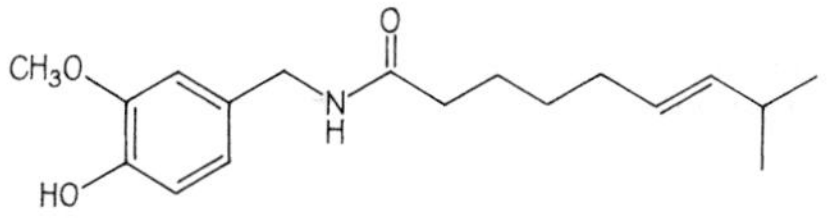

Capsaicin

Capsanthin: a carotenoid, M_r 584.85, m.p. 176 °C, $[\alpha]_{Cd}$ + 36° ($CHCl_3$). It possesses a terminal 5-membered ring. The secondary OH-groups have *R* configuration on C3, and *S* configuration on C3′. C. is the main red pigment of paprika *(Capsicum annum),* where it is accompanied by Capsorubin (see), cryptocapsin and capsanthin-5,6-epoxide. Ripe fruits contain 9–10 mg C. per 100 g fresh weight.

Capsid: see Viral coat protein.

Capsidiol: see Phytoalexins.

Capsorubin: a carotenoid, M_r 600.85, m.p. 218 °C. It contains 2 identical cyclopentanol rings. The secondary OH-groups have *R* configuration on C3, and *S* configuration on C3′. C. is present in paprika fruits *(Capsicum annum),* usually as esters. Ripe fruits contain about 1.5 mg per 100 g fresh weight.

Ca^{2+} pump: see Membrane transport.

Caran: see Monoterpenes, Fig.

Carbamate: see Carbamyl phosphate.

Carbamide: see Urea.

Carbamino group: see Buffer, section on buffers of body fluids.

6-Carbamoylglycyl purine nucleoside: see 6-Carbamoylthreonyl purine nucleoside.

Carbamoyl phosphate: H_2N–COO ~ PO_3H_2, an energy-rich phosphorylated carbamate and an important metabolic intermediate. Carbamic acid, NH_2COOH, is unstable in free form. Carbamate removed hydrolytically from carbamyl compounds, e.g. ureidopropionic acid (see Pyrimidine degradation), decomposes immediately into CO_2 and NH_3. C.p. is a specific precursor of arginine and urea (see Urea cycle), and of pyrimidines via orotic acid (see Pyrimidine biosynthesis).

C.p. is synthesized de novo or generated by phosphorolysis of ureido compounds. De novo synthesis involves 3 different enzymes.

1) Carbamoyl-phosphate synthetase (ammonia) (EC 6.3.4.16), present in vertebrate liver, catalyses the irreversible formation of C.p. from ammonium hydrogen carbonate at the expense of 2ATP, employing *N*-acetyl-L-glutamate (AGA) as cofactor:

$$NH_4HCO_3 + 2ATP \xrightarrow[AGA]{Mg^{2+}} H_2N\text{–COO–}PO_3H_2 + 2ADP + P_i.$$

AGA is required for the active enzyme conformation, not for activation of the carbon dioxide. The enzyme is localized in mitochondria, where it operates in the synthesis of arginine and urea.

2) Carbamoyl-phosphate synthetase (glutamine-hydrolysing) (EC 6.3.5.5) catalyses the irreversible synthesis of C.p. by transferring the amide nitrogen of L-glutamine:

$$HCO_3^- + \text{L-glutamine} + ATP + H_2O \xrightarrow{Mg^{2+}} \text{C.p.} + \text{L-glutamine} + ADP.$$

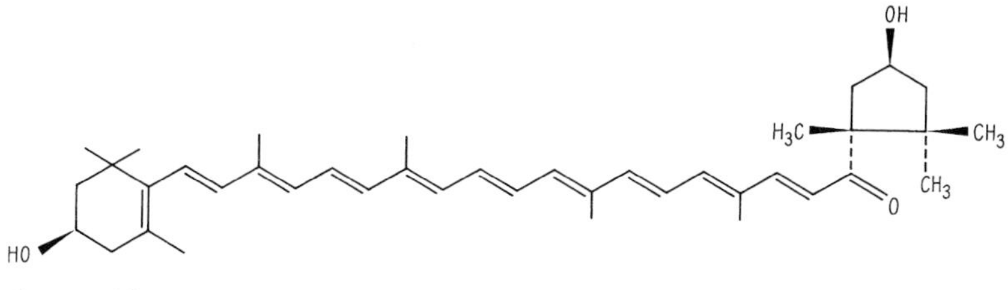

Capsanthin

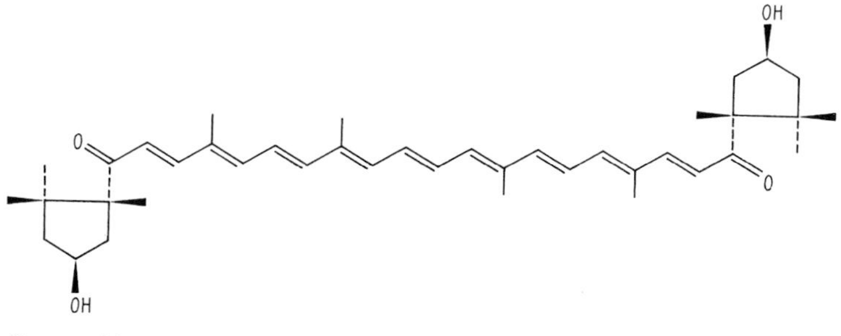

Capsorubin

The cytoplasmic enzyme is active in pyrimidine biosynthesis. Free ammonium ions can replace the glutamine, but only at higher than physiological concentrations.

3) Carbamate kinase (carbamyl phosphokinase) (EC 2.7.2.2) catalyses phosphorylation of carbamate by ATP: $NH_2–COO^- + ATP \rightleftharpoons NH_2–COO–PO_3H_2 +$ ADP. It is found in various microorganisms *(Streptococcus, Neurospora,* etc). Thermodynamically, ATP formation is favored, so the enzyme is thought to be involved in ATP generation rather than C.p. synthesis. Several microorganisms phosphorolyse allantoin and citrulline, forming C.p. which can then be used for ATP formation. Phosphorolysis of citrulline produces ornithine and C.p. In the course of allantoin fermentation by *Streptococcus allantoicus* and *Arthrobacter allantoicus,* carbamyloxamic acid (oxaluric acid) is formed then phosphorolysed: $NH_2CO–NH–CO–COOH$ (oxaluric acid) + $P_i \rightarrow NH_2CO–COOH$ (oxamic acid) + C.p. The formation of ATP from C.p. derived from citrulline or allantoin is an example of Substrate level phosphorylation (see).

All transcarbamylation reactions require C.p. as donor of the carbamyl group, i.e. transcarbamylation directly from ureido compounds does not occur. C.p. is probably also the carbamyl donor in the biosynthesis of *O*- and *N*-carbamyl derivatives, such as Albizziin (see). C.p. is a metabolically active form of ammonia used as the starting material for synthesis of other nitrogenous compounds.

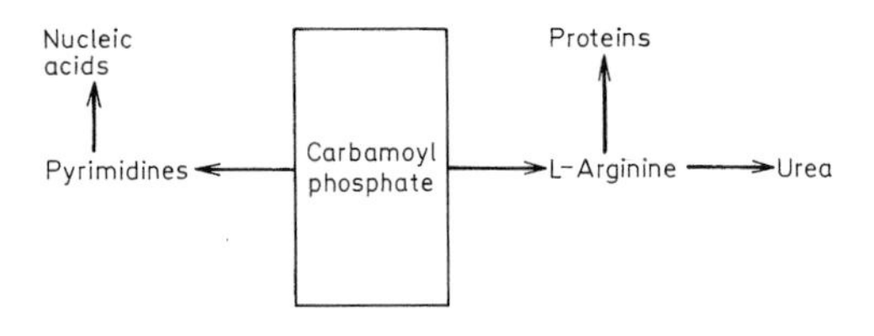

Carbamoyl-phosphate synthetase: see Carbamoyl phosphate.

6-Carbamoylthreonyl purine nucleoside, N-*(nebularin-6-ylcarbamoyl)-threonine:* a nucleoside containing one of the rare nucleic acid bases, present in certain transfer RNAs. The analogous compound, 6-carbamoylglycyl purine nucleoside has been isolated from yeast tRNA.

R	
R= —NH—CH_2—COOH	6-Carbamoylglycyl purine nucleoside
R= —NH—CH(—HCOH—CH_3)—COOH	6-Carbamoylthreonyl purine nucleoside

Carbamyl phosphokinase: see Carbamoyl phosphate.

Carbohydrases: see Glycosidases.

Carbohydrate metabolism: the constant formation, transformation and degradation of carbohydrates in living organisms. Major reactions in C.m. are: 1. interconversions of polymeric storage forms (glycogen and starch) and monomeric transport and substrate forms (glucose and glucose phosphates); 2 reactions of carbohydrate degradation and interconversion; and 3. reactions in which glucose is synthesized from noncarbohydrate precursors (glucogenic amino acids, fats). Glucose 6-phosphate has a central position in the entire C.m. Apart from minor pathways (see Glucuronate pathway, Entner-Doudoroff pathway, Phosphoketolase pathway), there are 4 main pathways of glucose 6-phosphate metabolism: 1. Glycolysis (see), 2. glycogen synthesis (see Glycogen metabolism), 3. Pentose phosphate cycle (see), and 4. enzymatic hydrolysis to free glucose. In the animal organism, the effectiveness of these pathways depends on the function of the tissue in question. A change in the activity of the tissue, as in illness, has a large effect on its C.m. (Fig.1).

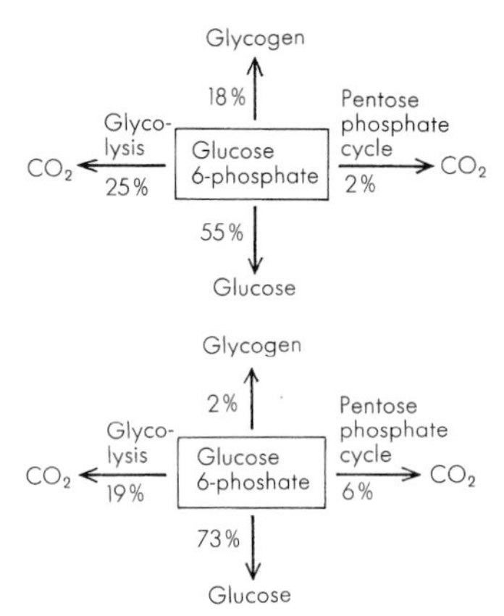

Fig. 1. *Turnover of glucose 6-phosphate.* Above: under normal conditions. Below:in diabetes mellitus.

Under certain conditions, carbohydrates may be resynthesized from degradation products of C.m. Starting materials for this Gluconeogenesis (see) are lactate and glucogenic amino acids.

There are 3 different phases of C.m.

1. Mobilization, in which poly-, oligo- and disaccharides are cleaved and phosphorylated to hexose phosphates, particularly to glucose 6-phosphate. In digestion, cleavage occurs by hydrolysis.

2. Interconverions. The mutual transformations of monosaccharides involve the following 3 types of reactions: a) epimerization, in which the steric arrangement on a C-atom is reversed by epimerases (e.g. in galactose metabolism); b) isomerization, in which aldoses and ketoses are reversibly interconverted by

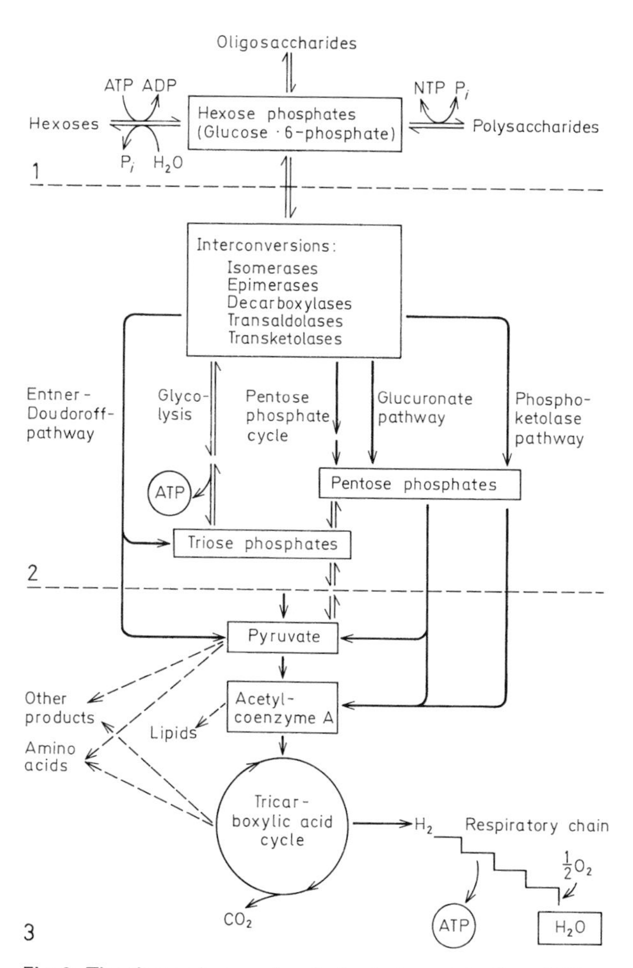

Fig. 2. *The three phases of carbohydrate metabolism.* NTP = nucleoside triphosphate

isomerases (e.g. conversion of glyceraldehyde 3-phosphate into dihydroxyacetone phosphate); c) transfer of C_3 (see Transaldolation) and C_2 fragments in the form of a dihydroxyacetone phosphate residue or "active glycolaldehyde" (see Transketolation); d) oxidation of an aldose to an acid and its subsequent decarboxylation. In this second phase of C.m., the intermediary products of the first phase are incompletely degraded (the main products are triose phosphates). About a third of the potential free energy content is released and partly used for ATP synthesis.

3. Amphibolic reaction chains. The degradation products of C.m. enter general metabolism as pyruvate and acetyl-CoA (Fig. 2).

Biosynthesis in C.m. Plants are the main producers of carbohydrates in nature. In photosynthesis, a series of enzymatic reactions (the "dark reactions" of photosynthesis) produces phosphorylated monosaccharide derivatives, which can be hydrolysed to free sugars or converted to Nucleoside diphosphate sugars (see).

In the biosynthesis of polysaccharides like starch, glycogen and cellulose, monosaccharides are activated by conversion to nucleotide derivatives, from which monosaccharide units are transferred to the nonactivated, growing end of the polysaccharide chain. The sugar-phosphate residue is removed by cleavage during the coupling reaction. Biosynthesis depends on the presence of a highly polymerized starter molecule. Uridine diphosphate glucose (UDP-glucose) and adenosine diphosphate glucose (ADP-glucose) provide activated sugars in the biosynthesis of

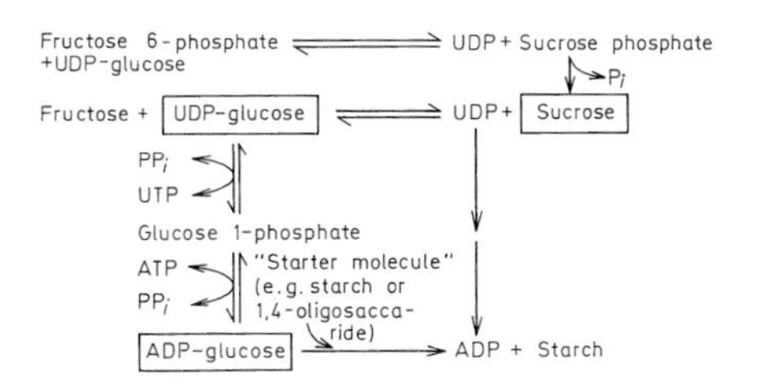

Fig. 3. *Reactions of starch and sucrose synthesis*

the disaccharide sucrose, and the polysaccharide starch (Fig. 3). UDP-glucose is probably the main precursor in the synthesis of the 1,4-glucosyl chain of cellulose, but guanosine diphosphate glucose (GDP-glucose) may also serve as a precursor.

Oligo- and polysaccharides are degraded in organisms by specific enzymatic hydrolysis (by hydrolases) or phosphorolysis (by phosphorylases).

Regulation of C. m. is characterized by tight interactions between the individual metabolic pathways, mediated by the metabolic products. Glycolysis is controlled by allosteric regulation (by the ATP/ADP ratio) of the enzymes Phosphofructokinase (see) and pyruvate kinase. Increased production of ATP by the respiratory chain leads to inhibition of phosphofructokinase (see Pasteur effect). On the other hand, an increased consumption of ATP stimulates the glycolytic turnover of glucose 6-phosphate, because the increased level of ADP stimulates phosphofructokinase. This in turn causes increased ATP formation by substrate level phosphorylation. Another regulatory quantity of this sytem is the amount of available oxygen (see Pasteur Effect). Glucose 6-phosphate activates glycogen synthetase, so that a high concentration of glucose 6-phosphate leads to increased production of the storage carbohydrate, glycogen. Intermediates of the pentose phosphate cycle decrease the rate of glycolysis by inhibiting the early enzyme of glycolysis, phosphoglucose isomerase. The NADPH produced in larger amounts by an active pentose phosphate cycle is used, inter alia, in fatty acid synthesis. Excessive fat synthesis is prevented by inhibition of glucose 6-phosphate dehydrogenase (key enzyme of the pentose phosphate cycle) by long-chain acyl-CoA compounds.

Carbohydrates: a large class of natural products, consisting of polyhydroxycarbonyl compounds and their derivatives. In general they have the composition $(C)_n(H_2O)_n$. They were originally characterized as hydrated forms of carbon and were named C. in 1844 by K. Schmidt. The name is retained, although it is chemically inaccurate, and compounds are now included that have other elemental compositions, e.g. aldonic acids, uronic acids and deoxysugars, or which contain additional elements, e.g. amino sugars, mucopolysaccharides. Mono- and oligosaccharides are also called *sugars.* Individual C. have trivial or systematic names ending in "-ose", e.g. glucose, fructose. The IUPAC Commission on the Nomenclature of Organic Chemistry and The IUPAC-IUB Commission on Biochemical Nomenclature published tentative rules for C. nomenclature in 1969.

C. are present in every plant or animal cell, and make up the largest portion by mass of organic compounds present on earth. They are formed in plants in the course of assimilation processes. Together with fats and proteins, they are the organic nutrients for animals including humans. C. are subdivided according to molecular size.

1. ***Monosaccharides*** (simple C.) cannot be further hydrolysed into simpler types of C. They can be formally regarded as primary oxidation products of aliphatic polyalcohols, usually with unbranched carbon chains. Monosaccharides carrying a terminal alde-

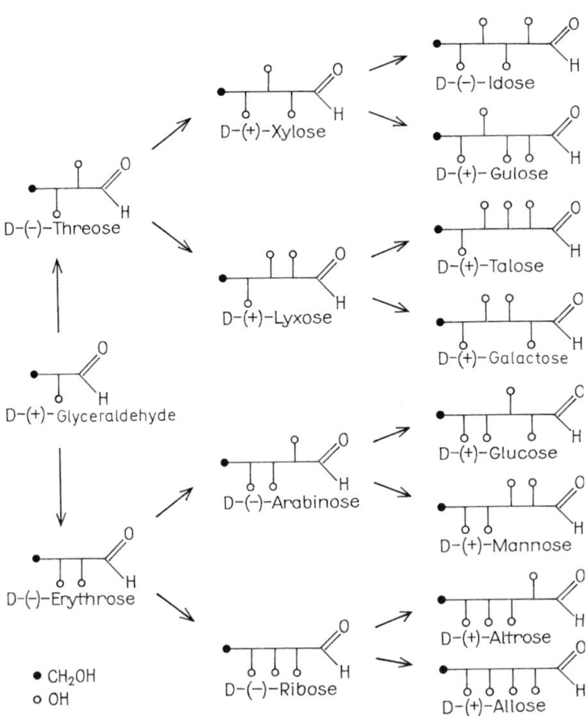

Fig. 1. *Schematic representation of the D-aldoses*

hyde group, i.e. polyhydroxy-aldehydes, are called *aldoses* (Fig. 1), whereas polyhydroxyketones (the keto group is usually on C2) are called *ketoses.*

a) *Structure.* Nearly all naturally occurring monosaccharides have unbranched carbon chains; some exceptions are hamamelose, apiose, streptose. Configuration is indicated by the prefix D or L (Fig. 2); this is not the optical rotation, which is indicated by (+) or (–) according to Wohl and Freudenberg, e.g. (+)-D-glucose. According to Rosanow, Wohl and Freudenberg, a D-sugar has the same configuration as D-glyceraldehyde, which was chosen arbitrarily as the reference compound. In this system, the asymmetric C-atom farthest from the carbonyl group has its hydroxyl group on the right in the Fischer projection; in an L-sugar the corresponding hydroxyl group is on the left. In general each C-atom carries a hydroxyl function or a derived functional group. Replacement of a hydroxyl group by hydrogen yields a deoxysugar; replacement by an amino group yields an amino sugar. C. usually possess a number of asymmetric centers. The number of stereoisomeric forms of a monosaccharide is given by the formula 2^n, where n is the number of asymmetric C-atoms.

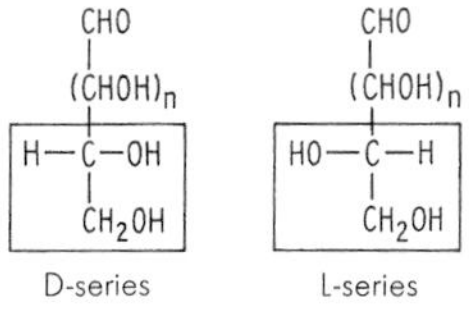

Fig. 2. *Fischer projection of D- and L-carbohydrates*

In Fischer projections (Fig. 3a), formulas are written vertically as chains, with the aldehyde group at the top and the hydroxymethyl group at the bottom. This representation is easy to comprehend, but it does not indicate the spatial structure. Furthermore, the open-chain form does not account for all the properties of a monosaccharide, e.g. sodium hydrogen sulfite or ammonia do not react with the aldehyde group of an aldose.

According to Tollens, this is because monosaccharides do not exist as open-chain structures (or only a small fraction of molecules are in this form). Instead, the carbonyl group forms a half-acetal bond with one of the hydroxyl groups, so that the molecule is an oxygen-containing ring. If the half-acetal bond links C1 to C4, the resulting 5-membered ring is a furanose; a 6-membered ring (C1 to C5) is called a pyranose. Most monosaccharides are pyranoses, but the furanose form is present in some oligosaccharides, e.g. sucrose, and in a few polysaccharides, e.g. arabans. The cyclic half-acetal form may be indicated according to Tollens (Fig. 3b), or better and more comprehensibly according to Haworth (Fig. 3c). Ring carbons may be omitted in the perspective representation, in which the heavy line between ring atoms indicates the side in the foreground. Substituents are drawn perpendicular to the plane of the ring. The furanose ring is nearly planar, but the pyranose ring is puckered because its C–O–C angle of 111° is very close to the tetrahedral angle of 109°28′.

The spatial relationships are therefore similar to those in cyclohexane. Due to the asymmetry caused by the hetero-atom, 2 chair and 6 boat forms are possible. However, pyranoses (e.g. D-glucose, D-galactose and D-mannose) spend most of their time in the energetically favorable chair form (Fig. 4).

Of the 10 substituents on the 5 ring atoms, 5 are axial and 5 equatorial, e.g. in β-D-glucose, all hydroxyl

Fig. 4. *Chair form of the pyranoses*

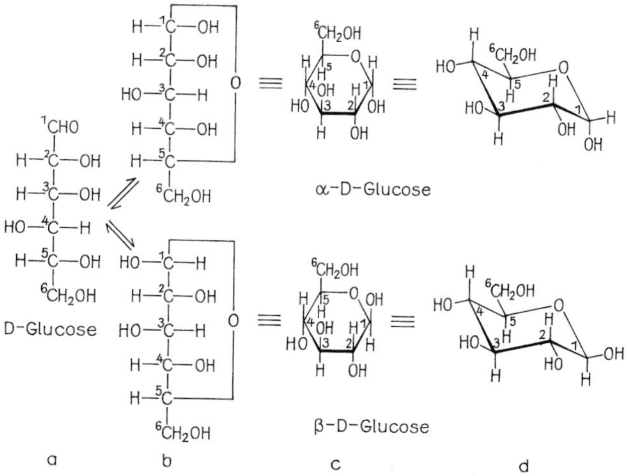

Fig. 3. *Conventions for structural representations of carbohydrates.*
a according to Fischer; *b* according toTollens; *c* according to Haworth; *d* chair conformation.

groups and the hydroxymethyl group are equatorial. The conformation formulas in Fig. 3d provide the best expression of the spatial arrangements of substituents, and enable a better understanding of the chemical, biochemical and physical properties of the molecule.

Ring formation creates a new asymmetric center on the atom originally carrying the carbonyl group (C1 in aldoses, C2 in ketoses), and thus 2 new series of isomers, called the α- and β-forms. These differ with respect to solubility, melting point, optical rotation, etc. According to Hudson, the diastereoisomer in the D-series with the more highly positive optical rotation is the α-form, while the more levorotatory form is the β-form; the reverse holds for the L-series. In Tollens' conformational formulas the hydroxyl group on C1 in the D-series is on the right side in the α-form and on the left side in the β-form. In the Haworth projection the corresponding hydroxyl group points down in the α-form and up in the β-form. In the L-series, hydroxyl groups have the opposite arrangement. The same holds for conformational formulas. It can be seen that in the α-form the hydroxyl groups on C1 and C2 are *cis* to each other, whereas in the β-form they are *trans*.

In solution the α and β isomers are in equilibrium via the open-chain form (oxo-cyclo tautomerism). This phenomenon, known as mutarotation, is based on the attainment of equilibrium between the α and β forms; thus, the optical rotation of a freshly prepared aqueous solution of a sugar changes until it reaches a constant value.

If 2 diasteromers differ only in the configuration about C1, they are called anomers, e.g. α- and β-glucose. Epimers are diastereomeric monosaccharides which have opposite configurations of a hydroxyl group at only one position, e.g. D-glucose and D-mannose at C2, or D-galactose and D-glucose at C4.

b) *Reactions.* The chemical properties of monosaccharides are due to the presence of reactive keto or aldehyde groups and the alcoholic hydroxyl groups. Aldoses give osazones with excess phenylhydrazine, while ketoses give oximes with excess hydroxylamine. Mild oxidation leads to aldonic acids, and stronger oxidation to aldaric acids. Oxidation of aldoses in which the sensitive carbonyl function is protected (by glycoside formation) yields uronic acids. Reduction of an aldose with uptake of 2 hydrogen atoms produces an alditol (Fig. 5). The hydroxyl group adjacent to the acetal bond (the glycosidic hydroxyl) is especially reactive, reacting with OH, NH or SH groups to form glycosides. The alcoholic hydroxyls

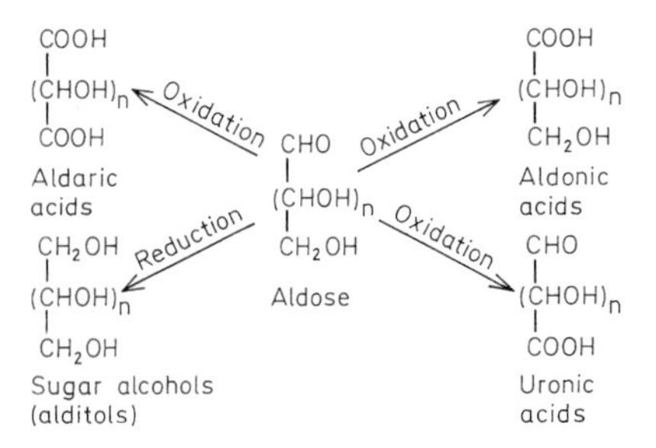

Fig. 5. *Types of carbohydrate*

can form esters or ethers. Sugar anhydrides are formed by intramolecular exclusion of water.

c) *Detection and determination.* Monosaccharides can be isolated and identified by chromatographic techniques, including paper, thin-layer, column and gas chromatography. The most modern techniques employ gas chromatography or high performance liquid chromatography in combination with mass spectroscopy. Many older methods of detection are based on color reactions with phenols, e.g. formation of a violet color with naphthol or a green color with resorcinol. When heated, pentoses lose water to form furfural, while hexoses form 5-hydroxymethylfurfural. The latter decomposes into levulinic acid, taking up water in the process. Furfural and levulinic acid condense easily with phenols to form dyes, which can be used for identification or quantitation. The ability of yeast to ferment a sugar constitutes an important biochemical test; since pentoses are generally not fermentable, they can be differentiated from hexoses. Reduction of metal salt solutions has long been a laboratory test for ***Reducing sugars*** (see).

d) *Occurrence.* Monosaccharides occur naturally in the free form, in particular D-glucose and D-fructose, and as components of many oligo- and polysaccharides. More than 100 naturally occurring monosaccharides are known. In the form of their phosphate esters, some play a crucial role in intermediary metabolism (see Carbohydrate metabolism, Glycolysis).

2. ***Oligosaccharides*** consist of between two and ten α- or β-glycosidically linked monosaccharides. They are classified as di-, tri-, tetrasaccharides, etc., according to the number of subunits. They can be hydrolysed by acid or enzymes into their subunits, which they resemble in physical and chemical properties. Oligosaccharides are widespread in plants and animals, occurring free or in bound form. They are biosynthesized from nucleoside diphosphate sugars. Certain disaccharides, which are composed of 2 glycosidically linked monosaccharide units of the same or different types, are particularly abundant. According to the nature of the glycosidic bond, they are classified as trehalose or maltose types. In the trehalose type (e.g. sucrose, trehalose), the hydroxyl groups on the C1 atoms of the two monosaccharides are linked with formal exclusion of a water molecule. Due to this 1,1 linkage, both monomeric units are full acetals, so that the disaccharide does not react like its component monosaccharides, e.g. it is not a reducing sugar, does not form a hydrazone or oxime, and does not display mutarotation. In the maltose type (e.g maltose, cellobiose, gentiobiose, melibiose), the glycosidic hydroxyl group of one monosaccharide is linked to an alcoholic hydroxyl group (usually on C4 or C6) of a second monosaccharide. Since there is a free reducing group, such disaccharides are reducing sugars, they display mutarotation, and they form hydrazones and oximes.

Some other oligosaccharides are the trisaccharides raffinose, gentianose and melezitose, the tetrasaccharide stachyose, and the pentasaccharide verbascose.

3. ***Polysaccharides.*** Numerous polysaccharides occur widely in plants and animals, serving as reserve substances or skeletal material. They consist of 10 or more monosaccharide units, linked α- or β-glycosidi-

cally according to the same structural principles as in oligosaccharides, forming branched or unbranched chains. The chains may be linear, or they may form spirals or spheres. The most common component monosaccharides are the hexoses D-glucose, D-fructose, D-galactose and D-mannose, the pentoses D-arabinose and D-xylose, and the amino sugar D-glucosamine. Polysaccharides containing only one type of unit are called homoglycans, those composed of different types of units, heteroglycans. Polysaccharides generally contain hundreds or thousands of monosaccharide units and have very high molecular masses. Polysaccharides differ not only in the types of units they contain, but also in their degree of polymerization and the nature of their glycosidic linkages. Their chemical and physical properties differ from those of their mono- or oligosaccharide components. Water solubility, reductive capacity and sweetness diminish with increasing molecular size. Skeletal C., e.g. cellulose and chitin are water-insoluble and resist enzymatic degradation. In contrast, reserve C. like starch, lichenin and glycogen are colloidally soluble in water and are more easily hydrolysed by enzymes. Acid hydrolysis of polysaccharides produces first oligosaccharides, then the monosaccharide subunits. Polysaccharides are not hydrolysed by yeast. They are optically active in colloidal solution and generally have a microcrystalline structure.

Carboline: see Indole alkaloids (Table).

Carbomycin: see Macrolide antibiotics.

Carbon dioxide assimilation, ***carbon fixation:*** the incorporation of CO_2 into larger organic compounds. C. d. a. is sometimes used as a synonym for photosynthesis, but strictly speaking photosynthesis is a series of reactions that generate the ATP and NAD(P)H required to drive the reactions of C. d. a. By far the greatest amount of C. d. a. occurs in green plants and cyanobacteria. Most of this occurs in the Calvin cycle (see), and a lesser amount by the Hatch-Slack-Kortschack cycle (see). C. can also be driven by chemical energy in autotrophic or chemoautotrophic organisms, in particular bacteria. Some of these use a process similar to the Calvin cycle. The methanogenic bacterium, *Methanobacterium thermoautotrophicum,* however, fixes CO_2 by a series of 5 major reactions which do not include the Calvin cycle (Fig.). There is also anaplerotic CO_2 fixation (see Carboxylation, Biotin enzymes).

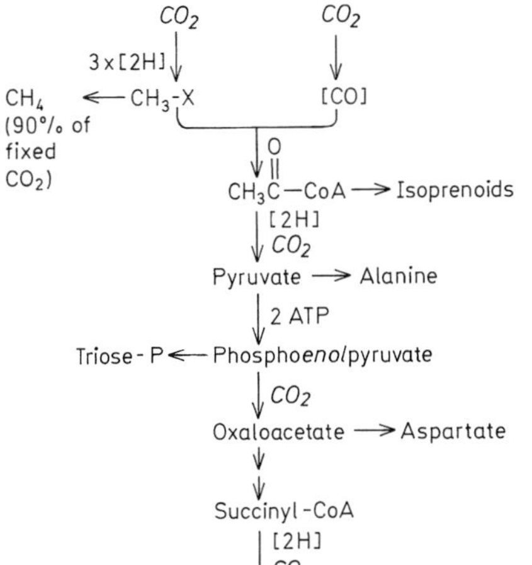

CO_2 —(3×[2H])→ CH_3-X → CH_4 (90% of fixed CO_2)
CO_2 → [CO]
CH_3-X + [CO] → CH_3C(=O)–CoA → Isoprenoids
CH_3C(=O)–CoA —([2H], CO_2)→ Pyruvate → Alanine
Pyruvate —(2 ATP)→ Phospho*enol*pyruvate → Triose-P
Phospho*enol*pyruvate —(CO_2)→ Oxaloacetate → Aspartate
Oxaloacetate → → Succinyl-CoA
Succinyl-CoA —([2H], CO_2)→ 2-Oxoglutaryl-CoA → Glutamate

Carbon dioxide assimilation in Methanobacterium thermoautotrophicum [adapted from M. Rühlmann et al. *Arch. Microbiol. 141* (1985) 399–406]

Carbonic acid anhydrase, carboanhydrase, carbonate dehydratase (EC 4.2.1.1): a widely occurring, zinc-containing enzyme, which is monomeric in most organs and organisms. It catalyses the reversible hydration of carbon dioxide ($CO_2 + H_2O \rightleftharpoons H^+ + HCO_3^-$) with one of the largest known turnover rates. C. a. a. also weakly catalyses hydration of aldehydes, and displays weak esterase activity. As a regulator of acid-base balance, C. a. a. plays an important role in respiration, CO_2 transport, and other physiological processes where the rapid conversion of CO_2 to hydrogen carbonate is vital. High concentrations of C. a. a. are found in erythrocytes, the mucous membrane of the stomach, the kidneys and eye lenses, as well as the gills, digestive glands and swim bladders of fish. It has also been found in all classes of nonvertebrates, higher and lower plants and bacteria. The most extensively studied erythrocyte C. a. a. is that of human origin, which exists as 3 isoenzymes: A, B and C. The A and B forms are very similar. B (256 amino acids, M_r 28,000) and C (259 amino acids, M_r 28,500) differ markedly with respect to primary sequence, chain conformation and catalytic activity. With the exception of the C. a. a. from parsley (M_r 180,000, 6 subunits of M_r 29,000, one Zn atom each), known C. a. a. consist of a single polypeptide chain (M_r 28,000–30,000) and contain one Zn atom essential for catalytic activity. The Zn is located in a hydrophobic pocket where 3 of its 4 ligand sites are occupied by histidine residues. Other characteristics common to the B and C forms are the absence of disulfide bridges and the relatively large amount of β-structure in the protein conformation, e. g. the C form contains 37 % β-structure and 20 % α-helix. Of the many inhibitors of C. a. a., the monovalent sulfide, cyanide and cyanate ions are the strongest (e. g. K_i for sulfide ions is 2×10^{-6} M). Of the sulfonamides, acetazolamide is the inhibitor most frequently used for therapeutic purposes (for treatment of glaucoma). [W. S. Sly & P. Y. Hu 'Human Carbonic Anhydrase and Carbonic Anhydrase Deficiencies' *Annu. Rev. Biochem.* **64** (1995) 375–401]

Carbon monoxide, ***CO:*** a colorless and practically odorless, combustible gas. CO is extremely poisonous, because its affinity for hemoglobin (Hb) is 300 times that of oxygen. The CO displaces O_2 from hemoglobin, which is then unable to transport oxygen to the tissues. The reaction between Hb and CO is reversible: $HbO_2 + CO \rightleftharpoons HbCO + O_2$. Only a large excess of O_2 can displace CO from Hb. Death by CO poisoning occurs after the following series of events: 1. the transport capacity of the blood for oxygen is decreased by formation of HbCO; 2. intoxication of oxygen-sensitive tissues occurs, especially in the brain (symptom: headache); 3. the respiratory center in the brain is incapacitated (symptom: unconsciousness);

4. the heart stops beating due to inadequate oxygen supply. CO can only enter the body through the lung alveoli. Concentrations >0.01 % are considered toxic. The maximal allowable concentration at the workplace is 55 mg CO/m^3 air.

CO is formed endogenously (mainly in spleen, liver and kidneys) during the degradation of protoheme (from hemoglobin and other hemoproteins), catalysed by microsomal heme oxygenase (decyclizing) (EC 1.14.99.3). In this process, the porphyrin ring system of protoheme is converted into the linear porphyrin structure of biliverdin, by the oxidation and release of an α-methene bridge carbon as CO.

Carbonyl cyanide-*p*-trifluoromethoxyphenylhydrazone, *FCCP*: an uncoupler of oxidative phosphorylation. For structure and mode of action, see Ionophore.

Carboxydismutase: see Ribulose *bis*phosphate carboxylase.

4-Carboxyglutamic acid, *γ-carboxyglutamic acid, Gla*: an amino acid residue found in certain proteins, and formed by posttranslational carboxylation of glutamate residues (see 4-Glutamyl carboxylase). It is present in the blood clotting factors, prothrombin, factor VII, factor X and factor IX, in low M_r protein isolated from the bones of several vertebrates (bovine bone protein M_r 6,800 contains 3 residues of Gla), in the calcium-binding protein from chick embryonic chorioallantoic membrane, in a plasma protein (protein C) not involved in blood coagulation, and in protein of the kidney cortex. The doubly charged side chain of Gla acts as a chelator of Ca^{2+}; many of the proteins known to contain Gla bind and/or transport Ca^{2+}.

Carboxylase: 1. see Carboxylation. 2. obsolete term for pyruvate decarboxylase.

Carboxylation: transfer of carbon dioxide, frequently in activated form. The C. of pyruvate to dicarboxylic acids (Table) was discovered by Wood and Werkman as a balanced reaction ($C_3 + C_1 = C_4$), and serves to renew the pool of oxaloacetate (anaplerotic CO_2 fixation). Oxaloacetate is used in various syntheses (see Tricarboxylic acid cycle). The C. of pyruvate occurs in 2 ways: 1. direct addition of "activated CO_2" (Wood-Werkman reaction) and 2. reductive C. by the "malic" enzymes (EC 1.1.1.38, 1.1.1.39 and 1.1.1.40) (Ochoa). In the latter process, pyruvate reduction by NAD(P)H and CO_2 fixation occur in a single step. The product is malate, which can then be converted into oxaloacetate.

Photosynthetic fixation of CO_2 is an important form of C. (see Calvin cycle), and C. also occurs in fatty acid and purine metabolism (Fig.). C. of 5-aminoimidazole-ribotide to 5-aminoimidazole-carboxamide-ribotide does not require preliminary activation of the CO_2.

Carboxyl carrier protein: see Biotin enzymes.

Carboxylic acid esterases: see Esterases.

Carboxylic acids: organic compounds containing the carboxyl group, COOH. Alkane and alkene monocarboxylic acids are also called fatty acids. In aqueous solution, C.a. dissociate into hydroxonium (H_3O^+) and carboxylate ions ($R\text{-}COO^-$). C.a. and their derivatives, esters and amides, are metabolically important. An ester is formally equivalent to a condensation product of an alcohol with a C.a. Although under physiological conditions the equilibrium favors the free components, esters such as acetyl-CoA and fats are present in large amounts in living cells. Amides are formed by replacing the OH group of the COOH with an NH_2 group. They are biosynthesized from activated acid derivatives, such as anhydrides or thioesters (e.g. acetyl-CoA). Other C.a. derivatives are Hydroxyacids (see) and Ketoacids (see). At physiological pH values, C.a. are present largely as anions.

Carboxylic esterases: 1. a subgroup of esterases acting on carboxylic esters, e.g. lipases. 2. Carboxylesterase (EC 3.1.1.1), a class of enzymes with wide specificity, usually for a short-chain acid and an alcohol

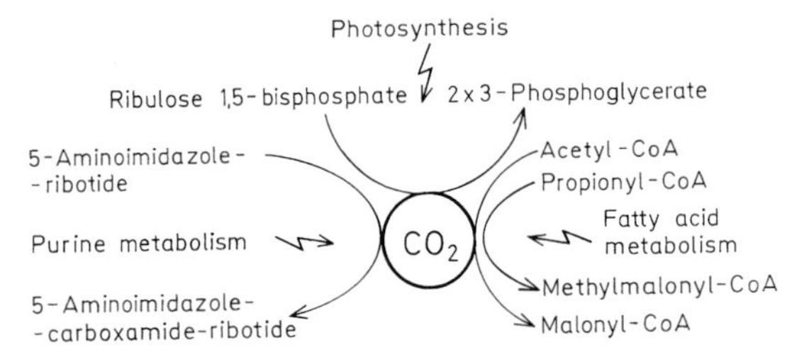

Metabolic carboxylations

Carboxylation reactions synthesizing oxaloacetate or malate

Enzyme	Reaction
Phosph*enol*pyruvatecarboxylase (EC 4.1. 1.31)	Phospho*enol*pyruvate + CO_2 → Oxaloacetate + P_i
Pyruvate carboxylase (EC 6.4. 1.1)	Pyruvate + CO_2 + ATP + $H_2O \xrightarrow{Mn^{2+}}$ Oxaloacetate + ADP + P_i
Malate dehydrogenase (decarboxylating) NADP) (EC 1.1. 1.39) (malic enzyme)	Pyruvate + CO_2 + NAD(P)H + $H^+ \rightleftharpoons$ Malate + NAD(P)

with only one OH group; these differ from lipases in that they are not activated by Ca^{2+} or taurocholate. This second group is widely distributed in vertebrate tissues, blood, serum and digestive juices, in the digestive juices of arthropods and mollusks, and in plant seeds, citrus fruits, mycobacteria and fungi. In mammals the highest activities of C. e. are found in the liver, kidneys, duodenum and in the testes and epididymis. The best studied liver and kidney C. e. are localized in the microsomes after cell disruption. In addition, the lysosomes of these organs contain an arylesterase with an acidic pH optimum. The physiological significance of these enzymes lies in their inactivation of pharmacological esters or amides (e. g. atropine, phenacetin), and their activity as aminoacyl group transferases.

Carboxyl transferases: see Biotin enzymes.

Carboxypeptidases: single-chain exopeptidases which contain zinc. They remove successive amino acids from the *C*-terminal ends of proteins. Animal C. contribute to protein digestion in the small intestine; their inactive precursors (zymogens) are secreted by the pancreas into the small intestine, then activated by trypsin. They are classified according to substrate specificity into 2 groups, C.A and C. B. C.A preferentially hydrolyse amino acids with aromatic or branched aliphatic side chains, whereas C.B attack peptide bonds in which the NH group belongs to a lysine or arginine residue. The structure and mechanism of bovine C.A have been extensively studied. The Zn atom is located in a pocket-like active center and participates in catalysis. The molecule contains 8 hydrophobic segments with β-structure, which account for 20 % of the amino acid residues. A further 35 % compose surface α-helices, while 20 % are part of random coil areas.

Pro-C.A (M_r 87,000) is converted into a main active C.A (M_r 34,409, 307 amino acids, known primary and tertiary structures), together with the active forms, C.Aβ (305 amino acids), C.Aγ and C.Aδ (300 amino acids each), formed by further enzymatic shortening of the amino terminal end. C.A and C.B (M_r 34,000, 300 amino acids, M_r of zymogen 57,400) have high esterase activity in addition to peptidase activity.

Other less specific C. have been isolated from citrus fruits and leaves (C.C), yeast (C.Y), *Pseudomonas* (M_r 92,000, two polypeptide chains), *Aspergillus* species, germinating barley and cotton seeds. Unlike C.A and C.B, these less specific C. are inhibited by the serine protease inhibitor, diisopropylfluorophosphate.

Carcinogen: see Cancer.

Cardenolides: see Cardiac glycosides.

Cardiac glycosides: poisonous vegetable glycosides with specific cardiac effects, and related compounds from animals, e. g. Toad toxins (see). The aglycones are steroids possessing an unsaturated lactone ring at position 17β, C/D-*cis* ring coupling, and a 14β-hydroxyl group. In addition to a 3β-hydroxyl, which is always present, there may be other oxygen substituents in positions 1, 11, 12, 16 and 19 (see Steroids, Fig. 2). C. g. with a 5-membered lactone ring are classified as cardenolides, e. g. Strophanthins (see) and *Digitalis* glycosides (see), while those with a 6-membered lactone ring are bufodienolides (bufogenins), e. g. Scillaren A (see). The sugar component is linked to the 3β-hydroxyl group. In addition to D-glucose, the sugar component is often a 2-deoxy sugar, e. g. D-digitoxose. C. g. have been found in 12 plant families, including *Apocyanaceae, Scrophulariaceae, Liliaceae, Moraceae* and *Ranunculaceae.* Some toad toxins, e. g. bufotoxin, are also bufadienolides. C. g. have also been found in insects: the grasshopper *Poekilocerus bufonius* and the butterfly *Danaus plexippus* acquire them from their food plants. Pregnenolone is a precursor in C. g. biosynthesis.

In therapeutic doses, C. g. have long been used to strengthen the contraction of heart muscle. Higher doses are toxic. They bind to specific membrane receptors, which are thought to be part of the Na^+/K^+-ATPase complex. Endogenous compounds binding to the same receptors ("ouabain-like compounds" or OLC) have been purified from several sources. [Y. Shimoni et al. *Nature* **307** (1984) 369–371; F. Abe et al. 'Presence of Cardenolides and Ursolic Acid from Oleander leaves in Larvae and Frass of *Daphnis nerii*' *Phytochemistry* **42** (1996) 51–60]

Cardiolipins: see Membrane lipids.

Cardiotoxins: in the widest sense, substances which, in toxic doses, cause heart damage and may lead to heart stoppage. They may interfere with the generation or conduction of stimuli or with the heart's own blood supply, or they may directly attack the heart muscle. In the narrower sense, they include the cardiac glycosides and their aglycones from plants, and a group of toad toxins.

Carminic acid: a red glucoside pigment (m. p. 130 °C) from the scale insect *Coccus cacti* L. which lives in Central American cacti of the genus *Opuntia.* C. a. is the principal component of cochineal, which was formerly a highly prized dye for wool and silk. Carmine is an insoluble complex of C. a. with alkaline earth or heavy metals, such as zinc. Carmine dyes are used in cosmetics, artist's colors, inks and food colors.

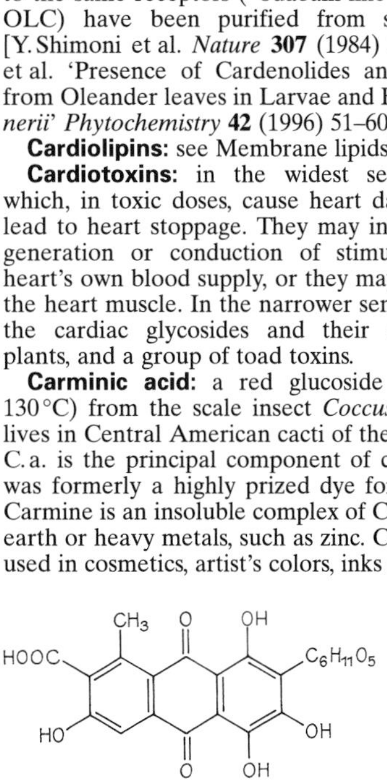

Carminic acid

Carnitine, *vitamin B_T*: a carrier of acetyl and acyl groups through the mitochondrial membrane (see Fatty acid degradation). C. is characteristically present in muscle tissue, and has been isolated in crystalline form from yeast. Mammals can synthesize C., but insects must obtain it in their diet, otherwise they are unable to complete metamorphosis. C. is biosynthesized from lysine and γ-butyrobetaine (Fig.) and degraded via glycine betaine and 2-methylcholine.

Carnosine: see Peptides.

Carotenes: isomeric unsaturated hydrocarbons with 9 conjugated *trans* double bonds, 4 branch methyl groups and usually a β-ionone ring at one end. Isomers differ in the arrangement at the other end of the chain (Fig.). C. with β-ionone rings are oxidatively cleaved to vitamin A in the wall of the small intestine, and are therefore important as provitamins. In this

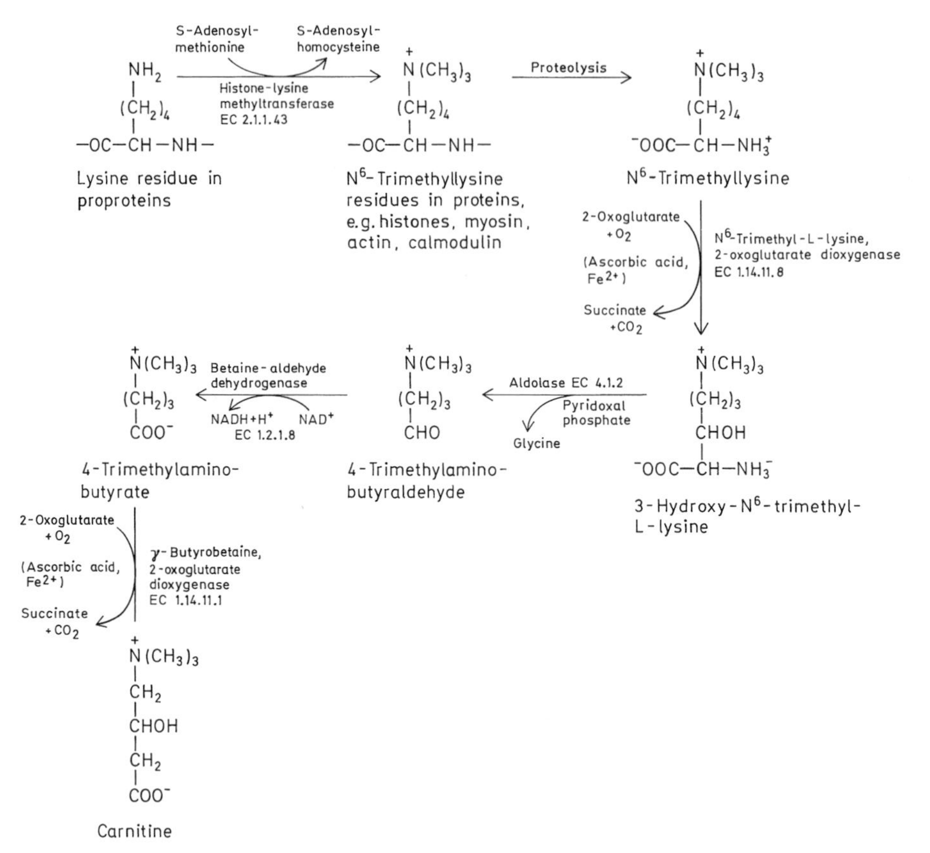

Biosynthesis of carnitine. γ-Butyrobetaine, 2-oxoglutarate dioxygenase (EC 1.14.11.1) occurs in liver, kidney and brain, whereas the other enzymes occur in most tissues. The aldolase cleaving 3-hydroxy-N^6-trimethyl-L-lysine may be identical with glycine hydroxymethyltransferase (EC 2.1.2.1). [W. A. Dunn et al. *J. Biol. Chem. 259* (1984) 10764–10770]

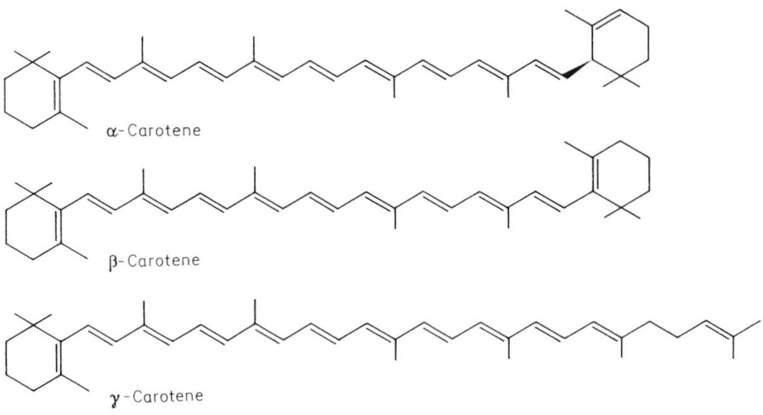

α-Carotene

β-Carotene

γ-Carotene

process, β-C. is cleaved to 2 molecules of vitamin A, whereas the α and γ isomers each yield 1 molecule of vitamin A (see Vitamins).

(+)-α-C., M_r 536.85, m.p. 188 °C (evacuated tube), $[\alpha]_D$ +385° (c = 0.08 in benzene), λ_{max} ($CHCl_3$) 485 and 454 nm, has a second α-ionone ring as the second terminus. It has the *R* configuration at C′6. α-C. makes up 15 % of the C. mixture in carrots. It is as widespread as β-C. but is present in smaller amounts. Its vitamin A activity is half that of β-C.

β-C., M_r 536.85, m.p. 183 °C (evacuated tube), λ_{max} ($CHCl_3$) 497 and 466 nm, has 2 terminal β-ionone rings and is optically inactive. Its intense yellow color is due to 11 *trans* conjugated double bonds. β-C. is the most common C. in plants and is also found in bacteria, fungi and animals (milk, fat, blood, serum, etc). The C. mixture of the carrot contains about 85 % β-C. It is now produced synthetically on a large scale and used as a food, pharmaceutical and cosmetic colorant.

γ-C., M_r 536.85, m.p. 178 °C (natural material crystallized from benzene/methanol), 153.5 °C (synthetic material crystallized from light petroleum), λ_{max} 437, 462 and 494 nm, has a β-ionone ring at one end and an open chain at the other, and is optically inactive. It has been found in bacteria, fungi and various higher plants, including the carrot (about 1 % of the C. mixture) but it appears to be less widespread than the α- and β-isomers. See also Carotenoids.

Carotenoids: a large class of yellow and red pigments, which are highly unsaturated aliphatic and alicyclic hydrocarbons and their oxidation products. C. are biosynthesized from isoprene units (C_5H_8) and therefore have the methyl branches typical of isoprenoid compounds. Most C. have 40 C atoms and are thus tetraterpenes (consisting of 8 isoprene units). A few C. contain 45 and 50 C atoms, particularly in nonphotosynthetic bacteria. C. with fewer than 40 C atoms are called nor-, seco- or apocarotenoids.

C. are subdivided into: 1. carotenes, which are hydrocarbons, e.g. lycopene, α-, β- and γ-carotene, neurosporene, phytofluene and phytoene; and 2. oxygen-containing, yellow xanthophylls, e.g. violaxanthin, zeaxanthin, fucoxanthin, lutein, neoxanthin, cryptoxanthin, astaxanthin, capsanthin, capsorubin, rubixanthin and rhodoxanthin. In the native state, xanthophylls often occur as carotenoproteins (the carotenoid is bound to a protein to form a soluble chromoprotein, which stabilizes the carotenoid against the effects of air and light), or esterified to fatty acids or glucose. C. contain 9–15 (usually 9 or 11) conjugated double bonds, usually *all* trans, and are therefore planar polyenes. Their usually intense yellow to red color is conferred by this chromophore system.

Most C. contain an unsaturated C_{22} chain with methyl branches and a 9-membered unit on each end, which may be cyclic as in α- and β-carotene, or acyclic as in lycopenes. When present, the terminal rings are usually 6-membered α- and β-ionone systems. The median 22 C atoms are generally constant, and oxidations to xanthophylls usually occur on the two C_9 end sections. Most of the 700 known C. are xanthophylls, with 1, 2 or many hydroxyl groups, and ether, aldehyde, ketone or acid substituents. Stereochemical differences arise from the *cis-trans* isomerism of the C = C double bonds, and from chiral centers that may be present. The prototype C. is Lycopene (see), an acyclic C_{40} compound from which other C. can be formally derived (Fig.).

C. represent a major class of natural pigments, occurring widely in plants and animals. In animals they occur mainly in surface tissues such as skin, shells, scales, feathers and beaks, but also in birds' egg yolks and as visual pigments. C. in animals are all of plant origin, as animals are incapable of their de novo synthesis. However, animals do convert plant C. into other forms, e.g. carotenes are oxidized to vitamin A. For C. biosynthesis, see Tetraterpenes.

In natural materials, the concentration of C. is usually in the order of 0.02–0.1 % of the dry mass. The C. content of the eye ring of the pheasant *Narcissus majalis* is extremely high, being 16 %, or about 10^4 times that in carrots. An estimated 10^8 metric tonnes of C. are produced per year by living organisms, the most abundant being fucoxanthin, lutein, violaxanthin and neoxanthin, followed by β-carotene, zeaxanthin, lycopene, capsanthin and bixin. More than 90 % of the C. in a plant is found in the leaves, usually as a mixture of 20–40 % carotenes (containing more than 70 % β-carotene) and 60–80 % xanthophylls like lutein, violaxanthin, cryptoxanthin and zeaxanthin.

C. contribute to energy transfer in photosynthesis, as well as providing protection against light. They are also important as precursors of vitamin A and the visual pigments. C. are isolated from plant material by extraction with organic solvents and adsorption chromatography. Partial and total syntheses of β-carotene are carried out on an industrial scale. Some C. (in particular β-carotene) are used as food, pharmaceutical and cosmetic colorants.

Carrageenan, *carrageen:* a mixture of polysaccharides, similar to agar-agar, obtained from red algae by hot-water extraction. C. contains about 45 % carragenin, a polysaccharide containing galactose and galactose sulfate. λ-Carragenin is a branched chain compound of 3,6-anhydro-α-D-galactose and D-galactose-4-sulfate, in which 1,3 and 1,6 links alternate. λ-Carragenin is composed of α-1,3-glycosidically linked D-galactose-4-sulfate residues. C. is used as a gelling agent in food, pharmaceuticals and cosmetics.

Carrier: see Biomembrane, Metabolic cycle.

Caryophyllane: see Sesquiterpenes, Fig.

Caryophyllenes: isomeric cyclic sesquiterpene hydrocarbons found in many essential oils. α-C. is the same as Humulene (see).

β-C. (4,11,11-trimethyl-8-methylenebicyclo[7.2.0]-undeca-4-ene), M_r 204.36, b.p.$_{10}$ 123–125 °C, $[\alpha]_D$ −9°, ρ_4 0.9074, n_D^{17} 1.4988, has a *trans*-linked cyclobutane-cyclononane ring system (the double bond of the unsaturated cyclononane ring is *trans),* shows a tendency for acid-catalysed cyclization, and forms a crystalline epoxide. It is present especially in clove oil and oils

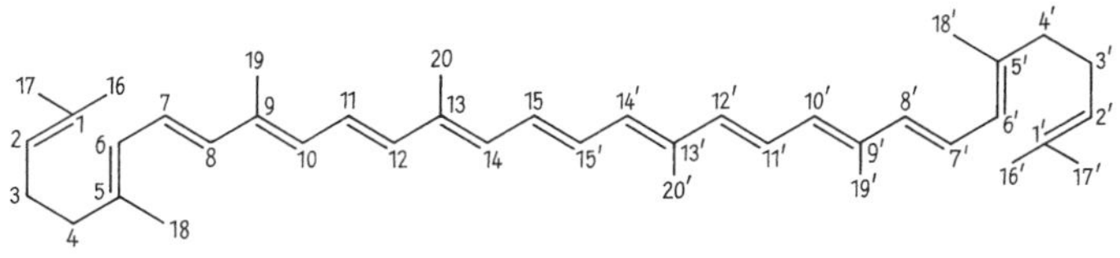

Lycopene

from the stem and flowers of *Syzygium aromaticum* (L.) Merril & Perry *(Jambrosa caryophyllus* Niedenzu; *Eugenia caryophyllata* Thunb.) *(Myrtaceae).*

γ-C. (isocaryophyllene), M_r 204.36, b.p.$_{14}$ 125 °C, $[\alpha]_D$ −26.2 °, differs from β-C. only in that the double bond of the unsaturated cyclononane ring is *cis.* For biosynthesis and structures, see Sesquiterpenes.

Caryoplasm: see Nucleus.

Caseins: see Milk proteins.

Cassaine: see Erythrophleum alkaloids.

Castaprenols: see Polyprenols.

Castoramine: see Nuphara alkaloids.

Catabolism, *dissimilation:* the sum of all degradative metabolic processes. Amino acids, purines, pyrimidines, etc., formed in the turnover of cell components, can be degraded, in some cases to inorganic compounds (water, carbon dioxide, ammonia, etc.). Catabolic processes are linked with the formation of ATP by substrate phosphorylation and respiratory chain phosphorylation. In a certain sense, C. is therefore identical to energy metabolism. Hydrolysis plays a major role in the degradation of biopolymers. The C. of carbohydrates during fermentation or respiration usually occurs by oxidation (usually dehydrogenation) followed by decarboxylation.

Catabolite repression: Inhibition of enzyme synthesis by increased concentrations of certain metabolic products. Enzymes subject to C.r. are formed in reponse to metabolic events (utilization of new nutrients by catabolic enzymes, synthesis of secondary products in certain developmental phases of microorganisms). C.r. is probably present in all organisms, although the molecular mechanisms are diverse. Examples of C.r. are glucose repression of catabolic enzymes in *E. coli* and the enzymes of secondary metabolism in microorganisms, e.g. those for the synthesis of penicillin, actinomycin and riboflavin.

A well-studied example of C.r. is the repression of β-galactosidase synthesis by glucose in *E. coli.* When β-galactosidase is induced, RNA polymerase binding is enhanced by the binding of a regulatory protein (catabolite-gene activator protein, CAP) just upstream (at a specific CAP binding site) from the promoter. In this process, cAMP acts as a signaling molecule by binding the CAP to induce a conformational change which enables it to bind to the CAP binding site. When glucose is present in excess (as well as the inducer lactose), the cAMP level is very low, so that the initiation complex at the promoter is less likely to be formed or cannot be formed. Conversely, glucose starvation induces an increase of intracellular cAMP, converting CAP to its active conformation, and enabling the bacterium to exploit alternative carbon sources. The ultimate pattern of gene expression depends on the levels of both Lac repressor protein (see) and CAP. X-ray data indicate that cAMP-CAP can bind tightly to a left-handed helix, rather than the normal right-handed helix. It is thought that by forcing the DNA of the binding site into a left-handed configuration the cAMP-CAP complex causes the nearby region to unwind, thus promoting transcription. Similar mechanisms operate for promoters of the genes encoding enzymes for the metabolism of maltose, galactose and other sugars.

Catabolite repressor protein, *CAP:* a protein controlling expression of a set of *E. coli* genes that become operative only in the absence of glucose (see Catabolite repression).

Catalase (EC 1.11.1.6): a tetrameric heme enzyme, M_r 245,000, which catalyses removal of the highly poisonous hydrogen peroxide from the cell: $H_2O_2 \rightarrow H_2O + {}^1/_2O_2$. Each subunit of C. ($M_r$ 60,000) contains a heme group in the form of ferriprotoporphyrin IX. C. is found in all animal organs, particularly liver (where it is present in peroxisomes) and erythrocytes, all plant organs and in nearly all aerobic microorganisms. The turnover number of C. (5×10^6 molecules H_2O_2 per min and C. molecule) is one of the highest known for an enzyme. At low concentrations of H_2O_2, C. behaves as a peroxidase. It is inhibited by hydrogen sulfide, hydrogen cyanide and azides, e.g. NaN_3.

Catalytic antibody: an antibody with the ability to (i) bind a particular compound with great specificity, and (ii) catalyse a reaction in which that compound is a substrate. The former ability is characteristic of both antibodies and enzymes but the latter is normally peculiar to enzymes. C.a. have both abilities and are therefore often called 'abzymes'. They are not natural products, but are generated in the laboratory by manipulation of normal physiological processes. An antibody binds very tightly to the normal, energetic ground state conformation of its antigen; in contrast, an enzyme binds relatively weakly to the normal, energetic ground state conformation of its substrate (forming an enzyme-substrate (ES) complex). On the other hand, as part of its catalytic function, an enzyme binds very tightly to the strained, relatively high-energy, transition-state conformation of its substrate. The change in substrate conformation from normal to transition state(s) (TS), following formation of the ES, is brought about by chemical groups at the active site of the enzyme, which interact electrostatically, hydrophobically or even covalently with appropriate groups in the substrate; similar interactions also bring about the subsequent conversion of the TS into the reaction products.

It was therefore conceivable that, if an antibody could be induced to bind to a molecule already in the TS conformation, it might act like an enzyme and catalyse the reaction to which the TS conformation is predisposed. The attraction of this to chemists is the prospect of creating catalysts with the specificity and efficiency of enzymes for thermodynamically-possible but kinetically-slow reactions for which there are no naturally occurring enzymes. In an enzyme-catalysed reaction the TS conformation of a substrate molecule has, by definition, a transient existence, i.e. it exists for a much shorter time than is necessary to elicit antibody formation. A further problem is that for many enzyme-catalysed reactions the structure of the TS conformation of the substrate is not known. However, it is known for some reactions, or can be guessed at with a considerable degree of certainty. For many reactions, knowledge of the TS conformation of the substrate has been provided by the discovery and design of 'TS-analog inhibitors'. In their normal, energetic ground state conformation, these inhibitors resemble the TS conformation of the substrate sufficiently closely to bind to the catalytic site almost as tightly as the TS conformation of the substrate, and many orders of magnitude more tightly than the substrate in its ground state conformation. It was

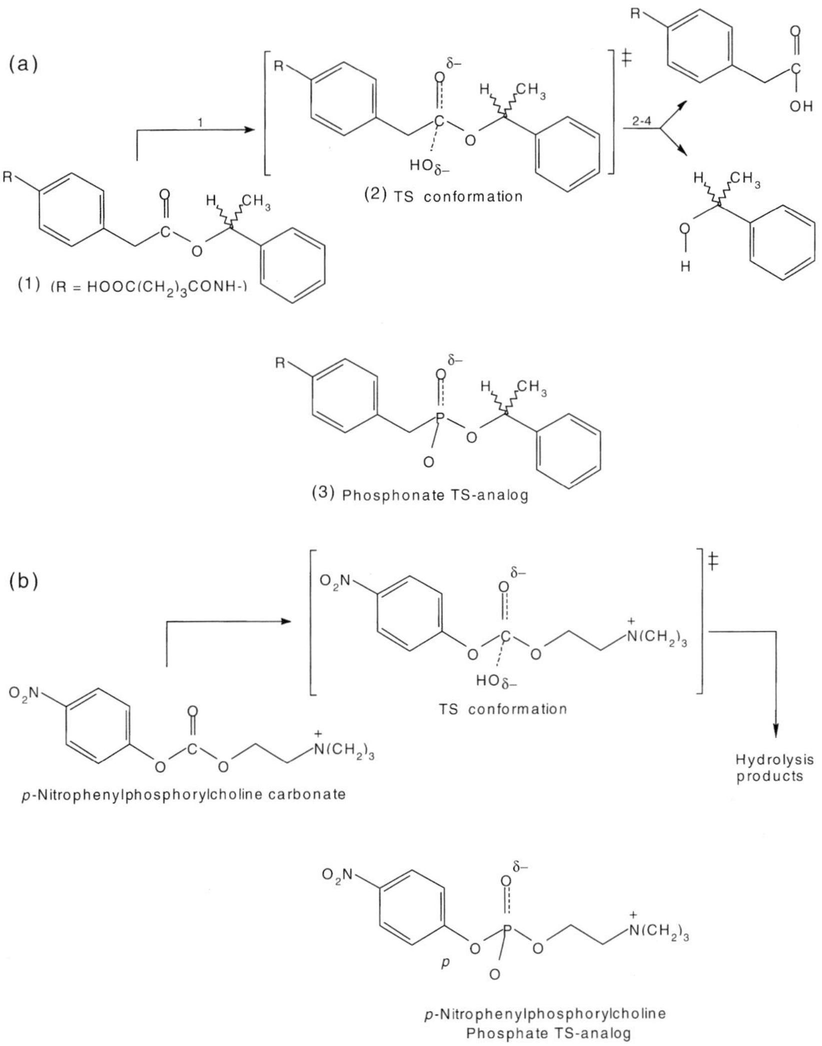

(a)

(1) (R = $HOOC(CH_2)_3CONH$-)

(2) TS conformation

(3) Phosphonate TS-analog

(b)

TS conformation

Hydrolysis products

p-Nitrophenylphosphorylcholine carbonate

p-Nitrophenylphosphorylcholine
Phosphate TS-analog

Fig. 1. ***Catalytic antibodies.*** **Examples of phosphonic- and phosphoric acid derivatives that have been used to generate catalytic antibodies for the hydrolysis of the specific ester and carbonate substrates shown. 1 & 2–4 refer to the steps in the hydrolysis of esters shown in Fig. 2; each step is effectively a group of electron shifts.**

then logical to propose that if a TS-analog of a reaction were used as an antigen, the resulting antibody might act as an enzyme for that reaction, i.e. the substrate of the reaction would be forced into its TS conformation when it bound to the antibody's binding site. This idea became practicable with the advent of monoclonal antibodies (see).

The first catalytic antibodies to be generated catalysed the hydrolysis of esters and carbonates. In both these reactions the attack of OH^- on the carbonyl carbon generates a tetrahedral TS conformation with a partial negative charge located on the carbonyl oxygen. Derivatives of phosphonic acid $[(H(HO)_2P=O]$ mimic the tetrahedral structure and charge distribution of the TS state in ester hydrolysis while derivatives of phosphoric acid $[(HO)_3P=O]$ mimic it in the hydrolysis of carbonates. Thus in Fig. 1 the phosphonate **3**, which is the structural and electronic mimic of the TS conformation (**2**) of the ester **1** (Fig. 1a), has been successfully used to generate a monoclonal antibody that catalyses the hydrolysis of the ester bond of **1** yielding 1-methylbenzyl alcohol and the phenylacetic acid derivative, and in Fig. 1b *p*-nitrophenylphosphorylcholine, the structural and electronic mimic of the TS conformation of *p*-nitrophenylphosphorylcholine carbonate, has been used

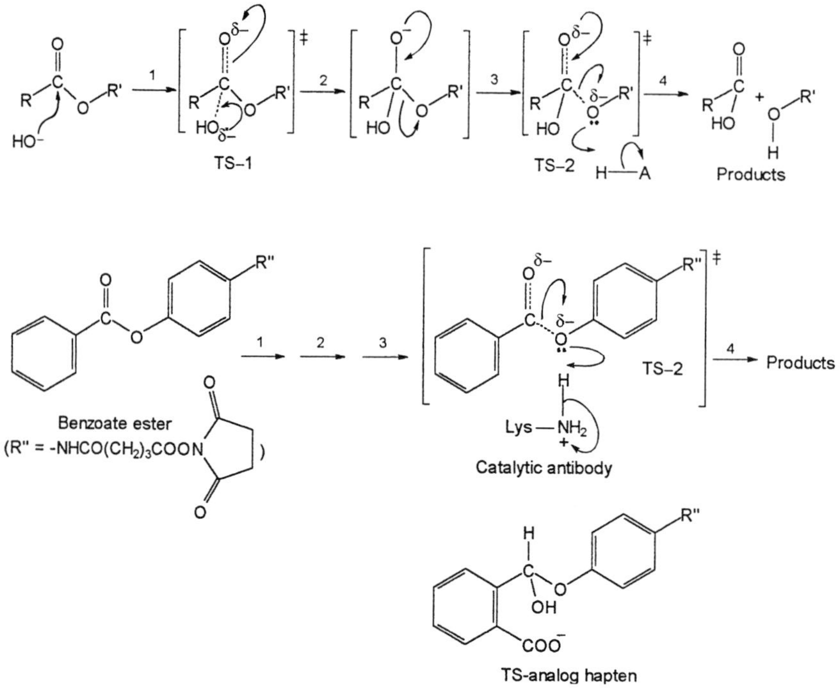

Fig. 2. *Catalytic antibodies.* Use of a TS-analog hapten with a charged group (COO^-) to generate a catalytic antibody with a complementary charge (H_3N^+) at its binding site. The final step in the hydrolysis of an ester is facilitated by the presence of an acidic group (H-A), which is close to the non-carbonyl oxygen of the ester bond, and which provides the mechanistically-required proton. In the catalytic antibody for benzoate ester cleavage, the proton required for this final step is provided by an appropriately positioned lysine (or arginine) residue, which is complementary to the -COO^- of the TS-analog hapten. This technique has been called 'antibody-hapten charge complementarity' or, more colloquially, 'bait and switch'. [K. D. Janda *J. Amer. Chem. Soc. 113* (1991) 5427–5434; S. J. Benkovic *Annu. Rev. Biochem. 61* (1992) 29–54]

to generate an abzyme that catalyses the hydrolysis of the latter.

Since 1986 abzymes capable of catalysing acyl transfer, β-elimination, C-C bond cleavage, C-C bond formation, peroxidation, porphyrin metallation and redox reactions have been described. Current work in this field is aimed at improving the catalytic efficiency of abzymes by using haptens which not only mimic the conformation of the TS of the reaction in question but also possess a charged group, whose purpose is to induce the introduction of an oppositely-charged group into the antibody, and which is so sited that it assists the conversion of TS to products by providing or abstracting a proton as required by the reaction mechanism; this is illustrated in Fig. 2 which shows that the fourth successive step in the hydrolysis of an ester is facilitated by the presence of an acidic group (H-A),which is close to the non-carbonyl oxygen of the ester bond, and which provides the mechanistically-required proton. In the abzyme for benzoate ester cleavage, the proton required for this final step is provided by an appropriately positioned lysine (or arginine) residue, which is complementary to the the analogous TS conformation (**2**) generated during hydrolysis of the benzoate ester (**1**) [K. D. Janda *J. Amer. Chem. Soc.* **113** (1991) 5427–5434]. This technique has been called 'antibody-hapten charge complementarity' or, more colloquially, 'bait and switch'. [S. J. Benkovic *Annu. Rev. Biochem.* **61** (1992) 29–54]

Catalytic center: see Active center.

Catalytic constants: see Michaelis-Menten equation.

Catechins: flavan-3-ols. (+)-Catechin and (–)-epicatechin are widespread in plants. Other C. have a more limited distribution (Fig.). C. have been implicated in the biosynthesis of condensed Tannins (see).

Catecholamines: alkylamino derivatives of pyrocatechol (*o*-dihydroxybenzene); a group of substances biosynthesized from L-tyrosine, and including the hormones Adrenalin (see), Noradrenalin (see) and Dopamine (see).

Catechol estrogens: 2-hydroxylated derivatives of estrogens. Estrogens are hydroxylated by the microsomal cytochrome P450 system of the liver. The 2-hydroxyl group is then methylated by the action of methyltransferase and *S*-adenosyl-L-methionine in the liver cytosol. The methoxy derivatives are excreted in the urine. C. e. and their methylated derivatives

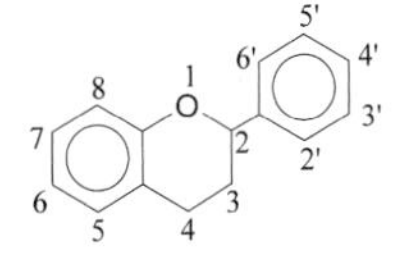

Structures of some catechins
3,5,7,3′,4′-Pentahydroxy-flavan:
(+)-*Catechin* (2*R*,3*S*); (–)-*Epicatechin* (2*R*,3*R*).
3,5,7,3′,4′,5′-Hexahydroxy-flavan:
(+)-*Gallocatechin* (2*R*,3*S*) *(Camellia sinensis);*
(–)-*Epigallocatechin* (2*R*,3*R*) *(Camellia sinensis).*
3,7,4′,5′-Tetrahydroxy-flavan:
(+)-*Fisetinidol* (2*R*,3*S*) *(Schinopsis quebracho-colorado);* (–)-*Fisetinidol* (2*R*,3*R*) (heartwood and bark of *Acacia mearnsii).*
3,7,3′,4′,5′-Pentahydroxy-flavan:
(–)-*Robinetinidol* (2*R*,3*R*) *(Robinia pseudoacacia, Acacia mearnsii).*
3,6,8,4′-Tetrahydroxy-flavan:
(+)-*Afzelechin* (2*R*,3*S*) *(Cochlospermum gillivraei, Desmoncus polycanthos, Eucalyptus calophylla);*
(+)-*Epiafzelechin* (2*S*,3*S*) *(Livinstoma chinensis);*
(–)-*Epiafzelechin* (2*R*,3*R*) (Afzelia wood, *Larix sibirica, Actinidia chinensis, Juniperis communis, Cassia javanica).*

can account for up to 50 % of excreted estrogen metabolites; they lack estrogenic activity and were discovered much later than other known estrogen metabolites (e.g. sulfates and glucuronides of estradiol and estrone), because they give no response in biological tests. Synthetic estrogens used in the contraceptive pill (e.g. ethinylestradiol, see Ovulation inhibitors) are also subject to 2-hydroxylation, often to a greater extent than natural estrogens. The resulting C.e. may delay inactivation of catecholamines by competing with them for methylation by methyl transferase. This may explain the hypertension that sometimes accompanies the use of oral contraceptives.

Since C.e. are nonestrogenic (i.e. nonuterotropic), they are generally regarded as inactivation products of the estrogens. There is, however, evidence that C.e. (in particular 2-hydroxyestrone) are hormones in their own right, and active in the suppression of prolactin secretion [J. Fishman & D. Tulchinsky *Science* **210** (1980) 73–74].

Naturally occurring catechol estrogens and methylated derivatives.
R_1 = H, R_2 = αHβOH: 2-Hydroxyestradiol-17β
R_1 = CH_3, R_2 = αHβOH: 2-Methoxyestradiol-17β
R_1 = H, R_2 = O: 2-Hydroxyestrone
R_1 = CH_3, R_2 = O: 2-Methoxyestrone

Catenane, ***catenated DNA:*** two or molecules of circular DNA interlocked like the links of a chain. The frequency of occurrence of catenated DNA is increased by inhibitors of protein synthesis. It has been found in cells infected with SV40 or ΦX174, in mouse mitochondria and in human leukocyte mitochondria. Catenation and decatenation (unlinking) of circular DNA is catalysed in vitro by Topoisomerases (see). The physiological function of catenated DNA is unknown, and it may sometimes occur as a result of errors in DNA replication. The highly catenated networks of DNA in the single kinetoplast of trypanosomes appear, however, to represent the normal state of the DNA. This kinetoplast DNA (kDNA) contains thousands of catenated DNA circles. Most of these are minicircles (0.8–2.5 kilobases, depending on the trypanosome species) with a few maxicircles (20–40 kilobases). For example, kDNA from *Trypanosoma brucei* consists of 6,000 catenated minicircles interlaced with 25–50 maxicircles. The diameter of the total network varies from about 5 fm in African trypanosomes (agents of human sleeping sickness) to about 15 fm in *Crithidia* (a trypanosome that parasitizes insects). [P. T. England et al. *Annu. Rev. Biochem.* **51** (1982) 695–726]

Catharanthus alkaloids: see Vinca alkaloids.

Cathepsins: see Proteolysis.

CCC: acronym of Chlorocholine chloride (see).

CCP: acronym of carboxyl carrier protein. See Biotin enzymes.

CD: acronym of cluster differentiation or cluster designation; a system of nomenclature for human leukocyte ‘differentiation antigens’ recognized by monclonal antibodies. Also the acronym of Circular dichroism (see).

cDNA: Complementary DNA (see). See Hybridization.

CDP: acronym of cytidine 5′-diphosphate. See Cytidine phosphates.

CDP-choline: cytidine diphosphate choline. See Membrane lipids (biosynthesis).

CDP-glyceride: cytidine diphosphate glyceride. See Membrane lipids (biosynthesis).

CDP-ribitol: cytidine diphosphate ribitol. See CDP-sugars.

CDP-sugars: cytidine disphosphate sugars, a metabolically activated form of sugars and sugar derivatives (see Nucleoside diphosphate sugars). CDP-ribitol, is a precursor in the biosynthesis of bacterial cell walls.

Cell adhesion molecules: cell surface glycoproteins which promote the adhesion between cells and between cells and their extracellular matrix that is necessary for tissue formation. The acronym CAM is usually reserved for cell adhesion molecules that are members of an immunoglobulin superfamily (they possess one or more immunoglobulin domains), e.g. neural (N-CAM), neural-glial (Ng-CAM) and liver (L-CAM) CAM. N-CAM exists in at least 2 forms, distinguished by the difference in the number of sialic acid residues that they carry. The embryonic form has 30 g sialic acid/100 g protein, while the adult form has 10 g sialic acid/100 g protein. Adhesion between neurons is specifically inhibited by antibodies to N-CAM, but adhesion between neurons and glial cells is not inhibited by purified anti-N-CAM. In chick embryos, both N- and L-CAM appear very early, before the formation of germ layers. Later, the distribution of the 2 types of CAM correlates with the fate map of blastodermal cells: N-CAM labels the future neural plate, notochord, somites and parts of the lateral plate mesoderm; while L-CAM labels non-neural ectoderm and endoderm. It is likely that embryological development is guided by the expression (sometimes transitory) of one or more types of CAM on the various cell types.

N-CAM is particularly well characterized. It exhibits homophilic binding (binds to N-CAMs on other cells), and it exists in several isoforms. The largest isoforms of N-CAM are transmembrane proteins (M_r 180,000 and 140,000). Smaller isoforms (M_r 120,000–125,000) are located on the membrane surface, where they are attached by a glycosylphosphatidylinositol anchor, whereas the smallest isoforms (M_r ~ 115,000) are found as soluble proteins.

Other cell adhesion molecules are:

1. *Cadherins* are transmembrane glycoprotein adhesion molecules (M_r ~ 124,000) involved in the formation of strong, Ca^{2+}-dependent cell-cell bonds. Three cadherins, each encoded by a different gene, have been studied in detail: N-cadherin (brain, skeletal muscle, cardiac muscle), E-cadherin (uvomorulin) (liver and other epithelial tissues), and P-cadherin (placenta and mesothelium). Ca^{2+} binds to the extracellular domain of the cadherin and stabilizes its structure. In its absence cadherins are rapidly degraded from the cell surface. Like N-CAM the cadherins bind by a homophilic mechanism. Binding specificity is relatively strict, and the different cadherin species do not appear to bind to each other or to other proteins. They are associated with the cytoskeleton via cytoplasmic proteins called catenins. Three catenins have been identified: α-, β- and γ-catenin (plakoglobin).

2. *Integrins* constitute a large superfamily of transmembrane heterodimers, which connect molecules (e.g. α-actinin, talin) of the intracellular cytoskeleton to molecules in the extracellular matrix (e.g. fibronectin, vitronectin, collagens, laminin), promoting both substrate adhesion and cell-cell adhesion. Sites of close cell-substrate contact mediated by integrins are known as focal adhesions. Migration of cells during embryonic development involves the alternate release and rebinding of fibronectin by integrin. In contrast to CAMs and cadherin, adhesion by integrins is heterophilic.

Eight β-chains are known, and they display 37–55 % sequence homology. More than 90 % of the β-chain is extracellular, and it displays a highly conserved pattern of Cys residues and four repeats of a Cys-rich domain. Its cytoplasmic domain (40–50 residues) often carries a phosphorylatable Tyr residue. β4 has an exceptionally long cytoplasmic tail of about 1000 amino acids, containing 4 fibronectin type III domains.

The α-subunits display sequence homology of 25–45 %, but their secondary and tertiary structures are highly conserved. They all possess the same pattern of Cys residues, a repeated (3 or 4 times) divalent cation-binding motif, and a relatively well conserved transmembrane domain.

The specificity of ligand binding is determined by the particular αβ combination, and some twenty different heterodimers have been identified so far; e.g. human integrin α8β1 functions as a receptor for tenascin, fibronectin and vitronectin. Integrins constitute one of the largest groups of C.a.m., and are found throughout the animal kingdom, in both vertebrates and invertebrates.

In addition to their role as C.a.m., integrins also appear to be involved in transmembrane signaling, e.g. in the activation of mitogen-activated protein kinases.

3. *Selectins* constitute a family of mammalian, carbohydrate-binding membrane glycoproteins present in leukocytes and endothelium. Three members of the family are well characterized: leukocyte homing receptor (LAM-1, L-selectin), endothelial leukocyte adhesion molecule (ELAM-1, E-selectin), and CD62 (P-selectin) of platelets. All three have an *N*-terminal extracellular domain of 117–120 amino acid residues, which is responsible for Ca^{2+}-dependent carbohydrate recognition. This is followed by an epidermal growth factor-like domain of 34–40 amino acids, in turn followed by variable numbers of 62-amino acid consensus repeats. There is then a transmembrane segment and a cytoplasmic domain.

Selectins are responsible for the recognition and binding between leukocytes and endothelium during the exit or homing of leukocytes from the blood stream, and migration of leukocytes into inflamed tissue. Binding by selectins is Ca^{2+}-dependent. [M.Grumet et al. *Science* **222** (1983) 60–62; G.M.Edelman *Ann. Rev. Biochem.* **54** (1985) 135–169; C.A.Otey et al *J.Cell Biol.* **111** (1990) 721–729; M.Takeichi *Science* **251** (1991) 1451–1455; M.B.Lawrence & T.A.Springer *Cell* **65** (1991) 859–873; N.Morino et al. *J.Biol. Chem.* **270** (1995) 269–273; K.Norgard-Sumnicht & A.Varki *J.Biol. Chem.* **270** (1995) 12012–12024; P.A.Sacco et al. J.Biol. Chem. 270 (1995) 20201–20206; K.Vuori & E.Ruoslahti *J.Biol. Chem.* 270 (1995) 22259–22262; L.M.Schnapp et al. *J.Biol. Chem.* **270** (1995) 23196–23202; M.H.Ginsberg et al. 'Integrins: Emerging Paradigms of Signal Transduction' *Annu. Rev. Cell Dev. Biol.* **11** (1995) 549–599]

Cell count determination: see Growth.

Cell cycle: a sequence of irreversible phases in the life of a eukaryotic cell, following mitosis and leading to the next mitosis. Occurring in the following order, they are known as as the M-, G_1-, S- and G_2-phases (G stands for "gap"). The M-phase is mitosis (30 min – several hours), culminating in cell division. DNA replication occurs during the S-phase (synthesis phase). G_1 and G_2 are the postmitotic and premitotic phases, respectively, and DNA synthesis (replication) is absent from both. The C.c. can last 12 hours or several weeks, depending on the species, cell line and environment (e.g. in the root meristem of broad bean, G_1 = 12 h, S = 6 h, G_2 = 8 h, M = 4 h). The widest variations occur in the duration of G_1, which may be almost too brief to measure (e.g. in the early frog embryo) or so extended that the cell appears to be quiescent. In the latter case, the cell is said to be in the G_0-phase. After mitosis, some cells (e.g. neurons) undergo such a marked differentiation that they can no longer divide, but remain viable for several years in G_0. Other cells (e.g. hepatocytes, lymphocytes) undergo considerable differentiation, but prepare again for cell division (i.e. they pass from G_0 to G_1) after a few weeks or months. Unfertilized eggs are arrested at some stage of the C.c., e.g. sea urchin eggs in G_1, frog eggs in M and clam eggs in G_2. Other cells divide regularly, i.e. they continually pass through the C.c. without pausing in any phase. Cell growth usually occurs during G_1. Also during the G_1-phase, RNA, tubulin, etc. are synthesized, together with enzymes of DNA and protein biosynthesis (e.g. DNA polymerase) which are required in the next phase (S-phase). The S-phase can be divided into S_1 (replication of euchromatin) and S_2 (replication of heterochromatin), which overlap one another. The quantity of chromatin is doubled during the S-phase, but the number of chromosomes is unchanged. After G_2, the nuclear membrane disintegrates and the chromosomes condense, marking the beginning of the prophase of mitosis. The numbers and specificities of cell receptors vary during the different phases of the C.c.

The C.c. of neoplastic (cancer) cells is no longer subject to control by the organism. Such cells divide autonomously, and usually pass through their C.c. more rapidly than normal cells; in some cases, however, they may take longer to divide than comparable normal cells.

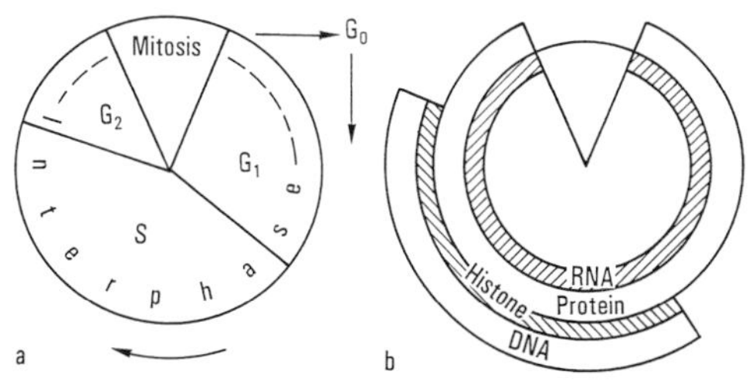

Cell cycle. a Schematic representation of the sequential phases of the cell cycle; the sections of the clock face represent the relative duration of each phase. *b* Periods of synthesis of different cytoplasmic and nuclear components during the cell cycle.

Advances in the understanding of molecular events during G_1 and G_2, which prepare the cell for entry into S and M, were made possible by the study of temperature-sensitive mutants of yeasts for DNA, RNA, or protein synthesis. Some of these mutations result in blockage at specific points in the cell cycle, and many such *cdc* (cell division cycle) mutants have been isolated. A structurally highly conserved protein kinase, which promotes transition from G_2 to M, has been found in all investigated eukaryotes, and is considered to be the master regulator of the C.c. Known as *maturation-promoting factor* (MPF), this protein kinase was first described as a protein fraction promoting the M-phase in amphibian eggs. Purified MPF consists of 2 proteins in equimolar amounts: a protein kinase, $p34^{cdc2}$ (product of gene $cdc2^+$), and a B-cyclin, whose potential biochemical role includes substrate specification and/or subcellular localization of the heterodimer. $p34^{cdc2}$-type protein kinases have been found in all investigated eukaryotes (e.g. P.C.L.John et al. *Aust. J.Plant Physiol* **20** (1993) 503–526 have reported the function of this protein as a molecular switch in the plant cell cycle). During interphase, $p34^{cdc2}$ is inactivated by phosphorylation of its ATP binding region. Entry into M-phase is marked by dephosphorylation of $p34^{cdc2}$, which then becomes an active phosphorylase. Substrates phosphorylated during M-phase include nuclear lamins (contributing to nuclear disassembly), vimentin and caldesmon (potential substrates of cytoskeletal rearrangement).

Commitment to enter S-phase is thought to occur during G_1, in a molecular event known as Start. A gene ($rum1^+$), which regulates G_1 progression and the onset of S-phase, has been identified in the fission yeast, *Schizosaccharomyces pombe.* Overexpression of $rum1^+$ leads to giant cells containing many times the haploid complement of DNA, whereas deletion of $rum1^+$ abolishes the interval between the end of mitosis and Start. $p34^{cdc2}$ possibly exists in two forms, an S form that allows progress through Start with subsequent replication of DNA, and an M form that allows progress through mitosis with subsequent chromosome segregation. The gene product of $rum1^+$ may play a critical role in the interconversion of these two forms, e.g. by maintaining $p34^{cdc2}$ in the inhibited G_1/S form until the critical cell mass for Start is attained.

An alternative explanation for the role of $rum1^+$ invokes the existing theory of a *licensing factor* that binds to DNA during mitosis, and is consumed during S-phase. A subsequent round of replication is not possible until the nucleus has disintegrated, enabling the chromosomes to bind fresh licensing factor. G_1 cannot be followed by mitosis, because licensing factor activates the gene product of $rum1^+$, which in turn inhibits mitotic MPF. During DNA synthesis, licensing factor is consumed and cytoplasmic licensing factor is excluded from the nucleus; the resulting activation of MPF programs the cell for entry into mitosis.

It has been shown that cyclin B has a short amino acid sequence near its *N*-terminus that targets the protein for destruction between metaphase and anaphase (destruction of cyclin B is mediated by ubiquitin-mediated proteolysis). Removal of this sequence (known as a destruction box) renders the cyclin indestructible, and shows that cyclin destruction is necessary for chromosome decondensation, nuclear envelope reformation and cytokinesis, but not for sister chromatid segregation.

It should be noted that replication and chromosome segregation are not mutually exclusive in bacteria, especially in rapidly growing forms, where DNA synthesis and chromosome segregation may occur at the same time. Thus, when applied to prokaryotes, the term C.c.has a different meaning from that discussed above. [C.N.Norbury & P.Nurse, *Ann. Rev. Biochem.* **61** (1992) 441–470; A.W.Murray, *Nature* **367** (1994) 219–220; S.Moreno & P.Nurse, *Nature* **367** (1994) 236–242; T.W.Jacobs 'Cell Cycle Control' *Annu. Rev. Plant. Physiol. Plant Mol. Biol.* **46** (1995) 317–339; A.Murray *Cell* **81** (1995) 149–152; J.T.Tyson et al. *Trends Biochem Sci.* **21** (1996) 89–96]

Cell mass determination: see Growth.

Cell membrane, *plasmalemma:* a Biomembrane (see) which serves as the outer boundary of the cell. In addition to a C.m., plant and bacterial cells also have a Cell wall (see).

Cellobiase: see Disaccharidases.

Cellobiose: a reducing disaccharide consisting of 2 molecules of D-glucose linked β1,4 (cf. maltose which has an α1,4 bond). C. is not fermented by yeast or hydrolysed by maltase. It represents the repeating disaccharide unit of cellulose and lichenin. It does not occur free in nature, except as an intermediate in the degradation of cellulose by cellulases. It is also present in certain glycosides.

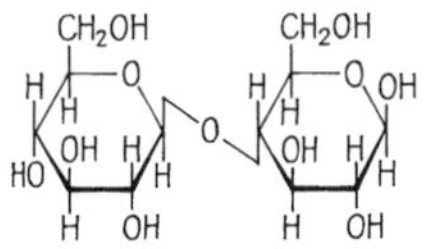

β-Cellobiose

Cell plate: see Cell wall.

Cell sap: see Vacuole.

Cellular enzymes: see Enzymes.

Cellular metabolism: see Primary metabolism.

Cellulases: enzymes which hydrolyse cellulose to cellobiose, and which are found in plants, bacteria and fungi, The C. of *Penicillium notatum* is well-studied; it consists of 324 amino acid residues (M_r 35,000) with a disulfide bridge and no free SH groups. C. are used in digestion tablets, to remove undesired cellulose in foods, and for the preparation of cellobiose from cellulose.

Cellulose: an unbranched plant polysaccharide, M_r 300,000–500,000, consisting of β1,4-linked glucose units. C. is enzymatically hydrolysed to the disaccharide cellobiose. It can be hydrolysed to D-glucose by treatment with concentrated acids, such as 40 % HCl or 60–70 % H_2SO_4 at high temperature. This process, called saccharification of wood, is used to produce fermentable sugar from wood.

C. is the main component of the plant cell wall. Certain plant fibers, such as cotton, hemp, flax and jute, are almost pure C. Wood, in contrast, is only 40–60 % C. In the cell wall, the C. is arranged in microfibrils which are 100–300 Å wide and about half as thick. The C. chains are parallel and stabilized by interchain hydrogen bonding. The microfibrils are embedded in a matrix of other polysaccharides, including pectin, hemicelluloses and lignin, and small amounts of a protein, extensin.

Microfibrils of C. are arranged transverse to the long axis of elongating cells, with a relatively large amount of angular dispersion. In thicker walls they are arranged in helical lamellae, in which the pitch of the spiral changes from one lamella to the next.

C. accounts for more of the earth's biomass than any other compound. The total amount is equivalent to about 50 % of the carbon dioxide in the atmosphere; about 100 billion metric tonnes are produced each year. It is degraded by organisms that possess Cellulases (see): lower plants, wood-destroying fungi and some bacteria. Termites, ruminants and some rodents harbor symbiotic bacteria in their digestive tracts which enable them to utilize cellulose. In animals that cannot digest it (e.g. humans and carnivores), C. is a ballast substance. C. is a very important industrial product, obtained primarily by acid (sulfite process) or alkaline (sulfate process) hydrolysis of wood or straw. It is used in the manufacture of paper, textiles, plastics, explosives, animal feeds and fermentation products. Some important industrial derivatives of C. are ethers, esters and xanthogenates.

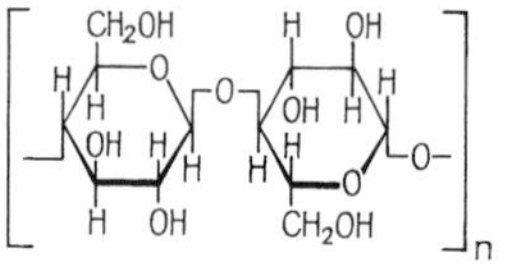

Cellulose

Cellulose ion-exchangers: see Ion-exchangers, Column chromatography.

Cell wall: a rigid structure external to the cell membrane of prokaryotes, green plants, fungi and some protists. It is synthesized by the protoplasm. Animal cells do not possess a C. w.

Bacterial C. w. Classification of bacteria as Gram-positive or negative on the basis of their reaction to the Gram stain corresponds to a fundamental difference in the structure of their C. w. Gram-positive bacteria have relatively simple walls, usually consisting of 2 layers. The outer layer is often a Teichoic acid (see), although in some species it may be a neutral polysaccharide or an acidic one called teichuronic acid. The inner layer is Murein (see). Gram-negative C. w. are more complicated. Under the electron microscope they appear to consist of at least 5 layers, comprising lipoproteins, lipopolysaccharides, proteins and murein; the murein again forms the innermost layer, or it may be separated from the cell membrane by an extra layer of protein.

Plant C. w. are exceedingly complex structures which have largely defied detailed analysis. The precursor of the C. w. is the cell plate, a non-cellulose structure formed between the 2 daughter nuclei during mitotic division of the cell. After the new cell membranes have formed on either side, the cell plate (a strongly hydrated structure) matures to a "middle lamella". As growth proceeds, the area of the new wall is increased by intussusception, i. e. incorporation of new material within the existing matrix. The thickness increases by apposition as new layers of wall material are added.

The primary C. w. is a complicated array of carbohydrates. The hemisubstances are hemicelluloses and polyuronides, plant gums, mucilages (e. g. fucoidin, laminarin, alginates, agar, carrageenans) and reserve carbohydrates (e. g. arabans, xylans, mannans and galactoarabans). [M. McNeil et al. *Ann. Rev. Plant Biochem.* **53** (1984) 625–663] These carbohydrates form the thick walls of certain plants, such as date palms and vegetable ivory. Fragments derived from them (see Oligosaccharins) may be potent effectors of cell function. The composition of the primary C. w. appears not to be random, and to require the expression of numerous carbohydrate-transferring enzymes to achieve the correct structures.

Secondary and tertiary C. w. contain mechanical and support materials. In most plant cells the chief structural material is cellulose (see), laid down as a network of submicroscopic microfibrils. In the C. w.

of Basidiomycetes and Phycomycetes, the main structural material is chitin. In plants, the interstices of the structural network contain incrustation materials, such as lignin, silicic acid (*Equisetum* and diatoms), calcium carbonate (stoneworts) or calcium oxalate (cypress).

Accrustation materials of the C.w. of boundary tissues (epidermis, periderm) are cutin and suberin. Cutin forms a cuticle of varying thickness and strength. Suberin is the material of cork, e.g. in the cork oak, *Quercus suber.* Some plant C.w. secrete wax, e.g. the Andean wax palm, *Copernica cerifera.* The walls of pollen grains and cryptogam (e.g. fern) spores consist of Sporopollenin (see).

Cembranes, ***cembranoids:*** monocyclic diterpenes isolated from several plants, in particular from the gum resins of pines. C. have also been found in marine coelenterates and insects.

The C. skeleton (1-isopropyl-4,8,12-trimethylcyclotetradecane, or octahydrocembrane) consists of a 14-membered carbocyclic ring with an isopropyl group in position 1, and 3 methyl groups at positions 4,8 and 12, but cembrane itself does not occur naturally. Cembrene (Fig.) was isolated from pine oleoresins, and eunicin (Fig.) from the Caribbean gorgonian, *Eunicea mammosa.* Cembrene A (or neocembrene) [$-C(CH_3) = CH_2$ at position 1] was isolated from the gum resin of *Commiphora mukul,* which is used in the Ayurvedic system of medicine. The same source also yielded mukulol (3,7,11-cembatriene-2-ol). Isocembrene (no 4,5-double bond, and $= CH_2$ at position 4) is present in *Pinus sibirica.* 2,7,11-Cembatriene-4-ol has been isolated from several pines.

Biosynthesis of crassin acetate (Fig.) of the Caribbean gorgonian, *Pseudoplexaura porosa,* appears to be carried out by its algal symbionts (zooxanthellae). Cell extracts of the zooxanthellae incorporate mevalonate and geranyl pyrophosphate into crassin acetate. [A.J.Weinheimer et al. *Progress in the Chemistry of Organic Natural Products* **36** (1979) 285–387]

Cembrene

Eunicin

Crassin acetate

Structures of some naturally occurring cembrane derivatives

Central dogma of molecular biology: the fact that genetic information can be transferred from DNA to protein, but not in the reverse direction (Fig.). The discovery that RNA can code for the synthesis of DNA (see RNA-dependent DNA-synthetase) does not alter the validity of the dogma.

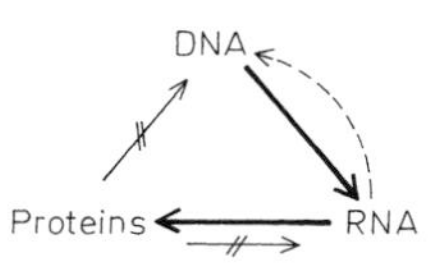

Schematic representation of the central dogma of molecular biology

Centomere: see Chromosomes.

Cephalins: see Membrane lipids.

Cephalosporin P$_1$: a tetracyclic triterpene antibiotic, M_r 574.73, m.p. 147 °C, $[\alpha]_D^{20}$ +28° (c = 2.7 in $CHCl_3$). C. differs from the related Fusidic acid (see) in that it lacks an 11α-hydroxyl group and has an additional 6α-acetoxy function. It is obtained from culture filtrates of *Cephalosporium* sp. and is effective against Gram-positive organisms.

Ceramides: see Membrane lipids.

Ceramide lactoside lipidosis: see Lysosomal storage diseases.

Ceramide 1-phosphorylcholine: see Membrane lipids.

Cerebrocuprein: see Superoxide dismutase.

Cerebronic acid, ***2-hydroxytetracosanoic acid, α-hydroxylignoceric acid:*** $CH_3-(CH_2)_{21}-CHOH-COOH$, a hydroxy fatty acid, M_r 384.63, m.p. 101 °C, present in various glycolipids.

Cerebrosides: see Membrane lipids.

Cerotinic acid, ***n-hexacosanoic acid:*** $CH_3-(CH_2)_{24}-COOH$, a fatty acid, M_r 396.7, m.p. 87.7 °C, present as an esterified acid in beeswax, wool grease, carnuba wax and montan wax, and in traces in plant fats.

Ceruloplasmin: a blue, copper-containing glycoprotein, which is present in mammalian blood plasma, and which is phylogenetically related to plant laccases and to ascorbate oxidase, both of which are blue. It is a transport and/or storage protein for copper, and it is also an oxidase. Substrates include unsaturated compounds, such as indole derivatives, amphetamine, adrenalin and dopamine. C. has a central role in copper metabolism; when C. is absent, as in Wilson's disease, copper is deposited in the tissues, causing death.

Early studies of C. suggested an octameric structure, then a tetrameric $\alpha_2\beta_2$ structure. However, gel filtration of reduced and alkylated C. in 6 M guanidine hydrochloride showed that the protein from several species is a single chain of M_r 120,000; the earlier results were probably proteolysis artifacts. The amino acid sequence consists of a twice repeated (i.e. 3 copies) 340-residue unit, which is 30 % homologous to the A subunit of blood clotting factor VIII. [L.Ryden in R.Lontie (ed.) *Copper Proteins and Copper Enzymes,* CRC Press, Boca Raton, 1984; R.M.Lawn *Cell* **42** (1985) 405–406]

Ceveratrum alkaloids: see Veratrum alkaloids.

Cevine: see Germine.

cGMP: acronym of cyclic guanosine 3′,5′-monophosphate. See Guanosine phosphates.

Chain conformation: see Proteins.

Chaksine: see Guanidine derivatives.

Chalcone isomerase (EC 5.5.1.6): a plant enzyme, catalysing the stereospecific isomerization of chalcones to the corresponding (–)(2*S*)flavanones (Fig.),

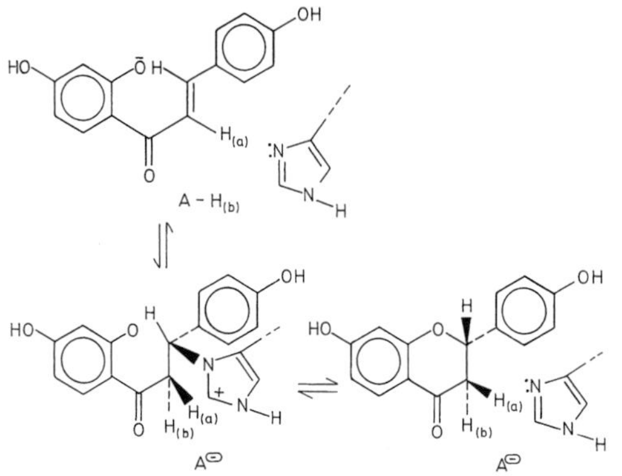

Proposed mechanism for the action of chalcone isomerase. [M. J. Boland & E. Wong *Bioorg. Chem. 8* (1979) 1–8]. Nucleophilic addition of an imidazole group at the active site of the enzyme to the double bond is followed by nucleophilic attack by the 2′-phenolate ion. A-H may be an acidic side chain or simply a water molecule.

an important early stage in the biosynthesis of Flavonoid (see) compounds.

Chalcones: flavonoids with the ring system shown (Fig.). C. are widely distributed in plants, particularly in the *Compositae* and *Leguminoseae,* and they contribute to flower color in certain members of the *Compositae, Oxalidaceae, Scrophulariaceae, Generiaceae, Acanthaceae* and *Liliaceae.* C. are the first detectable C_{15} precursors in flavonoid biosynthesis (see Flavonoids, Chalcone synthase). A. B. Bohm [in *The Flavonoids: Advances in Research,* J. B. Harborne & T. J. Mabry (eds.), Chapman and Hall, 1982] gives a checklist of 11 naturally occurring C. aglycons and 28 C. glycosides. Examples are ***butein*** (2′,4′,3,4-tetrahydroxychalcone from e. g. *Acacia),* **coreopsin** (4′-glucoside of butein from e. g. *Cereopsis),* **pedicin** (2′,5′-dihydroxy-3′,4′,6′-trimethoxychalcone from *Didymocarpus),* **ovalichalcone** (2′-hydroxy-3′-prenyl-4′,6′-dimethoxychalcone from *Milletia ovifolia)* and ***ψ-isocordein*** [2′,4′-dihydroxy-3′-(α,α-dimethylallyl)-chalcone from *Lonchocarpus* spp.].

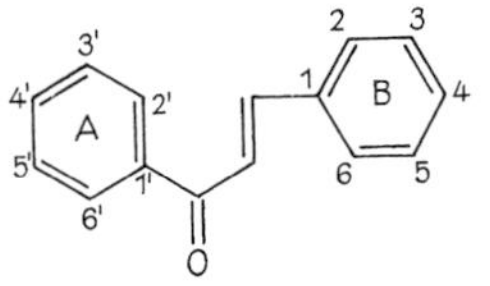

Chalcone ring system

Chalcone synthase, ***CHS:*** a plant enzyme (formerly called flavanone synthase) catalysing the synthesis of chalcones from one molecule of the CoA ester of a substituted cinnamic acid and 3 molecules of malonyl-CoA, a key reaction in flavonoid biosynthesis (see Stilbenes, Fig., and Flavonoid, Fig.). Specificity varies with the source of the enzyme, e. g. CHS from *Tulipa* stamens and *Cosmos* petals forms naringenin, eriodictyol or homoeriodictyol from 4-coumaroyl-CoA, caffeoyl-CoA or feruloyl-CoA, respectively, whereas CHS from *Petroselenum hortense* uses only 4-coumaroyl-CoA at pH 8.0, but also attacks caffeoyl-CoA at pH 6.0. The above products are flavanones, formed from the chalcone product by the action of chalcone isomerase, and to some extent spontaneously. Rigorous removal of chalcone isomerase during purification of CHS is necessary for the clear demonstration of chalcone formation. Formulas of these compounds can be found under Flavonoid, Flavanones, Lignin. M_r of CHS: 80,000 *(Phaseolus vulgaris),* 55,000 *(Tulipa, Cosmos),* 77,000 *(Petroselenum hortense, Brassica oleraea, Haplopappus gracilis).* The majority of CHS preparations appear to consist of 2 identical subunits. CHS closely resembles 3-oxoacyl-(acyl carrier protein) synthase *(syn* β-ketoacyl-ACP synthase) from type II (nonaggregated) fatty acid synthase, and it is suggested that CHS arose by gene duplication [F. Kreuzaler et al. *Eur. J. Biochem.* **99** (1979) 89–96]. CHS from all sources investigated so far catalyse formation of chalcones with phloroglucinol-type A-ring hydroxylation patterns, the three OH-groups originating from the CoA-esterified carboxyl groups of the malonyl-CoA. It is therefore necessary to postulate a separate 6′-deoxychalcone synthase to account for most of the isoflavonoid phytoalexins. It is suggested that synthesis of 5-deoxyisoflavonoids is catalysed by an enzyme system similar to 6-methylsalicylic acid synthase from *Penicillium patulum,* i. e. a multienzyme complex using acetyl-CoA primer and 3 molecules of malonyl-CoA, similar to type I (aggregated) fatty acid synthase from yeast and *P. patulum.* This suggestion is supported by the pattern of incorporation of [C^{13}]acetate into ring A of pisatin in *Pisum sativum* and phaseolin in *Phaseolus vulgaris* [M. Steele et al. *Z. Naturforsch* **37 c** (1982) 363–368; P. Elomaa et al. 'Transformation of antisense constructs of the chalcone synthase gene superfamily into *Gerbera hybrida:* differential effects on the expression of family members' *Molecular Breeding* **2** (1996) 41–50].

Chalinasterol, ***ostreasterol, 24-methylenyl cholesterol, ergosta-5,24(28)-dien-3β-ol:*** an animal sterol (see Sterols), M_r 398.66, m. p. 192 °C, $[\alpha]_D$ −35 ° ($CHCl_3$). C. is a characteristic sterol of pollen, and has also been found in sponges, oysters and mussels, and in honeybees.

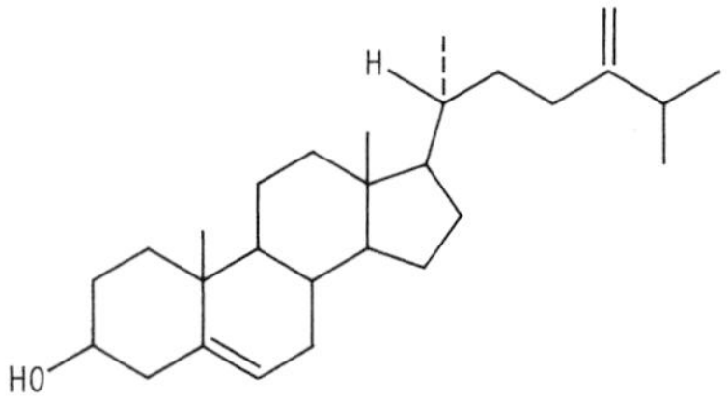

Chalinasterol

Chalones: antitemplate substances; tissue-specific, endogenous mitosis inhibitors. C. are allegedly produced by mature or differentiated cells, and it is claimed that they inhibit division of primordial cells by a type of negative feedback. The effect of C. is reversible and not species-specific. They have been isolated from various sources. C. from lymphocytes, granulocytes and fibroblasts are glycoproteins (M_r 30,000–50,000), while those from erythrocytes and liver are low M_r (~2,000) polypeptides. Thus, according to the C. theory, an epidermal C. is synthesized in the epidermis at a rate proportional to the thickness of the epidermis; it slows down mitosis in the basal layer, so that the rate of production of differentiated cells is matched to the required rate of replacement of the epidermis. C. are supposedly active in the late presynthetic G_1 phase and in the premitiotic G_2 phase of the Cell cycle (see). They are also thought to inhibit the decision phase E, in which the cell leaves the cell cycle and enters and the first postmitotic maturation phase A_1, leading to the aging phase A_2 and eventually cell death. Thus C. not only inhibit cell division, they also retard aging, thereby increasing the life expectancy of cells. C. are thought to play a central role in homeostasis (maintenance of metabolic balance in the body), regeneration and healing of wounds, and it is also suggested that they are involved in cancerous growth. Since the early reports of C. in the 1950s, there seems to have been little further confirmation of the above claims.

Chanoclavine: see Ergot alkaloids.

Chaperonins: see Molecular chaperones.

CHAPS: 3-[(3-cholamidopropyl)-dimethylammonio]-1-propanesulfonate, a nondenaturing, zwitterionic detergent, combining the properties of sulfobetaine-type detergents and bile salt anions. C. is more effective than sodium cholate or Triton X-100 at breaking protein-protein interactions. It is useful for the solubilization of membrane proteins without denaturation, e.g. solubilization of brain opiate receptors with retention of reversible opiate binding was first achieved with C., and it is still the detergent of choice for the isolation of many other membrane proteins, especially receptors. [L.M. Hjelmeland et al. *Anal. Biochem.* **130** (1983) 72–82; W.F. Simons et al. *Proc. Natl. Acad. Sci. USA.* **77** (1980) 4623–4627]

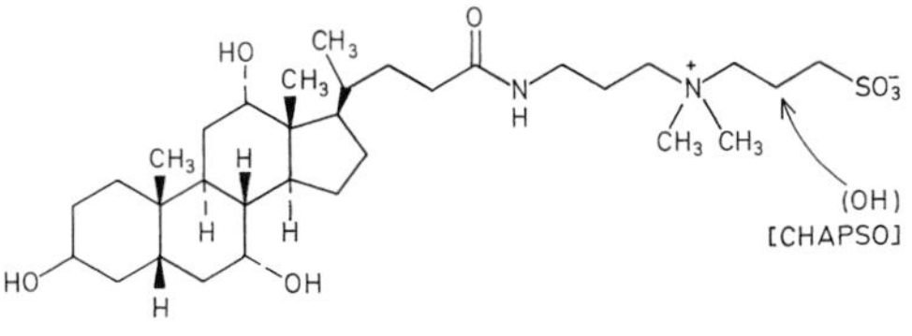

CHAPS and *CHAPSO*

CHAPSO: 3[(3-cholamidopropyl)-dimethylammonio]-2-hydroxy-1-propanesulfonate, a nondenaturing, zwitterionic detergent, chemically similar to CHAPS (see) and equally useful as a detergent in the isolation of membrane proteins.

Charge-relay system: a network of hydrogen bridges in the catalytic center of chymotrypsin and other serine proteases which is responsible for the high degreee of nucleophilicity of the Ser_{195} hydroxyl group. In addition to Ser_{195}, the system includes Asp_{102} in the hydrophobic interior of the molecule, and the imidazole ring of His_{57}, which lies between the other two. Electron flow is possible from the negatively charged carboxyl group of Asp_{102}, over the polarized imidazole ring of His_{57}, to the oxygen of Ser_{195} on the surface of the molecule. The oxygen is therefore highly nucleophilic, but only in alkaline medium (Fig.). In weakly acidic medium the Asp and His are ionically bonded rather than hydrogen-bonded. Thus the C-r.s. consists of a series of hydrogen bonds between the 3 amino acids that are catalytically active in serine proteases at pH 8.0.

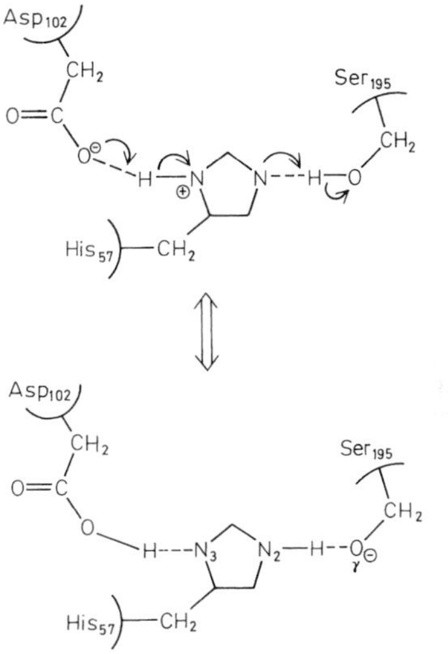

Charge-relay system is a serine protease

Chavicine: see Piperine.

Chebulic acid: see Tannins.

Chemical taxonomy: the use of the occurrence and distribution of chemical constituents to determine the taxonomic position and phylogenetic relationships of organisms. It is applied largely to plants, and is a valuable aid to systematic botany, e.g. the presence or absence of L-canavanine is a useful trait in the C.t. of the *Papilionaceae.* In the absence of chemical convergence, the presence of the same natural products indicates the presence of the same metabolic pathways, enzymes and genes, which can only have arisen through a common evolutionary history. There are, however, many cases of chemical convergence, so that the simple presence of the same natural product does not prove a taxonomic relationship. Such studies may therefore be supplemented by comparing the biosynthetic mechanisms of natural products, but it is difficult to compare enzyme patterns and reaction sequences in large numbers of organisms. Some examples of natural products employed in C.t. are alkaloids, flavonoids, proteins (comparison of amino acid sequences of homologous proteins) and nonprotein amino acids. Serology is also used in C.t., i.e. antiserum raised against a purified protein or protein extract from one plant is tested for cross reactivity with the proteins of other plants. Compared with the number of plant species (400,000 recent plants and 100,000 lower plants), the number of taxa studied by C.t. is small. Nevertheless, an enormous quantity of data has been gathered. Results from C.t. have partly confirmed the results of morphological comparisons (especially of flowering plants), and in some cases they have led to corrections in taxa where the relationships were uncertain. However, they have often brought more confusion than clarification, and the expectation that C.t. might transcend morphological studies and provide a more reliable and fundamental basis of taxonomy, has not been fulfilled. [*Chemotaxonomy of the Leguminosae,* edit. Harbourne, Boulter and Turner, Academic press, 1971; *The Biology and Chemistry of the Umbelliferae,* edit. Heywood, Academic Press, 1971; *The Biology and Chemistry of the Compositae,* edit. Heywood, Harbourne and Turner, Academic Press, 1977]

Chemiosmotic hypothesis: a mechanism proposed by Peter Mitchell [*Nature* **191** (1961) 144–148] to explain how free energy from the exergonic flow of electrons along an electron transport chain (in the mitochondrial inner membrane, the chloroplast thylakoid membrane and the prokaryotic plasma membrane) drives the endergonic phosphorylation of ADP, a reaction catalysed by the ATP synthase activity of (what came to be known as) the F_0F_1-complex (see F-type ATPases) in the same membrane. Until publication of the C.h., the mechanism linking electron transport to phosphorylation in energy-transducing membranes remained unresolved by experimentation suggested by other hypotheses (e.g. the chemical coupling hypothesis and the conformational coupling hypothesis; see Oxidative phosphorylation).

The C.h. postulates that as electrons flow exergonically down the chain of redox systems constituting the electron transport chain, from a more negative potential (e.g. –0.32V for NAD/NADH) to a less negative potential (e.g. +0.82V for O_2/H_2O), the free energy liberated by the successive redox reactions is used to transport protons (H^+) from the one side of the relevant membrane to the other, thereby generating an electrochemical gradient (i.e. electrical charge and concentration) across it. In the mitochondrion the protons are transported from the matrix to the intermembrane space (i.e. the space between the inner and outer membranes); in the chloroplast they are transported from the stroma into the thylakoid lumen, while in the case of prokaryotic cells they are transported from the cytoplasm to the extracellular side of the plasma membrane. The free energy liberated by electron flow is therefore conserved in the form of a proton electrochemical gradient. This gradient constitutes a force, for which Mitchell coined the name 'Proton motive force' (see), which, according to the present version of the C.h., drives protons back across the membrane (i.e. H^+ flowing from a higher concentration to a lower concentration and a region of greater positive charge to a region of lesser positive charge) by way of the F_0F_1-ATP synthase, which is forced to release the ATP that it has formed from ADP and P_i. Thus the free energy conserved in the electrochemical gradient is used to drive the endergonic phosphorylation of ADP (i.e. $ADP + P_i \rightarrow ATP + H_2O$; $\Delta G^{o'} = +30.5$ kJ.mol^{-1}). In this way the protons return to the side of the membrane from which they started and are available to be 'pumped' back across it once again by the operation of the electron transport chain.

It is a fundamental tenet of the C.h. that H^+ can traverse the participating membrane only by operation of the electron transport chain or via the F_0F_1-complex, and that the membrane is otherwise impermeable to H^+; moreover it must be structurally intact and completely enclose the compartment from which the H^+ are pumped. Ideally the membrane should also be impermeable to other major cellular cations (e.g. K^+) and anions (e.g. Cl^-), otherwise the generation of an electrical charge gradient across the membrane would be impossible (because the movement of H^+ would be balanced by the movement of anions in the same direction and/or cations in the opposite direction) and the only achievable gradient would be a H^+ ion (pH) gradient. The mitochondrial inner membrane has been found to fulfil all these requirements and a genuine electrochemical (electrical charge and concentration) gradient is achieved. However, the thylakoid membrane has been found to be rather more permeable to other ions and the electrochemical gradient that is achieved has a greater component of concentration than electrical charge. The requirement of the C.h. that these energy-transducing membranes be structurally intact and completely enclose the relevant cellular or organellar compartment has also been experimentally demonstrated; e.g. mitochondrial inner membrane fragments containing complete electron transport chains or parts thereof will not generate ATP even though they perform electron transport.

In addition to the above, the C.h. is supported by a wealth of evidence, the main items of which are: (i) oxidation of intramitochondrial NADH and $FADH_2$ (or substrates which cause their generation) by operation of the electron transport chain in the inner membrane causes a drop of about 1.4 pH units in the intermembrane space (seen experimentally in the medium outside the mitochondria because the outer membrane is freely permeable to H^+) relative to the matrix, and the appearance of an electrical potential gradient across the inner membrane, the outside being about 0.14V more positive than the inside; (ii) ATP is generated from ADP and P_i when a pH gradient is experimentally imposed upon mitochondria and chloroplasts even though electron flow down the relevant electron transport chains has been blocked; (iii) participation of the F_0F_1-complex in ATP generation and the dependence of this process on a distinct H^+-gradient-generating system was shown as follows: liposomes, containing in their membranes both bacteriorhodopsin and mitochondrial F_0F_1-complex, synthesized ATP when illuminated (bacteriorhodopsin is a purple membrane-bound protein from photosynthetic halobacteria, and it normally pumps H^+ from the cytoplasm to the outside of the cell when illuminated; this was incorporated into the liposome membrane with the opposite orientation so that it pumped H^+ from the external medium into the liposome lumen, while the F_0F_1-complex was oriented with its F_1-component, containing the ATP-synthase, on the outside of the liposome membrane); (iv) complexes I, III and IV of the mitochondrial electron transport chain (see Respiratory chain) have each been shown to be capable of catalysing transmembrane H^+ transport by incorporating them separately into the membranes of synthetic sealed vesicles (in the opposite orientation to that obtaining in the inner mitochondrial membrane), together with the F_0F_1-complex (oriented with its F_1-component on the outside of the vesicle membrane); ATP was synthesized when the vesicles were incubated with the appropriate oxidizable substrate; (v) compounds (e. g. 2,4-dinitrophenol) that can carry H^+ across the inner mitochondrial membrane have been shown to 'uncouple' ATP synthesis from electron transport. These 'uncoupling agents', by having dissociable protons and being able to traverse the membrane in both their protonated acid form (outside-to-inside) and their non-protonated conjugate base form (inside-to-outside), collapse the proton electrochemical gradient generated by electron flow down the electron transport chain; (vi) valinomycin, a natural antibiotic from *Streptomyces* sp. which carries no electrical charge and contains no ionizable groups, forms a positively charged complex with K^+ ions which has been shown to diffuse across mitochondrial, thylakoid and, to a lesser extent, bacterial plasma membranes, thereby markedly reducing the rate of ATP synthesis by collapsing the electrical component of the proton motive force, though leaving the pH component intact.

Though it is proven that the electron transport chains of the inner mitochondrial, thylakoid and prokaryotic plasma membranes act as transmembrane H^+ pumps when actively engaged in electron transport, the mechanism of this pumping action is not at all clear. Mitchell initially proposed the 'redox loop mechanism' in which the the participating redox systems were so arranged within the membrane that redox reactions that 'consume' H^+ ions, in addition to electrons, [e. g. $NADH + H^+ + FMN(FP_D) \rightarrow NAD^+ + FMNH_2(FP_D)$] take them from one side of the membrane (e. g. the mitochondrial matrix) and redox reactions that 'disgorge' H^+ ions [e. g. $FMNH_2(FP_D) + Fe\text{–}S\text{–}P_{ox} \rightarrow FMN(FP_D) + Fe\text{-}S\text{-}P_{red} + 2H^+$] do so on the other side of the membrane (e. g. the mitochondrial intermembrane space). The difficulty with this mechanism was that it demanded that the 'H^+ & electron' redox systems be spacially alternated with 'pure electron' redox sytems in each of the complexes (e. g. I, III and IV in mitochondria) constituting the chain. This was known not to be the case. Mitchell attempted to get round this problem in complex III by describing a way that ubiquinone (Q) (see) could participate twice in H^+ translocation in the so-called 'Q-cycle' (see). However because the problem could not be resolved with complex IV, which contains no 'H^+ & electron' redox system, the redox loop mechanism has been abandoned and replaced by the 'proton pump mechanism' which proposes that the transfer of electrons causes conformational changes within the proteins of a given complex which influence the pK_a values of ionizable groups in their amino acid side chains, thereby causing H^+ uptake on one side of the membrane and H^+ release on the other. There is little hard evidence for or against this mechanism at present. The mechanism by which the proton gradient generated by electron transport is currently believed to be 'translated' by the F_0F_1-complex into ATP synthesis is described in more detail in the entry on F_0F_1-ATP synthases (see). In essence the F_0-component of the complex extends across the thickness of the membrane and contains within it a 'pore' through which H^+ ions, driven by the proton motive force (electrochemical gradient), pass to the site of the ATP synthase located in the F_1-component. There it causes ATP synthesized from ADP and P_i, to be released from the catalytic site of the ATP synthase. Somewhat surprisingly, the H^+ electrochemical gradient is not required to drive the formation of ATP but only to release it from the ATP synthase.

Though the details of its mechanism remain to be elucidated, the C.h is generally believed to be the most accurate description yet available of the processes of oxidative phosphorylation in mitochondria and prokaryotes and photophosphorylation in chloroplasts. Mitchell received the Nobel Prize for Medicine and Physiology in 1978 for his work in this field. [P. Mitchell *Biochem. Soc. Trans.* **4** (1976) 399–430; *Science* **206** (1979) 1148–1159]

Chemolithoautotrophic metabolism: see Nutritional physiology of microorganisms.

Chemolithotrophy: see Chemosynthesis.

Chemostat: see Fermentation techniques.

Chemosynthesis: 1. Chemical synthesis. 2. Utilization of inorganic compounds or ions (ammonia, nitrite, hydrogen sulfide, thiosulfate, sulfite, iron(II) or manganese(II) ions) and of hydrogen or elemental sulfur to obtain reducing equivalents and ATP. Most organisms capable of C. (water and soil bacteria) fix CO_2 autotrophically. Substrates are oxidized by aerobic or anaerobic respiration. This autotrophic mode of existence was formerly called inorganic oxidation, and is now more appropriately called "chemolithotrophy".

Chemotherapy: C. was defined by its founder, Paul Ehrlich, as "the use of a drug to combat an invading parasite without damaging the host". The term is also used for the treatment of cancer with chemicals which are more damaging to dividing cells than to differentiating cells. In either case it is always possible that a resistant clone of bacteria or cancer cells will arise, and, since it has no competition from the wild type, multiply rapidly. This has made it necessary to develop a large arsenal of chemotherapeutic agents. The first were the sulfonamides, followed later by penicillin and other antibiotics.

Chenodeoxycholic acid, ***3α,7α-dihydroxy-5β-cholan-24-oic acid:*** the main bile acid [M_r 392.56, m. p. 119 °C, $[\alpha]_D + 11°$ (ethanol)] in the bile of chickens, geese and other fowl. It occurs in small amounts in the bile of ox, guinea pig, bear, pig and human.

Chicle: see Gutta.

Chinese gallotannin: see Tannins.

Chirality: the necessary and sufficient condition for optical activity (rotation of the plane of polarized light). C. means "handedness" (from the Greek κειρ = hand). Chiral molecules have no second order symmetry element (center, plane or axis of symmetry) and exist in two mirror-image forms (enantiomers) which cannot be rotated in such a way as to coincide. Most chiral compounds contain an asymmetrically substituted C-atom, i. e. a tetrahedral C-atom with 4 different substituents. [E. L. Eliel, S. H. Wilen & L. N. Mander *Stereochemistry of Organic Compounds,* Wiley & Sons, New York, 1994]

Chitin: a nitrogen-containing polysaccharide which forms a major component of the exoskeletons of arthropods. It consists of straight chains of *N*-acetyl-D-glucosamine residues linked β1,4. C. is also found in the cell walls of diatoms, fungi and higher plants. It usually occurs with other polysaccharides, proteins or inorganic salts (calcareous deposits). C.-hydrolysing enzymes such as chitinase (EC 3.2.1.14) and chitobiase (β-*N*-acetyl-D-glycosaminidase, EC 3.2.1.30) are widespread among microorganisms, animals and plants. Biosynthesis of C. from UDP-*N*-acetylglucosamine is catalysed by chitin synthase (EC 2.4.1.16).

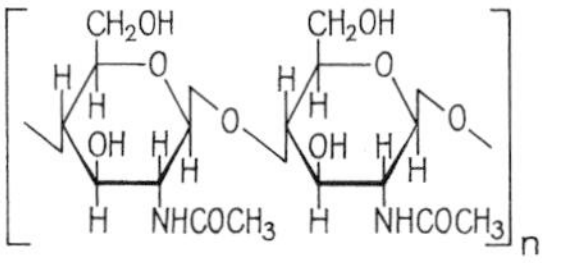

Chitin

Chitosamine: see D-Glucosamine.

Chloramphenicol: an antibiotic, M_r 323, from *Streptomyces venezuelae.* There are 4 stereoisomers, of which only D(–)-*threo*-C. (Fig.) is an antibiotic. C. inhibits protein synthesis on 70S ribosomes of prokaryotes, and on the mitochondrial ribosomes of eukaryotic cells. Protein synthesis on 80S eukaryotic ribosomes is not affected. C. inhibits peptide bond formation and peptidyl transferase activity on the 50S ribosomal subunit, by specifically binding to one of the 50S ribosomal proteins involved in these reactions. The protein in question is probably localized in the acceptor-donor region of the ribosome. C. is used as a broad-spectrum antibiotic in the treatment, e.g. of typhoid fever, paratyphus, spotted fever, infectious hepatitis, dysentery, diphtheria and "viral influenza". Because it inhibits protein synthesis in mitochondrial ribosomes, C. is relatively toxic. It is now produced entirely synthetically.

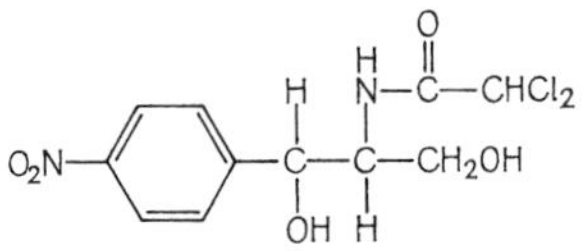

Chloramphenicol

Chlorin: one of the basic ring structures of the porphyrins. Using the numbering system of the Commission of Nomenclature of Biological Chemistry, 1960 (see Porphyrins), C. is 17,18-dihydroporphyrin. With the introduction of the term porphyrin, terms in the original Fischer nomenclature, such as C., porphin, phorbin, bacteriochlorin, etc., are now superfluous.

Chlormadinone acetate, ***6-chloro-17α-hydroxy-pregna-4,6-diene-3,20-dione acetate:*** a synthetic gestagen. Administered orally, C. has high progesterone activity and is used in oral contraceptives. It is also used in animal breeding to bring on heat and synchronize the ovulation cycle.

Chlormequat chloride: see Chlorocholine chloride.

Chlorobium chlorophylls: see Photosynthetic bacteria.

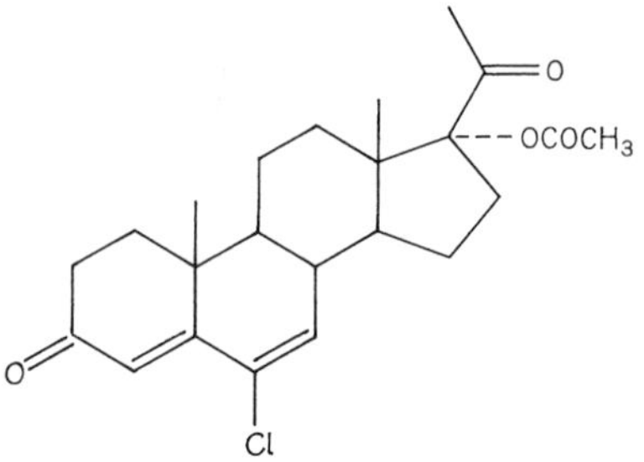

Chlormadinone acetate

Chlorocholine chloride, ***CCC, chlormequat chloride, 2-chloroethyltrimethylammonium chloride:*** $[ClCH_2–CH_2–N(CH_3)_3]^+Cl^-$, a synthetic plant growth regulator which inhibits cell elongation, leading to shortening and strengthening of the stem and a sturdier plant, as well as increased flowering and a larger harvest. It is used on a very wide variety of crop and ornamental plants, e.g. to increase resistance to lodging of wheat, rye, oats and triticale, for promoting lateral branching in azelias, fuchsias, vines, tomatoes, etc., for prevention of premature fruit drop in pears, apricots, etc., as well as being used on numerous other crops such as cotton and various vegetables. It blocks gibberellin synthesis, and may also increase chlorophyll synthesis and root development.

Chlorocruorin: a green respiratory protein containing heme(II) iron, found in the hemolymphs of some marine annelids. The C. of *Spirographis* (pI 4.3, M_r 3×10^6, twelve subunits of M_r 250,000) has a heme component more closely related to the heme of cytochrome oxidase than to that of hemoglobin.

Chlorocruorheme: the heme prosthetic group of Chlorocruorin (see).

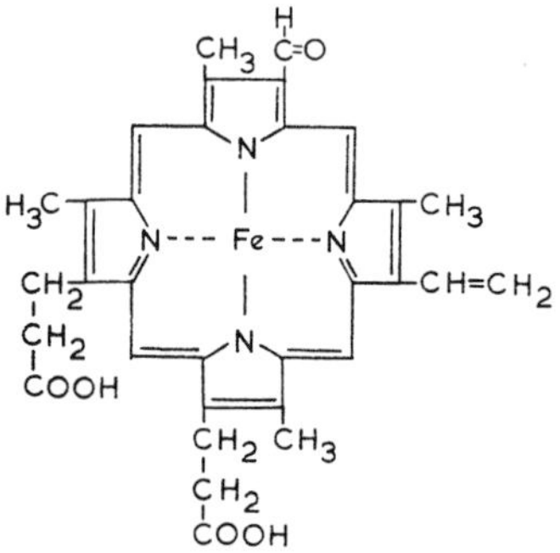

Chlorocruorheme

Chloroethyl choline chloride: see Chlorocholine chloride.

Chloroethylphosphonic acid, ***Ethrel:*** $Cl–CH_2–CH_2–PO(OH)_2$, a synthetic growth regulator, which generates ethylene (fruit ripening hormone). It is used for the loosening of *Prunus* fruits, and to stimulate the flowering of pineapple.

Chloroflurenol: see Morphactins.

Chlorofucin: see Chlorophyll.

Chlorohemin crystals: see Teichmann's crystals.

Chlorophylls: green photosynthetic pigments found in Chloroplasts (see) of all higher plants. C. are magnesium complexes of tetrapyrroles and can be considered as derivatives of protoporphyrins (see Porphyrins). They differ from other porphyrins in that: 1. they have a saturated rather than a double bond between C7 and C8; 2. they have a pentanone ring carrying a methylated carboxyl group fused to ring III of the pyrrole; 3. C7 carries an esterified propionic acid residue; in bacteriochlorophyll *a* and in C.*a,* this is esterified to phytol (Fig.), and in other bacteriochlorophylls it is esterified to farnesol (see Photosynthetic bacteria). This long-chain alcohol is responsible for the waxy properties of C. and makes it difficult to crystallize. It also provides a lipophilic anchor by which the molecule is held in place in the thylakoid membrane. The tetrapyrrole ring is hydrophilic.

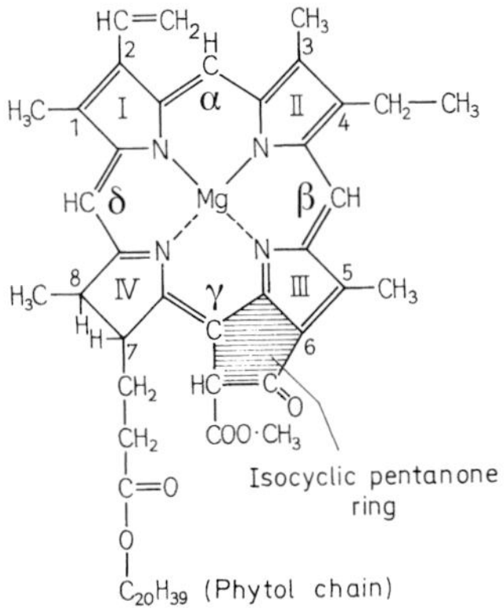

Fig. 1. *Chlorophyll a*

Removal of the Mg from C. generates pheophytin; hydrolysis of the phytol or farnesol ester (catalysed by chlorophyllase, EC 3.1.1.4) produces a water-soluble *chlorophyllide.* Removal of Mg from the chlorophyllide produces a *pheophorbide.* All these compounds are intensely colored and fluoresce strongly in strictly anhydrous solution in an organic solvent. The characteristic absorption spectra are used to identify and quantify C. and their derivatives. In vivo, the absorption spectra are strongly affected by binding to proteins. Some C.-protein complexes of the thylakoid membrane have been isolated and identified.

Only a very small part of the plant's C. (molecules of C.*a* or bacteriochlorophyll *a)* is directly active in Photosynthesis (see). Active centers consisting of C.a_I or C.a_{II} complexed with Plastoquinone (see) are absolutely necessary for the primary process of photosynthesis. More than 99 % of the C. serve as accessory pigments, as do the thylakoid Carotenoids (see) which trap and funnel light to the active centers of Photosystems I and II (see). Fig. 2 shows the possibilities for conversion of electronic energy resulting from the trapping of a photon by a C. molecule.

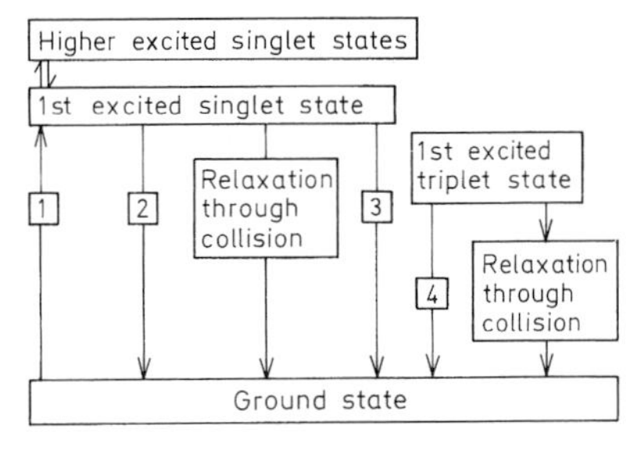

Fig. 2. *Chlorophyll.* Possible electronic energy conversions following the trapping of a photon.

Up to the level of protoporphyrin IX, C. biosynthesis is the same as that of the Porphyrins (see), bearing in mind that different mechanisms exist for the biosynthesis of 5-aminolevulinate, depending on the organism. Conversion of protoporphyrin IX into C.*a* is shown in Fig. 3. The final steps of biosynthesis appear to take place in situ in the thylakoid membrane. [*Biosynthesis of Heme and Chlorophylls,* H. A. Dailey (ed.) McGraw Hill, 1990; S. B. Brown et al. *J. Photochem. Photobiol., B: Biology* **5** (1990) 3–23]

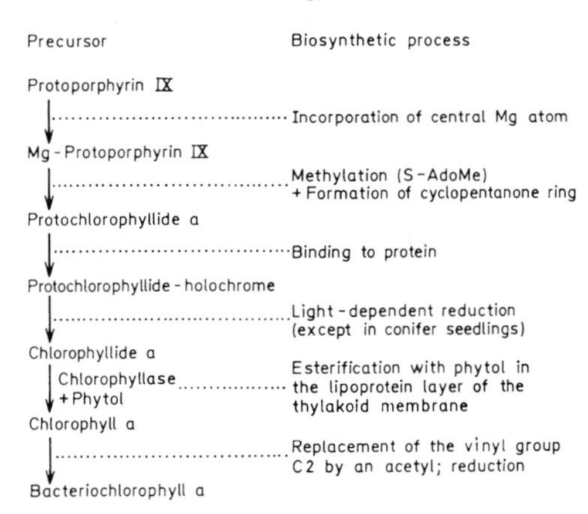

Fig. 3. *Chlorophyll.* Biosynthesis of chlorophyll *a* and bacteriochlorophyll *a.*

Chlorophyllase (EC 3.1.1.4): a plant carboxyesterase which catalyses the reversible transformation of chlorophyllides to chlorophyll in the last step of chlorophyll biosynthesis (Chlorophyllide *a* + phytol ⇌ chlorophyll + H_2O). The pH optimum for the reaction is 6.2. The enzyme is found in all plants, in both the green and nongreen parts such as roots. It is localized in the lipoprotein layer of the thylakoid membrane, which contains all the enzymes and pigments of the photosynthetic apparatus.

Chlorophyllide: see Chlorophyll.

Chloroplasts: the photosynthetic organelles of higher plants. The entire C. is lens-shaped and contains stacks of membranes (thylakoids) embedded in

Chlorophylls, structure and occurrence. Note that the macrocyclic ring displays different degrees of reduction, i.e. a chlorophyll may be a porphyrin, a dihydroporphyrin or a tetrahydroporphyrin. [Data from S.I. Beale & J.D. Weinstein in *Biosynthesis of Heme and Chlorophylls* (H.A. Dailey, ed.) pp. 287–391, McGraw-Hill Publishing Co., New York, 1990]

Chlorophyll	Occurrence	Structure (differences from chlorophyll *a*)
Chlorophyll *a*	All photosynthetic organisms except bacteria	Fig. 1. (a dihydroporphyrin)
Chlorophyll *b*	All higher plants (except the orchid *Neottia nidus avis*), green algae *(Euglenophyta, Chlorophyta)* and stoneworts *(Charophyta)*	–CHO replaces $-CH_3$ on C3.
Chlorophyll *c* (chlorofucin)	Diatoms *(Bacillariophyta), Dinophyta,* brown algae *(Phaeophyta)* and a few red algae *(Rhodophyta)*	Chlorophyll c_1: unesterified *(E)* acrylic acid residue on C7; ring IV has C7,C8 double bond. A porphyrin Chlorophyll c_2: unesterified *(E)* acrylic acid residue on C7; ethylidene residue on C4; ring IV has C7,C8 double bond. A porphyrin.
Chlorophyll *d*	Red algae (Rhodophyta)	CHO replaces vinyl group on C2.
Bacteriochlorophyll *a*	Purple sulfur bacteria *(Chromatiaceae,* formerly *Thiorhodaceae),* purple non-sulfur bacteria *(Rhodospirillacaeae,* formerly *Athiorhodaceae).*	Acetyl on C2; H & methyl on C3; H & ethyl on C4 (i.e. it is a dihydrochlorophyll); phytol or geranylgeraniol esterified to the C7 propionic acid residue. Both ring II and ring IV are more reduced than in chlorophyll *c.* A tetrahydroporphyrin
Bacteriochlorophyll *b*	Some purple non-sulfur bacteria *(Chromatiaceae),* e.g. *Rhodobacter viridis.*	Acetyl on C2; H & methyl on C3; ethylidene on C4; phytol, geranylgeraniol or $\Delta^{2,10}$-phytadienol esterified to the C7 propionic acid residue.
Bacteriochlorophyll *g*	Some strictly anaerobic purple photosynthetic bacteria	H & methyl on C3; ethylidine on C4; farnesol or geranylgeraniol esterified to the C7 propionic acid residue.
Bacteriochlorophyll *c* (formerly called a *Chlorobium* chlorophyll)	Green sulfur bacteria *(Chlorobiaceae)* and green non-sulfur bacteria *(Chloroflexaceae).*	α (*S* or *R*)-hydroxyethyl on C2; ethyl, *n*-propyl or *iso*-butyl on C4; methyl or ethyl on C5; no methylcarbonyl residue on C10; farnesol (or phytol in *Chloroflexus aurantiacus)* esterified to the C7 propionic acid residue; methyl on δC.
Bacteriochlorophyll *d* (formerly called a *Chlorobium* chlorophyll)	Green sulfur bacteria *(Chlorobiaceae)* and green non-sulfur bacteria *(Chloroflexaceae).*	α (*S* or *R*)-hydroxyethyl on C2; ethyl, *n*-propyl, *iso*-butyl or *neo*-pentyl on C4; methyl or ethyl on C5; no methylcarbonyl residue on C10; farnesol esterified to the C7 propionic acid residue.
Bacteriochlorophyll *e* (formerly called a *Chlorobium* chlorophyll)	Green sulfur bacteria *(Chlorobiaceae)* and green non-sulfur bacteria *(Chloroflexaceae).*	α (*S* or *R*)-hydroxyethyl on C2; CHO on C3; ethyl, *n*-propyl or *iso*-butyl on C4; ethyl on C5; no methylcarbonyl residue on C10; farnesol esterified to the C7 propionic acid residue; methyl on δC.
Protochlorophyll (biosynthetic precursor of chlorophyll *a)*		Double bond between C7 and C8; a porphyrin.
Chlorophyllide		No phytol residue, water-soluble

an aqueous phase (stroma). The chlorophyll is embedded in the thylakoids, where light energy is tranduced to the chemical energy of ATP and the reducing equivalents of NADPH; in a process called the light reaction (see Photosynthesis), hydrogen (as protons + electrons) is removed from water, and O_2 is released. In the dark reactions, which occur in the stroma, CO_2 is reduced to carbohydrate. The stroma also contains enzymes for the synthesis of starch, fatty acids and amino acids, as well as the genetic system of the C., including the components of transcription and translation.

C. of higher plants and green flagellates contain 10^{-15}–10^{-14}g of DNA which is organized in 20 or more redundant copies of a circular molecule. Each DNA molecule has M_r 90 × 10^6 and a length of 40 μm (peas, lettuce, *Euglena)*. The DNA and some of the 70S ribosomes are associated with the thylakoid membrane.

Chlorosomes: see Photosynthetic bacteria, Photosynthetic pigment systems.

4-Chlorotestosterone, ***4-chloro-17β-hydroxyandrost-4-en-3-one:*** an anabolic steroid obtained by partial synthesis from testosterone. Esters of 4-C. are used therapeutically, e.g. in convalescence, and for treatment of inflammation and tumors.

7′-Chlorotetracycline: see Tetracyclines.

Chlorotriazine dyes: dyes containing chlorotriazinyl groups. They react readily with polysaccharide matrices under alkaline conditions, forming stable, dyed products. A large number of such dyes is available, covering a wide range of colors. Two have found application in biochemistry: *Cibacron Blue F3GA* (Ciba-Geigy) and *Procion Red HE3B* (ICI). Cibacron Blue is the chromophore of *Blue Dextran,* a high M_r (average 2 million) polysaccharide used to measure the void volumes of gel filtration columns; it is completely excluded from all types of gel filtration columns. Some proteins (e.g. pyruvate kinase, phosphofructokinase, lactate dehydrogenase) bind to Blue Dextran and therefore emerge earlier than expected from gel filtration columns. This observation led to the discovery that the dye moiety (Cibacron Blue) of Blue Dextran interacts specifically with the nucleotide binding domain (see Dinucleotide fold) in lactate, malate and glyceraldehyde 3-phosphate dehydrogenases, and with the ATP-binding site in phosphoglycerate kinase. Columns of Blue Sepharose (Sepharose with bound Cibacron Blue) can therefore be used for the affinity chromatography of NAD^+- or ATP-dependent enzymes; the coenzyme is used as the competing ligand for protein elution. Procion Red has similar affinity properties to those of Cibacron Blue; immobilized Procion Red shows a higher specificity for binding $NADP^+$-dependent enzymes, and it has been used to advantage in the affinity chromatographic purification of glucose 6-phosphate dehydrogenase.

Cholane: see Steroids.

Cholanic acid: 5β-cholan-24-oic acid. See Bile acids.

Cholecalciferol: vitamin D_3. See Vitamins.

Cholecystokinin, ***pancreozymin:*** a tissue hormone consisting of 33 amino acids (porcine), M_r 3,838. It is formed in the mucosa of the upper intestine in response to the presence of chyme (acid mixture of partially digested food) or to a nervous stimulus, and promotes contraction of the gall bladder and secretion of pancreatic juice. [F. Ozcelebi & L. J. Miller *J. Biol. Chem.* **270** (1995) 3435–3441]

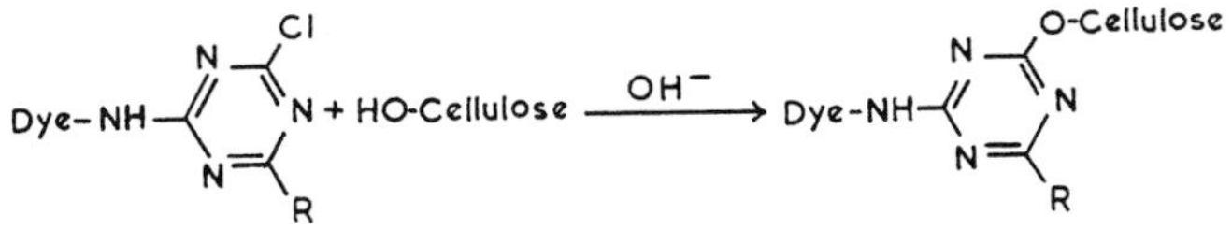

Coupling reaction of a chlorotriazine dye with cellulose

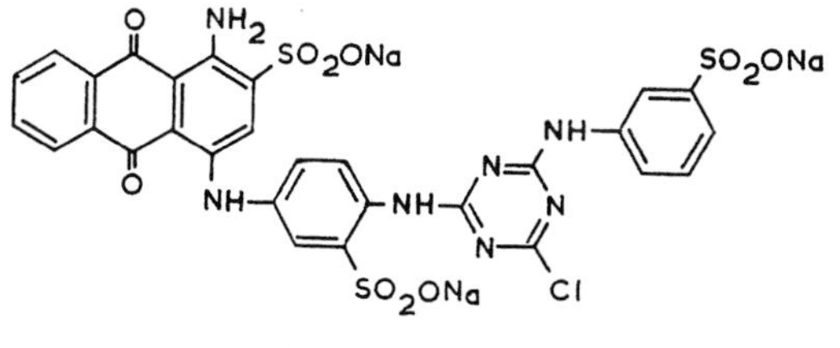

Cibacron Blue F3GA

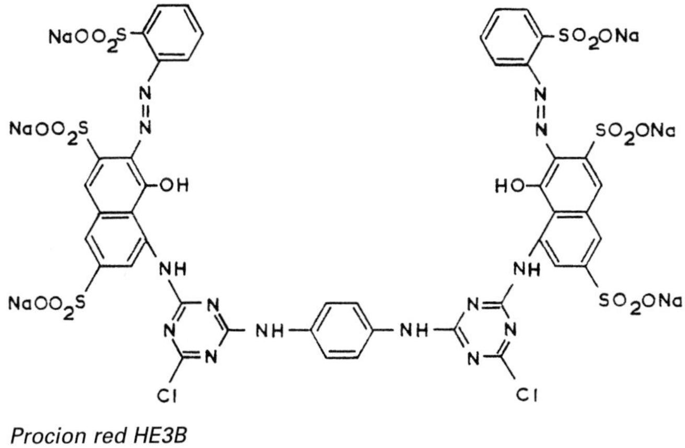

Procion red HE3B

Cholera toxins: see Toxic proteins.

Cholestane: see Steroids.

Cholestanol: 5α-cholestan-3β-ol, a zoosterol (see Sterols). It is a 5,6-dihydro-derivative of cholesterol, and occurs in small amounts with cholesterol in animal cells. In some sponges, C. is the main sterol. It differs from the stereoisomeric Coprostanol (see) in the configuration at C5 and the *trans* configuration of rings A and B.

Cholesterol: the main sterol of higher animals (see Sterols). It occurs free or esterified to fatty acids in all mammalian tissues, often together with phospholipids. It is especially abundant in brain (about 10% of dry matter), adrenals, egg yolk and wool grease. Blood contains about 2 mg/ml, bound to Lipoproteins (see). C. is a component of biomembranes and of the myelin sheaths around nerve axons. It has a detoxifying effect on hemolytically active Saponins (see), with which it forms insoluble complexes, thus protecting red blood cells from lysis. Human skin excretes up to 300 mg C. daily. Deposition of C. on the interior of artery walls (arteriosclerosis) and in gall stones is pathological (see Lipoproteins). C. was first isolated from the latter by Green in 1788.

C. has also been isolated in small amounts from plants, e.g. potato plants, from many pollens, isolated chloroplasts and from bacteria. It is a vitamin for many insects which require it as a precursor of ecdysone and related molting hormones. Commercially, C. is prepared from the spinal cords of cattle, or from wool grease. It was first synthesized chemically by Robinson and Woodward in 1951. For biosynthesis, see Terpenes.

C. is biosynthesized from acetyl-CoA via the triterpene lanosterol (see Steroids, Figs. 7 & 8) and zymosterol. In turn, it is a key intermediate in the biosynthesis of many other steroids, including steroid hormones, steroid sapogenins and steroid alkaloids.

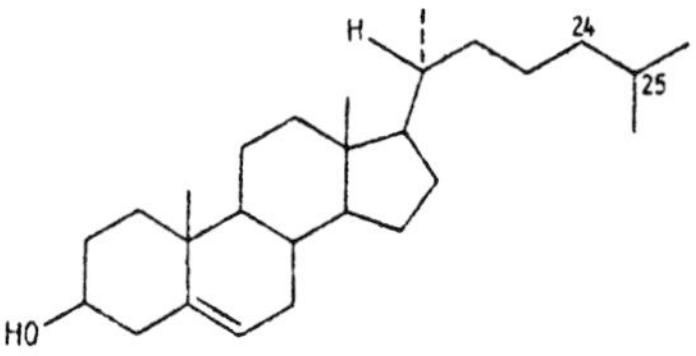

Cholesterol

Cholestyramine: a quaternary ammonium ion exchange resin, in which the basic groups are attached to a styrene-divinyl benzene copolymer by C-C bonds. It is a lyophilic solid, equivalent weight about 230. C. binds bile acids in the intestine, thereby preventing their reabsorption and return to the liver in the enterohepatic circulation. This results in an incease in the conversion of cholesterol to bile acids in the liver, and an increased uptake of cholesterol-containing phospholipids by the liver. Dietary administration of C. increases plasma cholesterol in cockerels, dogs and humans, and decreases aortic plaque formation in cholesterol-fed cockerels. It has been used to decrease plasma cholesterol levels in human heterozygotes for familial cholesterolemia. [J. W. Huff et al. *Proc. Soc. Exp. Biol. Med.* **114** (1963) 352–355; M. S. Brown & J. L. Goldstein *Science* **232** (1986) 34–47]

Cholic acid, *3α,7α,12α-trihydroxy-5β-cholan-24-oic acid:* a bile acid, M_r 408.56, m.p. 196–198 °C (anhydrous), $[\alpha]_D^{20}$ +37° (c = 0.6 in ethanol), found as its lysine or taurine conjugate in the bile of most vertebrates. It forms salts with fatty acids and other lipids, such as cholesterol and carotene. It is used as a starting material for the partial synthesis of therapeutically important steroid hormones.

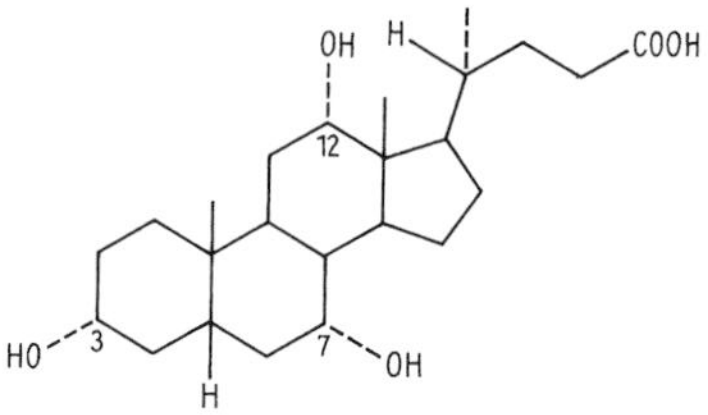

Cholic acid

Choline: $[(CH_3)_3N^+–CH_2–CH_2–OH]OH^-$, a component of some phospholipids (see Membrane lipids) and of Acetylcholine (see). It is a metabolic methyl group donor. It decreases the deposition of body fat, lowers the blood pressure, and causes the uterus to contract. Under certain conditions it may be required in the diet, and is therefore sometimes accorded the status of a vitamin.

Choline acetylase (EC 2.3.1.6): see Acetylcholine.

Cholinergic receptor: see Acetylcholine.

Cholinesterase, *pseudocholinesterase* (EC 3.1.1.8): an unspecific acylcholinesterase which hydrolyses butyroyl and propionoyl choline much faster than acetylcholine. It is found primarily in the serum (M_r of horse serum C. 315,000, 4 subunits of M_r 78,000; human serum C. M_r 348,000, 4 subunits of M_r 86,000), the liver and pancreas. C. is also found in cobra venom. Because it is secreted by the liver, the serum concentration of C. is measurably reduced in liver parenchyma damage. The enzyme is inhibited by the carbamate esters, physostigmine (eserine) and prostigmine, which have no effect at 10^{-4} M on type A and B carboxylesterases of serum and organs. Since the active center contains a catalytically important serine* residue (–Gly–Gly–Asp–Ser*–Gly–), the enzyme is stoichiometrically and irreversibly inhibited by organic phosphate esters (DFP, E600, E605, Wofatox and other nerve and tissue poisons). [D. M. Quinn et al. (eds.) *Enzymes of the Cholinesterase Family,* Plenum Press, New York & London, 1995]

Chondroitin sulfate: a water-soluble mucopolysaccharide, M_r ~ 250,000, found in animals. C. s. A and C consist of equimolar amounts of D-glucuronic acid and *N*-acetyl-D-galactosamine sulfate linked by alternating β-1,3 and β-1,4 bonds; they differ in the position of the esterified sulfate (Fig.). C. s. B (dermatan sulfate) contains L-iduronic acid instead of D-glucuronic acid (Fig.). C. ss. make up 40% of the dry weight of cartilage. They are also found in skin, tendons, umbilical cord, heart valves and other connective tissue. In vivo they are associated noncovalently with proteins.

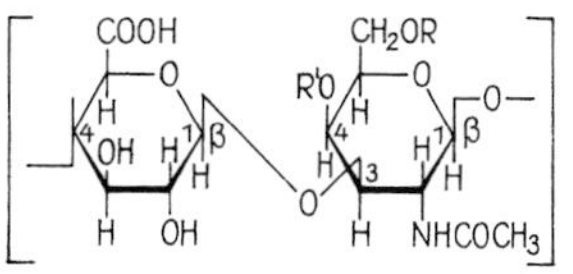

Chondroitin sulfate A: R = H, R' = SO_3H
C: R = SO_3H, R' = H

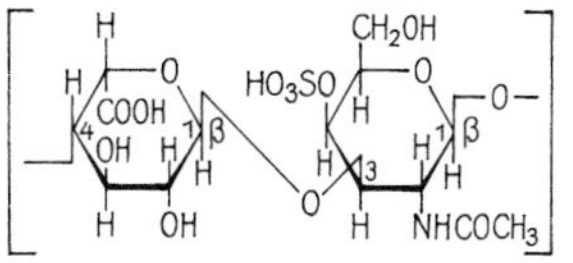

Chondroitin sulfate B

Chondrome: the genetic information contained in the mitochondria of a cell. Since the number of mitochondria per cell varies as widely as the amount of DNA in each mitochondrion, the size of the C. is highly variable.

Chondrosamine: see D-Galactosamine.

Choriogonadotropin, ***placental gonadotropin, human chorionic gonadotropin, hCG:*** an important hormone formed in the placenta during pregnancy. hCG is a glycoprotein, M_r 30,000, containing about 30% carbohydrate and 2 polypeptide chains: α-subunit, M_r 10,205, 92 amino acids; β-subunit, M_r 14,902, 139 amino acids. The α-subunit is common to other glycoprotein hormones (follicle-stimulating hormone, luteinizing hormone, thyrotropin) and is encoded by a single gene. The β-chain has 82% sequence homology with the β-chain of luteinizing hormone and probably evolved form it. hCG stimulates the ovaries to produce the steroid hormones necessary to maintain pregnancy. Using monoclonal antibodies, hCG is detectable in maternal urine or plasma as early as one week after implantation. Immunoassay of hCG is therefore used as a pregnancy test.

Choriomammotropin, ***placentalactogen, PL, human lactogen:*** a single chain polypeptide hormone of known primary structure (191 amino acid residues, M_r 22,308). It is synthesized in increasing amounts by the placenta during pregnancy, and secreted into the maternal circulation. Its action is similar to that of Somatotropin (see).

Chorismic acid: see Aromatic biosynthesis.

Christmas factor: see Blood coagulation.

Chromatid: see Chromosomes.

Chromatin: the stainable material of the interphase nucleus, consisting of DNA, RNA and several specialized proteins, which is dispersed randomly in the nucleus. The chromosomal DNA is intact in this phase; it is merely uncoiled or "relaxed". Immediately prior to cell division, C. condenses into dense bodies (chromosomes), which can be intensely stained. Heterochromatin is tightly coiled and densely packed chromosomal material which is not being transcribed. Euchromatin has a looser structure and is the site of transcription. See Nucleosomes. See Chromosome.

Chromatography: a type of method used for analytical or preparative separation of the components of mixtures. Paper and thin layer C. and simple column systems require only a modest amount of relatively simple equipment, whereas Gas chromatography (see), high performance liquid chromatography and automated ion exchange systems for amino acid analysis rely on advanced technology and electronic control systems. The lower limit for the scale of any chromatographic separation is determined by the sensitivity of the detection method. This group of techniques has revolutionized organic and biological chemistry by enabling the separation of closely related compounds which could not be separated by any other method.

All chromatographic methods employ a stationary and a mobile phase. The mixture to be separated is carried through or past the stationary phase by the

Definition of terms used in chromatography

Term	Definition
Equilibration	Equilibration of column material with the running solvent, or saturation of the paper to be used with the vapor of the solvent.
Elution	Emergence of material from a column, or washing out of material from a paper or thin layer chromatogram.
Eluant	The material which has been eluted.
Eluate	A solution emerging from a chromatography column, or a solution of material extracted from an area of a paper or thin layer chromatogram.
Developer, Eluent, or Elutant	The mobile phase.
Development	The process of chromatography, or treatment of a chromatogram with a detection reagent.
Solvent or solvent system	A pure liquid or a mixture of solvents.
Front	Position of the leading edge of the solvent
Running time	The time for which chromatography is performed.
Standard substance	An authentic substance used as an internal marker for the calibration of a system, or for comparison with and identification of a chromatographed substance.
Detection	Detection of material in a column eluate by its UV or visible absorption, fluorescence, etc. before or after treatment with other reagents; or by other methods such as detection of immunological activity, radioactivity, etc. Detection methods of gas chromatography include flame ionization. Detection on paper and thin layer chromatograms includes visualization by staining, and fluorescence under UV light.

mobile phase. Separation occurs because the components of the mixture are retarded to different degrees by the stationary phase. Different types of interactions between the components and the stationary phase are exploited. In *partition* C. the substances are distributed between two immiscible phases on the basis of their relative solubilities in the two phases. This is analogous to countercurrent distribution, but in C. the number of steps approaches infinity. *Adsorption* C. depends on the difference in degree of adsorption of the components to the solid stationary phase. The most elegant application of this principle is *affinity* C. (see Proteins), in which one component of a mixture is separated by its specific affinity to the column material. This stationary phase is prepared by linking a specific substrate (or antigen or biological binding partner) to an inert solid such as Sepharose. The desired enzyme (or antibody, hormone, repressor protein, etc.) is then selectively bound to its substrate while all others pass through. Other forms of interaction with the stationary phase occur in *Ion exchange* C. (see) and *Gel filtration* (see Proteins). C. is also named according to the type of stationary or mobile phase: 1. Paper C. (see), 2. Thin layer C. (see), 3. Column C. (see), and 4. Gas C. (see). Electrophoresis (see) is a related technique in which the components to be separated are moved through the stationary phase by electromotive force rather than by the mobile phase.

Chromatophores: 1. plastids of higher plants: Chloroplasts (see), Chromoplasts (see) and Leucoplasts (see). 2. The photosynthetic organelle of Photosynthetic bacteria (see). Bacterial C. are intraplasmatic membranes originating from the cell membrane. They may exist as closed vesicles or as flattened stacks, whose membranes contain the photosynthetic pigments, and the components of photosynthetic electron transport and photophosphorylation.

Chromium, *Cr*: an essential dietary constituent for animals. Very small amounts are required (at least 100 parts per billion in the diet of the rat), but the human dietary requirement has not been determined. Although relatively large amounts of Cr are associated with isolated RNA fractions, the relationship of Cr to RNA structure or function is unknown. Cr is a constituent of glucose tolerance factor (GTF), a water-soluble, relatively stable organic complex of Cr, M_r about 500, which is essential in animals and humans for normal glucose tolerance. The earliest symptom of Cr deficiency is impaired glucose tolerance. More severe deficiency leads to glycosuria, fasting hyperglycemia, impaired growth, shortening of life span. Diabetes refractory to insulin may be due to Cr deficiency. In such cases, children respond more readily than adults to infused Cr salts, suggesting that the ability to convert Cr into GTF decreases with age, although the ability of humans to convert Cr into GTF is not unequivocally proven. Many natural foods contain GTF, especially brewer's yeast, black pepper, liver, cheese, bread and beef. Foods highest in GTF are not necessarily highest in Cr. Only Cr^{3+} is physiologically active.

Chromogranins: proteins that specifically bind catecholamines (e.g. adrenalin, noradrenalin). C. are associated with catecholamines in storage vesicles within the cells that produce catecholamines.

Chromomeres: see Chromosomes.

Chromophores: see Pigments.

Chromoplast: a chromatophore filled with carotenoids and therefore red-orange to yellow in color. The pigments may be crystallized out of solution within the C., as in the carrot root. In forsythia petals the C. develop from chloroplasts. Pigment vesicles of red algae, which are colored red to violet by Biliproteins (see), are called Rhodoplasts (see).

Chromoproteins: proteins which contain a colored prosthetic group bound covalently or noncovalently. The group includes heme proteins and iron prophyrin enzymes, flavoproteins, chlorophyll-protein complexes, and the non-porphyrin iron and copper proteins in the blood of vertebrates (e.g. transferrin and ceruloplasmin), and invertebrates (e.g. hemerythrin and hemocyanin).

Chromosome: a subcellular structure for storage of genetic information and its transmission to the next generation. The term was originally applied to the stainable material in eukaryotic nuclei, but it has been extended to include the genetic material of any cell or organelle. Genes are sequences of nucleotide base pairs in a DNA molecule which forms the core of the C. (see Deoxyribonucleic acid). Genetic mapping shows that the genes of a C. form a single linear array, indicating that the C. contains only one very long molecule of DNA. Prokaryotes contain only a single, circular C., which is attached to the cell membrane (see Replicon). Prokaryotic C. are complexed with regulator proteins, but unlike eukaryotic C. they do not contain structural proteins (histones).

Eukaryotes have more than one C. per cell (each carrying a part of the eukaryotic genome), and the number and shapes of these C. are species-specific. As a prelude to nuclear division (mitosis or meiosis), eukaryotic C. form compact structures which are microscopically visible and can be intensely stained. At this time they consist of 2 identical longitudinal halves, the *chromatids,* which are joined at one point, the *centromere.* The centromere is also the point of attachment for the spindle fibers which pull apart the daughter chromatids during cell division. The nucleotide sequences of yeast centromere DNA have been analysed and inserted into Plasmids (see), which confers on the latter a mitotic stability almost as great as that of C. (segments of C. lacking centromeres are not evenly distributed between daughter cells and are rapidly lost from a dividing population).

During the prophase of meiosis, eukaryotic C. display dense, granular, heterochromatic regions, known as *chromomeres,* which are visible under the light microscope, and which contain densely packed DNA. They are separated by DNA-poor regions, forming a banding pattern that is specific for the chromosome and the species. The term, chromomere, is also applied to the condensed regions at the base of loops on Lampbrush C. (see), and the condensed bands in the polytene chromosomes of *Diptera* (see Giant C.).

Other specialized DNA sequences, the *Telomeres* (see), are found at the ends of linear C. Insertion of telomere sequences into a circular plasmid converts it to a linear structure. Another type of sequence (see Autonomously replicating sequence, ARS) confers on plasmids the ability to replicate independently of the C., and may well be the sites at which replication of the C. is initiated. [A.W. Murray & J.W. Szostak *Nature* 305 (1983) 189–193].

Eukaryotic C. are complexes of DNA (10–30 %), RNA (3–15 %) and protein (40–75 %), and the entire complex is known as *chromatin.* During interphase, when chromosomal DNA is transcribed and replicated, the chromatin becomes disperse, so that distinct C. are no longer recognizable. There are two forms of this disperse interphase chromatin: less densely packed *euchromatin* that is expressed genetically (i.e. its DNA is being transcribed to RNA), and more densely packed *heterochromatin* that is not being transcribed. The amount of DNA per C. is constant and species-specific, but the amount of RNA varies with the transcription activity in the cell (see Ribonucleic acid). In addition to the RNA involved in protein synthesis, there is also an organ-specific fraction called *chromosomal RNA,* which is probably part of the C. structure and may have a regulatory function in transcription.

There are 2 classes of chromosomal proteins, basic histones and more acidic nonhistone proteins. The Histones (see) form 5 major classes, but gel electrophoresis of nonhistone proteins produces very complicated patterns with hundreds of bands.

Histones are a group of basic proteins which form reversible complexes with DNA, called nucleohistones. They are divided into the following main classes: H1 (or I or f1) is lysine-rich; H2a and H2b (or IIb1 and IIb2 or f2a2 and f2b) are moderately lysine-rich; and H3 (or III or f3) and H4 (or IV or f2a1) are arginine-rich. Octomers containing two molecules each of H2a, H2b, H3 and H4 have been isolated in soluble form, and radioactive labeling indicates that these octomers segregate conservatively during chromatin replication [I.M. Leffak *Nature* **307** (1984) 82–85]. H1 is present in smaller quantities than the other H. and is thought to act as a link between "beads" (nu bodies) on the chromatin chain.

H2a, H2b, H3 and H4 are among the most highly conserved of all proteins, e.g. H4s from calf and pea differ at only two positions in their 102-residue sequences, and moreover the changes are conservative (Val60calf → Ile60pea; Lys77calf → Arg77pea). Such evolutionary constancy suggests that the structure of the four H. is critically adjusted to a crucial cell function, leaving very little, if any, latitude for structural variations. H1 is more variable than the other histones, i.e. it does not display such a high degree of conservation. H. undergo reversible posttranslational modification by methylation, acetylation and phosphorylation of specific Arg, His, Lys and Thr residues, all resulting in a decrease of the positive charge of the protein, and thereby altering the intensity of the H.-DNA interaction. The extent of posttranslational modification varies greatly according to the species, tissue and stage of the cell cycle. About 10 % of H2as are linked to Ubiquitin (see) via an isopeptide bond between the ε-amino group of Lys 119 and the terminal carboxyl of ubiquitin. [B.M. Turner *Cell* **75** (1993) 5–8; V. Ramakrishnan *Curr. Opin. Struct. Biol.* **4** (1994) 44–50]

Nucleosomes are spherical chromatin components containing DNA and histones. Electron microscopy of isolated chromatin (prepared by spreading) reveals smooth strands (20–30 nm diam.), a lesser proportion of thinner filaments (3 nm diam.), and stretches of nucleosomes resembling strings of beads. Brief digestion of chromatin with micrococcal nuclease (which cleaves double-stranded DNA) releases some of these bead-like nucleosomes, without attacking the DNA (about 200 bp) associated with them. At this stage, the nucleosomes are also complexed with some histone 1 (H1) and are also known as *chromatosomes.* Further digestion removes some of the DNA and releases the H1, producing a *nucleosome core particle.* The latter is a discrete particle, about 10 nm in diameter, consisting of a closely packed octamer of histones (H2A, H2B, H3, H4)2, encircled by 146 nucleotide pairs of B-DNA wound around the octomer cores in 1.8 turns of a flat left-handed superhelix of pitch 28 Å. This wound B-DNA does not describe a smooth superhelix, but bends sharply at several points along is length, so that large variations occur in the widths of its major and minor grooves. The DNA removed in the second digestion is responsible for linking adjacent nucleosomes, and is called *linker DNA.* It varies greatly in length according to its origin, but is usually about 55 bp. Histone 1 (H1) binds to the outside of the core particle and to the linker DNA, thereby promoting association between the nucleosomes in the nucleosome filament. Nucleosomes are also known as *ν* bodies.

300-Å Filaments represent a higher level of chromatin organization, in which the 100-Å nucleosome filament is wound into a solenoid with about 6 nucleosomes per turn and a pitch of 110 Å. The entire structure is thought to be stabilized by a helical polymer of H1 in the center of the solenoid. The single nucleosome filament represents a lower level of chromatin organization, which exists only at unphysiologically low salt concentrations. As the ionic strength is increased, nucleosome filaments begin to associate and fold, finally forming filaments of 300 Å diameter at physiological salt concentrations.

Looped domains represent an even higher level of chromatin organization observed in metaphase chromosomes, in which loops of 300-Å filaments extend radially from the central protein scaffold that forms the main chromosome axis. In the electron microscope, a cross section of a metaphase chromosome appears as a central fibrous region of protein surrounded by a halo consisting of loops of chromatin. Typical loops vary in size between 20,000 and 100,000 bp, corresponding to an average 300-Å filament length of about 0.6 μm. Since the filament is looped (i.e. bent back on itself), the average contribution to the diameter of the chromosome should be 0.3 μm on all sides. The central protein scaffold is 0.4 μm in diameter, so that the predicted diameter of the metaphase chromosome is 1.0 μm, in agreement with the value obtained from electron microscopic measurement. It is highly probable that the nonhistone proteins (which constitute 10 % of chromosomal proteins) are involved in the formation of looped domains. The mode of folding is unknown, but it has been shown by cytological labeling that the protein scaffold is highly enriched for DNA topoisomerase II.

A further level of organization is represented by Giant chromosomes (see) or polytene chromosomes, whose bands probably correspond to the looped domains of metaphase chromosomes. [J. Zlatanova & K. van Holde 'The Linker Histones and Chromatin Structure: New Twists' *Prog. Nucl. Acid Res. Mol. Biol.* **52** (1996) 217–259; C. Gruss & R. Knippers 'Structure of Replicating Chromatin' *Prog. Nucl. Acid Res. Mol. Biol.* **52** (1996) 337–365]

Chromosome crawling: an adaptation of the Polymerase chain reaction (PCR) (see) which enables DNA segments of unknown nucleotide sequence lying on either side of the PCR target DNA to be amplified. The name, coined by T. Triglia et al. [*Nucleic Acids Res.* **16** (1988) 8186], alludes to the fact that the procedure can be used to explore chromosomal sequences that are contiguous with a known segment of DNA; however, the procedure has also been called 'inverse PCR' [H. Ochman et al. *Genetics* **120** (1988) 621–623].

The procedure involves the initial digestion of the genomic DNA to completion with a restriction endonuclease (see) that has no cleavage site within the 'normal PCR' target DNA sequence. The fragment carrying the latter is then isolated; ideally it should be no more than 3,000 nucleotides long. Its two ends are then ligated to generate a circular DNA. The procedure can then continue in either of two ways, namely: (i) the circular DNA is denatured and then put through 25–30 cycles of PCR using the same two pimers as would have been used to amplify the 'normal PCR' target DNA sequence, or (ii) the circular DNA is firstly linearized by treatment with a restriction endonuclease that has a site within the 'normal PCR' target DNA sequence and then put through 25–30 cycles of PCR as before. Procedure (ii) has the advantage over (i) that *Taq* DNA polymerase works slightly better with linear DNA than it does with circular DNA. Whichever procedure is used, the major amplification product is a linear double-stranded DNA consisting of a head-to-tail arrangement of sequences that were originally on either side of the 'normal PCR' target DNA sequence, the junction between them being marked by a cleavage site for the restriction endonuclease that was used to digest the genomic DNA.

Chrysanthemin: see Cyanidin.

Chrysin: see Flavones (Table).

Chylomicrons: see Lipoproteins.

Chymopapain (EC 3.4.22.6): an enzyme (C.A and C.B both of M_r 35,000) from the latex of the papaya tree. In its substrate specificity and structure of its active center it resembles papain (see), but it differs in its acid stability and its very high isoelectric point (C.A pH 10, C.B pH 10.4; cf. papain pH 8.75).

Chymosin (EC 3.4.23.4): see Rennin.

Chymostatin: see Inhibitor peptide.

Chymotropic pigment: a pigment dissolved in the vacuole of a plant cell.

Chymotrypsin: a family of structurally and catalytically homologous serine proteases (see Proteases) whose precursors (zymogens) are formed and stored in the pancreas. Chymotrypsinogen A (pI 9.1, 245 amino acids, M_r 25,670) is cationic at pH 8, whereas chymotrypsinogen B (pI 5.2, 248 amino acids, M_r 25,760) is anionic at pH 8. The latter is lacking in swine pancreas, which contains chymotrypsinogen C (281 amino acids, Trp-rich, M_r 31,800). The activiated forms have different substrate specificities. All C. hydrolyse preferentially tyrosyl and tryptophanyl peptide and ester bonds with a pH optimum at pH 8–8.5. C.B also attacks other bonds (e.g. in glucagon), and C.C attacks leucyl and glutaminyl bonds.

Chymotrypsinogen A consists of a polypeptide chain with 5 intrachain disulfide bridges (1–122, 42–58, 136–201, 168–182, 191–221). Trypsin activates chymotrypsinogen A by hydrolysing the peptide bond between Arg^{15}-Ile^{16}, producing an active 2-chain structure known as π-C. In a process of autolysis, π-C. removes the dipeptides Ser_{14}-Arg_{15} and Thr_{147}-Asn_{148} from other molecules of π-C. to produce an equally active 3-chain structure known as α-C. The A, B and C chains of α-C. are held together by S-S bridges. The catalytically active amino acids His^{57}, Asp^{102} and Ser^{195} are located in the B and C chains (Fig. 1).

The tertiary structure of α-C. consists mostly of extended chain segments held together by hydrogen

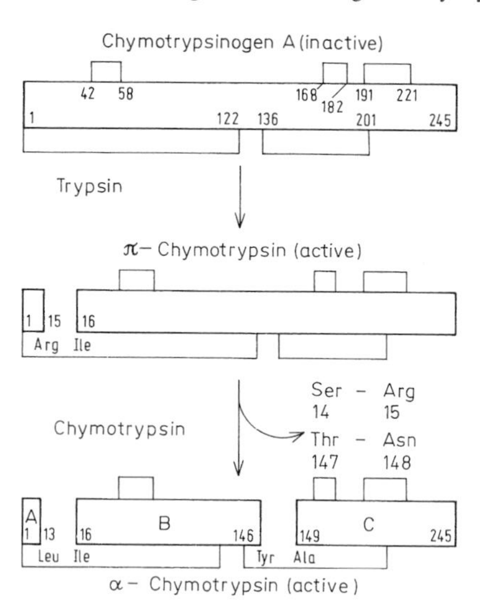

Fig. 1. *Chymotrypsin.* Activation of chymotrypsinogen.

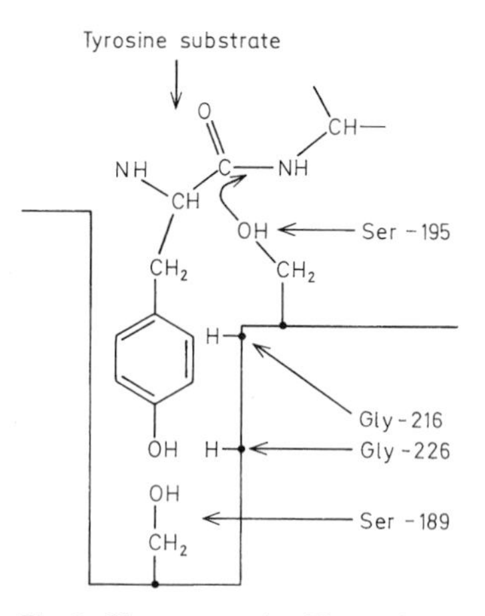

Fig. 2. *Chymotrypsin.* The substrate binding site of chymotrypsin.

bonds and 5 disulfide bridges. Except for the 11 amino acids at the carboxyl end of the C chain, the molecule contains no α-helices. As in trypsin, the active site is located in a loosely structured region of the molecule and has a pocket-like cleft for the side chain of the amino acid of the substrate (Fig. 2). This cleft determines the substrate specificify of the enzyme, and Ser_{189} is located deeply within it.

In the course of activation, chain segment 187–194 is rotated by 180°, bringing the catalytic amino acids to within 0.3 nm of the surface of the molecule. Substrate hydrolysis involves formation of an acyl-enzyme intermediate between the acid group of the peptide substrate and the hydroxyl of the Ser_{195}. The strongly nucleophilic character of this hydroxyl is due to the neighboring proton donor (or acceptor) His_{57}. The effect is amplified by Asp_{102} (see Charge relay system).

Cinchona alkaloids: a group of about 30 alkaloids from the bark of tropical trees, especially *Cinchona succiruba.* Since their precursors include tryptamine, they are considered to be indole alkaloids, although the main representatives contain a quinoline ring system The dried bark of *Cinchona* contains 7–10 % alkaloids, and 5–7 % of the main alkaloid, Quinine (see); secondary alkaloids include quinidine, epiquinine and epiquinidine. Other important C. a. are cinchonine and cinchonamine. In contrast to the above C. a., cinchonamine has an indole rather than a quinoline structure. Extracts of cinchona, as well as pure quinine and quinidine have several medical uses, in particular as antimalarial agents. The quinoline ring of the C. a. is biosynthesized from tryptophan via tryptamine, and the quinuclidine nucleus is derived from iridoid compounds. Cinchonamine is synthesized from a carboline-type compound by cleavage of the C ring and reaction of the N atom with the para side of the D ring. This is followed by oxidation of the primary alcohol group, hydroxylation and opening of the indole ring (compound I, Fig.).

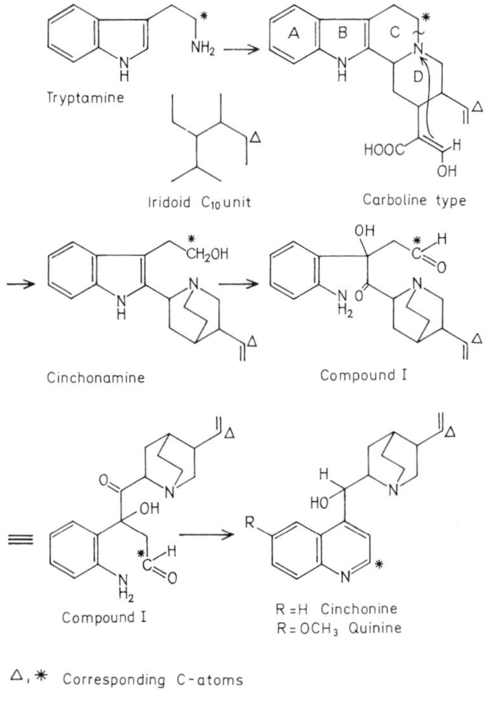

Tryptamine

Iridoid C_{10} unit

Carboline type

Cinchonamine

Compound I

R = H Cinchonine
R = OCH_3 Quinine

Δ, * Corresponding C-atoms

Biosynthesis of the Cinchona alkaloids

1,8-Cineole, *eucalyptol:* the main constituent of oil of eucalyptus. M_r 154.25, m. p. + 1.3 °C, b. p. 176–177 °C, ρ_{20} 0.9267, n_D^{20} 1.455–1.460. It is used in cough syrup.

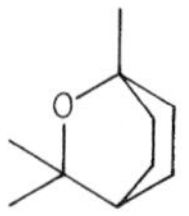

1,8-Cineole

Cinnamic acid 4-hydroxylase (EC 1.14.13.11): a mixed function monooxygenase, present in plants, which catalyses an early reaction in Flavonoid (see) biosynthesis, i. e. the insertion of an atom of oxygen into cinnamic acid to form 4-hydroxycinnamic acid (4-coumaric acid) with concomitant oxidation of one molecule of NADPH. The enzyme is a cytochrome P450 system associated with the microsomal fraction, and is specific for the *trans* isomer of cinnamic acid. During the hydroxylation, hydrogen at position 4 (experimentally tritium in position 4) is retained, i. e. there is an NIH shift (see). In vitro, a thiol, e. g. 2-mercaptoethanol, is required for activity. [P. R. Rich & C. J. Lamb *Eur. J. Biochem.* **72** (1977) 353–360]

Circular dichroism (CD): an optical property of a molecule that is indicative of asymmetric features of its molecular structure. CD spectra allow the secondary structure of proteins and nucleic acids to be characterized rapidly using 1–2 ml solution at a concentration of 0.05 – 0.5 mg/ml.

CD is based on the fact that left- and right-circularly polarized beams of light are absorbed to different extents by chiral molecules. CD spectra consist of a plot of a measure of this difference as ordinate against wavelength as abscissa. The difference may be expressed directly as ΔA, which is defined in equation 1, or indirectly as the molar ellipticity (θ_m or $[\theta]$).

$$\Delta A = A_L - A_R = \Delta\varepsilon.c.L \qquad (1)$$

where: A_L = absorbance of left-circularly polarized beam,
A_R = absorbance of right-circularly polarized beam,
$\Delta\varepsilon$ = the difference between the absorption coefficients of the two beams ($l\,mol^{-1}\,cm^{-1}$),
c = sample concentration ($mol\,l^{-1}$),
L = the length of the light path (cm).

Molar ellipticity arises from the fact that, although the incident left- and right-circularly polarized beams have the same amplitude (i. e. the combined beam is plane polarized), they have different amplitudes after passing though the solution of the sample; consequently the combined transmitted beam is elliptically polarized. Its ellipticity (θ_{obs}) is measured in degrees and is defined by equation 2.

$$\theta_{obs} = \tan^{-1}(b/a) \qquad (2)$$

where: b/a = the ratio of the two axes, minor to major, of the elliptically polarized beam.

Molar ellipticity (θ_m or $[\theta]$) is expressed in degrees $cm^2\ dmol^{-1}$ and is defined by equation 3, in which c and L have the meanings given earlier.

$$\theta_m \text{ (or } [\theta]) = (\theta_{obs} \times 10)/c.L \qquad (3)$$

It is related to $\Delta\varepsilon$ (see equation 1) by equation 4

$$\theta_m \text{ (or } [\theta]) = 3300\Delta\varepsilon \qquad (4)$$

Mean residue ellipticity is often used in the context of proteins; it is simply the ellipticity divided by the number of amino acid residues in the protein.

In protein molecules the main source of optical activity is the peptide bonds. Consequently CD spectra are taken at wavelengths shorter than 240 nm where the predominant absorption comes from peptide bonds. This arises from three electronic transitions, namely: (i) an $n \rightarrow \pi^*$ transition at ~210–220 nm, in which an electron in the non-bonding molecular orbital of the carbonyl oxygen is promoted to an antibonding π molecular orbital, (ii) a $\pi \rightarrow \pi^*$ transition at ~190 nm, and (iii) a probable $\pi \rightarrow \pi^*$ transition at ~160 nm; the latter, however, does not participate in CD spectra because of measurement difficulties in this wavelength range. Since the backbone of protein molecules is made up of amino acid residues linked together by peptide bonds, CD spectra taken in the wavelength range >180 nm – <240 nm reflect the secondary structure (i.e. α-helix, β-strand or random coil) of proteins. The CD spectral curves that are characteristic of these conformations were first identified using polypepeptides, such as poly-l-lysine, in known conformations (e.g. Fig. 1). Subsequently CD spectra were taken of a range of proteins whose type of secondary structure and the percentage that it constituted of the whole molecule was known from X-ray crystallography (see), and used to calculate a set of standard CD curves which can be used, in a semiquantitative way, to predict the percentage of α-helix, β-strand or random coil present in any new protein from its CD spectrum [N. Greenfield & G. D. Fasman *Biochemistry* **8** (1969) 4108–4116].

In nucleic acid molecules the main source of optical activity arises from the asymmetric positioning of the purine and pyrimidine residues. CD spectra, taken in the 220–320 nm wavelength range where the latter absorb but the sugar-phosphate backbone does not, therefore reflect the mode of *N*-base stacking, which is influenced not only by the conformation of the strand but also by the type of nucleic acid (i.e. DNA or RNA), the number of strands (i.e. single or double) and the nucleotide sequence. In spite of the latter three constraints, the right handed A-form and B-form of DNA can be recognized from their CD spectral curves, both of which consist of a peak (i.e. ΔA or $\theta_m > 0$) and a trough (i.e. ΔA or $\theta_m < 0$). For the A-form the areas of the peak and trough are roughly

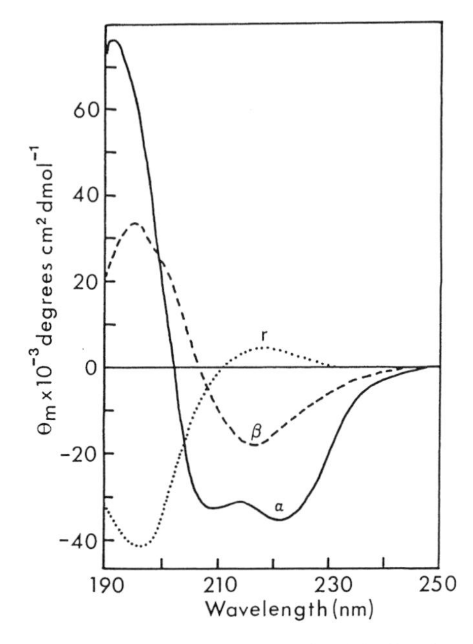

Circular dichroism spectra of polylysine: in α-helical (α), β-strand (β) and random coil (r) conformations. Adapted from N. J. Greenfield et al. *Biochemistry* 6 (1967) 1630–1637 and 8 (1969) 4108–4116.

equal with the λ_{max} of the peak occurring at 270–275 nm, the λ_{min} of the trough at 245–248 nm and the curve crossing zero at 257–259 nm. For the B-form the area of the peak (λ_{max} ~ 260 nm) is very much greater than that of the trough (λ_{min} ~ 210 nm) and the curve crosses zero at ~240 nm. Moreover, the use of artificial nucleotides has indicated that CD curves of nucleic acids in which the peak is at a shorter wavelength than that of the trough (i.e. the reverse of that just described) are indicative of a left-handed structure. [W. C. Johnson, Jr. *Annu. Rev. Biophys. Biophys. Chem.* **17** (1988) 145–166; I. Tinoco, Jr. & C. Bustamante *Annu. Rev. Biophys. Bioeng.* **9** (1980) 107–141]

Cisplatin, ***diamminedichloroplatinum, cis-DDP:*** a platinum-containing antitumor drug, first licensed in 1979. It is particularly effective in the control of testicular tumors, and is a common component of drug combinations against other tumors, notably ovarian tumors. Serious side effects include nephrotoxicity, and peripheral neuropathy has caused permanent disability in some patients. The cellular target of *cis*-DPP is thought to be DNA or chromatin. [D. C. H. McBrien & T. F. Slater (eds.) *Biochemical Mechanisms of Platinum Antitumor Drugs,* IRL Press, Oxford, 1986]

H_3N Cl
Pt(II)
H_3N Cl

Cisplatin

Cistron: a section of DNA which encodes the amino acid sequence of a polypeptide chain or a molecule of RNA (e.g. ribosomal RNA).

Citral: a doubly unsaturated monoterpene aldehyde, M_r 152.24. A mixture of *cis* and *trans* isomers is a component of many essential oils. C.A (*trans*-C., geranial): b.p.$_{12}$ 110–112 °C, ρ_{20} 0.8898, n_D^{17} 1.4894. C.B (*cis*-C., neral): b.p.$_{12}$ 102–104 °C, ρ_{20} 0.8888, n_D^{20} 1.4891. C. is a component of complex insect pheromone mixtures. When heated, it is converted to isocitral, and it undergoes photocyclization to *photocitral A*. Conversion of C. to pseudoionone with acetone is important as the first step in the industrial synthesis of vitamin A. In the perfume and food industries C. is the most important of the aliphatic monoterpenes.

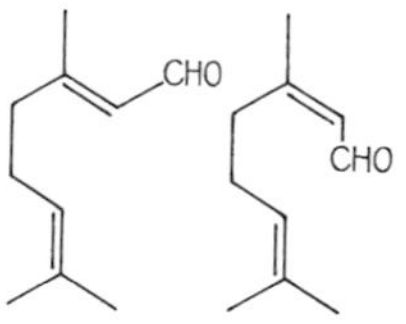

trans-Citral *cis-Citral*

Citrate cleavage enzyme: see ATP citrate (pro-3S)-lyase.

Citrate condensing enzyme: see Citrate *(si)*-synthase.

Citrate cycle: see Tricarboxylic acid cycle.

Citrate lyase: see Citric acid.

Citrate *(si)*-synthase, *citrate condensing enzyme, citrogenase* (EC 4.1.3.7): the tricarboxylic acid cycle enzyme which catalyses the aldol condensation of oxaloacetate and acetyl-CoA to form citrate. C.s. from *E. coli* (M_r 248,000) consists of 4 subunits (M_r 98,000). C.s. from pig or rat heart (M_r 98,000) consists of 2 subunits (M_r 49,000).

Citric acid: a key metabolic intermediate in the tricarboxylic acid cycle. Its concentration also coordinates several other metabolic pathways. Sufficiently high concentrations of C.a. allosterically activate acetyl-CoA carboxylase (EC 6.4.1.2), the key enzyme in fatty acid biosynthesis. C.a. is a negative allosteric effector of 6-phosphofructokinase (EC 2.7.1.11), the key enzyme of glycolysis. C.a. forms complexes with various cations, particularly iron and calcium. In animals, dietary C.a. improves the utilization of dietary calcium. In bacteria, C.a. can be hydrolysed by ATP citrate lyase (EC 4.1.3.8) and citrate pro-(3*S*)-lyase (EC 4.1.3.6) (Fig.).

C.a., m.p. 153–155 °C, was first isolated in 1784 from lemon juice by Scheele. As an intermediate of the tricarboxylic acid cycle, it is present in all aerobic organisms. It is found in relatively large quantities in many different plants, especially in fruits, but also in leaves and roots. C.a. is produced by industrial fermentation, using various microorganisms, e.g. *Aspergillus niger;* the usual substrate is molasses, and the yield is about 60 % of the sugar used.

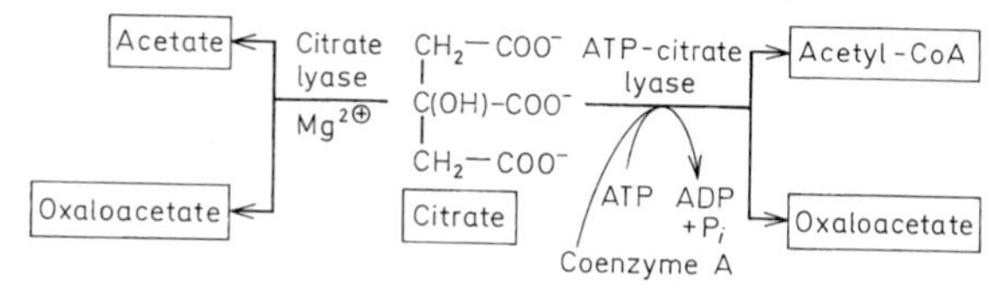

Cleavage of citrate

Citrinin: see Mycotoxins.

Citrogenase: see Citrate *(si)*-synthase.

Citronellal: an unsaturated monoterpene aldehyde, M_r 154.25. Both optical isomers and the isopropylidine form occur naturally. The most abundant form is (+)-C., b.p. 205–206 °C, $[\alpha]_D^{25}$ +11.5 °. C. is the main component of citronella oil and the essential oils of various eucalyptus species. It is an alarm pheromone of ants of the genus *Lasius*. Its tendency to cyclize is utilized in the synthesis of monocyclic monoterpenes such as menthol.

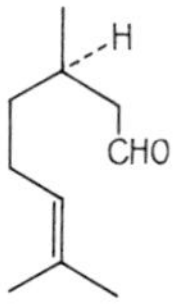

(+)-Citronellal

Citrostadienol, *4α-methyl-5α-stigmasta-7,24-(28)diene-3β-ol:* a phytosterol (see Sterols), M_r 426.7, m.p. 162 °C, $[\alpha]_D$ + 24 ° ($CHCl_3$), found in the oils of grapefruit and orange peel. It is a tetracyclic terpene and an intermediate in the synthesis of some plant sterols.

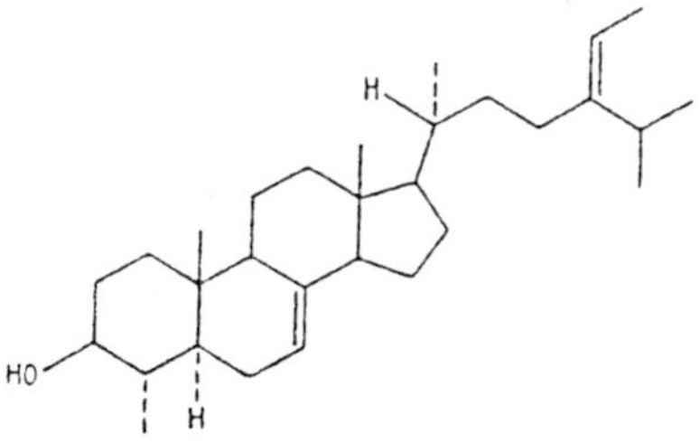

Citrostadienol

L-Citrulline, *N⁵-(aminocarbonyl)-L-ornithine, α-amino-δ-ureidovaleric acid:* $H_2N-CO-NH-(CH_2)_3-CH(NH_2)-COOH$, a nonprotein amino acid, M_r 175.2, m.p. 220 °C (d.), $[\alpha]_D^{25}$ +4.0 (c = 2 in water). It occurs free in plant and animals, and in large amounts in the sap of birches, alder and walnut trees. It is synthesized in the liver from carbamoyl phosphate and L-ornithine by the action of ornithine carbamoyl transferase (EC 2.1.3.3). The citrulline phosphorylase complex consists of ornithine carbamoyl transferase and carbamate kinase (EC 2.7.2.2) (see Carbamoyl phosphate) and catalyses the reaction: L-Citrulline + P_i + ADP ⇌ ATP + ornithine + HCO_3^- + NH_4^+. Arginine is synthesized from L-C. (see Urea cycle). L-C. was first isolated by Wada in 1930 from watermelon *(Citrullus vulgaris)* juice.

Citrullinemia: see Inborn errors of metabolism.

Clathrin: see Coated vesicle.

Clauberg test: see Progesterone.

Claviceps alkaloids: see Ergot alkaloids.

Clavine alkaloids: see Ergot alkaloids.

Cleland's short notation: a nomenclature for representing reaction mechanisms among several substrates. The symbols A, B, C ... are used for substrates; P, Q, R ... for products; I, J ... for inhibitors; E, F, G ... for stable enzyme forms; EA, EAB, FB ... for enzyme-substrate complexes; (EAB), (EPQ) ... for short-lived intermediate complexes. The molecularity (uni-, bi-, ter- ... molecular) of the reaction is determined by the number of reactants which are kinetically significant. The enzyme forms are written from left to right below solid horizontal lines, while reactants and products are indicated by vertical arrows (Fig.). In *sequential mechanisms,* all the substrates are associated with the enzyme before the first product is released. If the substrates must bind in a particular order, it is an *ordered mechanism;* otherwise, it is a *random mechanism.* In *ping-pong mechanisms,* one or more products dissociate from the enzyme before all the substrates have been bound.

Clinical chemistry: a subject bridging natural science and medicine, and now established as an independent discipline. It is defined by M.C.Sanz and P.Lous (IFCC News Letter, No.6, p.1) as follows: "Clinical chemistry encompasses the study of the chemical aspects of human life in health and illness and the application of chemical laboratory methods to diagnosis, control of treatment and prevention of disease." In practice, C.c. involves investigation of patient material with established analytical chemical or biochemical methods, for the purpose of diagnosis and therapy monitoring, as well as the development of new analytical methods and research into biochemical aspects of disease (pathobiochemistry). The clinical chemist usually operates from a centralized hospital laboratory, and must often also provide a medical interpretation of analytical results in consultation with physicians. Senior clinical chemists are also concerned with laboratory organization and management, and with the education of clinical chemists in training. For a discussion of the history of C.c. and the professional qualifications of clinical chemists, see Clinical Chemistry: "A Professional Field for Physicians and Natural Scientists in Europe", J.Büttner, *Eur. J. Clin. Chem. Clin. Biochem.* **29** (1991) 3–12; and "The Origin of Clinical Laboratories", J.Büttner, *Eur. J. Clin. Chem. Clin. Biochem.* **30** (1992) 585–593.

Clionasterol, ***poriferast-5-en-3β-ol:*** a marine zoosterol (see Sterols), M_r 414.7, m.p.138 °C, $[\alpha]_D$ – 42° ($CHCl_3$). C. differs from β-sitosterol (see Sitosterols) in its stereochemistry at C24. It is found in sponges, e.g. *Cliona celata* and *Spongilla lacustris.*

Clonal deletion theory: see Immunoglobulins.

Clonal selection theory: see Immunoglobulins.

Clone bank: see Recombinant DNA technology.

Clostripain (EC 3.4.22.8): an SH-dependent, trypsin-like protease *(M_r 50,000)* with endopeptidase and amidase-esterase activity, isolated from culture filtrates of *Clostridium histolyticum.* The endopeptidase acivity hydrolyses proteins, while the amidase-esterase activity cleaves synthetic amino acid amides and amino acid esters. It attacks only arginyl and lysyl residues, and is therefore used to isolate large peptide fragments without prior chemical modification of the substrate.

Cloverleaf model: see Transfer RNA.

Clupeine: see Protamines.

CMP: acronym of cytidine 5′-monophosphate.

CoA, ***CoA-SH:*** abb. of coenzyme A.

Coagulation vitamin: obsolete term for vitamin K. See Vitamins.

Coated pit: see Coated vesicle.

Coated vesicle:

1. Clathrin-coated vesicle: a transport vesicle, diameter about 800 Å, formed in the process of endocytosis (pinocytosis), and found in virtually all eukaryotic cells. A C-c.v. is therefore also an *endocytic* or *pinocytic vesicle.* A C-c.v. is formed by invagination of a *coated pit* in the plasma membrane. A coated pit is a highly specialized region of the plasma membrane carrying cell surface receptors for a variety of ligands that are normally taken up by the cell, e.g. serum proteins, insulin, lipoproteins, etc. The receptors become concentrated in the pits either before or after association with their ligands. The process of C-c.v. formation from the plasma membrane is therefore known as *receptor-mediated endocytosis.*

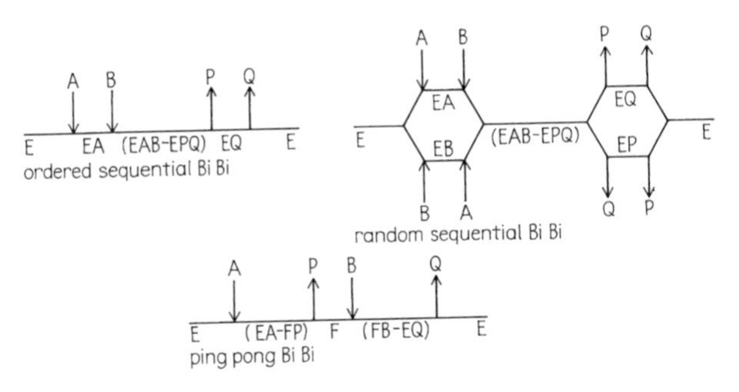

Cleland's notation

In cross section in the electron microscope, coated pits (which occupy about 2 % of the surface of a typical animal cell) appear as depressions or indentations of the plasma membrane with a thick layer of a protein called *clathrin* on their cytosolic face (see Fig. 4 of Liposome). As the coated pit invaginates, the clathrin forms a polyhedral framework (variously known as a lattice or basket) around it, eventually excising it from the membrane to form a C-c. v. Formation of C-c. v. is normally continuous during the life of the cell, e. g. cultured fibroblasts form about 2,500 C-c. v. per cell per minute. On average, a C-c. v. exists for only about 20 seconds. An ATP-dependent uncoating enzyme (uncoating ATPase, M_r 71,000), which removes clathrin from C-c.v., has been studied in vitro (3ATP are consumed for each triskelion (see below) removed); it is an abundant cytoplasmic protein and a member of the M_r 70,000 heat shock protein family (HSP 70). Shedding of the clathrin (which is recycled) is a key step in endocytosis, permitting the vesicles to fuse with endosomes, or to fuse with one another to form endosomes. Endosomes then fuse with each other to form vesicles of diameter 2,000–6,000 Å. It takes an average of about 20 seconds from the formation of a C-c. v. to its incorporation into an endosome. The interior of the endosomal compartment is acidic (pH 5–6), because ATP-dependent H^+ pumps in the endosomal membrane transport protons from the cytosol into the lumen of the endosome. This acidity is important for dissociating protein-receptor complexes, prior to the sorting and further transport of ligands to their intracellular destinations.

The polyhedral lattice of clathrin surrounding the C-c. v. is revealed by electron micrographs prepared by the rapid-freeze-deep-etch technique. Clathrin is a highly conserved protein, which can be reversibly dissociated into flexible three-legged protein complexes known as *triskelions*. Each triskelion consists of three large polypeptides (M_r 180,000; joined at their carboxyl termini at the vertex of the triskelion) and three smaller polypeptides (M_r 23–27,000) (Fig.). In all organisms studied, the large polypeptide is encoded by a single gene, whereas the other coat proteins are heterogeneous.

Under appropriate conditions, isolated triskelions spontaneously reassociate in vitro to form a range of open and closed baskets of various sizes. In vivo assembly to a clathrin basket of the correct size and structure involves participation of additional proteins known as *assembly proteins,* e. g. monomeric *clathrin assembly protein* (AP_{180}, M_r 180,000; different from the large triskelion polypeptide), monomeric *auxilin* (M_r 90,000), and a relatively small (M_r 20,000) clathrin assembly protein that is similar or identical to *myelin basic protein,* as well as the multimeric *adaptors.* All of these proteins promote the in vitro assembly of triskelions into homogeneous populations of clathrin cages. In vivo, assembly proteins are thought to be located between the membrane of the C-c. v. and its clathrin coat.

In solution, adaptors are heterotetramers consisting of a β-type adaptin (M_r 105,000) and an α- or a γ-adaptin (M_r 91,000–108,000) and 2 smaller proteins of M_r 50,000 and 20,000. In 3-dimensional structure, the adaptors appear as a large central unit with two small lateral domains connected by protease-sensitive

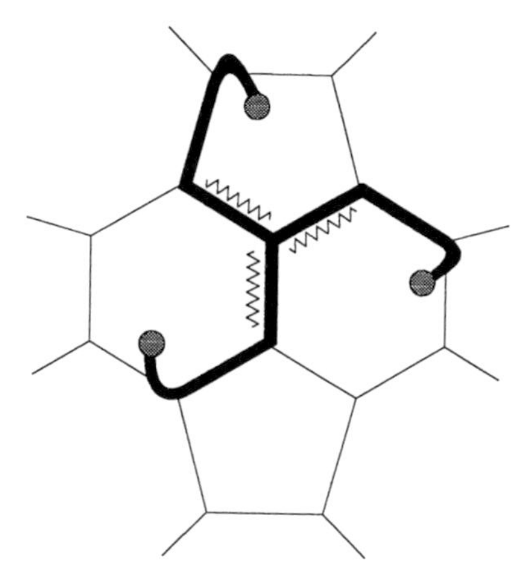

Diagrammatic representation of a triskelion, showing how it forms the sides of polyhedra in the clathrin lattice surrounding a clathrin-coated vesicle.
—— Large Subunit; ʌʌʌʌʌ Small subunit.

hinges. Two types of adaptors, both of $M_r \sim 270{,}000$, can be separated by hydroxyapatite chromatography: HA-I (or AP-1) is associated with the non-clathrin-coated vesicles of the *trans*-Golgi network, whereas HA-II (or AP-2) is associated with the clathrin-coated vesicles that arise from the plasma membrane. None of the coating or assembly proteins displays homology with other protein sequences at present stored in databases, although the three adaptins show some homology with each other.

Clathrin-coated transport vesicles also arise by budding of the membrane of the *trans* Golgi network. These are involved in extensive vesicular traffic between the Golgi apparatus, endolysosomes and other organelles. After uncoating (removal of clathrin) in the cytosol, the vesicle membrane fuses with the membrane of an endolysosome, and the vesicle contents ("cargo") are released into the lumen of the endolysosome. At the same time, it is thought that budding of the membrane of the endolysosome, followed by coating with clathrin, produces an empty C-c. v. which serves to recycle membrane and clathrin back to the Golgi membrane.

2. Non-clathrin-coated vesicle or ***COP (coat protein)-coated vesicle:*** a coated vesicle involved in transport from the endoplasmic reticulum to the Golgi apparatus, from one Golgi cisterna to another, and from the Golgi apparatus to the plasma membrane. Transport by such vesicles appears to be less specific than that performed by C-c. v., i. e. ligands bound to specific receptors are not selected for transport, so that COP-coated vesicles are involved in bulk flow of newly synthesized protein in secretory pathways. In the electron microscope, COP-coated vesicles are less structured than clathrin-coated vesicles. The coat of COP-coated vesicles consists of 4 major proteins: α-COP (M_r 160,000), β-COP (M_r 110,000), γ-COP (M_r 98,000) and δ-COP (M_r 61,000). The *N*-terminal half of β-COP displays considerable homology with that of β-adaptin. [T. Serafini et al. "Coatomer: a cytosolic protein complex containing subunits of non-clathrin-coated Golgi transport vesicles" *Nature* **349** (1991) 248–251; T. Kirchhausen "Coated pits and coated vesicles – sorting it all out" *Curr. Opin. Struct. Biol.* **3** (1993) 182–188; I. S. Trowbridge & J. F. Collawn 'Signal-Dependent Membrane Protein Trafficking in the Endocytic Pathway' *Annu. Rev. Cell Biol.* **9** (1993) 129–161; W. Ye & E. M. Lafer *J. Biol Chem.* **270** (1995) 10933–10939; K. Prasad et al. *J. Biol. Chem.* **270** (1995) 30551–30556; H. D. Blackbourn & A. P. Jackson 'Plant clathrin heavy chain: sequence analysis and restricted localisation in growing pollen tubes' *Journal of Cell Science* **109** (1996) 777–787]

Cobalamine: vitamin B_{12}. See Vitamins.

Cobalt, *Co:* an essential bioelement present in traces in plants, animals and microorganisms. It is important as a constituent of vitamin B_{12}. Traces of Co are required for microbial growth. It is a cofactor or prosthetic group of several enzymes, e. g. pyrophosphatases, peptidases, arginase, as well as certain enzymes involved in nitrogen fixation.

Cobamide coenzyme: see 5′-Deoxyadenosylcobalamine.

Cobramine: see Snake venoms.

Cobra toxin: see Snake venoms.

Coca alkaloids: see Tropane alkaloids.

Cocaine: a tropane alkaloid, the main alkaloid of the coca plant, *Erythroxylon coca,* and related forms growing in the tropics. (–)C. is a bitter white powder, m. p. 98 °C, b. p.$_{0.1}$ 187 °C, $[\alpha]_D^{20}$ –16° ($c = 4$, $CHCl_3$). Both C. and its secondary alkaloids are derivatives of the alkamine ecgonine. C. is prepared by extraction of coca leaves. The extract is hydrolysed and the ecgonine so obtained is easily converted to C. by esterification with methanol and benzoic acid. C. is sometimes used as a local anesthetic. Due to its euphoric or hallucinogenic (at higher doses) effects, the drug is a popular (though illegal) intoxicant. In South America coca leaves are chewed with lime to release the alkaloids. [E L. Johnson 'Alkaloid content in *Erythroxylum coca* Tissue During Reproductive Development' *Phytochemistry* **42** (1996) 35–38]

Cocarboxylase: obsolete term for thiamin pyrophosphate, and for the prosthetic group (also thiamin pyrophosphate) of pyruvate decarboxylase. The name is confusing, because this enzyme does not catalyse a carboxylation, but a decarboxylation.

Cochineal: see Carminic acid.

Cochliobolin B: see Sesterterpenes.

Cock's comb test: see Androgens.

CO_2-compensation point: the concentration of CO_2 at which the rate of photosynthesis (CO_2 incorporation) and the rate of respiration (CO_2 production) are balanced. The value varies with illumination and must be quoted for a given light intensity. The CO_2-c.p. for C3 plants is 40–60 ppm CO_2 at 25 °C, whereas for C4 plants it is often less than 10 ppm. In C3 plants the CO_2-c.p. increases with temperature, resulting in a loss of photosynthetic efficiency as the day temperature increases; the CO_2-c.p. of C4 plants, however, is not affected. Also, with increasing light intensity, the CO_2 concentration of the air decreases around growing plants, making a further contribution to the loss of efficiency. See Photorespiration, Light compensation point.

Code: see Genetic code.

Codehydrogenase I: see Nicotinamide adenine dinucleotide.

Codeine: an opium alkaloid present in several poppy species. Opium consists of about 4 % C. M_r 299.37, m. p. 154–156 °C, $[\alpha]_D^{20}$ –137.7° (ethanol). C. is morphine 3-methyl ether, and it is converted to morphine as the poppy ripens. About 80 % of the world production of morphine is methylated to the therapeutically more important C. In contrast to morphine, C. is only slightly analgesic, but it can increase the effects of other analgesics. It strongly inhibits coughing, and the danger of habituation or addiction is slight. For formula and biosyntheiss, see Benzylisoquinoline alkaloids.

Code triplet: see Codon.

Coding strand, *non-codogenic strand:* by convention (JCBN/NC-IUB Newsletter, 1989, reproduced in 'Biochemical Nomenclature & Related Documents – A Compendium', 2nd Edition 1992), the strand of a double-stranded stretch of DNA that has the same nucleotide sequence as that of the RNA transcript (e. g. mRNA) derived from that double-stranded DNA (save that T is in the place of U). It is therefore the DNA strand that does *not* act as the template for the RNA transcript and could thus be called the 'non-template' strand. Alternative, but in the opinion of the 1989 JCBN/NC-IUB Newsletter, less preferable names are 'sense strand' and 'transcribing strand'. See Nomenclatural conventions concerned with gene transcription.

Codogenic strand, *anticoding strand, sense strand:* the strand of double-stranded DNA that is transcribed into RNA, i. e. the strand that serves as a template for transcription.

Codon, *code triplet:* a linear sequence of 3 adjacent nucleotides in RNA which specify a particular amino acid. In the course of translation, the C. in the messenger RNA is paired with the anticodon in a tRNA which carries a specific amino acid (see Genetic code).

Coenzyme: in the narrow sense, the dissociable, low-molecular-mass active group of an enzyme which transfers chemical groups (see Group transfer) or hydrogen or electrons. C. in this sense couple two otherwise independent reactions, and can thus be regarded as transport metabolites. In a wider sense, a C. can be regarded as any catalytically active, low-molecular-mass component of an enzyme. This definition includes C. that are covalently bound to enzymes as prosthetic groups. A *holoenzyme* consists of a C. in combination with an *apoenzyme* (enzyme protein).

A C. in the narrow sense enters the reaction stoichiometrically, in that it reacts sequentially with two enzyme proteins, and therefore catalyses substrate turnover. An example is NAD, the active group of dehydrogenases and reductases. It first forms an active complex with a dehydrogenase (enzyme I) and accepts the hydrogen removed from the substrate. The resulting NADH then dissociates from enzyme I and associates with a reductase (enzyme II), then donates the hydrogen to the substrate of this enzyme (Fig.). Since the C. acts as a second substrate, it is sometimes called a *cosubstrate.* It must be able to react reversibly with 2 different apoenzymes. Many examples of NAD/NADH as a cosubstrate are found in Glycolysis, Alcoholic fermentation and the Tricarboxylic acid cycle.

Flavin, heme and pyridoxal phosphate are examples of C. in the wider sense. Metals are considered to be inorganic complements of enzyme reactions, and are called *cofactors* rather than C.

Many C. in the wider sense are synthesized from vitamins. The relationships of some C. to vitamins and metabolic function are listed in the table. Strictly speaking, ATP, which commands a special position in metabolism, does not fit the definition of a C. The C. of C_1-unit transfer are *S*-Adenosylmethionine (see), Tetrahydrofolic acid (see) and Biotin (see). The C. of C_2-transfer are Coenzyme A (see) and Thiamin pyrophosphate (see). Vitamin B_{12} is involved in various metabolic reactions, in free form, as methyl-vitamin B_{12} and as 5′-Deoxyadenosylcobalamine (see).

Nearly all C. contain a phosphate group. They often bind nonionized (uncharged) molecules or groups.

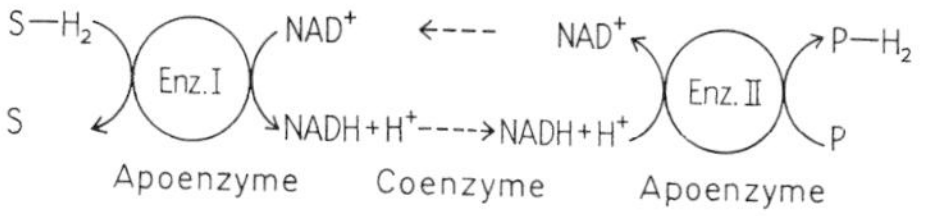

S—H₂ → S (Enz. I) NAD⁺ ← NADH + H⁺ ; NADH + H⁺ → NAD⁺ (Enz. II) P → P—H₂

Apoenzyme Coenzyme Apoenzyme

Role of NAD as a coenzyme or cosubstrate in dehydrogenation and hydrogenation. Enz. I = triosephosphate dehydrogenase, Enz. II = alcohol dehydrogenase; S and S-H_2 = oxidized and reduced substrate, respectively; P and P-H_2 = oxidized and reduced products.

Coenzyme I: see Nicotinamide adenine dinucleotide.

Coenzyme II: see Nicotinamide dinucleotide phosphate.

Coenzyme A, *CoA, CoA-SH:* the coenzyme of acylation, M_r 767.6, λ_{max} 257 nm at pH 2.5–11.0. Solutions of CoA are relatively stable between pH 2 and pH 6. CoA consists of adenosine 3′,5′-diphosphate linked via the 5′-phosphate to the phosphate of pantotheine 4′-phosphate (Fig. 1). The thiol group of the cysteamine is responsible for the biological function of CoA. Practically all the pantothenic acid in a cell is bound as CoA. The metabolically active form of CoA is acyl-CoA, which serves as an acyl group donor.

The metabolic significance of CoA rests on its ability to form high-energy thioester bonds. The methylene group next to the activated thioester tends to dissociate into a carbanion and a proton (Fig. 2), and the carbanion is subject to electrophilic attack. Thioester formation therefore activates both the carboxyl group (electrophilic form, nucleophilic reactions) and the neighboring position (nucleophilic form, electrophilic reactions). The following reactions of the carboxyl group are biochemically important: 1. reduction to the aldehyde; 2. transacylation in the formation of acetylcholine, hippuric acid, acetylated amino sugars, *S*-acetylhydrolipoic acid (see Pyruvate dehydrogenase) and acetylphosphate (Fig. 2a, see Phosphoroclastic pyruvate cleavage); 3. exchange of sulfhydryl com-

Classification, metabolic function and source of coenzymes

Coenzyme	Function	Vitamin source
1) *Oxidoreduction coenzymes*		
NAD	Hydrogen and electron transport	Nicotinic acid
NADP	Hydrogen and electron transport	Nicotinic acid
FMN	Hydrogen and electron transport	Riboflavin
FAD	Hydrogen and electron transport	Riboflavin
Ubiquinone (Coenzyme Q)	Hydrogen and electron transport	–
Lipoic acid	Hydrogen and acyl transfer	–
Heme coenzymes	Electron transport	–
Ferredoxins	Electron transport and hydrogen activation	–
Thioredoxins	Hydrogen transport	–
2) *Group transfer coenzymes*		
Nucleoside diphosphates	Transfer of phosphorylcholine (CDP) and sugars (UDP, GDP), TDP, CDP)	–
Pyridoxal phosphate	Transamination, Decarboxylation, etc.	Vitamin B_6
Phosphoadenosine phosphosulfate	Sulfate transfer	–
Adenosine triphosphate	Phosphorylation, pyrophosphorylation, transfer- of adenosyl and adenyl groups	–
S-Adenosyl-L-methionine	Transmethylation	(Methionine)
Tetrahydrofolic acid and conjugates	Transfer of formyl, hydroxymethyl and methyl groups	Folic acid
Biotin	Carboxylation, transcarboxylation, decarboxylation	Biotin
Coenzyme A	Transacylation, etc.	Panthothenic acid
Thiamin pyrophosphate	C_2-group transfer, C_1-group transfer	Thiamin (aneurin)
3) *Isomerization coenzymes*		
Coenzyme form of vitamin B_{12}	Isomerizations, e.g. methylmalonyl-CoA to succinyl-CoA. Also: ribonucleotide triphosphate reduction, homocysteine methylation, tRNA methylation, methane production by methanogenic bacteria, etc.	Vitamin B_{12}
Uridine diphosphate	Sugar isomerization	–

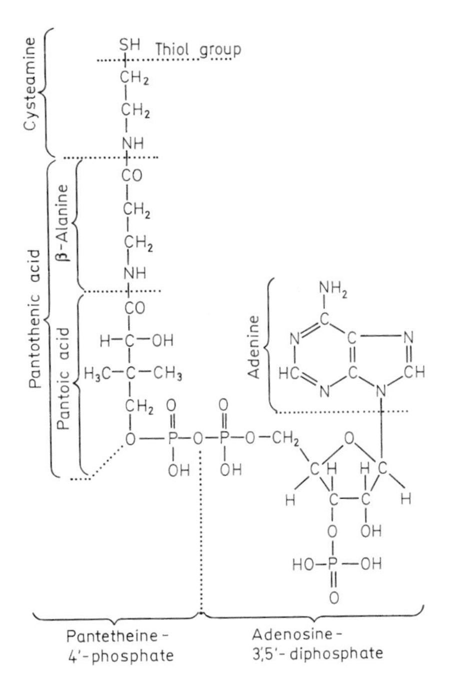

Fig. 1. *Coenzyme A*

$$R{-}CH_2{-}\overset{\delta^-O}{\overset{\|}{C}}{-}SCoA \rightleftharpoons R{-}\overset{\ominus}{C}H{-}\overset{O}{\overset{\|}{C}}{-}SCoA + H^\oplus$$

2a

$$H_3C{-}\overset{\delta^-O}{\overset{\|}{C}}{-}SCoA + {}^-O{-}P(=O)(OH)OH + H^+ \xrightleftharpoons[\text{(EC 2.3.1.8)}]{\text{Phosphate acetyltransferase}} H_3C{-}C(=O){-}O{-}P(=O)(OH)OH + CoASH$$

Acetylphosphate

2b

$$H^{\ominus\ominus}CH_2{-}\overset{O}{\overset{\|}{C}}{-}SCoA + O{=}C(COOH){-}CH_2{-}COOH \xrightarrow[+H_2O]{\text{Citrate synthase (EC 4.1.3.7 or 4.1.3.28)}} HO{-}C(COOH)(CH_2{-}COOH)_2 + CoASH$$

Oxaloacetic acid → Citric acid

Fig. 2. *Active forms of acyl-CoA (thioesters).* 2a Synthesis of acetylphosphate (activated carboxyl group). 2b Citrate synthesis (activated α-methylene group).

ponents in the thiophorase reaction, e.g. when CoA is transferred from succinyl-CoA to acetyl-CoA.

The α-methyl group in acetyl-CoA undergoes numerous condensation reactions, e.g. 1. carboxylation of acetyl-CoA to malonyl-CoA by biotin-dependent acetyl-CoA carboxylase (EC 6.4.1.2) in fatty acid biosynthesis, 2. aldol condensations, e.g. in citrate synthesis (Fig. 2b) in the tricarboxylic acid cycle. When acetoacetyl-CoA is synthesized from 2 molecules of acetyl-CoA (ester condensation), one molecule enters the reaction as an electrophile, the other as a nucleophile.

Derivatives of CoA occur as intermediates in the β-oxidation of fatty acids (see Fatty acid degradation) and in the synthesis of some alkaloids.

Coenzyme F: see Tetrahydrofolic acid.

Coenzyme Q: see Ubiquinone.

Coenzyme R: see Vitamins (vitamin H).

Colamine: see Ethanolamine.

Colamine cephalins: see Membrane lipids.

Colcemid, *demecolcine:* see Colchicum alkaloids.

Colchicine: an alkaloid extracted from *Colchicum autumnale* L. Because it binds specifically to tubulin, C. prevents "treadmilling" of microtubules (see Cytoskeleton), including those of mitotic and meiotic spindles. In plants, C. induces polyploidy by preventing separation of chromosomes during cell division. In small doses, C. relieves pain and suppresses inflammation, but it is highly toxic (lethal dose: 20 mg). It has been used against neoplastic growth.

Colchicum alkaloids: Isoquinoline alkaloids (see), in which the nitrogen is not heterocyclic but present as a substituted amino group on a tricyclic skeleton. The latter consists of an aromatic ring, a tropolone ring, and 7-membered ring. C. are synthesized by only a few genera of the *Liliaceae.* The name comes from *Colchicum autumnale* L., the meadow saffron. The main representatives of the group are Colchicine (see) and *demecolcine;* the former often occurs as the glucoside (colchicoside) (Fig. 1). In the light, the tropolone ring may rearrange to a C_4 and a C_5 ring *(lumicolchicines).*

C.a. are biosynthesized via a 1-phenylethyl-isoquinoline alkaloid, which is converted into androcymbine by hydroxylation, methoxylation, attack of phenol oxidases and oxidative couplings. Androcymbine is then converted to demecolcine and further to colchicine (Fig. 2).

Cold-sensitive enzymes: oligomeric enzymes which show decreased stability and loss of enzymatic activity with decreasing temperature. Low temperature causes these enzymes (about 25 are known) to dissociate into inactive subunits, due to weakening of

(Structure: R_2O, CH_3O, CH_3O, NH–R_1, O, OCH_3)

	R_1	R_2
Demecolcine	CH_3	CH_3
Colchicine	$COCH_3$	CH_3
Colchicoside	$COCH_3$	Glucose

Fig. 1. *The most common colchicum alkaloids*

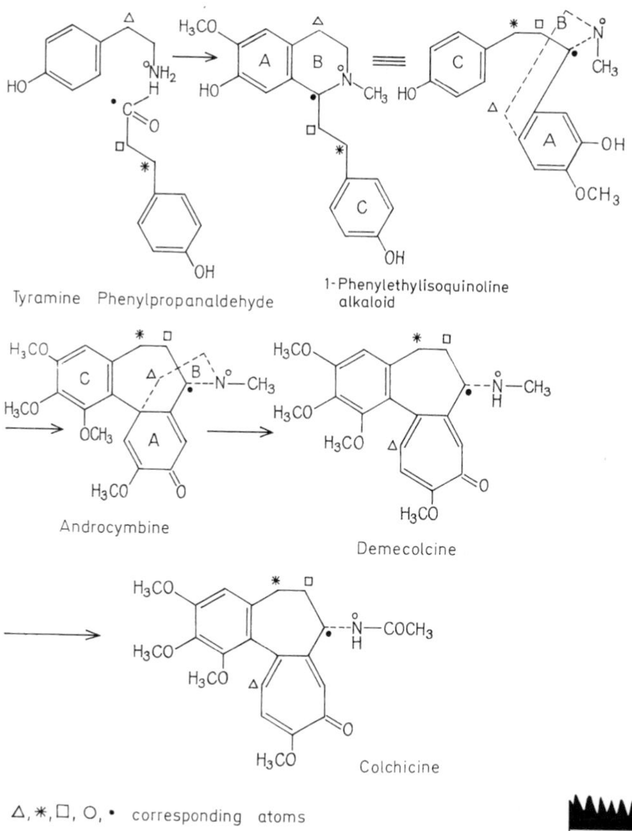

Tyramine Phenylpropanaldehyde

1-Phenylethylisoquinoline alkaloid

Androcymbine

Demecolcine

Colchicine

△, ✳, □, ○, • corresponding atoms

Fig. 2. *Biosyntheis of colchicine*

hydrophobic and/or electrostatic and ionic interactions. Examples are ATPase and pyruvate carboxylase from mitochondria, and glyceraldehyde phosphate dehydrogenase, yeast pyruvate kinase, fructose *bis*phosphatase and carbamoylphosphate synthetase from muscle.

Colicins: see Toxic proteins.

Collagen: an extracellular protein responsible for the strength and flexibility of connective tissue. It accounts for 25–30 % of the protein in an animal. Mature C. is insoluble under physiological conditions, although it is readily denatured by heat, alkali or weak acid.

Structure. Under the light microscope C. appears as fibrils, which under the electron microscope are seen to be composed of microfibrils. The latter display a charactersitic striation with a repeat distance of 670 Å, due to end-to-end alignment of the basic molecular unit, tropocollagen (Fig. 1). The tropcollagen molecule is a right-handed triple helix of 2 identical polypeptide chains (α_1) and one slightly different (α_2) chain. Each α-chain is itself a left-handed helix with a pitch of 9.5 Å, while the superhelix has a pitch of 104 Å. The helical and superhelical structures are stabilized by hydrogen bonds between the HN group of glycine in one chain and the O = C group of proline

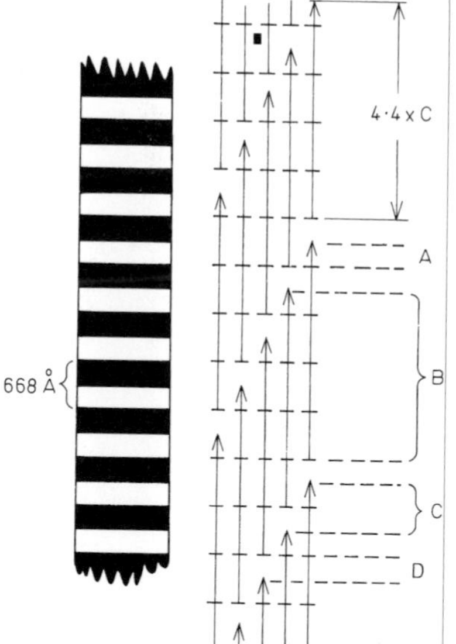

Fig. 1. *Schematic structure of the collagen microfibril.* A: a region of short overlap. B: a long overlap region. C: an overlap region corresponding to a single hole zone and one region of short overlap, and giving rise to the distance of 668 Å on the banded structure of the microfibril. D: a hole zone. Each single arrow (length 4.4 × C) represents a tropocollagen molecule.

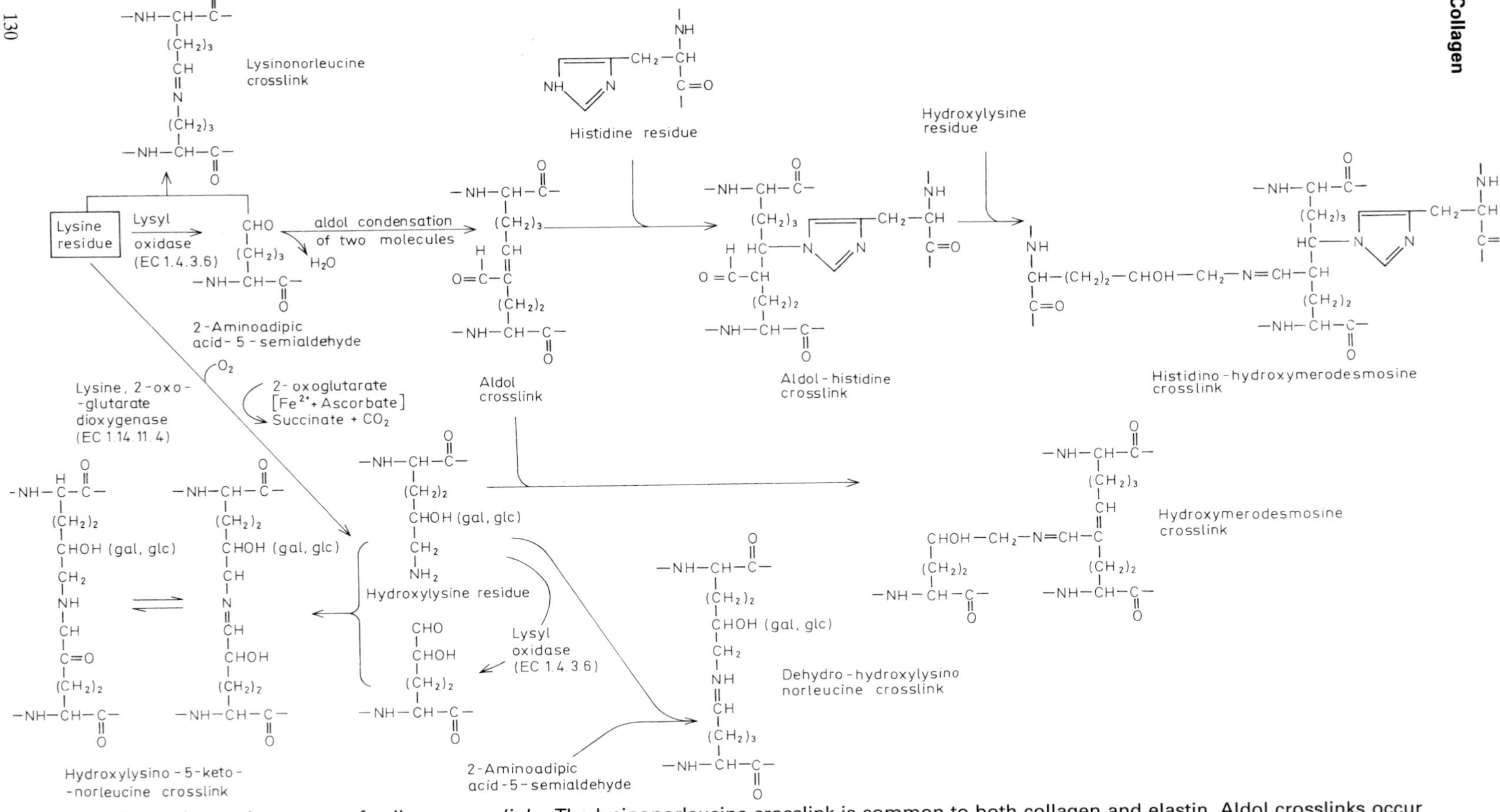

Fig. 2. *Formation and structure of collagen crosslinks.* The lysinonorleucine crosslink is common to both collagen and elastin. Aldol crosslinks occur only at the *N*-termini of collagen chains. Hydroxylysino-5-keto-norleucine crosslinks (with or without carbohydrate) are most abundant in mineralized collagens. Most collagens contain large amounts of dehydro- hydroxylysinonorleucine crosslinks. Aldol-histidine is only found in appreciable quantities in cow skin collagen. Histidinohydroxymerodesmosine crosslinks are common in most collagens. Hydroxymerodesmosine is analogous to merodesmosine (after reduction) in elastin. [M. L. Tanzer *Science 180* (1973) 551–566]

or other amino acid in an adjacent chain. The tropocollagen molecules are also cross-linked (see below). The unusual structural properties of collagen arise from its amino acid sequence. Each α-chain (M_r ~ 100,000) contains about 1,000 amino acid residues. The sequence can be summarized as $(Gly–X–Y)_n$, where X is frequently Pro, and Y is often Hyp (hydroxyproline).

Biosynthesis. C. is synthesized in fibroblasts as a precursor, procollagen, which also consists of 3 chains, each of M_r ~ 140,000. Hydroxyprolyl and hydroxylysyl residues are generated by post-translational modification of prolyl and lysyl residues, respectively, before the procollagen is extruded. Some hydroxylysyl residues are further modified intracellularly by addition of galactose or glucosylgalactose to the hydroxyl group. The chains are then extruded into the intracellular space, where *N*-terminal peptides of M_r 20,000 and *C*-terminal peptides of M_r 35,000 are removed by 2 procollagen peptidases. Tropocollagen then forms spontaneously. Specific lysyl and hydroxylysyl residues are oxidized by lysyl oxidase, resulting in loss of the side chain amino group and the formation of an aldehyde. These aldehydes spontaneously undergo Schiff's base and aldol condensations with neighboring side chains, forming a variety of crosslinks that contribute to the strength of C. (Fig. 2).

C. disorders. A number of hereditary and environmentally-caused disorders are due to impairment of C. synthesis. Hereditary enzyme deficiencies are responsible for the Ehlers-Danlos syndrome (hyperextensible skin), the Marfan syndrome (tendency of the aorta to rupture), osteogenesis imperfecta (very brittle bones), dermatosparaxis (brittle skin in cattle). Since ascorbic acid is required for the formation of hydroxyprolyl residues, a dietary deficiency of the vitamin (scurvy) blocks C. formation. Lathyrism results from ingestion by young animals of sweet peas (*Lathyrus odoratus*) or nitriles, or from copper deficiency; it is due to inhibition of lysyl oxidase and the resulting lack of crosslinks in the C. Normal C. fibers are degraded in rheumatoid arthritis, osteo-arthrosis, scleroderma and alkaptonuria. [W. G. Cole 'Collagen Genes: Mutations Affecting Collagen Structure and Expression' *Prog. Nucl. Acid Res. Mol. Biol.* **47** (1994) 29–80; D. J. Prockop 'Collagens: Molecular Biology, Diseases and Potentials for Therapy' *Annu. Rev. Biochem.* **64** (1995) 403–434]

Collagenase: a proteolytic enzyme, and the only enzyme capable of degrading native collagen to soluble, low M_r peptides. More than 20 C. have been described from bacteria, fungi, arthropods, amphibia and mammals. C. from *Clostridium histolyticum* and C. from tadpole tail have been extensively studied. Clostridial C. (C.A M_r 105,000, C.B M_r 57,000) attacks mainly the peptide bond preceeding the Gly-Pro sequence (–↓–Gly(16)–Pro–Ser–↓–Gly–Pro–); the bonds in front of Gly-Leu and Gly-Ala are attacked to a lesser extent, e. g. in the α_1-chain.

Colony stimulating factors, *CSFs:* a group of glycoproteins required for proliferation, differentiation and survival of hematopoietic progenitor cells, at concentrations of 10^{-11}-10^{-13} M. Functional subclasses are: M-CSF (macrophage-stimulating), G-CSF (granulocyte-stimulating), GM-CSF (granulocyte- and macrophage-stimulating), and multi-CSF (also called interleukin 3, burst-promoting activity, P-cell stimulating factor and hem(at)opoietic cell growth factor) which also stimulates proliferation of eosinophils, megakaryocytes, erythroid and mast cells, as well as neutrophilic granulocytes and macrophages. Some human CSFs are now produced in quantity as recombinant proteins, and used clinically to counteract leukocyte death during chemotherapy and to facilitate bone marrow transplantation. CSFs are probably produced by all animal tissues and by most normal cell types. Erythropoietin (see) may be considered as a circulating, hormone-type CSF. Pluripoietin (also called human pluripotent hematopoietic colony stimulating factor, or pluripotent CSF) is produced constitutively by human bladder carcinoma cell line 5637; it supports growth of human mixed colonies, granulocyte/macrophage colonies, and early erythroid colonies, and induces differentiation of human promyelocytic leukemic cell line HL-60 and the murine myelomonocytic leukemic cell line WEHI-35 (D+). Impure CSF material is generally called colony stimulating activity (CSA) or macrophage-granulocyte inducer (MGI).

Primary structures of several CSFs have been reported, based on partial protein sequencing and on sequence prediction from cloned cDNA: murine GM-CSF [M_r 23,000; N. M. Gough et al. *Nature* **309** (1984) 763–767]; murine multi-CSF [M_r 23,000–28,000; M. C. Fung et al. *Nature* **307** (1984) 233–237; T. Yokata et al. *Proc. Natl. Acad. Sci. USA* **81** (1984) 1070–1074]; human GM-CSF [M_r 22,000; G. G. Wong et al. *Science* **228** (1985) 810–815]; human M-CSF (M_r 45,000 of the homodimer; also called CSF-1; E. S. Kawasaki *Science* **230** (1985) 291–296]. The CSF amino acid sequences determined so far show no homologies, and they display considerable differences in predicted secondary and tertiary structures, which is surprising in view of their extensive functional overlap and the similar response of progenitor cells to different CSFs.

Human M-CSF is a disulfide-linked homodimeric glycoprotein, originally characterized by its stimulation of macrophage development from bone marow precursors; native M-CSF production is also associated with placental trophoblast function during pregnancy, and bone osteoclast survival. Recombinant human M-CSF supports growth, differentiation and defensive capabilities of macrophages in vitro, and is a chemotactic factor for monocytes [D. Munn & N-K. Cheung *Semin. Oncol.* **19** (1992) 395–407]. Significant concentrations of circulating M-CSF are found in healthy individuals, possibly reflecting its role in normal hematopoiesis. In contrast, serum levels of other cytokines increase during infection but are otherwise hardly detectable. M-CSF has been implicated in various diseases, e. g. atherosclerosis, lupus nephritis, mammary carcinoma, ovarian cancer and other cancers. In view of the possible role of M-CSF in tumorigensis, M-CSF antagonists may prove valuable for therapy and/or prophylaxis. Clinical trials suggest that recombinant M-CSF may be useful for treating fungal infections, and for lowering serum cholesterol.

Separate and specific receptors for each CSF are coexpressed on granulocytes and monocytes, and they show no cross competition for their respective ligands. However, a hierarchical down modulation has been observed, e. g. occupancy of multi-CSF receptor decreases the binding by all other receptors of their specific ligands; occupancy of GM-CSF receptors causes a similar down modulation of G-CSF and M-CSF receptors.

M-CSF receptor is the c-*fms* proto-oncogene product [C.J.Sherr et al *Cell* **41** (1985) 665–676], and it is structurally related to platelet-derived growth factor receptor and the c-*kit* proto-oncogene [A.Ullrich & J.Schlessinger *Cell* **61** (1990) 203–212]. This single high-affinity transmembrane M-CSF receptor consists of an extracellular protein tyrosine kinase domain linked to an intracellular protein tyrosine kinase domain by a single transmembrane helix. The extracellular part of the receptor consists of 5 immunoglobulin-like domains, but only a subset of these domains appears to be necessary for M-CSF-binding [E.R.Stanley *Methods Enzymol.* **116** (1985) 564–587; L.J.Guilbert & E.R.Stanley *J. Biol. Chem.* **261** (1986) 4024–4032; L.Coussens et al. *Nature* **320** (1986) 277–280]. Binding of M-CSF initiates covalent receptor dimerization, stimulation of receptor tyrosine kinase activity, and autophosphorylation of the cytoplasmic tyrosine kinase domain [C.J.Sherr *Trends Genet.* **7** (1991) 398–402; G.Vario & J.A.Hamilton *Immunol. Today* **12** (1991) 362–369]. Short, long and intermediate mRNA splicing variants of human M-CSF cDNA have been isolated from the M-CSF gene. Although post-translational modifications are extensive (glycosylation, chondroitin sulfate addition), these are not necessary for the expression of M-CSF activity in vitro. M-CSF does not increase phosphatidylinositol turnover, in contrast to related CSFs which promote hydrolysis of phosphatidylinositol 4,5-*bis*phosphate. Activation of the M-CSF receptor triggers the activation of phosphatidylinositol 3′-kinase, stimulation of phosphatidylcholine hydrolysis by phospholipase C or D, and phosphorylation (activation) of cytosolic phospholipase A_2 in human monocytes. Furthermore, addition of exogenous phosphatidylcholine-specific phospholipase C to the M-CSF-dependent cell line BACI.2F5 has the same effect as additon of the growth factor itself [M.Reedijk et al. *EMBO J.* **11** (1992) 1365–1372; Xiang-Xi et al. *J. Biol Chem.* **269** (1994) 31693–31700; E.W.Taylor et al. *J. Biol. Chem.* **269** (1994) 31171–31177]. The receptor for murine M-CSF has been isolated by cross-linking ^{125}I-M-CSF to its receptor, using disuccinimidyl suberate, followed by release from the membrane with detergent [C.J.Morgan & E.R.Stanley *Biochem. Biophys. Res. Commun.* **119** (1984) 35–41].

Human GM-CSF receptor is composed of at least 2 subunits; the β-subunit is identical with a subunit of the receptors for interleukin-3 and interleukin-5. Both subunits have domains that are structurally related to a fibronectin type III domain, a structure that is conserved in all members of the cytokine receptor superfamily. Genomic DNA clones containing the entire coding sequence of the α-subunit have been isolated and characterized [Y.Nakagawa et al. *J. Biol. Chem.* **269** (1994) 10905–10912].

Colophony: the residue from turpentine distillation. See Balsams.

Column chromatography: a chromatographic separation method, in which the carrier material is packed as a column inside a tube (usually of glass and known as a chromatography column). The term C.c. therefore indicates the mode of use of the chromatographic carrier, and does not indicate the type of chromatography, which may be partition, adsorption, ion exchange, gel filtration, affinity, etc., depending on the nature of the carrier. The column material is usually equilibrated with the running solvent or *elutant.* Sample solution is placed on top of the column and washed in with running solvent, followed by a continuous supply of the same solvent. Discrete samples of the effluent or *eluate* are collected from the bottom of the column (usually automatically with the aid of a fraction collector), then analysed for their contents. For column materials used in ion exchange C.c., see Ion exchangers. To increase the sharpness and speed of separations, gradient elution is often employed in ion exchange C.c.; a gradient mixer, supplied by two or more different buffer solutions, produces a continuous change in the pH and/or ionic strength of the running solvent entering to the top of the column. Such gradients may be linear, convex or concave.

Compactin, ***ML-236B, 6-demethylmevinolin, 1,2,6,7,8,8a-hexahydro-β,δ-dihydroxy-2-methyl-8-(2-methyl-1-oxobutoxy)-1-naphthalene-heptanoic acid δ-lactone:*** a fungal metabolite from culture media of *Penicillium citrinum* and *P. brevicompactum.* The parent hydroxyacid of C. is a potent competitive inhibitor (K_i 1.4 nM) of 3-hydroxy-3-methylglutaryl-CoA reductase (EC 1.1.1.34). C. acts like Mevinolin (see) in reducing plasma LDL cholesterol levels, but it is rather less potent than the latter. [A.Endo et al. *FEBS Letters* **72** (1976) 323–326; A.W.Alberts et al. *Proc. Nat. Acad. Sci. USA* **77** (1980) 3957–3961]

Compartment: a portion of the cell which is structurally or biochemically separate from the rest of the cell space. Compartmentation is the division of the cell into regions with different enzymatic equipment. Compartmentation facilitates the simultaneous operation of metabolic processes, which would otherwise mutually interfere, e.g. both the synthesis and degradation of metabolites.

Biomembranes (see) are particularly important for compartmentation. Serving both as barriers to free diffusion of metabolites and as a means of communication between separate C., they enable the directed transport of material, resulting in vectorial biochemical reactions.

Plasmatic C. (e.g. nucleoplasm, inner matrix of mitochondria and plastids) are surrounded by a double membrane, contain active nucleic acid, are the site of ATP formation and protein synthesis, and form α-glucans (glycogen or starch); these processes do not operate in nonplasmatic C. Nonplasmatic C. are, e.g. vacuoles, dictyosomes, lysosomes, uricosomes, peroxisomes, glyoxysomes, and the outer C. of mitochondria and plastids. If a cell is capable of forming β-glucans (cellulose, callose), these are formed in the nonplasmatic C.C. can only be partially demonstrated by light or electron microscopy, e.g. the lysosomal system of the cell can only be visualized histochemically using detection methods for lysosomal enzymes.

In a wider sense, compartmentation is possible through purely chemical means, e.g. embedding of enzymes in lipid layers and aggregation of enzymes to form multienzyme complexes.

Compartmentation: see Compartment, Biomembranes.

Competitive inhibitor: see Effectors.

Complementary DNA, *cDNA:* DNA complementary to a mRNA. It is prepared in the laboratory as a probe for hybridization studies by incubating the mRNA with dATP, dGTP, dTTP and dCTP in the presence of a reverse transcriptase (see RNA-dependent DNA-polymerase).

Complementary structures: two structures which define one another, e.g. the 2 polynucleotide chains in the DNA duplex. The base pairs adenine/thymine (or adenine/uracil in RNA) and guanine/cytosine are complementary, so that by base pairing the nucleotide sequence of one polynucleotide chain defines a unique sequence in the complementary strand.

Complement binding reaction: binding of the C1 component of complement to the Fc fragments (see Immunoglobulins) of antibodies which are bound to surface antigens or erythrocytes or bacteria. The Complement system (see) is activated by this reaction.

Complement system: a heat-labile (100% inactivation after 30 min at 56°C) cascade system in the serum of all vertebrates, and composed in mammals of at least 20 glycoproteins, 7 of which control localization of the effect. Each activated component (or complex) is a highly specific protease acting only on the next component of the cascade. Two pathways are recognized, the *classical* and the *alternative.* The classical pathway is activated by binding of C1 to immune complexes containing IgG or IgM [C1 binding sites are localized in the Fc regions of IgG and IgM (see Immunoglobulins), and binding is dependent on Ca^{2+}]. C1 consists of 3 subunits: C1q, C1r and C1s. The binding sites for IgG and IgM reside on C1q. Occupation of these binding sites confers proteolytic activity, which cleaves a single peptide bond of C1r. The resulting "activated" C1r in turn hydrolyses a peptide bond in C1s, thus yielding the fully active C1 complex, which initiates the sequential assembly of circulating components into a surface-bound protein complex (Fig.). The alternative pathway is initiated by a repertoire of activators, including antibodies (IgA and IgE, which do not activate the classical pathway when complexed with antigens), high M_r polysaccharides of bacteria and yeasts, fragments of plant cell walls and protozoa. This pathway is thought to provide the initial response to bacterial invasion, since it is activated in the absence of antibodies. Factor D of the alternative pathway is already proteolytically active in nonactivated serum, i.e. its formation from an inactive precursor is not part of the amplification system of the cascade.

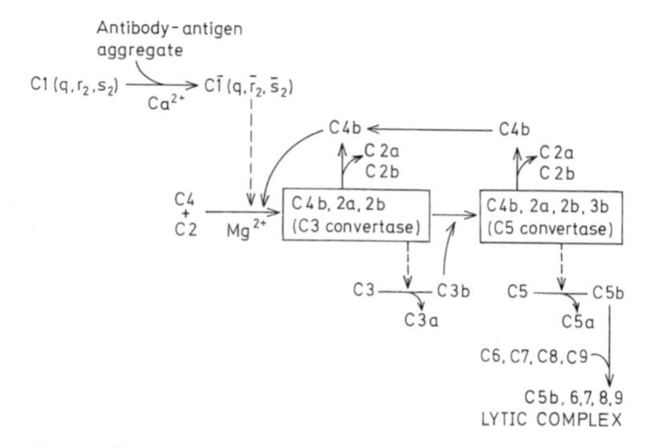

Fig. 1. *Classical pathway of complement activation.* A bar over the number or letter of a factor indicates that the factor is activated (i.e. proteolytically active). Activation by proteolysis occurs near to the *N*-terminus, producing a small (a) and a large (b) cleavage product, e.g. C3 → C3a + C3b.

Early components of each pathway are primarily concerned with the formation of 2 protein complexes, which function respectively as C3 and C5 convertases. Proteolytic activation of C5 is the final enzymatic event which triggers the spontaneous association of the late components (C6-C9) to form a nonenzymatic lytic complex capable of puncturing cell membranes (Figs. 1 & 2). Thus, the main function of the C.s. is lysis of foreign invasive cells. Also, through the anaphylatoxic and chemotactic properties of some components (notably C3a, C4a and C5a), foreign cells are rendered susceptible to phagocytosis. The C.s. is also involved in solubilization of immune complexes and in development of the cellular immune response. Deficiencies of C.s. components are associated with repeated bacterial infections, and are also implicated in certain autoimmune diseases.

The amino acid sequencies of most C.s. components have been determined by a combination of protein and cDNA sequencing techniques.

For experimental purposes, either fresh guinea pig serum or serum from patients genetically deficient in one component is used as a source of C.s. (hemolytic system). Sensitized sheep erythrocytes serve as immune aggregates for study of the classical pathway, and rabbit erythrocytes for the alternative pathway. [R.R. Porter, P.J. Lachmann & K.B.M. Reid (eds.)

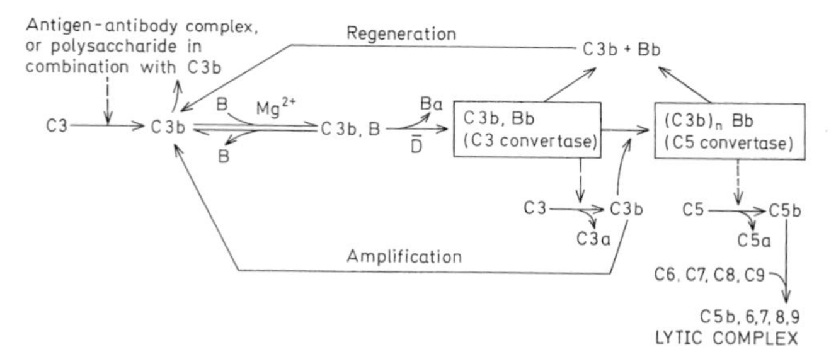

Fig. 2. *Alternative pathway of complement activation.* See legend to Fig. 1.

Biochemistry and Genetics of Complement, Cambridge University Press, 1986]

Conalbumin: see Siderophilins.

Concanavalin A: see Lectins.

Concentration variables, *fundamental variables, primary variables:* those substances in an enzymatic system whose concentrations can be directly controlled by the experimenter, e.g. substrates, products and effectors. They are therefore distinguished from the enzyme species, whose concentrations can be calculated from the kinetic equations at steady state for the given values of the C.v. Usually, in kinetic experiments, one C.v. is varied and the others are held constant.

Conchiolin: see Paleoproteins.

Concretion oils: see Essential oils.

Conessine: see Holarrhena alkaloids.

Coniine, *2-propylpiperidine:* the most important of the Conium alkaloids (see), and the toxic principle of the poison hemlock, *Conium maculatum,* which was used in ancient Athens to put Socrates to death. The lethal dose of C. in humans is 0.5–1 g. The largest quantities of C. are found in the unripe seeds. The synthesis of C. from α-picoline and paraldehyde by Ladenburg in 1886 was the first laboratory synthesis of any alkaloid. M_r 127.22, m.p. −2.5 °C, b.p. 166 °C, $[\alpha]_D^{17}$ ±16°.

Conium alkaloids: simple piperidine alkaloids found only in poison hemlock, *Conium maculatum.* The main alkaloids are Coniine (see) and γ-coniceine (M_r 125.22, b.p. 168 °C); the secondary alkaloids are *N*-methyl and hydroxy derivatives of coniine. In contrast to other piperidine alkaloids, the ring system of C.a. is synthesized from acetate rather than from lysine (Fig.).

Acetate → γ-Coniceine → Coniine

Biosynthesis of Conium alkaloids

Conotoxins: peptide neurotoxins from marine, fish-hunting snails of the genus *Conus.* They contain 13 to 29 amino acids, are strongly basic, and are highly cross-linked by disulfide bonds. ω-C. inhibit voltage-activated entry of Ca^{2+} into the presynaptic membrane and therefore the release of acetylcholine. α-C. inhibit the postsynaptic acetylcholine receptor. μ-C. prevent the generation of muscle action potentials. Their small size is probably an advantage in promoting rapid diffusion through the tissues of the prey; fish stung by these snails are paralysed in seconds. [B.M. Olivera et al. *Science* **230** (1986) 1338–1343]

Constitutive enzymes: enzymes which, in contrast to inducible enzymes, are constantly produced by the cell irrespective of the growth conditions.

Contig: a sequence of DNA assembled from overlapping cloned DNA fragments. The term is also applied to the section of a chromosome carrying the contiguous genes that are deleted in contiguous gene syndromes, e.g. in human thalassemias. These syndromes are associated with deletion of specific chromosome segments, and by implication a contiguous set of genes.

Contraceptives: see Ovulation inhibitors.

Contractile proteins: see Muscle proteins.

Convallatoxin: see Strophanthins.

Cooperative oligomeric enzymes, *allosteric enzymes:* enzymes composed of several subunits and displaying cooperativity.

Cooperativity: a phenomenon displayed by oligomeric or monomeric enzymes which possess more than one binding site for a particular ligand. Cooperative binding may be negative or positive, and it may occur for the same ligand (homotropic C.) or for a different ligand (heterotropic C.). Purely phenomenologically, this means that the dissociation constant for each successive ligand bond is lower (positive C.) or higher (negative C.) than the preceding one. C. is also involved if the binding of one substrate or effector molecule changes the configuration and thereby the reactivity or catalytic constant (see Michaelis-Menten equation) for other substrate molecules. The degree of C. is usually determined from a Hill plot (see).

In positive C. the binding curves (saturation curves) are sigmoidal (S-shaped). Various models have been developed to describe C. (see Cooperativity model). It is assumed that C. is caused by changes in the 3-dimensional structure of the enzyme protein, and that each subunit of an oligomeric enzyme can exist in at least 2 configurations which react differently with effector molecules. Further, the change in configuration of one subunit is thought to induce changes in the configurations of the other subunits in the same molecule. The most general sigmoidal rate equation is: $v^{-1} = a + bS^{-1} + cS^{-2} + \ldots$ This type of equation, as well as sigmoid binding curves can be derived from other mechanisms not included in the C. model.

Cooperativity model: a functional and structural model of cooperative oligomeric enzymes, which is intended to describe and explain Cooperativity (see) in the turnover of substrates or the binding of effectors. The C.m. can be used to derive binding potentials and equations, which can then be compared with experimental data. In order to arrive at and characterize a C.m. for a particular enzyme, a number of hypotheses must be made: 1. the number of subunits, 2. the geometric hypothesis, or arrangement of the subunits (tetrahedral, square planar, etc.), 3. the number of configurations of the subunits and the nature of their interactions with effectors (configurational hypothesis), and 4. the way in which the change in configuration of one subunit affects the other subunits (interaction hypothesis).

The C.m. for a tetrameric enzyme, for which the Monod and the Koshland models are limiting cases (Fig.), is often called the general C.m. It assumes only 2 subunit configurations (S and T), but even this is somewhat specialized. The configurational hypothesis of the Monod model (Monod-Wyman-Changeux model, MWC model, all-or-nothing model) presumes 2 configurations (the S and T states of the subunits), in which the effector binding of the second state can be exclusive or nonexclusive. The configurational hypothesis of the Koshland model (Adair-Koshland-Nemethy-Filmer model, AKNF model, induced fit model) assumes induced fit, i.e. the ligand-binding configuration is present only in negligible quantities in the ab-

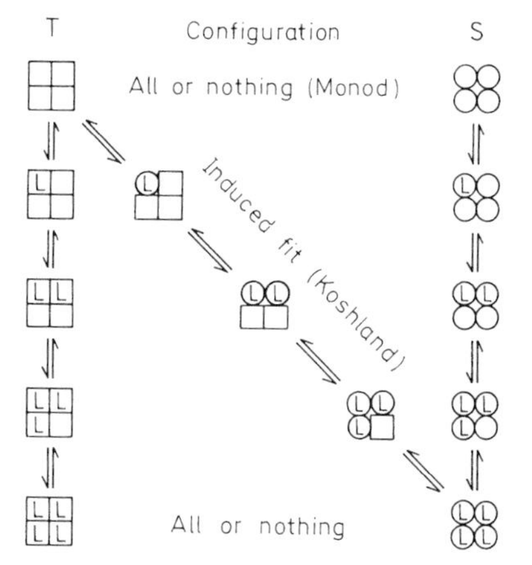

2-Configuration cooperativity model of a tetrameric enzyme. L = ligand (substrate or effector).

sence of ligands and is induced by the presence of the latter. The interaction hypothesis postulated by the Monod model is an all-or-nothing transition of the subunit configuration. This is also called a concerted transition. The Koshland model ascribes to each contact between subunits a free energy of interaction which is characteristic for the 2 configurations of the subunits.

Copolymer: a polymeric molecule containing more than one kind of monomer unit. In biochemistry it often refers to polynucleotides made by incubating two or more nucleoside di- or triphosphates with the appropriate polymerase. Using RNA polymerase, Khorana synthesized C. according to the following scheme (choice of the starter molecule was arbitrary and could be varied according to need):

$$\underset{\text{(starter)}}{\text{AUG}} + [\text{ATP} + \text{UTP} + \text{GTP}] \xrightarrow{\text{RNA polymerase}} \text{AUGAUGAUGAUG}\ldots$$

The synthesis of such C. with defined triplets played a major role in the cracking of the genetic code.

Copper, *Cu:* an important bioelement, involved in electron transport in mitochondrial membranes, required by plants for chlorophyll synthesis, and present in a number of enzymes (see Copper proteins). Large amounts of Cu are toxic, although tolerant strains of plants and microorganisms arise near copper mines and dumps. Threshold limits for exposure of workers in foundries and smelters in the USA are 0.1 mg/m^3 fumes and 1.0 mg/m^3 dusts and mists. The recommended daily intake for humans is 2 mg, and the amount in the body is 100–150 mg. The highest concentrations are found in liver and bones. Mammalian blood contains a number of Cu proteins and the synthesis of hemoglobin is dependent on Cu, although it does not contain the metal. The oxygen transport pigments in mollusk and crustacean blood contain Cu (see Hemocyanins).

Copper proteins: metalloproteins, often blue in color, which usually contain a mixture of mono- and mostly divalent copper in their molecules. Exceptions are the plastocyanins of chloroplasts (M_r 11,000, $2Cu^{2+}$), and Hemocyanins (see) (the oxygen transport proteins of arthropod and molluscan blood, which contain only Cu^{2+} or Cu^+). With the exception of the copper thiolate proteins (see Metallothioneins), known C.p. are either enzymes for which oxygen is a substrate, or they are oxygen transport proteins. Even Ceruloplasmin (see), which was long thought to be merely a transport protein, displays oxidase activity with unsaturated compounds, including indoles. Some of these oxidase reactions incorporate the entire O_2 molecule into hydrogen peroxide, e.g. the Cu-containing Amine oxidases (see) catalyse the reaction $RCH_2NH_2 + O_2 + H_2 \rightarrow RCHO + H_2O_2 + NH_3$, and galactose oxidase catalyses O_2 + galactose $\rightarrow H_2O_2$ + galactohexosedialdose. A more common type of Cu oxidase incorporates only one atom of the O_2 into the product, while the other is reduced to water. Some important examples are:

1) Dopamine β-monooxygenase, EC 1.14.17.1, which hydroxylates dopamine to noradrenalin (ascorbate donates the hydrogens for reduction of the second oxygen atom).

2) Monophenol monooxygenases, EC 1.14.18.1, which oxidize (hydroxylate) tyrosine and other phenols.

3) Laccases (EC 1.10.3.2) found in higher plants and fungi, especially white rot fungi, are involved in the metabolism of lignin.

4) Classical tyrosinase (catechol oxygenase, EC 1.10.3.1) catalyses the first step in the synthesis of Melanin (see) from tyrosine.

Cytochrome *c* oxidase (EC 1.9.3.1) and ascorbate oxidase (EC 1.10.3.3) reduce O_2 to 2 molecules of H_2O. Superoxide dismutase (see) (EC 1.15.1.1) has an unusual substrate, i.e. the superoxide radical ion. See Azurin. [R. Lontie (ed.) *Copper Proteins and Copper Enzymes,* CRC Press, Boca Raton, 1984; T. G. Spiro (ed.) *Copper Proteins,* Wiley, New York, 1981]

Coprogen: a Siderchrome (see) synthesized by *Penicillium* and *Neurospora* spp.

Coprostane: obsolete term for 5β-cholestane. See Steroids.

Coprostanol, *5β-cholestan-3β-ol:* a sterol alcohol, M_r 388.64, m.p. 101 °C, $[\alpha]_D$ + 28° ($CHCl_3$). C. is the main sterol of feces, where it arises by reduction of cholesterol by intestinal bacteria. It differs from its stereoisomer, Cholestanol (see), in the configuration at C5 and its A/B *cis* bonding.

Cordycepin, *3'*deoxyadenosine, adenine 9-cordyceposide: a purine antibiotic synthesized by *Cordyceps militaris* and *Aspergillus nidulans* (see Nucleoside antibiotics). M_r 251.24, m.p. 225–226 °C, $[\alpha]_D^{20}$ −47° (water). As an antimetabolite of adenosine, it inhibits purine biosynthesis, but is not very toxic. It is phosphorylated metabolically to the monophosphate.

Core particles: particles released from Chromatin (see) by partial enzymatic digestion. They contain about 140 base pairs of DNA and 2 molecules each of the "inner histones" H2A, H2B, H3 and H4. Within chromatin, C.p. are separated by about 40 base pairs of DNA which is associated with the "outer histones" H1 and H5. See Nu bodies.

Corepressor: see Enzyme repression.

Coriandrol: see Linalool.

Cori ester: see Glucose 1-phosphate.

Corilagin: see Tannins.

Coronatine: a chlorosis-inducing toxin produced by several species of *Pseudomonas*. These infect a variety of grasses, notably Italian ryegrass, as well as soybeans. The bioassay of C. is based on its ability to cause hypertrophic growth of potato tuber tissue. Structural and conformational studies of C. have all been performed on material from liquid cultures of bacteria, but evidence for isolation of C. from infected leaves of Italian ryegrass is fairly conclusive (chromatography, activity in the biological test). Culture filtrates of *Pseudomonas syringae* contain about equal quantities of C. and a structural analog, *N*-coronafacoylvaline, which is also a chlorosis-inducing toxin [R. E. Mitchell *Phytochemistry* **23** (1984) 791–793]

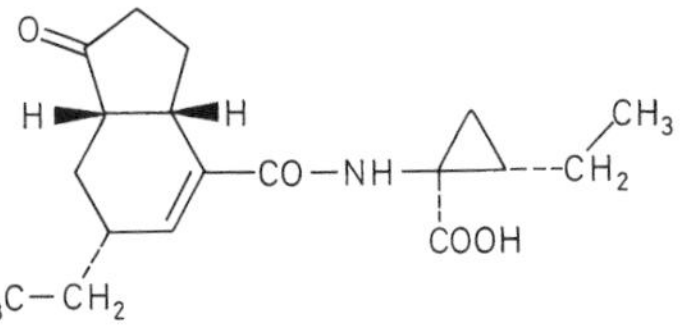

Coronatine [A. Ichihara et al. *Tetrahedron Lett.* No. 4 (1979) 365–368]

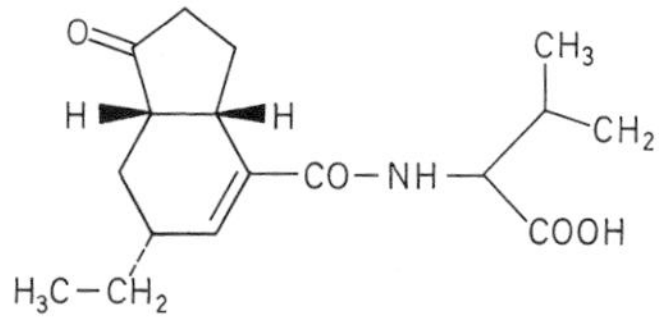

N-Coronafacoylvaline

Corpus luteum hormone: see Progesterone.

Corrinoids: compounds containing a corrin ring. Like the porphyrin ring system, the corrin system consists of 4 pyrrole rings linked in a large ring, but there are only 3 linking methenyl groups, the fourth link being a direct bond between 2 pyrroles. The pyrroles carry acetate, propionate and methyl substituents. The central atom is covalently bound cobalt. Vitamin B_{12} is a C. (see Vitamins).

Cortexolone, ***Reichstein's substance S, 11-deoxycortisol, 17α-hydroxy-11-deoxycorticosterone, 17α,-21-dihydroxypregn-4-ene-4,20-dione:*** a mineralocorticoid from the adrenal cortex. For structure and biosynthesis, see Adrenal corticosteroids.

Cortexone, ***Reichstein's substance Q, 11-deoxycorticosterone, DOC, 21-hydroxypregn-4-ene-3,20-dione:*** a mineralocorticoid from the adrenal cortex. The acetate or glucoside is used in the treatment of Addison's disease, and of shock. For structure and biosynthesis, see Adrenal corticosteroids.

Corticoids: see Adrenal corticosteroids.

Corticosteroids: see Adrenal corticosteroids.

Corticosterone, ***Reichstein's substance H, Kendall's substance B, 11β,21-dihydroxypregn-4-ene-3,20-dione:*** an adrenal cortex hormone and a glucocorticoid. C. is biosynthesized from progesterone, which is hydroxylated first to cortexone, then in position 11 to C. For structure and biosynthesis, see Adrenal corticosteroids.

Corticotropin, ***adrenocorticotropin, adrenocorticotropic hormone, ACTH:*** a polypeptide hormone secreted by the pituitary. The primary structure of the human hormone is Ser-Tyr-Ser-Met-Glu-His-Phe-Arg-Trp-Gly-Lys-Pro-Val-Gly-Lys-Lys-Arg-Arg-Pro-Val-Lys-Val-Tyr-Pro-Asn-Gly-Ala-Glu-Asp-Glu-Leu-Glu-Phe, M_r 4541. Only sequence 31–33 is species-specific. Biological activity is determined by the first (fixed) 20 amino acids. Sequence 1–13 is identical to that of α-melanotropin.

ACTH is the smallest of the hormones produced by the anterior pituitary. It is synthesized in the γ-cells when they are stimulated by corticotropin releasing hormone (see Releasing hormones). The target organ of ACTH is the adrenal cortex, which it stimulates to growth and increased production of glucocorticoids, an action mediated by adenylate cyclase.

ACTH is determined in blood by radioimmunological techniques; concentrations are in the range of ng/ml, and fluctuations are observed with daily and seasonal periodicities. The blood concentration of ACTH is controlled via the hypothalamus, in response to circulating concentrations of glucocorticoids.

The total synthesis of ACTH was reported in 1963 by Schwyzer.

Corticotropin-like peptide, ***CLIP:*** a peptide from the pars intermedia of rat and pig pituitaries. Its primary structure is idential to amino acid sequence 18–39 of ACTH. No definite function of CLIP is known. It may be a cleavage product of ACTH unavoidably formed during production of α-MSH. See Peptides, Fig. 3.

Corticotropin releasing hormone: see Releasing hormones.

Cortine: see Adrenal corticosteroids.

Cortisol, ***Reichstein's substance M, Kendall's substance F, 11β,17α,21-trihydroxypregn-4-ene-3,20-dione:*** a circulating glucocorticoid hormone produced by the adrenal cortex. For structure and biosynthesis, see Adrenal corticosteroids.

Cortisone, ***Reichstein's substance F, Kendall's substance E, Wintersteiner's substance F, 11-dehydro-17α-hydroxycorticosterone, 17α,21-dihydroxypregn-4-ene-3,11,20-trione:*** a glucocorticoid hormone from the adrenal cortex. It differs from Cortisol (see) in having a keto instead of a hydroxyl group in position 11. Like cortisol, it stimulates carbohydrate formation from protein, promotes glycogen storage in the liver, and raises the blood sugar level. It is biosynthesized from cortisol by enzymatic dehydrogenation of the 11β-hydroxyl group. C. can be obtained by partial synthesis from other pregnane compounds, bile acids and steroid sapogenins. The first total synthesis was reported in 1951 by Woodward. C. was first isolated from the adrenal cortex in 1935 simultaneously by Reichstein, Kendall and Wintersteiner. It is also present in blood and urine. *C. acetate* is used as a drug for rheumatoid arthritis and allergic skin reactions, but it is surpassed in these properties by synthetic derivatives such as prednisone (see Prednisolone) and triamcinolone. For structure and biosynthesis, see Adrenal corticosteroids.

Corynantheine: see Yohimbine.

COS-cells: various cell lines (COS-1, 3 and 7) derived from CV-1, an established line of simian cells, by transformation with an origin-defective mutant of

SV-40 virus which codes for wild type T antigen [Y. Gluzman, *Cell* **23** (1981) 175–182].

Cosmid: a DNA molecule made by fusing DNA from a phage with a bacterial plasmid, and used as a cloning vector (see Recombinant DNA technology). It contains the plasmid gene(s) for antibiotic resistance (for selection purposes), the plasmid replication origin, and the *cos* site of the phage. The *cos* site is required for packaging of the DNA into the phage protein coat. Many phage genes are lacking. After transfection, C. become packaged in multiple, infective, phage-like particles within the host bacterial cell, but cell lysis does not occur. C. are relatively small, consisting of about 8 kb, and they can carry more than 50 kb of foreign DNA by insertion.

Cosubstrate: a Coenzyme (see) which enters an enzymatic reaction as a second substrate.

C_0t (see also Hybridization): a measure of the degree of reassociation of single-stranded DNA, the latter having been produced by the heat denaturation of duplex DNA (see Hybridization). The two parameters affecting reassociation are: (i) the initial concentration (C_0) of single-stranded DNA, and (ii) the time (t) allowed for it to occur. C_0t is simply the product of these parameters, i.e. $C_0t = C_0$ (expressed in moles of nucleotide per litre) $\times$ t (expressed in seconds). A convenient way of comparing the reassociation rates of different DNAs is the $C_0t_{1/2}$ value, which is the C_0t value when half of the DNA has reassociated; the lower the $C_0t_{1/2}$ value the higher the rate at which the complementary strands of DNA reassociate. $C_0t_{1/2}$ values were instrumental in the discovery of tandemly repeated nucleotide sequences in eukaryotic DNA. If the genomic DNA of a given eukaryote is cleaved into fragments of 1,000 bp average length which are then denatured by heating to 92–94 °C to form single strands, the $C_0t_{1/2}$ values for the reassociation of the complementary pairs of the latter are not the same. About 10–15 % of the total DNA reassociates very rapidly ($C_0t_{1/2}$ values 0.01 or less), a further 22–40 % reassociates more slowly ($C_0t_{1/2}$ values of 0.01–10) while the rest reassociates very slowly ($C_0t_{1/2}$ values of 100–10,000). The first of these categories consists of several different types of DNA all of which are composed of many tandem repeats of a short nucleotide sequence, the nature of the latter being different for each type; this is often called 'simple sequence DNA' which is almost synonymous with the term Satellite DNA (see). ('Simple sequence DNA' is composed of 'satellite DNA' plus 'cryptic satellite DNA'. The latter is composed of 'simple sequence DNA' which has the same G + C content and therefore the same buoyant density as that of the main cellular DNA; consequently, it does not separate from the main cellular DNA during density gradient centrifugation and remains hidden within it; in contrast, 'normal' satellite DNA separates from the main cellular DNA during density gradient centrifugation). The second of the 3 categories consists of a few different types of DNA, all composed of thousands of tandem repeats of a long nucleotide sequence, which differs for each type; this is usually called 'intermediate repeat DNA'. The third category is composed largely of 'single copy DNA', which consists of nucleotide sequences that occur only once in the genome; it includes solitary genes that code for protein and 'spacer DNA' which has no known function. The reason for the differences in the $C_0t_{1/2}$ values of the three DNA categories is that the less 'random' the nucleotide sequence (i.e. the more short identical repeats it has) the easier it is for two complementary strands to reassociate.

Cotransport: see Membrane transport.

Cotylenins: leaf growth-promoting substances from *Cladosporium*. All C. are glycosides of cotylenol. The absolute stereochemistry of cotylenol is the same as that of the Fusicoccins (see), with which they probably share an essentially identical biosynthetic pathway. Formulas of known C. are shown in the Fig. Previously isolated C. B, D and G are now known to be artifacts. [T. Sassa et al. *Agric. Biol. Chem.* **39** (1975) 1729–1734 (A,C,E); T. Sassa & A. Takahama *ibid* (1975) 2213–2215 (C,F); A. Takahama et al. *ibid* (1979) 647–650 (H, I); A. Bottalico et al. *Phytopathol. Mediterr.* **17** (1978) 127–134 (structure-activity relationships)]

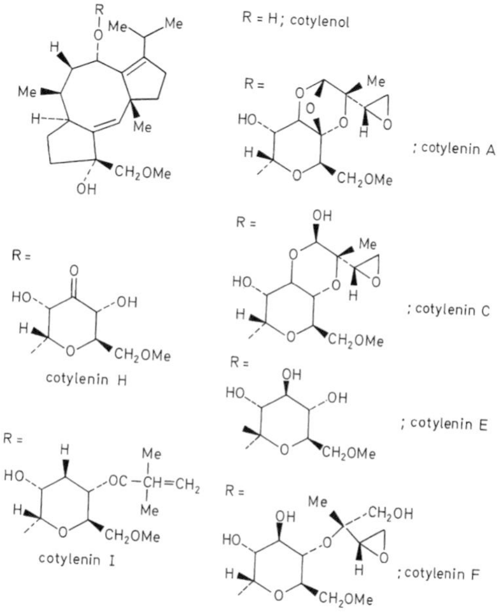

R = H; cotylenol

; cotylenin A

; cotylenin C

; cotylenin E

; cotylenin F

cotylenin H

cotylenin I

Cotylenins

Couepic acid: see Licanic acid.

4-Coumarate:CoA ligase, ***hydroxycinnamoyl-CoA ligase*** (EC 6.2.1.12): a plant enzyme catalysing a 2-step process:

$$1.\ E + R\text{–}CH = CH\text{–}COOH + ATP \overset{Mg^{2+}}{\rightleftharpoons} E[R\text{–}CH = CH\text{–}CO \cdot AMP] + PP_i;$$

$$2.\ E[R\text{–}CH = CH\text{–}CO \cdot AMP] + CoA\text{–}SH \rightleftharpoons R\text{–}CH = CH\text{–}CO \cdot S\text{–}CoA + AMP + E,$$

where E is the enzyme and R–CH = CH–COOH represents *trans*-4-hydroxycinnamic acid. From most sources the enzyme has M_r 55,000–67,000. Isoenzymes have been separated by ion exchange chromatography. The enzyme from all sources is inhibited competitively by AMP. In *Glycine max* and in cell suspension cultures of *Phaseolus vulgaris*, the enzyme increases in response to biotic elicitor from *Phytophthora megasperma* (see Phytoalexins). The catalysed reaction is an early step in Flavonoid (see) biosynthesis. [R. A. Dixon et al. *Advances in Enzymology* **55** (1983) 1–136]

Coumarins: lactones of *cis*-2-hydroxycinnamic acid derivatives, which are widely distributed in plants (Table), especially in the *Umbelliferae* and *Rutaceae*. The lactone ring can be opened by alkali treatment to give *cis*-2-hydroxycinnamic acids, which spontaneously recyclize in acid. Most C. are formal derivatives of umbelliferone, i.e. they carry an OH-group on C7. Some also possess a second OH-group (or alkoxyl group) on C5 or, more rarely, on C4. Coumarin itself (2H-1-benzopyran-2-one; 1,2-benzopyrone; *cis*-2-coumarinic acid lactone; *o*-hydroxycinnamic acid lactone) is a pleasant smelling compound.

Biosynthesis. C. are products of the shikimic acid pathway of Aromatic biosynthesis (see). The key intermediate is *trans*-cinnamic acid, which may be converted either to coumarin itself, or become hydroxy-

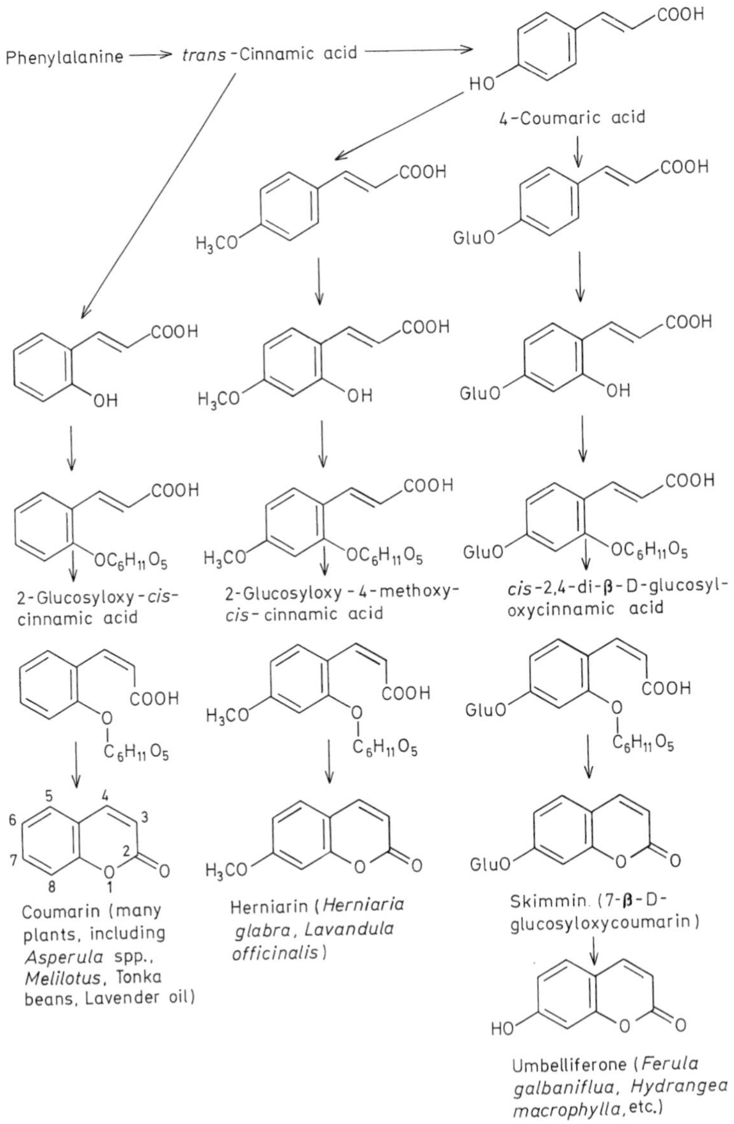

Phenylalanine → *trans*-Cinnamic acid → 4-Coumaric acid

2-Glucosyloxy-*cis*-cinnamic acid

2-Glucosyloxy-4-methoxy-*cis*-cinnamic acid

cis-2,4-di-β-D-glucosyloxycinnamic acid

Coumarin (many plants, including *Asperula* spp., *Melilotus*, Tonka beans, Lavender oil)

Herniarin (*Herniaria glabra, Lavandula officinalis*)

Skimmin (7-β-D-glucosyloxycoumarin)

Umbelliferone (*Ferula galbaniflua, Hydrangea macrophylla*, etc.)

Fig. 1. *Biosynthesis of coumarins. Trans*-Cinnamic acid is converted to 4-coumaric acid by Cinnamate 4-hydroxylase (see). Cyclization to the coumarin ring system occurs spontaneously after removal of the 2-glucosyl group by a specific glucosidase when the plant is damaged.

lated in the *para* position as a prelude to the synthesis of other C. (Fig. 1). The appropriately substituted *trans*-cinnamic acid precursor is then hydroxylated in the *ortho* position. Glycosylation of this hydroxyl group may be important for the subsequent *trans-cis* isomerization of the side chain, because strong intramolecular hydrogen bonding between the carboxyl group and the 2-hydroxyl group of the glucose residue is only possible in the *cis* form of the 2-glucosyloxycinnamic acid. There is evidence for the existence

Some naturally occurring coumarins

Trivial name	Substituents on coumarin ring system (see Fig. 1 for numbering)	Plant source
Angelicin	furano (2′:3′-8:7) [formula shown]	*Angelica archangelica* (roots)
Angustifolin	3-(1′,1′-dimethylallyl)-7-hydroxy [formula shown]	*Ruta angustifolia* (leaves)
Ayapin	6,7-methylenedioxy	*Eupatorium ayapana* (leaves)
Bergapten	5-methoxy-furano (2′:3′-6:7)	*Citrus bergamia* and many other plants
Braylin	6-methoxy-7,8-pyrano [formula shown]	*Flindersia brayleyana* (bark)
Calycanthoside	6,8-dimethoxy-7-glucosido	*Calycanthus occidentalis* (twigs)
Ceylantin	7,8-dimethoxy-5,6-pyrano [formula shown]	*Atalantia ceylanica* (heartwood)
Chicoriin	6-hydroxy-7-glucosido	*Chicorium intybus* (flowers)
Coumarin	see Fig. 1	
Dalrubone	see illustrated formula	*Dalea* spp.
Dalbergin	3-phenyl-6-hydroxy-7-methoxy	*Dalbergia sissoo* (heartwood)
Daphnetin	7,8-dihydroxy	*Euphorbia lathyris* (seeds)
Daphnin	7-glucosido-8-hydroxy	*Daphne alpina* (bark)
Esculetin	see Fig. 2	
Esculin	6-glucosido-7-hydroxy	*Aesculus hippocastanum* (bark)
Fraxetin	6-methoxy-7,8-dihydroxy	*Fraxinus intermedia* (bark)
Fraxin	6-methoxy-7-hydroxy-8-glucosido	*Fraxinus intermedia* (bark)
Fraxinol	5,7-dimethoxy-6-hydroxy	*Fraxinus excelsior* (bark)
Isobergapten	5-methoxy-furano (2′:3′-6:7)	*Heracleum lanatium* (roots)
Isofraxidin	6,8-dimethoxy-7-hydroxy	*Fraxinus intermedia* (bark)
Limettin	5,7-dimethoxy	*Citrus limetta* (fruit)
Osthenol	7-hydroxy-8-*iso*pent-2′-enyl, or 7-hydroxy-8-dimethylallyl	*Angelica archangelica* (roots)
Osthole	7-methoxy-8-*iso*pent-2′-enyl, or 7-methoxy-8-dimethylallyl	*Imperatorium ostruthium* (rhizomes)
Pereflorin	5-methyl-4-methoxy	*Perezia multiflora* (roots)
Psoralen	furano (2′:3′-6:7) [formula shown]	*Psoralea corylifolia* (seeds), *Xanthoxylum flavum* (wood)
Skimmin	see Fig. 1	
Suberosin	7-methoxy-6-*iso*pent-2′-enyl, or 7-methoxy-6-dimethylallyl [formula shown]	*Xanthoxylum suberosum* (bark), *Xanthoxylum flavum* (wood)
Trimethoxy-coumarin	6,7,8-trimethoxy	*Fagara macrophylla* (heartwood)
Umbelliferone	see Fig. 1	
Vellein	7-glucosido-8-*iso*pent-2′-enyl, or 7-glucosido-8-dimethylallyl	*Vellia discophora*
Xanthotoxin*	8-methoxy-furano (2′:3′-6:7)	*Fagara xanthoxyloides* (fruit)

* Xanthotoxin is used in indigenous medical centers in India for treatment of leucoderma.

of an isomerase, but it seems probable that the isomerization is largely photo-catalysed. Cinnamic acid 4-hydroxylase (see), present in the microsomal fraction of disrupted cells, converts *trans*-cinnamic acid to *p*-coumaric acid, whereas the *ortho* hydroxylation is catalysed by a chloroplast enzyme and stimulated by NADPH and 2-amino-4-hydroxy-6,7-dimethyl-5,6,7,8-tetrahydrobiopteridine [H. Kindl *Z. physiol. Chem.* **352** (1971) 78–84]. Coumarin itself and 7-hydroxylated C. rarely occur together in the same plant, i.e. 7-hydroxylated C. are normally synthesized, unless the membrane system catalysing *para* hydroxylation of *trans*-cinnamic acid is absent or of low activity.

Free C. do not occur in significant amounts in healthy plant tissue. When plant cells are damaged or the plant wilts, specific glucosidases remove glucose from the 2-glucosyloxy-*cis*-cinnamic acid precursors, which then spontaneously cyclize to C. Some C. may also be stored as glucosides, if a hydroxyl group is available for glucose attachment; thus, both skimmin and *cis*-2,4-di-β-D-glucosyloxycinnamic acid represent stored precursors, which are converted into umbelliferone when leaves of *Hydrangea macrophylla* are damaged [D. J. Austin & M. B. Meyer Phytochemistry **4** (1965) 255–262].

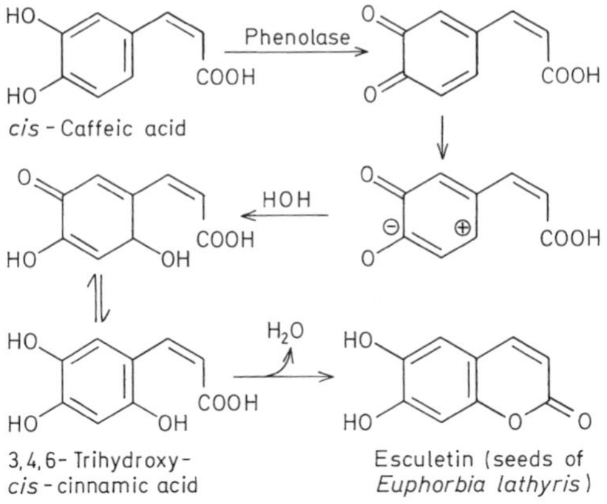

cis-Caffeic acid — Phenolase → ... HOH ... H_2O → Esculetin (seeds of *Euphorbia lathyris*); 3,4,6-Trihydroxy-*cis*-cinnamic acid

Fig. 2. *Coumarins.* Proposed pathway for the biosynthesis of esculetin from *cis*-caffeic acid, catalysed by phenolase from *Saxifraga stolonifera.*

Chloroplasts of *Saxifraga stolonifera* convert *cis*-caffeic acid into esculetin [M. Sato *Phytochemistry* **6** (1967) 1363–1373]. The enzyme responsible is a phenolase (ortho-diphenol: O_2 reductase), which has been solubilized by detergent treatment of chloroplasts. It is proposed that the *ortho*-quinone from *cis*-caffeic acid converts spontaneously to esculetin, following spontaneous hydroxylation by water in the *ortho* position (Fig. 2). During synthesis in the plant, the 3,4,6-trihydroxy-*cis*-coumaric acid is presumably trapped as its 2-glucosyloxy derivative. A different biosynthetic route has been proposed for dalrubone, and some C. may arise from the A ring of anthocyanins [D. L. Dreyer et al. *Tetrahedron* **31** (1975) 287–293].

The coumarin ring system in Novobiocin (see) from *Streptomyces spheroides* is derived from tyrosine, i.e.

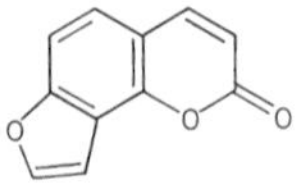

Angelicin — Angustifolin — Braylin — Ceylantin — Dalrubone — Pereflorin B — Psoralen — Suberosin

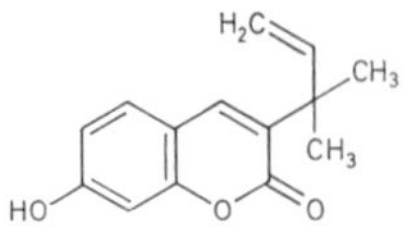

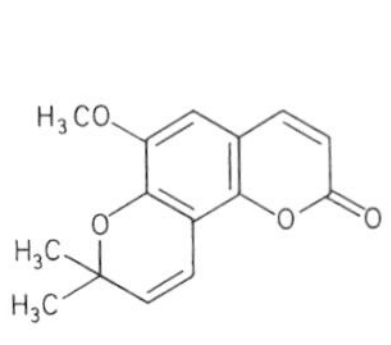

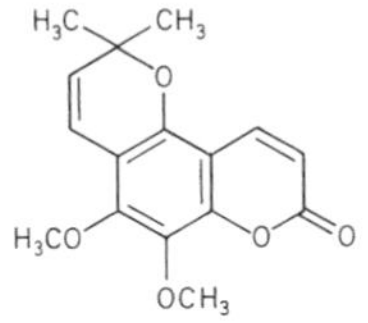

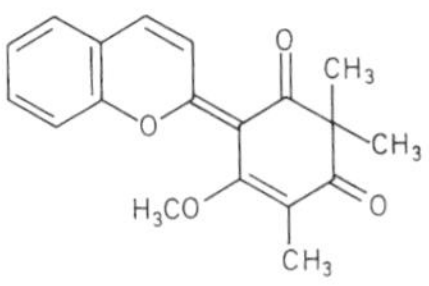

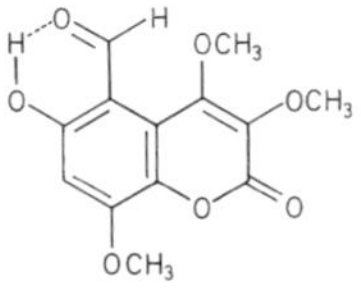

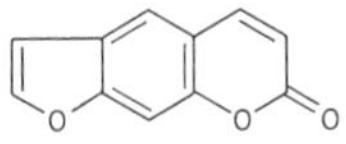

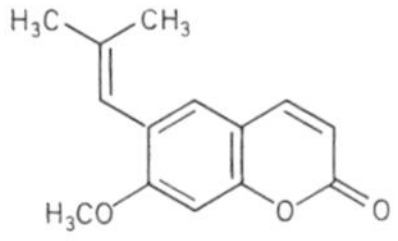

Structures of some coumarins (see Table for further description)

the hydroxyl group at C7 is introduced at an earlier stage in the bacterial than in the plant synthesis [K. B. G. Torsell *Natural Products Chemistry,* John Wiley, 1983; G. Billek (ed.) *Biosynthesis of Aromatic Compounds,* Academic Press, 1969]

Coumestans, ***coumaronocoumarins:*** compounds with the illustrated ring system (Fig.), which represents the highest possible oxidation level of the Isoflavonoid (see) skeleton. Like other isoflavonoids, C. are largely restricted to the *Leguminoseae* and they are accompanied by the corresponding 6a,11a-dehydropterocarpans (see Pterocarpans). Examples: ***lucernol*** (2,3,9-trihydroxycoumestan, *Medicago sativa),* **sativol** (4,9-dihydroxy-3-methoxycoumestan, *Medicago sativa),* **psoralidin** (3-hydroxy-9-methoxy-2-γ,γ-dimethylallylcoumestan, *Psoralea corylifolia),* 2-hydroxy-

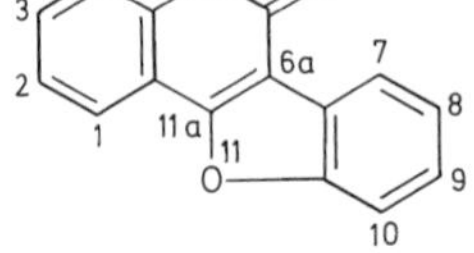

Coumestan ring system

1,3-dimethoxy-8,9-methylenedioxycoumestan *(Swartzia leiocalycina).* [J.B.Harborne, T.J.Mabry & H.Mabry (eds.) The Flavonoids, Chapman and Hall, 1975]

Coupling factors: see Respiratory chain.

Covalent modification of enzymes, ***enzyme modulation, enzyme interconversion:*** Oligomeric (i.e. multichain) enzymes may exist in two or more forms, which are interconvertible by enzyme-catalysed covalent modifications, and which differ in their catalytic properties, e.g. activity, substrate affinity and dependence on effectors. Usually the difference in activity is such that one form is active and the other inactive. The activities of the conversion enzymes are in turn regulated by other enzymes, metabolites and/or effectors. Covalent modifications are therefore important in physiological regulation, in addition to Allostery (see). Whereas allostery provides fine adjustment of metablic rates, C. provides an on/off switching of cellular functions, which is very sensitive to environmental influences.

The most common type of covalent modification appears to be a phosphoryation/dephosphorylation cycle. Many cell membrane receptor molecules are kinases, as are the products of some oncogenes.

The potential for such systems to respond radically to small changes in the concentration of effector molecules arises because their kinetic behavior can be described by the Michaelis-Menton equation (see). At steady state, the fraction of protein P in active form (P^*/P_{tot}) depends on the rate constants of the activation and inactivation reactions, $P \underset{k_b}{\overset{k_c}{\rightleftharpoons}} P^*$. If the concentration of substrate (the regulated protein) is less than the K_m values of the regulating enzymes, the kinetics are first order. For the forward reaction, $v_f = (V_{mf}/K_{mf})P$, and for the back reaction, $v_b = (V_{mb}/K_{mb})P^*$. If the two reactions are plotted as in Fig. 1, the steady-state fraction of activated protein is found at the intersection of the plots for the two enzymes. A minor change in the activity (V_m) of one of the two enzymes does not greatly affect the fraction of the substrate protein that is active.

In contrast, if the concentration of regulated protein is high enough to saturate the two modifying enzymes, giving zero-order kinetics, a relatively small change in the V_m value of one of them causes a large change in the fraction of the regulated protein that is in the active form (Fig. 1b). This makes the system very sensitive to changes in the concentration of an allostreric regulator for one of the modifying enzymes. An example of a system subject to zero-order regulation is isocitrate dehydrogenase of *E. coli* [D.C.LaPorte & D.E.Koshland, Jr. *Nature* **305** (1983) 286–290]. This enzyme determines the distribution of acetyl-CoA between the tricarboxylic acid cycle and the glyoxylate cycle. Isocitrate dehydrogenase, which is part of the TCA cycle, is inactivated by phosphorylation. The phosphate, which removes the phosphate groups and thus activates the dehydrogenase, is in turn activated by 3-phosphoglycerate, a metabolic precursor of acetyl-CoA. When 3-phosphoglycerate is abundant, the active isocitrate dehydrogenase allows the acetyl-CoA to enter the TCA cycle where it is oxidized to CO_2. However, when the bacteria are growing at the expense of acetate, the relative lack of 3-phosphoglycerate causes the phosphatase, and therefore the dehydrogenase, to be inactive. The TCA cycle is then shut down and the glyoxylate shunt provides carbon for the synthesis of cell material. The relationship between the K_m of the phosphatase and the concentration of isocitrate dehydrogenase is such that a small change in the concentration of 3-phosphoglycerate shifts the equilibrium of isocitrate dehydrogenase from nearly all active to nearly all inactive, and vice versa.

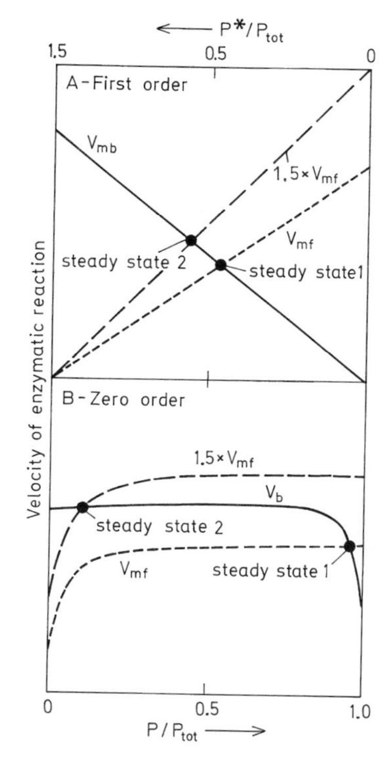

Fig. 1. *Steady state fractions of activated (P*) and inactive (P) protein.*
A. If the K_m values for the enzymes catalysing the forward and back reactions are greater than P_{tot}, the reaction kinetics are first order. Increasing the rate of the forward reaction by 50% does not shift the fraction of P in the active form (P*/P) very much.
B. In contrast, if P_{tot} is much larger than K_m, so that the reaction kinetics are zero order, a 50% increase in the rate of the forward reaction shifts the fraction of active protein from about 0.05 to 0.9 (in this example). [adapted from D.C.LaPorte & D.E.Koshland, Jr. *Nature 305* (1983) 286–290]

Phosphorylase, which catalyses glycogen breakdown, provides an example of covalent modification in response to the concentration of the "second messenger", cyclic AMP (see Hormones). Phosphorylase in its inactive *(b)* form is activated by phosphorylation of a serine residue. In muscle phosphorylase, this causes the dimeric *b* form of the enzyme to aggregate to a tetramer, the active *a* form (in liver, both *a* and *b* forms have the same M_r). The enzyme responsible for this activation, phosphorylase *b* kinase, must itself be activated by a specific kinase.

Some other enzymes subject to regulation by phosphorylation and dephosphorylation are pyruvate dehydrogenase, glycogen synthetase, phosphofructokinase and glutamate dehydrogenase. Covalent modification accompanied by a change of M_r, i.e. association and dissociation, affects glutamine phosphoribosylpyrophosphate amidotransferase, pancreatic lipase (F- and S-lipase), human glucose 6-phosphate dehydrogenase and rat kidney pyruvate kinase. See also ADP-ribosylation of proteins. [O.M. Rosen & E.G. Krebs (eds.) *Protein Phosphorylation.* Cold Spring Harbor Conferences on Cell Proliferation, Cold Spring Harbor Laboratory, Cold Spring Harbor, N.Y. (1982) vol. **8, A** & **B**]

Another well studied example of covalent modification is that of glutamine synthetase (EC 6.3.1.2) which catalyses L-glutamine synthesis: ATP + L-glutamate + $NH_3 \rightarrow$ ADP + P_i + L-glutamine. Control of this enzyme by covalent modification has been studied chiefly in *E. coli* and *Klebsiella* spp. (Fig. 2).

The enzyme consists of 12 subunits (each of $M_r \sim 50{,}000$) arranged as 2 hexagons attached face to face. The subunits are identical and contain a tyrosyl residue that can be adenylylated by glutamine synthetase adenylyltransferase (EC 2.7.7.42). A phosphoester linkage is formed between the phosphate of the AMP residue and the phenolic OH of the tyrosyl residue. When all 12 subunits are adenylylated (E_{12}) the enzyme is totally inactive, whereas full activity is expressed when no subunits are adenylylated (E_0). Between E_{12} and E_0 there are theoretically 382 possible forms of adenylylated glutamine synthetase, but the true arrangment of adenyl groups is not known for the intermediate levels of adenylylation. The plot of the number of AMP residues per molecule of enzyme (as extracted from cells grown under different conditions of nitrogen supply) against specific enzymatic activity is not a straight line; a plateau at about one half maximal activity is found between E_3 and E_7, suggestive of threshold values in a control system. Glutamine synthetase adenylyltransferase, purified from *E. coli,* contains 2 active centers, one adenylylating and the other deadenylylating the glutamine synthetase. The transferase therefore activates or inactivates the synthetase, depending on the relative activities of its 2 active sites. These, in turn, depend on a regulatory protein P_{II}, which consists of 4 identical subunits. Each subunit contains a tyrosyl residue that is subject to uridylylation by the action of a uridylyltransferase. The uridylylated form activates the deadenylylating activity, whereas the non-uridylylated form stimulates the adenylylating activity of the adenylyltransferase. The uridylyltransferase is activated by 2-oxoglutarate and inhibited by L-glutamine, so that the ratio 2-oxoglutarate/glutamine ultimately controls the intracellular activity of glutamine synthetase.

Cozymase: see Nicotinamide adenine dinucleotide.

C_3 plants, *Calvin plants:* green plants which produce C_3 compounds as the first product of CO_2 fixation. The CO_2 fixation reaction is catalysed by ribulose 1,5-*bis*phosphate carboxylase (EC 4.1.1.39): ribulose 1,5-*bis*phosphate + CO_2 + $H_2O \rightarrow 2 \times$ (3-phosphoglycerate). The products of the light reaction, ATP and NADPH, are used to reduce the 3-phosphoglycerate to 3-phosphoglyceraldehyde (see Carbon dioxide assimilation), which enters the Calvin cycle (see). Most green plants are C_3 plants.

C_4 plants: green plants in which the primary product of CO_2 fixation is not 3-phosphoglycerate (cf. C_3 plants) but a C_4 acid such as oxaloacetate, malate or aspartate. These plants possess two types of photosynthesizing cells. In *mesophyll cells* near the leaf surface, CO_2 is fixed into C4-compounds. This "prefixation" of CO_2 is due to the action of the cytosolic enzyme, phospho*enol*pyruvate carboxylase (EC 4.1.1.31), which carboxylates phospho*enol*pyruvate to oxaloacetic acid (see Hatch-Slack-Kortschak cycle). The Calvin cycle (see) operates in the the *vascular bundle cells* of C_4 plants, and CO_2 for the Calvin cycle is derived from the decarboxylation C_4 compounds rather than directly from the atmosphere. This "Kranz anatomy", i.e. photosynthetically active bundle sheath cells with a photosynthetically active layer

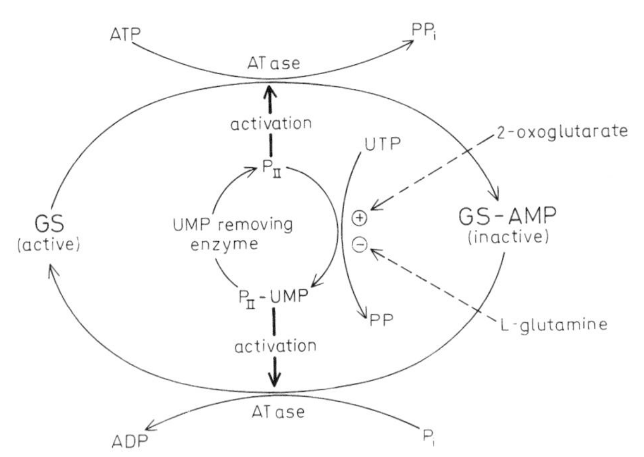

Fig. 2. *Control of the activity of glutamine synthetase by adenylylation and deadenylylation.* GS = glutamine synthetase; GS-AMP = adenylylated glutamine synthetase; ATase = adenylyltransferase.

of mesophyll cells on either side, is typical of the leaves of C_4 plants. In contrast, the vascular bundles of C_3 plants show very little photosynthetic activity. The C_4 reactions are found in many tropical plants and in unrelated dicotyledons which are adapted to hot, dry climates. These plants must close their stomata in the hot parts of the day to avoid excessive water loss, thus limiting the availability of atmospheric CO_2 to the mesophyll cells. However, the affinity of phospho*enol*pyruvate carboxylase for CO_2 is much higher than that of ribulose*bis*phosphate carboxylase (the first enzyme of the Calvin cycle), so that C_4 plants can fix scarce CO_2 within their leaves when light, and thus ATP, is abundant. Not all C_4 plants display this biochemical division of labor between vascular bundles and mesophyll cells (see Crassulacean acid metabolism).

Crabtree effect: a decrease in respiration rate after addition of glucose, when studied in isolated systems such as ascites tumor cells. This is probably due to a more effective competition by glycolysis for inorganic phosphate and NADH, leading to a deficiency of these materials for oxidative phosphorylation.

Crassulacean acid metabolism; *CAM:* Plants that exhibit CAM are succulents; many belong to the *Crassulaceae* (e.g. *Sedum* spp., *Kalanchoe* spp.) but others belong to a range of mono- and dicotyledonous families and even to the Pteridophyte family, *Polypodiaceae.* CAM plants have the following characteristics: (i) normally their stomata are open during the night (i.e. in the dark) and closed during the day (i.e. in the light); these stomatal movements are the reverse of those of non-CAM plants, (ii) CO_2 is fixed during the hours of darkness in the chloroplast-containing cells of photosynthetic leaf and stem tissue and considerable quantities of L-malic acid are synthesized, (iii) this malic acid is stored in the large vacuoles that are characteristic of the cells of CAM plants, (iv) during the hours of daylight the malic acid formed during the preceding night is decarboxylated and the resulting CO_2 is converted into sucrose and starch or some other storage glucan by a light-driven Calvin cycle (see), and (v) during the succeeding night some of the storage glucan is catabolized to provide an acceptor molecule for the dark CO_2-fixation reaction. The photosynthetic cells of CAM plants, therefore, exhibit a diurnal cycle in which the level of malic acid rises and that of storage glucan falls during the night and the converse occurs during the day.

The metabolism underlying these characteristics is as follows. During the night CO_2 enters the tissue through the open stomata and diffuses into the photosynthetic cells where it dissolves in the aqueous milieu and generates HCO_3^-, possibly with the assistance of carbonic anhydrase (EC 4.2.1.1). The HCO_3^- then reacts with phospho*enol*pyruvate (PEP) under the catalytic influence of PEP carboxylase (EC 4.1.1.31) to form oxaloacetic acid (OAA), which is then reduced to malic acid by NADH and NAD-malate dehydrogenase (EC 1.1.1.37). The malic acid is actively transported across the tonoplast membrane and accumulates in the cell vacuole. This keeps the concentration of malic acid low in the cytoplasm, which is important because it is an allosteric inhibitor of PEP carboxylase. Production of malic acid proceeds throughout the night but slackens off as dawn approaches. It has been suggested that the reason for this is that the turgor pressure of the vacuole, which has been progressively increasing through the night as malic acid and osmotically-attracted water have accumulated in it, exceeds a critical level and forces malic acid back into the cytoplasm where it inhibits PEP carboxylase. The source of the PEP for the carboxylation reaction is the storage glucan produced during the previous period of daylight; the glucan is catabolized by phosphorylase and debranching enzymes to D-glucose-1-phosphate (G-1-P) which is then converted to PEP via the glycolysis pathway. The NAD^+/NADH balance of the cell is maintained during the night by the reciprocal requirements of NAD-malate dehydrogenase and 3-phosphoglyceraldehyde dehydrogenase (EC 1.2.1.12).

During the ensuing period of daylight malic acid continues to pass back from the vacuole into the cytoplasm where it is decarboxylated to provide CO_2 for light-driven C_3 photosynthesis via the Calvin cycle. Two different decarboxylation reactions are used in CAM but they never occur together in the same plant species. CAM members of the *Crassulaceae, Cactaceae* and *Agavaceae* decarboxylate malic acid with the NADP-malic enzyme (EC 1.1.1.40), producing CO_2 and pyruvic acid (Pyr). CAM members of the *Liliaceae, Bromeliaceae* and *Asclepiadaceae* use PEP carboxykinase (EC 4.1.1.49) which means that the malic acid has to be converted back into OAA, the substrate for this enzyme; this conversion is catalysed by malate dehydrogenase. The PEP produced by PEP carboxykinase and the pyruvate produced by the NADP-malic enzyme are converted into 3-phosphoglyceric acid (3PGA), which is utilized in the Calvin cycle. The PEP $\rightarrow$ 3PGA conversion proceeds via 2-phosphoglyceric acid (2PGA) and requires the enzymes enolase (EC 4.2.1.11) and phosphoglyceromutase (EC 5.4.2.1). The Pyr $\rightarrow$ 3PGA conversion follows the same route but requires the initial pyuvate,orthophosphate dikinase (EC 2.7.9.1)-catalysed conversion of Pyr to PEP.

Following fixation of the CO_2 by ribulose bisphosphate carboxylase (EC 4.1.1.39) (see) D-fructose-6-phosphate (F-6-P) can be tapped off the Calvin cycle (as photosynthetic product) and used for the synthesis of storage glucan, e.g. starch, within the chloroplast. Additionally dihydroxyacetone phosphate can be tapped off the cycle and exported from the chloroplast to the cytoplasm where it can be used to synthesize sucrose (see Fig. 1 of the entry: Calvin cycle). The sucrose is then exported from the cell to the growing parts of the plant.

CAM is an adaptation, appearing late in evolution, that allows plants (e.g. cacti) to survive and grow in extremely arid environments. During the daytime, when it is extremely hot, cacti close their stomata. This, taken with their low surface-to-volume ratio and heavily waxed cuticle, prevents virtually all gaseous exchange with the atmosphere and thus cuts water loss down to almost zero. It also prevents any CO_2 uptake. However this does not matter because the CO_2 required by the light-driven Calvin cycle is obtained by decarboxylating the malic acid formed during the previous night. During the night, when it is much cooler and water loss by evaporation much less, the stomata may or may not open, depending on the availability of water. If there has been recent rainfall the stomata open and CO_2 can be taken in and used to generate malic acid which is used in the following period of daylight for the operation of the Calvin cycle; there is thus a net production of carbohydrate and the cactus can grow. If, however, there has been no rainfall for some time, as is usual in such regions, the stomata remain shut, because any evaporative water loss could not be made up by root absorption. Thus there can be no CO_2 uptake during the night and no net carbohydrate production during the following day. Under these conditions there can be no growth and the best that the cactus can do is survive. This it can do for months and years by utilizing a process that has been termed 'idling'. The main function of this process is to dissipate the light energy that the cactus cannot avoid absorbing during the daytime. The 'normal' way that a plant dissipates absorbed light energy is to use it productively to synthesize carbohydrate and grow, but this is denied to the cactus. It therefore dissipates this light energy nonproductively by employing a cycle of carboxylation and decarboxylation using CAM, photorespiration and dark respiration. During the day malic acid is decarboxylated to provide CO_2 for the light-driven Calvin cycle until the supply is depleted. When this happens photorespiration provides the CO_2. During the night the storage glucan formed during the preceding day is converted to CO_2 and PEP by the normal respiratory processes of Glycolysis (see) and the Tricarboxylic acid cycle (see); the CO_2 and PEP are then converted back to malic acid. This process of idling persists until the drought is broken, when the cactus takes up water, its tissues rehydrate, its stomata recommence their night-time opening and growth occurs once again as a consequence of the re-establishment of productive CAM.

Some CAM plants grow in regions where water availability is less of a problem. These plants appear to shift from CAM to normal C_3 photosynthesis during periods of water abundance and back to CAM when it is in short supply. Moreover there is evidence that in some plants CAM is an inducible adaptation. Thus there appear to be two types of CAM plants, obligate or constitutive CAM plants (e.g. cacti) and facultative or inducible CAM plants (e.g. *Mesembryanthemum crystallinium*).

Creatine, ***β-methylguanidoacetic acid, methylglycocyamine,*** **N-*amidosarcosine:*** a product of amino acid metabolism, which is readily converted into its cyclic anhydride, Creatinine (see). More than 90 % of the C. in an adult human is present in muscle. The concentration is particularly high in contracting muscle, i.e. when chemical energy is being converted into mechanical energy. The normal human plasma concentration of C. is higher for women (0.35–0.93 mg/100 ml) than that for men (0.17–0.50). C. is not a normal constituent of urine, except in pregnancy. It is also excreted in muscular dystrophy and other muscle wasting diseases. Creatinine, on the other hand, is one of the main nitrogenous compounds of normal urine.

For the role of C. phosphate as a phosphagen and for the formula and biosynthesis of C., see Phosphagens.

Creatine kinase, ***creatine phosphokinase*** (EC 2.7.3.2): a phosphotransferase found in brain and muscle. It is a dimeric enzyme of M_r 82,000. There are 3 different subunits: M (muscle), B (brains) and Mi (mitochondria), and 4 known isoenzymes (BB, MB, MM, MiMi). The BB dimer is found in brain, smooth muscle and embryonic skeletal muscle. As the latter develops there is a gradual change from BB through MB to MM dimers. Most C.k. is soluble but about 5 % of the MM dimer in striated muscle is located in the M-line (see Muscle proteins). MiMi-C.k. is found on the outer side of the inner mitochondrial membrane, where it is coupled to an ADP/ATP translocase.

C.k. catalyses ATP formation from ADP and creatine phosphate in a reversible reaction depending on Mg(II) or Mn(II). Muscle contraction consumes ATP, but during prolonged work ATP is not depleted, because C.k. continually catalyses the phosphorylation of ADP to ATP at the expense of the large amounts of stored creatine phosphate. The immobilized C.k. in the M-lines of muscle is sufficient to maintain the ATP level even during vigorous contraction. It is thought that the remaining 95 % of C.k., which is soluble, serves to rebuild the creatine phosphate supply rapidly after a period of depletion. Similarly, the MiMi-C.k. enables the efficient conversion of mitochondrial ATP to creatine phosphate. [T. Wallimann & H. M. Eppenberger in J. Shay (ed.) *Cell and Muscle Motility* **6,** Plenum Press, 1985, pp. 239–285]

Damaged skeletal or heart muscle releases C.k. into the serum. Increases in serum C.k. are therefore used in the early diagnosis of heart infarction, the detection of muscle wastage, and to distinguish heart infarction from lung embolism, in which there is no increase in serum C.k.

Creatine phosphate: see Phosphagens.

Creatinine: the internal cyclic anhydride of creatine. In healthy adult humans the normal plasma concentrations of C. are 0.95–1.29 mg/100 ml (males) and 0.77–0.98 (females). Normal adult urine contains an average of 2.15 g C. per 24 h, with wide variations depending mainly on total muscle mass. In the healthy human, urinary C. is of endogenous origin, formed continually from creatine in muscle, and hardly affected by changes of diet. In any individual it is therefore relatively constant, and is often used as an internal standard for reporting other urinary quantities.

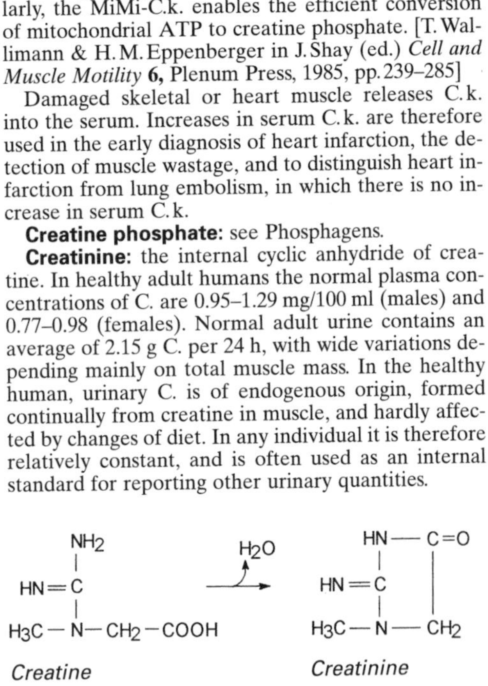

NH_2–C(=NH)–N(CH_3)–CH_2–COOH $\xrightarrow{-H_2O}$ HN=C–N(CH_3)–CH_2–C(=O)–NH (ring)

Creatine → *Creatinine*

CRH: acronym of corticotropin releasing hormone (see Releasing hormones).

Crigler-Najjar syndrome: see Inborn errors of metabolism.

Crocetin: a brick-red C_{20} dicarboxylic acid, M_r 328.39, m.p. 285 °C, with 7 conjugated *trans* double bonds, 4 methyl branches and 2 terminal carboxyl groups. It is a carotenoid oxidation product or apocar-

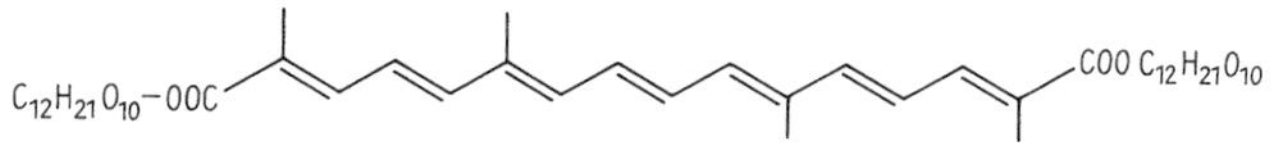

$C_{12}H_{21}O_{10}$–OOC ... COO $C_{12}H_{21}O_{10}$

Crocin

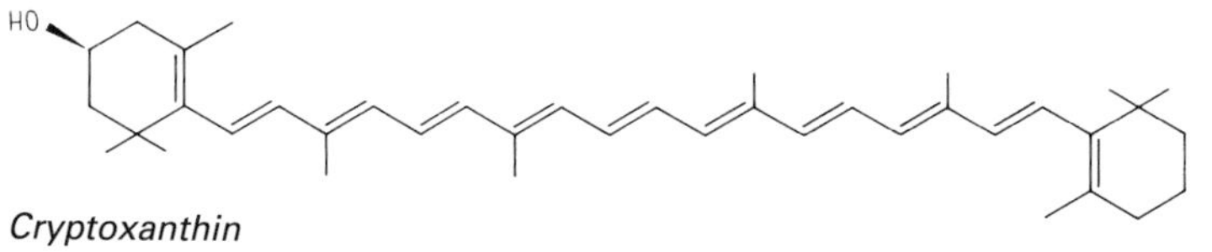

HO

Cryptoxanthin

otenoid. Its digentiobiose ester, crocin (M_r 976, m.p. 186 °C) (Fig.), occurs as a yellow pigment in crocuses and a few other higher plants.

Cross-link: a covalent bond between two polymers. DNA or soluble proteins are not naturally cross-linked, but C.l. can be artifically introduced into these polymers by reaction with bifunctional alkylating agents. C.l. are formed naturally (enzymatically) in insoluble structural proteins (see, e.g. Collagen) and in the cell walls of bacteria.

Crotactin: see Snake venoms.

Crotamine: see Snake venoms.

Crotoxin: see Snake venoms.

Crustecdysone: see Ecdysterone.

Cryptosterol: see Lanosterol.

Cryptoxanthin, ***(3R)-β,β-caroten-3-ol:*** a xanthophyll, M_r 552.85, m.p. 169 °C. It is a common plant pigment, found especially in fruits, e.g. bell peppers, oranges, tangerines and papayas, as well as in maize and egg yolk. As a provitamin, one molecule of C. yields one molecule of vitamin A.

Crystallins: soluble proteins which account for almost 90 % of the total vertebrate lens protein. Lens cells are not replaced, so they must survive the lifespan of the animal. Moreover, C. cannot be replaced, because lens cells lose their nuclei and other organelles; if these were present they would presumably cause discontinuities in refractive index. C. must therefore be exceptionally stable, withstanding all abuses which could lead to denaturation or aggregation. They must also have short range order in solution to insure a smoothly changing refractive index.

C. are classified into 3 major groups, α, β and γ, originally on the basis of their precipitability, but now according to their M_r range as determined by gel filtration. α-C. comprizes 2 subgroups, one of average M_r 800,000, the other much higher. All α-C. are oligomeric and differ only in the relative quantities of 4 subunits: αA_1, αA_2, αB_1 and αB_2. There are 2 primary gene products: αA_2 and αB_2, while αA_1 and αB_1 arise by post-translational modification. Lens mRNA from calves has been translated in vitro into α-C. and transcribed into cDNA [A.J.M. Berns & H. Bloemendal, *Methods in Enzymology* **30** (1974) 675–694].

β-C. comprise at least 7 gene products. All β-C. are oligomers, ranging from dimers to aggregates of up to 8 chains, and all share one major subunit, βBp, but they may differ considerably with respect to other subunits. γ-C are all monomeric, of M_r less than 28,000, representing 4 or 5 gene products. In addition to α, β and γ, reptile and bird lenses contain tetrameric δ-C.

C. are highly conserved, and have evolved relatively slowly since the vertebrate lens appeared about 500 million years ago [W.W. DeJong et al. *J. Mol. Evol.* **10** (1977) 123–135]. All β-and γ-polypeptides show a high degree of homology, and are probably derived from an ancestral protein by point mutation, insertion and deletion, and by *N*- and *C*-terminus extension of β-C. γ-C. contain an unusually large proportion of cysteine, and it is suggested that the free SH-groups serve to prevent oxidative cross-linking in other lens proteins.

The principal subunits of β-C (dimeric βBp), tetrameric avian δ-C. and several γ-C. have been crystallized. X-ray crystallography of calf γ-II-C. to a refinement of 1.6 Å reveals a symmetrical molecule with 2 lobes, constructed from 4 similar polypeptide domains, each containing about 40 amino acid residues. Many Met, Cys, Trp, Tyr and Phe residues lie close to one another (less than 6.5 Å), suggesting an interaction of aromatic side chains, Arg residues, the polarizable sulfur of Met residues and the SH of Cys. This may contribute to protein stability by delocalization of π electrons, and may also serve as an electron transfer system. [H. Bloemendal & W.W. DeJong *Trends in Biochemical Sciences* **4** (1979) 137–141; L. Summers et al. *Peptide and Protein Reviews* **3** (1984) 147–168].

Subunits of cephalopod C. are encoded in at least 6 genes, and have M_r 25,000–30,000; they display 52–57 % sequence similarity with cephalopod glutathione *S*-transferase (GST), and 20–25 % sequence similarity with mammalian GST. These similarities indicate that the genes for C. may have been recruited from those for detoxication enzymes, and that the octopus and squid lens GST-like C. gene familes expanded after the evolutionary divergence of mammals and cephalopods. [S.I. Tomarev et al. *J. Biol. Chem.* **266** (1991) 24226–24231]

CSF: acronym of Colony stimulating factor (see).

C-terminal amino acids: see Peptides.

C-toxiferin 1: calabash toxiferin 1. See Curare alkaloids.

CTP: acronym of cytidine 5′-triphosphate.

Cucumber hypocotyl test: see Gibberellins.

Cucurbitacins: tetracyclic triterpenes found as their glycosides in the *Cucurbitaceae* and *Cruciferae.* These toxic, bitter compounds are structurally related to the parent hydrocarbon, cucurbitane [19(10–9β)-abeo-5β-lanostane], which differs from lanostane (see Lanosterol) in the formal shift of the 10-methyl group to the 9β-position. Cucurbitacin E was formerly known as *elaterin.* C. have a laxative action; some serve as insect attractants, and a few have antineoplastic and antigibberellin activity.

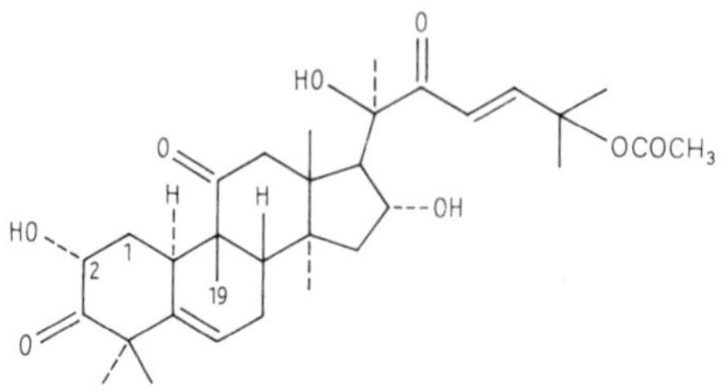

Cucurbitacin B

Cultivation of microorganisms, *culture techniques:* deliberate propagation of a microorganism. Cultivation may be discontinuous (static) or continuous. The technological or industrial culture of microorganisms is discussed under Fermentation techniques (see). In discontinuous culture, microorganisms are grown in a closed system until one factor in the nutrient medium becomes limiting. The culture conditions therefore change constantly with time. This growth can be compared to the genetically limited growth of a multicellular organism, whose life is divided into youth, maturity, aging and death. The corresponding growth phases in discontinuous culture are lag, exponential growth, stationary and post-stationary (decline) phases. These are represented graphically in a *growth curve* by plotting the logarithm of total cell numbers, or of the numbers of viable cells (see growth) against time. A semi-logarithmic representation is useful because exponential growth is then represented by a straight line, whose slope corresponds to the rate of cell division. Exponential growth is therefore also called logarithmic growth. The lag phase is the period between inoculation and the attainment of the maximal rate of cell division. The exponential phase merges slowly into the stationary phase, in which the cells no longer grow or divide. There may still be significant metabolic turnover during the stationary phase, and this may continue for a considerable time; it may also be artificially maintained (resting cells), as in the industrial production of Antibiotics (see). Transition to the post-stationary phase is also gradual.

Important growth parameters are the lag time, the growth rate (doubling time) and the yield. The ratio of the mass produced to the substrate consumed is called the coefficient of yield or economic coefficient. It can also be defined as the energy yield coefficient, if one compares the actual yield of ATP to the theoretically possible yield. As the substrate concentration decreases, cell density increases, and metabolic products accumulate or are excreted during discontinuous culture, the cells may actually begin the change during exponential growth.

There is a more or less clear relationship between the growth phase or growth rate and the synthesis of secondary metabolites. In many cases the latter begins during transition from the exponential to the stationary growth phase. The tropho or nutritional can therefore be distinguished from the idioproduction phase (Bu'Lock).

In continuous culture, a growing population of microorganisms is continuously supplied with fresh nutrient solution at the same rate that microorganism suspension is removed from the culture vessel. This open system (stabilized by limiting one growth factor) tends toward a steady state. Thus, the cells grow exponentially for an indefinite period under constant environmental conditions.

Cupressuflavone: see Biflavonoids.

Curare alkaloids: the toxic principles of arrow and bait poisons used by South American Indians. At the end of the 19th century the plant preparations were classified by Boehm as pot, tube or bamboo, and gourd or calabash curares, according to the vessels in which the Indians stored them. These terms have been retained because the materials from different sources are chemically and pharmacologically different, and the method of storage depends largely on the source. The C. a. are extracted from the plants with water, then concentrated by boiling.

Pot and *tube* curare are rather similar, and not highly poisonous. They are prepared from plants of the genus *Chonodendron* and are stored in clay pots or bamboo tubes. The alkaloids are isoquinolines, the most important being the quaternary *bis*benzyl-isoquinoline, tubocurarine.

Calabash curare, prepared from plants of the genus *Strychnos*, is packed in hollow gourds. It contains a large number of alkaloids and is extremely poisonous. The alkaloids contain an indole ring system and can be divided into 3 types: the yohimbine type (e. g. mavacurine), the strychnine type (e. g. Wieland-Gumlich aldehyde) and the *bis*indole type (e. g. calabash toxiferin-I).

In current scientific usage, curare refers only to those C. a. with muscle-relaxing (paralysing) effects, namely the dimeric indole alkaloids with a strychnine skeleton and two quaternary nitrogens. These compounds are extremely poisonous; calabash alkaloids E and G cause death by respiratory paralysis and are among the most toxic compounds known.

Pharmacology. C. a. interrupt nervous impulses at the end plates of motor nerves by displacing acetylcholine, leading to paralysis of striated muscles. Pharmacological effects are manifested only after injection (arrow or syringe). Because C. a. are absorbed very slowly from the intestine, animals killed by arrow poisons can be eaten with impunity. C. a. are used clinically as muscle relaxants in operations, and to relieve severe tetanus and nervous muscle cramps. Since the natural preparations vary in composition, and have unpleasant side effects, the pure alkaloids or synthetic or semisynthetic analogs (e. g. Alloferin) are used in modern medicine.

Curcumin, *turmeric yellow, diferuloylmethane:* a yellow pigment from the roots and pods of *Curcuma longa* L., which is cultivated in Southeast Asia. The dried root is used medicinally for liver and bile ailments, and is a component of curry powder. C. is used as a food color, as a textile dye, as a pH indicator (curcumin-boric acid paper), and as a test reagent

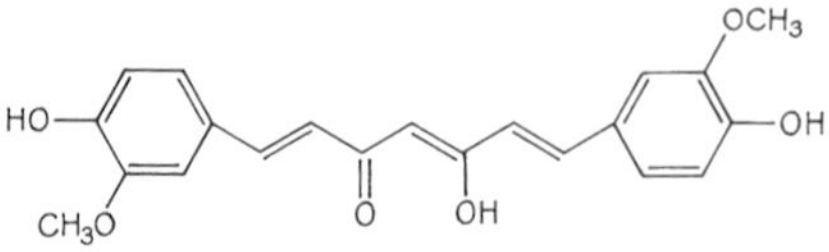

Curcumin

for beryllium. Phenylalanine and acetate/malonate are specific biosynthetic precursors.

Cuscohygrine: see Pyrrolidine alkaloids.

Cyanides: salts of hydrogen cyanide (HCN) with the general formula $Me^{I}CN$. Soluble C. are hydrolysed in water to the metal hydroxide and hydrogen cyanide, and are therefore highly poisonous. C. of heavy metals are generally insoluble and form complex ions with excess CN^-. HCN and alkali cyanides are often included as components of insecticides and other biocides. HCN is released by the enzymatic hydrolysis of Cyanogenic glycosides (see), which are the glycosides of α-hydroxynitriles (cyanhydrins). Nitriles ($R\text{–}C \equiv N$) do not occur widely in nature, but the nitrile group does occur in the alkaloid ricinine, and in the rare amino acid β-cyanoalanine. Assimilation of HCN by plants leads to formation of β-cyanoalanine which is then enzymatically hydrolysed to L-asparagine (Fig.).

C. reversibly inhibit cytochrome oxidase. In cases of cyanide poisoning, the venous blood is bright red because the oxygen has not been used. Hemoglobin does not react with C. The average lethal oral dose of C. for humans, is 60–90 mg HCN (200 mg KCN). An antidote is methemoglobin, which has a higher affinity for C. than does cytochrome oxidase. It forms cyanomethemoglobin which is slowly converted to normal hemoglobin as the CN^- is converted to thiocyanate (SCN^-) by rhodanese (EC 2.8.1.1). Extremely small quantities of C. are consumed in the formation of cyanocobalamin (vitamin B_{12}, see Vitamins).

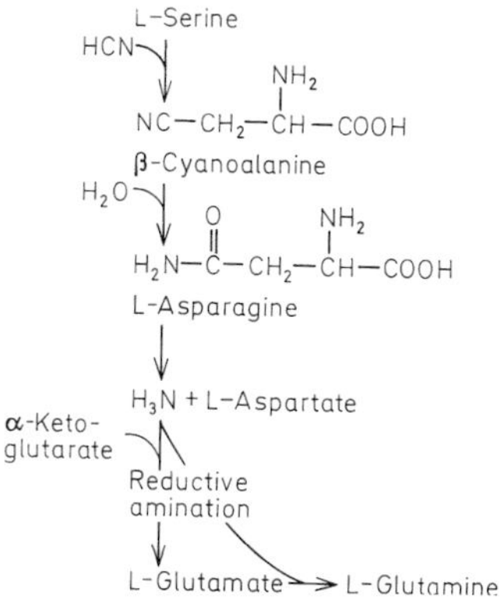

Cyanide assimilation in plants

Cyanidin, ***3,5,7,3′,4′-pentahydroxyflavylium cation:*** the aglycon of many Anthocyanins (see), m.p. 200 °C (d.). Glycosides of C. and a few acylated derivatives are found in many plants; the oxonium salts are responsible for the deep red color of many flowers and fruits such as red roses, geraniums, tulips, poppies and zinnias. Chelates with Fe(III) or Al(III) are deep blue in color. When bound to a polysaccharide carrier, they form chromosaccharides such as protocyanin, the blue pigment of cornflowers.

The structures of more than 20 different natural glycosides of C. are known, including *chrysanthemin* (3-β-glucoside) (red autumn leaves of some maples, wild strawberries and blackberries); *idaein* (3-β-galactoside) (apples and cranberries); *mekocyanin* (3-gentiobioside) (hibiscus blossoms and sour cherries); *keracyanin* (3-rhamnoglucoside) (snapdragons, canna lilies, tulips and cherries); *cyanin* (3,5-di-β-glucoside) (blue cornflowers, violets, dahlias, red roses, etc.).

Cyanin: see Cyanidin.

Cyanocobalamin: one of the B_{12} vitamins. See Vitamins.

Cyanogenic glycosides: O-glycosides formed from decarboxylated amino acids (Table). The cyano group arises from the α-C atom and the amino group. Hydrogen cyanide is generated from C. g. by β-glycosidases (e. g. emulsin) and oxinitrilases.

Cyanophycin granules: see Blue-green bacteria.

Cyasterone: a phytoecdysone (see Ecdysone), M_r 520.67, m.p. 164–166 °C. $[\alpha]_D + 64.5°$ (pyridine). It has been isolated from various plants, including *Cyathula capitata (Amaranthaceae)* and *Ajuga decumbens (Labiatae).* Unlike other ecdysones, C. has a stigmastane skeleton (see Steroids) and a γ-lactone group in the side chain.

Cyasterone

Cyanogenic glycosides

Amino acid ($R_2\text{–}CR_1H\text{–}CH(*NH_2)\text{–}COOH$)	Cyanogenic glycoside ($R_2\text{–}CR_1(O\text{-sugar})\text{–}CN*$)	R_1	R_2	Sugar
L-Valine	Linamarin	$-CH_3$	$-CH_3$	Glucose
L-Isoleucine	Lotaustralin	$-CH_2CH_3$	$-CH_3$	Glucose
L-Phenylalanine	Prunasin	$-C_6H_5$	–H	Glucose
L-Phenylalanine	Amygdalin	$-C_6H_5$	–H	Gentiobiose
L-Tyrosine	Dhurrin	$-C_6H_4OH$	–H	Glucose

Cyclic adenosine 3′,5′-monophosphate: see Adenosine phosphates.

Cyclic N^6,$O^{2'}$-dibutyryladenosine 3′,5′-monophosphate, *DBCAMP:* a synthetic derivative of cAMP. See Adenosine phosphates.

Cyclic guanosine 3′,5′-monophosphate: see Guanosine phosphates.

Cyclic inosine 3′,5′-monophosphate: see Inosine phosphates.

Cyclic nucleotides: see Nucleotides.

Cyclic uridine 3′,5′-monophosphate: see Uridine phosphates.

Cyclitols, *cyclohexitols, cyclohexanehexols, inositols:* the nine isomers of 1,2,3,4,5,6-hexahydroxycyclohexane (Table), all of which occur naturally. They have the same empirical formula ($C_6H_{12}O_6$, M_r 180.15) as the hexoses, and are biosynthetically related to them, but they have a hexane ring (with chair configuration) rather than a heterocyclic ring. The hydroxyl groups can lie either in the equatorial or the axial position, so that there are 8 possible *cis, trans* isomers; 7 of these are optically inactive, and another isomer pair results from the D and L forms of the optically active 8th *cis, trans* isomer. The most prevalent natural form is *Myo*-inositol (see) (*cis*-1,2,3,5-*trans*-4,6-cyclohexanehexol). Scyllitol (*scyllo*-inositol) has been found in the organs of various cartilaginous fishes (e.g. sharks, rays), in the urine of several mammals including humans, and in *Calycanthus occidentalis* and *Chimonanthus fragrans* (both plants of the family *Calycanthaceae).* Methyl ethers of *dextro*-inositol, *levo*-inositol and *myo*-inositol are widespread in plants, e.g. +-pinitol (3-*O*-methyl-*dextro*-inositol), (–)-pinitol (3-*O*-methyl-*levo*-inositol), (–)-quebrachitol (1-*O*-methyl-*levo*-inositol), bornesitol (1-*O*-methyl-*myo*-inositol), sequoyitol (5-*O*-methyl-*myo*-inositol), *meso*-dambonitol (1,3-di-*O*-methyl-*myo*-inositol), meso-ononitol (4-*O*-methyl-*myo*-inositol), liriodendritol (1,5-di-*O*-methyl-*myo*-inositol). See also Phytic acid.

[S.J. Angyal, *Quart. Reviews* **11** (1957) 212–226; S.J. Angyal & L. Anderson *Adv. Carbohydrate Chem.* **14** (1959) 135–212]

The nine isomeric cyclitols (see entry *Myo*-inositol for ring numbering system)

Name	*cis* OH-groups	*trans* OH-groups
cis-Inositol	1,2,3,4,5,6	–
epi-Inositol	1,2,3,4,5	6
allo-Inositol	1,2,3,4	5,6
neo-Inositol	1,2,3	4,5,6
myo-Inositol	1,2,3,5	4,6
muco-Inositol	1,2,4,5	3,6
scyllo-Inositol	1,3,5	2,4,6
dextro-Inositol	1,2,5	3,4,6
levo-Inositol	1,2,4	3,5,6

Cyclo-AMP: cyclic adenosine 3′,5′-monophosphate.

Cycloartenol, *9,19-cyclo-5α,9β-lanost-24-en-3β-ol:* a tetracyclic tripterpene alcohol, M_r 426.73, m.p. 115 °C, $[\alpha]_D$ + 54°. It is present in many plants, e.g. in the latex of members of the *Euphorbiaceae,* in the potato *(Solanum tuberosum)* and in nux vomica *(Strychnos nux vomica).* It is biosynthesized from squalene via 2,3-epoxysqualene (see Steroids, Fig. 6).

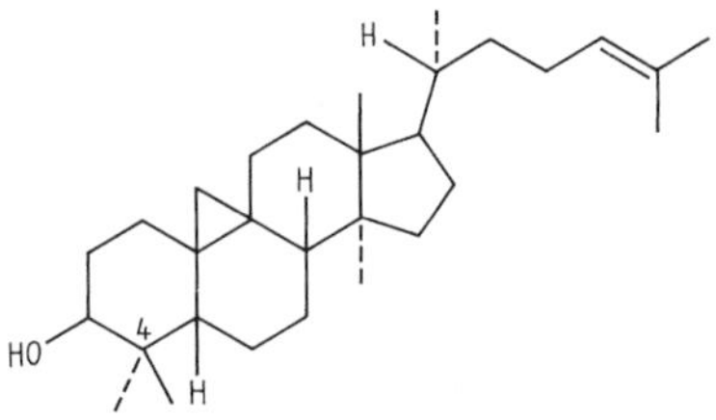

Cycloartenol

Cyclobuxamide H: one of the Buxus alkaloids (see).

Cyclo-GMP: cyclic guanosine 3′,5′-monophosphate. See Guanosine phosphates.

Cyclohexanehexols: see Cyclitols.

Cycloheximide, *actidione:* an antibiotic from *Streptomyces griseus.* Chemically, it is a derivative of glutarimide, and it is water-soluble. It inhibits protein biosynthesis on 80S ribosomes in eukaryotes by inhibiting peptidyl transferase on the large ribosomal subunit. Both initiation and elongation are prevented, and this can lead to an increase in the number of monosomes and a decrease in the number of polysomes. C. is used as a fungicide to control cherry leaf spot, turf diseases and rose powdery mildew.

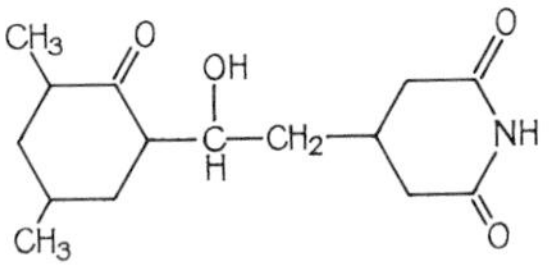

Cycloheximide

Cyclohexitols: see Cyclitols.

Cyclo-IMP: cyclic inosine 3′,5′-monophosphate. See Inosine phosphates.

Cyclopenase: see Viridicatine.

Cyclopentanoperhydrophenanthrene: see Steroids.

Cyclophosphamide, *Cytoxan, Endoxan:* an alkylating agent used as an immune suppressive. C. inhibits cellular and humoral immune reactions by alkylating the SH and NH_2 groups of proteins and the N^7 of guanine in nucleic acids. It leads to a long-lasting suppression of antibody synthesis, provided the C. is administered for 1 day before and 15 days after the antigen is introduced. It has also been proposed as an antineoplastic agent.

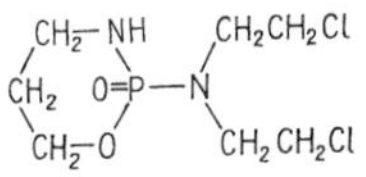

Cyclophosphamide

Cyclo-UMP: cyclic uridine 3′,5′-monophosphate. See Uridine phosphates.

Cyproterone acetate, ***17α-acetoxy-6-chloro-1α,2α-methylenepregna-4,6-diene-3,20-dione:*** an antiandrogen steroid containing a pregnane skeleton. It counteracts the effects of the male sex hormone, testosterone, and is used in cases of hypersexuality ("hormonal castration").

Cyproterone acetate

Cys: abb. of L-Cysteine.

Cystathioninuria: see Inborn errors of metabolism.

L-Cysteine, ***Cys, L-2-amino-3-mercaptopropionic acid:*** HS-CH_2-CH(NH_2)-COOH, a sulfur-containing, proteogenic amino acid, M_r 121.2, m.p. 240 °C (d.), $[\alpha]_D^{25}$ +6.5 (c = 2 in 5M HCl). Cys occupies a pivotal position in intermediary Sulfur metabolism (see) (Fig. 1), and is an important component of redox reactions in living cells. In neutral or alkaline solution, Cys is oxidized in the air to L-cystine. The SH-groups of Cys residues and the disulfide groups (–S–S–) of cystine residues are important for the tertiary structure and/or enzymatic activity of proteins.

Cys is synthesized in the course of sulfate assimilation (see) or from Methionine (see) by Transsulfuration (see) (Fig. 2).

Cysteine knot: a motif of protein structure consisting of two catenated (interlocking) rings. This arises

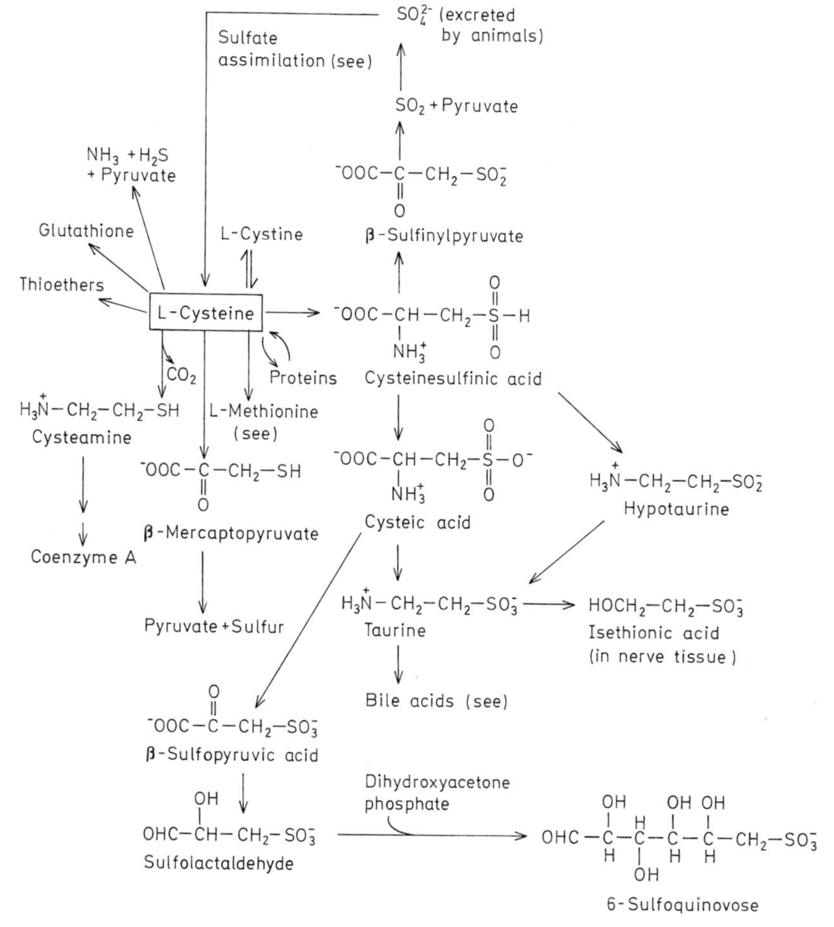

Fig. 1. *Position of L-cysteine in metabolism*

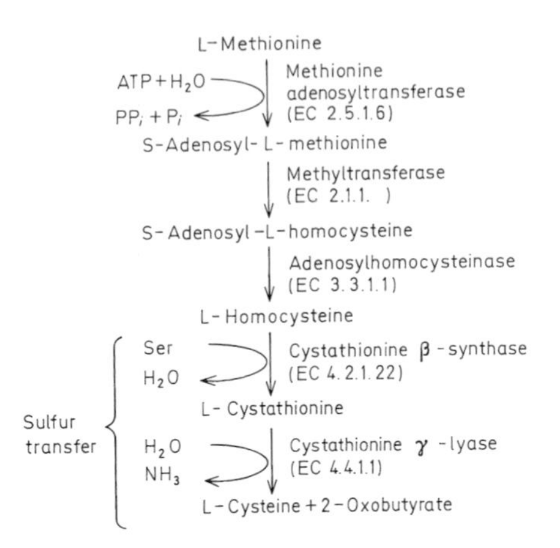

Fig. 2. *Formation of L-cysteine from L-methionine*

when a disulfide bridge connecting two β-strands passes through an eight-residue circle formed by a pair of disulfide bonds joining two other β-strands. A C. k. is present in the structure of human chorionic gonadotropin (hCG) and in several other protein growth factors, including platelet-derived growth factor (PDGF). [P. D. Sun & D. R. Davies 'Cystine Knot Growth Factor Superfamily' *Annu. Rev. Biophys. Biomol. Struct.* **24** (1995) 269–291]

Cysteinesulfinic acid: see Cysteine (Fig. 1).

Cysteine synthase: see Sulfate assimilation.

L-Cystine, ***dicysteine, 3,3'-dithiobis(2-aminopropionic acid):*** HOOC–(NH_2)CH–CH_2–S–S–CH_2–CH(NH_2)–COOH, a dimer of cysteine formed by oxidation of the SH-group to a disulfide (-S-S-). M_r 240.3, m.p. 258–261 °C (d.), $[\alpha]_D^{25}$ −232 ° *(c* = 1 in 5M HCl). All the disulfide bridges of proteins belong to cystine residues, and these are always formed by the post-translational oxidation of cysteine residues.

Cystine bridges: see Disulfide bridges.

Cyt: abb. of cytosine.

Cytidine, ***Cyd, cytosine riboside, 3-D-ribofuranosyl-cytosine:*** a β-glycosidic Nucleoside (see) consisting of D-ribose and the pyrimidine base cytosine. M_r 243.22, m.p. 220–230 °C (d.), $[\alpha]_D^{25}$ +34.2 °, *(c* = 2 in water). Cytidine phosphates (see) are metabolically important in all living organisms.

Cytidine diphosphocholine, ***CDP-choline, active choline:*** the activated form of choline, which is biosynthesized from phosphocholine and CTP. See Membrane lipids (biosynthesis).

Cytidine diphosphoglyceride, ***CDP-glyceride:*** see Membrane lipids (biosynthesis).

Cytidine phosphates: cytidine 5'-monophosphate (CMP, cytidylic acid, M_r 323.2), cytidine 5'-diphosphate (CDP, M_r 403.19) and cytidine 5'-triphosphate (CTP, M_r 483.16). For structure, see e. g. Pyrimidine biosynthesis. CTP is a precursor of RNA synthesis, while deoxy-CTP is a precursor of DNA synthesis. CDP may be regarded as the coenzyme of phospholipid biosynthesis (see Membrane lipids) (activated choline is CDP-choline). Glycerol and the sugar alcohol, ribitol, are also activated by bonding to CDP (see Nucleoside diphosphate sugars). Reduction of ribose to deoxyribose in the synthesis of deoxyribonucleotides usually occurs at the level of CDP (see Nucleotides).

Cytidylic acid, ***cytidine 5'-monophosphate, CMP:*** see Cytidine phosphates.

Cytimidine: see Amicetins.

Cytisine: see Lupin alkaloids.

Cytochalasins: a group of mold metabolites with cytostatic activity. Six chemically similar C. have been isolated from *Helminthosporium dematioideum, Metarrhizium anisopliae* and *Rosellina necatrix.* [M. S. Buchanan et al. *Phytochemistry* **41** (1996) 821–828; M. S. Buchanan et al. *Phytochemistry* **42** (1996) 173–176]

Structure of phomin, a typical cytochalasin

Cytochrome oxidase: the last member of the electron transport chain, catalysing the terminal reaction of electron transport and reducing oxygen to water. C. o. normally reacts with molecular oxygen, but can be inhibited with CN^- or CO. Its prosthetic group is hemin *a* (cytohemin), which has a lipophilic C_{12} side chain, and an aldehyde and a vinyl group on the porphyrin ring. Reaction with O_2 involves the redox systems of both the heme iron and copper (Fe^{2+}/Fe^{3+}; Cu^{2+}/Cu^+). Electron microscopic studies have shown that the protein part of the molecule undergoes conformational changes during the redox process; the oxidized form is crystalline, but the reduced form is amorphous. In the native state, C. o. is a component of a phospholipid-containing supermolecule (M_r 3×10^6) bound to the mitochondrial membrane. This complex can only be dissociated by treatment with detergent (2 % deoxycholate). Once isolated, C. o. (pI 4–5) forms a stable, active, noncovalent complex with cytochrome *c* (pI 10.1), from which it can be separated by treatment with detergent (0.1 % Emasol) and gel filtration.

Cytochrome aa_3 appears to be a single heme protein which exists in two functional states, rather than a complex of two proteins, as was thought earlier. The *a* state is not autoxidizable, and does not react with O_2, CO or CN^-; the a_3 form does, however, react with O_2, CO and CN^-.

Purified heart muscle C. o. is a tetrameric heme lipoprotein, which contains 4 heme and 4 copper chromophores per lipid-containing (M_r 440,000) or lipid-free (M_r 350,000) molecule. It is converted in the presence of low concentrations of guanidine (1 M) or dodecyl sulfate (0.5 %) to a dimeric form (M_r 190,000) which is 2 to 3 times as active as the tetramer. The monomer can be obtained by raising the pH, but, like the tetramer, it is not very active. Higher concentrations of dodecyl sulfate (1–5 %) cause the C. o. monomer (M_r 90,000 to 100,000) to dissociate into 4 to 6 nonidentical polypeptide chains (main species M_r 11,500, 14,000, 20,000, 39,000).

In contrast to the heart muscle C.o., the C.o. monomer from bacteria, e.g. *Pseudomonas,* the autoxidizable and copper-free cytochrome a_2 (M_r 120,000) contains only two subunits (M_r 58,000). It is not yet known which of the chains carries the prosthetic groups.

Cytochrome P450: a cytochrome whose reduced complex with carbon monoxide absorbs maximally at 450 nm. C.P450s function as terminal oxidases in C.P450-dependent monooxygenase systems (Figs. 1 & 2). The system comprises C.P450 reductase (a flavoprotein), C.P450, phospholipid and possibly nonheme iron. It is uncertain whether the phospholipid is directly involved in electron transfer, or whether it has only a structural role. In some systems electrons may be also transferred from cytochome b_5. The ultimate source of reducing power for the system operating via C.P450 reductase is NADPH. Prokaryotic C.P450s are soluble, whereas eukaryotic C.P450s are membrane-bound within the endoplasmic reticulum (e.g. in liver) or in the inner mitochondrial membrane. C.P450s in the mitochondria of the mammalian adrenal cortex are responsible for some of the hydroxylations in corticosteroid biosynthesis [S. Takemori and & S. Kominami *Trends. Biochem. Sci.* **9** (1984) 393–396]. In steroidogenic tissues, like adrenals, placenta and brain, the C.P450 responsible for cholesterol side chain cleavage (EC 1.14.15.6) (see Steroids) is present in the inner mitochondrial membrane [B. Walther et al. *Arch. Biochem. Biophys.* **254** (1987) 592–596]. Mammalian liver contains a superfamily of C.P450s with overlapping substrate specificities, which is responsible for the oxidative conversion of various endogenous metabolites (e.g. bile acid biosynthesis) and xenobiotics into more polar compounds. The latter are more easily excreted, either alone or conjugated with sulfate or glucuronate. Reactions catalysed by C.P450s are shown in the Table. C.P450s are also involved in fatty acid metabolism, prostaglandin biosynthesis, leukotriene biosynthesis, steroid biosynthesis, 1,25-dihydroxyvitamin D_3 biosynthesis and many other endogenous processes.

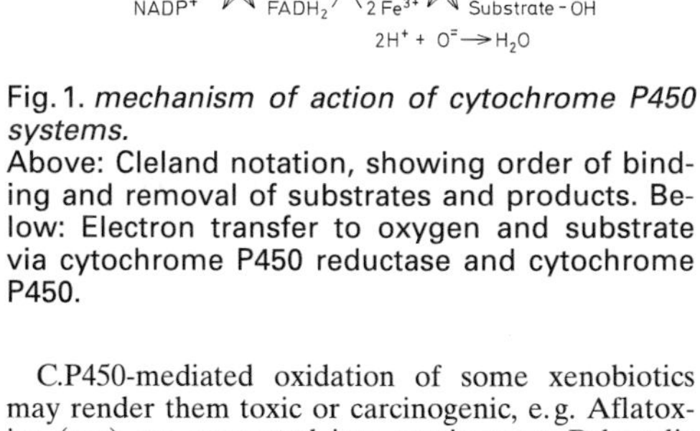

NADP⁺ ↑; AH ↓; NADPH + H⁺ ↓; O_2 ↓; AOH + H_2O ↑

E.FAD P_{450} — E.FAD.AH P_{450} — E.FADH$_2$.AH P_{450} — E.FADH$_2$.O$_2$AH ⇌ E.FAD.O$_2$.AH reduced P_{450}

NADPH + H⁺ / NADP⁺ — FAD / Reductase / FADH$_2$ — 2 Fe^{2+} / P_{450} / 2 Fe^{3+} — O_2 — Substrate-H / Substrate-OH

$2H^+ + O^= \rightarrow H_2O$

Fig. 1. *mechanism of action of cytochrome P450 systems.*
Above: Cleland notation, showing order of binding and removal of substrates and products. Below: Electron transfer to oxygen and substrate via cytochrome P450 reductase and cytochrome P450.

C.P450-mediated oxidation of some xenobiotics may render them toxic or carcinogenic, e.g. Aflatoxins (see) are converted into carcinogens. Polycyclic hydrocarbons and halogenated polycyclic hydrocarbons are also rendered carcinogenic by the action of C.P450s (see Table, see Bay-region theory of carcinogenesis). Administration of a xenobiotic to an animal promotes the hepatic synthesis of an appropriate C.P450. Thus phenobarbital feeding causes an increase in the hepatic C.P450 content in rats, and the increase is largely accounted for by C.P450b, which has a high specificity for the hydroxylation of phenobarbital. On the other hand, a different species (C.P450c), specific for the epoxidation of polycyclic hydrocarbons, is promoted by administration of benzpyrene or methylcholanthrene. Isosafrole promotes the synthesis of C.P450d. In these stimulation studies, several species of C.P450 are affected, but usually one species increases above the others. In contrast, Arochlor (a commercial mixture of halogenated polycyclic hydrocarbons, which is strongly carcinogenic) causes up to 40-fold increases in hepatic C.P450b, c, d and e, with little or no effect on C.P450a and other minor species.

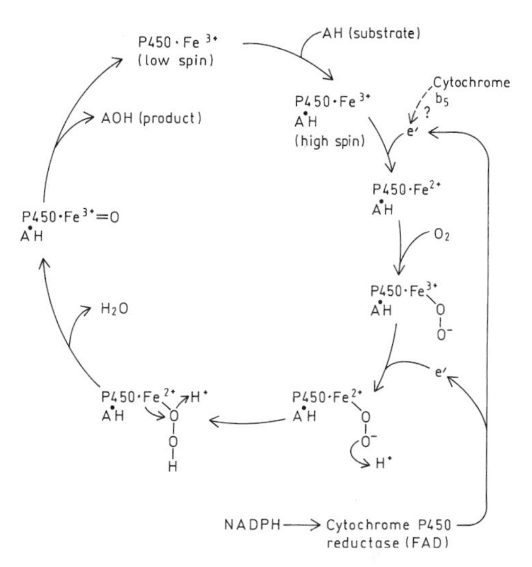

Fig. 2. *Mechanism of action of cytochrome P450.* AH, substrate. AOH, hydroxylated product. Binding of oxygen to the Fe(II) form of the hemoprotein, followed by delocalization of an electron, is an essential stage in all Fe-dependent oxygenase reactions, and in the binding of oxygen to carrier hemoproteins like hemoglobin.

Stimulation of C.P450 synthesis occurs at the transcriptional level, and different cytosolic receptors have been identified for benzpyrene, dioxin, isosafrole and 3-methylcholanthrene. The analogy with the mode of action of steroid hormones is striking. It must be remembered, however, that the existence of cytosolic steroid receptors in vivo is doubted [J. Gorski et al. *Molec. Cell. Endocrinol.* **36** (1984) 11–15]. Cytosolic receptors are probabl y artifacts of cell disruption, and steroid receptors are thought to reside only in the nucleus. This may also apply to cytosolic receptors of C.P450 substrates. [S. K. Yang et al. *Science* **196** (1977) 1199–1201, D. Pfeil & J. Friedrich *Pharmazie* **40** (1985) 217–221, J. Friedrich & D. Pfeil

Table. *Reactions catalysed by Cytochrome P450.*

1. *Hydroxylation of aromatic compounds,* e. g. salicylic acid, phenobarbital, acetanilide, synthetic and natural estrogens, diphenyl.

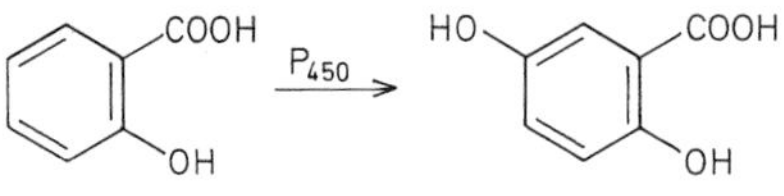

COOH, OH → P_{450} → HO, COOH, OH

Salicylic acid

2. *Hydroxylation of aliphatic compounds,* e. g. pentobarbital, antipyrine, tolbutamide, imipramine.

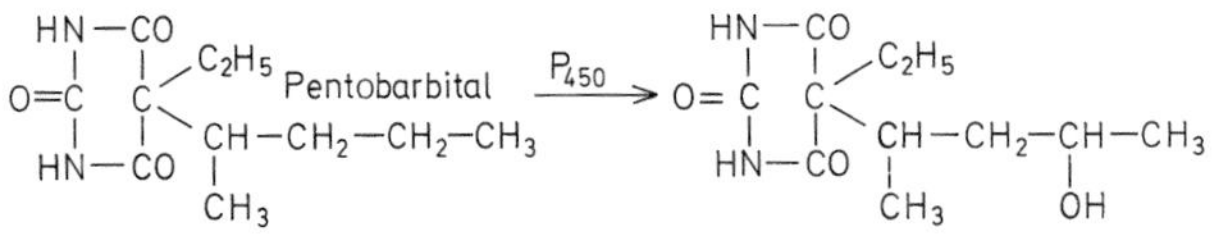

Pentobarbital $\xrightarrow{P_{450}}$ HN—CO, O=C, C, C_2H_5, $CH-CH_2-CH-CH_3$, HN—CO, CH_3, OH

3. *O-dealkylation,* e. g. phenacetin, griseofulvin, codeine.

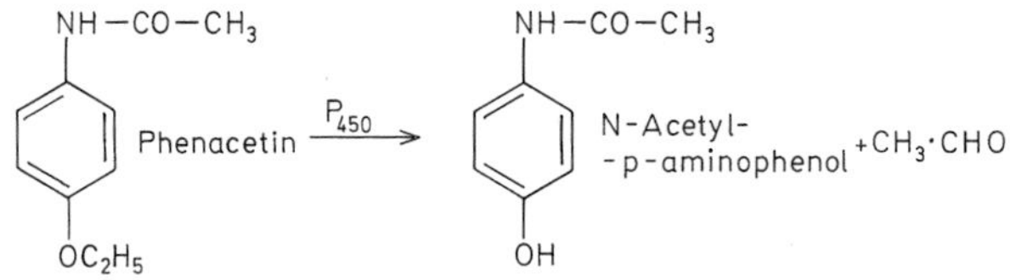

$NH-CO-CH_3$, OC_2H_5 Phenacetin $\xrightarrow{P_{450}}$ $NH-CO-CH_3$, OH N-Acetyl- -p-aminophenol $+CH_3 \cdot CHO$

4. *N-Dealkylation,* e. g. aminopyrine, chlorpromazine, ephedrine, morphine.

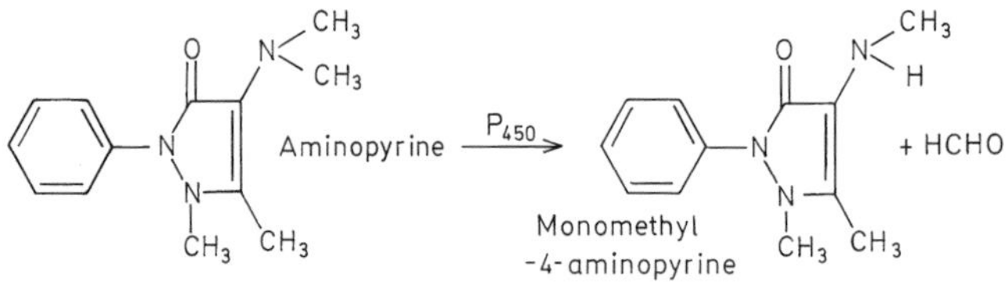

Aminopyrine $\xrightarrow{P_{450}}$ Monomethyl -4-aminopyrine + HCHO

5. *S-Dealkylation,* e. g. 6-methylmercaptopurine, methylthiobenzylthiazole, dimethylsulfide.

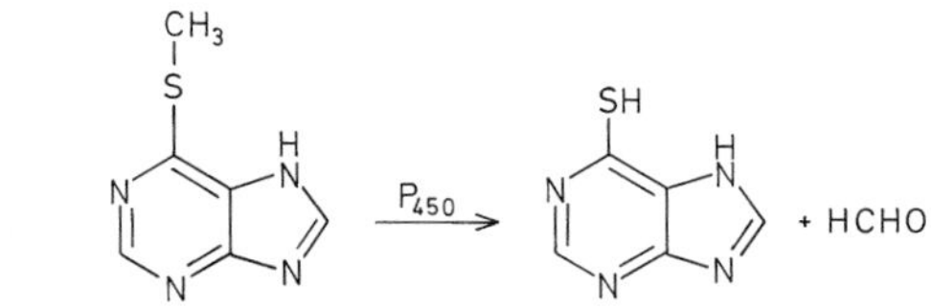

6-Methylmercaptopurine $\xrightarrow{P_{450}}$ 6-Mercaptopurine + HCHO

6. *N-Oxidation,* e. g. 2-acetylaminofluorene, nicotinamide, trimethylamine, guanethidine, chlorpromazine, imipramine.

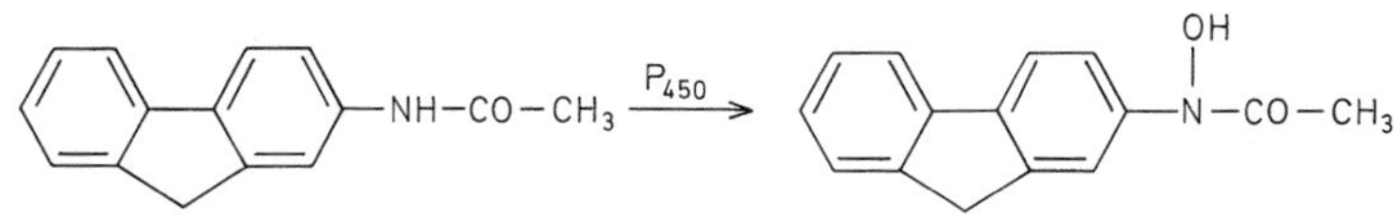

2-Acetylaminofluorene $\xrightarrow{P_{450}}$ N-Hydroxy-2-acetylaminofluorene

7. *Dehalogenation,* e.g. halothane, carbon tetrachloride, DDT, triiodothyronine.

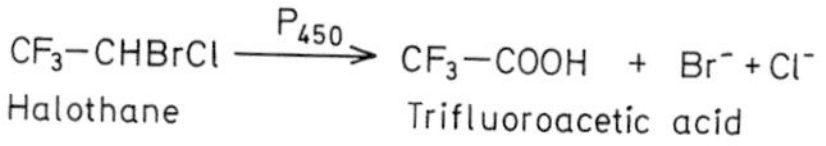

$$CF_3{-}CHBrCl \xrightarrow{P_{450}} CF_3{-}COOH + Br^- + Cl^-$$

Halothane → Trifluoroacetic acid

8. *Sulfoxidation,* e.g. chlorpromazine.

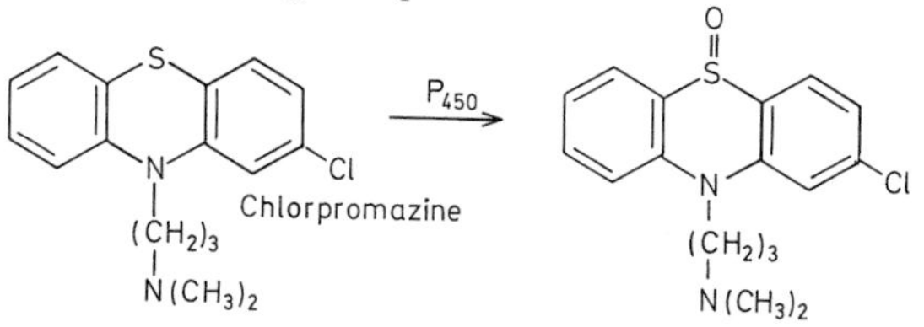

Chlorpromazine $\xrightarrow{P_{450}}$ sulfoxide

9. *Phosphothionate oxidation,* e.g. parathion.

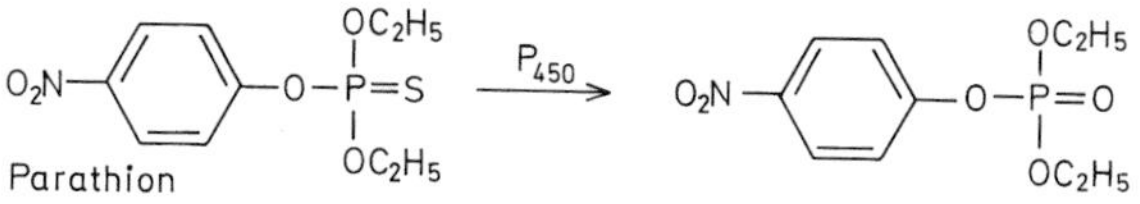

Parathion $\xrightarrow{P_{450}}$

10. *Deamination,* e.g. amphetamine, ephedrine.

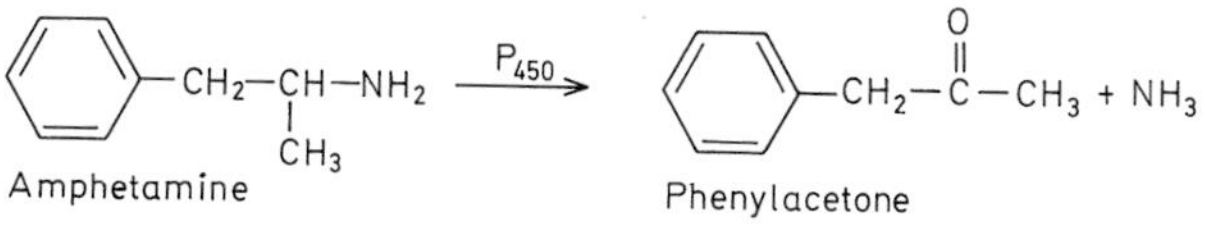

$$C_6H_5{-}CH_2{-}CH(CH_3){-}NH_2 \xrightarrow{P_{450}} C_6H_5{-}CH_2{-}CO{-}CH_3 + NH_3$$

Amphetamine → Phenylacetone

11. *Epoxidation,* e.g. benzpyrene, Aflatoxin (see).

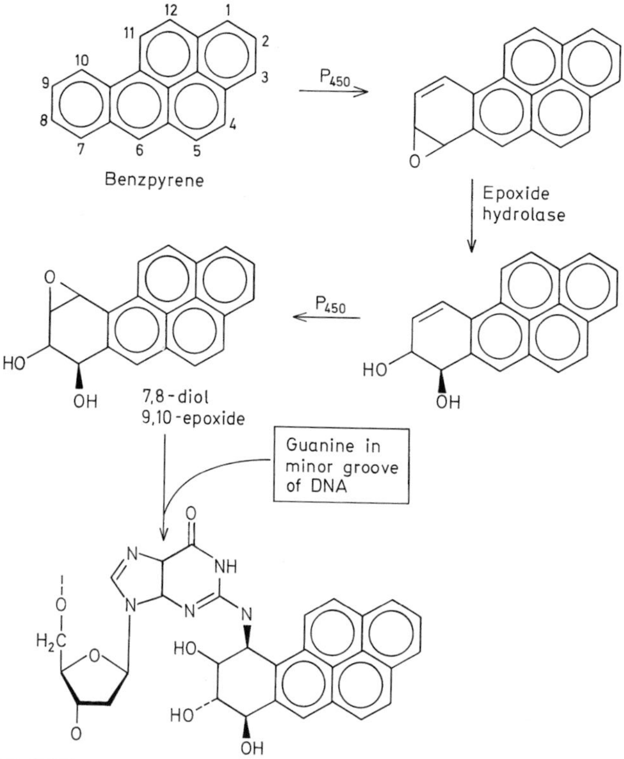

Benzpyrene $\xrightarrow{P_{450}}$ epoxide $\xrightarrow{\text{Epoxide hydrolase}}$ diol $\xrightarrow{P_{450}}$ 7,8-diol 9,10-epoxide → Guanine in minor groove of DNA

ibid. 228–232, H. V. Gelboin & P. O. P. Ts'o (eds.) *Polycyclic Hydrocarbons and Cancer* Vol. 1, *Environmental Chemistry and Metabolism* (Academic Press, 1978); S. Arnold et al. *Biochem. J.* **242** (1987) 375–381]

Cytochromes: heme proteins which serve as particle-bound redox catalysts in respiration, energy conservation, photosynthesis and some processes in anaerobic bacteria. They act as electron donors and acceptors by reversible valence changes in the iron atom at the center of their porphyrin complex:

$$Fe^{3+} \underset{-e^-}{\overset{+e^-}{\rightleftharpoons}} Fe^{2+}$$

The importance of the cytochromes can be seen from the fact that they are very old proteins, dating back more than 2 billion years, and their structures have been modified only insignificantly by point mutations. They are found in all organisms. On the basis of their structures and spectra, especially the α, β and γ bands, the C. fall into three main groups, *a, b* and *c*. About 30 different C. are known, and they are distinguished by subscripts, e.g. C.b_1. All three types are found in the mitochondria of higher plants and animals, where they are essential components of the Respiratory chain (see).

The cytochrome a/a_3 complex is identical with Cytochrome oxidase (see). The absorption spectra and behavior in the presence of inhibitors suggest the existence of 2 components (a and a_3), but these appear to be part of the same molecule. Another possibility is that cytochrome oxidase is a single species which oscillates between 2 very different configurations.

The prosthetic group of *cytochrome b* is the same as that of hemoglobin, iron(II)-protoporphyrin IX. It has the lowest redox potential of the respiratory chain C., and therefore lies between ubiquinone and C.*c*. C.*b*, a dimeric protein (M_r 60,000) with 1 heme group per monomer, is very firmly bound to the mitochondrial membrane, and can only be extracted with detergents. Its central iron atom, like that of C.*c*. is not autooxidizable and does not react with CO or cyanide. There is evidence that C.*b* exists in two forms, C.b_K and b_T, which have different redox potentials. C.b_7 is believed to function in the transfer of energy during electron transport.

C.b_5 from the microsomal fraction of bird and mammal livers is thought to deliver electrons to the fatty acid desaturase system of the endoplasmic reticulum. The heme group is noncovalently bound to the protein via histidine and does not react with O_2. It also protects the C.b_5 molecule from denaturation and proteolytic attack. C.b_5 solubilized by detergent treatment is an oligomer (M_r 120,000) of several monomers (M_r 16,000, 126 amino acids), while C.b_5 solubilized by protease or lipase treatment has only 82 to 98 amino acids, depending on the species. The primary sequences of these fragments are known. They have 50 % α-helix and 25 % β-structure.

Other C.b_5 are the soluble C.b_{562} from *E. coli* (110 amino acids, M_r 12,000, primary sequence and heme binding similar to myoglobin), which has the highest potential of the known C.*b*, and C.b_6 and C.b_{559} from the chloroplast grana of higher plants. These can only be solubilized by combined treatment with the

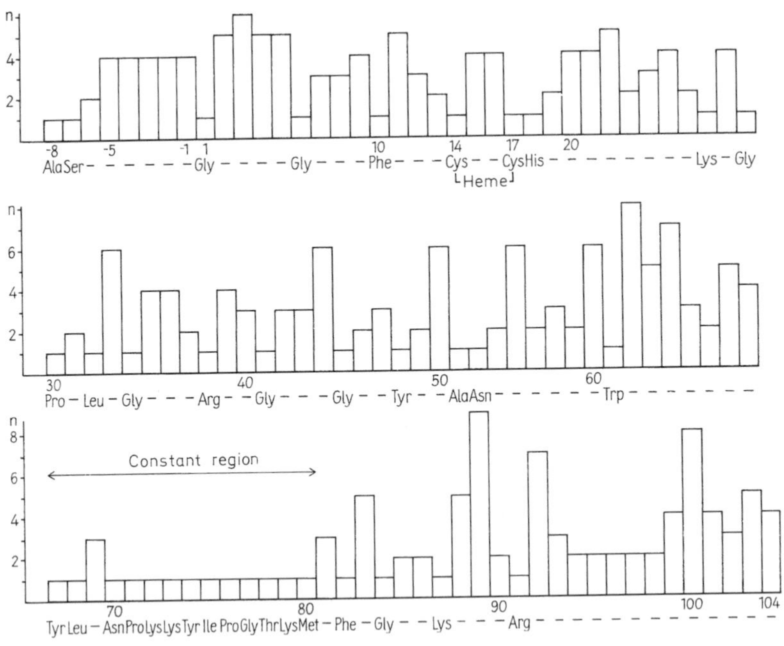

Fig. 1. *Primary structures of cytochrome c in 40 species (13 mammals, 5 birds, 2 reptiles, 1 frog, 5 fish, 1 snail, 4 insects, 5 higher plants, 4 fungi).* The names of the 35 or 37 invariable residues (in cytochromes containing 104 and 112 residues, respectively) are given. The number of different amino acids found in each position is given by the height of the column.

detergent Triton X-100 (0.1 %) and 4 M urea, and are obtained in disaggregated form. They are both part of the photosynthetic electron transport chain. C.b_1 and b_2, from *E. coli* and yeast, respectively, are both tetramers with dehydrogenase activity.

C.P-450 (see), a mixed-function mitochondrial oxidase, also belongs to the C.*b* group. The term P-450 refers to the unusual absorption maximum at 450 nm of the reduced CO-complex. This C. is involved in a number of steroid hydroxylation reactions in the adrenal cortex and is a prosthetic group in the monooxygenase which removes the side chain from cholesterol. The structure of the active oxygen on C.P-450 is unknown. The protein (M_r 850,000) has an unusually high partial specific volume of 0.765, is composed of 16 identical subunits (M_r 53,000 each), and contains 8 heme units.

Cytochrome c is the most widespread and best studied C. It is a central component of the Respiratory chain (see) in mitochondria. Because it is easy to extract and relatively small (C.c. of vertebrates, 104 amino acids, M_r 12,400; of higher plants, 111 amino acids, M_r 13,100). C.*c* has been used for phylogenetic studies (Fig. 1). The heme group (Fig. 2) is joined to the apoprotein by two thioether bonds to cysteine residues in the interior of the molecule. The iron atom is complexed with two other interior residues, methionine 80 and histidine 18, which prevent native C.*c* from reacting with O_2 or other heme-complexing agents, such as CO. However, aggregates of C.*c* are biologically inactive and autooxidizble. They are deaggregated and simultaneously reactivated by addition of urea or guanidine HCl. All C.c from higher organisms whose sequences are known have an *N*-acetylalanine or *N*-acetylglycine at the *N*-terminus. In invertebrates, there is a chain of at most 7 amino acids in place of the acetyl residue. Lysines 72 and 73 play a role in the interaction of C.*c* with cytochrome oxidase. The basicity of C.*c* (pI = 10) is due to lysines 27, 79, 87 and 100, and arginines 38 and 91.

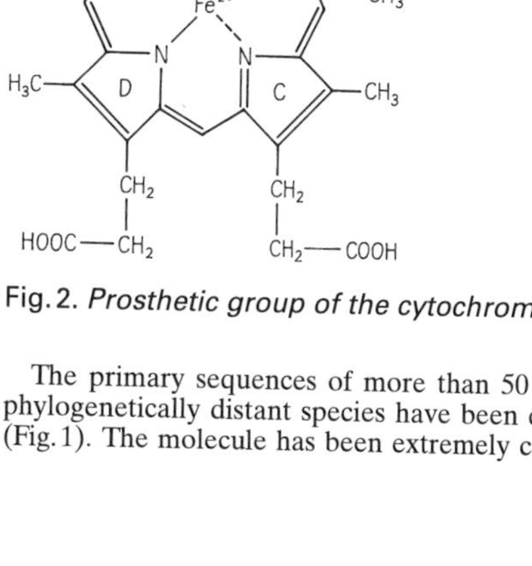

Fig. 2. *Prosthetic group of the cytochromes c*

The primary sequences of more than 50 C.*c* from phylogenetically distant species have been compared (Fig. 1). The molecule has been extremely conserved; 35 of the 104 to 112 residues are invariant, most of these invariant residues being located between positions 17 and 32, and 67 and 80. The configurations of the chains of four species studied, horse muscle, bonito, tuna and *Rhodospirillum*, were also found to be very similar. The molecule is unusual in undergoing a major structural change from the compact iron(III) form to the expanded iron(II) form. Some of the structural characteristics of C.*c* such as the large helix regions found only at the *N*- and *C*-termini, are shared by the prokaryotic C.c_2 and C.c_{550} (137 amino acids) from *Rhodospirillum* and *Micrococcus* (Fig. 3).

In addition to the classical C.*c*, eukaryotes also have a C.c_1, which is larger (M_r 37,000), insoluble, has a different amino acid composition, and forms part of cytochrome *c* reductase.

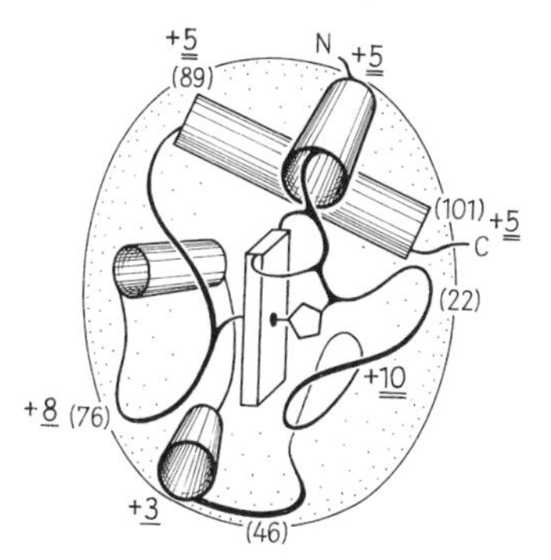

Fig. 3. *Folding of the polypeptide chains in vertebrate cytochrome c (104 residues), bacterial cytochromes c_2 (112 residues) and bacterial cytochrome c_{550} (137 residues).* The segments of α-helix are indicated by cylinders, and the heme by the rectangular plate. Numbers in parentheses show the numbering of the amino acid sequence in horse cytochrome *c*.

Bacterial C.*c* are similar to eukaryotic C.*c*, but they do not react with mammalian cytochrome *c* oxidase. The following bacterial C.*c* have been studied:

C.c_2 (112 amino acids, M_r 13,500, pI 6.3, primary and tertiary structure known) from *Rhodospirillum rubrum;* C.c_3 (102, 107 or 111 amino acids, with varying primary structure, 2 heme groups) from *Desulfovibrio* and other sulfate-reducing bacteria; C.c_4 (M_r 24,000, a dimer with 2 heme groups) and C.c_5 (M_r 24,400. a dimer) from *Azotobacter;* C.c_5 from *Pseudomonas* (87 amino acids, M_r 10,100, sequence known, an acidic protein); C.*cc'* (monomer obtained from guanidine treatment of the dimer, 125 amino acids, M_r 14,000, 2 heme groups) from *Chromatium* and C.*c'* (127 amino acids, primary structure similar to that of C.*cc'*) from *Alcaligenes,* both photosynthetic bacteria; C.c_{551} (82 amino acids, M_r 9,600, sequence known for 5 species, acidic protein), an electron carrier for the nitrite reductases from *Pseudomonas.* C.c_{553} (82 amino acids, M_r 9,600, basic, sequence known) and C.c_{555}, also called C.c_6 or C.*f*, are components of the sulfide reductases of green sulfur bacteria. C.*f* is also found in the chloroplast membranes of *Euglena gracilis* and in higher plants, where it is part of the photosynthetic electron transport chain, along with C.b_{559}. C.*f* from spinach (M_r 270,000) consists of 8 subunits (M_r 34,000), of which only 4 carry a heme group. C.*f* from *Euglena gracilis* (pI 5.5, M_r 13,500) is functionally similar to C.*c*.

The C. were discovered by MacMunn and rediscovered in 1925 by Keilin. The discovery that cellular respiration is based on iron catalysis was made by Otto Warburg.

Cytokines: a general term for a group of polypeptides, which are produced transiently by a range of different cell types, and which serve as short range intercellular signaling agents They usually act locally by binding to surface receptors, and they are involved in the promotion of cell division and the regulation of cell function, primarily by activating the expression of specific genes. About 40 different polypeptides are currently considered as C. Examples are: Interferons (see), Tumor necrosis factors (see), Epidermal growth factor (see), Insulin-like growth factors (see), Erythropoietin (see), Colony stimulating factors (see), Lymphokines (see), Transforming growth factors (see). Many C. have been cloned and are available in quantity as recombinant proteins. They share many properties with polypeptide hormones

Cytokinins, *phytokinins, kinins:* a group of plant hormones which promote cell division and generally stimulate plant metabolism, in particular RNA and protein synthesis. C. are generally *N*-substituted derivatives of adenine (Fig. 1). Together with other plant hormones (gibberellins and auxins), they mediate the plant's response to environmental factors, such as light.

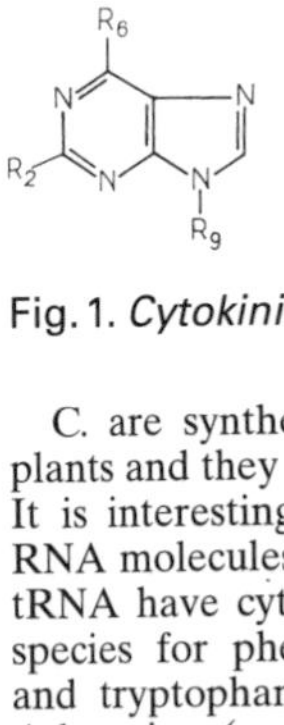

Fig. 1. *Cytokinins.* Substituted adenine

C. are synthesized mainly in the roots of higher plants and they are not subject to much translocation. It is interesting that they occur in certain transfer RNA molecules: 0.05 to 0.1 % of the purine bases in tRNA have cytokinin activity. In *E. coli,* the tRNA species for phenylalanine, leucine, serine, tyrosine and tryptophan contain C. N^6-(γ,γ -dimethylallyl)-Adenosine (see) also occurs in the free form. The most important C. are Kinetin (see), Zeatin (see) and Dihydrozeatin (see). The synthetic C., *N*-benzyladenine (see 6-Benzylaminopurine) and SD 8339 (Fig. 2) also have high cytokinin activity.

Biotests for the detection and quantitative measurement of C. measure one of the following: l. stimulation of cell division in tissue cultures, e. g. in tobacco or soybean callus tissue; 2. enlargement of cells in the leaf disc test, often with bean or radish leaves; 3. inhibition of the loss of chlorophyll from detached leaves; 4. promotion of seed germination in the dark.

The first C. to be discovered was kinetin, which was isolated from hydrolysed herring sperm DNA by Skoog and Miller in 1955. The first C. obtained from a plant was zeatin, which was isolated in 1964 from maize kernels.

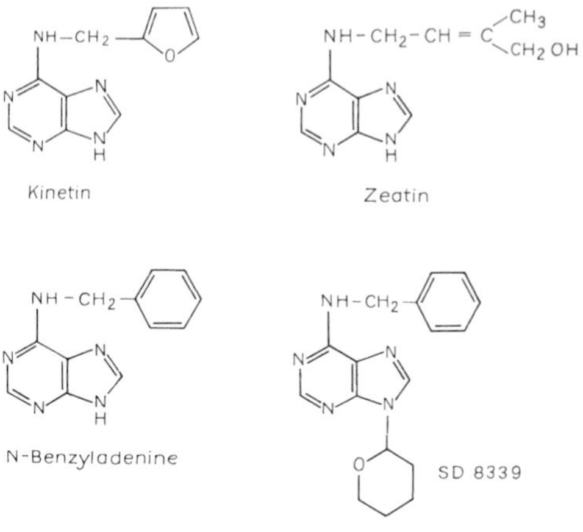

Kinetin

Zeatin

N-Benzyladenine

SD 8339

Fig. 2. *Some important cytokinins*

Cytolipin K: the same as globoside. See Lysosomal storage diseases (Sandhoff's disease).

Cytolysosomes: see Intracellular digestion.

Cytoplasm: see Cell, 2.

Cytoplasmic inheritance: transfer of genetic information in eukaryotic sexual reproduction, which is not carried by the chromosomes of the nucleus. C. i. is due to extrachromosomal genetic carriers, e. g. mitochrondrial and plastid DNA. C. i. does not obey Mendelian rules, and it permits mixing of cytoplasmic genetic factors during mitosis. Certain petite mutations of yeast, the killer property of certain strains of *Paramecium* and leaf pigmentation in *Antirrhinum majus* are examples of properties transmitted by C. i.

Cytoplasmon: the total extranuclear genetic information of a eukaryotic cell, excluding that in the mitochondria and plastids.

Cytosine, *C* or ***Cyt, 6-amino-2-hydoxypyrimidine:*** one of the pyrimidine bases of DNA and RNA. M_r 111.1, m. p. 320 to 325 °C (d.). C. is biosynthesized as its nucleoside triphosphate (see Pyrimidine biosynthesis). It is also a component of some nucleoside antibiotics.

Cytosine arabinoside: see Arabinosides.

Cytoskeleton: a three-dimensional network of fibrous proteins in the cytoplasm which provide structural support, motility and a scaffolding along which intracellular bodies can be moved. The C. has 3 distinct components: microtubules (25 nm diameter), microfilaments (6–8 nm) and intermediate filaments (10 nm).

Microtubules are polymers of the protein tubulin which, in vivo, are probably always associated with other proteins, the *Microtubule Associated Proteins* (MAPs). Microtubules are the main structural components of mitotic and meiotic spindles, and of cilia and flagella; in axons and dendrites, they serve as cables along which organelles are moved by kinesin and the retrograde translocator [R. D. Vale et al. *Cell* **43** (1986) 623–632]. Depending on the function of the microtubules, they are associated with different

MAPs. In addition, the tubulins are a family of related globular proteins which are the products of several genes. Although the sequences of amino acids have been highly conserved throughout metazoan evolution, there has been a divergence among different tubulin genes within a species, and these genes are differentially expressed during development.

Tubulins are amenable to isolation and purification because they bind specifically to the drug colchicine. Native tubulin is a dimer of α and β sub-units, of M_r 50,000–60,000, while a third subunit (γ) serves as a nucleus for the assembly of microtubules but is not incorporated into them. The *C*-terminal regions of both subunits are very acidic; the last 40 amino acid residues of the α-subunit include 16 Glu and 3 Asp. The tubulin dimer contains 2 bound guanosine nucleotides. One of these, bound to the E (E for exchangeable) site on the β-subunit, exchanges readily with exogenous GTP. The other, bound to the N (N for 'non-exchangeable') on the α-subunit site, does not exchange. After polymerization, neither site exchanges with the medium. Blocking or oxidation of the SH groups in tubulin prevents assembly of microtubules in vitro.

At relatively high concentrations, purified tubulin dimers will polymerize in vitro, and they are capable of hydrolysing GTP. Polymerization in vitro requires GTP, Mg^{2+} and EGTA to chelate Ca^{2+}, and occurs only at 35 °C. The resulting microtubules depolymerize when cooled. In the presence of MAPs, polymerization occurs at a lower tubulin concentration, and the resulting microtubules are more stable.

Although tubulin has GTPase activity, and although GTP hydrolysis accompanies microtubule assembly, the addition of a tubulin dimer to the polymer does not require GTP hydrolysis. Instead, the GTP appears to be hydrolysed after incorporation of the dimer into the polymer. A region of GTP-tubulin dimers forms near the (+)-end of the microtubule, while somewhat further back the E site of the dimers contains GDP. The dimers released from the (–)-end have one GTP and one GDP bound to them. It seems likely that the hydrolysis of GTP stabilizes the polymer.

Stages in the assembly of microtubules may be summarized as follows: (i) heterodimers associate 'α-tubulin end to β-tubulin end' to form short protofilaments; (ii) the protofilaments are unstable and associate side by side, with adjacent heterodimers slightly displaced vertically relative to each other so as to form a curved sheet; (iii) typically, when 13 protofilaments are present the curved sheet forms a 24 nm diam. hollow tube, in which horizontally-adjacent heterodimers form rows that appear to follow a spiral path because of the above-mentioned slight displacement; (iv) elongation of the tube by addition of more heterodimers at either end.

Assembly of microtubules is asymmetric. At one end, the (+)-end, tubulin dimers associate with the tubule faster than they dissociate, while at the other end, the (–)-end, dissociation occurs more rapidly than association. At equilibrium, the rate of addition of subunits at one end of the microtubule is equal to the rate of loss at the other end; the process is referred to as "treadmilling", because a given tubulin dimer will pass slowly from the plus to the minus end of the microtubule. In many cells, microtubules are stabilized by being anchored, presumably at the (–)-end, to microtubule organizing centers (MTOC). This is true of the neuronal axon, polar and mitotic microtubules, but kinetochore microtubules appear to be attached to the kinetochore by their plus ends. (This may be determined by decoration of the microtubules with Dynein (see) or C-shaped protofilament sheets under appropriate conditions).

The anti-cancer drugs, Colchicine (see), vincristine and vinblastine (see Vinca alkaloids) and Taxol (see) act by binding to tubulin and adversely affecting microtubule behavior, a consequence of which is the disruption of mitosis because the nuclear spindle is largely composed of microtubules. At high concentrations colchicine causes microtubule disassembly but at low concentrations it binds at the microtubule ends and prevents heterodimers being added to them or taken away from them; this 'poisoning' of the microtubule ends has the effect of blocking mitosis at metaphase. High concentrations of vincristine and vinblastine, which bind to a different site on tubulin from colchicine, cause the disassembly of microtubules and the precipitation of heterodimers in paracrystalline form; at low concentrations they inhibit the lengthening and shortening of microtubules. Taxol acts by hyperstabilizing microtubules.

MAPs may be defined as proteins with specific binding sites for tubulins. It is not known whether dynein, kinesin and the retrograde translocator (see Kinesin) are MAPs by this definition, because their affinity to microtubules may be mediated by other MAPs. The classic MAPs are listed below.

Protein	M_r	Properties
MAP_1	350,000	Projection on microtubule surface
MAP_2	270,000	Projection on microtubule surface
	70,000	Protein associated with MAP_2
Tau	55,000–62,000	Asymmetric protein
Type II cAMP-dependent protein kinase	54,000–39,000	Enzyme associated with MAP_2
Low M_r MAPs	28,000–30,000	Light chains of MAP_1

The biological functions of MAPs are not known. Although they promote microtubule assembly in vitro, they have not been observed to do so in vivo. MAP_2 binds to actin filaments, secretory granules, coated vesicles and neurofilaments. However, it is not certain that this binding is specific. It has been demonstrated by immunofluorescence that MAP_2 binds to intermediate filaments (vimentin). [T.W. McKeithan & J.L. Rosenbaum in J.W. Shay (ed). *The Cytoskeleton,* vol. 5 of *Cell and Muscle Motility* (Plenum Press, New York, 1985) pp. 255–288; R.B. Vallee, *ibid.* pp. 289–311; D.W. Cleveland & K.F. Sullivan, *Annu. Rev. Biochem.* **54** (1985) 331–365]

Microfilaments consist of Actin (see) and actin-binding proteins. Stress fibers, which were first observed by bright-field microscopy, are bundles of microfilaments with opposite polarities; they contain actin, α-actinin and myosin or myosin-like protein. Tropomyosin, filamen and vinculin are located in some domains of stress fibers. Stress fibers can contract in vitro, but in vivo they probably do not. Instead, they probably act as tensile elements which resist motion. Their absence from rapidly moving cells suggests this interpretation. Stress fibers probably anchor the cell to its substrate. *Vinculin,* M_r 130,000, is associated not only with the ends of stress fibers, but with substrate adhesion plaques and cell-cell contact regions. Fibronectin (see), which also mediates cell adhesion, is another organizer of microfilaments. Finally, the patches formed by cross-linking of cell-surface proteins by antibodies or lectins are associated with stress fibers. In addition to stress fibers, microfilaments make up polygonal nets within the cell; these have been dramatically demonstrated by immunofluorescence photographs of cells treated with anti-actin antibodies. Filamen (M_r 250,000) has been shown to cross-link microfilaments in vitro. Fascin and fimbrin are proteins which promote formation of actin bundles, while actin gelation protein and actin binding protein have been found to promote formation of an irregular mesh. [H. R. Byers et al in J. W. Shay (ed.) *The Cytoskeleton,* vol. 5 of *Cell and Muscle Motility* (Plenum Press, New York, 1985) pp. 83–137; D. J. DeRosier & L. G. Tilney, *ibid.* pp. 139–169]

Intermediate flaments (IF) are polymers of several different proteins which have very similar secondary structures and form fibers about 10 nm in diameter. They may be isolated because they are specifically bound by colcemid (demecolcine; see Colchicum alkaloids). Expression of these proteins is linked to differentiation of cell lineages, each of which has a characteristic complement of IF proteins when fully differentiated. The following 5 proteins have been recognized as components of IF for some time: *cyto-* (or *pre-)keratins* (from tonofilaments), subunit M_r 40,000–68,000; *neurofilament triplet proteins,* M_r 68,000, 160,000 and 200,000; *glial fibrillary acidic protein* (from glial filaments), M_r 51,000; *desmin,* M_r 53,000; and vimentin, M_r 53,500. (The subunit M_r values given for the first three were determined by SDS-PAGE, while the last two are derived from the known amino acid sequences.) In addition, there are several proteins which may be IF proteins, because they bind to colcemid, but which have not been polymerized in vitro: synemin (M_r 230,000, from avian muscle), paranemin (M_r 280,000, from embryonic skeletal muscle), a 66 kDa protein (rat brain, squid axoplasm, possibly identical to the 66 kDa heat-shock protein), a 68 kDa protein (associated with brain and spinal cord microtubules and skeletal muscle myofibrils, identical to the 68 kDa heat-shock protein), a 95 kDa protein (myofibrils, IF from various cultured cells), 50 kDa proteins (cold-labile neurofilaments), and 60–70 kDa proteins associated with IF in cultured cell lines.

IF usually form a network of insoluble protein throughout the cytoplasm. It may be revealed by dissolving the membranes in a detergent like Triton X-100, which allows soluble cytoplasmic components to diffuse away from the cytoskeleton. IF bind the MAPs of microtubules, and form end-to-side linkages with microfilaments (the pointed ends of the IF bind to the actin). Actin and the IF are also linked by 3-nm filaments which may be mde up of spectrin (fodrin). In pathological conditions like Alzheimer's disease, tangles of neurofilaments are found in degenerate neurons. On the basis of their distribution in the cell and their insolubility, IF have long been thought to have a structural function. P. Traub has proposed that in addition to this function, or instead of it, the IF may have a nuclear function. His hypothesis is based on the observation, firstly, that microinjection of antibodies against vimentin or cytokeratins or both causes the TF system to collapse into a dense ring around the nucleus. However, the treated cells survive for several days and may even divide once or twice, so that the structural function of the IF is not essential to cell survival. Secondly, IF are involved in receptor-mediated endocytosis, vectorial transport of the endosomes, and membrane capping, which occurs when cell-surface antigens are cross-linked by antibodies or lectins. (Capping precedes internalization of the antigen-antibody complexes and is a necessary step in lymphocyte activation.) Thirdly, vimentin, desmin, glial fibrillary acidic protein and the neurofibrillary triplet proteins bind both single-stranded DNA and ribosomal RNA. In the case of vimentin, the avidity of binding to DNA depends on the GC content of the latter. At neutral pH and physiological ionic strength, vimentin will bind to purified core histones, but not to H1 histone. A mixture of all 5 histones was resistant to hydrolysis by Ca^{2+}-activated proteinase, but they were hydrolysed in the presence of vimentin, suggesting that vimentin may loosen the chromatin structure and make DNA available for transcription.

Traub hypothesizes that the Ca^{2+}-activated proteinase associated with IF may convert the proteins to a form which does not readily polymerize, but which binds more readily to histones. The normal intracellular concentration of Ca^{2+} is far too low to activate the proteinase, but it is known that endocytosed vesicles and other membrane structures may release Ca^{2+}. Thus the local concentration could be high enough to activate the proteinase and to cause conversion of the IF protein from the fibrous to the "signal" form. [P. Traub, *Intermediate Filaments* (Springer, Heidelberg, 1985)]

Cytotactin: see Tenascin.

Cytoxan: see Cyclophosphamide.

D

Daidzein: 4′,7-dihydroxyisoflavone. See Isoflavone.

Daidzin: daidzein 7-glucoside. See Isoflavone.

Dalton (symbol Da), ***atomic mass unit*** (symbol u): the symbol, u, for this unit is possibly confusing, in that it also stands for unit; also multiples and submultiples, e. g. mu, could be misleading. The term, atomic mass unit, is also unwieldy. In 1981, the NCIUB and JCBN therefore asked the IUB and IUPAC to apply to the International Committee of Weights and Measures (Comité International des Poids et Mesures, CIPM) to approve the name "dalton" (symbol Da) as an alternative to the name "atomic mass unit". The dalton is one twelfth of the mass of the nuclide ^{12}C. It is equal to $1.6605655 \times 10^{-27}$ kg within about 6 parts per million.

Dalton complex: see Dictyosomes.

Dark reaction: see Photosynthesis.

Databases: computerized systems for storage, retrieval, analysis and modeling of biological data. In its simplest form a database is an electronically stored list of data in a particular field. Databases are particularly useful for handling the large number of data produced by research in molecular biology, such as protein and nucleic acid sequences, and they are also widely used for the collection, storage and dissemination of data in other areas, such as genome maps, cloning vectors, restriction endonucleases, carbohydrate sequences, experimental 3-dimensional atomic coordinates, collections of cell lines, hybridomas and microbial strains, enzyme EC numbers and reactions, *Escherichia coli* references, sequences and genetic map positions, etc., and there is a database of molecular biological databases. The field of activity concerned with the gathering of biological data, its storage by electronic means and its dissemination is known as ***bioinformatics***.

Databases are produced by national and international research centers, by chemical companies and by individual researchers. They are available on magnetic tape and on CD-ROM. Also many databases are available from file servers located at various sites around the world, which can be accessed by electronic mail requests. The present trend, however, is for databases to become accessible via the "net". Networked centralized major resources are clearly the answer to the problem posed by the exponential increase in the size of databases, and they also address the need for a more direct and interactive form of data retrieval. Networking also avoids the duplication of resources; for example, hundreds of different online versions of nucleotide and protein sequences are now in existence.

Databases of protein and nucleic acid sequences are invaluable to the molecular biologist and biochemist as a basis for sequence similarity searches. Thus, any nucleotide or protein sequence can be rapidly compared with all other known sequences stored in the database. Similar search procedures can also be used to identify promoters, signal sequences, etc., and to reveal sequence patterns in families of proteins.

DNA sequences are compiled in: 1. GenBank [Los Alamos National Laboratory, Los Alamos, NM 87545, USA (genbank@life.lanl.gov)]; 2. the Data Library of the European Molecular Biology Laboratory (EMBL) based at the European Bioinformatics Institute (EBI) Cambridge, UK (office@ebi.ac.uk); and 3. the DNA database of Japan (DDBJ) (ddbjupdt@ddbj.nig.ac.jp). These three international data banks, together with the Genome Sequence Database (GSDB) at Santa Fe, USA, are now cooperating by exchanging new nucleotide sequence data on a daily basis. The resulting database is called "DDBJ/EMBL/GenBank + GSDB international nucleotide sequence database", and is accessible through DDBJ, EBI and GenBank.

Protein and peptide sequences are stored in: 1. the Protein Identification Resources [National Biomedical Research Foundation, 3900 Reservoir Road, Washington, DC 200007, USA (pirmail@gunbrf.bitnet)]; and 2. in SwissProt (European Bioinformatics Institute).

Using parallel architecture computers, a 500 amino acid sequence can be compared with all presently known protein sequences in a few minutes, using programs such as PROSRCH, BLAZE or MPSRCH. For conventional architecture computers, programs such as BLAST and FAST have been written which reduce the run time by employing short cuts. The BLAST family of programs takes only a few minutes to compare all known DNA sequences with a probe of 1 kb, and it reports maximal segment pairs between the probe and database sequences. FAST takes about ten times longer then BLAST; it locates identities of fixed *k*-tuple between the probe and the database sequence, scores the best diagonal regions (as on a dot plot), and attempts to join these diagonal regions using an appropriate algorithm. [R. F. Doolittle (ed.) *Methods in Enzymology,* **vol 183,** *Molecular Evolution: Computer Analysis of Protein and Nucleic acid Sequences.* Academic Press, San Diego, New York, 1991; H. Mannila & K-J. Räihä *The Design of Relational Databases,* Addison-Wesley, 1992; C. Robinson *Trends in Biotechnology* **12,** No. 10, 1994; All papers pp. 2–22 in *The Biochemist (The Bulletin of the Biochemical Society)* December/January 1994/95; D. & M. Campbell *A Student's Guide to Doing Research on the Internet,* Addison-Wesley, 1995; R. A. Harper 'EMBnet: an institute without walls' *Trends Biochem. Sci.* **21** (1996) 150–152; G. Goos, J. Hartmanis & J. van Leeuwen (eds.) *Advances in Database Technology – EDPT '96. Proceedings of the 5th International Con-*

ference on Extending Database Technology (Lecture Notes in Computer Science 1057), Springer Verlag (Berlin, Heidelberg, New York) 1966]

Datura alkaloids: see Tropane alkaloids.

DBC coenzyme: see 5′-Deoxyadenosylcobalamin.

DCMU: see Dichlorophenyl-dimethyl urea.

ddATP, ddCTP, ddGTP and ddTTP: see Dideoxyribonucleotide triphosphates.

Deamination: removal of the amino group (-NH_2) from a chemical compound (usually an amino acid). Metabolically, D. may occur by: a) oxidative D. of amino acids to ketoacids and ammonia by Flavin enzymes (see) and pyridine nucleotide enzymes (see Amino acids, Table 3); b) Transamination (see) in which an amino group is transferred from an amino to a keto compound, and c) removal of ammonia from a compound, leaving a double bond, e.g. the D. of L-aspartate to fumarate, and the D. of histidine to urocanic acid. Transamination is important in the synthesis of amino acids from tricarboxylic acid cycle intermediates; the reverse reactions feed excess amino acids into the tricarboxylic acid cycle for oxidation.

Debranching enzyme: see Amylo-1,6-glucosidase.

Decarboxylases: enzymes which catalyse removal of a carboxyl group as CO_2 (see Decarboxylation) from α-ketoacids or from amino acids. Pyruvate decarboxylase (see) is an important enzyme of carbohydrate metabolism. For amino acid D., see Pyridoxal phosphate.

Decarboxylation: removal of a carboxyl group as CO_2, from a ketoacid or from an amino acid. D. of ketoacids occurs several times in the course of the Tricarboxylic acid cycle (see). The D. of β-ketoacids often occurs spontaneously. In biological systems the oxidative D. of α-ketoacids requires coenzymes such as thiamin pyrophosphate, lipoic acid, coenzyme A, flavin adenine dinucleotide or nicotinamide adenine dinucleotide. Oxidative D. of Pyruvate (see) to acetyl-CoA and of α-ketoglutarate to succinyl-CoA are nodes at which many metabolic pathways cross. D. of amino acids is catalysed by Pyridoxal phosphate (see) enzymes.

Decoyinin: see Angustmycin.

Defective organism: a term sometimes applied to an Auxotrophic mutant (see).

Defective viruses: viruses that lack genes for the synthesis of functional capsid or envelope proteins. For example, the Rous sarcoma virus is unable to synthesize one of its envelope proteins. In order to complete its replication cycle and form new virus particles, the missing gene must be provided by a *helper virus*. The helper virus is a member of a group of viruses often found in association with the Rous sarcoma virus, and known as Rous-associated viruses (RAV). In the absence of an RAV, the normal host cell is transformed into a tumor-forming cell. Most acute transforming Retroviridae are D.v. Deficient viruses (see) are often included with the D.v.

Deficiency mutants: see Auxotrophic mutants.

Deficient viruses: viruses which, through mutation, have lost the ability to synthesize certain essential enzymes, e.g. RNA-dependent RNA replicase. The replication cycle can be completed only if the cell is superinfected with a *helper virus*, which carries the gene for the missing protein. D.v. are often so dependent on their helper virus that they are only found in association with it. In such cases, the deficient virus is known as a satellite virus, e.g. the tobacco necrosis satellite virus (TNSV), which requires the RNA replicase of the tobacco necrosis virus (TNV) in order to complete its replication cycle. As a result, the replication of the TNV is retarded. The TNSV is therefore a virus parasite, because it parasitizes not only the host cell, but also the TNV. A similar relationship exists between the adeno-associated satellite viruses and certain adenoviruses.

Under certain conditions, especially in cell cultures, the missing enzyme can be supplied by the cells of a second organism, which are then known as *helper cells*. D.v. are sometimes grouped with the Defective viruses (see).

Dehydrobufotenin: a Toad poison (see) found in *Bufo marinus*. M_r 202. Minimal lethal dose in mice 6 mg/kg.

7-Dehydrocholesterol, ***provitamin D₃, cholesta-5, 7-dien-3β-ol:*** a zoosterol (see Sterols). M_r 384.6, m.p. 150 °C, $[\alpha]_D^{20}$ – 114° (*c* = 1 in chloroform). D. occurs in relatively high concentrations in animal and human skin, where it can be converted to vitamin D_3 by ultraviolet radiation. Prevention and cure of rickets by UV irradiation is due to this conversion.

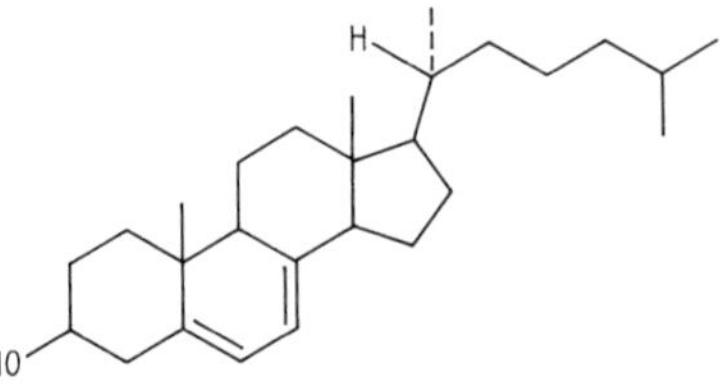

7-Dehydrocholesterol

24-Dehydrocholesterol: see Desmosterol.

11-Dehydrocorticosterone, ***Kendall's substance A, 21-hydroxypregn-4-ene-3,11,20-trione:*** a glucocorticoid hormone from the adrenal cortex. M_r 344.43, m.p. 178–180 °C, $[\alpha]_D^{25}$ + 258° (ethanol).

Dehydrogenases: redox enzymes which extract hydrogen (2 H^+ + 2 e^-) from a substrate (hydrogen donor) and transfer it to a second substrate (hydrogen acceptor). There are two main groups of D., those requiring a pyridine nucleotide (NAD^+ or $NADP^+$) and those requiring a flavin coenzyme (see Flavin enzymes) as the primary hydrogen acceptor.

Dehydrogenation: removal of hydrogen from a reduced substrate, which thereby becomes oxidized. Enzymatic D. is catalysed by dehydrogenases or oxidases. In general, 2H atoms (and 2 electrons) are removed at once. The opposite reaction is called hydrogenation. The following metabolic D. are common:

Saturated compounds	→	Unsaturated compounds
Alcohols	→	Carbonyls
Aldehyde hydrates	→	Carboxylic acids
Dihydropyridine nucleotides	→	Pyridine nucleotides
Reduced flavin nucleotides	→	Oxidized flavin nucleotides
Hydroquinones	→	Quinones
Amines	→	Imines

D. of amino acids and amines produces unstable imino compounds which spontaneously react with water, producing ammonia and the corresponding carbonyl compounds. Amino acids are converted in this way to the correspoding ketoacids.

Deletion: loss of one or more nucleotides of DNA, or of an entire segment of a chromosome; a form of mutation.

Delphinidin: 3,3′,4′,5,5′,7-hexahydroxyflavylium cation, the aglycon of many Anthocyanins (see). D. glycosides are widespread among plants and are responsible for the mauve and blue colors of many flowers and fruits. Some of the more important are tulipanin (3-rhamnoglucoside), the pigment of various tulips, violanin from pansies *(Viola tricolor)* and delphin (3,5-di-β-glucoside) from salvia and delphiniums.

Demecolcine: see Colchicum alkaloids.

Demissine: one of the Solanum alkaloids, a steroid found in wild potatoes *(Solanum demissum)*. It is a glycoalkaloid, consisting of the steroid base demissidine [50α-solanidan-3β-ol; M_r 399.67, m.p. 221–222 °C, $[\alpha]_D + 30°$ (chloroform)] and the tetrasaccharide β-lycotetrose (D-galactose, D-xylose and 2 D-glucose residues). The aglycon demissidine differs structurally from solanidine (see α-Solanine) in that it has no double bond at C5, and has the 5α-configuration. D. is repellant to the larvae of potato beetles, and protects the wild potato from predation by these insects.

Denaturation: structural change in biopolymers which destroys the native, active configuration. It is brought about by heat, pH changes or chemical agents. See Nucleic acids, Proteins.

Dendrobium alkaloids: a group of terpene alkaloids from various species in the orchid genus *Dendrobium*. The D. are sesquiterpenes with one heterocyclic nitrogen atom.

Denitrification: see Nitrate reduction.

De novo purine synthesis: see Purine biosynthesis.

De novo pyrimidine synthesis: see Pyrimidine biosynthesis.

Density gradient centrifugation: a method for separating macromolecules on the basis of their density. In a very high-speed centrifuge, a solution of CsCl will form a stable density gradient after a sufficient time (24–48 h). Macromolecules will come to rest in a layer, or isopycnic zone, which corresponds to their buoyant density. This parameter, expressed in g/cm, can be measured exactly after the centrifugation. The densities of the CsCI solution range from 1.3 to 1.8 g/cm^3, which covers the range of most biomolecules, e.g. DNA 1.7, RNA 1.6, proteins 1.35–1.4. Cell organelles, which have lower densities, can be separated in sucrose density gradients. In contrast to the above equilibrium method, sucrose gradient centrifugation can be used as a dynamic method in which the buoyant density is estimated from the rate at which macromolecules sediment in the sucrose gradient. In this case, the gradient is established by carefully layering sucrose solutions of linearly or exponentially decreasing density into the centrifuge tube. The macromolecules are layered onto the top of the solution, and the run is timed so that they do not have time to move all the way to the bottom. [G.B. Cline and R.B. Ryel, in *Methods in Enzymology* **22** (1971) 38–50; T.J. Bowen, *An Introduction to Ultracentrifugation,* Wiley (Interscience) New York, 1970.

5′-Deoxyadenosine: a β-glycosidic deoxynucleoside (see Nucleosides) which is important as a component of vitamin B_{12} (see 5′-Deoxyadenosylcobalamin).

S-(5′-deoxyadenosine-5′)-methionine: see S-adenosyl-L-methionine.

5′-Deoxyadenosylcobalamin, ***B_{12} coenzyme, DBC coenzyme (DBC = dimethylbenzimidazole cobamide):*** one of the coenzyme forms of vitamin B_{12} (see Vitamins). In this compound, the 6th coordination position of the cobalt atom in the center of the corrinoid ring is covalently bound to the 5′C atom of the deoxyadenosine. Other cobamide coenzymes contain an *N*-heterocyclic base other than dimethylbenzimidazole. D. is the coenzyme of certain isomerization reactions (Fig. 1). In the isomerization of L-glutamic acid to β-methylaspartic acid (I), there is a reversible transfer of the glycine portion of L-glutamate from the C2 to C3 of the propionic acid moiety, and at the same time, an H atom is shifted in the opposite direction. In the conversion of methylmalonyl-CoA to succinyl-CoA (II), the thioester group is shifted from the 2nd to the 3rd C of the propionic acid part of methylmalonyl-CoA. This reaction plays a part in the biological degradation of branched-chain amino acids and in the propionic acid metabolism of *Propionibacterium.*

Cobamide coenzymes are also involved in the degradation of L-lysine to fatty acids and ammonia in *Clostridium* and the conversion of 1,2-diols to aldehydes in various microorganisms (Fig. 2).

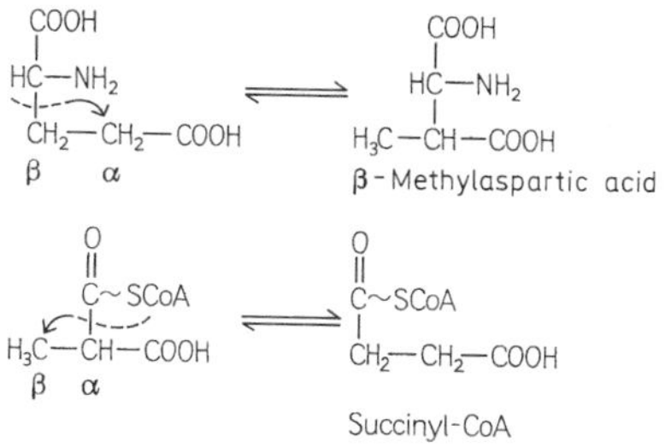

Fig. 1. *Isomerization reactions for which 5′-deoxyadenosylcobalamin is a cofactor*

Deoxycholic acid, ***3α,12α-dihydroxy-5β-cholan-24-oic acid:*** a bile acid, M_r 392.56, m.p. 176–177 °C, $[\alpha]_D^{20} + 55°$ (ethanol). D. is found in the bile of most mammals, including man, dog, ox, sheep and rabbit. It can be used as starting material for the partial synthesis of therapeutically important steroid hormones.

3-Deoxygibberellin C: see Gibberellins.

Deoxyhemoglobin: see Hemoglobin.

6-Deoxyhexoses: see 5-Methylpentoses, Deoxy sugars.

Deoxynucleoside: see Nucleosides.

Deoxynucleoside phosphorylases: see Nucleosides, Salvage pathway.

Deoxynucleotide: see Nucleotides.

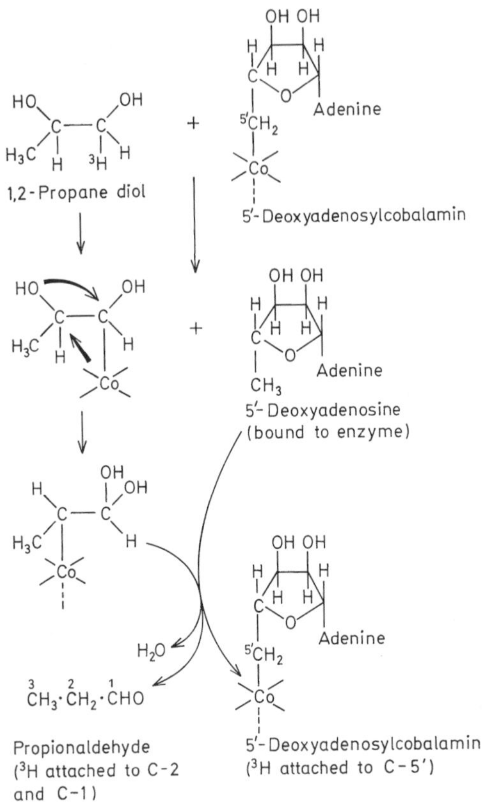

Fig. 2. *Mechanism of action of 5′-deoxyadenosylcobalamin as the cofactor of propanediol dehydratase* (EC 4.2.1.28). The mechanism is supported by the migration of ^3H from C1 of the substrate to C2 of the substrate and C5′ of the cofactor. The same enzyme also dehydrates ethylene glycol to acetaldehyde.

Deoxynucleotide biosynthesis: see Nucleotides.

Deoxyribonuclease I, *DNase I* (EC 3.1.21.1): an enzyme (normally obtained from bovine pancreas and therefore also known as *pancreatic deoxyribonuclease)* which catalyses random endonucleolytic cleavage of internucleotide bonds of double-stranded DNA, preferentially but not exclusively between adjacent purines and pyrimidines, producing 5′-phosphodi- and -oligonucleotides.

A divalent cation is required for activity (a requirement shared with all other enzymes catalysing phosphoryl transfer); Ca^{2+} is normally used for in vitro studies, but Mg^{2+} is probably more important in vivo. X-ray crystallographic studies reveal that the divalent cation is bound close to the cleaved phosphodiester bond, which is also very near to Glu_{75}, His_{131} and a bound water molecule, thus forming a Glu-His-H_2O triad (Fig.) very reminiscent of the Charge-relay system (see) in chymotryopsin and other serine proteases.

DNase I is used as a laboratory tool for distinguishing between regions of DNA that are bound to (and therefore protected by) specific proteins and those regions that do not display such binding (see Footprinting, DNase protection), e.g. DNase I-hypersensitive sites in chromatin are thought to represent DNA available for transcription. It also displays the interesting property of forming a 1:1 complex with G-actin (actin is therefore a natural and specific inhibitor of DNase I), which together with one Ca^{2+} ion and one molecule of ATP or ADP, can be crystallized. This enabled the X-ray analysis of G-actin, which is otherwise impossible to crystallize on account of its marked tendency to polymerize (see Actins).

Similar enzymes from sources other than pancreas are: streptococcal DNase (streptodornase), *E. coli* endonuclease, "nicking" enzyme of calf thymus, T_4 endonuclease II, T_7 endonuclease II, colicins E_2 and E_3.

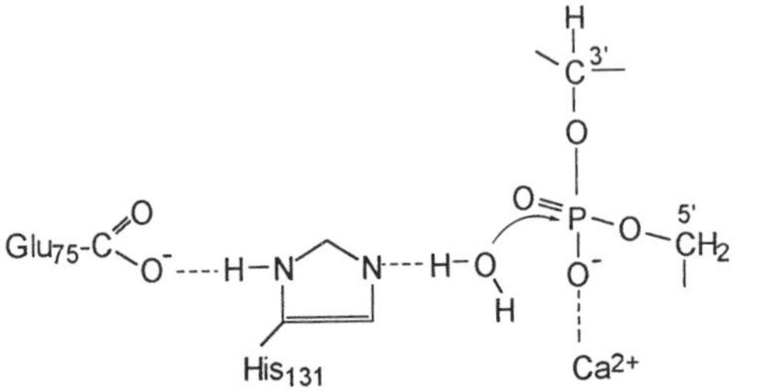

Catalysis of phosphodiester cleavage by the charge-relay system of DNase I. Transfer of a proton from water to His_{131} generates a positive charge on the imidazole side chain, which in turn is stabilized by the negative charge on the carboxyl group of Glu_{75}. The resulting hydroxide attacks the P atom, forming a pentacovalent intermediate stabilized by electrostatic interaction of the negatively charged O-atom and the divalent cation. With the subsequent departure of the 3′-OH group, the DNA strand is cleaved, leaving a terminal 5′-phosphate group.

Deoxyribonuclease II, *DNase II, pancreatic DNase II, acid DNase, calf thymus acid DNase, spleen acid DNase* (EC 3.1.22.1): a monomeric enzyme (M_r 31,000) which catalyses the endonucleolytic cleavage of double stranded DNA to 3′-phosphomononucleotides and 3′-phosphooligonucleotides. The earliest definitive studies were performed on the enzyme from hog spleen, which became known simply as "hog-spleen enzyme". But the enzyme (or enzymes with extremely similar properties) is widely distributed in animal cells, including, e.g. human pancreas, thymus, liver, gastric mucosa and cervix, as well as crab testis, snails and salmon testis. Although optimal activity occurs at pH 4–5, this optimum is not sharply defined, and considerable activity is retained even at neutral pH.

The enzyme is generally thought to be exclusively localized in lysosomes, although some authors claim that a small proportion (~ 10 %) of the total cellular activity resides in the nucleus. It is not unequivocally proven that the nuclear activity is not a contamination, or that the lysosomal and nuclear enzymes are identical. However, the reported 2–7-fold increase of nuclear DNase II activity in the S-phase of synchronized HeLa cells is interesting in that it suggests involvement of the enzyme in cellular DNA metabolism.

A natural protein inhibitor of DNase II, originally detected in mouse liver, has been purified to homogeneity from bovine liver (using affinity chromatography on insolubilized DNase II); it is a basic, monomeric protein, M_r 21,500, which forms a 1:1 complex with DNase II and causes the plot of enzyme activity against substrate concentration to become sigmoid. [G. Bernadi in *The Enzymes,* P. D. Boyer (ed.), Vol. **IV** (1971) pp. 271–287, Academic Press; D. Kowalski & M. Laskoski, Sr. in *Handbook of Biochemistry and Molecular Biology* 3rd edition, G. D. Frasman (ed.), Vol **II** *Nucleic Acids* pp. 491–535; H. Sierakowska & D. Shugar in *Progress in Nucleic Acid Research and Molecular Biology* Vol. **20,** W. E. Cohn (ed.), Academic Press, 1977, pp. 50–130]

Deoxyribonucleic acid, DNA: a polymer of deoxyribonucleotides found in all living cells and some viruses. It is the carrier of genetic information which is passed from generation to generation by an exact replication of the DNA molecule.

Structure. Each mononucleotide unit of the DNA polymer consists of phosphorylated 2-deoxyribose which is glycosidically linked to one of four bases: adenine, guanine, cytosine or thymine (abb. A, G, C or T, respectively). In the DNA of higher organisms, some of the cytosine is methylated in the 5 position, and in bacteriophage DNA, some is replaced by hydroxymethylcytosine. For the structures of the bases, see Nucleosides. The mononucleotides are linked by 3′,5′-diester bridges (see Nucleic acids, Phosphodiesters) in an unbranched polynucleotide chain. The base contents of DNA from different organisms vary widely (from 18 % A in tuberculosis bacteria to 30 % A in calf thymus), but the amount of A is always equal to the amount of T, and the amount of G is always equal to the amount of C (see GC content). This fact and the results of X-ray diffraction studies by Wilkins and Franklin led Watson and Crick to propose the double-helix model of the DNA structure [*Nature* **171** (1953) 737]. The life sciences were revolutionized by this concept of DNA structure. According to the Watson-Crick model, the DNA molecule consists of two complementary, but not identical strands which spiral around an imaginary common axis. The two spiral bands are made up of sugar phosphate chains from which the bases protrude at regular intervals into the interior of the helix. The two strands are held together by hydrogen bonds between bases on opposite strands. In order for the strands to fit together in the helix, a purine on one strand must always be opposed by a pyrimidine on the other. Hydrogen bonds can only form (within the constraints of the double helix) between adenine and thymine (A-T) or guanine and cytosine (G-C), so that the sequence of bases along one strand determines the sequence of the other (see Base pairing). The genetic information is encoded in the sequence of bases in the DNA (See Genetic code). The two strands are antiparallel, meaning that the phosphate diesters between the deoxyribose units read 3′ to 5′ in one chain and 5′ to 3′ in the other. In most organisms, only one of the two strands (the template strand, non-coding strand, anti-sense strand, non-codogenic strand) is transcribed, while its complementary non-template strand (coding strand, sense strand, codogenic strand) is not transcribed. The nomenclature of these complementary strands is clarified in the entry: *Nomenclatural conventions concerned with gene transcription.*

The double helix is not symmetrical; it has a broad and a narrow groove between the chains. These provide steric orientation for the processes of replication and transcription (Fig. 1). This right-handed double helix with 10 base pairs per helical turn is called B-DNA. It probably approximates closely the structure of relaxed (unstrained) DNA. It is generally accepted, however, that DNA is a dynamic molecule, with different conformations in equilibrium with one another. This equilibrium is affected by nucleotide sequence, ionic strength of the environment, presence of proteins [e. g. Histones (see) and other DNA binding proteins (see)] and the extent to which the molecule is under topological strain.

It has been shown that the DNA fragment d(CpGpCpGpCpG) crystallizes as a left-handed double helix [A. Wang, et al. *Nature* **282** (1979) 680–686]. This structure is known as Z-DNA, because an imaginary line joining the phosphate groups around the outer surface of the molecule describes a zig-zag course. (In B-DNA, the phosphate groups follow a smooth spiral) (Fig. 1). The two strands of Z-DNA are antiparallel, and complementary bases are hydrogen bonded as in B-DNA; however, the orientation of the bases to the backbone of the molecule is different from that in right-handed B-DNA. In Z-DNA the flat planes of the base ring systems are still more or less at 90° to the long axis of the molecule, and parallel with one another, but they are effectively rotated through 180° in comparison with bases in B-DNA. In the case of guanine this occurs by rotation about the glycosidic bond, so that guanine residues have a *syn* configuration; in the case of cytosine both the base and the deoxyribose are rotated, so that the cytosine residues retain the *anti* conformation. The result is an alternating orientation of adjacent sugars: the O1′ of the dG points down, and the O1′ of the dC points up. Because of this, the repeating unit of Z-DNA is a dinucleotide (in B-DNA it is a mononucleotide). See Table 1 for further comparisons.

A third type of helix is observed in X-ray studies of DNA made under conditions of low (75 %) humidity. A-DNA is right-handed, like B-DNA, but it has nearly 11 base pairs per helix twist, and they are inclined by 13° with respect to the plane perpendicular to the helix axis. A-DNA has a very deep major groove, in contrast to the B form. Since the 2′OH of the ribose moieties of RNA prevent RNA from adopting a B conformation, it has been suggested that DNA-RNA hybrids must have the A conformation.

The mean helix parameters of the three forms of DNA are given in the Table.

The "propeller twist" of a base pair is the angle between the planes of the two bases. (One can visualize the "rung" of the DNA ladder as being shaped like a two-bladed propeller rather than like a flat plank). The "base roll" and the "base inclination" refer to the average plane of the base pair. The base inclination is the angle between this average plane and the plane perpendicular to the helix axis, while the base roll is the angle between two successive base pairs. As can be seen from the Table, the standard deviations for propeller twist, base roll and base inclination are large. This is probably due to steric interac-

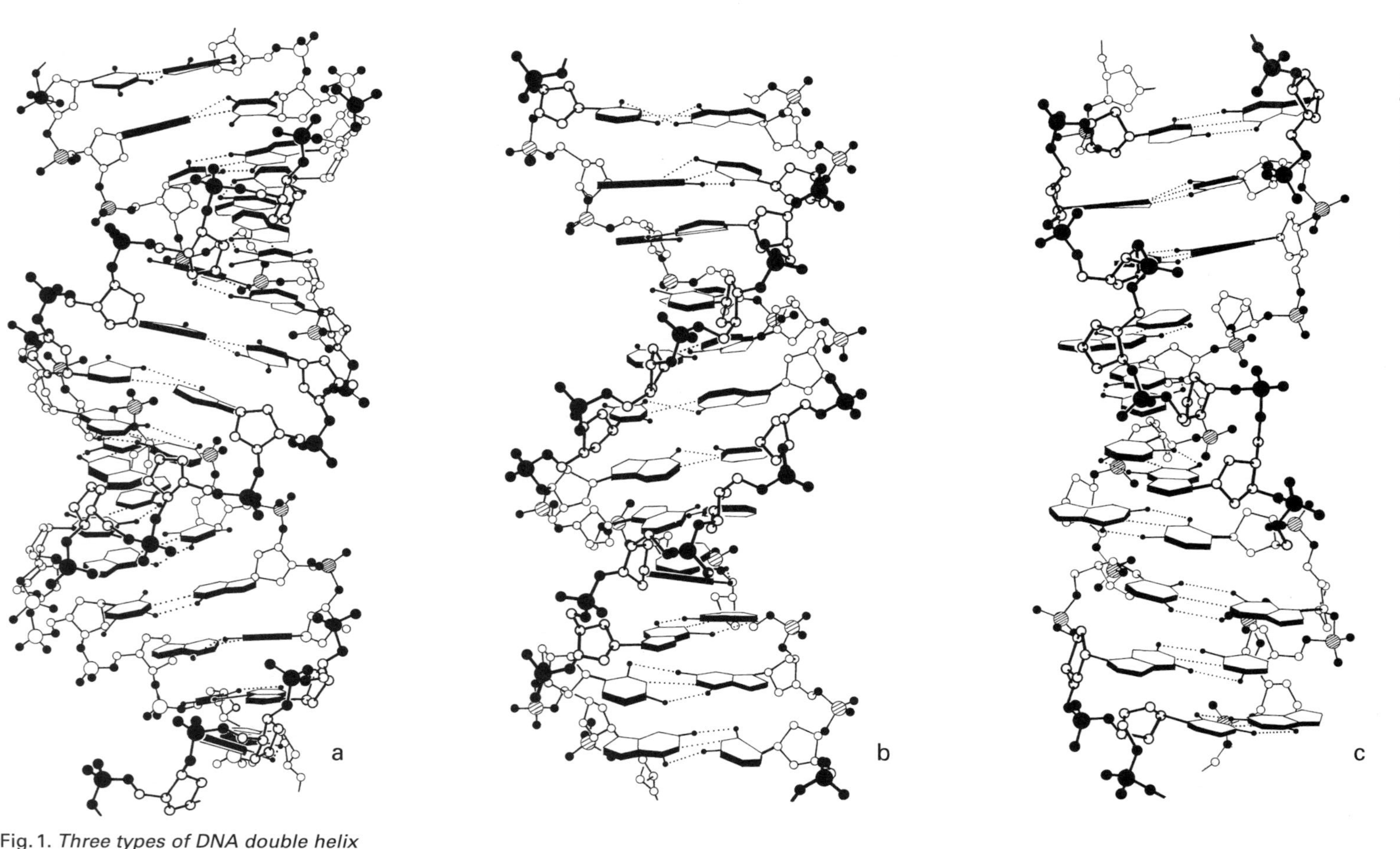

Fig. 1. *Three types of DNA double helix*
A-DNA (left) was generated by extension of the structure of the central six bases in the octamer GGTATACC, the crystal structure of which has been determined. B-DNA (center) was generated by repeating the central 10 base pairs of the dodecamer CGCGAATTCGCG. Z-DNA (right) is a left-handed helix of alternating guanines and cytosines; the structure here was generated by extension of the central 4 base pairs of CGCGCG. Hydrogen atoms are not shown. Bases are represented as laminae with black sides. All oxygen atoms bonded to phosphorus are black. Phosphorus atoms are black (front of helix) or hatched (rear of helix). The carbon atoms and ring oxygen of deoxyribose are all represented as open circles. For dimensions see Table 1.

Table 1. *Comparison of B-, A- and Z-DNA*

	B-DNA	A-DNA	Z-DNA
Handedness of helix	right	right	left
Residues per turn	10.4	10.9	12.0
Diameter of helix	23.7Å	25.5Å	18.4Å
Distance between adjacent base pairs along the axis (= rise per residue)	3.4Å	2.9Å	G-C 3.5Å C-G 4.1Å
Length of one helix turn (= helix pitch)	34Å	32Å	45Å
Rotation of adjacent base pairs relative to each other (= helical twist) (mean and observed range)	36° (16–44°)	33° (28–42°)	G-C, 51° C-G, 8.5°
Propeller twist	11.7 ± 4.8°	15.4 ± 6.2°	4.4 ± 2.8°
Base roll	−1.0 ± 5.5°	5.9 ± 4.7°	3.4 ± 2.1°
Base inclination	−2.0 ± 4.6°	13.0 ± 1.9°	8.8 ± 0.7°

tions between the bases; some combinations can be more tightly packed together than others. Van der Waals attraction between the bases insure that each pair will flt as closely as possible. The variations in the helix parameters are thus due to the base sequence, and can presumably be recognized by the proteins whose function demands recognition of specific sequences.

The axis of B-DNA passes through base pairs, whereas the axis of Z-DNA is practically empty, and a single deep, narrow groove extends into the center of the Z-DNA molecule. Part of the imidazole ring of guanine is exposed on the outer convex surface of the Z-DNA molecule. Thus, the double helix of Z-DNA resembles a ribbon of material with a serrated edge of phosphate groups wrapped around an imaginary central axis. In solution, poly(dG-dC)(dG-dC) exists in the B- or Z-conformation. Interconversion of the two forms can be measured from the inversion of the circular dichroism spectrum of the molecule [F.M. Pohl & T.M. Jovin, *J. Mol. Biol.* **67** (1972) 375–396]. Z-DNA and B-DNA are immunologically distinct. Thus, by using specific antibodies, it has been shown that the negatively supercoiled plasmid pBR322 contains a section of left-handed Z-DNA sequence, d(CpApCpGpGpGpTpGpCpGpCpApTpG) [A. Nordheim et al. *Cell* **31** (1982) 309–318]; the Z-conformation is therefore not restricted to poly(dG-dC) sequences. In this case, formation of the Z-structure is favored because it releases the strain caused by supercoiling. Both B- and Z-DNA probably exist as families of closely related structures, whose members differ from one another by slight modifications of conformation. Two such conformations of Z-DNA (Z_I and Z_{II}) have been described.

The DNA of certain viruses exists as a single-stranded coil rather than a double helix. Both single- and double-stranded DNA can form ring-shaped molecules (bacteria, mitochondria). The rings can be more tightly coiled (superhelices, hypertwisted configurations), the number of "supercoils" depending on the size of the molecule (33–44 for mitochondrial DNA). See Superhelix.

Because the double helix must unwind when the molecule is replicated, it was thought earlier by some authors that the double helix observed in crystalline DNA by X-ray diffraction might not exist in solution. In 1976, for example, Rodley et al. proposed the "side-by-side" model, in which the two strands are not twisted around each other. However, in 1983 Iwamoto and Hsu showed that the two chains of a 39-base-pair segment of double-stranded DNA in solution must have been twisted around each other 3–4 times in a right-handed direction; this is consistent with the double helix, but not the side-by-side model [S. Iwamoto & M.-T. Hsu, *Nature* **305** (1983) 70–72; F.H.C. Crick, *Proc. Natl. Acad. Sci. USA.* **73** (1976) 2639–2643]

On the basis of conformation studies on double-stranded DNA, Cyriax and Gäth concluded in 1978 that transitions are possible between double-helical and non-coiled structures. The non-coiled transition conformation was called the *cis-ladder conformation* because the sugar-phosphate chains have a *cis*-like position with respect to the base pairs, which are arranged like the rungs of a ladder. The DNA strands can change from the *cis*-ladder conformation into other conformations, including helices, without generating strains in the molecule, so this structure may be regarded as a transitional form between the double helix and the single-stranded state, and also between the double helix and other highly ordered structures. The molecular mass of DNA is difficult to measure, because the molecules break very easily during extraction. The highest measured M_r is 10^9 (about 2×10^6 base pairs), but the estimated M_r of *E. coli* DNA is 2.8×10^9 ($3–4 \times 10^6$ base pairs). (The ring-shaped molecule of *E. coli* DNA has been photographed in the electron microscope.) The M_r measured for mammalian DNA is much smaller (10^8), but this may reflect only the difficulty of separating it from chromosomal proteins. There is genetic evidence that the DNA of a single chromosome is one gigantic chain, which would then have to be 10 to 20 times as long as *E. coli* DNA. The *E. coli* DNA molecule is about 1000 μm long, while mitochondrial DNA is about 50 μm long.

Prokaryotic DNA is bound to the membrane as giant rings; sometimes the cells also contain smaller fragments, known as Plasmids (see) or episomes. About 95 % of the DNA of a eukaryotic cell is located in the nucleus, where it is complexed with proteins in the form of Chromosomes (see). Mitochondria and chloroplasts also contain DNA (extrachromosomal DNA); mitochondrial DNA accounts for 1–2 % of the total DNA of the cell, while chloroplast DNA may account for up to 5 %. The amount of DNA per cell varies widely among different species, but within a given organism it is constant (the germ cells of diploid organisms contain only half as much as the somatic cells).

The DNA of a cell can be fractionated according to its base composition (see GC content) by density gradient centrifugation in CsCl. The buoyant density, given as g $CsCl/cm^3$, is used to characterize the fractions. Mouse liver DNA, for example, consists of a main fraction (buoyant density 1.702) together with Satellite DNA (see) (1.691). The DNA fractions of *Euglena* have buoyant densities of 1.707 (nuclear DNA), 1.691 (mitochondrial DNA) and 1.686 (chloroplast DNA).

Replication of DNA. Each single strand of the parent duplex acts as a template for the synthesis of a new partner strand, i. e. replication is *semiconservative.* This was shown experimentally by the classical Meselson and Stahl experiment (see). Replication must therefore involve rotation of the entire parent molecule, or a mechanism which breaks and rejoins single strands, or both. Circular double strands cannot replicate without at least one breaking and rejoining event. When the strands are separated, each base binds a complementary base from the pool of nucleotides (base pairing), so that the sequence of nucleotides in the new strand is determined by that of the old strand. This mechanism allows information (the sequence of bases) to be reproduced and passed on to subsequent generations. The DNA of eukaryotic chromosomes is divided into replicative subregions *(replicons,* Fig. 2), but bacterial DNA is replicated as a single unit.

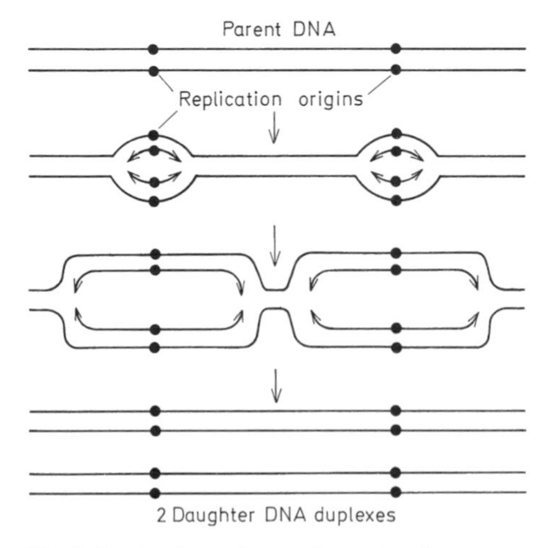

Fig. 2. *Repication of a eukaryotic chromosome.* Replication occurs simultaneously in different subregions, which eventually join with one another, producing two separate daughter duplexes.

The circular DNA remains attached to the cell membrane during the entire process, so that when the cell divides, each daughter cell receives a DNA molecule. However, the binding to the membrane could have a regulatory effect on initiation, especially since the attachment point lies near the replicator region. On the other hand, the DNA polymerase and the initiator proteins are associated with the membrane.

The replication process must occur at extremely high speed, since *E. coli* can divide every 20 minutes under optimal conditions. This requires the replication of at least 133,000 base pairs (133 kbases) or 50 µm DNA per min. DNA replication in eukaryotes occurs much more slowly. The process is catalysed by DNA polymerase (EC 2.7.7.7) which joins the four 5′-nucleoside triphosphates into the linear DNA molecule, using a single-stranded DNA as a template. Each nucleotide loses a pyrophosphate group (PP_i), and the energy stored in these bonds drives the reaction.

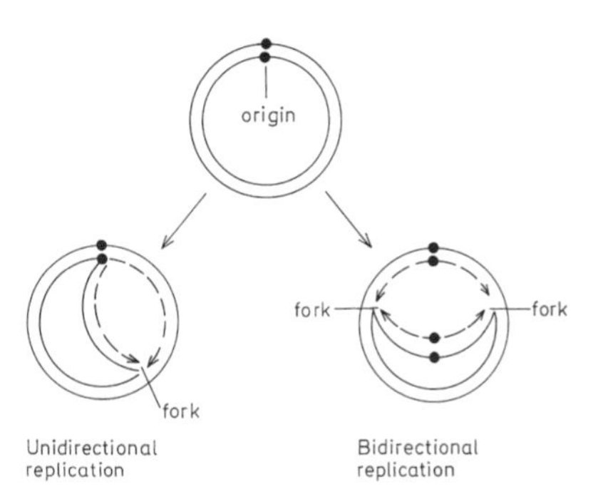

Fig. 3. *Unidirectional and bidirectional replication of prokaryotic and viral DNA.* The bidirectional mechanism is found in *E. coli* and other bacteria and many DNA viruses. The unidirectional mechanism has been observed in certain plasmids. Autoradiography indicates that the ring structure is retained during the entire process. Some viral DNAs are replicated by a unidirectional rolling circle mechnaism (Fig. 4).

DNA synthesis always occurs in the 5′→3′ direction; since association between strands is always antiparallel, the template strand is read in the 3′→5′ direction. Replication depends on the action of 20 or more enzymes and proteins. The several sequential steps of replication include recognition of the origin or starting point of the process, unwinding of the parent duplex (by helicases, see Topoisomerases), maintenance of strand separation in the replicating region, initiation of daughter strand synthesis, elongation of daughter strands, rewinding of the double helix (see Topoisomerases), and termination of the process. The complex of factors involved in replication is called the *DNA replicase system,* or *replisome* (Table 2; Fig. 6).

It is clear from autoradiographic studies that synthesis occurs simultaneously from the 3′ end of one parent strand and the 5′ end of the other, but all three of the known DNA polymerases work only in the

Table 2. *Some constituents of the replisome of* Escherichia coli

Protein	$M_r \times 1{,}000$	No. of subunits	Function	No. of molecules per cell
Single strand DNA binding protein	74	4	Binding to single strand after opening of helix	300
Protein i	66	3	Primosome assembly. Initiation of primosome action. Formation of primer RNA	50
Protein n	28	2		80
Protein n′	76	1		70
Protein n″	17	1		–
dnaC	29	1		100
dnaB	300	6		20
Primase	60	1		50
DNA polymerase III holoenzyme, which consists of:	760	2	Chain elongation	20
α	140	1		–
ε	25	1		–
θ	10	1		–
β	37	1		300
γ	52	1		20
δ	32	1		–
τ	83	1		–
DNA polymerase I	102	1	Excision of primer and replacement with DNA	300
Ligase	74	1	Ligation of extended Okazaki fragments	300
Gyrase, comprising:	400	4	Supercoiling	–
GyrA	210	2		250
GyrB	190	2		25
Helicase I (rep protein)	65	1	Helicase	50
Helicase II	75	1	Helicase	5,000
dnaA	48	–	Binds at the origin of replication	

5′→3′ direction. The 5′→3′ parent strand is replicated in short pieces (about 2,000 nucleotides in bacteria, less than 200 nucleotides in animal cells). These short pieces, known as *Okazaki fragments,* are later linked by a DNA ligase. Each Okazaki fragment is synthesized as an extension of a short RNA primer (about 10 nucleotides). The RNA primer is synthesized in the 5′→3′ direction on the template of the replicating DNA strand by the action of a specialized RNA polymerase, called a *primase.* The primase is associated with other proteins in a *primosome complex,* which may also be considered a part of the replisome complex (Table 2). DNA synthesis proceeds from the 3′ end of the primer. The RNA primer is then removed, one nucleotide at a time, by the 5′→3′ exonuclease activity of DNA polymerase I. As each ribonucleotide is removed, it is replaced by a corresponding deoxyribonucleotide by the polymerase activity of the same enzyme. Final splicing of the extended Okazaki fragments is due to *DNA ligase,* which catalyses phosphodiester bond synthesis between the 3′-phosphate group of the elongating DNA and the 5′-hydroxyl group of the newly synthesized Okazaki fragment. Ligation is coupled to pyrophosphate bond cleavage (in bacteria, NAD is cleaved to nicotinamide mononucleotide and AMP; in animal cells, ATP is cleaved to AMP and pyrophosphate) (Fig. 7).

Theoretically, discontinuous synthesis is not necessary for the progress of replication along the 3′→5′ replicating parent strand, but there is evidence that this may also be discontinuous. In addition to normal DNA-dependent DNA synthesis, RNA-dependent synthesis can occur in some cases (see RNA-dependent DNA polymerase). [A. Kornberg *DNA Replication,* W. H. Freeman & Co., 1980; A. Kornberg *1982 Supplement to DNA Replication*, W. H. Freeman & Co., 1982]

DNA degradation. In vital cells, DNA is normally not degraded (but see Apoptosis) but there are various enzymes in tissues which degrade the DNA of dead or broken cells (see Nucleases, Deoxyribonuclease, Phosphodiesterase). When bacteriophages enter bacterial cells, they produce nucleases which degrade the host's DNA. (The DNA of some bacteria

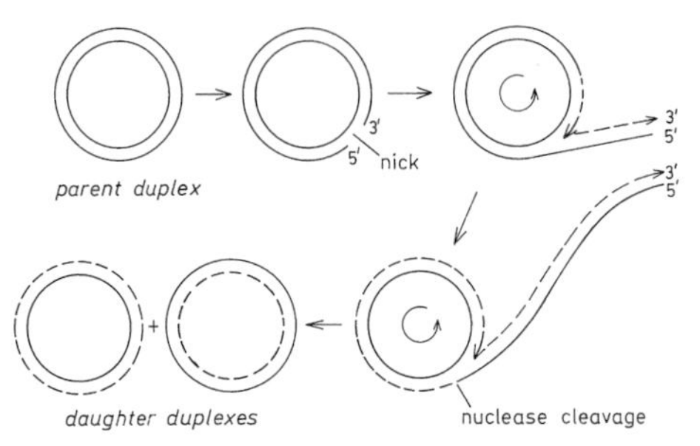

Fig. 4. *Rolling circle mechanism of replication of some viral DNAs.* The circular parent DNA is first cleaved ("nicked") enzymatically. New nucleotide units are added to the 3′ terminus of the broken strand, and the continuous growth of the new strand displaces the 5′ tail of the broken strand from the rolling circular template. Thus, the 5′ tail becomes a linear template for synthesis of a new complementary strand. The duplex originating from the 5′ tail is then cleaved from the other daughter duplex by a nuclease.

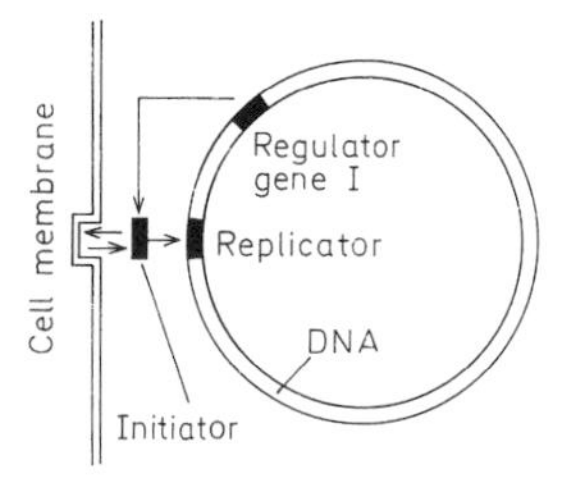

Fig. 5. *Regulon model of the regulation of replication in circular, bacterial DNA*

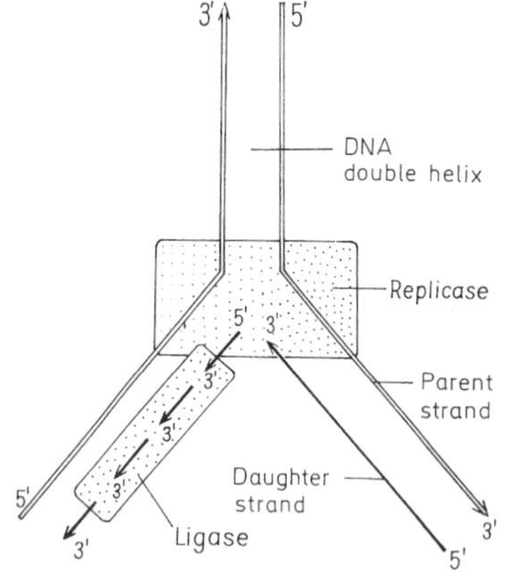

Fig. 6. *Replication fork of DNA.* The arrows indicate the direction of synthesis.

contains a significant number of methylated bases, e.g. 5-methylcytosine and 6-methyladenine, which appear to protect the DNA from viral degradation. The T-even bacteriophages, which attack *E. coli* contain

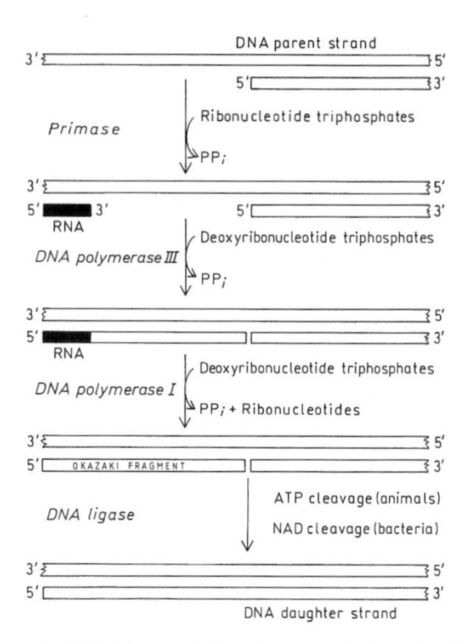

Fig. 7. *Role of RNA primer and Okazaki fragments in DNA replication*

hydroxymethylcytosine in place of cytosine, which protects their DNA from their own deoxyribonucleases.) In the laboratory, DNA can be degraded by acid hydrolysis. It is completely hydrolysed to phosphate, bases and deoxyribose by strong acid, but under milder conditions it can be hydrolysed to nucleosides or nucleotides.

Biological significance. The information for the synthesis of the cell's proteins is contained in the base sequence of its DNA (see Genetic code). This was proved directly by experiments on Transformation (see) and Transduction (see), and by demonstrating the role of phage DNA (see Phage development). Other DNA sequences do not encode proteins, but regulate their expression (see Operon, Intron, DNA-binding proteins). See DNA repair.

Deoxyribonucleotides: see Nucleotides.

2-Deoxy-D-ribose: a pentose lacking one hydroxyl group. M_r 134.13; α-form, m.p. 82 °C, $[\alpha]_D$ −58 ° (water); β-form, m.p. 98 °C, $[\alpha]_D^{20}$ −91 °→−58 ° (water). D. is the carbohydrate of deoxyribonucleic acids (DNA).

HOCH$_2$ O OH; H H H H; HO H

β-2-Deoxy-D-ribose

Deoxyribose phosphates: phosphorylated derivatives of the deoxypentose, 2-deoxy-D-ribose. They are biosynthesized by reduction of ribose phosphates in the course of nucleotide synthesis. There are two enzymes in *E. coli* which reduce cytidine diphosphate to deoxycytidine diphosphate. Deoxyribose 5- and 1-phosphate form an equilibrium mixture in the presence of phosphopentomutase (EC 2.7.5.6).

Deoxyribosyltransferases: enzymes which catalyse the transfer of deoxyribose from purine and pyrimidine deoxyribosides to free bases in the synthesis of deoxynucleosides.

Deoxyribotide: see Nucleotide.

Deoxy sugars: monosaccharides in which one or more hydroxyl groups have been replaced by hydrogen. There are two types, those with a methyl group in the terminal position, such as the 6-deoxy-hexoses L-fucose and L-rhamnose, and those with the hydroxyl missing from the middle of the molecule, such as the DNA component 2-deoxy-D-ribose. D. are often components of glycosides, e.g. D-digitoxose is the sugar component of many digitalis glycosides.

Deoxythymidine: see Thymidine.

Deoxythymidylic acid: see Thymidine phosphate.

Deposiston: see Ovulation inhibitors.

Depsipeptide: polypeptides which contain ester bonds as well as peptide bonds. Naturally occurring D. are usually cyclic peptides, also called peptolides, which generally have α- or β-hydroxyacids as heterocomponents. In the wider sense, this class also includes O-peptides and peptide lactones. The most important peptide lactones are the Actinomycins (see), Etamycin (see) and Echinomycin (see); the peptolides include the Enniatins (see), Valinomycin (see), Sporidesmolides (see), Serratamolide (see), Esperin (see), etc. D. are metabolic products of microorganisms which often have very high antibiotic activity. Many D. can be chemically synthesized by methods very similar to those used in the chemical synthesis of peptides.

Derepression: the release of an operon from repression of transcription. In prokaryotic cells it occurs by inactivation of a repressor, either by removal of a corepressor (see Enzyme repression) or by binding of an inducer (see Enzyme induction). D. in eukaryotic cells involves regulatory proteins and effectors, such as hormones (see Gene activation).

Dermatan sulfate: formerly called β-heparin and chondroitin sulfate B. It is a mucopolysaccharide containing L-iduronic acid linked α1,3 to *N*-acetyl-D galactosamine 4-sulfate; the latter is linked α1,4 to the next iduronic acid residue. See Chondroitin sulfate.

Desmin: a fibrous protein in muscle cells; a component of intermediate filaments (see Cytoskeleton). The subunit M_r is 52,000; the protein is structurally related to the α-keratins (see Keratins).

Desmoplakins: fibrous proteins found on the cytoplasmic side of desmosomes (cell-cell contact regions of the plasma membranes of epithelial cells). D. mediate attachment of tonofilaments (see Cytoskeleton) to the plasma membrane.

Desmosine: see Elastin.

Desmosterol, ***24-dehydrocholesterol, 5α-cholesta-5,24-diene-3β-ol:*** a zoosterol (see Sterols). D. differs from Cholesterol (see) in having an extra double bond between C24 and C25. It is an intermediate in the biosynthesis of cholesterol (see Steroids). It has been isolated from the barnacle *Balanus glandula,* chicken embryos and rat skin.

Desulfurase: see Desulfurication.

Desulfuricants: anaerobic bacteria of the genera *Desulfovibrio* and *Desulfotomaculum* whose Sulfate respiration (see) contributes to the process of Desulfurication (see). The most important of these bacteria is *Desulfovibrio desulfuricans. Desulfovibrio* are thought to be responsible for the hydrogen sulfide and other sulfides in the Black Sea.

Desulfurication: the anaerobic degradation of sulfur-containing organic compounds to inorganic sulfur. In these processes, the sulfhydryl groups are removed from proteins by *desulfurases* to yield hydrogen sulfide. D. is also the formation of hydrogen sulfide by desulfuricants.

Detergent degradation: the processes by which microorganisms digest synthetic detergents, thus removing them from the environment. Unbranched hydrocarbon chains can be degraded, while branched-chain compounds resist degradation. Thus it is possible to design biodegradable detergents.

DETPP: see Thiamin pyrophosphate.

Dexamethasone, ***9α-fluoro-16α-methylprednisolone, 9α-fluoro-11β,17,21-trihydroxy-16α-methylpregna-1,4-diene-3,20-dione:*** a synthetic pregnane derivative (see Steroids) which is highly antiinflammatory but has little mineralocorticoid activity. It is used to relieve arthritis. D. is synthesized from cortisol.

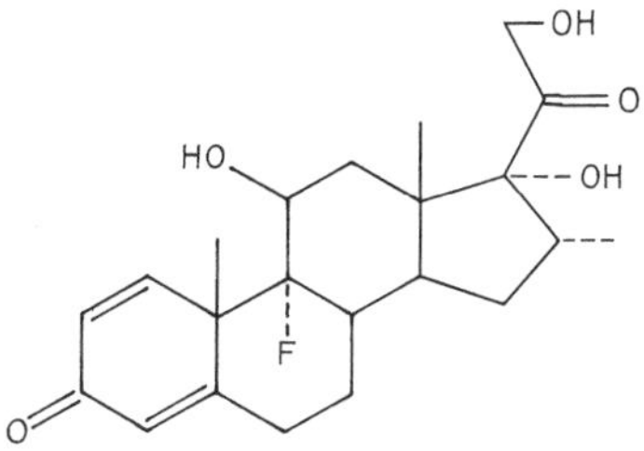

Dexamethasone

Dextranase: see Enzymes, Table 2.

Dextrans: high-molecular-mass polysaccharides synthesized by certain microorganisms. They consist of D-glucose linked α-glycosidically, primarily in 1,6 bonds, but with some 1,3 and 1,4. The M_r of the microbial product is several million. The colloidal osmotic pressure of D. with a M_r of about 75,000 corresponds to that of blood, so they are used as a substitute for blood plasma. These smaller D. are produced by controlled microbial synthesis or by partial hydrolysis of larger molecules. The organisms used are the lactic acid bacteria *Leuconostoc mesenteroides* and *Leuconostoc dextranicum,* which are grown anaerobically in sucrose-containing media. D. are the starting material for the dextran gels used for molecular sieves. The polysaccharides are cross-linked to form a three-dimensional network containing a large number of hydroxyl groups. The degree of cross-linkage determines the size of the pores and the capacity of the gel for water uptake. (More cross-linking means smaller pores and smaller water capacity). These cross-linked D. are sold under the trade name Sephadex.

Dextrin 6-α-D-glucanohydrolase, ***oligo-1,6-glucosidase*** (EC 3.2.1.10): a glycosidase present in the digestive juices of the small intestine, and formerly known as limit dextrinase, or isomaltase. It is specific for the hydrolysis of the α-1,6 bonds in isomaltose and in the oligosaccharides produced by the action of α-amylase on starch and glycogen. Products of the hydrolysis are glucose, maltose and unbranched oligosaccharides, which can then be degraded further by α-amylase.

Dextrins: water-soluble degradation products of starch. According to the color of their reaction product with iodine, they are classified as *amylodextrins* (blue reaction product), *erythrodextrins* (red), and low-molecular-mass *achroodextrins,* which have no colored reaction product with iodine. Heating starch with 3 % HCI or HNO_3 produces *acid dextrins. Schardinger dextrins* are the result of the action of *Bacillus macerans* on starch; these are α-1,4-linked rings of 6–8 glucose units. Depending on the size of the ring, they are called α-, β- or γ-D. D. (in particular British gum, starch gum or gommelin, produced by dry heating starch at 160 –200 °C) are used for dry extracts and pills, emulsifiers, thickeners for fabric dyes, sizing in paper and textiles, printing inks, glues, matches, fireworks and explosives.

Dextrose: see D-Glucose.

DHF: acronym of dihydrofolic acid.

Diabetes mellitus: a disease caused by partial or complete lack of Insulin (see), or by decreased numbers or sensitivity of cellular insulin receptors. D. m. is the most common human endocrine disorder (at least 1 % of the population of Europe and North America) except for obesity.

Insulin regulates the uptake of blood glucose, ions and other substances into cells. In D. m. very high blood glucose levels (8–60 mM) occur, with the result that glucose is lost through the kidneys. D. m. is characterized by copius, sweet urine. In uncontrolled D. m., the muscles and liver, which are unable to absorb glucose from the hyperglycemic blood, metabolize proteins and fats. The result is wasting similar to that in starvation. Two types of D. m. are known, juvenile onset (type I) and adult onset (type II). A propensity to type I is inherited; the gene or genes are located in the major histocompatibility complex. In type I D. m., the islets of Langerhans are destroyed, there is massive infiltration by lymphocytes, and the disease is thought to be autoimmune. Experimentally produced alloxan D. m. mimics this condition because alloxan preferentially destroys the pancreatic β-cells. Type II D. m. is strongly correlated with obesity, and may develop through the loss of fully active insulin receptors from cell membranes.

Mild D. m. can be controlled by diet, but in severe cases, insulin must be injected. Until the discovery of insulin early in this century, severe D. was invariably fatal. Milder D. is accompanied by a number of side effects, including damage to retinal capillaries, cataracts, dwindling and demyelination of neuronal axons, which lead to motor, sensory and autonomic dysfunction, arteriosclerosis and renal disease. The severity of these side effects correlates directly with the degree to which blood glucose exceeds normal levels. [M. Bliss, *The Discovery of Insulin,* University of Chicago Press, 1982; M. Brownlee & A. Cerami, *Ann. Rev. Biochem.* **50** (1981) 385–432; M. Hattori et al., *Science* **231** (1986) 733–735]

Diacetyl, butane-2,3-dione: CH_3–CO–CO–CH_3, a diketone produced as a byproduct of carbohydrate degradation, m. p. –3 °C, b. p. 88.8 °C. D. is a component of butter aroma, and has been found in many biological materials. It is produced by dehydrogenation from acetoin, the decarboxylation product of pyruvate. In microorganisms D. is also produced by reaction of active acetaldehyde with acetyl-CoA . Little is known about the further metabolism of D. It is used as an aroma carrier in the food industry.

Diacylglycerol: see Acylglycerols.

Diaminopimelic acid pathway: see L-Lysine.

6-Diazo-5-oxo-L-norleucine, ***DON:*** $N^- = N^+ =$ CH–CO–CH_2–CH(NH_2)–COOH, an antagonist of glutamine. It inhibits de novo purine synthesis in bacteria and mammals. It prevents the growth of experimental tumors, but is toxic for animals.

Dicarboxylic acid cycle: a cyclic pathway (Fig.) which enables the utilization of glyoxylic acid or one of its precursors (e. g. glycolic acid) as a carbohydrate source for the growth of microorganisms. The D. c. a. includes some of the reactions of the tricarboxylic acid cycle. It also involves malate synthase (EC 4.1.3.2; see Glyoxylate cycle). If the concentration of D. c. a. intermediates is too severely reduced by diversion to synthetic pathways, they can be replenished from the Glycerate pathway (see).

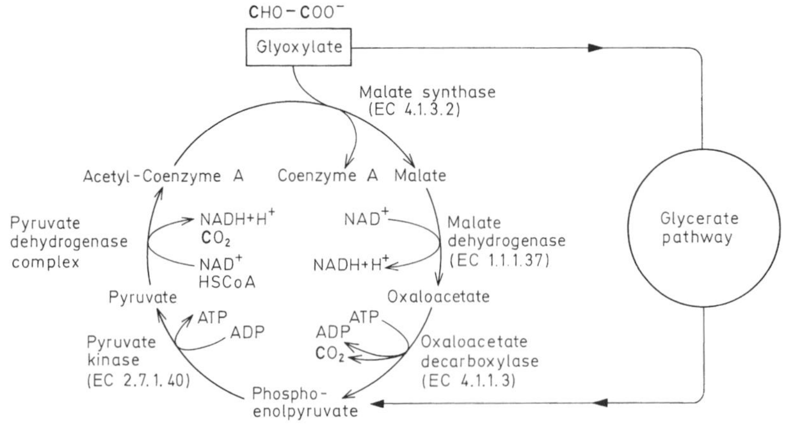

Oxidation of glyoxylate in the dicarboxylic acid cycle

Dichlorophenyl-dimethyl urea, ***DCMU, Diuron, 3-(3,4-dichlorophenyl)-1,1-dimethylurea, N^1-(3,4-dichlorophenyl)-N,N-dimethylurea:*** a systemic herbicide which blocks electron transport from photosystem II to photosystem I (see Photosynthesis). It is absorbed principally by the roots then translocated acropetally in the xylem.

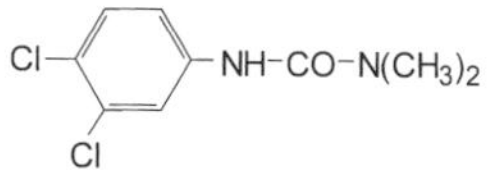

Cl–C6H3(Cl)–NH–CO–N(CH3)2

3-(3,4-dichlorophenyl)-1,1-Dimethylurea

Dichrostachinic acid, ***S-[(2-carboxy-2-hydroxyethylsulfonyl)methyl]cysteine:*** HOOC–CH(OH)–CH_2–SO_2–CH_2–S–CH_2–CH(NH_2)–COOH, a sulfur amino acid from seeds of *Dichrostachys glomerata* and *Neptunia oleracea.* [R. Gmelin *Hoppe Seyler's Z. Physiol. Chem.* **327** (1962) 186–194; L. Fowden et al. *J Chem. Soc.* (C) (1971) 833–840]

Dictyosomes: components of the Golgi apparatus, especially in plants. D. have also been called *lipochondria* and osmiophilic material. Sitte has suggested that the total of the dictyosomes in a plant cell should be called the Golgi apparatus (see) or the Dalton complex.

Dicumarol, ***dicoumarol, dicoumarin: 3,3'-methylenebis(4-hydroxy-2H-1-benzopyran-2-one); 3,3'-methylenebis(4-hydroxycoumarin):*** a vitamin K antagonist formed by the action of microorganisms on coumarin and/or coumarin precursors (see Coumarins) in improperly cured (spoiled) sweet clover (Melilotus) hay. Cattle eating this hay are subject to hemorrhage (sweet clover disease). D. is used clinically to prevent thromboses.

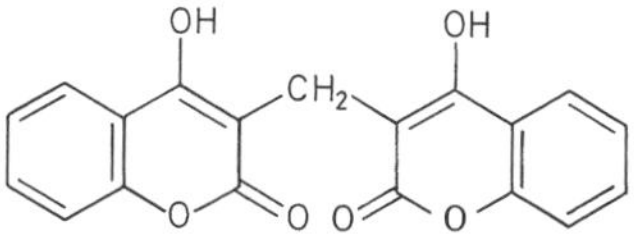

Dicumarol

Various organic syntheses have been reported. Of particular interest are those designed for synthesis of [^{14}C]D. for metabolic investigations, e.g. [2-^{14}C]4-hydroxycoumarin (from reaction of *o*-hydroxyacetophenone with [^{14}C]diethylcarbonate in the presence of sodium ethoxide) is converted to [2-^{14}C]D. by treatment with formaldehyde [H. R. Eisenhauer et al. *Can J. Chem.* **30** (1952) 245–250 (this is a useful reference source for other synthetic methods). Isolation and structural elucidation: H. A. Campbell & K. P. Link *J. Biol. Chem.* **138** (1941) 21–33; M. A. Stahmann et al. *J. Biol. Chem.* **138** (1941) 513–527]

Dicysteine: see L-Cystine.

Dideoxyadenosine triphosphate, ***Dideoxycytidine triphosphate, Dideoxyguanosine triphosphate, Dideoxythymidine triphosphate:*** see Dideoxyribonucleotide triphosphates.

Dideoxyribonucleotide triphosphates, ***terminating triphosphates:*** synthetic substrates of DNA polymerase l, which catalyses their incorporation into growing oligonucleotide chains in place of the normal deoxyribonucleotide triphosphate substrates. Incorporation of a terminating triphosphate results in chain termination, since the 3'-hydroxyl group is lacking. They are used in the Sanger method for DNA sequencing. In the reaction mixture of template DNA, DNA polymerase and the four deoxyribonucleotide triphosphate substrates, one terminating triphosphate is included at about 1 % of the concentration of its normal counterpart. This results in a family of variously extended DNA sequences, which can be sized by gel electrophoresis. By performing separate incubations for each terminating triphosphate, four patterns of bands are obtained from which the DNA sequence can be deduced. [F. Sanger et al. *Proc. Natl. Acad. Sci. USA.* **74** (1977) 5463–5467].

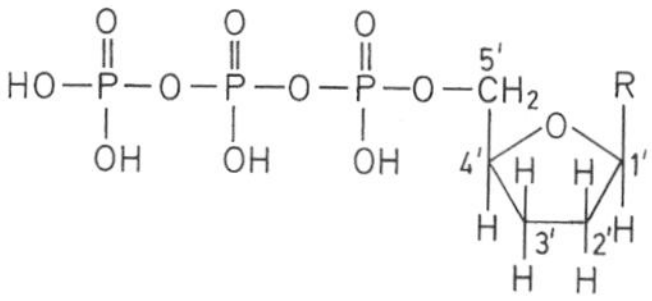

Dideoxyribonucleotide triphosphates

R = Adenine: Dideoxyadenosine triphosphate (ddATP)

R = Guanine: Dideoxyguanosine triphosphate (ddGTP)

R = Cytosine: Dideoxycytidine triphosphate (ddCTP)

R = Thymine: Dideoxythymidine triphosphate (ddTTP)

Didymocarpin: see Humulenes.

Differential gene activation: see Gene activation.

Differential gene expression: see Metabolic regulation.

Differential scanning calorimetry (DSC): see Membrane lipids.

Diffutin: 7-hydroxy-3',4'-dimethoxy-5'-*O*-β-D-glucosylflavan. See Flavan.

Digalactosyldiglycerides: see Membrane lipids.

Digestion: the totality of mechanical and chemical processes occurring in the digestive tract, which result in the degradation of foodstuffs to low M_r, absorbable, nonantigenic substances. The digestive tract, especially in mammals, shows considerable structural and biochemical adaptation to the nutritional physiology of the organism, e.g. in carnivores, herbivores and omnivores. Generally a distinction is made between buccal D., gastric D. and duodenal D. The chemical processes of buccal D. have little significance since salivary amylase occurs only in man, apes, pig and some rodents, and food is present for too short a time in the buccal cavity. The first main site of D. is the stomach. Despite the many different designs of this organ between individual species, it always serves as the site for the degradation of dietary proteins to peptones, and of starch (in animals with salivary amylase) to water-soluble dextrins. A special situation exists in ruminants and other herbivores, where the rumen, reticulum,

Digestive enzymes of vertebrates

Enzyme	M_r	Site of attack (↓)	pH-optimum
(I) Proteases			
1. Proteinases			
a) Gastric			
Pepsin A (alkali-labile)	34500	–Gly↓Try–Phe–, –Glu↓Phe	1.8
Pepsin B (Gelatinase)	36000	Hydrolyses only gelatin	
Pepsin C (Gastricsin)	31500	–Tyr↓Ser–, –Phe↓Ser	3.0
Rennin	30700	–Phe↓Met (in κ-casein)	4.8
b) Pancreatic			
Trypsin	23400	–Arg↓R, –Lys↓R	8.0
Chymotrypsin A (α and γ)	25170	–Tyr↓R, ↓Phe–R, ↓Try–R, –Met↓R	8.0
δ-Chymotrypsin	25400	as chymotrypsin A	
Chymotrypsin B	25400	as chymotrypsin A	
Chymotrypsin C	23900	as chymotrypsin A, plus –Leu↓R, –Glu(Asp)↓R	8.0
Elastase	25700	R–neutral amino acid↓R	8.0
Collagenase	?	Hydrolyses only collagen	5.5
c) Duodenal secretion Enterokinase	196000	H_2N–Val–$(Asp)_{2-5}$$Lys_6$↓$Ile_7$–Trypsin–COOH	8.0
2. Peptidases			
a) Pancreatic			
Carboxypeptidase A	34400	Peptidyl↓Phe, ↓Tyr, ↓Trp, ↓Leu	8.0
Carboxypeptidase B	34400	Peptidyl ↓Lys, ↓Arg	8.0
b) Duodenal secretion			
Leucine aminopeptidase	300000	H_2N–Leu↓peptide, or ↓polypeptide	8.9
Aminotripeptidase	300000	Ala↓dipeptide	8.0
Dipeptidases	100000	Gly↓Gly, Gly↓Leu, Cys↓Gly	7.8
Prolidase	?	Gly↓Pro	7.8
Prolinase	?	Pro↓Gly	7.8
(II) Glycosidases			
a) Salivary and pancreatic amylase, acting in buccal cavity, stomach and duodenum			
α-Amylase	50000	α-glycosidic 1 → 4 bonds	6.5
b) Duodenal secretion			
α-Glycosidases			
5 Specific maltases	~ 200000	α-glycosidic 1 → 4 bonds	7.0
A specific sucrase	~ 200000	α-glycosidic 1 → 4 bonds	7.0
A trehalase	~ 200000	α-glycosidic 1 → 1 bonds	7.0
α-1,3-Glycosidase	~ 200000	α-glycosidic 1 → 3 bonds	7.0
β-Galactosidase (Lactase)	~ 200000	β-glycosidic 1 → 4 bonds	6.0
Oligo-α(1 → 6)glucosidase	~ 200000	α-glycosidic 1 → 6 bonds in starch and glycogen	7.0
(III) Esterases			
a) Gastric			
Gastric lipase	35000	Ester bonds in triacylglycerols, especially milk fat	5.0
b) Pancreatic			
Pancreatic lipase	35000	Ester bonds in triacylglycerols	7.5
Phospholipase A + B	14000	Ester bonds in phosholipids	7.5
Cholesterol esterase	400000	Cholesterol fatty acid esters	7.5
c) Duodenal secretion			
Monoacylglycerol lipase	?	Ester bonds of monoacylglycerols	7.5
Carboxylic acid esterase	160000	Esters of aliphatic fatty acids	7.8
Alkaline phosphatase	140000	Phosphate ester bonds	9.0
(IV) Nucleases from pancreas			
Ribonuclease	13700	3′-Phosphate ester bonds	7.3
Deoxyribonuclease	31000	3′-Phosphate ester bonds	7.0

omasum and other compartments serve as bacterial fermentaion chambers. Rumen bacteria perform the anaerobic degradation of cellulose to absorbable end products; these are not glucose, but short chain fatty acids, like acetic, propionic, butyric and valeric acids. In adult mammals the stomach is not absolutely essential for life, but in young suckling animals, including man, HCl and rennin in the gastric juice are responsible the important process of milk coagulation. The most important stage of D. occurs in the small intestine, which contains all the Digestive enzymes (see) for the continuation and completion of D. Digestive enzymes are secreted by the walls of the small intestine (succus entericus) and by the pancreas (pancreatic juice). In herbivores, a small proportion of the digestive enzymes is derived from the food (dietary enzymes). Bile (secreted by the gall bladder in the liver) provides activator substances, in particular bile acid salts, which together with the $NaHCO_3$ of the pancreatic juice provide an optimal environment for the digestive processes. The absorbed endproducts of D. are L-amino acids, monosaccharides (glucose, fructose, galactose, mannose and pentoses), sodium salts of fatty acids, glycerol, monoglycerides (monoacylglycerols) and nucleosides. The remaining undigested and nonabsorbed material passes to the large intestine (colon) where it is concentrated by absorption of water. In the colon it is also subjected to various bacterial fermentation and putrifaction processes, which result in the production of lactic acid, acetic acid, various gases, poisonous amines and phenols. If passage through the intestinal tract takes too long, toxins may be absorbed in the latter half of the colon.

Degradation of the cell's own biopolymers is known as Intracellular D. (see), a process distinct from D. in the gastrointestinal tract.

Digestive enzymes: hydrolases present in the digestive tract of all animals, which catalyse hydrolysis of mechanically disrupted foodstuffs (proteins, carbohydrates, fats, nucleic acids) to their absorbable components. These low M_r components are absorbed as rapidly as they are formed, so that the equilibrium of digestive processes is continually displaced in favor of hydrolysis. With the exception of disaccharidases and certain peptidases, D.e. are Secretory enzymes (see), which are synthesized in high concentrations in accessory glands, like the pancreas or salivary glands, or in the gastric or intestinal mucosa. D.e. include some of the most thoroughly investigated enzymes. They are classified as 1. Proteases (see) and Peptidases (see); 2. Glycosidases (see) which cleave carbohydrates; 3. Esterases (see), especially lipases; and 4. Nucleases which cleave nucleic acids. See Table.

Digestive vacuole: see Intracellular digestion.

Digifolein: see Diginin.

Digifologenin: see Diginin.

Diginigenin: see Diginin.

Diginin: a digitanol composed of the pregnane derivative *diginigenin* (M_r 344.45, m.p. 115 °C, $[\alpha]_D$ −126 °) and the deoxysugar D-diginose. D. occurs together with cardiac glycosides in *Digitalis,* e.g. *Digitalis purpurea,* from which it was isolated in 1936 by Karrer.

Digifolein, which also occurs in *Digitalis* spp., has the aglycon *digifologenin* (M_r 360.45, m.p. 176 °C, $[\alpha]_D$ −269 °) which differs from diginigenin in having an extra 2β-hydroxyl group.

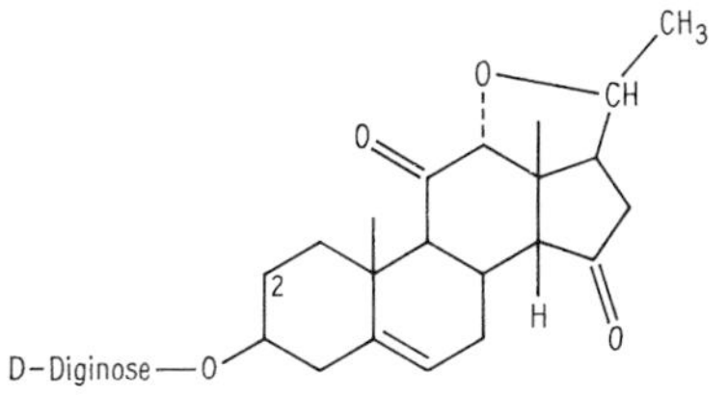

D-Diginose—O CH₃ CH O O 2 H O

Diginin

Digitalis glycosides: cardiac glycosides of the cardenolide group found in the leaves of foxgloves *(Digitalis purpurea* and *Digitalis lanata).* The three most important D.g. are *digitoxin* [M_r 764.92, m.p. 256–257 °C (anhydrous), $[\alpha]_D^{20}$ + 4.8 ° (c = 1.2 in dioxan)]; *digoxin* [M_r 780.92, m.p. 265 °C (d.), $[\alpha]_D$ + 11 °] and *gitoxin* [M_r 780.92, m.p. 285 °C (d.), $[\alpha]_D$ + 22 °]. The D.g. are secondary glycosides formed during preparation of the *Digitalis* leaves.

The native primary glycosides of the plants, the lanatosides, carry a D-glucose group and an acetate ester. The aglyca of digitoxin, digoxin and gitoxin are *digitoxigenin* [M_r 374.50, m.p. 253 °C, $[\alpha]_D^{17}$ + 19.1 ° (c = 1.36 in methanol)], *digoxigenin* [M_r 390.53, m.p. 222 °C (anhydrous), $[\alpha]_{546}^{20}$ + 27.0 ° (c = 1.77 in methanol)] and *gitoxigenin* [M_r 390.50, m.p. 234 °C, $[\alpha]_{545}^{20}$ + 38.5 ° (c = 0.68 in methanol)], respectively. The latter two differ from digitoxigenin in having an extra hydroxyl group at C12 or C16. The sugar component is always three molecules of D-digitoxose. D.g. are obtained by gentle extraction of fresh plant material with ethyl acetate or chloroform. Tannic acids are precipitated from the alcoholic solution with lead salts. The D.g. are released enzymatically from the lanatosides and separated chromatographically.

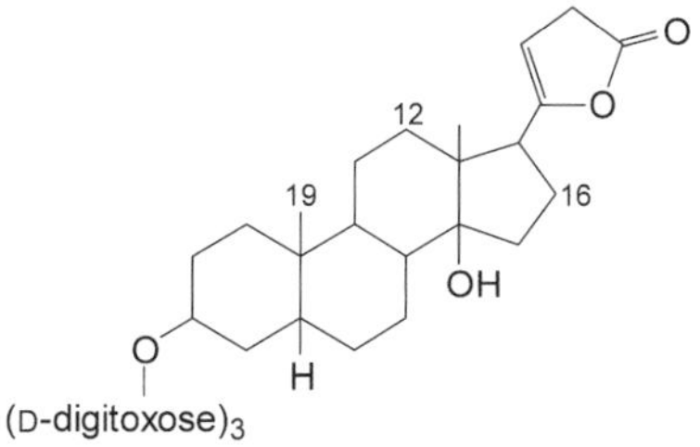

O O 12 19 16 OH O H (D-digitoxose)₃

Digitoxin

Some color reactions of D.g. are: red with sodium nitroprusside in sodium hydroxide solution, orange with alkaline picric acid solution, and blue-violet with alkaline *m*-dinitrobenzene solution. D.g. are indispensible cardiotonic agents, used for long-term treatment of chronic heart weakness and defective heart valves. The pure glycosides are now used instead of leaf powders or extracts.

Digitanols, *digitanol glycosides:* a group of plant glycosides with pregnane type Steroids (see) as aglyca, e.g. Diginin (see) and digifolein. D. occur together with cardiac glycosides, but have no cardiotonic activity themselves. They are biosynthesized from pregnenolone.

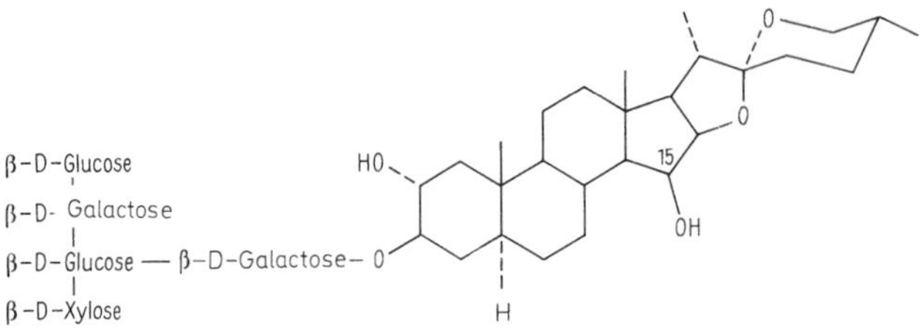

β-D-Glucose
β-D-Galactose
β-D-Glucose — β-D-Galactose–O
β-D-Xylose

Digitonin

Digitogenin: see Digitonin.

Digitonin: a mixture of four different steroid Saponins (see) from the seeds of the purple foxglove, *Digitalis purpurea.* The main component, also called D. [M_r 1229.36, m.p. 235°C, $[\alpha]_D^{20}$ −54° (c = 0.45 g in 15.8 ml methanol)], makes up 70–80 % of the mixture. The aglycon is digitogenin [(25R)-5α-spirostan-2α,3β,15β-triol, M_r 448.62, m.p. 296°C (d.), $[\alpha]_D^{10}$ −81° (c = 1.4 in $CHCl_3$)]. D. is a strong hemolytic poison, due to its affinity to blood cholesterol. It is used as a precipitating agent for cholesterol and other sterols.

Digitoxigenin: see Digitalis glycosides.

Digitoxin: see Digitalis glycosides.

Diglyceride: see Acylglycerols.

Digoxigenin: see Digitalis glycosides.

Digoxin: see Digitalis glycosides.

Dihydrofolic acid: see Tetrahydrofolic acid.

Dihydroorotate: an intermediate in Pyrimidine biosynthesis (see).

Dihydrouracil: an intermediate in Pyrimidine degradation (see). M_r 114.10, m.p. 274°C. 5,6-Dihydrouracil occurs as a rare base in some nucleic acids.

Dihydroxyacetone phosphate: see Triose phosphates.

20,22-Dihydroxycholesterol: see Cholesterol.

20,26-Dihydroxyecdysone: a steroid which acts as a molting hormone. It has been isolated together with Ecdysone (see) and ecdysterone from pupae of the tobacco hornworm, *Manduca sexta.*

Dihydrozeatin, *6-(4-hydroxy-3-methyl-butylamino)purine:* a cytokinin from corn *(Zea mays).* It is a derivative of zeatin and also occurs as a riboside and a ribotide.

Dimethazide: see Succinic acid mono-*N*-dimethylhydrazide.

N^6(-γ,γ-dimethylallyl)Adenosine, *N^6-isopentenyladenosine:* one of the Rare bases (see) in nucleic acids found in certain transfer RNAs, e.g. in serine tRNA. It also acts as a Cytokinin (see) and is found in free form in the culture media of *Coynebacterium* and *Agrobacterium.*

Bryokinin, a cytokinin found in the callus cells of moss sporophytes, is identical with the free base N^6-γ,γ-dimethylallyladenine.

Dimethylallylpyrophosphate: an intermediate in Terpene biosynthesis (see).

3,7-Dimethyloctane type: see Monoterpenes, Fig.

Dinucleotide fold: a characteristic folded protein structure constituting part or all of the structure of four NAD-dependent dehydrogenases, and certain other enzymes, some of which do not bind nucleotides. The D.f. was first identified in the tertiary structures of liver alcohol dehydrogenase (EC 1.1.1.1), glyceraldehyde 3-phosphate dehydrogenase (EC 1.2.1.12), lactate dehydrogenase (EC 1.1.1.28) and malate dehydrogenase (EC 1.1.1.37). All four of these dehydrogenases contain between 327 and 374 amino acid residues, which are folded into two distinct domains. One domain binds the NAD cofactor, while the other domain carries the binding and catalytic sites for the substrate. In each case, the NAD-binding domain has a fold consisting of a core of β-pleated sheet structure containing six parallel strands (strand order CBADEF), with the α-helical intrastrand loops above or below the sheet. A similar structure exists in phosphoglycerate kinase (EC 2.7.2.3), where the D.f. is responsible for binding ATP. Other enzymes with tertiary structures resembling the D.f. are phosphoglycerate mutase (EC 2.7.5.3), adenylate kinase (EC 2.7.4.3), phosphorylase a (EC 2.4.1.1) and pyruvate kinase (EC 2.7.1.40). Not all of these enzymes are known to bind nucleotides, and it is possible that the D.f. was present in an ancestral protein and was later exploited for nucleotide binding. On the other hand, the D.f. may be an especially stable structure, which has arisen in more than one enzyme family. It is noteworthy that enzymes so far shown to possess a D.f. bind either dinucleotides or 2-oxotrioses. There may in fact be an "oxotriose fold" sharing a common ancestry with the D.f. [C.C.F. Blake *Nature* **267** (1972) 482–483].

Dioscin: a steroid saponin (see Saponins), M_r 869.08, m.p. 275–277°C (d.), $[\alpha]_D^{13}$ −115° (c = 0.373 in ethanol). The aglycon of D. is diosgenin, (25*R*)-spirost-5-en-3β-ol, M_r 414.61, m.p. 204–207°C. D. is found in yams *(Dioscorea)* and trilliums. Diosgenin is an important starting material for the partial chemical synthesis of steroid hormones.

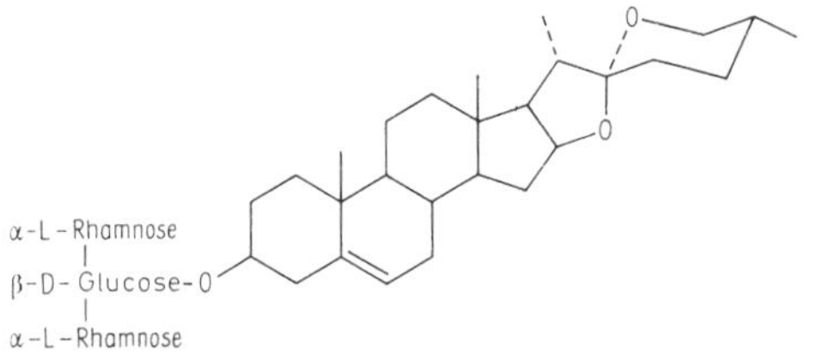

α-L-Rhamnose
β-D-Glucose-O
α-L-Rhamnose

Dioscin

Diosgenin: see Dioscin.

Dioxygenases: see Oxygen metabolism, Oxygenases.

Dipentene: see *p*-Menthadienes.

Diphosphatidylglycerols: see Membrane lipids.

2,3-Diphosphoglycerate, ***glycerate 2,3-bisphosphate, 2,3-bisphosphoglycerate:*** see Glycolysis, Hemoglobin, Rapoport-Luerbing shuttle.

Diphosphopyridine nucleotide: see Nicotinamide adenine dinucleotide.

Diptheria toxin: see Toxic proteins.

Disaccharidases: enzymes that hydrolyse disaccharides. They are most abundant in ripe fruits, microorganisms (yeasts) and the intestinal mucosa. Well studied examples are: 1) *β-D-fructofuranosidase* (invertase or saccharase) (EC 3.2.1.26) from yeast; 2) *α-1,4-glucosidases* (maltase) (EC 3.2.1.20) which hydrolyse α-D-glucosides like maltose, sucrose and turanose; 3) *β-galactosidase* (lactase) (EC 3.2.1.23) which hydrolyses lactose; 4) *β-1,4-glucosidases (β-D-glucoside glucohydrolases, gentiobiases, cellobiases)* (EC 3.2.1.21) which remove terminal nonreducing β-D-glucose residues from β-D-glucosides. β-D-Glucosidases display wide specificity, some of them attacking one or more other substrates such as β-D-galactosides, α-L-arabinosides, β-D-xylosides and/or β-D-fucosides.

Disaccharides: see Carbohydrates.

Disc electrophoresis: see Proteins.

Discontinuous process: see Fermentation techniques.

Dissimilation: see Catabolism.

Dissimilatory sulfate reduction: see Sulfate respiration.

Disulfide bridges, ***cystine bridges:*** a term referring to disulfide bonds, -S-S-, in proteins and peptides, formed by oxidation of two sulfhydryl groups: 2-SH →-S-S-. D. b. make a major contribution to the formation and maintenance of secondary structure in proteins. Proteins containing large numbers of D. b. are very resistant to denaturation by heat, acid or alkali, detergents, etc. and to hydrolysis by proteolytic enzymes. D. b. can be cleaved either reductively, e. g. with 2-mercaptoethanol, or oxidatively, e. g. with performic acid. In the living cell, formation and cleavage of D. b. is catalysed by enzymes, the best known being protein-disulfide reductase (glutathione) (EC 1.8.4.2).

Diterpene alkaloids: see Terpene alkaloids.

Diterpenes: terpenes derived from four isoprene units ($C_{20}H_{32}$). Phytol, an aliphatic D., is important as the ester component of chlorophyll and as a part of vitamins K and E. Other examples of naturally occurring D. are: Stevioside (see), vitamin A, retinol (chromophore of visual purple), certain alkaloids (cassaine, aconitine), and hormones such as gibberellins, trisporic acids and antheridiogens. Many cyclic D. are acids (e. g. the resinic acid, abietic acid; Fig.).

Biosynthesis. The starting compound is geranyl pyrophosphate (see Terpenes). Acyclic D. are formed by hydrolysis of the pyrophosphate residue (e. g. phytol). Geranylgeranylpyrophosphate is probably converted easily to geranyl linalool, which is then converted to bi- and tricyclic compounds (Fig.). In a few of the cyclic D., e. g. abietic acid, there is a migration of the substituents. The gibberellins are derived from the labdadiene type of D.

Dityrosine: a dimer of L-tyrosine found in acid hydrolysates of several biological materials: tussa silk, fibroin, insect cuticle resilin, spore coat of *Bacillus subtilis,* and the fertilization membrane of sea urchin egg. In *Saccharomyces cerevisiae,* D. is sporuation-specifc, being found only in spores and not in vegetative cells or non sporulating cells under sporulating conditions. It is now thought that dityrosyl residues exist in vivo, acting as cross-links in structural proteins. D. can be synthesized in vitro by the action of

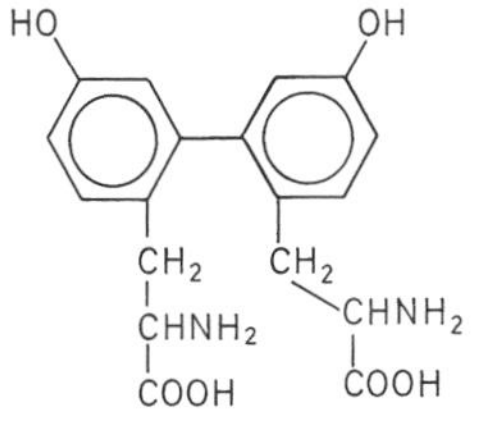

HO OH
CH_2 CH_2
$CHNH_2$ $CHNH_2$
COOH COOH

Dityrosine

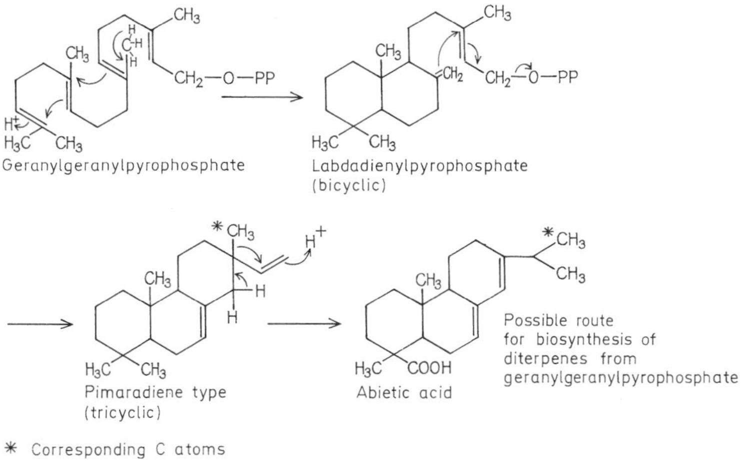

CH_3 CH_3 CH_2—O—PP
H_3C CH_3
Geranylgeranylpyrophosphate

CH_3 CH_3 CH_2 CH_2—O—PP
H_3C CH_3
Labdadienylpyrophosphate (bicyclic)

*CH_3 H^+ CH_3 H H
H_3C CH_3
Pimaradiene type (tricyclic)

*CH_3 CH_3 CH_3
H_3C COOH
Abietic acid

Possible route for biosynthesis of diterpenes from geranylgeranylpyrophosphate

* Corresponding C atoms

Possible route for biosynthesis of diterpenes from geranylgeranylpyrophosphate

horse radish peroxidase on L-tyrosine. NMR analysis shows that the earlier assigned structure (phenolic OH-groups *ortho* to the inter-ring bond) is wrong, and that the phenolic OH-groups are actually meta to the inter-ring bond (Fig.). [P. Briza *J. Biol. Chem.* **261** (1986) 4288–4294]

Diurnal acid rhythm: see Crassulacean acid metabolism.

Diuron: see Dichlorophenyl-dimethylurea.

Divided genome viruses: see Multipartite viruses.

DNA: acronym of Deoxyribonucleic acid (see).

DNA-binding proteins: structural and/or regulative proteins bound to DNA. Proteins which repress or induce transcription of particular genes (see Operon) must be able to recognize specific nucleotide sequences, while histones (see Chromosomes) may recognize more general features, such as areas of higher or lower GC content. Variations in the propeller twist and base roll (see Deoxyribonucleic acid) of the base pairs are highly correlated with the base sequences of synthetic DNA oligomers. It is likely that these variations affect the binding of hydrogen-bonding proteins, and thus are the basis of sequence recognition by the proteins. The Z structure should also be easily recognizable by proteins. It is usually a higher-energy configuration than A- or B-DNA but it is stabilized by alternating purine and pyrimidine sequences, or by proteins. In the ideal B-form of DNA, the minor groove (~5 Å-wide, ~8 Å-deep) is too narrow to accommodate protein structural elements (e.g. an α-helix), whereas the major groove (~12 Å-wide, ~8 Å-deep) is able to do so. It is also significant that alternating purine-pyrimidine sequences are clustered in the control regions of many genes.

According to X-ray and NMR analysis, a number of proteins which regulate gene expression possess a supersecondary structure, known as a *helix-turn-helix (HTH) motif.* The HTH motif consists of two symmetrically arranged α-helical structures, each containing about 20 residues with similar primary structures. The helices cross at about 120°, and are spaced in such a way that they can bind two sequential turns of the major groove of the DNA helix. Examples of regulatory proteins that possess a HTH motif are: (i) Cro, which is a repressor of the repressor maintenance promoter P-RM in the bacteriophage lambda; (ii) the lambda repressor, which can act as a repressor but can also stimulate expression of its own gene; (iii) CAP, the "catabolite gene activator protein", which promotes transcription of several genes in the presence of cyclic AMP, but in other circumstances can also act as a repressor; (iv) *lac* repressor; (v) *trp* repressor. X-ray analysis of complexes of repressor proteins with their target DNA show that the protein conforms closely to the DNA surface, interacting with bases and sugar-phosphate chains via hydrogen bonding, salt bridges and van der Waals contacts. [Y. Takeda et al. *Science* **221** (1983) 1020–1026; F. A. Jurnak & A. McPherson, eds. *Biological Macromolecules and Assemblies,* vol. 2, Wiley, New York, 1985; S. C. Harrison 'A Structural Taxonomy of DNA-binding domains' *Nature* **353** (1991) 715–719; P. S. Freemont et al. 'Structural Aspects of Protein-DNA Recognition' *Biochem. J.* **278** (1991) 1–23; R. E. Harrington & I. Winicov 'New Concepts in Protein-DNA Recognition: Sequence-directed DNA binding and Flexibility' *Prog. Nucl. Acid Res. Mol. Biol.* **47** (1994) 195–270]

DNA fingerprinting, ***DNA typing, DNA profiling:*** preparation of a pattern of DNA fragments that is characteristic of the genome and therefore of the individual from which the DNA is derived. The technique is based on the presence in the human genome, and probably that of all eukaryotic species, of short nucleotide sequences (11–60 bp, referred to subsequently as monomers) that are repeated tandemly many times at a considerable number of places (genetic loci) in many of the chromosomes constituting the particular genome. These stretches of tandemly repeated monomers have been called Minisatellite DNA (see), variable number tandem repeats (VNTRs), hypervariable loci and highly variable repeats (HVRs). The term 'variable' in these names refers to the fact that the number of times a monomer is tandemly repeated at a given genetic locus is frequently different in different individuals of that species; indeed it may be different in a pair of homologous chromosomes in the same individual. The monomers themselves appear to fall into different classes on the basis of their length and nucleotide sequence; however at a given locus the monomers typically have the same nucleotide sequence, though variations of this pattern are known. In humans the different classes of monomer are characterized by being G-rich and several have sequences within them that are sufficiently similar to allow the design of Probes (see) that will bind to all of them under low Stringency (see) conditions. The first minisatellite in the human genome was discovered by A. Wyman & R. White [*Proc. Natl. Acad. Sci. USA* **77** (1980) 6754–6758] when 15 alleles of different length, and therefore different numbers of monomer repeats, were found at the locus designated D14S1.

Although the functional significance of these hypervariable loci is not yet known, they have proved to be of considerable value in identifying a particular individual, in determining family relationships and in ecological and evolutionary studies. This stems from the coincidence of: (i) the variation in monomer number at a given hypervariable locus in the genome of a given species, (ii) the considerable number of hypervariable loci in the genome of a given species, and (iii) the inheritance of hypervariable loci in the same manner as Mendelian genes.

The way in which individuals differ at hypervariable loci is elucidated by restriction analysis of chromosomal DNA. This is made possible by the fact that restriction sites for many restriction endonucleases (see) do not occur in minisatellite DNA. This means that when the chromosomal DNA is digested with such an endonuclease all the species of minisatellite DNA remain intact, cleavages having occurred on either side of them. Thus individuals with different numbers of tandemly repeated monomers at specific hypervariable loci will produce restriction fragments of different length from those loci; these can be separated by gel electrophoresis and detected with an appropriate probe after Southern blotting (see).

The methology is as follows. The source of DNA, which should be sufficient to produce 60 ng or more, is incubated overnight with a mixture of sodium dodecyl sulfate (which lyses cell nuclei), proteinase K (which assists the subsequent recovery of DNA by digesting the protein that is present) and dithiothreitol (which assists the action of proteinase K by reducing disulfide bonds in proteins). For forensic or diagnostic purposes the DNA source may be whole blood or blood stains (i.e. the white cells therefrom), whole semen or semen stains (i.e. sperm cells), hair root cells, epithelial cells from the oral cavity or indeed cells from any body tissue that is available. The DNA is extracted then incubated with a restriction endonuclease which cleaves it into fragments. Of the many endonucleases that would be satisfactory for this purpose those that are usually chosen recognize a 4 bp sequence (e.g. *Hae*III which recognizes $^{5'}GG^{\downarrow}CC^{3'}$). The reason for this is that if the sequence of bases in the DNA were random a particular 4-base sequence (e.g. GGCC) would on average occur every $4^4 = 256$ bases which is considered to be a satisfactory frequency for cleavage. The DNA fragments, which include those with hypervariable loci, are then separated according to size by agarose gel electrophoresis. The fragments in each separated zone are then denatured *in situ* by soaking the gel in alkali; this breaks the H-bonds holding the two DNA strands of each fragment molecule together. The denatured DNA zones are transferred to a sheet of nitrocellulose paper by Southern blotting (see); this produces the same pattern of zones as was present on the gel. The nitrocellulose paper is then immersed in a solution containing a labeled probe for the hypervariable locus or loci being sought. The probe molecules only bind to those zones that are composed of denatured DNA fragments containing one or more monomers of the target hypervariable locus or loci; since the binding of a probe molecule involves the formation of H-bonds to a nucleotide sequence complementary to it in the DNA fragment, it is clear why prior denaturation of the latter is necessary. These zones are detected by monitoring the paper for the particular label present in the probe (e.g. autoradiography for a radioactively-labeled probe). When DNA fingerprinting was first used in the mid-1980s the labeled probes that were used detected several different types of hypervariable loci and were thus called multi-locus probes; the use of such a probe detects many zones on the Southern blot. However single-locus probes which detect one specific type of hypervariable locus are now also used. They detect far fewer zones on the Southern blot, a result which, taken in isolation, is less diagnostic than that obtained with a multi-locus probe. However it is usual to compensate for this by examining the restriction fragment sample separately with several different single-locus probes.

The diagnostic capabilities of DNA fingerprinting, first demonstrated in 1985 [A.J. Jeffreys et al. *Nature* (1985) **314**, 67–73], are demonstrated in Fig. 1 which depicts a specific hypervariable locus, bounded by a pair of restriction sites (indicated by the shaded boxes), in a homologous pair of chromosomes in four human individuals: A and B are the father and mother, respectively, of C and D, who are not identical twins. A and B are both heterozygous at the locus, A having four tandem repeats of the monomer (indicated by the arrow) in one homolog (labeled a1) and seven in the other (labeled a2) while B has two (b1) and five (b2). The children, C (with 4 & 2 repeats) and D (with 7 & 5 repeats) represent two (a1 & b1; a2 & b2) of the four possible genetic combinations at this locus in the offspring of A and B, the other two being a1 & b2 and a2 & b1). The lower part of Fig. 1 shows the zones on the Southern blot that are derived from this locus in the genome of each of these four individuals, clearly demonstrating that the technique has distinguished between them. However an examination of one hypervariable locus is insufficient to guarantee that C and D can be distinguished from each other; there is in fact a one in sixteen chance that C and D will have the same genetic make-up at the hypervariable locus in question. This chance, however, becomes exponentially smaller as the number of hypervariable loci under consideration is increased linearly; since probes, whether they be multi-site or a battery of single-site, often detect twenty or more hypervariable loci, the power of the technique as a diagnostic tool is apparent. The result is also consistent with the earlier statement that A and B are the parents of the two children because C and D each have one or other of the two zones of both A and B, but it is insufficient to prove parentage. However, when this criterion proves to be true at twenty or more different hypervariable loci the possibility that A and B are not the parents of C and D becomes vanishingly small. Track E on Fig. 1 shows the zones on the Southern blot derived from the hypervariable locus in the genome of a fifth individual, E, who is the child of B born in a previous marriage. The fact that neither of the fragments characteristic of A at this locus, namely a1 with 4 tandem repeats and a2 with 7, is present in track E rules out A as the father of E; this demonstrates the value of DNA fingerprinting in paternity suits.

The applicability of DNA fingerprinting to the forensic field was established by Gill [P. Gill et al. *Nature* **318** (1985) 577–579; P. Gill et al. *Electrophoresis* **8** (1987) 38–44]. The technique was used in a criminal case for the first time in 1986 [P. Gill & D.J. Werrett *For. Sci. Int.* **35** (1987) 145–148]; this resulted not only in the eventual conviction of a man for the rape and murder of two girls, one in 1983 the other in 1986, near Leicester (UK), but also in the earlier elimination from both investigations of a man who had confessed to the 1986 murder and who, on the basis of this, was believed by the police to have committed the 1983 murder. Fig. 2 shows the Southern blot after labeling with a hypervariable locus probe that could have been obtained in a hypothetical rape and murder investigation. The tracks show the fragments containing tandem repeats detected by the probe in the four samples of evidence. Track A is derived from DNA from the blood of the victim. Track B is derived from DNA from spermatozoa present in a vaginal swab obtained from the victim (spermatozoa are surrounded by sulfur-rich protein coat which renders them resistant to detergents that solubilise vaginal tissue cells, so enabling them and their DNA to be obtained free from the vaginal cells that are also present on the swab; DNA from vaginal cells would otherwise complicate the DNA fingerprint). Tracks C and D are derived from DNA from the blood of two men who, on the basis of other evidence, are police suspects. Comparison of the fragments patterns in tracks A and B shows that the rapist's DNA is not contaminated with that of his vic-

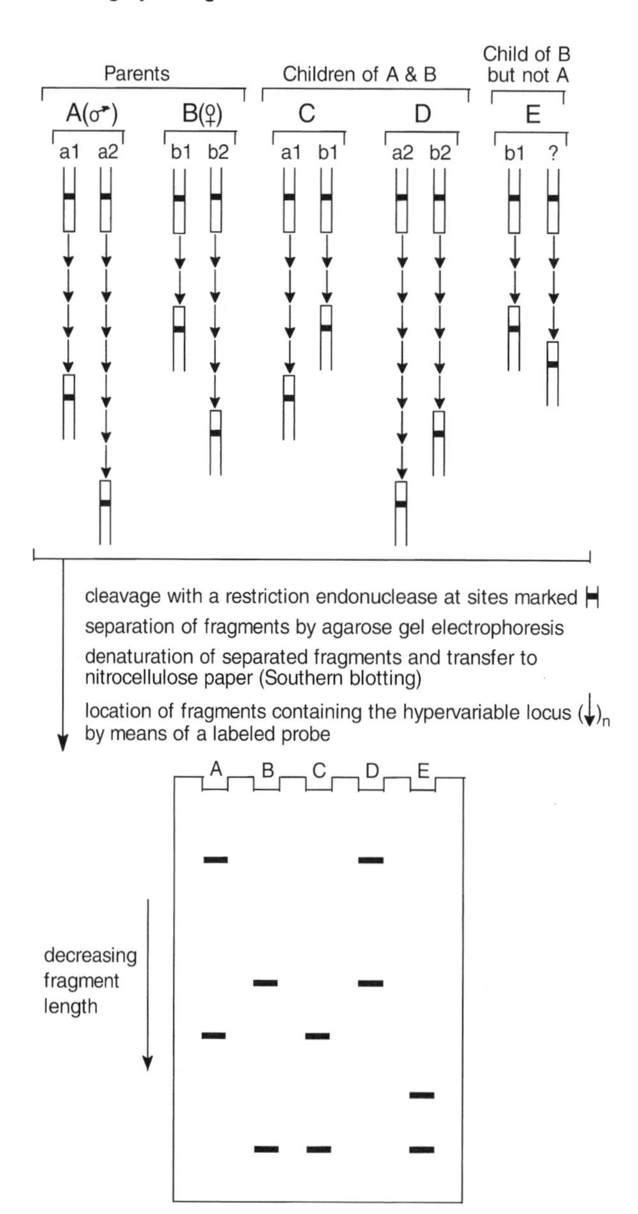

Fig. 1. *DNA fingerprinting.* Use of DNA fingerprinting in establishing family interrelationships. This example shows only the DNA fingerprint from one hypervariable locus.

tim. Similarly, comparison of tracks B and C reveals an exact coincidence of fragments while that of tracks B and D reveals a marked difference. This proves beyond doubt that suspect D was not the rapist and demonstrates a statistically high likelihood that suspect C is the rapist, since (i) suspect C and the rapist have 10 fragments in common, and (ii) Jeffreys [*Biochem. Soc. Trans.* **15** (1987) 309–317] quotes a fragment-sharing statistic of 0.25, the probability that two unrelated individuals (e.g. the rapist and an innocent individual) share the same fragment pattern is $0.25^{10} = 9.53 \times 10^{-7}$ or about 1 in a million. Although this is not absolute proof it would probably be sufficient to gain a conviction when added to all the other indicative evidence. Clearly the weight of the DNA fingerprint evidence increases considerably as the number of shared fragments increases.

The problem with some forensic samples is that too little DNA (i.e. < 60 ng) is available and that it is has become degraded into short pieces by age and/or the

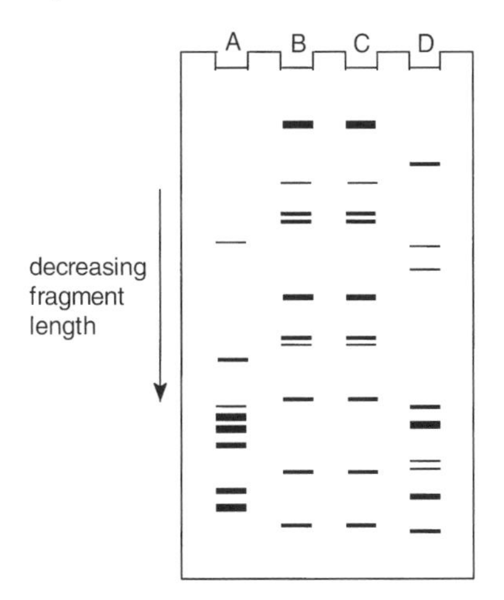

Fig. 2. *DNA fingerprinting.* The forensic use of DNA fingerprinting. A, B, C & D are DNA fingerprints from the victim, the evidence and two suspects, respectively, in a hypothetical murder/rape case.

conditions to which it has been exposed. This can be largely overcome by using the Polymerase chain reaction (PCR) (see) to amplify preselected, known hypervariable loci, which are then subjected to DNA fingerprint analysis and compared with equivalent loci in the suspect. However such is the power of PCR that great care has to be taken to avoid contamination of the DNA from scene of the crime with DNA from police, laboratory workers or anyone else. Strict guidelines have been instituted to avoid this possibility; for instance evidential DNA is analysed in a different laboratory from that of any suspect.

More recently a different DNA fingerprinting procedure has been described [A. J. Jeffreys et al. *Nature* **354** (1991) 204–209] based on sequence variation in hypervariable loci rather than the number of tandem repeats that are present. It has been applied successfully to hypervariable locus D1S8 and overcomes many of the limitations of conventional DNA fingerprinting.

DNA fingerprinting has become increasingly useful in the field of population genetics in non-human species, where its two most frequent applications are: (i) resolving questions of individual male reproductive success in particular locales amongst wild, mammalian, bird and reptile species, in which the female may mate with several different males during her fertile period or is able to store viable spermatozoa for a considerable length of time, e. g. in determining whether there is a significant relationship between the number of offspring fathered by a territory holder and the size of his territory [H. L. Gibbs *Science* **250** (1990) 1394–1397], and (ii) determining genetic relatedness, particularly in conservation genetics [B. Amos & A. R. Hoelzel *Biol. Conserv.* **61** (1992) 133–144] by measuring the 'bandsharing coefficient' [M. Lynch. *Med. Biol. Evol.* **7** (1990) 478–484]. The latter is an index of the similarity of two DNA fingerprint patterns and is calculated from the expression: $S_{xy} = 2n_{xy}/(n_x + n_y)$, where n_x and n_y are the number of bands in the DNA fingerprints of individuals x and y, and n_{xy} is the number of bands that the two patterns have in common; the higher its value the greater is the similarity of the two patterns (and therefore the genetic relatedness of the two individuals); a value of 1 indicates absolute similarity.

DNA gyrase: a type II topoisomerase. See Topoisomerases.

DNA-ligase: see Polynucleotide ligase.

DNA methylation: methylation of the adenine and cytosine residues of DNA, with the formation of N^6-methyladenine (m^6A), 5-methylcytosine (m^5C) and N^4-methylcytosine (m^4C), resulting in varying degrees and patterns of methylation that are specific to the species of origin. The methyl groups project into the major groove of B-DNA. In most eukaryotic DNA, m^5C is practically the only methylated base, and it occurs mostly in the CG dinucleotides of palindromic sequences. More than 30 % of C residues may be methylated in certain plants, and about 70 % of C residues in mammalian DNA. In all organisms, the methyl groups of m^6A, m^5C and m^4C are derived by transfer from *S*-adenosyl-L-methionine (see).

Restriction modification. In bacteria, species-specific patterns of DNA methylation serve to protect the cell's DNA from its own Restriction endonucleases (see). In *E. coli,* DNA methylation is largely the function of: 1. dam methyltransferase [product of *dam* (DNA-adenine-methylation) gene; methylates A in the palindromic sequence GATC], and 2. dcm methyltransferase [product of *dcm* (DNA-cytosine-methylation) gene; methylates both Cs in the palindromic sequence CC(A or T)GG]. The DNA methyltransferase (M.HhaI) from *Haemophilus haemolyticus* methylates one of the C residues in the sequence 5′-GCGC-3′ of duplex DNA, forming 5′-G-m^5C-GC-3′. The X-ray structure of the complex of M.HhaI with the self-complementary sequence d(TGATA**GCGC**TATC) shows that the DNA is bound in a cleft between two enzyme domains; the target cytosine flips out of the minor DNA groove and inserts into the active site of the enzyme, where it is methylated by *S*-adenosyl-L-methionine; as the target C flips out of position, it leaves an exposed G on the complementary strand, and the resulting gap is temporarily filled by the side chain of Gln 237 of the enzyme.

Mismatch repair. In prokaryotic systems, DNA methylation plays a role in the correction of mismatched base pairs, a process known as *mismatch repair.* In *E. coli,* a *mismatch correction enzyme* is encoded by genes *mutH, mutL* and *mutS.* The enzyme scans newly replicated DNA for mismatched base pairs, removes the single-stranded segment containing the wrong nucleotide, then DNA polymerase inserts a segment containing the correct base. But there must also be a mechanism for identifying the wrong base of a mismatched pair. This mechanism relies on the fact that methylation lags behind replication, so that for a brief period (seconds to minutes) a newly replicated daughter strand is temporarily less methylated than its parental strand. The correction enzyme monitors the degree of methylation of each strand by recognizing the unmethylated sequence GATC, but not its subsequent methylation product, G-m^6A-TC.

Eukaryotic gene regulation. Methylation of C residues in specific CG dinucleotides of palindromic sequences is important in eukaryotic gene regulation. In the eukaryotic genome, the frequency of occurrence of CG is about one fifth of that predicted for its random occurrence. In contrast, the upstream control regions of many genes contain 'islands' of 'normal' CG frequency. As shown by transfection, microinjection and cell-free transcription, specific methylation of C in the CpG dinucleotide residues in these upstream control regions (both the pattern and density of methylation are important) causes inhibition or inactivation of promoters, resulting in repression of transcription.

Repeat-induced gene silencing. The occurrence of this control phenomenon was first indicated by the observation that a marker construct in *Nicotiana tabacum* is reversibly methylated and inactivated following the introduction of a second recombinant gene that shares common homologies with the inactivated marker gene (Matzke et al., 1989). The phenomenon has since been observed in several other plants and in filamentous fungi.

Trans-inactivation of the marker gene is dependent on the presence of its homologous counterpart which is incorporated at a different position in the genome. When the two transgenes are separated by segregation, the *trans*-activated copy is reactivated in subsequent generations. This implies that the silenced transgene is epigenetically modified, and that this process is initiated by the homologous silencer locus. The susceptibility of transgenes to homology-induced silencing depends on the arrangement, modification, secondary structure and genomic location of transgene sequences. Clearly, this silencing mechanism has important implications for the control of gene expression in transgenic plants; in fact, an inverse correlation between transgene copy number and gene activity was reported as early as 1987.

A common feature of transcriptional silencing in plants and premeiotic gene inactivation in filamentous fungi is methylation of the cytosine residues in the repeated DNA sequences. In addition to the possible induction of DNA methylation by conformational changes, it is also likely that foreign sequences are specifically targeted for methylation. Methylation of repeated DNA sequences and of foreign DNA might be expected to confer an evolutionary advantage on the genome, as well as increasing the tolerance of the genome to incorporated foreign sequences. Thus, hypermethylation prevents somatic recombination between homologous sequences, generates silent epigenetic states (which may be reactivated under favorable environmental conditions), and it facilitates evolutionary sequence divergence by deamination of m^5C, resulting in the mutation $C \rightarrow T$. [R.L.P. Adams *Biochem. J.* **265** (1990) 309–320; A. Razin & H. Cedar *Microbiol. Reviews* **55** (1991) 451–458; A.P. Bird *Cold Spring Harbor Symp. Quant. Biol.* **58** (1993) 281–285; J.P. Jost & H.P. Saluz (eds.) *DNA Methylation: Molecular Biology and Biological Significance,* Birkhäuser (Basel, Boston, Berlin), 1993; E. Li et al. *Nature* **366** (1993) 362–365; G.L. Verdine *Cell* **76** (1994) 197–200; A. Razin & T. Kafoi 'DNA methylation from embryo to adult' *Prog. Nucl. Acid Res. Mol. Biol.* **48** (1994) 53–81; S.S. Smith 'Biological Implications of the Mechanism of Action of Human DNA (Cytosine-5)methyltransferase' *Prog. Nucl. Acid Res. Mol. Biol.* **49** (1994) 65–111; A.J.M. Matzke et al. 'Homology-dependent gene silencing in transgenic plants – epistatic silencing loci contain multiple copies of methylated transgenes' *Molecular and General Genetics* **244** (1994) 219–229; X. Cheng 'Structure and Function of DNA Methyltransferases' *Annu. Rev. Biophys. Biomol. Struct.* **24** (1995) 293–318; F. Radtke et al. *Biol. Chem. Hoppe Seyler* **377** (1996) 47–56; P. Meyer 'Repeat-Induced Gene Silencing: Common Mechanisms in Plants and Fungi' *Biol. Chem. Hoppe-Seyler* **377** (1996) 87–95; S. Prösch et al. *Biol. Chem. Hoppe-Seyler* **377** (1996) 195–201]

DNA nucleotidyltransferase: see DNA polymerase.

DNA polymerase, DNA nucleotidyltransferase (EC 2.7.7.7): an enzyme which catalyses the synthesis of DNA polynucleotide chains on a pre-existing DNA matrix (DNA replication). The precursors are the four 3′-deoxyribonucleotide triphosphates. In vitro, the enzyme can also synthesize homo- and copolymers from triphosphates. The D.p. from different organisms have different specificities, some for single and others for double strands of DNA as primers.

Three D.p. have been isolated from *E. coli.* Polymerase I (Kornberg enzyme) joins deoxyribonucleoside triphosphates to high-molecular-mass polynucleotides, simultaneously removing pyrophosphate. The chain grows from the 5′ phosphate to the 3′ end. In vitro the reaction requires an oligonucleotide primer; nucleotides are added to its 3′-hydroxyl end. A matrix DNA is also required to give the correct sequence. In 1967, Kornberg used this enzyme to totally synthesize the DNA of the single-stranded phage ΦX 174. However, it is probably not the enzyme responsible for replication of DNA in vivo; it is more likely to be a repair enzyme (see DNA repair). The function of D.p. II in the cell is still unclear. DNA replication is probably catalysed by D.p. III.

DNA-relaxing enzyme: a type I eukaryotic topoisomerase isolated from mammalian tissue culture cells [W. Keller, *Proc. Natl. Acad. Sci. USA.* **72** (1975) 2550–2554]. The term may also be loosely applied to all type I and type II Topoisomerases (see).

DNA repair: a variety of mechanisms for restoring the normal structure (i.e. the genetic integrity) of DNA after it has been damaged, e.g. by toxins, UV or ionizing radiation, and spontaneous bond cleavage (e.g. cleavage of glycosidic bonds; deamination of cytosine residues to uracil residues).

Replacement of pyrimidine dimers. UV irradiation of DNA causes dimerization of adjacent thymine residues on the same strand; intrastrand cytosine dimers and thymine-cytosine dimers are also formed, but at much lower rates. The resulting dimer cannot fit into a double helix, and therefore prevents normal transcription and replication.

This type of damage may be corrected by the action of *photoreactivating enzymes* or *DNA photolyases* (M_r 55,000–65,000), which possess a noncovalently bound chromophore of N^5,N^{10}-methenyltetrahydrofolate or of 5-deazaflavin, depending on the species of origin. The monomeric enzyme binds to the pyrimidine dimer. The chromophore absorbs light in the range 300–500 nm and transfers the excitation energy to noncovalently bound FADH⁻. The latter transfers an electron to the pyrimidine dimer and splits it into two monomers.

Reversal of alkylation by alkyltranserases. The bases of DNA can be alkylated by nonphysiological alkylating agents, such as N-methyl-N′-nitro-N-nitrosoguanidine (a powerful mutagen and carcinogen), or it may occur spontaneously in the untreated normal cell by the nonenzymatic methylating activity of *S*-adenosylmethionine. In *E. coli* and mammalian cells, O^6-methylguanidine and O^6-ethylguanidine are dealkylated by a protein known as *O^6-methylguanidine-DNA methyltransferase,* which transfers a single alkyl group to one of its own Cys residues, then becomes inactive. In *E. coli,* this O^6-methylguanidine-DNA methyltransferase activity is the property of the 178-residue *C*-terminal segment of the 354-residue Ada protein (the *ada* gene product)

Replacement of altered bases by nucleotide excision repair. As an alternative to the photoreactivation, a pyrimidine dimer may be excised and replaced by two monomers. In *E. coli,* this process is initiated by a complex of proteins known as the uvrABC excinuclease (encoded by the *uvrABC* genes) in response to the distortion generated by the dimer. The excinuclease cleaves the affected DNA strand 8 nucleotides away from the dimer on the 5′ side and 4 nucleotides away on the 3′ side. The excised oligonucleotide containing the offending dimer is lost. DNA polymerase I then synthesizes and replaces the missing segment, using the free 3′-end as the primer and the exposed stretch of intact complementary strand as the template. The 3′-end of the newly synthesized DNA is finally joined to the free 5′-end of the cleaved strand by DNA ligase.

The autosomal recessive hereditary human disease, *Xeroderma pigmentosum,* can arise from various defects in pyrimidine dimer repair. In one form of the disease, the excinuclease is deficient. The skin of homozygotes is sensitive to the ultraviolet component of sunlight, leading to atrophy of the dermis, keratosis, ulceration and skin cancer. In cultured normal human fibroblasts the half life of pyrimidine dimers is about 24 h, whereas practically no dimers are excised in 24 h in cultured fibroblasts from Xeroderma pigmentosa patients.

The cytosine and adenine residues of DNA are spontaneously deaminated respectively to uracil and hypoxanthine at a low but finite rates. Uracil pairs with adenine, so that replication of DNA containing a uracil residue would produce a daughter strand containing an AU base pair in place of the original GC. Similarly, hypoxanthine pairs with cytosine, and one of the daughter strands would contain CHyp in place of TA. These potential mutations are prevented by a repair process, initiated by hydrolytic removal of the offending base by a specific DNA glycosylase (e.g. *uracil-DNA glycosylase*). The resulting gap in the sequence of bases in an intact DNA molecule is known as an AP site (*ap*urine or *ap*yrimidine). An *AP endonuclease* cleaves the backbone of the DNA on one side of the AP site generated by removal of the base, leaving a deoxyribose phosphate residue. The latter is removed and replaced by a unit of deoxycytidine phosphate (determined by base-pairing with the guanine residue of the undamaged complementary strand) by DNA polymerase I, and the strand is finally sealed by DNA ligase.

Recombination repair. Replication may occur before the above repair mechanisms have time to operate. For example, replication of a DNA strand containing a pyrimidine dimer results in a daughter strand consisting of two sections. The two sections are bound by base complementarity to the faulty strand, and separated by a gap opposite the pyrimidine dimer. Excision repair is not possible, because this requires a perfect complementary strand. However, the sister duplex is perfect, and one of its strands contains the correct sequence of nucleotides required to fill in the gap. The gap is therefore filled by transfer of the appropriate sequence from the perfect sister duplex, i.e. by homologous Recombination (see). This leaves a gapped sister duplex, but since the nucleotide sequence opposite the gap is normal, it can be repaired by filling in and ligation. Following this recombination (or *postreplication repair*), one duplex still contains the pyrimidine dimer, but this can then be corrected by photoreactivation or nucleotide excision repair. Like homologous genetic recombination, recombination repair is also mediated by recA, and the two processes are mechanistically very similar.

SOS repair. See SOS response.

[L.C.Myers et al. *Biochemistry* **32** (1993) 14089–14094; B.Van Houtten & A.Snowden *BioEssays* **15** (1993) 51–59; K.Morikawa *Curr. Opin. Struct. Biol.* **3** (1993) 17–23; K.Tanaka & R.D.Wood *Trends Biochem. Sci.* **19** (1994) 83–86; M.H.Moore et al. *EMBO J.* **13** (1994) 1495–1501; DNA Repair: a special issue of *Trends in Biochemical Sciences* **20** (1995) issue 10; S.N.Guzder et al. 'Reconstitution of Yeast Nucleotide Excision repair' *J. Biol. Chem.* **270** (1995) 12973–12976; S.Griffin 'DNA damage, DNA repair and disease' *Current Biology* **6** (1996) 497–499]

DNA-RNA hybrids: double-stranded molecules, of which one strand is DNA, and the other the complementary RNA. They are formed during transcription (see Ribonucleic acids) and in the multiplication of oncogenic RNA viruses (see RNA-dependent DNA polymerase), but they can also be produced in vitro by Hybridization (see). They are resistant to ribonuclease.

DNA sequencing: see Nucleic acid sequencing.

DNA-swivelase: a type I Topoisomerase (see).

DNA synthesis: synthesis of oligodeoxyribonucleotides of specified base sequence, using chemical methods. In the original phosphodiester method, the 5′-phosphate of one nucleotide (with other functional groups protected) is condensed with the 3′-hydroxyl of another protected nucleotide (Fig. 1). Reaction times are long, and the yield decreases rapidly with the length of the synthesized chain. This method is historically important, since it was used to perform the first total synthesis of a gene, a biologically functional suppressor transfer RNA gene [H.G.Khorana *Science* **203** (1979) 614–625].

The phosphotriester method overcomes some of the disadvantages of the phosphodiester method by blocking each intermediate phosphodiester function during the synthesis of the required sequence. This method was used to synthesize 67 different oligonucleotides of chain length 10–20 which were then spliced to gen-

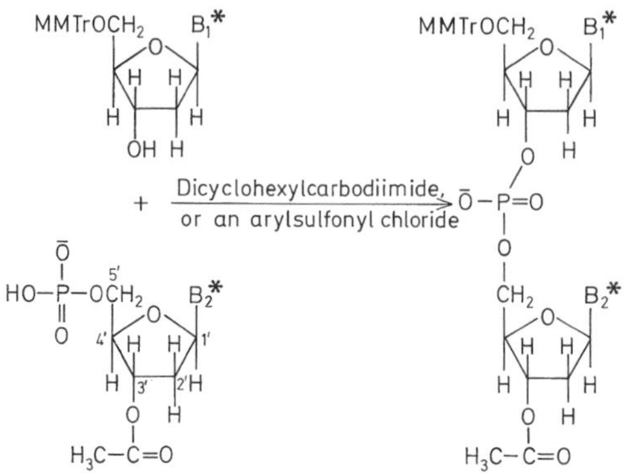

Fig. 1. *DNA synthesis.* Principle of the phosphodiester method of DNA synthesis. B* = protected base. For other abbreviations and formulas see Fig. 5.

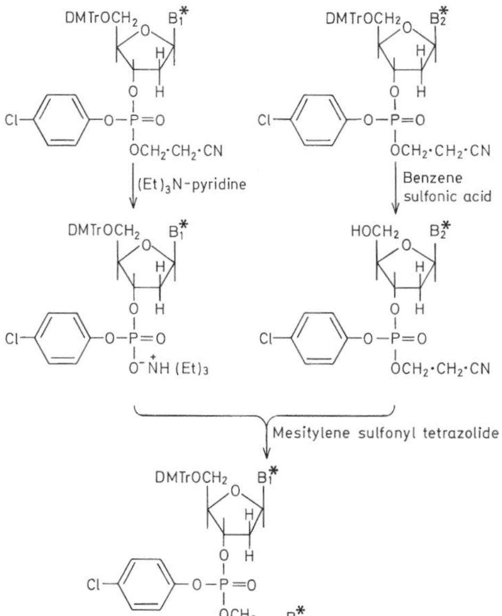

Fig. 2. *DNA synthesis.* The phosphotriester method of DNA synthesis. B* = protected base. For other abbreviations and formulas see Fig. 5.

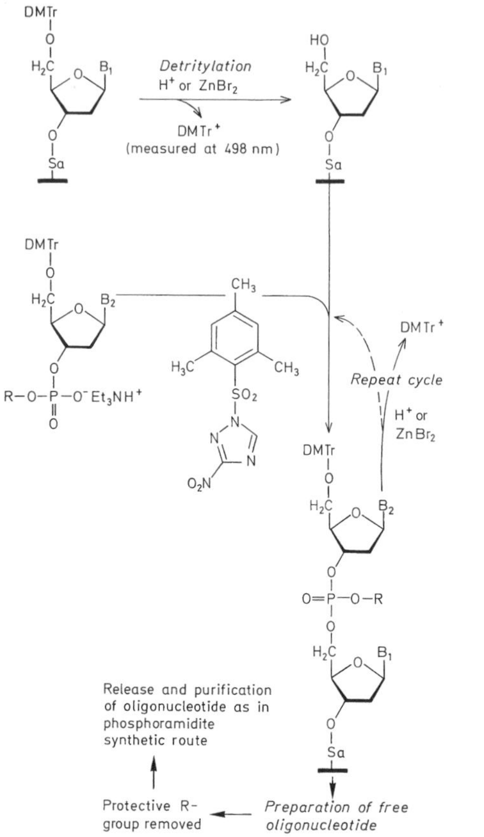

Fig. 3. *DNA synthesis.* Phosphotriester method of DNA synthesis, using a solid phase system. The brilliant orange color (λ_{max} 498 nm) of the DMTr cation can be used to monitor coupling efficiency between synthetic cycles. Sa = spacer arm. B_1, B_2 = protected bases. Condensation is activated by mesitylene-2-sulfonyl-3-nitro-1,2,3-triazole (formula shown). R is a phosphoryl protecting group (see Fig. 2).

erate a 517-base-pair α-interferon gene [M. D. Edge et al. *Nature* **292** (1981) 756–762]. The reactions may be performed in solution, as shown in Fig. 2, but solid phase synthesis is more efficient (Fig. 3).

In solid phase synthesis, the oligodeoxyribonucleotide is synthesized while covalently attached to a solid support, and excess soluble protected nucleotides and coupling reagents are used to drive each stage of the synthesis to completion. Solid phase synthesis is easily automated, and instruments for this purpose are technically advanced liquid-dispensing devices, in which the delivery of reagents to the solid support is metered and controlled by computer.

The modern method of choice (manual or automated) is a solid phase system, using phosphoramidite chemistry (Fig. 4). The most advanced automated systems employ β-cyanoethylamidites, rather than methylamidites. Cyanoethyl protection avoids the potential hazard of thymine methylation by internucleotide methyl phosphate, which is present when methylamidites are used; the cyanoethyl group is also more easily removed than methyl, so that deprotection of the phosphate, deprotection of the bases, and release from the solid support can be performed in a single stage.

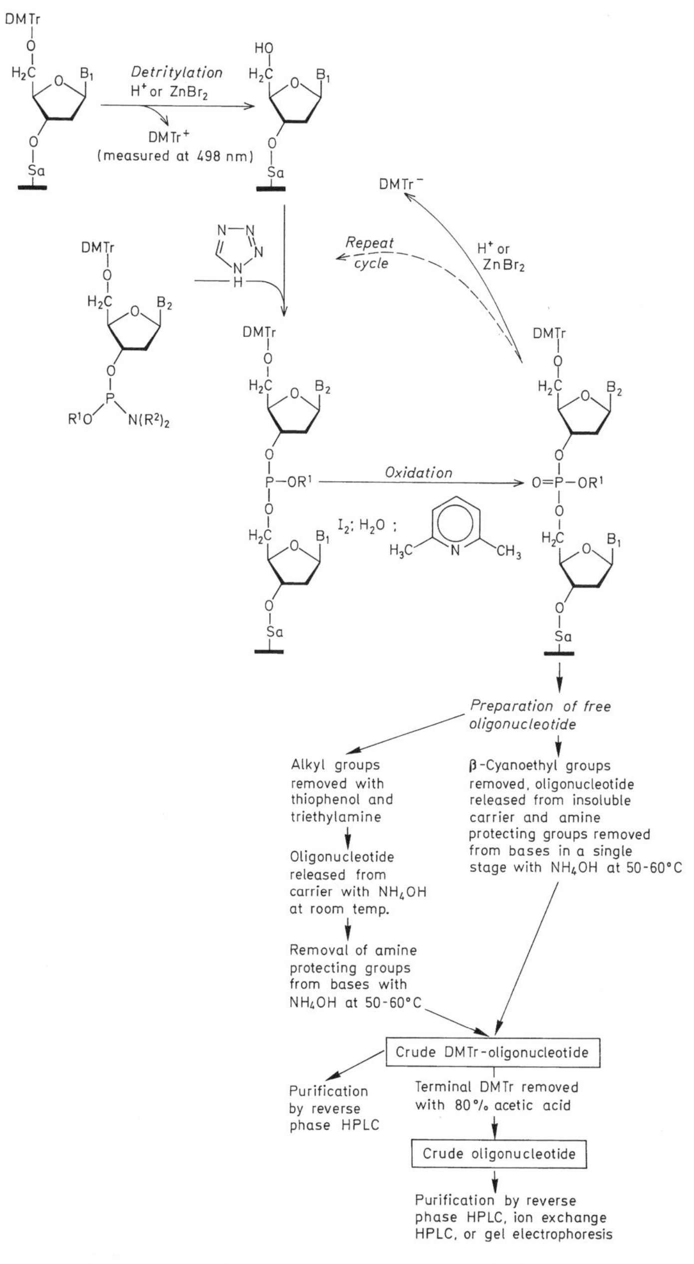

Fig. 4. *DNA synthesis.* Phosphoramidite method of DNA synthesis, using a solid phase system. Coupling efficiency is measured by monitoring the DMTr cation, as in Fig. 3. R^1 is a methyl or β-cyanoethyl group. R^2 is methyl, ethyl or isopropyl. Sa = spacer arm. The reactive phosphite is oxidized to a stable triester by aqueous iodine in the presence of lutidine. At the beginning of each new cycle, any unreacted, carrier-bound 5′-hydroxyl groups are acetylated with acetic anhydride, a process known as capping.

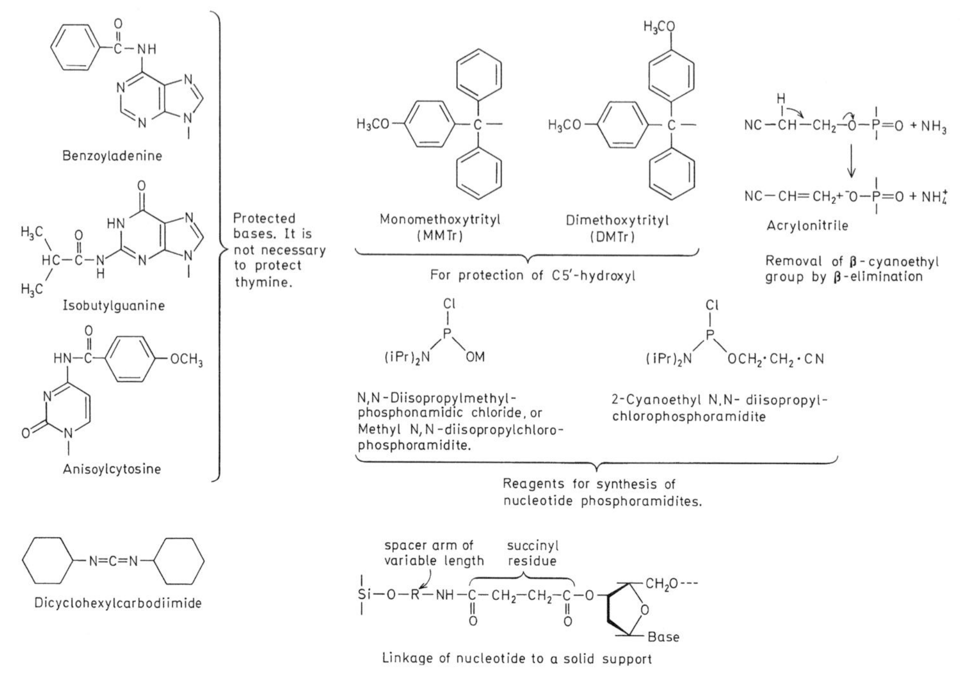

Fig. 5. *DNA synthesis.* Some structures and reagents in chemical DNA synthesis.

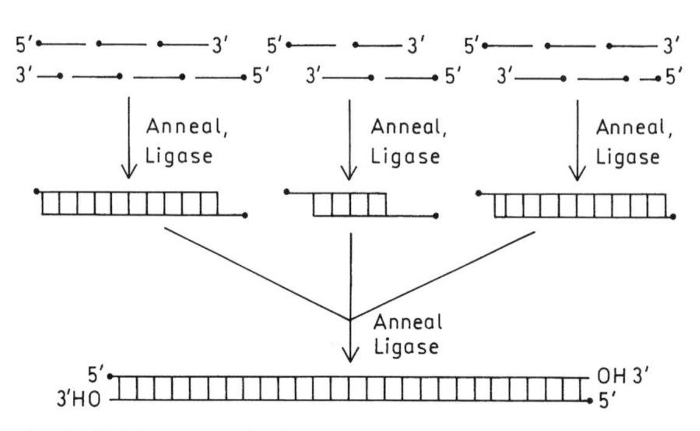

Fig. 6. *DNA synthesis.* Synthesis of double-stranded DNA from overlapping oligonucleotides of both strands. The 5'-end of each oligonucleotide is first phosphorylated by the action of polynucleotide kinase in the presence of ATP. —• = 5'-phosphate.

The solid support may be controlled pore glass or silica (phosphoramidite method), or polystyrene-divinylbenzene (phosphotriester method). Attachment is via a succinyl residue to the amino terminus of a spacer arm (Fig. 5). The product of chemical DNA synthesis is a single-stranded oligonucleotide. Short oligonucleotides can be converted to the double-stranded form in vitro by suitable enzymatic methods (see Recombinant DNA technology). For the synthesis of very long double-stranded DNA (e.g. a whole gene), overlapping oligonucleotides of both strands may be synthesized, then formed into the total polynucleotide by annealing and ligase action (Fig. 6). [R. L. Letsinger & W. B. Lunsford *J. Amer. Chem. Soc.* **98** (1976) 3655–3661; N. D. Sinha et al. *Nucleic Acids Res.* **12** (1984) 4539–4557; R. Newton, *Internat. Biotech. Lab.* **5** (1987) 46–53]

DNA technology: see Recombinant DNA.

DOC: see Cortexone.

Dolichol phosphate: membrane-bound polyprenol phosphates (derived from 13–20 isoprene units) which accept glycosyl units from soluble donors (uridine- or guanosine-diphosphate glycosides) and donate them in turn to membrane proteins or lipids. D. p. can therefore be regarded as coenzymes of pro-

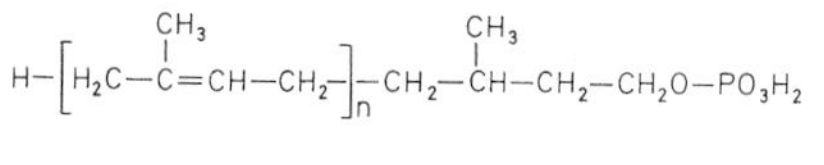

$$H-\left[H_2C-\overset{CH_3}{\overset{|}{C}}=CH-CH_2-\right]_n-CH_2-\overset{CH_3}{\overset{|}{C}H}-CH_2-CH_2O-PO_3H_2$$

n = 13 - 20 (mainly) in mammals

Dolichol phosphate

tein and lipid glycosylation. The highest concentrations of D.p. are found in nuclear, Golgi and rough endoplasmic reticulum membranes.

DON: acronym of 6-diazo-5-oxo-L-norleucine.

Donor position: the binding site for the peptidyl-tRNA on the ribosome during Protein synthesis (see).

Dopa: acronym of 3,4-dihydroxyphenylalanine. See Dopamine.

Dopamine, ***hydroxytyramine, 3,4-dihydroxyphenethylamine:*** one of the catecholamines, M_r 153.2. D. is formed by decarboxylation of 3,4-dihydroxy-phenylalanine (dopa), which in turn is formed by hydroxylation of tyrosine. D. is the precursor of the hormones Noradrenalin and Adrenalin (see). In the liver, lungs and intestines, it is the end product of tyrosine metabolism. In the central nervous system it serves as a neurotransmitter; the highest concentration of D. neurons occurs in the nigrostriatal system, which degenerates in Parkinson's disease. The blood-brain barrier is impermeable to D., but permeable to dopa (a D. precursor); the latter is therefore used to treat Parkinson's disease. [H.N.Wagner et al. *Science* **221** (1983) 1264–1266; E.S.Garnett et al. *Nature* **305** (1983) 137–138]. Agonists of D., such as amphetamines (see Antidepressants), can elicit psychotic symptoms similar to those of schizophrenia, while antagonists of D. (neuroleptics) are used to treat schizophrenia. It has been shown that post-synaptic D. receptors are functionally linked to an adenylate cyclase via an intrinsic membrane protein, the G/F protein. When the D. receptors are occupied, the G/F protein releases a molecule of GDP and binds a molecule of GTP. This is the rate-limiting step in the activation of adenylate cyclase. It has been suspected for many years that hypersensitivity to D. is a cause of schizophrenia, but comparison of brain tissue from schizophrenics and control persons has failed to reveal a difference in the activation of the adenylate cyclase by D. Two types of D. recognition sites have been described: the D_1 type mediates stimulation of the adenylate cyclase, and the D_2 type mediates inhibition. Both types are coupled to the cyclase via the G/F protein. It has been reported that 2,3,4,5-tetrahydro-7,8-dihydroxy-1-phenyl-1H-3-benzazepine, a selective D_1 agonist, stimulates more AMP formation in homogenates of the nucleus caudatus from schizophrenic brains than in homogenates from normal brains. This difference was observed only in the nucleus caudatus [M.Memo et al. *Science* **221** (1983) 1304–1306]. See also Ascorbate shuttle.

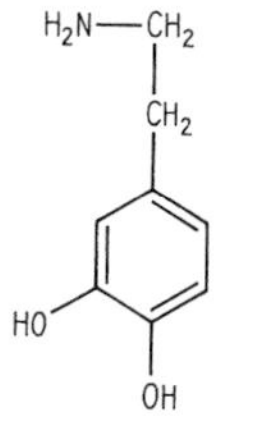

$H_2N-CH_2-CH_2-$(3,4-dihydroxyphenyl) (HO, OH)

Dopamine

Dopamine β-hydroxylase, ***3,4-dihydroxyphenylethylamine, ascorbate: oxygen oxidoreductase (β-hydroxylating)*** (EC 1.14.17.1): a mono-oxygenase catalysing the hydroxylation of dopamine to noradrenalin, and tyramine to octopamine. DβH is a copper protein, stimulated by fumarate and specifically inhibited by disulfiram. [Purification and physical properties: S.Friedman & S.Kaufman *J. Biol. Chem.* **240** (1965) 4763–4773]. Ascorbate, which is required as an electron donor, is converted into the free radical, Semidehydroascorbate (see) [T.Skotland & T.Ljones *Biochim. Biophys. Acta* **630** (1980) 30–35]. In vitro, two free radicals then dismute rapidly to form one molecule of ascorbate and one of dehydroascorbate. Under physiological conditions, e.g. during noradrenalin synthesis in the adrenal medulla, the semidehydroascorbate free radical is re-reduced to ascorbate. See Ascorbate shuttle.

Dormin: see Abscisic acid.

Dose: the amount of a pharmaceutical agent administered at a time. In animal experiments, the dose is given as the amount having a specified effect on a stated fraction of the individuals tested (Table). The D. should be given in mg per kg body weight; and the form in which it is administered should be stated explicitly, for example, LD_{50}, 20 mg/kg mouse, subcutaneously. The safety of an agent is indicated by the *therapeutic range* (therapeutic index), LD_{50}/ED_{50}.

Dose-effect relationships

Term	Abb.	Definition
Median effective D.	ED_{50}	D. at which 50 % of experimental subjects show an effect
Effective, therapeutic or curative D.	CD_{50}	Same as above; used with therapeutically applied compounds
Minimal lethal D.	LD_{05}	D. causing death of 5 % of subjects
Median lethal D.	LD_{50}	D. causing death in 50 % of experimental animals
Lethal D. Absolute lethal D.	LD_{100}	D. which kills 100 % of experimental animals

Dot plot: a graphic comparison of all the positions of two sequences, generated by drawing diagonals in the matching regions. For example, graphic comparison of a sequence with itself, using the same scale on both axes, produces a continuous diagonal line at 45°, representing total identity. Repeat sequences appear as patterns of shorter diagonals, occurring in mirror image on both sides of the central diagonal, and parallel to it. Many publications of DNA and polypeptide sequences contain Dot plots, e.g. the report by K.E.Sahr et al. (*J. Biol. Chem.* **265** (1990)

4434–4443) on the cDNA and polypeptide sequences of α-spectrin.

Double helix: see Deoxyribonucleic acid.

Double membrane: see Biomembrane.

Double strand break: a break in a double-stranded DNA molecule in which both strands are broken without separating from each other. A D.s.b. can be made by mechanical forces, radiation, chemicals or enzymes.

Downstream: a term strictly referring to any portion of the Coding strand (see) (otherwise known as the 'sense' or 'non-template' strand) of a stretch of double-stranded DNA that is on the 3′-side of the start site of transcription. The nucleotide at the start site is designated + 1 and the nucleotides downstream (i.e. on the 3′-side) of it are sequentially designated + 2, + 3, + 4 etc. The term is sometimes used more loosely to refer to any nucleotide or region of the coding strand of DNA that is on the 3′-side of some other nucleotide or region within the strand. See Nomenclatural conventions concerned with gene transcription.

DPN: acronym of diphosphopyridine nucleotide; see Nicotinamide adenine dinucleotide.

DPX 1840: 3,3a-dihydro-2-(*p*-methoxyphenyl)-8H-pyrazolo-[5,1-a]-isoindol-8-one, a synthetic growth regulator affecting auxin transport, ethylene production and root growth, etc. in cotton and soybeans.

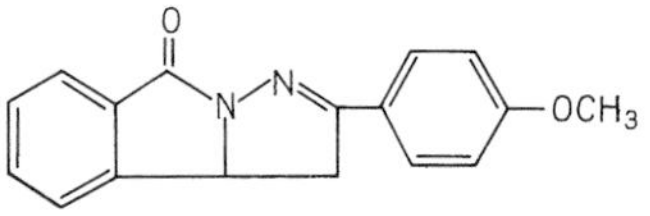

DPX 1840

D-RNA: see Messenger RNA.

Droplet countercurrent chromatography, *DCC chromatography, DCCC:* a separation method based on partition chromatography. Stationary and mobile phases are liquid and immiscible, and of different densities. Droplets of the mobile phase pass through a battery (normally about 300) of glass columns (2.0–3.0 mm internal diam., length 40 cm) containing the stationary phase. DCCC is very protective, and gives 100 % recovery of microgram or gram quantities of separated material. It has been successfully used for practically every type of natural product, including alkaloids, cardiac glycosides, saponins, anthraquinones, terpenoids, sugars, amino acid derivatives, peptides, etc.

Dry weight: the weight of material, in g or mg remaining after drying tissues, organisms, etc. at temperatures somewhat above 100 °C.

DSC: the acronym of Differential scanning calorimetry. See Membrane lipids.

dTDP: see Thymidine phosphates.

dTDP-sugars: sugars or sugar derivatives activated by bonding to deoxythymidine diphosphate sugars.

dThd: acronym of deoxythymidine. see Thymidine.

dTMP: see Thymidine phosphates.

dTTP: see Thymidine phosphates.

(–)-Duartin: (3S)-7,3′-dihydroxy-8,4′,2′-trimethoxyisoflavan. See Isoflavan.

Dulcitol, *galactitol:* an optically inactive, C_6 sugar alcohol derived from galactose. M_r 182.17, m.p. 189 °C. It is found in algae, fungi and the sap and bark of various higher plants. It is produced synthetically by reduction of galactose or by isolation from dulcite or Madagascar mannu *(Melampyrum nemorosum* L.).

Dwarf maize test: see Gibberellins.

Dwarf pea test: see Gibberellins.

Dynamic reciprocity: see SPARC, Tenascin, Thrombospondin.

Dynein: an ATPase responsible for motility in cilia and flagella. It causes the outer doublet microtubules in these organelles to slide with respect to one another. In cilia, the D.is firmly attached to the A subfiber of the outer doublet, and interacts transiently with the B subfiber of the adjacent fiber in such a way as to generate force; ATP is hydrolysed in the process. From the clearest electron micrographs available, the D. from cilia consists of 3 globular "heads", each of which is attached by a filament to a common, root-like base. The particle has M_r approx. 2×10^6. The three "heads" may not be identical. In cilia, the base is attached to the A subfiber of the microtubule, and the heads interact with the B subfiber. Micrographs of bovine brain microtubules show 7 D. molecules surrounding 14 protofilament microtubules, with the heads pointing toward the microtubule. [K. A. Johnson & J. S. Wall *J. Cell Biol.* **96** (1983) 669–678]

E

Eadie-Hofstee plot: see Kinetic data analysis.

EAG: acronym of electroantennogram. See Pheromones.

Early proteins: see Phage development.

Early RNA: see Phage development.

Ecdysone, ***α-ecdysone, molting hormone, (22R)-2β,3β, 14, 22, 25-pentahydroxy-5β-cholest-7-en-6-one:*** a steroid hormone which stimulates molting of caterpillars, pupa formation and emergence from the pupa. Its first recognizable effect is the activation of certain genes (see Gene activation). E. was the first insect hormone to be isolated in crystalline form. This was accomplished in 1954 by Butenandt and Karlson, who isolated it from pupae of the silkworm, *Bombyx mori* (25 mg/550 kg). It has also been detected in other insects. The structure was clarified in 1963 using X-ray crystallography. Phytoecdysones (compounds similar to E. and related insect and crustacean hormones) are synthesized by some plants, e.g. *Lemmaphyllum microphyllum* Presl. and *Polypodium vulgare* L. (see Ecdysterone). E. and related hormones are synthesized from cholesterol or phytosterols (see Sterols), which are required by the insects as vitamins.

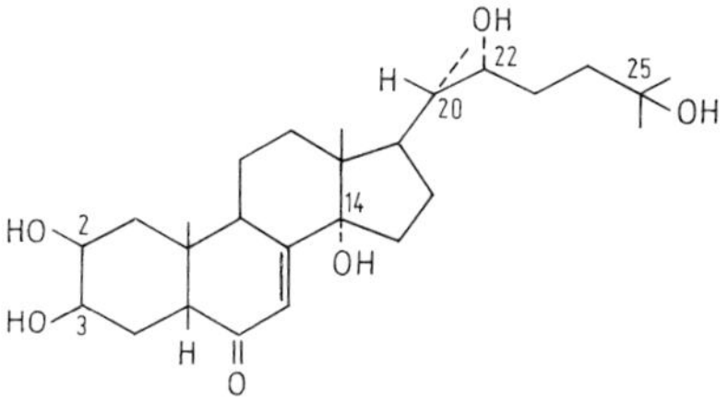

Ecdysone

β-Ecdysone: see Ecdysterone.

Ecdysterone, ***β-ecdysone, crustecdysone, 20-hydroxyecdysone:*** a molting hormone, M_r 480.65, m.p. 238 °C, found together with Ecdysone (see) in pupae of silkworms *(Bombyx mori)* in amounts of 2.5 mg/500 kg. It has also been isolated from other insects and crustaceans and, more recently, from many plants, including *Lemmaphyllum microphyllum* Presl., *Podocarpus elatus* and *Trillium smalli.* E. differs from ecdysone in having an extra hydroxyl group at position 20.

Ecgonine: the principal part of the cocaine molecule and the basis of many Coca alkaloids. M_r 185.22, (−) form, m.p. 205 °C (d.), $[\alpha]_D^{15}$ − 45.5 ° (c = 5 in water). E. has four chirality centers, and therefore exists in a number of naturally occurring stereoisomers. It is produced commercially by hydrolysis of the alkaloids of coca leaves, and converted to cocaine by esterification with methanol and benzoic acid. For formula and biosynthesis, see Tropane alkaloids.

Echinochromes: see Spinochromes.

Echinoderm saponins: see Echinoderm toxins.

Echinoderm toxins, ***echinoderm saponins:*** low M_r steroid toxins produced in the glands of echinoderms. Sea cucumbers *(Holothuria)* secrete highly toxic sulfated steroid glycosides, called holothurins. Starfish *(Asteroidea)* produce asteriotoxins or asteriosaponins, in which the main aglycon is pregnene diolone. See Asterosaponin A.

Echinomycin, ***quinomycin A:*** a depsipeptide antibiotic isolated from *Streptomyces echinatus* and effective against Gram-positive bacteria. It contains two quinoxalinoic acid residues as heterocomponents.

EC nomenclature: see Enzymes.

Ecological chemistry: 1. Investigation and optimization of the effects of man-made chemicals (e.g. pesticides, fertilizers) on the environment. E.c. also includes the development of chemicals and chemical processes that are not harmful to the environment, and the identification of harmful pollutants and their origins. 2. Study of the interactions of living organisms with each other, and with their environment, at a chemical level. This is therefore a wide subject, including such topics as the chemistry and biochemistry of pheromones, insect defense secretions, plant antifeeding substances, attraction and warning pigmentation in plants and animals, etc., and biochemical adaptation to drought, salinity, water-logging, high temperature, etc. Several names are used for this area of biochemistry: Ecological Chemistry, Chemical Ecology, Ecological Biochemistry, Biochemical Ecology.

Economic coefficient: see Cultivation of microorganisms.

Ectocarpene, ***all-cis-(1-cyclohepta-2′,5′-dienyl)-but-1-ene:*** a sexual attractant excreted by the female gametes of the brown alga *Ectocarpus stiliculosus.* b.p. 80 °C, $[\alpha]_D^{22}$ + 72 ° (c = 0.03 in $CHCl_3$).

Ectotoxins: see Toxic proteins.

Edestin: a hexameric, globular protein (M_r 300,000) from hemp seeds, *Cannabis sativa.* Each of the 6 subunits consists of two nonidentical polypeptide chains (M_r 27,000 and 23,000) joined by disulfide bridges. E. is very similar to arachin from peanuts and excelsin from brazil nuts.

EDTA: acronym of Ethylenediaminetetraacetic acid.

Effectors: chemical compounds, such as metabolites, hormones or cyclic AMP, which regulate the activity of a gene or enzyme, usually by allosteric interaction with a regulator protein or the enzyme protein. They may increase the rate of an enzyme reaction (activators) or decrease it (inhibitors). An inhibitor may occupy the active center of an enzyme and thus prevent the substrate from occupying it (competitive inhibition), or it may bind either the free enzyme or the enzyme-substrate complex (non-competitive inhibition), or it may bind only to the enzyme-substrate complex (uncompetitive inhibition). E. which bind to the enzyme at a site other than the catalytic center are called allosteric E.

EGF: acronym of Epidermal growth factor (see).

Ehlers-Danlos syndrome: see Inborn errors of metabolism.

EIDA: acronym of Enzyme immunodetection assay. See Recombinant DNA technology.

EL-531, ***α-cyclopropyl-α-(4-methoxyphenyl)-5-pyrimidine methanol:*** a synthetic growth retardant. It is a gibberellin antagonist which delays the growth of lettuce hypocotyls.

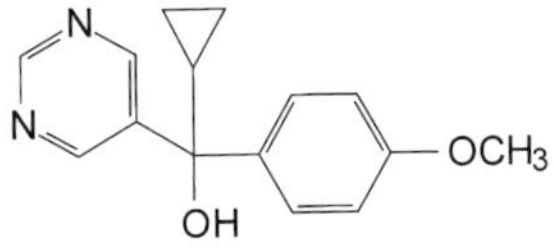

EL-531

Elaidic acid: see Oleic acid.

Elastase (EC 3.4.21.11): an endopeptidase specific for the Elastin (see) in animal elastic fibers. Its inactive precursor, *proelastase,* is formed in the vertebrate pancreas and converted in the duodenum to elastase by the action of trypsin. The natural substrate of E. is elastin, an insoluble protein rich in valine, leucine and isoleucine. E. attacks the peptide bond adjacent to a non-aromatic, hydrophobic amino acid. The best synthetic substrates are therefore acetyl-Ala-Ala-Ala-OCH_3 and benzoylalanine methyl ester. Benzoylarginine ester (a trypsin substrate), and acetyltyrosine ester (a chymotrypsin substrate) are not attacked by E.

The primary and tertiary structures of E. are very similar to those of the other pancreas proteinases. Of the 240 amino acids in E. (M_r 25,700), 52 % are identical to those in trypsin and chymotrypsin A and B. These include the catalytically important residues His_{57}, Asp_{102} and Ser_{195}, the ion pair Val_{16} Asp_{197} which is important for the conformation, and the 4 disulfide bridges. As might be expected, the 3-dimensional structure of E. is very similar to that of the other pancreatic serine proteases (see Chymotrypsin, Trypsin).

A second elastolytic enzyme (M_r 21,900) has been isolated from porcine pancreas. It shows higher activity than chymotrypsin in the hydrolysis of acetyltyrosine ester, which is used routinely to assay chymotrypsin. Another E.-like enzyme, α-lytic proteinase (M_r 19,900, 198 amino acids) has been isolated from the soil bacterium *Myxobacter 495.* This enzyme is remarkably similar to pancreatic E. both in structure (41 % homology, sequence in the active center Gly-Asp-Ser-Gly, 3 homologous disulfide bridges) and substrate specificity. Another E. (M_r 22,300) has been isolated from *Pseudomonas aeruginosa.*

Elastin: a Structural protein (see) which is the main component of the elastic fibers of tendons, ligaments, bronchi and arterial walls. It owes its elasticity to a high content of glycine, alanine and proline, and of valine (17 %), and of leucine and isoleucine (12 % together). The sequences -Gly-Val-Pro-Gly- and -Gly-Gly-Val-Pro- occur frequently, and the protein is cross-linked by two unusual, blue-fluorescing amino acids, *desmosine* and *isodesmosine.* The elasticity, yellow color, insolubility in water and sodium hydroxide solution, and resistance to denaturation and to proteases (except elastase) are all due the three-dimensional network created by these cross-links. E. is hydrolysed to the water-soluble *α-elastin* (M_r 70,000) by treatment with hot oxalic acid. The precursor of E., soluble tropoelastin, contains no desmosine or isodesmosine cross-links.

The complete nucleotide sequence of the human E. gene and its 5′ flanking regions has been determined. [M. M. Bashir et al. *J. Biol. Chem.* **264** (1989) 8887–8891; H. Yeh et al. *Biochemistry* **28** (1989) 2365–2370]

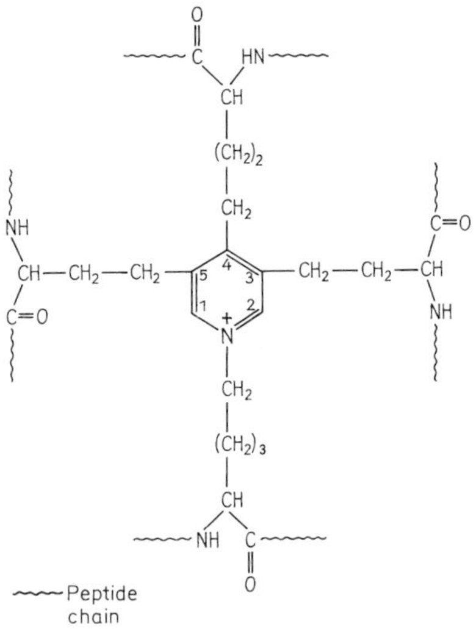

~~~ Peptide chain

Desmosine. In isodesmosine, a chain is attached to C2 of the pyridine ring instead of C4.

Elaterin: see Cucurbacitins.

Electroantennogram: see Pheromones.

Electroblot: see Southern blot.

Electrofocusing: see Proteins.

Electron spin resonance (ESR) spectroscopy: a technique used for the study of substances that are paramagnetic, a property caused by the presence of unpaired electrons. In biological materials paramagnetism is principally found in two main types of molecular species, free radicals and those that contain a transition metal ion. Some spectroscopists confine the term ESR spectroscopy to the study of free radicals, because they have an identifiable electron spin, and use the term electron paramagnetic resonance (EPR) spectroscopy for the study of transition metals, where paramagnetism arises from the distribution of electrons in the *d* orbitals. However the spectra are taken with the same instrumentation and most authors use the terms ESR- and EPR spectroscopy either interchangeably or according to personal preference.

ESR spectroscopy is analogous to nmr spectroscopy (see) in that it measures the resonant absorption
~~~

of electromagnetic radiation by entities (unpaired electrons in esr, the nuclei of certain atomic species in nmr) as they move from the lower to the higher of two energy levels induced by an externally-applied magnetic field and then relax back again. An unpaired electron, by virtue of its spin, which may be either $+^1/_2$ or $-^1/_2$, and its negative charge, has a magnetic moment and thus may take up one or other of two possible orientations when placed in an external magnetic field, each of which is associated with a particular energy. The orientation in which the magnetic moment is aligned with the external magnetic field has a lower energy than the orientation that is aligned against it. Thus an electron with the latter orientation is said to be at a higher energy level than one with the former orientation. The greater the external magnetic field the greater is the difference between these energy levels. In the absence of an external magnetic field the electron, regardless of its spin, is at a single intermediate energy level. Thus the external magnetic field effectively splits the latter energy level into two, one higher and one lower; this phenomenon is known as Zeeman splitting. An electron in the lower energy orientation may be flipped to the higher energy orientation by absorbing a photon of electromagnetic radiation equal to the energy difference between the two orientations. It may then revert to the lower energy orientation by loss of the same amount of energy in a non-radiative manner involving spin lattice interaction, known as relaxation. The magnitude of the energy difference, ΔE, is given by the equation: $\Delta E = g\mu B_o$, where g is the *Landé g-value* (a dimensionless quantity derived from the contributions of orbital and spin angular momentum to the angular momentum of the unpaired electron in the paramagnetic species under investigation; for a free electron it has the value 2.00232), μ is the Bohr magneton ($9.2741 \times 10^{-24} JT^{-1}$) and B_o is the stength of the external magnetic field (T; 1 telsa = 10^{-4} Gauss). Since ΔE is small ($\sim 10^{-24}$ J; *cf* $\sim 10^{-26}$ J in nmr spectroscopy, $\sim 10^{-19}$ J in absorption spectroscopy) the two energy levels are almost equally populated, with slightly more in the lower one; the ratio of the two populations may be calculated with the Bolzmann distribution equation (see Nuclear magnetic resonance spectroscopy). Thus irradiation with electromagnetic radiation of the appropriate frequency ($\sim 10^{10}$ Hz $\equiv$ wavelength, λ, of 0.03 m; *cf* $\sim 10^7$–10^8 Hz $\equiv \lambda \sim 3$–30 m in nmr, $\sim 10^{14}$ Hz $\equiv \lambda \sim 10^{-9}$ m for absorption spectroscopy) normally causes resonance (unless it is too powerful, causing a state termed saturation to be reached, when the population difference between the two orientations virtually disappears because energy cannot be lost fast enough by relaxation).

Since it is difficult to generate and transmit tuneable frequencies in the microwave region of the electromagnetic spectrum, the conventional continuous-wave ESR spectrometer uses a fixed microwave frequency (typically in the X-band, 8.5–10 GHz) generated by a klystron or Gunn diode, and varies the magnetic field, generated by a powerful electromagnet, linearly. The sample sits in the microwave cavity, which is positioned between the poles of the magnet. When resonance occurs a detector diode registers a change in the intensity of the microwave beam. The absorption signal (A), after electronic amplification, is plotted with its first derivative (i.e. dA/dB_o) as the ordinate against magnetic field strength (B_o) as the abscissa. Thus in an ESR spectrum the peak of an absorption band is marked by the point where the steep line in the middle of the first derivative curve crosses the horizontal (B_o) axis. Since most biological samples are aqueous and microwave absorption by water is high, sample volumes examined in ESR spectrometers are small, e.g. ~20 µl if in a capillary, slightly larger if in a flat cell; minimum detectable concentrations depend upon the nature of the paramagnetic material but are generally in the 0.1–100 µM range.

Two important features of the esr spectrum are, (i) the absolute magnetic field position of the lines, characterized by the g-value, and (ii) hyperfine splitting. The g-value in free radicals is often close to 2.00232, the theoretical value of the free electron, because the unpaired electron is delocalized over the whole molecule and thus behaves almost like a free electron. The g-values of transition metal ions differ considerably from 2, e.g. Fe^{3+} (low spin, $S = 1/2$) $g = 1.4$–3.1, Fe^{3+} (high spin, $S = 5/2$) $g = 2.0$–9.7, Cu^{2+} ($S = 1/2$) $g = 2.0$–2.4. The reason for this is that their unpaired electrons are in the d orbitals whose directionality contributes a strong angular momentum to the net electron magnetic moment. Hyperfine splitting is caused by an interaction between the magnetic moment of the unpaired electron and that of the nucleus of a nearby atom; it is thus analogous to the splitting caused by spin-spin coupling in nmr (see). Not all nuclei have a magnetic moment; those that do and are of particular biological importance are: ^1H ($I = 1/2$), ^{14}N ($I = 1$). Hyperfine splitting by ^1H causes two spectral lines while that by ^{14}N causes three; the former is shown diagrammatically in Fig. 1 and an example of the latter in Fig. 2.

Another important interaction affecting ESR spectra is that between the spins of unpaired electrons in the same molecule. This does not affect a free radical such as TEMPO (2,2,6,6-tetramethyl-1-piperidyloxy) shown in Fig. 2, which has only one unpaired electron, but it does affect molecules which have several unpaired electrons, such as cytochrome P-450 with its iron ion in the high spin ($S = 5/2$) Fe^{3+} state. In such molecules the unpaired electrons are strongly coupled, by exchange interactions, giving a complex set of energy levels which can give rise to spectral lines over a wide range of magnetic field.

ESR spectroscopy is a useful investigative tool in the biological context. It can be used to: (i) detect and identify radicals generated from enzymic substrates, co-enzymes, drugs and hormones during metabolic reactions, (ii) identify redox components and follow electron transfer in complex systems such as those operating in mitochondria and chloroplasts (iron-sulfur proteins, for example, were discovered by ESR), (iii) provide information on the structure and function of transition metal centers in metalloproteins such as Cytochromes (see), Plastocyanin (see) and Ferredoxin (see), (iv) investigate radiation damage and other abnormalities in nucleic acids and proteins, (v) investigate conformational changes in proteins, and (vi) investigate the structure and fluidity of membranes. Items (v) and (vi) require the use of 'spin-label probes'; these are small radicals like TEMPO which can be covalently attached to specific sites

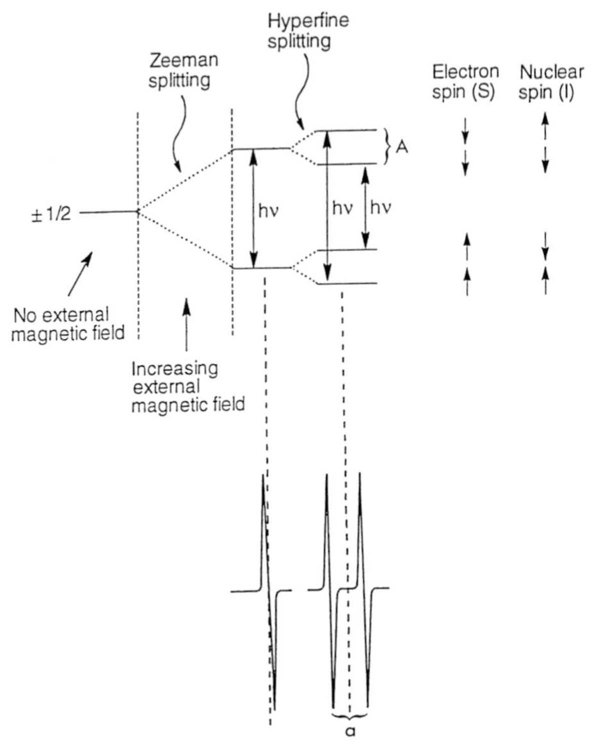

Fig. 1. *Electron spin resonance spectroscopy.* Zeeman splitting of the electron energy level corresponding to spin quantum numbers $\pm^1/_2$ and hyperfine splitting of the resulting energy levels as a result of interaction with a ^1H nucleus ($I = ^1/_2$).

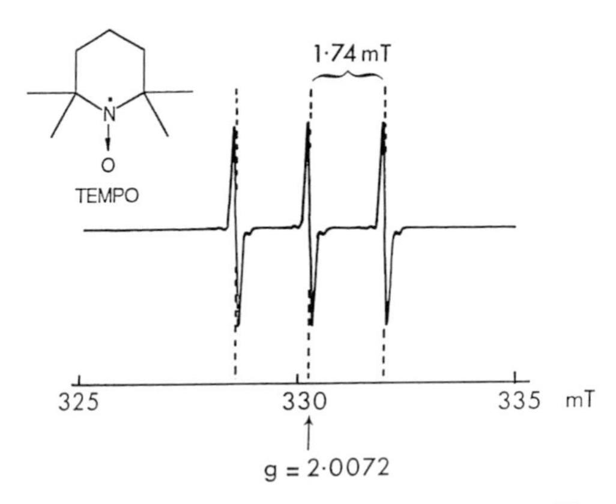

Fig. 2. *Electron spin resonance spectroscopy.* The ESR spectrum of TEMPO in aqueous solution, showing the three-way hyperfine splitting of the Zeeman energy levels as a result of interaction with the ^{14}N nucleus of the nitroxide moiety.

in molecules that are not otherwise paramagnetic. With proteins this allows the distances of various groupings, such as substrates and cofactors, from the spin-label reference point to be measured. With phospholipids, spin-labeled in a fatty acyl residue, it allows the rate of lateral diffusion of such molecules within the lipid bilayer to be estimated and the effect on the membrane fluidity of the incorporation of normal (e.g. cholesterol) and abnormal (e.g. 14α-methyl sterols) into the bilayer to be determined. An increase in membrane fluidity is indicated by a sharpening of the ESR spectral lines and a decrease by a broadening; this follows from the fact that a highly fluid membrane allows the spin-labeled phospholipid molecule to spin rapidly about its long axis thereby averaging out its anisotropy, the latter being a requirement for sharp ESR spectral lines.

Electron transfer flavins, *ETF:* flavoproteins (see Flavin enzymes) which mediate electron transport from reduced FADH to the cytochrome system. Flavoproteins oxidize those substrates of the Respiratory chain (see) whose oxidation does not involve pyridine nucleotides. However, the substrate must have a more positive redox potential than the $NADH + H^+/NAD^+$ system. ETF from pork liver has 6 molecules flavin per atom of iron, and copper is also present.

Electron transport chain: see Respiratory chain.

Electron transport particles, *ETP:* fragments of mitochondria, obtained by ultrasonic or detergent treatment, which contain the complete electron transport system of the Respiratory chain (see). In the electron microscope, ETP appear as membrane-enclosed vesicles, which are thought by some authors to be a giant molecule of defined composition. *Heavy*

ETP (ETP_H) also include succinate dehydrogenase. Extraction of the lipids destroys the activity of ETP, but it can be restored by addition of appropriate lipids. Further degradation of the ETP produces the separate complexes of the respiratory chain.

Electrophoresis: a method of separating charged particles or macromolecules by allowing them to migrate in an electric field. The method is used most frequently in biochemistry to separate delicate macromolecules, usually proteins or nucleic acids. In *free E.* (Tiselius, 1937), the substances to be separated are placed in a solution in a U-shaped tube. In *carrier E.*, the separation is performed on a carrier, e.g. paper, cellulose powder, glass powder, gels of starch, agar or polyacrylamide. These stabilize the separated bands against diffusion or convection, but they also interact with the materials under separation, so that additional effects such as molecular sieving or electrostatic interactions may contribute to the separation. E. can be automated, and the lower limit for the quantity of material separated depends only on the availability of a sufficiently sensitive detection method; thus µg, ng or even pg quantities of radiolabeled proteins and nucleic acids can be analysed by E. The separation of Plasma proteins (see) is an example of a relatively large scale, routine application of E. Special variants of E. are *immunoelectrophoresis, disc electrophoresis* and *isoelectric focusing,* which are discussed under Proteins (see).

Eledoisin: an undecapeptide, Pyr-Pro-Ser-Lys-Asp-Ala-Phe-Ile-Gly-Leu-Met-NH_2, from the salivary glands of a cephalopod *(Eledone moschata* and *E. Aldrovandi).* Its biological activities in vitro are similar to those of Substance P (see), with which it also shares certain sequence similarities. [V. Erspamer & A. Anastasi *Experentia* **18** (1962) 58–59; E. Sandrin & R. A. Boissonas *ibid.* 59–61]

(9*R*,11*S*)-Eleutherin: see Naphthoquinones (Table).

Elicitor: any factor, biotic or abiotic, which induces formation of Phytoalexins (see) in plant tissue, e.g. heavy metal salts like $HgCl_2$ and $CuCl_2$ are abiotic E. The term E. is often used specifically for a glucan fraction, released by heat treatment from the cell wall of a phytopathogenic fungus. Glucan E. is used experimentally to induce the synthesis of phytoalexins. E. activity does not appear to be species- or variety-specific, and is probably a general defensive response caused by any fungal wall. Specific Oligosaccharins (see) released from the cell wall of either the plant or the invading fungus have been identified as E. [A. R. Ayers et al. Plant Physiol **57** (1976) 751–759, 760–765, 766–774, 775–779; U. Zähringer et al. *Z. Naturforsch.* **360** (1981) 234–241; A. G. Darvill & P. A. Albersheim, *Ann. Rev. Plant Physiol.* **35** (1984) 243–275]

ELISA: acronym of Enzyme-Linked Immunosorbent Assay. See Immunoassays.

Elongation: the phase of Protein biosynthesis (see) in which the amino acid chain is extended by addition of new residues.

Elongation factors, *transfer factors:* proteins catalysing the elongation of peptide chains in Protein biosynthesis (see). Three have been isolated from bacteria: EFT, which is a mixed dimer of two proteins, Ts (M_r 42,000) and Tu (M_r 44,000), and EFG. Bacterial E.f. do not interact with the 80S ribosomes of eukaryotes. They can be obtained by saline treatment of 70S ribosomes.

Embden ester: a mixture of D-glucose-6-phosphate and D-fructose 6-phosphate, both of which are intermediates in glycolysis.

Embden-Meyerhof-Parnas pathway: see Glycolysis.

Embelin: see Benzoquinones.

Embryonal inducers: compounds which induce differentiation of organs in the course of embryonic development. A low-molecular-mass protein has been isolated from chick embryos which, when injected into the ectoderm of an amphibian gastrula, can induce the formation of kidney and muscle primordia (mesodermal factor) and of notochord tissue (neural factor).

Emerson effect, *enhancement effect:* an increase in the photosynthetic quantum yield from long-wave red light (700 nm) which occurs when a plant is simultaneously irradiated with shorter wavelengths (< 670 nm). There is a sharp fall in photosynthetic efficiency around 700 nm, but light of this wavelength can be utilized synergistically with light of shorter wavelengths. The effect indicates that two photosystems participate in the generation of oxygen, and that these have different light-collecting pigments, a conclusion supported by the action spectra of the E.e. See Red drop, Photosynthetic pigment systems, Photosynthesis (Fig. 1).

Emetine: a dimeric isoquinoline alkaloid [M_r 480.63, m.p. 74 °C, $[\alpha]_D^{20}$ – 50 ° (c = 2 in $CHCl_3$)] which is the principal alkaloid of ipecac, the ground roots of *Uragoga ipecacuanha.* E. is very poisonous, exerting a strong stimulus on mucous membranes. In large doses, it leads to vomiting. It is used in the treatment of amebic dysentery and its complications. The alkaloids of the E. group are biosynthesized by a Mannich condensation of two molecules of phenylethylamine with an iridoid C_9 body (Fig.).

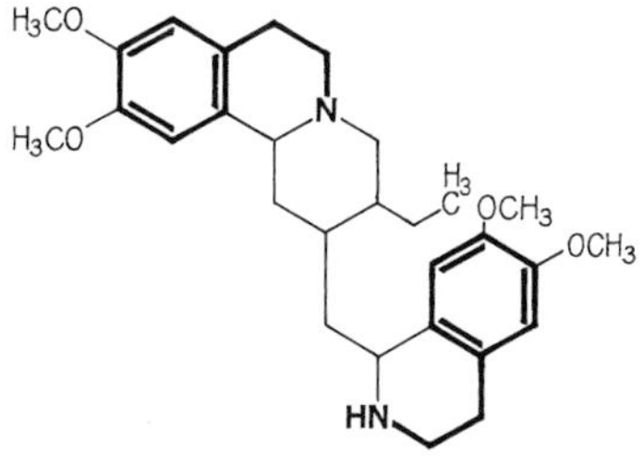

Emetine

Emodin: an orange anthraquinone pigment, m.p. 225 °C. It often occurs as a glycoside, or as the dimer, skyin. The 5,5′ dimer, (+)-skyrin, m.p. 380 °C, is found

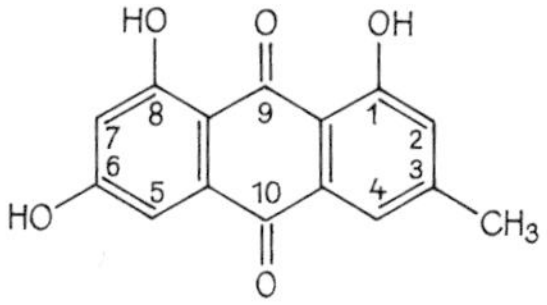

Emodin

in *Penicillium* species. E. is found in many higher plants including rhubarb root, alder buckthorn (*Rhamnus frangula* L.), *Cascara sagrada,* etc. It is used as a cathartic.

Encephalitogenic protein, *myelin protein A1:* the most important myelin protein of the mammalian central nervous system. On injection into guinea pigs, rabbits or rats, it induces allergic autoimmune encephalomyelitis (EAE), an inflammation of the brain and spinal column. The structure of E.p. from the myelin sheath of humans, cattle, rabbits and guinea pigs has been determined: M_r 18,000, 170 amino acids, including 11 % Arg, 8 % Lys and 6 % His, but no Cys. EAE can be induced by a relatively small active region of E.p., the location of which varies in the E.p. of different species.

Endocrine hormones: hormones produced by specialized cells in endocrine glands, and released into the blood stream. They stand in contrast to Tissue hormones (see), which are produced by individual cells in tissues specialized for other functions.

Endocytosis: see Coated vesicle.

Endogenous minimum: see Minimum protein requirement.

Endolysin: see Lysozyme.

Endomembrane system: see Endoplasmic reticulum.

Endonucleases: see Nucleases.

Endopeptidases: see Proteases.

Endoplasmic reticulum, *ER:* a net-like system of double membranes located in the cytoplasm of eukaryotic cells. The membranes form tubes or channels with cross sections of 50 to 500 nm. The ER appears to be continuous with the outer nuclear membrane and with the secretory system of the cell (the Golgi apparatus, see). If the ER is associated with ribosomes, it is called *rough ER;* otherwise it is called *smooth ER.* Rough ER is engaged in the synthesis of membrane proteins and proteins for export. The growing polypeptides pass through the membrane. When the cell is homogenized, the membranes of the ER form vesicles which can be isolated as the *microsome fraction.* Enzyme activities present in microsomes depend to some extent on the tissue of origin. They are usually characterized (e.g. liver microsomes) by the presence of the marker enzymes, glucose-6-phosphatase and the mixed-function oxidases. Other enzymes of the ER are responsible for the biosynthesis of triacylglycerols, glycerophospholipids, mucopolysaccharides and glucuronides. Various steps of steroid biosynthesis, including formation of mevalonic acid from hydroxymethylglutarate, synthesis of squalene from farnesyl pyrophosphate and cholesterol from lanosterol occur on or in the ER. (Other steps in cholesterol biosynthesis occur in the cytoplasm or mitochondria). Steroid conversions and degradation occur in the ER, and it is also involved in sulfur metabolism.

Endorphins: endogenous peptides with morphine-like effects (endogenous morphine), the natural ligands for the opiate receptors. See Opioid peptides.

Endotoxins: see Toxic proteins

Endoxan: see Cyclophosphamide.

End oxidation, *terminal oxidation:* the last step in catabolism. In aerobically respiring cells, E.o. is carried out via the tricarboxylic acid cycle.

End product: the last compound in a metabolic pathway, which is irreversible as written. The E.p. may either be the starting material for another metabolic pathway, or it may be accumulated or excreted. An E.p. can control its own rate of synthesis, either as an allosteric Effector (see) of one of the enzymes at the beginning of the pathway, or as a repressor of the operon coding for the enzymes.

End product inhibition: see Metabolic regulation.

End product repression: inhibition of the synthesis of the enzymes of a reaction sequence (see Enzyme repression) by the end product of that reaction sequence.

Energy charge: see Adenosine phosphates.

Energy metabolism: the totality of those reactions which serve to release energy by degradation of carbohydrates and fats. The most important link between the energy-producing and energy-consuming reactions is ATP,which is produced in large quantities by Respiration (see), and in smaller quantities by Glycolysis (see). In photosynthetic organisms, ATP is also produced in the light reactions (see Photosynthesis). Since the intermediate products of glycolysis and the tricarboxylic acid cycle are also starting materials for many syntheses there can be no sharp division between energy metabolism and synthesis.

Energy-rich bonds: see High energy bonds.

Energy-rich phosphates, *high energy phosphates:* phosphorylated compounds in which the phosphate ester displays a high free energy of hydrolysis (this is equivalent to the phosphoryl group having a high transfer potential). Chemical energy in biological systems is stored in energy-rich phosphates. These may be acid anhydrides of phosphoric acid (e.g. ATP), enol phosphates (e.g. see Phospho*enol*pyruvate) or amidine phosphates (e.g. Phosphagens). A high energy phosphate bond is represented as R ~ P (where P is a phosphate group) instead of the usual hyphen between groups. See High energy bonds.

The terms high energy phosphate and phosphate bond energy, and the so-called "squiggle" bond (~) are contrary to the purity and austerity of classical thermodynamics, but have proved useful in conveying the importance of phosphate groups in the storage and transfer of chemical energy in biochemical systems. These concepts were introduced in 1941 by Lipmann ("Metabolic Generation and Utilization of Phosphate Bond Energy", F.Lipmann *Advances in Enzymology* **1** (1941) 99–162).

Enhancement effect: see Emerson effect.

Enkephalins: Pentapeptides with affinity for the opiate receptors in the brain (see Opioid peptides). Met-E.has the structure Tyr-Gly-Gly-Phe-Met; Leu-E. has the sequence Tyr-Gly-Gly-Phe-Leu. These two sequences are also found at the *N*-termini of a number of opioid peptides. The physiological source of the E., however, is preproenkephalin A, a protein of 267 amino acid residues. Its sequence has been deduced from cloned DNA. [M.Noda et al *Nature* **295** (1982) 202–206; M.Comb et al. *ibid.* 663–666; V.Gubler et al. *ibid.* 206–208]

Enniatins: ring-shaped depsipeptide antibiotics produced by the fungus *Fusarium orthoceras* var. *enniatum* and other Fusaria. Enniatin A, cyclo-(-D-Hyv-MeIle-)$_3$ and enniatin B, cyclo-(-D-Hyv-MeVal-)$_3$

are found together (Hyv = hydroxyisovaleric acid, MeIle = methylisoleucine, MeVal = methylvaline). These compounds act as artificial pores in membranes, permitting potassium ions to penetrate them. See Ionophore.

Enolase (EC 4.2.1.11): an enzyme of Glycolysis (see) which catalyses the reversible dehydration of 2-phosphoglycerate to phospho*enol*pyruvate. It is a dimeric metalloenzyme, requiring $2Mg^{2+}$ per mol for the stabilization of structure, and $4Mg^{2+}$ per mol for activity. Zn^{2+} and Mn^{2+} also activate E., whereas F^- inhibits it. The M_r of some E. are (M_r of the subunit in parentheses): rabbit liver and muscle 82,000 (41,000); salmon 100,000 (48,000); yeast 88,000 (44,000); *E. coli* 90,000 (46,000). Like most glycolysis enzymes E. exists as several isoenzymes (see) which can be separated by electrophoresis.

Entactin, ***nidogen:*** a basement membrane glycoprotein, M_r 150,000, with a dumbell-shaped molecule which appears to be tightly bound to each Laminin (see) molecule at the meeting point of its three arms. It also binds to cells and to collagen IV. [A.E. Chung et al. *Kidney Int.* **43** (1993) 13–19; Li-J. Dong et al. *J. Biol. Chem.* **270** (1995) 15838–15843]

Enteroamine: see Serotonin.

Enterobacteriaceae: a family of bacteria capable of formic acid fermentation. They are Gram-negative rods which do not form spores. Motility is provided by peritrichal flagella. E. are facultatively anaerobic and can grow on simple synthetic media containing mineral salts, sugars (as carbon and energy sources) and ammonium salts (the nitrogen source). Some E., including *E. coli* (see), *Aerobacter aerogenes* and *Proteus vulgaris,* are intestinal bacteria. The family also includes the plant pathogens of the genus *Erwinia* (causes white rots), *Salmonella typhimurium* (causes food poisoning), *Salmonella typhi* (causes typhus), and *Shigella dysenteriae* (causes dysentery).

Enterodiol: 2,3-*bis*(3-hydroxybenzyl)butane-1,4-diol, a lignan which occurs as its glucuronide in primate urine, accompanied by the glucuronide of the related lignan, Enterolactone (see).

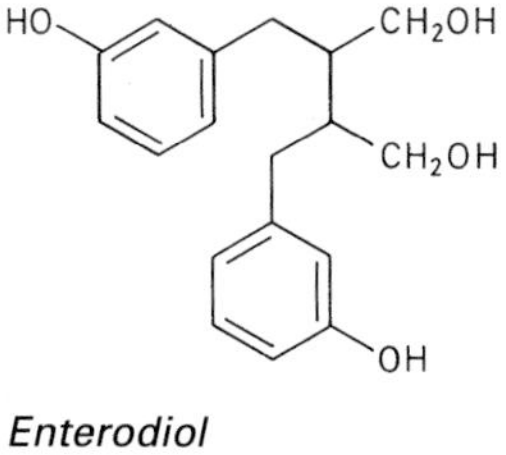

Enterodiol

Enterogastrone: see Gastric inhibitory peptide.

Enterohepatic circulation: see Bile acids.

Enterokinase: see Enteropeptidase.

Enterolactone: *trans*-3,4-*bis*[(3-hydroxyphenyl)-methyl]dihydro-2-(3H)furanone, or *trans*-2,3-*bis*(3-hydroxybenzyl)-γ-butyrolactone (Fig.), a lignan which occurs as its glucuronide in primate urine, accompanied by the glucuronide of the related lignan, Enterodiol (see). [S.R. Stitch, *Nature* **287** (1980) 738–740; D.R. Setchell et al. *Biochem. J.* **197** (1981) 447–458]

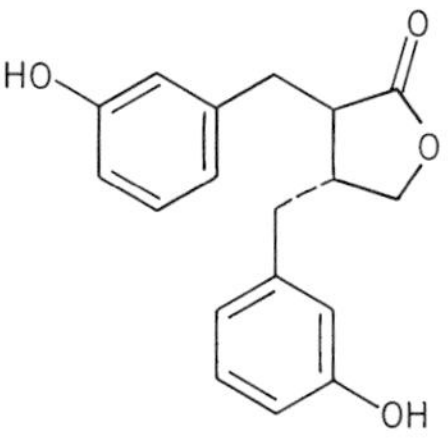

Enterolactone

Enteropeptidase, ***enterokinase*** (EC 3.4.21.9): a highly specific duodenal protease which acts only on trypsinogen, removing the *N*-terminal peptide (Val-$[Asp]_4$-Lys) to form trypsin. E. has M_r 196,000 (1,100 amino acids) and consists of two covalently bound glycopeptides (M_r 134,000 and 62,000). E. contains the most carbohydrate (37 %) of any digestive enzyme.

Enterotoxins: see Toxic proteins.

Entner-Douderoff pathway: an anaerobic degradation pathway (Fig.) for carbohydrates in some microorganisms, in particular *Pseudomonas* spp., which lack some glycolytic enzymes (e.g. phosphofructokinase and glyceraldehyde-3-phosphate dehydrogenase).

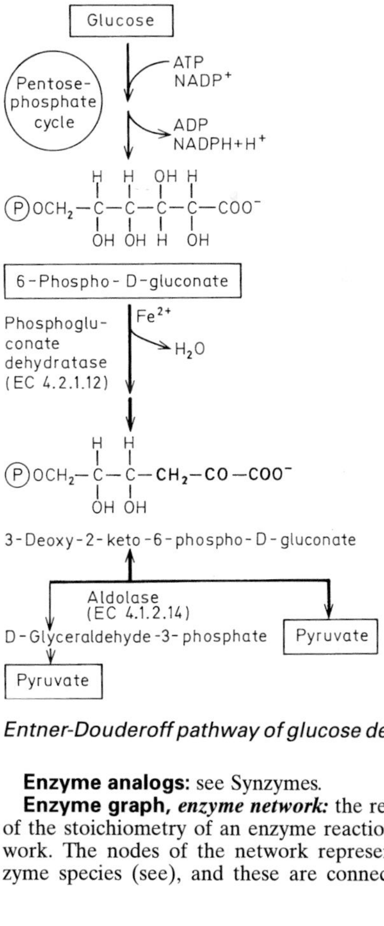

Entner-Douderoff pathway of glucose degradation

Enzyme analogs: see Synzymes.

Enzyme graph, ***enzyme network:*** the represention of the stoichiometry of an enzyme reaction as a network. The nodes of the network represent the Enzyme species (see), and these are connected by ar-

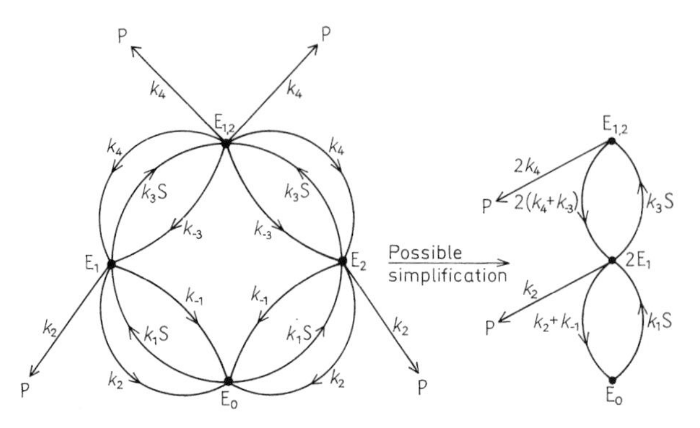

Enzyme graph for an enzyme with two active centers. E are enzyme species; P is product; S is substrate.

rows showing the direction of the reaction. Each arrow is labeled with the corresponding rate constant. An E.g. simplifies the writing of Rate equations (see) for the reactions. The Fig. shows an example of an E.g. for an enzyme with two active centers, where E_0, E_1 and E_2 represent enzyme species with zero, one and two substrate molecules bound in a Michaelis-Menten complex.

Enzyme immunodetection assay: see Recombinant DNA technology.

Enzyme induction: stimulation of enzyme synthesis by inducers. In bacteria, many catabolic enzymes are only synthesized when the appropriate substrate is available in the medium. The classical example is the induction of the enzymes of the lac operon in *E. coli* by their substrate, lactose (Fig.) (the true inducer is allolactose, formed in small quantities by rearangement of lactose). The regulator gene R codes for a specific repressor protein which, in the absence of an inducer, binds to the corresponding operator O and blocks transcription of the structural genes *z, y* and *a* (negative control). Since the mRNA for these genes is not produced, neither are the proteins. If lactose (or an artificial inducer) is added to the medium, the inducer binds to the repressor and inhibits its binding to the operator. Transcription of the structural genes is then possible, and the enzymes are synthesized. The concentration of β-galactosidase is about 1,000 times higher in induced cells than in those growing in the absence of an inducer. The presence of glucose prevents induction of β-galactosidase by lactose; this is due to Catabolite repression (see). E.i. also occurs in higher organisms, although the mechanism is not well understood. Hormones or substrates can cause a large increase in the synthesis of a particular enzyme, e.g. the hormone ecdysone induces dopa decarboxylase in insects, and nitrate (as substrate) induces nitrate reductase in higher plants. See Gene activation.

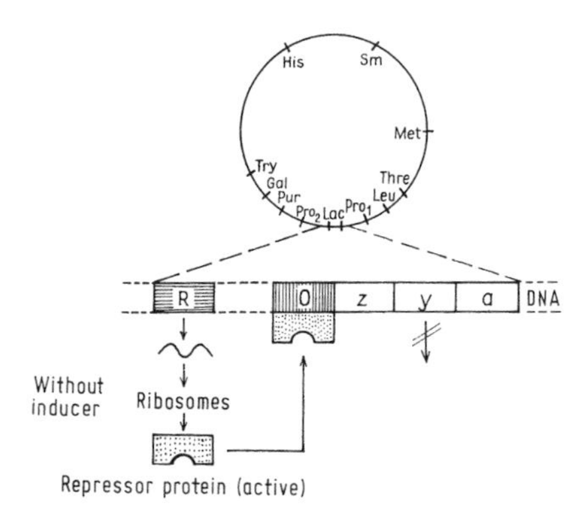

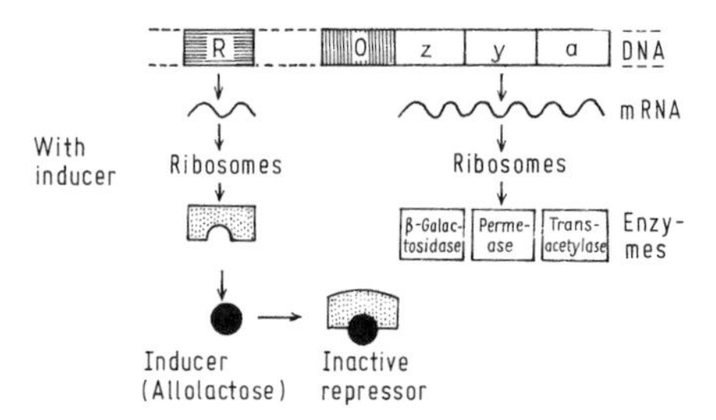

Induction of the lac operon. The top circle represents the *E. coli* chromosome map showing the lactose and other regions.

Enzyme interconversion: see Covalent modification of enzymes.

Enzyme isomerization: reversible changes in enzyme conformation in the course of a catalytic cycle. See Enzyme kinetic parameters.

Enzyme kinetic parameters, ***enzyme parameters:*** the parameters of enzyme rate equations which remain constant, provided that temperature, pressure, pH and buffer composition are constant. They are derived from the rate constants of the rate equations, and are frequently used to characterize the enzyme functionally. Some can be interpreted in physical terms, e.g. K_i as the dissociation constant for the enzyme-inhibitor complex, V_m as the maximal velocity at a saturating substrate concentration, K_s (substrate constant) as the concentration of substrate which half saturates the enzyme, K_m (Michaelis constant) as the substrate concentration for half-maximal velocity.

Enzyme kinetics: the mathematical treatment of enzyme-catalysed reactions. A great deal of information about reaction mechanisms can be obtained from kinetic experiments and evaluation of the data so obtained (see Kinetic data evaluation). A kinetic experiment consists of measuring the rate of disappearance of substrate, or appearance of product under controlled conditions of temperature, pH, substrate and enzyme concentration, buffer composition, etc. The simplest graphical representation of the data is a *progress curve,* a plot of $\Delta[S]$ against time (where [S] is the concentration of substrate). One is usually interested in the *instantaneous velocity,* $v = d[S]/dt$, which may not be easy to determine from the progress curve. In this case an *integrated rate expression,* or the time course of product formation may be useful. The *activity* of an enzyme is expressed as the amount of substrate consumed (or product formed) in a given time. In 1961 the Nomenclature Committee of the International Union of Biochemistry recommended the use of the *Unit (U),* which is the amount of enzyme required to turn over 1 micromole of substrate per minute under standard conditions. In accordance with the shift to the mks system, in 1972 they recommended the *katal,* the amount of enzyme turning over 1 mole substrate per second. This is a very large unit, so that in practice one uses the micro-, nano- and picokatal. *Specifc activity* is given in U/mg or kat/kg.

Single-substrate enzymes (see) display first order kinetics. The rate equation for such a *unimolecular* or *pseudounimolecular* reaction is $v = -d[S]/dt = k[S]$. The reaction is characterized by a *half-life* $t_{1/2} = \ln 2/k = 0.693/k$, where k is the first-order rate constant. The *relaxation time,* or the time required for [S] to fall to (1/e) times its initial value is τ; $\tau = 1/k = t_{1/2}/\ln 2$.

When there is more than one substrate (see Multisubstrate enzymes), the kinetics may be *second order* (or pseudo-second-order. See Cleland's short notation). The equation for a second order reaction is $A + B \xrightarrow{k_2} P$, where k_2 is the bimolecular rate constant, and $v = k_2[A][B]$. All chemical reactions are reversible andd eventually reach an equilibrium in which the rates of the forward and reverse reactions are equal.

At this point, $A + B \underset{k_2}{\overset{k_1}{\rightleftharpoons}} P$, where k_1 is the rate of the forward reaction and k_2 is that of the back reaction. The equilibrium constant is given by $K = [P]/[A][B] = k_1/k_2$.

The Michaelis-Menten treatment of E. assumes that the enzyme and substrate bind temporarily to form an *enzyme-substrate complex,* which may break down either to substrate or product: $E + S \rightleftharpoons ES \rightleftharpoons E + P$. For a short period of time, it can be assumed that the rate of change in [ES] is small compared with the change in [S] (the *steady state approximation*) because the rate at which it is formed is equal to the rate at which it breaks down. It follows that $[E][S] = \frac{(k_2 + k_3)}{k_1}[ES] = K_m[ES]$, where K_m is the *Michaelis constant. The Michaelis-Menten equation* for the initial rate of reaction (when [P] = O) is $v = \frac{V_{max}}{1 + K_m/[S]} = \frac{V_{max}[S]}{K_m + [S]}$. The quantities K_m and V_{max} are called the kinetic parameters of an enzyme (see Enzyme kinetic parameters). A plot of initial rate against substrate concentration is not linear (Fig.), so it is difficult to estimate the kinetic parameters from it (except by computer analysis).

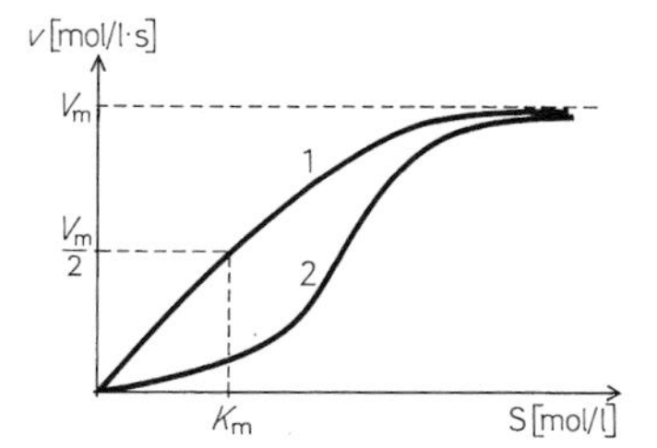

Plot of initial rate vs. substrate concentration, sometimes called a characteristic curve. 1 is a Michaelis-Menten hyperbola; 2 is a sigmoidal plot characteristic of a cooperative enzyme. For linear transformations see Kinetic data evaluation.

In the treatments discussed so far, it has been assumed that the back reaction could be neglected. The reactions catalysed by many enzymes are essentially irreversible or the products are immediately subject to further reaction, so that the assumption of irreversibility is valid. However, if the reaction is reversible, the Michaelis equation must be modified. Haldane suggested a notation in which V_f and V_r are the maximal velocities in the forward and reverse directions, and K_{mS} and K_{mP} are the Michaelis constants for the substrate and product. The *Haldane relationship* for a system with a single substrate and single product is then $K_{eq} = V_f K_{mP}/V_r K_{mS}$.

Multi-substrate enzymes (see) catalyse reactions of two or more substrates. Such enzymes can form a number of different complexes (known as *enzyme species*) with one or both substrates and/or products. The order in which these species are formed may be *random* or *ordered.* Cleland's short notation (see) is a convenient way of representing the possibilities. The kinetics of such reactions become extremely complicated; *enzyme networks* (see Enzyme graphs) provide a means of summarizing them. To evaluate the kinetic data for such systems, one must resort to a computer. Furthermore, the information gained from steady-state experiments may not be sufficient. A number of methods of very rapid measurement have been used to investigate the *pre-steady-state* condition of reactions, including stopped flow, temperature jump and flash methods.

Enzyme modulation: see Covalent modification of enzymes.

Enzyme network: see Enzyme graph.

Enzyme parameter: see Enzyme kinetic parameter.

Enzyme repression: blockage of the synthesis of enzymes of a biosynthetic pathway by the end product of the same pathway. This type of regulation is found in prokaryotes, in particular for operons of amino acid biosynthesis. If an amino acid is available in the growth medium, the synthesis of all the enzymes in the operon is turned off, but if it is in short supply, the operon is derepressed (see Derepression). See also Attenuation.

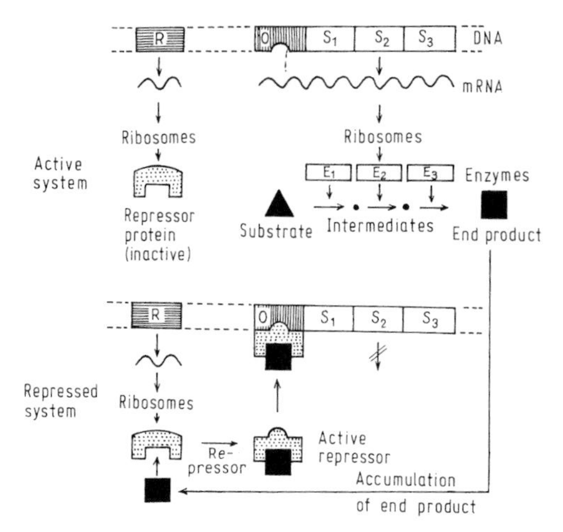

Repression of enzyme synthesis

The mechanism of E.r. is shown in the Fig. Regulator gene R encodes an aporepressor which cannot bind to the operator gene O unless it forms an active complex with the corepressor (the end product of the biosynthetic chain). In the absence of the corepressor the structural genes are transcribed and the enzymes synthesized. See Gene activation.

Enzymes: protein catalysts. About 90 % of cellular proteins are E., and some structural proteins (e.g. actin and myosin) are also E. A protein is classified as an E. if it is known to catalyse a reaction, but it is always possible that a given protein catalyses an unrecognized reaction. The Nomenclature Committee of the IUB lists well over 2,000 E. which have been well characterized; however, these are classifed according to the reaction catalysed, not the identity of the proteins. The same reaction may be catalysed by non-homologous proteins in different organisms, and thus the number of E. actually characterized is much higher than 2,000. The Nomenclature Committee publishes an enzyme catalog (EC) in which each E. is assigned four consecutive numbers. The first number indicates the general category of reaction (Table 1), the second and third numbers indicate the subgroup and sub-sub-group, while the fourth part number is assigned arbitrarily.

The committee also establishes a recommended trivial name for each E., to be used in text, and a systematic name as the basis for classification. The latter should be mentioned at least once in each publication.

Table 1. *The main enzyme groups in the enzyme catalog.*

Group	Type
1	Oxidoreductases
2	Transferases
3	Hydrolases
4	Lyases
5	Isomerases
6	Ligases (synthetases)

Since the properties of E. from different sources vary, often substantially, it is also necessary to indicate the source in a publication. The most striking difference between E. and chemical catalysts is the specificity of E. with regard both to their substrates and to the reactions catalysed. E. distinguish stereoisomers absolutely, due to the three-dimensional nature of the binding between substrate and enzyme. The lock and key model suggests that the E., like a complicated lock, is fitted only by a substrate (key) of exactly the correct shape. The *induced fit model* hypothesizes that when the substrate binds to the E., the conformation of the latter changes in such a way as to make the fit more complete. In either case, the difference between D- and L-forms of a substrate is analogous to the difference between right and left hands; the substrate binding site on the E. corresponds to a glove which will fit only one of the two hands. The substrate adheres to (and often reacts with) the E. because the arrangement of charged or hydrophobic groups in the enzyme precisely complements the arrangement of charges or hydrophobic sections of the substrate, or because hydrogen bonds between the two can form easily. The mechanisms by which E. catalyse reactions are usually fairly simple. In general, molecules do not react until they have been activated to a higher-energy electronic or vibrational state called the transition state. In spontaneous reactions, the necessary free energy is supplied by molecular collisions or light. However, when the molecule is activated in this way, it can usually rearrange itself, break down or react with another molecule in a number of different ways, so that there are usually a number of reaction products. Any catalyst increases the rate of a reaction by reducing its free energy of activation. A chemical catalyst may also promote the "right" reaction; E. invariably do. This is accomplished in several ways: by bringing the reactants together in the appropriate orientations; by providing acidic and/or basic groups at the right positions to promote acid or base catalysis of the "desired" reaction; by providing electrophilic or nucleophilic groups which form a temporary covalent bond with the substrate, and/or by providing for concerted mechanisms in which two catalytic groups on the E. attack the substrate molecule. Any of these forms of catalysis may be accompanied by conformational changes in the E. molecule which strain the substrate molecule and facilitate the breaking of bonds within it.

The amino acid residues which take part in catalysis and their immediate neighbors constitute the *active center* of the E. The *substrate binding site* is close to or includes the active center, but if the substrate is large, the binding site may extend beyond the active center. Allosteric E. (see Allostery) also have *effector binding sites* which are distinct from the substrate binding site.

Most E.are strictly intracellular, but some are excreted into the body fluids or the medium of unicellular organisms. Proteolytic E. are secreted in the form of inactive precursors which are only activated after they are safely out of the cell; those which are not secreted are sequestered into a special organelle, the lysosome. Some E. are specific for particular organs or tissues, and their presence in the blood can be used as a diagnostic test for damage to the tissue of origin. Within the eukaryotic cell there is a fair amount of compartmentation of the E. (and reactions) within or-

Table 2. *A selection of technical and medical applications of enzymes.*

Enzyme	Reaction	Application
Glucose oxidase from *Aspergillus niger* or *Penicillium*	Glucose $\xrightarrow{O_2}$ Gluconolactone	Preserving foods and drinks by removal of O_2. Prevention of brown discoloration (e.g. in dried eggs) by removal of glucose.
Catalase from microorganisms	$2H_2O_2 \rightarrow 2H_2O + O_2$	a) Food preservation, together with glucose oxidase. b) Removal of excess H_2O_2 in milk preservation by treatment with H_2O_2.
Glucose isomerase from yeast and other microorganisms	Glucose ⇌ Fructose	Fructose production to increase sweetness of drinks without adding carbohydrate. Preparation of fructose to add to paper to increase its plasticity.
Invertase (β-fructosidase) from yeast and *Aspergillus*	Sucrose (S.) → Glucose + Fructose (Invert sugar I.)	I. is sweeter and more easily digested than S., and is used in artificial honey, ice cream, chocolate creams, etc.
Lactase (β-galactosidase)	Lactose → Galactose + Glucose	Production of milk products for adults lacking the enzyme.
"Naringase"	Naringin → Naringenin + carbohydrate	Removal of the bitter taste of grapefruit juice. Naringenin is less bitter than naringin.
Lipase from pancreas or *Rhizopus nigricans*	Triacylglycerol → Glycerol + Fatty acids	Isolation of labile fatty acids, improvement of cheese aroma, cocoa processing, digestion aid.
L-Amino acid acylase from kidneys	DL-Acylamino acid → L-Amino acid + D-Acylamino acid	Production of essential amino aids for human and animal nutrition.
Penicillin amidase from microorganisms	Penicillin G → 6-Aminopenicillanic acid (6AP)	6AP is the starting material for semisynthetic penicillin.
α-Amylase (Endoamylase from pancreas, bacteria and *Aspergillus*)	Hydrolysis of α-1,4-glucans (e.g. starch, amylose, amylopectin) to dextrins and maltose	Digestion and baking aid; cleavage of starch in beverages; production of starch pastes and non-sweetening syrups; removal of starch in the textile industry.
α-Amylase + Glucoamylase	Removal of glucose from the nonreducing end of products of amylase action	Production of glucose from starch in high yield and purity.
Cellulase from *Aspergillus niger* and *Stachybotrysatra*	Hydrolysis of cellulose to cellobiose	Aid to extraction of pharmaceutically active principles; removal of cellulose from food for special diets; softening of cotton.
Pectin esterase from *Aspergillus niger*	Pectin methyl ester → Pectinic acid + CH_3OH	Removal of pectin sheaths from plant fibers; removal of cloudiness in fruit juices and beer; production of fruit juices and purees.
Lysozyme from microorganisms	Hydrolysis of the murein of bacterial cell walls	Removal of bacteria from cow's milk for infant formulas.
Proteases (trypsin, pepsin, elastase, papain, microbial enzymes, etc.)	Hydrolysis of peptide bonds	Digestion aids; removal of necrotic tissue; acne treatment; meat tenderizers; preparation of special diets and peptone media for microorganisms; removal of proteins from carbohydrates and fats; prevention of cloudiness in beer; laundry additive; leather tanning.
Rennet from calf stomach and similar enzymes from microorganisms	Hydrolysis of the Phe-Met bond in ϰ-casein	Cheese production.
Streptokinase from streptococci	Plasminogen → Plasmin	Removal of blood clots (fibrinolysis).

ganelles. When the cell is fractionated, these E. can be used as *markers* for the various fractions (mitochondria, cytosol, liposomes, etc.). In some cases, a given catalysis may be performed by two or more *isoenzymes* in different tissues or cell fractions.

In addition to their clinical value in the symptomology of tissue damage, E. are exploited as reagents for the specific qualitative and quantitative analysis of other metabolites, such as glucose, ATP, lactate, etc. Certain proteases are used medically to aid poor digestion, to prevent blood coagulation and to promote fibrinolysis.

With the development of methods for large-scale cultivation of microorganisms and isolation of E. in very large quantities, the industrial use of E. has become economically feasible, e.g. in food processing, in textile and paper manufacture, in laundry products and in pharmaceuticals. With the aid of genetic engineering it is potentially possible to develop artificial E. for the catalysis of reactions in organic and pharmaceutical chemistry.

Enzyme species, ***intermediates:*** all covalent and noncovalent complexes between an enzyme and a substrate and/or product or effector; the term also includes all the different conformations of an enzyme that occur during catalysis (see Enzyme isomerization). The concentrations of the E.s. cannot be measured directly by means of steady-state Enzyme kinetics (see), but their steady-state concentrations can be calculated from the kinetic equations for definite values of the concentration variables. In principle the concentrations of the E.s. might be determined by the methods of presteady-state kinetics.

Enzyme units: see Enzyme kinetics.

Ephedrin: see Antidepressants.

Epidermal growth factor, ***EGF:*** a mitogenic polypeptide, M_r 6,045, of known structure (Fig.) with growth stimulating activity for a wide variety of epidermal and epithelial cells in vitro and in vivo. Injected into newborn mice, EGF causes precocious opening of the eyelids and precocious tooth eruption. EGF accounts for about 0.5 % of the protein centent of adult male mouse submaxillary gland. A homologous polypeptide, human EGF [probably identical to Urogastrone (see)], has been isolated from human urine. EGF from mouse submaxillary gland is a high M_r complex, containing two molecules of EGF and 2 molecules of EGF binding protein (M_r 29,300). The binding protein has arginine esterase activity, and is thought to represent the processing enzyme that con-

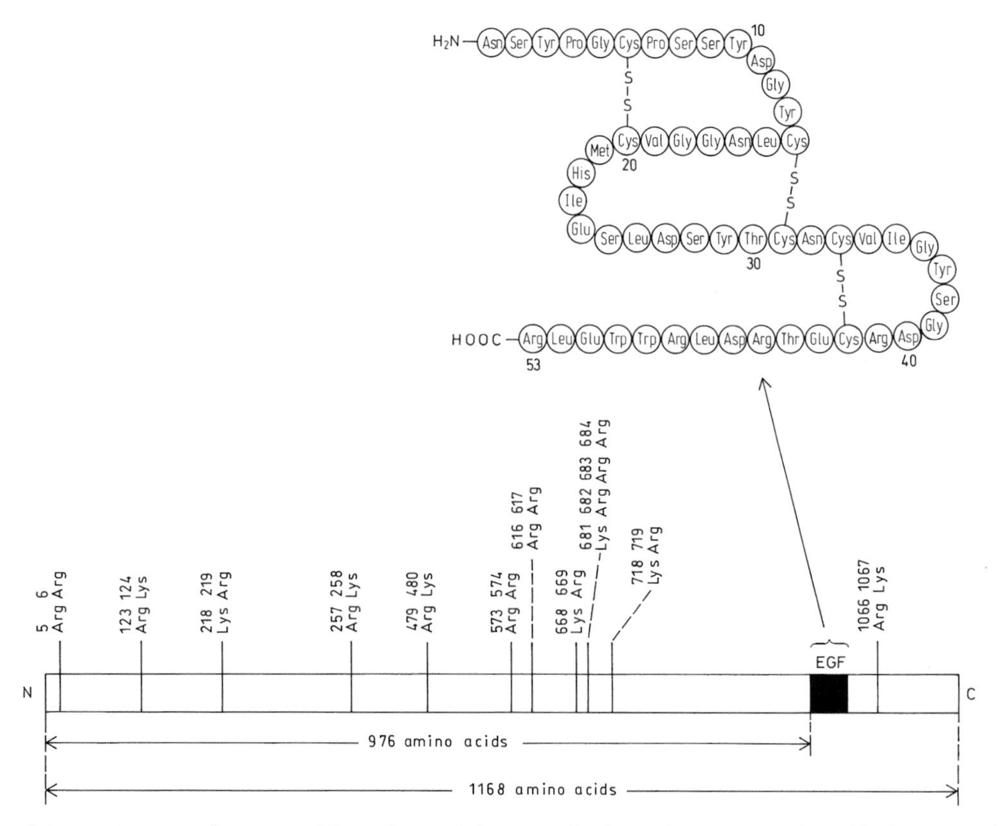

Primary structure of mouse epidermal growth factor, and schematic representation of its large protein precursor (pre-pro-EFG). The signal sequence is uncertain, but the amino terminus of pro-EGF is probably residue 23 (Ile), 26 (Val) or 29 (Trp). The indicated sequences of basic residues represent potential sites of proteolytic cleavage.

verts proEGF into EGF. The concentration of EGF in submaxillary gland is androgen-dependent, and the concentration is much lower in the glands of female mice. As determined by radioimmunoassay, mouse serum contains 1 ng/ml EGF. This may rise to 150 ng/ml when adrenergic receptors are stimulated by phenylephrine. Thus EGF seems to serve an as yet unidentified physiological function, and its secretion is regulated. Radioimmunoassay of human EGF shows a normal 24 h excretion of 63.0 ± 3.0 ng (2 SD) (males) and 52.0 ± 3.5 ng (females) with no diurnal or postprandial variation. Human saliva contains 6–17 ng/ml, human milk contains 80 ng/l, and it is undetectable in amniotic fluid. Circulating plasma levels are 2–4 ng/ml. Elevated levels are excreted by females taking oral contraceptives.

Analysis of a cDNA clone for mouse EGF suggests that a large protein precursor (1,168 amino acids) is first synthesized, then processed to EGF (Fig.). It seems unlikely that such a large polypeptide serves only as the precursor of EGF. There are several potential proteolytic cleavage sites within the large precursor, but if processing produces other physiologically active peptides, they have not yet been recognized. [J. A. Downie et al. *Ann. Rev. Biochem.* **48** (1979) 103–131; A. Gray et al. *Nature* **303** (1983) 722–725; S. R. Campion & S. K. Niyogi 'Interaction of Epidermal Growth Factor with Its Receptor' *Prog. Nucl. Acid Res. Mol. Biol.* **49** (1994) 353–383; P. A. Weernink & G. Rijksen *J. Biol. Chem.* **270** (1995) 2264–2267; S. Tomic et al. *J. Biol. Chem.* **270** (1995) 21 277-21 284]

Epimers: see Carbohydrates.

Epinephrin: see Adrenalin.

Epiphysis: see Pineal gland.

Episome: see Plasmid.

2,3-Epoxysqualene: see Squalene.

Equilenin, ***3-hydroxyestra-1,3,5(10),6,8-pentaen-17-one:*** an estrogen, M_r 266.32, m.p. 259 °C, $[\alpha]_D^{16}$ + 87 ° (12.8 mg made up to 1.8 ml in dioxane), found together with Equilin (see) in the urine of pregnant mares. It has 1/25 of the estrogenic activity of estrone. E. differs from equilin in having an aromatic ring B. It was the first natural steroid to be totally synthesized, in 1939.

Equilin, ***3-hydroxyestra-1,3,5(10),7-tetraen-17-one:*** an estrogen, M_r 268.34, m.p. 240 °C, $[\alpha]_D^{25}$ + 308 ° ($c = 2$ in dioxan), isolated from the urine of pregnant mares. It has 1/20 of the estrogenic acitivity of estrone.

Equilin

Equivalence point, ***equivalence zone:*** the region of the precipitation curve in which all the antibody binding sites are saturated with the antigenic determinants, and the antibodies have been quantitatively precipitated. Neither antigen nor antibody remains in the supernatant. At this point the antigen-antibody complex forms a complicated network structure (Fig.). Otherwise only smaller structures or soluble binary complexes are formed.

1 Antibodies in excess

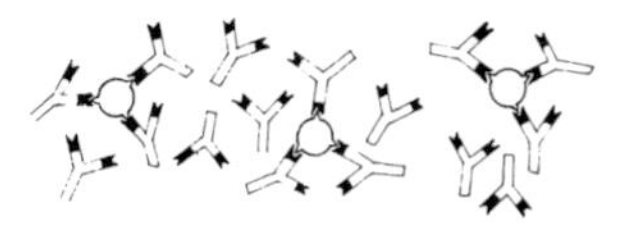

2 Antibodies and antigens mutually saturated (equivalence point); interlocking structure

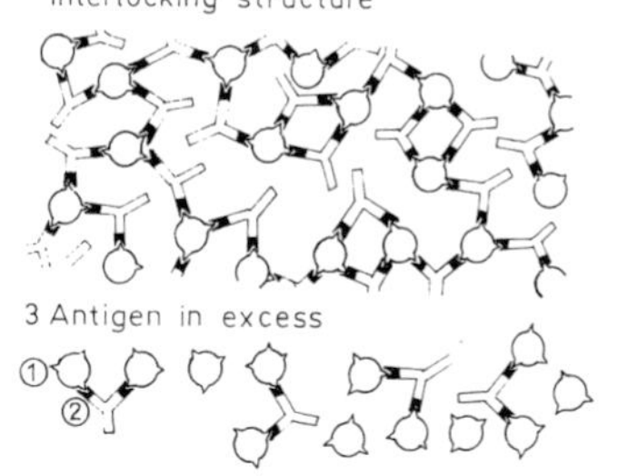

3 Antigen in excess

① Trivalent antigen
② Bivalent antibody

Basic types of antigen-antibody complex illustrated for the case of a trivalent antigen and a bivalent antibody. The variable part with the antigen-binding site is shown in black.

Equivalence zone: see Equivalence point.

Equol: see Isoflavone.

ER: acronym of Endoplasmic reticulum.

Erepsin: an outdated term for the amino- and dipeptidases secreted by the mucous membranes of the small intestine. Tripeptidases are not included in the erepsin complex, but are present in the mucous membrane itself.

Ergastoplasm: see Endoplasmic reticulum (rough).

Ergocalciferol: see Vitamins (vitamin D_2).

Ergochromes, ***secalonic acids:*** a group of weakly acidic, bright yellow natural pigments based on a dimeric 5-hydroxychromanone skeleton. They have been isolated from a number of molds and lichens and are poisonous. They are synthesized from acetate via the anthraquinone emodin. The most common is secalonic acid A; others are secalonic acid B, C and D, ergoflavin, ergochrysins A and B and the ergochromes AD, BD, CD and DD. The ergot mold con-

Structural system of ergochromes

tains two anthraquinone carboxylic acids, the red-orange endocrocin and the red clavorubin, which has an additional hydroxyl group on C5.

Ergoline alkaloids: see Ergot alkaloids.

Ergometrine: see Ergot alkaloids.

Ergosomes: see Polyribosomes.

Ergostane: see Steroids.

Ergosterol, ***provitamin D_2, ergosta-5,7,22-trien-3β-ol:*** the most abundant mycosterol (fungal sterol). In most fungi E. is an important component of cell or mycelium membranes. E. is also synthesized by some protozoa (notably *Trypanosomatidae* and soil amebas), *Chlorella,* and the primitive tracheophyte *Lycopodium complanatum.* It is not synthesized by all fungi, e.g. it is absent from *Pythium* and *Phytophthera,* which do not synthesize any sterols. In some fungi, E. is synthesized but is not the most abundant sterol, e.g. the predominant sterols in rusts *(Uredinales)* possess a C_2 substituent at C24. E. is converted by UV irradiation into vitamin D_2 (see Vitamins, cholecalciferol). The main commercial source of E. is yeast. For biosynthesis, see Steroids. [E. I. Mercer *Pestic. Sci.* **15** (1984) 133–155]

Ergot alkaloids, ***Claviceps alkaloids, ergoline alkaloids:*** a group of over 30 indole alkaloids, possessing the ergoline ring system. They are produced by various species of the fungal genus *Claviceps* (class Ascomycetes), which parasitize rye and wild grasses. Following infection by spores of *Claviceps purpurea,* the ears of rye develop 1–3 cm long dark violet (almost black), highly poisonous sclerotia, containing up to 1 % E.a. The same alkaloids have also been found in higher plants.

E.a. form 2 subgroups (lysergic acid derivatives and clavine alkaloids), depending on the oxidation state of the substituent at C8 of the tetracyclic ergoline ring system (Fig. 1). The first group includes simple amide derivatives (e.g. ergometrin) and cyclic tripeptides (e.g. ergotamine). The tripeptide is formed from three of the following: D-proline, L-leucine, L-valine, L-phenylalanine and L-alanine, and it is linked to the carboxyl group of lysergic acid via alanine (ergotamine type), or valine (ergotoxin type). All E.a. derived from lysergic acid are levorotatory. The dextrorotatory isomers are derived from isolysergic acid, and their trivial names carry the ending "-inine" (e.g. ergotaminine). In aqueous solution, the derivatives of lysergic acid slowly isomerize to isolysergic acid derivatives by a change in configuration at C8. E.a. are detected by fluorescence or by indole reagents.

Biosynthesis. In *Claviceps purpurea,* and presumably also in higher plants, E.a. are biosynthesized from tryptophan and isopentenylpyrophosphate (see Terpenes). Synthesis proceeds via 4-dimethylallyltryptophan, which is converted into the alkaloid chanoclavine (by hydroxylation, methylation, decarboxylation and formation of a new C-C bond). All the other clavine alkaloids and the lysergic acid derivatives are derived from chanoclavine. The peptide moieties of the ergotamine and ergotoxin alkaloids are formed by a multienzyme complex (Fig. 2). E.a. are prepared from the sclerotia of rye, previously inoculated with *Claviceps;* and they are also produced by culture of the fungus on artificial growth media.

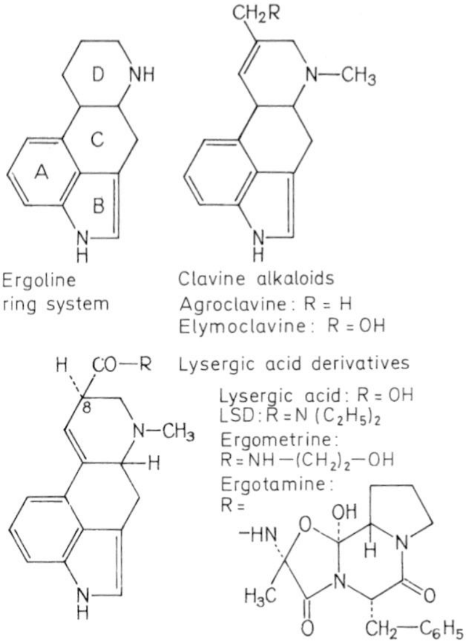

Ergoline ring system

Clavine alkaloids
Agroclavine: R = H
Elymoclavine: R = OH

Lysergic acid derivatives
Lysergic acid: R = OH
LSD: R = N $(C_2H_5)_2$
Ergometrine: R = NH$-(CH_2)_2-$OH
Ergotamine: R =

Fig. 1. *Ergot alkaloids*

Isolysergic acid and clavine alkaloids are physiologically inactive, whereas the lysergic acid derivatives (especially the peptide alkaloids) exhibit a variety of useful pharmacological properties. Extracts of ergot were used earlier, but owing to their variable alkaloid content, they have now been totally replaced by pure alkaloids (in particular ergotamine), or semisynthetic analogs. E.a. stimulate contraction of smooth muscle, especially of uterus and of arterioles in peripheral parts of the body; they are used to control of hemorrhage after childbirth. Owing to their wide spectrum of pharmacological activity E.a. are also used in combination preparations, e.g. with tropane alkaloids for the suppression of the autonomic nervous system. Whereas the natural E.a. have vasconstrictor activity, the semisynthetic compounds produced by hydrogenation of the Δ^9 double bond are vasodilatory.

Only a few new naturally occurring E.a. have been found in recent years. Of particular interest is ergoladinine (Fig. 3) of the *iso*-lysergic acid series; it is the first reported sulfur-containing, natural E.a., and the cyclic tripeptide moiety is derived from proline, valine and methionine.

Historical. The first description of the drug, ergot, with directions for its use dates from 1582. In the 19th century, the drug was introduced into many pharmacopeas. In 1816, Vauquelin started work on the identification of the active principle. 100 years later Stoll first succeeded in isolating crystalline ergotamine. For many centuries it was not realized that ergot was poisonous, and in Europe contamination of rye flour with ergot led to periodic widespread outbreaks of poisoning, known as "St. Anthony's fire", now called ergotism. Toxic symptoms are convulsions, permanent mental damage, limb gangrene, often followed by death. [L. Cvak et al. 'Ergoladinine, an ergot alkaloid' *Phytochemistry* **42** (1996) 231–233]

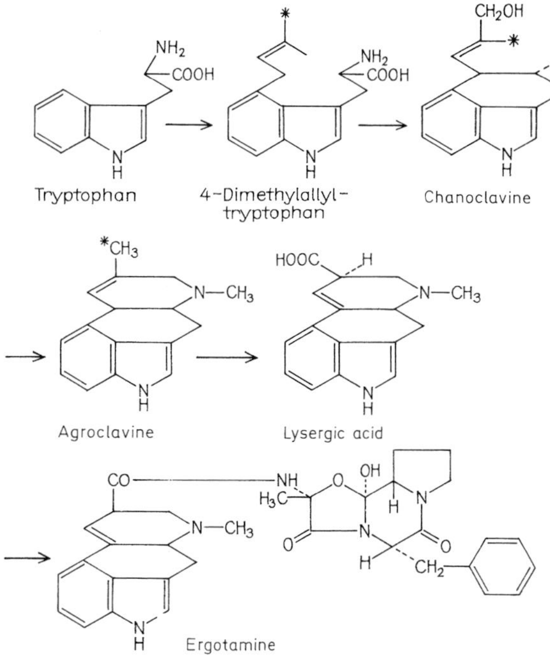

Fig. 2. *Ergot alkaloids.* Biosynthesis of clavine and lysergic acid alkaloids. C-atoms with an asterisk have the same origin.

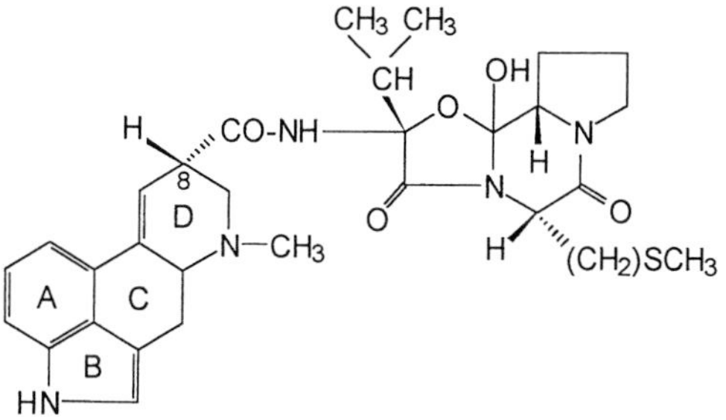

Fig. 3. *Ergoladinine from* Claviceps purpurea

Ergotamine: see Ergot alkaloids.

Ergotoxin: see Ergot alkaloids.

Eriodictyol: 5,7,3′,4′-tetrahydroxyflavanone. See Flavanone.

Erucic acid, *Z-13-docosenoic acid, Δ^{13-14}docosenoic acid:* CH_3–$(CH_2)_7$–CH = CH–$(CH_2)_{11}$–COOH, M_r 338.56, *cis*-form: m.p. 34 °C, b.p.$_{10}$ 254.5 °C, a fatty acid found esterified in the glycerides of many seed oils of the *Cruciferae* and *Tropaeolaceae.* It constitutes 40–50 % of the total fatty acids of rapeseed, mustard and wallflower seeds, and 80 % of the fatty acids of nasturtium seeds.

Erythrina alkaloids: isoquinoline alkaloids, usually tetracyclic, found exclusively in the legume genus *Erythrina.*

D-Erythritol: CH_2OH–CHOH–CHOH–CH_2OH, an optically inactive sugar alcohol (M_r 122.12, m.p. 122 °C), derived from D-erythrose. It is found in some algae, fungi, lichens and grasses.

Erythrocruorin: a hemoglobin-like protein found in many invertebrates. In some snails and worms (e.g. *Cirraformis)* it is a high-molecular-mass ($M_r = 3 \times 10^6$) extracellular respiratory pigment, consisting of 162 heme-bearing polypeptide chains of M_r 18,500. In sea cucumbers, mussels, polychaete worms and some primitive vertebrates, like the river lamprey, it occurs as an intramuscular, low-molecular-mass protein (M_r 16,700 to 56,500).

Erythrocuprein: see Superoxide dismutase.

Erythrocyte membrane: a typical plasma membrane, which because of its ease of preparation, ready availability and relative simplicity has often been chosen for study as a prototype membrane (see Fig. 1 of Biomembrane). However, it is unusual in being particularly tightly anchored to the cytoskeleton. Membranes of mature mammalian erythrocytes are simply prepared by osmotic lysis and washing. When returned to physiological media, these ‘ghosts’ reseal and retain the original shape of the erythrocyte. The E. m. consists of approximately 49 % protein, 44 % lipid and 7 % carbohydrate.

Solubilization of erythrocyte ghosts in 1 % sodium dodecyl sulfate, followed by SDS-polyacrylamine gel electrophoresis and staining with Coomassie blue reveals more than 10 well defined protein bands. When staining is performed with periodic acid-Schiff reagent (PAS), which stains carbohydrates, four bands (PAS bands) are revealed.

Some proteins can be extracted from the E. m. by altering the ionic strength or pH of the medium, indicating that they are peripheral proteins. Since these proteins are attacked when proteases are added to preparations of unsealed ghosts, but are unaffected in sealed ghosts or intact erythrocytes, they must be located on the cytoplasmic side of the membrane.

The peripheral proteins have been identified as follows (bands are numbered from the top of the SDS-polyacrylamide gel pattern; band 1 is therefore the slowest and has the highest apparent molecular mass). Bands 1 and 2: dimeric and monomeric form of spectrin, which forms a network and stabilizes and controls the shape of the erythrocyte by interacting with other proteins. Band 4.1: a protein that links spectrin to the membrane. Band 5: actin. Band 6: glyceraldehyde 3-phosphate dehydrogenase.

Bands 3 and 7 and all four PAS bands are integral membrane proteins, which can be dissociated from the E.m. only with detergents or organic solvents. Band 3 is the anion-transport protein (anion channel) that facilitates the exchange of bicarbonate and chloride that is necessary for the buffering of erythrocyte pH, and the import of phosphate and sulfate. The PAS bands represent different members of the glycophorin family.

Glycophorin A is a major sialoglycoprotein of the E.m., consisting of 60 % carbohydrate (16 oligosaccharide units, 15 *O*-linked, 1 *N*-linked) and 40 % protein. All the oligosaccharide units are located in the 72-residue *N*-terminal region on the extracellular surface of the membrane. This is followed by a central, 19-residue, α-helical, hydrophobic region spanning the nonpolar center of the lipid bilayer, while the 40-residue, *C*-terminal region, which has a high content of polar and ionized residues, is exposed on the cytosolic surface of the E.m. The PAS-1 band is a glycophorin A dimer, which is not disrupted by SDS, despite the fact that the two monomers are held together by noncovalent forces.

Spectrin constitutes about 75 % of the protein network ('skeleton') of the E.m. It consists of two closely similar subunits with a high degree of sequence homology. Band 1 is the α-subunit (M_r 280,000) and band 2 the β-subunit (M_r 246,000), each consisting of 106-residue repeating segments that are thought to fold into triple stranded α-helical coiled coils. In the electron microscope, the two subunits are seen to be intertwined to form a 1000 Å-long αβ heterodimer. Two such dimers are further associated head-to-head to form an $(\alpha\beta)_2$ tetramer. Each complete ghost contains about 100,000 such tetramers, cross-linked at both their ends by attachment to the proteins that constitute bands 4.1 and 5 (actin), forming a dense network on the cytosolic face of the E.m. In this network, spectrin also binds to Ankyrin (see), which in turn binds to the anion-transport protein (band 3).

This dense network of cross-linked spectrin is responsible for maintaining the concave-disc shape of the erythrocyte, and for allowing the extreme flexibility it needs to pass through capillaries.

Some hemolytic anemias arise from abnormalities in the protein skeleton of the E.m. One form of *Hereditary spherocytosis* (fragile, inflexible, spherical erythrocytes) is due to decreased synthesis of normal spectrin and the presence of an abnormal spectrin that binds band 4.1 protein with decreased affinity; another form is due to the absence of band 4.1 protein. Due to their inflexibility and consequent resemblance to aged erythrocytes, spherocytic erythrocytes are prematurely removed and destroyed from the circulation by the spleen *Hereditary elliptocytosis* or *Hereditary ovalcytosis* (elliptical erythrocytes), common in parts of Southeast Asia and Melanesia, arises from a defective anion-transport protein; the homozygotic condition is lethal, but heterozygotes are resistant to malaria. [R.T.Moon & A.P.McMahon 'General diversity in nonerythroid spectrins' *J. Biol. Chem.* **265** (1990) 4427–4433; K.E.Sahr et al. 'Complete cDNA and polypeptide sequence of human erythroid α-spectrin' *J. Biol. Chem.* **265** (1990) 4434–4443; J.A.Chasis & N.Mohandas 'Red blood cell glycophorins' *Blood* **80** (1992) 1869–1879; J.G.Conboy 'Structure, function, and molecular genetics of erythroid membrane skeletal protein 4.1 in normal and abnormal blood cells' *Semin. Hematol.* **30** (1993) 58–73; V.Bennett & D.M.Gilligan *Annu. Rev. Cell Biol.* **9** (1993) 27–66; P.O.Schischmanoff et al. 'Spectrin and actin-binding domains of 4.1' *J. Biol. Chem.* **270** (1995) 21243–21250; J.Whelan 'Selectin Synthesis and Inflammation' *Trends Biochem. Sci.* **21** (1996) 65–69 (plus corrigendum in same volume, p.160)]

Erythromycin: see Macrolide antibiotics.

Erythrophilic γ-globulin: predominantly IgG and a minor amount of IgM (see Immunoglobulins) which coats the surface of erythrocyte membrane, and is necessary for the integrity and normal survival of the erythrocyte. Normal plasma contains about 3,000 mg E.γ-g. per liter, and about 250 mg of bound E.γ-g is present in 1 liter of packed erythrocytes. After splenectomy, production of E.γ-g is markedly decreased, and the half life of the erythrocyte is decreased by up to 50 %. During the 4–8 month period after splenectomy, E.γ-g. levels gradually return to normal, as does the half-life of the erythrocytes.

Erythrophleum alkaloids: a group of terpene alkaloids from *Erythrophleum guineense* and *E. ivorense*. They are esters or amides of the diterpene cassainic acid with substituted ethanolamines. The most important is cassaine (Fig.), a cardiotonic agent equal in potency to the digitalis glycosides.

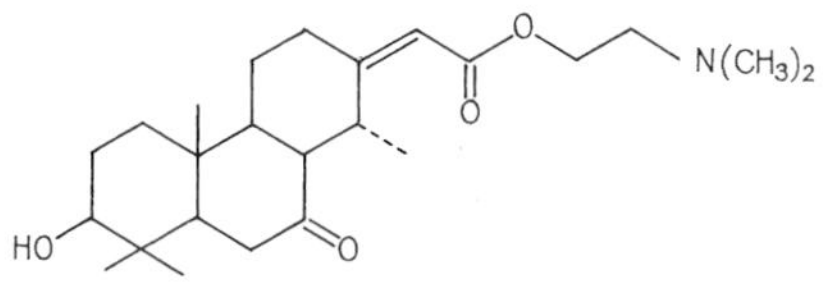

Cassaine

Erythropoietin: a glycoprotein hormone which stimulates production and release of erythrocytes in response to a lack of oxygen. E. is produced in the kidney and liver of adults, and in the liver of fetuses and neonates. It increases the rate of mRNA formation in bone marrow, resulting in increased heme synthesis and erythroblast production, and ultimately in larger numbers of circulating reticulocytes and erythrocytes. The hormone is quickly degraded and excreted by the kidneys. Human E. has 166 amino acid residues (M_r 18,398 without the carbohydrate), including 4 Cys and 3 potential sites of *N*-glycosylation (based on the consensus site -Asn-X-Ser(Thr), see Glycoproteins). It is

Chemical composition and biosynthetic capacity of Escherichia coli *in the logarithmic growth phase (modified from Lehninger).* DM = dry mass, $M_r$av = average molecular mass.

Material	$M_r$av	% DM	Molecules per cell	Molecules synthesized per second	Molecules ATP used per second	% Biosynthetic energy
DNA	2×10^9	5	4	0.033	6.0×10^4	2.5
RNA	1×10^6	10	15,000	12.5	7.5×10^4	3.1
Protein	6×10^4	70	1,700,000	1,400	2.1×10^7	88.0
Lipid	1×10^3	10	15,000,000	12,500	8.8×10^4	3.7
Polysaccharide	2×10^5	5	39,000	32.5	6.5×10^4	2.7

not known if *O*-linked glycans are present. The glycosylated protein has M_r 34,000–39,000, so that about one half of the mass of the native hormone is due to carbohydrate. The human E. gene has been cloned, and the cDNA has been expressed in COS-cells, giving a secreted product with biological activity. The amino acid sequence of E. and the nucleotide sequence of its cDNA show no significant homology with any published sequences. [K.Jacobs et al. *Nature* **313** (1985) 806–810]

D-Erythrose: CH$_2$OH–CHOH–CHOH–CHO, an aldotetrose, M_r 120. E. 4-phosphate is an intermediate in carbohydrate metabolism (see, e.g. Calvin cycle).

Escherichia coli, *E. coli:* an intestinal bacterium of the family *Enterobacteriaceae* and the most common experimental organism in molecular biology. The dimensions of a single cell are ~ 1 × 1 × 3 µm, volume ~ 2.25 µm^3, wet weight ~ 10^{-12} g, dry weight ~ 2.5×10^{-13} g. Under Optimal conditions, the organism doubles about every 20 minutes. The wild type can grow on a completely synthetic medium containing ammonium salts, and glucose. The chemical composition, biosynthetic capacity and energy expenditures are listed in the table. Most (70 %) of the dry mass is protein, but there is a much greater number of lipid molecules. When the cell is growing logarithmically, it must synthesize 12,500 lipid molecules and 1,400 protein molecules per second. Assuming an average of 400 peptide bonds per protein, this amounts to 560,000 peptide bonds per second, for which the cell expends 88 % of its biosynthetic energy; these estimates are based on certain simplifying assumptions; thus energy for synthesis of monomeric precursors was ignored, and it was assumed that there is no turnover of macromolecules.

The chemical composition figures given in the table are approximations. Such generalizations apply only to logarithmically growing bacteria, because in the stationary phase they accumulate reserve lipids, polysaccharides, polyphosphates, sulfur, etc., depending on the species; in addition, slime and capsule substances increase the proportion of polysaccharides during the stationary phase.

The nuclear material of *E. coli* is a single, circular DNA molecule; it is only 25 Å thick, but the single strand has an extraordinary total length of about 1 mm). This is called the bacterial chromosome (genome, lineome, genophore), and is transferred during bacterial conjugation. The number of DNA molecules per cell is given as 4, in accordance with Lehninger, although the number and molecular masses depend on how actively the cells are growing and dividing.

Eserine: see Physostigmine.

Esperin: a cyclic antibiotic depsipeptide from *Bacillus mesentericus.* It contains a long-chain α-hydroxy-fatty acid which is linked by an amide bond to the *N*-terminus of the sequence Glu-Leu-Leu-Val-Asp-Leu-Leu, and also forms a lactone with the γ-carboxyl group of the Asp residue.

Essential fatty acids: see Vitamins (vitamin F).

Essential oils: an extremely heterogeneous mixture of volatile, lipophilic plant products with characteristic odors. The International Standard Organization (ISO) has defined the E. o. in the strict sense as the steam distillates of plants, or of oils obtained by pressing out the rinds of certain citrus fruits. However, in practice the products of extraction with organic solvents, enfleurage or maceration of blossoms (flower oils), and the *resinoids* which can be extracted from other plant parts, and from resins and balsams are all embraced by the term E. o.

Ester alkaloids: see Steroid alkaloids.

Esterases: a large group of hydrolases acting on ester bonds: $R^1COOR^2 + H_2O \rightarrow R^1COOH + R^2OH$. The acid may be a carboxylic acid, phosphoric or sulfuric acid. The ester may be an alcoholic or a thiol ester.

Esters: see Carboxylic acids.

Estradiol, *oestradiol, 17β-estradiol, estra-1,3,5(10)-triene-3,17β-diol:* the most potent natural estrogen. E. occurs in high concentration in pregnancy urine, Graafian follicles and the placenta. In the organism, E. and Estrone (see) are interconvertible. The stereoisomeric 17α-E. has only 0.29 % of the activity of E. The E. used therapeutically is derived entirely from chemical synthesis (see Estrone). For structure, biosynthesis and therapeutic uses, see Estrogens.

Estrane: see Steroids.

Estriol, *oestriol, estra-1,3,5(10)-trien-3,16α,17β-triol:* an estrogen, M_r 288.39, m.p. 280 °C, $[\alpha]_D + 61°$ (CHCl$_3$). E. has been isolated from human female urine (especially during pregnancy) and placenta, from mare's urine, and from willow catkins. It is biosynthesized from estrone and estradiol, and differs from Estradiol (see) by the presence of an additional hydroxyl group (α-conformation) at position 16.

Estrogens, *oestrogens:* a group of female sex hormones. The ring system of E. is estrane (see Steroids). They have an aromatic A ring with a phenolic 3-OH group, and an oxygen function on C17. The chief E. are Estrone (see), Estradiol (see) and Estriol (see); see also Equilenin and Equilin. E. are produced in the Graafian follicles of the ovary, and in the corpus luteum; during pregnancy, they are also formed by the placenta. Small quantities of E. occur in male go-

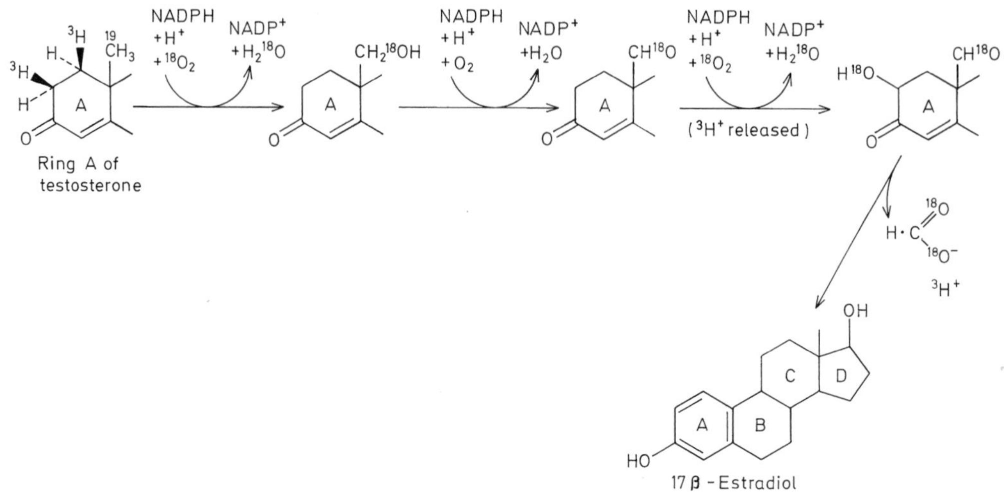

Fig. 1. *Estrogens.* Biosynthesis of estradiol from testosterone. C1β and C2β hydrogens are lost, and C19 is removed as formate. Oxygen incorporated in the first hydroxylation is retained during subsequent oxidations until it appears in the product formate. Oxygen incorporated in the final hydroxylation also appears in the product formate. [J. Fishman *Cancer Research* (Suppl.) *42* (1982) 3277s-3280s]

nads. In the female, they control the course of the menstrual cycle (in humans and apes), or the estrus or mating cycle (other animals). E., acting together with Progesterone (see) and the Gonadotropins (see), are responsible for proliferation of the uterine mucosa, growth of the mammary glands, and the appearance of secondary sexual characteristics. They are metabolized in the liver and kidney, and excreted in the urine in free form and as their sulfate and glucuronide esters. Total E. excretion: 11–47 μg/day in nonpregnant women; 20–40 mg/day in pregnant women; a maximum of 10 μg/day in men.

The Allen-Doisy test was a biological test for E., in which the response of ovarectomized mice to urinary or plasma E. was observed. It was replaced in clinical practice by a colorimetric test, the Korber method, in which the E. are treated with H_2SO_4 and hydroquinone to produce intensely fluorescing derivatives. However, the development of Immunoassays (see) for the steroid hormones has made the Korber method obsolete for routine clinical assays.

E. are biosynthesized from androgens by aromatization of ring A and loss of C19. Three sequential hydroxylations are involved. The first two occur at the C19 methyl group. The final and rate-limiting hydroxylation occurs at position 2β, followed by rapid nonenzymatic rearrangement to estradiol (Fig. 1). The aromatase system is in the endoplasmic reticulum, and has not yet been solubilized in active form. In the synthesis of E. during pregnancy there is an intricate interplay of fetus and placenta (Fig. 2). In particular, desulfatation of dehydroepiandrosterone 3-sulfate produced by the fetus can only be performed by the placenta. Estriol is not formed by direct hydroxylation of estradiol, but by hydroxylation of more oxidized precursors, followed by reduction, e. g. estrone →16-hydroxyestrone →estriol; 16α-hydroxyandrostenedione →estriol.

E. have also been isolated from plants, e. g. estrone from pomegranate seeds and date stones, and estriol from willow catkins. The roots of a Burmese tree, *Pueraria mirifica,* contain Mirestrol (see) which has the same E. potency as 17β-estradiol.

E. are used therapeutically for the treatment of menstrual disorders and menopausal problems, and as Ovulation inhibitors (see). Some synthetic steroids (see Ethinylestradiol; Mestranol) and nonsteroids (see Stilbestrol; Hexestrol) show E. activity. [Estrogen receptors (see also Hormones): J. H. Clark & E. J. Peck *Monogr. Endocrinol.* **14** (1979) 4–36; J. Norris et al. *J. Biol. Chem.* **270** (1995) 22777–22782]

Estrone, ***oestrone, 3-hydroxyestra-1,3,5(10)-trien-17-one:*** an estrogen, M_r 270.38, m. p. 259 °C, $[\alpha]_D$ + 160° ($CHCl_3$). E. occurs in human urine, ovaries and placenta, and it has also been found in pomegranate seeds (17 mg/kg) and palm oil. It was isolated independently by Doisy and by Butenandt in 1929 from pregnancy urine. E. differs from Estradiol (see) by the presence of an oxo-group at position 17. E. and estradiol are interconvertible in the organism. The total synthesis of E. was achieved in 1948 by Anner and Miescher. Nowadays some E. is obtained from horse urine, but most E. is prepared on a large scale by pyrolysis of androsta-1,4-diene-3,17-dione (obtained from cholesterol or sitosterol by microbial oxidation), or from diosgenin by the Marker synthesis (see). E. is the starting material for the chemical synthesis of other pharmacologically important steroids, e. g. estradiol and Ethynylestradiol (see).

Etamycin, ***viridogrisein:*** a cyclic peptide lactone antibiotic from *Streptomyces griseus* and related species. It is effective against Gram-positive bacteria and *Mycobacterium tuberculosis.*

Etephon: see Chloroethylphosphonic acid.

Ethanal: see Acetaldehyde.

Ethanaloic acid: see Glyoxylic acid.

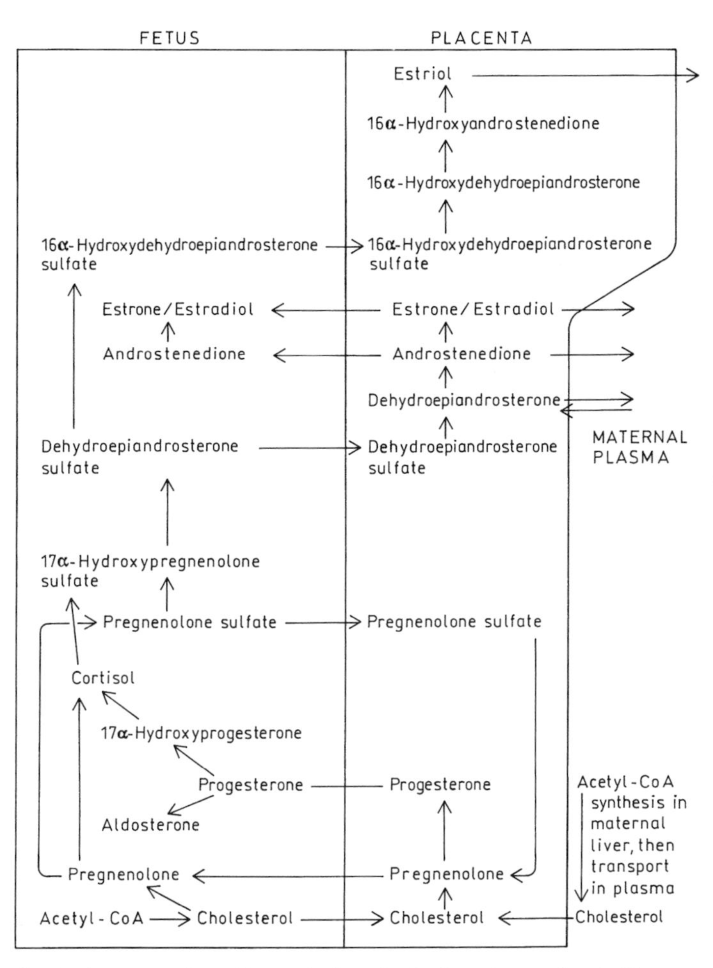

Fig. 2. *Estrogens.* Steroid metabolism in the fetoplacental unit. Plasma estriol levels are determined by fetal metabolism, so that maternal urinary or plasma estriol levels provide an index of fetal health.

Ethane dicarboxylic acid: see Succinic acid.

Ethanoic acid: see Acetic acid.

Ethanol, ***ethyl alcohol:*** CH_3–CH_2–OH, the end product of the Alcoholic fermentation of carbohydrates (see). E. is produced by decarboxylation of pyruvate, and is found in small amounts in many organisms. In humans the normal level is 0.002–0.005 % in the blood. It is degraded in the liver to acetaldehyde and acetate. In certain other tissues which lack the capacity for oxidative degradation of E., it is converted to fatty acid ethyl esters. E. causes short-term intoxication; when it is consumed in large amounts, the liver, and often the pancreas, heart and brain are damaged. Fatty acid ethyl esters are presumably responsible for the damage to the pancreas, heart and brain, which are not capable of oxidizing E. to acetaldehyde. [E. A. Laposata & L. G. Lange *Science* **231** (1986) 497–501]

Alcoholic fermentation is started by adding yeast *(Saccharomyces cerevisiae)* to any substrate containing starch or sugar, such as fruit, grain, potatoes, sugar cane or beets, or even wood hydrolysates. Fermentation stops at an alcohol content of 8 to 10 %, so the contents of the fermenter are distilled several times to obtain pure E. (see Fermentation techniques).

Ethanolamine, ***aminoethanol:*** H_2N–CH_2–CH_2–OH, a biogenic amine produced by decarboxylation of L-serine. It is a common constituent of phospholipids in which it is esterified to an acylglycerol phosphate moiety to form phosphatidyl ethanolamine (see One-carbon cycle).

Ether lipids: see Membrane lipids.

Ethrel: see Chloroethylphosphonic acid.

Ethyl alcohol: see Ethanol.

Ethylene, ***ethene, fruit-ripening hormone:*** a gaseous plant hormone found widely in plant tissues, which accelerates fruit ripening, leaf and fruit drop and plant aging. Exogenous E. accelerates the ripening and coloring of fruit, and induces blossoming and seed germination. E. biosynthesis increases as a prelude to developmentally programmed senescence, and in response to environmental stress. It is strongly suspected that the

E. receptor is a Zn^{2+}- or Cu^{2+}-containing metalloprotein, but it has not yet been characterized. The two carbon atoms of E. are derived biosynthetically from the two methylene groups of L-methionine via the intermediate 1-aminocyclopropane-1-carboxylate (ACC). Production of ACC is associated with a cycle of reactions, known as the 'Yang cycle', in which the methyl group and sulfur of L-methionine are retained and the remainder of the L-methionine molecule is continually resynthesized at the expense of the carbon atoms of ribose (Fig.); this continual replacement of L-methionine may be necessary because it is one of the less abundant amino acids. As five carbon atoms leave the cycle in the form of ACC they are replenished by incorporation of the adenosine moiety of ATP. The cycle is catalysed by a cytosolic enzyme, 1-aminocyclopropane-1-carboxylate synthase (ACC synthase). ACC synthesis is the rate-limiting step in E. synthesis, and factors promoting the production of E. (auxin, cytokinins + auxin, Ca^{2+} + cytokinins, ethylene itself, and stress factors such as wounding, anaerobiosis, heat, cold, Cd^{2+}, Li^{+}, UV light and pathogens) cause an increase in ACC synthase. Conversion of ACC to E., catalysed by a membrane-bound enzyme, ACC oxidase (likely membrane sites are the tonoplast and plasmalemma). The reaction requires molecular oxygen and ascorbate. CO_2 is stimulatory in vivo, and the purified enzyme shows a definite requirement for the presence of CO_2 in vitro; but no role has been found for this CO_2 in the stoichiometry of the conversion. Fe^{2+} is also required and is probably a normal constituent of the enzyme. The stoichiometry of the reaction is shown in the Fig.

ACC synthases have been cloned from several sources (485 amino acid residues in the tomato enzyme). They display a high degree of conservation, particularly with respect to 7 regions, one of which contains an active site lysine residue which binds pyridoxal phosphate and *S*-adenosyl-L-methionine. ACC oxidases have also been cloned from several sources (315 amino acid residues in the tomato enzyme), and they also display a high degree of conservation.

Genetic analysis of the E. signal transduction pathway has been facilitated by exploitation of the 'triple response' phenotype of *Arabidopsis.* Triple response phenotypes are mutants that synthesize increased quantities of E. The triple response consists of (i) inhibition of epicotyl and root elongation, (ii) radial swelling of epicotyl and root cells, and (iii) development of a horizontal (diageotropic) growth habit. By studying the action of inhibitors of E. perception or biosynthesis, and the behavior of mutants unable to respond to E., it has been shown that induction of the triple response depends entirely on the plant's ability to perceive and respond to E.

With aid of appropriate mutants of *Arabidopsis,* the existence of about 15 genes in the E. signal tranduction

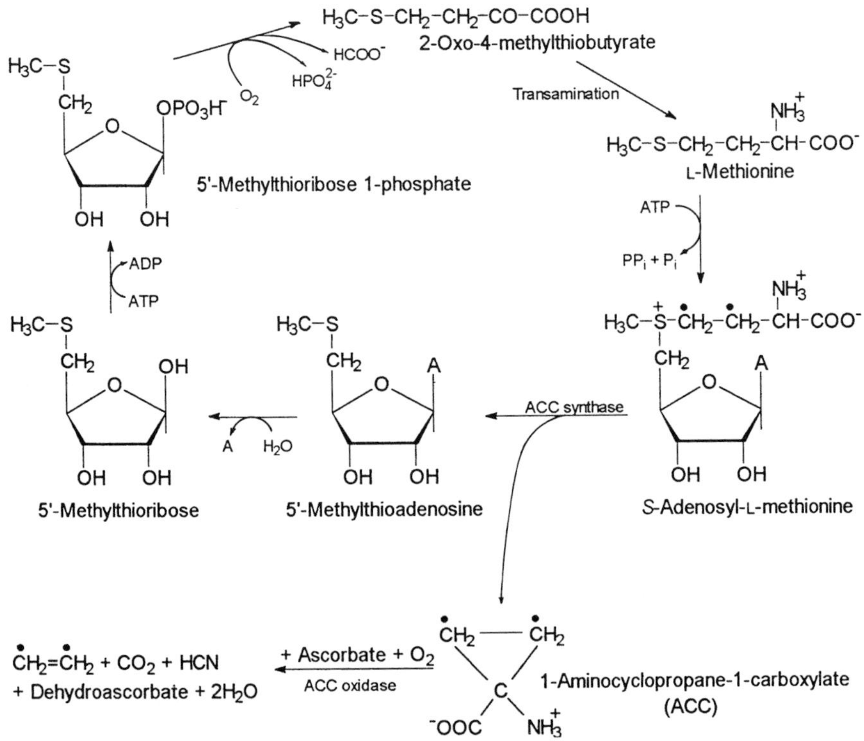

$H_3C\text{-}S\text{-}CH_2\text{-}CH_2\text{-}CO\text{-}COOH$
2-Oxo-4-methylthiobutyrate
$HCOO^-$
HPO_4^{2-}
O_2
OPO_3H^-
5'-Methylthioribose 1-phosphate
Transamination
$H_3C\text{-}S\text{-}CH_2\text{-}CH_2\text{-}CH(NH_3^+)\text{-}COO^-$
L-Methionine
ATP
$PP_i + P_i$
ADP
ATP
A
H_2O
ACC synthase
5'-Methylthioribose
5'-Methylthioadenosine
S-Adenosyl-L-methionine
$CH_2{=}CH_2 + CO_2 + HCN$
$+$ Dehydroascorbate $+ 2H_2O$
$+$ Ascorbate $+ O_2$
ACC oxidase
1-Aminocyclopropane-1-carboxylate
(ACC)

Biosynthesis of ethylene in plant tissue. The upper cycle is known as the 'Yang cycle'. The two methylene groups of *S*-adenosyl-L-methionine finally give rise, via the two methylene groups of ACC, to the two carbon atoms of ethylene (indicated by the solid circles). A = Adenine.

pathway has been demonstrated. A key component of the signaling pathway appears to be the product of a gene that is defective in *ctr1* mutants of *Arabidopsis*. The predicted product of the cloned CTR1 gene contains the typical domains of a serine-threonine protein kinase, and is structurally similar to mitogen-activating protein kinase kinase kinase (MAPKKK). Notably, in several eukaryotes, MAPKKK, MAPKK and MAPK function in phosphorylation cascades for various developmental and stress signals. [A. Theologis *Cell* **70** (1992) 181–184; H. Kende *Annu. Rev. Plant Physiol. Plant Mol. Biol.* **44** (1993) 283–307; J. R. Ecker *Science* **268** (1995) 667–675; C. Chang 'The ethylene signal transduction pathway in *Arabidopsis:* an emerging paradigm?' *Trends Biochem Sci.* **21** (1996) 129–133; D. J. Osborne et al. 'Evidence for a Non-ACC Ethylene Biosynthesis Pathway in Lower Plants' *Phytochemistry* **42** (1996) 51–60]

Ethylene generator: see Chloroethylphosphonic acid.

Ethylenediaminetetraacetic acid, *EDTA, ethylenedinitrolotetraacetic acid:* a chelating agent used in biochemical systems in vitro for the chelation of divalent metal ions. It is commonly used as the disodium salt (Na_2EDTA). Oral administration of EDTA may be used for the chelation of lead in cases of lead poisoning. In the older literature, EDTA was known by its trade name, Versene.

Ethynylestradiol, *19-nor-17α-pregna-1,3,5(10)-trien-20-yne-3,17β-diol:* a synthetic estrogen, M_r 296.41, m.p. 146 °C, $[\alpha]_D$ + 1 ° (dioxan). Administered subcutaneously, E. and estradiol have the same biological potency, but when administered orally, E. is much more active and is therefore used in oral contraceptives. It is synthesized by addition of ethyne (acetylene) to estrone in the presence of sodium in liquid ammonia. Small scale synthesis of E. (e.g. for synthesis of radiolabeled E.) is achieved by addition of lithium acetylide to estrone in dimethyl sulfoxide.

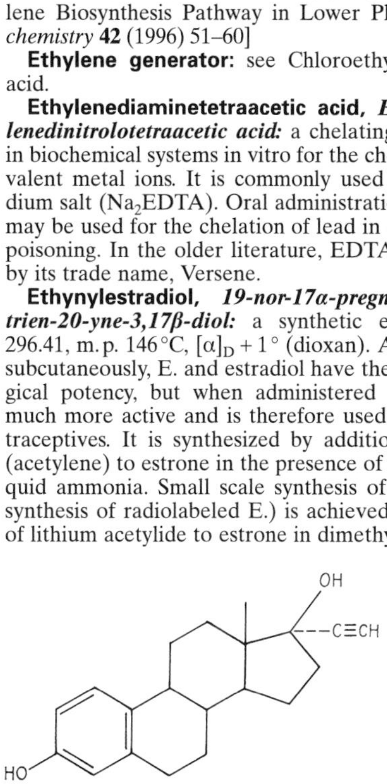

Ethynylestradiol

Etiocholane: see Steroids.

ETP: acronym of Electron transport particle (see). See also Mitochondria.

Eucalyptol: see 1,8-Cineol.

Eukaryote: an organism with eukaryotic cells. See Cell, 2.

Eumelanins: see Melanins.

Evolution (chemical and molecular): the processes by which biomolecules first arose, then gave rise to self-reproducing organisms. *Chemical evolution*, the process by which biomolecules arose from inorganic matter, is discussed under Abiogenesis (see).

It is not known how polynucleotides came to be self-replicating, but Kuhn has suggested that the alternation between single- and double-stranded forms must have been driven at first by periodic changes in the environment. Selection would have begun to operate as soon as there were systems capable of moderately accurate self-replication, and the traits presumably selected for at first were efficiency and accuracy of replication and stability of the nucleic acid molecules between replication cycles. The E. of nucleic-acid-directed protein synthesis (see Genetic code) also remains a riddle. Proteins could not be selected for their efficiency as enzymes or structural elements until their structures had been encoded in the genetic material of "proto-organisms". Furthermore, the entire system must by this time have been enclosed in some kind of membrane; otherwise the proteins would have diffused away from their nucleic acid "parents" and their stabilizing qualities would have no effect on the survival of their genes *("organic E.")*.

After the stage of nucleic-acid directed protein synthesis had been reached, the organisms must have multiplied rapidly. Selection would now favor the development of enzymes, first to synthesize nucleic acids and proteins from abiotically generated precursors, and later, as the *"primal soup"* was exhausted, to synthesize the immediate precursors of proteins and nucleic acids from available pre-precursors. The universal metabolic pathways would thus have arisen step by step, going backwards from the complex to the simple, as the growing populations of cells exhausted their immediate abiotic resources. The culmination of this process, called *"biochemical E."* was the E. of photosynthesis by blue-green bacteria, which finally produced an oxidizing atmosphere and put an end to abiotic formation of organic compounds. The oldest fossil bacteria and blue green bacteria are about 3 billion years old; the atmosphere became oxidizing roughly 2 billion years ago.

Mechanisms of molecular evolution: An organism with precisely the amount of DNA needed to encode a minimal complement of enzymes required for life would be very inflexible. Mechanisms which increase the complement of DNA per cell are thus essential for variation, adaptability and further E. Most important of these is gene duplication, which can be caused by an uneven crossing over or by a single locus duplication of a gene. At first the two duplicate genes are identical, but since mutation in one cannot be lethal if the other is functional, one at least begins to accumulate point mutations, deletions or additions (see Mutants). Selection begins to occur if the new protein assumes some function in the organism which better adapts it to its environment. A number of protein families have arisen in this way: the cytochromes *c*, myoglobins, α-, β-, γ- and δ-hemoglobin chains (Fig. 1), the fibrinopeptides, some protein hormones, the pancreatic serine proteases and probably many others.

Intragenal or *daughter-gene doubling* is a similar process, in which the duplication involves only part of a gene. Proteins which have arisen in this way have two or more regions of similar sequence along their length.

The most easily tolerated mutations are point mutations, in which a single nucleic acid base is exchanged. Since the genetic code is degenerate, both the original and the new codon may code for the same amino acid; in this case, nothing happens. If the new codon is specific for another amino acid the substitution may be either conservative, meaning that the new amino acid is chemically similar to the old (Lys for Arg, Glu for Asp, etc.), or radical (Lys for Asp). Conservative changes are of course more easily tolerated, and comparison of enzyme primary structures from different species reveals many such substitutions. Some amino acids cannot be changed without destroying the enzyme's activity, e.g. Ser_{195} in the serine proteases. Such a mutation, when it occurs, is generally lethal. Different families of proteins have accumulated substitutions at different rates. It is possible to calculate an average rate of mutation, given the number of substitutions between the primary sequences of enzymes from two species and the approximate time which has elapsed since their phylogenetic lines diverged. The table gives a few examples, including the extremely rapidly changing fibrinopeptides (90 point mutations per 100 amino acids and 100 million years) and the very conversative histones (2 mutations per 1.5 billion years). It is not that mutations are more frequent in the DNA coding for fibrinopeptides than for histones; rather, the structure of the histones is so rigidly determined by their function that no other structure will suffice, so that mutations are almost invariably lethal. Other forms of mutation observed in the laboratory, *deletions and additions*, can also be recognized in the sequences of existing proteins, such as the variable regions of the immunoglobulins, the α- and β-chains of hemoglobins, and the snake venom neurotoxins. Such radical changes seem not to have been tolerated in myoglobin, calcitonin and cytochrome c_{551}. *Hybridization* occurs

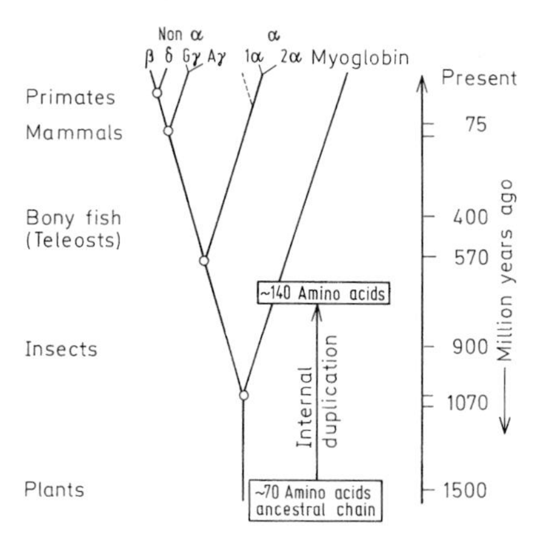

Fig. 1. *Evolution.* Evolution of human hemoglobin chains (α, β, γ, δ). Separation of horse and human α-chains is indicated by the dotted line.

Mutation rates of different protein families (abbreviated from Dayhoff: Atlas of Protein Sequence, 1972). n = successful point mutations per 100 amino acid residues in 100 million years.

Protein or peptide	n
Fibrinopeptides	90
Growth hormone	37
Pancreatic RNase	33
Immunoglobulins	32
Lactalbumin	25
Hemoglobin chains	14
Myoglobin	13
Pancreatic trypsin inhibitor	11
Animal lysosome	10
Trypsinogen	4
Cytochrome c	3
Histone IV	0.06

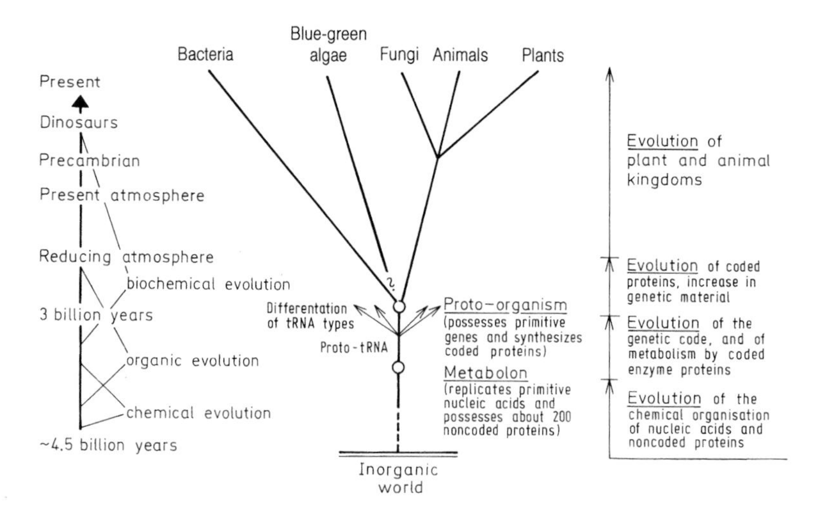

Fig. 2. *Evolution.* Phases of molecular evolution.

when there is crossing over between related but non-allelic genes, such as those for β- and δ-chains of hemoglobins. *Fusion* of two neighboring genes can produce bifunctional enzymes, such as tryptophan synthase. In *E. coli,* the α- and β-subunits of this enzyme are separate, but in *Neurospora* they are a single chain. In *Salmonella,* two enzymes of histidine biosynthesis have been fused to a single dimeric protein.

The processes of E. on a molecular level are fairly well understood. It is not known in detail how a particular mutation effects a particular change in morphology and behavior, although this is the level at which selection operates. (Selection also insures that all individuals are biochemically functional; a mutant lacking basic functions will generally die as an embryo.) We can assume, however, that the difference between man and monkey is the sum of a number of subtle differences in the actions of a relatively small number of development-regulating genes, enzymes and structural proteins. At both the molecular and the organism level, one can observe convergent and divergent E. Convergent E. occurs when similar environmental pressures favor development of similar traits in unrelated organisms, e.g. streamlining in fish and sea mammals, or enzymes with quite different sequences which catalyse the same reaction. Divergent E. is the process by which one species of organism or protein gives rise to two or more. The mechanisms discussed above (gene duplication, accumulation of mutations) account for molecular divergence. Divergence of organisms must begin with a change in one or several enzymes which, acting together with all the others in the organism, change it sufficiently so that it cannot interbreed with the members of the parent species.

Excision repair: see DNA repair.

Exon: see Intron.

Exonucleases: see Nucleases.

Exopeptidases: see Proteases.

Exotoxins: see Toxic proteins.

Extinction: see Absorbance.

Extinction coefficient: see Absorptivity.

Extracellular matrix: a network of proteins including Collagens (see) and Elastin (see), glycoproteins such as Laminin (see), Fibronectin (see) and Entactin (see), as well as various Proteoglycans (see; primarily heparan sulfates), which together constitute basement membranes and interstitial stroma. Numerous other protein components include Osteonectin (M_r 33,000; binds Ca^{2+}), Anchorin CII (M_r 34,000; binds to cells and collagen II), Epinectin (M_r 70,000; binds to epithelial cells and heparin), Thrombospondin (a trimer, M_r 140,000; binds to cells, proteoglycans, collagen, laminin, fibrinogen, Ca^{2+}), Chondronection (a trimer, M_r 56,000; binds to cells and collagen II), Vitronectin (M_r 75,000; binds to cells, heparin and collagens), Tenascin (cytotactin) (a hexamer, M_r 230,000; binds to cells and fibronectin).

Regulated matrix degradation is thought to be performed largely by matrix metalloproteinases.

See Cell adhesion molecules. [C. Ries & P. E. Petrides *Biol. Chem. Hoppe Seyler* 376 (1995) 345–355]

Extrachromosomal genes: DNA molecules located outside the nucleus, e.g. in Plastids (see) and Mitochondria (see). The corresponding E.g. in bacteria are the Plasmids (see).

Extrinsic factor: see Vitamins (vitamin B_{12}).

F

Fab fragment: see Immunoglobulins.

Fabry's disease: see Inborn errors of metabolism.

Facilitated diffusion: passive transport, the movement of specific compounds across a biomembrane from higher to lower concentration, but at a rate greater than simple diffusion. F. d. is saturable, meaning that above a certain concentration, the rate is not dependent on the substrate concentration. Furthermore, it is stereospecific and susceptible to competitive inhibition. Together, these properties indicate that the process is mediated by a "carrier" or pore protein in the membrane. F. d. differs from Active transport (see) in not requiring energy. A class of substances called Ionophores (see) mimic the carriers of F. d. by making membranes permeable to certain ions. Antibiotics that act in this way are called *transport antibiotics.*

Factor II: prothrombin. See Blood coagulation.

Factor VIII: antihemophilic factor. See Blood coagulation.

Factor XIII, ***fibrin-stabilizing factor:*** the last clotting factor to act in the blood coagulation cascade. It is an α2-plasma globulin of M_r 350,000 and contains 2α- and 2β-chains of M_r 100,000 and 77,000, respectively. It is activated by thrombin in the presence of Ca^{2+} to factor XIIIa, which catalyses the formation of γ-glutamyl-ε-lysine peptide bonds in a calcium-dependent transamidation reaction. These bonds serve to cross-link the fibrin chains into a 3-dimensional network, the clot.

FAD: acronym of Flavin adenine dinucleotide (see).

Farber's disease: see Lysosomal storage diseases.

Farnesol, ***3,7,11-trimethyl-2,6,10-dodecatrien-1-ol:*** an acyclic sesquiterpene alcohol. *Cis-trans* isomerism about two of the double bonds produces 4 isomeric forms. The most commonly occurring natural isomer is *trans-trans*-F. (M_r 222.36), an oil smelling of lilies of the valley, b. p.$_{10}$ 160 °C, ρ_4 0.895, n_D^{25} 1.4872. F. is unstable, is easily oxidized, and tends to cyclize. It occurs widely in essential oils, and it takes the place of phytol in the bacterial chlorophyll of the genus *Chlorobium.* It is also a pheromone for bumblebees. Farnesyl pyrophosphate is an important intermediate in the synthesis of many Terpenes (see).

OH

trans-trans-Farnesol

Farnesyl pyrophosphate: see Farnesol.

Farnoquinone: see Vitamins (vitamin K_2).

Fascin: see Cytoskeleton (microfilaments).

Fats: see Neutral fats.

Fatty acid biosynthesis: a process catalysed by fatty acid synthase, in which the fatty acid carbon chain is formed stepwise from 2-carbon units (derived from malonyl groups, with subsequent decarboxylation). The intermediates of F. a. b. are thioesters of Acyl carrier protein (see) (ACP) and not of coenzyme A as in fatty acid degradation.

Malonyl-CoA is synthesized in a biotin-dependent carboxylation of acetyl-CoA (see Biotin, under Vitamins):

ADP + P_i ⟵ ATP + HCO_3^-

Carboxybiotin ⟵ Biotin

CH_3–CO ~ SCoA ⟶ COO^-–CH_2–CO ~ SCoA

Acetyl-CoA ⟶ Malonyl-CoA

Acetyl-CoA carboxylase (Acetyl-CoA: carbon dioxide ligase [ADP-forming]) (EC 6.4.1.2)

The malonyl group of malonyl-CoA is transferred to ACP, where it forms a thioester linkage with the SH of covalently bound 4-phosphopantetheine. This phosphopantetheine arm serves as a carrier for substrates and intermediates, which are acted upon in turn by the other catalytic activities of fatty acid synthase shown in Table 1 and Fig. 1. Each newly formed saturated acyl group is transferred from the phosphopantetheine arm to the SH group of a cysteinyl residue ("peripheral" SH group) of β-ketoacyl synthetase. A further malonyl group is then attached to the freed central thiol, and another cycle of reactions (condensation, reduction, dehydration, reduction) extends the acyl group by two more carbon atoms. These cycles are repeated until a palmitoyl (C_{16}) residue is formed, which is then released as the free acid or as the fatty acyl-CoA, depending on whether the system possesses a thioesterase or CoA-transacylase (see Tables 1 and 2). For palmitate synthesis, the process can be represented stoichiometrically as: Acetyl-CoA + 7malonyl-CoA + 14NADPH + 14H^+ → Palmitate + 7CO_2 + 14$NADP^+$ + 8HSCoA + 6H_2O.

In most bacteria and in chloroplasts, the acyl carrier protein and the enzymes of F. a. b. are discrete proteins in a noncovalently associated multienzyme complex (type II fatty acid synthase), whereas the fatty acid synthase of animals (type I) is a dimer of a single multifunctional protein. Yeast fatty acid synthase is an intermediate type I/type II enzyme (see Table 2).

In animals, F. a. b. occurs in the cytoplasm, but the starting material, acetyl-CoA, is produced in the mi-

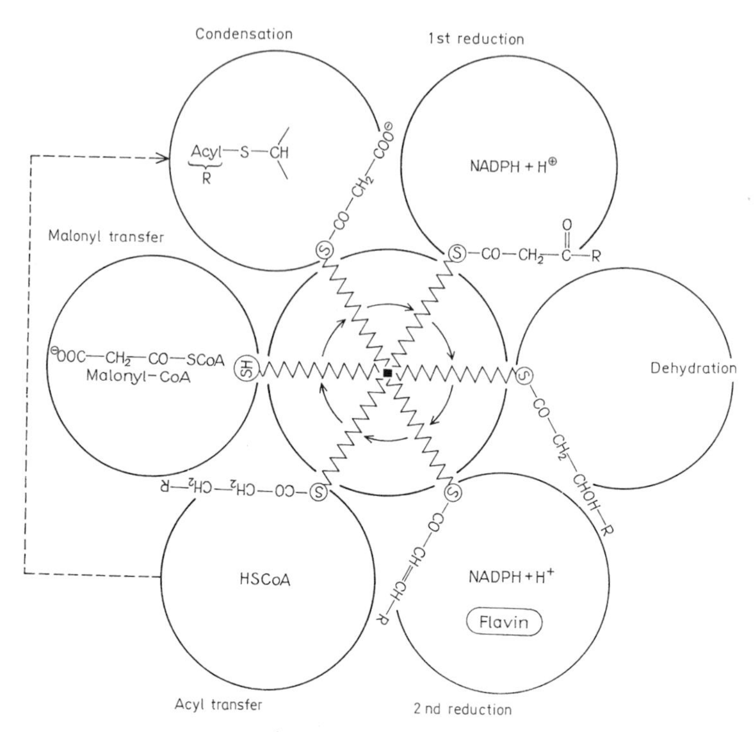

Fig. 1. *Fatty acid biosynthesis.* Cooperative action of the proteins of fatty acid synthase, which may be a multienzyme complex, or a single multifunctional protein (see Table 2). The zig-zags represent the movable pantetheine arm of the acyl carrier protein. The central circle represents acyl carrier protein; the other enzymes are represented by the six peripheral circles (see Table 1). This purely diagrammatic representation does not necessarily show the true juxtaposition of the proteins.

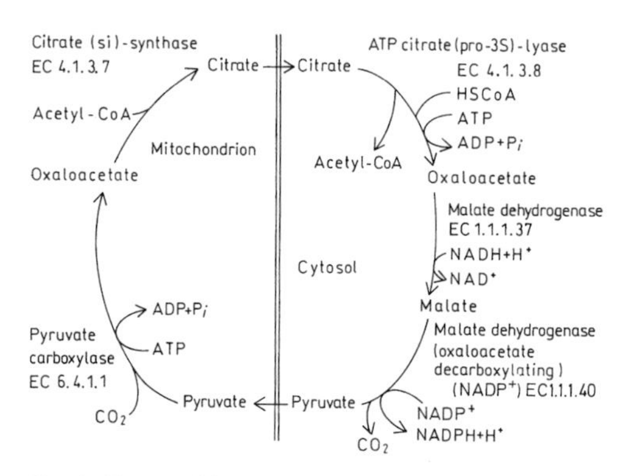

Fig. 2. *Fatty acid biosynthesis.* Production of cytosolic acetyl-CoA at the expense of mitochondrial acetyl-CoA by the formation, transport and breakdown of citrate.

tochondria by pyruvate dehydrogenase (see Multienzyme complex). The main route for provision of cytosolic acetyl-CoA is shown in Fig. 2.

For every acetyl-CoA transferred from the mitochondria to the cytosol, one NADPH is generated. Thus, in the conversion of 8 molecules of acetyl-CoA into palmitate, 8 of the required 14 NADPH molecules are provided by malate dehydrogenase (EC 1.1.1.40, Fig. 2). The additional 6 NADPH are provided by the Pentose phosphate cycle (see). The rate-limiting step in the biosynthesis of fatty acids from acetyl-CoA is the synthesis of malonyl-CoA, catalysed by acetyl-CoA carboxylase (see above). Active acetyl-CoA carboxylase from animal tissues is a polymer (M_r 4–8 × 10^6), which can be dissociated into inactive monomers or dimers of the M_r 230,000 subunit. Each subunit is a multifunctional protein containing the catalytic activities of biotin carboxylase, biotin-carboxy-carrier protein and transcarboxylase, as well as the regulatory allosteric site (see Biotin). Citrate and isocitrate activate the enzyme by promoting aggregation of the subunits. This effect of citrate is antagonized by palmitoyl-CoA, which also inhibits the carrier that transports citrate across the mitochondrial inner membrane (Fig. 2).

Elongation of the fatty acid carbon chain. The specificity of β-ketoacyl-ACP synthase is such that the enzyme normally binds all fatty acyl groups up to a chain length of C_{14} (tetradecanoyl). The hexadecanoyl (C_{16}, palmitoyl) cannot be bound, so that palmitate or palmitoyl-CoA is released as the end product of F. a. b. The chain length can be extended by elongation reactions which occur in the mitochondria and in the endoplasmic reticulum in animals. In mitochon-

Table 1. *Reactions of fatty acid synthesis.* The enzyme possesses a peripheral SH-group (which belongs to a cysteine residue) and a central SH-group (which belongs to pantetheine). With the exception of acetyl-CoA carboxylase, the catalytic activities listed below (together with acyl carrier protein) may belong to discrete proteins in a multienzyme complex, or they may all be present in a single multifunctional protein (see Table 2).

Enzyme		Function	Reaction
1) Acetyl-CoA carboxylase		Malonyl-CoA synthesis	$CH_3CO{\sim}SCoA + HCO_3^- + ATP \xrightarrow[\text{Biotin}]{H_2O,\ Mg^{2+}} {}^-OOC\text{-}CH_2\text{-}CO{\sim}SCoA + ADP + P_i$ (Acetyl-CoA → Malonyl-CoA)
2) Acetyl-transacylase		Acyl transfer (initiating reaction)	$CH_3CO{\sim}SCoA$ + Enzyme (HS, HS) ⇌ Enzyme (HS, $CH_3CO{\sim}S$) + HSCoA Acetyl-β-ketoacyl-ACP synthetase
3) Malonyl-transacylase	↑ Chain	Malonyl transfer	${}^-OOC\text{-}CH_2\text{-}CO{\sim}SCoA$ + Enzyme (HS, $CH_3CO{\sim}S$) ⇌ Enzyme (${}^-OOC\text{-}CH_2\text{-}CO{\sim}$S, $CH_3CO{\sim}S$) + HSCoA Acetyl-β-ketoacyl-ACP synthetase and Malonyl-ACP
4) β-Ketoacyl-ACP-synthetase	Chain	Condensation	Enzyme (${}^-OOC\text{-}CH_2\text{-}CO{\sim}$S, $CH_3CO{\sim}S$) → Enzyme ($CH_3\text{-}CO\text{-}CH_2\text{-}CO{\sim}$S, HS) + CO_2 Acetoacetyl-ACP
5) β-Ketoacyl-ACP-reductase	elon-	1st Reduction	Enzyme ($CH_3\text{-}CO\text{-}CH_2\text{-}CO{\sim}$S, HS) $\underset{NADP^+}{\overset{NADPH + H^+}{\rightleftharpoons}}$ Enzyme ($CH_3\text{-}CHOH\text{-}CH_2\text{-}CO{\sim}$S, HS) D-β-Hydroxybutyryl-ACP
6) Enoyl-ACP-hydratase (Crotonase)	gatio	Dehydration	Enzyme ($CH_3\text{-}CHOH\text{-}CH_2\text{-}CO{\sim}$S, HS) ⇌ Enzyme ($CH_3\text{-}CH{=}CH\text{-}CO{\sim}$S, HS) Crotonyl-ACP
7) Enoyl-ACP-reductase	n ↓	2nd Reduction	Enzyme ($CH_3\text{-}CH{=}CH\text{-}CO{\sim}$S, HS) $\underset{NADP^+\ (\text{Flavin})}{\overset{NADPH + H^+}{\rightleftharpoons}}$ Enzyme ($CH_3\text{-}CH_2\text{-}CH_2\text{-}CO{\sim}$S, HS)
8) Transacylase		Acyl transfer (followed by repeated malonyl transfer, i.e. reaction 3)	Enzyme ($CH_3\text{-}CH_2\text{-}CH_2\text{-}CO{\sim}$S, HS) ⇌ Enzyme (HS, $CH_3\text{-}CH_2\text{-}CH_2\text{-}CO{\sim}S$)
		Repeated cycles of chain elongation	C_4 6 8 10 12 14 16
9) Transacylase		Palmityl transfer (terminating reaction)	HSCoA + Enzyme ($CH_3\text{-}(CH_2)_{14}\text{-}CO{\sim}$S, HS) ⇌ Enzyme (HS, HS) + $CH_3(CH_2)_{14}\text{-}CO{\sim}CoA$ Palmityl-CoA — Alternative termination reactions
10) Thioesterase		Hydrolysis (terminating reaction)	H_2O + Enzyme ($CH_3\text{-}(CH_2)_{14}\text{-}CO{\sim}$S, HS) ⇌ Enzyme (HS, HS) + $CH_3(CH_2)_{14}\text{-}COOH$ Palmitate — Alternative termination reactions

Table 2. *Fatty acid synthases.*

Source	Description
Rat liver, rat adipose tissue, rat and rabbit lactating mammary gland, chicken liver, goose uropygial gland, *Ceratitis capitata* (an insect), *Crypthecodinium* (a marine dinoflagelate).	Type I fatty acid synthase. Enzyme activities are arranged as a series of globular domains in a single multifunctional protein, M_r 4–5 × 10^5, consisting of two identical subunits, each of M_r 1.8–2.5 × 10^5, i.e. α_2 structure. β-Ketoacyl synthase activity is only present in the dimer, since this activity requires juxtaposition of two thiols, one from each subunit. The enoyl reductase does not use a flavin coenzyme (NADPH is used directly for the reduction). Termination is by hydrolysis (thioesterase), and the products are chiefly palmitate with some stearate.

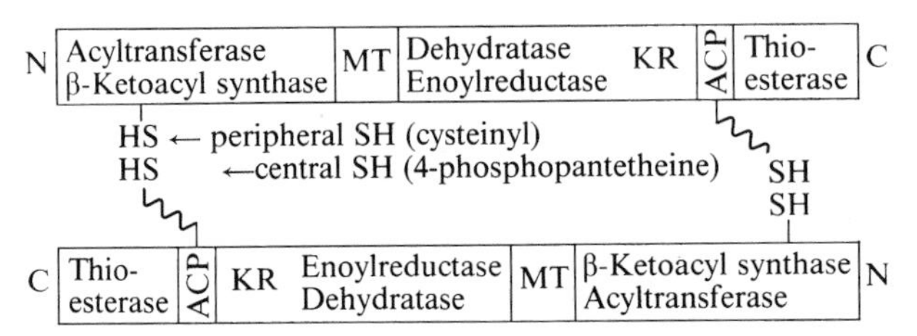

Proposed structure of the dimer of type I fatty acid synthase (based on studies of the chicken liver enzyme). ACP = acyl carrier protein, KR = β-ketoacyl reductase, MT = malonyl transacylase.

Source	Description
Mycobacterium smegmatis, Corynebacterium diphtheriae, Streptomyces coelicolor, Brevibacterium ammoniagenes.	Resembles type I enzyme, i.e. subunits are identical, and each is a multifunctional protein, M_r 2 × 10^6, consisting of six identical subunits, each of M_r 290000, i.e. α_6 structure. The enoyl reductase uses FMN. F.a.b. is dependent on both NADH (enoyl-acyl reduction) and NADPH (β-ketoacyl reduction). Palminate and tetracosanate are major products of *M. smegmatis* enzyme.
Saccharomyces cerevisiae (The fatty acid synthases from *Neurospora crassa, Penicillium patulum:* and *Pythium debaryanum* – all filamentous fungi – have M_r 2.2–4.0 × 10^6, and appear to be similar to the yeast enzyme)	Seems to be an evolutionary intermediate between types I and II. M_r 2.4 × 10^6, consisting of equal numbers of nonidentical subunits, M_r 213000 (α) and M_r 203000 (β), i.e. $\alpha_6\beta_6$ structure. The enoyl reductase in the β-subunit uses FMN. Products are palmitoyl-CoA or stearoyl-CoA.

α	β-Ketoacyl synthase	Acyl carrier	β-Ketoacyl reductase	
β	Enoyl reductase	Dehydratase	Acetyl trans-acylase	Malonyl and palmitoyl transacylase

Distribution of catalytic activities in the α and β-subunits of yeast fatty acid synthase

Source	Description
Euglena gracilis	Cytoplasm contains type I enzyme (α_2, M_r 200000), producing primarily palmitoyl-CoA. Chloroplasts contain type II system (distinct enzymes), producing primarily stearoyl-ACP.
Most bacteria, including *Escherichia coli, Clostridium butyricum, Bacillus subtilis, Pseudomonas aeruginosa, Phormium lunidum.* Plants, e.g. avocado, barley, spinach, safflower, parsley. *Chlamydomonas* (alga)	Type II fatty acid synthase. 6–7 discrete enzymes and an acyl carrier protein, associated noncovalently. In plants the enzymes are only in the plastids, and fatty acid synthesis does not occur in the cytoplasm. Similarity of plant and bacterial type II systems supports endosymbiont theory of origin of chloroplasts. Enoyl reductase uses FMN. Primary product is palmitate.

dria, elongation occurs by addition of acetyl-CoA units (not malonyl-ACP). This pathway closely resembles a reversal of β-oxidation (see Fatty acid degradation), except that the unsaturated bond at C2 is reduced by NADPH and not by $FADH_2$. In the endoplasmic reticulum, C_2 units for elongation are derived from malonyl-CoA, and all intermediates are in the form of their CoA esters (ACP esters are not involved) (Fig. 3). The microsomal elongation system is probably physiologically more important than the mitochondrial system. The mitochondrial system is only likely to operate when the [NADH]/[NAD^+] ratio is high in the mitochondria, i.e. under anaerobic conditions, or in liver when excessive quantities of ethanol are oxidized.

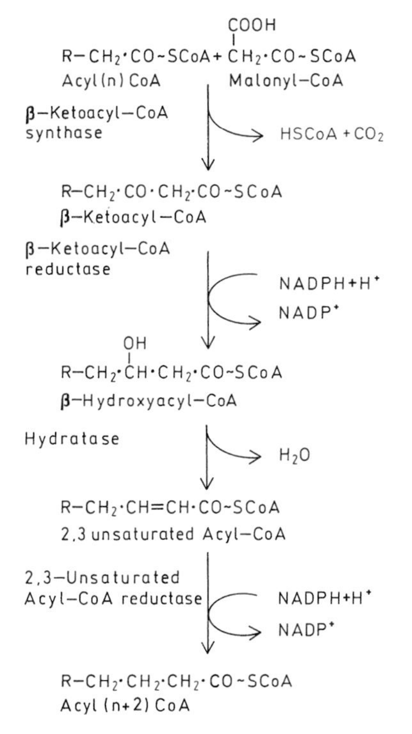

Fig. 3. *Fatty acid biosynthesis.* Endoplasmic reticulum system for the elongation of fatty acyl chains.

In bacteria and plants, elongation occurs by continuation of the reactions of F. a. b. beyond C_{16}, employing synthases of different specificity. Two β-ketoacyl synthases have been isolated from plants, the first producing primarily palmitate, while product distribution shifts primarily to stearate in the presence of the second synthase. Unsaturated acids are elongated in the same way as saturated acids.

Unsaturated fatty acid biosynthesis. Two apparently mutually exclusive pathways exist for the formation of unsaturated fatty acids, and together they account for essentially all naturally occurring unsaturated fatty acids. The *anaerobic pathway* operates only in bacteria. It can be considered as part of the pathway of saturated F. a. b., which branches at the level of a C_{10} intermediate to produce either palmitate or stearate on the one hand, or palmitoleate or vaccenate on the other. The branch point lies at D-β-hydroxy-decanoyl-ACP, a normal intermediate in the synthesis of palmitate. On the pathway of palmitate synthesis, this intermediate is dehydrated 2,3 (α,β) to produce a *trans* double bond, and the reactions of F. a. b. continue as described in Table 1. In the biosynthesis of unsaturated fatty acids, this intermediate is dehydrated to produce a *cis* double bond between C3 and C4 (β and γ) by the action of bacterial β-hydroxydecanoyl thioester dehydratase (Fig. 4). The *cis*-3 double bond is then carried through the subsequent chain lengthening steps of F. a. b. to yield palmitoleate (3 cycles) or vaccenate (4 cycles). *E. coli* possesses two distinct β-ketoacyl-ACP synthases (I and II). Enzyme I functions mainly in the elongation of *cis*-3-decenoate to palmitoleate, whereas enzyme II catalyses the condensations leading to palmitate and the chain extension of palmitoleate to vaccenate. *Brevibacterium ammoniagenes* produces oleate as the exclusive unsaturated fatty acid of the anaerobic pathway; this organism is also exceptional in using a type I fatty acid synthase complex for synthesis by the anaerobic pathway. In addition to palmitoleate and vaccenate, *Clostridium butyricum* also produces 7-hexadecenoate and oleate by chain extension of 3-dodecenoyl-ACP (Fig. 4).

The *oxyygen-dependent pathway of desaturation* operating in plants and animals produces a wide variety of unsaturated fatty acids differing in chain length, branching and position and number of double bonds. This is in marked contrast to the limited number and types of unsaturated fatty acids produced by bacteria in the anaerobic pathway. In animals, the oxygen-dependent desaturase is found in the endoplasmic reticulum; it is a mixed function oxygenase system containing cytochrome b_5 reductase, cytochrome b_5 and the desaturase (Fig. 5), and it is specific for fatty acyl-CoA substrates.

Animals are unable to synthesize linoleate or linolenate, which are therefore "essential" and must be supplied by the diet. Animals can desaturate stearoyl-CoA to oleoyl-CoA. Starting from linoleoyl-CoA, animals are able to produce a variety of polyunsaturated fatty acids by desaturation and chain elongation, including the prostaglandin precursor, arachidonate (Fig. 6).

Plant desaturase systems appear to be soluble, may contain ferredoxin in place of cytochrome b_5, and act upon fatty acyl-ACP or upon acyl groups already incorporated into membrane lipids (e.g. [1-oleoyl]-diacylgalactosylglycerol desaturated to [1-linoleoyl]-diacylgalactosylglycerol and then to [1-linolenoyl]-diacylgalactosylglycerol by spinach chloroplast desaturase).

Only a few bacteria operate the oxygen-dependent desaturase system, e.g. *Mycobacterium smegmatis* and *M. phlei, Alcaligenes faecalis, Bacillus megaterium.* The system from *M. smegmatis* is particulate, converts palmitoyl-CoA or stearoyl-CoA to the corresponding unsaturated Δ^9-derivatives, and requires NADPH. The substrate for desaturation by the *Bacillus megaterium* system is the acyl group of phosphatidylglycerol.

Synthesis of fatty acids with odd numbers of carbon atoms, and fatty acids with branched chains. Incor-

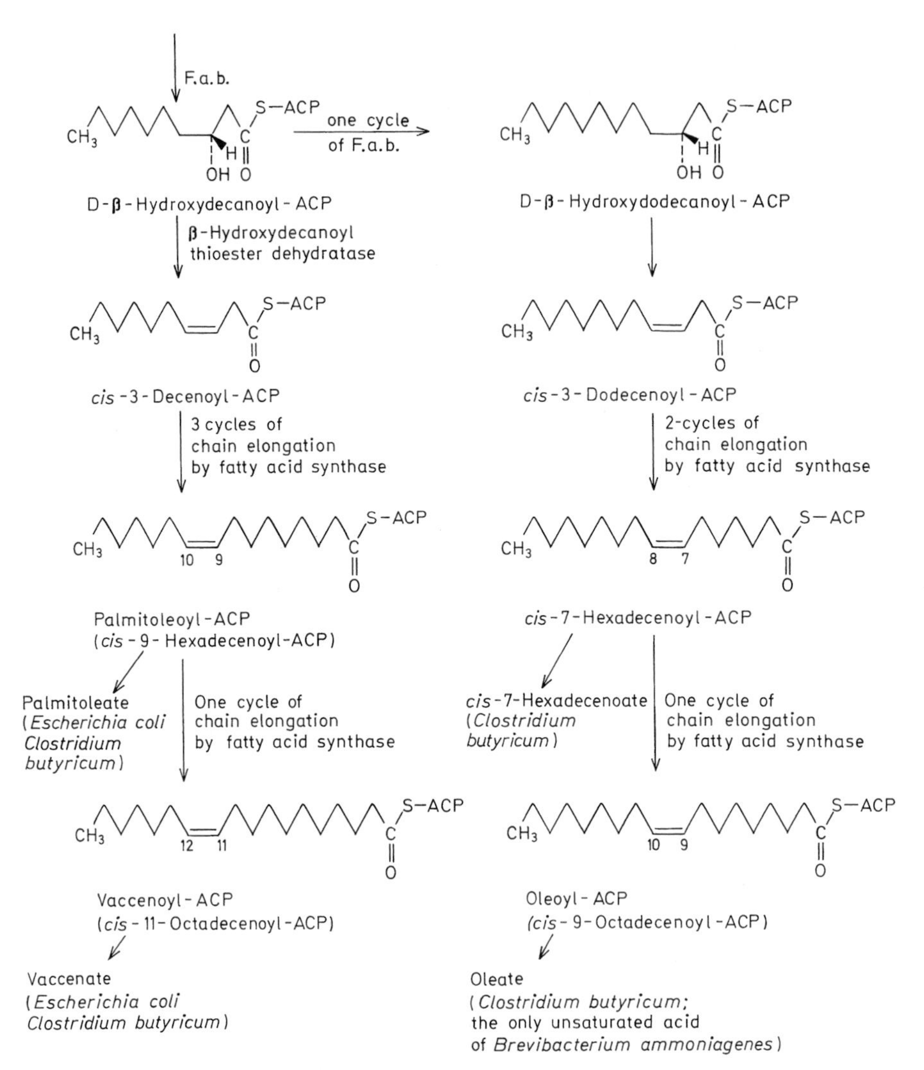

Fig. 4. *Fatty acid biosynthesis.* Synthesis of unsaturated fatty acids by the anaerobic pathway in bacteria.

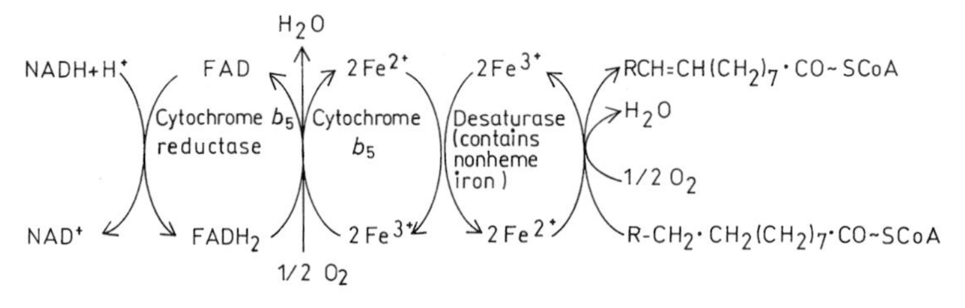

Fig. 5. *Fatty acid biosynthesis.* Desaturase system of rat liver endoplasmic reticulum.

poration of various primers or starter molecules in F. a. b. in place of acetyl-CoA results in odd-numbered carbon chains or in branching (Table 3).

Elongation of the branched starter molecules in Table 3 results in a branched fatty acid with branching distal to the carboxyl group. For the synthesis of branched fatty acids with branching adjacent to the carboxyl group, the starter molecule must be unbranched and the final stages of elongation must employ a branched substrate, e.g. methylmalonyl-CoA. The

Table 3. *Different starter molecules for fatty acid biosynthesis.* The acyl group of the CoA derivative is transferred to the peripheral SH-group (SH-group of β-ketoacyl-ACP synthase) (Table 1, reaction 2).

Starter molecule	Fatty acid product
$CH_3\text{-}\overset{O}{\overset{\|}{C}}\sim CoA$ *Acetyl-CoA*	Even numbered, straight carbon chain
$CH_3CH_2\overset{O}{\overset{\|}{C}}\sim CoA$ *Propionyl-CoA*	Odd numbered, straight carbon chain
$(CH_3)_2CH\overset{O}{\overset{\|}{C}}\sim CoA$ *Isobutyryl-CoA* (from valine)	Even numbered, branched chain (Iso series)
$(CH_3)_2CHCH_2\overset{O}{\overset{\|}{C}}\sim CoA$ *Isovaleryl-CoA* (from leucine)	Odd numbered, branched chain (Iso series)
$CH_3CH_2CH(CH_3)\overset{O}{\overset{\|}{C}}\sim CoA$ *2-Methylbutyryl-CoA* (from isoleucine)	Odd numbered, branched chain (Anteiso series)

mycocerosic acids are a family of multimethyl-branched acids produced by mycobacteria such as *Mycobacterium tuberculosis* var. *bovis Bacillus Calmette-Guérin.* They mostly possess a straight chain of 18–20 carbons with a multimethyl-branched region at the carboxy end:

$$CH_3(CH_2)_{19}\text{–}\underset{}{\overset{CH_3}{\overset{|}{C}}}H.CH_2.\overset{CH_3}{\overset{|}{C}}H.CH_2.\overset{CH_3}{\overset{|}{C}}H.CH_2.\overset{CH_3}{\overset{|}{C}}H.COOH$$

2,4,6,8-Tetramethyloctacosanic acid

$$CH_3(CH_2)_{19}\text{–}\overset{CH_3}{\overset{|}{C}}H.CH_2.\overset{CH_3}{\overset{|}{C}}H.CH_2.\overset{CH_3}{\overset{|}{C}}H.COOH$$

2,4,6-Trimethylhexacosanic acid

$$CH_3(CH_2)_{17}\text{–}\overset{CH_3}{\overset{|}{C}}H.CH_2.\overset{CH_3}{\overset{|}{C}}H.CH_2.\overset{CH_3}{\overset{|}{C}}H.COOH$$

2,4,6-Trimethyltetracosanic acid

} Mycocerosic acids

A novel fatty acid synthase, mycocerosic acid synthase, has been isolated from *Mycobacterium tuberculosis* var. *bovis BCG,* which elongates *n*-fatty acyl-CoA with methylmalonyl-CoA [D. L. Rainwater & P. E. Kolattukudy *J. Biol. Chem.* **260** (1985) 616–223]. It consists of two identical subunits (M_r subunit 238,000, M_r dimer 490,000). It elongates straight fatty acyl-CoA esters from *n*-C_6 to *n*-C_{20}, to generate the corresponding tetramethyl-branched mycocerosic acids, and it does not utilize malonyl-CoA.

Branch methyl groups located near the center of the fatty acid chain are derived by carbon transfer from *S*-adenosyl-L-methionine to an unsaturated fatty acyl derivative in a phospholipid, followed by a reduction, e. g. synthesis of 10-methylstearic acid (tuberculostearic acid) in mycobacteria, *Nocardia, Streptomyces* and *Brevibacterium:*

$$CH_3(CH_2)_7CH{=}CH(CH_2)_7\text{–}COO\text{–}X + S\text{-Adenosyl-L-methionine}$$
$$\downarrow$$
$$CH_3(CH_2)_7\underset{\underset{CH_2}{\|}}{C}\text{–}CH_2(CH_2)_7\text{–}COO\text{–}X + S\text{-Adenosyl-L-homocysteine}$$
$$\downarrow \; NADPH + H^+ \rightarrow NADP^+$$
$$CH_3(CH_2)_7\underset{\underset{CH_3}{|}}{C}H\text{–}CH_2(CH_2)_7\text{–}COO\text{–}X$$

Tuberculostearoyl group in a phospholipid

Cyclopropane fatty acids (Gram-negative bacteria, some green plants, zooflagellates) are also formed by carbon transfer from *S*-adenosyl-L-methionine to a *cis*-monounsaturated fatty acid derivative:

$$CH_3(CH_2)_nCH{=}CH(CH_2)_xCOO\text{–}X + S\text{-Adenosyl-L-methionine}$$
$$\downarrow$$
$$CH_3(CH_2)_n\overset{CH_2}{\overbrace{CH\text{–}CH}}(CH_2)_xCOO\text{–}X + S\text{-Adenosyl-L-homocysteine}$$

Hydroxy fatty acids may be intermediates of F. a. b which have been diverted (3-hydroxy fatty acids with D-configuration). They may also arise by oxygenation catalysed by cytochrome P-450 systems (especially ω-oxidation in eukaryotes; see Fatty acid degradation). Hydroxy fatty acids can also be biosynthesized by hydration of double bonds in unsaturated fatty acids; specific fatty acid hydratase systems catalysing this reaction have been characterized in bacter-

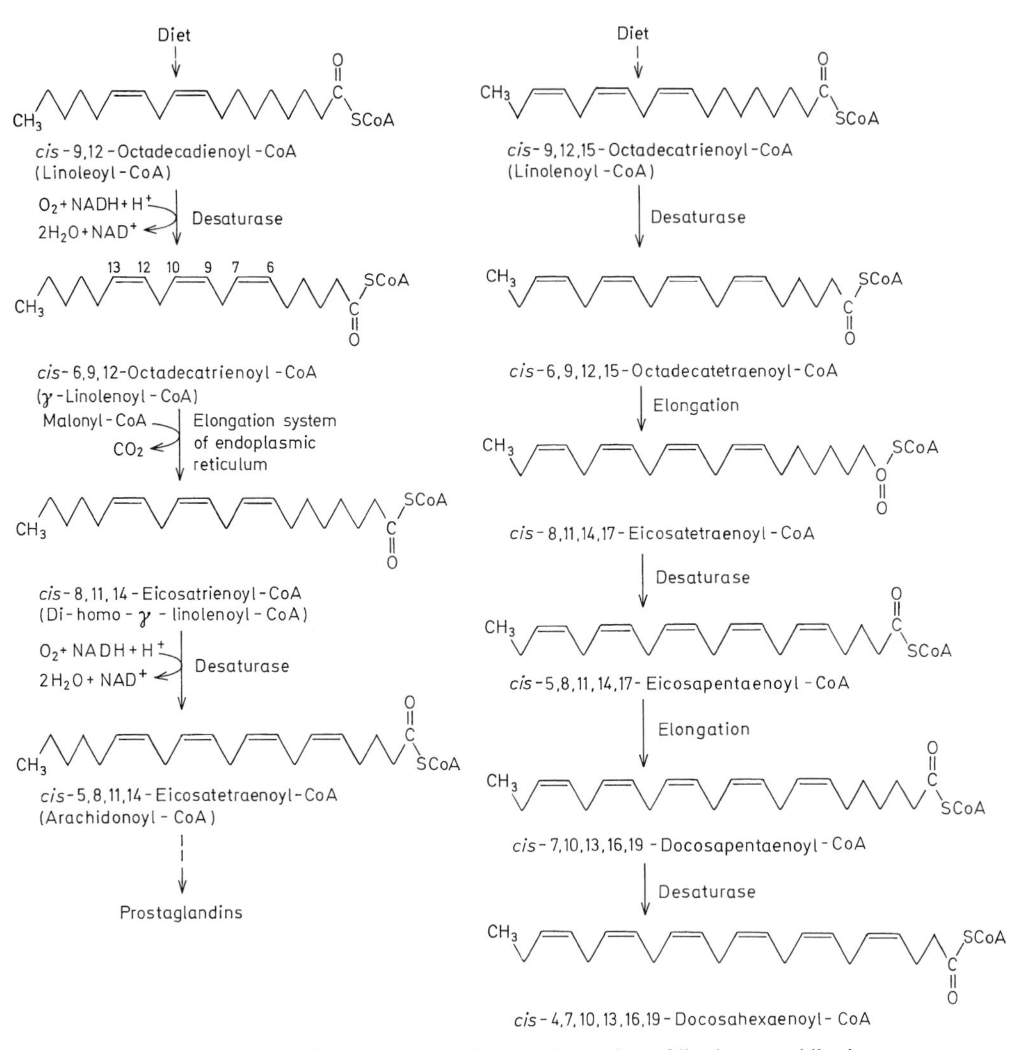

Fig. 6. *Fatty acid biosynthesis.* Desaturation and chain elongation of linoleate and linolenate

ia, especially *Pseudomonas* spp. [J. B. Ohlrogge *Trends Biochem. Sci.* **7** (1982) 386–387; S. J. Wakil et al. *Annu. Rev. Biochem.* **52** (1983) 537–579; A. J. Fulco *Prog. Lipid Res.* **22** (1983) 133–160; A. D. McCarthy & D. G. Hardie *Trends Biochem. Sci.* **9** (1984) 60–63]

Fatty acid degradation, ***fatty acid catabolism:*** This occurs mainly by the process of β-oxidation. The processes of α- and ω-oxidation represent minor pathways.

β-Oxidation is a cyclic (or spiral) process in which fatty acids are degraded stepwise from the carboxyl end. In each turn of the cycle, two carbon atoms are removed as acetyl-CoA, and the β-carbon atom is oxidized:

$$R\text{–}CH_2CH_2\text{–}CO \sim SCoA \xrightarrow{HS\text{–}CoA} R\text{–}CO\text{–}CoA + CH_3\text{–}CO \sim SCoA$$

Fig. 1 shows the "fatty acid spiral". The initiating reaction is the activation of the fatty acid to acyl-CoA by an acyl-CoA synthetase (EC 6.2.1.3), a process requiring ATP:

$$R\text{–}CH_2\text{–}CH_2\text{–}COOH + ATP + HS\text{–}CoA \rightarrow R\text{–}CH_2\text{–}CH_2\text{–}CO \sim SCoA + AMP + PP_i$$

The acyl-CoA is reduced by acyl-CoA dehydrogenase (EC 1.3.99.3) (an FAD-enzyme) to 2,3-dehydro-acyl-CoA. This compound is hydrated, then oxidized to the 3-keto-acyl-CoA. The latter is cleaved (thiolysed) to produce acetyl-CoA and a new acyl-CoA which is two carbons shorter than the original one. The new acyl-CoA is immediately subject to further β-oxidation. Thus each turn of this fatty acid spiral produces one molecule of acetyl-CoA, which is further oxidized in the tricarboxylic acid cycle. Complete oxidation of 1 molecule of Steroyl-CoA yields 148 molecules of ATP (18 C atoms yield 9 molecules of acetyl-CoA; 1 acetyl-CoA yields 12 molecules of ATP in the tricarboxylic acid cycle; in addition, 5 molecules ATP are formed in each of the 8 β-oxidation steps). The energy stored in the terminal phosphate bonds of 148 molecules of ATP is equivalent to 50% of the heat of combustion of the fatty acid.

Provided that the fatty acid contains an even number of C-atoms, it can be totally degraded to acetyl-CoA by β-oxidation. If the fatty acid contains an uneven number of C-atoms, however, the stepwise removal of two carbons at a time by β-oxidation eventually leads to a C_3 compound, i.e. propionyl-CoA, which must be metabolized by alternative pathways. These pathways are shown in Fig. 2; considerable quantities of propionyl-CoA also arise from the degradation of the branched-chain amino acids, isoleucine and valine (see L-leucine). The main pathway involves carboxylation of propionyl-CoA to methylmalonyl-CoA, followed by the vitamin B_{12}-dependent isomerization of methylmalonyl-CoA to succinyl-CoA; excretion of methylmalonyl-CoA is diagnostic of vitamin B_{12} deficiency.

Various pathways are employed for the degradation of branched-chain fatty acids, some of which arise from the metabolism of the branched-chain amino acids (see Leucine). Short-chain fatty acids are converted to their fatty acyl derivatives within the mitochondria, but long-chain fatty acids can be activated only by the endoplasmic reticulum and outer mitochondrial membrane. Long-chain acyl-CoA cannot penetrate the inner mitochondrial membrane, and must be transported into the mitochondria as acylcarnitine (Fig. 3).

α-Oxidation of fatty acids occurs in germinating plant seeds. A fatty acid peroxidase (EC 1.11.1.3) catalyses the decarboxylation and simultaneous formation of an aldehyde, in which H_2O_2 acts as hydrogen acceptor. The aldehyde can either be oxidized to a fatty acid or reduced to a fatty alcohol.

ω-Oxidation involves oxidation of the terminal methyl group of a fatty acid by enzymes localized in the

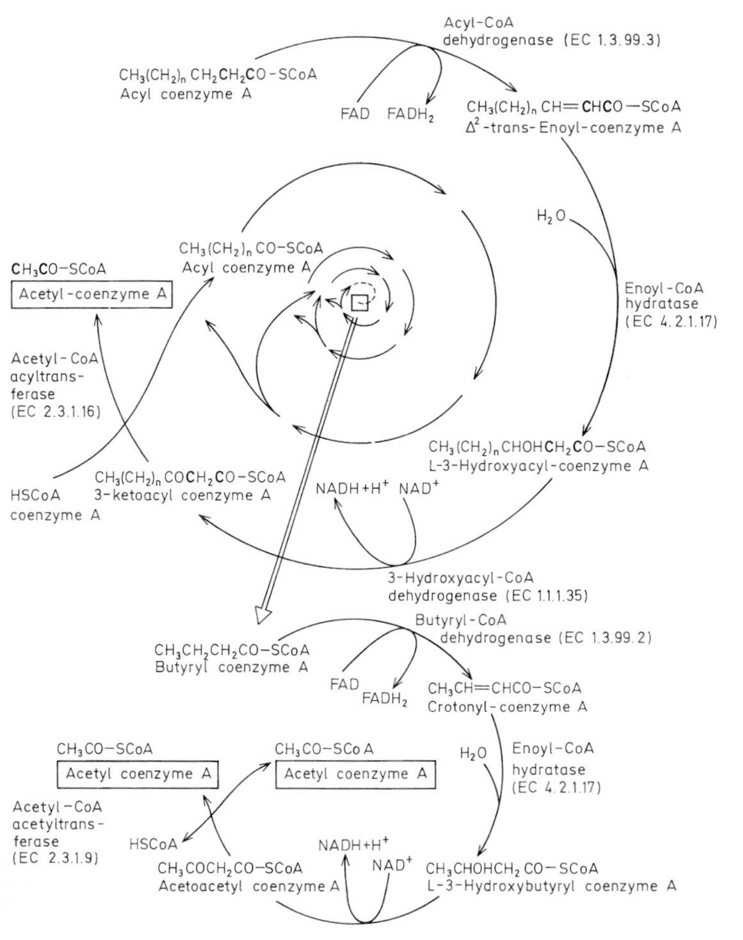

Fig. 1. *Fatty acid degradation.* Fatty acid spiral. In the degradation of fatty acids with an odd-number of C-atoms, the final step produces acetyl-CoA and propionyl-CoA (Fig. 2.).

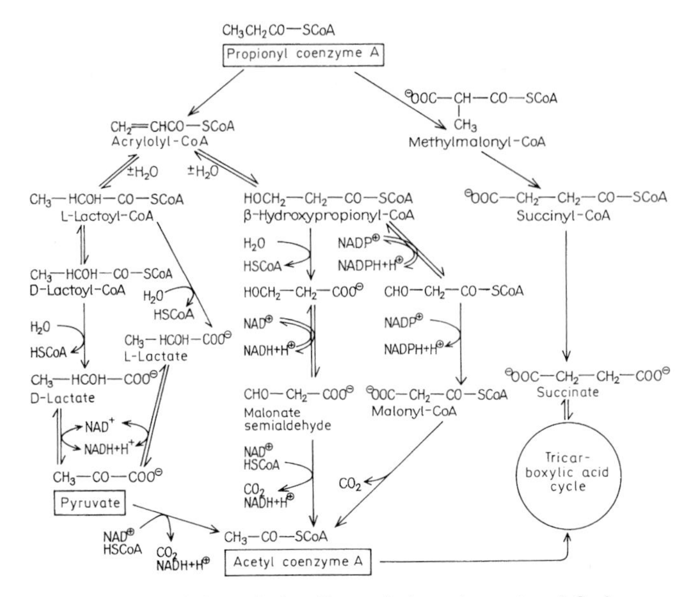

Fig. 2. *Fatty acid degradation.* Degradation of propionyl-CoA.
a) In the methylmalonyl pathway, propionyl-CoA is converted to methylmalonyl-CoA by an ATP-dependent carboxylation. *S*-Methylmalonyl-CoA mutase (EC 5.4.99.2), a cobamide-requiring enzyme, then converts methylmalonyl-CoA to succinyl-CoA.
b) In the lactate pathway, propionyl-CoA is dehydrogenated to acrylyl-CoA. α-Hydration produces L-lactoyl-CoA, which is hydrolysed to lactate.
c) In plant mitochondria, acrylyl-CoA is hydrated to 3-hydroxypropionyl-CoA, which is deacylated and oxidized to malonic acid semialdehyde. The latter is converted directly to acetyl-CoA by oxidative decarboxylation, or indirectly via malonyl-CoA.

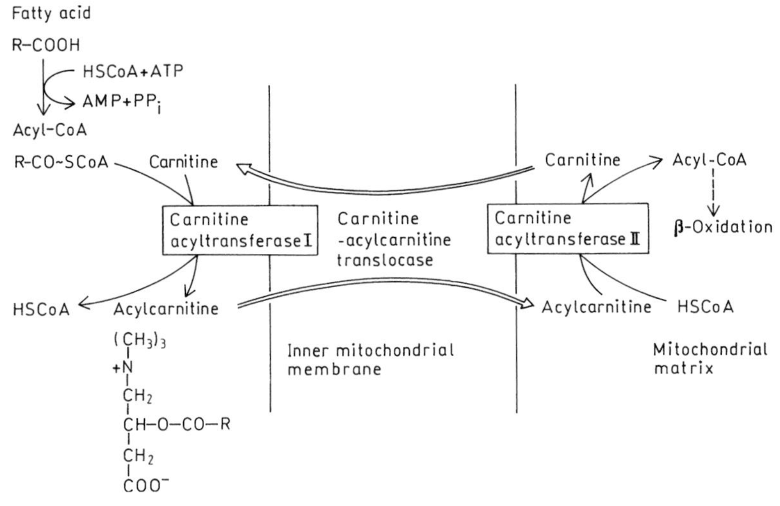

Fig. 3. *Fatty acid degradation.* Role of carnitine in the transport of long-chain fatty acids through the inner mitochondrial membrane. Carnitine-acylcarnitine translocase is an integral membrane exchange transport system. Carnitine acyltransferases I and II are located on the outer and inner surfaces, respectively, of the inner mitochondrial membrane.

microsomal fraction of animal and microbial cells. The substrate is usually a C_8 to C_{12} fatty acid, which is converted to the dicarboxylic acid in two steps: 1. hydroxylation of the ω-C to form an ω-hydroxy fatty acid, a process requiring oxygen and NADPH; 2. oxidation of the hydroxylated ω-C to a carboxyl group, catalysed by a soluble, nonmicrosomal enzyme, which is usually NAD^+-dependent.

Fatty acids: the esterified carboxylic acids of fats and oils. Saturated F.a. have the general formula $C_nH_{2n+1}COOH$. Unsaturated F.a. have double bonds between two or more carbons, and correspondingly fewer H atoms. Some common saturated F.a. are butyric, palmitic, lauric and stearic acids. Oleic, linoleic and linolenic acids are unsaturated or olefinic acids. *Essential F.a.:* see Vitamins (vitamin F).

Fatty acid synthetase complex: see Multienzyme complexes.

Fatty alcohols: unbranched, aliphatic monohydric alcohols with 10–20 C atoms. They occur naturally as esterified components of waxes, and are biosynthesized by reduction of fatty acids.

Fatty liver, ***fatty degeneration of the liver:*** see Lipotropic substances.

Favism: see Glucose 6-phosphate dehydrogenase.

FCCP: acronym of carbonyl cyanide-*p*-trifluoromethoxyphenylhydrazone. See Ionophore.

Fc fragment: see Immunoglobulins.

Fd: abb. for ferredoxin.

Febrifugine: a quinolazine alkaloid isolated from the comon hydrangea, and from the shrub *Dichroa febrifuga* which has been used in Chinese folk medicine since ancient times. m.p. 139–140 °C, $[\alpha]_D^{25}$ +6° (c = 0.5 in $CHCl_3$). F. is a powerful antimalarial and antipyretic, but is too toxic for human use.

Feedback mechanisms: see Metabolic regulation.

Feedforward mechanisms: see Metabolic regulation.

Fermentation: a form of metabolism producing incompletely oxidized end products. Per unit of substrate, F. yields far less energy than respiration, e.g. a yeast cell obtains 2 molecules ATP per molecule of glucose when it ferments glucose to ethanol, whereas complete respiration would yield 38 molecules of ATP (see Alcoholic fermentation). Strictly speaking, F. is an anaerobic process (Pasteur defined F. as "life without air") but the term is also widely and loosely applied to certain aerobic processes, such as acetic acid F., and to any industrial production process employing microorganisms in a fermentor (see Fermentation techniques).

Fermentation products: end products of anaerobic microbial metabolism. The products listed in the table are (or were) industrially significant, but in view of the finite amounts of fossil fuels, it seems likely that F. will become more important in the future.

Fermentation products

Product	Organism	Use
Ethanol	*Saccharomyces cerevisiae*	Industrial solvent; beverages
Glycerol	*Saccharomyces cerevisiae*	Production of explosives (nitroglycerin, dynamite)
Lactic acid	*Lactobacillus delbrueckii, L. bulgaricus*	Food and pharmaceuticals
Acetone and Butanol	*Clostridium acetobutylicum, Cl. butylicum*	Industrial solvents
2,3-Butylene glycol (2,3-butanediol)	*Bacillus polymyxa, B. subtilis, Aerobacter aerogenes*	

Fermentation techniques, ***microbial production techniques:*** techniques for large-scale production of microbial products. F.t. must both provide an optimal environment for the microbial synthesis of the desired product and be economically feasible on a large scale (see Cultivation of microorganisms). They can be divided into surface (emersion) and submersion techniques. The latter may be run in batches or continuously.

Surface techniques, in which microorganisms are cultivated on the surface of a liquid or solid substrate, are used for the production of enzymes on solid substrates and, in a few factories, for citric acid production; the microorganisms are grown in flat dishes which are stacked in large containers. A special surface technique, employing vertical containers of carrier material, is used in the production of acetic acid and in sewage treatment. The organisms grow on the carrier bed, and the medium or sewage flows past them.

In submersion processes, the microorganisms grow in a liquid medium. Except in traditional beer and wine fermentation, the medium is held in Fermentors (see) and stirred to obtain a homogeneous distribution of cells and medium. Most processes are aerobic, and for these the medium must be vigorously aerated. Most important industrial processes (production of biomass and protein, antibiotics and enzymes and some forms of sewage treatment) are carried out by submersion processes.

In batch techniques, the fermentor is filled with medium and inoculated; the cells multiply and synthesize the product, simultaneously consuming the nutrients in the medium. At the appropriate time, the culture is drained off and processed. Discontinuous processes are used, e.g. in the production of antibiotics and certain amino acids, which are synthesized only during particular growth phases of the organism. Their discontinuity of operation makes them less suitable for industrial production than continuous processes.

In continuous processes, the microorganisms are continuously supplied with new medium and the suspension of old medium and microbes is removed simultaneously by overflow. The overflow can be processed continuously. These techniques are used especially for the production of biomass protein and baker's yeast. A continuous process for brewing beer has been developed, and sewage treatment is also continuous. These processes have in common that the desired product is either the cells themselves (biomass, proteins) or a metabolic product which is continuously produced (fermentation products). However, continuous processes can also be adapted to products which are synthesized after cell growth has stopped, e. g. by setting up a series of fermentors to produce the desired physiological state in the cells. The main advantage of continuous processes is that they can be automated. In chemostats, the ratio of cell growth to synthesis of secondary products can be controlled by the rate at which new medium is added to the culture. Turbidostats have so far been used more in the laboratory for maintaining cultures at a defined cell densities. A light is shone through the fermenter, and when the intensity falling on a photocell on the other side has decreased below a preset threshhold, the culture is diluted with fresh medium. On the industrial scale continuous processes are difficult to keep sterile over long periods, so they are generally used only for non-sterile production.

Fermentors: large-scale vessels in which microorganisms are cultured to obtain some product of their metabolism (see Industrial fermentation). The main vessel is usually cylindrical; the requirements of the particular microorganism and the desired product dictate the accessories required for stirring, cooling, aeration, input of medium and harvesting. The simplest F. is a large tank like the vats originally used for wine and beer fermentation. A large vat must be cooled by a water jacket or heat exchanger so that the organisms' metabolic heat does not raise the temperature beyond their level of tolerance. Stirring is required in large F. to keep the organisms from settling and to provide a uniform supply of nutrients, including oxygen in aerobic fermentations. There are many types of stirrers, turbines, baffles and recirculating pumps. In an *air-lift* or *bubble F.*, stirring and aeration are provided by a stream of gas entering the bottom of the tank. In a *cyclone column F.*, a pump continuously circulates the culture from the bottom to the top of the vessel. The medium is ejected from a cyclone head tangentially to the wall of the F., so that it spirals down the wall as a thin film. Air or gas injected at the bottom of the vessel rises in a counter current to the culture flow. *Tower F.* are used for beer and other cultures in which the desired products are not produced continuously over the growth cycle. F. range in size from a few liters for laboratory use to more than 500 m^3 for industrial applications. The working volume of medium is usually about 2/3 of the total capacity of the F.

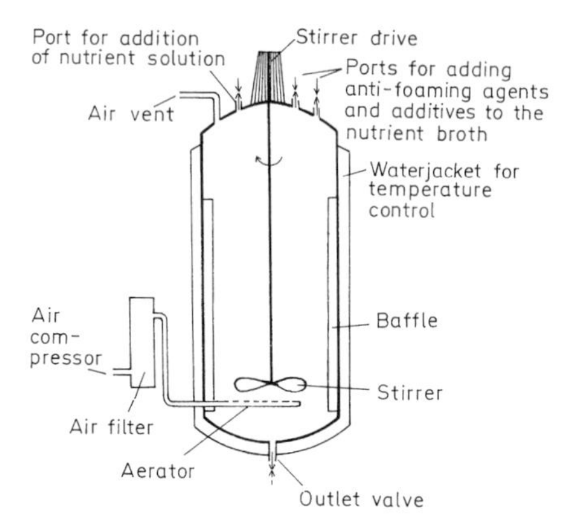

Cross section of a fermentor

Ferredoxins, ***Fd:*** low-molecular-mass iron-sulfur proteins which transfer electrons from one enzyme system to another, without possessing any enzyme activity themselves. The name was coined by Mortenson for iron-containing proteins from *Clostridium pasteuranum.* The *8-Fe-Fd* take part in many electron transport processes in organisms like Clostridia and photosynthetic bacteria (Table). The primary structures of many of these proteins have been determined. They consist of about 55 amino acids, including 8 cysteines, which occupy the same positions in each Fd. The molecule contains two identical 4Fe-4S clusters, each one forming a cube and covalently bonded to 4 cysteine residues in the peptide chain. Each 4Fe-4S center can transfer one electron.

4Fe-Fd have a single 4Fe-4S cluster as an active center. As in the clostridial Fd, the iron atoms are bound to the only 4 cysteine residues in the protein;

Ferredoxin-dependent metabolic reactions

Reaction	Ferredoxin type
Nitrogen fixation $N_2 + 6ATP + 6e^- \rightarrow 2NH_3 + 6ADP + 6P_i$	Nitrogenase complex: Molybdoferredoxin contains 18–28 FeS and 1Mo + Azoferredoxin (4Fe-4S center) + 8 Fe-Fd. Nitrogen-fixing bacteria and blue-green bacteria.
Hydrogen metabolism $2H^+ + 2e^- \rightleftharpoons H_2$	Hydrogenase (4 Fe-S center) + 4 Fe-Fd, 8 Fe-Fd or 2 Fe-Fd. Bacteria and blue-green bacteria
Phosphoclastic reaction $CH_3\text{–}CO\text{–}COO^- + P_i \rightarrow CH_3\text{–}CO\text{–}O\text{–}PO_3H_2 + CO_2$	8 Fe-Fd + complex Fe-S enzyme. *Clostridium* spp.
Synthesis of α-ketoacids (pyruvate, α-ketoglutarate) by reductive CO_2 fixation $CH_3\text{–}CO \sim CoA + CO_2 \rightarrow CH_3\text{–}CO\text{–}COO^- + CoA$	8 Fe-Fd. Photosynthetic and fermenting bacteria, e. g. *Chromatium, Clostridium.*
Photosynthetic NADP reduction $NADP^+ + H_2O \xrightarrow{Light} NADPH + H^+ + \frac{1}{2}O_2$	2 Fe-Fd. Higher plants and algae.

these Fd have been isolated primarily from bacteria, e.g. *Desulfovibrio gigas* and *Bacillus stearothermophilus*.

High potential iron-sulfur protein (HiPIP) is a special type of Fd which has been isolated from some photosynthetic bacteria and detected by ESR spectroscopy in other bacteria. HiPIP also contains a single 4Fe-4S cluster, but it differs from the other Fd in having a positive standard potential of about + 350 mV (most Fd have standard potentials in the range of the hydrogen electrode, about –420 mV). Furthermore, the HiPIP from *Chromatium* is paramagnetic in the oxidized state.

2Fe-Fd from blue-green bacteria, green algae and higher plants are highly homologous (96 to 98 amino acid residues with 4–6 cysteines). Four of the six cysteines are found in all plant Fd in positions 39, 44, 47 and 77, and are covalently bound to the Fe-S center. A proposed model of 2Fe-Fd is shown in the Fig. to Iron-sulfur proteins (see). 2Fe-Fd serve as electron-transfer catalysts in both cyclic (1) and noncyclic photophosphorylation (2):

$$NADP^+ + H_2O + ADP + P_i \xrightarrow{Fd/light} NADPH + H^+ + ATP + \tfrac{1}{2}O_2 \quad (1)$$

$$ADP + Pi \xrightarrow{Fd/light} ATP \quad (2)$$

The main metabolic reactions involving Fd are listed in the table.

A number of chemical structures for Fd have been suggested. The chelate structure [Fig.(1)] is based on chemical analysis, ESR and Mossbauer spectroscopy (see also Iron-sulfur proteins). Model experiments led to the construction of metal-atom island structures, or clusters. In this model, the "labile" sulfur atoms cause charge delocalization, and electrons are transferred from the sulfur to the iron which is bound between neighboring cysteine residues. The α-helical structure of the polypeptide prevents formation of a disulfide bond until the electron is transferred to another substrate [Fig.(2)].

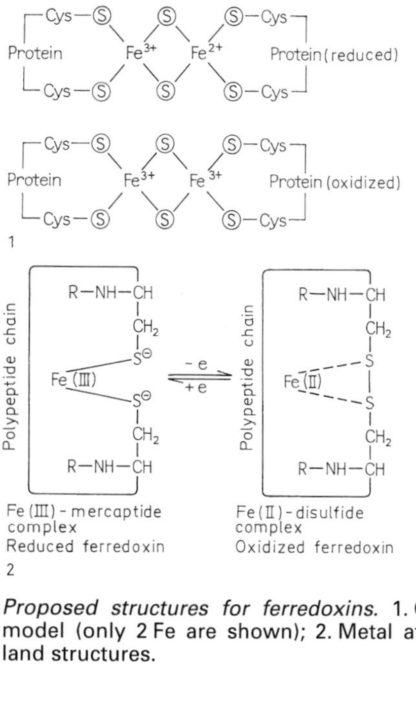

Proposed structures for ferredoxins. 1. Chelate model (only 2 Fe are shown); 2. Metal atom island structures.

Ferreirin: 5,7,6′-trihydroxy-4′-methoxyisoflavanone. See Isoflavanone.

Ferrichrome: see Siderochromes.

Ferrimycin: see Siderochromes.

Ferrioxamines: see Siderochromes.

Ferritin: the most important iron-storage protein in mammals. Together with the related Hemosiderin (see), it contains 25 % of the iron in the body. F. is found in the spleen, liver, bone marrow and reticulocytes, where excess iron is stored intracellularly and can be mobilized at need. F. has also been found in mollusks, plants and fungi. It consists of a shell-like protein, apoferritin, with a diameter of 12 nm. The interior is a chamber with a diameter of 7 nm in which the iron is deposited as iron-hydroxide-oxide micelles. One molecule of F. can hold up to 4,300 Fe(II) atoms (Fig. 1). The micellar iron has a composition of 8 $FeO(OH) \cdot FeO(PO_3H_3)$; but the phosphate is not always present. Iron-free apoferritin (isolated by chemical reduction of the iron or by various centrifugation techniques) consists of an oligomer (M_r 445,000) of 24 equal-sized, noncovalently bonded subunits (M_r 18,500). Apoferritin can be dissociated into subunits by treatment with 1 % sodium dodecylsulfate or 5 M guanidine hydrochloride, which disrupt the strong hydrophobic bonding between the subunits.

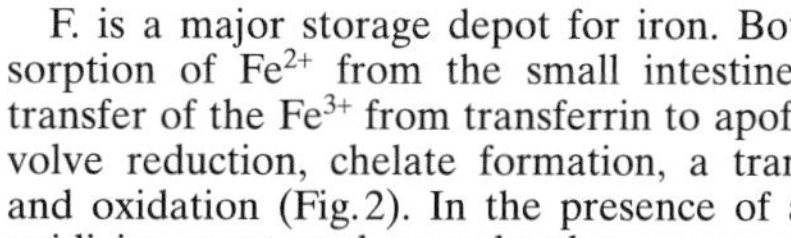

Fig. 1. *Ferritin.* Storage of iron in apoferritin in the presence of an oxidizing agent A, and mobilization of iron from ferritin, which requires NADH

F. is a major storage depot for iron. Both the resorption of Fe^{2+} from the small intestine and the transfer of the Fe^{3+} from transferrin to apoferritin involve reduction, chelate formation, a transfer step and oxidation (Fig. 2). In the presence of a suitable oxidizing agent, such as molecular oxygen, apoferritin catalyses the oxidation of Fe^{2+} to Fe^{3+}. The reverse reaction, the reductive mobilization of the ferritin iron, is catalysed by an NADH-dependent oxidoreductase which may be either membrane-bound or soluble.

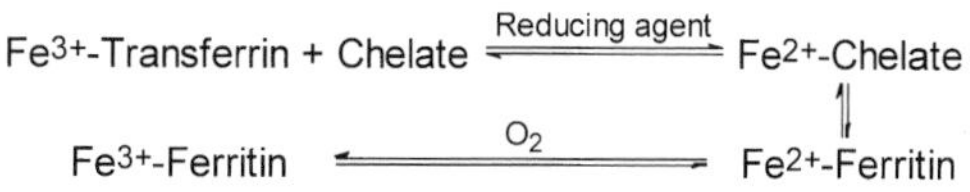

Fig. 2. *Ferritin.* Model for iron transfer from transferrin to ferritin (according to Crichton).

Fervenulin: 6,8-dimethylpyrimido-5,4-e-1,2,4-triazine-5,7-(6H,8H)-dione, M_r 193.17, m.p. 178–179 °C, a pyrimidine antibiotic synthesized by *Streptomyces* *fervens.* Its structure is analogous to that of toxoflavin, but F. is less toxic. It has a broad action spectrum, especially against cocci, Gram-negative and phytopathogenic bacteria and trichomonads.

Fervenulin

Fe-S-protein: see Iron-sulfur proteins.

α-Fetoprotein: the first α-globulin (M_r 70,000) formed in the serum of mammalian embryos. Synthesized first by the yolk sac and later by the fetal liver, it is replaced at birth by serum albumin. The presence of α-F. in adult serum (detected by radioimmunoassay) is symptomatic of liver carcinoma. Escape of α-F. into the amniotic fluid is symptomatic of spina bifida (failure of the neural fissure to close) and/or anencephaly; the condition can be detected by an immunoassay of amniotic fluid obtained by amniocentesis. [F. Jacob, "Expression of Embryonic Characters by Malignant Cells", *1983 Ciba Foundation Symposium 96: Fetal Antigens and Cancer* (Pitman, London) pp. 4–27]

FGAR: see *N*-Formylglycinamide ribotide.

FH_4: see Tetrahydrofolic acid.

Fibrin: the protein endproduct of blood coagulation. It is generated from the plasma protein Fibrinogen (see) by thombin in the presence of Ca^{2+} ions. Thrombin removes two fibrinopeptide pairs, A_2 and B_2, from fibrinogen. After the A peptide has been removed, the protein polymerizes in a pH-dependent, linear fashion. Thereafter, the B peptide is removed. At this point the fibrin bundle is still soluble in urea and has a cross-banding pattern visible in the electron microscope. It is stabilized by factor XIII, which must first be activated by thrombin and Ca^{2+}. Factor XIIIa removes the carbohydrate and generates two intermolecular ε-(γ-glutamyl)lysine bonds (see Isopeptide bonds) per molecule between the γ and α chains of the fibrin bundle. A cross-linked fibrin clot is insoluble both in water and in 8 M urea, and provides a stable wound closure (see Blood coagulation). It can be degraded to soluble cleavage products by Plasmin (see).

Fibrinogen: the precursor of Fibrin (see) and the only coagulable protein in the blood plasma of vertebrates and some arthropods. Human plasma contains 200 to 300 mg/100 ml. In the electrophoresis of the plasma proteins, F. migrates between the β and γ fractions; its M_r is 340,000, of which 15 % is carbohydrate. The accepted model of F. structure (Fig.) consists of 6 polypeptide chains, two each of Aα, Bβ and γ. These are mutually cross-linked by disulfide bonds in their *N*-terminal segments. A coiled coil made up of one of each kind of polypeptide emerges on each side of the central knot. In this region, the polypeptide chains are parallel and in register with one another. The coiled coil is separated by a second region of disulfide bonds ("disulfide swivel") from the terminal, globular region of the protein. The α-chains extend as "random coils" from this globular region and serve as cross-links in the fibrin clot. Thrombin first removes the A peptides from the Aα-chains by cleavage at Arg_{16}, forming fibrin I monomers. The second cleavage occurs at Arg_{14} in the Bβ chain after the fibrin I monomers have polymerized end-to-end to form protofibrils.

F. is also present in platelet α-granules and is secreted from stimulated platelets. It is a major regulator of platelet aggregation, binding to specific receptor sites on the platelet surface. In this respect, F. has features in common with other "adhesive" proteins: Fibronectins (see), von Willebrand factor (see) Thrombospondin (see). ["Molecular Biology of Fibrinogen and Fibrin" *Annals of the New York Academy of Sciences* **408** (1983)]

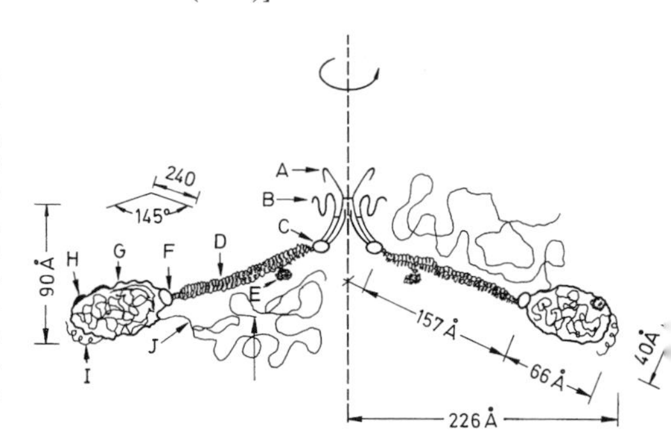

Detailed model of fibrinogen drawn to scale. A, fibrinopeptide A at amino- terminus of α-chain; B, fibrinopeptide B at amino-terminus of β-chain; C, first disulfide swivel holding all three chains together and in register; D, coiled-coil connection segment holding domains I and III together; E, carbohydrate attached to residue 52 of γ-chain; F, second disulfide swivel holding all three chains together; G, domain III which corresponds approximately to fragment D; H, carbohydrate on β-chain; I, carboxy-terminal segment of γ-chains which becomes involved in intermolecular cross-linking; J, plasmin attack point and α-chain cross-linking region; K, carboxy-terminus of α-chain. [from Doolittle et al. *Horizons Biochem. Biophys. 3* (1977) 164–191, with permission]

Fibrinolysin: see Plasmin.

Fibrinolysis: see Plasmin.

Fibrinopeptides: the two pairs of peptides (A and B) cleaved from the *N*-termini of the 2α and 2β chains of fibrinogen by thrombin. F. arise by cleavage of Arg-Gly bonds, so that Arg is the *C*-terminal end of the F. and Gly is the *N*-terminal end of the α and β chains of fibrin. Human F. A. is Ala-Asp-Ser-Gly-Glu-Gly-Asp-Phe-Leu-Ala-Glu-$(Gly)_3$-Val-Arg, and human F. B. is Pyr-Glu-Gly-Val-Asn-Asp-Asn-$(Glu)_2$-Gly-$(Phe)_2$-Ser-Ala-Arg. F. A. ranges in size from 14 amino acids (horse, lizard) to 19 (cattle), and F. B. from 9 (rhesus monkeys) to 21 (cattle, elk and kangaroo). The sequences of the F. have been used to establish a detailed phylogenetic tree for mammals which is very similar to the classical one. The F. have a vasoconstrictive effect which serves to keep the coagulation principles from being removed too quickly from an injury site.

Fibronectin: A family of closely related adhesion glycoproteins, M_r 200,000–250,000, which arise by the alternative splicing of the precursor mRNA. F. is found both as a cell surface protein of the extracellular matrix (also known as *Cell Surface Protein, Galactoprotein A, Large External Transformation Sensitive Protein, Surface Fibroblast Antigen,* and *Zeta* or *Z Protein*), and in plasma. Plasma fibronectin is incorporated into fibrin clots and is cross-linked to the fibrin, probably via a Gln-Lys transamidation reaction.

Cell surface F. functions in cell adhesion and cell migration (e.g. in wound healing and embryonic development). It has a very low mobility because it is anchored via integrin (a complex of proteins spanning the plasma membrane) to actin filaments on the other side of the plasma membrane. It exhibits strong affinity binding to fibrous proteins such as collagen and fibrin, and enables cells to interact with the extracellular matrix. Each F. molecule carries a linear series of domains (3 types of domain are recognized, designated I, II and III); thus, heparin binds to a group of 5 repeated type I domains, collagen binds to the sequence I-II-II-I-I-I, and cell binding is mediated by a group of 3 repeated type III domains. In tissue cultures, cellular F. is found in fibrillar arrays on cell surfaces, between cells and as part of the intracellular matrix. In vivo, it is found as a component of connective tissue matrices, i.e. in basement membranes and as thin fibers in loose connective tissue. F. also enables cells to migrate in developing embryos by alternately binding to and dissociating from integrin.

Transformed cells (which do not adhere to substrate) lack F., and addition of F. to some lines of transformed fibroblasts induces them to adhere to the dish and spread out. The adhesion-promoting activity can be mimicked by the tetrapeptide Arg-Gly-Asp-Ser, a sequence present in the cell attachment domain of F., or by Arg-Gly-Asp-Cys. The same tetrapeptides block adhesion of cells to F.-coated surfaces, which implies that the cell has a receptor specific for this sequence.

F. is a component of a fibrillar matrix underlying the blastocoel roof in the blastula stage of amphibian embryos, and has been shown to be necessary for mesodermal cell migration in gastrulation.

Fragments of F. have been isolated from normal human plasma and other sources. Enhanced levels of these fragments are characteristic of diseases with abnormally high rates of proteolysis, including cancer. This suggests that proteolysis of F. releases "activation peptides" which regulate cell division; evidence for this is the observation that proteolysis of F. by cathepsin D releases fragments which induce DNA synthesis in cultured normal fibroblasts. There is evidence that cell-surface F. is a "receptor" for binding of the infectious stage of trypanosomiasis parasites, and some bacterial pathogens. [E. Pearlstein et al. *Molec. Cell. Biochem.* **29** (1980) 103–128; M. J. Humphries & S. R. Ayad *Nature* **305** (1983) 811–813; J. C. Boucaut et al. *Nature* 307 (1984) 364–366; M. A. Ouaissi et al. *Nature* **308** (1984) 380–382; M. D. Pierschbacher & E. Ruoslahti *Nature* **309** (1984) 30–33; R. Hynes *Annu. Rev. Cell Biol.* **1** (1985) 67–90; M. Okada et al. *J. Biol. Chem.* 260 (1985) 1811–1820; E. J. H. Danen et al. *J. Biol. Chem.* **270** (1995) 21612–21618; L. M. Schnapp et al. *J. Biol. Chem.* **270** (1995) 23196–23202]

Fibrous proteins: see Structural proteins.

Ficaprenols: see Polyprenols.

Ficifolinol: 3,9-dihydroxy-2,8-di-γ-γ-dimethylallyl-pterocarpan. See Pterocarpans.

Field effect transistor, *FET*: an electronic device in which the conductivity of the semiconductor material is controlled by the electrical field at a particular part of its surface, known as the "gate". This electrical field may be varied by attachment (and subsequent reaction) of enzymes (ENZFET) or antibodies (IMMUNOFET) to the gate. The gate may also be sensitized to specific ions (ISFET) or other chemicals (CHEMFET).

Filaggrin: a basic protein, M_r 26,500 to 49,000 (depending on the animal species), from the stratum corneum of the epidermis. It is probably identical to the interfilamentous keratin matrix protein; its function is to organize prekeratin into filaments as the epidermal cells move outward from the basal layer and become keratinized. F. is synthesized as a highly phosphorylated precursor which cannot organize keratin. Later the phosphate-bearing region is removed by proteolysis; the remaining protein has a large number of Arg residues which presumably interact with the acidic regions of the keratins. Still later in its development, the Arg residues are converted to citrulline, and as this happens, F. loses the ability to aggregate pre-keratins. [P. Traub, *Intermediate Filaments,* Springer, Heidelberg, 1985)]

Filamen: see Cytoskeleton (microfilaments).

Fimbrin: see Cytoskeleton (microfilaments).

Finger print technique: see Proteins, DNA fingerprinting.

Fisetin: see Flavones (Table).

Flagellin: the main protein component of bacterial flagella. The filaments of F. have an α-keratin structure, which is reversibly converted to the β-keratin structure on stretching of the aggregate. The F. monomer (M_r 33,000 to 40,000) is dissociated by mild acid treatment (pH 3–4). It has 304 amino acid residues, but no cysteine or tryptophan and only traces of proline and histidine.

Flavan: a naturally occurring Flavonoid (see) with the ring system shown (Fig. 1).

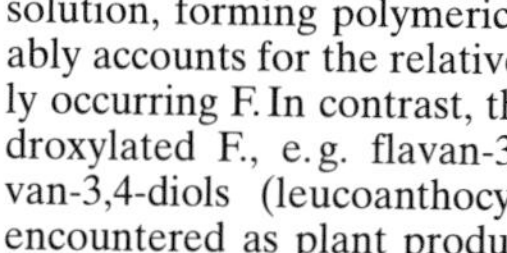

Fig. 1. *Flavan ring system*

F. unsubstituted at positions 3 or 4 are unstable in solution, forming polymeric products, which presumably accounts for the relatively few reports of naturally occurring F. In contrast, the more stable 3- or 4-hydroxylated F., e.g. flavan-3-ols (catechins) and flavan-3,4-diols (leucoanthocyanidins), are frequently encountered as plant products. Four reported F. glycosides are ***diffutin*** (7-hydroxy-3′,4′-dimethoxy-5′-*O*-β-D-glucosylflavan, *Canscora diffusa),* ***auriculoside*** (7,5′-dihydroxy-4′-methoxy-3′-*O*-β-D-glucosylflavan, *Acacia auriculiformis),* ***koaburarin*** ((2*R*)-5-hydroxy-7-*O*-β-D-glucosylflavan, *Enkianthus nudipes)* and 7,4′-dihydroxy-5-*O*-β-D-xylosylflavan *(Buckleya lanceolata).* F. have been found in 7 families: *Ericaceae,*

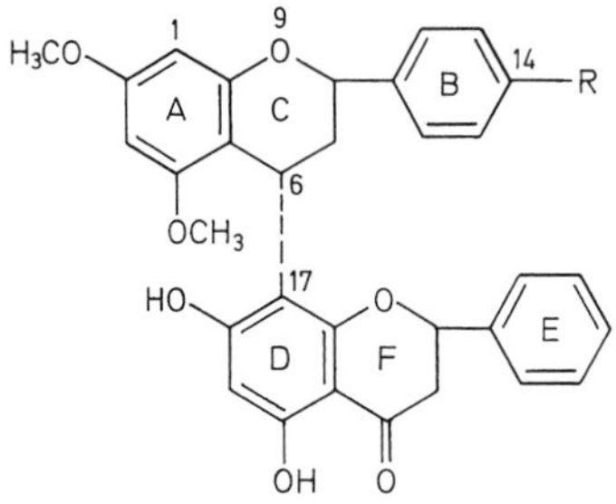

R =H: *Xanthorrhone*
R = OH: *14-Hydroxyxanthorrhone,*
both form *Xanthorrhoea (Liliaceae)*

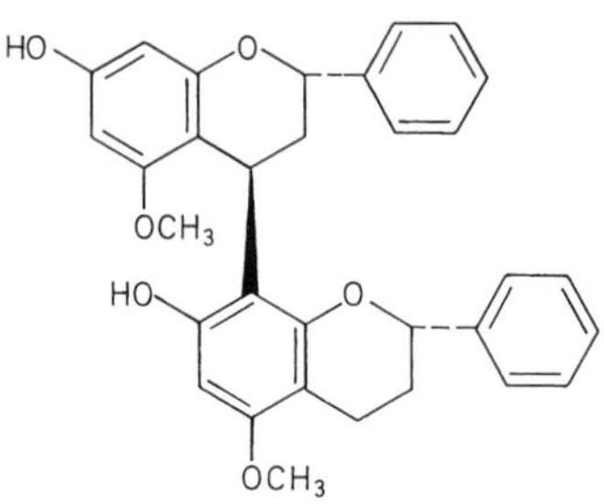

Biflavan 12 from "dragon's blood",
the resin of *Deamonorops draco (Palmae)*

Fig. 2. *Some naturally occurring flavans.*

Gentianaceae, Leguminoseae, Liliaceae, Myristicaceae, Santalaceae and *Amaryllidaceae.* Some F. have antimicrobial activity and some are also phytoallexins, e.g. ***broussin*** (7-hydroxy-4′-methoxyflavan) is formed by wounded xylem tissue of *Broussonetia papyrifera* (paper mulberry), and is absent from healthy tissue; it inhibits growth of the bacterium *Bipolaris leersiae* at 10^{-4}-10^{-5} M. Infection of *Narcissus pseudonarcissus* L. by *Botrytis cinerea* elicits production of 7-hydroxyflavan, 7,4′-dihydroxyflavan and 7,4′-dihydroxy-8-methylflavan, which inhibit growth of the fungus and are also active against fungal spore germination in liquid culture.

Diffutin (see above) has pronounced adaptogenic (anti-stress and anti-anxiety) activity, as well as being a mild CNS depressant (barbiturate potentiation). It also has a marked inotropic effect in perfused frog heart, and shows no arrhythmogenic properties. In addition, it potentiates the contractile response of guinea pig vas deferens to catecholamines, without inhibition of the uptake of adrenalin. At 500 mg/kg, diffutin is nontoxic to dogs. The use of *Canscora* in Indian medicine as a herbal remedy for certain mental disorders is supported by these observations.

Three dimers containing the F. ring system have also been isolated (Fig. 2).

[S. K. Saini & S. Ghosal *Phytochemistry* **23** (1984) 2415–2421]

Flavan-3,4-diol: see Leucoanthocyanidins.

Flavanone: a flavonoid with the ring system shown (Fig).

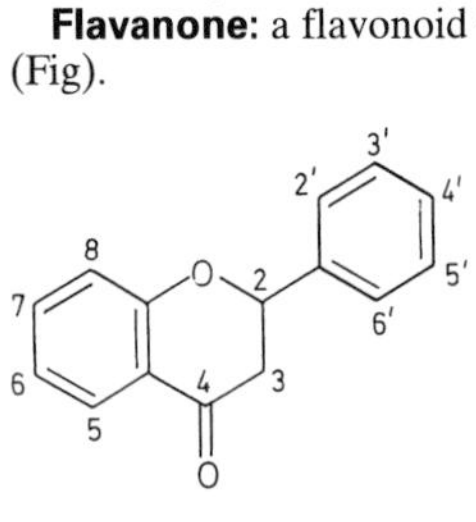

Flavanone ring system

F. hydroxylated at position 3 may be called flavanonols, but are more commonly known as dihydroflavonols. F. are amongst the earliest products of flavonoid biosynthesis (see Flavonoid, Fig.). Many F. have fungistatic or fungitoxic properties. They occur free or as glycosides in many angiosperm families and in many gymnosperms, but insufficient data are available for a useful taxonomic analysis. Examples are ***naringenin*** (5,7,4′-trihydroxyflavanone), found in several plant families and a growth inhibitor in dormant peach flowers; ***naringin*** (naringenin 7-neohesperidoside), a bitter constituent of grapefruit and bitter oranges; ***prunin*** (naringenin 7-glucoside) from *Prunus;* **pinostrobin** (5-hydroxy-7-methoxyflavanone) from *Prunus;* **pinocembrin** (5,7 dihydroxyflavanone) from *Pinus;* **bavachin** (7,4′-dihydroxy-6-prenyl-flavanone), from *Psoralea;* **eriodictyol** (5,7,3′,4′-tetrahydroxyflavanone) from *Eriodictyon;* ***hesperetin*** (5,7,3′-trihydroxy-4′-methoxyflavanone), from *Citrus;* hesperidin (hesperetin 7-rutinoside) from *Citrus.* For lists of known F., see B. A. Bohm in *The Flavonoids: Advances in Research,* J. B. Harborne & T. J. Mabry (eds.), Chapman and Hall, 1982, pp. 350–416.

Flavin: a compound containing the 6,7-dimethylisoalloxazine ring system (Fig.). Biologically important F. are Flavin adenine dinucleotide (see), Flavin mononucleotide (see) and Riboflavin (see under Vitamins). See Flavin enzymes.

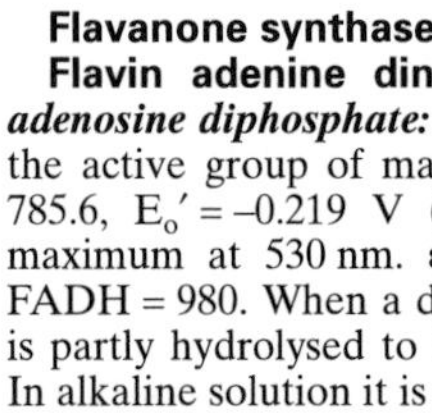

Flavin ring system

Flavanone synthase: see Chalcone synthase.

Flavin adenine dinucleotide, *FAD, riboflavin adenosine diphosphate:* a flavin nucleotide (Fig.) and the active group of many Flavin enzymes (see). M_r 785.6, E_o' = –0.219 V (pH 7.0, 30 °C), fluorescence maximum at 530 nm. ε_{450} of FAD = 11300, ε_{450} of FADH = 980. When a dilute solution is heated, FAD is partly hydrolysed to Flavin mononucleotide (see). In alkaline solution it is quickly converted to cyclic riboflavin 4′,5′-phosphate. FAD is less photolabile than flavin mononucleotide or riboflavin.

The isoalloxazine ring of the flavin acts as a reversible redox system (see Flavin mononucleotide). The colorless dihydro-compound is formed by addition of hydrogen to N1 and N10; oxidized FAD is yellow.

Flavin catalysis may involve three oxidation levels of the ring, and each of these may exist as a cation, anion or neutral compound, giving a total of 9 forms. The oxidation levels are flavoquinone (oxidized) ⇌ flavosemiquinone (half-reduced) ⇌ flavohydroquinone (reduced). Some catalytic mechanisms involve only one electron, i.e. the flavoquinone is converted to the semiquinone, or the semiquinone is converted to the hydroquinone. Other mechanisms involve two electrons, and the flavin shuttles back and forth between the quinone and the hydroquinone states. FAD is biosynthesized from flavin mononucleotide by the action of FAD pyrophosphatase (EC 3.6.1.8): FMN + ATP ⇌ FAD + PP_i.

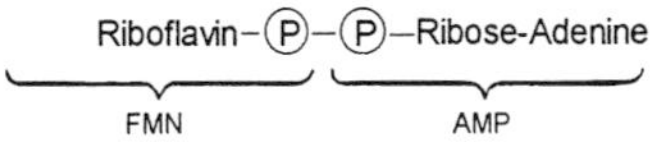

Riboflavin–Ⓟ–Ⓟ–Ribose-Adenine

FMN AMP

Structure of FAD

Flavin enzymes, *Flavoproteins, yellow enzymes:* a diverse group of more than 70 oxidoreductases of animals, plants and microorganisms which have Flavin adenine dinucleotide (FAD) (see) or Flavin mononucleotide (FMN) (see) as prosthetic groups or coenzymes. The coenzymes are reversibly reduced, either by transfer of hydrogen atoms from a substrate (e.g. succinate dehydrogenase) or from NAD(P)H. The properties of FAD and FMN depend strongly on the protein molecule around them. F.e. are structurally and functionally very heterogeneous, and there is no single generalized type. The metalloflavin enzymes (metalloflavoproteins) contain metals, such as Fe, Mg, Cu or Mo, which are involved in the redox reaction and in the binding of the F.e. to the mitochondrial membrane (see Molybdenum enzymes). F.e. can be divided into the following groups:

1. *Oxidases* use oxygen as electron acceptor and transfer two or four electrons. Those which transfer two electrons form H_2O_2. They include Glucose oxidase (see), the iron- and molybdenum-containing Xanthine oxidases (see), and the D-amino acid oxidases (containing FMN or FAD). The latter catalyse the irreversible formation of the corresponding α-ketoacid. Oxidases which transfer four electrons contain copper; they oxidize the substrate and form water, e.g. laccase, ascorbate oxidase and *p*-diphenol oxidase.

2. *Reductases* react primarily with cytochromes. They include cytochrome b_5 and *c* reductases and glutathione reductase (all contain FAD), GMP reductase and the Mo-containing nitrate reductase.

3. *Dehydrogenases;* the natural hydrogen acceptor of some of these is unknown. A well-known example of this group is Succinate dehydrogenase (see) of the tricarboxylic acid cycle. The respiratory chain NADH and NADPH dehydrogenases (FMN) and acyl-CoA dehydrogenase are further examples

In addition, there are more complex F.e., such as the hemoflavin enzymes (e.g. formic acid dehydrogenase from *E. coli)* which contain metals, sulfhydryl-disulfide systems and heme groups, in addition to the flavin component.

Flavin mononucleotide, *FMN, riboflavin 5'*-phosphate: the prosthetic group of various flavin enzymes. M_r 456.4, E_0' –0.219 V (pH 7.0, 30 °C), fluorescence maximum at 530 nm. FMN is composed of 6,7-dimethylisoalloxazine (flavin) and a ribitol residue linked to N9 (Fig.). It occurs as the free acid or the sodium salt and usually contains 2–3 molecules H_2O. The phosphate ester bond is hydrolysed in acid solution, and the isoalloxazine-ribitol bond is unstable in alkaline solution. The compound is photolabile over the entire pH range, but is particularly so in alkaline solution.

FMN is biosynthesized from riboflavin and ATP by a flavokinase. It is hydrolysed by acid and alkaline phosphatases.

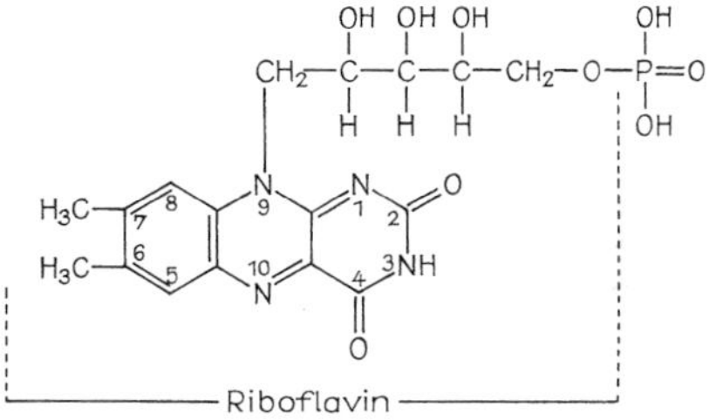

Riboflavin

Flavin mononucleotide

Flavin nucleotides: the coenzymes of the Flavin enzymes (see). Strictly speaking, they are Prosthetic groups (see), but they can be easily separated from the apoproteins of some flavin enzymes . The two F. are Flavin mononucleotide (see) and Flavin adenine dinucleotide (see).

Flavodoxins: low-potential, electron-transfer proteins (M_r ~ 20,000) containing one molecule of noncovalently bound Flavin mononucleotide (FMN) (see). They occur in a range of prokaryotes and a few eukaryotic algae and are generally only present when the environment is deficient in iron, whereupon they appear to be able to substitute for Ferredoxin (see) in all the reactions in which this iron-sulfur, electron-transfer protein participates. In nitrogen-fixing organisms flavodoxins are particularly favored as the reducers of Nitrogenase (see). FMN is the redox-active component of F. and is reduced in two one-electron steps via a semiquinone. The E'_0 of these two steps are usually widely separated; for instance the E'_0 value for reduction of the oxidized F. of cyanobacteria to the semiquinone ranges from –0.210 to –0.235V, whereas reduction of the semiquinone to the reduced F. has an E'_0 value of –0.414. F. can be crystallized in their oxidized, semiquinone or reduced forms.

Flavones, *flavone pigments:* a group of plant pigments containing the flavone ring system. This consists of two substituted phenyl rings (A and B) and the pyrone ring C, fused to ring A, which is responsible for the typical reactions of the F. The structures of about 300 natural F. are known; except for flavone, they all carry hydroxyl groups. Some also have methyl or methoxy substituents. Between 1 and 7 OH groups are usually present, the favored hydroxylation posi-

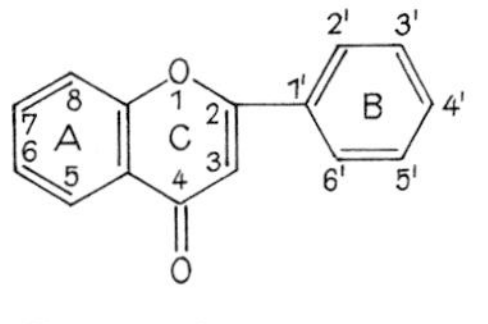

Flavone ring system

tions being 3,5,7,3′ and 4′. Those hydroxylated at position 3 are often called flavonols and are treated as a subgroup of the F. F. occur in the plant either in free form or as glycosides or esters. The most common positions for a glycosidic bond are 3 and 7.

Examples are: ***chrysin*** (5,7-dihydroxy-F., poplar buds, heartwood of many pines); ***primetin*** (5,8-dihydroxy-F., primroses); ***galangin*** (3,5,7-trihydroxy-F., wood of pine trees and roots of *Apinia officinarum*); ***apigenin*** (5,7,4′-trihydroxy-F., white and yellow blossoms); ***kaempferol*** (3,5,7,4′-tetrahydroxy-F., many plants, including 50 % of all angiosperms); ***fisetin*** (3,7,3′,4′-tetrahydroxy-F., wood and blossoms of many higher plants); ***luteolin*** (5,7,3′,4′-tetrahydroxy-F., flowering plants, e.g. mignonette, dahlia, broom); ***morin*** (3,5,7,2′,4′-pentahydroxy-F., various members of the *Moraceae,* e.g. old fustic); ***quercetin*** (3,5,7,3′,4′-pentahydroxy-F., wide occurrence in 56 % of all angiosperms); ***myricetin*** (3,5,7,3′,4′,5′-hexahydroxy-F., higher plants, e.g. bark of *Myrica nagi*).

F. and flavonols are vacuolar pigments with a high absorption from 240 to 270 nm and 320 to 380 nm, i.e. they are yellow; their frequent occurrence with anthocyanins leads to red and yellow flower colors. They used to be used in dying and printing, especially those from quercitron bark, old fustic (*Chlorophora tinctoria* L.), buckthorn berries and camomile.

F. are biosynthesized from a phenylpropane unit such as cinnamic acid, which contributes the aromatic ring B, and C-atoms 2, 3 and 4. The remaining C-atoms are added by head-to-tail condensation of acetate and malonate (see Stilbenes).

Flavonoids: natural products with a $C_6C_3C_6$ skeleton. In most F., this skeleton takes the form of a phenylchroman ring system, in which the phenyl group may be attached at position 2 (normal flavonoids), 3 (isoflavonoids) or 4 (neoflavonoids) of the pyran ring. Further classification is based on the degree of oxidation of the pyran ring; see Anthocyanidins, Flavan, Flavanone, Flavones, Isoflavonoid, Isoflavan, Isoflavanone, Isoflavone, Neoflavonoid, Leucoanthocyanidins, Catechins. The oxygen-containing ring may be contracted (see Aurones), or absent (see Chalcones; α-Methyldeoxybenzoins).

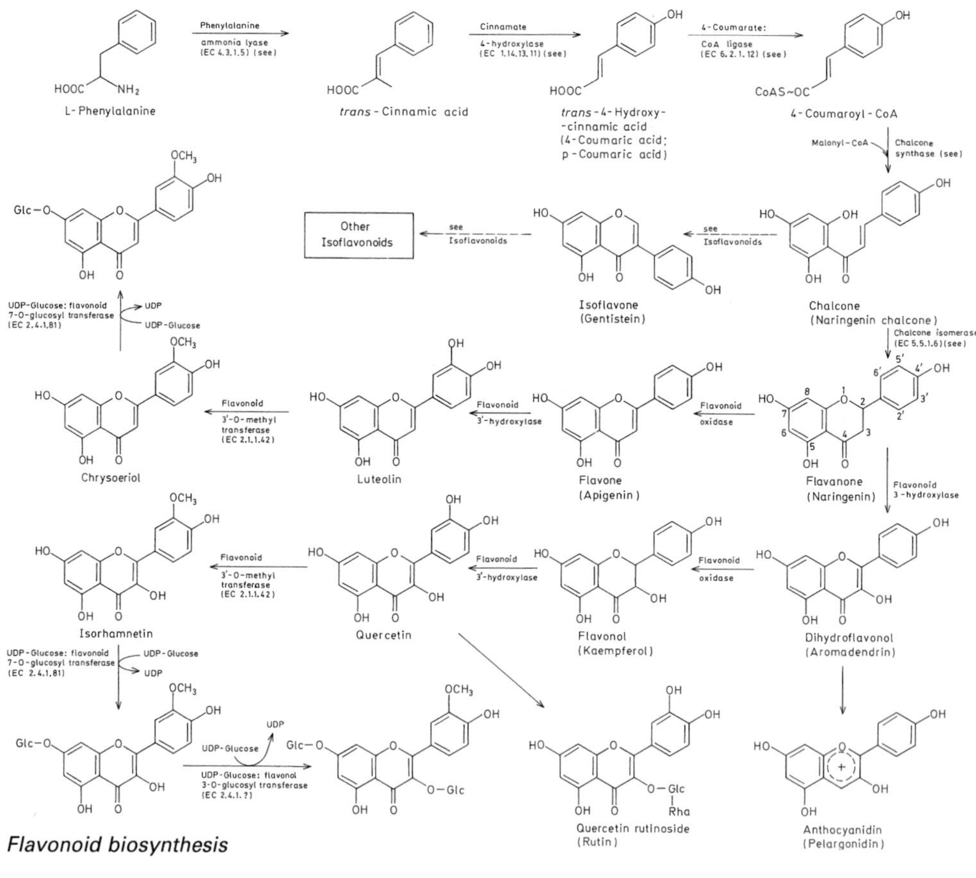

Flavonoid biosynthesis

Biosynthesis. It is well established by radiotracing that all classes of F. are biosynthetically closely related (Fig.). A chalcone is the latest intermediate in the biosynthetic sequence which is common to all F. The first stages of biosynthesis from phenylalanine also lead to the biosynthesis of other phenylpropyl compounds, e.g. coumarins, lignans, lignin, benzoic acid derivatives, aromatic esters, etc. Chalcones are also the precursors of stilbenes. The various patterns of hydroxylation and methoxylation in the F. may be partly established at an early stage, e.g. by hydroxylation of coumaroyl-CoA to caffeoyl-CoA, and *O*-methylation of caffeoyl-CoA to feruloyl-CoA; *S*-adenosyl-L-methionine:caffeoyl-CoA 3-*O*-methyltransferase (EC 2.1.1.104) has been characterized from plants (*S*-Adenosyl-L-methionine + caffeoyl-CoA → *S*-Adenosyl-L-homocysteine + feruloyl-CoA). Hydroxylation, *O*-methylation and glycosylation also occur at various stages after formation of the F. ring system. Some typical conversions are shown in the Fig., which also shows the biosynthetic relationships between the main classes of F. [J. Ebel & K. Hahlbrock in *The Flavonoids: Advances in Research* (ed. J. B. Harborne & T. J. Mabry) Chapman and Hall, 1982, pp. 641–675]

Flavonol: a flavone hydroxylated at position 3. F. and flavones are the two most abundant classes of flavonoids. For examples, see Flavones.

Flavoproteins: see Flavin enzymes.

Fluid mosaic model: see Biomembrane.

Fluorescamine: see Amino acid reagents.

Fluoride, F^-: an anion found in bone and tooth apatite. Small quantities are beneficial in lowering the incidence of caries, and this cariostatic effect of F^- has been clearly demonstrated in humans. Fluoridation of water is now a public health measure (optimal level in drinking water 1-2 ppm F^-). The role of F^- in the inhibition of osteoporosis is less certain. High levels of F^- are toxic (fluorosis). F^- affects several enzymes. Excess F^- decreases fatty acid oxidase in rat kidney, and partially inhibits intestinal lipase. Fatty acid utilization is generally impaired in fluorosis. Carbohydrate metabolism is also affected, probably due to inhibition of enolase and a shift of the NAD/NADH ratio in favor of NADH. Cow's milk contains 1-2 μg F^- per g. dry weight. Cereals contain 1–3 μg F^-/g. Tea (100 μg F^-/g.) is especially high in F^-. Sea foods contain 5-10 μg F^-/g.

Fluoroacetic acid: CH_2F-COOH, a very poisonous carboxylic acid which is converted in the tricarboxylic acid cycle to fluorocitrate, a strong inhibitor of aconitase. The Tricarboxylic acid cycle (see) therefore ceases to operate, with lethal results. Free F. a. occurs in the leaves of *Dichapetalum cymosum,* a poisonous African plant which is reponsible for cattle poisoning in South Africa. It is also present in other plants to which it confers toxicity. [L. P. Miller and F. Flemion *Phytochemistry* **3** (1973) 1–40]

Fluorocitric acid: see Fluoroacetic acid.

Flurenol: see Morphactins.

Fly agaric toxins: the toxins of *Amanita muscaria.* Poisonings by this mushroom are rarely fatal. The F. a. t. include Muscarin (see) and other quaternary ammonium bases, such as muscaridin, indole compounds, Ibotenic acid (see) and its easily formed derivatives, Muscimol (see) and Muscazone (see) (Fig.). Muscimol and ibotenic acid inhihit motor functions, and muscimol is psychotropic. This explains the use of the fly agaric as a psychedelic agent in some regions. The fly-killing power long ascribed to this fungus (hence its name) is due to the weak insecticidal action of ibotenic acid and muscimol, which, however, are only effective when consumed by the fly.

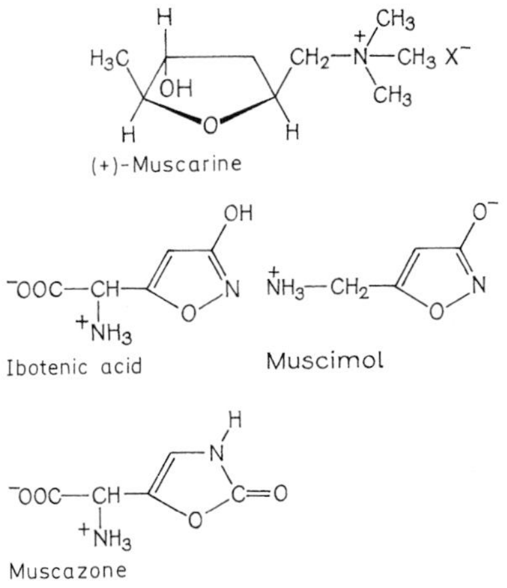

(+)-Muscarine

Ibotenic acid

Muscimol

Muscazone

Fly agaric toxins

FMN: see Flavin mononucleotide.

Folate-H_2: see Tetrahydrofolic acid.

Folate-H_4: see Tetrahydrofolic acid.

Folic acid: see Vitamins (vitamin B_2 complex).

Follicle stimulating hormone, *FSH, follitropin:* a gonadotropin secreted by the hypophysis. FSH is an acidic glycoprotein (M_r 32,000) containing a high proportion of glutamate, threonine and cysteine residues, and 27 % carbohydrate (consisting chiefly of sialic acid, galactose, mannose and glucosamine, with smaller amounts of galactosamine and fucose). The protein consists of an α-subunit (92 amino acids with carbohydrate attached to Asn_{52} and Asn_{78}) and a β-subunit, which displays microheterogeneity at the *N*- and *C*-terminals. The β-subunit contains between 108 and 115 amino acid residues and is unique to FSH. The α-subunit, however, shows close homologies with the α-chains of human chorionic gonadotropin, luteinizing hormone and thyrotropin.

FSH is synthesized in the adenohypophysis, and its secretion is regulated by follicle stimulating hormone releasing factor or folliliberin (which is identical with luteinizing hormone releasing factor) (see Releasing hormones). FSH is responsible for development and function of the gonads in both men and women. It stimulates spermatogenesis in the male testes, and it controls maturation of the female follicle. Together with luteinizing hormone, estradiol and progesterone, FSH is involved in the hormonal regulation of the monthly cycle. In women, the secretion of FSH is inhibited by estradiol, in men by testosterone.

Follicle-stimulating-hormone releasing hormone: see Releasing hormones.

Follitropin: see Follicle-stimulating hormone.

Footprinting: a procedure for identifying the sites in DNA that bind specific proteins [e.g. determination of promoter sequences of DNA that bind RNA polymerase (see Pribnow box)].

One strand of the DNA under investigation is labeled at its 5′ end with ^{32}P (techniques for end labeling are described in the entry Recombinant DNA technology) then incubated *in vitro* with the DNA binding protein; the binding protein may have been previously purified, or the end-labeled DNA may be incubated with a crude mixture of proteins containing the protein in question, e.g. a nuclear extract. Two alternative techniques are then possible: 1. the DNase I technique, based on the principle that the region of DNA bound by the protein is protected from attack by DNase I; and 2. the dimethyl sulfate (DMS) technique, based on the principle that the region of DNA bound by the protein is protected from methylation by DMS.

In the *DNase I technique* the protein-complexed, end-labeled DNA is incubated with DNase I under mild, controlled conditions to produce on average only one cleavage site ("nick") per DNA molecule. As a control, a sample of the same end-labeled DNA, which has not been complexed with protein, is treated similarly with DNase. Both samples of DNA are denatured and submitted to electrophoresis side-by-side in a sequencing gel (the fragments separate according to size). Autoradiography of the gels reveals that the control DNA has been cleaved randomly to produce an uninterrupted ladder of radioactive nucleotides. The autoradiogram of the DNase I-digested, protein-complexed DNA reveals a similar, but interrupted ladder (i.e. the ladder of nucleotides contains a gap or *footprint*), because certain potential cleavage sites were protected from DNase action by the bound protein (Fig.). All the bands in the complete ladder have the same 5′ terminus; they represent cleavage products ranging in steps of one nucleotide from the 5′ terminal mononucleotide to nucleotides containing *n*-1 and finally *n* residues, where *n* is the total number of nucleotide residues in a single strand of the DNA (or the number of base pairs in the original double-stranded DNA sample). Obviously, each DNA digest contains many more than *n* nucleotides, including all those retaining the 3′-terminus, as well as other fragments lacking both the 3′- and the 5′-terminus. However, only those nucleotides retaining the 5′-terminus are revealed by autoradiography. It is therefore possible not only to identify the relative position of the protein along the DNA, but also to determine the complete nucleotide sequence of the DNA and the identity of the particular nucleotide sequence protected by the protein (see Nucleic acid sequencing).

In the *dimethylsulfate (DMS) technique,* the end-labeled DNA segment is treated with DMS, cleaved chemically at its methylated G residues, and the resulting fragments separated electrophoretically; this entire procedure is described in detail in the entry Nucleic acid sequencing (see). As in the previous technique, gaps or *footprints* in the autoradiogram identify the position and nature of the DNA sequences that bind the protein in question.

Forbes' disease: see Glycogen storage disease.

Formaldehyde dehydrogenase, ***formaldehyde: NAD*⁺** **oxidoreductase** (glutathione formylating) (EC 1.2.1.1): an enzyme catalysing the NAD-dependent formation of *S* formylglutathione from glutathione and formaldehyde. This first stage in the conversion of formaldehyde to formate (Fig.) has been demonstrated in beef, chicken, human, monkey and rat liver, human and animal retinas, and in yeast. The second stage is catalysed by *S*-formylglutathione hydrolase. Both enzymes have been purified from rat liver [L. Uotila & M. Koivusalo *J. Biol. Chem.* **249** (1974) 7653–7663, 7664–7672]. Glutathione is also a substrate for Glyoxalase (see), but in the glyoxalase system the substrate is modified by an intramolecular hydride shift, whereas F.d. involves an NAD-linked

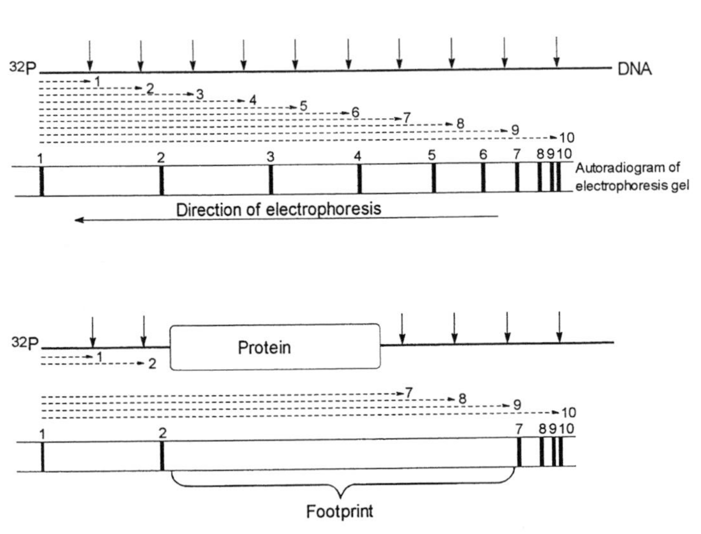

Diagrammatic representation of the DNase I technique of footprinting.
↓ site of cleavage by DNase I. ---→ 1, ----→ 2, -----→ 3 etc. represent nucleotides extending from the ^{32}P-labeled terminus to a cleavage site

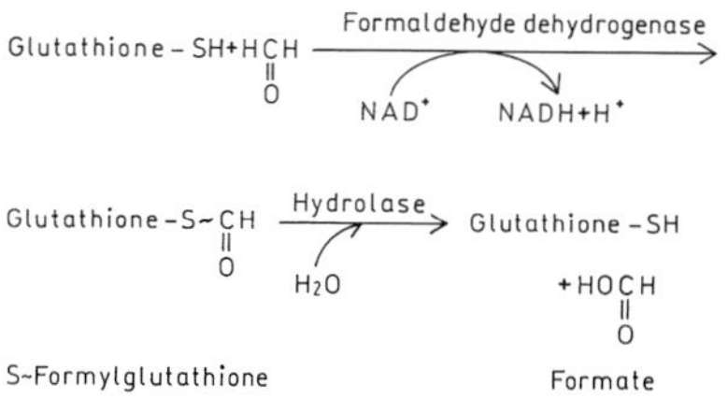

Glutathione – SH + HCHO —(Formaldehyde dehydrogenase; NAD⁺ → NADH + H⁺)→ Glutathione – S – CHO (S-Formylglutathione)

Glutathione – S – CHO + H_2O —(Hydrolase)→ Glutathione – SH + HOCHO (Formate)

Oxidation of formaldehyde to formate by the coupled action of formaldehyde dehydrogenase and S-formylglutathione hydrolase

oxidation (NADP also serves as a cofactor, but NAD is more efficient).

5-Formamidoimidazole-4-carboxamide ribotide: see Purine biosynthesis.

Formaminotransferase deficiency: see Inborn errors of metabolism.

Formononetin: ***7-hydroxy-4′-methoxyisoflavone.*** See Isoflavone.

Formycins: pyrimidine antibiotics (see Nucleoside antibiotics) synthesized by *Nocardia interforma. Formycin* itself is 3-β-D-ribofuranosyl-7-amino-pyrazolo-(4,3-d) pyrimidine. The biogenetic precursor is adenosine. F. is converted to formycin 5′-triphosphate, which acts as an ATP analog. *Mycobacterium* and *Xanthomonas oryzae* are particularly sensitive to formycin.

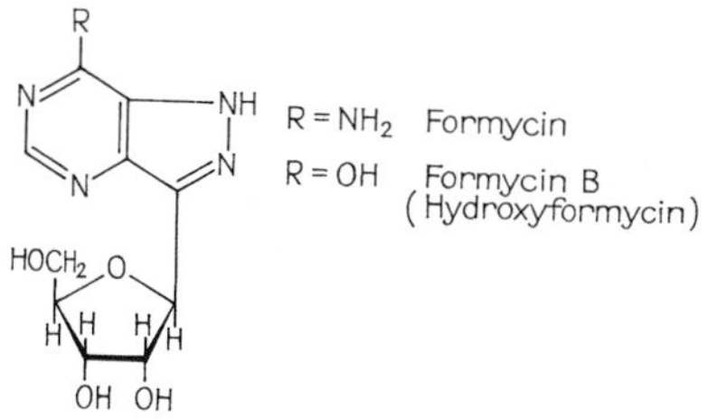

R = NH_2 Formycin
R = OH Formycin B (Hydroxyformycin)

Formycins

Formycin B or hydroxyformycin, is also synthesized by *Streptomyces lavendulae* and *S. roseochromogenes* var. *cyaensis.* Formycin B is the deamination product of formycin and is less toxic than the latter. Its effect is limited mostly to *Mycobacterium* spp. and a few viruses.

N*-Formylglycinamide ribotide, *FGAR: an intermediate in Purine biosynthesis (see).

***N*-Formylglycinamidine ribotide:** an intermediate in Purine biosynthesis (see).

Formylmethionyl-tRNA: see Initiation tRNA.

N^{10}-Formyltetrahydrofolic acid: see Active one-carbon units.

Fraction-1 protein: see Ribulose *bis*phosphate carboxylase.

Fragarin: see Pelargonidin.

Fragment reaction: a reaction used to assay the activity of peptidyl transferase. In a cell-free system containing 70S or 80 S ribosomes, the growing peptide chain is transferred to Puromycin (see) and released as peptidyl-puromycin.

FRH: acronym of follicle-stimulating hormone releasing hormone. See Releasing hormones.

5α-Friedelane: see Friedelin.

Friedelin: a pentacyclic triterpene ketone, M_r 426.70, m. p. 263 °C, $[\alpha]_D$ −27.8 ° ($CHCl_3$). F. is abundant in the bark of the cork oak (1 %), in grapefruit rinds and some lichens. The hydrocarbon skeleton of F. is 5α-friedelane.

Fructans: high-molecular-mass polysaccharides of 1,2- or 1,6-glycosidically linked D-fructose. They are common in plants, e. g. inulin and phlein, and the branched triticin, hordecin and graminin.

β-Fructofuranosidase: see Disaccharidases.

D-Fructose, *fruit sugar, levulose:* a ketohexose, M_r 180.16, m. p. 103–105 °C (d.), $[\alpha]_D^{20}$ −135 ° → −92 ° (c = 2, water). F. tastes sweeter than any other carbohydrate and is fermentable by yeast. It crystallizes as the β-pyranose, but is present in compounds as the furanose (see Carbohydrates). Chemical reduction yields a 1 : 1 ratio of D-sorbitol and D-mannitol. Its metabolically important derivatives are fructose 1,6-*bis*phosphate and the 1-phosphate.

F. is found together with glucose and sucrose in many sweet fruits and in honey. It is a component of many oligosaccharides, including sucrose, raffinose stachyose and gentianose, and of various polysaccharides, such as inulin and levan.

F. is used as a sweetener by diabetics, because it does not raise the blood sugar level, even in large amounts. *Metabolism:* F. is phosphorylated to fructose 1-phosphate by ketohexokinase (EC 2.7.1.3). Only a small amount is phosphorylated in position 6. In the liver, fructose 1-phosphate is split into dihydroxyacetone phosphate and glyceraldehyde; the former enters glycolysis directly, and latter is phosphorylated either to 2-phosphoglyceric acid (requiring ATP and NAD^+) or to glyceraldehyde 3-phosphate, both of which can enter general carbohydrate metabolism. Thus, F. degra-

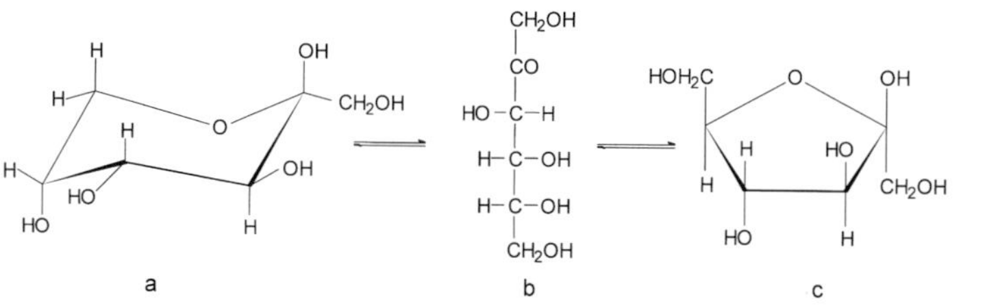

Fructose. Equilibrium in aqueous solution between the 6-membered ring structure of β-D-fructopyranose (a) the open chain structure (b; Fischer convention) and the 5-membered ring structure of β-D-fructofuranose (c).

dation and glucose degradation are different processes, and they are regulated independently. The liver can also convert F. to glucose, via the sugar alcohol sorbitol.

Fructose 1,6-*bis*phosphate, *fructose 1,6-diphosphate, Harden-Young ester:* a derivative of fructose in which the OH groups of C1 and C6 are esterified with phosphoric acid. It is an important intermediate of Glycolysis (see).

Fructose 2,6-*bis*phosphate, *F2,6P:* the most potent allosteric activator of animal phosphofructokinase and an inhibitor of fructose *bis*phosphatase. It plays an important role in the maintenance of blood glucose by the liver. Within the liver cell, the concentration of F2,6P is determined by the balance of its synthesis [by phosphofructokinase-2 (PFK-2)] and its degradation [by fructose *bis*phosphatase-2 (FBPase-2)], both enzyme activities residing on different domains of a ~ 100 kd homodimeric protein. Each of these enzyme activities is in turn regulated allosterically by phosphorylation/dephosphorylation by cAMP-dependent protein kinase and phosphoprotein phosphatase: phosphorylation of a specific Ser residue inhibits the PFK-2 activity and activates the FBPase-2 activity of the 100 kd homodimer. The following stages of regulation occur: low blood glucose → release of glucagon → increase in concentratioin of liver cAMP → decrease in concentration of liver F2,6P → decrease in PFK-2 activity → accumulation of glucose 6-phosphate which is diverted to glucose synthesis. At the same time, gluconeogenesis is stimulated by the increased activity of FBPase-2 (the concentration of its inhibitor, F2,6P, has decreased).

Heart muscle contains a different PFK-2/FBPase-2 isoenzyme, whose phosphorylation activates rather than inhibits PFK-2. Thus any hormone that promotes increased cAMP synthesis also causes an increase in the level of heart muscle F2,6P, leading to an increase in the rates of both glycogenolysis and glycolysis, so that glycogenolysis is followed by glycolysis and not by glucose secretion.

In skeletal muscle the isoenzyme lacks a phosphorylation site, and is not regulated by phosphorylation/ dephosphorylation.[L. Hue & M. H. Rider *Biochem J.* **245** (1987) 313–324; S. J. Pilkis & D. K. Granner *Annu. Rev. Physiol.* **54** (1992) 885–909; G. G. Rousseau & L. Hue *Prog. Nucleic Acid Res. Mol. Biol.* **45** (1993) 99–127]

Fructose-*bis*phosphate aldolase, *aldolase* (EC 4.1.2.13): a tetrameric lyase which reversibly cleaves fructose 1,6-*bis*phosphate into the two triose phosphates, dihydroxyacetone phosphate and D-glyceraldehyde phosphate. The reaction is analogous to the aldol condensation (CH_3CHO + CH_3CHO → CH_3–CHOH–CH_2–CHO), hence the name of the enzyme. The equilibrium concentrations are 89 % fructose *bis*phosphate and 11 % triose phosphate. The enzyme catalyses the condensation of a number of aldehydes with dihydroxyacetone phosphate, and can also cleave fructose 1-phosphate. Liver aldolase (aldolase B, M_r 156,000, 4 subunits of M_r 39,000) cleaves fructose 1,6-*bis*phosphate and fructose 1-phosphate at nearly the same rate. Muscle aldolase (aldolase A, M_r 160,000, 4 subunits of M_r 41,000, pI 6.1), however, is more active with the *bis*phosphate. Aldolase from yeast is inhibited by cysteine, and reactivated by Fe^{2+}, Zn^{2+} and Co^{2+}. Spinach leaf aldolase has a M_r of only 120,000 (M_r of subunits 30,000).

Of animal and human organs, skeletal muscle has the highest aldolase activity, 5 times that in brain, liver and heart muscle. For this reason the determination of aldolase in the serum is of diagnostic value in muscle diseases such as myoglobinuria, or progressive muscular dystrophy.

Fructose intolerance: see Inborn errors of metabolism.

Fructose 6-phosphate, *Neuberg ester:* an intermediate in Glycolysis (see) produced by isomerization of glucose 6-phosphate. It is also formed from erythrose 4-phosphate by transketolation.

β-h-Fructosidase: see Invertase.

Fructosuria: see Inborn errors of metabolism.

Fruit ripening hormone: see Ethylene.

Fruit sugar: see D-Fructose.

FSH: acronym of Follicle-stimulating hormone.

F-type ATPases: the ATPases present in the F_0F_1-complexes of the inner mitochodrial membrane, the thylakoid membrane of the chloroplast, and the plasma membrane of most bacteria. F-t.A. constitute one of 3 classes of Ion-motive ATPases (see). They consist of two major parts, each composed of several protein subunits, namely the roughly spherical (9 nm diam.) F_1-component, containing the catalytic site, and the transmembrane F_0-component, responsible for H^+ transport, which are linked together by a ~ 4.5 nm stalk. In the mitochondrion the F_1-component is located on the matrix face of the inner membrane, in the chloroplast it is located on the stroma face of the thylakoid membrane, and in bacterial cells it is located on the cytoplasmic face of the plasma membrane. The F_1-components of chloroplasts and thermophilic bacteria are often called CF_1 and TF_1 respectively. *In vitro* the F_1-components are relatively easily removed from their membrane-bound F_0-components and isolated; they are water soluble and catalyse the exergonic reaction ATP + H_2O → ADP + P_i. However, *in vivo* the mitochondrial and chloroplastidic F_1-components act as ATP synthases, catalysing the reverse, endergonic reaction, ADP + P_i → ATP + H_2O, at the expense of the electrochemical H^+ gradient generated by electron transport (see Proton motive force); however, the bacterial F_1-component *in vivo* appears to be capable of catalysing both these reactions according to the prevailing requirements of the cell, generating a transmembrane H^+ gradient by ATP hydrolysis or using a 'preformed' transmembrane H^+ gradient to generate ATP. The F_1-component can be re-attached *in vitro* to the F_0-component in its parent membrane in the presence of Mg^{2+}, with full recovery of all the functions of the original F_0F_1-complex.

All F_0F_1-complexes, regardless of source, have a similar subunit structure; moreover the amino acid sequence of several subunits exhibit a high degree of interspecies homology. The F_1-component of *E. coli* is composed of five subunits, namely: α (M_r 55,200; 513 amino acid residues), β (50,155; 459), γ (31,428; 286), δ (19,328; 177) and ε (14,920; 132), which occur as an $\alpha_3\beta_3\gamma\delta\varepsilon$ complex (38,1741; 3511). The F_1-component of beef heart mitochondria is also an $\alpha_3\beta_3\gamma\delta\varepsilon$ complex with α (55,164; 509), β (51,595; 480) and γ (30,141; 272) subunits that are very similar to those of *E. coli*; however, its '*E. coli*-equivalent' δ-subunit is usually called the 'oligomycin sensitivity-conferral protein' (OSCP) (20,967; 190), its '*E. coli*-equivalent' ε-subunit is labeled δ (15,056; 146) and its ε-subunit (5,652; 50) has no equivalent in *E. coli*. These nomenclatural problems only became apparent when the amino acid sequences of the F_1-subunits of *E. coli* and beef heart mitochondria were compared. In general it appears that the F_1-component of prokaryotes and chloroplasts (CF_1) resembles that of *E. coli* and that the F_1-component of eukaryotic mitochondria resembles that of beef heart mitochondria. The major structural feature of the F_1-component consists of three pairs of αβ-subunits arranged with 3-fold symmetry such that the α- and β-subunits alternate round a hexagon. It is believed that each αβ-subunit is a catalytic entity with catalytic sites for the binding of ADP, P_i and ATP on the β-subunit and non-catalytic sites that bind ADP or ATP at the interface of the two subunits.

The F_0-component of *E. coli* is composed of three subunits, namely: *a* (M_r 30,285; 271 amino acid residues), *b* (17,202; 156) and *c* (8,264; 79) which occur as an ab_2c_{9-12} complex. Subunit *a* is very hydrophobic and has at least seven transmembrane α-helical regions. Subunit *b* is rod-shaped and has a hydrophobic *N*-terminal tail anchored in the membrane and an extended, hydrophobic, α-helical domain which probably penetrates into the F_1-component where it cross-links with a β-subunit, thereby facilitating conformational interactions between the F_0 and F_1 components. Subunit *c* has two hydrophobic, membrane-spanning regions linked by a hydrophilic section which has the potential for interaction with the F_1-component. It also has an aspartate residue at position 61, flanked by hydrophobic residues, which reacts with dicyclohexyldiimide (DCCD). It is believed that the *c*-subunits form a transmembrane H^+-channel and it seems likely the carboxyl group of the side chain of the Asp-61 residue of each plays a role in the transport of H^+ through it; moreover the Asp-61 carboxyl groups evidently act cooperatively because reaction of just one of the *c*-subunits with DCCD blocks the H^+-translocating ability of the F_0-component. The F_0-component of beef heart mitochondria is similar to that of *E. coli*, having an *a* subunit (often called ATPase 6) (M_r 24,816; 226 amino acid residues) one or two *b* subunits, several *c* subunits (often called DCCD-binding protein or proteolipid) (7,402; 75) plus a subunit variously called A6L or *aap*1 (7,965; 66) and possibly others.

The stalk which links the F_0 and F_1 components together appears to be formed by interaction of subunits *b* (F_0) and β (F_1) (see above) and OSCP in mitochondria and yeast (δ in bacteria) and possibly others such as γ and ε.

The question of how a transmembrane flow of H^+ is capable of driving the endergonic phosphorylation of ADP has exercised the minds of biochemists for some years. Originally it was thought that the H^+ ions were directed by the F_0-channel to the catalytic site(s) in the F_1-component where they facilitated the removal of H_2O from orthophosphate, thereby shifting the equilibrium of the ATP + H_2O ⇌ ADP + P_i reaction towards ATP synthesis; but this became untenable when isotope-exchange experiments using $H_2{}^{18}O$ showed that F_1-bound ATP is produced in the absence of any proton motive force. However, when it was found that the ATP so formed remained tightly bound to the F_1-catalytic site unless H^+ flowed through the F_1-component, it was realized that the function of the proton motive force is not to force the synthesis of ATP but to force its release from the enzyme. This finding posed two questions, namely: (i) how can the phosphorylation of ADP, which is endergonic ($\Delta G^{o\prime}$ = +30.5 kJ mol^{-1}) in aqueous solution, be readily reversible (as shown by the isotope-exchange experiments) at the catalytic site of the enzyme?, and (ii) how does H^+ cause the release of ATP from its tight binding to the enzyme? Two answers have been advanced for the first question, namely that the catalytic site resides in a very hydrophobic pocket where, in consequence, the thermodynamics of hydrolysis in water are inapplicable, or that the tight, non-covalent binding of ATP to the enzyme supplies sufficient binding energy to render bound ATP as stable as ADP and P_i. The answer suggested for the second question is that H^+ ions arriving via the F_0-channel, cause conformational changes at the catalytic site(s) of the F_1-component which markedly alter the tightness of ATP binding.

The finding that there are multiple adenine nucleotide binding sites within the F_1-component and that they act co-operatively led Boyer to put forward the 'binding change mechanism' of ATP synthase action. According to this hypothesis the F_1-component has three nucleotide binding/catalytic sites, one for each αβ-subunit pair, which, though intrinsically identical, exist in different functional states at any given moment in time, one being in the open or O-state, which binds ADP and P_i very loosely, the second being in the L-state, which binds them loosely but is catalytically inactive, and the third being the T-state, which binds them tightly and is catalytically active. At the start of the catalytic cycle ATP is bound tightly at the site in the T-state, ADP and P_i are bound loosely at the site in the L-state and the site in the O-state is empty. Then the flow of H^+ through the F_0-channel, driven by the proton motive force, causes co-operative conformational changes within the F_1-component such that the three sites simultaneously undergo the following changes of state: $T \rightarrow O$, $O \rightarrow L$, and $L \rightarrow T$. These changes allow ATP to be released from the site that was in the T-state and is now in the O-state, ADP and P_i to form ATP at the site that was in the L-state and is now in the T-state, and ADP and P_i to bind very loosely to the site that was in the T-state and is now the O-state. More H^+ ions then flow though the F_0-channel and the process is repeated. Thus three ATPs are formed and released per turn of the cycle in response to three H^+ ion fluxes which drive the linked conformational changes in the three αβ-subunits. [P.L. Pedersen & E. Carafoli *Trends Biochem. Sci.* **12** (1987) 146–150, 186–189; A.E. Senior *Physiol. Rev.* **68** (1988) 177–231; P.D. Boyer *FASEB J.* **3** (1989) 2164–2178, *Biochim. Biophys. Acta* **1140** (1993) 215–250]

L-Fucose, ***6-deoxy-L-galactose:*** a component of the blood-group substances A, B and O and of various oligosaccharides in human milk, seaweed, plant mucilages. M_r 164, m.p. (α-form) 140 °C, $[\alpha]_D^{20}$ −153° → −76° (*c* = 9). It is also found in various glycosides and antibiotics. Some of the latter also contain D-fucose. L-F. is synthesized as its activated derivative, guanosine diphosphate fucose, from GDP-D-mannose by dehydrogenation, isomerization and reduction.

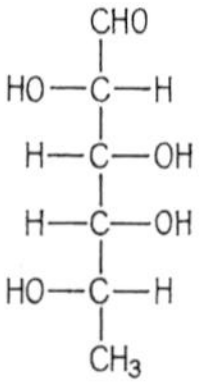

L-Fucose

Fucosidosis: see Inborn errors of metabolism.

Fucosterol, ***(24E)-stigmasta-5,24(28)-dien-3β-ol:*** a phytosterol (see Sterols), M_r 412.67, m.p. 124 °C, $[\alpha]_D^{20}$ −38.4° ($CHCl_3$). F. is characteristic of marine brown algae and has also been isolated from fresh-water algae.

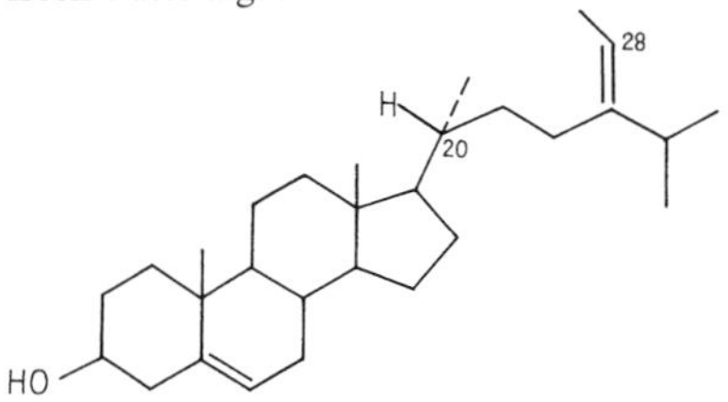

Fucosterol

Fucoxanthin: a carotenoid pigment with an allene, an epoxy and a carbonyl group, and three hydroxyl groups, one of them acetylated. M_r 658.88, m.p. 160 °C, $[\alpha]_D^{18}$ +72.5° ±9° ($CHCl_3$). The configurations of the chiral centers are 3*S*, 5*R*, 6*S*, 3′*S*, 5′*R*, 6′*R*. F. is found in many marine algae, especially brown algae (Phaeophyta), and it is the most abundant of the naturally occurring carotenoids.

Fugu poison: see Tetrodotoxin.

Fumarase, ***fumarate hydratase*** (EC 4.2.1.2): the tricarboxylic acid cycle enzyme which reversibly converts fumarate to malate by adding water to the double bond. In contrast to other hydrolyases, which require either pyridoxal phosphate or metal ions as cofactors, F. has no cofactor requirement. It is a tetramer (M_r 194,000, 1784 amino acids, no disulfide bridges) of identical subunits (M_r 48,500). It exists as a number of isoenzymes.

Fumaric acid, ***trans-ethylene dicarboxylic acid:*** an intermediate in the Tricarboxylic acid cycle (see), and the form in which the carbon skeletons of aspartate (see Urea cycle), phenylalanine and tyrosine (via fumarylacetoacetate) are fed into the tricarboxylic acid cycle. It occurs widely in the free form in plants and was first isolated from mushrooms in 1810 and from *Fumaria officinalis* in 1833. It melts at 286–287 °C in closed tube, and sublimes at 200 °C. The *cis*-isomer is maleic acid.

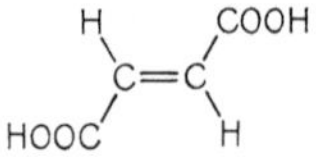

Fumaric acid

Fumigatin: see Benzoquinones.

Fundamental variable: see Concentration variable.

Funtumia alkaloids: a group of steroid alkaloids characteristic of the genus *Funtumia* of the dogbane *(Apocynaceae)* family. The F. a. are derived from pregnane (see Steroids) and have an amino or methylamino group on carbon 3 and/or 20. The most important representatives are *funtumine* (3α-amino-5α-pregnan-20-one) and *funtumidine* (3α-amino-5α-pegnen-20α-ol) from *Funtumia latifolia.* F. a. are biosynthesized from cholesterol and pregnenolone.

Furanoses: see Carbohydrates.

Fusel oil: unpleasant-tasting side product of alcoholic fermentation, consisting mainly of amyl, isoamyl, isobutyl and propyl alcohols. The compounds are formed from amino acids, in particular leucine, isoleucine and tyrosine by deamination and decarboxylation. Tyrosol, which is formed from tyrosine, is a component of beer.

Fusicoccin, ***fusicoccin A:*** the major toxin from culture filtrates of *Fusicoccum amygdali,* a pathogenic fungus responsible for a wilt disease of almond and peach. It is thought that F. specifically activates a single central transport system (possibly by interacting with the plasmalemma ATPase, thereby stimulating the conversion of phosphate bond energy into proton gradient energy), and that all other effects of the toxin are consequences of this fundamental process. F. has no effect on fungi, bacteria or animals. In higher plants, it generally causes cell enlargement, proton ef-

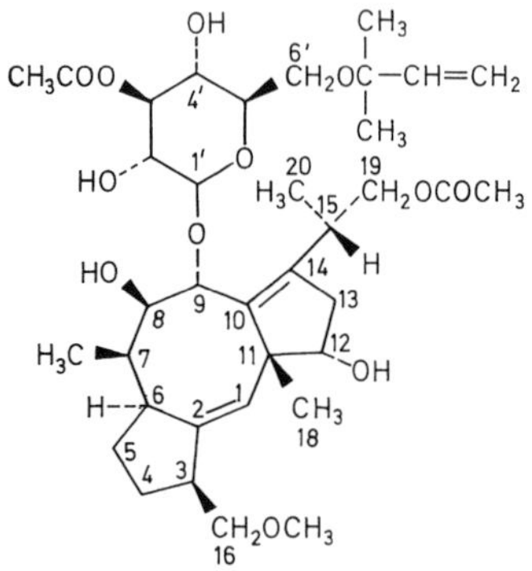

Fusicoccin [K. D. Barrow et al. *Chem Commun.* No 19 (1968) 1198–1200]

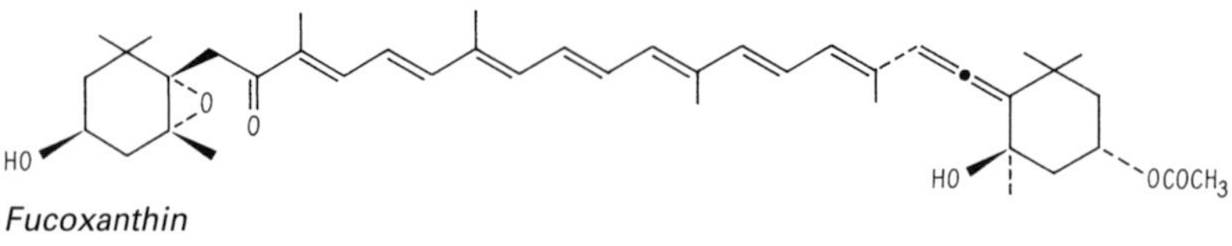

Fucoxanthin

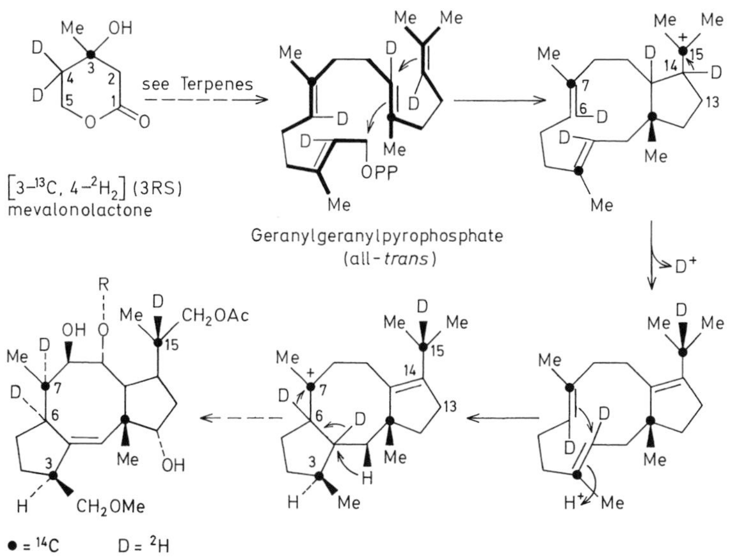

Biosynthesis of the aglycon ring of the fusicoccins

flux, K^+ efflux and stomatal opening. It also promotes seed germination in antagonism with abscisic acid.

There are two series of related fusicoccins, all present as co-metabolites in culture filtrates of *Fusicoccum amygdali:* fusicoccin A and 10 cometabolites differing only in the number and position of acetyl groups (probably resulting from nonenzymatic, *in vitro* migration of acetyl groups); and two 19-deoxyfusicoccins.

Labeling studies show that the fusicoccins are true diterpenes (note that the structurally similar ophiobolanes are sesterterpenes). Three of four possible 4-*pro-R* hydrogens of mevalonic acid are incorporated into the aglycon, including one at C6 and one at C15, but none at C3. One of the H-atoms at position 2 of mevalonic acid is incorporated at C8 (cf. ophiobolin, where this hydrogen migrates to C15). Six out of eight possible hydrogens at position 5 of mevalonic acid are incorporated into the aglycon, two of them at C9 and C13. Together with labeling patterns from ^{13}C- and ^{14}C-labeled mevalonic acid, these results are consistent only with synthesis via all-*trans*-geranylgeranylpyrophosphate. Final ring closure by two consecutive 1,2 hydride shifts was demonstrated by the incorporation of [3-^{13}C, 4-$^{2}H_2$]-(3*RS*)mevalonolactone; NMR analysis of the resulting F. showed that the signals due to $^{13}C7$ and $^{13}C15$ were strongly depressed, owing to the presence of ^{2}H on each of them.

The fusicoccins and Cotylenins (see) constitute a class of compounds with no other known naturally occurring representatives. The glycosidic sugar residue is also uncommon in fungal metabolites. [E. Marre *Annu. Rev. Plant Physiol.* **30** (1979) 273–278 (mode of action and physiology); A. Banerji et al. *J. Chem. Soc. Chem. Comm.* (1978) 843–845 (biosynthesis)]

Fusidic acid: a tetracyclic triterpene antibiotic, M_r 516.69, m.p. 192 °C, $[\alpha]_D^{20}$ −9° ($CHCl_3$). F. a. is isolated from culture filtrates of *Fusidium coccinium.* Like the structurally related antibiotics, Cephalosporin P_1 (see) and helvolic acid, it is effective against Gram-positive organisms. F. a. is biosynthesized from squalene via 2,3-epoxysqualene. It inhibits protein synthesis by preventing reaction of the elongation factor EFG with the small ribosomal subunit.

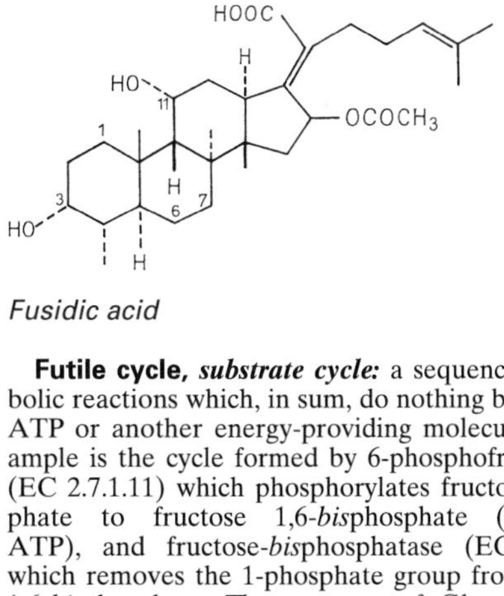

Fusidic acid

Futile cycle, ***substrate cycle:*** a sequence of metabolic reactions which, in sum, do nothing but degrade ATP or another energy-providing molecule. An example is the cycle formed by 6-phosphofructokinase (EC 2.7.1.11) which phosphorylates fructose 6-phosphate to fructose 1,6-*bis*phosphate (consuming ATP), and fructose-*bis*phosphatase (EC 3.1.3.11) which removes the 1-phosphate group from fructose 1,6-*bis*phosphate. The enzymes of Glycolysis (see) and Gluconeogenesis (see) provide another example. There are many others. It is believed that F. c. do not normally consume a large amount of cellular energy, because the enzymes of the opposing reactions are under tight Metabolic control (see), so that the overall reaction runs in only one direction at any given time. However, the fructose-phosphate cycle and the pyruvate → oxaloacetate → PEP → pyruvate cycle operate at measurable rates *in vivo* in liver. F. c. also provide heat in thermogenic tissues such as brown fat and insect thoracic muscle.

G

G: 1. a nucleotide residue (in a nucleic acid) in which the base is guanine. 2. Abb. for guanosine (e.g. GTP is acronym of guanosine triphosphate). 3. Abb. for guanine.

GA3: see Gibberellins.

GABA: see 4-Aminobutyric acid.

Gadoleinic acid, *Δ⁹-eicosenoic acid:* $CH_3-(CH_2)_9-CH=CH-(CH_2)_7-COOH$, a fatty acid, M_r 310.5, m.p. 39 °C. G.a. is a component of acylglycerols in plant and fish oils, and of phosphatides.

Galactans: polysaccharides of D-galactose found in plants. They are usually unbranched and have high molecular masses. Examples are agar-agar and carrageenan.

Galactitol: see Dulcitol.

Galactokinase deficiency: see Inborn errors of metabolism.

D-Galactosamine, *chondrosamine, 2-amino-2-deoxy-D-galactose:* an amino sugar, M_r 179.17, m.p. of the hydrochloride 185 °C, $[\alpha]_D^{20} + 125° \rightarrow + 98°$ (water). G. is biosynthesized from D-galactose by replacement of the OH group on C2 by an amino group. It usually occurs naturally as its *N*-acetyl derivative, which is present in a few mucopolysaccharides such as chondroitin sulfate, blood group substance A, etc. It is also found in mucoproteins.

Galactose epimerase deficiency: see Inborn errors of metabolism.

Galactose: an aldohexose, M_r 180.16, occurring naturally in its D- and L-forms. D-galactose, m.p. 167 °C, α-form, $[\alpha]_D^{20} + 151° \rightarrow + 80°$ (water), β-form, $[\alpha]_D^{20} + 53° \rightarrow + 80°$ (water). L-Galactose, m.p. 165 °C. G. is especially widespread in animals and is a component of oligosaccharides, such as lactose, and of cerebrosides and gangliosides in nervous tissues. In plants, it is a component of melibiose, raffinose, stachyose and the Galactans (see), and it occurs as the sugar component of some glycosides.

Metabolism. G. is synthesized as UDPgalactose from UDPglucose; epimerization on C4 is catalysed by UDPglucose 4-epimerase (EC 5.1.3.2). This reaction is reversible, and the degradation of UDPgalactose occurs via UDPglucose. G. enters general glucose metabolism via galactose 1-phosphate and UDP galactose (Fig.).

α-D-Galactose

Galactosemia: see Inborn errors of metabolism.

β-Galactosidase (EC 3.2.1.23): a disaccharidase which hydrolyses lactose to galactose and glucose (Fig. 1). The lactose operon (*lac* operon) of *E. coli* contains structural genes for β-G., galactoside permease and thiogalactoside transacetylase. Induction of transcription of the *lac* operon (manifested as induction of the synthesis of β-G. and the other two enzymes) enables the bacterium to use lactose as its sole source of carbon. The true inducer is 1,6-allolactose, formed from lactose by transglycosylation (Fig. 1). Nonmetabolizable or gratuitous inducers of the *lac* operon are also known, e.g. isopropylthiogalactoside (Fig. 2). Classical studies on the induction of β-G. led Jacob and Monod to propose the operon model (see Operon) for the regulation of protein synthesis. β-G. activity is conveniently measured with the colorless substrate, *o*-nitrophenyl-β-D-galactoside, which is hydrolysed to galactose and the colored product *o*-nitrophenol. The latter can be determined spectrophotometrically. See Enzyme induction, Enzyme repression. [J.H. Miller, *Experiments in Molecular Genetics* (Cold Spring Harbor, New York, 1972)]

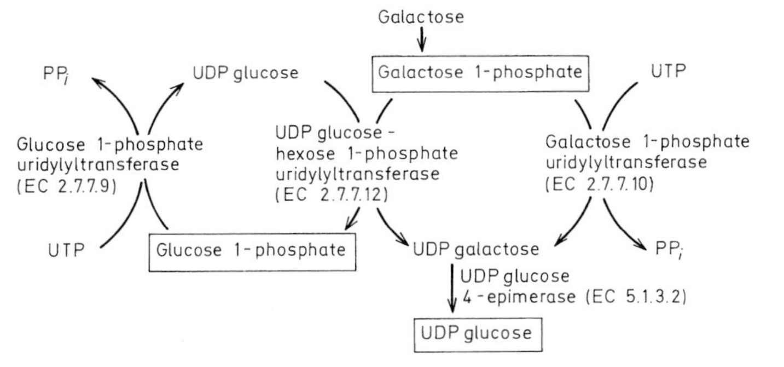

Relationship between galactose and glucose metabolism

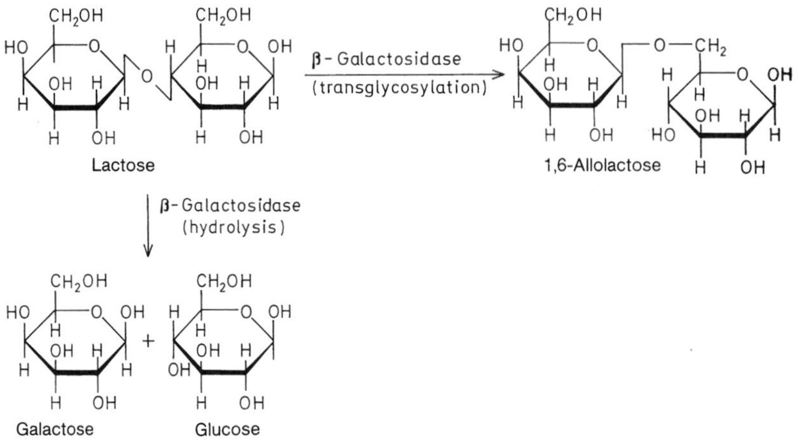

Fig. 1. *Hydrolysis of lactose to galactose and glucose, and transglycosylation of lactose to 1,6-allolactose.* Both reactions are catalysed by β-galactosidase.

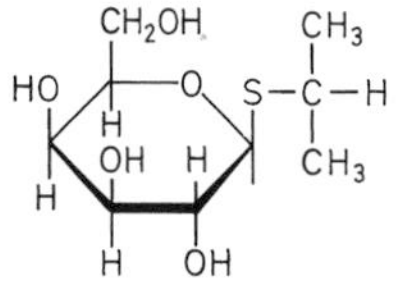

Fig. 2. *Isopropylthiogalactoside, a gratuitous inducer of β-galactosidase*

Galactosuria: see Inborn errors of metabolism.

Galactosylceramide lipidosis: see Lysosomal storage diseases.

D-Galacturonic acid: a uronic acid [M_r 194.14, m.p. 159 °C, α-form $[\alpha]_D^{20}+98° \rightarrow 50.8°$ (water), β-form $[\alpha]_D^{20}+27° \rightarrow 50.8°$ (water)] biosynthesized from D-galactose. Pectins contain 40–60 % G.a., which is also found in a few other plant polysaccharides.

Galangin: see Flavones (table).

Galanthamine: see Amaryllidaceae alkaloids.

Galegine, ***(3-methyl-2-butenyl)guanidine:*** a guanidine derivative found together with 4-hydroxygalegine in the seeds of goat's rue, *Galega officinalis.* G. is synthesized in the shoot and accumulated in the seeds. Its isoprenoid hydrocarbon chain does not arise from the mevalonic acid-isopentenyl pyrophosphate sequence of terpenoid biosynthesis. The guanidino group is added by a Transamidination (see).

Gallic acid: see Tannins.

Gamete attractants: see Sexual attractants.

Gamones: plant Sexual attractants (see).

Ganglioside: see Membrane lipids, Lysosomal storage diseases (Sphingolipidoses).

Gangliosidosis: see Lysosomal storage diseases.

GAR: see Glycinamide ribotide.

Gas chromatography: a separation technique based on the distribution of gaseous compounds between a mobile gaseous phase and a stationary adsorbant phase. The method was developed by Martin and James in 1952 [A. J. P. Martin in R. Porter (ed.) *Gas Chromatography in Biology and Medicine. A CIBA Foundation Symposium,* J. & A. Churchill, Ltd., London, 1969].

The GC system consists of five components: 1. the carrier gas supply, 2. the sample injection system, 3. the column and oven, 4. the detector, and 5. signal processing and control electronics. The carrier gas must be chemically inert with respect both to the sample and the stationary phase, and must not interfere with the detector function. H_2, He, N_2, O_2, Ar and CO_2 are commonly used. The sample must be vaporized before it enters the column. Thus for solid or liquid samples, the inlet is kept at a high temperature in order to vaporize the entire sample instantly. Liquid and solid samples are commonly introduced in solution; in this case the solvent must be carefully chosen so that it does not react chemically with either the sample or the stationary phase of the column at high temperature. It must also leave the column at a different time from the sample and not overload the column. If the sample is not volatile, it may be converted to a volatile chemical derivative, e. g. by reaction with trimethylchlorosilane or hexamethyldisilazane.

The column may be made of glass or metal, and it is either filled with a porous packing or it may be a capillary. Packed columns have inner diameters of 2 to 6 mm, and lengths of 1–3 m. Capillaries are 0.2–0.5 mm in inner diameter, and 10–100 m long. Materials to be used as stationary phase must be chemically inert, thermally stable and non-volatile. In gas-liquid chromatography (GLC), the liquid stationary phase is either adsorbed to the packing or to the capillary walls. Some common liquids used as stationary phase are Apiezon, SE 30 (methyl silicone), Carbowax 20 M, and mixtures of phenyl methyl silicone or trifluoropropyl silicone. In gas-solid chromatography (GSC), the packing is an active solid material, such as alumina, silica gel, activated carbon or zeolites. These packing materials can greatly increase retention times, and thus make possible the separation of very volatile substances which move too quickly through a GLC column.

The separation process involves an equilibrium partition of the solute between the stationary and mobile phases. The partition coefficient depends on the solute vapor pressure and is a function of temperature: $\ln(p) = \Delta H/RT + C$, where p is the solute vapor pressure, ΔH is the molar heat of solution and C is a constant. The higher the temperature, the less the solute is retained by the stationary phase, and thus the more quickly it leaves the column. This fact is utilized to drive compounds with longer retention times out of the column by gradually increasing the temperature.

Detection is based on a change in a physical property of the gas emerging from the column. The signal must be proportional to the amount of material eluted. For routine analyses, a flame ionization detector is used. It consists of a small H_2-air flame burning at a metal jet. The eluant gas mixes with the H_2 before the latter emerges from the jet. The combustion of H_2 produces free radicals, but no ions; however, the combustion of carbon produces positive ions in the flame as well as free radicals. The ions travel to the collector electrode, where they are measured by the current they induce. GC is of enormous value for the separation of small to medium sized molecules and is routinely used in clinical chemistry, e.g. to detect drugs or anesthetics in serum.

Gastric inhibitory peptide, *GIP:* a polypeptide hormone (for structure, see Secretin) purified from crude preparations of Cholecystokinin (see). GIP has potent enterogastrone activity, i.e. it inhibits secretion of acid and pepsin by the stomach, and inhibits gastric motility. It possesses no significant secretin or cholecystokinin activity. [J.C. Brown & J.R. Dryburgh *Canad. J. Biochem.* **49** (1971) 867–872]

Gastrin: a hexadecapeptide hormone from the gastric antrum. Human G.I is Pyr-Gly-Pro-Trp-Leu-$[Glu]_5$-Ala-Gly-Tyr-Trp-Met-Asp-Phe-NH_2, M_r 2116. Human G.II has an additional sulfate group on Tyr_{12}. Leu_5 is replaced by Met in porcine G. and by Val in sheep and bovine G. In bovine G. there is also an Ala instead of Glu in position 10. G. is homologous to the *C*-terminus of cholecystokinin. The protected *C*-terminal peptide, tetragastrin (Trp-Met-Asp-Phe-NH_2), is used to test secretion of digestive juice. A biologically active *big G.* with 34 amino acids can be detected by radioimmunoassay.

G. is formed in the antrum mucosa in response to alkaline pH, mechanical stimulation and vagus (acetylcholine) stimulation. Acidic stomach juice inhibits its secretion. The presence of G. in the blood stimulates the stomach mucosa to produce and secrete hydrochloric acid and the pancreas to secrete its digestive enzymes. Cholecystokinin is a competitive inhibitor of G.

Gated ion channel: see Ion channel.

Gaucher's disease: see Lysosomal storage diseases.

Gay-Lussac equation: see Alcoholic fermentation.

Gazaniaxanthin: see Rubixanthin.

GC content: the amount of guanine + cytosine in nucleic acids, expressed in mol % of the total bases. The GC content plus the AT (adenine + thymine) content of a nucleic acid is thus equal to 100 mole %, provided the molecule is double-stranded. The GC contents for higher organisms lie between 28 and 58 %, while those for prokaryotes range from 22 to 74 %. The density of a double-stranded molecular species depends on its GC content, as does its melting temperature, which increases almost linearly with increasing GC content. The flexibility of the DNA helix is also a function of GC content: poly(dG) · poly(dC) is much stiffer than random-sequence DNA, which in turn is stiffer than poly(dA) · poly(dT). This applies both to torsional and bending flexibility, and these differences are probably important in the supercoiling of the helices in Chromosomes (see). [M. Hogan et al. *Nature* **304** (1983) 752–754.]

DNA methylation (see) is a feature of many vertebrate genomes. There is some evidence that the sequence mCpG is susceptible to deamination of the methylcytosine to thymidine, so that the bulk DNA of vertebrates is has 4–5-fold fewer CpG than would be expected if the distribution were random; it is correspondingly enriched in TpG. This mechanism must contribute to the species differences in overall GC content. [A. Bird et al. *Cell* **40** (1985) 91–99].

GDP: acronym of Guanosine 5′-diphosphate. See Guanosine phosphates.

Gelatins: proteins obtained by extraction from tissues rich in Collagen (see).

Gel filtration: see Proteins.

Gene: a section of DNA coding for a single polypeptide chain (structural gene or *cistron*), a particular species of transfer or ribosomal RNA, or a sequence which is recognized by and interacts with regulator proteins (regulatory gene). In prokaryotes, several cistrons often comprise a single regulatory unit, the Operon (see). In eukaryotes, stretches of nucleotides which encode amino acid sequences (exons) are often interrupted by those that do not code for amino acid sequences (introns), and which are excised during Post-transcriptional modification of RNA (see). Such G. are sometimes called "split G".

Gene activation, *gene expression:* No organism continuously synthesizes all the proteins encoded in its genome. Even in prokaryotes, a cell (clone) may undergo many division cycles without producing the enzymes for catabolic or anabolic pathways it does not need, and other enzymes are made only at specific times during the cell cycle. In multicellular eukaryotes, the situation is further complicated by the existence of genes which are expressed only in specific tissues or at specific stages of development. In both types of cell, the rate of transcription of a gene may vary widely in response to environmental conditions. G.a. in prokaryotes is achieved largely by the binding of specific proteins (repressors or activators) to the DNA adjacent to the site of initiation of transcription (the promoter). See Operon.

In eukaryotes, chromosomal DNA is associated with a large number of proteins. The most abundant are the histones, very basic proteins of which there are five types: H1-H5. The DNA double helix is wound around a core structure formed by histones H2-H5; about 200 base pairs wrapped around such a core comprise a nucleosome. The H1 histones are associated with the DNA strand between nucleosomes. There is evidence that the nucleosomes play a role in G.a., in that RNA polymerase evidently cannot bind to DNA within the nucleosome. Thus if the initiation site of a gene is buried in a nucleosome, the gene cannot be transcribed.

Although operons have not been demonstrated in eukaryotic genomes, the expression of individual genes may be subject to a similar kind of regulation. Eukaryotic genes, like prokaryotic genes have promoters, which are untranslated stretches of DNA adjacent to the structural gene. These are involved in the binding of RNA polymerase to the DNA, and probably in causing it to locate the exact site at which transcription begins, i.e. the initiation site. In bacteria, specific proteins bound to a promoter can cause the associated gene (or operon) either to be transcribed or not transcribed; it seems very likely that such activator and repressor proteins also exist for eukaryotic genes. In fact the sites at which hormone-receptor complexes bind to DNA have been shown, for the prolactin gene family, to be located just upstream (on the 5′-side) of the structural gene.

In addition to their promoters, the genes of some viruses which infect eukaryotic cells are associated with enhancers. These can operate over long distances, enhancing transcription from a promoter up to 3,000 base pairs away, either upstream or downstream from the enhancer. They are effective when inserted into the genome in either the 5′ → 3′ or the 3′ → 5′ orientation relative to the direction of transcription. Enhancers have tissue and species specificity, but there is a conserved core sequence, $GGTGTGG\frac{AAA}{TTT}G$. Homologous structures have been found in the introns of immunoglobulin genes, and may exist in other cellular genes.

Another form of regulation of G.a. was discovered in yeast, which normally contains genes for both of the two mating types α and *a*. In vegetatively reproducing cells, neither mating type is expressed but under certain circumstances, either the Matα or the Mat*a* genes (there are two α and two *a* genes) are moved to a different location near the center of the chromosome which carries both sets of genes, and in this position they are expressed. The nontranslocated genes remain unexpressed. The flanking sequences of the genes are the same in both positions, out to 800 bp downstream (on the 3′ end) from the structural genes. Repression of the α genes in their non-active position requires activity of another gene, SIR, whose product presumably binds to the DNA at least 800 bp downstream from the gene. A possible mechanism whereby a protein bound to the DNA at such a distance from the structural gene can prevent transcription is that it causes the arrangement of the nucleosomes to shift in such a way that the initiation site of the gene is not available for RNA polymerase. [K.A.Nasmyth et al. *Nature* **289** (1981) 245–250]

The product of the Matα 2 gene of yeast, the Homeobox (see) sequence in *Drosophila* and homologous sequences in other organisms appear to function as control elements, similar to bacterial activators, for scattered genes which are induced simultaneously in the course of development. It is presumed that the genes induced by the products of the homeotic genes have sequences analogous to bacterial promoters or possibly to viral enhancers where these regulatory proteins can bind and stimulate transcription.

Mechanisms for transposing genes on a chromosome exist in many types of eukaryotic cell. For immunoglobulin and T-cell antigen receptor molecules, as for yeast mating type genes, rearrangement of the genes is a necessary step in activation of the respective lymphoid clones.

Finally, DNA methylation (see) of appears to be related to G.a. in eukaryotes. Stretches of DNA, in which a large proportion of the thymidine residues are methylated, are not transcribed; genes which are methylated (and inactive) in one tissue can be nonmethylated and expressed in another tisssue. [J.D.Hawkins in *Gene Structure and Expression,* Cambridge University Press, Cambridge, 1985; Benjamin Lewin, *Gene Expression,* Wiley, New York; A.Kumar (ed.) Eukaryotic Gene Expression, Plenum Press, New York, 1984].

Gene amplification: production of extrachromosomal copies of the genes for ribosomal RNA. In frog's egg cells, G.a. leads to the formation of numerous extrachromosomal nucleoli. G.a. is thus a special regulatory mechanism for RNA. It is not the same as Redundancy (see), which is the presence of multiple copies of the same gene on the chromosome.

Gene bank: see Recombinant DNA technology.

Gene cloning: see Recombinant DNA technology.

Geneserine: see Physostigmine.

Gene synthesis, ***cell-free gene synthesis:*** The term includes cell-free replication of a bacteriophage genome, which was first achieved in 1967 by Goulian, Kornberg and Sinsheimer with ΦX-174. Khorana and coworkers later synthesized the gene for alanine tRNA by a combination of chemical and enzymatic methods. They first synthesized by chemical means a series of polynucleotides of 8 to 12 bases. The free hydroxyl groups at the 5′ ends were then phosphorylated with polynucleotide kinase. Taken together, the sequences of these segments made up both strands of the DNA for the gene. Furthermore, they were designed so that each segment of one strand overlapped the ends of two segments on the other strand, each overlap extending for at least five bases. The double-stranded DNA was first organized, segment by segment, by hybridization; then the breaks in the strand were fused by a polynucleotide ligase. Using a similar approach, H.Koster et al. reported the total synthesis of the structural gene for the octapeptide angiotensin II (M_r 21800) in 1978. The gene for the peptide hormone Somatostatin (see) was synthesized in 1977 by chemical means. It was joined to the *E. coli* β-galactosidase gene in the plasmid pBR322 and introduced into the bacteria. The product was a chimeric polypeptide containing the amino-acid sequence of somatostatin. When subjected to cyanogen bromide cleavage, the polypeptide yielded biologically active somatostatin. Later the hormone Insulin (see) was synthesized analogously.

Gene therapy: the beneficial alteration of a phenotype by changing or normalizing defective genetic material. Currently, the term is normally taken to mean the curing of genetic defects in humans by genetic manipulation of specific cells in the body. There are basically three possible methods of G.t.

1. Cells are taken from the body of the genetically defective individual, treated, then returned to the body. An example of this 'ex vivo' method is given below for the correction of severe combined immunodeficiency. If blood cells are used, and since blood cells have a limited lifespan, periodic cycles of ex vivo treatment and reinfusion are necessary. It may be more expedient to target the stem cells of the bone marrow, because these are apparently immortal. Neonates with SCID have now been treated by inserting genes into the stem cells, and at 2 years of age these were still thriving without the benefit of further treatment.

2. Using carriers, the corrective genes are introduced directly into the individual, i.e. into the tissue where they are needed. This 'in situ' treatment has been explored for e.g. cystic fibrosis and muscular dystrophy (see below). The method may be used for the treatment of localized disorders, but clearly cannot be used to correct systemic conditions.

3. In the 'in vivo' approach gene carriers would be injected into the circulation, then home in on the appropriate target, to the exclusion of all other cell types. Conceptually this is the ideal method, but it does not yet exist.

The first experimental attempts at gene therapy were started in 1990 in the USA. The disorder in question was a rare immune deficiency disease, known as severe combined immunodeficiency (SCID) caused by the absence of adenosine deaminase (ADA). A disabled mouse retrovirus was used to carry healthy copies of the ADA gene into white blood cells taken from the patient; the treated white cells were then returned to the bloodstream. The patient is now (in 1995) nine years old, out of quarantine, and leading a full life. During the first 4 months after the beginning of treatment she received 4 infusions of her own genetically manipulated blood cells, and has received occasional follow-up infusions. A further seven children have been treated; all are now leading full lives and do not have be kept in a permanently sterile environment. Thus the apparent success rate of this particular gene therapy is 8/8, i.e. 100%. It has now been revealed, however, that these children are receiving regular injections of synthetic ADA as a precautionary measure. It is therefore difficult to assess the contribution of gene therapy to their improved condition.

In 1990, the National Institutes of Health (USA) approved three proposals to use genetically engineered cold viruses to carry healthy genes into the respiratory passages of cystic fibrosis patients. Cystic fibrosis is the commonest lethal inherited disease of Caucasians, affecting 1/2 000 to 1/2500 babies, whose life expectancy is about 30 years. It is characterized by the accumulation of dense, sticky mucus in the lungs and airway epithelia, leading to bacterial infection of respiratory passages. The airway epithelia of affected patients display defective cyclic AMP-regulated chloride conductance. Similar experimental therapies are planned for muscular dystrophy and severe melanoma.

In 1993, it was shown for the first time, that the 'in situ' method of gene therapy can restore normal function to an organism suffering from an inherited lethal mutation (S.C. Hyde et al., *Nature* **362** (1993) 250–255). These authors enclosed healthy copies of the human cystic fibrosis membrane conductance regulator (CFTR) gene in liposomes, which were then introduced into the lungs of mice with artificially induced cystic fibrosis (the liposomes fuse with cell membranes so that the healthy DNA passes into the cells). Normal ion transport was restored and the accumulation of mucus was prevented.

Despite the optimism generated by these experiments, and with the possible exception of SCID, all attempts to treat human genetic disorders with gene therapy have so far been disappointing. The US National Institutes of Health is now considering (November 1995) the diversion of funds from clinical trials back to basic research.

[J. Lyon & P. Gorner *Altered Fates: Gene Therapy and the Retooling of Human Life* W.W. Norton, 1995; K.W. Culver *Gene Therapy: A Handbook for Physicians* Mary Ann Liebert, Inc., Publishers, 1994; R.C. Boucher 'Current Status of Cystic Fibrosis Gene Therapy' *Trends Genet.* **12** (1996) 81–84]

Genetic code: the rules for translation of base sequences in nucleic acids into amino acid sequences in polypeptides (see Protein synthesis). The nucleic acid bases are read off as triplets. There are four bases, and thus 64 (4^3) different permutations ("words"). As there are only 20 amino acids specified by the code, many triplets can be, and are, redundant. The table lists the amino acids by triplet, or *codon*.

	U	**C**	**A**	**G** (middle nucleotide)		
U	Phe	Ser	Tyr	Cys	**U**	
1st position	Phe	Ser	Tyr	Cys	**C**	3rd position
(5′ end)	Leu	Ser	term	term	**A**	(3′ end)
	Leu	Ser	term	Trp	**G**	
C	Leu	Pro	His	Arg	**U**	
	Leu	Pro	His	Arg	**C**	
	Leu	Pro	Gln	Arg	**A**	
	Leu	Pro	Gln	Arg	**G**	
A	Ile	Thr	Asn	Ser	**U**	
	Ile	Thr	Asn	Ser	**C**	
	Ile	Thr	Lys	Arg	**A**	
	Met*	Thr	Lys	Arg	**G**	
G	Val	Ala	Asp	Gly	**U**	
	Val	Ala	Asp	Gly	**C**	
	Val	Ala	Glu	Gly	**A**	
	Val	Ala	Glu	Gly	**G**	

* AUG serves as part of the initiation signal, as well as encoding internal Met residues.

Three of the codons ("term" in the table) cause translation to stop (see Termination codon). Initiation of the chain is indicated by longer sequences which are recognized by the initiation factors (see Protein biosynthesis).

The code has no commas. The "reading frame", the way in which a sequence is divided into triplets, is determined by the precise point at which translation is initiated, e.g. the sequence CATCATCAT could be read CAT,CAT,CAT, or C,ATC,ATC,AT, or CA,T-

CA,TCA,T in the three possible reading frames. A mutation which changes the reading frame of a gene, e.g. the deletion of one or two nucleotides, is called a *frameshift mutation.* The code is normally not overlapping, but in the bacteriophage ΦX 174, two genes are entirely contained within other genes. These are translated in different reading frames, so the amino acid sequences of the proteins encoded by them are different.

Historical. The G.c. was cracked between 1961 and 1963, primarily by Khorana, Matthaei, Nirenberg and Ochoa, by the analysis of the translation products of synthetic copolymers of pyrimidine and purine nucleotides. Brenner and Crick showed that defined chemical changes in bacteriophage DNA produced certain types of mutations, which were in accordance with the proposed code. Comparison of the primary sequences of viral coat proteins (Wittmann) and bacterial lysozymes (Streisinger) from wild types and mutants confirmed the code. In 1954 Gamov tried to explain the G.c. by steric fitting of nucleotides and amino acids. In 1957, Crick suggested his Adaptor hypothesis (see), which was later confirmed experimentally.

Genetic code in mitochondrial mRNA: In mitochondrial mRNA, some of the codons are different from those listed in the table, which was established for cytoplasmic and bacterial mRNA. For example, in human mitochondrial mRNA, AGA and AGG are used as termination codons. Other differences are as follows:

Some codons in mitochondrial mRNA

mRNA	5′ CAA	AUA	CUA	UGA 3′
Yeast mitochondria	Gly	Ile	Thr	Trp
Human mitochondria	Gly	Met	Leu	Trp
Neurospora mitochondria	Gly	Ile	Leu	Trp

In addition, mitochondrial systems also use fewer species of tRNA (maximum 24) than nonmitochondrial systems (estimated 32–40). This would appear to be the result of a simpler decoding strategy in mitochondria, where tRNA with U in the wobble position pairs with any of the four bases (see Wobble hypothesis). Thus the eight amino acids which are encoded by at least four codons each can be served by only eight tRNAs.

Genetic code in ciliates. The G.c. in ciliates is similar to that described in other eukaryotes, except that UAA and UAG code for glutamine, and do not act as stop signals. Comparison of gene structure and protein sequence shows that TAA and TAG triplets in DNA (noncoding strand) correspond to glutamine in the surface antigen protein of *Paramecium,* α-tubulin of *Stylonchia,* and a histone of *Tetrahymena.* In ciliate mRNA, the only termination codon appears to be UGA. [F.Caron & E.Meyer *Nature* **314** (1985) 185–188; J.R.Preer et al. *Nature* **314** (1985) 188–190]

Genetic engineering: see Recombinant DNA technology.

Genetic material: the carrier of hereditary information. In higher organisms, bacteria and some viruses, it is double-stranded DNA; in other viruses it is single-stranded DNA, and in still other viruses, it is RNA. See Deoxyribonucleic acid, Chromosomes.

Genin: see Aglycon.

Genistein: 5,7,4′-trihydroxyisoflavone. See Isoflavone.

Genistin: genistein 7-glucoside. See Isoflavone.

Genome: the sum of all chromosomal genes in a haploid cell (including prokaryotes) or the haploid set of chromosomes in a eukaryotic cell.

Genomic library: see Recombinant DNA technology.

Genotype: the sum of an individual's genes, both dominant and recessive.

Gentamycin: see Streptomycin.

Gentiana alkaloids: terpene alkaloids with a pyridine ring system (they are therefore also pyridine alkaloids) found primarily in gentian *(Gentiana)* species. The biogenetic precursors of G.a. are probably bicyclic monoterpenes. This assumption is strengthened by the fact that Gentianine (see) can easily be obtained from gentiopicroside in the presence of ammonium ions (Fig.).

Gentianic acid: see Gentianine, 2.

Gentianine: 1. a terpene and pyridine alkaloid, M_r 175.18, m.p. 82 to 83 °C. G. occurs in many plants of the gentian family *(Gentianaceae).* Its existence was long disputed, because it is easily formed in the presence of ammonium ions from gentiopicroside. 2. ***Gentisin, gentianic acid, 1,7-dihydroxy-3-methoxy-9H-xanthene-9-one,*** a yellow pigment from the yellow gentian *(Gentiana lutea),* M_r 258.22, m.p. 266–267 °C. 3. An anthocyan pigment with a delphinidin structure from *Gentiana acaulis.*

Glucose $\xrightarrow{NH_4^{\oplus}}$

Gentiopicroside *Gentianine*

Gentianose: a nonreducing trisaccharide [m.p. 211 °C, $[\alpha]_D^{20}$ + 33.4 °] which serves as a storage compound in the roots of gentians. It is composed of one D-fructose and two D-glucose units. One glucose and the fructose are joined as in sucrose; the second glucose is linked β-1,6 to the first.

Gentiobiose: a reducing disaccharide consisting of two molecules of D-glucopyranose linked β-1,6. M_r 342.20, α-form, m.p. 86 °C, $[\alpha]_D^{22}$ + 16 °(3 min) → + 8.3 ° (3.5 h, c = 4); β-form, m.p. 190–195 °C, $[\alpha]_D^{22}$ – 5.9 ° (6 min) → + 9.6 ° (6 h, c = 3). G. differs from isomaltose only in having a β-glycosidic bond instead of an α-linkage. It is found naturally only in bound form, e.g. in glycosides such as amygdalin and gentiopicrin, and as the ester component of crocin.

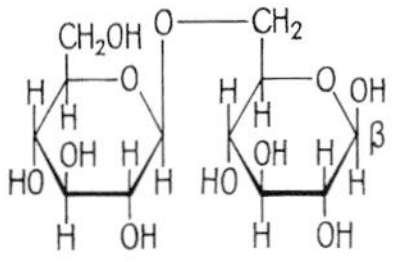

Gentiobiose

Gentiopicroside: see Gentianine 1).

Gentisin: see Gentianine 2).

Geranial: the *trans* isomer of Citral (see).

Geraniol, *2,6-dimethylocta-2,6-dien-8-ol:* an unsaturated monoterpene alcohol, M_r 154.24, m.p. –15 °C, b.p.$_{757}$ 299–230 °C, ρ^{20} 0.8894, n_D^{20} 1.4766, with a rose fragrance. The Δ^2 double bond has the *trans* configuration. *Nerol* is the double bond isomer with the *cis* configuration at position 2. In the structural isomer *linalool* the OH group is on C3 instead of C1. G. is a component of many essential oils, and constitutes up to 60 % of oil of roses. It is used in perfumery, primarily as the acetate (b.p. 242 °C)

The pyrophosphate ester of G. is an important intermediate in the biosynthesis of Terpenes (see). Head-to-tail coupling of two molecules produces geranylgeranylpyrophosphate, the precursor of the tetraterpenes.

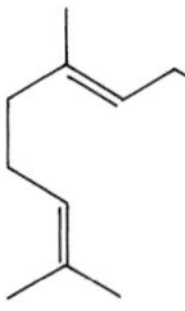

Geraniol 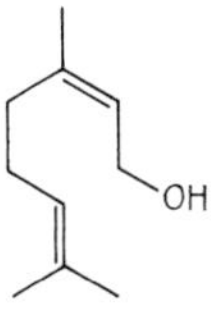 Nerol 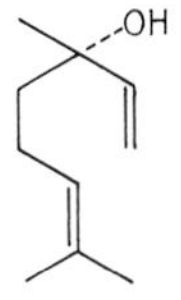 (+)-Linalool

Geranyl pyrophosphate: see Geraniol.

Germacrane: see Sesquiterpenes, Fig.

Germine: a Veratrum alkaloid with a C-nor-D-homo ceveratrum type structure. M_r 509.62, m.p. 221.5–223 °C, $[\alpha]_D^{25}$ + 4.5 ° (95 % ethanol), $[\alpha]_D^{16}$ + 23.1 ° (c = 1.13 in 10 % acetic acid). Esters of G. are present in *Veratrum album, V. viride, V. nigrum* and *Zygadenus venosus.* The most common acid components are acetate, angelate and tiglate. These ester alkaloids reduce blood pressure. The isomeric Veratrum base *veracevine* (m.p. 183 °C $[\alpha]_D$ – 33 °), which occurs as an ester alkaloid in *Veratrum sabadilla,* differs from G. in having a 12α hydroxyl group instead of the 15 hydroxyl. Alkaline isomerization of veracevine yields *cevine,* with a 3α-hydroxyl group.

Gestagens, *progestins, gestins:* a group of female gonadal hormones, including Progesterone (see) and other natural and synthetic steroids with progesterone-like effects, e.g. Norgestrel (see) and Chlormadione acetate (see). Oral G. are used to correct irregularities in the menstrual cycle, to treat repeated abortion, and as components of Ovulation inhibitors (see). They are also being used increasingly to regulate animal reproduction.

GH: acronym of growth hormone. See Somatotropin.

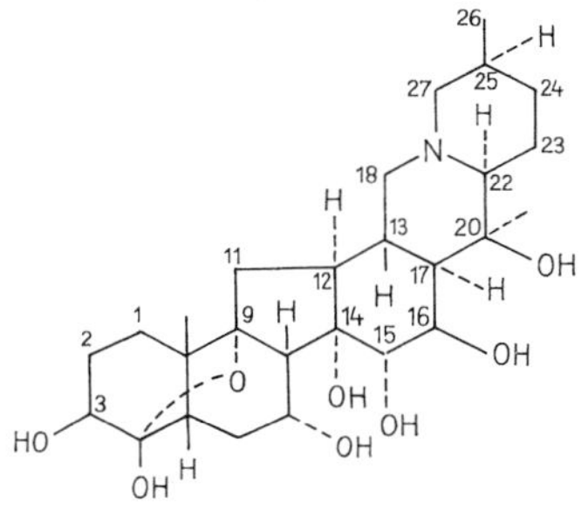

Germine

Giant chromosome, *polytene chromosome:* a particularly large type of chromosome present in members of the Diptera. G.c. in the salivary salivary glands of *Drosophila* have been extensively studied. They have a cable-like appearance, due to multiple endomitotic duplication of the chromosomes, without separation of the chromatids. Each diploid pair may replicate up to 9 times within the G.c., so that the G.c. finally contains up to $2 \times 2^9 = 1024$ DNA strands. Each of the 4 G.c. of *Drosophila* may attain a length of 0.5 mm and a thickness of 25 µm. The aggregate length is therefore about 2 mm; since the haploid genome contains 1.65×10^8 base pairs, the packing ratio of these G.c. is almost 30. The supercoiled DNA molecules are much more extended than in other chromosomes, and the duplicated molecules remain in register as they lie side by side. This gives rise to a banded structure (DNA-rich bands separated by DNA-poor bands) along the length of the G.c., which can be visualized by staining. The stainable, DNA-rich bands (chromomeres) contain about 95 % of the total chromosomal DNA. The banding or chromomere pattern is characteristic of each *Drosophila* strain.

G.c. have facilitated the study of chromosome morphology, and made possible the direct demonstration of the involvement of chromosomes in RNA synthesis. Individual bands (about 3,000 along the length of the G.c.) undergo periodic changes of shape: they become more loosely packed, and produce a swelling which encircles the axis of the G.c. This swollen structure is called a puff or Balbiani ring (Fig.). Formation of puffs (puffing) is accompanied by greatly increased RNA synthesis. DNA in the normal bands is tightly coiled and repressed by specific proteins, whereas DNA in the puffs is relaxed and derepressed, and it can function in the transcription of RNA. Specific incorporation of RNA precursors in the puffs can be demonstrated by autoradiography. Puff formation is reversible. The number, time of occurrence, duration and shape of puffs (i.e. the puff pattern) depend on the cell, the organ and developmental stage of the insect. Studies of G.c. have made an essential contribution to the undersatanding of differential Gene activation (see).

Giant messenger-like RNA: see Messenger RNA.

Gibbane: see Gibberellins.

Gibberellane: see Gibberellins.

Gibberellic acid: see Gibberellins.

Gibberellin antagonists: inhibitors of the effects of gibberellins on plants. Their effects can be at least

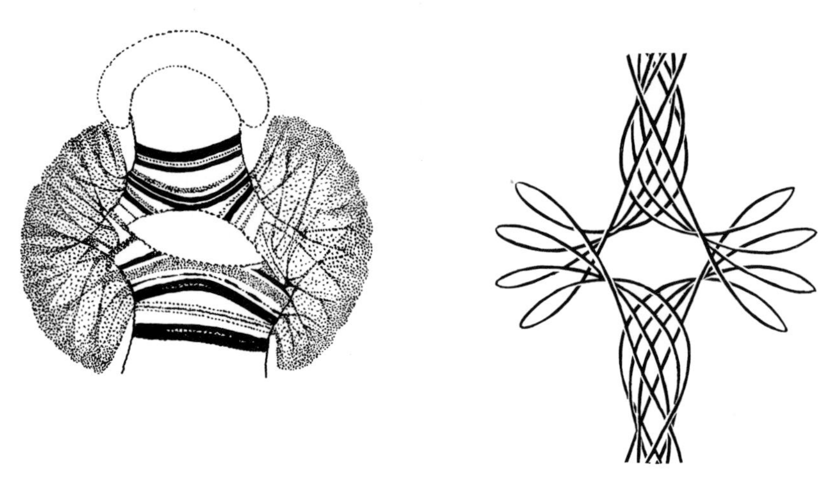

A puff in a giant chromosome (left) and a diagram of the course of the individual chromatids in the region of the puff (right).

partially overcome by gibberellins. The term is used independently of the mechanism of inhibition. Competitive inhibitors of gibberellins, compounds which bind to the same active sites as the hormones, are called *antigibberellins.* The G. a. include the phytohormone Abscisic acid (see) and a number of growth retardants, e. g. Chlorocholine chloride (see), Morphactins (see), AMO 1618 (see), EL 531 (see) and Succinic mono-*N*-methylhydrazide (see). Some natural products, e. g. tannins are also G. a.

Gibberellin glucosides: see Gibberellins.

Gibberellinic acid: see Gibberellins.

Gibberellins: a class of ubiquitous natural plant growth regulators, which stimulate extension growth. The first G. to be discovered was isolated in 1938 from *Gibberella fujikuroi (Fusarium moniliforme),* the pathogen of the rice disease Bakanae. The first pure G. was G.A_3 (gibberellic acid), crystallized in 1954. Since many fungal metabolites function as plant growth promoters, claims for G. production by other fungi, based only on biological activity, are unacceptable. G.A_4 has, however been unequivocally identified (mass spectrometry) as a toxin of pathogenic strains of *Sphaceloma manihoticola* [R. S. Ziegler et al. *Phytopathology* **70** (1980) 589–593]. All of the known ~90 G. contain the tricyclic gibbane ring system. The IUPAC recommends that nomenclature and numbering be based on the *ent-gibberellane* system (Fig. 1).

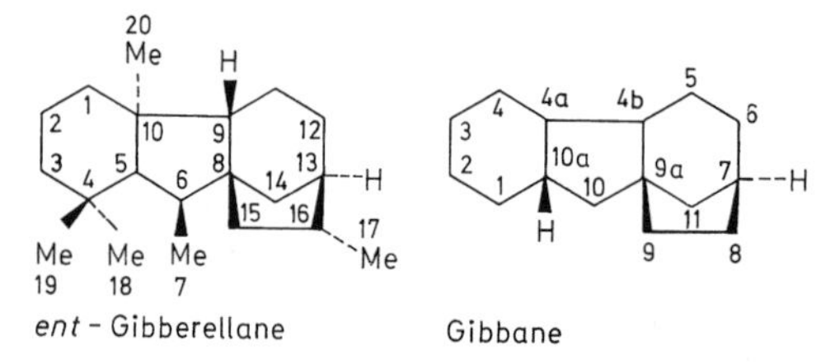

Fig. 1. *Numbering systems of ent-Gibberellane and Gibbane.*

The G. are referred to, in order of their discovery, as G.A_1 through A_{52}. There are two main groups: G. with 20 C-atoms *(ent-gibberellanes)* and those with 19-C atoms *(ent-20-nor-gibberellanes).* Other differences between G. lie in the presence or absence of hydroxyl groups at positions 1, 2, 3, 11, 12 and 13. The substance present in a given sample of plant material can often be characterized only by combined gas chromatography and mass spectroscopy of the methyl esters or trimethylsilyl ethers. G.A_3 (the best known of all the G.) is prepared commercially from culture filtrates of *G. fujikuroi.* Some of the known degradation products of G.A_3 are gibberellenic acid, allogibberic acid, gibberic acid, gibberene and gibberellin C. In addition to these free G., water-soluble G. glucosides and G. glucose esters have been isolated from plants. These are probably transport and storage forms of the biologically active free G.

The activities of G. in various biotests differ widely. Some appear to be physiologically unimportant, while others may be intermediates in the synthesis or degradation of active hormones. The most important effect of the G. is stimulation of extension growth and cell division. Dwarf mutants of peas and corn with genetically blocked G. synthesis are used in assays for G. which have a sensitivity of 1 ng (dwarf maize or pea test). The lettuce and cucumber hypocotyl tests are also based on the stimulation of shoot growth by G. In addition, G. inhibit root growth and influence seed dormancy and germination. Parthenocarpy can be induced in tomatoes, grapes and cucumbers by G. treatment. The acceleration of germination by G. is the basis of the seed germination test, and is used to obtain uniform and rapid germination in malting barley. G. also stimulates blossom formation and fruit growth, and both properties are utilized in commercial agriculture.

G.A_3 activates genes in barley from which the embryo has been removed, suggesting that the embryo releases G. into the aleurone layer, where it activates the genes for hydrolytic enzymes, particularly α-amylase. Another test for G., the α-amylase test, is based

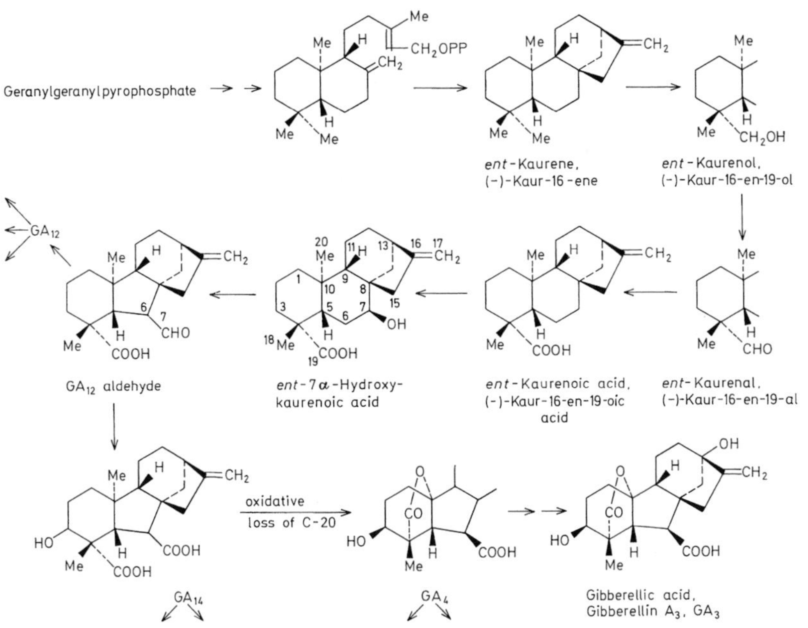

Fig. 2. *Biosynthesis of gibberellins*

on this property. A few G. produced by partial synthesis, e.g. 3-deoxygibberellin C and pseudogibberellin A_1, act as antigibberellins (see Gibberellin antagonists).

Biosynthesis. G. are diterpenes, synthesized from geranylgeranylpyrophosphate (Fig. 2). The pivotal, common intermediate in the biosynthesis of all G. is G.A_{12} aldehyde. Thereafter, the biosynthetic pathways diverge considerably. [A. Lang *Ann. Rev Plant Physiol.* **21** (1970) 537–570; J. E. Graebe & H. J. Ropers in *Phytohormones and Related Compounds – A Comprehensive Treatise* Vol 1, D. S. Letham, P. B. Goodwin & T. J. V. Higgins (eds.), North Holland, 1978, pp. 107–204; P. Hedden et al. *Annu. Rev. Plant Physiol.* **29** (1978) 149–192; P. Fuchs & G. Schneider 'Synthesis of glucosyl conjugates of [17-$^2H_2$]-labelled and unlabelled Gibberellin A_{34}' *Phytochemistry* **42** (1996) 7–10]

Gibberene: see Gibberellins.

Gibberic acid: see Gibberellins.

Gibbs effect: an apparently anomalous labeling of glucose following the brief photosynthetic assimilation of $^{14}CO_2$ by *Chlorella.* The resulting fructose *bis*phosphate is labeled symmetrically, as expected from the operation of the Calvin cycle (see). In contrast, glucose phosphates and the glucose moiety of starch are asymmetrically labeled, C4 containing significantly more ^{14}C than C3, and C1 and C2 containing significantly more than C5 and C6. [O. Kandler & M. Gibbs *Plant Physiol.* **31** (1956) 411–412]. The G. e. indicates that the two halves of the glucose molecule are not derived from the same pool of triose phosphates as fructose *bis*phosphate; it also suggests that fructose *bis*phosphate is not the precursor of glucose phosphates. Reactions in the revised version of the Pentose phosphate cycle (see) may explain the G. e.

Gitogenin: see Gitonin.

Gitonin: a steroid saponin, M_r 1051.21, m.p. 272 °C, $[\alpha]_D$ – 51° (pyridine). The aglycon is gitogenin, (25 *R*)-spirostan-2α,3β-diol, M_r 432.62, m.p. 272–275 °C (d.), $[\alpha]_D^{20}$ –70° (c = 1.02 in $CHCl_3$); the sugar chain consists of 2 galactose, 1 glucose and 1 xylose units. Gitogenin differs from digitogenin (see Digitonin) in lacking the 15β-hydroxyl group. G. has been isolated from *Digitalis purpurea* and *Digitalis germanicum.* Free gitogenin has been isolated from agave and yucca species.

Gitoxigenin: see Digitalis glycosides.

Gitoxin: see Digitalis glycosides.

Gla: abb. for L-4-carboxyglutamic acid.

Glabrene: see Isoflav-3-ene.

Gliadin: see Prolamines.

Gln: abb. for L-glutamine.

Globoid cell leukodystrophy: see Lysosomal storage diseases.

Globoside: see Lysosomal storage diseases (Fig. Catabolism of glycososphingolipids and sphingomyelin).

Globulins: a group of simple proteins which are insoluble in pure water but soluble in dilute salt solutions (salting in effect). They are found in all animal and plant cells and body fluids, including serum and milk. The group includes many enzymes and most glycoproteins. G. are precipitated by ammonium sulfate: fibrinogen at 20 to 25 %, the euglobulins at 33 % and the pseudoglobulins at 50 %. The most familiar are the serum G., which are separated by electrophoresis into the α_1, α_2, β_1, β_2 and γ-G. (see Plasma proteins).

Glomerine: a quinazoline in the defense secretion of the insect *Glomeris marginata.* The secretion also contains homoglomerine. These two quinazolines may be regarded as animal alkaloids.

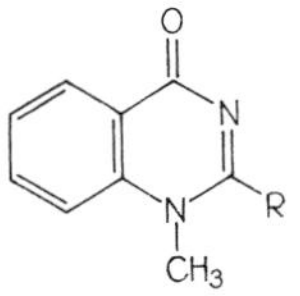

Glomerine ($R = C_2H_5$) and *Homoglomerine* ($R = CH_3$).

Glp: recommended abb. for Pyroglutamic acid (see).

< Glu: recommended abb. for Pyroglutamic acid (see).

⊏Glu: recommended abb. for Pyroglutamic acid (see).

Glu: abb. for L-glutamic acid.

Glucagon: a pancreatic polypeptide hormone consisting of a single chain of 29 amino acids (M_r 3485), which is produced in the A cells of the islets of Langerhans in the pancreas in response to a decrease in blood glucose concentration. It promotes the hydrolysis of glycogen and lipids and raises the blood sugar level. G. is degraded in the liver. It can be detected in blood serum (in the order of pg/ml) by radioimmunoassay.

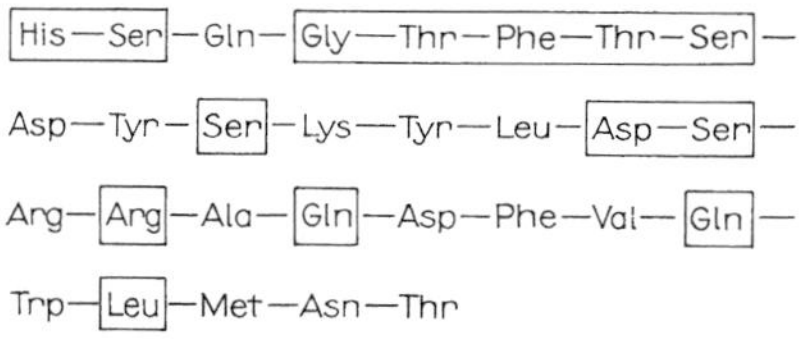

His—Ser—Gln—Gly—Thr—Phe—Thr—Ser—
Asp—Tyr—Ser—Lys—Tyr—Leu—Asp—Ser—
Arg—Arg—Ala—Gln—Asp—Phe—Val—Gln—
Trp—Leu—Met—Asn—Thr

Glucagon. The boxed amino acids are identical with the secretin sequence.

Glucans: linear or branched polysaccharides composed of D-glucose. The glycosidic linkages may be α-1,4 as in amylose and bacterial dextran, β-1,4 as in cellulose, β-1,3 as in leucosin and callose, or 1,6 as in pustulan. Branched glucans include amylopectin (α-1,4 and α-1,6 bonds), dextran, laminarin and lichenin.

Glucocorticoids: see Adrenal corticosteroids.

Glucokinase: see Kinases.

Gluconeogenesis: synthesis of glucose from pyruvate or Amino acids (see). G. cannot occur by a simple reversal of Glycolysis (see), because the equilibria of the glycolysis reactions are too unfavorable under physiological conditions. Instead, the pyruvate is carboxylated to oxaloacetate (directly, or indirectly via malate) (see Carboxylation). The oxaloacetate is then decarboxylated and phosphorylated simultaneously by phospho*enol*pyruvate carboxykinase (ATP) (EC 4.1.1.49) to form phospho*enol*pyruvate; in *E. coli* pyruvate is phosphorylated directly by ATP. Reversal of the glycolytic reactions then yields fructose 1,6-*bis*phosphate from the phospho*enol*pyruvate. The phosphofructokinase reaction is not reversible; instead fructose-*bis*phosphatase (EC 3.1.3.11) removes one phosphate group to form fructose 6-phosphate, which is readily converted to glucose 6-phosphate. If the blood sugar level is low, glucose 6-phosphate is hydrolysed to glucose by glucose-6-phosphatase (EC 3.1.3.9). Otherwise, the glucose 6-phosphate is used directly for glycogen synthesis. Overall reaction: 2pyruvate ($2CH_3COCOO^-$) + 2NADH + $4H^+$ + 6ATP → glucose ($C_6H_{12}O_6$) + $2NAD^+$ + 6ADP + 6 P_i. The energy required for G. can be obtained by oxidizing 20 to 30 % of the lactate to CO_2 and H_2O.

G. can proceed from the carbon skeleton of any amino acid that can be converted to a C4 carboxylic acid (glucoplastic amino acids), i.e. into any intermediate of the Tricarboxylic acid cycle (see) that can be converted to oxaloacetate.

G. in liver is promoted by Glucagon (see) and Adrenalin (see), the effects being mediated by cAMP. When the organism fasts, glucocorticoids (e.g. cortisol) are released from the adrenal glands and induce the synthesis of enzymes of G. in the liver; glucocorticoids also appear to make the cells more sensitive to cAMP, and thus to glucagon. The result is increased G. from amino acids in the fasting animal.

Glucoplastic amino acids: amino acids whose degradation products can contribute to Gluconeogenesis (see). See Amino acids.

Glucosamine, ***2-amino-2-deoxyglucose, chitosamine:*** a widely occurring aminosugar, M_r 179.17; α-form m.p. 88 °C, $[\alpha]_D^{20}$ + 100 ° → + 47.5 ° after 30 min (water); β-form m.p. 110 °C (d.), $[\alpha]_D^{20}$ + 28 ° → + 47.5 ° after 30 min (water). G. is a component of chitin, of mucopolysaccharides like heparin, chondroitin and mucoitin sulfate, and of blood group substances and other complex polysaccharides. It is usually present as *N*-acetylglucosamine.

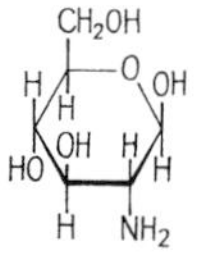

β-Glucosamine

Glucosaminoglycans: see Proteoglycans.

D-Glucose, ***dextrose, grape sugar, blood sugar:*** a hexose, M_r 180.16; α-form, m.p. 146 °C, $[\alpha]_D$ + 112.1 → + 52.7 ° (c = 10 in water); β-form, m.p. 148–155 °C, $[\alpha]_D$ + 18.7 → + 52.7 ° (c = 10 in water). The most stable configuration for the pyranose form is the chair, in which all the hydroxyl groups of the β-form are equatorial (see Carbohydrates, Fig. 4). In the α-form, the two hydroxyl groups at positions 1 and 2 are *cis.* D-G. undergoes various forms of anaerobic and aerobic fermentation to ethanol, lactate, acetate, or citrate. Metabolically, it is a very important animal monosaccharide, and it is also the most abundant natural organic compound. Free D-G. is found in many sweet fruits, honey and nectar, and blood (up to 0.1 %). It is also a component of many oligo- and polysaccharides (sucrose, lactose, maltose, starch, glycogen, cellulose, etc.) and glycosides. Phosphoric acid esters of D-G. are extremely important metabolic intermediates. The activated form, ADPglucose, is a biosynthetic precursor in starch synthesis in plants; UDP-glucose is the D-G. donor in the synthesis of many saccharides (see Nucleoside diphosphate sugars).

D-G. is produced commercially by acid or enzymatic hydrolysis of potato or corn starch or cellulose. It is used as a sweetener in beverages and as a nutrient in medicine.

α-D-Glucose

β-D-Glucose

Glucose 1,6-*bis*phosphate: a glucose derivative which is an important intermediate in glycolysis. It is synthesized in yeast, plants and muscle cells by the reaction glucose 1-phosphate + ATP → glucose 1,6-*bis*-phosphate + ADP; and in *E. coli* and muscle cells by the reaction 2glucose 1-phosphate ⇌ glucose 1,6-*bis*-phosphate + glucose. It is the cosubstrate of phosphoglucomutase (EC 2.7.5.1), which catalyses the interconversion of glucose 1- and 6-phosphates (Fig.) (see Glycolysis).

Glucose isomerase: see Enzymes, Table 2.

D(+)-Glucose oxidase (EC 1.1.3.4): a plant and bacterial flavin enzyme which oxidizes β-D-glucose in the presence of O_2 to gluconic acid and H_2O_2. G. o. is used for a specific enzymatic assay of glucose, e.g. in blood. G. o. from *Aspergillus* is a dimeric flavoglycoprotein (16% carbohydrate, M_r 160,000, 2FAD per molecule, 2 subunits of M_r 80,000 each) which is inhibited by *p*-chloromercuribenzoate.

Glucose 1-phosphate, *Cori ester:* the product of phosphorolysis of Glycogen (see) and Starch (see). It is converted to glucose 6-phosphate by phosphoglucomutase (EC 2.7.5.1).

Glucose 6-phosphate, *Robinson ester:* the key intermediate in Carbohydrate metabolism (see).

Glucose-6-phosphate dehydrogenase, *GPDH* (EC 1 1 1.49): the key enzyme of the Pentose phosphate cycle (see). It occurs widely in plants and animals and has been shown to be formed from inactive precursor subunits. GPDH is a tetramerie enzyme with M_r ranging from 206,000 in *Neurospora* to 240,000 in erythrocytes. Its dimers are held together by NADP. In humans, 50 hereditary variants of erythrocyte GPDH are known. It is lacking in some individuals who, after consuming certain legumes (e.g. fava beans), breathing bean pollen or taking certain pharmaceuticals, suffer a severe hemolytic anemia (favism). The condition is especially widespread in the Mediterranean region, Asia and America.

Glucose tolerance factor: see Chromium.

Glucosinolate, *mustard oil thioglucoside, mustard oil glucoside:* a natural plant product with the structure shown in Fig. 1. G. are particularly abundant in members of the *Cruciferae, Capparidaceae* and *Resedaceae.* Sinalbin was crystallized from white mustard in 1831, and sinigrin was isolated from black mustard in 1840. More than 70 different G. are now known. The carbohydrate residue is always a single glucose unit; all G. are 1-β-D-thioglucopyranosides. The term "glucosinolate" for the anion of the compound has been used since 1961. The cation is usually potassium, but in sinalbin it is the basic organic molecule, sinapine. An alternative and earlier nomenclature uses the prefix "gluco-" attached to an appropriate part of the Latinized binomial of the plant of origin, e.g. glucotropaeolin from *Tropaeolum majus,* etc.

Damage to the plant results in hydrolysis of G. and release of volatile, pungent, lacrimatory isothiocyanates, also known as mustard oils. This ability to form mustard oils has led to the adoption of several members of the *Cruciferae* (e.g. horseradish, mustard) as condiments. Mustard oils are normally absent from the plant, and are first formed when tissue damage permits interaction of G. and thioglucoside glucohydrolase (EC 3.2.3.1); activation of this enzyme by its cofactor, ascorbic acid, may also be promoted by tissue damage. Enzymatic hydrolysis of the *S*-glucoside bond is followed by moleular rearrangement of the aglycon with concomitant production of sulfate and isothiocyanate (Fig. 1). Small quantities of the corresponding thiocyanate and nitrile may also be produced.

Like the cyanogenic glycosides, G. are biosynthesized from amino acids. The pathway in Fig. 2 is strongly supported by isotopic tracing studies with ^{14}C,^{15}N-labeled amino acids or the corresponding aldoximes. The following G. are derived from proteogenic amino acids: methyl-G. (from alanine; not yet found in *Cruciferae,* but widely distributed and dominant in the *Capparaceae),* isopropyl-G. and sec-butyl-G. (from valine and isoleucine, respectively; both compounds found especially in the genera *Cardamine, Cochlearia, Lunaria* and *Sisymbrium),* isobutyl-G. (from leucine, found in *Conringia orientalis, Cochlearia officinalis* and 2 spp. of *Thelypodium),* 4-hydroxybenzyl-G. (from tyrosine; the salt with sinapine is sinalbin; especially abundant in, and rather restricted to, the genus *Sinapis),* benzyl-G. (from tryptophan; in several plant families, but restricted to seedlings and young vegetative tissue). The G. that would be derived from the remaining 13 proteogenic amino acids have not been encountered; owing to its cyclic structure, proline is not even a potential G. precursor.

G. are also known that arise from proteogenic amino acids after homologization (Fig. 3), which consists of a series of type reactions already well known in

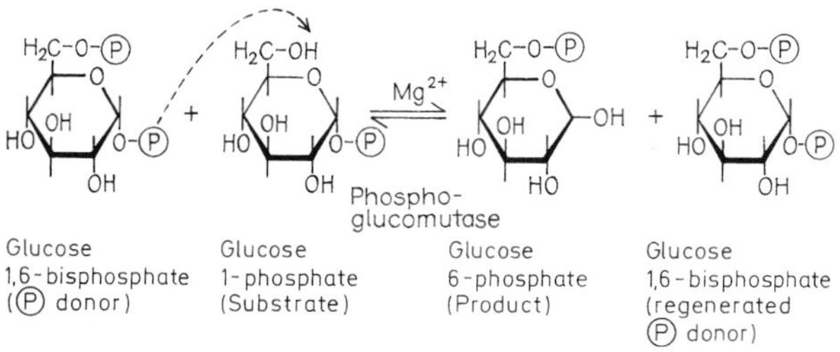

Phosphoglucomutase reaction

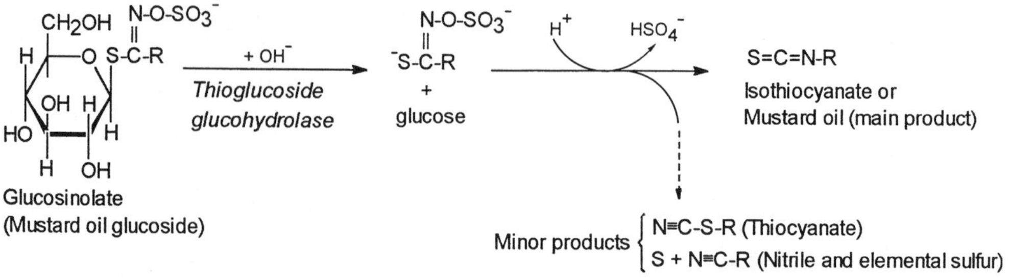

Fig. 1. *Glucosinolate.* Conversion of a glucosinolate to the corresponding isothiocyanate by the action of thioglucoside glucohydrolase.

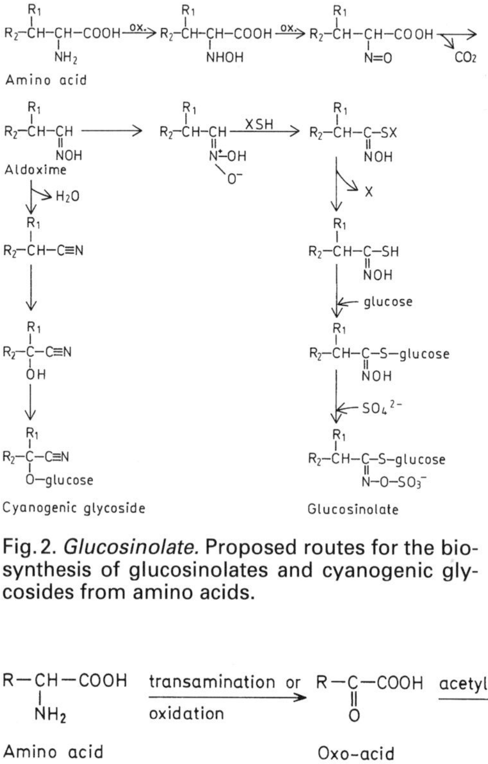

Fig. 2. *Glucosinolate.* Proposed routes for the biosynthesis of glucosinolates and cyanogenic glycosides from amino acids.

the Tricarboxylic acid cycle (see) and leucine biosynthesis, e.g. ethyl-G. (reported only in one species of *Lepidum)* is derived from alanine after one cycle of homologization. Similarly, 2(*S*)-methylbutyl-G. is derived from 2-amino-4-methylhexanoic acid, which in turn is derived from isoleucine by homologization. The *Cruciferae* have so far yielded 9 homologous ω-methylthioalkyl-G. (side chain CH_3-S-$(CH_2)_n$-, where n varies from 3 to 11); these are derived from methionine, following 1–9 cycles of homologization. The corresponding sulfoxides and sulfones are also known, e.g. 3-(methylsulfonyl)-propyl-G. from *Cheiranthus.* The biosynthetic origin of sinigrin is less obvious, but isotopic tracing studies have shown that the carbon atoms of the side chain are derived from methionine after one cycle of homologization (Fig. 4).

Allylisothiocyanate is highly toxic to the pathogen *Peronospora parasitica* (downy mildew). Varieties of cabbage bred for milder flavor (and hence lower sinigrin content) lack resistance to the pathogen. On the other hand, sinigrin is a feeding attractant for the cabbage butterfly larva *(Pieris brassicae)* and an oviposition stimulant for the adult female. It is also a feeding attractant to the cabbage aphid, which prefers mature leaves of medium sinigrin content, and avoids feeding on young leaves that are very rich in this G. [A. Kjaer

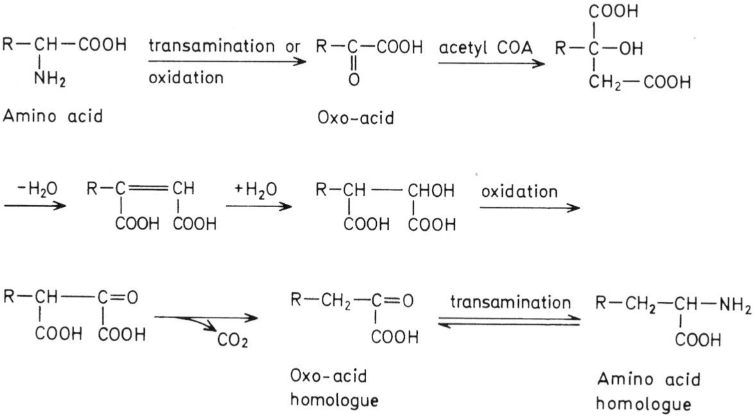

Fig. 3. *Glucosinolate.* Homologization. One cycle of this process extends the side chain of the precursor amino acid by one CH_2 group; it is analogous to the reactions of the tricarboxylic acid cycle and of leucine biosynthesis.

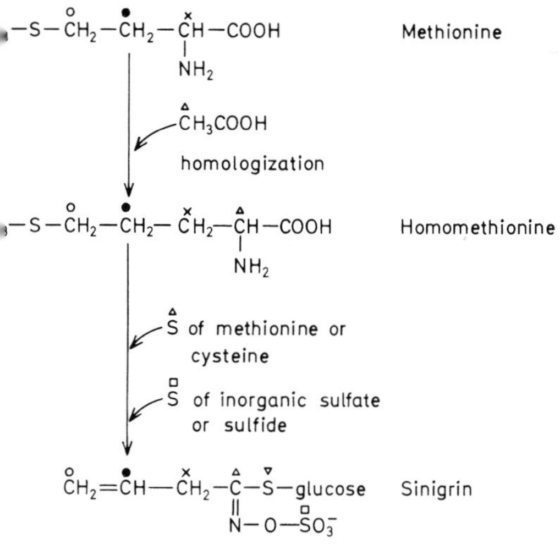

$-S-CH_2-CH_2-CH(NH_2)-COOH$ Methionine

↓ CH_3COOH homologization

$-S-CH_2-CH_2-CH_2-CH(NH_2)-COOH$ Homomethionine

↓ S of methionine or cysteine

↓ S of inorganic sulfate or sulfide

$CH_2=CH-CH_2-C(=N-O-SO_3^-)-S-glucose$ Sinigrin

Fig. 4. *Glucosinolate.* Summary of isotope (^{14}C and ^{35}S) tracing studies on the biosynthesis of sinigrin.

in *Progress in the Chemistry of Organic Natural Products* **XVIII,** L. Zechmeister (ed.), Springer, 1960, pp. 122–176.; *The Biology and Chemistry of the Cruciferae,* J. G. Vaughan et al. (eds.), Academic Press, 1976]

Glucuronate pathway, *glucuronate-xylose cycle, D-glucuronate-L-gulonate pathway:* a pathway in carbohydrate metabolism by which *myo*-inositol and ascorbate are synthesized and degraded (Fig. 1). Glucose is oxidized at position 6 to D-glucuronate, probably via UDPglucose (see Nucleoside diphosphate sugars). Glucuronate, which is also the product of *myo*-inositol oxygenase, is the starting material for the synthesis of glucuronides, and it is reduced to L-gulonate. Since C6 of glucuronate becomes C1 of gulonate, the latter belongs to the L-series of carbohydrates. L-Gulonate is diverted into the L-ascorbate pathway (see Vitamins: ascorbic acid) or it is oxidized to 3-keto-L-gulonate, which is decarboxylated to form L-xylulose. Xylulose is reduced to the sugar alcohol xylitol, which is reoxidized to D-xylulose. The change in configuration is again accomplished by an end-for-end shift in which C5 of xylulose is formed from C1 of xylitol. D-Xylulose is phosphorylated to xylulose 5-phosphate, which is a member of the Pentose phos-

Table. *Some examples of naturally occurring glucosinolates*

Glucosinolate	Source	Formula
Sinalbin, Sinapine glucosinalbate, Sinapine 4-hydroxybenzylglucosinolate	*Sinapis alba* (White or yellow mustard)	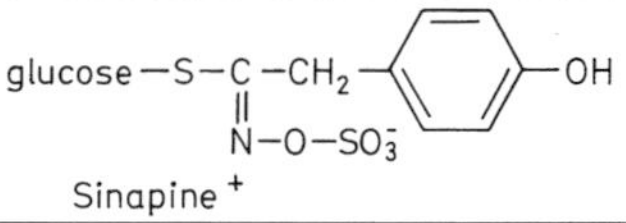 $glucose-S-C(=N-O-SO_3^-)-CH_2-C_6H_4-OH$ Sinapine$^+$
Sinigrin, Allylglucosinolate, Sinigroside, Potassium myronate	*Brassica nigra* (Black mustard), *Amoracia rusticana* (Horse radish)	$glucose-S-C(=N-O-SO_3^-\,K^+)-CH_2-CH=CH_2$
Benzylglucosinolate Glucotropaeolin	*Tropaeolum majus* (Nasturtium)	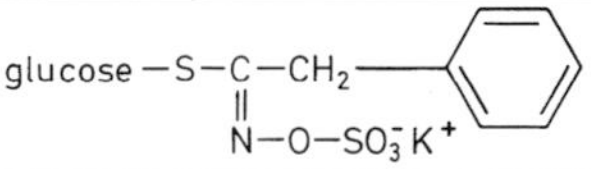 $glucose-S-C(=N-O-SO_3^-\,K^+)-CH_2-C_6H_5$
Phenylethyl glucosinolate, Gluconasturtiin.	*Nasturtium officinale* (Watercress)	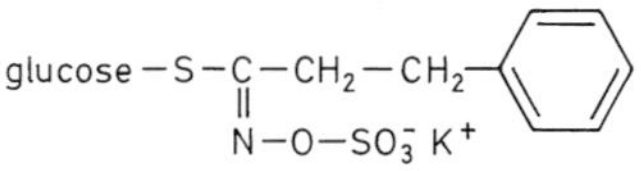 $glucose-S-C(=N-O-SO_3^-\,K^+)-CH_2-CH_2-C_6H_5$
3-(Methylsulfonyl)--propylglucosinolate Glucocheirolin	*Cheiranthus cheiri* (Wallflower)	$glucose-S-C(=N-O-SO_3^-\,K^+)-(CH_2)_3-SO_2-CH_3$
*2-Hydroxy-3-butenylglucosinolate, Progoitrin, Glucorapiferin	Various spp. of Brassica (esp. Yellow turnip)	$glucose-S-C(=N-O-SO_3^-\,K^+)-CH_2-CHOH-CH=CH_2$

* 2-Hydroxy-3-butenylisothiocyanate ($S=C=N-CH_2-CHOH-CH=CH_2$), the mustard oil from 2-hydroxy-3-butenylglucosinolate, cyclizes spontaneously to 5-vinyl-2-thiooxazolidone (goitrin), which is responsible for the goitrogenic action (see Goitrogens) of yellow turnips and rapeseed oil meal.

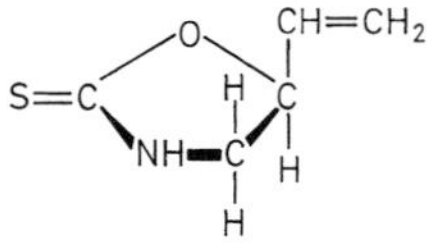

(S)-5-Ethenyl-2-oxazolidinethione (5-vinyl-2-thiooxazolidone, goitrin)

Glucuronate pathway

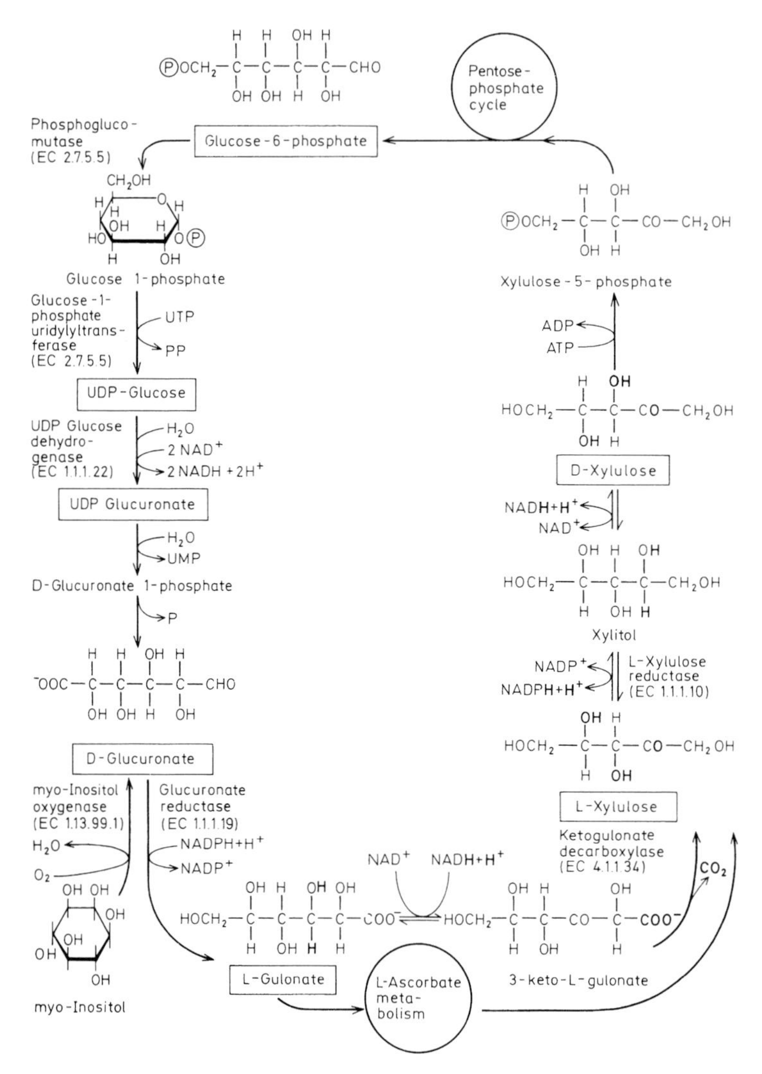

Fig. 1. *Glucuronate pathway*

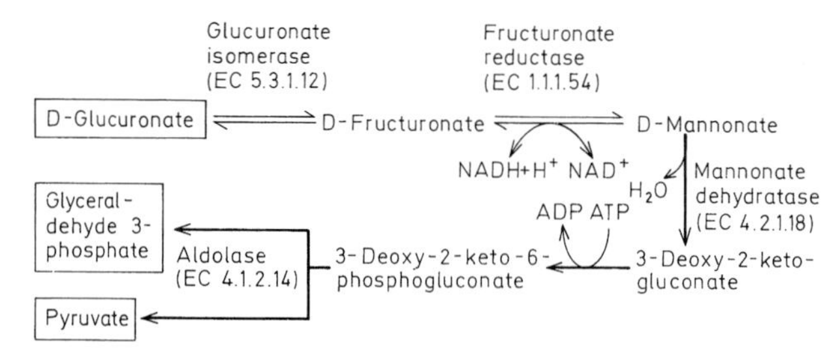

Fig. 2. *Alternative catabolic pathway for glucuronate in bacteria.*

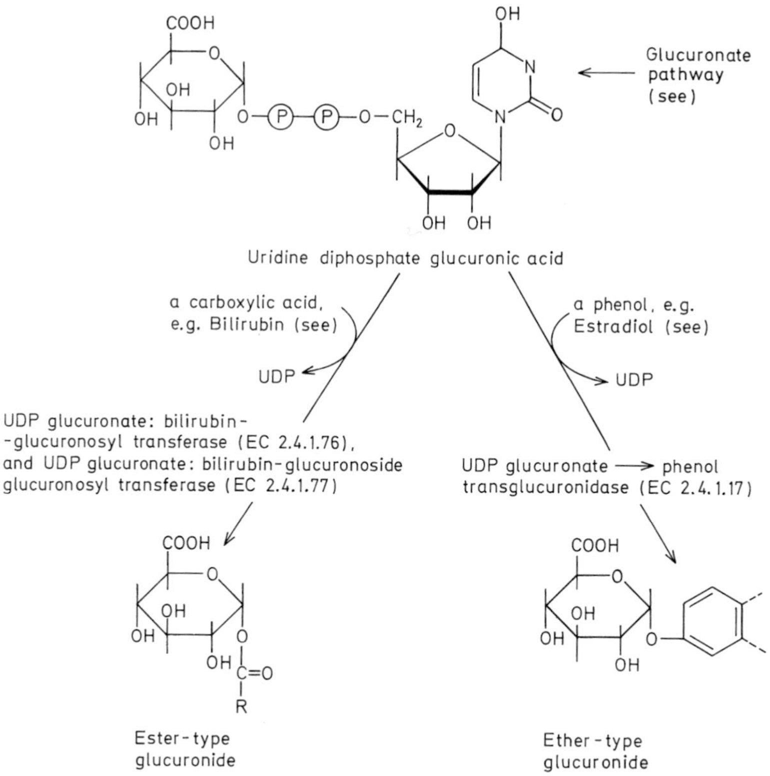

Glucuronate pathway (see)

Uridine diphosphate glucuronic acid

a carboxylic acid, e.g. Bilirubin (see)

UDP

a phenol, e.g. Estradiol (see)

UDP

UDP glucuronate: bilirubin-glucuronosyl transferase (EC 2.4.1.76), and UDP glucuronate: bilirubin-glucuronoside glucuronosyl transferase (EC 2.4.1.77)

UDP glucuronate ⟶ phenol transglucuronidase (EC 2.4.1.17)

Ester-type glucuronide

Ether-type glucuronide

Formation of glucuronic acid conjugates (glucuronides or glucuronosides). UDP glucuronate:phenol transglucuronidase (EC 2.4.1.17) catalyses glucuronidation of a wide range of phenols, alcohols and fatty acids. Glucuronidation of bilirubin is an important special case (see Crigler-Najjar syndrome in Inborn errors of metabolism).

phate cycle (see). Glucose 6-phosphate, the precursor of UDPglucose, is regenerated from xylulose 5-phosphate in the pentose phosphate cycle.

Bacteria possess an alternative pathway for the degradation of glucuronate to glyceraldehyde 3-phosphate and pyruvate (Fig. 2).

D-Glucuronic acid: a biosynthetic derivative of glucose found in mucopolysaccharides like hyaluronic acid and chondroitin sulfate. In animals it is an important conjugation partner for foreign and endogenous substances, especially phenols, which are excreted as G. a. conjugates (glucuronides). In most animals, G. a. is the starting material for ascorbic acid (see Vitamin C) synthesis.

L-Glutamic acid, *Glu, L-α-aminoglutaric acid:* a proteogenic amino acid with two carboxyl groups. Since only the α-carboxyl group forms peptide bonds in proteins, the remaining free carboxyl group gives the polypeptide an acid character. M_r 147.14, m.p. 247–249 °C (d.), $[\alpha]_D^{25} + 17.7°$ ($c = 2$ in water) or $+46.8°$ ($c = 2$ in 5 M HCl). Glu is present in nearly all proteins, especially seed proteins. It is easily cyclized to L-pyrrolidone carboxylic acid (Fig. 1) on heating; this compound is also formed as a post-translational modification of Glu residues in proteins (see Pyroglutamic acid). The biosynthesis of amino acids belonging to the 2-oxoglutarate (α-ketoglutarate) family begins with Glu (Fig. 2). Glu is involved in the transport of potassium ions in the brain, and also detoxifies ammonia in the brain by forming glutamine, which can cross the blood-brain barrier. In the central nervous system, Glu is decarboxylated to 4-aminobutyric acid by glutamate decarboxylase (EC 4.1.1.15).

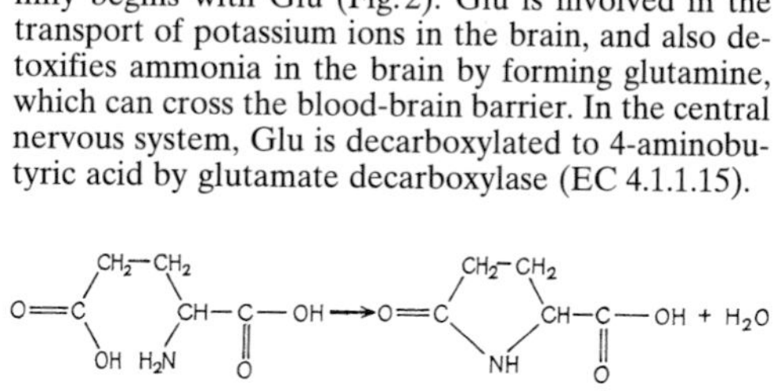

Fig. 1. *Cyclization of L-glutamic acid.*

Glu possesses flavor-enhancing properties (normally used as monosodium glutamate), and for this reason was the first amino acid prepared industrially; at first this was achieved by the acid hydrolysis of rich sources such as wheat gliadin (34.7 % Glu), but subsequently by industrial fermentation [A. L. Demain, *Naturwiss.* **67** (1980) 582–587].

Glu is a component of γ-glutamylpeptides, which are involved in a proposed mechanism for amino acid transport across cell membranes (see γ-Glutamyl cy-

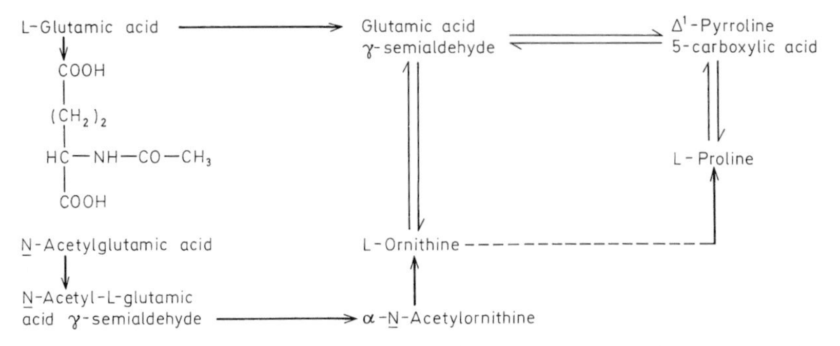

Fig. 2. *L-Glutamic acid.* Biosynthesis of L-ornithine and L-proline from L-glutamic acid.

cle). A few γ-glutamyl peptides are also found in the central nervous system, but they occur in quantity in plants, where they serve as storage compounds, and often contain rare amino acids. Glu is also a component of Glutathione (see). In animals Glu is biosynthesized by the transamination of 2-oxoglutarate, which in turn is derived directly from the tricarboxylic acid cycle. This transamination also occurs in green plants and microorganisms, but in these organisms Glu synthesis also occurs by the action of glutamate synthase or by the reductive amination of 2-oxoglutarate (see Ammonia assimilation). Glu is degraded, by transamination or dehydrogenation, to 2-oxoglutaric acid, or decarboxylated to 4-aminobutyric acid, then converted to succinic acid which enters the tricarboxylic acid cycle.

L-Glutamine, *Gln, Glu-NH₂*: a proteogenic amino acid, the 5-amide of L-glutamic acid, M_r 146.15, m.p. 185–186 °C (d.), $[\alpha]_D^{20}$ + 9.2° (c = 2 in water) or + 46.5° (c = 2 in 5 M HCl). In boiling, neutral aqueous solutions or in weak acid, Gln rapidly cyclizes to the ammonium salt of pyrrolidone carboxylic acid (see Glutamic acid).

Gln is synthesized from glutamic acid and ammonium in an endergonic reaction: Glu + NH_4^+ + ATP → Gln + ADP + P_i, catalysed by glutamine synthetase (EC 6.3.1.2). The synthesis is made thermodynamically favorable by coupling with the hydrolysis of ATP; there appear to be no free intermediates. The synthesis plays a role in both Ammonia detoxification (see) and the formation of nitrogen excretion products (Gln is the nitrogen donor in the synthesis of carbamoylphosphate and purines). In plants and microorganisms the synthesis of Gln by this mechanism plays a pivotal role in the assimilation of ammonium ions (see Ammonia assimilation). Gln synthesized by the same reaction is also an important nitrogen storage compound in plants.

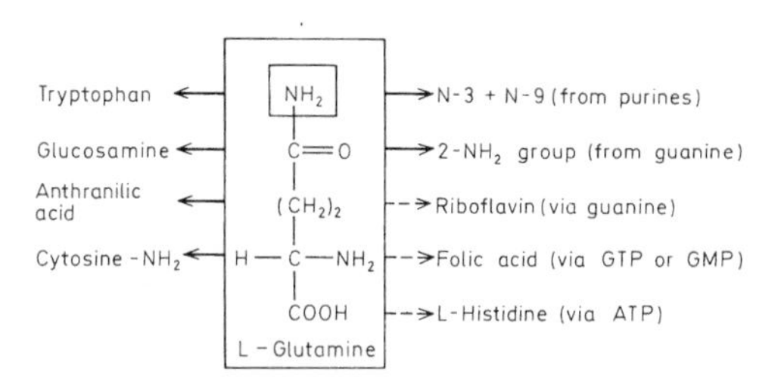

Gln is therefore a central intermediate in nitrogen metabolism, providing (by transamidation) the amide nitrogen required in the synthesis of a variety of other compounds (Fig.). It also plays an important role in the kidney, where it is hydrolysed to provide free ammonia, which aids resorption of potassium and sodium ions. Gln is an important nutrient for the brain (see Glutamic acid). Because its amide group is involved in the biosynthesis of hexosamines, Gln promotes the regeneration of mucoproteins and the intestinal epithelium.

Glutamine antagonists, *glutamine analogs*: structural analogs of L-glutamine which competitively inhibit glutamine-dependent enzyme reactions. Examples are: 6-Diazo-5-oxo-L-norleucine (see), Albizziin (see) and Azaserine (see).

Glutamine synthetase: see Ammonia assimilation, Covalent modification of enzymes, L-Glutamine.

Glutamino-carbamoylphosphate synthetase: see Carbamoyl phosphate.

4-Glutamyl carboxylase, *γ-glutamyl carboxylase, vitamin K-dependent γ-glutamyl carboxylase*: the enzyme responsible for the posttranslational carboxylation of glutamate to 4-carboxyglutamate residues in certain proteins. The reaction requires vitamin K, which explains the vitamin K requirement for the synthesis of blood clotting proteins, e.g. prothrombin, which contain residues of 4-carboxyglutamate. It is a microsomal enzyme, which can be solubilized with various detergents. There is an apparent ATP requirement for the microsomal system, but the solubilized enzyme does not require ATP. Biotin is not involved in this carboxylation.

$$\text{HC(CO)(NH)}-CH_2-CH_2-COO^- + HCO_3^- \xrightarrow{\text{vitamin K}} \text{HC(CO)(NH)}-CH_2-CH(COO^-)_2$$

L-Glutamate residue → L-Carboxyglutamate residue

γ-Glutamyl cycle: a cycle of reactions proposed for the transport of amino acids across cell membranes. At the outer surface of the membrane, amino acids are converted to their γ-glutamyl peptides by the action of membrane-bound γ-glutamyl transferase (EC 2.3.2.2), glutathione serving as a γ-glutamyl donor: amino acid + glutathione → Cys-

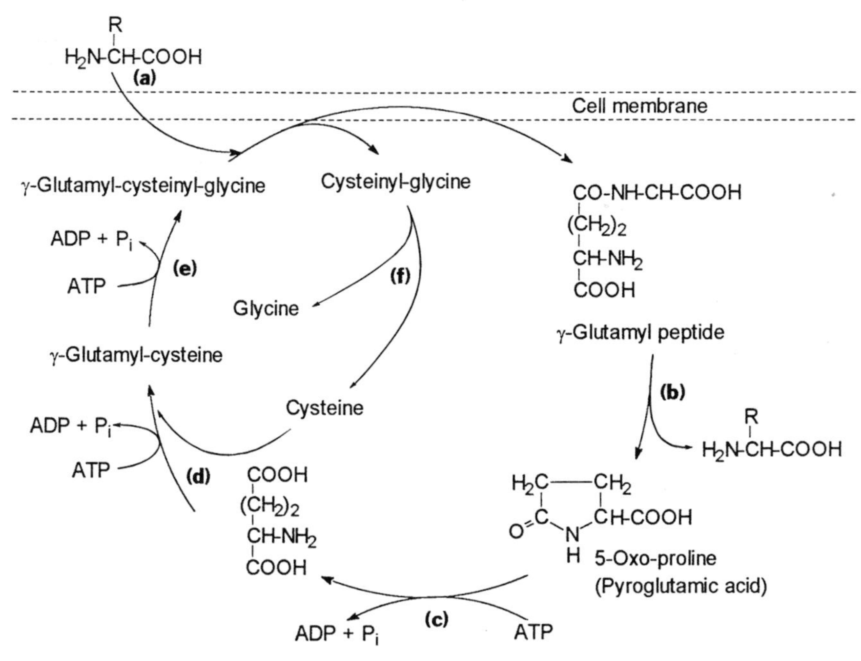

The γ-glutamyl cycle. (a) γ-Glutamyl transpeptidase (5-glutamyl)-peptide: amino acid 5-glutamyltransferase, EC 2.3.2.2). (b) γ-Glutamyl cyclotransferase (5-glutamyl)-L-amino acid 5-glutamyltransferase (cyclizing), EC 2.3.2.4). (c) 5-Oxo-prolinase (5-oxo-L-proline amido-hydrolase (ATP hydrolysing), EC 3.5.2.9). (d) γ-Glutamyl-cysteine synthetase (L-glutamate:L-cysteine γ-ligase (ADP-forming), EC 6.3.2.2). (e) Glutathione synthetase (γ-L-glutamate:L-cysteine: glycine ligase (ADP-forming), EC 6.3.2.3). (f) Dipeptidase.

Gly + γ-glutamyl amino acid. The products are transported into the cell and are cleaved, the Cys-Gly to Gly and Cys, and the γ-glutamyl amino acid to 5-oxoproline and the amino acid. 5-Oxoproline is converted to glutamic acid, so that the γ-glutamyl cycle can repeat. [A. Meister *Science* **220** (1983) 472–477]

Glutathione, ***L-γ-glutamyl-L-cysteinyl-L-glycine, GSH:*** a tripeptide found in animals, most plants and bacteria. It serves as a biological redox agent, as a coenzyme and cofactor, and as a substrate in certain coupling reactions catalysed by Glutathione *S*-transferase (see). The concentration of GSH in cells is 0.1 to 10 mM; it is also exported to the extracellular fluids in the course of the γ-Glutamyl cycle (see). As a redox agent, GSH scavenges free radicals and reduces peroxides ($2GSH + ROOH \rightarrow GSSG + ROH + H_2O$), and thus protects membrane lipids from these reactive substances. It is especially important in the lens and in certain parasites which have no catalase to remove H_2O_2.

Metabolism. Within the cell, GSH is formed in two steps: Glu + Cys → γ-Glu-Cys, and γ-Glu-Cys + Gly → GSH. These are catalysed, respectively, by γ-glutamylcysteine synthetase (EC 6.3.2.2) and GSH synthetase (EC 6.3.2.3). Some of the GSH is transported out of the cell, where it participates in the γ-Glutamyl cycle (see).

Glutathione S-transferase, ***GST, ligandin*** (EC 2.5.1.18): a group of enzymes of the liver cytosol, which catalyse reaction of glutathione (acting as a nucleophile) with a wide range of electrophilic, hydrophobic substrates: $RX + GSH \rightarrow HX + RSG$, where GSH is reduced glutathione, R may be an aliphatic, aromatic or heterocyclic radical, and X may be a sulfate, nitrite, halide, epoxy, or ethene group, or the cyanide group of a thiocyanate. Thus GST are involved in detoxication of numerous carcinogenic, mutagenic and toxic alkylating agents, as well as some pharmacologically active compounds and xenobiotics. The water-soluble glutathione conjugates are metabolized by cleavage of the glutamate and glycine residues, followed by acetylation of the free amino group of the cysteinyl residue, to produce water-soluble mercapturic acids, which are readily excreted. GST also bind strongly and noncovalently a wide variety of ligands, without catalysing further reaction, e.g. estrogens, steroid conjugates, bilirubin, probenecid, heme, penicillin, chloramphenicol. This latter property gave rise to the name, ligandin, before it was discovered that GST and ligandin are indentical. A third property of GST is its ability to bind covalently several electrophilic substrates of the transfer reaction in the absence of glutathione. Yet a further property of GST enzymes is their ability to catalyse certain isomerization reactions for which glutathione is a coenzyme, e.g. maleylacetone to fumarylacetone, and the positional isomerization of Δ^5- to Δ^4-unsaturated 3-ketosteroids.

All cytosolic GST from rat liver are basic proteins, M_r about 45,000. In addition there is a membrane-bound enzyme (present in the endoplasmic reticulum, outer mitochondrial membrane and peroxisomes) which is quite distinct from its cytosolic counterparts with respect to M_r, primary sequence and immunological properties.

See also Crystallins.

[W.B Jakoby & J.H.Keen *Trends in Biochemical Sciences* **2** (1977) 229–231; L.F.Chasseaud *Adv. Cancer Res.* **29** (1979) 175–274; R.Morgenstern & J.W.DePierre *Rev. Biochem. Toxicol.* **7** (1985) 67–104; B.Mannervik *Adv. Enzymol. Relat. Areas Mol. Biol.* **57** (1985) 357–417; *Glutathione Conjugation: Its Mechanism, and Biological Significance,* B.Ketterer & H.V.Sies (eds.), Academic Press Ltd. London, 1988]

Glutelins: a group of simple proteins from grain. They are generally insoluble in water, salt solutions and dilute ethanol, but at extreme pH values they do become soluble. They contain up to 45% glutamic acid. The best known are glutenin in wheat, orycenin in rice and hordenin in barley.

Gluten: a mixture of about equal parts of glutelins and prolamines. G. enables flour to form dough during bread making. Since wheat and rye flours contain G. they are suitable for preparing bread dough. Oat and rice grains, however, lack prolamines and their flour is not suitable for baking.

Glutenin: see Glutelins.

Gly: abb. for glycine.

Glyceollins: four phytoallexins produced by *Glycine max* L. in response to infection by *Phytophthora megasperma* var. *sojae,* or in response to treatment of soybean cotyledons or cell cultures with Elicitor (see). A particulate fraction from elicitor-treated soybean cotyledons or cell cultures contains dimethylallyl transferase activity, which catalyses the synthesis of 4- and 2-dimethylallyl-3,6a,9-trihydroxypterocarpan from (6a*S*, 11a*S*)-3,6a,9-trihydroxypterocarpan and dimethylallyl pyrophosphate. A biogenetic relationship between these pterocarpans (which also occur naturally in *Glycine max)* and the pterocarpan G. is strongly implied by these observations (Fig.). Microsomes from elicitor-challenged *Glycine* cell suspensions catalyse 6a-hydroxylation of 3,9-dihydroxypterocarpan to 3,6a,9-trihydroxypterocarpan. The reaction is dependent on NADPH and O_2. [U.Zahringer et al. *Z. Nalurforsch.* **36C** (1981) 234–241; M.-L. Hagmann et al. *Eur. J.Biochem.* **142** (1984) 127–131]

Glyceraldehyde 3-phosphate: see Triose phosphates.

Glycerate 2,3-*bis*phosphate, *2,3-diphosphoglycerate, 2,3-bisphosphoglycerate:* see Glycolysis, Hemoglobin, Rapoport-Luerbing shuttle.

Glycerate pathway: an anaplerotic pathway which enables the utilization of glyoxylate in plants and microorganisms. Two molecules of glyoxylate are converted to tartronate semialdehyde by tartronate

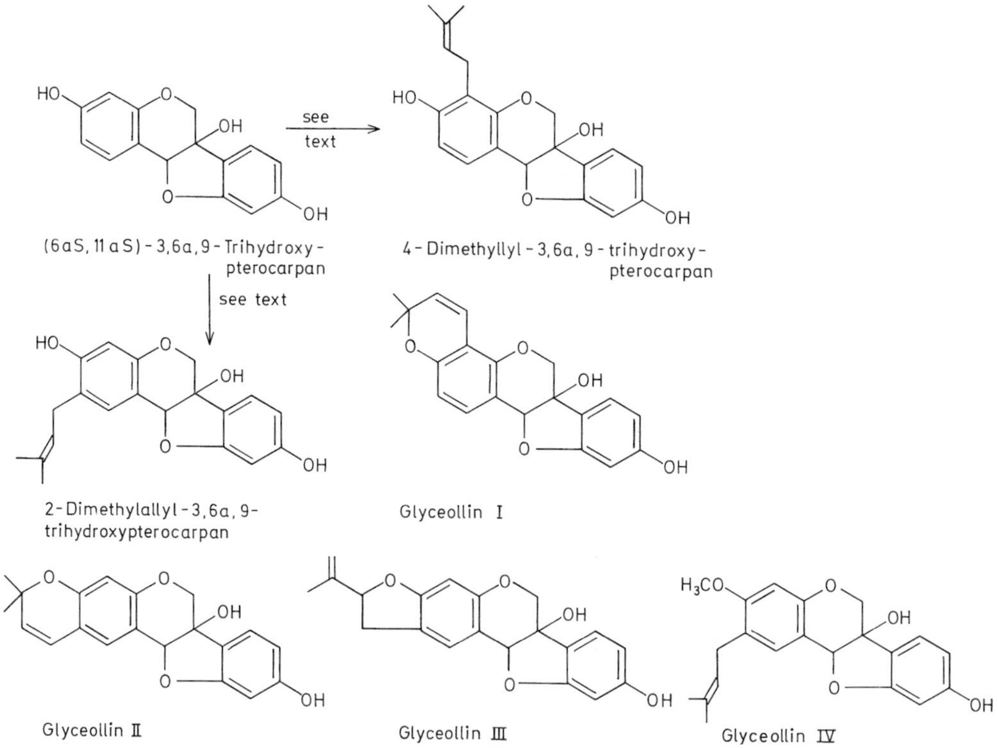

(6aS, 11aS)-3,6a,9-Trihydroxypterocarpan

4-Dimethyllyl-3,6a,9-trihydroxypterocarpan

2-Dimethylallyl-3,6a,9-trihydroxypterocarpan

Glyceollin I

Glyceollin II

Glyceollin III

Glyceollin IV

Pterocarpans from Glycine max L.

semialdehyde synthase (EC 4.1.1.47). The semialdehyde is then reduced to D-glycerate which is phosphorylated by glycerate kinase (EC 2.7.1.31) to 3-phosphoglycerate. This is converted by phosphoglyceromutase (EC 2.7.5.3) and enolase (EC 4.2.1.11) to phospho*enol*pyruvate, which enters general metabolism. Balance: 2glyoxylate + ATP + NAD(P)H + H^+ → phospho*enol*pyruvate + ADP + CO_2 + $NAD(P)^+$.

Glyceric acid: $HOCH_2$–CHOH–COOH, a dihydroxyacid which occurs in intermediary metabolism both free and phosphorylated. Glycerate is formed from glyoxylate via the Glycerate pathway (see) or from serine via hydroxypyruvate. The 1-, 2- and 3-monophosphates of G.a and 1,3-diphosphoglycerate are important intermediates in alcoholic fermentation, glycolysis and photosynthesis. Formation of 3-phospho-D-G.a. from glyceraldehyde 3-phosphate is linked to ATP formation.

Glycerides, ***acylglycerols:*** strictly speaking, any ester of glycerol. In normal usage, however, G. are esters of fatty acids with glycerol. The IUPAC-IUB Commission on Biochemical Nomenclature discourages this usage and proposes that the terms mono-, di- and triglyceride be replaced by mono-, di- and triacylglycerol. Single triglycerides (triacylglycerols) and mixtures of different triglycerides are also known as Neutral fats (see). Mono- and diacylglycerides usually occur only as metabolic intermediates.

Glyceroglycolipids: see Membrane lipids.

Glycerol, ***propan-1,2,3-triol:*** CH_2OH-CHOH-CH_2OH, a syrupy, sweet-tasting fluid. b.p. 290°C (d.). G. occurs widely as its esters in fats and fatty oils (see Glycerides, Neutral fats) and in phospholipids (see Membrane lipids). About 3% of G. is formed as a byproduct of alcoholic fermentation by reduction of dihydroxyacetone phosphate or glyceraldehyde 3-phosphate and hydrolysis of the phosphate group. It is produced commercially by adding sodium hydrogen sulfite to an alcoholic fermentation. Sodium hydrogen sulfite forms an addition compound with acetaldehyde; the latter is therefore no longer available for reduction, and dihydroxyacetone phosphate is reduced instead (2nd form of Neuberg fermentation).

Glycerol fermentation: see Neuberg's fermentation.

Glycerophospholipids: see Membrane lipids.

Glycinamide ribonucleotide, ***Gar:*** an intermediate of Purine biosynthesis (see).

Glycine, ***Gly, aminoacetic acid:*** H_2N-CH_2-COOH, the simplest proteogenic amino acid. M_r 75.07, m.p. 233–290°C (d.). Gly is not essential in the diet. Its amino nitrogen can easily be exchanged, so it is added to amino acid diets in large amounts to provide nitrogen for the synthesis of other amino acids. Gly is converted by transamination or oxidative deamination to glyoxylic acid, which is further metabolized to formic acid. It is synthesized from glyoxylate by transamination, or from serine by removal of the hydroxymethyl group. Glyoxylate and serine are relatively early products of photosynthesis (Fig. 1).

Reactions leading to glycine from glyoxylate are part of the *glycolic acid cycle.* The α-C atom of Gly is a source of Active one-carbon units (see), either directly or via glyoxylic acid. The direct cleavage is dependent on tetrahydrofolic acid (THF) and yields active formaldehyde: Gly + THF → $N^{5,10}$-methylene-THF + CO_2 + NH_3. The pathway starting with glyoxylic acid leads to active formic acid. The reaction 2Gly → L-serine + CO_2 + NH_3 requires that one of the Gly be converted to an active one-carbon unit, while the other serves as acceptor in the glycine-serine interconversion. Gly can be completely degraded, either via the Succinate-glycine cycle (see) or by the aminoethanol cycle (Fig. 2).

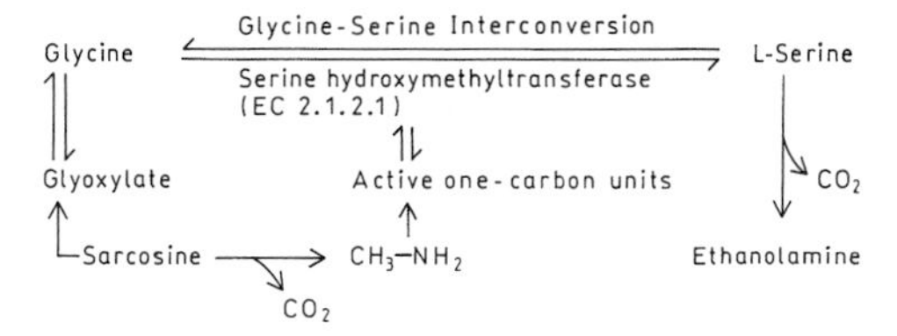

Fig. 2. *Glycine.* Aminoethanol (ethanolamine) cycle.

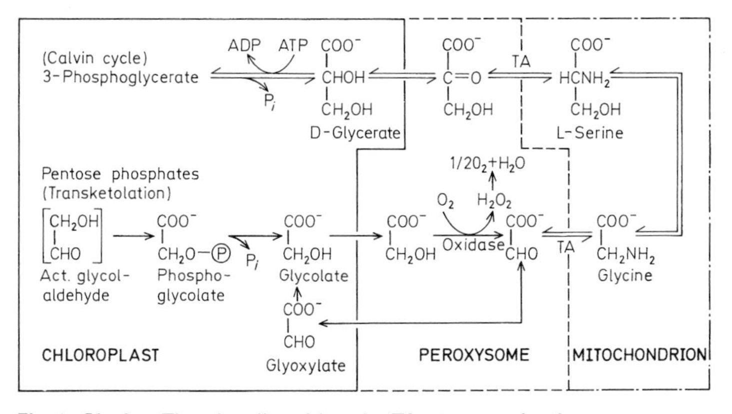

Fig. 1. *Glycine.* The glycolic acid cycle. TA = transamination.

The succinate-glycine cycle is the route by which the α-C atom and the amino nitrogen of Gly are incorporated into Porphyrins (see). The intact Gly molecule is incorporated into positions 4,5 and 7 of the purine ring, and also provides the C-atoms for positions 2 and 8 as one-carbon units (see Purine biosynthesis). Sarcosine and glycine betaine are formed by methylation of Gly. Glycocyamine, a precursor of creatine and creatinine, is synthesized by Transamidination (see) with L-arginine. Gly is also a component of glutathione. It was first isolated from gelatin in 1819 by Braconnot.

Glycine allantoin cycle, ***purine cycle:*** a series of reactions leading to the synthesis of urea (Fig.) in the lungfish *(Dipnoi)* and certain urea-accumulating plants. The glyoxylate and urea produced by purine degradation are reassimilated at different rates. Glyoxylate is converted to glycine, which can be reused for purine synthesis, thus closing the cycle. This cycle is particularly important, according to Krebs and Henseleit, for urea formation in lungfish (e.g. *Protopterus dolloi* and *P. aethiopicus*), which may accumulate up to 0.5 to 1 % of their dry weight in urea during estivation.

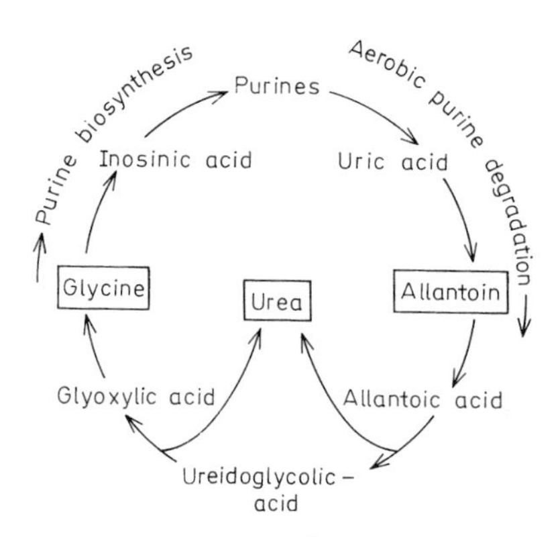

Glycine-allantoin cycle

Glycine cleavage system: an enzyme system accounting for a major proportion of glycine metabolism in vertebrate liver. It is composed of 4 proteins: 1. P-protein, which contains pyridoxal phosphate, 2. H-protein, which acts as a carrier and contains lipoic acid, 3. L-protein (lipoamide dehydrogenase) and 4. T-protein, a transferase dependent on tetrahydrofolate. The following reactions occur:

Sum: Glycine + THF + $NAD^+ \rightarrow N^{5,10}$-methylene-THF + $NH_3 + CO_2$ + NADH + H^+. [K. Fujiwara et al. *J. Biol. Chem.* **259** (1984) 10664–10668]

Glycine-succinate cycle: see Succinate-glycine cycle.

Glycinin: chief protein component of soybeans. It is stored in subcellular particles, known as "protein bodies". The dimer (M_r 350,000) consists of 12 subunits (M_r 28,500 each); 6 of these chains are acidic (pI 3.0–3.4) and 6 are basic (pI 8.0–8.5). G is structurally related to the protein Arachin (see).

Glycoalkaloids: see Steroid alkaloids, Saponins.

Glycocholic acids: see Bile acids.

Glycogen: an animal polysaccharide which consists of D-glucose units. Most of the glycosidic bonds are α-1,4, with α-1,6 linkages at branching points. The side chains consist of 6 to 12 glucose units. G. is therefore structurally similar to amylopectin, but is more (about 2-fold) highly branched than the latter. The M_r ranges from 1 to 16 million, which means a maximum of 10^5 glucose units. G. reacts with iodine to give a brown-violet color. It is most abundant in the liver (up to 10 %) and muscles (up to 1 %). Since it serves as a

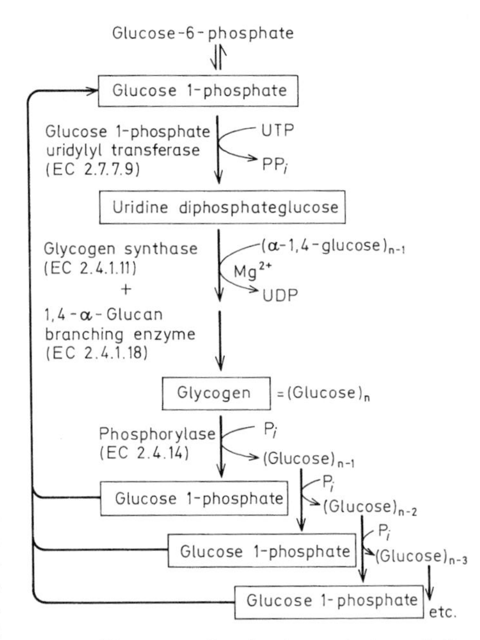

Fig. 1. *Glycogen.* Synthesis and degradation of glycogen.

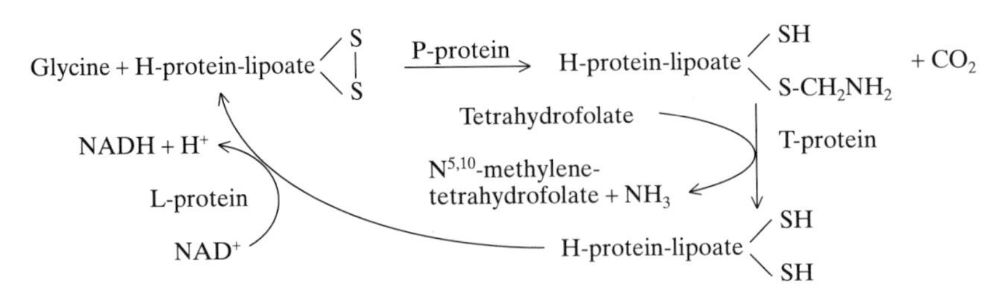

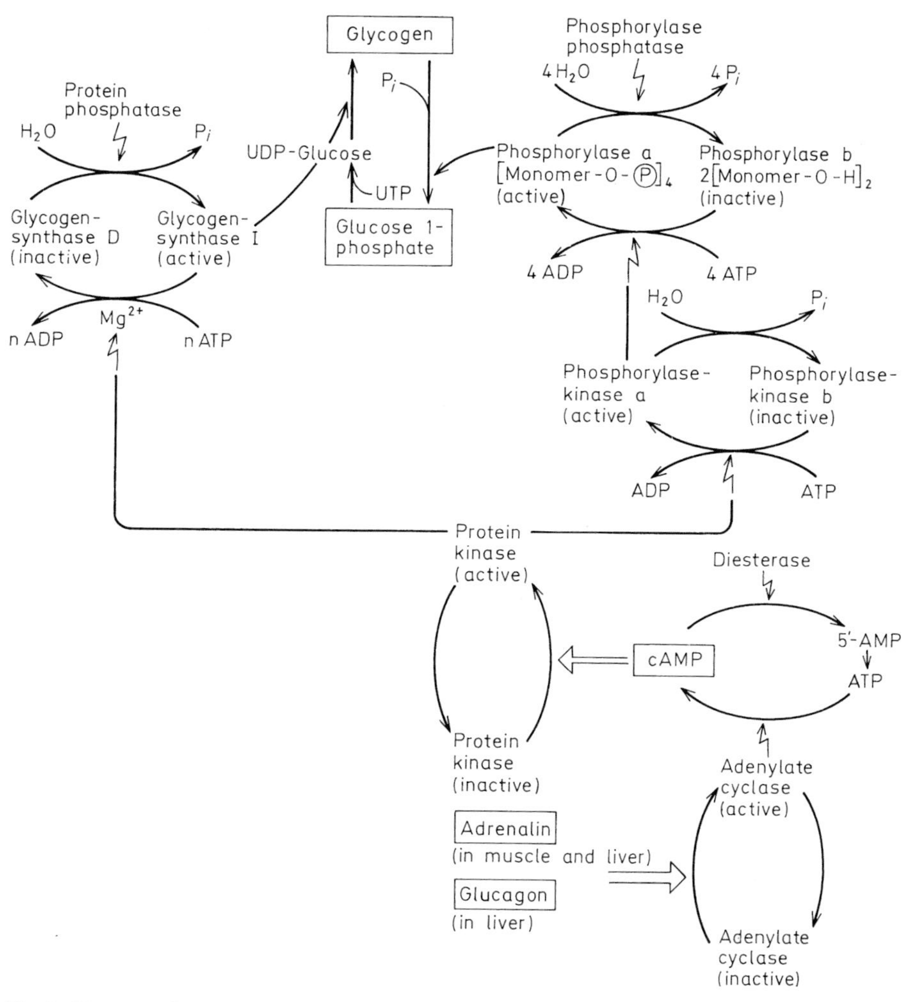

Fig. 2. *Glycogen.* Regulation of glycogen metabolism.

short-term storage material, it is subject to continuous synthesis and degradation.

Glycogen metabolism: the formation and breakdown of glycogen. The process has been well studied in mammalian muscle. Glycogen is synthesized from uridine diphosphate glucose (UDPG) and a starter molecule of the structure (α-1,4-glucosyl)$_n$ by glycogen synthase (EC 2.4.1.11). The branching of the molecule is accomplished by 1,4-α-glucan branching enzyme (EC 2.4.1.18), which transfers a segment of chain from a 4-OH to a 6-OH (the enzyme is sometimes called Q enzyme) (Fig. 1).

Glycogen degradation glycogenolysis) is started by the enzyme phosphorylase (EC 2.4.1.1) which transfers a glucose residue from the nonreducing end of the chain to inorganic phosphate. The product is glucose 1-phosphate, which is isomerized to glucose 6-phosphate, then enters Glycolysis (see). The 1,6 bonds are cleaved by a hydrolase rather than by phosphorylase. Complete degradation of glycogen therefore yields about 10 % free glucose and 90 % glucose 1-phosphate.

G.m. is controlled by interconversion of the enzymes involved (Fig. 2) (see Covalent modification of enzymes). Glycogen synthase is normally present in muscles in the active form, while phosphorylase is in its inactive form *(b).* However, phosphorylase *b* can be allosterically activated by a burst of AMP from muscle activity, so that phosphorylation can begin, although hormones and nervous stimulation are more important regulators of G. m. Adrenalin and glucagon activate adenylate cyclase (EC 4.6.1.1); the resulting cAMP activates a number of protein kinases, including one which simultaneously activates *phosphorylase kinase* (EC 2.7.1.38) and inactivates glycogen synthase; the inactive form of phosphorylase may also be allosterically activated by Ca^{2+} ions released as a result of muscle contraction. The activated phosphorylase kinase *a* in turn converts inactive phosphorylase *b* to active phosphorylase *a*. Phosphorylase *a* is sub-

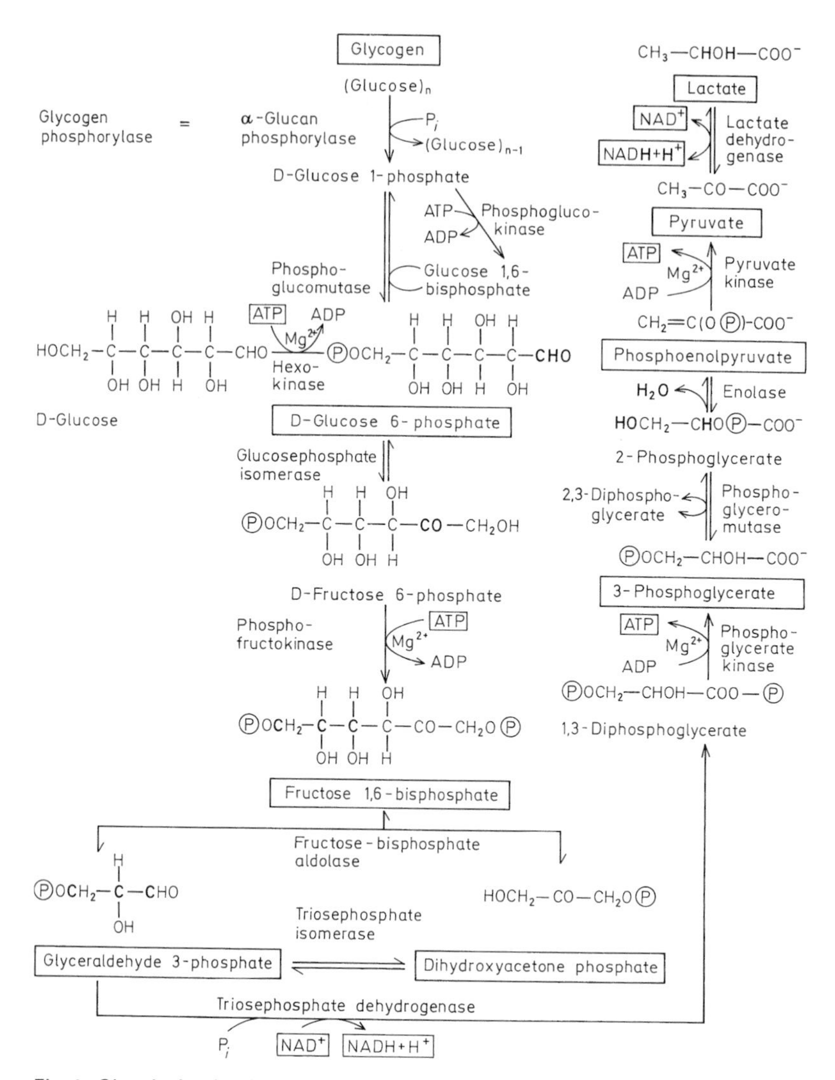

Fig. 1. *Glycolysis, the degradation of glycogen to lactate.*

ject to allosteric inhibition by glucose 6-phosphate, so that glycogen breakdown is slowed by a buildup of this product. In addition, glucose 6-phosphate allosterically activates glycogen synthase.

The system relaxes to the resting state as a number of phosphatases remove the phosphate groups added by the kinases. Less is known about these enzymes, but they must also be regulated. The total system is thus regulated both allosterically by substrates and products and by posttranslational modification of the enzymes in response to hormones and nerve impulses.

Glycogenolysis: see Glycogen metabolism.

Glycogen storage diseases: a group of conditions characterized by excessive storage of glycogen in the liver, muscles or other organs, due to hereditary lack of one of the enzymes of glycogen degradation.

Cori type I G. s. d. (von Gierke's disease). Glucose-6-phosphatase (EC 3.1.3.9) deficient. Enlarged liver and kidneys due to glycogen accumulation. Glycogen structure normal. No accumulation of glycogen in muscle. Hypoglycemia and lactic acidosis. Growth retarded. Prognosis favorable.

Cori type II G. s. d. (Pompe's disease). Amylo-1,4-α-glucosidase (EC 3.2.1.20) deficient. Glycogen structure normal. Glycogen accumulates in many tissues, including heart, which is enlarged. Most of the glycogen is segregated in large vacuoles (distended lysosomes) not seen in other forms of glycogenesis. The enzyme is lysosomal and glycogen enters lysosomes for this stage of degradation. Carbohydrate metabolism is not otherwise markedly affected. Death in first year from heart failure.

Cori type III G. s. d. (Forbes' disease).

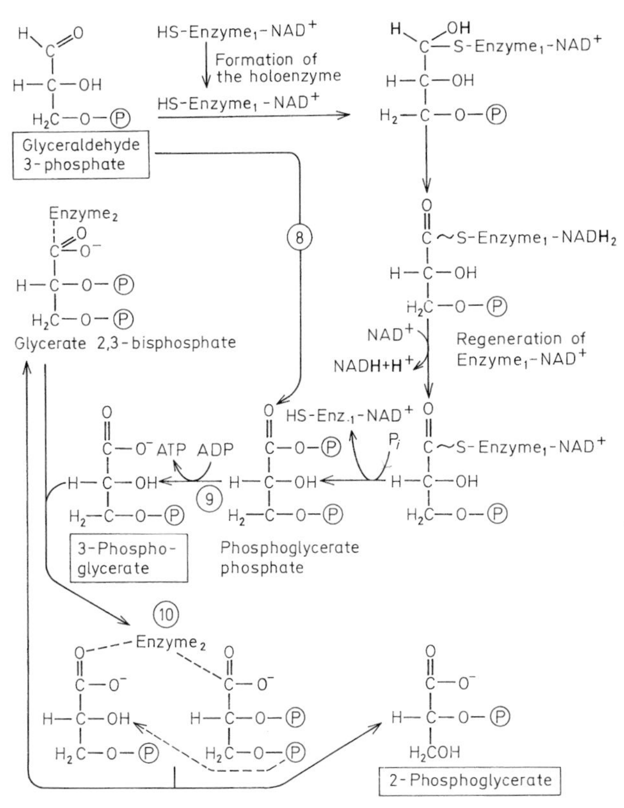

$Enzyme_1$ = triosephosphate dehydrogenase
$Enzyme_2$ = Phosphoglyceromutase

Fig. 2. *Mechanism of reactions 8 to 10 in glycolysis (see Table).*

Amylo-1,6-α-glucosidase, glycogen debranching enzyme (EC 3.2.1.33) deficient. Glycogen has short outer chains and is dextrin-like. Liver enlarged in infancy but may decrease in adolescence. Moderate hypoglycemia. Lactic acidosis mild or absent. Prognosis favorable.

Cori type IV G. s. d. (Andersen's disease, Amylopectinosis).

Amylo-(1,4 → 1,6)-transglucosidase, glycogen branching enzyme (EC 2.4.1.18) deficient. Glycogen has long inner and outer chains with very few branch points, and it is readily precipitated in the tissues. Liver cirrhosis. Fatal in early childhood.

Cori type V G. s. d. (McArdle's disease).

Muscle phosphorylase (EC 2.4.1.1) deficient. Glycogen structure normal. Glycogen accumulates only in muscle. Exercise causes muscle cramps; myoglobin from damaged muscle may appear in urine. Patients symptomless if they refrain from strenuous exercise. Prognosis favorable.

Cori type VI G. s. d. (Hers' disease).

Liver phosphorylase (EC 2.4.1.1) deficient. Glycogen structure normal. Enlarged liver. Moderate hypoglycemia and lactic acidosis. Prognosis favorable.

Type VII G. s. d. (not listed by Cori)

6-Phosphofructokinase (muscle isoenzyme) (EC 2.7.1.11) deficient. Glycogen structure normal. Glycogen accumulation in muscle. Abnormal muscular weakness and stiffness on vigorous or prolonged exercise. Fructose-6-phosphate and glucose-6-phosphate also accumulate. Half of enzyme activity in erythrocytes is lost (remaining 50 % is different isoenzyme). Mild hemolysis. Prognosis favorable.

Type VIII G. s. d. (not listed by Cori)

Phosphorylase kinase (EC 2.7.1.38) deficient. Glycogen structure normal. Liver enlarged. Prognosis favorable.

Glycolic acid, hydroxyacetic acid: $HOCH_2$-COOH, a hydroxycarboxylic acid (M_r 76.05, m.p. 80 °C) found in young plant tissue and green fruits, such as gooseberries grapes and apples. Glycolate is an intermediate in photosynthesis, where it is formed from active glycolaldehyde by Transketolation (see). G. a. is oxidized to glyoxylic acid by glycolate oxidase (EC 1.1.3.1). Glyoxylic acid may be converted to Glycine (see) or reduced back to G. a.

Glycolic acid cycle: see Glycine.

Glycolipids: see Membrane lipids.

Glycolysis, ***Embden-Meyerhof-Parnas pathway:*** the main pathway in all groups of organisms for the anaerobic degradation of carbohydrates (Fig. 1, Table). For each mole of glucose consumed, 150.7 kJ (36 kcal) energy are released. The organism ob-

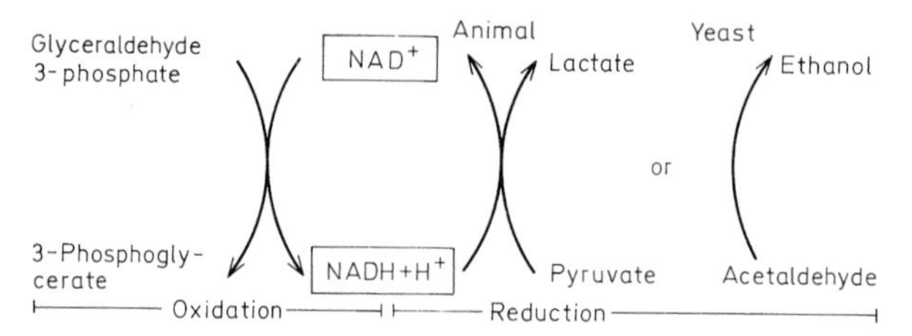

Fig. 3. *Oxidoreduction cycle of glycolysis.*

Reactions of glycolysis

No.	Equation	Enzyme	Inhibitors	$\Delta G^{o\prime}$ (free energy) kJ/mol (kcal/mol)
1	Starch or glycogen + nP_i $\rightarrow n$glucose-1-P	Phosphorylase (EC 2.4.1.1)		+3.06 (+0.73)
2	Glucose-1-P $\rightarrow$ glucose-6-P	Phosphoglucomutase (EC 2.7.5.1)	Fluoride, organic phosphates	−7.29 (−1.74)
3	Glucose + ATP $\rightarrow$ glucose-6-P + ADP	Hexokinase (EC 2.7.1.1)		−16.77 (−4.0)
4	Glucose-6-P $\rightleftharpoons$ Fructose-6-P	Glucosephosphate isomerase (EC 5.3.1.9)	2-Deoxyglucose-6-P	+1.68 (+0.4)
5	Fructose-6-P + ATP $\rightarrow$ Fructose-1,6-P_2	6-Phosphofructokinase (EC 2.7.1.11)	ATP, citrate	−14.2 (−3.40)
6	Fructose-1,6-P_2 $\rightleftharpoons$ dihydroxyacetone-P + glyceraldehyde-3-P	Fructose-bisphosphate aldolase (EC 4.1.2.13)	Chelating agents (only with bacterial enzymes)	+24.0 (+5.73)
7	Dihydroxyacetone-P $\rightleftharpoons$ glyceraldehyde-3-P	Triose phosphate isomerase (EC 5.3.1.1)		+7.66 (+1.83)
8	Glyceraldehyde-3-P + P_i + NAD^+ $\rightleftharpoons$ 1,3-P_2-glycerate + NADH + H^+	Glyceraldehyde phosphate dehydrogenase	Threose-2,4-P_2	+6.28 (+1.50)
9	1,3-P_2-Glycerate + ADP $\rightleftharpoons$ 3-P-glycerate + ATP	Phosphoglycerate kinase (EC 2.7.2.3)		−18.86 (−4.5)
10	3-P-Glycerate $\rightleftharpoons$ 2-P-glycerate	Phosphoglyceromutase (EC 2.7.5.3)		+4.44 (+1.06)
11	2-P-Glycerate $\rightleftharpoons$ Phospho*enol*pyruvate	Enolase (EC 4.2.1.11)	Ca^{2+}, F^-, P_i	+1.84 (+0.44)
12	Phosphoenolpyruvate + ADP $\rightleftharpoons$ pyruvate + ATP	Pyruvate kinase (EC 2.7.1.40)	Ca^{2+}, Na^+	−31.44 (−7.5)
13	Pyruvate + NADH + H^+ $\rightleftharpoons$ lactate + NAD^+	Lactate dehydrogenase (EC 1.1.1.27)	Oxamate	−25.1 (−6.00)

Note: For each mole of glucose, reactions 8 through 13 must be doubled, because two molecules of triose phosphate are produced from one mole of glucose. -P = phosphate; P_i = inorganic phosphate.

tains a net yield of 2 moles ATP per mole glucose.

The starting material is glycogen or starch, which is hydrolysed to glucose 1-phosphate or glucose monomers. G. can be divided into four phases: **1.** the formation of two molecules of triose phosphate (glyceraldehyde 3-phosphate and dihydroxyacetone phosphate) from one molecule of hexose; two molecules of ATP are consumed in this step. **2.** Dehydrogenation of the triose phosphates to 2-phosphoglycerate; NAD^+ is reduced in the process to NADH. One molecule of ATP is generated per triose phosphate, which makes up for the ATP invested in step 1. (Fig. 2). **3.** Conversion of 2-phosphoglycerate to pyruvate via phospho*enol*pyruvate; another ATP is generated here for each molecule of pyruvate. **4.** Reduction of pyruvate with regeneration of NAD^+. In muscle the pyruvate is converted to lactate, and in yeast it is reductively decarboxylated to ethanol (see Alcoholic fermentation). In an aerobically respiring cell, the NADH is oxidized ultimately by the respiratory chain (see Hydrogen metabolism), and the pyruvate is further oxidized in the Tricarboxylic acid cycle (see). Under anaerobic conditions G. can only be maintained continuously by the redox reactions shown in Fig. 3. Balance: Glucose ($C_6H_{12}O_6$) + $2P_i$ + 2ADP → 2lactate ($C_3H_6O_3$) + 2ATP.

If the starting material is glycogen, which is degraded to glucose 1-phosphate, the yield is 3ATP per glucose 1-phosphate consumed. The key enzyme of G. is 6-phosphofructokinase (EC 2.7.1.11), which is inhibited by high concentrations of ATP and is activated by ADP or AMP. Its product, fructose *bis*phosphate, activates pyruvate kinase. The Pasteur effect (see) is another form of regulation of G.

Glycophorin: see Erythrocyte membrane.

Glycoprotein hormones: a term applied to the two pituitary gonadotropins (luteinizing and follicle-stimulating hormones), choriogonadotropin and thyrotropin, all consisting of 2 peptide subunits (α and β), both of which are glycosylated and internally cross-linked by disulfide bonds. They are thus regarded as a family, and in the narrow sense other glycosylated protein hormones, e.g. erythropoietin, are excluded. It now seems appropriate to include inhibin (see) in the family of G. h. [J. G. Pierce & T F. Parsons *Annu. Rev. Biochem.* **50** (1981) 465–495]

Glycoproteins: proteins with covalently linked oligosaccharide. The protein of G. is translated from mRNA, and the addition of oligosaccharide represents a post-translational modification. This is in contrast to the peptidoglycans (see Murein), where the carbohydrate backbone is structurally dominant and cross-linked by relatively short polypeptide sequences synthesized independently of mRNA and ribosomes. G. are typically found in plants and animals, and not in bacteria (a notable exception is a G. from *Halobacterium,* see below). Most G. are either secreted into body fluids or are membrane proteins. They include many enzymes, most protein hormones, plasma proteins, all antibodies, complement factors, blood group and mucus components and many membrane proteins. The polypeptide chains of G. generally carry a number of short heterosaccharide chains, and these, in turn, almost always include *N*-acetylhexosamines and hexoses (usually galactose and/or mannose, less often, glucose). The last member of the chain is very often sialic acid or L-fucose. The oligosaccharide chains are often branched, and rarely contain more than 15 monomers (usual range: 2 to 10 monomers, corresponding to a M_r 540–3200). The number of saccharide chains on a polypeptide varies widely; ovalbumin contains only 3 % saccharide, whereas sheep submaxillary G. is 50 % carbohydrate. The sequences of sugar monomers are determined by the specificities of glycosyl transferases, and probably also by the concentrations of available activated sugars. Consequently, G. display microheterogeneity in their saccharide moieties.

Properties and functions. Due to their high viscosities, the G. serve as lubricants and protective agents, e.g. against proteolytic enzymes, bacteria and viruses. They play a role in cellular adhesion and contact inhibition of cell growth in tissue culture. They are also responsible for cellular recognition of foreign tissue.

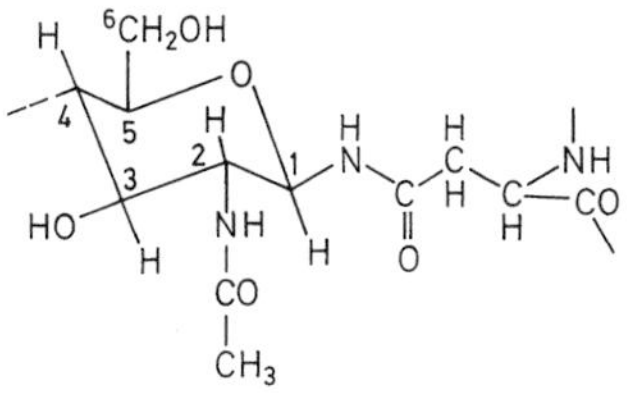

Fig. 1. *Glycoproteins.* Linkage of *N*-acetylglucosamine and an L-asparagine residue, representing the site of attachment of *N*-glycans in glycoproteins.

G. are essential components of receptors for virus and plant agglutinins, and of blood group substances. The siderophilins and ceruloplasmin are G. Some G. are membrane carrier proteins. The carbohydrate portion of gonadotropic hormones is essential to their biological activity; in many cases, selective removal of the terminal sialic acid residues inactivates the hormone. The carbohydrate evidently serves as a label for recognition by a receptor.

G. appear to be synthesized on the rough endoplasmic reticulum, the growing polypeptide chain being extended through the membrane from the cytoplasmic side. Sugar units are added to the chain in the cisternae of the endoplasmic reticulum (ER), preventing it from recrossing the membrane. In the case of *N*-glycans (see below), a standard complement of sugars is added in the ER, consisting of 2 N-acetylglucosamine, 9 mannose and 3 glucose units. Vesicles which bud off the ER carry the G. to the Golgi apparatus (see), where some of the sugar units are removed and others are added, to produce the structure appropriate to the protein. The processed G. may then be excreted, carried to one of the cell organelles, or incorporated into one of the cell membranes.

The covalent protein-carbohydrate linkage may be *N*-glycosidic or *O*-glycosidic. *N*-Glycosidically linked oligosaccharides are called *N*-glycans; *O*-glycosidically linked oligosaccharides are called *O*-glycans.

N-Glycosidic linkage. Only one type of *N*-glycosidic linkage is known in animal and plant G. This occurs between the amide nitrogen of an asparaginyl residue and C1 of an *N*-acetylglucosamine residue (GlcNAc) (Fig. 1). A sulfated G. in the cell wall of *Halobacterium* contains two types of carbohydrate: 1. a sulfated, repetitive, high-M_r saccharide resembling an animal glycosaminoglycan, and 2. a sulfated, low M_r saccharide with the composition 1 glucose:3 glucuronic acid:3 sulfate. In this compound, the glucose is linked *N*-glycosidically to the amide nitrogen of an asparaginyl residue. This is the first report of an asparaginylglucose linkage in any G. [F. Wieland et al. *Proc. Nat. Acad. Sci. USA* **80** (1983) 5470–5474]

Most N-glycans contain a common core (Fig. 3). Each terminal mannose of the core structure is regarded as an antenna for extension. If each terminal mannose carries at least one GlcNAc residue, the structure is known as *complex antennary.* If only one core mannose is substituted by GlcNAc, and the other is extended by mannose residues, the structure is known as *biantennary hybrid.* Attachment of more

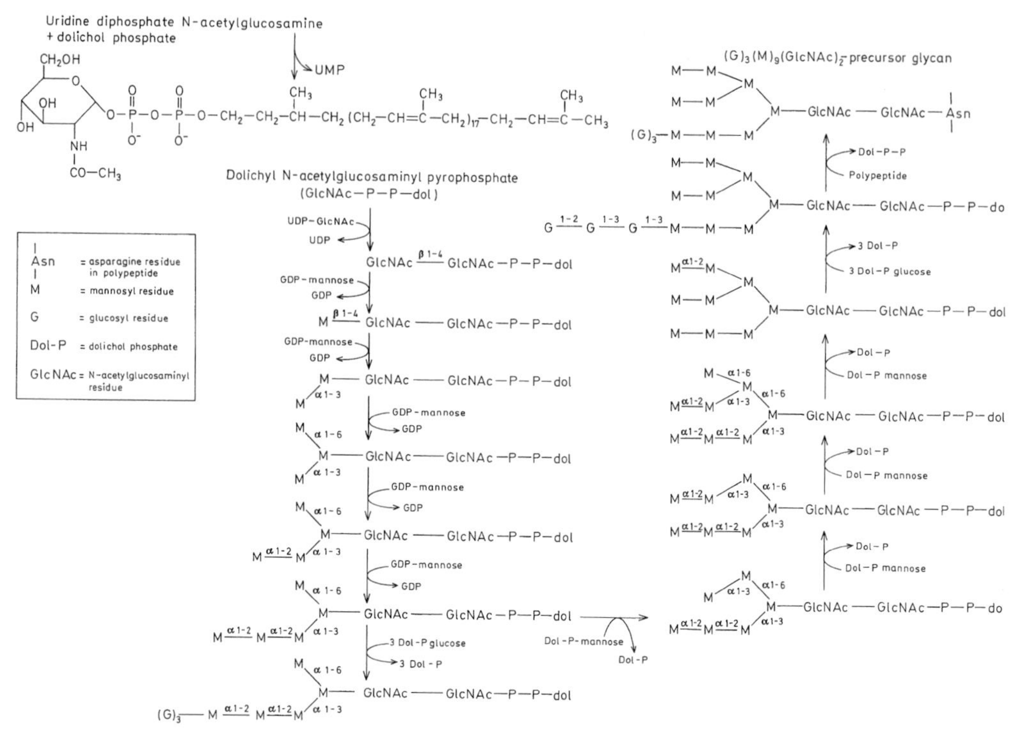

Fig. 2. *Glycoproteins.* Biosynthesis of the precursor of asparagine-linked *N*-glycans.

than one GlcNAc to a core terminal mannose results in *triantennary, tetra-antennary,* etc. structures. A *bisected antennary* structure possesses a GlcNAc residue linked $\beta 1 \rightarrow 4$ to the central mannose of the common core (Fig. 4).

Biosynthesis of N-glycans. A large precursor of the oligosaccharide is first synthesized on the lipid carrier, Dolichol phosphate (see). Glycan synthesis is initiated by interaction of dolichol phosphate with uridine diphosphate *N*-acetylglucosamine (UDPGlcNAc) to form dolichyl *N*-acetylglucosaminylpyrophosphate (Fig. 2). As illustrated in Fig. 2, subsequent attachment of α-mannosyl and finally glucose units is an ordered process, probably requiring separate, specific mannosyl and glucosyl transferases for each stage. The entire final product is transferred to a specific asparaginyl residue of a polypeptide acceptor (with release of dolichol pyrophosphate). Small peptides containing asparagine, as well as native proteins, act as acceptors; the enzyme is specific for the structure: -Asn-X-Ser (or Thr)-, where X may be any amino acid except proline and possibly aspartic acid, and the amino group of Asn and the carboxyl group of Ser (or Thr) must be in a peptide linkage or otherwise blocked.

Processing (in the Golgi apparatus) then commences with removal of the three glucose residues by specific α-glucosidases, followed by ordered removal of α-mannose residues by specific α-mannosidases, thereby accounting for all high-mannose structures encountered in G.

Complex and antennary *N*-glycans are derived from the $(man)_5$ high-mannose structure in a series of reactions initiated by transferase I (UDP-GlcNAc: α-D-mannoside(GlcNAc to Manα1–3) βl-2-GlcNAc transferase I) (Fig. 4). After attachment of the first GlcNAc to the $\alpha 1 \rightarrow 3$-linked core mannose of the $(man)_5$ structure, mannose is removed by α-mannosidases specific for $\alpha 1 \rightarrow 3$ and $\alpha 1 \rightarrow 6$ linkages (different enzymes from those involved in earlier processing). Further GlcNAc residues are introduced by specific GlcNAc transferases II, III, and IV, and fucose residues are attached to core GlcNAc residues by fucose transferase [GDP-Fuc:β-*N*-acetylglucosaminide (Fuc to Asn-linked GlcNAc) $\alpha 1 \rightarrow 6$-fucosyl transferase]. Galactose units are transferred from UDP-galactose to C4 of GlcNAc residues, producing the *N*-acetyllactosamine sequence present in many *N*-glycans. Sialic acid residues (which occur in C3 or C6 linkage to penultimate galactose residues) are transferred from CMP-sialic acid by the action of sialyl transferase. After incorporation into the glycan structure, sialic acid may be modified by hydroxylation of the *N*-acetyl group to *N*-glycolyl, or by acetylation.

O-Glycans. In the O-glycans, C1 of a sugar residue is linked glycosidically to the hydroxyl group of the side chain of serine, threonine, hydroxylysine or hydroxyproline. Arabinose linked glycosidically to hydroxyproline is found very commonly in plant G., but has not been identified in G. from any other type of organism. D-Galactose linked β glycosidically to

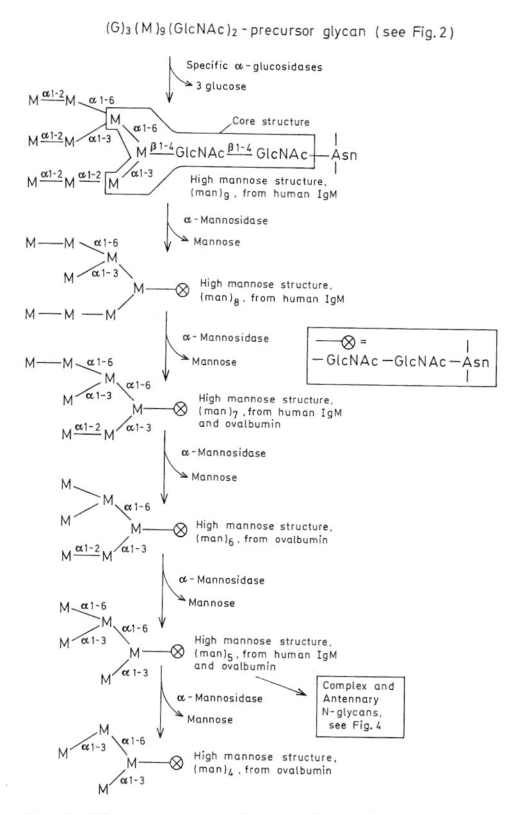

Fig. 3. *Glycoproteins.* Processing of the asparagine-linked precursor carbohydrate. Removal of the three glucose residues produces a high-mannose structure, designated $(man)_9$. Subsequent removal of mannose produces high-mannose structures $(man)_8$ to $(man)_4$.

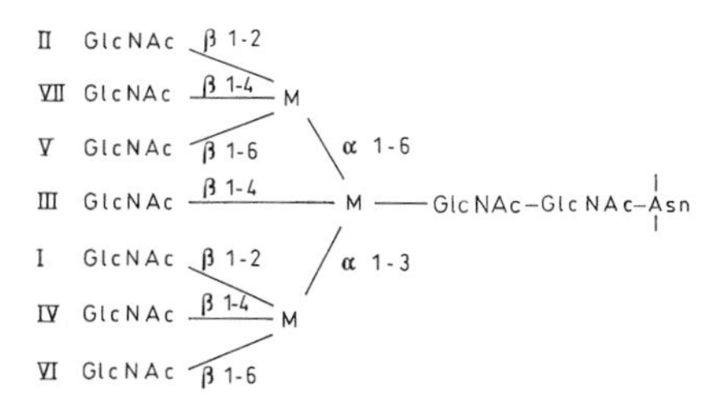

Fig. 4 a. Glycoproteins. Numbering system for antennas and GlcNAc transferases.

hydroxylysine has been found only in animal collagen. Collagen also contains 2-*O*-α-D-glucopyranosyl D-galactose linked to hydroxylysine. Galactose linked glycosidically to serine is found in the plant G. extensin, and in the collagen of the worms, *Lumbricus* and *Nereis.* In the G. of yeasts and fungi, carbohydrate chains are attached by a glycosidic linkage between mannose and serine or threonine, and this type of linkage also makes a minor contribution to mammalian G. structure. Some established reactions of *O*-glycan synthesis in submaxillary gland G. are shown in Fig. 5. [R. C. Hughes, *Glycoproteins (Outline Studies in Biology)* W. J. Brammer & M. Edidin, eds. (Chapman & Hall, 1983); H. Schachter et al. *Can. J. Biochem. Cell Biol.* **61** (1983) 1049–1066]

The disaccharide *O*-glycan (II) accumulates in sheep salivary gland, which contains only low activity of specific galactosyl transferase. II is not a substrate for other transferases.

Pig submaxillary gland contains high activities of various transferases, and longer glycans are present, e.g. III and IV are prominent.

Formation of the GlcNAc β1–6 linkage in V possibly initiates a series of reactions leading to synthesis of human blood group active products of secretory cells (see Blood group antigens). The GlcNAc transferase responsible for synthesis of V is highly specific for the presence of β1–3-linked unsubstituted Gal; the enzyme shows only very low activity toward VI. Glycan VIII is human blood group H-active. IX and X show human blood group A antigenic activity.

Glycosaminoglycan: see Mucopolysaccharide.

Glycosaminoglycan degradation: see Lysosomal storage diseases.

Glycosidases: a group of hydrolases which attack glycosidic bonds in carbohydrates, glycoproteins and glycolipids. G. are not highly specific. Usually they distinguish only the type of bond, e.g. *O*- or *N*-glycosidic, and its configuration (α or β). See Amylases, Taka amylase, Neuraminidase, Oligo-1,6-glucosidase, Amylo-1,6-glucosidase, Disaccharidases, Cellulases, Lysozyme and Invertase.

Glycosides: a group of compounds in which mono- or oligosaccharide units are linked by acetal bonds with hydroxyl groups of alcohols or phenols (*O*-glycosides) or with amino groups (*N*-glycosides). Acid hydrolysis cleaves these compounds into a sugar and a noncarbohydrate portion called the aglycon or genin. The Saponins (see), cardiac glycosides and cyanogenic glycosides are examples of *O*-glycosides; Nucleic acids (see) and Nucleosides (see) are *N*-glycosides.

Glycosylceramides: see Membrane lipids.

Glycyrrhetic acid, *glycyrrhetin:* a pentacyclic triterpene with carboxylic acid and ketone functions; M_r 470.7, m.p. 300 °C, $[\alpha]_D + 98°$ ($CHCl_3$). G. a. is the aglycon of glycyrrhizic acid, an extremely sweet principle from the root of licorice, *Glycyrrhiza glabra* L., and it is the aglycon of Saponins (see) from other plants, e.g. the bark of *Pradosia latescens* and rhizomes of the fern *Polypodium vulgare.* It differs structurally from β-amyrin in having an 11-keto and a 30-carboxyl group (see Amyrin, Fig.).

Glyoxalase, aldoketomutase: a system of enzymes found in many organisms. It consists of lactoyl-glutathione lyase (G.I) (EC 4.4.1.5) which catalyses condensation of methylglyoxal and glutathione to form *S*-lactoyl-glutathione, and hydroxyacylglutathione hydrolases (G.II) (EC 3.1.2.6) which catalyses hydrolysis of the condensation product to lactate and glutathione. The significance of the system is unknown.

Glyoxylate carboligase: see Tartronic semialdehyde synthase.

Glyoxylate cycle, *Krebs-Kornberg cycle:* an alternative to the tricarboxylic acid cycle found in micro-

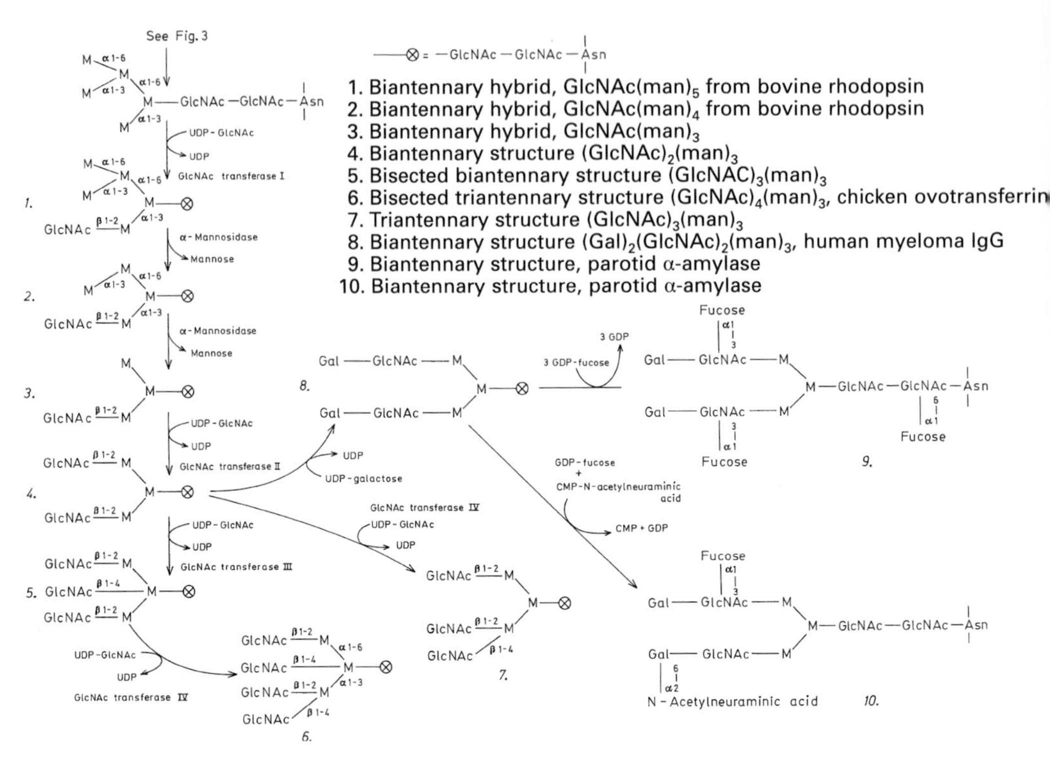

Fig. 4b. Glycoproteins. Reactions leading to formation of complex and antennary *N*-glycans.

organisms and plants. In the G.c., oxaloacetate is generated from acetyl-CoA (Fig.). The key enzymes are isocitrate lyase (isocitratase) (EC4.1.3.1) and malate synthase (EC 4.1.3.2). Balance: 2Acetyl-CoA + $NAD^+ + 2H_2O \rightarrow$ succinate + 2CoA + NADH + H^+. Intermediates of the G.c. serve as starting materials for various synthetic pathways. Succinate is especially important as the precursor for gluconeogenesis. The G.c. enables plant seedlings to utilize their fat reserves for the net synthesis of carbohydrate, and it enables microorganisms to grow on fatty acids or acetate as their sole carbon sources. Animals lack the G.c. (isocitrate lyase and malate synthase are absent) and are therefore unable to perform the net conversion of fat into carbohydrate.

Glyoxylate cycle

Glyoxylic acid, ***oxoacetic acid, glyoxalic acid:*** CHO-COOH, a carboxylic acid found in green fruits, seedlings and young leaves. In some plants (maple, borage, horse chestnut) G.a. is present in the form of allantoin and allantoic acid which arise in the course of purine catabolism. G.a. is synthesized by transamination or oxidative deamination of glycine, or from sarcosine. Decarboxylation of G.a. yields active formate. G.a. is the starting point of the Glyoxylate cycle (see).

GMP: acronym of guanosine 5′-monophosphate.

GMP reductase (EC 1.6.6.8): a flavin enzyme which catalyses the conversion of guanosine 5′-monophosphate to inosinic acid in a single step. It is part of the system for converting guanine to adenine compounds.

Goitrogens, ***antithyroid compounds:*** substances which inhibit iodine peroxidase (conversion of iodide to "active iodine": $H_2O_2 + 2I^- + 2H^+ \rightarrow 2$"I"$ + 2H_2O$), the iodination of tyrosine residues and the coupling of monoiodotyrosine and diiodo-tyrosine residues to form T_3 and T_4 (see Thyroxin). The resulting low plasma levels of T_3 and T_4 stimulate the release of thyrotropin from the anterior pituitary. This, in turn, results in a compensatory hypertrophy of the thyroid gland. Such an enlargement, without inflammation or malignancy, is called a goiter. Some G. are 2-

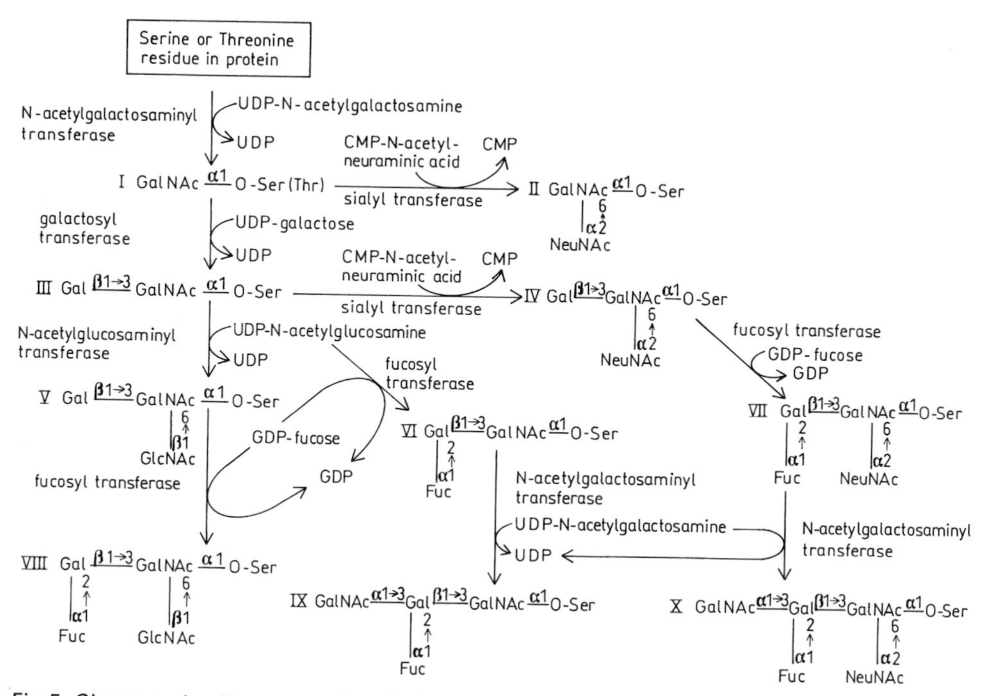

Fig. 5. *Glycoproteins.* Known reactions in the synthesis of submaxillary gland glycoproteins. The specificity of *N*-acetylgalactosaminyl transferase responsible for synthesis of I was studied using basic myelin protein [an unusual substrate which is not normally glycosylated; partial structure: Val(94) Thr-Pro-Arg-Thr-Pro-Pro-Pro(101)-]. Glycosylation is specific for Thr 98 (Thr 95 is not glycosylated); a proline-rich environment appears to be necessary around the glycosylated hydroxyamino acid residue.

thiouracil, thiourea, sulfaguanidine, propylthiourea, 2-mercaptoimidazole, 5-vinyl-2-thiooxazolidone (see Glucosinolate) and allylthiourea (in mustard).

Golgi apparatus: a complex of membranous structures in eukaryotic cells. The G.a. is the site of protein processing, in particular protein glycosylation. In 1898, Golgi observed a net-like structure which was made visible by staining with OsO_4 or $AgNO_3$, but this is not the same as the G.a seen in the electron microscope.

The G.a. in differentiated cells consists of three types of structure: cisternae (shaped like flattened balloons), vesicles which form on the edges of cisternae, and groups of vacuoles, which are also called "dictyosomes". The cisternae are stacked; the minimal number in a stack is 4, while in plant cells as many as 20 cisternae may comprise a stack. Mammalian cells typically have 5 or 6 cisternae per stack. There are at least three functionally different compartments in each stack. The cisterna(e) on the *cis* or protein-receiving side of the stack are characterized by the presence of mannosidase I; those in the middle, by mannosidase II and *N*-acetylglucosamine transferases; and those on the *trans* or protein-releasing side, by galactosyl and sialyl transferases (see Glycoproteins for discussion of these enzymes). Glycoproteins of the *N*-glycan type receive an oligosaccharide chain in the lumen of the endoplasmic reticulum. The same chain is added to all the proteins; it consists of 2 *N*-acetylglucosaminyl, 9 mannosyl and 3 glucosyl units. These proteins are carried in vesicles which bud off from the endoplasmic reticulum to the *cis* face of the G.a. Here the proteins destined for different cellular compartments, e.g. lysosomes, plasmalemma, or secretion are apparently recognized and processed differently, depending on their destinations. Lysosomal proteins are phosphorylated on the carbohydrate chains, but otherwise not greatly modified. Secretory and plasmalemma proteins typically lose all their mannose units and receive other carbohydrate units in a multistep process.

Proteins are transported from one cisterna of the G.a. to the next by the vesicles which form at the edges of the cisternae. Experiments with fused cells have indicated that proteins do not necessarily pass through all the steps of glycosylation in the same stack; thus proteins radioactively labeled in mutant cells (which are unable to complete processing) may be completed correctly if the mutant cells are fused with normal ones containing no radioactivity. This implies transit from the mutant to the normal G.a. stacks. At the end of processing, the glycoproteins may be stored in the dictyosomes, secreted, transported to the appropriate cell organelles, or incorporated into the plasmalemma. [W.G. Dunphy et al. ***Cell* 40** (1985) 463–472]

Gonadal hormones, *sex hormones:* steroid hormones produced in the male (testes) and female (ovaries) gonads. The G.h. determine the male or female character of an organism, in that they effect the normal development and function of the sex organs and the expression of the secondaly sexual traits. They are classified as *male* G.h. (see Androgens) and *female* G.h. The latter fall into two groups with different physiological effects, the Estrogens (see) and the Gestagens (see).

Gonadotropin releasing hormone: see Releasing hormones.

Gonadotropins: a group of glycoprotein hormones from the anterior lobe of the pituitary and the placenta. They stimulate the gonads to growth and production of sex-specific hormones (female, see Estrogens and Gestagens; male, see Androgens). The G. include Follitropin, or Follicle-stimulating hormone (FSH) (see), Lutropin, or Luteinizing hormone (LH) (see), Human menopausal gonadotropin (Urogonadotropin, hMG) (see), Prolactin (see), and Chorionic gonadotropin (Choriogonadotropin, hCG) (see). Follitropin, choriogonadotropin and urogonadotropin consist of α and β chains. The primary sequences of the α chains are nearly identical, while the β chains are hormone-specific, which suggests that the three hormones have evolved from a common ancestor. Lutropin and follitropin are produced in the anterior lobe of the pituitary, though not continuously. Choriogonadotropin is synthesized in the syncytiotrophoblasts of the placenta, maximal secretion occurring in the second month of pregnancy.

Synthesis of lutropin and follitropin is controlled by gonadotropin releasing hormone from the hypothalamus, which in turn is influenced by the sexual center in the brain. The sexual center, hypothalamus, anterior lobe of the pituitary and gonads or placenta form a feedback loop in which the sex hormones from the gonads exert negative feedback on the brain. Females display a cyclic pattern of variation in the hormone levels (menstrual cycle). In males, the variations are acyclic. In addition to these long-term variations, there are diurnal fluctuations in the hormone levels.

The G. interact with specific receptors in the theca and follicle cells of the ovary and corpus luteum and with the interstitial cells of Leydig in the testes. cAMP is released as a second messenger to stimulate the production of steroid sex hormones. G. can be determined by radioimmunoassay, the radioligand-hormone receptor method, or by biological assays (see Hormones).

Gossypol: an aromatic triterpene from cotton seed *(Gossypum hirsutum)*. M_r 518.54, m.p. 184 °C from ether, 199 °C from chloroform, 214 °C from light petroleum. G. is biosynthesized from mevalonic acid via neryl pyrophosphate and *cis,cis*-farnesyl pyrophosphate. It is somewhat poisonous, and has been suggested for use as an insecticide.

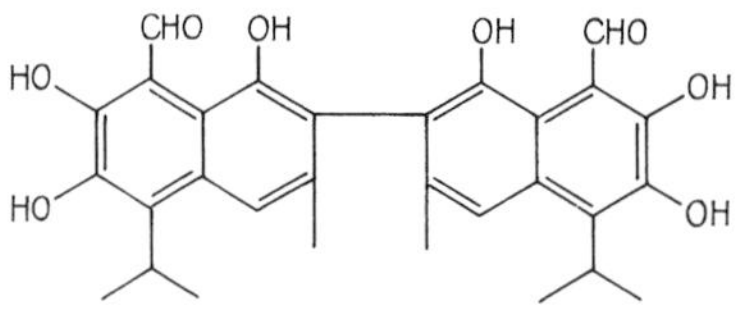

Gossypol

GOT: acronym of glutamate-oxaloactate transaminase. See Transaminases.

Gougerotin, *aspiculamycin, asteromycin:* 1-cytosinyl-4-sarcosyl-D-serylamino-1,4-dideoxy-β-D-glucopyranuramide, an important pyrimidine antibiotic (see Nucleoside antibiotics) synthesized by *Streptomyces gougeroti*. It inhibits protein biosynthesis on both eukaryotic and prokaryotic ribosomes.

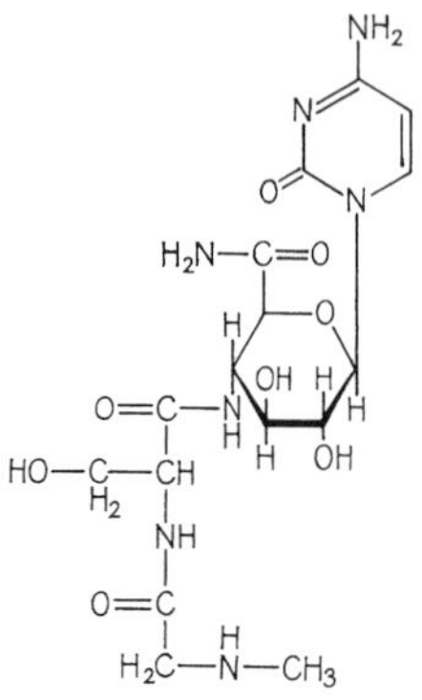

Gougerotin

G-proteins: see GTPases.

GPT: acronym of glutamate-pyruvate transaminase. See Transaminases.

Gradient elution: see Column chromatography.

Gramicidins: cyclic or linear peptide antibiotics produced by *Bacillus brevis*. Gramicidin S is cyclo (-D–Phe–L–Pro–L–Val–L–Orn–L–Leu–)$_2$. It is biosynthesized by small, soluble system of enzymes. The first enzyme activates L-Phe, and the second activates the other four amino acids by transferring them to the thiol groups of pantotheine molecules covalently bound to the enzyme. The activated phenylalanine, which at some point is converted to D-Phe, attacks the activated L-Pro to initiate synthesis. The Pro attacks the Val, and so on, to form the pentapeptide; two pentapeptides are then joined and cyclized. Gramicidin S is effective against Gram-positive but not Gram-negative bacteria. Its structure is thought to be an antiparallel pleated sheet.

Gramicidin D is a complex of four components, A, B, C and D. Gramicidin A is a linear peptide of alternating D and L amino acids: HCO-L-Val-Gly-L-Ala-D-Leu-L-Ala-D-Val-L-Val-D-Val-(L-Trp-D-Leu)$_3$-L-Trp-N-HCH_2CH_2OH [R. Sarges & B. Witkop, *J. Am. Chem. Soc.* **86** (1964) 1862.] A second series (the isoleucine gramicidins) has an isoleucine instead of valine in position 1. Gramicidin B has a Phe in position 11, while

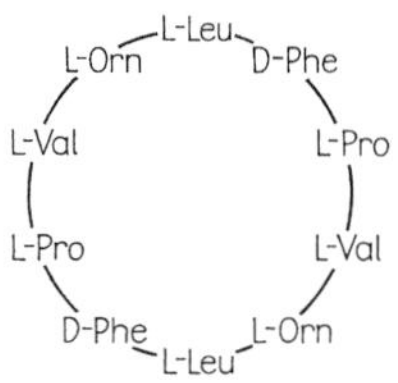

Gramicidin S

gramicidin C has Tyr at this position. The linear gramicidins act as ionophores, forming channels through biological membranes which allow passage of monovalent cations, thus allowing equilibration of intra- and extracellular Na^+ and K^+. The gramicidin A channel is a single-stranded helix comprising two gramicidin A molecules . The two CHO residues are adjacent in the middle of the membrane. [D.W. Urry et al. *Science* **221** (1983) 1064–1067]

Granatan: see Pseudopelletierine.

Granulocyte-macrophage colony stimulating factors: see Colony stimulating factors.

Granulomatous disease: see Inborn errors of metabolism.

GRH: see Releasing hormones.

Grisein: an iron-containing antibiotic synthesized by *Streptomyces griseus.* It is a cyclic polypeptide containing cytosine. The iron ions are strongly bound as hydroxamate-iron(III) complexes. G. has the same functional groups as Albomycin (see), and like the latter, G. is a sideromycin. It is especially effective against Gram-negative bacteria, but is not effective against fungi.

Griseofulvin: an antifungal agent synthesized by *Penicillin griseofulvi.* It is a polyketide synthesized from one molecule of acetyl-CoA and 6 molecules of malonyl-CoA (Fig.).

1 CH_3—CO~CoA + 6 CH_2(COOH)—CO~CoA → Benzophenone derivative → Griseophenone A radical → Griseofulvin

Biosynthesis of griseofulvin

Group transfer: enzymatic transfer of a functional group from one molecule to another (Table). Coenzymes (see) often serve as carriers of the groups, hence the term 'transport metabolites'. Hydrolysis and phosphorolysis can be regarded as special cases of G.t. in which the group is transferred to water or a phosphate ion.

Group transfer reactions

Group transferred		EC number
$-CH_3$	Methyl	2.1.1
$-CH_2OH$	Hydroxymethyl, formyl, etc.	2.1.2
$-C(=O)-OH$	Carboxyl and carbamoyl	2.1.3
$R-C(=O)-$	Aldehyde or ketone residues	2.2.1
$R-C(=O)-$	Acyltransferases	
	Acyl	2.3.1
	Aminoacyl	2.3.2
(glycosyl, CH_2OH)	Glucosyltransferases	
	Hexosyl	2.4.1
	Pentosyl	2.4.2
	Other glycosyl groups	2.4.3
$R-CH_2-$	Alkyl or aryl groups other than methyl	2.5.1
	Nitrogenous groups	
$-NH_2$	Amino-	2.6.1
$=N-OH$	Oximino-	2.6.3
H_2PO_3-	Phosphorus-containing groups	2.7
$-S$, $-SO_2$	Sulfur-containing groups	2.8

Growth: an irreversible increase in the mass of living material, usually accompanied by an increase in the size of a cell or organism, as well as an increase in the number of cells. Bacterial G. is measured as the increase of total cell material, or the increase of cell numbers, or both (see Cultivation of microorganisms); a strict relationship between cell numbers and cell mass does not necessarily exist (see Fermentation techniques).

In plant physiology, it is important to distinguish between G. by cell division and G. by cell stretching or extension. In cell extension, water is taken up powerfully by the cell, and the cell wall is physically stretched along the length of the cell. This phenomenon is particularly apparent in the sudden appearance of fruiting bodies of fungi, in the opening of flowers and in the thrusting of shoots.

Total cell numbers in cell suspensions can be determined microscopically in a counting chamber, or with the aid of a membrane filter. Cell mass may be determined directly by weighing, and quoted as the dry weight or fresh weight (dry material + cell water). Other suitable criteria for total cell material are the quantity of protein and the total nitrogen content. The concentration of cells in suspension can be measured indirectly from the absorption of light, i.e. turbidity (500–600 nm in a conventional spectrophotometer), or by the light scattering, i.e. by nephelometry. Such methods must be calibrated against one of the absolute methods of cell counting or mass determination. There may be a difference between *total cell count* and *viable cell count,* i.e. some cells may be dead.

The viable cell count of bacterial cultures is determined by plating out suitable dilutions of culture on solid growth medium and counting the number of colonies formed after incubation. In addition to a correct supply of nutrients, bacterial G. depends on the pH, temperature and osmotic pressure of the culture medium, and on the degree of aeration (aerobes) or the completeness of anaerobiosis (anaerobes).

Certain chemical substances cause a concentration-dependent inhibition of G. In bacteriology, these are classified as *bacteriostatic* and *bacteriocidal* substances. Bacteriostatic agents inhibit growth, but growth resumes after their removal. A bacteriocidal agent (or bacteriocide) kills the bacterium. The underlying mechanisms of G. inhibition are many and varied, depending on the inhibitor.

Growth curve: see Cultivation of microorganisms.

Growth factor: a growth-stimulating substance. The term is used in two distinct ways: 1. nutrients such as vitamins, essential amino acids, essential lipids, inorganic ions, etc.; or 2. a variety of non-nutrient, mitogenic substances.

Non-nutrient mitogenic growth factors are a group of multifunctional intercellular signaling polypeptides, which differ functionally from hormones in that they act locally within tissues, whereas hormones are generally long-range signaling molecules transported to their target by the circulation. They are usually active at very low concentrations (10^{-9}–10^{-11} M), and their activity is mediated by the specific high-affinity receptors of their target cells. The biological actions of G.f. include regulation of cell proliferation, differentiation, migration and gene expression

So far, some 80 genes have been reported that encode growth factors. Relatively few of these have been characterized; the following are listed separately: Platelet-derived growth factor, Epidermal growth factor, Insulin-like growth factor, Transforming growth factor β.

Growth hormone: see Somatotropin.

Growth medium: see Nutrient medium.

Growth parameters: see Cultivation of microorganisms.

Growth phases: see Culture methods.

Growth regulators: organic compounds, which in small quantities inhibit, accelerate or in some way influence physiological processes in plants. G.r. include natural (endogenous) substances, e.g. Phytohormones (see) and native inhibitors; and many synthetic compounds, especially herbicides and Growth retardants (see). The Auxins (see) include natural and synthetic compounds.

Growth retardants: synthetic plant growth inhibitors, often used in agriculture for the control of plant growth. In particular, they cause stalk shortening in grasses, and some G.r. are used for controlling stalk length in cereal crops. The following G.r. are listed separately: Chlorocholine chloride, AMO 1618, Succinic acid mono-*N*-dimethylamide, Succinic acid mono-*N*-dimethylhydrazide, EL-531, Maleic hydrazide, Phosphon D.

Growth substances: see Auxins.

Growth vitamin: vitamin A (see Vitamins).

GTF: acronym of glucose tolerance factor. See Chromium.

GTP: acronym of guanosine 5′-triphosphate. See Guanosine phosphates.

GTPases, *GTP-binding proteins:* a large superfamily of proteins that have the ability to bind GTP and catalyse its hydrolysis to GDP and P_i and which, by doing this, operate an 'on/off switching' mechanism in a particular cellular activity. The superfamily may be subdivided into several families, namely: (i) translational factors, which participate in ribosomal protein synthesis, e.g. EF-Tu; (ii) the heterotrimeric G-proteins, which mediate transmembrane signaling by hormones and light; (iii) the Ras proteins, which participate in the control of cell proliferation and differentiation; (iv) other small (20–35 kDa) GTPases ('small Gs'), which appear to be involved in regulating the intracellular transport of vesicles (e.g. Rab, ARF) and the cytoskeleton (e.g. Rho, Rac); (v) those involved in the targeting of nascent polypeptide chains to the endoplasmic reticulum (ER) (e.g. SRP, SRP-R). Not included in this superfamily are proteins that are known to bind GTP but which do not utilise it as an intracellular 'on/off' switch, such as Tubulin (see), guanylate cyclases and GTP-utilising kinases.

All the GTPases operate the same 'on/off' switching cycle, which involves two different conformational states depending on whether GTP or GDP is bound to the enzyme (E), i.e. E · GTP and E · GDP. E · GTP is the active conformation, which corresponds to the 'on' position of the switch and signals that the particular process being controlled by the switch can proceed; E · GDP is the inactive conformation, which corresponds to the 'off' position of the switch and signals the cessation of the process. Conversion of E · GDP to E · GTP occurs by exchange of GDP for GTP and is stimulated by a *guanine-nucleotide exchange-promoting protein* (GEP) which, by binding to E · GDP, alters the conformation of E in such a way that the affinity of the nucleotide binding site is simultaneously lowered for GDP and raised for GTP; thus GTP, which has a higher intracellular concentration than GDP, binds to E as the GDP dissociates. Conversion of E · GTP to E · GDP is brought about by the intrinsic GTPase activity of E; this activity is hugely stimulated by the binding of a *GTPase-activating protein* (GAP) to E · GTP. The nature of GEP and GAP varies according to the GTPase, e.g. in the case of the *E. coli* elongation factor, EF-Tu (see Protein biosynthesis) the role of GEP is played by EF-Ts and that of GAP by the ribosome, whereas for heterotrimeric G-proteins these two parts are played by the ligand-receptor complex and a ~ 133 amino acid domain of the α-subunit of the G-protein, respectively.

1. ***Translational factors.*** Included among the GTPases involved in protein biosynthesis (see) are the initiation factors IF-2 (prokaryotes) and eIF-2 (eukaryotes) and the elongation factors EF-Tu & EF-G (prokaryotes) and eEF-Tu (or EF-1α) & eEF-G (or EF-2) (eukaryotes).

The GTPase-catalysed cycle of the initiation of protein synthesis involves the formation of the small (30*S* or 40*S*) ribosomal subunit initiation complex and its subsequent binding to the large (40*S* or 60*S*) ribosomal subunit. In this process the initiation factor (IF-2 or eIF-2), in association with GTP, facilitates the appropriate association of the ribosomal subunit, mRNA, fMet- or Met-$tRNA^{Met}$ and other initiation factors to form the initiation complex which then binds to the large ribosomal subunit, an event that triggers the hydrolysis of GTP and the release of the initiation factors. Although it is not clear what plays the role of GEP in this cycle, the large ribosomal subunit appears to function as GAP.

The protein elongation process involves two GTPase-catalysed cycles, one with EF-Tu (or eEF-Tu) and one with EF-G (or eEF-G). The EF-Tu cycle of prokaryotes can be taken as starting with EF-Tu bound to GDP (and thus in the inactive 'switched off' conformation) as EF-Tu · GDP. EF-Ts, the GEP of the cycle, then binds to it and alters its conformation, thereby allowing the GDP to dissociate and be replaced by GTP. This produces a further conformational change which allows the dissociation of EF-Ts, giving the active *switched on* conformation EF-Tu · GTP, which subsequently binds to an aminoacyl-tRNA. The resulting GTP · EF-Tu · aminoacyl-tRNA complex binds to an mRNA-programed ribosome in such a way that the aminoacyl-tRNA is located at the A-site, thereby making possible the 23*S*-rRNA-peptidyltransferase-catalysed covalent bonding of the peptidyl residue on the tRNA at the P-site to the newly-introduced amino acid. However, before peptide bond formation can occur the GTP of the complex is hydrolysed (the ribosome acting as GAP), a reaction which allows the resulting EF-Tu · GDP to dissociate from the ribosome and begin another cycle. Hydrolysis of GTP in this cycle is believed to monitor and enhance the fidelity of mRNA-to-protein translation in a process termed *kinetic proofreading* which is based on the relationship between the duration of the codon-anticodon binding interaction with the time taken for GTP hydrolysis and the subsequent dissociation of EF-Tu · GDP. There are two proofreading steps. The first is based upon the premise that a GTP · EF-Tu · aminoacyl-tRNA with an 'incorrect' anticodon is more likely to dissociate from the mRNA before GTP is hydrolysed than one with the 'correct' anticodon. The second, which comes into play if the first fails, is based on the observation that the EF-Tu · GDP, although now in a dissociable conformation, nevertheless remains bound to its site on the ribosome for a brief period, thereby preventing the peptidyltransferase reaction and allowing further time for dissociation of an 'incorrect' anticodon. Thus the longer the time taken for GTP hydrolysis and the subsequent dissociation of EF-Tu · GDP the more likely it is that an incipient error will be corrected and the greater will be the fidelity of the translation process. However the longer these processes are the slower the translation process will be; clearly Nature has had to find a fine balance between accuracy and speed.

The EF-G cycle of prokaryotes can be taken as starting with EF-G bound to GDP (and thus in the inactive *switched off* conformation) as EF-G · GDP, whose conformation is such that it has no affinity for the ribosome and therefore readily dissociates from it. It then exchanges its GDP for GTP; it is not clear whether a GEP catalyses this exchange. However, the exchange brings about a conformational change which allows the resulting EF-G · GTP to bind to the ribosome, which, in turn, promotes release of the deaminoacylated tRNA from the P-site, translocation of the peptidyl-tRNA (elongated by one amino acid in the preceding EF-Tu-catalysed cycle) from the A-site to the P-site, and movement of the ribosome along the mRNA a distance of one codon in the $5' \rightarrow 3'$ direction. The hydrolysis of GTP is not required for this complex set of manoeuvres but it is required after their completion for the release of EF-G from the ribosome; the GTPase activity of EF-G is stimulated by the ribosome, which thus acts as the GAP.

In the protein elongation process the EF-Tu and EF-G cycles themseves interact with the mRNA-programed ribosome cyclically and since (i) EF-Tu has to be released from the ribosome before EF-G can bind to it and *vice versa* and (ii) GTP hydrolysis is required for the release of both factors, there is a stoichiometric relationship between the number of GTPs hydrolysed and the number of amino acids incorporated into the protein synthesized. This contrasts with the other GTPases, e.g. heterotrimeric G-proteins, Ras proteins, which continue to transmit a signal as long as they remain in the E · GTP form, a key feature of the signal amplification process.

2. ***Heterotrimeric G-proteins*** constitute a numerous (e.g ~ 16 identified in mammals alone) group of proteins that mediate transmembrane signaling and which consist of three distinct subunits, namely α (39–52 kDa), β (35–36 kDa) and γ (7–10 kDa), of which the β and γ are so tightly associated that they can only be dissociated in vitro by treatment with denaturants and function as a dimeric unit in vivo. Individual G-proteins are presently identified by their α-subunit which gives them their specificity; thus there are G-proteins with $G_s\alpha$-, $G_i\alpha$-, $G_{olf}\alpha$-, $G_t\alpha$-, $G_q\alpha$-, and $G_o\alpha$-subunits of which there are several species of the first four types, usually differentiated by the addition of a numeral, e.g. $G_s1\alpha$-, $G_s2\alpha$-, $G_i1\alpha$-, $G_i2\alpha$-. These α-subunits appear to fall into three families on the basis of amino acid sequence homologies, namely $G_s\alpha$, $G_i\alpha$ and $G_t\alpha$. Each α-subunit appears to be composed of two domains, one of ~ 170 amino acid residues containing a single high affinity guanine nucleotide (GDP or GTP) binding site and having a low intrinsic GTPase activity, and one of ~ 133 amino acid residues which functions like a built-in GAP when the α-subunit binds to its effector protein and assumes the appropriate conformation. Gα · GDP binds tightly to the βγ-subunit and is the inactive, *switched off* conformation. Gα · GTP dissociates from βγ-subunit and is the active, *switched on* conformation. AlF_4^-, in the presence of Mg^{2+}, can interact with Gα · GDP to mimic GTP and activate Gα. Although the βγ-subunits of all G-proteins except G_t are interchangeable and were originally thought to be the inactive partner in the G-protein signal transduction cycle, it is now clear that, though this is generally true, there are exceptions; for instance all the physiological responses triggered by stimulation of the pheromone receptors of yeast are mediated by Gβγ not Gα · GTP. G-proteins are bound to the cytoplasmic surface of plasma membranes; this binding is facilitated by the presence of isoprenyl groups on the γ-subunit and myristoyl goups on the $G_i\alpha$-subunits which, being hydrocarbon in nature, readily associate with the lipid bilayer and act as anchors.

The G-protein GTPase-catalysed cycle can be taken as starting with the conversion of the inactive GDP · Gα-Gβγ complex to the active Gα · GTP form by interaction with a GEP which promotes the exchange of GDP for GTP and thereby lowers the affinity of Gα for Gβγ. The nature of the GEP varies according to the G-protein. Usually it is a ligand-receptor protein complex but in the case of the G_t-proteins (the transducins), which participate in the transduction of visual signals in the rods ($G_t1\alpha$) and cones ($G_t2\alpha$) of the retina of the eye, it is an activated opsin (see Visual process). The receptor of the ligand-receptor protein complex is a Seven-spanning receptor protein (see), so-called because its amino acid sequence contains seven stretches of ~ 23 hydrophobic residues that form seven α-helixes traversing the plasma membrane, whilst the ligand is usually a hormone, e. g. Adrenalin (see), Glucagon (see), or odorant. The $G\alpha \cdot GTP$ binds to its specific effector protein and modulates its activity. Thus $G_s\alpha \cdot GTP$ and $G_{olf}\alpha \cdot GTP$ bind to adenylate cyclase, thereby altering its conformation and converting it from its catalytically inactive form to its catalytically active form; similarly $G_i\alpha \cdot GTP$ binds to adenylate cyclase and inactivates it, $G_t\alpha \cdot GTP$ binds to cGMP-phosphodiesterase and activates it, and $G_o\alpha \cdot GTP$ and $G_q\alpha \cdot GTP$ bind to phospholipase C and activate it. The binding of the $G\alpha \cdot GTP$ to its effector protein also alters the conformation of Gα in such a way that its intrinsic GTPase activity is stimulated; this conformational change is believed to involve the GAP-like ~ 133 amino acid domain mentioned previously. Hydrolysis of the bound GTP to GDP alters the conformation of Gα so that its affinity for its effector protein is removed and its affinity for Gβγ is restored. Thus the $G\alpha \cdot GDP$ dissociates from its effector protein, and binds to Gβγ so reconstituting the $GDP \cdot G\alpha$-Gβγ complex with which the cycle began. The dissociation of the $G\alpha \cdot GDP$ from the effector reverses the effect that the binding of $G\alpha \cdot GTP$ had on the effector protein, which was stimulation by α-subunits of the 's', 'olf', 't', 'q' and 'o' types and inhibition by those of the 'i' type. G_{olf} is peculiar to olfactory epithelial cells; it is switched on by the binding of an odorant molecule to a specific receptor protein located in the apical plasma membrane of the receptor cell (it is currently believed that each olfactory epithelial cell has odorant receptors specific to one species of odorant) and, in turn, activates adenylate cyclase which markedly increases the intracellular concentration of cAMP. The latter binds to cAMP-gated Na^+-channels and causes them to open; this results in the depolarization of the cell membrane and the sending of an electrical signal to the olfactory cortex of the brain.

Although G_o and G_q both activate phospholipase C (though not phospholipase C_γ), they are distinct entities. This is evident from the fact that Pertussis toxin, secreted by the whooping cough bacterium *Bordetella pertussis*, catalyses the transfer of the ADP-ribose moiety of NAD^+ to $G_o\alpha$ but not to $G_q\alpha$. This prevents the exchange of GDP for GTP and keeps G_o in the inactive $GDP \cdot G\alpha$-Gβγ form and thus prevents the activation of phospholipase C. Pertussis toxin acts in the same way on a G_i, thus preventing the inhibition of adenylate cyclase when the appropriate receptor is stimulated.

Cholera toxin interrupts the GTPase-switching cycle of the G_s of intestinal epithelial cells, thereby causing the massive water loss from the body that is characteristic of cholera victims. It is composed of an A-subunit, consisting of an A_1-peptide (23 kDa) linked to an A_2-peptide by an -S-S- bridge, and five B-subunits. The latter bind specifically to the G_{M1} ganglioside located on the luminal surface of the plasma membrane of intestinal epithelial cells. This binding allows the A-subunit to enter the cell, where it undergoes proteolytic cleavage and reduction of the -S-S- bridge. The A_1-peptide, so released, catalyses the transfer of ADP-ribose from NAD^+ to Arg_{201} of the α-subunit of the $GTP \cdot G_s\alpha$-adenylate cyclase complex, thereby inhibiting its GTPase activity. Thus the adenylate cyclase remains in its active form and keeps the intracellular cAMP level high. The latter causes the continuous secretion of ions (Na^+, Cl^- & HCO_3^-) and water from the cells, and ultimately the other bodily tissues, into the intestinal lumen. Cholera toxin also catalyses the ADP-ribosylation of Arg_{174} of the α-subunit of G_t, with similar loss of GTPase activity.

3. ***Ras proteins.*** This class of proteins was originally discovered as the product of an oncogene (see) in tumor-causing ***ra****t* ***s****arcoma* viruses and termed the *ras* gene. It is now known that this gene is just one of a family of closely related *ras* genes which occur, not only in viruses, but also in the cells of probably all eukaryotic organisms. Moreover it is known that most of these genes are not cancer-causing but encode highly conserved (~ 90 % amino acid identity) 21 kDa Ras proteins which are GTPases that participate in the regulation of cell proliferation and differentiation. However because mutations in them can cause the production of 'mutant' GTPases (which function abnormally and allow the cells in which they occur to proliferate in an uncontrolled manner causing tumors), they are called proto-oncogenes. Introduction of a *ras* oncogene into the cell's genome can occur by spontaneous or induced mutation of the *ras* proto-oncogene, or as a result of infection by a virus containing the *ras* oncogene.

Mammalian cells have three Ras proteins, Ha-Ras, Ki-Ras and N-Ras, which are present in most cell types. They are bound to the cytoplasmic surface of the plasma membrane by (i) a farnesyl moiety bound by a thioether linkage to the *C*-terminal cysteine residue whose carboxyl group has been methylated by SAM (see) and (ii) a palmitoyl residue bound by a thioester linkage to a cysteine residue 2–5 residues from the *C*-terminus in the case of Ha- and N-Ras, or a polybasic domain containing six lysine residues in the case of Ki-Ras.

Ras proteins are composed of ~ 170 amino acids and have a guanine nucleotide-binding site and a 3D-structure similar to that of the guanine nucleotide-binding domain of the α-subunit of G-proteins. $Ras \cdot GDP$ is the conformationally inactive form while $Ras \cdot GTP$ is the active form. The Ras GTPase-catalysed cycle can be taken as starting with the conversion of $Ras \cdot GDP$ to $Ras \cdot GTP$ by interaction with a GEP which promotes the exchange of GDP for GTP. The GEP is a cytosolic protein termed Sos, which itself has to be activated in order to carry out its exchange-promotion function. The activation of Sos is a multistep process, beginning with the binding of a growth factor [e.g. Epidermal growth factor (EGF) (see)] to its specific Receptor tyrosine kinase (RTK) (see) located in the plasma membrane. This causes dimerization of two such ligand-RTK complexes and the autophosphorylation of specific tyrosine residues in their cytosolic domains. A phosphotyrosine and its immediately adjacent amino acid residues act as a binding site for the SH2 domain of a cytosolic protein known as GRB2. GRB2 also has two SH3 domains which bind to proline-rich sites in Sos, thereby activating it. The resulting membrane-bound 'ligand-RTK-GRB2-activated Sos' binds to the Ras · GDP and promotes the GDP/GTP exchange. The Ras · GTP typically remains in the active state for about a minute before its GAP-induced GTPase activity hydrolyses the GTP to GDP. While in the active state the Ras · GTP participates in a multistep process involving several kinases which leads to the phosphorylation of many substrates (e.g. nuclear transcription factors) whose combined activity causes cell proliferation or differentiation. The first step in this process is the binding of a cytosolic serine/threonine kinase termed Raf, by its *N*-terminal domain, to the membrane-bound Ras · GTP. The second step is the binding of the inactive form of a cytosolic, dual-specificity (serine and tyrosine) kinase called MEK to the *C*-terminal domain of the Ras · GTP-bound Raf. Raf then uses ATP to phoshorylate MEK and thereby activate it. The third step is the MEK-catalysed, ATP-driven phosphorylation of the hydroxyl groups of serine and tyrosine residues in the inactive form of another cytosolic serine/threonine kinase, called MAP (**m**icrotubule **a**ssociated **p**rotein or **m**itogen-**a**ctivated **p**rotein kinase), thereby activating it. The cytosolic, activated-MAP now catalyses the ATP-driven phosphorylation of the hydroxyl groups of serine and/or threonine residues in the cocktail of proteins, cytosolic and nuclear, involved in cell proliferation and/or differentiation; this serves to modulate their activity in such a way that their collective action brings about the latter effect.

It has been found that one mutation in the *ras* gene causing the replacement of glycine$_{12}$ by valine in Ras protein is sufficient to convert it into an oncogene. The mutant Ras-Val$_{12}$ has a markedly lower GTPase activity than normal; thus the cell proliferation signaling cascade remains permanently on and tumor formation occurs.

4. ***Other small GTPases (Rab & ARF; Rho & Rac).*** Members of the Rab and ARF families participate in the intracellular transport of vesicles. Rab proteins, of which ~15 are presently known, participate in the targeting of transport vesicles to their correct cellular destination. They have ~200 amino acid residues, of which a central segment of ~145 is common to all, whereas the ~20 amino acid *N*-terminal and ~35 amino acid *C*-terminal sequences are peculiar to each Rab. The *C*-terminal segment contains the membrane binding site and determines which membrane it binds to. Membrane binding is facilitated by a geranylgeranyl moiety, attached by a thioether linkage to the *C*-terminal cysteine residue, which inserts itself into the lipid bilayer. Cytosolic Rab contains bound GDP; this is exchanged with GTP when Rab binds to its receptor protein in the membrane of a vesicle which is just about to bud off its parent membrane. The GDP/GTP exchange causes a conformational change in Rab which facilitates its binding to its correct acceptor organelle; recognition of the correct receptor organelle depends on the specific *C*-terminal sequence of Rab. This binding initiates the vesicle fusion process. The latter requires the participation of several other proteins, and when it is complete the intrinsic GTPase activity of Rab is activated and its bound GTP is hydrolysed to GDP, thereby triggering the release of Rab · GDP into the cytosol. The identity of GEP and GAP in the Rab GTPase cycle is not yet clear. Rab-1 & -2 direct vesicles from the ER to *cis*-Golgi cisternae. Rab-3 is involved in the direction of synaptic vesicles in neural tissues. Rab-4 & -5 are localized in early endosomes (derived from clathrin-coated vesicles soon after they bud off the plasma membrane) and are involved, along with Rab-7 of late endosomes, in their route to lysosomes. Rab-6 directs vesicles from medial to *trans* Golgi cisternae. Rab-9 directs vesicles from *trans*-Golgi cisternae to lysosomes. Rab-10 directs vesicles to the ER.

ARF proteins participate in the inititiation of vesicle formation, which begins with a Golgi-bound (GEP) enzyme promoting the exchange of GDP for GTP on cytosolic ARF. The conformation of the resulting ARF · GTP now allows it to bind to ARF-receptors on the cytosolic face of the Golgi membrane. This is followed by the binding to ARF, and adjacent membrane-bound proteins, of a complex composed of several proteins (α-, β-, γ-COP plus four others) known as *coatomer,* which causes the budding of the vesicle. Fission of the completed vesicle from the Golgi cisterna requires fatty acyl-CoA, though how it functions is not yet known. This is accompanied by hydrolysis of the GTP and release of ARF · GDP to the cytosol.

Members of the Rho and Rac families participate in the organization of the cytoskeleton and the activation of its motility by acting as trans-membrane signal transducers. Growth factors such as PDGF, EGF and insulin, when they bind to their receptors on fibroblasts, initiate a Rac-mediated signaling train which causes the polymerisation of actin filaments and the later formation of tight adhesions to the substratum which cause the ruffling of the leading membranes of the cells. Other growth factors initiate a Rho-mediated signaling train which causes the assembly of stress fibers and adhesion plaques. The intermediate steps in these trains are not yet known.

5. ***GTPases involved in the targeting of nascent polypeptide chains to the ER (e.g. SRP, SRP-R).*** Polypeptides destined for the ER membrane or the ER lumen rather than the cytosol are synthesized with an *N*-terminal sequence of ~ 30 predominantly hydrophobic amino acids which acts as a signal directing them to the ER, and which is enzymically removed once this has been done. Since the ribosomal translation of mRNA causes polypeptides to be synthesized in the *N*-terminus → *C*-terminus direction, the signal sequence emerges from the mRNA-ribosome complex first. When the polypeptide chain is ~ 70 amino acids long a cytosolic signal recognition particle (SRP) binds to the signal sequence and the resulting complex then binds to an SRP-receptor (SRP-R) in the ER membrane. This allows an interaction with another ER membrane component, signal sequence binding protein (SSBP), which causes the signal sequence to insert itself into and across the ER membrane in such a way that its *N*-terminus remains at the cytosolic side and the growing end is at the luminal side. During or immediately after the insertion step SRP and SRP-R dissociate from the SRP · SRP-R · ribosome/mRNA/polypeptide/SSBP

complex; SRP returns to the cytosol and SRP-R remains as an integral ER membrane protein. This is followed by (i) aggregation of several other ER membrane proteins to form a transmembrane channel through which the progressively elongating polypeptide chain can pass, and (ii) cleavage of the signal sequence. The insertion step and the dissociation step both require the hydrolysis of bound GTP; since both SRP and SRP-R are known to have components with GDP/GTP binding and GTPase activity it is presumed that they are responsible for this, although the details of the two GTPase-cycles are not yet known.

SRP is composed of six different protein subunits, known as P72, P68, P54, P19, P14 & P9 (the numerals refer to their approximate size in kDa) and a 7SL RNA of 300 nucleotides. P54 has the GDP/GTP binding site in its *N*-terminal domain. It also has the signal sequence binding site in its *C*-terminal domain; the latter is rich in methionine residues which bind to the side chains of the hydrophobic amino acids of the signal sequence. P19 binds P54 to the RNA, P72 and P68 form a dimer which binds SRP to SRP-R, while P14 and P9 prevent elongation of the growing polypeptide chain beyond a length of ~ 70 amino acids until the SRP-mediated docking of the signal sequence with the ER membrane has occurred.

SRP-R is composed of two proteins of ~ 640 and ~ 300 kDa, the larger of which contains the GDP/GTP binding site and is anchored to the ER membrane by a hydrophobic sequence near its *N*-terminus.

[L. Birnbaumer *Cell* **71** (1992) 1069–1072; M. S. Boguski & F. McCormick *Nature* **366** (1993) 643–654; H. R. Bourne et al. *Nature* **348** (1992) 125–132; S. Collins et al. *Trends Biochem. Sci.* **17** (1992) 37–39; W. J. Fantl et al. *Annu. Rev. Biochem.* **62** (1993) 453–481; J. R. Hepler & A. G. Gilman *Trends Biochem. Sci.* **17** (1992) 383–387; Y. Kaziro et al. *Annu. Rev. Biochem.* **60** (1991) 349–400; E. Nishida & Y. Gotoh *Trends Biochem, Sci.* **18** (1993) 128–131; G. Panayotou & M. D. Waterfield *BioEssays* **15** (1993) 171–177; C. D. Strader et al. *Annu. Rev. Biochem.* **63** (1994) 101–132; C. W. Emala et al. *Prog. Nucl. Acid Res. Mol. Biol.* **47** (1994) 81–111; C. M. Fraser et al. *Prog. Nucl. Acid Res. Mol. Biol.* **49** (1994) 114–156]

Gua: see Guanine.

Guanidine: see Guanidine derivatives.

Guanidine derivatives: compounds containing the strongly basic guanidine group (Fig. 1). Guanidine itself, $H_2N\text{-}C(=NH)\text{-}NH_2$, is found in free form only in a few plants. The guanidino group is synthesized de novo in the course of arginine biosynthesis (see Urea cycle); other G. are formed by Transamidination (see) from arginine (Fig. 2).

L-Arginine and γ-guanidobutyric acid are probably present in all living organisms. Some G. d. are found only in plants, e. g. L-Canavanine (see) and Galegine (see); other G. d., such as Phosphagens (see), only occur in animals. The G. d., streptidin, is a component of the antibiotic streptomycin.

G.d. are degraded by various enzymes: 1. Transamidinase can act as a catabolic enzyme, if the G. d. is further degraded by other enzymes and the resulting amino compound is catabolized (Fig. 3); 2. Arginase and heteroarginase catalyse the hydrolysis of G. d. to produce urea. Arginase (L-arginine-ureo-hydrolase) cleaves L-arginine, L-canavanine and γ-hydroxyarginine, and it probably occurs as isoenzymes that possess heteroarginase activity. Heteroarginases differ markedly from classical arginase with respect to substrate specificity; they cleave G. d. with chain lengths less than 6 C-atoms, e. g. γ-guanidobutyric acid, and they show wider specificity in that they cleave certain special G. d., e. g. γ-guanidobutyric acid (also cleaved by γ-guanidobutyrase from *Streptomyces griseus),* arcain, agmatine and streptomycin.

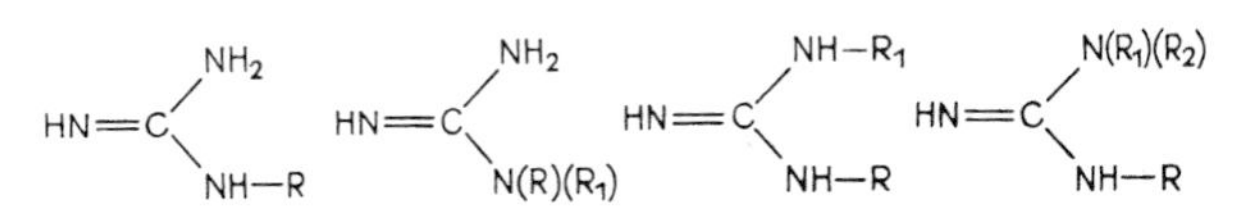

Fig. 1. *Guanidine derivatives.* Different types of guanidine derivatives.

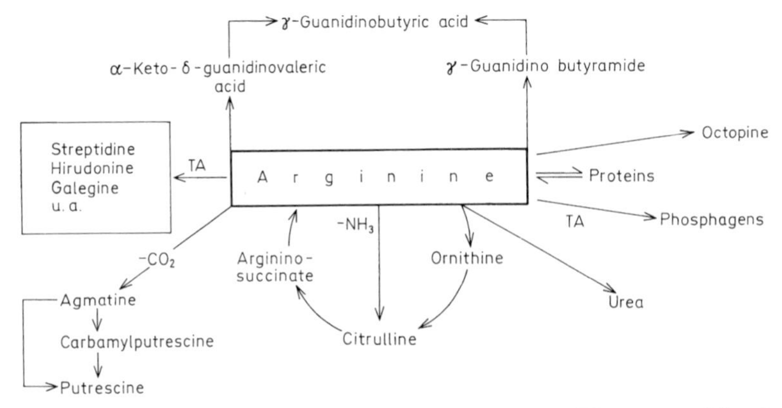

Fig. 2. *Guanidine derivatives.* Metabolic fates of arginine. TA = transamidination.

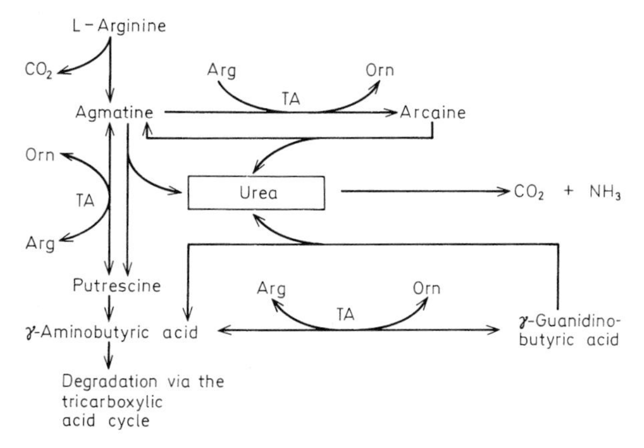

Fig. 3. *Guanidine derivatives.* Arginine degradation in a mushroom. TA = transamidination, Arg = L-arginine, Orn = L-ornithine.

Arginine decarboxy-oxidase, hitherto described only from *Streptomyces griseus* and the pond snail *(Limnaea stagnalis),* degrades arginine to γ-guanidobutyramide, canavanine to β-guanido-oxypropionamide, and homoarginine to δ-guanidovaleramide. Other reactions for the degradation of L-canavanine are found in bacteria.

Guanine, *G., Gua, 2-amino-6-hydroxypurine:* one of the four nucleic acid bases. M_r 151.13, m.p. 360°C (d.). G. is also a component of nucleotide coenzymes and the starting material for the biosynthesis of many natural products, including pterins and the vitamins folic acid and riboflavin. Free G. is rare in nature. It was first discovered in 1844 in Peruvian guano. It is the nitrogen excretion compound of spiders. G. is deaminated by guanine deaminase (EC 3.5.4.3) to xanthine; this reaction is the first step in purine degradation.

Guanine

Guanosine, *Guo:* 9-β-D-ribofuranosylguanine, a β-glycosidic nucleoside containing D-ribose and guanine. M_r 283.2, m.p. 237–240°C, $[\alpha]_D^{25}$ – 72° *(c* = 1.4 in 0.1M NaOH). Guanosine phosphates (see) are metabolically important in all organisms.

Guanosine diphosphate sugars, *GDP-sugars:* guanosine-diphosphate-activated forms of various sugars. Their synthesis is analogous to that of other Nucleoside diphosphate sugars (see), i.e. the condensation of a sugar phosphorylated at C1 and guanosine triphosphate, with release of pyrophosphate. Guanosine diphosphate mannose is of particular importance; glucose, fucose and rhamnose also occur as GDP derivatives.

Guanosine phosphates: phosphoric acid esters of guanosine, which are metabolically important nucleotides. The biologically important derivatives are esterified on C5′ of the ribose. According to the number of phosphoric acid residues, G.p. are classified as guanosine mono-, di- and triphosphates.

1. *Guanosine 5′*-monophosphate, GMP, guanylic acid, M_r 363.2, m.p. 190–200°C (d.) is synthesized in the purine pathway from xanthosine monophosphate and is the starting material for the synthesis of the other guanosine phosphates. GMP is used as a flavoring and aroma substance, and is extracted for this purpose from yeast nucleic acids or produced on a large scale by mutants of certain microorganisms, such as *Corynebacterium glutamicum.*

2. *Guanosine 5′* diphosphate, *GDP, M_r* 443.2, is produced by phosphorylation of GMP by a kinase or by dephosphorylation of guanosine triphosphate. Certain sugars, e.g. mannose, are activated by binding to GDP (see Nucleoside diphosphate sugars).

3. *Guanosine 5′*-triphosphate, GTP, M_r 523.2, like adenosine triphosphate (ATP), can provide energy for biochemical reactions. The energy released by dehydrogenation of α-ketoglutarate (2-oxoglutarate) in the Tricarboxylic cycle (see) flows into the GDP/GTP system, from which it can be transferred to the ADP/ATP system. GTP can also provide the phosphate group for the synthesis of phospho*enol*pyruvate from oxaloacetate in the course of gluconeogenesis. GTP is an important energy source in Protein biosynthesis (see).

4. *Cyclic guanosine 3′*-monophosphate, cyclo-GMP, cGMP, M_r 345.2, a structural analog of cyclic adenosine 3′,5′-monophosphate (see Adenosine phosphates) which occurs in similar concentrations in many tissues. cGMP is synthesized by a guanylate cyclase which is highly specific for GTP. The biological significance of the cGMP-guanylate cyclase system lies in its mediation of the action of certain hormones and neurohumoral transmitters, such as acetylcholine, prostaglandins and histamine.

Guanylic acid: see Guanosine phosphates.

(+)-Guibourtacacidin: 7,4′-dihydroxyflavan-3,4-diol. See Leucoanthocyanidins.

Guo: abb. for Guanosine

Gutta: a rubber-like polyterpene of about 100 isoprene units in which the double bonds are in the *trans* configuration (see Polyterpenes, Fig.). G. is produced on the Malayan peninsula and the Indonesian islands from the latex of *Palaquium gutta.* G. is less elastic than rubber, but since it is more resistant to chemicals and environmental influences it serves well as an insulating material. Depending on its source, G. occurs in mixtures with other terpenes. The mixture with resins is called guttapercha, and that with triterpene alcohols is chicle (starting material for manufacture of chewing gum).

Guvacine: see Areca alkaloids.

Guvacoline: see Areca alkaloids.

Gyrase: a type II topoisomerase. See Topoisomerases.

H

Haem: British spelling of heme. See Hemes.

Haginin A: 7,4′-dihydroxy-2′,3′-dimethoxyisoflav-3-ene. See Isoflav-3-ene.

Haginin B: 7,4′-dihydroxy-2′-methoxyisoflav-3-ene. See Isoflav-3-ene.

Hallucinogens: a group of drugs which cause changes in mood, perception, thinking and behavior. The group does not include addictive drugs, but some users become dependent on H. Those H. with the strongest action are called psychedelics. The H. have been divided into four groups: derivatives of indole alkaloids (tryptamine, harmine, ibogaine, LSD and psyilocybin), derivatives of piperidine (belladonna, atropine, scopolamine, hyoscyamine and phencyclidine), phenylethylamines (mescaline, amphetamine and adrenochrome), and cannabinols from Cannabis plants.

Most H. are sympathomimetics and cause a rise of blood pressure, higher pulse rate, dilation of the pupils, sweating, palpitation and increased tendon reflexes. Marijuana (the dried leaves of the Cannabis plant) is the weakest, and remains in the body for 2 to 3 hours if smoked, longer if eaten. LSD is the most potent of this group, 25 μg being an effective dose for an adult. It remains in the body for 6 to 8 hours, with the greatest psychological effects at 3 to 4 hours, by which time most of it has disappeared from the central nervous system.

There are legitimate medical and psychiatric uses for the H., and the nonmedical use of several of them is widespread. For a review see Marihuana Research Findings: 1976, (Robert C. Peterson, ed.) National Institute on Drug Abuse Research Monograph # 14 (1977) from the US Department of Health, Education and Welfare.

Haloopsin: see Bacteriorhodopsin.

D-Hamamelose: a branched monosaccharide, m. p. 111 °C, $[\alpha]_D^{20}$ − 7.1 ° (water). D-H. occurs in higher plants, for example *Hamamelis* (witch hazel) bark. Its biosynthesis, which is similar to that of apiose, involves an intramolecular rearrangement of an unbranched hexose.

Hanes-Wilkinson plot: see Kinetic data evaluation.

Haptens: partial or incomplete antigens. H. are either chemically defined molecules, e. g. dinitrophenol, or part of the molecular structure of an antigen. They bind specifically to the corresponding antibodies (see Immunoglobulins), but they cannot elicit an immune response (i. e. antibody formation) unless coupled to a carrier protein. After parenteral application of a protein-coupled H., the body produces two specific antibodies, one against the protein and one against the bound H. (hapten-specific antibodies). Half-haptens react with their antigen, but without precipitation.

One of the most effective H. is a glycoplipid from animal organs known as the Forsman H. Forsman H. from horse is *N*-acetyl-α-galactosaminyl-*N*-acetyl-β-galactosaminyl-(1–3)-[galactosyl-(1–4)]$_2$-glucosyl ceramide. Enzymatic removal of the terminal GalNAc group destroys the Forsman antigenicity. The Forsman antigen, which is composed of the Forsman H. and a specific protein, induces hemolysin formation.

Haptoglobin, *Hp:* an acid α_2-plasma glycoprotein, which binds specifically to free plasma oxyhemoglobin to form a high molecular mass complex, M_r 310,000, which cannot be filtered by the kidneys. The associated conformational change in the hemoglobin allows the heme α-methenyl oxygenase of the liver to remove the heme porphyrin ring. Thereafter the globin is degraded by the trypsin-like protease action of the β-chain of the Hp. Hp. is a tetramer consisting of two nonequivalent chain pairs, 2α and 2β, held together by disulfide bridges. Human Hp. exists in three genetic variants Hp 1–1, 2–2 and 2–1 which differ in their electrophoretic patterns. While Hp 1–1 moves as a single band, Hp 2–2 and 2–1 display several (up to 14) discrete bands, which represent stable oligomers of the monomer (one β, one α_1 and one α_2-chain in Hp 2–1; one β and one α_2-chain in Hp 2–2). The M_r of the Hp 2–2 monomer is 57,300. This heterogeneity is due to the existence of two different α-chain types, α_1 and α_2. While α_1 contains 82 amino acids (M_r 9,000), α_2 is nearly twice as large, with M_r 17,300 and 142 amino acids. The α_2-chain, which is only found in human Hp 2–2 and 2-l, is a union of 71 amino acids each from the *N*- and the *C*-terminal ends of the α_1-chain. It is the product of a second autosomal allele, i. e. a homologous gene pair which arose from the α_1 structural gene through unequal crossing over. Unlike the α-chain, the carbohydrate-carrying and much larger β-chain (M_r 40,000; without sugar residues 35,000) is the same size in all 3 Hp types. The Hp 1–1 ($2\alpha_1 + 2\beta$) has M_r 98,000; 20 % carbohydrate; pI 4 (19 % of the amino acids are aspartate or glutamate). The main component of Hp 2–2 $(\alpha_2 + \beta)_2$ has M_r 114,000. The M_r values of the Hp 2–2 polymers are even and odd multiples of the monomer $(\alpha_2 + \beta)_n$ (n = 2–14).

On the basis of sequence homologies between the Hp α-chain and the immunoglobulins, Hp appears to be a natural, preformed antibody against hemoglobin.

Har: abb. for Homoarginine.

Harden-Young ester: see Fructose 1,6-*bis*phosphate.

Harmaline: see Harman alkaloids.

Harman alkaloids: a group of indole alkaloids with a β-carboline ring system. H. a. are biosynthesized from tryptophan and a carbonyl component. The same reaction can be performed *in vitro* using conditions designed to simulate those in the cell. The ubiquitous occurrence of the precursors and the simple synthesis probably account for the wide occurrence of H.a. in many different plant groups. The most abundant H.a. (Fig.) are harmine, harman, 3,4-dihydroharmine (harmaline), and 1,2,3,4-tetrahydroharmine. H.a. are occasionally used to treat encephalitis and Parkinson's disease. Some H.a. cause hallucinations and intoxication.

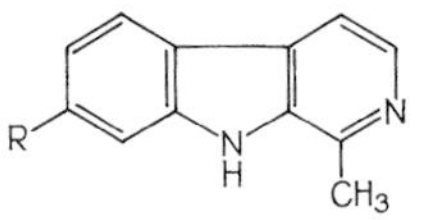

Harman: R=H
Harmine: R=OCH_3

Harman alkaloids

Harmine: see Harman alkaloids.

Hashish: the dried resin from the glandular hairs of the female hemp plant *(Cannabis sativa* L.). Due to its content of psychoactive $\Delta^{9,10}$-tetrahydrocannabinol (2–8 %), it is one of the most common Narcotics (see). Marihuana (often used synonymously with hashish) is the dried and chopped tips of the shoots of the female hemp with a content of 0.5 to 2 % $\Delta^{9,10}$-tetrahydrocannabinol. Both hemp drugs have been used for millenia in folk medicine for their intoxicating action. Today they are, along with alcohol, the most widely used drugs.

The amount of Δ^{9-10}-tetrahydrocannabinol and of structurally related, nonpsychotropic compounds (e.g. cannabinol, cannabidiol and cannabidiolic acid) varies considerably. In contrast to the tropical culture forms, H. from European hemp contains little Δ^{9-10}-tetrahydrocannabinol and much cannabidiol and cannabidiolic acid.

Hatch-Slack-Kortschak cycle, ***HSK cycle, C_4 acid cycle, C_4 pathway:*** The C_4 pathway was discovered in two tropical grasses, maize and sugar cane, but is now known to operate in many other species of *Gramineae,* some species of *Cyperaceae* and in some species of several dicotyledonous families. Plants operating the C_4 pathway (C_4 plants) grow optimally at high light intensities and day temperatures of 30–35 °C. They characteristically have: (i) high photosynthetic rates (40–80 mg CO_2 fixed per dm^2 of leaf surface per hour, compared with 15–40 for C_3 plants), (ii) high growth rates (4–5 g of dry wt. produced per dm^2 of leaf surface per day, compared with 0.5–2 for C_3 plants), (iii) low rates of water loss in relation to dry wt production (250–350 g H_2O lost per g dry matter produced, compared with 450–950 for C_3 plants), (iv) low photorespiration rates, and (v) an unusual leaf structure, typified by an open internal structure giving atmospheric CO_2 ready access to many photosynthetic cells, a double spiral of cells tightly packed around the vascular bundles (the so-called 'bundle sheath cells'), loosely packed spirals of mesophyll cells around the bundle sheath cells with many plasmodesmata connecting the two cell types, and a marked structural difference between the chloroplasts of the mesophyll cells, which have 'normal' grana, and those of the bundle sheath cells, which have very small grana or no grana at all.

The first indication that photosynthetic CO_2 fixation in these plants was different from that in other plants (which subsequently became known as C_3 plants) came with Kortschak's discovery in 1965 that when sugar cane leaves were allowed to photosynthesize in $^{14}CO_2$ the first compounds to be labeled were the C_4-dicarboxylic acids, D-malic acid and L-aspartic acid, and there was a distinct time lag before label appeared in 3-phosphoglyceric acid (3PGA), the product of the carboxylation reaction of the Calvin cycle (see), and in other participants of the Calvin cycle. Hatch and Slack then showed that: (i) oxaloacetic acid (OAA) was also labeled rapidly, (ii) the ^{14}C label was located only in the C4 carboxyl group of oxaloacetic, malic and aspartic acids, (iii) 3PGA was labeled only in its carboxyl group (i.e. C1), and (iv) the labeling pattern of the carbon chains of the hexose phosphates was consistent with their formation from triose phosphates derived from 3PGA (i.e. ^{14}C was present in C3 & C4). These findings suggested that CO_2 initially reacted with a C_3 compound to form a C_4 dicarboxylic acid which then underwent a 'C4 transcarboxylation reaction' with an acceptor molecule to regenerate a C_3 compound and yield [1-^{14}C]3PGA, which was then converted into triose and hexose phosphates. The transcarboxylation reaction,

Cannabidiolic acid

Cannabidiol

Cannabinol

$\Delta^{9,10}$-Tetrahydrocannabinol

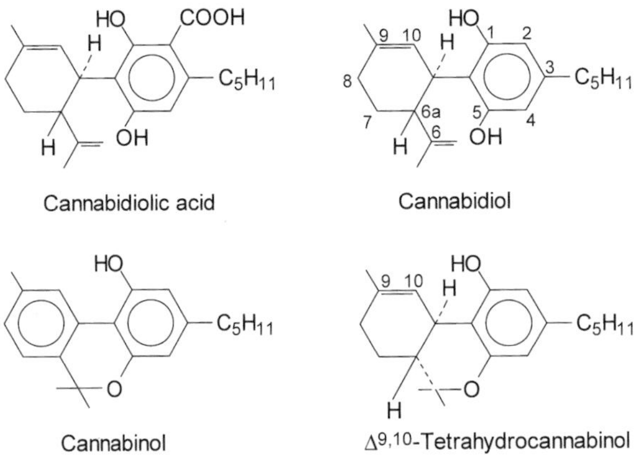

Some components of hashish

however, could not be found and it was eventually discovered that the C4 carboxyl group of the C_4 dicarboxylic acid was removed in a decarboxylation reaction as CO_2, which was then fixed into 3PGA by ribulose bisphosphate carboxylase (Rubisco) (see). The two carboxylation reactions that this finding indicated, were found to be physically separated in C_4 plants; the first one, the phospho*enol*pyruvate (PEP) carboxylase-catalysed conversion of CO_2 and PEP to OAA, takes place in the cytoplasm of the mesophyll cells, while the second one, the Rubisco-catalysed conversion of CO_2 and D-ribulose-1,5-bisphosphate to 3PGA, takes place in the chloroplast stroma of the bundle sheath cells. The decarboxylation reaction also occurs in the bundle sheath cells.

The mechanism of the C_4 pathway is broadly similar in all C_4 plants; however three variants, based mainly on the nature of the decarboxylation reaction, are known. These are: (i) C_4 plants that use the NADP-malic enzyme (EC 1.1.1.40) to catalyse the decarboxylation reaction (eq. 1); these include sugar cane (*Saccharum officinarum*), maize (*Zea mays*) and sorghum (*Sorghum sudanense*), (ii) C_4 plants that use the NAD-malic enzyme (EC 1.1.1.39) to catalyse the decarboxylation reaction (eq. 2); these include *Atriplex spongiosa, Portulaca oleracea* and *Amaranthus edulis*, and (iii) C_4 plants that use PEP carboxykinase (EC 4.1.1.49) to catalyse the decarboxylation reaction (eq. 3); these include *Panicum maximum, Chloris gayana* and *Sporobolus fimbriatus.*

$$\text{L-Malic acid} + NADP^+ \rightarrow CO_2 + \text{pyruvic acid} + NADPH + H^+ \quad (eq. 1)$$

$$\text{L-Malic acid} + NAD^+ \rightarrow CO_2 + \text{pyruvic acid} + NADH + H^+ \quad (eq. 2)$$

$$\text{Oxaloacetic acid} + ATP \rightarrow CO_2 + \text{phosphoenolpyruvic acid} + ADP \quad (eq. 3)$$

The C_4 pathway in the plants of group (i) (NADP-malic enzyme; eq. 1) is shown in the Fig. Air passes through the open stomata of the leaf into the extensive intercellular spaces and bathes the mesophyll cells. CO_2 passes into the cytoplasm of these cells where it dissolves and ionizes; the latter process is probably facilitated by carbonic anhydrase (EC 4.2.1.1) (A, Fig.). The resulting HCO_3^- is then used by PEP carboxylase (EC 4.1.1.31) (B, Fig.) to carboxylate PEP and so form OAA (eq. 4). Note that, unlike Rubisco, PEP carboxylase uses HCO_3^- and not CO_2 as its carboxylation substrate.

$$HCO_3^- + PEP + H^+ \rightarrow OAA + H_3PO_4 \quad (eq. 4)$$

Thus far the pathway is the same in all three groups of C_4 plants and in all of them rapid interconversion of OAA, malic acid and aspartic acid now occurs. In the plants of group (i) the OAA passes into the mesophyll chloroplasts where the NADP-dependent malate dehydrogenase (EC 1.1.1.82) (C, Fig.), a light activated enzyme, catalyses its reduction to malic acid using NADPH generated by the light phase of photosynthesis. The malic acid then passes from the chloroplasts of the mesophyll cells to those of the bundle sheath cells probably via the cytoplasm of the plasmodesmata that connect the two types of cell. It is then decarboxylated by the NADP-malic enzyme (eq. 1; D, Fig.). The resulting CO_2 becomes the substrate for Rubisco (E, Fig.), enters the Calvin cycle, and is converted to photosynthetic product. The pyruvate from this reaction passes back from the bundle sheath chloroplasts to the mesophyll chloroplasts, probably via the cytoplasm of the connecting plasmodesmata. It is then converted back into PEP by the light-activated enzyme, pyruvate,orthophosphate dikinase (EC 2.7.9.1), using ATP generated in the light phase of photosynthesis (eq. 5; I, Fig.). This reaction is pulled in the direction of PEP synthesis by the rapid, exergonic hydrolysis of pyrophosphate to orthophosphate by pyrophosphatase (EC 3.6.1.1) (J, Fig.). The AMP formed in eq. 5 is phosphorylated at the expense of ATP by adenylate kinase (EC 2.7.4.3) (eq. 6; K, Fig.), and the resulting two molecules of ADP converted back to ATP in the light phase of photosynthesis. The NADPH, formed in the bundle sheath chloroplasts when malic acid is decarboxylated (eq. 1; D, Fig.), is used to drive the Calvin cycle. This explains the observation that bundle sheath chloroplasts of C_4 plants using the NADP-malic enzyme are largely deficient in PSII, without which they cannot operate non-cyclic photophosphorylation and so generate NADPH; however ATP is generated by cyclic photophosphorylation, which uses only PSI (see Photosynthesis).

$$\text{Pyruvic acid} + ATP + P_i \rightarrow PEP + AMP + PP_i \quad (eq. 5)$$

$$AMP + ATP \rightarrow 2\ ATP \quad (eq. 6)$$

The photosynthetic product of the Calvin cycle, dihydroxyacetone phosphate (DiHOAcP), is transported from the bundle sheath chloroplasts to the cytoplasm where it is converted into sucrose (see Fig. 1 of the entry Calvin cycle) then exported to the growing parts of the plant via the phloem of the adjacent vascular bundle.

The metabolic capabilities of the chloroplasts of the mesophyll and bundle sheath cells of C_4 plants are quite different. The PSII-deficient bundle sheath chloroplasts of C_4 plants of group (i) were mentioned earlier. That apart, the chloroplasts of both cell types generate NADPH and ATP by non-cyclic photophosphorylation. However, only the bundle sheath chloroplasts of all three groups of C_4 plants have a functional Calvin cycle. Rubisco and several other Calvin cycle enzymes are absent from the mesophyll chloroplasts.

It is believed that C_4 photosynthesis is an evolutionary adaptation to allow plants to grow well in hot, dry conditions. This belief is based on the following reasoning. To grow under such conditions C_4 plants must reduce water loss through their stomata. They have done this by increasing the resistance of their stomata to gaseous diffusion. However the adverse consequence of this is that, compared with C3 plants, there is a much steeper CO_2 concentration gradient between the air outside the leaf and that in the intercellular spaces surrounding the photosynthetic cells. Thus the CO_2 concentration in C_4-mesophyll cells is far lower than the ~ 10 μM concentration present in C_3-mesophyll cells. At such a low CO_2 concentration, Rubisco, $[K_m(CO_2) \sim 10\ \mu M]$, is not an adequate catalyst for the carboxylation reaction in C_4-mesophyll cells. This problem has been solved by the use of PEP carboxylase, which has a very high affinity for HCO_3^- and catalyses a reaction (eq. 4) that is essentially irreversible because the cleavage of its enolic phosphate bond is highly exergonic ($\Delta G^{o\prime}$ of PEP hydrolysis = -61.9 kJ mol^{-1}). By this means CO_2 is trapped in a C_4 dicarboxylic acid which is then transferred

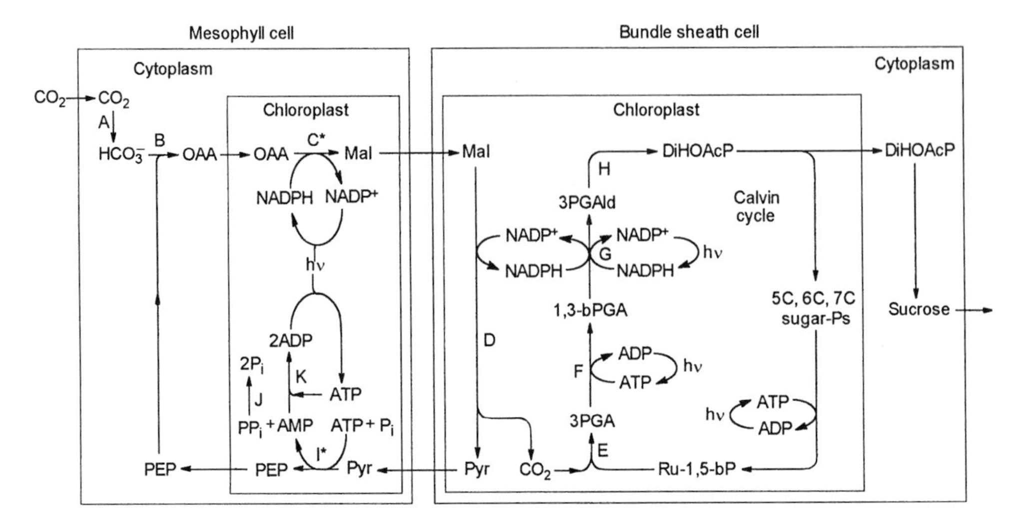

Hatch-Slack-Kortschak cycle. The C_4 pathway in those plant species that utilize the NADP-malic enzyme to catalyse the decarboxylation reaction.
OAA = oxaloacetic acid; mal = malic acid; pyr = pyruvic acid; PEP = phospho*enol*pyruvic acid; 3PGA = 3-phosphoglyceric acid; 1,3-bPGA = 1,3- *bis*phosphoglyceric acid; 3PGAld = 3-phosphoglyceraldehyde; DiHOAcP = dihydroxyacetone phosphate; P_i = inorganic orthophosphate; PP_i = inorganic pyrophosphate; A = carbonic anhydrase; B = PEP carboxylase; C = NADP-malate dehydrogenase; D = NADP-malic enzyme; E = ribulose *bis*phosphate carboxylase; F = phosphoglycerate kinase; G = 3-phosphoglyceraldehyde dehydrogenase; H = triose phosphate isomerase; I = pyruvate,orthophosphate dikinase; J = pyrophosphatase; K = adenylate kinase; asterisks indicate light- activated enzymes.

to the bundle sheath cell. Within the bundle sheath cell, CO_2 is then released by decarboxylation and concentrated to a level that is sufficiently high to allow Rubisco to function adequately. There is another benefit attached to this concentration of CO_2 in the bundle sheath chloroplasts, namely that the oxygenase activity of Rubisco is depressed (CO_2 is a competitive inhibitor of the oxygenase activity of Rubisco) which, in turn, markedly lowers the rate of the wasteful process of Photorespiration (see). The latter more than compensates for the apparently lower efficiency of C_4 photosynthesis (5ATP & 2NADPH are required to fix one CO_2) relative to that of C_3 photosynthesis (3ATP & 2NADPH are required to fix each CO_2).

Hb: abb. for hemoglobin.

HbS: abb. for sickle-cell hemoglobin.

hCG: abb. for human chorionic gonadotropin; same as chorionic gonadotropin.

Hcy: abb. for homocysteine.

HDL: acronym of high density lipoprotein. see Lipoproteins.

Heat-shock proteins: ***Hsp:*** a small number of proteins synthesized in response to hyperthermia or other noxious conditions, such as ethanol or sodium arsenite. They are presumably required for the recovery of gene expression following the deleterious effects of hyperthermia, and they are the first major proteins transcribed from the embryonic genome in the mouse embryo.

Heat shock causes changes in gene transcription, as well as changes in the specificity of mRNA translation. Certain species of mRNA present before heat shock may be degraded, or they may be preserved and translated later when the heat shock response is switched off.

The changes in transcription are due to the presence of heat shock elements (HSE) in heat shock gene promoters. In *E. coli,* the consensus sequence for the HSE is CTGCCACCC at nucleotide positions –44 to –36 relative to the transcription initiation site. In eukaryotes, the HSE consist of contiguous repeats (in alternating orientation) of the 5-base pair sequence NGAANN (N = any nucleotide), upstream of the TATA box.

The RNA polymerase of *E. coli* recognizes the HSE only when the enzyme possesses a special regulatory sigma subunit (σ^{32}; M_r 32,000; product of the *rpoH* gene). The usual sigma subunit (in the absence of the heat shock response) is σ^{70} (M_r 70,000; product of *rpoD* gene). In eukaryotes, the HSE is the binding site for a transcription activating factor, which becomes operative in the heat shock response.

There are four broad categories of Hsp: Hsp90 (M_r 80,000–90,000), Hsp70 (M_r 68,000–74,000), Hsp60 (M_r 58,000–60,000) and a number of relatively small Hsp (e.g. Hsp10), with their homologous counterparts in bacteria (*E. coli*) and eukaryotes.

Many appear to act as chaperonins (see Molecular chaperones) in bacterial cells, the cytoplasm of eukaryotic cells, and mitochondria. The Hsp 70 family in particular appears to chaperone the insertion of proteins into membranes. [O. Bensaude et al. *Nature* **305** (1983) 331–332; E. A. Craig 'The heat shock response' *CRC Crit. Rev. Biochem.* **18** (1986) 239–280; C. Wu 'Heat Shock Transcription Factors: Structure and Regulation' *Annu. Rev. Cell Dev. Biol.* **11** (1995) 441–469]

Heat-stable enzymes: see Thermostable enzymes.

Table 1. *Heavy metals in biomolecules.*

Heavy metal	Type of biomolecule	Biological function
Iron	Hemoglobins; cytochromes	O_2-Transport; electron transport
	Flavin enzymes (metalloflavo-enzymes) e.g. xanthine oxidase	Oxidation, dehydrogenation, and/or reduction
	Iron-sulfur protein e.g. ferredoxin, complexes of the respiratory chain	Electron transfer
	Nitrogenase	Reduction of N_2 to ammonia
	Ferritin, conalbumin, transferrin (siderophilin)	Fe-storage; Fe-transport
	Siderochromes, mycobactin; enterobactin, etc.	Fe-transport in microorganisms
Cobalt	Vitamin B_{12} and its coenzyme forms	Reduction and methyl group transfer (methionine synthesis); isomerization
Copper	Laccase, cytochrome oxidase	Oxidation
	Cupreine, ceruloplasmin	Storage and transport of copper
Manganese	Arginase	Arginine hydrolysis, urea cycle
	Decarboxylases and other enzymes	Release of CO_2, etc.
Molybdenum	Nitrogenase	Binding and activation of molecular nitrogen (subsequent reduction also requires iron)
	Nitrate reductase	Nitrate reduction
	Xanthine oxidase	Purine oxidation, etc.
Zinc	Carbonic anhydrase, peptidases, phosphatases, pyridine nucleotide enzymes and other proteins	Zn functions in substrate binding (e.g. for hydride transfer in ternary complexes with pyridine nucleotide enzymes), and in protein structure (e.g. alcohol dehydrogenase)
	Insulin	Aggregation of the polypeptide

Table 2. *Ligands of heavy metals in biomolecules.*

Heavy metal	Ligand	Example
Iron	Porphyrin, imidazole	Myoglobin
	Sulfur	Ferredoxin
	Phenolate	Transferrin
Cobalt	Corrin, benzimidazole	B_{12} and derivatives
Copper	> N bases	Cupreine
Manganese	Carboxylate, phosphate, imidazole	Glycolysis and proteolysis enzymes
Vanadium	No ligand yet identified	Vanadium proteins of tunicates
Zinc	Imidazole (His), carboxyl groups (Glu)	Carboxypeptidases
	R-S^-	Dehydrogenases
	Imidazole (His)	Carbonic anhydrase

Heavy metals: all metals with a density greater than 5. In living organisms they are usually present in stable organic complexes. Their biological functions are listed in Table 1. Chemical ligands of H.m. are listed in Table 2. Iron (see) is present in a particularly large number of biomolecules. Vanadium is essential for tunicates (a group of strictly marine filter-feeding animals related to the vertebrates) and for lower plants. It is accumulated from sea water by some species of tunicate, notably the *Ascidiidae* and *Perophoridae* (order Phlebobranchia), and is found chiefly in the vacuoles of certain types of blood cell, called vanadocytes or morula cells. Vanadocytes of *Ascidia ceratodes* contain approximately 1.2 M V(III), representing a concentration factor over sea water greater than 10^7. The mechanism of this highly selective accumulation process is unknown.

Other H.m., e.g. lead, mercury, cadmium and copper are toxic (see Lead). Toxicity is usually attributed to the ability of these metals to react irreversibly with free SH-groups of proteins.

Heavy water: see Water.

Heidelberger curve: see Precipitation curve.

Heinz bodies: irregular, refractile inclusions in erythrocytes, attached to the inner surface of the membrane. They may be up to 3 μm in diameter, and are stained by vital dyes. Unstained, they are detectable by their green autofluorescence. H.b. are insoluble aggregates of degraded hemoglobin admixed with fragments of lipid and other protein. Some authors maintain that H.b. formation requires initial formation of methemoglobin, but this is contradicted by the observation that some H.b.-producing drugs do not produce a measurable increase in methemoglobin. H.b. formation is promoted by the same drugs and environmental pollutants that also cause hemolytic anemia, e.g. phenylhydrazine, *O*-methyl-, *O,N,*-dimethyl and trimethyl hydroxylamine. [H. Martin et al. *Klin. Wochenschr.* **42** (1964) 725–731; G. Rentsch *Biochem. Pharmacol.* **17** (1968) 423–427; E. Beutler *Pharmacol. Rev.* **21** (1969) 73–103; C.C. Winterbourn & R.W. Carrell *Brit. J. Haematol.* **25** (1973) 585–592]

Heliangin: a sesquilactone and plant growth inhibitor first isolated from the leaves of the Jerusalem artichoke *(Helianthus tuberosus).* It is a gibberellin antagonist and inhibits the growth of oat coleoptiles; but it stimulates root growth in beans (*Phaseolus* spp.).

Helix: the spiral arrangement of a biopolymeric compound, e.g. starch, some proteins and DNA (see Deoxyribonucleic acid, Fig.).

Helix-random-coil transition: see Denaturation.

Helminthosporal: a product of the phytopathogenic fungus *Helminthosporium sativum* (= *Bipolaris sarokiniana,* Shoemaker), which has a physiological effect on plants similar to that of gibberellins.

Helvolic acid: a tetracyclic triterpene antibiotic, M_r 568.7, m.p. 215 °C. It differs structurally from Fusidic acid (see) in having a Δ^1-3,6-diketo and a 7α-acetoxy function, and in lacking the 11α-hydroxyl group. H.a. is produced by *Aspergillus fumigatus* and related fungal species. Like the structurally related triterpene antibiotics, fusidic acid and cephalosporin P_1, it is effective against Gram-positive organisms.

Hemagglutination: see Agglutination.

Hemagglutinins: see Immunoglobulins.

Heme: 1. a metalloporphyrin which acts as the prosthetic group of a hemoprotein, e.g. cytochrome, hemoglobin, nitrite reductase, etc. The iron atom is coordinately bound to the four pyrrole nitrogen atoms of the porphyrin ring. Biosynthesis of iron porphyrin starts from protoporphyrin IX (see Porphyrins). Iron is added by the action of ferrochelatase (EC. 4.99.1.1) (Fig.). See Protoheme, Chlorocruoroheme, Siroheme.

2. A complex of iron in which the organic component is not a porphyrin, but a related tetrapyrrole structure, e.g. verdoheme, biliverdin heme.

3. An iron complex of a Chlorin (see).

Heme iron: iron which is coordinately bound in porphyrins. See Heme, Hemoproteins.

Heme c: the heme prosthetic group in cytochromes *c*, b_4 and *f*

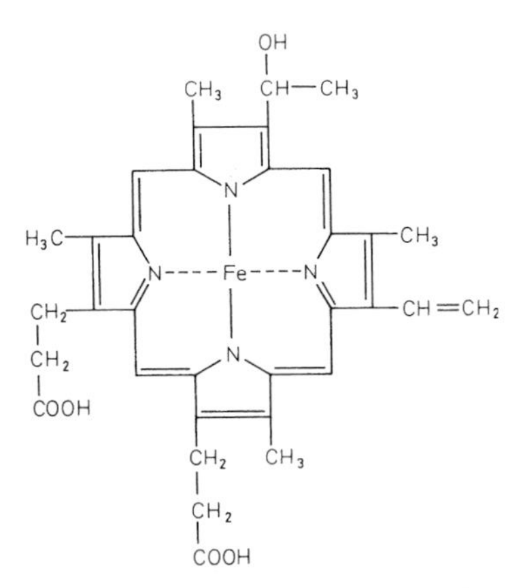

Heme d, formerly *heme* a_2*:* a heme prosthetic group in the terminal oxidase system of many bacteria.

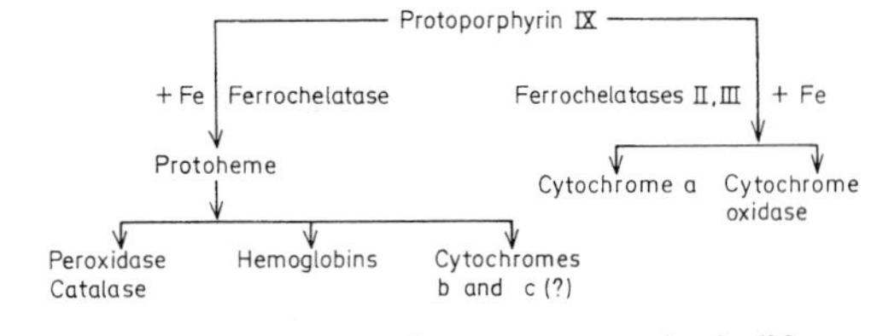

Derivation of hemes from protoporphyrin IX

Heme a: the heme prosthetic group in cytochromes a/a_3.

Hemerythrin: a red-brown, oxygen-transporting chromoprotein, which contains iron but not porphyrin, and which serves as a respiratory pigment in the blood cells of certain marine invertebrates, such as sipunculoid worms, polychete worms and lamp shells *(Brachiopoda)*. The H. of sipunculoid worms is an octamer (M_r 108,000, subunit M_r 13,500, 113 amino acids). Both the iron (2 Fe^{2+}/chain or 16 Fe^{2+}/mol. H.) and the oxygen (1 molecule O_2/chain) are bound directly to the protein.

Hemicelluloses: high molecular mass polysaccharides of β-1,4-linked hexose and pentose residues, often also containing uronic acid. H. occur in woody parts of plants together with cellulose, serving as structural compounds and sometimes as reserve substances. They are insoluble in water, but soluble in dilute alkali. Humans and animals cannot digest them. Important H. are arabans, xylans, glucans, galactans, fructans and mannans.

Hemisubstances: see Cell wall.

Hemiterpenes: terpenes composed of a single isoprene unit (C_5H_8). The group is represented by only a few members, the most important being isoprene, which is formed by removal of pyrophosphate from isopentenylpyrophosphate ("active" isoprene) (Fig.).

$H_2C{=}C(H_3C)-CH_2-CH_2-O-PP \longrightarrow H_2C{=}C(H_3C)-CH{=}CH_2$

Isopentenylpyrophosphate → Isoprene

Formation of isoprene from isopentenyl pyrophosphate

Hemocyanins: copper proteins responsible for oxygen transport in arthropod and molluscan hemolymph. They do not contain porphyrin and are not particle-bound. Oxidized H. are blue. Snail blood contains 2–4 % H., which has a copper content of 0.24 %. Two copper (I) ions are required to bind one molecule of oxygen. Molluscan H. are cylindrical oligomers with subunits of M_r 400,000, each subunit containing 8 oxygen-binding sites, i.e. 16 copper atoms. The H. from arthropods consist of hexamers, with subunits of M_r 75,000, which can aggregate to form multi-hexamers; each subunit carries one pair of copper atoms. The oxygen-binding centers of these two classes of H. are believed to be similar, and may resemble the active sites of other copper proteins, such as oxytyrosinase, ascorbate oxidase, laccase and ceruloplasmin. The structure of spiny lobster *(Panulirus interruptus)* H. has been determined to 3.2 Å resolution [W. P. J. Gaykema et al. *Nature* **309** (1984) 23–29].

Hemoglobin, *Hb*: a vertebrate oxygen-transporting protein, responsible for the red color of blood. It is present as a 34 % solution in the erythrocytes (red blood cells) and it carries oxygen from the lungs to the other tissues. Hb is a tetramer of two pairs of polypeptide chains and 4 heme groups; M_r 64,500. With a Fe^{2+} content of 0.334 %, the total of 950 g Hb in the human body represents 3.5 g or 80 % of the total body iron. Adult human Hb consists of 96.5–98.5 % HbA_1 ($\alpha_2\beta_2$) and 1.5–3.5 % HbA_2 ($\alpha_2\delta_2$). At least 7 different structural genes specify globin; these are clustered on chromosomes 11 and 16 (Fig. 1).

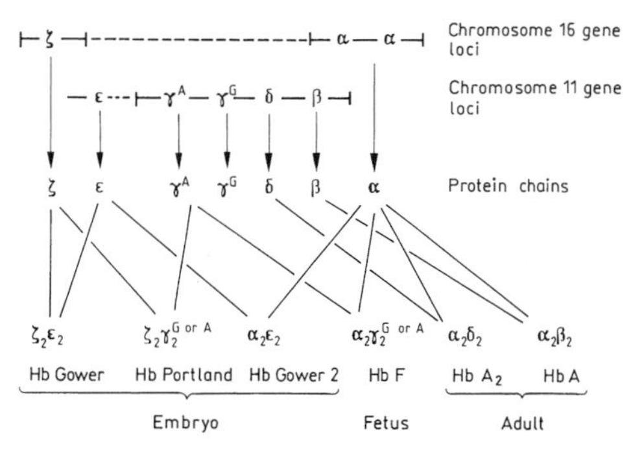

Fig. 1. *Hemoglobin.* Arrangement of subunits of different hemoglobins, and their genetic origin. Hb = hemoglobin.

There are two α-genes per haploid genome, i.e. a total of 4 α-genes. Three forms of Hb are found very early in embryonic development: $\zeta_2\varepsilon_2$ (Hb Gower), $\zeta_2\gamma_2$ (Hb Portland) and $\alpha_2\varepsilon_2$ (Hb Gower 2). After 3 months gestation, ζ and ε polypeptides are no longer synthesized (Fig. 2), and all 3 embryonic Hbs (HbEs) disappear to be replaced by fetal Hb (HbF, $\alpha_2\gamma_2$). The ζ-chain can be considered as an embryonic α-chain; it differs from the α-chain by about 60 residues. Most of these differences are conservative, but in particular the ζ-chain contains 4 more Arg residues and 3 fewer His residues than the α-chain, the *N*-terminus is acetylated (affecting the Bohr shift), and His(H5) is present.

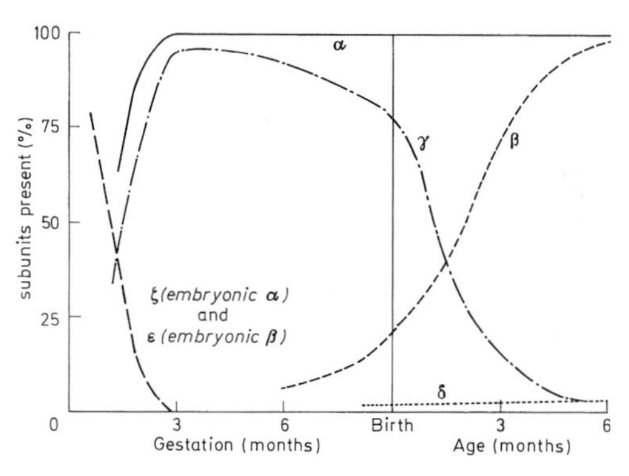

Fig. 2. *Hemoglobin.* Changes in the concentration of hemoglobin subunits during human gestation and early neonatal development. Since all subunits are part of a hemoglobin tetramer, these changes in turn reflect the developmental changes in the types of hemoglobin present (see Fig. 1).

Functionally, the ε-chain is an embryonic β-like chain; it differs from the β-chain by about 32 residues and from the γ-chain by about 25 residues; most of these differences are conservative, but in particular ε77 (E20) is Asn, ε116(G18) is Thr, and there is an extra Lys at ε87(F3) (cf. amino acid sequences in Fig. 3). HbE and HbF differ from adult Hb in their higher O_2 affinity, making possible the exchange of gas between

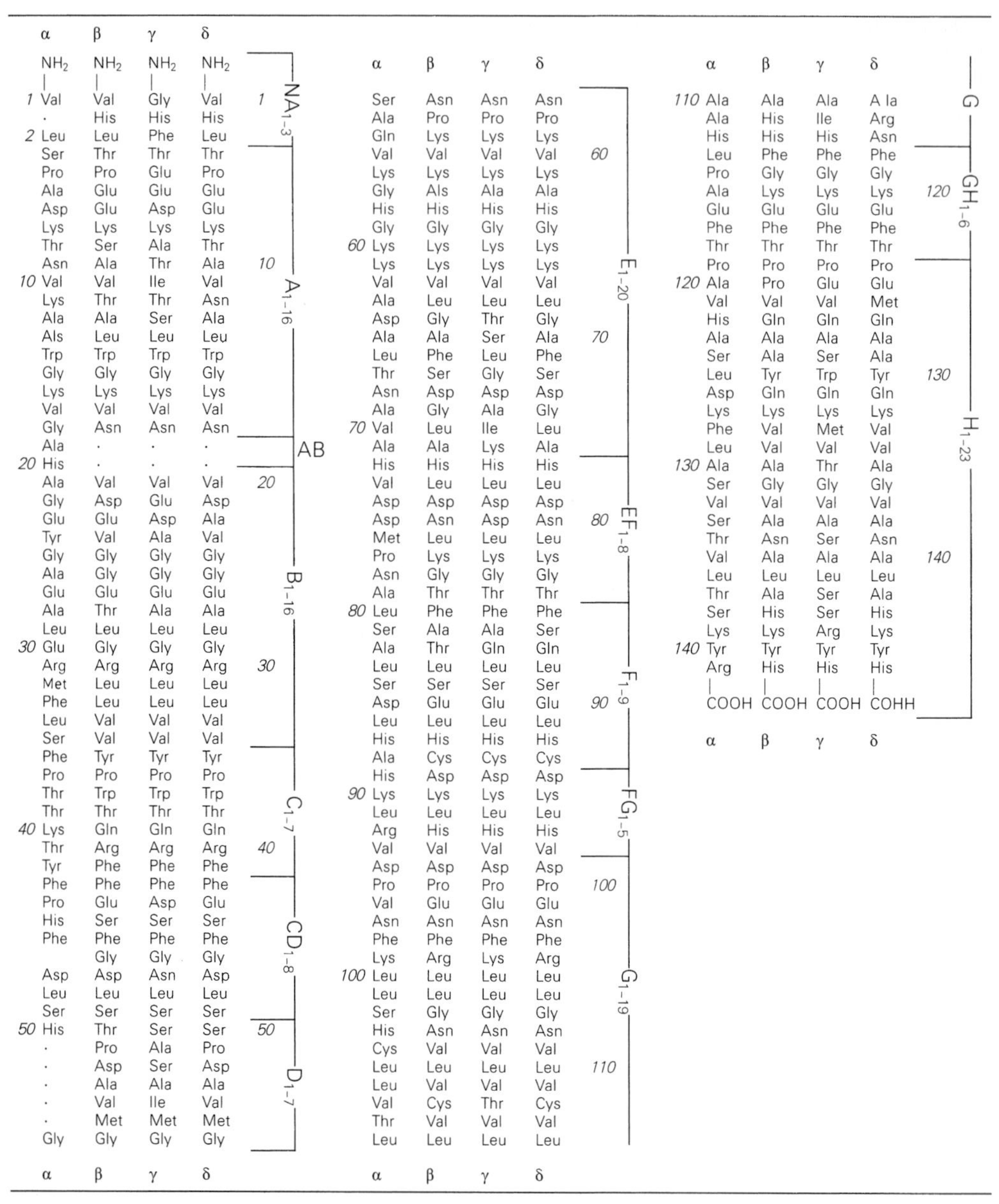

No. (α)	α	β	γ	δ	No. (β, γ, δ)	Region
	NH2	NH2	NH2	NH2		
1	Val	Val	Gly	Val	1	NA_{1-3}
	·	His	His	His		
2	Leu	Leu	Phe	Leu		
	Ser	Thr	Thr	Thr		A_{1-16}
	Pro	Pro	Glu	Pro		
	Ala	Glu	Glu	Glu		
	Asp	Glu	Asp	Glu		
	Lys	Lys	Lys	Lys		
	Thr	Ser	Ala	Thr		
	Asn	Ala	Thr	Ala	10	
10	Val	Val	Ile	Val		
	Lys	Thr	Thr	Asn		
	Ala	Ala	Ser	Ala		
	Als	Leu	Leu	Leu		
	Trp	Trp	Trp	Trp		
	Gly	Gly	Gly	Gly		
	Lys	Lys	Lys	Lys		
	Val	Val	Val	Val		
	Gly	Asn	Asn	Asn		
	Ala	·	·	·		AB
20	His	·	·	·		B_{1-16}
	Ala	Val	Val	Val	20	
	Gly	Asp	Glu	Asp		
	Glu	Glu	Asp	Ala		
	Tyr	Val	Ala	Val		
	Gly	Gly	Gly	Gly		
	Ala	Gly	Gly	Gly		
	Glu	Glu	Glu	Glu		
	Ala	Thr	Ala	Ala		
	Leu	Leu	Leu	Leu		
30	Glu	Gly	Gly	Gly		
	Arg	Arg	Arg	Arg	30	
	Met	Leu	Leu	Leu		
	Phe	Leu	Leu	Leu		
	Leu	Val	Val	Val		
	Ser	Val	Val	Val		
	Phe	Tyr	Tyr	Tyr		C_{1-7}
	Pro	Pro	Pro	Pro		
	Thr	Trp	Trp	Trp		
	Thr	Thr	Thr	Thr		
40	Lys	Gln	Gln	Gln		
	Thr	Arg	Arg	Arg	40	
	Tyr	Phe	Phe	Phe		
	Phe	Phe	Phe	Phe		CD_{1-8}
	Pro	Glu	Asp	Glu		
	His	Ser	Ser	Ser		
	Phe	Phe	Phe	Phe		
		Gly	Gly	Gly		
	Asp	Asp	Asn	Asp		
	Leu	Leu	Leu	Leu		
	Ser	Ser	Ser	Ser		
50	His	Thr	Ser	Ser	50	D_{1-7}
	·	Pro	Ala	Pro		
	·	Asp	Ser	Asp		
	·	Ala	Ala	Ala		
	·	Val	Ile	Val		
	·	Met	Met	Met		
	Gly	Gly	Gly	Gly		
	Ser	Asn	Asn	Asn		E_{1-20}
	Ala	Pro	Pro	Pro		
	Gln	Lys	Lys	Lys		
	Val	Val	Val	Val	60	
	Lys	Lys	Lys	Lys		
	Gly	Als	Ala	Ala		
	His	His	His	His		
	Gly	Gly	Gly	Gly		
60	Lys	Lys	Lys	Lys		
	Lys	Lys	Lys	Lys		
	Val	Val	Val	Val		
	Ala	Leu	Leu	Leu		
	Asp	Gly	Thr	Gly		
	Ala	Ala	Ser	Ala	70	
	Leu	Phe	Leu	Phe		
	Thr	Ser	Gly	Ser		
	Asn	Asp	Asp	Asp		
	Ala	Gly	Ala	Gly		
70	Val	Leu	Ile	Leu		
	Ala	Ala	Lys	Ala		
	His	His	His	His		EF_{1-8}
	Val	Leu	Leu	Leu		
	Asp	Asp	Asp	Asp		
	Asp	Asn	Asp	Asn	80	
	Met	Leu	Leu	Leu		
	Pro	Lys	Lys	Lys		
	Asn	Gly	Gly	Gly		
	Ala	Thr	Thr	Thr		
80	Leu	Phe	Phe	Phe		F_{1-9}
	Ser	Ala	Ala	Ser		
	Ala	Thr	Gln	Gln		
	Leu	Leu	Leu	Leu		
	Ser	Ser	Ser	Ser		
	Asp	Glu	Glu	Glu	90	
	Leu	Leu	Leu	Leu		
	His	His	His	His		
	Ala	Cys	Cys	Cys		
	His	Asp	Asp	Asp		FG_{1-5}
90	Lys	Lys	Lys	Lys		
	Leu	Leu	Leu	Leu		
	Arg	His	His	His		
	Val	Val	Val	Val		
	Asp	Asp	Asp	Asp		G_{1-19}
	Pro	Pro	Pro	Pro	100	
	Val	Glu	Glu	Glu		
	Asn	Asn	Asn	Asn		
	Phe	Phe	Phe	Phe		
	Lys	Arg	Lys	Arg		
100	Leu	Leu	Leu	Leu		
	Leu	Leu	Leu	Leu		
	Ser	Gly	Gly	Gly		
	His	Asn	Asn	Asn		
	Cys	Val	Val	Val		
	Leu	Leu	Leu	Leu	110	
	Leu	Val	Val	Val		
	Val	Cys	Thr	Cys		
	Thr	Val	Val	Val		
	Leu	Leu	Leu	Leu		
110	Ala	Ala	Ala	A la		
	Ala	His	Ile	Arg		
	His	His	His	Asn		
	Leu	Phe	Phe	Phe		GH_{1-6}
	Pro	Gly	Gly	Gly		
	Ala	Lys	Lys	Lys	120	
	Glu	Glu	Glu	Glu		
	Phe	Phe	Phe	Phe		
	Thr	Thr	Thr	Thr		
	Pro	Pro	Pro	Pro		H_{1-23}
120	Ala	Pro	Glu	Glu		
	Val	Val	Val	Met		
	His	Gln	Gln	Gln		
	Ala	Ala	Ala	Ala		
	Ser	Ala	Ser	Ala		
	Leu	Tyr	Trp	Tyr	130	
	Asp	Gln	Gln	Gln		
	Lys	Lys	Lys	Lys		
	Phe	Val	Met	Val		
	Leu	Val	Val	Val		
130	Ala	Ala	Thr	Ala		
	Ser	Gly	Gly	Gly		
	Val	Val	Val	Val		
	Ser	Ala	Ala	Ala		
	Thr	Asn	Ser	Asn		
	Val	Ala	Ala	Ala	140	
	Leu	Leu	Leu	Leu		
	Thr	Ala	Ser	Ala		
	Ser	His	Ser	His		
	Lys	Lys	Arg	Lys		
140	Tyr	Tyr	Tyr	Tyr		
	Arg	His	His	His		
	COOH	COOH	COOH	COHH		

Fig. 3. *Hemoglobin.* Amino acid sequences of human hemoglobin subuniuts α, β, γ and δ. Numbers on the left of each column refer to residues in the α-subunit only. Numbers on the right refer to the residues in subunits β, γ and δ. Single letters, A, B, C, etc. refer to helices according to Kendrew's nomenclature (see Figs. 4 & 5 below, and Figs. 9 & 10 in Proteins). Paired letters, AB, CD, etc. refer to nonhelical, interhelical stretches. The number of residues in each region is given as a subscript in the designating letter.

fetal and maternal blood. The Hb of all human races and even of chimpanzees is identical. Anomalies are pathological and arise through point mutation, which leads to the substitution of amino acids, or, more rarely, to their absence. Of the presently known 153 abnormal Hb, 87 are variations of the β-chain. The most frequent and best known is Sickle cell Hb (see). Although only a few of the variants lead to disease, such as hemolytic anemia, the absence of an entire chain represents a severe condition for the carrier. In β-thalassemia, which occurs in Mediterranean countries, no β-chains are formed (HbA_1 is missing). In its place the blood of these patients contains HbF (85–95 %) and HbA_2 (5–15 %). Human α-chains contain 141 amino acid residues, and β-chains 146 residues; the sequences of human Hb and of many other vertebrate Hb have been determined. The tertiary structures of the Hb chains and the quaternary structure of the entire tetrameric molecule were determined by Perutz, using X-ray analysis. Aside from slight differences due to variations in primary structure, the Hb chains are folded in a manner very similar to that of the myoglobin molecule. Attachment of the heme group by bonding of its iron(II) atom to two histidine residues and the hydrophobic interactions in the heme pocket of Hb also correspond to myoglobin. However, the primarily hydrophobic interactions between the individual Hb chains, which are not present in myoglobin, are much more complicated. There are large regions of hydrophobic contact, especially between the α- and β-chains, and these are the basis for interactions of the four spatially separated heme groups in the reversible, cooperative binding of four oxygen molecules to each molecule of Hb.

The hydrophobic interactions also make possible the gliding motions of the two αβ-dimers with respect to one another during O_2 loading and release. This allosteric effect is responsible for the sigmoidal O_2 binding curve of Hb. The lack of covalent bonds between the Hb chains can be shown by the reversible dissociation of Hb. At pH 4, the dissociation is asymmetric, forming $\alpha_2 + \beta_2$; at pH 11; in 1 M NaCl, it is symmetric, forming 2 (αβ) units; in the presence of *p*-chloromercuribenzoate, dissociation into monomers occurs. It is medically significant that the affinity of Hb to the poisonous CO is 325 times greater than its affinity to O_2.

The single-chain, Hb-like respiratory pigments in invertebrates (M_r about 16,000) are similar in primary and tertiary structure to myoglobin and to the α- and β-chains of Hb; they are thus regarded as ancestral to the tetrameric Hb. The single-chain hemoglobin in yeast (M_r 50,000) is unique, because it contains FAD in addition to heme.

Conformational states of hemoglobin. In deoxyHb, the 4 subunits are relatively tightly associated, and the chains are linked by several noncovalent (ionic) bonds, as well as hydrophobic interactions. DeoxyHb is therefore known as the T-form (taut, tight or tense). The more important ionic bonds (salt bridges) of the T-form are: free terminal NH_2 of Val-1 (α_2) with the free (terminal) COOH of Arg-141, similarly Val-1 (α_1) with Arg-141 (α_2); guanidinium of Arg-141 (α_1) with side chain COOH of Asp-126 (α_2), similarly Arg-141 (α_2) with Asp-126 (α_1); free (terminal) COOH of His-146 (β_1) with ε-NH_2 of Lys-40 (α_2), similarly His-146 (β_2) with Lys-40 (α_1); imidazole of His-146 (β_1) with side chain COOH of Asp-94 (β_1), similarly His-146 (β_2) with Asp-94 (β_2) (Fig. 4)

In deoxyHb the penultimate Tyr residues of all 4 chains lie in hydrophobic pockets between helices F and H. The Fe atom of heme is displaced about 0.08 nm from the plane of the porphyrin ring system; the hemes lie in V-shaped pockets formed by helices F and E. In the α-subunits the heme groups are in open pockets, permitting access of O_2, whereas the

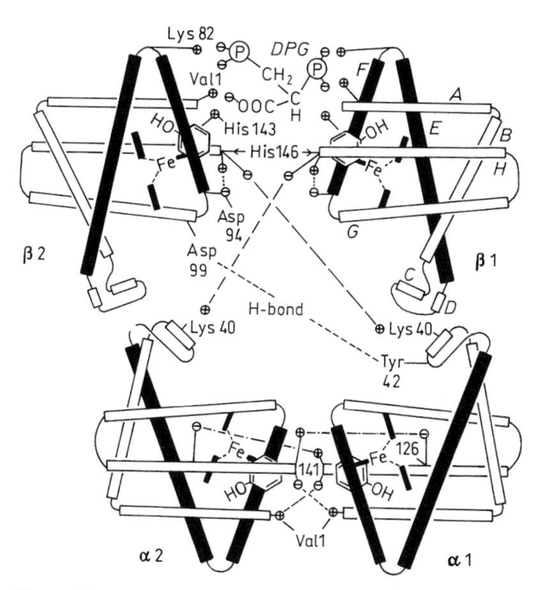

Fig. 4. *Hemoglobin.* Schematic representation of the four subunits of deoxyhemoglobin. Helical regions are labeled A, B, C, etc., the same designation as in Fig. 3. For the true perspective of the tetramer, see Fig. 6 below, and Figs. 9 & 10 in proteins. DPG = 2,3-diphosphoglycerate.

heme pockets of the β-subunits are more compressed, preventing entry of O_2. 2,3-Diphosphoglycerate is present and forms salt bridges with positively charged groups of the two β-chains (Fig. 4).

In oxyHb the subunits are less tightly associated, so that this form is known as the R-form (relaxed). In the R-form, the dissociation constants of ionizable groups are changed, and salt bridges present in deoxyHb are broken; the most important of these seem to be associated with His-146 (β) and Val-1 (α). Also in oxyHb, the β-subunits move closer together; the heme pockets of the β-subunits become wider and admit O_2; as the iron of each heme group binds oxygen, it moves 0.08 nm into the plane of the porphyrin ring, pulling His-92 with it. 2,3-Diphosphoglycerate is not present (Fig. 5).

In the transition from the T to the R-form, one pair of subunits ($\alpha_1\beta_1$) rotates through 15° relative to the other pair ($\alpha_2\beta_2$). The axis of rotation is eccentric, so that the $\alpha_1\beta_1$ pair also moves slightly towards the axis (Fig. 6).

Bohr effect. a reversible shift in the O_2-binding curve of Hb, which permits O_2 binding and CO_2 release in the alveolar capillaries of the lungs, and the reverse process in respiring tissues (Fig. 7). Carbonic anhydrase in erythrocytes promotes the rapid formation of carbonic acid from dissolved CO_2 in respiring tissue, and each molecule of H_2CO_3 spontaneously dissociates into bicarbonate and a proton. The protons are absorbed by deoxyHb, which thereby acts as a buffer. OxyHb does not bind protons, but for every 4 O_2 molecules lost, the resulting deoxyHb binds 2 protons. Also, about 15 % of CO_2 carried in the blood

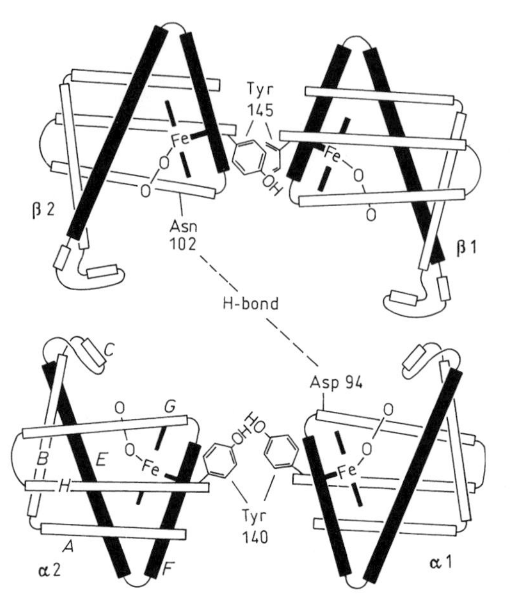

Fig. 5. *Hemoglobin.* Schematic representation of the four subunits of oxyhemoglobin. Helical regions are labeled A, B, C, etc., the same designation as in Fig. 3. 2,3-Diphosphoglycerate is no longer present and the two β-subunits have moved closer together (cf. Fig. 4). For the true perspective of the tetramer, see Fig. 6 below, and Figs. 9 & 10 in proteins.

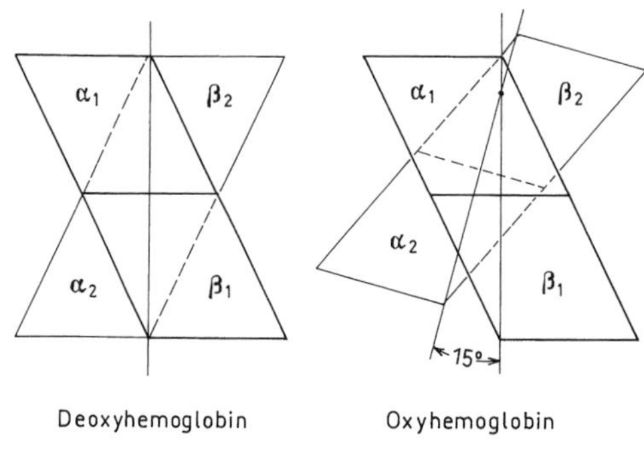

Fig. 6. *Hemoglobin.* Relative positions of the subunits in deoxyhemoglobin (T-form) and oxyhemoglobin (R-form).

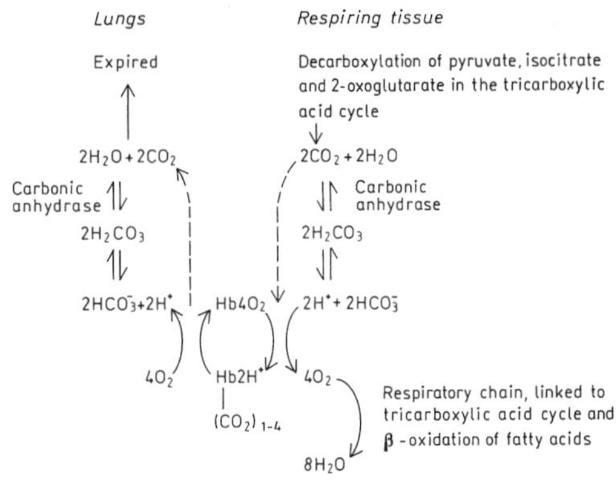

Fig. 7. *Hemoglobin.* The Bohr effect. Binding of protons and CO_2 in respiring tissue causes a shift in the oxygen-binding curve of hemoglobin (Hb) so that oxygen is readily lost. In the capillaries of the lung alveoli, the reverse process occurs, and the high oxygen tension promotes oxygen binding with concomitant loss of protons and CO_2.

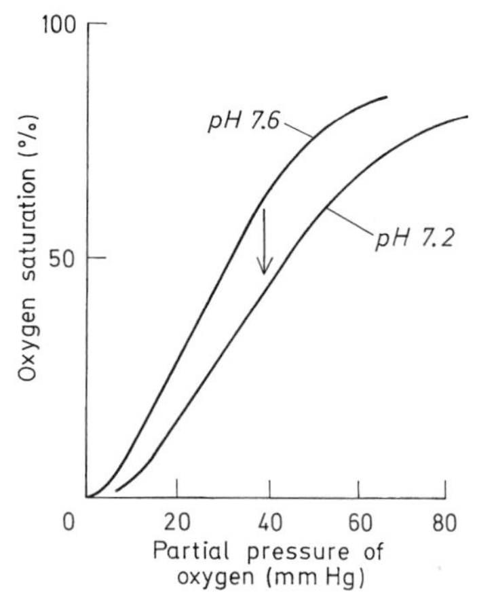

Fig. 8. *Hemoglobin.* Effect of pH on the oxygen saturation curve of hemoglobin. As acidity increases, i.e. a decrease in pH from 7.6 to 7.2, oxygen release is favored (arrow) because percentage saturation is less at any give oxygen partial pressure.

is bound to Hb as carbamino groups. The process is reversed in the lungs, where O_2 binds to deoxyHb with release of protons, i.e. O_2 binding promotes exhalation of CO_2 from the lungs, whereas proton binding promotes release of O_2 in respiring tissues.

This shift in the O_2-binding properties of Hb is caused by the acidity (Fig. 8) and high CO_2 tension (CO_2 alone is effective at constant pH) in respiring tissue, and the low acidity and low CO_2 tension in the lungs. The carbamino groups, formed by reaction of CO_2 with the non-ionized terminal α-amino groups of each of the 4 chains, form salt bridges which help to stabilize the deoxyHb: $RNH_2 + CO_2 \rightleftharpoons RNHCOO^- + H^+$.

In the transition from oxyHb to deoxyHb, 3 pairs of negatively charged (i.e. proton-binding) groups are moved to more negative environments: His-146 (*C*-terminal) on each β-chain; terminal NH_2 groups of the α-chains, and His-122 on each α-chain. The resulting increases in pK of these groups renders them available for binding protons.

The β-chain *C*-terminal His-146 in oxyHb can rotate freely, but in deoxyHb the β His-146 becomes involved in several interactions, and in particular it comes into close proximity with negatively charged Asp-94 on the same β-chain. In deoxyHb the terminal amino group of one α-chain interacts with the carboxyl terminal of the other α-chain, thereby raising the pK of the amino group and increasing its affinity for H^+.

2, 3-Diphosphoglycerate (DPG), glycerate 2, 3-bisphosphate. DPG binds noncovalently to deoxyHb but not to oxyHb. In deoxyHb, one mol DPG is associated with the charged α-amino groups of the *N*-terminal valine residues of the two β-chains. Other β-chain groups possibly contributing to DPG binding in deoxyHb are His-2, Lys-82 and His-143 (Fig. 3). DPG pulls the equilibrium between oxyHb and deoxyHb + O_2 to the right. The molar concentration of DPG in erythrocytes is about the same as that of Hb; this is sufficient to shift the dissociation curve to the right at all times (Fig. 9): $Hb(O_2)_4 + DPG \rightleftharpoons Hb \cdot DPG + 4O_2$.

The erythrocyte DPG concentration may change in response to defective oxygen delivery to the tissues. For example, if airflow in the bronchioles is restricted, as in obstructive pulmonary emphysema, the O_2 pressure of the arterial blood is decreased. This is compensated by a shift in the O_2 dissociation curve, due to an increase in the erythrocyte concentration of DPG from 4.5 mM to as high as 8.0 mM (Fig. 9). DPG may also play a part in high altitude adaptation; transfer from sea level to 4,500 meters results, after two days, in an increase of the erythrocyte DPG concentration to 7.0 mM. The O_2 affinity of blood increases during storage, due to loss of DPG (DPG decreases to 0.5 mM in 10 days when blood is stored in acid-citrate-dextrose). Although transfused blood can regain DPG (half normal levels attained after 24 h, following total depletion), this may not be rapid enough in critical cases. Addition of DPG to the blood has no effect, because it cannot cross the erythrocyte membrane. Addition of Inosine (see), however, maintains the DPG level of stored red cells, because inosine crosses the erythrocyte membrane and is converted into DPG, a conversion involving reactions of the Pentose phosphate cycle (see). See Rapoport-Luerbing shuttle. [R. E. Dickerson & I. Geis *The Structure and Action of proteins* (W. A. Benjamin, 1969); M. F. Perutz *Annu. Rev. Biochem.* **48** (1979) 327–386, *Nature* **228** (1970) 726–739, *Sci. Amer.* **239** (6) (1978) 92–125; J. V. Kilmartin *Brit. Med. Bull.* **32** (1976) 209–222; A. Arnone *Nature* **237** (1972) 146–149]

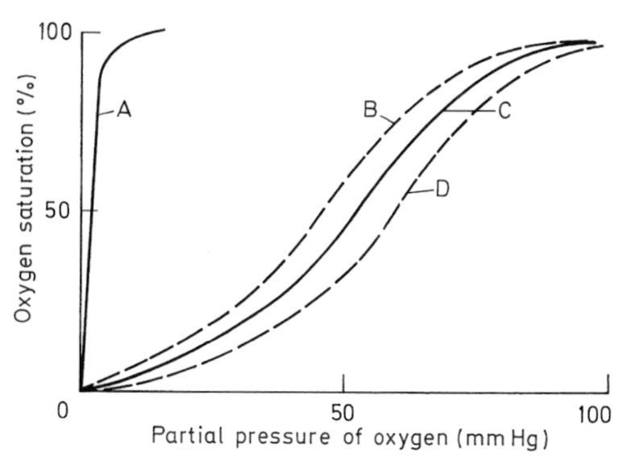

Fig. 9. *Hemoglobin.* Effect of 2,3-diphosphoglycerate on the oxygen saturation curve of hemoglobin.
A, 2,3-diphosphoglycerate absent.
B, 2,3-diphosphoglycerate concentration low.
C, 2,3-diphosphoglycerate concentration normal.
D, 2,3-diphosphoglycerate concentration high.

Hemoglobinopathy: an inherited abnormality of hemoglobin structure, usually the substitution of a single amino acid in the α- or β-chains as a result of a point mutation in the α or b gene. H. due to amino acid deletions are also known. H. are Inborn errors of metabolism (see). Thalassemias (see) may also be classified as H. Many structurally abnormal hemoglobins are clinically unimportant. Others may show decreased or increased affinity for oxygen, or display instability leading to Heinz body (see) formation, sickling, decreased erythrocyte survival and hemolytic anemia. Other functional defects include lack of subunit cooperativity, decreased Bohr effect and decreased effect of 2 3-diphosphoglycerate (see Hemoglobin). The main initial screening techniques for abnormal

Table. *Abnormal hemoglobins.* The names of abnormal hemoglobins usually refer to the geographical location where the hemoglobin was first discovered. The term, unstable, refers to those hemoglobins associated with accelerated erythrocyte destruction *in vivo* and precipitation of hemoglobin *in vitro* when warmed to 50 °C. For residue numbering system and comparison with normal hemoglobin A, see Hemoglobin.

A. ***Amino acid replacements affecting contact with heme***

A-1. *Hemoglobins M (M for Methemoglohin).* The heme iron is irreversibly oxidized, i. e. permanently in the Fe(III) state. The resulting methemoglobin cannot be reduced by methemoglobin reductase. Five types are known. Only heterozygotes survive.

Boston: α58(E7)His → Tyr. Permanently in deoxy form. Low O_2 affinity. Low cooperativity. Bohr effect decreased.
Saskatoon (=Emory): β63(E7)His → Tyr. Unstable. High O_2 affinity. Decreased cooperativity. Bohr effect normal.
Milwaukee: β67(E11)Val → Glu. Properties as for Saskatoon.
Iwate: α87(F8)His → Tyr. Permanently in deoxy form. Low O_2 affinity. Cooperativity absent. Bohr effect absent.
Hyde Park: β92(F8)His → Tyr. High O_2 affinity. Decreased cooperativity. Bohr effect normal.

N—Fe^{2+}—O_2 ⟶ O—Fe^{3+}—OH_2

His F8 Tyr F8

Replacement of heme-binding His F8 by Tyr destroys the ability of heme to bind oxygen.

A-2. *Other substitutions in the heme binding region*

Fort de France: α45(CD3)His → Arg. High O_2 affinity.
Hirosaki: α43(CDI)Phe → Leu. Unstable.
Torino: α43(CD1)Phe → Val. Low O_2 affinity.
J Buda: α61(E10)Lys → Asn. Low O_2 affinity. Slightly unstable.
Moabit: α86(F7)Leu → Arg. Low O_2 affinity. Slightly unstable.
Bibba: α136(H19)Leu → Pro. Unstable.
Hammersmith: β42(CD1)Phe → Ser. Low O_2 affinity. Unstable. Slightly decreased cooperativity. Polar Ser probably permits water to enter heme pocket.
Zürich: β63(E7)His → Arg. High O_2 affinity. Unstable. Bohr effect normal. Decreased cooperativity.
Shepherd's Bush: β74(E18)Gly → Asp. High O_2 affinity. Unstable. Bohr effect normal. Decreased cooperativity. Effect of 2,3-diphosphoglycerate decreased.
Sabine: β91(F7)Leu → Pro. Unstable. Heme group absent.
Casper: β106(G8)Leu → Pro. Unstable. Heme group absent.
Bryn Mawr: β85(F1)Phe → Ser. High O_2 affinity. Unstable.

B. ***Amino acid replacements affecting contact between α1 and β1** subunits.*

Khartoum: β124(H2)Pro → Arg. Unstable.
Fannin Lubbock: β119(GH2)Gly → Asp. Unstable.
Madrid: β115(G17)Ala → Pro. Unstable.
San Diego: β109(G11)Val → Met. High O_2 affinity.
Philly: β35(C1)Tyr → Phe. High O_2 affinity. Unstable.
Hemoglobin E: β26(B8)Glu → Lys. Purified HbE has normal O_2 affinity. In erythrocytes homozygous HbE has low O_2 affinity due to increased 2,3-diphosphoglycerate in cells. Cells containing HbA and HbE (heterozygotes) have normal O_2 affinity. See section G.
Heathrow: β103(G5)Phe → Leu. High O_2 affinity.
Prato: α31(B12)Arg → Ser. No abnormal properties.
Chiapas: α114(GH2)Pro → Arg. No abnormal properties.

C. ***Amino acid replacements affecting contact between α_1 and β_2 subunits.***

Hiroshima: β146(C-term His) → Asp. Bohr proton cannot be donated. High O_2 affinity. Bohr effect decreased.
Richmond: β102(G4)Asn → Lys. No abnormal properties.
Kempsey: β99(G1)Asp → Asn. H-bond between Asp(βG1) and Tyr(αC7) absent. High O_2 affinity. DeoxyHb less stable.
Setif: α94(G1)Asp → Tyr. Unstable.

D. ***Amino acid replacements in cavity between like chains***

Manitoba: α102(G9)Ser → Arg. Slightly unstable.
Jackson: α127(H10)Lys → Asn. Abnormal properties not reported.
Surenes: α141(HC3)Arg → His. High O_2 affinity.
Helsinki: β82(EF6)Lys → Met. Low O_2 affinity. Bohr effect decreased. Lys_{82} is diphosphoglycerate binding site.
Altdorf: β135(H13)Ala → Pro. High O_2 affinity. Unstable.
Syracuse: β143(H21)His → Arg. High O_2 affinity. Bohr effect decreased. His_{143} is diphosphoglycerate binding site.

E. ***Amino acid replacements in the interior of subunits.*** Replacement of a nonpolar by a polar residue in the hydrophobic interior, or a small residue by a large one can cause instability. Insertion of Pro into a helix causes distortion (Pro is a "helix breaker") and instability.

Port Phillip: α91(FG3)Leu → Pro. Unstable.
Perth: β32(B14)Leu → Pro. High O_2 affinity. Unstable.
Riverdale-Bronx: β25(B6)Gly → Arg. Unstable. Helices B and E closely packed in this region. Arg too large.

F. ***Amino acid deletions.*** All these Hb are unstable.

Freiburg: β23(B5)Gly deleted. High O_2 affinity.
St. Antoine: β74–75(E18-E19)Gly-Leu deleted. O_2 affinity normal.
Gun Hill: β91–95(F7-FG2)Leu-His-Cys-Asp-Lys deleted. Essential contacts with heme absent. β-chains contain no heme. Hemolytic anemia.

G. ***Amino acid replacements on the outer surface of the molecule.***

Hemoglobin C: β6(A3)Glu → Lys. Relatively common in West Africans and people of West African extraction. Heterozygotes have 30–40 % HbC (+ about 60 % HbA) and are healthy. Homozygotes may have mild anemia and their lifespan is normal.
Hemoglobin D: β121(GH4)Glu → Gln. Relatively prevalent among Negroes (0.4 %), Algerians (2.0 %) and Sikhs of north and central India, with sporadic occurrence in other groups. Homozygotes show very minor symptoms (anemia?) and the term "hemoglobin D disease" is probably too strong.
Hemoglobin E: β26(B8)Glu → Lys. See also section B. The second most common abnormal Hb in the world, in people of southeast Asian origin. Homozygotes show mild anemia.
Hemoglobin J: α115(GH3)Ala → Asp. Found in Melanesia (New Hebrides) New Guinea. Seems to be of no pathological significance.
Hemoglobin O Arab: β121(GH4)Glu → Lys. Enhances sickling when heterozygous with HbS. Low O_2 affinity of erythrocytes in homozygotes, which show a sickling condition. Probably originates from non-Arab peoples of presemitic Egypt. Reservoir is possibly Sudanese, spreading through Ottoman Empire. Also found in Jamaica, Roumania, Bulgaria, Hungary.
Korle Bu: β73(E17)Asp → Asn. Described in only one West African family. Homozygotes normal with no pathology.
Hemoglobin S (sickle cell hemoglobin). The most common pathologically abnormal hemoglobin. See separate entry.

hemoglobins are starch gel and paper electrophoresis (Fig.). Electrophoresis cannot differentiate all mutant hemoglobins. Evidence for the presence of an abnormal hemoglobin is often physiological or clinical, e.g. abnormal oxygen affinity, instability, etc. Ultimately, the α- and β-chains of the suspect hemoglobin must be purified and analysed by conventional techniques of protein sequence analysis to identify the site of amino acid substitution. Lettering of abnormal hemoglobins (see Fig. and Table, section G) was at first an initial system (F=fetal, S=sickle, M=methemoglobin), after which letters were assigned in order of discovery. The table does not list every known H., and new examples are continually reported in the literature. About 350 different H. are known; about 150 of these are discussed by E.R.Huehns in *Blood and Its Disorders* (R.M.Hardisty & D.J.Weatherall, eds. 2nd edition, Blackwell Scientific Publications, Oxford, 1982). The International Hemoglobin Information Center publishes updated lists of abnormal hemoglobins in the journal *Hemoglobin.* See also R.G.Schneider "Methods for Detection of Hemoglobin Variants and Hemoglobinopathies in the Routine Clinical Laboratory" *RC Critical Review in Clinical Laboratory Sciences* (November 1978).

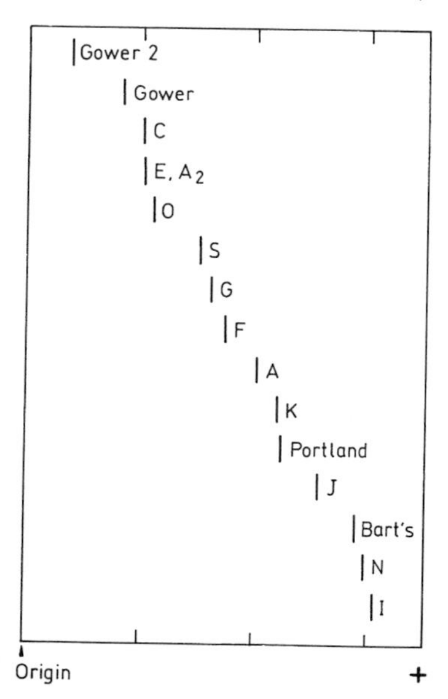

Electrophoresis of normal and variant hemoglobins. Relative migration distances in starch gel electrophoresis in Tris-EDTA-borate buffer, pH 8.6. Hemoglobin I is the same as H or β_4. Hemoglobin Bart's is hemoglobin γ_4.

Hemopexin: a single-chain, heme-binding β1-plasma glycoprotein, M_r 57,000. H. contains 22% carbohydrate. In contrast to Haptoglobin (see), H. binds neither hemoglobin nor cytochrome *c*, but only their prosthetic group, heme. 100 ml human plasma contains 80 to 100 mg H.

Hemopoietic cell growth factor: see Colony stimulating factor.

Hemoproteins: ubiquitous chromoproteins which, as respiratory pigments, are involved in oxygen transport (see Hemoglobin, and Hemerythrin) and in oxygen storage (see Myoglobin). Catalase and peroxidase are responsible for reduction of peroxides, while cytochromes are involved in electron transport between dehydrogenases and terminal acceptors. Their prosthetic group, iron porphyrin IX or heme, is tightly bound to the protein component. While 4 of the iron ligands are occupied by the porphyrin ring, the other 2 are used for binding to the protein (via histidine) and to O_2 (in the respiratory H.) or also to the protein (via Cys, Met, Trp, Lys or Tyr). In cytochrome *c* the porphyrin ring is additionally bound to the protein by covalent bonds between two SH groups and the two vinyl groups of the heme. One of the characteristic properties of the H. in the reduced state (Fe II) is a spectrum with 3 intense bands in the visible range: the α-band, λ_{max} 550–565 nm; the β-band, λ_{max} 520–535; and the γ or Soret band, λ_{max} 400–415 nm.

Hemosiderin: an iron storage protein of the mammalian organism, functionally related to Ferritin (see). H. is deposited in the liver and spleen (hemosiderosis), particularly in diseases associated with increased blood destruction, such as pernicious anemia, or with increased iron resorption (hemochromatosis), or even in hemorrhages. Most of the deposits are located in the liver, which may contain up to 50 g H., compared with the normal content of 120 to 300 mg H.H. from horse spleen consists of 26–34% iron(III), and up to 35% protein (aposiderin). The rest is made up of octasubstituted porphyrin, mucopolysaccharides and fatty acid esters.

Heparin: a dextrorotatory acid mucopolysaccharide (M_r ~ 16,000) from animal tissues which prevents blood clotting. H. consists of equal amounts of α-1,4-glycosidically linked D-glucosamine and D-glucuronic acid, and it contains *O*- and *N*-sulfate residues. It is acidic and therefore forms salts The effective protein component in blood is called heparin complement. H. prevents the clotting of blood by preventing the conversion of prothrombin to thrombin and of fibrinogen to fibrin. Clinically it is applied parenterally for the treatment of thrombosis, phlebitis and embolism.

R=H or SO_3H

Heparin

Heparitin sulfate, *heparan sulfate:* a monosulfate ester of an acetylated (*N*-acetyl) heparin. As isolated from animal tissues (liver), H.s. is probably a mixture of mucopolysaccharides with varying degrees of sulfatation or amino group acetylation.

HEPES: acronym of *N*-2-Hydroxypiperazine-*N*-2-ethanesulfonic acid, a buffering compound used in the range pH 6.8–8.2.

Heptoses: monosaccharides containing 7 C-atoms. The 7-phosphates of D-mannoheptulose and D-sedoheptulose are important in carbohydrate metabolism.

Heroin, ***diacetylmorphine, diamorphine:*** one of the most dangerous narcotics, m.p. 173 °C, $[\alpha]_D^{20}$ −166 ° (methanol). H. is not very stable and decomposes in boiling water. It is synthesized chemically by acetylation of the two hydroxyl groups of morphine with acetyl chloride. This increases the analgesic effect by a factor of six. Due to the extreme danger of addiction, the therapeutic use of H. is forbidden in most countries.

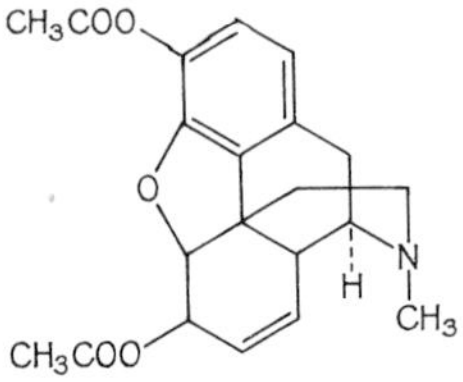

Heroin

Hers' disease: see Glycogen storage diseases.

Hershberg test: see Anabolic steroids.

Hesperetin: 5,7,3′-trihydroxy-4′-methoxyflavanone. See Flavanone.

Hesperidin: hesperetin 7-rutinoside, a Flavanone (see) glycoside which makes up 8 % of the dry weight of orange peel.

Heteroauxin: see Auxins.

Heteroglycans: polysaccharides composed of two or more different monosaccharide residues, e.g. pectins, plant mucilages, plant gums and mucopolysaccharides.

Heterophagy: see Intracellular digestion.

Heteropolar bond: see Noncovalent bonds.

Heteropolypeptides: see Proteinoids.

Heteroside: a compound of one or several carbohydrate residues and a component belonging to a different class of substance and known as the aglycon or genin. Thus glycosides are H.

Heterotrophy, ***heterotrophic nutrition:*** nutritional dependence on organic compounds. In carbon H. organic carbon compounds serve as sources of carbon and energy for the synthesis of body substituents and ATP. The degree of H. varies widely among the various heterotrophic organisms (all animals, including humans, and most microorganisms), and may include a dependence on externally supplied essential amino acids, fatty acids and vitamins. Auxotrophic mutants (see) have special nutritional requirements. Parasitism (see), Saprophytism (see) and Symbiosis (see) are special forms of heterotrophic feeding. The terms autotrophy and H. do not suffice to describe the widely varying forms of microbial nutrition, since here the nature of the carbon source and the energy source, as well as the chemical nature of the reducing agent used for reductive syntheses must be taken into account. The converse of H. is Autotrophy (see).

HETPP: acronym of hydroxyethylthiamin pyrophosphate. see Thiamin pyrophosphate.

Hexabrachion: see Tenascin.

Hexestrol, ***meso-hexestrol:*** a synthetic compound with estrogenic activity. It is not a steroid, but is used therapeutically in the same way as natural estrogens.

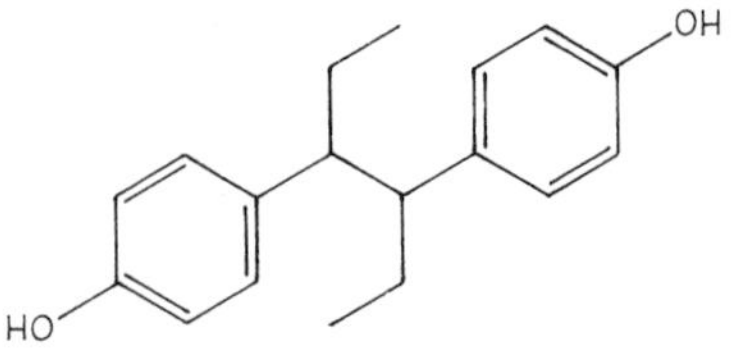

Hexestrol

Hexitols: sugar alcohols with 6 C-atoms. Of the 10 possible isomers, D-sorbitol, dulcitol, D-mannitol, iditol and allitol occur naturally.

Hexokinase (EC 2.7.1.1): an enzyme catalysing the transfer of a phosphoryl group from ATP to the C6 oxygen of glucose or another hexose, such as mannose or D-glucosamine; the physiological substrate is MgATP. The reaction with glucose is the first step in Glycolysis (see). H. binds specifically to porin, a protein of the outer mitochondrial membrane which permits ADP and saccharides to penetrate this membrane. Although H. is a soluble enzyme, it may therefore be associated with the mitochondrion under physiological conditions. This notion is supported by the observations that mitochondrion-associated H. has a higher K_m for MgATP (0.25 mM vs. 0.12 mM for non-associated H.) and is more susceptible to product inhibition by glucose 6-phosphate.

X-ray studies of H. crystals grown in the presence and absence of glucose indicate that when glucose is bound, one lobe of the molecule rotates through 12 °, thereby closing the substrate binding cleft around the glucose molecule. In this conformation, water is excluded from the cleft. The same conformational change also occurs in H. in aqueous solution, and analogs of glucose which are too bulky to allow this change are not substrates of the enzyme, but some are inhibitors. Although water is able to bind to the substrate pocket in the same position as the 6-OH group of the sugar, it does not induce the conformational change, nor is it readily phosphorylated (this would amount to hydrolysis of ATP; H. is not very active as an ATPase). Interestingly, the enthalpy and heat capacity of the enzyme are nearly unchanged by the binding. [B. I. Kurganov in G. R. Welch (ed.) *Organized Multienzyme Systems* (Academic Press, Orlando, 1985) pp. 241–268; K. Takahashi et al. *Biochem.* **20** (1981) 4693–4697, W. S. Bennett & T. A. Steitz, *Proc. Natl. Acad. Sci. USA* **75** (1978) 4848–4852]

Hexosans: high-molecular-mass plant polysaccharides, which are homoglycans composed of hexoses. Examples are glucans, fructans, mannans and galactans.

Hexose monophosphate pathway: see Pentose phosphate cycle.

Hexoses: aldoses containing 6 C-atoms, an important groups of monosaccharides (see Carbohydrates). All possible stereoisomeric aldohexoses (there are four asymmetric C atoms) have been isolated or synthesized. D-Glucose, D-mannose, D-galactose and L- and D-talose are widespread in nature, both as free sugars and in bound form. Some phosphorylated H. are very important metabolic intermediates. The two 6-deoxy-sugars, L-rhamnose and L-fucose, are also H. The ketohexoses corresponding to the aldohexoses are called hexuloses. They include the naturally occurring monosaccharides D-fructose and L-sorbose.

Hexuloses: see Hexoses.

Hibbert's ketones: see Lignin.

High energy bonds, *energy-rich bonds:* chemical bonds which release more than 25 kJ/mol on hydrolysis. They are usually esters (enol, thio and phosphate esters), acid anhydrides, or amidine phosphates.

In biological systems, the energy released is used to transfer the hydrolysed residue to other metabolic compounds (group transfer). H. e. b. are symbolized by ~ instead of the usual hyphen between groups (e. g. $CH_3CO \sim SCoA$). The free energy of hydrolysis of H. e. b. is at the same time a measure of the potential for group transfer (Table). See Energy-rich phosphates.

ΔG_0 of hydrolysis

Compound	ΔG_0 (kJ/mol)	(kcal/mol)
Creatine phosphate	42.7	10.2
Phospho*enol*pyruvate	53.2	12.7
Acetylcoenzyme A	34.3	8.2
Aminoacyl-tRNA	29.0	7.0
ATP $\rightarrow$ ADP + P_i	30.5	7.3
ATP $\rightarrow$ AMP + P ~ P	36.0	8.6
P ~ P $\rightarrow$ $2P_i$	28.0	6.7

High-yielding strains: see Production strains.

Hill plot: a graphic method for the determination of the degree of cooperativity of an enzyme (see Cooperativity). The plot of log $Y_s/(1-Y_s)$ versus log α is a curve with a slope of 1 for large or small α and a finite energy of interaction between the substrate-binding sites. Y_s is the saturation function, i. e. the fraction of the enzyme in the enzyme-substrate complex, and $\alpha = S/K_m$. For the values of α usually obtained experimentally, an approximate straight line is obtained with the maximal slope h (Fig.), the Hill coeffcient. This serves as a measure of the cooperativity, and is not usually identical to the number n of subunits of an enzyme, but is only a minimal estimate of n. If the energy of interaction between the substrate-binding sites is infinite, the Hill plot degenerates into a straight line with a slope equal to the number of subunits. The segment of the ordinate AB, constructed

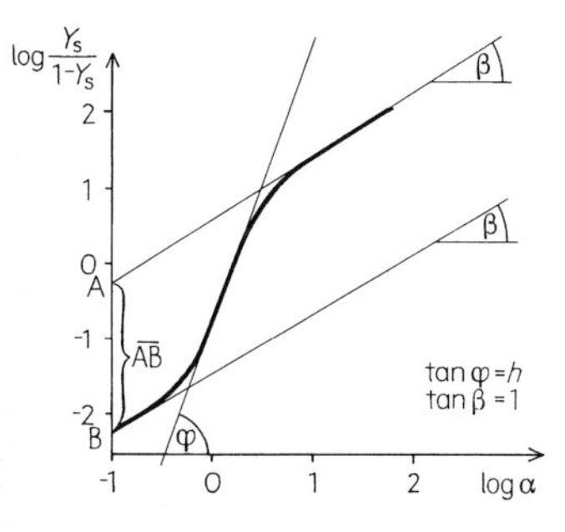

Hill-Plot

as shown in the Fig., can be used to determine the total energy of interaction ΔG_w between substrate or effector-binding sites: $\Delta G_w = 2.303\ RT \cdot AB$, where R is the general gas constant and T the absolute temperature. The saturation functions for effectors can also be determined. Y_E is the fraction of the enzyme in the enzyme-effector complex. For enzymes in equilibrium with their enzyme-substrate complexes, the saturation function Y_s can be replaced by the kinetic saturation v/V_m, so that log $v(V_m\text{-}v)$ can be plotted against log α. Correspondingly, for an effector, log $(v\text{-}v_o)/(V\text{-}v)$ must be plotted against log α; v is the measured rate of reaction at constant S, v_o is the rate in the absence of effector at the same S, and V is the rate at a saturating concentration of effector for the chosen S; S is the substrate concentration.

Hill reaction: light-dependent production of oxygen by the photosynthetic system in the presence of an artifical oxidizing agent (electron acceptor). R. Hill first observed this reaction in illuminated isolated chloroplasts, in the absence of CO_2, and using iron(III) oxalate as oxidizing agent. Iron(III) oxalate (Fe^{3+} is reduced to Fe^{2+} in the reaction) can be replaced by potassium ferricyanide, quinone and other compounds (Hill reagents). Spinach chloroplasts catalyse the following H.r: $4K_3Fe(CN)_6 + 2H_2O + 4K^+ \rightarrow$ 4K4 $Fe(CN)_6 + 4H^+ + O_2$. The "natural" Hill reaction is the photolysis of water, and the "natural" Hill reagent is oxidized NADP.

Hinekiflavone: see Biflavonoids.

Hippuric acid: C_6H_5-CO-NH-CH_2-COOH, the *N*-benzoyl derivate of glycine. Mammalian herbivores detoxify benzoic acid by converting it to H. a.

Hircinol: see Orchinol.

His: abb. for L-Histidine.

Histamine, *β-imidazol-4(5)-ethylamine:* a biogenic amine, M_r 114.14. H. is formed by enzymatic decarboxylation of L-histidine. It stimulates the glands in the fundus of the stomach to secrete digestive juices, dilates the blood capillaries (important for increasing blood flow and decreasing blood pressure), increases permeability (urtication and reddening after local application of histamine), and causes contraction of the smooth muscles of the digestive tract, the uterus and the bronchia (in bronchial asthma). H. is catabolized by diamine oxidases and aldehyde oxidases to imidazolylacetic acid. It is widely distributed in plants and animals, occurring e. g. in stinging nettles, ergot, bee venom and the salivary secretions of biting insects. As a tissue hormone, H. is present in the liver, lungs, spleen, striated muscles, mucus membranes of stomach and intestine, and it is stored with heparin in mast cells. The amounts found in tissue are on the order of μg/g fresh weight.

Histamine

L-Histidine, *His, imidazolylalanine:* a weakly glucoplastic, proteogenic amino acid, M_r 155.2. It is a component of the catalytic centers of many enzymes,

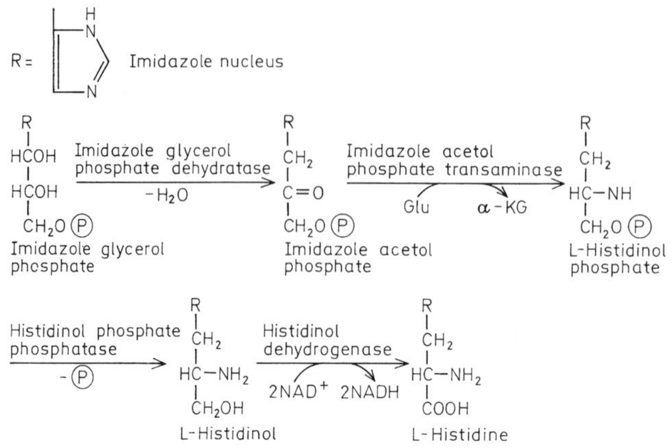

R = Imidazole nucleus

Imidazole glycerol phosphate (R–HCOH–HCOH–CH_2OⓅ) —[Imidazole glycerol phosphate dehydratase, $-H_2O$]→ Imidazole acetol phosphate (R–CH_2–C=O–CH_2OⓅ) —[Imidazole acetol phosphate transaminase; Glu → α-KG]→ L-Histidinol phosphate (R–CH_2–HC–NH–CH_2OⓅ)

—[Histidinol phosphate phosphatase, -Ⓟ]→ L-Histidinol (R–CH_2–HC–NH_2–CH_2OH) —[Histidinol dehydrogenase; $2NAD^+$ → 2NADH]→ L-Histidine (R–CH_2–HC–NH_2–COOH)

Fig. 1. *Histidine.* Terminal reactions in histidine biosynthesis.

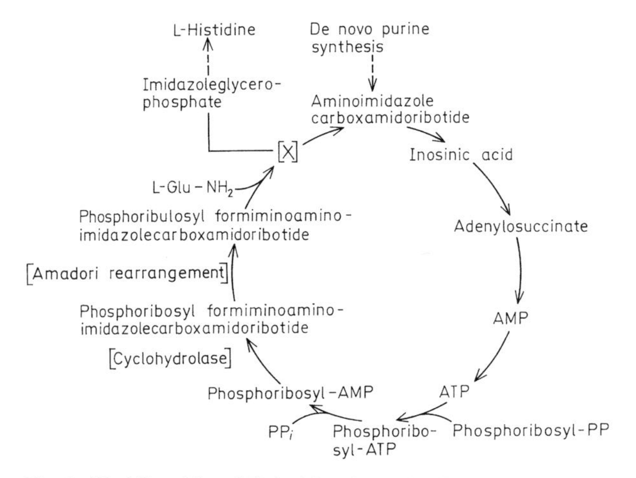

Fig. 2. *Histidine.* The ATP-imidazole cycle. The latter part of this cycle is shown in greater detail in Fig. 3.

and is also a component of carnosine and anserine. In the absence of dietary His, adult animals can maintain their nitrogen balance for a short time, but the amino acid is essential for growing animals. The imidazole ring cannot be synthesized by mammals. In bacteria His is formed via imidazole glycerophosphate in the last part (Fig. 1) of the ATP-imidazole cycle (Fig. 2). In this biosynthesis, phosphoribosyl-formimino-aminoimidazole carboxamidoriboride is synthesized by an Amadori rearrangement, which is relatively rare in cellular metabolism. It is significant that ATP is the substrate. Only the C2 atom and the N1 atom of the purine ring (Fig. 3) are incorporated into the His molecule. His catabolism is shown in Fig. 4; the formimino group of formiminoglutamate is a metabolic source of Active one carbon units (see). His is used in the treatment of allergies and anemias, and it is a useful laboratory buffer in the physiological pH range. [R. G. Martin et al. Methods in Enzymology **XVII B** (1971) 3–44; M. Brenner & B. N. Ames "The Histidine Operon and Its Regulation" in Greenberg (ed.) *Metabolic Pathways* (3rd. edn) **5** 349–387]

Histidinemia: see Inborn errors of metabolism.

Histocompatibility antigens: see Major histocompatibility complex.

Histones: see Chromosome.

Histopine: see D-Octopine.

hMG: acronym of human menopausal gonadotropin.

HMTPP: acronym of 2-hydroxymethylthiamin pyrophosphate. see Thiamin pyrophosphate.

Holarrhena alkaloids, *kurchi alkaloids:* a group of steroid alkaloids which are characteristic constituents of plants of the dogbane family *(Apocyanaceae)* genus *Holarrhena.* Those so far isolated (about 50) contain a pregnane ring system (see Steroids) substituted in positions 3 and 20 with amino or methylamino groups, e.g. the widespread conessine (Fig.) and its 12β-hydroxy-derivative holarrhenine; in both these

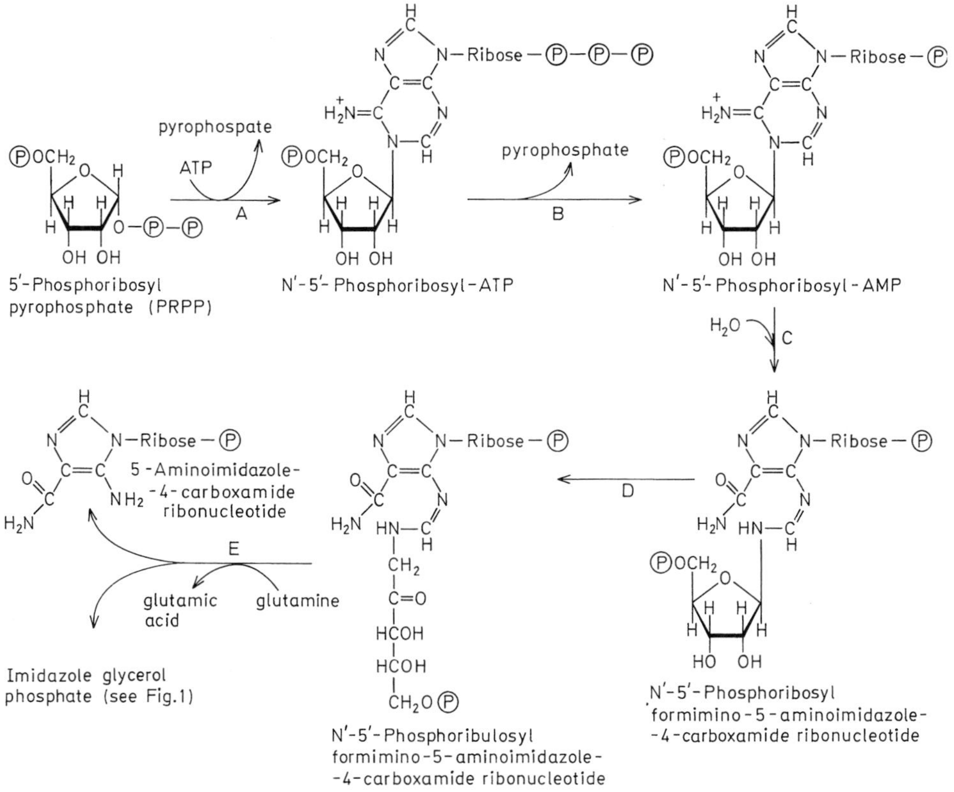

5'-Phosphoribosyl pyrophosphate (PRPP)
N'-5'-Phosphoribosyl-ATP
N'-5'-Phosphoribosyl-AMP
N'-5'-Phosphoribosyl formimino-5-aminoimidazole-4-carboxamide ribonucleotide
N'-5'-Phosphoribulosyl formimino-5-aminoimidazole-4-carboxamide ribonucleotide
5-Aminoimidazole-4-carboxamide ribonucleotide
Imidazole glycerol phosphate (see Fig.1)

Fig. 3. *Histidine.* First stages of histidine biosynthesis.
A. *N*-1-(5'-phosphoribosyl)adenosine triphosphate:pyrophosphate phosphoribosyl transferase, or ATP phosphoribosyltransferase (EC 2.4.2.17).
B. Phosphoribosyl ATP pyrophosphohydrolase (EC 2.4.2.17). In *Neurospora* this enzyme is trifunctional, also catalysing reaction A and the conversion of histidinol to histidine (Fig. 1).
C. 1-*N*-(5'-phospho-D-ribosyl)-AMP 1,6-hydrolase, or phosphoribosyl-AMP cyclohydrolase (EC 3.5.4.19).
D. Amadori rearrangement of the phosphoribosyl residue.
E. Amidotransferase and cyclase. Presumably an intermediate is produced by transfer of nitrogen from glutamine, followed by cyclization to form the imidazole ring of imidazole glycerol phosphate. The intermediate is not known, and the amidotransferase and cyclase activities have not been separated.

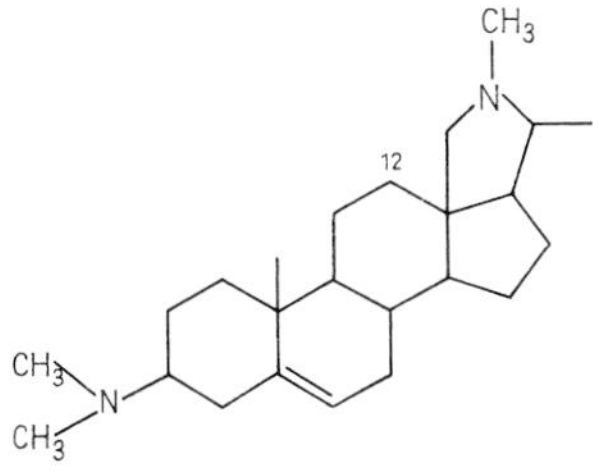

Conessine

alkaloids the 20 amino group is bound to C18 to form a pyrrolidine ring. Conessine is used as a starting material for the chemical synthesis of aldosterone. H.a. have hypotensive, curare-like, diuretic and narcotic properties. The alkaloid-rich bark of the shrub, *Holarrhena antidyserenterica,* is used for treating dysentery in India. H.a. are biosynthesized via cholesterol and pregnenolone.

Holarrhenine: see Holarrhena alkaloids.

Holoenzyme: see Coenzyme.

Holosides: compounds consisting only of glycosidically linked sugar residues, e.g. oligo- and polysaccharides.

Holothurins: a group of highly neurotoxic compounds from sea cucumbers *(Holothurioidea).* H. are water-soluble, sulfated steroid glycosides, which have saponin properties and possess greater hemolytic activity than saponins of plant origin. The Cuviers glands of the Caribbean sea cucumber *Actinopyga agassiz* produce a mixture of holothurins and are a major source of these compounds. Holothurin A carries a chain of 4 sugar residues: D-glucose, D-xylose, 3*(O)*-methylglucose and D-quinovose; the sulfate is attached to the xy-

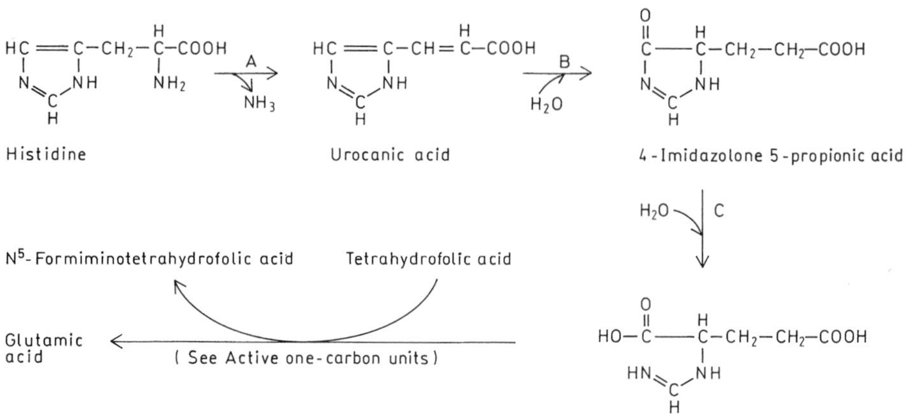

Histidine — Urocanic acid — 4-Imidazolone 5-propionic acid

N5-Formiminotetrahydrofolic acid — Tetrahydrofolic acid — Glutamic acid (See Active one-carbon units)

Fig. 4. *Histidine.* Biodegradation of histidine.
A. L-Histidine ammonia lyase (histidase) (EC 4.3.1.3).
B. 4-Imidazole-5-propionate hydro-lyase (urocanase) (EC 4.2.1.49).
C. 4-Imidazolone-5-propionate amidohydrolase (EC 3.5.2.7).

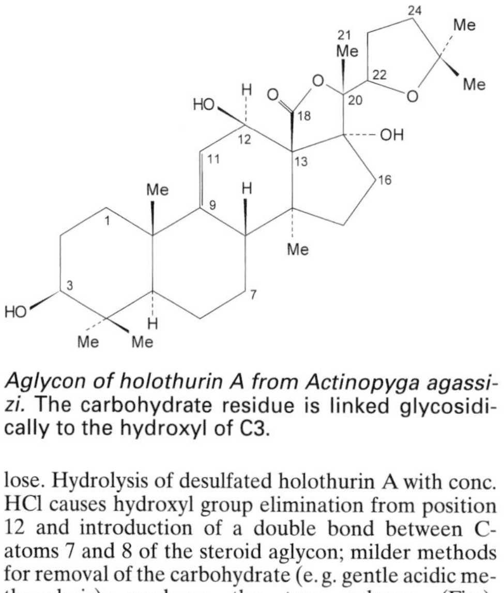

Aglycon of holothurin A from Actinopyga agassizi. The carbohydrate residue is linked glycosidically to the hydroxyl of C3.

lose. Hydrolysis of desulfated holothurin A with conc. HCl causes hydroxyl group elimination from position 12 and introduction of a double bond between C-atoms 7 and 8 of the steroid aglycon; milder methods for removal of the carbohydrate (e. g. gentle acidic methanolysis) produces the true aglycon (Fig.). [J. D. Chanley & C. Rossi *Tetrahedron* **25** (1969) 1911–1920; J. S. Grossert *Chem. Soc. Rec.* **1** (1972) 1–25]

Holothurinogenins: see Holothurins.

Homeo box: A sequence of base pairs in DNA about 180 kilobases in length. The sequence is highly conserved, and has been identified in insects, other arthropods, annelids and chordates, including *Homo sapiens.* The term is derived from "homeosis", the replacement of one body structure by an homologous one, e. g. an antenna by a leg in certain mutant flies. In insects, a number of genes which control development of body segments contain the H. b.; and all genes known to contain the box are related to embryonic development. The functions of non-insect genes containing the H. b. are not known. Conceptual translation of the H. b. (i. e. determination of the polypeptides it would encode if it were translated *in vivo*) suggests that it would encode a large number of basic amino acids, i. e. the encoded proteins would possibly contain DNA-binding domains. [W. J. Gehring *Cell* **40** (1985) 3–5]

Homeoviscous adaptation: see Membrane lipids.

Homoarginine, *Har:* a higher homolog of arginine with an additional methylene group in the side chain.

Homocysteine, *Hcy:* a higher homolog of cysteine with an additional methylene group in the side chain.

Homocysteinemia: see Inborn errors of metabolism.

Homocystinuria: see Inborn errors of metabolism.

Homoferreirin: 5,7-dihydroxy-4′,6′dimethoxy-isoflavanone. See Isoflavanone.

Homoglycans: straight or branched chain polysaccharides containing only one kind of monosaccharide residue. H. are widespread in the vegetable kingdom. They include arabans, xylans, glucans, fructans, mannans, galactans, the starch components amylose and amylopectin, cellulose and glycogen.

Homologization: see Glucosinolate.

Homologous proteins: proteins which have arisen through divergent evolution from a common ancestor. They usually have very similar primary and tertiary structures. Examples are the cytochromes, hemoglobin and myoglobin, the ferredoxins, fibrin peptides, immunoglobulins, peptide hormones (e. g. insulin and pituitary hormones), snake venom toxins and enzymes like the serine proteases of the pancreas (trypsin, chymotrypsin, elastase) or the blood-clotting enzymes (e. g. plasmin, thrombin) and lactate dehydrogenase. As examples of homologies in primary structure, the partial sequences around the catalytically important amino acid residues of the serine proteases are given in the table. Comparison of the primary structures and the location of disulfide bridges in trypsin, chymotrypsin, and elastase is shown in Fig. 1.

Fig. 2 shows that the structural homologies are also reflected in the conformation of the polypeptide chains, taking chymotrypsin and elastase as an example. Although only about 40 % of the amino acids in

*Partial sequences of homologous proteolytic (A) and esterolytic (B) enzymes from the region about the active serine residue**

Enzyme	Sequence									
A Trypsin (beef, sheep, pig, dogfish, shark, shrimp)	Asp	Ser	Cys	Glu	Gly	Asp	Ser*	Gly	Gly	Pro
Chymotrypsin A and B (beef)	Ser	Ser	Cys	Met	Gly	Asp	Ser*	Gly	Gly	Leu
Elastase (pig)	Ser	Gly	Cys	Glu	Gly	Asp	Ser*	Gly	Gly	Pro
Thrombin (beef)	Asp	Ala	Cys	Glu	Gly	Asp	Ser*	Gly	Gly	Pro
"Trypsin" *(Streptomyces griseus)*	Asp	Thr	Cys	Glu	Gly	Asp	Ser*	Gly	Gly	Pro
B Acetylcholinesterase			Phe	Gly	Glu		Ser*	Ser	Glu	Gly
Pseudocholinesterase (horse)				Gly	Glu		Ser*	Ala	Gly	Gly
Liver esterase (pig, horse, sheep, chicken)				Gly	Glu		Ser*	Ala	Gly	Gly
Pancreatic lipase (pig)				Leu			Ser*	Gly	His	
Alkaline phosphatase *(Escherichia coli)*	Asp	Tyr	Val	Thr	Asp		Ser*	Ala	Ala	Ser

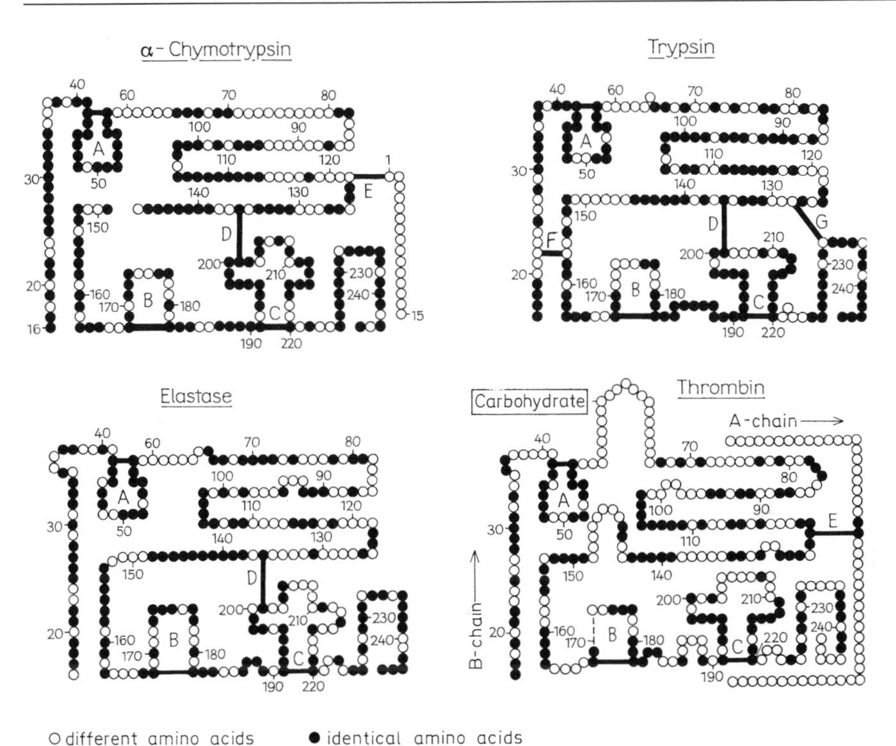

Fig. 1. *Homologous proteins.* Comparison of the primary structures and the positions of the disulfide bridges of four serine proteases. A to G indicate the homologous disulfide bridges.

the two proteins are homologous, i.e. identical, their spatial folding is similar. Positions 57, 102 and 195 are occupied by the important amino acids histidine, aspartic acid and serine.

Homopolymer: a polymer consisting of identical monomers, e.g. amylose and polyphenylalanine. In a narrower sense, H. are synthetic polynucleotides in which all the nucleotides contain the same base, e.g. polyadenylic acid, polyuridylic acid, polydeoxyadenylic acid. Homopolynucleotides (usually single-stranded) are synthesized *in vitro* from nucleoside di- or triphosphates using the appropriate polymerases without a matrix. An oligonucleotide is needed as a primer. Homopolynucleotides, in particular poly(A) sequences, occur naturally in some eukaryotic RNA (see Messenger RNA).

Homopterocarpin: 3,9-dimethoxypterocarpan. See Pterocarpans.

Homosteroids: see Steroids.

Hopanoids: naturally occurring derivatives of the pentacyclic triterpene, hopane (Fig.).

Eukaryotic H. can be subdivided into two classes: (i) those with an oxygen-containing group at C3, which are formed biosynthetically by the cyclization of 2,3-epoxysqualene, and (ii) those that do not and which are generated by cyclization of squalene

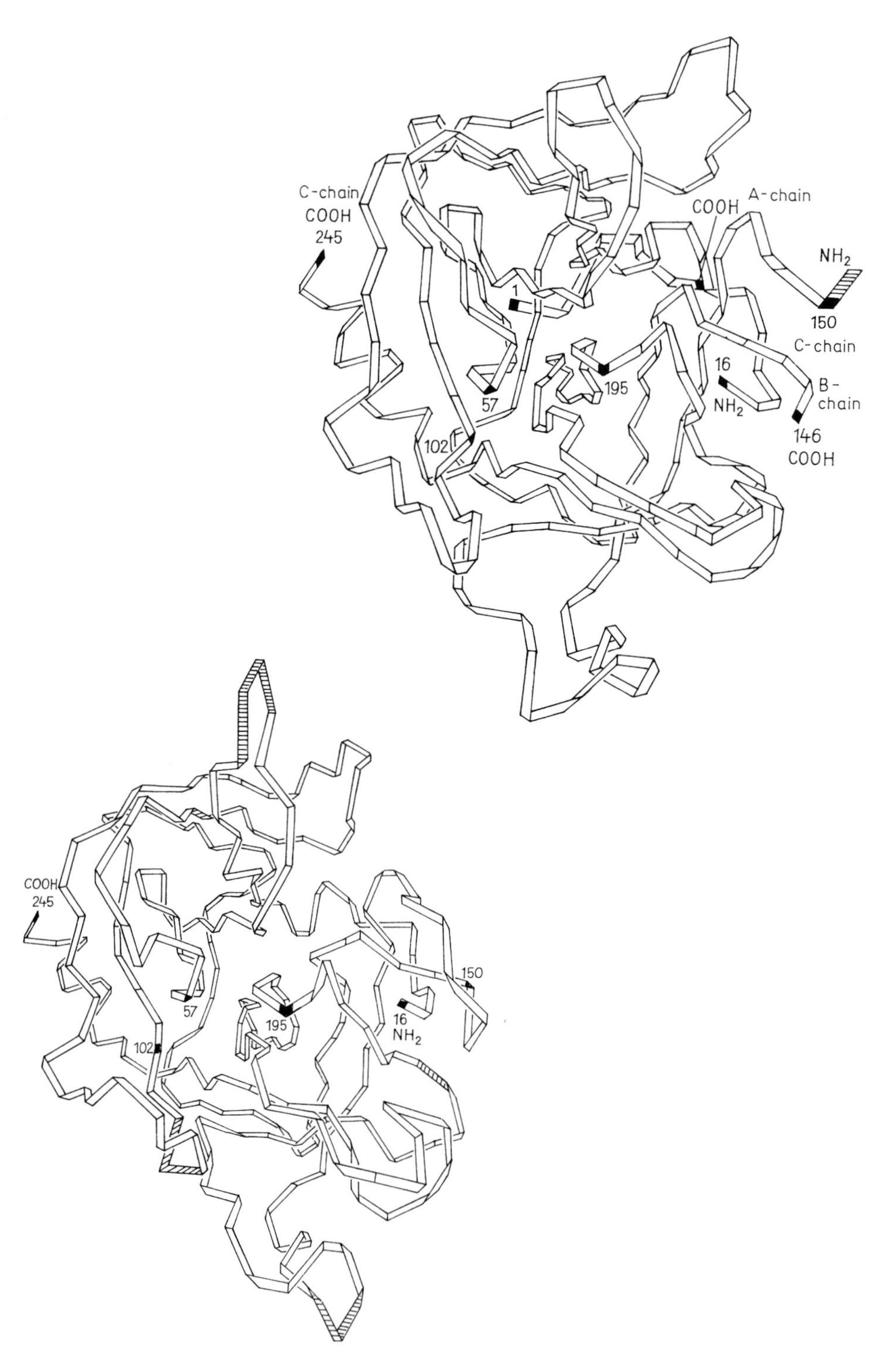

Fig. 2. *Homologous proteins.* Conformation of the polypeptide chains of the two homologous pancreas enzymes, α-chymotrypsin (above) and elastase (below).

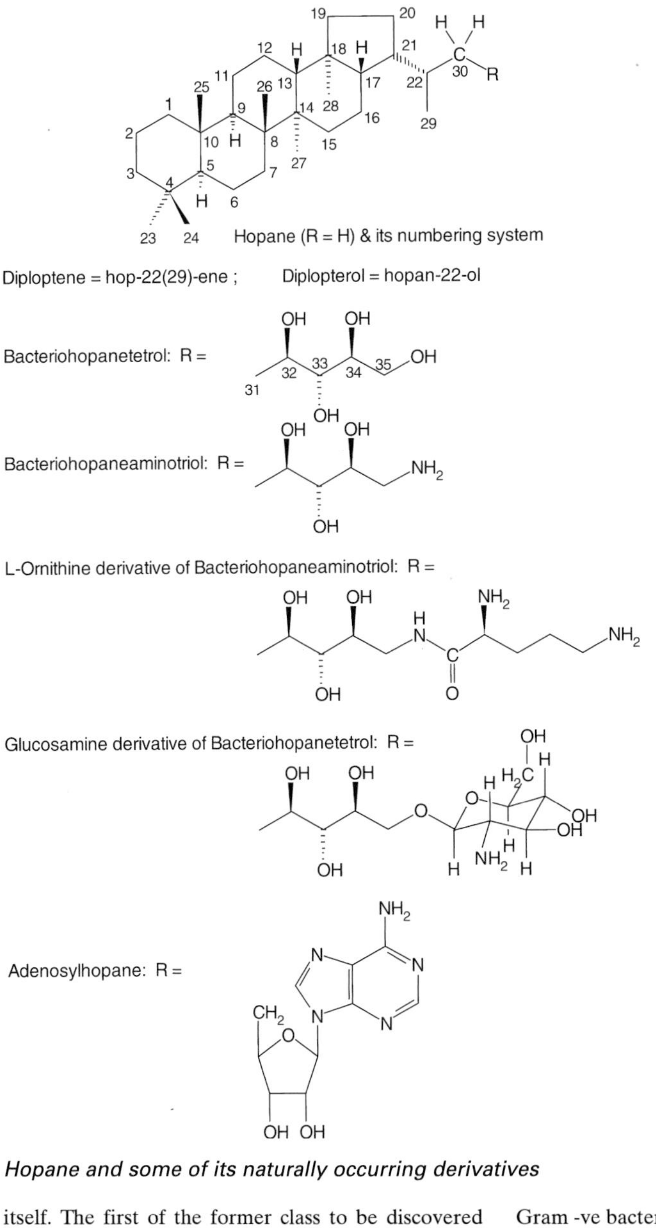

Hopane (R = H) & its numbering system

Diploptene = hop-22(29)-ene ; Diplopterol = hopan-22-ol

Bacteriohopanetetrol: R =

Bacteriohopaneaminotriol: R =

L-Ornithine derivative of Bacteriohopaneaminotriol: R =

Glucosamine derivative of Bacteriohopanetetrol: R =

Adenosylhopane: R =

Hopane and some of its naturally occurring derivatives

itself. The first of the former class to be discovered was 22-hydroxyhopan-3-one, which was isolated from the dammar resin of *Hopea* sp. *(Dipterocarpaceae);* other 3-hydroxy- and 3-keto-H. have been found in various higher plant taxa. Members of the latter class, e.g. diploptene, diplopterol (Fig.) have so far only been found in ferns, a few fungi and protozoa.

Prokaryotic H. (bacteriohopanoids), discovered later than their eukarytoic counterparts, are widely distributed amongst cyanobacteria, methylotrophic bacteria, acetic acid bacteria, *Rhodospirillaceae* (purple non-sulphur bacteria) and in various Gram + ve and Gram -ve bacteria, where they are present at levels similar to those of sterols in eukaryotes (i.e. 0.1–2 mg/g dry wt.), but do not appear to be present in archaebacteria, *Chromataceae* (purple sulphur bacteria) or enterobacteria. Structurally they are like the 3-desoxyhopanoids of eukaryotes save that they have a C_5 *n*-alkyl polysubstituted chain attached to C30; the most widespread bacteriohopanoids are bacteriohopanetetrol and bacteriohopaneaminotriol (Fig.). Free polyols and aminopolyols such as these have only been found in *Acetobacter* spp. and methylotrophs respectively. Usually the polysubstituted C_5 alkyl moiety of bacteriohopanoids is linked to a further polar moiety (Fig.) which may be an amino acid (e.g. L-ornithine, L-tryptophan), a sugar (e.g. D-glucosamine) or a nucleoside (e.g. adenosine). The hopane residue of prokaryotic H. is formed biosynthetically by the cyclization of squalene while the C_5 residue appears to be derived from ribose or a ribose-containing compound. Bacteriohopanoids and compounds readily recognizable as being derived from them are abundant and widespread in geological sediments and have been termed geohopanoids; they have their origin in the H. present in the cells of successive populations of marine and lacustrine microorganisms that have lived and died during the last 1.5×10^9 years. The main interest in prokaryotic H. stems from the increasing likelihood that they perform the same stabilizing role in the membranes of prokaryotes as sterols do in those of eukaryotes. In this respect it is interesting that those prokaryotes that do not have H. have alternative membrane-stabilising molecules (e.g. the phytanyl ethers of archaebacteria). [G. Ourisson et al. *Annu. Rev. Microbiol.* **41** (1987) 301–333]

Hordein: see Prolamines.

Hordenine, ***anhaline:*** *N,N*-dimethyltyramine, a widely distributed biogenic amine, m.p. 117–118 °C, b.p. 173–174 °C. As a derivative of phenylethylamine, H. increases blood pressure, but it has a low physiological activity.

Hormones: organic compounds produced and interpreted as intercellular signals by nearly all organisms. Very low concentrations of H. are usually encountered, e.g. some 10^{-12} –10^{-15} mole per mg tissue protein in the case of hypothalamic releasing H., pituitary-like H. and gastrointestinal H. in animals. As nervous systems evolved, some H. came to serve as neurotransmitters, e g. acetylcholine and some peptide H. The existence of a separate endocrine system is phylogenetically more recent, occurring only in the vertebrates. There are close parallels between neurons which secrete neurotransmitters and neurohormones and the endocrine gland cells which secrete the specialized glandular H.

H. have a variety of chemical structures: there are Steroid H. (see), amino acid derivatives (e.g. see Adrenalin, Thyroxin, Auxins), Peptide H. (see), Protein H. (see) and fatty acid derivatives (see Prostaglandins). In plants, the Cytokinins (see) are mostly derivatives of adenine. Insect hormones (see) are discussed in a separate entry.

All H. act by binding to receptors, either on the membrane of the target cell (e.g. insulin, adrenalin) or within the cell (e.g. steroid H.). What happens next depends on the H. and the cell. In many cases, the H. serves as a *first messenger,* which activates a *second messenger* (often cyclic AMP, see Adenosine phosphates) within the cell. The cAMP may activate a protein kinase (or the receptor may itself be a protein kinase activated by its H.) which in turn activates some other key enzyme and thus alters the metabolism of the cell (see Post-translational modification of proteins), or the permeability of its membrane to ions or other molecules. Steroid H. bind to their intracellular receptors, forming a complex which interacts with nuclear chromatin, leading to specific gene activation. Peptide H. may also affect gene expression, in some cases apparently through the action of histone kinases activated by the H. The H. itself is usually subject to rapid inactivation and/or degradation and excretion.

Biosynthesis, storage and secretion. In animals there are two types of endocrine cells: those synthesizing protein and peptide H., and those synthesizing steroids. The former are stimulated to produce their H. by neurotransmitters (including those released by the neurons synapsing onto the cells), other H., metabolic products or dietary substances. The H. or their precursors (e.g. proinsulin) synthesized on the ribosomes enter the Golgi apparatus of the cell and are packed there into small vesicles and stored in the cytosol. On demand, which is signaled by various stimuli (e.g. a high blood glucose level), the vesicles move to the cell membrane and release the H. into the blood stream.

The enzymes required for steroid hormone biosynthesis must be synthesized or activated when the cell is stimulated; e.g. corticotropin causes cholesterol to be converted into glucocorticoids in the adrenals. Steroid H. are not stored; the appropriate enzyme system can be rapidly activated and deactivated. Cells which produce steroid H. have large Golgi apparatuses, much smooth endoplasmic reticulum, lipid droplets and lysosomes.

H. are transported in free form or bound to specific or unspecific proteins, e.g. oxytocin is bound to neurophysin, transported within the axon of a nerve from the hypothalamus to the posterior lobe of the hypophysis, and stored there. This binding is loose, noncovalent and easily dissociated. H. are transported in the blood, bound noncovalently to plasma proteins. Thus the albumins carry somatotropin, steroid binding globulin possesses a relatively high affinity for steroid H., and some steroid H. have specific transport proteins, e.g. cortisol is transported by transcortin. However, serum albumin, which possesses a relatively low affinity for steroid H. is present in such high concentration that it binds the major fraction of circulating steroid H. The H. is partly inactivated by this binding, but is also stabilized and protected against enzymatic attack. Any method for H. determination in the blood must take into account the fact that the H. is present in both bound and free forms.

Inactivation of H. The action of a H. is immediately stopped by: 1. enzymatic hydrolysis of a cyclic nucleotide second messenger to a mononucleotide (e.g. cAMP to 5′-AMP); or 2. enzymatic degradation of the H. Peptide and protein H. are inactivated by proteolytic enzymes, catecholamines by monoamine oxidases, steroid H. by oxidation or reduction (e.g. about 50 % of estrogens in humans are oxidized to nonestrogenic catechols) and by conversion to readily excreted (in urine and bile) glucuronides or sulfates.

Regulation of H. effects. In every H. effect, two points must be kept in mind: 1. there is no such thing as an isolated H. effect. H. have a definite, but not the dominant role in the regulation of intracellular metabolism and of total metabolism; 2. H. effects are usually recognizable as a change in enzyme activity. Every H. effect is subject to fine tuning by a feedback mechanism, involving the H. itself, the metabolic products dependent on it, and the nervous system. In most cases synthesis and secretion of the first H. in a system is inhibited by the H. whose production it stimulates. A classic example of the coordination of various regulatory circuits is the system composed of hypothalamus (releasing hormones), anterior lobe of the pituitary (hypophyseal H.) and target organ, in which there is feedback at every level.

Pathobiochemistry. Due to the close coordination of the H. and the nervous system with metabolism, any disturbance in the synthesis, secretion or transport of H., or the lack of receptors, or a disturbance of H. catabolism is reflected by a pathology of the entire metabolism. Lack of mineralocorticoids, for example, leads to a disruption of the mineral and water balance; too much or too little somatotropin (growth hormone) produces gigantism or dwarfism; imbalances in the thyroid H. system (see Thyroxin) cause a derangement of energy metabolism and are reflected in hyper- or hypothyreosis; sexual functions and the normal course of pregnancy are severely affected when Gonadotropins (see) and Gonadal hormones (see) are under- or overactive; and there is some evidence that certain psychiatric illnesses are due to imbalances in peptide H. or neurotransmitters.

Methods of determination. H. can be assayed by biological, chemical or immunological methods. The traditional assays are biological, e.g. the pregnancy test in which gonadotropins in the urine cause ripening of the follicles in mouse ovaries. Assays of biological activity are also required to show that a laboratory synthesis has duplicated the natural H. structure. Structurally simple H. can be detected by chemical methods, such as gas chromatography for detection of prostaglandins. Most H. are now determined by immunoassays, because the very small amounts of H. in tissue require the sensitivity of these methods. An indirect method of demonstrating H. synthesis in cells is to use a cloned, radioactive DNA probe for the gene in question. If mRNA for the (peptide or protein) H. is present, it will hybridize with the DNA and bind it to the preparation, which is then subjected to radioautography. [K. Talmadge et al. *Nature* **307** (1984) 37–40]. See Choriogonadotropin.

Hp: abb. for Haptoglobin.

HSK cycle: acronym of Hatch-Slack-Kortschak cycle.

Human chorionic gonadotropin: see Choriogonadotropin.

Human genome project: an international initiative aimed at "acquiring complete knowledge of the organization, structure and function of the human genome – the master blueprint of each of us" *(Human Genome,* 1991–1992 Program report, U.S. Department of Energy, Office of Energy Research, Office of health and Environmental research, Washington, D.C. 20585. The U.S Department of Energy Human Genome Initiative was announced in 1986).

A series of international Human Gene Mapping Workshops was initiated in 1973, in response to the increasing quantity of data generated by somatic cell genetics. With the advent of in vitro Recombinant DNA technology (see), new initiatives became necessary to coordinate the mapping activities and results from the greatly increased number of markers and genes. This resulted in the development of computer systems for the collection, storage, analysis and reporting of results. Data are now continuously entered into an international database (central node: Human Genome Database (GDB), John Hopkins University, Maryland, USA) permitting the immediate sharing of results and continuous (day to day) updating in database nodes throughout the world (see below). Single Chromosome Workshops, annual Chromosome Coordination Meetings and biannual Human Genome Mapping Workshops are now organized and promoted by the Human Genome Organization (international).

The first aim of the project is to develop genetic maps of all the human chromosomes, i.e. to chart the positions of all loci on the 22 autosomes and the two sex chromosomes, bearing in mind that the term locus embraces not only functional genes but also segments of DNA of no or no known function. This basically means producing a series of descriptive maps of ever increasing resolution as the project progresses. The Human genome project also aims to sequence the DNA of each human chromosome (see Nucleic acid sequencing). It is accepted that the capacity, speed and cost of presently available methods of DNA sequencing do not meet the requirements of the project (it is estimated that the human genome contains 3×10^9 nucleotides), and that it will be necessary to develop new sequencing technologies.

Each chromosome has a characteristic length. After staining with certain dyes, commonly quinacrine or Giemsa, each chromosome also shows a characteristic pattern of transverse, lightly and heavily stained bands. Each band represents about 5–10 % of the chromosome length, and corresponds to about 10^7 base pairs of DNA. It is now possible to apply banding techniques to chromosomes at the less condensed prophase stage, revealing even more bands (as many as 3,000 can be identified). After pretreatment (e.g. with trypsin), followed by staining with adenine/thymine (AT)-specific or guanine/cytosine (GC)-specific dyes, distinct patterns of bands are discernable, which indicate the distribution of GC- and AT-rich regions. Allocation of genetic markers to specific chromosome bands is an important stage in the mapping of gene loci, providing important reference points along the chromosome for the orientation and localization of genes and markers.

Studies of gene distribution in families have long been used to map gene loci, especially those of disease genes, by investigating genetic linkage, i.e. the extent to which genes on the same chromosome before meiosis are passed on together to offspring (equivalent to the frequency with which they are still located on the same chromosome after the completion of meiosis). If genes are widely separated on a chromosome or located on different chromosomes, there is a 50 % probability that they will be passed on together; such genes are said to be unlinked. The greater the proximity of two genes on a chromosome, the smaller the probability that they will be separated at meiosis, and the greater the frequency of their appearance together in offspring, i.e. the greater their degree of linkage. The resolution of maps based on linkage studies is relatively coarse.

A gene can be mapped by physical methods, provided it can be cloned or has a distinctive property that can be monitored at the cellular level. Somatic cell genetics have been used widely for the physical mapping of human genes. This involves the construction of a heterokaryon by fusing the cell of interest (in this case a human cell) with another cell (e.g. mouse tumor cell) which lacks the gene in question. Eventually, the heterokaryon proceeds to mitosis and produces a hybrid cell (somatic cell hybrid), in which the two separate nuclear envelopes disassemble and allow both sets of chromosomes to come together in a single large nucleus. The initial hybrid cells tend to be unstable and lose chromosomes. Human-mouse hybrid cells randomly lose human chromosomes or parts thereof. Thus, a human-mouse hybrid useful for human gene mapping contains all the mouse chromosomes with a subset of human chromosomes, single human chromosomes, or parts of human chromosomes. Individual hybrid cells are propagated and maintained as cell lines containing specific human chromosomes. Provided a retained human gene has a product detectable at the cellular level, then in cell culture it can be attributed to a particular chromosome or even part of a chromosome. Of course, it is necessary to known which human chromosomes or parts of chromosomes are represented in the cell hybrid. This is usually performed by labeling hybrid cell DNA (e.g. with biotin) and hybridizing it with normal human metaphase DNA (a technique known as chromosome painting).

A human gene in a somatic cell hybrid can also be detected and mapped without being expressed at the cellular level, e.g. by restriction enzyme digestion of DNA from the hybrid, followed by Southern blotting and detection of the human gene with a radioactively labeled probe, or by identifying primers which can be used in the Polymerase chain reaction (see) to amplify specifically the human and not the mouse gene.

As an alternative to somatic cell hybrids, human chromosomes can be separated from each other by a process known as flow sorting, a process pioneered at Los Alamos National Laboratory, which uses flow cytometry to separate chromosomes according to size. As the chromosomes flow singly past a laser beam they are differentiated according to their DNA content, then directed into appropriate collection vessels. The high-speed chromosome sorter analyses chromosomes at the rate of up to 20,000/s and reliably produces 250–1000 ng of sorted chromosome DNA per day.

DNA sequence differences between individuals, i.e. regions of polymorphism (see DNA fingerprinting) can be precisely characterized, are therefore useful markers, and are also included on genetic maps. Such polymorphic regions are due to: 1. Restriction fragment length polymorphisms (see), and 2. variable numbers of tandem repeat sequences. On average, polymorphic regions occur every 300 to 500 base pairs.

cDNA maps show the positions of DNA that is expressed, i.e. DNA that is transcribed into mRNA. Laboratory-synthesized cDNA is used as a probe to locate the site of synthesis on the chromosome of its counterpart mRNA. When the location of a disease gene is known approximately from linkage analysis, a cDNA map of the corresponding region may provide a set of candidate genes for further testing.

Genetic maps have proved useful for locating the positions of genes associated with inherited diseases, even before the molecular basis of a disease is known or the responsible gene identified. Thus, the combination of linkage analysis with polymorphic markers has greatly facilitated the location and isolation of the genes for cystic fibrosis and Duchenne muscular dystrophy (see Gene therapy).

High resolution mapping may produce a macro-restriction map (top-down mapping) or a contig map (bottom-up mapping). In *top-down mapping,* a single chromosome is cleaved into large fragments (up to several megabases) using rare cutter restriction enzymes. The fragments are separated by pulsed field gel electrophoresis and mapped for genes. The resulting macro-restriction map shows the order of, and distance between sites at which rare cutter enzymes cleave. Such maps show more continuity and fewer gaps between fragments than contig maps, but their resolution is lower. In *bottom-up mapping,* the chromosome DNA is cut into small pieces, then each piece is cloned and isolated in Cosmids (see) or Yeast artificial chromosomes (see). Overlapping clones can then be ordered into a physical map or Contig (see) including all the genes in that region.

Human genome data can be accessed via the following nodes.

Main node (USA):
GDB Human Genome Database, John Hopkins University School of Medicine, 2024 E. Monument St., Baltimore, Maryland 21205–2100 USA.
User support registration: (410) 955–9705.
User support Fax line: (410) 614–0434.
User support and registration (e-mail): help@gdb.org
FTP server address: ftp.gdb.org
WWW (world-wide-web) site: http://gdbwww.gdb.org/

United Kingdom:
Administration, UK Human Genome Mapping Program Resource Center, Hinxton Hall, Hinxton, Cambridge CB10 1RQ, U.K.
Phone: 44–1223–494–511
Fax: 44–1223–494–512
e-mail: admin@hgmp.mrc.ac.uk

Germany:
Dr. Otto Ritter, Deutsche Krebsforschungszentrum, German Cancer Research Center, Department of Molecular Biophysics, Im Neuenheimer Feld 280, D-6900 Heidelberg 1, FRG.
Phone: 49–6221–42–2372
Fax: 49–6221–42–2333
e-mail: dok261@cvx12.dkfz-heidelberg.de

Australia:
Dr. Alex Reissner, Australian National Genomic Information Service (ANGIS), Dept. Electrical Engineering, Building JO3, The University of Sydney, N.S.W. 2006, Australia.
Phone: 61–2-692–2948
Fax: 61–2-692–3847
e-mail: reissner@gis.su.oz.au

France:
Dr. Philippe Dessen, Service de Bioinformatique, CNRS-INSERM, 7 rue Guy Moquet – BP8, 94801 VILLEJUIF Cedex, France.
Phone: 33–1-45–59–52–41
Fax: 33–1-45–59–52–40
e-mail: gdb@genome.vjf.inserm.fr

Sweden:
GDB User Support, Biomedical Center, Box 570, S-75123 Uppsala, Sweden.

Phone: 46–18–17–40–57
Fax: 46–18–52–48–69
e-mail: help@gdb.embnet.se
Netherlands:
CAOS/CAMM Center, Faculty of Science, University of Nijmegen, P.O. box 9010, 6500 GL NIJMEGEN, The Netherlands.
Phone: 31–80–653–391
Fax: 31–80–652–977
e-mail: post@caos.caos.kun.nl
Israel:
Dr. Jaime Prilusky, Bioinformatics Unit, Weizmann Institute of Science, 76100 Rehovot, Israel.
Phone: 972–8–343456
Fax: 972–8–344113
e-mail: Isprilus@weizmann.weizmann.ac.il
Japan:
Mika Hirakawa, JICST GDB Center, Numajiri Sangyo Building, 783–12, Enokido, Tsukuba City, Ibaraki 305, Japan.
Phone: 81–298–38–2965
Fax: 81–298–38–2956
e-mail: mika@gdb.gdbnet.ad.jp
[A.J. Cuticchia (ed.) *Human Genome Mapping – A Compendium,* John Hopkins University press, Baltimore & London, 1995]

Human lactogen: see Placenta lactogen.

Human menopausal gonadotropin, ***HMG, castration gonadotropin:*** a glycoprotein (M_r 31,000, 30 % carbohydrate) of the anterior lobe of the pituitary. Its action is similar to that of follicle stimulating hormone. Increased quantities of HMG are formed in the pituitary of women in the menopause, or women who have been ovarectomized. The increased synthesis (accompanied by an increased excretion) of HMG is explained by the absence of the negative feedback by sex hormones from the ovary on the hypothalamus and the pituitary.

Humulane type: see Sesquiterpenes (Fig.).

Humulenes: isomeric monocyclic sesquiterpene hydrocarbons, M_r 204.36, found in many essential oils. α-H. (α-caryophyllene) (2,6,6,9-tetramethyl-1,4,8-cycloundecatriene), b.p.$_{10}$ 123 °C, ρ_4 0.8905, n_D 1.5508, is present especially in oil of hops *(Humulus lupulus* L.) *(Moraceae)* and leaves of *Lindera strychnifolia* (F.) Will *(Lauraceae).* β-H. [(E,E)-1,4,4-trimethyl-8-methylene-1,5-cycloundecadiene], ρ_4 0.8907, n_D 1.5012, also present in hop oil, has the same 11-membered ring system as the α-isomer, but one of the 3 double bonds is exocyclic. Hop oil also contains oxygenated derivatives of H. and β-caryophyllene. For biosynthesis and formulas, see Sesquiterpenes.

Hunter's syndrome: see Lysosomal storage diseases.

Hurler's syndrome: see Lysosomal storage diseases.

Hyaloplasm: see Cell 2).

Hyaluronic acid: an unbranched mucopolysaccharide, M_r 200,000–400,000. The repeating subunit is a disaccharide, *N*-acetyl-D-glucosamine glycosidically linked β-1,4 to D-glucuronic acid, the latter being linked β-1,3 to the next disaccharide unit. H. a. occurs in various animal tissues and joint fluids. Its aqueous solution is highly viscous, which explains its biological function as a lubricant. It is synthesized from D-glucose in the fibroblasts.

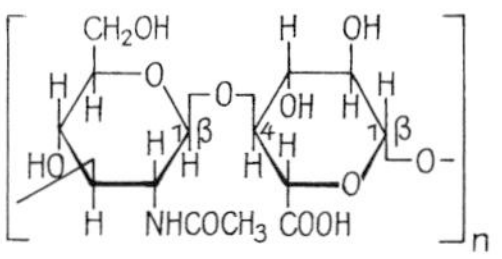

Hyaluronic acid

Hybridization: formation of a hybrid nucleic acid duplex by association of single strands of DNA and RNA (DNA:RNA hybrid) or single strands of DNA not previously associated with each other in a natural duplex (DNA:DNA hybrid). RNA:RNA hybrids are also possible. H. is used to detect and isolate specific nucleotide sequences, and to measure the extent of homology between nucleic acids. In principle, the two nucleic acids are denatured by heating above the T_m (see Melting point of DNA) then allowed to hybridize at about 25 °C below the T_m; single strands must also be denatured by heating to remove intrastrand base pairing. Usually one component (RNA or cDNA) of the H. mixture is present in relatively low concentration and is radioactive (of known specific radioactivity, labeled with ^{32}P or ^{3}H), while the other component (cellular DNA or fragments thereof) is unlabeled and present in excess. H. is determined by separating single from double-stranded nucleic acids and measuring the specific activity of the latter.

The filter or gel technique is used widely for the measurement of H. Here, DNA is thermally denatured and rapidly cooled so that dimers cannot form; DNA can also be denatured by adjusting its pH to 12 with NaOH, followed by HCl to pH 7. The single-stranded DNA is then embedded in agar or polyacrylamide gel, or adsorbed on a disk of cellulose nitrate. The DNA or RNA to be tested for complementarity is radioactively labeled then applied to the gel or cellulose nitrate filter and incubated at about 25 °C below the T_m of the native material. Unbound radioactive material is removed by digestion with ribonuclease (DNA:RNA hybrids are resistant to the enzyme) or a deoxyribonuclease which preferentially attacks single-strand DNA. The amount of radioactivity remaining on the filter or gel after washing is a measure of the extent of complementarity between the sequences of the two samples. Cellulose nitrate disks are more convenient than other supporting substances; the original method of Gillespie and Spiegelman [*J. Mol. Biol.* **12** (1965) 829–841] is still used with little modification. H. may also be performed in solution (see DNA:DNA hybrids, below), in which case chromatography on hydroxyapatite is used to separate single and double-stranded material. The three methods, filter, gel and solution, have been critically compared by D. Kennel and A. Kotoulas [*J. Mol. Biol.* **34** (1968) 71–84].

DNA:RNA hybrids. H. of rRNA with DNA has been used extensively to study gene multiplicity. As increasing quantities of radioactive RNA are incubated with cellular DNA, the formation of the DNA:RNA hybrid shows saturation kinetics. The limiting value for the number of RNA binding sites on the DNA is equivalent to the number of RNA cistrons per genome.

DNA:RNA H. is also used for gene isolation. In this technique, the DNA is sheared into shorter lengths (mechanical breakage or treatment with restriction endonucleases) then hybridized with RNA under conditions that permit only about 10 % renaturation of the DNA. Double- and single-stranded molecules are separated by hydroxyapatite chromatography, and the double-stranded fraction is repeatedly denatured and renatured until single-stranded DNA is effectively removed. Many genes have been partially purified in this way, e.g. rRNA cistrons of *Salmonella typhimurium* [A. Udvardy & P. Venetianer *Eur. J. Biochem.* **20** (1971) 513–517].

Another technique is cytological H., in which radioactive RNA is hybridized with chromosomal DNA in a histological preparation. This indicates the chromosomal location of the gene(s) for the RNA in question. Conversely, cloned DNA probes can be used to show the histological distribution of mRNA corresponding to the cloned gene. Similarly, ^{32}P-labeled RNA or cDNA is used to locate complementary DNA sequences in Southern blots (see) of electrophoretically separated restriction fragments.

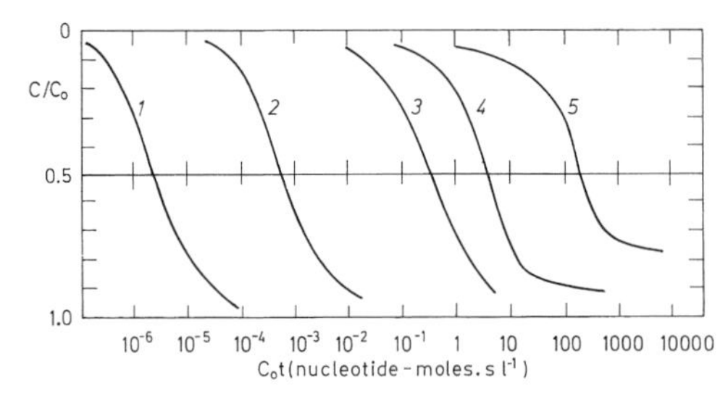

Hybridization plots or "Cot curves" for different DNA samples.
1. Poly U + Poly A.
2. Mouse satellite DNA.
3. Phage T4 DNA.
4. *Rhizobium* DNA.
5. Excess soybean DNA with radioactive root nodule leghemoglobin cDNA (single- and double-stranded DNA separated on hydroxyapatite, and radioactivity of double-stranded DNA monitored as an index of the fraction (C/C_0) of cDNA annealed).

DNA: DNA hybrids. If heat-denatured DNA (see Melting point of DNA) is allowed to cool slowly, double-stranded molecules may be reformed, i.e. the single strands reassociate or become annealed. When two different DNAs are denatured in the presence of each other, the annealed mixture will include hybrid molecules, providing the two DNAs have some base sequences in common. Reassociation and H. (both can be called annealing) are therefore kinetically and mechanistically identical. The following conditions are required for efficient reassociation or hybridization of denatured DNA:

1) An adequate concentration of cations; reassociation is virtually nonexistent below 0.01 M Na^+, very dependent on [Na^+] below 0.4 M NaCl, and almost independent of [Na^+] above 0.4 M NaCl.

2) An optimal temperature about 25 °C below T_m. The reaction rate increases as the temperature decreases below T_m, reaching a broad, flat maximum between (T_m-15) °C and (T_m-30) °C, then decreases with further decrease in temperature.

3) Incubation time and DNA concentration must be sufficient to permit an adequate number of collisions.

4) The statistical probability that the very long complementary strands of denatured eukaryotic DNA will become correctly aligned by random collision is extremely low. The rate of reassociation is therefore conveniently controlled by shearing the DNA and working with relatively small fragments (200–500 base pairs) of known size. On the other hand, certain bacterial genomes can be denatured and reassociated to double-stranded, biologically active DNA, e.g. *E. coli* DNA (4.5 million base pairs) is often used to standardize measurements of the kinetics of reassociation.

In many studies of H., it is advantageous to use cDNA (see Complementary DNA) in place of mRNA; using the mRNA as a template, cDNA of high specific radioactivity can be synthesized in relatively large quantities with efficient use of the radioactive nucleotide precursors; moreover, mRNA is constantly at risk from attack by RNAses.

The literature describes many varied applications of DNA:cDNA H., e.g. determination of the number of copies of globin genes in mouse DNA [P. R. Harrison et al. *J. Mol. Biol.* **84** (1974) 539–554]; and the demonstration that the globin of root nodule leghemoglobin is encoded by the plant genome and not by that of the bacterial symbiont [R. Sidloi-Lumbroso et al. *Nature* **273** (1978) 558–560]. In this type of experiment, sheared cellular DNA and cDNA are fractionated to obtain preparations of the same average nucleotide length (200–500 nucleotides or base pairs), and the cellular DNA is present in large excess (about 10^7-fold) over the radioactive cDNA. As a taxonomic tool, DNA DNA H. in solution, followed by determination of the T_m of the resulting hybrid, represents a powerful tool of biochemical taxonomy. The difference in T_m (ΔT_m) between conspecific (same species) and heterospecific (different species) DNA hybrids is a measure of the evolutionary closeness of the two organisms. A 1 % difference in nucleotide sequence results in a ΔT_m of about 1 °C. Species of the same genus show ΔT_m up to 4 °C. For different genera of the same family, ΔT_m is in the range of 4–11 °C; for different families within the same order, 11–20 °C; and for members of the same class but distantly related orders, about 25 °C [J. M. Diamond *Nature* **305** (1983) 17–18].

Kinetics of H. Association of the two strands follows second-order kinetics. Consider the equilibrium: $A + B \underset{K_2}{\overset{K_1}{\rightleftharpoons}} AB$, where AB represents the hybrid or reassociated molecule. Let $\boldsymbol{a}$ and $\boldsymbol{b}$ represent the initial concentrations of the two strands (A and B) respectively, and $\boldsymbol{x}$ the concentration of AB at time t. The rate of association, $dx/dt = K_1(\boldsymbol{a}-\boldsymbol{x})(\boldsymbol{b}-\boldsymbol{x})-K_2\boldsymbol{x}$. Since K_2 is much smaller than K_1, this can be rewritten: $dx/dt = K_1(\boldsymbol{a}-\boldsymbol{x})(\boldsymbol{b}-\boldsymbol{x})$. Although one H. partner is often present in relatively low concentration, (i.e. $\boldsymbol{a} >> \boldsymbol{b}$), there are circumstances in which A and B are present in equal concentrations, e.g. in the reassociation of melted DNA, and in the H. of equal quantities of DNA from two different species. The above equation then becomes $dx/dt = K_1(\boldsymbol{a}\text{-}\boldsymbol{x})^2$, and integration gives $\boldsymbol{x}/\boldsymbol{a} = \boldsymbol{a}K_1t/(1 + K_1\boldsymbol{a}t)$. This equation represents the mathematical basis of the C_0t (see) method, and is usually expressed as $C/C_0 = 1/(1 + KC_0t)$, where C = concentration of DNA (mol/l) remaining at time t (seconds), and C_0 = initial concentration of denatured DNA (mol/l). H. or association is expressed graphically by plotting C_0t against C/C_0 (Fig.). Since the observed rates range over at least 8 orders of magnitude, it is necessary to plot C_0t on a logarithmic scale; the resulting second-order curve is conveniently symmetrical. The units of C_0t are moles of nucleotide × seconds per liter ($mol \cdot s \cdot l^{-1}$). $C_0t_{1/2}$ (the half period of reassociation, i.e. the half period for 50 % annealing) can be predicted [J. G. Wetmur & N. Davison *J Mol. Biol.* **31** (1968) 349–370]: Let N = complexity of the DNA, i.e. the number of base pairs of nonrepeating sequences; and let L = average number of nucleotides per single strand of denatured DNA. The second-order rate constants for all DNAs are then about $3 \times 10^5 \times L^{0.5}/N$ $l \cdot mol^{-1} \cdot sec^{-1}$. The reaction rate also increases slightly with GC content and is affected by the viscosity of the matrix.

Reassociation of eukaryotic DNA (especially when fragmented) is often faster than predicted [R. J. Britten & D. R. Kohne *Science* **161** (1968) 529–540; also a valuable reference for the principles and experimental basis of DNA reassociation]. This observation gave rise to the hypothesis that certain sequences are repeated, sometimes hundreds or even thousands of times. It is now known that virtually all eukaryotic DNAs contain families of repetitive sequences ranging in length from 130 to 300 base pairs.

Hybridoma: an immortal cell line prepared by fusing lymphocytes with an appropriate line of transformed cells. H. are the usual source of Monoclonal antibodies (see Immunoglobulins).

Hydroazulene: see Proazulene.

Hydrocarbon degradation, ***microbial hydrocarbon degradation:*** degradation of hydrocarbons by certain microorganisms which utilize these compounds as their sole source of carbon and energy. This ability is important for the microbial production of protein and for the elimination of environmental contamination by mineral oils. H.d. depends greatly on the hydrocarbon structure; unbranched alkane chains of 10 to 18 carbons are most readily degraded. The most important pathway involves oxidation of one end of the chain to $-CH_2OH$ by Cytochrome P450 (see), followed by oxidation to –CHO then –COOH by NAD-linked dehydrogenases. The resulting fatty acids are then further degraded by β-oxidation. Aromatic hydrocarbons are less readily degraded than aliphatic structures. Ring cleavage is always preceded by the formation of a phenol by a monooxygenase; a dioxygenase then cleaves the ring next to the hydroxyl function: pyrocatechol → *cis,cis*-muconate → α-ketoadipate → acetate + succinate.

Hydrocyanic acid, ***hydrogen cyanide, HCN:*** a highly poisonous compound found widely in nature. It is released from Cyanogenic glycosides (see) by the action of β-glucosidases (e.g. emulsin) and oxinitrilases. A number of plants, especially those containing cyanogenic glycosides, can metabolize HCN, usually by reaction with serine or cysteine to form cyanoalanine. Addition of water converts the latter to asparagine.

Hydrogen: see Bioelements, Hydrogen metabolism.

Hydrogenase: see Hydrogen metabolism.

Hydrogenation: see Reduction.

Hydrogen bonding: see Noncovalent bonds.

Hydrogen metabolism: metabolic redox reactions, involving pyridine nucleotide and flavin coenzymes; alternatively all metabolic reactions involving hydrogen, i.e. hydrogenation, dehydrogenation, transhydrogenation, activation and formation of molecular hydrogen.

In anaerobic and aerobic respiration, substrates are oxidized by removal of hydrogen (dehydrogenation), a process that does not involve transfer of oxygen. Dehydrogenations are catalysed by dehydrogenases (NAD^+ and $NADP^+$ act as coenzymes) and by oxidases. FAD and FMN act as redox prosthetic groups or cofactors of the flavoenzymes. Hydrogen atoms ($H = H^+ + e^-$) removed from a substrate are transferred to the active group of the dehydrogenase or oxidase. Hydrogen is transferred to NAD^+ or $NADP^+$ as a hydride ion, i.e. one proton (H^+) and an electron pair ($2e^-$) are transferred from the substrate hydrogen (2[H]), leaving one proton free (see Nicotinamide-adenine-dinucleotide). In anaerobic carbohydrate degradation the reduced coenzyme (NADH) is reoxidized by coupling its oxidation to the reduction of an endproduct of glycolysis; this is an internal oxidoreduction system in which molecular oxygen plays no part. In the complete, aerobic oxidation of glucose via glycolysis, the tricarboxylic acid (TCA) cycle and the respiratory chain, NADH is reoxidized by the respiratory chain with the uptake of molecular oxygen.

NAD and NADP serve different metabolic functions. In aerobic metabolism, most of the NADH produced by the cell is reoxidized by the respiratory chain for the purposes of energy production (see Oxidative phosphorylation). Because NAD^+ is the acceptor for most of the hydrogen produced in catabolic reactions, the term "catabolic reduction charge" (CRC) has been proposed for the ratio $[NADH]/([NADH] + [NAD^+])$. NADPH, on the other hand, does not transfer hydrogen to the respiratory chain; it functions as a reducing agent in reductive biosyntheses. Thus the ratio $[NADPH]/([NADPH] + [NADP^+])$ is the "anabolic reduction charge" (ARC). The CRC and the ARC are important quantities in the regulation of cellular metabolism. [K. B. Anderson & K. von Meyenburg *J. Biol. Chem.* **252** (1977) 4151]

In the cytoplasm, the CRC is normally maintained at a value much less than 1, i.e. there is much more NAD^+ than NADH. In the mitochondria, however, there is much more of the reduced form, in spite of the fact that this is the site of oxidation of NADH to NAD^+. This suggests that NADH produced in the cytoplasm is efficiently oxidized by the mitochondria, but this cannot occur directly because the inner mitochondrial membrane is impermeable to NAD^+ and NADH in both directions. Instead, reducing equivalents are transferred from the cytoplasm to the mitochondria, a process accomplished by the following *shuttle systems:*

1. *β-Hydroxybutyrate/acetoacetate shuttle.* Cytoplasmic acetoacetate is reduced to β-hydroxybutyrate by an NADH-dependent reductase (EC 1.1.1.30). β-Hydroxybutyrate enters the mitochondrion, where it is oxidized to acetoacetate by an NAD-dependent dehydrogenase The resulting NADH is oxidized by the respiratory chain, and the acetoacetate leaves the mitochondrion to be reduced by more cytoplasmic NADH. The reality of this system is in some doubt. Certainly it could not operate in mitochondria that lack β-hydroxybutyrate dehydrogenase, i.e. hepatic mitochondria of ruminants. Vertebrate red muscle contains high levels of mitochondrial β-hydroxybutyrate dehydrogenase, so the shuttle might operate in this tissue.

2. *Dihydroxyacetone phosphate/α-glycerolphosphate shuttle.* Dihydroxyacetone phosphate is reduced in the cytoplasm by glycerol-3-phosphate dehydrogenase (EC 1.1.1.8) and NADH. The resulting α-glycerolphosphate is oxidized in the mitochondria by an efficient glycerol 3-phosphate dehydrogenase (an FAD-flavoprotein, EC 1.1.99.5), and the dihydroxyacetone phosphate returns to the cytoplasm. Whereas the oxidation of NADH by the respiratory chain gives rise to 3 ATP, this shuttle system produces 2 ATP per NADH oxidized, because the mitochondrial dehydrogenase is FAD-dependent. This is an active shuttle system in blowfly flight muscle. The component enzymes are also present in mammalian muscle in amounts compatible with the operation of the shuttle, but it seems doubtful that this particular shuttle is operative in liver.

3. *Malate/oxaloacetate shuttle.* Oxaloacetate is reduced at the expense of NADH by the action of a cytoplasmic malate dehydrogenase (EC 1.1.1.37). Malate enters the mitochondria, where it is dehydrogenated by mitochondrial malate dehydrogenase. Oxaloacetate does not leave the mitochondria, but it is transaminated with glutamate to form aspartate and 2-oxoglutarate. Aspartate crosses the mitochondrial membrane and is transaminated to oxaloacetate in the cytoplasm. This shuttle does not cause an imbalance of charge across the mitochondrial membrane, because the entry of malate is coupled with the exit of 2-oxoglutarate (both dicarboxylic acids), and the entry of glutamate is coupled with the exit of aspartate (both acidic amino acids).

Many other shuttles can be contrived theoretically, but they lack experimental support. Liver cells are probably served by several shuttles, rather than one major system. In contrast to the CRC, the ARC in the cytoplasm is low. The high concentration of NADPH relative to that of $NADP^+$ helps to drive biosyntheses forward by mass action; among them are the reduction of glycerate 3-phosphate to glyceraldehyde 3-phosphate in the photosynthetic assimilation of CO_2 (see Calvin cycle), reductive synthesis of glutamate by glutamate synthase (see Ammonia assimilation), and Fatty acid biosynthesis.

In heterotrophic organisms, NADPH is formed in the oxidative phase of the Pentose phosphate cycle (see), and in photosynthetic organisms (except photosynthetic bacteria) in the light reaction of Photosynthesis (see). Another important source is the cytoplasmic oxidative decarboxylation of malate by $NADP^+$-linked malate dehydrogenase (EC 1.1.1.40). Pyruvate enters the mitochondria and undergoes ATP-dependent carboxylation to oxaloacetate, which is hydrogenated by NADH-dependent malate dehydrogenase (EC 1.1.1.37). The resulting malate leaves the mitochondrion and is decarboxylated to pyruvate in the cytoplasm by $NADP^+$-linked malate dehydrogenase (EC 1.1.1.40). Operation of this cycle results in the export of reducing equivalents from the mitochondrion to the cytoplasm.

In contrast to the oxidation of NADH, there is no mechanism for the direct oxidation of NADPH by the respiratory chain. In fact the physiological problem faced by animal cells in particular is how to maintain an adequate supply of NADPH for reductive biosynthesis, and how to supplement this by exploiting the reducing equivalents of NADH, e.g. by the malate-pyruvate cycle described above. Nevertheless, some of the hydrogen of NADPH is probably oxidized by the respiratory chain. For example, NADH and NADPH might become equilibrated by any enzyme that can use both, and such an equilibration has been shown to occur. Enzymes that use both cofactors with more or less equal facility are glycerol dehydrogenase (pig and rat liver, *E. coli*), glutamate dehydrogenase (muscle, liver, yeast) and 3β-hydroxysteroid dehydrogenase (liver). Liver and heart mitochondria perform a transhydrogenation in which hydrogen is transferred from NADH to $NADP^+$. The process is ATP-dependent and is catalysed by an enzyme in the inner mitochondrial membrane. The equilibrium is so strongly in favor of the formation of NADPH that the reverse process is not easily measured. Clearly this reaction will not lead to the net production of NADH from NADPH; rather it serves to supply NADPH required for intramitochondrial reductive biosynthesis. Both the donor (NADH) and the acceptor ($NADP^+$) interact with transhydrogenase on the M-side of the mitochondrial membrane. During transhydrogenation, there is no exchange of hydrogen with protons of water. The hydrogen atom is transferred from the A side of NADH to the B side of $NADP^+$, indicating that the planes of the nicotinamide rings of the two cofactors must become closely associated on the enzyme surface. Cytoplasmic transhydrogenases have also been reported, which catalyse transhydrogenation with an equilibrium constant of about unity.

In the majority of living organisms, molecular hydrogen has no significant metabolic role. However, certain enzymes called hydrogenases have been found in a wide range of organisms, including bacteria, plants and animals. It has been suggested (Krebs) that hydrogenase serves to release H_2 from excess NADH, when biosyntheses have a prevalence of oxidative steps and therefore disturb the normal redox balance of the cell, e.g. in microorganisms growing anaerobically, excess reducing power may be used to synthesize reduced products that are excreted, or it may simply be released as molecular hydrogen by the agency of hydrogenase.

Some hydrogenases are membrane-bound, and they are often linked to formate dehydrogenase ($HCOOH + NAD^+ \rightarrow CO_2 + NADH + H^+$; $NADH + H^+ \rightarrow NAD^+ + H_2$). In strictly anaerobic bacteria, hydrogenases are linked to ferredoxin, and they catalyse an oxidoreduction reaction between hydrogen and ferredoxin: $H_2 + 2Fd_{ox} \rightarrow 2H^+ + 2Fd_{red}$, where Fd_{ox}

and Fd_{red} represent oxidized and reduced ferredoxin, respectively. Certain bacteria, e.g. *Hydrogenomonas, Pseudomonas* and *Alcaligenes,* can oxidize H_2 with O_2; a normal electron transport chain operates and 3 molecules of ATP are generated.

Ferredoxin-dependent hydrogen production catalysed by hydrogenase, e.g. in *Clostridium,* is inhibited by carbon monoxide, and is independent of ATP. It therefore differs from the ATP-dependent CO-sensitive hydrogen production by Nitrogenase (see). Evolution of hydrogen gas during nitrogen fixation is due to competition between nitrogen and protons for electrons at the reducing center of the enzyme. In addition to ferredoxin, the Redoxins (see) are important electron transferring agents in hydrogen metabolism.

Hydrogen transfer: see Hydrogen metabolism, Pyridine nucleotide coenzymes.

Hydrolases: see Enzymes, Table 1.

Hydrophobic bonds: see Noncovalent bonds.

Hydroxamic acids: derivatives of carbonic acid containing the tautomeric group R–CO–NHOH ⇌ R–C(OH) = N–OH. H. a. form stable, five-membered rings with metal ions, and they are especially important in the iron metabolism of many organisms. Well-known examples of H. a. are aspergillic acid (Fig.), synthesized from leucine and isoleucine by *Aspergillus flavus,* and the Siderochromes (see).

Aspergillic acid

Hydroxyacetic acid: See Glycolic acid.

Hydroxyacid: a carboxylic acid, in which one or more hydrogen atoms of the alkyl moiety is replaced by a hydroxyl group. The position of the OH group in the alkyl chain is indicated by α, β, γ, δ, etc., or by 2, 3, 4, 5, where the C-atom of the COOH is No. 1; thus lactic acid is 2-hydroxypropionic or α-hydroxypropionic acid. Some important H. are glyceric, malic, lactic and citric acids.

2′-Hydroxy-3-arylcoumarins: a group of naturally occurring Isoflavonoids (see), e.g. pachyrrhizin from *Pachyrrhizus erusus* and *Neorautanenia* spp. [L. Crowbie & D. A. Whiting *J. Chem. Soc.* (1963) 1569–1579] (Fig.).

Pachyrrhizin

3-Hydroxy-2-butanone: see Acetoin.

***o*-Hydroxycinnamic acid lactone:** see Coumarin.

Hydroxycinnamoyl-CoA ligase: see 4-Coumarate:CoA ligase.

9-Hydroxy-trans-2-decenoic acid: see Queen substance.

N*-2-Hydroxyethylpiperazine-*N*-2-ethanesulfonic acid, *Hepes: a compound (M_r 238.3) used for the preparation of Buffers (see) in the pH range 6.8–8.2.

Hydroxyisovaleric aciduria: see Inborn errors of metabolism.

Hydroxylases: see Oxygenases.

Hydroxylation: see Oxygenases.

Hydroxylubimin: see Phytoalexins.

5-Hydroxymethylcytosine: one of the rare nucleic acid bases, M_r 141.1, decomposes above 200 °C without melting. H. does not arise by modification of cytosine in nucleic acids; rather 5-hydroxymethyldeoxycytidylic acid is formed de novo in the course of pyrimidine biosynthesis. It is found in place of cytosine in the DNA of bacteriophages of the T2, T4, T6 series.

Hydroxymethyl glutarate cycle: see Ketogenesis.

Hydroxynervon: see Glycolipids.

Hydroxynervonic acid, *Δ¹⁵-2-hydroxytetracosanoic acid:* a hydroxylated, unsaturated fatty acid, CH_3–$(CH_2)_7$–CH = $CH(CH_2)_{12}$–CHOH–COOH, M_r 382.5, m. p. 65 °C. It is a major component of cerebrosides.

2-Hydroxy-3-oxoadipate synthase, *2-hydroxy-3-oxoadipate glyoxylate-lyase (carboxylating)* (EC 4.1.3.15): An enzyme in bacteria and mammalian liver, which catalyses the decarboxylative condensation of 2-oxoglutarate with glyoxylate. The product, 2-hydroxy-3-oxoadipate, is important in mammals for diverting glyoxylate metabolism from oxalate synthesis (see Oxalic acid; Oxalosis under Inborn errors of metabolism). *In vitro,* the enzyme catalyses a variety of related decarboxylation and condensation reactions (Fig.), which are analogous to Acetoin (see) formation from pyruvate and acetaldehyde. [M. A. Schlossberg et al. *Biochemistry* **9** (1970) 1148–1153].

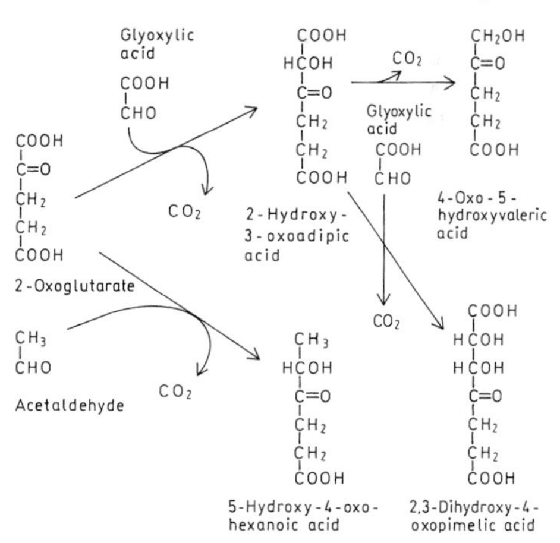

Some reactions catalysed by 2-hydroxy-3-oxoadipate synthase.

3α-Hydroxy-5α-pregnan-20-one: a catabolite of progesterone. Like its stereoisomers, 3β-hydroxy-5α-pregnan-20-one and 3α-hydroxy-5β-pregnan-20-one, 3α-H appears in human pregnancy urine.

Hydroxyproline, *Hyp* or ***Pro(OH):*** an amino acid residue in collagen formed by the post-translational modification of proline residues (see Post-translational modification of proteins). Hydroxylation occurs at the 4 or the 3 position. 4Hyp is more common in collagen. After enzymatic degradation of collagen, 4Hyp is degraded by reductive ring cleavage to 4-hydroxy-2-oxoglutarate, then to pyruvate and glyoxylate (see L-Proline). 3Hyp is also a constituent of Amanita toxins (see Amatoxins).

5-Hydroxytryptophan: An intermediate in the synthesis of Melatonin (see) and Serotonin (see). It is used clinically in the treatment of depression and myoclonus [M. A. A. Namboodiri *Science* **221** (1983) 659–660].

Hydroxyxanthorrhone: see Flavan.

Hygrine: see Pyrrolidine alkaloids.

Hyocholic acid: 3α,6α,7α-trihydroxy-5β-cholan-24-oic acid, one of the bile acids. H. is a component of pig and rat bile. Like the main component of pig bile, hyodeoxycholic acid (3α,6α-dihydroxy-5β-cholan-24-oic acid), it is important as a starting material for the chemical synthesis of steroid hormones.

Hyodeoxycholic acid: see Hyocholic acid.

Hyoscine: see Scopolamine.

Hyoscyamine: see Atropine.

Hyp: abb. for Hypoxanthine (see) or Hydroxyproline (see).

Hyperammonemia: see Inborn errors of metabolism.

Hyperbilirubinemia: see Inborn errors of metabolism.

Hyperchromic effect: increase in the absorbance of a solution at a particular wavelength due to structural changes in the solute molecules. The H. e. is a useful experimental index of DNA denaturation, since the A_{260} of a DNA solution increases when the double helix is transformed by heating into a disordered random coil (see Hybridization).

Hyperlysinemia: see Inborn errors of metabolism.

Hyperornithinemia: see Inborn errors of metabolism.

Hyperprolinemia: see Inborn errors of metabolism.

Hypersarcosinemia: see Inborn errors of metabolism.

Hypersarcosinuria: see Inborn errors of metabolism.

Hypertensin: see Angiotensin.

Hypervitaminosis: see Vitamins.

Hypochloremic alkalosis: see Alkalosis.

Hypochromes: see Pigment colors.

Hypochromic effect: an optical phenomenon in molecules with several chromophores, in which the sum of the absorbance of the individual components is greater than the absorbance of the whole molecule. The absorbance of the nucleic acids at 260 nm, for example, is less than the calculated sum of the absorbances of the component bases. The H. e. depends on the content of adenine and thymine and is therefore greater in DNA than in RNA. Double-stranded polynucleotides have a greater H. e. than single-stranded, because the effect is intensified by hydrogen bonds.

Hypophosphatasia: see Inborn errors of metabolism.

Hypophysis, ***pituitary gland:*** a vertebrate endocrine gland at the base of the brain, and connected to the midbrain by the hypophyseal stalk. In humans, it weighs 0.7 g. The H. is composed of two parts with different ontogenies. The adenohypophysis, which includes the anterior and middle parts, arises from the roof of the embryonic mouth. The neurohypophysis, or posterior lobe is formed from an outgrowth of the base of the midbrain. In the anterior lobe, releasing hormones from the hypothalamus stimulate the formation of somatotropin and prolactin in the acidophilic (α) cells; follicle-stimulating hormone, luteinizing hormone and thyreotropin in the basophilic β-cells; and corticotropin in the chromophobic γ-cells. Melanotropin is synthesized in the middle part of the H. The neurohormones, oxytocin and vasopressin, are stored in the posterior lobe and released on demand into the circulation. See separate entries for each of the above-mentioned hormones.

Hypotaurine: see Cysteine.

Hypothalamus: the lowest part of the midbrain, which also includes the thalamus and epithalamus. The hypophyseal stalk with the neurohypophysis arises from the underside of the H. As part of the limbic system, the H. is a "gateway to consciousness". Afferent (incoming) stimuli from the breast and abdominal areas and from the circulatory system are processed by the thalamus and passed on to the cerebrum. Conversely, efferent stimuli from the cerebrum flow to the thalamus and H. A series of nuclear regions controlling important vegetative functions, such as blood pressure, temperature, sweat production, water balance, motions of the digestive tract organs, and sexual function are located in the H. The entire metabolism can be greatly influenced by the appetite and eating center of the H. In addition, in certain nuclear areas, nervous excitation is transformed into hormonal signals (releasing hormones and neurohypophyseal hormones), and conversely, hormones from the hypophyseal hormone glandular system exert negative feedback on the formation of hormones in the nerve cells.

Hypovitaminosis: see Vitamins.

Hypoxanthine, ***Hyp:*** 6-hydroxypurine, a widely distributed purine derivative in plants and animals (formula, see Inosine), M_r 136.11, m.p. 150 °C (d.). It is produced during aerobic Purine catabolism (see) by deamination of adenine compounds or by hydrolysis of inosine compounds. It is found as a rare base in certain transfer RNAs.

Hypoxanthinosine: see Inosine.

Hypusine, ***N^ε-(4-amino-2-hydroxybutyl)lysine:*** an amino acid present in eukaryotic initiation factor 4D

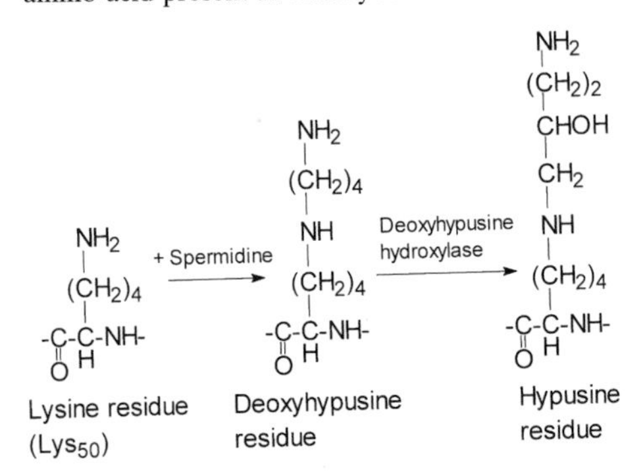

Biosynthesis of the hypusine residue in eukaryotic initiation factor 4D.

(eIF-4D) and formed by con- or post-translational modification of Lys_{50} (Fig.). Radiotracing shows that the terminal four carbons of hypusine are derived from the butyl carbons of spermidine, but the enzymology of the process is not known. Deoxyhypusine hydroxylase requires only sulfhydryl reagents for activity; absence of requirements for Fe^{2+} (which inhibits), 2-oxoglutarate and ascorbate indicates that the enzyme is not a 2-oxoacid dehydrogenase. The variant of eIF-4D, in which residue 50 remains as unmodified Lys, is biologically inactive, i.e. it does not stimulate synthesis of methionyl-puromycin or globin *in vitro,* and it does not significantly inhibit the action of native eIF-4D *in vivo.* [Myung Hee Park et al. *J. Biol. Chem.* **259** (1984) 12123–12127; A. Abbruzzese et al. *J. Biol. Chem.* **261** (1986) 3085–3089; Z. Smit-McBride et al. *J. Biol. Chem.* **264** (1989) 18527–18530]

I

IAA: see Auxins.

Ibotenic acid: α-amino-3-hydroxy-5-isoxazole-acetic acid, M_r 158, m.p. 145 °C (d.), a psychotropic, weakly insecticidal substance. I. a. and its decarboxylation product, Muscimol (see), are amanita toxins. I. a. is found only in a few species of *Amanita* (fly agaric), at concentrations averaging 0.05 % of the fresh weight. It is pharmacologically very active, but less so than muscimol in most tests. Another I. a. derivative is Tricholomic acid (see).

I-cell disease: see Inborn errors of metabolism.

ICSH: see Luteinizing hormone.

Idaeine: see Cyanidin.

IDP: acronym of Inosine 5′-diphosphate.

Ig: abb. for Immunoglobulin.

Ile: abb. for L-Isoleucine.

Imidazole alkaloids: alkaloids of sporadic occurrence, possessing the imidazole ring system. Their most important representative is Pilocarpine (see). The biosynthesis of the I. a. is coupled to histidine metabolism.

Immobilized enzymes: soluble enzymes bound to an insoluble organic or inorganic (e. g. porous glass) matrix, or encapsulated within a membrane in order to increase their stability and make possible their repeated or continual use. The five most commonly used methods of enzyme immobilization are: 1. Adsorption onto an inert or electrically charged carrier (cross-linked dextrans); 2. Covalent binding to a carrier polysaccharide, e. g. Sepharose; 3. Entrapment in a three-dimensional cross-linked carrier, e. g. polyacrylamide gel; 4. Cross linkage, i. e. condensation of several enzyme molecules with bi- or polyfunctional agents, like glutaraldehyde and epichlorhydrin; 5. Microencapsulation. Covalent binding to a carrier is widely used. Since the enzyme is chemically fixed, it is not washed out and can be used repeatedly. Of the organic carriers, Sepharose and large-pore cross-linked dextrans come close to meeting all the requirements of an enzyme carrier, i. e. minimal solubility, high mechanical strength, suitable particle size and shape, high binding capacity for the enzyme, but no adsorption of the substrates and products, and resistance to attack by microorganisms. Being organic, however, they are subject to microbial attack, and they swell and shrink, depending on pH and other environmental conditions. Since 1969, silanized porous glass, which has neither of these disadvantages, has been used increasingly for both adsorptive and covalent attachment of enzymes. For covalent attachment, functional groups must be present on the carrier. A frequently used activation procedure, especially with dextran gels, consists of reaction with cyanogen bromide (CNBr), which produces reactive iminocarbonate groups (Fig. 1).

At pH 7–9, these react with the free amino groups of the enzyme protein, with the formation of substituted imidocarbonates (=C=N-protein). Many other coupling techniques have been reported for the covalent attachment of enzymes to agar, agarose and Sephadex supports, and to the silanized surface of porous glass. Details of these techniques and other aspects of immobilized enzymes are comprehensively treated in *Methods in Enzymology,* **XLIV,** 1976, Klaus Mosbach, ed. Academic Press

The quantities of covalently bound enzymes are generally low (1–5 %); in exceptional cases, especially when carrier and protein are oppositely charged, 10 % or more may be bound.

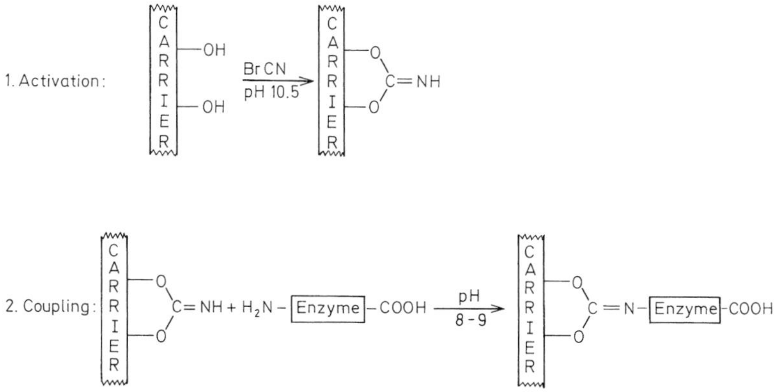

1. Activation: CARRIER(–OH)(–OH) $\xrightarrow[\text{pH 10.5}]{\text{BrCN}}$ CARRIER(–O–)(–O–)C=NH

2. Coupling: CARRIER(–O–)(–O–)C=NH + H$_2$N–Enzyme–COOH $\xrightarrow[\text{8-9}]{\text{pH}}$ CARRIER(–O–)(–O–)C=N–Enzyme–COOH

Fig. 1. *Immobilized enzymes.* Covalent binding of an enzyme to an unsubstituted polysaccharide (e. g. Sepharaose) by the cyanogen bromide method.

Properties. In general, the free and the immobilized enzyme catalyse the same reaction, but depending on the supporting material and nature of the binding, there may be differences in pH and temperature optima (the latter is usually increased), K_m value and specific and maximal activities (the latter is usually decreased) (see Enzyme kinetics). Chief reasons for these alterations are the decreased flexibility and mobility of the coupled enzyme, and steric factors which interfere with access of the substrate to, and diffusion of product from, the active center. These changes are usually more than compensated by increased stability of the enzyme. They can be avoided or reduced by attaching the enzyme to the support by a side chain, or spacer, which allows greater mobility and unhindered contact with substrates.

Enzymes may also be immobilized by microencapsulation. In this technique, which has medical applications, enzymes are enclosed by various types of semipermeable membrane, e. g. polyamide, polyurethane, polyphenyl esters and phospholipids. Microcapsules of phospholipids are also called *liposomes.* The microencapsulated enzymes and proteins inside the microcapsule cannot pass the membrane envelope, but low M_r substrates can pass into it, and products can leave. Such encapsulated proteins do not elicit an antigenic response, and they are not attacked by proteases outside the microcapsule. They are therefore suitable for the delivery of enzymes for therapeutic purposes. This area of application is still at an early stage of development, but positive results have been reported from animal experiments and clinical studies, e. g. treatment of inherited catalase deficiency with encapsulated catalase. There are various methods of administration: intramuscular, subcutaneous or intraperitoneal injection. However, their major area of application is outside the body. For example, microencapsulated urease can be employed as an artificial kidney in hemodiffusion (Fig. 2).

Examples of I. e.used commercially are: proteases (in detergents, and for making cheese and other food products), carbohydrases (for hydrolysing starch, converting glucose to fructose, etc.), lipases (in food processing); various other I. e. are used in analysis, and in the production and development of pharmaceuticals [*Industrial Enzymology,* Nature Press, New York, 1983]

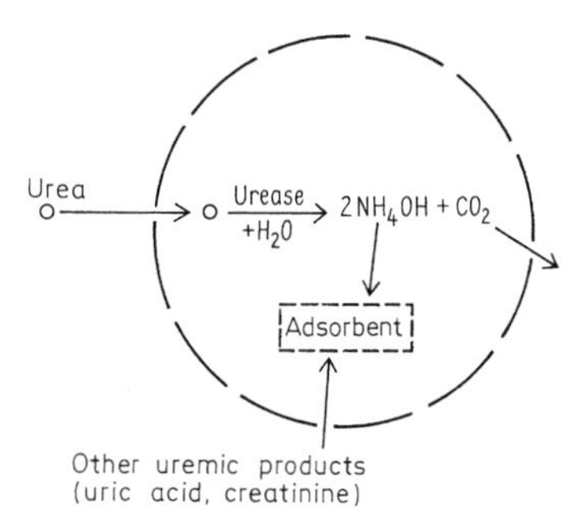

Fig. 2. *Immobilized enzymes.* Schematic representation of an artificial cell, containing urease and albumin-coated active charcoal as an absorbent for uric acid, ammonia and creatinine. A 10 ml suspension of these 20 μm-diameter urease capsules corresponds to a surface area of 20,000 cm², which is larger than that of the conventional artificial kidney.

Immune response: a specific protective or defense reaction against foreign substances, called Antigens (see). In the humoral I. r., Immunoglobulins (see), which are also called antibodies, are secreted into the blood, lymph or mucosal secretions. Immunoglobulins (Ig.) are produced by lymphocytes derived from bone marrow in mammals or from the Bursa of Fabricius in birds (B cells). B cells display on their surfaces membrane-bound Ig. called surface (s-)Ig. When they are incubated with a fluorescent-labeled antigen (see Immunofluorescence), those B cells with sIg. complementary to the antigen can be seen to undergo "capping". The antigen-sIg. complexes gather together in one area of the cell membrane, and are eventually ingested. This appears to be a necessary step in the activation of B cells to multiply and secrete Ig. Some antigens, including bacterial lipopolysaccharides, can stimulate purified cultures of B cells to produce antibodies, but most antigens require the presence of macrophages and T (thymus) lymphocytes. Particulate antigens, such as bacteria, are engulfed by phagocytes, which then "present" the antigen to the T and B cells. B-cells, when stimulated by "helper" T-cells, multiply and secrete antibodies into the blood or lymph. Multiplication of stimulated B-cell clones is also regulated by "suppressor" T-cells In many cases, expression of antibody by B cells requires both "presentation" by a macrophage and activation by helper T-cells. The chemical mediators of the interaction are called Lymphokines (see).

As an infection progresses, many B cells switch from the production of IgM antibodies (see Immunoglobulins) to IgG ("gamma globulins"). The latter usually have higher affinity for the antigen. After the infection has subsided, the serum levels of IgG drop, but do not fall to zero. Some of the B cells are transformed into "memory cells"; in the event of a secondary challenge with the same antigen, these are capable of dividing and producing IgG much more quickly than was possible in the primary response.

In *cellular immunity,* some T cells, rather than regulating B cells, directly attack foreign cells or tissue (as in graft rejection). These are called cytotoxic T cells; populations of such cells in apparently naive animals are called "natural killer" cells.

Immune serum: see Antiserum.

Immune suppression: the unspecific suppression of the Immune response (see) by corticosteroids, antimetabolites (purine analogs, folic acid antagonists), alkylating substances like Cyclophosphamide (see), ionizing radiation or antilymphocyte serum (see). Although the latter is directed against the cells of the immune system, the other agents have an antiproliferative effect, i. e. they hinder multiplication of the plasma cells. At the molecular level, these agents directly or indirectly inhibit the synthesis of DNA and RNA. I. s. is an important form of therapy for autoimmune diseases and for preventing rejection of transplanted organs. However, suppression of antibody formation results in higher susceptibility to infection and to tumor formation. The pathological I. s. observed in Aquired Immune Deficiency Syndrome (AIDS) is often (perhaps always) fatal, due to adventitious infection or tumors. A true alternative to I. s. is specific I. s., achieved by means of antigen-specific immune sera or by the induction of an immune tolerance.

Immunization: artificial stimulation of antibody production for protection against pathogens and other antigens. *Active I.* is performed by the injection of live, weakened pathogens (e. g. Sabin poliomyelitis vaccine, smallpox vaccine) or killed pathogens, or purified fractions of them. *Passive I.* is performed by injecting antiserum or antibodies raised against the pathogen in another organism and extracted from it. While repeated active I. confers life-long immunity, passive I. is effective for only a few weeks, and cannot be repeated with antiserum or antibodies from the same species of animal, because the foreign antibodies act as antigens. Passive I. is therefore used for short-term prophylaxis or therapy, to bridge the time until the body can produce its own antibodies (e. g. in tetanus simultaneous immunization).

Immunoassays: assays which exploit the specific interaction between an antigen (Ag) and its antibody (Ab) to determine the presence or the concentration of either. Detection and quantitation of an Ab are most easily performed with the aid of the pure or nearly pure Ag. If only impure Ag is available it may be necessary to supplement the assay with a secondary technique such as immunoprecipitation or immunoblotting to increase the ability of the Ab (particularly if it is polyclonal) to distinguish the correct Ag from the background. Similarly, a suitable Ab is required for the detection and quantitation of an Ag; this Ab may be polyclonal (ideally one that is affinity-purified), or monoclonal. If the Ag to be assayed is a Hapten (see), a corresponding Ab can only be generated by first rendering it immunogenic by conjugating it to a carrier protein.

Immunoassays can be divided into three stages, namely: (i) the Ab-Ag reaction, (ii) the partitioning and separation of the Ab-Ag complex from other components of the reaction mixture, particularly unbound Ab or Ag, and (iii) measurement of the response. There are two alternative configurations for the Ab-Ag reaction: (a) precipitation of one reactant with excess of the other, or (b) competition among known amounts of Ab, Ag and the material being assayed (an unknown amount of Ag or Ab). Thus, Ab may be assayed (a) by using excess Ag, or (b) by allowing a known amount of labeled Ab to compete for a fixed amount of Ag with the unknown amount of unlabeled Ab; similarly, Ag may be assayed (a) by using excess Ab, or (b) by allowing a known amount of labeled Ag to compete for a fixed amount of Ab with the unknown amount of unlabeled Ag.

Since the first appearance of immunoassays in about 1960 many methods have been used to partition and separate the Ab-Ag complex from other components of the reaction mixture. These include salting out with $(NH_4)_2SO_4$ or Na_2SO_4, protein denaturation/precipitation by solvent (e. g. CH_3OH, C_2H_5OH, acetone) or precipitation by polyethylene glycol, followed by collection of the precipitated Ab-Ag by filtration or centrifugation. Nowdays the method of choice involves immobilisation of the Ab (for assay of the Ag) or protein Ag (for assay of the Ab) to a solid support material; the Ab-Ag complex is therefore also bound to the support, thereby allowing the rest of the reaction mixture to be washed away. The solid support material may be the wall of the reaction vessel (e. g. the well of a microtiter plate), or it may take the form of sheets or beads; it may be composed of nitrocellulose, polyvinyl chloride or polystyrene (which noncovalently bind up to 100 μg, 300 ng & 300 ng of protein respectively per cm^2), diazotized paper (which binds >10 mg protein/cm^2 by covalent bonds formed with free protein amino groups), agarose or polyacrylamide beads (which, when subjected to the CNBr activation procedure, bind up to 10 mg protein/ml by forming covalent bonds with free protein amino groups) and *Staphylococcus aureus* Protein A beads (which have a high affinity for the Fc region of some Ab and can therefore bind noncovalently to them; they have an Ab protein capacity of 20 mg/ml).

Several methods have been used to measure the response of the Ab-Ag interaction; they all involve the use of either a labeled Ab, Ag or secondary reagent (the latter is often an Ab to the Ab or Ag of the Ab-Ag reaction, and is added to the reaction mixture after the formation of the Ab-Ag complex). Radioactive labels were the first to be used in immunoassays and are still in use. The γ-emitter ^{125}I ($t_{1/2}$ = 59.6 d) is frequently used to label the phenyl group of tyrosine residues in Ab and proteinaceous Ag (1–5 μg of protein and [^{125}I]iodide in slightly alkaline solution are treated with chloramine T which yields the mild oxidizing agent hypochlorous acid which is presumed to promote iodination by generating I^+ ions). However, Ab and Ag labeled chemically or biosynthetically with β-emitters such as ^3H ($t_{1/2}$ = 12.4 y), ^{14}C ($t_{1/2}$ = 5730 y), ^{35}S ($t_{1/2}$ = 87.4 d) and ^{32}P ($t_{1/2}$ = 14.3 d) have also been used. More recently enzymes and fluorescent compounds have been used to label Ab, Ag and secondary reagents. The enzymes used most commonly are horseradish peroxidase, alkaline phosphatase and β-galactosidase which are covalently coupled to Ab and proteinaceous Ag using glutaraldehyde which acts as a linker between enzyme-NH_2 groups and Ab/Ag-NH_2 groups. Horseradish peroxidase is often assayed by incubating it with 3,3′,5,5′-tetramethylbenzidine and H_2O_2, acidifying and then measuring the absorbance of the yellow product at 450 nm. Alkaline phosphatase is often assayed by incubating it with *p*-nitrophenyl phosphate at pH 9.5 then measuring the absorbance of the yellow product at 400 nm. β-Galactosidase is often assayed by incubating it with *o*-nitrophenyl-β-D-galactopyranoside then measuring the absorbance of the yellow product at 410 nm. The most commonly used fluorescent compounds are fluorescein (excitation 495 nm, emission 525 nm), tetramethylrhodamine (Rhodamine B) (excitation 552 nm, emission 570 nm), sulforhodamine 101 acid chloride (Texas red) (excitation 596 nm, emission 620 nm) and phycoerythrins (excitation ~545 nm, emission ~575 nm). The isothiocyanate derivatives of fluorescein and tetramethylrhodamine label Ab and Ag by forming thiocarbamido linkages with free amino groups. Fluorescently-labeled compounds are usually assayed spectrofluorimetrically, using an appropriate excitation wavelength and measuring the intensity of the emitted fluorescence. Small Ag, however, may sometimes be measured by fluorescence polarisation. The rationale of fluorescence polarization is that the conjugate of a small Ag and a fluorescent compound tumbles so rapidly in aqueous solution that when it is excited by polarized light, the emitted fluorescence is unpolarized; however, when an Ab binds to the Ag of the conjugate the new complex, being much larger, tumbles more slowly, so that a significant fraction of the emitted light is polarized. Thus the appearance of polarization in the emitted light is an indication of Ab-Ag binding.

Regardless of the label used, the response of the Ab-Ag interaction can be augmented by making use of the affinity of the protein Avidin (see) or Streptavidin (see) for biotin (see Vitamins). In this procedure the Ab or proteinaceous Ag is covalently coupled to biotin; this is accomplished by treating the Ab or Ag with the succinimide ester of biotin, which couples to free amino groups, usually those of lysine residues, in the protein; thus each Ab or Ag molecule is likely to be conjugated to several biotin residues. Additionally the (strept)avidin is labeled, either radioactively with ^{125}I or by covalently coupling it to one of the enzymes or fluorescent compounds mentioned earlier. Because each biotinylated Ab or Ag molecule carries several biotin residues it will bind several labeled (strept)avidin molecules. This markedly augments the label on the Ab or Ag so that when it participates in the Ab-Ag reaction the detection and assay of the Ab-Ab complex is considerably easier; it is estimated that a biotin-avidin enhanced immunoassay is 1,000-fold more sensitive than the equivalent assay without the the participation of the biotin-avidin couple.

The extensive range of immunoassays now available can be classified in several ways. For the researcher wishing to choose the procedure best suited to the constraints (e.g. type and purity of Ab or Ag available) of the measurement, a particularly useful classification is that of E. Harlow & D. Lane (*Antibodies: a laboratory manual* 1988, pp. 553–612, Cold Spring Harbor Laboratory); this is based on methodology and divides immunoassays into three classes: (i) Ab capture assays, (ii) Ag capture assays, and (iii) two-Ab sandwich assays, each of which is then divided into four subclasses on the basis of whether it is carried out in 'Ab excess' or in 'Ag excess' or as an 'Ab competition' or an 'Ag competition'. The commonest system of classification, however, is based on the labeling method used in the measurement of the Ab-Ag binding response, and this system is used in the present account; examples of some of these classes are described briefly below.

Immunoassays using a radioisotopic label

1. *Competitive binding radioimmunoassay.* This assay procedure is also known as *Inhibition radioimmunoassay* and is regarded by many researchers as the 'real' radioimmunoassay (RIA) technique; certainly it was the first to be developed (by R. S. Yalow & S. A. Berson in the late 1950s for the detection and quantitation of insulin in human blood plasma, work for which Rosalyn Yalow shared the 1977 Nobel Prize in medicine; her Nobel lecture [*Science* **200** (1978) 1236–1245] summarizes this work) and has been used to quantitate a wide range of compounds. For the quantitative assay of compound 'X' by this procedure it is necessary to have a pure sample of unlabeled X, a sample of radioisotopically labeled X (i. e. *X) and an Ab to X (i. e. anti-X) which ideally should be of high specificity and avidity. It is then necessary to determine the appropriate concentration of anti-X for use in the competitive binding assay. This is accomplished by making serial dilutions of the anti-X antiserum and incubating them for ~ 2 h with the fixed amount of *X that will be used in the subsequent preparation of a standard curve and in the assay of unknown samples of X. The anti-X-*X complex formed in each dilution is then separated from unbound *X and its radioactivity measured. From a plot of 'radioactivity in anti-X-*X' versus 'log anti-X concentration' the concentration of anti-X that binds ~ 70 % of the maximum *X bound is determined. This concentration is used, along with the fixed amount of *X mentioned previously, in the preparation of the standard curve and the assay of unknown samples of X; this insures that anti-X is in limiting concentration in the competitive assay. The standard curve used to measure the concentration of X in unknown samples is prepared by incubating each of a series of different, known concentrations of unlabeled X with the fixed concentration of anti-X and the fixed amount of *X determined previously. In each incubation mixture the unlabeled X competes with *X for the anti-X binding sites. The greater the concentration of unlabeled X the less *X will bind to the anti-X; consequently the radioactivity in the generated Ab-Ag complex (a mixture of anti-X-X and anti-X-*X) decreases as the concentration of unlabeled X increases. After the incubation period is complete, the Ab-Ag complex is separated from unbound *X and its radioactivity measured. The standard curve consists of a plot of either 'ratio of bound radioactivity to total radioactivity present' (i. e. radioactivity in anti-X-*X / radioactivity in the fixed amount of *X added to each incubation mixture) or '% of *X bound as anti-X-*X' (as ordinate) versus 'log concentration of unlabeled X' (as abscissa). Within the resulting curve, the region that closely approximates to a straight line is used to determine the concentration of X in an unknown sample. The latter is accomplished by: (i) incubating known dilutions of the unknown sample with the fixed concentration of anti-X and the fixed amount of *X used in the generation of the standard curve, (ii) isolating the resulting anti-X-*X, (iii) determining the radioactivity of the latter, (iv) calculating the appropriate ordinate value, (v) selecting the dilution with an ordinate value on the straight part of the standard curve, (vi) reading off the corresponding abscissa value, then determining the concentration of X by taking the antilog and correcting for the dilution.

2. *Immunoradiometric assay (IRMA).* IRMA is a modification of RIA, in which the Ab is radioactively labeled rather than the Ag under assay (i.e. X). It is often used when X cannot be obtained in radiolabeled form or when its immunological properties are altered by radiolabeling. IRMA can be carried out using either a single, radiolabeled Ab (i. e. anti-*X_1) or two Ab, each of which binds to a different site on the X molecule and one of which (the second to be used in the assay) is radiolabeled (i. e. anti-X_1 & anti-*X_2); the latter procedure is called a 'two-site IRMA' and is an example of a 'two-Ab sandwich as-

say'. Two-Ab sandwich assays, whether they use radioisotopic, enzymic or fluorometric labels, have the advantage of giving greater chemical specificity, because they use two distinct Ag sites; they have the disadvantage of requiring two Ab, which may be either two monoclonals or a polyclonal (as anti-X_1) and a monoclonal (as anti-*X_2). Both IRMA procedures nowadays usually utilize a solid phase method (e.g. beads or microtiter plate) for separation of the labeled Ab-Ag complex from unbound, labeled Ab.

The two-site IRMA procedure for the assay of X begins with the binding of anti-X_1 to the wells of a microtiter plate. This is achieved by placing in each well the same volume (e.g. 50 µl) of an appropriate concentration (e.g. 10 µg/ml) of anti-X_1 in phosphate-buffered saline (PBS), then incubating overnight at 4 °C; after thorough washing, the remaining protein binding sites are blocked with an irrelevant protein such as hemoglobin (Hb), by incubation with PBS-Hb. A standard curve is then generated by adding a different, known concentration of X to each well and incubating for 4–16 h to allow formation of anti-X_1-X; the anti-X_1 in each well must be in excess of the amount of X in the most concentrated sample used. The wells are then washed thoroughly with PBS-Hb to remove unbound X, then incubated with the same volume of a solution of anti-*X_2 (e.g. $1–2 \times 10^5$ cpm) at room temperature for ~ 2 h. This is followed by thorough washing to remove unbound anti-*X_2; the wells are then cut out and their radioactivity, which is due to the bound 'sandwich' of anti-X_1-X-anti-*X_2, is determined. The standard curve consists of a plot of 'bound cpm' (i.e. radioactivity in anti-X_1-X-anti-*X_2) as ordinate versus 'log concentration of X' as abscissa. Within this curve the region that closely approximates to a straight line is used to determine the concentration of X in an unknown sample. The latter is accomplished by repeating the procedure used to generate the standard curve, using known dilutions of the unknown sample of X in place of the known concentrations of X; the dilution that has a 'bound cpm' on the straight part of the standard curve is selected; the corresponding abscissa value is read off, and the concentration of X is determined by taking the antilog and correcting for the dilution.

Immunoassays using an enzymic label

1. *Enzyme-multiplied immunoassay.* This is a competitive binding assay which operates similarly to RIA and is widely used in the field of therapeutic drug monitoring. An assay for a compound 'X' requires the generation of an antibody to it (i.e. anti-X) and the synthesis of a sample of X linked covalently to an enzyme (i.e. X-Enz) in such a way that the enzyme retains its activity and X retains its ability to act as an Ag of anti-X. A further requirement is that the enzyme loses its activity totally when X-Enz binds to anti-X to form the Ab-Ag complex, anti-X-X-Enz. The assay procedure, like that of RIA, depends on the use of a standard curve which is prepared by incubating, for an appropriate time, each of a series of different, known concentrations of X with previously determined fixed amounts of anti-X and X-Enz, then determining the enzyme activity of the reaction mixture by adding an appropriate substrate and measuring the resulting product. As the concentration of X increases, enzyme activity also increases, because X competes with the constant amount of X-Enz for the constant number of anti-X binding sites, thereby keeping progressively more of the X-Enz in the catalytically active, unbound form. The standard curve consists of a plot of 'enzyme activity' as ordinate versus 'log concentration of X' as abscissa; the region of this plot that closely approximates to a straight line is used to determine the concentration of X in an unknown sample as described earlier.

2. *Enzyme-linked immunosorbent assay (ELISA).* This is a 'two-Ab sandwich' assay that uses an enzyme as the label. It can be used to detect or measure either an Ab or an Ag. When applied to an Ag the assay operates in a manner similar to that of the 'two-site IRMA' described earlier, the only difference being that the second antibody used in the assay of X (i.e. anti-X_2) is labeled by covalently bonding it to an enzyme (i.e. anti-X_2-Enz) rather than by a radioisotope. A nonquantitative version of such an assay is used in human pregnancy testing kits, when the Ag under assay is Human chorionic gonadotrophin (hCG) (see) which is present in maternal urine about one week after embryo implantation. Although different manufacturers configure the test in different ways in an attempt to combine reliability with ease of use, it essentially involves the following sequence: (i) addition of a small urine sample to an Ab to hCG (i.e. anti-hCG_1) bonded to a solid support; bound Ab-Ag complex (i.e. anti-hCG_1-hCG) is then formed if the urine contains hCG, (ii) washing the support free of urine, (iii) adding a second anti-hCG, that is enzyme-labeled and specific to a different site on the hCG molecule (i.e. anti-hCG_2-Enz), so as to form a bound Ab_1-Ag-Ab_2 sandwich (i.e. anti-hCG_1-hCG-anti-hCG_2-Enz), provided bound anti-hCG_1-hCG had been formed earlier, (iv) washing the support free of unbound anti-hCG_2-Enz, and (v) adding a solution of a colorless substrate of the enzyme and looking for the appearance of the colored product.

When used to assay an Ab (e.g anti-X), the Ag (i.e. X) is bonded to a solid support and the solution containing the unknown concentration of anti-X added to it, so as to generate bound X-anti-X. The latter is then detected by adding an enzyme-labeled Ag that binds to the Fc region of anti-X, so forming a bound Ag-Ab-anti-Ab sandwich (i.e. X-anti-X-anti-[anti-X]-Enz), followed by a solution of a colorless substrate of the enzyme, so producing a colored product. The latter can be measured spectrophotometrically, if required, and the assay quantitated with the aid of a previously prepared standard curve.

Immunoassays using a combination of radioisotopic and enzymic labels

1. *Ultrasensitive enzymic radioimmunoassay (USERIA).* The combination of RIA and ELISA has on occasion been used to create an assay for a given Ag (e.g. X) that is more sensitive by up to three orders of magnitude than either of these techniques used by themselves. For this assay three different Ab are required: (i) an Ab to X (i.e. anti-X_1), which is bonded to a solid support, (ii) a second Ab to X which binds to a different site on the X molecule (i.e. anti-X_2), and (iii) an alkaline phosphatase-labeled Ab that binds to the Fc region of anti-X_2 (i.e. anti-[anti-X_2]-alkaline phosphatase), along with [^{3}H]adenosine-5′-monophosphate, a substrate of alka-

line phosphatase which is hydrolysed to [^{3}H]adenosine and orthophosphate. The assay consists of the following sequence of steps: (i) the solution containing an unknown concentration of X is added to the solid support-bound anti-X_1, so forming bound anti-X_1-X, (ii) unbound X is washed away, (iii) a solution containing excess anti-X_2 is added, so forming bound anti-X_1-X-anti-X_2, (iv) unbound anti-X_2 is washed away, (v) a solution containing excess anti-[anti-X_2]-alkaline phosphatase is added, so forming bound anti-X_1-X-anti-X_2-anti-[anti-X_2]-alkaline phosphatase, (vi) unbound anti-[anti-X_2]-alkaline phosphatase is washed away, (vii) a solution containing excess [^{3}H]adenosine-5′-monophosphate is added and the mixture incubated for an appropriate time, during which [^{3}H]adenosine is formed, (viii) the soluble phase of the incubation mixture, which now contains [^{3}H]adenosine, orthophosphate and residual, unhydrolysed [^{3}H]adenosine-5′-monophosphate, is transferred to a chromatography column and the [^{3}H]adenosine separated from the other components, collected and radioassayed. A standard curve of 'radioactivity in [^{3}H]adenosine' versus 'log concentration of X' is prepared and used in essentially the same way as for the previously described assays. [T.J.Ngo & H.M.Lenhoff, eds. *Enzyme-Mediated Immunoassay* (Plenum, New York, 1985); S.B.Pal, ed. *Immunoassay Technology,* vol. **1** (1985) and **2** (1986) (Walter de Gruyter, Berlin); D.W.Chan & M.T.Perlstein, eds. *Immunoassay: A Practical Guide* (Academic Press, New York, 1987); L.J.Krickla *J. clin. Immunoassay* **16** (1993) 267–271]

Immunoelectrophoresis: see Plasma proteins.

Immunofluorescence: a sensitive technique for detection of antigens or antibodies in which the antibody is coupled to a highly fluorescent compound such as rhodamine or fluorescein isothiocyanate. Tissues, cells or electrophoresis gels containing the antigen are incubated with the fluorescent antibody; after thorough washing, the label indicates the presence and position of the antigen, and can be detected with the aid of a fluorescence microscope. In direct I., the specific antibody is labeled. In indirect I., antibody against the specific antibody is labeled (e.g. the specific antibody might have been raised in a rabbit, and the fluorescent anti-antibodies could then be goat anti-rabbit serum). This allows an amplification of the fluorescence: whereas only one specific antibody molecule can bind to the antigenic site, several anti-antibodies can bind to different sites on the specific antibody.

Immunoglobulins, ***Ig, antibodies:*** specific defense proteins found in blood plasma, lymph and many body secretions of all vertebrates. Cell-bound hemagglutinins, which are phylogenetic precursors of the Ig, are also found in invertebrates, including annelids, crustaceans, spiders and mollusks. The most salient features of the system are its ability to respond to the presence of any foreign antigen (primary response), to respond more quickly to a previously encountered foreign antigen (secondary response), and under normal circumstances not to respond to components of the animal's own body (self:nonself discrimination).

Lymphocytes derived from bone marrow in mammals or the Bursa of Fabricius in birds (B-cells) display Ig on their surfaces. Upon exposure to antigen, those cells whose surface Ig can bind the antigen are stimulated to proliferate and differentiate into plasma cells which secrete antibodies of the same specificity as the ancestor of the clone. The process of stimulation is complex and involves cooperation with other lymphocytes (T-cells) and macrophages; but B-cells can also be stimulated to proliferate by polyclonal activators such as concanavalin A (see Lectins) or bacterial lipopolysaccharide. When bound to antigen,

Properties of human immunoglobulins

	IgG	IgM	IgA (serum)	IgA (secretions)	IgD	IgE
Sedimentation constant	6.5–7S	19S	7S	11S	6.8–7.9S	8.2S
M_r, of which the L-chain contributes 23000	155000	940000 (pentamer)	170000	380000 (dimer)	185000	196000
H-chain type and M_r	γ 1–4 50000 to 60000	μ 71000	α 64000	α 64000	δ 60000 to 70000	ε 75000
Chain formula (L = ϰ or λ)	$L_2\gamma_2$	$(L_2\mu_2)_5$	$L_2\alpha_2$	$(L_2\alpha_2)_2$	$L_2\delta_2$	$L_2\varepsilon_2$
Carbohydrate	2–3 %	10–12 %	8–10 %	8–10 %	12.7 %	10–12 %
Fraction of the serum Ig	70–75 %	7–10 %	10–22 %	10–22 %	0.03–1 %	0.05 %
Serum concentration (mg/100 ml)	1300 (800–1800)	140 (60–280)	210 (100–450)		3 (1–40)	0.03 (0.01–0.14)
Valence of bonding	2	5(10)	1	2	?	2
Biological half-life (days)	8(IgG3) or 21	5.1	5.8		2.8	2–3
Complement binding	yes	yes	no		no	no

some Ig can fix the C1q component of the Complement system (see), thus facilitating lysis of the foreign cell to which they are bound. Because they are multivalent, Ig also cross-link soluble antigens and facilitate their clearance from the blood or lymph by macrophages.

Structure. Ig is tetrameric, consisting of two identical light (M_r 22,000 to 24,000) chains and two identical heavy, carbohydrate-containing chains (M_r 50,000 to 73,000) (Fig.). There are two types of light chain, κ and λ, each of which can be associated with any of five types of heavy chain, μ, γ, δ, α or ε. The Greek letter designating the type of heavy chain corresponds to the class of Ig (A, D, E,G or M); these can be distinguished by electrophoresis or serologically.

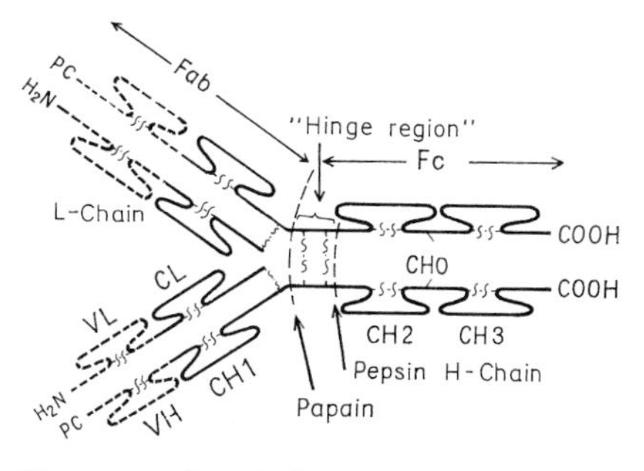

Structure of an IgG molecule. Variable (VL, VH) and constant (CL, CH) parts (domains) of the light and heavy chains, respectively. The molecule shown represents an IgG or IgA secreted Ig. Other classes have different numbers of heavy-chain domains (2 to 4), and the hinge region may not be present (IgM or IgE). The C-terminal sequence of membrane-bound Ig is different (and slightly longer) than that of the secreted form. CHO = carbohydrate chain. VL and VH form the antigen-binding site. Papain cleavage produces two Fab and one Fc fragment; pepsin cleavage forms a $(Fab')_2$ fragment and a smaller Fc′ fragment. PC = pyrrolidone carboxylic acid.

Papain treatment of Ig releases monovalent antigen-binding fragments, Fab (M_r 50,000) and a complement-binding, Fc (M_r 60,000) fragment. Pepsin cleavage, on the other hand, releases a bivalent antigen-binding $F(ab')_2$ fragment and a somewhat smaller complement-binding fragment Fc′. (The terms Fc and Fc′ are also applied to the corresponding region of Ig which does not bind complement.) The $F(ab')_2$ can be dissociated by thiol reagents into monovalent fragments, indicating that the two H-chains in the intact molecule are held together by one or more disulfide bonds between the sites of pepsin and papain cleavage (Fig.).

The Y shape of the Ig molecule has been confirmed by electron microscopy and X-ray diffraction. The former shows that the molecule is flexible at the "hinge" region, as already postulated from the susceptibility of this region to proteolytic attack. Fluorescence polarization shows a correlation between the flexibility of the hinge region, which varies from one class of Ig to the next, and the ability of the Ig to fix complement [V. T. Oi et al. *Nature* **307** (1984) 136–140].

Historically, on the basis of their electrophoretic behavior, Ig were divided into five classes, IgA, IgD, IgE, IgG and IgM. It was later realized that there are several genetically independent IgG subclasses in mice, rats and humans, while humans also possess subclasses of IgA. The classes and subclasses (or isotypes) are determined by the constant regions of the heavy chains, which are given the corresponding Greek letters (α, δ, ε, μ and γ1, γ2,etc.) Each type of heavy chain can be associated with either of the two types of light chain, κ or λ, but in a given molecule, both light chains are of the same class.

Because of the extreme heterogeneity of normal Ig it is impractical to attempt to purify them for amino acid sequencing. However, *monoclonal antibodies,* the products of a single clone of B cells, are available from several sources. Paraproteins (see), including Bence-Jones proteins (see), are found in the plasma of patients with certain types of lymphomas. In these cancers, the transformed cells retain enough of their B-cell characteristics to secrete whole or partial Igs. Rapid advances in the understanding of Ig structure and function were made possible by hybridomas, which are cell lines resulting from the fusion of antibody-secreting B cells and myeloma cells maintained in tissue culture. The B-cells are taken from the pancreas of recently immunized animals, or they may be from the peripheral circulation. The hybridoma clones can be selected for production of Ig of the desired antigenic specificity and isotype.

Comparison of the DNA and RNA from hybridomas or myelomas with that from embryonic cells reveals that genes for the constant region, one or two joining regions and the variable region become spliced together during differentiation of plasma cells. The intervening sequences of DNA are lost. [M. M. Davis et al. *Nature* **283** (1980) 733–739; F. R. Blattner & P. W. Tucker "The molecular biology of Immunoglobulin D" *Nature* **307** (1984) 417–422].

Immunology: the science of the biological and chemical bases of immunity, or the defense mechanisms of the human and animal organism, which are activated by the invasion of antigens. I. encompasses cellular and humoral immunity and the connexions between the immune reaction and physiological and biochemical processes in the organism, in particular those which control the Immune response (see).

Immunochemistry is an essential component of I. which deals with the antigen-antibody reaction, the Immunoglobulins (see), Agglutination (see), the Complement-binding reaction (see), etc. Immunological and serological techniques are particularly important in clinical chemistry. They are also applied in genetics, molecular biology and chemotaxonomy.

Immunosensor: a Biosensor (see) which makes use of an antibody-antigen reaction to generate an electronic signal. Enzyme I. are based on competition between antigen and enzyme-linked antigen, so that sensor operation is equivalent to conventional enzyme immunoassay. See Field effect transistor.

Immunotolerance: the lack of an immune response to: a) the body's own substances or antigens with which the body has had contact since, before or shortly after birth (natural I.), and b) larger or smaller amounts of certain antigens (tolerogens), which the body does not recognize as foreign and therefore tolerates (acquired I.). Acquired I. can only be maintained when the tolerogens are constantly present; otherwise the immune response to this antigen arises again. Immunosuppressive measures facilitate the induction of I. in adults.

Immunotoxins: Synthetic compounds made by coupling plant or bacterial toxins to monoclonal antibodies specific for a particular cell type. Toxins coupled to antibodies against cancer antigens should be highly specific chemotherapeutic agents; I. based on specific types of anti-lymphocyte antibodies are expected to be useful in modulating the immune response for therapeutic purposes. [E. Vitetta et al. *Science* **219** (1983) 644–650.]

IMP: acronym of inosine 5′-monophosphate (see Inosine phosphates).

cIMP: acronym of cyclic inosine 3′,5′-monophosphate (see Inosine phosphates).

Inactivases: see Regulation of metabolism.

Inborn errors of metabolism: the title of a book by Archibold Garrod published in 1902, in which the author recognized the relationship between genes and enzymes. Many metabolic disorders caused by the absence of a protein or the synthesis of a biologically inefficient form of a protein are genetic in origin. I. e. m. is therefore a biochemical and genetic concept synonymous with inherited metabolic block, inherited metabolic disorder, heritable disorder of metabolism, enzymopathy, and other similar terms.

The *one gene one enzyme* hypothesis (now more correctly called one cistron one polypeptide), which developed later from work on auxotrophic mutants of *Neurospora* [G. W. Beadle and E. L. Tatum *Proc. Natl. Acad. Sci. USA* **27** (1941) 499–506] and on pigmentation mutants of insects (see Ommochromes), confirms the original concept of inborn errors. Clinically, an I. e. m. may result in the dangerous accumulation of unmetabolized material before the site of the metabolic block (e. g. phenylketonuria) and/or failure to produce an essential metabolite (e. g. albinism). The concept now embraces nonenzymatic proteins, e. g. abnormal hemoglobins (see Hemoglobinopathy) are also the result of inborn errors. Known human and animal I. e. m. are clearly of medical interest, and a selection of these is listed in the table. Mutant microorganisms, on the other hand, have made a major contribution to the study of intermediary metabolism and molecular biology (see Auxotrophic mutants, Mutant technique). [H. Harris, *The Principles of Human Biochemical Genetics (Frontiers of Biology, Vol 19),* 2nd ed., North Holland Publishing Co., 1975; Stanbury et al. (eds.) *The Metabolic Basis of Inherited Disease,* 5th ed., McGraw-Hill, 1983; H. Galjaard *Genetic Metabolic Diseases (Early Diagnosis and Prenatal Analysis),* Elsevier/North Holland, 1980]

Inborn errors of metabolism in humans (with deficient enzyme(s) and clinical findings)

Acatalasia.
Catalase (EC 1.11.1.6) deficient in all tissues. In some individuals activity is less than 1 % of normal with no ill effects; in others there may be ulceration of nasal and buccal mucosae, and oral gangrene.

Adrenal hyperplasia I.
Cholesterol monooxygenase (side-chain-cleaving) (EC 1.14.15.6). Deficient conversion of cholesterol to pregnenolone; therefore impaired synthesis of mineralocorticoids, glucocorticoids and sex hormones.

Adrenal hyperplasia II, or ***Nonvirilizing congenital adrenal hyperplasia.***
3β-Hydroxysteroid dehydrogenase (EC 1.1.1.51). Decreased synthesis of all 3 types of adrenal corticosteroids, especially aldosterone, cortisol, and testosterone. Severe adrenal insufficiency manifested shortly after birth. Males show pseudohermaphroditism.

Adrenal hyperplasia III.
Steroid 21-hydroxylase (EC 1.14.99.10). Mild form known as simple virilizing adrenal hyperplasia. Decreased cortisol synthesis induces increased ACTH secretion, leading to overproduction of cortisol precursors and sex steroids. Excessive excretion of pregnane-triol. Over 20 other metabolites lacking 21-OH in urine. Female external genitalia masculinized, but internal genitalia normal (pseudohermaphroditism). Virilization in males. Short stature in both sexes (premature fusion of bony epiphyses). Severe form known as salt-losing congenital adrenal hyperplasia; all features of mild form present plus severe aldosterone deficiency with salt and water loss.

Adrenal hyperplasia IV, or ***Hypertensive congenital adrenal hyperplasia.***
Steroid 11β-Hydroxylase (EC 1.14.15.4). Decreased cortisol synthesis induces increased ACTH secretion with resulting overproduction of deoxycorticosterone (a potent salt-retaining hormone). Virilization in both sexes and female pseudohermaphroditism. Not all patients show hypertension

Adrenal hyperplasia V.
Steroid 17α-hydroxylase (EC 1.14.99.9). Decreased secretion of glucocorticoids and sex steroids. Excessive secretion of mineralocorticoids. Hyperkalemia and hypertension. Sexual infantism in females. Pseudohermaphroditism in males.

Aldosterone deficiency.
Steroid 18-hydroxylase (corticosterone 18-hydroxylase) (EC 1.14.15.5). Deficient aldosterone synthesis. Normal synthesis of cortisol and sex hormones. Severe salt loss.

Albinism. See Melanins.

Alkaptonuria. See Phenylalanine.

Drug-induced apnea.
Pseudocholinesterase (EC 3.1.1.8). Enzyme present in serum but shows atypical kinetics. Condition first discovered with introduction of suxamethonium (succinyl dicholine) as muscle relaxant in electroconvulsion therapy. This drug is normally rapidly hydrolysed by pseudocholinesterase, and its effects last only a few minutes. Affected subjects (1 in 2,000 Europeans) develop prolonged muscular paralysis and apnea (up to 2 hours) after normal drug dose. Condition screened for by measuring inhibition of serum cholinesterase by dibucaine; percent inhibition is called dibucaine number (80 % for normal enzyme, 20 % for atypical enzyme at 10^{-5}M dibucaine). Dibucaine numbers of about 62 % also occur in 4 % of Europeans, who possess about equal amounts of normal and atypical forms.

Argininemia.
Arginase (EC 3.5.3.1). Elevated blood arginine and ammonia. Neurological damage and mental retarda-

tion. Primary defect in liver, but as in other inherited defects of urea cycle, there is damage to central nervous system, probably due to toxicity of elevated ammonia.

Argininosuccinic aciduria.

Argininosuccinase (argininosuccinate lyase) (EC 4.3.2.1). Argininosuccinate (normally present in only trace amounts) increased in serum and excreted in large quantities. Blood ammonia rises significantly after protein meal. In normal subjects, the enzyme does not appear to be saturated, but in affected subjects the decreased (less than 5 % normal activity) enzyme is saturated and therefore catalyses a higher rate of arginine synthesis. Urea production is consequently not significantly decreased, while the concentrations of intermediates preceding argininosuccinate cleavage are altered. Some degreee of mental retardation and sometimes other neurological disorders, e. g. fits.

Citrullinemia.

Argininosuccinate synthetase (EC 6.3.4.5). Elevated blood citrulline and ammonia, and large amounts of urinary citrulline. Urea synthesis still occurs. Neurological damage and mental retardation.

Crigler-Najjar syndrome, or ***Constitutional nonhemolytic hyperbilirubinemia.***

Glucuronyl transferase (UDP-glucuronate: bilirubin-glucuronyltransferase) (EC 2.4.1.76). Enzyme completely absent. Severe jaundice and brain encephalopathy. Often death in infancy. Conjugation is obligatory for excretion of bilirubin. In a milder form of the disease some enzyme is present, and treatment is possible by administration of phenobarbital, which stimulates hepatic uptake, conjugation and biliary secretion (as glucuronide) of bilirubin.

Cystathioninuria.

Cystathionase (homoserine dehydratase, cystathionine γ-lyase) (EC 4.4.1.1). High urinary cystathionine. Increased concentrations of cystathionine in serum and tissues. See Cysteine.

Ehlers-Danlos syndrome.

A group of heritable disorders. Clinical features include hyperelastic skin, hyperextensible joints, easy bruising and poor wound healing. Collagen has decreased content of hydroxylysine. Deficient enzyme may be *Lysyl oxidase* (EC 1.4.3.6), *Lysyl protocollagen hydroxylase* (EC 1.14.11.4) or *Procollagen peptidase* (*C*-endopeptidase EC 3.4.24.19, or *N*-endopeptidase EC 3.4.24.14).

Formaminotransferase deficiency syndrome.

Formaminotransferase (EC 2.1.2.5). Increased urinary formamino-glutamate after oral histidine load, despite adequate serum folate. Mental and physical retardation. Neurological abnormalities. See Histidine.

Fructose intolerance.

Fructose bisphosphate aldolase (isoenzyme B, M_r 156,000) (EC 4.1.2.13). Fructosemia, fructosuria and hypoglucosemia after intake of fructose. Intracellular accumulation of fructose 1-phosphate. Hyperuratemia. Hepatomegaly. Renal tubular dysfunction. Intraocular bleeding. Patients symptom-free and healthy if fructose avoided. Aldolase A (muscle and most other tissues) and aldolase C (brain and heart) present and fully active.

Fructosuria, or ***Essential fructosuria.***

Ketohexokinase (EC 2.7.1.3). Fructosemia and fructosuria after intake of fructose. No pathological consequences.

Galactokinase deficiency, or ***Galactosuria,*** or ***Galactose diabetes.***

Galactokinase (EC 2.7.1.6). Galactosemia and galactosuria after intake of galactose or lactose (i.e. when milk is fed). Galactitol also produced. Severe lens cataracts from an early age. Affected individuals may otherwise be normal, but some evidence of neurological disturbance. Treated with galactose-free diet throughout life.

Galactose epimerase deficiency.

UDPglucose 4-epimerase (UDPgalactose 4-epimerase) (EC 5.1.3.2). Enzyme activity absent only in erythrocytes and leukocytes (intermediate activity in heterozygotes). Liver activity normal. Benign. Treat with high milk intake.

Galactosemia.

UDPglucose-hexose-1-phosphate uridylyl transferase (galactose 1-phosphate-uridylyl-transferase) (EC 2.7.1.12). Specific inability to metabolize galactose. High concentrations of galactose and galactose 1-phosphate in tissues and body fluids. Consequences generally severe. Failure to thrive. Retarded mental development. Hepatomegaly, eventually cirrhosis. Renal tubular dysfunction. Often early death, but if recognized in first days, an entirely galactose-free diet (rigorous exclusion of milk) may permit normal growth and development.

Chronic granulomatous disease.

Phagocytes fail to produce O_2^- or H_2O_2, owing to lack of cytochrome b_{245}, flavoprotein or protein kinase C, or production of defective (low affinity) NADP oxidase. Catalase-positive bacteria not killed. Granulomata in lymph glands. Chronic or recurrent lymph adenitis and respiratory disease. Usually death in childhood, unless maintained on antibiotics.

Histidinemia.

Histidine ammonia lyase (histidase) (EC 4.3.1.3). Failure to form urocanic acid from histidine. Histidine elevated in blood and cerebrospinal fluid. Increased urinary histidine, imidazolepyruvate, imidazolelactate and imidazoleacetate. Formiminoglutamate not excreted after histidine load. Usually benign. Mental retardation rare.

Homocystinuria, or ***Homocysteinemia.***

Cystathionine β-synthase (EC 4.2.1.22). Failure to form cystathionine from homocysteine and serine. Elevated homocysteine and methionine in serum. Urine contains homocystine and homocysteine-cysteine disulfide. Cystathionine virtually absent from brain, where it is normally present in significant quantities. Mental retardation. Lens detachment. Skeletal abnormalities (tall stature, arachnodactyly). Arterial and venous thrombosis.

3-Methylcrotonyl-glycinuria, or ***β-Hydroxyisovaleric aciduria.***

Methylcrotonyl-CoA carboxylase (EC 6.4.1.4). Urinary excretion of 3-methylcrotonylglycine and 3-hydroxyisovalerate. Acidosis in early childhood, and failure to thrive. Hypotonia and muscle atrophy. Treatment is low leucine diet. See Leucine.

Hydroxyprolinemia.

4-Hydroxyproline dehydrogenase (EC 1.5.99.?). Failure to convert 4-hydroxyproline to (*S*)-1-pyrroline-3-hydroxy-5-carboxylate (see L-Proline). Serum 4-hydroxyproline 30–50 times normal. Urinary 4-hy-

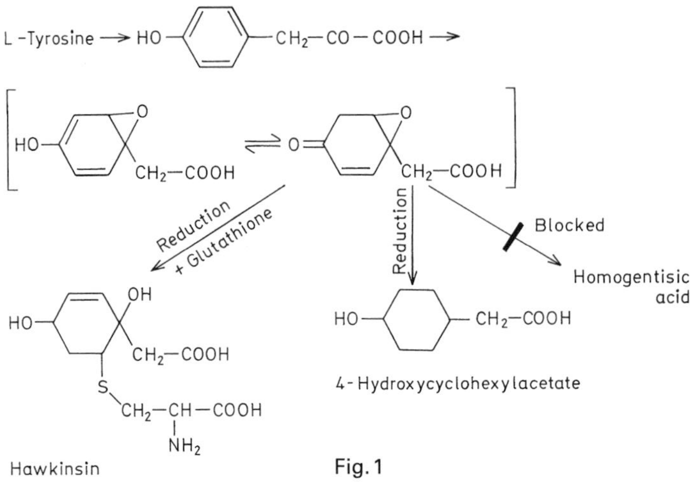

L-Tyrosine → HO–C_6H_4–CH_2–CO–COOH →

Reduction + Glutathione

Reduction

Blocked

Homogentisic acid

Hawkinsin

4-Hydroxycyclohexylacetate

Fig. 1

droxyproline elevated. No abnormality in collagen metabolism. Severe mental retardation. Otherwise benign.

Hawkinsinuria (see Fig. 1).

4-hydroxyphenylpyruvate dioxygenase (EC 1.13.11.27). Urinary excretion of (2-L-cystein-*S*-yl-1,4-dihydroxycyclohex-5-en-1-yl)acetic acid, also known as hawkinsin, and of hydroxycyclohexylacetic acid. There is mild hypertyrosinemia and acidosis. Treated with low protein diet and ascorbic acid supplement. Defective enzyme appears to release reactive intermediates before completion of the conversion of 4-hydroxyphenylpyruvate to homogentisate (see Phenylalanine). [B. Wilcken et al. *New Engl. J. Med.* **305** (1981) 865–869]

Hyperammonemia type I, or ***Carbamoylphosphate synthase deficiency.***

Mitochondrial *carbamoylphosphate synthase (ammonia)* (EC 2.7.2.5). Extreme hyperammonemia. Elevated glutamine in plasma and cerebrospinal fluid. Urea excretion low. Coma and death usually in postnatal period

Hyperammonemia type II, or ***Ornithine carbamoyltransferase deficiency.***

Ornithine carbamoyltransferase (ornithine transcarbamylase) (EC 2.1.3.3). Gross elevation of blood ammonia. Elevated glutamine in plasma and cerebrospinal fluid. Urea excretion low. Urinary orotic acid increased. Uracil and uridine in urine. Gene for enzyme X-linked. Condition severe in boys (0–0.2 % of normal enzyme activity in liver), who die in postnatal period (some cases of late onset have been reported). Girls have 5–10 % normal liver enzyme activity. Some girls have died in late infancy or childhood, and others have survived with restricted protein intake. Abnormal EEG, mental retardation, brain atrophy and hepatomegaly.

Hyperlysinemia.

Saccharopine dehydrogenase (NADP$^+$, L-lysine-forming) (EC 1.5.1.8) and *Saccharopine dehydrogenase (NAD$^+$,* L-glutamate-forming) (EC 1.5.1.9). Elevated plasma lysine. (0.2–1.5 mmol/l). Increased urinary lysine, *N*-acetyllysine, homocitrulline and homoarginine. Usually no symptoms. Some patients mentally retarded, but this may not be due to the metabolic disorder. See Lysine.

Hyperprolinemia type I.

Proline dehydrogenase (EC 1.5.99.8). Elevated serum and urinary proline. Urine also contains 4-hydroxyproline and glycine (probably due to saturation by proline of common transport system for these amino acids in renal tubules). Benign. Mental retardation in some cases. See Proline.

Hyperprolinemia type II.

1-Pyrroline-5-carboxylate dehydrogenase (EC 1.5.1.12). Elevated serum and urinary proline. Serum pyrroline-5-carboxylate concentration 10–20 times normal. Increased urinary pyrroline-5-carboxylate, proline, 4-hydroxyproline and glycine. Mental retardation and convulsions. Low proline diet only partly effective in control. See Proline.

Hypophosphatasia.

Alkaline phosphatase (EC 3.1.3.1). Defective ossification and skeletal abnormalities. Increased urinary *O*-phosphoethanolamine and inorganic pyrophosphate, which are also abnormally high in plasma. In "pseudohypophosphatasia", the enzyme is present but shows decreased affinity for phosphoethanolamine. [Osteoblasts normally secrete extracellular vesicles containing alkaline phosphatase, which is responsible for production of inorganic phosphate by hydrolysis of pyrophosphate and organic phosphates such as *O*-phosphoethanolamine. This inorganic phosphate forms apatite crystals by reaction with calcium, leading to bone mineralization]

Isovaleric acidemia.

Isovaleryl-CoA dehydrogenase (EC 1.3.99.10). Defective conversion of isovaleryl-CoA to β-methylcrotonyl-CoA (see Leucine). Elevated isovalerate in plasma and urine; also increased urinary isovalerylglycine, isovalerylcarnitine and sometimes 3-hydroxyisovalerate. Ketoacidotic crises, sometimes with fatal coma. Slight mental retardation in survivors. Treated with low leucine diet and supplements of glycine and/or carnitine to increase excretion of isovaleryl conjugates. Peritoneal dialysis in crises.

Lactose intolerance.
Lactase (intestinal) (EC 3.2.1.23). Lactase is present in infancy when it is required for digestion of lactose in the mother's milk. The enzyme then decreases markedly with age, so that abdominal pain and diarrhea result from drinking milk, due to failure to hydrolyse lactose in the intestinal mucosa and lactose malabsorption. Condition prevalent throughout the world, and hereditary persisitence of high intestinal lactase activity prevails only on Northern European populations (and those derived from them) and certain Arab and Hamitic races.

Pancreatic lipase deficiency.
Pancreatic lipase (EC 3.1.1.3). Lipase of pancreatic juice decreased to about 10 % of normal. Triacylglycerols in feces. Growth normal. Fat absorption about 70 % of normal. Very rare condition.

Maple syrup urine disease, or ***Leucinosis.***
Branched-chain-oxoacid dehydrogenase complex, which is responsible for the oxidative decarboxylation of the oxoacids derived from leucine, isoleucine and valine. Increased concentrations of all 3 branched chain amino acids and their oxoacids in urine, plasma and cerebrospinal fluid. Serum also contains alloisoleucine (probably derived from isoleucine). Urine has characteristic odor. Marked cerebral degeneration apparently shortly after birth. Usually fatal within weeks or months of birth.

Methemoglobinemia.
NAD-methemoglobin reductase (EC 1.6.2.?). Affected individuals grayish blue and cyanotic in appearance, due to large proportion of circulating methemoglobin, which cannot transport oxygen. No serious incapacitation. In a clinically similar congenital condition, hemoglobin is abnormal and cannot react with the enzyme. Very rare condition. [Methemoglobin is continually formed by oxidation of hemoglobin in erythrocytes. Normally it is reduced to hemoglobin, largely by NADH-dependent reductase (67 %) and to a lesser extent by NADPH-dependent reductase, as well as by nonenzymatic interaction with glutathione and ascorbate]

Methylmalonic aciduria, or ***Methylmalonic acidemia.***
Methylmalonyl-CoA mutase (EC 5.4.99.2). Failure to convert (*R*)-methylmalonyl-CoA into succinyl-CoA. Large quantities of methylmalonic acid appear in plasma and urine. Affected children fail to thrive and show pronounced ketoacidosis. Often fatal in early life. Hyperammonemia and intermittent hyperglycinemia are also typical. Restricted protein intake and synthetic diets are helpful, in particular low intakes of leucine, isoleucine, valine, threonine and methionine. A similar condition may arise from a congenital deficiency of *methylmalonyl-CoA epimerase* (EC 5.1.99.1). Both conditions unresponsive to vitamin B_{12}. Another type of methylmalonyl aciduria is thought to result from an hereditary deficiency of *deoxyadenosyl transferase* (transfers the 5′-deoxyadenosyl group in cobalamin synthesis), which provides the coenzyme of methylmalonyl-CoA mutase. This condition responds to injection of B_{12}. Dietary B_{12} deficiency also results in methylmalonic aciduria.

Ornithinemia, or ***Hyperornithinemia type I.***
Ornithine-oxoacid aminotransferase (EC 2.6.1.13). Ornithine concentration increased in plasma, urine and cerebrospinal fluid. Progressive loss of vision and usually blindness before 40th year. Treated with large pyridoxine supplements and by restriction of arginine intake. Type II hyperornithinemia is due to an unidentified defect (possibly defective ornithine transport into mitochondria); it is characterized by hyperornithinemia, hyperammonemia, homocitrullinemia and homocitrullinuria; the eyes are not affected, but there is lethargy, coma and mental retardation.

Orotic aciduria.
Orotidine-5′-phosphate pyrophosphorylase (EC 2.4.2.10) and *orotidine-5′*-phosphate decarboxylase (EC 4.1.1.23). Abnormally high urinary orotic acid. Severe megaloblastic anemia. Marked retardation of growth and development, and slight mental retardation. [A gross deficiency of 2 enzymes of pyrimidine biosynthesis]

Oxalosis type I.
2-Hydroxy-3-oxoadipate synthase (EC 4.1.3.15). High urinary oxalic acid and glycolic acid. Calcium oxalate crystals in many body tissues. Nephrocalcinosis, urolithiasis, with progressive renal insufficiency and usually death before 20 years. See Oxalic acid.

Oxalosis type II.
Glycerate dehydrogenase (EC 1.1.1.29). High urinary oxalic acid and L-glyceric acid. Clinically similar, but milder than type I.

Pentosuria.
NADP-specific xylitol oxidoreductase (L-xylose reductase) (EC 1.1.1.10). L-Xylulose is continuously excreted in large amounts in the urine. Administration of glucuronic acid further increases xylulose excretion. Benign. No treatment needed. See Glucuronate pathway.

Hereditary tyrosinemia type II, or ***Tyrosinosis type II,*** or ***Hypertyrosinemia type II,*** or ***Richner-Hanhart syndrome.***
Cytosolic tyrosine aminotransferase (EC 2.6.1.5). Elevated tyrosine in blood and cerebrospinal fluid. Increased urinary tyrosine, 4-hydroxyphenylpyruvate, 4-hydroxyphenyllactate and 4-hydroxyphenylacetate. Slight to moderate mental retardation. Blistering and hyperkeratosis of palms and soles of feet. Photophobia. No hepatorenal dysfunction (cf. Hereditary tyrosinemia type I). Controlled with diet low in phenylalanine and tyrosine.

Hereditary tyrosinemia type I, or ***Tyrosinosis,*** or ***Hereditary hepatorenal dysfunction*** (see Fig. 2).
Fumarylacetoacetase (EC 3.7.1.2). General aminoaciduria, glucosuria, proteinuria and hypokalemia. Hypoprothrombinemia and jaundice. Death usual in early childhood. Survivors develop cirrhosis and hepatorenal dysfunction; often also malignant hepatoma, acidosis and vitamin D-resistant rickets. [Failure to metabolize fumarylacetoacetate by normal route (see Phenylalanine) results in reduction of accumulated fumarylacetoacetate (or maleylacetoacetate or both) to succinylacetoacetate, which is decarboxylated to succinylacetone. The latter inhibits *4-hydroxyphenylpyruvate:oxygen oxidoreductase (hydroxylating, decarboxylating)* (EC 1.13.11.27), causing tyrosinemia (blood methionine often also elevated) and increased urinary tyrosine, 4-hydroxyphenylpyruvate, 4-hydroxyphenyllactate and 4-hydroxyphenylacetate. *Porphobilinogen synthase* (EC 4.2.1.24) also inhibited, re-

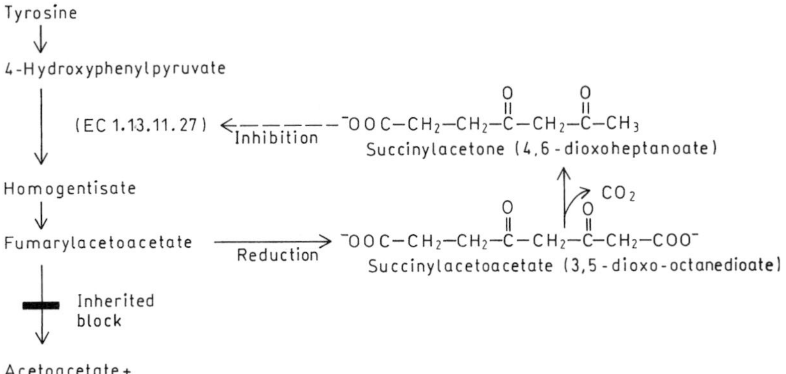

Fig. 2

sulting in high excretion of 5-aminolevulinic acid and symptoms of acute hepatic porphyria. B. Lindblad et al. *Proc. Natl. Acad. Sci. USA* **74** (1977) 4641–4645]

Refsum's disease, or ***Phytanic acid storage disease,*** or ***Heredopathia atactica polyneuritiformis*** (see Fig. 3).

Phytanic acid α-hydroxylase. Phytanic acid accumulates in liver and kidneys, and it may represent over 50 % of total liver fatty acids. Plasma phytanic acid concentrations of 200–3,100 mg/l have been reported (normal <2 mg/l). Peripheral neuropathy and ataxia, retinitis pigmentosa and skin and bone abnormalities. Treatment by plasma exchange and low phytol intake. [Phytanic acid is normally formed in the body from the plant alcohol, phytol, present as an ester in chlorophyll. The presence of a branch methyl group at position 3 of phytanic acid means that the normal process of β-oxidation is blocked. Oxidation of fatty acids one carbon at a time (α-oxidation) is common in plants, but also occurs to some extent in animals, especially in the brain, where it serves to initiate degradation of phytanic acid. The resulting pristanic acid is then degraded by β-oxidation]

Lecithin-cholesterol acyltransferase deficiency, or ***Norum's disease.***

Phosphatidylcholine-sterol acyltransferase (EC 2.3.1.43). Plasma cholesterol and triacylglycerols increased. Lysophosphatidylcholine and cholesterol esters decreased. Turbid or milky plasma. Multiple lipoprotein abnormalities. Corneal opacities. Normochromic anemia and proteinuria, due to renal damage. Therapy by enzyme replacement. [The enzyme catalyses formation of cholesterol esters by transfer of an unsaturated fatty acid from position 2 of lecithin to the 3-OH of cholesterol]

Sarcosinemia.

Sarcosine dehydrogenase (EC 1.5.99.1). Hypersarcosinemia and hypersarcosinuria. Some cases with mental retardation. Probably benign.

Sulfite oxidase deficiency, or ***Sulfituria.***

Sulfite oxidase (sulfite dehydrogenase) (EC 1.8.3.1). Urinary excretion of *S*-sulfo-L-cysteine, sulfite and thiosulfate, and virtual absence of urinary sulfate. Progressive neurological abnormalities, mental retardation, lens dislocation. Treatment with diet low in sulfur amino acids. Death in postnatal period possi-

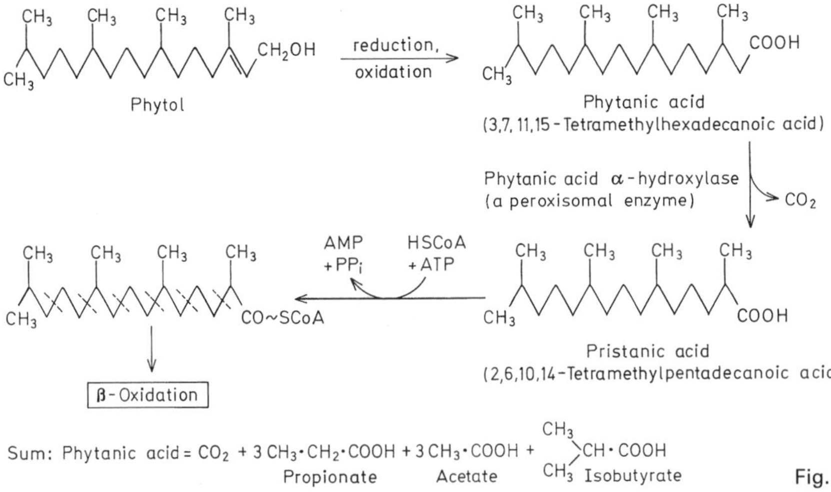

Fig. 3

ble. [Cysteine sulfinic acid, an intermediate in the metabolism of cysteine, can be transaminated to β-sulfinylpyruvate, which readily loses SO_2 in a reaction analogous to the decarboxylation of oxaloacetate. The sulfite is oxidized to sulfate by sulfite oxidase]

Xanthurenic aciduria.

Kynureninase (EC 3.7.1.3). Failure to convert 3-hydroxykynurenine to 3-hydroxyanthranilic acid, and kynurenine to anthranilic acid. Increased urinary xanthurenic acid, kynurenine and 3-hydroxykynurenine, especially after ingestion of tryptophan. Some patients mentally retarded, others symptomless. Enzyme may not be absent, but structurally abnormal, with decreased affinity for coenzyme (dietary vitamin B_6 temporarily corrects metabolic disturbance, and addition of pyridoxal phosphate to liver biopsy material increases enzyme activity to near normal). See Tryptophan.

Xanthinuria type I.

Xanthine oxidase (EC 1.2.3.2). Failure to convert xanthine to uric acid (see Purine degradation). Xanthine therefore replaces uric acid as end product of purine metabolism. Urinary xanthine greatly increased. Urinary uric acid abnormally low. Xanthine calculi tend to form in renal tract.

Xanthinuria type II.

Malabsorption of molybdenum. Activities of *xanthine oxidase, sulfite oxidase* and *aldehyde oxidase* are therefore deficient (see Molybdoenzymes). Urinary xanthine and uric acid high and low, respectively (see Purine degradation). Mental retardation, seizures, cerebral atrophy, lens dislocation.

Other I.e.m. include ***Glycogen storage diseases*** (see), ***Lysosomal storage diseases*** (see) and ***Porphyrias*** (see).

Indican: a glucoside composed of indoxyl and one molecule of glucose, m.p. 57 °C, $[\alpha]_D^{15}$ −66 ° (water). It occurs in *Indigofera* species, e.g. the indigo plant *(Indigofera tinctoria),* and several other higher plants. It is the precursor of natural Indigo (see).

Indicator methods: see Methods of biochemistry.

Indicaxanthin: a yellow betaxanthine from the prickly pear *(Opuntia ficusindica).* I. is the best known representative of the betaxanthines. It differs from betanin by replacement of the cyclodopa residue by L-proline.

Indicaxanthin

Indigo: a dark blue vat dye, known in ancient times and particularly valued in the middle ages. I. was formerly extracted from a few *Indigofera* species, like *I. tinctoria* and from dyer's woad *(Isatis tinctoria).* The latter contains the colorless ester, indoxyl-5-oxofuranogluconate, whereas *Indigofera* contains the colorless indoxyl-β-D-glucoside (indican). When plants are damaged both of these compounds become enzymatically hydrolysed to the corresponding sugars and indoxyl (3-hydroxyindole). Indoxyl is converted to I. by oxidation in the air. Natural I. contains up to 95 % I. accompanied by indirubin and indigo brown. Today I. is produced synthetically. It is fast against light, acids and bases, and is particularly suitable for dyeing cotton and wool. I. was formerly the most important organic dye, but now halogenated derivatives and other indigoid dyes have become more important. [E. Epstein et al. *Nature* **216** (1967) 547–549]

Indigo

Indole alkaloids: one of the largest groups of alkaloids, with more than 600 representatives. Biogenetically, almost all I.a. are derived from tryptophan; the majority also contain a monoterpene component (iridoid I.a.), which is responsible for the large variety of forms. It is advantageous to classify the I.a. according to their sources (Table), because many plant genera are characterized by particular structural forms of the alkaloids.

Classification of the indole alkaloids

Structural type	Typical representatives
Carboline	Harman alklaoids
Pyrrolidinoindole	Physostigmine
Ergoline	Ergot alkaloids
Iridoid indole alkaloids	Alkaloids of *Rauwolfia,* Calabash curare, *Vinca, Strychnos* and *Cinchona*

The cinchona alkaloids are also included in the table because they are derived from indole precursors, although their main alkaloids have a quinoline skeleton. Most of the I.a. have pronounced pharmacological properties.

In addition to the I.a., there are many natural indole compounds which are not alkaloids, e.g. the melanins, betalaines, sporidesmins and gliotoxin.

Induced fit: see Cooperativity model.

Inducer: a chemical compound which stimulates the synthesis of enzymes (see Enzyme induction). The I. may be the substrate of the enzyme or another effector, such as a hormone.

Induction: see Enzyme induction.

Infectious nucleic acid: viral nucleic acid which enters the host cell, where its genetic information is translated into viral products by the synthetic apparatus of the host cell (see Viruses, Phage development).

Informofers: protein particles generated by removal of RNA from nucleoprotein particles (containing mRNA or its precursor) (see Messenger RNA).

Informosomes: according to Spirin, nucleoprotein particles containing mRNA which can be isolated from the cytoplasm of eukaryotic cells. They can be separated from ribosomes and polyribosomes by density-gradient centrifugation. I. are thought to be transport particles for mRNA. They pick up the mRNA formed in the cell nucleus, probably at the nuclear membrane, and make it available in the cytoplasm for the formation of polyribosomes. See Messenger RNA.

INH: acronym of isonitinic hydrazide (see Nicotinic acid).

Inhibin: a glycoprotein, M_r 31,000, consisting of 2 subunits (α or A, M_r 18–20,000; β or B, M_r 14–15000) cross-linked by one or more disulflde bridges. There are two types of β-subunit (βA and βB), giving rise to inhibin A and inhibin B, respectively. Both heterodimers are secreted by the gonads, and are specific and potent inhibitors of FSH production and release by cultured pituitary cells. In view of its effect on the secretion of follicle-stimulating hormone, inhibin appears to be a key hormone in the control of folliculogenesis and spermatogenesis. It may be the missing link in the mechanism controlling differential secretion of pituitary gonadotropins, and it is of interest as a potential contraceptive. It is present in seminal plasma and follicular fluid, and the latter has proved to be the most successful source for purification purposes.

The β-subunits of inhibin also combine to form dimers (βAβA, βAβB, βBβB) known as *activins,* which stimulate FSH production by cultured pituitary cells, as well as steroidogenesis in granulosa cells. Human leukemia cell-derived erythroid differentiating factor is identical with the homodimeric βA activin. The β-subunits also show 30 % sequence similarity with the β_1 and β_2 chains of transforming growth factor β, indicating an evolutionary relationship between their genes.

Each subunit of inhibin (and activin) represents the *C*-terminal region of a larger precursor, from which it is released by proteolytic processing. The cDNAs (derived from porcine and bovine ovarial mRNA) encoding the biosynthetic precursors of the α and β subunits have been cloned and analysed. The α and β subunits show considerable homology, suggesting a common origin from distantly related genes; this is also a feature of the α and β subunits of pituitary and placental Glycoprotein hormones (see). [S.-Y. Ying *Proc. Soc. Exp. Biol. Med.* **186** (1987) 253–364; S. Cheifetz et al. *J. Biol. Chem.* **263** (1988) 17225–17228]

Two proteins from human seminal plasma have been called α-inhibin (92 amino acids) and β-inhibin (94 amino acids); they are, however, not of gonadal origin. α-Inhibin is formed in the prostate, and β-inhibin in the seminal vesicles. They are structurally unrelated to the dimeric molecule described above that is generally accepted as inhibin [A. J. Mason et al. *Nature* **318** (1985) 659–663; R. G. Forage et al., *Proc. Natl. Acad. Sci.* **83** (1986) 3091–3095; T. K. Woodruff & J. M. Mather 'Inhibin, Activin and the Female Reproductive Axis' *Annu. Rev. Physiol.* **57** (1995) 219–244]

Inhibitor peptides: low molecular mass oligopeptide-fatty acid compounds of microbial origin which irreversibly inactivate plant and animal proteases. The inhibition is stoichiometric, i. e. 1 molecule I. p. inhibits 1 molecule enzyme. Examples are: *Leupeptin* [acetyl-(or propionyl-)L-Leu-L-Leu-arginal; the L-leucine can also be replaced by L-isovaline or L-valine], from *Streptomyces* species, inhibits cathepsin B, papain, trypsin, plasmin and cathepsin D, the effectiveness of the inhibition decreasing in that order. *Pepstatin* (isovaleryl-L-Val-L-Val-β-hydroxy-γ-NH_2-ε-CH_3-heptanoyl-L-Ala-β-hydroxy-γ-NH_2-ε-heptanoic acid), from actinomycetes, inhibits pepsin and cathepsin D. *Chymostatin* inhibits all known chymotrypsin types, cathepsin A, B, and D and papain. *Antipain* inhibits papain trypsin and plasmin.

Inhibitor proteins: for the most part, low molecular mass, resistant proteins with compact structures and a lack of species specificity. At the molecular level, I. p. reversibly or irreversibly inhibit certain anabolic or catabolic processes. At the cellular level, they reversibly or irreversibly inhibit growth and maturation processes of normal and malignant cells. The best known group of I. p. are the protease inhibitors, a group of protease-insensitive, disulfide-rich polypeptides widely distributed in animals and plants. They are especially common in the nutritional protein (egg white) of many eggs, and in plant seeds. Many animal protease inhibitors are secretory proteins, like the trypsin inhibitor of the mammalian pancreas, blood and seminal plasma, milk, colostrum, salivary and snail mucus. The M_r of most proteinase inhibitors lies between 5,000 and 25,000. However, the carbohydrate-containing I. p. of human blood plasma, which are α-globulins, have much higher M_r, e. g. antitrypsin (M_r 54,000), and antichymotrypsin (M_r 68,000), inter-α-trypsin inhibitor (M_r 160,000) and α-macroglobulin (M_r 820,000). The proteinase inhibitors form temporary or permanent inactive enzyme-inhibitor complexes with the proteases of the digestive tract, the blood-clotting system and the blood pressure regulatory system, and with cell and tissue proteases. There are also I. p. for other enzymes, including a special inhibitor of cytochrome oxidase which has been characterized. Another group of I. p., including thc toxalbumins and bacterial toxins, inhibit protein synthesis. In contrast, repressor proteins, antitumor proteins and chalones block the synthesis or function of DNA or RNA, which manifests itself in the case of chalones in the inhibition of mitosis. Interferons inhibit virus growth by inducing a new RNA coding for an anti-virus protein. Other surface-active I. p., such as snake venom toxins, some bacterial toxins and lectins, block certain receptor proteins on the synapses or the erythrocytes, producing severe disruption of the nervous system or cell agglutination. Examples of I. p. with narrowly limited effects are the apotransferrins, which inhibit microbial growth, troponin I. p., which inhibits actomyosin-ATPase, and the peptones that inhibit blood clotting.

Initial rate technique: a graphic method to determine the initial reaction rate from the reaction kinetic curves. The slopes of the tangents passing through the origin $t = 0$ for the reaction curves at various substrate concentrations are the initial rates at those concentrations.

Initiation codon, ***start codon:*** a sequence of three nucleotides in mRNA which is recognized by the anticodon of formylmethionyl-tRNA (prokaryotes) or methionyl-tRNA (eukaryotes) and thus serves as the start signal for polypeptide synthesis. The I.c. has the sequence 5′-AUG and is apparently localized in a sterically favorable position of the mRNA. See Genetic code.

Initiation factors, ***IF:*** catalytic proteins required for the initiation of RNA synthesis (see Ribonucleic acids) and of Protein synthesis (see). Three structurally and functionally characterized IF of *E. coli* are: IF 1 (M_r 9,000), IF 2 (M_r 97,000) and IF 3 (M_r 22,000). They appear to be loosely associated with the ribosomes, but can be dissociated from them with 0.5 M NH_4Cl or KCl, or isolated from the cytoplasm after homogenization. Some eukaryotic systems possess more than 10 IF (eIF, "e" for eukaryotic), none of which can functionally substitute for the bacterial IF. There is thus strict class specificity in the initiation of protein synthesis.

An IF is also required for the specific initiation of RNA synthesis at the promoter. In prokaryotes this is called the sigma (σ) factor, and has M_r 90,000.

Initiation tRNA, ***starter tRNA:*** a species of tRNA, specific for the methionine that is introduced as the first amino acid residue in the biosynthesis of all proteins. This particular tRNA differs in its primary and tertiary structure from the tRNA which supplies the methionine incorporated into the middle of a polypeptide chain.

In *prokaryotes,* the initiating residue is not methionine itself, but *N*-formylmethionine. Formylation occurs after esterification of the methionine residue with the tRNA, and it is catalysed by a formyltransferase which transfers the formyl group from formyltetrahydrofolic acid. The complete tRNA derivative responsible for introduction of the first *N*-terminal residue is therefore fMet-$tRNA_f^{Met}$ (formylmethionyl-tRNA), where the subscript "f" indicates that this species of RNA carries a formylated methionine.

In *eukaryotes* (i.e. in the initiation of protein synthesis on 80S ribosomes) the initiating tRNA derivative is Met-$tRNA_f^{Met}$, where the subscript "i" distinguishes this species of tRNA from the that of prokaryotes.

N-terminal fMet or Met is removed from most proteins during their biosynthesis (see Post-translational modification of proteins), and therefore does not appear in the mature protein.

Initiator complex: see Ribonucleic acids, Protein biosynthesis.

Inosine, ***Ino, hypoxanthinosine, hypoxanthine riboside, 9β-D-ribofuranosylhypoxanthine:*** a β-glycosidic nucleoside of D-ribose and hypoxanthine, M_r 268.23, m.p. 218 °C (d.), $[\alpha]_D^{25}$ −73.6° ($c = 2.5$, 0.01 M NaOH). It occurs free, is particularly abundant in meat and yeast, and is formed by dephosphorylation of inosine phosphates. It fulfills a specific function as a component of the anticodon of certain tRNA species (see Rare nucleic acid bases).

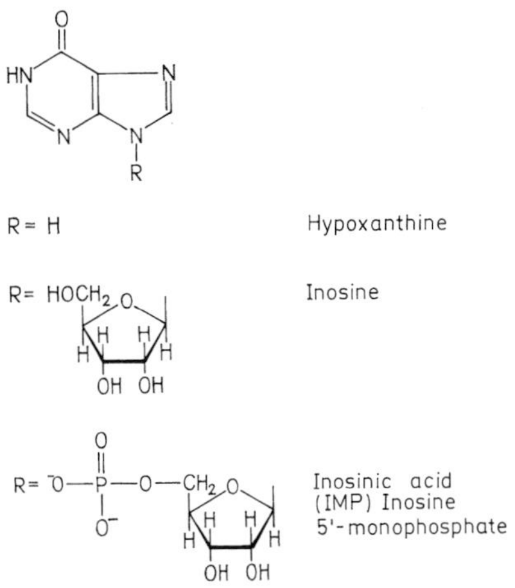

R = H Hypoxanthine

R = $HOCH_2$ (ribose) Inosine

R = $^-O-P(O)(O^-)-O-CH_2$ (ribose) Inosinic acid (IMP) Inosine 5′-monophosphate

Hypoxanthine and its derivatives

Inosine phosphates: nucleotides, phosphoric acid esters of inosine. 1. ***Inosine 5′-monophosphate*** (IMP, inosinic acid, or hypoxanthine riboside 5′-phosphoric acid; for formula, see Inosine) consists of the purine base hypoxanthine, D-ribose and phosphoric acid, M_r 348.22, $[\alpha]_D^{25}$ −36.8° ($c = 0.87$, 0.1 M HCl). IMP is the biosynthetic precursor of all the other purines (see Purine biosynthesis). Together with guanylic acid, IMP serves as a flavoring principle and it is isolated for this purpose either from meat extract or from hydrolysates of yeast nucleic acids, or it is produced on a large scale by mutants of certain microorganisms, e.g. *Corynebacterium glutamicum.* 2. ***Inosine 5′-triphosphate*** (ITP) (M_r 508.19) can serve as an energy-rich phosphate, replacing ATP in certain metabolic reactions (carboxylations). It is formed by phosphorylation of IMP via ***inosine 5′-diphosphate*** (IDP, M_r 428.2). 3. ***Cyclic inosine 3,5′-monophosphate*** (cyclo-IMP, cIMP) is a structural analog of cyclic adenosine 3′,5′-monophosphate (see Adenosine phosphates), and like the adenosine derivative, it inhibits the growth of certain transplantable tumors.

Inorganic bulk elements: see Mineral nutrients.

Inosinic acid: see Inosine phosphates.

Inositols: see Cyclitols.

Inositol phosphates: phosphate esters of the cyclic alcohol *myo*-inositol. They are found in the plasma membrane as components of phospholipids known as phosphatidyl inositols. Inositol 1,4,5-*tris*phosphate ($InsP_3$) is released from phosphatidylinositol-4,5-*bis*-phosphate ($PtdInsP_2$) in the plasma membrane by a variety of membrane receptors for hormones or neurotransmitters when the receptors are occupied. $InsP_3$ appears to be the active species responsible for the release of Ca^{2+} which follows stimulation of the cell; it is thus a second messenger (see Hormones, Calcium) for such agents as acetylcholine, vasopressin, substance P and epidermal growth factor (see Peptide hormones).

$InsP_3$ is soon dephosphorylated, in three steps, to *myo*-inositol (Fig.). The last of these reactions is catalysed by *myo*-inositol 1-phosphatase (EC 3.1.3.25) which is inhibited by lithium. The de novo synthesis of *myo*-inositol yields *myo*-inositol-1-P, rather than free *myo*-inositol, so that inhibition of this enzyme interrupts the cycle shown in the figure, and prevents formation of phosphatidyl inositol. It is likely that this effect of lithium is responsible for its pharmaceutical effects in manic depressive illness.

Phosphatidyl inositol is far more abundant than PtdIns4P or PtdIns4,5P, but the three species are in equilibrium, so that PtdIns4,5P removed from the membrane by cleavage is rapidly replaced. The diacylglycerol released from PtdIns4,5P by cleavage is, like $InsP_3$, a second messenger. It activates, inter alia, the membrane C kinase, which in turn activates a number of other proteins by phosphorylation (see Calcium). Interestingly, the diacylglycerol portion of PtdIns4,5P tends to have arachidonic acid in its 2 position; when released from the diacylglycerol by hydrolysis, the arachidonic acid may also serve as a cell activator.

Thus in the form of phosphatide esters, the inositol phosphates play an essential role in the activation of many types of cell. Cleavage of PtdIns4,5P releases at least three different second messengers: $InsP_3$, which in turn generates a surge in intracellular Ca^{2+}, diacylglycerol and arachidonic acid.

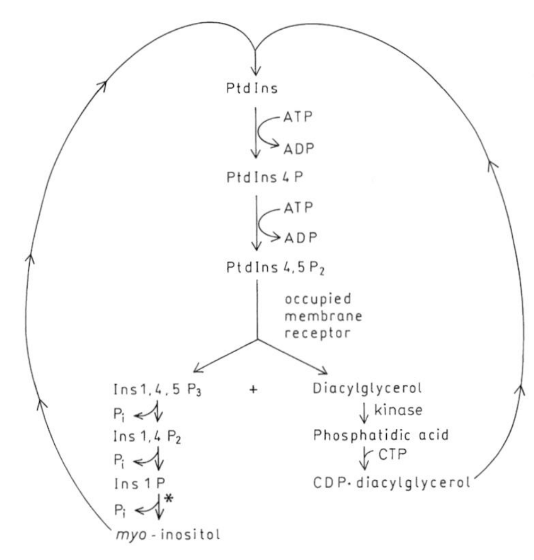

Metabolism of inositol phosphates. The reaction with an asterisk is inhibited by lithium. [M.J.Berridge *Biochem. J. 220* (1984) 345–360; D.H.Carney et al. *Cell 42* (1985) 479–488]

Insect attractants: see Pheromones.

Insect hormones: mostly low molecular mass substances, which are responsible for directing the life cycle of insects. Three hormones regulate the post-embryonic development of insects: 1. activation hormone (brain hormone, adenotropic factor), 2. molting hormone (see Ecdysone), and 3. Juvenile hormone. Each molt is induced by the activation hormone, a polypeptide produced in the brain. It causes an increase in RNA synthesis in the molting glands. This is accompanied by the secretion of a molting hormone from the prothoracic gland. The type of molt is decided by the juvenile hormone, a sesquiterpene from the corpora allata. High concentrations of this hormone produce a larval molt, low concentrations a pupal molt. An imaginal molt occurs when the juvenile hormone is acting alone.

Ecdysone or close structural relatives are widespread in the plant kingdom, often in high concentrations (phytoecdysones). For example, 15 kg fresh yew leaves yield 300 mg ecdysterone, an amount which could only be obtained from 6,000 kg insect pupae.

In addition to the natural compounds, many synthetic analogs have high activities in biological tests, especially juvenile hormone activity. Such environmentally non-damaging biological substances can be used for the control of insect pests.

Insecticyanin: a blue biliprotein from the hemolymph and integument of the tobacco hornworm *(Manduca sexta)*. With the yellow color of carotenoids, which are also present, I. is responsible for the green color of the insect larvae. I. is produced only by the larvae, but persists throughout pupation into the hemolymph of the adult female, and is then sequestered in the egg. It is a tetramer of identical subunits, each of M_r 21,378, containing 189 amino acid residues of known sequence, and 2 disulfide bridges. The chromophore of I. is biliverdin IX, which is tightly bound and is only removable under denaturing conditions. High resolution X-ray crystallographic studies have been reported. [H.M.Holden et al. *J. Biol. Chem.* **261** (1986) 4212–4218]

Insect viruses: viruses that cause disease in insects. Mostly affected are insects with complete metamorphosis, especially *Lepidoptera, Hymenoptera, Diptera* and *Coleoptera.* Larval and pupal stages are affected, although adult insects may carry the virus without manifesting a disease state. In certain infections, the cells of afflicted insects contain light microscopically visible, polygonal to ellipsoid, paracrystalline, often strongly refractive *inclusion bodies,* containing virus particles embedded in their matrices; in such structures the virus particles may remain infective for many years, even after the death and disintegration of the host. Inclusion bodies may be nuclear or cytoplasmic. When inclusion bodies are present, the virus is classified as a *polyhedrose* or *polyhedrosis virus.* Best known of the cytoplasmic polyhedroses is that responsible for the "yellowing disease" of the silkworm. In a second group, known as *granuloses* or *granulosis viruses,* geometrically shaped inclusion bodies are absent, and the infection is characterized by an accumulation of small granular inclusions called *capsules* in the infected cells, each capsule containing a single rod of viral DNA [e.g. infections of the Cabbage white butterfly *(Pieris brassicae)* and the Turnip dart moth *(Agrotis segetum)*]. Sacbrood, which kills honey bee larvae, is caused by a noninclusion virus, i.e. neither capsules nor inclusion bodies are present in infected cells.

Some I.v., in particular polyhedroses, are used for biological pest control, e.g. against the Pine sawfly *(Diprion pini)* and the Fir sawfly *(Pritisphora abietina),* two of the most important European forestry pests. The American Environmental Protection Agency has approved the use of viruses for the control of the Cotton bollworm *(Heliothis zea),* Gypsy moth *(Lymantria dispar)* and Tussock moths *(Hemerocampa).*

Insertion: see Mutation.

Insertion sequence: see Transposon.

Insulin: a polypeptide hormone, M_r 5,780 (bovine), synthesized in, and secreted by, the B cells of the islets of Langerhans. The first protein primary sequence ever to be elucidated was that of I. (Fig. 1) [F. Sanger et al. *Biochem. J.* **59** (1955) 509–518]. I. is the only hormone that decreases the blood glucose concentration. It affects the entire intermediary metabolism, especially of the liver, adipose tissue and muscle. I. increases the permeability of cells to monosaccharides, amino acids and fatty acids, and it accelerates glycolysis, the pentose phosphate cycle, and, in the liver, glycogen synthesis. It promotes the biosynthesis of fatty acids and proteins. These indirect effects on various enzymes and metabolic processes are listed in the tables.

Most species possess only one type of I., but three rodents (laboratory rat, mouse, spiny mouse) and two fish (tuna, toadfish) have two distinct hormones. Accordingly, the rat possesses two I. genes, whereas only one is present in humans. Rat and human I. genes have been sequenced [G. I. Bell et al. *Nature* **284** (1980) 26–32]. The mature mRNA transcript encodes preproinsulin, which carries an amino terminal hydrophobic, 16-residue signal sequence (see Signal hypothesis) (Figs. 2 & 3). As preproinsulin traverses the membrane into the lumen of the endoplasmic reticulum, the signal sequence is removed, leaving proinsulin. Proinsulin is transported to the Golgi, where proteolysis starts to remove an internal sequence known as connecting peptide (C-peptide). Proteolysis continues in storage granules; in addition to other structural homologies, C-peptide from different species contains Arg-Arg at its amino end, and Lys-Arg at its carboxyl end, representing proteolytic cleavage sites for attack by trypsin-like enzymes. Disulfide bonds are formed immediately after translation, and are present in proinsulin. For the amino acid sequence of proinsulin, see Insulin-like growth factor. Fusion of the membranes of the mature storage granules with the plasma membrane of the cell results in release (secretion) of I. Secretion of I., which occurs in response to respirable metabolites (e.g. glucose) and certain hormones (e.g. acetylcholine), is triggered by an increase in the concentration

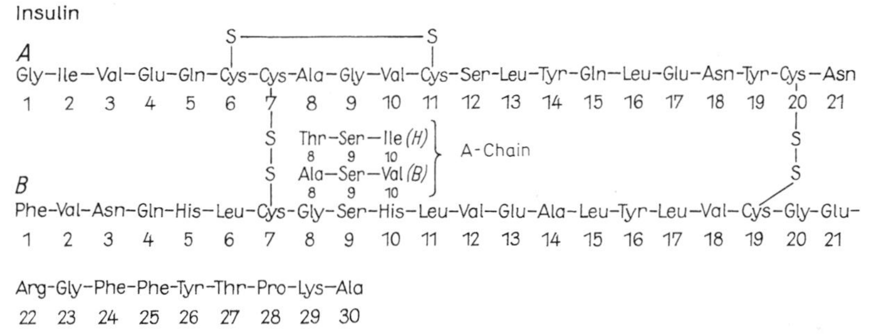

Fig. 1. *Insulin.* Primary structure of sheep insulin. Human (H) and bovine (B) insulin differ from sheep insulin in the sequence region A8 to 10; in addition, in human insulin, the *C*-terminal alanine of the B-chain is replaced by threonine.

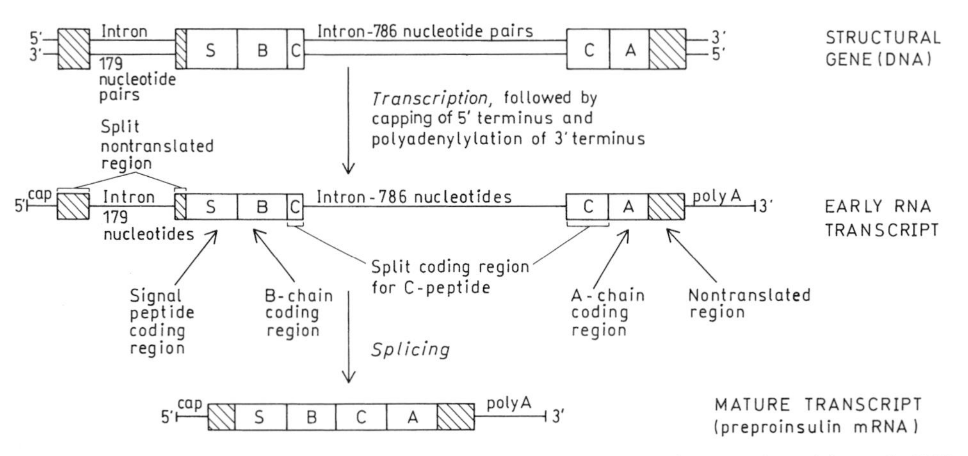

Fig. 2. *Insulin.* Transcription of the human insulin structural gene, and processing of the early RNA transcript. All processes shown here occur in the nuclei of the B-cells of the islets of Langerhans. The mature RNA transcript then passes through pores in the nuclear membrane and enters the cytoplasm, where it is translated.

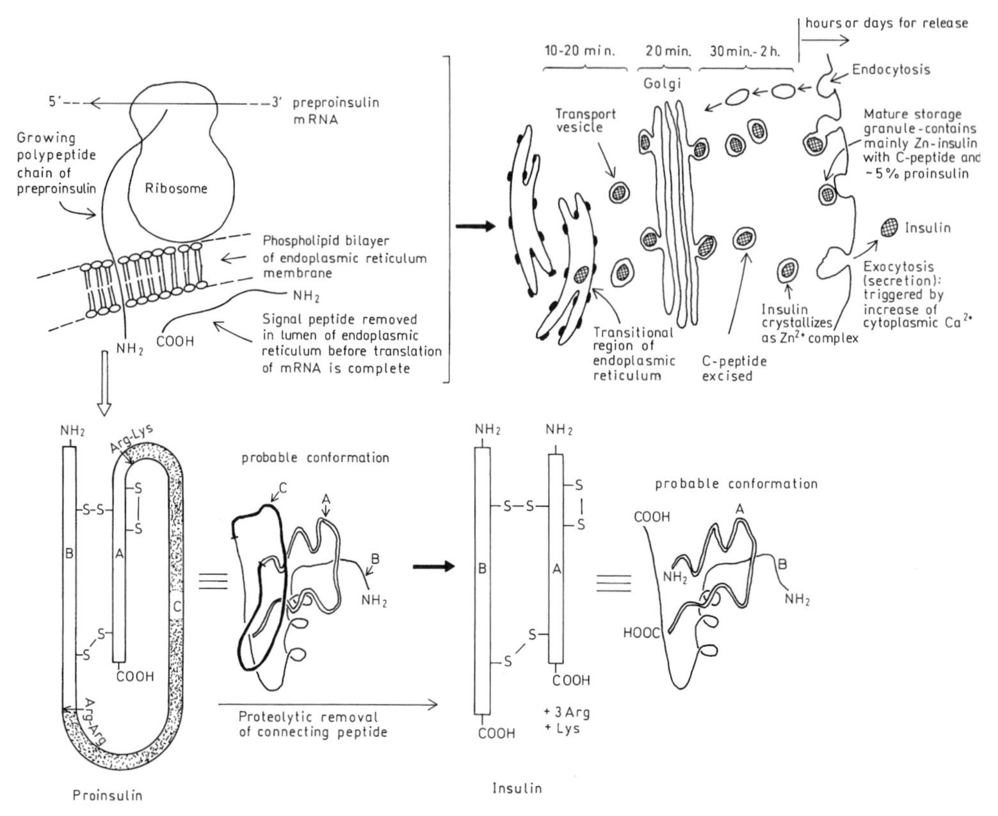

Fig. 3. *Insulin.* Post-translational modification (processing) of proinsulin in the lumen of the endoplasmic reticulum, Golgi apparatus and transport vesicles. These processes occur in the B-cells of the islets of Langerhans. Secretion of insulin occurs by exocytosis. Membranes may be recycled by endocytosis.

of free cytosolic Ca^{2+}, caused by an increased entry of Ca^{2+} into the cell via voltage-dependent Ca^{2+} channels, and by the mobilization of intracellular pools of Ca^{2+}.

Inositol triphosphate may serve as a messenger to release Ca^{2+} from internal stores (see Inositol phosphates) and there is now considerable evidence that this event may be universal in the stimulation of various secretory cells. [S. K. Joseph et al. *J. Biol. Chem.* **259** (1984) 12952-12955; C. B. Wollheim & G. W. G. Sharp, *Physiol. Rev.* **61** (1981) 914–973]

An important physiological stimulus for I. secretion is a high concentration of blood glucose, e.g. the hyperglycemia following a meal promotes I. secretion. Other primary physiological stimuli of I. secretion are mannose, leucine, arginine, lysine, short chain fatty acids, long chain fatty acids, acetoacetate, β-hydroxybutyrate. Secondary physiological stimuli (do not promote I. release directly, but alter response to primary stimuli) include glucagon, secretin, pancreozymin, gastrin, acetylcholine, prostaglandins E_1 and E_2. I. secretion is inhibited by somatostatin, adrenalin and noradrenalin. A distinction must be drawn between biosynthesis and secretion of I. Thus, glucose concentrations higher than 2–4 mM stimulate I. synthesis, while concentrations higher than 4–6 mM are required for stimulation of I. secretion. I. biosynthesis is also promoted by mannose, dihydroxyacetone, glyceraldehyde, leucine, *N*-acetylglucosamine, α-ketoisocaproate, glucagon, methylxanthines. Adrenalin inhibits I. biosynthesis, whereas galactose has no effect. The sulfonylureas, which are used pharmacologically to stimulate I. secretion in type 2 diabetes (defective B-cell secretory response), do not stimulate I. biosynthesis.

I. is determined radioimmunologically; the normal concentration of circulating I. in human blood is 1 ng/ml.

I. is rapidly removed from the circulation and degraded. The major site of degradation (inactivation) is the liver, but most peripheral tissues contain specific insulin-degrading enzymes. Such enzymes are localized in the cells, so I. must be internalized before destruction. This occurs by receptor-mediated endocytosis, after which I. dissociates from the receptor and is degraded, while the receptor is probably recycled to the plasma membrane. A soluble I. protease is capable of attacking both the I. molecule and its separated A and B chains. In addition, the cell contains an I.-glutathione transhydrogenase, associated with in-

N ——→ C

HPI (1-30) F V N Q H L C G S H L V E A L Y L V C G E R G F F Y T P K T ←B-chain

IGF-I (1-29) G P E T L C G A E L V D A L Q F V C G D R G F Y F N K P T

IGF-II (1-32) A Y R P S E T L C G G E L V D T L Q F V C G D R G F Y F S R P A

N ——→ C

HPI (31-65) R R E A E D L Q V G Q V E L G G G P G A G S L Q P L A L E G S L Q K R ←C-peptide

IGF-I (30-41) G Y G S S S R R A P Q T

IGF-II (33-40) S R V S R R S R

N ——→ C

HPI (66-86) G I V E Q C C T S I C S L Y Q L E N Y C N ←A-chain

IGF-I (42-70) G I V D E C C F R S C D L R R L E M Y C A P L K P A K S A

IGF-II (41-67) G I V E E C C F R S C D L A L L E T Y C A T - - P A K S E

Comparison of the primary structures of IGFs and human proinsulin (HPI). For the one letter code see Amino acids.

Effects of insulin on phosphorylation and activity of various enzymes and proteins. The effects are indirect, i.e. insulin binds to its receptor on the cell membrane, initiating a chain of events, culminating in the phosphorylation (kinase reaction) or dephosphorylation (phosphatase action) of the enzyme.

Enzymes of carbohydrate metabolism	Activity	Phosphorylation
Fructose 6-phosphate 2-kinase	↑	↓
Glycogen synthase	↑	↓
Phosphorylase	↓	↓
Phosphorylase kinase	↓	↓
Phosphoprotein phosphatase inhibitor 1	↓	↓
Pyruvate dehydrogenase	↑	↓
Pyruvate kinase	↑	↓
Enzymes of lipid metabolism		
Acetyl-CoA carboxylase	↑	↑
ATP-citrate lyase	No change	↑
Diacylglycerol acyltransferase	↑	↓
Glycerol phosphate acyltransferase	↑	↓
Hydroxymethylglutaryl-CoA reductase	↑	↓
Hydroxymethylglutaryl-CoA reductase kinase	↓	↓
Triacylglycerol lipase	↓	↓
Other		
Insulin receptor (β-subunit)	?	↑
Ribosomal protein S6 (in 40S subunit)	?	↑
Cyclic AMP phosphodiesterase (low K_m type)	↑	↑
Ca-ATPase of plasma membrane	↓	↓ ?
Na/K-ATPase of plasma membrane	↑ ?	↑

↑ = phosphorylation or increase in activity; ↓ = dephosphorylation or decrease in activity

Effects of insulin and other hormones on carbohydrate and lipid metabolism in muscle, adipose tissue and liver. Insulin stimulates amino acid uptake and protein synthesis in all three tissues.

Tissue	Process	Insulin	Adrenalin	Glucagon
Muscle	Glucose uptake	↑	↑	–
	Glycolysis	↑	↑	–
	Glycogenolysis	↓	↑	–
	Glycogen synthesis	↑	↓	–
Adipose tissue	Glucose uptake	↑	↓	(↓)
	Lipogenesis	↑	↓	(↓)
	Lipolysis	↓	↑	(↑)
Liver	Fatty acid synthesis	↑	↓	↓
	Fatty acid oxidation	↓	↑	↑
	Gluconeogenesis	↓	(–)	↑
	Glycogenolysis	↓	↑	↑
	Glycogen synthesis	↑	↓	↓
	Ketone body formation	↓	(–)	↑

↑ = increase in activity; ↓ = decrease in activity

tracellular membranes, which cleaves the disulfide bonds of I., producing separate A and B chains. I. induces synthesis of the degrading enzymes; the I. degrading activity of the liver decreases in starvation, and increases again on refeeding. Excessive degradation of I. may be one cause or contributing factor of diabetes mellitus (see). Two major types of this disease are recognized clinically: Type 1 or I.-dependent diabetes (IDD) is caused by a lack of functional B cells, resulting in I. insufficiency. IDD classically appears during childhood or adolescence, but can appear at any age; patients are totally dependent on exogenous I. for survival, and are prone to ketosis. Type 2 or non-I.-dependent diabetes (NIDD) is due to a relative I. deficiency, often related to defective secretion, although synthesis of I. may be adequate. NIDD is more characteristic of middle age and older, and is also known as maturity onset diabetes; exogenous I. is not required, and control by diet alone is often possible. [W. Montague, *Diabetes and the Endocrine Pancreas – A Biochemical Approach* (Croom Helm, London, 1983); M. P. Czech, ed., *Molecular Basis of Insulin Action* (Plenum Press, New York 1985); L. M. Graves & J. C. Lawrence, Jr. 'Insulin, Growth Factors, and cAMP. Antagonism in the Signal Transduction Pathways' *Trends Endocrinol.* **7** (1996) 43–50]

Insulin-like growth factor, ***IGF, non-suppressible insulin-like activity NSILA:*** a family of polypeptides present in vertebrate blood that share considerable structural and functional similarity with insulin, but are distinguishable from it by a lack of immunological cross reactivity. Insulin has long-term growth effects, but is more remarkable for its potent short-term action, whereas IGFs show potent long-term effects. Two IGFs (IGF-I and IGF-II) are present in human serum; they possess growth-promoting activity on chick embryo fibroblasts at a concentration of 10^{-9} M.

IGF-I is also called somatomedin C. IGF-II is similar to Multiplication stimulating activity (see) of rat. The structures of human IGF-I and IGF-II and proinsulin show obvious homologies (Fig.). The number of sequence differences between the IGFs and proinsulin suggest that duplication of the common ancestral gene occurred before the appearance of the vertebrates. The three-dimensional structures of insulin and the IGFs are probably similar, since all half-Cys and Gly residues and most of the nonpolar core residues of the insulin monomer are conserved.

IGF-I receptor and Insulin receptor (see) are immunologically, structurally and functionally related. IGF-I receptor is a disulfide-linked glycoprotein heterotetramer, M_r about 400,000, which can be resolved into α- and β-subunits, M_r 130,000 and 98,000, respectively. It binds IGF-I with high affinity, and IGF-II, multiplication stimulating activity and insulin with lower affinity; it displays Protein-tyrosine kinase (see) activity. IGF-II receptor is a single chain glycoprotein, M_r about 250,000, with a high affinity for IGF-II and multiplication stimulating activity, and no affinity for IGF-I or insulin. Insulin can cause a reversible desensitization of the induction of tyrosine aminotransferase by IGF-I or IGF-II, possibly by affecting a post-binding step of IGF action, which may be common to both insulin and IGF. There is evidence that Ca^{2+} has a role in the binding and subsequent cellular processing of IGF-II in pancreatic acini. [K.-T. Yu & M. P. Czech, *J. Biol. Chem.* **259** (1984) 3090–3095; C. Thibault et al. *J. Biol. Chem.* **259** (1984) 3361–3367; J. Mössner et al. *J. Biol. Chem.* **259** (1984) 12350–12356; W. S. Cohick & D. R. Clemmons 'The Insulin-like Growth Factors' *Annu. Rev. Physiol.* **55** (1993) 131–153; R. Yamamoto-Honda *J. Biol. Chem.* **270** (1995) 2729–2734; D. E. Jensen et al. *J. Biol. Chem.* **270** (1995) 6555–6563]

Insulin receptor: The short-term metabolic effects and the long-term growth-promoting activity of insulin are initiated by its binding to specific, high-affinity, cell surface receptors. The I. r. is an integral membrane glycoprotein, consisting of 2α and 2β subunits linked by disulfide bonds. This heterotetrameric structure is derived from a single polypeptide precursor (Fig. 1), the structure of which was deduced from a cDNA clone. Removal of the signal sequence, partial glycosylation, folding of the polypeptide, and formation of the disulfide linkages (destined to link the α and β subunits) occur in the endoplasmic reticulum. Further glycosylation and proteolytic cleavage into α

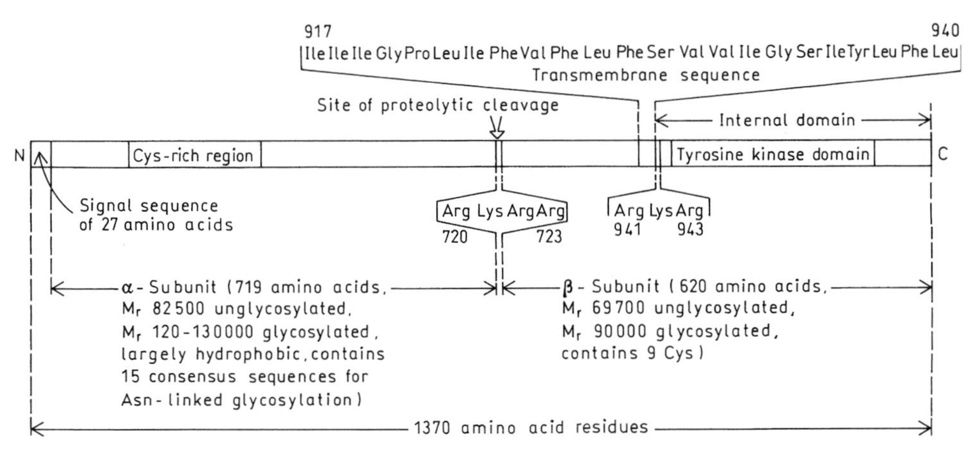

Fig. 1. *Insulin receptor.* Structure of the precursor polypeptide of the insulin receptor. A sequence of basic amino acids (in this case Arg_{941} Lys_{942} Arg_{943}) at the junction of the transmembrane sequence and the cytoplasmic domain is a common feature of transmembrane proteins. It is thought that they interact with polar groups of phospholipids on the membrane surface.

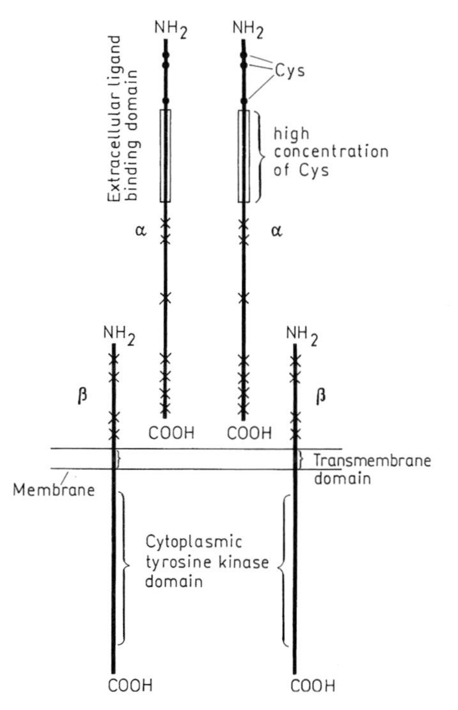

Fig. 2. *Insulin receptor.* Proposed organization of the heterotetrameric insulin receptor complex. Single Cys residues that may be involved in disulfide linkages between units are represented by X.

and β subunits occur in the Golgi apparatus, followed by transport to the plasma membrane. Analysis of the binding of ^{125}I-labeled insulin to the I. r. shows a curvilinear Scatchard plot and a binding constant of 1 nM. Mild reductive cleavage (dithiothreitol) of Triton X-100-solubilized, affinity-labeled I. r. (M_r about 440,000) produces identical dimers (M_r about 220,000) by cleavage of disulfide bonds between α-subunits. Complete reduction produces separate α and β subunits (M_r about 120,000 and 90,000, respectively) with high activity of ^{125}I-insulin attached to the α-subunit. Thus each α-subunit binds a molecule of insulin externally to the membrane surface. Only the β-subunits traverse the membrane (Fig. 2). Insulin binding causes the appearance of tyrosine kinase activity (interpreted as a large increase in the V_{max} of an existing active center) in the intracellular domain of the β-subunit. This insulin-dependent kinase catalyses phosphorylation by ATP of Tyr residues in the β-subunit itself, as well as in other peptides and proteins (see Protein-tyrosine kinase, Receptor tyrosine kinases). Serine residues of the I. r. also become phosphorylated, but the serine kinase responsible is not intrinsic to the I. r.

There are many similarities between the β-subunit of the I. r., the receptors for other growth factors, and certain oncogene products. Thus tyrosine kinase activity is also present in epidermal growth factor receptor, which shows considerable sequence homology with the β-subunit of the I. r., and also possesses similar domains (extracellular transmembrane, cytoplasmic). Homologies of amino acid sequence also exist between the cytoplasmic domain of the I. r. β-subunit and products of a family of viral oncogenes, commonly known as the *src* family, which possess tyrosine kinase activity. Despite this structural similarity, no oncogene counterpart of the I. r. has yet been identified. Similarly, the receptors for platelet-derived growth factor and for insulin-like growth factor are structurally related to the I. r. β-subunit, and they possess tyrosine kinase activity, but transforming proteins derived from them by mutation have not been found. In contrast, part of the epidermal growth factor receptor is known to be a proto-oncogene product. [M. P. Czech, *Recent Progress in Hormone Research* **40** (1984) 347–377; A. Ullrich et al. *Nature* **313** (1985) 756–761]

Integrated rate equation: an equation which represents the concentration of the substrate or product as a function of time. The corresponding plots are called the progress curves, which can also be obtained by direct measurement (see Enzyme kinetics). By integrating the Michaelis-Menten equation, one obtains for example, with $(S_0 - S) = P$, the integrated velocity equation $P(t) = V_m t + K_m \ln S/S_0$, where P is the product concentration, S the substrate concentration, S_0 the substrate concentration at $t = 0$, and V_m and K_m are the maximal velocity and Michaelis constant, respectively.

Integrins: see Cell adhesion molecules.

Intercalation: a special interaction between dye molecules (e.g. proflavin) and DNA in which the dye molecule inserts itself between two neighboring base pairs of the DNA double helix. This stretching of the DNA can cause errors in transcription, leading to defective ("mutated") mRNA, whose translated protein contains the wrong amino acid sequence.

Interconversion: in biochemistry in general, the conversion of one intermediary metabolic product into another; in enzymology, in particular, the transformation of the active form of an enzyme into the inactive form, and vice versa, as a mechanism of metabolic regulation.

Interconversion of enzymes: see Covalent modification of enzymes.

Interconvertible enzymes: see Covalent modification of enzymes.

Interferons. ***IFNs:*** species-specific glycoproteins, first described as agents produced and secreted by vertebrate cells in response to penetration by viral (or synthetic) nucleic acid (IFN synthesis is induced by double-stranded RNA, which is probably generated by DNA viruses, as well as by RNA viruses, during infection). They leave the infected cell to confer resistance on other cells of the same organism, i.e. by binding to surface receptors, they produce an antiantiviral state in other cells, which prevents the replication of many different RNA and DNA viruses. IFNs therefore represent a major defense against viral infection (they are active at concentrations as low as 3×10^{-14} M). Since some cancers are induced by viruses, IFNs have also attracted interest as potential anti-cancer agents. They are now used against certain malignant tumors as well as virus infections, and the IFNs needed for such treatments are produced by molecular cloning.

Three families are recognized:

(i) ***Leukocyte IFN*** or ***IFN-α*** is produced principally by T and B lymphocytes and monocytes, but there are many other cell sources. This is a large family of closely related, nonglycosylated proteins encoded by a supergene family (about 20 different genes all located on human chromosome 9). Human IFN-αs have 166–172 amino acid residues (M_r 16,000–27,000). The human IFN-α family is divided into two subfamilies: IFN-αI and IFN-αII (also called IFN-ω).

(ii) ***Fibroblast IFN*** or ***IFN-β*** (formerly IFN-β1) is produced principally by fibroblasts and leukocytes, but there are many other cell sources. The human protein has 166 amino acid residues ($M_r \sim 20{,}000$), encoded by a single gene on human chromosome 9. The protein previously known as IFN-β2, encoded on chromosome 7, is now known as interleukin-6.

(iii) ***Lymphocyte IFN, immune IFN*** or ***IFN-γ*** is normally a glycosylated dimer ($M_r \sim 50{,}000$). The human protein has 143 amino acid residues per monomer, encoded by a single gene on human chromosome 12. It is produced by T-lymphocytes, which have been stimulated by interleukin. It enhances the cytotoxic activity of T-lymphocytes, macrophages and natural killer cells, and thus has antiproliferative effects. The crystal structure of recombinant rabbit IFN-γ has been reported at 2.7 Å resolution [C.T. Samudzi et al. *J. Biol. Chem.* **266** (1991) 21791–21797].

All IFNs generally impair virus proliferation by inhibiting the protein synthesis of infected cells. This occurs in two distinct ways: (i) induction of the synthesis of *double-stranded RNA-activated inhibitor* which, in the presence of double-stranded RNA, inhibits initiation of protein synthesis on the ribosome by phosphorylating the α-subunit of eukaryotic initiation factor-2; or (ii) induction of the synthesis of *(2′,5′)*-oligoadenylate synthetase, which, in the presence of double-stranded RNA, catalyses synthesis of the oligonucleotide $pppA(2'p5'A)_n$ ($n = 1$–10), which in turn activates RNase L. The latter degrades mRNA, thereby halting protein synthesis. Rapid degradation of $pppA(2'p5'A)_n$ by (2′,5′)-phosphodiesterase means that this effect can only be maintained by continual synthesis of the oligonucleotide.

All IFNs also protect cells against other intracellular parasites, as well as inhibiting proliferation of some normal and transformed cells. IFNs also appear to be involved in the regulation of cell differentiation, and they enhance the expression of class I major histocompatibility antigens.

IFN-γ displays certain properties not shared by INFs α and β: (i) it also increases the production of antibodies in response to antigens, possibly by enhancing the antigen-presenting function of macrophages; (ii) it activates natural killer cells; (iii) it induces some cytokines (e.g. interleukin-1, tumor necrosis factor, colony-stimulating factor); (iv) it induces or enhances the expression of class II major histocompatibility antigens; (v) it regulates the expression of receptors for other cytokines and for the Fc portion of IgG; (vi) it acts synergistically with interleukin-2 in the stimulation of proliferation of B lymphocytes and production of IgG immunoglobulins.

The genes for I. contain no introns (the only other vertebrate structural genes known to lack introns are those encoding histones).

Interlaboratory survey, ***collaborative interlaboratory survey, quality control survey, multicenter evaluation:*** a system for evaluating the accuracy and precision of reference methods conducted by individual clinical chemical laboratories, and for assessing the performance of analytical instruments. Samples of the same specimen are analysed by different laboratories, and the range of interlaboratory scatter is determined for the particular analytical technique. In some countries, it is now mandatory for clinical chemical laboratories to participate is such quality assurance surveys, and those failing to fulfil requirements for a particular analyte may not be recognized by medical insurance companies. Similar surveys are conducted to test the reliability, accuracy, precision, running costs and ease of operation of new instruments. Reports of I.s. appear regularly in the *European Journal of Clinical Chemistry and Clinical Biochemistry.* [Symposium on Reference Methods in Clinical Chemistry. *Eur. J. Clin. Chem. Clin. Biochem.* **29** (1991) 221–279; W.J. Geilenkeuser & G. Röhle, *Eur. J. Clin. Chem. Clin. Biochem.* **32** (1994) 369–375]

Interleukin 3: see Colony-stimulating factors.

Interleukins: growth factors for blood cells and their precursors. See Lymphokines.

Intermediary metabolism: a term from early physiological chemistry, signifying all those metabolic reactions occurring between the uptake of foodstuffs and the formation of excretory products. In modern usage, I. m. is essentially identical with Primary metabolism (see).

Intermediate: any compound of Intermediary metabolism (see). In the narrow sense, an I. is a compound in a metabolic chain of reactions, excluding the starting compound and the endproduct.

Intermediate filaments: see Cytoskeleton.

Interphase: the phase between two mitoses, comprizing the G_1, S and G_2 phases of the Cell cycle (see). During I., the chromosomes, bounded by the nuclear membrane, are completely unwound and present as randomly distributed, uncondensed chromatin.

Interphase nucleus: See Nucleus.

Interstitial cell stimulating hormone: see Luteinizing hormone.

Intervening sequence: see Intron.

Intracellular digestion: digestion within the cell, in which the lysosomal system plays an important role. Macromolecular substances are taken up by pinocytosis and phagocytosis in vacuoles called phagosomes, which merge with primary lysosomes to form digestive vacuoles (secondary lysosomes). I. d. affects exogenous materials brought into the cell from outside, and endogenous cell components, the respective processes of I. d. being known as heterophagy and autophagy. Autolysis (see) is also a form of autophagy, although it occurs in dying or dead cells. In living cells, autophagy may be expressed as the I. d. of entire regions of the cytoplasm. Cytoplasmic regions bounded by membranes which are later digested, are called cytolysosomes. The fragments produced by I. d. must be removcd from the digestive vacuoles and cytolysosomes so that they can be further degraded. I. d. has been most thoroughly investigated in protozoa, fibroblasts in tissue culture and polymorphonuclear leukocytes.

Intrinsic factor: see Vitamins (vitamin B_{12}).

Intron: an intervening sequence in a eukaryotic gene. The same term is sometimes applied to the corresponding intervening sequence in the RNA transcript, but strictly speaking this should be called the *intron transcript* (IT). The term, *intervening sequence* (IVS), is also used for both the I. and IT. I. and the coding sequences *(exons)* are transcribed together; the ITs are then removed to produce functional RNA. An I. therefore makes no contribution to the final gene product of the flanking exons. I. have been found in eukaryotic mRNA, tRNA and nuclear rRNA, and in mitochondrial mRNA and rRNA. Prokaryotic genes generally do not contain I., and they are less common in yeast genes than in those of "higher" eukaryotes. It might be thought that the lack of I. is a more primitive condition. However, analysis of the pyruvate kinase genes of the chicken (13 kilobases, 9 or 10 I.) and yeast (1.5 kilobases, no I.) suggests that the gene ancestral to both was first assembled from smaller blocks of DNA, presumably before the divergence of pro- and eukaryotes. The subsequent loss of I. from the yeast and prokaryote genes may be seen as an adaptive response to selection for more rapid reproduction [N. Lonberg & W. Gilbert *Cell* **40** (1985) 81–90]. Genes containing I. are known as *split genes.* The process of removal of an I. from an RNA transcript and the joining of the neighboring exons is known as *splicing.* The base sequence of I. in mRNA begins with 5′GU and finishes with AG3′. These sequences act as recognition sites for *splicing enzymes.* The 5′GU ... AG3′ pattern is not found in tRNA precursor molecules, suggesting the existence of at least two splicing enzymes, one for mRNA, the other for tRNA. I. are often longer than exons, and I. may account for a greater proportion of a gene than exons, e.g. the gene for ovalbumin contains seven I. and 7,700 base pairs, whereas the corresponding spliced mRNA contains 1,859 bases.

There are possibly several different mechanisms of splicing, depending on the organism and the type of precursor RNA. A model for the splicing of fungal mitochondrial RNA is shown in Fig. 1.

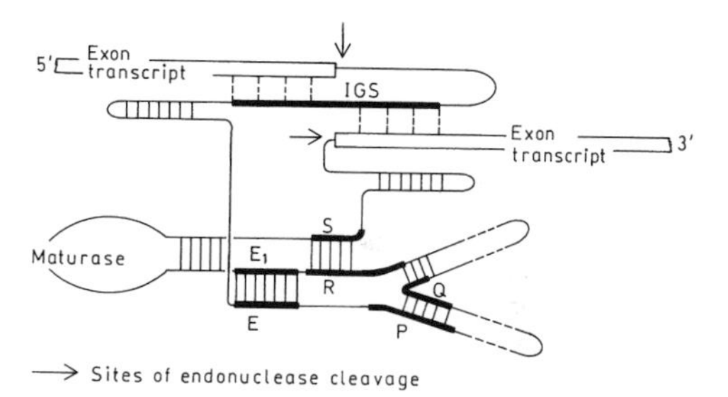

Fig. 1. *Intron.* Secondary structure of the intron transcript (IT) of fungal mitochondrial precursor RNA [adapted from R. W. Davies et al. *Nature 300* (1982) 719–724]. Nucleotide sequence analysis of four mitochondrial introns from *Aspergillus nidulans* and five from *Saccharomyces cerevisiae* shows that all the corresponding ITs are able to form the same secondary structure. P, Q, R and S represent highly conserved regions. E and E_1 are not highly conserved, but are always complementary. This secondary structure brings the ends of the two flanking exon transcripts close to one another; precise alignment of the splicing sites is guaranteed by an internal guide sequence (IGS) in the IT. The IGS consists of two regions in tandem. which base-pair with the terminal regions of the respective flanking exon transcripts. The maturase loop represents a region that translates for a mRNA maturase, a protein essential for excision; this is, however, quite exceptional; most ITs do not code for any protein. The diagram is not to scale; tertiary structure, not represented here, is also present.

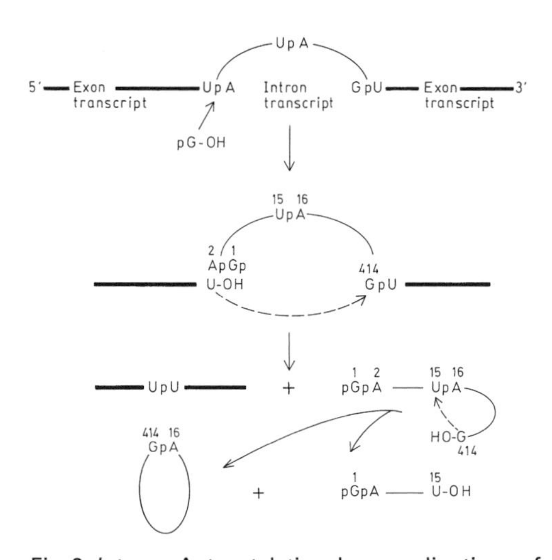

Fig. 2. *Intron.* Autocatalytic cleavage-ligation of *Tetrahymena* pre-ribosomal RNA [adapted from A. J. Zaug et al. *Nature 301* (1983) 578–583]. The process is initiated by insertion of guanosine between the 3′ end of the exon transcript (U) and the 5′ end (A) of the IT. The -UpA- phosphodiester bond is thereby cleaved, and the guanosine becomes attached as the 5′-terminus of the IT, extending the IT from 413 to 414 nucleotides. Guanosine can be replaced by GMP, GDP or GTP. For experimental purposes ^{32}P-labeled GMP is used, thereby conveniently labeling the 5′ end of the IT. The free terminus (-U-OH) of the exon transcript ligates with the 5′-terminal U of the other exon transcript (forming a new -UpU- phosphodiester bond) with concomitant release of the IT at its 3′-terminal. The 3′- terminal G (nucleotide 414) cleaves the phosphodiester bond between nucleotides 15 (U) and 16 (A), forming a cyclic RNA fragment (nucleotides 6–414) and a linear fragment (nucleotides 1–15). Thus, two phosphodiester bonds are cleaved and two are formed. Cyclization probably serves to prevent this unwanted RNA fragment from taking part in the reverse reaction. The entire process can be described as an autocatalytic cleavage-ligation reaction, and the RNA is known as *self-splicing RNA.* The term, *ribozyme,* has also been suggested for self-splicing RNA. Removal of ITs by splicing is one type of Post-transcriptional modification of RNA (see).

Splicing of *Tetrahymena* pre-ribosomal RNA occurs independently of ATP or protein (Fig. 2). [L. D. Hurst & G. T. McVean *Current Biology* **6** (1996) 533–536]

Inulin: a high molecular mass vegetable reserve carbohydrate, $[\alpha]_D$ −40°, m.p. 178 °C. I. is a fructan consisting of 20–30 β-1,2-linked fructofuranose units. Probably the non-reducing end of the chain terminates with glucose. I. is found as a reserve substance in the tubers and roots of many members of the *Compositae,* like dahlia and Jerusalem artichoke tubers. It is used in food for diabetics.

Inulin

Invertase, ***β*-D-*fructofuranosidase*** (EC 3.2.1.26): a saccharide-cleaving hydrolase from yeast, fungi and higher plants. Yeast I. is a dimeric glycoprotein (M_r 270,000), *Neurospora* invertase a tetrameric glycoprotein (M_r 210,000, M_r of the subunit, 51,000). Aniline (fungal I.), pyridoxal (potato I.) and zinc ions (all I.) are inhibitors of I.

Invert sugar: a mixture of equal parts D-glucose and D-fructose which, in contrast to dextrorotatory sucrose, is levorotatory (see Mutarotation). I. s. is generated by acid or enzymatic hydrolysis of sucrose. Because bees have the corresponding invertase, honey is 70–80 % I. s. The sweet taste is essentially due to the fructose.

In vitro methods: see Methods of biochemistry.

In vivo methods: see Methods of biochemistry.

Ion channel: a transmembrane protein or protein complex that forms a hydrophilic pore, thus enabling ions to migrate between the aqueous environments on either side of the membrane. I. cs. are specific for either cations or anions, and specificity may be even stricter, so that only one ionic species is accepted. An I. c. my be a permanently open channel, allowing free passage of an ion at any time (e. g. the I. c. that permits passage of K^+ during generation of the resting membrane potential), but most I. cs. open only in response to a stimulus, and are therefore known as *gated ion channels.* The latter can be broadly classified as: (i) ligand-gated I. cs. which open in response to the binding of a specific ligand, e. g. the nicotinic receptor of Acetylcholine (see); (ii) voltage-gated I. cs. which open in response to changes in membrane potential. [W. A. Catterall 'Structure and Function of Voltage-gated Ion Channels' *Annu. Rev. Biochem.* **64** (1995) 493–531]

Ion-exchange chromatography: a liquid-solid phase chromatographic method for the analytical and preparative separation and purification of mixtures of substances. I. c. is based on electrostatic binding between cations or anions of the substance to be studied or separated and the corresponding Ion exchanger (see). The components of an electrolyte solution are separated by successive adsorption and desorption through ion exchange. I. c. can be run as a front, replacement or elution process (see Chromatography). Ion exclusion and ion retardation are special techniques of I. c. Ion exclusion can be used to separate ionized from nonionized substances or substances ionized to different extents, e. g. inorganic from organic acids. Ions appear in the eluate sooner than non-ions, because the latter penetrate the interstitial and the interior resin phases and thus pass through the column more slowly. Ion retardation is a delayed flow of ions over a mixed bed of cation and anion exchangers. The ions are eluted with water. I.c. can be used to separate closely related ions and compounds, e.g. inorganic ions (particularly rare earth metals), amino acids, alkaloids and optically active compounds from their racemates.

Ion exchangers: all natural and artificial substances, mostly solids, which are able to exchange bound ions for ions from the surrounding liquid medium. The structure of the solid I.e. is not significantly changed in the process. The exchange depends on the properties of the ions involved; purely absorptive binding can also occur. I.e. are high molecular mass, insoluble polyelectrolytes capable of swelling. There are acidic (solid acids, macropolyacids) and basic (solid bases, macropolybases) types of widely varying chemical nature. The exchange of ions occurs stoichiometrically; accordingly, a chemical equilibrium can be established: $IG_1 + G_2 \rightleftharpoons IG_2 + G_1$, where I is the ion exchanger, and G_1 and G_2 are the counter ions of the exchanger and the milieu. Depending on the type of ions on the exchanger, it is called a *cation exchanger,* which has negative functional fixed ions with positive counter-ions, or an *anion exchanger,* with positively charged fixed ions and negative counter-ions (anions) to equalize the charge. The anchor groups (exchange-active components) in commercially available cation exchangers are usually $-C_6H_5O^-$, $-SO_3^-$, $-COO^-$, $-PO_3^{2-}$ or $-AsO_3^{2-}$; those of anion exchangers are usually $-N^+H_3$, $-N^+H_2R$, $-N^+HR_2$, $-N^+R_3$ (R = organic residue). Amphoteric I.e. have both acidic and basic fixed ions, while metal-specific ion exchange resins have specific fixed ions arranged in pairs so they can complex certain metal ions (specific I.e.). I.e. have large surface areas, so that their exchange capacity is usually very high, e.g. 500 m^2/g, with an exchange capacity of 0.5 to 10 mval/g, which can lead to an ion concentration equivalent to that of a 10 M solution. – The different types of I.e. can be classified as follows.

1. *Synthetic ion exchange resins* consist of a hydrophobic matrix capable of limited swelling. Their properties depend on their chemical type, degree of cross-linking and the number and kind of fixed ions. In the synthesis of artificial resin exchangers, the basic substance is cross-linked by bridge-forming groups, and at the same time, the anchor groups are attached. Synthetic resin exchangers consist of a carbohydrate network with many ionizable groups, either acidic or basic, which are covalently anchored to the basic substance. There are two types of resin, polycondensation and polymerization resins. Polymerization resins, consisting of polystyrene cross-linked with divinylbenzene, are the most common. Dowex resins and the various Wofatit resins are of this type. The chemical and physical properties of the exchange resin are changed by modification of the matrix. Polymerization resins are classified according to particle size and degree of branching (X-number). The ion-exchange equilibrium is more quickly reached with smaller resin particles, because the speed of diffusion of the ions is inversely proportional to the third power of the particle diameter. The degree of cross-linking determines the degree of swelling and porosity of the I.e. and thus affects its molecular sieve effect and the rate of flow through the bed of resin, as well as the exchange capacity. The degree of cross-linking is defined as the number of intermolecular bonds in the matrix, referred to 1 g I.e. The capacity of an exchanger can be expressed either as the total capacity (total number of fixed ions per gram dry exchanger, expressed in mval/g) or the useful capacity, which is the number of counter-ions which can be exchanged. The capacity depends on several factors, and there are sometimes considerable differences between strong and weak I.e. Thus, a strongly acidic resin with sulfonate groups as fixed ions is highly ionized, both in the acid and the salt form, so that ion exchange can occur over the entire pH range, while a weakly acidic, carboxyl exchanger is active only at pH 7.

2. *Cellulose ion exchangers* carry active groups near the surface of the hydrophilic network of cellulose which are introduced by substitution or impregnation. They are produced as paper or powder, and have the principal advantage that even large molecules can be exchanged, which would not be able to penetrate the network of an artificial resin exchanger. They are therefore of particular importance in protein chemistry. DEAE-cellulose (diethylaminoethylcellulose) is a substituted cellulose, an anion exchanger which contains $-C_2H_4-N(C_2H_5)_2$ groups. Carboxymethylcellulose is a cation exchanger containing $-CH_2-COO^-$ groups.

3. *I.e. based on dextran gels* have active groups which are bound to the glucose residues in the hydrophobic matrix.

4. *Inorganic I.e.* are the silicic permutites (formerly used for softening water) and zeoliths. The latter are natural crystalline aluminosilicates with a network of SiO_4 and AlO_4 tetrahedra. They are now the most commonly used water-softening I.e., because their interstitial spaces contain not only water, but alkali and alkaline earth metal ions, which can be exchanged.

Applications. I.e. are primarily used for water softening and desalting, for simple ion-exchange processes, e.g. separation of neutral salts and salt-salt exchange, for enrichment, isolation and determination of trace elements, for separation of ions with similar properties (see Ion-exchange chromatography), and for catalysis (e.g. sugar inversion). In biochemistry they are widely used for the separation and determination of charged molecules, e.g. nucleotides, proteins, amino acids (see Proteins).

Ion filtration chromatography: see Proteins.

Ion-motive ATPases: membrane-bound enzymes which, as part of their catalytic cycle, couple the transport of one or more ionic species across the membrane in which they are located, to either the hydrolysis of ATP to ADP and P_i or to the synthesis of ATP from ADP and P_i. They comprise 3 major classes: P-type ATPases (see), F-type ATPases (see) and V-type ATPases (see).

The 3 types of I-m.A. are coupled in a 'master-slave' relationship in many cells, particularly those of higher eukaryotes. For instance, in higher animal cells the F-type ATPase of mitochondria operates solely in the direction of ATP synthesis, driven by the Proton motive force (see) generated by electron flow in the Respiratory chain (see); it is therefore the 'master' supplying the 'slave' P- and V-type ATPases of the extramitochondrial membranes, which hydrolyse it and use the resulting free energy to drive endergonic transmembrane ionic flow.

Ionone ring: see Tetraterpenes.

Ionophore: a compound which increases the permeability of membranes to ions. I. act by delocalizing the charge of the ion and shielding it from the hydrophobic region of the lipid bilayer. M_r of I. are typically in the range 500–2,000. They possess both hydrophobic (confers lipid solubility) and hydrophilic (binds the ion) regions. There are two types: a) Mobile carriers, which diffuse within the membrane and catalyse the transport of up to 1,000 ions per second; they show high specificity for specific ions. b) Channel-formers, which create a channel in the membrane; these are less specific, but may catalyse the transport of up to 10^7 ions per second, i.e. passage of the ion through the membrane may be limited only by the rate of diffusion.

Examples of I.: *Valinomycin* (see) is a mobile carrier with a high specificity for the transport of K^+; it discriminates between K^+ and Na^+ in the ratio of 10,000:1. *Gramicidin* (see) is a channel-former with low discrimination between protons, monovalent cations and NH_4^+. *Nigericin* is a mobile carrier, which loses a proton when it binds a cation; it thus forms a neutral complex which diffuses through the membrane, catalysing an overall electroneutral exchange of K^+ for H^+. *Uncoupling agents* (see Oxidative phosphorylation) are also I. which are specific for protons, e.g. carbonyl cyanide-*p*-trifluoromethoxyphenylhydrazone (FCCP) contains an extensive conjugated system with delocalized π electrons; it is therefore lipid-soluble and can also form an anion (Fig.).

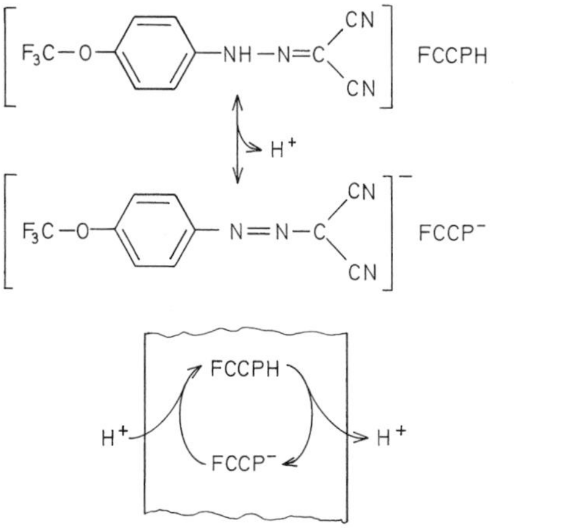

$$\left[F_3C{-}O{-}C_6H_4{-}NH{-}N{=}C(CN)_2\right] \quad \text{FCCPH}$$

$$\updownarrow H^+$$

$$\left[F_3C{-}O{-}C_6H_4{-}N{=}N{-}C(CN)_2\right]^- \quad \text{FCCP}^-$$

Structure of FCCP and its mode of action as an ionophore in the transport of protons across a membrane

Ion pumps: metabolic cycles within cell membranes which can transport ions against the prevailing concentration gradient. The bioelectric membrane potential of the nerves, for example, is based on different distributions of Na^+ and K^+ ions. The high concentration of K^+ in the inside and the predominance of Na^+ on the outside of the nerve produces a normal potential of –60 mV. The energy required to maintain the ionic disequilibrium is provided by the ATP obtained from oxidative phosphorylation. When the nerve is stimulated, there is a change in the permeability of the membrane and a reversal of polarity. Na^+ ions diffuse into the cell and K^+ ions out of it. After a few milliseconds, the original condition is restored (the Na^+ ions are pumped out of the cell). The sudden change in permeability is probably due to structural changes in the membrane proteins induced by released acetylcholine.

IPA: 6-Δ^2-isopentenylaminopurine; 6-(3-methyl-2-butenyl)aminopurine; N^6-γ,γ-dimethylallyladenine; also known as bryokinin (see N^6-[γ,γ-Dimethylallyl]-adenosine). IPA is a Cytokinin (see).

Ipecacuanha alkaloids: iridoid Isoquinoline alkaloids (see). The most important representative of the group is Emetine (see).

Ipomoeamarone: a Phytoalexin (see) with a sesquiterpene structure. I. is formed when sweet potatoes are infected by phytopathogenic microorganisms. For synthesis, see Fig.

Iridine: see Protamines.

Iridodial: a representative of the Iridoids (see), m.p. 90–92 °C, n_D^{19} 1.4782. The dialdehyde form is in equilibrium with the half-acetal form (Fig.). I. was first identified in the defensive secretion of various ants.

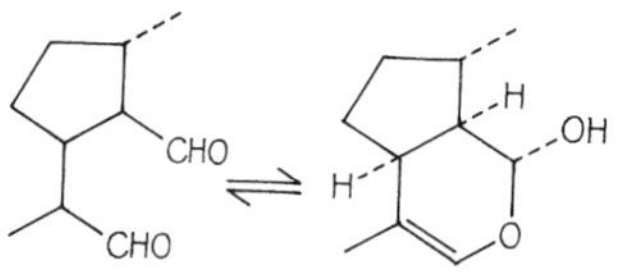

Iridodial

Iridoid alkaloids: see Terpene alkaloids.

Iridoid indole alkaloids: see Indole alkaloids.

Iridoid isoquinoline alkaloids: see Isoquinoline alkaloids.

Iridoids: a group of natural products characterized by a methylcyclopentanoic monoterpene skeleton. I. used to be called, incorrectly, pseudoindicans, because some of them can be converted into intensely blue compounds in the presence of air and acids. This process has not been elucidated, but it is not chemically related to the transformation of indican glycosides into indigo. The name I. is derived from iridodial, which was discovered in the defense secretions of ants. The I. are widespread in higher plants and are often present as glycosides. In addition to the C_{10} I., e.g. verbenalin and loganin, there are compounds with 9 C-atoms, e.g. aucubin, and 8 C-atoms. A few I. have additional structural elements, e.g. acetate in plumieride. The Valepotriates (see) are a complete subgroup of the I. The secoiridoids, which include bitter substances like gentiopicroside, arise from the I. by cleavage of the methylcyclopentane ring. Some I. are easily converted into monoterpene alkaloids, and are therefore possible biogenetic precursors.

Biosynthesis of the I. starts from geranyl pyrophosphate, which is first hydroxylated, then isomerized to neryl pyrophosphate. A number of intermediate reactions (oxidations, reductions, hydroxylations or decarboxylations) lead both to the loganin type of compound (aucubin, usnedoside, actinidine), and to the gentiopicroside type (which also involves cleavage of ring A and recyclization (see Gentiana alkaloids). [A. Bianco et al. *Phytochemistry* **42** (1996) 81–91; A. J. Chulia et al. *Phytochemistry* **42** (1996) 139–143]

Iridomyrmecin: a monoterpene lactone, one of the iridoids, M_r 162, m.p. 59–60 °C, $[\alpha]_D^{17}$ +205 °. I. was first isolated from a pheromone mixture from ants of the genus *Iridomyrmex*. Treatment with alkali produces the isomeric isoiridomyrmecine, m.p. 58–59 °C, $[\alpha]_D^{17}$ −62 °, which is a pheromone for other ant species.

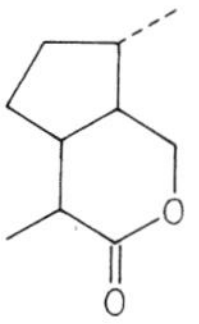

Iridomyrmecin

Iristectorigenin: 7,5,3′-trihydroxy-6,5′-dimethoxyisoflavone. See Isoflavone.

Iristectorin A: iristectorigenin 7-glucoside. See Isoflavone.

Iron, *Fe:* a bioelement found in all living cells. The human body contains 4–5 g Fe, of which 75 % is in hemoglobin. In living organisms Fe occurs in the II and III oxidation states; in higher animals it is stored bound to protein. It is transported in the blood as a complex with transferrin (see Siderophilins), from which it is transferred enzymatically to metal-free porphyrin molecules (see Heme iron). Non-heme iron (see) is also found in a number of compounds, e.g. Iron-sulfur proteins (see). The Fe metabolism of microorganisms is mediated by a group of natural products called Siderochromes (see).

Fe catalyses most redox reactions in the cell (see Cytochromes). It is involved in the reduction of ribonucleotides to deoxyribonucleotides, is a coenzyme for aconitase (EC 4.2.1.3) in the tricarboxylic acid cycle, and is a component of a number of metalloflavoproteins. It has a regulatory role in many microorganisms, e.g. as an inhibitor of citrate synthesis in *Aspergillus niger,* and as a promoter of antibiotic synthesis by *Streptomyces* species.

Iron metabolism: see Ferritin.

Iron porphyrin: see Heme.

Iron-sulfur proteins, *Fe-S-proteins:* a group of proteins found in all organisms. They contain *iron-sulfur centers* (iron-sulfur clusters) and take part in electron transfer processes. They are involved in H_2 metabolism, nitrogen and carbon dioxide fixation, oxidative and photosynthetic phosphorylation, mitochondrial hydroxylation and nitrite and sulfite reduction. The iron in the active centers is coordinated with the sulfur atoms of cysteine residues. In addition, all Fe-S-proteins except for Rubredoxins (see) contain the same number of "labile" or inorganic sulfur atoms as iron atoms, and both are covalently bound in iron-sulfur clusters. Since the iron is not bound in a porphyrin ring, this group of proteins is included in the Non-heme iron proteins (see). It can be subdivided into the *simple Fe-S-proteins* [Rubredoxins, Ferredoxins (see) and others] and the *conjugated Fe-S-proteins,* like Fe-S-flavoproteins, Fe-S-molybdenum proteins, Fe-S-molybdenum flavoproteins, Fe-S-heme proteins, Fe-S-heme flavoproteins, etc. A number of Fe-S-proteins operate in cooperation with other biological electron transport systems (cytochromes, flavoproteins and other oxidoreductases). The NADH dehydrogenase complex contains 28 Fe per molecule in seven Fe-S clusters, in addition to FMN; it catalyses the following reaction in the Respiratory chain (see): $NADH + H^+ + CoQ \rightarrow NAD^+ + CoQH_2$.

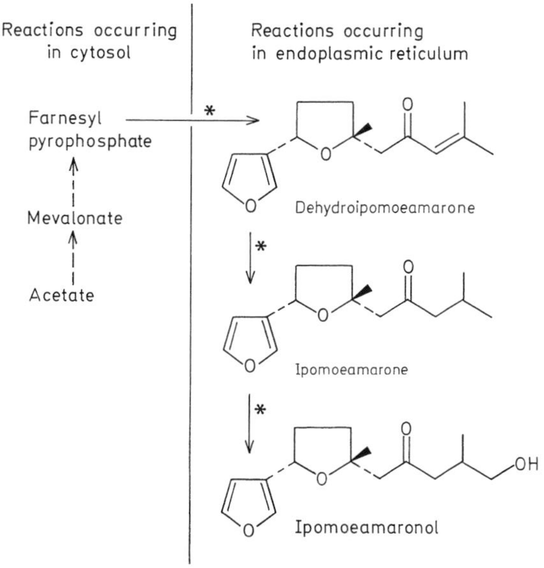

Reactions occurring in cytosol | Reactions occurring in endoplasmic reticulum

Acetate → Mevalonate → Farnesyl pyrophosphate →* Dehydroipomoeamarone →* Ipomoeamarone →* Ipomoeamaronol

Biosynthesis of ipomoeamarone and related compounds in sweet potato root (Ipomoea batatas Lam.) infected with Ceratocystis fimbriata. Conversion of acetate to farnesyl pyrophosphate (see Terpenes) occurs in the uninfected plant, and is stimulated by infection. All reactions marked with an asterisk are absent before infection. [P. A. Brindle & D. R. Threlfall *Biochem. Soc. Trans. 11* (1983) 516–522]

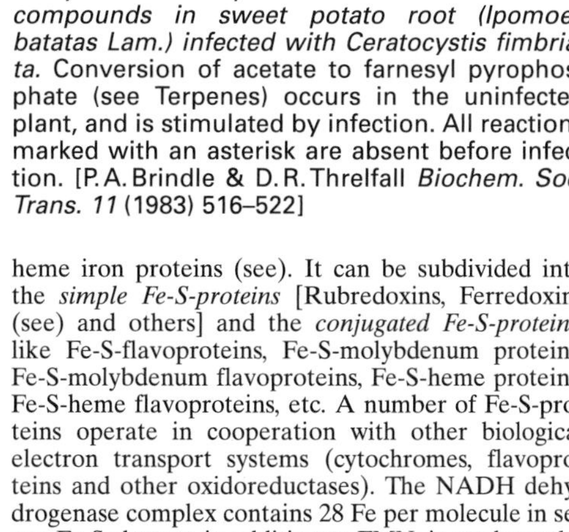

Fe^{3+} Fe^{3+} (S = 0) Oxidized $\xrightarrow{e^-}$ Fe^{2+} Fe^{3+} (S = 1/2) Reduced

Arrangement of iron atoms in a 2Fe-2S center

Succinate dehydrogenase, which contains FAD, contains 8 Fe atoms per molecule, in two 2Fe-2S and one 4Fe-4S centers. Xanthine oxidase contains FAD, Mo and 8 Fe per molecule in two 2Fe-2S clusters.

The ubiquitous occurrence of Fe-S-proteins suggests that they arose early in the evolution of life. The active center of the Fe-S-proteins can be extracted easily by chemical means, and the iron and sulfur can be reintroduced into the apoprotein without loss of biological activity.

The apoproteins always contain at least four cysteine residues per Fe-S cluster. The type of cluster and some of its chemical and biological properties depend on the total length of the amino acid chain the location of the cysteine residues and the amino acids between and around the cysteines.

The iron atoms in the clusters can be replaced by ^{57}Fe (which has a nuclear spin), then studied by Mossbauer spectroscopy. The Fe atoms have a nearly tetrahedral coordination, which means they are in the "high spin" state, irrespective of whether they are oxidized (Fe^{3+}) or reduced (Fe^{2+}). The Fe atoms are close to each other, so that their spins are coupled. When both are in the Fe^{3+} state, the spin coupling produces a nonmagnetic ground state (no ESR signals), but when an electron is added to the cluster it becomes paramagnetic (total spin $S = {}^1/_2$). Due to the antiferromagnetic coupling, the g-value of this signal is lower than that of a free electron: 1.96 instead of 2.0023.

Isethionic acid: see Cysteine.

Isocitrate dehydrogenases: enzymes that catalyse dehydrogenation of the secondary alcohol group of isocitrate, with simultaneous decarboxylation, to form 2-oxoglutarate. The NAD-specific form (EC 1.1.1.41) is found only in mitochondria, where it is an enzyme of the TCA cycle. This form is activated by ADP (animal tissues) or AMP (yeast, molds), and inhibited by ATP; it catalyses the reaction only in the direction of 2-oxoglutarate and does not decarboxylate added oxalosuccinate. The NADP-specific form (EC 1.1.1.42) is found in both mitochondria and cytoplasm; it requires magnesium or manganese, also decarboxylates added oxalosuccinate, does not take part in the TCA cycle, and serves to produce reducing power for biosyntheses.

Isocitric acid: HOOC–CH_2–CH(COOH)–CHOH–COOH, a monohydroxy tricarboxylic acid, an isomer of citric acid, which is widely distributed in the plant kingdom and occurs in free form especially in plants of the stone-crop family *(Crassulaceae),* and in fruits. The salts of I.a., isocitrates, are important metabolically as intermediates in the Tricarboxylic acid cycle (see), where they are formed from citrate by the enzyme aconitase, then oxidized to 2-oxoglutarate. In the Glyoxylate cycle (see), isocitrate is cleaved to succinate and glyoxylate.

Isoelectric focusing: see Proteins.

Isoelectric point, *pI:* the pH value of a solution at which the net charge on the dissolved ampholyte is zero, i.e. the sum of the cationic charges is equal to the sum of the anionic charges. The pI of electrolytes, e.g. amino acids, peptides or proteins, may lie in the range from pH 1 (pepsin) to pH 11.8 (protamine), and is characteristic for each ampholyte. Certain characteristic properties appear at the pI, e.g. a minimal solubility and viscosity. Electrophoretic methods of separation are based on the differences in the pI of the individual components. The pI. can be determined either electrophoretically at various pH values or by electrofocusing on an ampholine pH gradient.

Isoenzymes, *isozymes:* multiple forms of an enzyme with the same substrate specificity, but genetic differences in their primary structures. If there are no differences in primary structure, one speaks of Pseudoisoenzymes (see). I. often differ in their isoelectric points (charge isomers), and sometimes in their M_r (size isomers, e.g. glutaminase I. from *Pseudomonas,* glutamate dehydrogenase I. from *Chlorella*). In oligomeric I., these differences are localized in their subunits. Other differences are found in catalytic properties, e.g. in the K_m values (see Enzyme kinetics), in the pH and temperature optima and heat lability, response to effectors, immunological behavior, patterns of distribution in different organs and cell components. I. may be developmental Isoforms (see), or they may perform similar tasks within different tissues of an adult organism. They may be genetically independent products of different genes, e.g. the mitochondrial and cytosolic malate dehydrogenases, or they may arise from differences in the control sequences in the DNA or RNA which govern transcription or RNA processing (see Post-transcriptional modification of RNA). In the case of proteins which consist of non-identical subunits, I. arise through formation of hybrid forms, e.g. heart and muscle Lactate dehydrogenase (see). I. also arise from genetic variation of the enzyme (alleles), e.g. human glucose-6-phosphate dehydrogenase, of which more than 50 genetic variants are known. The term I. is also used for enzymes of the same catalytic activity which can be separated by suitable methods, such as electrophoresis. Due to their differences in charge, I. can usually be separated by electrophoresis, isoelectric focusing or ion exchange chromatography. If in addition there are differences in size, the I. can be separated by zonal centrifugation in density gradients (e.g. isocitrate dehydrogenase I.), or by gel filtration. I. with different binding affinities for inhibitors, e.g. carboanhydratase I., can be separated by affinity chromatography. Likewise, affinity methods employing monoclonal antibodies can be used to separate and identify I.

Isoflavan: a compound containing the isoflavan skeleton (Fig.), which is the most reduced of the Isoflavonoid (see) ring systems. 3*R* and 3*S* configurations both occur naturally. Where pterocarpans and I. are present in the same plant, both have the same configuration, thus indicating that they are biogenetically related. The first reported natural I. was the animal metabolite, equol (see Isoflavone). All other known natural I. are plant products, e.g. ***(–)-duartin*** [(3*S*)-7,3′-dihydroxy-8,4′,2′-trimethoxyiisoflavan, wood of *Machaerium* spp.], ***(–)-mucronulatol*** [(3*S*)-7,3′-dihydroxy-4′,2′-dimethoxyisoflavan, wood of *Machaerium* spp.], ***(+)-vestitol*** [(3*S*)-7,2′-dihydroxy-4′-methoxyisoflavan, wood of *Machaerium vestitum, Dalbergia variabilis*], ***(–)-vestitol*** [(3*R*) configuration, *Cyclolobium claussenii, C. vecchi*], ***(+)-laxifloran*** [(3*R*)-7,4′-dihydroxy-2′,3′-dimethoxyisoflavan, *Lonchocarpus laxflorus*], ***(+)-lonchocarpan*** [(3*R*)-7,4′-dihydroxy-2′,3′,6′-trimethoxy-isoflavan, *Lonchocarpus laxiflorus*], ***Phaseolin isoflavan*** [(3*R*)-7,2′-dihydroxy-3′,4′-dimethylchromenyl-isoflavan, *Phaseolus vulgaris*], ***(+)-licoricidin*** [5,4′,2′-trihydroxy-7-methoxy-6,3-di-γ,γ-dimethylallylisoflavan, *Glycyrrhiza glabra*], and (3*S*)-2′-hydroxy-7,4′-dimethoxyisoflavan (*Dalbergia ecastophyllum*).

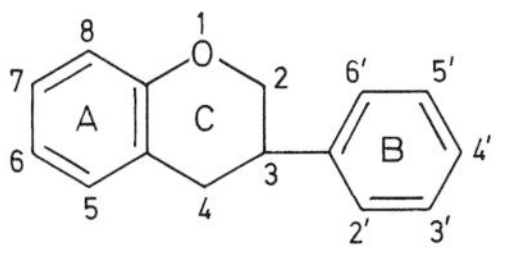

Isoflavan ring system

Compared with the corresponding isoflavones and isoflavanones, I. are relatively effective fungicides.

[Krämer et al. *Phytochemistry* **23** (1984) 2203–2205. General reference: *The Flavonoids,* J.B. Harborne T.J. Mabry & H. Mabry, eds. (Chapman and Hall, 1975)]

Isoflavanone: an isoflavonoid with the ring system shown (Fig.). Relatively few naturally occurring I. are known. Like other Isoflavonoids (see), I. are restricted to various subfamilies of the *Leguminoseae.* Examples: ***padmakastein*** (5 4′-dihydroxy-7-methoxyisoflavanone, bark of *Prunus puddum*), ***ferreirin*** and ***homoferreirin*** (5,7,6′-trihydroxy-4′-methoxy- and 5,7-dihydroxy-4′,6′-dimethoxyisoflavanone, respectively, from heartwood of *Ferreirea spectabilis*), ***sophorol*** (7,6′-dihydroxy-3′,4′-methylenedioxyisoflavanone, *Maakia amurensis*). [*The Flavonoids,* J.B. Harborne, T.J. Mabry & H. Mabry, eds. (Chapman and Hall, 1975)]

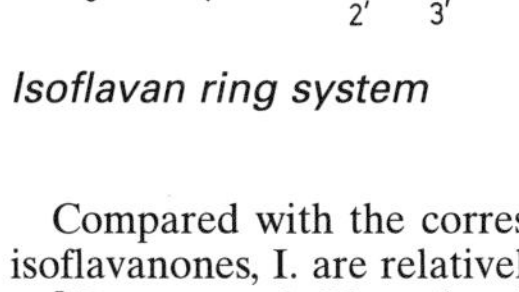

Isoflavanone ring system

Isoflav-3-ene: an Isoflavonoid (see) with the ring system shown (Fig. 1). I. have been known chemically for many years as dehydration products of isflavonols; the first naturally occurring I. was reported in 1974. I. are probably biosynthetic precursors of coumestans. Examples: ***haginin B*** and ***haginin A*** (7,4′-dihydroxy-2′-methoxyisoflav-3-ene and 7,4′-di-hydroxy-2′,3′-dimethoxyisoflav-3-ene, respectively, both from *Lespedeza cyrtobotrya*), ***sepiol*** (7,2′,3′-trihydroxy-4′-methoxyisoflav-3-ene) and ***2′-methylsepiol*** (both from *Gliricidia sepium*), ***neorauflavene*** (Fig. 2) and ***glabrene*** (Fig. 2). [Dewick in *The Flavonoids: Advances in Research,* J.B. Harborne & T.J. Mabry, eds. (Chapman and Hall, 1982)]

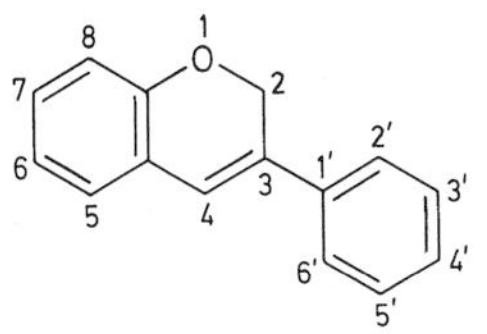

Fig. 1. *Isoflav-3-ene ring system*

Isoflavone: one of the naturally occurring Isoflavonoids (see) based on the isoflavone ring system (Fig.). I. are biosynthesized from the corresponding chalcones by migration of the B-ring from C2 to C3 during formation of the oxidized chroman ring system (see Isoflavonoid).

Genistein (5,7,4′-trihydroxyisoflavone) was isolated as early as 1899 from the coloring matter of dyer's broom (*Genista tinctora*), then subsequently identified in subterranean clover (*Trifolium subterraneum*), soybean, and the fruits of *Sophora japonica.* Genistein possesses estrogenic activity, and is responsible for infertility in Australian sheep grazing on pastures of subterranean clover, and for the "spring flush" of milk production in dairy cows in Britain. A major metabolite of genistein in animals is equol (Fig.) [M.N. Cayen et al. *Biochim. Biophys. Acta* **86** (1964) 56–64]. Slight estrogenic activity (may increase uterine weight when administered to immature mice) is shown by ***biochanin A*** (5,7-dihydroxy-4′-methoxyisoflavone, *Cicer arietinum, Ferreirea spectabilis, Trifolium* spp.) and ***prunetin*** (5,4′-dihydroxy-7-methoxyisoflavone, *Prunus puddum, P. avium, Pterocarpus angolensis*). Definitive proof of estrogenic activity is lacking for all other I.

Further examples of I. with simple substituents are ***daidzein*** (4′,7-dihydroxyisoflavone, *Pueria* spp. and other legumes), ***orobol*** (3′,4′,5,7-tetrahydroxyisoflavone, roots of *Lathyrus montanus*), formononetin (7-hydroxy-4′-methoxyisoflavone, soybean, *Trifolium subterraneum, T. pratense*), ***muningin*** (6,4′-dihydroxy-5,7-dimethoxyisoflavone, *Pterocarpus angolensis*), ***afrormosin*** (7-hydroxy-6,4′-dimethoxyisoflavone, *Afrormosia elata*), ***tlatlancuayin*** (5,2′-dimethoxy-6,7-methylenedioxyisoflavone, Mexican "tlatlancuaya" *Iresine celosioides*). As shown by the foregoing list, I. are largely restricted to the *Papilionoideae,* a subfamily of the *Leguminoseae.* Notable exceptions are the I. of *Iris* spp. (*Iridaceae*): 4′,5,7-trihydroxy-3′,6-dimethoxyisoflavone (*I. germanica*), 7,5,3′-trihydroxy-6,5′-dimethoxyisoflavone (trivial name, ***iristectorigenin,*** *I. tectorum, I. spuria*), 5,7-dihydroxy-6,2′-dimethoxyisoflavone (*I. spuria*) [A.S. Shawl et al. *Phytochemistry* **23** (1984) 2405–2406]. ***Orobol*** and ***pratensein*** (5,7,3′-trihydroxy-4′-methoxyisoflavone) have been reported in *Bryum capillare,* where they exist as 7-O-glucosides and predominantly as 7-(6″-malonyl)-glucosides; this is the first report of I. in a bryophyte [S. Anhut et al. *Phytochemistry* **23** (1984) 1073–1075].

Like other flavonoids, I. exist in the plant as glycosides, the most frequent carbohydrates being glucose

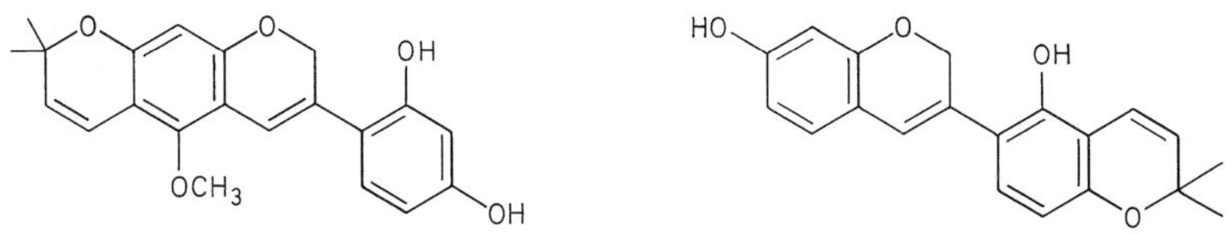

Fig. 2. *Isoflav-3-enes* [Neorauflavene (left) and Glabrene (right)]

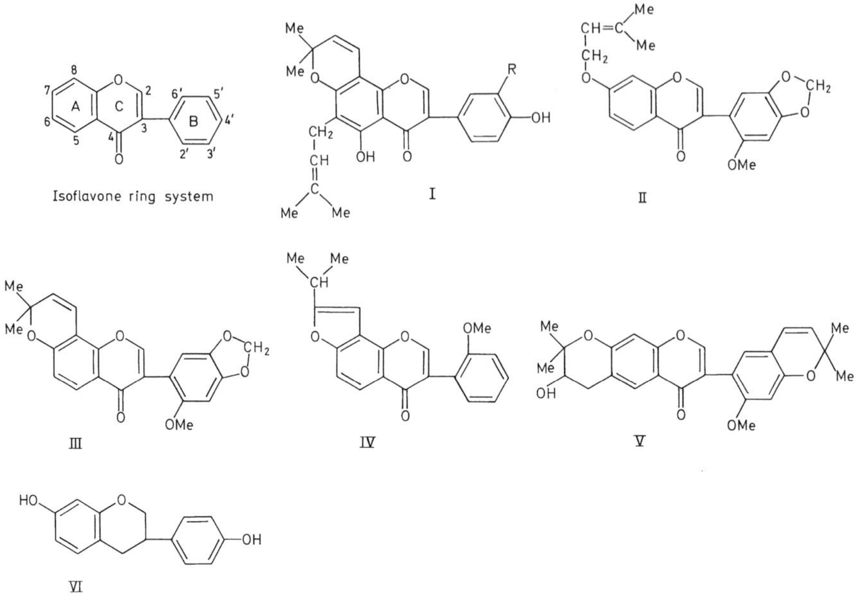

Isoflavone ring system

I

II

III

IV

V

VI

Isoflavones
I: R = H, Osajin; R = OH, Pomiferin (Osaje orange tree, *Maclura pomifera).*
II: Maxima substance C *(Tephrosia maxima).*
III: Jamaicin *(Piscidia erythrina).*
IV: Munetone *(Mundulea suberosa).*
V: Mundulone *(Mundulea sericea).*
VI: Equol *(metabolite of genistein in animals).*

and rhamnose. Examples are ***daidzin*** (daidzein 7-glucoside), ***ononin*** (formononetin 7-glucoside), ***genistin*** (genistein 7-glucoside), ***sophoricoside*** (genistein 4′-glucoside), *sophorabioside* (genistein 4′-rhamnoglucoside), ***prunetrin*** (prunetin 7-glucoside), ***iristectorin A*** (iristectorigenin 7-glucoside).

The formulas of some complex T. are shown (Fig.).

Isoflavonoid: a flavonoid with the branched $C_6C_3C_6$ skeleton shown (Fig.).

Most I. contain the 3-phenylchroman skeleton, in which the C_3 chain is cyclized with oxygen.

I. include isoflavones, isoflavanones, isoflavans, isoflav-3-enes, rotenoids, pterocarpans, coumestans, 3-aryl-4-hydroxycoumarins, 2′-hydroxy-3-arylcoumarins, 2-arylbenzofurans, α-methyldeoxybenzoins and individual compounds like lisetin and ambanol (see separate entries). I. have a restricted botanical distribution, occurring chiefty in the subfamily *Papilionoideae* of the *Leguminoseae,* and sometimes in the subfamily *Caesalpinioideae.* Their occasional presence in other families has also been reported (*Rosaceae, Moraceae, Amaranthaceae, Podocarpaceae, Chenopodiaceae, Cupressaceae, Iridaceae, Myristicaceae, Stemonaceae*). I. have also been reported from microbial cultures (e.g. T. Hazato et al. *J. Antibiot.* Tokyo **32** (1979) 217–222), but in all cases the culture media contained soybean meal, so the possibility remains that these I. are of plant origin.

Biosynthesis. All Flavonoids (see) are biosynthesized via chalcones. The 1,2 aryl migration which leads to the characteristic I. skeleton occurs during conversion of the chalcone, and is accompanied by net oxidation. In contrast, biosynthesis of all other flavonoids involves conversion of the chalcone to a flavanone of the same molecular formula. An attractive theory for the mechanism of aryl migration involves phenolic oxidation via a spirodienone intermediate (Fig. 2) [A. Pelter et al. *Phytochemistry* **10** (1971) 835–850]. In agreement with this proposed mechanism, 4-methoxychalcones do not act as substrates for aryl migration. In fact, it is possible that only two chalcones act as substrates, i.e. 2′,4′,4-trihydroxychalcone and 2′,4′,6′,4-tetrahydroxychalcone, which would give rise to daidzein and genistein, respectively. Formononetin and biochanin A can arise by methylation of daidzein and genistein, respectively, but there is strong evidence that 4′-methylation occurs mainly during conversion of the chalcone, i.e. during aryl migration. These four I. could then act as precursors of virtually all other known natural I. Although of rare

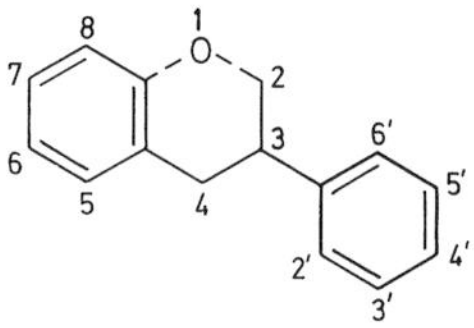

Fig. 1. *Isoflavonoids.* Isoflavonoid ring system.

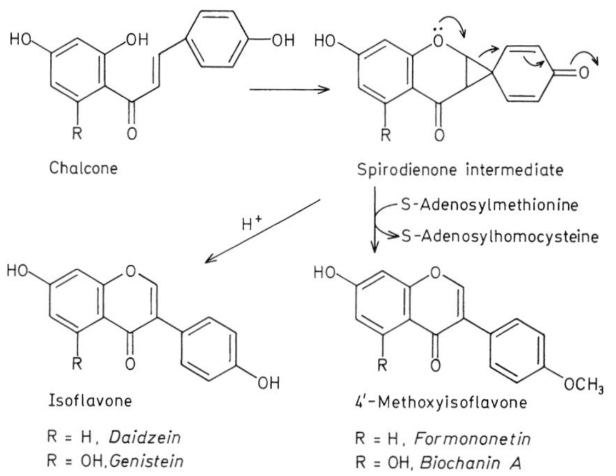

Chalcone

Spirodienone intermediate

H^+

S-Adenosylmethionine

S-Adenosylhomocysteine

Isoflavone

R = H, *Daidzein*
R = OH, *Genistein*

4′-Methoxyisoflavone

R = H, *Formononetin*
R = OH, *Biochanin A*

Fig. 2. *Isoflavonoids.* Proposed mechanism for cyclization and oxidation of chalcones to isoflavones.

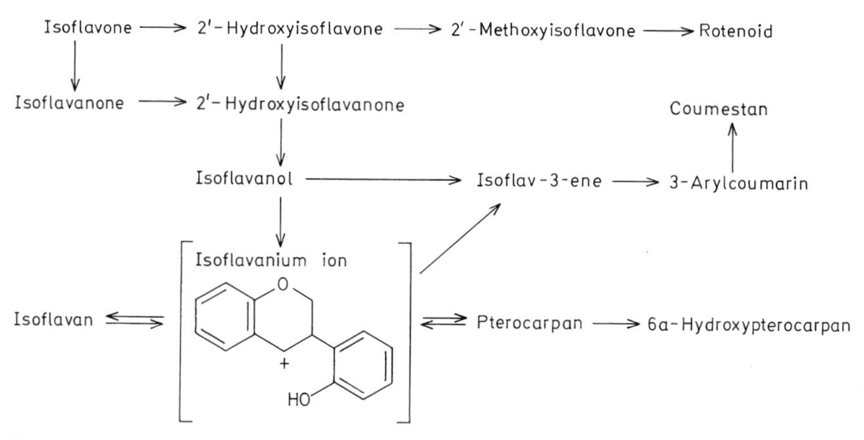

Fig. 3. *Isoflavonoids.* Biosynthetic relationships of the main groups of isoflavonoids. For formulas see individual entries.

occurrence, I. lacking an oxygen function at 4′ are also known. These may be formed by an analogous phenolic oxidation of 2-hydroxychalcones, leading to 2′-hydroxy (or methoxy) flavones. The B-ring substitution pattern appears to be determined in the isoflavone by hydroxylation sequences of 4′ → 2′,4′ and 4′ → 4′,5′ → 2′,4′,5′. A-ring substitution at 7 or 5,7 is determined by the chalcone, but further hydroxylation at 6 is also possible in the isoflavone. The biosynthetic relationships of the main groups of I., elucidated chiefly by isotopic tracing (^{14}C and ^{3}H), are shown in Fig. 2. [P. M. Dewick in *The Flavonoids: Advances in Research* J. B. Harborne & T. J. Mabry, eds. (Chapman and Hall, 1982) 535–640; J. L. Ingham, "Naturally Occurring Isoflavonoids (1855–1981)" in *Progress in the Chemistry of Organic Natural Products* **43** (1983) 1–266]

Isoforms: Different forms of cells or macromolecules which arise and replace each other sequentially during ontogeny. In mammals, for example, a myosin light chain present in skeletal and cardiac tissue is replaced by adult light chains which are specific for muscle type. In addition, the heavy chain of fetal myosin is replaced by a transient form of chain, neonatal heavy chain, which in turn is replaced by adult heavy chain. [A. I. Caplan et al. *Science* **221** (1983) 921–927.]

28-Isofucosterol: see Avenasterol.

Isohydric principle: see Buffer (section on Buffers of body fluids).

L-Isoleucine, ***Ile:*** L-α-amino-β-methylvaleric acid, CH_3–CH_2–CH(CH_3)–CH(NH_2)–COOH, an aliphatic, neutral amino acid found in proteins. Ile is found in relatively large amounts in hemoglobin, edestin, casein and serum proteins, and in sugar beet molasses, from which it was first isolated in 1904 by F. Ehrlich. It is an essential dietary amino acid, and is both glucoplastic (degradation via propionic acid) and ketoplastic (formation of acetate) (see Leucine). The biosynthesis of Ile starts with oxobutyrate and pyruvate. Oxobutyrate is synthesized by deamination of L-threonine by threonine dehydratase (threonine de-

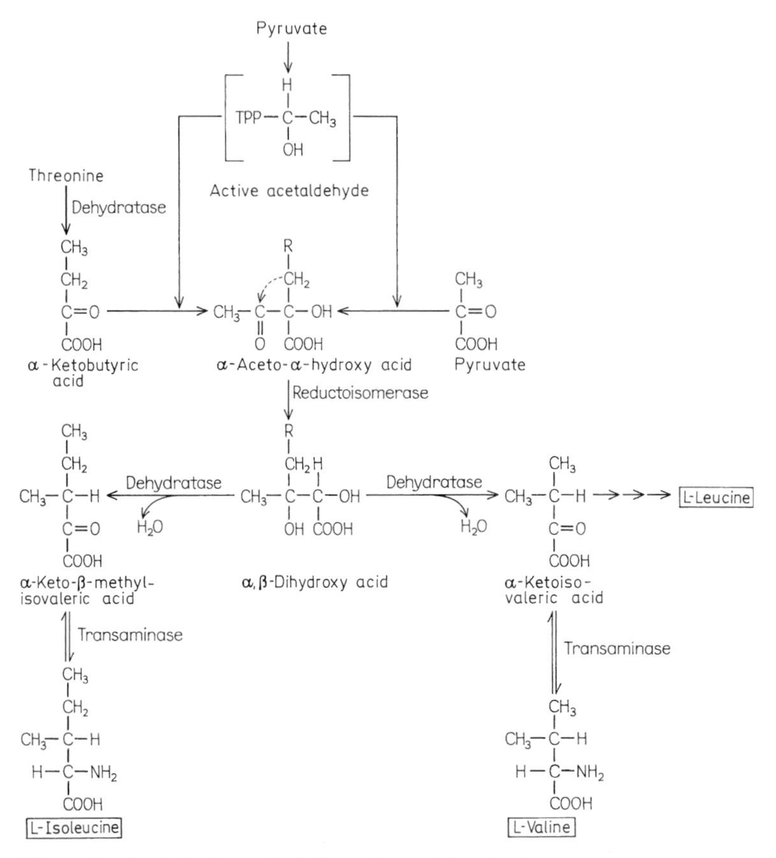

Biosynthesis of the branch-chain amino acids, L-isoleucine, L-valine and L- Leucine. TPP = thiamin pyrophosphate.

aminase). Ile and L-valine are synthesized by parallel pathways. The individual reaction steps (Fig.) are catalysed by the same enzyme (see Auxotrophic mutants). Biosynthesis of Ile diverges from that of the branched-chain amino acids at the level of the valine precursor, oxoisovalerate.

Isolysergic acid: see Lysergic acid.

Isomagnolol: see Neolignans.

Isomaltose: a reducing disaccharide, composed of two molecules of D-glucopyranose linked 1,6-glycosidically. I. is formed by the enzymatic degradation of branched polysaccharides, e.g. amylopectin. I. is a stereoisomer of gentiobiose.

Isomerases: see Enzymes (Table 1).

Isonicotinic acid hydrazide: see Nicotinic acid.

Isopelletierine: see Punica alkaloids.

Isopentenylpyrophosphate: an intermediate in the biosynthesis of Terpenes (see).

Isopentylacetate: $(CH_3)_2CH–CH_2–CH_2–O–CO–CH_3$, M_r 130, the most effective alarm pheromone (see Pheromones) of the honey bee. It is synthesized in the glandular tissue of the sting palps, and is released when the bee stings. Its odor attracts other bees.

Isopeptide bond: a covalent cross-linking bond between the ε-amino group of lysine and the side-chain carboxyl group of glutamate or aspartate, formed by condensation: $H_2N–CH(COOH)–CH_2–CH_2–COOH + H_2N–CH_2–CH_2–CH_2–CH_2–CH(NH_2)–COOH \rightarrow N^{\varepsilon}$-(γ-glutamyl)-lysine + H_2O. The I. b. has been found in polymerized fibrin and native wool. It is not hydrolysed by the body's own digestive proteases, but only by the bacteria in the large intestine. Its presence in food proteins therefore reduces their nutritional value.

Isoprene: see Terpenes, Hemiterpenes.

Isoprene rule: see Terpenes.

Isoprenoids: see Terpenes.

Isopycnic zone: see Density gradient centrifugation.

Isoquinoline alkaloids: a large group of alkaloids occurring widely in the plant kingdom. The heterocyclic skeleton is usually synthesized in vivo by a Mannich condensation between a phenylethylamine derivate and a carbonyl component. The resulting tetrahydroisoquinoline derivatives are converted by dehydrogenation into isoquinoline derivatives (Fig.). There are various structural types distinguished by the nature of the carbonyl group, which are typical for certain kinds of plants and are therefore named after them (Table). Compounds with more complica-

Classification of the isoquinoline alkaloids

Structural type	Typical examples
Tetrahydroisoquinoline alkaloids	Anhalonium (cactus) alkaloids
Phenylisoquinoline alkaloids	Amarylidae alkaloids
Benzylisoquinoline alkaloids	Poppy *(Papaver)* alkaloids, Erythrina alkaloids
Phenylethylisoquinoline alkaloids	Colchicum alkaloids
Bis-benzylisoquinoline alkaloids	Some curare alkaloids
Iridoid isoquinoline alkaloids	Ipecacuanha alkaloids

Phenylethylamine + Aldehyde → Tetrahydro-isoquinoline → Isoquinoline

Biosynthesis of isoquinolines

ted ring structures may arise by secondary cyclization due to phenol oxidation, and by rearangements and ring cleavages, and these can only be identified as I. a. by their common biosynthetic pathways.

Isorubijervine: a Veratrum alkaloid of the jerveratrum type (see Veratrum alkaloids). M_r 413.65, m.p. 273°C, $[\alpha]_D$ + 6.5° (alcohol). I. occurs in hellebores (*Veratrum album, V. escholtzii and V. viride*) and differs from solanidine (see α-solanidine) in having an additional 18-hydroxyl group. The glycoalkaloid ***isorubijervosine,*** from *V. escholtzii,* consists of I. linked to D-glucose.

Isosteviol: see Stevioside.

Isotope technique, ***tracer technique:*** the use of radioactive and stable isotopes (more exactly, nuclides) in biological, chemical and physical research, and in technology. Since atoms, groups of atoms or molecules are labeled by addition or chemical incorporation of indicator atoms, this technique is regarded as one of the indicator methods (see Methods of biochemistry). Both unstable radionuclides, which decay with a fixed half life and emit α, β or γ radiation, and stable nuclides may be used as markers. When a stable nuclide is used, it must be added in sufficient amount so that it can be detected above its natural background level, i.e. the rarer isotope, which is to serve as a marker, must be enriched. The term tracer technique for I. t. refers to the fact that one can follow the labeled atom in a mixture with unlabeled atoms of the same element, using suitable techniques for detection and measurement. The labeling guides the researcher just as a black ball among a number of white balls of the same size can always be recognized.

Radioactive or stable nuclides of all bioelements are available (Table) for use in biochemistry. There

Selected nuclides of the bioelements

Nuclide	Symbol	Half-life	Type of radiation
Hydrogen			
(Deuterium)	^{2}H	Stable	–
(Tritium)	^{3}H	10.46 years	β, very soft
Carbon	^{13}C	Stable	-
	^{14}C	5568 years	β, soft
Nitrogen	^{13}N	10.05 min	β
	^{15}N	Stable	–
Phosphorus	^{32}P	14.3 days	β
Sulfur	^{35}S	87.1 days	β, soft

are three main advantages of the I. t. for biological research: 1. the specific behavior of atoms and molecules can be followed; 2. when isotope effects are excluded, exact and very sensitive studies are possible; 3. the normal (physiological) behavior of a biological system is not, as a rule, affected by isotopes.

Radionuclides can be detected and their amounts determined by several methods (Geiger-Müller counter, thin-layer scanner, scintillation counter, etc.). The practical unit of measurement is counts per minute (cpm), which can be converted with the help of standards and the known efficiency of the counter into the actual number of disintegrations per minute (dpm). The Curie unit (Ci) is equivalent to 3.7×10^{10} disintegrations per second. For chemical and biochemical purposes, the millicurie (mCi) (2.22×10^{9} disintegrations per minute) and the microcurie (μCi) (2.22×10^{6} disintegrations per minute) are more convenient units. The Curie and its multiples should now be replaced by the Bequerel (Bq), which is equivalent to one disintegration per second. The megabequerel (10^{6} disintegrations per second) is more appropriate for biochemical purposes. Specific radioactivity is the number of disintegrations or impulses per time interval divided by the mass unit, e.g. cpm/mmol, cpm/g, mCi/mol, MBq/mol, etc. The specific incorporation, expressed in per cent (%), is the quotient of the specific radioactivity of the product, multiplied by 100, and the specific radioactivity of the isotopically labeled precursor. Stable nuclides are most easily determined by mass spectroscopy. For studies of biosynthesis, or precursor-product relationships, it is often not sufficient to determine the specific incorporation, especially if it is relatively low (1 % and lower), for the following reasons: an applied labeled compound can enter and be degraded by various metabolic pathways, so that the specific incorporation of the precursor into the metabolic product can only be proven by determination of the site of labeling within the product. This is achieved by chemical structural analysis, which must be developed and tested on a microscale. If the isotope can be located at a specific position in the product, the possibility of unspecific metabolic labeling or "smearing" of the isotope is excluded. For example, a compound labeled with ^{14}C could become metabolized to $^{14}CO_2$, which is then incorporated unspecifically by various carboxylation reactions into the reaction product.

Radioactive compounds are obtained in various ways: 1. by synthesis in the laboratory; 2. biosynthetically, using the specific synthetic capacities of organisms; 3. by radiochemical methods, e.g. tritiated compounds by Wilzbach tritiation. The second and third technique often yield unspecific, uniform labeling, but not necessarily an equal degree or labeling of all atoms. For example, in Wilzbach labeling, the C-T bonds are stable, but the O-T and N-T bonds are labile.

I.t. is applied in biochemistry as follows: 1. for the localization of metabolites, enzymes and metabolic reactions in the organism and in the cell; 2. to follow processes of uptake, transport and accumulation; 3. to determine physiological functions, e.g. thyroid activity with the aid of ^{131}I; 4. to measure the turnover of biomolecules and cell components, e.g. protein synthesis.

The combination of I.t. with chromatographic and histochemical or cytochemical techniques represents a powerful experimental technique known as autoradiography. When X-ray film is laid directly over a preparation containing radionuclides, their positions are indicated by darkening of the film. The technique has been extended to electron microscopic radioautography at the supermolecular and molecular level. Elucidation of the dark reactions of photosynthesis by M.Calvin and coworkers relied heavily on paper autoradiography, in which the radioactive products of $^{14}CO_2$ assimilation were separated by paper chromatography, and their positions determined by exposure of the chromatograms to X-ray film.

Isovaleraldehyde: $(CH_3)_2CH–CH_2–CHO$, a naturally occurring aldehyde. ρ 0.7977, b.p. 92.5 °C. I. is a colorless, highly reactive, sharp-smelling liquid which polymerizes easily in the presence of acid. It occurs in many aromatic oils, in particular in eucalyptus oils, and is obtained synthetically by oxidation of isoamyl alcohol.

Isovaleric acidemia: see Inborn errors of metabolism.

Isozymes: see Isoenzymes.

ITP: acronym of Inosine 5′-triphosphate.

IUB: acronym of International Union of Biochemistry.

IUPAC: acronym of International Union of Pure and Applied Chemistry.

IVS: acronym of intervening sequence. See Intron.

J

JAK/STAT signaling pathway: see Signal transducers and activators of transcription.

Jamaicin: see Isoflavone.

Jasmonic acid: 3-oxo-2-(2′-*cis*-pentenyl)-cyclopropane-1-acetic acid, $C_{12}H_{18}O_3$, M_r 210.272, a viscous oil, $bp_{0.001}$ 125 °C, $[\alpha]_D$ −83.5 ° (c = 0.97, $CHCl_3$). The volatile methyl ester ($bp_{0.001}$ 81–84 °C) is also present in plants and displays similar properties. The two compounds are jointly referred to as "jasmonate". Both occur in their readily interconvertible 1*R*,2*S*- and 1*R*,2*R*-isomeric forms, with the former predominating. Both have high biological activity as a signaling molecules, hormones or growth regulators, and the ester is the more potent; both are present in most organs of most plant species. Concentrations of J. a. range from 10 ng to 3 µg/g fresh weight, depending on the tissue and species, and it is particularly abundant in young organs. Its biosynthesis from α-linolenic acid involves lipoxygenase-mediated oxygenation, followed by additional modifications. Jasmonate initiates the de novo transcription of genes involved in the chemical defense mechanisms of plants, e. g. phenylalanine ammonia lyase. In the absence of elicitors it induces the synthesis of specific Phytoalexins (see), and it appears to have an integral role in the intracellular signal cascade that starts with the interaction of elicitor molecules with the plant cell surface and concludes with the accumulation of phytoalexins. After treatment with nmole quantities of J. a. and/or methyl jasmonate, cotyledons and leaves of various *Brassica* species show a sustained systemic increase (up to 20-fold) in the concentration of species-specific indole glucosinolates; mechanical wounding or damage by feeding insects elicit the same response. This effect is specific to jasmonate; other compounds associated with plant stress responses (e. g. abscisic acid, auxin, cytokinins) have no influence on indole glucosinolate production. Certain proteinase inhibitors induced by wounding in tomato and potato are also induced by jasmonate. It also induces various seed storage proteins, and oil body membrane proteins (oleosins) in developing embryos of *Brassica napus,* as well as promoting tendril coiling, ethylene synthesis and β-carotene synthesis. Some inhibitory effects of exogenously applied jasmonate may be physiologically irrelevant and due to the toxic effects of high doses (e. g. inhibition of callus growth, root growth, pollen germination, chlorophyll production). Although exogenous jasmonate has been reported to increase the rate of senescence of foliage leaves and to promote de novo synthesis of senescence-specific proteins, its greater abundance in young organs is inconsistent with a role in senescence.

The fragrant methyl ester is present in the essential oils of some flowers (e. g. jasmine) and fruits, and it is synthesized and used in large quantities by the perfumery industry. It is also an active component of the sex pheromone released by male oriental fruit moths.

[R. K. Hill et al. *Tetrahedron* **21** (1965) 1501–1507; F. Johnson et al. *J. Org. Chem.* **47** (1982) 4254–4255; P. E. Staswick *Plant Physiol.* **99** (1992) 804–807; H. Grundlach et al. *Proc. Natn. Acad. Sci. U. S. A.* **89** (1992) 2389–2393; R. P. Bodnaryk *Phytochemistry* **35** (1994) 301–305]

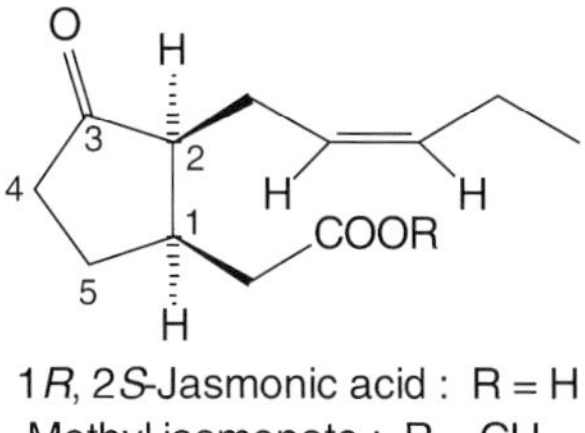

1*R*, 2*S*-Jasmonic acid : R = H
Methyl jasmonate : R = CH_3

Jasmonic acid and its methyl ester

Jerveratrum alkaloids: see Veratrum alkaloids.

Jervine: a jeveratrum type of Veratrum alkaloid with a C-nor-D-homo structure. M_r 425.62, m. p. 238 °C, $[\alpha]_D$ −147 °. J. is the main alkaloid of white and green hellebore (*Veratrum album and V. viride*).

Juglone: see Naphthoquinones (Table).

Jumping genes: see Transposon.

Juniperic acid: 16-hydroxypalmitic acid, $HOCH_2$–$(CH_2)_{14}$–COOH, a fatty acid, M_r 272.42, m. p. 95 °C. J. a. is a typical wax acid in the waxes of many gymnosperms, e. g. juniper (*Juniperus communis*).

Juvabione, ***paper factor:*** a monocyclic sesquiterpene ester from the wood of the North American balsam fir (*Abies balsamea*). (+)-J. is an oil, M_r 266, $[\alpha]_D^{20}$ +79.5 ° (c = 3.5, $CHCl_3$). J. was the first juvenile hormone isolated from a plant, and its structure was the first elucidated. It is specific for *Pyrrhocoris apterus* and only the dextrorotatory form is biologically active. The search for J. began when it was observed that filter paper made from the wood of the balsam fir contained a factor which produced developmental anomalies in the larvae of *Pyrrhocoris.* In addition to

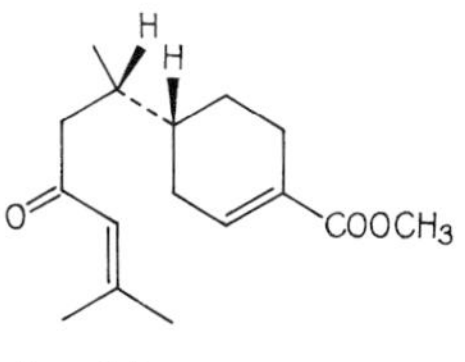

Juvabione

J., the wood of balsam fir from other areas also contains, dehydro-J., M_r 264, $[\alpha]_D^{20}$ +102.5° ($c = 3.6$, $CHCl_3$). In both compounds the steric arrangement of the side chain differs from the bisabolane type.

Juvenile hormone, ***larval hormone, status-quo hormone:*** insect hormone responsible for the control of molting. The first J. h. to be isolated and structurally elucidated was obtained from the abdomens of the male silk worm moth (*Hyalophora cecropia*) in 200 µg quantities. Homologs of this compound were discovered later, and other J. h. were postulated. Juvabione (see) and farnesol derivatives are among the natural compounds with J. h. activity. Some synthetic substances are more active than the J. h., and they are potentially useful in biological strategies for controling insect pests.

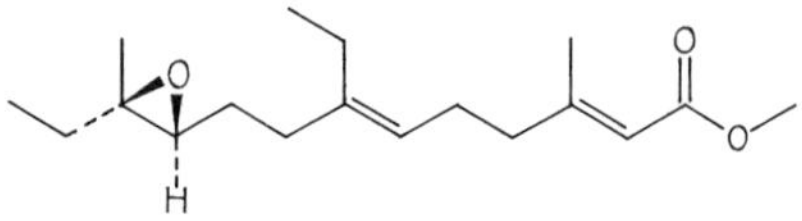

Cecropia juvenile hormone

K

Kachirachirol-B: see Neolignans.

Kaempferol: see Flavones (Table).

Kainic acid: an analog of the cyclic form of Glutamic acid (see) extracted from the seaweed *Digenea simplex.* It is an anthelminthic, and has become important as a selective neurotoxin in neurobiology. It destroys neurons, leaving axons and synapses intact. [E. G. McGeer et al. *Kainic Acid as a Tool in Neurobiology* (Raven, New York, 1978)]

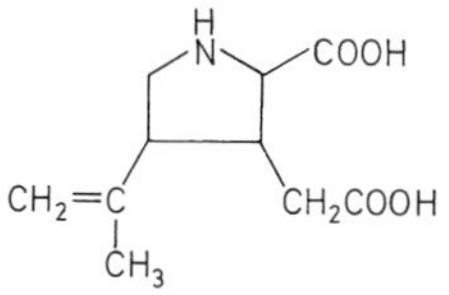

Kainic acid

Kairomones: see Pheromones.

Kallidin 10: see Bradykinin.

Kallikrein, *kininogenin, kininogenase:* a proteolytic enzyme (EC 3.4.21.8) which preferentially cleaves Arg and Lys peptides. There are at least three types. Plasma K. activates the Hageman factor and kininogen (see Blood coagulation) and releases Bradykinin (see) from kininogen. Pancreatic and submandibular K. release lysyl-bradykinin (kallidin) from kininogen, as well biologically active peptides other than kinins from high molecular mass plasma proteins. Urinary K. releases kallidin from kininogen and is antagonistic to the renin-angiotensin system (see Angiotensin).

Kanamycin: an aminoglucoside antibiotic. See Streptomycin.

Kappa-casein: see Milk proteins.

Kasugamycin: an aminoglucoside antibiotic. See Streptomycin.

Kaurene: ent-kaur-16-ene, a tetracyclic diterpene found in the plant kingdom in both the (+) and (–) forms. (–)-K. M_r 272, m.p. 50 °C, $[\alpha]_D^{20}$ –75 °. There are various stereoisomers of K., some of them occurring naturally. The formula and significance of K. as an intermediate in the biosynthesis of the gibberellin phytohormones are given under Gibberellins.

Kava: a narcotic drink made from the roots of the kava plant (*Piper methysticum* L.) in the Pacific islands. K. has a mild pain-killing and euphoric effect (see Narcotics). The active constituents have not all been elucidated, but those which have been isolated are α-pyrones, including dihydromethysticin and dihydrokawain.

kDNA: kinetoplast DNA. See Catenane, Kinetoplast.

Kendall's compounds: see 11-Dehydrocorticosterone, Corticosterone, Cortisone, Cortisol.

Keracyanin: see Cyanidin.

Keratan sulfate: an acidic mucopolysaccharide composed of *N*-acetyl-D-glucosamine 6-sulfate and D-galactose, linked alternately by β-1,3 and β-1,4 glycosidic bonds. K. s. is found in the cornea of the eye, in cartilage, in the aorta and in intervertebral discs. For formula, see Lysosomal storage diseases (Mucopolysaccharidoses).

Keratins: fibrous intracellular proteins which are components of intermediate filaments (IF) in epithelial tissues, and the proteins of hair, scales, silk, etc. At least three classes of proteins are known to form IF: acidic (Type I) keratins, neutral-basic (Type II) keratins, and vimentin, desmin and glial fibrillary acidic protein. Neurofilament proteins make up at least one other type. Around 30 different K. are known, with M_r 40,000–70,000. From amino acid sequence data, it can be seen that although IF subunits are diverse with respect to size and molecular properties, they all contain a central α-helical rod domain of 311–314 amino acids. The secondary structure of this domain is highly conserved; it is made up of repeated heptads $(a\text{-}b\text{-}c\text{-}d\text{-}e\text{-}f\text{-}g\text{-})_n$ of amino acids, where the residues a and d are usually apolar. The heptads coil into an α-helix with a stripe of apolar residues spiraling around the coil; two such helices associate by apposition of the apolar stripes. Assembly of K.IF in vivo and in vitro requires equimolar amounts of Type I and Type II subunits, and it seems probable that each pair of coils has one subunit of each type. The rod domains are 45–48 nm long; values reported for the diameter range from 7 to 15 nm. It seems plausible that the end domains of the polypeptide chains protrude from the dense core formed by the α-helical rod domains. Early evidence suggested that the basic structure of the protein was a left-handed helix of three α-helices twisted around each other, but more recently this has been shown to be a four-chain structure consisting of two double helices.

In contrast to the α-helical structure of the α-K. discussed above, the β-K. have β-pleated sheet structure. The most prominent representative of this class is silk fibroin (M_r 365,000, 2 subunits). Here the chains run antiparallel rather than parallel, and form a zig-zag structure. The formation of hydrogen bonds between the –CH(=O) and –NH– groups of neighboring chains stabilizes the pleated sheet structure. Together with weak hydrophobic interactions, the hydrogen bonds link pairs of polypeptides into a three-dimensional protein complex. These are additionally stabilized, in silk, by a water-soluble protein, sericin. The resultant fiber is very resistant and flexible, but only slightly elastic. The amino acid sequence which repeats over long stretches of the chain is, for silk fibroin, $(\text{Gly-Ser-Gly-Ala-Gly-Ala-})_n$.

Kermesic acid: an anthraquinone, m.p. 250 °C (d.), which occurs naturally as a bright red insect dye. It is structurally closely related to Carminic acid (see); it possesses the identical structure of a tetrahydroxylated methylanthraquinone carboxylic acid, but there is no C-glycosidic glucose on C2. K.a. makes up 1–2 % of kermes, the dried bodies of female scale insects *Kermococcus ilicis*. Kermes is one of the oldest known dyes and was used in ancient times as a scarlet mordant dye (Venetian scarlet). It was supplanted in the 16th century by cochineal.

Ketoacid, *oxoacid:* a carboxylic acid containing a carbonyl group (–C=O) in addition to the carboxyl group (COOH). Depending on the position of the carbonyl with respect to the carboxyl, the acid is referred to as an α-, β- or γ-ketoacid (2-, 3- or 4-oxoacid). If the two groups are adjacent, the compound is called an α-K.a. or a 2-oxoacid (e.g. pyruvic acid and α-ketoglutaric acid); if they are separated by one CH_2 group, the compound is a β-K.a. or a 3-oxoacid (e.g. acetoacetic acid).

α-Ketoacid dehydrogenase complex: see Multienzyme complex.

Ketoeleostearic acid: see Licanic acid.

Ketogenesis: the formation of ketone bodies. The primary product of K. is acetoacetate, which is synthesized in the liver from acetyl-coenzyme A via acetoacetyl-CoA and β-hydroxy-β-methylglutaryl-CoA. β-Hydroxybutyrate dehydrogenase (located in mitochondria) catalyses the conversion of acetoacetate to β-hydroxybutyrate. Acetone is formed by the spontaneous decarboxylation of acetoacetate (Fig. 1). Acetoacetate is also produced by degradation of the ketoplastic amino acids, leucine, isoleucine, phenylalanine and tyrosine.

In normal liver, only relatively small amounts of ketone bodies are formed. Their concentration in the blood is 0.5–0.8 mg per 100 ml plasma. The acetoacetate produced by this physiological K. is degraded in the peripheral musculature. Coenzyme A from succinyl-CoA is transferred to the acetoacetate by acetoacetate:succinyl-CoA transferase. Direct activation of acetoacetate by coenzyme A and ATP can also occur (Fig. 2). The acetoacetyl-CoA produced in either case is thioclastically cleaved into two molecules of acetyl-CoA, consuming a CoA molecule in the process. In carbohydrate deficiency (starvation, ketonemia in ruminants), or deficient carbohydrate utilization (diabetes mellitus), K. is greatly increased. The cause of this pathological K. is a disturbance of the equilibrium between the degradation of fatty acids to acetyl-CoA and its utilization in the tricarboxylic acid cycle. The several-fold increase in the oxidation of the fatty acids leads under these conditions to an increase in the intracellular acetyl-CoA concentration. This leads to the condensation of 2 molecules of

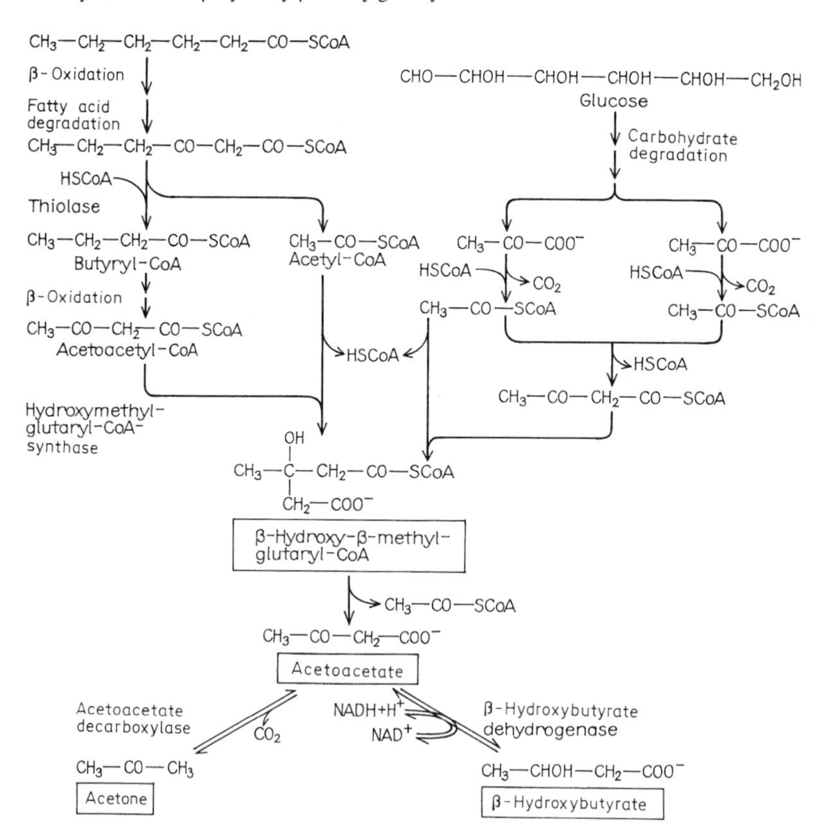

Fig. 1. *Ketogenesis.* Biosynthesis of the ketone bodies, acetoacetate, β-hydroxybutyrate and acetone.

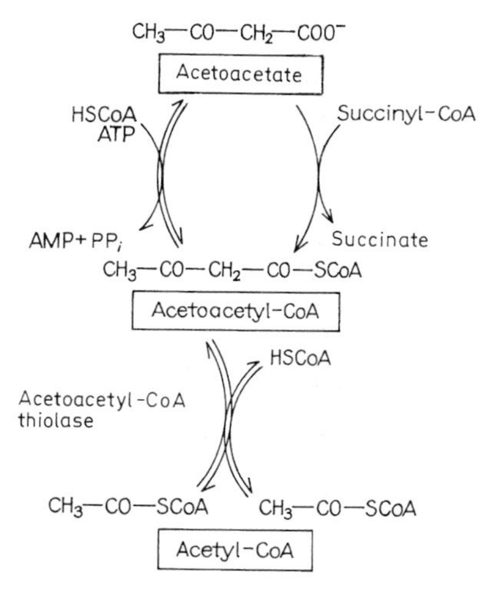

Fig. 2. *Ketogenesis.* Conversion of acetoacetate to acetylcoenzyme A.

acetyl-CoA to acetoacetate via the hydroxymethylglutarate cycle (Fig. 3), with the release of coenzyme A. Under pathological conditions, this cycle regenerates the coenzyme A which is required for the β-oxidation of fatty acids. This series of reactions thus takes over the role of the citrate synthase in the tricarboxylic acid cycle.

In an alternative pathway for K., acetoacetyl-CoA, which is a normal intermediate in the β-oxidation of fatty acids, acts as a precursor of acetoacetate; the enzyme thiolase is active in this process. Acetoacetyl-CoA reacts with acetyl-CoA to form hydroxymethylglutaryl-CoA, which is converted into acetoacetate.

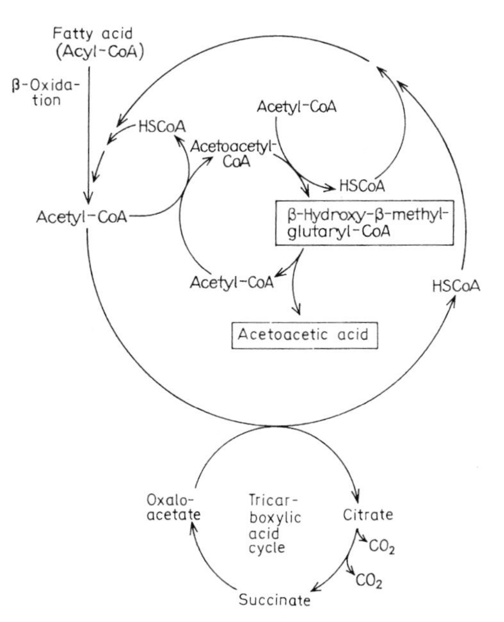

Fig. 3. *Ketogenesis.* Hydroxymethylglutarate cycle for regeneration of coenzyme A for the β-oxidation of fatty acids.

Ketogenic amino acids: see Amino acids.

α-Ketoglutarate dehydrogenase complex: see Multienzyme complex.

α-Ketoglutaric acid, ***2-oxoglutaric acid:*** HOOC–CO–CH_2–CH_2–COOH, a keto-dicarboxylic acid which represents an important branching point in the tricarboxylic acid cycle. α-K. a. is formed as its anion (α-ketoglutarate) by oxidative decarboxylation of isocitrate, by transamination of glutamate in amino acid metabolism, and by degradation of lysine via glutarate and α-hydroxyglutarate. Oxidative decarboxylation of α-K. yields succinyl-coenzyme A. Reductive amination of α-K leads to glutamate.

Ketone bodies: organic compounds produced by ketogenesis in living organisms. K. b. are acetoacetate and the compounds formed from it, β-hydroxybutyrate and acetone. Increased production of K. b. under certain pathological conditions, e. g. diabetes mellitus, leads to acidosis, because acetoacetate and β-hydroxybutyrate are present as anions, which reduce the concentration of HCO_3^- in the blood. Other consequences are excretion of the K. b. by the kidneys, production of acid urine, and damage to the central nervous system.

Ketoses: polyhydroxyketones, a subgroup of the monosaccharides (the other is aldoses). A characteristic feature of K. is their non-terminal –C=O group, which is given the lowest possible number in systematic numbering. The K. are formally derived from dihydroxyacetone. In all known natural K., the carbonyl group is in position 2. Depending on the number of carbons in their chain, they are classified as tetruloses, pentuloses, etc. They are either indicated by the ending "-ulose", e. g. ribulose, or have trivial names, like D-fructose.

Kidney: a paired mammalian organ with excretory and metabolic functions. Human K. are bean-shaped, weigh about 150 g. each, and lie in the posterior part of the abdominal cavity, on either side of the vertebral column. The functional unit of the K. is the nephron, consisting of a glomerular capsule and a long tubule. Each K. contains about a million nephrons. K. are responsible for maintaining the constancy of the internal milieu of the body. About 1.2 liters of blood pass through the two kidneys per minute, giving rise to about 120 ml of glomerular ultrafiltrate, i. e. protein-free primary urine, containing a wide range of low molecular mass, nonvolatile and inorganic and organic substances, including unmetabolized drugs. Many components of this ultrafiltrate are then passively or actively reabsorbed in the tubule: water (up to 99 %), glucose, amino acids and electrolytes. The daily urinary output is about 1.5 liters.

As a metabolic organ, the K. is responsible for the constancy of the acid-base balance of the body. The normal pH of the blood is 7.4, and it is important that this value is not increased (see Alkalosis) or decreased (see Acidosis). The K. contains amine oxidases and glutaminase, which catalyse the production of free ammonia. In acidosis, these reactions are brought into play and excess H^+ ions are consumed by formation of NH_4^+. In addition to the natural secretory processes, foreign substances, such as pharmaceuticals, can be removed from the blood by active transport. To support the large number and intensity of these active processes, the K. has a high oxygen consumption and high ATP production. K. function is regulated by the nervous system, and by three hormones: vasopressin increases water resorption in the distal part of the tubule by acting on the adenyl cyclase system; the mineralocorticoids, especially aldosterone, promote the resorption of Na^+; and parathormone favors the excretion of phosphate by the tubule itself.

The K. is the site of synthesis of Erythropoietin (see) and Renin (see). Renin is synthesized in the juxtaglomerular cells; it releases Angiotensin (see), which in turn stimulates the release of aldosterone by the adrenal cortex. 25-Hydroxycholecalciferol (produced in the liver from vitamin D_3 or cholecalciferol) is converted by the kidney into 1,25-dihydroxycholecalciferol, which promotes calcium uptake by the intestine and calcium mobilization in bone.

Kinases: enzymes which catalyse the transfer of a phosphate residue (strictly speaking a *phosphoryl* group is transferred) from ATP to another substrate, in particular to the alcoholic hydroxyl groups of monosaccharides. Some important K. are hexokinase, which phosphorylates several hexoses at the C6 position, glucokinase, which is responsible for glucose 6-phosphate formation in the liver, and phosphofructokinase.

Kinesin: a soluble protein isolated from the axoplasm of squid giant axons; apparent M_r 600,000. In the presence of the non-hydrolysable ATP analog, adenylyl imidodiphosphate, it binds tightly to microtubules, but it is released from them by ATP. K. apparently consists of subunits of M_r 110,000, 70,000 and 65,000. Polypeptides from bovine brain with similar binding properties and M_r 120,000 and 62,000 have also been isolated. In the presence of ATP, K. causes axoplasmic organelles to move along microtubules, but in one direction only, from the minus to the plus ends of the tubules (in nerve axons, microtubules are usually oriented with their minus ends pointing toward the cell body and their plus ends toward the nerve terminal). The axoplasm contains a further soluble factor which induces movement of organelles in the opposite direction; the motion induced by this factor differs from the K.-induced motion in its sensitivity to *N*-ethylmaleimide and vanadate. [R. D. Vale et al. *Cell* **42** (1985) 39–50; R. D. Vale et al. *Cell* **43** (1985) 623–632]

Kinetic data evaluation: Data evaluation by computer is performed by a nonlinear regression analysis, using an objective procedure, in which the sums of the squares of the differences between calculated and experimental values are minimized (method of least squares). If a reaction follows Michaelis-Menten kinetics, a plot of initial velocity ($\boldsymbol{v}$) against substrate concentration (S) gives a rectangular hyperbola of the form: $\boldsymbol{v} = V_{max}\,S/(K_m + S)$. For a first approximation in the laboratory, this equation can be rearranged, forming the basis of several linear transformations for the determination of V_{max} (maximal velocity) and K_m (Michaelis constant). All such graphical forms involve a certain degree of subjectivity.

The Lineweaver-Burk, or double reciprocal plot is widely used, but it is the least reliable of the possible linear transformations; small errors of $\boldsymbol{v}$ for small values of $\boldsymbol{v}$ result in very large errors in $1/\boldsymbol{v}$; these same small errors in large values of $\boldsymbol{v}$ become almost negligible in $1/\boldsymbol{v}$. The authors of this method pointed out the inherent error of high $1/\boldsymbol{v}$ values, and the necessity to apply higher weightings to the low values of $1/\boldsymbol{v}$, but this has often been ignored by subsequent users. It has been suggested that the Lineweaver-Burk plot should be abandoned as a method for the determination of K_m values.

Linear transformations of the equation $v = V_{max}S/(K_m + S)$, where v = initial velocity, V_{max} = maximal velocity when enzyme is saturated with substrate, K_m = Michaelis constant, S = starting substrate concentration. Also included is the Scatchard linear equation for ligand binding, where b = concentration of bound ligand, f = concentration of free ligand, K_d = dissociation constant, b_{max} = maximal concentration of b when ligand is saturating.

Name and No. of illustrated plot	Equation	Ordinate	Abscissa	Ordinate intersected at	Abscissa intersected at	Slope
Lineweaver-Burk, or double reciprocal plot (1.)	$\frac{1}{v} = \frac{K_m}{V_{max}} \times \frac{1}{S} + \frac{1}{V_{max}}$	$\frac{1}{v}$	$\frac{1}{S}$	$\frac{1}{V_{max}}$	$\frac{-1}{K_m}$	$\frac{K_m}{V_{max}}$
Eadie-Hofstee (2.)	$v = -K_m \frac{v}{S} + V_{max}$	v	$\frac{v}{S}$	V_{max}	$\frac{V_{max}}{K_m}$	$-K_m$
Hanes-Wilkinson (3.)	$\frac{S}{v} = \frac{1}{V_{max}} S + \frac{K_m}{V_{max}}$	$\frac{S}{v}$	S	$\frac{K_m}{V_{max}}$	$-K_m$	$\frac{1}{V_{max}}$
Eisenthal-Cornish Bowden, or direct linear plot (4.)	$V_{max} = v + \frac{v}{S} K_m$	v	$-S$	lines joining $S_1 - v_1$, $S_2 - v_2$, $S_n - v_n$; intersection at V_{max} and K_m		
Scatchard (5.)	$\frac{b}{f} = \frac{1}{K_d}(b_{max} - b)$	$\frac{b}{f}$	b	$\frac{b_{max}}{K_d}$	b_{max}	$\frac{1}{K_d}$

A more satisfactory distribution of error is found for S/v versus S, known as the Hanes plot, or the Hanes-Wilkinson plot (Table & Fig.). A further linear transformation is represented by v versus v/S, known as the Eadie-Hofstee plot. The error increases with v/S, but since v is a component of both coordinates, the errors vary with respect to the origin rather than the axis, i.e. all error bars converge on the origin (Table & Fig.).

A most satisfactory treatment of kinetic data is the direct linear plot of Eisenthal and Cornish Bowden (1974). Axes are drawn with –S on the abscissa and v on the ordinate, but instead of making the usual hyperbolic plot of S/v (see Enzyme kinetics), corresponding points (each reading of –S and its related v value) are joined by straight lines. The point of intersection of this family of lines gives the values of V_{max} and K_m. Mathematically, this plot corresponds to a rearrangement of the general equation: $V_{max} = v + vK_m/S$, in which V_{max} and K_m appear to be variables. In practice, however, only one value for V_{max} and K_m will satisfy all pairs of v and S. Owing to experimental error, the point of intersection of the lines is not always clearly defined but the best point is easily found where most of the lines crowd closest together. The advantage lies in the fact that each observation stands alone, and it is revealed as a bad observation if its line does not conform to the majority (Table & Fig.).

The Scatchard plot is frequently used for the determination of binding constants; the other plots (1, 2, 3 and 4) are also appropriate for this purpose. Such measurements are important, e.g. for determination of the strength of association between hormones and cell membranes, between control enzymes and effector molecules, and between steroid hormones and their high affinity receptor proteins in the cells of the target organ. The Scatchard plot is a special case of the Hill equation (see), where $n_H = 1$. The concentration of one binding partner is kept constant; this will usually be a protein, perhaps an organelle or even a cell. The concentration of the smaller ligand is then varied (Table & Fig.).

Kinetic equations: a system of differential equations which describes the changes with time of the concentrations of enzyme species as a function of the reaction rate constants, the concentrations of enzyme spe-

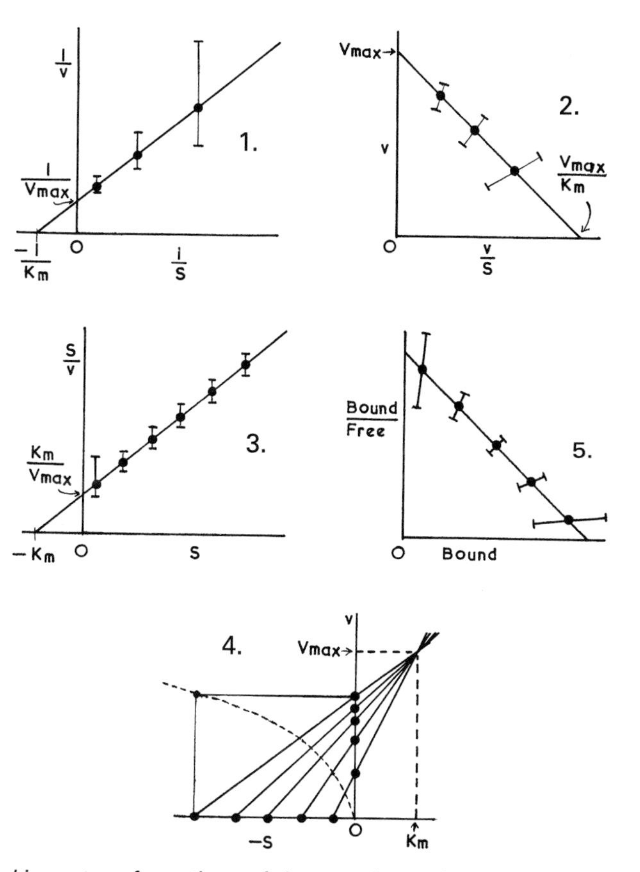

Linear transformations of the equation $v = V_{max}S/Km + S$. Error bars are also shown. 1. Lineweaver-Burk, or double reciprocal plot; 2. Eadie-Hofstee plot; 3. Hanes-Wilkinson plot; 4. Eisenthal-Cornish Bowden, or direct linear plot; 5. The Scatchard plot which is used for determination of ligand binding constants.

cies and the concentration variables. K. e. for the Michaelis-Menten scheme are: $dE/dt = -k_1S \cdot E + k_{-1}ES$; $dES/dt = k_1S \cdot E - (k_1 + k_2)ES$ (see Michaelis-Menten equation).

Kinetin: 6-furfurylaminopurine, the model substance for cytokinins. K., in conjunction with other factors, such as auxins, induces renewed cell division in resting plant tissue. It influences the nucleic acid and protein metabolism of the plant. In addition to many other physiological effects, it prevents yellowing of isolated leaves and promotes protein synthesis at its site of application. K. is obtained by hydrolysis of deoxyribonucleic acid. The 2-deoxyribose of the DNA provides the furfuryl residue of the K. It can also be made synthetically from 6-mercaptopurine and furfurylamine.

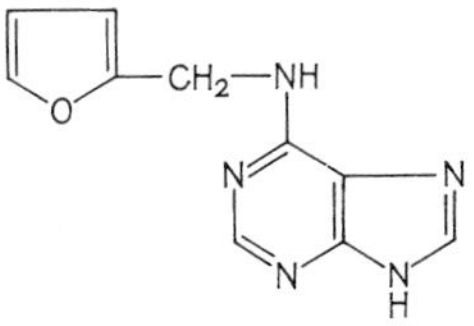

Kinetin

Kinetoplast: a structure at the base of the single flagellum of a trypanosome. The K. has a high affinity for basic dyes, and was first observed early in the 20th century by light microscopy. Electron microscopy shows that the K. is a disc-like structure in the matrix of the cell's single mitochondrion. The K. has attracted biochemical interest because it contains an unusual form of highly catenated DNA, known as kinetoplast DNA (kDNA). See Catenane.

Kinetoplast DNA: see Catenane.

King-Altman method: a method for derivation of rate equations according to simple rules. These rules come from the application of the determinant theory to the solution of inhomogeneous systems of linear equations. One first draws all possible geometric figures of the enzyme graphs which transform the various enzyme forms (enzyme species) into one another. The number of lines (edges) is 1 less than the number of enzyme forms. Circles and cycles are forbidden and are eliminated. The edges are assigned the appropriate reaction rate constants or the products of rate constants and concentration variables of the corresponding step of the reaction (e.g. $E \xrightarrow{k_1S} ES$) (edge analysis). According to the King-Altman rules, the following distribution equation then holds: Enzyme form/E_t = sum of the products of the edge analysis of all pathways leading to this enzyme form, divided by Σ. Here E_t is the total enzyme concentration, Σ the sum of the numerators of all distribution equations of the enzyme graph. The rate equation is then obtained by multiplication of the product-producing enzyme forms, such as EP, with the associated catalytic constant: $v = k_{cat}EP$, where k_{cat} is the catalytic constant, and EP the enzyme product complex. If there are several product-producing enzyme forms, the partial rates are added.

Kinins: see Cytokinins.

Kjeldahl technique: see Proteins.

Klenow fragment: a single polypeptide chain, M_r 76,000, produced by cleavage of *E. coli* DNA polymerase I with subtilisin. It carries the 5′→3′ polymerase activity and the 3′→5′ exonuclease activity, but the 5′→3′ exonuclease activity is lacking. K. f. is now available as a cloned protein. [H. Jacobsen et al. *Eur. J. Biochem.* **45** (1974) 623–627].

Koaburarin: (2*R*)-5-hydroxy-7-*O*-β-D-glucosyl-flavan. See Flavan.

Kornberg enzyme: see DNA polymerase.

Koshland model: see Cooperativity model.

Krabbe's disease: see Lysosomal storage diseases.

Kranz anatomy: see C_4 plants.

Krebs cycle: see Tricarboxylic acid cycle.

Krebs-Henseleit cycle: see Urea cycle.

Krebs-Kornberg cycle: see Glyoxylate cycle.

***k*-Tuple:** an oligonucleotide or peptide sub-sequence of length *k*.

Kurchi alkaloids: see Holarrhena alkaloids.

Kwashiorkor: a chronic form of malnutrition, occurring chiefly in the second year of life. The name is from the Ga language, and was introduced into modern medicine in 1933 by Cicely Williams [*Archs. Dis. Childh.* **8** (1933) 423]. The word means "deposed child", i. e. deposed from the breast by the advent of another pregnancy. The earlier literature emphasizes the relationship between K. and various tribal customs and taboos, which result in the administration of a low protein, high carbohydrate diet to children. It has now been shown, however, that there are no essential differences in the dietary protein/energy ratio of marasmus and K. victims [C. Gopolan, in R. A. McChance and R. M. Widdowson, eds. *Calorie deficiencies and Protein Deficiencies,* Churchill Livingstone (1968) pp. 49–58]. Like all other forms of protein-energy malnutrition, K. is often precipitated and exacerbated by microbial infections and intestinal worms. It is more common in rural than in urbanized areas of the developing world. K. victims show retarded growth and anemia. Plasma albumin concentrations fall below 20 or even 10 g/l (in marasmus, plasma albumin is usually about 25 g/l). The resulting decrease in the osmolarity of the blood is thought to be partly responsible for the accumulation of fluid in the body, and the watery, bloated state of the tissues; edema is a common feature of K. The hair becomes sparse, and especially in negro children, it may show patches or streaks ("flag sign") of red, blonde or gray. These hair lesions are probably due to a specific tyrosine deficiency, since the periodic administration of tyrosine to phenylketonurics causes very similar alternating bands of deeply pigmented hair. The skin shows a very characteristic dermatosis, with areas of pigmentation, depigmentation and peeling, the most severely affected areas being the lower limbs and buttocks. Muscles are always wasted, so that walking or crawling may not be possible. Fatty liver is also very characteristic of K.; an average lipid content of 390 g/kg liver has been reported (cf. 35 g/kg in marasmus); this increase is due to triacylglycerols, the level of phospholipids being relatively unaffected. In K., but not in marasmus, plasma levels of triacylglycerols and cholesterol are low. Subcutaneous fat is retained in K., in contrast to marasmus, in which it is severely depleted. Changes in plasma amino acids are similar in most types of protein-energy malnutrition. In more than 50 % of cases, a mixture of all the essential amino acids effects a complete cure of K. In the remaining cases the response is partial or negative.

K. is common in hot, humid areas, but not in dry, hot areas. It is especially prevalent in the wet season. Furthermore, it is never found in temperate regions. These observations suggest the contribution of other factors, in addition to protein deficiency. There is now strong evidence that K. is caused by a combination of malnutrition and aflatoxin poisoning [R. G. Hendrickse, *British Medical Journal* **285** (1982) 843–846; S. M. Lamplugh & R. G. Hendrickse, *Annals of Tropical Paediatrics* **2** (1982) 101–104]. In hot, humid countries of the developing world, many market foods contain aflatoxins, especially the groundnut oil used for cooking. Well nourished infants can degrade and excrete these relatively small quantities of toxins, but this ability is weakened by protein deficiency. In the resulting vicious cycle, accumulation of aflatoxins causes further liver damage and a marked decrease in the protein synthesizing ability of the liver.

Kynureninase deficiency: see Inborn errors of metabolism.

Kynurenine: 3-anthraniloylalanine, M_r 208.2. An intermediate in Tryptophan (see) degradation.

L

Labdadienyl pyrophosphate: an intermediate in the biosynthesis of the Diterpenes (see).

Labeling site: see Isotope technique.

Laccase: see Monophenol monooxygenase.

Lac repressor protein: the first repressor substance (product of a repressor gene) to be isolated and characterized. L.r.p. is an acidic (pI 5.6) allosteric protein of M_r 152,000. It consists of four identical subunits (M_r 38,000, 347 amino acids, sequence and 3-dimensional structure known). Its primary structure has recognizable similarities to histones or to β-galactosidase. L.r.p. is transcribed from a particular gene in *E. coli* which regulates the synthesis of three enzymes of lactose metabolism (the *lac* region). It inhibits the transcription of the lactose operon (see Operon) by specifically binding to the operator gene. Allolactose, which is formed from the substrate lactose, binds the L.r.p., and causes it to be released from the operator and inactivated. See Enzyme induction. [R.T. Sauer 'Lac Repressor at last' *Structure* **4** (1996) 219–222]

***Lac* system:** the region of the genome of *E. col* and other enterobacteria which controls the ability to utilize lactose and other β-galactosides. It consists of the structural genes *lacZ* for β-galactosidase, *lacY* for galactoside permease and *lacA* for thiogalactoside transacetylase, an operator, a promoter and a regulator gene which is responsible for the synthesis of the *lac* repressor (see Repressor).

The enzymes of the L.s. are inducible, i.e. they are synthesized only when β-galactosides are present in the medium (see Enzyme induction).

The *lac* operon comprises less than 0.1 % of the *E. coli* DNA. The gene *lacZ* was the first ever to be isolated in pure form, which was accomplished in 1969 by Beckwith. This section of DNA, together with the short operator region (about 27 nucleotide pairs), is about 1.4 µm long and contains about 4,000 nucleotide pairs.

α-Lactalbumin: see Milk proteins.

Lactate dehydrogenase, *LDH, lactic acid dehydrogenase, (S)-lactate: NAD⁺* oxidoreductase (EC 1.1.1.27): an oxidoreductase and a much studied isoenzyme. LDH catalyses an NAD- or NADH-dependent side reaction of glycolysis: lactic acid ⇌ pyruvic acid. The enzyme is absolutely specific for L(+)-lactate, and shows no activity toward the D(−)-isomer. The highest LDH activities are found in heart muscle and liver. LDH (M_r 140,000) consists of four subunits of equal size (M_r 35,000). There are two types of subunits, differing in charge, catalytic properties and organ specificity, the heart-muscle (H) and muscle (M) types. The five-fold isomerism of LDH is due to the five possible combinations of the two types of subunit in a tetramer: H_4, H_3M, H_2M_2, HM_3 and M_4 (the latter two in skeletal muscle). Because they have different pI values, the 5 forms can be separated electrophoretically. The H_4 form has the largest negative charge. The two subunit types differ in their susceptibility to inhibition by pyruvate (H-type is strongly inhibited, M-type weakly), which serves to regulate glycolysis and the tricarboxylic acid cycle. H_4 and H_3M are, in addition, heat-labile, so that they are 100 % inactivated after 5 min at 65 °C. In addition to these 5 bird and mammal LDHs, the sperm cells of these animals contain a sixth type of isoenzyme, LDH X. This LDH, which is also tetrameric, contains a third type of polypeptide chain and has a wider substrate specificity than the other LDHs. An LDH of the same size (M_r 140,000), but without subunit structure or isoenzyme properties, has been demonstrated in the shrimp *Artemia salina*.

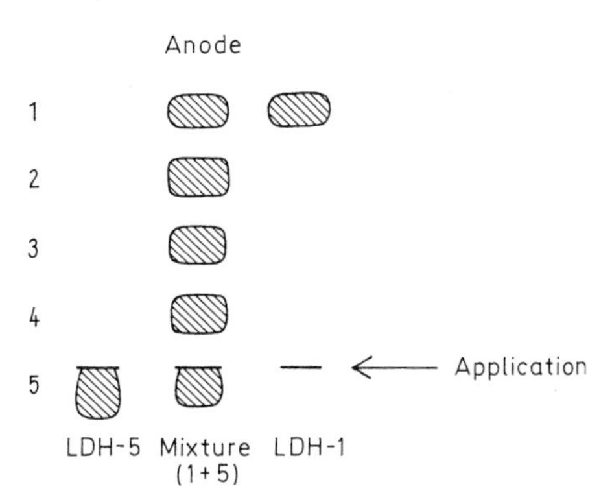

The 5 molecular variants (allomers, isoenzymes) of LDH. The diagram shows separation of the variants by starch gel electrophoresis. 1 to 5 (center) were produced artificially by random recombination of the subunits of the homogeneous tetramers LDH-5 (M-type, left) and LDH-1 (H-type, right).

LDH is used for diagnosis of heart infarction and hepatitis, since its level in serum is considerably elevated in both these diseases. Purified LDH is used in coupled optical tests for the assay of other enzymes, e.g. pyruvate kinase, enolase, transaminases, and for the enzymatic determination of many metabolites, such as ADP, ATP, L-lactate and pyruvate.

Lactic acid: CH_3–CHOH–COOH, an aliphatic hydroxy acid, widely distributed in plants, especially seedlings.

DL-form: m.p. 18 °C, b.p. 119 °C at 12 Torr. L-(+)-form and D-(−)-form: m.p. 25–26 °C. The salt (lactate) of the dextrorotatory L(+)L.a. is the end product of anaerobic Glycolysis (see), and a substrate of Gluconeogenesis (see). Increased degradation of glycogen in contracting muscle can lead to an increase of blood L(+)L.a. from 5 mg% to100 mg%. In the subsequent rest period, most of this L. a. is used for the synthesis of glycogen by the liver. In microbial fermentation, the DL-form of L. a. is produced. [A. Weltman *The Blood Lactate response to Exercise.* Current Issues in Exercise Science, Monograph No. 4, Human Kinetics 1995; ISBN 0-87322-769-7]

Lactoferrin: see Milk proteins.

Lactoflavin: see Riboflavin.

β-Lactoglubulin: see Milk proteins.

Lactose, *milk sugar:* a reducing disaccharide. M_r 324.3, α-form m.p. 223 °C, $[\alpha]_D^{20}$ + 89.4 ° → + 55.5 ° (water), β-form m.p. 252 °C, $[\alpha]_D^{20}$ + 34.9 ° → 55.3 ° (water). L. crystallizes from water as the β-form above 93 °C, and as α-lactose monohydrate below 93 °C. It consists of galactose β-1,4 glycosidically linked to glucose, both monosaccharide residues in the pyranose form. L. is not fermented by ordinary yeasts, but it is fermented by yeasts such as kefir. The souring of milk consists of the conversion of L. into lactic acid by lactic acid bacteria.

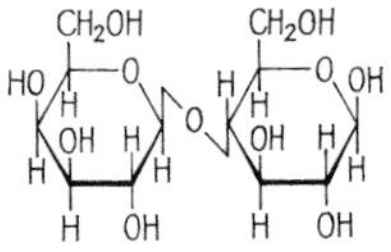

β-Lactose

L. is the most important carbohydrate in the milk of all mammals. Human milk contains 6–8 %, cow's milk 4–5 %. L. is a component of a few oligosaccharides, and also occurs in plants, e. g. in fruits and pollen. In L. synthesis in the milk glands of mammals, the galactose residue of uridine diphosphate β-galactose is transferred and linked to the OH on C4 of D-glucose, a process catalysed by lactose synthetase.

L. can be cleaved by the action of β-galactosidase or by acid hydrolysis. It is prepared by evaporation of the whey from cheese production; first the lactalbumin and then L. precipitates. L. is used as a starting material in pharmaceutical preparations, and as a carbon source in the culture of microorganisms, e. g. in the production of penicillin.

Lactose intolerance: see Inborn errors of metabolism.

Lactosyl ceramidosis: see Lysosomal storage diseases.

Lactotropin: see Prolactin.

Lactucerol: see Taraxasterol.

Laki-Lorand factor: see Blood coagulation (Table).

Laminin: a large, 3-chain, adhesive glycoprotein (M_r 850,000–1,000,000) of the extracellular matrix. It is found specifically in the basal lamina, and enables epithelial cells to bind to underlying connective tissue. It consists of a number of functional domains, one binding with high affinity to collagen IV of the basement membrane, one to the proteoglycan, heparan sulfate, and others binding to laminin receptor proteins on cell surfaces. The 3 chains are held together by disulfide bonds, and, as seen in the electron microscope, the entire molecule is arranged in the shape of a cross with one long arm and three short arms. The three chains (A, B_1, B_2) form a long α-helical domain, then separate in different directions to form the three shorter arms of the cross.

At least 7 isoforms of laminin are known, each consisting of 3 different chains. Most extensively characterized of these is laminin-1 (M_r 900,000) from mouse Engelbreth-Holm-Swarm tumor. [A. Utani et al. *J. Biol. Chem.* **270** (1995) 3292–3298; M. Nomizu et al. *J. Biol. Chem.* **270** (1995) 20583–20590; C. Matsui *J. Biol. Chem.* **270** (1995) 23496–23503; A. R. E. Shaw et al. *J. Biol. Chem* **270** (1995) 24092–24099]

Lamp-brush chromosomes: very large chromosomes, up to 1 mm long, which occur in the meiotic prophase in newts and a few other animals. They form loops along the sides, which give them a brush-like appearance. The loops are not fixed structures; they represent unwound, individual chromomeres, consisting of DNA, protein and RNA. They appear at the sites of active RNA synthesis, thus indicating increased physiological activity in that particular chromosome section (see Gene activation).

Lanatosides: see Digitalis glycosides.

Lanolin, *wool fat, wool wax:* the fatty or more correctly waxy substance secreted by the skin of the sheep, m. p. 36–42 °C. L constitutes up to 50 % of the weight of raw wool. It is a complicated mixture of fatty acids, alcohols, fats and waxy substances. The latter are chiefly esters of steroids (cholesterol and lanosterol) and long chain aliphatic alcohols with higher fatty acids, which are δ-hydroxylated or carry a terminal isopropyl or isobutyl residue. L. is obtained from raw wool by extraction with organic solvents or soap solutions. It forms water-in-oil suspensions, and is used widely in the pharmaceutical and cosmetics industries (as Adeps Lanae) as an ointment base.

5α-Lanostane: see Lanosterol.

Lanosterol, *kryptosterol:* 5α-lanosta-8(9),24-dien-3β-ol, a tetracyclic triterpene alcohol. M_r 426.7, m. p. 140 °C, $[\alpha]_D$ + 60 ° ($CHCl_3$). L. is also a zoosterol (see Sterols) present in large amounts in the wool fat of sheep. Structurally, it is based on the hydrocarbon 5α-lanostane ring system. It is biosynthesized from squalene, via 2,3-epoxysqualene, and is an important intermediate in the synthesis of all further tetracyclic triterpenes of the lanostane type and of the steroids.

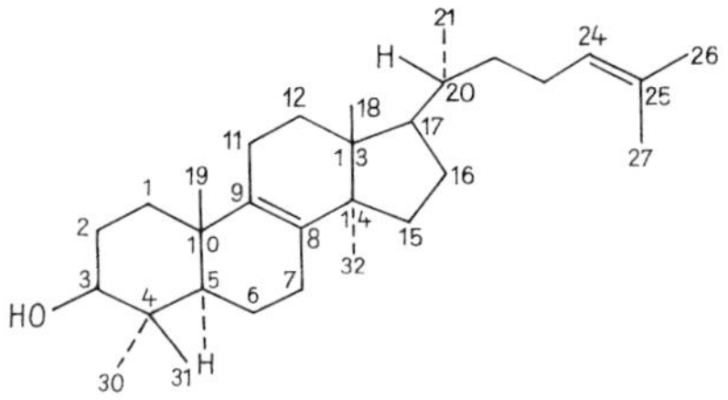

Lanosterol

Lanthanide ion probes: The trivalent ions of the rare earth metals (lanthanides) can be monitored in biological systems, due to their magnetic and spectroscopic properties. For example, external Eu(III) and Pr(III) shift the ^{1}H-NMR resonance of the -$N(CH_3)_3$ head groups of lecithins in the external layer of phospholipid bilayers. This property has been used to monitor the action of local anesthetics on phospholipid bilayers, and to study the transport of lanthanide ions by ionophores. Lanthanide-induced NMR shifts have also been used to determine the conformation of 3′,5′-cyclic AMP and other nucleotides in solution.

Lanthanide ions have been used notably as replacement ions for Ca(II), which itself lacks useful physical properties for the study of its behavior in biological systems. The four bound Ca(II) ions of thermolysin can be replaced by three lanthanide ions (two of these Ca(II) ions are only 3.8 Å apart and bridged by three carboxylate groups; with a small rearrangement, a single lanthanide ion is accommodated in place of the two closely paired Ca(II) ions). This exchange causes no significant change in the conformation of the polypeptide backbone, and the bound lanthanide ions serve as X-ray heavy atoms for crystallographic study of the protein. Similarly, the two bound Ca(II) ions of parvalbumin can be replaced by Eu(III) or Tb(III). which then serve as X-ray heavy atoms, or can be investigated by laser-induced luminescence. Lanthanide (III) ions have been employed as X-ray heavy atoms in many other biological macromolecules, e.g. concanavalin A, lysozyme, tRNA, and bacterial ferredoxin, but in these cases it is not established that the lanthanide ion occupies a Ca(II) binding site. Tb(III) and Eu(III) have useful luminescent properties with excited state lifetimes in the convenient range of 100–3,000 μs. Sensitized luminescence of protein-bound Tb(III) has been observed in several proteins. The excitation spectrum is diagnostic of the aromatic residues responsible for sensitization (excitation maxima: 259 nm for Phe, 280 nm for Tyr, 295 nm for Trp). Gd(III) has isotropic magnetic properties and a long electronic spin-lattice relaxation time, which makes it ideal as a nuclear relaxation probe. Most of the other lanthanide ions have shorter relaxation times and fairly large magnetic anisotropies, which makes them more suitable as dipolar shift probes. Gd(III) also has a room temperature EPR spectrum, but this property has not been widely exploited. The magnetic circular dichroism of Nd(III) is quite intense and has been used to determine Nd(III) binding to thermolysin. [W. Horrocks *Adv. Inorg. Bichem.* **4** (1982) 201–261]

Lapachol: see Napthoquinones (Table).

Larval hormones: see Juvenile hormones.

Late proteins: see Phage development.

Late RNA: see Phage development.

Latex: see Caoutchouc.

Lathyrin: see Guanidine derivatives.

Lathyrinogenic amino acids: nonproteogenic amino acids occurring in the seeds of some species of vetch (*Lathyrus*). They include diaminobutyric acid $H_2N–(CH_2)_2–CH(NH_2)–COOH$ (neurolathrinogenic effect), β-aminopropionitrile, which occurs as the glutamyl peptide in the seeds of *Lathyrus odoratus,* and presumably the *N*-oxaloyl-α,β-diamino-propionic acid $HOOC–CH(NH_2)CH_2–NH–CO–COOH$. The disease caused in humans and animals by L. a. a. is called lathyrism, and takes various forms, e. g. neuro-(nerve)- and osteo(bone)-lathyrism; e. g. β(*N*γ-glutamyl)aminopropionitrile causes skeletal abnormalities in rats.

Lauric acid: n-dodecanoic acid, $CH_3–(CH_2)_{10}–COOH$, one of the most widespread fatty acids, a typical wax fatty acid, M_r 200.3, m.p. 44 °C, b.p.$_{100}$ 225 °C. L. a. is present esterified in the seed fats of the laurel family (*Lauraceae*), and makes up 52 % of the fatty acids in palm seed oil, 48 % in coconut fat, 4–8 % in butter. It is an esterified acid of spermaceti.

Lawsone: see Naphthoquinones (Table).

(+)-Laxifloran: (3R)-7,4′-dihydroxy-2′,3′-dimethoxyisoflavan. See Isoflavan.

LDH: acronym of lactate dehydrogenase.

LDL: see Lipoproteins.

Lead, *Pb:* a highly toxic cumulative element in man and animals. It affects adversely nearly all steps in heme synthesis; it inhibits the mitochondrial enzyme 5-aminolevulinic acid synthase, but inhibits even more strongly 5-aminolevulinic acid dehydrase (see Porphyrins). The result of this inhibition is an increase in the blood level of 5-aminolevulinic acid, which is also detectable in the urine. Other enzymes inhibited by absorbed Pb are cytochrome P450 (liver), adenyl cyclase (brain and pancreas), enzymes of collagen synthesis, some ATPases, and lipoamide dehydrogenase. Clinical symptoms of Pb poisoning (plumbism) include anemia, lead-line stippling of the gums, alimentary pain, muscle weakness, and encephalopathic manifestations that occur mainly in children (convulsions, delirium, loss of memory, hallucinations). Most blood Pb is present in red cells and one clinical characteristic of plumbism is the appearance, in the bone marrow and circulating blood, of basophilic stippled red cells (resembling reticulocytes in that they contain mitochondria; stippling is caused by clumped ribosomes).

Learning: see Memory.

Lecithin: see Membrane lipids.

Lecithin-cholesterol acyltransferase deficiency: see Inborn errors of metabolism.

Lectin, *phytohemagglutinin:* as defined by the Nomenclature Committee of the IUB, "sugar-binding protein or glycoprotein of nonimmune origin which agglutinates cells and/or precipitates glycoconjugates." This definition may be broadened to include similar proteins which, although they specifically bind complex saccharides, do not precipitate or agglutinate them. Such proteins can be called "monovalent L.". L. are found in almost every major taxon of flowering plants, and in some nonflowering plants as well. Vertebrate and microbial L. have also been identified. They bind to erythrocytes, leukemia cells, yeast and several types of bacteria. As the binding is saccharide-specific, L. will not agglutinate cells which do not carry the appropriate surface saccharides. It may be expected that as more kinds of surface oligosaccharide are used in screening assays, more L. will be found. Indeed they may be ubiquitous. L. may account for as much as 10 % of the soluble protein in extracts from mature seeds; they are also present in other plant tissues at lower concentrations. The L. present in vegetative tissues are often different from the seed L. in the same plant. The physiological function of plant L. is unknown. It has been suggested that they promote infection of legume root hair tips by *Rhizobium,* or that they inhibit pathogenic microorganisms. The vertebrate L. may function in development; they have a role in receptor-mediated endocytosis.

Relative molecular masses and subunit structures of some lectins from plant seeds

Source	M_r	Subunits No.	M_r	Remarks
Jack beans (Concanavalin)	at pH > 7 110,000	4	27,000	238 amino acids, n.c.
	at pH < 6 54,000	2	27,000	Primary structure known; not a glycoprotein but a lipoprotein
Lima beans	247,000	4	62,000	Glycoprotein, n.c.
	124,000	2	62,000	
	247,000	8	31,000	Glycoprotein, c.
	124,000	4	31,000	
Green beans *(Phaseolus)*	140,000	2α	35,000	Glycoprotein
		1β	35,000	
		1β	36,500	
	126,000	4	31,000	
Wheat hemaglutinin	34,000	2	17,000	Not a glycoprotein, n.c.
Castor beans *(Ricinus)*	125,000	1	33,000	Glycoprotein, c.
		1	30,000	
		1	27,500	
Potato	95,000	2	46,000	Glycoprotein, n.c. (50 % carbohydrate)

c. = covalent; n.c. = noncovalent binding of subunits

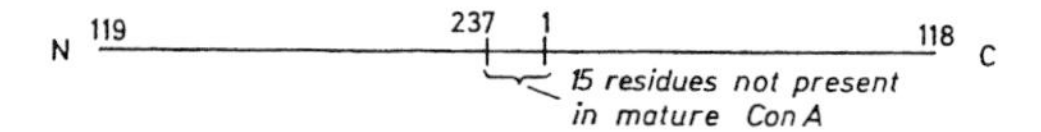

Con A precursor (Structure of Con A translation product as deduced from cloned cDNA). The numbers refer to residues in mature Con A

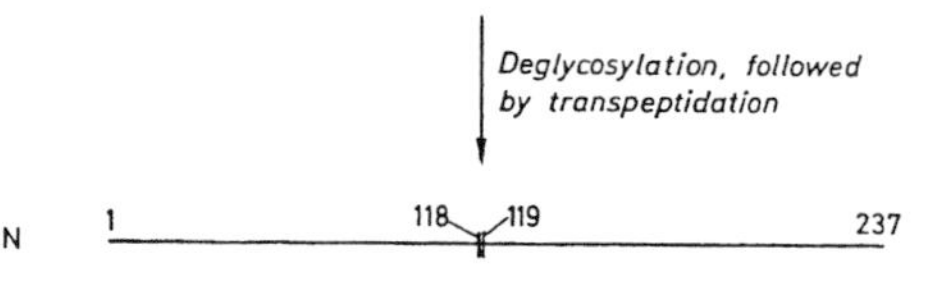

Mature Con A

Maturation of ConA precursor. Pulse-chase experiments with radioactive amino acids strongly support a mechanism of transpeptidation, rather than proteolytic cleavage followed by ligation. The only other known precedent for such a mechanism is the last step of peptidoglycan synthesis in bacteria (see Murein). Inspection of possible three-dimensional structures of Con A and its precursor show that maturation by transpeptidation is possible without unfolding of the polypeptide chain.

The best known plant L. are Concanavalin A (Con A) from jack beans (*Canavalia ensiformis*), which was crystallized in 1919 by Sumner, and the agglutinins from wheat germ (WGA), lima beans, green beans (*Phaseolus*), castor beans (*Ricinus*) and potatoes. WGA is very well characterized, and L. with similar specificities and structure are present in rye and barley. Seeds from 90 other members of the *Triticeae* tribe of the grass family have L. which are immunochemically identical to WGA, although their specificities are not all the same. The seed L. in the grasses are found only in the embryos. Legume seeds are very rich in L., and complete amino acid sequences are known for a number of them, including Con A. There are extensive homologies among the L. from related legumes. Studies on the structure of Con A have revealed a new type of protein maturation [D. J. Bowles et al. *J. Cell Biol.* **102** (1986) 1284–1297]. Con A (consisting of 4 identical subunits, each of M_r 27,500) shows maximal sequence similarity with the L. of lentil, soybean and fava bean when its *N*-terminus is positioned near the middle of the sequences of the other L. It is now known that the primary translation product undergoes transpeptidation (Fig.). [M. E. Etzler, *Ann. Rev. Plant Physiol.* **36** (1985) 209–234; T. C. Bøg-Hansen & E. van Driessche, eds. *Lectins: Biology Biochemistry, Clinical Biohemistry* vol. **5** (de Gruyter, Berlin, 1986); see also preceeding volumes of this series]

Leghemoglobin, ***Legoglobin:*** an autoxidizable hemoprotein, present in the root nodules of leguminous plants. L. is structurally and functionally related to hemoglobin and myoglobin. Amino acid sequence of L. shows homology with that of animal myoglobin. X-ray crystallography shows similar topology and three-dimensional structure of L. and animal myoglobins. L. is essential for symbiotic nitrogen fixation in the root nodules of leguminous plants, where it is responsible for the rapid flow of oxygen to the bacteroids (cf. role of myoglobin in transport of oxygen to respiratory enzymes of muscle). Synthesis of the globin is under the genetic control of the macrosymbiont (host plant); the heme is synthesized by the microsymbiont (*Rhizobium* bacteroids). The concentration of L. shows a positive correlation with the N_2-fixing capacity of the nodule tissue. There may be more than one L., depending on the species of the host plant. L. from soybean can be resolved into at least four components a-d, of which a and c are the main ones; these have molecular masses 16,800 and 15,950; both have been crystallized: Component c has been further resolved into c_1 and c_2, which differ only in their *C*-termini (lysine in c_1; phenylalanine in c_2). L. contains no cysteine or methionine, and it shows no immunological cross reaction with hemoproteins from *Rhizobium.* L. is unique to the legume-*Rhizobium* symbiotic system; it is absent from free-living Rhizobia and from uninfected leguminous plants.

It is located between the bacteroids and the membrane envelope. (See Nitrogen fixation).

Legumin: an oligomeric storage protein of legume seeds, M_r about 328,000, consisting of 6 pairs of subunits. Each subunit pair consists of an α-subunit (M_r about 36,000) and a β-subunit (M_r about 20,000) linked covalently by a disulfide bond. L. is synthesized in vitro as a single polypeptide chain, M_r about 60,000, which is cleaved in vivo into α- and β-subunits. The precursor polypeptide already contains the disulfide bond, so the subunit pairs are specific. Study of *Vicia faba* L. shows that at least two different gene families (which may have arisen from a single gene) code for L. precursor polypeptide, giving rise to type A and type B subunit pairs. Type Aα- and β-subunits both contain Met, which is absent from type B subunits. Since Met is an essential amino acid in animal nutrition, and is a limiting factor in the nutritional value of legume seed protein, it should be possible to improve the quality of this protein by plant breeding to increase the type A/B ratio. Further heterogeneity of amino acid composition within the subunit pairs has been detected and may result from mutation of the common ancestral gene. [C. Horstman *Phytochemistry* **22** (1983) 1861–1866]

Lettuce hypocotyl test: see Gibberellins.

L-Leucine, *Leu:* L-2-amino-4-methylvaleric acid, $(CH_3)_2$–CH–CH_2–CH(NH_2)–COOH, an aliphatic, neutral, proteogenic amino acid, which is both essential and ketogenic. It is particularly abundant in serum albumins and globulins. Leu is degraded to isovaleric acid by deamination and decarboxylation, then further to acetate via acetoacetate (Fig.). Leu biosynthesis follows the scheme for branched amino acids (see L-Isoleucine) and branches off at the level of 2-oxo-3-methylvaleric acid, which undergoes condensation with an acetyl group. The subsequent reactions are analogous to those in the tricarboxylic acid cycle, and result in 2-oxo-4-methylvaleric acid, which is transaminated to Leu.

Leucine aminopeptidase: see Aminopeptidases.

Leucine zipper: a structural motif of DNA-binding proteins, which mediates both homodimerization and heterodimerization of these proteins, but is not itself a DNA-binding motif. Thus, several DNA-binding proteins contain heptad repeats, in which every seventh position is occupied by a leucine residue, e.g. rat liver transcription factor C/EBP (CCAAT/ enhancer binding protein) and yeast transcriptional activator GCN4, as well as several DNA-binding proteins encoded by oncogenes. It was therefore predicted that these proteins form coiled coils, aided by interdigitation of the leucine side chains, like the teeth of a zipper. The prediction was confirmed by X-ray analysis of the leucine zipper region (33 residues) of GCN4. The first 30 residues of the zipper region contain about 3.6 heptad repeats; these form an α-helix of about 8 revolutions, which dimerizes to form a quarter turn of a parallel left-handed coiled coil, i.e. the dimer takes the form of a twisted ladder, every second rung formed by side-to-side contacts (not interdigitated as originally predicted) of leucine residues. The alternate rungs are formed by side-to-side contact of other hydrophobic residues (mostly valine residues). [W. H. Landschulz et al. *Science* **240** (1988) 1759–1764; E. K. O'Shea et al. *Science* **254** (1991) 539–544; T. E. Ellenberger et al. *Cell* **71** (1992) 1223–1237; T. E. Ellenberger *Curr. Opin. Struct. Biol.* **4** (1994) 12–21; K. Gramatikoff et al. *Biol. Chem. Hoppe Seyler* **376** (1995) 321–325]

Leucinosis: see Inborn errors of metabolism.

Leucoanthocyanidins, *flavan-3,4-diols:* a class of Proanthocyanidins (see). L. are common plant constituents, especially in wood, bark, and the rind of fruits. Examples are ***(+)-guibourtacacidin*** (7,4′-dihydroxyflavan-3,4-diol, absolute configuration 2*R*:3*S*:4*S*, from *Guibourtia coleosperma* and *Acacia cultriformis;* the configuration 2*R*:3*S*:4*R* is also present in *A. cultriformis*), ***(–)-leucofisetinidin*** (7,3′,4′-trihydroxyflavan-3,4-diol, abs. config. 2*S*:3*R*:4*S*, from *Schinopsis lorentzii*), ***(+)-leucorobinetinidin*** (7,3′,4′,5′-tetrahydroxyflavan-3,4-diol, abs. config. 2*R*:3*S*:4*R*, from *Robinia pseudoacacia*), and ***(+)-mollisacacidin*** (7,3′,4′-trihydroxyflavan-3,4-diol, abs. config. 2*R*:3*S*:4*R*, from heartwood of *Acacia baileyana* and sapwood of *A. mearnsii;* see Fig.). L. have been implicated in the biosynthesis of condensed tannins and condensed proanthocyanidins (see Tannins).

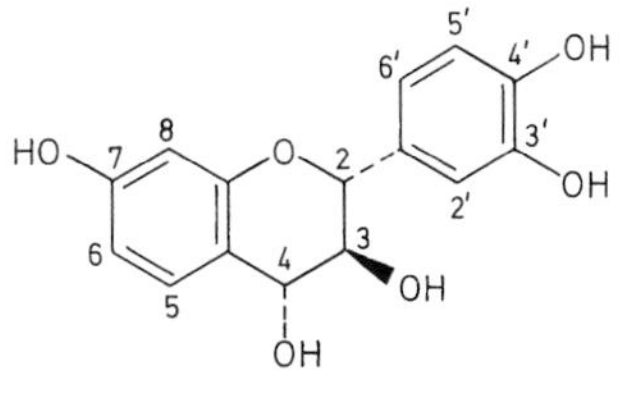

Mollisacacidin

(–)-Leucofisetinidin*:* 7,3′,4′-trihydroxyflavan-3,4-diol. See Leucoanthocyanidins.

Leucomycin: see Macrolide antibiotics.

Leucoplasts: colorless plastids in plant cells, generally those which are not exposed to light. The L. include the starch-storing amyloplasts, the protein-storing aleuroplasts, the fat-storing elaioplasts and the etioplasts, which are the colorless chloroplasts of sprouts and stolons (e.g. potato shoots) which have been kept in the dark (etiolated). Amyloplasts are found in the non-green storage tisues, such as the endosperm of grain or the cotyledons of legumes. Reserve starch is formed and stored in them in the form of starch grains. Although the amyloplasts are colorless, they are sometimes considered to be chromoplasts.

Leucopterin: 2-amino-5,8-dihydro-4,6,7(1H)-pteridinetrione, a white pigment, m.p. 350 °C, found in

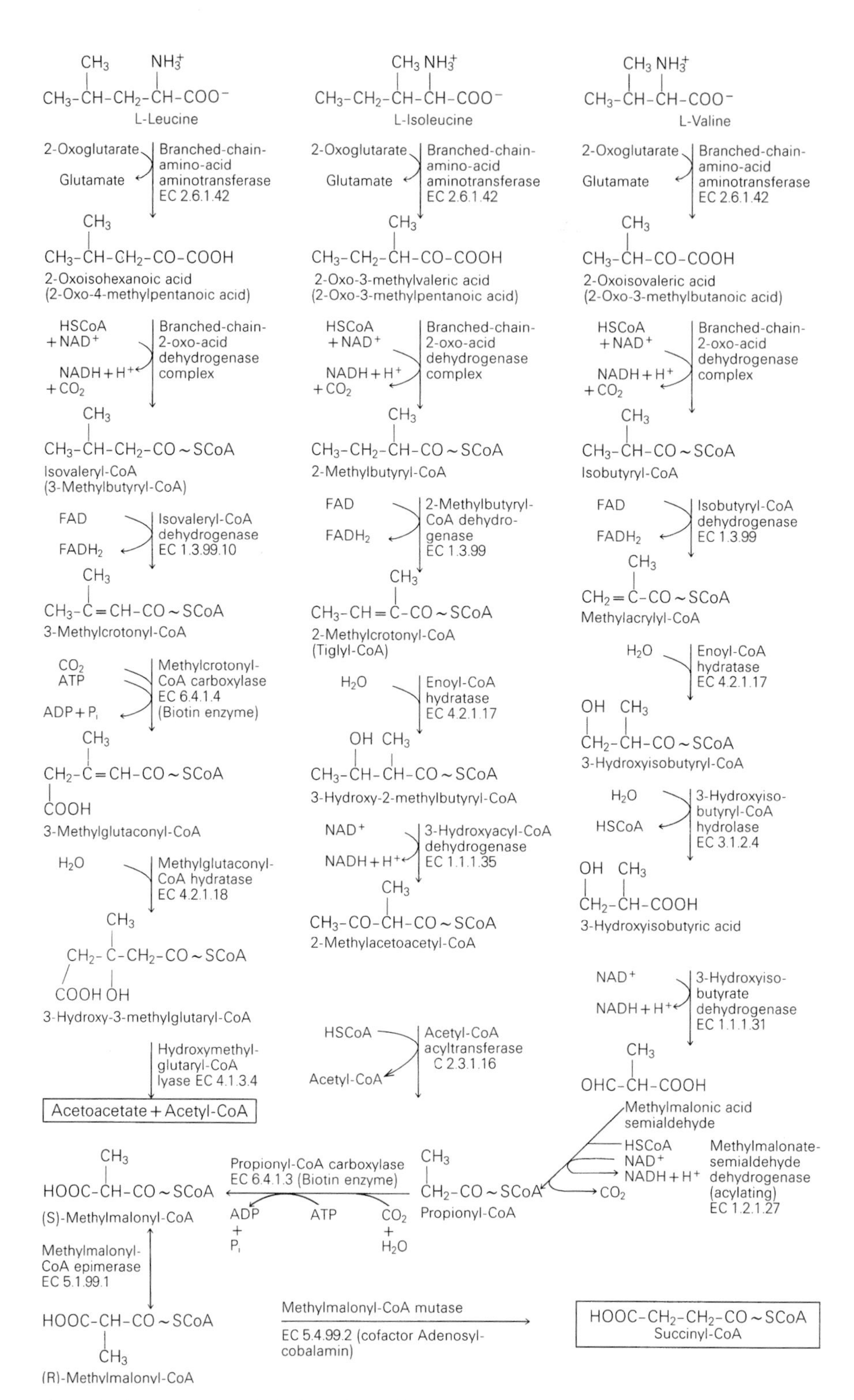

Degradation of branched chain amino acids

the wings of cabbage white butterflies and other butterflies. It is biosynthesized from guanine and two C-atoms of a pentose. It was isolated in 1926 from 200,000 cabbage white butterflies by H. Wieland and C. Schopf, and its structure was established in 1940 by R. Purrmann.

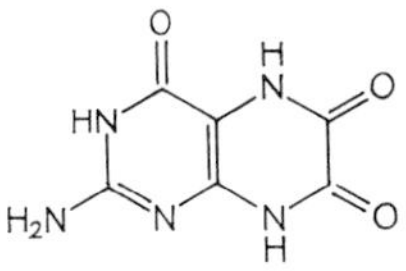

Leucopterin

(+)-Leucorobinetinidin: 7,3′,4′,5′-tetrahydroxy-flavan-3,4-diol. See Leucoanthocyanidins.

Leukokinin: the specific leukophilic γ-globulin which binds to the leukocyte membrane and acts as the precursor of Tuftsin (see).

Leukophilic γ-globulin: predominantly γG-globulin and minor amounts of γA-globulin and γM-globulin, which coat the surface of the leukocyte membrane. See Tuftsin.

Leukotrienes: lipid hormones derived from Arachidonic acid (see). Leukocytes are important sources of L. (hence the "leuko-" prefix), and all of them contain three conjugated double bonds (hence "triene"). The major L. are actually tetra-unsaturated, but the fourth double bond is not conjugated with the others.

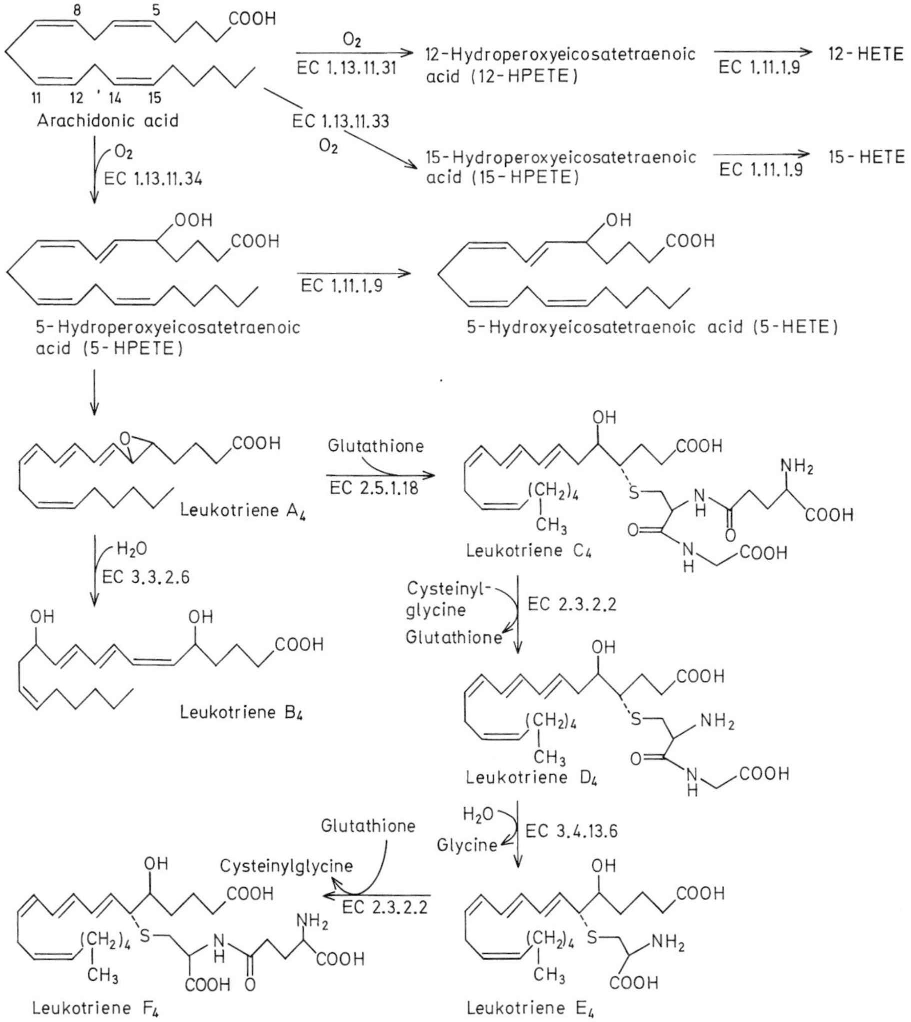

Lipoxygenase pathway of arachidonate metabolism. EC 1.11.1.9: Glutathione peroxidase. EC 1.13.11.31: Arachidonate 12-lipoxygenase. EC 1.13.11.33: Arachidonate 15-lipoxygenase. EC 1.13.11.34: Arachidonate 5-lipoxygenase. EC 2.3.2.2: γ-Glutathione transferase. EC 3.3.2.6: Leukotriene-A_4 hydrolase. EC 3.4.13.6: Cysteinyl-glycine dipeptidase.

The L. are the same as the "slow-reacting substances of anaphylaxis", which mediate immune hypersensitivity. They also potentiate inflammation. All L. cause powerful contraction of the lungs (they are hundreds to thousands of times more potent than the histamines), and they stimulate release of thromboxanes and prostaglandins. A lipoxygenase acting at C5 of arachidonic acid produces 5-hydroperoxyeicosatetraenoic acid. Conversion of this compound to the 5,6-epoxide produces L.A_4, which is very unstable; in buffer at pH 7.4 and at 25 °C its half-life is less than 10 sec. However, it is stabilized by alkaline conditions or by albumin. The structures of L.A_4 and other L. are given in the figure. The physiological effects of different L. are not identical. L.C_4, L.D_4 and L.E_4 are mostly myotropic, stimulating contraction of smooth muscle. L.B_4 is a chemotactic agent for macrophages and eosinophils; the former are caused to aggregate and to release superoxide and lysosomal enzymes. L.B_4 induces transformation of T lymphocytes into suppressor T cells, and may have other effects which modulate the immune response. [P. Borgeat et al. "Leukotrienes. Biosynthesis, Metabolism and Analysis" *Adv. Lipid Res.* 21 (1985) 47–77; P. Sirois, "Pharmacology of the Leukotrienes" ibid. 79–101].

Leupeptin: see Inhibitor peptides.

Leurocristine: see Vincristine.

Levulose: see D-Fructose.

LH: acronym of luteinizing hormone.

Liberin: a Releasing hormone (see); a suffix (-liberin) used in the nomenclature of releasing hormones.

Licanic acid, *couepic acid, ketoeleostearic acid, oxoeleostearic acid:* 4-oxo-9,11,13-octadecatrienoic acid, $CH_3(CH_2)_3(CH{=}CH)_3(CH_2)_4CO(CH_2)_2$ COOH, from the seed fat or oil of *Licania rigida* (see Oiticica oil). The original source of the oil was erroneously thought to be *Couepia grandiflora,* hence the name couepic acid. L. a. exists in two forms: α-L.a. is *cis*-9, *trans*-11, *trans*-13 (m.p. 74–75 °C); β-L.a. is all-*trans* (m.p. 99.5 °C). α-L.a. readily isomerizes to β-L.a. by the action of light and in the presence of traces of sulfur or iodine. It is the only unsaturated oxoacid that has been isolated from a natural fat, and it is also present in the seed fats of other *Licania* species, e.g. *L. arborea* (Mexico), *L. crassifolia* (East Indies) and *L. venosa* (Guyana), and the seed fats of several species of *Parinarium.*

The glyceride oils of L. a. are commercially important in the manufacture of alkali and water resistant coatings in the paint industry.

Licensing factor: see Cell cycle.

Lichenin, moss starch: a polysaccharide serving as both a storage and a structural compound, M_r 25,000 to 30,000, $[\alpha]_D + 120°$. L. is composed of 150–200 D-glucose units linked β-1,4 with about 25 % β-1,3 glucosidic linkages distributed at random in the molecule. It is found in many lichens and has antineoplastic properties.

(+)-Licoricidin: 5,4′,2′-trihydroxy-7-methoxy-6,3-di-γ,γ-dimethylallylisoflavan. See Isoflavan.

Liebermann-Burchard reaction: a reaction used for colorimetric determination of sterols. The substance to be tested is dissolved in chloroform and treated with sulfuric acid and acetic anhydride. A color change from pink to blue to green indicates unsaturated sterols and was formerly used for the quantitative determination of cholesterol in blood.

Ligandin: see Glutathione *S*-transferase.

Ligases: see Enzymes.

Light compensation point: the light intensity at which the rate of photosynthesis (CO_2 incorporation) and the rate of respiration (CO_2 production) are balanced. See CO_2-compensation point.

Lightening hormone: Pyr-Leu-Asn-Phe-Ser-Pro-Gly-Trp-NH_2, M_r 930, a neurohormone produced by the eye-stalk glands of crustaceans. The hormone is released from nerve endings in the gland in response to visual stimuli. It controls the distribution of pigment granules in the hypodermal chromatophores, enabling the animal to adjust its color to match the surroundings.

Light-harvesting protein: a strongly hydrophobic, integral membrane protein, isolated from the thylakoids of many angiosperms, gymnosperms and green algae. It is chiefly associated with photosystem II but some activity with photosystem I has also been demonstrated. Chlorophylls *a* and *b* are bound in equimolar amounts, together with lutein and β-carotene. The chlorophyll/carotenoid molar ratio is 3–7/1. M_r of protein 27,000–35,000, depending on the species. Complexes are also known, which contain two dissimilar subunits. Up to 6 mol Chlorophyll (see) are bound per mol protein (assuming M_r of 30,000). The protein serves in the transfer of light energy from chlorophyll *b* to *a,* and it is involved in (although not essential for) the stabilization or stacking of the grana. Only very small amounts of L. h. p. are found in the bundle sheath cells of C-4 plants.

Light reactions: see Photosynthesis.

Lignans: plant products (L. have also been found in primate urine; see Enterolactone, Enterodiol) formally equivalent to two n-propylbenzene (phenylpropane) residues linked at the central carbons of their side chains:

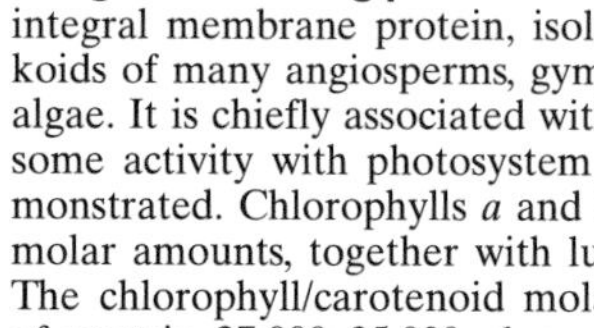

5″ 4″ 6″ 9″ 7 2 1″ 8″ 1 3″ 8 3 2″ 7″ 9 6 4 5

The two benzene rings are usually identically substituted, and the type of substitution is similar to that present in ring B of the C_6C_3 residues of flavonoids. It is therefore generally accepted that L. are biosynthesized by dimerization of a C_6C_3 precursor, but direct experimental evidence for this is lacking. Linear L. can be classified in four groups, according to the type of structure between the two aromatic rings:

1. ***Lignans*** (derivatives of butane)

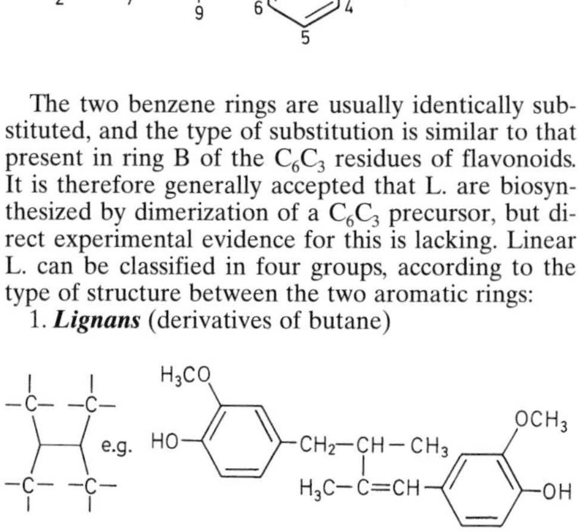

e.g. H_3CO HO– CH_2–CH–CH_3 H_3C–C=CH– OCH_3 –OH

Guaiaretic acid (constitues about 10 % of the resin of *Guaiacum officinale* L.)

2. ***Lignanolides*** (derivatives of butanolide)

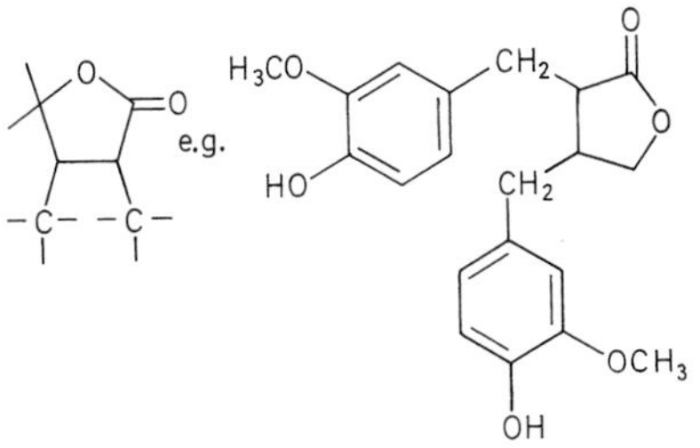

Matairesinol (from *Podocarpus spicatus,* or New Zealand "matai", and many different woods)

3. ***Monoepoxylignans*** (derivatives of tetrahydrofuran)

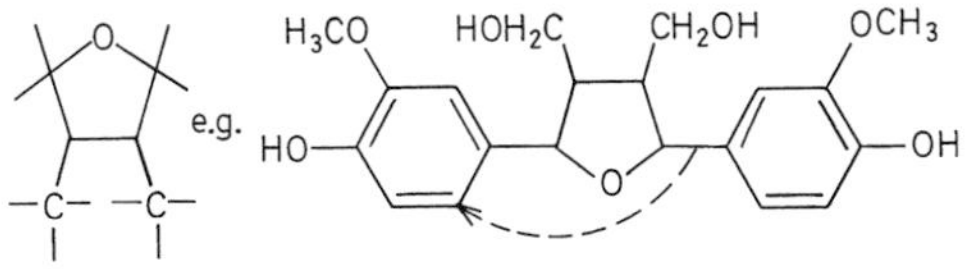

(–)-Olivil (from resin of the olive tree, *Olea europea).* The broken arrow represents cyclization to iso-olivil in the presence of mineral acid (see text).

4. ***Bisepoxylignans*** (derivatives of 3,7-dioxabicyclo[3.3.0]-octane)

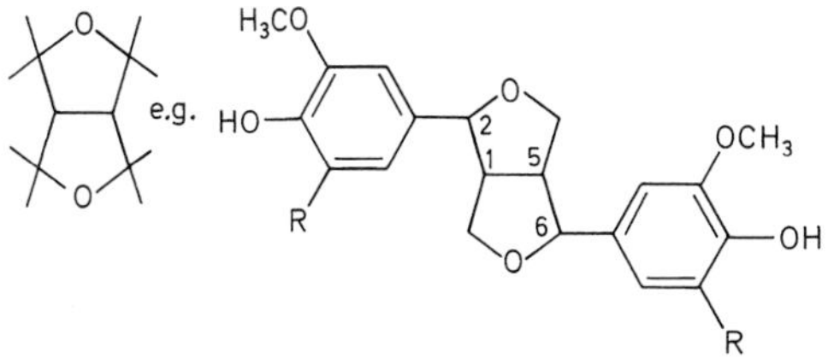

R = H: *Pinoresinol* (from exudate of *Pinus lavico* and other pines).
R = OCH_3: *Syringiaresinol* (1*R*,2*R*,5*R*,6*R*-form); also called *Lirioresinol A* or *Episyringiaresinol;* from *Artemesia absinthium, Lirodendron* spp. and *Magnolia grandiflora;* from degradation of birch lignin). Other configurational forms occur naturally.

Further cyclization (C7 to C6'') produces ***cyclolignans:*** Cyclolignans containing a tetrahydronaphthalene ring system:

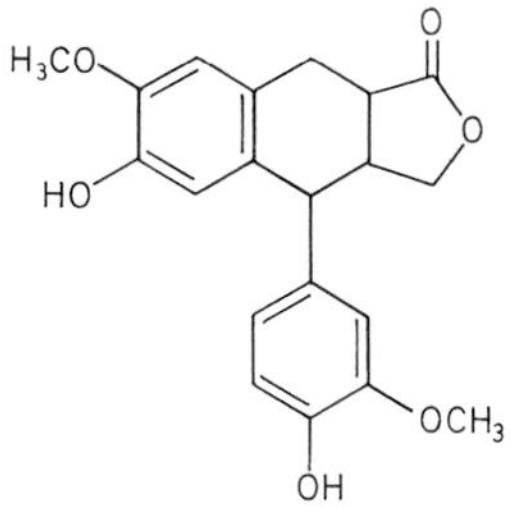

Conidendrin (from spruce; present in large amounts in waste sulfite liquor from paper making)

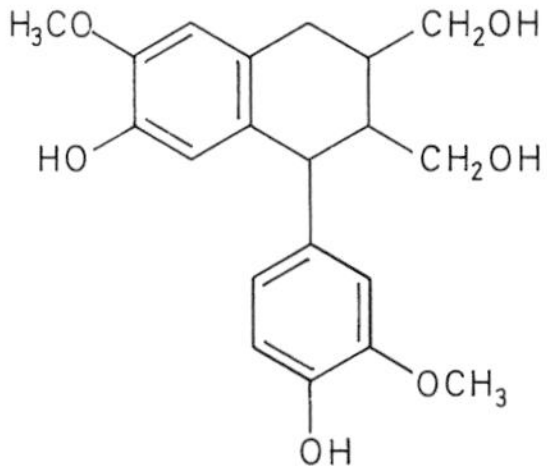

iso-Olivil (from the Australian olive, *Olea cunninghamii*)

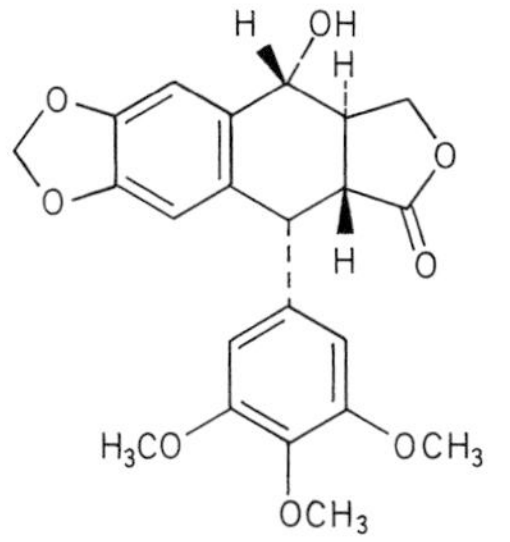

Podophyllotoxin (from the mayapple, *Podophyllum* spp.)

Cyclolignans containing the naphthalene ring system:

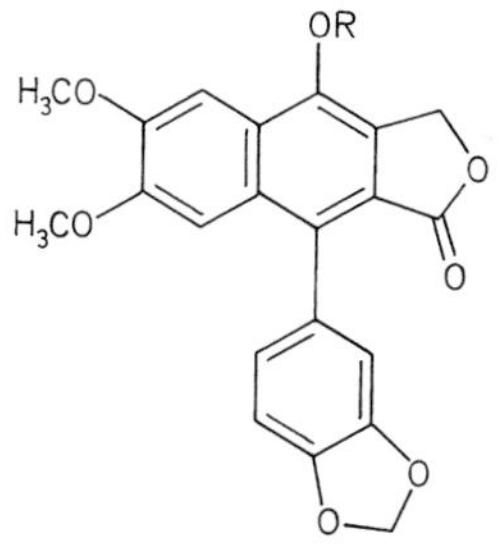

R = CH_3: *Justicidin A*
R = H: *Justicidin B* (piscicidal constituents of *Justicia hayatai*)

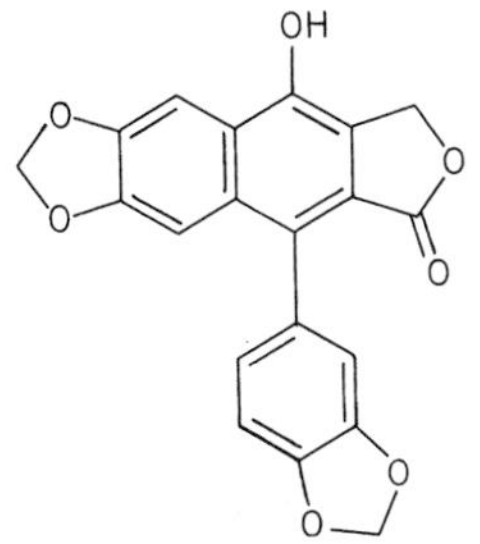

Diphyllin (from roots of *Diphylleia grayi;* also a piscicidal constituent of *Justicia hayatai*)

According to Freudenberg, L. represent stages in the biosynthesis of Lignin (see); enzymatic oxidation of coniferyl alcohol with a fungal laccase at pH 5 (i.e. conditions similar to those for formation of spruce lignin in vitro) produces dehydroconiferyl alcohol and pinoresinol. Lignin formally resembles a polymer of propylbenzene (phenylpropane) units, and the *bis*epoxylignan structure occurs repeatedly in the lignin polymer.

Treatment of (–)-olivil with mineral acid breaks one of the benzyl ether linkages, forming a carbonium ion. The positive center then attacks the opposite guaiacol residue by electrophilic substitution, to produce iso-olivil. This type of reaction has been observed for many other L. It is probably also biologically important, because iso-olivil occurs naturally in the Australian olive.

Representatives of 55 families of vascular plants have been found to contain L. Gymnosperms feature prominently as L.-containing plants. L. occur in all parts of the plant, wood and resin being especially rich sources. L. have been isolated which possess antitumor, antimitotic, antiviral, insecticidal, piscicidal, antibacterial and fungistatic properties. Some L. inhibit mammalian cyclic AMP phosphodiesterase, and others have been shown to possess cathartic activity, cardiovascular activity, or the ability to damage DNA and inhibit nucleic acid synthesis. Podophyllotoxin and related compounds are particularly active, showing antimitotic activity and the ability to bind to purified tubulin preparations, probably at the same site as colchicine; they also display antiviral and antitumor activity, and they inhibit DNA and RNA synthesis in mammalian cell culture. L. have been classified by the U.S. National Cancer Institute as compounds of "high interest" as potential anticancer agents, and tumor trials have been conducted on several L.

A useful source reference with literature list covering all aspects of L.: W. D. MacRae & G. H. N. Towers *Phytochemistry* **23** (1984) 1207–1220.

Lignification: see Wood.

Lignin: a polymer responsible for the thickening and strengthening of plant cell walls. The properties associated with wood are due to the incrustation of plant cell walls with L. Chemically, L. cannot be exactly defined. According to Freudenberg, L. is a highly cross linked, macromolecular, branched polymer, formed irreversibly by dehydrogenation and condensation. According to Adler and Gierer, L. is an essen-

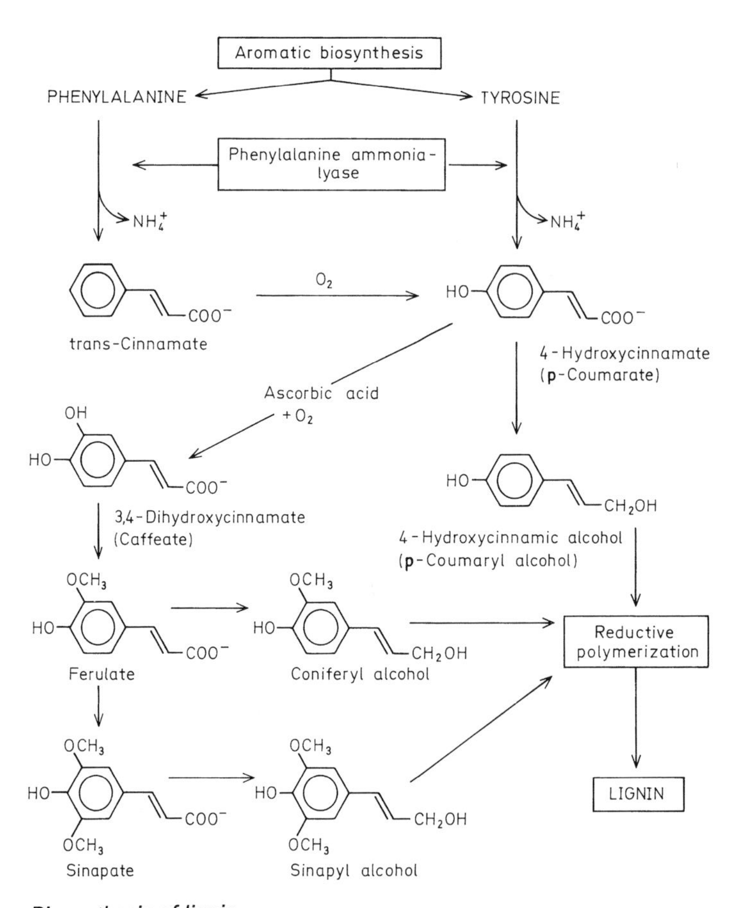

Biosynthesis of lignin

tially acid-resistant, polymorphic, amorphous incrustation material found in wood, consisting of methoxylated phenyl-propane units linked by ether linkages and C-C bonds. It has also been described as a "random polymer of hydroxyphenylpropane units". The M_r of L. is greater than 10,000, and it is insoluble in hot 70 % sulfuric acid. The chemical composition of L. differs according to the plant species. Beech L. is the most extensively studied.

Biosynthesis. The primary precursors of L. are coniferyl, sinapyl and *p*-coumaryl alcohol, which are derived from 4-hydroxycinnamic acid. L. from conifers (i.e. from softwood) is derived chiefly from coniferyl alcohol with variable but small proportions of sinapyl and *p*-coumaryl alcohol. L. from dicotyledonous angiosperms (i.e. from hardwood), particularly deciduous trees, is formed chiefly from sinapyl (~44 %) and coniferyl (~48 %) alcohol, with about 8 % *p*-coumaryl alcohol. L. in grasses is formed from *p*-coumaryl (~30 %), coniferyl (~50 %) and sinapyl (~20 %) alcohol. These primary L. precursors are formed from the aromatic amino acids L-phenylalanine and L-tyrosine by a series of reactions shown (Fig.). The first reaction is catalysed by L-Phenylalanine ammonia-lyase (EC 4.3.1.5) (see); this enzyme is induced by light in a process involving phytochrome, and it is of general importance in the synthesis of plant phenolic compounds from phenylalanine and tyrosine.

D-Coniferin (glucoside of coniferyl alcohol), D-glucocoumaryl alcohol (glucoside of *p*-coumaryl alcohol) and D-syringin (glucoside of sinapyl alcohol) are storage forms of L. precursors. β-Glucosidases in cells between the cambium and the mature wood hydrolyse the glucosides and thereby release the alcohols for L. synthesis.

The total biosynthesis of L. represents the interplay of enzymatic phenol dehydrogenation and the non-enzymatic coupling of radicals generated by loss of an electron from a phenolate ion, a process referred to as *reductive polymerization.*

The process can be demonstrated by the synthesis of artificial L in a model system, in which coniferyl alcohol (80 mol %), *p*-coumaryl alcohol (14 mol %) and sinapyl alcohol (6 mol %) are incubated with a phenol oxidase (e.g. fungal laccase) under strongly aerobic conditions. The product is identical with spruce L. Presumably a peroxidase or similar oxidation system is responsible for L. biosynthesis in vivo. When isotopes are used to study L. biosynthesis strict criteria are applied to the identification of L., which must be clearly distinguished from other cell wall polymers. One method is to degrade the L. in situ, followed by isolation of degradation products known unequivocally to be derived from L. For example, Hibbert showed that wood refluxed with ethanol containing 2 % HCl gave a mixture of water-soluble aromatic ketones. These compounds, known as "Hibbert's ketones", have the structures: R–CO–CO–CH_3, R–CH_2 CO–CH_3, R–CO–CH(OC_2H_5)–CH_3 and R–CH(OC_2H_5)–CO–CH_3, where R is guaiacyl or syringyl. These compounds represent the intact phenylpropane structure of the original L. [D. Fengel & G. Wegener *Wood – Chemistry, Ultrastructure, Reactions,* DeGruyter (Berlin, New York, 1989); A.M. Boudet & J. Grima-Pettenati 'Lignin genetic engineering' *Molecular Breeding* **2** (1996) 25–39]

Lignoceric acid: n-tetracosanoic acid, CH_3–$(CH_2)_{22}$–COOH, a fatty acid M_r 368.6, m.p. 84 °C. L.a. occurs as an esterified component of glycerides (usually less than 3 %) in many seed oils, such as ground nut and rape seed oil. It is an esterified component of certain cerebrosides (e g kerasin), phosphatides and waxes.

Limonene: see *p*-Menthadienes.

Linalool, ***coriandrol:*** a doubly unsaturated monoterpene alcohol found in essential oils. Both optical isomers occur naturally. L. is an oil with a scent resembling that of lily of the valley. M_r 154.25, (+)-L. b.p. 194–198 °C, $[\alpha]_D^{20}$ – 19.4 °. The pure alcohol and its esters are used in perfumery. For formula, see Geraniol.

LINE(s): the acronym of Long INterspersed DNA sequence Element(s) which constitute one of the two main classes of 'intermediate repeat DNA' sequences (see C_0t) occurring in the mammalian genome, the other being SINEs (see). They consist of a single nucleotide sequence 6,000–7,000 bp long that occurs as a non-tandemly repeated element at thousands of different places in the genome, e.g. there are about 50,000 copies of the most abundant LINE in the human genome, the L1 family, constituting ~5 % of the total human DNA. They are capable of moving (transposing) to new sites within the genome and therefore fall into the category of 'mobile DNA elements' (see) and the sub-subdivision of the latter known as 'non-viral retrotransposons' (see). They appear to be derived from RNA polymerase II-generated transcripts.

Lineweaver-Burk plot: see Kinetic data evaluation.

Linking difference: see Linking number.

Linking number, ***topological winding number, α:*** a number specifying the number of times two strands of the DNA double helix are intertwined. This is always an integer. In solution, DNA has a pitch of 10.4 base pairs per helical turn. Closed circular duplex DNA is therefore considered to be relaxed if α approximates to the number of base pairs divided by 10.4. The totally "relaxed" value for α is represented by α^0 or β. A L.n. smaller than α^0 means that the DNA is negatively supercoiled (i.e. underwound), whereas DNA with a L.n. greater than α^0 is positively supercoiled. In both cases the DNA is under torsional strain. This strain is partitioned between twist (altered pitch of the double helix) and writhe (contortion of the double helix). Writhe is equivalent to supercoiling. The ***linking difference*** ($\Delta\alpha = \alpha - \alpha^0$) is an expression of strain, but is dependent on the length of the DNA. The ***specifc linking difference*** ($\Delta\alpha/\alpha^0$), being independent of molecular length, is a more convenient measure of the strain on the DNA duplex. See also Superhelix density.

Linoleic acid: $\Delta^{9,12}$-octadecadienoic acid, CH_3–$(CH_2)_4$–CH = CH–CH_2–CH=CH–$(CH_2)_7$–COOH, an essential fatty acid, M_r 280.44, m.p. 5 °C, b.p.$_{14}$ 202 °C. L.a. is widely distributed in plants and animals. It occurs as an esterified component in many fats and oils, and it is found in phosphatides. It is an essential dietary constituent for mammals.

Linolenic acid: $\Delta^{9,12,15}$-octadecatrienoic acid, CH_3 –CH_2–CH=CH–CH_2–CH=CH–CH_2–CH=CH–CH_2–$(CH_2)_6$–COOH, an essential fatty acid, M_r 278.44, m.p. −11.2°C, b.p.$_{17}$ 232°C. L.a. occurs in animals and plants, and is an especially common esterified component of plant fats and glycerophosphatides. L.a. is not synthesized by mammals. The above compound is also known as *α-linolenic acid.* $\Delta^{6,9,12}$-Octadecatrienoic acid, $CH_3(CH_2)_3CH_2$–CH=CH–CH_2–CH =CH–CH_2–CH=CH–$(CH_2)_4$–COOH, is called *γ-linolenic acid.*

Lipase deficiency: see Inborn errors of metabolism.

Lipases: a group of carboxylesterases, which preferentially hydrolyse emulsified neutral fats to fatty acids and glycerol or monoacylglycerols. Calcium ions are required for activity. Pancreatic L. also requires taurocholate. Pancreas and certain plant seeds (e.g. *Ricinus*) contain particularly high activities of L. In addition, high activities are found in adipose tissue, in the stomach (especially in unweaned infants) and in the liver. Pancreatic L. has M_r 35,000. Rapid removal of the third and last fatty acid residue from mixed triacylglycerols is catalysed by a specific monoacylglycerol lipase (EC 3.1.1.23) produced by the intestinal mucosa.

Lipidosis: see Inborn errors of metabolism.

Lipids: a heterogeneous group of biological compounds which are sparingly soluble in water, but very soluble in nonpolar solvents. They can be extracted from animal and plant tissues with a variety of organic solvents, e.g. benzene, chloroform, trichloroethene. As a class, L. are defined by their solubility; they include such chemically diverse compounds as Triacylglycerols (see), Waxes (see), and Terpenes (see; these include monoterpenes, diterpenes, carotenoids, steroids, etc.). The more complex L., such as Glycolipids (see) and phospholipids (see Membrane lipids), are also called lipoids. Triacylglycerols and waxes are known as saponifiable L., whereas the terpenes are called nonsaponifiable L.

Lipochondria: see Dictyosomes, Golgi apparatus.

Lipofuscin, ***age pigment, wear and tear pigment:*** intracellular clusters of yellowish granules of universal occurrence in humans. L. is deposited continuously throughout life in nerve, heart and liver cells, as well as other organs. L. is very prominent in the pyrimidal cells of the cerebral cortex, in spinal ventral horn cells and the cell bodies of the hypoglossal nuclei in elderly persons. In the neurons of the olivary nucleus, deposition of L. commences earlier and is prominent by middle age. L. appears to be derived from lysosomes, which have accumulated indigestible material. Most of these congested lysosomes, or residual bodies, are removed by exocytosis, but some persist within the cell and their constituent lipids become oxidized. The resulting mixture of proteins and partly oxidized polyunsaturated fatty acids constitutes L. It is stained intensely by Sudan black and by periodic acid-Schiff reagent.

Lipofuscinosis: see Lysosomal storage diseases.

Lipoic acid, ***6-thioctic acid, (+)-5[3-(1,2-dithiolanyl)] pentanoic acid:*** a coenzyme of hydrogen transfer and acyl group transfer reactions. L.a. is a component of the pyruvate dehydrogenase and the 2-oxoglutarate dehydrogenase complexes (see Multienzyme complexes), which catalyse the oxidative decarboxylation of the corresponding 2-oxoacids (see also Tricarboxylic acid cycle). The natural form of L.a. is α-(+)-L.a., M_r 206.3, m.p. 47.5°C, $[\alpha]_D^{25}$ +96.7° (*c* = 1.88, benzene), E_0' −0.325 V (pH 7.0, 25°C). The asymmetric carbon has the *R* configuration (Fig.).

S—S
COOH

(3R)-1,2-dithiolane-3-pentanoic acid (the oxidized form of lipoic acid)

L.a. and dihydrolipoic acid form a biochemically important redox pair. Dihydrolipoic acid (reduced L.a.) is represented as $Lip(SH)_2$ and the oxidized form as $Lip(S_2)$. L.a. is frequently present in the cell as the carboxylic acid amide (lipoamide). When serving as a coenzyme, L.a. is bound covalently through an amide bond to the ε-amino group of a lysyl residue in the enzyme. When acting in the generation and transfer of acyl groups, the acyl group becomes attached to one of the sulfurs by a thiol ester linkage. In the oxidation of pyruvate by *E. coli.* the intermediate is 6-*S*-acetyl-6,8-dithioloctanoic acid (Fig.); presumably all other acyl groups are carried in the 6-*S*-position, since all synthesized 8-thiolacyl derivatives of dihydrolipoic acid are biologically inactive.

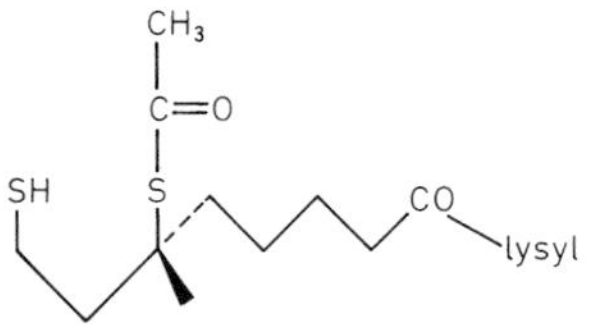

CH_3
C=O
SH S CO
lysyl

6-S-acetyl-6,8-dithioloctanoic acid. This acetyl derivative of the reduced form of lipoic acid is shown here attached to an enzyme protein by an amide linkage with the ε-amino group of a lysine residue

Lipoproteins: lipid-protein conjugates found in cellular membranes, blood plasma, cell cytoplasm and egg yolk. Blood plasma L. are responsible for the transport and distribution of lipids (hormones, dietary lipids from the intestine, fat-soluble vitamins) via the blood and lymph systems. The largest L. are the microscopically visible chylomicrons (diameter 500 nm); these are responsible for the lipemia (milky turbidity) of the blood, which follows the digestion of a fatty meal, and which disappears after about 5 hours. The other L. are always present in the blood; the apoproteins in them are produced in the liver.

In starch block electrophoresis, blood L. migrate with the β_1-, α_1- and α_2-globulin fractions. Ultracentrifugation in a high concentration of NaCl, or in a density gradient, produces a better separation and is suitable for the preparation of the various fractions. Owing to their lipid component, the L. float rather than sediment (with the exception of the very high density L. fraction, VHDL), i.e. they move centripetally towards the surface of the ultracentrifuged suspension. Thus, the flotation coefficient S_f is a characteristic of the L. L. contain between 1 and 2 % carbohydrate.

The L. content of the blood is affected by both genetic and non-genetic factors, such as age, sex, diet, hormone balance, exercise and occupation. Chylomicrons, VLDL, IDL and LDL are involved in cholesterol transport and deposition, and the correlation between high blood cholesterol levels and atherosclerosis has focused attention on these L. Nascent VLDL are produced in the liver, and they contain large amounts of triacylglycerols and cholesteryl esters. Nascent chylomicrons are produced by the small intestine, and they also contain large amounts of triacylglycerols and cholesteryl esters. The dietary cholesterol, which eventually controls endogenous cholesterol synthesis and LDL receptor synthesis, emerges from the intestine as nascent chylomicrons. The VLDL and chylomicrons transport triacylglycerols to adipose tissue and muscle; a lipase (lipoprotein lipase, clearing factor lipase, EC 3.1.1.34) on the membranes of the capillary endothelial cells hydrolyses about 90 % of the triacylglycerols. The resulting fatty acids diffuse into the tissue cells (see Acylglycerols). Some phospholipid is also lost, together with some apoproteins, which are transferred to HDL. The chylomicron remnants and VLDL remnants (IDL) continue in circulation (Figs. 1 & 2). HDL appears to function largely as a transport vehicle by transferring components between the other L., and by receiving and esterifying free cholesterol from the tissues (Fig. 3). IDL contain two apoproteins, E and B-100. The former is present in several copies, and has a high affinity for cell-surface receptors called LDL receptors (a slight misnomer, as their affinity for IDL is higher). The LDL receptors congregate in clathrin-coated pits (see Coated vesicles), and are soon taken into the cell by endocytosis (Fig. 4), whether or not they have bound IDL or LDL. Those IDL which are not internalized by cells

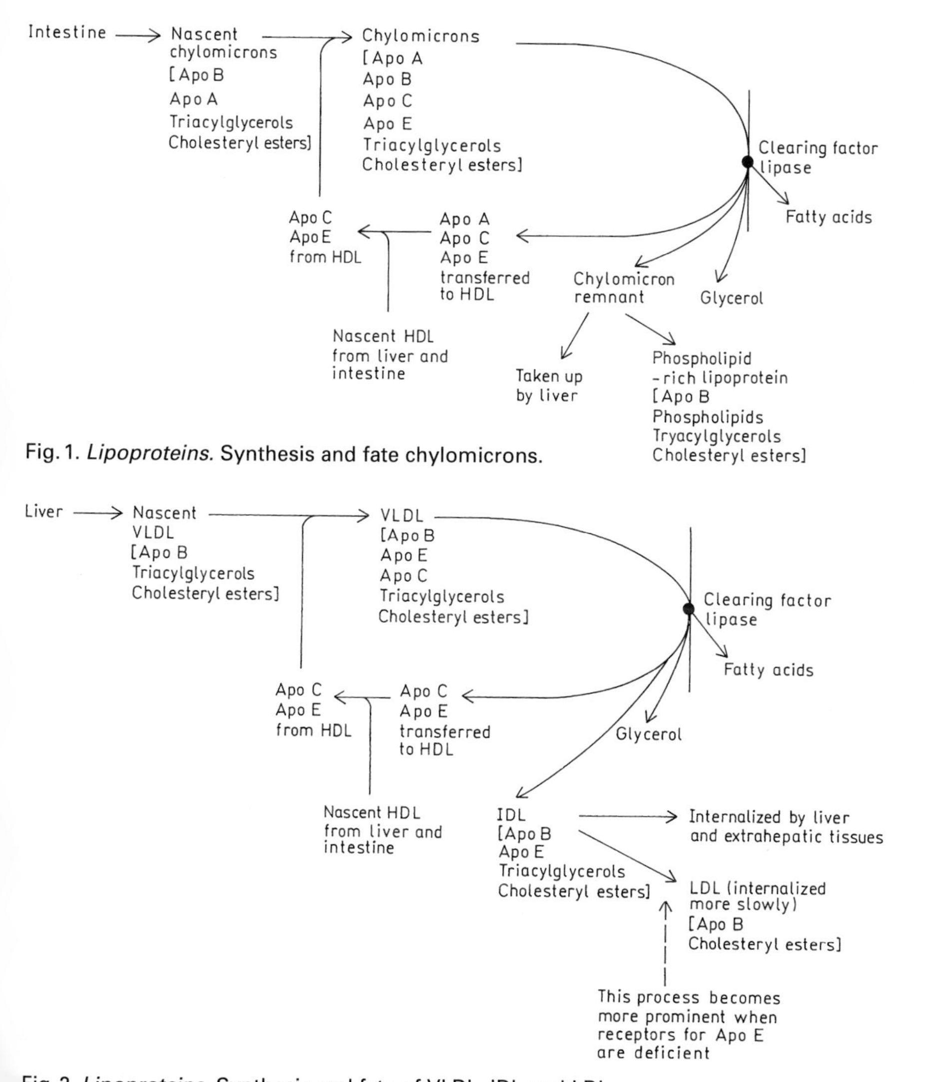

Fig. 1. *Lipoproteins.* Synthesis and fate chylomicrons.

Fig. 2. *Lipoproteins.* Synthesis and fate of VLDL, IDL and LDL.

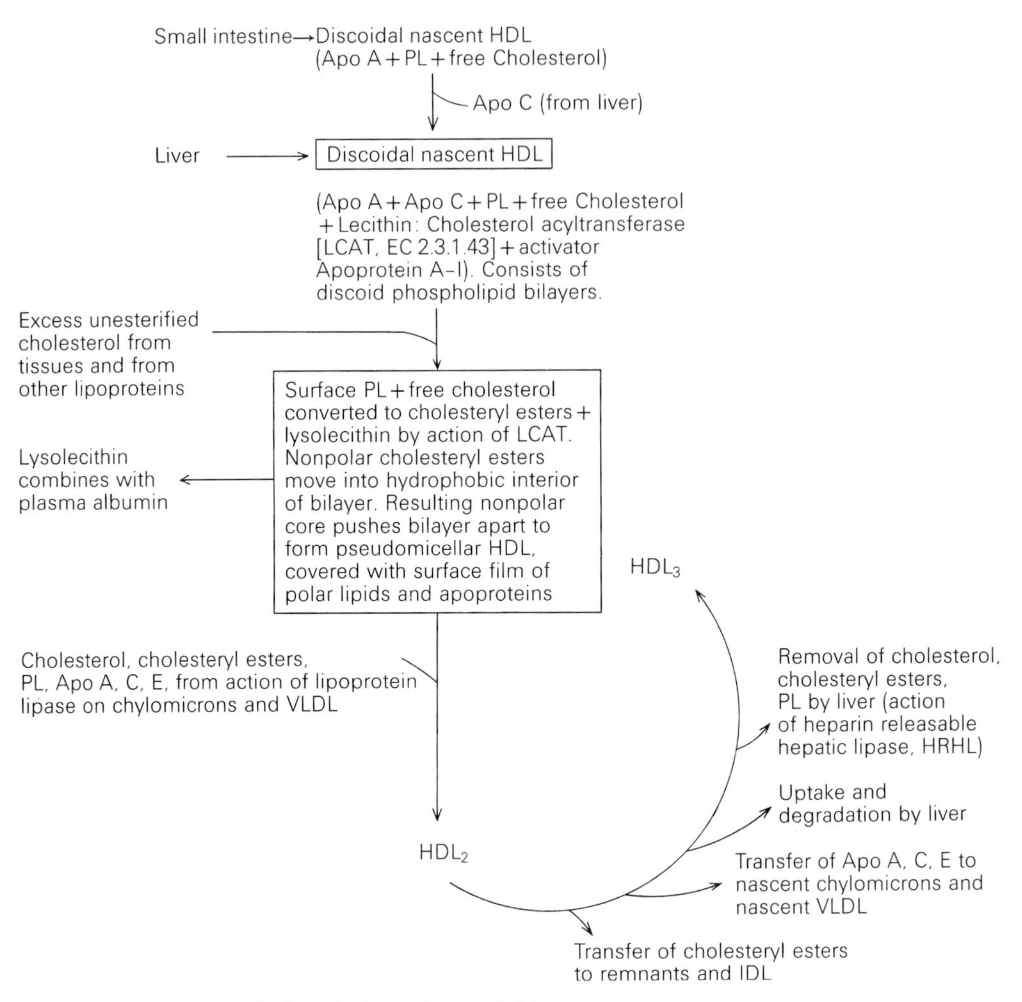

Fig. 3. *Lipoproteins.* The LDL cycle. PL = phospholipid.

are degraded by loss of the apoprotein E, leaving LDL, which contain only a single molecule of apoprotein B-100 and a high content of cholesteryl esters. Apoprotein B-100 also binds the LDL receptor, so that LDL are normally cleared from the blood, albeit more slowly than IDL. Only a small proportion of chylomicron remnants is destroyed by extrahepatic tissues. Chylomicron remnants are more readily internalized than IDL by the liver, and cholesterol derived from chylomicron remnants serves to control liver cholesterol biosynthesis and LDL receptor synthesis. The liver appears to possess specific receptors for chylomicron remnants, in addition to LDL receptors. In an animal model for familial hypercholesterolemia (Watanabe rabbit), LDL receptors are defective, but the uptake of chylomicron remnants by the liver is unaffected. The reasons for the different receptor-binding behavior of chylomicron remnants and IDL are not clear; the phospholipid concentration may be important, and it may be significant that chylomicron remnants possess apoprotein B-48, which differs from B-100 of the IDL.

LDL are kept in suspension in the blood by a small amount of unesterified cholesterol, which presents its hydrophilic OH-group on the outside of the particle. This cholesterol is susceptible to loss from the particle, however, and can then settle on the inside of arteries as atherosclerotic plaque; this is thought to be the reason for the correlation between high concentrations of LDL in the blood and atherosclerosis.

LDL receptors are present in many tissues, with the highest concentrations in adrenal glands and the largest total numbers in the liver. These tissues have especially high requirements for cholesterol, which is the precursor of the adrenal steroid hormones and of the liver bile acids. The number of LDL receptors in any given cell depends on the cell's requirement for cholesterol. In familial hyperlipidemias, the genes for the LDL receptors are defective and the IDL and LDL are removed from the blood slowly or not at

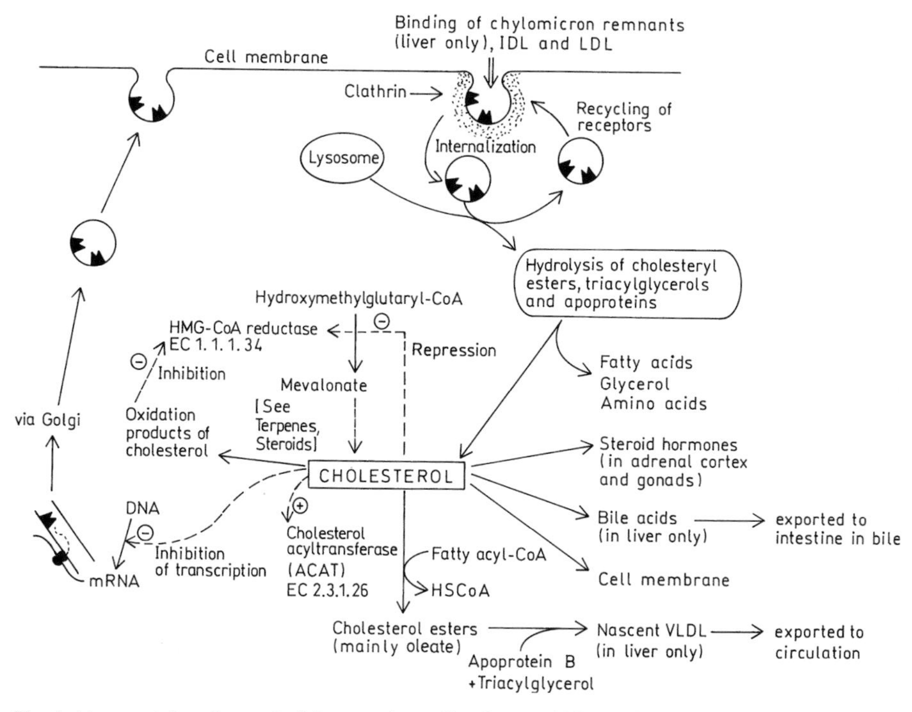

Fig. 4. *Lipoproteins.* Control of the uptake, utilization and biosynthesis of cholesterol by hepatic and extrahepatic tissues.

Classification and properties of the lipoproteins of human blood plasma. [F. T. Lindgren et al. in G. J. Nelson (ed.) *Blood Lipids and Lipoproteins: Quantitation, Composition and Metabolism* (Wiley-Interscience, New York, 1972) pp. 181–274]

Electrophoretic fraction	Ultracentrifuge fraction	Density	Flotation (S_f)	M_r (x 10^{-6})	mg/100 ml plasma	% protein	High content of
Chylomicrons		<0.96	10^3–10^5		0–50	1	Triacyl-
Pre-β	VLDL	0.960–1.006	20–400	5.0–20	150–250	7	glycerols
α_2, β_1	IDL	1.006–1.019	12–20	3.4	50–100	11	Cholesterol
β	LDL	1.019–1.063	0–12	2.0–2.7	315–385	21–23	esters
α_1	HDL	1.063–1.210		0.375	270–380	35–50	Phospholipids
α_1	$VHDL_1$	>1.210	Sediment.	0.145	?	65	Free fatty
Albumin	$VHDL_2$	1.210	constant 2–10S	0.280	?	97	acids

VLDL = Very low density lipoproteins; IDL = Intermediate density lipoproteins; LDL = Low density lipoproteins; HDL = High density lipoproteins; VHDL = Very high density lipoproteins

all. Hence a larger fraction of IDL remains in circulation long enough to be converted to LDL; high concentrations of LDL are the hallmark of this disease. Most (95 %) cases of atherosclerosis are not, however, due to defective LDL receptors.

Oxidized derivatives of cholesterol, rather than free cholesterol, control cholesterol biosynthesis by inhibiting 3-hydroxy-3-methylglutaryl-CoA reductase (HMG-CoA reductase; EC 1.1.1.34) [A. A. Kandutsch et al. *Science* **201** (1978) 498–501]. Prolonged high intracellular concentrations of cholesterol also lead to decreased synthesis of HMG-CoA reductase. Endogenous cholesterol synthesis, however, cannot be totally abolished. Intracellular cholesterol also suppresses synthesis of LDL receptors (as demonstrated in fibroblast culture) (Fig. 4). High intracellular cholesterol therefore ultimately prevents uptake of LDL and IDL by the tissues, so that plasma LDL and IDL persist at abnormally high concentrations; this probably explains the connexion between dietary cholesterol and atherosclerosis. LDL receptors (but not chylomicron remnant receptors) of most inhabitants of the developed world are permanently down-regulated, owing to high dietary cholesterol. On a low-cholesterol diet, the liver of a normal healthy adult synthesizes about 800 mg cholesterol per day, which replaces that lost in the feces. Free cholesterol and bile salts (derived from cholesterol) are excreted in the bile. Some of this is returned in the enterohepatic circulation, but about 250 mg/day of bile salts and about 550 mg/day of free cholesterol are lost in the feces. On a low cholesterol diet, a total of about 1,300 mg of cholesterol are returned to the liver each day by the enterohepatic circulation and by HDL, which carry cholesterol from peripheral tissues to the liver. See Cholestyramine, Compactin, Mevinolin. [M. S. Brown & J. L. Goldstein, *Science* **232** (1986) 34–47; C. R. Scriver, A. L. Beaudet, W. S. Sly & D. Valle (eds.) *The Metabilic and Molecular Bases of Inherited Disease,* 7th edn., McGraw-Hill, New York, 1995; papers by: R. J. Havel & J. B. Kane (pp. 1841–1851), J. B. Kane & R. J. Havel (pp. 1853–1885), J. D. Brunzell (pp. 1913–1932), J. L. Goldstein et al. (pp. 1981–2030), J. L. Breslow (pp. 2031–2052); R. Hampton et al. 'The biology of HMG-CoA reductase: the pros of contra-regulation' *Trends Biochem. Sci.* **21** (1996) 140–145]

Liposome, *phospholipid bilayer vesicle:* an aqueous compartment enclosed by a completely sealed lipid bilayer. L. are formed by ultrasonic irradiation (sonication) of phosphoglycerides (or other suitable lipids) and water. Alternatively, phospholipids are dispersed in the aqueous medium with a detergent, then the concentration of detergent is gradually decreased by dialysis until L. are formed. More simply, phospholipids are allowed to swell in an aqueous medium. Egg lecithin is often used as the phospholipid. L. are spherical or slightly elongated, and may be up to 1–2 μ in diameter. The smallest L. (about 25 nm diameter) are formed by sonication of larger ones. L. may be graded according to size by gel filtration. Rather than a simple vesicle, L. often consist of several concentric vesicles, each with a bilipid membrane, enclosing a layer of aqueous medium. If L. production is performed in the presence of salts, proteins, or other water-soluble components, these also become entrapped in the aqueous phase of the L., where they are isolated from the external environment. L. have attracted considerable interest, firstly as a model for the structure of biological membranes, and secondly as a vehicle for the encapsulation of therapeutic agents. The rationale behind the use of L. for the transport of drugs is based on the fact that L. injected into the blood stream are taken up rapidly by cells of the reticuloendothelial system. This approach is under development for the therapy of leishmaniasis (a parasitic disease of the tropics and subtropics, caused by a hemoflagellate protozoan, *Leishmania,* which invades chiefly the phagocytic cells of the reticuloendothelial system). Injection of L. containing antimonial drugs (meglumine antimonate, or sodium stibogluconate) into leishmaniasis-infected hamsters is about 1,000 times more effective against the disease than injection of the drug alone. [R. M. Straubinger *Methods in Enzymology* **221** (1993) 361–376; K. D. Lee et al. 'Delivery of Macromolecules into Cytosol Using Liposomes Containing Hemolysin from *Listeria monocytogenes*' *J. Biol. Chem.* **271** (1996) 7249–7252]

β-Lipotropic hormone, *β-LPH:* a melanotropic peptide from the pituitaries of several species. The complete structure of β-melanotropin is contained within residues 41–58 of β-LPH (see Peptides, Fig. 3). Human β-MSH (see Melanotropin) is an artifact formed from β-LPH during extraction of the pituitary. [P. J. Lowry & A. P. Scott *Gen. Comp. Endocrinol.* **26** (1975) 16]

Lipotropic substances: compounds directly or indirectly involved in fat metabolism, which can prevent or correct fatty degeneration of the liver. They serve as substrates of phosphatide biosynthesis, or contribute (e.g. by methylation) to the synthesis of these substrates. Thus choline and any substance capable of contributing methyl groups for choline synthesis (e.g. methionine) are L.s. Liver is the major site of synthesis of plasma phosphoglycerides; when the availability of choline is restricted, the rate of phosphatidylcholine synthesis decreases, and the rate of removal of fatty acids from the liver falls below normal. If the rate of supply of fatty acids (free and esterified) to the liver remains normal, the resulting accumulation of fat gives rise to the condition of *fatty liver,* or *fatty degeneration of the liver.*

Lipotropin, *lipotropic hormone, adipokinetic hormone, LPH:* a polypeptide hormone from the hypophysis, that stimulates mobilization of lipids, especially fatty acids, from lipid depots. β-LPH contains 91 amino acid residues, M_r 9,894 (porcine). γ-LPH contains 58 amino acid residues, identical in sequence to residues 1–58 of β-LPH. Corticotropin, melanotropin, ACTH and LPH are all derived from a common precursor (see Peptides); they contain an identical heptapeptide in their structures, and they are grouped together to form the ACTH family of peptide hormones. [J. Bogard et al. *J. Biol. Chem.* **270** (1995) 23038–23043]

Lisetin: an Isoflavonoid (see) and the only known naturally occurring coumaronochromone (Fig.). L. was isolated from the heartwood of *Piscidia erythrina* [Falshaw et al. *Tetrahedron* (1966) Suppl. No. 7, 333–348].

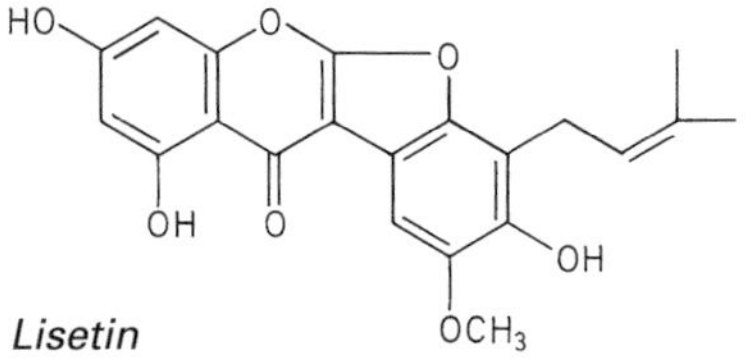

HO O O OH O OH OCH_3

Lisetin

Lithocholic acid: 3α-hydroxy-5β-cholanoic acid, or 3α-hydroxy-5β-cholan-24-oic acid, M_r 376.58, m.p. 185 °C, $[\alpha]_D + 32°$ (ethanol). L. a. is a monohydroxylated steroid carboxylic acid, and one of the bile acids. It has been isolated from the bile of man, cow, rabbit, sheep and goat, and is normally prepared from bovine bile. L. a. is formed from chenodeoxycholate by intestinal bacteria. It is absorbed and returned to the liver for secretion. It is not readily conjugated, and it is relatively toxic to the liver. L. a. may be important in the pathogenesis of liver damage following biliary stasis.

Liver: the largest metabolic gland of vertebrates. L. and pancreas arise from the midgut in the course of embryonic development The human L. weighs 1.5 kg and lies in the abdominal cavity in the right subphrenic space. 1–2 l blood per minute is supplied by the portal vein and the liver arteries, and leaves via the liver veins. The L. secretes bile, which flows out into the duodenum through a system of vessels, of which the gall bladder is a side arm. The functional unit of the L. is the liver cell or *hepatocyte,* which is surrounded by blood vessels and bile capillaries. Metabolically, the L. is the most versatile organ in the body. Nearly all the products of digestion absorbed by the intestine are carried to the liver by the portal vein, there to be transformed or degraded.

Together with the muscles, the L. is the most important site of carbohydrate metabolism. The L. cell is freely permeable to glucose, in contrast to other organs whose permeability to glucose is regulated by insulin. The L. evens out the discontinuous supply of glucose provided by food by storing it as glycogen and releasing it as needed to maintain a constant blood sugar level. A number of hormones regulate these processes. Glucagon stimulates glycogenolysis, while the glucocorticoids promote gluconeogenesis, or synthesis of glucose from precursors such as lactate (which is produced in the muscles) and from glucogenic amino acids. In addition, glucose is converted into other monosaccharides in the liver, including pentoses which are produced by the pentose phosphate cycle. These sugars are needed for the synthesis of nucleic acids. The NADPH produced in the course of pentose synthesis is used in the synthesis of fatty acids. Fructose (from sucrose) and galactose (from lactose) are also metabolized in the L.

The close relationship between lipid and carbohydrate metabolism is shown by the interaction of L. and fat tissue. Fatty acids from food or the body's fat depots are degraded by β-oxidation in the L. Excess fatty acids lead to the formation of ketone bodies. When needed, fatty acids can be synthesized from carbohydrates and transported by the blood as phosphatides to the fat tissue. Here they are converted to neutral fats. The L. is able to synthesize cholesterol and to export it to other tissues, or convert it into bile acids.

The L. plays an important part in nitrogen metabolism. Amino acids absorbed from the digestive tract or produced by internal metabolism are degraded in the L. by transamination and deamination. They can also be synthesized from α-keto acids. The L. is the site of synthesis of the blood plasma proteins (albumin, some of the globulins and fibrinogen) and the hepatogenic blood-clotting factors (prothrombin, accelerator globulin, proconvertin, Christmas factor). The nitrogen from amino acid degradation is converted in the L. into urea, and excreted as such, via the kidneys, in the urine. The L. is also capable of synthesizing uric acid and creatine. It also degrades hemoglobin to bile pigments, which are excreted in the bile.

Many substances, both natural (e.g steroid hormones) and foreign (e.g. drugs) are transformed or degraded in the L. and transported to the kidneys in the blood, in the form of glucuronides or sulfates, where they are excreted in the urine. Insulin is inactivated in the L. by reduction. Vitamin D_3 (cholecalciferol) is enzymatically transformed into 25-hydroxycholecalciferol. In the kidney, this compound is further hydroxylated to 1,25-dihydroxycholecalciferol, which acts as a hormone in the regulation of calcium metabolism.

Given the number and variety of metabolic reactions occurring in the L., it follows that any L. disorder will result in a more or less severe metabolic disturbance.

Living matter: in biochemistry, the body substance of living organisms and cells. It has been determined by elemental analysis that about 40 chemical elements (see Bioelements) occur in L.m., but of these, about 10 account for more than 99 % of the body substance. The dominant inorganic compound is water, the milieu in which life processes take place. Of the organic molecules of L.m. (biomolecules), the quantitatively most important are the carbohydrates, proteins and lipids, which are therefore the main classes of nutrients for humans and animals. Qualitatively, the proteins, as determinants of biological structure, function and specificity, the nucleic acids as carriers of genetic information, and the lipids as structural components of proteolipid membranes (biomembranes) are the most significant classes. Table 1 shows the average water, mineral and biomolecule contents of plant and animal bodies. The large differences in the compositions of L.m. of plants and animals reflect the basic difference in the way these organisms obtain their nutrients.

The carbon-autotrophic green plant is the primary producer of carbohydrates on the earth and uses carbohydrates extensively as storage and cell-wall substances (starch, cellulose, etc.). The nitrogen-heterotrophic animal usually consumes a rather protein-rich diet and uses proteins as supportive materials. The approximate chemical composition of the human body is shown in Table 2. Most of the mass of bacteria (see *Escherichia coli*) is made up by proteins, which may account for 70 % of the dry weight. Numerically,

Table 1. *Water, mineral and organic contents (percent fresh weight) of animal and plant body substance.* The values are rough approximations.

Class of substance	Animal (%)	Plant (%)
Water	60	75
Minerals	4.3	2.5
Carbohydrates	6.2	18
Lipids	11.7	0.5
Proteins	17.8	4.0

Table 2. *Chemical composition of the human body (adapted from Rapoport).* The values are rounded off.

Class of substance	%	Kg (fresh weight)
Water	60	42.0
Minerals	4	2.8
Carbohydrates	1	0.7
Lipids	15	10.5
Proteins	19	13.3
Nucleic acids	1	0.7

however, the lipid molecules predominate. The number of molecules of each class of substance was calculated from the mass using Avogadro's number (6.02×10^{23} = number of molecules in one mole). The biosynthetic capacity of a bacterial cell is given under *Escherichia coli.*

LLD-factor: see Vitamins (vitamin B_{12}).

Loading effect: see Pheromones.

Loading technique: see Methods of biochemistry.

Lobelia alkaloids: an extensive group of 2,6-disubstituted piperidine alkaloids of the genus *Lobelia,* especially from the medicinal plant, *Lobelia inflata,* cultivated in some European countries and in the USA. There are three structural types, depending on the functional groups of the substituents: *lobelidiols, lobelionols* and *lobelidiones* (Fig.). R_1 and R_2 may $-CH_3$, $-C_2H_5$ or $-C_6H_5$, and the configuration of the side chains may be *cis* or *trans.* Compounds lacking the *N*-methyl group, and 2-monosubstituted L. a. also occur.

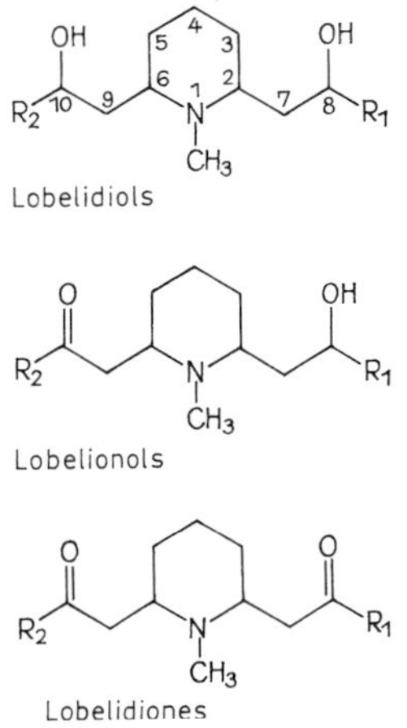

Lobelidiols

Lobelionols

Lobelidiones

Lobelia alkaloids

The chief alkaloid is Lobelin (see). In attempts to imitate the biological synthesis of L. a., lobelidiones have been synthesized under mild conditions from glutardialdehyde, methylamine and acylacetic acid.

Lobelidiols: see Lobelia alkaloids.

Lobelidiones: see Lobelia alkaloids.

Lobelin: (–)-lobelin, *cis*-8,10-diphenyl-lobelionol, the main Lobelia alkaloid (see). Structurally, it is a lobelionol, in which both R_1 and R_2 are phenyl ($-C_6H_5$) groups. L. crystallizes as colorless needles, m. p. 130–131 °C, $[\alpha]_D^{15} - 43°$ ($c = 1$, ethanol). It is isolated from *Lobelia inflata,* and is used medicinally as a respiratory analeptic. On account of its nicotine-like properties, it is also used in the treatment of smoking addiction. Simultaneous administration of nicotine and L. has an additive effect, leading to nausea and aversion.

Lobelionols: see Lobelia alkaloids.

Loganin: an ester glucoside belonging to the iridoid group, m. p. 222–223 °C. Cleavage with emulsin produces the aglycon, loganetin. L. and the free acid, loganic acid, are found in *Strychnos* and *Menyanthus* spp. L. is a key compound in the biosynthesis of iridoids and many alkaloids. For the biosynthesis of L., see Iridoids.

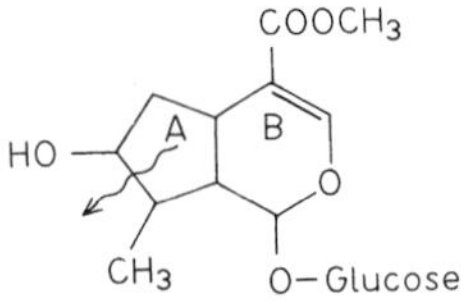

Loganin

Lohmann reaction: see Phosphagens.

Lomatiol: see Naphthaquinones.

Lophenol: 4α-methyl-5α-cholest-7-en-3β-ol, a zoo- and/or phytosterol (see Sterols), M_r 400.69, m. p. 151 °C, $[\alpha]_D + 5°$ ($CHCl_3$). L. has been isolated from rat tissues and feces, and from the cactus *Lophocereus schottii.* It has the structure of a tetracyclic triterpene with a 31,32-*bis*demethyl-lanostane skeleton. Its structure is therefore intermediate between lanosterol and the sterols derived from lanosterol.

Lophophorin: see Anhalonium alkaloids.

Lowry method: see Proteins.

LPH: see Lipotropin

LRH: see Releasing hormones.

LSD: see Lysergic acid diethylamide.

LTH: see Prolactin.

Lubimin: see Phytoalexins.

Luciferase: a low molecular mass oxidoreductase, which catalyses the dehydrogenation of luciferin in the presence of oxygen, ATP and magnesium ions. During this process, 96 % of the energy released appears as visible (mostly blue) light. This is the basis of Bioluminescence (see). All L. so far purified are oligomeric, low molecular mass, thiol proteins. L. of the American firefly (*Photinus pyralis*) has M_r 95,000, with 2 subunits (M_r 50,000). L. of *Photobacterium* has M_r 80,000, with subunits of M_r 38,000 and 41,000. L. of the phosphorescent coral (*Renilla reniformis*) has M_r 34,000, with 3 subunits (M_r 12,000). [T. O. Baldwin 'Firefly luciferase: the structure is known, but the mystery remains' *Structure* **4** (1996) 223–228; E. Conti et al. 'Crystal structure of firefly luciferase throws light on a superfamily of adenylate-forming enzymes' *Structure* **4** (1996) 287–398]

Luciferin: a collective name for the substrates of luciferases. By the action of the enzyme and in the presence of oxygen, L. gives rise to bioluminescence.

Electron-excited states of the oxidation product of L. (thought to be peroxides) are responsible for light emission. *Photinus* L. (for reaction mechanism and formula, see Bioluminescence) is derived from fireflies of the genus *Photinus,* and it represents the only known naturally occurring benzthiazole derivative. Both the D- and the L-form are chemiluminescent, but only the D-compound is biologically active. It crystallizes as yellowish needles, m. p. 190 °C (d.) $[\alpha]_D^{24} - 29°$ (formamide). Different species of *Photinus* emit different colors of yellow to green light, but these differences are due to the different types of luciferase, and the L. is always the same. *Latia* L. from the limpet, *Latia neritoides,* has an unusual sesquiterpene structure. When stimulated, the animal produces a slime, which has a powerful fluorescence due to the interaction of L., luciferase, O_2 and a "purple protein". In the ostracod crustacean, *Cypridina hilgendorfi,* the L. and a specific luciferase are stored in separate glands. When stimula-

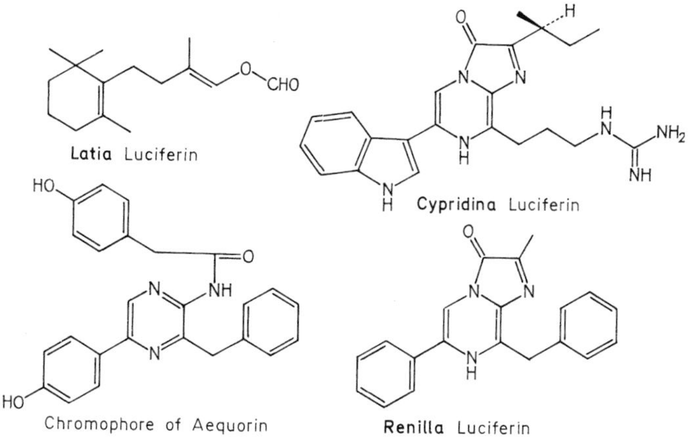

Latia Luciferin

Cypridina Luciferin

Chromophore of Aequorin

Renilla Luciferin

Luciferins

ted, the animal secretes both components simultaneously into the surrounding, oxygen-containing sea water, where they produce a blue fluorescence. *Cypridina* L. is biosynthesized from tryptophan, arginine and isoleucine. *Renilla* L. from the coelenterate *Renilla reniformis* (sea pansy) is an unstable compound, which is stored as its sulfate ester. L. is released by an L. sulfokinase. Electrical or mechanical stimulation results in bioluminescence, which spreads over the surface of the animal in concentric waves of green light.

Bacterial L. has not yet been characterized. The corresponding luciferase produces luminescence in the presence of FMNH and straight chain aldehydes with more than 7 C-atoms. Structural elucidation of L. was delayed on account of their low natural concentration. 30,000 fireflies were required for the isolation of 15 mg L., and 40,000 sea pansies yielded only 0.5 mg of the *Renilla* L. Many L. and synthetic analogs show a spontaneous luminescence in proton-free solvents, such as dimethyl sulfoxide, but the quantum yield is lower than in bioluminescence. See also Photoproteins.

Lumazine: see Pteridines.

Lumicolchicine: see Colchicum alkaloids.

Lumisterol: see Vitamins (vitamin D).

5α-Lupane: see Lupeol.

Lupanine: see Lupin alkaloids.

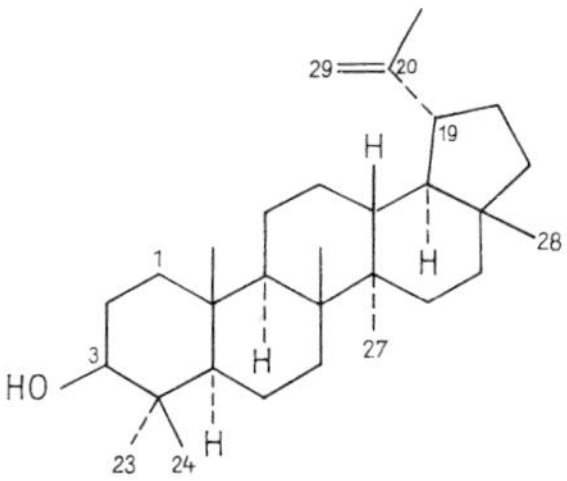

Lupeol

Lupeol: a pentacyclic, triterpene alcohol, M_r 426.73, m.p. 215 °C, $[\alpha]_D^{20}$ + 27.2° (c = 4.8 in $CHCl_3$). L. has a 5α-lupane ring system. It occurs free, esterfied and as the aglycon of triterpene Saponins (see) in many plants. It has been found, e.g. in the latex if *Ficus* spp., in the seed coats of the yellow lupin (*Lupinus luteus*) and in the leaves of mistletoe. It has also been detected in the cocoons of the silk-worm, *Bombyx mori.*

Lupin alkaloids: a group of quinolizidine alkaloids, containing a ring system variously known as quinolizidine, octahydropyridocoline, norlupinane, or l-azabicyclo [0,4,4] decane. This ring system may be

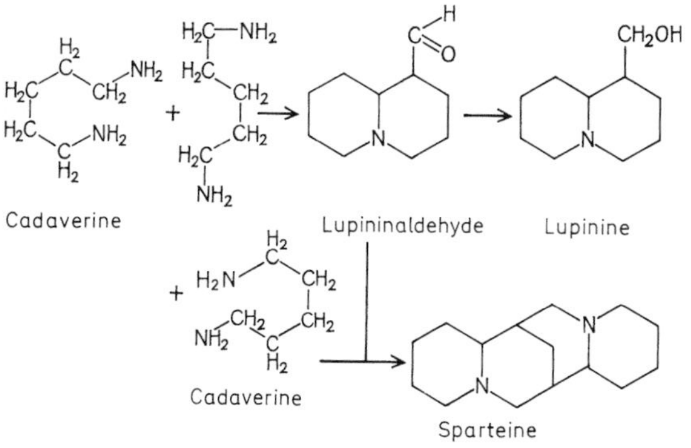

Cadaverine

Lupininaldehyde

Lupinine

Cadaverine

Sparteine

Biosynthesis of the lupin alkaloids, lupinine and sparteine

further condensed with other N-containing ring systems, so that tri- and tetracyclic, as well as bicyclic L. are known. Chief representatives are lupine, lupanine, sparteine and cytisine. L. are derived biosynthetically from lysine, via its decarboxylation product, cadaverine. Two molecules of cadaverine give rise to lupinine, whereas sparteine is derived from three molecules of cadaverine (Fig.). L. occur in plants of the genus *Lupinus* (lupins), in broom (*Sarothamnus scoparius* Koch), laburnum (*Laburnum anagyroides* Medic.) and gorse (*Ulex europaeus.* L.). Bitter lupins cannot be used directly as animal feed, owing to the bitter and toxic L., but these can be removed by steaming, soaking and extraction. Alternatively, selectively bred alkaloid-poor sweet lupins may be used.

Lupinine: see Lupin alkaloids.

Lutein, ***xanthophyll:*** (3*R*,3′*R*,6′*R*)-β,ε-carotene-3,3′-diol, or 3,3′-dihydroxy-α-carotene (Fig.), a carotenoid of the xanthophyll group, M_r 568.85, m.p. 193 °C, $[\alpha]^{20}_{Cd}$ + 160 ° (chloroform), + 145 ° (ethyl acetate). L. contains the same chromophore as α-carotene, and it is isomeric with zeaxanthin. It is a yellow pigment present, together with carotene and chlorophyll, in all green parts of plants. It is also present in many yellow and red flowers and fruits, and is also found in animals, e.g. in bird feathers, egg yolk and the corpus luteum. It may be present in free form, or as an ester; it has no vitamin A activity. L. was first crystallized in 1907 by R. Willstätter from stinging nettles, then later from chicken egg yolk. The structure was elucidated in 1951 by P. Karrer.

Luteinizing hormone, ***LH, lutropin, gonadotropin II, prolan B, metakentrin, corpus luteum ripening hormone, interstitial cell-stimulating hormone, ICSH:*** an important gonadotropin. LH is a glycoprotein (M_r 26,000) containing 22 % carbohydrate. It consists of an α-chain (96 amino acid residues, M_r 10,791) and a β-chain (119 amino acid residues, M_r 12,715) (bovine LH). The primary structures of the chains are known. The α-chain is identical with that of follicle stimulating hormone, thyrotropin and chorionic gonadotropin. The β-chain is hormone-specific. LH acts with follicle stimulating hormone to stimulate growth of the gonads and the synthesis of sex hormones.

Luteinizing hormone releasing hormone: see Releasing hormones.

Luteolin: see Flavones.

Luteolinidine: see Anthocyanins.

Luteotropic hormone: see Prolactin.

Luteotropin: see Prolactin.

Lutropin: see Luteinizing hormone.

Lyases: see Enzymes, Table 1.

Lycopene: one of the carotenoids. L. is an unsaturated, aliphatic hydrocarbon of isoprenoid origin (for formula see Carotenoids), M_r 536, m.p. 173 °C. L. is a red plant pigment, widely distributed and especially plentiful in fruits and berries. For biosynthesis of L. see Tetraterpenes. L. was crystallized from tomatoes in 1876 by A. Millardet. Its empirical formula was determined in 1910 by R. Willstätter and H. H. Escher. The final structural elucidation was performed from 1930 to 1932 by P. Karrer et al. and by R. Kuhn and C. Grundmann.

β-Lycotetraose: see α-Tomatin.

Lymphocytes: colorless, nucleated blood cells, approximately the same size as erythrocytes. L. are immunologically competent, i.e. they recognize antigens, and they play a central role in the Immune response (see). An average human possesses about 10^{12} L., which make up 1.3 kg. Only 3 g are present in the blood, the majority being found in bone marrow, spleen, lymph glands and lymph. L. are classified, according to their origin and function, into ***thymus-dependent L. (T-L.)*** and ***thymus-independent L. (B-L.).*** The T-L. include "killer cells" which form the basis of cellular immunity (e.g. T-L. are responsible for the rejection of transplants), "helper" and "suppressor" T-L. The latter interact with B-L. to promote or inhibit their response to antigens. Clonal multiplication of B-L. gives rise to plasma cells, which produce antibodies (humoral immunity). Each plasma cell produces 2,000 molecules of a single type of antibody per second. Organs rich in lymphocytes are spleen, tonsils, aggregated lymphatic nodules (Peyer's patches) of the ileum, and the vermiform appendix.

Lymphokines: proteins released from activated cells of the immune system which coordinate the immune response. ***Interleukin I*** is a protein of M_r 12,000–16,000, which is released from macrophages stimulated by contact with an antigen. It enhances the response of T-cells which have been primed with antigen, by stimulating their production of interleukin 2. It also stimulates the proliferation of B-cells, in conjunction with B-cell growth factor. Interleukin 1 or a very similar protein is also produced by the keratin-producing cells of the skin corneal cells and cells lining the mouth. It has a range of effects on a variety of non-lymphatic cells, all of which contribute to inflammation. Interleukin 1 may be identical to endogenous pyrogen, a macrophage product which elevates body temperature by interaction with the appropriate region of the hypothalamus.

Interleukin 2 is a protein of M_r 15,500 produced by T-cells, which induces the proliferation of T-cells, but only those which have been antigenically stimulated. ***B-cell growth factor (BCGF)*** is produced by activated T-cells. It stimulates proliferation of B-cells, but not their differentiation to the antibody-producing state. For this, additional L. are required. [J. L. Marx *Science* **221**(1983) 1362–1364.]

Interleukin 3 (Il-3) regulates growth and differentiation of the pluripotent stem cells which are precur-

OH

HO

Lutein

sors to all types of blood cells. It appears to be identical with the following: multi-colony stimulating factor CSF, hematopoietic growth factor, burst-promoting activity, P-cell stimulating factor, mast cell growth factor, histamine-producing cell stimulating factor, Thy 1-inducing activity.

Il-3 is a glycoprotein with an estimated M_r of about 30,000. The murine gene for Il-3 has been cloned and sequenced. It encodes a polypeptide of 166 amino acids, of which the *N*-terminal 32 are removed during maturation. [M.C.Fung et al. *Nature* **307** (1984) 233–237]

Lymphotoxin: see Tumor necrosis factor.

Lys: abb. for Lysine.

Lysergic acid: a tetracyclic indole derivative, M_r 368.32. The D-form, m.p. 240°C (d.), $[\alpha]_D^{20}+40°$ ($c = 0.5$, pyridine), is one of the Ergot alkaloids (see). Change of configuration at C8 produces biologically inactive isolysergic acid. L. is a close structural relative of LSD (see Lysergic acid diethylamide).

Lysergic acid diethylamide, *LSD:* a synthetic derivative of D-lysergic acid. It is the most potent psychotomimetic substance known. Hallucinogenic derivatives of lysergic acid occur in the Mexican ritual drug, ololiuhqui (seeds of *Rivea corymbosa*); these are lysergic acid amide (ergine) and lysergic acid hydroxyethylamide. LSD can be prepared semisynthetically from ergot alkaloids, which also occur in *Rivea* seeds (e.g. ergometrine). The hallucinogenic action of LSD, which is characterized by a state resembling schizophrenia, was discovered accidentally by A.Hofmann during the recrystallization of a sample of LSD tartrate.

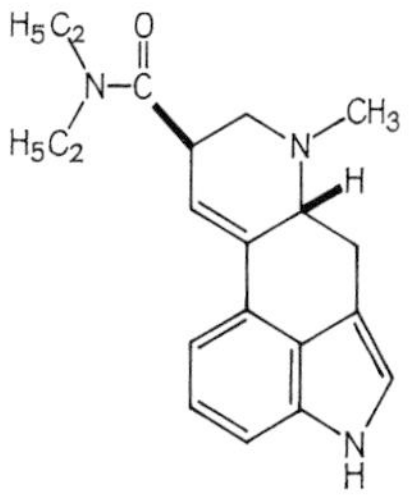

D-Lysergic acid diethylamide (LSD-25)

L-Lysine, *Lys:* 2,6-diaminocaproic acid, $H_2N-(CH_2)_4-CH(NH2)-COOH$, a basic proteogenic, essential amino acid, M_r 146 2, m.p 224°C (d), $[\alpha]_D^{25}$ +25.9° ($c = 2$ in 5 M HCl), + 13.5° ($c = 2$, water). The proteins of cereals (wheat, barley, rice) and other vegetable foodstuffs are rather poor in Lys. Children and young growing animals have a particularly high requirement for Lys, since it is needed for bone formation. Like threonine, Lys does not take part in reversible transamination.

In rat liver, Lys degradation (Fig. 1) occurs primarily in the mitochondria. Some features of the degradation pathway appear to be a reversal of the reactions of Lys synthesis (Figs. 4 & 5). Saccharopine is formed, offering a direct route to 2-aminoadipic-6-semialdehyde, and bypassing the cyclic piperideine and piperidine intermediates. The semialdehyde is also produced by a second pathway, initiated by L-amino acid oxidase. The resulting oxoacid cyclizes spontaneously to Δ^1-piperideine 2-carboxylic acid.

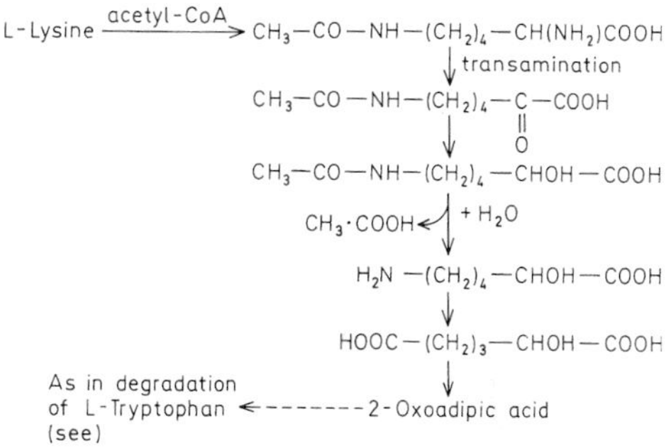

L-Lysine —(acetyl-CoA)→ $CH_3-CO-NH-(CH_2)_4-CH(NH_2)COOH$
↓ transamination
$CH_3-CO-NH-(CH_2)_4-C(=O)-COOH$
↓
$CH_3-CO-NH-(CH_2)_4-CHOH-COOH$
↓ + H_2O, → $CH_3 \cdot COOH$
$H_2N-(CH_2)_4-CHOH-COOH$
↓
$HOOC-(CH_2)_3-CHOH-COOH$
↓
2-Oxoadipic acid - - - → As in degradation of L-Tryptophan (see)

Fig. 2. *L-Lysine.* Degradation of L-lysine by yeasts.

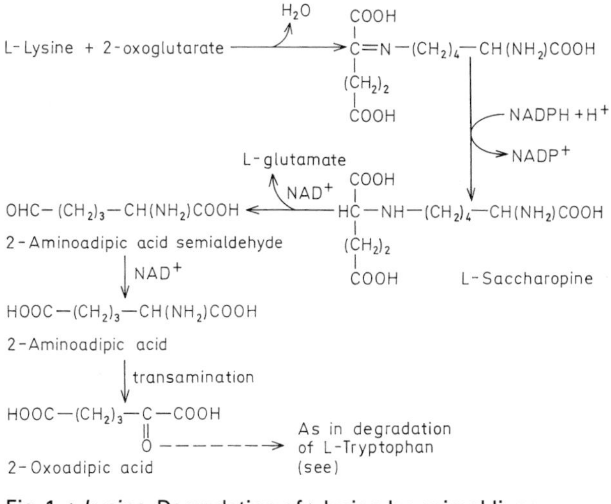

L-Lysine + 2-oxoglutarate —(− H_2O)→ $HOOC-(CH_2)_2-C(COOH)=N-(CH_2)_4-CH(NH_2)COOH$
↓ NADPH + H^+ → $NADP^+$
L-Saccharopine: $HOOC-(CH_2)_2-CH(COOH)-NH-(CH_2)_4-CH(NH_2)COOH$
↓ NAD^+ → L-glutamate
$OHC-(CH_2)_3-CH(NH_2)COOH$
2-Aminoadipic acid semialdehyde
↓ NAD^+
$HOOC-(CH_2)_3-CH(NH_2)COOH$
2-Aminoadipic acid
↓ transamination
$HOOC-(CH_2)_3-C(=O)-COOH$
2-Oxoadipic acid - - - → As in degradation of L-Tryptophan (see)

Fig. 1. *L-Lysine.* Degradation of L-lysine by animal liver.

The semialdehyde is subsequently oxidized to 2-aminoadipic acid, which is transaminated to 2-oxoadipic acid. Further degradation, via glutaryl-CoA to acetyl-CoA is identical with the final stages of L-tryptophan degradation (see L-Tryptophan). An alternative pathway (Fig. 2) of Lys degradation in mammals (also in yeast) involves acetylated intermediates; cyclization is thus prevented. A further pathway exists in bacteria (Fig. 3). D-Lysine may also enter this pathway by conversion to L-lysine by a pyridoxal phosphate-dependent racemase. Lys is converted to 5-aminovaleramide by L-lysine oxygenase.

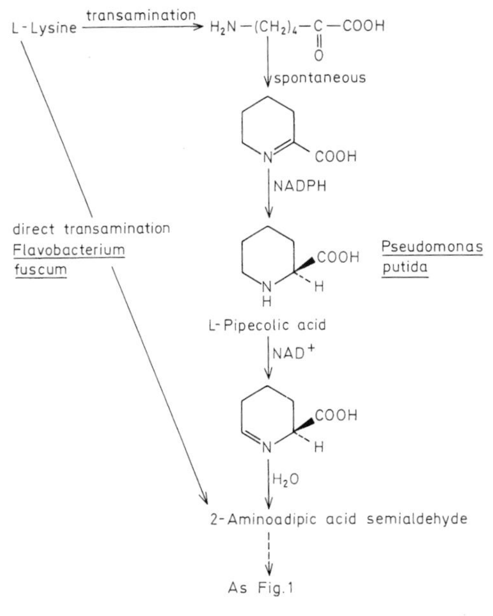

Fig. 3. *L-Lysine.* Bacterial degradation of L-lysine.

There are two pathways of Lys biosynthesis in plants and microorganisms: 1. The *aminoadipate pathway* (Fig. 4) is found in *Neurospora crassa* and *Saccharomyces cerevisiae,* and in other fungi and yeasts. The reactions of this pathway are analogous to those of the Tricarboxylic acid cycle (see). 2. The *diaminopimelate pathway* (Fig. 5) occurs in bacteria, cyanobacteria, green algae, higher plants and certain fungi.

About 20,000 tons of Lys per year are produced in Japan by a specific fermentation process (Kyowa fermentation industry). This Lys is used as a supplement to foodstuffs, and to increase the biological value of low value plant dietary proteins. The growth rate of poltry and pigs is appreciably increased by the addition of 0.1–0.3 % Lys to their feed. Also of technical importance is the enzymatic synthesis of Lys from DL-2-aminocaprolactam, using microbial L-aminocaprolactam hydrolase. The remaining D-2-aminocaprolactam is racemized enzymatically, then converted entirely to L-Lys. Lys was first isolated by Drechsel from casein in 1889.

Lysocephalins: see Membrane lipids.

Lysogeny: see Phage development.

Lysolecithins: see Membrane lipids.

Lysophosphatidic acid: see Lysosomal storage diseases (mucolipisosis IV).

Lysopine: see D-Octopine.

Lysosomal storage diseases: Inborn errors of metabolism (see), in which specific hydrolases are absent from the lysosomes, leading to accumulation

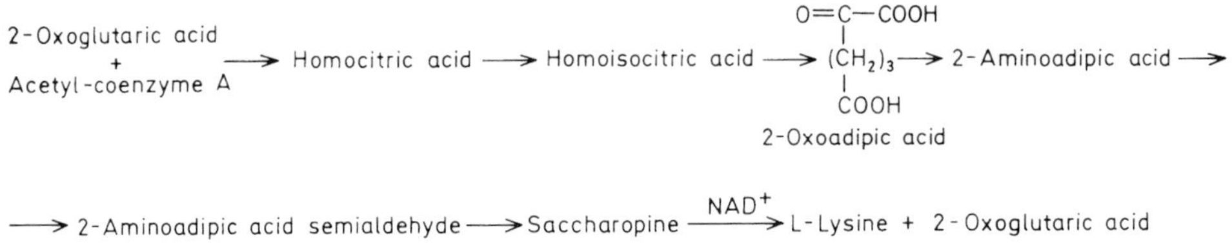

2-Oxoglutaric acid + Acetyl-coenzyme A ⟶ Homocitric acid ⟶ Homoisocitric acid ⟶ $O{=}C(COOH)-(CH_2)_3-COOH$ (2-Oxoadipic acid) ⟶ 2-Aminoadipic acid ⟶

⟶ 2-Aminoadipic acid semialdehyde ⟶ Saccharopine $\xrightarrow{NAD^+}$ L-Lysine + 2-Oxoglutaric acid

Fig. 4. *L-Lysine.* 2-Aminoadipic acid pathway of L-lysine biosynthesis.

Pyruvic acid + L-Aspartic acid $\xrightarrow{-H_2O}$ 2,3-Dihydropicolinic acid $\xrightarrow{NADPH+H^+}$ Δ^1-Piperideine-2,6-dicarboxylic acid ⟶ N-Succinyl-2-oxo-L-6-aminopimelic acid ⟶

⟶ N-Succinyl-L-2,6-diaminopimelic acid ⟶ L-2,6-Diaminopimelic acid ⟶ $HOOC-CH(NH_2)-(CH_2)_3-CH(NH_2)-COOH$ (Meso-2,6-Diaminopimelic acid) $\xrightarrow{-CO_2}$ L-Lysine

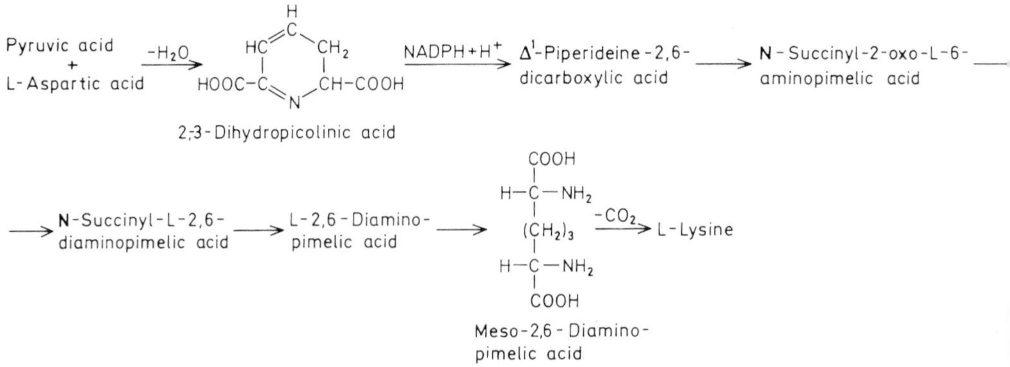

Fig. 5. *L-Lysine.* Diaminopimelic acid pathway of L-lysine biosynthesis.

or "storage" of the enzyme substrate. A genetic defect may result in total failure to synthesize a lysosomal hydrolase, absence of an activator protein, synthesis of a less stable enzyme, or failure to target the synthesized enzyme to lysosomes. L.s.d. are usually classified and named according to the nature of the accumulated material, e.g. mucopolysaccharidoses, oligosaccharidoses, mucolipidoses, sphingolipidoses, etc.

Some human lysosomal storage diseases (with deficient enzyme and clinical findings)

Acid phosphatase deficiency.
Lysosomal acid phosphatase (EC 3.1.3.2). Fatal in infancy.

Glycogenosis type II, or ***Pompe's disease.***
See Glycogen storage diseases. Of the several glycogen storage diseases, only Pompe's disease is also a L.s.d.

Cholesterol ester storage disease.
Acid lipase (EC 3.1.3.2). Hepatic lipase about 25 % of normal. Cholesterol esters deposited in liver, spleen, intestinal mucosa, lymph nodes, aorta. Hepatomegaly, leading to hepatic fibrosis. Sometimes jaundice and/or splenomegaly. Relatively benign. Autosomal recessive. Wolman's disease (below) is probably the expression of a different mutant allele at the same locus. [J.M.Hoeg et al. *Amer. J.Hum. Gen.* **36** (1984) 1190–1203]

Wolman's disease.
Acid lipase (EC 3.1.3.2). Cholesterol esters and triacylglycerols deposited in adrenals, liver, spleen, bone marrow, capillaries, endothelium, ganglion cells of mesenteric plexus and mucosa of small intestine. Plasma lipids mainly normal. Hepatosplenomegaly. Adrenal calcification and enlargement. Failure to thrive in infancy, rapid deterioration and death. Autosomal recessive. See Cholesterol ester storage disease.

Aspartylglycosaminuria (an oligosaccharidosis).
N^4-(β-N-Acetylglucosaminyl)-L-asparaginase (EC 3.5.1.26). Impaired degradation of some glycoproteins. Severe mental retardation, motor impairment, and large amounts of urinary aspartylglycosamine (Fig. 1) and other glycoasparagines.

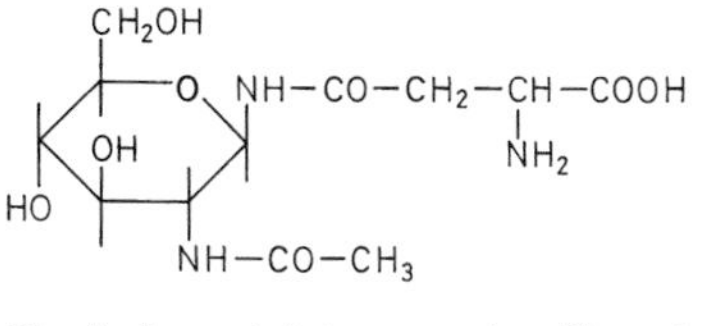

Fig. 1. *Aspartylglucosamine [2-acetamido-1-(β-aspartamido)-1,2-dideoxyglucose]*

Mannosidosis (an oligosaccharidosis).
α-Mannosidase (EC 3.2.1.24). Accumulation of mannose-rich, glucosamine-containing oligosaccharides. Brain damage, bone abnormalities, opaque cornea and cataracts, hepatosplenomegaly. Skeletal involvement and facial features similar to Hurler's syndrome (below).

Fucosidosis (an oligosaccharidosis).
α-L-Fucosidase (EC 3.2.1.51). Accumulation of fucose-rich glycoproteins, sphingolipids and glycosaminoglycans. Severe progressive cerebral degeneration in infancy (rapid in type I, slow in type II). Skeletal involvement similar to Hurler's syndrome (below). Hepatosplenomegaly (type I shows cardiomegaly). Facial features different from Hurler's syndrome.

Sialidosis, or ***Mucolipidosis I*** (an oligosaccharidosis and a mucolipidosis).
Glycoprotein sialidase (neuraminidase) (EC 3.2.1.18). Other lysosomal activities in liver sometimes increased. Sialyloligosaccharides accumulate. Mental retardation, skeletal abnormalities and Hurler-type facial features may be present in varying degrees, or absent. Hepatosplenomegaly, renal involvement and hydrops fetalis sometimes present.

Sialic acid storage disease (an oligosaccharidosis).
Unidentified defect. possibly a defect of lysosomal membrane transport. Free acetylneuraminic acid in urine. Impaired motor function and ataxia. In severe form (type II) there is progressive neurological deterioration and early death. Type II involves moderate coarsening of features, growth retardation (rickets) and hepatosplenomegaly; these effects are absent from type I, which shows some skeletal involvement (curved tibiae).

Mucolipidosis I.
The same as Sialidosis (above).

Mucolipidosis II, or ***I-cell disease.***
UDP-N-acetyl-D-glucosamine-lysosomal-enzyme N-acetylglucosamine phosphotransferase (EC 2.7.8.17). Glycoaminoglycans and glycolipoids accumulate in fibroblasts, hepatocytes and Schwann cells. Clinically similar to Hurler's syndrome (below). Death in childhood.

Mucolipidosis III, or ***Pseudo-Hurler polydystrophy.***
UDP-N-acetylglucosamine-lysosomal-enzyme N-acetylglucosamine phosphotransferase (EC 2.7.8.17). Similar to, but milder than mucolipidosis type II. Survival to adulthood.

Mucolipidosis IV.
Exo-α-sialidase (EC 3.2.1.18). Enzyme hydrolyses α-2,3-, α-2,6- and α-2,8-glycosidic linkages of terminal sialic acid residues. Gangliosides, glycolipids, lipofuscin and lyso*bis*phosphatidic acid accumulate in lysosomes of many tissues. Corneal opacities. Dementia. No skeletal effects. No hepatomegaly.

Type I_H mucopolysaccharidosis, or ***Hurler's syndrome,*** or ***Gargoylism*** (see Fig. 2).
α-L-Iduronidase (EC 3.2.1.76). Large skull, depressed nose bridge, hypertrichosis, short neck, projecting forehead, large tongue and lips, widely spaced teeth, and gum hypertrophy. Thick, hairy skin, and usually clouded cornea. Coronary valves, vessels and heart muscles often affected, leading to death from heart failure before 20th year. Hepatosplenomegaly and skeletal abnormalities (dwarfism and kyphosis). Severe and progressive mental retardation. Heparan sulfate and dermatan sulfate accumulation in tissues.

Type I_S mucopolysaccharidosis, or ***Scheie's syndrome*** (see Fig. 2).
α-L-Iduronidase (EC 3.2.1.76). Heart involved only in some cases. No mental retardation. Skeleton affected less than in type $I_{H/S}$. Facial features similar to type I_H.

Type II mucopolysaccharidosis, or ***Hunter's syndrome*** (see Fig. 2).

Iduronate-2-sulfatase (EC 3.1.6.13). Heparan sulfate and dermatan sulfate accumulate. No clouding of cornea. Heart involvement rare. Mental retardation less than in type I_H. Facial features similar to type I_H. Type II_B slightly less severe than II_A.

Type III_A mucopolysaccharidosis, or ***Sanfilippo A*** (see Fig. 2).

Heparan N-*sulfatase.* Heparan sulfate accumulates. No clouding of cornea. Heart not affected. Skeleton only slightly affected. Facial features similar to type I_H, but less severe.

Type III_B mucopolysaccharidosis, or ***Sanfilippo B*** (see Fig. 2).

α-N-*Acetylglucosaminidase* (EC 3.2.1.50). Heparan sulfate and glycososphingolipids accumulate. Facial features similar to type III_A.

Type III_C mucopolysaccharidosis, or ***Sanfilippo C.***

Glucosamine N-*acetyltransferase* (EC 2.3.1.3). Heparan sulfate accumulates. Facial features similar to type III_A.

Type III_D mucopolysaccharidosis, or ***Sanfilippo D*** (see Fig. 2).

N-*Acetylglucosamine-6-sulfatase* (EC 3.1.6.14). Heparan sulfate accumulates. Mild osteochondrodystrophy, and presence of hypoplastic odontoid process. Facial features similar to type III_A.

Type IV_A mucopolysaccharidosis, or ***Morquio-Brailsford syndrome*** (variant 1) (see Fig. 2).

N-*Acetylgalactosamine-6-sulfatase* (EC 3.1.6.4). Keratan sulfate and sometimes chondrotin sulfate peptide accumulate. No mental retardation. Cornea sometimes clouded. Heart involvement rare. Marked stunting of growth. Severe and distinctive changes in ribs, sternum, vertebrae and bones of hands and feet. Thin enamel.

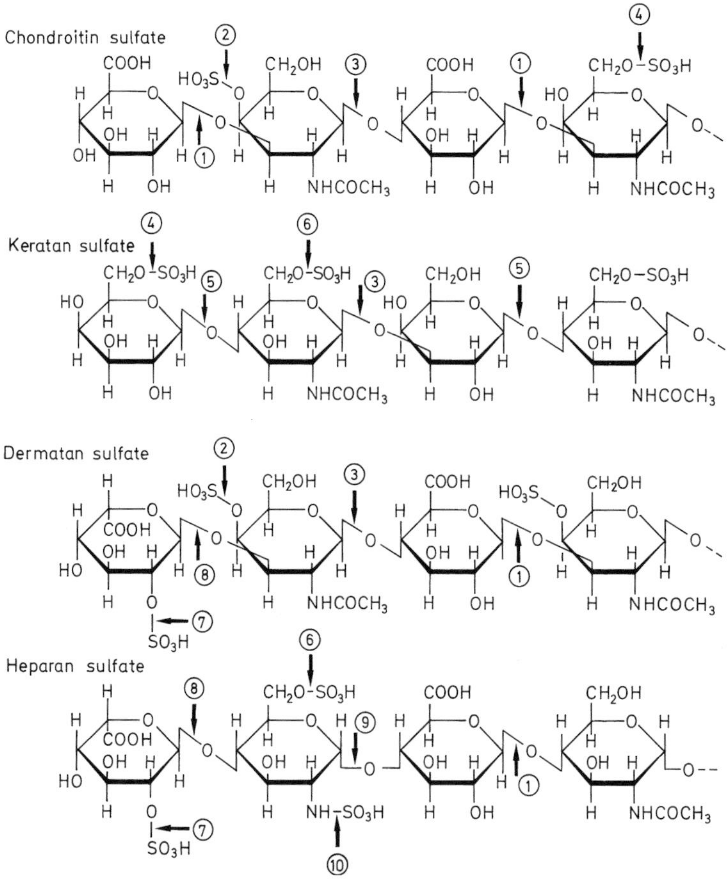

Fig. 2. *Glycosaminoglycans (mucopolysaccharides) and sites of attack by degradative enzymes.* Absence or defective function of any one of these enzymes leads to accumulation of incompletely degraded mucopolysaccharide, i.e. mucopolysaccharidosis. 1. β-Glucuronidase; 2. *N*-Acetylgalactosamine-4-sulfatase; 3. β-*N*-Acetylhexosaminidase; 4. *N*-Acetylgalactosamine-6-sulfatase; 5. β-Galactosidase; 6. *N*-Acetylglucosamine-6-sulfatase; 7. Iduronate-2-sulfatase; 8. L-Iduronidase; 9. α-*N*-Acetylglucosaminidase; 10. Heparan *N*-sulfatase.

Type IV$_B$ mucopolysaccharidosis, or ***Morquio-Brailsford syndrome*** (variant 2) (see Fig. 2).

β-Galactosidase (EC 3.2.1.23). Enzyme has high K_m for keratan sulfate and normal K_m for gangliosides. Keratan sulfate accumulates. No mental retardation. Enamel normal. Only mild bone changes. Heart involvement rare. Cornea clouded. Facial features normal and unaffected.

There is no type V mucopolysaccharidosis.

Type VI$_A$, VI$_B$, VI$_C$ mucopolysaccharidosis, or ***Maroteaux-Lamy syndrome*** (see Fig. 2).

N-*Acetylgalactosamine-4-sulfatase* (EC 3.1.6.12). Dermatan sulfate accumulates. No mental retardation. Heart affected in some cases. Facial features less severely affected than in type I_H. In type VI_A skeletal involvement is similar to type I_H without vertebral deformation. VI_B and VI_C show moderate and mild skeletal effects, respectively.

Type VII mucopolysaccharidosis, or ***Sly syndrome*** (see Fig. 2).

β-Glucuronidase (EC 3.2.1.31). Chondroitin 4-sulfate and chondrotin 6-sulfate accumulate. Dermatan sulfate and/or heparan sulfate sometimes also accumulate. Mental retardation not severe. Cornea sometimes clouded. Facial appearance may resemble type I_H, or may be less severely affected, or normal. Skeletal deformities usually very marked. Heart not affected

Gangliosidosis, or ***Generalized gangliosidosis,*** or ***Type I G$_{M1}$ gangliosidosis,*** or ***Neurovisceral lipidosis,*** or ***Pseudo-Hurler syndrome,*** or ***Maladie de Landing*** (a sphingolipidosis) (see Fig. 3).

β-Galactosidase A_1, A_2 and *A_3* (EC 3.2.1.23). A sphingolipidosis, involving accumulation of G_{M1} ganglioside and desialo-G_{M1} ganglioside in neurons. Keratan sulfate-related glycosaminoglycan accumulates in spleen, liver, epithelial cells of kidney glomeruli, and bone marrow. Mental and motor deterioration. Hepatosplenomegaly. Invariably fatal by end of 2nd year.

Juvenile gangliosidosis, or ***Type II G$_{M1}$ gangliosidosis*** (a sphingolipidosis) (see Fig. 3).

β-Galactosidase A_2 and A_3 (EC 3.2.1.23). Only two of the three liver β-galactosidases are absent, and sphingolipids do not accumulate in the liver. Progress of disease is slower than that of generalized gangliosidosis, but the clinical picture is otherwise similar. Death usually at 3–10 years. G_{M1} deposited exclusively in central nervous system.

Adult gangliosidosis, or ***Type III G$_{M1}$ gangliosidosis*** (a sphingolipidosis) (see Fig. 3).

β-Galactosidase (EC 3.2.1.23). Enzyme activity about 5 % of normal. Motor deterioration. Little intellectual impairment. [A. T. Hoogeveen et al. *J. Biol. Chem.* **259** (1984) 1974–1977]

G$_{M3}$ Gangliosidosis, or ***G$_{M3}$ Sphingolipodystrophy*** (a sphingolipidosis) (see Fig. 3).

(N-*Acetylneuraminyl)-galactosylglucosyl-ceramide* N-*acetylgalactosaminyl transferase* (EC 2.4.1.92).

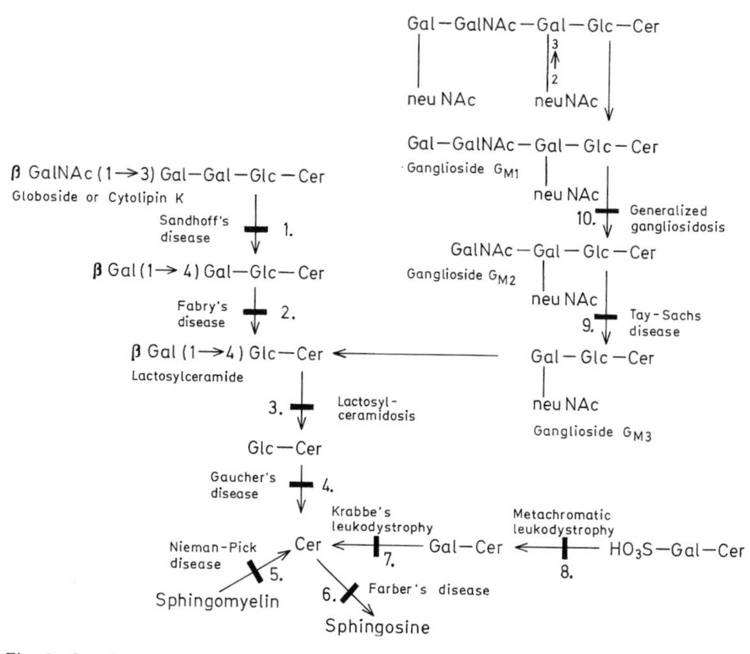

Fig. 3. *Catabolism of glycososphingolipids and sphingomyelin in humans.* The sites of metabolic defects are shown by solid bars. Cer, ceramide; Gal, galactosyl; Glc, glucosyl; GalNAc, *N*-acetylgalactosaminidyl; neuNAc, *N*-acetylneuraminidyl; 1. β-*N*-Acetylhexosaminidase, EC 3.2.1.20; 2. Ceramide trihexosidase, EC 3.2.1.22; 3. Ceramide lactoside β-galactosidase, EC 3.2.1.?; 4. Glucosylceramidase, EC 3.2.1.45; 5. Sphingomyelin phosphodiesterase, EC 3.1.4.12; 6. Acylsphingosine deacylase, EC 3.5.1.23; 7. Galactosylceramidase, EC 3.2.1.46; 8. Arylsulfatase, EC 3.1.6.8; 9. β-*N*-Acetylhexosaminidase, EC 3.2.1.30; 10. β-Galactosidase, EC 3.2.1.23.

Synthesis of G_{M1} and G_{M2} is deficient, and G_{M3} accumulates in brain and liver. Clinically similar to generalized gangliosidosis. Death in first year.

Adult G_{M2} gangliosidosis (a sphingolipidosis) (see Fig. 3).

β-N-Acetylhexosaminidase A (EC 3.2.1.30). G_{M2} accumulates intraneuronally in white and gray matter of brain; small quantities accumulate in liver and spleen. Manifested at 18–40 years (sometimes 2–4 years, followed by slow progression). Mental and motor deterioration. Death in few years after first manifestation.

Lactosyl ceramidosis, or ***Ceramide lactoside lipidosis*** (see Fig. 3).

Ceramide lactoside β-galactosidase. Accumulation of ceramide lactoside. Hepatomegaly. Splenomegaly. Progressive brain damage and neurological impairment.

Tay-Sachs disease (a sphingolipidosis) (see Fig. 3).

β-N-Acetylhexosaminidase A (EC 3.2.1.30). In type B, the enzyme is absent. In type AB, the enzyme is present but an activator protein is absent. In type $A^{M}B$, the enzyme is present but defective. G_{M2} accumulates intraneuronally in white and gray matter of the brain, with small quantities in spleen and liver. Clinical picture and prognosis similar to infantile form of Sandhoff's disease, with manifestation at 4–6 months (ganglioside accumulation begins in utero).

Fabry's disease, or ***Angiokeratoma corporis diffusum*** (a sphingolipidosis) (see Fig. 3).

Ceramide trihexosidase (α-galactosidase) (EC 3.2.1.22). Sphingolipidosis due to accumulation of ceramide trihexoside. Skin lesions (purple macules and papules). Corneal opacities, cataracts, retinal edema. Cardiovascular, neurological and gastrointestinal disorders. Characteristic severe burning pains in extremities. X-chromosome-linked.

Gaucher's disease (a sphingolipidosis): (a) Chronic (adult) form (at least 2 subtypes); (b) Infantile neuronopathic form; (c) Juvenile neuronopathic form (Norbottnian); (d) Prenatal neuronopathic form. See Fig. 3.

Glucosylceramidase (EC 3.2.1.45). Enzyme activity about 15 % of normal in (a). Enzyme totally absent in (b) and (d), and about 2.5 % of normal in (c). Glucosyl ceramide accumulates in liver, spleen, bone marrow and leukocytes; it accumulates in the brain in (b), (c) and (d), and in the lungs in some cases of (b). Hepatosplenomegaly. Anemia. Pancytopenia. Ostealgia and osteoporosis. Purpura. Cerebral degeneration in (b) and (c). Type (a) is manifested between 1 and 60 years, and may result in death at any time from infection or liver dysfunction. Type (b) is manifested in the first year, and is fatal by the end of the first or second year. Type (c) is manifested at 6–20 years, and is fatal in adolescence or early adulthood. Type (d) results in death in utero (ascites hydrops fetalis).

Sandhoff's disease (a sphingolipidosis) (see Fig. 3).

β-N-Acetylhexosaminidases A and *B* (EC 3.2.1.30). G_{M2} and G_{A2} accumulate intraneuronally in white and gray matter of the brain, and in spleen and liver. Globoside (cytolipin K) accumulates in kidneys and spleen. *N*-Acetylglucosaminyloligosaccharides appear in tissues and urine. The infantile form is manifested between 4 and 6 months (ganglioside accumulation begins in utero), with marked hyperacusis, a cherry red spot in the macular region, progressive cerebral degeneration, and usually death before 5 years. The juvenile (late infantile) form is manifested in the first year, progresses slowly, and shows no red macular spot.

Metachromatic leukodystrophy (a sphingolipidosis) (see Fig. 3).

Arylsulfatase A (EC 3.1.6.8). Accumulation of galactosyl sulfate ceramide and lactosyl sulfate ceramide in kidneys and gall bladder, and galactosyl sulfate ceramide alone in brain white matter and peripheral nerve sheaths. Cerebral degeneration and motor disturbances. Death 5–10 years after onset, or at 3–7 years of age (infantile form).

Krabbe's disease, or ***Krabbe's leukodystrophy,*** or ***Galactosylceramide lipidosis,*** or ***Globoid cell leukodystrophy*** (a sphingolipidosis) (see Fig. 3).

Galactosylceramidase (EC 3.2.1.46). Failure to hydrolyse galactosylceramide to galactose and ceramide. Galactosylceramide accumulates in axons, but not in myelin sheaths. Myelin almost totally disappears. Globoid cells infiltrate brain white matter. Rapid neurological deterioration. Always fatal in infancy.

Mucosulfatidosis (a form of metachromatic leukodystrophy, and a sphingolipidosis).

Nine distinct *sulfatases* acting on sulfatides, steroids and glycosaminoglycans, and *β-Galactosidase.* Accumulated metabolites include gangliosides, sulfatides, cholesterol sulfate and dehydroepiandrosterone sulfate in neurons and liver. Glycosaminoglycans excreted in the urine. Moderate hepatosplenomegaly. Progressive, with motor disturbances, ataxia and convulsions. Cerebral degeneration. Onset at 1–4 years. Death before 12th year.

Nieman-Pick disease (a sphingolipidosis) (see Fig. 3).

Sphingomyelin phosphodiesterase (EC 3.1.4.12). Sphingomyelin and cholesterol accumulate in neurons, liver, spleen, bone marrow, lymphatic tissue and lungs. Five types of the disease are recognized (types A-E), differing in severity. Type D possesses normal, and type E normal or lowered sphingomyelin phosphodiesterase activity, and the genetic defect has not been identified. Type E is benign. The other types involve hepatosplenomegaly and mental, physical and motor retardation, except type B, which has no neurological involvement. In type A, deposition of sphingomyelin begins in utero, and death usually occurs at 2–3 years. Type C is characterized by excessive accumulation of unesterified cholesterol, accompanied by relatively less accumulation of sphingomyelin. Cholesterol accumulation appears to be due to an error of cholesterol processing and homeostasis, in which cellular uptake of cholesterol from low density lipoproteins is excessive, and its subsequent esterification is deficient. Types C (onset during first 6 months) and D (onset at birth) show neurological disturbances at 5–10 years, with death usually at 5–15 years. [H. S. Kruth et al. *J. Biol. Chem.* **261** (1986) 16769–16774; P. G. Pentchev et al. *J. Biol. Chem.* **261** (1986) 16775–16780]

G_{M1} Gangliosidosis with sialidosis (a sphingolipidosis) (see Fig. 3).

β-Galactosidase (EC 3.2.1.23) and *Sialidase* (neuraminidase) (EC 3.2.1.18). Both enzymes lack a protective glycoprotein and are unstable. G_{M1} and sialyloligosaccharides accumulate in neurons, spleen, liver, bone marrow and kidneys. The infantile form (type I) is manifested in early infancy and is clinically similar to generalized gangliosidosis. The juvenile form (type II) is manifested at 3–20 years, shows progressive neurological deterioration with ataxia and moderate mental retardation, coarse facial features, bone deformities and cherry-red macular spots, and does not involve visceromegaly. Life expectancy is possibly normal.

Juvenile G_{M2} gangliosidosis (a sphingolipidosis) (see Fig. 3).

β-N-Acetylhexosaminidase (EC 3.2.1.30). Late infantile and juvenile forms are recognized. Loss of enzyme activity shows wide variations. Site of G_{M2} accumulation as in Tay-Sachs disease. Clinically similar to Tay-Sachs, but often no red macular spot, and progress of the disease is slower. Death in childhood.

Farber's disease, or ***Disseminated lipogranulomatosis*** (a sphingolipidosis) (see Fig. 3).

Acylsphingosine deacylase (ceramidase) (EC 3.5.1.23). Ceramide, glycolipids and sometimes dermatan sulfate accumulate subcutaneously, in tendons (especially at joints), and in neurons. Swollen joints and subcutaneous nodules. Growth retardation. Death usually at age 7–22 months, but some survivors.

Neuronal ceroid lipofuscinosis, or ***Familial amaurotic idiocy*** (a sphingolipidosis). Types: ***Santavuori*** (infantile), ***Jansky-Bielschowsky*** (late infantile), ***Batten-Spielmeyer-Vogt*** (juvenile), ***Kufs*** (adult).

Defect unidentified. Possibly involves metabolism of dolichol-linked oligosaccharides. Arachidonic acid-containing phosphoglycerides increased in brain and serum. Dolichol, dolichyl phosphate, ceroid and lipofuscin accumulate in neurons. Cerebral degeneration with optic atrophy. Mental and motor retardation with blindness. Late forms do not involve blindness.

Lysosomes: organelles, 0.2–2 nm diameter, found in the cytoplasm of eukaryotic cells. L. are bounded by a single lipoprotein membrane, but otherwise show no fine structure. Under the light or electron microscope, L. are markedly polymorphic. They can be characterized biochemically or histochemically, but not morphologically. L. are sites of Intracellular digestion (see), particularly of biological macromolecules, such as proteins, polynucleotides, polysaccharides, lipids, glycoproteins, glycolipids, etc. Approximately 40 different lysosomal hydrolases are responsible for this degradative activity; they all show optimal activity at acidic pH values. Marker enzyme for L. is acid phosphatase. Under anaerobic conditions, L. are destroyed, and the lysosomal enzymes are released into the cytoplasm; subsequent degradation of the cell contents by the lysosomal enzymes is known as autolysis. Autolysis is a characteristic post mortem process.

Primary L. are formed in the Golgi apparatus of the cell. Fusion of L. with phagocytosing and pinocytosing vacuoles (phagosomes) produces digestive vacuoles, known as secondary L. Excess or old cell parts, including mitochondria, may be digested by L. (phagolysosomes). In ameba, L. apparently provide the digestive enzymes. L. were discovered in 1959 by De Duve (Nobel prize 1974).

Lysozyme, *endolysin, muramidase, N-acetylmuramide glycanohydrolase* (EC 3.2.1.17): a widely occurring hydrolase, found in phages, bacteria, plants, invertebrates and vertebrates. In the latter, it is found particularly in egg white, saliva, tears and mucosas. L. acts as a bacteriolytic enzyme by hydrolysing the β-1,4 linkage between *N*-acetylglucosamine and *N*-acetylmuramic acid in the proteoglycan of the bacterial cell wall. L. therefore affords protection against bacterial invasion. All animal L. consist of a single chain of 129 amino acid residues, with homologous sequence (M_r 14,200–14,600). There are four disulfide bridges, and the chain is folded into a known tertiary structure (42 % α-helix, with hydrophobic tryptophan residues on the outer surface), with prominent hydrogen bonding between the side chains of Ser, Thr, Asn and Gln. Like other hydrolases (e.g. papain, ribonuclease), the molecule of L. has a cleft which houses the active center of the enzyme and serves for the attachment of the substrate, in this case a hexasaccharide unit of the proteoglycan molecule. There is a remarkable correspondence between the primary and tertiary structures of L. and α-lactalbumin (123 residues). It is thought that both proteins arose from a common precursor protein with lysozyme activity, an example of divergent evolution by gene duplication. On the other hand, there is no structural relationship between animal L. and bacteriophage L. The latter contain 157 residues (λ phage endolysin, M_r 17,873), or 164 residues (T4- and T2-phage L., M_r 18,720), and they are either endoacetylmuramidases (T4, T2), or endoacetylglucosaminidases (Streptococcal L.).

Lysylbradykinin: see Bradykinin.

M

mμ: millimicron, an obsolete symbol for nm.

Maackiain: 3-hydroxy-8,9-methylenedioxyptero-carpan. See Pterocarpans.

Macdougallin: 14α-methyl-5α-cholest-8(9)-en-3β,6α-diol, a phytosterol (see Sterols), M_r 416.69, m.p. 173 °C, $[\alpha]_D$ +72 ° ($CHCl_3$), from the cacti, *Peniocerius fosteriunus* and *P. macdougalli.* It is a tetracyclic triterpene with a 30,31-*bis*demethyllanostane ring system, and it is intermediate in structure between lanosterol and those sterols derived from lanosterol.

Macroelements: see Mineral nutrients.

α_2-Macroglobulin, *α_2-antiplasmin:* an α_2-plasma protein. The first reported M_r (by sedimentation diffusion) was 820,000. Later determination by sedimentation equilibrium gave values of 725,000. These results, together with studies of subunit composition, indicate that the true M_r lies in the range 650,000–725,000. α_2-M. is a glycoprotein containing 8.2 % carbohydrate. The carbohydrate moiety contains mannose, fucose, *N*-acetyl-glucosamine and sialic acid. Electron microscope studies of the protein reveal a structure resembling two beans facing each other; these two identical subunits are bound noncovalently. Each subunit consists of two peptide chains linked covalently by a disulfide bridge. α_2-M. binds tightly and inhibits a number of proteases of varying specificity and origin, e. g. trypsin, plasmin, thrombin, kallikrein and chymotrypsin. It is therefore a natural inhibitor of plasmin. Unlike other protease inhibitors, it does not block the active centers of the enzymes, so that α_2-M.-proteinase complexes are almost fully active toward low molecular mass synthetic substrates. The bound protease acts upon the α_2-M. subunit (M_r about 350,000) and cleaves one protein chain near its midpoint. All the products of this cleavage are still linked covalently by disulfide bridges. Treatment of the α_2-M.-trypsin complex with urea and dithiothreitol reveals two proteins derived from α_2-M., M_r 185,000 (intact subunit chain, originally linked to a similar chain by disulfide bridge), and M_r 85,000 (proteolytic derivative of a 185,000 M_r chain). α_2-M. selectively inhibits the growth of tumors in cell cultures and in rats. α_2-M. and ceruloplasmin are two plasma proteins with a specific binding affinity for zinc. In humans, the normal plasma concentration of α_2-M. is 220–380 mg per 100 ml.

Macrolides, *macrolide antibiotics:* a group of antibiotics from various strains of *Streptomyces,* all with the same complex macrocyclic structure. M. inhibit protein synthesis by blocking transpeptidation, and translocation on the 50S ribosomal subunit (similar to Chloramphenicol, see). Examples of M. are erythromycin (Fig.), spiramycin, oleandomycin, carbomycin, angolamycin, leucomycin, picromycin. Almost all M. are used therapeutically as broad spectrum antibiotics.

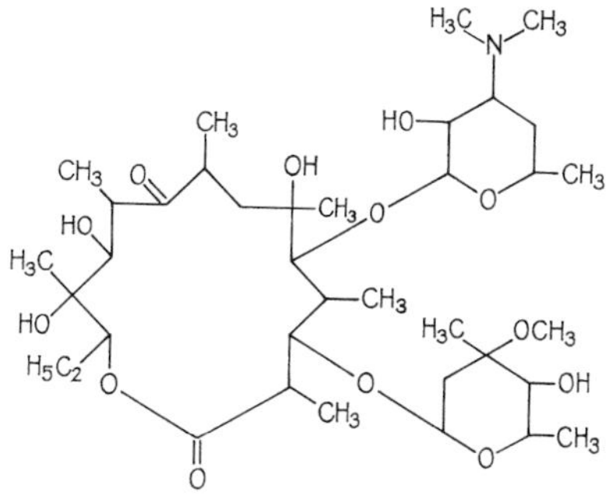

Erythromycin

Madder dyes: plant dyes from madder (*Rubia tinctorum*) and other members of the madder family (*Rubiaceae*). Important representatives are the glycosidically bound components, alizarin and purpurin. M. d. were chiefly used to make madder enamel, a colored paint with outstanding fastness to light. Now the product is made almost exclusively from synthetic alizarin.

Magnesium, *Mg:* a cation widely distributed in biological systems, with many different biological functions. Mg is the fourth most abundant cation in vertebrates, and it has great biological significance as a constituent of the porphyrin system of chlorophyll. Mg is an essential nutrient, and it plays an important part in metabolism by acting as a cofactor of many different enzyme systems. In particular, it is involved in the reactions of energy metabolism, including the enzymatic cleavage of phosphate esters and the transfer of phosphate groups. It is an activator of phosphatases and a cofactor of practically all ATP-dependent phosphorylation reactions, e.g hexokinase, phosphofructokinase and adenylate kinase; in every case an ATP-Mg complex (1 : 1) is formed, and this (not free ATP) is a substrate. Intracellularly, Mg is present chiefly as Mg^{2+} and $MgOH^+$.

Maize factor: see Zeatin.

Major histocompatibility complex, *MHC:* several closely linked genetic loci in vertebrates which encode cell-surface glycoproteins and serum proteins known as *histocompatibility antigens.* In addition to the MHC, which have been found in all vertebrates examined for them, there are other cell surface antigens which have been little studied and are referred to as minor histocompatibility antigens (complexes).

The MHC in humans and mice are by far the best studied. They were discovered as factors which cause rejection of organ grafts between genetically nonidentical individuals, hence the term "histocompatibility". Their significance to the organism, however, lies in their function as moderators of the immune response. Three classes of MHC antigens have been characterized in humans and mice; other MHC genes in the mouse have been mapped to the H-2 region.

Class I includes the HLA-A, HLA-B, HLA-C and HLA-D gene products in humans, and the *H2K* and *H2D* gene products in mice. The corresponding loci in rats are Rt-1, and in chickens, B. The products of these highly polymorphic genes are components of cell-surface glycoproteins which consist of two polypeptide chains each. Only one of these polypeptides is encoded by the MHC gene; the other is a β_2-microglobulin. In the mouse, the MHC chain has M_r about 44,000, while the β_2-microglobulin chain has M_r 12,000. Class I antigens are found on the surfaces of all cells except sperm. They are important in the cell-mediated immune response, in which macrophages, cytotoxic T lymphocytes (CTL) and granulocytes are induced to kill body cells which bear specific foreign (viral or cancer) antigens on their surfaces. However, these specifically "programmed" killer cells attack only cells which display both the foreign antigen and the same class I antigens as the killer cells themselves. Activated CTL from one mouse cannot kill cells from a second mouse infected with the same virus unless the second mouse has the same set of class I antigens as the first. The actual function of the class I antigens in this process is still unknown.

Class II MHC antigens are found only on certain lymphoid cells. These are products of (mouse) *Ia* and *Ir* (for immune response) genes.

The *Ia* proteins are membrane-bound dimers of α- (M_r 55,000) and β- (M_r 28,000) subunits; they are responsible for activation of helper T lymphocytes. The first step of this process is ingestion of an antigen by accessory cells (macrophages), which partially digest proteins and "present" peptides to the precursors of helper T cells. Helper T lymphocytes enhance antibody production by those B lymphocytes which recognize the same antigen as the helper cells; thus in order to be antigenic, a peptide must stimulate both T and B lymphocytes; such peptides are actually bound to the *Ia* molecules. The ability of a peptide to bind to a given *Ia* molecule is closely correlated to its antigenicity. Peptides which do not bind to the *Ia* molecules of a given mouse strain are not antigenic in that strain. [S. Buus et al. *Science* **235** (1987) 1353–1358]

The known immune response genes influence antibody production, delayed type hypersensitivity and proliferation of T cells in vitro. Enhancing *Ir* genes map to the A, B, C or E subregions of the *I* part of the mouse *H-2* (MHC) complex; suppressing *Ir* genes map to the J subregion of the *I* region.

Class III MHC genes code for a serum protein consisting of three covalently linked polypeptides, known as the α-, β- and γ-chains (M_r 87,000, 78,000 and 33,000, respectively). This protein is the C4 component of the classic Complement system (see). Other complement components encoded by MHC-linked genes are C3 in mouse, Bf in guinea pig, man and mouse, and C2 in man. [J. Klein "The Major Histocompatibility Complex of the Mouse" *Science* **203** (1979) 516–521; H. L. Ploegh et al. *Cell* **24** (1981) 287–299]

Major inorganic elements: see Mineral nutrients.

Malate: see Malic acid.

Maleic hydrazide, ***MH:*** 1,2-dihydro-3,6-pyridazinedione, a synthetic plant growth retardant, used as a herbicide against grasses. It causes a unique depression of growth, inhibition being marked but temporary. It inhibits seed germination, suppresses growth of roots and terminal shoots, and retards flower and bud development. It prevents the formation of suckers in tobacco and tomatoes, and prevents the sprouting of onions and potatoes. The effects are limited to green plants and there is little or no effect on any other organism. The reason for this specificity is not known. Animals are very tolerant to MH; the LD_{50} for rats is 4 g per kg, and rats are unaffected by a diet containing 1 % MH throughout their lives. There is nevertheless some evidence for carcinogenic activity and MH is listed as a suspected carcinogenic agent.

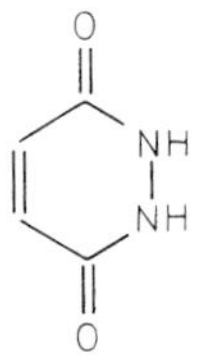

Maleic hydrazide

Malformin A: a heterodetic cyclic pentapeptide with antibiotic activity, from *Aspergillus niger.* It causes malformation of the roots of cereals. The primary structure was revised in 1974, and confirmed by total synthesis.

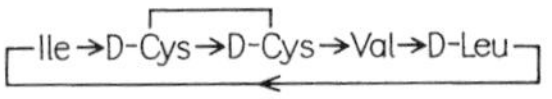

Ile→D-Cys→D-Cys→Val→D-Leu (cyclic; disulfide bridge between the two Cys)

Malformin A

Malic acid: monohydroxysuccinic acid, HOOC–CHOH–CH_2–COOH, a dicarboxylic acid found in many plant juices, usually in the L(+)-form, m.p. 100 °C, b.p. 140 °C (d.). The malate ion is formed in the Tricarboxylic acid cycle (see) and in the Glyoxylate cycle (see). Malate plays an important role in the Diurnal acid rhythm (see) of the *Crassulaceae* (stonecrop family).

Malic enzyme(s), ***L-malate-NADP oxidoreductase, decarboxylating*** (EC 1.1.1.40): an important enzyme found in most organisms, which catalyses the decarboxylation of L-malate to pyruvate and CO_2, with concomitant reduction of $NADP^+$ to NADPH (or the synthesis of malate by the reverse reaction): HOOC–CH_2–CHOH–COOH + $NADP^+$ $\rightleftharpoons$ CH_3–CO–COOH + CO_2 + NADPH + H^+. M. e. has various metabolic roles: 1. Synthesis of malate by the action of M. e. may serve as an Anaplerotic reaction (see) of the TCA-cycle; 2. An important route for the total combustion of any TCA-cycle intermediate is conversion to malate followed by decarboxylation of malate to pyruvate and CO_2 by M. e. Animal mitochondria contain two M. e., one specific for NADP, the other utilizing both NADP and NAD. Regulation is complex; when glycolysis is low, free CoA activates M. e., thereby promoting oxidation of malate, whereas rapid glycolysis increases NADH which inhibits M. e.; 3. In plants operating the C4 pathway of photosynthesis (see Hatch-Slack-Kortschack cycle), mesophyll cells export malate to bundle sheath cells, where it is decarboxylated by M. e.; the resulting CO_2 is assimilated by the Calvin cycle while the pyruvate returns to the mesophyll cells; 4. M. e. appears to be important in a cycle for the generation of cytoplasmic NADPH: malate exported from mitochondria is decarboxylated by cytoplasmic M. e., with the formation of NADPH. The pyruvate enters the mitochondria and is converted to oxaloacetate, followed by hydrogenation to malate by NADH. Thus, reducing power (NADH) from mitochondria is converted into reducing power (NADPH) in the cytoplasm, where it is available for NADPH-dependent biosynthetic reactions (see Hydrogen metabolism).

Malonic acid: HOOC–CH_2–COOH, a dicarboxylic acid, m. p. 135.6 °C, which has been found in the free form in plants, but is of only sporadic occurrence. At the pH of the cell M. a. is present as its anion (malonate), which is a known competitive inhibitor of succinate dehydrogenase in the tricarboxylic acid cycle. A metabolically important derivative of M. a. is malonyl-CoA, an intermediate of Fatty acid biosynthesis (see).

Maltase: see Disaccharidases.

Maltose, *maltobiose, malt sugar:* 4-O-α-D-glucopyranosyl-D-glucose, a reducing disaccharide, M_r 342.3, $[\alpha]_D^{20}$ +112° → + 130.4° (c = 4). The monohydrate of M. (m. p. 102–103 °C), crystallized from water or dilute ethanol, does not lose water of crystallization when dried at room temp. in vacuo over P_2O_5. M. consists of two D-glucopyranose residues linked by an α-1,4-glycosidic bond. M. is a stereoisomer of cellobiose. With lactose and sucrose, it is one of the three most common naturally occurring disaccharides. It can be hydrolysed to 2 molecules of D-glucose by dilute mineral acids, or enzymatically by α-glucosidase (known previously as maltase) of yeast, malt and digestive secretions. M. is the fundamental structural unit of starch and glycogen; it also occurs free in higher plants. The monohydrate of M. is prepared technically in about 80 % yield by degradation of starch with amylases (diastase). It serves as a fermentable substrate in brewing, as a component of prepared bee food, as a substrate in microbiological growth media, and generally in food and pharmaceuticals as a nutrient and sweetener (one third as sweet as glucose).

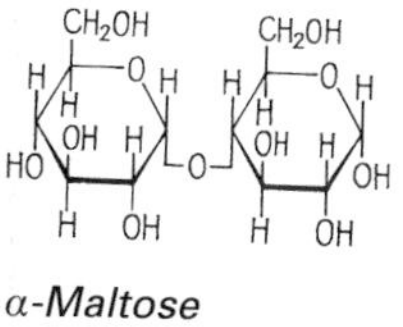

α-Maltose

Malt sugar: see Maltose.

Malvidin: 3,5,7,4′-tetrahydroxy-3′,5′-dimethoxyflavylium cation, the aglycon of various anthocyanins. 10 natural glycosides of M. are known, e. g. oenin (syn. primulin, ligulin), the 3-β-glucoside of M., is the pigment in the skins of black grapes and the flowers of *Primula* spp.; and malvin (3,5-di-β-glucoside of M.) from common mallow (*Malva sylvestris*) and other flowering plants.

Malvin: see Malvidin.

Mammotropin: see Prolactin.

Manganese, *Mn:* a bioelement present in all living cells, which usually contain less than 1 ppm on a dry weight basis, or less than 0.01 mM in fresh tissue. Bone contains 3.5 ppm. Bacterial spores contain high levels of Mn(II) (0.3 % dry weight). Mn(II) is necessary for sporulation in *Bacillus subtilis,* and these bacteria can maintain an intracellular Mn concentration of 0.2 mM against an external concentration of 1 μM. Mn is an essential nutrient for animals and plants. Mn deficiency in animals leads to degeneration of the gonads and to skeletal abnormalities; the characteristic skeletal abnormality in chickens is called slipped tendon disease, or perosis. In plants, Mn deficiency results in chlorosis and mottling.

Many glycosyl transferases (in particular galactosyl and *N*-acetylgalactosaminyl transferases) require Mn for activity, which explains the impairment of mucopolysaccharide metabolism associated with the symptoms of Mn deficiency. Mn is required by the enzyme farnesyl pyrophosphate synthetase, which catalyses a stage in cholesterol synthesis. Mn is also required at an earlier point in cholesterol synthesis probably at some stage in the conversion of acetate into mevalonate. Lactose synthetase requires Mn. Pyruvate carboxylase contains four atoms of Mn (i. e. one for each biotin molecule); Mn(II) is essential for the transcarboxylation, and the initial ATP-dependent carboxylation of biotin requires either Mn(II) or Mg(II). Superoxide dismutase from *E. coli* (M_r 39,500) contains two atoms of Mn(III). Yeast mitochondrial superoxide dismutase is a tetramer containing one atom of Mn per subunit (M_r 24,000). Similar enzymes are present in chicken liver mitochondria. Presumably the Mn of superoxide dismutases alternates between the III and II states during catalysis. Manganin, a protein present in peanuts (M_r 56,000), contains one atom of Mn. Avimanganin, a protein of unknown function from avian liver (M_r 89,000) contains one atom of Mn(III). Concanavalin A (M_r 190,000) from jack bean contains one atom Mn(II).

Mn may also be important in the regulation of enzyme activity, e. g. nonadenylylated glutamine synthetase requires Mg(II), but the adenylylated form binds Mn(II). Also the specificity of nucleases and DNA polymerases is changed when Mg(II) is replaced by Mn(II). The physiological significance of these differences is not clear.

The reaction center of photosystem II of the chloroplast contains 2–4 Mn atoms. The effect of Mn deficiency on photosynthesis resembles poisoning by DCMU (see), i. e. evolution of O_2 by photosystem II is inhibited, while photosystem I is unaffected. It is thought that the Mn is intimately involved in the primary event of the photolysis of water, probably alternating between the III and II states.

Mannans: polysaccharides widely distributed in plants as reserve material, and in association with cellulose as hemicellulose. M. of plants consists of D-mannose predominantly in α-1,4-glycosidic linkage. Yeast cell walls contain M., consisting of a backbone of α-1,6-linked mannose with short (1–3 mannose units) branches attached by α-1,2- and α-1,3-linkages.

D-Mannitol, *mannite, manna sugar:* HOH_2C–CHOH–CHOH–CHOH–CHOH–CH_2OH, M_r 182.17, m.p. 166 °C, $[\alpha]_D$ −2.1° (water). M. is a hexitol structurally related to D-mannose. It is found widely in plants and plant exudates, and in fungi and seaweeds. The exudate of the manna ash (*Fraxinus ornus*) contains 75 % M. M. is used as a sugar substitute for diabetics.

D-Mannosamine: 2-amino-2-deoxy-D-mannose, an amino sugar related to mannose (the hydroxyl group on C2 of mannose is replaced by an amino group). M_r 179.17, m.p. of the hydrochloride 180 °C, $[\alpha]_D$ −3°. M. is a constituent of neuraminic acids, animal mucolipids and animal mucoproteins.

D-Mannose, *seminose, carubinose:* a monosaccharide hexose, M_r 180.16. α-Form: m.p. 133 °C, $[\alpha]_D^{20}$ +29.3° → +14.2° (water). β-Form: decomposes at 132 °C, $[\alpha]_D^{20}$ −17.0° → +14.2° (c = 4). D-M. is a C2 epimer of D-glucose. In plants, free D-M. is found only occasionally, but it is a constituent of many high molecular mass polysaccharides in algae, yeasts and higher plants. Manna (the exudate of the manna ash, *Fraxinus ornus*) contains particularly high levels of D-M. In the metabolism of D-M., the activated form of the sugar is the GDP derivative (not the UDP derivative which occurs for many other sugars). D-M. is degraded via its 6-phospho-derivative, which is converted to glucose 6-phosphate.

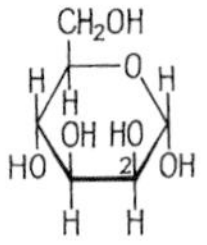

D-Mannose

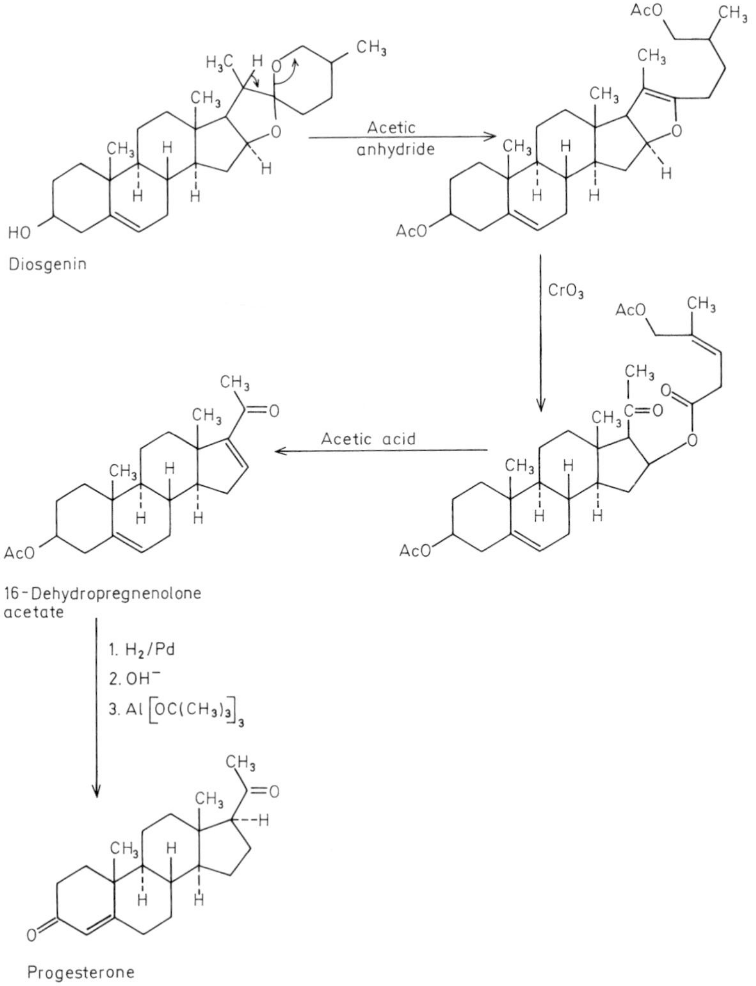

Diosgenin

Acetic anhydride

CrO_3

Acetic acid

16-Dehydropregnenolone acetate

1. H_2/Pd
2. OH^-
3. Al$[OC(CH_3)_3]_3$

Progesterone

Marker synthesis of progesterone from diosgenin

Mannosidosis: see Lysosomal storage diseases.

D-Mannuronic acid: a uronic acid, M_r 194.14, derived from D-mannose. M. is a constituent of the polyuronide, alginic acid.

Maple syrup urine disease: see Inborn errors of metabolism.

Marasmus: a nutritional deficiency disease, mostly in children under one year, due to lack of both dietary energy and protein. The condition is usually compounded by deficiencies of minerals and vitamins, and by gastrointestinal infections. Body weight is less than 60 % of the standard for the age. Dehydration is common, there is little or no subcutaneous fat, and plasma albumin levels are about 25 g/l (normal levels are above 40 g/l). The characteristic hair and skin lesions of kwashiorkor are absent. M. victims often have a history of abrupt and early weaning, followed by dilute (and often unhygienic) artificial feeds that are low in both energy and protein. Subsequent gastroenteritis may be treated by withholding food, thus precipitating and/or exacerbating the disease. A distinction is made between nutritional M. (resulting from inadequate food alone) and M. produced by infection (although inevitably against a background of poor nutrition). M. is more common in urbanized than in rural areas of the developing world. Unlike kwashiorkor, M. shows no dependence on climate, and it was once well known in industrial areas of Europe and North America. M. in children is equivalent to starvation in adults. See Kwashiorkor, Protein-energy malnutrition.

Marihuana: see Hashish.

Marker enzymes: see Enzymes.

Marker rescue: see Recombinant DNA technology.

Marker synthesis: a laboratory synthesis developed by R.E.Marker for the conversion of diosgenin into progesterone (Fig.). From 1939 to 1942, Marker undertook a series of plant collecting expeditions to find a high yield source of the starting material, diosgenin, which was already known from a species of *Dioscorea* in Japan. Two good sources were found in Mexico, i.e. *Dioscorea composita* ("barbasco") and *D. macrostachya* ("cabeza de negro"), otherwise known as "Mexican yams". By the 1950s virtually all steroid hormones had been made in the laboratory from diosgenin. Other synthetic methods are now available, but the Marker synthesis is still important. As a result of the Marker synthesis the price of progesterone fell in the late 1940s from over 80 dollars per gram (obtained by extraction from animal ovaries) to less than 50 cents per gram.

Maroteauz-Lamy syndrome: see Lysosomal storage diseases.

Mastich: see Resins.

Matrix: see Stroma.

Maturation of RNA: see Post-transcriptional modification of RNA.

Maturation-promoting factor: see Cell cycle.

Mavacurin: one of the Curare alkaloids (see).

Maximal rate: see Michaelis-Menten kinetics.

Maxima substance C: see Isoflavone.

McArdle's disease: see Glycogen storage disease.

Mecocyanin: see Cyanidin.

Melanins: high molecular mass, amorphous polymers of indole quinone, empirical formula $(C_8H_3NO_2)_x$, containing 6–9 % nitrogen. M. are natural pigments occurring predominantly in the animal kingdom in vertebrates and insects, and occasionally in microorganisms, fungi and higher plants. Mammalian colors are determined chiefly by two types of M.: the black/brown insoluble, nitrogenous Eumelanins (see), and the lighter colored, sulfur-containing, alkali-soluble Phaeomelanins (see). In addition, the low molecular mass, yellow, red and violet Trichochromes (see) are usually classified with the M., since they also serve as pigments, and are biogenetically closely related to M., i.e. they are derived from the oxidation of tyrosine (Fig.).

M. are synthesized in melanosomes, which are found in cells called melanocytes. M. are also synthesized in the retina of the eye. All M. are biosynthesized from L-tyrosine (I in Fig.), which is converted to indole 5,6-quinone (XI), via dopa (II), dopaquinone (III), leucodopachrome (VIII), dopachrome (IX) and 5,6-dihydroxyindole (X). XI is oxidatively polymerized to eumelanin. Alternatively, III interacts with cysteine to form cysteinyldopa (IV), which is then oxidized to cysteinylquinone (V). Cyclization of V leads to VI, which is reduced to benzothiazine (VII) by IV. Formation of V from IV requires only catalytic amounts of oxygen, and thereafter the reaction proceeds spontaneously. The dihydrobenzothiazine (VII) acts as the precursor of both the phaeomelanins and the trichochromes. In the native state, M. are often bound to proteins, forming melanoproteins with a protein content of 10–15 %. In humans and other mammals, the pigments of skin, hair and eyes are almost exclusively M. The pigments of many bird feathers, the skins of reptiles and fish, the exoskeletons of insects and the colored component (sepia melanin) from the ink sac of the cuttlefish (*Sepia officinalis*) are all M. M. are sometimes distributed diffusely, or they may be present in granules. In humans, the degree of coloration of the skin, from pale through brown to almost black, depends entirely on the concentration of melanin. Moles and freckles represent areas of especially high M. concentration. Sunlight causes increased M. synthesis (sun tan). M. acts as a protective agent against excess ultraviolet irradiation of the body surface. The color change of the chameleon and other color-adaptive animals is the result of a hormonally controlled change in the distribution of M. (see Melanotropin).

Oxidative conversion of dopa into M. is catalysed by the enzyme Tyrosinase (EC 1.14.18.1) (see). Absence of tyrosinase (usually an autosomal inherited defect in the ability to synthesize the enzyme) results in albinism; most groups of mammals occasionally produce albino individuals, which completely lack any M. pigmentation of eyes, skin, hair, feathers, etc. Albinism may also result from 1. deficient melanin polymerization, 2. failure to synthesize the protein matrix of the melanin granule, 3. lack of tyrosine, and 4. presence of tyrosinase inhibitors.

Melanotropin, *melanocyte stimulating hormone, MSH:* a peptide hormone or neuropeptide produced by opiomelanotropinergic cells of the pars intermedia of the pituitary and by neurons in the central nervous system. In species lacking the pars intermedia (e.g. chickens, porpoise, whale), α-MSH is produced by the neurohypophysis. Production of α-MSH by the pituitary is under control of MRH and MIH (see Re-

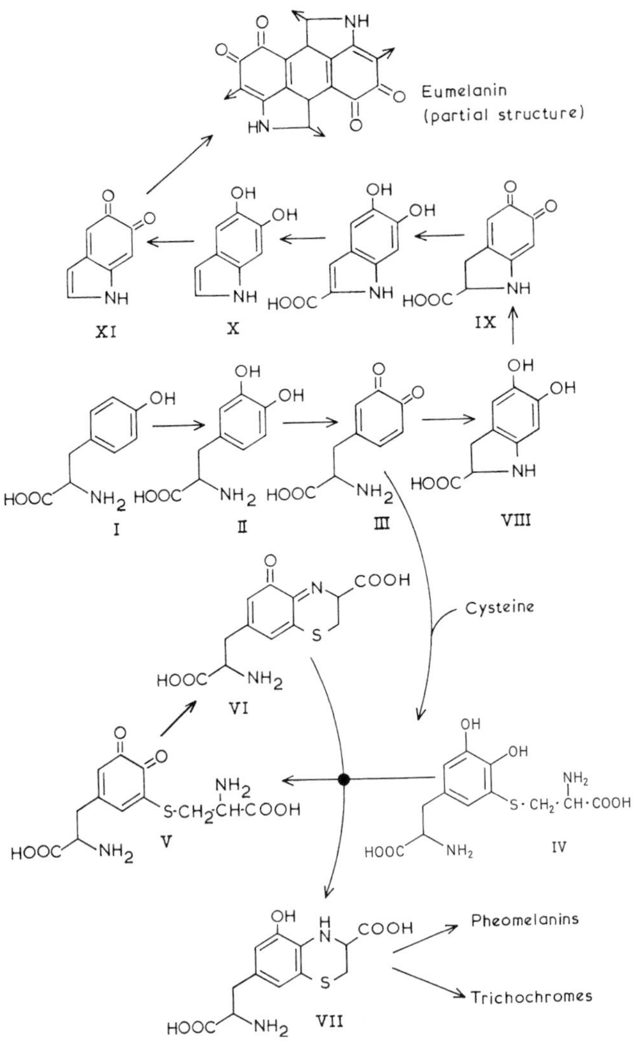

Biosynthesis of melanins

leasing hormones) of the hypothalamus. α-MSH belongs to a family of peptide hormones, including ACTH, β-lipotropin, endorphins, enkephalin and others, all of which are derived from a common precursor protein, pro-opiomelanocortin (see Peptides, Fig. 3). α-MSH was first recognized and named for its ability to cause dispersion of melanin in the melanophores of the skin of cold-blooded vertebrates; it also causes increased deposition of melanin in mammals, but its role in pigmentation in humans is doubtful. MSH-like activity of blood and urine increases during pregnancy in humans (the frog skin test described below was originally designed for detection of human pregnancy). This MSH-like activity may be of fetal origin (fetal pituitary is thought to produce α-MSH, which may play a role in fetal development), but its source and identity have not been established.

Adult humans do not appear to possess a circulating MSH. α-MSH is, however, present extensively throughout the central nervous system of humans and other animals, where the deacetylated form is more abundant than the acetylated form. It is synthesized and secreted by an opiomelanotropinergic multineuronal transmitter system, the specific neurons of which have been located in the brain by immunological methods. Particularly high concentrations of α-MSH are present in the axons and terminals of the hypophysiotropic area of the hypothalamus, and the cell bodies containing α-MSH are localized in the arcuate nucleus and in the dorsolateral region of the hypothalamus. Behavioral effects of α-MSH in mammals include arousal, increased motivation, longer attention span, memory retention and increased learning ability. α-MSH has been shown to promote soma-

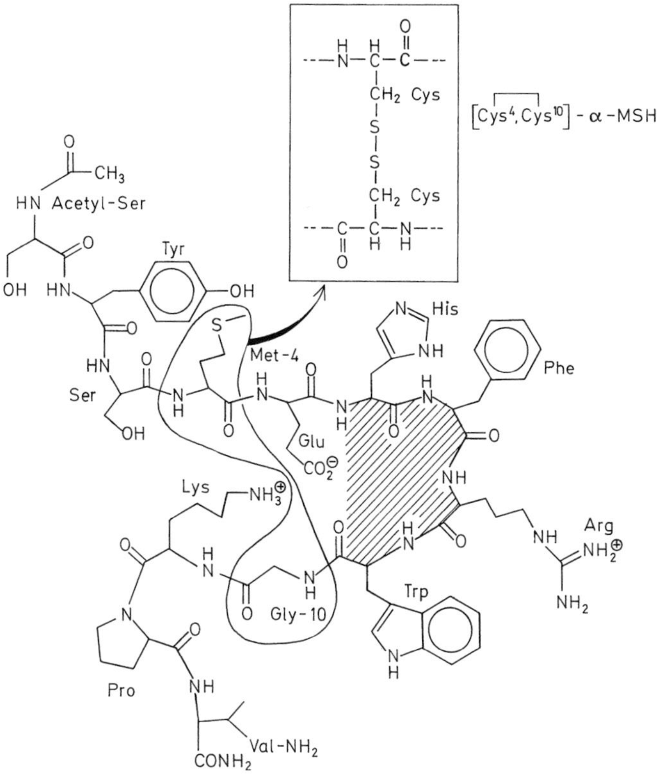

Two-dimensional representation of the reverse turn conformation of α-MSH. The shaded area represents the messenger sequence or active site. Replacement of Met-4 and Gly-10 by oxidatively coupled Cys residues results in [Cys⁴, Cys¹⁰]-α-MSH, a superpotent MSH.

totropin secretion, and to inhibit the secretion of prolactin and luteinizing hormone.

The same neurons produce both α-MSH and β-endorphin (which stimulates prolactin secretion). α-MSH may therefore be an endogenous antagonist of β-endorphin, thereby preventing hyperprolactinemia, but the true interplay of these two opposite effects is not clear. The classical assay for α-MSH is the in vitro frog (*Rana pipiens*) skin bioassay, which is based on the centrifugal dispersion of melanin granules in the dendritic processes of dermal melanophores, leading to skin darkening. It is capable of detecting minimal concentrations of α-MSH of 10^{-11} M, and gives dose-response curves in the range $2.5 \times 10^{-11} - 4 \times 10^{-10}$ M. Lizard skin may also be used, but is less sensitive. A later method measures the increase in the conversion of [α-^{32}P]ATP into [^{32}P]cAMP by plasma membranes from mouse S-91 (Cloudman) melanoma cells in response to α-MSH. Activation of tyrosinase or production of cAMP in whole melanoma cells have also been used as assay systems.

Mammalian α-MSH is an acetyltridecapeptidamide, identical with amino acid sequence 1–13 of ACTH (see Corticotropin): Ac-Ser-Tyr-Ser-Met-Glu-His-Phe-Arg-Trp-Gly-Lys-Pro-Val-NH_2. The deacetyl form is less potent and acetylation/deacetylation is probably a method of α-MSH potency control in mammals (α-MSH and β-endorphin are the only known *N*-acetylated peptides, although many *N*-acetylated proteins have been described). In salmon, only the deacetylated form has been found. Shark (*squalus acanthias*) α-MSH is also a tridecapeptide, and differs from mammalian α-MSH in only two aspects: the *N*-terminal serine is never acetylated and the *C*-terminus is Met (which may be amidated or possess a free carboxyl group). The major storage and release form in rat pituitary is *N,O*-diacetyl-α-MSH, i. e. the amino and hydroxyl groups of the *N*-terminal Ser are both acetylated.

Certain fragments of α-MSH have melanotropic activity with decreased potency. This has led to the concept of an active site or messenger sequence. Two such messenger sequences have been claimed for α-MSH: -Met-Glu-His-Phe-Arg-Trp-Gly- (residues 4–10 of α-MSH and ACTH) and -Lys-Pro-Val-NH_2 (residues 11–13 of α-MSH and ACTH), both independently capable of triggering the hormone receptor site responsible for melanin dispersion. Other authors have been unable to confirm the activity of the latter sequence, but there is general agreement that the 4–10 heptapeptide sequence represents a true messenger sequence of α-MSH.

β-MSH contains the same messenger heptapeptide sequence, but with different flanking amino acids: (bovine) Asp-Ser-Gly-Pro-Tyr-Lys-Met-Glu-His-Phe-Arg-Trp-Gly-Ser-Pro-Pro-Lys-Asp. Human β-MSH has 4 additional residues (Ala-Glu-Lys-Lys-) at the *N*-terminus. α- and β-MSH are derived from different regions of the same precursor, pro-opiomelanocortin. β-MSH has biological activity similar to that of α-MSH in the frog skin test; it probably occurs naturally, but is thought to have no peripheral physiological function, and to represent a byproduct of β-lipotropin degradation. β-MSH from pituitaries is now known to be a fragment of β-lipotropin formed during extraction. Earlier reports of circulating human β-MSH were mistaken.

A third region of pro-opiomelanocortin contains the sequence: -Met-Gly-His-Phe-Arg-Trp-Asp-, which is closely similar to the heptapeptide messenger sequence. This suggested the existence of a third MSH. The peptide Tyr-Val-Met-Gly-His-Phe-Arg-Trp-Asp-Arg-Phe-Gly has now been found naturally and named γ-MSH, but nothing is known of its biological function, if any.

Structure-activity studies of α-MSH and its fragments strongly suggest that the active molecule possesses a reverse turn configuration in the central tetrapeptide sequence: -His-Phe-Arg-Trp (Fig.). The plausibility of this hypothesis is greatly increased by the synthesis of a conformationally restricted α-MSH analog, in which Met-4 and Gly-10 are replaced by Cys residues. Oxidative cross linkage of these Cys residues maintains the molecule in a permanent reverse turn configuration. The resulting [Cys^4, Cys^{10}]-α-MSH has a minimal potency in the frog skin test which is 10,000 times that of α-MSH. [V. J. Hruby et al. *Peptide and Protein Rev.* **3** (1984) 1–64; O. Khorram et al. *Proc. Natl. Acad. Sci. USA* **81** (1984) 8004–8008]

Melanotropin release inhibiting hormone: see Releasing hormones.

Melanotropin releasing hormone: see Releasing hormones.

Melanocyte stimulating hormone: see Melanotropin.

Melatonin: *N*-Acetyl-5-methoxytryptamine, M_r 232.3, a hormone of the pineal gland and retina. It inhibits development of gonadal function in young animals and humans and the action of gonadotropins in mature animals. Synthesis of M. is suppressed by light, acting via the eyes and nervous system, and it peaks in the dark. A corresponding circadian rhythm has been observed in the activity of serotonin *N*-acetyltransferase (see below) in rodent pineal glands and the retinas of frogs and chickens. It seems likely that M. is an effector of mammalian or avian behavioral rhythms, although the molecular mechanisms are not known [J. Redman et al. *Science* **119** (1983) 1089–1091]. Rats kept in continuous light are in continuous estrus, due to the lack of the antigonadotrophic activity of M. In amphibians, M. mediates the response of skin pigmentation to light; it reverses the darkening effect of melanotropin by causing the melanin granules within the melanocytes to aggregate instead of disperse. M. appears also to influence the shedding of photoreceptor disks in the vertebrate retina [J. Besharse & P. M. Iuvone *Nature* **305** (1983) 133–135].

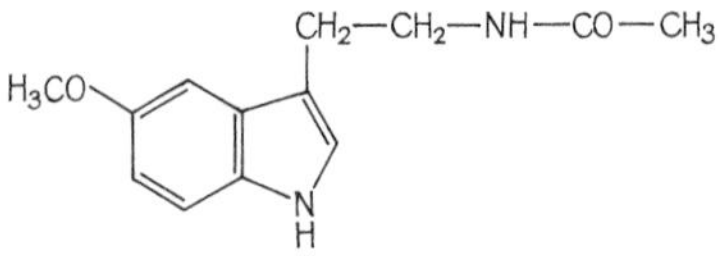

H_3CO … $CH_2—CH_2—NH—CO—CH_3$ … N H

Melatonin

M is synthesized from serotonin by acetylation to *N*-acetylserotonin, followed by methylation (catalysed by acetylserotonin methyltransferase, EC 2.1.1.4) to M. In the pineal gland M. biosynthesisis is controlled by epinephrine, which stimulates adenylate cyclase (EC 4.6.1.1), which in turn stimulates protein kinase. The first three steps of M. synthesis (L-tryptophan → 5-hydroxytryptophan → serotonin → *N*-acetyl-5-hydroxytryptamine) are promoted in this way, but the methyltransferase is not affected. However, the rate-limiting step is catalysed by serotonin *N*-acetyltransferase. M. is inactivated and excreted as 6-hydroxymelatonin, or as 5-methoxyindoleacetic acid.

Melezitose: *O*-α-D-glucopyranosyl-(1 → 3)-*O*-β-D-fructofuranosyl-(2 → 1)-D-glucopyranoside, a trisaccharide consisting of a molecule of D-glucose linked to the diasaccharide turanose (an isomer of sucrose). High concentrations of M. are found in manna formed on the surfaces of pine trees, and in honey manufactured by bees which collect M. from this manna in times of drought.

Melibiose: 6-*O*-α-D-galactopyranosyl-D-glucose, a reducing disaccharide consisting of galactose and glucose linked α-1,4 glycosidically, both sugars being in the pyranoid form. M. is present in plant juices. A M. residue is present in the trisaccharide raffinose.

Melissic acid: n-triacontanoic acid, CH_3-$(CH_2)_{28}$-COOH, a long-chain monocarboxylic fatty acid, M_r 452.8, present as an esterified acid in beeswax and montan wax.

Melissyl alcohol, *myricyl alcohol:* 1-triacontanol, or 1-hydroxytriacontane, CH_3–$(CH_2)_{28}CH_2OH$. M. is present in plant cuticle waxes. Beeswax consists principally of the palmitate ester of M. (myricin).

Melittin: a linear, toxic (hemolytic) polypeptide amide, and the chief component of bee venom (about 50 % of dried venom, and at least in 50-fold molar excess over other venom constituents). The primary product of M. biosynthesis is prepromelittin, consisting of 70 amino acid residues. Removal of the *N*-terminal signal sequence of 21 residues (see Signal hypothesis) leaves promelittin. The protease required for removal of the signal sequence is present in many animal cells and is not species-specific, e. g. promelittin and not prepromelittin is the first detectable product when melittin mRNA is injected into oocytes of *Xenopus laevis*.

The proteolytic release of M. from promelittin probably occurs outside the venom gland cells. Since M. disrupts phospholipid membranes, it is not surprising that the promelittin inside the secretory cells is less potent than M.

Primary structure of M.: Gly-Ile-Gly-Ala-Val-Leu-Lys-Val-Leu-Thr-Thr-Gly-Leu-Pro-Ala-Leu-Ile-Ser-Trp-Ile-Lys-Arg-Lys-Arg-Gln-Gln-NH_2. There is a very unequal distribution of hydrophobic, neutral (1–

20) and hydrophilic, mostly basic (21–26) amino acid residues. The resulting tenside character probably accounts for the pharmacological and biochemical effects of M.

Melting point, T_m t_m: the temperature, in °C, at which a double-stranded nucleic acid becomes 50% denatured to the single-stranded form. A DNA solution is heated and its absorbance at 260 nm is plotted against temperature. Transition from double to single-stranded DNA occurs over a relatively narrow temperature range, and is characterized by an increase in absorbance at 260 nm (melting of DNA is also accompanied by a marked reduction in viscosity, alteration of optical rotation, and an increase in density). T_m is taken as the temperature at the midpoint (half the final increase of absorbance at 260 nm) of the S-shaped curve. The sharpness of the transition indicates a cooperative alteration of structure throughout the molecule (T_m is well above the temperature required to unstack the bases and destroy a single helix, but these are preserved by hydrogen bonding between the two helices; separation of the two strands occurs when the hydrogen bonds between base pairs finally break). Single-stranded RNA, on the other hand, shows only a gradual increase of absorbance over a wide temperature range and has no T_m.

Under standard conditions of pH and ionic strength, T_m of DNA is proportional to the stability of the molecule. Since the base pair guanine-cytosine (see Base-pairing) has 3 hydrogen bonds, and adenine-thymine has only 2 hydrogen bonds, there is a linear relationship between the GC content of DNA and its T_m value: $T_m = 69\,°C + 0.41$ (molar % GC), i.e. the higher the degree of hydrogen bonding, the higher the temperature required to separate the strands of the double helix.

Measurement of Tm of hybrid nucleic acids (see Hybridization): 1. Cellulose nitrate discs containing known quantities of the radioactive hybrid (e.g. DNA: ^{32}P-RNA) are incubated for 15 min at various elevated temperatures, cooled, treated with a nuclease specific for single-stranded nucleic acid (in this case an RNase), washed and assayed for radioactivity. The melting profile is determined from the loss of radioactivity. 2. A hydroxyapatite column can be used for both large and small scale procedures. Double-stranded nucleic acids are retained by hydroxyapatite and single-stranded material is removed by washing. The column (water-jacketed) is eluted at various elevated temperatures (e.g. in the range 60–120 °C) and the effluent monitored (radioactivity or A_{260}) for single-stranded material.

Membrane: see Biomembrane.

Membrane enzymes: enzymes present on the surfaces or integrated into the phospholipid bilayer of the many different types of Biomembranes (see), e.g. glucose 6-phosphatase of the endoplasmic reticulum, galactosyl transferase of the Golgi apparatus, oligomycin-sensitive ATPase of the inner mitochondrial membrane, monoamine oxidase and rotenone-insensitive NADH-cytochrome *c*-reductase of the outer mitochondrial membrane, and the sodium-independent ATPase and 5′-nucleotidase of the cell membrane. Very many reactions occur on membranes, and M.e. play a central part in cell metabolism. M.e. are useful as markers in the isolation and identification of different types of membrane from cell homogenates.

Membrane lipids: the lipids that consititute the lipid bilayer of biomembranes (see). They are amphipathic molecules, having within their overall structure both hydrophilic and hydrophobic regions. The hydrophilic region associates with hydrophilic environments such as the aqueous milieu of intracellular and extracellular compartments segregated by biomembranes, while the hydrophobic region associates with hydrophobic environments such as the internal region of biomembranes themselves. The hydrophilicity of the hydrophilic regions of M.l. is due to their polarity (possession of electrical asymmetry within their structure). This results either from the presence of full positive or negative charges on ionized groups, e.g. NH_2 ($\rightarrow {}^+NH_3$), COOH ($\rightarrow COO^-$) and $(R.O)_2P(=O)OH$ ($\rightarrow (R.O)_2P(=O)O^-$) or from partial positive or negative charges (δ+, δ–) arising from the inductive effect due to the permanent, unequal sharing of the electron pair constituting a sigma (σ) bond linking together two unlike atoms. The latter is typified by the bond linking the O and H in a hydroxyl group which, though normally regarded as a covalent bond, has 39% ionic character because the oxygen atom is much more electronegative than the hydrogen atom and therefore has a disproportionately greater share of the bonding electron pair. This bond, of which there are two in the water molecule making an angle of 104.8° to each other, renders the hydroxyl group a distinctly polar group; thus the more hydroxyl groups there are in a molecule or residue the more hydrophilic it is. The hydrophobicity of the hydrophobic regions of M.l. is due to their lack of polarity. Typically these regions are composed of hydrocarbon chains which are not polar because they do not ionize and the inductive effect is minimal (the % ionic characters of the C–C and C–H bonds are 0 and 4 respectively). The hydrophilic and hydrophobic regions of a M.l. are usually referred to as the "head" and "tail" of the molecule respectively.

Classification and Structure: M.l. can be classified structurally in several different ways. The present classification (Fig. 1) divides them into phospholipids, glycolipids and Sterols (see). The phospholipids are divided into glycerophospholipids and sphingophospholipids, the glycolipids into glyceroglycolipids and sphingoglycolipids; each of these subclasses is then divided into subsubclasses. Phospholipids and glycolipids can be described as acylated-M.l. because they have at least one fatty acyl residue (R.C = O where R is a hydrocarbon chain) in their structure; in this they differ from sterols which occur in biomembranes in non-acylated form.

a) *Phospholipids* are defined, according to this classification, as any M.l. containing a phosphoric acid (Fig. 2a) residue as a diester in which one alcoholic moiety (X in Fig. 2b) is polar and the other (Y in Fig. 2b) is nonpolar. The third and remaining acidic hydrogen of the phosphoric acid residue is ionized at physiological pH.

(i) *Glycerophospholipids* (recommended IUPAC-IUB nomenclature), sometimes also called *phosphoglycerides.* Here the non-polar alcoholic residue (Y) is a glycerol (Fig. 2c) derivative, namely 1,2-diacyl-*sn*-glycerol or 1-*O*-(1-alkenyl)-2-*O*-acyl-*sn*-glycerol;

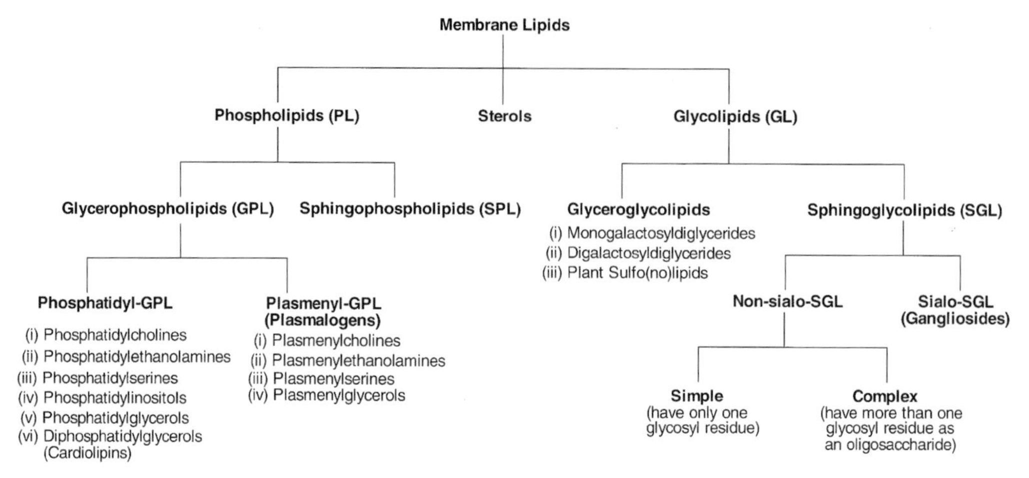

Fig. 1. *Membrane lipids.* Classification of membrane lipids.

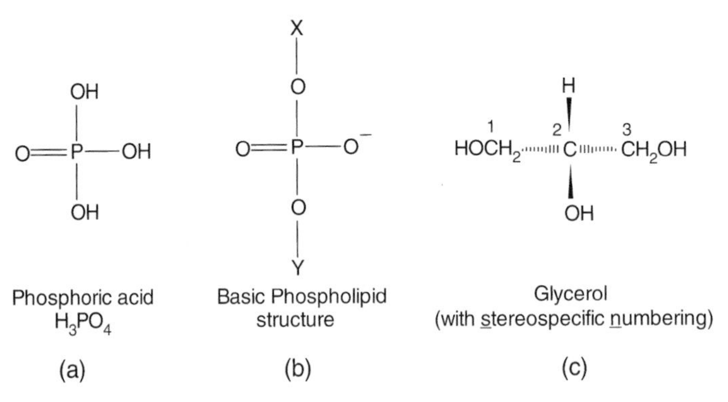

Fig. 2. *Membrane lipids.* Structure of the basic components of phospholipids.

in both cases it is the *sn*-3 hydroxyl group of the glycerol moiety that is esterified to phosphoric acid. Glycerophospholipids of the first type (Fig. 3 a) are derivatives of *L-α-phosphatidic acid* (1,2-diacyl-*sn*-glycerol-3-phosphoric acid) (Fig. 3 b) and are named as such (e.g. phosphatidylethanolamine) while those of the second type (Fig. 3 c) are derivatives of *plasmenic acid* (1-*O*-(1-alkenyl)-2-*O*-acyl-*sn*-glycerol-3-phosphoric acid) (Fig. 3 d) and may also be named as such (e.g. plasmenylethanolamine) although they are collectively known as plasmalogens. Note that the substituent at the *sn*-1 position of the plasmalogens is ether-linked to the glycerol carbon rather than ester-linked; plasmalogens are therefore often referred to as ether lipids and may be classified as such with the phytanyl ethers of the Archaebacteria (see later). The phosphatidyl and plasmenyl glycerophospholipids subsubclasses are further subdivided according to the polar alcohol esterified to their phosphoric acid moiety as the X component of Fig. 3 a & 3 b. Table 1 shows the structure of these polar alcohols and gives the names of the final glycerophospholipid subdivisions. It should be noted, however, that within each such subdivision a considerable variety of molecular species is possible, because many different fatty acids can be esterified to the glycerol residue within the parent structure; this is also true of all other acylated phospholipid subdivisions. Fig. 4 shows the structure of some of the glycerophospholipids. Notice that the head groups of ethanolamine- and choline-containing glycerophospholipids are electrically neutral because each has the same number of full positive and negative charges due to ionization, although, of course it is the local electrical asymmetry associated with each of the these charges that gives rise to the polarity. The head groups of all the other glycerophospholipid species are electrically negative because of the predominance of groups that ionize to give full negative charges. The polarity of some of these net negatively charged species (e.g. phosphatidylinositol and the phosphatidylglycerols) is augmented by the presence of hydroxyl groups. A number of derivatives of phosphatidylinositol (PI) are also present in biomembranes; for instance much of the PI of animal plasma membranes consists of its 4-phosphate (PIP) and 4,5-*bis*phosphate (PIP_2) derivatives, while some has been shown to anchor a range of proteins to the outer surface of various single-celled eukaryotes (e.g. *Trypa-*

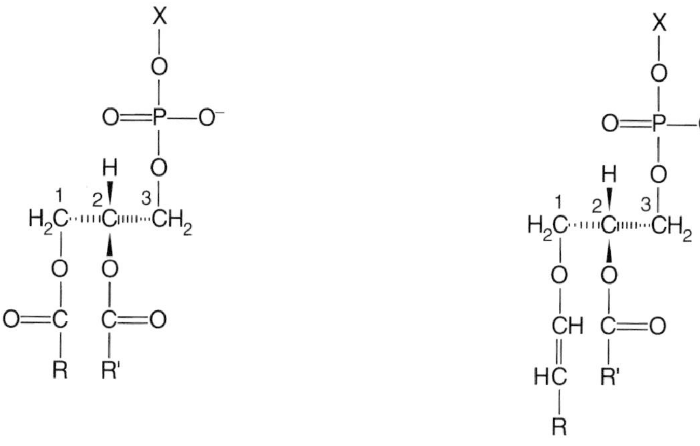

(a) A Glycerophospholipid
(X = a polar alcohol residue)

(b) L-α-Phosphatidic acid (X = H)

(R and R' = hydrocarbon chains)

(c) A Plasmalogen
(X = a polar alcohol residue)

(d) Plasmenic acid (X = H)

(R and R' = hydrocarbon chains)

Fig. 3. *Membrane lipids.* Basic structures of phosphatidyl- and plasmenyl glycerophospholipids along with those of phosphatidic acid and plasmenic acid.

nosoma spp.) and mammalian cell types, including lymphocytes, by means of an ethanolamine-containing oligosaccharide; these are consequently known as glycosyl-PI (GPI) anchors, and their basic structural pattern is shown in Fig. 4.

(ii) *Sphingophospholipids.* Here the non-polar alcoholic residue (Y) is derived from an *N*-acylated derivative of the long chain aminoalcohol, sphinganine [(2*R*,3*R*)-2-amino-1,3-octadecanediol], or one of its derivatives, sphingosine [*trans*-4-sphingenine; (2*S*,3*R*,4*E*)-2-amino-4-octadecene-1,3-diol] or phytosphingosine [4-D-hydroxysphinganine; (2*S*,3*S*,4*R*)-2-amino-1,3,4-octadecanetriol]; such *N*-acylsphinganine derivatives are called ceramides. When the C1 hydroxyl group of a ceramide is esterified with phosphoric acid the resulting structure resembles that of phosphatidic and plasmenic acids, in that two hydrocarbon chains are effectively linked to adjacent carbons on a glycerol-like three-carbon chain; this structural similarity is emphasized in Fig. 5. A sphingophospholipid is generated when the phosphoric acid residue of a ceramide-1-phosphate forms a second ester linkage with a polar alcohol. In animal tissues the best known sphingophospholipid is *sphingomyelin* (Fig. 5), in which *N*-acylsphingosine-1-phosphate is esterified to choline; the fatty acyl moiety is commonly derived from lignoceric (C24:0) or nervonic (C24:1, *cis* Δ^{15}) acid. The sphingophospholipids of plant membranes utilise phytosphingosine and have a complex oligosaccharide as the polar alcoholic moiety; in the latter, *myo*-inositol (see cyclitols) is often the component that is linked, via its C1 hydroxyl group, directly to the phosphoric acid residue of the ceramide-1-phosphate.

b) *Glycolipids* are defined, according to the classification used here, as any non-phosphoric acid-containing M.l. whose non-polar alcoholic moiety (Y in Fig. 6a) is linked to a polar mono- or oligosaccharide residue by means of a glycosidic bond.

(i) *Glyceroglycolipids.* Here the non-polar alcoholic moiety is an *sn*-1,2-diacylglycerol residue, the *sn*-3 oxygen of which is glycosidically linked to the the polar mono- or oligosaccharide residue as shown in Fig. 6b. The best known of all the glyceroglycolipids are the 1,2-diacyl-3-β-D-galactosyl-*sn*-glycerols (often called monogalactosyldiglycerides) which are the major lipids of chloroplast membranes, constituting up to 40 % of their dry weight; this makes them the most abundant M.l. in the biosphere. They are accompanied in chloroplast membranes by the 1,2-diacyl-3-[α-D-galactosyl (α1 → 6)-β-D-galactosyl]-*sn*-glycerols (often called digalactosyldiglycerides) and the 1,2-diacyl-3-(6-sulfo-α-D-quinovosyl)-*sn*-glycerols (often called plant sulfolipids, in spite of the fact that plant sulfonolipids is a better name because it emphasizes the sulfonic acid nature of the sulfur-containing moiety). The structures of these compounds are shown in Fig. 7.

(ii) *Sphingoglycolipids.* Here the non-polar alcoholic residue is an *N*-acylsphinganine derivative (a ceramide), the C1 oxygen of which is glycosidically linked to the polar mono- or oligosaccharide residue as shown in Fig. 6c. These lipids are often called gly-

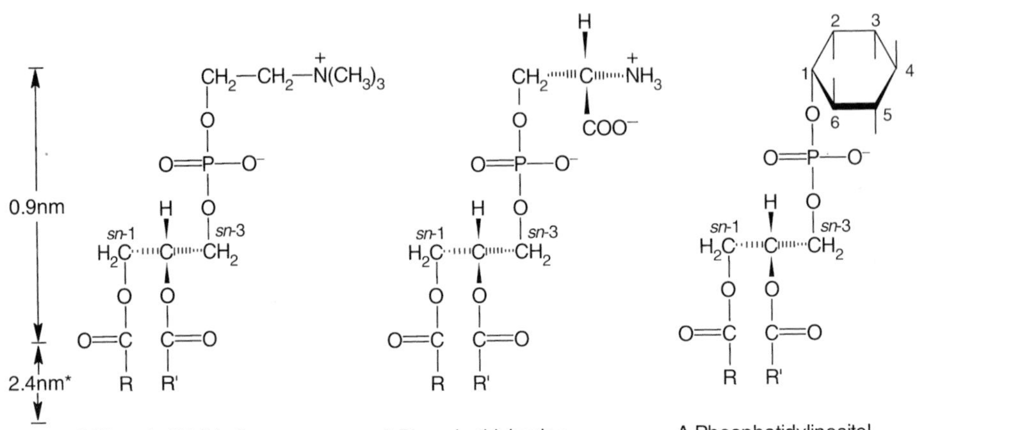

A Phosphatidylcholine

A Phosphatidylserine

A Phosphatidylinositol

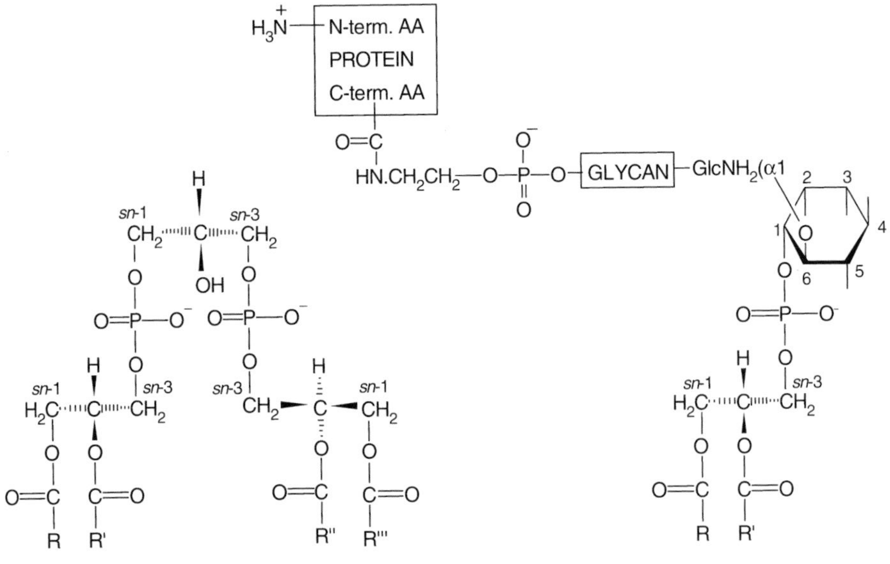

A Diphosphatidylglycerol (A Cardiolipin)

A Glycosyl-phosphatidylinositol anchor-protein complex

Fig. 4. *Membrane lipids.* Structure of some glycerophospholipids. R, R′, R″ & R‴ are hydrocarbon chains whose structure may be the same or different; * for the purposes of the dimensions shown for this structure R & R′ are assumed to be derived from C_{18} fatty acids).

cosylceramides. They can be subdivided into those (usually called gangliosides) which possess one or more sialic acid (see neuraminic acids) residues, and those which do not. The non-sialo sphingoglycolipids may be further divided into simple and complex categories. The former, often called cerebrosides, possess only one glycosyl residue which is sometimes sulfated at the 3-hydroxyl group, and are typified by 1-β-D-galactosylceramide (Fig. 8a) and its glucosyl analogue; the latter have an oligosaccharide (see examples in Table 2) which is β-glycosidically-linked to the ceramide C1 via the lactosyl residue at its reducing end. Lactosylceramide (Fig. 8b) also participates in the structure of gangliosides, whose oligosaccharide moieties (see examples in Table 3) are made more complex by the presence of one or more sialic acid residues. The latter are usually derived from *N*-acetylneuraminic acid (NANA) although some may come from *N*-glycolylneuraminic acid (NGNA) (Fig. 9); they are linked to glycosyl residues in the oligosaccharide by $\alpha 2 \rightarrow 3$ glycosidic bonds or to each other by $\alpha 2 \rightarrow 8$ glycosidic bonds.

Table 1. *Structure of the polar alcohols esterified to the phosphoric acid residue of glycerophospholipids*

Polar alcohol	Formula (at physiological pH)	Name of glycerophospholipid*
L-Serine	↓ $HOCH_2—C(H)(COO^-)—\overset{+}{N}H_3$	Phosphatidylserine Plasmenylserine
Ethanolamine	↓ $HOCH_2—CH_2—\overset{+}{N}H_3$	Phosphatidylethanolamine Plasmenylethanolamine
Choline	↓ $HOCH_2—CH_2—\overset{+}{N}(CH_3)_3$	Phosphatidylcholine Plasmenylcholine
myo-Inositol (1,2,3,5/4,6-Hexahydroxy-cyclohexane)	Cyclohexane ring, carbons numbered 1–6; arrow at C-1	Phosphatidylinositol
Glycerol	↓ *sn*-1 $HOCH_2—C(H)(OH)—CH_2OH$ *sn*-3	Phosphatidyglycerol
Phosphatidylglycerol	↓ *sn*-1 $HOCH_2—C(H)(HO)—CH_2$ *sn*-3 $—O—P(=O)(O^-)—O—$ *sn*-3 $CH_2—C(H)(O—C(=O)—R)—CH_2$ *sn*-1 $—O—C(=O)—R'$	Diphosphatidylglycerol (Cardiolipin)

* Strictly the prefix (3-*sn*)- should precede the name of phosphatidylglycerophospholipids so as to specify the position of the polar alcoholic residue
↓ = the oxygen atom that is ester-linked to the phosphoric acid residue of phosphatidic acid or plasmenic acid
R and R′ = hydrocarbon chains

c) *Fatty acid residues present in acylated M.l.* In animals these are predominantly straight-chain, even-numbered (mainly C_{16}–C_{22}), saturated, monounsaturated (*cis*-Δ^9) or polyunsaturated; the latter have a methylene-interrupted system of 2–6 *cis*-double bonds which tends to occur towards the carboxyl end of the molecule. In spite of the fact that plants use a much greater range of fatty acid residues in their M.l. than animals, the majority are straight-chain, even-numbered (mainly C_{12}–C_{18}), saturated, monounsaturated (*cis*-Δ^9) or polyunsaturated; the latter have a methylene-interrupted system of 2–3 *cis*-double bonds which, in contrast to those of animals, tends to occur towards the methyl end of the molecule. A small proportion of plant membrane fatty acid residues follow this structural pattern but have either a shorter or longer carbon chain, while others have unusual features, such as triple bonds, conjugated double bond systems, branched chains, cyclopropene or cyclopentene rings,

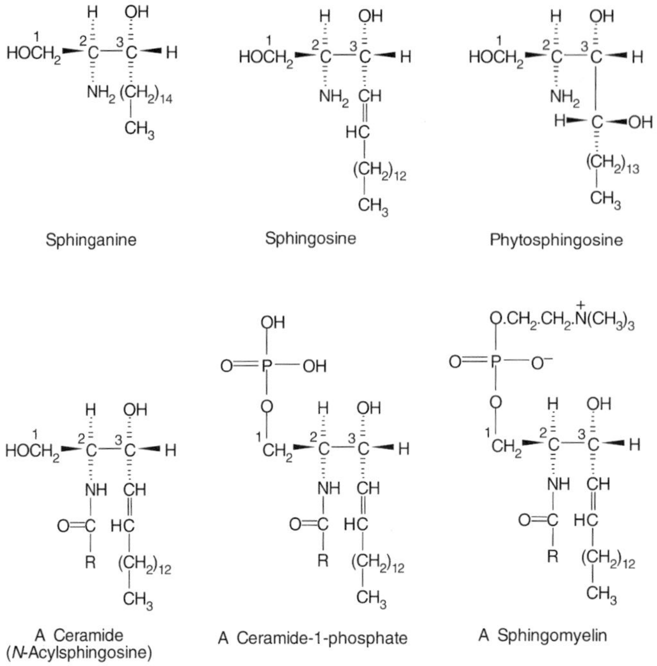

Sphinganine | Sphingosine | Phytosphingosine

A Ceramide (*N*-Acylsphingosine) | A Ceramide-1-phosphate | A Sphingomyelin

Fig. 5. *Membrane lipids.* Structure of sphingomyelin and sphingophospholipid components. R = a hydrocarbon chain.

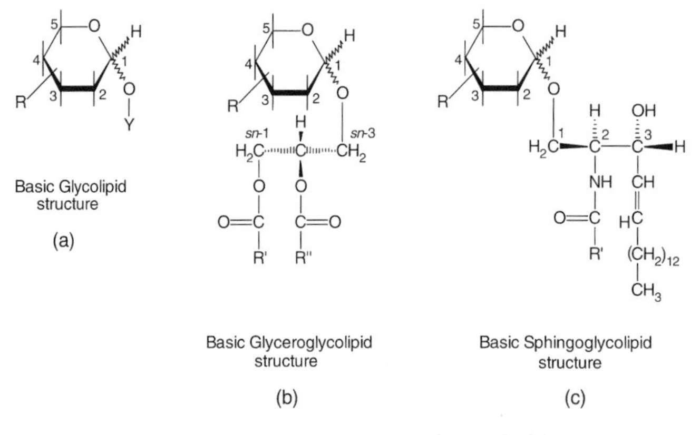

Fig. 6. *Membrane lipids.* Basic structures of glycolipids and their subclasses, the glyceroglycolipids and the sphingoglycolipids. R = rest of the glycosidic polar moiety. R′ & R″ = hydrocarbon chains. Y = nonpolar residue.

and hydroxy, epoxy or keto groups. Chloroplast thylakoid M.l. are unique in having *trans*-Δ^3-hexadecenoic acid residues. The fatty acid residues of bacterial M.l. are predominantly straight-chain, even-numbered (mainly C_{12}-C_{20}), saturated or monounsaturated but not polyunsaturated; less common are those with an odd number of carbon atoms, a branched chain or cyclopropyl, 2-hydroxy or 4-hydroxy moieties.

d) *Sterols* (see) are the least amphipathic of the M.l.; their only hydrophilic moiety, the 3β-hydroxyl group, is small compared with their large hydrophobic, hydrocarbon, cyclopentanoperhydrophenanthrene nucleus (19C) and 17β side chain (8–10C). Cholesterol (see) is the main sterol of animal membranes, while various combinations of Sitosterol (see), Stigmasterol (see) and Campesterol (see) are to be found in plant membranes. The $\Delta^{5,7}$-sterol, Er-

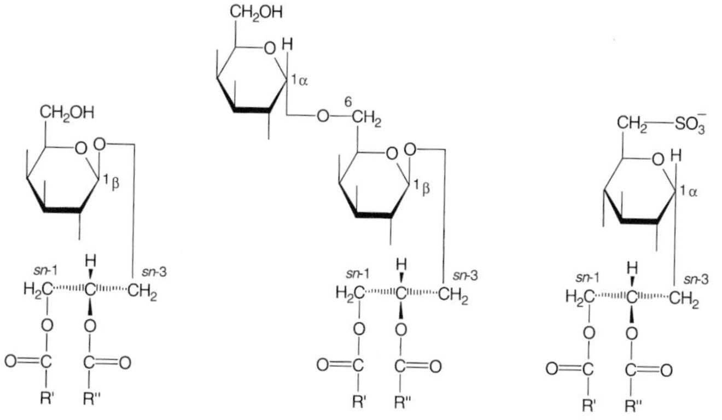

CH2OH O O 1β H sn-1 sn-3 H2C C CH2 O O O=C C=O R' R''

CH2OH O H 1α 6 O CH2 O O 1β H sn-1 sn-3 H2C C CH2 O O O=C C=O R' R''

CH2 SO3− O H 1α H sn-1 sn-3 H2C C CH2 O O O=C C=O R' R''

A Monogalactosyldiglyceride A Digalactosyldiglyceride A Plant Sulfo(no)lipid

Fig. 7. *Membrane lipids.* Structures of some glyceroglycolipids. R & R′ = hydrocarbon chains.

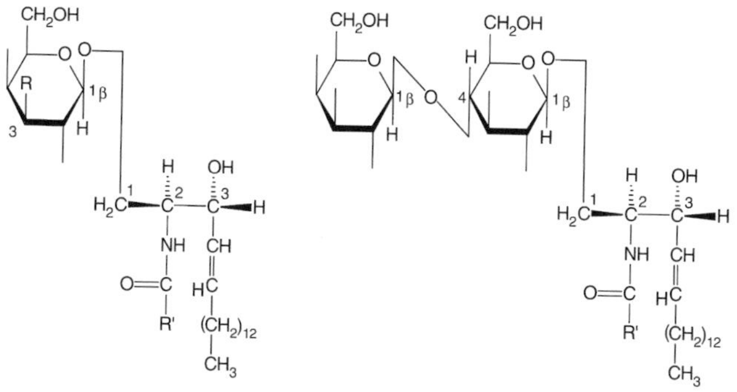

CH2OH O O R 1β 3 H H OH 1 2 3 H2C C C H NH CH O=C HC R' (CH2)12 CH3

CH2OH CH2OH O H O O 1β O 4 1β H H H OH 1 2 3 H2C C C H NH CH O=C HC R' (CH2)12 CH3

A 1-β-D-Galactosylceramide (R = OH) and its sulfate ester (R = O−SO_3^-) (R′ = a hydrocarbon chain)

(a)

A Lactosylceramide (R′ = a hydrocarbon chain)

(b)

Fig. 8. *Membrane lipids.* (a) A galactosylceramide (a simple, non-sialo sphingoglycolipid. (b) A lactosyl ceramide (a component of complex, non-sialo sphingoglycolipids and of gangliosides).

Table 2. *Structure of some complex, non-sialo sphingoglycolipids* (note that all have a lactosyl [D-Gal-(β1 → 4)D-Glc] residue glycosidically-linked to C1 of the ceramide)

Structure	Trivial name of oligosaccharide	Name of sphingoglycolipid
Gal(α1 → 4)Gal(β1 → 4)Glc(β1 →)Cer	Globotriaose	Globotriaosylceramide
GalNAc(α1 → 4)Gal(α1 → 4)Gal(β1 → 4)Glc(β1 →)Cer	Globotetraose	Globotetraosylceramide
Gal(α1 → 3)Gal(β1 → 4)Glc(β1 →)Cer	Isoglobotriaose	Isoglobotriaosylceramide
GalNAc(α1 → 4)Gal(α1 → 3)Gal(β1 → 4)Glc(β1 →)Cer	Isoglobotetraose	Isoglobotetraosylceramide
Gal(β1 → 4)Gal(β1 → 4)Glc(β1 →)Cer	Mucotriaose	Mucotriaosylceramide
Gal(β1 → 4)Gal(β1 → 4)Gal(β1 → 4)Glc(β1 →)Cer	Mucotetraose	Mucotetraosylceramide
GalNAc(β1 → 3)Gal(β1 → 4)Glc(β1 →)Cer	Lactotriaose	Lactotriaosylceramide
Gal(β1 → 3)GalNAc(β1 → 3)Gal(β1 → 4)Glc(β1 →)Cer	Lactotetraose	Lactotetraosylceramide
Gal(β1 → 4)GalNAc(β1 → 3)Gal(β1 → 4)Glc(β1 →)Cer	Neolactotetraose	Neolactotetraosylceramide

Cer = Ceramide; Gal = D-Galactopyranose; Glc = D-Glucopyranose; GalNAc = 2-*N*-Acetyl-D-galactopyranose.

gosterol (see), is the main sterol of the membranes of most, but by no means all, fungi; some fungal membranes have Δ^5- or Δ^7-sterols instead. The membranes of prokaryotes typically contain no sterols at all. Within the eukaryotic membrane the sterol molecule has a cross-sectional area of 0.35–0.4 nm^2 (cf. 0.52 nm^2 for a phospholipid) and has a length of 2.1 nm (cf. 3.3 nm for phosphatidylcholine, see Fig. 4) of which 1.2 nm is taken up by the nucleus and 0.9 nm by the side chain (Fig. 10). Interestingly, the point at which the rigidity of the nucleus gives way to the flexibility of the side chain coincides with the position within the biomembrane that is occupied by the *cis*-Δ^9-double bond of the unsaturated fatty acid (e.g. oleic acid) residues of acylated M.l. The position of the sterol hydroxyl group within the biomembrane occurs at the same level as that of the ester linkages of the acylated M.l.

e) *Phytanyl ether lipids* predominate in the membranes of the Archaebacteria, a primitive bacterial family, whose members occupy a number of harsh ecological niches (e.g. low pH, high salt concentration, high temperature). Structurally (Fig. 11) they most resemble the plasmalogens; however the residues ether-linked to the *sn*-1 and *sn*-2 carbons are formally derived from phytanol, which is composed of four saturated isoprene units linked together in a head-to-tail fashion. In some molecules the two phytanyl residues are joined together by head-to-head isoprenoid linkages while in others they are joined, also by head-to-head isoprenoid linkages, to those of an equivalent molecule in the other monolayer of the membrane lipid bilayer. It is tempting to speculate that such molecules, spanning the bilayer, serve to strengthen the membrane and thereby give it greater stability in the testing conditions to which it exposed.

2. ***Biosynthesis:*** The glycerol moiety of both glycerophospholipids and glyceroglycolipids is generated by the biosynthetic pathway: D-glucose → dihydroxyacetone phosphate → *sn*-glycerol-3-phosphate → phosphatidic acid → *sn*-1,2-diacylglycerol. The phosphatidylinositols and -glycerols are tapped off this pathway at the level of phosphatidic acid, which reacts with CTP to yield CDP-diglyceride which in turn reacts with inositol, glycerol or phosphatidylglycerol, with the elimination of CMP. Phosphatidylcholines (*lecithins*), phosphatidylethanolamines and phosphatidylserines (these latter two groups are called *cephalins* in the older literature) and the glyceroglycolipids are derived from *sn*-1,2-diacylglycerol, the latter by the transfer of one or more glycosyl residues from the appropriate nucleoside diphosphate (usually UDP) sugar and the others by the transfer of the appropriate phosphorylalcohol (e.g. phosphorylcholine) from the corresponding CDP-alcohol to the *sn*-3 position. *Lysophosphatidic acids* (*lysolecithins* and *lysocephalins*) contain an unesterified hydroxyl group; they are produced from glycerophospholipids by the action of phospholipase A_2 (EC 3.1.1.4) and A_1 (EC 3.1.1.32). The ceramide moiety of both sphingophospholipids and sphingoglycolipids is generated by the biosynthetic pathway: palmityl-CoA + L-serine → 3-ketosphinganine → sphinganine → sphingosine → cera-

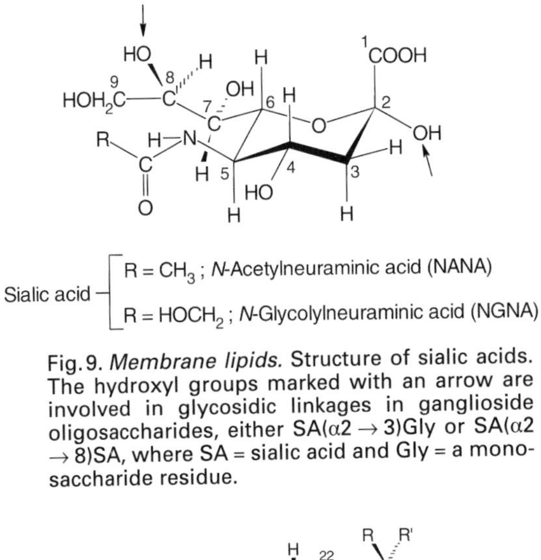

Sialic acid: $R = CH_3$; *N*-Acetylneuraminic acid (NANA)
$R = HOCH_2$; *N*-Glycolylneuraminic acid (NGNA)

Fig. 9. *Membrane lipids.* Structure of sialic acids. The hydroxyl groups marked with an arrow are involved in glycosidic linkages in ganglioside oligosaccharides, either SA(α2 → 3)Gly or SA(α2 → 8)SA, where SA = sialic acid and Gly = a monosaccharide residue.

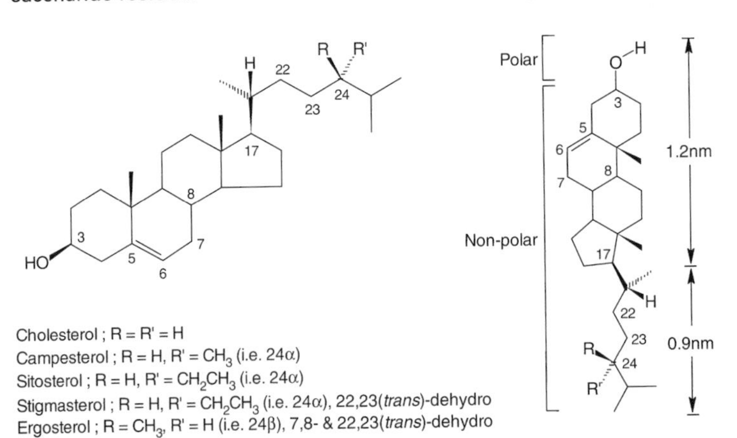

Fig. 10. *Membrane lipids.* Structure of the main membrane sterols and their orientation and dimensions in a monolayer of the lipid bilayer of a biomembrane.

Table 3. *Structure of some gangliosides* (note that all have a lactosyl [D-Gal(β1 → 4)D-Glc] residue glycosidically-linked to C1 of the ceramide)

Structure	Designation*
Gal(β1 → 4)Glc(β1 →)Cer 3 ↑ α2 NANA	GM3
GalNAc(β1 → 4)Gal(β1 → 4)Glc(β1 →)Cer 3 ↑ α2 NANA	GM2
Gal(β1 → 3)GalNAc(β1 → 4)Gal(β1 → 4)Glc(β1 →)Cer 3 ↑ α2 NANA	GM1
Gal(β1 → 3)GalNAc(β1 → 4)Gal(β1 → 4)Glc(β1 →)Cer 3 3 ↑ ↑ α2 α2 NANA NANA	GD1a
Gal(β1 → 3)GalNAc(β1 → 4)Gal(β1 → 4)Glc(β1 →)Cer 3 ↑ α2 NANA(α2 → 8)NANA	GD1b
Gal(β1 → 3)GalNAc(β1 → 4)Gal(β1 → 4)Glc(β1 →)Cer 3 3 ↑ ↑ α2 α2 NANA NANA(α2 → 8)NANA	GT1b

* The nomenclature is that of Svennerholm (*J. Neurochem. 10* (1963) 613–623) in which G = ganglioside, M = monosialo-, D = disialo-, T = trisialo-, and the arabic numerals indicate the sequence of migration on thin layer chromatograms.

mide. *Sphingomyelins* are generated by the transfer of the appropriate phosphorylalcohol (e.g. phosphorylcholine) from the corresponding CDP-alcohol to the ceramide C1 hydroxyl group. Non-sialo sphingoglycolipids are generated by the transfer of glycosyl residues, in the correct sequence, from the appropriate UDP-sugar initially to the ceramide C1 hydroxyl group and thereafter to the terminal glycosyl residue of the growing oligisaccharide residue. Gangliosides are derived from lactosylceramide, formed by the process just described, by the transfer of glycosyl and sialic acid residues, in the correct sequence, to its galactosyl residue and thereafter to the terminal glycosyl or sialic acid residue of the growing molecule. The biosynthesis of sterols is described in the entry on steroids.

The enzymes catalysing the early stages in the biosynthesis of all these M.l. are located in the soluble phase of the cell, whereas those catalysing the final stages are located in the membrane itself. This means that the early biosynthetic intermediates are present in the soluble milieu of the cell while the later ones are built into the membrane. More specifically, they are built into the monolayer of the lipid bilayer that is adjacent to the soluble phase of the cell. Evidence is accumulating for the existence of specific enzymes, which have been called "flipases", located in the membrane, whose function is to transfer an appropriate proportion of the newly synthesised lipid molecules from the inner monolayer to the outer monolayer. Such a transfer, termed a "flip-flop" movement, is a rare event in artificial lipid bilayers, because passage of the polar head group through the nonpolar core is thermodynamically unfavorable, but it has been shown to occur readily in natural membranes. In eukaryotic cells, M.l. are known to be transferred between different membrane types in the form of membrane fragments. For instance, sections of the endoplasmic reticulum, a site of active membrane lipid synthesis, are known to be incorporated into Golgi

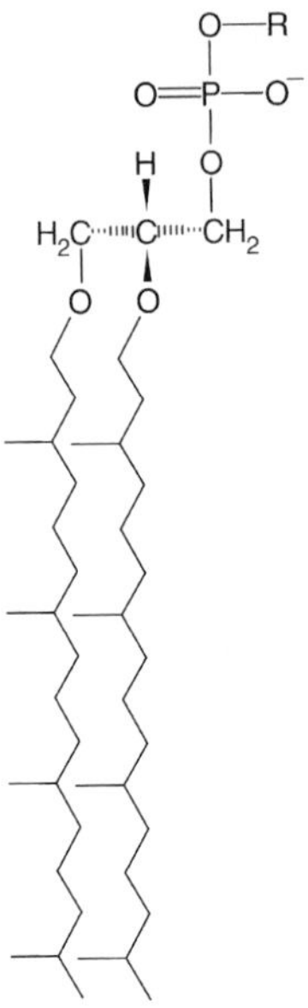

R = – $CH_2.CHOH.CH_2OR'$ where R' = H or SO_3^- or PO_3^{2-}
or
– Glc.Man.Gal.OR' where R' = H or SO_3^-

Fig. 11. *Membrane lipids.* Structure of the phytanyl ether membrane lipids of Archaebacteria. Sometimes the ends of the phytanyl residues of a particular molecule are linked to each other or to those of an equivalent molecule in the other monolayer of the lipid bilayer of the biomembrane.

cisternae, from which small membrane vesicles, containing secretory products, are continually pinched off. These vesicles then become part of the plasma membrane when they fuse with it during the secretion of their contents by exocytosis. It seems likely that there are similar processes transporting M.l. to other cellular organelles.

3. ***Function:*** At present our knowledge of the function(s) of the various species of M.l. is far from complete. However it is probable that all participate in the "collective" function of maintenance of fluidity, the object of which is to attain and maintain a degree fluidity that best suits the normal functioning of the membrane under the prevailing environmental conditions; this fluidity roughly corresponds to that possessed by olive oil at room temperature. The hydrocarbon chains of the phospho- and glycolipids play the dominant role in determining fluidity, but this is modulated by the size and charge of the polar head groups, as can be seen from their effect on the transition temperatures (T_c) of a selection of these lipids (Table 4). Below the T_c a bilayer composed of a given molecular species behaves as if it were a solid; its X-ray diffraction pattern indicates that its molecules are packed closely in hexagonal array. This is usually referred to as the gel state. When the temperature is gradually raised, a point (the T_c) is reached when the X-ray diffraction pattern changes suddenly and dramatically, the spacing increasing from 0.42 nm to 0.46 nm, indicating that the hexagonal array has broken down and that the molecules are further apart. This is the liquid crystalline state, in which the bilayer continues to hold together but individual molecules and their component parts have considerable freedom of movement. This phase change and its T_c are analogous to the melting of an organic solid and its melting point. It is frequently monitored by differential scanning calorimetry (DSC), a technique which follows the flow of heat into the bilayer sample as its temperature is raised; at the T_c there is a peak in the DSC heat uptake trace which marks the intake of the energy required to disrupt the crystal lattice of the solid gel state. Although the T_c is quite sharp for bilayers composed of a single lipid species, it is not at all sharp for mixtures, particularly those of the complexity seen in biomembranes, where the phase transition ("melting") extends over a range of 10–25 °C. Within this "melting" range an equilibrium mixture of gel-state and liquid crystalline-state regions co-exist within the bilayer. It is believed that biomembranes can only function properly within a fairly narrow range of bilayer fluidities; the fluidity must be between a minimum value, which appears to correspond to that of an equilibrium mixture containing 90 % of the lipid in the gel state, and a maximum value, which appears to correspond to the fluidity that would obtain at a temperature just above the T_c. Clearly, for the membranes of an organism to perform normally, the mix of its bilayer lipids must be such that this "functional fluidity range" is provided over the range of environmental temperatures, including both diurnal (day = warm/night = cool) and seasonal (summer = hot/winter = cold) variations. This is not a problem for homeothermic animals, but poikilothermic organisms, which include cold-blooded animals as well as plants and microorganisms, have no ability to control their internal temperature. Several of the latter have been shown to be capable of adjusting the "functional fluidity range" of their membranes in response to long term changes in environmental temperature, by altering the fatty acyl moieties of their phospho- and glycolipids; a rise in temperature causes a shift to longer chain, more saturated fatty acyl residues while a fall causes a shift to shorter chain, more unsaturated or branched fatty acyl residues. This process has been termed homeoviscous adaptation and is accomplished while the lipid remains in the membrane by the action of specific lipases and acyltransferases. Homeoviscous adaptation is however unsuited to coping with diurnal temperature changes and it is in this context that sterols help. Sterols have two quite opposite effects on membrane fluidity depending upon the circumstances. When sterols are incorporated into a bilayer which is in its liquid crystalline state they decrease fluidity; this is called the condensing effect of sterols. On the other hand when sterols are introduced into a bilayer which is in its gel state they increase fluidity; this is called the liquifying effect of sterols. The condensing effect of sterols is

due to their hindrance of the easy movement of the phospho- and glycolipids when they are in the liquid crystalline state; they have a liquifying effect because they intercalate between the phospho- and glycolipids, thereby disrupting the crystal lattice in the gel state. The combined effects of the condensing and liquifying abilities of sterols significantly extends the temperature range over which the membrane is functionally fluid. Both extremes of diurnal temperature may fall within such an extended range, thereby allowing the membrane to function normally throughout the day.

Table 4. *Phase transition temperatures of some glycerophospholipids*

Glycerophospholipid	Fatty acyl residues	T_c °C
Phosphatidylcholine	14:0/14:0	23
	16:0/16:0	40.5
	18:0/18:0	58
	18:1(*sn*-1)/18:0(*sn*-2)	15
	18:0(*sn*-1)/18:1(*sn*-2)	3
	16:0(*sn*-1)/18:1(*sn*-2)	–5
	18:1/18:1	–22
Phosphatidylcholine	16:0/16:0	40.5
Phosphatidylglycerol	16:0/16:0	41
Phosphatidylserine	16:0/16:0	53
Phosphatidylethanolamine	16:0/16:0	64

T_c = temperature of transition from a solid gel state to a flexible, fluid-like liquid crystalline state

In addition to their collective function of maintaining membrane fluidity, some M.l. have specific functions. For instance, many glycolipids with large oligosaccharide head groups act as receptor sites, which participate in immunological defense, cellular differentiation and organellar and cellular recognition. They are suited to this purpose because the structural variety that is possible in their oligosaccharides allows them to carry information and serve as signaling molecules. The phosphatidylinositol derivative, PIP_2, on the other hand, participates in the transduction of hormonal and neurotransmitter messages by liberating the second messengers, Inositol-1,4,5-*tris*phosphate (IP_3) (see) and *sn*-1,2-diacylglycerol within the target cell in response to the appropriate extracellular signal.

Membranochromic pigments: plant pigments which impregnate the cell wall, e.g. the phenols and quinones that give color to heart wood.

Memory: the ability of a nervous system to store information. A functional distinction can be made between short-term and long-term M: the former lasts hours or days, while the latter is more or less permanent.

Studies on the marine snail *Aplysia* have indicated some of the molecular events associated with facilitation and depression of synaptic transmission from sensory to motor neurons. Habituation, a reduction in intensity and duration of a reflexive response, occurs when a weak stimulus is repeated but not coupled with either a positive or negative reinforcement. In the snail synapses, habituation is associated with a reduction in the number of calcium ions entering the cell as a result of each action potential spike. The calcium influx determines the release of neurotransmitter from the neuron; thus a decrease in calcium influx causes fewer molecules of neurotransmitter to be released into the synaptic gap and weakens the signal transmission.

In contrast, facilitation of a response occurs when a weak stimulus to one region is coupled with a strong stimulus to another. (When facilitation becomes permanent, it is equivalent to classical conditioning). In these snails, facilitation has been traced to an ion channel for K^+ which normally assists in the return to normal membrane voltage after a voltage spike has arrived. If this channel is partially blocked, the action potential is prolonged, more Ca^{2+} can enter the neuron, and this causes more neurotransmitter to be released. Both serotonin and cAMP facilitate synaptic transmission between sensory and motor neurons in *Aplysia.* Kandel and Schwartz have proposed a mechanism which would account for this: the serotonin enhances cAMP synthesis; cAMP in turn activates a protein kinase, which phosphorylates one or more proteins associated with the K^+ channel, and thus reduces the efficiency of the channel in transporting K^+ outward. These authors speculate that long-term memory might be the result of permanent facilitation of the same synapses, possibly due to the synthesis of a site-specific, highly cAMP-sensitive regulatory protein. This is consistent with the observation that learning in vertebrates is prevented by inhibitors of protein synthesis. [E.R.Kandel & J.H.Schwartz *Science* **218** (1982) 433–443]

Another model for M. is long-term potentiation (LTP), which occurs in the hippocampus of vertebrate brains. LTP is a stable facilitation of synaptic responses resulting from very brief trains of high-frequency nervous impulses. It has been reported that injection of EGTA (a calcium chelator) into postsynaptic neurons inhibits LTP; possibly, calcium activates a proteinase which in turn increases the number of glutamate binding sites on the postsynaptic membrane in these synapses. By binding the calcium which enters the neuron, the EGTA would prevent activation of the proteinase and thus block LTP. [G.Lynch et al. *Nature* **305** (1983) 719–721]

Menadione: vitamin K_3 and provitamin of the vitamin K group (see Vitamins).

Menaquinone-6: vitamin K_2 (see Vitamins).

***p*-Menthadienes:** doubly unsaturated, monocyclic monoterpene hydrocarbons, M_r 136.24. Of the 14

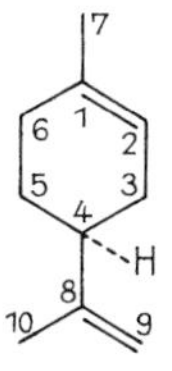

(+)-Limonene

Some common menthadienes

Name	Δ	b. p. (°C)	$[\alpha]_D$	Occurrence
(+)-Limonene	1,8	177–178	+126.8	Orange and lemon oil
(–)-Limonene	1,8	177–178	–122.6	Peppermint oil
(±)-Limonene (Dipentene)	1,8	175–176		Pine oil
α-Terpinene	1,3	173–175		*Elettaria* spp.
β-Terpinene	1(7),3	75.5 (22 mm)		*Pittosporum* spp.
γ-Terpinene	1,4	183		Widely distributed
(+)α-Phellandrene	1,5	58–59 (16 mm)	+45	Eucalyptus oil
(–)α-Phellandrene	1,5	173–175	–17.7	Fennel oil
(+)β-Phellandrene	1(7),2	173	+62.5	Fennel oil
(–)β-Phellandrine	1(7),2		–74.4	Eucalyptus oil

structural isomers of *p*-M., 9 are found naturally in volatile oils. The commonest *p*-M. are shown in the table.

Extensive groups of oxygen-containing, monocyclic monoterpenes are derived from the *p*-M.

***p*-Menthane:** see Monoterpenes, Fig.

Menthol: *p*-menthan-3-ol (M_r 156.27), commercially the most important of the monocyclic monoterpene alcohols. M. contains 3 asymmetric carbon atoms; it can exist as 4 racemates and in 8 stereoisomeric forms. The naturally occurring form, (–)-M. (m. p. 43 °C, b. p. 216 °C, $[\alpha]_D$ –49.4°, ρ_5^{15} 0.904, n_D^{20} 1.4609), is the chief constituent of peppermint oil. It is used in large quantities in the preparation of flavorings, tooth paste, etc. Oxidation of M. produces (–)-menthone, M_r 154.25, m. p. –6 °C, b. p. 204 °C, $[\alpha]_D$ –29.6, which is also a component of peppermint oil.

(–)-Menthol

(–)-Menthone: see Menthol.

Mercaptans: see Sulfur compounds.

Mercapto group: see Thiol group.

Mercapturic acids: see Glutathione *S*-transferase.

Meromyosin: see Muscle proteins.

Merrifield synthesis: see Peptides.

Mescaline: 3,4,5-trimethoxy-phenylethylamine, the principal hallucinogenic component of the drug Peyotl or Peyote, used as a ceremonial intoxicant by Mexican Indians living near the Mexico-USA border. Peyote is the cut and dried parts of the cactus *Anhalonium lexinii.* The synthetic M. derivative, 2,5-dimethoxy-4-methylamphetamine (DOM), is a more potent hallucagen than M.

Mescaline

Meselson and Stahl experiment: the first definitive experiment showing that DNA replication is semiconservative [M. Meselson & F. W. Stahl *Proc. Natl. Acad. Sci. USA* **44** (1958) 671–682]. DNA in which all the nitrogen atoms of the bases are ^{15}N is slightly heavier than "normal" DNA, in which all the bases contain only ^{14}N. The difference in buoyant density is sufficient to permit separation of ^{14}N- and ^{15}N-labeled DNA by centrifugation in a density gradient of cesium chloride. The DNA sample (<100 mg/l) is centrifuged in a centrifugal field exceeding 100,000 × g in a solution of 8 M CsCl. Redistribution of the CsCl occurs, so that the concentration increases toward the bottom of the cell, forming a density gradient from about 1.6 to 1.8 g/ml. The DNA, which has a buoyant density of about 1.7 g/ml, sediments from the region of low density and moves centripetally from the region of high density, collecting as a narrow band where its buoyant density equals that of the matrix.

E. coli was grown for several generations in a medium containing isotopically pure $^{15}NH_4^+$ as the sole nitrogen source. Under these conditions, ^{14}N-containing compounds of the original inoculum are diluted out, and all the nitrogenous compounds, including DNA, of the bacterial cells contain ^{15}N. ^{15}N-labeled DNA has a buoyant density of 1.744 g/ml, whereas ^{14}N-labeled DNA has a buoyant density of 1.704 g/ml. If replication were conservative (i. e. the parent duplex remains intact in one daughter cell, while the other daughter cell contains a duplex of two newly synthesized strands), then two species of DNA, buoyant densities 1.704 and 1.744 g/ml, would be present after one phase of cell division in a growth medium containing $^{14}NH_4^+$. If replication is semiconservative (i. e. if each strand of the parent duplex acts as a template for the synthesis of its partner), then after one phase of cell division in the presence of $^{14}NH_4^+$, all the daughter DNA will be a ^{14}N:^{15}N hybrid of intermediate buoyant density. After one phase of cell division in the presence of $^{14}NH_4^+$, the band of ^{15}N-labeled DNA is replaced by a single band of the ^{15}N:^{14}N hybrid (buoyant density 1.725 g/ml) and the ^{14}N-labeled DNA (buoyant density 1.704 g/ml) does not appear until a second phase of cell division (Fig.). See Deoxyribonucleic acid.

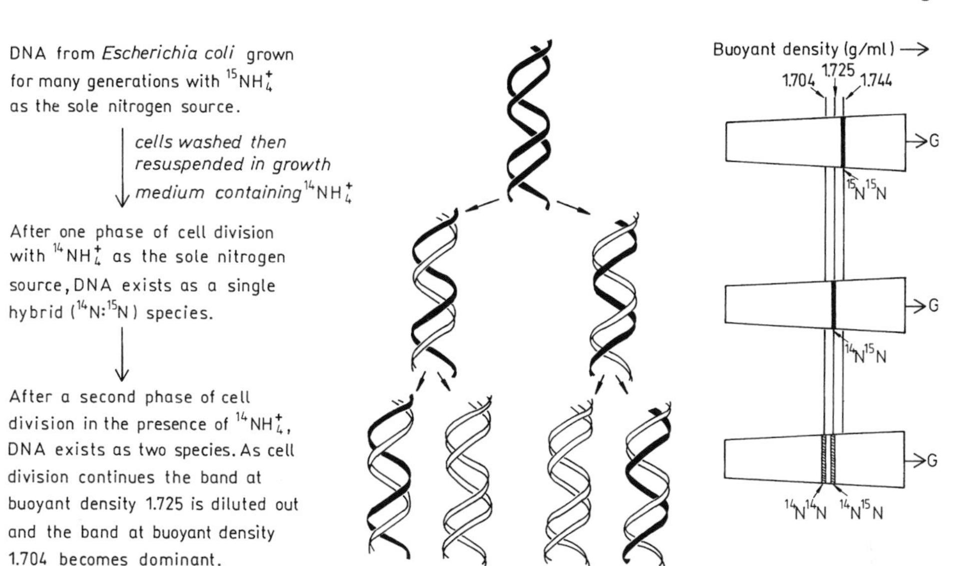

DNA from Escherichia coli grown for many generations with $^{15}NH_4^+$ as the sole nitrogen source.
cells washed then resuspended in growth medium containing $^{14}NH_4^+$
After one phase of cell division with $^{14}NH_4^+$ as the sole nitrogen source, DNA exists as a single hybrid (^{14}N:^{15}N) species.
After a second phase of cell division in the presence of $^{14}NH_4^+$, DNA exists as two species. As cell division continues the band at buoyant density 1.725 is diluted out and the band at buoyant density 1.704 becomes dominant.
Buoyant density (g/ml)
1.704
1.725
1.744
G
$^{15}N^{15}N$
G
$^{14}N^{15}N$
G
$^{14}N^{14}N$
$^{14}N^{15}N$

The Meselson and Stahl experiment. Bands of DNA in the cell of the analytical ultracentrifuge are detected by their absorption of UV light. →G indicates the direction of acceleration due to gravity (centrifugal field).

Mesobilirubin: see Bile pigments.

Mesobilirubinogen: see Bile pigments.

Messenger RNA, *mRNA, template RNA:* RNA which is translated into a polypeptide on the ribosome (see Protein biosynthesis). One molecule of mRNA may carry information for the synthesis of more than one protein, having been transcribed without interruption from several neighboring cistrons of the DNA. This polycistronic mRNA has so far been found only in prokaryotes. The polypeptides translated from polycistronic mRNA usually have related functions, e.g. 10 enzymes of histidine biosynthesis are encoded in and translated from a polycistronic mRNA containing about 12,000 nucleotides, M_r 4×10^6. Viral RNA is functionally very similar to mRNA. The entire length of viral RNA or mRNA is not translated, the start codon (always AUG) is located some distance in from the 5′-end of the molecule. For example, the start codon for β-galactosidase in the Lac mRNA of *E. coli* occurs at position 39. The untranslated sequence of nucleotides at the 5′-end of prokaryotic mRNA includes nucleotides complementary to a sequence in the 16S rRNA of the ribosome and this is believed to help bind the mRNA to the ribosome.

Functional mRNA is single-stranded. In prokaryotic cells, transcription and translation are usually coupled: mRNA becomes bound to ribosomes and translation begins before transcription is complete. The messenger is usually translated by several ribosomes at once, and thus several to many protein molecules are made from it. In prokaryotes, however, mRNA lifetimes are short, with half-times of several minutes. Eukaryotic mRNA is normally stable for hours or days.

In eukaryotic cells, the synthesis of mRNA and its subsequent translation are more complex. The first transcriptional product is a heterogeneous, very long RNA (*giant messenger-like RNA, mlRNA, heterogeneous nuclear RNA, HnRNA*). Its M_r in animal cells is $1–15 \times 10^6$. HnRNA is synthesized in the nucleoplasm (in contrast to ribosomal mRNA, which is synthesized in the nucleolus), and it is the precursor of active, polysomal mRNA. HnRNA contains both nucleotide sequences which are ultimately translated into polypeptides (exons) and large tracts of sequences which are not translated. The non-translated sequences located between exons are called "introns" (or intervening sequences); there are also repetitive sequences at the 5′-end (see Redundancy), which are not translated. The 3′-end of HnRNA and eukaryotic mRNA carries a poly(adenine) sequence (see Post-transcriptional modification of RNA), and the extreme 5′-end of most messengers carries a "cap" of 7-methylguanosine triphosphate linked 5′ to 5′ to the first "normal" nucleotide.

Capped messengers are more readily translated than uncapped ones in eukaryotic cell-free protein synthesis systems; removal of the caps interferes with the binding of the mRNA to the ribosome. It is not known when the cap is applied in vivo, but it may be part of the initiation of transcription, as capped mRNA fragments can be isolated from cells in which elongation of the messenger has been inhibited. [A. J. Shatkin et al. in *Messenger RNA and Ribosomes in Protein Synthesis,* C. F. Phelps & H. R. V. Arnstein, eds. (The Biochemical Society, London, Symposium 47) 1982.]

Although the 3′-poly(A) tail is regarded as a feature of eukaryotic mRNA, as distinct from prokaryotic mRNA, there are some notable exceptions, including histone mRNA, some HeLa cell mRNA and some early mRNA of the sea urchin embryo, which have no 3′-poly(A) tail.

Directly after synthesis, all HnRNA becomes bound to protein particles, which can be isolated from the nucleus. The protein components have been shown to include poly(A) polymerases and endonucleases (for processing of the HnRNA). The protein particle, without the RNA, is sometimes called an *informofer.*

HnRNA is processed in the nucleus by removal and degradation of the exons and some of the untranslated end sequences. The poly(A) sequences are largely unaffected, but some may be partially or completely degraded. The degradation is so extensive that in some cases as much as 90 % of the HnRNA never enters the cytoplasm. The resulting mRNA molecules (in animal cells, M_r 0.05×10^6-1.5×10^6) are transported into the cytoplasm, where they first appear as ribonucleoprotein particles (RNP), sometimes called informosomes. The mRNA then leaves the RNP and becomes associated with ribosomes to form active polysomes. Fully processed monocistronic eukaryotic mRNA still possesses some non-translated nucleotides at the 5′-end, in addition to the 3′-poly(A) tail. Storage and release of mRNA from cytoplasmic RNP may be important in the regulation of translation. Gene expression is also regulated at other steps in the process (see Gene activation, Operon, Enzyme induction; Enzyme repression, Attenuation). See also Protein biosynthesis.

Mesterolone: 1α-methyl-3β-hydroxy-5α-androstan-3-one, a synthetic androgen. M. shows high activity when administered orally. Like Methyltestosterone (see), M. is used for the therapy of male gonadal insufficiency and endocrine disorders.

Mestranol: 17α-ethinyl-3-methoxyestra-1,3,5(10)-trien-17β-ol, a synthetic estrogen. M. has high biological activity when administered orally, and it is used as a component of Ovulation inhibitors (see). M. differs from Ethinylestradiol (see) by the presence of a 3-methoxy group.

Met: abb. for L-methionine.

–Met–NH$_2$: the abbreviation for an amidomethionyl residue at the *C*-terminus of a peptide or protein.

Metabolic acidosis: see Acidosis.

Metabolic alkalosis: see Alkalosis.

Metabolic block: see Mutant technique, Auxotrophic mutants.

Metabolic bypass: see Metabolic shunt.

Metabolic control, *metabolic regulation:* Metabolism is subject to control, and an analogy may be drawn with electronic and mechanical regulation processes used in technology. In a purely formal way, living systems can be regarded as cybernetic machines. Control is a fundamental principle in the organization of living organisms, and depending on the nature of the signal or method of information transfer, there are four broad types:

1. *Neural (nervous) control.* The nerve impulse is an electrical signal, and the regulatory response may also be electrical (e. g. further nerve impulse to a muscle) or chemical (e. g. production of a hormone). The nervous system may be considered as a physiological broadcasting system.

2. *Hormonal (humoral) control.* Hormones (see) act as chemical signals in a regulation system that is superimposed on the more basic levels of M. c. Cyclic AMP acts as a second messenger for many hormones. Hormones are synthesized at specific sites (endocrine glands), then transported to the target tissue or organ. Neural and hormonal regulation represent intercellular M. c.

3. *Differential gene expression.* The signals (or triggers) of differentlal gene expression may be chemical (hormones) or environmental (e. g. light). Differential gene expression is responsible for regulation, at a molecular level, of differentiation and development.

4. *Feedback and feedforward mechanisms,* in which metabolites themselves act directly as signals in the control of their own breakdown or synthesis. Feedback is negative or positive. *Negative feedback* results in inhibition of the activity or synthesis of an enzyme or several enzymes in a reaction chain by the endproduct. Inhibition of the synthesis of enzymes is called *Enzyme repression* (see). Inhibition of the activity of an enzyme by an endproduct is an allosteric effect (see Allostery, Aromatic biosynthesis); this type of feedback control is well known for amino acid biosynthesis in prokaryotic organisms, and is variously known as *endproduct inhibition, feedback inhibition* and *retroinhibition.* In positive feedback, or *feedback activation,* an endproduct activates an enzyme responsible for its production, e. g. thrombin activates factors VIII and V during blood clotting, thus contributing to the speed of the cascade system and the rapid formation of a clot. An example of *feedforward enzyme activation* is found in the activation of glycogen synthetase by glucose 6-phosphate, i. e. a metabolite activates an enzyme involved in its utilization. *Enzyme induction* (see) is a positive feedforward process.

Table 1. *Control of enzymes in metabolism*

Control mechanism	Type of control
Chemical modification by attachment of covalently bound groups by specific enzymes.	Enzyme activity
Physical modification by noncovalent interactions: allosteric control (feedback inhibition, precursor activation).	Enzyme activity
Induction (derepression), repression.	Enzyme concentration controlled by enzyme synthesis
Exposure of active center by removal of peptides.	Activation of zymogen by limited proteolysis
Association of enzyme proteins; shift of balance between de novo enzyme synthesis and enzyme degradation by group-specific proteases (inactivases).	Enzyme concentration

Mechanisms discussed under 4. are all intracellular mechanisms of M. c., which have been studied chiefly in prokaryotic organisms.

4 a. *Chemical modification of enzymes.* This occurs by forming or breaking covalent bonds. Two mechanisms have been studied in detail: phosphorylation-dephosphorylation by protein kinases and protein phos-

Table 2. *Differences between feedback inhibition and repression of enzyme synthesis.*

Feedback inhibition	Repression
Inhibition of enzyme activity.	Inhibition of enzyme synthesis.
Allosteric interaction of enzyme and endproduct, which acts as a negative allosteric effector.	Endproduct acts as corepressor and activates a repressor protein, which prevents protein synthesis at the level of transcription.
Epigenetic regulation.	Genetic or transcriptional regulation.
The allosteric enzyme of a reaction chain (usually the first specific enzyme) is inhibited.	If the relevant enzymes are controlled by an operon, coordinate regulation results in decreased synthesis of all enzymes in the reaction chain.
Rapid, fine regulation.	Slow, coarse regulation, in which the enzymes in question are diluted out by turnover and by cell growth and division.
A reversible inhibition, depending on the endproduct concentration and the nature of the sigmoid relationship between enzyme activity and substrate concentration.	Reversible by removal of the inhibitory metabolite, i.e. the repressed system can be derepressed.
Can be demonstrated in vitro, i.e. with purified enzymes.	Cannot be demonstrated with the purified enzymes; it depends on the protein-synthesizing system of the cell. In whole cells derepression is prevented by inhibitors of transcription and translation, and by inhibitors of nucleic acid and protein metabolism.

Table 3. *Intracellular metabolic regulation*

Regulation system	Mechanism and consequences
Compartmentation	Differential location of enzymes in organelles, membranes, etc.
Membrane barriers	Membranes regulate the selective exchange of materials, separate cytoplasmic and noncytoplasmic compartments, and create concentration and pH gradients, which drive osmosis and phosphorylation.
Multienzymes	Association of enzymes into multienzyme systems and complexes, i.e. metabolic compartmentation.
Metabolic pools	Metabolic compartmentation.
Enzyme kinetic regulation	Competition between enzymes for common substrates and cosubstrates on the basis of their Michaelis constants; product inhibition; stoichiometric effects of metabolites on equilibria and reaction rates; changes in the type and quantity of enzyme effectors.
Regulation of enzyme activity:	
Isostery	Competitive inhibition by structural analogs.
Allostery	Physical modification of enzyme and therefore enzyme activity by allosteric effectors.
Chemical modification.	Attachment or removal of covalently bound groups which influence the biologically active quaternary structure of enzyme proteins.
Limited proteolysis	Exposure of active centers of peptide hormones and digestive enzymes by removal of a peptide sequence, i.e. a form of processing.
Regulation of enzyme concentration:	
Regulation of gene activity	Induction (derepression) and repression of enzyme synthesis by transcriptional regulation on the DNA template, or the RNA polymerase.
Regulation of protein biosynthesis	Regulation at various stages of translation.
Regulation of proteolysis	Change in protein turnover by shift of balance between de novo enzyme synthesis and enzyme degradation.

phatases (see Adenosine phosphates, Glycogen metabolism, Regulation), and adenylylation-deadenylylation (see Covalent modification of enzymes).

4b. *Physical modification of enzymes* is a feature of allostery.

Both 4a and 4b result in a change in the active conformation of the enzyme protein. Chemical modification causes a change in the equilibrium: protomers $\rightleftharpoons$ oligomers, leading to the establishment or abolition of quaternery structure.

M.c. is also achieved by competition of enzymes for common substrates and cosubstrates, cofactor stimulation, regulation of coenzyme synthesis, product inhibition and stoichiometrie inhibition by metabolites. An important principle of M.c. in eukaryotes appears to be the regulation of active enzyme concentrations by a change in the turnover of enzyme proteins. This type of control therefore depends on the regulation of Proteolysis (see), e.g. the active concentration of tryptophan synthase in yeast is controlled by group-specific proteases, called inactivases; further regulation is achieved by an inactivase inhibitor, which is also a protein. Group-specific proteases probably initiate a cascade reaction, in which they prepare proteins for further degradation by unspecific proteases. [For further details of enzyme regulation by proteolysis, see R.T. Schimke, "On the Roles of Synthesis and Degradation in Regulation of Enzyme Levels in Mammalian Tissues", pp. 77–120 in *Current Topics in Cellular Regulation,* B.L. Horecker, and E.R. Stadtman (eds.), Vol. 1, 1969, Academic Press; and all contributions to the Symposium: *Intracellular Protein Turnover,* R.T Schimke. & N. Katunuma (eds.) 1975, Academic Press]

Metabolic cycle: a catalytic series of reactions in which the product of one bimolecular reaction is regenerated: $A + B \rightarrow \rightarrow \rightarrow C + A$. Thus A acts catalytically, is required only in small amounts, and can be considered as a carrier of B. The catalytic function of A and other members of the M.c. insure economic conversion of B into C. B is the substrate of the M.c. and C is the product. If intermediates are withdrawn from the M.c., e.g. for biosynthesis, the stationary concentrations of the M.c. intermediates must be maintained by synthesis. Replenishment of depleted M.c. intermediates is called *anaplerosis.* Only one anaplerotic reaction is necessary, since the resulting intermediate is in equilibrium with all other members of the cycle. Anaplerosis may be served by a single reaction, which converts a common metabolite into an intermediate of the M.c. (e.g. pyruvate to oxaloacetate in the tricarboxylic acid cycle), or it may involve a metabolic sequence of reactions, i.e. an *anaplerotic sequence* (e.g. the glycerate pathway which provides phospho*enol*pyruvate for anaplerosis of the dicarboxylic acid cycle).

M.c. are *anabolic, catabolic* or *amphibolic.* The Calvin cycle (see) is an anabolic (synthetic) cycle. A truly catabolic cycle, which does not supply intermediates for biosynthesis, probably does not exist. The Tricarboxylic acid cycle (see) is an important central metabolic pathway, serving both the terminal oxidation of substrates and the provision of intermediates for biosynthesis (e.g. biosynthesis of porphyrins and certain amino acids); it is therefore an amphibolic M.c. Similarly, the Pentose phosphate cycle (see) has a catabolic function and provides ribose phosphate for the synthesis of nucleic acids and certain coenzymes.

The first M.c. to be recognized was the Urea cycle (see), described by Krebs and Henseleit in 1932. This may be considered as an anabolic cycle since it results in the energy-dependent synthesis of urea; but with respect to its metabolic role in the degradation of protein and detoxication of ammonia, it is catabolic. However, under certain conditions, it also provides arginine for protein synthesis, and anaplerosis occurs by the synthesis of ornithine from glutamate.

Metabolic shunt, ***metabolic bypass:*** a metabolic pathway which bypasses some reactions and exploits others of a primary metabolic pathway, e.g. the Aminobutyrate pathway (see) and Glyoxylate cycle (see). The term is sometimes used to mean secondary metabolism (see Secondary metabolites).

Metabolism: the sum total of chemical (and physical) changes that occur in living organisms, and which are fundamental to life. Nutrients from the environment are used in two ways by living organisms; they may serve as constituents in the synthesis of components of the organism (assimilation), or they may be oxidatively degraded for energy production (dissimilation). All constituents of living organisms are subject to a process of continual breakdown and resynthesis, normally referred to as Turnover (see). Processes of breakdown are collectively called Catabolism (see), and all reactions concerned with synthesis are called anabolism. An Amphibolic pathway may serve both degradation and synthesis. M. represents a cybernetic network with self-regulatory properties (see Metabolic control), containing many cyclic processes (see Metabolic cycle).

Since M. is an open system, its multistage processes continually approach a state of equilibrium that is never quite achieved, at least not while the organism is alive. M. is therefore better described as a steady state, rather than an equilibrium. Let S. be a starting substrate (e.g. a dietary component), which is converted to endproduct E via intermediates $I_1, I_2, I_3 \ldots I_n$, then $S \xrightarrow{1} I_1 \xrightarrow{2} I_2 \xrightarrow{3} I_3 \xrightarrow{n} I_n \rightarrow E$. Reactions 1, 2, 3 ...... n are enzyme-catalysed conversions, leading to the endproduct, according to the principle of "organization by specificity" (Dixon). E may be accumulated or excreted. Per unit time, the quantity of S converted is proportional to the quantity of E produced (it is assumed that the quantity of available S is so large that its decrease per unit time can be ignored). The steady state concentrations of the intermediates $I_1, I_2, I_3 \ldots I_n$ are constant, i.e. the rate of change of concentration is zero:

$$-\frac{ds}{dt} = +\frac{dE}{dt}; \quad \frac{dI_{1,2,3\ldots n}}{dt} = 0$$

The quantity of I_2 formed per unit time is equal to the quantity of I_1 converted, and is proportional to the concentration of I_1:

$$+\frac{dI_2}{dt} = \frac{dI_1}{dt} = k_1 I_1;$$

$$-\frac{dI_2}{dt} = k_2 I_2; \quad (k = \text{rate constant})$$

At equilibrium, $k_1 I_1 = k_2 I_2$, or $\frac{I_1}{I_2} = \frac{k_2}{k_1}$.

This leads to the following conclusions: 1. Concentrations of intermediates depend on the rate constants; the larger the rate constant of a reaction, the smaller the concentration of the intermediate. 2. The rate of the conversions S → E is determined by the rate of the slowest step; Krebs called this reaction the *pacemaker* reaction. The situation is usefully illustrated by analogy with a sluice: above the sluice gate (pacemaker), the water level is high (intermediates are present in high saturating concentrations). The sluice gate determines the rate of water flow (substrate conversion). A sudden increase in the concentration of an intermediate would be self-regulated by an overflow and diversion of the intermediate into alternative metabolic channels, which occur earlier than the pacemaker reaction. After the pacemaker reaction, such an increase leads to increased synthesis of endproduct, which is stored or excreted. The endproduct may also regulate its own synthesis by actually inhibiting the synthetic process, so that the rate of production of endproduct is finely adjusted to its rate of utilization by the cell (i.e. endproduct inhibition, see Metabolic control). See also Primary metabolism, Secondary metabolism, Metabolic shunt, Intermediary metabolism, Energy metabolism, Carbon dioxide assimilation, Respiration.

Metabolite: a substance produced or consumed by metabolism. Biopolymers are not included in this definition. The precursors and degradation products of biopolymers are, however, true M. All small molecules produced or converted by enzymes during metabolism are M. According to the nomenclature of enzymology, substances (including biopolymers) that are attacked by enzymes are called substrates.

Metachromatic leukodystrophy: see Lysosomal storage diseases.

Metalloflavoproteins: see Flavin enzymes.

Metalloproteins: proteins containing complexed metals. In the metalloenzymes, the metals are functional components. In the metal-transporting M. (e.g. the blood proteins, transferrin and coeruloplasmin) and the metal-storage depot proteins (e.g. ferritin), the metal binding is reversible and the metal is a temporary component. Important iron-containing M. are the cytochromes, oxygen transport pigments (e.g. hemoglobin) and enzymes (e.g. catalase). Zinc is present in insulin and in carbonic anhydrase. Some glycolytic enzymes and proteases contain Manganese (see). Cu is present in certain oxidoreductases (see Copper proteins). The catalytic M. also include enzymes that require specific cations for activity, e.g. the sodium-dependent membrane ATPases. See also Molybdoenzymes.

Metallothioneins: highly conserved, sulfhydryl-rich, cytoplasmic metalloproteins, present in all vertebrates, invertebrates and fungi. Mammalian cells synthesize two isoforms, M.I and M.II. Each M. consists of 61 amino acid residues, of which 20 are Cys, and there are no disulfide bridges. Accumulation of M. is induced by Zn, Cu, Hg and especially Cd, as well as other transition metals. This induction is due to transcription of M. genes, which are regulated by metals, glucocorticoids and interferons. M. bind and detoxify several metals, and are involved in Zn and Cd resistance in mammalian cells. Gold compounds (e.g. Auranofin) used in the treatment of rheumatoid arthritis are potent activators of M. genes in Chinese hamster ovary cells. [B.P. Monia et al. *J. Biol. Chem.* **261** (1986) 10957–10959]

Methemoglobin: a hemoglobin in which the iron is trivalent (Fe III). M. are unable to transport oxygen.

Methemoglobinemia: see Inborn errors of metabolism.

L-Methionine, ***Met:*** α-amino-γ-methylmercaptobutyric acid, a sulfur-containing, essential proteogenic amino acid, M_r 149.2, m.p. 281 °C (d.), $[\alpha]_D^{25}$ +23.2° (c = 0.5–2.0, 5 M HCl), or −10.0° (c = 0.5–2.0, water). The nutritional value of many plant proteins is limited by their low content of Met. In the first stage of Met biosynthesis in *E. coli,* cysteine and homoserine condense to form cystathionine. The latter is cleaved to give homocysteine, which is methylated to Met by transfer of a methyl group from N^5-methyl-tetrahydrofolic acid (Fig.). Vitamin B_{12} is a coenzyme in this methyl group transfer. In green plants, Met can also be synthesized by transamination of 2-oxo-4-methylthiobutyrate (see Ethylene). The active form of Met, i.e. *S*-Adenosyl-L-methionine (see), is the methyl group donor in Transmethylation (see). Degradation of Met proceeds via L-Cysteine (see).

Homoserine (CH_2OH–CH_2–$HCNH_2$–COOH) + Cysteine (SH–CH_2–$HCNH_2$–COOH) → Cystathionine (HOOC–$HCNH_2$–CH_2–S–CH_2–CH_2–$HCNH_2$–COOH) →

→ Homocysteine (CH_2SH–CH_2–$HCNH_2$–COOH) + $CH_3COCOOH$ + NH_3

Homocysteine + N^5-Methyltetrahydrofolic acid (+ Vitamin B_{12}) → Methionine (CH_2–S–CH_3, CH_2, $HCNH_2$, COOH) + Tetrahydrofolic acid

L-Methionine biosynthesis in Escherichia coli

DL-Met is prepared on the industrial scale (more than 100,000 tons annually) by a Strecker synthesis, using β-methylmercaptopropionaldehyde prepared from acrolein and methylmercaptan. DL-Met is used to supplement poltry feed. Both D- and L-forms are effective, so that no prior separation of enantiomers is necessary. [F. Takusagawa et al. 'Crystal Structure of *S*-Adenosylmethionine Synthetase' *J. Biol. Chem.* **271** (1996) 136–147]

Methods of biochemistry: usually methods for the study of metabolic processes, but in the widest sense including isolation, identification and characterization of natural substances. Methods for the study of metabolism are classified into 3 types:

1. *In vivo methods* employ whole organisms, their organs or cells, or populations of cells of microorganisms. In balance studies, substances are administered to the organism and the time course of their conversion to various products is determined by analysis of body materials or excretory products (feces, urine, expired gases) (see, e.g. Nitrogen balance). Balance studies are also performed on isolated organs, e.g. perfused liver. The load test is also a form of balance study, in which excess of a substance is administered to the organism to test the ability of an organ or organ system to deal with the substance in question. Such tests are used clinically to investigate organ function, e.g. to test kidney function by measurement of the rate of urinary excretion of injected phenol red. The most important in vivo methods are the indicator methods. Classical among these is the use, by Dakin and Knoop, of "non-combustible" aromatic residues on fatty acids, which led to the discovery of the β-oxidation of fatty acids (see Fatty acid degradation). This and similar chemical labeling methods have the drawback that the administered material is different from the natural material under investigation. These difficulties are now overcome by the use of isotopes; a compound in which one or more of the constituent atoms is present as a stable or radioactive isotope, is chemically and biochemically identical with the unlabeled compound (see Isotope technique). Thus isotopically labeled natural metabolites can be administered and their natural fate within the organism can be determined by isotopic tracing. A further important in vivo method is the Mutant technique (see).

A combination of isotope and mutant techniques has made a considerable contribution to the exponential growth of biochemical knowledge.

2. *In vitro methods* are performed outside the whole organism. They are essentially "test tube" methods, employing crude cell homogenates, subcellular fractions thereof, or purified enzymes. For methods of cell disruption, subcellular fractionation and enzyme purification, see Proteins, Density gradient centrifugation.

The techniques of histochemistry and cytochemistry are also in vitro methods: the sites of metabolites, enzymes or metabolic reactions are identified in organ or tissue slices and in cells by characteristic chemical reactions, e.g. specific color reactions.

A large proportion of in vitro studies is concerned with the activity and behavior of enzymes, coenzymes and substrates. Such studies require enriched or purified enzymes. The degree of purity necessary for the reliable determination of kinetic parameters varies according to the nature of the enzyme and the possible impurities, but all such investigations are preferably performed with a homogeneous enzyme protein. The completely pure enzyme is necessary for the further investigation of physical and chemical properties, involving e.g. photometric measurements, and for determination of amino acid composition. Ultimately, the crystalline enzyme is required for the study of the detailed mechanism of enzyme catalysis, and for the determination of the three dimensional structure of the enzyme and enzyme-substrate complexes.

3. *Synthetic methods* include the construction and reconstruction of e.g. multienzyme complexes and biochemical systems; and modeling and simulation, e.g. Synzymes (see) and computer simulation of glycolysis.

Methotrexate: see Aminopterin.

Methoxymellein: see Phytoalexins.

Methyl-accepting chemotaxis proteins, ***MCP:*** cell membrane proteins in *E. coli* involved in the initiation and control of chemotactic behavior. There are at least 3 different MCP in *E. coli,* each responsive to stimuli from different types of chemoreceptor, including both attractants and repellants. Attractants elicit counterclockwise rotation of the flagella, which results in smooth swimming. Repellants elicit clockwise rotation, which changes the bacterium's direction of motion. Methylation of MCPs results in adaptation, or lessened responsiveness to stimulation. [D. Sherris & J. S. Parkinson *Proc. Natl. Acad. Sci. USA* **78** (1981) 6051–6055]

γ-*N*-Methylasparagine: an amino acid residue in the β-subunit of *Anabena variabilis* allophycocyanin, formed by post-translational modification of an asparaginyl residue. [A. V. Klotz et al. *J. Biol. Chem.* **261** (1986) 15891–15894]

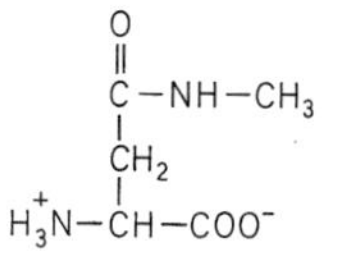

$$H_3\overset{+}{N}-CH(COO^-)-CH_2-C(=O)-NH-CH_3$$

γ-N-Methylasparagine

Methylated xanthines: *N*-methyl derivatives of xanthine, biosynthesized by the enzymatic methylation of free xanthine (N-1, 3 and 7) with *S*-adenosyl-L-methionine. Caffeine, theobromine and theophylline (Fig.) occur in certain plants and are known as purine alkaloids.

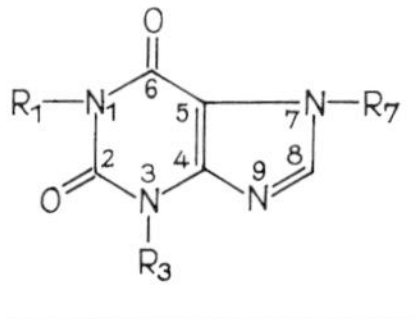

	R_1	R_3	R_7
Xanthine	H	H	H
Theophylline	CH_3	CH_3	H
Theobromine	H	CH_3	CH_3
Caffeine	CH_3	CH_3	CH_3

Structures of methylated xanthines

Caffeine (syn. thein, coffeine, guarine) acts as a central stimulant, and is present in tea, coffee, maté leaves, guarana paste and cola nuts. Theobromine is the principal alkaloid of cacao beans (1.5–3 %) and is also present in cola nuts and tea. Theobromine is usually prepared from cacao bean hulls, which contain 0.7–1.2 %. It acts as a diuretic, smooth muscle relaxant, cardiac stimulant and vasodilator. Theophylline is present in small amounts in tea. It has similar pharmaceutical properties to theobromine. The water solubility of M.x. can be increased by formation of molecular compounds with diethanolamine or isopropanolamine, and their solubility is also greatly increased by the presence of alkali benzoates, cinnamates, citrates or salicylates. For therapeutic purposes synthetic derivatives of M.x. are often used; these have improved water solubility compared with the natural M.x., e.g. 7-theophylline-acetic acid (1,2,3,6-tetrahydro-1,3-dimethyl-2,6-dioxopurine-7-acetic acid) and 1-theobromineacetic acid (2,3,6,7-tetrahydro-3,7-dimethyl-2,6-dioxo-1H-purine-1-acetic acid), which have the same therapeutic properties as theophylline and theobromine. The salt of 1-theobromineacetic acid with bromocholine phosphate is used as an antihypertensive. The 1-hexyl derivative of theobromine, called pentifylline, is used as a vasodilator; it has increased lipid solubility, which favors absorption.

***O*-Methylbufotenin:** a toxin from the toad, *Bufo alvarius* (see Toad poisons); it has also been found in plants. In addition to its general properties as a toad poison, *O*-M. also has a psychotropic effect. The lowest fatal dose for mice is 75 mg per kg.

Methylcrotonylglycinuria: see Inborn errors of metabolism.

α-Methyldeoxybenzoins: Only two naturally occurring representatives of this group are known: angolensin (Fig.) from the heartwood of *Pterocarpus* spp. and the wood of *Pericopsis* spp. (teak), and 2-*O*-methylangolensin (wood of *Pericopsis*). In *Pericopsis,* the α-M. are accompanied by the isoflavones afrormosin and biochanin A. It is therefore assumed that the α-M. are also of isoflavonoid origin. [W.D. Ollis et al. *Aust. J. Chem.* **18** (1965) 1787–1790; M.A. Fitzgerald et al. *J. Chem. Soc. Perkin I* (1976) 186–191]

HO, OH, CH_3, O, H, OCH_3

(–)-Angolensin

Methylglyoxal: an intermediate of carbohydrate degradation in certain organisms. In some bacteria (*Pseudomonas* spp.) glyceraldehyde 3-phosphate does not enter the normal glycolytic pathway; instead it is dephosphorylated to glyceraldehyde, followed by dehydration to M., which is converted to lactate (precursor of pyruvate) by a catalytic cycle involving glutathione (Fig.).

N^6 cis-γ-Methyl-γ-hydroxymethylallyladenosine: 6-(4-hydroxy-3-methyl-but-*cis*-2-enyl)-aminopurine, an adenine derivative, and the *cis* isomer of Zeatin (see). It is a rare nucleic acid component in certain tRNA, and it is biosynthesized by modification of an adenosine residue in the nucleic acid. The free compound shows cytokinin activity.

Methylmalonic acidemia: see Inborn errors of metabolism.

Methylmalonic aciduria: see Inborn errors of metabolism.

Methylnissolin: 3-hydroxy-9,10-dimethoxypterocarpan. See Pterocarpans.

Methylotrophy: the use of one-carbon compounds more reduced than CO_2 as the sole source of carbon and energy. [C. Anthony, *The Biochemistry of Methylotrophs* Academic Press, London, 1982]

Methyltestosterone: 17α-methyltestosterone, 17α-methyl-17β-hydroxyandrost-4-ene-3-one, a synthetic androgen. M. shows high biological activity when administered orally, and it is used especially for the therapy of hypogenitalism, hormonal impotence, and peripheral circulatory disturbances. It is the 17α-derivative of Testosterone (see).

Mevaldic acid: an intermediate in Terpene (see) biosynthesis.

Mevalonic acid: an intermediate in Terpene (see) biosynthesis.

Mevinolin: 1,2,6,7,8,8a-hexahydro-β,δ-dihydroxy-2,6-dimethyl-8-(2-methyl-1-oxobutoxy)-1-naphthalene-heptanoic acid δ-lactone (Fig.), a fungal metabolite from the culture medium of *Aspergillus terreus.* The parent hydroxyacid (mevolinic acid) is a potent competitive inhibitor (K_i 0.6 nM) of 3-hydroxy-3-methylglutaryl-CoA reductase (EC 1.1.1.34). Oral administration of M. produces a 30 % decrease of plasma LDL cholesterol and a moderate increase in the number of LDL receptors in human heterozygotes for familial hypercholesterolemia (see Lipoproteins). When M. is administered with Cholestyramine (see),

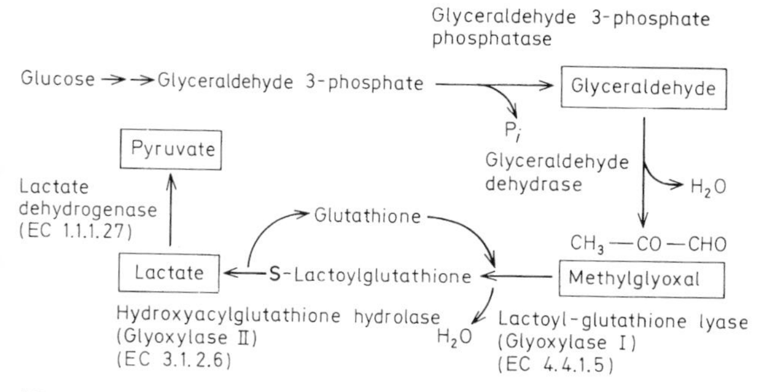

Metabolism of glyceraldehyde 3-phosphate in Pseudomonas. Synthesis and metabolism of methylglyoxal.

plasma LDL cholesterol decreases by 50–60 % and the increase of LDL receptors is even greater. See Compactin. [A. W. Alberts et al. *Proc. Natl. Acad. Sci. USA* **77** (1980) 3957–3961]

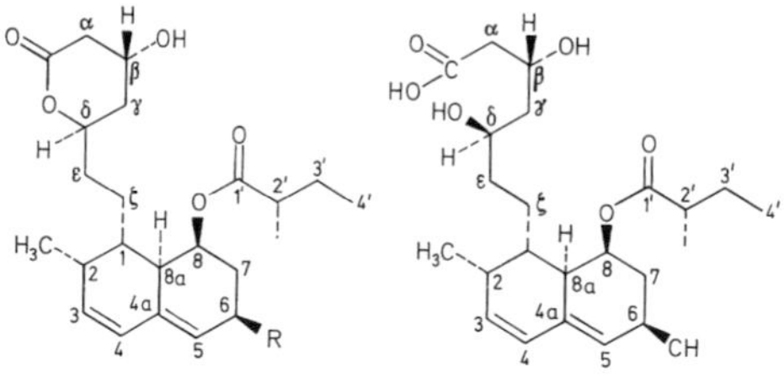

Compactin (see) (R = H) *Mevinolinic acid*
Mevinolin (R = CH_3)

Meyerhof quotient: see Pasteur effect.

MF: acronym of maize factor (see Zeatin).

MH: acronym of Maleic hydrazide.

Michaelis constant: see Michaelis-Menten kinetics.

Michaelis-Menten kinetics:

1. The Michaelis-Menten stoichiometric model shows the relationship between free enzyme (E), substrate (S), enzyme-substrate complex (Michaelis complex, ES) and product (P):

$$E + S \underset{k_{-1}}{\overset{k_1}{\rightleftharpoons}} ES \xrightarrow{k_2} E + P,$$

where k_1 and k_2 are rate constants (k_2 is also known as the catalytic constant, or k_{cat}). $(k_{-1} + k_2)/k_1$ is a kinetic constant, known as the Michalis constant and represented by K_m. If $k_2 << k_{-1}$, then $K_m \approx K_s = k_{-1}/k_1$ (Michaelis condition), and the equilibrium constant K_s is known as the substrate constant. If $k_{-1} << k_2$, then $K_m \approx k_2/ k_1$ (Briggs-Haldane condition).

2. The Michaelis-Menten rate equation shows the relationship between v (rate of reaction), V_m (maximal rate when enzyme is saturated with substrate), and S (substrate concentration): $v = V_m S/(K_m + S)$. When $S << K_m$, the reaction is first order, and $v = (V_m/K_m)S$. When $K_m << S$, the reaction is zero order, and $v = V_m$ = constant. V_m and K_m are enzyme kinetic parameters. The Michaelis-Menten equation is often valid for other cases, where the derivation of the kinetic parameters from the rate constants is more complicated.

Microbiological conversions, ***microbiological transformations:*** conversions of materials occurring in one or more stages, and catalysed by microorganisms. M. c. are the result of microbiological enzyme action, and often have no importance for the microbial cell. Several M. c. are important in the pharmaceutical industry. Examples are the stereospecific conversions of steroids, oxidation of sorbitol to sorbose by *Acetobacter suboxydans* (in the production of vitamin C), and the addition of acetaldehyde to benzaldehyde by *Saccharomyces cerevisiae.* The product of this last reaction is phenylacetylcarbinol, a precursor for D-ephedrine synthesis.

Microbiological industry, ***fermentation industry:*** a branch of industry in which materials are produced or converted by the action of microorganisms. It includes fermentation industry (production of alcoholic beverages and organic acids), production of antibiotics, production of enzymes and biochemicals, production of baking yeasts and yeasts for foodstuff manufacture. See Industrial microbiology, Industrial biochemistry.

Microelements: see Trace elements.

Microfilaments: see Cytoskeleton.

β_2-Microglobulin: The smallest known plasma protein, M_r 11815 (100 amino acid residues of known primary structure) [P. A. Peterson et al. *Proc. Natl. Acad. Sci. USA.* **69** (1972) 1697; B. A. Cunningham et al. *Biochemistry* **12** (1973) 4811]. Increased quantities of β_2-M. are found in the urine in Wilson's disease and in cadmium poisoning. There are sequence homologies between the constant region of the γG1-immunoglobulins and β_2-M., which suggest an evolutionary relationship between the two proteins. The β_2-M. gene may have evolved directly from the immunoglobulin precursor gene before its duplication.

Micronutrients: see Trace nutrients.

Microsomes: a heterogeneous fraction of submicroscopic vesicles, 20–200 nm diameter, formed during disruption of the cell by the resealing of fragments of the endoplasmic reticulum (and to some extent of the plasma membrane). Under the electron microscope, Ribosomes (see) can be seen attached to the outside of the M. M. are prepared by differential centrifugation of disrupted, homogenized cells; following the sedimentation of larger fragments, M. are sedimented at 100,000 × g.

Microtubule-associated proteins, ***MAPs:*** see Cytoskeleton.

Microtubules: see Cytoskeleton.

Middle lamella: see Cell wall.

MIH: acronym of melanotropin release inhibiting hormone. See Releasing hormones.

Milk proteins: soluble proteins present in milk, consisting of caseins and whey proteins. The chief caseins are α_s, β and κ-casein. The most important whey proteins are β-lactoglobulin, α-lactalbumin and lactoferrin. In addition, milk contains several enzyme proteins, e. g. lactoperoxidase, xanthine oxidase and immunoglobulins. IgG, IgA and IgM. These immunoglobulins are absorbed directly and without cleavage by the intestine of the infant, and provide it with passive immunity against those pathogens to which the mother is immune.

Millimicron, ***mμ:*** a unit of length, the same as a Nanometer (see).

Mineral nutrients, ***major inorganic elements, inorganic bulk elements, macroelements:*** inorganic nutrients required by living organisms in greater quantity than the Trace elements (see). They are absorbed as cations (Na^+, K^+, Ca^{2+} etc.) and anions (Cl^-, I^-, NO_3^-, SO_4^{2-} etc.). In some cases they may occur in the organism in large concentrations, e. g. skeletal structures such as the bones of vertebrates, or the silicic acid exoskeleton of diatoms; the epidermis of grasses, sedges, horsetails (*Equisetum*) is sometimes so completely impregnated with silicic acid that if the tissue is burned, a complete siliceous skeleton of the cells is left behind; the cell walls of the stoneworts (order *Charales*) are incrusted with calcium carbonate. Generally, however, very high local concentrations of inorganic elements do not occur in living organisms.

The very mobile cations K^+ and Na^+ are important in membrane transport (see ion pumps), in which K^+ is the intracellular and Na^+ the extracellular cation of the active transport mechanism. Cl^- is an intra- and extracellular anion, and it is the counterion of H^+ in the production of H^+Cl^- in the stomach. Calcium is a structural component of bone; it also occurs as calcium pectinate in the middle lamella (primordial membrane) of the plant cell wall; it acts as an oxalate trapping agent in plants that store oxalate as crystals of calcium oxalate; it is also an important ion in muscle contraction. Magnesium (see) is required for the activity of many enzymes; it is involved in bone formation, and is a component of chlorophyll.

Mineralocorticoids: see Adrenal corticosteroids.

Minimum protein (or nitrogen) requirement: the amount of complete protein required daily to compensate the nitrogen lost by excretion. Adults require 25–35 g complete (containing optimal amounts of essential amino acids) protein per day. The absolute M.p.r. is the amount of nitrogen excreted on a protein-free but calorically adequate diet, i.e. about 2.4 g N = 15 g protein per day for adults. The United States Dept. of Agriculture's recommended daily allowance for adults has been revised downward in recent years from 70 g to 40 g protein per day for adults.

Minisatellite DNA: a subset of Satellite DNA (see). In humans and other mammals some of the satellite DNA (also known as simple sequence DNA) has $C_0t_{1/2}$ values of 0.01 or less and, with the exception of 'cryptic' satellite DNA, a buoyant density significantly different from that of the main bulk of the genomic DNA. It consists of relatively short (typically 1,000–5,000 bp) lengths composed of up to 50 or so tandemly repeated (i.e. linked head-to-tail) nucleotide sequences, each typically composed of 10–60 bp. These regions are called minisatellites or minisatellite DNA to distinguish them from the rest of the satellite DNA which consists of tandemly repeated sequences that are 10^5-10^6 bp long. Within the genome there are different minisatellites which differ from one another in the nucleotide sequence of their repeating unit. Generally the nucleotide sequence of the repeating unit of the various minisatellites is constant, though variations resulting from base change, deletion or addition are known. However there is considerable variation in the number of repeat units in a given minisatellite among the genomes of individuals of a particular species; this is believed to be caused by unequal crossing over during meiosis. Because of this, minisatellites have been called 'hypervariable loci', variable number tandem repeats (VNTRs) and highly variable repeats (HVRs). It is the variability in the number of tandem repeats possessed by given minisatellite from individual to individual that has led to their use in the technique of DNA fingerprinting (see).

Minor bases: see Rare nucleic acid components.

Minor nucleic acid components: see Rare nucleic acid components.

Minus strand: see Codogenic strand.

Miraculin: a taste-modifying glycoprotein from the miraculous berry (*Synsepalum dulcificum,* family *Sapotaceae*) native to West Africa. M. does not itself taste sweet, but prior exposure of the tongue to M. causes sour substances to taste sweet; this activity may persist for several hours. Some work has been reported on amino acid composition and possible subunit structure of M. [R.H.Cagan *Science* **181** (1973) 32–35].

Miraxanthins: yellow Betaxanthins (see) found in *Mirabilis jalapa,* and consisting of conjugates of betalamic acid.

M.-I: betalamic acid conjugated with methionine sulfoxide.

M.-II: betalamic acid conjugated with aspartic acid.

M.-IV: betalamic acid conjugated with tyramine.

M.-VI: betalamic acid conjugated with dopamine.

Mirestrol, ***miroestrol:*** a potent plant estrogen from the tubers of *Pueraria mirifica* found in Thailand. The tubers of the plant are known in folk medicine for their rejuvenating and oral contraceptive properties. [N.E.Taylor et al. *J. Chem. Soc.* (1960) 3685; M.C.L. Kashemsanta et al. *Kew Bull.* (1952) 549]

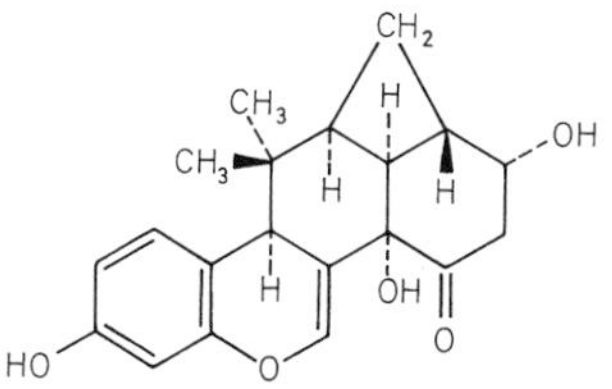

Mirestrol

Mitchell hypothesis: see Respiratory chain phosphorylation.

Mitochondria, ***chondriosomes*** (obsolete): organelles present in all eukaryotic cells. M. are 0.3–0.5 µm long, and they show a wide range of shapes and sizes. The average M. is rather elongated and about the same size as a cell of *E. coli.* M. are on the threshold of visibility under the light microscope, and they are made more readily observable by staining with Janus green B. M. of muscle cells are known as *sarcosomes.* In the differential centrifugation of cell homogenates, M. are sedimented by centrifugation at 10,000 × g for 10-15 min, i.e. their sedimentation properties are intermediate between those of the nuclear and microsomal fractions. M. are isolated in hypertonic or isotonic solutions of nonelectrolytes, such as mannitol or sucrose.

Electron microscopy of M. shows a characteristic internal structure. There are 2 concentric membranes, each 5–7 nm thick (see Biomembranes). Between the *outer* and *inner membranes* lies the *intermembrane space* (also called *external matrix* or *outer mitochondrial space*) (Figs. 1 & 2). These 2 membranes have different submicroscopic structures, their biogenesis is different, and they are functionally distinct. The outer membrane can be removed by osmotic rupture.

The density of the outer membrane is about 1.1 g/cm^3 and it is permeable to most substances of M_r 10,000 or less. It contains a high proportion of phospholipid (phospholipid to protein ratio is about 0.82 by weight); after extraction of 90 % of mitochondrial phospholipid with acetone, the inner membrane re-

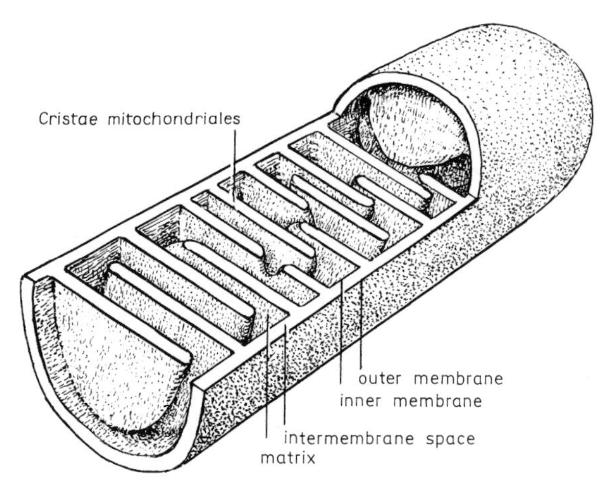

Fig. 1. *Mitochondria.* Diagram of a cristae-type mitochondrion. Each membrane consists of a bilayer

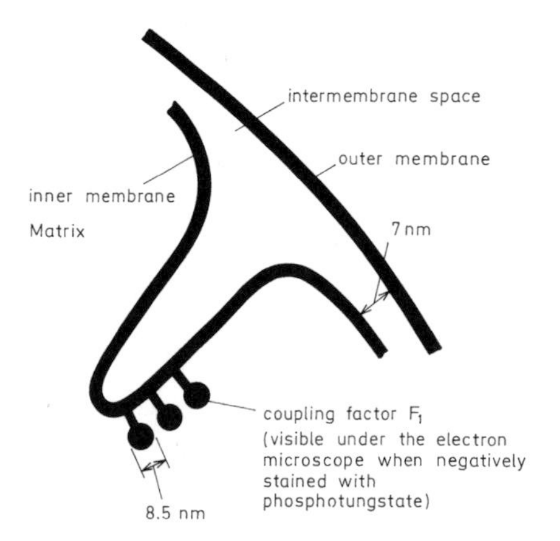

Fig. 2. *Mitochondria.* Diagrammatic representation of part of a mitochondrion, showing an infolding of the inner membrane to form a crista, and the arrangement of knob-like coupling factors (F_1) facing into the matrix.

mains intact and its double layered structure is retained, whereas the outer membrane is destroyed. The lipid fraction contains a low concentration of cardiolipin, a high concentration of phosphoinositol and cholesterol, and no ubiquinone. The density of the inner membrane is about 1.2 g/cm^3. Neutral substances, e.g. sugars, of M_r less than 150 seem to cross the inner membrane freely, but the passage of all other substances is subject to tight control. There is a low content of phospholipids (phospholipid to protein ratio is about 0.27 by weight), containing about 20 % cardiolipin. The components of the respiratory chain, including ubiquinone, are present in the inner membrane.

The inner membrane encloses the *matrix* or *stroma,* which is a contractile network of structural proteins embedded in an aqueous phase. The inner membrane also shows characteristic folds, known as *cristae,* which extend into the matrix. In some cases these structures may more closely resemble tubes which are known as *tubuli.* The terms cristae space (matrix) and intracristae space (intermembrane space) have been used, but they are equivocal and should be avoided. The precise shape and internal structure of the M. depends on its functional state (which varies from cell to cell) and may also be influenced by the methods of isolation, fixation and staining. The number of M. per cell depends on the cell type. Thus a sperm contains 20–24 M. whereas the protozoan *Chaos chaos* contains 500,000. Liver parenchyma cells contain about 500 M.

Disruption of M. produces smaller fragments known as submitochondrial particles (SMP). SMP consist chiefly of fragments of inner membrane, which become resealed to form vesicles; these are sometimes referred to as "inside-out-particles", because the outer surface (i.e. exposed to the surrounding medium) corresponds to the inner surface of the membrane in the intact M. (i.e. exposed to the matrix). The method of disruption of M. (sonication, mechanical shear, detergents) and the intensity of its application determine the nature of the resulting SMP. The capacity for oxidative phosphorylation may be lost, but the particles may still actively respire (electron transport particles, ETP). On the other hand, careful and mild disruption of M. produces SMP that are still able to carry out oxidative phosphorylation.

In addition to respiration, M. are the sites of other metabolic activities, e.g. glucuronate and ascorbic acid synthesis. The inner membrane contains all the components of the Respiratory chain (see), succinate dehydrogenase (EC 1.3.99.1), NADH dehydrogenase (EC 1.6.99.3), the enzymatic apparatus of Oxidative phosphorylation (see), ATPase (EC 3.6.1.8), 3-hydroxybutyrate dehydrogenase (EC 1.1.1.30), ferrochelatase (EC 4.99.1.1), δ-aminolevulinate synthase (EC 2.3.1.37), carnitine palmitoyltransferase (EC 2.3.1.21), fatty acid elongation system (C_{10}), NAD(P) transhydrogenase (EC 1.6.1.1), choline dehydrogenase (EC 1.1.99.1).

The intermembrane space contains adenylate kinase (EC 2.7.4.3), nucleosidediphosphate kinase (EC

2.7.4.6) and creatine kinase (heart muscle M.) (EC 2.7.3.2).

Marker enzymes for the outer membrane are flavin-containing amine oxidase (monoamine oxidase) (EC 1.4.3.4) and NADH dehydrogenase (EC 1.6.99.3) (also known as cytochrome *c*-reductase, or NADH-cytochrome b_5-reductase; this enzyme is insensitive to rotenone, amytal or antimycin A, and it is similar to the NADH-cytochrome b_5-reductase of microsomes). In addition, the outer membrane contains kynurenine 3-monooxygenase (EC 1.14.13.9), ATP-dependent fatty acyl-CoA synthetase (EC 6.2.1.3), glycerophosphate acyltransferase (EC 2.3.1.15), acylglycerophosphate acyltransferase (EC 2.3.1.51, or EC 2.3.1.52), lysolecithin acyltransferase (EC 2.3.1.23), cholinephosphotransferase (EC 2.7.8.2), phosphatidate phosphatase (EC 3.1.3.4), phospholipase A_2 (EC 3.1.1.4), nucleosidediphosphate kinase (EC 2.7.4.6), and a fatty acid elongation system (C_{14}, C_{16}). It is noteworthy that the outer membrane contains many enzymes of phospholipid metabolism.

The matrix contains all the enzymes of the TCA-cycle (see), with the exception of succinate dehydrogenase. Also present are glutamate dehydrogenase (EC 1.4.1.3), pyruvate carboxylase (EC 6.4.1.1), aspartate aminotransferase (EC 2.6.1.1), carbamoyl phosphate synthetase (EC 6.3.4.16) (utilizing ammonia and concerned in urea synthesis; unlike a similar enzyme in the cytosol which utilizes glutamine and is involved in the synthesis of pyrimidines), ornithine carbamoyltransferase (EC 2.1.3.3), GTP- and ATP-dependent fatty acyl-CoA synthetases (EC 6.2.1.10 and 6.2.1.3, respectively), and enzymes for the β-oxidation of fatty acids.

On the matrix side of the inner membrane are attached small particles (M_r about 85,000), visible under the electron microscope with negative staining (Fig. 2); these constitute the ATP-synthesizing system of oxidative phosphorylation, and are described in detail in the entry F-type ATPases (see).

Coenzymes such as NAD^+, $NADP^+$ and coenzyme A are concentrated in the aqueous phase of the matrix. The matrix also contains the protein synthesizing system of the M. Vertebrate M. contain Ribosomes (see) of size 50–55S, whereas the M. of other eukaryotic organisms contain 70S ribosomes. The DNA of M. is circular, histone-free, and attains a length (in mammals) of 5 μm. It contains cistrons for ribosomal RNA and tRNA, and the genes for ATPases, for 2 of the 3 subunits of cytochrome *b*, and for 4 of the 7 subunits of cytochrome oxidase. The biosynthesis of all the other mitochondrial constituents appears to be under the genetic control of the cell nucleus. The genetic code and tRNA complement of M. are different from those of the cytoplasm (see Genetic code).

M. are formed by the fission of M., or from promitochondria. The latter are smaller, with practically no internal structure, and no enzymes of the respiratory chain or oxidative phosphorylation. Evidence is lacking for a de novo formation of M., or the biogenesis of M. from precursor components in the cell. The 2 compartments of M., one inside the other, appear to be of different evolutionary origin.

Mitomycin C: an aziridine antibiotic, M_r 334, first isolated in 1956 from *Streptomyces caespitosus.* M. leads to an inhibition of DNA synthesis. In the natural, oxidized state, M. is inactive. In the cell it undergoes intramolecular rearrangement and reduction to form an unstable bifunctional alkylating agent. The accompanying formation of carbonium ions at positions 1 and 10 enables the compound to bind covalently to DNA and irreversibly cross link the two strands of the double helix, thus preventing the action of DNA polymerase. M. is a bacteriocide and a cytotoxin. The aziridine antibiotics include other mitomycins and porfiromycins.

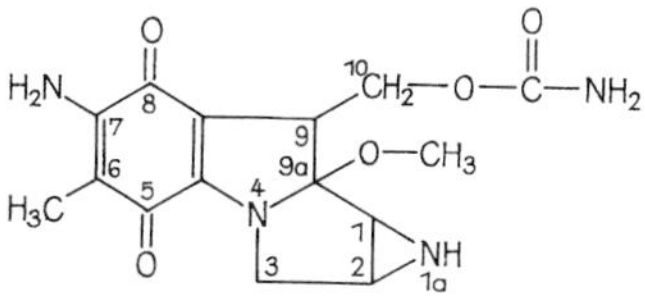

Mitomycin C

Mitosis: nuclear division in the somatic cells of eukaryotic organisms, producing two identical daughter cells with the same genetic constitution. M. occurs in 4 stages, called prophase, metaphase, anaphase and telophase. Each daughter cell resulting from M. contains the same number of chromosomes as the parent cell.

Mitosis cycle: see Cell cycle.

Mixotrophy: see Nutritional physiology of microorganisms.

mlRNA: giant messenger-like RNA. See Messenger RNA.

Mobile DNA elements, ***transposable genetic elements:*** nucleotide sequences that are capable of moving, or transposing, to new sites within the genome of both prokaryotic and eukaryotic cells. Prokaryotic M. D. e. are subdivided into (i) insertion sequences (IS), which are <2,000 bp long, (ii) bacterial transposons, which are >2,000 bp long, and (iii) bacteriophages such as Mu and D108 which replicate by transposition. Their mechanism of transposition is either conservative, in which the transposing element is excised from the donor site of the genome and inserted into the target site, or replicative, in which the transposing element remains at the donor site and a replicate of it is inserted at the target site. Eukaryotic M. D. e. are subdivided into (i) eukaryotic transposons, which transpose directly through DNA, and (ii) retrotransposons, which transpose via an RNA intermediate transcribed from the mobile element in question by an RNA polymerase, and are then converted back to DNA by reverse transcriptase (see). Retrotransposons are subdivided into 'viral retrotransposons', so-called because they are similar to the genomes of retroviruses, and 'non-viral retrotransposons', which are in turn subdivided into LINEs (see) and SINEs (see).

Moderator protein: see Calmodulin.

Modification: see Restriction endonucleases.

Modulation:

1. A change in the kinetics of an enzyme, resulting from the reversible alteration of its protein conformation by the binding of a modulating or allosteric effector (see Allostery, Covalent modification of enzymes).

2. Regulation of the rate of translation of mRNA by a modulating codon, which codes for a rare tRNA known as modulator tRNA. This is important in the regulation of the synthesis of series of enzymes from polycistronic mRNA, where specific modulator tRNA is responsible for the continuity of translation between the individual enzymes, e.g. in the biosynthesis of the 10 enzymes of histidine biosynthesis.

Molar activity: see Enzyme kinetics.

Molar mass (symbol *M*): numerically the same as the Relative molecular mass (see), but with units, e.g. the molar mass of glycine is 75 g mol^{-1}. It is sometimes needed instead of M_r, e.g. in the Svedberg equation: $M = R \times T \times s/D\ (1\text{-}v\rho)$.

Molecular chaperones: "a family of unrelated classes of proteins that mediate the correct assembly of other polypeptides but that are not components of the functional assembled structures" [R.J. Ellis & S.M. Hemmingsen *Trends Biochem. Sci.* **14** (1989) 339–342]. The classic theory of self-assembly [C.B. Anfinsen *Science* **181** (1973) 223–230] postulates that the total information for the folding of a polypeptide chain to a conformation of lower free energy is inherent in its amino acid sequence. While this may be true, it does not follow that all of the polypeptide chain will spontaneously fold into the functional polypeptide, since alternative interactions within the chain and with other molecules are also possible, leading to a certain proportion of "incorrect" (biologically useless or even deleterious) structures. M.c. prevent incorrect interactions by binding noncovalently to specific parts of the polypeptide chain that are exposed in the early stages of folding or assembly. At the same time, the principle of self-assembly is not violated, because M.c. do not provide steric information for polypeptide folding or for the association of folded polypeptides into oligomers.

M.c. also guard against folding errors when interactive regions of proteins become temporarily exposed during certain cellular processes, e.g. transport of unfolded or partly folded proteins across the membranes of organelles. Also, during the normal function of some oligomeric proteins, the sites of interaction between polypeptides may change, and former sites of interaction may become exposed (e.g. in DNA replication, recycling of clathrin, assembly of microtubules); such transiently exposed regions are then prevented by M.c. from undergoing undesirable interactions. Proteins within developing organelles are probably restrained by M.c. until the full complement of proteins is present, thus preventing incorrect interactions. Many heat shock or stress proteins (Hsps) are also M.c., and such proteins are now known to be abundant even in the absence of stress. Thus, the group known as *heat shock proteins 70* is represented by several homologs in the nucleus, cytosol, endoplasmic reticulum of eukaryotes, and in bacteria, where they chaperone the assembly and disassembly of newly synthesized, transported and oligomeric proteins.

Interaction of M.c. with a polypeptide is normally considered to be noncovalent. However, the definition may be extended to include special cases of covalently bound M.c. such as the pro-sequence of prosubtilisin and alpha-lytic protease, and the ubiquitin sequence at the aminoterminus of 2-ribosomal precursor proteins in eukaryotes. In each of these cases, correct self-assembly depends the presence of a covalently bound sequence that is subsequently removed. By the same token, the connecting peptide of proinsulin (see Insulin) insures the correct alignment of the A and B chains of the active hormone. Proteins that are transported across membranes as they are synthesized are possibly prevented from adopting a nontransportable conformation by the presence of their pre-sequence, which may therefore be regarded as a M.c.

Some other examples of M.c. are:

1. The widely distributed *chaperonins* (chaperonin 60, chaperonin 10, TCP1, TF55) which function in the folding of proteins in bacteria, plastids, mitochondria and eukaryotic cytosol. Chaperonins are large, multisubunit, ring-structured proteins that prevent nonspecific aggregation of unfolded proteins by enclosing them (an ATP-dependent process) in a protective environment, thereby increasing the yield, but not the rate of formation, of the native configuration. Two families of chaperonins are recognized: (i) Heat shock proteins 60 (Hsp60) (GroEL in *E. coli;* Cpn60 in chloroplasts) consisting of two apposed rings, each of the same 7 identical subunits ($M_r \sim 60{,}000$), and (ii) Heat shock proteins 10 (Hsp10) (GroES in *E. coli;* Cpn10 in chloroplasts) consisting of single heptameric rings of identical subunits ($M_r \sim 10{,}000$).

2. *Lim protein* of pseudomonads, which chaperones the folding of bacterial lipase.

3. *Signal recognition particle* in bacterial cytoplasm and eukaryotic cytosol, which insures correct folding of transported precursor proteins.

The term M.c. was first coined by R.A. Laskey et al. [*Nature* 275 (1978) 416–420] to describe the action of nucleoplasmin, an acidic nuclear protein that mediates the in vitro assembly of nucleosomes from histones and DNA. In solutions of physiological ionic strength in vitro, DNA and histones interact to form a precipitate. If the histones are first mixed with a molar excess of nucleoplasmin, the positive charges of the histones are decreased by the nucleoplasmin; subsequent addition of DNA results in the formation of soluble nucleosome cores and release of the nucleoplasmin. Further development of the M.c. concept was initiated by the work of R.J. Ellis on ribulose *bis*phosphate carboxylase-oxygenase (rubisco). It has been suggested that prions may be rogue M.c. [R.J. Ellis & S.M. van der Vies *Annu. Rev. Biochem.* **60** (1991) 321–347; R.J. Ellis et al. (eds.) *Phil. Trans. Roy. Soc. Lond. B* 338 (1993) 255–373; D. Wall et al. *J. Biol. Chem.* **270** (1995) 2139–2144; J.S. Weissman et al. *Cell* **84** (1996) 481–490; A.A. Antson et al 'Circular Assemblies' *Curr. Opin. Struct. Biol.* **6** (1996) 142–150; P.A. Cole 'Chaperone-assisted protein expression' *Structure* **4** (1996) 239–242]

Molecular genetics: a component of genetics and of Molecular biology (see). Classical genetics defined the gene (on the basis of its phenotypic expression and the statistical analysis of its distribution) as a hypothetical unit of inheritance. M.g. recognizes genes as defined segments of DNA, and shows the molecular basis for the causal relationship between the gene and its phenotypic expression.

Molecular mass: the mass of a molecule measured in Daltons (see).

Molecular weight: see Relative molecular mass.

(+)-Mollisacacidin: 7,3′,4′-trihydroxyflavan-3,4-diol. See Leucoanthocyanidins.

Molting hormone: see Ecdysone.

Molybdenum, *Mo:* an important trace element for plants and bacteria, and for the symbiotic fixation of nitrogen. Mo has not been demonstrated to be an essential nutrient in animals, but it is known to be a constituent of at least 3 animal enzymes (see Molybdoenzymes). As a constituent of various molybdoenzymes, Mo plays an important part in the nitrogen metabolism of plants and bacteria. In biological systems there is a partial antagonism between Mo and Cu.

Molybdoenzymes: At present, 6 oligomeric oxidoreductases are known, which contain Mo as an essential constituent: 1. Nitrogenase (see); 2. Nitrate reductase, EC 1.6.6.3 (see); 3. Xanthine oxidase, EC 1.2.3.2 (see), from animals and bacteria; 4. Aldehyde oxidase, EC 1.2.3.1 from animal liver, which catalyses the reaction $R\text{-}CHO + H_2O \rightarrow R\text{-}COOH + 2H^+ + 2e^-$; 5. Sulfite oxidase, EC 1.18.3.1 from mammalian and bird liver (M_r 114,000; 2 subunits), which catalyses the reaction $SO_3^{2-} + H_2O \rightarrow SO_4^{2-} + 2H^+ + 2e^-$; this enzyme also contains a b_5-like cytochrome and passes electrons directly to cytochrome *c* in the respiratory chain; and 6. Formate dehydrogenase, EC 1.2.1.2, a membrane-bound protein from *E. coli,* containing one atom each of molybdenum and selenium, one heme group and nonheme iron-sulfur centers. It is NAD^+-dependent and catalyses the reaction $HCOO^- + NAD^+ \rightarrow CO_2 + NADH$.

Those molybdoenzymes that are also flavoproteins, e.g. nitrate reductase, xanthine oxidase, and aldehyde oxidase, are also called molybdoflavoproteins. In molybdoenzymes, Mo is part of the redox system for the transfer of electrons.

Monellin: an intensely sweet tasting dimeric protein; the A-chain contains 44, the B-chain 50 amino acid residues. M. tastes 3,000 times (weight basis) or 90,000 times (molar basis) sweeter than sucrose. It was discovered in the fresh fruits of an African berry, *Dioscoreophyllum cumminsii* ("wild red berry") of the *Menispermaceae.* Another sweet protein is Thaumatin (see).

Monoacylglycerols: see Acylglycerols.

Monoclonal antibodies: see Immunoglobulins.

Monod model: see Cooperativity model.

Monod-Wymann-Changeux model: see Cooperativity model.

Monogalactosyldiglycerides: see Membrane lipids.

Monoglyceride: see Acylglycerols.

Monohydroxysuccinic acid: see Malic acid.

Monomer: see Subunit.

Monophenol monooxygenase, *laccase* (EC 1.14.18.1): see Oxygen metabolism, Lignin.

Monosaccharide: see Carbohydrates.

Monosomes: see Ribosomes.

Monoterpenes: aliphatic mono-, di-, or tricyclic terpenes, formed from two isoprene units ($C_{10}H_{16}$). Their occurrence is limited chiefly to plants. Most M. are volatile liquids, and they are present in volatile

Geranyl pyrophosphate → Neryl pyrophosphate → Iridoids

Type	3,7-Dimethyl-octane	*p*-Methane	Thujane	Carane	Pinane	Bornane
Hydrocarbons	Ocimene Myrcene	Terpinene Limonene	Sabinene	Δ^3-Carene	α-Pinene β-Pinene	
Alcohols	Linalool Citronellol	Terpeneol Menthol				Borneol
Carbonyl compounds	Citral Citronellal	Menthone Carvone				Camphor

Types of monoterpene structures

oils and balsams. For their separation and identification from volatile oils, M. are converted into nitrosochlorides by treatment with nitrosyl chloride. Cantharidin (see) and the Pyrethrins (see) are solid M., which possess an unusual linkage between the two isoprene components. The many different structural types of the remaining M. arise from different types of cyclization and the introduction of substituents (Fig.). The monoterpene alkaloids (e.g. Nuphara alkaloids) are M. that also contain nitrogen; M. also participate in the synthesis of the iridoid indole alkaloids (see Iridoids).

Apart from their intermediate role in the biosynthesis of other terpenes, little is known about the biological function of M. They are important volatile components of olfactorally active pheromone mixtures (e.g. citral, iridodiol). The bitter principle, Gentiopicroside (see), is also a M. In pure form or as mixtures, e.g. balsams, oil of turpentine, valtratum, camphor), M. are used widely in pharmacy, the food industry, perfumery and paint manufacture. Nerol and Geraniol are used in perfumery.

M. are biosynthesized from geranyl pyrophosphate (see Terpenes) via its isomer neryl pyrophosphate (Fig.). Aliphatic M. arise by hydrolysis of the phosphate bond, or the elimination of pyrophosphate (synthesis of 3,7-dimethyloctane type). Cyclic M. usually arise by nucleophilic substitution at C1 of the neryl pyrophosphate with loss of pyrophosphate. Iridoid compounds arise from the reductive cyclization of neryl pyrophosphate.

Montanic acid: n-octacosanoic acid, CH_3-$(CH_2)_{26}$-COOH, M_r 424.73, m.p. 91 °C. Esterified M.a. is present in various waxes, e.g. montan wax, beeswax, Chinese wax.

MOPS: acronym of 2-(*N*-Morpholino)-propane sulfonic acid (see).

Morbus Addison: see Adrenal corticosteroids.

Morin: see Flavones (Table).

Morindone: 1,2,5-trihydroxy-6-methylanthraquinone, an orange-red anthraquinone, m.p. 285 °C (from glacial acetic). M. occurs in the roots, bark and wood of *Morinda* species. In some Asian countries, particularly India, it was earlier used extensively as a natural dye.

Morphactins: a group of highly potent synthetic plant growth regulators. M. are derivatives of flurenol (9′-hydroxy-flurene-[9]-carboxylic acid, R = H), chiefly by esterification of the carboxyl group and/or substitution in the ring system, e.g. chloroflurenol (R = Cl). Over a wide concentration range, M. inhibit and modify plant growth by acting as gibberellin antagonists. M. act synergistically with certain herbicides, e.g. 2,4-D, which enables their use as broad spectrum herbicides. Among the many physiological properties of M. are: shortening of shoot internodes (generation of compact bushy growth habit), reversible inhibition of mitosis, effects on root growth, geotropism and phototropism, delayed induction of flowering, production of seedless fruit, and an influence on the sexual development of bisexual flowers.

HO COOH

R

Flurenol (R=H)

Chloroflurenol (R=Cl)

Morphactins

Morphinane: see Benzylisoquinoline alkaloids.

Morphine: medically the most important, and quantitatively the chief member of the opium alkaloids. M. [M_r 285.35, m.p. 254 °C, $[\alpha]_D^{20}$ −130.9° (methanol)] is isolated in large quantities (over 1,000 tons per year) from opium. Turkish and Macedonian opium contain 15–20 % M. Most commercial M. is converted into codeine by methylation. M. acts as an anesthetic without decreasing consciousness, and it is the most powerful analgesic known. High doses cause death by respiratory failure. Owing to its dangerous properties as a hypnotic and narcotic, the use of M. is restricted. The diacetyl derivative of M., Heroin (see), is an even more dangerous narcotic. M. was isolated in 1806 by Serturner as the soporific principle of poppy. It was the first alkaloid to be isolated.

For formula and biosynthesis, see Benzylisoquinoline alkaloids.

2-(*N*-Morpholino)-propane sulfonic acid, *MOPS*: a buffering agent (M_r 209.3, pK_a 7.2 at 25 °C) used for the preparation of buffers in the pH range 6.5–7.9.

Morquio-Brailsford syndrome: see Lysosomal storage diseases.

MPF: acronym of Maturation-promoting factor. See Cell cycle.

MRH: acronym of melanotropin releasing hormone. See Releasing hormones.

mRNA: see Messenger RNA.

MSH: acronym of melanocyte stimulating hormone. See Melanotropin.

Mucins: see Glycoproteins.

Mucolipidoses: Lysosomal storage diseases (see) arising from defects in glycoprotein biosynthesis.

Mucopolysaccharide degradation: see Lysosomal storage diseases.

Mucopolysaccharides: heteroglycans of animal connective tissue. Each of these acidic polysaccharides consists of an acetylated hexosamine (*N*-acetylglucosamine, or *N*-acetylgalactosamine) and a uronic acid (usually glucuronic acid, sometimes iduronic acid as in dermatan sulfate), which form a characteristic repeating disaccharide unit. The repeating structure of the polymer involves alternate 1,4 and 1,3 linkages. Many M. also contain sulfate. Examples of M. are Hyaluronic acid (see), Chondroitin sulfate (see), Dermatan sulfate (see), Keratan sulfate (see), Heparin (see), Heparitin sulfate (see). Some blood group substances and some bacterial polysaccharides are also M. In animals, M. function as supportive and protective materials and as lubricants.

M. are synthesized from the UDP-*N*-acetylhexosamine and UDP-D-glucuronic (or iduronic) acid. Sulfate is transferred directly from adenosine-3′-phosphate-5′-phosphosulfate after formation of the polysaccharide chain. Synthesis takes place in the endoplasmic reticulum.

Mucopolysaccharidoses: Lysosomal storage diseases (see), arising from defects in the degradation of chondrotin sulfate, keratan sulfate, dermatan sulfate and/or heparan sulfate.

Mucosulfatidosis: see Lysosomal storage diseases.

(–)-Mucronulatol: (3*S*)-7,3′-dihydroxy-4′,2′-dimethoxy-4′-methoxyisoflavan. See Isoflavan.

Müllerian inhibiting substance, *MIS:* a homodimeric protein of 70–72 kDa glycosylated subunits, and a highly specialized member of the TFG-β family of cytokines (see Transforming growth factor-β). MIS occurs exclusively in mammalian testes, where it causes regression of the Müllerian ducts during development of the male embryo. Its highly conserved *C*-terminal domain displays marked homology with human transforming growth factor-β and the β-chain of porcine inhibin. Analysis of cDNA clones reveals a protein of 575 amino acids containing a 24 amino acid leader peptide. The human gene has 5 exons that encode a protein of 560 amino acids. [R. L. Cate et al. *Cell* **45** (1986) 685–698]

Multienzyme complex: an ordered association of functionally and structurally different enzymes which catalyse successive steps in a chain of metabolic reactions. The known M. c. consist of 2–7 different catalytic units associated non-covalently (M_r 160,000 to a few million), with no associated lipids, nucleic acids or enzymatically inactive proteins. Some can be partly dissociated into smaller active groups of enzymes and half molecules, or even into their fundamental subunits (usually inactive) by changes of pH, temperature and ionic strength, or by treatment with neutral or anionic detergents. In many cases, dissociated M. c. have been reassociated to active forms similar to the physiological complex. M. c. represent an important form of subcellular organization (compartmentation), which is much simpler than the localization of enzymes within an organelle (e. g. tricarboxylic acid cycle). Owing to the close proximity of the active centers of the enzymes within the M. c., and their high affinity for substrates, the reactions of M. c. are rapid, but controlled. Intermediates are passed directly from one enzyme to the next, and they do not dissociate from the complex. The following M. c. have been particularly well studied biochemically, physically (hydrodynamic methods) and by electron microscopy.

2-Oxoacid dehydrogenases. Both the pyruvate and 2-oxoglutarate dehydrogenase complexes can be separated into 3 components: a dehydrogenase (or decarboxylase) with the dissociable cofactor thiamin pyrophosphate; a flavoprotein, dihydrolipoyl dehydrogenase; and a core enzyme containing lipoic acid which acts as a dihydrolipoyl transacetylase or transsuccinylase. The lipoic acid is bound covalently to the ε-NH_2 group of a lysine residue in the transferase. This transferase also has a special structural role; it must be present as a core for the association of the other enzymes. The two oxoacid dehydrogenases have analogous reaction mechanisms: the lipoic acid oxidizes the thiamin-bound active aldehyde, and the resulting acyl group becomes bound to the 1.5 nm long arm of the dihydrolipoic acid. Carrying the acyl group, this arm moves to the bound CoA on the transferase. The acyl-CoA is released, and the dihydrolipoic acid arm moves to the dihydrolipoyl dehydrogenase, where it becomes reoxidized. The $FADH_2$ is then reoxidized by NAD^+. For equations showing the case of 2-oxoglutarate dehydrogenase (α-ketoglutarate dehydrogenase), see Tricarboxylic acid cycle, Fig. 2.

The composition of the oxoacid dehydrogenase M. c. from *E. coli* is shown in the table. The corresponding M. c. of mammalian origin are larger, e. g. bovine pyruvate dehydrogenase contains a single central transacetylase molecule (M_r 3.12×10^6, consisting of 60 subunits), with 60 dimeric (αβ) dehydrogenase molecules ($M_r\alpha = 42{,}000$; $M_r\beta = 37{,}000$) and 5–6 dimeric flavoprotein molecules (M_r 110,000). Pig heart 2-oxoglutarate dehydrogenase consists of a central succinyl transferase molecule with 6 molecules each of 2-oxoglutarate dehydrogenase and dihydrolipoyl dehydrogenase. Under the electron microscope, pyruvate dehydrogenase from *E. coli* shows a polyhedral structure.

Fatty acid synthase. Type II fatty acid synthase is a M. c. (see Fatty acid biosynthesis). 6-Methylsalicylic

Composition of the 2-oxoacid dehydrogenase complex of Escherichia coli

Enzyme	Oligomeric enzyme molecules		Subunits per enzyme molecule		Total subunits
	n	M_r	*n*	M_r	*n*
1. *Pyruvate dehydrogenase complex*					
Pyruvate-DH	12	192,000	2	96,000	24
DL-Transacetylase	1	1.7×10^6	24	67,500	24
DL-DH (Flavoprotein)	6	112,000	2	56,000	12
Complex	19	4.67×10^6			60
2. *α-Ketoglutarate dehydrogenase complex*					
α-Ketoglutarate-DH	12	190,000	2	90,000	24
DL-Transacetylase	1	1.1×10^6	24	46,000	24
DL-DH (Flavoprotein)	6	112,000	2	60,000	12
Complex	19	2.94×10^6			60

DH = Dehydrogenase, DL = Dihydrolipoyl, *n* = Number.

acid synthetase (M_r 1.3 × 10^6; 7 constituent enzymes) from *Penicillium patulinum* closely resembles fatty acid synthase.

Other examples of M.c. are anthranilate synthase and tryptophan synthase, which are involved in microbial Aromatic biosynthesis (see), and citrate lyase. The definition of M.c. may be extended to include the membrane-bound respiratory chain, the contractile protein complexes of muscle, and the ribosome (which also contains RNA), etc.

Multipartite viruses, *coviruses:* plant viruses whose genome is divided and packed into more than one virion, often in unequal sized portions. Divided viral genomes packaged in two or three different virions, all required for infectivity and production of new virus progeny, are called dipartite and tripartite, respectively. Thus, the dipartite genome of *comoviruses* (e.g. cowpea mosaic virus) is divided between two similarly sized icosahedral virions, containing RNAs of M_r 2.3 and 1.4 × 10^6, respectively. *Almoviruses* (e.g. alfalfa mosaic virus) consist of 4 nonhelical, bacilliform rods of different lengths, which according to their positions in density gradient centrifugation are classified as bottom (B), middle (M) and top (T) components. B particles are the largest (length 58 nm) and they consist of two types: those containing RNA component 1 (M_r 1.1 × 10^6) and those containing RNA component 2 (M_r 1.0 × 10^6). The RNA of the M particles has M_r 0.8 × 10^6. Some T particles (Tb) contain no RNA, while the Ta particles contain RNA of M_r 0.3 × 10^6. Only the 3 largest RNA components are required for infection and production of progeny, so the genome is tripartite. The genomes of *bromoviruses* (e.g. brome mosaic virus) and cucumoviruses (e.g. cucumber mosaic virus) consist of 4 RNA components (2 molecules of M_r about 10^6 and one each of M_r about 0.7 and 0.3 × 10^6) separately packaged in icosahedral virions. Viral infection and production of progeny require the presence of only the three largest RNAs, so these genomes are also tripartite.

In ***divided genome viruses,*** portions of the genome are not packaged in freely separable virions, but within the same virion. For example, the genome of influenza viruses consists of 8 different RNA molecules. Within the influenza virion, each RNA molecule is weakly encapsulated in a delicate helical, filamentous nucleocapsid, about 10 nm in diameter and of varying length. The frequency of recombination of this multiplicity of genome components accounts for the frequent changes of the influenza virus, leading to new strains, some with increased virulence and pathenogenicity, and different serological properties. The only other animal viruses displaying such a high degree of physical segmentation are the *Orthomyxoandreoviridae.* The reoviruses and several plant viruses (e.g. wound tumor mosaic virus) are not multipartite, but their genomes consist of several elements of double-stranded RNA, all packaged in a single virion.

Multisubstrate enzymes: enzymes that require two or more substrates in order to catalyse a particular reaction. Accordingly, the enzyme forms a ternary (two substrates), quaternary (three substrates), etc. complex. Many enzymes are of this type, e.g. NAD-dependent dehydrogenases must bind both the substrate and NAD^+. See Cleland's short notation.

Mundulone: see Isoflavone.

Munetone: see Isoflavone.

Muningin: 6,4′-dihydroxy-5,7-dimethoxyisoflavone. See Isoflavone.

Muramic acid: *N*-acetyl-3-*O*-carboxyethyl-D-glucosamine, or *O*-lactyl-*N*-acetylglucosamine, m.p. 151 °C, $[\alpha]_D$ 144 °, an aminosugar derived from glucose, and an important constituent of mucopolysaccharide-peptide complexes of bacterial cell walls. For the biosynthesis of M.a., see Murein.

Muramidase: see Lysozyme.

Murein, *peptidoglycan:* a cross-linked polysaccharide-peptide complex of indefinite size, of the inner cell wall layer of all bacteria. It constitutes 50 % of the cell wall in Gram-negative, and 10 % in Gram-positive bacteria. M. is tough and fibrous, and it enables the cell to withstand high internal osmotic pressure. It forms an immense bag-shaped molecule or "sacculus", which gives shape and rigidity to the bacterial cell. In Gram-positive bacteria the M. layer may be up to 10 nm thick, consisting of about 20 layers of crosslinked peptidoglycan, whereas in Gram-negative bacteria, the M. layer is probably rather more flexible and less tough. M. consists of linear parallel chains of up to 20 alternating residues of β-1,4-linked residues of *N*-acetyl-glucosamine and *N*-acetylmuramic acid, extensively cross-linked by peptides.

In the M. of *Staphylococcus aureus* (Gram-positive), the *N*-acetylmuramic acid residues carry a branched peptide of 9 amino acids. The *N*-terminal branch of this peptide is linked to the *C*-terminal branch of a similar unit in a different polysaccharide chain. The branched peptide is L-alanyl-D-isoglutaminyl-L-lysyl-D-alanine, with an "extending arm" of pentaglycine attached to the ε-NH2 group of the lysine. The isoglutaminyl residue is so called because the α-carboxyl group is substituted with an amino group; linkage to the lysine occurs via the γ-carboxyl group. In the M. of *E. coli* (Gram-negative) the lysine is replaced by mesodiaminopimelic acid, and there is no pentaglycine "extending arm"; cross-linkage occurs between the D-alanine and the terminal NH_2 group of diaminopimelic acid. Also, a specific D-alanyl carboxypeptidase removes the terminal D-alanine after cross-linkage has occurred through the neighboring diaminopimelic acid. Thus the scope for cross-linking is limited. Only 15–30 % of the peptide side chains of *E. coli* M. are involved in cross-links; the remainder are free or (one in every 10–12 peptide side chains) they lose the D-alanine and become attached to the ε-NH_2 group of a lysine in a lipoprotein, via the carboxyl at the L-center of the mesodiaminopimelic acid.

Considerable variation is found in the composition of the cross-linking peptide, depending on the organism: the D-isoglutaminyl residue may be replaced by D-isoglutamyl or 3-hydroxy-D-isoglutaminyl; the alanyl residue may be replaced by L-seryl or glycyl; the lysyl residue by mesodiaminopimelyl, L,L-diaminopimelyl, ornithyl, diaminobutyryl, or homoseryl.

Biosynthesis of M. can be considered in 4 stages:

1. Synthesis of a water-soluble precursor, the UDP derivative of *N*-acetylmuramyl pentapeptide: *N*-acetylglucosamine 1-phosphate + UTP → UDP-*N*-acetylglucosamine + pyrophosphate.

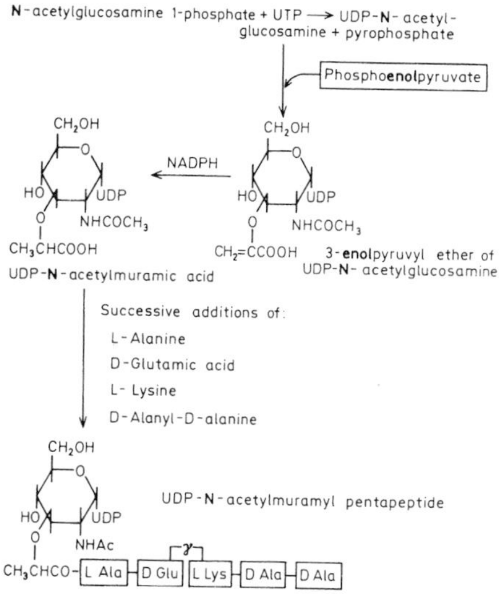

N-acetylglucosamine 1-phosphate + UTP → UDP-N-acetylglucosamine + pyrophosphate

Phosphoenolpyruvate

UDP-N-acetylmuramic acid ← NADPH — 3-enolpyruvyl ether of UDP-N-acetylglucosamine

Successive additions of:
L-Alanine
D-Glutamic acid
L-Lysine
D-Alanyl-D-alanine

UDP-N-acetylmuramyl pentapeptide

CH_3CHCO–L Ala–D Glu–(γ)–L Lys–D Ala–D Ala

The sequence of the pentapeptide residue is a function of the specificity of the enzymes concerned; there is no involvement of mRNA, ribosomes or tRNA; addition of each amino acid and of the D-alanyl-D-alanine is accompanied by cleavage of one ATP to ADP and phosphate.

2. Attachment of the precursor to the cell membrane:

The pentapeptide derivative is transferred, with loss of UMP, and becomes attached by a pyrophosphate linkage to a lipophilic carrier embedded in the membrane. The carrier is undecaprenyl phosphate, consisting of 11 isoprene units. A residue of *N*-acetylglucosamine, derived from UDP-*N*-acetylglucosamine, is added to the complex in 1,4-β-glycosidic linkage with the *N*-acetylmuramic acid. This is followed by conversion of the γ-glutamyl residue to isoglutaminyl (addition of NH_4^+, accompanied by cleavage of ATP to ADP and phosphate). A "spacer arm" of 5 glycine residues is then formed on the ε-NH_2 group of the lysine, by the addition of one glycine residue at a time from glycyl-tRNA. Ribosomes or mRNA are not involved, and in contrast to protein synthesis, formation of the pentaglycine chain starts with the *C*-terminus and finishes with the *N*-terminus. The product represents the fundamental unit of M. At this stage it is still bound to the inside surface of the cell membrane by a pyrophosphate linkage to an undecaprenyl residue:

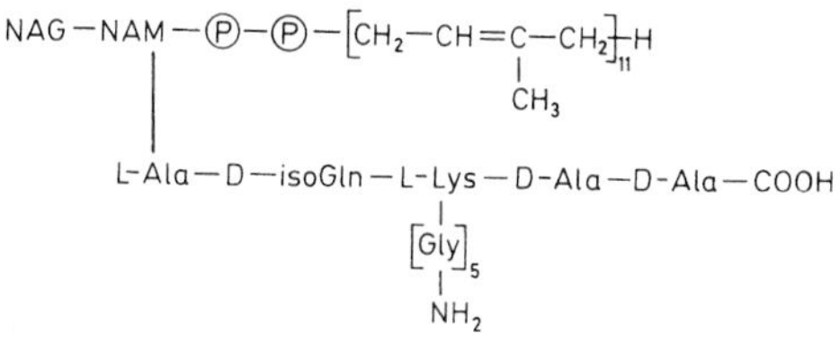

NAG—NAM—Ⓟ—Ⓟ—$[CH_2—CH=C(CH_3)—CH_2]_{11}$—H

L-Ala—D—isoGln—L-Lys—D-Ala—D-Ala—COOH, with $[Gly]_5$—NH_2 on L-Lys

NAG = N-acetylglucosamine
NAM = N-acetylmuramic acid
Ⓟ = phosphate

3. The complex is transferred to the outside surface of the cell membrane, then released, leaving undecaprenyl pyrophosphate. a) The monophosphate of undecaprenol is regenerated by the action of a specific phosphatase, and the energy of this hydrolysis appears to be utilized in moving the undecaprenyl monophosphate back to the inside surface of the membrane. The antibiotic activity of Bacitracin (see) is due to its inhibition of this pyrophosphate cleavage. b) Complexes combine into linear polymers.

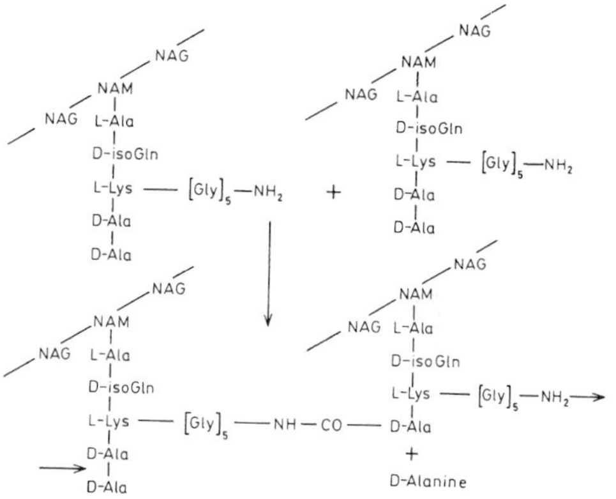

NAG–NAM–L-Ala–D-isoGln–L-Lys($[Gly]_5$—NH_2)–D-Ala–D-Ala + NAG–NAM–L-Ala–D-isoGln–L-Lys($[Gly]_5$—NH_2)–D-Ala–D-Ala

↓

NAG–NAM–L-Ala–D-isoGln–L-Lys—$[Gly]_5$—NH—CO—D-Ala–L-Lys($[Gly]_5$—NH_2)–D-isoGln–L-Ala–NAM–NAG + D-Alanine

4. Cross-links are formed by transpeptidation, the necessary energy being derived from the cleavage of the terminal D-alanine, i.e. one peptide bond is formed at the expense of another. This stage is inhibited by penicillin.

Lysozyme (see) destroys the cell wall structure of bacteria by hydrolysing 1,4-glycosidic linkages between C4 of *N*-acetylglucosamine and C1 of *N*-acetylmuramic acid in M. The glycosidic bond between C1 of *N*-acetylglucosamine and C4 of *N*-acetylmuramic acid is not cleaved. The products of cleavage are "muropeptides", e.g.

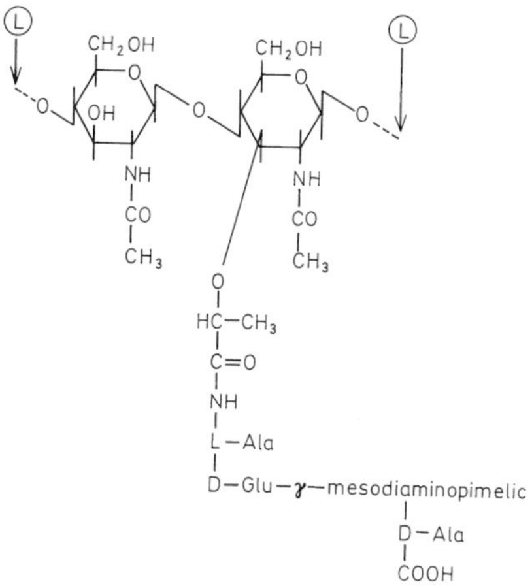

Ⓛ CH_2OH … NH–CO–CH_3 … O–HC(CH_3)–C=O–NH–L-Ala–D-Glu–γ–mesodiaminopimelic–D-Ala–COOH Ⓛ

Ⓛ = site of action of Lysozyme

Muscaaurins: orange Betalains (see) present in the cap of the fly agaric. All M. are derivatives of Betalamic acid (see), which in M.I is linked to Ibotenic acid (see), and in M.II to glutamic acid. In addition to M.I to VII, the fly agaric also contains yellow muscaflavin (an isomer of betalamic acid), violet muscapurpurin and red-brown muscarubrin.

Muscarin: a biogenic amine and one of the amanita toxins. M. is a quaternary ammonium base. (+)-M. from the fly agaric (*Amanita muscaria*) is a salt of 2*S*,4*R*,5*S*-(4-hydroxy-5-methyltetrahydrofurfuryl)-trimethylammonium. It is concentrated in the skin of the cap of the fruiting body. (+)-M. is present in high concentrations in species of the fungal genera *Inocybe* and *Clitocybe*. The symptoms of experimental muscarin poisoning are not the same as those of fly agaric poisoning; other water-soluble compounds of fly agaric are responsible for its primary toxicity, namely Ibotenic acid (see), Muscimol (see) and Muscazone (see).

Muscarinic receptors: see Acetylcholine.

Muscazone: an α-amino acid, M_r 144, m.p. 190°C (d.), with a heterocyclic substituent of 2(3H)-oxazolone. M. occurs in the fly agaric (*Amanita muscaria*). It is readily formed from ibotenic acid by UV-irradiation of dilute aqueous solutions, whereby the isoxazole ring of 3-hydroxyisoxazole is converted into the 2(3H)-oxazolone system of M. M. is much less toxic than Ibotenic acid (see).

Muscimol: the enol betaine of 5-aminomethyl-3-hydroxyisoxazole. M. [M_r 100, m.p. 155–156°C (hydrate), 174–175°C (anhydrous)] is strongly polar. It is probably not a native constituent of the agaric (*Amanita muscaria*), but an artifact formed from ibotenic acid during processing of the fungus. M. is pharmacologically very potent, and in most tests it is more active than ibotenic acid. The primary pharmacological effect is inhibition of motor function. Experimentally, M. causes euphoric and dysphoric ill humor, and leads to a condition resembling a model psychosis, but without hallucinations.

Muscle adenylic acid: see Adenosine phosphates.

Muscle proteins: proteins present in muscle and constituting 20 % of muscle tissue. The insoluble contractile proteins are organized in myofibrils which consist of organized arrays of thick and thin myofilaments. Thick filaments (about 1.5 µm long, 12–16 nm diam.) contain 200–400 molecules of the protein myosin (Fig. 1).

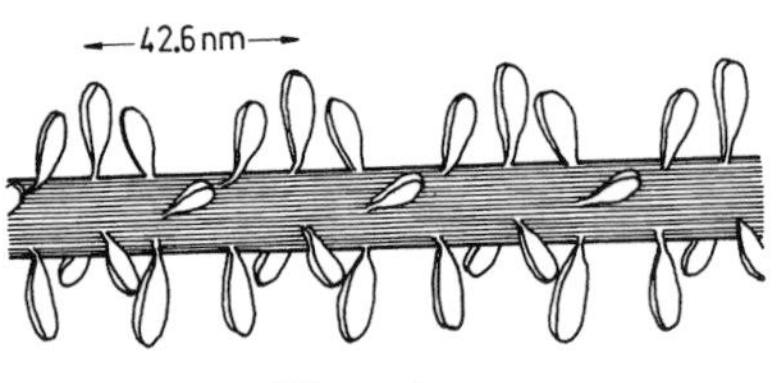

Fig. 1. *Muscle proteins.* Diagram of a thick filament, consisting of numerous parallel myosin molecules. During contraction, the globular heads of the myosin molecules, protruding from the side of the filament, make contact with actin molecules of the neighboring filaments.

Thin filaments (8 nm diam.) consist of polymerized Actin (see) (300–400 actin molecules per 1.0 µm length), each chain being accompanied by thread-like tropomyosin molecules and, in striated muscle, globular molecules of troponin. In smooth muscle troponin is replaced by caldesmon. The actin and myosin are responsible for muscle contraction, while the tropomyosin and troponin or caldesmon are regulatory proteins.

In striated muscle, the myofibrils are aligned so that alternating dense and less dense bands are visible with a repeating distance of about 2.5 µm (Fig. 2). The less dense *I bands* consist mainly of actin filaments which appear to be anchored at a central, darker *Z line*. The denser *A bands* consist of interdigitated myosin and actin filaments; each actin filament extends from the Z line into the A band. The center of the A band is a weakly staining line, the *M line*, which is usually visible only by electron microscopy. α-Actinin, a protein which serves as an anchor for actin in other types of cells, is found in the Z line. Among other proteins, the M lines contain Creatine kinase (see). When muscle contracts, the I bands shorten and may nearly disappear; this suggests that contraction involves sliding of the thick and thin filaments past one another.

Contraction in smooth muscle appears to occur by a similar mechanism; however, here the myofibrils are not aligned and striations are not visible. Another difference between striated and smooth muscle is that actomyosin (see below) can be extracted from striated muscle at high ionic strength, but from smooth muscle at low ionic strength. [J. V. Small & A. Sobieszek in Newman & Stephens (eds.) *Biochemistry of Smooth Muscle* (CRC Press, Boca Raton, 1983) pp. 85–140]

Myosin accounts for one third of M.p. and two thirds of the contractile M.p. It consists of a long, slender stem (about 135 nm long, 2 nm diam.) and a globular head region (designated S-1, about 10 nm long). The stem is formed from two polypeptide chains of 2,000 amino acid residues each; these are twisted together in an α-helix. The two chains are parallel with re-

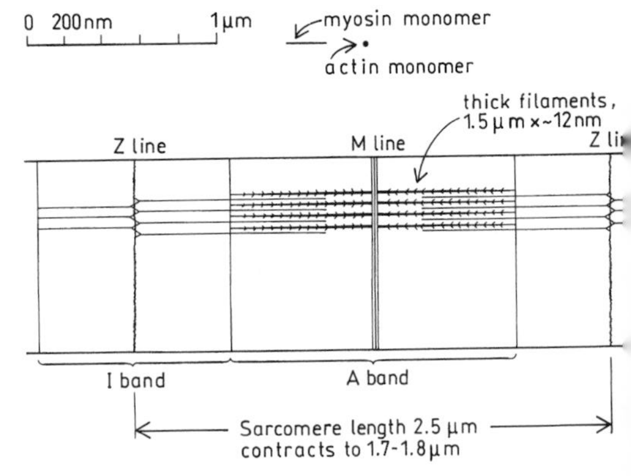

Fig. 2. *Muscle proteins.* Interdigitation of thick and thin filaments in striated muscle. [used with permission from D. E. Metzler, *Biochemistry. The Chemical Reactions of Living Cells,* Academic Press, 1977]

spect to their termini, and the *C*-termini are distal to the head end. Morphologically, there are two globular S-1 "heads", each approximately $15 \times 4 \times 3$ nm. This region contains an ATPase (part of the system producing energy for contraction) and two essential SH-groups. Treatment of extracted myosin (M_r 480,000) with guanidine or dodecyl sulfate causes the molecule to dissociate into two main subunits of the tail region (each M_r 212,000) and four light chains of the head region. Two of these light chains (each M_r 18,000) carry essential SH-groups; the remaining two chains are designated A_1 (M_r 20,700; 190 amino acids) and A_2 (M_r 16,500; 148 amino acids). Myosin from mollusks contains only three light chains.

Trypsin or papain cleaves myosin into two water-soluble fragments, the meromyosins (MM). Under controlled conditions, the cleavage occurs chiefly at one point part way along the helical stem; this point is known as the "hinge" region. Light meromyosin (LMM, M_r 140,000) is fibrillar and consists of two polypeptides; it represents a long *C*-terminal section of the original helical stem. Heavy meromyosin (HMM, M_r 340,000) represents the head and part of the tail of myosin; it is globular, contains the ATPase and consists of 9 polypeptides (7 derived from the head region). HMM has an affinity for actin and is often used to "decorate" actin filaments to reveal their orientation. Decorated actin is also used as a crystalline model in X-ray studies of the actin-troponin-myosin complex (Fig. 3).

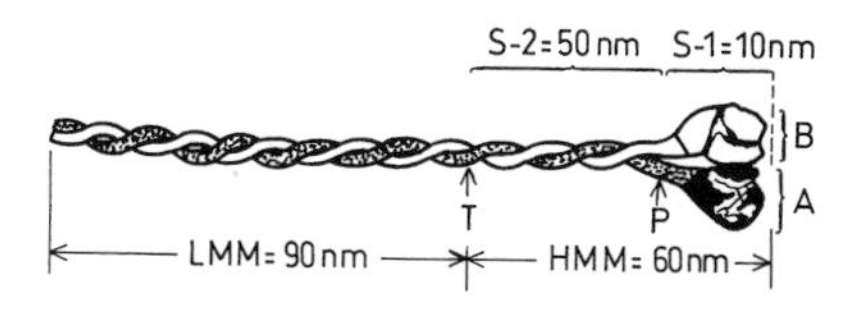

Fig. 3. *Muscle proteins.* Diagram of a myosin molecule.
T = site of attack by trypsin. P = site of attack by low concentrations of papain. LMM + S-2 = tail. S-1 = head region.

The globular S-1 heads of myosin protruding from the side of the thick filament become temporarily cross-linked with thin actin filaments to form the actomyosin complex. It is thought that the cross-bridges formed by the myosin S-1 "heads" move the actin molecules relative to the myosin molecules, but it is not known whether the myosin heads are rigidly or flexibly bound to the actin during the "power stroke" which generates tension. The difference may be visualized as follows: if the bond to actin is flexible, the myosin heads would act as rigid paddles moved by a conformational change elsewhere in the molecule; but if they are rigidly bound to the actin, the conformational changes needed to propel the actin past the myosin would include bending of the paddle-like head.

The actomyosin complex can be isolated directly from muscle, and it can also be formed in vitro from its two separate components. The complex consists of three myosin molecules and one fibrillar actin molecule (F-actin), and it has marked ATPase activity which is activated by magnesium and calcium ions. ATP causes the complex to dissociate into actin and myosin, which then reassociate when the ATP is totally cleaved. In vitro, actomyosin fibrils can be induced to contract by addition of ATP; this phenomenon, called superprecipitation, clearly demonstrates the cooperative interaction of actin and myosin, but its significance as an in vitro model of muscle contraction is disputed.

Actin (see) constitutes 14 % of total M. p. The naturally occurring thin actin filaments of striated muscle myofibrils are associated with two regulatory M. p., tropomyosin B and troponin. Tropomyosin B (M_r 68,000) is 40 nm long and has no ATPase activity. It contains about 90 % α-helix. Acting directly or indirectly via the troponin I component (see below), tropomyosin inhibits the ATPase activity of actomyosin, and thereby slows or stops contraction. In the actin filament, one tropomyosin molecule is in contact with 7 actin monomers and one troponin molecule. This complex of three components is known as α-actin.

Troponin (TN, M_r 80,000) is the central regulatory protein of skeletal muscle contraction in higher vertebrates. TN is absent from mollusks, where myosin acts as the regulatory protein. TN consists of three functionally distinct components: TN-1 (M_r 24,000), inhibitor of actomyosin ATPase; TN-T (M_r 37,000), which contains the binding site for tropomyosin; and TN-C (M_r 18,000), which binds calcium ions. The binding of calcium by TN-C abolishes the inhibitory action of TN on the actin filaments; thus when calcium floods the sarcoplasmic reticulum in response to nervous stimulation of the muscle, ATP hydrolysis can occur and is accompanied by contraction.

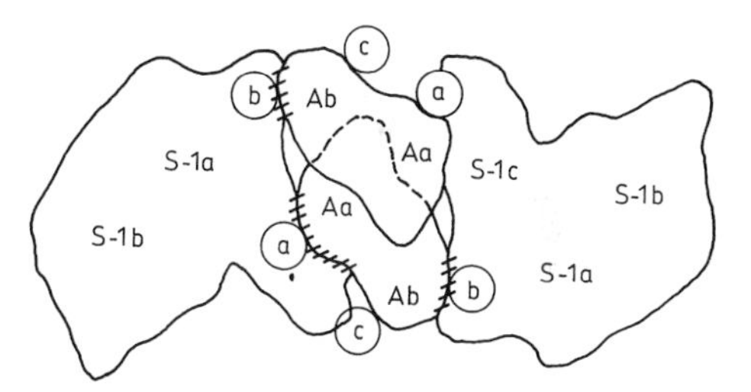

Fig. 4. *Muscle proteins.* Positions of myosin heads (S-1), actin monomers (A) and troponin (circles) in active and relaxed states shown in relationship to a cross section of decorated actin. S-1a, S-1b and S-1c indicate the domains of the HMM. Aa and Ab are the domains of the actin monomers. The circles labeled "a" show troponin fibrils in the active position; "b" and "c" are possible positions in the relaxed state. This drawing is not a helical projection but a schematic view of two actin monomers (the one on top is 27 Å above the other one and rotated by about 167° around the filament axis) and their associated S1s. The S1 on the left is therefore 27 Å above the S1 on the right hand side of the drawing. In three dimensions, additional S1s would be in contact with both of the actin subunits shown. Therefore it only appears in this two-dimensional drawing that the tropomyosin and S1 do not make equivalent contacts with the two actins. [Used with permission from E. H. Egelman *J. Musc. Res. Cell. Motil. 6* (1985) 129–151]

In lower vertebrates, TN-C is absent, and the calcium-binding function is performed by structurally similar proteins, M_r 12,000, known as Parvalbumins (see). The parvalbumins characteristically contain no cystine, cysteine, tyrosine or tryptophan, and therefore have no absorption maximum at 280 nm. Higher vertebrates also contain these proteins. Parvalbumins have been isolated from the muscle of chicken, turkey, rabbit and human.

Other fibrillar elements, the paramyosins, are present in the slow contracting fibrils of smooth muscle. They are especially prominent in the 'catch' muscles of mollusks, including those which hold the shells of bivalves closed without expenditure of energy by the animal. The main component of paramyosins is tropomyosin A, which has a high α-helix content and no ATPase activity.

Other M.p. are filamin, M_r 250,000, which binds actin; vinculin, M_r 130,000, which is part of the Z line, and titin (or connectin), M_r 250,000, found in cardiac and skeletal muscle. The giant, single molecule of Titin forms a filament extending from the M-line to the Z-line in the striated muscle sarcomere. In smooth muscle, α-actinin, vinculin and filamin anchor the thin filaments to the cell membrane. [R.M.Bagby in Newman & Stephens (eds.) *Biochemistry of Smooth Muscle* (CRC Press, Boca Raton, 1983) pp.1–84; R.M.Dowben & J.W.Shay (eds.) *Cell and Muscle Motility* Vol.4 (Plenum, New York, 1983); S.B.Marston & C.W.J.Smith *J. Muscle Res. Cell Motil.* **6** (1985) 669–708; K.Wang in *Cell and Muscle Motility* (ed. J.W.Shay) Vol.6 (Plenum, New York, 1985) pp.315–369; J.-P.Jin *J. Biol. Chem.* **270** (1995) 6908–6916; B.J.Agnew et al. *J. Biol. Chem.* **270** (1995) 17582–17587; A.S.Rovner et al. *J. Biol. Chem.* **270** (1995) 30260–30263]

Mustard oil: see Glucosinolate.

Mustard oil glucoside: see Glucosinolate.

Mustard oil thioglucoside: see Glucosinolate.

Mutagen, ***mutagenic agent:*** an agent which causes an increase in the rate of mutational events (see Mutation). In practice, the following M. are used: X-rays, UV-radiation at 260 nm (absorption maximum of DNA), nitrosomethylguanidine, methyl- and ethylmethane sulfonic acid, nitrite, hydroxylamine, base analogs, acridine dyes like proflavin, etc. The various physical and chemical M. have different mechanisms of action, some of which are imperfectly understood. The base analogue 5-bromouracil (BU for the keto form, BU* for the enol form) is a structural analogue of the DNA base, thymine. It is incorporated in place of thymine during DNA synthesis, resulting initially in replacement of some A-T base pairs by A-BU pairs (see Base pairing). BU, however, shows a marked tendency to enolize; its enol form, BU*, behaves similarly to cytosine, and base pairs with guanine. Thus A-T pairs become replaced by BU-G; effectively A is replaced by G, resulting in a change (transition) of the base (i.e. nucleotide) sequence of the DNA. Nitrite deaminates DNA bases that possess an amino group (i.e. adenine, guanine, cystosine), e.g. cytosine is converted into uracil, which is not a normal DNA component. Deamination of a DNA base by nitrite always results in a false base partner, and thus leads to a different base pair in subsequent replication. Proflavin intercalates between neighboring DNA bases, so that the base sequence and therefore the instructional content of the genetic code is changed.

Mutants: organisms, which as a result of Mutation (see), show characteristic differences from the parent or wild type organism, e.g. morphological differences (changes or defects in cell wall formation), physiological differences (e.g. change in temperature sensitivity), or new nutritional requirements (see Auxotrophic mutants). Auxotrophic M. are obtained in pure culture by suitable enrichment and selection techniques, e.g. the Penicillin selection technique (see). The precise growth requirements of M. are determined by growth studies on supplemented minimal media. Further characterization includes identification of the site of the metabolic block, and analysis of any effects on metabolic regulation. This is achieved by analysis of excretory products and the sizes of metabolic pools, by measurement of the activities or concentrations of individual enzymes, and by serological tests (see Immunology).

The mutation rate represents the probability of a mutation occurring in the generation of one cell; its value lies between 10^{-6} and 10^{-10}.

The mutational frequency represents the proportion of M. in a cell population; its value lies between 10^{-4} and 10^{-11}, depending on the mutation in question.

Auxotrophic M. and M. with regulatory defects have proved very effective in the study of metabolic pathways (see Auxotrophic M., Mutant technique), the elucidation of control mechanisms (e.g. see Repressor) and in the mapping of the bacterial genome.

Mutant technique: an important area of biochemical methodology, employing Mutants (see), in particular Auxotrophic mutants (see) of microorganisms. An auxotrophic mutant has a metabolic block at a certain point in the biosynthetic pathway of an essential product (e.g. an amino acid or coenzyme). Such a block is caused by the Mutation (see) of a gene, which (via transcription and translation) is responsible for production of an enzyme:

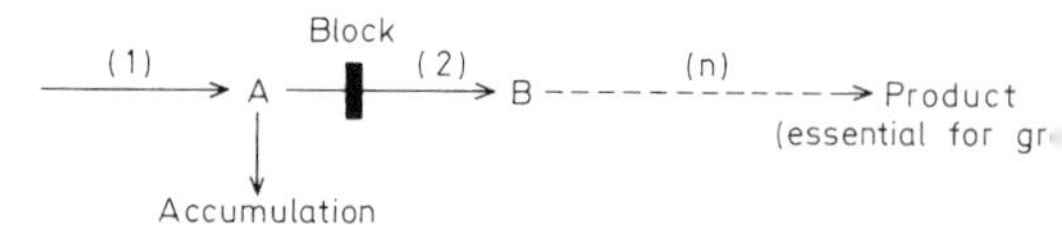

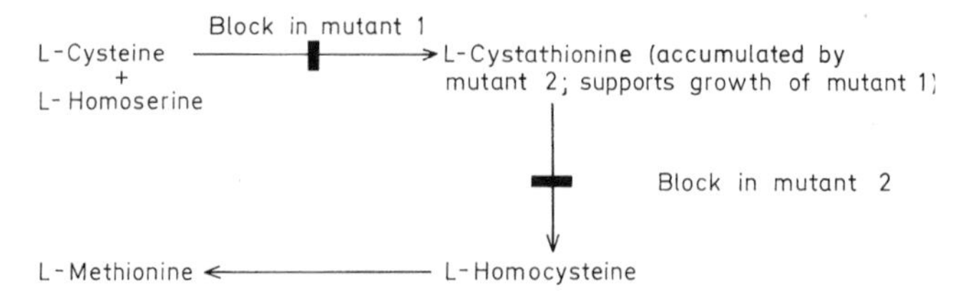

Thus in the absence of enzyme 2, the intermediate A cannot be converted into B, and synthesis of the essential end product is blocked. Intermediate A accumulates and is excreted, or its higher concentration may result in its conversion into other products by reactions that are normally insignificant in the nonmutant organism, e.g. phosphorylated intermediates of histidine biosynthesis are excreted in a dephosphorylated form by appropriate mutants. In order for growth to occur, the auxotrophic mutant must be supplied with an external source of the essential product, or with any biosynthetic intermediate between (and including) B and the product. Thus two types of analysis are possible: *Accumulation analysis,* in which intermediate A, or its metabolic product(s), is isolated and identified; and *Supplementation analysis,* in which biosynthetic intermediates occurring after the metabolic block are identified by their ability to support growth (i.e. to be converted into the end product) when added to the growth medium of the auxotrophic mutant. Growth may be measured by determination of dry weight (e.g. growth of fungal mycelia), measurement of turbidity (e.g. suspension cultures of bacteria or yeasts), or by total protein determination.

Example of accumulation analysis. Two groups of L-methionine-requiring (met⁻) *E. coli* were isolated. Both could be supplemented with L-homocysteine, whereas only one could use L-cysteine in place of L-methionine for growth. One mutant accumulated a compound, which satisfied the growth requirement of the other. This compound was identified as L-cystathionine.

This test is performed by placing the mutants close to each other on the surface of a solidified (with Agar-agar) growth medium, containing a very low concentration of L-methionine to support suboptimal growth. After about 24 hours at 30 °C, cystathionine diffuses from mutant 1 to mutant 2, which then begins to grow strongly. This phenomenon is known as cross feeding or syntrophism. With the aid of this technique, it is possible to test a large number of mutants auxotrophic for the same end product, and arrange them in the order of the metabolic blocks that they carry, e.g. if mutant 3 feeds mutant 2 feeds mutant 1 (mutant 3 will also feed mutant l), then mutant 3 is blocked later in the pathway than mutant 2, which is blocked later than mutant 1. Such an analysis can be performed with no knowledge of the nature of the intermediates involved. In the wild type organisms, biosynthetic intermediates are often present in such low concentrations that their analysis is very difficult. In contrast, they may be produced in exceptionally large quantities by auxotrophic mutants. This is probably due to the absence (or very low concentration) of the end product which in the wild type organism is responsible for the feedback control of the biosynthetic pathway. In the absence of the end product, the pathway runs out of control, at least up to the site of the metabolic block, and processes an unusually large quantity of material.

Example of supplementation analysis. The biosynthesis of L-arginine was elucidated with the aid of arginine-requiring mutants of *Neurospora crassa.* From studies on perfused liver and tissue slices, the following pathways had already been proposed: precursors → L-ornithine → L-citrulline → L-arginine. Arginine auxotrophic mutants (arg⁻) were isolated on a minimal agar medium supplemented with ornithine, citrulline and arginine. The supplementation test showed the existence of three groups:

Mutant group	Growth with Ornithine	Growth with Citrulline	Growth with Arginine	Site of block
I	–	–	+	Cit → Arg
II	–	+	+	Orn → Cit
III	+	+	+	Precursors → Orn

The detailed and complete reaction sequence was finally elucidated by the use of further mutants (blocked at various stages between glutamate and ornithine), by the additional use of isotopic methods, and by the characterization of the individual enzymes.

The site of the metabolic block can be further ascertained by measurement of the appropriate enzyme activity. This will be absent or greatly decreased, compared with that in the wild type organism. Sometimes, serologically similar, but enzymically inactive protein is produced, i.e. the mutation has not prevented transcription and subsequent translation of a protein, but small changes (sometimes replacement of a single amino acid) have occurred which destroy catalytic activity. Such proteins are known at CRiM proteins (Cross Reacting Materia; the "i" is added for pronunciation purposes).

Elucidation of the biosynthetic pathway of the aromatic amino acids (see Aromatic biosynthesis) was greatly aided by the use of mutants of *E. coli* and *Aerobacter aerogenes.* Many bacterial mutants were known to have a simultaneous requirement for phenylalanine, tyrosine, tryptophan, *p*-aminobenzoic acid and hydroxybenzoic acid, thus indicating a common biosynthetic pathway for all these compounds. This multiple requirement was satisfied by the single compound, shikimic acid, which can be isolated from the leaves of various plants. Shikimic acid was later found to be excreted by certain auxotrophs with multiple aromatic growth requirements. Subsequently, auxotrophic mutants for each step of aromatic biosynthesis were isolated, the accumulated material was identified and shown to support growth of mutants with earlier blocks, and the appropriate enzymes isolated and studied. The key branch point compound, chorismic acid, was identified with the aid of a triple mutant with metabolic blocks in the conversion of chorismic acid into prephenic acid, prephenic acid into *p*-hydroxyphenylpyruvic acid, and anthranilic acid into indole 3-glycerol phosphate. In the presence of L-tryptophan (to repress the formation of enzymes converting chorismic acid into anthranilic acid), washed suspensions of this mutant (*A aerogenes* 62–1) excreted chorismic acid.

The use of mutants, and in particular the supplementation test, are difficult when a single mutation of a fundamental metabolic reaction gives rise to multiple growth requirements, i.e. polyauxotrophy. This must be clearly distinguished from polyauxotrophy resulting from polygenetic mutations, i.e. single mutations in several distinct pathways.

The mutant technique has also been applied to naturally occurring mutations in animals, e.g. in the study of the degradation of phenylalanine and tyrosine (see Phenylalanine).

Mutarotation: a change in the optical rotation of an optical isomer, usually a carbohydrate, in aqueous solution. The carbohydrate molecule can exist in anomeric forms, designated α and β. These two diastereomers differ in chemical and physical behavior, such as melting point, solubility, and especially optical activity. In aqueous solution an equilibrium is gradually established between the two diastereomeric half acetal forms and the open chain form, with a consequent change in the optical rotation. Interconversion of the two diastereomers occurs via the intermediate open chain form. Attainment of equilibrium is accelerated by acids or bases. The starting value of $[\alpha]_D^{20}$ for α-D-glucose in water is +113°, that of β-D-glucose is + 19.7°. After a few hours, when equilibrium between α and β forms is established, the value $[\alpha]_D^{20}$ is +52.3°, representing a mixture of 37% α- and 63% β-glucose. A monosaccharide showing M. is often characterized by measurement of the starting and final values for rotation, e.g. $[\alpha]_D$ +113° $\rightarrow$ 52°.

Mutation: chemical or physical changes in the genetic material of a cell or organism. A single M. represents a change in a gene, which is a defined segment of DNA (or RNA). Chemical changes in the DNA include substitution of one nucleotide for another, due to an error in copying (a base pairing error) or to a change such as dimerization of adjacent bases which prevents accurate copying. Physical changes include breakage and loss of part of a DNA molecule, or rearrangement of the molecule.

A point mutation affects a single nucleotide; it may consist of a) substitution of a different nucleotide, b) loss of a nucleotide (deletion) or c) insertion of an extra nucleotide. In each case the nucleotide sequence is altered; in b and c, the "reading frame", or division of the sequence into triplets to be translated into amino acid (see Genetic code, Protein biosynthesis) is shifted, with severe effects on the protein product of the gene if it is a structural gene.

Substitution of one base for another may or may not change the interpretation of a base triplet or a non-translated stretch of DNA. M. which have no phenotypic effects are called "silent M." Most of the DNA in a multicellular organism appears never to be translated. Some portions of this untranslated DNA regulate the expression of structural genes (see Gene expression), some apparently serve to make rearrangements such as crossing over of sister chromatids and movement of Transposons (see) possible, some may serve to promote pairing of homologous chromosomes at meiosis, and others may have no function at all or serve merely as spacers between genes (see Introns). M. in such regions are likely never to be detected.

Detectable M. result in changes in structural or regulatory genes, and thus are reflected by the absence of, or changes in, enzymes, structural proteins or metabolic reactions, or by changes in the overall performance of a metabolic pathway. At the morphological level, the results may be, e.g. changes in pigmentation, body structure or response to environmental changes.

M. may be random, spontaneous or induced, and in a multicellular organism they may occur either in the germ-line or in somatic cells; naturally only the former are passed on to subsequent generations. A random M. may be selected by environmental pressure for preservation, although it is far more likely to be deleterious. Spontaneous M. occurs because DNA replication is subject to occasional errors, and for reasons which are not understood, some stretches of DNA are more subject to such errors than others ("hot spots"). Genetic variability within a species confers the advantage of adaptability to changing environmental conditions, and thus evolution appears to have selected DNA polymerases with a low but finite tendency to err in replication. Induced M. is caused by Mutagens (see), which are defined as agents which increase the low rate of spontaneous M.

Mutational frequency: see Mutants.

Mutation rate: see Mutants.

MWC model: see Cooperativity model.

Myasthenia gravis: see Acetylcholine.

Mycobactin: a Siderochrome (see) synthesized by *Mycobacterium* spp.

Mycocerosic acids: see Fatty acid biosynthesis.

Mycorrhiza: symbiotic association between the roots of a higher plant (forest trees, orchids, etc.) and a fungus. The fungal mycelium covers the root tips, and the root hairs become reduced or disappear. The fungal hyphae take up nutrients from the soil. The hyphae may penetrate the intercellular spaces of the root (ectotrophic M.), or actually penetrate the plant cells (endotrophic M.). Endotrophic M. is necessary for the germination of orchid seeds. Many forest trees possess ectotrophic M.

Mycosterols: see Sterols.

Mycotoxins: metabolic products of certain fungi and other microorganisms, which are harmful to other organisms, especially vertebrates, including man. The same M. may be produced by more than one fungal species. Of about 100,000 described species of fungi, 50 are known to produce M.; these may damage the host directly (e.g. plant pathogenic fungi), or indirectly by causing illnesses in animals and man when M. are consumed in the diet. Ergotism is a classical example of mycotoxicosis (see Ergot alkaloids). M.-producing organisms frequently develop on improperly stored foodstuffs, leading to food poisoning. Such M. are e.g. botulinus toxins (see Toxic proteins), Aflatoxins (see) and Ochrotoxins (see). Other M. include the nephritic toxin citrinin from *Penicillium citrinum,* notatin from *Penicillium notatum* and sporidesmin from *Pithomyces chartarum* (formerly *Sporidesmin bakeri*). Other important M. producers are *Penicillium islandicum, Penicillium rubrum. Paecilomyces varioti, Fusarium sporotrichioides* and *Stachybotrys atra.* M. also include the bacterial toxins, which are subdivided into endo- and exotoxins (see Toxic proteins).

Mydriatic alkaloids: see Tropane alkaloids.

Myelin protein A1: see Encephalogenic protein.

Myoglobin: a single chain heme protein (M_r 17,200; 153 amino acid residues) of skeletal muscle. The function of M. is to store and transfer oxygen (i.e. from hemoglobin to respiratory enzymes). The affinity of M. for oxygen is higher than that of hemoglobin. Muscle has a high content of M. especially the cardiac muscle of diving mammals, such as whale and seal, which contains up to 8% M., compared with 0.5% M. in the cardiac muscle of dog. High levels of M. are also found in the flight muscles of birds. M. and hemoglobin were the first globular proteins to be structurally elucidated by X-ray diffraction analysis. M. possesses no disulfide bridges or free SH-groups. It contains 8 variously sized right-handed helical regions, joined by non-ordered or random coil regions. 121 Amino acids are involved in helix formation, representing a total α-helix content of 77%. These 8 helices (A,B,C,D,E,F,G,H) are folded back on top of one another, and the heme is situated between helices E and F (see Proteins, Fig. 9). The molecule has the shape of a flattened ball with a pocket for the heme; the heme is almost totally buried, with just one edge exposed (the edge that carries the two hydrophilic propionic acid groups). The heme is held in position by a coordination complex between the central iron (II) atom and two histidine residues (on helices E and F, respectively). One of these histidines binds to the oxygen of the water molecule, which is bound to the heme. The position and the functional competence of the heme also depend upon the hydrophobic amino acids that line the inside of the heme pocket. Slight changes in the tertiary structure of M. destroy the oxygen-binding function of the heme. Metmyoglobin, i.e. with Fe(III), does not bind oxygen.

The primary and even tertiary structure of M. and hemoglobins from many different species show striking similarities. M. appears to be phylogenetically the oldest known heme protein, from which hemoglobin evolved as an independent molecule 600 million years ago. M. also appears to be phylogenetically related to Leghemoglobin (see).

Myo*-inositol, *meso-inositol, i-inositol, bios I, dambose, nucite, phaseomannite, meat sugar, meso-inosite, rat antispectacled eye factor, mouse antialopecia factor, cis-1,2,3,5-trans-4,6-cyclohexanehexol: an optically inactive, sweet-tasting Cyclitol (see) (m.p. 225–227 °C) found free and combined in animals and plants. It was first isolated in 1850 from muscle tissue, and is now prepared commercially by the hydrolysis of Phytic acid (see). It is an essential nutrient for yeasts and human cell lines, and deficiency symptoms have been described in animals. Relatively large amounts are normally present in brain cells, lens, thyroid gland, muscle, lung and liver, where it is largely present in phospholipids. Phospholipids containing M. play a role in the response of mammalian cells to external stimuli such as hormones (see Inositol phosphates, Membrane lipids). M. is also generally referred to as *inositol,* but it should be borne in mind that inositol is also a generic term for all 9 isomeric cyclitols. M. is biosynthesized from D-glucose, and the configuration of C-atoms 1–5 is preserved during the conversion: a cyclase (EC 5.5.1.4)

Myo-inositol

catalyses conversion of glucose 6-phosphate into inositol 1-phosphate, which is dephosphorylated by a phosphatase (EC 3.1.3.25). Inositol oxygenase converts M. into D-glucuronate, an intermediate of the Glucuronate pathway (see).

Inositol niacinate (inositol nicotinate; hexanicotinoyl *cis*-1,2,3,5-*trans*-4,6-cyclohexane; myo-inositol hexa-3-pyridine-carboxylate) is a synthetic derivative of M., in which each hydroxyl group is esterified with a molecule of nicotinic acid; it causes peripheral vasodilation and is used clinically for the symptomatic relief of severe intermittent claudication and Raynaud's phenomenon.

Myokinase: see Adenylate kinase.

Myosin: see Muscle proteins.

Myrcene: a triply unsaturated acyclic monoterpene hydrocarbon. M is a pleasant smelling liquid, M_r 136.24, b.p.$_{12}$ 55–56 °C, ρ^{15} 0.8013, n_D^{19} 1.470. It is a component of many essential oils, and it is prepared for the perfumery industry by pyrolysis of β-pinene (from oil of turpentine). M. is also used commercially for the preparation of isomeric acyclic monoterpene alcohols and their acetates.

Myrcene

Myricetin: see Flavones (Table).

Myristic acid: n-tetradecanoic acid, CH_3-$(CH_2)_{12}$ COOH, a fatty acid, M_r 228.4, m.p. 58 °C, b.p.$_{100}$ 250.5 °C. Glycerol esters of M.a. occur naturally in nearly all plant and animal fats, e.g. coconut oil, ground nut oil, linseed oil, rapeseed oil, milk fat and fish oils. Oil of nutmeg (*Myristicaceae*) is particularly rich in M.a. Esters of M.a. are also found in waxes.

Myrosin: see Thioglucoside glucohydrolase, Glucosinolate.

Myrosinase: see Thioglucoside glucohydrolase, Glucosinolate.

N

NAD: acronym of Nicotinamide adenine dinucleotide (see). NAD^+ represents the oxidized form, NADH the reduced form.

NADP: acronym of Nicotinamide adenine dinucleotide phosphate (see). $NADP^+$ represents the oxidized form, NADPH the reduced form.

Na^+K^+-ATPase: see Transport.

Nalidixic acid: an antibiotic and a naphthyridine derivative, M_r 220, used therapeutically against Gram-negative bacteria. It inhibits DNA replication in growing bacteria.

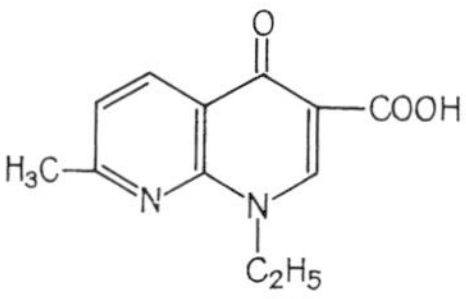

Nalidixic acid

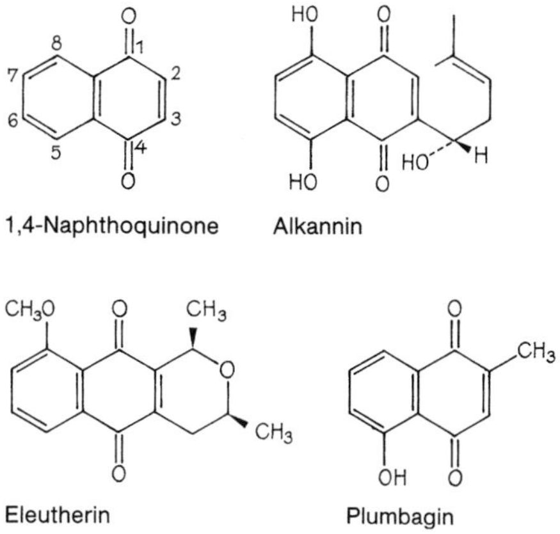

1,4-Naphthoquinone Alkannin

Eleutherin Plumbagin

Naphthoquinones

Nanometer, ***nm:*** unit of length equal to 10^{-9} m or 10^{-7} cm, and widely used for the wavelength of light.

Naphthoquinones: naturally occurring derivatives of 1,4-naphthoquinone. N. are widely distributed, and over 120 different N. are known in higher plants, bacteria and fungi. Examples of N. from plants are alkannin, eleutherin, juglone, lapachol, lawsone, lomatiol plumbagin and shikonin. In the animal kingdom, N. have been found in echinoderms, especially sea urchins (see Spinochromes). Other important N. are the K vitamins. Fungal N. are generally synthesized from acetate and malonate by the polyacetate pathway, whereas in higher plants and bacteria N. are derived from shikimic acid and a C3 compound.

Narcotics, ***narcotic drugs:*** substances which act predominantly on the central nervous system. De-

Naphthoquinones found in plants

Name	m.p. (°C)	Color	Occurrence
Alkannin*	147–149	brown-red	Roots of *Boraginaceae,* e. g. *Alkanna tinctora*
(9*R*,11*S*)-Eleutherin	175	yellow	*Eleutherina bulbosa*
Juglone (5-Hydroxy-1,4-N.)	165	yellow to brown-red	Shells of unripe walnuts. Leaves and roots of the walnut tree
Lapachol (2-Hydroxy-1,4-N., with a C_5 isoprene side chain)	142	yellow	Members of the *Bignonaceae,* e. g. *Tecoma aralia ceaea*
Lawsone (2-Hydroxy-1,4-N.)	195	yellow to orange	Leaves of henna (*Lawsonia inermis*)
Lomatiol	128	yellow	The Australian plant, *Lomatia ilicifolia*
Plumbagin (5-Hydroxy-3-methyl-1,4-N.)	78	yellow	Roots of *Plumbago* species
Shikonin (an enantiomer of alkannin)	149	brown-red	*Lithospermum* species

N. = Naphthoquinone. * Alkannin is the only one of these pigments still used as a dye and an indicator.

pending on the dose, N. show different phases of activity: small doses have a sedative action, somewhat higher doses are stimulatory, while even higher doses cause loss of consciousness (narcosis).

According to the 1964 WHO classification there are 7 groups: 1. alkaloids (LSD, mescalin, opium, etc.), 2. barbiturates and other sleeping drugs, 3. alcohol, 4. cocaine, 5. hashish and marihuana, 6. hallucillogens, 7. stimulants or antidepressants (e.g. amphetamines).

N. may be used in psychiatry and psychotherapy, but misuse can lead to acute and chronic physical and mental deterioration, and to addiction. Regular use of N. leads to tolerance, i.e. an increased rate of breakdown of N. by the body, so that 10–20 times the normal dose may be necessary to achieve the desired effect.

Naringenin: 5,7,4′-trihydroxyflavanone. See Flavonoids (Fig.), Flavanone.

Naringin: naringenin 7-neohesperidoside. See Flavanone.

Natural pigments, *biochromes:* Colored organic compounds found very widely in plants and animals. Their color is due to their chemical structure, which absorbs light in the visible spectrum between 400 and 800 nm, and reflects or transmits the unabsorbed wavelengths. If all wavelengths of the visible spectrum are absorbed more or less equally, the substance appears grey to black, while the color white results from the nonabsorption and the reflection of all wavelengths in the visible spectrum. These pigment colors are distinct from Structural colors (see) which arise from light reflection and refraction by the physical structure of surfaces. Natural (and synthetic) organic pigments are unsaturated compounds containing a system of conjugated double bonds. The chromophoric groups (or chromophores) are structures such as –CH=CH–, =CO, –N=O, or –N=N–, which, when present in the conjugated system, are responsible for absorption bands in the visible region. Auxochromic groups (or auxochromes), e.g. $-NR_2$, $-NH_2$, –OH do not produce color themselves, but they intensify the color effect of existing chromophores. The color may be shifted to longer or shorter wavelengths by bathochromic or hypochromic groups, respectively. Most N.p. also show some degree of fluorescence and/or phosphorescence. In some N.p., UV light may excite fluorescence at a wavelength near that of the visible color, e.g. ribitylflavin; this coincidence may increase the normal intensity of the visible color.

N.p. may be classified: 1.according to their occurrence, i.e. as plant pigments (flower, leaf, wood pigments, etc.), as animal pigments (blood, skin, hair, insect, eye, wing pigments, etc.), or as fungal, bacterial, algal or lichen pigments. Animal pigments are called zoochromes. The term, Phytochrome (see), however, has a restricted meaning and it does not refer to all plant pigments; 2. according to their chemical structure, i.e. as carotenoids, pteridines, tetrapyrroles, quinones, melanins, flavonoids, ommochromes, betalains, ergochromes indigoid pigments, etc.

Many N.p. act as attractant, warning or camouflage pigments, and are therefore important for the survival of the species. Others are protective, e.g. against UV light (melanin), or against fungal attack (some flavonoids). Some are involved in the trapping and utilization of light energy in plants, while animal pigments like hemoglobin are important for the transport of oxygen. Often, however, N.p. are endproducts of metabolism and have no other apparent function.

N.p. were used in antiquity as dyes. Among the oldest of these are alizarin, indigo, tyrian purple, safran, kermes, cochineal and many flavonoid-containing colored woods. In the dyeing industry they have now been largely replaced by superior synthetic compounds, but natural coloring materials are still used in the food industry.

Nearest neighbor frequency: the frequency with which a given pair of bases are immediately adjacent in the sequence of a polynucleotide chain. Since there are 4 bases, there are 16 possible pairs of nearest neighbors. Although it is now possible to determine the sequence of DNA, the parameter still gives useful information about the structure of the molecule.

Nebularine: 9-β-D-ribofuranosylpurine [M_r 252.23, m.p. 181–182 °C, $[\alpha]_D^{25}$ −48.6° (c = 1, water)], a purine antibiotic (see Nucleoside antibiotics) synthesized by the mushroom *Agaricus (Clitocybe) nebularis,* and by *Streptomyces* spp. N. is selectively active against mycobacteria It has marked cytostatic properties, and in animals it is among the most poisonous of the purine derivatives.

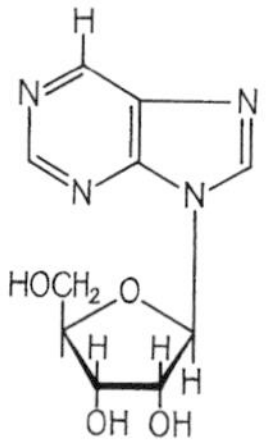

Nebularine

Necines: see Pyrrolizidone alkaloids.

Necinic acid: see Pyrrolizidone alkaloids.

Negative control: inhibition of transcription by repression or by inhibition of enzyme activity by metabolites.

Neoflavones: see Isoflavones.

Neoflavonoids: Flavonoids (see) with a 4-phenylchroman (neoflavan) skeleton, or a corresponding structure in which the C_3 chain (C2, C3, C4) is not cyclized with oxygen.

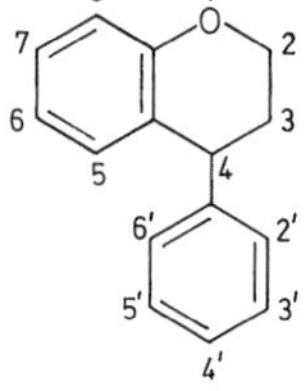

Neoflavan ring system

N. include: ***4-arylcoumarins,*** e.g. calophyllolide from *Calophyllum inophyllum* (nuts),

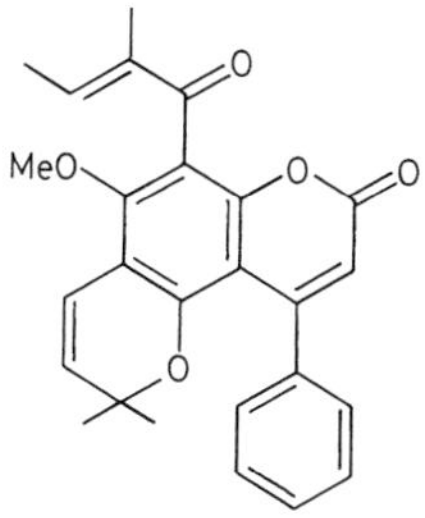

Calophyllolide

4-arylchromans, e.g. hematoxylin from *Haematoxylon campechianum* (wood),

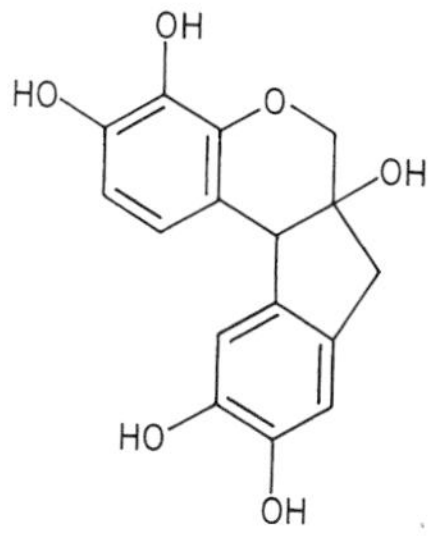

Hematoxylin

neoflavenes, e.g. 6-hydroxy-7-methoxy-3,4-dehydroneoflavan from *Dalbergia sissoo* and *D. latifolia* (bark),

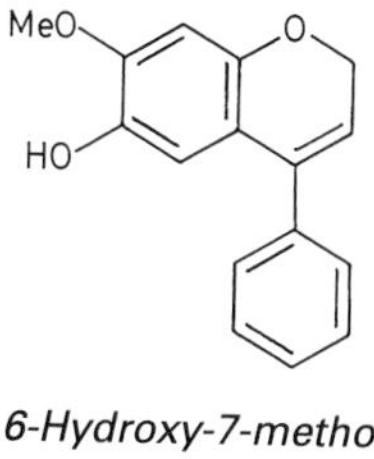

6-Hydroxy-7-methoxy-3,4-dehydroneoflavan

dalbergiquinols (3,3-diarylpropenes), e.g. latifolin from *Dalbergia latifolia* (bark, wood),

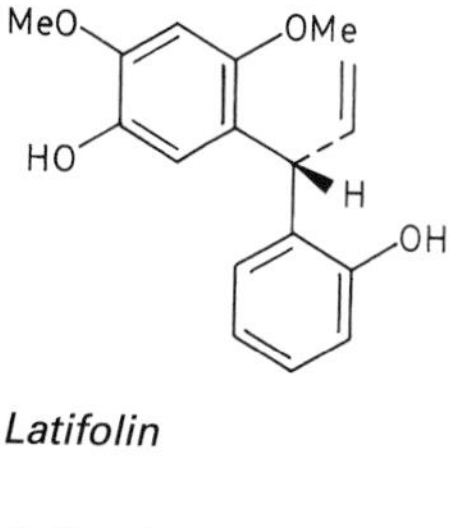

Latifolin

dalbergiones, e.g. 4-methoxydalbergione from *Dalbergia* spp. (bark, wood) (both configurations occur naturally),

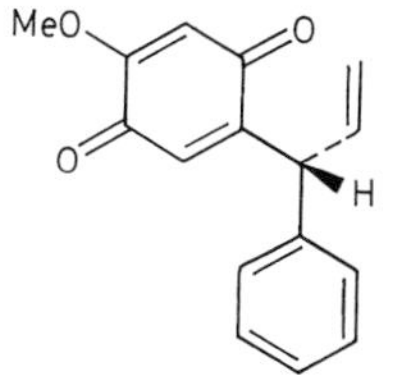

4-Methoxydalbergione

coumarinic acids, e.g. calophyllic acid from *Caesalpinia inophyllum* (nuts).

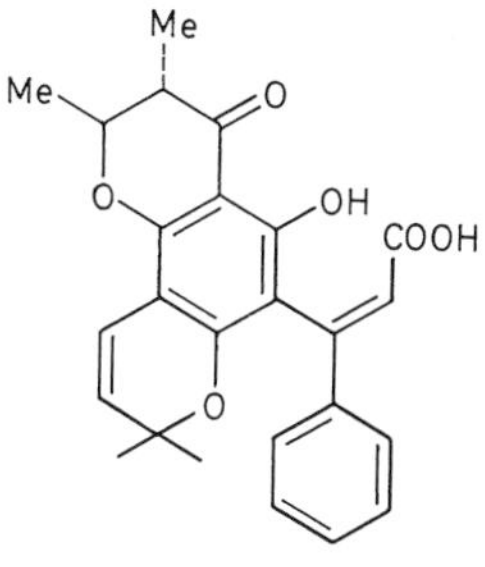

Calophyllic acid

Biosynthesis. Radioactivity from [3-^{14}C]phenylalanine is incorporated chiefly into C4 of calophyllolide by young shoots of *Calophyllum inophyllum,* which indicates that no aryl shifts are involved, and that C2, C3 and C4 of the N. are derived from C1, C2 and C3, respectively, of phenylalanine. Other aspects of the biosynthesis are less certain and are discussed in *The Flavonoids* (ed. J.B.Harborne, T.J.Mabry & H.Mabry) (Chapman and Hall, 1975).

Neolignans: *bis*-arylpropanoids, in which the inter-aryl linkage is other than C8, C8″ (see Lignans, in which the linkage is C8, C8″). N. are less numerous and have a more restricted phylogenic distribution than lignans. Some examples are shown (Fig.). For speculative mechanisms of N. biosynthesis, see O.R.Gottlieb *Phytochemistry* **11** (1972) 1544–1570. For comprehensive review of N., see O.R.Gottlieb, *Prog. Chem. Org. Natural Products* **35** (1978) 2–72.

Neomycin: an aminoglucoside antibiotic (see Streptomycin).

Neoplasm: see Cancer research.

Neorauflavene: see Isoflav-3-ene.

Neoretinal b: see Vitamins (vitamin A).

Neoxanthin: a xanthophyll containing an allene group, two secondary and one tertiary hydroxyl group. The chiral centers are 3*S*, 5*R*, 6*R*, 3′*S*, 5′*S*, 6′*S*. N. is present in the green parts of all higher plants, and together with lutein and violaxanthin it is one of the commonest carotenoids. 9′-*cis*-N. and probably all-*trans*-N. are intermediates in the biosyntheis of abscisic acid. For formulas see Abscisic acid.

Neral: the *cis*-isomer of Citral (see).

Nerol: a doubly unsaturated, acyclic monoterpene alcohol, M_r 154.25, b.p. 225–226 °C, ρ^{15} 0.8813, which

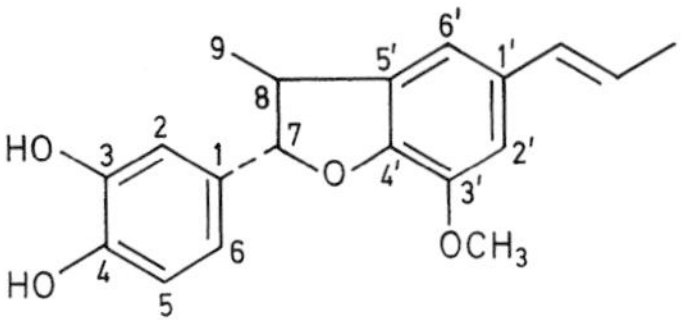

Kachirachirol-B (Magnolia kachirachirai) (8–5′ linkage)

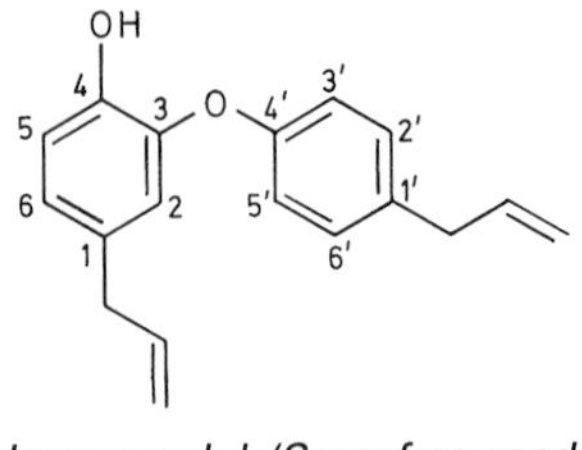

Isomagnolol (Sassafras randaiense) (2-O-4′ linkage)

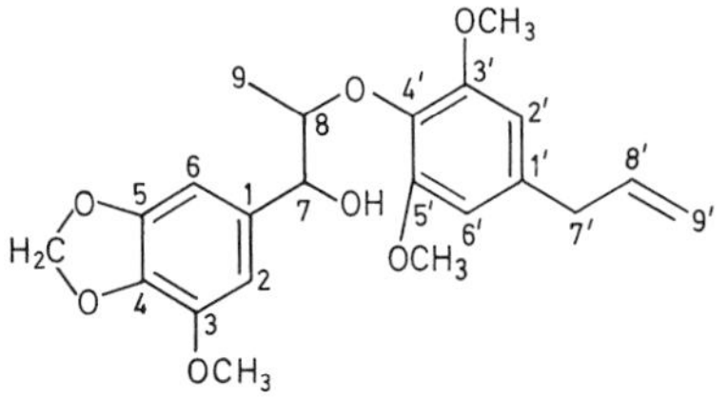

A neolignan from members of the *Myristicaceae (8-O-4′* linkage)

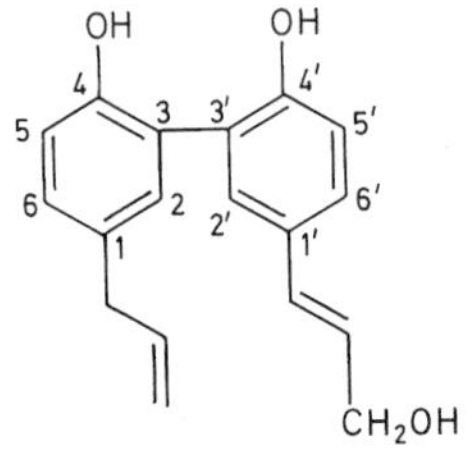

Randainol (Sassafras randaiense) (3-3′ linkage) [F. S. El-Feraly *Phytochemistry 23* (1984) 2329–2331]

Examples of neolignans

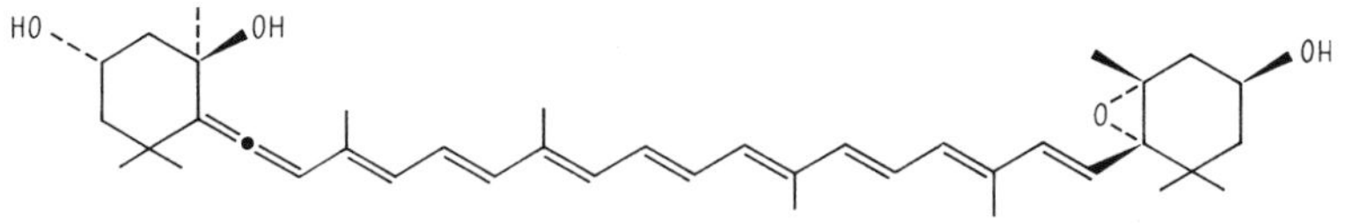

Neoxanthin

is a structural isomer of linalool and a *cis-trans*-isomer of Geraniol (see). N. is an oil with an odor of roses. It is the most valuable acyclic monoterpene alcohol used in perfumery. The pyrophosphate ester of N., neryl pyrophosphate, is an intermediate in the biosynthesis of the Monoterpenes (see).

Nerve growth factor, ***NGF:*** a tightly (noncovalently) associated dimer of identical polypeptides, each of M_r 13,259 (Fig.), which stimulates division and differentiation of sympathetic and embryonic sensory neurons in vertebrates. The mitogenic effect of NGF is restricted to early embryonic life; it has no effect after 9 days post partum in mice. A sensitive assay for NGF relies on the stimulation of the outgrowth of neurites from cultured ganglia.

NGF occurs in high concentration in adult mouse submaxillary gland, and in some snake venoms, but it is unlikely that these high local concentrations play any part in the stimulation of nerve growth. The first reported source of NGF was mouse sarcoma. It cannot be detected in innervated tissues, but denervation leads to the appearance of measurable NGF in the target tissue; when nerve regeneration is complete, NGF again becomes undetectable. NGF is therefore thought to carry information from the innervated end organs or tissues to the cell bodies of neurons by retrograde axonal transport.

NGF is isolated from mouse submaxillary gland as a high M_r complex (designated 7S), with subunit structure $\gamma 2\alpha 2\beta$, containing 1–2 g atoms Zn^{2+} per mole. The β-subunit is the hormonally active NGF dimer (also called β-NGF). The γ-subunit (also called γNGF, but with no mitogenic activity) has esteropeptidase activity with a marked preference for Arg, and it is thought to be the processing enzyme which converts proNGF to NGF. It belongs to a family of proteases which includes the binding protein for Epidermal growth factor (see). The α-subunit (also called αNGF, but with no mitogenic activity) is structurally homologous with the γ-subunit, but it lacks proteolytic activity and appears to inhibit the activity of the γ-subunit.

The structures of human and mouse preproNGF, predicted from the analysis of cDNA clones, are highly homologous (Fig.). [R. H. Angeletti & R. A. Bradshaw, *Proc. Natl. Acad. Sci. USA* **68** (1971) 2417–2420; J. Scott et al. *Nature* **302** (1983) 538–540; A. Ullrich et al. *Nature* **303** (1983) 821–825; P. J. Isackson & R. A. Bradshaw, *J. Biol. Chem.* **259** (1984) 5380–5383; D. Johnson et al. *Cell* **47** (1986) 545–554; T. H. Large et al. *Neuron* **2** (1989) 1123–1134; D. R. Kaplan et al. *Science* **252** (1991) 554–558; R. Klein et al. *Cell* **65** (1991); S. B. Woo et al. *J. Biol. Chem.* **270** (1995) 6278–6285]

Nervone: see Glycolipids.

Nervonic acid: Δ^{15}-tetracosenoic acid, CH_3–$(CH_2)_7$–CH=CH–$(CH_2)_{13}$–COOH, a fatty acid, M_r 336.6, m.p. 42 °C. N. a. is an esterified component of cerebrosides (see Glycolipids).

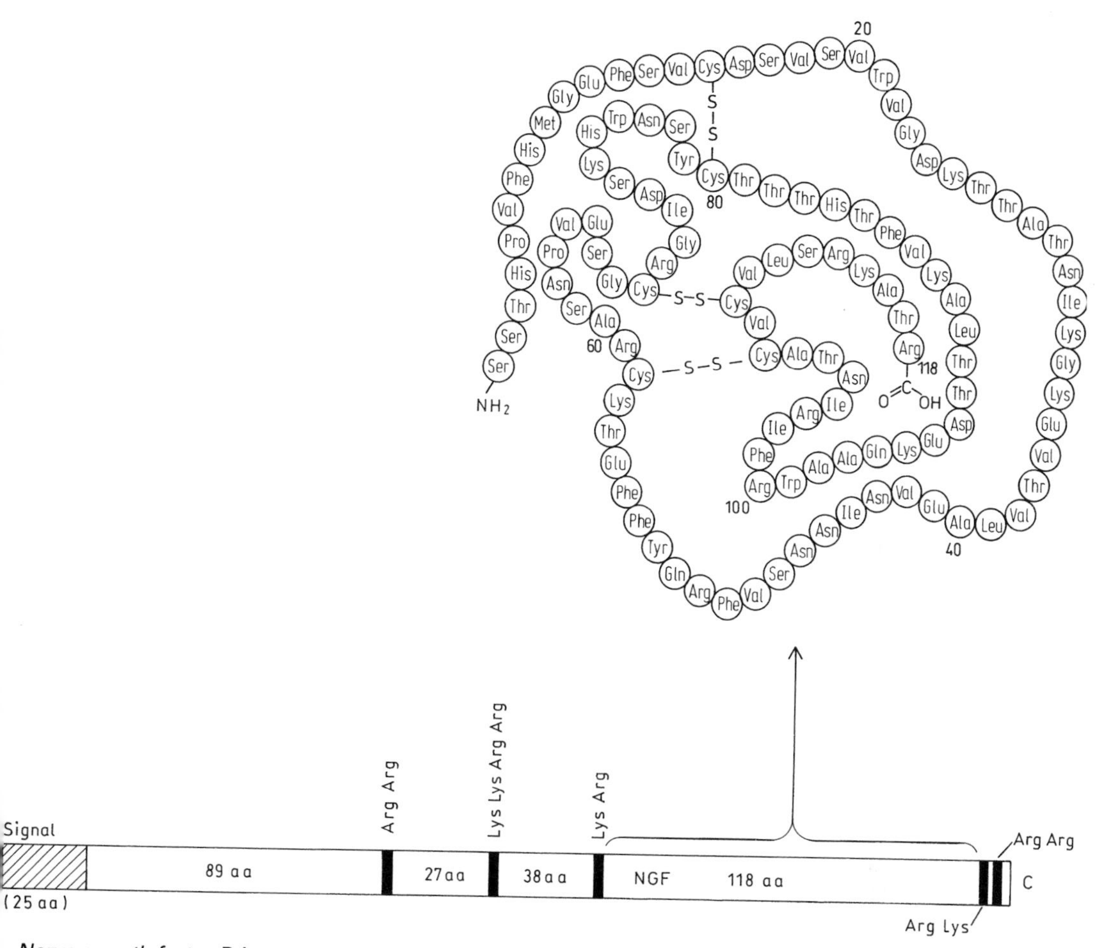

Nerve growth factor. Primary structure of mouse nerve growth factor, and schematic representation of its submaxillary precursor protein (prepro-NGF) predicted from analysis of cDNA clones.

Neryl pyrophosphate: see Nerol.

Neuberg ester: old name for fructose 6-phosphate.

Neuberg fermentation: Neuberg's first form of fermentation is the normal "unsteered" fermentation by yeasts, i.e. all the reactions and side reactions of the anaerobic degradation of carbohydrate, with the production of ethyl alcohol (see Alcoholic fermentation).

Neuberg's second form of fermentation (or sulfite fermentation) occurs in yeasts in the presence of sodium hydrogen sulfite. The acetaldehyde, which would normally act as a hydrogen acceptor, is converted to its bisulfite addition compound. The resulting excess NADH reduces dihydroxyacetone phosphate to glycerol 1-phosphate (or 3-phosphate), a reaction catalysed by glycerol 1-phosphate (or 3-phosphate) dehydrogenase (Baranowski enzyme) (EC 1.1.1.8). The oxido-reduction cycle (i.e. reduction of NAD and its regeneration from NADH) is thus maintained. The glycerol 1-phosphate is dephosphorylated by a phosphatase (EC 3.1.3.21), and free glycerol is produced. The process is therefore also known as a glycerol fermentation.

In the liver, free glycerol is dehydrogenated to glyceraldehyde, with formation of NADH. The reactions of Neuberg's second form of fermentation are also important in insect flight muscle, where there is only a low activity of lactate dehydrogenase, and dihydroxyacetone phosphate serves as the normal hydrogen acceptor. In the liver, glycerol 1-phosphate is a precursor of phospholipids, and it is important for the transport of reducing power from the cytoplasm to the mitochondria (see Hydrogen metabolism). Glycerol 1-phosphate can also be produced from glycerol and ATP by the action of glycerol kinase.

Neuberg's third form of fermentation proceeds under alkaline conditions. Two molecules of glucose are converted into two molecules of glyceraldehyde and two molecules of acetaldehyde (via pyruvate by decarboxylation). The glyceraldehyde is reduced to glycerol, and the acetaldehyde undergoes a dismutation to produce equivalent amounts of ethanol and acetate.

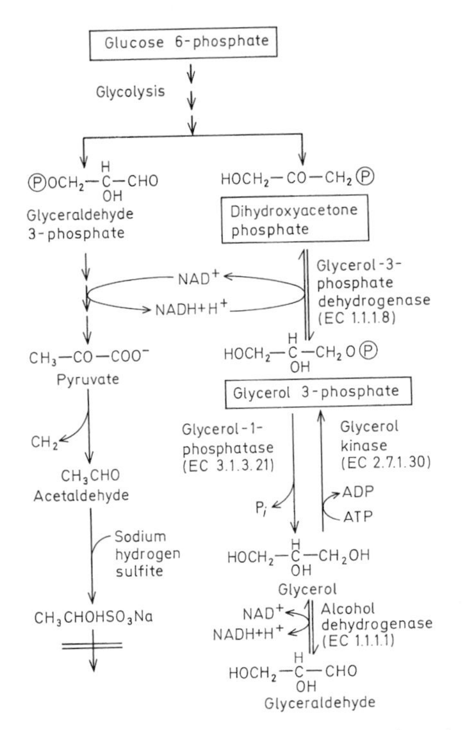

Significance of dihydroxyacetone phosphate in the 2nd Neuberg fermentation

Neuraminic acid: 5-amino-3,5-dideoxy-D-glycero-D-galactononulosonic acid, a C_9-aminosugar biosynthesized from mannosamine and phospho*enol*pyruvate. It is widely distributed in the animal kingdom in mucolipids, mucopolysaccharides, glycoproteins and the oligosaccharides of milk. The *N*-acetyl, *N*-glycoloyl and *O*-acetyl derivatives are called sialic acids. Only *N*-acetylneuraminic acid appears to be present in human glycoproteins.

$\boxed{HOOC\text{-}CO\text{-}CH_2}\text{-}CHOH\text{-}CH(NH_2)\text{-}(CHOH)_3\text{-}CH_2OH$

Neuraminic acid. The boxed part of the molecule is derived biosynthetically from phospho*enol*pyruvate.

Neuraminidase (EC 3.2.1.8): a hydrolase which cleaves *N*-acetylneuraminic acid from the nonreducing end of heterosaccharide chains. N. occurs in mixoviruses, various bacteria, blood plasma, and the lysosomes of many animal tissues. The richest source of N. is the culture filtrate of the cholera organism, *Vibrio cholerae*. M_r of the dimeric N. from *Vibrio* is 90,000; N. from influenza virus has M_r 130,000. Most neuraminic acid-containing proteohormones and some enzymes are inactivated by treatment with N.; on the other hand, the electrophoretic mobility of many glycoproteins, e.g. plasma proteins, may be altered after treatment with N., but their activity is unaffected. The N. of influenza virus destroys the protective mucus layer of the attacked organ.

Neuroendocrinology: the study of the interaction of nervous and hormonal systems in the regulation of metabolism, and in the adjustment of the individual organism to its environment.

Neurohormones: a group of hormones which, in conjunction with the nervous system and endocrine systems, play an important part in the regulation of somatic function, and in the adjustment of the individual organism to its environment. Examples are Releasing hormones (see), Neurohypophyseal hormones (see) and Neurotransmitters (see).

Neurohypophysial hormones: a group of hormones produced in the hypothalamus (not, as the name suggests, in the neurohypophysis; see Hypophysis). The action spectrum of these phylogenetically ancient hormones, Oxytocin (see) and Vasopressin (see), extends from an effect on the smooth muscle of the uterus, mammary gland and blood vessels, to an alteration of the permeability of the skin, urinary bladder and kidney tubules. The carrier protein of N.h. is Neurophysin (see). N.h. are small molecules (nonapeptides) and they represent an ideal model for the study of the mechanism of hormone action at a molecular biological level.

Neuronal ceroid lipofuscinosis: see Lysosomal storage diseases.

Neurophysins: small proteins associated with the hormones oxytocin and vasopressin. In bovines, oxytocin-like arginine vasopressin and N.II are synthesized from a single polypeptide precursor, as are oxytocin and N.I. Most species have more than two N., but usually there are two major N. The middle sequence of N. is conserved, and the N. differ in their terminal sequences. The N. bind to the neurohormones with which they are synthesized and are secreted with them into the blood. The biological function, if any, of the N. is not known. [H. Land et al. *Nature* **302** (1983) 342–344; A. G. Robinson in D. T. Krieger & J. C. Hughes (eds.) *Neuroendocrinology* (Sinauer Assoc., Sunderland, Mass., 1980)]

Neurosporene: a hydrocarbon carotenoid, M_r 538, m.p. 124 °C, containing 12 double bonds, nine of them conjugated. N. is a direct biosynthetic precursor (see Phytoene) of lycopene, and it is found in *Neurospora crassa*.

Neurotensin, *NT*: a tridecapeptide (Pyr-Leu-Tyr-Glu-Asn-Lys-Pro-Arg-Arg-Pro-Tyr-Ile-Leu) first discovered in and isolated from acid/acetone extracts of bovine hypothalamus. The same peptide has been isolated and unequivocally identified from extracts of calf and human small intestine. Subsequent studies on the distribution and concentratiion of NT have relied

Neurosporene

on immunoassay. Antisera against the *C*-terminal and *N*-terminal regions of NT show varying degrees of cross reactivity with other naturally occurring peptides and give different quantitative results in immunoassays. When NT is measured by immunoassay it is therefore preferable to refer to "neurotensin-like immunoreactive material" (NTLI). The following approximate concentrations of NTLI (pmol/g fresh weight) were found in rat tissues: hypothalamus and mucosa of small intestine (50–60); thalamus, spinal cord and brainstem (12–16); posterior pituitary (30); anterior pituitary (24); cerebral cortex (2); cerebellum, esophagus, stomach and pancreas (less than 1). NT is thus a typical brain-gut peptide and a member of the Amine precursor uptake decarboxylase system (see). NTLI has been found in the gut of members of most chordate phyla, including Uro- and Cephalochordata. NT immunoreactive cells have been found in avian thymus, and NTLI has been reported in cat adrenal medulla. NT displays such a wide variety of activities in vivo that it is doubtful whether these represent true physiological functions. It seems more likely that NT acts locally, affecting only those cells near the site of release, and is then rapidly degraded (half-life of NT when injected into rats is 30 sec). It is rigidly excluded by the blood-brain barrier, so that its peripheral and cerebral effects are distinct. Peripheral injection of NT into rats has the following effects: increased vascular permeability, hypotension, cyanosis, vasodilation, peripheral blood stasis, greater than 30-fold increase in plasma histamine, increased secretion of ACTH, LH, FSH, GH and prolactin, hyperglycemia, inhibition of gastric secretion of acid and pepsin, inhibition of gastric motility. Central injection causes hypothermia, enhanced phenobarbital action (NT does not itself induce sleep), and analgesia. In vitro effects include contraction of guinea pig ileum, rat fundus strip and guinea pig atrium, relaxation of rat duodenum, and release of histamine from isolated mast cells.

The amino acid sequence of NT is not present within any other known peptide or protein, but it does have certain similarities to other biologically active peptides. Thus *N*-terminal pyroglutamate is also present in xenopsin, thyrotropin releasing hormone, and luteinizing hormone releasing hormone, where it possibly protects against *N*-terminal degradation. The structures of vasopressin, LRH and NT suggest a distant evolutionary relationship for these three peptides. A similarity between the *C*-terminal sequence of NT (-Arg-Pro-Tyr-Ile-Leu-OH) and angiotensin I (-His-Pro-Phe-His-Leu-OH) should also be noted.

In the mucosa of the mammalian jejuno-ileum, NT is localized in endocrine-like cells called N- or NT-cells, which make contact with the intestinal lumen via a narrow apical process. It is suggested that a receptor on the cell surface responds to intraluminal lipid and signals release of NT from storage granules. Plasma NT increases after a high fat meal. Normal concentrations of plasma NTLI are 15–25 fmol/ml.

Several peptides cross-reacting with antiserum against the NT *C*-terminus, but not with antiserum against the *N*-terminus, have been isolated from chicken small intestine. Two have been analysed: Chick I: Pyr-Leu-Tyr-Glu-Asn-Lys-pro-Arg-Arg-Pro-Tyr-Ile-Leu-OH; Chick II: H-Lys-Asn-Pro-Tyr-Ile-Leu-OH.

The amphibian peptide, Xenopsin (see) and other xenopsin-like peptides also appear to belong to the NT family.

A detailed survey of work on N. up to 1982 is given in *Annals of the New York Academy of Sciences* **400** (1982), C. B. Nemeroff & A. J. Prange Jr. (eds).

Neurotoxins: toxins that act specifically on nervous tissue. They may act as antagonists of neurotransmitters, e. g. snake venom N. block acetylcholine receptors; strychnine is an antagonist of glycine (an inhibitory transmitter in the spinal cord); or they may interfere with cation transport across the nerve membrane, e. g. batrachotoxin. See Snake venoms, Bicuculline, Batrachotoxin, Curare alkaloids, Picrotoxin, Saxitoxin, Strychnine.

Neurotransmitters: small, diffusible molecules, such as acetylcholine, noradrenalin, amino acids, amino acid derivatives and peptides (e. g. substance P, enkephalin, etc.), which are released from synaptic vesicles (storage vesicles) at nerve endings, synapses and motor end plates in response to electrical stimuli. They act as chemical messengers, which transmit a signal from a nerve cell to its target organ. Adrenergic N., e. g. Noradrenalin (see) and Adrenalin (see), are formed in sympathetic postganglionic synapses. Cholinergic N., e. g. Acetylcholine (see) are found in pre- and postganglionic synapses of the parasympathetic nervous system. The central nervous system contains additional N., which may be excitatory, e. g. L-Glutamic acid (see), or inhibitory, e. g. γ-Aminobutyric (see) and Glycine (see), or able to show both effects, e. g. Dopamine (see) and Serotonin (see). N. are often considered to be synonymous with neurohormones, but this is debatable.

Neutral fats, ***fats, triacylglycerols:*** triesters of glycerol with saturated or unsaturated fatty acids. The fatty acid residues may be different or the same. Almost without exception, the fatty acids of natural N. f. are unbranched and have an even number of carbon atoms, usually between 4 and 26. In plant N. f. the primary hydroxyl groups at positions 1 and 3 of the glycerol are usually occupied by saturated fatty acids, whereas position 2 carries an unsaturated acid. In animal N. f. the reverse situation is often found. More than 50 fatty acids have been identified as components of N. f. The most common are the saturated palmitic (C_{16}) and stearic (C_{18}) acids and the unsaturated oleic, linoleic and linolenic (all C_{18}) acids. The double bonds in naturally occurring fatty acids are always in the *cis* configuration. Branched or hydroxylated fatty acids, or those with an uneven number of C-atoms are found occasionally, such as ricinolic, cerebronic and hydroxy-nervonic acids.

Hydrolysis of N. f., by alkali (saponification) or by lipases, produces glycerol and fatty acids. The alkali salts of fatty acids produced by saponification are *soaps*.

The melting point of a N. f. depends on the nature of the fatty acids in it. Shorter hydrocarbon chains and unsaturation confer lower melting points. Those which are liquid at room temperature are called fatty oils. These are further subdivided, according to their tendency to undergo autocatalytic oxidation in the presence of oxygen, into drying oils, semi-drying oils and non-drying oils. Drying depends on polymerization and cross-linking by oxygen, peroxy- and hydrocarbon bridges between the fatty acids. Only polyunsaturated acids can undergo these reactions. Linseed and poppy seed oils are examples of drying oils; peanut and rapeseed oils are semi-drying, and olive oil is a non drying oil.

Plant and animal N.f. can be distinguished analytically because they are associated with different sterols, e.g. phytosterol in plant fat and cholesterol in animal fat. They are of equal nutritional value, however, provided that the same vitamins and essential fatty acids are present. Plant N.f. are most abundant in seeds (40–45 % in rapeseed, poppy seed and linseed). Olives contain up to 25 % N.f. The most important fruit oils are palm and olive oils; the most important solid seed N.f. are coconut and palm seed fats and cocoa butter. The seed oils of cotton, corn, sunflowers, peanuts, soy, almonds, sesame, flax (linseed), poppies, rape, mustard and *Ricinus* are economically important.

Animal N.f. are stored in subcutaneous and omentum tissue, in the peritoneum and the region around the kidneys. The body N.f. of pigs, cattle, sheep and geese, and of sea animals (whale, seal, fish liver) and the milk N.f. of cattle, goats and sheep are economically significant.

Liquid oils are converted to solids (e.g. margarine) by hydrogenation of their double bonds, transesterification and fractionation, and by removal of the lower-melting fraction. N.f. rather easily become rancid, i.e. the fatty acids are released by hydrolysis, and if unsaturated they are then easily oxidized to aldehydes and ketones. Antioxidants (vitamin E, butylated hydroxyanisole [BHA], butylated hydroxytoluene [BHT], propyl gallate) can slow this process.

N.f. provide more metabolic energy than proteins and carbohydrates: 37.7 kJ/g (9.0 kcal/g). They are also important as body insulation and cushioning materials for organs, and as components of cell membranes. In addition to their nutritional role, N.f. are used commercially in the production of fatty acids, glycerol, soaps, salves, candles, fuel and lubricants. Drying oils are used in paints, varnishes and textile dyes. For biosynthesis and degradation of N.f. see Acylglycerols.

NGF: acronym of Nerve growth factor.

NHI-proteins: acronym of Nonheme iron proteins.

Niacin: inclusive term for nicotinic acid and nicotinamide, which are members of the vitamin B_2 complex. See Vitamins.

Niacinamide: a member of the vitamin B_2 complex. See Vitamins.

Nickel, *Ni:* a metal present only in traces in living systems. In particular, it seems to be associated with RNA. A nickel metalloprotein, named "nickeloplasmin" has been isolated from human and rabbit serum, but its function is not known. Ni protects the structure of the ribosome against heat denaturation, and it restores the sedimentation characteristics of *E. coli* ribosomes that have been denatured by EDTA. Ni can activate some enzymes in vitro, e.g. deoxyribonuclease, acetyl CoA synthetase and phosphoglucomutase. Ni deficiency causes changes in the ultrastructure of the liver and alters the level of cholesterol in the liver membranes. It may be important in the regulation of prolactin.

Nicking-closing enzymes: eukaryotic type I Topoisomerases (see).

Nick translation: a procedure for preparing ^{32}P-labeled DNA probes for use in hybridization tests. In this context the word "translation" is not related to the translation process of protein synthesis, but to the translatory movement of a cleavage point or "nick" along a duplex DNA molecule. Nicks are introduced at widely separated sites in DNA by very limited treatment with DNAase. The resulting nicks expose free 3′OH groups. *E. coli* DNA polymerase I is then used to remove concomitantly the 5′-mononucleotide terminus (i.e. 5′ → 3′ exonuclease activity of polymerase I), and to incorporate appropriate nucleotides from ^{32}P-labeled deoxynucleoside triphosphates (α^{32}P-dNTP, where N = G, A, T or C). In the presence of all 4 radiolabeled deoxynucleoside triphosphates, label is progressively incorporated into the duplex at random, so that the DNA effectively becomes uniformly labeled. If only one of the deoxynucleoside triphosphates is labeled, the pattern of labeling becomes nonuniform, especially with respect to homopolymer regions (Fig.). [P.W.J Rigby et al. *J Mol Biol* **113** (1977) 237–251]

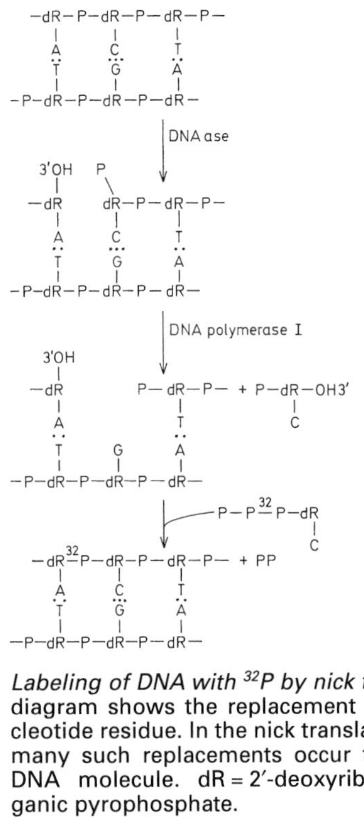

```
  -dR-P-dR-P-dR-P-
   |     |     |
   A     C     T
   :     :     :
   T     G     A
   |     |     |
-P-dR-P-dR-P-dR-
          |
          | DNA ase
          v
 3'OH  P
  |     \
 -dR    dR-P-dR-P-
  |      |    |
  A      C    T
  :      :    :
  T      G    A
  |      |    |
-P-dR-P-dR-P-dR-
          |
          | DNA polymerase I
 3'OH     v
  |
 -dR        P-dR-P- + P-dR-OH3'
  |           |           |
  A           T           C
  :           :
  T      G    A
  |      |    |
-P-dR-P-dR-P-dR-
          |
          |<---- P-P-32-P-dR
          v                |
                           C
 -dR-32-P-dR-P-dR-P- + PP
  |      |    |
  A      C    T
  :      :    :
  T      G    A
  |      |    |
-P-dR-P-dR-P-dR-
```

Labeling of DNA with ^{32}P by nick translation. The diagram shows the replacement of a single nucleotide residue. In the nick translation procedure many such replacements occur throughout the DNA molecule. dR = 2′-deoxyribose. PP = inorganic pyrophosphate.

Nicotiana alkaloids, *tobacco alkaloids:* a group of pyridine alkaloids occurring mainly in the tobacco plant (*Nicotiana*). N.a. contain a pyridine ring system substituted at position 3 by pyrrolidyl or piperidyl residues. Many varieties of tobacco plant contain more than 10 % N.a. Nicotine (see) is usually the chief N.a., but in *Nicotiana glauca* this is Anabasine (see) and in *N. glutinosa* the main and apparently only al-

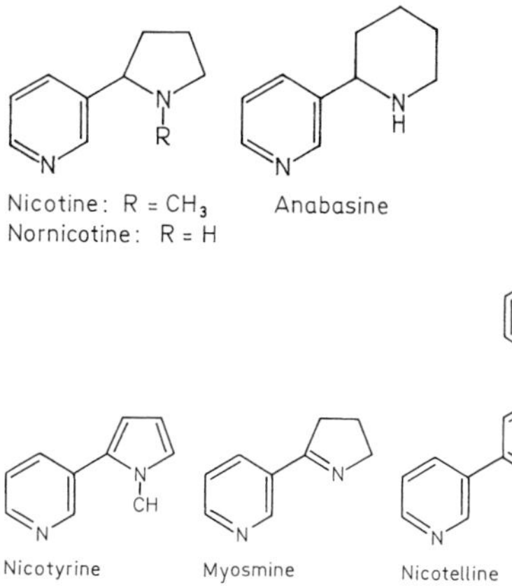

Nicotine: R = CH_3
Nornicotine: R = H
Anabasine

Nicotyrine
Myosmine
Nicotelline

Some Nicotiana alkaloids

kaloid is nornicotine. Other N.a. occurring in small quantities are nicotyrine, nicotelline and myosmine. All N.a. are very poisonous.

The pyridine ring system of N.a. is biosynthesized from nicotinic acid (or a closely related derivative; see Pyridine nucleotide cycle). The pyrrolidine ring is derived from ornithine or a closely related compound in the ornithine-proline-glutamic acid family of amino acids. During the biosynthesis of nicotine, the C3 carboxyl group of nicotinic acid is lost as CO_2, and the pyrrolidine ring is inserted in its place; this is accompanied by a labilization of the hydrogen at position C6 of the pyridine ring of nicotinic acid. In the biosynthesis of anabasine, the precursor of the piperidine ring system is Δ^1-piperideine, which is derived from lysine (probably by oxidation to α-oxo-ε-amino-caproic acid, followed by ring closure to the acid, which is then decarboxylated). The *N*-methyl group in the pyrrolidine ring is derived from methionine (see Transmethylation).

Nicotinamide: a member of the vitamin B_2 complex. See Vitamins.

Nicotinamide-adenine-dinucleotide, *NAD, diphosphopyridine nucleotide, DPN, codehydrogenase I, coenzyme I, cozymase:* a pyridine nucleotide coenzyme involved in many biochemical redox processes. It is the coenzyme of a large number of oxidoreductases, which are classified as pyridine nucleotide-dependent dehydrogenases. Mechanistically, it serves as the electron acceptor in the enzymatic removal of hydrogen atoms from specific substrates.

In the oxidized form of NAD, the pyridinium cation of nicotinamide is bound by an *N*-glycosidic linkage to C1 of D-ribose. This nicotinamide riboside moiety is linked to adenosine via a pyrophosphate group. NAD therefore has the structure of a dinucleotide (Fig. 1). M_r of the oxidized form (NAD^+) = 663.4.

NAD occurs in two forms, distinguished by the configuration of the glycosidic linkage of the nicotinamide, hence α-NAD and β-NAD. Only β-NAD is enzymatically active. In both α- and β-NAD, the glycosidic linkage of the adenine has the β-configuration. $[\alpha]_D^{23}$ −34.8° (β-NAD); + 14.3° (α-NAD).

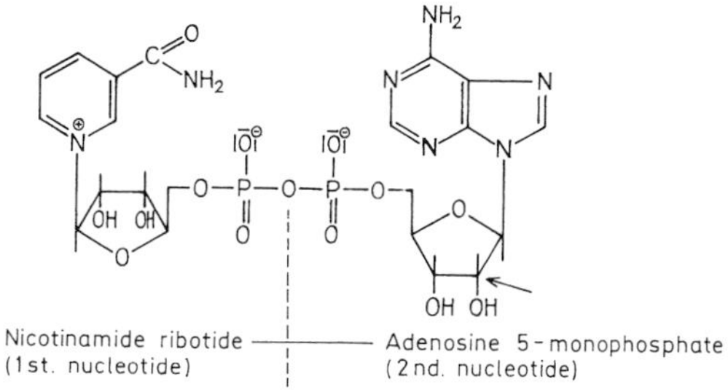

Nicotinamide ribotide (1st. nucleotide)
Adenosine 5-monophosphate (2nd. nucleotide)

Fig. 1. *Structure of nicotinamide-adenine-dinucleotide (NAD).* In nicotinamide-adenine-dinucleotide phosphate (NADP) an additional phosphate group is present on C2 of the 2nd nucleotide, as indicated by the arrow.

The hydrogen reversibly carried by NAD is attached to the nicotinamide residue (Fig. 2). In view of the positive charge on the coordinated pentavalent nitrogen of the pyridinium ion, oxidized NAD is represented as NAD^+, and in accordance with the guidelines of the International Commission on Nomenclature, the reduced form is represented as NADH. When 2[H] and a pair of electrons are transferred from a substrate, the pyridinium cation loses its aromaticity and becomes reduced, and a proton is released: Substrate-H_2 + NAD^+ ⇌ Substrate + NADH + H^+.

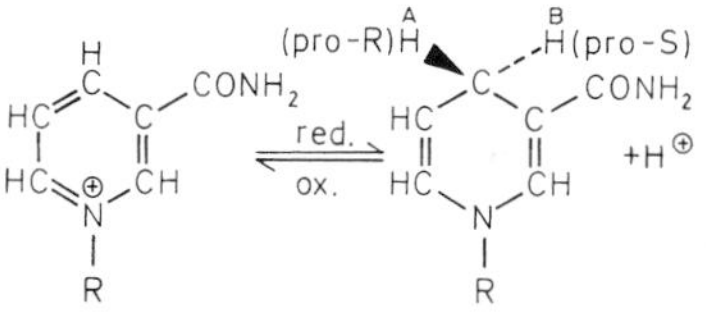

Fig. 2. *Reversible reduction of the pyridine ring of NAD or NADP, showing the prochiral center at C4 of the pyridine ring*

Thus, of the two hydrogens transferred, one becomes covalently bound to the NAD, while the other becomes a proton in equilibrium with the protons of the aqueous medium.

Reduction of NAD introduces a prochiral center at C4 of the pyridine ring. Transfer of the hydride ion (a proton with an electron pair) to NAD^+ is stereospecific, i.e. the newly introduced hydrogen is either *pro-R* or *pro-S* (in older nomenclature, it becomes attached to either the A(α) or B(β) side of the reduced nicotinamide ring). For example, ethanol, lactate and malate dehydrogenases transfer hydrogen to and from the *pro-R* position, whereas 3-phosphoglyceraldehyde and glucose dehydrogenases are *pro-S*-specific.

The oxidized and reduced forms have different spectrophotometric properties. Both show an intense absorption band in the region of 260 nm, due to the adenine; this is slightly higher in NAD^+ due to an additional contribution from the pyridine ring. NADH exhibits a broad absorption band at 340 nm which is entirely absent from NAD^+, and is due to the quinoid structure of the reduced nicotinamide ring. Thus reduction or oxidation of NAD is relatively easy to monitor by the change in light absorption at 340 nm. Since the absorption band is relatively wide, it may also be determined at wavelengths other than the maximum, e.g. the mercury lines 334 nm and 366 nm are suitable. This difference in absorption is exploited in the Optical test (see). λ_{max} NAD^+ (pH 7.0) = 260 nm (ε = 18,000, i.e. 1 mole in 1 liter corresponds to an aborbance of 18,000, when the light path is 1 cm). 1st λ_{max} NADH (pH 10.0) = 259 nm (ε = 14,400). 2nd λ_{max} NADH (pH 10.0) = 340 nm (ε = 6,200); at 334 nm, ε = 6,000; at 366 nm, ε = 3,300.

In living cells NAD is present predominantly as NAD^+, e.g. in rat liver cells $NAD^+/NADH + H^+$ = 2.6/1. A considerable proportion of this intracellular NAD is, however, not free but bound to dehydrogenases.

The nicotinamide moiety of NAD is biosynthesized from quinolinic acid (see Pyridine nucleotide cycle, Tryptophan). NAD is degraded by a pyrophosphatase, and by nucleosidases that cleave the glycosidic bonds to nicotinamide and adenine.

For the metabolic functions of NAD as a hydrogen carrier, see Hydrogen metabolism.

In addition to its function in hydrogen transfer NAD^+ is the donor of the ADP-ribosyl group in the ADP-Ribosylation of proteins (see).

Nicotinamide-adenine-dinucleotide phosphate, ***NADP, triphosphopyridine nucleotide, TPN, coenzyme II:*** a pyridine nucleotide coenzyme, which differs from NAD by the presence of an extra phosphate residue in the 2′-position of the adenosine moiety (see Fig. 1 in Nicotinamide-adenine-dinucleotide). M_r 743.4, E_0' = –0.317 V (pH 7.0, 30 °C), λ_{max} of the oxidized form ($NADP^+$) at pH 7.0 = 260 nm (ε = 18,000). 1st λ_{max} of the reduced form (NADPH) at pH 10.0 = 259 nm (ε = 14,100). 2nd λ_{max} of the reduced form = 340 nm (ε = 6,200).

NADP is the coenzyme of dehydrogenases and hydrogenases. It is important as the agent of hydrogen transfer in reductive syntheses, e.g. in Photosynthesis (see) and in Ammonia assimilation (see). It is a fair generalization that NAD is concerned in the transfer of hydrogen in oxidative, energy-yielding processes (e.g. oxidation in the respiratory chain), whereas NADP provides reducing power in synthetic reactions. Most intracellular NADP is present in the reduced form (NADPH). In analogy with Nicotinamide-adenine-dinucleotide (see), the oxidized form is represented as $NADP^+$. Excess NADPH can be removed by transhydrogenation (see Hydrogen metabolism).

The spectrophotometric properties of $NADP^+$ and NADPH are completely analogous to those of NAD^+ and NADH. The change in extinction at 340 nm, due to oxidation and reduction, is exploited in the Optical test (see).

NADP is biosynthesized by phosphorylation of NAD in an ATP-dependent kinase reaction.

For the metabolic functions of NADP, see Hydrogen metabolism.

Nicotine: chief representative of the *Nicotiana* alkaloids (see). Oxidative degradation of N. produces nicotinic acid. N. is present in all parts of the tobacco plant (*Nicotiana*), being biosynthesized in the roots of the plant, then transported to the aerial parts. Depending on the variety, the N. content of the leaves varies between 0.05 and 10 %. It is also found in several plants of other families.

N. has many different physiological effects. Owing to its toxicity, it is not used therapeutically, although it is the most widely consumed addictive alkaloid in the world. Between 50 and 100 mg N.(the content of half a cigar) is fatal in man, causing respiratory paralysis. The concentration in smokers' blood plasma is 5 to 50 ng/ml. At this concentration, N. enhances the chemotactic effect of several other substances on polymorphonuclear neutrophils, which are involved in the inflammatory process. At 5 µg/ml, a concentration which may be reached in lung fluid, N. is strongly chemotactic for neutrophils. Thus N. is probably a primary etiological factor in emphysema. [N.Totti III et al. *Science* **223** (1984)169–171]

Nicotinic acid: pyridine 3-carboxylic acid, M_r 123.11, m.p. 234–237 °C, first obtained in 1867 from the oxidation of nicotine. It is biologically important as a member of the vitamin B_2 complex (see Vitamins). Isonicotinic hydrazide (INH), M_r 137.14, m.p. 163 °C, is an antituberculosis agent, which shows antivitamin properties against both nicotinic acid and vitamin B_6.

Nicotinic receptors: see Acetylcholine.

Nieman-Pick disease: see Inborn errors of metabolism.

NIH-shift: an anionotropy, whereby the cation intermediate produced in the hydroxylation of an unsaturated or aromatic substance by an oxygenase is stabilized. It is named after the ***N****ational* ***I****nstitute of* ***H****ealth*. A group (usually a hydride ion) in the position of the incoming hydroxyl group is shifted to the neighboring position.

Ninhydrin: see Amino acid reagents.

Nissolin: 3,9-dihydroxy-10-methoxypterocarpan. See Pterocarpans.

Nitrate ammonification: see Nitrate reduction.

Nitrate assimilation, ***assimilatory nitrate reduction:*** see Nitrate reduction.

Nitrate dissimilation, ***dissimilatory nitrate reduction:*** microbial reduction of nitrate to various products, which are not assimilated. See Nitrate reduction.

Nitrate reductase: enzymes which catalyse the reduction of nitrate to nitrite. All N.r. studied so far contain iron and molybdenum. In the sequence of electron transfer, molybdenum appears to be the ultimate acceptor, which then transfers electrons to nitrate; during this process the molybdenum alternates between Mo(V) and Mo(VI). Dissimilatory N.r. from *E. coli* (also called respiratory N.r.) (EC 1.7.99.4) is a transmembrane protein, containing Mo, inorganic sulfur and nonheme iron, approximate M_r 220,000, consisting of two subunits of M_r 150,000 and 60,000. Higher M_r, up to 10^6, have been reported; this reflects the problems inherent in the M_r determination of membrane proteins, which must first be solubilized, chiefly by the use of detergents. The higher M_r material probably resulted from aggregation and/or association with other proteins. During several stages of the purification procedure, the N.r. is accompanied by a small polypeptide containing cytochrome b_{556}; this is probably the natural electron donor for N.r. within the membrane. The preferred substrates for nitrate reduction in *E. coli* are NADH or formate. Formate dehydrogenase is itself a molybdenum enzyme (it also contains selenium), donating electrons to another *b*-type cytochrome. Formate dehydrogenase and N.r. are intimately associated in the membrane, and the electron flow proceeds: formate → [Mo,Se] → Cytochrome *b* → Quinone pool → Cytochrome b_{556} → Fe/S → Mo → NO_3^-. This results in a transmembrane proton translocation, serving to generate ATP by chemiosmosis (see Oxidative phosphorylation). N.r. of *E coli* is induced by nitrate and repressed by oxygen; it may represent up to 15 % of the total protein of the inner bacterial membrane. Similar respiratory N.r. have been found in several other bacterial genera. In the absence of an alternative nitrogen source for the organism, the nitrite produced by nitrate reduction can be reduced to ammonia (see Nitrite reductase), then assimilated (see Ammonia assimilation). Fungi, green algae and higher plants possess assimilatory N.r. (EC 1.6.6.1–3), which are high M_r (200,000–500,000) multimeric enzyme complexes, containing FAD, cytochrome b_{557} and molybdenum. The natural electron donor is NADH or NADPH, and most N.r. react with both. All these pyridine nucleotide-dependent N.r. also reduce cytochrome *c* (referred to as "diaphorase" activity) in vitro, but this property probably has no physiological significance. The sequence of electron transfer is apparently:

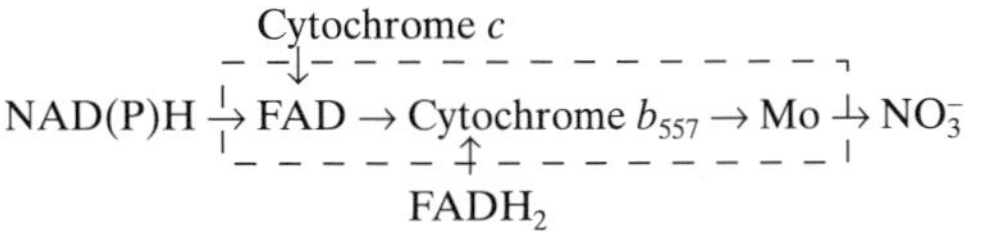

Cytochrome *c*

NAD(P)H → FAD → Cytochrome b_{557} → Mo → NO_3^-

$FADH_2$

N.r. of *Neurospora crassa* has M_r 228,000. A N.r. mutant (*Neurospora crassa nit-3*) produces a subunit of N.r. containing cytochrome b_{557} and Mo (M_r 160,000); this does not react with NAD(P)H, but transfers electrons to nitrate from exogenous $FADH_2$ or reduced methyl viologen. The other component is produced by *Neurospora crassa nit-1;* it is unable to reduce nitrate, but catalyses the reduction of cytochrome *c* by NAD(P)H.

Assimilatory N.r. from bacteria (*Azotobacter chroococcum, Ectothiorhodospira shaposhnikovii*) and cyanobacteria (*Anabaena cylindrica, Anacystis nidulans*) are specific for reduced ferredoxin, and they do not react with NAD(P)H.

The assimilatory N.r. have been described as soluble, cytoplasmic enzymes. They may, however, be attached to membranes or particles. Thus the N.r. of higher plants can be isolated in association with chloroplasts, or as an apparently soluble enzyme, depending on the technique of cell homogenization. Also, the nitrate-reducing system of *Nostoc muscorum, Anabaena cylindrica* and *Anacystis nidulans* (cyanobacteria) has been found tightly bound to photosynthetic particles, or as a soluble enzyme, depending on the technique of cell disruption. These N.r.-containing, photosynthetic particles catalyse the photoreduction of nitrate to nitrite, accompanied by the stoichiometric evolution of oxygen. In higher plants also the reducing power from the light reaction of photosynthesis may be consumed in the reduction of nitrate to nitrite (see Nitrate reduction). A detailed study of N.r. of bacteroids from leguminous root nodules (see Nitrogen fixation, Nitrogenase) shows that the enzyme consists of two subunits: a consitutive unit containing molybdenum and nonheme iron, and an inducible unit containing iron. The latter is inducible by nitrate, whereas the constitutive subunit appears to be identical with the molybdenum-containing subunit found in xanthine oxidase (from milk, mammalian intestine and bird liver), and with fraction I of nitrogenase from *Azotobacter vinelandii* and soybean nodule bacteroids. It has been suggested that competition between the inducible subunit of N.r. and fraction II of Nitrogenase (see) for the Mo-containing subunit is the basis of a control mechanism, whereby nitrogenase activity is manifested only when nitrate is in short supply. In the presence of nitrate, the inducible N.r. subunit is formed in such quantity that it binds all the Mo-Fe protein and renders it unavailable for the formation of nitrogenase. This may not be the only basis of control, and inhibition of nitrogenase by traces of nitrite appears to be an important factor in the suppression of nitrogenase activity by the presence of nitrate.

Nitrate reduction: In the narrow sense, N.r. is the reduction of the nitrate ion (NO_3^-) to nitrite (NO_2^-), catalysed by Nitrate reductase (see). In the broad sense, it is the reduction of nitrate to gaseous products (nitrous oxide or nitrogen), or to ammonia.

N.r. may be assimilatory or dissimilatory. Assimilatory N.r. is found in green plants, green algae and fungi, where it involves the reduction of nitrate to ammonia, and operates in conjunction with the assimilation of the ammonia. It occurs in two stages: reduction of nitrate to nitrite by Nitrate reductase (see), and the reduction of nitrite to ammonia by Nitrite reductase (see). In green plants. N.r. may occur in the roots or in the green tissue. In some species (e.g. the field pea, *Pisum arvense*), nitrate is reduced and assimilated in the roots. The reducing power for N.r. is provided by respiration, and the xylem contains high levels of organic nitrogen relative to free nitrate. Other species (e.g. *Xanthium pennsylvanican* [Cocklebur], *Stellaria, Trifolium*) transport high levels of nitrate in the xylem from the roots to the green tissue; in such plants most reactions of nitrogen metabolism occur within the chloroplast, and the necessary ATP and reducing power can be provided directly by the light reaction of photosynthesis. Thus reduced ferredoxin in the chloroplast transfers electrons directly to nitrite reductase and to glutamate synthase, and ATP is utilized by glutamine synthetase. Nitrate reductase is remarkably sensitive to inhibition by cyanide. 10^{-9} M CN^- totally and reversibly inhibits nitrate reductase from *Chlorella* (cf. 10^{-5} M CN^- for the inhibition of mitochondrial respiration). N.r. and assimilation in the chloroplast may be controlled by the production of HCN [L.P. Solomonson & A.M. Spehar *Nature* **265** (1977) 373–375]. Nitrate accumulation is typical of plants growing in nitrate-rich soils, e.g. on rubbish dumps in areas of human occupation. Control of N.r. always appears to be exerted at the early stage of reduction of nitrate to nitrite, so that nitrite never accumulates. Green plants also show a light-dependent cycle of nitrate accumulation and depletion; nitrate increases in the dark and decreases during daylight. The mechanism of the light effect is probably complicated: photosynthesis supplies reducing power and ATP for N.r. and ammonia assimilation; the carbohydrate produced by photosynthesis is used for the synthesis of the carbon skeletons of amino acids; and respiration of carbohydrate also produces reducing power and ATP for N.r. and ammonia assimilation.

There are two main types of dissimilatory N.r.: 1. reduction of nitrate to ammonia, which is not assimilated (ammonification of nitrate), and 2. reduction of nitrate to gaseous compounds such as nitric oxide, nitrous oxide and molecular nitrogen (denitrification). Ammonification and denitrification represent forms of respiration, in which nitrate replaces oxygen as the electron acceptor in an electron transport chain (see Respiratory chain). These forms of dissimilatory N.r. are therefore called nitrate respiration. Ammonification and denitrification have only one step in common, i.e the reduction of nitrate to nitrite. In ammonification the nitrite is then reduced to ammonia, e.g. in *Clostridium perfringens, Achromobacter fischeri, Bacillus filaris, E. coli.* In denitrification the nitrite is reduced to various gaseous products, according to the sequence: $NO_3^- \rightarrow NO_2^- \rightarrow NO \rightarrow N_2O \rightarrow N_2$.

Many bacterial genera contain denitrifying species: *Achromobacter, Alcaligenes* (*Alcaligenes odorans* denitrifies nitrite), *Bacillus, Chromobacterium, Corynebacterium, Halobacterium, Hyphomicrobium, Moraxella, Paracoccus, Pseudomonas, Spirillum, Thiobacillus* and *Xanthomonas.* In some species of *Pseudomonas* and *Corynebacterium,* N_2O is the final denitrification product. All these bacteria are aerobes that are able to respire (denitrify) nitrate under anaerobic conditions. The only true anaerobe able to carry out denitrification is *Propionibacterium.* There is no evidence for other intermediates in the above denitrification pathway, but in the formation of nitrous oxide (N_2O) an NN bond must be formed, and there may exist transient enzyme-bound intermediates that have not yet been identified. The enzymology of denitrification from nitrite is poorly understood. It seems likely that each stage is linked to electron transport via a cytochrome system, but sites of ATP synthesis have not been unequivocally located.

Nitrate respiration: see Nitrate reduction.

Nitric oxide, *nitrogen monoxide, NO:* a highly reactive, naturally occurring gas, produced and utilized by a wide range of animals. Synthesis of NO has been demonstrated in arthropods (e.g. king crabs) and in several mammals, including humans. It is speculated that the biological production of NO arose in response to increased levels of atmospheric oxygen early in the earth's history. Animals have subsequently exploited NO in two ways: 1. as an assassin molecule (macrophages produce NO when stimulated by bacterial lipopolysaccharide or γ-interferon from immune cells), and 2. as a signaling or messenger molecule in both the nervous system and the vascular system.

NO carries an unpaired electron and is therefore highly reactive. After it has been released by a cell, it rapidly combines with oxygen to form nitrogen dioxide (NO_2), then nitrite and nitrate ions, or it binds to hemoglobin. It therefore has short half-life, and can act only over a relatively short range. There appear to be no specific membrane receptors for NO. The gas gains access to the cell interior by rapid diffusion across the membrane. Inside the cell it stimulates guanylate cyclase, which catalyses the production of cyclic GMP from GTP. Cyclic GMP is a messenger that activates many processes, such as muscle relaxation and changes in brain chemistry.

The family of NO synthases (L-arginine, NADPH:oxidoreductase, NO-forming, EC 1.14.13.39) (abbreviated here as NOS) comprizes three mammalian isoforms, all of which have been purified, cloned and expressed. All three enzymes differ in their primary sequence, chromosomal localization and regulation, but they share a common catalytic mechanism, i.e. in the presence of NADPH they catalyse a two-stage, 5-electron reduction of L-arginine, yielding NO and L-citrulline (molecular oxygen is incorporated into both NO and L-citrulline) (Fig.).

Neuronal constitutive NOS [ncNOS, M_r (human) 161,000; 1433 amino acids], originally identified in neuronal tissues, is expressed constitutively, is strictly Ca^{2+}-dependent, and is activated by small Ca^{2+} transients in the physiological range of Ca^{2+} concentration. Although ncNOS has been described as a cytosolic enzyme, ncNOS antigens have been demonstrated immunologically in the sarcolemma of type II fast fibers of skeletal muscle [L. Kobzik et al. *Nature* **372** (1994) 546–548], and the enzyme also contains a sequence motif [known as a discs-large homologous region (DHR) or GLGF repeat] which is thought to play a role in targeting proteins that contain it to the submembranous cytoskeleton [W. Hendriks *Biochem J.* **305** (1995) 687–688]. As a neurotransmitter, NO occurs in certain nerve networks, e.g. it is active in the dilation of arteries supplying the penis, and in the relaxation of muscles of the corpora cavernosa. NO released from stomach nerves causes the stomach mus-

$H_2N(C{=}N^+H_2)-NH-(CH_2)_3-CH(H_3^+N)(COO^-)$ $\xrightarrow[O_2]{NADPH}$ $H_2N(C{=}NOH)-NH-(CH_2)_3-CH(H_3^+N)(COO^-)$ $\xrightarrow[O_2]{1/2\ NADPH}$ $H_2N(C{=}O)-NH-(CH_2)_3-CH(H_3^+N)(COO^-)$ + NO

L-Arginine — N^G-Hydroxy-L-arginine — L-Citrulline

Two-stage reduction of L-arginine to L-citrulline and nitric oxide

cles to relax in order to accommodate food. Intestinal nerves also cause relaxation of the intestinal muscles by releasing NO. In addition, nervous activity in the cerebellum is increased by NO, which acts by promoting the release of cyclic GMP. There is evidence that NO is an important neurotransmitter in nervous processes concerned with memory.

Endothelial constitutive NOS [ecNOS, M_r (human) 133,000; 1203 amino acids] is also strictly Ca^{2+}-dependent and activated by small Ca^{2+} transients. It was first shown to be constitutive in vascular endothelial cells, and is the enzyme responsible for synthesis of "endothelium-derived relaxing factor" discovered by R. Furchgott et al. (State University, New York) in the early 1980s, then identified as NO in 1987 by S. Moncada et al. at the Wellcome Research Laboratories, UK. Synthesis of NO in blood vessel epithelia occurs in response to the distortion of blood vessels by blood flow. The gas then rapidly diffuses into the surrounding muscle layers, causing them to relax. One possible cause of high blood pressure may be the defective production of NO by blood vessel epithelia. Through the relaxation of vascular smooth muscle by NO, ecNOS therefore represents a system for regulation of tissue conductance, blood pressure control and regulation of blood flow. It also leads to inhibition of platelet aggregation and reactivity. EcNOS is *N*-myristoylated, indicating that the enzyme is membrane-associated.

Inducible NOS [iNOS, M_r (human) 131,000; 1153 amino acids] was identified as an enzyme in macrophages and liver cells that is inducible by endotoxins and cytokines. (Under certain conditions, ncNOS and ecNOS are also induced and are probably transcriptionally controlled, but the term 'inducible' is reserved for the cytokine-inducible, macrophage-like type of NOS). iNOS forms a very tight, noncovalent complex with calmodulin at extremely low Ca^{2+} concentrations and therefore appears to be Ca^{2+}-independent; the binding is so tight that calmodulin is not separated from the enzyme by boiling in the presence of sodium dodecyl sulfate. Native iNOS is a homodimer containing FAD, FMN, tetrahydrobiopterin, calmodulin and heme, whereas the purified subunits contain only FAD, FMN and calmodulin; subunit dimerization is promoted by heme, tetrahydrobiopterin and L-arginine; this requirement for two prosthetic groups and the substrate for subunit association may be unique among multimeric proteins.

The iNOS of macrophages enables macrophages to kill parasites and tumor cells by the poisonous action of NO. Septic shock is thought to be due to the induction of NO synthase, e.g. by bacterial lipopolysaccharide, resulting in overproduction of NO and excessive dilation of blood vessels.

All three enzymes share only 51–57 % sequence similarity, but they all possess consensus sequences for the binding of NADPH, FAD, FMN and calmodulin and for protein kinase A phosphorylation. The same isoenzymes from different species display considerable homology, e.g. the cloned genes for iNOS from mouse and human display 80 % homology in the coding region and 55 % in the promoter region. [M. A. Marletta et al. *Biochemistry* **27** (1988) 8706–8711; D. J. Stuehr et al. *J. Exp. Med.* **169** (1989) 1011–1020; J. B. Hibbs et al. *Biochem. Biophys. Res. Commun.* **157** (1988) 87–94; Lowenstein et al. *Proc. Natl. Acad. Sci. USA.* **90** (1993) 9730–9734; Chartrain et al. *J. Biol Chem.* **269** (1994) 6765–6772; *The Biochemist (The Bulletin of The Biochemical Society* **16** (1994) contains six review articles on nitric oxide; K-D. Kröncke et al. *Biol. Chem. Hoppe-Seyler* **376** (1995) 327–343; Five papers on the special topic 'Nitric Oxide' in *Annu. Rev. Physiol.* **57** (1995) pp. 659–790]

Nitrite reductase: Dissimilatory N. r. (see Nitrate reduction), found in various bacteria, catalyses the reduction of nitrite, probably to nitric oxide; some authors claim that the reduction product is nitrous oxide. N. r. from *Pseudomonas aeruginosa, Paracoccus denitrificans* and *Alcaligines faecalis* contains both a *c*- and a *d*-type heme. The *Pseudomonas* N. r. consists of two similar polypeptides, each of M_r 63,000, and each containing two heme groups. Electrons are donated either by cytochrome c_{551} or the blue copper protein, azurin. N. r. from *Achromobacter cycloclastes* and *Pseudomonas denitrificans* are both copper proteins. The structure, mechanism of action and true intracellular location of these bacterial N. r. are poorly understood.

Assimilatory N. r. (see Nitrate reduction), found in fungi, green algae and higher plants, catalyse the reduction of nitrite to ammonia. This reduction is equivalent to the transfer of 6 electrons. There are no intermediates in the true sense, but in the presence of high concentrations of ammonium ions, the enzyme exhibits abnormal kinetics and releases traces of hydroxylamine. NADPH-dependent assimilatory N. r. (EC 1.6.6.4) from *Neurospora crassa* has M_r 290,000. It is assumed to be a flavoprotein but FAD appears to be removed during purification and must be added in vitro to achieve catalysis. The enzyme also contains Siroheme (see) as a prosthetic group, and the siroheme component interacts with the nitrite. The sequence of electron transfer is: NADPH $\rightarrow$ FAD $\rightarrow$ Siroheme $\rightarrow$ Nitrite.

Assimilatory ferredoxin-dependent N. r. (EC 1.7.7.1) from spinach has M_r 61,000, contains one molecule of siroheme and one inorganic (Fe_2–S_2)-center. In vivo, reducing power is supplied by reduced ferredoxin from the light reaction of photosynthesis. The sequence of electron transfer is: Ferredoxin $\rightarrow$ (Fe_2S_2) $\rightarrow$ Siroheme $\rightarrow$ Nitrite.

Nitrogen: see Bioelements.

Nitrogenase (EC 1.18.2.1): the enzyme system responsible for biological nitrogen fixation. All N. systems are a complex of two proteins, namely (i) one of three possible proteins: a MoFe protein, a VFe protein or a Fe-Fe protein, and (ii) an Fe protein, which is similar, if not identical in all N. systems. It is therefore possible to classify the various N. systems as molybdenum nitrogenase, vanadium nitrogenase or iron nitrogenase, according to the metal that characterizes component I. In *Azotobacter,* the prevailing complex depends on the availability of the respective metals. In the presence of sufficient Mo, only the MoFe protein is synthesized. If Mo is deficient and V is available, the VFe protein is synthesized. The FeFe protein is synthesized only if both Mo and V are deficient. Before V-N. and Fe-N. were discovered, considerable progress had already been made in the elucidation of N. structure, the mechanism of N. catalysis, the genetic analysis of N. and its metabolic control, based entirely on the study of Mo-N.

Molybdenum nitrogenase. As mentioned above N. consists of two proteins, both of which are required for activity. One of these proteins contains iron, molybdenum and acid-labile sulfur (Mo-Fe protein, component I, molybdoferredoxin, azofermo) and the other contains iron and labile sulfur (Fe-protein, component II, azoferredoxin. azofer). The two components are present in the ratio 1 Mo-Fe protein: 2 Fe-proteins. Both protein components have been isolated and characterized from various nitrogen fixing organisms, and the separated components can be reconstituted to active N.

The Mo-Fe component has M_r in the range 200,000–270,000, and is tetrameric ($\alpha_2\beta_2$). Each Mo atom forms part of a polynuclear cluster containing Fe, S^{2-} and homocitrate (*R*-2-hydroxy-1,2,4-butanetricarboxylic acid); the cluster takes the form of a distorted octahedron in which the Mo is coordinated by three S atoms, three Fe atoms and three O, N or C atoms; all the available evidence indicates that these Fe-Mo coordination centers are the sites of N_2 binding and reduction. The FeMo coordination center or cluster can be removed from the denatured protein without causing essential changes in its structure; it can then be used to restore activity to inactive MoFe protein from mutants unable to synthesize the FeMo cluster. It is therefore also referred to as the FeMo cofactor or FeMoco.

The Fe protein has M_r about 60,000, and is dimeric. It possesses a single 4Fe-4S redox center (E_m –400 mV), which appears to bridge the subunits, two Cys residues from each subunit providing ligands to the cluster. Binding of MgADP or MgATP to the Fe protein causes a conformational change accompanied by a shift in the mid-point potential of the Fe-S center to about –500 mV, thereby generating an extremely powerful reducing agent with the unique property of being able to transfer electrons to the Mo-Fe protein. N. is therefore regarded as having two active centers: the electron-activating (on the Fe protein), and the substrate-activating (on the Mo-Fe protein) center. Electrons are transferred one at a time from the Fe protein to the MoFe protein. Since the Fe protein is a one electron donor, the MoFe protein acts rather like an electrical capacitor, receiving and storing electrons until sufficient reducing power has accumulated. Under optimal conditions, dinitrogen reduction can be represented as: $N_2 + 8H^+ + 8e^- + 16MgATP \rightarrow 2NH_3 + H_2 + 16MgADP + 16P_i$; thus the system appears to display a certain inefficiency, in that molecular hydrogen is also produced at the expense of MgATP hydrolysis. In *Azotobacter* and in *Rhizobium* bacteroids, hydrogen produced by N. is taken up again by hydrogenase, then oxidized with consequent synthesis of ATP.

N. catalyses the MgATP-dependent reduction of several substrates that are of similar molecular size to, and are isoelectronic with molecular nitrogen, e.g. azide, nitrous oxide, cyanide, acetylene. Attempts to produce models for the substrate activating center of N. have resulted in the preparation of organic complexes of transition metals. These are of potential interest in chemical industry, where more efficient catalysts are sought for industrial ammonia synthesis. Such catalysts might also be exploited in the synthesis of hydrazine which is a high-energy and environmentally clean fuel, but prohibitively expensive. Combustion of hydrazine produces nitrogen and water, and the heat yield is about 544 kJ/mol (130 kcal/mol).

N. from different sources are very similar. Fe protein from one organism (e.g. *Azotobacter*) will often cross react with Fe-Mo protein from another organism (e.g. *Klebsiella*) to give a functional enzyme. On the other hand, heterologous N. components may generate catalytically inactive complexes, e.g. N. of *Clostridium* is inhibited by Mo-Fe protein of *Azotobacter,* because *Clostridium* Fe protein forms a tight, inactive complex with *Azotobacter* Mo-Fe protein, even in the presence of excess *Clostridium* Mo-Fe protein.

Vanadium nitrogenase. VFe proteins (e.g. from *Azotobacter chroococcum* and *A. vinelandii*) are hexameric ($\alpha_2\beta_2\delta_2$), but are otherwise structurally and mechanistically similar to MoFe proteins. The VFe cluster is similar to the MoFe cluster, i.e. the V is in a very similar environment to that of Mo. The similarity is sufficient to allow the isolated VFe cluster to activate inactive MoFe protein from mutants unable to synthesize the MoFe cluster. The resulting protein does not, however, reduce N_2, although protons and C_2H_2 still serve as substrates. Vanadium N. are active at lower temperatures than are molybdenum N., and the reduction of N_2 by vanadium N. is characterized by the production of small quantities of N_2H_4. Also, the reduction of C_2H_2 leads to production of small quantities of C_2H_6 in addition to C_2H_4.

Iron nitrogenase. FeFe proteins (e.g. from *Azotobacter vinelandii* and *Rhodopseudomonas capsulatus*) are hexameric ($\alpha_2\beta_2\delta_2$) and they actually contain low levels of Mo or V. It is assumed that the redox center is similar to those of MoFe and VFe proteins, since the amino acid sequences in these regions are conserved, and genes involved in synthesis of the FeMo cluster are also necessary for synthesis of a functional FeFe protein.

Component I of the nitrogenases of* Azotobacter *– subunit composition and content of metals and inorganic sulfur (contents expressed in g atom/mol)

Protein (subunit composition) and native M_r	Subunit M_r	Mo	V	Fe	S^{2-}
MoFe ($\alpha_2\beta_2$) 229,494	α 55,288 β 59,459	2	0.05	32	30
VFe ($\alpha_2\beta_2\delta_2$) 240,048	α 53,877 β 52,775 δ 13,372	0.06	2	21	19
FeFe ($\alpha_2\beta_2\delta_2$) 249,996	α 58,326 β 51,153 δ 15,369	0.01	0.08	24	18

Study of gene expression and mapping in *Klebsiella pneumoniae* has revealed 20 nitrogen fixation (*nif*) genes arranged in 8 coordinately regulated operons. Most of these genes are probably present in all nitrogen-fixing organisms. Thus, when *Klebsiella nif* genes are transferred on RP4 (a plasmid first isolated from

Pseudomonas aeruginosa) to mutant *nif⁻ Azotobacter*, nitrogen fixation is restored despite the different physiology of the donor and recipient organisms. Transfer of the *nif* genes from *Klebsiella pneumoniae* to *E. coli* by R-factor mediated conjugation was the first case of the transfer of nitrogen fixing ability to a non-nitrogen fixing organism.

Regulation of *nif* gene expression occurs mainly at the level of transcription, and depends on the products of 2 genes: *rpoN* and *nifA*.

Many promoters engaged in the control of *nif* gene expression contain the highly conserved consensus sequence, $TGGCAC[N]_5TTGC$, between –26 and –12 (in relation to the start of transcription) which characterizes prokaryotic genes that employ a minor RNA polymerase containing sigma factor σ^{54}. The *rpoN* gene encodes sigma factor σ^{54}.

All σ^{54}-dependent promoters require an activator protein which binds to an activator sequence about 100 bp upstream of the region defined by –26 and –12. In the case of *nif* gene expression, this activator protein (*nif*-specific activator protein) is the product of *nifA*. It has 3 structural domains: (i) the *N*-terminal domain which possibly interacts with environmentally sensitive proteins, (ii) a central domain which is present in all σ^{54}-dependent activators, and which is thought to interact with RNA polymerase or with σ^{54}, and (iii) the *C*-terminal domain which contains a typical DNA-binding motif (a helix-turn-helix) responsible for binding to the upstream activator sequences. *nifA* is widely distributed in numerous free-living and symbiotic nitrogen fixing organisms. In cyanobacteria and archaebacteria *nif* gene regulation has not yet been analysed, but *nifA* and *rpoN* do not appear to be involved.

In enteric bacteria, transcription of the *nif* operon is promoted by a transcriptional activator (NtrC), which is active in its phosphorylated form. Some stages in this control mechanism are reminiscent of the control of glutamine synthetase activity (see Covalent modification of enzymes). Thus the nitrogen status of the cell is reflected by the ratio of the concentrations of 2-oxoglutarate and glutamine. When this ratio is high in favor of 2-oxoglutarate (i. e. metabolism and growth are limited by the availability of nitrogen), uridylyltransferase is stimulated to catalyse the uridylylation of a strategic tyrosine residue on each subunit of the tetrameric protein P_{II}. The resulting P_{II}-UMP promotes the kinase activity of a regulatory protein known as NtrB, which in turn promotes phosphorylation (i. e. activation) of NtrC (a His residue is first phosphorylated, and the phosphate group is then transferred to an Asp residue in the *N*-terminal region). When nitrogen is in plentiful supply (reflected intercellularly by a relatively high concentration of glutamine and a low concentration of 2-oxoglutarate) uridylyltransferase acquires uridylyl-removing activity and converts P_{II} to its nonuridylylated form. NtrB then promotes dephosphorylation (i. e. inactivation) of NtrC, and NtrC-dependent promoters are switched off. In *Klebsiella pneumoniae* and *Rhizobium capsulatus*, NtrC governs the expression of *nifA*, so that the nitrogen status of the cell exerts control at the level of nitrogenase synthesis.

As discussed in the entry 'Nitrogen fixation' (see), N. is affected in various ways by molecular oxygen. It is synthesized by strict anaerobes such as *Clostridium*, by very powerfully respiring aerobic organisms such as *Azotobacter*, and by organisms with moderate oxygen consumption such as *Klebsiella* and *Rhizobium*. Different types of oxygen regulation can therefore be anticipated at the level of translation, transcription and/or activity control.

NifA proteins from *Rhizobium* and *Rhodobacter* share two common features: they are directly inhibited by oxygen and they both possess a –Cys–X–X–X–X–Cys– motif. Oxygen sensitivity is possibly due to the binding of a metal ion by this structural motif. In *Rhizobium*, two gene products (FixL and FixJ) are responsible for regulating *nifA* expression. FixL

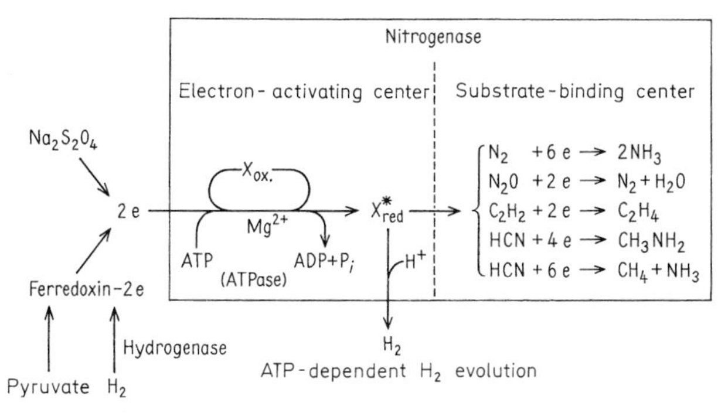

Schematic representation of the transfer of reducing power during the operation of nitrogenase. The left hand box (electron activating center) represents the Fe protein which interacts with MgATP to become a powerful reducing agent. The right hand box (substrate binding center) represents the MoFe protein (or the VFe or FeFe protein) which stores electrons from the Fe protein, finally passing them to the substrate. A variety of substrates is shown, all known to be reduced by nitrogenase. X_{ox} represents the metal center before association with MgATP. X^*_{red} represents a powerful reducing agent, possibly a metal hydride. In vivo, electrons (derived variously from aerobic respiration, anaerobic phosphoroclastic pyruvate cleavage, hydrogenase, etc.) are transferred to nitrogenase via a ferredoxin. In vitro, in cell-free systems, electrons are passed directly to nitrogenase from sodium bisulfite.

is a membrane-bound haemoprotein with histidine protein kinase activity. In response to low atmospheric oxygen tension FixL phosphorylates (i.e. activates) FixJ, which in turn activates *NifA* expression. As a general observation, anaerobiosis in many prokaryotes induces negative DNA supercoiling. DNA supercoiling is known to be necessary for transcripton of several prokaryotic promoters, including that of the nifA operon in *Klebsiella pneumoniae.* Thus oxygen also appears to exert control at the level of DNA supercoiling in this organism.

Activity of N. preparations is usually measured by colorimetric assay of ammonia, following incubation with molecular nitrogen. Nitrogen fixation in general can be determined from the incorporation of $^{15}N_2$ into cell material, but the method is relatively expensive and tedious. The use of acetylene (ethyne) as an alternative substrate of N. has revolutionized studies of nitrogen fixation. Assays are performed in a closed system. The product ethylene (ethene) is not assimilated and is easily assayed by gas chromatography.

See Nitrogen fixation for further information on N. [*New Horizons in Nitrogen Fixation,* R. Palacios et al. (eds), Kluwer Academic, Dordrecht, 1993; R.H. Burris 'Breaking the N ≡ N Bond' *Annu. Rev. Plant Physiol. Plant Mol. Biol.* **46** (1995) 1–19]

Nitrogen balance: difference between the total nitrogen intake of an organism and its total nitrogen loss. Young growing animals are in positive N.b., i.e. they retain more nitrogen (as protein added during growth) than they excrete. Mature, healthy adults show a zero N.b., i.e. nitrogen intake is exactly balanced by nitrogen excretion. A negative N.b. results from a deficiency of an essential amino acid; a decrease in the concentration of any proteogenic amino acid in the body's amino acid pool impairs total protein synthesis, so that the concentration of all other free amino acids in the pool is increased; this leads to an exaggeration of degradative pathways and an increase in urea formation. The classical deletion method for the determination of the essential (indispensable) amino acid requirements of an animal involves the measurement of N.b. in adult animals receiving a complete diet, except for the omission of the amino acid under test. The daily food intake is determined, and the nitrogen content of an identical food sample is measured. Strictly, in determining the daily loss of nitrogen, dropped hair, sloughed skin and perspiration should be taken into account, but it is usually sufficient to omit these minor contributions, and use only the nitrogen content of feces and urine in the deterrnination of N.b.

Nitrogen catabolite repression: repression by ammonia of the synthesis of various enzymes of nitrogen metabolism. See Ammonia assimilation.

Nitrogen cycle: see Nitrogen fixation.

Nitrogen excretion: see Ammonia detoxification.

Nitrogen fixation: a process in which atmospheric nitrogen is converted into ammonia. Activation of molecular nitrogen and its reduction to ammonia depend on the catalytic activity of the enzyme Nitrogenase (see). The ammonia is then incorporated into the various nitrogenous compounds of the cell by the processes of Ammonia assimilation (see). Nitrogenase is a very unstable enzyme, especially in anaerobic organisms. N.f. is fundamentally important for the nitrogen economy of soils and waters, and it forms an essential stage in the nitrogen cycle of the biosphere. Certain free living soil microorganisms, especially of the genera *Clostridium* and *Azotobacter,* are capable of N.f. Other microorganisms fix nitrogen in symbiosis with higher plants. notably the *Leguminosae.* Many instances of N.f. by symbiotic associations between microorganisms and nonleguminous plants are also known, e.g. an actinomycete (*Frankia* spp.) has been isolated from the nitrogen fixing root nodules of Alder (*Alnus*). In water, especially in the ocean, the most important nitrogen fixers are the blue green bacteria or cyanobacteria (also commonly known as blue green algae, a term that is preferably avoided, because these nitrogen fixing organisms are prokaryotes, whereas the algae are eukaryotes). N.f. by cyanobacteria is of practical importance for the cultivation of rice in the tropics. The lichens (symbiotic associations between a blue-green bacterium and a fungus), and the symbiotic system *Nostoc-Gunnera* (i.e. symbiosis between a cyanobacterium and an angiosperm) are ecologically very important, since they are able to colonize habitats that have extreme climates or are poor in nutrients. The carbon and nitrogen requirement of the lichens are met by photosynthesis and N.f. Such symbiotic systems are therefore sustained largely by the atmosphere, and their nutritional demands on the remaining environment are relatively small. Lichens pioneer the exploitation of barren environments and pave the way for later colonization by plants with more exacting nutritional requirements. In poor soils, the *Nostoc-Gunnera* system can fix about 70 g atmospheric nitrogen per m^2 per year.

Diazotrophy or the ability to perform N.f. is a special characteristic of relatively few prokaryotic organisms (diazotrophs); it has never been detected in a eukaryote. In *Clostridium pasteurianum,* which contributes to the nitrogen enrichment of agricultural soils, both the reducing power and the ATP required for N.f. are derived from phosphoroclastic pyruvate cleavage. In cell-free enzyme preparations pyruvate can be replaced by ATP or an ATP generating system, and a reducing agent (hydrogen or electron donor). Suitable reducing agents include sodium dithionite and potassium borohydride. Nitrogenase will also catalyse the transfer of electrons from molecular hydrogen to nitrogen in the presence of a ferredoxin-dependent hydrogenase. In most nitrogen-fixing systems, the natural electron donor is a ferredoxin; in certain cases, this is replaced by other electron transferring proteins, e.g. flavodoxin, or rubredoxin. Four molecules of ATP are required for the transfer of each pair of electrons. The stepwise reduction of nitrogen on the surface of the nitrogenase may take place via enzyme-bound intermediates, but free intermediates between ammonia (the product) and N_2 (substrate of N.f.) have not been observed.

So far, the most extensively studied N.f. system is the symbiotic association between members of the *Leguminosae* and *Rhizobium;* the legume often chosen for these studies is *Glycine max* (soybean). Infection of the plant roots by virulent Rhizobia leads to the formation of root nodules, which have the capacity to fix nitrogen. The Rhizobia living free in the soil do not fix nitrogen. Under laboratory conditions, however, pure cultures of *Rhizobium* will fix nitrogen, providing a pentose (e.g. arabinose) and a dicarboxylic acid (fumarate or succinate) are present in the culture medium. During the infection process the *Rhizobium* cells lose their rod-like shape and eventually become globular-shaped bacteroids. Reduction of nitrogen to ammonia and the assimilation of the ammonia occur in these bacteroids; carbon compounds for the respiration of the bacteroids and for the assimilation of the ammonia are provided by the plant; amino acids are exported to the host plant tissues. Leghemoglobin is necessary for N.f. by legume root nodules, and the leghemoglobin concentration in root nodules is an index of the nitrogen fixing capacity. Leghemoglobin does not occur in nonlegume N.f. systems. Within the cells of the root nodule the bacteroids are bathed in a solution of leghemoglobin, which is enclosed by a membrane envelope. The rate of oxygen transport through an unstirred solution of leghemoglobin is eight times higher than its rate of diffusion through water. This facilitated diffusion of oxygen to the bacteroids permits a high respiration rate, which is necessary to produce the relatively large quantities of ATP required by the nitrogenase. In contrast, oxygen interferes during the laboratory preparation of active bacteroids, due to the presence of phenols and polyphenol oxidases from the host plant tissue. These can be inactivated by adsorption onto polyvinyl pyrrolidone in the presence of ascorbic acid. Nitrogen fixing bacteroid suspensions can therefore be isolated from homogenized root nodules by using strictly anaerobic conditions, e.g. centrifugation of the homogenate under argon, or by abolishing polyphenol oxidase activity. The bacteroids can then be treated like any other bacterial source of nitrogenase. Subsequent disruption of the cells and purification of the enzyme by selective precipitation and column chromatography must be performed under strictly anaerobic conditions, because nitrogenase is irreversibly inactivated by oxygen; this is especially critical at later stages of purification, as the oxygen sensitivity of nitrogenase increases with purification. The separate protein components of nitrogenase are both inactivated by oxygen, and the Fe-protein is the more sensitive.

Nitrogen storage: see Ammonia detoxification.

nm: nanometer.

Nodule bacteria: see Rhizobia.

Nomenclatural conventions concerned with gene transcription: The term 'transcription unit' is given to a segment of double-stranded DNA (containing a single gene or group of genes) that is transcribed into a single molecule of RNA. Of the two strands of DNA constituting the transcription unit one acts as the template that defines the sequence of nucleotides in the RNA transcript produced by RNA polymerase action while the other, the non-template strand, does not. Convention requires that when the nucleotide sequences of a transcription unit, or any section of double-stranded DNA, are written down horizontally, the sequence of the template strand has its 3′-end at the left of the page such that normal left-to-right reading proceeds in the $3' \rightarrow 5'$ direction, while that of the non-template strand, being antiparallel, has its 5′-end at the left so that normal reading proceeds in the $5' \rightarrow 3'$ direction (Fig.). This convention means that, since RNA synthesis takes place (a) on the template DNA strand and (b) in the $5' \rightarrow 3'$ direction, the RNA transcript has a nucleotide sequence, complementary to that of the template, that is written with its 5′-end at the left of the page so that normal reading proceeds in the $5' \rightarrow 3'$ direction (Fig.). The consequence of this convention is that the nucleotide sequence of the RNA transcript is identical to that of the non-template DNA strand, save that U-containing nucleotides replace T-containing nucleotides (Fig.). Because the genetic code (see) is expressed as RNA codons (i.e. nucleotide triplets read in the $5' \rightarrow 3'$ direction) which, in mRNAs, are translated into proteins (starting with the amino acid at the *N*-terminal end) in the $5' \rightarrow 3'$ direction, it is the nucleotide sequence in the RNA transcript that best describes the relevant genetic information. Furthermore, because the nucleotide sequence of the RNA transcript is identical to that of the non-template DNA strand (except for the U/T replacement) it is the latter that best expresses the coding information in the DNA transcription unit. Thus, by convention, the non-template strand (i.e. the one that is not transcribed into RNA) of the double-stranded DNA transcription unit is called the 'coding strand' or the 'sense strand'. By the same token the template strand (i.e. the one that is transcribed into RNA) is called the 'non-coding strand', the 'non-sense strand' or the 'anti-sense stand' (Fig.).

There has been a great deal of confusion in student textbooks over the use of the terms template/non-template, coding/non-coding, sense/non-sense or anti-sense in the description of the two strands of DNA. The JCBN/NC-IUB Newsletter, 1989 (reproduced in 'Biochemical Nomenclature and Related Documents – A Compendium; 2nd Edition, 1992) recognized this confusion and recommended the use of the convention described above, on the grounds that "as the word 'coding' refers to the relationship between nucleic acids and proteins, rather than the mere transcription of DNA into RNA, it is logical to call the strand with the mRNA sequence the coding strand". The newsletter also expressed a preference for 'coding/non-coding' over 'sense/non-sense', although in practice the latter pairing is frequently used. It also recommended that when DNA sequences are described by giving the sequence of only one of the two strands of a double-stranded DNA, this should normally be the strand with the same sequence (save for the T/U replacement) as that of the RNA transcript (i.e. the strand designated above as the coding strand).

RNA polymerase that is about to transcribe a particular transcription unit binds to double-stranded DNA at a position (the promoter) which is said to be 'upstream' of the nucleotide at which transcription will ultimately begin. The term 'upstream' strictly refers to any portion of the 'coding strand' that is on the 5′-side of the start site of transcription. The nucleotide at the start site is designated +1 and the nucleotides upstream (i.e. on the 5′-side) of it are sequentially designated –1, –2, –3 etc, no nucleotide being designated 0. The promoter of a transcribed operon, which includes the Pribnow (see) or TATA box (–10 region) and the –35 region, is, for instance, upstream of the initiation site of transcription. The term is sometimes used more loosely to refer to any nucleotide or region of the 'coding strand' of DNA

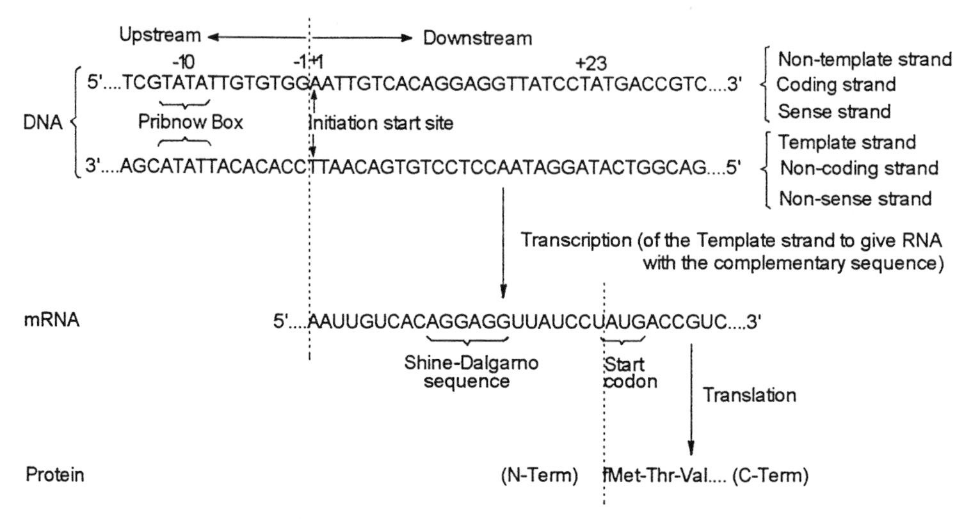

Nomenclatural conventions. Transcription and translation of a hypothetical segment of double-stranded, prokaryotic DNA, showing the structural relationships of the participants and their accepted nomenclature.

that is on the 5′-side of some other nucleotide or region within the strand. Similarly, the term 'downstream' strictly refers to any portion of the 'coding strand' of double-stranded DNA that is on the 3′ side of the start site of transcription. The nucleotides downstream (i.e. on the 3′-side) of it are sequentially designated +2, +3, +4 etc. The term is sometimes used more loosely to refer to any nucleotide or region of the 'coding strand' of DNA that is on the 3′-side of some other nucleotide or region within the strand.

Other aspects of nucleic acid nomenclature may be found in 'Biochemical Nomenclature and Related Documents – A Compendium; 2nd Edition, 1992, pp. 109–126, or in *Eur. J. Biochem.* **15** (1970) 203–208, **131** (1983) 9–15 & **150** (1985) 1–5.

Non-coding strand: by convention (JCBN/NC-IUB Newsletter, 1989, reproduced in 'Biochemical Nomenclature & Related Documents – A Compendium', 2nd Edition 1992), the strand of a double-stranded stretch of DNA that has the complementary nucleotide sequence to that of the RNA transcript (e.g. mRNA) that is derived from that double-stranded DNA. It is therefore the DNA strand that acts as the template for the RNA transcript and could thus be called the 'template' strand. Alternative, but in the opinion of the 1989 JCBN/NC-IUB Newsletter, less preferable names are 'non-sense strand' and 'non-transcribing strand'. Another alternative name is 'anti-sense strand'. See Nomenclatural conventions converned with gene transcription.

Noncompetitive inhibition: see Effectors.

Noncovalent bonds: various types of noncovalent bond are responsible for maintaining chain conformation and quarternary structure of proteins, and they are also important in the structure and function of nucleic acids.

1. *Hydrogen bonds* are formed between neighboring peptide bonds (separation distance 0.28 nm), between tyrosyl and carboxyl or imidazole, and between seryl and threonyl residues. Hydrogen bonding in particular is an important factor in nucleic acid structure, and in template recognition during replication, transcription and translation (see Base pairing, Ribonucleic acid, Protein biosynthesis, Deoxyribonucleic acid).

2. *Heteropolar (electrostatic) bonds* in proteins are formed between residues of opposite charge, e.g. lysyl and glutamyl.

3. *Apolar (hydrophobic) bonds* in proteins are formed between very close, uncharged groups, e.g. $-CH_3$ and $-CH_2OH$, or between more widely separated, uncharged groups, e.g. phenyl and leucyl. The effective strength of these hydrophobic bonds is increased by the entropy effect of the repulsion of the surrounding water, and they contribute to the stability of protein conformation especially at elevated temperatures.

4. *Van der Waals forces* act only at very short distances, and they represent the weak attraction between the positively charged nucleus of one atom and the negatively charged electrons of another. They are important in base stacking in the double helix of DNA (see Deoxyribonucleic acid).

Nonheme iron proteins, *NHI-proteins:* proteins containing iron that is not bound in a heme system. In these proteins, the iron is bound by the sulfur of cysteine residues, and it is often also associated with inorganic sulfur. They are also called iron-sulfur proteins. See Ferredoxin, Rubredoxin.

Nonhistone chromatin proteins: "acidic chromatin proteins", a highly heterogeneous group of tissue-specific proteins, which are bound to certain DNA sequences. Their M_r in detergents is 30,000–70,000, i.e. markedly higher than the M_r of Histones (see). As gene derepressors, they play a part in the regulation of gene expression in mammalian cells, especially in cell proliferation.

Nonordered conformation: see Proteins.

Non-reducing end (of an oligo- or polysaccharide): the terminal monosaccharide residue which does *not* have an underivatized hydroxyl group on its anomeric carbon atom (i.e. C1 in aldoses, C2 in ketoses), because it has been replaced, as the result of glycosidic bond formation, by OR (where R is the rest of the oligo- or polysaccharide molecule), thereby preventing the residue from fulfilling the requirements of a reducing sugar (see). All linear oligo- and polysaccharides (e.g. amylose, cellulose) have one N-r.e. while all branched-chain oligo- and polysaccharides (e.g. amylopectin, glycogen) have many because each exterior chain terminates in one. Several enzymes that participate in either the breakdown of oligo- and polysaccharides (e.g. β-amylase EC 3.2.1.2; phosphorylase, EC 2.4.1.1) or their biosynthesis (e.g. starch [glycogen] synthases, EC 2.4.1.11 & EC 2.4.1.21) recognize N-r.es. and progressively remove glycose moieties from or add glycose moieties to the N-r.e. of glycan chains respectively.

Nonsense codon: an Amber codon (see), or Ochre codon (see).

Non-sense strand: an alternative name for the Non-coding strand (see) of double-stranded DNA, which, by convention, has the complementary nucleotide sequence to that of the RNA transcript (e.g. mRNA) derived from that double-stranded DNA (save that T is in the place of U). See Nomenclatural conventions concerned with gene transcription.

Non-template strand: the strand of a double-stranded stretch of DNA that does not act as the template for the generation of an RNA transcript (e.g. mRNA). It therefore has the same nucleotide sequence as that of the RNA transcript, save that T is in the place of U. Alternative names are Coding strand (see) and Sense strand (see). See Nomenclatural conventions concerned with gene transcription.

Nopaline: see D-Octopine.

Nopalinic acid: see D-Octopine.

Noradrenalin, *norepinephrin*, *arterenol*: dihydroxyphenylethanolamine, a hormone and adrenergic Neurotransmitter (see) in the sympathetic nervous system, M_r 169.2. N. causes contraction of blood vessels, with the exception of coronary vessels; it therefore causes an increase in the blood pressure. It also relaxes smooth muscle, but stimulates cardiac muscle. Comparison of the effects of N. and Adrenalin (see) has led to classification of postsynaptic receptors as α or β receptors. The former are more responsive to adrenalin, and are generally excitatory, except in the intestinal smooth muscle, while the latter respond to N. and are generally inhibitory. N. is a catecholamine synthesized from L-tyrosine via dopa (see Dopamine) in the adrenal medulla and in adrenergic neurons in the nervous system. It is converted, in part, to Adrenalin (see) in these same tissues by the enzyme noradrenalin *N*-methyltransferase (E.C.2.1.1.28).

N. is deactivated by *O*-methylation; a monoamine oxidase then removes the NH_2 group to produce 3-methoxy-4-hydroxymandelic acid (vanillylmandelic acid, VMA). VMA is the major metabolite of N. in the peripheral parts of the body and in the urine. The amount of VMA in the urine is an index of parasympathetic nervous function, and is used for the diagnosis of tumors which produce N. or adrenalin. See ascorbate shuttle.

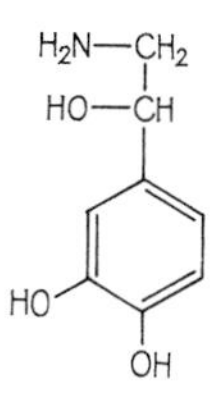

$H_2N{-}CH_2$, HO—CH, HO, OH

Noradrenalin

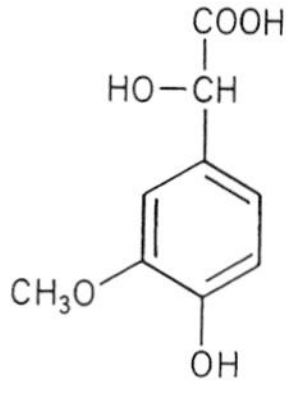

COOH, HO—CH, CH_3O, OH

Vanillymandelic acid

Norepinephrin: see Noradrenalin.

Norgestrel: a synthetic gestagen. N. is used in oral contraceptives, and it is the most potent of the orally active gestagens.

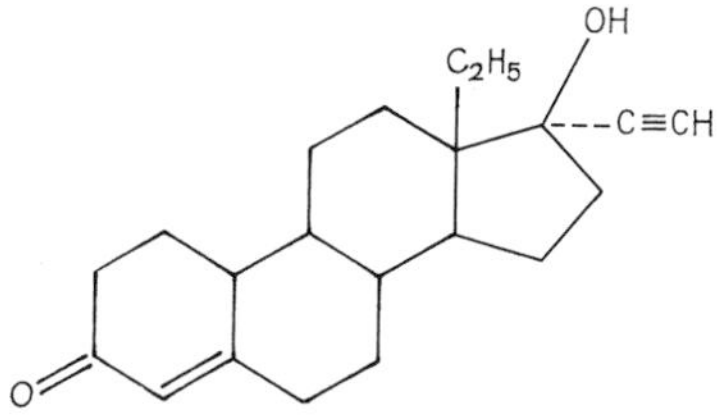

OH, C_2H_5, C≡CH, O

Norgestrel

Norlaudanosine: see Benzylisoquinoline alkaloids.

Nornicotine: see Nicotine.

Norsteroids: see Steroids.

Northern blot: see Southern blot.

Norum's disease: see Inborn errors of metabolism.

Notatin: see Mycotoxins.

Nu bodies, *nucleosomes*: subunits of Chromatin (see) containing 180 to 200 base pairs of DNA and an approximately equal weight of histone. They are released by partial digestion of chromatin by staphylococcal nuclease.

Nuclear magnetic resonance (nmr) spectroscopy: a technique used in the determination of structure of molecules at atomic resolution. The phenomenon of nmr was first observed in 1946 and nmr spectroscopy has been routinely used by organic chemists since about 1960. With the advent of increasingly powerful instrumentation, not least in the field of computers, the size of the molecules that can be successfully examined has increased to the point where nmr is of enormous value to biochemists and molecular biologists since it can be applied to small proteins (M_r < 25,000) and other biopolymers. Because nmr spectra are taken of compounds in solution the atomic structural information revealed complements that of X-ray crystallography (see) which, in contrast, requires the presence of a crystal lattice; indeed the application of the two techniques to identical proteins has shown that, in general, proteins in solution have the same 3D structure as those in the crystalline state.

The basis of nmr spectroscopy is the fact that the nucleus of some species of atoms within a molecule spin about an axis. For a nucleus to have a spin it must contain either an odd number of protons, or an odd number of neutrons, or both of these conditions.

Over sixty elements have one or more isotopes whose nuclei fulfil these requirements and so have spin and can therefore be studied by nmr. They include several that are particularly useful to the biochemist and molecular biologist, namely ^{1}H (99 · 9844 %), ^{13}C (1 · 108 %), ^{15}N (0 · 365 %), ^{19}F(100 %) & ^{31}P (100 %) where the figure in parenthesis denotes their natural abundance; ^{12}C (98 · 892 %) and ^{16}O (99 · 759 %) have no nuclear spin and therefore do not exhibit nmr.

The spin of these nuclei is associated with a circulation of electric charge. The latter arises from the presence of one or more positively-charged protons in the nucleus of each isotopic species. Since circulating electrical charges give rise to magnetic fields, each of these spinning nuclei also has a magnetic moment. Thus when they are put in an external magnetic field, each turns so as to take up a particular orientation, rather like the magnetized needle of a compass in the earth's magnetic field. Since all of the nuclei named above have a spin *(I)* of $^{1}/_{2}$ only two orientations are possible, one aligned with the external magnetic field and one against it. However, the magnetic field of the nucleus does not lie at exactly 90° to the faces of the pole pieces of the external magnet because it is spinning. This spin makes it behave like a gyroscope and so it precesses about its axis; thus the two orientations, one aligned with and the other against the external magnetic field, may be represented by Figs. 1 a and 1 b respectively.

Since nuclei obey quantum laws, each of the two orientations is associated with a particular energy; the orientation aligned with the external magnetic field has the lower energy and is said to be the ground (or α) state whilst the orientation aligned against the external magnetic field has a higher energy and is said to be the excited (or β) state. Since the difference between these energy levels is very small (about five orders of magnitude lower than the average thermal energy of the molecule in which the nucleus resides) they are almost equally populated with nuclei, with fractionally more in the lower one; this is exemplified by ^{1}H nuclei where, in a typical nmr experiment, the lower energy level is more populated by a factor of 1.00001 (i.e. in a population of 2000000 nuclei, 999995 are present in the higher energy orientation and 1000005 in the lower energy orientation). The ratio of these two populations may be calculated by equation 1, known as the Boltzmann distribution:

$$\frac{N_\beta}{N_\alpha} = e^{-\frac{\Delta E}{kT}} \qquad (1)$$

where: N_α = the number of nuclei in the ground (or α) state,
N_β = the number of nuclei in the excited (or β) state,
ΔE = the difference in energy (J) between the α & β states,
k = the Boltzmann constant = 1.380626×10^{-23} JK^{-1},
T = the temperature in degrees Kelvin (= 273.15 + °C).

A nucleus may change from the lower energy orientation to the higher energy orientation by absorbing a photon of electromagnetic radiation with an energy equal to that of the energy difference (ΔE) between the two orientations. The magnitude of this energy difference, expressed in Joules (J) is given by equation 2:

$$\Delta E = \frac{h\gamma B_o}{2\pi} \qquad (2)$$

where: h = Planck's constant = 6.6262×10^{-34} Js,
γ = the magnetogyric ratio (a proportionality constant, which is different for each nuclear species and is in essence a measure of the magnetic strength of the particular nucleus) = 2.675197×10^{8} $s^{-1}T^{-1}$ for ^{1}H,
B_o = the strength of the external magnetic field (T) (1 tesla, T = 10^{4} Gauss).

The frequency (ν), expressed in Hertz (Hz, cycles s^{-1}), and the wavelength (λ), expressed in metres (m), of the photon carrying energy equal to ΔE can be calculated using the equations $E = h\nu$ and $E = hc/\lambda$ respectively, where h is Planck's constant and c is the speed of light (2.99792459×10^{8} ms^{-1}). Because ΔE is so small for all nuclei exhibiting nmr the photons effecting the transitions from the α-state to the β-state fall in the radiofrequency range of the electromagnetic spectrum (i.e. ν runs into MHz and λ into meters).

The frequency of the radiation necessary to effect the transitions from the α-state to the β-state is directly proportional to the strength of the external magnetic field experienced by the nucleus and the nuclear species in question. This follows from equa-

a

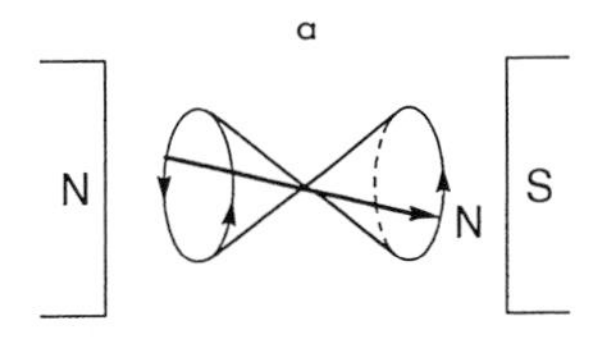

Proton aligned with the external magnetic field (ground state or α-state) showing the precesson of its axis

b

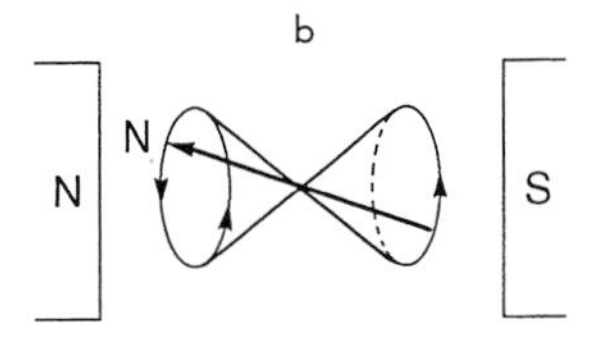

Proton aligned against the external magnetic field (excited state or β-state) showing the precesson of its axis

Fig. 1. *Nuclear magnetic resonance spectroscopy.* Precession of the axis of the magnetic moment of a ^{1}H nucleus aligned with (a) and against (b) the magnetic field of an external magnet whose poles are indicated by N and S.

tion 3, in which ν, γ and B_o have the same meanings as given in earlier equations.

$$\nu = \frac{\gamma B_o}{2\pi} \quad (3)$$

Table 1 shows this effect by giving relevant values for the ^{1}H and ^{13}C nuclei.

The frequency at which the 'α-state → β-state' transition typically occurs is different for different nuclei. For instance, at an external magnetic field strength of 2.3487 T it occurs at the following frequencies (MHz) for the nuclei mentioned earlier: ^{1}H, 100; ^{13}C, 25.14; ^{15}N, 10 · 13; ^{19}F, 94 · 08; ^{31}P, 40 · 48. [At this point it is perhaps worth mentioning that, because of the widespread use of ^{1}H nmr spectroscopy (often called 'proton nmr' or simply 'pmr'), it has become common practice to refer to an nmr spectrometer with a magnetic field strength of 2.3487T as a 100 MHz instrument regardless of the species of nucleus it is being used to examine.]

The absorption of such radiofrequency energy temporarily increases the population of nuclei in the higher energy orientation; however, the population distribution mentioned earlier is re-established by loss of energy to the surroundings by a combination of processes not involving the radiation of radiofrequency energy, collectively known as 'relaxation'. Hence, as nuclei resonate between the α and β-states there is a net absorption of energy which can be measured.

The nmr spectrum can be measured in either of two ways, known as 'continuous wave' (CW) and 'Fourier transform' (FT). In the CW method the frequency range of the nucleus being studied is scanned steadily from one extreme to the other, either by varying the frequency of the radiotransmitter while keeping the external magnetic field constant or, (more usually) by varying the magnetic field while keeping the radiofrequency constant, and picking up the response (equivalent to the net energy absorption) on a receiving coil. The spectrum is plotted as absorption as the ordinate against frequency (but see later) as the abscissa (increasing in value from right to left, as in UV, visible & IR spectroscopy). In the FT method the sample is irradiated for a fixed time (typically 5 μs for ^{1}H & 10 μs for ^{13}C) by a pulse of radiofrequency characteristic of the nucleus under investigation. The response is picked up by the receiving coil during an acquisition period (typically a few seconds) as the oscillation dies away to zero. This signal, known as the free induction decay (FID), is a complex wave pattern which can be converted by Fourier transformation into an FT spectrum which is then plotted in the same form as a CW spectrum. The FT method has the advantage over the CW method of being capable of much greater sensitivity. This stems from the fact that, since all the data are handled in digital form by computer, as soon as the FID from one radiofrequency pulse has been acquired another identical pulse can be administered and its FID acquired and added to the first. In fact the cycle of 'pulse followed by acquisition' can be repeated many times and the summation of all the FIDs subjected to Fourier transformation. This increases sensitivity by increasing the absorption signal and decreasing the noise signal, which being random is largely cancelled out; relative to a single spectrum the summation of n spectra improves the signal to noise ratio by $\sqrt{n}$. This feature is particularly necessary for nuclei like ^{13}C which have a low natural abundance (1.108 %). Another feature that is necessary for nuclei like ^{13}C is a magnet with a high and uniform magnetic field; this follows from the fact that the ^{13}C nucleus is considerably less sensitive to nmr than the ^{1}H nucleus, as is evident from Table 1 which shows that at any given field strength the energy difference between its α and β-states is about a quarter of that of the ^{1}H nucleus. Errors that may result from slight non-uniformy of the magnetic field are minimized by rotating the tube containing the sample about its vertical axis at 30 rps. The sample size for a ^{1}H CW spectrum is 50 mg or more whilst that for an FT spectrum is 1–10 mg. The sample size for a routine ^{13}C FT spectrum is 50–100 mg, although if time is available for a large number of 'pulse-FID cycles' this can be cut to 1 mg. Samples are dissolved in a solvent that does not itself give rise to nmr signals; CCl_4 and $CDCl_3$ are commonly used for lipid-soluble materials and D_2O for water soluble materials. An internal reference compound, tetramethylsilane (TMS) with organic solvents or 3-trimethylsilyl-1-propane sulfonate (Tier's salt) with D_2O, is added to the solution.

From the explanation of nmr given so far it would be expected that all the nuclei of the particular species (e.g. ^{1}H, ^{13}C) under examination in the molecules of the sample would resonate at the same frequency (i.e. at 2.3487 T, ^{1}H at 100 MHz and ^{13}C at 25.144 MHz). However, this is not so. The reason for this is that each nucleus is surrounded by electrons. These electrons, when placed in a magnetic field, behave like perfectly conducting shells and weak electric currents are induced in them. These currents, in

Table 1. *Relationship between external magnetic field strength and the frequency of the absorbed radiation*

B_o (T)	^{1}H nucleus			^{13}C nucleus		
	ν (MHz)	λ (m)	E_{photon} (J)	ν (MHz)	λ (m)	E_{photon} (J)
1.4092	60	4.997	3.97×10^{-26}	15.087	19.871	0.99×10^{-26}
2.3487	100	2.998	6.62×10^{-26}	25.114	11.923	1.67×10^{-26}
4.6974	200	1.499	13.3×10^{-26}	50.288	5.962	3.33×10^{-26}
7.0460	300	0.999	19.9×10^{-26}	75.432	3.974	4.99×10^{-26}
9.3947	400	0.750	26.5×10^{-26}	100.577	2.981	6.66×10^{-26}
11.7434	500	0.600	33.1×10^{-26}	125.720	2.385	8.33×10^{-26}

turn, generate a magnetic field which opposes the externally-applied magnetic field. The consequence of this is that the nucleus at the centre of the electron shells experiences a magnetic field that is fractionally smaller than the externally-applied one; the electrons, in effect, partially shield the nucleus from the external magnetic field. Thus the energy difference between the α and β-states is slightly less than it would have been if no electrons were present, so the radiofrequency necessary to make the nucleus resonate is lower. Because the electron distribution around chemically different nuclei of the same species in a molecule is different, these nuclei experience slightly different external magnetic fields and are therefore brought into resonance by slightly different radiofrequencies; the greater the shielding effect of the electrons the lower the radiofrequency required. The usefulness of this effect is that, at a given external magnetic field strength, the radiofrequencies at which resonance occur are diagnostic of nuclei in particular chemical environments (e.g. ^{1}H nuclei in a methyl group, ^{13}C in a benzene ring). Because of this, samples are irradiated with an appropriate range of radiofrequencies. For instance in a 100 MHz (i.e. 2.3487 T) instrument the range of frequencies used when taking ^{1}H spectra is about 1000 Hz in the region of the ^{1}H resonant frequency of 100 MHz, while that used when taking ^{13}C spectra is about 5000 Hz in the region of the ^{13}C resonant frequency of 25.144 MHz; each of these frequency ranges is wide enough to bring each of the chemically different types of ^{1}H or ^{13}C nuclei found in most organic molecules into resonance.

Because the frequency at which ^{1}H or other nuclei resonate depends upon the strength of the externally-applied magnetic field (see Table 1) and therefore varies with the nmr spectrometer used, the parameter that is plotted on the abscissa of the nmr spectrum is not actually frequency (ν) expressed in Hz but 'chemical shift units' (δ) expressed in parts per million (ppm), which are, in effect, frequencies relative to that of the internal reference standard (e.g. TMS) and thus independent of the instrumentation used. They are defined by equation 4.

$$\delta = \frac{\nu_s - \nu_{TMS}}{\nu_{instrument}} \qquad (4)$$

where: δ = chemical shift units (ppm),
ν_s = frequency (Hz) of the resonance of nuclei in a particular chemical environment in the molecules of the sample,
ν_{TMS} = frequency (Hz) of the resonance of the ^{1}H nuclei of the methyl groups of TMS,
$\nu_{instrument}$ = operating frequency (Hz) of the nmr spectrometer used.

By definition the δ value of TMS, which is used as the reference for ^{1}H and ^{13}C nmr spectra, is zero; δ values determined using Tier's salt as the reference compound are not significantly different from d_{TMS} values. TMS is used as a reference compound primarily because: (i) its twelve ^{1}H nuclei have identical chemical environments and so resonate at the same frequency, thereby giving a single intense peak that is easily recognized, and (ii) the ^{1}H nuclei of its four methyl groups are better shielded by their planetary electrons than those of almost any other organic compound [because silicon has a lower electronegativitiy (1.8) than hydrogen (2.1) or carbon (2.5), with the result that electron density is pulled from the Si atom towards the methyl groups], thereby requiring a lower radiofrequency to bring them into resonance; thus the radiofrequency required to bring into resonance ^{1}H nuclei in virtually all other chemical environments is higher than that for the ^{1}H nuclei of TMS. Such ^{1}H nuclei are said to be *deshielded* relative to those of TMS and to give an nmr signal that is *downfield* of them (i.e. at a higher frequency and to the left of them in the standard representation of the nmr spectrum).

Deshielding, and thus the downfield shift of the nmr signal, relative to TMS is caused by differences in the electronegativities of adjacent atoms and by multiple bonds. The effect of electronegativity is exemplified by the δ values of the ^{1}H nuclei of the methyl groups in the series CH_3I (2.10), CH_3Br (2.70), CH_3Cl (3.06) & CH_3F (4.27) where the progressive increase in the electronegativity of the halogen (I = 2 · 5; Br = 2 · 8; Cl = 3 · 0; F = 4 · 0) exerts an increasing withdrawal of electron density from the methyl groups and thus an increased deshielding of the ^{1}H nuclei. The effect of multiple bonds is more complex. By virtue of their π-bonds (one and two respectively for double and triple bonds) they are regions of high electron density and can thus set up magnetic fields. However, these fields are stronger in one direction than another; they are *anisotropic*. Consequently functional groups that have multiple bonds have regions of greater and lesser shielding around them. A double bond (e.g. >C = O, >C = C<) has regions of greater shielding above and below its general plane and a region of lesser shielding (relative deshielding) around its plane; thus ^{1}H nuclei attached to the carbon atoms of these groups, being located in the plane of the double bond and therefore relatively deshielded, have higher δ values than would be expected simply from electronegativity considerations. A triple bond (e.g. –C≡C–), on the other hand, has is long axis lying in a cylindrical region of greater shielding; thus ^{1}H nuclei attached to the carbon atoms, being located along the same axis, are relatively strongly shielded and therefore have lower δ values than would be expected simply from electronegativity considerations.

Analysis of the nmr spectra of a large number of known compounds has enabled extensive tables of the δ values of ^{1}H, ^{13}C and other nuclei in a wide variety of structural circumstances to be prepared.

In addition to the δ value of a particular peak in the nmr spectrum, two other parameters that are of importance in the determination of the structure of the compound under examination are: (i) the area under the peak, (ii) spin-spin coupling, and (iii) the nuclear Overhauser effect (NOE).

For CW spectra and FT spectra of ^{1}H nuclei the area under each absorption peak is proportional to the number of ^{1}H nuclei present in it; this is not true for ^{13}C nuclei however. Thus knowledge of the ratio of the peak areas in a pmr spectrum helps in the identification of the chemical structure responsible for each of the peaks. Because of this, all nmr spectrometers have the ability to integrate the area under each peak and to plot this information as an integration trace which takes the form of a horizontal line drawn from left (high ν, $\delta > 0$ end) to right (low ν, $\delta = 0$ end) rising at each peak to a height proportionate to the area of the peak and which can be superimposed on the usual nmr spectrum; this is shown in Fig. 2 for the two clusters of peaks in the ^{1}H nmr spectrum of 1,1-dichloroethane.

Spin-spin coupling is the name given to the phenomenon that arises from the fact that the magnetic monent generated by the spinning of one nucleus can affect the magnetic field experienced by a nearby nucleus. This effect is transmitted through the electrons of the covalent bonds that connect the atoms of the nuclei in question by the induction of magnetic moments in them. Because these induced magnetic moments are weak the effect does not extend far; for instance ^{1}H nuclei can usually only interact with other ^{1}H nuclei that are up to three bonds away (i.e. H–C–C–H). This interaction is called coupling and is manifested by the splitting of a low resolution nmr absorption peak into a cluster of two or more sharp, equidistant peaks, whose δ value is always given as the δ value at its central point. The distance between adjacent peaks in the cluster is called the coupling constant *(J)*; it is always expressed in Hz (i.e ν rather than δ ppm) and is characteristic of the nuclei involved. Coupling only occurs between nuclei that are in different chemical environments.

The pmr spectrum of 1,1-dichloroethane ($Cl_2.CH.CH_3$; Fig.2) provides a simple example of spin-spin coupling. The spectrum contains two clusters: (i) a doublet with peaks of equal area and a separation of about 7 Hz (i.e. $J = 7$ Hz), centered at δ 1.95, which is due to the three chemically equivalent ^{1}H nuclei of the methyl group, and (ii) a quartet with peak areas in the ratio 1:3:3:1 and peak separations also of about 7 Hz (i.e. $J = 7$ Hz), centered at δ 5.6, which is due to the single methine ^{1}H nucleus. The signal for the methyl ^{1}H nuclei appears as a doublet because of the influence on them of the single methine ^{1}H nucleus. At any given time about half of the methine ^{1}H nuclei in the sample have their spins aligned with the externally-applied magnetic field and half against it. Thus the magnetic field experienced by the ^{1}H nuclei of the methyl group is slightly stronger or weaker than the externally-applied magnetic field. Consequently their absorption of radiofrequency energy occurs at two slightly different (by ~7 Hz) values. The two peaks are of approximately the same area (i.e. intensity) because of the approximate equality of the numbers of two oppositely spinning methine ^{1}H nuclei in the sample. The signal for the methine ^{1}H nucleus is a quartet with a peak area ratio of 1:3:3:1 because of the influence on it of the three ^{1}H nuclei of the methyl group. The latter have four different ways in which they can be oriented with respect to the externally-applied magnetic field, which may be represented as ↑↑↑, ↑↑↓, ↑↓↓ & ↓↓↓ where ↑ and ↓ represent single methyl ^{1}H nuclei with spins aligned with and against the externally-applied magnetic field respectively. The abundance in the sample of these four combinations of alignments occurs in the ratio 1:3:3:1 because both of the center two are effectively the sum of three possibilites, each of which is just as likely to occur as the two outer combinations (↑↑↑ and ↓↓↓), i.e. ↑↑↓ represents ↑↑↓ + ↑↓↑ + ↓↑↑ and ↑↓↓ represents ↑↓↓ + ↓↑↓ + ↓↓↑.

The splitting patterns for ^{1}H nuclei of hydrogen atoms that are on adjacent carbon atoms and in different chemical environments are determined by the number of hydrogen atoms on the neighboring carbon atoms. The number of peaks in the cluster is $N + 1$, where N is the number of neighboring hydrogen

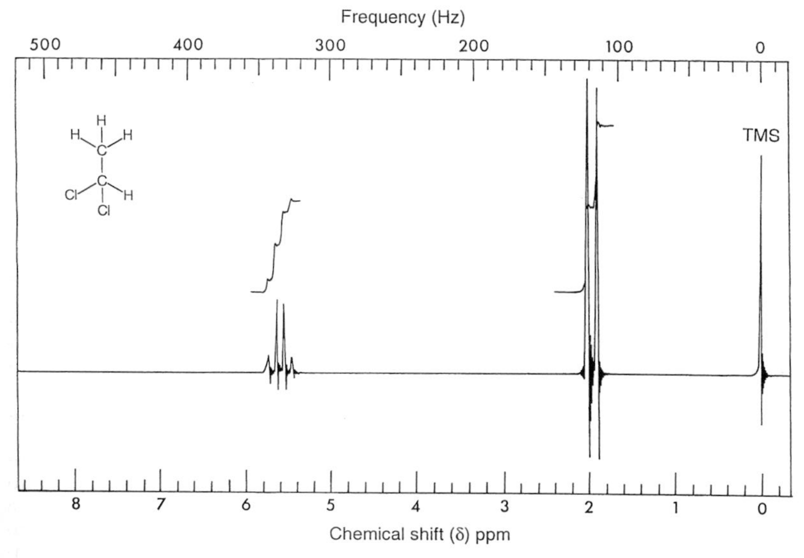

Fig. 2. *Nuclear magnetic resonance spectroscopy.* The ^{1}H nmr spectrum of 1,1-dichloroethane showing spin-spin coupling.

atoms that are in identical chemical environments or that have equivalent coupling constants. This, along with the relative intensities of the peaks in the cluster, is shown in Table 2.

Table 2. *Splitting patterns resulting from spin-spin coupling of N neighboring structurally equivalent ¹H nuclei*

N	No. of peaks ($N+1$)	Relative peak intensities
0	Singlet (1)	1
1	Doublet (2)	1:1
2	Triplet (3)	1:2:1
3	Quartet (4)	1:3:3:1
4	Quintet (5)	1:4:6:4:1
5	Sextet (6)	1:5:10:10:5:1
6	Septet (7)	1:6:15:20:15:6:1

Thus the shape of the peak clusters and their coupling constants, as well as their δ values, give useful information about the chemical environment of ^1H nuclei in the molecules of the sample under investigation and therefore of the chemical structure of the molecules themselves.

The nuclear Overhauser effect (or enhancement) (NOE) was first described in 1953 [A.W. Overhauser *Phys. Rev.*, (1953) **91**, 476] and is given by nuclei that interact through space (not through the electrons of intervening bonds as in spin-spin coupling). The effect is only seen when the nuclei in question are close to one another, generally within 0.2–0.4 nm, because it falls off rapidly as the inverse sixth power of the distance between them. It is revealed in the following way. Having taken a normal nmr spectrum of the sample which has revealed, for example, the presence of two non-chemically equivalent ^1H nuclei (or groups of chemically-equivalent ^1H nuclei), H_A and H_B, a second spectrum is taken in which the sample is irradiated with the resonant frequency of H_A. The integrated forms of the two spectra are then compared. If H_B is close to H_A in space (i.e. within 0.2–0.4 nm) the rise in the integration curve corresponding to H_B will be greater (i.e. enhanced) in the second spectrum. If, however, H_A and H_B are separated by more than ~0.4 nm no such enhancement will be seen. The cause of the enhancement is that the irradiated H_A provides another relaxation pathway for H_B by dipolar interaction. The importance of NOE is that it provides information about the three dimensional structure of the molecule under examination (i.e. its conformation or arrangement in space) by showing which of its parts are close to each other and which are not.

Two dimensional nmr. As the size of the molecules examined by nmr increases so does the complexity of the spectrum, so much so that 'through-bond' (i.e. spin-spin coupling) and 'through-space' (i.e. NOE) interactions become difficult to see. This problem has been markedly lessened in recent years by the development of two dimensional (2D) nmr, the theory of which has been well described by R. Benn and H. Gunther [*Angew. Chem. Int. Ed. Engl.* **22** (1983), 350–380] and A. Bax and L. Lerner [*Science* **232** (1986) 960–967]. The most commonly used 2D nmr technique applied to 'through-bond' interactions is termed (J-) correlated spectroscopy (COSY), others are total correlation spectroscopy (TOCSY), which allows somewhat longer-range 'through-bond' connectivities to be observed than with COSY, spin echo coherence transfer spectroscopy (SECSY), relayed coherence transfer spectroscopy (RELAY), double quantum spectroscopy (DQNMR) and homonuclear Hartmann-Hahn spectroscopy (HOHAHA). The most commonly used 2D nmr technique applied to 'through-space' interactions is termed nuclear Overhauser effect spectroscopy (NOESY); also used is the closely related rotating-frame NOESY (ROESY).

2D nmr data are gathered using a special pulse sequence in an FT instrument which allows the chemical shift (δ ppm) values to be determined on two different frequency axes (usually called F_1 and F_2) which are plotted as a 2D spectrum. This is exemplified in Fig. 3 which shows diagrammatically the COSY nmr spectrum of ethyl 4-methylbenzoate. The spectrum takes the form of a square, the sides of which are the two frequency axes. The chemical shifts of the ^1H nuclei are plotted on both axes, resulting in the generation of small, approximately circular or oval zones (shown by black dots in Fig. 3) which fall either on a diagonal of the square or symmetrically on either side of it. Each of these zones, in fact, represents the contor taken horizontally through an absorption peak (equivalent to the ordinate parameter of the normal one- dimensional pmr spectrum) at that position. The zones that lie on the diagonal correspond to the one-dimensional pmr spectrum whilst the off-diagonal zones (also known as correlation or cross signals) indicate which ^1H nuclei in the molecule are coupled together. To determine whether or not a given ^1H nucleus is coupled to one or more other ^1H nuclei in the molecule a line is projected from its signal (zone) on the diagonal parallel to the horizontal axis. If this line crosses one or more off-diagonal zones, a line from each is projected parallel to the vertical axis. The signal (zone) on the diagonal that each of these lines crosses indicates that the ^1H nucleus corresponding to it is coupled to the first-mentioned ^1H nucleus. In Fig. 3 it is clear that in ethyl 4-methylbenzoate the three chemically equivalent ^1H nuclei of methyl group 'a' are coupled to the two chemically equivalent ^1H nuclei of methylene group 'b' and that the two chemically equivalent pairs of aromatic ^1H nuclei 'c' and 'd' are coupled to each other. The lack of any other off-diagonal signals (zones) also makes it clear that the ^1H nuclei of the methyl group 'e' are not coupled to any other ^1H nuclei in the molecule and that there are no other couplings of 'a', 'b', 'c' or 'd'.

Determination of protein structure by nmr. Proton nmr is the most appropriate tool for this purpose because of the large number of hydrogen atoms present in proteins and the high natural abundance of the ^1H isotope; ^{13}C and ^{15}N are generally of too low an abundance to be of use in this context unless extra atoms of these isotopes have been purposely incorporated into the protein during their biosynthesis. All the ^1H nuclei in a protein can be observed by pmr except those of the hydrogen atoms of $-NH-$, $-NH_2$, $-OH$

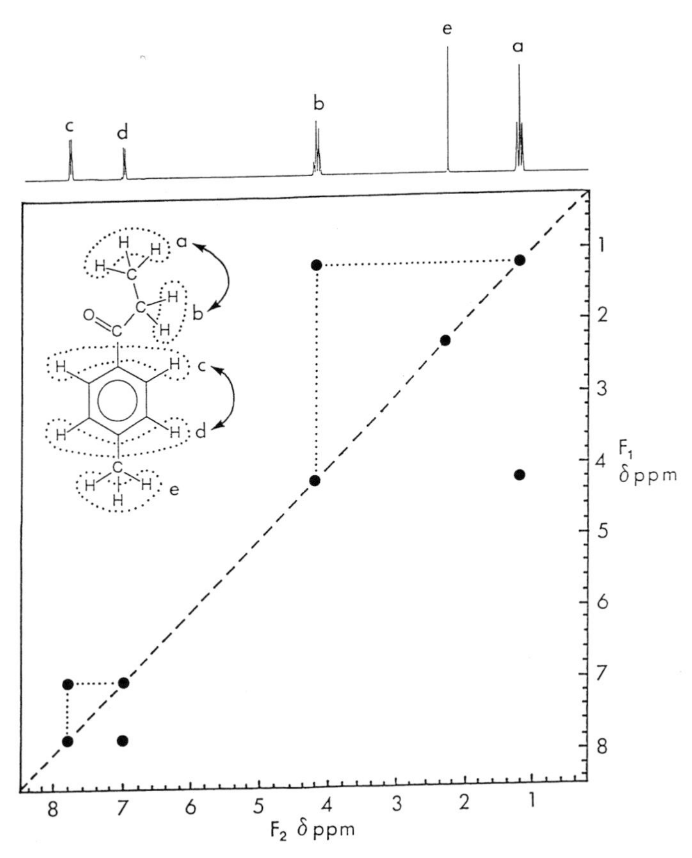

Fig. 3. *Nuclear magnetic resonance spectroscopy.* The COSY ^{1}H nmr spectrum of ethyl 4-methylbenzoate.

and –SH groups which exchange rapidly with the aqueous solvent. This difficulty can be overcome by making the protein solution slightly acidic (pH 4–5), which slows the rate of exchange markedly. Theoretically it is possible to get a unique signal (i.e. chemical shift, δ) for each ^{1}H nucleus in the protein molecule, with the exception of those that are chemically equivalent. However, in practice ^{1}H nuclei are so numerous in the protein that the signals of many of them overlap in the conventional one-dimensional spectrum. This difficulty has largely been solved by the use of two-dimensional pmr spectra, e.g. COSY, TOCSY, NOESY, particularly as a result of the pioneering efforts of K. Wüthrich [e.g. *J. Mol. Biol.* **180** (1984) 715–740; *Science* **243** (1989) 45–50; *Acc. Chem. Res.* **22** (1989) 36–44; *J. Biol. Chem.* **265** (1990) 22059–22062] who introduced the method of 'sequential assignment'.

It is important to appreciate that nmr cannot be used to determine the amino acid sequence (primary structure) of a protein. However, provided that the amino acid sequence is known, 2D nmr techniques can be used to deduce the secondary and tertiary structure of a protein (i.e. the way in which the polypeptide chain is coiled to give α-helical, β-stranded or random coil regions and the way in which these regions themselves are arranged in space). To this extent nmr is similar to X-ray crystallography which also requires that the primary structure of the protein be known in order to determine its 3D-conformation. However, getting this information is not a great problem nowadays; it can be obtained either directly using a protein sequenator if sufficient protein for structural studies is readily available or indirectly from the nucleotide sequence of its gene if it is necessary to use recombinant DNA technology to generate the amount required.

The first step in 'sequential assignment' is the use of COSY and TOCSY to identify the signals in the 2D spectrum of the protein according to the particular species of amino acid giving them; thus one set of signals may be identified as coming from all the alanine resides in the protein and another set from all the valines, etc. This can be done because, using COSY and TOCSY together, most of the amino acid species found in proteins give unique 'through-bond' connectivities (Fig. 4).

The second step is to use the NOESY 2D spectrum of the protein and the known amino acid sequence to decide which signals are due to amino acids that are next to one another in the primary structure. This can be done because certain NOESY cross signals

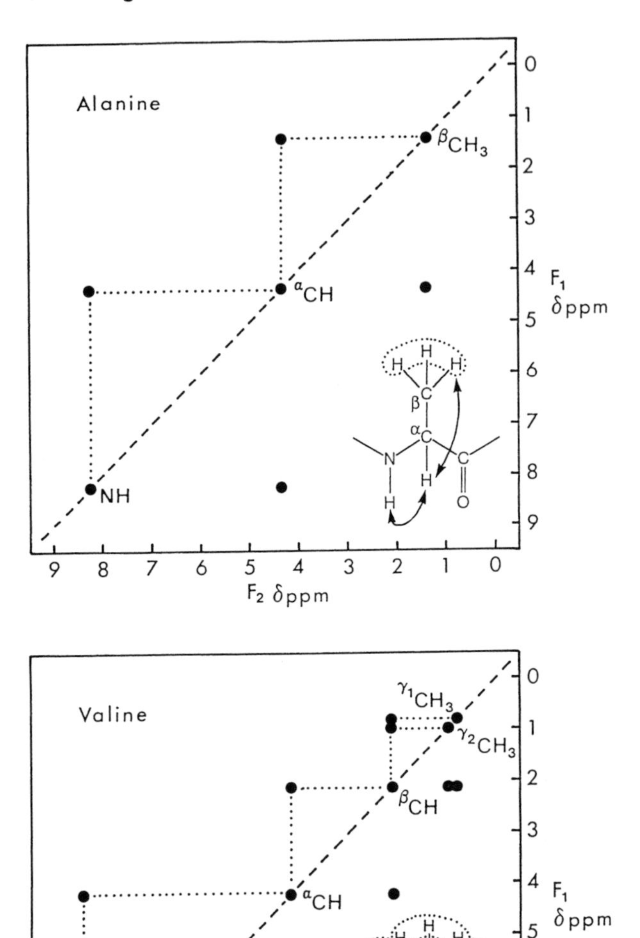

Fig. 4. *Nuclear magnetic resonance spectroscopy.* The COSY ^{1}H nmr spectra of alanine and valine.

come from 'through-space' connectivities between two peptide bond-linked amino acids regardless of the protein backbone conformation. This is shown in Fig. 5 where the 'through-space' distance between the NH ^{1}H nucleus of residue $i+1$ is sure to be ≥0.3 nm from at least one of the NH, $^{\alpha}$C or $^{\beta}$C ^{1}H nuclei of residue i and therefore have a strong NOE connectivity with it.

The third step is to use NOESY connectivities that, by elimination, are known not to arise from amino acids that are next to each other in the primary structure to infer the conformation of the protein. These connectivities fall into to two categories, resulting from (i) secondary structure, and (ii) tertiary structure, 'through-space' ^{1}H nuclei proximities. Stretches of α-helix are indicated by strong NOE connectivities between NH_i & NH_{i+1} and $^{\beta}CH_i$ & NH_{i+1} but not between $^{\alpha}CH_i$ & NH_{i+1}; β-strands, however, are indicated by strong NOE connectivities between $^{\alpha}CH_i$ & NH_{i+1} but not between NH_i & NH_{i+1}. Tertiary struc-

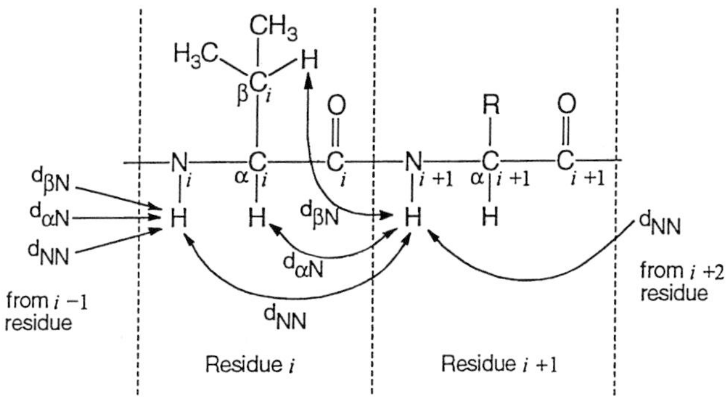

Fig. 5. *Nuclear magnetic resonance spectroscopy.* Nuclear Overhauser effect (NOE) connectivities between ^{1}H nuclei of two peptide bond-linked amino acids in a polypeptide chain that are always present regardless of the conformation of the polypeptide backbone. $d_{\alpha N}$, $d_{\beta N}$ & d_{NN} are the 'through space' distances (≥0.3 nm) between ^{1}H nuclei of the α-C, β-C & peptide bond-N atoms respectively of amino acid residue *i* and that of the peptide bond-N atom of amino acid residue *i* + 1.

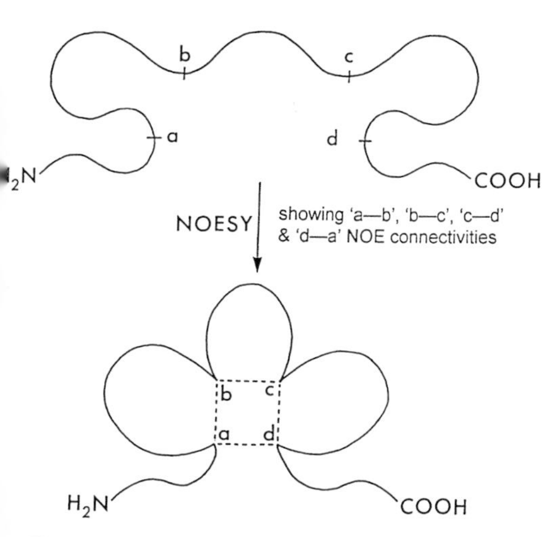

Fig. 6. *Nuclear magnetic resonance spectroscopy.* Foldings of a polypeptide chain that can be detected from Nuclear Overhauser effect (NOE) connectivities between ^{1}H nuclei that are distant in the primary structure and are not part of an α-helix or β-strand.

ture is deduced from NOE connectivities between ^{1}H nuclei of amino acids that are more distant in the primary structure than these. For instance, in an antiparallel β-pleated sheet there are strong connectivities between the ^{1}H nuclei of the $^{\alpha}$C atoms of the amino acid residues of adjacent polypeptide chains as well as between those of the peptide NHs of adjacent chains, while in a parallel β-pleated sheet there are strong connectivities between the ^{1}H nuclei of the $^{\alpha}$C atoms of the amino acid residues of one chain with those of the peptide NHs of the adjacent chain and also between those of $^{\alpha}C_i$ and peptide N_{i+1} atoms in the same chain. Other foldings of the polypeptide chain can be deduced from NOE connectivities between ^{1}H nuclei of amino acids that are distant in the primary structure but are not part of an α-helix or a β-strand (Fig. 6).

The correlation of all the 2D nmr spectral data is carried out by sophisticated computer graphics which ultimately produces a structural conformation, or more frequently a set of closely similar conformations, for the protein that fits the data.

Use of nmr to follow metabolic processes. Some metabolic processes occurring in skeletal muscle, heart and brain can be explored non-invasively by nmr. A classic example of this is the work of G.K. Radda [*Science* **233** (1986) 640–645] showing the effect of exercise on the levels of creatine phosphate (see Creatine), ATP (see Adenosine phosphates) and orthophosphate in human forearm muscle using ^{31}P nmr. In the ^{31}P nmr spectrum of resting muscle (Fig. 7 a) there are five peaks, a large one due to the phosphorus atom in creatine phosphate, a small one due to orthophosphate and three others due to the α, β and γ phosphorus atoms of ATP. Other compounds possessing phosphorus atoms (e.g. ADP) do not contribute sharp peaks to the spectrum because they are either present in too low a concentration or because their ^{31}P nuclei are not free to rotate rapidly, as in nucleic acids and membrane phospholipids. Fig. 7 b shows the ^{31}P nmr spectrum of the same muscle after 19 minutes of exercise. It is apparent that the creatine phosphate and orthophosphate peaks have markedly decreased and increased respectively, whilst those of ATP have remained roughly the same. Since peak area in this instance is an indication of the compound's concentration in the muscle, it is clear that exercise has caused a decrease in creatine phosphate, an increase in orthophosphate and little change in ATP. This is what would be expected from previous research that has shown that whilst the free energy that drives muscle contraction is derived directly from the hydrolysis of ATP to ADP & P_i, it is indirectly derived from the large buffer store of creatine phosphate, which is used to rephosphorylate ADP and thereby restore the ATP level; the reduced creatine phosphate level is restored during the ensu-

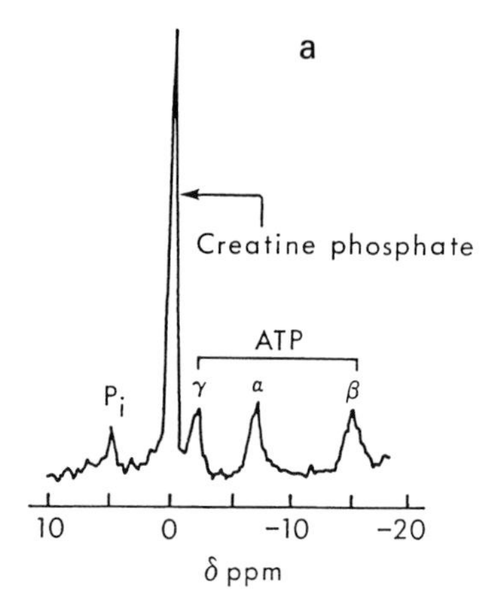

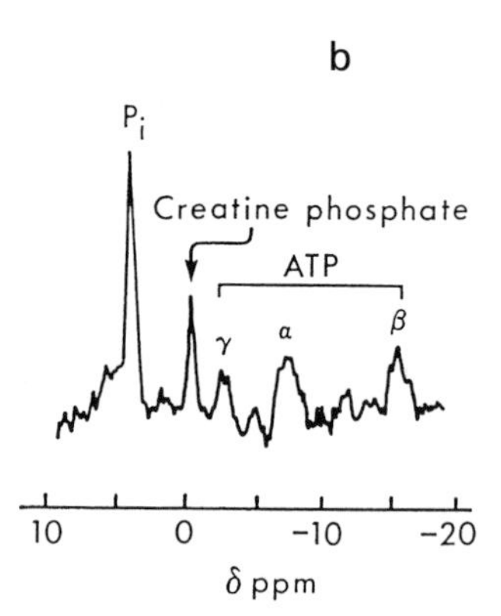

Fig. 7. *Nuclear magnetic resonance spectroscopy.* Use of ^{31}P nmr to show the effect of exercise on the levels of creatine phosphate, ATP and orthophosphate in human forearm muscle; ^{31}P nmr spectra of (a) resting muscle and (b) muscle after 19 minutes of exercise. Adapted from G. K. Radda *Science* 233 (1986) 640–645.

ing period of muscle inactivity. The importance of nmr investigations of this type is that they can be used to reveal abnormalities of metabolism which may be indicative of an underlying pathology.

Use of nmr in body scanning. ^1H nmr is widely used to visualize the different tissues in the body. This is possible because hydrogen atoms, and thus ^1H nuclei, are so numerous in the major components (e.g. water, lipids) of the soft tissues of the body that they give an intense signal, thereby enabling soft tissues to be 'seen' more easily than bone. Moreover, differences in their environments from tissue to tissue produces different signals thereby allowing an appreciable differentiation amongst the soft tissues. Nmr images can be taken from a limb, the head or the whole body in any plane and the information derived from them complements that of the hard tissues obtained by X-ray scanning.

Nucleases: a group of hydrolytic enzymes which cleave nucleic acids. Exonucleases attack the nucleic acid molecule at its terminus, whereas endonucleases are able to catalyse a hydrolytic cleavage within the polynucleotide chain. Deoxyribonucleases (DNAases) are specific for DNA, ribonucleases (RNAases) for RNA. All N. are Phosphodiesterases (see); they catalyse the hydrolysis of either the 3′ or 5′ bond of the 3′,5′-phosphodiester linkage. Ribonuclease (see) has been extensively studied.

The *N*-glycosidic bond of nucleic acids is cleaved by Nucleosidases (see).

Nucleic acid bases: constituent bases of nucleic acids. N.a.b. are fundamental to the storage and transfer of genetic information by nucleic acids. They are Adenine (see), Guanine (see), Cytosine (see), Thymidine (see), Uracil (see) and others that occur less frequently (see Rare nucleic acid components). See also Nucleic acids, Genetic code, Base pairing.

Nucleic acids: polymerized nucleotides found in all cells and viruses. N.a. were first isolated in 1869 from the white blood cells of pus by Miescher, who called the material nuclein. The term N.a. was introduced in 1889 by Altman, in recognition of their acidic properties. There are two main classes of N.a., distinguished by their carbohydrate component: *Ribonucleic acid* (RNA) (see), which contains ribose, and *Deoxyribonucleic acid* (DNA) (see) which contains 2-deoxy-D-ribose. Both types have certain common structural features, but they have different biological functions. DNA stores genetic information; it is replicated during cell division, so that each daughter cell receives DNA that is identical in structure and informational content (see Genetic code). RNA is intimately involved in protein synthesis, and is primarily responsible for translating the information of the DNA into the primary structure of specific proteins (see Protein biosynthesis).

N.a. are polymers with M_r between 20,000 and about 10^9. They contain three structural components: the purine and pyrimidine bases, a pentose (either D-ribose in RNA or 2-deoxy-D-ribose in DNA) and esterified phosphate. The five major bases are uracil (RNA only), thymine (DNA only), adenine, guanine and cytosine (RNA and DNA); in addition, there are over 30 Rare nucleic acid components (see) which occur in various N.a. These rare components are formed by modification of existing structures within the N.a., e.g. by methylation, hydrogenation or rearrangement of normal bases.

Each mononucleotide unit is linked to its neighbor by a phosphate group, which forms an ester linkage with position 3′ of one sugar and 5′ of the neighboring sugar. This 3′,5′-linkage results in a linear chain of phosphate-linked sugar residues; a base is attached to C1 of each sugar by an *N*-glycosidic linkage (Fig.). The linear order of the bases is statistically irregular, representing the informational code of the N.a.

The reactive $-NH_2$, -OH and -NH groups of purine and pyrimidine bases are responsible for certain properties of N.a., e.g. formation of specific hydrogen bonds between purines and pyrimidines, leading to secondary structures. Thus complementary linear chains can form a double helix (see DNA), or a linear strand can fold on itself, forming alternate linear and helical regions (RNA). Other forces involved in the spatial conformation of N. a. (see Noncovalent bonds) are homopolar cohesive forces (Van der Waals forces), hydrophobic interactions between bases and solvent, and electrostatic interactions (ionic bonds).

Physical properties and analytical methods. The conjugated double bonds in the heterocyclic rings of the bases absorb UV light in the region of 260 nm; UV spectrophotometric analysis is therefore used for characterization and quantitative determination of N. a. and lower M_r related compounds, such as mono- and oligonucleotides. The intensity of UV absorption depends on the conformation, i. e. the secondary structure of the N. a. Thus optical methods can also be used in the structural elucidation of N. a. and for detecting and monitoring structural changes. Denaturation results in an increase of absorption at 260 nm (see Hyperchromic effect). From the course of the heat denaturation curve, called the melting point curve, it is possible to assess the helical content of a N. a. and to determine the GC content of a DNA sample (see T_m value). If the heat-denatured N. a. is cooled slowly, the original structure is largely reformed (renaturation). Bacterial DNA renatures more extensively than nuclear DNA from higher organisms. Heat-denatured RNA can also be largely renatured to its original form.

Other physical methods used in the investigation of N. a. structure are Hybridization (see), electron microscopy, analytical ultracentrifugation, CsCl density gradient centrifugation, X-ray diffraction analysis, infrared spectroscopy, optical rotatory dispersion, light scattering photometry and viscosity measurements.

N.a. are purified by column chromatography (e. g. methylalbumin-silicic acid columns), by electrophoresis (e. g. polyacrylamide gels) and by density gradient centrifugation (e. g. sucrose or CsCl). A N. a. may be determined quantitatively from the UV absorption of its bases, by determination of the phosphate content, or by specific color reactions for ribose or deoxyribose (Dische reagent for DNA, Dische-Schwarz reagent for RNA). The Fuelgen reaction is used for histochemical detection of DNA.

Nucleic acid sequencing: determination of the order of nucleotides in a nucleic acid or a polynucleotide segment, i. e. determination of the primary structure. Both RNA and DNA can be sequenced, but most contemporary analyses are confined to the latter. However, the first biologically significant nucleic acids to be sequenced were small RNA species, transfer RNAs (the first being the 76-nucleotide alanine tRNA from yeast completed by Holley in 1965) and 5S ribosomal RNAs. The procedure involved the initial cleavage of the RNA into quite short fragments by partial digestion with endonucleases, e. g. ribonuclease T_1 from *Aspergillus oryzae* (cleaves 3′,5′-phosphodiester linkages after guanine(G)-containing ribonucleotide residues, producing oligonucleotides terminating in Gp) and pancreatic ribonuclease (cleaves 3′,5′-phosphodiester linkages after pyrimidine-containing nucleotide residues, producing oligonucleotides terminating in Cp or Up). These fragments were separated then sequenced by partial digestion with one of two exonucleases, namely snake venom diesterase, which successively cleaves the terminal nucleotide residue from the 3′ end yielding 5′-mononucleotides and a mixture of n–1, n–2 etc. oligonucleotides, and spleen phosphodiesterase, which successively cleaves the terminal nucleotide residue from the 5′-end yielding 3′-mononucleotides and a mixture of n–1, n–2 etc. oligonucleotides. The oligonucleotide mixtures resulting from exonuclease action were then separated by chromatography or electrophoresis and the base composition of each component determined. Comparison of the base composition of pairs of components that differed in length by one nucleotide (deduced from their relative chromatographic or electrophoretic mobilities) then established the identity of the 3′- or 5′-terminal nucleotide

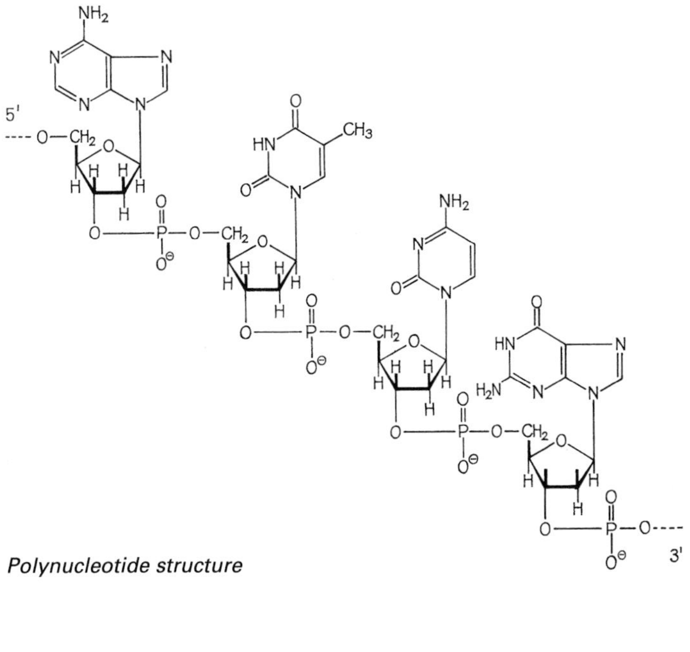

Polynucleotide structure

(depending upon the exonuclease used) of the larger component. When this had been done for all the possible oligonucleotide pairs, the nucleotide sequence of the RNA fragment from which they were derived was apparent. When this information was available for all the endonuclease-derived RNA fragments, the nucleotide sequence of the original RNA species could be determined by looking for overlapping sequences amongst them. The sequencing of RNA using this technique is painstaking and time consuming, but its use nevertheless enabled Fiers in 1976 to publish the sequence of the 3,569 nucleotide genome of bacteriophage MS2.

In 1977 two methods for sequencing DNA were published which, when allied to the increasing availability of Restriction endonucleases (see) (which enable duplex DNA to be cleaved at specific nucleotide sequences) and the development of molecular cloning (see Recombinant DNA technology) (which allows the faithful amplification of the quantity of any identifiable DNA segment), opened the way to DNA sequencing and enabled it to be carried out far more efficiently and rapidly than that of RNA as described above. They are known as the Chemical Cleavage Method [A.M.Maxam and W.Gilbert, *Proc. Natl. Acad. Sci. USA.* **74** (1977) 560–564; *Methods Enzymol.* **65** (1980) 499–560] and the Chain Termination Method (or Dideoxy Method) [F.Sanger et al. *Proc. Natl. Acad. Sci. U.S.A.* **74** (1977) 5463–5467].

The Chemical Cleavage Method is carried out on a single-stranded fragment of DNA which has been radiolabeled at one end, often the 5′-end. This may be accomplished by incubating the DNA segment with [γ-^{32}P]ATP in the presence of polynucleotide 5′-hydroxyl-kinase which results in addition of [^{32}P]phosphate to the 5′-hydroxyl group at the 5′-end; if the latter already carries a phosphate group, as in native DNA, it must first be removed by incubating the DNA with *E. coli-* or calf intestinal alkaline phosphatase. Aliquots of the labeled DNA preparation are then subjected to at least four different chemical procedures each of which is designed to cleave the DNA by removing, in effect, a specific nucleoside (i.e. a N-base and its *N*-glycosidically-linked 2-deoxy-D-ribose) or nucleoside type from the polynucleotide chain. The conditions under which each of these cleavage procedures is carried out are carefully chosen to insure that each labeled DNA molecule in the DNA sample is cleaved at an average of just one randomly located, susceptible nucleoside. Thus, if the DNA in question is labeled at the 5′-end with ^{32}P then cleaved with a procedure that specifically removes guanosine residues, each DNA molecule is typically cleaved at one position, namely where a G-containing nucleotide is located, so generating two cleavage products: one with a [^{32}P]phosphate at its 5′-end and unlabeled phosphate at its 3′-end, and one with an unlabeled phosphate at its 5′-end and either a hydroxyl or an unlabeled phosphate group at its 3′-end. Of these two products only the labeled one will be detected by subsequent autoradiography. For example, if this procedure were applied to $^{(5')}$CGATGGCAGTCT$^{(3')}$ the ^{32}P-CGATGGCAG-TCT molecules formed would be cleaved so as to produce the following set of 5′-labeled fragments: ^{32}P-C, ^{32}P-CGAT, ^{32}P-CGATG, ^{32}P-CGATGGCA. This set of fragments is then subjected to electrophoresis in one lane of a polyacrylamide sequencing gel, with the analogous sets of fragments from the three or more other cleavage procedures in parallel lanes alongside. This process separates the fragments in each lane by size and is capable of separating fragments that differ in length by just one nucleotide residue; the gel contains ~ 8 M urea and is electrophoresed at ~ 70 °C to suppress hydrogen bonding interactions. The positions of the separated fragments in each lane are then determined by autoradiography, and from the pattern created by the darkened bands of the parallel lanes on the resulting autoradiogram almost all the nucleotide sequence of the DNA under investigation can be deduced, as explained below.

Four of the cleavage procedures commonly used in the Chemical Cleavage Method cleave the radiolabeled DNA by removing from it the following *N*-bases or types of *N*-bases along with their attendant 2-deoxy-D-ribose residues: (i) guanine (G), (ii) purines (G & A), (iii) cytosine (C), and (iv) pyrimidines (C & T). Other cleavage procedures have also been used and have been reviewed by B.J.B.Ambrose and R.C.Pless [*Methods Enzymol.* **152** (1987) 522–538]. Cleavage at G involves treatment of the labeled DNA with dimethyl sulfate in aqueous pH 8 buffer which methylates the N-7 of guanine but not that of adenine, thereby rendering the C8-C9 bond susceptible to base-catalysed cleavage and the 2-deoxy-D-ribose residue susceptible to removal with piperidine as shown in Fig.1. Cleavage at G & A makes use of the fact that acidic conditions (pH 2) protonate the purine N atoms of the labeled DNA thereby weakening the glycosidic bond of G and A and opening the way to their removal and that of their 2-deoxy-D-ribose residues by piperidine treatment. Cleavage at C & T involves treatment of the labeled DNA with aqueous hydrazine which opens the pyrimidine rings by conjugate nucleophilic addition to their carbonyl group; treatment with piperidine then partially removes C & T as a 5-membered *N*-heterocycle and subsequently removes their ring-remainders as urea, along with their attendant 2-deoxy-D-ribose residues (Fig.2). Cleavage at C alone involves a procedure identical to that for cleavage at C & T save that hydrazine is used in the presence of 1.5 M NaCl to prevent ring-opening of T.

Figure 3 shows the set of labeled fragments generated from the DNA mentioned above, i.e. $^{(5')}$CGATGGCAGTCT$^{(3')}$, by each of the four cleavage methods, and their separation by electrophoresis in parallel lanes of a sequencing gel. The uncleaved, labeled DNA, which is 12 nucleotides long, runs the least distance and appears as a dark band at the same position in all four lanes. Moreover a dark band corresponding to fragments with 11, 10, ... 2 and 1 nucleotides occurs in at least one or other of the lanes. For instance the fragments with 10 and 6 nucleotides occur in the C cleavage lane and indicate that there must have been cytosine-containing nucleotides at positions 11 and 7 in the original DNA, which have been removed by the C cleavage procedure. These two fragments obviously also occur in the C & T cleavage lane but are accompanied by fragments with 11, 9 and 3 nucleotides which must have resulted specifically from T cleavage, since they do

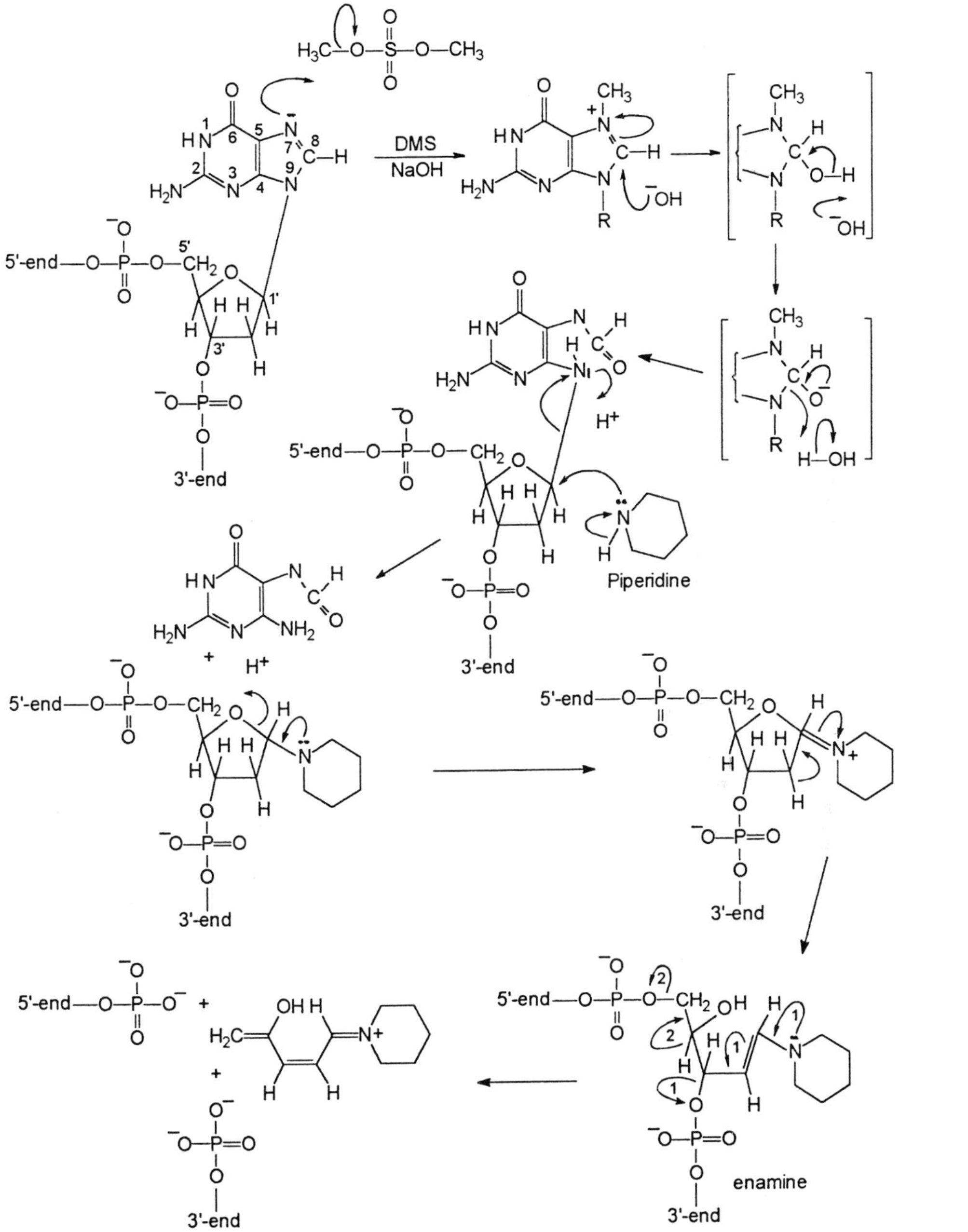

Fig. 1. *Nucleic acid sequencing.* Postulated mechanism of the reactions involved in the cleavage of DNA by removal of a 2-deoxyguanosine residue. Two fragments are produced, one from the 5′-side, the other from the 3′-side. NB: the sequence of electron shifts labeled "1" in the enamine occur prior to and distinct from that labeled "2". DMS = dimethyl sulfate.

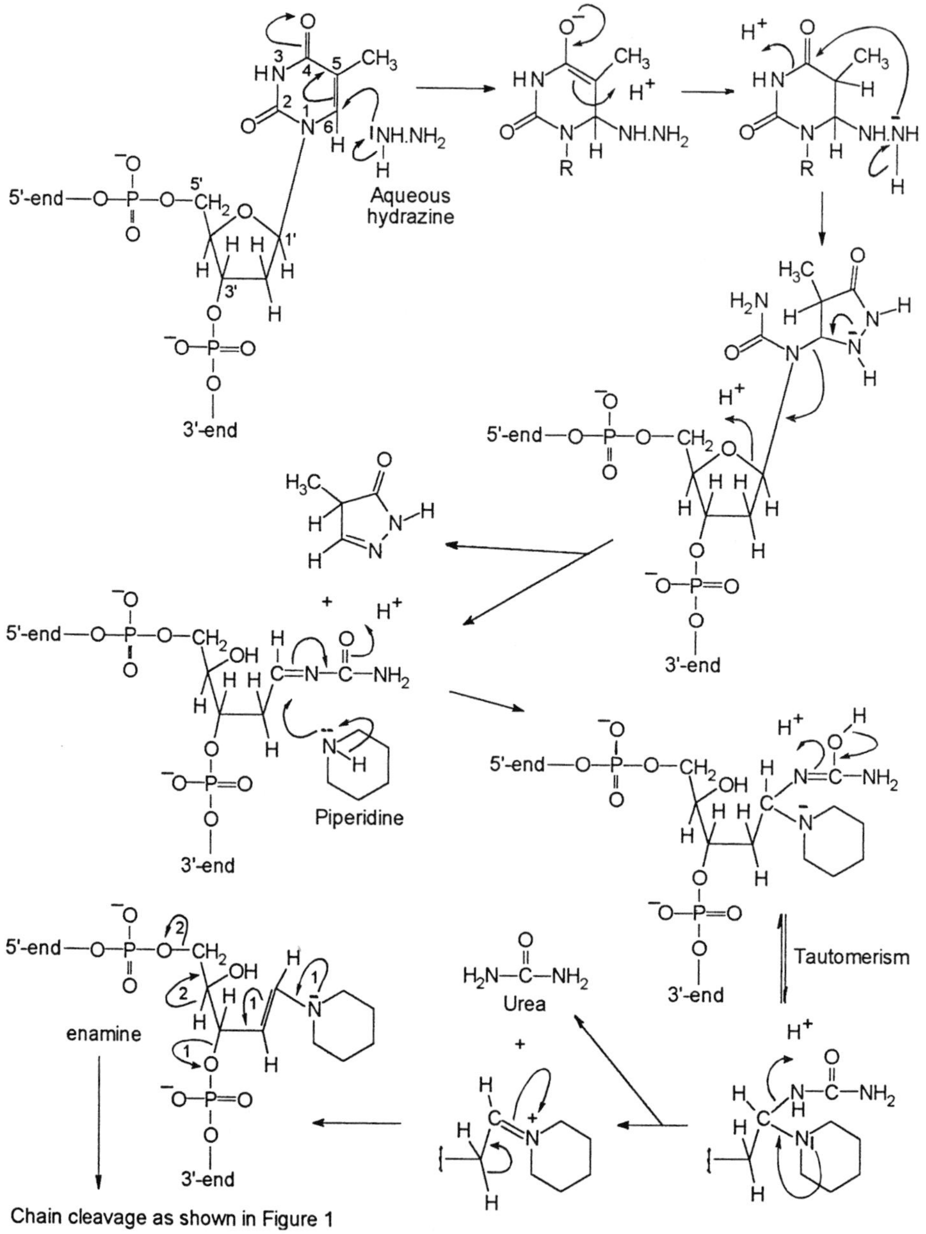

Fig. 2. *Nucleic acid sequencing.* Postulated mechanism of the reactions involved in the cleavage of DNA by removal of a 2-deoxythymidine residue. Two fragments are produced, one from the 5′-side, the other from the 3′-side. NB: cleavage of the enamine produced by the reaction sequence shown is mechanistically identical to that shown in Fig. 1.

not occur in the C cleavage lane; this indicates that there must have been thymine-containing nucleotides at positions 12, 10 and 4 in the original DNA, which have been removed by the C & T cleavage procedure. Application of analogous reasoning to the bands seen in the G and G & A cleavage lanes indicate that guanine- and adenine-containing nucleotides occur at positions 9, 6, 5 & 2 and at 8 & 3, respectively. Notice that because each of the cleavage methods removes particular nucleotide species from the DNA there

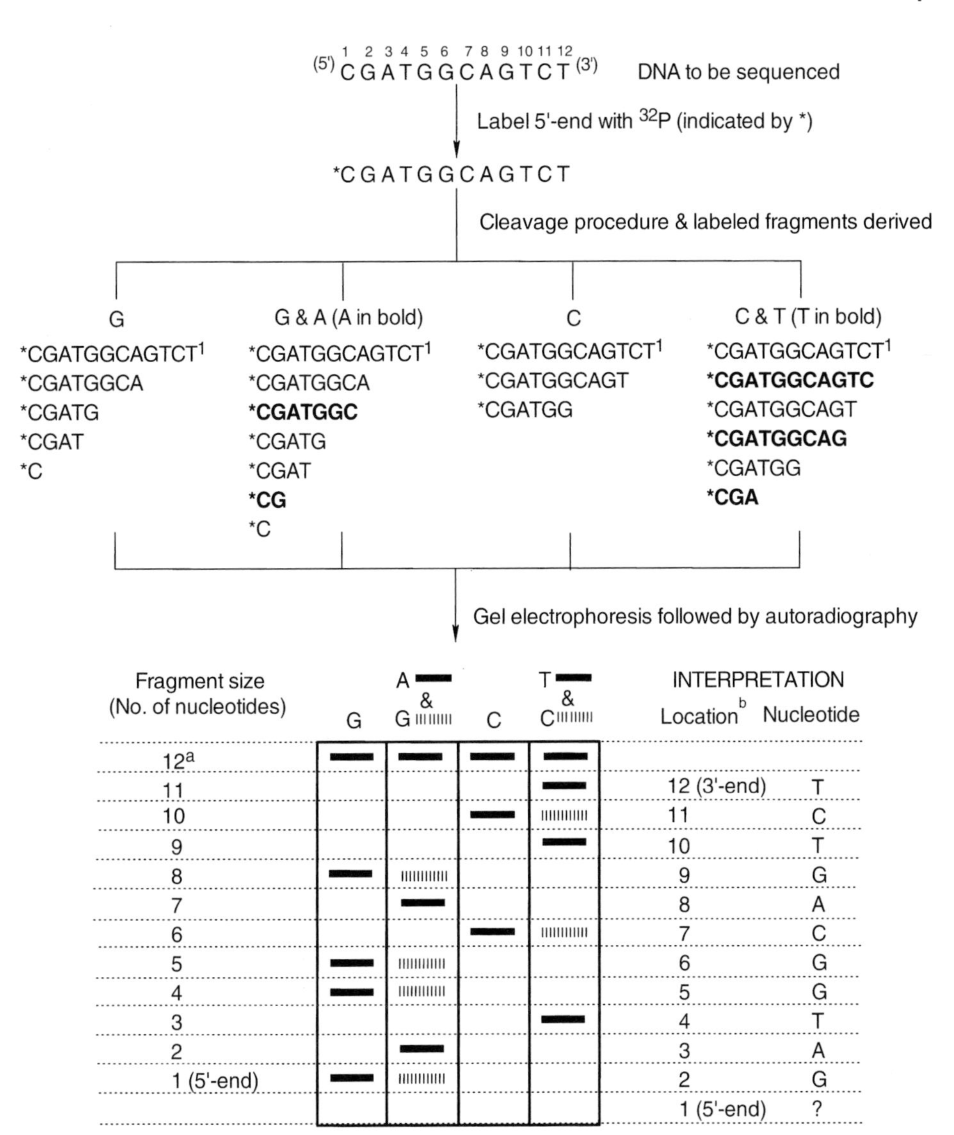

Fig. 3. *Nucleic acid sequencing.* Scheme showing the chemical cleavage method of Maxam and Gilbert applied to the determination of the nucleotide sequence in a hypothetical length of DNA. a = original DNA, i.e. the uncleaved but ^{32}P-labeled DNA being sequenced. b = location of each nucleotide in the DNA being sequenced, numbering from the 5′-end.

cannot be a fragment corresponding to the 5′-nucleotide, so that the Chemical Cleavage Method does not identify the nucleotide at the 5′-end of the DNA. In practice the mononucleotide identifying the second nucleotide from the 5′-end (*C in Fig. 3) is frequently not detected on the gel. This is of no great importance, because the identities of these two nucleotides can be found by sequencing the complementary strand of the DNA, an operation which also confirms the sequence of the original strand. Although Fig. 3 shows the sequencing of only a short DNA sequence, the Chemical Cleavage Method can be successfully used to sequence DNAs of up to 250 nucleotides.

For a number of years, the Chemical Cleavage Method was more extensively used than Sanger's Chain Termination Method, because it did not require such specialized reagents, e.g. specific oligonucleotide primers and appropriate DNA polymerases. The latter method, however, is now the one of choice for most workers, largely because the specialized reagents are now readily available and the procedure has been automated.

The Chain Termination Method uses a DNA polymerase to catalyse the synthesis of complementary copies of the single strand of DNA that is to be sequenced, and it is the nucleotide sequence of the complementary copy that is directly determined; that of the original DNA is deduced from base pairing. Thus the DNA to be sequenced is used as a template for the generation of complementary copies. The template is incorporated, as an insert, at a specific point into a longer piece of DNA whose nucleotide sequence is already known; the latter is derived from appropriate phage or plasmid DNA; template insertion involves the use of restriction endonucleases and standard molecular biological procedures. To initiate the replication process DNA polymerase requires that a primer, a short segment of DNA, be stably base-paired to a complementary nucleotide sequence in the phage/plasmid DNA, starting at the nucleotide on the 3′-side of the 3′-end of the template insert and extending away from it. Primers with a nucleotide sequence enabling them to bind with such precision can be synthesized because the nucleotide sequences of the phage/plasmid DNAs used in this work are known, as are the sites within them where template DNA may be inserted; such primers are about 20 nucleotides long and are now available commercially; the Chain Termination procedure requires that the appropriate primer be annealed to the phage/plasmid DNA containing the template insert. Because of the antiparallel pairing of complementary DNA strands, the nucleotide at the 5′-end of the primer is base-paired with the nucleotide on the 3′-side of the template insert. When the template/primer duplex is incubated with the four deoxyribonucleoside 5′-triphosphates (dATP, dGTP, dCTP, dTTP) in the presence of the appropriate DNA polymerase, the primer is progressively elongated by the successive addition, to the terminal 3′-hydroxyl group, of deoxyribonucleoside 5′-monophosphate residues in a sequence governed by that of the template and the rules of base-pairing. In this process, which proceeds in the $5' \rightarrow 3'$ direction, each new phosphodiester linkage is formed by condensation of the α-phosphate residue on the 5′-hydroxyl group of the incoming deoxyribonucleotide with the 3′-hydroxyl group of the terminal deoxyribonucleotide of the elongating chain. Elongation is therefore absolutely dependent on the presence of a 3′-hydroxyl group on the terminal deoxyribonucleotide of the elongating chain.

When the Chain Termination Method was first introduced, four incubations were carried out, each containing the template/primer duplex, the appropriate DNA polymerase (at that time the Klenow fragment of *E. coli* DNA polymerase I), equal quantities of the four deoxyribonucleoside 5′-triphosphates, at least one of which (usually dATP) was labeled with ^{32}P in the α-phosphate residue, and a small quantity of the 2′,3′-dideoxy derivative of one of the four deoxyribonucleoside 5′-triphosphate species (i.e. one out of ddATP, ddGTP, ddCTP and ddTTP). In each incubation, whenever a 2′,3′-dideoxy analog was incorporated into the growing polynucleotide chain in the place of the corresponding 2′-deoxyribonucleotide, chain elongation ceased because the new terminal nucleotide lacked a 3′-hydroxyl group. The small quantity of 2′,3′-dideoxyribonucleotide present in each incubation mixture, relative to that of each of the 2′-deoxyribonucleotides, insured that a set of ^{32}P-labeled chains of different length were generated, each component of which had a 3′-terminal 2′,3′-dideoxyribonucleotide which corresponded to the position of its base-paired partner in the template DNA. Thus if ddGTP were used and the template DNA had the sequence $^{(3')}$TCTGACGGTAGC$^{(5')}$ the set of ^{32}P-labeled chains released from the enzyme would contain the following: Primer-$^{(5')}$AddG$^{(3')}$, Primer-$^{(5')}$AGACTddG$^{(3')}$ and Primer-$^{(5')}$AGACTGCCATCddG$^{(3')}$. This and the three other sets of chains generated by using the other 2′,3′-dideoxyribonucleotides were then run in parallel lanes on a polyacrylamide sequencing gel which separates their components according to their number of nucleotide residues. The positions of the separated components in each lane were then determined by autoradiography. From the pattern created by the darkened bands in the parallel lanes on the resulting autoradiogram the nucleotide sequence of the DNA strand complementary to the template DNA was deduced in much the same way as described for the Chemical Cleavage Method. Fig. 4 shows the application of this procedure to the template DNA mentioned above, i.e. $^{(3')}$TCTGACGGTAGC$^{(5')}$.

In the succeeding years several variations and improvements have been made to this methodology but the basic concept remains the same. These fall into five general areas, namely: (i) improvements in the applicability and ease of use of the cloning vectors into which the DNA to be sequenced is inserted, (ii) improvements in the method of labeling the sets of 2′,3′-dideoxyribonucleotide-terminated chains that are generated, (iii) use of analogs of 2′-deoxyribonucleotide 5′-triphosphates to eliminate the so-called "compression" of bands on the sequencing gel, (iv) improvements in the DNA polymerase used, and (v) ability to sequence double-stranded DNA templates as well as single-stranded ones.

Improvement (i) came with the development of phage (e.g. M13 and its constructs such as M13mp19), plasmid (e.g. the pUC series) and phagemid (e.g. pBluescript® II KS +/–) DNAs with high copy numbers, resistance to specific antibiotics, multiple restriction endonuclease cleavage sites and known coding sequences into which the DNA to be sequenced could be inserted with precision.

Improvement (ii) came initially with the use of [^{35}S]dATPαS instead of [^{32}P]dATP in the incubation mixtures; this compound has a ^{35}S atom attached by a double bond to the P atom of the α-phosphate residue of the 2′-deoxyribonucleotide instead of the double-bonded O atom present in dATP. Thus, when a dAMP residue is incorporated by DNA polymerase into the growing DNA chain the latter becomes labeled with ^{35}S because an atypical phosphodiester linkage is formed, namely R-O-P$^{(35}$S)(OH)-o-R′. Since the β-particles emitted by ^{35}S have about 10 times less energy than those of ^{32}P, there is less radiological damage to the labeled DNA chains and the bands on the autoradiograms of the sequencing gels are smaller and less diffuse.

An alternative, more recent, improvement in labeling has been the use of dyes which fluoresce at different wavelegths. These dyes may be used in two differ-

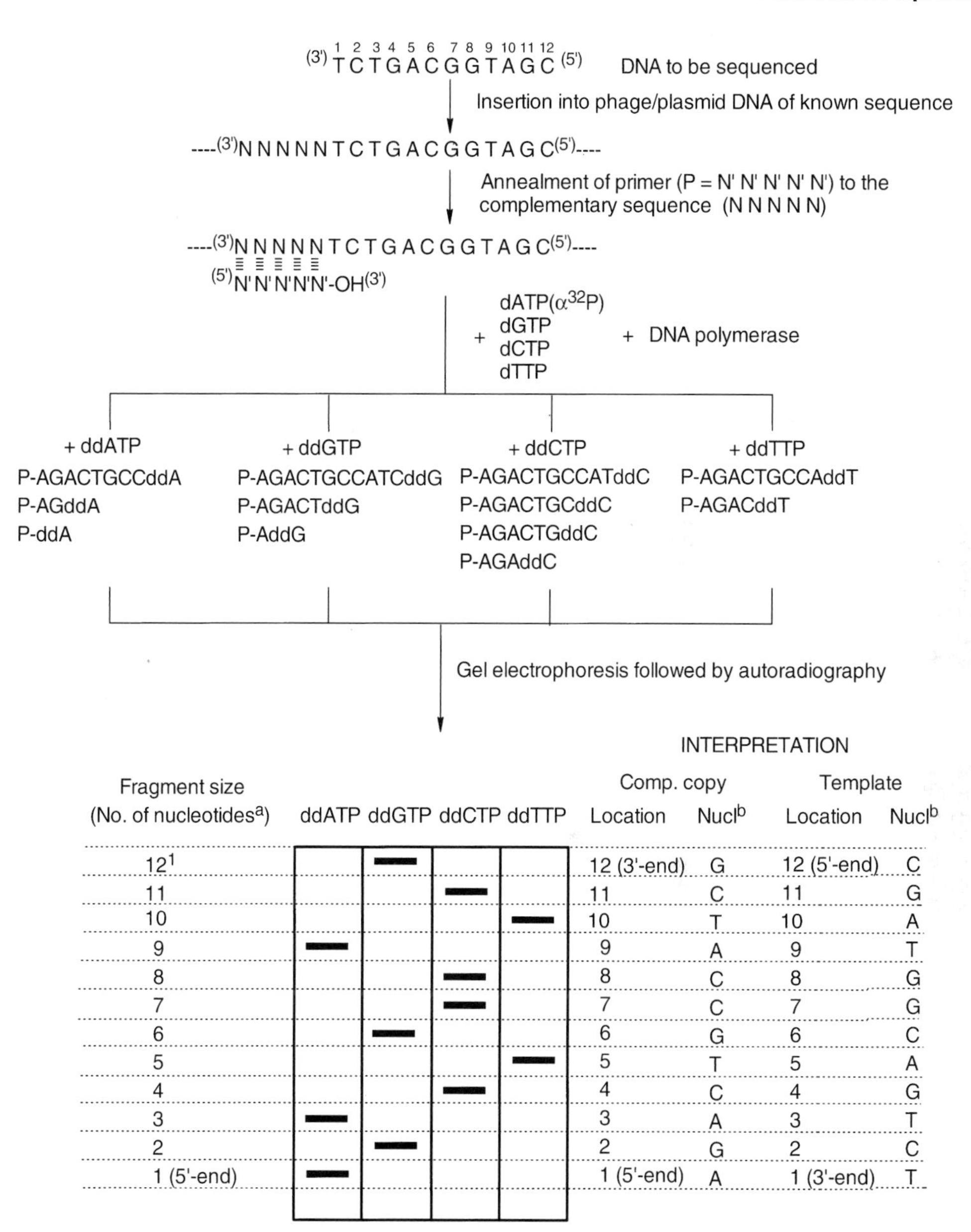

Fig. 4. *Nucleic acid sequencing.* Scheme showing the chain termination method of Sanger applied to the determination of the nucleotide sequence of a hypothetical length of DNA. a = the number shown in the column below is the number of nucleotides in the DNA chain that is complementary to the original DNA; it takes no account of the constant number of nucleotides in the primer, P, attached at the 5′-end. b = nucleotide. Comp = complementary.

ent ways, namely: (a) to label the primer, by covalently bonding a dye molecule to its 5′ end, and (b) to label the 2′,3′-dideoxyribonucleotides by covalently bonding a different dye molecule to each. The use of these dye-labeled materials gives rise to two slightly different DNA sequencing procedures known as 'dye primer labeling' and 'dye terminator labeling'. Both of these procedures have now been automated.

In 'dye primer labeling' four differently fluorescing dye-primer species, synthesized by reacting each of four samples of the same primer with one of four differently fluorescing dyes, are used; these will be called the blue, green, red and yellow primers, although the color difference of their specrophotometrically distinct fluorescence is not as marked as these names imply. Four incubations are carried out, each containing a differently 'colored' primer annealed to the DNA to be sequenced, the DNA polymerase, the four different dNTPs and a small amount of one of the ddNTPs, such that each ddNTP is arbitrarily assigned to a differently 'colored' primer (e.g. ddGTP with the red primer etc.). Each incubation therefore generates a set of DNA chains of different length, each component of which will be labeled at its 5′-end with a primer of a particular color and have at its 3′-end the 2′,3′-dideoxynucleotide assigned to that primer; thus if ddGTP were assigned to the red primer and the DNA under investigation had the sequence $^{(3')}$TCTGACGGTAGC$^{(5')}$ the set of primer-labeled chains released from the enzyme would contain the following: red primer-$^{(5')}$AddG$^{(3')}$, red primer-$^{(5')}$AGACTddG$^{(3')}$ and red primer-$^{(5')}$AGACTGCCATCddG$^{(3')}$. The four sets of primer-labeled DNA chains are pooled and separated electrophoretically, according to length (i.e. nucleotide number), in a single lane of a polyacrylamide sequencing gel. The nucleotide sequence is then obtained by determining the sequence of 'colors' (i.e. dye fluorescences) of the bands in the electrophoretogram as the lane is passed through a spectrofluorimetric detector.

In 'dye terminator labeling' four differently-fluorescing dye-ddNTP species, synthesized by reacting ddATP, ddGTP, ddCTP and ddTTP each with a different dye, are incubated, in small amounts, in the same reaction vessel with the primer annealed to the DNA to be sequenced, the four dNTPs and the DNA polymerase. This generates a mixture of four sets of DNA chains, the components of each of which have the primer at their 5′-end and a differently-fluorescing 2′,3′-dideoxyribonucleotide at their 3′-end. This mixture is then processed and interpreted in the same way as that of the dye primer labeling method.

Improvement (iii) can, if necessary, be applied to any of the variants of the Chain Termination methodology. Its purpose is to overcome the phenomenon of 'compression', in which adjacent bands on the sequencing gel are so close to each other (i.e. compressed) that it becomes difficult to read off the nucleotide sequence; this difficulty is particularly apparent when four parallel lanes are being used. Compression arises when the DNA chains generated during incubation have dyad symmetry and form intra-strand, H-bonded, secondary structures that are not fully denatured during gel electrophoresis, in spite of the precautions taken to prevent this occurring. This most commonly happens when the DNA chains have a high GC content because G binds to C more tightly than A to T (due to the involvement of three H-bonds rather than two). The DNA strands with intra-strand secondary structure migrate distances that are different from those of fully denatured strands of identical length during gel electrophoresis, thereby causing the reading problems described. This difficulty can often be alleviated by replacing dGTP in the incubation mixtures with 2′-deoxyinosine 5′-triphosphate (dITP) or 7-deaza-2′-deoxyguanosine 5′-triphosphate (7-deaza-dGTP). This results in the incorporation of dIMP or 7-deaza-dGMP residues into the generated DNA chains wherever dGMP residues would normally have been. Because neither of the N-bases of dIMP or 7-deazadGMP H-bond as tightly to cytosine as does guanine, the formation of DNA chains with intra-strand, secondary structures that are resistant to the denaturing conditions of the sequencing gel is much less likely, and compression of bands is decreased. Hypoxanthine, the N-base of inosine, does not H-bond to cytosine so well as guanine, because its structure, which is 'guanine minus the 2-amino group', allows it to form only two H-bonds instead of three; similarly the replacement of the 7-N atom of guanine by carbon alters the electronic configuration of 7-deazaguanine sufficiently to prevent its forming three H-bonds with cytosine.

Improvement (iv) has been in the DNA polymerases used to catalyse the synthesis of the DNA chains. Initially the Klenow fragment (see) of *E. coli* DNA polymerase I was used but this has several disadvantages, namely: (a) low processivity (processivity is the ability of the enzyme to continue to catalyse polynucleotide synthesis without dissociating from the template after the addition of each nucleotide), (b) weak acceptance of ddNTPs, and (c) intolerance of dNTP analogs. Consequently it has now been largely replaced by DNA polymerases from thermophilic bacteria (e.g. Taq DNA polymerase from *Thermus aquaticus*) and from bacteriophage T7. Taq DNA polymerase has high processivity, is tolerant of ddNTPs and dNTP analogs and is particularly useful in determining the sequence of DNA whose complementary DNA chains form extensive, intra-strand, secondary structures at 37°C (since the enzyme is from a thermophilic organism, it can be used at incubation temperatures of 70–75°C at which H-bonded secondary structures are precluded). The most successful T7 DNA polymerases are those that have been modified to reduce (e.g. sequenase™ 1) or even eliminate (e.g. sequenase™ 2) the enzyme's 3′ → 5′-exonuclease activity; they have high processivity, a high tolerance of ddNTPs and dNTP analogs, a high elongation rate (~300 nucleotides added per second; cf. Klenow fragment 30–45 and Taq DNA polymerase 35–100) and can use short primers. Taq DNA polymerase and the sequenases can often be used to determine a DNA sequence several hundred nucleotides long from a single set of reactions.

Improvement (v) is the introduction of methods for sequencing double-stranded DNA templates. The double-stranded DNA to be sequenced is incorporated into plasmid or phage DNA and amplified using standard molecular biological techniques. The circular, double-stranded plasmid or phage DNA, containing the double-stranded template insert, is isolated and then denatured by heating in the presence of NaOH followed by rapid cooling, a procedure that ruptures the H-bonds holding the strands together. A primer with a nucleotide sequence complementary to the sequence on the 3′-side of the template insert in one of the two strands is then annealed to it. The resulting primer/template duplex is then sequenced in the usual manner. The sequence of the other strand of the DNA template can be deduced from base-pairing rules or can be sequenced directly, providing a primer for the sequence on the 3′-side of the template in that strand is available. This technique requires longer primers (25–30 nucleotides long) than are used for the sequencing of single-stranded template DNA, and can reliably sequence 200 nucleotide-long stretches of DNA.

Although relatively rarely required nowadays, RNA sequencing is carried out by using a slight modification of the DNA sequencing procedures. A method which operates in principle like the Chemical Cleavage Method using reactions that cleave at specific ribonucleosides or types of ribonucleosides has been developed. Alternatively the RNA can be transcribed into a complementary strand of DNA (cDNA) using the enzyme known as Reverse Transcriptase or RNA-directed DNA-polymerase (see). This is then sequenced by either the Chemical Cleavage Method or the Chain Terminator Method, and the RNA sequence deduced from base-pairing rules.

Nucleocidin: a purine antibiotic synthesized by *Streptomyces calvus* (see Nucleoside antibiotics), M_r 392, $[\alpha]_D^{25}$ −33 3°. N. is active against bacteria and fungi, and it is used therapeutically against trypanosomes. It inhibits protein synthesis.

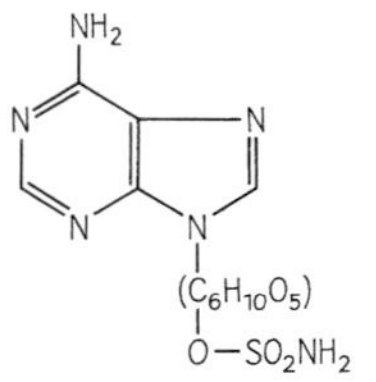

NH_2 ... $(C_6H_{10}O_5)$ ... $O-SO_2NH_2$

Nucleocidin

Nucleolus (plur. Nucleoli): a compartment of the nucleus. The N. contains the N. organizer (see). The number of N. per nucleus varies widely. The main components of the N. are proteins (over 80 % of the dry weight), RNA (over 5 %) and DNA. The N. contains the following recognizable structures: a ground material of amorphous protein, ribonucleoprotein granules, ribonucleoprotein fibrils, and the chromatin fibrils of the N. organizer. The N. is the site of biosynthesis of ribosomal RNA (see Ribosomes). During nuclear division, the N. temporarily disappears as a visible structure.

Nucleolus organizer: a specific region on one or more eukaryotic chromosomes, where the nucleolus is formed. The DNA in this region contains genetic information for the synthesis of ribosomal RNA.

Nucleoplasm: see Nucleus.

Nucleoprotein: heteropolar complexes of nucleic acids (in particular, nuclear DNA) with basic, acid-soluble proteins (histones or protamines), and with acidic, base- or detergent-soluble non-histone chromatin proteins. N. occur mainly in the chromatin of the cell nucleus in its quiescent state, and in the chromosomes when the nucleus is active, i.e. dividing. Many viruses consist entirely of N., but N. are absent from bacteria. N. are concerned in DNA replication, and in the control of gene function during protein biosynthesis.

The protein-RNA complexes of the ribosome are also N.

Nucleosidases: enzymes that catalyse the cleavage of the bond between the sugar residue and the base of a Nucleoside (see). The reaction is usually a phosphorolysis (not a hydrolysis), involving orthophosphate.

Nucleoside antibiotics: purine or pyrimidine nucleosides with antibiotic activity. They act as antimetabolites of natural substrates, and inhibit the growth of microorganisms by blocking the metabolism of purines, pyrimidines and proteins. Some N.a. (e.g. showdomycin) contain an analog base, others (e.g. gougerotin) contain an analog sugar, or both moieties may be modified (e.g. puromycin) (see Table).

The analog components are formed by modification of primary metabolites. The sugars or sugar derivatives, such as cordycepose, psicose, angustose and glucuronamide are derived from D-glucose or D-ribose by various reactions, e.g. epimerization, isomerization, oxidation, reduction and decarboxylation. Methyl groups occur frequently in N.a., and are derived by transmethylation. A normal nucleoside may be modified to a N.a. without prior cleavage of the *N*-glycosidic bond, e.g. the synthesis of tubercidin. Alternatively, the free base may combine with the analog sugar, as in the synthesis of psicofuranin. Unusual

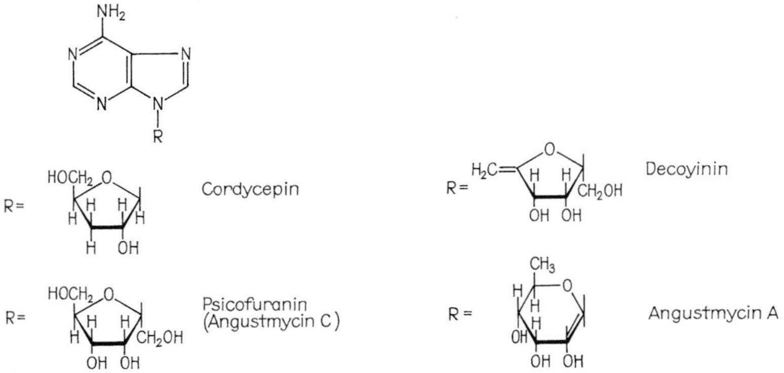

R = Cordycepin

R = Psicofuranin (Angustmycin C)

R = Decoyinin

R = Angustmycin A

Fig. 1. *Nucleoside antibiotics.* Structures of cordycepin, psicofuranin, decoyinin and angustmycin A.

Nucleocidin

Nucleoside antibiotics. Naturally occurring nucleoside analogs with antibiotic activity. For mode of action, see separate entry for each compound.

Antibiotic	Produced by	Base	Sugar	Antimetabolite of
3′-Acetamido-3′-deoxyadenosine	*Helminthosporium* spp.	Adenine	3-Acetamido-3-deoxyribose	Adenosine
Amicetin A	*Streptomyces fasciculatus, S. vinacius-drappus*	Cytosine	Amicetose	Cytidine
Amicetin B (Plicacetin)	*Streptomyces plicatus*	Cytosine	Amicetose	Cytidine
3′-Amino-3′-deoxyadenosine	*Cordyceps militaris, Helminthosporium* spp.	Adenine	3-Amino-3-deoxyribose	Adenosine
Angustmycin A (Decoyinin)	*Streptomyces hygroscopicus*	Adenine	L-2-Ketofuco-pyranose	Adenosine Guanosine
Angustmycin C	*Streptomyces hygroscopicus*	Adenine	Psicose (Psicofuranose)	Adenosine Guanosine
Arabinofuranosyl-adenine	*Streptomyces antibioticus*	Adenine	Arabino-furanose	Adenosine
5-Azacytidine	*Streptoverticillus lakadamus*	Azacytosine	Ribose	Cytidine
Bamicetin	*Streptomyces plicatus*	Cytosine	Amicetose	Cytidine
Blasticidin S	*Streptomyces griseochromogenes*	Cytosine	4-Deoxy-4-amino-2,3-hexen-uronic acid	Cytidine, Acyl-tRNA
Cordycepin	*Cordyceps militaris, Aspergillus nidulans*	Adenine	Cordycepose (3-Deoxyribose)	Adenosine
Formycin	*Nocardia interforma*	7-Aminopyrazolo-pyrimidine	Ribose	Adenosine
Formycin B	*Nocardia interforma, Streptomyces lavendulae, S. roseochromogenes*	7-Hydroxypyrazolo-pyrimidine	Ribose	Adenosine
Gougerotin	*Streptomyces gougeroti*	Cytosine	4-Dideoxy-glucopyran-uronamide	Cytidine, Acyl-tRNA
Nebularin	*Agaricus (Clitocybe) nebularis, Streptomyces* spp.	Purine	Ribose	Adenosine
Nucleocidin	*Streptomyces calvus*	Adenine	Hexose of unknown structure	Adenosine
Puromycin	*Streptomyces albo-niger*	Dimethylamino-purine	3-Amino-3-deoxyribose	Acyl-tRNA
Sangivamycin	*Streptomyces* spp.	4-Amino-5-carbox-amide-7-pyrrolo-pyrimidine	Ribose	Adenosine
Showdomycin	*Streptomyces showdoensis*	Maleimide	Ribose	Uridine
Toyocamycin	*Streptomyces toyocaensis, S. rimosus*	4-Amino-5-cyano-7-pyrrolopyrimidine	Ribose	Adenosine
Tubercidin	*Streptomyces tubercidicus*	4-Amino-7-pyrrolo-pyrimidine	Ribose	Adenosine

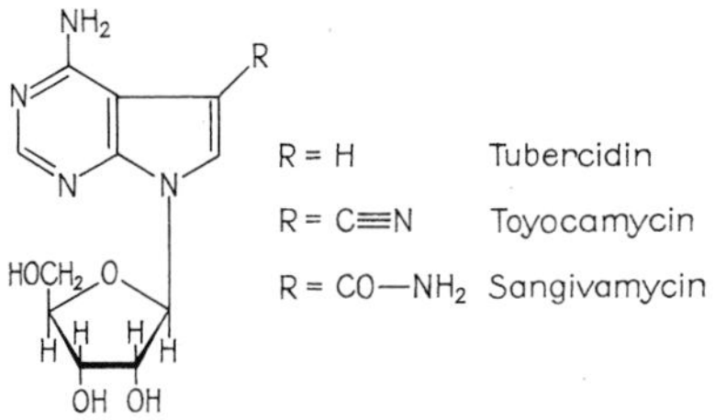

NH2, R, N, N, N, HOCH2, O, H, H, H, H, OH OH

R = H Tubercidin
R = C≡N Toyocamycin
R = CO—NH2 Sangivamycin

Fig. 2. *Nucleoside antibiotics.* Structures of tubercidin, toyocamycin and sangivamycin.

amino acids occur in some N.a., e.g. amicetin contains α-methyl-D-serine. Some N.a. contain unusual bond types, such as the aza bond in 5-azacytidine, or unusual functional groups, such as the CN-group in toyocamycin. The same N.a. may be formed by systematically unrelated organisms, e.g. cordycepin from *Cordyceps* and *Aspergillus.* One organism may produce several N.a., e.g. psicofuranin and decoyinin from *Streptomyces hygroscopicus.*

Nucleoside diphosphate compounds: compounds containing a nucleoside diphosphate grouping. This grouping has an activating effect, so that the molecule has a high group transfer potential. Examples of N.d.c. are Nucleoside diphosphate sugars (see) and Cytidine diphosphate choline (see).

Nucleoside diphosphate sugars, ***nucleotide sugars:*** energy-rich nucleotide derivatives of monosaccharides. The activating group is a nucleoside diphosphate. Uridine diphosphate glucose (UDP-glucose, UDPG, "active glucose") is of widespread general importance in carbohydrate metabolism. It is synthesized from glucose 1-phosphate by reaction with uridine triphosphate (Fig.). Other nucleoside diphosphate groups found in N.d.s. are listed in the table.

The activated sugar can take part in various metabolic reactions. Of particular importance is the transfer of the sugar moiety to the OH-group of another molecule, e.g. in the synthesis of oligo- and polysaccharides (see Carbohydrate metabolism).

Glucose 6-phosphate
Phosphoglucomutase (EC 2.7.5.1)
Glucose 1,6-diphosphate
Glucose 1-phosphate
Glucose 1-phosphate uridylyltransferase (EC 2.7.7.9)
Uridine triphosphate (UTP)
Pyrophosphate
OH, N, O, N, O, CH2—O—Ⓟ~Ⓟ—O, CH2OH, O, OH, OH, OH, OH OH
Uridinediphosphate glucose

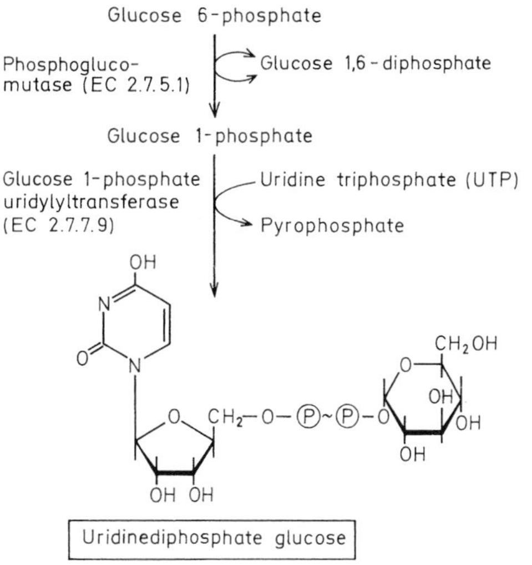

Biosynthesis of uridinediphosphate glucose

Nucleoside diphosphate derivatives of uronic acids are also metabolically important, e.g. uridine diphosphate glucuronic acid is formed from uridine diphosphate glucose, and it is a precursor of glucuronides (see Glucuronate pathway).

Nucleosides: *N*-glycosides of heterocyclic nitrogenous bases. The *N*-glycosides of purines and pyrimidines with pentoses are of particular biological importance. The sugar component is either D-ribose or D-2-deoxyribose, both in the furanose form (Table). C1 of the pentose residue is linked to N9 of the purine or N1 of the pyrimidine by an *N*-glycosidic linkage (C-N bond). To distinguish between the numbering systems of the base and sugar, the numbers of the sugar atoms are characterized by a prime, i.e. C-atoms 1′ to 5′. Deoxynucleosides contain D-2-deoxyribose, whereas ribonucleosides contain D-ribose. N. have trivial names derived from the component base. Pyrimidine N. end in "-idine", purine N. in "-osine".

Nucleoside diphosphate sugars

Activating nucleoside diphosphate group	Activated molecule	Function
Uridine diphosphate	Glucose	Involved generally in carbohydrate metabolism. Glycogen (see) synthesis. Murein (see) synthesis.
	Galactose	Galactose (see) metabolism
	Glucuronate	Glucuronate pathway (see). Glucuronate synthesis
	N-Acetylglucosamine	Metabolism of aminosugars. Chitin synthesis.
Adenosine diphosphate	Glucose	Starch (see) synthesis.
Guanosine diphosphate	Mannose	Synthesis of L-fucose, D-rhamnose and 6-deoxy hexoses.
Deoxythymidine diphosphate	Glucose	Synthesis of L-rhamnose.
Cytidine diphosphate	Ribitol, Glycerol	Synthesis of Teichoic acids (see).

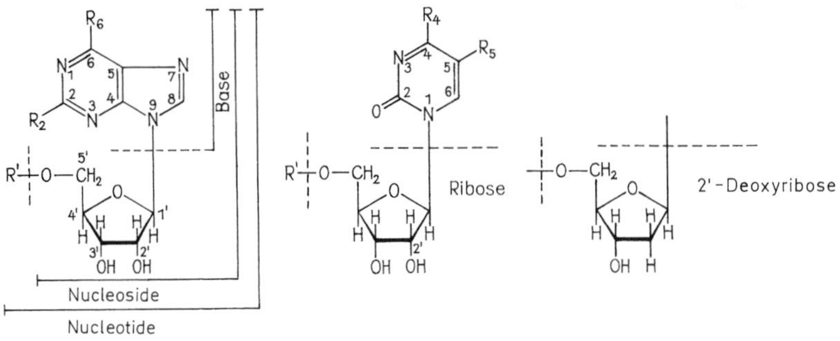

Structure of purine and pyrimidine bases, nucleosides and nucleotides

Purine base		$R^1 = H$ Nucleoside or Deoxynucleoside	Pyrimidine base		$R^1 = H$ Nucleoside or Deoxynucleoside
$R_2 = H$ $R_6 = NH_2$	Adenine	Adenosine Deoxyadenosine	$R_4 = OH$ $R_5 = H$	Uracil	Uridine Deoxyuridine
$R_2 = NH_2$ $R_6 = OH$	Guanine	Guanosine Deoxyguanosine	$R_4 = NH_2$ $R_5 = H$	Cytosine	Cytidine Deoxycytidine
$R_2 = H$ $R_6 = OH$	Hypoxanthine	Inosine (Deoxyinosine)	$R_4 = OH$ $R_5 = CH_3$	Thymine	Ribothymidine Thymidine*
		R^1 = Ⓟ	R^1 = Ⓟ ~ Ⓟ		R^1 = Ⓟ ~ Ⓟ ~ Ⓟ
Nucleotides:		Adenosine, Guanosine, Inosine, Uridine, Cytidine, Thymidine } mono-phosphate	Adenosine, Guanosine, Inosine, Uridine, Cytidine, Thymidine } di-phosphate		Adenosine, Guanosine, Inosine, Uridine, Cytidine, Thymidine } tri-phosphate

* The sugar component of thymidine is 2′-deoxyribose

The Rare nucleic acid components (see) represent nucleoside moieties in which the base or sugar is chemically modified.

N. and deoxynucleosides can be synthesized via a Salvage pathway (see). They are also produced by hydrolysis of nucleic acids and nucleotides. Nucleoside phosphorylases and deoxynucleoside phosphorylases catalyse the reversible, phosphate-dependent cleavage of N. and deoxyribonucleosides, forming ribose 1-phosphate or deoxyribose 1-phosphate and the free base. N. and deoxyribonucleosides can be converted into their corresponding nucleotides by the action of specific kinases.

Strictly speaking, the term nucleoside is reserved for base-sugar combinations present in nucleic acids, but the term is often applied to any base-sugar compound.

Nucleosomes: see Chromosome.

Nucleotide coenzyme: a coenzyme containing a nucleotide structure. N.c. are Pyridine nucleotide coenzymes (see), and the nucleoside diphosphate moieties of Nucleoside diphosphate sugars (see) and Coenzyme A (see). The Flavin nucleotides (see) are also N.c. Strictly speaking FMN is not a nucleotide but the term nucleotide can be generally applied to any base-sugar-phosphate group.

Nucleotides, *nucleoside phosphates:* phosphoric acid esters of Nucleosides (see). *o*-Phosphoric acid is esterified with a free OH-group of the sugar. If the sugar is D-ribose, the N. is called a ribonucleotide or ribotide. If the sugar is D-2-deoxyribose, the N. is a deoxyribonucleotide, deoxynucleotide or deoxyribotide. The phosphate may be present on position 2′, 3′ or 5′ (in deoxyribonucleotides, only the 3′- and 5′-phosphate are possible). 5′-nucleoside phosphates are metabolically very important; they may be mono-, di- or triphosphorylated, e.g. guanosine 5′-monophosphate, cytidine 5′-diphosphate and adenosine 5′-triphosphate.

The cyclic N. (cyclic 3′,5′-monophosphates) have important regulatory properties (see Adenosine phosphates, Guanosine phosphates, Inosine phosphates, Uridine phosphates).

Nucleoside monophosphates are synthesized de novo in the course of Purine biosynthesis (see) and Pyrimidine biosynthesis (see). They are then phosphorylated stepwise by the action of kinases, to produce nucleoside di- and triphosphates. The 2-deoxyribose moiety is formed by reduction of the ribose in ribonucleotides (see Ribonucleotide reductase). Reduction of free ribose to 2-deoxyribose does not occur in vivo.

N. and deoxynucleotides are the monomeric components of Oligonucleotides (see) and Polynucleotides (see). Enzymatic degradation of oligo- and polyribonucleotides (but not the corresponding deoxyribonucleotides) produces cyclic 2′,3′-nucleoside phosphates. N. are cleaved hydrolytically to nucleosides by the action of 5′- or 3′-nucleotidases, which function as phosphomonoesterases. N. are cleaved to free bases and phosphoribosylpyrophosphate in a pyrophosphate-dependent reaction catalysed by N. pyrophosphorylases. Certain coenzymes contain nucleotide structures (see Nucleotide coenzymes); thus NADP contains the structure of adenosine 2′,5′-diphosphate; and coenzyme A the structure of adenosine 3′,5′-diphosphate. In all living cells, N. (especially adenosine 5′-triphosphate) act as high energy compounds in the storage and transfer of chemical energy.

Nucleotide sugar: see Nucleoside diphosphate sugars.

Nucleus: a large structure (~ 5 μm diam.) in eukaryotic cells, containing the bulk of the cellular DNA, and representing the chief site for the storage, replication and expression of genetic information. In the period between cell divisions, i.e. at interphase, the nucleus is densely and uniformly packed with DNA and shows few distinct structures, even under the electron microscope.

Distinguishable features are Chromatin (see), nucleoplasm, nuclear membrane and the Nucleoli (see). At nuclear division, the highly structured Chromosomes (see) are formed from the chromatin.

Chromatin and chromosomes are composed mainly of DNA, RNA and numerous proteins. Important enzyme proteins are DNA-polymerase for DNA-replication, and RNA-polymerase for transcription. Other chromosomal proteins (histones, protamines and acidic proteins) are engaged in the regulation of replication and transcription (see Chromosomes).

Other important metabolic processes occur in the nucleoplasm, e.g. glycolysis and tricarboxylic acid cycle. NAD is synthesized only in the nucleus. The nuclear Na^+ concentration is ten times higher than that of the cytoplasm. The nucleus also contains a complete protein synthesizing system. The nuclear membrane is important in the transfer of high and low M_r compounds between nucleus and cytoplasm.

Isolated nuclei are highly permeable to histones, protamines and other biological macromolecules, whereas ATP and Na^+ become tightly bound. The chief components of isolated and disrupted nuclei are DNA-histone complexes (nucleohistones), ribonucleic acids and poorly soluble acidic proteins (residual proteins). Nuclei also contain high concentrations of an arginase and an adenosine 5′-phosphatase of unknown function.

Nuphara alkaloids: a group of alkaloids possessing a piperidine or quinolizidine ring system, which are found in various species of water lily (*Nuphar* spp.). All N.a. contain a sesquiterpene skeleton, which is cyclized by the inclusion of nitrogen, oxygen or sulfur. The chief N.a. are nupharidine, nupharamine and thiobinupharidine. Castoreum or castor, the secretion from the preputial follicles of the beaver, contains castoramine (M_r 247, m.p. 65–66 °C) which is related to the N.a.; it is not known whether it is synthesized by the animal, or derived from water lilies in its diet. For the biosynthesis of N.a., see Terpene alkaloids.

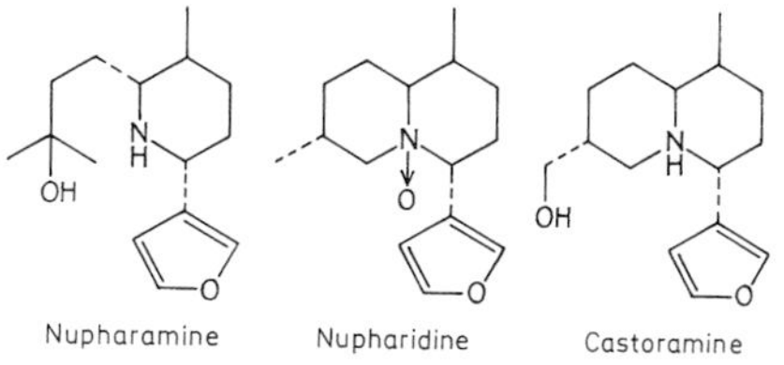

Nupharamine Nupharidine Castoramine

Nuphara alkaloids

Nutrient medium, ***growth medium:*** a medium, liquid or solid, for the cultivation of microorganisms, cells, tissues or organs. Solid media are prepared from liquid media by the addition of a gelling agent, e.g. gelatin (nowdays used only in special cases), silicic acid (when exclusion of organic compounds is necessary), or Agar-agar (see) (used widely in bacteriology, usually at a concentration of 1.5–2 %). N.m. contain fairly large amounts of mineral elements, together with trace elements. Sufficient quantities of certain trace elements are often already present as impurities in the other components of the medium. The mineral constituents, or inorganic nutrients, must be correctly balanced in order to avoid competitive effects of ions, to achieve an appropriate pH value, and to establish the correct oxido-reduction status of the N.m.

The composition of a complex N.m. is more or less ill defined. This may arise when the nutritional requirements of the culture are not exactly known, and the requirements can only be satisfied by the inclusion of, e.g. yeast extract, yeast autolysate, meat extract, peptone, coconut milk, or other complicated natural mixtures. The constant aim in the use of N.m. is the definition of the minimal growth requirements for the system in question; this enables the use of synthetic N.m. (i.e. made by mixing defined chemicals of known purity), or minimal N.m. (synthetic media of exactly known composition containing only those components that are required). Some growth systems have rather complicated and exacting growth requirements, so that it is necessary to add Growth factors (see). Auxotrophic organisms also show requirements for growth factors, which are not required by the parent, wild-type prototrophic organism (see Mutant technique).

Glucose often serves as the source of carbon and energy in synthetic N.m. In the presence of glucose, the utilization of other, less efficient carbon sources

Table 1. *A simple synthetic culture medium for microorganisms* (after Schlegel)

Component	Amount
Glucose	10.0 g
NH_4Cl	1.0 g
K_2HPO_4	0.5 g
$MgSO_4 \cdot 7H_2O$	0.2 g
$FeSO_4 \cdot 7H_2O$	0.01 g
$CaCl_2$	0.01 g
Trace element solution	1 ml
Water	1,000 ml

Table 2. *Vitamin solution used for the preparation of culture media for soil and water bacteria* (after Schlegel). 2–3 ml of this solution are added to 1,000 ml of culture medium.

Biotin	0.2 mg
Vitamin B_{12}	2.0 mg
Nicotinic acid	2.0 mg
p-Aminobenzoic acid	1.0 mg
Thiamin	1.0 mg
Pantothenic acid	0.5 mg
Pyridoxamine	5.0 mg
Distilled water	1,000 ml

Table 3. *The A-Z solution, or trace element solution of Hoagland*

$Al_2(SO_4)_3$	0.055 g
KI	0.028 g
KBr	0.028 g
TiO_2	0.055 g
$SnCl_2.2H_2O$	0.028 g
LiCl	0.028 g
$MnCl_2.4H_2O$	0.389 g
$B(OH)_3$	0.614 g
$ZnSO_4$	0.055 g
$CuSO_4.5H_2O$	0.055 g
$NiSO_4.7H_2O$	0.059 g
$Co(NO_3)_2.6H_2O$	0.055 g
Distilled water	100 ml

is usually prevented by catabolite repression. The source of nitrogen depends on the growth system; it may be inorganic, e.g. nitrate or ammonium, or organic, e.g. urea. Table 1 shows the composition of a simple bacteriological growth medium. Tables 2 & 3 give the composition of a vitamin solution and a trace element solution used in the preparation of some N.m.

Nutritional physiology of microorganisms: The terms autotrophy and heterotrophy are too broad to distinguish between all the different forms of microbial nutrition. Different types of nutritional physiology are classified according to: 1. the nature of the carbon source; 2. the source and mechanism of formation of ATP; 3. the source of reducing power for the synthesis of cell constituents.

Phototrophs use light energy (see Photosynthesis), and if they obtain all their energy from light and all their carbon from CO_2, they are called *photoautotrophs*. Some phototrophs can use organic carbon sources, obtaining all or some energy from light; these are called *photoheterotrophs* (i.e. they grow under mixotrophic conditions, see below).

Chemotrophs obtain ATP by the oxidation of inorganic or organic substrates, and assimilate CO_2 at the expense of the resulting oxidation energy (see Chemosynthesis).

Bacteria that obtain energy from the oxidation of inorganic compounds (i.e. inorganic hydrogen donors) are called *lithotrophs* or *chemolithotrophs*, e.g. *Nitrosomonas*, which oxidizes ammonium to nitrite and nitrate; *Hydrogemonas*, which oxidizes gaseous hydrogen with oxygen; *Thiobacillus*, which oxidizes sulfide, elemental sulfur, thiosulfate or sulfite to sulfate. If all of their carbon is derived from CO_2, they are called *chemoautotrophs*. Most of these bacteria possess an electron transport chain, which is similar to that in other bacteria and mitochondria, and ATP is synthesized by "oxidative" phosphorylation during oxidation of the inorganic source of reducing power. The potentials of the reactions involved may, however, be lower than in the aerobic respiration of organic substrates, so that P/O ratios are also lower.

The use of organic sources of reducing power is known as *organotrophy*. Most microorganisms, like animals, are *chemoorganotrophs*. Cyanobacteria and purple sulfur bacteria, which carry out photosynthesis, are *photolithotrophs*.

Some autotrophs can also grow heterotrophically (facultative autotrophs), using organic energy sources. Obligate autotrophs are unable to grow heterotrophically, e.g. cyanobacteria, some species of *Thiobacillus*, and *Nitrosomonas*. On the other hand, most obligate autotrophs can assimilate organic compounds as carbon sources, but not as energy sources. This ability makes growth possible under conditions that are best described as *mixotrophic*. Under these conditions, the energy source is needed only for the generation of ATP; the organic compound provides a source of reduced carbon and, if necessary, reducing power. *Beggiatoa*, a sulfur bacterium, is unable to grow on a completely inorganic medium (reduced sulfur compounds and CO_2), and in order to utilize reduced sulfur compounds, it requires an organic carbon source such as acetate, i.e. it exhibits mixotrophy. Similarly, in the mixotrophic (photoheterotrophic) culture of green algae or euglenoids, the growth medium contains an organic energy and carbon source (e.g. glucose) and the culture is illuminated; the cells become green, and growth is supported at least in part by photosynthesis.

O

Oat coleoptile test, ***Avena test:*** a biotest for the quantitative determination of auxins, which is carried out as follows. The tip of the coleoptile is cut off, and the cotyledon within it is removed by pulling on its base. An agar block with the test substance is set on one side of the coleoptile stump. Auxin diffuses into the side of the coleoptile covered by the agar and, due to the one-sided stimulation of growth, causes it to bend. The angle of the bend is a function of the auxin concentration.

Ochnaflavone: see Biflavonoids.

Ochratoxins: mycotoxins produced by *Aspergillus ochraceus* during food spoilage. In rats and mice O. cause pronounced liver damage. O.B is the dechloro derivative of O.A (Fig.). O.C is the ethyl ester of O.A.

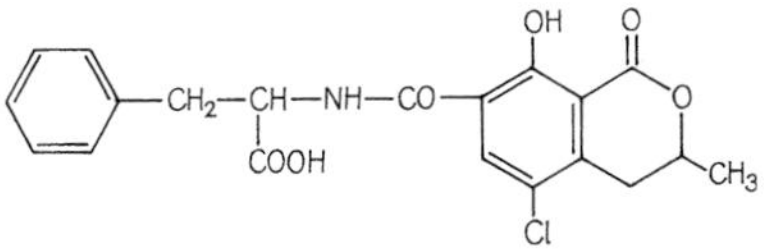

Ochratoxin A

Ochre codon: the UAA sequence in mRNA. Like the amber codon, it signals the end of protein biosynthesis. The synthesized polypeptide chain is released after the incorporation of the amino acid encoded immediately before the O.c. Ochre is probably the natural termination codon, and the one most widely employed by all living systems. See Ochre mutants.

Ochre mutants: bacterial mutants, which, as a result of a point mutation, possess a UAA codon in their mRNA (see Ochre codon). There are specific Suppressors (see) of ochre mutations.

Ocimene: a triply unsaturated monoterpene hydrocarbon. O. is an oily liquid, M_r 136.24, b.p.176–178 °C, ρ^{15} 0.8031, n_D^{18} 1.4857. It is a double bond positional isomer of Myrcene (see), and a component of many essential oils.

D-Octopine: *N*-α-(1-carboxyethyl)-arginine, N^2-(D-1-carboxyethyl)-L-arginine, M_r 246.3, m.p. 262–263 °C (d.), $[\alpha]_D^{24}$ +20.6 (c = 1, water). It is found in the muscles of certain invertebrates, e.g. *Octopus, Pecten maximus, Sipunculus nudus,* where it serves as a functional analog of lactic acid, i.e. the NADH produced by glycolysis is oxidized to NAD^+ during the synthesis of D-O., and NAD^+ can be reduced to NADH by the reversal of the same process. D-O. is biosynthesized from pyruvate and arginine by a reductive condensation catalysed by an unspecific NADH-dependent dehydrogenase.

D-O. is also found in certain plant tumors induced by *Agrobacterium tumefaciens.* This bacterium induces tumors in dicotyledenous plants by transferring a large bacterial plasmid (called the T_i plasmid) to the eukaryotic cell. In the transformed tissue, the T_i plasmid determines the synthesis of novel amino acids, which serve as specific substrates for the bacterium. These may be D-O. and related compounds (the "octopine family"), or nopaline and nopalinic acid (the "nopaline family"), but not both. [F. Marincs & D. W. R. White 'Divergent Transcription and a Remote Operator Play a Role in Control of Expression of a Nopaline Catabolism Promoter in *Agrobacterium tumefaciens*' *J. Biol. Chem.* **270** (1995) 12339–12342]

Octopinic acid: see D-Octopine

Oenin: see Malvidin.

Oestradiol: see Estradiol.

Oestriol: see Estriol.

Oestrogen: see Estrogen.

Octopine family:

R–CH–COOH / NH / CH_3–CH–COOH

R = NH_2–C(=NH)–$NH(CH_2)_3$–,	Octopine.
R = NH_2–$(CH_2)_3$–,	Octopinic acid.
R = NH_2–$(CH_2)_4$–,	Lysopine.
R 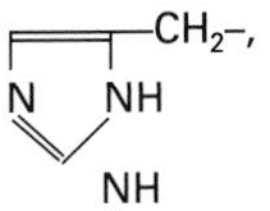 = imidazolyl–CH_2–,	Histopine

Nopaline family:

R–CH–COOH / NH / HOOC–$(CH_2)_2$–CH–COOH

R = NH_2–C(=NH)–$NH(CH_2)_3$–,	Nopaline.
R = NH_2–$(CH_2)_3$–,	Nopalinic acid or Ornaline.

Oestrone: see Estrone.

Oils: water-insoluble, liquid organic compounds. They are combustible, lighter than water, and soluble in ether, benzene and other organic solvents. Naturally occurring O. may be acylglycerols, e. g. the O. stored in certain seeds, or fish liver oils (see Fats); or they may be nonsaponifiable lipids, e. g. Essential oils (see).

Oiticica oil: the seed fat of *Licania rigida,* a tree native to the semiarid areas of northeast Brazil. Licanic acid (see) represents 50–80 % of the total esterified fatty acids of O. o. In addition, O. o. contains esterified palmitic, stearic, oleic and eleostearic acids.

Okazaki fragments: see Deoxyribonucleic acid.

Oleandomycin: a Macrolide (see) antibiotic.

5α-Oleane: see Amyrin.

Oleanolic acid: a monounsaturated, pentacyclic Triterpene (see), with a carboxylic acid group. M_r 456.71, m. p. 310 °C, $[\alpha]_D$ +80 ° (methanol). It differs structurally from β-amyrin by the presence of a carboxyl group in place of the 28-methyl group (see Amyrin). O. a. occurs free, esterified with acetic acid, or as the aglycon of triterpene saponins (see Saponins) in many plants, e. g. sugar beet, bilberry, mistletoe, cloves and cacti.

Oleic acid: Δ^9-octadecenoic acid, $CH_3-(CH_2)_7-CH=CH-(CH_2)_7-COOH$, the most widely distributed naturally occurring unsaturated fatty acid, M_r 282.45, m. p. 13 °C, b. p.$_{10}$ 223 °C. The double bond has the *cis* conformation. The *trans* isomer is called elaidic acid. O. a. is present in practically all the acylglycerols of depot and milk fats, and it is a component of phospholipids.

Oleoresins: see Balsams.

Oligo-1,6-glucosidase: see Dextrin 6-α-D-glucanohydrolase.

Oligonucleotides: linear sequences of up to 20 nucleotides, joined by phosphodiester bonds. Position 3′ of each nucleotide unit is linked via a phosphate group to position 5′ of the next unit. In the terminal units, the respective 3′ and 5′ positions may be free (i. e.-OH groups) or phosphorylated. O. are named according to chain length, i. e. di-, tri-, tetra-, pentanucleotides, etc. Linear sequences of more than 20 nucleotide units are called Polynucleotides (see and compare).

Oligopeptides: see Peptides.

Oligosaccharides: see Carbohydrates.

Oligosaccharidoses: inherited diseases affecting the degradation of asparagine-linked oligosaccharides. See Lysosomal storage diseases.

Oligosaccharins: oligosaccharides of specific composition which act as modulators of cell behavior in plants. They are thought to control growth, development, reproduction and defense. O. are present in enzymatic and acidic digests and heat-treated preparations of plant and fungal cell wall polysaccharides. Such preparations contain small quantities of active material in the presence of large amounts of inactive substances of similar monotonous structure, so that progress in the isolation and characterization of O. has been slow. O. are specific, in contrast to other plant hormones, which are pleiotropic. Hormones, such as auxin and gibberellin, may function by activating enzymes which release O. from cell walls. For example, auxin stimulates growth of pea stems and increases 50-fold an enzyme which releases active material from cell wall xyloglycan. In turn, this material inhibits auxin-stimulated growth. Thus as auxin stimulates growth at the shoot tip, O. may be transported down the stem and inhibit the growth of lateral buds, thereby causing apical dominance.

O. have been shown to inhibit flowering, promote vegetative growth and control organ development in plant tissue culture. Effective concentrations of O. are 100–1000-fold lower than those of other plant hormones, such as auxins, cytokinins and gibberellins.

O. are also thought to be involved in elicitation (see Elicitor) of Phytoalexin (see) synthesis. In elicitation, O. may be released: 1. from the cell wall of the invading organism by host enzymes, 2. from the host cell walls by enzymes of the invading organism, and 3. from host cell walls by enzymes released from the host itself by cell damage.

A hexaglucopyranosylglucitol, released from the cell wall of *Phytophthora megasperma* f.sp. *glycinea* by acid hydrolysis, has been characterized (Fig.) and shown to be highly active (10 ng applied to 1 g of plant tissue is effective) in promoting transcription of mRNA encoding the enzymes of phytoalexin synthesis in *Glycine max.* This is the first regulatory plant oligosaccharide to be completely characterized. It is presumed that similar O. are released enzymatically.

Other O. active in elicitation appear to be linear oligosaccharides of galacturonic acid, derived from host (e. g. *Ricinus*) cell wall by fungal enzymes. An enzyme from the plant pathogen *Erwinia carotovoris* also releases oligogalacturonides from the cell walls of *Glycine max* and other hosts. [A. G. Darvill & P. Albersheim *Ann. Rev. Plant Physiol.* **35** (1984) 243–275; P. Albersheim & A. G. Darvill *Scientific American* **253** (1985) 58–73; D. A. Smith & S. W. Banks *Phytochemistry* **25** (1986) 979–995]

Ommatins: see Ommochromes.

Ommins: see Ommochromes.

Ommochromes: a class of natural pigments containing the phenoxazone ring system. Their colors range from yellow through red to violet. They are especially common in, but not limited to, the *Arthropoda,* and were named for their occurrence in the om-

β-D-Glcp-(1 → 6)-β-D-Glcp-(1 → 6)-β-D-Glcp-(1 → 6)-β-D-Glcp-(1 → 6)-Glucitol
with β-D-Glcp-(1 → 3) branches on the second and fourth Glcp residues:
3 ↑ | β-D-Glcp (on 2nd residue); 3 ↑ | β-D-Glcp (on 4th residue)

Heptaglucoside from the cell wall of Phytophthora megasperma. This oligosaccharide, which stimulates phytoalexin synthesis in *Glycine max,* was isolated from acid digests of the fungal cell wall. [J. K. Sharp et al. *J. Biol. Chem. 259* (1984) 11321–11336]

matidia of the insect eye. They are divided into two groups: low M_r, alkali-labile, dialysable ommatins; and high M_r, alkali-stable ommins. In the organism they are often bound to proteins as chromoprotein granules.

Crystalline dihydroxanthommatin was first prepared from the meconium (post pupal secretion) of the small tortoiseshell butterfly (*Vanessa urticae*). It is found universally in insects as an eye pigment, accompanied in most orders by a greater amount of ommin. In the eyes of many *Diptera* (*Calliphora erythrocephala, Syrphus pyrastri, Musca domestica, Drosophila melanogaster*), xanthommatin is the only O. Rhodommatin (the O-glucoside of dihydroxanthommatin) and ommatin D (the sulfate ester of dihydroxanthommatin) have only been found in the *Lepidoptera*; they are present in the wings of the *Nymphalidae,* where they contribute greatly to pigmentation, but they are absent from the eyes and all other ectodermal structures. They are also present in meconium as their water soluble ammonium salts, and the first isolates were made from this source. The large amounts of xanthommatin found in meconium were probably derived from ommatin D, and it is doubtful whether xanthommatin occurs in fresh secretions. Xanthommatin has also been identifed in the eggs of the marine worm *Urechis caupo* (*Echiuridae*), and very small amounts, together with ommin, have been found in some crustaceans.

Ommin has been found in all investigated orders of the *Insecta* and *Crustaceae.* It is an especially common eye pigment in crabs, spiders, insects and cephalopods, and it is also present in the epidermis of *Cragnon, Limulus* and *Gryllus,* but not *Carcinus* and *Portunus.* Ommin is present in the eyes and skin, but not in the ink, of *Sepia officianalis* (*Cephalopoda*). About 75 % of the pigment known as ommin contains a violet-black component called ommin A.

O. are biosynthesized from 3-hydroxykynurenine, an intermediate of L-Tryptophan (see) metabolism. The phenoxazine ring system is formed by the oxidative coupling of two molecules of 3-hydroxykynurenine; it is a general reaction of *o*-aminophenols that they can be oxidized, chemically or enzymatically, to phenoxazones. Further cyclization of one side chain produces the quinoline ring system that is also present in the ommatins. Mutants of many insects are known, in which the capacity for O. synthesis is impaired. Such mutations are usually recognized from the abnormal eye color. Mutation may affect the conversion of L-tryptophan to *N*-formylkynurenine (e.g. *white eye* mutation of *Periplaneta americana*), *N*-formylkynurenine to kynurenine (e.g. *a* mutation of *Ephestia kühniella*), kynurenine to 3-hydroxykynurenine (e.g. *cinnabar* mutation of *Drosophila melanogaster*) 3-hydroxykynurenine to O. (e.g. *white-2* mutation of *Bombyx mori*), or the synthesis of the protein which binds the O. may be impaired (e.g. *wa* mutation of *Ephestia kühniella*).

Our knowledge of the structure and biochemistry of O. is due largely to the work of Butenandt et al.

OMP: acronym of Orotidine 5′-monophosphate.

Oncogenes: genes which are capable, under certain conditions, of inducing neoplastic transformation of cells. O. were first identified as nucleic acid sequences which are necessary for oncogenicity of certain viruses; strains of the viruses in which these genes were altered or deleted were found incapable of transforming cells in vitro or inducing tumors in vivo. Retrovirus O. (retroviruses are RNA viruses whose genomes are transcribed into DNA in the infected cells. The resulting single-stranded DNA is

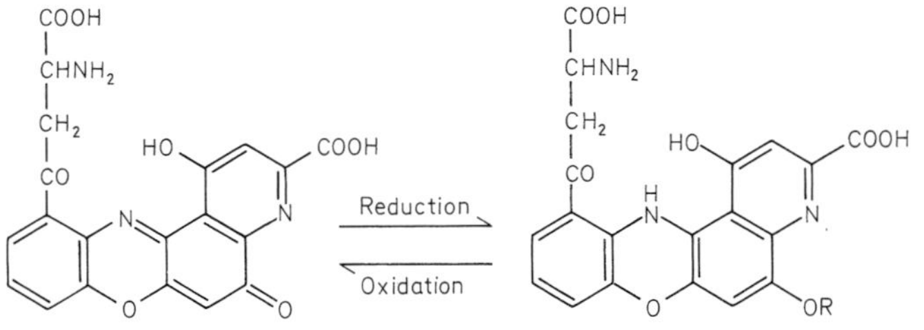

Xanthommatin (yellow-brown)

R = H, Dihydroxanthommatin (red).
R = SO_3H, Ommatin D (red).
R = 1-glucosyl, Rhodommatin (red).

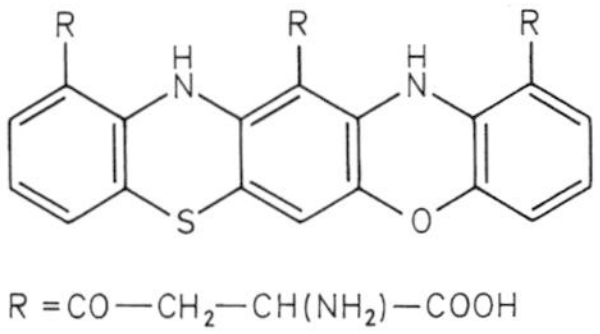

R = CO—CH_2—$CH(NH_2)$—COOH
Ommin A (violet-black)

Ommochromes

replicated to form double-stranded DNA, which is then integrated into a chromosome, where it is known as a provirus) were soon found to be homologous with cellular genes which have been highly conserved in evolution. The latter are called proto-oncogenes; apparently a virus strain can become oncogenic by incorporating one or more proto-oncogenes into its genome. As a rule, the process does not conserve the entire proto-oncogene. Deletions and fusion with viral genes are common, as are point mutations. DNA tumor viruses, such as polyoma, adenovirus and SV40, also contain O., and homologs have not been discovered in the vertebrate genome.

Cellular O. have also been identified in tumors by testing tumor DNA for its ability to transform cells in culture. (It is customary to refer to a viral O. as v-*onc;* and to the corresponding cellular gene as c-*onc,* e.g. v-*myc* and c-*myc.*) A c-*proto-onc* does not cause tumors; therefore, a c-*onc* must have been modified in some way through mutation or a change in the regulatory mechanism controlling its expression, or both. Activation of c-*onc* has been postulated as the mechanism by which chemical carcinogens and radiation induce cancers; another possibility is that the mutagen activates a genomic provirus which had been unexpressed.

Activation of *proto-onc* may occur by mutation, rearrangement of DNA or insertion of a segment of viral DNA into the chromosome adjacent to the c-*onc.* Either of the latter events brings the c-*onc* under the control of a foreign promoter. For example, in Burkitt's lymphoma and human chronic myeloid leukemia, chromosome exchange between non-homologous chromatids is often observed to have occurred. In some B-cell leukemias, the O. has come under the control of the immunoglobulin promoter and enhancer. It is probable that both mutation and change in the level of expression are required in most cases, as transformation in vivo usually appears to be a multistep process.

In some cases, two or more O. must be present in a virus for it to be able to transform cells in vitro (especially in the case of primary cells, i.e. cells taken directly from an animal and cultured briefly in vitro); this is usually true in vivo. Some O. are able by themselves, to induce transformation of certain cell lines in vitro. The standard test of transforming ability for DNA is transfection of NIH 3T3 cells (an immortalized line of mouse cells). Some authors argue that these cells have already taken the first step towards neoplasticity, and that the ability of an O. to transform NIH 3T3 cells in vitro does not prove that it can induce a cancer in vivo in the absence of other oncogenic factors.

The normal function of most proto-oncogenes is unknown. It is believed that elucidation of the normal functions of the proto-oncogene products will suggest the mechanism(s) by which O. can induce cancers. It is thought that proto-O. are essential to cell growth and differentiation; mutations or inappropriate expression of them might thus lead to cellular immortality and/or failure to differentiate.

A brief description of the more commonly studied O. is given below. In accordance with convention, the products of these genes are denoted by the letters "p", "gp" or "pp" to indicate "protein", "glycoprotein", and "phosphoprotein", respectively, followed by the approximate M_r (in kilodaltons). A superscript to the right of the symbol indicates the gene encoding the protein, e.g. $pp60^{src}$ is a phosphoprotein of M_r 60,000 encoded by the *src* gene. The products of fused genes are indicated by a hyphenated superscript, e.g. $gp180^{gag\text{-}fms}$ is a glycoprotein of M_r 180,000 which is the product of a fused viral *gag* gene and the proto-oncogene *fms.* In the following list, the virus from which the v-*onc* was first isolated and the species infected by it are given in parentheses after the name of the gene.

src (Rous sarcoma virus; infects chickens). The protein product, $pp60^{src}$, is a Protein tyrosine kinase (see). The 19 *C*-terminal amino acids of $pp60^{c\text{-}src}$ are replaced by 12 new ones in the viral gene; most of these are encoded by a sequence about 1 kb downstream from *c-src.* The 3′ end of *c-src* inhibits transcription of the gene. Other O. which encode Protein tyrosine kinases (see) are listed under that entry. pp60 is a myristylated protein, which is anchored in the plasma membrane by its myristic acid residue. Loss of myristylation by mutation also results in loss of oncogenicity.

mos (Moloney sarcoma virus; mice) shows homology to *src,* but its product appears not to be a tyrosine kinase. c-*mos* does not initiate transformation; it differs from v-*mos* at several points and its transcription is also inhibited by upstream sequences. v-*mos* requires viral LTR (long terminal repeat) sequences for transforming activity; it is present at 1–10 copies per cell, while expression of c-*mos* has not been detected.

fms (McDonough feline sarcoma virus) encodes a glycoprotein which is related or identical to the receptor for colony stimulating factor [R.L. Mitchell et al. *Cell* **45** (1986) 497–504]. It has tyrosine kinase activity in vitro, but this has not been demonstrated in vivo. The protein co-purifies with intermediate filament proteins, and its location in the cell (by double immunofluorescence) corresponds closely to that of vimentin and keratin, but not actin. c-*fms* is expressed in detectable amounts in the extra-embryonic tissues of mice, but not in the fetus or adult mouse.

v-yes (avian sarcoma virus Y73), ***v-fps*** (Fujinami sarcoma virus), ***v-fgr*** (Rasheed feline sarcoma virus), ***v-abl*** (Abelson murine leukemia virus) and ***v-ros*** (UR2 avian sarcoma virus) show a certain degree of homology with one another. All encode tyrosine kinases, and there is a cellular homolog of each. Each is capable of transforming cells (in vitro) independently. In the case of v-*fps* and the homologous v-*fes* from feline sarcoma virus, it has been shown that the tyrosine kinase activity of their products is necessary for the initiation of transformation by the viruses carrying them. The p130 product of v-*fps* is not an integral membrane protein, but appears to be linked to the membrane or cytoskeleton. Infection of chicken macrophage progenitors with retroviruses carrying v-*fps* enables them to differentiate in the absence of macrophage colony stimulating factor in the medium.

myc (avian myelocytomatosis virus). Other avian tumor viruses which contain v-*myc* cause leukemia, carcinomas and sarcomas. These have in common a 1.6 kb sequence at the 3′ end of the gene which is shared with the c-*myc*. However, the 5′ ends of the viral and cellular genes vary. In the human genome, c-*myc* is normally on chromosome 8, near the break-point at which part of chromosome 8 is transferred to chromosome 14 in Burkitt's lymphoma. c-*myc* is expressed in normal avian and mammalian cells and the gene product is located in the nucleus. Excessive expression in transgenic mice does not prevent normal development but does make the animals highly susceptible to a number of different kinds of tumor. [A. Leder et al. *Cell* **45** (1986) 485–495]

myb (avian myeloblastosis virus, which causes chicken leukemia). The viral gene encodes a protein of M_r 48,000 (located in the nucleus), while the cellular proto-oncogene encodes a p110 and is expressed in hematopoietic cells. *myb* is distantly related to *myc* and the adenovirus oncogene E1A, with homologies of 15 to 20 % among the proteins.

v-mht (avian retrovirus MH2, which contains both this O. and *myc*).

erb (avian erythroblastosis (acute leukemia) virus) encodes two proteins (A and B). $p75^{erb\ A}$ is found in transformed cells; the *N*-terminal 1/3 of this protein is the product of the viral *gag* gene. *erb* B is a glycoprotein found in membranes; it is nearly identical to the tyrosine kinase part of the epidermal growth factor receptor molecule. However, it has no kinase activity in vitro.

v-sis (simian sarcoma virus); the encoded protein is structurally related to the β-chain of platelet-derived growth factor. c-*cis* is located on chromosome 22 of the human genome; it is transferred to chromosome 9 in the Philadelphia translocation associated with human myelogenous leukemia.

rel (avian reticuloendotheliosis virus). c-*rel* is expressed in normal chicken cells; there are a number of mutations in v-*rel* with respect to c-*rel*.

fos (murine sarcoma virus). The *N*-terminal 332 amino acid residues Of $p^{v\text{-}fos}$ (located in the nucleus) are identical to the corresponding part of the c-*fos* protein; however, there is a deletion in the viral gene which shifts its reading frame and causes the C-terminal 49 amino acids to be entirely different from those of the proto-oncogene. c-*fos* is expressed in the placenta and extra-embryonic membranes of fetal mice. It is also expressed in neonatal mouse bones, muscles, skin and connective tissues. It is transiently expressed in resting cells which have been challenged with mitogen.

There are three forms of the ***ras*** gene: Ha-*ras* (Harvey murine sarcoma virus), Ki-*ras* (Kirsten murine sarcoma virus), and N-*ras* from some tumor cell DNA. The human genome contains cellular versions of all three genes, and the transforming genes isolated from human tumors are often derived from the *ras* family. The proteins encoded by the genes consist of 189 amino acid residues, and are called "$p21^{ras}$". $p21^{N\text{-}ras}$ binds GDP and phosphorylates itself; membrane proteins with these properties are often involved in the transduction of cellular signals. Injection of either c-Ha-*ras* or v-Ha-*ras* proteins into fibroblasts in vitro induces membrane ruffling and pinocytosis; however, the effects of the cellular gene product are much shorter-lived than those of the oncogene product. The effect depends on calcium and apparently involves the Inositol phosphate (see) signal pathway. [D. Bar-Sagi & J. R. Feramisco *Science* **233** (1986) 1061–1068; G. F. Vande Woude et al. (eds.) *Oncogenes and Viral Genes,* vol.**2** of *Cancer Cells* (Cold Spring Harbor Laboratory, 1984)]

Oncovin: see Vincristin.

One-carbon cycle: a cycle of methyl transfer and methyl oxidation involving glycine, sarcosine, dimethylglycine, betaine and choline, first proposed in 1958 from observations on the oxidation of dimethylglycine and sarcosine by rat liver mitochondria [Mackenzie & Frisell, *J. Biol. Chem.* **232** (1958) 417–427]. The existence of a O-c.c. is now confirmed; it involves choline oxidation in mitochondria and phosphatidylcholine synthesis in the endoplasmic reticulum (Fig.). [A.J . Wittwer & C.Wagner *J. Biol. Chem.* **256** (1981) 4102–4108, 4109–4115]

Dehydrogenation of sarcosine and dimethylglycine is closely linked to the conversion of glycine to serine via the cycle of $N^{5,10}$-CH_2-THF utilization and regeneration. When isolated from rat liver, sarcosine and dimethylglycine dehydrogenases possess a tightly (but noncovalently) bound THF derivative with a chain of 5 glutamate residues This $THF(Glu)_5$ may replace THF in the reactions shown below. In the absence of THF, oxidation of dimethylglycine and sarcosine still occurs, but the products are glycine and formaldehyde, possibly formed by hydrolysis of enzyme-bound oxidation products:

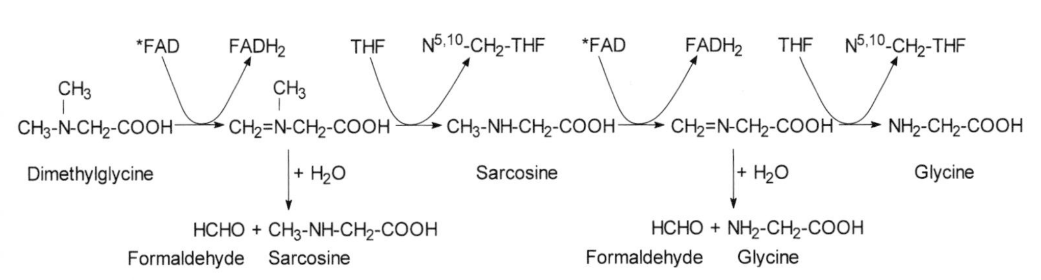

* The FAD of sarcosine and dimethylglycine dehydrogenases is covalently bound (position 8α of the isoalloxazine ring is linked to the imidazole (N3) of a histidine residue) [R. J. Cook et al. *J. Biol. Chem. 260* (1985) 12998–13002].

Other products are CO_2 and formate, resulting from the breakdown of THF derivatives which fail to contribute to the conversion of glycine to serine:

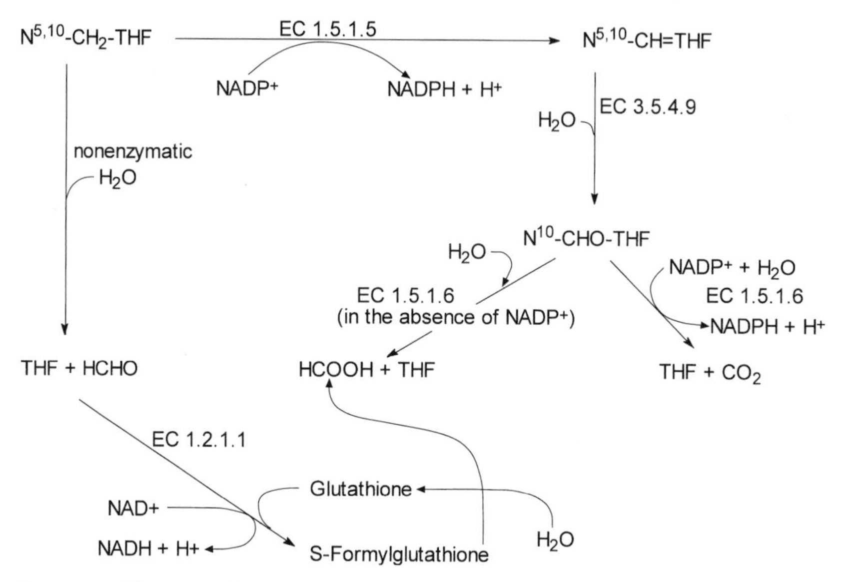

Enzymes: EC 1.2.1.1, Formaldehyde dehydrogenase (see); EC 1.5.1.5, Methylene-THF dehydrogenase ($NADP^+$); EC 1.5.1.6, Formyl-THF dehydrogenase; EC 3.5.4.9, Methylene-THF cyclohydrolase.

Onic acids: see Aldonic acids.

Ononin: formononetin 7-glucoside. See Isoflavone.

Oogoniols: a family of sex hormones secreted by the male mycelium of *Achlya* spp. in response to stimulation by antheridiol (see), a sex hormone secreted continuously by the female mycelium of *Achlya* spp. when in the vegetative state. O. stimulate the female *Achlya* mycelium to produce branches upon which oogonia are borne. The oogonial initials are believed to secrete even larger amounts of antheridiol than the vegetative female *Achlya* mycelium; this acts as a chemotropic agent, directing the growth of the antheridial hyphae towards the developing oogonium and ultimately leading to conjugation and the production of diploid oospores. The O., like antheridiol, are biosynthetically derived from fucosterol (see), the major sterol of *Achlya* spp.

Oogoniol	R = H
Oogoniol-1	R = $(CH_3)_2CHCO$
Oogoniol-2	R = CH_3CH_2CO
Oogoniol-3	R = CH_3CO

Dehydro-oogoniols corresponding to the four Oogoniols above save that they have a double bond between C-24 and C-28(or 24^1) also occur

Oogoniols. Dehydro-oogoniols also occur naturally; they posses a double bond between C24 and C28 (or 24^1), and otherwise correspond to the four oogoniols shown above.

Open system: 1. a system in dynamic equilibrium (see Steady state) with its surroundings, i.e. there is a continual exchange of material, energy and information with the environment. Application of the theory of O.s. to living systems (by Bertalanffy) involves the thermodynamics of irreversible processes.

2. In biology, plants are considered to be O.s., whereas animals are closed systems. The plant is theoretically capable of unlimited growth; certain cells remain embryonal and able to divide and differentiate, so that growth occurs from vegetative sites, such as the meristematic regions of the shoot and root tips, intercalary meristems, etc. In the animal, however,

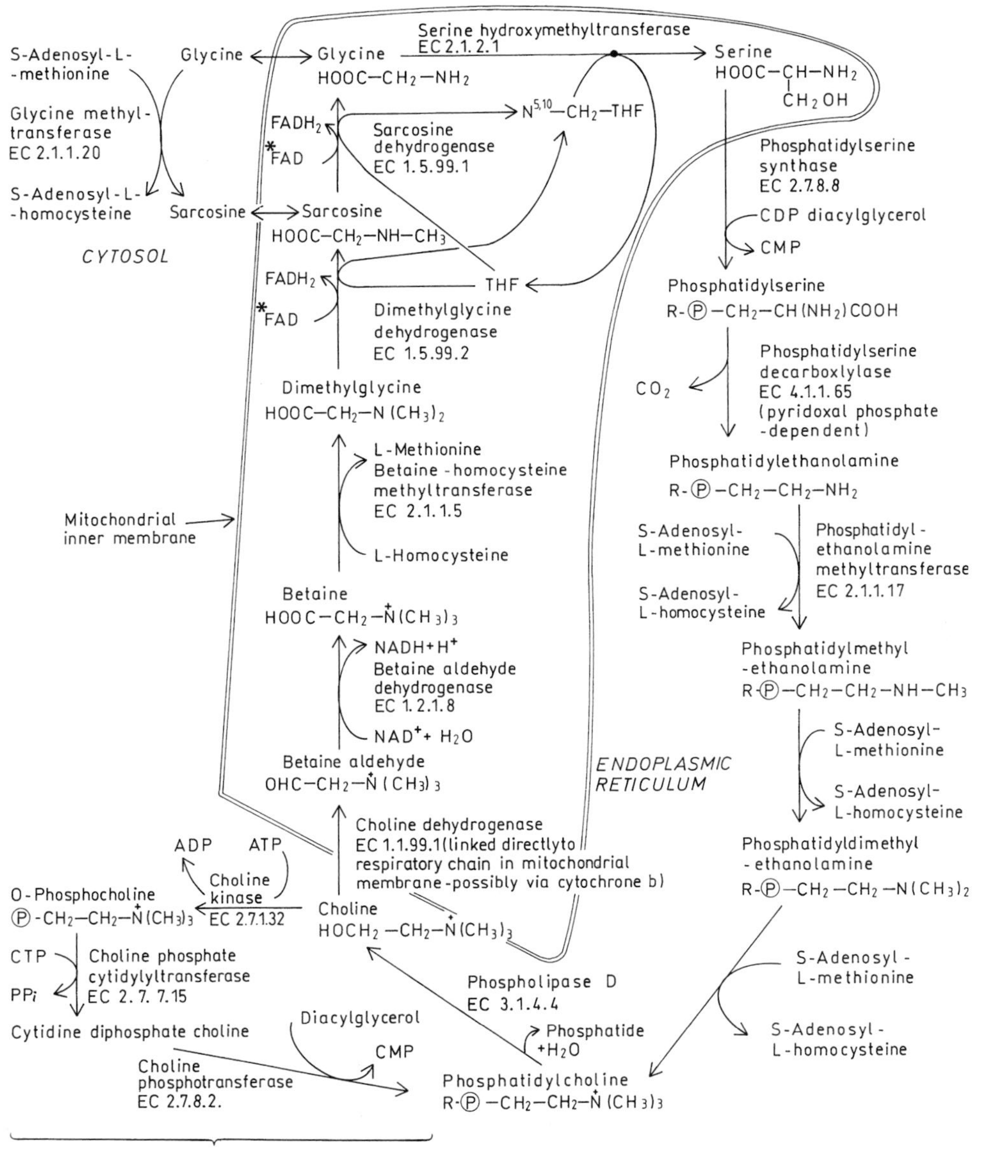

One carbon cycle in rat liver. THF = tetrahydrofolic acid (see Active one-carbon units).
R = diacylglycerol.

differentiation is essentially complete after the conclusion of embryonal development.

Operator: see Operon.

Operon: a group of neighboring genes, which represent a functional unit. An operon contains the following.

1. Structural genes (S_1 to S_4 in the Fig.), which code for the primary structures of enzyme proteins catalysing successive steps in a metabolic pathway, e.g. the enzymes in the biosyntheisis of an amino acid. The primary transcription product of this group of genes is a polycistronic mRNA; therefore, during the control of transcription, all the structural genes are affected equally.

2. The promoter (P in the Fig.) is the starting point of transcription. This section of the DNA is "recog-

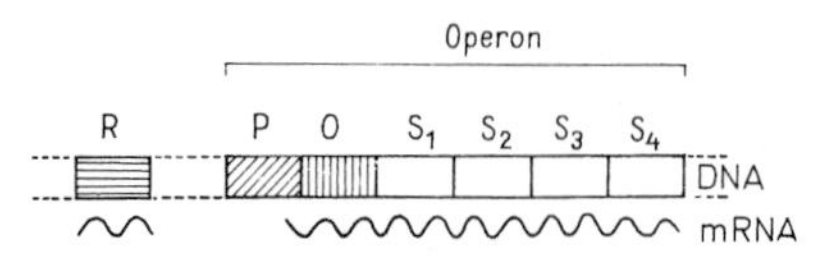

Schematic representation of an operon

nized" by RNA polymerase with the aid of the sigma factor. The affinity of the promoter for RNA polymerase (apparently determined by the structure of the promoter) is one of the factors that regulate the transcriptional frequency of the operon. When the RNA polymerase has bound to the promoter, it must then pass through the operator region in order to reach the structural genes.

3. The operator (O in the Fig.) is the control gene for the function (i.e. the transcription) of the structural genes. It is able to bind a repressor protein, which is the product of the regulator gene (R in the Fig.). R. is not a part of the operon, and it is located in a different region of the chromosome. If the specific repressor protein binds to the operator, transcription of the structural genes is blocked, i.e. the RNA polymerase cannot pass to the structural genes. If the operator is unoccupied, transcription of the structural genes can proceed. The details of these control mechanisms are described under Enzyme induction (see) and Enzyme repression (see).

The nucleotide sequence of the lactose operator was elucidated in 1973 by Gilbert and Maxam; this operator is double stranded and consists of the following base pairs:

5′ TGGAATTGTGAGCGGATAACAATT 3′
3′ ACCTTAACACTCGCCTATTGTTAA 5′

The operon model was developed by Jacob and Monod, and it has only been demonstrated in prokaryotic systems. It is the basis for the explanation of Enzyme induction (see) and Enzyme repression (see). See also Attenuation.

Ophiobolanes: sesterterpene phytotoxins present in various species of *Helminthosporum* and *Cochliobolus.* Ophiobolin A (ophiobolin, cochliobobin, cochliobobin A) is thought to be responsible for some of the symptoms of corn blight caused by *Helminthospor-*

Mevalonic acid x 5
(see Terpenes)
• = Carbon from [2-^{14}C] mevalonic acid
T = ^{3}H from [4R-^{3}H] mevalonic acid

all-*trans*-Geranylfarnesylpyrophosphate
(via 2-*cis*-Geranylfarnesylpyrophosphate?)

Ophiobolin F
(from *Cochliobolus heterostrophus*)

Ophiobolin C (Zizanin A)
(from *Helminthosporum zizaniae*)

Oxygen at C-14 derived from atmospheric O_2

Ophiobolin A (Cochliobolin A)
(from *Helminthosporum maydis*)

Ophiobolin B (Cochliobolin B, Ophiobolisin A, Zizanin B)
(from *Cochliobolus heterostrophus Helminthosporum zizaniae, H. oryzae*)

Biosynthesis of ophiobolanes

ium maydis. Biosynthesis occurs by head-to-tail linkage of 5 isoprene units, forming all-*trans*-geranylfarnesylpyrophosphate (Fig.). The bioconversions and incorporation of labels shown in the figure were studied in fungal cultures. Other ophiobolins have been isolated, differing chiefly with respect to the presence or absence of oxygen at C14 and the oxidation level of C21. [A. Stoessel in *Toxins in Plant Disease,* R. D. Durbin (ed.) (Academic Press, 1981) pp. 178–181]

Opioid peptides: a group of peptides found in the brain, hypophysis, adrenal medulla, intestines, blood and urine. They act as opiates, causing, e. g. a specific reduction in the contraction threshhold of the mouse vas deferens and the guinea pig ileum. The two *Enkephalins* (see) are pentapeptides first isolated from pig brain: Met-enkephalin (Tyr-Gly-Gly-Phe-Met) and Leu-enkephalin, (Tyr-Gly-Gly-Phe-Leu). The term "endorphin" (a contraction of "endogenous morphine") was coined to include all O. p. However, it is often used to refer specifically to several larger peptides: β-endorphin, α-neo-endorphin, dynorphin, adrenorphin, etc. Each of these peptides contains the Met- or Leu-enkephalin sequence at its *N*-terminus; the pentapeptide sequence is therefore thought to be responsible for the specific effect of the O. p., while the remaining amino acids give the "address" at which the peptide is most effective.

Three precursor proteins of O. p. have been identified and sequenced. The first, *preproopiomelanocortin,* is the source of corticotropin (ACTH) and β-lipotropin (see Hormones). These hormones are cleaved to smaller peptides; β-endorphin is released from β-lipotropin. The second and third precursors are preproenkephalins A and B (PPE A and PPE B). PPE A contains four copies of Met-enkephalin, one of Leu-enkephalin, and two extended Met-enkephalins with six and seven amino acids, respectively. It is also the source of adrenorphin, and other active peptides. PPE B is the precursor of α-neo-endorphin, dynorphin, and PH-8P (a predominant O. p. in brain).

In general, the points at which the precursor proteins are hydrolysed to release O. p. are marked by pairs of basic amino acids, such as Arg-Arg, Arg-Lys, etc. In the case of adrenorphin, the *C*-terminal NH_2 is donated by a Gly residue immediately preceding the signal. In this case, however, and in that of PH-8 P, the signal is Arg-Pro. [*Nature* **278** (1979) 423–427 ; *ibid.* **295** (1982) 202–206, 206–208, 663–666 ; *ibid.* **298** (1982), 245–249; *ibid.* **305** (1983) 721–723] The opiate antagonist naloxon(e) relieves the O. p.-induced inhibition of contraction, suggesting a specific interaction between the O. p. and the opiate receptors. Unfortunately, the early hopes that O. p. might be used as non-addictive analgesics have not been fulfilled. The physiological function of the O. p. is not known. The enkephalins may be neurotransmitters, and the longer-chained O. p. could be neurohormones. It is of interest that large amounts of enkephalins are stored with adrenalin in the chromaffin cells of the adrenal medulla and released together with adrenalin during stress. [*Science* **221** (1983) 957–960] O. p. are implicated as mediators of the suppressive effect of stress on natural killer cell cytotoxicity. [Y. Shavit et al. *Science* **223** (1984) 188–190.] Dynorphin-A-(1–8) is contained in the same neurosecretory vesicles in the hypophysis as vasopressin and neurophysin, and may therefore be involved in the antidiuretic response. [M. Whitnall et al. *Science* **222** (1983) 1137–1139] They may also be modulators of behavior, and a connection with acupuncture analgesia has been suggested. They may even be involved in the pathogenesis of schizophrenia, hallucinations, etc.

The endorphin and the opiate receptor system are also apparently involved in growth and development. Human infants and laboratory animals exposed to opiates like heroin and methadone are retarded in somatic and neurobiological growth and development. Conversely, newborn rats continuously maintained on the opiate antagonist naltrexone gain weight and pass developmental milestones earlier than litter-mate controls. [*Science* **221** (1983) 1179–1180; *Science* **222** (1983) 1246–1248; F. Meng et al. (κ & δ receptors) *J. Biol. Chem.* **270** (1995) 12730–12736; J.-C. Xue et al. (μ receptor) *J. Biol. Chem.* **270** (1995) 12977–12979]

Opium: the congealed or dried latex of *Papaver somniferum.* The chief active constituents of O. are the Opium alkaloids (see). O. is smoked, or preparations from it may be ingested or injected. Uncontrolled use leads to addiction and deterioration of personality. The action of O. is essentially similar to, but weaker than that of its constituent Morphine (see).

Opium alkaloids: the *Papaveraceae* alkaloids (see) that occur in Opium (see). Main representatives of the approximately 40 different O. a. are Morphine (see), Codeine (see) and Papaverine (see). More than 1,000 tons of morphine are produced annually. Nowdays the straw of the poppy plant is also used for the isolation of O. a. This new source accounts for about one third of the annual production of morphine.

Opsin: see Vitamins (vitamin A).

Opsonin: the name given by Wright and Douglas in 1903 to the thermostable material present in serum, which stimulates phagocytosis of bacteria. It was shown to act directly on bacteria and not on the phagocytes. It is probably identical to C3 b of the complement system (see Opsonization), although other complement components may also be active to a lesser extent. In addition to its role in the opsonization of immune complexes, C3 b binds to various structures, such as foreign erythrocytes and bacterial cells, and renders them more readily phagocytosed by immune adherence to phagocytes.

Opsonization: the ability of serum to render an immune complex more readily phagocytosed. O. is a property of the Complement system (see). Although immune complexes are subject to phagocytosis, interaction with complement greatly increases the rate. Component C3 b (activated C3 of the complement system) has a labile binding site(s), which permits it to bind tightly to antigen-antibody complexes (opsonic adherence). Other stable binding sites on C3 b enable this opsonized immune complex to bind to polymorphonuclear leukoytes, monocytes and macrophages (immune adherence); this binding results in enhanced phagocytosis. The phagocytosis can be abolished by metabolic inhibitors, whereas the immune and opsonic adherence (due to the physical chemical interaction of receptor and binding sites) are unaffected.

Optical density: see Absorbance.

Optical test: a method introduced in 1936 by Otto Warburg for the determination of the enzymatic activity of NAD and NADP-dependent dehydrogenases. Absorbance at 340 nm (or another suitable wavelength in that region) is measured as an index of the degree of reduction of NAD or NADP (see Nicotinamide-adenine-dinucleotide). The principle of this method is now widely used for measuring enzyme activities and the concentration of metabolites. A coupled enzyme system may be used, in which the reaction of interest does not produce or consume NAD(P)H, but is linked to one that does, e.g. the concentration of glucose can be determined from the increase in absorbance at 340 nm, when the unknown glucose concentration is incubated in the presence of excess ATP and $NADP^+$ and the enzymes hexokinase and glucose 6-phosphate dehydrogenase:

Glucose + ATP → Glucose 6-phosphate + ADP.

Glucose 6-phosphate + $NADP^+$ → 6-Phosphogluconate + NADPH + H^+.

Determinations of dehydrogenase activities by the O. t. are now conveniently performed in an automatic recording spectrophotometer, so that the change in absorbance at 340 nm is monitored continuously with time, e.g. the activity of malate dehydrogenase is measured from the rate of increase of absorbance at 340 nm due to the production of NADH in the reaction: NAD^+ + Malate → Oxaloacetate + NADH + H^+.

Orchinol: 9,10-dihydro-2,4-dimethoxy-7-phenanthrol, a phytoalexin. O. is formed by the orchid *Orchis militaris* as a defense against infection by the fungus *Rhizoctonia repens.* The structurally related

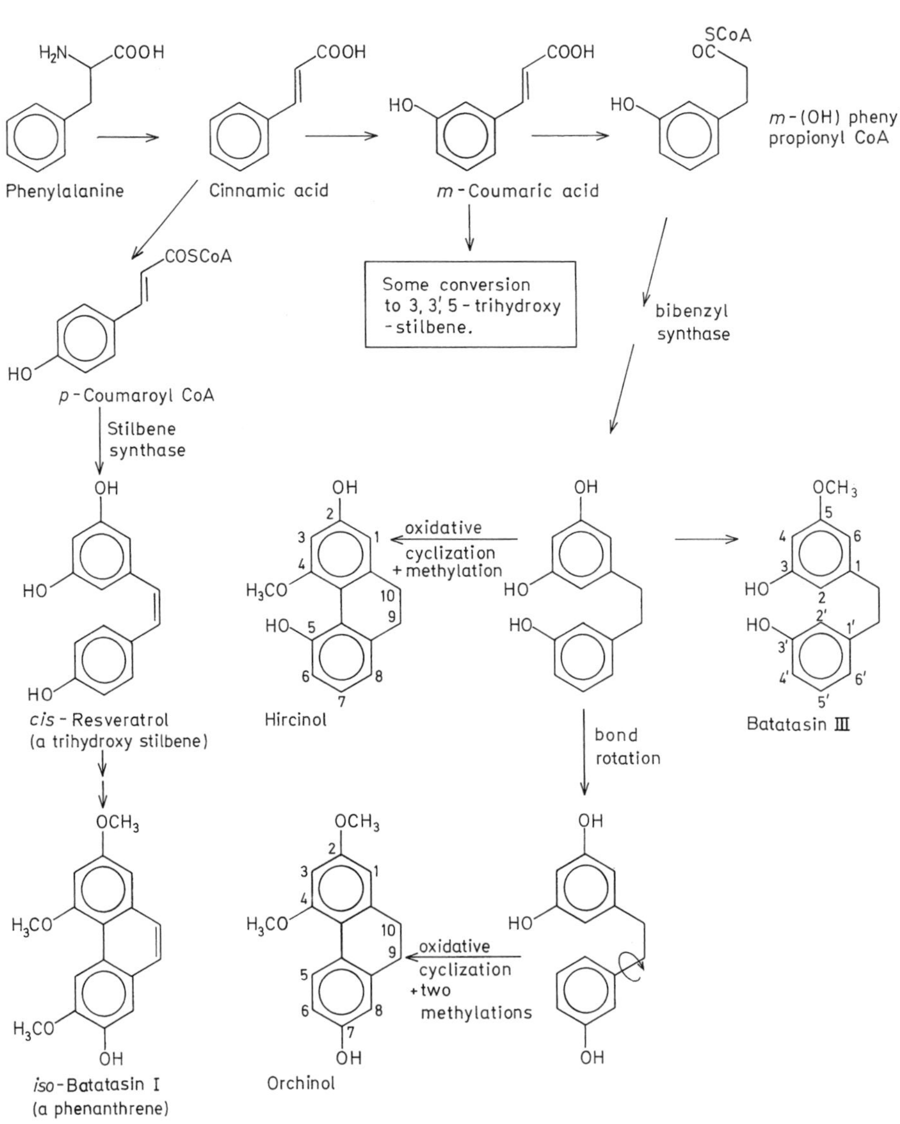

Biosynthesis of orchinol, hircinol and batatasins

hircinol (9,10-dihydro-4-methoxyphenanthrene-2,5-diol) is formed by the orchid *Himantoglossum (Loroglossum) hircinum,* against infection by *Rhizoctonia.* Hircinol and O. are biosynthesized via phenylalanine and *m*-hydroxyphenylpropionic acid by the "m-coumaric acid pathway" (Fig.). Hircinol has also been found in the peel of *Dioscorea rotondata,* together with related compounds, the batatasins. The latter have been identified as the dormancy-inducing principle in bulbils of the yam, *Dioscorea batatas.* Batatasins, with hircinol, may also protect against fungal attack, since they inhibit the growth of *Cladosporium cladosporioides* and certain yam soft root pathogens. Several batatasins are dihydrostilbenes (e.g. batatasin III), which share a common precursor (dihydrostilbene) with hircinol and O. (Fig.). Batatasin I, a phenanthrene, is present in tubers of many species of *Dioscorea,* and has also been found in *Tamus communis (Dioscoreaceae).* See Stilbenes. [K.H. Fritzmeier & H. Kindl *Eur. J. Biochem.* **133** (1983) 545–550; K.H. Fritzmeier et al. *Z. Naturforsch.* **39c** (1984) 217–221].

ORD: acronym of Optical Rotatory Dispersion.

Ord: abb. of Orotidine.

Organotrophy: see Nutritional physiology of microorganisms.

Orn: abb. of L-Ornithine.

Ornaline: see D-Octopine.

L-Ornithine, *Orn:* α,δ-diaminovaleric acid, 2,5-diaminopentanoic acid, a nonproteogenic amino acid. In mammals, L-Orn is an intermediate in the Urea cycle (see), and it is the precursor of the Polyamines (see). It is also an intermediate in the biosynthesis of Arginine (see) in all arginine-synthesizing organisms. In a few microorganisms, L-Orn is formed from citrulline by the action of citrulline phosphorylase. Various antibiotics contain L-Orn, e.g. the peptide gramicidin S.

2-*N*-Acetyl-L-ornithine is an intermediate in the biosynthesis of arginine from glutamic acid in microorganisms, and it is an important allosteric effector of carbamoyl phosphate synthetase. 5-*N*-acetyl-L-ornithine, which is found in various plants, is a nonproteogenic amino acid, and a structural analog of citrulline.

Ornithine cycle: see Urea cycle.

Ornithinemia: see Inborn errors of metabolism.

Oro: abb. of Orotic acid.

Orobol: 3′,4′,5,7-tetrahydroxyisoflavone. See Isoflavone.

Orosomucoid, *α_1-seromucoid, α_1-acid glycoprotein, α_1AGp:* a plasma protein of mammals and birds. O. contains about 38 % carbohydrate, and is the most carbohydrate-rich and water-soluble of all the plasma proteins. The seromucoid fraction of blood consists of O. together with two other carbohydrate rich proteins (C1-inactivator and hemopexin). Increased blood levels of O. and of the total seromucoid fraction are associated with inflammation, pregnancy and various disease states, such as cancer, pneumonia, rheumatoid arthritis. After major surgery, increased levels of O. are produced until the wound is healed. O. also binds certain steroids, especially progesterone, but the binding affinity is lower than that for corticosteroid binding globulin. Membranes of human platelets contain considerable quantities of tightly bound O. Human O. is a single chain glycoprotein: M_r 39,000 by sedimentation equilibrium, 41,600 by sedimentation diffusion; isoionic pH 3.5. It has pronounced anion binding capacity so that the isoelectric point depends on the type and concentration of anions. The protein chain of O. contains 181 amino acid residues (primary sequence: K. Schmid et al. *Biochemistry* **12** (1973) 2711–2724). Five oligosaccharide units are attached to the β-carboxyl groups of asparaginyl residues at positions 15, 38, 54, 75 and 85. The native protein contains two disulfide bridges, located between cysteine residues at 5 and 147, and at 164 and 72. The *N*-terminus is pyroglutaminyl, and the *C*-terminus is serine. The C-terminal half of O. shows great similarities of sequence with the α-chain of haptoglobin and the H-chain of immunoglobulin G. Analysis of the carbohydrate of O. [O.H.G. Schwarzmann et al. *J. Biol. Chem.* **253** (1978) 6983–6987] shows 14 residues of *N*-acetylneuraminic acid (sialic acid), about 34 residues of neutral hexoses (D-galactose/D-mannose in an average ratio of 1.4); 31 residues of *N*-acetylglucosamine; 2 residues of L-fucose. *N*-Acetylneuraminic acid is always located terminally via a 2-ketosidic bond; the galactose occupies a penultimate position, linked β-glycosidically to the third sugar, *N*-acetylglucosamine. O. is synthesized in the liver. Synthesis of the carbohydrate moiety is initated by transfer of an *N*-acetylglucosaminyl residue to an asparaginyl residue when the nascent polypeptide chain is still attached to the ribosome (see Post translational modification of proteins).

Orotic acid, *Oro:* uracil 4-carboxylic acid, M_r 156.1, m.p. 344 347 °C (d.). Oro is an intermediate in Pyrimidine biosynthesis (see), and it is secreted in large quantities by mutants of *Neurospora crassa* that show a growth requirement for uridine, cytidine or uracil.

Orotic aciduria: see Inborn errors of metabolism.

Orotidine, *Ord:* orotic acid-3-β-D-ribofuranoside, a β-glycosidic Nucleoside (see) of D-ribose and the pyrimidine, orotic acid. M_r 288.21, m.p. > 400 °C. Cyclohexylamine salt: m.p. 184 °C, $[\alpha]_C^{23}$ +14.3° ($c = 1$, water). O. does not appear to be a normal intermediate of pyrimidine metabolism, but it is produced by mutants of *Neurospora crassa.*

Orotidine 5′-monophosphate, *OMP:* a nucleotide of orotic acid. M_r 368.2. OMP is an intermediate in Pyrimidine biosynthesis (see). Orotidine 5′-phosphate pyrophosphorylase catalyses the synthesis of OMP from orotic acid and 5-phosphoribosyl 1-pyrophosphate.

Orthonil, *PRB 8:* 2-(β-chloro-β-cyanoethyl)-6-chlorotoluene, a synthetic, stimulatory growth regulator. It is claimed that O. causes an increase in the sugar content of sugar beet.

Orthonil

Orycenin: see Glutelins.

Osajin: see Isoflavone.

Osmosis: a phenomenon associated with semipermeable membranes, especially Biomembranes (see). When two solutions are separated by a membrane which is permeable only to selected components of the solution (e.g. water), the component which can cross the membrane will flow from the side on which its partial pressure is higher to the side on which its partial pressure is lower. Cytoplasm is a concentrated aqueous solution of salts, sugars and other small molecules, most of which cannot diffuse across the plasma membrane; the water of the solution, however, can cross the membrane. As a result, the cells of freshwater organisms, and of plant root hairs, experience an inflow of water until the pressure within the cell increases to the point where the partial pressure of water is equal inside and outside the cell. Fresh-water unicellular animals expel the excess water by means of contractile vacuoles; multicellular animals excrete it through their kidneys. Plant cells are surrounded by rigid Cell walls (see), which enable them to withstand the high internal pressure caused by O. without bursting. Higher land plants utilize the pressure gradient created by O. to help drive their sap up from the roots. Salt-water organisms live in an environment in which the salt concentration outside the plasma membranes is higher than inside. O. thus tends to dehydrate the cell rather than to cause it to burst, and the organism must expend metabolic energy to excrete salt in order to maintain the lower internal concentration.

Osteocalcin: A small protein (M_r 5,800) which comprises 10–20 % of bone protein. O. from a variety of vertebrates has γ-carboxyglutamate (Gla) residues at positions 17, 21 and 24, and a disulfide bond between cysteine residues at positions 23 and 29. In the presence of Ca^{2+}, the protein undergoes a conformational change so that all the Gla residues are on the same side of an α-helix, and 5.4 Å apart. This distance is close to the spacing between adjacent Ca atoms in hydroxyapatite crystals, which make up the mineral part of bone. Thus O. is presumed to have a role in mineralization of bone. It is synthesized in bone; small amounts are also found in blood. [V. Geoffroy et a. 'A PEBP2α/AML-1-related Factor Increases Osteocalcin Promoter Activity through Its Binding to an Osteoblast-specific cis-Acting Element' *J. Biol. Chem.* **270** (1995) 30973–30979]

Osteonectin: see SPARC.

Ostreasterol: see Chalinasterol.

Ouabagenin: see Strophanthins.

Ouabain: see Strophanthins.

Ouchterlony technique: see Precipitation.

Ovalbumin: see Albumins.

Ovosiston: an oral contraceptive (see Ovulation inhibitors), consisting of a mixture of the progestin, Chlormadinone acetate (see), and the estrogen, Mestranol (see).

Ovulation inhibitors: a group of steroids, which inhibit ovulation by feedback inhibition of the production of Luteinizing hormone (see) and/or Follicular stimulating hormone (see). They are used extensively as oral contraceptives (otherwise known as "the pill"). Most commercially available O.i. preparations contain a combination of a progestin and an estrogen. The progestin is the actual O.i., whereas the estrogen is included to prevent breakthrough (midcycle) bleeding. Combination O.i. consist of a progestin-estrogen combination, which is taken daily throughout the cycle. A sequential O.i. was introduced in 1965 to mimic more closely the normal rise and fall of estrogen and progestin in the female monthly cycle; the first 15 pills contain estrogen only, and the last 5 contain a progestin-estrogen combination. Since these were less effective than the combination products, and showed undesirable side effects, they were removed from the US. market in 1976. The "minipill" contains a low dose of progestin (0.075 mg norgestrel, or 0.35 mg norethindone) and no estrogen; it is free from many side effects but users sometimes have irregular bleeding. Trials are also being conducted on the use of once-a-month and once-a-week O.i., and on injectable depot O.i. Since the natural progestin and estrogens show poor gut absorption, all the progestins and estrogens used in oral O.i. are synthetic products which show efficient gut absorption. Synthetic progestins used as O.i. are norethindone, norgestrel, norethynodrel, ethynodiol diacetate, dimethisterone. Synthetic estrogens used in O.i. preparations are Ethinylestradiol (see), Mestranol (see) ethinylestradiol 3-cyclopentyl ether (quingestanol). The quantities used are 0.5–5 mg progestin, and 0.05–0.1 mg estrogen per tablet. There are several different commercial oral contraceptives, containing various combinations of the above synthetic progestins and estrogens. See also Chlormadinone acetate. O.i. are also used for the treatment of dysmenorrea, endometrioses, cycle-dependent migraines and sterility.

Oxalic acid: HOOC–COOH. O.a. occurs widely in plants as its calcium, magnesium and potassium salts. By forming insoluble calcium salts in the intestine, it hinders the absorption of calcium. It is not metabolized by animals, and in large quantities it is poisonous. Humans normally excrete 10–30 mg O.a. daily; higher levels may lead to kidney damage (formation of oxalate kidney stones).

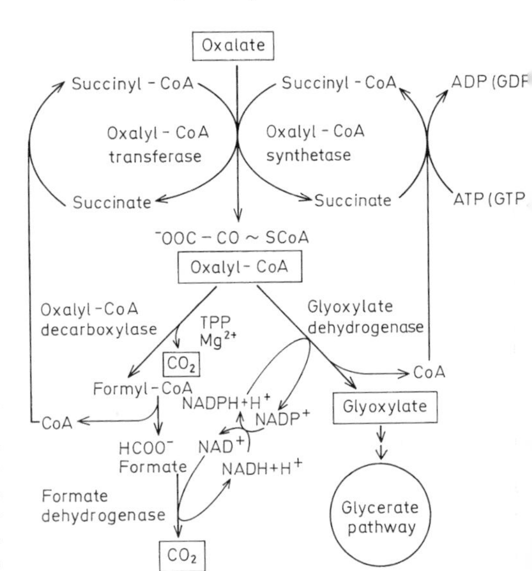

Fig. 1. *Oxalic acid.* Oxalate metabolism in *Pseudomonas oxalaticus.*

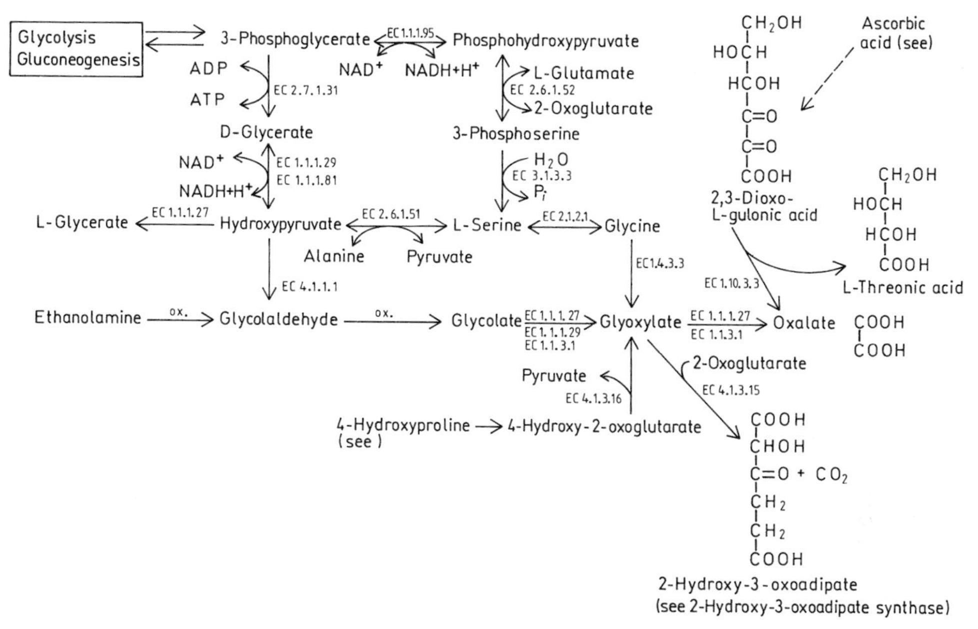

Fig. 2. *Oxalic acid.* Metabolic sources of oxalate. 2-Hydroxy-3-oxoadipate is a normal excretory product in humans, which diverts glyoxylate from oxalate production (see Inborn errors of metabolism, Oxalosis).
EC 1.1.1.27, L-Lactate dehydrogenase; EC 1.1.1.29, Glycerate dehydrogenase; EC 1.1.1.81, Hydroxypyruvate reductase; EC 1.1.1.95, Phosphoglycerate dehydrogenase; EC 1.1.3.1, Glycolate oxidase; EC 1.4.3.3, D-Amino acid oxidase; EC 1.10.3.3, Ascorbate oxidase; EC 2.1.2.1, Glycine hydroxymethyltransferase; EC 2.6.1.51, Serine-pyruvate aminotransferase; EC 2.6.1.52, Phosphoserine-oxoglutarate aminotransferase; EC 2.7.1.31, Glycerate kinase; EC 3.1.3.3, Phosphoserine phosphatase; EC 4.1.1.1, Pyruvate decarboxylase; EC 4.1.3.15, 2- Hydroxy-3-oxoadipate synthase; EC 4.1.3.16, 4-Hydroxy-2-oxoglutarate aldolase.

The O. a. content of fruits and vegetables is usually less than 10 mg/100 g fresh weight; however, rhubarb leaves contain 700 mg, rhubarb stems 300 mg, celery stems up to 600 mg, and spinach 60–200 mg/100 g fresh weight.

O.a. was first prepared in 1773 from wood sorrel (*Oxalis acetosella*). It is synthesized in plants by the oxidation of excess glyoxylate. It is also produced by oxidative cleavage of phenylpyruvate to benzaldehyde and O. a. (a reaction in the conversion of phenylalanine to hippuric acid).

In the bacterium *Pseudomonas oxalaticus,* O. a. is metabolized (via oxalyl-CoA) by an oxidative and a reductive pathway, which are coupled and operate simultaneously (Fig. 1); the balance is 2 $^{-}OOC–COO^{-} \rightarrow 2\ CO_2 + CHO–COO^{-} + H_2O$.

Oxaloacetic acid: $HOOC–CO–CH_2–COOH$, an oxodicarboxylic acid present in fairly high concentrations in plants, e. g. red clover, peas. The anion (oxaloacetate) is an intermediate in the Tricarboxylic acid cycle (see), and O. a. is the oxoacid derived from Aspartic acid (see) by Transamination (see). The enol form of O. a. may be *cis* or *trans:* hydroxymaleic acid (*cis*, m. p. 152 °C), and hydroxyfumaric acid (*trans*, m.p 184 °C [d.]).

Oxalosis: see Inborn errors of metabolism.

Oxalosuccinic acid: a β-ketotricarboxylic acid (2-oxotricarboxylic acid):

$$HOOC–CH_2–\underset{}{\overset{COOH}{\overset{|}{C}}}H–CO–COOH$$

The anion (oxalosuccinate) is an intermediate in the isocitrate dehydrogenase reaction of the Tricarboxylic acid cycle (see). Isocitrate dehydrogenase catalyses both the oxidation of isocitrate to oxalosuccinate, and its decarboxylation to 2-oxoglutarate.

Oxidases: oxidoreductases (see Enzymes), which use molecular oxygen as an electron acceptor. See also Flavin enzymes.

Oxidation: the loss of electrons. Classically, O. was defined as combination with oxygen or removal of hydrogen. The electrons are transferred to the oxidizing agent, which becomes reduced. Therefore O. is always coupled to Reduction (see), so that any O. or reduction is part of an oxidoreduction process. In metabolism there are different mechanisms of enzyme catalysed O., i. e. dehydrogenation, electron transfer, introduction of oxygen, and hydroxylation (Table).

The oxidizing or reducing power of a substance is indicated by its redox (reduction-oxidation) potential. The redox potential is related to the potential of the

Oxidative metabolism

Oxidation reactions in metabolism

Type of reaction	Enzymes	General reaction
Dehydrogenation	Oxidoreductases	$SH_2 \rightleftharpoons S$ (D → DH_2)
Electron transfer	4 and 2 electron transferring oxidases	$SH_2 \rightarrow S$ ($1/2O_2$ → H_2O or O_2 → H_2O_2)
Oxygen transfer	Dioxygenases	$S + O_2 \rightarrow SO_2$
Hydroxylation	Hydroxylases	$S + DH_2 + O_2 \rightarrow SOH + D + H_2O$

S = substrate, D = Hydrogen carrying cofactor

hydrogen electrode, and it is a quantitative index of electron affinity (i.e. oxidizing power) or tendency to lose electrons (i.e. reducing power). The defined standard potential in physical chemistry, or Normal potential, E_0 (pH = O, $pH_2 = 1$ atmosphere) is, however, inappropriate for biological purposes. Instead, the Normal potential E'_0, at pH 7.0 is taken as the reference point. E'_0 has a value of 0.42 volt on the E_0 scale. The E_0 value is only valid under standard reaction conditions, i.e. all reactants at unit activity. Such conditions do not obtain in the living cell, and the ratio of concentrations of oxidized and reduced forms of the redox pair is included in the calculation. The actual redox potential E′ is therefore expressed as: $E' = E'_0 + (0.06/n) \log (Ox/Red)$ (30 °C). The redox potential is related to free enthalpy (Gibb's potential) as follows: $\Delta G^0 = -nFE_0$, where n is the change of valency, and F the Faraday constant. As oxidizing potential increases E'_0 becomes more positive, so that strong reducing agents are characterized by high negative E'_0 values, and strong oxidizing agents by high positive values. See Table 3 under the entry for Respiratory chain; here the members of the respiratory chain and other biological redox systems are arranged in order of their redox potentials. Each component in the table is a more powerful oxidizing agent than those above it.

Oxidative metabolism: see Respiration.

Oxidative phosphorylation, *respiratory chain phosphorylation:* formation of ATP coupled with the operation of the Respiratory chain (see). Energy available from the flow of electrons from substrate to oxygen via the respiratory chain drives the synthesis of ATP from ADP and inorganic phosphate. Oxidation of one molecule of reduced nicotinamide adenine dinucleotide ($NADH + H^+$) generates three molecules of ATP, while oxidation of one molecule of reduced flavin adenine dinucleotide ($FADH_2$) yields two ATP. Complete oxidation of one molecule of glucose yields 38 ATP (2 from glycolysis and 36 from O.p.).

The mechanism of O.p., i.e. the nature of the energy transducing mechanism that converts the energy of electron flow into the chemical energy of ATP, has always been a controversial area of biochemistry. The earliest hypothesis, chemical coupling, proposes the existence of an energy-rich intermediate generated by electron flow and consumed in the phosphorylation of ADP. A second hypothesis, conformational coupling, proposes that the energy is stored in a protein conformational change caused by electron transfer; return to the original conformation is linked to ATP synthesis. Whatever the mechanism of O.p. eventually accepted, there seems little doubt that it will be based on the mechanism of chemiosmosis proposed by Nobel laureate P. Mitchell. This is described in detail under Chemiosmotic hypothesis (see).

22,25-Oxidoholothurinogenin: see Holothurines.

Oxidoreductases: see Enzymes (Table 1).

Oxidoreduction: see Oxidation.

9-Oxo-*trans*-2-decenoic acid: see Queen bee substance.

2-Oxoacid dioxygenases: Oxygenases (see) which catalyse substrate hydroxylation linked to the oxidation of a 2-oxoacid (usually 2-oxoglutarate) with formation of CO_2. They all require Fe^{2+} and ascorbate, the latter being necessary to maintain the Fe^{2+} in its reduced form. The reaction can be generalized as:

$$>C{-}H + O_2 + HOOC{-}CO{-}CH_2{-}CH_2{-}COOH \xrightarrow{\text{Ascorbate} + Fe^{2+}} >C{-}OH + HOOC{-}CH_2{-}CH_2{-}COOH + CO_2,$$

Substrate; 2-Oxoglutarate; Hydroxylated product; Succinate

where one atom of the substrate molecular oxygen is incorporated into a carboxyl group of succinate, and the other forms the hydroxyl oxygen of the hydroxylated product.

The following 2-oxoacid dioxygenases are known: *4-Trimethylaminobutyrate, 2-oxoglutarate:oxygen oxidoreductase (3-hydroxylating),* or *γ-Butyrobetaine, 2-oxoglutarate dioxygenase* (EC 1.14.11.1). It catalyses the hydroxylation of 4-trimethylaminobutyrate to L-3-hydroxy-4-*N*-trimethylaminobutyrate, or Carnitine (see).

Prolyl-glycyl-peptide, 2-oxoglutarate:oxygen oxidoreductase (4-hydroxylating), or *Proline, 2-oxoglutarate dioxygenase,* or *Protocollagen hydroxylase* or *Proline hydroxylase* (EC 1.14.11.2). It catalyses the hydroxylation of proline residues adjacent to glycine residues in procollagen, forming residues of 4-*trans*-hydroxyproline (see Collagen).

Thymidine, 2-oxoglutarate:oxygen oxidoreductase (2′-hydroxylating), or *Thymidine, 2-oxoglutarate dioxygenase,* or *Thymidine* 2′-hydroxylase, or *Pyrimidine deoxyribonucleoside* 2′-hydroxylase (EC 1.14.11.3). It catalyses hydroxylation of C2′ of the deoxyribose moiety of thymidine.

Peptidyllysine, 2-oxoglutarate:oxygen 5-oxidoreductase, or *Lysine, 2-oxoglutarate dioxygenase,* or *Lysine hydroxylase* (EC 1.14.11.4). It catalyses hydroxylation of lysine residues in procollagen, forming residues of 5-hydroxylysine (see Collagen).

Thymine, 2-oxoglutarate:oxygen oxidoreductase (7-hydroxylating), or *Thymine, 2-oxoglutarate dioxygenase,* or *Thymine 7-hydroxylase* (EC 1.14.11.6). (The dioxygenase of EC number 1.14.11.5 has now been deleted from the EC list).

4-Hydroxyphenylpyruvate:oxygen oxidoreductase (hydroxylating, decarboxylating), or *4-Hydroxyphenylpyruvate dioxygenase* (EC 1.13.11.27). It catalyses the oxidation and decarboxylation of 4-hydroxyphenylpyruvate to 2,5-dihydroxyphenylacetate (homogentisate), a reaction in the degradation of Phenylalanine (see) and tyrosine. The conversion is accompanied by migration of the side chain, and the substrate serves as its own 2-oxoacid electron donor, i.e. the enzyme may be described as an internal 2-oxoacid-dependent dioxygenase. [S. Lindstedt & M. Rundgren *J. Biol. Chem.* **257** (1982) 11922–11931]

N^6,N^6,N^6-Trimethyl-L-lysine, 2-oxoglutarate:oxygen oxidoreductase (3-hydroxylating), or *Trimethyllysine hydroxylase* (EC 1.14.11.8). It catalyses the hydroxylation of 6-*N*-trimethyl-L-lysine to 3-hydroxy-6-*N*-trimethyl-L-lysine, a reaction in Carnitine (see) biosynthesis. [R. Stein & S. England *Arch. Biochem. Biophys.* **217** (1982) 324–331]

Prolyl-glycyl-peptide,2-oxoglutarate:oxygen oxidoreductase (3-hydroxylating) or *Procollagen-L-proline, 2-oxoglutarate:oxygen oxidoreductase (3-hydroxylating),* or *Procollagen-proline 3-dioxygenase* (EC 1.14.11.7). It catalyses the hydroxylation of proline residues in procollagen to 3-*trans*-hydroxyproline (3-Hyp) residues in Collagen (see). 3-Hyp appears to occur only in the sequence -Gly-3-Hyp-4-Hyp- [K. I. Kivirikko & R. Myllylä *Methods in Enzymology* **82** (1982) 245 304]

Oxoeleostearic acid: see Licanic acid.

2-Oxoglutaric acid: see α-Ketoglutaric acid.

5-Oxoproline: see Pyroglutamic acid.

Oxygen: see Bioelements.

Oxygenases: enzymes that catalyse the incorporation of the oxygen of molecular oxygen into their organic substrates, i.e. the oxygen atom(s) appearing in the product is (are) derived from atmospheric O_2, and not from water. Dioxygenases (oxygen transferases) catalyse the introduction of both atoms of molecular oxygen. Monooxygenases or hydroxylases catalyse the introduction of one atom from molecular oxygen; the other atom is reduced to water. Monooxygenases therefore require a second substrate, which serves as an electron donor; for this reason they are also called mixed function oxygenases. They catalyse the following type of reaction: $AH + O_2 + DH_2 \rightarrow AOH + H_2O + D$, where AH is the substrate, AOH the hydroxylated substrate, DH_2 an electron donor, and D the oxidized electron donor.

Flavoprotein hydroxylases are found primarily in bacteria, e.g. 4-hydroxybenzoate hydroxylase has been crystallized from 4 different specis of *Pseudomonas.* The prosthetic group is FAD, and hydroxylation of the substrate to 3,4-dihydroxybenzoate is coupled to the oxidation of NADPH. Pteridine-dependent hydroxylases are a class of monooxygenases with a pteridine prosthetic group, e.g. L-phenylalanine-4-monooxygenase, tyrosine-3-monooxygenase of the adrenal medulla, and tryptophan-5-monooxygenase of the brain.

Heme-coupled monooxygenases contain cytochrome P-450. They are present in microsomes and are responsible for many hydroxylation reactions, e.g. 11β-hydroxylation of steroids in the adrenal cortex, 2-hydroxylation of estrogens in the liver; the liver system is especially important in the hydroxylation of drugs and xenobiotics, thus rendering them water soluble, capable of conjugation and easily excretable. A cytochrome P-450 system responsible for the hydroxylation of camphor (a 5-*exo*-hydroxylase) has been purified from *Pseudomonas putida,* and named putidaredoxin; it contains FAD, an $Fe_2S_2Cys_4$ center, and a P-450 cytochrome; substrate hydroxylation is coupled to the oxidation of NADPH.

Certain monooxygenases contain copper (copper-containing hydroxylases), e.g. dopamine-β-hydroxylase; this enzyme is associated with the chromaffin granules of the adrenal medulla, and is responsible for the oxidation of dopamine to noradrenalin; the second substrate (electron donor) is ascorbic acid. Various phenolases are also Cu-containing monooxygenases. The mechanism of action is unclear, but it appears that a monophenol is hydroxylated to an *o*-diphenol, coupled to the oxidation of a diphenol (i.e. the second substrate) to an *o*-quinone.

Monooxygenases are responsible for the hydroxylation of proline and lysine residues in collagen. These hydroxylases contain ferrous iron and have a specific requirement for 2-oxoglutarate; the latter acts as an electron donor by becoming oxidatively decarboxylated to succinate and CO_2 (for this reason, these enzymes are also classified as dioxygenases, and they are described in greater detail in the entry: 2-Oxoacid dioxygenases). 4-Hydroxyphenylpyruvate hydroxylase from liver and kidney is also an Fe-containing decarboxylating monooxygenase; hydroxylation is accompanied by migration of the side chain, while oxidative decarboxylation of the side chain serves to consume the other oxygen atom; the product is homogentisic acid (see L-Phenylalanine).

All known dioxygenases contain iron, either in heme groups or as Fe-S centers. Some also contain copper. Examples are Tryptophan-2,3-dioxygenase (see), homogentisate oxidase (see L-Phenylalanine), 3-hydroxy-anthranilate oxidase (see L-Tryptophan).

Mono and dioxygenase are employed extensively by bacteria in the degradation of aromatic compounds and are therefore fundamentally important in the carbon cycle of the biosphere. See 2-Oxoacid dioxygenases.

Oxygen electrode: an amperometric sensor for detecting oxygen concentration changes in solution. In the polarographic O. e., a polarizing voltage of approximately + 0.6 V, from an external source, is established between a platinum cathode and a silver/silver chloride anode bathed in saturated KCl; the current passing is proportional to the oxygen concentration in solution. In the galvanic O. e., the current is produced between a silver cathode and a lead anode bathed in saturated lead acetate; in this case the potential is generated by the electrode system and does not need to be supplied externally. In both types of O. e., the electrode is normally separated from the solution by an oxygen-permeable plastic membrane.

Oxygen metabolism: metabolic reactions involving oxygen, including: 1. General oxidation of metabolites by Dehydrogenation (see), removal of electrons (oxidases, see Flavin enzymes), and addition of oxygen to the substrate (see Oxygenases). 2. Oxidation (see) of reduced coenzymes via the Respiratory chain (see).

Oxyhemoglobin: see Hemoglobin.

Oxysomes: see Mitochondria.

5′-Oxytetracycline: see Tetracyclines.

Oxytocin: a neurohypophysial, peptide hormone, M_r 1,007. O. causes contraction of the smooth muscle of the uterus and of the mammary gland (milk ejection). It is structurally related to the other neurohypophysial hormone, Vasopressin (see); these two hormones have a common ancestry and there is an overlap in their physiological activities. In lower vertebrates, four other neurohypophysial hormones have been identified, which are variants of O. and vasopressin, with different amino acid residues in positions 4 or 8, or both. [Arg^8]Oxytocin is a hybrid analog, in which the tripeptide tail of vasopressin is attached to the ring of O.; it occurs naturally in chicken pituitary, and is the probable ancestor of all other related neurohypophysial hormones. [Ser^4, Ile^8]Oxytocin (also called isotocin) occurs in teleost fish; [Ser^4, Glu^8]Oxytocin (also called glumitocin) occurs in elasmobranchs. [Ile^8]Oxytocin (also called mesotocin) is present in amphibians and reptiles.

Cys–Tyr–Ile–Gln–Asn–Cys–Pro–Leu–Gly–NH_2 (disulfide bridge between Cys and Cys)

Oxytocin

O. is synthesized in the hypothalamus in the Nucleus paraventricularis, and transported to the posterior lobe of the pituitary (hypophysis) via the Tractus paraventriculo-hypophyseus, in combination with the transport protein, Neurophysin I (see). It is released into the blood stream in response to psychological and tactile stimulation of genitalia, or suckling stimulation of the mammary gland, and it acts on its target organ via the adenylate cyclase system. Methods for the assay of O. are chiefly biological, based on the milk ejecting activity in lactating animals, or the behavior of perfused sections or strips of mammary tissue. Radioimmunoassay is also possible. O. is inactivated and degraded in the organism by proteolysis. Over 300 analogs of O. have been synthesized, some of which have higher biological activity than the natural product. The role of O. in parturition is unclear. Although O. is probably involved in uterine contraction, it is not generally accepted that the onset of parturition is determined by the release of O. Structural elucidation and the first synthesis of O. were reported in 1953/54 by V. duVigneaud et al. The most active analog obtained so far by amino acid substitution is [Thr^4]Oxytocin. [M.-Y. Ho et al. Bovine Oxytocin Transgenes in Mice *J. Biol. Chem.* **270** (1995) 27199–27205]

Oxytocinase: see Aminopeptidases.

P

Ⓟ: The letter P in a circle denotes inorganic phosphate (see P_i). If the circle is attached to another group by a bond (R–Ⓟ), the encircled P may represent $-PO_3H_2$ (the group that is transferred in phosphorylation reactions), or it may represent phosphate ($-O-PO_3H_2$). If the bridge oxygen atom is shown, then the encircled must be $-PO_3H_2$. The symbol is no longer recommended, but is still used.

~Ⓟ, ~*P*: a high energy or labile phosphate group. See High energy bonds, Energy-rich phosphates.

***P*:** the currently recommended abbreviation or symbol for bound phosphate and all its ionized forms. Attached to one other symbol, it denotes PO_3H_2 (and its ionized forms). Attached to two other symbols, it denotes PO_2H (and its ionized form). The bridge oxygen atom(s) to attached groups is (are) therefore not included. [*Eur. J. Biochem. 1* (1967) 259–266]

P_i: inorganic phosphate, any anion of orthophosphoric acid, (PO_4^{3-}, HPO_4^{2-}, $H_2PO_4^-$)

Pacemaker reaction: see Metabolism.

Padmakastein: 5,4′-dihydroxy-7-methoxyisoflavanone. See Isoflavanone.

Paeonidin: 3,5,7,4′-tetrahydroxy-3-methoxyflavylium cation. The 3,5-β-diglucoside (paeonin) is the chief pigment of peony.

Pahutoxin: the main toxin of the tropical box fish *Ostracion lentiginosus.* P. is a choline derivative, highly poisonous for fish, but not for warm blooded animals.

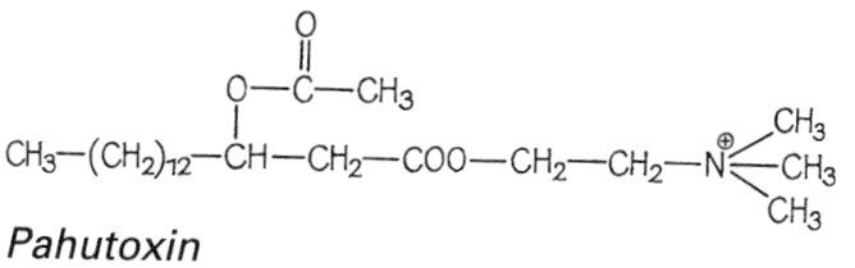

$$CH_3-(CH_2)_{12}-CH(O-CO-CH_3)-CH_2-COO-CH_2-CH_2-N^{\oplus}(CH_3)_3$$

Pahutoxin

Paleoproteins: a group of proteins from fossils, in particular the exoskeleton of mollusks (snail and mussel shells, and cuttle bone). The most studied P. is conchiolin from the hard parts of fossil mussels from the Lower Tertiary (25–58 million years), the Jurassic (150 million years) and the Silurian (440 million years). Conchiolin has been characterized chemically and by electron microscopy; it is a fibrillar scleroprotein, rich in glycine, alanine and serine.

Palindrome: a nucleic acid sequence that is identical to its complementary strand (when both are read in the same 5′-3′ direction). In the region of a P. there is therefore a twofold rotational symmetry. Perfect P., e.g. GAATTC, often occur as recognition sites for restriction enzymes. Imperfect palindromes, e.g. TACCTCTGGCGTGATA, often act as binding sites for proteins such as repressors. Interrupted P., e.g. GGTTXXXXXAACC, make possible the formation of a stem with a loop (hairpin structure) as in tRNA.

Palmitic acid: *n*-hexadecanoic acid, $CH_3-(CH_2)_{14}-COOH$, a fatty acid, M_r 256.4, m.p. 63 °C, b.p.$_{100}$ 271.5 °C, b.p.$_{15}$ 215 °C. Together with stearic acid, P.a. is one of the most widely distributed natural fatty acids, and is present in practically all natural fats, e.g. 36 % in palm oil 29 % in bovine carcass fat, 15 % in olive oil; it is also found in phosphatides and waxes. P.a. is the raw material for the manufacture of candles, soap, wetting agents and antifoams.

Palmitoleic acid: Δ^9-hexadecanoic acid, $CH_3(CH_2)_5-CH=CH-(CH_2)_7-COOH$, an unsaturated fatty acid, M_r 254.4, m.p. 1 °C, b.p.$_{15}$ 220 °C. P.a. is present esterified in the acylglycerols of many plant and animal fats, and in phosphatides.

Pancreas: a vertebrate organ producing a digestive secretion which enters the adjacent duodenum in response to the hormones Secretin (see) and Cholecystokinin (see). It also contains about 1 million islets of Langerhans, each with a diameter of 150 μm; these have a rich blood supply and are innervated with unmyelinated nerve fibers. The islets contain various types of hormone-producing cells, the A cells which produce Glucagon (see), the B cells which make Insulin (see) and the D cells which manufacture Gastrin (see).

Pancreatic enzymes: a group of at least 12 digestive enzymes, including some of the most investigated of all enzymes. Autolysis of the pancreas does not occur, because the proteolytic enzymes, trypsin, chymotrypsin A and B, elastase and carboxypeptidase A and B, and phospholipase A_2 are synthesized and stored in the pancreas as inactive zymogens. The other P.e. require effectors for optimal activity, which are present in the duodenum. Trypsin inhibitors in the pancreatic tissue and secretion afford additional protection against proteolytic destruction by active P.e. With the exception of cholesterol esterase (M_r 400,000), the M_r of P.e. lie between 13,700 (ribonuclease) and 50,000 (α-amylase).

Pancreatic lipase deficiency: see Inborn errors of metabolism.

Pancreatin: defatted, powdered preparations of pancreas. Acetone-dried pancreas powders can be stored for long periods without loss of activity. P. contains all the pancreatic enzymes in active form, and it serves as a starting material for their laboratory purification. P. is also used in the pharmaceutical industry for the preparation of enzyme tablets, which are used in cases of secretory malfunction of the pancreas.

Pancreozymin: see Cholecystokinin.

Pantetheine 4′-phosphate, *phosphopantetheine:* the phosphate ester of *N*-(pantothenyl)-β-aminoethanethiol. A residue of P. is present in the molecule of Coenzyme A (see). P. is the prosthetic group of acyl carrier proteins in certain multienzyme complexes, e.g. fatty acid synthetase and gramicidin S synthetase, where it serves as a "swinging arm" for attachment of activated fatty acid and amino acid groups.

Pantherine: see Ibotenic acid.

Pantothenic acid: a vitamin of the B_2 group. See Vitamins.

PAP: acronym of 3′-phosphoadenosine 5′-phosphate. See Sulfotransferase.

Papain (EC 3.4.22.2): a thiol enzyme from the latex and unripe fruit of *Carica papaya* (tropical melon or papaw). P. is unusually stable at high temperatures and in high concentrations of denaturing agents, e.g. 8 M urea or organic solvents such as 70 % ethanol or 15 % dimethylsulfoxide. P. is a carbohydrate-free, basic, single chain protein (M_r 23,350; 212 amino acid residues; methionine absent; pI 8.75) with 4 disulfide bridges, and catalytically important cysteine (position 25) and histidine (position 158) residues. The molecule is a rotational ellipsoid, divided by a cleft, and containing a predominance of antiparallel β-structures.

For activity, P. requires a free SH group and an operating pH of 5–5.5. SH blocking agents, like iodoacetic acid, are powerful P. inhibitors, whereas SH compounds, like mercaptoethanol, are potent activators. P. has a broad specificity, embracing endopeptidase, amidase and esterase activities. It catalyses the cleavage of a wide variety of peptide bonds, indicating a fairly low specificity for peptide bond cleavage. Systematic studies with model peptides have, however, revealed a high degree of specificity for certain groupings. The active site of P. can be divided into 7 "subsites", each accommodating one amino acid residue of the substrate. Four of these subsites are on one side of the catalytic site, three on the other:

$$H_2N{-}R_1{-}R_2{-}R_3{-}R_4{-}R_5{-}R_6{-}R_7{-}COOH$$

↑ cleavage

The site corresponding to R_3 specifically interacts with the side chain of phenylalanine (Phe). The presence of Phe in position 3 or further from the *C*-terminus increases the susceptibility of the peptide to hydrolysis, so that cleavage occurs at the peptide bond one residue removed from the Phe towards the *C*-terminus:

↓

H_2N–Ala–Phe–Ala–Lys–Ala–$CONH_2$

For the same reason, peptides containing Phe as the second residue from the *C*-terminus are inhibitors of P., e.g. Ala-Ala-Phe-Ala, or Ala-Ala-Phe-Lys.

P. is used medically for the treatment of necrotic tissue and eczema, and in protein chemistry for cleaving proteins into large peptides.

Papaveraceae alkaloids, *Papaver alkaloids, poppy alkaloids:* a group of Benzylisoquinoline alkaloids (see), occurring especially in species of poppy (*Papaver*). They include the important Opium alkaloids (see).

Papaverine: an opium alkaloid occurring with morphine in various species of poppy. M_r 339.39, m.p. 147–148 °C. Physiologically it acts peripherally, and causes relaxation of smooth muscle; it is therefore used for the treatment of spasms in the gastrointestinal tract. P. is not addictive. For formula and biosynthesis, see Benzylisoquinoline alkaloids.

Paper autoradiography: see Isotope technique.

Paper chromatography: a chromatographic separation method, which employs a high quality filter paper (chromatography paper) as the carrier. P.c. is almost pure partition chromatography on cellulose. Under certain conditions, the separation deviates from that predicted by the Nernst distribution, owing to slight adsorption and ion exchange effects. The stationary phase is a film of water or adsorbed hydration layer on the cellulose fibers of the paper. The mobile phase, an organic solvent or mixture of solvents, migrates over the stationary phase. In a defined solvent system, each substance has a rate of migration, which is expressed as the R_f value (ratio to the solvent front): R_f = (migration distance of substance)/(migration distance of the solvent). The migration distance of the solvent is the distance from the point of application of the substance to the front of the advancing solvent. Under standardized conditions (composition of solvent system, temperature, degree of vapor saturation in the chromatography tank), the R_f value is a constant.

In practice, P.c. consists of the physical separation, followed by special detection methods for the location of the individual substances. It may be performed on an analytical or preparative scale. With respect to the movement of the solvent system and position of the paper, P.c. can be ascending, descending or horizontal. Single or two dimensional procedures, and radial P.c. (circular filter paper technique) are commonly used. Additional useful techniques are multiple development (the first solvent is removed by drying and a different solvent is run in the same dimension) and flow-through or run-off (the solvent is allowed to run for a prolonged period by dripping from the edge of the paper).

For separation of lipophilic substances, such as fatty acids and steroids, a reversed phase system of P.c. can be used. In this technique, the chromatography paper is made hydrophobic by impregnation with silicone or paraffin oil, or by acetylation. The phases are therefore reversed, i.e. the hydrophilic solvent system is repelled by the cellulose fibers, so that it forms the mobile phase; the more strongly hydrophobic organic solvent becomes the stationary phase. Zaffaroni systems are important for the separation of steroids; these are usually water-free, and the paper is impregnated with formamide or propylene glycol, which acts as the stationary phase.

Paper factor: see Juvabione.

PAPS: acronym of Phosphoadenosinephosphosulfate.

Paramyosin: see Muscle proteins.

Paraproteins: abnormal immunoglobulins, or normal proteins which are found in increased quantities in blood plasma in various hematological disturbances, known as paraproteinemias.

Known P. are Bence-Jones proteins, amyloid proteins, Waldenström-type IgM and cryoglobulin (a 7S globulin). Owing to their homogeneous character and relative ease of isolation on a preparative scale, P. are among the best studied immunoglobulins.

Parapyruvate: a dimer which accumulates during storage of pyruvate.

$$^{-}OOC-\underset{CH_3}{\overset{OH}{C}}-CH_2-CO-COO^{-}$$

Parapyruvate

Parathormone, *parathyrin:* a hormone produced by the parathyroid gland, which influences the metabolism of calcium and phosphate. It is a single chain proteohormone with 84 amino acid residues of known primary structure [R.T. Sauer et al. *Biochemistry* **13** (1974), 1994–1999]. M_r 9,402 (porcine). P. influences the cells that degrade bone (osteoclasts) by activation of membrane-bound adenylate cyclase and by increasing the entry of Ca^{2+} into these cells. The resulting mobilization of Ca^{2+} causes an increase in blood calcium. This is necessarily accompanied by the release of free phosphate which is excreted via the kidneys. Thus P. favors phosphate secretion in the distal part of the kidney tubule, and inhibits phosphate resorption in the proximal tubule. P. promotes calcium absorption by the intestine. The action of P. is therefore opposite to that of Calcitonin (see). P. is degraded by the liver, and some is excreted in the urine. Absence of P. leads to a decrease of blood calcium, accompanied by neuromuscular overexcitability (tetany).

Parathyrin: see Parathormone.

Parathyroid glands, *Glandula parathyreoidea, Sandstroem's body:* endocrine organs, which develop from the endodermal lining of the third and fourth pharyngeal pouches of the embryo. P. g. are present in all vertebrates, excluding fishes. As a rule there are four P. g., but the location and number of individual P. g. may vary considerably. Parathyroid tissue is sometimes found in the mediastinum. In humans the P. g. are small, oval bodies about 5 mm long and 4 mm wide, embedded in the posterior surface of the lobes of the thyroid gland (2 in the superior, and 2 in the inferior lobes). In non-mammals, the P. g. are not embedded in the thyroid. Although the P. g. are usually physically associated with the thyroid, there is no functional relationship between the two glands. P. g. secrete parathormone, which raises the concentration of blood calcium and causes resorption of bone calcium.

Paromomycin: an aminoglucoside antibiotic. See Streptomycin.

Partition chromatography: see Chromatography.

Parvalbumins: water soluble, acidic, monomeric proteins, M_r approx. 12,000. P. were originally thought to be present only in skeletal muscle of fish and amphibia, but have now been found in mammalian muscle. They have two high-affinity sites for Ca^{2+} ions ($K_{diss} = 0.1–4 \times 10^{-6}$), which also bind Mg^{2+} competitively. In relaxed muscle (Ca^{2+} approx. 10^{-8} M), P. bind $2Mg^{2+}$ and no Ca^{2+}. In contracting muscle, intracellular Ca^{2+} increases to $10^{-6}–10^{-5}$M, and Ca^{2+} becomes bound with displacement of Mg^{2+}. Thus exchange of Ca^{2+} with Mg^{2+} seems to be important in the physiological activity of P. From rabbit, rat, chicken, carp and frog, P. are isostructural. They have 6 α-helical regions, A to F. Loops between helices C and D and between E and F are the binding sites for the 2 Ca^{2+} ions. P. have also been found in some neurons of the CNS, in Leydig cells of the testis, and in the ovary. It seems probable that P. are involved in the relaxation of fast-twitch muscle fibers in mammals. See also Muscle proteins. [C. W. Heizmann *Experentia* **40** (1984) 910–921]

Passive hemagglutination: see Agglutination.

Passive transport: see Transport.

Pasteur effect: an inverse relationship between the rate of glucose utilization and the availability of oxygen. This was first observed in 1860 by L. Pasteur, who noted that yeasts decompose more sugar under anaerobic than under aerobic conditions. In the anaerobic glycolysis of glucose, there is a net yield of 2 molecules of ATP per molecule of glucose, compared with 36 ATP for the complete aerobic respiration of glucose. From the ratio 36/2 = 18, it is evident that 18 times more glucose must be consumed under anaerobic conditions than under aerobic conditions, in order to obtain an equivalent amount of energy. Thus the P. e. represents the regulation of glucose consumption to match the energy needs of the cell. Naturally, the effect is only observed in facultative cells, i. e. cells which can adjust their metabolism to either aerobic or anaerobic conditions, e. g. yeast cells, muscle cells.

In the absence of oxygen, animal cells perform anaerobic glycolysis and produce lactate, yeast cells ferment glucose to various fermentation products, e. g. glycerol and notably ethanol. When oxygen is admitted, production of lactate quickly ceases and lactate rapidly disappears; at the same time, the rate of glucose uptake is markedly decreased. The P. e. can largely be explained by the allosteric properties of phosphofructokinase (EC 2.7.1.11), the rate-limiting ("pace maker") enzyme of glycolysis. This enzyme is inhibited by ATP and activated by ADP and AMP. The ATP inhibition is overcome by AMP, cAMP, P_i, fructose 1,6-*bis*phosphate and fructose 6-phosphate, but it is increased by citrate. The citrate effect is significant in yeast and in aerobic muscle, which also use fatty acids and ketone bodies as major sources of energy. The effect of cAMP may be important in adipose tissue, where it overcomes the citrate inhibition.

The P. e. is especially important in skeletal muscle; the quantity of ATP in muscle (5–7 μmoles per g fresh weight) is sufficient to support contraction for 1–5 seconds. ATP synthesized from creatine phosphate (see Phosphagens) supports further short term contraction. During this early period the muscle becomes relatively anaerobic as the oxygen stored in the myoglobins is consumed and contraction restricts the blood supply by squeezing the blood vessels. Rephosphorylation of ADP to ATP then depends upon the anaerobic glycolysis of glycogen to lactate; the activity of phosphofructokinase is markedly increased, with a corresponding increase in the rate of glycolysis. However, this effect cannot simply be explained by the change in level of ATP; in fact, the steady state concentration of ATP in contracting muscle is only slightly lower than in resting muscle; for example in insect flight muscle a 10 % decrease in the ATP level results in a 100-fold increase in the rate of glycolysis.

Two complementary mechanisms have been proposed for this amplification: 1. Adenylate kinase (EC 2.7.4.3) catalyses the interconversion, 2ADP ⇌ ATP + AMP with an equilibrium constant of approximately 0.5; under the steady state conditions in the cell, the ATP concentration is much higher than those of ADP and AMP. Thus a small change in ATP concentration causes a much larger percentage change in AMP concentration. 2. Simultaneous operation of phosphofructokinase and fructose 1,6-*bis*phosphatase (EC 3.1.3.11) would constitute a Futile cycle (see). In resting muscle phosphofructokinase is largely inhibited by the ATP/AMP ratio. Furthermore, much of the fructose 1,6-*bis*phosphate that is formed is converted back to fructose 6-phosphate by fructose 1,6-*bis*phosphatase, i.e. the futile cycle does operate at a low rate, and the net flow rate of glycolysis is equal to the difference between the activities of phosphofructokinase and fructose 1,6-*bis*phosphatase. When the muscle contracts, the small decrease in ATP causes a large relative rise in AMP; the activity of phosphofructokinase increases, that of fructose 1,6-*bis*phosphatase decreases, and there is a marked increase in the rate of production of fructose 1,6-*bis*phosphate. The first reaction of glucose utilization is catalysed by hexokinase, (EC 2.7.1.1) which is inhibited by high concentrations of glucose 6-phosphate. In cells containing a reserve carbohydrate (e.g. glycogen in muscle cells), glucose 6-phosphate is also formed via glucose 1-phosphate. Glucose 6-phosphate accumulates during the inhibition of phosphofructokinase in sufficient quantity to inhibit hexokinase, thus preventing even the first stage of glucose utilization. A quantitative index for the effect of oxygen on glycolysis and fermentation is given by the Meyerhof oxidation quotient ($R_{an.f.}$ = rate of anaerobic fermentation; $R_{a.r.}$ = rate of aerobic respiration; RO_2 = rate of O_2 uptake): $(R_{an.f.} - R_{a.r.})/RO_2$, which is usually about 2. Consumption of one molecule of oxygen generally inhibits the production of two molecules of lactate or ethanol.

The P.e. is not observed, or only slightly, in malignant tumors, intestinal mucosa, red blood cells (mammalian) and retina (mammalian and especially avian). These tissues obtain their energy by the anaerobic degradation of glucose, even in the presence of oxygen.

Uncoupling agents (e.g. dinitrophenol) abolish the P.e., so that full anaerobic carbohydrate utilization is maintained in the presence of oxygen.

Pathocidin: see 8-Azaguanine.

Pathological proteins: see Paraproteins.

PCA: acronym of pyrrolidone carboxylic acid. See Pyroglutamic acid.

P-cell stimulating factor: see Colony stimulating factors.

P700 Chlorophyll a-protein: a strongly hydrophobic, integral membrane protein isolated from the thylakoids of many angiosperms, gymnosperms and green algae. It is a component of photosystem I. The ratio chlorophyll *a*/P700 is in the range 40–45/1, and preparations are also associated with cytochromes *f* and b_6. The complex contains 14 mol chlorophyll per mol protein (M_r 110,000). Thus only one in three of the complex molecules contains P700, and the total pigment system must contain at least 3 molecules of the protein. Photosynthesis-deficient mutants are known, which lack P700 Chlorophyll a-protein, e.g. mutants of *Scendesmus* and *Antirrhinum*. See Photosynthesis, Photosynthetic pigment systems.

PCM: acronym of Protein-calorie malnutrition.

Pectin esterase: see Enzymes. Table 2.

Pectin lyase: see Enzymes, Table 2.

Pectins: high M_r polyuronides, consisting of α1,4 glycosidically linked D-galacturonic acid residues. Some of the carboxyl groups are present as their methyl esters. The free acids are called pectinic acids. Varying extents of esterification and polymerization give rise to a wide variety of P. They are distributed widely in plants, and are found in association with cellulose; they are structurally important as cement and support substances, particularly in the middle lamellae and primary walls of plant cells. P. are especially plentiful in fleshy fruits, roots, leaves and green stems. They are prepared from sliced sugar beet and from apple and citrus residues (following juice extraction) by gentle extraction with hydrochloric, lactic or citric acids.

Due to the presence of hydrophilic groups, P. have a high capacity for binding water. They are therefore powerful gelling agents, a property which is widely exploited in the food industry (P. are used as setting agents in jams, etc.), and in the preparation of pharmaceuticals and cosmetics.

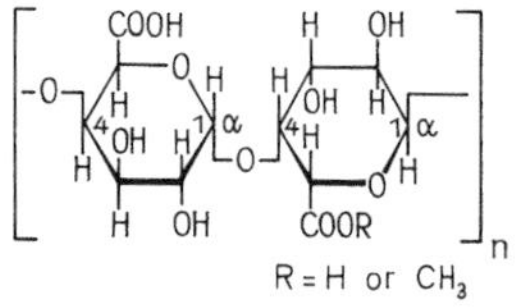

R = H or CH_3

Pectin

Pelargonidin: 3,5,7,4′-tetrahydroxyflavylium cation, an aglycon of many Anthocyanins (see). M.p. >350 °C. P. is widely distributed in higher plants as its various glycosides, which are responsible for the rose, orange-red and scarlet colors of petals and fruits. In addition to some acylated derivatives, the structures of about 25 natural glycosides are known, e.g. callistephin (3-β-glucoside) of the red aster; fragarin (3-β-galactoside) of the strawberry; and pelargonin (3,5-di-β-glucoside) of various *Pelargonium* species and garden dahlias.

Pelargonin: see Pelargonidin.

Pellagra: see Vitamins (nicotinic acid).

Pellotine: see Anhalonium alkaloids.

PEM: acronym of Protein-energy malnutrition.

Penicillinamidase: see Enzymes, Table 2.

Penicillinases (EC 3.5.2.6): enzymes which catalyse the hydrolysis of the β-lactam ring of penicillin; the resulting penicilloic acids have no antibiotic activity. Production of P. is adaptive, and P.-producing (i.e. penicillin-resistant) strains can be isolated by treating bacterial populations with penicillin. All P. are single chain polypeptides. The primary structures of soluble P. from *Staphylococcus aureus* (257 amino acid residues, M_r 28,800) and *Bacillus licheniformis* (265 amino acid residues, M_r 29,500) are known [R. P. Ambler & R. J. Meadway *Nature* **222** (1969) 24; R. J. Meadway *Biochem. J.* **115** (1969) 12P–13P]. *Ba-*

cillus licheniformis possesses a membrane-bound P. (M_r about 33,000) which is a precursor of the soluble exopenicillinase [S. Yamamoto & J. O. Lampen *J. Biol. Chem.* **251** (1976) 4095–4101].

Penicillins: sulfur-containing antibiotics produced by fungi of the genera *Penicillium, Aspergillus, Trichophyton* and *Epidermophyton.* All P. contain a condensed β-lactam-thiazolidine ring system, whereas the acyl group (R) is variable.

```
               S    CH3
R-NH-CH-CH       C
     |  |        |  CH3
   O=C--N-------CH-COOH
```

Name	R
Benzylpenicillin (penicillin G)	$C_6H_5-CH_2-CO-$
Pentenylpenicillin (penicillin F)	$CH_3-CH_2-CH=CH-CH_2-CO-$
n-Heptylpenicillin (penicillin K)	$CH_3-(CH_2)_6-CO-$
Penicillin N	$HOOC-CH(NH_2)(CH_2)_3-CO-$
6-Aminopenicillanic acid ("6 APS")	H–
Penicillin X	$HO-C_6H_4-CH_2-CO-$
Penicillin V	$C_6H_5O-CH_2-CO-$
Ampicillin	$C_6H_5-CH(NH_2)-CO-$

The growth inhibitory properties of P. were first observed in 1928 in a *Staphylococcus* culture by Alexander Fleming, and its use against bacterial infections in humans was first tested in 1941. Against sensitive bacteria, penicillin G is active at a dilution of 1:50,000,000; one mg of penicillin G corresponds to 1670 international units. The P. are, however, unstable to acids, and they are readily hydrolysed to inactive penicilloic acids by the action of Penicillinases (see). Commercially, P. are prepared by large scale fermentation in steel vats (up to 100 m^3), using high yielding strains. These commercial strains produce about 20,000 times more P. than the original isolates, and the yield is further increased by the addition of precursors, e.g. phenylacetic acid for penicillin G.

6-Aminopenicillanic acid (6 APS) is an important precursor for the organic synthesis of new P. The compound itself has no antibiotic activity; it is isolated as a fermentation product from cultures of *Pencillium chrysogenum,* or prepared by the enzymatic hydrolysis of benzylpenicillin. Thousands of new P. have been prepared by the acylation of 6 APS, but only a few of these are therapeutically useful, e.g Penicillin V is relatively stable to acid and is not hydrolysed in the stomach, so that it may be administered in tablet form; Ampicillin (the aminophenylacetyl derivative of 6 APS), has a wider spectrum of activity than most other P., including activity against various Gram-negative bacteria (*Typhus, E. coli,* etc.).

Penicillin V serves as the starting material for the semisynthesis of a group of antibiotics related to cephalosporin C, which are active against penicillin-resistant strains.

P. are biosynthesized from the amino acids α-aminoadipic acid, cysteine and valine, which become linked by peptide bonds. The residue of α-aminoadipic acid is usually subsequently lost and replaced by a different acyl residue, but it is retained in the molecule of penicillin N.

P. act by inhibiting cross linkage of the muropeptide in the Murein (see) layer of the cell wall. The cell wall is thus weakened and cannot withstand the high internal pressure of the bacterial cell (about 30 atmospheres). P. are the most widely and intensively used antibiotics; they are well tolerated by the animal organism and have a relatively broad spectrum of activity.

Penicillin selection technique: an aid to the isolation of chosen auxotrophic mutations in a bacterial population. It is based on the fact that penicillin kills growing bacterial cells, but does not affect nongrowing cells.

In a medium containing only growth requirements for the wild type organism, the wild type will grow, whereas nutritional auxotrophs will fail to grow (at a suitable concentration of cells over a suitable period of time, to avoid cross feeding of auxotrophs by wild cells). Addition of penicillin then kills the wild cells. After washing away the penicillin, the cells are cultured on a minimal medium supplemented by the auxotrophic requirement to favor growth of the auxotrophs (e.g. histidine in the isolation of histidine auxotrophs). The alternating cycles of penicillin treatment and growth in the presence of the auxotrophic requirement may be repeated several times, resulting in a considerable enrichment of auxotrophs. Cultures are finally plated on solid media, and individual colonies tested for their auxotrophic requirements.

Pentamethylenediamine: see Cadaverine.

Pentitols: C5-sugar alcohols. Naturally occurring P. are D- and L-arabitol, ribitol and xylitol.

Pentosans: polysaccharides consisting of pentose residues, e.g. arabans and xylans. P. are widely distributed in the plant kingdom, and they are important as cell wall and storage materials.

Pentose metabolism: see Pentose phosphate cycle, Phosphoketolase pathway.

Pentose phosphate carboxylase: see Ribulose 1,5-diphosphate carboxylase.

Pentose phosphate cycle, ***hexose monophosphate shunt, Warburg-Dickens-Horecker pathway, phosphogluconate pathway:*** an oxidative pathway of carbohydrate metabolism, in which glucose 6-phosphate (derived from glucose by phosphorylation; see Kinases) is totally degraded to carbon dioxide, accompanied by reduction of $NADP^+$ to $NADPH + H^+$. Overall equation: $C_6H_{12}O_6 + 7H_2O + 12NADP^+ + ATP \rightarrow 6CO_2 + 12NADPH + 12H^+ + ADP + P_i$. The importance of the P.p.c. lies in the production of reduced NADPH which is required for biosynthesis (e.g. of fatty acids), and in the production of pentoses required for the synthesis of nucleosides, nucleotides and nucleic acids. By the action of a transhydrogenase system, or NADPH cytochrome *c* reductase, the NADPH can be reoxidized to produce energy (36 molecules ATP per molecule glucose), but operation of the P.p.c. for the sole purpose of energy production is unusual.

Pentose phosphate cycle

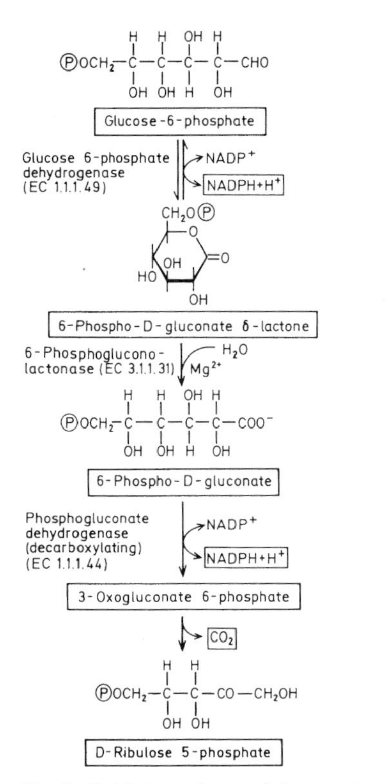

Fig. 1. *Oxidative phase of the pentose phosphate cycle.*

The P. p. c. consists of an oxidative phase (resulting in the production of NADPH, CO_2 and D-ribulose 5-phosphate) and a nonoxidative phase (a fairly complex series of sugar interconversions). The experimental evidence for the oxidative phase (Fig. 1) belongs to classical biochemistry, and the reactions leading from glucose 6-phosphate to D-ribulose 5-phosphate are well proven. Glucose 6-phosphate dehydrogenase (EC 1.1.1.49) ("Zwischenferment" in the older literature) was discovered in 1931 by Otto Warburg. In 1934, Warburg and Christian discovered $NADP^+$ as a result of their investigation of the oxidation of glucose 6-phosphate by erythrocytes. Further contributions to the study of the conversion of glucose 6-phosphate into D-ribulose 5-phosphate were made by Dickens, Horecker, Racker and Kornberg.

In the scheme shown in Fig. 2, tentatively proposed by Horecker in 1954, the nonoxidative phase results in the conversion of 3 molecules of D-ribulose 5-phosphate into 2 molecules of fructose 6-phosphate and one molecule of glyceraldehyde 3-phosphate. Evidence for this nonoxidative phase was obtained from buffered extracts of acetone-dried rat liver, pea root and pea leaf. The pathway became generally accepted and it has been published widely in biochemical texts. However, it has long been known that the results of certain isotopic labeling studies are inconsistent with this scheme.

In the light of subsequent evidence [J. F. Williams et al. *Biochem. J.* **176** (1978) 241–256; J. F. Williams et al. *ibid.* 257–282], it became necessary to revise the accepted scheme of sugar interconversions in the nonoxidative phase of P. The new scheme (Figs. 3 & 4) is consistent with all the isotopic labeling data. Evidence

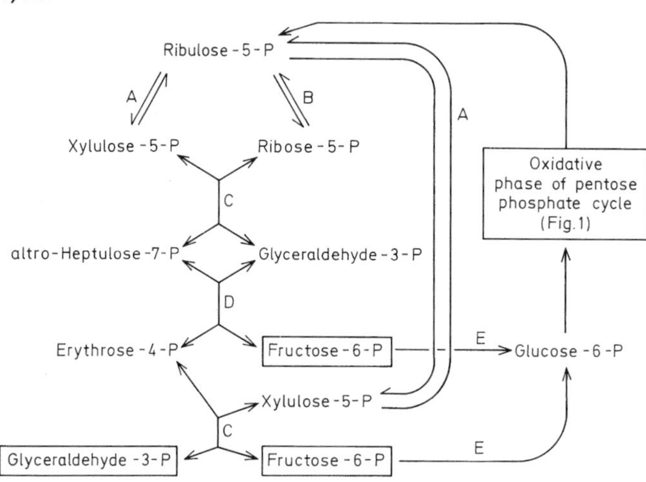

Sum of reactions in nonoxidative phase:
3 Ribulose-5-P ⟶ Glyceraldehyde-3-P + 2 Fructose-6-P

Fig. 2. *The nonoxidative phase of the pentose phosphate cycle proposed in 1954 and widely accepted up to 1978.*
A: Ribulose phosphate 3-epimerase (xylulose phosphate 3-epimerase) (EC 5.1.3.1).
B: Ribose phosphate isomerase (EC 5.3.1.6).
C: Transketolase (EC 2.2.1.1).
D: Transaldolase (EC 2.2.1.2).
E: Glucose phosphate isomerase (EC 5.3.1.9).
P = phosphate.

for the revised pathway was obtained from buffered extracts of acetone-dried rat liver. This later work resulted in the discovery of five additional intermediates of pentose phosphate metabolism: D-manno-heptulose 7-phosphate, D-altro-heptulose 1,7-*bis*phosphate, D-glycero-D-ido-octulose 1,8-*bis*phosphate, D-glycero-D-altro-octulose 1,8-*bis*phosphate, and D-arabinose. A new enzyme, arabinose phosphate 2-epimerase was also tentatively identified, which catalyses the reaction: D-arabinose 5-phosphate ⇌ D-ribose 5-phosphate.

Sum of reactions:
3 Ribulose-5-P ⟶ Dihydroxyacetone-P + Fructose-6-P + Glucose-6-P

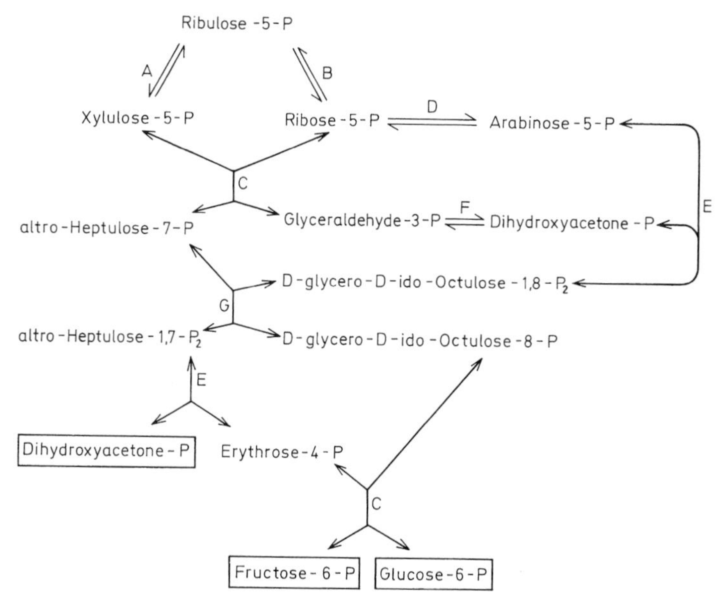

Fig. 3. *Current scheme (since 1978) for the nonoxidative phase of the pentose phosphate cycle* (see also Fig. 4).
A: Ribulose phosphate 3-epimerase (xylulose phosphate 3-epimerase) (EC 5.1.3.1).
B: Ribose phosphate isomerase (EC 5.3.1.6).
C: Transketolase (EC 2.2.1.1).
D: Arabinose phosphate 2-epimerase.
E: Fructose-*bis*phosphate aldolase (EC 4.1.2.13).
F: Triose phosphate isomerase (EC 5.3.1.1).
G: A phosphotransferase.
P = phosphate.

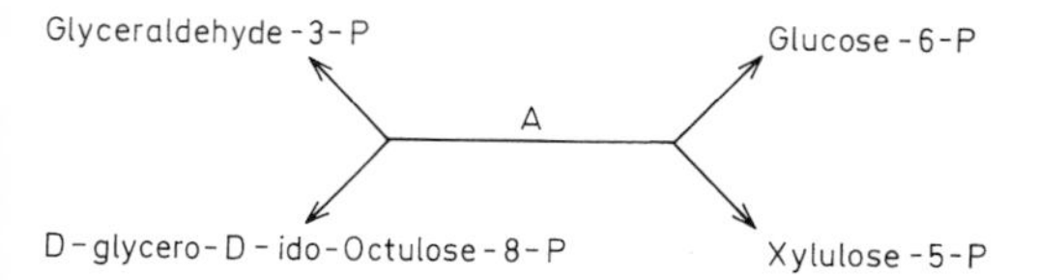

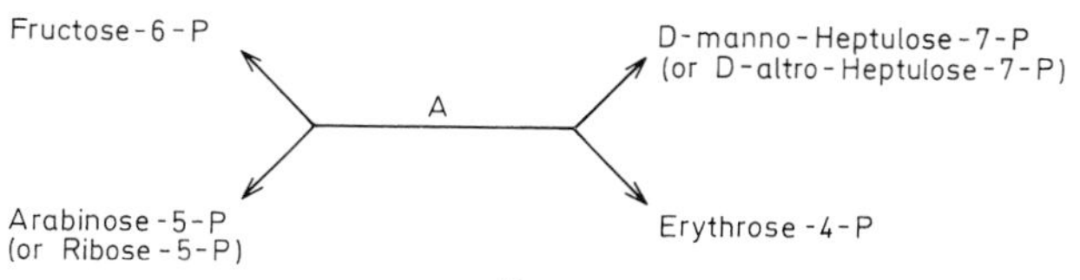

Ribose-5-P + Dihydroxyacetone-P $\overset{B}{\rightleftharpoons}$ D-glycero-D-altro-Octulose-1,8-P_2

Fig. 4. *Additional reactions that contribute to the nonoxidative phase of the pentose phosphate cycle shown in Fig. 3.*
A: Transketolase (EC 2.2.1.1).
B: Fructose-*bis*phosphate aldolase (EC 4.1.2.13).
P = phosphate.

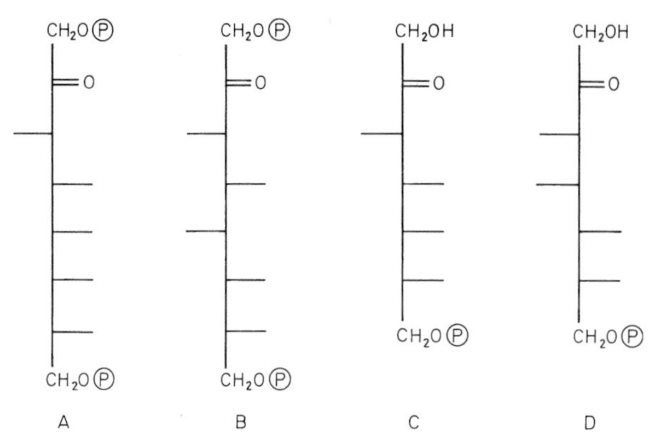

Fig. 5. *Structures of 7- and 8-carbon sugars involved in the pentose phosphate cycle.*
A: D-glycero-D-altro-octulose 1,8-*bis*phosphate.
B: D-glycero-D-ido-octulose 1,8-*bis*phosphate.
C: altro-heptulose 7-phosphate (or sedoheptulose 7-phosphate).
D: manno-heptulose 7-phosphate.

The new scheme requires high aldolase activity and has no transaldolase, whereas the old scheme has no aldolase and depends on transaldolase. D-Arabinose 5-phosphate inhibits transaldolase, but if the transaldolase is present in sufficient quantity, some activity probably remains in the presence of D-arabinose 5-phosphate.

Clearly it is incorrect to consider the nonoxidative phase of the P.p.c. as a fixed, albeit fairly complex mechanism for the conversion of 3 molecules of pentose phosphate into a triose phosphate and 2 molecules of hexose phosphate. There exists a network of possible reactions, which can change in emphasis, depending on the tissue, and possibly on the physiological state of the tissue. Thus, in rat epididymal fat pad, isotopic labeling suggests that the older scheme of Horecker operates for the metabolism of pentose phosphates. Operation of the new scheme in plants would provide an explanation of the Gibbs effect (see). Other workers [T. Wood & A. Gascon *Archives of Biochemistry and Biophysics* **203** (1980) 727–733] have reported their failure to demonstrate the interconversion of D-arabinose 5-phosphate and D-ribose 5-phosphate, or the role of D-arabinose 5-phosphate as an acceptor for transketolase in baker's yeast, *Candida utilis,* or rat liver.

The rate at which C1 of glucose, as compared with C6, is converted to CO_2 is used as an index for the quantitative importance of different pathways of carbohydrate metabolism. In Glycolysis (see), C1 and C6 become equilibrated at the level of the triose phosphates. Subsequent oxidation and decarboxylation in the tricarboxylic acid cycle therefore results in equal rates of production of $^{14}CO_2$ from [1-^{14}C]glucose and [6-^{14}C]glucose.

$$\text{i.e.}\ \frac{^{14}CO_2 \text{ from } [1\text{-}^{14}C]\text{glucose}}{^{14}CO_2 \text{ from } [6\text{-}^{14}C]\text{glucose}} = 1.$$

In the oxidative phase of the P.p.c. C1 of glucose is removed totally by decarboxylation of 3-oxogluconate 6-phosphate, so that a ratio greater than 1 is evidence for the functioning of the P.p.c. In the Glucuronate cycle (see), however, decarboxylation of 3-oxo-L-gluconate represents the removal of C6 of glucose; thus ratios less than 1 indicate that the glucuronate pathway is operating.

In mammals the P.p.c. is especially important in the cornea, lens, liver and lactating mammary gland. It is also a common pathway in invertebrates, bacteria and plants. The P.p.c. is completed when glucose 6-phosphate (some of which is derived from fructose 6-phosphate by the action of glucose phosphate isomerase) is converted into ribulose 5-phosphate by the oxidative phase (Fig. 1) of the cycle. [W. Martin et al. 'Microsequencing and cDNA cloning of the Calvin cycle/oxidative pentose phosphate pathway enzyme ribose-5-phosphate isomerase (EC 5.3.1.6) from spinach chloroplasts' *Plant Molecular Biology* **30** (1996) 796–805]

Pentoses: aldoses containing five carbon atoms. P. are an important group of monosaccharides (see Carbohydrates). Naturally occurring P. include D- and L-arabinose, L-lyxose, D-xylose, D-ribose and 2-deoxy-D-ribose, and the ketopentoses (pentuloses) D-xylulose and D-ribulose. P. occur chiefly in the furanose form. They are not fermented by the usual yeasts. By distillation with dilute acids, P. are converted into furfural, a reaction which serves for the detection of P. and their differentation from hexoses.

Pentosuria: see Inborn errors of metabolism.

Pepsin (EC 3.4.23.1): a protease in the stomach of all vertebrates with the exception of stomachless fish (e.g. carp). Purified P. shows maximal activity at pH 1–2, but in the stomach the optimal pH is 2–4. Above pH 6, P. is inactivated by denaturation. It preferentially catalyses hydrolysis of peptide bonds between two hydrophobic amino acids (Phe-Leu, Phe-Phe, Phe-Tyr). With the exception of protamines, keratin, mucin, ovomucoid and other carbohydrate-rich proteins, most proteins are attacked by P. The products of P. action are peptone, i.e. mixtures of peptides in the M_r range 300–3,000. P. is a highly acidic (pI 1), single chain phosphoprotein (327 amino acid residues of known primary sequence, M_r 34,500), which is released from its zymogen (pepsinogen, M_r 42,500) by autocatalysis in the presence of hydrochloric acid.

The largest of the cleavage peptides (M_r 3,000) produced in this activation process is a P. inhibitor. Further peptic degradation of this inhibitor to inactive products is therefore necessary for the full realization of pepsin activity. At least 4 different active pseudoisoenzymes of porcine and human P. are known. These are P.A (EC 3.4.23.1) (the classical alkali-labile P.), B (EC 3.4.23.2) (the alkali-stable gelatinase, M_r 36,000, 332 amino acid residues), C (EC 3.4.23.3) (the alkali-stable gastricsin, M_r 31,500, 298 amino acid residues) and D. Probably B and D are autolysis products of the zymogen of the predominant A-form, since no pepsinogens of B and D have been detected. Pepsinogen C, the zymogen of P.C, has M_r 41,000.

Pepstatin: see Inhibitor peptides.

Peptidases: see Proteases.

Peptide antibiotics: oligopeptides with antibiotic activity. They are usually cyclic. In addition to L-amino acids, they often contain nonproteogenic D-amino acids and unusual amino acids, as well as other components like branched hydroxy-fatty acids. The cyclic structures are closed by amide or (in depsipeptide antibiotics) ester bonds. Owing to their ring structures and high content of nonproteogenic constituents, P.a. are resistant to proteolysis. Most P.a. are relatively toxic, and only a few are used clinically (notably Penicillins, see), particularly for treatment of Gram-negative infections (*Proteus*, *Pseudomonas*). Most known P.a. are of bacterial origin; a few are produced by *Streptomycetes* or lower fungi. Their antibiotic action is exerted in different areas of metabolism, e.g. cell wall biosynthesis, nucleic acid and protein biosynthesis, energy metabolism, nutrient uptake. Biosynthesis of P.a. does not involve ribosomes or mRNA (see Gramicidins). The following important P.a. are listed separately: Penicillins, Gramicidins, Bacitracins, Tyrocidins, Polymyxins, Actinomycins.

Peptide bond: the most important type of covalent bond between amino acid residues in peptides and proteins. Formally, a P.b. is an amide bond between the carboxyl group of one amino acid and the amino group of another. Shortening of the C-N bond by mesomerism of the P.b. has been confirmed by X-ray crystallographic analysis. Since free rotation around the bond axis is not possible, two planar conformations of the P.b. are possible, i.e. *cis* and *trans* (Fig.). In native peptides and proteins, the trans P.b. predominate, but P.b. to the nitrogen atom of proline are always *cis*. The relative rigidity of the P.b. strongly affects secondary structure in Proteins (see). Cyclic dipeptides (2,5-dioxopiperazines) and polyproline contain only *cis* P.b.

Peptide hormones: a group of hormones with the chemical structure and physical properties of peptides. The smallest P.h., thyrotropin releasing hormone, is a tripeptide. Most P.h. are oligopeptides with 5 to 30 residues; see Oxytocin, Vasopressin, Releasing hormones, Opioid peptides. Although P.h. are considered separately from Proteohormones (see), there is a range of hormonally active compounds extending from P.h. to protein hormones. There are, however, differences as well as similarities: some P.h. are synthesized from amino acids by specific synthetases (e.g. releasing hormones), and many are formed by proteolysis of protein precursors (e.g. Angiotensin, see). P.h. act chiefly via the adenylate cyclase system (see Adenosine phosphates) and they are degraded by proteolytic enzymes.

Sequence analysis of P.h. reveals that many are capable of adopting stable amphiphilic secondary structures at membrane/aqueous interfaces. That such structures are important to the action of the P.h. is indicated by studies with synthetic analogs with similar or dissimilar amphiphilicity but otherwise different primary structures from the natural P.h. [E.T. Kaiser & F.J. Kezdy *Science* **223** (1984) 249–255]

Peptides: organic compounds consisting of two or more amino acids joined covalently by peptide bonds (see Peptide bond). The number of amino acid residues in a peptide is indicated by a Latin prefix, e.g. dipeptide, tripeptide. *Oligopeptides* contain 10 or fewer residues; more than 10 residues constitute a *polypeptide*. P. with M_r 10,000 (about 100 amino acid residues) lie on the borderline between polypeptides and proteins. P. dialyse through natural membranes, whereas proteins do not.

In the names of P., an amino acid contributing its carboxyl group to the peptide bond is given the ending -yl, and only the amino acid retaining its free carboxyl group retains its usual name, e.g. glycyl-histidyl-lysyl-alanine. Using the accepted abbreviations for Amino acids (see), the name of this tetrapeptide becomes Gly-His-Lys-Ala. It is customary to write the formula horizontally, always starting with the *N*-terminal amino acid (the one with the free α-amino group) on the left and finishing with the *C*-terminal amino acid on the right. The use of this abbreviated notation presupposes that amino acids with extra functional groups (e.g. lysine or glutamic acid) are linked by α-peptide

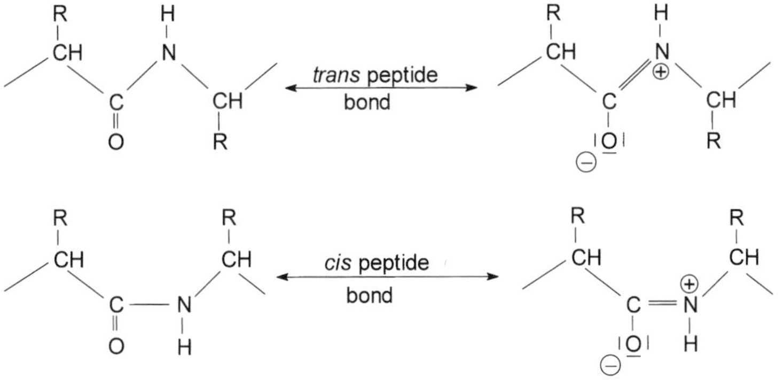

trans peptide bond

cis peptide bond

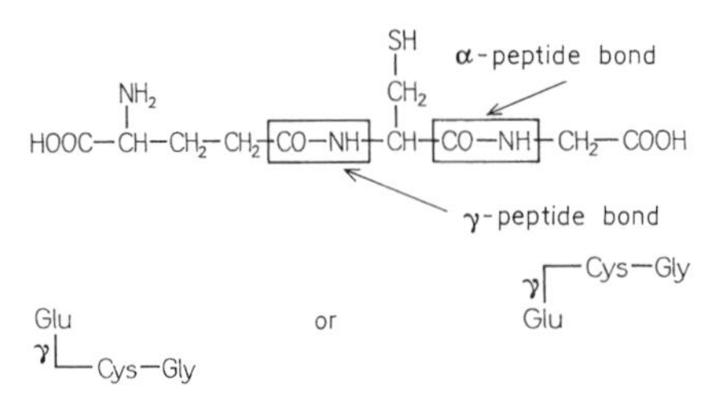

Fig. 1. *Glutathione*

bonds. Fig. 1 shows the methods of representing the α- and γ-peptide bonds in the tripeptide glutathione.

The complete amino acid sequence of a P. is shown by linking the 3-letter amino acid symbols by hyphens. If only part of a sequence is shown, the terminal symbols must also carry a second hyphen, e.g. -Ala-Ile-Val-Lys-. If part of a sequence is unknown the symbols are enclosed in parentheses and separated by commas, e.g. Ser-Phe-Gly-(Tyr, Asn, Val)-Pro-Ala. Using the one-letter notation (see Amino acids), this peptide would be represented as SFG(Y,B,V)PA, where the absence of punctuation between letters indicates a known sequence.

P. which contain only amino acid residues are called *homomeric P.,* whereas P. with non-amino acid constituents are known as *heteromeric P.* Each of these classes is further subdivided into *homodetic P.,* which contain only peptide bonds, and *heterodetic P.,* which contain other bonds, e.g. ester, thioether or disulfide. Heterodetic P. possessing one or more ester bonds as hetero link are called *Depsipeptides* (see). The classification of depsipeptides does not accord entirely with the homomeric/heteromeric system; it has arisen from the fact that the synthesis of these peptides presents rather special problems. There are two types of homomeric depsipeptides: *O-peptides* contain ester-linked hydroxyamino acids, and they may be linear or cyclic (i.e. lactones); in *S-peptides* or *thiodepsipeptides,* the mercapto group of a cysteine residue is acylated with an amino acid. Heteromeric depsipeptides contain hydroxy acids as the hetero component, and they are also known as *peptolides.* P. with non-amino acid components covalently bound to terminal amino or carboxyl groups, or to side chains, are called *peptoids.* Important representatives of this group are glyco-, nucleo-, lipo-, phospho- and chromopeptides.

In *linear P.,* it is not usually considered necessary to denote the direction of the peptide bond (from -CO to NH-), but in cyclic P., the arrow should always be used.

Synthetic analogs of naturally occurring P. are named according to the following IUPAC-IUB rules: If one or more amino acid residues are replaced, the full name of the new amino acid(s) and its(their) sequence position(s) are written in square brackets before the trivial name of the peptide, e.g. [7-alanine]-oxytocin, [3-leucine, 7-alanine]oxytocin, or in abbreviated notation, [Leu^3, Ala^7]oxytocin.

If the peptide chain is extended at the terminus, the additional amino acid is represented in the normal way, e.g. valyl-bradykinin (abb. Val-bradykinin) has valine added to the *N*-terminus of bradykinin, while bradykinyl-valine (abb. bradykinyl-Val) has valine added to the *C*-terminus.

When an extra amino acid is introduced into the chain, this is indicated by the prefix *endo,* e.g. endo-6a-glycine-bradykinin (abb. endo-Gly^{6a}-bradykinin) has a glycyl residue inserted between residues 6 and 7 of bradykinin.

Omission of amino acids is shown by the prefix *des,* e.g. des-3-proline-bradykinin (abb. des-Pro^3-bradykinin).

P. representing partial sequences of larger P. are represented by placing the sequence numbers of the first and last amino acids in brackets after the trivial name, followed by the Greek root for the number of residues in the partial sequence, e.g. bradykinin-(3–9)-heptapeptide.

P. occur widely and have many different biological functions. Many are hormones, e.g. Corticotropin, Melanotropin, Oxytocin, Vasopressin, Releasing hormones, Insulin, Glucagon, Gastrin, Secretin, Angiotensin, Bradykinin, Endorphins, Opioid peptides (see entries for each of these). Many microorganisms produce P., often with antibiotic activity (see Peptide antibiotics). Some P. are very toxic, e.g. Phallotoxins (see), Amatoxins (see) and Melittin (see). Very simple P. are

Homomeric homodetic peptides

H—Ala—Val—Glu—Ile—Phe—Leu—OH
Linear peptide

H—Ala—Val—Glu—Ile—Phe—Leu—OH
└Leu—Val—OH
Linear branched peptide

Ala→Val→Glu
↑ ↓
Leu←Phe←Ile
Homodetic cyclic peptide

Ala←Val←Lys←Glu←Ile←Phe←Leu—H
↓ ↑
Leu→Val→Ile
Homodetic cyclic branched peptide

Homomeric heterodetic peptides

H—Ala—Val—Ser—Ile—Phe—Leu—OH
O—Leu—Val—H
Linear branched peptide

H—Ala—Val—Cys—Ile—Phe—Leu—OH
S—Leu—Val—H
Linear branched S-peptide

Ala→Val→Ser→Ile
↑ ↓
H—Cys—S — S — Cys—OH
Herodetic cyclic peptide
(cyclic disulfide)

H—Ala→Asp→Ser→Ile
↓ ↓
Val—CO—O—Thr—OH
Heterodetic cyclic branched peptide
(peptide lactone)

Fig. 2. *Classification of homomeric peptides*

found in muscle, e.g. carnosine (β-Ala-His), anserine (β-Ala-MeHis). A wide variety of biologically active P. have been isolated from plants, e.g. γ-glutamyl-β-cyanoalanine (a neurotoxin from *Vicia sativa*). Some P. inhibit enzymes, e.g. pancreatic Trypsin inhibitor (see).

Some P. are synthesized by reaction of an acyl phosphate (e.g. glutamate + ATP → γ-glutamyl phosphate + ADP) with the amino group of the next amino acid (γ-glutamyl phosphate + cysteine → glutathione; see Glutathione). In a second type of P. synthesis, activated aminoacyl groups (amino acid + ATP → aminoacyl-AMP + PP_i) are transferred to –SH groups to form intermediate thioesters (see Gramicidins for details. Thirdly, some P. are formed by the controlled hydrolysis of proteins (see Protein biosynthesis, Post translational modification of proteins). Thus peptide hormones (e.g. see Insulin, Bradykinin) are formed by enzymatic degradation of protein precursors. Stepwise degradation of the original protein precursor may produce a family of hormones, in which one active hormone acts as the precursor of another, e.g. Releasing hormone (see), Pro-Leu-Gly-$CONH_2$, is produced from oxytocin by a membrane-bound exopeptidase. The ACTH family is derived from a single precursor molecule (Fig. 3).

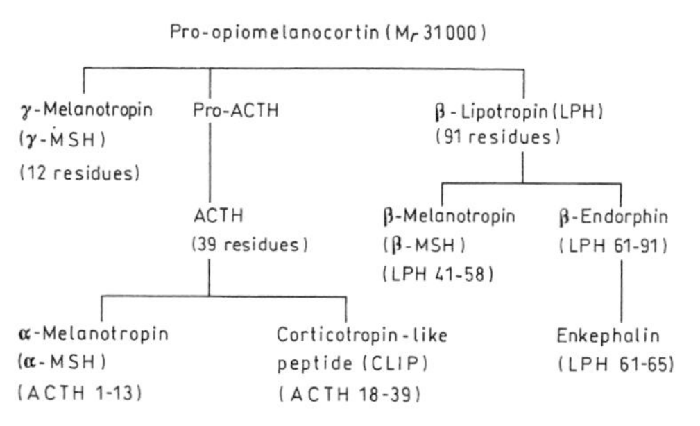

Fig. 3. *Derivation of the ACTH family of peptides from a single precursor of M_r 31,000*

The half-life of most peptide hormones is less than 30 minutes; rapid inactivation results from the concerted action of various endo- and exopeptidases.

The primary structure of P. is elucidated by the standard methods of sequence analysis (see Proteins). The conformation of P. is an important determinant of their biological activity. It is stabilized by peptide and disulfide bonds. In P. containing unusual bonds and/or constituents, other factors also stabilize the conformation. Although X-ray crystallography (see) was used in the elucidation of the three-dimensional structures of insulin and gramicidin S, the conformational analysis of P. is now by preference carried out in solution, using the spectroscopic methods ORD, CD, IR, NMR and ESR.

With respect to their general properties, P. are intermediate between amino acids and proteins (the following generalizations do not always apply to cyclic peptides). The acid-base behavior and solubility of linear P. depend on the amino acid sequence; the ampholite properties depend on the number and distribution of basic and acidic groups, and solubility is influenced by the relative numbers of hydrophobic and hydrophilic side chains. Compared with amino acids, the acidic character of P. is more pronounced. As a rule, the amount of P. can be determined by alkalimetric titration in 50% alcohol. Like amino acids, P. produce a blue to blue-violet color with ninhydrin, which can be used for their detection in chromatography and electrophoresis. P. can be differentiated from amino acids by the biuret test. P. are hydrolysed to their constituent amino acids by proteases, acids or bases, and their quantitative amino acid composition can be determined by conventional methods (see Proteins).

Chemical synthesis of P. serves a number of purposes: 1. Confirmation of the primary structure of natural P. and proteins. Some of these, e.g. releasing hormones, exist in such low concentrations that structural elucidation must be performed in conjunction with trial syntheses. 2. Investigation of the relationship between structure and biological activity. 3. Modification of the activity and specificity of naturally occurring, biologically active P. 4. Preparation of model P. for conformational analyses, e.g. polyamino acids. 5. Preparation of test P. for immunological studies.

The basic principle of P. synthesis is the acylation of the amino group of one amino acid with the carboxyl function of a second amino acid. This must be designed to give the correct sequence of the two amino acids, and to ensure that there is no racemization at the optically active center of either amino acid. For the synthesis of a defined P., it is therefore necessary temporarily to block (protect) all the functional groups which would take part in undesired side reactions, including the carboxyl, amino and sulfhydryl groups of side chains. Some protective groups are listed in the table. In the formation of a peptide bond, the carboxyl C-atom of one amino acid undergoes nucleophilic attack by the amino N-atom of the other

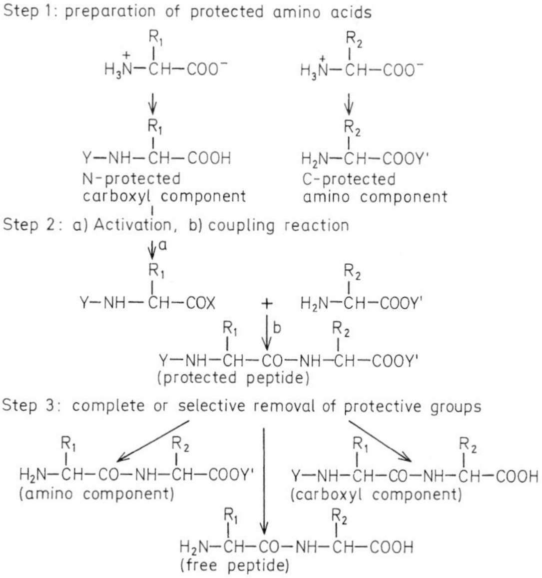

Step 1: preparation of protected amino acids

$H_3\overset{+}{N}-CH(R_1)-COO^-$ → $Y-NH-CH(R_1)-COOH$ (N-protected carboxyl component)

$H_3\overset{+}{N}-CH(R_2)-COO^-$ → $H_2N-CH(R_2)-COOY'$ (C-protected amino component)

Step 2: a) Activation, b) coupling reaction

a: $Y-NH-CH(R_1)-COX$ + $H_2N-CH(R_2)-COOY'$

b: $Y-NH-CH(R_1)-CO-NH-CH(R_2)-COOY'$ (protected peptide)

Step 3: complete or selective removal of protective groups

$H_2N-CH(R_1)-CO-NH-CH(R_2)-COOY'$ (amino component)

$Y-NH-CH(R_1)-CO-NH-CH(R_2)-COOH$ (carboxyl component)

$H_2N-CH(R_1)-CO-NH-CH(R_2)-COOH$ (free peptide)

Fig. 4. *Steps in the chemical synthesis of a peptide*

amino acid, i.e. the unbonded electrons on the N atom are attracted to the electrophilic (slightly positive) C nucleus. The process is facilitated by the presence of a substituent (X) on the C which draws electrons away from the C nucleus; in this case the reaction can take place under mild conditions. The last step of P. synthesis consists of the removal of protective groups, either from all the reactive groups or from only one of them, if the protected peptide is to serve as an intermediate in the synthesis of a higher P. (Fig. 4) Fig. 5 shows some of the available methods for activation. For the synthesis of higher P., two different strategies are usually applied. In order to avoid racemization, C-protected P. are extended one residue at a time by reaction with N-protected amino acids. At a later stage, fragment condensation may be used; N-protected P. are condensed with C-protected P. This has the advantage that the product of the condensation reaction is markedly different from the starting materials, whereas stepwise synthesis results in a product which is rather similar to the precursor P. Purification and characterization of the product is therefore much easier after fragment condensation.

By exploiting a number of different protective groups and methods of activation, it is possible to attain maximal blocking of side chain functions and a minimum of undesirable reactions. This synthetic strategy is limited by problems of solubility, especially for P. with more than about 30 amino acid residues. However, its success has been shown by the total synthesis of insulin, glucagon, corticotropin and other P.

The difficulties of conventional P. synthesis are theoretically overcome by synthesis on carrier polymers (Merrifield synthesis), introduced in 1962. In principle, an amino acid is bound via its carboxyl group to an insoluble, easily filtered, polymeric carrier, then the P. chain is built up stepwise from the *C*-terminal end (Fig. 6). Thus the laborious purification steps of the conventional P. synthesis are reduced to simple filtration and washing. With complete reaction at each stage of the synthesis, cleavage of the final P. from the carrier should give a pure product. The technical operation of the method is therefore very simple. In addition, the total synthesis can be automated, and machines for this purpose are commercially available. In practice, however, 100 % yields are not realized, and the final product is accompanied by incomplete and wrong P., making the final purification especially difficult. Assuming a conversion rate of 99 % at each coupling stage, polypeptides with 100 to 200 amino acid residues can still be synthesized, albeit in low yield. At the present state of development, despite many variations in the technique and attempts at optimization, an unequivocal synthesis is only possible for P. or fragments containing up to 10 to 20 amino acids. On the other hand, the Merrifield syn-

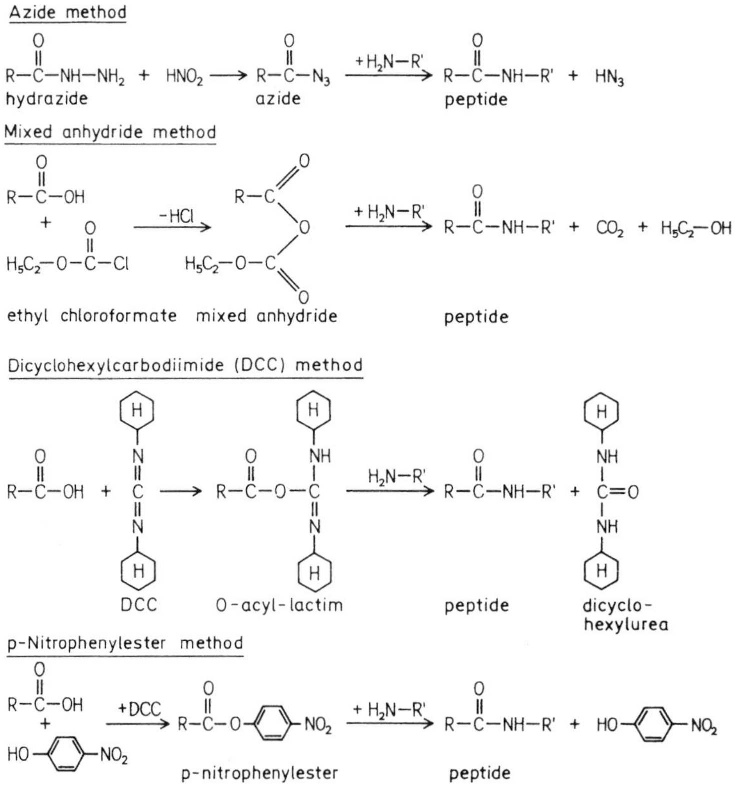

Azide method

$$R{-}CO{-}NH{-}NH_2 + HNO_2 \longrightarrow R{-}CO{-}N_3 \xrightarrow{+H_2N-R'} R{-}CO{-}NH{-}R' + HN_3$$

hydrazide — azide — peptide

Mixed anhydride method

$$R{-}CO{-}OH + H_5C_2{-}O{-}CO{-}Cl \xrightarrow{-HCl} R{-}CO{-}O{-}CO{-}O{-}C_2H_5 \xrightarrow{+H_2N-R'} R{-}CO{-}NH{-}R' + CO_2 + H_5C_2{-}OH$$

ethyl chloroformate — mixed anhydride — peptide

Dicyclohexylcarbodiimide (DCC) method

$$R{-}CO{-}OH + C_6H_{11}{-}N{=}C{=}N{-}C_6H_{11} \longrightarrow R{-}CO{-}O{-}C({=}N{-}C_6H_{11}){-}NH{-}C_6H_{11} \xrightarrow{H_2N-R'} R{-}CO{-}NH{-}R' + C_6H_{11}{-}NH{-}CO{-}NH{-}C_6H_{11}$$

DCC — O-acyl-lactim — peptide — dicyclohexylurea

p-Nitrophenylester method

$$R{-}CO{-}OH + HO{-}C_6H_4{-}NO_2 \xrightarrow{+DCC} R{-}CO{-}O{-}C_6H_4{-}NO_2 \xrightarrow{+H_2N-R'} R{-}CO{-}NH{-}R' + HO{-}C_6H_4{-}NO_2$$

p-nitrophenylester — peptide

Fig. 5. *Some methods of peptide synthesis*

Some protective groups in general use

	IUPAC abb.	Cleavage reactions
Amino-protective groups		
Benzyloxycarbonyl	Z	HBr/glacial acetic acid; H_2/Pd; Na/liquid NH_3
tert-Butyloxycarbonyl	Boc	M HCl/glacial acetic acid; trifluoroacetic acid
Trifluoroacetyl	Tfa	M piperidine; NaOH/acetone
2-(*p*-Biphenylyl)-iso-propyloxycarbonyl	Bpoc	5 % trifluoroacetic acid/CH_2Cl_2
Carboxyl-protective groups		
Alkylesters	–OMe –OEt	NaOH/acetone
Benzyl ester	–OBzl	H_2/Pd; Na/liquid NH_3; NaOH
tert-Butyl ester	OBu^t	trifluoroacetic acid
Side chain-protective groups		
S-Benzyl	Bzl–	Na/liquid NH_3
S-Acetamidomethyl	Acm–	Hg^{2+} (pH 4)
O-tert-Butyl	Bu^t	trifluoroacetic acid; HCl/diethyl ether
O-Benzyl	Bzl	NH_3

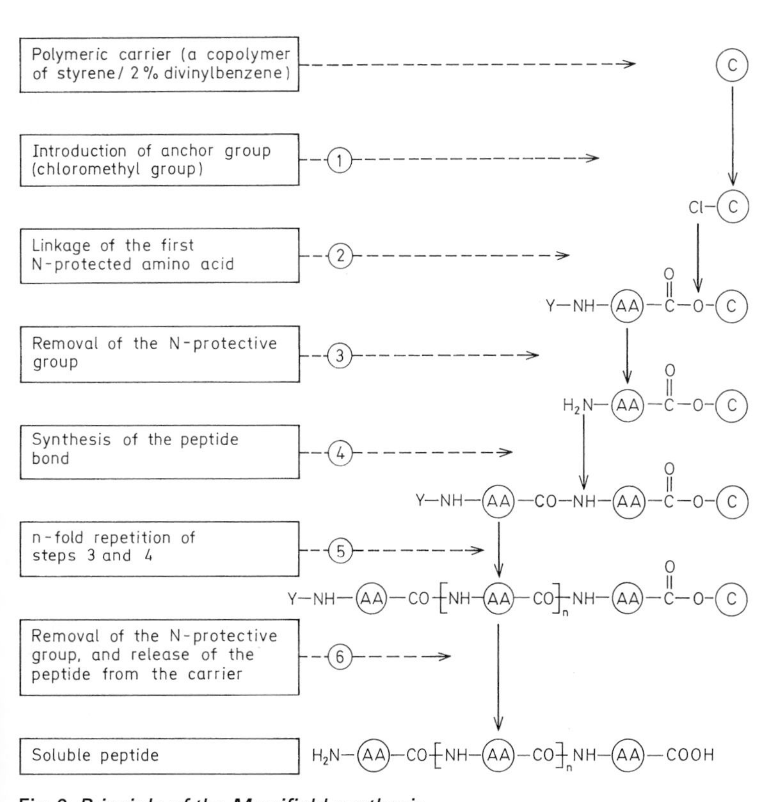

Fig. 6. *Principle of the Merrifield synthesis.* Ⓒ carrier; Ⓐ amino acid.

thesis has been used for the total synthesis of ribonuclease A with high enzymatic activity and the same specificity as the native enzyme, and for the synthesis of an analog of human growth hormone with 188 amino acids. The success of these syntheses was, however, assessed on the basis of biological activity, and it was not possible to apply classical criteria of protein or P. purity.

Peptide synthesis: see Peptides.

Peptidoglycan: see Murein.

Peptidyltransferase (EC 2.3.2.12): an integral enzymatic activity of the large ribosomal subunit. P. t.

catalyses peptide bond formation between the NH_2 group of the amino acid of an aminoacyl-tRNA and the COOH group of the terminal amino acid of the peptidyl-tRNA. The reaction appears to be sterically favored by the relative positions of the two reaction partners at the acceptor and donor sites of the ribosome. It is thought that P.t. is located on the surface of the 50S subunit, where it makes contact with the 30S subunit.

Peptolides: see Depsipeptides.

Peptones: mixtures of polypeptides produced by the partial degradation of protein. P. are prepared by the acid or enzymatic (pepsin, trypsin) hydrolysis of dietary proteins, like albumins, caseins and muscle proteins. Autolysed and trypsin-treated yeast contains vitamins and other growth factors in addition to P. Owing to their low M_r of 600–3,000, P. are not coagulable and cannot be salt-precipitated. They inhibit blood coagulation and act as lymphagogues (i.e. they increase the flow of lymph). P. are used in the formulation of many microbiological growth media.

Perezone: see Benzoquinones.

Permeability factor: see Vitamins (vitamin P).

Permeases: a term applied to carrier proteins involved in the transport of materials across membranes. P. is a misnomer and should be avoided.

Peroxidase (EC 1.11.1.7): an oligomeric oxidase which utilizes hydrogen peroxide as an oxidant in the dehydrogenation of various substrates. The H_2O_2 is reduced to water: $AH_2 + H_2O_2 \rightarrow A + 2H_2O$

In the cell, P. is often accompanied by catalase and localized in the peroxisomes. The prosthetic group of plant P. is ferriprotoporphyrin (a red hemin); animal P. contain a green hemin. The most widely studied P. is the crystallizable horse radish P., which can be separated into apoenzyme and hemin under mild conditions by treatment with acetone-hydrochloric acid. The nonenzymatic, heat-stable P. activity of hemoproteins is called pseudoperoxidase, e.g. as shown by hemoglobin and its degradation products. The P. reaction is diagnostically important, because the P. activity of myelotic granulocytes is greatly increased, whereas lymphocytes, reticular cells, carcinoma, sarcoma and myeloma cells have no P. activity. The P. of thyroid gland (M_r 200,000) consists of 3 subunits (M_r 67,000).

Peroxisomes, ***microbodies:*** structurally similar, but functionally heterogeneous cytoplasmic organelles, which are present in all eukaryotic cells, and which are the sites of certain oxidative metabolic reactions. Oxidative reactions in the P. give rise to hydrogen peroxide, which is converted into water and oxygen by the action of catalase. Catalase therefore serves as a detoxifying agent by preventing the accumulation of toxic hydrogen peroxide. This catalase constitutes up to 40 % of total peroxisomal protein, and serves as a marker enzyme for P., i.e. for the histochemical demonstration of P. The various oxidative reactions of P. are not accompanied by ATP synthesis, since P. possess no phosphorylation mechanism; the resulting energy is dissipated as heat.

P. were first identified in 1954 in kidney tubules. They are approximately spherical in shape (diameter 0.5–2 mm), bounded by a single membrane, often contain a crystal of urate oxidase, and do not contain DNA or ribosomes. P. are thought to be formed in the same way as chloroplasts and mitochondria, i.e. by selective protein and lipid import from the cytosol. They were formerly thought to arise by abstriction from the endoplasmic reticulum, which they resemble in being a self-replicating organelle without a genome. Subsequent studies, however, indicate that P. always arise by growth and fission of existing P.

Two types of P. have been studied in plants. The type present in leaves catalyses the oxidation of glycolate to glyoxylate (the glycolate is derived from phosphoglycolate, a byproduct of CO_2 fixation in C3 plants; see Photorespiration). The other type is present in germinating seeds, where it converts fatty acids (from stored triacylglycerols) into carbohydrates. This latter type is called a glyoxysome, because the conversion of fatty acids into sugars involves the glyoxylate cycle.

P. are particularly abundant in liver and kidney cells, e.g. a single rat liver cell contains 350–400 P. The half-life of P. is 1.5–2 days. Other cell types usually contain smaller P. with homogeneous contents (microperoxisomes). In addition to catalase, P. contain, inter alia, D-amino acid oxidase, α-hydroxyacid oxidase and urate oxidase. The latter enzyme is often present as a large crystal in the otherwise homogeneous matrix. These enzymes are particularly important in the oxidative degradation of metabolic intermediates (e.g. purine bases) and in the formation of carbohydrates from amino acids and other materials. [L.J. Olsen & J.J. Harada 'Peroxisomes and Their Assembly in Higher Plants' *Annu. Rev. Plant Physiol. Plant Mol. Biol.* **46** (1995) 123–146]

Pestalotin: a metabolic product from the culture filtrate of the fungus *Pestalotia cryptomeriaecola* Sawada, M_r 214.25, m.p. 88–88.5 °C. P. was isolated in 1971 by Japanese workers. It behaves as a synergist of the gibberellins.

Pestalotin

Petite mutants: spontaneous mutants, chiefly yeasts, with chemical or physical defects in the respiratory chain. P.m. grow very slowly and form small ("petite") colonies on nutrient agar. The same phenotype can be produced by a chromosomal mutation (segregational petite), or a mutation in the mitochondrial DNA (vegetative or neutral petite). In the latter case, mitochondrial structure is considerably altered, largely due to changes in the amino acid composition of the structural proteins of the inner mitochondrial membrane. Since these structural proteins are important for the correct arrangement and conformation of the respiratory chain enzymes, the effect of petite mutation on the respiratory chain is probably secondary.

Petroselinic acid: *cis*-Δ^6-octadecenoic acid, $CH_3–(CH_2)_{10}–CH=CH–(CH_2)_4–COOH$, an unsaturated fatty acid, M_r 282.5, m.p. 33 °C, b.p.$_{18}$ 238 °C. P.a. occurs esterified in the seed fats of many aromatic plants, e.g. parsley, celery, caraway.

Petunin: a blue anthocyanin in the flowers of the blue garden petunia. M.p. 178 °C (3,5,7,4′,5′-pentahydroxy-3′-methoxyflavylium cation). The structures of more than 10 other naturally occurring glycosides of petunidin have been determined.

PGE_1: see Prostaglandins.

$PGF_{1\alpha}$: see Prostaglandins.

pH-activity profile: see Enzyme kinetics.

Phaeophorbide: see Chlorophyll.

Phaeophytin: see Chlorophyll.

Phaeoplasts: photosynthetic organelles in the brown algae and in diatoms. P. are brown due to the presence of carotenoids like fucoxanthin and β-carotene.

Phage, *bacteriophage* [Greek, *phagein* = to devour]: viruses that attack bacteria. As a generalization, they all consist of a protein coat (the capsid or capsomer) surrounding the genetic material, which is DNA or RNA. The nucleic acid is usually double stranded, but ΦX 174 (phi-chi, but usually called phi-ex) contains circular, single stranded DNA.

Length and M_r of circular DNA from some coliphages

Phage	Length of DNA (μm)	$M_r \times 10^6$
ΦX 174	1.77	1.7
λ	17	32
T2	56	130
T3	11.6	23
T4	50	125
T7	12.5	25

P. have various shapes and sizes, e.g. ΦX 174 and Qβ (both coliphages) are spheroidal particles, and apical knob-like structures are also discernible on ΦX 174; coliphage fd is elongated, best described as a flexuous strand; PM2 of *Pseudomonas* is spheroidal with apical knobs and enclosed by an additional envelope. The most complicated P. structures are found among certain well studied coliphages (i.e. P. that attack *E. coli;* these have been given arbitrary names based on letters and numbers, such as T_1 T_2, P_1 λ (lambda), etc. (T_2, T_4 and T_6 are called the T-even P.).

The structure of coliphage T_2 (Fig.) may be taken as an example of the structure of T-even P. In the electron microscope, the head and tail parts can be distinguished, together with thread-like appendages. The head contains DNA plus putrescine and spermidine, which neutralize about 30 % of the acidic groups of the DNA. The tail (a hollow cylinder of contractile protein) serves as an injection tube for the DNA. The collar and the base plate of the tail have a hexagonal symmetry, which is related to the symmetry of the head (a deformed icosahedron). The base plate contains a special lysozyme and 6 molecules of the coenzyme, 7,8-dihydropteroylhexaglutamate. The fibers, which serve to attach the P. to the bacterial surface (a process called adsorption), are appendages of the base plate. Adsorption represents a specific interaction of the tail fibers with molecular structures on the bacterial surface; this process closely resembles an immunochemical reaction, and infection cannot

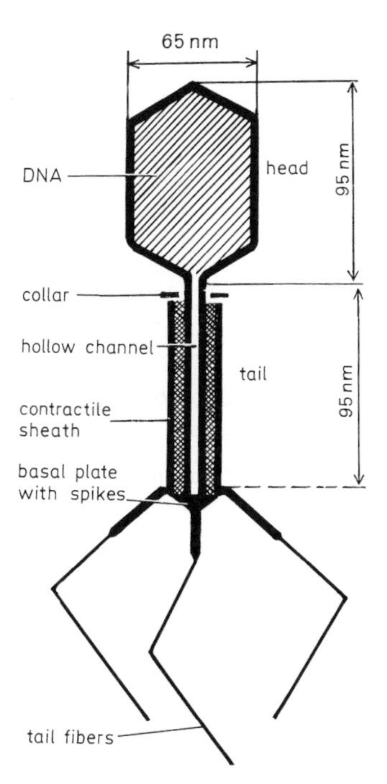

Schematic representation of coliphage T_2

occur unless specific receptors are present on the bacterial surface. There is therefore a high degree of specificity between a P. and its host bacterium. With attachment of the tail fibers, the base plate is brought to the cell surface. Lysozyme in the base plate catalyses a local dissolution of the bacterial cell wall, the protein of the tail contracts, and the DNA is injected into the bacterial cell.

T_1, T_3, T_5, T_7 and λ have similar structures to those of the T-even phages, but the tail is less rod-like (i.e. it is more flexuous), and it does not contract during the infection process.

Phage development: entry of phage DNA (or RNA) into the host cell initiates a series of processes leading to new phage progeny. P.d. can be divided into three phases: 1. synthesis of early phage RNA and early phage proteins, and termination of the synthesis of all host nucleic acids and proteins; 2. synthesis of late RNA and late proteins; 3. morphogenesis of new phages. This complicated interaction of host cell and phage is regulated largely on the level of gene expression.

Development of virulent T-phages (e.g. T7). The protein coat of the phage remains outside the bacterial cell, and only phage DNA is injected. Upon entry of the phage DNA, synthesis of host DNA, RNA and protein immediately stops. Phage nucleic acids and proteins are then synthesized, using phage DNA and the transcription und translation apparatus of the host cell. The progress of P.d. can be represented as follows (all genes are transcribed from the 5′ end of the phage genome): Synthesis of early RNA is catalysed by the host RNA polymerase, which recognizes the operator at the 5′ end of the phage genome.

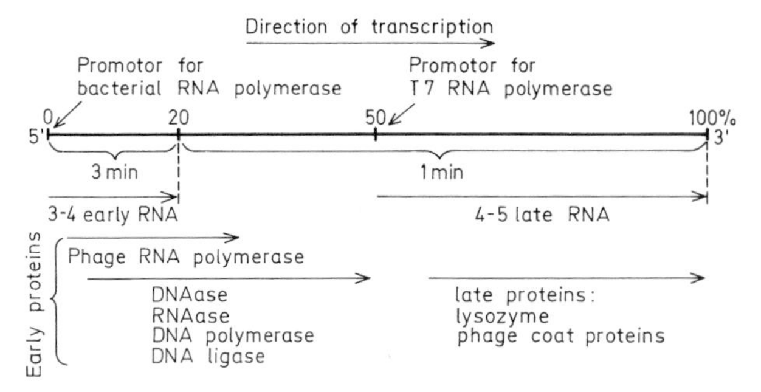

Simplified scheme for transcription of the T7 phage genome

Owing to the relatively unspecific binding of the host RNA polymerase, synthesis of early RNA is relatively slow, so that 75 % of the total time of P.d. is required for the transcription of only 20 % of the phage genome. The products of early RNA, the early proteins, are responsible for the termination of host nucleic acid synthesis. The first protein is an RNA polymerase specific for the phage, possibly accompanied by a specific sigma factor. The phage RNA polymerase permits the transcription of further phage genes for DNAase and RNAase (for the degradation of host nucleic acids), and DNA polymerase and DNA ligase (for replication of phage DNA). Late RNA, transcribed from the second half of the phage genome, codes for late protein, e.g. lysozyme and proteins of the head and tail. All these phage components are then assembled into new, viable phage progeny, a process which may require some enzymatic control, but which also involves a high degree of cooperative self assembly. Finally, the lysozyme lyses the bacterial cell wall. Depending on the particular strains of phage and host, and on the culture conditions, 100–200 viable phage particles are released per infected bacterial cell.

Development of the temperate coliphage λ. Infection can lead to immediate production of new phage progeny and lysis of the bacterial cell, as described for T7 (see above); this is called a nonlysogenic reaction. Usually, however, λ phage initiates a lysogenic reaction, i.e. P.d. is repressed and the phage DNA is integrated into the bacterial chromosome. In lysogeny, the host DNA and the integrated phage DNA are replicated synchronously, and the phage DNA is present in the DNA of all subsequent daughter cells. This integrated phage DNA is called a prophage. A regulator gene in the prophage controls the synthesis of a specific repressor, which prevents the independent replication of phage DNA. P.d. can be induced by various chemical or physical stimuli (e.g. temperature rise, UV irradiation, osmotic shock, chemical mutagens). Under the influence of an appropriate stimulus, synthesis of the repressor ceases and certain genes of the prophage are derepressed. Transcription of these derepressed genes requires a sigma factor from the host cell, which "recognizes" the promoter of the phage (prophage) DNA. The first products of these derepressed genes are an "antirepressor" protein, then an "antitermination" factor. The latter exerts a Positive control (see) by enabling the transcription (dependent on the bacterial RNA polymerase) of two further λ genes. The product of one of these genes modifies the host DNA polymerase, so that it becomes specific for the replication of λ DNA. The product of the other gene is a phage-specific sigma factor, which enables the RNA polymerase to transcribe further phage genes to polycistronic late RNA. The latter is translated into late proteins (head and tail proteins, lysozyme, etc.) by the translation apparatus of the host. The scheme described above is simplified. In reality several operons of the phage genome share one promoter; genes are transcribed both leftward and rightward of the promoter during the early stages of induction; several genes have overlapping sequences; and there are various postive and negative controls.

The breakage and resealing of DNA during phage integration is mediated by a phage-specific *integrase* (product of the *λint* gene). In vitro studies show that integrase nicks and reseals one strand of the DNA duplex (i.e. it acts as a type I topoisomerase), as well as resolving synthetic Holliday structures (see Recombination). Integrase operates in conjunction with a basic histone-like protein known as *integration host factor,* which binds to attachment sites on the phage and host DNA. Subsequent excision of the prophage is performed by *excisionase* (product of of the *λxis* gene), working in conjunction with integrase, integration host factor and a DNA-binding protein known as Fis (factor for inversion stimulation), which also stimulates gene inversion in *E. coli.* [W.C. Earnshaw & S.C. Harrision *Nature* **268** (1977) 598–602; H. Echols *Trends Genet.* 2 (1986) 26–30; A. Landy *Annu. Rev. Biochem.* **58** (1989) 913–949; H. Murialdo *Annu. Rev. Biochem.* **60** (1991) 125–153; J.D. Karam (ed.) *Molecular Biology of Bacteriophage T4,* American Society for Microbiology, Washington DC., 1994; V. Ellison et al. *J. Biol. Chem.* **270** (1995) 3320–3326]

Phagosomes: see Intracellular digestion.

Phallotoxins: heterodetic, cyclic heptapeptides, present in *Amanita phalloides.* Together with the Amatoxins (see), P. are the main toxic components of this fungus. The chief P. are phalloidin, phalloin and phallacidin (Fig.). Phallacidin is similar to phalloidin, but the D-threonine-alanine grouping is replaced by valyl-D-erythro-β-hydroxyaspartic acid. P. are not as toxic as the amatoxins, but they act more quickly. The structural requirements for toxicity

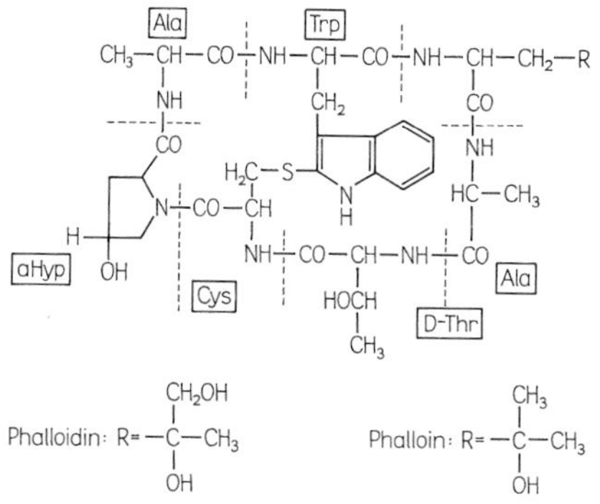

Phallotoxins

are the cycloheptapeptide structure and the characteristic thioether bridge between the indole residue of the tryptophan and the mercapto group of the cysteine. LD_{50}-values in white mice: phalloin 1.35 mg, phalloidin 1.85 mg, phallacidin 2.5 mg per kg body weight.

Phaseolin, ***phaseollin:*** 3-hydroxy-9,10-dimethylchromenylpterocarpan (Fig.) (see Pterocarpans). P. is a Phytoalexin (see) produced by various bean species in response to infection by phytopathogenic microorganisms, e.g. *Phytophthora, Monilinia,* and to other forms of stress like treatment of the plant with heavy metal salts. P. production can also be induced by a water-soluble polypeptide (monilicolin A, M_r 8,000, 65 amino acids) isolated from *Monilinia fructicola.* The polypeptide is not fungi- or phytotoxic, and it induces P. at concentrations as low as 2.5×10^{-9} M; it appears to be a specific inducer of P. and does not, for example, induce pisatin production in pea plants. Induction of P. by any means is always accompanied by an increase in the measurable catalytic activity of phenylalanine ammonia lyase.

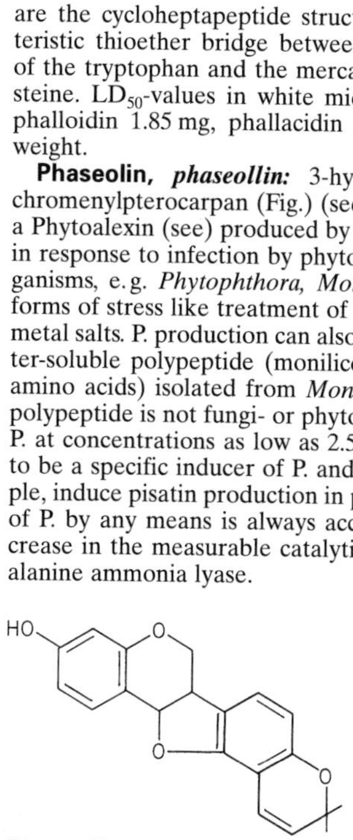

Phaseolin

Phaseolin isoflavan: (3*R*)-7,2′-dihydroxy-3′,4′-dimethylchromenylisoflavan. See Isoflavan.

Phe: abb. for L-phenylalanine.

Phellandrene: see *p*-Menthadienes.

Phenazines: compounds based on the phenazine ring system (Table). All known naturally occurring P. are produced only by bacteria, which excrete them into the growth medium. Both six-membered carbon rings of P. are biosynthesized in the shikimate pathway of aromatic biosynthesis, via chorismic acid (not from anthranilate, as reported earlier). The earliest identified biosynthetic intermediate after chorismate is phenazine 1,6-dicarboxylate, which has been isolated from *Pseudomonas phenazinium* and from nonpigmented mutants of other organisms. Phenazine 1,6-dicarboxylate is thought to be the key branch point compound at the beginning of a least two pathways accounting for the biosynthesis of a wide variety of different P. [G. S. Byng *J. Gen. Microbiology* **97** (1976) 57–62; G. S. Byng & J. M. Turner *Biochem. J.* **164** (1977) 139–145].

Phenosulfate esters: see Sulfate esters.

Phenotype: see Mutation

L-Phenylalanine, ***Phe:*** L-α-amino-β-phenylpropionic acid, an aromatic proteogenic amino acid, M_r 165.2. Phe is essential in the animal diet, and it is both glucogenic and ketogenic The first stage in the catabolism of Phe is hydroxylation to L-tyrosine, which is the precursor of Melanin (see), the neurotransmitter Dopamine (see), the hormones Adrenalin, Noradrenalin and Thyroxin (see separate entries), and other compounds. This first step is catalysed by Phe hydroxylase (a monooxygenase), EC 1.14.16.1. Excess L-tyrosine is broken down to fumarate and acetoacetate (Fig. 1).

Classical phenylketonuria is an hereditary defect in the synthesis of Phe hydroxylase (the enzyme may be absent or inactive), which affects about 1 infant in 10,000. These individuals are unable to convert Phe into tyrosine, and the major route of Phe metabolism is thus blocked. Phenylpyruvate and phenylacetic acid are excreted in the urine. The condition is accompanied by defective pigmentation and, if untreated, by severe mental retardation (hence the other name, phenylpyruvic oligophrenia, also known as Fölling's syndrome). The urine of newborn infants is now routinely tested (Guthrie test) for the presence of phenylketones; the condition can be compensated by a diet low in phenylalanine, and the typical mental retardation is thereby avoided. Other types of phenylketonuria are due to defective reduction or synthesis of dihydrobiopterin (see Inborn errors of metabolism).

A further hereditary defect results in the absence of homogentisate oxidase (a dioxygenase), EC 1.13.11.5. This condition is known as alcaptonuria. It is harmless, but the urine of affected individuals turns black on standing, due to the autoxidation of the excreted homogentisic acid. In plants and bacteria, Phe and L-tyrosine are synthesized by the shikimate pathway of Aromatic biosynthesis (see). Chorismic acid, the branch compound of aromatic biosynthesis, is converted into prephenic acid, which is the precursor of both L-tyrosine and Phe (Fig. 2). Pretyrosine dehydrogenase, pretyrosine dehydratase, prephenate dehydrogenase and prephenate dehydratase may serve as taxonomic markers for microorganisms. Thus, in Cyanobacteria, some Coryneform bacteria, *Hansenula henrici, Halobacterium halobium, Trichococcus, Sulfolobus acidocaldarius, Methanosarcina barkeri, Methanococcus voltae, Micrococcus luteus, Leptothrix* spp., *Sphaerotilus* spp., *Euglena gracilis* and *Zea mays,* tyrosine is synthesized exclusively via pretyrosine. In Mung bean and *Pseudomonas* spp., however, tyrosine is synthesized by two pathways, i.e. via pretyrosine and via 4-hydroxyphenylpyruvate. In some organisms, pretyrosine may be converted into Phe by pretyrosine dehydratase, but this is the exception and the majority of Eubacteria, Cyanobacteria, Archaebacteria and eukaryotes (yeasts, green plants)

Table. *Structures of naturally occurring phenazines*

(Phenazine ring numbering: positions 1–4 on one ring, N10 and N5, positions 6–9 on the other ring)

Name	Positions of substituents (C8 always carries hydrogen in known phenazines) 1	2	3	4	5	6	7	9	10	Source and comments
Aeruginosin A	COOH				CH_3		NH_2			*Pseudomonas aeruginosa* (red strain)
Aeruginosin B	COOH		SO_3^-		CH_3		NH_2			*Pseudomonas aeruginosa* (red strain)
Iodinin	OH				→O	OH			→O	First identified from *Chromobacterium iodinum* (renamed *Brevibacterium iodinum*), later from many other genera. Purple crystals with coppery sheen from $CHCl_3$, m.p. 236 °C (d.).
Pyocyanine	OH				CH_3					*Pseudomonas aeruginosa*
Chlororaphin	$CONH_2$									*Pseudomonas chlororaphis*. 1:1 mixture of dihydro- and dehydro-forms. Green crystals from acetone, m.p. 228–235 °C. Sublimes at 210 °C in absence of oxygen. Rapidly forms oxychlororaphin in air.
Griseolutein A	COOH					$CH_2OC{=}O$ / $HOCH_2$ / CH_2OH		OCH_3		Broad spectrum antibiotics from *Streptomyces griseoluteus*. Relatively low toxicity against mammals. Especially active against *Staphylococcus aureus*.
Griseolutein B	COOH							OCH_3		
Griseoluteic acid	COOH				$=\overset{+}{N}_5-$ / CH_2-O-	$HO-C_6=$ / C / CH_2OH		OCH_3		*Streptomyces griseoluteus*
Lomofungin	$COOCH_3$			OH		CHO	OH	OH		*Streptomyces lomondensis*
Myxin	OH				→O	OCH_3			→O	*Sorangium*. Red needles from acetone, m.p. 120–130 °C; also given as 149 °C (d.).
2(OH)Phenazine 1-carboxylate	COOH	OH								Many different species of *Pseudomonas*. Colors ranging yellow to orange.
Phenazine 1-carboxylate	COOH									
2(OH)Phenazine		OH								
Phenazine 1,6-dicarboxylate	COOH					COOH				

synthesize Phe exclusively via phenylpyruvate [F. Lingens & E. Keller *Die Naturwissenschaften* **70** (1983) 115–118]. The detailed enzymology of tyrosine synthesis in *Pseudomonas aeruginosa* has been reported [Patel et al. *J. Biol. Chem.* **252** (1977) 5839–5846].

L-Phenylalanine ammonia-lyase, ***PAL*** (EC 4.3.1.5): a fungal and plant enzyme catalysing the conversion of L-phenylalanine into *trans*-cinnamic acid and ammonia by nonoxidative deamination (Fig.), an early key reaction in Flavonoid (see) and Lignin (see) biosynthesis. From most plant sources, PAL is a tetrameric protein, M_r about 330,000, consisting of 4 identical subunits of M_r 83,000. Exceptions include PAL from mustard cotyledons (4 subunits, each of M_r 55,000) and PAL from wheat leaves (2 subunits of M_r 75,000 plus 2 of M_r 85,000). PAL possesses two

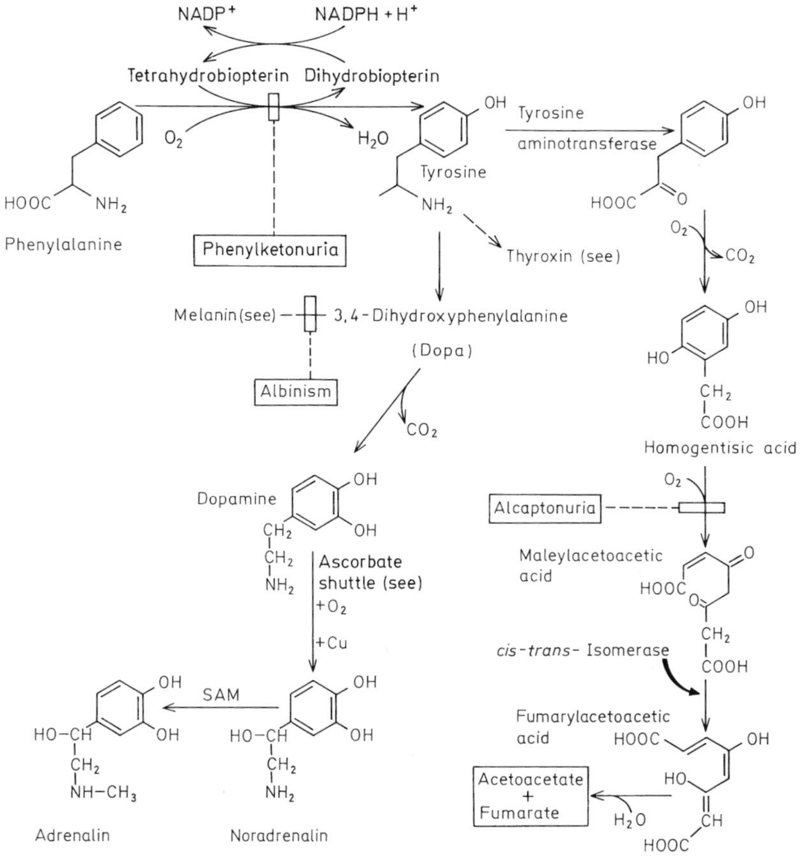

Fig. 1. *Metabolism of L-phenylalanine*

functionally active sites per tetramer, and, like histidine ammonia-lyase, the active sites contain a dehydroalanine residue (not part of an orthodox peptide chain, but present in a Schiff's base linkage at the active site). It is thought that the amino group of Phe adds to the β-position of the dihydroalanyl double bond, resulting in formation of enzyme-ammonia and enzyme-cinnamate intermediates. This would account for the correct stereochemical elimination of the pro-3*S* hydrogen of Phe, leading to *trans* elimination of ammonia, which has been proven by using Phe stereospecifically labeled with isotopic hydrogen at C3. PAL from many sources shows negative rate cooperativity (Hill coefficient < 1.0) with respect to Phe. Since PAL is competitively inhibited by *trans*-cinnamic acid, the enzyme in vivo is probably more sensitive to cinnamic acid concentration than to the pool size of Phe. Cinnamic acid therefore appears to be important in regulating the flux through the phenylpropanoid pathway. PAL from monocots and microorganisms also catalyses deamination of Tyr to 4-coumaric acid and ammonia. PAL increases rapidly when biosynthesis of flavonoid Phytoalexins (see) is initiated; it therefore represents an ideal system for investigation of de novo gene regulation in plants. [R.A. Dixon et al. *Advances in Enzymology* **55** (1983) 1–136; G.B. D'Cunha et al. 'Purification of phenylalanine ammonia lyase from *Rhodotorula glutinis*' *Phytochemistry* **42** (1996) 17–20; T. Fukasawa-Akada et al. 'Phenylalanine ammonia-lyase gene structure, expression, and evolution in *Nicotiana*' *Plant Molecular Biology* **30** (1996) 687–695].

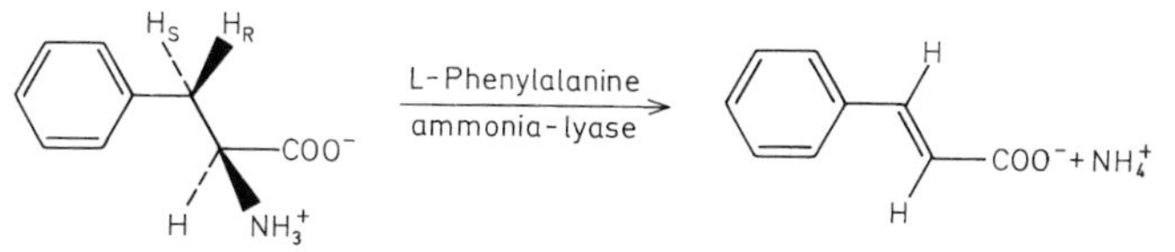

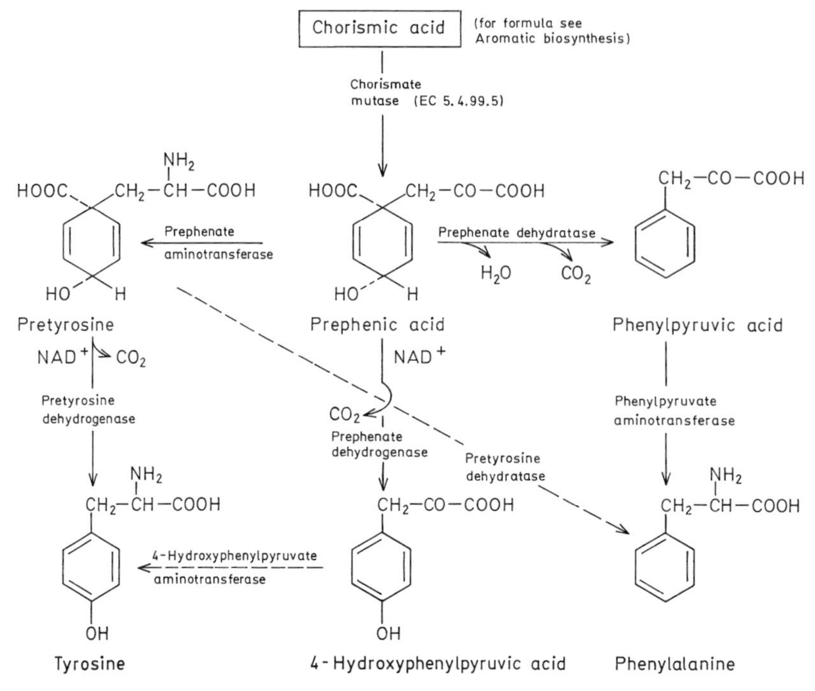

Fig. 2. *Prephenate branch of aromatic biosynthesis*

Phenylketonuria: see L-Phenylanine

Phenylpyruvic oligophrenia: see L-Phenylalanine.

Pheophorbide: see Chlorophyll.

Pheophytin: see Chlorophyll.

Pheromones: predominantly low M_r substances produced by animals, especially insects (insect attractants), and secreted outside the body for purposes of communication (chemical biocommunication) with members of the same species. The main structural types are lower terpenes and higher unbranched fatty acids, and there are also ali- and heterocyclic representatives.

Depending on their mode of perception (susception), P. are classified as oral or olfactory. An animal may show a direct reaction to P., which ceases when the P. disappears (i.e. a releasing effect), or the response may constitute a long term physiological change (i.e. a priming effect). In most cases, a mixture of P. is necessary, together with appropriate biotic und abiotic environmental factors.

For individual members of a species, the most important P. are the Sexual attractants (see). For social insects, e.g. honey bees, ants and termites, aggregation P., alarm P. and trail P. are also important. The activity of P. can be studied by electrophysiological measurement of the nerve impulse of isolated receptor organs (e.g. electroantennograms, abb. EAG); known quantities of P. (10^{-6}-10^{-2} μg) are blown over the isolated antennae in a stream of air, and the resulting cell potential is fed to a recorder with the aid of microelectrodes. The olfactory cells of the silkworm moth or the cockroach produce a measurable nerve impulse in response to one molecule of the appropriate P. Generally the quantity of P. produced by one animal is less than 1 μg. The quantity of Bombykol (see) (the sexual attractant of the silkworm moth) produced by one female moth is sufficient to attract all male members of the species in existence, if they were within range. The activity of a sexual attractant is expressed in attractant units (AU). One unit is the concentration of material (μg/ml) in light petroleum (b.p. 30–50 °C), so that when a glass rod is dipped in the solution and dried it will attract 50 % of experimental males from a distance of 1 m, and stimulate them to vigorous wing buzzing. One unit is in the range of 10^{-6} μg and lower.

Scientifically, P. are useful for the investigation of biological information transfer, receptor theories and structure-activity relationships. Some P. are being used increasingly in the integrated control of insect pests; their potency serves the interest of science and efficiency, while they have no harmful effect on the environment.

P. must be distinguished from allelochemicals, which serve as signal substances between individuals of different species. These are subdivided into allomones (of benefit to the producer, e.g. warning secretions), and kairomones (e g. flower scents). A strict subclassification is, however, difficult, especially for substances with multiple functions.

Phlein: a high M_r reserve carbohydrate in plants. P. is a straight chain polymer of fructofuranose units joined by 2,6-glycosidic linkages. There is probably a D-glucose unit at the non-reducing end of the chain.

Phleomycins: see Bleomycins.

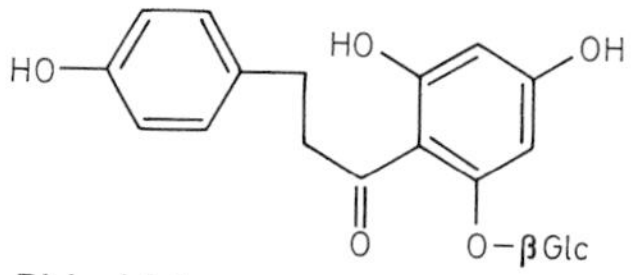

Phlorhizin

Phlorhizin, *phloridzin:* a dihydrochalcone found in the root bark of pears, apples and other members of the *Rosaceae.* P specifically blocks resorption of glucose by kidney tubules, thus inducing glucosuria. It therefore finds use in experimental physiology. Its activity may be due to inhibition of mutarotase.

Phlorin: 5,22-dihydroporphyrin.

Phomine: see Cytochalasins.

Phorbol esters: esters of phorbol, $C_{20}H_{28}O_6$. Some are tumor-promoting compounds originally discovered in oil expressed from the seeds of *Croton tiglium,* but not all P.e. are tumorigenic [E. Hecker & R. Schmidt *Fortschr. Chem. Org. Naturst.* **31** (1974) 377–467]. There is a direct correlation between tumorigenicity and ability to stimulate an amiloride-sensitive Na^+/H^+ exchange across the plasma membranes of cultured human leukemia cells [J. M. Besterman & P. Cuatrecasas *J. Cell. Biol.* **99** (1984) 340–343]. A similar exchange appears to be associated with initiation of proliferation or differentiation in cultured cells.

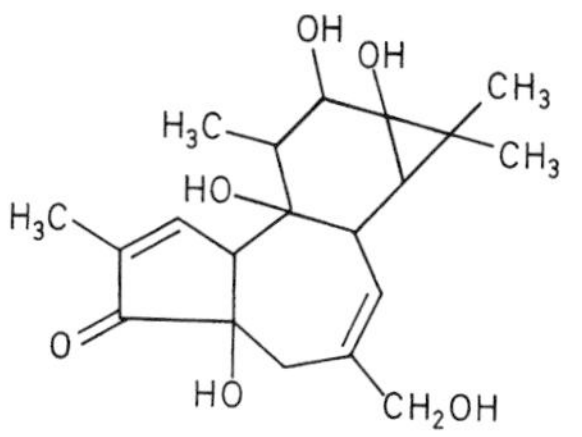

Phorbol

Phosphagens: energy-rich guanidinium or amidine phosphates, which function as storage depots for high energy phosphate in muscle. Excess energy-rich phosphate (i. e. ATP) is transferred to P., from which ATP can be regenerated when required. In invertebrates, arginine phosphate (phosphoarginine, M_r 254.2) is the commonest P. The phosphate group is attached to the guanidine nitrogen of the amino acid:

$$\text{ATP} + \text{guanidine derivate} \rightleftharpoons \text{ADP} + \text{phosphagen } (\underset{\text{H}}{-\text{N}} \sim \text{Ⓟ})$$

The reaction is catalysed by arginine kinase (EC 2.7.3.3) (Fig.). There are several other invertebrate P., particularly in worms, e. g. lombricine phosphate and taurocyamine phosphate. The P. of vertebrate muscle is creatine phosphate (phosphocreatine: M_r 211.1), which is formed from creatine and ATP by the action of creatine kinase (EC 2.7.3.2) (Lohmann reaction, Fig.): reversal of the reaction regenerates ATP. The system, creatine phosphate + creatine kinase + ADP, is often used in vitro for the continual generation of ATP, if the enzyme under investigation is inhibited by substrate levels of ATP.

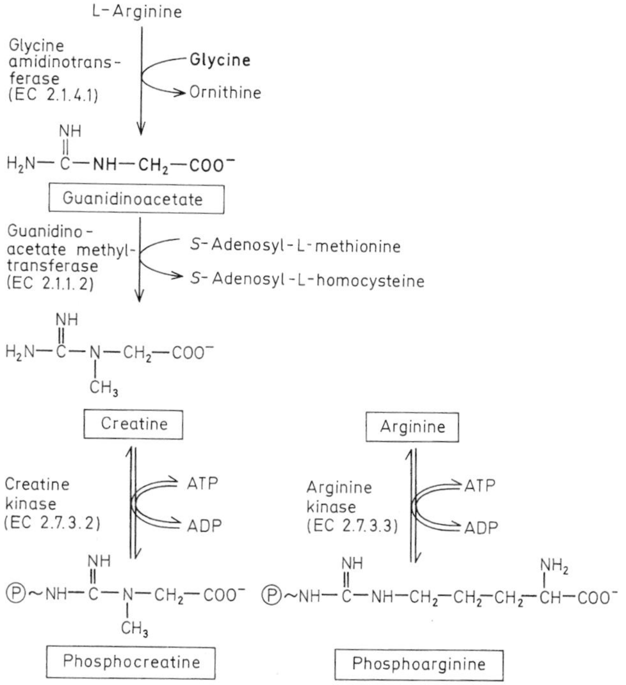

L-Arginine
Glycine amidinotransferase (EC 2.1.4.1)
Glycine
Ornithine
Guanidinoacetate
Guanidinoacetate methyltransferase (EC 2.1.1.2)
S-Adenosyl-L-methionine
S-Adenosyl-L-homocysteine
Creatine
Arginine
Creatine kinase (EC 2.7.3.2)
ATP
ADP
Arginine kinase (EC 2.7.3.3)
ATP
ADP
Phosphocreatine
Phosphoarginine

Biosynthesis of phosphocreatine (creatine 5-phosphate) and phosphoarginine (arginine phosphate)

Phosphatases. ***phosphoric monoester hydrolases*** (EC sub-sub-group 3.1.3): esterases that catalyse the hydrolysis of monophosphate esters. P. are widely distributed in living organisms, and they are mostly dimeric proteins with a catalytically important serine residue in the active center. They are classified on the basis of their pH optima as acid P. (EC 3.1.3.2), e.g. the single chain P. of liver (M_r 160,00) and erythrocytes (M_r 10,000), and prostate P. (M_r 102,000; 2 chains, each of M_r 50,000), or alkaline P. (EC 3.1.3.1), e.g. intestinal mucosa P. (M_r 140,000; 2 chains, each of M_r 69,000), placenta and bone P. (M_r 120,000; 2 chains, each of M_r 60,000) and the P. of *E. coli* (M_r 85,000; 2 chains, each of M_r 43,000). The alkaline P. contain one or two essential zinc atoms per subunit, and also require Mg(II) ions for full activity; these cations have no effect on acid P. In contrast to other esterases, P. are only slightly inhibited by diisopropylfluorophosphate. On the other hand, the alkaline P. are strongly inhibited by ethylenediaminetetraacetic acid (EDTA), inorganic phosphate and L-phenylalanine, while the acid P. are inhibited by fluoride. D(+)-Tartaric acid is a specific and highly potent inhibitor of prostate acid P.

In addition to the relatively unspecific P., some P. show a high specificity, e.g. 5′-nucleotidase from snake venom and 3′-nucleotidase from rye, which catalyse the hydrolysis of 5′-nucleotides or 3′-nucleotides to their respective nucleosides (see Purine degradation, Pyrimidine degradation).

Determination of the acid tartrate-sensitive P. in serum is important in the diagnosis of prostate cancer, which is accompanied by a marked increase in the serum level of this enzyme. Measurement of alkaline P. in serum is used in the diagnosis of bone disease, especially bone tumors, and diseases of the liver (hepatitis) and the gall bladder (e.g. obstructive jaundice), all of which result in a several fold increase in serum P. The isoenzymes of both acid and alkaline P. differ in their neuraminic acid contents. The most widely used substrate for P. assay is *p*-nitrophenyl phosphate; the released *p*-nitrophenol can be determined directly by photometry (λ_{max} 405 nm).

Phosphatidases: see Phospholipases.

Phosphatide degradation: the hydrolytic cleavage of phosphatides (see Phospholipases).

Phosphatides: see Membrane lipids.

Phosphatidic acid: see Membrane lipids.

Phosphatidylcholines: see Membrane lipids.

Phosphatidylcholine-sterol acyltransferase deficiency: see Inborn errors of metabolism.

Phosphatidylethanolamines: see Membrane lipids.

Phosphatidylglycerols: see Membrane lipids.

Phosphatidylinositols: see Membrane lipids, Inositol phosphates.

Phosphatidylserines: see Membrane lipids.

Phosphoadenosinephosphosulfate, ***3′-phosphoadenosine-5′-phosphosulfate, adenosine-3′-phosphate-5′-phosphosulfate, PAPS, active sulfate:*** the product of sulfate activation. PAPS is a key intermediate in the reduction of sulfate and the formation of sulfate esters by sulfokinases. It is produced in a two stage reaction:

ATP + SO_4^{2-} → APS + PP_i (ATP sulfurylase) (EC 2.7.7.4)
APS + ATP → PAPS + ADP (APS kinase) (EC 2.7.1.25)
Sum: 2ATP + SO_4^{2-} → PAPS + ADP + PP_i

Synthesis of PAPS (Fig.) is known as Sulfate activation (see). In the first stage, the terminal pyrophosphate of ATP is replaced by sulfate, forming adenosine phosphosulfate (APS); this reaction is catalysed by sulfate adenylyltransferase (ATP sulfurylase (EC 2.7.7.4). In the second stage, the 3′ position of the adenosine residue of APS is phosphorylated by ATP (catalysed by adenylsulfate kinase, EC 2.7.1.25). Formation of the anhydride bond between adenylic acid and the sulfate anion is a strongly endergonic reaction. The energy balance of the overall process becomes weakly negative only after hydrolysis of the pyrophosphate and consumption of two molecules of ATP. As mixed acid anhydrides of sulfate and adenylic acid, PAPS and APS have high group transfer potentials.

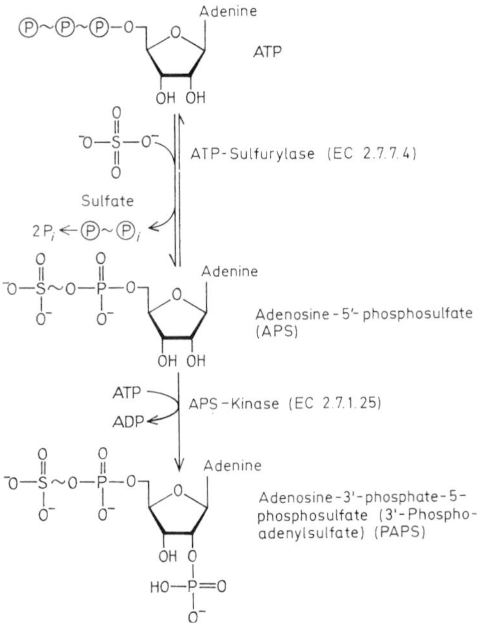

Sulfate activation

Phosphoarginine: see Phosphagens.

Phosphocreatine: see Phosphagens.

Phosphodiester: a phosphate ester in which two hydroxyl groups of the phosphoric acid are esterified with organic residues: RO-PO_2H-OR′. For example, R and R′ may be nucleosides. All polynucleotides and nucleic acids are P., in which the 3′ and 5′ positions of neighboring pentose units are linked by esterification with a phosphate residue.

Phosphodiesterases: enzymes that catalyse the hydrolytic cleavage of phosphodiesters, e.g. endonucleases, Ribonuclease (see) and Deoxyribonuclease (see), and the less specific exonucleases. The latter degrade both DNA and RNA stepwise in the 3′ → 5′ direction, producing 5′-mononucleotides (snake venom P., EC 3.1.4.1), or in the 5′ → 3′ direction, producing 3′-mononucleotides (spleen P., EC 3.1.16.1). P. have been used for the sequence determination of nucleic acids (especially RNA). 3′:5′-Cyclic-nucleotide P. (EC 3.1.4.17) catalyses the hydrolysis of cyclic AMP (see Adenosine phosphates).

Phospho*enol*pyruvate: see Pyruvate.

Phospho*enol*pyruvate carboxylase: see Photosynthetic carboxylation.

Phosphofructokinase, ***6-phosphofructokinase*** (EC 2.7.1.11): an oligomeric phosphotransferase, and a key control enzyme of glycolysis. P. is induced by insulin, and its activity is subject to allosteric control; it is activated by AMP, fructose 6-phosphate, fructose 1,6-*bis*phosphate, magnesium, potassium and ammonium ions, and inhibited by ATP and citrate (see Pasteur effect). P. catalyses the phosphorylation of fructose 6-phosphate by ATP in the presence of magnesium ions to form fructose 1,6-*bis*phosphate. The reaction is irreversible, and the conversion of fructose 1,6-*bis*phosphate to fructose 6-phosphate is catalysed by the fluoride-sensitive enzyme, fructose *bis*phosphatase, EC 3.1.3.11 (fructose diphosphatase). P. of yeast, muscle, liver and erythrocytes have been isolated and studied. Yeast P. consists of six subunits, 3α and 3β, which are immunologically unrelated. Subunit M_r is 130,000 in the native protein, and 96,000 in the protein modified by partial degradation by yeast proteases. The hexameric enzyme has M_r 755,000 (native), or 570,000 (proteolytic modification). In 6 M guanidinium chloride, these subunits dissociate into two equal sized chains of M_r 63,000 (native) or 59,000 (modified).

The degree of association of muscle P. depends on its concentration. At 7 mg protein/ml it exists as an active hexameric aggregate (M_r 2×10^6); at 0.5 mg protein/ml as an active monomer (M_r 340,000). In 6 M guanidinium chloride and 0.1 M mercaptoethanol, the latter dissociates into its 4 inactive chains (M_r 80,000). Similar behavior has been described for the octameric erythrocyte P. (M_r 500,000). Chicken liver P. (M_r of the smallest active form 400,000) is cold-sensitive; it dissociates reversibly at 0 °C into four inactive protomers (M_r 100,000), each consisting of two identical chains. *Clostridium* species contain a P. of M_r 144,000 (4 subunits, M_r 35,000).

Phosphogluconate pathway: see Pentose phosphate cycle.

Phosphoglyceric acids: monophosphate esters of Glyceric acid (see).

Phosphoglycerides: see Membrane lipids.

Phosphohexoketolase pathway: see Phosphoketolase pathway.

Phosphoketolase pathway: a pathway of carbohydrate degradation found in various microorgan-

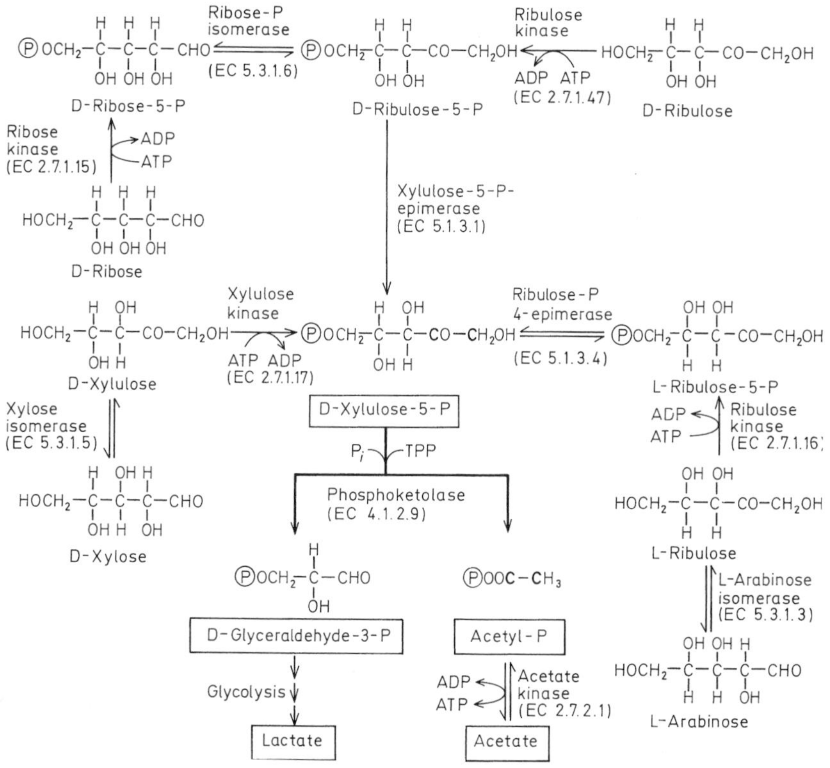

D-Ribose-5-P ⇌ (Ribose-P isomerase, EC 5.3.1.6) D-Ribulose-5-P ← (Ribulose kinase, EC 2.7.1.47; ATP → ADP) D-Ribulose

D-Ribose → (Ribose kinase, EC 2.7.1.15; ATP → ADP) D-Ribose-5-P

D-Ribulose-5-P → (Xylulose-5-P-epimerase, EC 5.1.3.1) D-Xylulose-5-P

D-Xylulose → (Xylulose kinase, EC 2.7.1.17; ATP → ADP) D-Xylulose-5-P ⇌ (Ribulose-P 4-epimerase, EC 5.1.3.4) L-Ribulose-5-P

D-Xylose ⇌ (Xylose isomerase, EC 5.3.1.5) D-Xylulose

L-Ribulose → (Ribulose kinase, EC 2.7.1.16; ATP → ADP) L-Ribulose-5-P

L-Arabinose ⇌ (L-Arabinose isomerase, EC 5.3.1.3) L-Ribulose

D-Xylulose-5-P + P_i (TPP) → (Phosphoketolase, EC 4.1.2.9) D-Glyceraldehyde-3-P + Acetyl-P

D-Glyceraldehyde-3-P → Glycolysis → Lactate

Acetyl-P + ADP → (Acetate kinase, EC 2.7.2.1) Acetate + ATP

The position of phosphoketolase in pentose metabolism. P = phosphate. TPP = thiamin pyrophosphate.

isms, especially *Lactobacillus*, in which a ketopentose phosphate undergoes phosphorolytic cleavage to triose phosphate and acetyl phosphate (Fig.). The key enzyme is phosphoketolase (EC 4.1.2.9) which catalyses the TPP-dependent, irreversible cleavage of D-xylulose 5-phosphate to D-glyceraldehyde 3-phosphate and acetyl phosphate. The balance of pentose metabolism in *Lactobacillus* species is: Pentose ($C_5H_{10}O_5$) + 2ADP + 2P_i → Acetate ($C_2H_4O_2$) + L-Lactate ($C_3H_6O_3$) + 2ATP. There is a net yield of 2ATP per molecule of pentose; one molecule of ATP is required for the phosphorylation of the pentose which is ultimately converted to D-xylulose 5-phosphate (Fig.); 2ATP are derived from the Glycolysis (see) of glyceraldehyde 3-phosphate, and one from the acetyl kinase reaction with acetyl phosphate.

In *Acetobacter xylinum*, carbohydrate is degraded by a phosphohexoketolase pathway; a specific phosphohexoketolase catalyses the phosphorolytic cleavage of fructose 6-phosphate to D-erythrose 4-phosphate and acetyl phosphate.

Phosphokinases: see Kinases.

Phospholipases, *phosphatidases:* a collective name for the carboxylic acid esterases, P.A_1, A_2 and B, and the phosphodiesterases, P.C and D, which are specific for lecithins. P.A_1 (EC 3.1.1.32) catalyses the release of the fatty acid at C1 of the glycerol, producing a lysophosphatide (a 2-acylglycerophosphocholine) which hemolyses erythrocytes. P.A_2 (EC 3.1.1.4) removes the unsaturated fatty acid at C2. P.B (EC 3.1.1.5) also removes the unsaturated fatty acid, but only from lysophosphatides. P.C (EC 3.1.4.3) releases the base in its phosphorylated form. P.D (EC 3.1.4.4) cleaves on the other side of the phosphoryl group and releases the nonphosphorylated base (Fig.). P. activity is particularly high in liver and pancreas (P.A_1), in bee and snake venom (P.A and P.A_2), in bacteria (P.C) and plants (P.D). Due to their compact structures (6–15 disulfide bridges), P.A_2 and P.C display unusual heat stability (5 min at 98 °C) and are insensitive to diethyl ether, chloroform and 8 M urea. P.A_2 from porcine pancreas is synthesized as a zymogen (130 amino acid residues of known sequence, M_r 14,660), which is converted to active P.A_2 (123 amino acid residues) by tryptic removal of the *N*-terminal heptapeptide, Pyr-Glu-Gly-Ile-Ser-Ser-Arg-. P.A_2 from snake venom (*Crotalus atrox*) (M_r 14,500) and bee venom (M_r 14,550, 129 amino acid residues) are about the same size as pancreas P.A_2. Both P.A_2 isoenzymes from the venom of *Crotalus adamanteus* (266 amino acid residues, M_r 29,865) are twice the size of other P.A_2. Bacterial P.C have been isolated from *Clostridium welchii* and *Bacillus cereus*.

Sites of attack by lecithin-cleaving enzymes. R_1 saturated fatty acyl residue. R_2 mono- or multiunsaturated fatty acyl residue.

Phospholipid biosynthesis: see Membrane lipids.

Phospholipids: see Membrane lipids.

Phosphon D: 2,4-dichlorobenzyltri-*n*-butylphosphonium chloride, a synthetic plant growth retardant; e.g. it inhibits the growth of chrysanthemum stems and induces flowering.

Phosphopantetheine: see Pantetheine 4′-phosphate.

Phosphoproteins: conjugated proteins containing phosphate esterified with the hydroxyl groups of serine or (less often) threonine residues. Well known P. are casein (see Milk proteins) and ovalbumin (see Albumins). The latter can contain one or two phosphate groups; this microheterogeneity (see Heterogeneity) is due to variations in the production and activity of the phosphoprotein phosphokinase in the hen oviduct. Other P. are phosphovitin and vitellin of egg yolk, and pepsin of gastric juice.

Phosphopyruvate kinase: see Pyruvate kinase.

5-Phosphoribose 1-diphosphate: see 5-Phosphoribosyl 1-pyrophosphate.

5-Phosphoribosylamine, *PRA:* an intermediate in Purine biosynthesis (see).

5-Phosphoribosyl 1-pyrophosphate. *5-phosphoribose 1-diphosphate, PRPP:* an energy-rich sugar phosphate, M_r 390.1, formed by transfer of a pyrophosphoryl residue from ATP to ribose 5-phosphate. PRPP is concerned in various biosynthetic reactions, e.g. biosynthesis of purines, pyrimidines and histidine.

Phosphoroclastic fission of pyruvate: a special mechanism for the cleavage of pyruvate found only in saccharolytic Clostridia. It is responsible for the synthesis of ATP during nitrogen fixation. The first stage is the synthesis of acetyl phosphate, with production of CO_2 and hydrogen:

$$\underset{\text{Pyruvate}}{CH_3{-}\overset{O}{\overset{\|}{C}}{-}COOH} + P_i \rightarrow \underset{\text{Acetyl phosphate}}{CH_3{-}\overset{O}{\overset{\|}{C}}{-}O{-}PO_3H_2} + CO_2 + H_2$$

followed by synthesis of ATP from acetyl phosphate catalysed by acetokinase:

Acetyl phosphate + ADP ⇌ ATP + Acetate. The first stage requires several enzymes (in the form of a multienzyme system analogous to pyruvate dehydrogenase, together with phosphotransacetylase and hydrogenase) and cofactors (thiamin pyrophosphate, ferredoxin and coenzyme A).

Phosphorus: see Bioelements.

Phosphotransacetylase: see Acetyl phosphate.

Phosphotransferase system: see Active transport.

Phosphovitin: an egg yolk protein containing 10 % phosphate (M_r 180,000). Serine constitutes 50 % of the total amino acid content, and all the serine residues are phosphorylated. P. is synthesized in the liver of the laying hen and transported in the blood to the developing egg.

Photocitral: a cyclization product of Citral (see).

Photoheterotrophism: see Nutritonal physiology of microorganisms.

Photolysis of water: cleavage of water by the light reaction of Photosynthesis (see). P. is a property of photosystem II. It is not a simple photodissociation of water, but is the physiological counterpart of the

Hill reaction (see). Electrons are withdrawn from water or OH^- ions, then transported to $NADP^+$ via an electron transport chain; this results in the production of molecular oxygen with the formation of NADPH + H^+. Since Hill reagents oxidize water by withdrawal of electrons, $NADP^+$ is the natural Hill reagent.

Photophosphorylation: synthesis of ATP in Photosynthesis (see).

Photoproteins: proteins responsible for luminescence in many light-emitting coelenterates. Light emission by P. does not involve a luciferin-luciferase system (see Luciferin), and the reaction proceeds in the absence of oxygen. Aequorin, the P. of the jellyfish *Aequorea,* contains a substituted 2-aminopyrazine as chromophore; light production (λ_{max} 469 nm) is activated specifically by Ca^{2+}. A similar Ca^{2+}-activated P., obelin, has been isolated from *Obelia geniculata* (λ_{max} of emitted light 475 nm). Both of these proteins have been employed as sensitive probes for measuring intracellular Ca^{2+} concentrations, with a sensitivity of at least 10 nM Ca^{2+}. [A. K. Campbell & J. S. A. Simpson "Chemi- and bioluminescence as an analytical tool in biology", in *Techniques in Metabolic Research* **B213**, pp. 1–56, Elsevier (1979)]

Photoreactivation: see DNA repair.

Photorespiration: light-enhanced respiration in photosynthetic organisms. Illumination of C3-plants markedly increases the rate of oxygen utilization; this increase in respiration can be as high as 50 % of the net photosynthetic rate. P. thus results in a loss of yield in the photosynthesis of C3-plants. In C4-plants, P. is either absent or extremely low. P. is largely due to the oxygenase activity of Ribulose-*bis*phosphate carboxylase (see), which oxidatively cleaves ribulose 1,5-*bis*phosphate into phosphoglycolate and 3-phosphoglycerate. Glycolate (derived from the phosphoglycolate) leaves the chloroplasts and enters the peroxisomes, where it is oxidized (by a flavoprotein oxidase) to glyoxylate. Hydrogen peroxide from the action of the flavoprotein oxidase may oxidize some of the glyoxylate to formate and CO_2, but the majority is destroyed by peroxidases and catalase. Most of the glyoxylate is transaminated to glycine, which enters the mitochondria. Glycine may be decarboxylated and/or converted into serine, some of which may reenter the peroxisomes and become oxidized to hydroxypyruvate and D-glycerate. Thus various reactions occur that result in loss of carbon as CO_2. The process depends on light because light is required for the operation of the Calvin cycle, which supplies the ribulose 1,5-*bis*phosphate. Some of these reactions are illustrated in the diagram of the glycolate cycle shown under Glycine (see), but here the phosphoglycolate is shown as coming from active glycolaldehyde. See also CO_2-compensation point, Light compensation point. [S. Krömer 'Respiration During Photosynthesis' *Annu. Rev. Plant Physiol. Plant Mol. Biol.* **46** (1995) 45–70]

Photosynthesis: Although the term photosynthesis can be used to mean any light-dependent synthesis, it is usually used to describe the light-dependent, reductive conversion of CO_2 into carbohydrate carried out by organisms that possess chlorophyll. These comprise members of the plant kingdom, ranging from algae (including the Prochlorophyta, e. g. *Prochloron* and *Prochlorothrix* spp.) to higher plants, cyanobacteria (formerly classified as blue-green algae) and bacteria of the order Rhodospirillales (see Photosynthetic bacteria).

By convention P. is divided into two phases, namely (i) the light phase, in which light energy is absorbed by one or two chlorophyll-containing pigment systems and used to generate ATP (in all photosynthetic organisms) and a reducing agent (in most photosynthetic organisms) and (ii) the dark phase, in which the product(s) of the light phase are used, as soon as they are formed, to convert CO_2 into carbohydrate. The terms 'light' and 'dark' in this context indicate the participation or non-participation respectively of one or more photochemical reactions in the particular phase, and have their origin in the 'light-' and 'dark reactions' proposed by Blackman in 1905 after it had been discovered that an increase in temperature, within the limits of enzyme stability, caused an increase in the rate of photosynthesis, despite the fact that photochemical reactions are temperature-independent. However, the term 'dark phase of photosynthesis' should be used with caution because some of the enzymes catalysing reactions in it have been shown to be activated by light (see e. g. Calvin cycle). Perhaps the time has come to replace it with a term that is less open to misinterpretation, such as 'CO_2 fixation phase'.

Photosynthesis in members of the plant kingdom and in cyanobacteria occurs in the chloroplast. The components of the light phase are built into the chloroplast thylakoid membranes and are organized into an electron transport chain, composed of proteinaceous and lipid redox systems interlinking two pigment systems (PSI & PSII), whose function is to use electrons derived from H_2O to generate ATP (from ADP & P_i) and NADPH (from $NADP^+$ & H^+). This process, termed noncyclic photophosphorylation, is highly endergonic ($\Delta G^{o\prime}$ + 220 kJ mol^{-1}) because it involves (i) the forcing of electrons up a potential gradient of 1.14V, from the very positive redox potential of water (E'_0 + 0.82V) to the negative redox potential of NADPH (E'_0 –0.32V), and (ii) the phosphorylation of ADP ($\Delta G^{o\prime}$ + 30.5 kJ mol^{-1}). It therefore requires the input of energy and it is the function of PSI and PSII to provide this energy. The removal of electrons from H_2O leads to the formation of O_2 as a byproduct; consequently organisms operating the process are said to carry out 'oxygenic photosynthesis'. A modified version of it, termed cyclic photophosphorylation and involving only PSI and about half the usual redox systems, can also operate in these organisms. It does not use up H_2O, generate NADPH or evolve O_2 and its sole function appears to be to boost the intrachloroplastidic level of ATP as and when required. The enzymes catalysing the dark phase in these organisms are not membrane components. Those of the Calvin cycle (see), which operates in the chloroplasts of (i) organisms that carry out C_3 photosynthesis, (ii) the bundle sheath cells of C_4 plants (see Hatch-Slack-Kortschak cycle), and (iii) plants that are able to carry out Crassulacean acid metabolism (CAM) (see), are located in the stroma. However some of the other enzymes, e. g. phospho*enol*pyruvate carboxylase, that participate in the dark phase of photosynthesis in C_4 and CAM plants occur outside the chloroplast, usually in the cytoplasm of the cell.

Photosynthetic bacteria (see) do not have chloroplasts. The components of the light phase occur in the plasma membrane, either as integral proteins or as extrinsic structures projecting into the cytoplasm, and are organized into an electron transport chain, composed of proteinaceous and lipid redox systems containing a single pigment system, the bacterial pigment system (BPS). Only one pigment system is required because photosynthetic bacteria do not use H_2O as the electron donor for noncyclic photophosphorylation; instead they use electron donors that have much less positive E'_0 values [e.g. H_2S (E'_0 – 0.23V) in the *Chromatiaceae,* SO_3^{2-} (E'_0 – 0.32V) in the *Chlorobiaceae;* or succinate (E'_0 + 0.03V) and malate (E'_0 – 0.17V) in the *Rhodospirillaceae*] and consequently do not have to force electrons up such a large potential gradient as do members of the plant kingdom and cyanobacteria. Because photosynthetic bacteria do not use H_2O as an electron donor they do not produce O_2 as a byproduct and are thus said to carry out anoxygenic or nonoxygenic photosynthesis. All photosynthetic bacteria are able to carry out cyclic photophosphorylation, thereby generating ATP. However, collectively they use a variety of ways of generating the reductant, NADH (not NADPH), which is also necessary for the dark phase. This variety probably reflects stages in the evolutionary route to the more advanced oxygenic photosynthesis of the plant kingdom. The enzymes catalysing the reactions of the dark phase are located in the cytoplasm. All photosynthetic bacteria use the Calvin cycle to fix CO_2, with the exception of the *Chlorobiaceae* which use a Reductive citrate cycle (see).

Light phase of photosynthesis:

1. ***In members of the plant kingdom and the cyanobacteria.*** The process of non-cyclic photophosphorylation (nc-p/p) begins with the absorption of photons of light by the antenna pigments of PSI and PSII and the transfer of their energy by inductive resonance to P700 and P680 respectively at the PSI and PSII reaction centers (see Photosynthetic pigment systems), so raising them from the ground state electronic energy level (S_0) to the first excited state singlet (S_1) electronic energy level, represented by P700* and P680* in Fig. 1. The latter constitute the reduced components of two redox systems that have very negative redox potentials, $P700^+/P700^*$ ($E'_0 \sim -1.2V$) and $P680^+/P680^*$ ($E'_0 \sim -0.8V$); consequently they are very powerful reducing agents (i.e. electron donors in redox reactions). Once formed the P700* and P680* rapidly (i.e. within 10 psec) reduce the oxidized component of the redox systems that constitute the immediate electron acceptors of PSI and PSII. The resulting $P700^+$ and $P680^+$ constitute the oxidized components of two redox systems that have very positive redox potentials, $P700^+/P700(S_0)$ (E'_0 + 0.49 V) and $P680^+/P680(S_0)$ ($E'_0 \sim +1.1V$); consequently they are good oxidizing agents (i.e. electron acceptors in redox reactions). Once formed the $P700^+$ and $P680^+$ rapidly (20–30 nsec) oxidize the reduced component of the redox systems that constitute the immediate electron donors to PSI and PSII. By participating in these redox reactions P700 and P680 drive electrons along a chain of redox systems from H_2O (the ultimate electron donor of nc-p/p) to $NADP^+$ (the ultimate electron acceptor). This generates two required products: ATP and NADPH; the latter is the reduced component of the $NADP^+$/NADPH redox system (E'_0 – 0.32V). The byproduct of this process is O_2, which is the the oxidized component of the O_2/H_2O redox system (E'_0 + 0.82V). As electrons are driven along the nc-p/p electron flow pathway, protons are 'pumped' across the thylakoid membrane from the chloroplast stroma into the thylakoid lumen; thus part of the energy of the absorbed photons is used to generate a proton motive force (PMF) (see). The protons then pass back to the stroma via the membrane-spanning CF_0CF_1 complex, thereby promoting the ATP-synthase-catalysed phosphorylation of ADP and the release of the resulting ATP into the stroma (see Chemiosmotic hypothesis). Thus nc-p/p electron flow generates both NADPH and ATP.

Since two electrons (and one H^+) are needed to reduce $NADP^+$ to NADPH the stoichiometry of the nc-p/p electron flow pathway requires that two photons of light be absorbed by each pigment system, thereby generating 2P700* and 2P680*, per H_2O oxidized and per $^1/_2O_2$ generated ($H_2O \rightarrow 2e^- + 2H^+ + {}^1/_2O_2$). However the operation of the water-oxidizing enzyme complex (also known as the oxygen-evolution complex) requires that water molecules are oxidized in pairs, yielding four electrons, four protons and one molecule of O_2.

Starting at the electron donor end of the nc-p/p electron transport chain, electrons are removed one at a time from two molecules of H_2O (from the thylakoid lumen) by the membrane-bound water-oxidizing enzyme complex. There is still considerable uncertainty about this enzyme. It is known that it is intimately associated with the membrane-spanning PSII complex and that for activity it requires Cl^-, Ca^{2+} and a cluster of four manganese ions. The latter is intimately concerned with the stepwise transfer, via Tyr_z, of four electrons from $2H_2O$ to the $P680^+$ form of a single P680 molecule (probably one of a 'special pair') as it goes through the $P680(S_0) \rightarrow P680^* \rightarrow P680^+$ cycle driven by the absorption of four photons. The Mn cluster was thought to be located on one of the three extrinsic proteins (17, 23, & 33 kDa) bound to the luminal surface of the PSII complex and known to be required for O_2 evolution, but this has now been shown to be unikely. Tyr_z, the immediate recipient of electrons from the Mn cluster and immediate donor of electrons to $P680^+$, is a tyrosine residue at position 161 of the D_1 polypeptide (32 kDa) of the PSII complex. The electron flow sequence from H_2O to $P680^+$ is thus $H_2O \rightarrow$ Mn cluster $\rightarrow Tyr_z \rightarrow P680^+$ and represents a fall in redox potential (E'_0) from + 0.82V to ~ + 1.1V, rendering it exergonic to the extent of ~ 54 kJ mol^{-1} of electron pairs under standard conditions. The H^+ ions liberated by the oxidation of H_2O remain on the luminal side of the thylakoid membrane and contribute to the proton gradient that nc-p/p generates across it.

The P680* formed by PSII light absorption reduces the oxidized form of its immediate electron acceptor, pheophytin (Pheo) (E'_0 – 0.61V), located in the PSII complex. The resulting $Pheo^-$ then reduces Q_A, a plastoquinone (PQ) molecule tightly bound to protein D_2 of the PSII complex. The receipt of one electron from $Pheo^-$ converts Q_A to Q_A^-; this is followed by the receipt of a second, forming Q_A^{2-}. At neither the Q_A^- stage nor the Q_A^{2-} stage are H^+ ions taken up to form the semiquinone or the quinol. The Q_A^{2-}

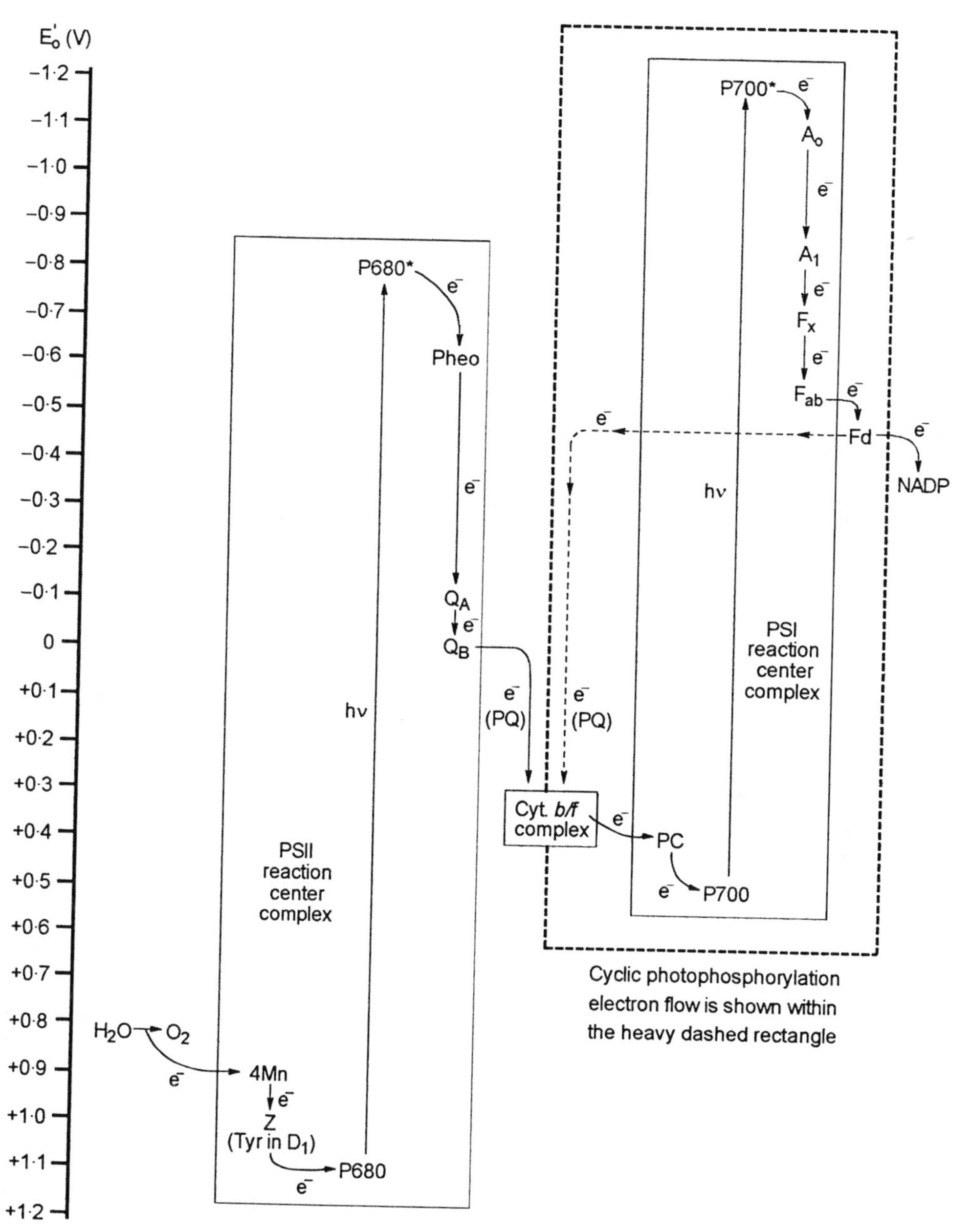

Fig. 1. *The electron flow pathways of non-cyclic- and cyclic photophosphorylation in the plant kingdom and the cyanobacteria.* Electron flow in non-cyclic photophosphorylation extends from H_2O to NADP and involves PSI and PSII, whereas that in cyclic photophosphorylation (shown within the heavy dashed rectangle) involves only PSI and follows a cyclic path. This cyclic path is composed of the 'P700→Fd' and '*bc*-complex→P700' (which also occur in non-cyclic photophosphorylation) plus the 'Fd→*bc*-complex' section (shown with a dashed arrow) that links them; Fd = ferredoxin; Pheo = pheophytin; (PQ) = plastoquinone, mobile in the lipid bilayer and exchanging with PQ at the P_B site; PC = plastocyanin.

then reduces Q_B, a PQ molecule loosely bound to protein D_1 of the PSII complex, along with two H^+ ions taken from the stroma, to form the quinol, Q_BH_2. The E'_0 of the two PQ redox systems is about 0V, with Q_B probably a little more positive than Q_A. Once formed, the Q_BH_2 rapidly (~5 msec) exchanges with a PQ molecule from the pool of mobile PQ molecules in the hydrophobic central region of the membrane lipid bilayer.

The exchanged Q_BH_2 (i.e. PQH_2), having been released from the stromal end of the transmembrane PSII complex into the membrane bilayer, now diffuses to the transmembrane cytochrome *bc* complex. This contains three redox components, cytochrome b_6 (E'_0 ~ –0.1V), the Rieske 2Fe-2S protein (E'_0 ~ +0.25V) (see) and cytochrome *f* (E'_0 ~ +0.34V), a typical *c*-type cytochrome (the *f* comes from *frons*, the Latin for leaf). The *bc* complex accepts electrons from Q_BH_2 (thereby oxidizing it and allowing it participate in the PQ/PQH_2 exchange again) and transfers them to plastocyanin (PC) (see), a soluble copper-containing redox protein (E'_0 ~ +0.37V) present in the aqueous milieu of the thylakoid lumen. The way in which it does this is controversial. One theory is that it operates in a manner similar to that outlined in Fig. 3, which shows the *bc* complex of photosynthetic bacteria participating in the Q-cycle (see). This process (explained below) involves all three redox systems of the complex and results, not only in the exergonic transfer of electrons down a potential drop of 0.37V, but also the 'pumping' of H^+ across the thylakoid membrane from the chloroplast stroma into the thylakoid lumen, thereby contributing to the ATP-generating PMF. The reduced PC then diffuses to the transmembrane PSI complex where it is oxidized by $P700^+$ of the 'special pair' located at the luminal end, thereby regenerating P700(S_0) and readying it for the absorption of another photon; the oxidized PC diffuses back to the *bc* complex to pick up another electron. The electron flow sequence from P680* to $P700^+$ is thus P680* → Pheo → Q_A → Q_B → $PQ_{(mobile)}$ → *bc* → PC → $P700^+$ and represents a fall in redox potential (E'_0) from ~ –0.8V to +0.49V, rendering it exergonic to the extent of ~249 kJ mol^{-1} of electron pairs under standard conditions.

The P700* formed by PSI light absorption reduces the oxidized form of its immediate electron acceptor, A_0, believed to be a monomeric chlorophyll *a* (E'_0 ~ –1.0V) that is functionally analogous to the Pheo of PSII, located in the PSI complex. This then passes the electron exergonically on through a sequence of redox systems with progressively less negative E'_0 values that are components of the PSI complex, namely A_1 (probably phylloquinone), F_x (a 4Fe-4S protein with an E'_0 of –0.7V) and F_{ab} (a complex of two 4Fe-4S proteins with E'_0 values of –0.59V and –0.54V). The reduced form of the latter then reduces the oxidized form of a soluble ferredoxin (Fd) (E'_0 0.42V) (see) in the chloroplast stroma. The enzyme ferredoxin:NADP oxidoreductase (EC 1.6.99.4), an FAD-containing flavoprotein bound to the stromal surface of the thylakoid membrane (particularly that of non-appressed thylakoids), then catalyses the reduction of $NADP^+$ by the reduced ferredoxin. This reaction requires the uptake of one H^+; this comes from the chloroplast stroma and serves to accentuate the proton gradient across the thylakoid membrane created by the 'pumping' of protons mentioned earlier. The electron flow sequence from P700* to NADPH is thus P700* → A_0 → A_1 → F_x → F_{ab} → Fd → FAD → NADPH and represents a drop in redox potential (E'_0) from ~ –1.2V to –0.32V, rendering it exergonic to the extent of ~170 kJ mol^{-1} of electron pairs under standard conditions.

The number of molecules of ATP generated as a result of the combined activities of the proton gradient and the CF_0CF_1 complex is controversial. However, it seems unlikely that nc-p/p generates sufficient ATPs per $NADP^+$ reduced (or per electron pair transported) to satisfy the demands of the Calvin cycle (which requires 1.5 ATPs per NADPH) and the other ATP-requiring processes of the chloroplast (e.g. synthesis of starch, lipids, pigments, proteins, nucleic acids). It is probably for this reason that the evolutionarily more primitive process of cyclic photophosphorylation (c-p/p) persists in chloroplasts; c-p/p generates only ATP and therefore assists nc-p/p in supplying the total ATP requirements of the chloroplast.

The path of electron flow in c-p/p is shown in Fig. 1 (within the heavy dashed rectangle). Electrons are driven around this cyclic path by light photons absorbed by PSI only. P700*, generated by photon absorption, transfers electrons to soluble Fd via the same chain of redox systems as operate in nc-p/p. The resulting reduced Fd then transfers electrons to the cytochrome *bc* complex, instead of to $NADP^+$ as occurs in nc-p/p. PQ is the 'go-between' in this transfer, but there are a number of uncertainties about the way it accomplishes this task; for instance it is not clear whether the transfer of electrons from Fd to PQ is catalysed by a specific ferredoxin:PQ oxidoreductase or by the ferredoxin:NADP oxidoreductase that participates in nc-p/p. However, it is clear that interaction of the PQ/PQH_2 redox system and the *bc* complex leads to the 'pumping' of protons across the thylakoid membrane from the stroma to the lumen, probably by a mechanism involving the Q-cycle and reminiscent of that operating in photosynthetic bacteria (Fig. 3). The resulting proton gradient then drives the CF_0CF_1 ATP-synthase-catalysed phosphorylation of ADP as in nc-p/p. From the *bc* complex the electrons are transferred to PC, which then diffuses through the luminal milieu and reduces $P700^+$ to P700(S_0) at the luminal face of the PSI complex, thereby completing the cycle. Although c-p/p probably occurs in all photosynthetic cells in members of the plant kingdom and the cyanobacteria, it is the only or main light phase mechanism in certain specialized cells, such as the heterocysts of cyanobacteria, which require ATP for nitrogen fixation, and the bundle sheath cells of some C_4 plants (the NADP-malic enzyme users – see Hatch-Slack-Kortschak cycle), because they either do not possess a PSII complex or have a much less abundant one.

2. ***In the purple sulfur-, purple non-sulfur- and green non-sulfur bacteria.*** In these three photosynthetic bacterial families, the *Chromatiaceae*, *Rhodospirillaceae* and *Chloroflexaceae* respectively, ATP is generated by c-p/p. NADH, the reductant needed, along with ATP, for the dark phase of photosynthesis, is generated in different bacterial species in different ways. These, however, appear to fall into three gener-

al categories, namely: (i) the use of H_2 and the enzyme hydrogenase, (ii) the use of electron donors with E'_0 values that are more negative than Q_B of the bacterial pigment system (BPS) complex, and (iii) the use of electron donors with E'_0 values that are more positive than Q_B of the BPS complex.

The electron flow pathway of c-p/p in these bacteria is shown in Fig. 2(a). The process is initiated by the absorption of photons of light by the antenna pigments of the BPS and the transfer of their energy by inductive resonance to the BPS reaction centre; this generates P870*, the reduced component of the B870*/B870$^+$ redox system ($E'_0 \sim -0.9$V). P870* rapidly reduces its more electropositive neighbors in BPS, so generating BPheo$^-$, the reduced form of bacteriopheophytin. BPheo$^-$ then reduces the quinone at the Q_A site of BPS, which is ubiqinone (UQ) in some bacteria (e.g. *Rhodobacter sphaeroides*) and menaquinone (MQ) in others (e.g. *Rhodobacter viridis, Chloroflexus*). The resulting Q_A^- does not take up a proton but immediately reduces the quinone at the Q_B site, which is again UQ in some bacteria (e.g. *Rb. sphaeroides, Rb. viridis*) and MQ in others (e.g. *Chloroflexus*), forming Q_B^-. The latter then accepts a second electron from newly re-reduced Q_A^- and the resulting Q_B^{2-} takes up two H^+, thereby becoming a quinol, QH_2. The two H^+ come from the cytosol because the Q_A and Q_B sites are at the cytosolic end of the transmembrane BPS complex and participate in the generation of a H^+ gradient across the bacterial plasma membrane. The QH_2 formed at the Q_B site then exchanges with Q, which is mobile in the lipid bilayer of the plasma membrane. It moves to the periplasmic end of the transmembrane cytochrome *bc* complex and participates in a rather complicated and imperfectly understood Q-cycle which results in the

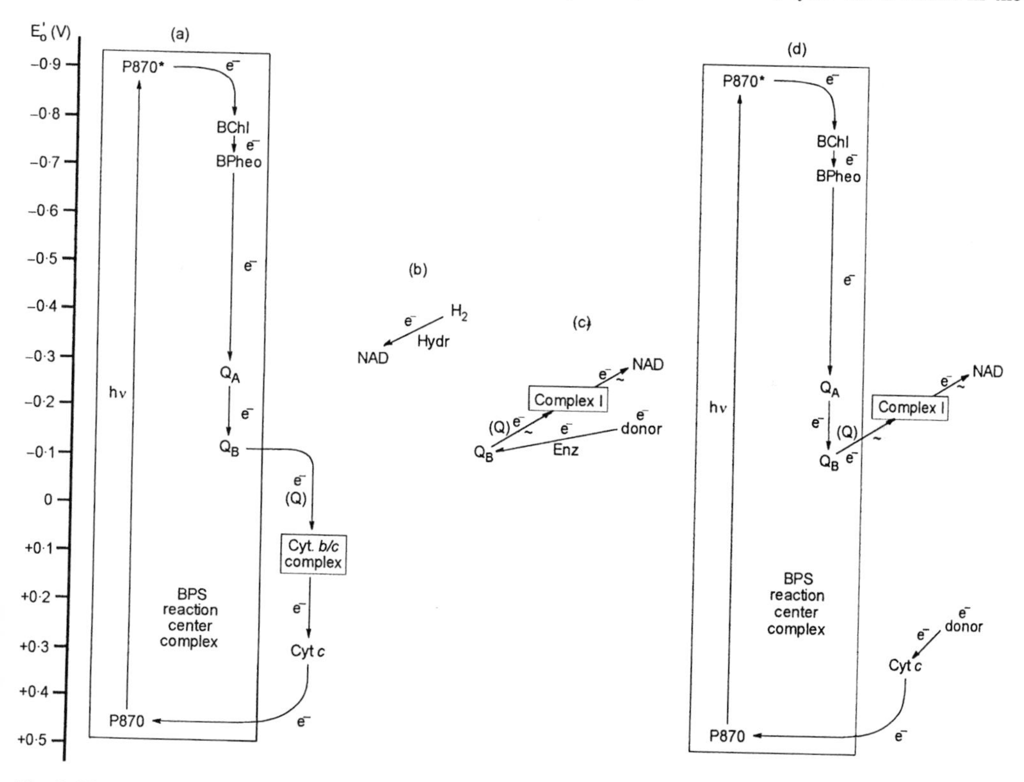

Fig. 2. *The electron flow pathways of cyclic photophosphorylation in purple sulfur-, purple non-sulfur- and green non-sulfur bacteria, including the main strategies that these organisms have collectively evolved for the generation of reductant for use in the dark phase of photosynthesis.* (a) cyclic photophosphorylation, (b) use of H_2 and hydrogenase for the generation of NADH, (c) use of electron donors with E'_0 values that are more negative than Q_B for the generation of NADH, (d) use of electron donors with E'_0 values that are more positive than Q_B for the generation of NADH; ~ in diagrams (c) and (d) indicates the use of PMF; Complex I in diagrams (c) and (d) is NADH dehydrogenase (quinone) (EC 1.6.99.5); P870 = P870 at the ground state energy level; P870* = P870 at its excited S_1 energy level after absorbing a photon of light (hv) of the appropriate wavelength; BChl and BPheo = bacteriochlorophyll *a* or *b* and bacteriopheophytin *a* or *b* respectively, depending upon the bacterial sp.; Q_A and Q_B are ubiquinone and/or menaquinone, depending upon the bacterial sp.; Cyt. *c* = cytochrome *c* or c_2, depending upon the bacterial sp.; in (a) and (d) the proton gradient generated across the bacterial plasma membrane that accompanies electron flow is used to phosphorylate ADP.

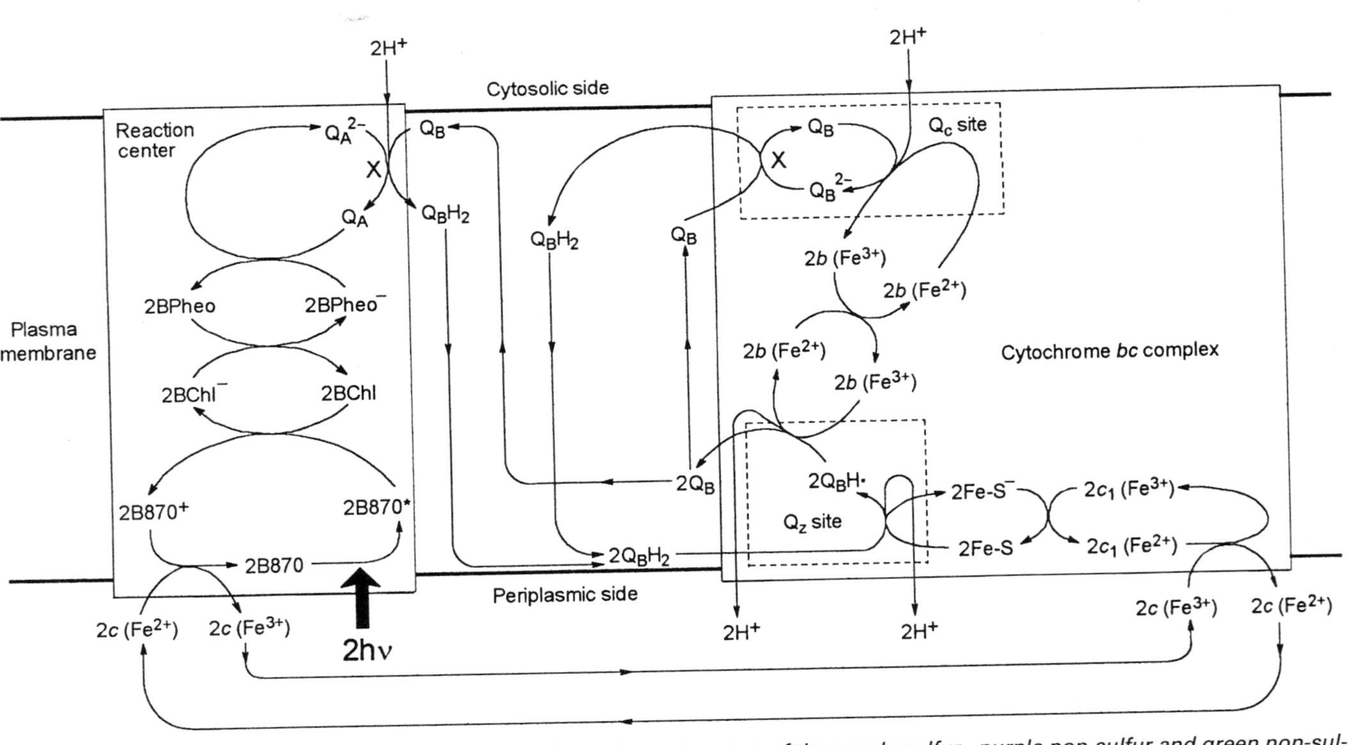

Fig. 3. *The Q-cycle operating with the reaction center of the pigment system of the purple sulfur-, purple non-sulfur and green non-sulfur bacteria.* B870 = 'special pair' of bacteriochlorophyll (BChl) *a* or *b*; B870* = B870 in the S_1 excited state; BChl = bacteriochlorophyll *a* or *b* molecules directly reduced by B870*; BPheo = bacteriopheophytin *a* or *b*; Q_A = bound quinone [ubiquinone (UQ) in *Rhodobacter sphaeroides*; menaquinone (MQ) in *Rhodobacter viridis* and *Chloroflexus* spp.]; Q_B = mobile quinone [UQ in *Rb. sphaeroides* and *Rb. viridis*; MQ in *Chloroflexus* spp.] which exchanges with Q_BH_2 at the sites marked with an X; Q_BH_2 = fully reduced (quinol) form of Q_B; $Q_BH\cdot$ = semiquinone form of Q_B; Fe-S = Rieske iron-sufur center; *b*(Fe^{2+}) or *c*(Fe^{2+}) = reduced forms of cytochromes *b* or *c*; *b*(Fe^{3+}) or *c*(Fe^{3+}) = oxidized forms of cytochromes *b* or *c*; X = Q_B/Q_BH_2 exchange site.

transfer of its two electrons, via the components of the the *bc* complex, to a soluble *c*-type cytochrome in the periplasmic fluid.

A possible form of this Q-cycle is shown in Fig. 3. Its stoichiometry is best understood if its description starts with the binding of two QH_2 molecules at the Q_Z site of the *bc* complex. Here they each give up one electron and one H^+ in reducing the Rieske Fe-S center (see) and in doing so become a semiquinone, QH. The Rieske Fe-S center passes the two electrons, one at a time, to cytochrome c_1 of the *bc* complex, which in turn uses them to reduce two molecules of soluble cytochrome *c* in the periplasmic fluid. The two H^+ ions pass into the periplasmic fluid and contribute to the H^+ gradient generated by the Q-cycle. The two QH· radicals then each reduce a *b*-type cytochrome of the *bc* complex, thereby generating two H^+, which also pass into the periplasmic fluid and contribute to the H^+ gradient, and two oxidized Q molecules, which pass into the lipid bilayer. These two Qs have different fates. One returns to the Q_B site on BPS where it exchanges with a QH_2 newly generated by the activities of BPS; the other moves to the Q_C site at the cytosolic end of the *bc* complex where it exchanges with bound QH_2. The resulting two QH_2 molecules return to the Q_Z site to start the next Q-cycle. The exchange of Q for QH_2 at the Q_C site requires that QH_2 be generated there by reduction of a molecule of Q already bound at the site; this involves the receipt of two electrons and two H^+. The latter are taken from the cytosol and also participate in the generation of the H^+ gradient. The two electrons are derived from the two *b*-type cytochromes of the *bc* complex that were reduced by the two QH· formed at the Q_Z site. These two cytochromes firstly reduce two other *b*-type cytochromes of the *bc* complex, which then reduce the Q bound at the Q_C site. The latter process involves one electron from each reduced cytochrome *b* and the two H^+ from the cytosol. It may take one of two routes, namely: (i) $Q \rightarrow Q^- \rightarrow Q^{2-} \rightarrow QH_2$ (cf. the interaction of Q at the Q_A and Q_B sites of BPS), or (ii) $Q \rightarrow Q^- \rightarrow QH\cdot \rightarrow QH^- \rightarrow QH_2$. With the formation of the quinol product, QH_2, the Q/QH_2 exchange can occur. As a result of the operation of this Q-cycle four H^+ are 'pumped' across the bacterial plasma membrane from the cytosol to the periplasm per electron pair transferred from Q to soluble cytochrome *c*. The resulting proton gradient drives the F_0F_1 ATP-synthase-catalysed phosphorylation of ADP as in plant c-p/p and nc-p/p; the H^+-conducting F_0 component spans the bacterial plasma membrane and the ATP-synthase-containing F_1 component projects from it into the cytosol.

The c-p/p process is completed by the soluble cytochrome *c* molecules, reduced as a result of the activities of Q-cycle, which diffuse through the periplasmic fluid to the BPS complex where their E'_0 value is sufficiently negative at ~ +0.3V to enable them to reduce the $P870^+$ of the $P870^+/P870(S_0)$ redox system (E'_0 ~ +0.45V).

Fig. 2(b), (c) & (d) show the different methods used by these bacteria to generate NADH for use in the dark reaction. Some bacteria (e.g. *Chromatium*) are able to use hydrogen gas for this purpose [Fig. 2(b)] and can thus grow well in the presence of light, H_2 and CO_2. They possess the enzyme, hydrogenase (EC 1.12.1.2), which catalyses the reaction shown in eq. 1, in which H_2 reduces NAD^+.

$$H_2 + NAD^+ \rightarrow H^+ + NADH \qquad (1)$$

This reaction is possible because the H^+/H_2 redox system (E'_0 −0.42V) is a more powerful reducing system than the NAD^+/NADH redox system (E'_0 −0.32V) and is exergonic to the extent of 19.3 kJ mol^{-1} under standard conditions. Thus NADH generation in *Chromatium* growing on H_2 is a light-independent process; however light is required to drive c-p/p and generate the ATP that is required, in addition to NADH, for CO_2 fixation.

Other bacteria are able to use as electron donors compounds that constitute the reduced components of redox systems with E'_0 values more negative than that of the quinone/quinol redox system (E'_0 ~ 0V) which binds at the Q_B site of BPS [Fig. 3(c)]. Examples of such electron donors are H_2S (E'_0 of S/H_2S = −0.23V), which is used by some purple sulfur bacteria (including the versatile *Chromatium*), and malic acid (E'_0 of oxaloacetate/malate = −0.17V), which is used by some purple- and green non-sufur bacteria. Organisms using such electron donors have an enzyme that catalyses the generalized reaction shown in eq. 2; this enzyme is the functional equivalent of the succinate:UQ oxidoreductase of complex II of the mitochondrial electron transport chain.

$$\text{Electron donor} + Q \rightarrow \text{oxidized form of } e^- \text{ donor} + QH_2 \qquad (2)$$

The activity of this enzyme results in the exergonic transfer of an electron pair from the electron donor to the quinone. The resulting QH_2 is then used to reduce NAD^+. This is an endergonic process because the E'_0 of the Q/QH_2 redox system is less negative than that of the NAD^+/NADH redox system. In fact, in mitochondrial aerobic respiration this reaction proceeds in the opposite direction with electrons flowing exergonically from NADH to Q under the catalytic influence of NADH dehydrogenase (quinone) (EC 1.6.99.5) in complex I; moreover much of the free energy 'liberated' in the redox reaction is used to 'pump' H^+ across the inner mitochondrial membrane and is thus conserved as a proton gradient, which is then used by the F_0F_1 complex to generate ATP. However in these bacteria the electrons are driven in the endergonic direction by part of the energy of the proton gradient generated by c-p/p operating as shown in Fig. 3(a) with the assistance of a membrane-bound complex I-like enzyme system.

Still other bacteria are able to use as electron donors compounds that constitute the reduced components of redox systems with E'_0 values more positive than that of the quinone/quinol redox system which binds at the Q_B site of BPS [Fig. 3(d)]. These organisms probably have an enzyme, located in the periplasmic space, that catalyses the generalized reaction shown in eq. 3.

$$\text{Electron donor} + \text{cyt } c\ (Fe^{3+}) \rightarrow \text{oxidized form of } e^- \text{ donor} + \text{cyt } c\ (Fe^{2+}) + H^+ \qquad (3)$$

The activity of this enzyme results in the exergonic transfer of electrons from the electron donor to the soluble cytochrome *c* in the periplasmic fluid. The reduced cyt. *c* then diffuses to the BPS reaction center complex where it reduces $P870^+$. The absorption of light by BPS leads to the formation of QH_2 at the Q_B site as described earlier. The UQ_2 then exchanges with Q in the lipid bilayer and diffuses to the membrane-bound complex I-like enzyme system which uses it to reduce NAD^+ as before. Thus the energy barrier for the cyt $c \rightarrow$ Q step is overcome by light energy absorbed by BPS and that for the $QH_2 \rightarrow NAD^+$ step by the light-generated proton gradient.

3. ***In green sulfur bacteria.*** The *Chlorobiaceae,* unlike the other photosynthetic bacteria, are able to carry out nc-p/p (Fig. 4). However, they cannot use H_2O as their electron donor because the E'_0 of the $B840^+/B840(S_0)$ system in the BPS reaction center complex is far too negative, at ~ +0.3V, to be reduced by water. Thus they have to use electron donors with E'_0 values that are of the same order as those used by the other families of photosynthetic bacteria, namely ~ −0.3V to ~ +0.1V; being sulfur bacteria these electron donors are sulfur compounds, such as H_2S and SO_3^{2-}. The reason why the green sulfur bacteria can carry out nc-p/p whereas the others cannot is that the primary mobile reductant generated by the BPS reaction center complex is the reduced form of the Fe-S protein, ferredoxin (Fd), rather than QH_2. The bacterial Fd/Fd^- redox system has an E'_0 of −0.45V which is 0.13V more negative than the $NAD^+/NADH$ redox system; thus Fd^- can reduce NAD^+ in a reaction that is exergonic to the extent of ~ 25 kJ mol^{-1} under standard conditions. This contrasts with the Q/QH_2 redox system whose more positive E'_0 of ~ 0V does not allow the exergonic reduction of NAD^+. The $Fd^- \rightarrow NAD^+$ electron transfer is analogous to that by which NADPH is generated in plant nc-p/p as a result of the operation of PSI (Fig. 1). Thus the electron flow pathway of nc-p/p in green sulfur bacteria is believed to start with the exergonic sequence: electron donor $\rightarrow$ Q? $\rightarrow$ cyt *bc* complex $\rightarrow$ cyt *c*-553 (mobile in periplasmic fluid) $\rightarrow P840^+$; as electrons flow through this sequence a Q-cycle is presumed to generate a proton gradient across the plasma membrane and lead to the phosphorylation of ADP. The absorption of a photon by BPS then generates the powerful reductant P840* (E'_0 ~ −1.2V) which sets in train the exergonic sequence: P840* $\rightarrow$ BChl *a* $\rightarrow$ BPheo(?) $\rightarrow$ [?] (probably Fe-S centers like F_x & F_{ab} of PSI) $\rightarrow$ soluble Fd $\rightarrow NAD^+$. Green sulphur bacteria are also able to operate c-p/p, with the Fd^- reducing Q rather than NAD^+ to complete the cyclic electron flow.

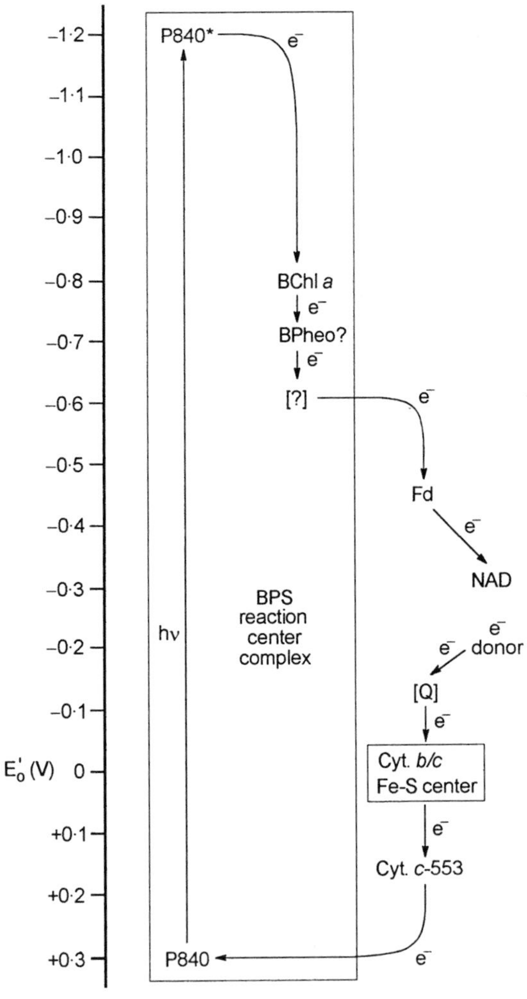

Fig. 4. *The electron flow pathway of non-cyclic photophosphorylation in green sulfur bacteria.* P840 = P840 at the ground state energy level; P840* = P840 at its excited S_1 energy level after absorbing a photon of light (hν) of the appropriate wavelength; BChl = bacteriochlorophyll *a;* Bpheo = bacteriopheophytin *a;* Fd = ferredoxin; [Q] = probably menaquinone.

4. ***Evolutionary aspects of cyclic- and non-cyclic photophosphorylation.*** There is evidence to indicate that when life appeared on earth the atmosphere contained hydrogen gas but no oxygen. One can therefore envisage the evolution of a primitive life form that could trap light energy from the sun and produce ATP by c-p/p. With this ATP and atmospheric H_2 to generate a reductant it would be able to fix CO_2 into carbohydrate. This would suggest that the most primitive form of photosynthesis is that carried out today by bacteria such as *Chromatium* when they grow in the light on H_2 and CO_2. This type of photosynthesis would have gradually used up the atmospheric H_2 and the next stage in the evolutionary process came when mutations allowed electron donors with E'_0 values less negative than NAD to be used. Such mutants would have become increasingly dominant as the level of atmospheric H_2 approached zero. It is possible that these mutants initially had BPS reaction center complexes which produced QH_2 as the primary reductant and so had to use the H^+ gradient generated by c-p/p to form NADH by processes analogous to those shown in Fig. 3, (c) & (d). If this were the case then the next evolutionary advance could be envisaged as the appearance of mutants with a ferredoxin as the primary reductant generated by the BPS reaction center complex. This would have allowed them to operate nc-p/p in a manner analogous to present day green sulfur bacteria (Fig. 4).

However all these forms of photosynthesis would have had two major disadvantages, namely: (i) the organisms that operated them were tied to a source of their particular electron donor, and (ii) the supplies of these electron donors were finite. What was needed was an electron donor that was both abundant and widespread in the biosphere and was a renewable commodity. The obvious choice was water. However, its major disadvantage as an electron donor was its very positive redox potential. Thus it could not be used until mutations within existing photosynthetic organisms (these were probably already capable of nc-p/p of the type carried out by present day green sulfur bacteria) led to the development of a second pigment system capable of being exergonically reduced by water. The system that eventually appeared was, of course, PSII, which operates in tandem with PSI, the functional equivalent of the BPS, in the oxygenic nc-p/p seen in the most evolutionarily advanced of today's photosynthetic organisms. The beauty of the use of water as an electron donor for nc-p/p was that the oxygen produced as a byproduct opened the way to the evolution of respiration, a process that generates water. Thus water ultimately became the renewable commodity.

It is interesting to note that the primitive process of c-p/p is still retained by present day members of the plant kingdom and cyanobacteria, even though they operate oxygenic nc-p/p. There are two possible reasons for this, namely: (i) nc-p/p does not generate a sufficiently high ATP/NADPH ratio to support the Calvin cycle and all the other intrachloroplastidic ATP-requiring processes, and so c-p/p is required to provide a supplementary supply of ATP; from this it could argued that the nc-p/p operating today is no more than an evolutionary step in the direction of 'perfect nc-p/p', a process that would supply all the ATP and NADPH needs of the chloroplast, and that when this ideal is attained c-p/p will be redundant and will be lost, or (ii) the chloroplast finds it advantageous to be able to generate ATP without simultaneously generating NADPH; from this it could be argued that c-p/p will be retained no matter what improvements in ATP/NADPH ratio evolution might bring to nc-p/p.

Dark phase (CO_2 fixation phase) of photosynthesis. The various aspects of this phase of photosynthesis are described in the following separate articles, which are best read in the following sequence: (i) *Calvin cycle*, which can be regarded as the basic CO_2 fixation process, (ii) *Hatch-Slack-Kortschak cycle*, which deals with CO_2 fixation in C_4-plants, such as sugar cane, (iii) *Crassulacean acid metabolism*, which deals with CO_2 fixation in plants, such as cacti, growing in arid climates, and (iv) *Reductive citrate cycle*, which describes CO_2 fixation in green sulfur bacteria.

Photosynthetic bacteria: bacteria that carry out Photosynthesis (see). They fall into two quite different groups, namely: (i) members of the order Rhodospirillales, and (ii) Cyanobacteria.

The former are Gram-negative bacteria, all of which can carry out photosynthesis without producing oxygen (non-oxygenic photosynthesis) under anaerobic conditions, though some can also grow chemolithotrophically under aerobic or microaerobic conditions. They are divided into two suborders, namely: (i) Rhodospirillineae (purple photosynthetic bacteria), and (ii) Chlorobiineae (green photosynthetic bacteria). Rhodospirillineae are divided into two families, namely: (i) *Chromatiaceae* (purple sulfur photosynthetic bacteria; formerly known as *Thiorhodaceae*) which are obligate anaerobes and use sulfur compounds (typically H_2S) as the photosynthetic electron donor, and *Rhodospirillaceae* (purple non-sulfur photosynthetic bacteria; formerly known as *Athiorhodaceae*) which generally do not use sulfur compounds as their photosynthetic electron donor (though some can use H_2S if its concentration is low), preferring instead to use organic compounds (though many can use H_2 as an alternate electron donor). Chlorobiineae are divided into two families, namely: (i) *Chlorobiaceae* (green sulfur photosynthetic bacteria, formerly known as *Chlorobacteriaceae*) which, like the *Chromatiaceae,* are obligate anaerobes and use sulfur compounds as the photosynthetic electron donor, and (ii) *Chloroflexaceae* (green non-sulfur photosynthetic bacteria) which are filamentous facultative anaerobes usually living in algal mats in hot springs where they grow aerobically in an unpigmented form.

All members of the Rhodospirillales have one functional type of photosynthetic pigment system, the equivalent of PSI of green plants, built into the plasma membrane. Its basic unit is composed of a reaction center and antenna chromoproteins. The latter may constitute either an array of integral membane proteins or a membrane-bound extrinsic structure projecting into the cytosol. Rhodospirillineae have the former, and their plasma membrane proliferates to form extensive folded invaginations projecting into the cytosol; when the cell is homogenized these invaginations are torn apart but then reseal to form vesicles known as chromatophores which are inside-out with respect to the plasma membrane. Chlorobiineae have extrinsic antennae assemblies rich in bacteriochlorophyll *c* (BChl *c*, formerly known as Bacterioviridin or Chlorobium chlorophyll), BChl *d* or BChl *e*, bounded by a lipid monolayer envelope, which are known as *chlorosomes.* The reaction centers of *Chromataceae, Rhodospirillaceae* and *Chloroflexaceae* are very similar, having a cytochrome *c* and ubiquinone (UQ) as the electron donor and electron acceptor respectively and a BChl *a* (P870) as the pigment component; the only difference appears to be that the purple bacteria have four BChl *a* and two bacteriopheophytin *a* (BPh *a*) molecules attached to the L and M protein subunits of the complex, whereas the *Chloroflexaceae* have three of each and lack the H protein subunit. The reaction center of *Chlorobiaceae* is significantly different from that of the other three families, having cytochrome *c*-553 and ferredoxin as the electron donor and electron acceptor respectively, and a different bacteriochlorophyll *a* dimer (P840) as the pigment component. In all the Rhodospirillales, the reduced coenzyme generated in the light phase of photosynthesis that drives the fixation of CO_2 in the dark phase is NADH.

Cyanobacteria (formerly known as Blue-green algae) are prokaryotes that characteristically carry out photosynthesis with the production of oxygen (oxygenic photosynthesis), though some are capable of non-oxygenic photosynthesis; the oxygen is a byproduct from the H_2O used as the photosynthetic electron donor. Their photosynthetic process is therefore much more closely related to that of green plants than that of the Rhodospirillales. Like green plants they have two photosynthetic pigment systems, PSI and PSII, built into the membrane of flattened vesicles (thylakoids) located in the cytosol. The basic unit of PSI and PSII consists of a reaction center and antenna chromoproteins. Both reaction centers use chlorophyll *a* (Chl *a*), P700 for PSI and P680 for PSII. The antenna chromoproteins are not organized into the light-harvesting complexes that are characteristic of green plants, but are present in an extrinsic assembly of phycobilin-protein complexes, known as phycobilisomes (see), which are attached to the cytosolic surface of the thylakoid membrane. The phycobilins, principally the blue phycocyanobilin (λ_{max} 620 nm), are covalently bound to the protein of the phycobilisome, a water-soluble heterodimer of α (16 kDa) and β (20 kDa) polypeptide chains, which forms disks of tri- and hexameric aggregates, each containing 300–800 pigment molecules. In the Cyanobacteria, the reduced coenzyme generated in the light phase of photosynthesis to drive the fixation of CO_2 in the dark phase is NADPH, as it is in green plants.

Photosynthetic carboxylation: the enzymatic fixation of carbon dioxide in photosynthesis. In C-3 plants, the photosynthetic carboxylation enzyme is ribulose *bis*phosphate carboxylase (EC 4.1.1.39). In C-4 plants it is phospho*enol*pyruvate carboxylase (EC 41.1.31). P.c. is the first step of carbon dioxide assimilation in photosynthesis, and one of the dark reactions.

Photosynthetic cycle: see Calvin cycle.

Photosynthetic experimental organisms and systems: for technical reasons, certain systems are preferred for the investigation of photosynthesis, e.g. green algae such as *Chlorella*, photosynthetic bacteria and isolated chloroplasts. Green algae and euglenoids (e.g. *Euglena gracilis*) can be cultured under defined conditions in an illuminated chemostat (see Fermentation techniques), and they can also be grown in Synchronous culture (see). It is relatively easy to obtain chlorophyll-deficient mutants of these organisms, which must grow heterotrophically (see Mutant technique). Plastid formation can be prevented by culture in the dark; illumination induces formation of the photosynthetic apparatus. Such systems are therefore especially suited to the study of the regulation of autotrophism and heterotrophism.

Photosynthetic pigments: pigments that take part in the trapping and utilization of light in Photosynthesis (see). Seed plants (*Spermatophyta*), ferns (*Pteridophyta*), mosses (*Bryophyta*), green algae (*Chlorophyta*, e.g. *Chlorella*), euglenoids (*Euglenophyta*, e.g. *Euglena*) and stoneworts (*Charales*) contain both chlorophylls *a* and *b*, and carotenoids, but no biliproteins. The latter are found in red algae (see Rhodoplasts) und Blue-green bacteria (see). Certain algae lack chlorophyll *b* (*Chrysophyta, Pyrrophyta, Cryptophyta*). Table 1 lists the P.p. of various organisms, and table 2 shows the thylakoidal (see Thylakoids) carotenoids of the red beech (*Fagus silvatica*) and the green alga, *Chlorella pyrenoidosa*.

All P.p. are either hydrophobic, or they possess a strongly hydrophobic grouping, e.g. the phytol residue of chlorophyll. A simple model, in which P.p. are associated with the lipid layer of the thylakoid membrane, is however, unsatisfactory. It is necessary to propose a certain degree of ordered structure for P.p., and this is not possible if P.p. are subject to the random mobility of the lipid membrane components, as demanded by the fluid-mosaic model for membrane structure. Binding of a P.p. molecule to a protein would also be an unsatisfactory model, because the various P.p. would then be too widely separated for the efficient transfer of photons or resonance energy. A more feasible model would involve binding of several P.p. molecules to one protein, and there is much evidence for a system of this kind; e.g. several chlorophyll-binding proteins have been isolated from

Table 2. *Percentage composition of thylakoid carotenoids (from Wiessner)*

	Red beech (*Fagus sylvatica*) %	*Chlorella pyrenoidosa* %
α-Carotene	–	4
β-Carotene + Lycopene	34	15
Sum of carotenes	34	19
Lutein	45	50
Violaxanthin	14	10
Neoxanthin	7	12
Sum of xanthophylls	66	72

Ratio xanthophylls/carotenes 1.95 *(Fagus)*, 4.3 *(Chlorella)*.

Table 1. *Photosynthetic pigments and their occurrence in the plant kingdom*

Organism	Chlorophylls						Biliproteins		Carotenoids	
	a	*b*	*c*	*d*	B*a*	B*c*/*d*	Per	Pcy	Carotenes	Xanthophylls
Higher plants*	+	+							+	+
Green algae	+	+							+	+
Brown algae	+		+						+	+
Diatoms	+		+						+	+
Red algae	+		+	+			+	+	+	(+)
Blue-green bacteria	+						+	+	+	+
Green sulfur bacteria					+	+			+	(+)
Purple sulfur bacteria					+				+	+

Footnotes to Table 1:
* seed plants, ferns and mosses.
B*a* = bacteriochlorophyll *a*, B*c*/*d* = bacteriochlorophylls *c* and *d*.
Per = phycoerythrin, Pcy = phycocyanin, (+) = a trace.

thylakoid membranes, in particular P700-Chlorophyll *a*-protein (see) and Light-harvesting protein (see).

Photosynthetic pigment systems: complex aggregations of membrane-associated, pigment-protein complexes (chromoproteins) whose function is to absorb photons of light and transfer their energy by inductive resonance to a reaction center. The reaction center then uses the energy to reduce the oxidized component of a redox system by transferring an electron to it, thereby initiating the electron flow sequences of cyclic and noncyclic photophosphorylation (see Photosynthesis). Photosynthetic bacteria (see) have one P.p.s., termed the 'bacterial pigment system' (BPS) whereas members of the plant kingdom have two P.p.s., termed 'pigment system I' (PSI) and 'pigment system II' (PSII). The P.p.s. of all types of photosynthetic organisms are composed of two functional units, namely: (i) light-harvesting or antenna structures or complexes, and (ii) reaction center complexes.

The former can be subdivided into extrinsic structures and intrinsic complexes. The extrinsic structures are attached to the surface of the relevant membrane, whereas the intrinsic complexes are built into the membrane, usually spanning its thickness. Organisms that use extrinsic light-harvesting structures are the green sulfur bacteria *(Chlorobiaceae)*, which have chlorosomes, and the Cyanobacteria and the red algae *(Rhodophyceae)*, which have phycobilisomes (see). Organisms that have intrinsic light-harvesting complexes are the purple sulfur bacteria *(Chromatiaceae)*, the purple non-sulfur bacteria *(Rhodospirillaceae)*, green non-sulfur bacteria *(Chloroflexaceae)* and all members of the plant kingdom except the red algae. The light-harvesting complexes of PSI and PSII are usually referred to as LHCI and LHCII respectively. The reaction center complexes of all photosynthetic organisms are intrinsic membrane-spanning structures.

Light harvesting structures.

1. *Chlorosomes* are flattened discoid structures attached to the outer surface of the plasma membrane. Within their outer envelope of proteins are the light absorbing units, 120 rods carrying about 10,000 non-covalently bound bacteriochlorophyll *c* (BChl *c*) molecules and some of γ-carotene. Each rod is composed of 12 identical proteins, each of which has 7 bound BChl *c* molecules. The rods lie on a baseplate composed of water-soluble, globular proteins to which BChl *a* molecules are non-covalently bound.

2. *Phycobilisomes* (see) are attached to the stromal surface of thylakoid membranes and pass the light energy they absorb to both PSI and PSII.

3. *Purple sulfur-, purple non-sulfur- and green sulfur bacteria.* The basic unit of the light-harvesting complexes of purple bacteria are pairs of two different polypeptides, α and β, each made up of ~ 50 amino acid residues and carrying non-covalently bound Bchl *a*; the α carries one, the β two. Both have a central membrane-spanning α-helical section with the hydrophilic ends projecting into the aqueous milieu on either side, the *N*-terminal end into the cytoplasm and the *C*-terminal end into the periplasmic fluid. The light-harvesting complexes consist of aggregates of these αβ-heterodimers. The light-harvesting complex of green sulfur bacteria appears to be built on similar lines.

4. *LHCII of PSII.* LHCII of higher plants appears to be composed of several very similar proteins. Each of these proteins carries noncovalently bound chlorophyll (4 Chl *a* & 3 Chl *b*) and carotenoid (2 lutein or neoxanthin or 1 of each) molecules, and each possesses three thylakoid membrane-spanning α-helical regions. They may be organized into trimeric functional units. LHCII proteins of the various algal families carry the chlorophyll and carotenoid species that are characteristic of them, e.g. Chl. *c* and fucoxanthin in brown algae *(Phaeophyceae)*.

5. *LHCI of PSI.* The chromoproteins of higher plant LHCI are difficult to detach from those of the reaction center which they surround, because they span the thylakoid membrane. They appear to be in the molecular mass range 19–24 kDa and to be bound noncovalently to Chl *a* and Chl *b*.

Reaction center complexes.

1. *Purple sulfur-, purple non-sulfur- and green non-sulfur bacteria.* A breakthrough in our knowledge of reaction center structure came in the first half of the 1980s when the reaction center of the purple non-sulfur bacterium *Rhodopseudomonas viridis* (now renamed *Rhodobacter viridis*) was crystallized [H. Michel *J. Mol. Biol.* **158** (1982) 567–572] and its structure then elucidated by X-ray crystallography [J. Deisenhofer et al. *J. Mol. Biol.* **180** (1984) 385–389 & *Nature* **318** (1985) 618–624], work that was awarded the 1988 Nobel Prize for chemistry. Since then the reaction centers of other purple bacteria have been shown to be almost identical to that of *Rb. viridis*, and that of PSII to be similar.

The bacterial reaction center is composed of three protein subunits, L, M and H. These designations originally indicated their relative molecular masses, based on their SDS-PAGE migration, as being low, medium and heavy. However it was subsequently shown that H has a lower molecular mass (28.5 kDa; 258 amino acid residues) than L (31.5 kDa; 273 aa) or M (34.5 kDa; 323 aa) and that its SDS-PAGE behavior resulted from its being much less hydrophobic than L and M. These proteins are accompanied by a cytochrome containing 4 heme groups, 4 BChl (*b* in *Rb. viridis*; *a* in most other species), two of which constitute a 'special pair' which receives photon energy directly from the light-harvesting complex, and two that have an accessory role (sometimes called 'voyeur' BChl), 2 BPheo (*b* in *Rb. viridis; a* in most other species), one quinone (MQ in *Rb. viridis*; UQ in *Rb. sphaeroides*) at a site designated Q_A, one quinone (UQ in *Rb. viridis & Rb. sphaeroides*) at a site designated Q_B, and a non-heme Fe^{2+}.

The core of the reaction center consists of the L and M proteins, both of which have five long, membrane-spanning α-helices, designated A-E (hence LA-LE & MA-ME), and two short α-helices linking C to D (designated LCD & MCD) and D to E (designated LDE & MDE). The non-α-helical *N*-termini and *C*-termini occur on the cytoplasmic and periplasmic sides of the membrane respectively. As is shown in the figure, the helices of the L and M proteins fit together in the membrane to give approximately two-fold symmetry. This symmetry is maintained by the positions occupied by the BChl, BPheo and quinone molecules, its axis running perpendicular to the membrane surface, between the BChl special pair at

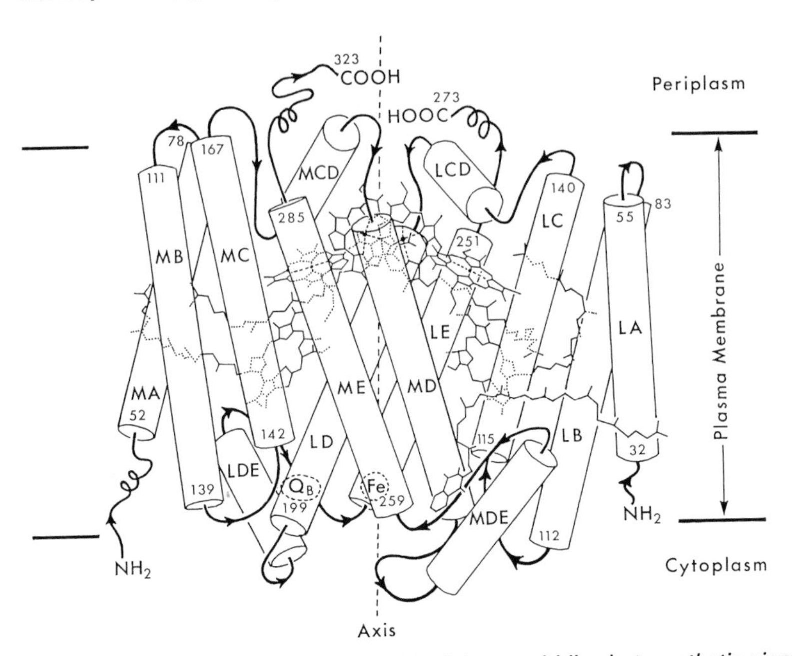

The core of the reaction center of the Rhodobacter viridis photosynthetic pigment system. The α-helical sections of the two protein subunits, L and M, are represented by cylinders. The position of the amino acid at each end of the transmembrane α-helices, LA-LE & MA-ME, in the primary structure of the proteins is indicated by a number; the numbers that have been omitted due to lack of space are: MD, 197–225; LD, 170; LE, 225. The axis of symmetry runs between the overlapping pyrrole ring As of the porphyrin rings of the bacteriochlorophyll *b* 'special pair' (BChl b_{sp}) at the periplasmic side (top) of the complex through the non-heme iron at the cytoplasmic side (bottom). Parts of chemical structures drawn with dotted lines are behind the α-helix cylinder at that location. Looking from the periplasmic side (top) towards the cytoplasmic side (bottom), the pigment and quinone components occur in the sequence: (i) BChl *b* 'special pair', (ii) two 'voyeur' BChl *b* molecules (drawn without their phytyl side chains), (iii) two bacteriopheophytin *b* (BPheo *b*) molecules, and (iv) menaquinone (MQ) at the Q_A site; the position of the ubiquinone (UQ), most of which is lost during the isolation and crystallization of the reaction center, is indicated by an encircled Q_B. Electron transfer through the core appears to follow the route BChl $b_{sp} \rightarrow$ BPheo $b \rightarrow$ MQ on the right hand side of the axis as drawn; electrons are then transferred 'laterally' to the exchangeable UQ at the Q_B site.

the periplasmic end of the reaction center towards the non-heme Fe^{2+} at the cytoplasmic end. The isoprenoid side chains of the BChl special pair 'overlap' the porphyrin rings of the two BPheo which lie beneath the special pair and further from the axis of symmetry. The two 'voyeur' BChl lie slightly below and on either side of the BChl special pair. The binding site for the quinone ring of Q_A is located at the cytoplasmic end of the M protein and involves interactions with the MD and ME α-helices. The binding site for the quinone ring of Q_B is located in an equivalent position on the L protein and is thus at the same level in the membrane as the Q_A site.

The reaction center core serves to effect a charge separation across the membrane by the rapid transfer of an electron from the BChl special pair to the quinone at the Q_B site. The process is initiated by the receipt by the BChl special pair of the energy of a light photon absorbed by the light harvesting complex. This causes an electron in a π-bonding molecular orbital of a BChl at its ground state electronic energy level to be raised to a π-antibonding molecular orbital (termed a $\pi \rightarrow \pi^*$ transition) thereby making the BChl (usually depicted as BChl*) into a very powerful reducing agent. The special pair BChl* then reduces BPheo. This electron transfer is extremely rapid; it has been shown to occur within 4 picoseconds (4×10^{-12} s). The $BPheo^-$ then reduces the quinone at the Q_A site; this electron transfer is also rapid, occurring within 200 ps. Up to this point the electron has moved across the thickness of the membrane from the periplasmic side toward the cytoplasmic side. However, the next electron transfer step, from Q^- at the Q_A site to the quinone at the Q_B site, is lateral and takes rather longer, occurring within ~ 10 μs. By this time a molecule of reduced cytochrome *c* has diffused in the periplasmic fluid to the $BChl^+$ of the special pair at the periplasmic end of the reaction center core; the reduced cytochrome *c* then transfers an electron to the $BChl^+$, which therefore returns to the ground state, readying the special pair for the receipt of the energy of another photon. It now appears that the 4-heme-cytochrome, tightly bound to the periplasmic side of the reaction center, is reduced by the soluble, mobile cytochrome

c and then acts as the immediate electron donor to $BChl^+$. The rapidity of the reactions and the physical separation of charge across the membrane thickness is essential to prevent either the loss of the energy of BChl* by the processes of internal conversion (effectively loss as heat) or fluorescence (loss as light) which occur within about 10^{-10} s, or back reactions that would return the electron to $BChl^+$. At the cytoplasmic side of the reaction center core the quinone at the Q_B site accumulates two electrons, derived from Q^- at the Q_A site in two steps, and then takes up two H^+ from the cytoplasm to form QH_2. This then exchanges with an oxidized molecule of Q that is part of a mobile pool of Q in the hydrophobic region of the membrane lipid bilayer. After its release from the Q_B site the QH_2 diffuses through the lipid bilayer until it reaches the trans-membrane cytochrome *bc* complex, to which it transfers its electrons (see Photosynthesis). The function of the two 'voyeur' BChl molecules in the reaction center is not fully understood.

Although the components of the reaction center core are paired and exhibit two-fold symmetry, there appears to be only one electron transfer route rather than two, namely the one that uses the BChl and BPheo molecules that are more closely associated with the L protein.

The reaction center of the green non-sulfur bacterial species *Chloroflexus* is very similar to that of the purple bacteria, except that it has no H protein, and it has three BChl and three BPheo instead of four and two respectively.

2. *PSII.* The reaction center core of PSII shows considerable similarity to that of the purple bacteria, which is remarkable since it evolved much later to fulfil the different and taxing role of abstracting electrons from water (see Photosynthesis). It has two 32 kDa proteins designated D_1 and D_2, the equivalents of proteins L and M respectively, which traverse the thylakoid membrane. It also has two Chl *a* molecules that are believed to constitute a P680 special pair, two Pheo *a* molecules which are the immediate acceptors of electrons from P680, and two plastoquinone (PQ) binding sites, Q_A and Q_B, which are located at the stromal end of D_2 and D_1 respectively. The flow of electrons from P680* through these components is believed to parallel that seen in the reaction center core of purple bacteria, with the end product PQH_2 formed at the Q_B site exchanging with mobile PQ in the lipid bilayer.

The $P680^+$ formed by this electron flow is reduced by an electron derived from water in the thylakoid lumen. The transfer of this electron from water to $P680^+$ requires the participation of a cluster of four manganese ions and a tyrosine residue (designated Tyr_z) at position 161 of the D_1 protein, the sequence being $H_2O \rightarrow$ Mn cluster $\rightarrow Tyr_z \rightarrow P680^+$. The precise location of the Mn cluster is not known; it was originally thought to be on one of three extrinsic proteins (17, 23 & 33kDa) bound to the luminal surface of the reaction center complex, but this is now known not to be so. However, it is logical to suppose that it is at the luminal end of the reaction center core, because its function is to take electrons from water in the thylakoid lumen. The positioning of the P680 special pair, the two Pheo and the Q_A and Q_B sites relative to each other and to proteins D_1 and D_2 are believed to be very like those of their equivalents in the purple bacterial reaction center core, with the possible proviso that the special pair may not be as close to the membrane surface, so as to allow for the operation of the Mn cluster.

3. *PSI and green sulfur bacteria.* The reaction centers of PSI and green sulfur bacteria are very similar. They both contain two membrane-spanning ~ 65 kDa proteins which carry a large number of light-harvesting pigments in addition to the components that bring about charge separation. They both have special pairs, P700 in PSI and P840 in green sulfur bacteria, in which the P^*/P^+ redox system has an E'_o of ~ −1.2V, i.e. about 0.3V more negative than that of PSII of the purple or green nonsulfur bacteria. Their immediate electron acceptor is a (B)Chl *a* (often designated A_0), from which the electron passes through several bound Fe-S centers (variously designated as F_x and F_{ab} or F_a & F_b). The latter are quite characteristic of PSI and the green sulfur bacterial reaction centers, and they allow these centers to reduce soluble ferredoxin, which has a sufficiently negative redox potential to reduce NAD^+ exergonically.

Phototrophism: see Nutritional physiology of microorganisms.

Phrenosine: see Glycolipids.

Phycobilins: open-chain tetrapyrroles that function as photosynthetic light-harvesting pigments when covalently linked to a specific protein. The resulting phycobilin-protein complex is called a phycobiliprotein. The two major classes of phycobiliproteins are phycocyanins (blue) and phycoerythrins (red), and these are largely responsible for the color of the organisms that contain them. See Phycobilisome.

Phycobiliproteins: see Phycobilisome.

Phycobilisome: a light-gathering structure in cyanobacteria and red algae. P. are directly attached to photosynthetic membranes, but are not part of them. They consist of Phycobiliproteins (see) (also known as biliproteins). According to the nature of the protein-bound Phycobilin (see), phycobiliproteins are classified as phycoerythrin (PE), phycocyanin (PC) and allophycocyanin (APC). The rods of PE and PC are thought to be held together by noncolored proteins with M_r around 30,000. The P. are somewhat larger than ribosomes, and may form a regular 2-dimensional array adjacent to the photosynthetic membrane.

Light energy is passed efficiently from PE to PC to APC to photosystem I or II (Fig. 1). The amounts of PE and PC in the cell may vary, depending on the light conditions. Absorbance maxima of PE are in the range 498–568 nm, while PC absorbs around 625 nm, and APC at 618–673 nm. The relative amounts of PE and PC are determined by the spectral characteristics of the available light; in redder light, more PC is present. The total phycobiliproteins may amount to 60 % of the soluble protein in the cell.

The open-chain tetrapyrrole chromophores of phycobiliproteins (Fig. 2) share a common biosynthetic pathway with heme and chlorophyll from 5-aminolevulinate to protoporphyrin IX (a pathway also shared by phytochrome of higher plants). Each phycobiliprotein consists of α- and β-chains. The M_r of these polypeptides varies among species, but the α-chain (M_r

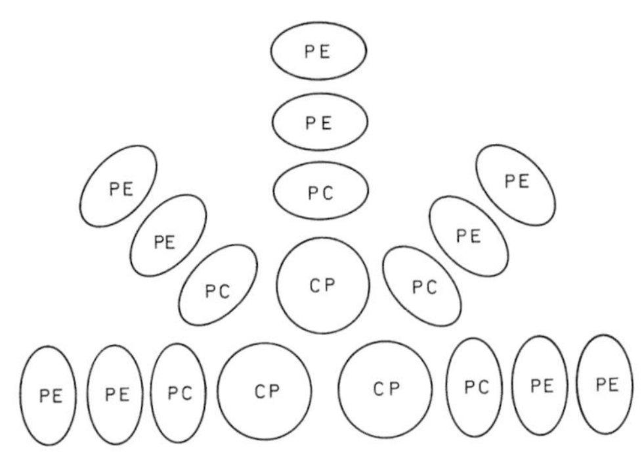

Fig. 1. *Model of phycobilisome structure*
CP = core protein, probably allophycocyanin
PC = phycocyanin
PE = phycoerythrin

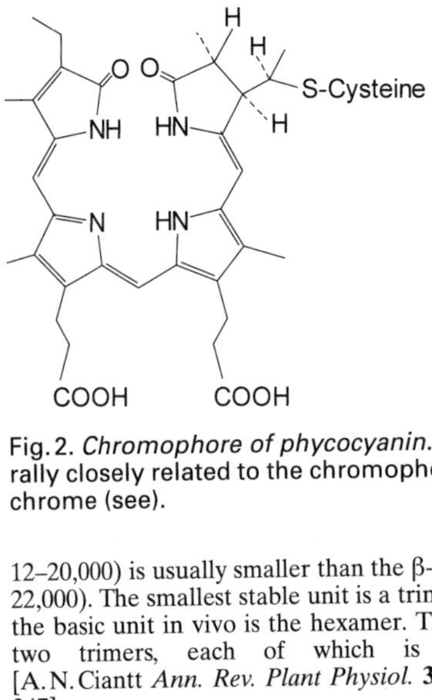

Fig. 2. *Chromophore of phycocyanin.* It is structurally closely related to the chromophore of Phytochrome (see).

12–20,000) is usually smaller than the β-chain (M_r 15–22,000). The smallest stable unit is a trimer, $(\alpha\beta)_3$, but the basic unit in vivo is the hexamer. This consists of two trimers, each of which is disk-shaped. [A. N. Ciantt *Ann. Rev. Plant Physiol.* **32** (1981) 327–347]

Phylloquinone: see Vitamins (vitamin K).

Physalaemin: Pyr-Ala-Asp-Pro-Asn-Lys-Phe-Tyr-Gly-Leu-Met-NH_2, an amphibian tachykinin from the skin of the South American frog *Physalaemus fuscumaculatus* [V. Erspamer et al. *Experentia* **20** (1964) 489–490]. P. has also been identified as a tumor-related peptide in extracts of human lung small cell carcinoma [L. H. Lazarus et al. *Science* **219** (1983) 79–81]. It affects blood vessels and extravascular smooth muscle, stimulates exocrine secretion, and has a distinct neurotropic effect. The *C*-terminal hexapeptide from P. elicits full biological response in guinea pig ileum, while the presence of the five *N*-terminal amino acids is necessary for full expression of sialogogic activity. Conformational studies have been reported. [J.-L. Bernier et al. *Eur. J. Biochem.* **142** (1984) 371–377]

Physostigmine, *eserine:* an indole alkaloid from calabar beans, the ripe seeds of *Physostigma venenosum* (a woody vine indigenous to the west coast of

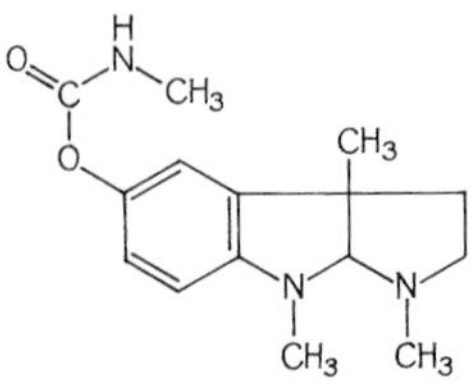

Physostigmine

Africa). P. contains a pyrrolidinoindole ring structure and a urethane group (Fig.), and it exists in an unstable form, m. p. 87 °C. M_r 275.35, $[\alpha]_D$ −82 ° ($CHCl_3$) and a stable form, m. p. 105–106 °C. It occurs with its *N*-oxide, geneserine, m. p. 129 °C, $[\alpha]_D$ −175 ° (acetone).

P. is used in opthalmic practice in the same way as pilocarpine, for pupil contraction and for the reduction of intraocular pressure. It is an inhibitor of acetylcholinesterase, a property shared by certain other basic urethanes, such as neostigmine (prostigmine); these urethanes presumably acylate, and therefore block, the enzyme. P. and neostigmine have been used in surgery to counteract the action of curare, and both have been used for the relief of Myasthenia gravis, a disease characterized by muscular weakness associated with a rapid breakdown of acetylcholine. These have now been largely replaced by other synthetic compounds.

The fatal human dose of P. is about 10 mg. It was first isolated in 1864, when forty-six children in Liverpool were poisoned by eating calabar beans thrown on a rubbish heap from a West African cargo ship.

Phytanic acid: see Inborn errors of metabolism (Refsum's disease).

Phytanic acid storage disease: see Inborn errors of metabolism (Refsum's disease).

Phytanylether lipids: see Membrane lipids.

Phytic acid, myo-*inositol hexakis (dihydrogen phosphate),* cis-*1,2,3,5*-trans-*4,6-cyclohexanehexol-hexaphosphate:* a major phosphate storage compound in plants, which is especially abundant in oil seeds, legumes and cereal grains. It is the hexaphosphate of *Myo*-inositol (see), in which each OH-group of *myo*-inositol is esterified with phosphoric acid. Calcium and magnesium salts of P. a. are known as phytin. The commercial preparation of *myo*-inositol involves extraction of P. a. from corn (maize) steep liquor, hydrolysis of the P. a. to *myo*-inositol and inorganic phosphate, and crystallization of the *myo*-inositol from water.

Phytin: see Phytic acid.

Phytoalexins, *phytoncides, stress compounds:* substances with antibiotic activity produced by plants in response to injury or stress, e. g. infection with fungi, bacteria or viruses, mechanical wounding, UV irradiation, cold and treatment with phytotoxic chemicals (e. g. heavy metals) or Elicitors (see). They function as growth inhibitors of pathogens, chiefly fungi. They have also been defined as novel post-infectional metabolites produced by plants in response to fungal infection, and which, because of their antifungal activity, protect the plant from attack by fungi. Under certain conditions P. may also be antibacterial. P. do not include antifungal compounds already present in the uninfected plant, e.g. protocatechuic acid and catechol (onion bulbs) or chlorogenic acid (many plants). The original concept of P. included compounds formed from immediate precursors, e.g. the antifungal dihydroxymethoxybenzoxazine released from its glucoside, which is present in cells of wheat and maize. The modern definition of P. excludes such compounds, and is restricted to compounds synthesized by a remote pathway, which is first activated by infection or stress (see e.g. Fig. 2). Since there are no immediate precursors, the appearance of P. is delayed until the biosynthetic pathway has been activated.

Most known P. have been discovered in the subfamily *Papilionoideae* of the *Leguminoseae,* followed by the *Solanaceae,* while many examples are known from other families. P. do not belong structurally or biosynthetically to any one class of compounds. Many are isoflavonoids and several of these are pterocarpans, e.g. pisatin (see Pterocarpans), Phaseolin (see), Glyceollins (see). There are also terpenes (see Fig. 2; see Ipomoeamarone), acetylenic compounds (see Safynol, Wyerone acid), α-hydroxydihydrochalcones, stilbenes and polyketides (Fig. 1).

In orchids P. are formed only when the roots and shoot of the germinating orchid seed are colonized by fungi. The fungi penetrate the plant, but are prevented from spreading by P. (see Orchinol). In most plants producing P., infection or stress cause the formation of a number of different P., i.e. there is a multicomponent response. For example, medicarpin (a pterocarpan) and a variety of compounds related to wyerone acid are all produced by *Vicia faba,* while rishitin, lubimin and phytuberin are produced by potato. [J. L. Ingham "Phytoalexins and Other Natural Products as Factors in Plant Disease Resistance" *The Botanical Review* (New York Botanical Garden) **38** (1972) 343–424; R.A Dixon et al. *Advances in Enzymology* (ed. Alton Meister) **55** (1983) 1–136; J. A. Bailey & J. W. Mansfield (eds.) *Phytoalexins,* Blackie & Son Ltd. 1983; P. A. Brindle & D. R. Threlfall *Biochemical Society Transactions* **11** (1983) 516–522; G. A. Cooper-Driver, T. Swain & E. E. Conn (eds.) *Chemically Mediated Interactions between Plants and Other Organisms* (Recent Advances in Phytochemistry) **19** (Plenum Press, 1985); D. A. Smith & S. W. Banks, *Phytochemistry* **25** (1986) 979–995;

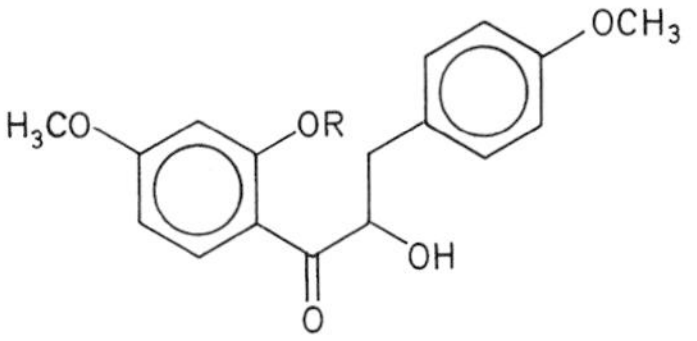

H_3CO, OR, OCH_3, OH, O

R = H: *Odoratol,*
R = CH_3: *Methylodoratol,* two α-hydroxydihydrochalcones from pods and cotyledons of *Lathyrus odoratus* infected with *Phytophthora megasperma* [A. Fuchs et al. *Phytochemistry 23* (1984) 2199–2201]

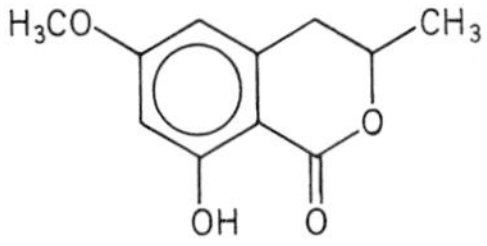

H_3CO, CH_3, O, OH, O

6-Methoxy-mellein, a polyketide from carrot infected with *Ceratocystis fimbriata* [J. A. Bailey & J. W. Mansfield (eds.) *Phytoalexins* (Blackie & Son Ltd., 1982)]

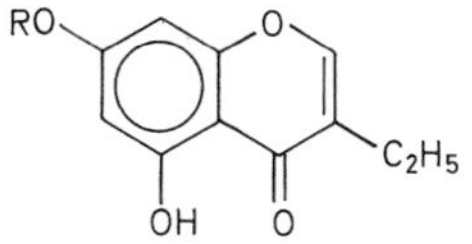

RO, O, C_2H_5, OH, O

R = H: *Lathodoratin*
R = CH_3: *Methyl-lathodoratin,* two chromones from *Lathyrus odoratus* and *L. hirsutus* infected with *Botrytis cinerea* or *Helminthosporum carbonum* [D. J. Robeson et al. *Phytochemistry 19* (1980) 2171–2173

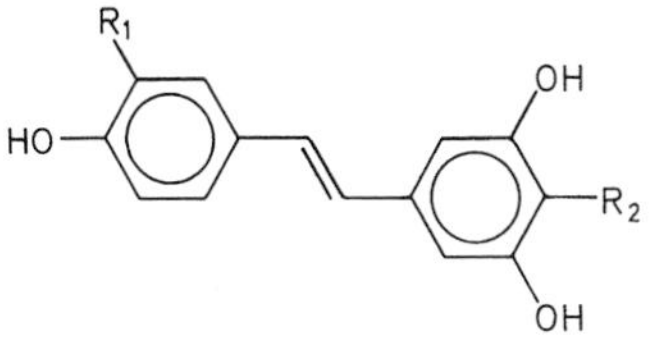

R_1, HO, OH, R_2, OH

R_1 = OH; R_2 = $(CH_3)_2$CH–CH=CH–,

4-(3-methylbut-1-enyl)-3,5,3′,4′-tetrahydroxystilbene

R_1 = H; R_2 = $(CH_3)_2$CH–CH=CH–,

4-(3-methylbut-1-enyl)-3,5,4′-trihydroxystilbene

R_1 = H; R_2 = $(CH_3)_2$C=CH–CH_2–,

4-(3-methylbut-2-enyl)-3,5,4′-trihydroxystilbene (or 4-issopentenyl resveratrol)

Stilbenes from *Arachis hypogaea* seeds, sliced and allowed to become infected with natural microflora. The compounds are active against *Cladosporium cucumerinum.* [G. Aguamah et al. *Phytochemistry 20* (1981) 1381–1384]

Fig. 1. *Examples of phytoalexins.* Other phytoalexins are named in Fig. 2 and cross-referenced in the text as individual entries.

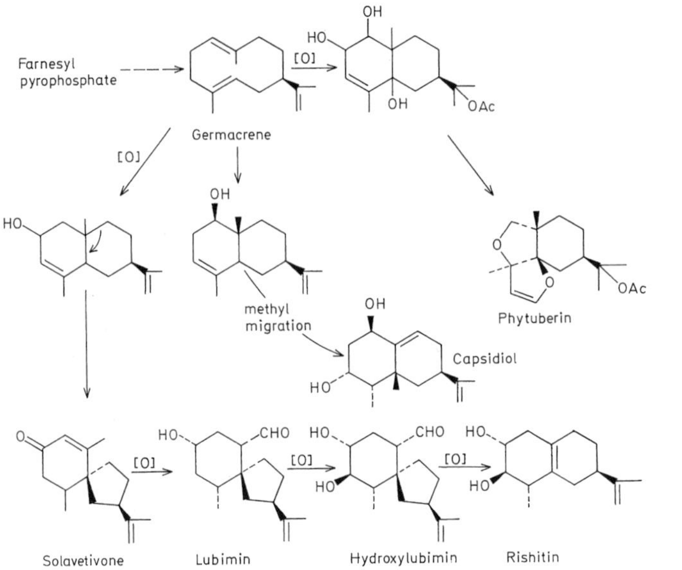

Fig. 2. *Biosynthesis of some sesquiterpene phytoalexins.* Phytuberin, solavetivone, lubimin, hydroxylubimin and rishitin are produced by *Phytophthora*-infected potatoes *(Solanum tuberosum)* and by potato tuber disks treated with Elicitor (see). Capsidiol is a phytoalexin from *Capsicum,* [P.A. Brindle & D.R. Threlfall *Biochem. Society Transactions 11* (1983) 516–522]

M.S. Kemp & R.S. Burden *Phytochemistry* **25** (1986) 1261–1269]

Phytochelatin: a plant peptide produced in response to heavy metals, e.g. cadmium, copper, mercury, lead and zinc. The structure is $(\gamma\text{-Glu-Cys})_n\text{-Gly}$ (n = 3–7). Like Metallothionein (see), P. form metal-thiolate bonds and thus sequester toxic metal ions. They are probably derived from glutathione rather than RNA-directed protein synthesis.

Phytochemistry: the chemistry of natural products from plants (see Natural product chemistry). It is part of plant biochemistry, and it is concerned chiefly with secondary metabolites.

Phytochromes: A family of ubiquitous plant pigments, localized in the cytosol, which mediate the light sensitivity of many growth and developmental processes. They are composed of two polypeptides, each of M_r 120,000–127,000, and each carrying a covalently linked tetrapyrrole chromophore (Fig.) in the *N*-terminal domain and dimerization determinants in the *C*-terminal domain. The chromophore is related to c-phycocyanin, which is found in blue-green bacteria. P. exist in two forms, one with a light-absorption maximum around 660 nm (P_r), and one which absorbs maximally around 730 nm (P_{fr}).

P. are synthesized as P_r, which is thought to be the inactive form. Exposure to red light of 660 nm wavelength converts P_r to P_{fr}, the active and more labile form. Exposure of P_{fr} to far-red light (730 nm) converts it back to P_r. Conversion to the P_{fr} form activates signaling pathways, which in turn lead to the changes in gene expression underlying the physiological and developmental responses to light [P.H. Quail *Curr. Opin. Genet. Dev.* **4** (1994) 652; P.H. Quail *Annu. Rev. Genet* **25** (1991) 389]. Dark-grown tissues accumulate Pr, but because of the faster rate of destruction or inactivation of P_{fr}, the total P. content of light-grown tissues is only 1–3 % of that of dark-grown tissues.

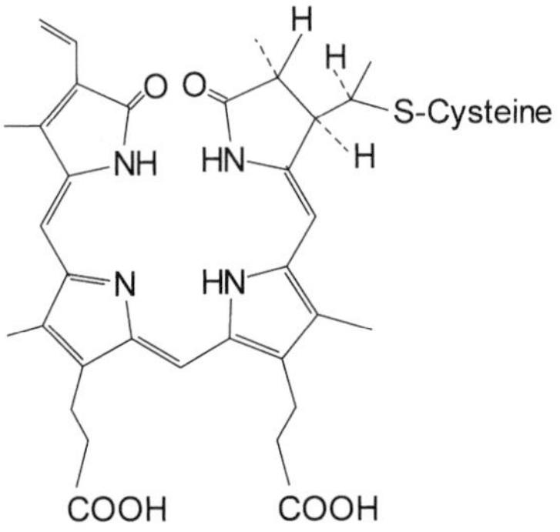

Chromophore of phytochrome. It is structurally closely related to the chromophore of phycocyanin (see Phycobilisome).

The diversity of processes promoted or inhibited as a result of light absorption by P. (e.g. shade avoidance, seed germination, seedling deetiolation) is difficult to reconcile with the existence of a single species of P. It is becoming increasingly apparent, however, that various members of the P. family fulfil different and specialized functions. For example, in *Arabidopsis* seedlings, phytochrome A (PA) is necessary for continuous far-red light (FRc) perception (the FR-high irradiance response: FR-HIR), whereas PB is neither necessary nor sufficient for FRc perception. On the other hand,

PB is necessary for continuous red light (Rc) perception, whereas PA is neither necessary nor sufficient for this process. PC, PD and PE are insufficient for both processes. Mutually antagonistic signals from PA and PB in response to both Rc and FRc light largely account for 'shade avoidance' by green plants. Thus when PB is activated by Rc it induces deetiolation (inhibits hypocotyl elongation), but absorption of FRc by $P_{fr}B$ converts it back to the inactive form. On the other hand, absorption of FRc by PA induces deetiolation (FH-HIR), but this induction is suppressed by absorption of Rc. In the shade of other vegetation an emerging seedling receives light depleted in RC which has been absorbed by the chlorophyll of plants in the higher canopy. In contrast, a seedling exposed to open sunlight receives plentiful Rc. Consequently, seedlings emerging into open sunlight use PB for deetiolation, whereas seedlings in the shade of other vegetation use PA until they attain the light, then switch to PB.

No striking sequence similarities have been found between P. and other proteins in databases. It is therefore not possible to predict possible mechanisms of primary signal transduction. Studies on the structure of *Arabidopsis* PA and PB (both with 1,200 residues) and their mutant counterparts have revealed separate domains and residues involved in assembly, photochemical activity and regulation. The tetrapyrrole chromophore is attached to a target Cys in the *N*-terminal domain, and this attachment occurs autocatalytically when recombinant PA, PB or PC are exposed in vitro to the appropriate tetrapyrrole. The *N*-terminal domain also contains subdomains the insure the fidelity of photoreception and the interconversion of the P_r and P_{fr} forms. Two regions of the polypeptide, one near the center and the other near the *C*-terminus, appear to be involved in dimerization. Mutagenesis has revealed regions at both ends of the polypeptide, within the terminal 110 residues, which are involved in regulatory activity of the photoreceptor. Subsequent transmission of the perceived signal to other components of the signal transduction pathway depends on a segment in the *C*-terminal region, consisting of residues 681–840. In both PA and BP, mutation in this region critically affects transmission of the effects of light signals to dowstream transduction factors; a stretch of 18 residues within this region appears to be especially crucial.

Mutants defective in components downstream of P. have been identified, but these have not yet led to the elucidation of the pathway of signal transduction. Certain factors are, however, strongly implicated, i. e. hetrotrimeric GTP-binding proteins, calcium calmodulin, cGMP, and the COP-DET-FUS class of master regulators. [P. H. Quail *Annu. Rev. Genet.* **25** (1991) 389–409; P. H. Quail et al. *Science* **268** (1995) 675–680]

Phytoecdysone: see Ecdysone.

Phytoene: an aliphatic, colorless hydrocarbon carotenoid, M_r 544. P. is a polyisoprenoid containing six branch methyl groups, two terminal isopropylidene groups and nine double bonds, three of them conjugated. Only the Δ^{15} double bond has *cis* configuration. Biosynthetically, P. is derived from two molecules of geranylgeranyl pyrophosphate, and it serves as a C40-starter molecule in the biosynthesis of other carotenoids: phytofluene, carotene, neurosporene and lycopene are formed by the stepwise dehydrogenation of P. It is found widely in plants, and is especially plentiful in tomatoes and carrot oil. It was isolated in 1946 from tomatoes by J. W. Porter & F. P. Zscheile, and its structure was elucidated in 1956 by W. J. Rabourn & F. W. Quackenbush. For structure and biosynthesis, see Tetraterpenes.

Phytofluene: an aliphatic polyisoprenoid hydrocarbon carotenoid, M_r 548. P. contains ten double bonds (five of them conjugated), six branch methyl groups and two terminal isopropylidene groups. As in phytoene, the central double bond between C15 and C16 has *cis* configuration. P. is found widely in plants, e. g. tomatoes and carrots, and it is an intermediate in lycopene biosynthesis (see Tetraterpenes).

Phytohemagglutinins: see Lectins.

Phytohormones, *plant hormones:* The classic P. are endogenous regulators of plant growth and development. They are analogous to animal hormones in that the sites of their synthesis may be remote from the sites of their action, but in contrast to animal hormones, P. have multiple activities and low action specificities. Oligosaccharins (see), however, are much more specific and are effective at concentrations similar to those at which animal hormones are effective (10^{-8}–10^{-9} M), but they may not be transported from their site of synthesis to the site of their action.

The classic P. are Auxins (see), Gibberellins (see), Antheridiogen (see), Cytokinins (see), Abscisic acid (see), Flowering hormone (see) and Fruit ripening hormone (see).

Phytokinins: see Cytokinins.

Phytol: see Chlorophyll.

Phytoncides: see Phytoalexins.

Phytosphingosine: see Membrane lipids.

Phytosterols: see Sterols.

Phytotoxin: a compound produced by a fungal or bacterial plant parasite, which is toxic to the plant. The term should not be confused with Plant toxin (see). Compared with the large number of plant parasites with demonstrable phytotoxic activity, only relatively few phytotoxins have been identified. See separate entries: *Alternaria alternata* toxins, Coronatine, Fusicoccin, Gibberellins, Ophiobolanes Stemphylotoxins, Tabtoxins.

Phytuberin: see Phytoalexins.

Picromycin: see Macrolides.

Picrotoxin, *cocculin:* a molecular compound of one molecule picrotoxin and one molecule picrotin. P. is a neurotoxin, which occurs in the seeds of *Anamirta coculus,* and is also found in *Tinomiscium phi-*

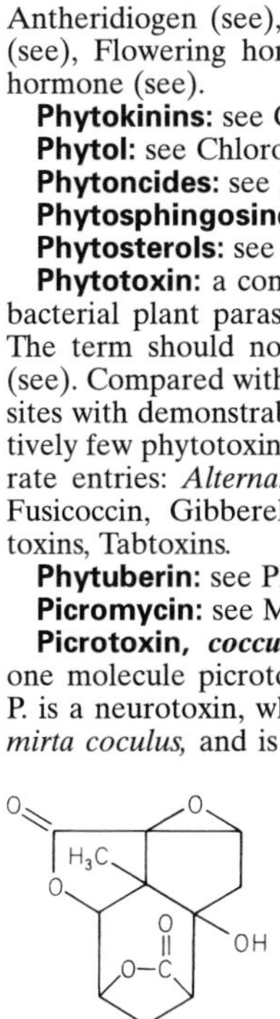

Picrotoxinin, R: $H_3C{-}C{=}CH_2$

Picrotin, R: $H_3C{-}C(CH_3){-}OH$

lippinense. As a specific antagonist of 4-Amino-butyric acid (see), it acts as a central and respiratory stimulant, and is an antidote to barbiturates. It is extremely toxic to fish. Picrotoxinin is the active component of the molecular compound; picrotin is physiologically inactive.

Piezoelectric sensors: Quartz crystals may be induced to resonate under electrical control. In this state they are sensitive to the mass of absorbing material. Piezoelectric sensors make use of this effect for the detection of very small amounts of inorganic or biological material, e. g. by use of surface immobilized enzymes or antibodies.

Pigment 700, *P700:* a chlorophyll with an absorption maximum at 700 nm. It is a component of photosystem I and serves as an energy sink or trapping center in the light reaction of this system (see Photosynthetic pigment systems, Photosynthesis). The reaction center of photosystem I consists of about 500 molecules of closely packed P700 in a quasicrystalline state resembling the crystal of a semiconductor.

Pigment systems: see Photosynthetic pigment systems.

PIH: acronym of prolactin release inhibiting hormone (see Releasing hormones).

Pilocarpine: an imidazole alkaloid, and the chief alkaloid from the leaves of Brazilian *Pilocarpus* species. M_r 208.26, m.p. 34 °C, b.p.$_5$ 260 °C, $[\alpha]_D^{20}$ –100.5 ° ($CHCl_3$). P. is used therapeutically as a diaphoretic, i. e. to induce sweating, and especially in nephritis to relieve the kidneys and remove toxic metabolites. It is also used in opthalmology as an antagonist of atropine, and for regulating the intraocular pressure in glaucoma.

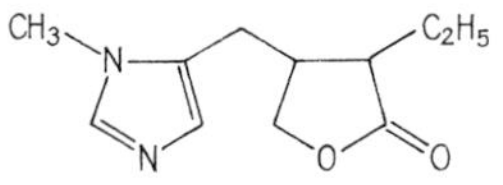

Pilocarpine

Pimaradiene type: see Diterpenes (Fig.).

Pinane type: see Monoterpenes (Fig.).

Pineal gland, *pineal body, epiphysis, Corpus pineale:* a small, cone-shaped, unpaired organ situated between the cerebral hemispheres on the roof of the third ventricle of the mammalian brain. Phylogenetically, the P. g. is a vestigial parietal eye, the light sensitive organ of reptiles. It produces the hormone Melatonin (see).

Pinecembrin: 5,7-dihydroxyflavanone. See Flavanone.

Pinestrebin: 5-hydroxy-7-methoxyflavanone. See Flavanone.

Ping-Pong mechanism: see Cleland notation.

Pinitol: see Cyclitols.

L-Pipecolic acid: piperidine-2-carboxylic acid, a nonproteogenic amino acid. It is formed from L-lysine, either by α-deamination followed by cyclization and reduction, or as a normal intermediate in the degradation of lysine to α-aminoadipic acid. The 4- and 5-hydroxy derivatives of L-P.a. are found especially in mimosas and palms.

L-Baikiain (1,2,3,6-tetrahydropyridine-α-carboxylic acid), a rare nonproteogenic amino acid first isolated trom the wood of *Baikiaea plurijuga,* is structurally related to L-P.a.

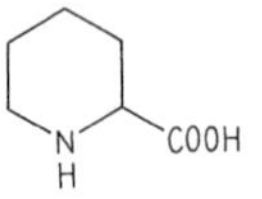

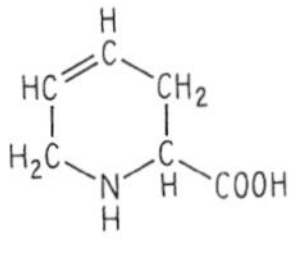

L-Pipecolic acid *L-Baikiain*

Piper alkaloids: a group of alkaloids occurring in various species of *Piper,* especially black pepper (*Piper nigrum*). Structurally, P. a. consist of an aromatic carboxylic acid with an unsaturated side chain (e. g. piperic acid, sinapic acid) in amide linkage with a basic component, usually piperidine. The chief representative is Piperine (see).

Piperidine alkaloids: a group of alkaloids containing the piperidine ring system. Simple P. a. are the alkyl substituted piperidines which occur sporadically. The other P. a. are classified according to their origin, e. g. Conium alkaloids (see), Punica alkaloids (see), Sedum alkaloids (see) and Lobelia alkaloids (see). These various groups are structurally different and have different mechanisms of biosynthesis. Other P. a. are found in water lilies, and are biosynthesized from mevalonic acid (see Nuphara alkaloids). A dehydropiperidine structure is present in the Areca alkaloids (see) and the Betalains (see).

Piperine: piperic acid piperidide, a Piper alkaloid (see), the chief alkaloid of black pepper (*Piper nigrum*) and responsible for its sharp taste. M_r 285.35, m.p. 16 °C. Both double bonds are *trans* (Fig.). The *cis-cis* isomer, earlier called chavicine, does not occur naturally.

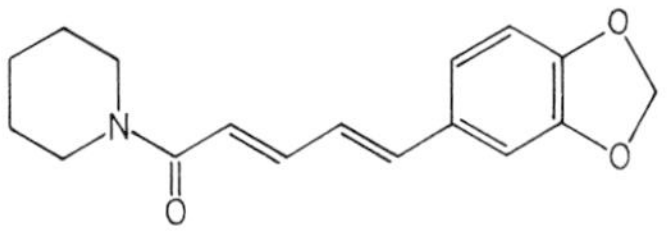

Piperine

Pisatin: 3-methoxy-6a-hydroxy-8,9-methylenedioxypterocarpan. See Pterocarpans.

Pituitary gland: see Hypophysis.

pK: see Buffers.

PL: acronym of placentalactogen.

Placentagonadotropin: see Choriogonadotropin.

Placentalactogen: see Choriomammotropin.

Plant hormones: see Phytohormones.

Plant mucilages: high M_r, complex, colloidal polysaccharides, which form gels and have adhesive properties. They are widely distributed in the plants, being found as secondary membrane thickening and as intercellular and intracellular material. They occur in root, bark, cortex, leaves, stalks, flowers, endosperm and seed coat. Some bulbs contain special mucilage cells. Some P. m. may function as food reserves. On account of their high affinity for water certain P. m. may be used as water reservoirs (i. e. as antidesiccants) by plants that live under very dry conditions; mucilagenous seed coatings may have a similar function. Together with the structurally related plant gums, P.m. form an ideal material for sealing damaged tissue; owing to their often heterogeneous carbohydrate composition, P.m are relatively resistant to microbial attack.

P.m. can be roughly classified into three groups: 1. Neutral, containing one or more types of sugar, but no uronic acids, e.g. a linear polymer of 1,4-linked mannopyranose isolated from the tubers of certain orchids. 2. Acidic, resembling the plant gums, but usually containing D-galacturonic acid as the acidic residue. In this type of P.m., the ratio of uronic acid residues to neutral sugar residues is usually about 1:3, e.g. a P.m. from the bark of *Ulmus fulva* (slippery elm), which contains D-galactopyranose, D-galacturonic acid, L-rhamnopyranose and 3-methyl-D-galactose. 3. P.m. present in algae (notably seaweeds), often of highly complicated structure and very high M_r and often containing esterified sulfate. This group is well exemplified by the agars from seaweeds. The agars range from a neutral species, agarose (see Agar-Agar), to the highly acidic, sulfated Carrageenans (see). The primary sequence of the entire agar family is based on a repeating disaccharide unit of galactose derivatives. Other seaweed P.m. are Alginic acid (see), laminarin (β-1,3 linked glucose residues with some β-1,6 branch points; isolated from *Laminaria*) and fucoidin (contains sulfated fucose residues; isolated from *Laminaria* and *Fucus* spp.).

Plant pigments: see Natural pigments, Photosynthetic pigments, Photosynthetic pigment systems.

Plant toxin: a compound produced by a plant which adversely affects man or animals, e.g. ricin from castor bean, amanitin from the death cap fungus. See also Toxic proteins.

Plaque: a transparent area in a lawn of bacteria on the surface of a solidified growth medium. P. is caused by lysis of bacteria in that area by bacteriophage. Under controlled conditions, each P. represents a center of infection initated by one infective bacteriophage particle. The number of P. produced after evenly spreading a known volume of phage suspension over the surface of the bacterial culture is used as a simple assay of the number of infective phage particles.

The term is also used in a general sense for an area of lysed cells in a lawn. In immunology, a lawn of red blood cells is used to detect immunoglobulin-producing cells in a P. assay. When complement is added to the system, those erythrocytes that have bound to the immunoglobulin molecules are lysed, leaving a transparent P. Since most antigens can be artificially bound to erythrocyte membranes, the technique can be used for practically any antibodies capable of fixing complement.

Plasma albumin: see Albumins

Plasma factors: see Blood coagulation.

Plasmakinins: physiological, highly active oligopeptides, with hormone-like properties. P. act on the smooth muscle of blood vessels, gastrointestinal tract, uterus and bronchi. Important representatives are Bradykinin (see), kallidin and methionylysyl-bradykinin.

Plasmalemma: the cell membrane. See Biomembrane.

Plasmalogens: see Membrane lipids.

Plasma proteins: a complex of predominantly conjugated proteins in the blood plasma of vertebrates. The number of P.p. is estimated to be more than 100. In mammalian plasma the concentration of P.p. is 6–8 %. Serum proteins lack Fibrinogen (see) and Prothrombin (see), but are otherwise essentially the same as P.p. Approximately 60 P.p. have been isolated and characterized. Of these, only albumin, prealbumin, retinol binding protein and a few trace proteins (e.g. lysozyme) do not possess bound carbohydrate. The remaining P.p. are glycoproteins e.g. Orosomucoid (see), Hemopexin (see), Haptoglobin (see), C1-inactivator, Immunoglobulins (see); some may also contain lipids (see Lipoproteins). P.p. help to regulate the pH-value and osmotic pressure of the blood; they transport ions, hormones, lipids, vitamins, metabolic products, etc. P.p. are also responsible for blood coagulation, for defense against foreign proteins or microorganisms (see Immunoglobulins), and for certain enzyme reactions. With the exception of immunoglobulins, the P.p. are synthesized in the liver. The 5 main groups of P.p., i.e. albumin, α_1-, α_2-, β- and γ-globulin (in order of increasing migration rate from the anode), can be separated by electrophoresis. The earliest separations were performed in the Tiselius carrier-free buffer system. Nowdays a carrier such as filter paper or cellulose acetate may be used, and an even better separation is achieved by exploiting the additional effect of electroosmosis (in agarose or starch gel), or sieving (polyacrylamide). The best single stage separation of P.p. into over 30 fractions is achieved by immunoprecipitation. Two dimensional immunoelectrophoresis after Laurell gives an even higher resolution; in this method the gel for the second electrophoretic direction contains the antibody mixture. See Fig. illustrating various, separation methods. The smallest known P.p. is β_2-Microglobulin (see).

P.p. are usually isolated on a preparative scale without loss of biological activity, by precipitation with ammonium sulfate, ethanol or rivanol. To an increasing extent, however, these methods are being replaced, even on an industrial scale, by efficient column methods, such as gel filtration, ion exchange, affinity and immunoadsorption chromatography, and by ultrafiltration.

Standardization of the measurement of plasma and serum proteins presents a particular problem in clinical biochemistry. Thus, quality control surveys in Western Europe and the United States during the early 1990s showed that the reported values for individual proteins in serum may vary by as much as 100 per cent, depending on the reference material used.

In 1989 the International Federation of Clinical Chemistry (IFCC) established a Committee for Plasma Protein Standardization to address this problem by preparing, characterizing and calibrating a new international secondary reference preparation for plasma proteins. Forty thousand 1 ml vials of a lyophilised preparation were produced and calibrated [S. Baudner et al, 'Manufacture and characterization of a new reference preparation for 14 plasma proteins/CRM 470 = RPPHSS Lot 5' *J. Clin. Lab. Anal.* **8** (1994) 177–190]. The material was released in 1993, in Europe, by the Community Bureau of Reference of the Commission of the European Communities (BCR) as the Certified Reference Material for Immunochemical Measurements of 14 Human Serum Proteins (CRM 470) and in 1994 in USA by the College of

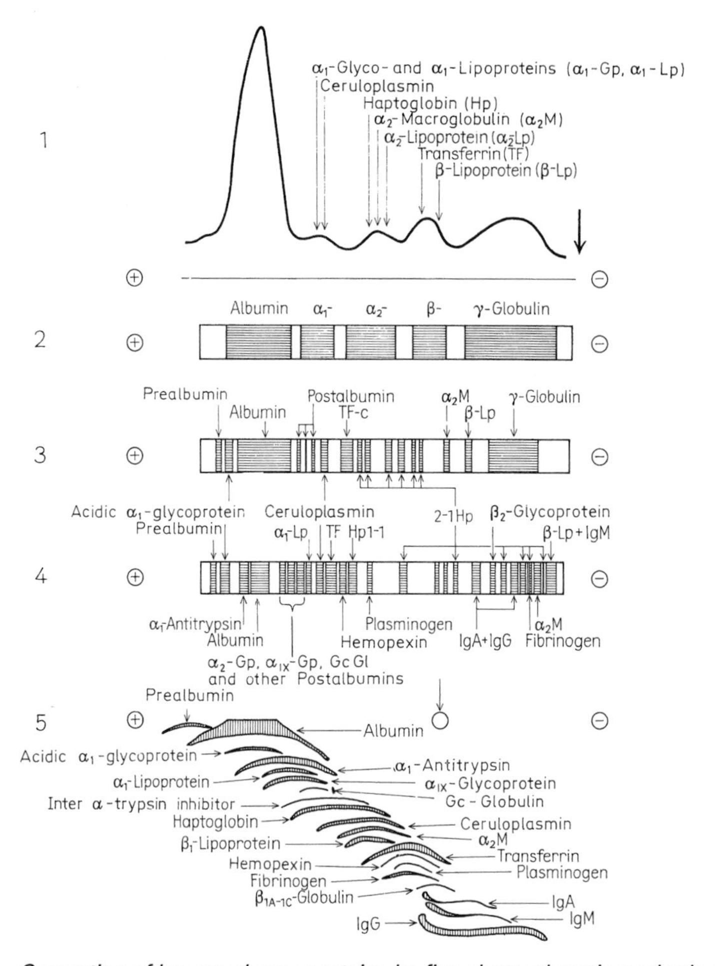

Separation of human plasma proteins by five electrophoretic methods at pH 8.
1. Carrier-free (Tiselius) electrophoresis. 2. Paper electrophoresis. 3. Starch gel electrophoresis. 4. Polyacrylamide gel electrophoresis (pH 9). 5. Immunoelectrophoresis (in practice the precipitation bands against polyvalent antihuman serum overlap one another, but have been drawn widely separated for legibility). Arrows in 1. and 5. indicate the site of application of the plasma sample.

American Pathologists (CAP) as the Reference Preparation for Proteins in Human Serum Lot No. 5 (RPPHS 5). The new reference material is a secondary reference material for albumin, ceruloplasmin, haptoglobin, α2-macroglobin, C3, C4, IgG, IgA, IgM and C-reactive protein, values for which have been assigned against existing international reference materials. It is however a primary material for transthyretin (prealbumin), α1-acid glycoprotein (orosomucoid), α1-antitrypsin (α1-protease inhibitor) and transferrin, values for which have been assigned against new purified protein preparations. A further project has been completed to assign values for α1-antichymotrypsin. The material is intended to be used for transfer of values to tertiary reference materials, commercial calibrants and controls via a well defined protocol.

Use of the new reference material results in significant changes in reference values for some proteins. The introduction of new reference ranges for several of the 14 serum proteins after calibration with CRM 470/RPPHS 5 is now urgently needed. The establishment of reference ranges from populations worldwide will require a huge international effort which is now being started by the new IFCC Committee on Plasma Proteins. Clearly such a project will take several years and an interim solution to the use of reference ranges for some proteins is required. In order to address this problem several professional societies and diagnostic companies have agreed to use interim consensus reference ranges derived from data produced by a number of recent studies in various parts of the world. They will be used immediately by most diagnostic companies in their package inserts and will be applied for all immunochemical methods independently of the system or instrument used.

Interim Consensus values have been reported by F. Dati et al. 'Consensus of a group of professional bodies and diagnostic companies on guidelines for interim reference ranges for 14 proteins in serum based on standardization against the IFCC/BCR/CAP reference material (CRM 470)' *J. Clin. Chem. Clin. Biochem.* **34** (1996) in press May 1996. [N. H. Packter et al. 'Characterization of Human Plasma Glycoproteins Separated by Two-Dimensional Gel Electrophoresis' *Biotechnology* **14** (1996) 66–70]

Plasmenic acid: see Membrane lipids.

Plasmid: extrachromosomal DNA in the bacterial cell. P. carry genetic information and replicate independently of the bacterial chromosome. Episomes are P. that can become reversibly integrated into the host DNA. P. are circular molecules of duplex, supercoiled DNA, ranging in size from M_r about 1.5×10^6 to 1.5×10^8. P. DNA differs physically and chemically from DNA of the bacterial chromosome, e.g. in its GC-content. P. DNA represents 1 to 3 % of total cell DNA, but in exceptional cases it may reach 20 %. P. genes are responsible for various nonessential properties of bacteria. Resistance to drugs or antibiotics is determined by P. called drug resistance factors or R-factors (see). Conjugation in bacteria is determined by P. called bacterial sex factors, fertility factors or F-agents. Bacterial virulence also depends upon P. Colicinogenic factors are P. in *E. coli,* which determine the synthesis of colicins (see Toxic proteins).

Plasmin, *fibrinolysin* (EC 3.4.21.7): a trypsin-like enzyme containing two polypeptide chains. Its inactive precursor or zymogen (plasminogen, profibrinolysin) occurs in blood at a concentration of 50–100 mg/100 ml serum. P. is responsible for fibrinolysis, i.e. dissolution of blood clots by proteolytic degradation of fibrin to soluble peptides. In addition, the products of fibrinolysis inhibit the conversion of fibrinogen to fibrin, thus forming a delicately balanced mechanism for the prevention of clotting. The natural inhibitor of P. is α_2-Macroglobulin (see).

Plasminogen (a single chain β-globulin, M_r 81,000 containing 8–10 % carbohydrate) is converted into P. (i.e. "activated") by hydrolysis of a specific arginyl-valine bond. The resulting P. consists of two chains held together by a single disulfide bond. The proteolytic activity of P. depends on the light chain of 233 amino acid residues; this contains the catalytically important serine and histidine residues characteristic of the serine proteases, and it also contains sequences homologous with those of trypsin. P. further resembles trypsin in that it splits arginyl linkages, and it is also inhibited by soybean and pancreatic trypsin inhibitors. Plasminogen activation is catalysed by specific tissue activators, which are released under various conditions, e.g. emotional stress, during exercise, following the injection of adrenalin, and it is released from blood vessel walls on vascular injury. The best known activator is the kidney enzyme, urokinase (EC 3.4.21.31) (M_r 53,000). Streptokinase (M_r 47,000, 416 amino acid residues, cysteine-free), produced by β-hemolytic streptococci, is of medical interest; although it is not generally considered to be a proteolytic enzyme, it is one of the most potent exogenous activators of human (not animal) plasminogen; it forms a complex with plasminogen, which then releases active P.

Plasminogen: see Plasmin.

Plasmochromic pigments: plant pigments contained in plastids, e.g. Chlorophyll (see) and Carotenoids (see).

Plasmon: the total extrachromosomal hereditary complement of a eukaryotic cell. See Chondrome, Plastome, Cytoplasmon.

Plast: see Cell, 2.

Plastids: organelles in the cells of eukaryotic plants. P. contain DNA, and are self-replicating. Division of P. and replication of P.-DNA are not synchronized with nuclear division and replication of nuclear DNA. P. are usually ellipsoid, 1–10 μm long. They can, however, be larger and have various shapes, e.g. in green algae. Chloroplasts (see) are P. Chromoplasts (see) contain various pigments responsible for plant (especially flower) coloration. P. devoid of pigments are called Leucoplasts (see); these act as storage sites for starch in the root and shoot. P. are derived biogenetically from Proplastids (see).

Plastocyanin: a blue protein (M_r 10,500) composed of a single polypeptide chain of 99 amino acids with a single copper ion bound to His-37, Cys-84, His-87 and Met-92 in a distorted tetrahedral arrangement. All its hydrophobic amino acid residues project inwards from a β-barrel-type structure giving it great solubilty in water. It is located in the aqueous milieu of the thylakoid lumen of chloroplasts where it carries electrons from cytochrome *f* of the b_6f complex to $P700^+$ of PSI. It is a one-electron redox system (Cu^+/Cu^{2+}) with an E'_0 of +0.37V; it is thus ideally suited to its electron transfer function, being susceptible to exergonic reduction by cytochrome *f* (E'_0 + 0.34V) and exergonic oxidation by $P700^+$ (E'_0 of the $P700^+/P700$ redox system + 0.43V). P. was originally discovered in *Chlorella ellipsoidea* [S. Katoh *Nature* **186** (1960) 533–534] but has since been found to be widespread in the chloroplasts of species capable of oxygenic photosynthesis. The amino acid sequence of P. from many species has been determined and 64 of its 99 amino acids found to be conserved. P. is encoded in the nucleus (the gene from several species has been sequenced), synthesized in the cytoplasm and translocated to the chloroplast. Its 3D structure has been determined by X-ray crystallography at 0.16 nm resolution [J. M. Guss & H. C. Freeman *J. Mol. Biol.* **169** (1983) 521–563]. Copper deficiency in the growth medium results in the replacement of P. by the soluble cytochrome *c*-552 in some algae (e.g. *Scenedesmus, Chlamydomonas*).

Plastome, *plastidom:* the total genetic information contained in the DNA of the plastids of a eukaryotic plant cell.

Plastoquinone: a polyisoprenoid quinone. P. is structurally similar to ubiquinone, but contains a methyl group (not a methoxy group) in the aromatic ring. P. can be isolated from chloroplasts; it acts as a reversible redox component in photosynthetic electron transport (see Photosynthesis).

Platelet activating factor, *PAF:* a phospholipid released by IgE-sensitized leukocytes in the presence of antigen, and possibly endogenous to platelets as well. It is present during anaphylactic shock, and appears to mediate inflammation and allergic responses. In vitro, at concentrations of 10^{-11} to 10^{-10} M, PAF causes platelets to change shape, aggregate and release the contents of their granules. PAF is structurally similar to an antihypertensive polar renomedullary lipid (APRL) from kidneys, which reduces the blood pressure of artificially hypertensive animals.

Platelet-derived growth factor, *PDGF:* a hydrophobic, heat-stable (100 °C), cationic (pI 9.8–10.2) polypeptide (2 homologous chains linked by disulfide bonds), M_r 32,000 contained in the α-granules of platelets and released into serum during blood clotting. PDGF is a potent mitogen, which stimulates growth of fibroblasts, glial cells, monocytes, neutrophils and smooth muscle cells; it represents the major growth factor of human serum. PDGF is the product of a proto-oncogene, c-*sis*. The oncogene counterpart is v-*sis*, an insertion sequence carried in the genome of simian sarcoma virus (SSV). The v-*sis* sequences are essential for the transforming activity of the virus, and they give rise to a M_r 28,000 oncogene product known as $p28^{sis}$. PDGF and $p28^{sis}$ compete for the same receptor, possess striking sequence homology, and are immunologically cross-reactive.

Binding of PDGF to its specific, saturable cell surface receptor stimulates Protein-Tyrosine kinase (see; see also Receptor tyrosine kinase) activity in the receptor molecule, leading to phosphorylation of the receptor itself (autophosphorylation) as well as other cell proteins. PDGF receptor (M_r about 170,000) has been isolated by affinity cross linking to ^{125}I-labeled PDGF (using disuccinimidyl tartrate or disuccinimidyl suberate) followed by precipitation with a monoclonal phosphotyrosine antibody from detergent (Triton)-treated cells after stimulation with PDGF. [M. D. Waterfield et al. *Nature* **304** (1983) 35–39; R. F. Doolittle et al. *Science* **221** (1983) 275–277; J. Y. J. Wang & L. T. Williams *J. Biol. Chem.* **259** (1984) 10645–10648; A. R. Frackelton *J. Biol. Chem.* **259** (1984) 7909–7915; J.-C. Yu et al. *J. Biol. Chem.* **270** (1995) 7033–7036; S. Mori et al. *J. Biol. Chem.* **270** (1995) 29447–29452]

Plicacetin: see Amacetins.

Plumbagin: see Naphthoquinones.

Plumierid: an Iridoid (see).

Pluripoietin: see Colony stimulating factors.

PMS: see Pregnant mare serum gonadotrophin.

Pmr: abbr. of Proton nmr. See Nuclear magnetic resonance spectroscopy.

Poisons, *toxins:* compounds which damage an organism in relatively small amounts, and which may kill it at sufficiently high doses. *Plant poisons* are found in all parts of certain plants; they are not produced by specialized cells. They tend to affect the heart and circulation. Many types of compound are represented: proteins, alkaloids, steroids, amines, unusual amino acids, etc. Since plants synthesize about 100 times as many natural products as animals, it is not surprising that the number of plant poisons is very large. Of those whose structures are known, the most important include the proteins ricin and the viscotoxins, nearly all alkaloids, especially the Aconitum, Colchicum, Conium, Curare, Strychnos and Tropane alkaloids, the cardiac glycosides, the unusual amino acids like djencolic acid and lathryin, and mushroom poisons. The toxins of lower fungi and bacteria (mycotoxins and antibiotics) are particularly diverse in structure. Plant poisons tend to be less toxic than animal poisons.

Animal poisons are produced by the animals under normal conditions. They are structurally diverse, including peptides and proteins, acids, amines, betaines, quinones, alkaloids, terpenes and steroids. The defensive secretion of the water beetle *Dytiscus marginalis* contains 0.4 mg deoxycorticosterone, which is also a vertebrate hormone. One thousand animals would be needed to obtain this amount from bovine adrenal glands!

Most animal poisons are mixtures. They occur in almost every group of animals, excepting perhaps some mollusks, tunicates and birds. Passive poisons are contained in the tissues or organs of the animal and are only effective if the flesh is eaten. Actively poisonous animals, however, usually produce the poison in special glands and have various means of delivering it: secretion from skin glands (see Amphibian toxins), ejection from the mouth or abdominal region by muscle pressure on the poison glands, or injection through teeth (see Snake venoms), sting (see Bee toxin) or hairs (caterpillars). Some snakes and all skunks can spray their poison in a chosen direction. The bombadier beetle's spray is ejected by a chemical explosion which occurs when the animal adds catalase to a mixture of quinones and 23 % hydrogen peroxide. The heat of reaction brings the mixture almost to boiling point, and the pressure of the released oxygen causes it to spray out. Animal poisons have various effects, but the most common are heart, blood and nerve poisons. Some, like snake toxins, are used therapeutically, or to produce antisera.

Comparison of the toxicities of various poisons

Substance	Occurrence or significance	LD_{50} (μg/kg mouse)
Inorganic poisons		
Potassium cyanide (KCN)		3,000
Arsenic (As_2O_3)		10,000
Mercuric chloride ($HgCl_2$)		23,000
Plant poisons		
Strychnine	Strychnos alkaloid	500
C-Toxiferin I	Curare alkaloid	25
Colchicine	Colchicum alkaloid	3,000
Nicotine	Tobacco alkaloid	16,000
Coniine	Conium alkaloid	60,000
k-Strophanthine	Cardiac glycoside	15,000
Digitonin	Cardiac glycoside	150,000
Fungal poisons		
α-Amanitin	*Amanita phalloides*	200
Ergotoxin	Ergot alkaloid	100,000
Bacterial toxins		
Botulin toxin	*Clostridium botulinum*	0.00003
Tetanus toxin	*Clostridium tetani*	0.0001
Diphtheria toxin	*Corynebacterium diphtheriae*	0.3
Animal poisons		
Batrachotoxin	Arrow-poison frog	2
Tetrodotoxin	Puffer fish	8
Cobra toxin	Cobra	0.3
Crotalus toxin	Rattlesnake	0.2
Bufotoxin	Toad poison	400
Melittin	Bee venom	4,000
Salamandarine	Salamander alkaloid	1,500

The toxicity of natural poisons, expressed by their LD_{50} (see Dose), is much higher than that of conventional inorganic poisons (see Table).

Modern toxicology is concerned with the relationships between the structure and effects of poisons and drugs. The site and mode of action, and detoxifying reactions are of particular interest. Many poisons inhibit vital enzymes, e. g. cyanide ions bind to the iron of cytochrome oxidase and block electron transport to oxygen, while carbon monoxide blocks the oxygen-transport function of hemoglobin by binding more tightly to the oxygen-binding sites than oxygen does. Other poisons interfere with the basic metabolism of the cell and affect its structure. Snake venom enzymes, for example, remove specific fatty acid components from phospholipids. The lysolecithins and lysocephalins formed in this way emulsify the lipids of the erythrocyte membranes and cause the cells to lyse. Melittin, the main component of bee venom, has the same effect. The amanitins and phalloidins irreversibly destroy the endoplasmic reticulum of liver cells. The organism responds to toxins by attempting to detoxify them. Foreign proteins are attacked by specific antibodies and most poisons are modified structurally in the liver. The simplest form of inactivation consists of conjugation to form glycosides, which are excreted. An animal may become tolerant of a P. if the detoxifying enzyme system is induced, so that the P. is quickly inactivated.

Pollinastanol: 4,4-desmethylcycloartanol, a plant constituent related structurally to the sterols and to cycloartenol. Particularly good sources of P. are pollen from members of the *Compositae,* the fern *Polypodium vulgare,* and roots of sarsaparilla (*Smilax medica*). In these plants, P. is an important intermediate in the biosynthesis of cholesterol.

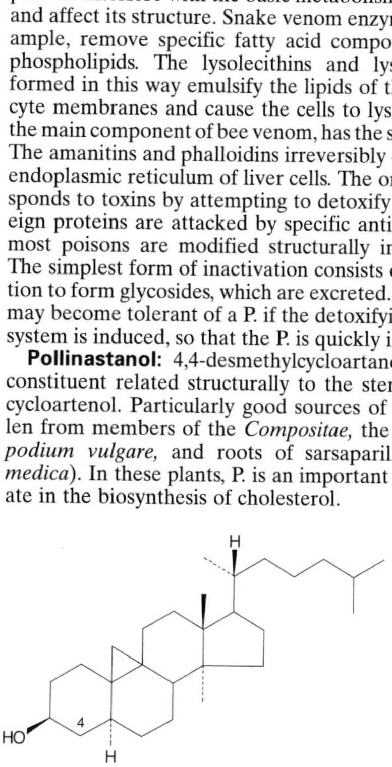

Pollinastanol

Poly A: abb. for Polyadenylic acid (see).

Polyacrylamide gel electrophoresis: see Proteins.

Polyadenylic acid, *Poly A:* a Homopolymer (see) consisting entirely of residues of adenylic acid. Poly A sequences of varying lengths are found at the 3′ end of many eukaryotic mRNA molecules (see Post-transcriptional modification of RNA, Poly A polymerase). The length of the poly A sequence depends on the source of the RNA and the physiological state of the cell, and is inversely related to the half-life of the mRNA.

The function of poly A sequences is not clear. They may be required for transport of mRNA from nucleus to cytoplasm or for binding to the ribosome. However, mRNA is only attacked by hydrolytic enzymes after the loss of all poly A; thus the metabolic half-life of a mRNA molecule may be regulated by the length of its poly A sequence. Degradation of a poly A sequence may be controlled by proteins which are actually bound to it.

The histone mRNA of higher eukaryotes does not have a poly A tail. A short sequence of oligo A has been found in the 3′-region of bacteriophage T7 RNA, and a short oligo A sequence has also been detected at the 3′-end of RNA from *E. coli.* The mRNAs of the slime mold *Dictyostelium* and of HeLa cells contain two poly A segments separated by a short sequence of other nucleotides.

The presence of the poly A sequence has been exploited for the isolation and purification of eukaryotic mRNA, by using affinity columns of oligo-dT-cellulose.

Poly ADP-ribose: see ADP-ribosylation of proteins.

Polyamines: a group of aliphatic, straight-chain amines derived biosynthetically from amino acids. They include spermine, spermidine, cadaverine and putrescine. The latter two are diamines produced by decarboxylation of lysine or ornithine, respectively. In plants, putrescine can also be synthesized by decarboxylation of arginine to agmatine, followed by hydrolytic removal of urea. Animals do not have arginine decarboxylase (EC 4.1.1.19), but in plants it is often more abundant than ornithine decarboxylase (EC 4.1.1.17). Putrescine is converted to spermidine and spermidine to spermine, by the addition of an aminopropyl group. This group is provided by decarboxylated *S*-adenosyl-L-methionine (Fig.).

P. are required for cell division, and probably for differentiation. Spermine apparently stabilizes the DNA which is tightly packed in the heads of sperm cells. When putrescine is witheld from putrescine auxotrophs of Chinese hamster ovary cells, the cells are found to lack 90 % of their actin filament bundles and to have essentially no microtubules. It may be inferred that P. are essential for the stability of these structures. P. stimulate growth of higher plants and ap-

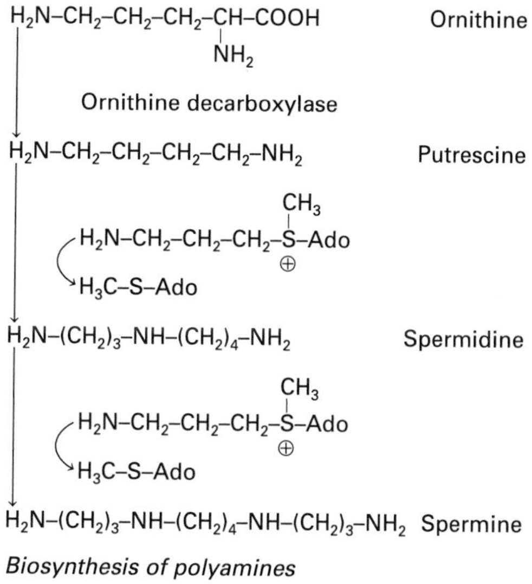

$H_2N-CH_2-CH_2-CH_2-CH(NH_2)-COOH$ Ornithine

↓ Ornithine decarboxylase

$H_2N-CH_2-CH_2-CH_2-CH_2-NH_2$ Putrescine

↓ $H_2N-CH_2-CH_2-CH_2-S^{\oplus}(CH_3)-Ado$ → $H_3C-S-Ado$

$H_2N-(CH_2)_3-NH-(CH_2)_4-NH_2$ Spermidine

↓ $H_2N-CH_2-CH_2-CH_2-S^{\oplus}(CH_3)-Ado$ → $H_3C-S-Ado$

$H_2N-(CH_2)_3-NH-(CH_2)_4-NH-(CH_2)_3-NH_2$ Spermine

Biosynthesis of polyamines

pear in some cases, e.g. tomatoes, to be required for formation of fruit. Putrescine accumulates in plants subjected to stress, especially deficiency of K^+ and Mg^{2+}, feeding of NH_4^+, acidification and high salinity.

In organisms lacking arginine decarboxylase, putrescine can be synthesized only by decarboxylation of ornithine. Many inhibitors of ornithine decarboxylase are known, and these have a number of medical applications against tumors, viruses, trypanosomes (e.g. *Trypanosoma brucei,* which causes African sleeping sickness) and plasmodia (e.g. *Plasmodium falciparum,* the agent of malaria). [*Polyamines in Biology and Medicine,* D.R.Morris & L.J.Marton, eds. (Dekker, New York, 1981); A.E.Pegg & P.P. McCann *Am. J. Physiol* **243** (1982) C212–221; T.A. Smith *Ann. Rev. Plant Physiol.* **36** (1985) 117–143]

Polyamino acids naturally occurring or synthetic polymers consisting of identical amino acids linked by peptide bonds. P.a. in the capsular substance of anthrax bacteria contain residues of D-glutamic acid linked by γ-peptide bonds. Poly-γ-D-glutamic acid precipitates antibodies against anthrax, a property not shared by poly-γ-L-glutamic acid. P.a. of M_r 103–106 are easily prepared by polymerization of aminocarboxylic acid anhydrides. P. are structurally similar to Sequence polymers (see).

Poly A polymerase: an enzyme which specifically catalyses the synthesis of poly A sequences, a process which does not require DNA. P. from calf thymus has M_r about 150,000. It is localized in the nucleus, and is responsible for the covalent attachment of poly A sequences to RNA (see Polyadenylic acid).

Polyisoprenes: see Polyterpenes.

Polyisoprenoids: see Polyterpenes.

Polyketides, ***acetogenins:*** natural products containing several recurring two-carbon units, formally equivalent to the condensation products of several molecules of acetate. Biosynthesis of P. occurs on a multienzyme complex. The first stage involves interaction of a starter molecule with a peripheral SH-group of the enzyme complex. A common starter molecule is acetyl-CoA (e.g. synthesis of anthraquinones and griseofulvin), i.e. HSCoA is lost and the acetyl group becomes attached via a thiol ester linkage to the peripheral acyl carrier protein (HSp). Other starter molecules are propionyl-CoA, cinnamoyl-CoA (synthesis of flavonoids and stilbenes), malonic acid amide-CoA (synthesis of tetracyclines), the CoA derivatives of the oxoacids of leucine, isoleucine and valine, nicotinyl-CoA, etc. Following the introduction of the starter group, a malonyl residue (from malonyl-CoA) is attached to the SH-group of a pantetheine residue, which forms the prosthetic group of a protein situated centrally in the multienzyme complex (HS_C). The starter group (e.g. acetyl) is then transferred with simultaneous decarboxylation of the malonyl residue, forming an acetoacetyl group on HS_C. This acetoacetyl group is then transferred to HSp, and another malonyl residue is introduced to HS_C. There is thus an extremely close analogy with the inital stages of fatty acid synthesis, but the two reduction steps of fatty acid synthesis do not occur in P. synthesis. The synthesis of P. proceeds by the stepwise introduction of malonyl groups. The recurring two carbon units evident in the structure of all P. are therefore derived biosynthetically by the incorporation and decarboxylation of malonyl-CoA. The β-polyketone intermediate does not accumulate; it cyclizes by ester or aldol condensation, forming the ring systems found in various P., and carbonyl groups are usually present in the resonance-stabilized enol form. Further modifications include introduction of extra oxygen by hydroxylation, methylation of hydroxyl groups to form methoxy groups, or direct methylation of carbon atoms (the methyl donor is *S*-adenosyl-L-methionine), reduction, substitution with polyisoprenoid residues, and chlorination of an aromatic ring.

Examples of P. are Tetracyclines (see), Griseofulvin (see), Macrolide antibotics (see), Cycloheximide (see), and various fungal products such as orsellinic acid, 6-methylsalicylic acid and cyclopaldic acid. ["The Biosynthesis of Acetate-Derived Phenols (Polyketides)" by N.M.Packter pp.535–570, in *The Biochemistry of Plants,* Vol 4, 1980 (Edit. P.K. Stumpf), Academic Press; S.Sahpaz et al. *Phytochemistry* **42** (1996) 103–107]

Polymerase chain reaction (PCR): a process, developed in 1984 by scientists at the Cetus Corporation, that is used to amplify a specific segment of DNA (the target DNA), which lies between two regions of known nucleotide sequence within a longer DNA template. The latter, the starting material, may be as little as a single molecule of DNA; moreover it may be present in a complex mixture of other DNA species, e.g. in a crude cell lysate. The target DNA is usually 200–500 nucleotides long, 2,000 nucleotides being the present practical upper limit. In addition to the template DNA, the process requires two oligonucleotide primers and a DNA polymerase. The primers are 20–25 nucleotides long, have different sequences and are produced synthetically. If the template is a single strand of DNA (see Fig.) one primer has a nucleotide sequence that is complementary to that of the known sequence at the 3′-end of the target DNA, while the other has a sequence that is complementary to the equivalent sequence in the complementary template strand. If the template is double-stranded the primers have nucleotide sequences that are complementary to known sequences that lie on opposite strands, each at the 3′-end of the target DNA. In the early days of PCR the DNA polymerase used was the Klenow fragment (see) of *E. coli* DNA polymerase I [R.K.Saiki et al. *Science* **230** (1985) 1350–1354; K.B.Mullis & F.A.Faloona *Methods Enzymol.* **155** (1987) 335–350], but this has now been replaced by the *Taq* DNA polymerase [R.K.Saiki et al. *Science* **239** (1988) 487–493] from the thermophilic bacterium *Thermus aquaticus* which, being capable of withstanding extended incubation at 95 °C, allows the process to be automated.

PCR occurs in a series of cycles, each of which essentially doubles the original quantity of target DNA present in the incubation mixture; the latter also contains a large molar excess of the two primers and the four 2′-deoxyribonucleotide 5′-triphosphates, dATP, dGTP, dCTP and dTTP. The first step in each cycle is a denaturation whose purpose is to render the template DNA single-stranded (note that the only occasion that this is not required, although it is carried out for the sake of convenience, is the first cycle which starts with a single-stranded template; see Fig.); this is accomplished by raising the temperature

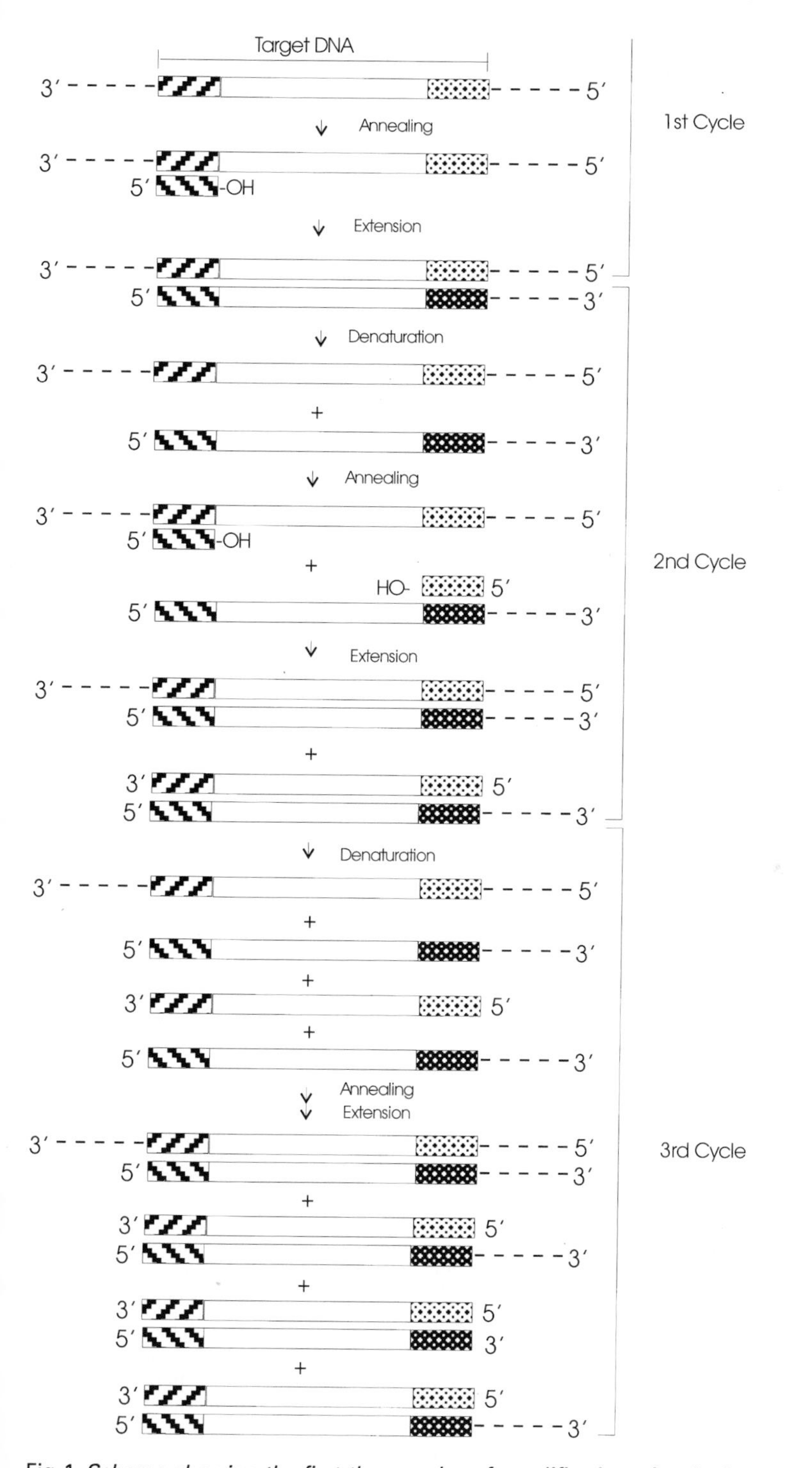

Fig. 1. *Scheme showing the first three cycles of amplification of a single-stranded target DNA by the polymerase chain reaction.*
/ & / represent pairs of complementary sequences of about 20 nucleotides.

of the incubation mixture to 92–96 °C for about 60 seconds. The second step is the annealing step, in which the primers anneal to their complementary sequences in the single-stranded DNA template molecules; this is accomplished by lowering the temperature to 55–60 °C for about 30 seconds. The third and final step is the extension step, in which the DNA polymerase extends the primer H-bonded to each template DNA molecule, in the 5′ → 3′ direction, by successively adding 2′-deoxyribonucleotide 5′-monophosphate residues to the 3′-hydroxyl group at its 3′-end in a sequence dictated by that of the template; this is accomplished by raising the temperature to 72 °C for 1–3 minutes. The step 3 incubation time depends upon the length of the target DNA being copied; using *Taq* DNA polymerase, which has an extension rate in excess of 50 nucleotides per second at 70 °C, 1 minute is satisfactory for target DNAs of 500 nucleotides or less. This cycle of three steps is repeated many times, resulting in an exponental amplification of the target DNA, because each cycle theoretically doubles the amount of template available for the succeeding cycle. Examination of the three depicted cycles (Fig.) shows that three species of DNA strands are present at the end of all cycles except the first, namely: (i) the original template, (ii) those that have one of the two primers at their 5′-end and have a 3′-end that extends as far as the 5′-end of the original template, and (iii) those that are the exact length of the target DNA since they have one of the two primers at their 5′-end and have a 3′-end that extends only as far as the end of the sequence that is complementary to the other primer. The Figure also shows that the relative proportions of these three species, (i), (ii) & (iii) at the end of cycles 1, 2 and 3 are 1:1:0, 1:2:1, 1:3:4 respectively. It is easy to show that the proportions for cycles 4, 5 and 6 are 1:4:11, 1:5:26 and 1:6:57, respectively, from which it is obvious that the PCR causes no amplification of the original template species, an arithmetic amplification of species (ii) and an exponential amplification of species (iii), which is the target DNA. In fact after 20 cycles the relative proportion of species (i), (ii) & (iii) in the amplification product is 1:20:1,048,555, showing that 99.998 % of it is the target DNA. Essentially the same degree of amplification occurs when the original template DNA is double-stranded.

The typical PCR starting incubation mixture has a volume of 100 μl and contains up to 2 μg of template DNA of which ≤ 1 amol may be target DNA, 100 pmol of each of the two primers, 20 nmol of each of the dNTPs, 2.5 Units of *Taq* DNA polymerase and an amplification buffer that gives 50 mM KCl, 10 mM Tris HCl, 1.5 mM $MgCl_2$ and 0.01 % gelatin concentrations in the reaction mixture and a pH of 7.2 at 72 °C. This mixture is overlaid with 100 μl of light mineral oil to prevent evaporation and sealed in a 0.5 ml microfuge tube. The tube is then typically subjected to 25–30 of the thermal cycles described above in an automated thermocycler. A recently developed thermocycler [L. Haff et al. *Biotechniques* **10** (1991) 102–112], which is capable of very rapid changes of temperature, removes the need for a mineral oil layer by heating the whole sample tube including the cap. The number of thermal cycles required depends upon the concentration of target DNA in the incubation mixture; at least 25 cycles are needed to amplify a single-copy target DNA in mammalian genomic DNA to an amount which can be detected by direct examination of agarose or polyacrylamide gels. However the efficiency of amplification, which starts off at 100 % (i.e. every template molecule in the incubation mixture is fully replicated during a given thermal cycle), falls off and plateaus as the number of cycles increases above 30. There are at least two reasons for this, namely: (i) as the concentration of product DNA becomes high there is increased competition between the annealing of the primers to their templates (i.e. the product DNA) and the reannealing of the complementary strands of the latter, and (ii) the amount of DNA polymerase in the incubation mixture eventually becomes insufficient to extend all the primer-template complexes in the time allocated for the extension step of the cycle. The average efficiency of a series of PCR cycles can be calculated from the equation: $N = n(1 + E)^c$ where N = the amount of target DNA produced by the series of cycles, n = the initial amount of target DNA, E = the amplification efficiency (where $1 \equiv 100\,\%$) and c = the number of cycles [N. Arnheim & H. Erlich *Annu. Rev. Biochem.* **61** (1992) 131–156]. Because amplification efficiency falls off after 30 cycles the most effective way of obtaining further amplification, should it be required, is to dilute the product of 25 cycles 10^3- to 10^4-fold and use appropriate aliquots to act as the starting material for further 25-cycle bouts of PCR synthesis.

The high specificity with which PCR amplifies only the target DNA depends on the designing and synthesizing of two primers that are of such a length that their sequences are unique within the complex mixture of DNAs that may be used as the PCR starting material, and which as a consequence allow them to H-bond only to the complementary sequence at the 3′-end of one or other of the target DNA strands. Failure in this respect would allow the the primers to anneal to non-target DNA sequences and could result in the amplification of non-target DNA. The magnitude of this specificity problem is apparent from the fact that a 1,000-base pair target constitutes only a millionth of the DNA in a typical mammalian genome of 10^9 nucleotides. Two approaches that have been used to minimize the problem and ensure that only the target DNA is amplified are known as 'nesting' [K. B. Mullis & F. A. Faloona *Methods Enzymol.* **155** (1987) 335–350] and 'hemi-nesting' [H. Li et al. *Proc. Natl. Acad. Sci. USA* **87** (1991) 4580–4584]. Both these strategies involve two bouts of PCR using two sets of primers, with a small aliquot of the amplification product of the first bout with one set of primers being used as the starting material for the second bout with the second set of primers. In the former procedure the sequences in the target DNA that are complementary to the first-acting pair of primers are external to those that are complementary to the second-acting pair, whose separation distance define the size of the final amplified product. In the second procedure only three different primers are used, with one of the first-acting pair being replaced by the third for the second bout of amplification. A recent modification of the hemi-nesting procedure obviates the need to carry out the two bouts of PCR in separate tubes; all three primers are present throughout but the thermal cycling conditions are adjusted so as to favor the annealing of the external pair during the first bout and the nested primer during the second [H.A Erlich et al. *Science* **252** (1991) 1643–1651]. These nesting strategies work because any non-target sequences that were amplified in the first bout cannot possibly be amplified with a pair of internal, target-specific primers or or even a single one.

Although PCR produces a remarkably accurate amplification of target DNA it is not totally error-free. The commonest type of mistake to occur is that of 'misincorporation' in which the wrong 2′-deoxyribonucleotide residue is incorporated into the extending DNA chain. This is most likely to occur with enzymes, like the *Taq* DNA polymerase, that have no 'proofreading' 3′ → 5′-exonuclease activity. Estimates of the misincorporation with this enzyme range from 5×10^{-6} to 1.7×10^{-4} nucleotides per cycle and appears to vary with the concentration of some of the components of the incubation mixture and the annealing temperature used.

The search for new thermostable DNA polymerases that may be useful in PCR continues [N. Arnheim & H. Erlich *Annu. Rev. Biochem.* **61** (1992) 131–156]. For instance, enzymes from *Thermoplasma acidophilum*, *Thermococcus litoralis* and *Methanobacterium thermoautotrophicum* have been reported to have 3′ → 5′-exonuclease activity and would therefore be expected to have lower misincorporation rates than *Taq* DNA polymerase. A DNA polymerase from the thermoacidophilic archebacterium *Sulfolobus acidocaldarius* catalyses primer extension at 100 °C which could prove useful in the amplification of target DNAs with regions of high secondary structure. A DNA polymerase with Reverse transcriptase (see) activity in the presence of $MnCl_2$ has been discovered in *Thermus thermophillus.* Since the DNA polymerase activity of this enzyme can be stimulated by chelating the Mn^{2+} and adding $MgCl_2$, it may be possible to perform cDNA synthesis followed by PCR amplification in a single-enzyme, single tube reaction.

Although PCR is a relatively new technique it is being used in a variety of ways in molecular cloning and DNA analysis, such as: (i) generation of specific sequences of cloned, double-stranded DNA for use as probes, (ii) generation of probes for uncloned genes by the selective amplification of specific segments of cDNA, (iii) generation of libraries of cDNA from small amounts of mRNA by amplifying first-strand cDNA generated from the mRNA templates by reverse transcriptase and an oligo-dT primer, (iv) generation of large amounts of DNA for sequencing, (v) analysis of mutations, and (v) Chromosome crawling (see). It has also found application in the analysis of mutations in oncogenes, the genetic identification of forensic samples, the detection of DNA sequences derived from pathogenic organisms in clinical samples and in the diagnosis of genetic disorders [J. Sambrook et al. *Molecular Cloning* 2nd Edn., Vol. 2 (1989) 14.2–14.21].

Polymerases: a collective term for enzymes that catalyse the formation of macromolecules from simple components, e. g. see separate entries for each of the following: DNA-polymerase, RNA-polymerase, RNA-synthetase, RNA-dependent DNA-polymerase, Polynucleotide phosphorylase, Poly A-polymerase.

Polymerization: see Ribonucleic acid.

Polymorphism: a genetically determined Heterogeneity (see) of proteins, especially enzymes. P. occurs when the frequency of a genetic variant in a population is greater than 1 %. Frequencies of this order develop by positive selection or by the effect of incidental genetic drift on rare mutations that have a heterozygotic advantage. The resulting allelomorphs (gene pairs) of a protein differ from each other by substitution or deletion of an amino acid at one or more sites in the peptide chain. P. is shown by, e. g. the β-chain of hemoglobin, haptoglobin, transferrin, adenosine deaminase, and glucose 6-phosphate and 6-phosphogluconate dehydrogenases. P. is found especially in enzymes not involved in primary metabolic processes, and in enzymes with broad in vitro substrate specificity

Polymyxins: heteromeric, homodetic, cyclic, branched peptides, produced by *Bacillus polymyxa,* possessing antibiotic activity against Gram-negative bacteria. P. are used to treat infections of the intestinal and urinary tracts, sepsis and endocarditis. Various P., designated A, B_1, B_2, C, D, E and M are known. P.B_1 is a cyclic decapeptide, consisting of a ring of 7 amino acids with a tripeptide side chain linked via the γ-amino group of a diaminobutyric acid residue. The *N*-terminus (D-α,γ-diaminobutyric acid) is acylated by (+)-6-methyloctanoic acid (MOA, isopelargonic acid).

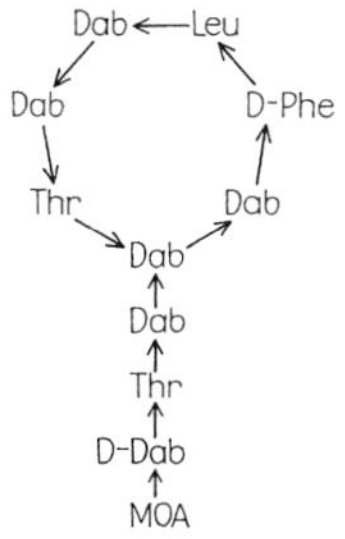

Dab ← Leu
Dab   D-Phe
Thr   Dab
Dab
↑
Dab
↑
Thr
↑
D-Dab
↑
MOA

Polymyxin B_1

Polynucleotide ligase, *DNA-ligase:* an enzyme that joins two DNA fragments by catalysing the formation of an internucleotide ester bond between phosphate and deoxyribose (see Single strand breakage). P. are active in the repair of damaged DNA, and in the linkage of Okazaki fragments during DNA replication. They have been used in vitro in the cell-free synthesis of genes (DNA) (see Gene synthesis).

Polynucleotide-methyltransferase, *polynucleotide methylases:* specific enzymes that catalyse the methylation of purine and pyrimidine bases, or sugars in the intact polynucleotide chain by transfer of methyl groups from *S*-adenosyl-L-methionine.

Methylated bases and/or sugars (see Rare nucleic acid components) are found in tRNA, mRNA, rRNA and DNA. 5-Methylcytosine and 6-methyladenine are particularly common. tRNA (see) is notable for its high content of modified components, many of them methylated. The 5′-terminal cap structures of eukaryotic and viral mRNA are methylated after transcription, and N^6-methyladenosine is typically found internally between the 5′ and 3′ ends of the mRNA molecule (see mRNA). Methylation of rRNA is essential for ribosome maturation. Methylation of DNA serves in the recognition of specific regulatory proteins, and it protects DNA from the action of restriction endonucleases producted by its own cell. 1–1.5 % of mammalian DNA and 5–6 % of plant DNA is methylated. The degree of methylation is often organ-specific. Phage DNA is also methylated. DNA from T2, T4, T7 and P1 contains 6-methyladenine, whereas that from T3 and T5 contains methylthymine; the latter, however, arises by methylation of deoxy-Tpp during DNA replication, and not by the action of P. m. on the intact DNA. See DNA methylation.

Polynucleotide phosphorylase: an enzyme catalysing the synthesis in vitro of polyribonucleotides. 5′-Nucleoside diphosphates are added to oligonucleotide starter molecules (primers) with the release of phosphate. The resulting sequence depends on the availability of components for the reaction. Thus it is possible to synthesize homopolymers (poly A, poly U, etc.) or Copolymers (see) (e. g. poly AU). The function of P. p. in the cell is not clear. Since it also acts as an exonuclease and catalyses the reverse reaction, it is possible that the enzyme functions in the synthesis and remobilization of storage polynucleotides.

Polynucleotides: linear sequences of at least 20 nucleotides, in which the 3′-position of each monomeric unit is linked to the 5′ of the neighboring unit via a phosphate group. The valency angles between phosphate and sugar residues are such that the sugar-phosphate backbone forms a helix, with the bases projecting sideways. The sugar may be D-ribose (in ribopolynucleotides) or 2-deoxy-D-ribose (in deoxyribopolynucleotides). The Nucleic acids (see) are specific high M_r P. Various types of P. structure are found e. g. single-stranded, double-stranded, or internally folded (alternating regions of single and double strands). The base sequence of P. is denoted by the base letters, A, G, C, U, T. The letter p is used before the base letter for 5′-substitution, and after the base letter for 3′-substitution, e. g. pGpApU represents 5′-phosphate → 3′-hydroxyl.

Polynucleotide thioltransferases: thiolases catalysing the specific thiolation of purine and pyrimidine bases in the synthesis of Rare nucleic acid components (see).

Polypeptides: see Peptides.

Polyporenic acid, ***polyporenic acid A:*** a tetracyclic triterpene carboxylic acid, M_r 486.74, m. p. 200 °C, $[\alpha]_D + 74°$. P a. is structurally related to 5α-lanostane (see Lanosterol). It occurs in the fungus *Polyporus* spp. growing on birch trees.

Polyporinic acid: see Benzoquinones.

Polyprenols, ***polyprenyl alcohols:*** acyclic, polyisoprenoid alcohols. P. occur free or esterified with higher fatty acids in microorganisms, plants and animals. The natural source and the number of isoprene units is usually indicated in the names of P., e. g. betulaprenol-8 is formed from 8 isoprene units and is found in *Betula verrucosa.*

P. also differ from each other in the order of double bonds and by the fact that some are partially hydrogenated. Dolichol-24 has the highest M_r of known P.

Polyprenols

Name (orgin)	No. of isoprene units	Conformation of double bonds
Solanesol (Tobacco)	9	all *trans*
Castaprenols (Chestnut)	11–13	3 *trans*, remainder *cis*
Ficaprenols (Ornamental rubber tree)	10–12	3 *trans*, remainder *cis*
Betulaprenols (Silver birch)	6–13	3–4 *trans*, remainder *cis*
Undecaprenol (Bacteria)	11	2 *trans*, 9 *cis*
Dolichols (Mammals, Microorganisms)	14–24	3–4 *trans*, remainder *cis*

Undecaprenol (bactoprenol) from *Salmonella* contains eleven isoprene units, and two *trans* and nine *cis* double bonds. In the form of undecaprenyl phosphate, it acts as a carrier of carbohydrate residues in the biosynthesis of bacterial antigenic polysaccharides; synthesis of Murein (see) also depends on undecaprenyl phosphate. In eukaryotes the Dolichol phosphates (see) function in the transfer of carbohydrate residues in the synthesis of glycoproteins and glycolipids. Probably the long lipid chains of these P. serve to anchor them in membranes, while the phosphate group acts as a carrier by protruding into the cytoplasm. It is not known whether all P. function as carbohydrate carriers. The structural relationship between solanesol and plastoquinone-9 and ubiquinone-9, and the joint occurrence of these compounds suggest a precursor role for P. Biosynthesis of P. proceeds from mevalonic acid and the conformation of all double bonds is predetermined in early precursors.

Polyribosomes, ***polysomes:*** the structural unit of Protein biosynthesis (see), consisting of several to many Ribosomes (see) attached along the length of a strand of mRNA. P. occur free or attached to the endoplasmic reticulum. The individual ribosomes are separated by 60–90 nucleotide units of the mRNA, i. e. the distance between them is 20–30 nm. Each ribosome covers about 35 nucleotide residues. Length of the P. is approximately proportional to the size of the synthesized polypeptide. P. are formed by association of mRNA with ribosomal subunits (produced at the termination of translation) under the influence of initiation factor IF 3, or by direct association of mRNA with newly formed ribosomes.

Polysaccharides: see Carbohydrates.

Polysaccharide sulfate esters, ***polysaccharide sulfates:*** sulfate esters of polysaccharides, synthesized by the action of sulfokinases. There are two possible mechanisms of biosynthesis: 1. Sulfation of the polysaccharide chain by phosphoadenosine phosphosulfate (PAPS); 2. Polymer formation from sulfated sugar nucleotides.

P.s.e. are found especially in algae, where they act as cell wall components. They serve as supportive material in the vegetative forms of the *Thallophyta,* and protect them from desiccation; this is particularly important in terrestial forms and in those that grow in intertidal zones. Many Plant mucilages (see) are P.s.e., e.g. Agar-agar (see) and Carrageenans (see). Other examples are Chondroitin sulfates (see) and Heparin (see).

Polysaccharide sulfates: see Polysaccharide sulfate esters.

Polysomes: see Polyribosomes.

Polytenic chromosomes: see Giant chromosomes.

Polyterpenes, ***polyisoprenes, polyisoprenoids:*** acyclic, unsaturated, terpene hydrocarbons or alcohols, of the general formula $(C_5H_8)_n$, and consisting of a large number of isoprene units.

Classification of polyterpenes

Compound	No. of isoprene units	Double bonds
Polyprenols	6–24	*cis/trans*
Gutta, Balata	about 100	*trans*
Natural caoutchouc	10,000	*cis*

All P. are unbranched molecules, and the double bonds may be *cis* or *trans* (Fig.). For the biosynthesis of P. see Terpenes. With the exception of certain Polyprenols (see), P. are found only in plants, in latex and latex cells.

0.810 nm
Natural caoutchouc (*cis*)
0.472 nm
Gutta Balata (*trans*)

Polyterpenes, showing conformation of double bonds.

Poly U: abb. for polyuridylic acid.

Polyuridylic acid, ***poly U:*** a Homopolymer (see) of uridylic acid, containing an indeterminate number of nucleotide units. In a cell-free, ribosomal protein biosynthesis system, poly U acts as synthetic mRNA, and its translation product is polyphenylalanine. It is often used to determine the synthetic capacity of such systems. Poly U played an important part in early work on deciphering the genetic code.

Polyuronides: macromolecular compounds found in plants, consisting of units of uronic acids in the pyranose form (see Carbohydrates). The main components are D-glucuronic acid, D-galacturonic acid and D-mannuronic acid. P. contain free carboxyl groups and are consequently more strongly hydrated than polysaccharides formed from neutral monosaccharides. Examples are pectins, alginic acid and plant mucilages.

Pomiferin: see Isoflavone.

Pompe's disease: see Glycogen storage disease.

Ponasterone A: 2β,3β, 14α, 20α(*R*),22β(*R*)-pentahydroxy-5β-cholest-7-en- 6-one, a phytoecdysone (see Ecdysone), M_r 464.65, m.p. 260 °C, $[\alpha]_D$ +90° (methanol). P. A was isolated from the fern *Podocarpus nakaii* Hay, and it has molting hormone activity in insects. The closely related ponasterone B and ponasterone C were also isolated from *Podocarpus;* they differ from P. A in that P. B has the opposite configurations at C2 and C3, while P. C possesses an additional hydroxyl group at C24.

OH
HO
24
HO
2
3
OH
HO
H
O

Ponasterone A

Poppy alkaloids: see Papaveraceae alkaloids.

P/O ratio, ***P/O quotient:*** see Respiratory chain.

Porifersterol: Poriferasta-5,22-dien-3β-ol [formerly (24S)-5α-stigmasta-5,22-dien-3β-ol] a marine zoosterol (see Sterols), M_r 412.7, m.p. 156 °C, $[\alpha]_D$ −49° ($CHCl_3$). P. is a characteristic sterol of sponges, and has been isolated from, e.g. *Haliclona variabilis, Cliona celata* and *Spongia lacustris.* It differs from Stigmasterol (see) by an altered configuration at C24.

Porphin: the parent tetrapyrrole of the Porphyrins (see). Porphin is a term from the original Fischer nomenclature. It is synonymous with porphyrin. See also Chlorin.

Porphobilinogen: see Porphyrins.

Porphodimethene: 5,10,15,22-tetrahydroporphyrin.

Porphomethene: 5,15-dihydroporphyrin

Porphyrias: clinical conditions resulting from genetic defects in heme biosynthesis. For the pathway of heme biosynthesis, see Porphyrins. Inborn errors have been described for 7 of the 8 enzymes in this pathway. Although no major genetic defect has been described for the first enzyme of the pathway, 5-aminolevulinate synthase (EC 2.3.1.37), low activity has been reported in a case of congenital sideroblastic anemia [G. R. Buchanan et al. *Blood* **55** (1980) 109–115]. Heme is an essential constituent of many important enzymes and hemoproteins. Absence of heme synthesis is therefore incompatible with life, and homozygotes of inherited autosomal dominant disorders of heme synthesis are not viable, unless there is residual activity of the enzyme concerned. P. are classified as erythropoietic or hepatic, depending on whether the defect is located mainly in the erythroid cells or the liver.

Human porphyrias (with deficient enzyme and clinical findings)

Congenital erythropoietic porphyria, or ***Gunther's disease.***

Uroporphyrinogen III synthase (EC 4.2.1.75). This enzyme converts hydroxymethylbilane into uroporphyrinogen III. Large amounts of uroporphyrin I and coproporphyrin I are deposited in tissues and bone marrow, and excreted in urine and feces. Hemolytic anemia and photosensitivity. Pink coloration of deciduous teeth. Early death is common. Treatment by protection from UV light. splenectomy and blood transfusion. Autosomal recessive. About 100 cases known.

Acute intermittent porphyria, or ***Pyrroloporphyria,*** or ***Swedish hepatic porphyria.***

Porphobilinogen deaminase (EC 4.3.1.8). Enzyme activity about half normal in liver and erythrocytes. Excess formation of 5-aminolevulinate and porphobilinogen, resulting from decreased feedback inhibition of 5-aminolevulinate synthase. High urinary porphobilinogen and 5-aminolevulinate. Peripheral neuritis, paralysis and general demyelination. Neurosis, psychosis and abdominal colic. No photosensitivity. Mortality rate high. Onset after puberty. Treatment with intravenous hematin and high dietary carbohydrate. Many drugs are known to precipitate acute attacks. e.g. barbiturates, methyldopa, sulfonamides, sulfonylurea, pyrazinamide, griseofulvin, glutethimide, meprobamate. Other precipitating factors are ethanol, hormonal changes, starvation, mental stress and infection. Frequency 0.015–0.1 %. Autosomal dominant. Asymptomatic in about 90 % of carriers.

Variegate porphyria, or ***Porphyria variegata,*** or ***Protocoproporphyria,*** or ***South African hepatic porphyria.***

Protoporphyrinogen III oxidase (EC 1.3.3.4). Ferrochelatase activity may also be decreased. Impaired feedback inhibition of 5-aminolevulinate synthase results in excessive porphyrin production. Increased urinary porphobilinogen, 5-aminolevulinate, protoporphyrin and coproporphyrin during acute attacks. Fecal protoporphyrin and coproporphyrin constantly elevated. Fecal porphyrin-peptide conjugates increased. Mild photodermatoses; clinical picture otherwise similar to that of acute intermittent porphyria. Treatment as for latter. Rare, except in white South Africans, where frequency is 0.4 %. Autosomal dominant.

Porphyria cutanea tarda. Severe form: ***Hepatoerythropoietic porphyria.***

Uroporphyrinogen decarboxylase (EC 4.1.1.37). The enzyme is unstable. Increased fecal excretion of uroporphyrin, porphyrins possessing 7 carboxylic acid groups, coproporphyrin and isocoproporphyrin. Photodermatosis. Hepatic siderosis. Risk of primary hepatocellular carcinoma greatly increased. Treatment by phlebotomy. Autosomal dominant. The condition may also be induced by exposure to halogenated aromatic hydrocarbons (e.g. polychlorinated biphenyls, hexachlorobenzene).

Erythropoietic protoporphyria.

Ferrochelatase (EC 4.99.1.1). Activity greatly decreased in bone marrow and liver. High levels of protoporphyrin in erythrocytes, normoblasts and feces. Mild photodermatosis. Erythema and pruritis. Mild edema. Usually cholelithiasis. Sometimes liver failure. Treatment by oral administration of β-carotene, and protection from UV light. Autosomal dominant.

Porphobilinogen synthase deficiency, or ***Acute congenital porphyria.***

Porphobilinogen synthase (5-amino-levulinate dehydrase) (EC 4.2.1.24). Enzyme activity in erythrocytes is about 1 % of normal. High urinary 5-aminolevulinate and coproporphyrin III. Porphyrin excretion in feces is normal. Neuropathic symptoms like those of acute intermittent porphyria. Autosomal recessive (?)

Hereditary coproporphyria.

Coproporphyrinogen:oxygen oxidoreductase (decarboxylating) (Coproporphyrinogen III decarboxylase) (EC 1.3.3.3). High urinary and fecal coproporphyrin III. High porphyrin content in liver. Harderoporphyrin found in feces in one variant. Resembles mild form of acute intermittent porphyria, and is often symptomless. Treatment as for latter. Autosomal dominant.

Porphyrinogen: 5,10,15,20,22,24-hexahydroporphyrin. See Porphyrins.

Porphyrins: cyclic tetrapyrroles which can be considered as derivatives of the parent tetrapyrrole, por-

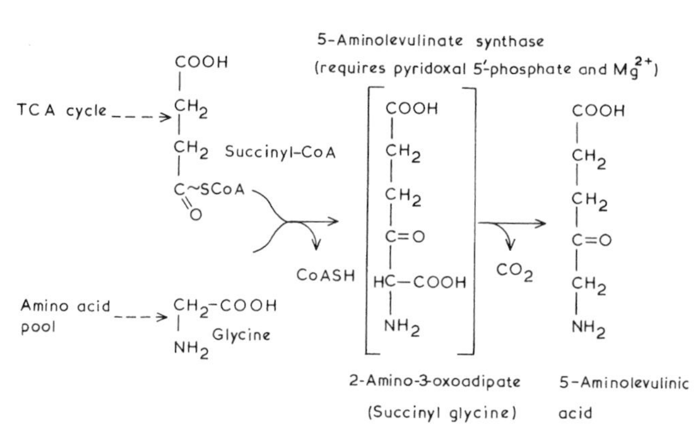

Fig. 1. *The rate-controlling step of porphyrin synthesis, catalysed by 5-aminolevulinate synthase.* This reaction occurs in liver mitochondria, and is allosterically inhibited by heme and hemin. The intermediate shown in brackets ("succinyl glycine") has not been unequivocally demonstrated; the condensation and decarboxylation may occur simultaneously.

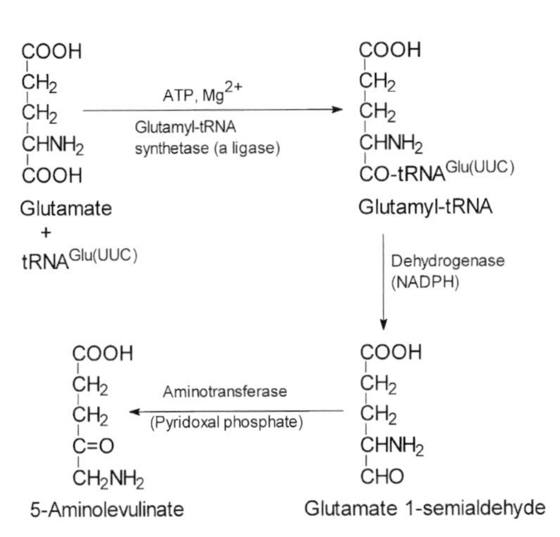

Fig. 2. *Biosynthesis of 5-aminolevulinate in some photosynthetic bacteria, in cyanobacteria, higher plants and algae.* The tRNA involved has also been called δ-aminolevulinic acid-RNA (δ-ALA-RNA), but it appears to be identical to the tRNA used by chloroplasts for translation of chloroplast mRNA. Chloroplast glutamic acid tRNA ligases (glutamyl-tRNA synthetases) have been purified to homogeneity from several sources. The final reaction is an *intramolecular* transamination, which, like conventional *intermolecular* transaminations, requires pyridoxal phosphate.
Although it is widely held that glutamate semialdehyde is the substrate of the final step (with diaminovalerate as a possible enzyme-bound intermediate), some authorities claim that 4,5-dioxovalerate is an intermediate, and that this is converted into 5-aminolevulinate by 4,5-dioxovalerate aminotransferase. The latter enzyme is present in plants and bacteria, but does not always display high activity.

Table. *Porphyrins*

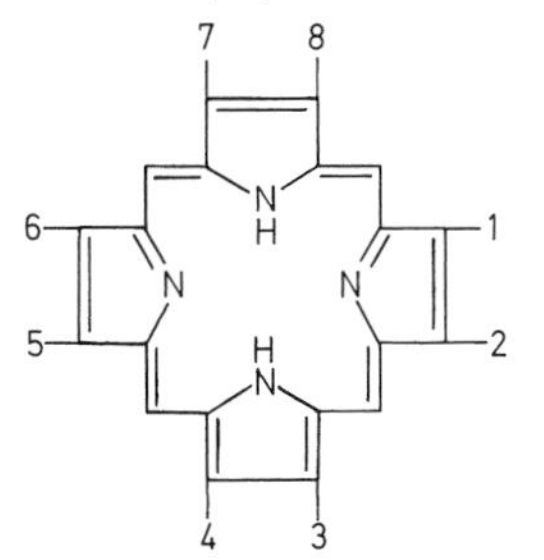

	Substituent No.							
	1	2	3	4	5	6	7	8
Coproporphyrin I	M	P	M	P	M	P	M	P
Coproporphyrin III	M	P	M	P	M	P	P	M
Deuteroporphyrin I	M	H	M	H	M	P	M	P
Deuteroporphyrin III	M	H	M	H	M	P	P	M
Etioporphyrin I	M	E	M	E	M	E	M	E
Etioporphyrin III	M	E	M	E	M	E	E	M
Hematoporphyrin I	M	HE	M	HE	M	P	M	P
Hematoporphyrin III	M	HE	M	HE	M	P	P	M
Mesoporphyrin I	M	E	M	E	M	P	M	P
Mesoporphyrin III	M	E	M	E	M	P	P	M
Protoporphyrin I	M	V	M	V	M	P	M	P
Protoporphyrin III	M	V	M	V	M	P	P	M
Uroporphyrin I	A	P	A	P	A	P	A	P
Uroporphyrin III	A	P	A	P	A	P	P	A
Isocoproporphyrin	M	E	M	P	P	P	P	M

M = methyl, $-CH_3$; P = propionic acid, $-CH_2CH_2-COOH$; H = hydrogen; HE = hydroxyethyl, $-CH_2CH_2OH$; V = vinyl, $-CH{=}CH_2$; E = ethyl, $-CH_2CH_3$; A = acetic acid, $-CH_2-COOH$. The corresponding porphyrinogens (see Fig. 2 porphyrinogen ring system) have identical pattterns of substitution.

phin. The eight β-hydrogens of the parent porphin are completely or partly substituted by side chains, e.g. alkyl, hydroxyalkyl, vinyl, carbonyl or carboxylic acid groups. The different P. are classified on the basis of these side chains, e.g. protoporphyrin, coproporphyrin, etioporphyrin, mesoporphyrin, uroporphyrin. In all naturally occurring P., the two side chain substituents of each pyrrole ring are dissimilar; therefore when two types of side chain substituents are present (e.g. etioporphyrin, coproporphyrin, uroporphyrin) there are 4 possible isomers (designated I to IV). With three different side chains (e.g. protoporphyrin) the number of possible isomers increases to 15. Protoporphyrin occurs naturally only as the IX isomer. Since it can be considered as a derivative of coproporphyrin III, protoporphyrin IX is also called protopor-

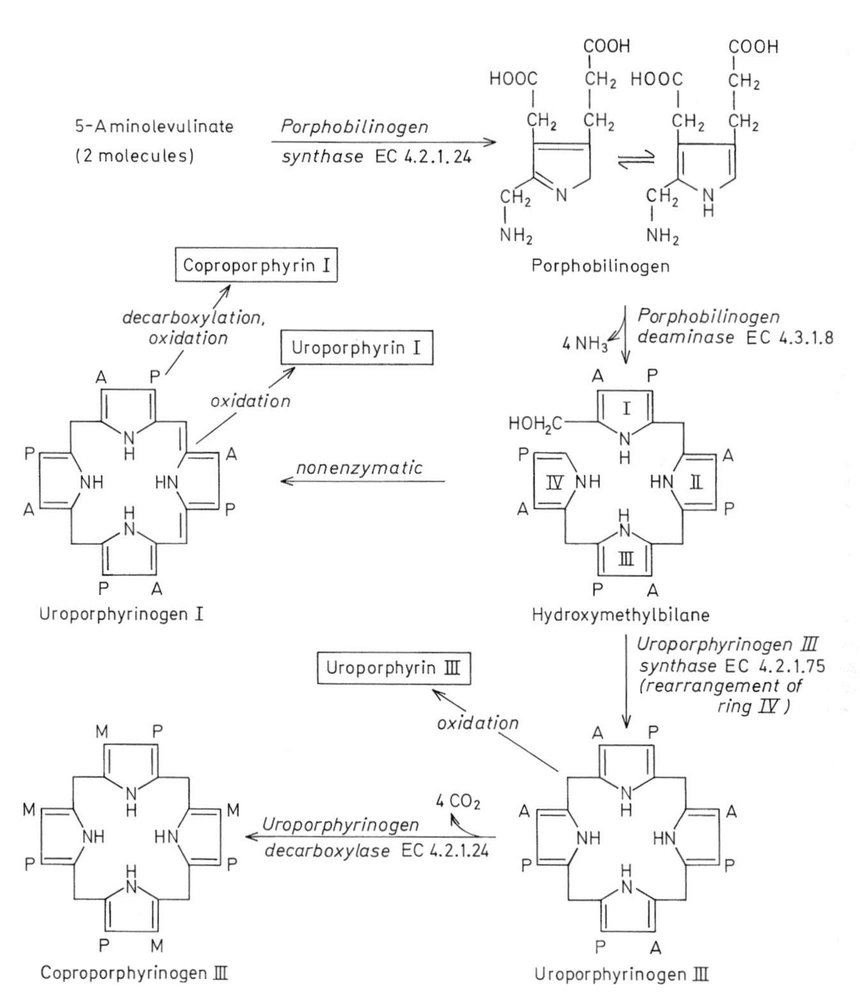

Fig. 3. *Stages of porphyrin synthesis which occur in the liver cytoplasm.* Coproporphyrin I is excreted in constant but small amounts in the urine (40–190 mg/day) and feces (300–1,100 mg/day); its rate of excretion increases during hemolytic disorders. The further conversion of coproporphyrinogen III occurs in the mitochondria.

phyrin III. Of the many possible P. isomers, only types I and III occur naturally. Type I P. have no known useful function and are excreted. Protoporphyrin IX is particularly abundant in nature, occurring as the corresponding heme in hemoglobin, myoglobin and most cytochromes. Many metal ions are complexed by P., forming metalloporphyrins. The chelate complex of protoporphyrin IX with Fe(II) is called protoheme or Heme (see). With Fe(III), the complex is called hemin or hematin. Chlorophylls (see) are magnesium complexes of various P. In cobalamin (see under Vitamins), cobalt is complexed by a corrin ring structure, which is structurally and biosynthetically related to the P. The Bile pigments (see) are derived metabolically from P. Porphyrinogens (which are intermediates in P. biosynthesis) represent P. in which both pyrrolenine nitrogens and all the bridge methene carbons are hydrogenated.

The biological metalloporphyrins share a common pathway of biosynthesis from 5-aminolevulinate to protoporphyrin IX (Figs. 3 & 4).

Two distinct pathways exist for the biosynthesis of 5-aminolevulinate. In some photosynthetic bacteria, yeast and animal mitochondria, 5-aminolevulinate arises by condensation of glycine and succinyl-CoA, a process catalysed by the single enzyme, 5-aminolevulinate synthase (Fig. 1). In several other photosynthetic bacteria and in cyanobacteria, higher green plants and algae, the intact carbon skeleton of glutamate is converted to 5-aminolevulinate by a multienzyme system. The interesting first stage of this conversion (also known as the 'five carbon pathway') involves ligation of the α-carboxyl group of glutamate to a $tRNA^{Glu}$ (Fig. 2).

In the continuation of P. synthesis, two molecules of 5-aminolevulinic acid condense to form the substitu-

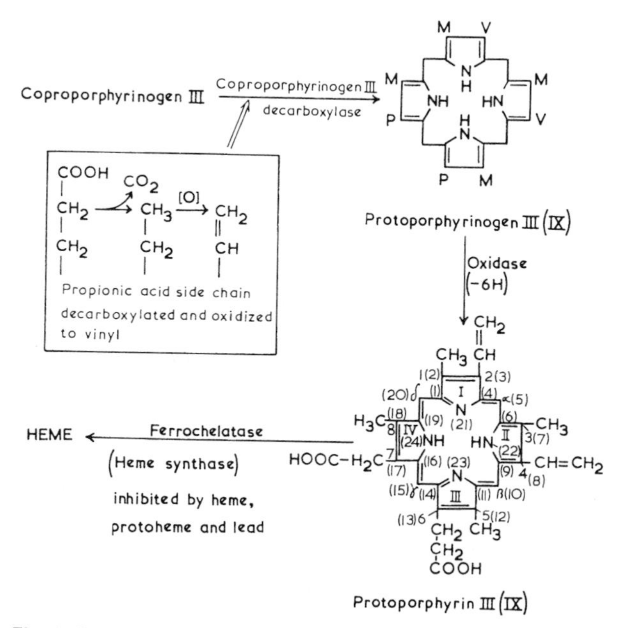

Fig. 4. *Final Stages in the synthesis of protoporphyrin III (IX) in liver mitochondria, and its conversion into heme.* The bracketed numbers and Greek letters correspond to the Fischer system. The unbracketed numbers represent the 1–24 porphyrin numbering system.

ted monopyrrole, porphobilinogen. The enzyme catalysing this reaction, 5-aminolevulinate dehydrase (porphobilinogen synthase, EC 4.2.1.24), is very sensitive to lead; in lead poisoning there is a significant increase in the urinary excretion of 5-aminolevulinic acid, and the activity of 5-aminolevulinate synthase in the blood is decreased. Four molecules of phorphobilinogen react via their aminomethyl side chains, with loss of ammonia, to form the tetrapyrrylmethane intermediate, hydroxymethylbilane. During metabolism of hydroxymethylbilane to uroporphyrinogen III, ring IV undergoes rearrangement (compare positions of ring IV propionate and acetate side chains in hydroxymethylbilane and uroporphyrinogen III, Fig. 3). In contrast, no rearrangement of ring IV occurs in the nonenzymatic conversion of hydroxymethylbilane to uroporphyrinogen I. The resulting uroporphyrinogen III is decarboxylated to coproporphyrinogen III. Further decarboxylation and dehydrogenation lead to the key substance of P. synthesis, protoporphyrinogen IX.

There is a lack of uniformity in the nomenclature of P. All types of nomenclature are current, and appear to depend on individual editorial policy. Naming and numbering systems were proposed by the Commission of Nomenclature of Biological Chemistry in 1960, and in the recommendations of the IUPAC/IUB Commission for Biochemical Nomenclature and the IUPAC Commission on Nomenclature in Organic Chemistry [*Pure and Applied Chemistry* **51** (1979) 2251–2304]. The original system of numbering and trivial names introduced by Fischer is, however, still widely favored. [D. Dolphin (ed.) *The Porphyrins* Vol. **1**. *Structure and Synthesis* Part A, Academic Press, 1978; H. A. Dailey (ed.) *Biosynthesis of Heme and Chlorophylls,* McGraw Hill, 1990]

Positive control: activation of transcription, or the activity of enzymes by activators or metabolites.

Post-transcriptional modification of RNA, ***maturation of RNA, processing of RNA:*** Most eukaryotic RNA is modified after transcription. The genes for transfer RNAs are about 100 nucleotides long, whereas the mature molecules contain only about 80, the remainder having been removed during RNA maturation. Eukaryotic ribosomal RNA (rRNA) consists of two species, with Sedimentation coefficients (see) of 18 and 28S. It is transcribed, however, as a 45S molecule. Nearly half of this is excised and degraded, leaving the two smaller molecules of mature rRNA. Prokaryotic rRNA is less severely pruned, losing only about 10 % of its nucleotides during P.

The mRNA of bacteria is probably not processed, as translation begins before the mRNA has been completely transcribed. In eukaryotes, however, P. of the precursors of mRNA (heteronuclear RNA, or hnRNA) involves at least three operations: capping, addition of a "tail" of polyadenosine, and excision of Introns (see). The "cap" is a 7-methyl-guanosine residue linked via 3 (rather than the usual 1) phosphate groups to the 5′-OH group of the 5′-terminal ribose of the RNA. This 5′-terminal ribose is also methylated at its 2′-OH group. The cap is added by nuclear enzymes very early in transcription. Shortly after transcription is complete, another enzyme locates a signal sequence which includes AAUAAA, cuts the RNA about 20 base pairs beyond the signal and adds a length of about 200 base pairs of polyadenosine. This process does not occur on all mRNA precursors, but mRNA lacking the tail (i.e. histone mRNA) has a very short lifetime in the cytoplasm. Finally, the non-coding sequences (introns) are removed from the RNA. Both the introns and the sequences downstream from the polyadenylation site are degraded. In some cases, only about 10 % of the original hnRNA appears in the cytoplasm as mRNA. [C. Montell et al. *Nature* **305** (1983) 600–605]

At least some RNA maturation is catalysed by Small nuclear riboproteins ("snurps") (see). [*Nature* **307** (1984) 412–413]

Post-translational modification of proteins, ***processing:*** in a general sense, any difference between a functional protein and the linear polypeptide sequence encoded between the the initation and termination codons of its structural gene can be regarded as a P. t. m. Thus, the folding of the polypeptide chain, stabilized by weak, noncovalent interactions, and the association of the subunits of an oligomeric protein both represent a modification that occurs after translation. It is more usual, however, to restrict the meaning of P. t. m. to modifications that involve the making or breaking of covalent bonds. Examples of P. t. m. by proteolysis are:

1. The first amino acid introduced during initiation of Protein biosynthesis (see) is *N*-formylmethionine in prokaryotes and methionine in eukaryotes. Few final proteins contain the formyl or even the methionine residue, e.g. no completed protein isolated from *E. coli* contains an *N*-terminal *N*-formylmethionine, and only 30 % have *N*-terminal methionine.
2. Removal of the signal sequence responsible for the transport of secretory proteins through the membrane of the rough endoplasmic reticulum (see Signal hypothesis).
3. Activation of proenzymes (zymogens) (see Pepsin, Trypsin, Chymotrypsin).
4. Activation of the Complement (see) system.
5. Proteolysis in the cascade system of Blood coagulation (see) (see Prothrombin).
6. Activation of proteo- and peptide-hormones (see Insulin).
7. Proteolysis of larger protein precursors, e.g. proteolysis of procollagen.

Another type of P. t. m. is the attachment of prosthetic groups, e.g. the insertion of heme into hemoproteins, and the attachment of carbohydrates to produce glycoproteins.

A further important form of P. t. m. is the modification of amino acid residues. Only 20 amino acids are encoded during translation, yet well over 100 different amino acid residues are known from various proteins. Examples are: hydroxylation of proline and lysine residues to form the hydroxyproline and hydroxylysine residues in Collagen (see); phosphorylation of serine to phosphoserine residues (see Phosphoproteins); adenylylation of tyrosine residues (see Metabolic control); carboxylation of glutamate to γ-carboxyglutamate residues (see γ-Glutamylcarboxylase); amide groups are sometimes found on *C*-terminal residues, e.g. glycinamide; the ε-amino group of lysine residues may be modified by methylation, acetylation or phosphorylation, and by the formation of peptide linkages with, e.g. biotin or lipoic acid; lysine residues are the precursors of desmosine (see Elastin). Clearly, the difference between the modification of an amino acid residue and the attachment of a prosthetic group is sometimes purely semantic.

Most amino acid modifications occur after release of the polypeptide from the ribosome, and a large number of processing enzymes (in particular those responsible for glycosylation, disulfide bridge formation and iodination) are found in the endoplasmic reticulum and Golgi apparatus. Limited proteolysis occurs in the extracellular space and in secretory granules. Cross linking reactions (e.g. cross linking of lysine and hydroxylysine residues in collagen; formation of isopeptide cross-links between carboxyl groups of glutamate or aspartate and the ε-amino group of lysine; formation of ether cross-links involving iodotyrosine) are extracellular.

Modification sometimes occurs at an earlier stage. Formylmethionyl-tRNA is formed by the formylation of methionyl-tRNA before it is incorporated as the *N*-terminal amino acid in prokaryotic translation. It appears that Pyroglutamic acid (see) and possibly certain acetylated amino acids are also formed at the aminoacyl-tRNA stage. Some modifications occur in the nascent polypeptide chain, i.e. while the polypeptide is still attached to the ribosome. Hydroxylation of lysine and proline occur in the nascent chain of the collagen precursor protein. *N*-terminal formylmethionine and methionine are removed when the nascent chain is about 50 amino acid residues long, as demonstrated in the synthesis of bacteriophage f2 coat protein in *E. coli,* and in the synthesis of rabbit hemoglobin. Some disulfide bridge formation, some *N*-terminal acylation and some glycosylation also occur in the nascent polypeptide chain. Many different kinds of P. t. m. are exemplified in the synthesis of Collagen (see).

PP-factor: pellagra preventative factor; biologically equivalent to nicotinic acid and nicotinamide. See Vitamins.

PP_i: inorganic pyrophosphate; the anion of pyrophosphoric acid, $P_2O_7^{4-}$.

Ⓟⓟ: inorganic pyrophosphate (see PP_i)

PRA: acronym of phosphoribosylamine.

Pratensein: 5,7,3′-trihydroxy-4′-methoxyisoflavone. See Isoflavone.

PRB 8: see Orthonil.

Precalciferol: see Vitamins (vitamin D).

Precipitation: formation of an insoluble, inactive antigen-antibody complex from a soluble antigen and a bivalent specific antibody (the precipitin). Monovalent or incomplete antibodies do not precipitate the soluble antigens; the resulting complexes cannot form cross-linkages and therefore remain soluble. In the quantitative determination of the precipitation reaction (due to Heidelberger), increasing quantities of antigen are added to a constant amount of antibody. At the Equivalence point (see) the supernatant above the precipitate contains neither antigen nor antibody. In most studies, only the antigen titer is determined, i.e. the dilution of antigen at which P. is still just visible. For this purpose the Ouchterlony technique is used, in which both reaction partners are allowed to diffuse into one another in an agar gel, forming precipitation lines in their area of contact; this is an economical method requiring only very small quantities of antigen and antibody.

Precipitation curve, ***Heidelberger curve:*** a plot of the quantity of precipitate formed during titration of an antibody with an antigen, or vice versa. The antibody must be at least bivalent. The P. c. attains a max-

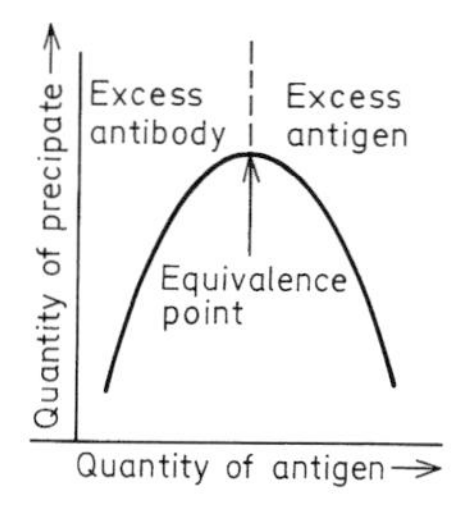

Precipitation curve. An increasing quantity of antigen is added to a constant amount of antibody. The resulting precipitate is removed by centrifugation, and antigen and antibody are determined in the supernatant.

imum at the Equivalence point (see) then decreases as the concentration of one component exceeds its optimum (Fig.).

Prednisolone: 1,2-dehydrocortisol, 11β,17α,21-trihydroxy pregna-1,4-diene-3,20-dione, a synthetic steroid prepared by chemical or microbiological dehydrogenation of cortisol. It has powerful antiinflammatory and antiallergic activity, very similar to that of prednisone (1,2-dehydrocortisone, the corresponding dehydro derivative of cortisone); both compounds are used in the treatment of disorders such as asthma, arthritis and eczema.

Prednisolone

Prednisone: see Prednisolone.

Pregnancy hormones: see Progesterone.

Pregnane: see Steroids.

5β-Pregnane-3α-diol, ***pregnanediol:*** a biological degradation product of progesterone. M_r 320.52, m.p. 224 °C, $[\alpha]_D$ + 27 °. It occurs as the glucuronide, particularly in pregnancy urine.

5β-Pregnane-3α,20α-diol-glucuronide

Pregnant mare serum gonadotropin, ***PMS:*** a hormone produced by the uterine endometrium. M_r 28,000. PMS is poorly excreted by the kidneys, and it accumulates in the blood. Its action is similar to that of Follicle stimulating hormone (see).

Pregnenolone: 3β-hydroxy-pregn-5-en-20-one, an intermediate in the biosynthesis of progesterone from cholesterol, and in the biosynthesis of androgens in the adrenal cortex. See Steroids, Androgens.

Prephenic acid: see L-Phenylalanine.

Pre-proprotein: a precursor in the biosynthesis of a secretory protein (see Signal hypothesis). The P-p. possesses an *N*-terminal signal peptide sequence in addition to the proprotein sequence, e.g. pre-proinsulin, pre-protrypsin, pre-proparathormone, pre-promellitin, etc.

Preprotein: the biosynthetic precursor of a secretory protein containing a very short-lived signal peptide sequence (see Signal hypothesis), e.g. preovomucoid and prelysozyme.

Pretyrosine, ***arogenic acid:*** see L-Phenylalanine.

PRH: acronym of Prolactin Releasing Hormone (see Releasing hormones).

Pribnow box: a sequence of seven nucleotide pairs, which is the same or very similar in all promoters. It is located five to seven nucleotide pairs from the initiation point of RNA transcription. The structure of the P.b is:

5′ T A T Pu A T G 3′
3′ A T A Py T A C 5′

where Pu is a purine and Py a pyrimidine.

Primary metabolism: those metabolic processes (see Metabolism) that are basically similar in all living cells, and are necessary for maintenance and survival. P.m. includes the fundamental processes of growth (synthesis of biopolymers and their constituents, synthesis of macromolecular superstructures of cells and organelles), energy production and transformation, and the Turnover (see) of body and cell constituents.

Primary structure: see Proteins.

Primer: 1. a small polymer required as a starter for the synthesis of a larger biopolymer, e.g. in nucleic acid synthesis, an oligonucleotide serves as a primer, and it is extended by the enzyme-catalysed addition of further nucleotide units from nucleoside triphosphates. The P. molecule is therefore incorporated into the final product. A P. or starter must not be confused with a Template (see) which is not incorporated into the product. 2. a Pheromone (see) that causes a long term physiological change.

Primetin: see Flavones (Table).

Primin: see Benzoquinones.

Primobolane: 1-methyl-17β-acetoxy-5α-androst-1-en-3-one, a synthetic Anabolic steroid (see). It is a less potent (one tenth) androgen, but a more potent (five times) anabolic agent than testosterone propionate (see Testosterone).

Prion: an agent generally responsible for a group of diseases known as spongiform encephalopathies (SE). Transmissible spongiform encephalopathies (TSE) are infectious prion diseases, rather than those that are inherited. Under appropriate conditions, familial (inherited) prion diseases can also be transmitted. All known prion diseases cause neurodegeneration and ultimately death. At autopsy the brain displays spongiform degeneration, often with highly ordered amyloid plaques that stain with Congo red and react with PrP (prion protein; see below) antibodies. The name prion is derived from *pro*teinaceous *in*fectious agent,

Known human prion diseases are Creutzfeldt-Jacob disease (CJD), Gerstmann-Sträussler-Schienker syndrome (GSS) and fatal familial insomnia (FFI). GSS and FFI are inherited, whereas CJD may be inherited, sporadic or infectious. Another infectious form of CJD (iatrogenic CJD) arises from inadequately sterilized surgical instruments, dura mater grafts, and from human growth hormone isolated from cadavers. Kuru is a classical example of an infectious human prion disease; transmitted by the ritual cannibalism of human brains, this disease was formerly common in the Foré tribe of New Guinea.

Scrapie, a prion disease of sheep and goats, has been known for at least 200 years. Consequently, "Scrapie-like disease" has become a synonym for prion disease or spongiform encephalopathy, and the superscript "Sc" in PrP^{Sc} (see below) is short for scrapie. Other known prion diseases are bovine spongiform encephalopathy (BSE, "mad cow disease"), transmissible mink encephalopathy, chronic wasting disease of the mule deer and elk. TSE have also been observed in zoo animals and pets, possibly due to feeding of infected meat. Thus cases have been reported in cats (70), cheetahs (3), pumas (2) an ocelot, as well as the following members of the *Bovidae:* Nyala, Eland, Kudu, Gemsbock, Oryx and Ankole cattle. Prion diseases (in particular scrapie) have been transmitted experimentally to laboratory mice and hamsters, and mice are used for the typing of TSE. One primate, a marmoset, has been infected experimentally. Pigs cannot be infected orally, and infection is only achieved with large doses of prion-containing material introduced intracranially.

The origin of SE is not always clear. Thus, it is widely thought that bovine spongiform encephalopathy (BSE, "mad cow disease") in British cattle arose because inadequately treated, scrapie-infected sheep products were incorporated into cattle feed. However, research by the US Department of Agriculture shows that calves infected experimentally with sheep scrapie develop a SE that is different from BSE; the animals do not display the behavioral abnormalities of BSE and the postmortem brain pathology is different. There still remains the possibility that this experimental "cow scrapie" and BSE are due to different strains of scrapie. The other possibility is that BSE is a new disease that first arose in a few British cows, and was then repeatedly recycled in the cattle food chain by the use of rendered cattle carcasses in cattle feed, possibly building up for years at a non-symptomatic low level, then emerging as an epidemic. The current widespread and intense interest in prion diseases is stimulated by this pressing question of the true nature of BSE and its transmissibility.

At present much attention is focused on cases of CJD in Britain that are phenotypically different from previously described cases, e.g. the vacuolation and pattern of amyloid deposition in the brain are different, and all cases are younger than 42, whereas the usual age of onset is greater than 65. The possibility is being explored that this may be a transmissible form of the disease contracted from eating infected beef products. This investigation depends on laborious strain testing with the aid of laboratory mice. Thus, different strains of the disease develop at different rates in defined strains of experimental mice; each test requires several months.

Prion diseases seem to arise from an aberrant isoform of a normal cellular prion protein, designated PrP^C. The aberrant isoform, designated PrP^{Sc}, differs from the normal protein in its conformation, and the weight of evidence now indicates that a transmissible prion consists exclusively of PrP^{Sc}. No nucleic acid is present, as shown by the fact that infectivity is unaffected by all procedures that specifically modify nucleic acids; moreover, no candidate TSE-specific polynucleotide has ever been found. PrP^C and PrP^{Sc} have been purified from the brains of healthy and scrapie-infected Syrian hamsters, respectively. PrP^{Sc} is also called the SAF isoform from its ability to aggregate into Scrapie-Associated Fibrils.

Prion diseases may be genetic or infectious. Nineteen mutations of the PrP gene are associated with inherited human prion disease, all distinguishable by their phenotypes, e.g. patients with Val-129 on the mutant allele develop familial CJD, whereas those with Met-129 develop FFI. The ability of prion diseases to cross species barriers depends, at least in part, on the structural similarity between the respective PrP amino acid sequences.

The amino acid sequences of different mammalian PrP^C are very similar, and the conformation of the protein is virtually the same in all mammalian species, including humans, mouse, sheep and cow. Mature PrP^C is a glycoprotein (209 amino acid residues; M_r 33,000–35,000). Most PrP molecules carry bi-, tri- and tetra-antennary neutral and *N*-linked sialylated oligosaccharides at asparagine 181 and asparagine 197. A disulfide bond links the only two cysteines (Cys-179, Cys-214). Both PrP^C and PrP^{Sc} are encoded by the same exon of the chromosomal gene, and translated as a 254-residue protein. An *N*-terminal 22-residue signal peptide is then removed posttranslationally, and a 23-residue *C*-terminal peptide is also lost upon addition of a GPI anchor. The above numbering of amino acid positions refers to the codons of the gene, and therefore to the immature translation product before posttranslational removal of the *N*- and *C*-terminal sequences.

Although PrP^C possesses a highly conserved hydrophobic domain, it is not an integral membrane protein, but it is anchored to cell surface plasma membranes by a glycosylphospatidylinositol (GPI) (see Membrane lipids) attached to its *C*-terminal amino acid. This GPI is unusual in that it is sialylated. PrP^C (the normal isoform of the protein) is completely degraded by proteases.

Under the same conditions of proteolysis, PrP^{Sc} (the abnormal, disease-causing isoform of the protein) is only partly degraded, leaving a protease-resistant core (M_r 27,000–30,000; the first 67 amino acids of the mature PrP^C protein are removed), which is designated PrP27–30, and which is also infective.

Despite the sequence identity of PrP^C and PrP^{Sc} from the same species, they display quite different physical properties. Fourier transform infrared (FTIR) and circular dichroism (CD) spectroscopy reveal that PrP^C is essentially α-helical and devoid of β-pleated sheet structure, whereas the β-pleated sheet content of PrP^{Sc} is ~40 %, while PrP27–30 contains more than 50 % β-pleated sheet. Procedures that destroy the infectivity of PrP27–30 and PrP^{Sc} (e. g. purification by SDS-polyacrylamide gel electrophoresis or treatment with alkali) also substantially reduce the β-pleated sheet content. It therefore appears that the difference between the normal and disease-causing isoforms of PrP is entirely conformational and related to the respective contents of α-helical structure and β-pleated sheet structure.

Most of the point mutations of PrP^C that result in SE represent amino acid replacements predicted to increase β-pleated sheet formation. With model structures it can be predicted that a Pro → Leu mutation at codon 102 or at codon 105 will greatly increase the tendency of PrP^C to lose much of its α-helical structure and convert to PrP^{Sc} with a high content of β-pleated sheet; accordingly, mutation of codon 102 causes spontaneous TSE in transgenic mice. It is notable that most of the known point mutations resulting in spontaneous SE lie within the sequence of PrP27–30 and are located in or adjacent to the four putative helices. However, an octa repeat near to the *N*-terminus is not included in the sequence of PrP27–30, and insertion of multiple additional copies of this octa repeat also cause inherited SE.

Less is known about the significance of quaternary structure for infectivity. Although ordered fibrils or rods of PrP^{Sc} are a common histological feature in the postmortem brains SE-infected animals, and although purified preparations of these rods are infective, aggregation of PrP^{Sc} into rods is not actually necessary for infectivity; in fact scrapie infectivity is increased by dispersion of the rods in detergent-protein-lipid complexes. On the other hand, inactivation of prions by ionizing radiation suggests a target size of M_r 55,000, implying that the infective unit is a dimer.

The progression of infectious prion diseases appears to involve interaction between PrP^C and PrP^{Sc}, which induces a conformational change of the α-helix-rich PrP^C to the β-pleated sheet-rich conformer of PrP^{Sc}. Similar interactions have been demonstrated in vitro. Differences in the amino acid sequence of PrP appear to present a partial barrier to interspecies transmission of prion diseases, partly due to the decreased efficiency of interaction between the endogenous PrP^C and the exogenous (infective) PrP^{Sc}. For example, the PrP sequences of mouse and human differ in 28 residues, and only about 10 % of mice develop TSE when inoculated intracerebrally with human CJD prions, and then only after incubation for longer than a year. Studies on the susceptibility of mice expressing the human PrP gene suggest the existence of a host factor ("protein X", possibly a chaperone) that promotes conversion of PrP^C into PrP^{Sc}. Comparison of PrP sequences shows that the *N*-terminal region of PrP27–30, including the putative helix region H1, plays a crucial role in determining species barriers.

Despite its apparently strategic location as a cell surface molecule, and the fact that it is widely expressed in embryonic and adult tissues, PrP^C appears to have no essential function. Alternatively, its function may be shared by other molecules that are able to replace it. After deletion of the PrP^C gene by homologous recombination, embryonic murine stem cells still develop into fertile chimaeric mice; their PrP^C null mutant offspring are resistant to infection by scrapie prions, but are otherwise normal. [Richard Lacey, *Mad Cow Disease: History of BSE in Britain,* St. Helier, Jersey: Cypsela Publications. 1994; *Transmissible spongiform encephalopathies: a summary of present knowledge and research,* Spongiform Encephalopathy Advisory Committee, London HMSO 1995; *Bovine spongiform encephalopathy in Great Britain: a progress report,* Great Britain Ministry of Agriculture, Fisheries and Food, 1995; H. M. Schätzl et al. *J. Mol. Biol.* **245** (1995) 362–374; M. A. Baldwin et al. "Prion Protein Isoforms, a Convergence of Biological and Structural Investigations" *J. Biol. Chem.* **270** (1995) 19197–19200]

Pristanic acid: see Inborn errors of metabolism (Refsum's disease).

Pro: abb. for L-proline.

Proaccelerin: see Blood coagulation (Table).

Proanthocyanidins: colorless substances from plants, which are converted into anthocyanidins by heating with acid. This conversion is chemical and does not imply a biogenetic relationship. P. are subdivided into: 1. monomeric flavan-3,4-diols, also known as Leucoanthocyanidins (see), and 2. dimers and higher oligomers of flavan-3-ols, known as condensed P. (also classified as condensed vegetable tannins; see Tannins). The four major, naturally occurring dimeric flavan-3-ols are all configurational isomers (Fig. 1), representing dimers of catechin and/or epicatechin (see Catechins). These compounds are also known as procyanidins, because cyanidin is the product of their treatment with acid. Early proposals for procyanidin biosynthesis implicated a symmetrical flav-3-en-3-ol intermediate [E. Haslam et al. *J. Chem. Soc.* (Perkin 1) (1977) 1637–1643]. According to later work, however, the biosynthesis is catalysed by multienzyme complexes, a flav-3-en-3-ol is not involved, and the 2,3-*cis* stereochemistry of (–)-epicatechin arises by the action of an epimerase on the precursor (+)-dihydroquercetin (Fig. 4) [H. A. Stafford *Phytochemistry* **22** (1983) 2643–2646].

Less widely distributed are the prodelphinidins, which yield delphinidin with acid, e. g. a prodelphinidin from *Ribes sanguineum* (Fig. 2). A profisetinidin has been reported from the heartwood of *Acacia baileyana* (Fig. 3).

A new, convenient system of nomenclature (used in Figs. 1, 2 & 3) has been proposed for P. [L. J. Porter et al. *J. Chem. Soc. (Perkin I)* (1982) 1209–1216], in which the names of common catechins with the most frequently encountered 2*R* configuration (see Catechins) are retained for the individual units of the dimer. Where 2*S* configuration occurs, it is denoted by the prefix enantio. The configuration at C4 is denoted by α or β. [M. Hör et al. *Phytochemistry* **42** (1996) 109–119]

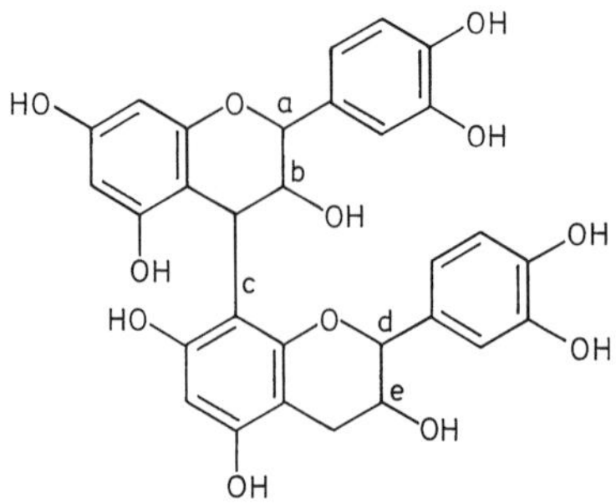

Name and source	Bond configuration				
	a	b	c	d	e
Epicatechin-(4β → 8)-epicatechin (Apple, hawthorn, cocoa bean, *Cotoneaster*, quince, cherry, horse chestnut)	R ---	R ---	β ◄	R ---	R ---
Epicatechin-(4β → 8)-catechin (Grape, cranberry, sorghum)	R ---	R ---	β ◄	R ---	S ◄
Catechin-(4α → 8)-epicatechin (Raspberry, blackberry)	R ---	S ◄	α ---	R ---	R ---
Catechin-(4α → 8)-catechin (Willow and poplar catkins, strawberry, hops, rose hips)	R ---	S ◄	α ---	R ---	S ◄

Fig. 1. *Configurations and sources of naturally occurring procyanidins*

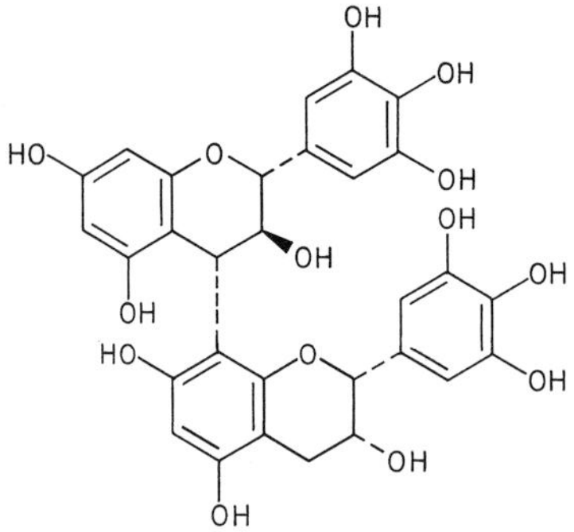

Fig. 2. *Gallocatechin-(4α → 8)-epigallocatechin* (a prodelphinidin from *Ribes sanguineum)*

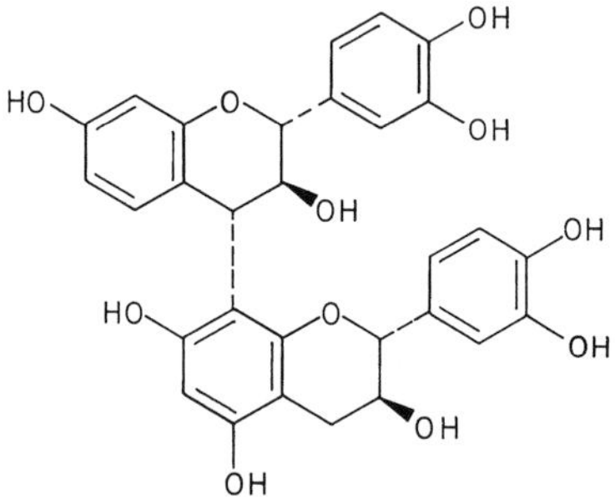

Fig. 3. *Fisetinidol-(4α → 8)-catechin* (A profisetinidin from *Acacia baileyana).* [L. Y. Foo *Phytochemistry 23* (1984) 2915–2918]

Proazulenes, ***azulenogens, hydroazulenes:*** a group of natural cyclic sesquiterpenes, which can be terminally dehydrogenated or dehydrated to Azulenes (see). P. are chiefly compounds of the guaiane type, e.g. guaiol.

Proazulene type: see Sesquiterpenes (Fig.).

Probe: a molecule used (i) to search for a particular gene, gene product or protein, or (ii) to monitor a particular cellular environment. A P. of class (i) is a molecule that (a) binds specifically to a particular nucleotide sequence in the target DNA or RNA or to a characteristic structural feature of the target protein and (b) can be produced in a radioactively or chemically labeled form. Thus when binding of the P. and its target takes place the latter is identified. Such P. may be mRNAs (which bind to the DNA template strand of their gene), cDNAs (see; which bind to the DNA sense strand of their gene), stretches of DNA (which bind to DNA or RNA with complementary nucleotide sequences) or antibodies (Immunoglobu-

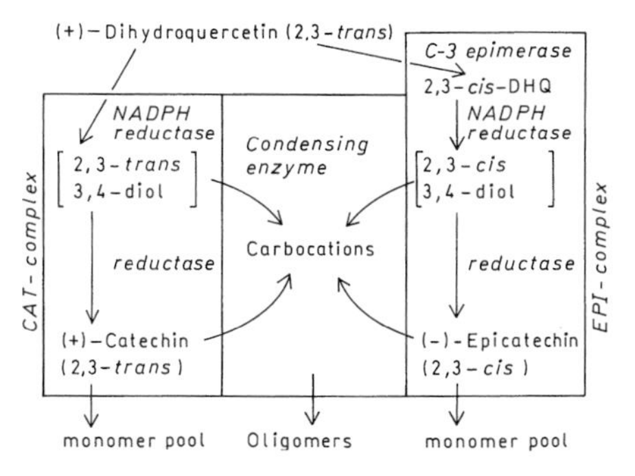

Fig. 4. *Suggested multienzyme complexes in procyanidin biosynthesis.* DHQ = dihydroquercetin. After H. A. Stafford *Phytochemistry 22* (1983) 2643–2646.

lins, see), particularly the monoclonal variety (which bind to their protein antigens). P. can be designed to detect the gene for a particular protein provided that at least part of the amino acid sequence of that protein is known; this is possible because (a) the codons for each protein amino acid are known, and (b) technology is available for the synthesis of DNA oligonucleotides of known sequence. Such a P. is usually a short stretch of DNA (20–36 nucleotides long) that has the same nucleotide sequence, save that T replaces U, as that of the mRNA that codes for a 7–12 amino acid sequence present in the protein. The short segment of the protein chosen for the generation of the P. is one where there is least codon degeneracy amongst the amino acids present, thereby minimizing the number of possible nucleotide sequences. Even so, several P., each of slightly different nucleotide sequence, may have to be synthesized to insure that one will bind to the relevant sequence of the template strand of the gene. Fortunately this problem is eased by the finding that a P. need not have exact sequence complementarity for this binding.

P. of class (ii) are typified by the electron spin resonance (esr) and fluorescence polarisation probes that are inserted into biological membranes to get a measure of their fluidity. The former are often sterol-like or phospholipid-like molecules with an attached paramagnetic nitroxide-containing moiety which give a broad esr signal when they cannot rotate freely about their long axis within the lipid bilayer (low membrane fluidity) and a sharp esr signal when they can rotate freely (high membrane fluidity). The latter are often long thin molecules like diphenylhexatriene (DPH) which, when irradiated with plane-polarized light of the appropriate wavelength, emit light with the same plane of polarization provided that they have not rotated about their long axis within the lipid bilayer during the life-time of their excited state (i. e. ~10^{-9}sec), as is the case at low membrane fluidity. A change in the polarization of the emitted light thus indicates an increase in membrane fluidity.

Process control: a term used in industrial biochemistry, particularly with reference to control of production of microbial fermentation products, e. g. by control of nutrient supply, air supply, etc., and choice of continuous or batch culture. See Fermentation techniques.

Processing: modification of protein molecules after translation (see Post-translational modification of proteins), or modification of RNA after transcription (see Post-transcriptional modification of RNA).

Prochirality: A molecule (or atom) is prochiral if it becomes chiral (asymmetrical or dissymmetrical) by the replacement of one point ligand by a new point ligand. A prochiral carbon atom possesses two identical (a, a) and two different (b, c) substituents. It has a one-fold alternating axis, and a mirror plane of symmetry, i. e. the plane through C, b and c cuts the molecule into mirror image halves. A classical example is the prochirality of citric acid, which is discussed in greater detail under Tricarboxylic acid cycle (see). NADH and NADPH are also prochiral, even though these molecules have perfect bilateral symmetry. Enzymes distinguish between the two sides of NAD(H) and NADP(H), so that hydrogen is added to or removed from the 4*R* or 4*S* position of C4 of the nicotinamide ring.

The *R* and *S* system [R. S. Cahn, et al. *Angew. Chem. Int. Ed. Engl.* **5** (1966) 385–415] is used to specify configuration at prochiral centers. Each substituent is given a certain priority (based on atomic number). The model of the molecule is viewed from the side furthest from the lowest priority group. If the sequence of groups in decreasing priority is clockwise, the configuration is *R* (rectus: right handed); if the sequence is anticlockwise the configuration is *S* (sinister: left handed). Naming of pairs of hydrogen atoms at prochiral centers is particularly important. Clearly any atom or group replacing a hydrogen will have a higher atomic number than the hydrogen it replaces. If the replacement results in *S*-chirality, then the replaced hydrogen is designated H_S; similarly, the other hydrogen must be H_R. [T. W. Goodwin "Prochirality in Biochemistry" in *Essays in Biochemistry* **9** (1973) 103–160; R. Bentley *Molecular Asymmetry in Biology* 2 Vols Academic Press (1970)].

Proconvertin: see Blood coagulation

Prodigiosin: a red pigment and secondary metabolite of the bacterium, *Serratia marcescens.* In the biosynthesis of P., L-proline enters intact, and contributes a greater number of carbon atoms than any other amino acid. All the carbon atoms of ring A and the associated carbon atom in ring B are derived directly from L-proline. The biosynthetic origin of the remainder of the molecule is less well understood. Carbon atoms 3 and 2 of alanine contribute the methyl group and its associated ring carbon atom in ring C.

Prodigiosin

Progesterone, ***corpus luteum hormone, luteohormone:*** pregn-4-ene-3,20-dione, Δ^4-pregnane-3,20-dione, M_r 314.47, m. p. 128 °C (ethanol/water), 121 °C (pentane/hexane), $[\alpha]_D$ +192 ° (ethanol). P. is structurally related to the parent hydrocarbon, pregnane (see Steroids). It is the natural progestin, and an antagonist of the Estrogens (see). It promotes prolifera-

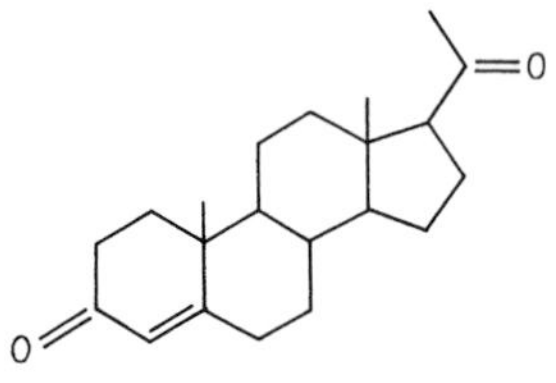

Progesterone

tion of the uterine mucosa. Having thus prepared the uterus, P. promotes implantation and further development of the fertilized ovum in the uterine mucosa (secretion phase). During pregnancy, it prevents further maturation of follicles, and stimulates development of the lactatory function of the mammary gland. In mammals, including humans, P. is produced by the corpus luteum; during pregnancy the production of P. by the corpus luteum is markedly increased, and some is also produced by the placenta. Absence of, or failure to produce P. causes abortion.

P. can be determined biologically by the Clauberg test: young female rabbits (600–800 g) receive estrogens subcutaneously daily for 8 days, which results in proliferative changes in the uterine mucosa. The P. under test is then injected subcutaneously in oil. One rabbit unit is the smallest quantity of P. required to change the endometrium to the secretory phase (detected histologically). An immunoassay is available, and in some laboratories, P. and other steroids are determined by combined gas chromatography and mass spectroscopy.

P. is degraded in the liver and kidneys; the chief products are hydroxylated pregnanes, e.g. 5β-pregnane-3α, 20α-diol.

P. is biosynthesized from cholesterol via pregnenolone. By conversion to 17α-hydroxyprogesterone, P. acts as a precursor of androgens and adrenocortical hormones.

P. was first isolated in 1934 from corpus luteum by Butenandt et al. It was later shown to present in the plant *Holarrhena floribunda.* It is used in veterinary medicine to correct irregularities of estrus and to control habitual abortion.

Programmed cell death: see Apoptosis.

Progress curve: in enzyme kinetics, a plot of the concentration of a reactant, or several reactants (e.g. substrates, products, enzyme-substrate complexes) or an enzymatic reaction as a function of the time for which the reaction has progressed. A P.c. may be derived theoretically, preferably with the aid of a computer (e.g. simulated from the Michaelis-Menten scheme, making certain assumptions about the values of rate constants), or obtained experimentally. The experimental determination may be discontinuous (analysis of discrete samples at different time intervals) or continuous (continuous monitoring of parameters in a single sample, e.g. by spectrophotometric or polarographic methods). Detailed mathematical treatment and design of computer programs for theoretical P.c. have been reported [W.W. Cleland *Biochim. Biophys. Acta.* **67** (1963) 104, 173, 188; W.W. Cleland *Nature* **198** (1963) 463–465]. None of these publications contains a graphical representation of theoretical P.c. The originals of such plots first appeared in Cleland's lecture notes, and have been reproduced, e.g. in all editions of *Biological Chemistry* by Mahler and Cordes, published by Harper and Row.

Prokaryote, ***prokaryotic organism:*** see Cell.

Prolactin, ***lactotropin, lactogenic hormone, mammotropin, luteomammotropic hormone, luteotropic hormone, LTH, luteotropin:*** a gonadotropin. Phylogenetically, P. is one of the oldest adenohypophysial hormones. It acts primarily on the mammary gland by promoting lactation in the postpartal phase, and in rodents it also acts on the ovary. Its activity in males is not clear. Bovine P. is a single chain polypeptide (198 amino acid residues, M_r 22,500, three disulfide bridges), of known primary structure.

P. synthesis in the α-cells (eosinophils) of the anterior pituitary (adenohypophysis) is under the control of the hypothalamus hormones PRH and PIH (see Releasing hormones). It is produced in increasing amounts during pregnancy and during suckling; it can be detected in the blood (ng/ml range) by radioimmunological assay. P. promotes metabolism and growth, and it influences osmoregulation, pigment metabolism and parental behavior, and it suppresses reestablishment of the menstrual cycle post partum. P. acts on its target cells via the adenylate cyclase system.

Prolactin release inhibiting hormone: see Releasing hormones.

Prolactin releasing hormone: see Releasing hormones.

Prolamines: a group of simple (unconjugated) proteins, soluble in 50–90 % ethanol. They occur in cereals, and contain up to 15 % proline and 30–45 % glutamic acid, but they have only low contents of essential amino acids. The chief representatives are gliadin (wheat and rye), zein (maize; contains no tryptophan or lysine) and hordein (barley; contains no lysine). Oats and rice do not contain P.

L-Proline, ***Pro:*** pyrrolidine-2-carboxylic acid, a proteogenic amino acid. Pro is very soluble in water, but is also soluble in ethanol, so it can be separated from other amino acids by ethanol extraction. Being an imino acid, it forms a yellow color with ninhydrin, rather than the purple color characteristic of α-amino acids.

Together with hydroxyproline (Hyp), Pro is an essential component of collagen, gliadin and zein. Collagen yields 15 % Pro on hydrolysis.

Pro is glucogenic. It does not participate in α-helix formation, and therefore has special importance in protein tertiary structure (see Proteins). Pro is biosynthesized chiefly from L-glutamate via glutamic-γ-semialdehyde. Some may also be formed from exogenous ornithine via pyrroline carboxylic acid. It is a nonessential amino acid in mammals, but is essential for the growth of chickens. Azetidine-2-carboxylic acid is a Pro antagonist.

Proline derivatives: trans-4-hydroxy-L-proline (Hyp) is an important component of animal supportive and connective tissues. Free all-*cis*-4-hydroxy-L-proline occurs in *Santalum album* and other plants. Small quantities of 3-hydroxy-L-proline are also present in collagen. 4-Hydroxy-L-proline is formed mainly by hydroxylation of ribosome-bound peptidylprolyl-RNA, a reaction requiring ascorbic acid and catalysed by proline hydroxylase (see Oxygenases). Free 4-hydroxy-L-proline is formed by cyclization of γ-hy-

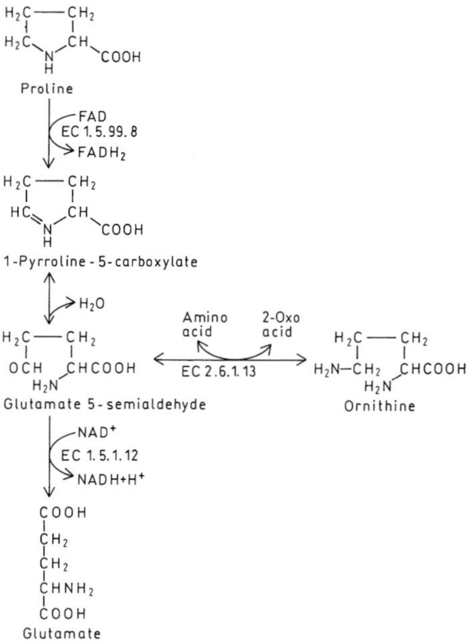

Proline
FAD / EC 1.5.99.8 / $FADH_2$
1-Pyrroline-5-carboxylate
H_2O
Glutamate 5-semialdehyde
Amino acid / 2-Oxo acid / EC 2.6.1.13
Ornithine
NAD^+ / EC 1.5.1.12 / $NADH+H^+$
Glutamate

Metabolism of L-ornithine and L-proline. EC 1.5.1.12, l-pyrroline-5-carboxylate dehydrogenase. EC 1.5.99.8, proline dehydrogenase. EC 2.6.1.13, ornithine-oxoacid aminotransferase.

4-Hydroxyproline
FAD / EC 1.5.99 / $FADH_2$
1-Pyrroline-3-hydroxy-5-carboxylate
H_2O
4-Hydroxyglutamate 5-semialdehyde
NAD^+ / EC 1.5.1.12 / $NADH+H^+$
4-Hydroxyglutamate
2-Oxo acid / Amino acid / EC 2.6.1.23
4-Hydroxy-2-oxoglutarate
EC 4.1.3.16
Glyoxylate + Pyruvate

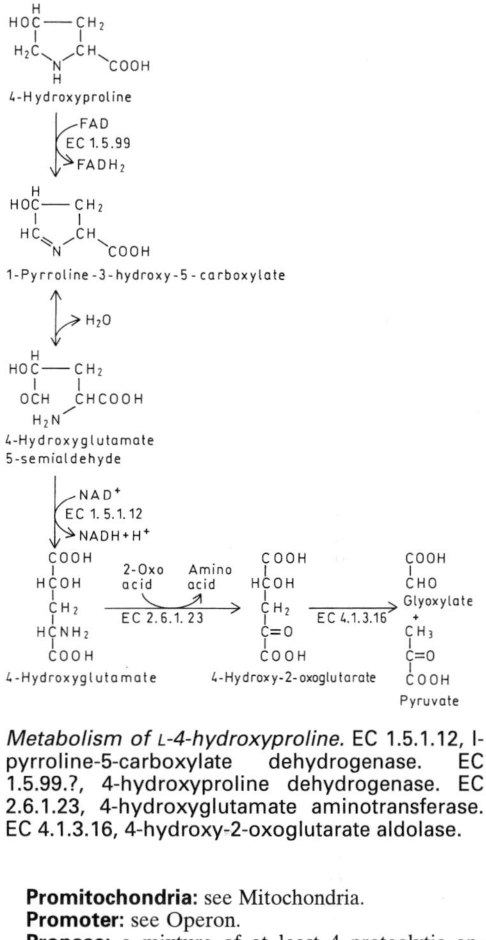

Metabolism of L-4-hydroxyproline. EC 1.5.1.12, l-pyrroline-5-carboxylate dehydrogenase. EC 1.5.99.?, 4-hydroxyproline dehydrogenase. EC 2.6.1.23, 4-hydroxyglutamate aminotransferase. EC 4.1.3.16, 4-hydroxy-2-oxoglutarate aldolase.

droxy-L-glutamate. Other derivatives of Pro are 4-methylproline (found in certain antibiotics) and 4-hydroxymethylproline (apple skin).

Proline oligopeptidase, ***post proline cleaving enzyme, prolyl endopeptidase, proline-specific endopeptidase*** (EC 3.4.21.26): a new type of serine oligopeptidase with specificity restricted to small peptides. It is the only proline-specific endopeptidase presently known in mammals, and it was first discovered as an oxytocin degrading enzyme [R. Walter et al. *Science* **173** (1971) 827–829]. It cleaves on the carboxyl side of Pro, provided this residue is preceded by at least 2 residues. Only small peptides are attacked, and the enzyme is therefore thought to be involved in the metabolism of neuropeptides; indeed, it displays high activity toward oxytocin, bradykinin, substance P and angiotensin. There is increasing evidence that the enzyme is involved in the renin-angiotensin system and may therefore be important in hypertension. It has been purified from *Flavobacterium* and from human tissues, and the gene encoding the human enzyme has been cloned and sequenced from lymphocytes; it shows no homology with other peptidases of known structure. [A. J. Barret & N. D. Rawlings *Biol Chem. Hoppe Seyler* **373** (1992) 353–360; G. Vanhoof et al. *Gene* **149** (1994) 363–366; G. Vanhoof et al. *FASEB J.* **9** (1995) 736–744]

The activity concentration of the serum enzyme correlates with different stages of depression [M. Maes et al. *Biol. Psychiatary* **35** (1994) 545–552] and inhibition of the enzyme prevents scopolamine-induced amnesia [T. Yoshimoto et al. *Pharmacobio-Dyn.* **10** (1987) 730–735].

Promitochondria: see Mitochondria.

Promoter: see Operon.

Pronase: a mixture of at least 4 proteolytic enzymes from *Streptomyces griseus.* Two peptide esterase components resembling chymotrypsin and trypsin have been separated and further characterized. Both are inhibited by chicken ovoinhibitor.

Pro-opiomelanocortin, ***pro-opiocortin:*** a protein (M_r 31,000) from the pars distalis and pars intermedia of the pituitary. P. contains the sequences of ACTH and β-LPH (see Peptides, Fig. 3). In the pars distalis, ACTH is released from P. to regulate adrenocortical function. In the pars intermedia, P. is cleaved to unprocessed α-MSH, which then undergoes amidation and acetylation to α-MSH (see Melanotropin).

Prophage: see Phage development.

Propionic acid: CH_3–CH_2–COOH, a simple fatty acid, m. p. –22 °C, b. p. 140.9 °C. P. a. occurs as its salts (propionates) and esters in many plants. It is especially important in the metabolism of propionic bacteria, which perform a propionic acid fermentation. *Propionibacterium shermanii* synthezises P. a. from pyruvate (Fig.). Propionylcoenzyme A (see) is an important metabolic derivative.

Propionyl-coenzyme A, ***propionyl-CoA:*** activated propionic acid, formed by attachment of coenzyme A to propionic acid by a thioester linkage. Propionyl-CoA is important in fatty acid biosynthesis and in fatty acid degradation; it is formed during the β-oxidation of fatty acids with odd numbers of carbon atoms, or with branched chains.

Proplastids: rounded (0.2–1 μm diam.), colorless and largely structureless organelles in the meristematic tissues of higher plants, or in unicellular algae cultured in the dark. P. are the biogenetic precursors of Plastids (see). They can be converted into chloroplasts by illumination. During this transformation, the contents of nucleic acids, ribosomes and enzymes of transcription and translation (already present at low levels in P.) are increased, while thylakoid membranes and the enzymes of photosynthesis are synthesized de novo. Like plastids, P. cannot be formed de novo, and they reproduce by division.

Proproteins: inactive protein precursors, which are activated by the removal (a highly specific reaction) of a peptide sequence. Examples are various Secretory enzymes (see) (procarboxypeptidase, proelastase, prothrombin, etc.) hormone precursors (proinsulin, proparathormone, etc.), peptide toxins (promellitin, etc.) and Zymogens (see).

Prostacyclin, *PGI₂*: see Prostaglandins.

Prostaglandins, *PG*: a group of biologically active unsaturated C20 fatty acids derived primarily from Arachidonic acid (see). The PG are related structurally and metabolically to the Leukotrienes (see) and Thromboxanes (see). All PG can be considered as formal derivatives of prostanoic acid (which does not occur naturally). The conventional abbreviation is PG, with an additional letter and number. The E series are β-hydroxyketones, the F series are 1,3-diols; the A type are α,β-unsaturated ketones. All PG have a double bond at C13 and an OH group at C15. The series number indicates the number of double bonds, which depends on the fatty acid precursor. Series 1 are biosynthesized from 8,11,14-eicosatrienoic acid, series 2 from arachidonic acid, and series 3 from 5,8,11,14,17-eicosapentaenoic acid. Series 2 compounds are most abundant. Some PG have little or no biological activity and are presumably metabolic products of the active species. The active species are very unstable, and it is often difficult to determine whether a given compound is active in its own right or is merely a precursor of a more active PG. However, it appears that the most active species are PGE_2, $PGF_{2\alpha}$, PGD_2, PGG_2, PGH_2, PGI_2 and the thromboxanes.

Biosynthesis of the PG is catalysed by a multienzyme complex (Fig.), prostaglandin synthase. The endoperoxides PGG_2 and PGH_2 are very active, but are also rapidly metabolized. PGH_2 is the precursor for both prostacyclin (PGI_2) and the thromboxanes. Very generally, the PG prevent aggregation of platelets and clotting (at high concentrations) while the thromboxanes promote aggregation and clotting. Since PGI_2 and thromboxanes are formed from a limited amount of common precursor, the branch point is an important site for homeostatic control. In mammals, there are receptors for PGI_2 on platelets, vascular muscle and other cells; these receptors are coupled to adenylate cyclase. The endothelial cells of mammalian blood vessels produce PGI_2 in response to injury or irritation. In birds, however, the thrombocytes have receptors for PGE_2, and it is PGE_2 rather than PGI_2 which is produced by their endothelial cells [J.M. Ritter et al. *Lancet* (1983) 1–317]. In mammals, PGI_2 inhibits aggregation of platelets and causes elevation of platelet cAMP. At concentrations high enough to activate the adenylate cyclase, PGI_2, PGE, or PGD_2 will inhibit clotting, but at concentrations which are too low to activate the cyclase, these PG activate factor X and thus initiate blood clotting [A.K. Dutta-Roy et al. *Science* **231** (1986) 385–388]. The effects of different PG in various tissues are not identical, e.g. PGI_2 does not cause diarrhea, but PGE_2 does. PGE_1 and PGI_2 stimulate adenylate cyclase in platelets, while thromboxane A_2 and PGH_2 inhibit the stimulation. Most PG cause contraction of the smooth muscle of blood vessels, gastrointestinal tract and uterus, but PGH_2 and PGI_2 are generally vasodilatory. The relation of the PG to inflammation and pain is not clear, but it is very likely that the anti-inflammatory and analgesic effects of aspirin are due to its inhibition of PG synthesis. Aspirin inhibits the

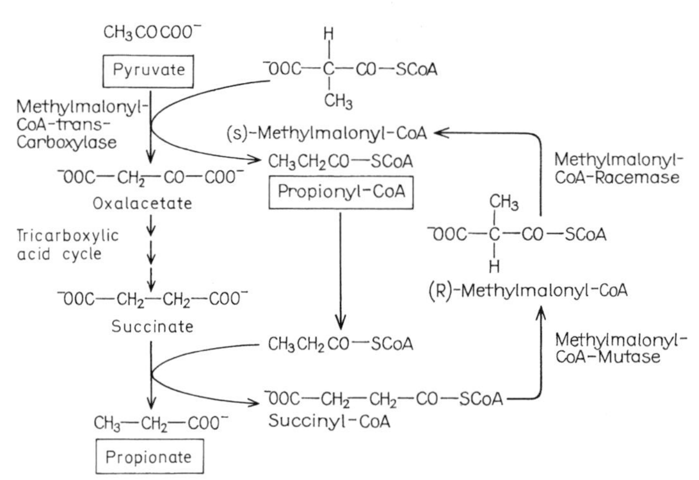

Biosynthesis of propionate from pyruvate by Propionibacterium shermanii

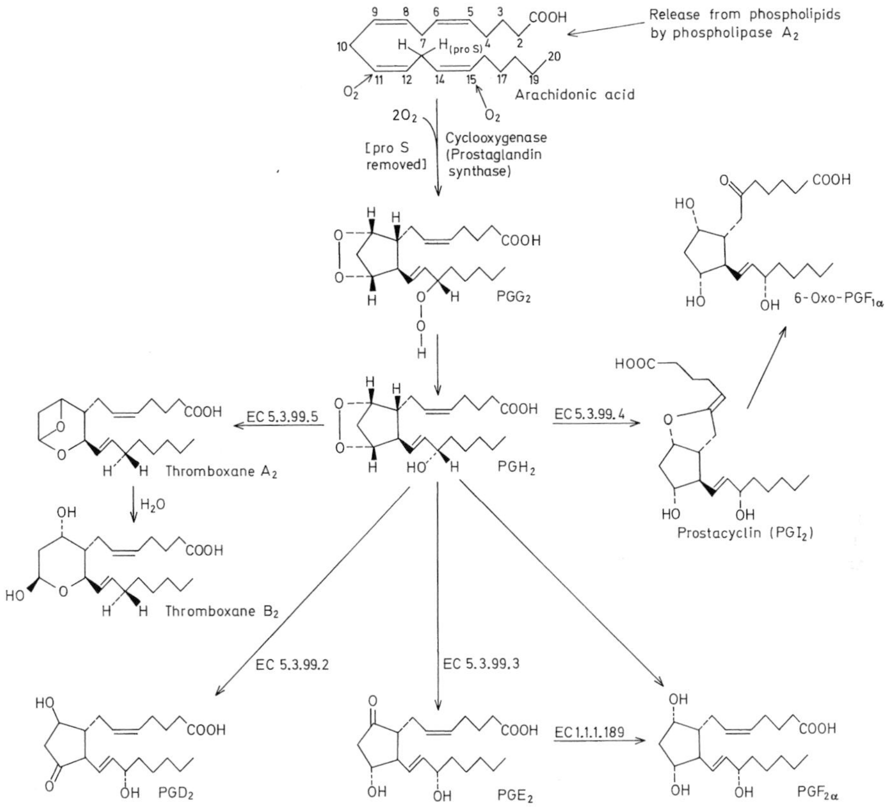

The cyclooxygenase pathway of arachidonate metabolism.
EC 1.1.1.189: Prostaglandin-E_2 9-oxidoreductase.
EC 5.3.99.2: Prostaglandin-H_2D-isomerase.
EC 5.3.99.3: Prostaglandin-H_2E-isomerase.
EC 5.3.99.4: Prostacyclin synthase.
EC 5.3.99.5: Thromboxane synthase.

synthesis of thromboxanes at lower concentrations than are required to inhibit PG synthesis; thus low concentrations of aspirin enhance bleeding by shifting the metabolism of PGH_2 toward PGI_2 and away from thromboxanes. [S. Moncada & J. R. Vane *Pharmacological Rev.* **30** (1979) 293–331; P. B. Curtis-Prior *Prostaglandins, an Introduction to their Biochemistry, Physiology and Pharmacology* (North-Holland, 1976)]

PG exhibit a wide variety of pharmacological properties. Of particular importance are: bronchoplasmolytic activity (treatment of acute asthma); control of gastric secretion (possible ulcer therapy); antagonism of the diuretic action of angiotensin (treatment of essential hypertension and cardiovascular disorders); initiation of ovulation (e.g. in cows, pigs and sheep); relief of pain during parturition; and finally, very small amounts of PG cause abortion.

Prostanoic acid: see Prostaglandins.

Prosthetic group: a nonproteogenic, low M_r group present in a conjugated protein. In an enzyme, a P. g. is a catalytically active group attached to the enzyme protein (apoenzyme). In the wide sense, P. g. are therefore Coenzymes (see), but many authors make a clear distinction, i.e. P. g. are covalently bound (e.g. heme, biotin phosphopantetheine), whereas coenzymes are freely dissociable and can be removed by dialysis (e.g. pyridoxal phosphate). The flavins (FAD and FMN) of flavoproteins and flavoenzymes are usually regarded as P. g.

Protamines: a group of strongly basic, low M_r, simple (unconjugated) proteins associated with DNA in the cell nucleus. They replace the somatic Histones (see) in sperm, at least during spermogenesis. P. have been prepared from fish and bird sperm; they contain 80–85 % arginine, the remaining amino acids being alanine, glycine, proline, serine and valine (or isoleucine). M_r of P. is between 4,050 (clupein, 30 amino acid residues) and 4,420 (iridin, 33 amino acid residues). The sequence of salmin A-I is Pro-Arg_4-Ser_3-Arg-Pro-Val-Arg_5-Pro-Arg-Val-Ser-Arg_6-Gly_2-Arg_4. Protamine-like proteins containing 40–80 amino acid residues have been isolated from invertebrate sperm.

Protease inhibitors: see Snake venom.

Proteases: all enzymes that catalyse the exergonic hydrolysis of peptide bonds in proteins and peptides (Fig.). P. are divided into two groups, according to their site of attack on the polypeptide chain:

$$\begin{array}{c} \begin{array}{cc} R_1 & R_2 \\ -\underset{H}{N}-\underset{H}{C}-\underset{O}{\overset{\|}{C}}- & -\underset{H}{N}-\underset{H}{C}-\underset{O}{\overset{\|}{C}}- \end{array} \\ \downarrow + H_2O \\ \begin{array}{ccc} R_1 & & R_2 \\ -\underset{H}{N}-\underset{H}{C}-\underset{O}{\overset{\|}{C}}-OH & + \; H_2N- & \underset{H}{N}-\underset{H}{C}-\underset{O}{\overset{\|}{C}}- \end{array} \end{array}$$

1. *Endopeptidases (proteinases)* catalyse the hydrolysis of bonds within the peptide chain, forming variously sized cleavage peptides. They can be further subdivided into acidic, neutral and basic endopeptidases. Neutral and basic types can each be divided into Serine proteases (see) and thiol proteinases (see Thiol enzymes). Examples of animal endopeptidases are Pepsin (see), Rennet enzyme (see), Trypsin (see), Elastase (see), Thrombin (see), Plasmin (see) and Renin (see). For examples of plant and bacterial endopeptidases, see Papain, Subtilisin, Bromelain. Endopeptidases have also been isolated from yeast and fungi.

2. *Exopeptidases* catalyse the hydrolytic removal of only terminal amino acids from the polypeptide chain. They can therefore be classified into *N*-terminal exopeptidases (aminopeptidases, e.g. leucine aminopeptidase) and *C*-terminal exopeptidases (e.g. carboxypeptidases A, B, Y etc.). Tri- and dipeptidases are also classified as exopeptidases.

Owing to their central importance in protein degradation and turnover, P. are widely distributed in the organism. Especially high concentrations are found in the digestive tract and in lysosomes, where they catalyse total degradation of dietary and cellular proteins, respectively, to amino acids. Most known P. are unconjugated metalloenzymes, containing or requiring a metal (e.g. zinc) as an activator or stabilizer. Self digestion (autolysis) is avoided by synthesis as inactive precursors (see Zymogens), by the presence of specific protease inhibitors, or by storage in special cell organelles called lysosomes.

Substrate specificity of P. may be high, e.g. rennet enzyme, carboxypeptidase B, enterokinase, etc., or low, e.g. pronase, pepsin and intracellular proteinases. Often P. are specific for certain amino acid residues, e.g. trypsin catalyses hydrolysis of arginyl and lysyl bonds. Pancreatic P. are among the most extensively studied of all enzymes.

Proteasomes, ***prosomes:*** large (20S and 26S) multimeric protease complexes that degrade specifically targeted intracellular proteins. Most of the proteins selectively degraded by P. are targeted by ubiquitination (see Ubiquitin). P. are important in cellular regulation, e.g. they are responsible for the degradation of cyclins at key stages of the cell cycle, and for the degradation of enzymes of intermediary metabolism. In addition they remove abnormal proteins during the stress response, and they are involved in the immune response by degrading intracellular proteins to the shorter cell surface fragments that initiate the immune response.

20S ($M_r \sim 700{,}000$) proteasomes have been found in the cytoplasm and nucleus of all investigated eukaryotes and in the archaebacterium *Thermoplasma acidophilum.* Electron microscopy and X-ray diffraction reveal a hollow cylinder (length 148 Å, diam. 113 Å) of 4 stacked rings, each ring containing 7 low molecular mass (M_r 20,000–35,000) subunits. Eukaryotic 20S proteasomes contain several different types of subunits, whereas the *Thermoplasma* 20S complex contains only 2 types of subunits, α and β. In vitro studies show that 20S proteasomes attack artificial chromogenic peptides and the A and B chains of insulin. It is uncertain whether 20S proteasomes are responsible for the degradation of specific intracellular proteins, or whether they exist only as precursors of 26S proteasomes.

26S proteasomes ($M_r \sim 1{,}700{,}000$) perform the ATP-dependent degradation of ubiquitinated proteins. In the electron microscope, the complex is seen as a 20S core with additional 19S cap complexes at each end. The cap complexes consist of about 15 different subunits of M_r 25,000–110,000; they appear to be necessary for recognition, unfolding and transport of ubiquitinated proteins to the proteolytic core. [K. Scherrer & F. Bey *Prog. Nucl. Acid Res. Mol. Biol.* **49** (1994) 1–64; A. Ciechanover *Cell* **79** (1994) 13–21; S. Jentsch & S. Schlenker *Cell* **82** (1995) 881–884; J. J. Monaco & D. Nandi 'The Genetics of Proteasomes and Antigenic Processing' *Annu. Rev. Genetics* 29 (1995) 729–754; W. Hilt & D. H. Wolf *Trends Biochem. Sci.* **21** (1996) 96–102]

Proteid: an obsolete name for a conjugated protein (see Proteins).

Protein: a naturally occurring polymer of high M_r, consisting predominantly of amino acids linked by peptide bonds. P. account for more than 50 % of the organic constituents of protoplasm. By weight, they are the major component of the dry material of living organisms, and they are among the most important functional components of the living cell. All enzymes are P., and enzymes catalyse the many biochemical reactions that constitute the metabolism of the living cell; these reactions are controlled by modification of the activity and/or quantity (i.e. rate of synthesis) of appropriate enzymes. Some of the regulators and transmitters in these control processes are also P., e.g. the proteohormones (see Hormones), repressor and activator molecules (see Gene activation, Operon), and the membrane P. which determine intracellular concentrations of many enzyme substrates and products (see Active transport). Structural P. contribute to the mechanical structure of organs and tissues (e.g. elastin, collagen), or they may constitute the bulk of a natural structure (e.g. the silk of insect cocoons and spiders' webs is the protein fibroin; feathers, hair, nails and hooves are mainly α-keratin). Contractile P. (e.g. actin and myosin of muscle, and dynein of cilia and flagella) are responsible for active movement in living organisms. Examples of storage P. are ovalbumin (from egg white), casein (from milk), gliadin (wheat seeds) and zein (maize seeds). Examples of transport P. are hemoglobin (transport of oxygen in vertebrate blood) and serum albumin (transport of fatty acids and many other substances, including hormones, drugs, etc.). P. are also involved in biochemical defense, e.g. antibodies, complement, interferons and various P. of the blood-clotting system. Certain bacterial and plant toxins are P. (see

Toxic proteins). Snake venoms (see) contain various enzymes and toxic P. P. exposed on the surfaces of cells permit recognition of one cell type by another, and thus have a role in morphogenesis and the recognition of foreign tissue (as in graft rejection).

The behavior and properties of P. are determined by their structures Theoretically, there is no limit to polypeptide chain length, and all permutations and combinations of the 20 constituent amino acids are possible (see Genetic code). Further possibilities for structural variation are offered by Post-translational modification of proteins (see), attachment of nonprotein prosthetic groups, and different levels of quaternary structure, so that the possible diversity of structure (and therefore function) is almost limitless.

P. in solution are neither rigid nor motionless. The bonds between the carbon atoms in the protein backbone and amino acid side chains allow considerable rotational and torsional flexibility, and the thermal motions of the individual atoms therefore produce writhing motions of the chain. These motions are almost certainly important to the enzymatic activity of the P.; and mutations which affect the flexibility of the chain could also be expected to affect the kinetics of the enzyme catalysis. (See Allostery) [R.H. Pain *Nature* **305** (1983) 581–582]

Classification. A logical and systematic classification of the many thousand (about 40,000 in the human) P. from all living species would only be possible on the basis of their primary and tertiary structures. A simple division into enzyme P. and nonenzyme P. is meaningful and practical. A further useful classification is: simple (pure, or unconjugated) P. and conjugated P. The simple P. can be further subdivided, according to their solubility properties and molecular shape, into water-soluble globular and water-insoluble fibrous P. (Table). Alternatively, P. may be classified according to their origin, e.g. viral, bacterial, plant and animal P. Different P. within one organism may be classified as blood, milk, cerebrospinal, secretory, muscle or structural P., etc. Subcellular localization is also used for classification, e.g. ribosomal, microsomal, mitochondnal, lysosomal, nuclear, cytosol and membrane P. Since the middle of the 20th century, electrophoresis has been an especially valuable tool for the differentiation of the P. of serum and cerebrospinal fluid in clinical biochemistry and diagnosis. The different electrophoretic migration of the P. of these fluids permits their classification into prealbumin, albumin, α_1-, α_2- β_1-, β_2- and γ-globulins.

Classification of proteins

I. *Simple or unconjugated proteins*

1. Globular proteins: soluble in water or dilute salt solutions; rotational ellipsoids with frictional coefficients $f/f = 1.1–1.5$. Representatives are Albumins (see), Globulins (see), Protamines (see), Histones (see), Prolamines (see) and Glutelins (see).

2. Fibrous proteins: insoluble in water and salt solutions; greatly extended molecular structures with very high degree of molecular asymmetry; structural proteins, very stable to acids, alkalis and proteolytic enzymes.

II. *Conjugated proteins:* In addition to amino acids they contain a nonprotein moiety, called the prosthetic group, which is usually essential and is attached by covalent, heteropolar or co-ordinate linkage. Prosthetic groups include carbohydrates, lipids, nucleic acids, metals, chromogens, heme groups and phosphate residues. P. containing the above are called glyco-, lipo-, nucleo-, metallo-, chromo-, hemo- and phospho-P., respectively.

Physical chemical properties:

1. *Ampholyte properties.* The ampholyte character of P. is determined by the presence of free acidic and basic groups in the P. molecule. Depending on the pH of the solvent, P. may have acidic or basic properties. The excess positive or negative charge results in an increase in the hydration and solubility of P. The degree of hydration depends upon the size of the net charge, irrespective of whether the charge is negative or positive. On the other hand, the direction of migration in electrophoresis does depend on the sign of the charge. At its isoelectric point (pI) a P. has no net charge (i.e. it exists as a zwitterion) so that its solubility and degree of hydration are minimal. The important buffering action of P. is due to their ampholyte character.

2. *Solubility.* This is determined by amino acid composition, the distribution of polar and nonpolar amino acids on the surface of the molecule, the molecular shape, and the surrounding milieu (pH, ionic strength, temperature). Hydrophilic P. are surrounded by a layer of water (hydration layer) and they are able to occlude hydrophobic substances and prevent them from separating from aqueous solution. This protective colloid function is responsible for the stability of many body fluids. Addition of polar solvents such as alcohol or acetone, or high concentrations of neutral salts, results in loss of the hydration layer and causes precipitation (salting out) of P. Many P, however, require a low salt concentration to prevent their precipitation. This salting-in effect is due to the suppression of the ordered (association) or random (aggregation) assembly of P. molecules by accumulation of electrolyte ions on the protein surface. The solubility diagram is a sensitive criterion of protein purity: for a homogeneous protein, the plot of the quantity of added versus dissolved protein is linear up to a sharply defined saturation point.

3. *Denaturation.* Loss of native structure and biological activity of P. by destruction of tertiary and quaternary structure, with the formation of random coil or metastable forms, is known as denaturation. In denaturation, noncovalent bonds are broken, in particular hydrogen bonds. Disulfide bridges are only broken in the presence of reducing agents (reductive denaturation). (As a prelude to amino acid sequence determination, disulfide bridges are often cleaved oxidatively with performic acid, producing two residues of cysteic acid). P. can be denatured with 6–8 M urea, 6 M guanidinium chloride, 1 % sodium dodecylsulfate and other detergents, extreme acidic or alkaline pH-values, heat and by UV- or X-irradiation. Denaturation is often reversible (renaturation), even after treatment with reducing agents.

Fig. 1 is a diagrammatic representation of a) the reductive denaturation of ribonuclease in 8 M urea with thioethanol (2-mercaptoethanol), b) its reoxidation in 8 M urea to one of the 105 possible disulfide bridge isomers, and c) its renaturation to the enzymically active form in urea-free medium, by disulfide exchange in the presence of traces of thioethanol. Formation of completely disordered structures results in irreversible denaturation, e.g. heat denaturation of ovalbu-

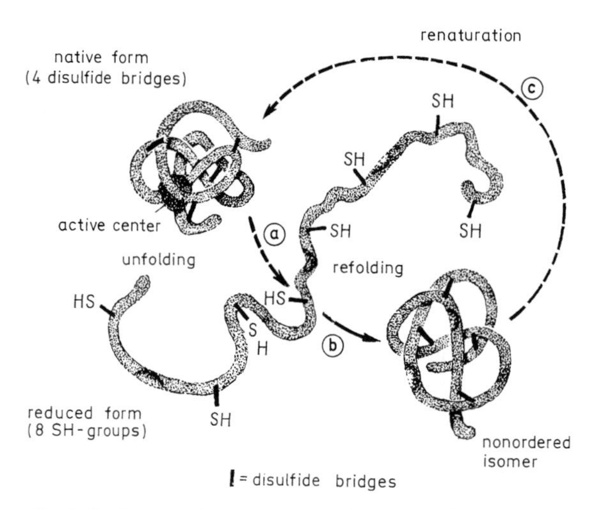

Fig. 1. *Reductive denaturation and renaturation of pancreatic ribonuclease*

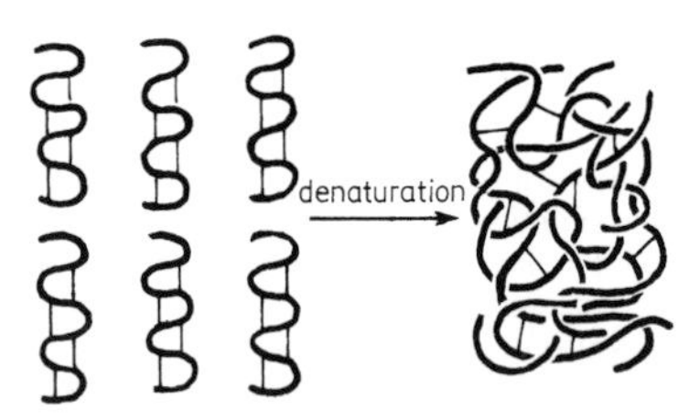

Fig. 2. *Irreversible heat denaturation of ovalbumin.* Weak intramolecular forces (not disulfide bridges) are broken, and the unfolded polypeptide chains become tangled to form a gel, as in meringue or cooked egg white.

min (Fig. 2), where the conformation of the polypeptide chain is irreversibly destroyed.

Determination of M_r: Several physical chemical methods are available for the determination of the M_r of P. They can be classified into two main types, i.e. kinetic and equilibrium methods. Kinetic methods depend on the measurement of particle transport; they include techniques for the measurement of rates of diffusion, rates of sedimentation in the ultracentrifuge, viscosity (e.g. in the Ubbelohde viscometer), rates of migration in electrophoresis, and rates of migration in gel chromatography on polyacrylamide or dextran gels of known porosity (molecular sieve, or gel chromatography). Equilibrium methods are applied to P. solutions that are in thermodynamic equilibrium. Such methods include the measurement of osmotic pressure in a membrane osmometer, measurement of light scattering (the Tyndall effect, which increases markedly with increasing particle size, is measured in a light scattering photometer), and measurement of small angle, X-ray diffraction (which depends on the mass radius of the protein) with a special camera. Electron microscopy represents a further specialized method for the determination of molecular shape and subunit composition; it is particularly suited to the determination of molecular shape and subunit composition of multienzyme complexes and P. with quaternary structure; the preparation is usually negatively stained with osmium tetroxide or other heavy metals, and the resolution can be as high as 1.5 nm. If the amino acid composition and the number of tryptic peptides, or the primary structure of a P. are known, the absolute M_r can be calculated. Currently, the most preferred methods for M_r determination of native and dissociated P. are the various ultracentrifuge methods, gel filtration and polyacrylamide gel electrophoresis.

1. *Ultracentrifuge methods; M_r* can be determined in the ultracentrifuge by measurement of the rate of sedimentation, or of the sedimentation equilibrium. In the rate methods, very high centrifugal forces are used (e.g. rotor speeds of 40–50,000 rpm, representing forces on the order of 100,000 g). Under these conditions the colloidal protein molecules have a higher density than the solvent, their rate of diffusion is high in comparison with the rate of back diffusion, and they are completely sedimented. The sedimentation rate is directly proportional to M_r, i.e. each protein migrates to the bottom of the ultracentrifuge cell at a different rate, depending on its size. The resulting changes in protein concentration along the length of the cell during ultracentrifugation are monitored by optical methods, e.g. with schlieren optics or Rayleigh interference optics, and by UV absorption. The mass unit for sedimentation per unit time is the Svedberg (S). The S-values of P. lie in the range 1.2S (insulin) to 185S (tobacco mosaic virus). P. mixtures become crudely separated into their major components. For example, serum is separated into the most rapidly sedimenting macroglobulins (M_r around 10^6), globulins (M_r about 160,000) and albumin (M_r 67,500) or prealbumin (M_r 61,000) (Fig. 3). It should be borne in mind that at the start of ultracentrifugation, the ultracentrifuge cell contains a homogeneous solution of P. Thus the peaks shown in Fig. 3 are not peaks of protein concentration; they represent boundaries of P. concentration, which are formed as P. become concentrated toward the bottom of the ultracentrifuge cell at different rates; the schlieren optical system, employ-

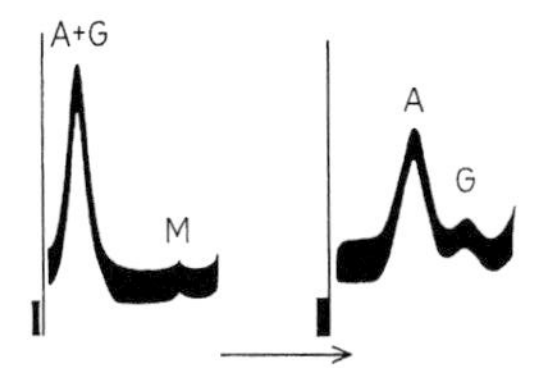

Fig. 3. *Sedimentation pattern of normal human serum in the ultracentrifuge, measured with a schlieren optical system.* Left hand picture taken 51 min after start of centrifugation; right hand picture after 125 min. Centrifugation at 59,800 rpm. A = albumin, 4.5S; G = globulin, 7S; M = macroglobulin, 19S.

ing cylindrical lenses, shows the rate of change of concentration at each boundary. M_r is calculated from the Svedberg equation, $M_r = RTS/D(1 - v\rho)$, where S is the sedimentation constant corrected for viscosity and density of water at 20 °C, R is the gas constant, T is the absolute temperature (K), D is the diffusion constant, v is the partial specific volume of the protein (= reciprocal of the density of the molecule; for most P. it is between 0.71 and 0.74) and ρ is the density of the solvent system. *D.* is determined separately by measurement of peak (boundary !) broadening caused by diffusion during low speed centrifugation.

In equilibrium methods, the P. solution is subjected to a centrifugal force of only 10,000–15,000 g. Under these conditions, sedimentation is slow and the effects of back diffusion are no longer negligible. After a fairly long period (hours to days) a stationary state is achieved, in which the net flow of protein molecules is zero. From the resulting concentration gradient between the meniscus and the bottom of the cell, the M_r can be calculated without knowing the diffusion coefficient. The time required for the low speed method can be shortened by using the approach-to-equilibrium method of Archibold (1947), in which the concentration gradient is measured at the top of the cell when the P. zone is only partly pulled away from the meniscus. The centrifugation time can be further shortened to 2–4 hours by using the meniscus depletion method of Yphantis (1964) under high speed conditions (high speed equilibrium method): only small volumes of P. solution are placed in the ultracentrifuge cell (0.1 ml of a 5 % P. solution, forming a liquid column of 3 mm). The sedimentation equilibrium can then be determined at high rotor speeds (up to 40,000 rpm) when the mensicus becomes depleted of P. Sedimentation equilibrium methods are not suitable for P. mixtures, because overlapping concentration gradients would be formed. The Yphantis method is particularly suitable for the investigation of the dissociation and association of oligomeric P.

In density gradient centrifugation (Martin & Ames, 1961), P. migrate in a sucrose gradient of increasing density, and each P. comes to rest in a discrete zone where particle density matches that of the solvent. Since the distance moved by a P. in the gradient is inversely proportional to its M_r, an approximate M_r of an unknown P. can be determined by comparison with known standards. The same method can be used to characterize subcellular fractions and these can be isolated in preparative quantities by applying the principle to large rotors (zonal centrifugation).

2. *Gel filtration,* or *molecular sieve chromatography.* Gels of chosen pore size are prepared from beads or granules of cross-linked dextrans, agarose, or polyacrylamide. Analyses are performed on columns, or by thin-layer chromatography. The gel particles contain pores of defined size. The solvent, salts and other low M_r material enter these pores and permeate the gel particles, but macromolecules above a certain size are excluded and they can only migrate in the solvent between the gel particles. The method is cheap, only simple apparatus is required, gels may be reused many times, and the separation times are short. The thin-layer version of the method is especially economical in the use of gel material and P. The quantity for M_r determination is the ratio of the elution volume, V_e, of the P. and the exclusion volume, V_0 (the void, or dead volume, i.e. the volume of the interstitial liquid between the gel particles, which is the same as the elution volume of a substance that is completely excluded from the gel). In thin-layer methods, the distance moved on the gel plate is compared with that of standards. A calibration curve is prepared with the aid of several standard P.: V_e versus log M_r or V_e/V_0 versus log M_r. Depending on the type of gel, the method measures approximate M_r in the range 500 to several million, with an accuracy of 5–10 %. Gel chromatography is also useful for the M_r determination of P. subunits: the gel consists of a cross-linked dextran or agarose, and it is equilibrated with the dissociating medium which contains 0.1–1 % dodecylsulfate, or 4–6 M guanidinium chloride. The analytical M_r range is then limited to 5,000 to 70,000.

3. *Polyacrylamide gel electrophoresis, PAGE.* A gel with the desired pore size is prepared by the simultaneous polymerization and cross-linking of acrylamide with *bis*-acrylamide. Electrophoresis is performed in tubes, or on slabs of gel, with a gel length of 7–10 cm. In the methods of Fergusun (1964) and of Hedrick and Smith (1968), the rate of migration of P. is measured in gels of different concentrations (i.e. different pore size). Migration distance is plotted against gel concentration (%) for a series of standard proteins. A linear calibration is obtained by plotting the logarithms of the M_r. Alternatively, electrophoresis is performed in the presence of sodium dodecylsulfate (SDS). The method was introduced by Shapiro (1967) and is commonly known as SDS-PAGE. The anionic detergent molecules form a layer around the polypeptide chain with their negative charges exposed to the surrounding medium. Globular P. are unfolded (it may be necessary to add 2-mercaptoethanol to cleave disulfide bridges), forming structures that approximate to rigid rods. The SDS molecule binds to the polypeptides with a constant weight ratio, corresponding to about one SDS molecule per three peptide bonds, so that the charge per unit length is constant (charges on the native protein are smothered and can be ignored). Thus in any given electrical field, the acceleration is the same for every SDS-coated protein molecule. Separation depends solely on the size of the molecule and the sieving effect of the gel. Log M_r versus relative distance of migration is linear, and the system can be calibrated with P. of known M_r. Since most oligomeric P. are dissociated by SDS, the

method is often used for the M_r determintion of subunits. It is the most rapid method (2–4 hours), and the most economical in use of test material (10–50 μg P. per analysis), with an accuracy of 5–10 %.

Single-chain P. have M_r between 10,000 and 100,000; oligomeric (multichain) P. have M_r between 50,000 and several million.

Determination of molecular symmetry. Molecular symmetry of P. can be determined from measurements of viscosity, streaming birefringence, rates of sedimentation and diffusion, or directly by electron microscopy. For a known M_r, the frictional coefficient can be calculated from ultracentrifugal measurements, e.g. from the sedimentation: $f = [M_r(1 - v\rho)]/S$, where v is the partial specific volume, ρ is the density and S is the sedimentation constant. The axial ratio a/b of a P. can be derived from the frictional ratio f/f_0 where f_0 is the f of a spherical molecule. The value of a/b for most globular P. is between 2 and 20, and greater than 20 for fibrous P., e.g. the axial ratio of fibrinogen is 30.

Purification and isolation. Biological material usually contains a great variety of P., together with carbohydrates, lipids and nucleic acids. Purification is therefore necessary as a prelude to the detailed characterization of any P. The insoluble fibrous structural P. and the soluble P. that occur in high concentrations (e g. hemoglobin from erythrocytes, casein from milk, trypsin from pancreatic juice) can be isolated relatively easily. For most globular P., however, separation from impurities requires multi-step purification procedures. Artifacts caused by the action of proteases are avoided by using protease inhibitors (e.g. diisopropylfluorophosphate) and/or by working at 4 °C so that proteases are relatively inactive. With the exception of P. in serum, cerebrospinal fluid, milk, egg white and various secretions, P. are localized inside cells and tissues. The first stage of P. purification therefore consists of extraction from biological material, following disruption of cell walls and/or membranes by mechanical methods (high speed rotary cutting blade; abrasion between closely fitting glass, or glass and teflon surfaces, e.g. Potter-Elvehjem homogenizer), by physical methods (ultrasound; alternate freezing and thawing; grinding of frozen tissue with fine Al_2O_3 granules; shaking with glass beads; change of pH; osmotic shock; treatment with detergent; forcing a frozen paste of cells between closely fitting steel surfaces, e.g. Hughe's press for disrupting bacteria; forcing a cell suspension at high pressure through a fine orifice, e.g. French press), or by enzymatic methods (treatment with proteases such as trypsin, papain, lipases, neuraminidase, hyaluronidase, or lysozyme). The resulting cell homogenate can be extracted with various reagents, such as salt solutions glycerol, dilute acids, detergents and a wide variety of buffers. Cellular debris is removed by coarse filtration or low speed centrifugation. The second step is usually a preliminary separation by ammonium sulfate fractionation. If necessary, nucleic acids are removed by precipitation with protamine sulfate, or they can be hydrolysed to low M_r products by addition of ribonuclease and deoxyribonuclease. Subsequent purification steps include chromatography (gel filtration, ion exchange, adsorption), and possibly preparative electrophoresis with or without carrier, or isoelectric focusing. Some rapid and high performance methods are affinity chromatography (1968) and ion filtration chromatography (1972), protective crude separation by ultrafiltration membranes (1964), preparative electrophoresis (1959), isoelectric focusing (1962), ammonium sulfate gradient stabilization chromatography (1972), gradient sieve chromatography (1973) and fractional precipitation with polyethylene glycol (1973).

Affinity chromatography is a column technique with wide application. It exploits the reversible and biospecific interaction between the functional groups of the protein under investigation and a ligand (either a small molecule or an antibody, receptor or other specific protein) which is covalently bound to an insoluble, inert carrier. If the ligand is too close to the matrix, interaction with the protein may be sterically hindered. For high binding capacity, small ligands are usually attached to the matrix via a hydrocarbon chain (flexible arm, or spacer group) of variable length. For maximal stability of the protein-ligand complex, a modified substrate or specific inhibitor (usually competitive) is used as the ligand; for the purification of antibodies, the appropriate antigens are used as ligands. Alternatively, antibodies specific for the desired P. may be immobilized on the matrix. Owing to the high selectivity of the protein-ligand interaction, it is often possible to isolate a P. from a mixture in a single step of affinity chromatography; all other substances, including P., pass unretarded through the column. The desired P. is released (i.e. the protein-ligand complex is dissociated) by changing the pH or ionic strength, or by adding a competing ligand to the elution solvent. Columns can be regenerated and used repeatedly. One of the many examples of affinity chromatography is the purification of NAD-dependent dehydrogenases, e.g. glyceraldehyde 3-phosphate dehydrogenase, using NAD as the affinity ligand covalently bound to Sepharose (matrix) via a spacer arm of 6-aminohexanoic acid. After washing a crude enzyme mixture through the column, dehydrogenases are eluted with buffers containing NAD (see Chlorotriazine dyes). In addition to the purification of a large number of enzymes, the method has also been used for the purification of antibodies and antigens, hormones, macromolecular inhibitors, transport and receptor P. (e.g. the hitherto intractable insulin receptor was purified 250,000-fold), and even cell populations.

Ion filtration chromatography is a time-saving combination of ion exchange and gel chromatography without a salt gradient. P. are separated on diethylaminoethyl-(DEAE)-dextran (cross-linked) at a constant pH within 2 to 3 hours.

Ultrafitration is a rapid and protective technique, by which a P. solution is simultaneously concentrated and freed from small molecular contaminants. It is usually operated with excess, rather than reduced pressure. By using the appropriate membrane filter (pore sizes range from 15 μm upwards), P. mixtures can also be crudely fractionated according to M_r. Biologically active P. can be prepared on an industrial scale by using tubular membranes with large surface areas and a daily throughput capacity of several thousand liters.

High performance liquid chromatography (HPLC) has now largely replaced other methods (e.g. thin layer chromatography) for separating and identifying PTH-amino acids from sequencing studies (see below). HPLC of whole proteins on hydroxyapatite is possible, but recovery is often very low. Silane-coated silica is used in reversed phase HPLC systems for whole proteins (silica alone is unsuitable, but its low compressibility makes it ideal as a packing for HPLC columns). Low compressibility organic polymers suitable for HPLC of proteins are under development.

With new carrier systems and refrigerated apparatus with voltages in the thousands, electrophoretic methods can now be used for preparative separation of P. Carrier-free systems (*continuous electrophoresis*), or zone electrophoresis with a carrier (*starch block* or *polyacrylamide gel electrophoresis*) are used. Continuous electrophoresis is usually performed on paper; P. move vertically with a downward flow of buffer, and are at the same time subjected to horizontal separation by a potential difference across the paper. In starch block and continuous electrophoresis, P. are mostly negatively charged (buffer pH 7–9), and they migrate according to their net charge, whereas migration in polyacrylamide gel electrophoresis includes the additional effect of molecuar size (molecular sieve effect). After zone electrophoresis, the starch block or gel is sectioned, and more or less pure proteins are eluted from the individual slices. In *isoelectric focusing,* P. are separated electrophoretically in a pH-gradient stabilized by a sucrose gradient, so that separation depends upon their different isoelectric points. During isoelectric focusing, each P. migrates to a position corresponding to its own pI. The current is switched off and the column is carefully emptied to collect each band of P. directly and in preparative quantity.

Discontinuous electrophoresis (abb. disc electrophoresis; the term "disc" also conveniently describes the sharp, disc-shaped bands of P. that are formed in the tubular column version of this method) exploits a discontinuity of gel pore size and pH. It has a very high resolving power in comparison with other methods of P. separation. Nowadays any claim for P. homogeneity should be supported by evidence from disc electrophoresis. The principle was first applied to separations in tubular columns of polyacrylamide gel. Later it was made even more incisive by the introduction of a slab gel technique, in which several P. samples can be run alongside each other on the same gel. As described here, the conditions were originally designed for the investigation of serum P. A short length of large pore gel buffered at pH 6.7 (known as the stacking gel, containing about 3 % acrylamide) is layered on top of a small pore gel buffered at pH 8.9 (known as the running gel, containing 7.5 % acrylamide). Both gels contain a highly mobile ion (the leading ion), which is usually chloride. The buffer (pH 8.3) in the anode and cathode compartments contains glycine (the trailing ion), which has a net negative charge, but is less mobile than chloride. The P. are (is) incorporated into a gel mixture identical to that of the stacking gel, and layered on top of the latter. Conditions of pH are chosen so that the P. have a mobility between the leading ion and the trailing ion. As the glycine enters the stacking gel, the P. become sandwiched between the two ions and concentrated into very tight bands, one stacked on top of the other in order of decreasing mobility, with the last one followed by glycine. In this state they enter the running gel and are thus subjected to a discontinuity of both pore size and pH. The mobility of the fastest protein now falls, glycine overtakes all the P. and runs directly behind the chloride, and each thin starting zone of P. is in a linear voltage gradient. Subsequent migration depends upon both the charge and the size of each P. molecule.

The criterion of P. purity is homogeneity. A "pure" P. preparation (i.e. only one P. is present) may contain inorganic salts, other small organic molecules (e.g. substrates, coenzymes) and water. Even a crystallized P. will contain much water and possibly other small molecules and ions. A special question of P. homogeneity arises in the study of quaternary structure, when it is necessary to determine whether subunits are identical or nonidentical. Homogeneity can be monitored by analytical disc electrophoresis (one P. band), analytical ultracentrifugation (a single symmetrical boundary), solubility diagram and pI. The amino acid composition of a homogeneous P. shows all the constituent amino acids in quantities of more or less the same order, and in sensible ratios; the presence of a contaminating P. of different amino acid composition is therefore easily recognized, especially if the amino acid composition of the pure P. is already known. Detection of more than one *N*-terminal amino acid (by Edman degradation, by dansylation, or by 2,4-dinitrophenylation) is indicative of contaminating P. Similarly, heterogeneity is indicated by the presence of more than one *C*-terminal amino acid (determined by treatment with carboxypeptidase). If a P. is known to contain x lysine and arginine residues (i.e. trypsin-sensitive sites), the theoretical maximal number of tryptic peptides is $x + 1$. Inhomogeneity is indicated if this number is exceeded. The ability of a P. to crystallize shows that it is fairly pure. For enzymes, further evidence of purity is given by pH and temperature optima, substrate specificity, kinetic parameters and specific catalytic activity.

Detection and measurement. P. can be detected qualitatively by a variety of denaturation methods: precipitation with trichloroacetic acid, perchloric acid, picric acid and heavy metal salts (Cu, Fe, Zn, Pb), or by warming at their pI. There are also color reactions for certain amino acid residues (aromatic, phenolic, SH and guanido groups) and for the peptide bond. The biuret color reaction, which is specific for peptide bonds, has been combined with the Folin reagent, which is specific for tyrosyl and tryptophanyl residues, to produce the sensitive Lowry method for the colorimetric assay of P. (lower detection limit 5–10 µg P. per ml); the blue color complex has an absorption maximum at 750 nm and is stable for several hours. The method is usually standardized against human or bovine serum albumin. The oldest quantitative procedure for P. is the Kjeldahl method, in which the sample is digested with sulfuric acid to convert the P. nitrogen into ammonium sulfate; after alkalization of the digest, the ammonia is steam-distilled into 0.1 or 0.01 M acid, then determined by back titration. The quickest and most suitable method for routine measurements (e.g. in a flow-through spectrophotometer cell) is the measurement of UV absorption at 280 nm, which depends chiefly on the presence of tyrosyl residues in the P. Using the measured absorption coefficient, $A_{280,\ 1cm}^{1\%}$, of a purified P. (e.g. the value for trypsin is 15.4), the absorption at 280 nm can be used to determine unknown concentrations of the same P.

```
          |  CH3  CH3  |              |        (C6H5)   |      O=C-NH2
  CH3     |     \ /    |              |         CH2     |        CH2
   |      |     CH     |              |          |      |         |
H2N—CH—CO—|—NH—CH—CO—|—NH—CH2—CO—|—NH—CH—CO—|—NH—CH—CO
   Ala   ———   Val   ———   Gly   ———   Phe   ———   Asn
```

Fig. 4. *Primary structure of the N-terminal end of a protein chain*

Composition and structure. The sequence of amino acids in a P. is genetically determined. The correct order or sequence of all the amino acid residues is known as the *primary structure* (Fig. 4).

Determination of primary structure is usually preceeded by the determination of amino acid composition, i.e. total hydrolysis of the P., followed by quantitative determination of all the amino acids in the hydrolysate. Several methods are available for the separation and quantitative determination of amino acids in a mixture, e.g. paper chromatography, high voltage electrophoresis and gas-liquid chromatography of volatile amino acid derivatives. However, the separation of all the P. amino acids and several other substances, using a 100 cm column of sulfonated polystyrene (Dowex 50-X8) and buffers of progressively increasing pH from 3.4 to 11 (Stein and Moore, 1951) represented a major advance in P. chemistry The automated version of this method (together with other improvements, such as micropowder in place of bead resin, use of 8 % cross-linked Amberlite IR-120, and use of two columns, one for neutral plus acidic, and one for basic amino acids) revolutionized the study of P. structure and composition. The machine automatically collects column effluent, adds ninhydrin, heats for color development, records color intensity, and plots the intensity on a graph (Fig. 5) (see Amino acid reagents).

From 1 mg P., the quantity of each amino acid (corresponding to the area under each peak) can be determined with an accuracy of a few percent. Analysis is complete in about 20 hours. Analysis is further facilitated by on-line computation of results with direct readout of amino acid composition. The most common method of hydrolysis consists of heating the P. in 6 M HCI at 105–110 °C for 20–70 hours in a sealed tube. A drawback of acid hydrolysis is that some amino acids are altered: tryptophan is totally destroyed, asparagine and glutamine are hydrolysed to aspartic and glutamic acid, respectively, and cysteine is oxidized to cystine. There is also some loss of serine, threonine, cystine and tyrosine; many of these destructive processes are encouraged by the presence of carbohydrate and may be at least partly avoided by performing the hydrolysis under nitrogen and in the presence of stannous chloride. Hydrolyses are performed for varying periods and the values for labile amino acids are extrapolated to zero time. Alkaline hydrolysis (5 M NaOH or $Ba(OH)_2$ at 110 °C for 20 hours) is especially destructive and causes racemization, but it is particularly suitable for the determination of tryptophan alone. Most of these problems can now be avoided by hydrolysing the P. with 3 M *p*-toluene-sulfonic acid plus 2 % thioglycolic acid in a sealed tube. Unfortunately, the residual *p*-toluene-sulfonic acid is nonvolatile and must be loaded onto the analytical column with the amino acids; it emerges early and tends to interfere with the separation of the acidic amino acids. Tryptophan may also be determined from the UV absorption of the P., after making allowance for the tyrosine content. Some P. can be hydrolysed for analytical purposes by a mixture of proteolytic enzymes, such as papain or subtilisin (both fairly nonspecific) in conjunction with leucine aminopeptidase and proline iminopeptidase, but such hydrolyses are often incomplete.

The next stage in primary structure determination is the reductive (2-thioethanol) or oxidative (performic acid) cleavage of disulfide bridges, followed by enzymatic (e.g. with trypsin) and/or chemical (e.g. with cyanogen bromide) cleavage of the polypeptide chain. Peptides from this cleavage are separated by ion-exchange chromatography or by the finger print technique (a two-dimensional separation on pa-

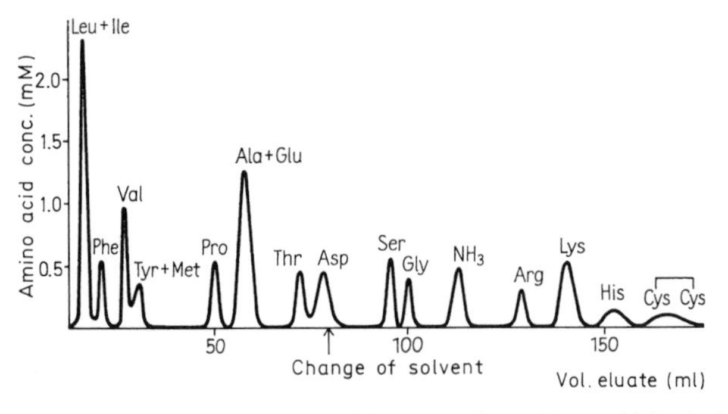

Fig. 5. *Column chromatographic separation of an acid hydrolysate of serum albumin in the automatic amino acid analyser* (after Moore & Stein).

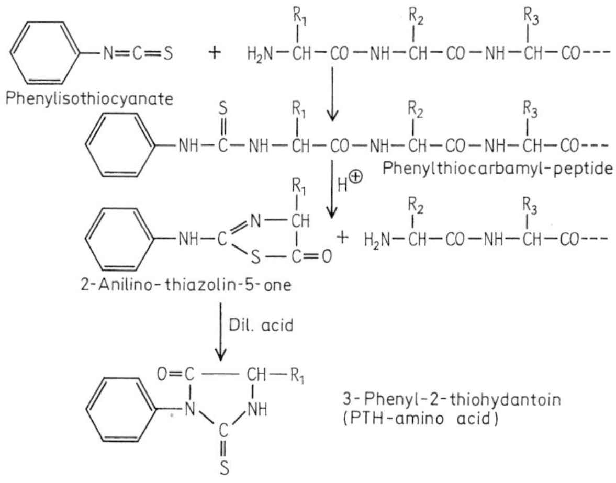

Phenylisothiocyanate

Phenylthiocarbamyl-peptide

2-Anilino-thiazolin-5-one

Dil. acid

3-Phenyl-2-thiohydantoin (PTH-amino acid)

Fig. 6. *Edman degradation of a peptide*

per or thin-layer plates, in which electrophoresis is used in the first dimension and partition chromatography in the second). Finally, the amino acid composition and the *N*- and *C*-terminal amino acids of each cleavage peptide are determined.

In the Edman degradation, the peptide is reacted with phenylisothiocyanate at pH 8–9 at 40 °C; acid treatment removes the *N*-terminal amino acid (R_1) as a substituted phenylthiohydantoin (PTH-amino acid), which is identified chromatographically (Fig. 6); the next amino acid (R_2) becomes the *N*-terminus of the new peptide. The method has been automated, and with the aid of a specially built sequenator, it is possible to determine the sequence of amino acids in a peptide, starting from the *N*-terminus. Sequences of more than 100 amino acids can be determined in this way, using only 1 nmole of protein. Each operational cycle (i.e. derivatization and separation of each residue) requires 20–30 min. In the original liquid phase method, reagents washed over a thin film of reactants adsorbed on the inside surface of a spinning glass cup. In the later gas-phase sequencers, the protein is dried on a glass fiber disc, and some reagents (triethylamine, trifluoroacetic acid) are passed as vapors over the adsorbed reactants, thus avoiding build-up of reagent impurities and permitting higher sensitivity. In both methods, the protein is adsorbed noncovalently to a glass support, and wash-out is always a danger (especially with less hydrophilic membrane proteins). Sequenators have been developed, in which the protein is attached covalently to an activated glass. Identification of the *N*-terminus may also be performed with Sanger's reagent or dansyl chloride. Once the amino acid sequence of each peptide is known, their order in the original P. must be established. Peptides at the *N*- and *C*-termini can be identified by comparison with the termini of the original P. The positions of interior peptides are established by cleavage of the P. by more than one method, e.g. proteolytic enzymes of different specificities can be used. Comparison of the amino acid sequences of the two (or more) sets of peptides permits alignment because the termini of one set are overlapped by peptides of the second set. Positions of disulfide bridges are determined after the linear sequence has been established. Tryptic digestion is performed with the disulfide bridges intact, and cystine-containing peptides are isolated. Their disulfide bridges are then cleaved; the resulting two peptides are identified by comparison with the tryptic peptides of the linear sequence. During this procedure, any free SH-groups are blocked with iodoacetate while the disulfide bridges are still intact.

An early method of sequence determination employs partial hydrolysis of the P. with acid or with nonspecific endopeptidases. The resulting hydrolysate contains a large number of relatively small peptides, representing many overlapping regions of the sequence. By analysis of a sufficient number of peptides, and comparison of their amino acid sequences, it is possible (but very laborious) to construct the linear sequence of the original P. The classical determination of the primary structure of the A chain of Insulin (see) was performed in this way.

Comparative studies on the primary structures of homologous P. from different species (e.g. hemoglobin from vertebrates, see Homologous proteins) or analogous P. (e.g. subtilisin from *Bacillus subtilis* and mammalian trypsin) have made a valuable biochemical contribution to questions of divergent and convergent evolution. However, for an explanation of P. function and behavior, especially the mechanism of enzyme action, the primary structure alone is insufficient, and a knowledge of secondary and tertiary structure is needed.

The first primary P. structure to be determined was that of insulin in 1953. By the end of 1980, over 700 complete primary P. structures had been reported, using the chemical methods of protein analysis described above. Although these methods are still used, in particular for smaller polypeptides and partial sequences, primary P. sequences are now mostly determined from the structure of the cloned gene encoding the P. in question (see Recombinant DNA technology, Nucleic acid sequencing). In 1995, the number of known primary P. sequences, determined by either chemical analysis or by the interpretation of gene structure, was nearly 100,000.

Secondary structure refers to the way in which the polypeptide chain is folded. Folding is due to the formation of hydrogen bonds between peptide bonds in close juxtaposition to one another (separated by 0.28 nm). These bonds involve the carbonyl oxygen of one peptide bond and the amide hydrogen of the other. In helical structures, hydrogen bonds exist within the peptide chain, whereas pleated sheets or β-structures result from hydrogen bonding between two lengths of polypeptide chain (which may or may not be part of the same chain). The most common helical structure is the α-helix, with 3.6 amino acid residues per revolution (Fig. 7).

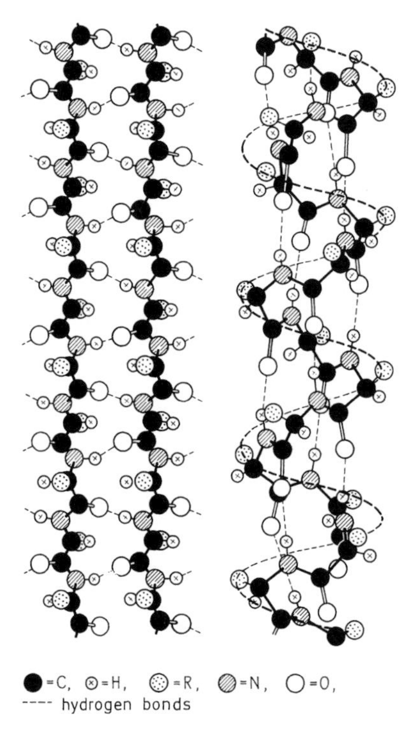

Fig. 7. *Secondary structure of proteins.* Left, β-pleated sheet structure of a fibrous protein; right, α-helix structure of a polypeptide chain.

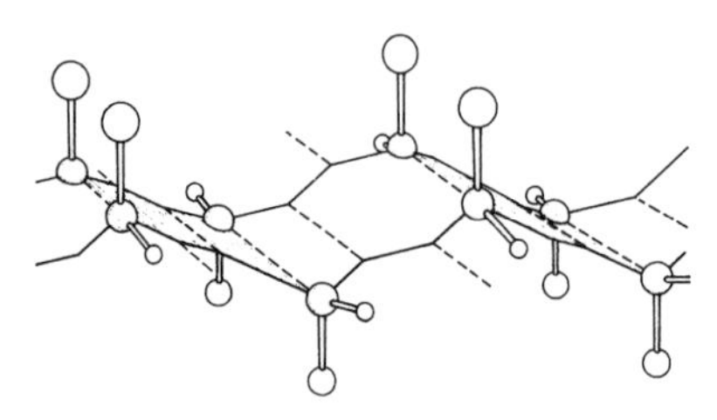

Fig. 8. *β-Pleated sheet structure; antiparallel arrangement of two polypeptide chains.*

Collagen (see) has a specialized structure containing interchain, hydrogen-bonded, left-handed helices. Otherwise, all known P. helices are right-handed, but the possibility remains that left-handed helices might be found in P. which have not yet been analysed by X-ray diffraction, e. g. membrane P. In the pleated sheet structure, the polypeptide chain is more or less stretched out, and neighboring lengths of chain can be parallel or antiparallel (with respect to their *N*- and *C*-termini), e. g. the pleated sheet structure of silk fibroin (Fig. 8) consists of antiparallel polypeptide chains.

Most P. also contain a large number of amino acid residues which cannot be assigned to any regular structure. Only certain peptide conformations are possible; others would bring neighboring unbonded atoms too close together (see Ramachandran plot). The fully denatured state of a polypeptide is referred to as a random coil; this completely disordered structure is also found in helix-coil transitions and in synthetic polyamino acids.

The percentage of α-helical structure in a P. can be estimated by measurements of optical rotatory dispersion (ORD) and circular dichroism (CD), because the helices are themselves optically active. The slow rate of exchange of P. hydrogen with deuterium oxide or tritium oxide depends inversely on the amount of intramolecular hydrogen bonding; estimates of α-helix based on this method often disagree with the amounts obtained by optical methods. This could be due to additional forms of hydrogen bonding (e. g. in β-sheets or loops), or the existence of left-handed helices (which have optical activity of opposite sign from right-handed helices), or the discrepancy may arise because hydrogen exchange is retarded by factors other than hydrogen bonds, e. g. areas so surrounded by hydrophobic bonds that hydrogen exchange is difficult. X-ray crystallography, which is used for the determination of tertiary structure, also provides detailed information about secondary structure. Hemoglobin and myoglobin have about 75 % helical content as determined by X-ray crystallography and ORD, but these two methods show poor agreement on the α-helical content of other P., such as cytochrome *c* and carbonic anhydrase. This probably signifies differences in the secondary structures of dissolved and crystalline P.

Tertiary structure defines the overall molecular shape and gives detailed information about the spatial arrangement of reactive amino acid residues, e. g. at the active sites of enzymes, or at the antigen-binding site of antibodies. With a knowledge of tertiary structure it is possible to visualize the three-dimensional shape of a P. to a high resolution (0.2 nm or better) and to observe the changes in molecular architecture that accompany the formation of an enzyme-substrate or enzyme-inhibitor complex. Tertiary structure is determined from the X-ray diffraction of isomorphic crystalline heavy metal derivatives of the P. The resulting diffraction diagram, and the electron density map constructed from it, provide information on the type and position of amino acid residues; at higher resolution (0.15 nm) it is even possible to determine the distance between the atoms in the P. molecule. Despite the very specialized and demanding nature of this technique, the spatial structures of numerous P. have been determined (Fig. 9).

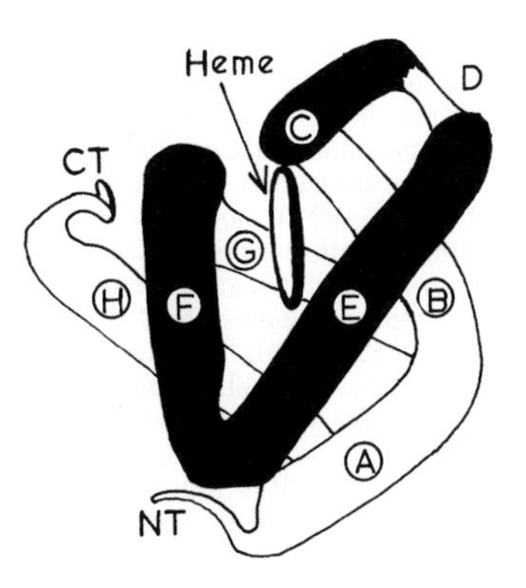

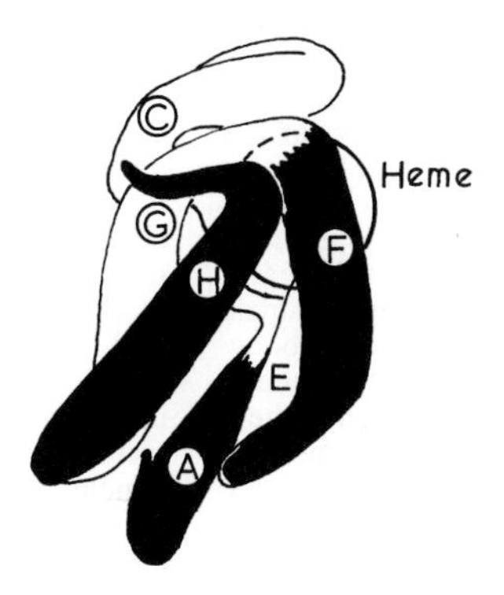

Fig. 9. *Chain conformation (tertiary structure) of myoglobin.* The molecule consists of eight stretches of α-helix, lettered A to H. NT is the *N*-terminus, containing two nonhelical amino acids. CT is the *C*-terminus, containing four nonhelical amino acids. The number of amino acid residues in each helical stretch is A(16), B(16). C(7), D(7), E(20), F(10), G(19), H(26). The nonhelical regions between the helical stretches are AB (1 amino acid residue), CD(8), EF(8), FG(4), GH(5).

The occurrence and stabilization of the three-dimensional P. structure is the result of several forces: hydrogen bonds between tyrosyl and carboxyl or imidazole groups, and between seryl and threonyl residues; disulfide bridges, which have a primary function in the stabilization of conformation; Van der Waals forces; noncovalent mutual attraction between the uncharged ($-CH_3$, $-CH_2OH$) or hydrophobic (phenyl, leucyl) residues separated by about 0.3 nm; electrostatic attraction between polar side groups (e.g. $COO^- \ldots {}^+NH_3$), which are also involved in the solvation of the molecule; interaction of P. and aqueous solvent, which favors the formation of hydrophobic bonds (interatomic distance 0.31–0.41 nm) in the nonpolar interior of the molecule, and thus makes an important contribution to natural P. conformation. Secondary and tertiary structures are referred to jointly as chain conformation. In fact, it is sometimes difficult to make a clear distinction between secondary and tertiary structure. According to nuclear magnetic resonance (NMR) studies, the chain conformation of a protein is changeable within certain limits, so that the conformation determined by X-ray crystallography represents one of several possible states "frozen" by crystallization.

Aggregation or association of two or more identical or different polypeptide chains by noncovalent interaction leads to stable oligomeric (or multimeric) P. These ordered associations represent quaternary structure, and the individual polypeptide chains are called subunits (Fig. 10). In rare cases, quaternary structure may also be maintained by disulfide bridges. P. with quaternary structure are of widespread occurrence. Most of the known multimeric P. contain either 2 or 4 similar sized subunits. Far less common are P. with uneven numbers of subunits with different sized subunits, with subunits capable of independent enzymatic activity, or with both regulatory and catalytic subunits. Possession of quaternary structure appears

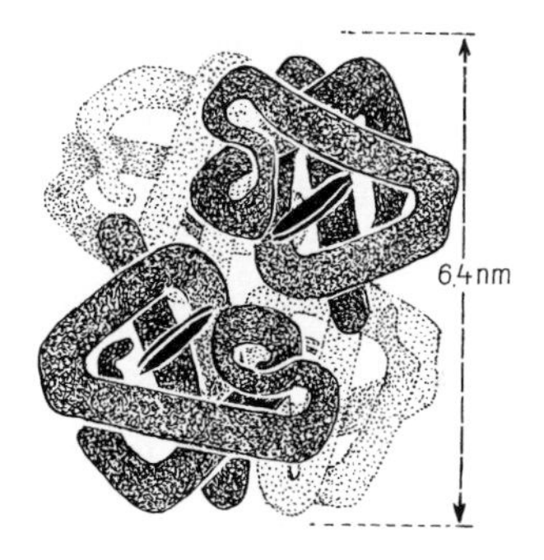

Fig. 10. *Quaternary structure of hemoglobin, showing spatial arrangement of the α- and β-chains.* Black disks represent heme groups.

to confer a flexibility of shape and activity, which is necessary for the physiological role of the P. Monomers derived from multimeric enzymes are usually inactive. The subunit composition of oligomeric P. can be determined by dissociation of the aggregate and investigation of the separate subunits by ultracentrifugation, polyacrylamide disc electrophoresis, gel filtration, ion exchange chromatography or viscometry. Alternatively, the structure of the intact molecular aggregate can be studied by electron microscopy, or by low angle X-ray or neutron diffraction. Aggregates are dissociated by treatment with sodium dodecyl sulfate (1 %), 8 M urea or 6 M guanidium chloride, by changes of pH, temperature or protein concentration, or by chemical modification (succinylation, maleylation, removal or attachment of cofactors). See Nuclear magnetic resonance (nmr) spectroscopy, X-ray crystallography.

Protein A: a protein produced by most strains of *Staphylococcus aureus.* It is mainly covalently linked to cell wall peptidoglycan, and a small proportion of the free protein is secreted into the growth medium. Certain methicillin-resistant strains of *S. aureus* produce only free P. A. One of the richest sources of P.A is the cell wall of *S. aureus* strain Cowan I, which contains 6.7 % by weight of the protein. P.A is usually isolated by digestion of bacteria with lysostaphin, followed by ion exchange, gel filtration, and finally affinity chromatography on IgG-Sepharose. It is a single polypeptide, M_r 42,000, with little or no carbohydrate, *C*-terminal lysine, and a blocked *N*-terminus. Tryptophan, cysteine and cystine are absent. There are 4 highly homologous domains (each containing a tyrosine residue, which serves as an iodination site for labeling purposes), and a *C*-terminal domain which is bound to the cell wall.

P.A induces synthesis of polyclonal antibodies in B lymphocytes of human and mouse origin, and it is probably a T-cell-regulated polyclonal activator of human B cells. It is chemotactic, blocks heat-labile and heat-stable opsonins, activates or inhibits complement fixation (depending on the dose) and has hypertensive activity when injected in some animals and humans.

P.A binds IgG of all mammals. It does not bind avian IgG, and ruminant IgG is bound only weakly. IgA and IgM of some species are also bound. This property is exploited in the isolation and purification of IgG subclasses, which are eluted separately from P.A-Sepharose with a stepwise pH gradient. ^{125}I-labeled P.A is especially useful as a revealing agent for bound antibodies (see, e. g. Recombinant DNA technology, subsection on Screening). [J. W. Goding *J. Immunol. Methods* **20** (1978) 241–253; A. Surolia et al. *Trends Biochem. Sci.* **7** (1982) 74–78]

Protein biosynthesis: a cyclic, energy-requiring, multistage process, in which free amino acids are polymerized in a genetically determined sequence to form polypeptides. P b. represents the translation of genetic information carried by mRNA. It requires the presence of mRNA, Ribosomes (see), tRNA, amino acids and a number of enzymes and protein factors, some of which are integral components of the ribosome. In addition, certain cations are required, and ATP and GTP are necessary as sources of energy. The mechanism and the individual steps of P. b. were elucidated mainly by careful preparation of the functional components, by the use of synthetic polynucleotides, and in particular by the use of specific inhibitors in cell-free systems. The process is similar in prokaryotes and eukaryotes, although it involves more factors in the latter; differences are mentioned in the text where appropriate. A standardized nomenclature exists for the "factors" involved in initiation (IF-1, IF-2, etc.), elongation (EF-1, etc.) and termination (release factor, RF). To distinguish them from prokaryotic factors, eukaryotic factors are designated "eIF-1", etc.

There are four phases of protein biosynthesis.

Activation of amino acids. Each amino acid is esterified with a specific tRNA (see Aminoacyl-tRNA synthetase). This phase does not require the presence of polysomes. The next phases occur successively on each ribosome of the polysome.

Initiation must start wiith a small ribosomal subunit; 80S or 70S ribosomes are inactive and their spontaneous dissociation into subunits under physiological conditions is slow. In *E. coli*, factor IF-1 promotes the dissociation of the ribosomes. The protein IF-3 (or eIF-3) prevents the association of the subunits, and thus makes the small one available for initiation. In eukaryotes, the actual first step is the binding of initiator tRNA (Met-tRNA$_f$) along with eIF-2 and GTP to the 40S subunit. In prokaryotes, this may be the first step, or it may come after the binding of mRNA. The assembled complex of 30S subunit, IF-1, 2 and 3, GTP and mRNA with the fMet-tRNA$_f$ is called the 30S-initiation complex. The corresponding initiation complex in eukaryotes involves another factor, eIF-4, and ATP, which are required for binding the mRNA. How much ATP is required is not known. The 40S initiation complex is thus 40S:eIF-3:Met-tRNA$_f$:eIF-2:GTP:eIF-4:mRNA. (eIF-4 consists of several proteins; see Table.)

The initiation codon for both pro- and eukaryotes is AUG, but the mechanism by which the ribosome locates it differs. Prokaryotes seem to recognize sequences in the messengers to the left (toward the 5′-end) of the AUG codon; these sequences are complementary to part of the 16S rRNA molecule. The 70S ribosomes can start either at the first AUG in the messenger or at an internal AUG, and can even translate circular messengers. Eukaryotic ribosomes, however, must start with the AUG nearest the 5′-end of the messenger. The m7G5′pppX "cap" (see Messenger RNA) appears to guide the system, but some uncapped messengers can also be translated. The role of the cap-binding protein (see Table) in this context is not clear. It is thought that the ribosome must "walk" down the messenger until it comes to the first AUG. It is possible that ATP provides the energy for the "walk".

In prokaryotes and eukaryotes, the initiation complex is prepared for the addition of the large ribosomal subunit by the release of initiation factor 3. In bacteria, the 50S subunit appears simply to replace IF-3, while IF-1 and IF-2 leave the complex afterward. In eukaryotes, another factor, eIF-5, catalyses the departure of the previous initiation factors and the joining of the 60S subunit. In both cases, release of initiation factor 2 involves hydrolysis of the GTP bound to it. Also, the Met-tRNA$_f$ is bound to the P site of the large ribosomal subunit.

Elongation. The ribosome can accommodate two tRNA molecules at once. One of these carries the Met-tRNA or the peptide-tRNA complex, and is thus called the P site; the other accepts the incoming aminoacyl-tRNA and is therefore called the A site. What binds to the A site is actually a complex of GTP, elongation factor TU, and aminoacyl-tRNA. The tRNA, of course, must be aligned with the next codon on the messenger which is to be read; the elongation factor is presumably responsible in some way for guiding it to precisely the right nucleotide triplet. The GTP is then hydrolysed to GDP, and the EF-TU :GDP complex leaves the ribosome. The GDP is released from the factor when the latter forms a complex with elongation factor TS: EF-TU:EF-TS. Although the affinity of EF-TU for GDP is higher than its affinity for GTP, the complex with EF-TS has a higher affinity for GTP, which replaces the EF-TS. The EF-TU-GTP is then ready to pick up another aminoacyl tRNA and to recycle. Meanwhile, on the ribosome, a reaction is catalysed between the carbox-

Prokaryotic and eukaryotic protein synthesis factors

Factor	M_r	Function
IF-1		Equilibration of 70S ⇌ 50S + 30S units. Stabilization of initiation complex.
IF-2		Binding of fMet-$tRNA_f$ to 30S subunit. Process may or may not require mRNA.
IF-3		Prevents association of 30S and 50S subunits
eIF-1	15,000	Stabilization of initiation complex.
eIF-2	α-subunit 32,000–38,000, β-subunit 47,000–52,000, γ-subunit 50,000–54,000	Binding of Met-$tRNA_f$ to 40S subunit; process requires GTP and occurs before mRNA is bound.
eIF-2A	50,000–96,000	Binding of Met-$tRNA_f$ to 40S subunit; process requires mRNA but not GTP.
eIF-3	500,000–750,000 (complex of 7–11 polypeptides)	Prevents association of ribosomal subunits, stabilizes initiation complex.
eIF-4A	48,000–53,000	Binding of mRNA to 40S initiation complex.
eIF-4B	80,000–82,000	Binding of mRNA to 40S initiation complex.
eIF-4C	19,000	Stabilization of initiation complex.
eIF-4D	17,000	Stabilization of initiation complex.
eIF-5	125,000–160,000	Release of eIF-2 and eIF-3 from initiation complex; binding of the 60S subunit to the 40S complex.
Cap recognition protein	24,000	Not clear; binds to mRNA cap.
EF-TU	43,000	GTP-EF-TU binds aminoacyl-tRNA to ribosomal A site.
EF-TS	35,000	Displaces GDP from EF-TU-GDP which has been released from the ribosome; EF-TU:EF-TS complex reacts with GDP to regenerate EF-TU-GTP.
EF-G	80,000	Involved in translocation of peptidyl-tRNA from the A to the P site; GTPase.
eEF-TU (EF-1α)	53,000	GTP-eEF-TU binds aminoacyl-tRNA to ribosomal A site.
eEF-TS (EF-1β)	30,000	Displaces GDP from eEF-TU-GDP which has been released from the ribosome; eEF-TU:eEF-TS complex reacts with GTP to regenerate eEF-TU-GTP.
eEF-G (EF-2)		Involved in translocation of peptidyl-tRNA from the A to the P site.
RF-1	47,000	Recognizes UAA and UAG termination codons; releases peptide from ribosome-bound tRNA.
RF-2	35,000–48,000	Recognizes UAA and UGA termination codons; releases peptide from ribosome-bound tRNA.
RF-3	46,000	Stimulates RF-1 and RF-2 activities.
eRF	56,000–105,000	Recognizes all three termination codons; has ribosome-dependent GTPase activity.

yl of the P-site occupant and the (free) amino group of the A-site occupant. The peptidyl transferase activity which catalyses this is intrinsic to the ribosome.

The final step of elongation is the motion of the ribosome relative to the mRNA, which is accompanied by the translocation of the peptidyl-tRNA from the A to the P site. Elongation factor G is involved in this step, although translocation can occur slowly in the absence of this factor. A complex of EF-G and GTP binds to the ribosome, and GTP is hydrolysed in the

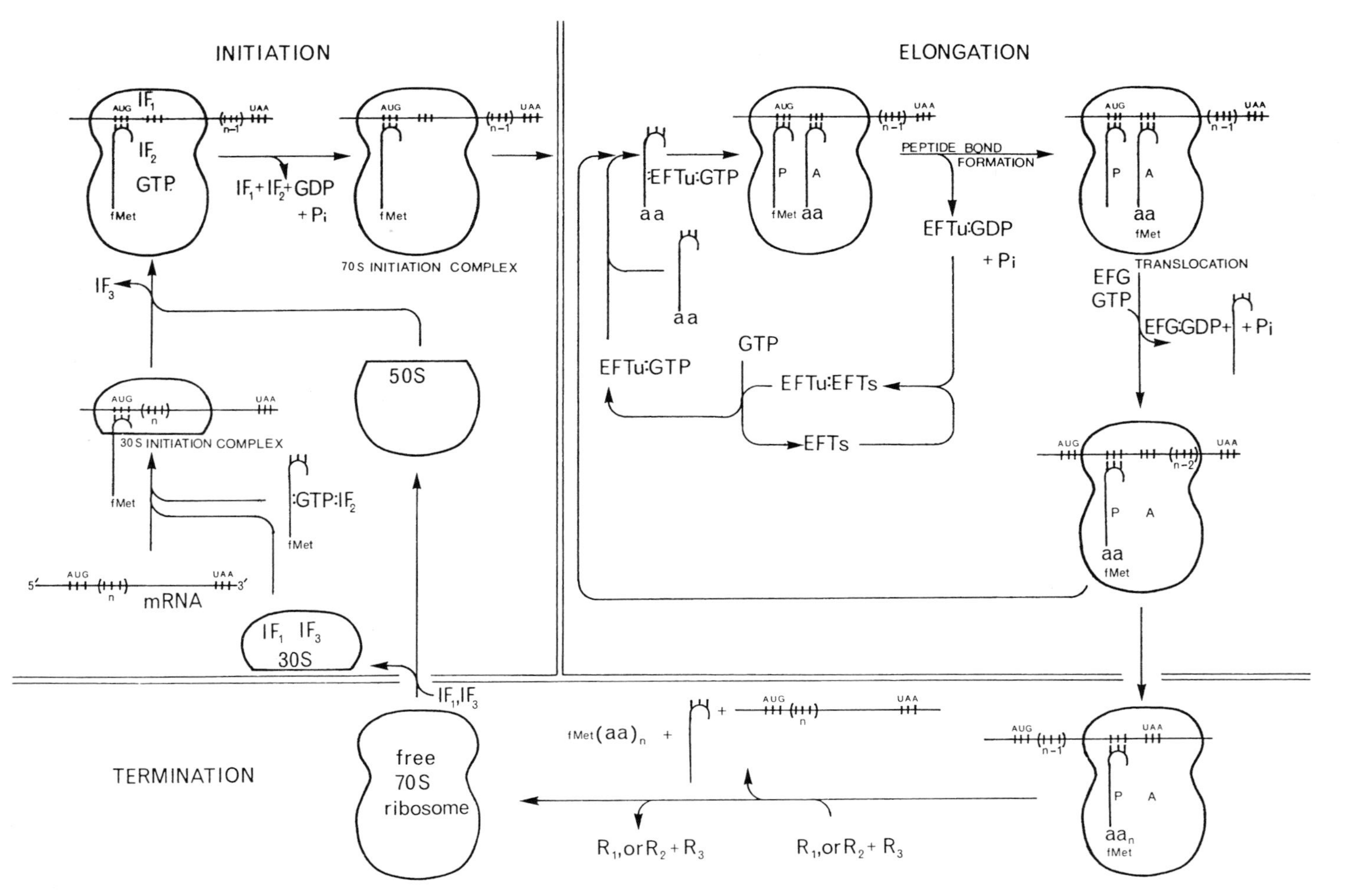

Diagrammatic representation of translation on prokaryotic ribosomes. The elongation cycle starts by interaction of the 70S initiation complex with fMet- tRNA:EFT_U:GTP. In all subsequent rounds of the cycle, fMet-tRNA:EFT_U:GTP interacts with the mRNA:ribosome complex carrying the growing polypeptide chain. Termination occurs when n amino acids have been incorporated, where n represents the number of codons between the initiation

course of the reaction. The deacylated tRNA is also released from the P site at this time.

It is interesting that IF-2 and EF-TU have analogous functions: IF-2 recognizes only Met-$tRNA_f$, while EF-TU recognizes all other aminoacyl-tRNAs; but each serves to introduce an aminoacyl-tRNA to a specific site on the ribosome. Moreover, the two proteins also have a stretch of homology about 100 amino acids in length. It must also be pointed out that hydrolysis of the GTP bound to EF-TU or IF-2 does not provide the energy, *per se* for the binding of the aminoacyl-tRNA to the ribosome. Instead, the GTP hydrolysis provides for the release of the tightly bound EF-TU or IF-2 from the ribosome. The energy for peptide bond formation is invested at the stage of loading the amino acids onto the tRNAs, when two phosphate bonds of ATP are hydrolysed in the course of forming the ester bond between the amino acid and the 2′ or 3′OH of the tRNA. All the other ATPs and GTPs hydrolysed in the course of P.b. apparently serve to increase the accuracy of the process.

Termination. The end of a polypeptide synthesis is signaled by a Termination codon (see) at the A site. Three prokaryotic release factors are known: RF-1 is specific for termination codons UAA and UAG, while RF-2 is specific for UAA and UGA. RF-3 stimulates RF-1 and RF-2, but does not itself recognize the termination codons. RF-3 also has GTPase activity; it is likely that it accelerates termination at the expense of GTP. Only one eukaryotic release factor is known, and this has GTPase activity. It is possible that eIF-3 interacts with the ribosome as it leaves the mRNA, causing the two subunits to separate. This is suggested by the observation that eukaryotic cells may have a large pool of 80S ribosomes which are not active in protein synthesis, and do not exchange with the pool of actively cycling ribosomal subunits. The preceding account describes P.b. for one ribosome, but it should be pointed out that the functional system is the Polysome (see). At any one time, several ribosomes are positioned along the mRNA; those nearest the 3′-end carry the longest newly synthesized polypeptide chains, whereas those at the 5′-end have translated fewer codons and therefore carry a relatively shorter length of peptide. Thus initiation, elongation (at various stages) and termination proceed simultaneously on the same length of mRNA.

The tertiary structure of the final protein begins to appear during P.b., before completion of the polypeptide. In many cases, the protein is subjected to further reactions, which convert it into its biologically active form, e.g. by covalent attachment of certain groups, or by removal of certain sequences of amino acids (see Post-translational modification of proteins). The initial translation product of secretory proteins contains a metabolically short-lived *N*-terminal peptide sequence, which functions in the attachment of the ribosome to the membrane of the rough endoplasmic reticulum and in the transfer of these proteins across the membrane into the tubules of the endoplasmic reticulum (see Signal hypothesis).

Proteinase inhibitors: see Inhibitor proteins.

Proteinases: a group of Proteases (see).

Protein-calorie malnutrition: see Protein-energy malnutrition.

Protein deficiency: see Protein-energy malnutrition, Kwashiorkor, Marasmus.

Protein degradation: see Proteolysis.

Protein-energy malnutrition, *PEM, protein-calorie malnutrition, PCM:* a spectrum of nutritional deficiency states, occurring characteristically in children under five years, although no age is immune. Marasmus and kwashiorkor are regarded as the two extremes of this spectrum. For many years, kwashiorkor was attributed to protein deficiency with adequate energy intake, in contrast to marasmus, which was attributed to deficiencies of both energy and protein. It is now known that there is no essential difference in the dietary protein:energy ratio of marasmus and kwashiorkor victims, and the characteristic clinical presentation of kwashiorkor is thought to be due to aflatoxin poisoning (see Kwashiorkor for further details). Three intermediate forms of PEM have been defined: marasmic kwashiorkor, nutritional dwarfism and "underweight child". PEM is often complicated by vitamin and mineral deficiencies, and by infection. Plasma albumin concentrations below 35 g/l are characteristic of protein deficiency and of PEM. Gamma globulin, on the other hand, may be raised in response to the presence of infection. There is always a decrease in metabolic rate, but this is roughly in proportion to the decrease in cell mass. Plasma concentrations of branched chain amino acids and tyrosine are lower than normal, whereas the concentrations of some nonessential amino acids may be increased; similar abnormal patterns of plasma amino acids appear after only 4 days on a protein-free diet. Therapeutic administration of protein causes an increase in the concentrations of plasma amino acids, sometimes accompanied by overflow aminoaciduria.

Loss by diarrhea is probably largely responsible for the low plasma potassium often observed in PEM; values less than 2.5 mmol/l have been recorded. Plasma magnesium may be low for the same reason. Plasma sodium is usually normal. Total body water increases to 65–80 % of body weight (60 % is normal).

Endocrine hypofunction is not a feature of PEM. Increased levels of certain hormones are sometimes observed, e.g. increased growth hormone has been reported in kwashiorkor; plasma cortisol and other adrenocorticosteroids may be raised in PEM. Thyroid function is usually normal, but fasting concentrations of plasma insulin may be low. For a review of the effect of malnutrition on circulating hormones, see R.D.G. Milner *Mem. Soc. Endocr.* **18** (1970) 191.

In PEM there appear to be no adaptive changes toward a more economic utilization of energy. Decreased protein intake results in increased levels of amino acid activating enzymes and decreased rates of urea synthesis, but normal amino acid and protein metabolism is rapidly restored when adequate dietary protein becomes available (except in kwashiorkor, where aflatoxin poisoning causes liver damage). Rapid recovery therefore proceeds immediately on receipt of a balanced diet; during recovery, energy utilization may be nearly 40 % higher than normal until the correct body weight for the age is achieved. See Kwashiorkor, Marasmus.

Protein engineering: the alteration of a protein by genetic or chemical means, or the direct chemical synthesis of proteins with novel properties. P.e. is performed in order to 1. study the effects of alterations on protein function, thereby defining the role of different amino acid residues and sequences in catalytic activity, binding properties and protein 3-dimensional structure, and 2. produce novel proteins with changed or optimized properties.

Ultimately, the aim of P.e. is to tailor proteins for specific roles in technology and medicine. This includes the production of enzymes with 1. increased stability to heat, extreme pH values, oxidizing atmospheres and organic solvents (a chemically engineered bacterial protease used in washing powders is stable at alkaline pH, stable up to 70 °C, resistant to denaturation by detergent, and resistant to oxidation by the chlorine used in bleaches), 2. improved or novel substrate specificity, 3. altered properties that facilitate recovery in downstream processing. Since the conformation and properties of proteins are determined by their primary sequence (see Protein folding), P.e. consists basically of the planned alteration of amino acid sequences of existing proteins. Alternatively, small proteins can be synthesized chemically (see Peptides). Chemical synthesis provides further control over the secondary structure of the protein by enabling the incorporation of unnatural amino acids, such as 2,2-dimethylglycine which is conformationally restricted and therefore imposes a stable secondary structure. Chemical modification of native enzymes can also be applied, with a view to retaining normal properties (e.g. enzymatic activity) while conferring greater stability. Thus, tissue plasminogen activator, used for the treatment of thrombosis, is made resistant to proteolytic degradation in the body by modification of its surface lysine residues by reaction with an acid anhydride.

A commonly used approach is to synthesize genes (see Recombinant DNA technology, DNA synthesis) encoding a chosen polypeptide sequence. Synthetic genes of up to about 100 nucleotides can be synthesized chemically. Hybrid genes can be chemically synthesized by joining segments of natural genes to chemically synthesized DNA sequences. Alternatively, new primary sequences can be programmed by using synthetic DNA sequences to extend natural genes, or to replace segments of natural genes. The final synthetic or hybrid DNA is then inserted into a plasmid for synthesis of the planned protein (see Recombinant DNA technology).

Engineered enzymes have application in the resolution of stereoisomers, especially in the pharmaceutical industry for the preparation of pure isomers of synthetically prepared drugs. For example, lactate dehydrogenase from *Bacillus stearothermophilus* has been engineered to accept substrates with a long branched side chain, without loss of its stereospecificity (the native enzyme reduces pyruvate to L-lactate).

Clinical applications of P.e. include the production of engineered antibodies. DNA encoding the specific antigen-binding site of a mouse monoclonal antibody is inserted into DNA from a human antibody gene, which is then expressed in cultured cells to produce a mouse-human hybrid antibody. This has the advantage that the human immune system does not recognize the antibody as foreign and therefore does not neutralize it, because most of the surface amino acid sequence is that of a human antibody. It is also possible to express synthetic antibody genes in bacteria, and it can be anticipated that all engineering of specific antibodies will eventually be performed independently of the immunization of animals. The potential for antibody production by P.e. is also important because antibodies are also used in biosensors and in diagnostic methods, as well as finding use as biocatalysts (see Catalytic antibody).

The visualization of native protein structure and the structure of engineered proteins by 3-dimensional computer graphics is widely used in P.e. (see e.g. van Aalten et al.). [D. A. Oxender & C. F. Fox (editors) *Protein Engineering. Tutorials in Molecular and Cell Biology,* Alan R. Liss, Inc., New York, 1987; H. M. Wilks et al. *Biochemistry* **29** (1990) 8587–8691; G. Winter & C. Milstein *Nature* **349** (1991) 293; D. M. F. van Aalten et al. *Protein Engineering* **8** (1995) 1129–1135; J. L. Cleland & C. S. Craik (eds.) *Protein Engineering, Principles and Practice* Wiley-Liss (A John Wiley & Sons, Inc., Publication) New York, Chichester, Brisbane, Toronto, Singapore, 1996 (ISBN 0-471-10354-3)]

Protein folding: the folding of a random coil polypeptide into its *native* structure, i.e. its 3-dimensional, biologically functional structure, also known as the native *conformation.* Loss of this native structure is known as *denaturation,* and the re-establishment of native structure is known as *renaturation.* (For the reductive denaturation and oxidative renaturation of pancreatic ribonuclease, see Protein).

In the living cell, a newly synthesized protein folds rapidly and spontaneously into its native 3-dimensional structure. Biological macromolecules and the components of macromolecular complexes generally exhibit an ability to fold and aggregate into appropriate tertiary and quaternary structures, despite the existence of many theoretically possibilities, a phenomenon known as *cooperative self-assembly.* The fact that folding to the native state occurs without the aid of any form of template strongly implies that the native state is dictated by information inherent in the amino acid sequence of the protein. However, in contrast to the rapid and accurate folding of protein chains in the living cell, the renaturation of many unfolded proteins in vitro under simulated physiological conditions occurs relatively slowly and with low efficiency (many non-native forms and unspecific aggregates may be formed). In fact, mechanisms do exist in the living cell to assist and accelerate correct folding, largely by enabling more efficient exploitation of the folding information already present in the amino acid sequence. For example, *folding accessory proteins* assist and accelerate correct folding and correct assembly to quaternary structures. In addition, disulfide bond formation in a newly synthesized protein may dictate the structure of the post-translationally modified protein (see Insulin).

The role of the primary sequence of a protein in determining its native conformation and ease of renaturation can be investigated by replacing or modifying amino acid residues, by studying the effects of mutations, and by comparing evolutionarily related and homologous proteins. Such studies have been performed largely on globular, nonmembrane proteins, which generally have an hydrophilic exterior surface and an hydrophobic interior. Different results should be anticipated for membrane proteins, which exist in an hydrophobic environment. In general, it can be stated that altered residues on the exterior surface of the folded nonmembrane protein are more readily tolerated than modified internal residues. It seems, therefore, that P.r. is driven by hydrophobic forces, related to the hydrophobic amino acids that largely occupy the interior of the protein. It is also observed that the proportion of α-helices and β-pleated sheets increases with the degree of compaction of the protein (equivalent to the number of intrachain contacts), indicating that while relatively powerful hydrophobic forces determine the general internalization and compaction of the folded protein, weaker (shorter-range) but more specific forces (hydrogen bonding, ionic pairing, Van der Waals forces) determine its unique internal structure.

It can be calculated that a small protein chain of 100 residues would require billions of years to explore all possible conformations, yet most proteins fold within seconds to their native conformation. P.f. cannot therefore occur by random selection, and must be ordered in some way, probably by the occurrence of sequential stages: 1. Short stretches of α-helices and β-turns are first formed, and these serve as nuclei (scaffolding) for stabilizing other ordered regions of the protein; 2. Such nuclei involve only 8–15 residues and are not necessarily permanent, being formed and lost in milliseconds. Those that belong to the native conformation persist and grow cooperatively, eventually forming a domain. A domain usually contains no more than about 200 residues, possibly reflecting the fact that the ordering of greater numbers of residues would take too long; 3. In the folding of multidomain proteins, there is an intermediate stage with extensive secondary structure, but disordered tertiary structure, in which the hydrophobic amino acids are partially exposed to the aqueous solvent; 4. This intermediate multidomain protein (or 'molten globule') undergoes small conformational changes and attains the compact tertiary structure of the monomeric protein; 5. In oligomeric proteins, monomers associate to a precursor of the quaternary structure; 6. Small conformational changes produce the native quaternary structure.

Once the native conformation is established, further subtle but crucial changes are induced by the specific binding of ligands (e.g. enzyme substrates, ligands of carrier proteins, ligands of receptors). Such changes can be calculated and monitored by computer simulation, using e.g. the molecular dynamics method and the essential dynamics method (Amadi et al.). A typical example of this approach is found in the study of retinol binding by retinol-binding protein (D.M.F. van Aalten et al.).

Folding accessory proteins. The following three types of folding accessory proteins are recognized. 1. *Protein disulfide isomerases (PDI)* catalyse the interchange of disulfide bonds (the random cleavage and reformation of Cys-Cys linkages) in the protein, until the most thermodynamically favorable conformation is attained and native Cys-Cys pairs are established. The mammalian enzyme contains 3 Cys residues, and one of these must have a free -SH group for enzymatic activity. The translation product of certain structural genes carries a Cys-containing *N*-terminal sequence, which has PDI-like activity, e.g. bovine pancreatic trypsin inhibitor (BPTI) is synthesized as a propeptide carrying a 13-residue, Cys-containing *N*-terminal sequence, which promotes correct folding of BPTI. This 13-residue sequence is excised *after* the BPTI has attained its native conformation. The attached *N*-terminal sequence and PDI may act synergistically to increase the rate and efficiency of folding. Both appear to serve as thiol-disulfide reagents. 2. *Peptidyl prolyl* cis-trans *isomerases (PPI or rotamases)* promote formation of the *cis* peptide bonds that precede Pro residues, by catalysing the *cis-trans* interconversion of X-Pro peptide bonds, where X is any other amino acid. 3. *Molecular chaperones* (see) are discussed separately.

Since the native conformation of a protein is ultimately determined by its primary sequence and its intracellular environment, it is theoretically possible to predict this native 3-dimensional structure from the amino acid sequence. Basically, this amounts to determining the minimum energy conformation of a protein, an approach that is still in its early stages of development, and which is not sufficiently advanced to allow reliable predictions of native protein structures.

An alternative and hitherto more successful approach is based on the comparison of known protein structures with their amino acid sequences. This empirical method has proved particularly successful in predicting secondary structure. Thus, amino acids display different propensities for occurring in an α-helix or a β-pleated sheet. The propensity to occur in an α-helix is defined as $P_\alpha = f_\alpha/<f_\alpha>$, where f_α is the frequency with which a residue occurs in α-helices in a number of proteins ($f_\alpha = n_\alpha/n$, where n_α is the number of times the amino acid occurs in α-helices, and n is the total number of times it occurs in all the studied proteins) and $<f_\alpha>$ is the average value of f_α for all 20 amino acids. P_β is derived similarly, based on the frequency of occurrence in β-pleated sheets. With respect to α-helix formation or β-pleated sheet formation, each amino acid residue can therefore classified as a strong former (*H*), former (*h*), weak former (*I*), indifferent former (*i*), breaker (*b*), or strong breaker (*B*). Using these propensity values, the positions of α-helices and β-pleated sheets can be predicted with up to 80 % reliability, although the average reliability is about 50 %. Further empirical considerations can be used to predict the occurrence of reverse turns, which are, for example, favored by stretches of low hydrophobicity and high hydrophilicity on the protein surface. See also Ramachandran plot. [E.R.Blout et al. *J. Amer. Chem. Soc.* **82** (1960) 3787–3789; P.Y.Chou & G.D.Fasman *Annu. Rev. Biochem.* **47** (1978) 251–276; G.D.Fasman (ed.) *Prediction of Protein Structure and the Principles of Protein Conformation,* Plenum Press, 1989; T.E.Creighton (ed.) *Protein Folding,* Freeman, 1992; M.Blaber et al. *Science* **260** (1993) 1637–1640; C.R.Matthews *Annu. Rev. Biochem.* **62** (1993) 653–684; Amadi et al. *Protein Stuct. Funct. Genet.* **17** (1993) 412–425; D.M.F. van Aalten et al. *Protein Engineering* **8** (1995) 1129–1135; 'Protein Stability and Folding: Theory and Practice' *Methods in Molecular Biology* (ed. B.A.Shirley) **40** (1995); E.Freire 'Thermodynamics of Partially Folded Intermediates in Proteins' *Annu. Rev. Biophys. Biomol. Struct.* 24 (1995) 141–165; J.T.Pederson & J.Moult 'Genetic algorithms for protein structure prediction' *Curr. Opin. Struct. Biol.* **6** (1996) 227–231; P.Bamborough & F.E.Cohen 'Modeling protein-ligand complexes' *Curr. Opin. Struct. Biol.* **6** (1996) 236–241]

Protein kinase C: a membrane protein apparently involved in several types of cellular signal transduction. It phosphorylates serine and threonine residues and depends on calcium and phospholipid for activity. At physiological calcium concentrations it is activated by diacylglycerol, a product of the action of phospholipase C on inositol phospholipids (see Inositol phosphates).

P. purified from bovine brain has M_r 81,000; the primary structure of the molecule has been deduced from cloned cDNA [P.J. Parker et al. *Science* **233** (1986) 853–859]. Hybridization of cDNA to libraries of human, bovine and rat genomes revealed that there are several distinct forms of the enzyme, which in the human genome are located on separate chromosomes. It is postulated that the existence of a family of similar P. may explain the diversity of effects of the enzyme and also account for differentiation of its functions in different tissues. [L. Coussens et al. *Science* **233** (1986) 859–866; Y. Nishizuka *Nature* **334** (1988) 661–665]

Binding of a number of growth factors, hormones and extrinsic agents to their receptors causes an increase in cellular calcium and cleavage of inositol phospholipids, which thus serve as second messengers. P. is one of the targets of these messengers, presumably regulating the activity of the proteins it phosphorylates and thus affecting the cellular activities. P. is evidently the site of binding of Phorbol esters (see) and other tumor promoters. It is thought that activation of P. by these esters mimics the effects of growth or proliferation signals, but is not subject to the same controls as the latter. [S. Jaken 'Protein kinase C isoenzymes and substrates' *Current Opinion in Cell Biology* **8** (1996) 168–173]

Proteinoids: heteropolyamino acids; artificially prepared polypeptides ($M_r > 1,000$) formed in 20–40 % yield by heating a mixture of several amino acids for 16 h at 170 °C (thermal condensation). P. show many similarities with globular proteins, e.g. relative quantities of individual amino acids, solubility, spectral properties, denaturation, degradation by proteases, catalytic activity (e.g. esterase, ATPase, decarboxylase activities) and hormone action (MSH activity). They can be regarded as models for the first information-carrying molecules. In water, they become organized into microsystems with a typical ultrastructure (microspheres), which share a number of properties with living cells, i.e. they possess a bilayer membrane, which exhibits a certain degree of semipermeability; they can also multiply in the absence of nucleic acids, by budding.

Protein-tyrosine kinase, *PTK* (see also Receptor tyrosine kinases): an enzyme that catalyses the ATP-dependent phosphorylation of tyrosyl residues in proteins:

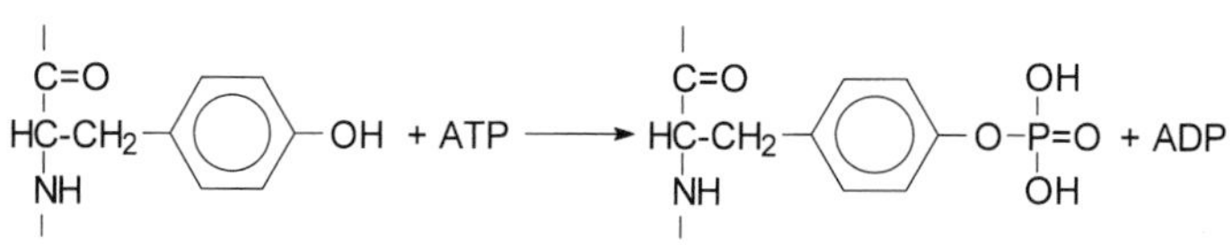

Tyrosyl residue (C=O / HC-CH$_2$-C$_6$H$_4$-OH / NH) + ATP → Phosphotyrosyl residue (C=O / HC-CH$_2$-C$_6$H$_4$-O-P(=O)(OH)$_2$ / NH) + ADP

Some normal cellular proteins and oncogene products which possess protein-tyrosine kinase activity

Normal protein	Normal gene	Oncogene product	Oncogene	Oncogene carrier
Insulin receptor	Insulin receptor gene	–	–	–
Insulin-like growth factor (IGF-1)	IGF-1 gene	–	–	–
Platelet-derived growth factor	c-*sis*	p28^{sis}	v-*sis*	Simian sarcoma virus
Epidermal growth factor receptor	c-*erb*-B	gp68/74$^{v\text{-}erb\text{-}B}$	v-*erb*-B	Avian erythroblastosis virus
pp60$^{c\text{-}src}$	c-src	pp60$^{v\text{-}src}$	v-*src*	Rous sarcoma virus
p98$^{c\text{-}fps}$	c-*fps*	p140$^{gag\text{-}fps}$	v-*fps*	Fujinami avian sarcoma virus
p92$^{c\text{-}fes}$	c-*fes*	p85$^{gag\text{-}fes}$	v-*fes*	Many feline sarcoma viruses
p-150$^{c\text{-}abl}$	c-*abl*	p120$^{gag\text{-}abl}$	v-*abl*	Abelson murine leukemia virus
–	c-*ros*	p68$^{gag\text{-}ros}$	v-*ros*	Avian sarcoma virus UR2
–	c-*yes*	p90$^{gag\text{-}yes}$	v-*yes*	Y73 Avian sarcoma virus
–	c-*fgr*	p70$^{gag\text{-}fgr}$	v-*fgr*	Gardner-Rasheed feline sarcoma virus
Colony-stimulating factor (CSF-1) receptor	c-*fms*	gp140$^{v\text{-}fms}$	v-*fms*	McDonough feline sarcoma virus

PTK activity is characteristic of certain membrane proteins, and about one third of known oncogenes and their proto-oncogene counterparts code for PTKs (see Table). The phosphotyrosine content of cell proteins may increase up to 10-fold in response to transformation by viral PTK-containing oncogenes, with the exception of v-*erb*, v-*fms* and v-*ros*. Among the substrates which become phosphorylated are: the glycolytic enzymes, enolase, phosphoglycerate mutase and lactate dehydrogenase three cytoskeletal proteins (vinculin of adhesion plaques; p36, which is part of a Ca^{2+}-sensitive complex in the submembranous cortical cytoskeleton and perhaps in ribonucleoprotein particles; p81 which, like p36, is submembranous, but present in microvillar cores), a 50,000 M_r protein associated with $pp60^{v\text{-}src}$, and some membrane glycoproteins. PTKs become autophosphorylated by acting as substrates to their own PTK activity; in some cases autophosphorylation appears to be necessary to sustain PTK activity towards other substrates. In the case of membrane receptor PTKs (see Insulin receptor), binding of the specific ligand on the outer surface of the cell promotes the appearance of PTK activity in the cytoplasmic domain. See Oncogene. [T. Hunter & J. A. Cooper *Ann. Rev. Biochem.* **54** (1985) 897–930; J. A. Cooper, *BioEssays* **4** (1986) 9–15]

Proteoglycans: high M_r compounds of carbohydrate and protein, found in animal structural tissues, e. g. the ground substance of cartilage and bone. The ground substance and gel fluids of these tissues owe their viscosity, elasticity and resistance to infective organisms to the presence of P. Each P. contains 40–80 acidic mucopolysaccaride chains (glucosaminoglycans) bound to protein via *O*-glycosidic linkages to serine or threonine. In contrast to the Glycoproteins (see), the prosthetic group of P. has M_r 20,000–30,000, consisting of many (100–1000) unbranched, regularly repeating disaccharide units. The disaccharides are composed of *N*-acetylhexosamine (which may or may not be sulfated) linked to a uronic acid or to galactose. In the chondroitin sulfates, the linkage region between polysaccharide and protein contains xylose linked *O*-glycosidically to serine; this is followed by two galactose residues and one glucorunic acid, to which the first repeating disaccharide is attached. In corneal keratan sulfate the polysaccharide chain is linked to the protein via a glucosamine residue, which is joined to an asparagine side chain by a glycosylamine linkage. In cartilage keratan sulfate, most of the carbohydrate-protein linkages are *O*-glycosidic bonds between *N*-acetylglucosamine and the hydroxyl of serine or threonine.

The chondroitin sulfates, together with collagen, form the major component of cartilage. Mammalian skin contains proteodermatan sulfate, and the intestinal mucosa contains protein-bound heparin. Heterogeneity of P. is due to differences in polypeptide chain length, and to the number and distribution of the attached polysaccharide chains. Microheterogeneity also exists, due to small differences in the chain lengths of the polysaccharide chains, and the distribution of sulfate residues.

P. can be extracted from cartilage under mild conditions with 4 M guanidinium chloride. The resulting P. subunit has $S_w = 16S$ (see Sedimentation coefficient), and M_r 1.6×10^6. In the tissue, P. exist as giant molecular aggregates ($S_w = 70S$ and 600S), which are formed by noncovalent association of P. subunits with a glycoprotein.

Proteohormones, ***protein hormones:*** proteins (often glycoproteins) with hormonal function. Like other proteins, they are synthesized by the translation of appropriate mRNA, and degraded by proteolysis. M_r of P. are between 5,000 and 25,000 for the monomers and correspondingly higher for the dimers and polymers. See Choriogonadotropin, Follicle stimulating hormone, Luteinizing hormone, Thyrotropin. Although a close relationship exists between P. and Peptide hormones (see), a distinction is made between these two.

Proteolysis, ***protein degradation:*** hydrolysis of proteins by the action of proteolytic enzymes, or nonenzymatically by acids (e. g. 6 M HCl at 110 °C for 24 h or longer) or alkalis. Ultimate products of P. are amino acids. Dietary proteins are hydrolysed to L-amino acids by P. in the intestine. After absorption, these amino acids are used in the synthesis of new proteins specific to the organism; these, in turn, are eventually hydrolysed to amino acids by P. as part of the continual process of synthesis and degradation of cellular constituents (see Turnover). A distinction is

Formation of biologically active proteins from inactive precursors by limited proteolysis AA, amino acids; Chy, chymotrypsin; CP, carboxypeptidase

Inactive protein	M_r	No. of AA	Active fragment	M_r	No. of AA
Pepsinogen	42,500	362	Pepsin	34,500	327
Prorennin	36,200	321	Rennin	30,700	272
Trypsinogen	24,000	229	Trypsin	23,400	223
Chy-ogen	25,666	245	Chy A	25,170	241
Pro-CP A	90,000	850	CP A	34,300	307
Prothrombin	72,000	560	Thrombin	39,000	309
Proinsulin	9,100	84	Insulin	6,000	51
Fibrinogen	340,000	3,400	Fibrin monomer	327,000	~3,270

In limited P. only certain peptide bonds of a protein are hydrolysed; this results in the production of biologically active (e. g. enzymes or hormones) or inactive (e. g. para-ϰ-casein) proteins or peptides. Limited P. occurs in digestion, blood coagulation and milk clotting; it is responsible for the activation of zymogens and for the release of certain peptide hormones, e. g. insulin, angiotensin, vasopressin, oxytocin and various kinins (see table).

drawn between extracellular P. (e.g. digestion and blood coagulation) and intracellular P. The latter occurs at neutral and acidic pH values, and the relevant endopeptidases (called cathepsins) are localized chiefly in the lysosomes. The cathepsins from protease-rich organs, like liver, spleen and kidney, can be separated into cathepsins A, B, C, D, E and L. Cathepsins A to E have pH optima in the range 2.5–6, and (with the exception of D and E) they also hydrolyse synthetic, low M_r substrates. Other cathepsins are active only at neutral pH, and they attack only proteins. M_r of cathepsins are between 25,000 (cathepsin B) and 100,000 (cathepsin E). Cathepsins B, B_2, C and some neutral cathepsins are SH-enzymes. In the intact cell P. is controlled and it occurs in the lysosomes (autophagy). In damaged cells, the same cathepsins are released from the ruptured lysosomes and are responsible for autolysis, i.e. the uncontrolled total degradation of the cell. See Proteasomes.

Prothrombin, ***factor II:*** the precursor of Thrombin (see) in the Blood coagulation (see) system. The protein is a single polypeptide chain, M_r 72,000 (582 amino acid residues), with a carbohydrate content of 14.7% (bovine) or 11.8% (human). Synthesis of P. occurs in vertebrate liver and requires vitamin K for the synthesis of γ-carboxyglutamate (Gla) residues in the *N*-terminal region of the chain. During blood coagulation, P. is converted to thrombin by factor X_a (EC 3.4.21.6). This reaction is accelerated 20,000-fold by active factor V and acidic phospholipid; the latter is normally sequestered on the inside of cell membranes, and its presence in plasma would thus signal injury. However, the reaction is accelerated by a factor of 100,000 in the presence of platelets and factor Va.

P. binds to the phospholipids by its 10 Gla residues; the binding is mediated by Ca^{2+}. The region (called F1) containing these residues is removed by thrombin in vitro; however, in vivo, this cleavage is suppressed. The first cleavage by factor X_a removes the segment called F1·2 (M_r of 32,000). In addition to the Gla region, this segment contains the F2 region, which mediates binding to factor V_a. Finally, the Arg-Ile bond between peptides A and B is cleaved to produce mature thrombin.

Prothromboplastin: see Blood coagulation.

Protoalkaloids: see Biogenic amines.

Protocyanin: see Cyanidin.

Protoheme, ***heme, ferroheme, ferroprotoporphyrin, protoheme IX:*** [7,12-diethenyl-3,8,13,17-tetramethyl-21H,23H-porphine-2,18-dipropanoate (2-)-N^{21},N^{22},N^{23},N^{24}]-iron; or 1,3,5,8-tetramethyl-2,4-divinylporphine-6,7-dipropionic acid ferrous complex, $C_{32}H_{32}FeN_4O_4$, M_r 616.48. Protoheme crystallizes as fine brown needles with a violet sheen; $\varepsilon_{572} = 5.5 \times 10^3$; in phosphate buffer pH 7.0, absorption maxima occur at 575 nm and about 550 nm.

It is the prosthetic group of a number of hemoproteins, e.g. hemoglobins, erythrocruorins, myoglobins, some peroxidases, catalase and cytochromes *b*. The four coordinate bonds of the iron lie in the plane of the nearly planar porphyrin ring structure, while the two unoccupied sites of the iron are perpendicular to it.

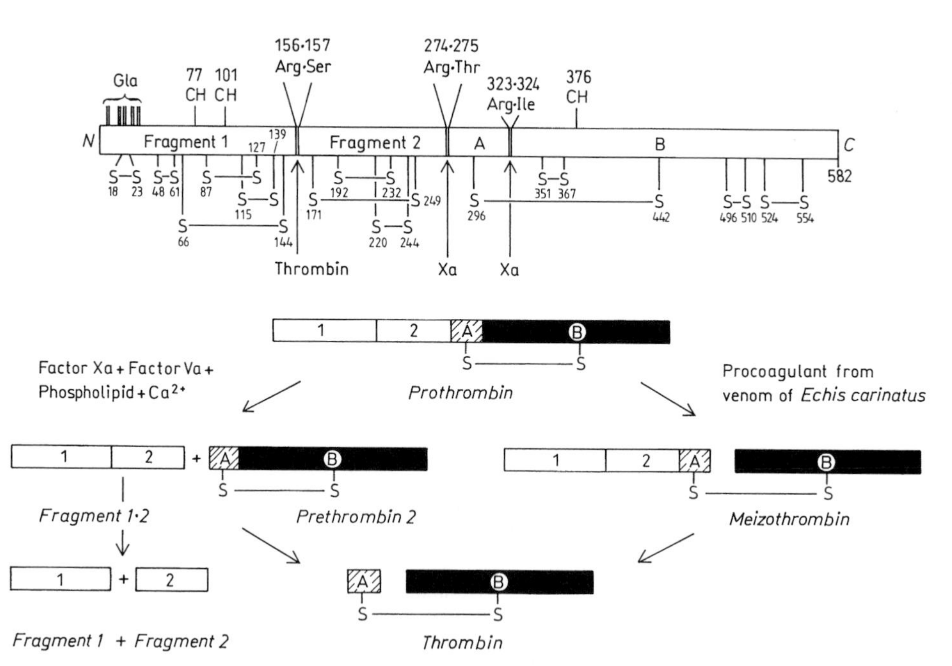

Structure of prothrombin and its conversion into thrombin. The pathway on the left operates during normal blood coagulation in response to injury, whereas the right-hand pathway is promoted by the procoagulant of snake venom. Fragments 1 and 2, formed from fragment 1·2 by the action of thrombin, have no known physiological function. Gla = 4-carboxyglutamate residues.

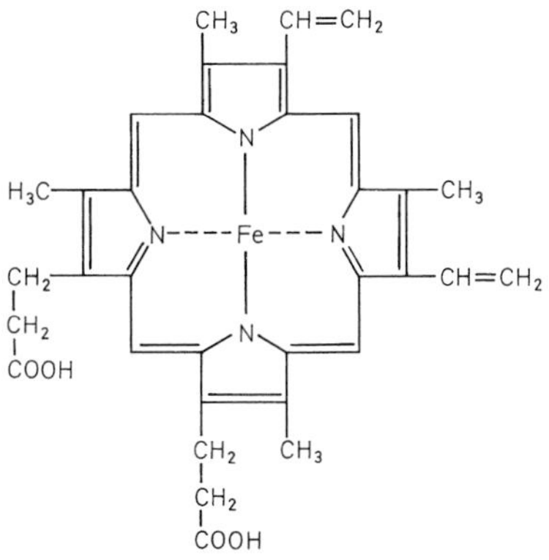

Protoheme

Protokaryote: see Cell.

Protomer: see Subunit.

Proton motive force: the proton (H^+) equivalent of electromotive force (EMF). The term was used by Mitchell in his Chemiosmotic hypothesis (see) to describe the 'force' of the proton electrochemical gradient which is generated across the relevant energy-transducing membrane when electrons flow exergonically through the electron transport chains of mitochondria (inner membrane), chloroplasts (thylakoid membrane) and bacteria (plasma membrane), and which is used to drive the endergonic, ATP-synthase (F_0F_1 complex)-catalysed formation of ATP by the reaction ADP + P_i → ATP + H_2O. PMF (also referred to by the terms $\Delta\mu_{H^+}$ and Δp), technically defined as 'the electrochemical potential difference for protons between two bulk phases separated by a membrane (in this case the energy-transducing membrane)', has dimensions of volts and can be described by the following expression:

$$\text{PMF (or } \Delta\mu_{H^+} \text{ or } \Delta p) = [(2.303RT/F).\Delta pH] + \Delta\psi \quad \text{(i)}$$

where: R = Gas constant = 8.314 $JK^{-1}mol^{-1}$;
T = Absolute temperature = 273.15 + °C;
F = Faraday = 96487 $Cmol^{-1}$;
$\Delta pH = pH_{in} - pH_{out}$ (where 'in' and 'out' refer to the inside and ouside of the relevant cellular compartment);
$\Delta\psi$ = the electrical potential difference (V) between the two bulk phases separated by the relevant membrane, with the outside being more positive than the inside (i.e. $\Delta\psi = \psi_{out} - \psi_{in}$).

This expression is derived in the following way. The reaction in question is:

$H^+_{in} \rightarrow H^+_{out}$ (where 'in' and 'out' have the meanings given above, e.g. in the context of the mitochondrion they refer to the matrix and the intermembrane space respectively).

In this reaction two endergonic processes occur, namely: (i) the formation of a concentration (i.e. a chemical potential) gradient across the membrane with $[H^+]_{out}$ greater than $[H^+]_{in}$, and (ii) the formation of an electrical potential gradient across the membrane with the outside more positive than the inside, because H^+ is a charged species. These two processes combine additively to give a 'proton electrochemical gradient'. The magnitude of the 'chemical' (concentration) component of this proton electrochemical gradient is derived in the following way. The equation for the free energy change, ΔG, in a chemical reaction that converts a reactant, R, into a product, P, is:

$$\Delta G = \Delta G^{o\prime} + 2.303RT.\log([P]/[R]) \quad \text{(ii)}$$

When the reaction is simply the transport of a solute between two compartments in which its concentration is different (as in the reaction in question), no chemical bonds are broken or made and the standard free energy change, $\Delta G^{o\prime}$, is zero. Thus equation (ii), for the '$H^+_{in} \rightarrow H^+_{out}$' reaction becomes

$$\Delta G = 2.303RT.\log([H^+]_{out}/[H^+]_{in}) \quad \text{(iii)}$$

in which ΔG is the difference in chemical potential rather than the difference in free energy, because H^+ has been treated as an uncharged species.

The logarithmic term of equation (iii) can be modified as follows:

$$\log([H^+]_{out}/[H^+]_{in}) = \log[H^+]_{out} - \log[H^+]_{in} \quad \text{(iv)}$$

but $pH = -\log[H^+] \therefore \log[H^+] = -pH$

Substituting for log $[H^+]$ in (iv) gives:

$= (-pH_{out}) - (-pH_{in})$
$= -pH_{out} + pH_{in}$
$= pH_{in} - pH_{out}$
$= \Delta pH$ (i.e. ΔpH in this context is defined as $pH_{in} - pH_{out}$)

Thus equation (iii) becomes:

$$\Delta G = 2.303RT.\Delta pH \quad \text{(v)}$$

The magnitude of the 'electrical' component of the proton electrochemical gradient is given by the expression

$$\Delta G = ZF\Delta\psi \quad \text{(vi)}$$

where Z = the electrical charge on the ion (i.e. 1 for H^+);
F = Faraday = 96487 $Cmol^{-1}$;
$\Delta\psi$ = the electrical potential difference (V) between the H^+_{out} bulk phase and the H^+_{in} bulk phase, the former being more positive than the latter;
ΔG = the difference in electrical potential rather than the difference in free energy because H^+ has been treated purely as a positive charge.

The true free energy change, ΔG, of the '$H^+_{in} \rightarrow H^+_{out}$' reaction is the sum of the 'chemical' and 'electrical' potential components, which is obtained by adding equations (v) and (vi) to give:

$$\Delta G = 2.303RT.\Delta pH + F\Delta\psi \quad \text{(vii)}$$

(Z is omitted because it equals 1)

Equation (i) can be obtained by dividing equation (vii) by F, as follows

$$\Delta G/F = [(2.303RT/F).\Delta pH] + \Delta\psi \quad \text{(viii)}$$

where $\Delta G/F$ has dimensions of volts (i.e. [J/mol]/[C/mol] = J/mol × mol/C = J/C = V) and is the proton motive force (PMF or $\Delta\mu_H$ + or Δp), thus giving equation (i); a similar line of reasoning shows that division by F has given the 'chemical' potential term the dimensions of volts.

At 30 °C equation (viii) becomes PMF = 0.06.ΔpH + Δψ.

In the mitochondrion ΔpH is about 1.4, thereby giving a value of 0.06 × 1.4 = 0.084V for the 'chemical' potential component, and Δψ is about 0.14V; thus the PMF has a value at 30 °C of about 0.224V. This corresponds to a ΔG value of about 21.6 kJ per mole of H^+ transported, as can be calculated by insertion of the ΔpH and Δψ values into equation (vii). Thus it can be deduced that at least two moles of H^+ have to be transported from the mitochondrial matrix to the intermembrane space by exergonic flow of electrons down the electron transport chain to drive the endergonic generation of one mole of ATP (ADP + P_i → ATP + H_2O; $\Delta G^{o'}$ = + 30.5 kJ.mol^{-1}).

Proton nmr: see Nuclear magnetic resonance spectroscopy.

Protopectin(s): a ground substance in plant cell walls. P. consists of insoluble Pectins (see) and are probably not pure homoglycans. They are present in the cell wall as salts of calcium and magnesium. The constituent polygalacturonic acid chains of P. are linked to one another by salt linkages, phosphate bonds and esterification with arabinose.

Protoplast: see Cell, 2.

Prototrophism: the property of being able to grow at the expense of usual or common nutrients (see Nutrient medium), with no special requirement for Growth factors (see). P. is a property of prototrophic organisms.

Provitamin D_2: see Ergosterol.

Provitamin D_3: see 7-Dehydrocholesterol.

Provitamins: inactive precursors of Vitamins (see). P. are mostly of vegetable origin, and are converted into active vitamins after absorption from the diet.

PRPP: acronym of 5-phosphoribosyl 1-pyrophosphate.

Prunetin: 5,4′-dihydroxy-7-methoxyisoflavone. See Isoflavone.

Prunetrin: prunetin 7-glucoside. See Isoflavone.

Prunin: naringenin 7-glucoside. See Flavanone.

Pseudoalkaloids: a group of alkaloids earlier assigned to other groups (e.g. some were grouped with the terpenes) with which they show a close structural relationship. At the time, their nitrogen content seemed incidental.

Pseudobaptigen: see Pterocarpans.

Pseudogibberellin A_1: see Gibberellins.

Pseudohermaphroditism: see Inborn errors of metabolism (Adrenal hyperplasia).

Pseudoindicans: an old name for Iridoids (see).

Pseudo-isoenzymes: multiple forms of an enzyme, which catalyse the same reaction. They have similar properties to isoenzymes, but do not have genetically determined differences of primary structure. Their multiplicity is the result of enzymatic or nonenzymatic modification of one original primary sequence, either in vivo or in vitro (i.e. during isolation). Examples of P. formed in vivo are the different chymotrypsins, trypsins, pepsins and carboxypeptidases, each group being derived from a single zymogen. Also in this category are the different degrees of aggregation shown by oligomeric enzymes that consist of identical subunits, e.g. glutamate dehydrogenase. P. formation in vitro is responsible for the occurrence of the 4 α-amylases, the 2 yeast phosphofructokinases, as many as 13 heart muscle lipoyl dehydrogenases, and the numerous forms of phosphoglucose isomerase.

Pseudopelletierine, ***Ψ-pelletierine, pseudopunicine, 9-methyl-3-granatanone:*** 9-methyl-9-azabicyclo-[3,3,1] nonan-3-one, the most important representative of the Punica alkaloids, present in the root bark of *Punica granatum*. M_r 153.21, m.p. 54 °C, b.p. 246 °C. Its structure is based on the meso form of granatane (9-azabicyclo [3,3,1] nonane). For biosynthesis, see Punica alkaloids.

Pseudotropine: see Tropane alkaloids.

Pseudouridine, ***5-β-D-ribofuranosyluracil, 5-ribosyluracil, Ψ:*** a structural analog of uridine containing a C-C bond between C5 of uracil and C1 of ribose. M_r 244.2, m.p. 223–224 °C. Ψ is a Rare nucleic acid component (see) found in tRNA. Despite earlier conflicting evidence, it is now clear that Ψ is formed by rearrangement of uridine after assembly of the tRNA chain, i.e. by post-transcriptional modification of the tRNA. An enzyme that can modify specific uridine residues in the anticodon region of many species of tRNA has been purified from *Salmonella typhimurium* and *E. coli*.

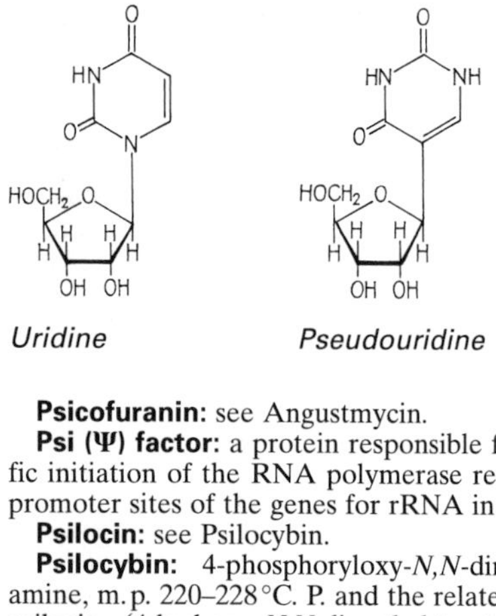

Uridine *Pseudouridine*

Psicofuranin: see Angustmycin.

Psi (Ψ) factor: a protein responsible for the specific initiation of the RNA polymerase reaction at the promoter sites of the genes for rRNA in bacteria.

Psilocin: see Psilocybin.

Psilocybin: 4-phosphoryloxy-*N,N*-dimethyl-tryptamine, m.p. 220–228 °C. P. and the related compound psilocin (4-hydroxy-*N,N*-dimethyl-tryptamine, m.p. 173–176 °C), are jointly responsible for the psychotropic action of the fruiting body of the Mexican hallucinogenic fungus Teonanacatl (*Psilocybe mexicana*). P. was the first naturally occurring phosphorylated indole derivative to be isolated. It can be hydrolysed to psilocin. Both compounds are slightly poisonous. Administered by mouth or by intramuscular injection, they cause hallucinations similar to those caused by LSD; the latter is, however, about 100 times more potent.

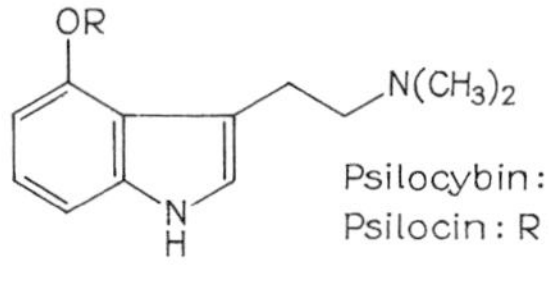

Psilocybin : R = H_2PO_3
Psilocin : R = H

Psychodelic drugs: see Hallucinogens.

Psychodysleptic, Psycholitic and Psychotomimetic drugs: see Hallucinogens.

Psychotropic agents: chemical compounds that influence the human psyche. P. a. are used in psychiatry. They include Narcotics (see) and Hallucinogens (see), and are almost exclusively of vegetable origin. About 50 different modes of psychotropic activity are found among about 20,000 natural plant products. In addition, there are semisynthetic (chemically modified natural products) and synthetic P. a., some of which are used in medicine. Some P. a. are also used as narcotics. All P. a. eventually produce a personality change. Use and possession of P. a. are therefore subject to legal controls, designed to prevent misuse. P. a. have scientific as well as medical uses; psychotropic natural products serve as model substances in pharmacy, pharmacology and toxicology, and P. a. in general are used in the biochemical investigation of the function of the central nervous system.

Pteridines: a group of compounds containing the pteridine ring system (Fig.). The majority of naturally occurring P. are chemically related to pterine (Fig.). A smaller number are derived from lumazine (Fig.). Both folic acid [see Tetrahydrofolate; Vitamins (B_2 complex)] and Tetrahydrobiopterin (see) are P. and serve as hydrogen transfer cofactors. Folic acid is a vitamin for mammals, but they are able to synthesize tetrahydrobiopterin; in spite of their chemical similarity, these compounds are synthesized by different pathways.

Pteridine

Pterine

Lumazine

Because of their role as enzyme cofactors, P. are ubiquitous. The cofactors are metabolized and excreted or deposited as pigments, e. g. as Xanthopterin (see), Leucopterin (see), sepiapterin, etc. in the wings of insects. The compounds were discovered by G. Hopkins in 1890, in butterfly wings. The name is derived from *pteron,* the Greek for wing. Mammals excrete bio-, xantho-, neopterin and others in their urine.

Elevated excretion of neopterin is correlated with certain malignant diseases, viral infection and graft rejection, apparently because neopterin is secreted by macrophages in the course of T-lymphocyte activation [I. Ziegler, in *Biochemical and Clinical Aspects of the Pteridines* **4** (1985) 347–361]. Activation of lymphocytes in vitro is stimulated by sepiapterin, dihydro- and tetrahydropterins, but repressed by xantho- and isoxanthopterins. These P. are thus lymphokines.

Pterins: see Pteridines.

Pterocarpans, ***coumaranochromans:*** isoflavonoids with the ring system shown (Fig).

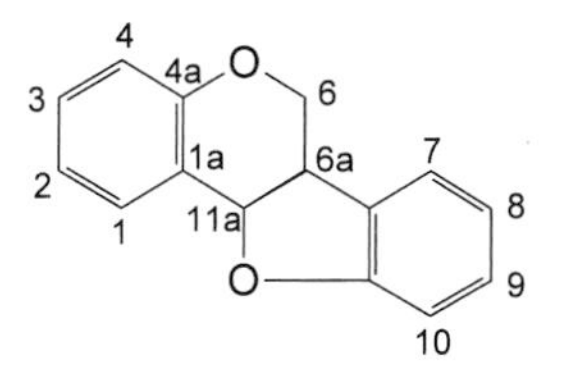

Pterocarpan ring system (systematic name: 6a,11a-dihydro-6H-benzofuro[3,2c][1]benzopyran; the systematic Ring Index numbering is shown)

Examples: ***pterocarpin*** (3-methoxy-8,9-methylenedioxypterocarpan, *Sophora japonica*), ***homopterocarpin*** (3,9-dimethoxypterocarpan), ***ficifolinol*** (3,9-dihydroxy-2,8-di-γ,γ-dimethylallylpterocarpan, *Neorautanenia ficifolia*). Some P. are Phytoalexins (see), e. g. ***Phaseolin*** (see), ***Glyceollins*** (see), ***pisatin*** (3-methoxy-6a-hydroxy-8,9-methylenedioxypterocarpan, *Pisum sativum*), ***maackiain*** (3-hydroxy-8,9-methylenedioxypterocarpan, *Lathyrus* spp.), ***variabilin*** (3,9-dimethoxy-6a-hydroxypterocarpan, *Lathryus* spp.), ***medicarpin*** (3-hydroxy-9-methoxypterocarpan, *Lathyrus* spp.), ***nissolin*** (3,9-dihydroxy-10-methoxypterocarpan, *Lathryus nissolia*), ***methylnissolin*** (3-hydroxy-9,10-dimethoxypterocarpan, *Lathyrus nissolia*) [D. J. Robeson & J. L. Ingham *Phytochemistry* **18** (1979) 1715–1717]. Many P., however, have been isolated from healthy, unstressed plants, especially from the heartwood of tropical genera of the *Leguminoseae.*

The relative configuration of the 6a,11a chiral center is *cis.* The configuration of (–)-P. is 6a*R*,11a*R*, and that of (+)-P. is 6a*S*,11a*S* [K. G. R. Pachler & W. G. E. Underwood *Tetrahedron* **23** (1967) 1817–1826].

Like other isoflavonoids, P. are largely restricted to the *Leguminoseae.* Structural relationships governing the phytoalexin (i. e. fungicidal) properties of P. have been studied [H. D. Van Etten *Phytochemistry* **15** (1976) 655–659]. 6a,11a-dehydro-P. (which have a fundamentally different shape from that of P.) are also active fungicides. A common 3-dimensional molecular shape, therefore, does not seem to be necessary.

Biosynthesis. Investigation of the biosynthesis of phytoalexin P. is greatly helped by the fact that their synthesis can be induced by treatment of the plant with fungi, Elicitor (see), UV light or heavy metal salts. Isotopically labeled precursor can be added at the time of maximal synthesis, and a high incorporation of isotope is observed. It has been shown, e. g. that 2′,4′,4-trihydroxychalcone and formononetin are excellent precursors of medicarpin in lucerne seedlings. In the same system, medicarpin and the isoflavan vestitol are interconvertible, presumably via an isoflavanium ion (see Isoflavonoids). [1,2-$^{13}C_2$]Acetate has been administered to $CuCl_2$-treated *Pisum sativum,* and the ^{13}C-^{13}C coupling in the resulting pisatin analysed by ^{13}C-NMR. This shows that C-atoms 1 and 1a, 2 and 3, and 4 and 4a are incorporated as intact C2 units, i. e. the carbons are not randomized by rotation of the acetate-derived ring, and the absence of oxygen on C1 is due to

Chalcone —see Isoflavonoids→ Formononetin → Calycosin → Pseudobaptigen → 7,2′-Dihydroxy-4′,5′-methylenedioxyisoflavone → Maakiain → 6a-Hydroxymaakiain → Pisatin

Biosynthesis of pisatin in Pisum sativum

loss of oxygen before cyclization. In contrast, other flavonoids (e.g. apigenin, kaempferol) are synthesized with free rotation of the acetate-derived ring (see Chalcone synthase). The biosynthetic sequence shown (Fig.) is strongly supported by the observation that all the represented compounds act as efficient precursors of pisatin in *Pisum sativum*. [P.M. Dewick in J.B. Harborne & T.J. Mabry, eds. *The Flavonoids: Advances in Research* (Chapman and Hall, 1982) pp. 535–640]

Pterocarpin: 3-methoxy-8,9-methylenedioxypterocarpan. See Pterocarpans.

Pteroylglutamic acid: see Vitamins (folic acid).

P-type ATPases: membrane-bound ATPases that generate a covalent phosphorylated intermediate (hence the designation 'P') during their catalytic cycle. P-t.A. constitute one of 3 classes of 'ion-motive ATPases' (see), and they are readily distinguished from the other two classes, V-type (see) and F-type (see), by their susceptibility to inhibition by vanadate (VO_4^{3-}, a transition state analog of phosphate). They include: (i) the H^+-transporting ATPases of the plasma membranes of lower eukaryotes (yeast, fungi) and of higher plants (plasmalemma) which pump H^+ from the cytosol into the extracellular compartment adjacent to and including the cell wall; within this compartment the lowered pH activates hydrolases believed to be concerned in cell wall formation, (ii) the H^+/K^+-transporting ATPase of the plasma membrane of animal gastric mucosal epithelial cells which pump, per 1ATP hydrolysed, $1H^+$ from the cytosol into the stomach lumen in exchange for $1K^+$, (iii) the Na^+/K^+-transporting ATPases of the plasma membrane of higher animal cells which pump, per 1ATP hydrolysed, $3Na^+$ from the cytosol to the extracellular compartment in exchange for $2K^+$, thereby generating the lower Na^+/higher K^+ intracellular concentration and 50–70 mV transmembrane potential difference (+ ve outside) characteristic of these cells, and (iv) the Ca^{2+}-transporting ATPases of the sarcoplasmic reticulum (s.r.) which pump, per ATP hydrolysed, 1–2 Ca^{2+} from the cytosol into the lumen of the s.r., thereby keeping the cytosolic Ca^{2+} concentration low.

Typically they consist of a single peptide (α) of 70–110 kDa, which has two domains, one spanning the membrane and containing several α-helices, the other projecting into the cytoplasm and containing the phosphorylation site, which is an Asp residue in a sequence that is typically -Ile-Cys-Ser-Asp-Lys-Thr-Gly-Thr-Leu-Thr-, and the ATP binding site; however the Na^+/K^+-ATPases have a second peptide (β) of ~55 kDa whose function is not clear.

During their catalytic cycle, P-t. A. can exist in two different conformational states, E_1 and E_2, which correspond respectively to the nonphosphorylated and phosphorylated states, and which have different affinity for the ion(s) to be transported across the membrane. The change from E_1 to E_2, by the reaction ATP + -Asp- → ADP + -Asp ~ P- causes the transport of one ionic species from one side of the membrane to the other while the change from E_2 to E_1, by -Asp ~ P- + H_2O → -Asp- + P_i, causes either the re-establishment of the original state or the transport of the second ionic species in the opposite direction, depending upon the species of P-t. A.

In addition to VO_4^{3-}, the Na^+/K^+-ATPases are inhibited by ouabain (see), and the plasma membrane H^+-ATPases are inhibited by dicyclohexyldiimide (DCCD) and diethylstilbestrol.

[P.L. Pederson & E. Carafoli *Trends Biochem. Sci.* **12** (1987) 146–150, 186–189]

Puff: see Giant chromosomes.

Punctuation codon: see Termination codon.

Punica alkaloids: a group of piperidine alkaloids, originally isolated from the bark of the pomegranate tree (*Punica granatum* L., official drug Cortex granati), and subsequently isolated from other plant families. Decoctions of the drug, or the isolated alkaloids,

Lysine or Cadaverine + 2 CH_3COOH → Isopelletierine / Pseudopelletierine

Biosynthesis of the Punica alkaloids, isopelletierine and pseudopelletierine

have some use as vermifuges. Biosynthesis of the predominant P.a., pseudopelletierine and isopelletierine, are shown in the Figure. *N*-Methylisopelletierine is also an important member of the group. P.a. are higher homologs of the tropane or pyrrolidine alkaloids, which have an analogous mode of biogenesis.

Purine: a heterocyclic compound with a condensed pyrimidine-imidazole ring system. M_r 120.1, m.p. 217 °C. P. was prepared from uric acid by Emil Fischer in 1884. The free compound is not known to occur naturally, but the otherwise unsubstituted P. ring system is found in combination with ribose in the nucleoside antibiotic, Nebularine (see).

Various substituted and oxidized purine derivatives occur naturally, and are of considerable biological importance. The purine derivatives, Adenine (see) and Guanine (see), are present in DNA and RNA, and they are commonly referred to as purine bases. Certain Rare nucleic acid components (see) are formed by modification of the P. bases in the polynucleotide chain. P. analogs, like 8-Azaguanine (see), can also be incorporated in place of natural P. bases during nucleic acid biosynthesis. P. bases are found in some low M_r Nucleotide coenzymes (see), and they are also components of other biologically active compounds, e.g. Nucleoside antibiotics (see) alkaloids (see Methylated xanthines), vitamins (see Vitamin B_{12}) and Cytokinins (see). The purine nucleotides, ATP (see Adenosine phosphates) and GTP (see Guanosine

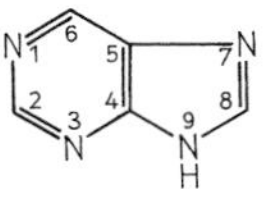

Purine

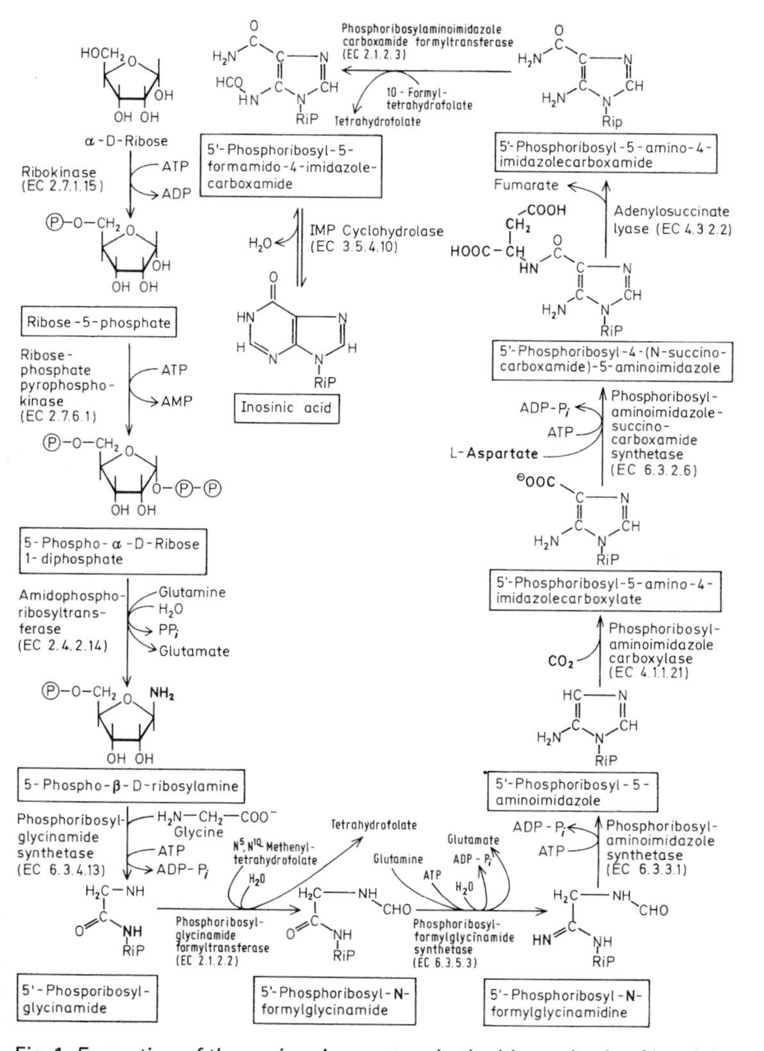

Fig. 1. *Formation of the purine ring system in the biosynthesis of inosinic acid*

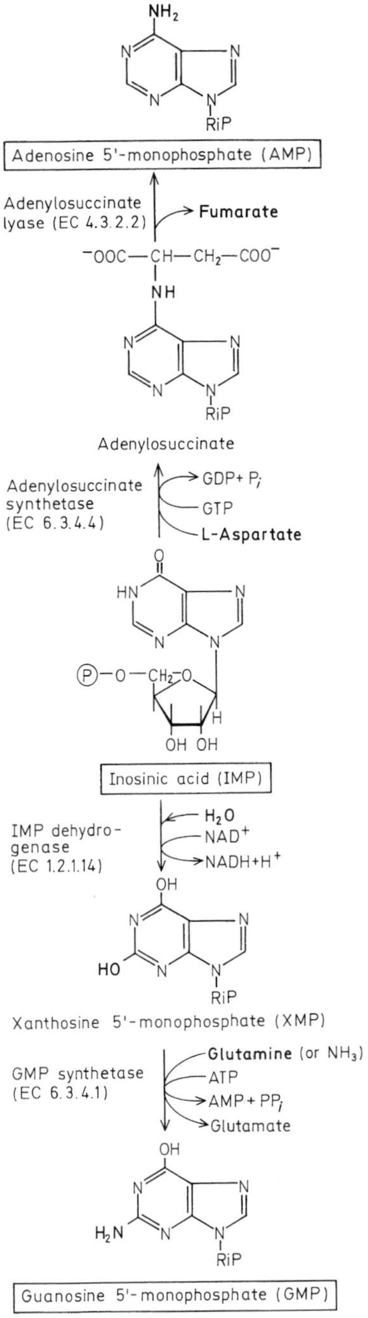

Fig. 2. *Biosynthesis of adenosine 5'*-monophosphate and guanosine 5'-monophosphate from inosinic acid

phosphates), are key compounds in biological energy metabolism. In some animals (birds, reptiles, insects), the chief route of nitrogen excretion is via purine synthesis; common excretory products are the purine oxidation product, Uric acid (see), and its further oxidation product, Allantoin (see). Spiders excrete guanine. See also Purine degradation, Purine biosynthesis. Purines are weak bases with specific light absorption between 230 and 280 nm. They display lactam-lactim and/or enamine-ketimine-tautomerism.

Purine antibiotics: purine derivatives with antibiotic activity. They occur as nucleosides (see Nucleoside antibiotics), polypeptides (see Viomycin), or free bases (see 8-Azaguanidine).

Purine bases: see Purine, Adenine, Guanine.

Purine biosynthesis, ***de novo purine biosynthesis:*** a common pathway for the biosynthesis of the purine ring system found at all levels of evolutionary development. α-D-Ribose 5-phosphate is pyrophosphorylated to 5-phosphoribosyl 1-pyrophosphate. The pyrophosphate group is then replaced by an amino group, which is transferred from the amide group of L-glutamine. The nitrogen of this amino group is destined to become N-9 of the purine ring system. The ribose phosphate remains attached throughout the successive enzyme-catalysed steps, which eventually lead to the complete purine ring system. Thus the purines are synthesized as their nucleoside monophosphates. The first product with a complete purine ring system is inosinic acid, which serves as the precursor of the other purine nucleotides. All the stages of P. b. are shown in Fig. 1, and the subsequent interconversions leading to the synthesis of AMP and GMP are shown in Fig. 2.

The nucleoside monophosphates are converted to the triphosphates (the direct precursors of RNA) by two kinase reactions. These kinases have a low specificity, and they catalyse the phosphorylation of nucleotides of adenine, guanine and the pyrimidines (Fig. 3). An alternative route for the synthesis of purine nucleotides is the Salvage pathway (see).

P.b. is regulated by both end products, AMP and GMP, which jointly inhibit phosphoribosylpyrophosphate amidotransferase (EC 2.4.2.14). GMP also inhibits IMP-dehydrogenase (EC 1.2.1.14); AMP inhibits adenylsuccinate synthetase (EC 6.3.4.4). Further control is exerted by the requirement for GTP in AMP synthesis (Fig. 2). ATP inhibits GMP reductase, which converts GMP into IMP in one step.

The foundation work on P. b. was performed by Buchanan and Greenberg with pigeon and chicken liver extracts. This led to the formulation of the scheme of P. b. by Greenberg in 1953. The postulated intermediates were isolated from mutant microorganisms.

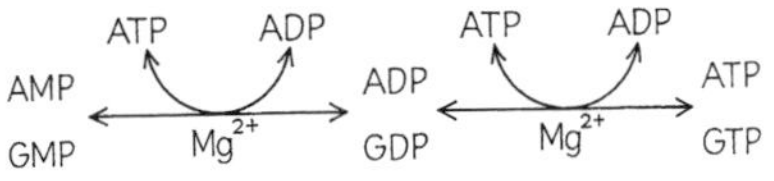

Fig. 3. *Interconversion of nucleoside mono-, di- and triphosphates*

Purine catabolism: see Purine degradation.

Purine cycle: see Glycine-Allantoin cycle.

Purine degradation, ***purine catabolism:*** a series of reactions in which purines are degraded by cleavage of the purine ring. P.d. is usually aerobic, but anaerobic P.d. occurs in certain microorganisms.

Aerobic P.d. The amino groups of adenine and guanine are removed hydrolytically by specific deaminases, which attack the free bases, the nucleosides or the nucleotides (Fig.). Uric acid is then produced by the action of xanthine oxidase (EC 1.2.3.2), which is the key enzyme of aerobic P.d. In humans and apes, the uric acid is excreted largely unchanged. In most reptiles and mammals, it is oxidized to allantoin by uricase (EC 1.7.3.3) (uricolysis).

In other organisms, including most fish and amphibians, allantoin is converted into allantoic acid, which is further degraded in two stages to yield 1 molecule of glyoxylic acid and 2 molecules of urea.

The inherited metabolic disease, xanthinuria, is caused by the absence of xanthine oxidase; xanthine and hypoxanthine are excreted instead of uric acid. Gout is caused by an increase in the rate of purine biosynthesis. The resulting increase in the concentration of blood uric acid leads to the deposition of crystalline uric acid in the joints.

Anaerobic P.d., anaerobic xanthine degradation. In certain microorganisms (*Micrococcus* and *Clostridium*) the substrate of nonoxidative P.d. is xanthine. Hydrolysis between C6 and N1 of the 6-membered ring of xanthine produces ureidoimidazolyl-carboxylic acid. Further hydrolytic removal of ammonia and CO_2 produces aminoimidazolecarboxylic acid. This is decarboxylated to aminoimidazole. The ring of aminoimidazole is opened by the simultaneous loss of ammonia and addition of 2 molecules of water, to produce formiminoglycine. The latter is hydrolysed to glycine, ammonia and formate.

Purine interconversion: see Purine biosynthesis.

Puromycin: a nucleoside antibiotic (M_r 472) from *Streptomyces alboniger.* As a structural analog of the 3′-terminal end of aminoacyl-tRNA (Fig.), P. inhibits protein biosynthesis on 70S and 80S ribosomes. It replaces the 3′-terminus of aminoacyl-tRNA during the elongation phase of protein biosynthesis, and a pep-

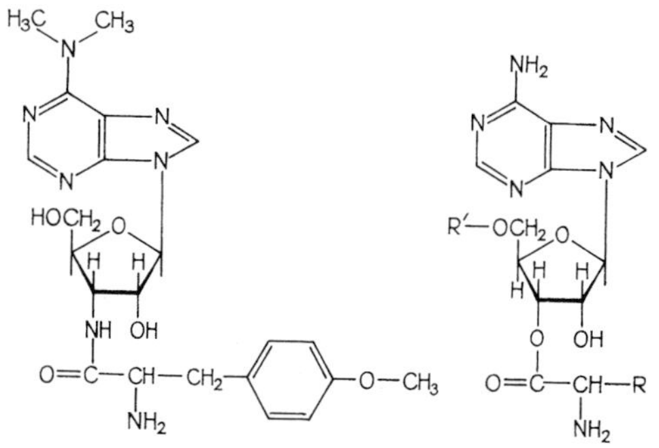

Comparison of the structures of puromycin (left) and the 3′-terminal end of an aminoacyl-tRNA (right)

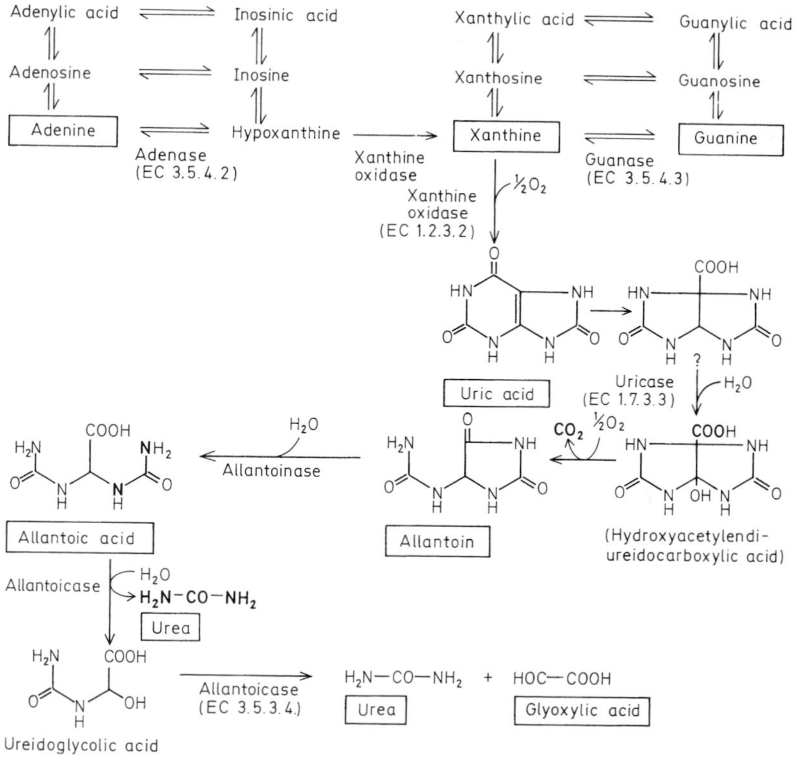

Aerobic purine degradation

tide bond is formed between the free amino group of P. and the COOH group of the *C*-terminal amino acid of the preceding peptidyl-tRNA (see Fragment reaction). Further elongation of the polypeptide is prevented, and the polypeptide fragment with the attached P. becomes separated from the ribosome. Other aminoacylnucleoside antibiotics, e.g. gougerotin and blasticidin S, have a similar action mechanism.

Purpurin: 1,2,4-trihydroxyanthraquinone, a red anthraquinone pigment, m.p. 263 °C. The glycoside of P. occurs in madder root (*Rubia tinctorum*) (accompanied by alizarin), and in other members of the *Rubiaceae.* P. is formed from its glycoside during storage, and there is no appreciable quantity of P. in the fresh root. It is used as a reagent for the detection of boron, for the histological detection of insoluble calcium salts, and as a nuclear stain. It forms colored lakes with various metal salts, and is used as a fast dye in cotton printing.

Putrescine: tetramethylenediamine, a ubiquitous Polyamine (see), formed by decarboxylation of ornithine and, in some organisms, by decarboxylation of arginine to agmatine, followed by cleavage to P. and urea. It is the precursor of spermine and spermidine in ordinary metabolism, and is essential for cell division. It accumulates during bacterial degradation of arginine. Increased protein decomposition (e.g. in cholera) leads to the appearance of P. in urine and feces.

Pyr: abb. for Pyroglutamic acid (see for other recommended abbreviations).

Pyranose: see Carbohydrates.

Pyrethrins: diterpene insecticides present in the flowers of *Chrysanthemum cinerariaefolium* (syn. *Pyrethrum cinerariaefolium*). The dried flowers, known as "pyrethrum", also have insecticidal activity and serve as starting material for the preparation of P. P. are esters of chrysanthemic acid (giving series I compounds) or pyrethric acid (giving series II compounds) with the alcohol pyrethrolone (giving pyrethrins), cinerolone (giving cinerins) or jasmololone (giving jasmolins).

Structures of naturally occurring pyrethrins

R_1	R_2	Name
$-CH_3$	$-CH_3$	Cinerin I
$-CH_3$	$-C_2H_5$	Jasmolin I
$-CH_3$	$-CH{=}CH_2$	Pyrethrin I
$-CO_2CH_3$	$-CH_3$	Cinerin II
$-CO_2CH_3$	$-C_2H_5$	Jasmolin II
$-CO_2CH_3$	$-CH{=}CH_2$	Pyrethrin II

Pyridine alkaloids: a group of alkaloids containing the pyridine ring system, which occur in various unrelated plants, and as metabolic products of microorganisms. Important examples are Nicotiana alkaloids (see), Areca alkaloids (see), Gentiana alkaloids (see), and Valeriana alkaloids (see). They are biosynthesized either from nicotinic acid, or as products of terpene synthesis.

Pyridine nucleotide coenzymes: Nicotinamide-dinucleotide (see), and Nicotinamide-adenine-dinucleotide phosphate (see).

Pyridine nucleotide cycle: a salvage pathway in which nicotinamide produced by degradation of

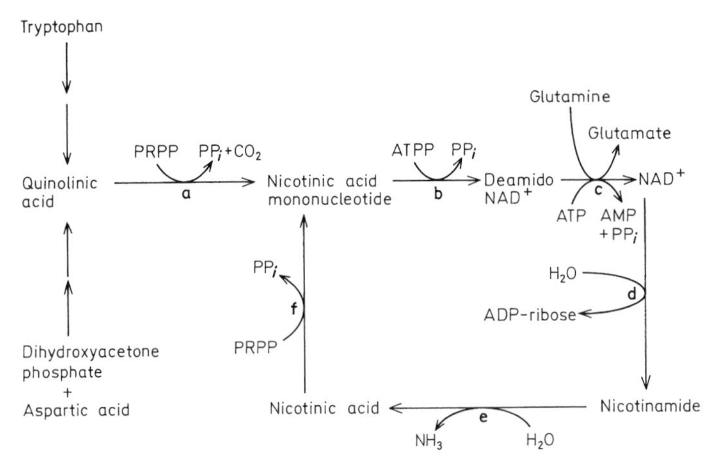

The pyridine nucleotide cycle. For formulas of intermediates, see L-Tryptophan and Nicotinamide adenine dinucleotide.

a Nicotinate-nucleotide-pyrophosphorylase (carboxylating) (EC 2.4.2.19) (also called quinolinate transphosphoribosylase).

b Nicotinate mononucleotide adenylyltransferase (EC 2.7.7.18).

c NAD$^+$ synthetase (glutamine hydrolysing) (EC 6.3.5.1).

d NAD$^+$ nucleosidase (EC 3.2.2.5).

e Nicotinamidase (EC 3.5.1.19).

f Nicotinate phosphoribosyltransferase (EC 2.4.2.11).

PRPP = 5-phosphoribosyl 1-pyrophosphate, or 5-phospho-α-D-ribose 1- diphosphate.

NAD$^+$ is reutilized to synthesize more NAD$^+$. The P. n. c. probably operates in all organisms, whether or not they are capable of synthesizing the pyridine ring system, and irrespective of the pathway of synthesis (from L-tryptophan in animals, *Neurospora* and *Xanthomonas pruni;* from aspartate and dihydroxyacetone phosphate in plants and most bacteria).

Pyridine nucleotide transhydrogenase: see Hydrogen metabolism.

Pyridoxal phosphate, *PalP:* the coenzyme form of Vitamin B_6 (see Vitamins for formula of PalP and related compounds). M_r of PalP 247.1. PalP is stable in aqueous solution when kept refrigerated and protected from light. It is particularly sensitive to photodecomposition in the solid state and in alkaline solution. PalP plays an important central role in amino acid metabolism, acting as the coenzyme in many different metabolic conversions of amino acids. It is formed from pyridoxal by a Mg^{2+}-dependent kinase reaction: pyridoxal + ATP $\rightarrow$ pyridoxal 5-phosphate + ADP. With amines and amino acids, PalP forms Schiff's bases (azomethines). The substrate of a pyridoxal phosphate enzyme is the Schiff's base of the amino acid with PalP; the action specificity, i. e. transamination, decarboxylation, racemization etc. is determined by the apoenzyme. The initial reaction of PalP with an enzyme involves the formation of a Schiff's base between the ε-amino group of a lysine residue in the active center; an exchange process then results in the formation of a Schiff's base between the substrate amino acid and PalP, with release of the free ε-amino group of lysine (see Transamination). In the Schiff's base, the electrophilic form of the positive pyridine nitrogen favors the formation of a mesomeric structure, but this is only possible by removal of a substituent from the α-C atom as a cation, e. g. $^+CH_2OH$ in the L-serine hydroxymethyl transferase reaction, H^+ in racemization and transamination, or CO_2 in amino acid decarboxylation. In another group of reactions (catalysed by L-serine dehydratase, L-cysteine desulfhydrase, L-tryptophan synthase, and tryptophanase) labilization of the α-hydrogen is followed by removal of the β-substituent (Fig.).

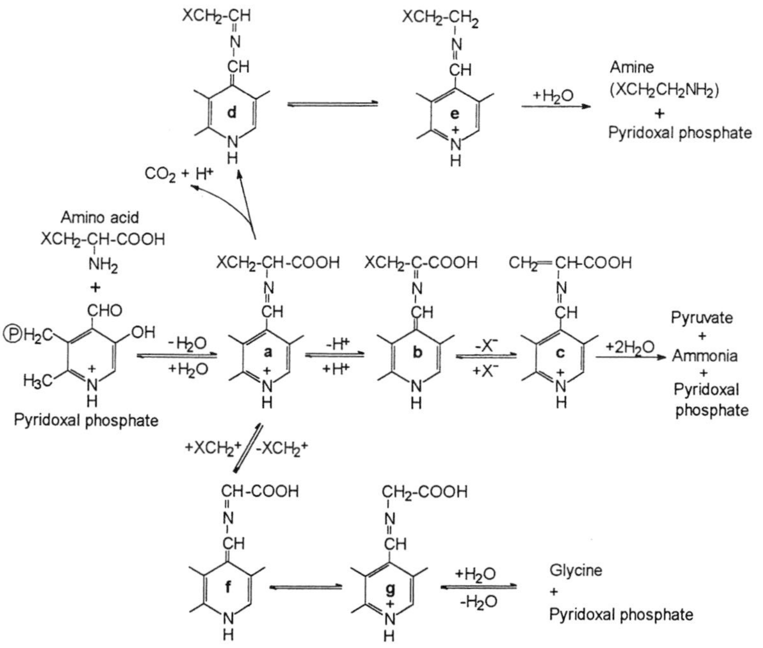

Coenzyme role of pyridoxal phosphate (PalP) in various reactions of amino acid metabolism.
In all reactions, the first stage is formation of Schiff's base *a* by condensation of PalP and the amino acid. Schiff's bases *a* and *b* represent part of transamination, but for the complete mechanism see Transamination. *Racemization: a* $\rightarrow$ *b,* followed by *b* $\rightarrow$ *a* $\rightarrow$ amino acid + PalP, with addition of the proton in the opposite configuration. *Amino acid decarboxylation: a* $\rightarrow$ *d* $\rightarrow$ *c* $\rightarrow$ amine + PalP. *Serine hydroxymethyltransferase* (EC 2.1.2.1): X = OH; L-serine + PalP $\rightarrow$ *a* $\rightarrow$ *f* $\rightarrow$ *g* $\rightarrow$ glycine + PalP; reversal of these reactions leads to L-serine synthesis from glycine; the hydroxymethyl group is carried by tetrahydrofolic acid. *Cysteine desulfhydrase* (EC 4.4.1.1): X = SH; cysteine + PalP $\rightarrow$ *a* $\rightarrow$ *b* $\rightarrow$ *c* $\rightarrow$ hydrogen sulfide + pyruvate + ammonia + PalP. *Serine dehydratase* (EC 4.2.1.13): X = OH; L-serine + PalP $\rightarrow$ *a* $\rightarrow$ *b* $\rightarrow$ *c* $\rightarrow$ water + pyruvate + ammonia + PalP. *Tryptophanase* (EC 4.1.99.1): X = indole; L-tryptophan + PalP $\rightarrow$ *a* $\rightarrow$ *b* $\rightarrow$ *c* $\rightarrow$ indole + pyruvate + ammonia + PalP. *Tryptophan synthase* (EC 4.2.1.20): 1st stage: X = OH; L-serine + PalP $\rightarrow$ *a* $\rightarrow$ *b* $\rightarrow$ *c;* 2nd stage: X = indole; *c* $\rightarrow$ *b* $\rightarrow$ *a* $\rightarrow$ L-tryptophan + PalP.

Pyridoxamine phosphate: see Transamination.

Pyridoxine: see Vitamins (vitamin B_6).

Pyridoxol: see Vitamins (vitamin B_6).

Pyrimidine: 1,3-diazine, a heterocyclic compound, consisting of a six-membered ring with 2 nitrogen atoms (Fig. 1). M_r 80.1, m.p. 20–22 °C, b.p. 124 °C. The P. ring system is present in many natural compounds, e.g. antibiotics (nucleoside antibiotics), pterins, purines and vitamins, it is especially important in the pyrimidine bases, Cytosine (see), Uracil (see) and Thymine (see), which are constituents of nucleic acids. Pyrimidine itself does not occur naturally. Pyrimidine analogs (see) can also be incorporated into nucleic acids.

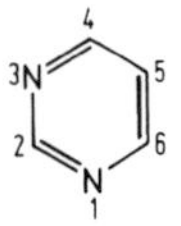

Fig. 1. *Numbering system of pyrimidine.* An older system uses the numbering of the pyrimidine ring of Purine (see).

The pyrimidine bases of nucleic acids possess an amino or hydroxyl group at position 6, and always an oxygen function at position 2. This gives rise to tautomeric structures, in which hydrogen is bound to oxygen or to ring nitrogen (shown for uracil in Fig. 2).

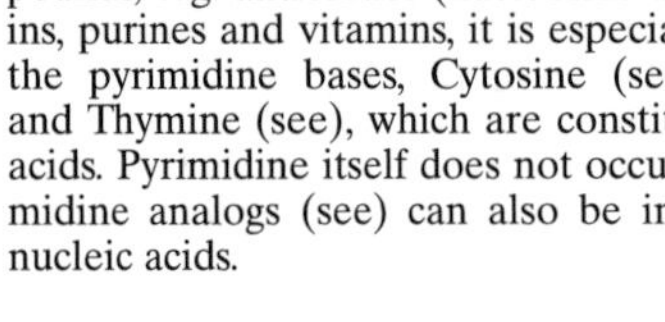

Lactim Lactam

Fig. 2. *Tautomers of uracil*

Pyrimidine analogs, ***antipyrimidines:*** pyrimidines and pyrimidine nucleosides structurally related to, but different from the natural compounds. They therefore act as antimetabolites and selectively inhibit certain biochemical pathways, especially nucleic acid synthesis. Most P. a. are modified bases or their nucleosides, but there are also pyrimidine nucleoside analogs with modified sugar components. The most common chemical modifications are the introduction of substituents (e.g. halogens on C5 of uracil and cytosine), replacement of an OH with an SH group (e.g. 2-thiouracil), and replacement of a ring carbon with nitrogen (5-azauracil). Arabinonucleosides (see) (steric inversion of the OH at C2 of the ribose) and Xylosylnucleosides (see) (inversion at C3) of natural pyrimidine bases are also active P. a. Incorporation of 5-bromouracil (an analog of thymine) into DNA was first reported in 1952. The 5-fluorouracil compounds were developed in 1957; 5-fluorouracil is incorporated into RNA in place of uracil. [*Antimetabolites of Nucleic Acid Metabolism. The Biochemical Basis of their Action with Special Reference to their Application in Cancer Therapy* by Peter Langen 1975 Gordon and Breach (London New York, Paris)]

Pyrimidine antibiotics: structurally modified pyrimidine derivatives with antibiotic activity. They occur as nucleosides (see Nucleoside antibiotics), polypeptides (e.g. albomycin and grisein), or free bases (e.g. bacimethrin). The P. a., Toxoflavin (see) and Fervenulin (see), are biosynthesized from purines.

Pyrimidine bases: see Pyrimidine.

Pyrimidine biosynthesis, ***de novo pyrimidine biosynthesis:*** total synthesis of the pyrimidine ring of uracil, thymine, cytosine and their derivatives from carbamoyl phosphate and aspartate in all living cells. The pyrimidine ring of thiamin (vitamin B_1) has a different biosynthetic origin (see below).

Biosynthesis of uridine and cytidine nucleotides. This is shown in Fig. 1. The first pyrimidine nucleotide to appear de novo in this pathway is uridine 5′-monophosphate (UMP, uridylic acid). This is phosphorylated to uridine 5′-triphosphate (UTP). By donation of an amino group from ammonia or glutamine (in animal tissues), UTP is converted to cytidine 5′-triphosphate (CTP).

Biosynthesis of thymine nucleotides. This is shown in Fig. 2. Since thymine is a constituent of DNA, the corresponding nucleotides contain 2-deoxyribose. Thymidylic acid (TMP) is therefore more correctly dTMP (deoxythymidine 5′-monophosphate). The reaction sequence is: CMP → CDP → dCDP → dCMP → dUMP → TMP (dTMP) → TDP (dTDP) → TTP (dTTP). Methylation of dUMP to TMP is catalysed by thymidylate synthase (EC 2.1.1.45). The cofactor, N^5,N^{10}-methylenetetrahydrofolic acid, transfers the active C1 unit to C5 of dUMP, and it also functions as a reducing agent in the formation of the methyl group from the active C1 unit.

Biosynthesis of ribothymidylic acid. The unit occurs as a minor component (see Rare nucleic acid components) in many species of tRNA. It is formed by methylation (from *S*-adenosyl-L-methionine) of C5 of uracil in the existing nucleic acid molecule.

Biosynthesis of S-hydroxymethyldeoxycytidylic acid. This component of the DNA of T-even phages is biosynthesized from deoxycytidine 5′-monophosphate (Fig. 3). Pyrimidine nucleotides may also be produced by the Salvage pathway (see).

Regulation of P.b. In *E. coli* carbamoyl phosphate synthetase is activated by the purine nucleotides, IMP and XMP, and it is inhibited by the pyrimidine nucleotides, UMP and UDP. The key control point is the synthesis of *N*-carbamoylaspartic acid, catalysed by aspartate carbamoyl-transferase (aspartate transcarbamylase, EC 2.1.3.2). In *E. coli* and *Aerobacter aerogenes* this enzyme is inhibited by CTP, and the inhibition is prevented by ATP. In *Pseudomonas fluorescens,* the enzyme is inhibited by UTP, whereas in higher plants the regulatory inhibitor is UMP.

Aspartate carbamoyltransferase from *E. coli* is one of the most thoroughly studied allosteric enzymes. It has M_r 310,000, and can be dissociated into two identical catalytic subunits (each of M_r 100,000, and containing three polypeptide chains, called C-chains, of M_r 34000), and three identical regulatory subunits. Each regulatory subunit contains two R-chains, M_r 17,000; each R-chain binds one molecule of CTP. The catalytic subunits are active in the absence of the regulatory subunits, but regulation by CTP only occurs in the complete oligomeric enzyme. The con-

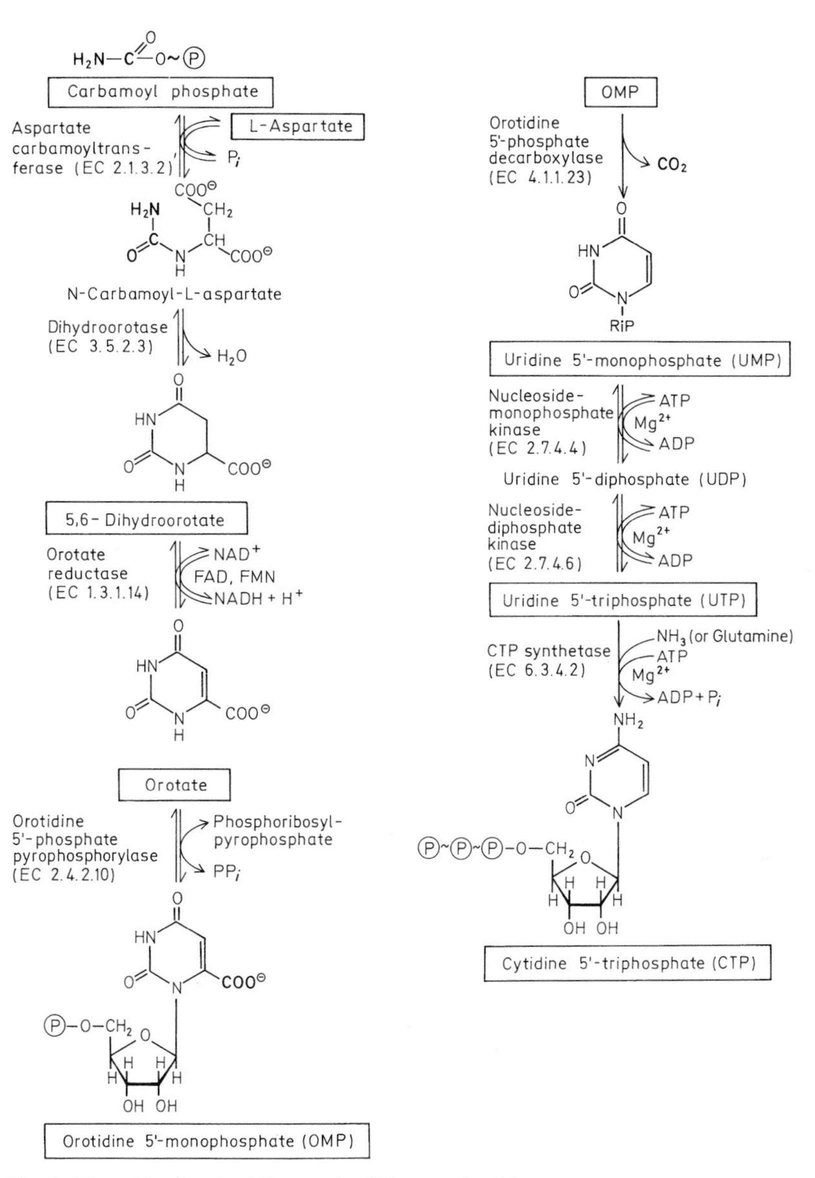

Fig. 1. *Biosynthesis of uridine and cytidine nucleotides*

trol mechanism is explained by the Cooperativity model (see) of allosteric enzymes. In some cases, uracil may also repress the synthesis of aspartate carbamoyltransferase and dihydroorotate oxidase (EC 1.3.3.1).

Biosynthesis of the pyrimidine ring of thiamin (vitamin B_1) from aminoimidazoleribonucleotide. The 2-methyl-4-amino-5-hydroxymethyl-pyrimidine ring present in thiamin is synthesized from aminoimidazoleribomlcleotide, which is an intermediate in purine biosynthesis (Fig. 4).

Pyrimidine degradation, ***pyrimidine catabolism:*** reductive or (in special cases) oxidative reactions leading to the cleavage of the heterocyclic ring of natural pyrimidines.

1. *Reductive P.d.* (Fig.). To a certain extent, this process represents a reversal of Pyrimidine biosynthesis (see). The pyrimidine ring is partially hydrogenated, and the resulting dihydro-compound is cleaved hydrolytically. Cytosine is converted to uracil by deamination, and uracil is degraded to β-alanine. Thymine is degraded to β-aminoisobutyrate. These endproducts are transaminated and metabolized to common metabolic intermediates (Fig.).

2. *Oxidative P.d.* In *Corynebacterium* and *Mycobacterium,* uracil is oxidized to barbituric acid, which

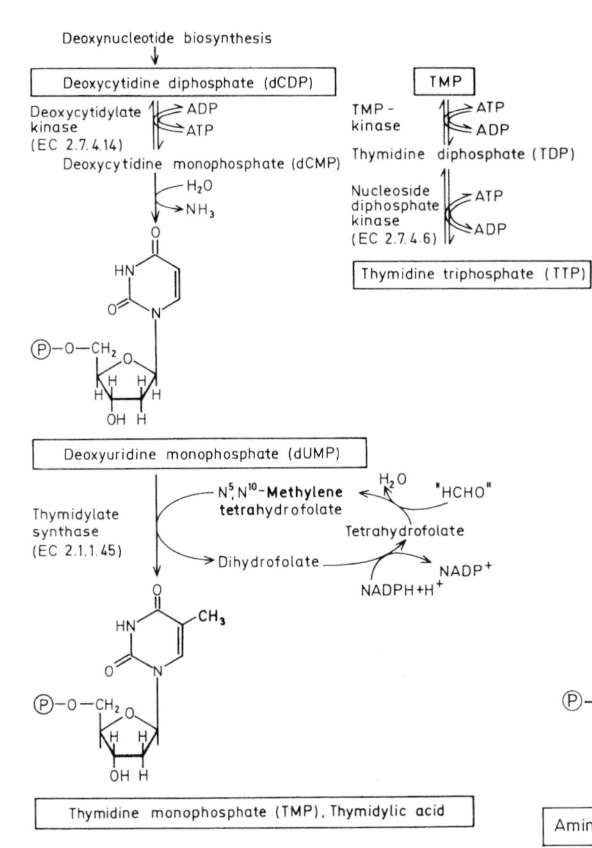

Fig. 2. *Biosynthesis of thymidine nucleotides*

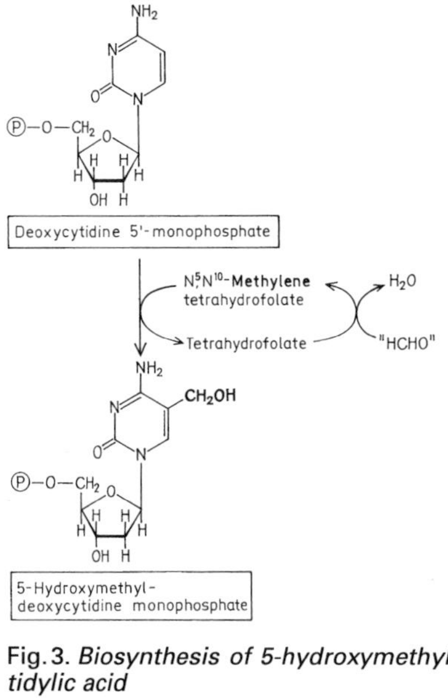

Deoxycytidine 5'-monophosphate

N^5N^{10}-**Methylene** tetrahydrofolate

H_2O

Tetrahydrofolate

"HCHO"

CH_2OH

5-Hydroxymethyl-deoxycytidine monophosphate

Fig. 3. *Biosynthesis of 5-hydroxymethyl-deoxycytidylic acid*

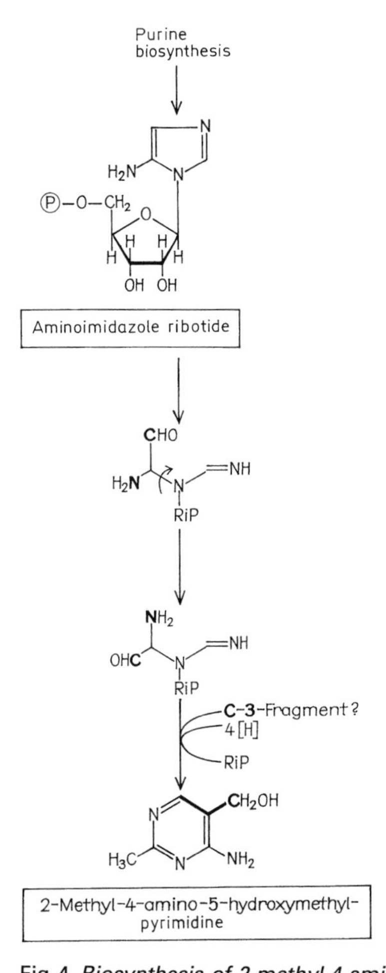

Fig. 4. *Biosynthesis of 2-methyl-4-amino-5-hydroxymethyl-pyrimidine from aminoimidazole ribotide* (RiP = ribose 5-phosphate).

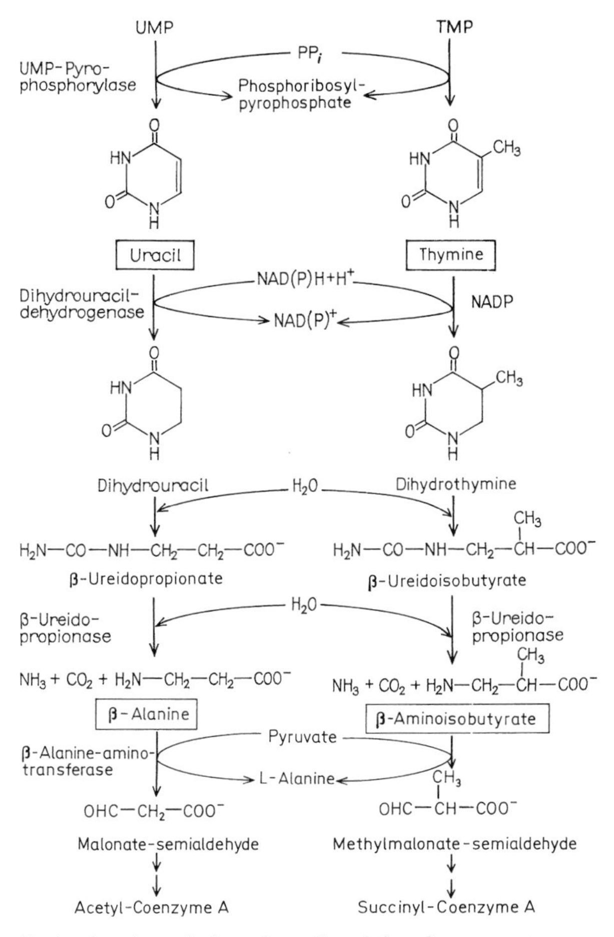

Reductive degradation of uracil and thymine

is cleaved hydrolytically to urea and malonic acid. Thymine is oxidized to 5-methylbarbituric acid, followed by hydrolysis to urea and methylmalonic acid.

Pyrimidine dimers: see DNA repair.

Pyroglutamic acid, ***pyrrolidone carboxylic acid, 5-oxoproline, pyrrolid-2-one-5-carboxylic acid, Glp, Pyr, PCA, <Glu,⌞G:*** an internal cyclic lactam of glutamic acid, representing a condensation of the α-amino with the γ-carboxyl group. Pyroglutamic acid occurs as the *N*-terminal residue of some proteins. It was first reported as an *N*-terminus in the heavy chain of rabbit IgG [Wilkinson et al. *Biochem J.* **100** (1966) 303–308]. The abbreviation PCA was used in this early work, but the IUPAC-IUB Commission on Biochemical Nomenclature (Recommendations, 1971) recommend the use of <Glu or ⌞G. Pyr is frequently used as a nonstandard abbreviation.

Pyrroles, ***pyrrole derivatives:*** compounds containing the pyrrole ring. They are subdivided into mono-, di-, tri- and tetrapyrroles. Tetrapyrroles may be noncyclic or cyclic. Bile pigments (see) and the chromophores of Biliproteins (see) are linear, while Porphyrins (see) and Corrinoids (see) are cyclic tetrapyrroles.

Pyrrolidine-2-carboxylic acid: see L-Proline.

Pyrrolidine alkaloids: a group of Alkaloids (see) with simple structures. P. a. are either derivatives of proline (e.g. stachydrin and its diastereoisomer, betonicin), or they are derived from a *N*-methyl-2-alkylpyrrolidine (e.g. hygrin and cuskhydrin). The latter occur together with the tropane alkaloids, with which they share the same biogenetic precursors, ornithine and acetate.

Hygrin

Pyrrolid-2-one-5-carboxylic acid: see Pyroglutamic acid.

Pyrrolizidine alkaloids, ***Senecio alkaloids:*** a group of ester alkaloids, in which amino alcohols (necines) are esterified with necic acids. The necines are derivatives of the pyrrolizidine ring system (also

Retronecine

known as 1-azabicyclo[0,3,3]octane) (see Alkaloids, Table) and they possess one or two alcoholic hydroxyls, e.g. retronecine (Fig.).

The necic acids (esterified with the hydroxyls of the necines) are branched aliphatic, mono- or dibasic acids, containing 5–10 carbon atoms, e.g. angelic, monocrotalic, senecic and tiglic acids. The various combinations of necines and necic acids, together with amine oxides and isomers gives rise to a very large number of P.a. Free necines also occur naturally. P.a. occur chiefly in species of *Senecio,* the largest genus of the *Compositae.* They are hepatotoxic, and can cause liver cirrhosis in grazing animals.

Pyrrolo quinoline quinone, ***PQQ:*** 2,7,9-tricarboxy-1*H*-pyrrolo [2,3-*f*] quinoline-4,5-dione, the cofactor of methanol dehydrogenase (EC 1.1.99.8) from *Hyphomicrobium X* and *Methylophilus methylotrophus,* and of glucose dehydrogenase (EC 1.1.99.17) from *Acinebacter calcoaceticus.* [J. A. Duine et al. *Eur. J. Biochem.* **108** (1980) 187–192]

Pyrrolo quinoline quinone

Pyruvate: the anion of pyruvic acid. P. is an important metabolic intermediate in aerobic and anaerobic metabolism (Fig.).

P. synthesis: 1. P. is synthesized from phospho*enol*pyruvate in Glycolysis (see). Phospho*enol*pyruvate is an enol ester and an energy-rich compound with a free energy of hydrolysis of 50.24 kJ (12 kcal) per mol. During catalysis by pyruvate kinase, this free energy is exploited for the transfer of the phosphate group to ADP, resulting in the synthesis of ATP and P. 2. P. is produced in the metabolism of certain amino acids, in particular transamination of alanine, dehydration of serine, and desulfhydration of cysteine.

P. metabolism. 1. P. is reduced to lactate in anaerobic glycolysis. 2. It is converted to ethanol in anaerobic Alcoholic fermentation (see). 3. By the action of the pyruvate dehydrogenase complex, under aerobic conditions, P. is oxidatively decarboxylated to acetyl coenzyme A. The latter is an important metabolite in various other biosynthetic and biodegradative processes. Equation for oxidative decarboxylation of P.: CH_3COCOO^- (pyruvate) + HSCoA (coenzyme A) + $NAD^+ \rightarrow CH_3CO$-SCoA (acetyl coenzyme A) + CO_2 + NADH + H^+. Complete oxidation of one molecule of P. via the TCA-cycle results in the production of 15 molecules of ATP (14 from the respiratory chain + 1 from the substrate level phosphorylation in the conversion of succinyl-CoA to succinate). 4. P. is carboxylated to oxaloacetate (see Carboxylation), representing the first stage in Gluconeogenesis (see). 5. During Nitrogen fixation (see) in *Clostridium,* P. undergoes phosphoroclastic fission to acetyl phosphate and CO_2.

Pyruvate carboxylase (EC 6.4.1.1): a biotin-dependent ligase, in animals and plants, which catalyses the addition of CO_2 to pyruvate: Pyruvate + CO_2 + ATP + $H_2O \rightleftharpoons$ Oxaloacetate + ADP + P_i. The enzyme is Mn^{2+}-dependent, and it is practically inactive in the absence of its positive allosteric effector, acetyl-CoA. This reaction is an important early stage of Gluconeogenesis (see), and is an example of CO_2 fixation in the animal organism. For the mode of attachment of the coenzyme, biotin, and the mechanism of CO_2 transfer, see Biotin enzymes. The active form of P.c. is a tetramer, M_r 600,000 (yeast), 650,000

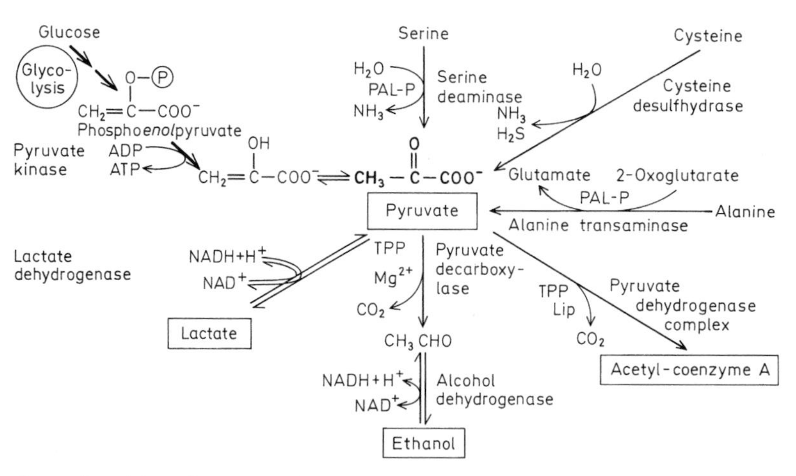

Central position of pyruvate in metabolism.
TPP Thiamin pyrophosphate, PAL-P pyridoxal phosphate, Lip lipoamide.

(chicken liver), or 520,000 (porcine liver), which is in equilibrium with the corresponding dimers and monomers. The dimers and monomers (M_r 130,000, 3 chains, each of M_r 47,000) of the porcine enzyme are also enzymatically active. Avian P.c. is cold sensitive, and reversibly dissociates into inactive monomers (M_r 160,000) at 0 °C. Like all biotin enzymes, P.c. is inactivated by avidin, due the binding of the coenzyme as an avidin $(biotin)_4$-complex. On the basis of their structural homologies, P.c. and acetyl-coenzyme-A-carboxylase are thought to have evolved from a common enzyme.

Pyruvate decarboxylase, ***carboxylase*** (EC 4.1.1.1): a thiamin pyrophosphate (TPP)-dependent lyase, absent from animals, and present in high activity in yeast and wheat seedlings. P.d. is a specific enzyme of alcoholic fermentation, catalysing the cleavage of pyruvate (via active acetaldehyde) into acetaldehyde and CO_2. The cofactors are Thiamin pyrophosphate (see) and magnesium ions. In the plant cell, P.d. competes with the pyruvate dehydrogenase complex for pyruvate. M_r of P.d. from yeast and *E. coli* is 190,000 (two indentical subunits, M_r 95,000).

Pyruvate dehydrogenase: a Multienzyme complex (see) responsible for the formation of acetyl-CoA from pyruvate, one of the central metabolic reactions (see Pyruvate, Acetyl-coenzyme A). It is subject to three types of control: 1.The enzyme complex is inhibited by acetyl-CoA and NADH; the transacetylase is inhibited by acetyl-CoA, and NADH inhibits the dihydrolipoyl dehydrogenase. These inhibitions are reversed by CoA and NAD^+, respectively. 2. Enzyme activity is influenced by the energy state of the cell; the complex is inhibited by GTP and activated by AMP. 3. The complex is inhibited when a specific serine residue in the pyruvate decarboxylase is phosphorylated by ATP. This phosphorylation is inhibited by pyruvate and ADP. The complex is reactivated by removal of the phosphoryl group by a specific phosphatase. [S. S. Mande et al. *Structure* **4** (1996) 277–286]

Pyruvate kinase, ***phosphopyruvate kinase*** (EC 2.7.1.40): a widely distributed, metal-ion dependent phosphotransferase, present in yeast, muscle, liver, erythrocytes and other organs and cells. It catalyses the last reaction of glycolysis: Phospho*enol*pyruvate (PEP) + ADP → Pyruvate + ATP (substrate level phosphorylation). Each subunit of P.k. forms an intermediate, cyclic, ternary metal bridge complex:

P.k.–Mn–ADP
\ /
PEP

in which the PEP and ADP are bound to the enzyme via a manganese (II) ion. Tetrameric P.k. from muscle and erythrocytes (M_r 230,000) shows Michaelis-Menton type kinetics (plot of initial velocity against substrate concentration is a rectangular hyperbola), whereas the yeast enzyme (M_r 190,000, 4 or 8 subunits) is an allosteric enzyme, showing sigmoid kinetics. The nucleotide base sequence of the P.k. gene from chicken and a discussion of the evolutionary implications of the gene structure has been published by N. Lonberg & W. Gilbert [*Cell* **40** (1985) 81–90].

Pyruvate phosphate dikinase: see Hatch-Slack-Kortschak cycle.

Pyruvic acid: CH_3–CO–COOH, the simplest and most important α-ketoacid (2-oxacid), m.p. 11.8 °C, b.p. 165 °C (d.). For the role of P.a. in metabolism, see Pyruvate.

Pythocholic acid: 3α,7α,16α-trihydroxy-5β-cholan-24-oic acid, a bile acid possessing three hydroxyl groups. M_r 408.58, m.p. 187 °C. P.a. is a characteristic component of the bile of many snakes, and has been isolated from the bile of the tiger snake, python and boa-constrictor, among others.

Q

Q: abb. for coenzyme Q. See Ubiquinone.

Q-cycle: a cycle devised by P. Mitchell [*FEBS Lett.* **56** (1975) 1–6; **59** (1975) 137–139] to overcome the requirement of the 'redox loop mechanism' (see Chemiosmotic hypothesis) for a 'H^+ & electron' carrier in the cytochrome bc_1-containing Complex III of the mitochondrial electron transport chain. The Q. c. proposed that ubiquinone (coenzyme Q), the only mobile, hydrophobic redox component of the chain, participates in electron transfer from cytochrome b to cytochrome c_1 within Complex III by one-electron steps involving the fully reduced quinol-form (QH_2), a stabilized free-radical semiquinone-form ($QH\cdot$) and the fully oxidized quinone-form (Q). It also made use of the observation that cytochrome b appears to be a dimer composed of b_T (b_{566}) and b_K (b_{562}), which is buried deeply in the membrane with b_T probably on the cytosolic side and b_K on the matrix side. In the Fig., outlining the proposed mechanism, it can be seen that two protons are 'pumped' across the membrane (steps 1 & 9 for uptake from the matrix and steps 3 & 7 for discharge into the intermembrane space) per electron transported (i. e. $4H^+$ per electron pair).

Although the 'redox loop mechanism' is no longer believed to explain how electron flow through an electron transport chain causes the generation of a trans-membrane H^+ gradient, there is considerable evidence that the Q. c. operates in mitochondria, bacteria and possibly chloroplasts under low light conditions.

Quantosome: the smallest structural unit of photosynthesis; small elementary units of the thylakoid measuring $18 \times 15 \times 10$ nm, M_r 2 million, containing 230 chlorophyll molecules, cytochromes, copper and iron. Q. are obtained by ultrasonic disintegration of isolated chloroplasts, and they can be visualized in the electron microscope. They can also be observed as granular units in the chloroplast lamella. The functional status of Q. is not clearly defined; they may be involved in both electron transport and photophosphorylation, and therefore analogous to the electron transport particles of the respiratory chain.

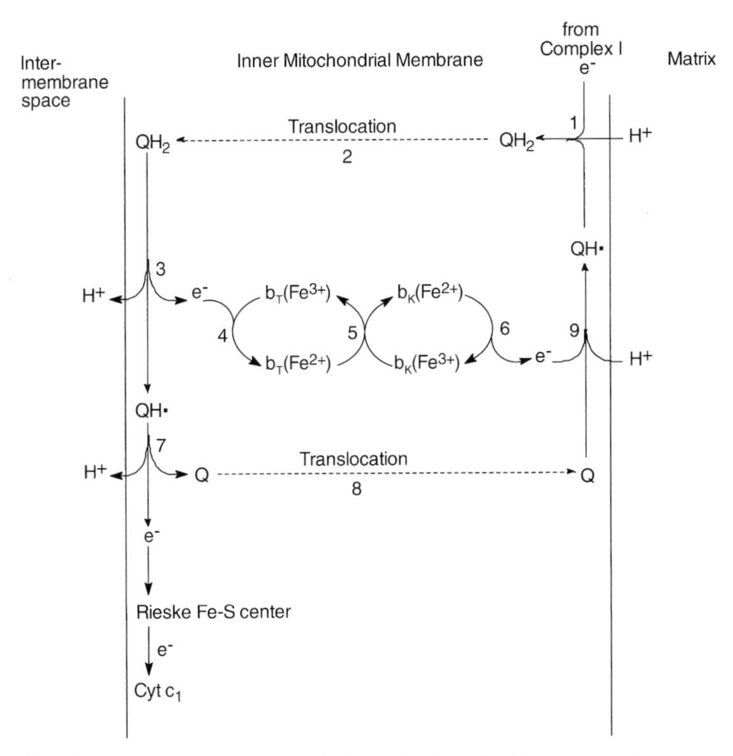

The Q-cycle as it may operate in the mitochondrial inner membrane

Quantum efficiency, ***quantum yield:*** see Quantum requirement.

Quantum requirement: the number of light quanta required for the formation of one molecule of O_2 in Photosynthesis (see). Two quanta are required per electron. The theoretical value of Q.r. is eight, since the production of one molecule of O_2 proceeds according to the following equation, with transfer of four electrons from water (see Photolysis of water) to $NADP^+$: $2H_2O \rightarrow O_2 + 4H^+ + 4e^-$. The experimentally determined value of Q.r. lies between 8 and 10 for leaves, and between 10 and 14 for isolated chloroplasts. It is influenced by the physiological state of the experimental system.

The inverse of Q.r. is the quantum yield (or quantum efficiency). The quantum yield therefore represents the number of molecules transformed (i.e. CO_2 molecules liberated) per quantum of light absorbed.

Quantum yield, ***quantum effciency:*** see Quantum requirement.

Quaternary structure: see Proteins.

Queen substance: originally a term for the entire mandibular gland secretion of the queen bee, which contains about 30 substances. It is now the trivial name for 9-oxo-trans-2-decenoic acid. This compound, together with 9-hydroxy-*trans*-2-decenoic acid, is very important as a pheromone for the maintenance of the division of labor within the hive. In the course of caring for the young, the worker bees lick the pheromone mixture off the queen. This causes their ovaries to shrink, and they are inhibited from building queen cells. Larvae in queen cells are not fed honey, but royal jelly, a mixture of pollen and secretions, which does not contain Q. Royal jelly is recommended as a health product, but its effectiveness is disputed.

R=O Queen substance
R=H, OH 9-Hydroxy-*trans*-2-decenoic acid

Quebrachitol: see Cyclitols.

Quercetin: see Flavones (Table).

Quercitrin: see Flavones (Table).

Quinazoline alkaloids: a group of about 30 alkaloids, which occur in higher plants (in families taxonomically very distant from one another), animals and bacteria. They are derived biosynthetically from anthranilic acid. The simplest representative of the Q.a. is Glomerine (see) which is a very rare animal Q.a. Of the plant Q.a., Febrifugine (see) has some significance. In the wide sense, tetrodotoxin from the puffer fish can be included among the Q.a., as it is a zwitterionic polyhydroxy-2-imino-perhydroquinazoline.

Quinidine: see Quinine.

Quinine: the most important of the cinchona alkaloids. M_r 324.21, m.p. 57 °C (trihydrate), 174–175 °C (dehydrated crystals). $[\alpha]_D^{17}$ − 284.5 (0.05 M H_2SO_4). Structurally, Q. consists of a quinoline ring system connected via a secondary hydroxyl on C4 to a quinuclidine structure (see Cinchona alkaloids, Fig.). It occurs naturally in association with its stereoisomers, quinidine [the C-9 epimer, m.p. 172.5 °C, $[\alpha]_D^{15}$ + 334.1 (0.05 M H_2SO_4)], and epiquinine and epiquinidine (the C 8′ epimers). Q. forms bitter-tasting salts and has many physiological effects. It is used therapeutically as a drug against malaria and bacterial influenza. By reducing the rate of tissue respiration, it has an antipyretic effect. It also acts as an analgesic and inhibits heart excitation, although quinidine surpasses it in this last respect. Q. is toxic (fatal dose about 10 g), leading to deafness and blindness.

Quinoline alkaloids: a group of alkaloids based on the quinoline skeleton. They are found both in microorganisms (see Viridicatine) and in higher plants. The most important therapeutically are the Cinchona alkaloids (see). The starting material for the biosynthesis of some Q.a. is anthranilic acid (see Viridicatine); for others it is tryptophan (see Cinchona alkaloids).

Quinolizidine alkaloids: a group of alkaloids containing the quinolizidine (norlupinane) ring system. The most important Q.a. are the Lupin alkaloids (see), which are synthesized from lysine via cadaverine. The Nuphara alkaloids (see), in contrast, also possess a Q.a. ring system, but are synthesized by the terpene pathway.

Quinones: aromatic dioxo compounds derived from benzene or multiple-ring hydrocarbons such as naphthalene, anthracene, etc. They are classified as Benzoquinones (see), Naphthoquinones (see), Anthraquinones (see), etc. on the basis of the ring system. The C = O groups are generally ortho or para, and form a conjugated system with at least two C = C double bonds; hence the compounds are colored, yellow, orange or red. This type of chromophore is found in many natural and synthetic pigments.

Q. are a large and varied group of natural products found in all major groups of organisms. Those with long isoprenoid side chains, such as plastoquinone, ubiquinone and phylloquinone are involved in the basic life processes of photosynthesis and respiration. Q. are biosynthesized from acetate/malonate via shikimic acid. A few Q. are used as laxatives and worming agents, and others are used as pigments in cosmetics, histology and aquarell paints.

R

Radioimmunoassay: see Immunoassays.

Raffinose, ***melitose:*** a nonreducing trisaccharide, m.p. 120°C, $[\alpha]_D^{20}+123°$ (water). R. contains units of D-galactose, D-glucose and D-fructose. The galactose and glucose are linked by an α-1,6-glycosidic bond, and the fructose is linked to the glucose by an a β-1,2-glycosidic bond. R. is easily fermented by yeasts. Yeast enzymes hydrolyse R. to D-fructose and melibiose; emulsin hydrolyses R. to sucrose and D-galactose. R. is found widely in many higher plants, where it is the second (the first is sucrose) most commonly occurring free sugar. Sugar beet, molasses and many seeds, e.g. cotton seeds are especially rich in R. In plants, R. may function in place of sucrose as a transport carbohydrate.

Ramachandran plots, ***conformational maps:*** plots of rotation about the αC-carbonyl-C bond (Ψ) in a peptide linkage against rotation about the αC-amino-N bond (Φ). A general R.p. is constructed with the aid of models and computers. Using the accepted atomic radii of C, N, O and H, possible combinations of the two angles (i.e. regions of no steric hindrance) are indicated by blocked out areas on the graph or map. With only a small decrease in contact distance, these permissible regions become larger and a new permissible region appears. The permissible conformations include antiparallel β-pleated sheets, parallel β-pleated sheets, polyproline helix, collagen supercoil, right and left handed α-helices, right handed ω-helix, 3_{10} threefold helix, and the π-helix (4.4 residues per turn). An R.p. for a given protein can be constructed from Ψ and Φ values determined experimentally by X-ray diffraction and model building. [G.N. Ramachandran in *Aspects of Protein Structure* Academic Press (1963) p.39]

Randainol: see Neolignans.

Random coil: see Proteins.

Rapanone: see Benzoquinones.

Raphanatin: 7-glucosylzeatin. R. is formed from the cytokinin, zeatin, and it has no cytokinin activity. It is a storage form of zeatin, present in radish seedlings. Glucosylation at position 7 of the purine ring probably serves to protect zeatin from enzymatic degradation.

Rapoport-Luerbing shuttle: part of the Embden-Meyerhof pathway of glycolysis, constituting a self-regulating system which maintains the concentrations of 2,3-diphosphoglycerate and ATP at the expense of each other in the erythrocyte. A decrease in pH increases the activity of 2,3-diphosphoglycerate phosphatase, so that the concentration of 2,3-diphosphoglycerate decreases. The activity of diphosphoglycerate mutase therefore increases due to relief of the inhibition by 2,3-diphosphoglycerate and extra provision of its cofactor 3-phosphoglycerate. The net result is an increased flux through 2,3-diphosphoglycerate in the conversion of 1,3-diphosphoglycerate to 3-phosphoglycerate, and less conversion of ADP to ATP by phosphoglycerate kinase (Fig.). Hypoxia increases the 2,3-diphosphoglycerate concentration by increasing pH, and at the same time an increased amount of the ester is bound to deoxyhemoglobin, thereby decreasing the pool of the free ester. See Hemoglobin, Glycolysis. [E. Gerlach & J. Duhm *Scand. J. Clin. Lab. Invest.* **29** (1972) *Suppl.126,* 5.4a-5.4h].

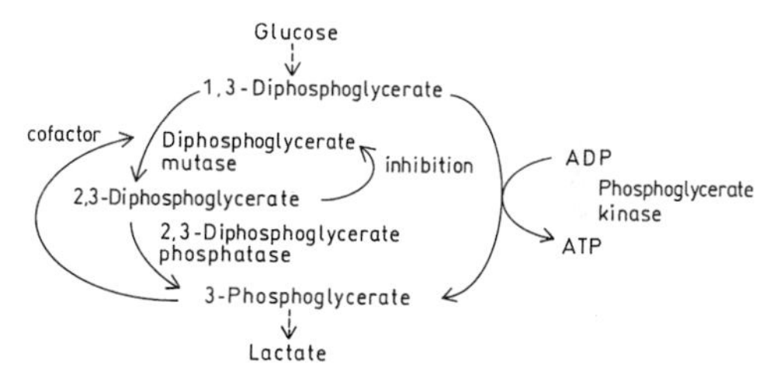

Rapoport-Luerbing shuttle of the erythrocyte

Rare nucleic acid components, ***unusual nucleic acid components, minor nucleic acid components:*** nucleic acid components of relatively infrequent occurrence, formed by the enzyme-catalysed modification of either the base or sugar of the usual nucleic acid constituents, i.e. modification of adenine, guanine, cytosine, uracil, thymine or ribose. With the exception of 5-hydroxymethyldeoxycytidylic acid (see Pyrimidine biosynthesis), all R.n.c. are formed by modification of residues in the intact polynucleotide chain of the nucleic acid. Modified nucleic acid bases are also called *minor bases.* Enzymic derivatization of free purines and pyrimidines does not occur, except in the formation of purine alkaloids in plants (see Methylxanthines).

The main types of modification are acetylation with acetyl-CoA (formation of N^4-acetylcytidine and 5-acetyluridine), glucosylation with UDPG (glucosylation of 5-hydroxymethyl-cytidine to 5-glucosylhydroxymethylcytosine), isoprenylation with γ,γ-dimethylallylpyrophosphate (conversion of adenosine to N^6-isopentenyladenosine), reduction (uridine to 5,6-dihydrouridine), thiolation with cysteine (formation of 2-thiouridine), cleavage of a C-C bond (conversion of uridine to pseudouridine), and methylation, which is particularly common. During methylation, methyl groups are transferred from *S*-adenosyl-L-methionine (SAM) to C, N, or O-atoms on bases (e.g. 5-methyluridine) or sugar moieties (e.g. 2-O-methyluridine). Other important R.n.c. are inosine and ribothymidine. About 40 R.n.c. are known. The enzyme-catalysed modifications are species-specific. Different types of nucleic acids differ significantly with respect to their contents of R.n.c. Transfer RNA contains an especially large number of R. n. c., which occur in specific positions, largely in the single-stranded regions, e. g. within and directly adjoining the anticodon loop. DNA and rRNA contain methylated nucleotides. Prokaryotic and phage mRNA appear to contain no methylated residues, i. e. if present their concentration is lower than the detection limit (less than 1 in 3,500 residues). Eukaryotic and viral mRNA contain unique 5-terminal cap structures with methylated residues, and some internal 6-methyladenylic acid residues (see mRNA).

In the methylation of DNA, specific methylases catalyse transfer of methyl groups from SAM to the 6-amino groups of adenine residues and C5 of cytosine. A specific pattern of methylation serves to protect DNA from the cell's own Restriction endonucleases (see); these enzymes destroy the DNA of invading viruses. The DNA of viruses that are able to replicate within a particular host are protected from the endonucleases by methylation of bases at the endonuclease-sensitive sites.

Carcinogenic alkylating agents alkylate chiefly guanine residues on the N7 atom. Some of the R. n. c. in tRNA are active cytokinins in the free state, e. g. N^6-(γ,γ-dimethylallyl)-adenosine.

Rate equation: in Enzyme kinetics (see), an equation expressing the rate of a reaction in terms of rate constants and the concentrations of enzyme species, substrate and product. When it is assumed that steady state conditions obtain, the Michaelis-Menten equation (see) is a suitable approximation. R. e. are represented graphically (see Enzyme graph); they may be derived by the King-Altman method (see).

Rauwolfia alkaloids: a group of about 50 structurally related indole alkaloids from the roots and rhizomes of various species of *Rauwolfa, Aspidosperma* and *Corynanthe.* All R. a. contain a β-carbolene skeleton; they are classified into 3 types: 1. yohimbine (corynanthine), 2. ajmaline, 3. serpentine. The large number of R. a. is due to the existence of stereoisomers. Thus *Rauwolfia* contains seven stereoisomers of yohimbine.

Some R. a. have valuable pharmacological properties. They may act centrally (e. g. see Reserpine) or peripherally (e. g. see Yohimbine, Ajmaline). In addition to the pure alkaloids or their synthetic analogs, extracts of the drug Radix Rauwolfiae and combination preparations are also used. The drug has been known since early times in Indian folk medicine, and its systematic investigation began in 1930.

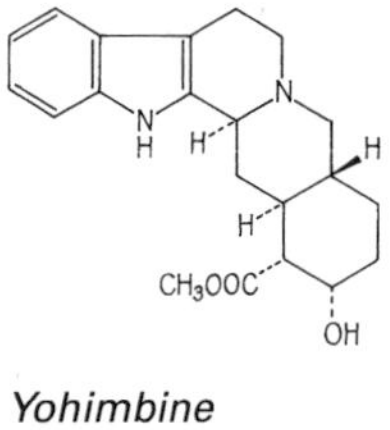

Yohimbine 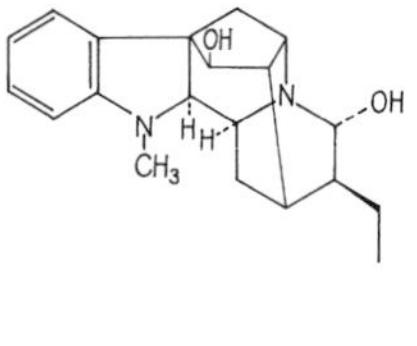 *Ajmaline*

Reagin: see Immunoglobulins.

RecA: see Recombination.

Receptor: see Hormones.

Receptor-mediated endocytosis: see Coated vesicle.

Receptor tyrosine kinases, *RTKs* (see also Protein tyrosine kinase): a large class of cell-surface receptors which is currently subdivided into four subclasses on the basis of their amino acid sequence homology, their 3D-structure and the similarity of their ligands. Subclass 1 includes the epidermal growth factor receptor (EGFR), subclass 2 includes the insulin receptor and the insulin-like growth factor-1 (IGF-1) receptor, subclass 3 includes the platelet-derived growth factor receptors A & B and, according to some authorities, the fibroblast growth factor receptors (FGFR) 1–4, and subclass 4 includes the Eph, Elk, Eek, Eck & Erk receptors.

Subclasses 1, 3 and 4 consist of an extracellular glycosylated domain containing a ligand-binding site, a single transmembrane α-helix composed mainly of hydrophobic amino acids, and a cytosolic domain containing one or two regions of tyrosine kinase activity. Subclasses 1 and 4 both have one region of tyrosine kinase activity but differ in the number of cysteine-rich regions in their extracellular domain (two in type 1; one in type 4). Subclass 3 has two regions of tyrosine kinase activity and a variable number of immunoglobulin-like loops in the extracellular domain. Subclass 2 receptors consist of two extracellular, glycosylated α-subunits (each with a single cysteine-rich region), which are linked to each other and to two transmembrane β-subunits by -S-S- bridges; the cytosolic domains of the β-subunits each have a single region with tyrosine kinase activity.

In subclasses 1, 3 and 4, ligand binding to the receptor site causes two ligand-RTK complexes to bind together (dimerize), whereupon the tyrosine kinase, located in the cytosolic region of each, catalyses autophosphorylation, i. e. the ATP-dependent phosphorylation of the hydroxyl groups of a distinct set of tyrosine residues in the cytosolic region of the other member of the dimer. The phosphorylated tyrosines resulting from this *trans* autophosphorylation process plus their immediately adjacent amino acids now act as binding sites for cytosolic proteins that possess SH2 domains (SH2 stands for **s**rc **h**omology region **2**, a region in the cytosolic tyrosine kinase encoded by the *src* gene). The 3D-structure of the SH2 domains of different proteins is very similar and yet sufficiently different to allow it to bind to a specific 'phosphotyrosine plus its immediately adjacent amino acids' site in the activated RTK; for instance the SH2 domain of the Src tyrosine kinase binds to the core sequence -Phosphotyr-Glu-Glu-Ile-. Proteins that possess SH2 domains and bind to activated RTKs include: (i) adapter proteins, such as GRB2 (see GTPases) that participate in a signaling train from the activated RTK to the ultimate recipient of the signal, and (ii) enzymes such as phospholipase C_γ, phosphadidylinositol-3 kinase (IP-3 kinase), a tyrosine kinase termed Syp, and the Ras-associated GAP.

Ligand binding to the receptor site of RTKs of subclass 2 initiates a somewhat different signaling train from that of the other RTK subclasses. For instance the binding of insulin to the binding site located in the α-subunit region of the insulin receptor, does not cause dimerization, presumably because this RTK is effectively already a dimer. However it does cause autophosphorylation of specific tyrosine residues in the cytosolic domains of both β-subunits. It also causes the phosphorylation (presumably catalysed by the tyrosine kinase of the activated insulin-RTK) of specific tyrosine residues in a 130 kDa, cytosolic protein called the 'insulin receptor substrate-1' (IRS-1). Phosphorylated IRS-1 molecules, which remain unbound to the activated insulin-RTK, then bind to several cytosolic proteins which have SH2 domains. These include GRB2 (see GTPases), PI-3 kinase and Syp; this constitutes another difference from the signaling train of the other RTK subclasses, where these proteins bind to the phosphorylated RTK. [P. van der Geer & T. Hunter 'Receptor Protein-Tyrosine Kinases and their Signal Transduction Pathways' *Annu. Rev. Cell Biol.* **10** (1994) 251–337]

Recombinant DNA technology, ***genetic engineering:*** the isolation and study of single genes, and the reintroduction of these genes into cells of the same or different species. *Gene cloning* is essential to R. D. t.

If a single gene is cloned in sufficient quantity, it can be sequenced (see Deoxyribonucleic acid) and the amino acid sequence of the gene product can be predicted from the nucleotide sequence of the DNA. For example, the amino acid sequences of the insulin receptor, nicotinic and muscarinic cholinergic acetylcholine receptors, epidermal growth factor precursor and many (thousands) other proteins have been predicted in this way.

Factors controlling gene expression can be studied by gene cloning. For example, eukaryotic genes for tyrosine aminotransferase (TAT), together with flanking nucleotides, including initiation and promotion elements, can be incorporated into a plasmid adjacent to, and upstream of, the bacterial gene for chloramphenicol acetyltransferase (CAT). After initiation, transcription proceeds through the structural TAT gene and continues through the adjacent CAT gene. Thus, factors affecting TAT transcription exert a similar influence on transcription of the CAT gene, and they can be studied by measurement of the CAT catalytic activity of the bacterial host. With the aid of this system, a glucocorticoid control region of the TAT gene has been identified and shown to consist of two glucocorticoid receptor binding sites.

The ultimate goal of the genetic engineer is the expression of purified genes in commercially exploitable, rapidly growing species. Human insulin, ovalbumin, fibroblast interferon, somatostatin, human growth hormone and other eukaryotic proteins can all be produced by expression of their cloned genes in *E. coli.* Stages in the isolation and cloning of a gene are outlined in Fig 1, and the strategy for cloning human growth hormone is shown in Fig. 2.

R. D. t., together with hybridization and blotting techniques, can be used to screen for a variety of genetic defects (see Hemoglobinopathy, Inborn errors of metabolism). Fetal cells, obtained by amniocentesis of the amniotic fluid during the 16th week of pregnancy, are cultured and their DNA is analysed. Mutation often destroys a restriction site or creates a new one, so that the pattern of nucleotides obtained after restriction analysis may change, thereby revealing the existence of the mutation. Thus, the gene for normal hemoglobin contains GAG for the glutamate residue at position 6 of the β-chain. This forms part of

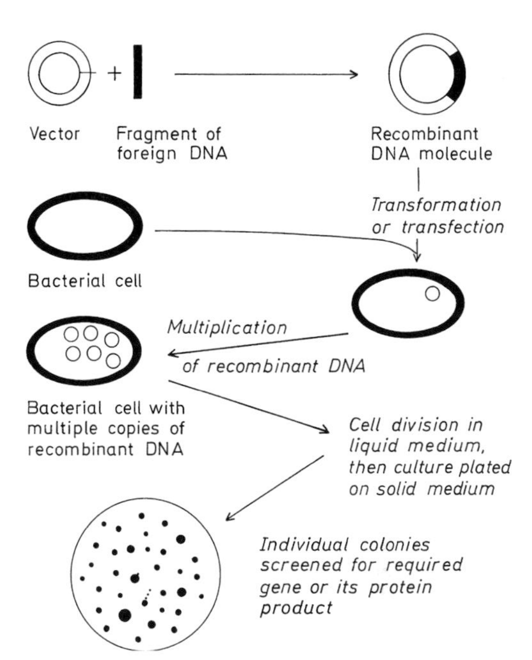

Fig. 1. *Essential stages in gene cloning using a bacterial plasmid vector*

a restriction site for Dde (which attacks $\frac{\text{CTNAG}}{\text{GANTC}}$) and for Mst II (which attacks $\frac{\text{CCTNAGG}}{\text{CCANTCC}}$). The mutation responsible for sickle cell anemia (see Sickle cell cell hemoglobin) eliminates this restriction site by converting GAG to GTG, resulting in valine at position 6 of the β-chain. The gene for sickle cell anemia can therefore be detected early in fetal development by digesting embryo-derived DNA with Dde I or Mst II and performing a Southern blot (see), using cloned normal β-globin DNA as the hybridization probe.

DNA vectors. These may be Plasmids (see), viruses (see Phage, Phage development) or Cosmids (see). Plasmids used in gene cloning usually carry one or more genes which confer characteristics on their host cells, enabling them to be distinguished from cells not carrying the plasmid (i. e. selection markers). For example, antibiotic resistance of a bacterium is often due to the presence of a plasmid carrying the genes for the resistance. All plasmids carry at least one replication initiation site, so that they can multiply independently of the host chromosome. Most plasmids are bacterial. Some strains of *Saccharomyces cerevisiae* contain a 2 μm circular plasmid, which has facilitated the construction of cloning vectors for this industrially important organism. Plasmids have not been found in other eukaryotes. Cassette vectors are the most advanced generation of vectors, which carry all the signals needed for gene expression (promoter, terminator, ribosome binding site) in the form of a cassette. The foreign gene is inserted into a unique restriction site in the cluster of expression signals. To construct a cassette vector, an

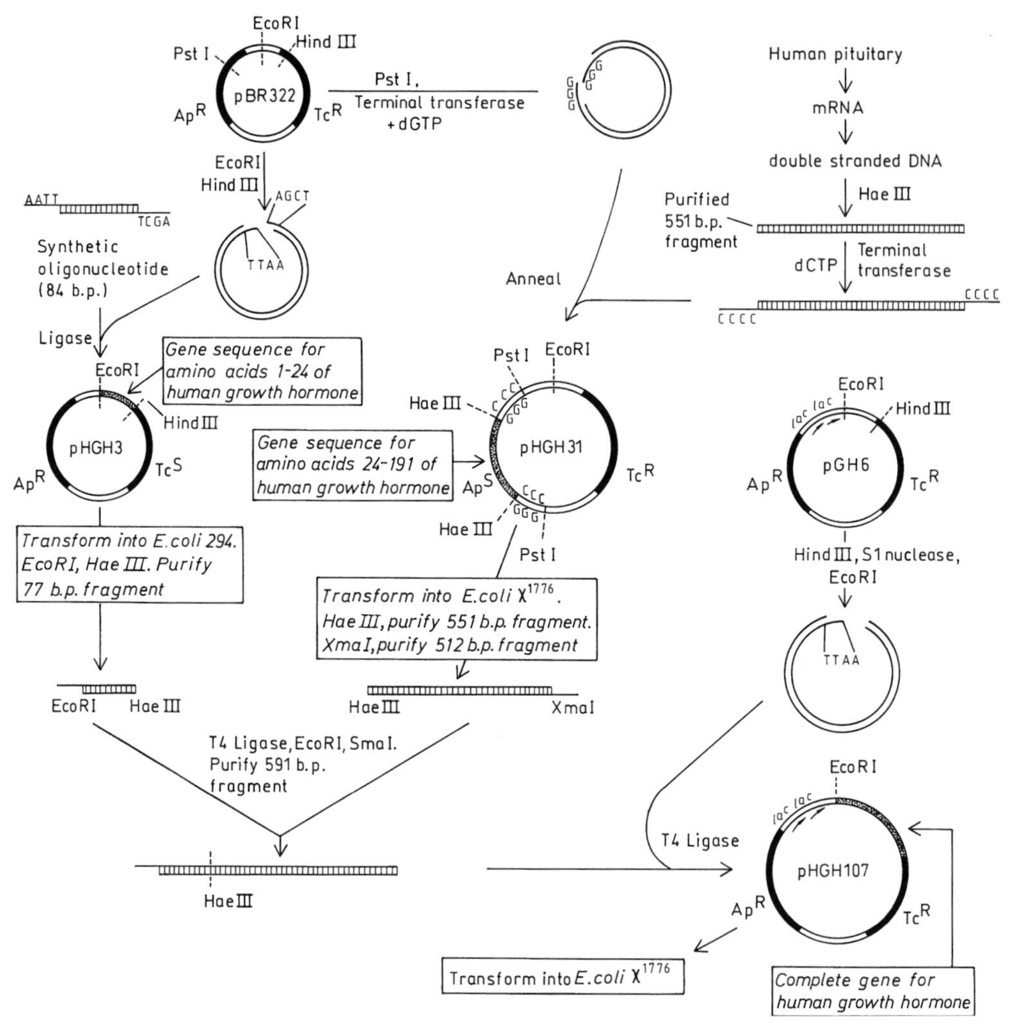

Fig. 2. *Cloning of human growth hormone.* Growth hormones are species-specific, and human cadavers were previously the only source of human growth hormone for clinical use, e.g. correction of dwarfism. One liter of a culture of *E coli* χ^{1776} carrying pHGH107 plasmids produces about 2.5 mg of human growth hormone; previously about 200 human cadavers would have been needed to prepare this quantity. The protein is not secreted, and it must be isolated and purified from the harvested cell paste. Rigorous purification is necessary to avoid the presence of antigenic bacterial proteins, if the product is to be injected into humans.

Hae III restriction sites are present on the 3′ noncoding region of the gene, and in the sequence coding for amino acids 23 and 24. Hae III restriction therefore cleaves the structural gene into 2 fragments, which are cloned separately and later ligated. The smaller cleavage fragment was discarded and replaced by a chemically synthesized oligonucleotide containing an ATG initiation codon. In ligation, restriction enzymes are also used to cleave unwanted dimerization products, e.g. in the formation of the 591 b.p. fragment, Eco I cleaves dimers formed from the Eco I sites, and Sma I cleaves Xma I dimers. *E. coli* χ^{1776} is a "safe" host, specially constructed to meet the safety requirements of genetic engineering; it can only survive under specially controlled laboratory conditions, and it cannot colonize, or survive in, the intestinal tract of warm blooded animals. ApR and TcR represent the genes for resistance to ampicillin and tetracyclin, respectively. If resistance is lost by inactivation of the gene, the same region is denoted by ApS or TcS (S = sensitive). b.p. = base pair. [D.V. Goeddel et al. *Nature 281* (1979) 544–548]

entire *E. coli* gene with its expression signals is inserted into a vector. The reading frame is then removed, leaving the expression signals intact. Initial difficulties in the construction of cassette vectors are now overcome by separate oligonucleotide synthesis of promoter, terminator and ribosome binding signals, which are ligated to form a cassette, then inserted into a plasmid. Imperfect cassette vectors (i.e. vectors retaining some of the original *E. coli* reading frame) are still widely used and have made a significant contribution to studies on the production of recombinant proteins.

Sources of DNA for cloning. *cDNA clone bank.* To produce a cDNA clone bank, mRNA is isolated from an organ, tissue or organism, then used as a template to synthesize cDNA. In multicellular eukaryotes, any one type of specialized cell does not express its total genome. Relatively few proteins are produced, and the corresponding mRNA is therefore present in relatively high proportion. For example, mRNA from pancreas contains a high proportion of mRNA for preproinsulin, legume root nodules contain a high level of leghemoglobin mRNA, and the mRNA for silk fibroin is predominant in the silk-synthesizing glands of the silk worm. An additional advantage of starting with mRNA is that it represents the end product of processing, in which intron transcripts (see Intron) may have been removed. The cDNA nucleotide sequence therefore corresponds to the amino acid sequence of the gene product. Eukaryotic recombinant genes containing introns may not be processed after being genetically engineered into a prokaryotic environment. Since most eukaryotic mRNA carries a 3'-polyA tail (see Messenger RNA), it can be isolated by affinity chromatography on oligo-dT cellulose.

mRNA used for cDNA synthesis can be enriched with respect to the message of interest by sucrose density gradient centrifugation, or by high performance liquid chromatography. Each mRNA fraction is tested in an in vitro translation system (reticulocyte lysate or wheat germ systems are used for eukaryotic mRNA), using radioactive amino acid precursors. Usually the translation product is analysed by precipitation with specific antibody, followed by sodium dodecyl sulfate gel electrophoresis and subsequent fluorography.

The first stage of cDNA synthesis is performed with reverse transcriptase (see RNA-dependent DNA polymerase), resulting in a mRNA-cDNA hybrid (Fig.3). Such hybrids have been successfully inserted into bacterial plasmids, but usually the mRNA template is removed by alkaline hydrolysis, and the single strand of cDNA is converted to the double stranded form. A region of self complementarity at the 3' end of the cDNA results in the formation of a small hairpin structure. Further action of reverse transcriptase or DNA polymerase I in the presence of deoxyribonucleotide triphosphates results in continued $3' \rightarrow 5'$ elongation, forming a double stranded structure with a closed end. Cleavage of the hairpin loop with S1 nuclease produces double stranded cDNA suitable for cloning. DNA polymerase I also possesses nuclease activity, which may partially destroy the synthesized DNA. It is therefore usual to use the Klenow fragment (see) of DNA polymerase I, or a different polymerase, such as bacteriophage T4 DNA polymerase.

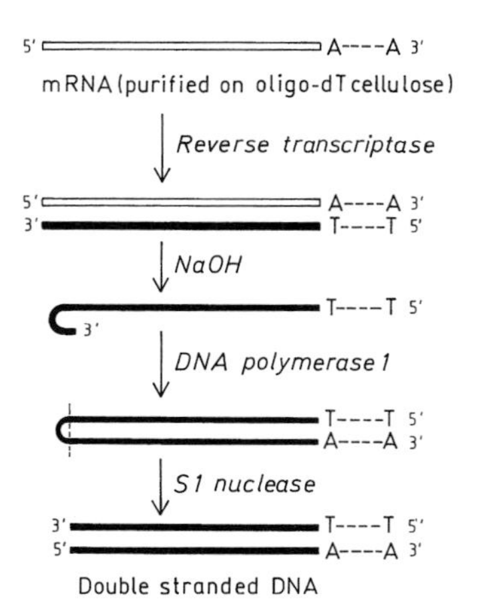

Fig.3. *In vitro synthesis of single-stranded and double-stranded cDNA.* The small hairpin structure at the 3' end of the initially synthesized single strand primes the synthesis of the second DNA strand.

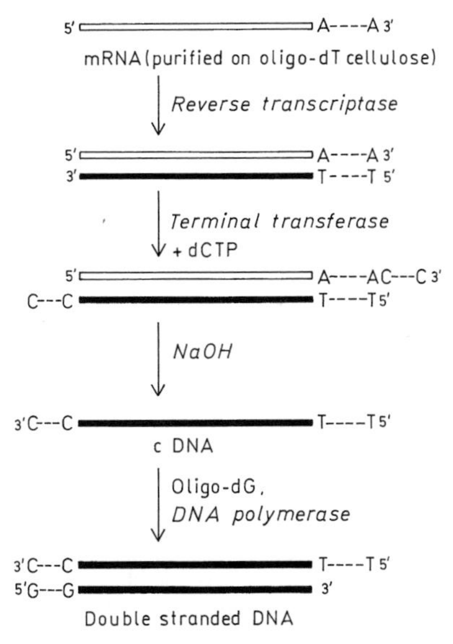

Fig.4. *In vitro synthesis of single-stranded and double-stranded cDNA.* Synthesis of the second strand of DNA is primed by oligo-dG, which associates with a 3' oligo-dC tail added at an earlier stage by the action of terminal transferase and dCTP.

Cleavage of the hairpin loop with S1 nuclease necessarily removes part of the DNA sequence corresponding to the 5' end of the mRNA. An improved method for full length duplex cDNA synthesis overcomes this problem. The first cDNA strand is tailed with oligo-dC so that second strand synthesis can be primed with oligo-dG (Fig.4).

Genomic library or gene bank. A genomic library is prepared from total cellular DNA, which is fragmented by cleavage with restriction endonucleases or by mechanical shear. Long, thin DNA molecules are easily broken by shearing forces. Intense irradiation with ultrasound produces DNA fragments of about 300 nucleotide pairs, while treatment in a high speed blender (1,500 rev/min for 30 min) produces fragments of about 8 kb pairs. To establish a genomic library, the total complement of DNA fragments is cloned in a suitable vector, usually a phage or cosmid. This approach is sometimes referred to as a shot gun method, since it is nonselective and relies on the statistical probability that the required gene is contained within at least one of the DNA fragments. The number of clones needed to insure that a genomic library contains all the genes of the cellular genome can be calculated from the formula: $N = [\ln(1 - P)]/[\ln(1 - \boldsymbol{a}/\boldsymbol{b})]$, where N is the number of clones required, P is the probability that any gene is present (e.g. this value can be set variously at 95 %, 80 %, etc.), ***a*** is the average size of the DNA fragment inserted into the vector, and ***b*** is the size of the total genome.

If bacteriophage DNA is used as a cloning vector, it is first purified then treated with a restriction endonuclease. The fragmented cellular genome and phage DNA restriction fragments are mixed, annealed and ligated, producing a population of recombinant DNA phage molecules, in which all the restriction fragments of the cellular genome are randomly distributed. This hybrid phage DNA is transfected into a host bacterium (most studies have been performed with bacteriophage λ and *E. coli*), eventually producing a phage population carrying the genomic library. Bacteriophages may act as *insertion vectors* or *replacement vectors.* Insertion vectors are opened at a single restriction site, so that insertion of the foreign DNA increases the size of the vector. Replacement vectors are cleaved at two restriction sites and the intervening sequence is replaced by the fragment of foreign DNA. The insertion vector, phage λNM607, can carry up to 9 kb of new DNA, whereas the replacement vector, phage λEMBL4, can carry up to 23 kb fragments. This compares with a maximal insert of 5 kb or less for most plasmid vectors. Cosmids, however, possess a much higher capacity, with a maximal limit of about 52 kb and a working limit of about 40 kb. They can therefore carry DNA fragments that are too large to be handled by plasmid or viral vectors, and they are useful for the establishment of genomic libraries of eukaryotic organisms which have very large genomes. Cellular genome fragments may be screened at an earlier stage, thereby avoiding the need to screen all the clones of a genomic library (see below: Blotting and hybridization techniques).

Cleavage and rejoining of DNA. These processes are fundamental to R.D.t. and gene cloning. With the aid of restriction endonucleases of different specificities, the DNA vectors and cellular genomes can be cleaved at chosen sites. Large DNA molecules may also be cleaved by mechanical shear. Most restriction enzymes generate cohesive or "sticky" ends of 1 to 4 nucleotides, so that when vector and donor DNA are cleaved by the same enzyme, the DNA termini can be annealed, then covalently linked by DNA ligase (DNA ligase seals single strand nicks between adjacent nucleotides in a duplex chain (Fig. 5). Such simple rejoining of the two DNAs is not always possible,

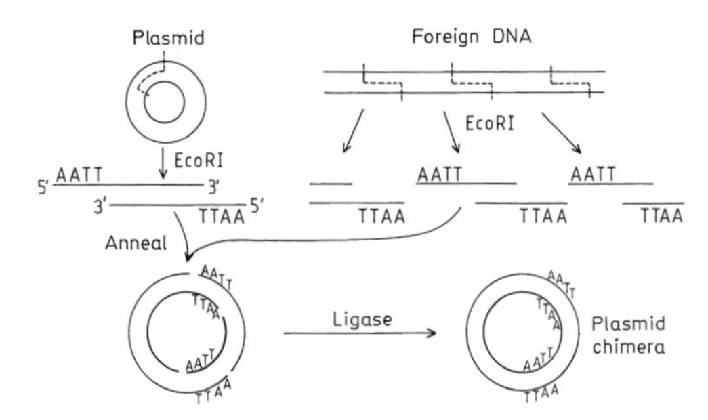

Fig. 5. *In vitro generation of a chimeric plasmid containing foreign DNA.* The required sticky ends are created by cleaving both the plasmid and the foreign DNA with the same restriction endonuclease (in this example, EcoRI).

because 1. the restriction enzyme may leave a "blunt" end, 2. it may be necessary to use different restriction enzymes, so that the potentially sticky ends are noncomplementary, 3. DNA fragments produced by mechanical shear, enzymatic cDNA synthesis or chemical DNA synthesis do not have sticky ends. However, sticky ends can be generated by addition of nucleotide "tails" to the 3′ end of DNA chains (a process known as *tailing*), by the action of terminal transferase from calf thymus (Fig. 6). Alternatively, synthetic DNA *linkers* may be ligated to the vector and/or donor DNA. The linker is a short DNA fragment containing the recognition sequence for one or more restriction enzymes, which is not present in the DNA receiving the linker. After ligation of the linker to blunt ended DNA, it is cleaved with an appropriate restriction enzyme to generate a sticky end (Fig. 7). Clearly, the two sticky ends produced by restriction are potentially capable of reuniting with each other, rather than participating in the intended hybridization. This tendency is overcome partly be performing hybridization at a relatively high DNA concentration to increase the frequency of intermolecular reactions. Also, treatment of linearized vector DNA with alkaline phosphatase removes 5′-terminal phosphate and prevents recircularization and dimer formation by plasmids; circularization of the vector can then occur only by insertion of non-phosphatase-treated DNA, which brings with it one 5′-terminal phosphate to each junction. The resulting nick is repaired by the host after transformation.

Introduction of recombinant DNA into a host cell. Plasmids enter the bacterial cell by Transformation (see). It is usually necessary to make the bacterial cell receptive to transformation, i.e. the cells must be competent. For *E. coli,* this is usually achieved by treatment with a solution of 50 mM $CaCl_2$. This causes the DNA to bind more effectively to the cell surface. Movement of DNA into the competent cells is then stimulated by raising the temperature briefly (2 min) to 42 °C, a process known as heat shock. Phage and cosmid DNA enter the bacterial cell in a process equivalent to transformation, but usually referred to as *transfection.*

Probes. ^{32}P-labeled mRNA, cDNA or synthetic oligonucleotides are used as probes in Southern blotting

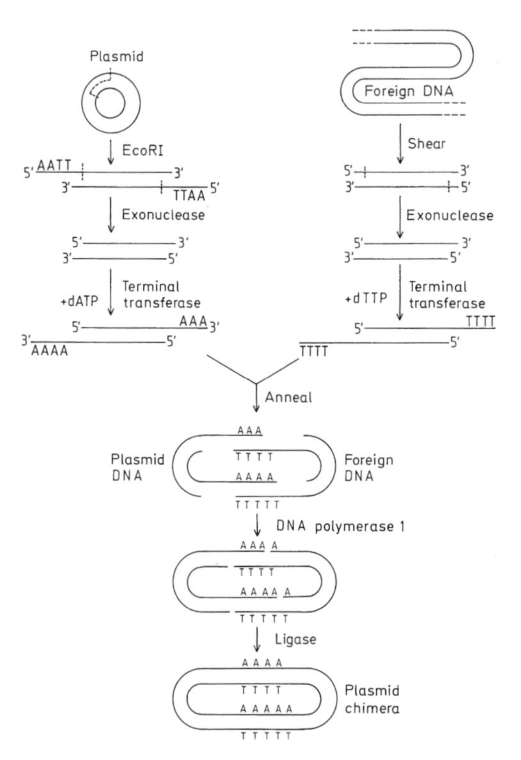

Fig. 6. *In vitro generation of a chimeric plasmid containing foreign DNA.* Sticky ends are created by the action of terminal transferase and an appropriate deoxyribonucleotide triphosphate.

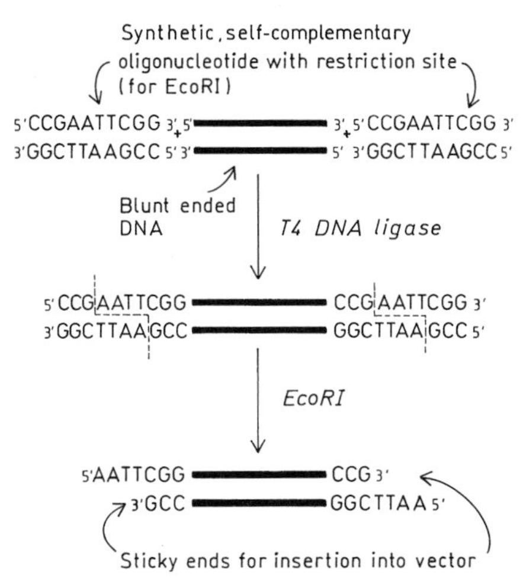

Fig. 7. *Generation of sticky ends on double-stranded, blunt-ended DNA by the use of linkers.* Linkers are self-complementary, synthetic oligonucleotides containing appropriate restriction sites.

analysis. Double-stranded DNA can be labeled with ^{32}P by Nick translation (see), by 3′ end labeling (Fig. 8), by the random priming process (Fig. 9), or by 5′ end labeling (Fig. 10). Synthetic oligonucleotide probes need contain only about 15–20 nucleotides (about 6 gene codons). Short sequences (about 6 amino acids) of the gene product are selected. If the total sequence is not known, sequences are chosen from those of the peptides produced by proteolysis. The corresponding oligodeoxyribonucleotide is synthesized chemically (see DNA synthesis). Owing to the degeneracy of the genetic code, there are several possible oligonucleotide structures for one amino acid sequence. Oligopeptide sequences containing tryptophan or methionine are therefore favored, because these two amino acids have single codons, thereby decreasing the possible number of probe structures. In practice, all of the possible oligonucleotide sequences are not synthesized individually. During each polymerization step where the sequence is degenerate, a mixed addition reaction is performed, resulting in a mixed probe, which is radiolabeled and used directly. If the gene structure is known, however, single oligonucleotide probe sequences can be selected and synthesized.

Blotting and hybridization techniques. DNA fragments from restriction digests are separated according to their size by agarose gel electrophoresis. The electrophoretogram is screened for the presence of specific nucleotide sequences by performing a Southern blot (see), followed by hybridization with ^{32}P-labeled specific RNA or DNA probes, then autoradiography. In this way, the total fragment mixture from a cellular genome can be screened for those fragments carrying a particular gene; this dispenses with the need later to screen a total genomic library (Fig. 11). A screening procedure for detection of DNA sequences in transformed bacterial colonies by hybridization in situ with radioactive probe RNA was developed by Grunstein and Hogness in 1975. A single colony among thousands can be detected by this technique. Colonies to be screened are replica plated onto a nitrocellulose filter disk on the surface of an agar culture plate, and allowed to grow. The nitrocellulose filter is then treated with alkali to lyse the bacterial colonies and denature their DNAs. Protein is removed with a proteinase, leaving denatured DNA bound to the nitrocellulose. Single-stranded, but not double-stranded, DNA binds very strongly to nitrocellulose. The DNA is then fixed firmly to the nitrocelluose by heating at 80 °C. To identify the original sites of colonies carrying the required transforrned gene, the nitrocellulose disk is incubated with ^{32}P-labeled, specific RNA. Only the specified DNA hybridizes with the RNA, and this is located by autoradiography. Comparison of the original culture plate with the autoradiogram permits identification of any colony carrying the gene of interest (Fig. 12).

Gene expression. Synthesis of a functional protein depends on transcription of the gene, translation of the mRNA, often processing of the mRNA, and often post-translational processing of the initial translation product (see Post-translational modification of proteins). Transcription of a cloned insert requires the presence of a promoter which is recognized by the host RNA polymerase. Translation requires a ribo-

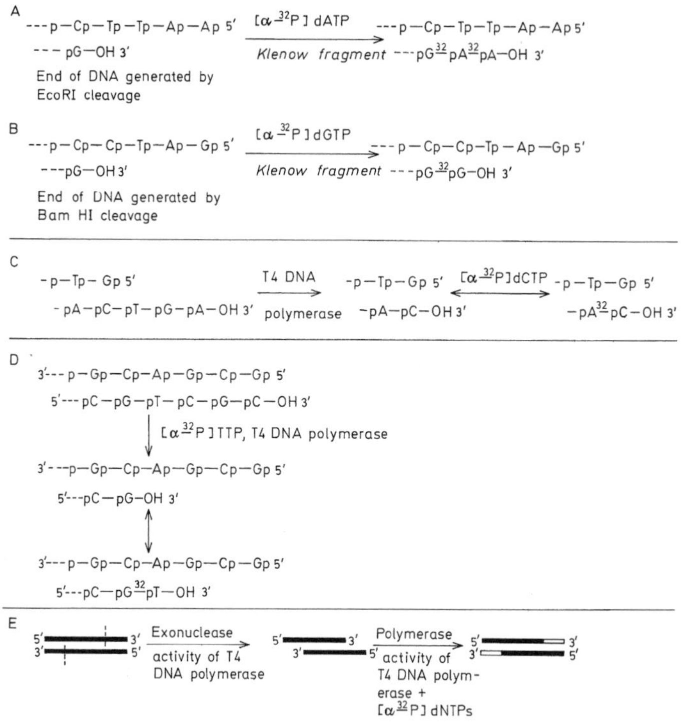

A

---p—Cp—Tp—Tp—Ap—Ap 5′ —[α-^{32}P] dATP, Klenow fragment→ ---p—Cp—Tp—Tp—Ap—Ap 5′

--- pG—OH 3′ → ---pG$\overset{32}{-}$pA$\overset{32}{-}$pA—OH 3′

End of DNA generated by EcoRI cleavage

B

---p—Cp—Cp—Tp—Ap—Gp 5′ —[α-^{32}P] dGTP, Klenow fragment→ ---p—Cp—Cp—Tp—Ap—Gp 5′

---pG—OH 3′ → ---pG$\overset{32}{-}$pG—OH 3′

End of DNA generated by Bam HI cleavage

C

-p—Tp—Gp 5′ —T4 DNA polymerase→ -p—Tp—Gp 5′ ←[α-^{32}P]dCTP→ -p—Tp—Gp 5′

-pA—pC—pT—pG—pA—OH 3′ → -pA—pC—OH 3′ → -pA$\overset{32}{-}$pC—OH 3′

D

3′---p—Gp—Cp—Ap—Gp—Cp—Gp 5′

5′---pC—pG—pT—pC—pG—pC—OH 3′

↓ [α-^{32}P] TTP, T4 DNA polymerase

3′---p—Gp—Cp—Ap—Gp—Cp—Gp 5′

5′---pC—pG—OH 3′

↕

3′---p—Gp—Cp—Ap—Gp—Cp—Gp 5′

5′---pC—pG$\overset{32}{-}$pT—OH 3′

E

5′ 3′ / 3′ 5′ —Exonuclease activity of T4 DNA polymerase→ 5′ 3′ / 3′ 5′ —Polymerase activity of T4 DNA polymerase + [α-^{32}P] dNTPs→ 5′ 3′ / 3′ 5′

Fig. 8. *Methods for the 3′ end labeling of double-stranded DNA.* The reactions are similar to those used for filling recessed ends of double-stranded DNA after restriction cleavage, except that only one [α-^{32}P]deoxyribonucleotide triphosphate is used.
A and B: Klenow fragment (see) possesses only polymerase activity, and it efficiently catalyses addition of the appropriate nucleotide residue at the recessed 3′ end of DNA.
C: T4 DNA polymerase also possesses exonuclease activity. The protruding 3′ arm is degraded one nucleotide at a time until a residue is encountered corresponding to the nucleotide substrate present in the incubation. The polymerase activity of the enzyme is then manifested, and an exchange reaction occurs.
D: The reactions shown in C may also be applied to blunt-ended DNA.
E: The combined exonuclease and polymerase activities of T4 DNA polymerase are exploited in the O'Farrell method for the general labeling of double-stranded DNA. dNTP = deoxyribonucleotide triphosphate.

some binding site on the mRNA. If the mRNA is translated in *E. coli,* the ribosome binding site includes the translation start codon (AUG or GUG) and a ribonucleotide sequence complementary to the bases at the 3′ end of the 16S ribosomal RNA. These are known as S-D sequences (named after Shine and Dalgarno, who first postulated these sequences in 1975), and they are present in almost all *E. coli* mRNAs. S-D sequences vary in length from 3 to 9 nucleotides, and they are situated 3 to 12 bases before the translation start codon. Although a foreign protein product may be synthesized from its own *N*-terminus under the control of an *E. coli* promoter (e.g. human growth hormone), usually a fusion hybrid or chimeric protein is produced. For example, the eukaryotic gene for β-endorphin was attached to the bacterial gene for β-galactosidase and expressed in *E. coli.* The resulting protein was therefore a hybrid of β-galactosidase and β-endorphin, from which the β-endorphin (a polypeptide of 31 amino acids) could be cleaved proteolytically: 1.lysine residues were protected by reaction with citraconic anhydride; 2.citraconylated endorphin was released by treatment of the hybrid with trypsin, which attacks an arginyl bond immediately preceding the β-sequence (β-endorphin contains no arginyl residues); 3.native endorphin was prepared by removal of citraconyl groups at pH 3.0. Several genes have been cloned as partners of the β-galactosidase gene, resulting in their initial expression as fusion proteins, e.g. somatostatin and human insulin. Several strategies can be employed for final cleavage of the chimeric protein. For example, if the gene is entirely synthetic (see DNA synthesis), an additional gene codon for *N*-terminal methionine can be included, thereby creating a CNBr-sensitive cleavage site in the chimeric protein. A particular problem was encountered in the expression of the human insulin gene. Proinsulin contains a 35-residue C-peptide

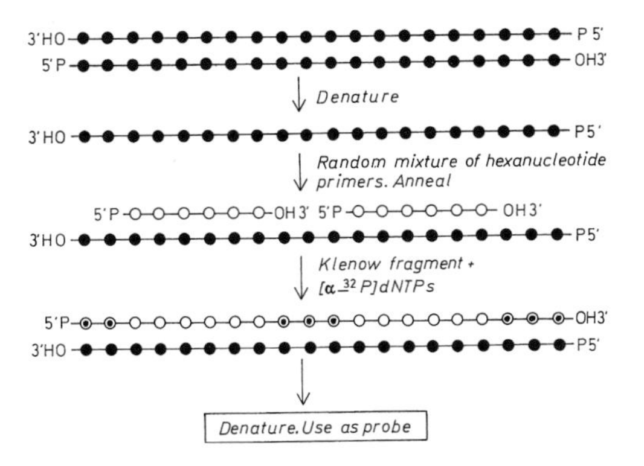

Fig. 9. *Random priming for the in vitro labeling of DNA probes.* Single-stranded DNA is incubated with a random mixture of hexanucleotides. Some of these will be complementary to regions of the DNA, and they act as primers for the synthesis of a complete complementary strand, using the Klenow fragment of DNA polymerase I and [α-^{32}P]deoxyribonucleotide triphosphates (dNTPs). ● = Nucleotide residues of the original DNA. ○ = Nucleotide residues of the hexanucleotide primers. ⦿ = ^{32}P-labeled nucleotide residues. Random priming gives a relatively low concentration of probe, but the specific activity is high. It is therefore suitable for preparing probes for single copy gene detection in a Southern blot. By using 200 μCi of [α-^{32}P]dNTPs at 6,000 Ci/mol, probes with specific activities greater than 5×10^9 dpm/μg can be prepared. To achieve the equivalent probe specific activity by Nick translation (see), it would be necessary to use 1 mCi labeled dNTP per μg DNA. [A. P. Feinberg & B. Vogelstein *Anal. Biochem. 132* (1983) 6–13; A. P. Feinberg & B. Vogelstein, Addendum *Anal. Biochem. 137* (1984) 266–267]

which must be removed after translation (see Insulin). A gene was constructed encoding an analog of proinsulin, in which the C-peptide sequence is replaced by a group of 6 amino acids (Arg-Arg-Gly-Ser-Lys-Arg), which is easily removed by proteolysis in vitro.

Oocytes are able to transcribe exogenous DNA, and *Xenopus* oocytes in particular have been used experimentally as a surrogate genetic system for expression of cloned DNA, e.g. genes for ovalbumin, sea urchin histones, and various proteins of phage SV40 have been expressed in this way. The oocyte is a relatively large cell with a large nucleus, so that microinjection of DNA into the nucleus is technically easy, e.g. 20–40 nanoliters of DNA solution can be injected via a fine glass capillary.

Mouse embryos have also been used as expression systems for foreign DNA. Microinjection of SV40 DNA into preimplantation blastocysts, followed by implantation into the uteri of foster mothers, resulted in progeny which were mosaics, i.e. only a proportion of cells in each tissue contained integrated SV40 DNA; but the subsequent generation produced genetically defined substrains, showing that

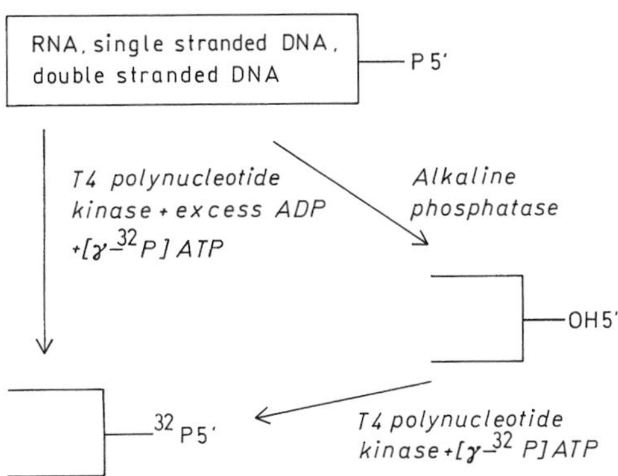

Fig. 10. *5′ End labeling of DNA and RNA.* Labeling is most efficient with single- stranded molecules and at the 5′ ends of double-stranded DNA with recessed 3′ ends. Blunt-ended DNA and DNA with recessed 5′ ends react more slowly. After removal of the 5′ phosphate with alkaline phosphatase, the 5′OH group is rephosphorylated by the action of polynucleotide kinase and ATP. Alternatively, in the presence of excess ADP, the reverse reaction is encouraged, effectively leading to exchange between then 5′ phosphate and the γ-phosphate of ATP.

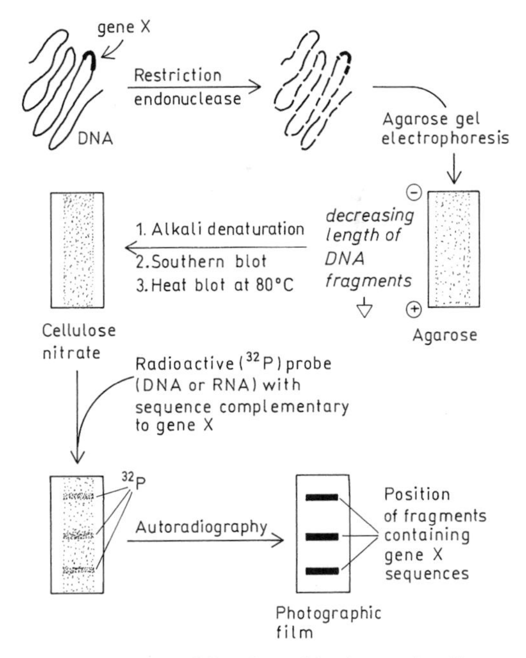

Fig. 11. *The use of Southern blotting and radioactive probes for screening fragmented DNA*

SV40 DNA had become integrated into germ line cells. To insure integration into a host chromosome at an early stage of development, and to avoid mosaicism, the DNA can be injected into the male pro-

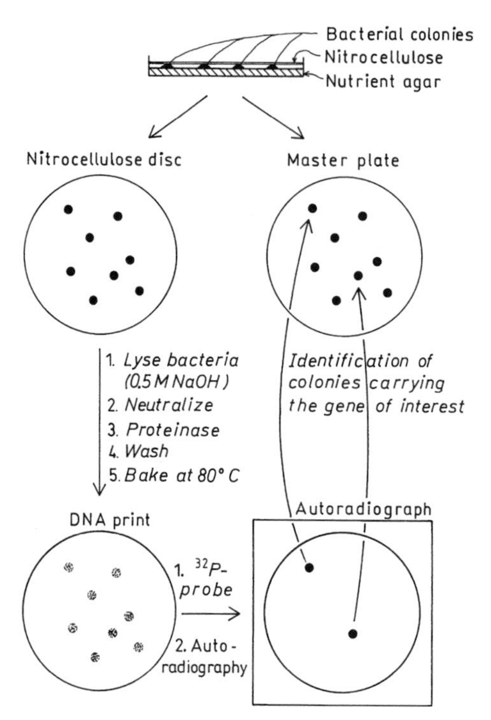

Fig. 12. *Screening procedure of Grunstein and Hogness*

nucleus of the newly fertilized egg. Between 3 % and 40 % of animals from these embryos are transgenic mice, i.e. their DNA contains the integrated foreign gene. Mouse metallothionein (MMT) gene has been fused with other genes and reintroduced into mouse embryos. Induction of the MMT gene with heavy metals (e.g. zinc or cadmium) results in expression of the hybrid partner as well. Thus, transgenic mice carrying fused (hybrid) genes for [MMT-*Herpes simplex* thymidine kinase], [MMT-rat growth hormone] and [MMT-human growth hormone] have been reared.

Selection. In direct selection methods, only the desired recombinants grow after plating transformants on an agar growth medium. For example, if a gene specifying antibiotic resistance is cloned in a sensitive host, then only transformants carrying the resistance gene will survive and form colonies on growth medium containing the antibiotic. In a direct selection technique known as *marker rescue,* auxotrophic mutants are used as hosts for the vector carrying a biosynthetic gene, e.g. if the A gene for tryptophan synthase is cloned in a *trp*A⁻ auxotroph, only those cells containing a plasmid-borne copy of the A gene are able to form colonies on growth media lacking tryptophan. Selection for the desired gene can be performed by hybridization, either on the initial digest of the cellular genome, or on the plated colonies of the final transformed population (see Blotting and hybridization techniques, above). Enrichment of the desired mRNA prior to cDNA synthesis is also a form of screening.

If the cloned gene is expressed, screening can be performed by exploiting some property of the synthesized protein. For example, in the immunochemical method of Broome and Gilbert, the transformed cells are first plated in the conventional way on nutrient agar, and a replica plate is prepared. The colonies on one of the plates are lysed (by exposure to chloroform vapor, by spraying with an aerosol of virulent phage, or by thermo-induction of an inducible prophage already present in the host) to release antigen from the positive colonies. The surface of the plate is then placed against a sheet of polyvinyl coated with antibody to the gene product. The washed polyvinyl sheet, carrying bound gene product, is exposed to ^{125}I-labeled IgG, which reacts with the bound antigen at different sites from those involved in the initial binding. Positive areas are detected by autoradiography. Peptide acetyltransferase was cloned in phage λ from a cDNA library derived from rat brain mRNA, and selected by a novel procedure known as *enzyme immunodetection assay* (EIDA) [J. H. Eberwine et al. *Proc. Natl. Acad. Sci. USA* **84** (1987) 1449–1453]. The t-butoxycarbonyl (Boc) derivative of [Leu^5]enkephalin hydrazide was reacted with paper carrying diazotized aminophenylthioether groups. After removal of the Boc groups with acid, the resulting substrate paper possessed covalently bound [Leu^5]enkephalin with a free amino-terminus. Bacterial plates with patterns of phage plaques (derived from λ phage carrying the cDNA library) were placed in contact with the substrate paper. The covalently bound enkephalin became acetylated at the site of any plaque expressing peptide acetyltransferase activity. After incubation of the paper with antisera specific for the acetylated form of enkephalin the acetylated areas retained bound IgG. Finally the IgG was revealed by incubation with ^{125}I-labeled Protein A (see) and autoradiography.

[T. Shine et al. *Nature* **285** (1980) 456–461; R. Wetzel et al. *Gene* **16** (1981) 63–71; H. Yoshioka et al. *J. Biol. Chem.* **262** (1987) 1706–1711; H. M. Jantzen et al. *Cell* **49** (1987) 29–38; R. W. Old & S. B. Primrose *Principles of Gene Manipulation* 3rd edtn. (Blackwell 1985); M. Grunstein & D. S. Hogness, *Proc. Natl. Acad. Sci. USA* **72** (1975) 3961–3965; S.Broome & W. Gilbert *Proc. Natl. Acad. Sci. USA* **75** (1978) 2746–2749; T.Moniatis, E. F. Fritsch & J.Sambrook, eds. *Molecular Cloning* (Cold Spring Harbor Laboratory, 1982)]

Recombination, *genetic recombination:* the totality of processes that result in new gene combinations, during meiosis or mitosis. R. occurs frequently and in all organisms, so that all DNA isolated from natural sources is recombinant DNA. There are two different forms of R.: *General R.* and *Specialized* or *Site-specific R.* The latter includes those forms of R. that do not require homologous regions in the parent DNA molecules, e.g. rearrangements of chromosomes (internal R.) and of the genes for vertebrate immunoglobulin and T-cell receptors, as well as transposition (see Transposon) and the generation of a lysogenic state by incorporation of bacteriophage DNA into the host chromosome (see Phage development). The present account deals exclusively with General R. Functionally, General R. can be defined as an exchange between any pair of homologous sequences on parental DNA molecules.

1) ***Meiotic (sexual) R.***

a) *Unlinked genes.* The allele pairs of the genes in question (which belong to different linkage groups) can recombine freely, due to the random orientation of bivalents and the random distribution of chromosomes during meiosis, and the random union of gametes during fertilization. The process obeys Mendelian laws.

b) *Linked genes.* The alleles of the genes in question belong to the same linkage group and behave as a unit in meiosis, i.e. they cannot freely recombine. R. between paired chromosomes in meiosis may be intergenic (nonallelic) or intragenic (allelic). In intergenic (nonallelic) R. there is a reciprocal exchange of segments between the chromatids of the paired chromosomes. The exchange involves functional genes (cistrons), and it is reciprocal and symmetrical, i.e. each crossing over during meiosis results in a cell tetrad containing two cells with exchange chromatids and two cells with nonexchange chromatids. Intragenic (allelic) R. occurs between the subunits of a functional gene (cistron), which is present in two different allelic forms in the homologous chromosomes. In this case, R. is nonreciprocal (asymmetric).

Thus in reciprocal R., parent types a+ and +b give rise to a tetrad of cells with equal proportions of genotypes a+, +b (parent combinations), ab and ++ (recombinants), provided crossing over occurs between the two genes. In nonreciprocal, intragenic R., the four genotypes do not occur with equal frequency, i.e. many cells are wild-type recombinants, and the double mutants expected from R. are absent (or vice versa). A further difference is that intergenic R. generally shows positive interference, whereas intragenic R. usually shows negative interference.

2) ***Somatic*** or ***parasexual R.*** This occurs in the mitosis of somatic cells, and not during meiosis. Like meiotic R., it may be intergenic or intragenic. Recombinable linked structures are found in diploid, heterokaryotic and heterogenotic cells. Somatic R. is rare in higher organisms, but it is the normal process of R. in certain fungi and in bacteria and viruses. The heterogenetic state of bacteria, which is a precondition for genetic R., is attained by conjugation, transduction or transformation.

In Genetic engineering (see Recombinant DNA technology), R. is performed in vitro, in order to prepare DNA containing new gene combinations.

As indicated above, R. is a result of crossing over. In the accepted model of the crossed-over intermediate, the nicked strands of two aligned homologous DNA duplexes cross over and pair with their complementary counterparts, followed by sealing of the nicks. Since models of the resulting 4-stranded, crossed-over structure of the heteroduplex (the Holliday structure) show that all the bases are paired without steric strain, the Holliday structure is apparently stable. Two possible mechanisms for resolution of the Holliday structure, i.e. for cleavage of the DNA chains followed by formation of two DNA duplexes without loss of continuity, are shown in the Fig.

In *E. coli,* recA protein (352 amino acid residues; product of the *recA* gene) plays a major role in R. RecA binds specifically and cooperatively to single-stranded DNA, forming a filamentous complex. As a prelude to the exchange of homologous segments, this nucleofilament then associates with double-stranded DNA. The duplex DNA is then rapidly scanned for a sequence complementary to that of the single-stranded DNA. As scanning progresses, the duplex is partly unwound in an ATP-dependent process catalysed by recA. When a complementary sequence is encountered, it pairs with the recA-coated single-stranded DNA. Strand exchange then occurs, and unidirectional branch migration results in the formation of a new heteroduplex molecule, as well as a displaced strand of single-stranded DNA. The displaced resident strand forms a D-shaped loop (actually known as a *D loop*) extending from the main body of the DNA.

Electron microscopy of the filamentous complex of recA-DNA reveals bound filaments of polymerized recA in the form of a right-handed helix (~6.2 recA monomers per turn, with a pitch of 95 Å). Each recA monomer binds to a region of DNA equivalent to 3 base pairs (or 3 nucleotide residues). The structure of recA determined by X-ray crystallography shows a major central domain flanked by smaller *N*- and *C*-terminal domains. The monomers are associated into a ~120 Å-wide helical filament (6 monomers per turn, pitch 82.7 Å), which is very similar to the right-handed recA helix of the recA-DNA complex. The helical recA polymer possesses a large groove which, from its size, is capable of accommodating either 3 or 4 DNA strands, i.e. the question arises as to whether the key stages of R. between two DNA duplexes involve a 3-stranded or a 4-stranded intermediate The weight of evidence is now in favor of a 3-stranded intermediate.

The single-stranded nicks necessary for the binding of recA are made by the ATP-dependent enzyme recBCD (M_r 330,000, product of the SOS genes *recB, recC* and *recD*). RecBCD both winds and unwinds duplex DNA, but unwinding is faster than winding, so that two single-stranded loops are formed. These two loops grow larger as the enzyme migrates along the duplex, until it encounters the sequence 3′GCTGGTGG5′, starting at its 3′ end. Known as the *Chi sequence,* this occurs approximately once in every 10,000 bases in the *E. coli* chromosome. RecBCD then cleaves the single-stranded loop 4–6 nucleotide residues to the 3′ side of the Chi sequence.

A variety of other catalytic proteins also participate in R. in *E. coli.* These include: Topoisomerases (see) (which maintain an appropriate degree of supercoiling for the R. process), Single-strand binding protein (SSB; modulates recA function, and helps to keep DNA single-stranded), RuvB Protein (M_r 27,000; a DNA-dependent ATPase that drives branch migration), RuvA protein (M_r 22,000; binds specifically to Holliday junctions and to RuvA), RuvC protein (M_r 19,000; a nuclease catalysing the cleavage that separates the two recombinant duplexes), recJ (a 5′- to 3′-exonuclease which degrades the single-stranded DNA displaced during strand exchange, thereby removing a potential competitor for strand pairing), DNA ligase (ligates the DNA strands at the resolved cross-over junctions).

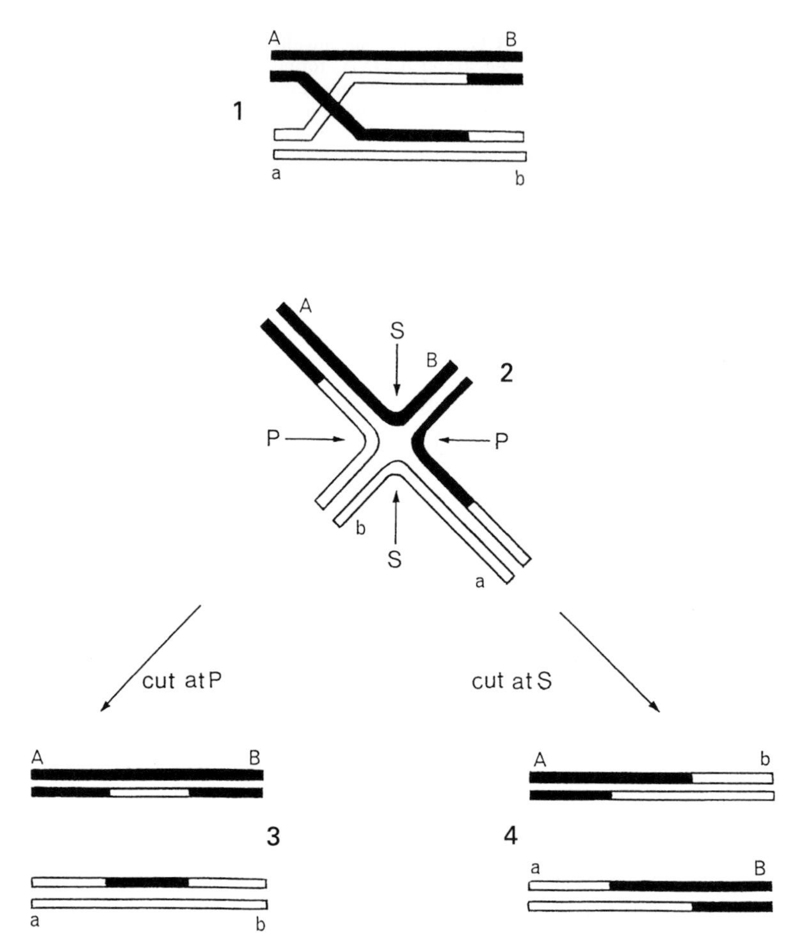

Resolution of crossovers in homologous recombination. 1. The Holliday structure: crossover between DNA strands has occurred and the point of crossover has moved by branch migration). 2. Alternative diagrammatic representation of the 4-way DNA junction of the Holliday structure, which may be cut at P then rejoined to give a "patch" recombination of recombinant molecules (3), or cut at S then rejoined to give a "splice" recombination of recombinant molecules (4).

The mechanism of R. in eukaryotes has not yet been as thoroughly investigated as that in *E.coli.* Yeast RAD51 protein, which displays 30 % homology with recA, also catalyses ATP-dependent DNA repair and recombination, and its complex with DNA is practically identical to that of recA in the electron microscope. Homologs of RAD1 are found in a variety of animals, including chickens, mice and humans, suggesting that the basic mechanism of R. is the same in all organisms. [R.Holliday *Genet. Res.* **5** (1964) 282–304; S.C.West *Annu. Rev. Biochem.* **61** (1992) 603–640; A.F.Taylor *Cell* **69** (1992) 1063–1065; R.M.Story et al. *Nature* **355** (1992) 318–325 (see also page 367); E.H.Engelman *Curr. Opin. Struct. Biol.* **3** (1993) 189–197; M.Takahashi & B.Norden *Adv. Biophys.* 30 (1994) 1–35; S.C.West *Cell* **76** (1994) 9–15; J.Lee & M.Jayaram *J. Biol. Chem.* **270** (1995) 4042–4052; S.E.Corrette-Bennett & S.T.Lovett *J. Biol. Chem.* **270** (1995) 6881–6885; M.M.Cox *J. Biol. Chem.* **270** (1995) 26021–26024; F.Maraboeuf et al. *J. Biol. Chem.* **270** (1995) 30927–30932]

Red drop: a sharp fall in the quantum efficiency of photosynthesis at wavelengths greater than 680 nm, although chlorophyll still absorbs light from 680 to 700 nm. However, the quantum efficiency of light above 680 nm is increased by the simultaneous presence of shorter wavelength light. This Emerson effect (see) led to to the proposal that photosynthesis depends on the interaction of two light reactions (i.e. two photosystems), both driven by light less than 680 nm, but only one by light of longer wavelengths. See Photosynthesis, Photosynthetic pigment systems.

Redoxases: see Ferredoxin.

Redoxin: an electron-transferring protein. R. that contain iron, bound as a functional group to S, N or O-ligands of the protein, are known as Ferredoxins (see). An R. containing no metal is Thioredoxin (see). See also Rubredoxin.

Redox potential: see Oxidation.

Reducing end: The R.e. of an oligo or polysaccharide is the terminal monosaccharide residue which has an underivatized hydroxyl group on its anomeric carbon atom (i.e. C-1 in aldoses, C-2 in ketoses), thereby fulfilling the requirements of a Reducing sugar (see). Although all polysaccharides necessarily have one reducing end, its presence among the large number of non-reducing monosaccharide residues that constitute the rest of the molecule is insufficient to give a positive result when polysaccharides are subjected to the Fehling's and Benedict's tests for reducing sugars.

Reducing sugars: monosaccharides or water-soluble, short-chain oligosaccharides which, when boiled with '$CuSO_4/OH^-$'-based reagents (e.g. Fehling's, Benedict's), reduce the Cu^{2+} ions to Cu^+ ions with the production of a brick-red precipitate of Cu_2O. For a sugar to give this reaction it must already have, or be capable of being converted by dilute alkali (0.2 M OH^-) into a compound that has, a free aldehyde (CHO) group. It is this aldehyde group that reduces the Cu^{2+} and in doing so is oxidized to a carboxyl (COOH) group. Sugars that fulfil this structural requirement are: (i) open-chain aldotrioses (e.g. D-glyceraldehyde), (ii) open-chain ketotrioses (e.g. dihydroxyacetone) and ketotetroses (e.g. D-erythulose), (iii) monosaccharide aldoses and ketoses that exist predominantly in aqueous solution in either their furanose or pyranose ring forms *provided that* they have an underivatized hydroxyl (OH) group on their anomeric carbon atom (i.e. C1 in aldoses, C2 in ketoses) (e.g. D-glucose and D-fructose, but *not* their glycosides), and (iv) water-soluble, short-chain oligosaccharides which have a monosaccharide residue (aldose or ketose) with an underivatized hydroxyl group on its anomeric carbon atom (e.g. maltose but *not* sucrose). Sugars that fall into category (i) clearly already possess a free aldehyde group but those falling into category (ii) do not, but they develop one by undergoing a reversible, base-catalysed, tautomeric isomerisation in the presence of dilute alkali (0.02 M OH^- is sufficient) known as the Lobry de Bruyn – Alberda van Ekenstein rearrangement (Fig. 1). Sugars in category (iii) exist solely as either the α- or β-anomer of one of their ring forms in the crystalline state, but undergo Mutarotation (see) in aqueous solution, so that they exist as an equilibrium mixture of the α- and β-anomers of one or more of their ring forms and the open-chain form. Attainment of this equilibrium is slow in pure water but is almost instantaneous in the presence of OH^- ions. Mutarotation of aldoses therefore produces the free aldehyde, open-chain form of the sugar that is necessary for the reduction of Cu^{2+}. On the other hand, mutarotation of ketoses produces the free keto, open-chain form of the sugar which only becomes capable of reducing Cu^{2+} after undergoing the Lobry de Bruyn – Alberda van Ekenstein rearrangement, in which its free-aldehyde, open-chain isomeric forms are reversibly generated. However the relatively high concentration of OH^- ions in Fehling's and Benedict's reagents causes the free-aldehyde, open-chain sugar (derived from a ring-form aldose or ketose) to rearrange to yield not only 1,2-enediols but also 2,3- and 3,4-enediols, all of which are unstable under the prevailing conditions, and fragment to yield a series of lower aldehydes, such as formaldehyde (HCHO), glycolaldehyde (CH_2OH–CHO) and glyceraldehyde (CH_2OH–CHOH–CH_2OH) (Fig. 2). It is this complex mixture of aldehydes that is reponsible for the reduction of the Cu^{2+}; this also explains why pentoses and hexoses have reducing power similar to that of the lower M_r aldehydes. The ring forms of sugars in which the OH group on the anomeric carbon atom is replaced by OR (where R is any group except H) are not reducing sugars, because they do not mutarotate and consequently do not form a free-aldehyde (or free keto), open-chain form in aqueous solution. Sugars in category (iv) have a single monosaccharide residue at one end of the molecule (known as the reducing end) that has an underivatized OH on its anomeric carbon atom. This terminal monosaccharide residue can undergo mutarotation in aqueous solution to generate an equilibrium mixture containing the free-aldehyde (or free keto), open-chain form of the residue, which in turn can form the complex mixture

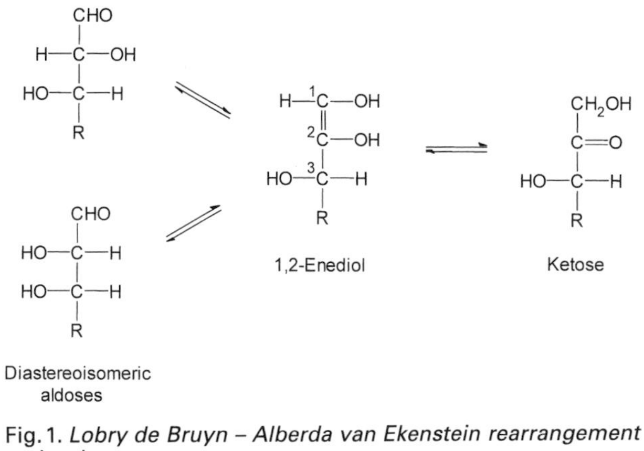

Diastereoisomeric aldoses

1,2-Enediol

Ketose

Fig. 1. *Lobry de Bruyn – Alberda van Ekenstein rearrangement of sugars in dilute alkali.* R = rest of the molecule.

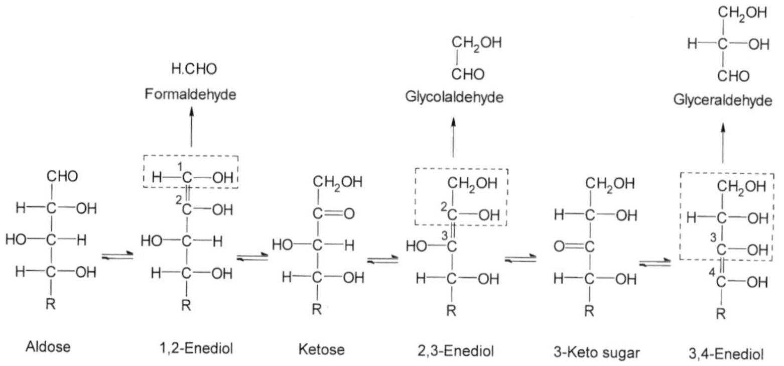

Fig. 2. *Formation of 1,2-, 2,3- and 3,4-enediols and their breakdown to yield reducing aldehydes in 0.5 M alkali.* R = rest of the molecule.

of reducing aldehydes in the presence of the OH^- ions of the Fehling's and Benedict's reagents as described above.

Reductases: see Flavin enzymes.

Reduction: the addition of electrons. R. is the converse of Oxidation (see). In biochemical systems R. may involve electron transfer only (e.g. R. of cytochromes and ferredoxin), but the majority of biochemical R. involve the addition of hydrogen (hydrogenation). The pyridine nucleotide coenzymes, NAD^+ and $NADP^+$, play an important part in R., and hydrogen transfer in reductive biosynthesis is usually mediated by NADPH. In green plants, the reducing power from the light reaction of photosynthesis appears as NADPH, which is then utilized in various reductive biosyntheses, notably the conversion of CO_2 to carbohydrate; in photosynthetic bacteria, the reductant formed in photosynthesis is NADH. R. must be accompanied by a corresponding oxidation, so that enzymes catalysing these reactions are generally known as oxidoreductases (see Enzymes Table 1). See also Dehydrogenation.

Reductive citrate cycle, *reductive carboxylic acid cycle:* the metabolic cycle used by photosynthetic bacteria of the family *Chlorobiaceae* (green suphur bacteria) to perform the photoautotrophic fixation of CO_2. In contrast, all other photosynthetic organisms employ the reductive pentose phosphate cycle (Calvin cycle; see). The R.c.c. (Fig.) consists in essence of a modified Tricarboxylic acid (TCA) cycle, driven in the opposite direction to that of the true TCA cycle by ATP and reducing power generated in the light phase of photosynthesis. The cycle fixes two molecules of CO_2 per turn and generates one molecule of acetyl-CoA. As in the Calvin cycle, the final product of the overall CO_2 fixation process is triose phosphate. This triose phosphate is formed from the acetyl-CoA (generated by the R.c.c), by driving it to pyruvate and then along the latter half of a reversed and slightly modified Glycolysis (see) pathway, again using ATP and reducing power generated by operation of the light phase of photosynthesis; this causes a further molecule of CO_2 to be fixed (in the reductive carboxylation of acetyl-CoA to pyruvate), so that a total of $3CO_2$ are eventually fixed for each turn of the cycle.

In the R.c.c., two of the reactions of the TCA cycle are replaced by alternative reactions catalysed by non-TCA cycle enzymes: (i) the citrate (*si*)-synthase (EC 4.1.3.7)-catalysed formation of citrate from acetyl-CoA and oxaloacetate is replaced by the ATP citrate (*pro*-3*S*)-lyase (EC 4.1.3.8)-catalysed, ATP-driven cleavage of citrate to acetyl-CoA and oxaloacetate (eq. 1; J, Fig.), and (ii) the α-ketoglutarate dehydrogenase complex-catalysed oxidative decarboxylation of α-ketoglutarate to CO_2 and succinyl-CoA is replaced by the α-ketoglutarate synthase (EC 1.2.7.3)-catalysed reductive carboxylation of succinyl CoA (eq. 2; O, Fig.), in which the reductant is reduced ferredoxin (Fd_{red})

$$\text{Citrate} + \text{CoA} + \text{ATP} \rightarrow \text{acetyl-CoA} + \text{oxaloacetate} + \text{ADP} \quad (1)$$

$$\text{Succinyl-CoA} + CO_2 + Fd_{red} \rightarrow \alpha\text{-ketoglutarate} + \text{CoA} + Fd_{ox} \quad (2)$$

All the other reactions of the TCA cycle, catalysed by the usual TCA cycle enzymes, operate in reverse in the R.c.c. These physiologically reversible reactions are forced to operate in an oxidative direction in the TCA cycle by the directionality of the physiologically irreversible α-ketoglutarate dehydrogenase complex-catalysed reaction; once that constraint has been removed and the necessary ATP and reductant are available (as is the case in the R.c.c.), they readily operate in the CO_2 fixing/acetyl-CoA forming direction.

Conversion of acetyl-CoA to triose phosphate (glyceraldehyde 3-phosphate and dihydroxyacetone phosphate) requires that two of the reactions of the catabolic pathway be replaced by alternative reactions catalysed by different enzymes; these are: (i) the pyruvate dehydrogenase complex-catalysed oxidative decarboxylation of pyruvate to acetyl-CoA, which is replaced by the pyruvate synthase (EC 1.2.7.1)-catalysed reductive carboxylation of acetyl-CoA (eq. 3; I, Fig.), in which the reductant is reduced ferredoxin (Fd_{red}), and (ii) the pyruvate kinase (EC 2.7.1.40)-catalysed, ATP-generating conversion of phospho*enol*pyruvate to pyruvate, which is replaced by the pyruvate, orthophosphate dikinase (EC 2.7.9.1)-catalysed, ATP-driven conversion of pyruvate to phospho*enol*pyruvate (eq. 4; F, Fig.), a reaction that is greatly assisted by the rapid exergonic hydrolysis of pyrophosphate to orthophosphate by pyrophosphatase (EC 3.6.1.1) (eq. 5; H, Fig. 1).

Acetyl-CoA + CO_2 + Fd_{red} → pyruvate + CoA + Fd_{ox} (3)

Pyruvate + P_i + ATP → phospho*enol*pyruvate + AMP + PP_i (4)

Pyrophosphate + H_2O → 2 orthophosphate (5)

Regeneration of ATP from AMP, the other byproduct of eq. 4, requires the expenditure of another ATP. This occurs in the adenylate kinase (EC

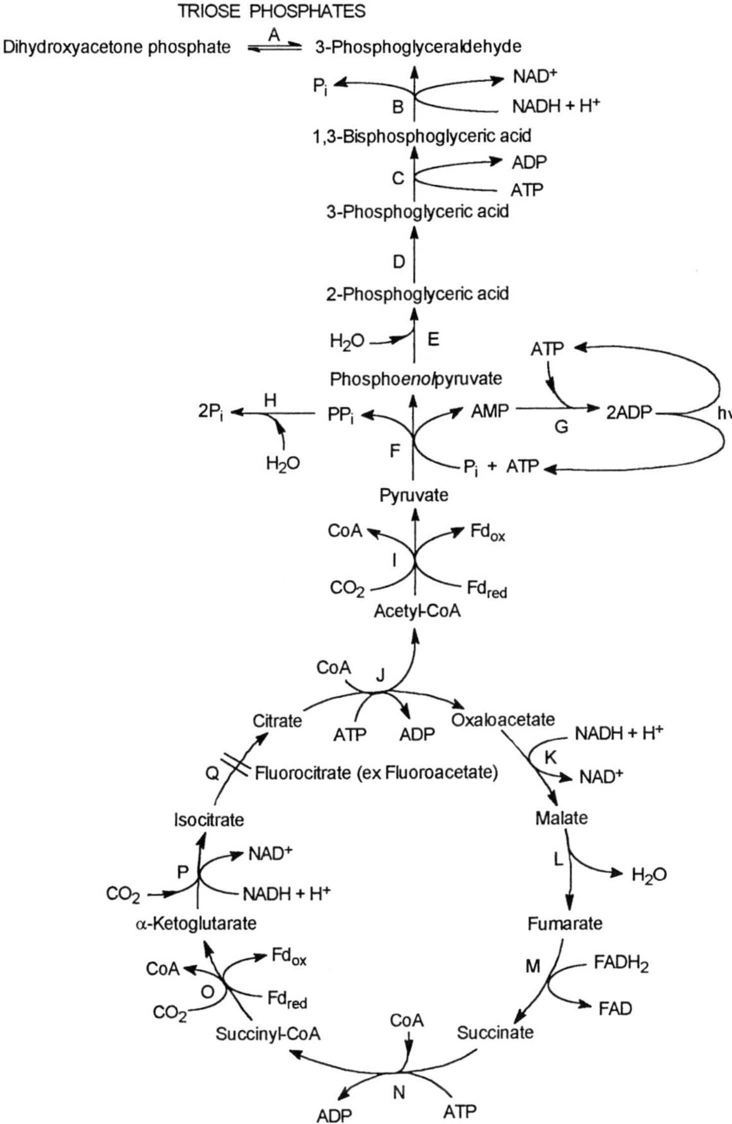

The reductive citrate cycle in the Chlorobiaceae *(purple sulfur bacteria).*
A = triose phosphate isomerase, EC 5.3.1.1; B = 3-phosphoglyceraldehyde dehydrogenase, EC 1.2.1.12; C = phosphoglycerate kinase, EC 2.7.2.3; D = phosphoglycerate mutase, EC 2.7.5.3; E = enolase, EC 4.2.1.11; F = pyruvate, orthophosphate dikinase, EC 2.7.9.1; G = adenylate kinase, EC 2.7.4.3; H = pyrophosphatase, EC 3.6.1.1; I = pyruvate synthase, EC 1.2.7.1; J = ATP citrate (pro-3S)-lyase, EC 4.1.3.8; K = malate dehydrogenase, EC 1.1.1.37; L = fumarase, EC 4.2.1.2; M = succinate dehydrogenase, EC 1.3.99.1; N = succinyl-CoA synthetase, EC 6.2.1.5; O = α-ketoglutarate synthase, EC 1.2.7.3; P = isocitrate dehydrogenase, EC 1.1.1.41; Q = aconitase, EC 4.2.1.3.

2.7.4.3)-catalysed reaction shown in eq. 6 (G, Fig.); the two molecules of ADP from this reaction are then converted back to ATP in the light phase of photosynthesis. Thus the conversion of pyruvate to phospho*enol*pyruvate effectively uses two molecules of ATP.

$$AMP + ATP \rightleftharpoons 2\ ADP \qquad (6)$$

All the other reactions of the triose phosphate-pyruvate segment of the glycolysis pathway operate in reverse to complete the conversion of acetyl-CoA to triose phosphate; these physiologically reversible reactions are forced to operate in the oxidative direction in glycolysis by the directionality of the physiologically irreversible pyruvate kinase-catalysed reaction.

The key to the reversal of the pyruvate → acetyl-CoA and α-kg→succinyl-CoA steps (these are physiologically irreversible when catalysed by NAD^+-dependent α-ketoacid dehydrogenase complexes) is the use of reactions catalysed by Fd_{red}-dependent α-ketoacid synthases; the increased reducing power of Fd_{red} ($E'_0 \sim -0.42V$) over that of NADH (E'_0 −0.32V), which is equivalent to an input of an extra ~ 20 $kJ.mol^{-1}$, is evidently sufficient to overcome the hitherto adverse thermodynamics.

The overall R.c.c. process is more energy-efficient than the Calvin cycle, as can be seen from: (i) a comparison of the number of reducing equivalents (expressed as $[2e^-]$ where $2e^- \equiv NAD(P)H$, $FADH_2$ or Fd_{red}; note that most bacterial ferredoxins are two-electron redox systems, in contrast to plant chloroplast ferredoxin, which is a one-electron redox system), (ii) a comparison of the number of ATPs required for the fixation of three molecules of CO_2 and the generation of one molecule of triose phosphate. Eq. 7 gives the stoichiometry of R.c.c. and eq. 8 that of the Calvin cycle.

$$3CO_2 + 6[2e^-] + 5ATP \rightarrow C_3H_6O_3\text{–}P + 5ADP + 3P_i \qquad (7)$$

$$3CO_2 + 6[2e^-] + 9ATP \rightarrow C_3H_6O_3\text{–}P + 9ADP + 8P_i \qquad (8)$$

The greater energy efficiency of R.c.c. is particularly appropriate for the *Chlorobiaceae* because their typical habitat has low light intensity.

The main evidence for the operation of the R.c.c. in *Chlorobiaceae,* which was first proposed by M.C.W. Evans et al. [*Proc. Natl. Acad. Sci. USA* **55** (1966) 928–934], is: (i) the finding [R. Sirevåg et al. *Arch. Microbiol.* **112** (1977) 35–38] that the way in which *Chlorobium thiosulphatum (Chlorobiaceae)* discriminates during photosynthesis against ^{13}C (atmospheric CO_2 contains 1.108 % ^{13}C, the rest being ^{12}C), is quite different from that of other photosynthetic bacteria, namely *Rhodospirillum rubrum (Rhodospirillaceae), Chromatium (Chromatiaceae)* and the green alga *Chlamydomonas reinhardii*, which behave similarly in this respect and are known to fix CO_2 in the Calvin cycle; this shows that the enzyme(s) catalysing CO_2 fixation in *C. thiosulphatum* is (are) different from ribulose bisphosphate carboxylase (EC 4.1.1.39) (see) and indicates, therefore, that the Calvin cycle is not operative in this organism, (ii) the finding [R. Sirevåg & J.G. Ormerod *Science* **169** (1970) 186–188; *Biochem. J.* **120** (1970) 399–408] that low concentrations of fluoroacetate (see), a known TCA cycle inhibitor, inhibit photosynthetic CO_2 fixation in *C. thiosulphatum*, and (iii) the finding that *C. thiosulphatum* possesses α-ketoglutarate synthase and pyruvate synthase, enzymes that are peculiar to R.c.c., but lacks ribulose *bis*phosphate carboxylase and phosphoribulokinase (EC 2.7.1.19), enzymes that are unique to the Calvin cycle.

Redundancy: 1. The occurrence (frequent in eukaryotes, only in isolated instances in prokaryotes) of linearly arranged, largely identical, repeated sequences of DNA. Between 50 and 107 such repeated base sequences have been demonstrated, depending on the source of the DNA. The degree of R. in a sample of DNA can be determined from the rate of renaturation following heat denaturation: the rate of reannealing increases with the degree of R. Repetitive DNA may constitute 20–80 % of the total DNA. Satellite DNA has especially high R., whereas genes coding for proteins normally have low R.; an exception is found in the high R. of the histone genes of the sea urchin. The nontranscribed part of giant messenger-like RNA also contains repetitive sequences, but their function is unknown. Genes for tRNA, 5S-RNA and rRNA (but not mRNA) show R., the R. value for rRNA lying between 100 and 7,500.

In prokaryotes, gene R. has only been detected in isolated cases, and the values are low. In *Bacillus subtilis* only 0.4 % of the genome is redundant (e.g. 9–10 genes for rRNA and 40 genes for tRNA).

2. *Terminal redundancy:* the existence of identical genetic information at each end of a viral chromosome. In λ-phages this constitutes up to 20 nucleotide pairs, whereas up to 6,000 nucleotide pairs may be terminally redundant in T-even phages.

Refsum's disease: see Inborn errors of metabolism.

Regulator gene: see Operon.

Regulon: a group of structural genes, whose gene products (enzymes) are involved in the same reaction pathway, and which are regulated together. The individual genes are in different regions of the chromosome, i.e. they do not lie on neighboring sequences of DNA, as in an operon. In *E. coli* the 8 genes for arginine synthesis constitute a R. In *Neurospora crassa,* the genes for histidine biosynthesis are found on four different chromosomes, but they are still subject to a concerted regulation, and therefore constitute a R.

Reichstein's substances F, H, M, Q and S: see Cortisone, Corticosterone, Cortisol, Cortexone, Cortexolone.

Relative molecular mass, M_r: the mass of a molecule relative to the Dalton (see). M_r has no units. It is equivalent to "molecular weight". a term used widely in physical and biological sciences. The word "weight" is, however, dimensionally incorrect, so that "molecular weight" should be avoided and replaced by relative molecular mass (M_r).

Relaxation protein: a type I eukaryotic topoisomerase (see Topoisomerases) isolated from the nuclei of LA9 mouse and HeLa cells, and characterized by its ability to remove superhelical turns from closed, circular DNA. [H-P. Vosberg et al. *Eur. J. Biochem.* **55** (1975) 79–93]

Relaxin: a female sex hormone, formed in mammals during pregnancy. R. is a heterodetic, cyclic polypeptide, M_r 12,000; the A-chain contains 22, the B-chain 26 amino acid residues. It is probably produced by the ovary and placenta and/or uterus under the influence of progesterone. It causes relaxation of collagenous connective tissue (sensitized by estrogens) in the symphysis and ileosacral joints, thus enabling enlargement of the pelvic girdle during pregnancy and at birth.

Release factors, *termination factors:* catalytically active proteins necessary for the termination step of RNA synthesis and protein biosynthesis. For further details, see Ribonucleic acid, Protein biosynthesis (subheading Termination).

Release inhibiting hormone: see Releasing hormones.

Releaser: see Pheromones.

Releasing factors: see Releasing hormones.

Releasing hormones, *releasing factors, liberins, statins:* a group of peptide neurohormones, synthesized in various distinct nuclei of the hypothalamus. They are released into the capillaries of the portal vessels in the median eminence of the hypothalamus, then carried to the anterior pituitary (adenohypophysis) where they regulate production and secretion of tropic hormones (e.g. thyrotropin, somatotropin, gonadotropins). R.h. are produced in nerve cells, and their synthesis represents the conversion of an electrical stimulus into hormonal signals. In the median eminence of the hypothalamus, R.h. are stored in nanogram quantities. In the anterior pituitary, R.h. act via the adenylate cyclase system, and are then degraded by proteolysis.

Thyrotropin releasing hormone, thyrotropic hormone releasing factor(TRF), thyroliberin: Pyr-His-Pro-NH_2 (Pyroglutamyl-L-histidyl-L-prolinamide), M_r 262. The identical hormone is found in all hitherto investigated species. Secretion of TRF is promoted by neurotransmitters, e.g. noradrenalin, and inhibited by serotonin. TRF stimulates the anterior pituitary to synthesize and secrete thyrotropin, which in turn stimulates the thyroid gland to secrete thyroxin and triiodothyronin. The latter two hormones exert a negative feedback on the secretion of TRF and thyrotropin. TRF also stimulates the secretion of prolactin and acts as a neurotransmitter in the central nervous system.

Luteinizing hormone releasing factor (LRH), luliberin: Pyr-His-Trp-Ser-Tyr-Gly-Leu-Arg-pro-Gly-NH_2, a decapeptide, M_r 1182, which stimulates synthesis and secretion of luteinizing hormone (LH) and follicle stimulating hormone (FSH) by the anterior pituitary. The names, gonadotropin releasing factor (GRF) and gonadoliberin have therefore been suggested for LRH. Synthesis of LRH is regulated by the central nervous system and is subject to environmental influences (time of year, light, olfactory stimuli, sexual stimulation). LH and FSH stimulate production of sex hormones by the gonads. LRH acts as a neurotransmitter in the central nervous system; it can be measured by radioimmunological techniques.

Follicle stimulating hormone releasing factor (FRH), folliliberin: this is chemically identical with LRH.

Corticotropin releasing factor (CRF), corticotropin releasing hormone: a polypeptide of 41 amino acids [J. Spiess et al. *Proc. Natl. Acad. Sci. USA.* **78** (1981) 6517–6521] which stimulates both β-endorphin and corticotropin secretion in the anterior pituitary and in some areas of the central nervous system. Corticotropin stimulates synthesis and secretion of glucocorticoids by the adrenals; the glucocorticoids inhibit the secretion of CRF or the response of the anterior pituitary to the releasing hormone (negative feedback). β-Endorphin, in turn, inhibits secretion of LRH in neurons which control sexual behavior.

Three of the seven anterior pituitary hormones do not act on endocrine glands, but influence tissue metabolism directly. There is no hormone feedback mechanism for the control of these three hormones, and control is exerted by the action of release-inhibiting hormones.

Somatotropin releasing factor (SRF), somatotropin releasing hormone, somatoliberin. The identity of the native hormone is unknown. Val-His-Leu-Ser-Ala-Glu-Gln-Lys-Glu-Ala, a decapeptide, M_r 1112 (porcine), has been found to stimulate synthesis and secretion of somatotropin and is therefore called SRF.

Somatotropin release inhibiting hormone (SIH), somatostatin: a polypeptide, M_r 1638, which inhibits secretion of somatotropin, thyrotropin, insulin, glucagon, gastrin and cholecystokinin. SIH has a broad activity spectrum; it acts outside the hypothalamus as a neurotransmitter in the central nervous system, and as a hormone in the intestinal tract; it has also been found in the thyroid and pancreas.

Ala-Gly-Cys-Lys-Asn-Phe-Phe-Trp-Lys-Thr-Phe-Thr-Ser-Cys

Prolactin releasing factor (PRF), prolactin releasing hormone, prolactoliberin: a hormone thought to be identical with thyrotropin releasing factor. The structure of *Prolactin release inhibiting hormone (PIH) (prolactostatin)* is unknown. PRF promotes, whereas PIH inhibits the synthesis and secretion of prolaction by the anterior pituitary.

Melanotropin releasing factor (MRF), melanotropin releasing hormone, melanoliberin: a hormone which corresponds structurally to the open hexapeptide ring of Oxytocin (see), without the tripeptide side chain, i.e. Cys-Tyr-Ile-Gln-Asn-Cys.

Melanotropin release inhibiting hormone (MIH), melanostatin: a hormone structurally equivalent to the peptide side chain of oxytocin, Pro-Leu-Gly-NH_2 (bovine, rat), or to Pro-His-Phe-Arg-Gly-NH_2 (bovine). MRF stimulates, whereas MIH inhibits synthesis and secretion of melanotropin by the anterior pituitary, leading respectively to darkening or lightening of amphibian skin. [Special section incorporating 4 papers on Gonadotropin-Releasing Hormone *Trends Endocrinol.* **7** (1996) 55–68]

Renaturation: conversion of a denatured protein or nucleic acid into its native configuration. See Nucleic acids, Proteins.

Renin (EC 3.4.99.19): an endopeptidase, M_r 43,000, formed in the juxtaglomerula cells (cells next to the glomerula) of the kidney. In the plasma, it releases angiotensin I (a decapeptide) from the *N*-terminal sequence of angiotensinogen (an α_2-plasma globulin). Another enzyme (M_r 210,000) in lung (3 subunits, each M_r 70,000), kidney (8 subunits, each M_r 25,000) and plasma removes histidine and leucine from the *C*-terminus of angiotensin I, releasing the highly hypertensive octapeptide, angiotensin II.

Rennet enzyme, ***rennin, chymosin:*** a pepsin-like proteinase probably present in the stomach of all nursing mammals. R.e. is prepared from the abomasum (fourth stomach) of young ruminants. It is formed as inactive prorennin, which is converted to active rennin by the action of pepsin or by autocatalysis. In the process its M_r is reduced from 36,200 (prorennin) to 30,700 (rennin), by removal of an activation peptide (M_r 5,500, 49 amino acids) from the *N*-terminal end. R. is the milk-coagulating enzyme of young mammals and requires Ca^{2+} for its action (pH optimum 4.8). It specifically attacks a single Phe-Met bond in κ-casein (-His-Leu-Ser-***Phe-Met***-Ala-), and cleaves κ-casein into insoluble para-κ-casein and a *C*-terminal glycopeptide (M_r 8,000). The Ser residue next to this Phe-Met bond serves as a binding site for R., while the His (3 residues from the Phe-Met bond) serves as a proton acceptor and donor. This process also occurs in the absence of proteases when the protein is heated at slightly acid pH. In both cases, the protective colloid function of κ-casein in the milk is destroyed.

Repair enzymes: enzymes that catalyse stages in the repair of DNA. They include exo- and endonucleases, DNA-polymerases and ligases. See Deoxyribonucleic acid (Repair of DNA).

Repeatability: In accordance with the International Organization for Standardization (Precision of test methods – Determination of repeatability and reproducibility. Draft International Standard ISO/Dis 5725, October 1977), *repeatability* refers to measurements in one laboratory over a short time period whereas *reproducibility* refers to measurements over longer periods of time and/or in different laboratories.

Replication: DNA replication. See Deoxyribonucleic acid.

Replication site: the site of DNA replication in vivo. In bacteria, the R.s. is on the cell membrane. The circular DNA is bound to the cell membrane at its initiation region, together with DNA-polymerase and initiation proteins. See Deoxyribonucleic acid.

Replicator: see Replicon.

Replicon: a term proposed by Jacob and Brenner for the replication unit. In prokaryotes or viruses, it is the complete circular or linear DNA (or RNA in RNA viruses), representing the total chromosome of the organism. If a cell contains more than one R., they are replicated independently. R. are subject to positive control, i.e. they contain a gene for the production of an initiator molecule, which initiates replication by interacting with a replicator (the initiation site on the DNA).

Bacterial chromosomes, episomes or plasmids, viruses, DNA molecules of chloroplasts and mitochondria are R. Eukaryotic chromosomes consist of a large number of R., i.e. one DNA molecule has several replication sites. See Deoxyribonucleic acid.

Repression: see Enzyme repression.

Repressor: an allosteric protein, which regulates the transcription of structural genes, and is encoded by a regulator gene. R. binds to the operator region of DNA, thereby preventing synthesis of the mRNA for the structural genes of an operon or regulon (see Enzyme repression). The ***Lac repressor,*** which regulates transcription of the Lac operon in *E. coli* (see Enzyme induction), is a tetrameric protein (M_r 152,000) consisting of 4 identical subunits (each of M_r 38,000, containing 347 amino acid residues). It was the first R. to be purified and structurally elucidated (Bayreuther et al. 1973).

In an inducible enzyme system, the R. is inactive in the presence of the effector (inducer); binding to the inducer apparently changes the conformation of the R., so that it no longer binds to the operator (see Enzyme induction). Synthesis of mRNA can therefore proceed only when the inducer is present. In enzyme repression, the situation is reversed; R. is activated by a corepressor (the endproduct of a biosynthetic pathway, e.g. an amino acid) so that it can bind to the operator. In this case, synthesis of mRNA proceeds only in the absence of corepressor (see Enzyme repression). See also Derepression.

Reproducibility: This is not the same as Repeatability (see).

Reserpine: therapeutically the most important Rauwolfia alkaloid. M_r 608.9, m.p. 265 °C, $[\alpha]_D^{25}$ – 123° ($CHCl_3$). Hydrolysis of R. with alcoholic KOH produces reserpinic acid (structurally similar to yohimbine), trihydroxybenzoic acid and methanol. R. is found widely in the genus *Rauwolfia* and is responsible for the sedative properties of these plants. After a latent period, R. causes long lasting sedation, with decrease of blood pressure and decreased pulse rate. It is used as a powerful neuroleptic drug in psychiatry.

Reserve cellulose, ***storage cellulose:*** see Cell wall.

Resilin: a Structural protein (see) from the exoskeleton of arthropods, especially insects. R. has a high glycine content, and no cystine. It is located between the chitin lamellae, and endows the arthropod exoskeleton with a certain elasticity. A notable component of R. is trityrosine (Fig.), which is formed (after translation of the protein) by cross linking the tyrosine side chains of one or more polypeptide chains. This results in an irregular three dimensional lattice, which is responsible for the rubber-like properties of R.

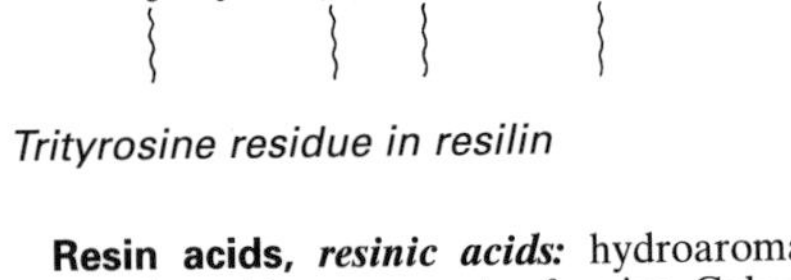

Trityrosine residue in resilin

Resin acids, ***resinic acids:*** hydroaromatic diterpenes, the acid components of resins. Colophony (rosin) consists of up to 90 % R.a. The most important representatives are abietic acid, neoabietic acid, dextro-pimaric acid and neopimaric acid. Salts and esters of the R. are called resinates. The alkali salts are also called resin soaps.

Resinates: salts and esters of resin acids.

Resinoids: see Essential oils.

Resinols: see Resins.

Resinotannins: see Resins.

Resins: largely amorphous, solid or half-solid, transparent, odorless and tasteless organic substances, usually of vegetable origin. Tree R. are classified according to age into fossil R., such as amber, recent fossil R. (several years to centuries old), e.g. copal R., and recent R., which occur mostly as balsams fresh from injured trees. Caoutchouc (see) is included with the R. Herbaceous plants produce R., e.g. mastic, but not in any considerable quantity. Mixtures of R. with mucin are called gum R. Solutions of R. are referred to as balsams. The most important animal R. is shellac, produced by the female East Asian scale insect (*Tachardia lacca*).

R. are supercooled melts, soluble in nonpolar solvents. Like essential oils, they are complex mixtures in which terpenes and aromatics predominate. The structures of resin components are not well clarified, but they can be classified according to their chemical properties into: *Resinotannols:* aromatic phenylpropane resin alcohols allied to the tannins, occurring partly free, but more generally combined with aromatic acids or with umbelliferone in the form of esters (tannol R.); *Resinols:* crystalline, colorless, resin alcohols, e.g. terpene alcohols, occurring partly free, partly as esters; *Resin acids:* partly crystalline and mostly free; they combine with alkalis to form soaps, and they form crystalline salts or esters called resinates (see Resin acids); *Resenes:* indifferent substances that are neither esters nor acids; *Resines:* esters, e.g. coniferyl benzoate.

Crude R. and refined resin components are widely used in the production of paints, varnishes, textile conditioners, cosmetics and pharmaceuticals.

Resin soaps: see Resin acids.

Respiration, *oxidative metabolism:* a process by which cells derive energy in the form of ATP from the controlled reaction of hydrogen with oxygen to form water. The hydrogen is derived from the degradation of organic substrates by oxidases or dehydrogenases, which in turn release it to a system of redox catalysts (see Respiratory chain) located in the inner mitochondrial membrane, or in the cell membrane of bacteria. The electron is stripped from the hydrogen atom and passed along the respiratory chain; three of the transfers from one redox catalyst to the next power the formation of ATP from ADP and P_i (see Oxidative phosphorylation). The last redox catalyst passes the electrons to a terminal acceptor, which is usually oxygen, but may be nitrate (nitrate respiration). The hydrogen ions produced at the beginning of the chain then join the O^{2-} to form H_2O.

Respiratory acidosis: see Acidosis.

Respiratory alkalosis: see Alkalosis.

Respiratory chain, *electron transport chain:* a series of redox catalysts which transport electrons from respiratory substrates to oxygen. The energy of this electron flow may be used in the synthesis of ATP. The coupling of ATP synthesis with electron transport in the R.c. is known as Oxidative phosphorylation (see). The R.c. complex is located in the inner mitochondrial membrane of eukaryotes, and in the cell membrane of prokaryotes. It is functionally closely associated with the enzymes of the Tricarboxylic acid cycle (see) which supply reducing equivalents, mostly in the form of NADH, with some $FADH_2$. De-

Table 1. *E_0' values for some biological redox pairs*

Redox system	E_0' in volts
H^+/H_2	− 0.420
$Ferredoxin_{ox}/Ferredoxin_{red}$	− 0.420
$NAD^+/NADH + H^+$	− 0.320
$Glutathione_{ox}/Glutathione_{red}$	− 0.230
$FMN/FMNH_2$	− 0.122
Cytochrome *b* (Fe^{3+}/Fe^{2+})	+ 0.075
Fumarate/succinate	+ 0.031
Ubiquinone/ubihydroquinone	+ 0.100
Cytochrome *c* (Fe^{3+}/Fe^{2+})	+ 0.254
Cytochrome *a* (Fe^{3+}/Fe^{2+})	+ 0.290
Fe^{3+}/Fe^{2+}	+ 0.770
$^1/_2O_2/O^{2-}$	+ 0.810

Table 2. *Components of the mitochondrial respiratory chain*

Complex	Component	Functional group	Activity
	Pyridine nucleotide-dependent dehydrogenases	NAD^+ or $NADP^+$	
I	$Flavoprotein_D$	FMN, Non-heme iron	NADH:ubiquinone oxidoreductase
II	$Flavoprotein_S$	FAD, Non-heme iron	Succinate:ubiquinone oxidoreductase
-	Ubiquinone (Coenzyme Q)	Reversibly reducible quinone structure	
III	Iron-sulfur protein Cytochrome *b* (b_K and b_T) Cytochrome c_1	Iron-sulfur centers Heme (noncovalent) Heme (covalently bound)	Ubiquinone:cytochrome *c* oxidoreductase
-	Cytochrome *c*	Heme (covalently bound)	
IV	Cytochrome *a* Cytochrome a_3	Heme *a* Copper protein, Heme *a*	Cytochrome oxidase

pending on the tissue and its metabolic state, other pathways, e.g. Fatty acid degradation (see) may be more important than the tricarboxylic acid cycle in providing NADH or $FADH_2$ for the R.c.

The overall reaction is $NADH + H^+ + {}^1/_2O_2 \rightarrow H_2O + NAD$, $\Delta G^\circ = -221.7$ kJ/mol $= -53$ kcal/mol. The standard electrochemical potentials of individual steps of the R.c. are given in Table 1. In three of the steps (Fig.), the potential difference is large enough to supply the energy required for the phosphorylation of ADP; these are recognized as the three sites of Oxidative phosphorylation (see). The first site is the transfer of electrons from NADH to ubiquinone. Since electrons from $FADH_2$ enter the chain at the level of ubiquinone, they promote the formation of only two molecules of ATP per electron pair. The amount of ATP generated by electrons from any given substrate can be expressed as the P/O ratio, i.e. the number of molecules of ATP generated per atom of oxygen consumed. For dehydrogenations involving NAD as the coenzyme, the P/O ratio is 3. For substrates oxidized by flavin nucleotide enzymes, the P/O ratio is 2. Electrons flow singly along the chain of cytochromes, yet each phosphorylation requires that 2 electrons pass the site, and 4 electrons are needed to reduce one molecule of oxygen. The mechanisms for the accommodation of these requirements are not clear. Since the R.c. is a membrane-bound system, its components can only be purified after disruption of the membrane (see Electron transport particle, Mitochondria). Using sonic oscillation, E. Racker produced submitochondrial particles capable of oxidative phosphorylation coupled to electron transport. Treatment with urea removes a component F_1 from the particles and prevents further phosphorylation, but electron transport, i.e. respiration, continues. Since isolated F_1 structures have ATPase activity, it seems likely that they form part of the ATP synthe-

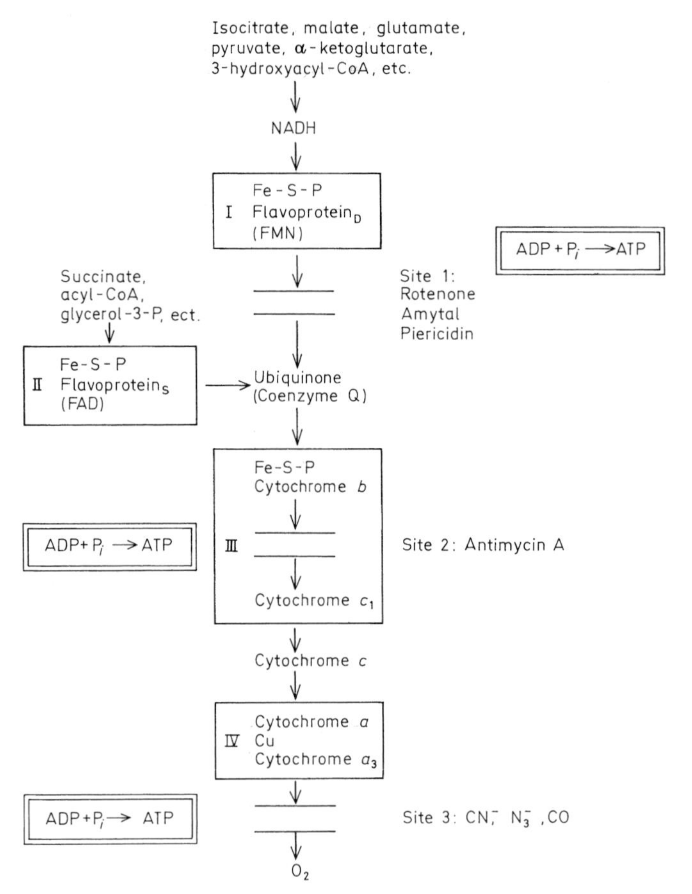

The respiratory chain. Boxes indicate the composition of complexes I to IV. Electron flow is shown by arows. Sites of action of some inhibitors are labeled 1, 2 and 3, and indicated by horizontal bars. Sites 2 and 3 are also coupled with ATP synthesis, i.e. according to the chemiosmotic theory, each of these stages of electron transfer (from cytochrome b to cytochrome c_1, and from cytochrome a_3 to oxygen) provides enough energy to generate a proton gradient for the synthesis of one molecule of ATP. The first site of ATP synthesis may not be identical with inhibition site 1 as shown here, but it is known to exist on the ubiquinone side (rather than the substrate side) of complex I.

sizing apparatus in situ (see Ion-motive ATPases, F-type ATPases). Treatment with detergents has been used to isolate four enzyme complexes from mitochondrial membranes, designated I, II, III and IV. The components and activities of these complexes are shown in Table 2. The best studied components of the R.c. are ubiquinone and cytochrome *c*, which are soluble and easily removed, and thus accessible. They are relatively small molecules that appear to serve as carriers transporting electrons between immobilized components.

Because the mitochondrial membrane is impermeable to ATP/ADP and NAD/NADH, two transport systems are necessary for the function of the R.c. One is the ATP/ADP carrier, which effects facilitated exchange diffusion, and the other is a metabolic shuttle (see Hydrogen metabolism) which allows reducing equivalents generated in the cytoplasm to enter the mitochondrion.

Some organisms use hydrogen acceptors whose redox potentials are higher than that of NADH (e.g. sulfide, thiosulfate, nitrite). The thermodynamic work required to reduce these substrates is derived from participation of ATP in Reversed electron transport (see) along a cytochrome chain. In general, such organisms are obligate anaerobes, and their electron transport chains lack cytochrome oxidase.

Respiratory chain phosphorylation: see Oxidative phosphorylation.

Respiratory control: see Respiratory chain, Oxidative phosphorylation.

Respiratory inhibitor: a compound which interferes in some way with the respiratory chain. There are three types: 1. Uncouplers (see) which prevent the synthesis of ATP without stopping the flow of electrons; 2. inhibitors of oxidative phosphorylation in the narrower sense; and 3. inhibitors of electron transport along the respiratory chain. The second type inhibits both electron transport and the generation of ATP. The prototype of this group is oligomycin, which interferes with the use of energy-rich intermediates for ATP synthesis. Atractyloside, a plant poison, and the antibiotic bongkrekic acid inhibit the carrier which mediates the exchange of ADP and ATP across the mitochondrial membrane. Oxidative phosphorylation is then stopped for lack of ADP in the mitochondria. Another form of inhibition is exercised in the presence of certain monovalent cations by ionophores like valinomycin, nigericin, nonactin and gramicidin, etc. which divert the energy of respiration to mitochondrial ion transport. The points of attack of the 3rd type of inhibitor are shown in the Fig. to Respiratory chain (see). The order of the electron carriers in the respiratory chain was deduced by using this type of inhibitor. The last member of the chain in aerobic organisms, cytochrome oxidase, is inhibited by cyanides, azides and carbon monoxide. The toxicity of the latter in mammals, however, is due to its affinity to hemoglobin, not to its inhibition of the respiratory chain.

Respiratory poison: see Respiratory inhibitor.

Restriction endonucleases (EC 3.1.23.1 to EC 3.1.23.45): a large group of enzymes, all with different specificities (Table), present in a wide variety of prokaryotic organisms, where they serve to cleave foreign DNA molecules (e.g. phage DNA). R.e. recognize specific palindromic sequences in double-stranded DNA, and they represent a powerful set of tools for the analysis of chromosome structure. Host DNA is protected from the activity of the host R.e. by methylation (by *S*-adenosyl-L-methionine). This methylation process is known as *modification* (Table). The phenomenon whereby foreign DNA introduced into a prokaryotic cell becomes ineffective owing to the action of R.e. is known as *restriction.*

Some palindromic sequences of double-stranded DNA recognized by specific restriction endonucleases (↓ indicates cleavage site) *and/or modification enzymes* (* indicates methylated base)

Enzyme	Target palindromic sequence
Eco RI (from *Escherichia coli* strain R)	* ↓ 3′-CTTAAG-5′ 5′-GAATTC-3′ ↑ *
Eco RI′ (*E. coli* strain R)	↓ 3′-PyPyTAPuPu-5′ 5′-PuPuATPyPy-3′ ↑
Eco RII (*E. coli* strain R)	* ↓ 3′-GGACC-5′ 5′-CCTGG-3′ ↑ *
Eco (PI)	* 3′-TCTAGA-5′ 5′-AGATCT-3′ *
Hin DII (*Hemophilus influenzae*)	* ↓ 3′-CAPuPyTG-5′ 5′-GTPyPuAC-3′ ↑ *
Hpa I (*Haemophilus parainfluenzae*)	↓ 3′-CAATTG-5′ 5′-GTTAAC-3′ ↑
Hae III (*Haemophilis aegypticus*) and Bsu × 5 (*Bacillus subtilis*)	↓ 3′-CCGG-5′ 5′-GGCC-3′ ↑
Hpa II (*Haemophilus parainfluenzae*)	↓ 3′-GGCC-5′ 5′-CCGG-3′ ↑

Two types of *restriction-modification* system have been found in bacteria. In type I systems, the methylase and R.e. are both associated with a complex containing three different polypeptide chains: an α-chain with R.e activity, a β-chain with methylase activity and a γ-chain with the recognition site for the DNA sequence. Type I systems require *S*-adenosyl-L-methionine and ATP for both R.e. and methylase activities; they are less specific, and cleavage sites may be random and far removed (1,000 base pairs) from the 5′ side of the recognition site. In type II systems, methylases and R.e. are separate, *S*-adenosyl-L-methionine donates the methyl group but does not take part in DNA cleavage. Type II systems do not require ATP, and they are also highly specific.

Restriction fragment length polymorphisms: variations in the lengths of nucleic acid fragments produced by endonuclease digestion, resulting from polymorphism in the genome. After digestion of DNA with an appropriate restriction endonuclease, the electrophoretic pattern of DNA fragments is analysed by the Southern blot procedure. R.f.l.p. have been used for assessing the structural heterogeneity of genes. They are also used in linkage studies as markers for the detection of disease, as tracers for the identification of alleles in pedigree analysis, and for determining the clonal origin of tumors. In linkage studies, the polymorphism may not exist within the gene in question, but is closely linked to it, e.g. the normal human globin gene and the sickle cell globin gene are linked to different *Hpa* I sites (*Hpa* I is a restriction endonuclease from *Haemophilus parainfluenzae,* with the target sequence GTT↓AAC). Thus *Hpa* I digestion of DNA containing the sickle gene produces a 13 kb fragment that is absent from digests of normal DNA, and the digest of normal DNA contains a 7.6 kb fragment that is absent from digests of DNA containing the sickle gene (Fig.). [D. Botstein et al. *Am. J. Hum. Genet.* **32** (1980) 314–331; B. Vogelstein et al. *Science* **227** (1985) 642–645]

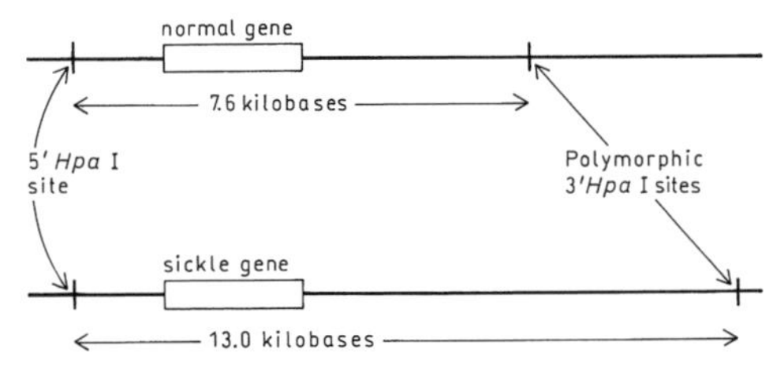

Polymorphic 3′Hpa I sites linked to normal and sickle β-globin genes

Restriction-modification system: see Restriction endonucleases.

Reticuline: see Benzylisoquinoline alkaloids.

Retinal: see Vitamins (vitamin A).

Retinoic acid: see Vitamins (vitamin A).

Retinol: see Vitamins (vitamin A).

Retronecine: see Pyrrolizidine alkaloids.

Reversed electron transport: reversal of Oxidative phosphorylation (see) in which NAD^+ is reduced by an *ATP-dependent* reverse transport of electrons. R.e.t. occurs in organisms that oxidize hydrogen donors whose redox potential (see Oxidation) is more positive than that of the pyridine nucleotide coenzymes, and it operates in the oxidation of substrates not specific for NAD (see Respiratory chain), e.g. Succinate + NAD^+ → Fumate + NADH + H^+. The redox system: succinate/fumarate (E'_0 = +0 · 031 V) is 325 mV more positive than the redox system: NAD^+/NADH + H^+ (E'_0 = −0.32 V); electrons are passed from succinate to flavoprotein in the respiratory chain, then via NADH-dehydrogenase to NAD^+. R.e.t. has been shown in nitrate bacteria (*Nitrobacter*), insect flight muscle mitochondria and kidney mitochondria under anaerobic conditions. It is a feature of bacterial Photosynthesis (see).

Reversed phase chromatography: see Paper chromatography.

Reverse transcriptase: see RNA-dependent DNA-polymerase.

R_f-value: see Paper chromatography.

L-Rhamnose: 6-deoxy-L-mannose, a deoxyhexose M_r 164.16, m.p. 122 °C, $[\alpha]_D^{20}$ + 38 ° → + 8.9 °. L-R. is a component of many glycosides, e.g. anthocyanins, and of plant mucilages. It is biosynthesized from glucose.

Rhein: 1,7-dihydroxy-3-carboxyanthraquinone, a yellow anthraquinone, m.p. 321 °C, present free or as a glycoside in the roots of many higher plants, particularly rhubarb. R. has a purgative action, and various R.-containing drugs (Radix Rhei, Folia Sennae) are used therapeutically.

Rhesus factor, *Rh-factor*: several closely related, blood group-specific erythrocyte antigens present in 85 % of Europeans (Rh-positive). The antigen is absent from the remaining 15 %, who are Rh-negative. The natural antibody does not normally occur in humans, but is formed in rabbits or guinea pigs after immunization with rhesus monkey erythrocytes. If Rh-negative individuals come into contact with Rh-antigen, e.g. by blood transfusion, or from an Rh-positive fetus (the antigen crosses the placenta), Rh-antibodies are formed. Repeated transfusion may then lead to hemolysis, or an Rh-positive fetus may suffer hemolytic damage (erythroblastosis of the newborn).

Rh-factor: see Rhesus factor.

Rhizobia: bacteria of the genus *Rhizobium,* which can live free in the soil, or enter into a symbiotic relationship with leguminous plants. As leguminous symbionts, R. are responsible for the formation of root nodules and the fixation of atmospheric nitrogen (see Nitrogen fixation). In the nodule of the host plant, R. are present as bacteroids, which differ morphologically (ill defined globular shape) and biochemically (high rate of nitrogen fixation, and absence of ribosomes) from the free living form. Under defined laboratory conditions, i.e. in the presence of a dicarboxylic acid (e.g. succinate or fumurate) and a pentose (ribose, xylulose, arabinose), pure cultures of free living R. show low rates of nitrogen fixation, but the free form probably does not fix nitrogen in the soil. Host specificity is fairly strict, and species are named accordingly, e.g. *Rhizobium leguminosarum, R. meliloti, R. trifolii, R. lupini, R. japonicum,* etc.

Rho (ρ) factor: see Ribonucleic acid.

Rhodoplasts: photosynthetic organelles of the red algae (*Rhodophyta*). R. are red or red-violet, due to the presence of important light-trapping photosynthetic pigments, called Biliproteins (see). R. are responsible for the characteristic color of these marine algae.

Rhodopsin: see Vitamins (vitamin A).

Rhodoxanthin: 3,3′-dioxo-β-carotene, a xanthophyll, M_r 562, m.p. 219 °C, found as a pigment in brown-red ("copper") leaves, in the needles of various conifers (e.g. yew) and in bird feathers.

RIA: acronym of Radioimmunoassay. See Immunoassays.

Ribitol: an optically inactive, C_5-sugar alcohol, biosynthesized by reduction of ribose. M.p. 102 °C. R. is a component of the flavin molecule, e.g. riboflavin (see Flavin-mononucleotide).

Riboflavin: see Vitamins (vitamin B_2).

Riboflavin-adenosine-diphosphate: see Flavin-adenine-dinucleotide.

Riboflavin 5′-phosphate: see Flavin-mononucleotide.

Ribonuclease (EC. 3.1.27.5): a pancreatic phosphodiesterase specific for RNA. R. catalyses hydrolysis of the phosphate ester bond between pyrimidine nucleoside 3-phosphate residues and the 5-hydroxyl group of the neighboring ribose residues. The cleavage products are 3′-ribonucleoside monophosphates and oligonucleotides possessing a terminal pyrimidine nucleoside 3′-phosphate. Although R. is present in most vertebrates, high activities are only found in the pancreas of ruminants, certain rodents and some herbivorous marsupials. In ruminants R. is required for digestion of the nucleic acids present in the rumen microflora, which account for a large proportion of the dietary nitrogen and phosphorus. Bovine pancreas contains four types of R. (A, B, C and D). ***R.A,*** which contains no carbohydrate, is the predominant form. The other R. are glycoproteins, which differ only in their sugar composition. Variations in the carbohydrate moiety gives rise to microheterogeneity and to the existence of pseudo-isoenzymes.

Primary structure (single chain, basic protein, 124 amino acid residues, M_r 13,700), secondary structure (4 disulfide groups, about 15 % α-helix, about 75 % β-structures), tertiary structure (active center containing His_{12}, His_{119}, Lys_7 and Lys_{41}, located in a cleft, which divides the molecule into two halves, Fig.), and the mechanism of action of R.A. are known. The pH optimum is 7.0–7.5, and the isoelectric point is 7.8. It is inhibited by penicillin, basic dyes (e.g. acridine), EDTA and divalent cations (Cu^{2+}, Zn^{2+}). R.A. is relatively heat stable (heating to 85 °C causes no loss of activity). In the presence of 2-mercaptoethanol in 8 M urea, it is denatured to a random coil, which can be reoxidized under mild conditions to reform the fully active R. (reversible denaturation). Limited proteolysis with subtilisin forms a peptide containing residues 1–20 (S-peptide); the remaining protein, known as S-protein or ***R.S*** (residues 21–124), has full enzymatic activity, providing it is still associated with S-peptide. Removal of S-peptide, by treatment with urea or acid, causes a reversible loss of activity. Only residues 2 and 14 of the S-peptide are necessary for maintaining the normal configuration and function of R. Even Val_{124} is unnecessary for R. activity. The only known dimeric R. was isolated from bovine seminal fluid; it consists of two identical polypeptide chains (each of M_r 14,500 and containing 124 amino acid residues identical in sequence with pancreatic R.) joined by two disulfide bridges.

R. have also been isolated from fungi and bacteria. ***R.T1*** (EC 3.1.27.3) and ***R.T2*** (EC 3.1.27.1) are present as impurities in *Aspergillus* Taka-diastase. RT.1 (primary structure known, 104 amino acid residues, M_r 110,902, pI 2.9, pH optimum 7.4) catalyses hydrolysis of internucleotide bonds between 3′-guanylic acid and 5′-hydroxyl groups. R.T2 catalyses hydrolysis between 3′-adenylate and 5′-hydroxyl groups of neighboring nucleotides. R.T1 and R.T2 were used in the sequence elucidation of tRNA. The bacterial nuclease of *Staphylococcus aureus* (primary and tertiary structures known, 149 amino acid residues, M_r 16,800) is an extracellular, unspecific phosphodiesterase, which degrades both DNA and RNA to 3′-phosphomononucleotides and dinucleotides.

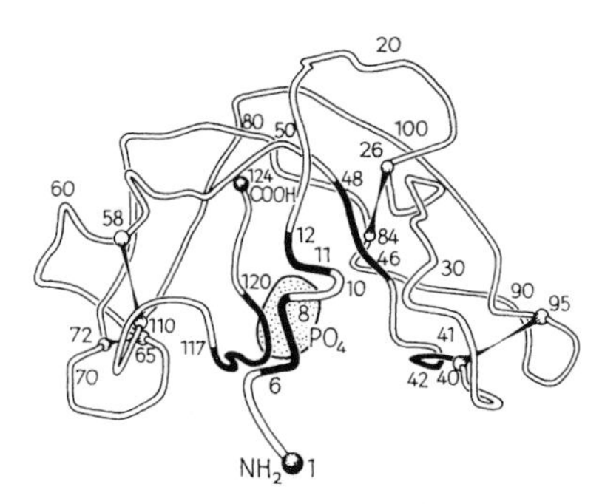

Schematic representation of the tertiary structure of the ribonuclease molecule.
The continuous band represents the conformation of the α-atoms of the peptide chain. Numbers show the localization of amino acid residues. Blacked-in parts of the peptide chain represent catalytically important regions. PO_4 = position of the phosphate residue between His_{119} and His_{12}.

As a model protein, R. has made a considerable contribution to the understanding of protein chain conformation.

Ribonucleic acid, *RNA:* a biopolymer of ribonucleotide units, present in all living cells and some viruses.

Structure. The mononucleotides of RNA consist of ribose phosphorylated at C3, and linked by an N-glycosidic bond to one of four bases: adenine, guanine, cytosine or uracil. Many other bases (chiefly methylated bases) also occur, but are less common (see Rare nucleic acid components). The mononucleotide units form a linear chain via 3′,5′ phosphodiester bonds (see Nucleic acids). Sequence analysis of RNA has become a standard technique. In many cases, the amino acid sequences of proteins are predicted from the sequence of the corresponding mRNA [or DNA] because it is much easier to clone the nucleic acid than to isolate the protein.

RNA does not form a double-stranded α-helix. Single chains, however, show partial folding into α-helical regions, probably by hydrogen bonding between complementary bases. These α-helical regions are separated by regions of single-stranded, unordered RNA.

Types of RNA and their occurrence. There are three main types of RNA, classified on the basis of their function: Messenger RNA (mRNA) (see), ribosomal RNA (rRNA) (see Ribosomes) and Transfer RNA (tRNA) (see); they also have different secondary and tertiary structures. Viral RNA is structurally and functionally very similar to mRNA. In eukaryotic cells, RNA is found in the nucleus, cytoplasm and in the cytoplasmic organelles (ribosomes, mitochondria, chloroplasts). The nucleus is the chief site of RNA synthesis. rRNA is synthesized by the nucleoli, whereas high M_r, polydisperse RNA (precursor of cytoplasmic mRNA) is transcribed on the DNA of the chromatin. Low M_r RNA is also synthesized in the nucleus, consisting partly of tRNA, and partly of RNA which has a regulatory function in gene activation.

In addition to tRNA, the cytoplasm also contains rRNA (in the ribosomes) and mRNA (in polysomes). Other nucleoprotein particles have also been demonstrated, which represent transport forms of mRNA (see Informosomes). All cytoplasmic RNA is synthesized in the nucleus. Mitochondria and plastids also contain mRNA, tRNA and rRNA which are transcribed from the DNA of the organelle.

The bacterial cell also contains tRNA (cytoplasm), rRNA (ribosomes) and mRNA (polysomes).

Function. In all cells, RNA functions in the transfer of genetic information from DNA to the site of protein biosynthesis (mRNA) and in the translation of this information during protein biosynthesis (mRNA, rRNA and tRNA). In addition, the RNA in *E. coli* and *Bacillus subtilis* ribonuclease P has been shown to be responsible for the catalytic activity of the riboprotein [C. Guerrier-Takada & S. Altman *Science* **223** (1984) 285–286]. The RNA in Small nuclear ribonucleoproteins (see) also appears to be catalytically active; other examples of catalytically active RNA are known, e.g. the autocatalytic cleavage-ligation of *Tetrahymena* pre-ribosomal RNA (see Intron).

Biosynthesis.

1) *DNA-dependent RNA synthesis (transcription).* With the exception of viral RNA, all types of RNA are synthesized on a template of DNA. According to the principle of Base pairing (see), the base sequence of the DNA determines the synthesis of a complementary base sequence in the RNA (Fig.). The growing RNA chain is released from the template, so that the process can start again, even before synthesis of the previous molecule is complete. Nucleoside triphosphates, positioned next to each other by complementary pairing of their bases along the transcribing strand of DNA, become linked to form a polynucleotide by the removal of pyrophosphate from each triphosphate group. This reaction is catalysed by RNA polymerase (EC 2.7.7.6):

DNA template: pA pC pT pG pC pT

Substrates: UTP + GTP + ATP + CTP + GTP + ATP

$\longrightarrow$ DNA template + UpGpApCpGpAp + $6PP_i$ (RNA)

Subsequent hydrolysis of the pyrophosphate helps to shift the equilibrium of the reaction in favor of RNA synthesis. The process is well understood in bacteria, and can be considered in three stages:

a) *Initiation.* RNA becomes bound to a specific site on the DNA, known as the promoter (see Operon), which contains high concentrations of cytosine and thymidine. Sigma factor, a subunit of RNA polymerase (see), is required for the start of synthesis. Sigma factor also insures that only the codogenic strand of the DNA is transcribed (strand selection). A short section of the double-stranded DNA comes apart and the first nucleotide (always ATP or GTP) be-

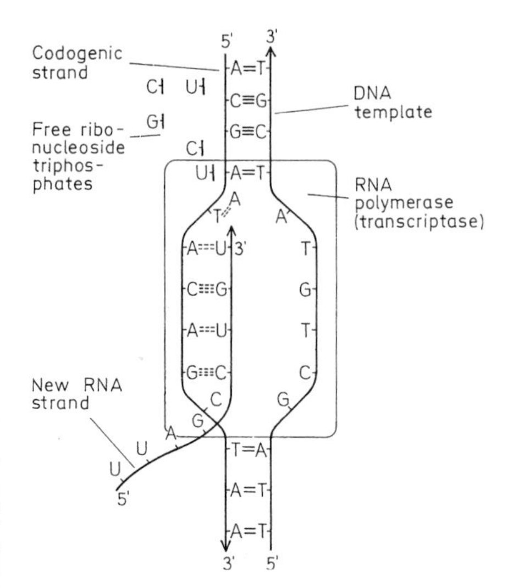

Schematic representation of transcription on a DNA template.
A = adenine. C = cytosine. G = guanine. U = uracil. T = thymine.

comes bound. An initiation complex, consisting of DNA, RNA polymerase and the first nucleotide is formed.

b) *Polymerization.* RNA polymerase moves along the DNA. As it progresses, it opens the double helix and links the ribonucleoside monophosphate units in the 5′-3′ direction in the order dictated by the base of the codogenic DNA strand. As bases in the synthesized RNA strand leave their complementary sites on the DNA, the DNA base pairs reform and the double helix closes again.

c) *Termination.* The end of RNA synthesis and complete release of RNA from the DNA template are signaled by termination codons in the DNA, which are recognized by a termination factor, known as rho-(ρ)-factor.

There are several specific inhibitors of RNA synthesis, e.g. actinomycin, rifamycin, α-amanitin. Often, especially in eukaryotes, the primary transcription products are not identical with the functional, mature RNA molecules. The primary transcription products are usually high M_r and subject to successive stages of degradation, known as maturation or processing (see Post-transcriptional modification of RNA).

Transcription of each gene is regulated according to the requirements of the cell (see Gene activation, Operon, Enzyme induction, Enzyme repression, Messenger RNA).

2) *RNA-dependent RNA synthesis.* Multiplication of RNA viruses depends upon an RNA-dependent reaction, in which viral RNA acts as a template for the synthesis of new RNA. The process is catalysed by RNA synthetase (see).

3) *Starter-dependent RNA synthetase.* In vitro, polynucleotides can be synthesized by polymerization of 5′-nucleoside diphosphates with the release of phosphate. The process requires the presence of an oligonucleotide, which serves as a starter (primer),

and is catalysed by polynucleotide phosphorylase. The base sequence of the resulting polynucleotide is not encoded by a template, and it depends on the relative concentrations of the different nucleoside diphosphates in the reaction mixture.

RNA degradation. RNA is continually degraded in the cell. It is cleaved by various ribonucleases, polynucleotide phosphorylases and phosphodiesterases. In strong acid, RNA is hydrolysed completely to bases, phosphate and ribose; alkaline hydrolysis produces 2′- and 3′-nucleoside monophosphates.

Ribonucleotide reductase (EC 1.17.4.1 or 1.17.4.2): an enzyme system that catalyses reduction of ribonucleotides to 2-deoxyribonucleotides. This is a stage in the biosynthesis of DNA precursors, and is the only metabolic route for the reduction of ribose to deoxyribose. R.r. is subject to a complicated control mechanism, in which an excess of one deoxyribonucleotide inhibits reduction of all other ribonucleotides. This means that DNA synthesis can be inhibited by an excess of any one deoxyribonucleotide or deoxyribonucleoside (subsequently phosphorylated in the cell), and the enzyme has been widely investigated as a possible target for anticancer drugs. R.r. from *E. coli* and mammals catalyse reduction of nucleoside diphosphates; in *Lactobacillus* and *Euglena,* the enzyme requires vitamin B_{12} and reduction occurs at the nucleoside triphosphate level. The oxygen at C2 is reduced and removed as water; the immediate reducing agent is Thioredoxin (see), which is itself reduced by NADPH and thioredoxin reductase (a flavoprotein). During reduction of the ribose moiety, the hydrogen on C2 exchanges with protons from water without loss of configuration.

Ribonucleotides: see Nucleotides.

D-Ribose: a monosaccharide pentose, M_r 150.13, m.p. 87 °C, $[\alpha]_D^{20}$ − 23.7 °C, not fermented by yeasts. The dry solid occurs as the pyranose form. At 35 °C in aqueous solution it forms an equilibrium mixture: 6 % α-furanose, 18 % β-furanose, 20 % α-pyranose and 56 % β-pyranose. It is a component of RNA, some coenzymes, vitamin B_{12}, ribose phosphates and various glycosides. D-R. is prepared by acid hydrolysis of yeast nucleic acids, or by the chemical conversion of arabinose.

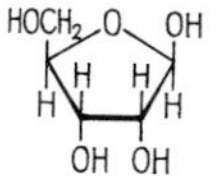

HOCH2 O OH
H H
H H
OH OH

β-D-ribose

Ribose phosphates: phosphorylated derivatives of ribose. Ribose is phosphorylated in position 5 by the action of ribokinase (EC 2.7.1.15) and ATP; ribose 5-phosphate is also produced in the Pentose phosphate cycle (see), and in the Calvin cycle (see) of photosynthesis. Phosphoribomutase catalyses the interconversion of ribose 5-phosphate and ribose 1-phosphate, and the cosubstrate of this reaction is ribose 1,5-*bis*phosphate. 5-Phosphoribosyl 1-pyrophosphate donates a ribose 5-phosphate moiety in the de novo biosynthesis of purine and pyrimidine nucleotides (see Purine biosynthesis, Pyrimidine biosynthesis), in the Salvage pathway (see) of purine and pyrimidine utilization, in the biosynthesis of L-Histidine (see) and L-Tryptophan (see) and in the conversion of nicotinic acid into nicotinic acid ribotide (see Pyridine nucleotide cycle). Ribose 1-phosphate can also take part in nucleotide synthesis (see Salvage pathway).

Ribosomal proteins: integral proteins of ribosomes. Prokaryotic ribosomes contain 35–40 % protein, and eukayotic ribosomes contain 48–52 % protein. The most extensively studied R.p. are those of *E coli.* The 50S-subunit contains 34 different L-proteins (L = large subunit), and the 30S-subunit contains 21 different S-proteins (S = small subunit) (Fig.). All 55 R.p. are immunologically distinct, except for L7 and L121. Each protein has been isolated and characterized with respect to M_r, amino acid composition, pK-value, stoichiometry within the ribosome, and specific interaction with rRNA. With the exception of protein S1 (M_r 65,000), the M_r are all in the range 9,000–28,000. The pK-values are basic (pK > 9), except for S6 (pK 4.9), L7 (pK 4.8) and L12 (pK 4.9). The strong basic character of R.p. is due to their high contents of lysine and arginine. L7 and L12 both contain 120 amino acid residues, and differ only by an *N*-terminal acetyl group in L7. The chemical and physical properties of L7 and L12 closely resemble those of contractile proteins, such as myosin and flagellin. 50S-Ribosomal subunits lacking L7 and L12 have no GTPase activity, but EFG-de-

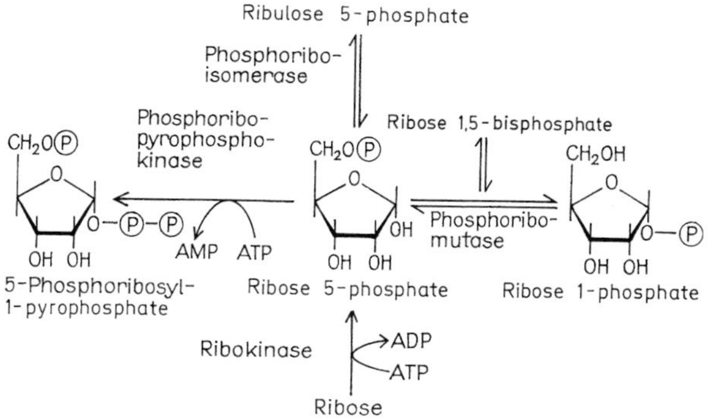

Ribulose 5-phosphate
Phosphoribo-isomerase
Phosphoribo-pyrophosphokinase
Ribose 1,5-bisphosphate
CH2O(P)
O
O—(P)—(P)
OH OH
AMP ATP
CH2O(P)
O
OH
OH OH
Phosphoribo-mutase
CH2OH
O
O—(P)
OH OH
5-Phosphoribosyl-1-pyrophosphate
Ribose 5-phosphate
Ribose 1-phosphate
Ribokinase
ADP
ATP
Ribose

Biosynthesis of ribose phosphates

pendent GTPase is restored by the addition of either protein. When antibodies to L7 and L12 are added to a reconstitution mixture (see Ribosomes), the resulting particles lack GTPase activity. Antibodies to L7 and L12 also prevent formation of a complex between EFG, GTP, 50S-subunit and Fusidic acid (an inhibitor of translocation). It is therefore suggested that L7 and L12 are involved in GTP hydrolysis at translocation by acting as the binding site for EFT_U, EFT_S, and EFG on the 50S-ribosomal subunit (see Protein biosynthesis); they may even serve as contractile protein in the physical movement of the ribosome along the mRNA during translocation. The use of specific antibodies is a powerful tool in the study of the function and location of R.p. within the ribosome. All the S-proteins and most of the L-proteins are accessible to antibody, which indicates that these proteins are at least partly exposed on the surface of the ribosome. They are either bound covalently to rRNA (primary binding proteins), or they interact strongly with it. Each protein is probably dispersed within a matrix of RNA, so that there is very little, if any surface interaction between the different R.p. (see ribosome, Fig. 2). This model is supported by studies with cross-linking reagents; when the intersite distance of the reagent is 5 Å there is practically no cross-linkage, relatively little cross-linkage when the distance is 9 Å, and much more with an intersite distance of 12 Å. In order to study the function of ribosomal proteins, self assembly has been performed in the absence of the protein in question, or in the presence of its mutant form. Spatial relationships have also been investigated by attempting reconstitution with cross-linked proteins. Studies on the effect of the order of addition of each protein during reconstitution have been particularly informative. The grouping of proteins has also been studied by mild ribonuclease digestion of the intact 30S-subunit. This produces ribonucleoprotein fragments (Brimacombe fragments) of varying size, representing subsets of the original ribosomal subunit. For example, one such fragment contains S7, S9, S10, S13, S14 and S19, while another contains S4, S5, S6, S8, S11, S15, S16, S17, S18 and S20, representing two clusters on opposite sides of the 30S-subunit. Particles reconstituted in the absence of S4, S7, S8, S9, S16 or S17 are nonfunctional and show a grossly altered sedimentation rate. Particles reconstituted in the absence of S3, S5, S10, S11, S14 or S19 may show small alterations in sedimentation behavior, but have greatly impaired function. S1, S2, S6, S12, S13, S18, S20 and S21 are not required for assembly, but they are necessary or stimulatory for function.

S4, S8, S15 and S20 are bound particularly tightly to the 16S rRNA. As far as it is possible to attribute function to the various R.p., the following relationships have been shown: S1 (mRNA binding); S2, S3, S10, S14, S19, S21 (fMet-tRNA binding); S3, S4 S5, S11, S12 (codon recognition); S1, S2, S3, S10, S14, S19, S20, S21 (function of A and P sites); S9, S11, S18 (binding of aminoacyl-tRNA); S2, S5, S9, S11 (close proximity to GTPase). Functional ribosomes can be reconstituted from rRNA and R.p. from different organisms, e.g. 16S rRNA from *Bacillus stearothermophilus* and S-proteins from *E. coli* form active hybrid 30S-subunits.

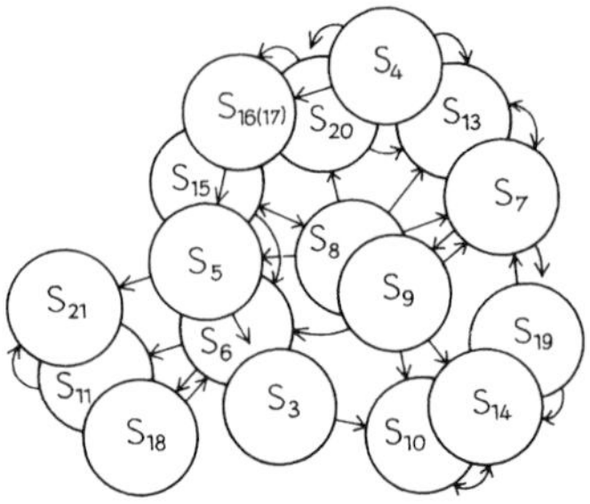

Three dimensional arrangement of the 21 S-proteins of the 30S subunit of the E. coli *ribosome.* Arrows indicate the nature and intensity of the interactions between individual subunits. 16S RNA provides the framework for the assembly process. [M. Nomura *Science 179* (1973) 869]

Proteins involved in binding the 50S- to the 30S-subunit are S20 (actually binds to 50S), S5 and S9 (both restore ability of 30S to combine with 50S, and antibodies to S9 prevent combination of 30S with 50S), S16 (becomes cross linked to 50S by cross linking reagents), S12 (binds to 23S rRNA), S11 (antibodies to S11 prevent combination of 30S with 50S, and S11 binds to 23S rRNA).

The replicase of virus Qβ (an RNA-phage) consists of four subunits: one is encoded by the viral genome, while the other three are the host proteins EFT_U, EFT_S and R.p. S1. The Fig. shows a three dimensional model for the arrangement of R.p. in the 30S-subunit. Many antibiotics act by combining with R.p., e.g. streptomycin interacts with S12, and bacterial mutants resistant to streptomycin have been shown to have an altered S12. Erythromycin interacts with L22, spiramycin with L4.

Ribosomal RNA: see Ribosomes.

Ribosomes, *monosomes:* the sites of Protein biosynthesis (see) in the cell. R. resemble giant multienzyme complexes. They are spherical to ellipsoid, highly hydrated cell organelles, 15 to 30 nm in diameter, normally present in the cytoplasm as Polysomes (see). They were first described in 1953 by Palade (Nobel Prize 1974).

The number of R. in a cell is directly correlated with its capaclty for protein synthesis. There are two main types of R., depending on size and origin (Table): 80S-R. from the cytoplasm of eukaryotic cells; and 70S-R. from prokaryotic cells, plastids and some mitochondria. The mitochondria of vertebrates contain 55S-R.

The subunit composition of R. depends essentially on the ionic concentration of the suspension medium, especially the Mg^{2+} concentration. If this is less than 0.001 M, the R. dissociate into two morphologically and functionally dissimilar subunits (Fig. 1). Thus 70S-R. consist of a large 50S and a small 30S subunit, whereas 80S-R. consist of 60S and 40S subunits. In the complete absence of Mg^{2+} (or in the presence of about 1 M monovalent cations, such as Li, Cs or K), the subunits dissociate into still smaller discrete ribonucleoprotein particles, known as core par-

Comparison of some properties of 70S- and 80S-ribosomes

	70S *(Escherichia coli)*	80S (Mammals)
M_r ($\times 10^6$)	2.7	4.0
S-values of subunits	50 + 30	60 + 40
% RNA	65	50
S-values of high M_r rRNA	23 + 16	28 + 18
M_r of high M_r rRNA ($\times 10^6$)	1.1 + 0.56	1.7 + 0.7
GC content of high M_r rRNA (%)	54 + 54	67 + 59
Number of ribosomal proteins	34 + 21	about 70
Initiation of protein biosynthesis by	formyl-Met-$tRNA_F$	Met-$tRNA_{Met}$
Inhibition of protein biosynthesis:		
by chloramphenicol	+	–
by cycloheximide	–	+

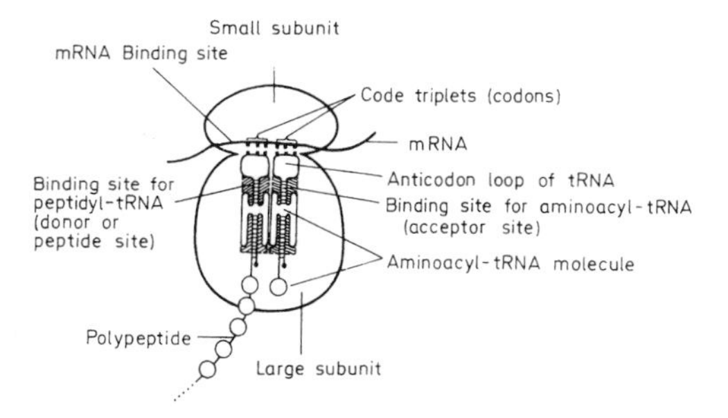

Fig. 1. *Schematic representation of a ribosome*

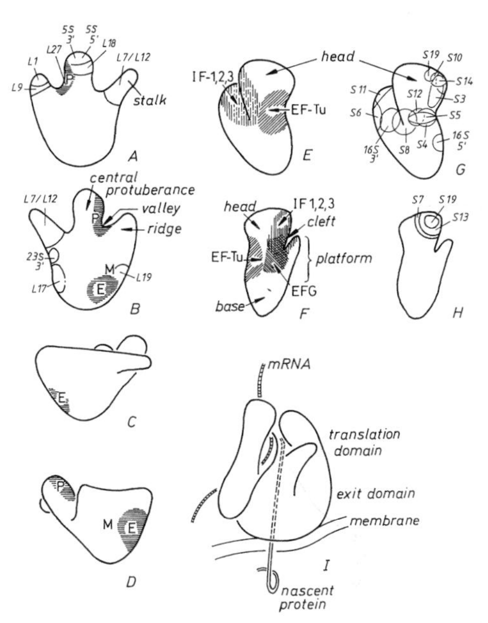

Fig. 2. *Consensus structure of the E. coli 70S ribosome and its subunits.* A, B, C and D are different orientations of the large (50S) subunit. E, F, G and H are two orientations of the small (30S) subunit. On the large subunit, E, M and P represent the nascent protein exit site, the membrane binding site, and the peptidyl transferase site, respectively. 23S 3′ indicates the position of the 3′ terminus of 23S rRNA. On the small subunit, IF-1,2,3 represents the probable location of initiation factors 1, 2 and 3. EF-Tu represents the binding site of the EF-Tu:GTP:aminoacyl-tRNA complex (see Protein biosynthesis). EF-G represents the binding site of elongation factor G (see Protein biosynthesis) near the interface area with the large subunit. 16S 3′ and 16S 5′ indicate the positions of the 3′ and 5′ termini of 16S rRNA. Numbers preceded by S and L represent ribosomal proteins of the small and large subunits, respectively, which have been mapped by electron microscopic visualization of subunit-antibody complexes. *I* is a diagrammatic representation of the whole ribosome, showing the probable location of mRNA and newly synthesized polypeptide, and the position and orientation of the ribosome with respect to the membrane of the endoplasmic reticulum during synthesis of secreted proteins.

ticles; at the same time, certain proteins (called split proteins) are removed. Core and split proteins are inactive in protein biosynthesis, but they can reassociate into functionally competent ribosomal subunits.

Compositon and fine structure of R.: The three-dimensional structure of R. has been studied by electron microscopy (whole R., subunits, and subunits cross-linked by antibodies against individual ribosomal proteins); by low angle diffraction of X-rays and neutrons; and by Fourier reconstruction of electron micrographs of crystallized subunits. Repeated confirmation of the main features of R. structure by different authors has led to a "consensus" structure, in particular for the R. of *E. coli* (Fig. 2). Both subunits are asymmetric. The large (50S) subunit consists of a central protuberance (head) and two side arms or protrusions (the ridge and the L7/L12 stalk) inclined at about 50°on either side of the head. The small (30S) subunit possesses a cleft or indentation which divides the structure into unequal parts, resulting in regions termed platforrn, head and base. In whole R. the platform and much of the base of the 30S subunit is in contact with the 50S subunit.

Using antibodies against purified ribosomal proteins, subunits can be cross-linked by bivalent antibody attachment. Examination of electron micrographs of the resulting subunit dimers permits identification of the site of attachment of the antibody molecule, a site which presumably represents an exposed part of the antigenic protein. In this way, several ribosomal proteins have been mapped on subunit surfaces (Fig. 2).

R. structure has been highly conserved during evolution, and it is similar for the most distantly related organisms. Nevertheless, electron microscopy reveals small but distinct differences in the shapes of R. from eubacteria, archebacteria, eocytes and eukaryotes (cytoplasmic R.), which have been used to interpret phylogenetic relationships between these groups. [J. A. Lake *Ann. Rev. Biochem.* **54** (1985) 507–530]

R. contain only RNA and proteins, 65 % RNA and 35 % protein in the case of 70S and 50 % each in the case of 80S R. Ribosomal RNA (rRNA) contains helical regions (about 70 % of the total rRNA) which alternate with nonhelical regions. The latter are linked to ribosomal proteins by specific ionic interactions and hydrogen bonding between nucleotides and amino acids (Fig. 3).

Total reconstitution of each ribosomal subunit from its separate RNA and protein components was first achieved by Nomura et al. between 1969 and 1972. Reassociation is spontaneous, and proceeds by a process of cooperative self assembly, i. e. all the information needed for the correct assembly of a ribosome is contained in the structure of its components. Reconstitution of the 30S subunit proceeds at 40 °C at high KCl concentrations and takes about 10 min. Reconstitution of the 50S subunit is slower and requires higher temperatures. Reconstitution of eukaryotic ribosomal subunits has not been reported. The various rRNAs can be separated by chromatography or electrophoresis. Prokaryotic rRNA consists of 3 fractions: M_r 0.56×10^6 (≈ 16S) from the 30S subunit; M_r 1.1×10^6 (≈ 23S); and M_r 50,000 (≈ 5S) from the 50S subunit. Eukaryotic rRNA consists of 4 fractions: M_r 0.7×10^6 (≈ 18S) from the small subunit, and from the large subunit, 5S, 5.8S and a fraction whose size depends on its origin, i. e. M_r 1.3×106 (≈ 25S) to 1.75×10^6 (≈ 29S). 5.8S rRNA has no size counterpart in prokaryotes, and it is generally thought to be characteristic of eukaryotes. Its sequence corresponds to that of the first 150 (approx.) nucleotides of prokaryotic 23S rRNA. The parasitic eukaryote *Vairimorpha necatrix* (Microsporidia), however, has no 5.8S rRNA. Microsporidia ribosomes are also prokaryotic in size (70S; subunits 30S and 50S, containing 16S- and 23S-like rRNAs). 23S-like *V. necatrix* rRNA, 25S rRNA of *Saccharomyces cerevisiae,* and *E. coli* 23S rRNA all have a certain degree of structural and sequence homology. [C. R. Vossbrinck & C. R. Woese *Nature* **320** (1986) 287–288.]

rRNA sequences are published by V. A. Erdmann et al. in supplements to the journal *Nucleic Acids Research.* These rRNA sequences are part of the Berlin Databank, which is online accessible worldwide, and is continually updated as soon as a new RNA sequence is available.

Biogenesis and processing of rRNA: 28S, 18S and 5.8S eukaryotic rRNA are transcribed as a single large 45S RNA bound to protein in the Nucleolus (see). The 45S RNA is rapidly degraded by specific endonucleases, in several stages, to produce mature 28S, 18S and 5.8S rRNA (Fig. 4). Thus after processing only about 45 % of the transcribed 45S precursor RNA appears in R. The other cleavage products are apparently destroyed.

In prokaryotic cells, 23S and 16S rRNA are formed in tandem; the precursor molecule is only 10 % longer than the two mature rRNAs, and processing consists of the removal of 200 to 250 nucleotides from the 5′ ends of the 23S and 16S precursors. Prokaryotic and eukaryotic 5S rRNA is always transcribed separately from the other rRNA. Methylated bases are characteristic components of all rRNA.

At every stage of maturation, the various RNA fractions are associated with proteins (see Ribosomal proteins), most of which are basic. Some of these proteins may be endonucleases involved in the maturation process, but the nature and function of these proteins is as yet poorly understood. During the maturation process, proteins are lost and exchanged, the true ribosomal proteins appearing at the end of ma-

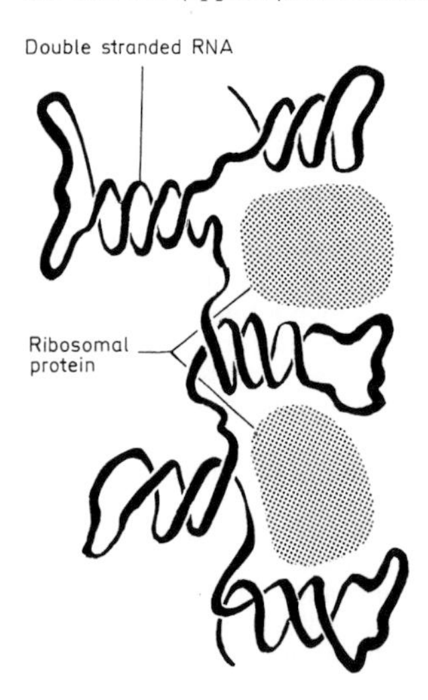

Fig. 3. *Schematic representation of the arrangement of ribosomal RNA and ribosomal protein within the ribosome*

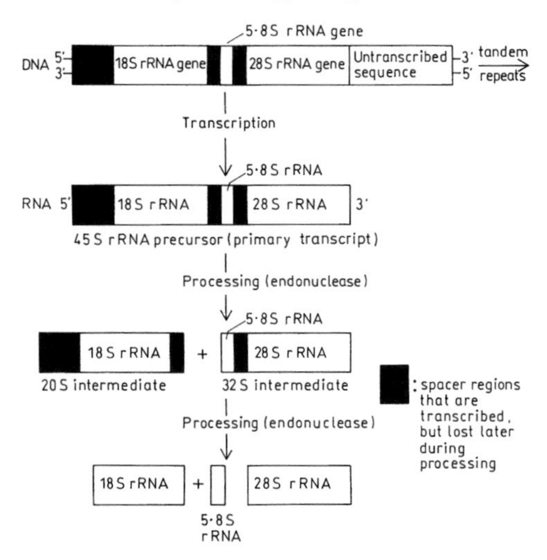

Fig. 4. *Transcription and processing, leading to mature mammalian 5.8S, 18S and 28S rRNA*

turation. It is unclear to what extent the ribosomal proteins are synthesized within the nucleus, or transported into the nucleus after synthesis in the cytoplasm. It is certain, however, that ribosomal subunits are first assembled in the nucleolus, then exported into the cytoplasm. [D.C. Eichler & N. Craig 'Processing of Eukaryotic Ribosomal RNA' *Prog. Nucl. Acid Res. Mol. Biol.* **49** (1994) 197–239]

Ribosylthymine, *2'-hydroxythymidine, ribothymidine:* 5-methyluridine, a pyrimidine derivative, and a Rare nucleic acid component (see) found in tRNA. See Pyrimidine biosynthesis, Thymidine phosphates.

Ribothymidylic acid: the monophosphate of Ribosylthymine (see).

Ribotides: see Nucleotides.

Ribozyme: see Intron.

D-Ribulose: a monosaccharide pentulose, M_r 150.13, $[\alpha]_D^{21} + 16°$. The 5-phosphate and 1,5-diphosphate of D-R. are important intermediates of carbohydrate metabolism. Ribulose 1,5-*bis*phosphate is the CO_2 acceptor in the dark reaction of photosynthesis (see Calvin cycle), and ribulose 5-phosphate is an intermediate in the Pentose phosphate cycle (see).

Ribulose *bis*phosphate carboxylase, *ribulose bisphosphate carboxylase/oxygenase, Rubisco, carboxydismutase, Fraction-1 protein* (EC 4.1.1.39): the enzyme responsible for catalysing photosynthetic CO_2 fixation in all photosynthetic organisms except green sulfur bacteria *(Chlorobiaceae)*. In plants it occurs in the chloroplast stroma where it constitutes about 50 % of the total protein; it should be noted, however, that it is absent from the chloroplast stroma of the mesophyll cells of C_4 plants, although it is present in the chloroplast stroma of the bundle sheath cells (see Hatch-Slack-Kortschak cycle). The abundance of choroplast-containing tissue in Nature makes it likely that Rubisco is the most abundant protein in the biosphere. In photosynthetic bacteria Rubisco is present in the cytoplasm. As well as catalysing the carboxylation reaction in the Calvin cycle (see), it has a second catalytic activity, that of an oxygenase, and it functions as such in the process of photorespiration (see).

The mechanism of the carboxylase reaction catalysed by Rubisco is shown in Fig. 1. It is believed that the enzyme initially induces the tautomerism of D-ribulose-1,5-*bis*phosphate (Ru-1,5-bP) by abstracting a proton from C-3 to form an enediol, thereby facilitating a nucleophilic attack by C2 on CO_2 which results in the generation of a six-carbon 2-carboxy-3-keto species which has now been identified [J. V. Schloss & G. H. Lorimer *J. Biol. Chem.* **257** (1982) 4691–4694] as 2-carboxy-3-keto-D-arabinitol-1,5-*bis*phosphate (i.e. the configuration at C2 is *R*). This intermediate is rapidly attacked at C3 by a water molecule leading to cleavage of the C2-C3 bond and the production of a molecule of 3-phosphoglyceric acid (3PGA), from C-atoms 3, 4 & 5 of the original molecule of Ru-1,5-bP, and a molecule of the *aci*-form of 3PGA, from C-atoms 1 & 2 plus the CO_2 carbon, which is promptly protonated to yield 3PGA. Thus two molecules of 3PGA are formed from one molecule each of Ru-1,5-P and CO_2.

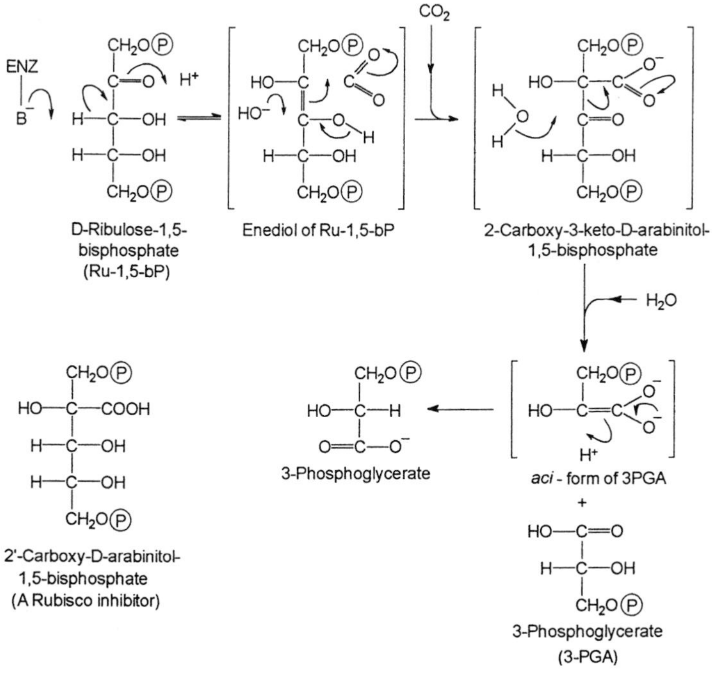

D-Ribulose-1,5-bisphosphate (Ru-1,5-bP)

Enediol of Ru-1,5-bP

2-Carboxy-3-keto-D-arabinitol-1,5-bisphosphate

2'-Carboxy-D-arabinitol-1,5-bisphosphate (A Rubisco inhibitor)

3-Phosphoglycerate

aci - form of 3PGA

3-Phosphoglycerate (3-PGA)

Fig. 1. *The carboxylase reaction catalysed by ribulose bisphosphate carboxylase, and the structure of 2'-carboxy-D-arabinitol-1,5-bisphosphate, a potent inhibitor of the enzyme*

The mechanism of the oxygenase reaction catalysed by Rubisco not well understood but can be represented in the manner shown in Fig. 2. It is believed that the enediol, formed as before, reacts with oxygen and a proton to generate a hydroperoxide which is hydrolytically cleaved between C-atoms 2 & 3 to yield 3PGA (from C-atoms 3, 4 & 5 of the original molecule of Ru-1,5-bP) and 3-phosphoglycolic acid from C-atoms 1 & 2. Evidence compatible with this representation is the observation that the use of [^{18}O]oxygen leads to the formation of [^{18}O]phosphoglycolic acid and unlabeled 3PGA. The main difficulty with the reaction is that it is not clear how Rubisco succeeds in activating oxygen; since oxygen is a ground state triplet, it does not react with the vast majority of organic molecules that are ground state singlets, e. g. molecules like Ru-1,5-bP. Most oxygenases overcome this difficulty by using a transition metal (e. g. Cu, Fe) ion or organic cofactor (e. g. flavin, pterin), to form a complex with O_2 but the Rubisco oxygenase does not seem to adopt this approach and it is likely that the reaction proceeds by a free radical mechanism [G. H. Lorimer *Annu. Rev. Plant Physiol.* **32** (1981) 349–383].

In most photosynthetic organisms Rubisco is a soluble protein of M_r 560,000 composed of eight identical large subunits (LSU) of M_r ~ 55,000 and eight identical small subunits (SSU) of M_r ~ 15,000; it is thus a L_8S_8 structure. However, Rubisco from the purple non-sulfur bacterium *Rhodospirillum rubrum* is an L_2 structure; other members of the *Rhodospirillaceae* (e. g. *Rhodobacter* spp.) have two types of Rubisco, an L_8S_8 and a smaller homo-oligimer, L_x where x is a small even number. The L subunits of plant and bacterial Rubisco are similar, all having three distinct regions, namely: an *N*-terminal domain, a central barrel and a short helical *C*-terminal domain. They are arranged in pairs with the carboxylase and oxygenase catalytic sites located at the interface between the components of the pair. The crystal structure, at 3Å resolution, of *R. rubrum* and chloroplast Rubisco has been determined [M. S. Chapman et al. *Nature* **329** (1987) 354–356]. The function of the small subunits is not known. In eukaryotes the LSU is encoded in the chloroplastidic DNA and synthesized in the chloroplast whilst the SSU is encoded in the nuclear DNA, synthesized on cytoplasmic ribosomes and imported into the chloroplast where chaperone-assisted assembly of the L_8S_8 complex takes place.

Up to 1966 it was believed that the carboxylase activity of Rubisco was too low to account for observed photosynthetic CO_2 fixation rates. This belief was based on the incompatibility of measured $K_m(CO_2)$ values of 200–500 μM with the fact that the concentration of CO_2 in the aqueous milieu of the chloroplast is ~ 10 μM (the conc. of CO_2 in water in equilibrium with air containing 0.03 % CO_2 by vol., at 1 atmosphere pressure and 25 °C is 10 μM). However it was then found that if the protein concentration were kept high during its isolation, Rubisco had a $K_m(CO_2)$ near to 10 μM [R. G. Jensen & J. A. Bassham *Proc. Natl. Acad Sci. USA* **56** (1966) 1095–1101], indicating that at the intrachloroplastidic CO_2 concentration the Rubisco-catalysed carboxylation reaction would be operating at about half its maximum velocity, an acceptable rate. It is now apparent that Rubisco exists in two forms, the high $K_m(CO_2)$ form and the low $K_m(CO_2)$ form, and that these two forms are interconvertible and constitute the basis of a regulatory mechanism. The low activity [high $K_m(CO_2)$] form is converted to the high activity [low $K_m(CO_2)$] form by carbamylation of the ε-amino group of a lysine residue at position 210 in LSU in the presence of Mg^{2+} and ATP and an enzyme termed Rubisco activase:

$LSU\text{-}Lys\text{-}NH_3^+ + CO_2 + ATP + Mg^{2+} \rightarrow LSU\text{-}Lys\text{-}NH\text{-}COO^- \ldots Mg^{2+} + ADP + P_i + H^+$ (At high levels of CO_2 this reaction occurs nonenzymatically).

Rubisco is also activated by the rises in pH (from ~ 7.0 to ~ 8.5) and Mg^{2+} concentration (from ~ 1 mM to ~ 5 mM) that occur as a result of the electrochemical gradient generated when the chloroplast is illuminated. Activation of the carboxylase activity of Rubisco also causes activation of its oxygenase activity.

Rubisco is inhibited by 2′-carboxy-D-arabinitol-1-phosphate, which accumulates in the chloroplast in the dark [S. Gutteridge et al. (1987) in *Progress in Photosynthesis Research* (J. Biggins, ed.) Proc. 7th Int. Congr. Photosynth., Vol. 3, pp. 395–398] and is rapidly destroyed by light. It has been called the 'pre-dawn inhibitor' and insures that during the hours of darkness the Calvin cycle is switched off; the latter is vital because the Calvin cycle requires ATP and this would have to come from carbohydrate oxidation in the absence of light, a process that would nullify the photosynthesis of previous day. It is believed that 2′-carboxy-D-arabinitol-1-phosphate is inhibitory because its structural similarity to 2-carboxy-3-keto-D-

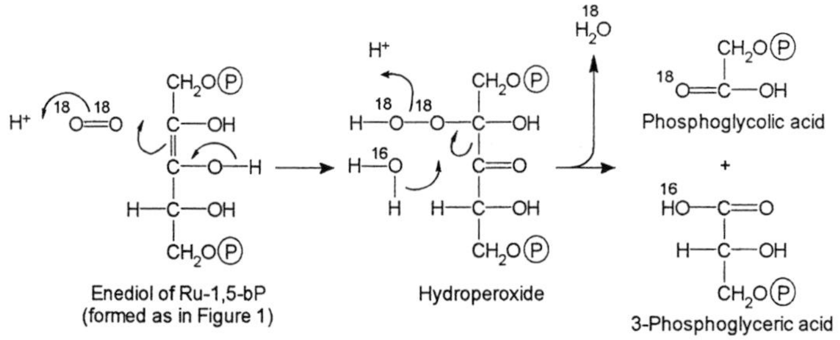

Enediol of Ru-1,5-bP (formed as in Figure 1) → Hydroperoxide → Phosphoglycolic acid + 3-Phosphoglyceric acid

Fig. 2. *The oxygenase reaction catalysed by ribulose bisphosphate carboxylase*

Receptor
Nucleus
Cell with surface receptors
Toxin

Hypothetical scheme for the interaction of a toxin (ricin or abrin) and a eukaryotic cell. A = A-chain. B = B-chain.

arabinitol-1,5-bisphosphate, the intermediate generated in Rubisco-catalysed carboxylation (see Fig. 1), enables it to bind to the catalytic site of the enzyme. For the same reason, 2′-Carboxy-D-arabinitol-1,5-bisphosphate (Fig. 1), a compound that does not occur naturally, is an irreversible inhibitor of Rubisco; it binds very tightly to the enzyme ($K_d \sim 300$ fM). The carboxylase activity of Rubisco is competitively inhibited by O_2 [$K_i(O_2) = 200$–$400\ \mu$M], while its oxygenase activity is competitively inhibited by CO_2 [$K_i(CO_2) = 20$–$40\ \mu$M].

Richner-Hanhart syndrome: see Inborn errors of metabolism.

Ricin: a toxalbumin phytotoxin from *Ricinus* seeds. M_r 66,000, 493 amino acid residues. R. inhibits protein biosynthesis (causes dissociation of polysomes) and has antitumor properties. It consists of an A-chain (M_r 32,000) and a B-chain (M_r 34,000) joined by disulfide bridges. After reductive separation by 2-mercaptoethanol, both chains show increased inhibitor activity but markedly decreased toxicity. Toxic activity is carried by the A-chain (effectomer), while the B-chain (haptomer) binds the toxin to the cell surface. Similar action and structure are possessed by abrin (M_r 65,000; A-chain 30,000, B-chain 35,000), a toxalbumin from the red seeds of *Abrus precatorius.* Abrin is used in opthalmology.

Entry of R. or abrin into the cell occurs in two stages (Fig.). In the first stage, the toxin becomes bound by its B-chain to the terminal galactose of the receptor on the cell surface. In the second stage, a disulfide cleaving system of the cell releases the A-chain which enters the cell by endocytosis. The A-chain binds to the 60S ribosomal subunit, which is then incapable of reacting with elongation factor EF2, and protein synthesis stops. Other toxic proteins are thought to act by an analogous two-stage mechanism.

Ricinin: a poisonous pyridine alkaloid from seeds of *Ricinus communis.* M_r 164.17, m.p. 201 °C, b.p.$_{20}$ 170–180 °C. R. is an exceptional alkaloid, in that it occurs in only one type of plant and is not accompanied by other alkaloids. It is biosynthesized from nicotinic acid; biosynthetic precursors of nicotinic acid are aspartic acid and a 3-carbon compound (probably hydroxyacetone phosphate). Administration of radioactive precursors to *Ricinus* shows a high incorporation of ^{14}C from aspartate and glycerol into R.

OCH_3 CN O N CH_3

Ricinin

Ricinoleic acid: 12-hydroxyoleic acid, CH_3–$(CH_2)_5$–CHOH–CH_2–CH = CH–$(CH_2)_7$–COOH, a fatty acid, M_r 298.45, m.p. (α-form) 7.7 °C, (β-form) 16.0 °C, b.p.$_{15}$ 250 °C. R. a. is present in the acylglycerols of castor oil (*Ricinus* oil) where it accounts for 80–85 % of the total esterified fatty acids. It is also present in maize oil, wheat oil and various other vegetable oils.

Rickets: vitamin D deficiency disease. See Vitamins.

Rieske protein/center: an iron-sulphur protein first isolated from Complex III of the mitochondrial electron transport chain, in which it occurs with cytochromes *b* and c_1 [J. S. Rieske et al. *Biochem. Biophys. Res. Commun.* **15** (1964) 338–344], but which has now been found in the equivalent cytochrome *bc* complexes in the bacterial plasma membrane and the chloroplast thylakoid membrane. The latter, known as the cytochrome b_6f complex, participates in cyclic and noncyclic electron flow in the light phase of photosynthesis (see Photosynthesis). All Rieske proteins are one-electron redox systems with a standard redox potential in the + 0.2 to + 0.3V range and have a [2Fe-2S] center, a single membrane-spanning α-helix, and a characteristic electron spin resonance (ESR) spectrum. The chloroplastidic R.p/c, with a M_r of ~ 20,000, is smaller than that of the mitochondrion. It is encoded in the nucleus, synthesized in the cytoplasm and translocated to the chloroplast, where it is inserted into the thylakoid membrane. Within the thylakoid membrane its [2Fe-2S] redox centre (near to its *C*-terminus) can readily pass electrons to cytochrome *f*, a *c*-type cytochrome that projects from the luminal surface; cytochrome *f* then passes electrons to plastocyanin (see) dissolved in the aqueous milieu of the thylakoid lumen.

Rifamycins: a group of antibiotics produced by *Streptomyces mediterranei.* They contain a naphthalene ring system bridged between positions 2 and 5 by an aliphatic chain. Rifamycin SV and rifampicin inhibit DNA-dependent RNA synthesis in prokaryotes, chloroplasts and mitochondna, but not in the nuclei of eukaryotes. Inhibition is due to the formation of a stable complex between RNA polymerase and R.; binding of the enzyme to DNA still occurs, but incorporation of the first purine nucleotide into RNA is prevented. Thus R. specifically inhibit initiation of RNA synthesis, but not chain elongation. Some R. also inhibit eukaryotic and viral RNA-polymerases.

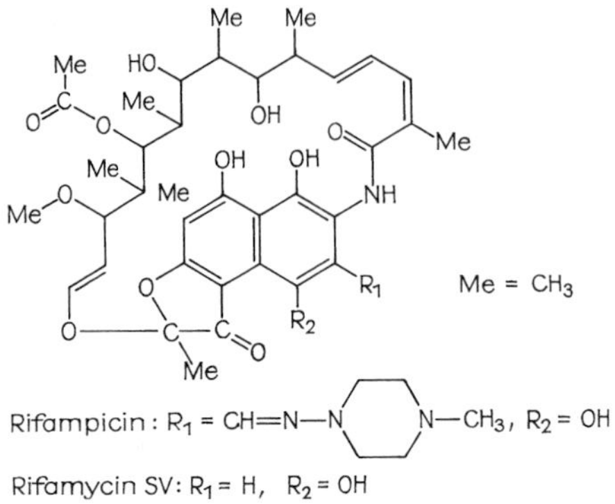

Me = CH_3

Rifampicin: R_1 = CH=N–N(piperazine)N–CH_3, R_2 = OH

Rifamycin SV: R_1 = H, R_2 = OH

Structures of Rifamycins

Rishitin: see Phytoalexins.

RNA: acronym of Ribonucleic acid (see).

RNA-dependent DNA-polymerase, *reverse transcriptase:* an enzyme present in retroviruses, some of which cause cancer, e.g. avian myeloma virus and various leukemia viruses. These are RNA viruses, and the enzyme catalyses the synthesis of the provirus DNA, using the viral RNA as a template. The resulting DNA is then incorporated into the genome of the infected cell. Study of the process of RNA-dependent DNA synthesis, involving several virus-specific enzymes, may therefore contribute to the understanding of malignant transformation of cells by RNA viruses and the problem of cancer in general. Occurrence of the enzyme in cells or viruses can be used for the diagnosis of oncogenic viruses. DNA synthesis is analogous to transcription or replication, i.e. RNA acts as a template for the formation of a base-complementary molecule of single-stranded DNA. The latter is replicated by the action of a DNA-dependent DNA-polymerase, thus forming double-stranded DNA, which can be further replicated (this is the provirus). Host DNA is cleaved by a viral endonuclease, and the provirus DNA is inserted by the action of a ligase.

RNA-dependent RNA-Polymerase: see RNA synthetase.

RNA-dependent RNA synthesis: see RNA synthetase.

RNA polymerase, *DNA-dependent RNA polymerase, nucleoside triphosphate:RNA nucleotidyltransferase, transcriptase* (EC 2.7.7.6): an enzyme which catalyses the synthesis of RNA on a DNA template. The base sequence of the resulting RNA is complementary to that of the DNA template (see Ribonucleic acid).

There are at least four types of RNA-P. from eukaryotic cells: RNA-P.I is found in the nucleolus and preferentially catalyses the synthesis of rRNA. RNA-P.II is present in the nucleoplasm, is mainly responsible for the synthesis of mRNA, and is specifically inhibited by α-amanitin. RNA-P.III transcribes tRNA and other classes of small RNA. The fourth RNA-P. is smaller than the other three and transcribes RNA from mitochondrial DNA; however, it is encoded by a nuclear gene. The RNA-P. of chloroplasts and mitochondria show similarities with prokaryotic RNA-P., e.g. inhibition by rifamycins.

There is extensive sequence homology between the largest subunits of RNA-P.II and III from yeast; these subunits also display great homology with the largest subunit (β') of prokaryotic RNA-P [L.A. Allison et al. *Cell* **42** (1985) 599–610]. The largest subunit of *Drosophila* RNA-P.II is involved with chain elongation and may, like the homologous β' subunit of *E. coli,* interact with promoters of genes [J. Biggs et al. *Cell* **42** (1985) 611–621].

Native bacterial RNA-P. has M_r 500,000 and is a complex of five subunits: $\alpha_2\beta\beta'\sigma$; M_r of subunits: α_2 40,000, β 155,000, β'165,000, σ 90,000. The enzyme has different functional parts: the initiation site binds nucleoside triphosphate, the polymerization site catalyses the formation of internucleotide linkages. The β'-subunit is responsible for binding the enzyme to the template DNA. The σ factor contains a binding site for specific nucleotide sequences in the codogenic DNA strand (promoter region); it therefore acts as the initiation factor for the synthesis of RNA exclusively on the codogenic strand. Termination of RNA synthesis, signaled by specific regions of the DNA, depends on the rho (ρ) factor, a protein which is not considered to be part of the enzyme.

Other extra protein factors appear to be necessary for the specific function of bacterial RNA-P., e.g. psi (Ψ) factor for promoter initiation in ribosomal RNA synthesis, or kappa (κ) factor as a further termination factor.

Infection of cells with DNA phage may lead to the production of a new, phage-specific RNA-P. (e.g. T7 phage; RNA-P. is a single-chain protein, M_r 107,000), or the host RNA-P. may be modified with the addition of new protein subunits encoded by the phage (e.g. T4 and λ phage); in each case the resulting RNA-P. is specific for transcription of phage DNA.

RNA-replicase: see RNA-synthetase.

RNA-synthetase, *RNA-replicase, RNA-dependent RNA-polymerase:* an enzyme that appears in bacterial, animal and plant cells, following infection with RNA viruses. Using a template of single-stranded (viral) RNA, the enzyme catalyses the synthesis of a complementary strand of RNA. The substrates are 5'-nucleoside triphosphates, which become linked by a phosphodiester bond between positions 3' and 5' of neighboring ribose units, with the release of pyrophosphate. The newly synthesized RNA is hydrogen-bonded by base-pairing to the template RNA, thus forming double-stranded RNA (replicative form). This consists of a minus strand (which serves for the synthesis of a new plus strand), and a plus strand (the original viral RNA). At least some of the subunits of the RNA-S. are encoded by the viral genome (see Phage development).

Robinin: see Flavones (Table).

Robinson ester: see Glucose 6-phosphate.

Robustaflavone: see Biflavonoids.

Rocellic acid: (*S*)-2-dodecyl-3-methylbutanedioic acid; (*S*)-2-dodecyl-3-methylsuccinic acid, a branched chain dicarboxylic acid found in lichens. M.p. 132 °C, $[\alpha]_D$ + 17.4° (ethanol). R.a. is biosynthesized from itaconic acid.

Rotenoids: chromanochromanones present in plants (especially roots) of the *Papilionaceae,* a subfamily of the *Leguminoseae,* notably in the related genera, *Derris, Lonchocarps, Tephrosia* and *Mundu-*

lea. R. are powerful insecticides, and preparations of R-containing plants are used as garden sprays and dusts (derris dust). R. are toxic to humans and animals only in large doses, and they are absorbed more efficiently by the lungs than the alimentary canal. R. are also active piscicides. R.-containing plants are among those used by native fisherfolk for paralysing river fish (*Derris* in Asia, *Lonchocarpus* in South America, *Tephrosia* on all four continents).

The most important R. is rotenone (Fig.), which gives its name to the group.

Biosynthesis. Tracing studies (^{14}C) with *Derris elliptica* plants and germinating seeds of *Amorpha fruticosa* have shown that ring A of rotenone is derived from the aromatic ring of phenylalanine, and that Cl and C2 of the phenylalanine side chain provide C12 and C12a of rotenone, respectively; this indicates that an aryl migration occurs from C3 to C2 of phenylalanine. Structural comparison suggests that R. are derived from isoflavonoids by addition of the methylene carbon C6, which closes ring B. C6 of rotenone is derived from the methyl group of methionine. [L. Crombie *J Chem. Soc* (C) Org. (1963) 3029–3032]

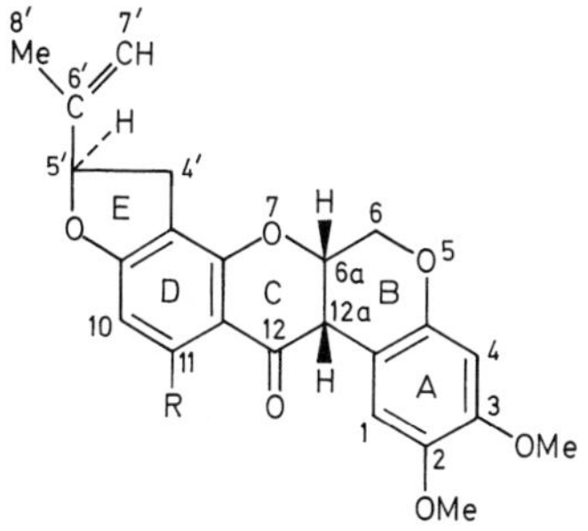

They are always accompanied by isoflavones. All R. so far examined have positive Cotton effects, and therefore form one stereochemical series with respect to the 6a and 12a chiral centers (6a*S*, 12a*S*) [J. Claisse et al. *J. Chem. Soc.* (1964) 6023–6036.]

Rotenone is used experimentally as an inhibitor of mitochondrial respiration. It has little effect on the mitochondrial oxidation of succinate, but it powerfully inhibits all oxidations which operate via NADH dehydrogenase. The site of inhibition by rotenone has been located on the oxygen side of the nonheme iron of NADH dehydrogenase (see Respiratory chain). Piericidin and amytal appear to act at or very close to the same site. [W. W. Wainio *The Mammalian Mitochondrial Respiratory Chain* (Academic Press, 1970); T. P. Singer & M. Gutman *Adv. Enzmol.* **34** (1971) 79–153; J. B. Harborne, T. J. Mabry & H. Mabry eds. *The Flavonoids* (Chapman & Hall, 1975)]

R = H, Rotenone *(Derris* spp., *Tephrosia* spp., *Lonchocarpus* spp., *Piscidia erythrina, Neorautanenia ficifolia, Pachyrrhizus erosus).*
R = OH, Sumatrol *(Derris malaccensis, Piscidia erythrina).*

rRNA: ribosomal RNA (see Ribosomes).

Rubber: 1) vulcanized caoutchouc. 2) the water-soluble components of rubber resin. (See Caoutchouc).

Rubijervine: 12α-hydroxysolanidine; solanid-5-ene-3β,12α-diol, a Veratrum alkaloid of the jerveratrum type. M_r 413.65, m. p. 242 °C, $[\alpha]_D$ + 19 ° (ethanol). It occurs in hellebore (*Veratrum album, V. nigrum* and *V. viride*) and differs structurally from solanidine (see α-Solanine) by the presence of a 12α-hydroxyl group.

Rubisco: see Ribulose *bis*phosphate carboxylase.

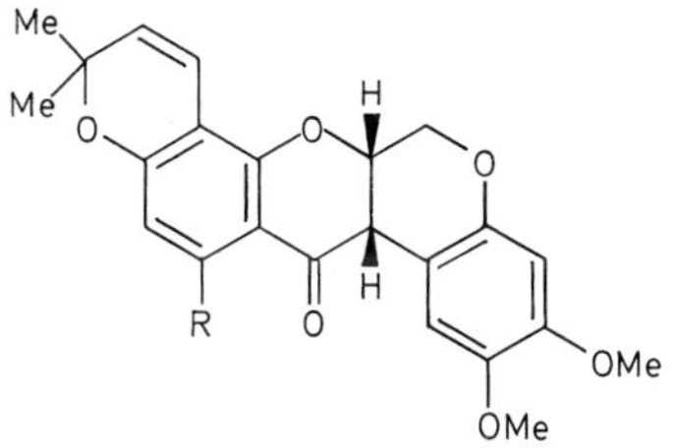

R = H, Deguelin (*Tephrosia* spp., *Derris* spp., *Lonchocarpus nicosa*)
R = OH, α-Toxicarol (*Tephrosia* spp., *Derris* spp.)

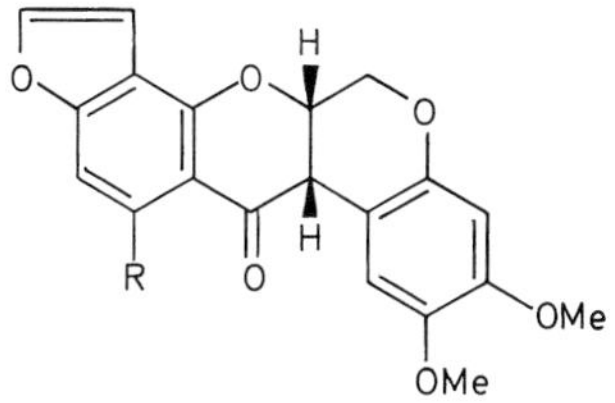

R = H, Elliptone *(Derris elliptica)*
R = OH, Malaccol *(Derris malaccensis)*

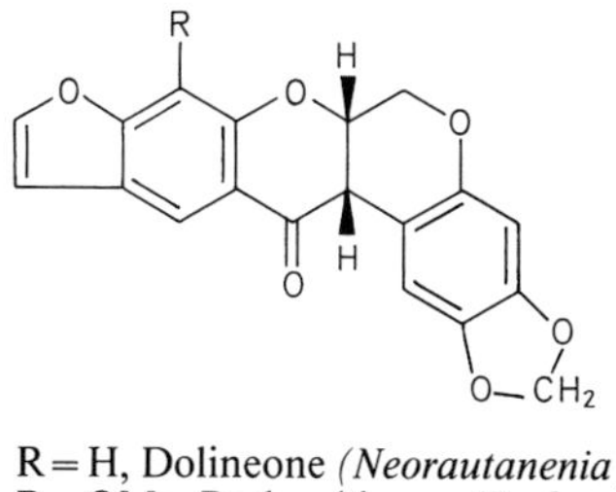

R = H, Dolineone *(Neorautanenia pseudopachyrhiza)*
R = OMe, Pachyrrhizone *(Pachyrrhizus erosus)*

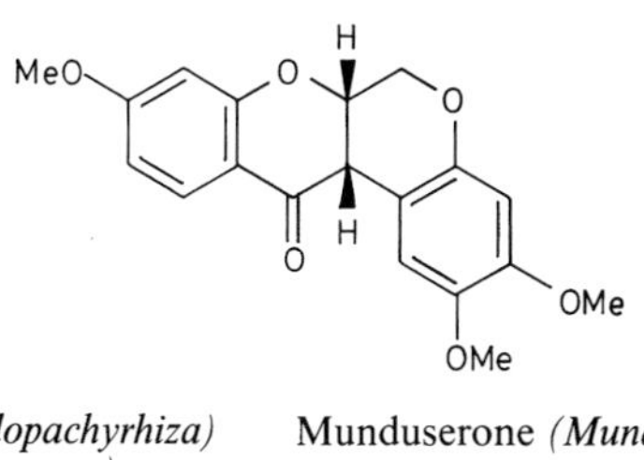

Munduserone *(Mundelea sericea)*

Some typical rotenoids

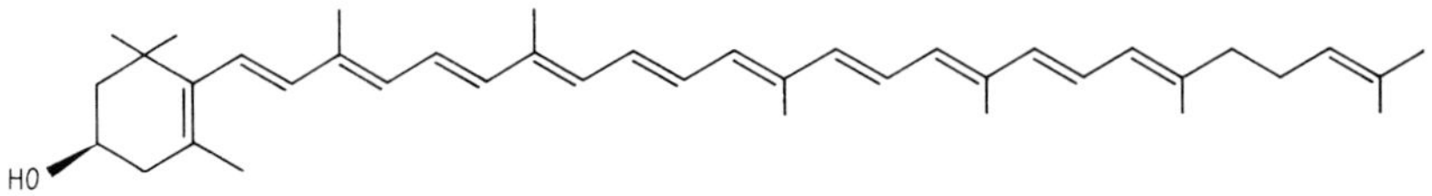

Rubixanthin

Rubixanthin: 3(*R*)-β,Ψ-carotene-3-ol; 3(*R*) hydroxy-γ-carotene, a xanthophyll, M_r 552, m.p. 160 °C. R. is a copper-red pigment of various rose species and some other higher plants. The 5′-*cis*-isomer, gazaniaxanthin, is the pigment of various *Gazania* species.

Rubredoxin: a redoxin, functionally similar to Ferredoxin (see). M_r 6,000. R. was isolated from *Clostridium pasteurianum;* synthesis of R. appears to be promoted by a relative deficiency of iron. It contains one Fe atom/molecule of protein, which is less than the Fe content of ferredoxin. Under acid conditions it is more stable than ferredoxin, and it has a more positive redox potential (E'_0 −0.057 V); thus when R. replaces ferredoxin in a ferredoxin-dependent reaction the reaction rate is decreased. The iron is bound by coordination with 4 cysteinyl residues; other possible ligands are tyrosine and lysine. Redoxins similar to R. have been isolated from *Peptostreptococcus elsdenii* and other bacteria. R. from *Micrococcus aerogenes* contains 53 amino acid residues of known sequence.

Rubrosterone: 2β,3β,14α-trihydroxy-5β-androst-7-ene-6,17-dione, a plant steroid, M_r 334.42, m.p. 245 °C, $[\alpha]_D$ + 119 ° (methanol). R. was isolated, together with ecdysterone, from the roots of *Amarantha obtusifolia* and *A. rubrofusca,* and is considered to be a biogenetic degradation product of ecdysone.

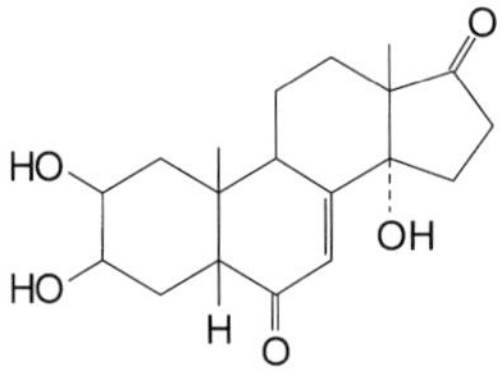

Rubrosterone

Rutaceae alkaloids: alkaloids from common rue (*Ruta graveolens* L.) and other members of the *Rutaceae.* They include quinoline, furanoquinoline, pyranoquinoline and acridine compounds, all biosynthesized from anthranilic acid. The furanoquinolines have spasmolytic activity, and the drugs are sometimes used therapeutically.

Rutin: see Flavones (Table).

S

S, *Svedberg unit:* see Sedimentation coefficient.

Sabininic acid: 12-hydroxylauric acid, $HOCH_2-(CH_2)_{10}-COOH$, a fatty acid, M_r 216.31, m.p. 84 °C, present as a typical esterified wax acid in the wax of many pines.

Safynol: *trans,trans*-3,11-tridecadiene-5,7,9-triine-1,2-diol, a Phytoalexin (see). S. and Δ^3-dehydrosafynol are formed by safflower (*Carthamus tinctorius*) following infection by *Phytophthora.* ED_{50} is 12 µg/ml for S. and 1.7 µg/ml for the dehydro derivative.

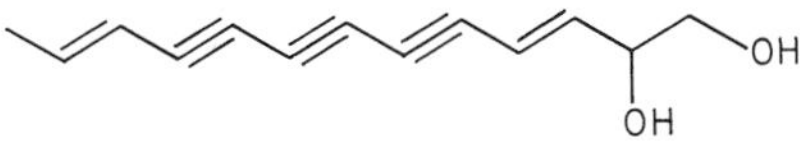

Safynol

Sakaguchi reaction: see L-Arginine.

Salamander alkaloids: toxic steroid alkaloids secreted by the skin glands of salamanders, e.g. *Salamandra maculosa* (European fire salamander). S.a. are modified steroids, in which the A-ring is expanded to a seven membered ring by a nitrogen between C2 and C3 (A-azahomosteroids). They excite the central nervous system and cause paralysis. The chief representative, *samandarin* (1α,4α-epoxy-3-aza-A-homoandrostane-16β-ol) (Fig.) also causes hemolysis (lethal dose for mice 1.5 mg/kg). *Samandarone* possesses a keto group in place of the hydroxyl group at position 16. *Samandaridine* contains a lactone bridge ($-CH_2-CO-O-$) between C16 and C17, and also possesses local anesthetic properties.

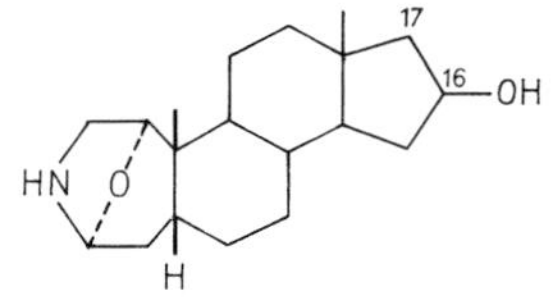

Samandarin

Salamander toxins: toxins secreted by the skin glands of *Salamandra maculosa* (European fire salamander) and *Salamandra atra* (alpine salamander). S.t. include the Salamander alkaloids (see), biogenic amines (tryptamine, 5-hydroxytryptamine), and high M_r substances that cause skin irritation and hemolysis.

Salmine: see Protamines.

Salsola alkaloids: a group of simple isoquinoline alkaloids, occurring in *Salsola* spp. Chief representative is salsoline (1-methyl-6-hydroxy-7-methoxy-1,2,3,4-tetrahydroisoquinoline), which occurs in the D-form, m.p. 215–216 °C, $[\alpha]_D + 40°$ (water) and in the DL-form.

Salsoline: see Salsola alkaloids.

Salting in/Salting out: see Proteins.

Salvage pathway: utilization of preformed purine and pyrimidine bases for nucleotide synthesis. In addition to de novo synthesis, the S.p. represents an alternative pathway for formation of purine and pyrimidine nucleotides. In mutant microorganisms lacking de novo purine and pyrimidine synthesis, the S.p. is the only route for nucleotide synthesis following administration of exogenous purine and pyrimidine bases.

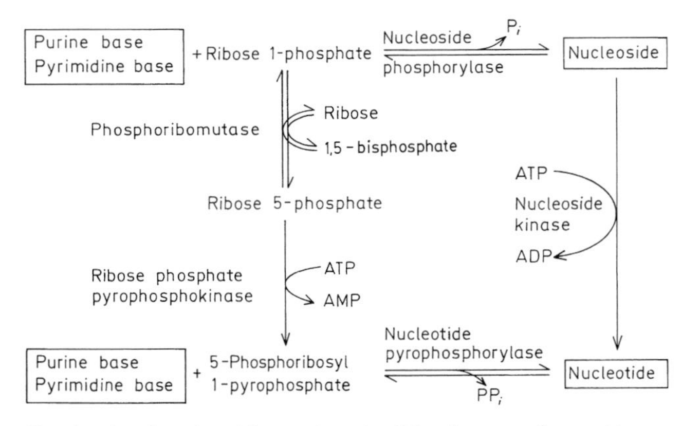

Synthesis of nucleosides and nucleotides from preformed bases.

In liver, specific pyrophosphorylases catalyse the synthesis of nucleotides from free bases and 5-phosphoribosyl 1-pyrophosphate; alternatively, nucleosides may be formed first from bases and ribose 1-phosphate by the action of a nucleoside phosphorylase (Fig.). Deoxynucleosides can be formed from bases and deoxyribose 1-phosphate by the action of deoxynucleoside phosphorylase.

SAM: acronym of *S*-adenosyl-L-methionine.

Samandaridine: see Salamander alkaloids.

Samandarine: see Salamander alkaloids.

Samandarone: see Salamander alkaloids.

Sandhoff's disease: see Lysosomal storage diseases.

Sanfilippo syndrome: see Lysosomal storage diseases.

Sangivamycin: 4-amino-5-carboxamide-(D-ribofuranosyl)-pyrrolo-(2,3-d)-pyrimidine, a deazaadenine-type antibiotic from *Streptomyces* (see Nucleoside antibotics). The antibiotic activity of S. is similar to that of Toyocamycin (see).

Sapogenins: see Saponins.

Saponins: a large and widely distributed group of plant substances, named for their ability to form strongly foaming, soap-like solutions with water. They are all glycosides, and are classified according to the nature of the aglycon, i. e. steroid, triterpene and steroid-alkaloid saponins; the latter are also known as glycoalkaloids. Aglycons of the steroid saponins are also called sapogenins; these are spirostane-type, C27-steroids, all possessing a 3β-hydroxyl group (spirostanols), which forms a glycosidic linkage with the sugar, e. g. D-glucose, D-galactose or L-arabinose. The triterpene saponins are less well studied; their aglycons are chiefly tetra and pentacyclic triterpenes.

S. are powertul surfactants, cause hemolysis and are potent plasma toxins and fish poisons, but they have no toxic effects when ingested by humans, because they are not absorbed. Many S. have antibiotic activity, mainly against lower fungi. With steroids, e. g. cholesterol, S. form poorly soluble 1:1 molecular compounds, which can be used for the analytical separation of S. or steroids.

S. are biosynthesized by cyclization of 2,3-epoxysqualene (see Squalene); in the case of the steroid S., cholesterol is a subsequent intermediate in the biosynthesis of the spirostane skeleton. The stage at which the sugar becomes attached and the mechanism of this attachment are not known.

S. have been found in over 100 plant families; they occur, e. g. in soapwort, rape, soybean, foxglove, sycamore. Important representatives are Digitonin (see) and Dioscin (see). S. are used as detergents and foaming agents, and have been used since antiquity as fish poisons.

Saprophytism: a heterotrophic mode of nutrition, in which dead, organic material serves as the substrate. Many bacteria and most fungi are saprophytic; it is nevertheless possible to grow some fungi and to achieve fruiting body formation on entirely synthetic media.

Many carbohydrates can serve as carbon sources for saprophytes, and in some cases, alcohols, fats, organic acids or hydrocarbons are used. Proteins are also used as a carbon source.

Biological decay and decomposition are due to S., which is therefore important in the cycling of elements in the biosphere. Such processes may be primarily concerned with the incorporation of nutrients for growth, or with energy production, e. g. fermentation. For utilization for growth of the saprophyte, all nutrients must ultimately be converted to primary metabolites, like glucose, or intermediates of glycolysis or the TCA-cycle.

There are many different nitrogen sources. Some molds, yeasts and bacteria are able to utilize inorganic nitrogen, e. g. nitrate or ammonium salts. Certain soil bacteria can even assimilate atmospheric nitrogen. Other saprophytes require organic nitrogen in the form of amino acids, peptones or proteins.

Many intermediate stages exist between S. and Parasitism (see).

Sarcine: an obsolete name for hypoxanthine.

Sarcosine, *N-methylglycine:* CH_3–NH–CH_2–COOH, an intermediate in the metabolism of choline in liver and kidney mitochondria (see One-carbon cycle). It has also been isolated from starfish and sea urchins, where it appears to be a major metabolite.

Sarcosine dehydrogenase (EC 1.5.99.1): a mitochondrial flavoprotein (the flavin is covalently bound) catalysing conversion of sarcosine to glycine and a one-carbon unit at the oxidation level of formaldehyde (or bound –CH_2–). The metabolic fate of the one-carbon unit depends on the availability of tetrahydrofolate (THF). In the absence of THF, the products are glycine and formaldehyde. See One-carbon cycle.

Sarcosinemia: see Inborn errors of metabolism.

Sarcosomes: see Mitochondria.

Sargasterol: (20S)-stigmasta-5,24(28)-diene-3β-ol, a phytosterol (see Sterols), M_r 412.7, m p 132 °C, $[\alpha]_D$ –48 ° ($CHCl_3$), in brown algae, e. g. *Sargassum ringolianum.* It differs from Fucosterol (see) by the opposite configuration at C20.

Satellite DNA: DNA fractions that can be separated from the main DNA by CsCl-gradient centrifugation. S.DNA has been demonstrated in the nuclei of many eukaryotes, accounting for about 1 % of nuclear DNA in man, and about 10 % in the mouse. It consists of a linear double helix, and differs markedly from the rest of the nuclear DNA with respect to base composition and density. It contains more 5-methylcytosine than nuclear DNA. The DNA of the nucleolus is S.DNA, containing the cistrons for ribosomal RNA. With the aid of cytological hybridization experiments, it has also been demonstrated in various parts of the chromosomes, where its function is unknown. S.DNA often shows a high level of redundancy: mouse S.DNA contains about 10^6 copies each of 150–300 nucleotide pairs; in guinea pig S.DNA, there are about 10^7 copies of the 6 base sequence, GGGAAT. This high level of redundancy results in a rapid rate of renaturation, following heat denaturation.

Extrachromosomal DNA is also considered to be S.DNA.

Saxitoxin: a neuromuscular blocking agent which prevents nerve transmission by blocking sodium pores in postsynaptic membranes. S. is produced by dinoflagellates of the genus *Gonyaulax,* found in "red tides". S. accumulates in shellfish that ingest the dinoflagellates, hence cases of poisoning from eating the Californian sea mussel *(Mytilus californianus)*, the Alaskan butterclam *(Saxidomus giganteus)* and the scallop.

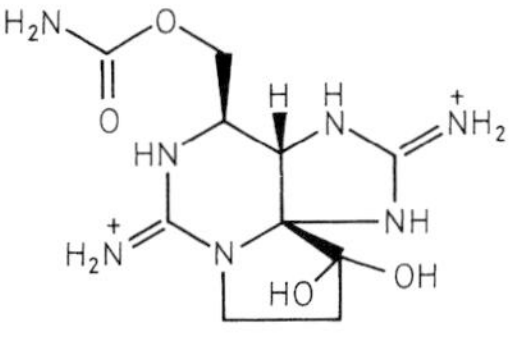

Saxitoxin

Schardinger enzyme: see Xanthine oxidase.

Scheie's syndrome: see Lysosomal storage diseases.

Schemochromes: see Structural colors.

Schiff's bases: see Pyridoxal phosphate.

Schottenol: 5α-stigma-7-ene-3β-ol, a widely distributed phytosterol (see Sterols), M_r 414.72, m.p. 151 °C, isolated, e.g. from the cactus *Lophocereus schottii.* S. is an essential dietary constituent for the insect *Drosophila pachea,* which lives on this plant.

Scillabiose: see Scillaren A.

Scillaren A, *glucoproscillaridin A, transvaalin:* a bufadienolide cardiac glyoside, M_r 692.78. m.p. 184–186 °C (prisms), 208–211 °C (leaflets), $[\alpha]_D^{23} - 71.9°$ (c = 1.011, methanol). The aglycon, *scillarenin* has M_r 384.52, m.p. 232–238 °C, $[\alpha]_D^{20} - 16.8°$ (c = 0.357, methanol), + 17.9° (c = 0.39, $CHCl_3$). The carbohydrate residue is the disaccharide, scillabiose (6-deoxy-4-*O*-β-D-glucopyranosyl-L-mannose; 4-*O*-β-D-glucopyranosyl-L-rhamnose) attached glycosidically at C3 of the aglycon. S.A. is the chief active component of *Scilla maritima* (squill), used since antiquity as a diuretic, cardiac stimulant and mouse poison.

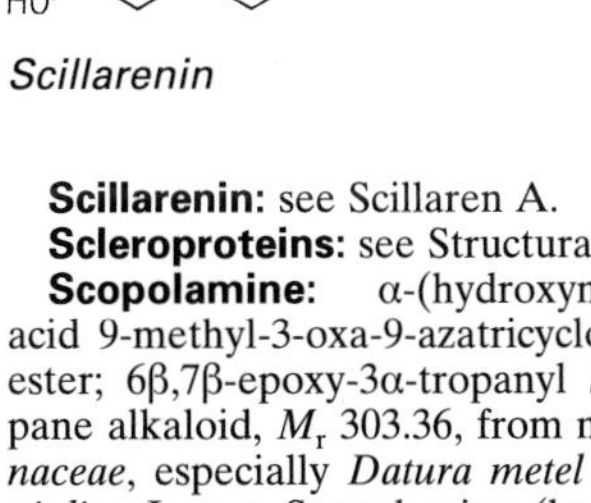

Scillarenin

Scillarenin: see Scillaren A.

Scleroproteins: see Structural proteins.

Scopolamine: α-(hydroxymethyl)benzeneacetic acid 9-methyl-3-oxa-9-azatricyclo $[3.3.1.0^{2,4}]$ non-7-yl ester; 6β,7β-epoxy-3α-tropanyl *S*-(−)-tropate, a tropane alkaloid, M_r 303.36, from members of the *Solanaceae*, especially *Datura metel* L. and *Scopola carniolica* Jacq. L-Scopolamine (hyoscine), a viscous liquid, $[\alpha]_D^{20} - 28°$ (water), −18° (ethanol), is soluble in water at 15 °C, forming a crystalline monohydrate, m.p. 59 °C. DL-Scopolamine (atroscine), forms an efflorescent hydrate, m.p. 55–57 °C (also reported as 82–83 °C). S. has similar pharmacological activity to that of hyoscamine, but has comparatively less activity on the peripheral nervous system. It is a highly toxic, anticholinergic agent. The hydrobromide has been used to sedate mental patients, as a preanesthetic, and to control motion sickness. For formula and biosynthesis, see Tropane alkaloids.

Scopoletin: see Coumarin.

Scorpamines: see Scorpion venoms.

Scorpion venoms: secretions of the scorpion stinging apparatus. Active principles of S.v. are the neurotoxic scorpamines, which are similar to cobra toxins (see Snake venoms) with respect to M_r (6,800–7,200, 4 disulfide bridges, 63–64 amino acid residues of known sequence), amino acid composition (high contents of basic and aromatic amino acids) and activity (both peripheral and central nervous system). The toxin from the North African scorpion, *Androctonus australia,* is one of the most potent known nerve poisons.

Scotophobin: Ser-Asp-Asn-Gln-Gln-Gly-Lys-Ser-Ala-Gln-Gln-Gly-Gly-Tyr-NH_2, a pentadecapeptide isolated from the brains of rats trained to avoid the dark. It induces dark avoidance in untrained animals. Isolation, biological activity and proof of structure of S. have been the subject of controversy: G. Ungar et al. "Isolation, Identification and Synthesis of a Specific-Behaviour-Inducing Brain Peptide" *Nature* **238** (1972) 198–202; W.M. Stewart "Comments on the Chemistry of Scotophobin" *Nature* **238** (1972) 202–210.

Scurvy: Vitamin C deficiency disease. See Vitamins (ascorbic acid).

Scyllitol: see Cyclitols.

Scymnol: see Bile alcohols.

SD 8339: see Cytokinins.

S-D sequence: see Recombinant DNA technology.

Second messenger: see Hormones.

Secondary metabolism: see Secondary metabolites.

Secondary metabolites: Substances such as pigments, alkaloids, antibiotics, terpenes, etc., which occur only in certain organisms, organs, tissues or cells, and are the products of secondary metabolism. They are thus distinct from primary metabolites (products of primary or general metabolism), which are concerned in the energy metabolism, growth and structure of all, or at least very large groups of, organisms, e.g. glycolysis and TCA cycle intermediates, amino acids and their biosynthetic precursors, proteins, purines and pyrimidine bases, nucleosides, nucleotides, nucleic acids, sugars, polysaccharides, fatty acids, triacylglycerols, etc.

Many S.m. have no apparent biological function; some, however, have been exploited during the course of evolution (Table 2), and have even become fundamentally important to the life of their producing organisms (e.g. plant and animal hormones). Others have ecological importance, by acting as attractants (scents, colors), antifeedants and toxic defense or attack substances, like salamander alkaloids, cardiac glycoside toad poisons, and physiologically active compounds produced by insects, e.g. HCN, formic acid, *p*-cresol, *p*-benzoquinone.

Table 1. *Relationship of secondary metabolism to total metabolism*

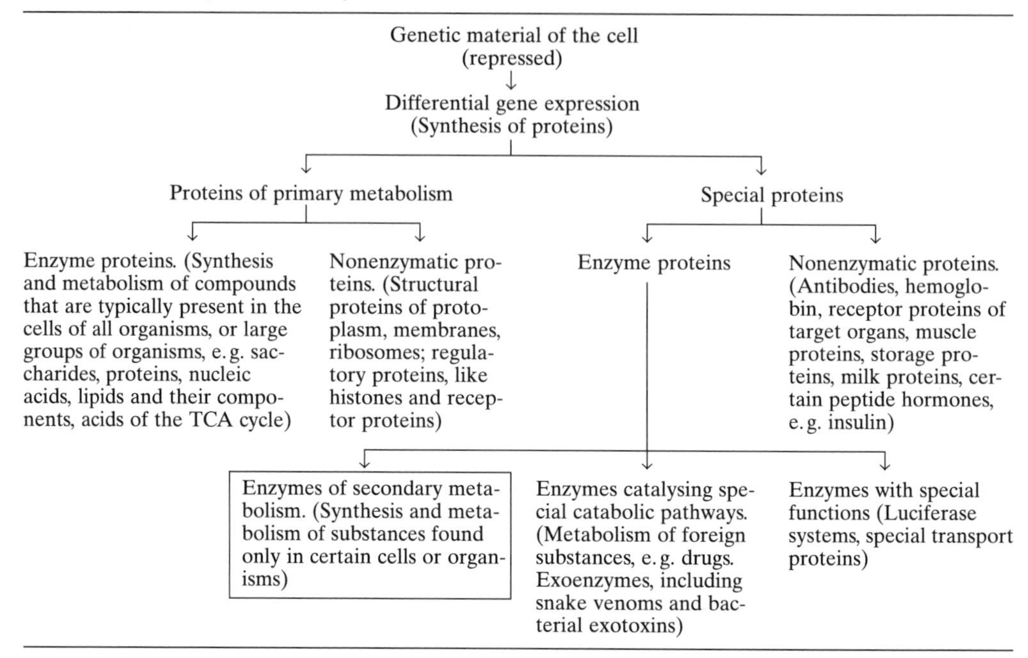

Despite the great chemical diversity of S. m., they are assembled from relatively few precursors, e. g. acetate, shikimate, isopentenyl pyrophosphate, etc. (Table 3), which usually occupy a key branchpoint position in primary or general metabolism. The S. m. of animals are not always synthesized de novo by the organism, but may be derived from the diet. The milkweed (*Asclepias curassavica*) produces several cardiac glycosides (e. g. calotropin) within its tissues as a defense against insect feeding. These substances are bitter and toxic, but certain insects, notably the caterpillar of the monarch butterfly, have become adapted to them and feed upon the plant. The glycosides are sequestered during feeding and subsequently passed on to the tissues of the adult butterfly. If a bird (commonly a blue jay) eats the butterfly, it is caused to vomit by the bitter cardiac glycosides; thereafter it avoids feeding on this butterfly, which it recognizes from its typical wing pattern. Not only the cardiac glycosides, but the butterfly wing pigments are S. m.; moreover, other butterflies, which contain no cardiac glycosides, have achieved protection against bird attack by evolving a wing pattern and coloration similar to that of the monarch butterfly. Thus S. m. may be intimately involved in complicated ecological interactions. In multicellular organisms, S. m. are produced by specific organs, tissues or cells, which contain the appropriate enzymes (see, e. g. Salamander alkaloids); they are usually formed only during certain periods of development or differentiation of the producing organism or cell. In microorganisms, S. m. (e. g. antibiotics, acetate-derived compounds) are usually synthesized at the end of the exponential growth phase (trophophase) or at the beginning of the stationary phase (idiophase), and their formation is usually repressed during rapid growth. This is probably due to catabolite represssion of enzyme synthesis during rapid growth on readily assimilated carbon sources. The risk of "metabolic suicide" is thus diminished, since growth and cell division are complete before toxic secondary products (e. g. antibiotics) are produced.

Table 2. *Biological functions of secondary natural products*

Effectors within the synthesizing cell, i. e. intracellular messengers.
Effectors of other cells of the same organism, i. e. intercellular messengers (plant and animal hormones, neuroendocrine transmitters).
Effectors of other organisms (blood pigments, flower scents, pheromones, antibiotics, insecticides, phytoalexins, toxins, antifeeding agents, sexual attractants).
Factors for the exploitation of specific ecological situations (chelating agents, e. g. siderochromes).
Storage forms of waste products from primary metabolism.

The majority of known S. m. are synthesized by plants. More than 5,000 plant alkaloids have been identified, compared with about 50 animal alkaloids. This difference may be related to excretory metabolism. Animals are able to remove from their bodies the endproducts and byproducts of metabolism, whereas plants employ "metabolic excretion", i. e. products are accumulated in vacuoles, cell walls and

Table 3. *Relationships between primary and secondary metabolites*

Primary metabolites	Secondary products
Sugars	Unusual sugars (amino, deoxy and methyl sugars, and sugars with branched chains). Reduction products (sugar alcohols, cyclitols, streptidine). Oxidation products (uronic acids, aldonic acids, sugar dicarboxylic acids).
Acetate/malonate	Fatty acid derivatives (*n*-alkanes, acetylene derivatives). Polyketides (anthracene derivatives, tetracyclines, griseofulvin, phenolcarboxylic acids from fungi and lichens, pyridine derivatives).
Isopentenyl pyrophosphate	Hemiterpenes (isoprene). Monoterpenes (iridoid components of volatile oils). Sesquiterpenes (bitter principles, components of volatile oils). Diterpenes (components of resins, gibberellins, phytol). Triterpenes (squalene, sterols, etc.). Tetraterpenes (carotenoids, xanthophylls). Polyterpenes (caoutchouc, gutta percha).
Propionate	Methyl fatty acids. Macrolide antibiotics.
Acids of the TCA and glyoxylate cycles	Alkyl citric acids.
Shikimate pathway of aromatic biosynthesis	Naphthoquinones, anthraquinones, quinoline and quinazoline alkaloids, phenazines.
Amino acids	Amines, methylated amino acids, betaines, cyanogenic glycosides, mustard oils, alkaloids, glycine conjugates, glutamine and ornithine, *S*-alkylcysteine derivatives, dioxopiperazines, peptides (penicillins), hydroxamic acids.
Phenylpropane amino acids	Cinnamic acid, coumarins, lignin, lignans, flavan derivatives, stilbenes, phenolcarboxylic acids, phenols, components of volatile oils.
Porphyrins	Bile pigments.
Purines	Methylated purines, purine antibiotics, pteridines, benzopteridines, pyrrolopyrimidines.

(lipophilic substances) in special excretory cells or spaces (volatile oil cells, resin ducts, etc.). Sites of synthesis and accumulation of S.m. are therefore often different. In animals, S.m. are usually stored in special organs (e.g. the salamander alkaloids and toad poisons are stored in skin glands). They may also be present in body fluids (e.g. cantharidine in insects is present in the lymph), or in hair and skin (e.g. melanins).

"Metabolic excretion" is only one of several theories put forward to explain the synthesis of S.m., especially those with no known function, like plant alkaloids and acetate-derived compounds (see Polyketides) from microorganisms. Any explanation must embrace the fact that so-called "degenerate" mutants can be isolated which do not produce the S.m., yet grow and divide as usual: nonalkaloid-producing strains of alkaloid plants are known (e.g. nicotineless tobacco plants), and pigment, antibiotic and polyketide production by microorganisms shows great strain variation with no apparent effect on viability. More recent explanations, especially relevant to microorganisms, place emphasis on the production of S.m. rather than S.m. themselves, i.e. it is thought that the operation of secondary metabolism is advantageous by keeping metabolism running at a low rate, rather than closing it down completely after growth ceases. The unbalanced growth hypothesis put forward by Bu'lock (see H.B.Woodruff "The Physiology of Antibiotic Production: The Role of the Producing Organisms" in *Biochemical Studies of Antimicrobial Drugs, 16th Symposium of the Society For General Microbiology,* 1966, pp.22–46 Cambridge University Presss) suggests that mechanisms controlling primary metabolism are not adequate to prevent overproduction of some compounds when balanced growth ceases. Since these compounds may be toxic to the cell, secondary metabolism diverts synthesis to the production of harmless products which are excreted. According to this theory, secondary metabolism should increase long-term viability; there is some evidence that *Pseudomonas aeruginosa* loses viability when grown under conditions that prevent secondary metabolism.

Secondary structure: see Proteins.

Secosteroids: see Steroids.

Secretin: a polypeptide hormone, M_r 3,050, containing 27 amino acid residues. S. shows considerable sequence homology with Glucagon (see), Vasoactive intestinal peptide (see) (VIP) and Gastric inhibitory peptide (see) (GIP), and it is thought that these four hormones evolved from a common ancestral protein by a process of gene multiplication.

Production of S. by the duodenal mucosa is stimulated by the acidic pH of the chyle, by large peptides ("secretogogues") from the incomplete hydrolysis of

					5					10					
Secretin:	His-	Ser-	Asp-	Gly-	Thr-	Phe-	Thr-	Ser-	Glu-	Leu-	Ser-	Arg-	Leu-	Arg-	
Glucagon:	His-	Ser-	Gln-	Gly-	Thr-	Phe-	Thr-	Ser-	Asp-	Tyr-	Thr-	Lys-	Tyr-	Leu-	
VIP:	His-	Ser-	Asp-	Ala-	Val-	Phe-	Thr-	Asp-	Asn-	Tyr-	Thr-	Arg-	Leu-	Arg-	
GIP:	Tyr-	Ala-	Glu-	Gly-	Thr-	Phe-	Ile-	Ser-	Asp-	Tyr-	Ser-	Ile-	Ala-	Met-	

	15					29					25				
Secretin:	Asp-	Ser	Ala-	Arg-	Leu-	Gln-	Arg-	Leu-	Leu-	Gln-	Gly-	Leu-	Val-	NH_2	
Glucagon:	Asp-	Ser-	Arg-	Arg-	Ala-	Gln-	Asp-	Phe-	Val-	Gln-	Trp-	Leu-	Met-	Asp-	Thr
VIP:	Lys-	Gln-	Met-	Ala-	Val-	Lys-	Lys-	Tyr-	Leu-	Asn-	Ser-	Ile-	Leu-	Asn-	NH_2
GIP:	Asp-	Lys-	Ile-	Arg-	Gln-	Gln-	Asp-	Phe-	Val-	Asn-	Trp-	Leu-	Leu-	Ala-	Gln- Gln

Amino acid sequences of secretin, glucagon, VIP and GIP. Amide groups are present at the *C*-termini of secretin and VIP. The sequence of GIP is shown incomplete; it continues: -Lys-Gly-Lys-Lys-Ser-Asp-Trp-Lys-His-Asn-Ile-Thr-Gln (total 43 residues).

dietary protein, and by fat and alcohol, i. e. it is generally stimulated by food with the exception of carbohydrate. S. is secreted into the blood. It stimulates formation and secretion of $NaHCO_3$-rich pancreatic juice and $NaHCO_3$-rich bile, and inhibits HCl production by the stomach. [V. Mutt et al. *Eur. J. Biochem.* **15** (1970) 513–519; M. Bodansky et al. *Proc. Natl. Acad. Sci USA* **70** (1973) 382–384].

Secretory enzymes: see Secretory proteins.

Secretory proteins: proteins synthesized intracellularly, often in specialized secretory organs (e. g. digestive glands), then secreted. S. p. that are also enzymes are called *secretory enzymes.* In cells actively engaged in synthesizing S. p., the rough endoplasmic reticulum (RER) is highly developed. As a generalization, it can be stated that proteins retained by the cell are synthesized on polysomes that are not attached to membranes, whereas S. p. are synthesized on polysomes bound to the endoplasmic reticulum. Our concept of S. p. synthesis and secretion are due largely to the work of Palade (see G. Palade *Science* **189** (1975) 347–358); his studies on the synthesis and secretion of digestive enzymes by the guinea pig pancreatic exocrine cell provided a conceptual model for this process in all secretory cells. During synthesis on the RER, S. p. pass throught the reticular membrane into the lumen; they then pass to the Golgi apparatus, where they become condensed into secretory granules, which leave the cell by exocytosis. In other cells (e. g. hepatocytes), the smooth endoplasmic reticulum (SER) may function in the transfer of S. p. from the lumen of the RER to the Golgi apparatus. Synthesis and secretion into the lumen of the RER depend on the formation and ultimate removal of a signal peptide sequence (see Signal hypothesis). Other modifications, such as glycosylation (see Post-translational modification of proteins) are also associated with the intracellular production of S. p. (many S. p. are glycoproteins). Most proteolytic S. p. are synthesized as inactive precursors (e. g. trypsinogen), thus avoiding self digestion (autolysis) by the producing cell. S. p. are secreted into the blood (e. g. serum albumin, serum cholinesterase and blood coagulation enzymes are synthesized in the liver), or into ducts from their producing glands (e. g. salivary and pancreatic amylases).

Sedamine: see Sedum alkaloids.

Sedimentation coefficient, *sedimentation constant:* a measure of the rate of sedimentation used in the determination of M_r of macromolecules by ultracentrifugation. The sedimentation coefficient (s) is equal to the rate of sedimentation of a macromolecule per unit centrifugal field; specifically,

$s = \frac{dx/dt}{\omega^2 \cdot FEx}$, where s is the sedimentation coefficient,

ω is the angular velocity of the centrifuge rotor (radians/sec), x is the distance from the center of rotation, dx/dt is the velocity of sedimentation. The sedimentation coefficient has the dimensions of time per unit force, and usually lies between 1×10^{-13} and 200×10^{-13}; the factor 1×10^{-13} is called the Svedberg unit (S), i. e. $1S = 10^{-13}s$. Sedimentation is monitored in the centrifuge cell by schlieren optics (Philpot-Svensson method), or by UV-absorption. Ideally, sedimentation constants are determined at a number of different macromolecule concentrations and the s-values are extrapolated to zero concentration, where the activity coefficient becomes unity. In addition, sedimentation coefficients are corrected to a standard state with respect to solvent viscosity, which is taken as that of water at 20 °C; this gives the *standard sedimentation coefficient,* or $S°_{20w}$-value. $S°_{20w}$ values for most proteins and nucleic acids lie between 4 and 40S (Svedberg units), for ribosomes and ribosomal subunits between 30 and 80S and for polysomes above 100S.

D-Sedoheptulose, *D-altro-2-heptulose:* a monosaccharide from *Sedum* (stonecrop), M_r 210.19, m. p. (monohydrate) 102 °C. The 7-phosphate is an intermediate of carbohydrate metabolism (see, e. g. Pentose phosphate cycle); aldolase reactions of sedoheptulose 7-phosphate give rise to D-erythrose 4-phosphate, which is a precursor in Aromatic biosynthesis (see).

```
      CH2OH
      |
      CO
      |
HO—C—H
      |
 H—C—OH
      |
 H—C—OH
      |
 H—C—OH
      |
      CH2OH
```

D-Sedoheptulose

Sedum alkaloids: a group of piperidine alkaloids from *Sedum* spp. They are 2- or 2,6-substituted piperidine derivatives, similar in structure and biosynthesis to the Punica and Lobelia alkaloids. Chief representative is *sedamine:* *N*-methyl-2-(β-hydroxy-β-phenylethyl)-piperidine, M_r 219; the L-form has m.p. 61–62 °C, $[\alpha]_D$ –82 ° (methanol).

Seed germination test: see Gibberellins.

Selectins: see Cell adhesion molecules.

Selenium, *Se:* an element toxic in large quantities, but an essential micronutrient for mammals, birds, many bacteria, probably fish and other animals. A requirement by higher plants is uncertain. Se is an essential component of the enzyme glutathione peroxidase, which is important in the protection of red cell membranes and other tissues from damage by peroxides: 2GSH + H_2O_2 (or R–OOH) → GSSG + $2H_2O$ (or H_2O + R–OH).

Normal sheep muscle contains a low M_r selenoprotein of unknown function, which is absent from the muscle of Se-deficient sheep suffering from dystrophic white muscle disease.

Residues of the selenoamino acid, selenocysteine, are present in bacterial formate dehydrogenase (*E. coli*), and glycine reductase (*Clostridium*). [T.C. Stadtman *Advances in Enzymology* **48** (1979) 1–28]

Selenoamino acids: amino acids containing selenium (Se) in place of sulfur (S), e.g. Se-methylselenocysteine. They are formed in plants growing on Se-rich soils. See Selenium.

Self-splicing RNA: see Intron.

Semidehydroascorbate: a free radical formed from ascorbate when the latter serves as a reducing agent. The free radical has a highly unstable electronic configuration, and it is able to act as an oxidant in the NADH system. It is also re-reduced to ascorbate by cytochrome b_{561} (see Ascorbate shuttle). Formation of the free radical can be demonstrated by EPR spetroscopy [I. Yamazaki & L. H. Piette *Biochim. Biophys. Acta* **50** (1961) 62–69] and by a scavenger method involving one electron transfer to oxidized cytochrome *c* [I. Yamazaki *J. Biol. Chem.* **237** (1962) 224–229]. In the absence of an efficient trap, two free radicals dismute to form one molecule of ascorbate and one of dehydroascorbate.

Senecio alkaloids: see Pyrrolizidine alkaloids.

Sense strand: an alternative name for the Coding strand (see) of double-stranded DNA, which, by convention, has the same nucleotide sequence as that of the RNA transcript (e.g. mRNA) derived from that double-stranded DNA (save that T is in the place of U). See Nomenclatural conventions concerned with gene transcription.

Sephadex: a trade name for a series of polydextrans used in gel filtration chromatography. See Dextrans, Proteins.

Sequence: see Proteins.

Sequence polymers: synthetic amino acid polymers, consisting of multiple repeats of one short sequence. In contrast to Polyamino acids (see), S.p. contain more than one type of amino acid residue. They are prepared by the self condensation of activated peptides (di to hepta). *p*-Nitrophenyl esters (see Peptides) and other activated esters can be used; ring closure is avoided by using high concentrations of reactants in polar solvents in the presence of organic bases. The average M_r of the resulting S.p. are considerably lower than those of the synthetic polyamino acids. They are used to investigate molecular aspects of antigenicity and antibody specificity, as model enzymes, and as models in the study of protein structure.

Sequential mechanism: see Cleland short notation.

Sequential therapy: see Ovulation inhibitors.

Ser: abb. for L-serine.

L-Serine, *Ser:* L-α-amino-β-hydroxypropionic acid, $HOCH_2$-$CH(NH_2)$-COOH, a proteogenic, glucogenic amino acid, M_r 105.1, m.p. 223–228 °C, $[\alpha]^{25}_D$ – 7.5 ° (*c* = 2, water), + 15.1 ° (c = 2 in 5 M HCl). Ser is a major component of silk fibroin. In phosphoproteins, phosphate is esterified chiefly with the hydroxyl groups of Ser residues. During acid hydrolysis of proteins, a large proportion of Ser is destroyed. It is converted quantitatively into formaldehyde by periodate oxidation. Metabolically, Ser and glycine are interconvertible by the action of tetrahydrofolate-5,10-hydroxymethyltransferase; in this reaction, the β-C atom of Ser is transferred as an active hydroxymethyl group, which is a very important metabolic source of one carbon units. L-Serine dehydratase (a pyridoxal phosphate enzyme, EC 4.2.1.13) catalyses the conversion of Ser into pyruvate and ammonia. Ser is synthesized from glycine or from 3-phosphoglycerate (a glycolytic intermediate). In liver and in *E. coli,* 3-phosphoglycerate is dehydrogenated (NAD^+-dependent) to phosphohydroxypyruvate, which is transaminated to 3-phosphoserine. The latter is dephosphorylated by a specific phosphatase. In plants 3-phosphoglycerate (a photosynthetic product) is dephosphorylated to glyceric acid, which is dehydrogenated to hydroxypyruvate; the latter transaminates with L-alanine, forming Ser and pyruvate. Transsulfuration of Ser with L-homocysteine produces L-cysteine.

Serine cephalins: see Membrane lipids.

Serine hydrolases: hydrolases which have a catalytically active serine residue in their active center, e.g. trypsin, chymotrypsin A, B and C, thrombin and B-type carboxylic acid esterases. See Serine proteases.

Serine hydroxymethyltransferase: see Active one-carbon units.

Serine phosphoglycerides: see Membrane lipids.

Serine proteases: a group of well studied animal and bacterial endopeptidases (see Proteases) which have a similar action mechanism, and a catalytically active serine residue in their active centers (serine residue 195 in chymotrypsin). In all S.p., catalysis involves formation of an ester between the hydroxyl group of the catalytically active serine and the carboxyl group of the cleaved peptide bond (acyl-enzyme intermediate); this is hydrolysed in the deacylation stage of the reaction, restoring the free hydroxyl group of the serine and releasing the cleavage peptide. The active serine residue is selectively and irreversibly acylated by organic phosphate esters, like diisopropylfluorophosphate (DFP), or phenylmethanesulfonylfluoride (PMSF), which therefore inhibit S.p. Trypsin, chymotrypsin A, B and C, pancreatic elastase, invertebrate trypsins and chymotrypsins, thrombin, plasmin and kallikrein of the blood, the elastase-like α-lytic protease from Myxobacter and the trypsin-like S.p. from *Streptomyces griseus* are all thought to be homologous proteins, derived from a common ancestral protein by gene multiplication.

Subtilisin (see) from *Bacillus subtilis* is a S.p.; it resembles chymotrypsin in the hydrogen bonding of the charge-relay system, but is otherwise structurally dissimilar; it is thus an example of convergent evolution of a catalytic center in two different groups of proteins.

See Homologous proteins (Table).

Serine sulfhydrase: see Sulfate assimilation.

Serotonin: 5-hydroxytryptamine, M_r 176.2, a plant and animal hormone. It is produced by hydroxylation of L-tryptophan to 5-hydroxytryptophan, followed by decarboxylation. The synthesis occurs in the central nervous system, lung, spleen and argentaffine "light" cells of the intestinal mucosa. S. is stored in thrombocytes and mast cells of the blood. It acts as a Neurotransmitter (see), stimulates peristalsis of the intestine, and causes a dose-dependent constriction of smooth muscle. It stimulates the release from arterial endothelium of a dilator substance which counteracts its primary constricting effect [T.M. Cocks & J.A. Angus *Nature* **305** (1983) 627–630]. S. is a precursor of the hormone Melatonin (see). It is inactivated and degraded by monoamine oxidases and aldehyde oxidases to 5-hydroxy-indoleacetic acid.

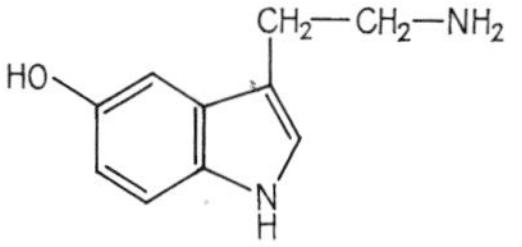

Serotonin

Serratomolide: a cyclic depsipeptide produced by *Serratia marcescens*. Chemically, it is the cyclic dimer of serrataminic acid (D-β-hydroxydecanoyl-L-serine).

Serum albumin: see Albumins.

Serum proteins: see Plasma proteins.

Sesquiterpenes: aliphatic, mono-, di- or tricyclic terpenes, formed from three isoprene units ($C_{15}H_{24}$). About 100 structural types are known, and about 1,000 natural representatives, forming the largest class of terpenes. Most are found in the volatile oils of plants. Little is known of the physiological significance of S.; some compounds have an ecological role (Table). Some are isolated for use in perfumery.

Some sesquiterpenes and their functions

Function	Sesquiterpenes
Juvenile hormones	Juvabione, farnesyl derivatives
Phytohormones	Abscisic acid
Plant sex hormones	Sirenin
Pheromones	Farnesol
Antibiotics	Trichothecin
Proazulenes	Guaiol
Alkaloids	Nupharidine
Scents	Santalols, cedrenes
Bitter principles	Cnicin
Phytoalexins	Ipomeamarone

S. are biosynthesized from farnesylpyrophosphate (see Terpenes). Acyclic S., e.g. farnesol, are formed by hydrolytic removal of the pyrophosphate group. The various types of cyclic S. are formed by elimination of the pyrophosphate residue to form an unstable cation, which stabilizes by loss of a proton.

Sesterterpenes: terpenes formed from five isoprene units ($C_{25}H_{40}$). They have a tricyclic skeleton, and have been isolated from insect secretions and lower fungi. See Ophiobolanes.

Seven-spanning receptors, ***seven-transmembrane-domain receptors:*** receptors in the plasma

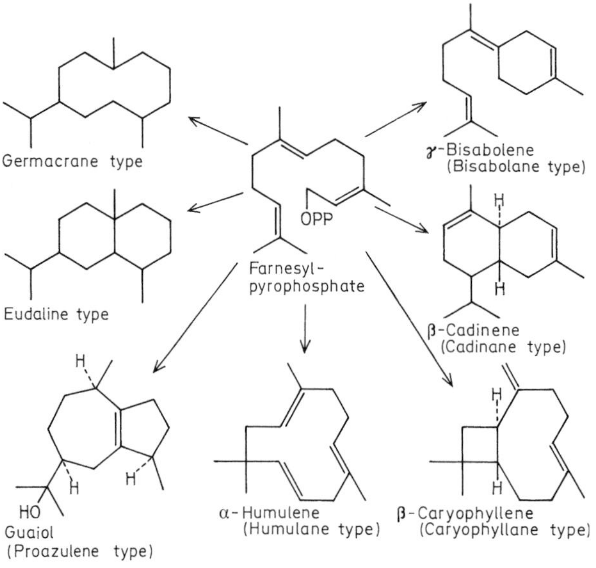

Some important sesquiterpene structural types

membranes of mammalian and other eukaryotic cells which, when bound to their specific ligand, activate a signal-transducing G-protein (see GTPases). Their structure is reminiscent of that of bacteriorhodopsin (see) in that it is composed of seven transmembrane α-helixes, each composed of 20–25 largely hydrophobic amino acids, labeled H_1-H_7 which are linked together on the cytosolic side of the membrane by polypeptide loops labeled C_1 (H_1–H_2), C_2 (H_3–H_4) & C_3 (H_5–H_6) and on the exterior side by loops labeled E_2 (H_2–H_3), E_3 (H_4–H_5) & E_4 (H_6–H_7); additionally E_1 is an exterior polypeptide chain extending from H_1 to the *N*-terminus and C_4 is a cytosolic chain extending from H_7 to the *C*-terminus. Chain C_4 and loop C_3, which is much longer than the other cytosolic loops, are believed to contribute to the G-protein binding site. The location of the ligand binding site, which must be accessible from the exterior face of the receptor, depends on the identity of the ligand. Judging from the way the β-adrenergic agonist, isoproterenol, binds to the β_2-adrenergic receptor, it is believed that catecholamine (see) hormones make the following binding interactions: (i) the H_3N^+ forms an ionic bond with the COO^- of Asp_{113} in H_3, (ii) the aromatic ring engages hydrophobically with that of Phe_{290} of H_6, and (iii) the two OHs form hydrogen bonds with the OHs of Ser_{204} and Ser_{207}. It is further believed that the COO^- of Asp_{113} is involved in the binding of other small hormones with an amino group, such as histamine (see), serotonin (see) and acetylcholine (see). Large protein hormones like luteinizing hormone (LH, 26 kDa) are thought to bind to E_1 which is very much longer (333 amino acids for the LH receptor) in the receptors for proteins than in the receptors for small ligands such as the catecholamines and peptide hormones. In fact the latter difference is used to divide the class of seven-spanning receptors into two subclasses, namely (i) those with a short E_1, which include the receptors for adrenalin (α_1, α_2, β_1 & β_2), serotonin, acetylcholine (muscarin), angiotensin, bradykinin and bombesin, and (ii) those with a long E_1, which include the receptors for LH and possibly follicle stimulating hormone and thyroid stimulating hormone.

Sexual attractants: natural products involved in sexual interaction. The S.a. of insects (see Pheromones) are particularly numerous and well studied examples of animal S.a. They are usually produced by the sexually mature female in order to attract and predispose the male to copulation.

Plant S.a. are called gamones, or plant sex hormones. They occur when at least one of the gametes involved in fertilization has a free existence, i.e. in many algae, lower fungi, mosses and ferns. The gamones that have been investigated, e.g. sirenin and ectocarpene, are produced by the female gametes and act as chemotactic agents to attract the male gametes. In contrast to the pheromones, gamones only influence interaction of the gametes (i.e. they act at the cellular level) and do not affect the behavior of the whole organism.

SF: acronym of Sulfation factor (see Somatomedin).

SH domains: see Src homology domains.

Shellac: see Resins.

Shemin cycle: see Succinate-glycine cycle.

SH-enzymes: see Thiol enzymes.

Shikimic acid: see Aromatic biosynthesis.

Shikonin: see Naphthoquinones (Table).

Showdomycin: 2-β-D-ribofuranosylmaleinimide, a C-substituted Nucleoside antibiotic (see) from *Streptomyces showdoensis*, structurally related to uridine and pseudouridine. M.p. 153 °C, $[\alpha]_D^{23}$ + 50 ° (*c* = 1, water). It selectively inhibits enzymes of uridine and orotic acid metabolism; the maleinimide moiety reacts with sulfhydryl groups of the affected enzymes. S. is especially active against *Streptococcus haemolyticus*.

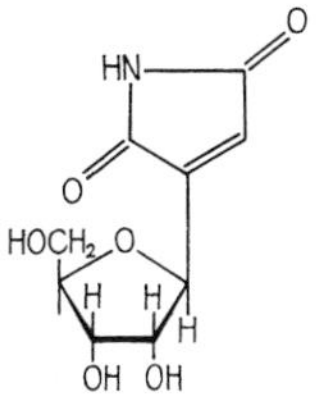

Showdomycin

Sialic acids: see Neuraminic acid.

Sialic acid storage disease: see Lysosomal storage diseases.

Sickle cell hemoglobin, *HbS*: one of the most frequently occurring abnormal hemoglobins, especially in negroids. As a result of a point mutation, the glutamic acid residue at position 6 in the β-chain (normal hemoglobin) is replaced by valine. The α-chain is normal ($\alpha_2\beta_2^{6Glu \rightarrow Val}$). There are no marked differences in the conformations of HbS and normal hemoglobin. DeoxyHbS undergoes self association and forms a liquid crystalline phase which distorts the erythrocytes into a sickle shape. In this phase, deoxy-HbS monomers are in eqilibrium with polymers. The polymers consist of 6–8 helical deoxy-HbS strands lying side by side to form a tubular shape of diameter 140–148 Å. Deformation of the erythrocytes into sickle shapes ("sickling") leads to their aggregation and to a decrease in blood circulation. Clinical symptoms are anemia and acute ischemia, tissue infarction and chronic failure of organ function. In heterozygotic carriers, the condition known as sickle cell trait may be without serious clinical effects; it may even go unrecognized, but sickling occurs when the individual is subjected to abnormally low oxygen tension. Negroid trainee airline or airforce pilots must therefore be screened for heterozygotic sickling. Homozygotes die from extensive hemolytic anemia.

Sideramines: see Siderochromes.

Siderochromes: iron-containing, red-brown water soluble secondary metabolites produced by microorganisms. S. include a series of antibiotics, the *sideromycins* (albomycin ferrimycin, danomycin, etc.) and a class of compounds with growth factor properties for certain microorganisms, the *sideramines* (ferrichrome, coprogen, ferrioxamine, ferrichrysin, ferrirubin, etc.). The sideramines competitively inhibit the antibiotic activity of the sideromycins.

S. are specific ligands for iron. Their synthesis and secretion by microorganisms is increased under conditions of iron deficiency. S. of the catechol type are produced by anaerobic microorganisms, whereas the hydroxamate type are produced by aerobic microorganism. They typically contain a central iron(III) trihydroxamate complex (Fig.), and specifically bind Fe^{3+}.

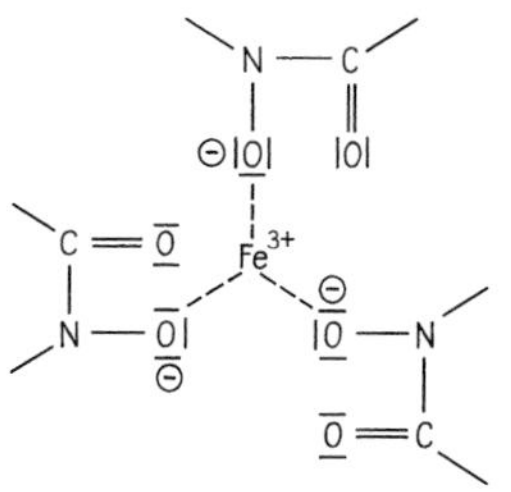

Central iron(III)-trihydroxamate complex of a hydroxamate-type siderochrome

As metal chelating agents, sideramines fulfil two functions: they transport iron into the microbial cell and/or make the chelated iron available for heme synthesis. In analogy with animal transferrin, the iron of microbial ferrichrome is transferred enzymatically into the porphyrin molecule during heme synthesis.

Other microbial chelating agents, e.g. mycobactin, aspergillic acid and schizokinen, are sometimes classified with the sideramines.

Sideromycins: see Siderochromes.

Siderphilins: nonheme, iron-binding, single chain animal glycoproteins, M_r about 77,000, carbohydrate content about 6%. On the basis of their occurrence, they are classified as transferrin (vertebrate blood), lactoferrin (mammalian milk and other body secretions) and conalbumin or ovotransferrin (avian blood and avian egg white). S. differ in their physical, chemical and immunological properties, but each possesses two binding sites for iron(III). The iron is bound less firmly than in Ferritin (see). Transferrin (a β-globulin) is the best studied S.; 15 genetic variants are known, the most common being transferrin A, B and C. Its main function is the transport of absorbed dietary iron(III) to iron depots (liver and spleen), or to the reticulocytes and their precursors in the bone marrow. Transferrin becomes bound to surface receptors of the reticulocytes, enters the cell and releases its iron, then is returned to the blood as iron-free apotransferrin. Of the total 7–15 g transferrin in the body, only one third is complexed with iron(III). By virtue of their ability to chelate iron, all S. inhibit bacterial growth, an important property in avian eggs and in milk. [W.A. Jeffries et al. *Trends Cell Biol.* **6** (1996) 223–228]

SIF: acronym of Somatotropin release inhibiting factor. See Releasing hormones.

Sigma (σ) factor: see Ribonucleic acid, Initiation factors.

Signal hypothesis: A mechanism proposed by Blobel for the segregation of secretory proteins during their biosynthesis. It has since been found applicable to most proteins which are either inserted into membranes or transported across them, including organelle proteins which are synthesized in the cytosol. Such proteins are transcribed with a signal or leader peptide of 15 to 30 amino acid residues. About 70 amino acid residues must be incorporated into a peptide before it can protrude from the ribosome; at this point the translation of secretory proteins is arrested until the polypeptide chain binds a soluble signal recognition particle (SRP). The complex of SRP and nascent peptide subsequently binds to an SRP receptor or "docking protein" in the appropriate membrane: endoplasmic reticulum, mitochondrial membrane, bacterial plasma membrane, etc. The signal peptide is then released from the complex of SRP and docking protein, and translation of the polypeptide resumes (dissociation of signal sequence, SRP and docking protein is accompanied by the binding and subsequent hydrolysis of GTP to GDP). The polypeptide is extruded through or into the membrane. In many cases, the signal peptide is cleaved on the other side of the membrane (the intralumenal space, in the case of the endoplasmic reticulum) by a specific signal peptidase. However, this is not always the case; some membrane-spanning proteins retain their signal peptide sequences. The chains of some membrane-spanning proteins cross the membrane several times, these are supposed to contain internal signal sequences which promote spontaneous insertion into the lipid bilayer. It is also possible for membrane proteins to be translated completely before insertion into the membrane. [W.T. Wickner & H.F. Lodish *Science* **230** (1985) 400–407; R.Gilmore & G. Blobel *Cell* **42** (1985) 497–505]

Proteins with cleavable signal peptides are called *preproteins* prior to cleavage of the signal. Since many secretory proteins are synthesized as inactive *proproteins*, the corresponding precursors are called *preproproteins*.

Signal transducers and activators of transcription, *STATs*: a family of DNA-binding proteins. Biological effects of STATs range from antiviral responses to cell transformation, and they appear to be particularly involved in signaling pathways activated by cytokines. So far, 6 STATs and their corresponding genes have been identified in mammals, and 2 in *Drosophila*.

STATs display sequence similarities over virtually their entire length of some 700 amino acid residues, and maximal sequence alignment reveals marked homology in residues 600–700, a stretch which matches the SH2 domains of other proteins. Residues 500–600 show a distinct sequence similarity to SH3 domains (see Src homology domains). DNA binding site specificity is determined by the sequence of residues 400–500. All STATs have a single Tyr residue in the region of residue 700, which becomes phosphorylated during cytoplasmic activation of the protein. The resulting activated protein then binds to DNA in a sequence-specific manner. In the region of residue 727, some STATs (STAT1a, STAT3, STAT4) possess a Ser residue which can be phosphorylated (possibly by a mitogen-activated kinase), representing another level of regulation of STAT activity.

STATs are components of the JAK/STAT signaling pathway, which was identified by study of the transcriptional activator response to certain cytokines and growth factors. JAK proteins (janus kinases, a family of tyrosine kinases) are bound to the membrane-proximal domain of cytokine receptors. Cytokine binding induces receptor dimerization, which brings the associated JAKs sufficiently close to one another to permit activation by transphosphorylation. In turn, these activated JAKs phosphorylate a distal tyrosine on the receptor. The phosphotyrosyl residue of the receptor is then recognized by the SH2 domain of the STAT. A complex is therefore formed, in which the STAT is activated by phosphorylation of its strategic Tyr by the action of the JAKs. Activated STATs then undergo hetero- or homo-dimerization, and are translocated to the nucleus, where they activate gene transcription.

STATS were first recognized from studies on signaling pathways initiated by binding of interferon (IFN) by cells. Two members of the STAT family (STAT1 and STAT2) are activated by IFNα, whereas only STAT1 is activated by IFNγ. The role of the janus family of tyrosine kinases was established by the observation that cells lacking JAK1 are unable to respond to IFNα, and that mutant cells lacking either JAK1 or JAK2 are unresponsive to IFNγ. [C. Schindler & J.E. Darnell *Annu. Rev. Biochem.* **64** (1995) 621–651; J.N. Ihle & I.M. Kerr *Trends Genet.* **11** (1995) 69–74; J.N. Ihle *Cell* **84** (1996) 331–334; X.S. Hou et al. *Cell* **84** (1996) 411–418; R. Yan et al. *Cell* **84** (1996) 419–430; M.A. Meraz et al. *Cell* 84 (1996) 431–442; J.E. Durbin et al. *Cell* 84 (1996) 443–450]

SIH: acronym of Somatotropin release inhibiting hormone. See Releasing hormones.

Silicon, *Si:* an essential trace element in human nutrition [E.M. Carlisle *Science* **178** (1972) 619–612; E.M. Carlisle *Fed. Proc. Fed. Amer. Soc. Exp. Biol.* **32** (1973) 930]. Si is a cross-linking agent in connective tissue. It is thought that Si is bound via oxygen to the C-skeleton of mucopolysaccharides, thus linking parts of the same polysaccharide, or linking acidic mucopolysaccharides to proteins. Si may also serve a matrix or catalytic role in bone mineralization. High levels of Si (as SiO_2) are present in plants and diatoms (see Mineral elements).

Silk fibroin: see Keratins.

Silybin: a flavanolignan from *Silybum marianum* Gaert. (milk thistle). S. protects animals against poisoning by phalloidin (see Phallotoxins). Crystallographic studies of S. indicate that the spacing and alignment of the aromatic rings A and B (Fig.) are almost identical with those of the Phe residues 9 and 10 in Antamanide (see), which also protects against phalloidin poisoning. It is also suggested that the correct arrangement of these two aromatic rings is essential for attachment to a target receptor on the liver cell membrane, thus preventing entry of phalloidin into the cell. [H.L. Lotter, *Zeitschrift für Naturforschung* **39c** (1984) 535–542]

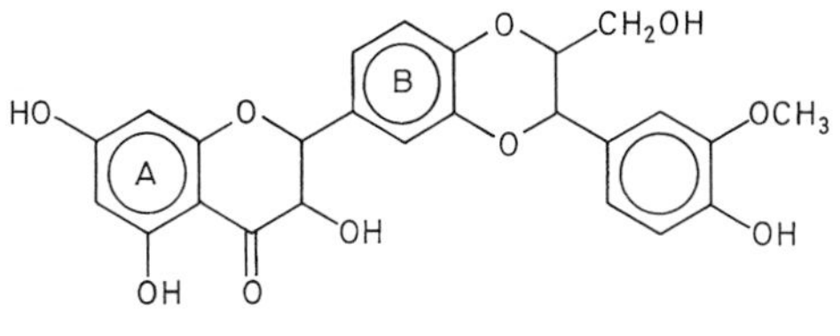

Silybin

Sinalbin: see Glucosinolate.

Sinapine, *sinapic acid choline ester:* 2{[3-(4-hydroxy-3,5-dimethoxyphenyl)-1-oxo-2-propenyl]oxy}-N,N,N-trimethylethanammonium hydroxide, an alkaloidal base occurring widely in the *Cruciferae,* first isolated in 1825 from black mustard seeds. It serves as the cation of sinalbin (see Glucosinolate). The component alcohol, choline, is a common plant metabolite involved in phospholipid synthesis and transmethylation; sinapic acid is also a widely distributed plant constituent implicated in Lignin (see) biosynthesis. The combination of these two compounds to form S., however, appears to be a peculiarity of the *Cruciferae.* S. behaves as a storage compound; during germination of mustard seeds, it is hydrolysed to choline and sinapic acid, which are then further metabolized. Two weeks after germination S. is no longer detectable. The esterase responsible for S. hydrolysis has been purified 20-fold from white mustard seedlings. [A. Tzagoloff *Plant Physiology* **38** (1963) 207–213]

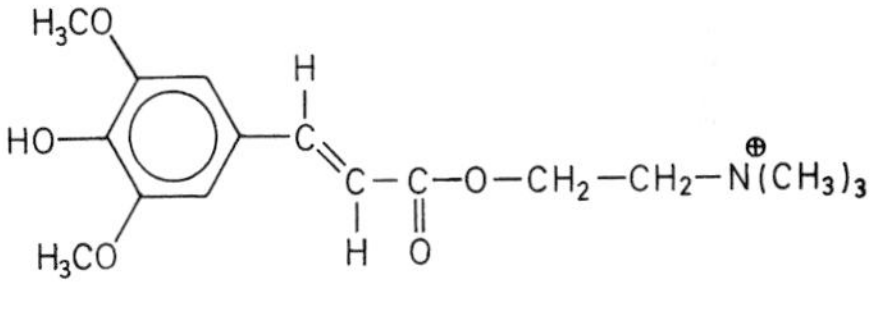

Sinapine

SINE(s): acronym of Short INterspersed DNA sequence Element(s) which constitute one of the two main classes of *intermediate repeat DNA sequences* (see C_0t) occurring in the mammalian genome, the other being LINEs (see). They consist of a single nucleotide sequence 130–200 bp long that occurs as a non-tandemly repeated element at hundreds of thousands of places in the genome; usually there is at least one per 5,000 bp. They make up ~ 5 % of the total human DNA. The most common SINEs in mammals constitute the *Alu* family, so-called because each contains a site for the restriction endonuclease (see) *Alu*I; they have a close sequence similarity to the 294-nucleotide 7SL RNA. SINEs are capable of moving (transposing) to new sites within the genome and therefore fall into the category of 'mobile DNA elements' (see) and the sub-subdivision of the latter known as 'non-viral retrotransposons' (see). They appear to be derived from RNA polymerase III-generated transcripts.

Single-strand break: a break in a double-stranded DNA molecule which involves only one of the two strands, so that the molecule remains together. One S.s.b. is required to initiate unwinding of the double helix during replication. S.s.b. are caused by endonucleases, physical conditions or chemicals. Such breaks are repaired by the polydeoxyribonucleotide synthetases (EC 6.5.1.1 and 6.5.1.2).

Single-substrate enzymes: enzymes which catalyse reactions involving only one substrate. They are usually isomerases or hydrolytic enzymes; in the latter case, the water involved in the reaction is regarded as a constant, and there is frequently no special enzyme-water complex formed.

Sinigrin: see Glucosinolate.

Sirenin: the first plant sexual attractant or gamone to be structurally elucidated. S. is a sesquiterpene, M_r 236. It occurs naturally in the L-form, $[\alpha]_D^{23}$ – 45° (c = 1.0, $CHCl_3$), but the DL-form is also biologically active. S. is produced by the female gametes of the fungus *Allomyces,* which lives in damp soils. Gametes swim from the mycelium, and S. acts as a chemotactic stimulus to attract the male to the female gametes. It is active at a concentration of 10^{-10} M.

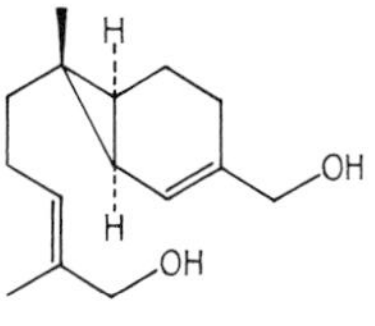

L(–)-Sirenin

Siroheme: the heme prosthetic group found in sulfite reductase of *E. coli* and nitrite reductase of green plants.

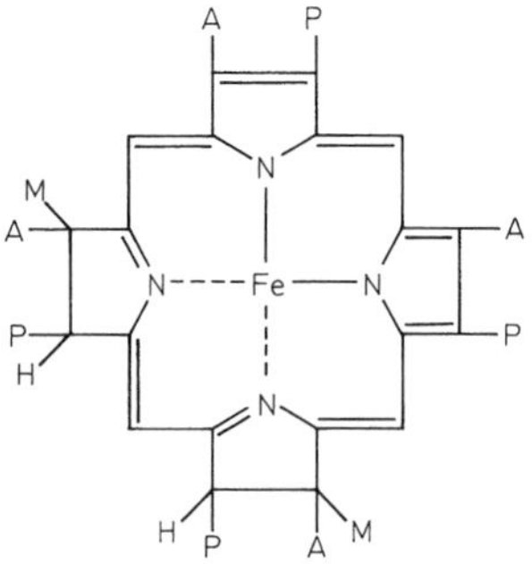

Siroheme. A = CH_2COOH. P = CH_2CH_2COOH. M = CH_3.

Site of labeling: see Isotope technique.

Sitosterol: stigmast-5-en-3β-ol, M_r 414.7: m.p. 140 °C: $[\alpha]_D$ –37 °($CHCl_3$), a sterol formerly known as β-sitosterol. It is widely distributed in higher plants where it commonly occurs with stigmasterol (see) and campesterol (see) in the plasma membrane and the endoplasmic reticulum. Sometimes it is also present as esters of fatty acids (e.g. palmitic, oleic, linoleic & α-linolenic acid), as β-*O*-glycosides or β-*O*-acylglycosides (in the latter a non-anomeric hydroxyl group of the sugar residue, commonly that on C6, is esterified to a fatty acid).

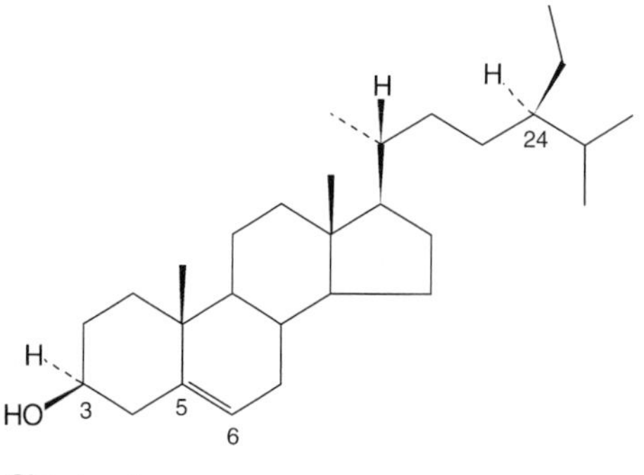

Sitosterol

(+)-Skyrin: see Emodin.

Slow reacting substance A: abb. for slow reacting substance of anaphylaxis. It is produced by sensitized cells as part of the immune response to antigens. See Leukotrienes.

Slow reacting substances: substances that cause smooth muscle to contract slowly in vitro. See Leukotrienes.

Sly syndrome: see Lysosomal storage diseases.

Small nuclear ribonucleoproteins: see U snRNP. Not all S.n.p. belong to the U class, which are the best characterized.

Snake venoms: a mixture of toxins produced in the venom glands (parotid gland, or salivary gland of the upper jaw) of venomous snakes (asps or hooded snakes, e.g. the cobra; sea snakes; vipers, e.g. puff adder, rattlesnake). They consist of highly toxic, antigenic polypeptides and proteins (which cause paralysis and death of the prey), and enzymes (which facilitate the spread of the toxins, and initiate digestion of undivided swallowed prey). The enzymes include hyaluronidase (promotes spread of toxins), ATPase and acetylcholine esterase (paralysis), phospholipases (hemolysis), proteinases and L-amino acid oxidases (tissue necrosis and blood clotting).

Important toxins are cobramine A and B from cobra toxin; and crotactine and crotamine from crotoxin, the toxin of the North American rattlesnake. The toxic proteins are classified according to their mode of action; cardiotoxins, neurotoxins and protease inhibitors (with inhibitory activity toward chymotrypsin and trypsin). *Cardiotoxins* (heart muscle poisons) cause an irreversible depolarization of the cell membranes of heart muscle and nerve cells. *Neurotoxins* (nerve poisons) show curare-like activity; they prevent neuromuscular transmission by blocking the receptors for the transmitters at the synapses of autonomic nerve endings and at the motor end plate of skeletal muscle. *Protease inhibitors* inhibit acetylcholine esterase and similar enzymes involved in nerve transmission.

The cardiotoxins and protease inhibitors consist of 60 amino acid residues (M_r 6,900), whereas the best studied neurotoxins (cobra) can be subdivided into long chain (71–74 amino acids; M_r 8,000) and short chain (60–62 amino acids, M_r 7,000) toxins. Despite their structural and pathophysiological differences, the cardio- and neurotoxins of the asps show sequence homologies. Neurotoxins so far isolated from the vipers have high M_r (e.g. rattlesnake crotoxin, M_r 30,000); they may even show subunit structure, e.g. the most potent S.v., taipoxin, consists of two nonidentical polypeptide chains. Owing to the presence of disulfide bridges (4 at M_r 7,000, 5 at M_r 8,000 and 7 at M_r 13,500), the neurotoxins are extremely stable. They suffer no loss of activity when treated with 8 M urea for 24 hours 25 °C, or for 30 min at 100 °C. But they are rapidly inactivated by strong alkali, probably by disulfide exchange or desulfuration. In addition to the disulfide groups, a tryptophan and a glutamate residue are also necessary for the neurotoxic activity of cobra toxins. Cobra toxins are similar to Scorpion venoms (see).

The yearly death rate from snake bites is estimated at 30,000–40,000, about 50 being registered in Europe, the remainder chiefly in Asia. Many animals (hedgehog, ichneumon) have a natural immunity against S. v. Large quantities of S. v. are used for the preparation of specific antisera (by the periodic active immunization of horses) for immunization against snake bite. Some S. v. are used therapeutically for the treatment of neuralgic pain, rheumatic diseases and epilepsy.

Snakes are farmed for the production of S. v. The animals are allowed to bite a membrane and inject the venom into a glass container, or the venom is expressed by pressure on the venom glands. This produces between 10 (vipers) and a few hundred mg (Asiatic snakes) S. v. from each snake.

snRNA: see U snRNP.

snRNP: see U snRNP.

Snurp: see U snRNP.

α-Solamargine: see α-Solasonine.

α-Solamarine: a Solanum glycoalkaloid first isolated from woody nightshade (*Solanum dulcanamara*). The aglycon is tomatidenol (22*S*,25*S*)-spirosol-5-ene-3β-ol, M_r 413.67, m.p. 239 °C, $[\alpha]_D$ – 37.8 ($CHCl_3$); the carbohydrate residue is a trisaccharide of one molecule of D-glucose and two molecules of L-rhamnose. Tomatidenol differs from tomatidine (see α-Tomatine) by the presence of a double bond at position 5; it can be used as a starting material for the laboratory synthesis of steroid hormones. β-S. inhibits growth of mouse Sarcoma 180.

Solanesol: see Polyprenols.

Solanidane: see Solanum alkaloids.

Solanidine: see α-Solanine.

α-Solanine, *solanine:* a Solanum alkaloid, the chief toxic alkaloid of the potato (*Solanum tuberosum*), also present in other *Solanum* spp. It is a glycoalkaloid containing the aglycon solanidine (solanid-5-ene-3β-ol), M_r 397.65, m.p. 218 °C, $[\alpha]_D$ –27 ° ($CHCl_3$), and the trisaccharide β-solatriose (Fig.). Potato tubers contain no more than 0.01 % α-S., which is harmless in this quantity. Potato shoots contain 0.5 % α-S. and therefore cannot be used as food.

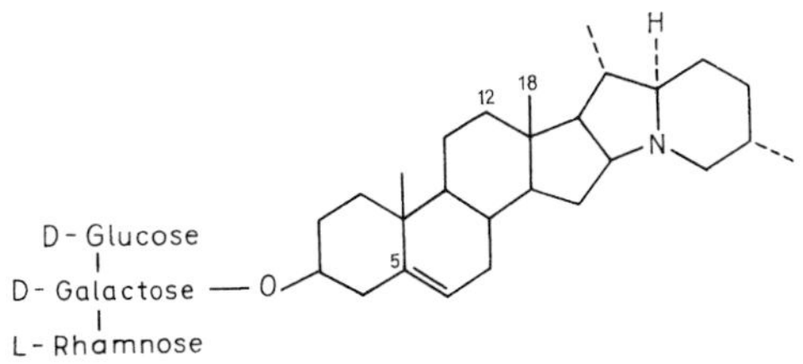

D-Glucose
D-Galactose — O
L-Rhamnose

α-Solanine

Solanum alkaloids: steroid alkaloids that occur in plants of the nightshade family (*Solanaceae*) of the genera *Solanum, Lycopersicon, Cyphomandra* and *Cestrum.* S. a. are structurally related to the parent hydrocarbon, cholestane (see Steroids).

In addition, they contain nitrogen, representing structural derivatives of either spirosolane (contains a secondary amino group), or solanidane (contains a tertiary amino group) (Fig.). In the plant S. a. occur mostly as glycoalkaloids (e. g. see α-Solanine, α-Tomatine); the free aglycons can be released by acid hydrolysis. These glycoalkaloids are also Saponins (see); they form insoluble 1:1 addition products with cholesterol. The aglycons of S. a. are biosynthesized from cholesterol. Certain S. a. serve as starting material for the laboratory synthesis of steroid hormones.

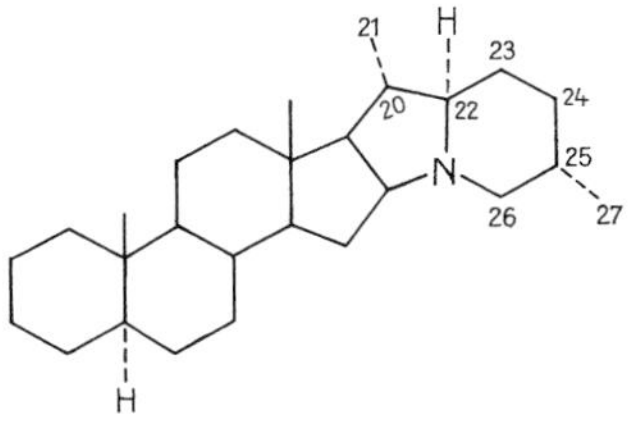

5α-Solanidine

Solasodine: see α-Solasonine.

α-Solasonine, *solasonine:* a Solanum steroid alkaloid occurring in many species of *Solanum*, e. g. *S. sodomeum, S. aviculare, S. laciniatum* and *S. nigrum.* It is a glycoalkaloid containing the aglycon, *salasodine* [(22*R*:25*R*)-spirosol-5-ene-3β-ol], M_r 413.67, m. p. 201 °C, $[\alpha]_D$ –107 ° ($CHCl_3$); and a branched trisaccharide.

Other solasodine glycosides are *β-solasonine* (with L-rhamnose and D-glucose), and *α-solamargine,* the first glycoside isolated from *Solanum marginatum* (with two molecules of L-rhamnose and one molecule D-glucose). Solasodine is used as starting material for the laboratory synthesis of steroid hormones.

β-Solasonine: see α-Solasonine.

β-Solatriose: see α-Solanine.

Solavetivone: see Phytoalexins.

Soluble RNA: see Transfer RNA.

Solvent accessible area, *solvent accessible surface:* that part of the surface of a macromolecule which is accessible to a solvent molecule, i. e. that part of the van der Waals surface which can be in contact with a probe sphere representing a solvent molecule. [M. L. Connolly *Science* **221**(1983) 709–713]

Somatomedin: a collective term for a group of peptides. S. C is the same as insulin-like growth factor (IGF) I, and S. A is a mixture consisting largely of IGF II and some IGF I (see Insulin-like growth factor). S. A causes an increase in sulfur incorporation into cartilage and is therefore called *sulfation factor.* [E. M. Spencer, ed. *Insulin-Like Growth Factors, Somatomedins* (de Gruyter, Berlin, 1983)]

Somatostatin: see Releasing hormones.

Somatotropic hormone: see Somatotropin.

Somatotropin, *somatotropic hormone, STH, growth hormone, GH:* a fundamentally important hormone, which in conjunction with other hormones (insulin, thyroxin, etc.) controls growth, differentiation and the continual renewal of body substances. GH is a single chain polypeptide containing 190 amino acid residues and two disulfide bridges (for primary sequence, see Fig.). It is synthesized in the anterior pituitary in response to SRF and SIH (see

```
NH2-Phe-Pro-Thr-Ile-Pro-Leu-Ser-Arg-Leu-Phe-Asp-Asn-Ala-Met-Leu-Arg-Ala-His-Arg-Leu-
                    5                   10                  15                  20
His-Gln-Leu-Ala-Phe-Asp-Thr-Tyr-Gln-Glu-Phe-Glu-Glu-Ala-Tyr-Ile-Pro-Lys-Glu-Gln-
                    25                  30                  35                  40
Lyc-Tyr-Ser-Phe-Leu-Gln-Asn-Pro-Gln-Thr-Ser-Leu-Cys-Phe-Ser-Glu-Ser-Ile-Pro-Thr-
                    45                  50                  55                  60
Pro-Ser-Asn-Arg-Glu-Glu-Thr-Gln-Lys-Ser-Asn-Leu-Gln-Leu-Leu-Arg-Ile-Ser-Leu-Leu-
                    65                  70                  75                  80
Leu-Ile-Gln-Ser-Trp-Leu-Glu-Pro-Val-Gln-Phe-Leu-Arg-Ser-Val-Phe-Ala-Asn-Ser-Leu-
                    85                  90                  95                  100
Val-Tyr-Gly-Ala-Ser-Asn-Ser-Asp-Val-Tyr-Asp-Leu-Leu-Lys-Asp-Leu-Glu-Glu-Gly-Ile-
                    105                 110                 115                 120
Gln-Thr-Leu-Met-Gly-Arg-Leu-Glu-Asp-Gly-Ser-Pro-Arg-Thr-Gly-Gln-Ile-Phe-Lys-Gln-
                    125                 130                 135                 140
Thr-Tyr-Ser-Lys-Phe-Asp-Thr-Asn-Ser-His-Asn-Asp-Asp-Ala-Leu-Leu-Lys-Asn-Tyr-Gly-
                    145                 150                 155                 160
Leu-Leu-Tyr-Cys-Phe-Arg-Lys-Asp-Met-Asp-Lys-Val-Glu-Thr-Phe-Leu-Arg-Ile-Val-Gln-
                    165                 170                 175                 180
Cys-Arg-Ser-Val-Glu-Gly-Ser-Cys-Gly-Phe-COOH
                    185                 190
```

Complete amino acid sequence of somatotropin

Releasing hormones), which are produced in the hypothalamus. Production of these latter horrnones is in turn affected by low blood sugar levels, increased levels of blood amino acids and by a variety of neural influences (stress situations). The growth stimulating activity of GH is especially apparent in pituitary dwarfism. GH acts directly on cellular metabolism, without the intermediate participation of other glandular hormones; as a prerequisite of growth stimulation, it promotes nucleic acid and protein biosynthesis. Treatment with GH results in increased growth of muscle, skeleton and organs, and an increased functional capacity of kidneys and liver.

Somatotropin release inhibiting hormone: see Releasing hormones.

Somatotropin releasing factor: see Releasing hormones.

Somatotropin releasing hormone: see Releasing horrnones.

Sophorabioside: genistein 4′-rhamnoglucoside. See Isoflavone.

Sophoricoside: genistein 4′-glucoside. See Isoflavone.

Sophorol: 7,6′-dihydroxy-3′,4′-methylenedioxyisoflavanone. See Isoflavanone.

D-Sorbitol: a C6-sugar alcohol found widely in plants. M_r 182.17, m.p. 97 °C, $[\alpha]_D^{20}$ – 2° (water). It can be prepared by catalytic or electrochemical reduction of the configurationally related D-glucose, D-fructose or L-sorbose. Technically, it is prepared by catalytic hydrogenation of D-glucose. D-S. is the starting material for the technical synthesis of ascorbic acid; it is used in the food industry as a preservative and as a softening agent in sweets. Since it is well tolerated, it is used as a sweetner in diabetic diets.

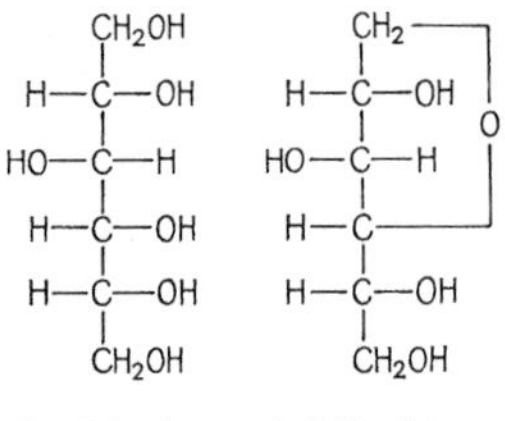

```
     CH2OH              CH2 ─────┐
     |                   |       |
  H─C─OH             H─C─OH      |
     |                   |       O
 HO─C─H             HO─C─H       |
     |                   |       |
  H─C─OH             H─C ────────┘
     |                   |
  H─C─OH             H─C─OH
     |                   |
     CH2OH              CH2OH
```

Sorbitol *1,4-Sorbitan*

By heating in the presence of acid catalysts, D-S. undergoes an intramolecular loss of water with formation of internal ethers. One such ether, 1,4-sorbitan, is converted commercially into dispersants and emulsifying agents by partial esterification with fatty acids and reaction with ethylene oxide.

L-Sorbose: a monosaccharide hexulose, M_r 180.16, m.p. 162 °C, $[\alpha]_D^{20}$ – 42.9°, present in certain plant juices, e.g. rowan berries, and biosynthesized from D-sorbitol. L-S. is an intermediate in the commercial synthesis of ascorbic acid.

SOS response: a collective response to DNA damage observed in bacteria. It includes the appearance of new, error-prone DNA repair activity, so that DNA repair is associated with mutagenesis. It also includes massive synthesis of recA protein, lytic development of certain prophages in lysogenic strains, and inhibition of cell division, leading to filamentous growth in nonlysogenic strains. The total collective SOS response can be completely abolished by mutation at either the *rec*A or *lex*A loci. Alternatively, mutation at these loci may result in temperature-sensitive (42 °C) gratuitous expression of the SOS response (*tif* at the *rec*A locus, and *tsl* at the *lex*A locus).

Incubation of nonlysogenic *tif* or *tsl* strains at 42 °C results in a total SOS response, including formation of long, multinucleate filaments. Filamentous growth is abolished by *sfi*A mutations, but all other components of the SOS response are unaffected, thereby defining a factor involved in the inhibition of cell division.

The hypothesis that *rec*A and *lex*A jointly control inducible functions related to DNA damage was originally known as the *SOS hypothesis* (SOS for the international distress signal).

Mechanism. LexA protein normally blocks the expression of fifteen or more genes involved in DNA repair. RecA protein specificially binds to the single-strands produced by physical or chemical damage to DNA (see Recombination for details of recA structure and DNA binding). The single-strand DNA-recA complex hydrolyses an Ala-Gly bond in lexA protein, thereby inactivating it as a repressor of DNA repair genes.

Chemical or physical DNA damage can be monitored by the *SOS function,* which is measured by the *SOS chromotest.* This test therefore serves as a screen for potential mutagens in prokaryotes, which are inferred to be potential carcinogens in eukaryotes.

The SOS chromotest uses a strain of *E. coli* in which the structural gene for β-galactosidase (see) is linked to the promoter of the *sfi*A operon [by operon fusion, using phage Mud(ap, *lac*)]. In the proper orientation, the *lac* genes of the prophage are expressed from the adjacent bacterial promoter. DNA damage induces a high level of expression of the *sfi* gene (i.e. part of the SOS response) leading to synthesis of β-galactosidase, which can be monitored colorimetrically using the chromogenic substrate *o*-nitrophenyl-β-D-galactoside. [E.M.Witkin *Bacteriological Reviews* **40** (1976) 869–907; O.Huisman & R.D'Ari *Nature* **290** (1981) 797–799; J.W.Little & D.W.Mount *Cell* **29** (1982) 11–22]

Southern blot: a method for transferring fragments of DNA from electrophoresis gels to cellulose nitrate membranes, first described by E.M.Southern [*J. Mol. Biol.* **98** (1975) 503–517]. Similar procedures for replicating electrophoretograms of RNA or proteins have been capriciously named Northern or Western blots, respectively. As yet, there is no Eastern blot.

The gels used for electrophoresis (agarose, polyacrylamide) are often unsuitable for further study of the resolved substances, whereas the replica (or blot) can be submitted to a variety of analytical procedures. For example, extremely small quantities of DNA can be located by hybridization with radioactive RNA, and proteins can be located and identified with antibodies. In the original method of Southern, the nitrocellulose membrane is layered on top of the gel (DNA is first denatured by immersion of the gel in buffered ethidium bromide), followed by moist filter paper and then by several layers of dry filter paper. DNA fragments move from the gel by capillarity and are trapped in the nitrocellulose. For the replication of protein separations, zones may be transferred to nitrocellulose, or to diazobenzyloxymethyl (DBM) paper or diazophenylthioether (DPT) paper. Transfers can be performed more rapidly by electrophoresing the zones from the gel into the receiving layer; the replica is then called an electroblot or "blitz" blot.

Soybean trypsin inhibitor, *STI:* the best known plant trypsin inhibitor. With bovine trypsin, at pH 8.3, it forms a stoichiometric, enzymatically inactive, stable complex with an association constant of 5×10^9 per mol STI. It also inhibits other vertebrate and invertebrate trypsins and plasmin. Chymotrypsin is inhibited to a small extent, and other endopeptidases not at all. STI is a single polypeptide chain, M_r 21,100, 181 amino acid residues (of known sequence) and two disulfide bridges. The molecule is compact and has a low α-helix content due to the presence of proline residues. It is consequently resistant to proteases and to denaturation. The reactive center contains a specific peptide bond Arg_{63}-Ile_{64}, which is hydrolysed when the trypsin-STI complex is formed. The ensuing interaction with the active site of trypsin, which also involves Arg_{65} of STI, results in firm binding of STI and inactivation of the trypsin.

Soybean also contains the Bowman-Birk inhibitor, which consists of 78 amino acid residues (M_r 8,000) and is much smaller than STI. It is a "double headed" inhibitor, i.e. in addition to inhibiting trypsin, it can simultaneously inhibit a second protease, e.g. chymotrypsin.

Since the nutritional value of protein-rich soybean products may be reduced by the presence of STI and other inhibitors, these are inactivated by heating the milled beans.

SPARC, *osteonectin:* an acidic, cysteine-rich protein of the extracellular matrix. SPARC displays a high degree of interspecies sequence conservation. It occurs in zones of mineralization in fetal growth plates, and it exhibits an affinity for type I collagen, hydroxyapatite and calcium, indicating a role in the mineralization of bone and cartilage. During embryo development high levels of SPARC mRNA are found in limb buds and in invasive cells of extraembryonic tissue. In the adult, SPARC appears to be restricted to cells that are actively proliferating or differentiating, as well as nonproliferating steroidogenic cells. With Thrombospondin (see) and Tenascin (see), it is considered to be a member of a family of proteins that modulate cell-matrix interactions. All three proteins influence embryogenesis and morphogenesis, and there is a positive correlation between their synthesis and cell proliferation. Reasoning that cells that produce an extracellular matrix are also influenced by it, Sage & Bornstein coined the concept of "dynamic reciprocity", i.e. agents such as SPARC may regulate cell development by influencing the cell's synthesis of its own extracellular matrix. [E.H.Sage & P.Bornstein *J. Biol. Chem.* **266** (1991) 14831–14834]

Specific activity: see Enzyme kinetics.

Specific incorporation rate: see Isotope technique.

Specificity constant, *physiological effectivity:* a measure of the turnover of a substrate. S.c. is the ratio of the catalytic constant and the Michaelis constant: k_{cat}/K_m. It is equal to the rate constant of a reaction for the rate equation: $v = k_{cat}E_0S/(K_m + S)$, where E_0 is the total enzyme concentration, and S is the substrate concentration, when $S << K_m$.

Specific linking difference: see Linking number.

Specific radioactivity: see Isotope technique.

Spectinomycin: see Streptomycin.

Spectrin: see Erythrocyte membrane.

Spermaceti: a solid animal wax, m.p. 45–50 °C, obtained from the head of the sperm whale, *Physeter macrocephalus.* An oily, liquid, crude sperm oil is secreted in a special large cylindrical organ in the upper region of the huge jaw and above the right nostril of the whale. After capture, this cavity is emptied of its oil, which, on cooling, deposits crystalline S. This is separated by pressure and purified by remelting and washing with dilute NaOH to remove the last traces of oil. S. consists chiefly of cetyl palmitate, accompanied by smaller quantities of cetyl laurate and myristate. It is used in the pharmaceutical and cosmetic industries as a basis for creams.

Spermidine: see Polyamines.

Spermine: see Polyamines.

Spheroidine: see Tetrodotoxin.
Sphinganine: see Membrane lipids.
Sphingoglycolipids: see Membrane lipids.
Sphingolipidoses: Lysosomal storage diseases (see) resulting from defective catabolism of glycosphingolipids.
Sphingomyelin: see Membrane lipids.
Sphingophospholipids: see Membrane lipids.
Sphingosine: 2-amino-4-octadecene-1,3-diol, M_r 299.48, m.p. 67 °C, a long-chain amino alcohol which is a component of sphingomyelins and glycolipids (see Membrane lipids). It is not found in free form in plants or animals.
Spider toxins: toxic substances produced in the venom glands of many spiders. They serve to paralyse and kill prey, and are dangerous to humans only in rare cases, e.g. the toxin of the South European *Latrodectus tredecimguttatus,* or the American black widow (*Latrodectus mactans*). The active principles of S.t. are proteinaceous and related to those of snake and scorpion venoms. They contain hyaluronidase and proteolytic activity, but phospholipases and hemolytic or blood clotting activities are absent.
Spinasterols: a group of very similar phytosterols (see Sterols), found in higher plants. Chief representative is *α-spinasterol* (5α-stigmasta-7,22-diene-3β ol; Fig), M_r 412.7, m.p. 172 °C, isolated e.g. from spinach, senega root and lucerne. α-, β-, γ- and δ-S. differ in the position of the side chain double bond.

α-Spinasterol

Spinochromes: derivatives of 1,4-naphthoquinone (see Naphthoquinones), responsible for the red or orange color of sea urchin shells. Over 20 different hydroxylated S. are known. In the native state, they are present as calcium and magnesium salts. They differ from the echinochromes, which occur in the eggs, perivisceral fluid and internal organs of the sea urchin. Echinochrome A is a red pigment in the eggs and skeleton of the sea urchin; it is a pentahydroxy-1,4-naphthoquinone with an ethyl group at C6, m.p. 223 °C.
Spinulosine: see Benzoquinones.
Spiramycin: a Macrolide (see) antibiotic.
Spirographis porphyrin: see Chlorocruoroporphyrin.
Spirosolane: see Solanum alkaloids.
Spirostane: the oxygen-containing parent structure of the steroid saponins (see Saponins). The S. system is formally derived from the parent hydrocarbon cholestane (see Steroids). The name S. embraces the configuration of all asymmetric centers with the exception of positions 5 and 25.
Spirostanols: see Saponins.
Splicing: see Intron.

(25S)-5α-Spirostane

Splicing enzymes: see Intron.
Split gene: see Intron.
Split proteins: see Ribosomes.
Spongonucleosides: see Arabinosides.
Spongosine: 9-β-D-ribofuranosyl-2-methoxy-adenine, a nucleoside with a modified base, isolated from sponges.
Spongosterol: (24*R*)-5α-ergost-22-ene-3β-ol, a marine zoosterol (see Sterols), M_r 400.66, m.p. 153 °C, $[\alpha]_D$ + 10° ($CHCl_3$), occurring as a typical sterol of sponges (*Spongia*) and isolated, e.g. from *Suberitis domuncula* and *S compacta.*
Spongothymidine: see Arabinosides.
Spongouridine: see Arabinosides.
Sporidesmolides: cyclic depsipeptides from the fungus *Pithomyces chartarum.* Sporidesmolide I is cyclo-(-Hyv-D-Val-D-Leu-Hyv-Val-MeLeu-); sporidesmolide II contains D-allo-isoleucine in place of D-valine, while sporidesmolide III contains L-leucine in place of L-*N*-methylleucine. (Hyv represents a residue of α-hydroxyisovaleric acid).
Sporopollenin: the material of the outermost cell wall layer (exine) of pollen grains and spores of pteridophytes, and also present in small amounts in fungal zygospore walls (e.g. zygospore wall of *Mucor mucedo* contains 1 % S.). S. is extremely resistant to physical, chemical and biological dcgradation. Pollen grains are therefore well preserved in geological strata, and have proved archeologically useful as quantitative and qualitative markers of previous plant life and agriculture. P. is an intimate mixture or complex of 10–15 % cellulose, 10 % of an ill-defined xylan fraction, 10–15 % of a lignin-like material, and 55–65 % of a lipid. It is now generally accepted that the lipid material is formed by oxidative polymerization of carotenoids; a virtually identical material can be synthesized in the laboratory by the catalytic oxidation of β-carotene. ["Sporopollenin" by G. Shaw, pp. 31–58 in *Phytochemical Phylogeny* (edit. J. B. Harborne) 1970, Academic Press]
S-protein: a cleavage product of ribonuclease, representing amino acid residues 21–124 of the ribonuclease primary sequence. It is produced, together with S-peptide (positions 1–20), by the action of subtilisin on ribonuclease.
Squalene: biochemically the most important aliphatic triterpene. M_r 408, b.p.$_{10}$ 262–264 °C, ρ 0.8584. For formula see Triterpenes. S. was first isolated from fish liver oils, and later found in many plant oils. It is the intermediate in the biosynthesis of all cyclic triterpenes. Cyclization of S. is catalysed by a mixed function oxygenase, and proceeds via 2,3-epoxysqualene (squalene epoxide).

Src homology domains, ***SH domains:*** conserved ~100-residue domains first recognized in tyrosine kinases related to Src protein (an oncogene product of Rous sarcoma virus). Many non-receptor tyrosine kinases that are activated by tyrosine kinase-associated receptors are members of the Src family. These ~500-residue, membrane anchored (by myristoylation) proteins contain at least one SH domain (SH2), and some possess a second SH domain known as SH3. Certain membrane-associated proteins possess only the SH3 domain.

SH2, which specifically binds phosphotyrosine residues with high affinity, is a hemispherical domain with a central 5-stranded antiparallel β-pleated sheet, sandwiched between two nearly parallel α helices, so that its *N*- and *C*-terminal residues are in close proximity.

SH3, which shows no homology with SH2, binds short (9 or 10 residues) Pro-rich sequences. It consists of two 3-stranded, closely packed, antiparallel β-pleated sheets. As in SH2, the *N*- and *C*-terminal residues of SH3 are in close proximity. The role of SH3, which occurs in both receptor and nonreceptor tyrosine kinases, is less clear. Its deletion from the proto-oncogenes *Src* and *Abl* converts them to oncogenes, suggesting that SH3, like SH3, plays a key role in the interaction between kinases and regulatory proteins. [T. Pawson & J. Schlessinger *Curr. Biol.* **3** (1993) 432–434; B. J. Mayer & D. Baltimore *Trends Cell Biol.* **3** (1993) 8–13; C-H. Lee et al. *Structure* **2** (1994) 423–438; A. Musacchio et al. *Nature Struct. Biol.* **1** (1994) 546–551]

SRH: acronym of somatotropin releasing hormone. See Releasing hormones.

sRNA: soluble RNA, now known as Transfer RNA (see) or tRNA.

SRS: acronym of slow reacting substance. See Leukotrienes.

Stachydrine: see Pyrrolidine alkaloids.

Stachyose: a nonreducing tetrasaccharide (see Carbohydrates) found in plants. M.p. 170 °C, $[\alpha]_D^{20}$ + 149 °. The four sugar residues are linked in the order D-galactose-D-galactose-D-glucose-D-fructose.

Staphylococcal protein A: see Protein A.

Starch: a high M_r polysaccharide, formula $(C_6H_{10}O_5)_n$, the chief storage carbohydrate in most

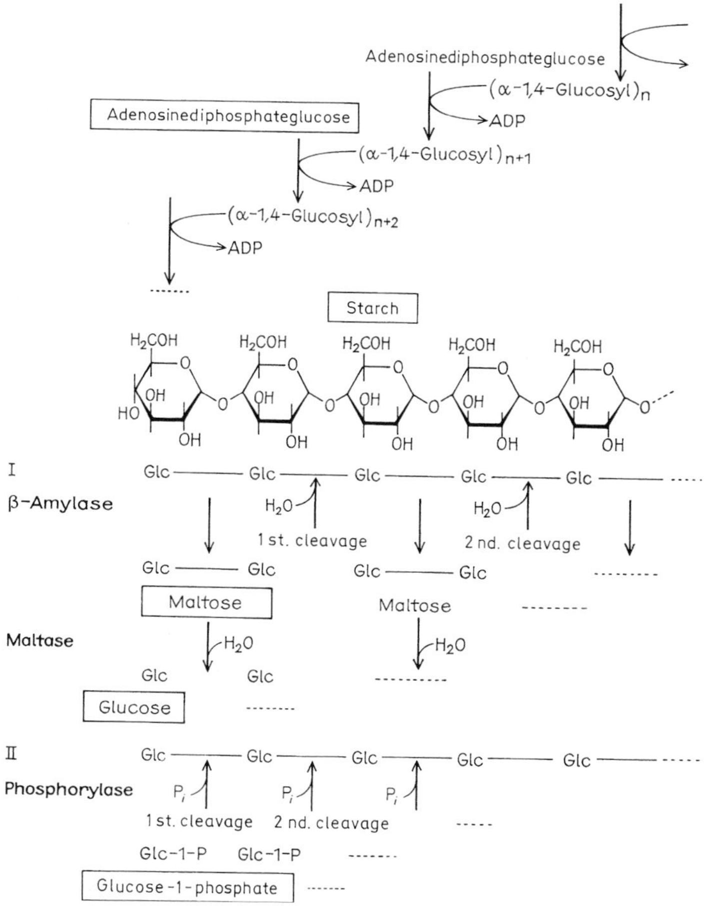

Adenosinediphosphateglucose
(α-1,4-Glucosyl)$_n$
ADP
Adenosinediphosphateglucose
(α-1,4-Glucosyl)$_{n+1}$
ADP
(α-1,4-Glucosyl)$_{n+2}$
ADP
Starch
I
β-Amylase
Glc — Glc — Glc — Glc — Glc — ·····
H_2O 1 st. cleavage
H_2O 2 nd. cleavage
Glc — Glc Glc — Glc
Maltose Maltose
Maltase H_2O H_2O
Glc Glc
Glucose
II
Phosphorylase
Glc — Glc — Glc — Glc — Glc — ·····
P_i P_i P_i
1 st. cleavage 2 nd. cleavage
Glc-1-P Glc-1-P
Glucose-1-phosphate

Biosynthesis and degradation of starch. I, hydrolytic degradation. II, phosphorolytic degradation. Glc = glucose. P_i = inorganic phosphate.

higher plants, consisting of about 80 % water-insoluble Amylopectin (see) and 20 % water-soluble Amylose (see). In plant metabolism, S. first appears as an assimilation product in the chloroplasts. It is then degraded, the products of degradation are translocated, and S. is resynthesized as storage S. (S. grains or granules) in storage organs, e.g. roots, tubers or pith. S. grains are classified as compound, simple, centric or acentric, depending on their characteristic shapes and stratifications. On the basis of these characteristic forms, flour from different sources (e.g. maize, rice, wheat, rye) can be identified microscopically.

S. is biosynthesized from adenosine-diphosphate-glucose (Fig.). During animal digestion, S. is hydrolysed by amylases. α-Amylase (EC 3.2.1.1) hydrolyses α(1 → 4) linkages at random, forming a mixture of glucose and maltose. Maltose is hydrolysed to glucose by α-D-glucosidase (maltase, EC 3.2.1.20). Amylose is thus completely degraded to glucose. α-Amylase cannot hydrolyse α(1 → 6) linkages at the branch points of amylopectin; the product of α-amylase action on amylopectin is therefore a large, highly branched core or limit dextrin, representing 40 % of the original amylopectin. The small intestine also contains an oligo-α(1 → 6)-glucosidase which hydrolyses the 1 → 6 linkages, and completes the total degradation of amylopectin. β-Amylases (EC 3.2.1.2) are found especially in germinating seeds (e.g. malt); they act on the nonreducing end of the polysaccharide chain, removing successive maltose units. β-Amylase is also unable to attack α(1 → 6) branch points. In plant cells, storage S. is remobilized by phosphorolysis to glucose 1-phosphate.

S. is very important in human nutrition, supplying most of the dietary carbohydrate requirement (humans require about 500 g carbohydrate per day). Potatoes, cereals and bananas are particularly rich in S. Metabolism of 1 g S. supplies 16.75 kJ (4 kcal). S. is prepared commercially from plant sources, in particular potatoes, wheat rice and maize, and it has many uses in the food industry and in technology.

Start codon: see Initiation codon, Protein biosynthesis.

Starter: see Primer.

Starter tRNA: see Initiation tRNA.

Statins: see Releasing hormones.

STATs: see Signal transducers and activators of transcription.

Status-quo hormones: see Juvenile hormones.

Steady state: a chain of chemical reactions is in a steady state when the concentration of all intermediates remains constant, despite a net flow of material through the system, i.e. the concentration of intermediates remains constant, while a product is formed at the expense of a substrate. The term is used in two ways:

1. *Kinetic S.s.* is part of the kinetic response of an enzyme-catalysed reaction during transition from the initial state to the equilibrium state (steady state region) (Fig.). It is characterized by a slow change in the concentration of intermediates (see Enzyme species), while the concentration variables alter with the rate of the overall reaction. In the S.s. region, $(dES/dt)/(dS/dt) << 1$, and in the kinetic equation dES/dt can be approximated to zero.

2. In the *induced S.s. (true S.s.)* the concentrations of all reactants remain constant, since the rates of formation and removal are equal. For the simple case of a chain of metabolic reactions:

$$\xrightarrow{v} X_1 \xrightarrow{k_1} X_2 \xrightarrow{k_2} \cdots \xrightarrow{k_{n-1}} X_n \xrightarrow{v},$$

it can be stated that $dX_i/dt = 0$ and $k_iX_i = v$, where $i = 1, \cdots, n-1$.

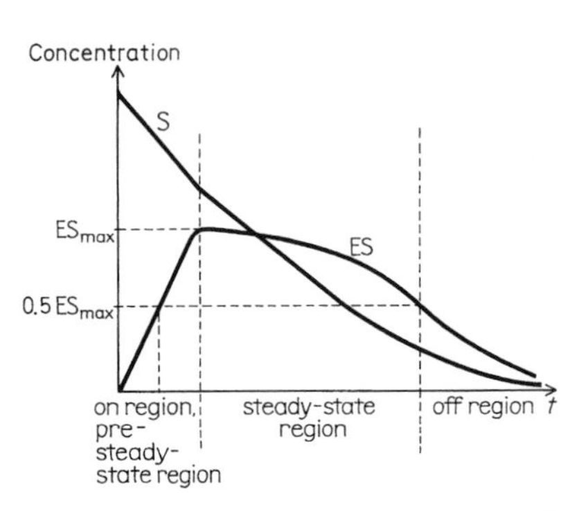

Time course of the concentrations of substrate and enzyme-substrate complex during an enzymatic reaction. S = substrate concentration. ES = concentration of enzyme-substrate complex.

Stearic acid: n-octadecanoic acid, CH_3-$(CH_2)_{16}$-COOH, a fatty acid, M_r 284.5, m.p. 71.5 °C, b.p.$_{15}$ 232 °C. Together with palmitic acid, S.a. is one of the most plentiful and most widely distributed fatty acids, occurring esterified in practically all animal and plant oils and fats, e.g. 34 % in cocoa butter, 30 % in mutton fat, 18 % in beef fat, 5–15 % in milk fat. It is used in the manufacture of candles, soaps, detergents, antifoams, pharmaceuticals and cosmetics. Stearin is the triacylglycerol of S.a.

Stemphylotoxins: phytotoxins (Fig.) isolated from culture filtrates of *Stemphylium botryosum* Wallr. f.sp. *lycopersici*, the causative agent of leaf spot and foliage blight disease of tomato. The symptoms of these diseases are caused by external application of S. Toxicity of S. is measured by their inhibition of incorporation of ^{14}C-amino acids into protein in exponentially growing tomato cell suspensions. S. also inhibit rootlet elongation of tomato, and they inhibit growth of *Spirodella oligorrhiza* (duckweed). S.I is about 100 times more toxic than S.II. Conversion of I into II occurs readily by acid or base catalysis. S. show a high affinity for ferric but not for ferrous iron [apparent stability constants for the Fe^{3+} complexes: 1.7×10^{24} (I) and 1.6×10^{24} (II);

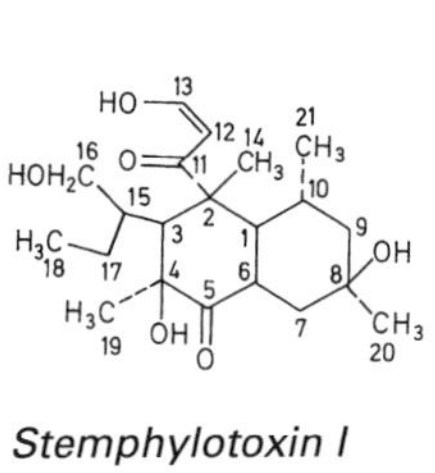

Stemphylotoxin I

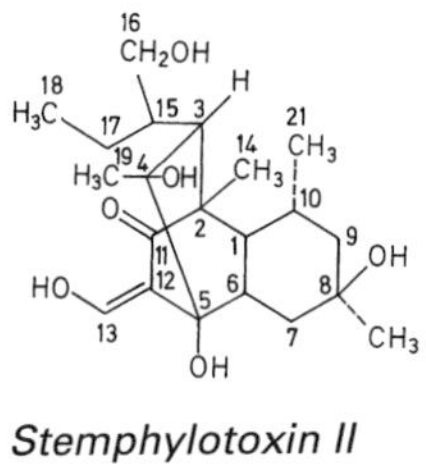

Stemphylotoxin II

cf. 10^{30}-10^{34} for hydroxamate Siderophores (see)]. S. may therefore function as siderophores by sequestering ferric iron from the host plant for use by the fungus. Biosynthesis of S. is regulated by iron, the optimal concentration for S. production being 2 mg/*l* (Fe^{2+} or Fe^{3+}).

Sterane: an earlier name for gonane. See Steroids.

Stercobilin: see Bile pigments.

Stercobilinogen: see Bile pigments.

Steroid alkaloids: nitrogen-containing steroids present in plants. S. a. are especially common in the plant families *Solanaceae* (nightshades), *Liliaceae* (lilies), *Apocyanaceae* (periwinkles) and *Buxaceae* (boxwoods), where they often occur as glycoalkaloids, or esterified as ester alkaloids. They are classified, according to their origin, into Solanum alkaloids (see), Veratrum alkaloids (see), Funtumia alkaloids (see) Holarrhena alkaloids (see) and Buxus alkaloids (see), but each group may contain different structural types. The Salamander alkaloids (see) are of animal origin and represent a special group. S. a. are formal derivatives of either cholestane or pregnane, and they are further subdivided on the basis of structure, e. g. Solanum alkaloids comprise spirosolanes and solanidanes. Some S. a., e. g. certain Veratrum alkaloids, are used pharmacologically.

Steroid alkaloid saponins: see Saponins.

Steroid hormones: groups of steroids which function as hormones. They comprise Adrenal corticosteroids (see), sex hormones (see Gonadal hormones), Ecdysone (see) and related molting hormones, and the plant sex hormone Antheridiol (see). 1,25-Dihydroxycholecalciferol (an active metabolite of the seco-steroid vitamin D_3) is also considered to be a S.

It was earlier thought that the S. bind to specific receptors in the cytosol of target tissues, followed by entry of the receptor-steroid complex into the nucleus. This widely held hypothesis appears to be no longer tenable. The receptor is bound in the nuclear matrix. Cytosolic receptors represent a reproducible artifact, due to release of the nuclear receptor by homogenization and dilution of the cell contents with homogenizing buffer.

Sex steroid-binding globulin (SBG) may play an active role in the entry of S. into target cells. Thus, ^{125}I-labeled SBG (purified from retroplacental blood serum) specifically binds to plasma membranes of human decidual endometrium only when it is carrying estradiol. [J.Gorski et al. *Mol. Cell. Endocrinol.* **36** (1984) 11–15; P.J.Sheridan et al. *Nature* **282** (1979) 579–582; H.R.Walters et al. *J. Biol. Chem.* **255** (1980) 6799–6905; K.Liimpaphayon et al. *J. Obstetr. Gynecol.* III, No.8 (1971) 1064–1068; N.I.Zhuk et al. *Biochemistry (USSR)* **50** (No.7, part 1) (1986) 936–939]

Steroids: a large class of of terpenoids (see Terpenes) including many biologically important compounds, e.g. Sterols (see), Steroid hormones (see), Bile acids (see), Cardiac glycosides (see), Steroid alkaloids (see) and steroid saponins (see Saponins). Synthetic S., e.g. Ovulation inhibitors (see), Anabolic steroids (see) and structurally modified steroid hormones, are pharmacologically important. More than 20,000 S are known, of which about 2 % have some medical significance.

Structurally, S. are derivatives of the hydrocarbon cyclopentanoperhydrophenanthrene, which contains 6 asymmetrical centers (marked with asterisks in Fig.1). There are thus $2^6 = 64$ theoretically possible stereoisomers, but relatively few of these are known among the S. The nomenclature of the tetracyclic ring structure (often referred to as the steroid 'nucleus') and the numbering of its carbon atoms are shown in Fig.2. It should be noted that the IUPAC-IUB Joint Commission on Biochemical Nomenclature recommended in 1989 [*Eur. J. Biochem.* **186** (1989) 429–458] a new system of numbering of the carbon atoms, replacing that recommended previously [*Eur. J. Biochem.* **10** (1969) 1–19; amendments *Eur. J. Biochem.* **25** (1972) 1–3]. This involved the change of the numbering of the carbons formerly numbered 28, 29, 30, 31 and 32 (shown in bold type in Fig.2) to 24^1, 24^2, 28, 29 and 30 (shown within ellipses in Fig.2) respectively. The purpose of this change is to bring steroid numbering into line with that of triterpenoids since some tetracyclic triterpenoids, e.g. lanosterol, can be regarded as trimethyl steroids. Coincidentally it also fits in with the biosynthetic origins of the carbons; C-atoms 1–30 of the 1989 numbering system are all derived from either the methyl or the carbonyl carbon of acetyl-CoA via isopentenyl pyrophosphate (IPP), the isoprenoid precursor of all terpenes (see), while the carbons of the 1C (methylene

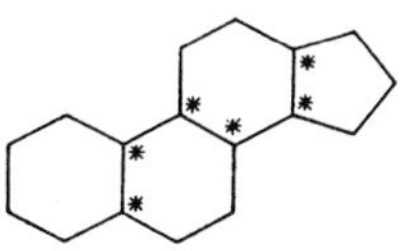

Fig. 1. *Cyclopentanoperhydrophenanthrene*

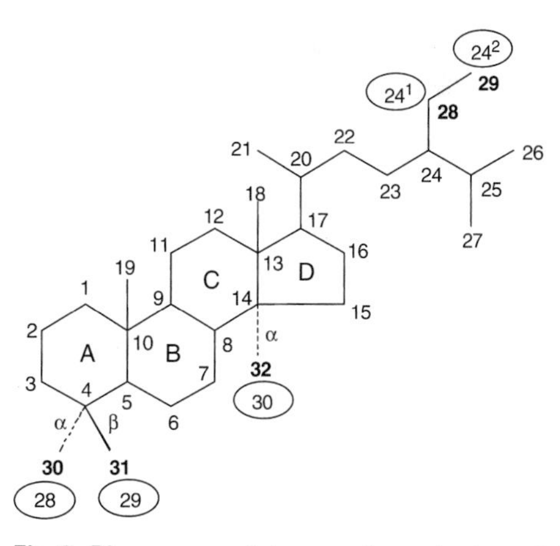

Fig. 2. *Ring nomenclature and numbering of the carbon atoms in the steroid molecule.* The IUPAC-IUB Joint Commission on Biochemical Nomenclature recommended in 1989 [*Eur. J. Biochem. 186* (1989) 429–458] that the numbers shown within the ellipses should replace the numbers shown in bold type; the latter were assigned in the 1969 recommendations [*Eur. J. Biochem. 10* (1969) 1–19] and the amendments to them [*Eur. J. Biochem. 25* (1972) 1–3]; all other numbers are unchanged.

Table. *Parent hydrocarbon structures of the steroids*

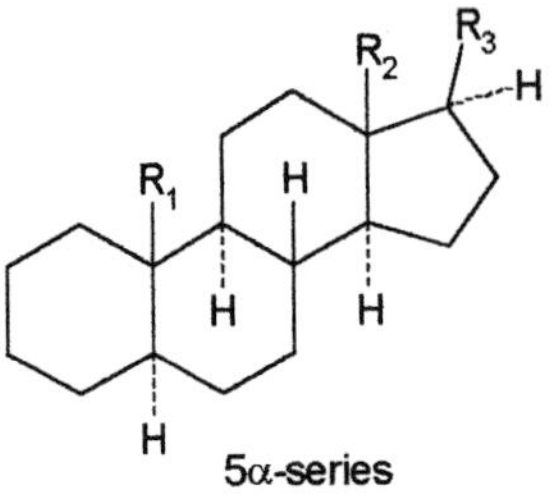

5α-series

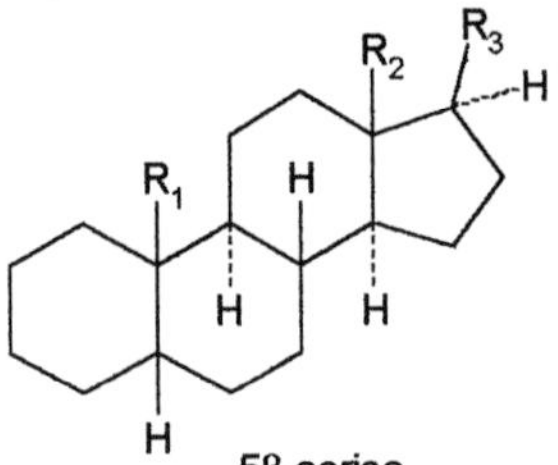

5β-series

Name	R_1	R_2	R_3
5α-Gonane (formerly Sterane) 5β-Gonane	H	H	H
5α-Estrane 5β-Estrane	H	CH_3	H
5α-Androstane (formerly Testane) 5β-Androstane (formerly Etiocholane)	CH_3	CH_3	H
5α-Pregnane (formerly Allopregnane) 5β-Pregnane	CH_3	CH_3	20*R* CH_2CH_3
5α-Cholane (formerly Allocholane) 5β-Cholane	CH_3	CH_3	20*R* $CH(CH_3)CH_2CH_2CH_3$
5α-Cholestane 5β-Cholestane (formerly Coprostane)	CH_3	CH_3	20*R* $CH(CH_3)CH_2CH_2CH_2CH(CH_3)_2$
5α-Ergostane 5β-Ergostane	CH_3	CH_3	20*R* 24*S* $CH(CH_3)CH_2CH_2CH(CH_3)CH(CH_3)_2$
5α-Campestane 5β-Campestane	CH_3	CH_3	20*R* 24*R* $CH(CH_3)CH_2CH_2CH(CH_3)CH(CH_3)_2$
5α-Poriferastane 5β-Poriferastane	CH_3	CH_3	20*R* 24*S* $CH(CH_3)CH_2CH_2CH(C_2H_5)CH(CH_3)_2$
5α-Stigmastane 5β-Stigmastane	CH_3	CH_3	20*R* 24*R* $CH(CH_3)CH_2CH_2CH(C_2H_5)CH(CH_3)_2$

or methyl) or 2C (ethylidene or ethyl) groups commonly found attached to C24 of some steroids, e. g. ergosterol (see), sitosterol (see), are non-isoprenoid in origin, being derived from the *S*-methyl group of L-methionine via the transmethylating agent *S*-adenosylmethionine (see). Both numbering conventions are now used in textbooks and learned journals. Although the recommended changes are logical, confusion will reign until authors indicate whether they are using the old (1967) or new (1989) convention. The present text uses the old convention.

There are several fundamental type-hydrocarbons, defined by the substituents at C-atoms 10, 13 and 17 and latterly, according to the 1989 nomenclatural recommendations, at C24 (Table). Each of these hydrocarbon types is divided into two sub-types on the basis of the way in which adjacent rings are fused to each other. There are two possible methods of fusion

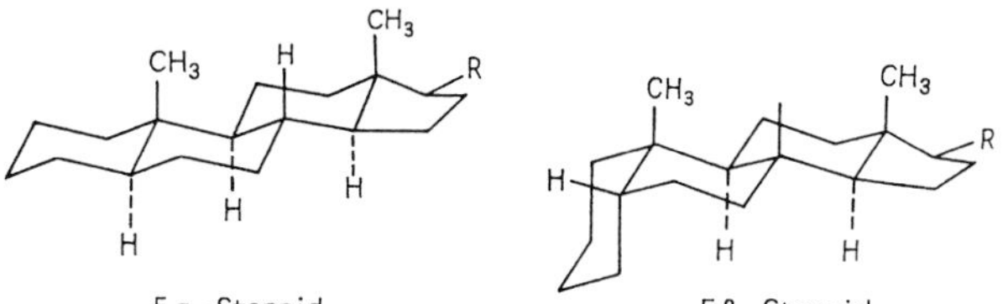

5α-Steroid
(A/B-trans)

5β-Steroid
(A/B-cis)

Fig. 3. *Conformation of the steroid molecule*

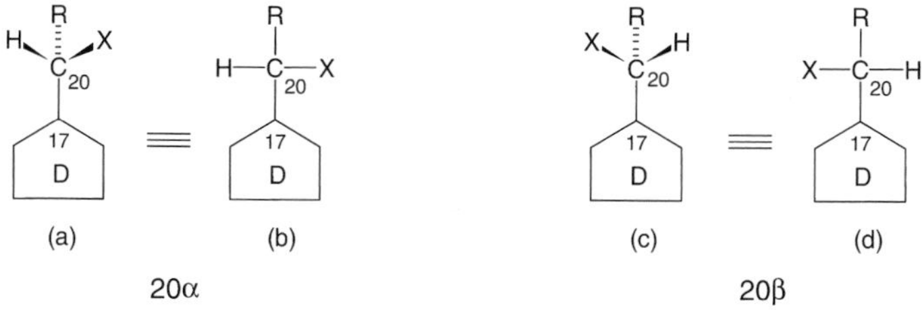

R = C-20 substituent with the longest carbon chain (CH_3 in the case of 20-hydroxypregnane)
X = CH_3 (OH in the case of 20-hydroxypregnane)

Fig. 4. *Use of the α/β system to specify configuration at the steroid C20*

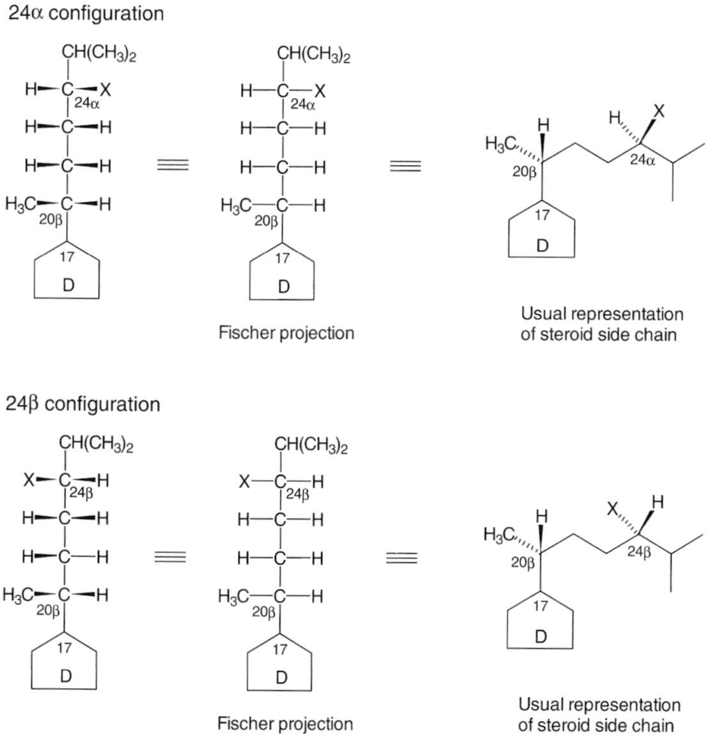

(X = methyl or ethyl)

Fig. 5. *Use of the α/β system to specify configuration at the steroid C24*

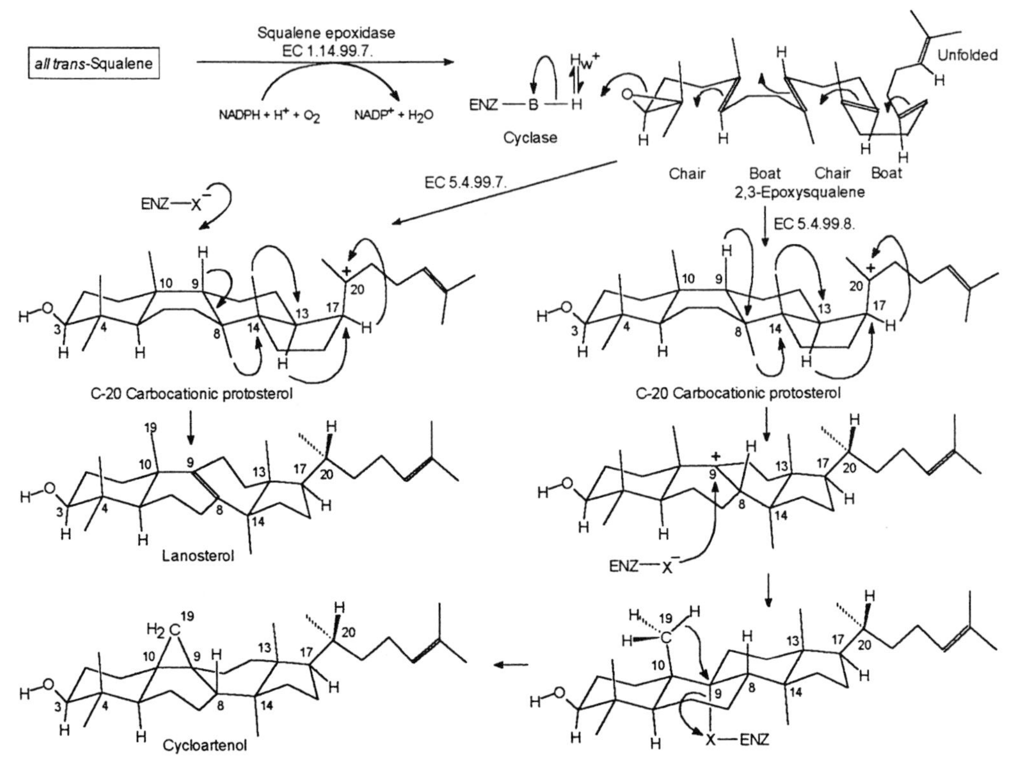

Fig. 6. *Biosynthesis of the steroid ring system from squalene.* Squalene epoxidase, an FAD-requiring flavoprotein monooxygenase, catalyses the oxidation of *all trans*-squalene to *all trans*-(3*S*)-2,3-epoxysqualene, which then undergoes cyclization, catalysed by different cyclases, to yield lanosterol or cycloartenol. Both cyclases first catalyse a forward cyclization so generating the same transient C20 carbocationic protosterol by initiating a protonic attack on the 'epoxy O-C2' bond and causing the wave of electron shifts shown. They then cause the backward rearrangement of this carbocation by promoting a series of Wagner- Meerwein shifts (i.e. sequentially *trans* 1,2 shifts of H^- or H_3C^-). In the case of the lanosterol-forming cyclase there are four such shifts: (i) 17α-H (not 17β-H as previously thought) to C20, which takes up the *R* configuration, (ii) 13α-H to C-17, where it becomes the 17α-H, (iii) 14β-CH_3 to C13, where it becomes the 13β-CH_3, (iv) 8α-CH_3 to C14, where it becomes the 14α-CH_3; the enzyme then stabilizes the resulting C8 carbocation by removing the 9β-H as a proton and generating an 8,9 double bond. The cycloartenol-forming cyclase catalyses the same four shifts as described above, then adds a fifth, namely: (v) 9β-H to C8, where it becomes the 8β-H. The enzyme then stabilizes the resulting C9 carbocation by removing one of the C19 hydrogens as a proton and using the newly available electrons to form a 'C9-C19' bond, thereby generating a 9β,19- cyclopropane ring. For steric reasons this step cannot be concerted with the 9β-H shift; it is therefore conceived as being composed of two substeps, namely stabilization of the C9 carbocation by a nucleophilic group (X^-) at the catalytic site of the cyclase followed by the *trans* elimination of the C19 hydrogen and the Enz- X^- and concomitant C-C bond formation.

which are exemplified by *cis*-decalin and *trans*-decalin; in the former the non-ring atoms (Hs) attached to the two carbon atoms common to both rings project in roughly the same direction and are said to be *cis* with respect to each other, whereas in the the latter they project in opposite directions and are said to be *trans.* When these two substituents are *cis* the two rings are said to be *cis*-fused and when they are *trans* the two rings are said to *trans*-fused. Among the naturally occurring steroids the ring A/B junction is *cis*- or *trans*-fused while the ring B/C and C/D junctions are *trans*-fused, so giving the two sub-types for each hydrocarbon type, namely 'A/B *trans*, B/C *trans*, C/D *trans*' and 'A/B *cis*, B/C *trans*, C/D *trans*'. This gives rise to the two series of steroids shown in Fig. 3 and the Table. Steroids with the 'A/B *trans*, B/C *trans*, C/D *trans*' ring system give rise to the 5α-series whilst those with the 'A/B *cis*, B/C *trans*, C/D *trans*' ring system give rise to the 5β-series. The terms 5α and 5β refer to the orientation of the H atom on C5 relative to that of the substituents on C-atoms 10 and 13 (usually methyl groups and referred to as the 'angular methyls') which project upwards from the general plane of the ring system when the latter is depicted in the

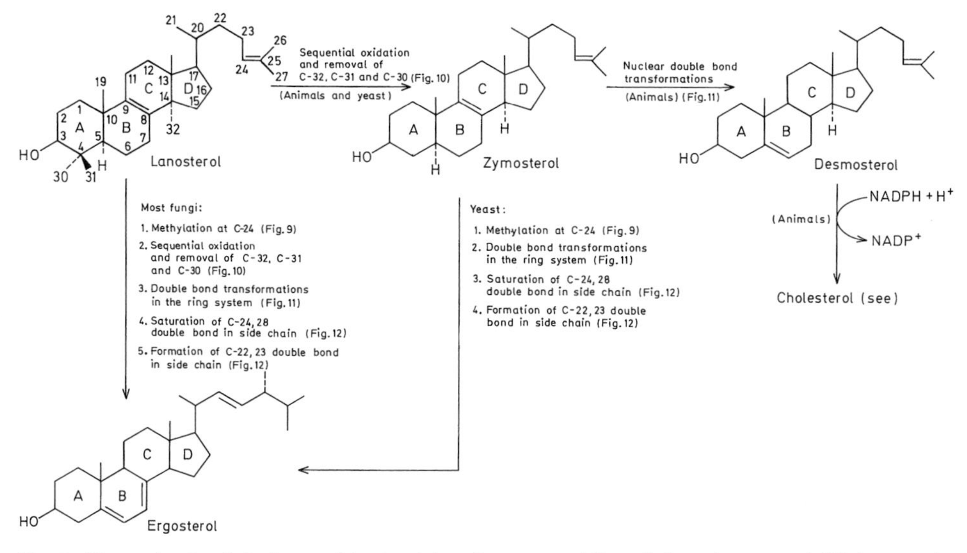

Fig. 7. *Biosynthesis of cholesterol (animals) and ergosterol (fungi) from lanosterol.* All the reactions listed (see Figs. 9, 10, 11, 12) are catalysed by enzymes of the endoplasmic reticulum.

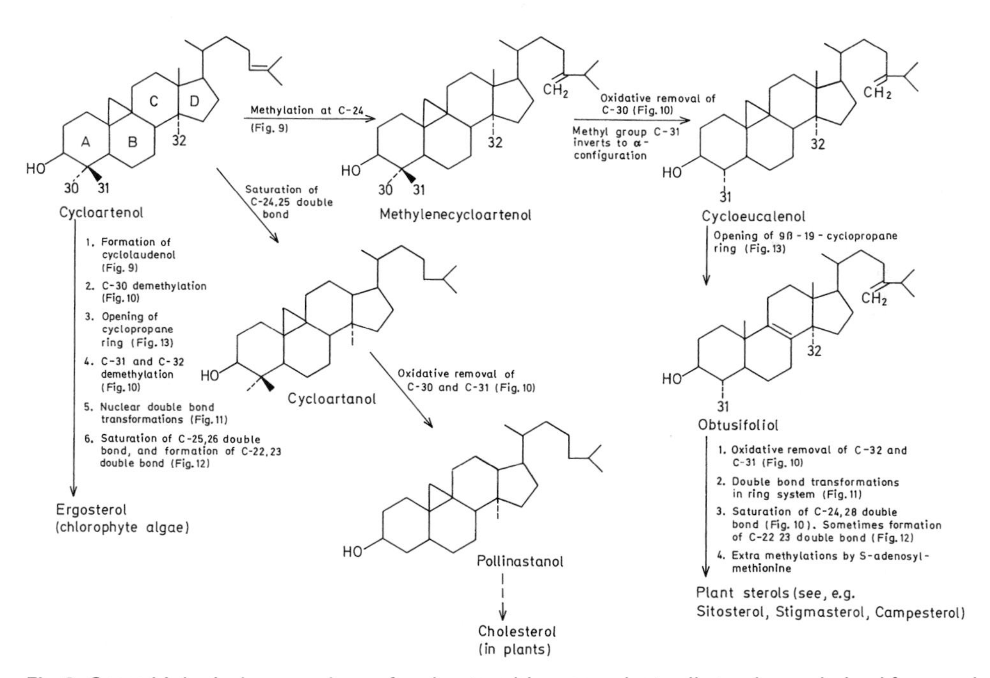

Fig. 8. *Some biological conversions of cycloartenol.* In green plants all sterols are derived from cycloartenol. Cholesterol, the predominant animal steroid, is also widespread as a very minor sterol in plants; however, in some plant tissues, e.g. *Compositae* pollen, it is abundant. In plants cholesterol is the biological precursor of phytoecdysones, cardenolides and steroid hormones, which may play a part in defense mechanisms against herbivorous animals.

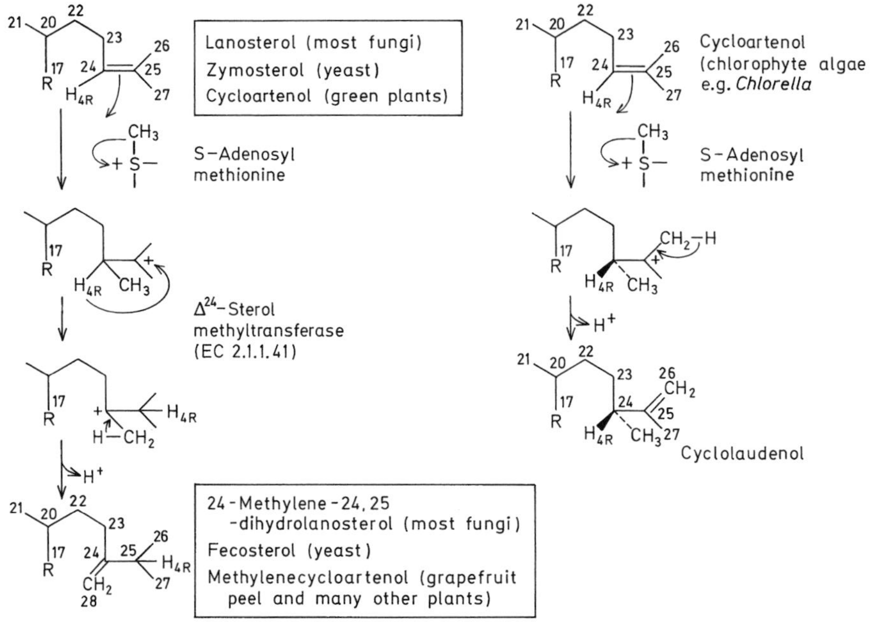

Fig. 9. Methylation at C24 of the C24,25-unsaturated side chain of various sterols. R represents the ring system of the appropriate sterol. Δ^{24}-Sterol methyltransferase has been purified from the microsomal fraction of yeast grown aerobically at the expense of glucose. When ethanol serves as the carbon source, the enzyme is found in the mitochondria. If mitochondrial development is represssed by anaerobiosis in the presence of 10% glucose, the enzyme is present in the promitochondrial fraction.

The substrate must possess a C24,25 double bond. Evidence for the mechanism depicted above was obtained in several fungi, using [*methyl*-$^{2}H_3$]methionine and [2-^{14}C,(4*R*)-4-$^{3}H_1$]mevalonic acid as sterol precursors. The methylation process in those chlorophyte algae which synthesize ergosterol differs from that in fungi. Both cycloartenol and cyclolaudenol are present in *Chlorella*, and studies with [*methyl*-^{14}C]-*S*-adenosylmethionine (SAM) and [*methyl*-$^{2}H_3$]methionine support the mechanism shown above. Confirmatory studies with labeled mevalonic acid are not possible because green algae are impermeable to mevalonic acid.

In the biosynthesis of plant sterols, the side chain may be subject to further methylations, all dependent on SAM as the methyl donor. The 24α-ethyl groups found in sitosterol and stigmasterol are formed by the transfer of a methyl group to a 24-methylene sterol, which is generated as shown above; the resulting *Z*-24-ethylidene sterol intermediate is isomerized to a Δ^{24}-24-ethyl sterol whose C-24(25) double bond is subsequently stereospecifically reduced by an NADPH-dependent sterol Δ^{24}-reductase.

usual manner (i.e. as shown in Fig. 3) and which, by convention, are said to be β-orientated. All substituents of the ring system that are *cis* with respect to the angular methyl groups are also said to be β-orientated while those that are *trans* are designated as α. Since C-atoms 5 and 10 are the atoms common to rings A and B and since the C10 methyl is β-orientated, the presence of a 5α-H clearly indicates a *trans* A/B-fusion and therefore a steroid of the 5α-series (i.e. with the 'A/B *trans*, B/C *trans*, C/D *trans*' ring system). Similarly the presence of a 5β-H indicates a *cis* A/B-fusion and therefore a steroid of the 5β-series (i.e. with the 'A/B *cis*, B/C *trans*, C/D *trans*' ring system).

When steroid ring systems are represented as projections onto the plane of the paper (i.e. as in the Table) substituents of the ring atoms that have the α-orientation are drawn with broken bond lines (---- or ┆┆┆┆┆┆┆┆), while those that have the β-orientation are drawn with solid bond lines (——) which may be thickened (▬▬) or wedge-shaped (◀). When the orientation of a substituent is not known it is drawn with a wavy (〰) bond line and is designated by the lower case Greek letter *xi* (ξ).

Structural differences among S. arise from the type, number, position and orientation of the substituents, as well as the number and position of double bonds. A substituent containing oxygen is denoted by adding to the name of the parent hydrocarbon the appropriate prefix, e.g. hydroxy, oxo, or suffix, e.g. -ol, -one, preceded by the number of the carbon to which it is attached and its orientation if appropriate; if hydroxyl and keto groups occur in the same steroid the former is denoted by a prefix and the latter by a suffix, e.g. 3α-hydroxy-5α-androstan-17-one. The presence of double bond(s) is indicated by changing the terminal

Oxidative removal of 14α-methyl (C-32) as formate

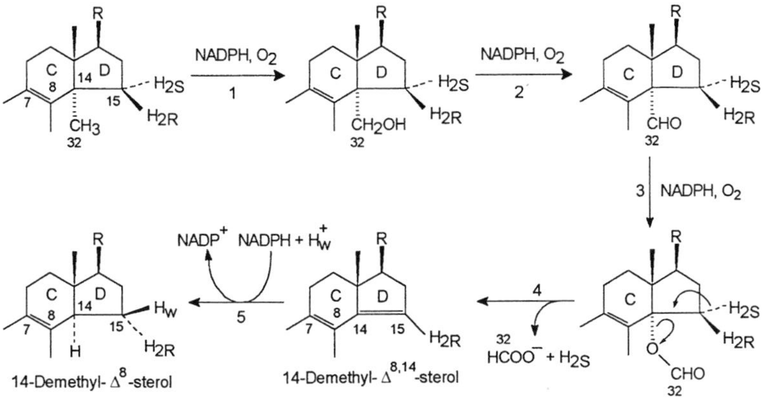

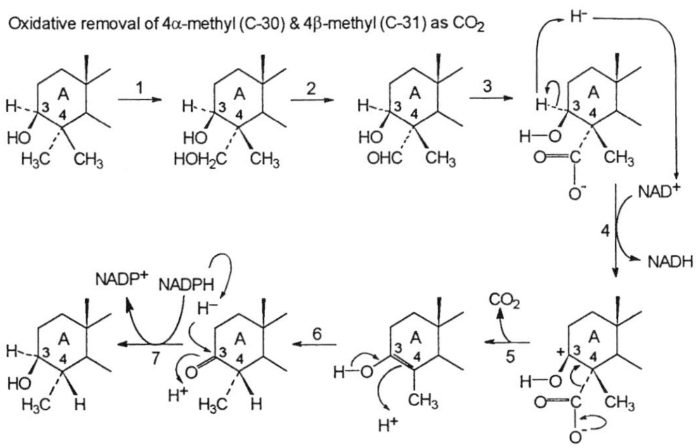

Oxidative removal of 4α-methyl (C-30) & 4β-methyl (C-31) as CO_2

Fig. 10. *Sequential oxidation and removal of C32, C31 and C30 of sterols.* In fungi and animals, C32 (14α-methyl group) is removed first. The overall process involves two enzymes, namely: (i) sterol 14-demethylase, a cytochrome P450 (often referred to as cyt. $P450_{14DM}$) which catalyses steps 1–4 above, consuming three molecules of NADPH and O_2 per sterol molecule, and eliminating C32 as formate and the 15α-H (*ex* 2-*pro-S* H of MVA) as H^+, so producing a C14,15 double bond; and (ii) sterol Δ^{14}- reductase, an NADPH-dependent oxidoreductase, which catalyses step 5 above, the saturation of the C14,15 double bond. The next carbon to be removed is C30 (4α-methyl). Its removal causes the remaining methyl on C4 (C31, originally 4β) to take up the 4α orientation, thereby allowing the same sterol 4-demethylase complex to catalyse both demethylations at C-4. The 4-demethylase complex is microsomal and consists of (i) a non-cytochrome P450 monooxygenase called 4-methylsterol oxidase, which uses three molecules of NADPH and O_2 per sterol molecule to catalyse steps 1–3 above and generates a 4α-carboxy sterol, (ii) an NAD^+-dependent sterol 4-decarboxyase, which catayses the removal of the carboxyl group as CO_2 (steps 4 & 5) so generating a Δ^3-3β- hydroxy sterol, which then spontaneously tautomerizes (step 6) to a 3-ketosteroid, and (iii) an NADPH-dependent 3-ketosteroid reductase, which catalyses the stereospecific reduction of the 3-keto group to a 3β-hydroxy group (step 7).

In photosynthetic organisms the same enzyme-catalysed reactions are involved, but the sequence of demethylations is different. The 4α-methyl group is removed first, producing a 4α,14α-dimethyl sterol (4-demethylcyclolaudenol in chlorophyte algae, and cycloeucalenol in tracheophytes). This is followed by the opening of the 9β,19-cyclopropane ring to give the corresponding Δ^8-sterols (Fig. 13) followed by isomerization to Δ^7-sterols (Fig. 11). The 14α-methyl group is then removed, yielding a Δ^7-4α-methyl sterol. The last methyl group to be removed is the one that was originally 4β.

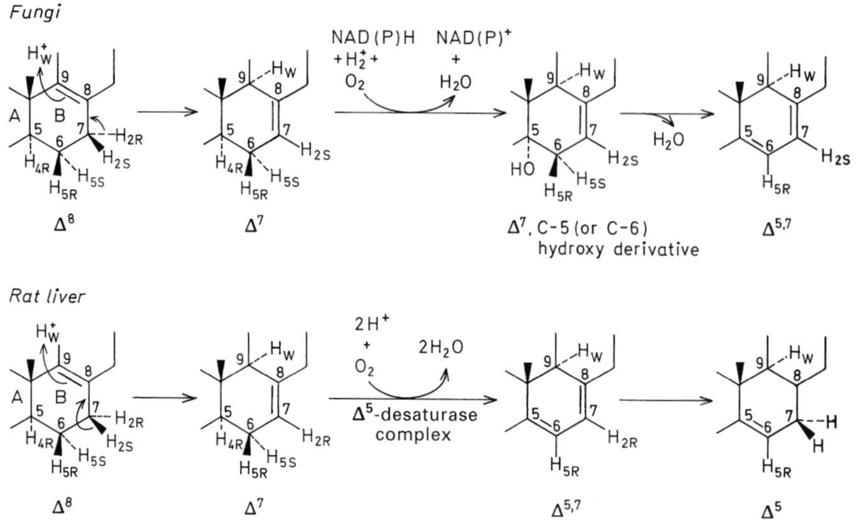

Fig. 11. *Double bond transformations in the sterol ring system.* The sequence of nuclear double bond transformations is well established in animal sterol biosynthesis, and the same sequence appears to occur in green plants. This sequence is $\Delta^8 \rightarrow \Delta^7 \rightarrow \Delta^{5,7} \rightarrow \Delta^5$. The last step does not occur in the biosynthesis of ergosterol, which retains the UV-absorbing (λ_{max} 282 nm) $\Delta^{5,7}$-dienoid system.

'ane' of the name of the parent hydrocarbon to 'ene, diene' etc. or 'en, dien' etc., and immediately preceding it with the number given to the lower-numbered of the two carbon atoms joined by the double bond, e.g. androst-5-ene, cholest-5,7-dien-3β-ol. When a double bond is denoted by a single number, as in the two preceding examples, it means that it connects the numbered carbon atom to the carbon atom with the next highest number, i.e. 8-en denotes a double bond between C8 and C9. A double bond that does not link two carbon atoms with consecutive numbers is denoted by the lower number followed immediately by the other one in parenthesis, e.g. 8(14)-en denotes a double bond between C-8 and C-14. The position of double bonds is also denoted by use of the numbers described above as superscripts immediately following the upper case Greek letter *delta* (Δ), e.g. $\Delta^{5,8(14)}$-cholestadien-3β-ol denates a S. with one double bond between C5 and C6 and another between C-8 and C-14. The latter system is however discouraged in the 1989 IUPAC-IUB recommendations except for generic terms such as Δ^5-steroids; however it appears so widely in scientific literature that it cannot be ignored.

The 17β-substituent, containing 2–10 carbon atoms, that is present in many naturally occurring S. is usually referred to as the 'side chain'. Depending on its length it has one or two asymmetric centres, C20 and C24. Two methods have been used to designate the configuration at these centers, namely the α/β-system [see L.F.Fieser & M.Fieser (1959) *Steroids* 4th edn, pp. 337–340] and the *R/S* (Cahn-Ingold-Prelog) convention [see R.S.Cahn *J. Chem. Educ.* **41** (1964) 116–125]. The α/β-system is peculiar to the steroid side chain and therefore unrelated to the α/β nomenclature described earlier to designate the orientation of substituents of the carbon atoms of the steroid nucleus. It is based on an arbitrary convention for the definition of the configuration at C20 of 20-hydroxypregnane, a steroid with -CHOH-CH_3 as its 17β substituent, and then widened to the more general structure shown in Fig. 4, in which the OH and CH_3 of 20-hydroxypregnane become X and R respectively, where X can be OH or CH_3, and R is the C20 substituent with the longest carbon chain (which is CH_3 in the case of 20-hydroxypregnane). When this structure is arranged as shown in Fig. 4, with R projecting into the plane of the paper and the other two substituents, H and X, projecting outward, the two possible configurations at C20 (and thus the two C20 stereoisomers) can be depicted by changing the positions of H and X [Fig. 4, (a) & (c)], thus allowing the generation of two distinct Fischer projection formulas [Fig. 4, (b) & (d)]. Structure 4(b), with X to the right of the vertical carbon chain (composed of C-atoms 17, 20 and the first carbon of R), was arbitrarily given the α-configuration at C20 while structure 4(d), with X to the left, was given the β-configuration. The extension of this convention to C24 is shown in Fig. 5; with the side chain arranged as a Fischer projection the 24α and 24β configurations have the X substituent (CH_3 or CH_2CH_3) to the right and left of the carbon chain respectively (i.e. 24α-X is on the same side of the carbon chain as 20α-X and the opposite side to 20β-X; similarly 24β-X is on the same side of the carbon chain as 20β-X and the opposite side to 20α-X). The use of the α/β system is discouraged in the IUPAC-IUB recommendations in the naming of specific compounds but allowed in the names of enzymes, e.g. 20α-hydroxysteroid dehydrogenase. However the α/β system has the great advantage over the *R/S* system in being independent of the substituents on neighboring atoms and appears so widely in scientific literature that it cannot be ignored. The *R/S* system, though accurate

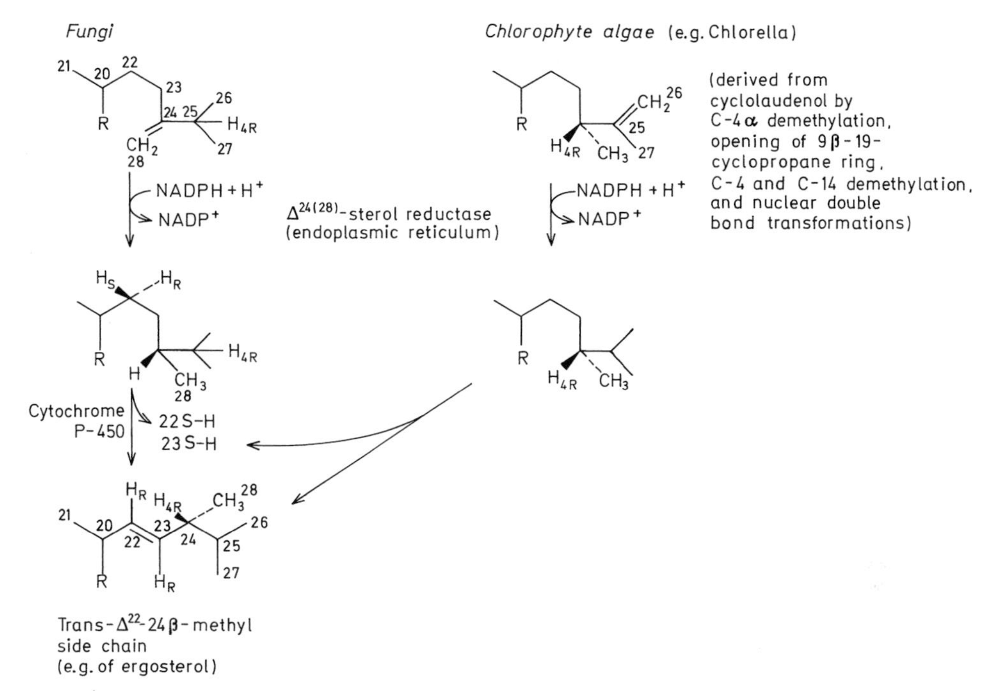

Fig. 12. *Saturation of the C24,28 double bond (and the C25,26 double bond) and formation of the C22,23 double bond in the sterol side chain.* Saturation of the C24,28 double bond, catalysed by the microsomal, NADPH-dependent sterol $\Delta^{24(28)}$-reductase, only occurs in sterol biosynthesis in fungi and certain members of the plant kingdom, typified by the Chrysophyte algae, which form 24β-alkylated sterols by directly reducing the $\Delta^{24(28)}$-double bond of 24-methylene- or 24-ethylidene sterols generated by transmethylations at C24 and C28 (Fig. 9). Saturation of the C25,26 double bond, catalysed by the NADPH-dependent sterol Δ^{25}-reductase, occurs in sterol biosynthesis in Chlorophyte algae and higher plants that form 24β-alkylated sterols by the Δ^{25}-sterol route (Fig. 9). Animals have no need of either of these enzymes because they do not carry out sterol 24-alkylation; instead they have an NADPH-dependent sterol Δ^{24}-reductase, which catalyses the saturation of the C24,25 double bond. Sterol Δ^{22}-desaturation, catalysed by a specific cytochrome P450 (cyt. $P450_{22Desat}$), occurs during sterol biosynthesis in fungi and green plants but not in animals.

Fig. 13. *Opening of the 9β,19-cyclopropane ring during sterol biosynthesis in photosynthetic organisms.* This is catalysed by a microsomal enzyme that appears to require no cofactors and generates a sterol product that possesses a 10β-methyl group and a Δ^8-double bond.

and wholly admirable because of its applicability throughout organic chemistry, can lead to confusion amongst the unwary when used to specify configuration at the steroid C24, because the designation 24*R* or 24*S* depends on the priorities of the atoms or groups of atoms covalently attached to C24. For example the introduction of a Δ^{22}-double bond into the side chain of 5α-ergostane, which has the 24*S* configuration (see Table), confers upon C24 the *R* configuration, in spite of the fact that there has been no change in the relative spacial locations of the substituents of C24; the change in configuration from 24*S* to 24*R* is simply due to the fact that, under the Priority Rule of the *R/S* convention, C25 has greater priority than C23 in the saturated side chain, whereas the reverse obtains in the Δ^{22}-side chain. This problem does not exist with the α/β system. The relationship of the α/β system to the *R/S* convention at C24 with respect to the absence or presence of a Δ^{22}-double bond can be summarized as follows: 24α ≡ 24*R* (Δ^{22}

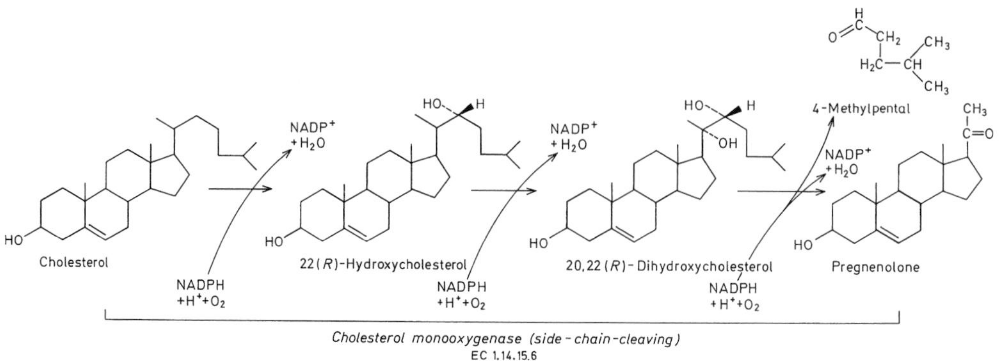

Fig. 14. *Conversion of cholesterol to pregnenolone by oxidative side chain cleavage.* This is an essential first stage in the biosynthesis of progesterone, androgens, estrogens and adrenal cortical steroids from cholesterol.

absent) ≡ 24*S* (Δ^{22} present) and 24β ≡ 24*S* (Δ^{22} absent) ≡ 24*R* (Δ^{22} present).

Ring contraction by loss of a CH_2 is indicated by the prefix 'nor'. The CH_2 lost is considered to be the one with the highest numbered carbon of that particular ring and this number precedes 'nor' in the name of the steroid, e.g. 4-nor-5α-androstane (C-4 is missing from ring A). Ring expansion by the inclusion of a CH_2 is indicated by the prefix 'homo' preceded by a number/letter combination specifying the formal location of the new carbon atom, e.g. 4a-homo-5α-androstane. Ring cleavage, with the addition of a hydrogen atom at each terminal group so created, is indicated by the prefix 'seco' preceded by the numbers of the carbons previously joined together, e.g. 2,3-seco-5α-cholestane.

S. are biosynthesized ultimately from acetyl-CoA. The early stages of this biosynthesis, leading to the non-cyclic triterpene squalene, is discussed under Terpenes (see). Squalene is oxidized to all-*trans*-(3*S*)-2,3-epoxysqualene, which is then cyclized to lanosterol in non-photosynthetic organisms, and to cycloartenol in photosynthetic organisms (Fig. 6). Lanosterol is the biosynthetic precursor of zoosterols, e.g. cholesterol, and mycosterols, e.g. ergosterol, (Fig. 7), whereas cycloartenol is the biosynthetic precursor of the phytosterols, e.g. sitosterol (Fig. 8). Details of reactions involved in the biosynthesis of sterols from lanosterol and cycloartenol are shown in Figs. 7–13 [see also E. I. Mercer *Prog. Lipid Res.* **34** (1993) 357–416]. Further biological conversion of cholesterol to a variety of important animal steroids commences with side chain cleavage to pregnenolone (Fig. 14). Dietary cholesterol is also the biosynthetic precursor of ecdysones (see) in carnivorous insects; ecdysones are biosynthesized in phytophagous insects from dietary phytosterols, a process that involves dealkylation at C24.

In fungal sterol biosynthesis the $\Delta^8 \rightarrow \Delta^7$-double bond isomerisation involves the loss of the 7α-hydrogen (derived biosynthetically from the 2-*pro-R* hydrogen of (+)-*R*-mevalonic acid [MVA]), whereas in animal, higher plant and chlorophyte algal sterol biosynthesis the 7β-hydrogen (*ex* 2-*pro-S* H of MVA) is lost.

Insertion of the C5,6 double bond ($\Delta^7 \rightarrow \Delta^{5,7}$) involves removal of the 5α (*ex* 4-*pro-R* H of MVA) and 6α (*ex* 5-*pro-S* H of MVA) hydrogens. In rat liver and yeast microsomal preparations the reaction requires aerobic conditions and NADH or NADPH and both of the eliminated hydrogens are found in H_2O. This suggests a mechanism in which hydroxylation at C5 or C6 is followed by dehydration across these two carbons or, alternatively, a fatty acid-type desaturation. There is evidence of the former in yeast, whereas the latter seems more likely in rat liver where the multienzyme complex catalysing it consists of the Δ^5-desaturase itself, cytochrome b_5 and an NAD(P)H-dependent flavoprotein; the Δ^5-desaturase is a monooxygenase which uses electrons derived from NAD(P)H, via the flavoprotein and cytochrome b_5, and O_2 to effect the removal of the two hydrogens from the sterol.

The insertion of the *trans* Δ^{22}-double bond found in myco- and phytosterols involves the stereospecific removal the 22-*pro-S* (*ex* 2-*pro-S* H of MVA) and 23-*pro-S* (*ex* 5-*pro-S* H of MVA) hydrogens and is probably catalysed by a cytochrome P450 monooxygenase.

Steroid saponins: see Saponins.

Sterols: a group of naturally occurring steroids possessing a 3β-hydroxyl group and a 17β-aliphatic side chain. S. are structural derivatives of the parent hydrocarbons, cholestane, etc. (see Steroids, Table). They occur in animal and plant cells as free S. and as glycosides and esters. According to their origin, they are classified as zoosterols (animals), phytosterols (plants), mycosterols (fungi) and marine S. (marine animals and plants). More recently, S. have also been found in bacteria. Some important S. are Cholesterol (see), Stigmasterol (see) and Ergosterol (see). S. are essential dietary components for insects. With digitonin, S. form insoluble addition compounds, and inhibit the hemolytic action of saponins. For biosynthesis, see Steroids. S. are isolated from the nonsaponifiable fraction of neutral fat. They are separated and identified by adsorption chromatography on silicic acid (column and thin layer) and by gas chromatography. Mass spectrometry is widely used for identification of S. With strong acid, S. give typical color reactions, which are used for their quantitative determination, e.g. Liebermann-Burchard reaction (see).

Steviol: the aglycon of Stevioside (see).

Stevioside: a tetracyclic diterpene, M_r 804, m.p. 196–198 °C, $[\alpha]_D$ −39.3° (c = 5.7, ethanol), from the bush, *Stevia rebaudiana,* which is native to Paraguay. S. is both a glycoside and a Glucose ester (see). Enzymatic hydrolysis releases the aglycon, steviol ($R_1 = R_2 = H$), M_r 318, m.p. 215 °C, $[\alpha]_D$ −94.7° (ethanol). Acid hydrolysis causes a Wagner-Meerwein rearrangement of rings C and D. to produce isosteviol, m.p. 234 °C, $[\alpha]_D$ −78° (ethanol). S. is 300 times sweeter than sucrose and would be an ideal sweetening agent; commercial exploitation is difficult on account of its low concentration (about 6 g S. per kg dried leaves). It possesses a gibberellin-like growth stimulating activity.

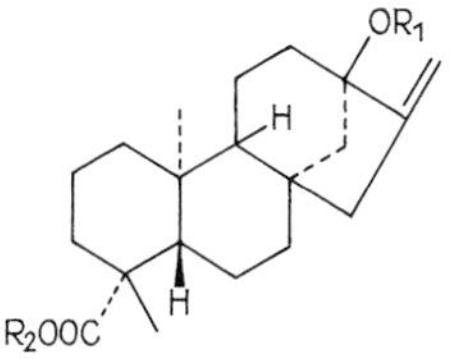

R_1 = Glucose-Glucose
R_2 = Glucose

Stevioside

STH: acronym of somatotropic hormone.

STI: acronym of soybean trypsin inhibitor.

Stigmastane: see Steroids.

Stigmasterol: stigmasta-5,22-diene-3β-ol, M_r 412.7, m.p. 170 °C, $[\alpha]_D$ −49° ($CHCl_3$), a widely distributed phytosterol (see Sterols), first isolated from calabar beans, and later from many other sources, e.g. soybeans, carrots, coconut oil and sugar cane wax. It is used as a starting material in the technical synthesis of steroid hormones.

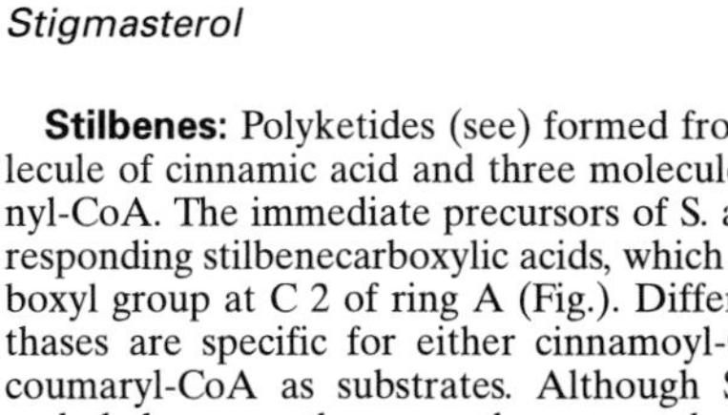

Stigmasterol

Stilbenes: Polyketides (see) formed from one molecule of cinnamic acid and three molecules of malonyl-CoA. The immediate precursors of S. are the corresponding stilbenecarboxylic acids, which have a carboxyl group at C 2 of ring A (Fig.). Different S. synthases are specific for either cinnamoyl-CoA or *p*-coumaryl-CoA as substrates. Although S. synthase and chalcone synthase use the same substrates, they are distinct enzymes (antibodies to one do not cross-react with the other). S. synthase has been purified from cell suspension cultures of *Arachis hypogaea* (peanut). It is a dimer of M_r 90,000 (monomer M_r 45,000), which converts 1 mol *p*-coumaroyl-CoA and 3 mol malonyl-CoA into 3,4′,5-trihydroxystilbene (resveretrol) (see Orchinol). [A. Schoeppner & H. Kindl *J. Biol. Chem.* **259** (1984) 6806–6811]

Stilbestrol, *stilboestrol, trans-stilbestrol:* a synthetic, nonsteroid Estrogen (see), used in estrogen therapy.

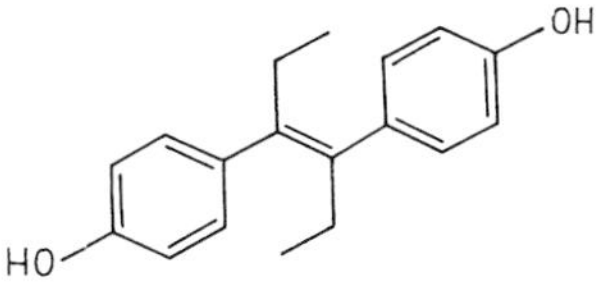

Stilbestrol

Stimulant amines: see Antidepressants.

Stoichiometric model, *stoichiometric reaction scheme:* in enzyme kinetics, a chemical reaction equation in which molecules involved in the reaction, including the enzyme, are represented as letters. These also indicate the molar concentrations. The numbers in front of the letters are the stoichiometric coefficients, which indicate how many moles or molecules reactant are involved in the corresponding reaction step (the coefficient is usually omitted). The simplest example of a S.M. in enzyme kinetics is: $E + S \rightleftharpoons ES \rightarrow E + P$. More complicated reactions are more conveniently expressed by enzyme graphs.

Stop codon: see Termination codon.

Strand polarity, *antiparallel conformation:* the polarity of nucleotide chains, with reference to the sequence of 3′,5′-phosphodiester bonds. Polynucleotide chains have a 3′-end (the terminal sugar residue is linked to the preceding residue via its 5′-hydroxyl, and the 3′-hydroxyl is free or phosphorylated) and a 5′-end (the 5′-hydroxyl is free or phosphorylated). In DNA and other double-stranded nucleic acids, the two strands always lie antiparallel to one another. During replication and transcription, the newly synthesized strand is always antiparallel to its template. The polarity of a nucleotide strand is indicated by $3' \rightarrow 5'$ or $5' \rightarrow 3'$.

Strand selection: the ability of DNA-dependent nucleic acid polymerases to choose the codogenic strand of the double-stranded DNA.

Streptavidin: a tetrameric protein from *Streptomyces avidinii* that binds biotin. It is used in various laboratory techniques, e.g. Immunoassays (see), to detect biotinylated molecules. It is preferred to Avidin (see) for this purpose because it has a more favorable isoelectric point.

Streptogenin peptides: natural products (e.g. liver extracts, peptones, partial hydrolysates of proteins) or synthetic peptides, which stimulate the growth of microorganisms, especially lactic acid bacteria. The unknown growth stimulant is called streptogenin. The growth stimulation of *Lactobacillus casei* by 1 mg standard liver extract is equivalent to one streptogenin unit. In synthetic peptides streptogenin activity depends on the presence of cysteine and high activity is guaranteed by a neighboring or *N*-terminal leucine residue.

Streptokinase: see Plasmin.

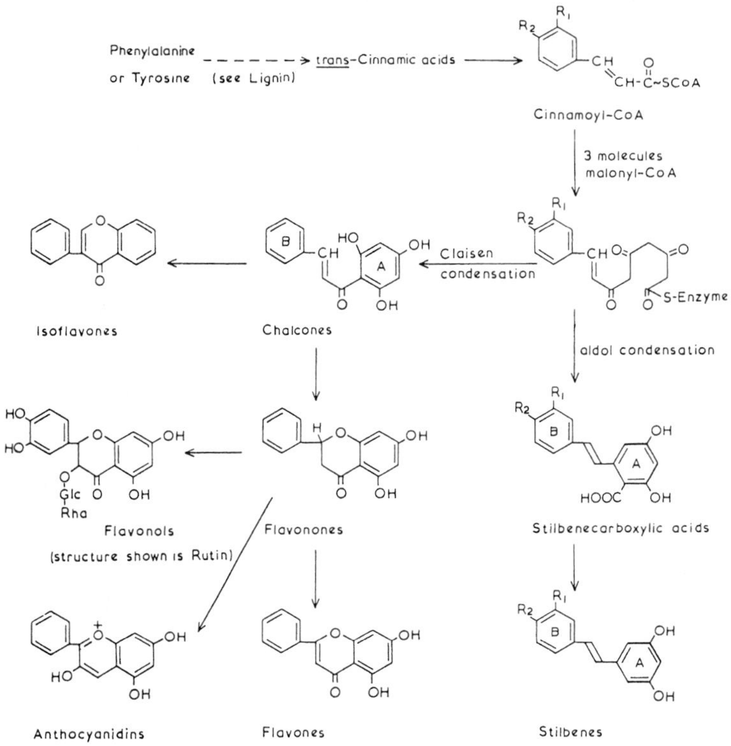

Phenylalanine or Tyrosine (see Lignin) → trans-Cinnamic acids → Cinnamoyl-CoA

3 molecules malonyl-CoA

Claisen condensation

S-Enzyme

aldol condensation

Isoflavones

Chalcones

Flavonols (structure shown is Rutin)

Flavonones

Stilbenecarboxylic acids

Anthocyanidins

Flavones

Stilbenes

Biosynthesis of stilbenes and related compounds

Cinnamic acid precursor of ring B	R_1	R_2	Stilbene
Cinnamic acid	H	H	Pinosylvin
p-Coumaric acid	H	OH	Resveratrol
Caffeic acid	OH	OH	Piceatannol
Isoferulic acid	OH	OCH_3	Rhapontigenin

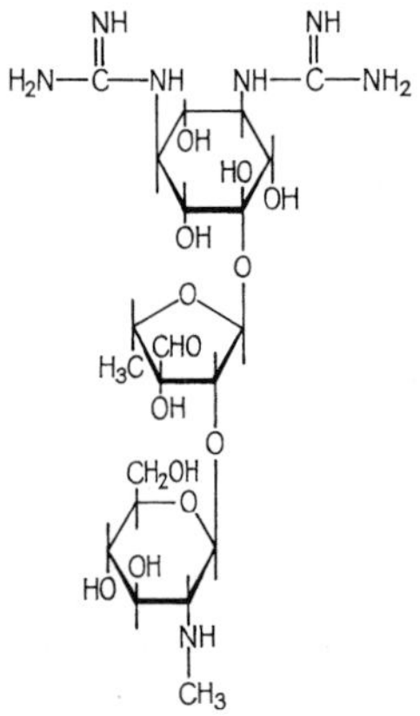

Streptomycin 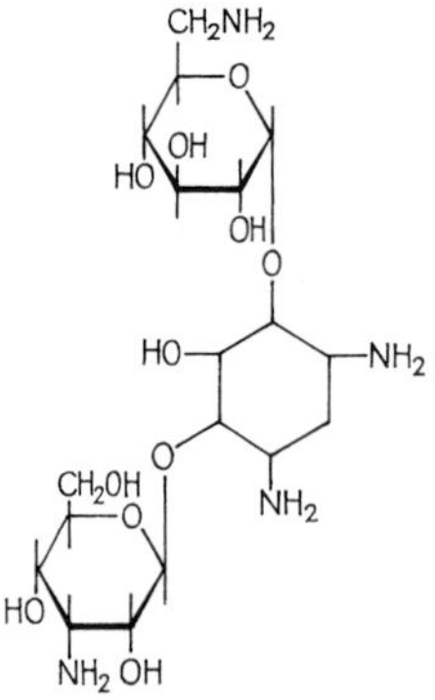 *Kanomycin* 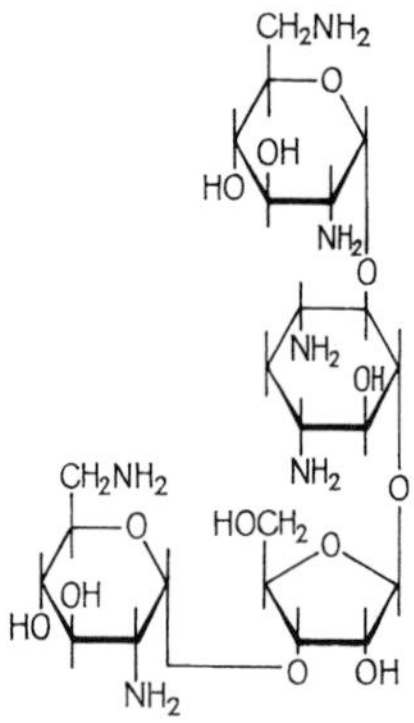 *Neomycin C*

Streptomycin: an antibiotic from *Streptomyces griseus.* M_r 581.6. S. is an aminoglucoside, in which streptidine is linked glycosidically to the disaccharide, streptobiosamine (Fig.). S. inhibits protein biosynthesis on 70S ribosomes. It becomes bound to the 23S core protein of the 30S ribosomal subunit. This protein appears to be responsible for the binding of mRNA, which is prevented by S. Similar modes of action are shown by the aminoglucoside antibiotics, kanamycin and neomycin (Fig.) and by paromomycin, kasugamycin, spectinomycin and gentamycin; all are used therapeutically. S. was one of the first antibiotics used against tuberculosis. However, it has been shown to have toxic side effects, e.g. damage to the auditory nerves.

L-Streptose: 5-deoxy-3-formyl-L-lyxose, a monosaccharide, M_r 162.14, with a branched carbon chain. In combination with 2-deoxy-2-methylamino-L-glucose, L-S. forms streptobiosamine, the disaccharide component of Streptomycin (see). L-S. is biosynthesized by rearrangement of an unbranched hexose.

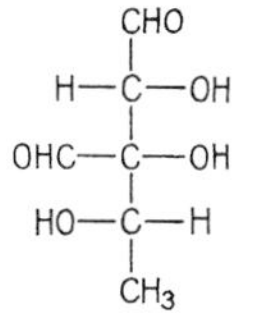

CHO
H—C—OH
OHC—C—OH
HO—C—H
CH_3

L-Streptose

Stringency: the *conditions* under which two strands of nucleic acid bind together in an antiparallel manner (i.e. one strand runs 5′ → 3′ whilst the other runs 3′ → 5)' by means of hydrogen bonding between complementary base pairs (G/C and A/T(U) to form a DNA:DNA or DNA:RNA duplex, a process that has been variously called annealing and hybridization. Conditions of high stringency require a perfect or near perfect match of complementary nucleotide sequences. Conditions of low, or relaxed, stringency allow the binding of nucleic acid strands with considerably less than perfect sequence complementarity. Stringency can be decreased by decreasing the temperature (e.g. from 60 °C to 45 °C) or by increasing the salt concentration, and *vice versa.* Reducing stringency is useful when searching for homologous sequences in the DNA of other species, and in allowing the use of primers in the Polymerase chain reaction (see), and probes in DNA fingerprinting (see), that do not have the precise sequence complementarity of their target DNA.

Stringent control: the tight control of a factor, such as a plasmid copy number, to a limited number (usually one) per genome. Its converse is 'relaxed control'.

Stringent response: a mechanism present in bacteria such as *E. coli*, but not in eukaryotes, by which ribosomes are able to synthesize certain guanosine polyphosphates (pppGpp and ppGpp) in response to growth conditions in which amino acids are limiting [G. Edlin & P. Broda *Bacteriol. Rev.* **32** (1968) 206–226; J. A. Gallant *Annu. Rev. Genet.* **13** (1979) 393–415]. The guanosine polyphosphates, in a manner that is largely unknown but which probably involves the regulation of gene transcription, cause the selective inhibition of the synthesis of rRNA and tRNA, with less effect on mRNA synthesis, thereby avoiding the wasteful synthesis of more ribosomes and tRNA and conserving the amino acids generated by protein turnover for the synthesis of essential proteins. The net result is a decrease in a wide range of cellular activities until growth conditions improve.

Stroma, ***matrix:*** colorless, homogeneous (by light microscopy) ground substance of cell organelles, like Chloroplasts (see) and Mitochondria (see).

Strophanthidin: see Strophanthins.

Strophanthins, ***strophanthosides:*** cardenolide cardiac glycosides from *Strophanthus* spp., e.g. g-strophanthin (ouabain) from *Strophanthus gratus,* and k-strophanthin from *Strophanthus kombe.* g-Strophanthin contains the aglycon, ouabagenin (g-strophanthidin), M_r 438.52, m.p. 255 °C, $[\alpha]_D + 11°$, linked glycosidically at position 3 with L-rhamnose. Ouabain is a specific inhibitor of the membrane-bound Na^+,K^+-ATPase, which is responsible for maintaining high intracellular concentrations of K^+ and low intracellular concentrations of Na^+. Ouabain therefore inhibits sodium ion transport out of the cell and potassium ion transport into the cell, and simultaneously inhibits entry of glucose and amino acids. The aglycon of k-strophanthin is strophanthidin, M_r 401.51, m.p. 136 and 235 °C, $[\alpha]_D + 41°$, which is linked to one molecule each of D-glucose and D-cymarose. Strophanthidin is the aglycon of other cardiac glycosides, e.g. convallatoxin of *Convallaria majalis* (lily of the valley); it differs structurally from digitoxin (see Digitalis glycosides) by the presence of an 18-aldehyde group. S. are highly potent toxins, which have been used In Africa since antiquity as arrow poisons.

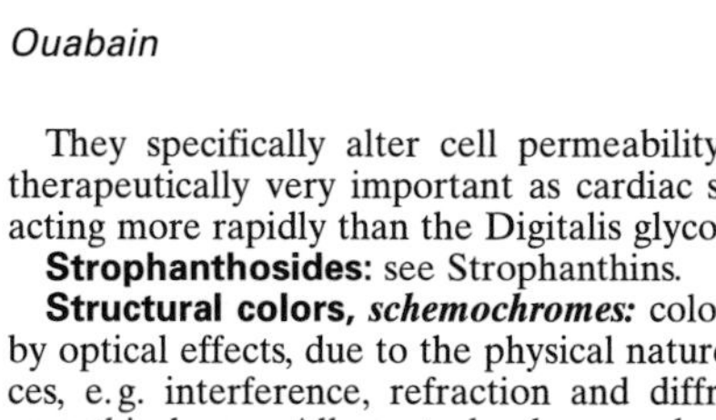

HO
HO
$HOCH_2$
18
OH
3
L-Rhamnose-O
OH
O
O

Ouabain

They specifically alter cell permeability, and are therapeutically very important as cardiac stimulants, acting more rapidly than the Digitalis glycosides.

Strophanthosides: see Strophanthins.

Structural colors, ***schemochromes:*** colors created by optical effects, due to the physical nature of surfaces, e.g. interference, refraction and diffraction on very thin layers. All spectral colors can be produced in this way, including white (total reflection) and black (total absorption). A change of color with the viewing angle gives rise to iridescence. S.c. occur frequently in Nature, e.g. the color effects of pearls and the inner layers of sea shells, which result from light interference at thin calcium carbonate layers. Bird feather and butterfly wing colors are due to chemical pigments with a contribution from S.c.

Structural genes: see Operon.

Amino acid composition of some structural proteins

Amino acid residues	Skin collagen (round worm)	Skin collagen (calf)	Cuticulin (round worm)	Elastin (Lig. nuchae)	Fibroin (*B. mori*)	Keratin (wool)	Resilin (locust)
Glycine	274	330	150	360	445	85	376
Gly + Ala	346	441	317	597	739	146	487
Pro + Hypro	312	221	301	124	3	85	79
Hypro	16	98	0	12	0	0	0
Lys + Hylys + Arg + His	82	88	36	12	9	104	45
Asp + Glu	136	117	146	39	23	173	152
Cys/2	27	0	24	0	0	106	0
Trp	0	0	0		2	12	

Structural proteins, ***skeletal proteins, scleroproteins, fibrous proteins:*** simple animal proteins with structure and support functions. They are generally insoluble in water and salt solutions. The best known S.p. are the cystine-rich Keratins (see). Others are Collagen (see), Elastin (see), Crystallins (see), silk fibroin, chondrin, spongin, etc. They are subdivided on the basis of chain conformation into: 1. S.p. with α-helical structure, e.g. α-keratins; 2. S.p. with β-pleated sheet structure, e.g. β-keratins, silk fibroin; 3. S.p. with triple helical structure, e.g. collagen. Conformation is related to amino acid composition. The amino acid composition of S.p. with β-pleated sheet structure shows 90% of the simple amino acids, glycine, alanine and serine, whereas α-helical S.p. contain a much higher percentage of bulky side chains; α-keratin also contains a large number of cystine residues. Collagen characteristically contains a high concentration of the nonhelix forming amino acids, proline and hydroxyproline. Owing to their unusual amino acid composition and fibrous structure, and their resulting insolubility in water, S.p. are not attacked by most proteolytic enzymes. Poor digestibility and absence of essential amino acids make them unsuitable as dietary proteins.

Strychnine: a Strychnos alkaloid, M_r 334.42, m.p. 286–288 °C, b.p.$_5$ 270 °C, $[\alpha]_D^{20}$ – 139° ($CHCl_3$). S. is the chief alkaloid in the seeds of tropical *Strychnos* spp., from which it is obtained commercially; the total synthesis by Woodward is uneconomic and only of scientific interest. S. has many therapeutic properties, but its high toxicity (see Neurotoxins) precludes its extensive use in medicine; 5 mg in children and 30–100 mg in adults are sufficient to cause death from muscle rigor. It is used as a rodent poison (mixed with red-dyed wheat grains). For formula and biosynthesis see Strychnos alkaloids. S. causes a blockade of central inhibition by selectively antagonizing the effect of glycine (glycine is an inhibitory transmitter of 4-aminobutyrate receptors in the central nervous system of vertebrates, particularly in the spinal cord). S. is thought to interact with receptor sites on postsynaptic membranes, thus preventing access of glycine.

Strychnos alkaloids: a group of indole alkaloids from the tropical plant genus *Strychnos.* The highly toxic main alkaloids, Strychnine (see) and Brucine (see), contain a heptacyclic ring structure (Fig.).

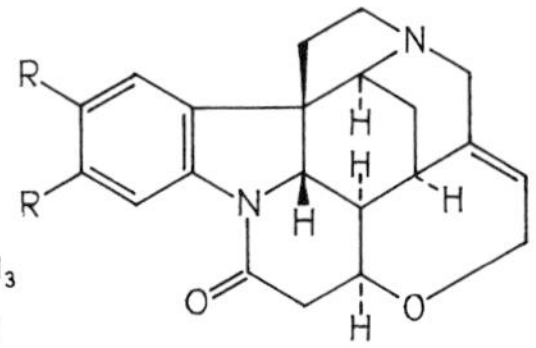

Brucine R=OCH_3
Strychnine R=H

Strychnos alkaloids

They are accompanied by Yohimbine (see)-type alkaloids. S.a. are biosynthesized from tryptophan and a terpenoid C_{10}-unit.

Suberic acid: octanedioic acid, HOOC–$(CH_2)_6$–COOH, a higher saturated dicarboxylic acid. m.p. 142 °C, b.p. 300 °C (d.). S.a. is formed (together with its homologs azelaic and sebacic acids) by oxidation of cork or ricinus oil with nitric acid. It is obtained in higher yields from cyclooctene (technical synthesis).

Submersed culture: see Fermentation techniques.

Submitochondrial particle: see Electron transport particle, Mitochondria.

Substance P, ***SP:*** an undecapeptide: Arg-Pro-Lys-Pro-Gln-Gln-Phe-Phe-Gly-Leu-Met-NH_2, present in brain and intestine, and a member of the Amine precursor uptake decarboxylase system (see). Studies were first performed on dried, powdered acid-alcohol extracts of equine brain and intestine; the active principle was called substance P (P for powder), a name that is now widely accepted. The highest concentrations of substance P-like immunoreactivity (SPLI, measured by radioimmunoassay) in the brain are found in the substantia nigra, habenula, hypothalamus, spinal cord sensory ganglia, dorsal roots and substantia gelatinosa. It is now established that SP acts as a neurotransmitter, probably throughout the central nervous system, and certainly in the sensory neurons of the spinal cord. SPLI has also been found in one type of enterochromaffin cell (EC_1) present in the gastrointestinal tract and bile duct. SP stimulates contraction of isolated rabbit jejunum and causes transient hypotension when injected intravenously into anesthetized rabbits. It shows close structural similarities to Physolaemin (see) and Eledoison (see). [*Substance P in the Nervous System, Ciba Foundation Symposium 91,* R. Porter and M. O'Connor (eds.), Pitman, 1982]

Substrate binding center: see Active center.
Substrate binding site: see Active center.
Substrate constant: see Michaelis-Menten kinetics.
Substrate level phosphorylation: adenosine triphosphate (ATP) synthesis not involving photosynthetic phosphorylation or oxidative phosphorylation in the respiratory chain. S.l.p. occurs in Glycolysis (see) and in the conversion of succinyl-CoA into succinate in the Tricarboxylic acid cycle (see). In some bacteria, S.l.p. can also occur from carbamoyl phosphate, formed in the degradation of L-citrulline or allantoin. The fine mechanism of S.l.p. is best studied for the glycolytic enzymes, glyceraldehyde 3-phosphate dehydrogenase and 3-phosphoglycerate kinase. The enzyme-substrate complex is oxidized to an energy-rich acylthioester by NAD^+; the acyl group is transferred to phosphoric acid to form 1,3-diphosphoglycerate, which contains a low energy alcohol-phosphate ester and a high energy acyl (carboxyl) phosphate. By the action of 3-phosphoglycerate kinase, the energy-rich phosphate on C1 is transferred to ADP, forming ATP and 3-phosphoglycerate (for formulas, see Fig. 2, Glycolysis). A thioester (i.e. succinyl-CoA) is also formed in the oxidative decarboxylation of 2-oxoglutarate in the TCA-cycle. In this case the first product of S.l.p. is GTP, which can give rise to ATP by transphosphorylation: GTP + ADP → ATP + GDP. Synthesis of ATP from phospho*enol*pyruvate and ADP in glycolysis is also S.l.p. There is no unifying general mechanism of S.l.p.
Substrate mediated transhydrogenation: see Hydrogen metabolism.
Substrate specificity: the ability of an enzyme to recognize and specifically bind its substrate. S.s. is a function of protein structure. *High* or *strict* S.s. indicates the ability to bind and convert only one or a limited number of substrates. An enzyme with broad S.s. converts a wide range of related substrates.
Subtilisin (EC 3.4.21.4): an extracellular, single chain, alkaline serine protease from *Bacillus subtilis* and related species. S. are known from four different species of Bacillus: *S. Carlsberg* (274 amino acid residues, M_r 27,277), *S. BPN'* (275 amino acid residues, M_r 27,537), *S. Novo* (identical with S.BPN') and *S. amylosacchariticus* (275 amino acid residues, M_r 27671). The observed sequence differenees between different S. represent conservative substitutions and are limited to the surface amino acids. Like the pancreatic proteinases, S. has catalytic Ser_{221}, His_{64} and Asn_{32} residues, but it is structurally very different from the other serine proteases, e.g. the active center of S. is -Thr-Ser-Met-, whereas that of the pancreatic enzymes is -Asp-Ser-Gly-; pancreatic enzymes contain 4–6 disulfide bridges, whereas S. contains none; S. contains 31 % α-helical structure and 3 spatially separated domains, whereas the pancreatic enzymes have 10–20 % α-helical structure and a high content of β-structures; in both types, the active center is a substrate cleft. S. also have a broader substrate specificity than the pancreatic enzymes. This is a notable example of the convergent evolution of catalytic activity in two structurally completely different classes of proteins. S. is used in the structural elucidation of small peptides; and on account of its stability to anionic detergents, it is used in biological washing powders.
Subunit: 1. In protein chemistry the smallest protein or polypeptide unit that can be separated from an oligomeric protein without breaking covalent bonds. S. may be identical or nonidentical. In allosteric enzymes, e.g. aspartate transcarbamylase, nonidentical S. can be further classified into regulatory and catalytic S. Sometimes, the smallest active combination of S. is called a *monomer,* and the smallest identical subgrouping of S. in an oligomer is called a protomer (e.g. the αβ-protomer of hemoglobin), producing the sequence: subunit-monomer-protomer-oligomer (quaternary protein). According to this definition polypeptides linked by disulfide bridges are not S., e.g. in insulin, chymotrypsin, fibrinogen and immunoglobulins, and such proteins are not considered to have quaternary structure (see Proteins). There are several ways of dissociating quaternary proteins. Usually the protein is treated with 1 % sodium dodecylsulfate (SDS), 6 M guanidine hydrochloride, or 8 M urea. Other treatments for dissociation of quaternary proteins into individual S. include increase or decrease of ionic strength, pH, protein concentration or temperature, removal or addition of a cofactor, or chemical modification (reaction with succinic or maleic anhydride). Demonstration and M_r determination of S. are performed by polyacrylamide gel electrophoresis, gel filtration and various other physical chemical methods (see Proteins) in the presence of the appropriate denaturing agent.
2. The term is also applied to the two subunits of Ribosomes (see).
Succedaneaflavanone: see Biflavonoids.
Succinate dehydrogenase (EC 1.3.99.1): an oligomeric flavoenzyme [M_r (bovine heart) 100,000] of the TCA-cycle, which catalyses the oxidation of succinate to fumarate. S.d. consists of two nonidentical, iron-containing subunits (M_r 70,000 and 30,000). Coenzyme of S.d. is FAD, which receives hydrogen directly (without the participation of NAD or NADP) and transfers it to ubiquinone or cytochrome *b* of the respiratory chain, or (in vitro) to redox dyes, e.g. methylene blue. S.d. occurs only in mitochondria (and prokaryotic cell membranes), where it is firmly bound in the membrane in association with the respiratory chain. It has a regulatory role in the TCA cycle, and can be isolated from the succinoxidase complex.
Succinate-glycine cycle, *glycine-succinate cycle, Shemin cycle:* a bypass of the TCA-cycle of particular importance in the metabolism of red blood cells. It converts succinyl-CoA and glycine into 5-aminolevulinate, which is the biosynthetic precursor of the Porphyrins (see). Alternatively, 5-aminolevulinate is deaminated to 2-oxoglutarate semialdehyde, and the cycle is completed by formation of succinyl-CoA via 2-oxoglutarate. One turn of the cycle converts glycine into 2 molecules CO_2 and 1 molecule NH_3. 2-Oxoglutarate semialdehyde can also be converted into succinate and a C1-unit.
Succinic acid, *ethane dicarboxylic acid:* $HOOC–CH_2–CH_2–COOH$, m.p. 185 °C, b.p. 235 °C (d.). Succinate is an important intermediary metabolite gener-

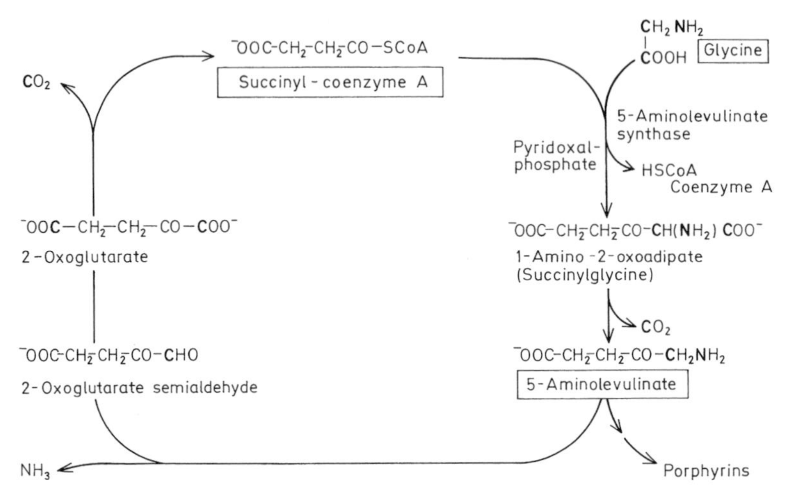

Synthesis of 5-aminolevulinate via the succinate-glycine cycle

ated both in the Tricarboxylic acid cycle (see) and in the Glyoxylate cycle (see). S. a. can be used via the reactions of the tricarboxylic acid cycle for the synthesis of amino acids and carbohydrates. Succinyl-coenzyme A is the starting point for the synthesis of porphyrins from intermediates of the Succinate-glycine cycle (see). Succinyl-coenzyme A is also formed by carboxylation of propionyl-coenzyme A in the degradation of valine and isoleucine, and from 2-methylmalonyl-coenzyme A in the degradation of fatty acids with an odd number of carbon atoms. Conversion of succinyl-coenzyme A to succinate in the tricarboxylic acid cycle generates one molecule of GTP, which is equivalent to 1 ATP.

Succinic acid 2,2-dimethylamide, ***Alar 85, B9:*** $HOOC-(CH_2)_2-CO-N(CH_3)_2$, a synthetic growth regulator, used to stimulate blossoming in apples and to accelerate the development of fruit color and fruit loosening in cherries.

Succinic acid 2,2-dimethylhydrazine, ***dimethazide, B-995:*** $HOOC-(CH_2)_2-CO-NH-N(CH_3)_2$, a synthetic growth regulator. It acts as a gibberellin antagonist and growth retardant, and is especially active in dicotyledonous plants. It increases the number of blossoms on fruit trees and prevents loss of immature fruit.

Succinylacetoacetate: see Inborn errors of metabolism (hereditary tyrosinemia type I).

Succinylacetone: see Inborn errors of metabolism (hereditary tyrosinemia type I).

***N*-Succinyladenylate:** adenylosuccinate. See Purine biosynthesis.

Succinylcholine: see Acetylcholine.

Succinyl-coenzyme A: see Succinic acid.

Sucrase, ***invertase:*** see Disaccharidases.

Sucrose, ***cane sugar, beet sugar:*** α-D-glucopyranosyl-β-D-fructofuranoside, M_r 342.31, decomposes at 160–186 °C with charring, $[\alpha]_D^{20}$ + 66.5 °. S. is a trehalose-type, nonreducing disaccharide (see Carbohydrates) biosynthesized from fructose 6-phosphate and UDP-glucose (Fig.). It does not give typical sugar reactions, i.e. does not form an osazone or oxime, or show mutarotation. Hydrolysis with dilute acid or enzymes, e.g. α-glucosidase (maltase, EC 3.2.1.20), or β-D-fructofuranosidase (invertase, EC 3.2.1.26) cleaves S. into equal parts of D-glucose and D-fructose (see Invert sugar). As a weak acid ($K \approx 10^{-13}$), S. (and saccharides derived from S.) forms salts (saccharates) with alkali and alkaline earth metal hydroxides. S. is widely distributed in plants, where it represents a transport form of soluble carbohydrate. It is an important metabolic product in all chlorophyll-containing plants, and is not synthesized by animals.

S. is obtained commercially from two economically important plants: the expressed juice of sugar cane (*Saccharum officinarum*), which grows in the tropics, contains 14–21 % S.; the expressed juice of sugar beet (*Beta vulgaris*), grown in temperate climates, contains 12–20 % S. The chemical synthesis of S. was reported in 1953, but is economically unimportant. S. is used as a food and sweetener. It is fermentable, but in high concentrations it inhibits growth of microorganisms, and is therefore used as a preservative. It is used as the substrate in the industrial fermentative production of ethanol, butanol, glycerol, citric and levulinic acids, and as a preservative, sweetener and demulcent in pharmaceutical preparations. Esters, ethers and other chemical modification products of S. are used in the preparation of detergents, soaps and plastics.

Sucrose density gradient centrifugation: see Density gradient centrifugation.

Sugar: in the narrow sense a general name for commercially available Sucrose (see). In the wider sense a term used for Carbohydrates (see), in particular mono- and oligosaccharides.

Sugar alcohols: polyhydric alcohols, which occur widely as metabolic reduction products of monosaccharides. They are named by replacing the '-ose' ending of the parent monosaccharide by '-itol'. According to the number of carbon atoms they are classified as pentitols, hexitols, etc. S. a. show only low optical activity, are not fermented by yeasts, and do

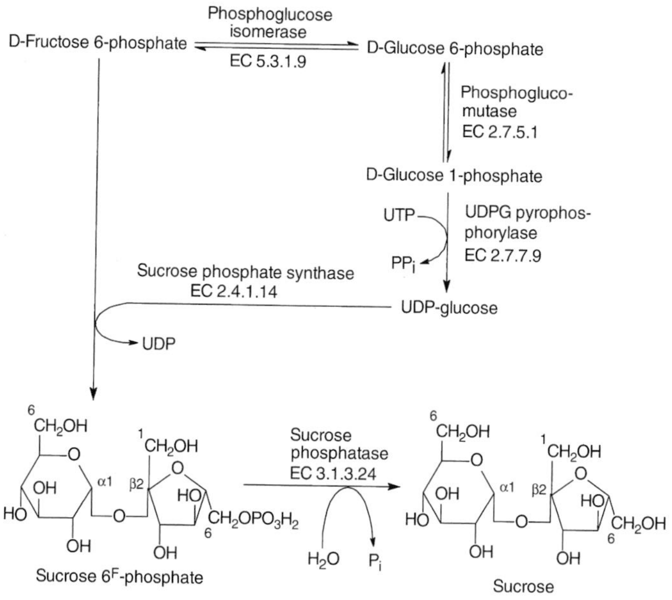

Biosynthesis of sucrose

not react with Fehling's solution or phenylhydrazine. They are biosynthesized by reduction of the corresponding monosaccharide with NADH or NADPH. Important natural representatives are glycerol, erythritol, ribitol, xylitol, D-sorbitol, D-mannitol, dulcitol.

Sugar anhydrides: internal acetals formed by intramolecular removal of water from a sugar molecule. They are chemically similar to glycosides and can be hydrolysed to the corresponding sugars by water or dilute acids.

Sugar esters: esters of mono- or oligosaccharides with organic or inorganic acids. The phosphate esters, e.g. glucose 6-phosphate, are fundamentally important in intermediary carbohydrate metabolism. Chondroitin and mucoitin sulfates, and heparin are typical animal sulfate esters.

Sulfamates: see Sulfur compounds

Sulfatases: see Esterases.

Sulfate activation: formation of the active sulfates, Phosphoadenosinephosphosulfate (see), (PAPS) and adenosine 5′-phosphosulfate (APS). PAPS is the substrate of sulfotransferases in the synthesis of sulfate esters. APS is the substrate of sulfate assimilation and sulfate respiration.

Sulfate assimilation, ***assimilatory sulfate reduction:*** reductive assimilation of oxidized inorganic sulfur (see Sulfate reduction), leading to the biosynthesis of L-cysteine. S. a. is a property of plants and bacteria, whereas protozoa and animals can only perform the first step of S. a., i.e. sulfate activation (see Phosphoadenosinephosphosulfate). The sulfate ion SO_4^{2-} is reduced to the oxidoreduction level of the sulfide ion, S^{2-}, or the mercapto group of L-cysteine. According to one view, Sulfate reduction (see) proceeds via free intermediates to hydrogen sulfide, H_2S, which is incorporated into the SH-group of cysteine by the action of serine sulfhydrase (cysteine synthase, EC 4.2.1.22):

$$\underset{\text{L-Serine}}{H_2S + HOCH_2CH(NH_2)COOH} \xrightarrow{PalP} \underset{\text{L-Cysteine}}{HSCH_2CH(NH_2)COOH + H_2O}$$

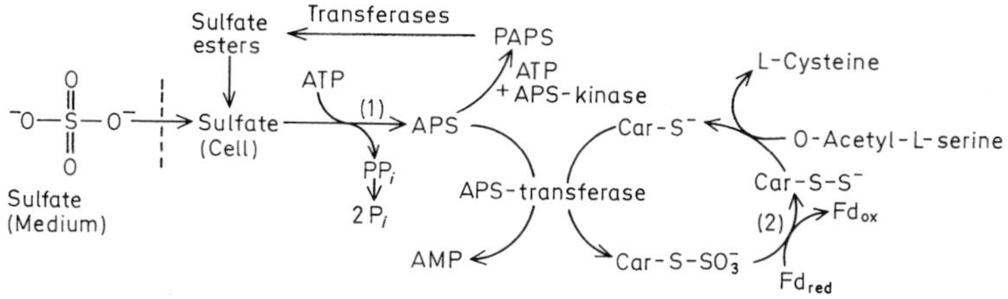

Fig. 1. *Sulfate assimilation via enzyme- and carrier-bound intermediates.* (1) ATP-sulfurylase. (2) APS-reductase.

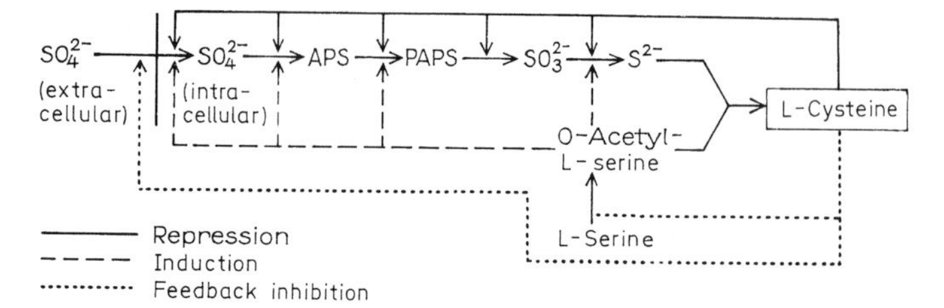

Fig. 2. *Sulfate assimilation in* E. coli *via free intermediates, with control mechanisms.*

Serine sulfhydrase is a pyridoxal phosphate enzyme. According to a different view of this process, the free intermediates of S.a. are byproducts, and S.a. proceeds via enzyme- and carrier-bound intermediates (Fig. 1). Sulfate is converted to adenosine 5′-phosphosulfate (APS), which is phosphorylated to PAPS. The latter takes part in the synthesis of sulfate esters, or it can be converted back to APS by a specific 3′-nucleotidase. The substrate of S.a. is not PAPS, but APS (according to work on *Chlorella,* yeasts and spinach chloroplasts). An APS-transferase transfers sulfate from APS to a hypothetical carrier, Car-S⁻, which is possibly identical with thioredoxin, although only one thiol group is apparently required. This carrier is possibly a coenzyme of the APS-transferase. The product of the transferase reaction is probably Car-S-SO_3^-; this is reduced to Car-S-S⁻ by a ferredoxin-dependent reductase. Car-S⁻ (or the coenzyme of APS-transferase) is regenerated by reaction of Car-S-S⁻ with *O*-acetyl-L-serine, with the formation of L-cysteine. Free intermediates could arise from this reaction pathway by reductive cleavage of bound intermediates. Free sulfide might arise from reductive dismutation of Car-S-S⁻ or some similar enzyme-bound intermediate. Exogenous sulfite might enter the system by combining with Car-S⁻ to form Car-S-SO_3^-, which would agree with results from defective mutants of *Salmonella, E. coli, Chlorella,* yeast, *Aspergillus* and *Neurospora.* It is, however, possible that two alternative pathways of S.a. exist: via free intermediates (see Sulfate reduction), and a pathway via enzyme- and carrier-bound intermediates (see Fig. 1), in which organic or bound thiosulfate, -S-SO_3^-, plays a central role.

S.a. is regulated by enzyme induction and repression, and by feedback inhibition (Fig. 2). In *E. coli,* L-cysteine represses all steps of S.a. (assuming S.a. to proceed via free intermediates) starting with the uptake of sulfate by the cell. *O*-Acetyl-L-serine induces uptake of sulfate, synthesis of APS and PAPS and synthesis of sulfite reductase, but not of sulfate reductase. Acetylation of L-serine to *O*-acetyl-L-serine is feedback-inhibited by L-cysteine.

Sulfate esters: products formed by the transfer of a sulfuryl group to the oxygen function of organic compounds. Donor for the sulfuryl transfer is Phosphoadenosinephosphosulfate (see) (PAPS). Naturally occurring S.e. are Polysaccharide sulfate esters (see), and sulfate esters of phenols, steroids, choline cerebrosides and flavonoids. Synthesis of phenol sulfate esters in mammalian liver was the first biological sulfate ester synthesis to be recognized, and the first to be performed in vitro. It is an important reaction for the detoxication of phenols in animals and man. Transfer of the sulfuryl group from PAPS to the acceptor is catalysed by phenolsulfotransferase. Ascorbic acid 2-sulfate may also function as a donor of sulfuryl groups.

Sulfate reduction: reduction of the sulfate ion SO_4^{2-} to the sulfide ion S^{2-}, in which the hexavalent, positive sulfur of the sulfate is converted to the divalent, negative form. S.r. must be preceded by Sulfate activation (see). The substrate of enzymatic S.r. is therefore either adenosine phosphosulfate (APS) or phosphoadenosinephosphosulfate (PAPS). The enzymology of S.r. has been studied in particular in enzyme preparations from baker's yeast (*Saccharomyces cerevisiae*) and the anaerobic bacterium *Desulfovibrio desulfuricans.* The former organism performs assimilatory S.r. (see Sulfate assimilation), the latter dissimilatory S.r. (see Sulfate respiration).

Sulfate respiration, *dissimilatory sulfate reduction:* a form of respiration in which the sulfate ion replaces oxygen as the terminal electron acceptor (see Sulfate reduction). The sulfate ion must first be activated (see Sulfate activation). S.r. is an anaerobic process in which sulfate is reduced to hydrogen sulfide, which is excreted. Ecologically, S.r. contributes to desulfurication, and is important for the sulfur cycle of the biosphere.

S.r. is limited to the members of two bacterial genera: *Desulfovibrio* (Fig.) and *Desulfotomaculum.* The process is described by the equation: $8[H] + SO_4^{2-} \rightarrow H_2S + 2H_2 + 2OH^-$. Hydrogen is donated by alcohols, organic acids (e.g. pyruvate), or molecular hydrogen (many strains have a constitutive hydrogenase). With pyruvate, the process can be represented as: 4pyruvate + H_2SO_4 → 4acetate + $4CO_2$ + H_2S. Since acetic acid appears as a fermentation product,

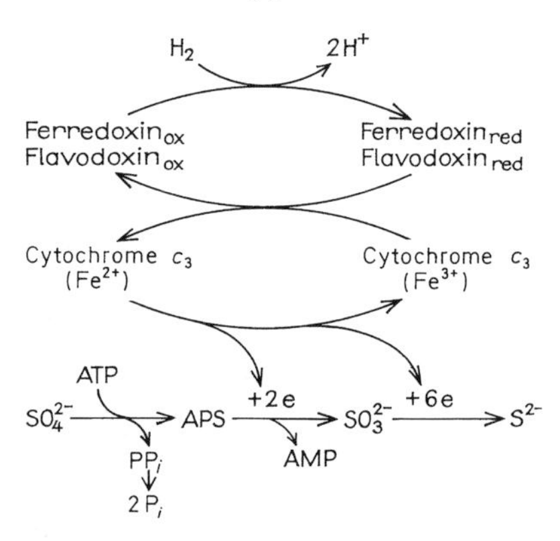

Dissimilatory sulfate reduction in Desulfovibrio.

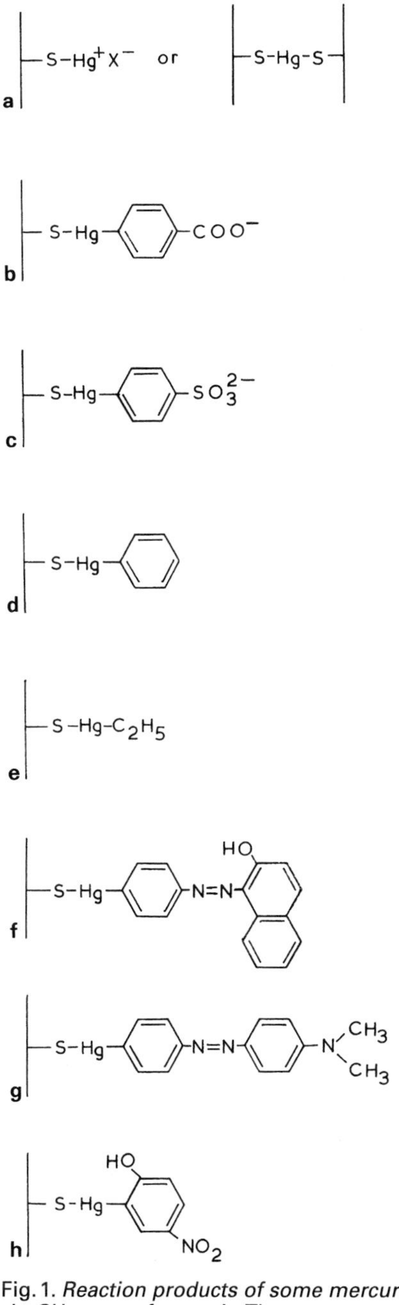

this S.r. can be considered as an anaerobic acetic acid fermentation.

Sulfatides: see Glycolipids.

Sulfation factor: see Somatomedin.

Sulfhydryl group: see Thiol group.

Sulfhydryl reagents, *SH-reagents:* substances that react with thiol groups (syn. sulfhydryl or SH-groups) of proteins. In vivo they cause metabolic changes or alterations of function and they are generally toxic. In vitro they are used to detect, titrate and characterize SH-groups. If the catalytic activity of an enzyme is inhibited by SH-reagents, the enzyme is eligible for designation as an SH-enzyme or Thiol enzyme (see). The affected SH-groups may be in the active center of the enzyme, or they may be elsewhere, so that inhibition of activity results from a change in the conformation or solubility of the enzyme.

Three types of cell constituents react with SH-reagents: 1. low M_r thiols, such as lipoic acid, coenzyme A, glutathione and cysteine; 2. non-enzyme proteins, such as actomyosin, membrane proteins and structural proteins; 3. enzymes. The toxic action of SH-reagents in vivo is due chiefly to their effect on SH-enzymes.

The most potent SH-reagents are the mercurials, which react with SH-groups according to the equation: Protein-SH + $^{+}$Hg-X → Protein-SHg-X + H^+ (see Fig. 1).

4-Chloromercuribenzoate is the most commonly used mercurial for the spectrophotometric titration of SH-groups; reaction with SH-groups at pH 7 results in an increase in absorbance at 250 nm. 4-Chloromercuriphenylsulfonic acid has the advantage of greater water solubility, but its poor lipid solubility makes it unsuitable for general biological purposes. Phenylmercury chloride and ethylmercury chloride show good lipid solubility and are therefore able to penetrate into the interior of subcellular particles and into cells. The various colored mercurials which possess azo- or nitro-groups, are useful for the histochemical detection of SH-groups (e.g. f and g, Fig. 1), and some are also used for the spectrophotometric titration of SH-groups (e.g. h, Fig. 1). Formation of mercuric derivatives has been used for the purification and crystallization of enzymes, e.g. crystalline Hg complexes of enolase from yeast, lactate dehydrogenase from Jensen sarcoma and rat muscle, and papain have been prepared. These complexes are enzymatically inactive, but Hg^{2+} can be removed and activity restored by dialysis against cyanide.

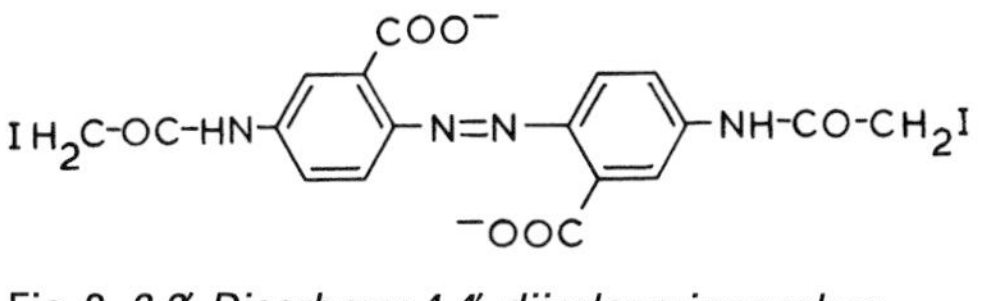

a: –S–Hg^{+}X^{-} or –S–Hg–S–

b: –S–Hg–C$_6$H$_4$–COO$^-$

c: –S–Hg–C$_6$H$_4$–SO$_3^{2-}$

d: –S–Hg–C$_6$H$_5$

e: –S–Hg–C$_2$H$_5$

f: –S–Hg–C$_6$H$_4$–N=N–(HO-naphthyl)

g: –S–Hg–C$_6$H$_4$–N=N–C$_6$H$_4$–N(CH$_3$)$_2$

h: –S–Hg–(HO, NO$_2$-phenyl)

Fig. 1. *Reaction products of some mercurials with the SH-group of a protein.* The reagents are: mercuric salts (giving product *a*), 4-chloromercuribenzoate *(b)*, 4-chloromercuriphenylsulfonic acid *(c)*, phenylmercury chloride *(d)*, ethylmercury chloride *(e)*, 4-chloromercuriphenylazo-β-naphthol *(f)*, 4-chloromercuri-4′-dimethylaminobenzene *(g)* and 2-chloromercuri-4-nitrophenol *(h)*. These reagents are usually employed as their chlorides (as in the above names) or as their acetates.

IH$_2$C-OC-HN–C$_6$H$_3$(COO$^-$)–N=N–C$_6$H$_3$(COO$^-$)–NH-CO-CH$_2$I

Fig. 2. *2,2′-Dicarboxy-4,4′-diiodoaminoazobenzene*

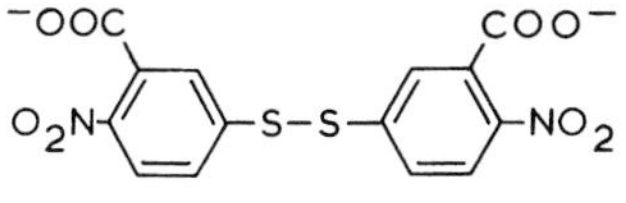

O$_2$N–C$_6$H$_3$(COO$^-$)–S–S–C$_6$H$_3$(COO$^-$)–NO$_2$

Fig. 3. *Ellman's reagent*

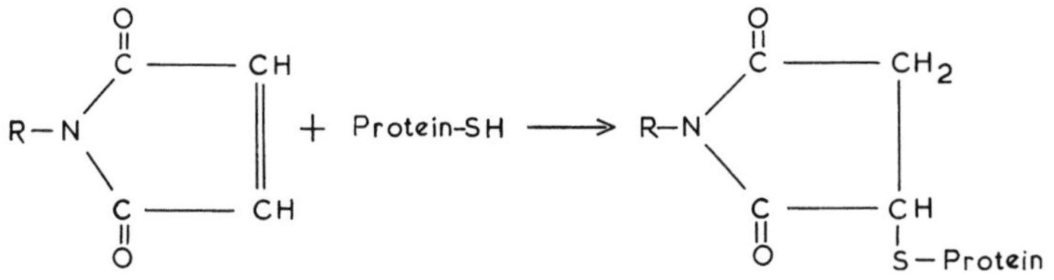

$$R{-}N\begin{matrix}C(=O){-}CH\\ \| \\ C(=O){-}CH\end{matrix} + \text{Protein-SH} \longrightarrow R{-}N\begin{matrix}C(=O){-}CH_2\\ | \\ C(=O){-}CH{-}S{-}\text{Protein}\end{matrix}$$

Fig. 4. *Reaction of a substituted maleimide with the SH-group of a protein.*

Compared with the mercurials and heavy metals, other SH-reagents are much less potent. On the other hand, they are often very useful for their selective activity against certain enzymes (e.g. the classical inhibition of glyceraldehyde 3-phosphate dehydrogenase, and therefore glycolysis, by iodoacetate), and as reagents in protein chemistry (e.g. ^{14}C-labeled iodoacetate is used to label cysteine residues in proteins: $I^{14}CH_2$-$^{14}COOH$ + Protein-SH → Protein-S-$^{14}CH_2$-^{14}-COOH + HI). Two reactive iodine groups are present in 2,2'-dicarboxy-4,4'-diiodoaminoazobenzene, so that this reagent is able to cross-link SH-groups.

Certain disulfides are also active SH-reagents, reacting with SH-groups by a disulfide-sulfhydryl interchange: X-SS-X + Protein-SH → Protein-SSX + XSH. A well known compound of this type is 5,5'-dithio-*bis*(2-nitrobenzoate), or Ellman's reagent. The substituted maleimides are used for both labeling and inhibition studies (Fig. 4).

The most widely used of these is *N*-ethylmaleimide ($R = C_2H_5$ in Fig. 4), which can be very selective for certain SH-groups of an enzyme. It is used as a general SH-reagent for detection of SH-functions in purified proteins and in biological processes, such as hormone release, metabolite uptake, membrane transport, etc.

Alloxan, well known for its ability to specifically damage the insulin secreting β-cells in the pancreatic islets, is an SH-reagent. It shows high specificity, being quite inactive towards many known thiol enzymes.

Arsenite and the arsenicals are among the earliest known SH-reagents, but the exact mechanism of their reaction with SH-groups is still unclear. The reaction often involves formation of a cross-link between two SH-groups, and may therefore depend on the correct juxtaposition of groups within the protein.

Sulfides: see Sulfur compounds.

β-Sulfinylpyruvic acid: see Cysteine.

Sulfite dehydrogenase deficiency: see Inborn errors of metabolism.

Sulfite fermentation: see Neuberg fermentation.

Sulfite oxidase: see Molybdoenzymes.

Sulfite oxidase deficiency: see Inborn errors of metabolism.

Sulfituria: see Inborn errors of metabolism.

Sulfokinase: see Sulfotransferase.

Sulfo(no)lipids: see Membrane lipids.

Sulfonic acids: see Sulfur compounds.

Sulfonolipids: see Capnine.

Sulfotransferase, *sulfokinase:* an enzyme that catalyses the transfer of sulfuryl groups from phosphoadenosinephosphosulfate (PAPS) to oxygen and nitrogen functions of suitable acceptors. Transfer to an oxygen function (alcoholic and phenolic hydroxyl groups) produces sulfate esters. Sulfated nitrogen functions are found in mustard oil glycosides and arylsulfamates; PAPS is the sulfuryl donor in the formation of sulfamates. S.t. are very important in the synthesis of polysaccharide sulfate esters, and in the detoxication of phenols by sulfate ester formation in animals and man. The other product of the sulfuryl transfer reaction is 3'-phosphoadenosine 5'-phosphate (PAP): PAPS + acceptor → PAP + product. Ascorbic acid 2-sulfate may also act as a biological donor of sulfuryl groups. [K. Kresse & E. F. Newfield *J. Biol. Chem.* **247** (1972) 2164–2170]

Sulfoxides: see Sulfur compounds.

Sulfur: see Bioelements.

Sulfur compounds: compounds containing reduced or oxidized sulfur, seldom both (Table). Biochemically important S.c. are the sulfur amino acids (see L-Cysteine, L-Methionine), biotin (thiophane ring) and thiamin (thiazole ring) (see Vitamins), sulfatides (complex lipids of the nervous system see Glycolipids), and thiol peptides (see Glutathione, Vasopressin, Oxytocin, Insulin). S.c. also include Penicillin (see), and the sulfonamides which are important synthetic therapeutic agents. The mustard oil glycosides contain both oxidized and reduced sulfur. Sulfate esters (see) are excreted by animals.

Sulfur cycle: see Sulfur metabolism.

Sulfuretin: 6,3',4'-trihydroxyaurone. See Aurones.

Sulfur metabolism: the metabolism of sulfur-containing compounds in living organisms. Sulfur in various forms (see Sulfur compounds) is needed by all living organisms for the synthesis of biomolecules. Many microorganisms can utilize inorganic sulfur sources, like sulfide, sulfite, sulfate, thiosulfate, and in some cases even elemental sulfur. Plants assimilate inorganic sulfate (see Sulfate assimilation), and to a limited extent they can also incorporate atmospheric sulfur dioxide into cysteine and methionine. Sulfur dioxide cannot, however, satisfy the total sulfur requirement of plants – higher concentrations cause damage, as frequently observed in woodlands in industrial areas. Sulfur dioxide is oxidized in the leaf to sulfate, which can be transported and assimilated. Sulfate is taken up from the soil, reduced (see Sulfate reduction), then assimilated. The chief source of sulfate in the soil for the nutrition of most plants is gypsum ($CaSO_4.2H_2O$) and anhydrite ($CaSO_4$). Sulfides, like FeS and FeS_2 (pyrites) can be oxidized to sulfur by purely chemical processes in the soil; the sulfur is then oxidized to sulfate by sulfur bacteria. Colorless sulfur bacteria (e.g. *Beggiatoa, Thiobacillus, Thiothrix*) oxidize reduced forms of sulfur, and thus play an important part in the circulation of sulfur in the biosphere (i.e. the sulfur cycle, see Fig.). Reduction of sulfate is specific to microorganisms (see Sulfate

Naturally occurring sulfur compounds

Compound	General structure	Examples
Thiols (mercaptans)	RSH	L-Cysteine; coenzyme A
Sulfides	RSR_1	L-Methionine
Sulfoxides	$RSOR_1$	Allicin; formation from alliin by alliinase when onions *(Allium cepa)* are grated
Methylsulfonium compounds	$(CH_3)_2S^+R$	*S*-Adenosyl-L-methionine; dimethyl-β-propiothetin
Sulfate esters	$R{-}O{-}S(=O)_2{-}O^-$	Phenol sulfates; polysaccharide sulfates
Sulfamates	$R{=}N{-}O{-}S(=O)_2{-}O^-$	Aryl sulfamates; mustard oil glycosides
Sulfonic acids	$R{-}C{-}S(=O)_2{-}O^-$	Glucose 6-sulfonate; cysteic acid; taurine and methyltaurine

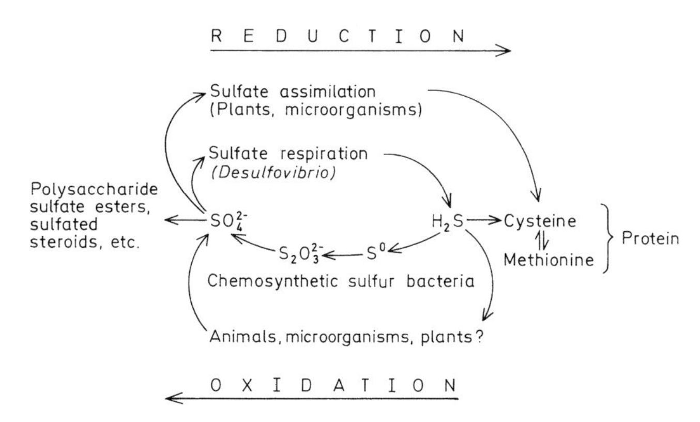

Oxidoreduction of sulfur in different organisms

respiration) and plants (see Sulfate assimilation), i.e. these organisms perform inorganic S.m. In contrast, animals depend on a dietary source of organically bound sulfur, chiefly in the form of the proteogenic amino acids, L-cysteine and L-methionine. Cysteine can also be synthesized from dietary methionine by animals. In plants and animals the same mechanism is used for the synthesis of methionine from cysteine (see L-Methionine). Some coenzymes also contain sulfur (see Sulfur compounds). Sulfate respiration and sulfate assimilation involve the initial formation of active sulfate (see Phosphoadenosinephosphosulfate), a process which also occurs in animals (see Sulfotransferase). Oxidized organic sulfur compounds arise chiefly from the degradation of L-cysteine or secondary sulfur-containing metabolites.

Supercoil: see Superhelix, Topoisomerases.

Superhelix: 1. a tertiary DNA structure formed by further helical coiling of the DNA double helix (see Deoxyribonucleic acid) (secondary structure is represented by the double helix, primary structure by the linear sequence of nucleotides). There is an even higher level of organization, in which the superhelix is further coiled to form a super-superhelix. Supercoiling of DNA enables this large molecule to be packed into a relatively small space, and is essential for the formation of the DNA-histone complexes of chromatin. (Chromosomal DNA is compacted about 8,000-fold in length, while DNA in a nucleosome is folded to give a 6.8-fold contraction of length, with formation of a unit fiber of length 100 Å.) In addition, the superhelix appears to serve in the control of gene expression.

The superhelix is left-handed, whereas the double helix is right-handed. This can be compared to the Mobius strip, in which a strip of paper is twisted before joining the ends, thus producing a higher order of twist, or a kink, in the completed circle. In conformity with this model, unwinding of the superhelix also causes unwinding of the double helix. Supercoiled, circular DNA denatures more easily than nonsupercoiled DNA, i.e. supercoiled DNA is under strain. Enzymes called Topoisomerases (see) are responsible for converting nonsupercoiled DNA into a superhelix and for regulating the degree of supercoiling.

Studies in vitro show that replication and transcription do not occur unless the DNA is supercoiled. It appears that uncoiling of the superhelix promotes separation of the strands of the double helix, thus providing access for DNA- or RNA-polymerases. The strain of the DNA supercoil can be released by "nicking" with gamma rays. Such "relaxed" supercoiled DNA shows a decreased ability to promote RNA transcription, or to bind RNA polymerase. It is therefore suggested that gene expression is controlled by the degree of supercoiling of DNA; transcribed genes are supercoiled to the correct degree, so that an RNA-polymerase can cause local unwinding, whereas untranscribed genes are coiled to the wrong degree.

Binding of one molecule of ethidium bromide unwinds the helix by 26°; the quantity of bound ethidium can be determined fluorimetrically, thus forming a basis for the determination of the degree of superhelicity. Supercoiling can also be measured by velocity sedimentation or equilibrium sedimentation in the presence of intercalating dyes, and by agarose gel electrophoresis (supercoiled DNA migrates more rapidly than relaxed DNA). For other information on superhelices and for references, see Topoisomerases, Linking number.

2. See also Collagen

Superhelix density, ***σ:*** according to W. Bauer & J. Vinograd [J. Mol Biol. **33** (1968) 141–172], the number of superhelical turns per 10 base pairs of the DNA circular duplex. It is therefore a measure of the strain on the DNA due to supercoiling. Negative S.d. ranging from 0.04 to 0.09 have been reported.

According to A. Nordheim et al. [*Cell* **31** (1982) 309–318], superhelical density (denoted by these authors also as σ) is equivalent to specific linking difference (see Linking number).

Superoxide dismutase, ***SOD, hyperoxide dismutase, superoxide:superoxide oxidoreductase*** (EC 1.15.1.1): When it was found that the copper- and zinc-containing proteins hemocuprein (bovine erythrocytes), hepatocuprein (bovine liver) and cerebrocuprein (human brain) have a common identity, they were given the single name, cytocuprein [R. J. Carrico & H. F. Deutsch *J. Biol. Chem.* **245** (1970) 723–727] Cytocuprein is now known to be identical with SOD.

SOD catalyses the dismutation (disproportionation) of superoxide: $O_2^- + O_2^- + 2H^+ \rightarrow O_2 + H_2O_2$. There are two main types of SOD: 1. cyanide-sensitive, Cu- and Zn-containing, eukaryotic enzymes, M_r 31,000–33,000 (2 subunits, M_r 16,000); and 2. cyanide-insensitive, Fe- or Mn-containing, prokaryotic enzymes, M_r about 40,000 (2 subunits, M_r 20,000). SOD from liver mitochondria contains Mn and has M_r 80,000 (4 subunits, M_r 20,000). The designations "prokaryotic" and "eukaryotic" are not applicable in every case: Mn-type SOD has been demonstrated in eukaryotes, while cyanide-sensitive SOD is absent from many algae. In contrast to the view widely held in the early 1970s, there is also a lack of correlation between the presence of SOD and aerobicity. Many anaerobes possess SOD, while some aerobes do not (however, these organisms do have high levels of catalase). There has been debate over the biological significance of SOD, due to the realization that the superoxide radical is not highly reactive in aqueous solution. Some authors have suggested that SOD is primarily a metal storage protein. However, studies on the structure and mechanism of Cu-Zn-SOD have shown that its structure has been highly conserved through evolution [J. A. Tainer et al. *Nature* **306** (1983) 284–286; E. D. Getzoff et al. *ibid* 286–290]. Others have demonstrated that SOD specifically catalyses dismutation of superoxide, and that this radical, if its concentration is elevated, may react with hydrogen peroxide to form a much more reactive species, possibly the hydroxyl radical. [B. Halliwell in R. Lontie, ed. *Copper Proteins and Copper Enzymes, Vol. 2* (CRC Press, Boca Raton, 1984) pp. 63–102; I. Fridovich 'Superoxide Radical and Superoxide Dismutases' *Annu. Rev. Biochem.* **64** (1995) 97–112].

Supplementation test: see Mutant technique.

Suppressor: With reference to Amber mutants (see) and Ochre mutants (see), a S. mutation is a secondary mutation in a tRNA cistron, which restores the ability of the tRNA to recognize the nonsense codon. The resulting new tRNA is called suppressor tRNA. The nonsense codon then becomes a sense codon and codes for a specific proteogenic amino acid during translation.

In a general sense, a S. mutation is any mutation that partly or completely restores a genetic function lost by another mutation. By definition, a S. mutation must be at a different site from that affected by the first mutation. It may be in the same gene (intragenic S. mutation) or a different gene (intergenic S. mutation). A S. gene reverses the effect of a mutation in another gene.

Suprarenin: see Adrenalin.

Svedberg unit: see Sedimentation coefficient.

Swivelase: a type I topoisomerase. See Topoisomerases.

Symbiosis: close spatial coexistence of different species of advantage to all partners. It is therefore also called mutualistic S. The distinction between S., Commensalism (see) and Parasitism (see) is not always clear. Biochemically relevant symbiotic systems are root nodule bacteria – leguminous plants (see Nitrogen fixation); alga – fungus in lichens; rumen flora – ruminants; fungus – higher plant (see Mycorrhiza). Partners showing large differences in size are called macrosymbionts and microsymbionts.

Symport: see Transport.

Synchronous culture: a synchronized cell population, in which all cells divide and pass through subsequent phases of the cell cycle at the same time. Synchronization can be achieved in various ways, e.g. by nutrient limitation, light stimulation, temperature change, treatmennt with antimetabolites of nucleic acid metabolism. In S.c. the cell count increases stepwise. Synchrony is usually lost after a few synchronous divisions, i.e. the cell count reverts to a continual increase. S.c. techniques have been applied to various bacteria, *Chlorella, Euglena gracilis,* etc. Light synchronization is important in the study of endogenous biological rhythms.

Syndein: see Ankyrin.

Synergists: substances or factors that increase the biological activity of another substance or factor, but are inactive alone in the same quantity.

Synthases: see Enzymes (Table).

Synthetases: see Enzymes (Table).

Synzymes, ***enzyme analogs:*** synthetic macromolecules with enzymatic activity. S. may be prepared by polymerization of amino acids or their derivatives, or by the attachment of catalytic groups to nonprotein materials. Examples of amino acid-derived S. are: a glutamic acid-phenylalanine copolymer (Glu: Phe = 9: 1) with one third of the activity of natural lysozyme; a copper(II)-poly-ε-carbobenzoxy-L-lysine (M_r 440,000) with specific alcohol dehydrogenase activity. For non-amino acid S., methylenimidazole (nucleophilic, catalytically active) and dodecyl (apolar substrate binding site) side chains are attached to a polymethylenimine carrier. This globular macromolecule has a high esterase activity for phenylsulfate esters, which is 100 times greater than that of natural aryl sulfatase. The cyclodextrins are S. with phosphatase activity. Low M_r peptides with enzymatic activity are also S. The decapeptide, Glu-Phe-Ala-Ala-Glu-Glu-Ala-Ala-Ser-Phe shows glycosidase activity for dextran and chitin.

T

T: 1. a nucleotide residue (in a nucleic acid) in which the base is thymine. 2. Abb. for thymidine (e.g. TMP is acronym of thymidine monophosphate). 3. Abb. for thymine.

T_3: see Thyroxin.

T_4: see Thyroxin.

T_m, t_m: the temperature at the mid-point of a temperature-dependent transition. T_m is commonly used to denote the Melting point (see) of DNA.

Tabtoxins: chlorosis-inducing dipeptides produced by several species of phytopathogenic *Pseudomonas;* e.g. "wild fire", a highly infectious and destructive leafspot disease of tobacco, is caused by *Pseudomonas tabaci* which produces T. T. consists of threonine linked to tabtoxinine β-lactam [2-amino-4-(3-hydroxy-2-oxo-azocyclobutan-3-yl)butanoic acid]. In [2-serine]T., the threonine is replaced by serine (Fig.). Tabtoxinine β-lactam is the actual phytotoxin; it is released from T. by plant proteases, and acts as an inhibitor of glutamine synthetase. [P.A. Taylor et al. *Biochim. Biophys. Acta* **286** (1972) 107–117; T.F. Uchytil & R.D. Durbin *Experentia* **36** (1980) 301–302]

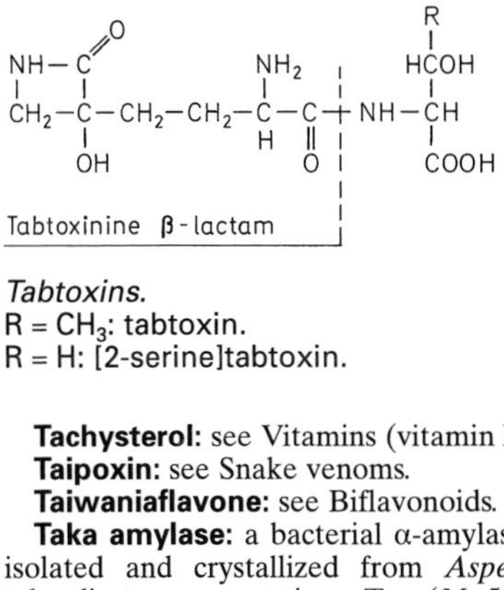

```
         O                                R
        //                                |
NH — C                NH2    |          HCOH
|    |                 |     |            |
CH2—C—CH2—CH2—C—C—+ NH—CH
     |                 H  ||  |           |
     OH                   O   |          COOH
                              |
Tabtoxinine β-lactam          |
```

Tabtoxins.
R = CH_3: tabtoxin.
R = H: [2-serine]tabtoxin.

Tachysterol: see Vitamins (vitamin D).

Taipoxin: see Snake venoms.

Taiwaniaflavone: see Biflavonoids.

Taka amylase: a bacterial α-amylase (EC 3.2.1.1) isolated and crystallized from *Aspergillus oryzae* taka diastase preparations. T.a. (M_r 50,000) is a calcium-containing, single-chain protein with *N*-terminal alanine and *C*-terminal serine. Like the tetrameric *Bacillus subtilis* α-amylase (M_r 96,000), T.a. is resistant to sodium dodecylsulfate but is reversibly denatured by 6 M guanidine or 8 M urea. T.a. must not be confused with Adenosine deaminase (see), which is also present in taka diastase.

Tannic acid: Chinese gallotannin. See Tannins.

Tannins: originally substances of vegetable origin capable of converting animal skin into leather. A more modern and appropriate definition is naturally occurring compounds of M_r 500–3,000, containing a sufficient number of phenolic hydroxyl groups (about 2 groups per M_r 100) to form cross links between macromolecules, such as proteins, cellulose and/or pectin.

By cross-linking proteins, T. can inhibit the activities of plant enzymes and organelles. Polyvinylpyrrolidone is therefore often added to adsorb T. during the isolation of plant enzymes. The chief reactive centers responsible for cross-linking and complex formation by T. are ortho-dihydroxy phenolic groups; isolated phenolic hydroxyl groups do not seem to make a significant contribution to these reactions. There are two classes of vegetable T.:

1. Hydrolysable T., which can be hydrolysed to glucose (or another polyhydric alcohol) and gallic acid (gallotannins) or ellagic acid (ellagitannins). The simplest known gallotannin is 1-*O*-galloyl-β-D-glucopyranose from *Rheum offcinale* (Chinese rhubarb). In contrast, Chinese gallotannin ("tannic acid") (widely distributed in *Hamamalidaceae, Paeonaceae, Aceraceae* and *Anacardiaceae,* and sporadically in the *Ericaceae*) may contain up to 8 galloyl groups (Fig.).

Ellagitannins are derivatives of hexahydroxydiphenic acid, which lactonizes to ellagic acid during hydrolysis. The simplest known ellagitannin is corilagin from *Caesalpina coriaria* and other plants (Fig.). Other phenolic components are sometimes found in place of gallic or hexahydroxydiphenic acid in T., e.g. chebulic acid (in myrobalans T.) and brevifelin carboxylic acid (in algarobilla T.).

2. Condensed T. are polymers in which the monomeric unit is a phenolic flavan, usually a flavan-3-ol, and in which the flavan units are linked by 4:8 (C-C) bonds. Many higher oligomers and polymers of Proanthocyanidins (see) are therefore condensed vegetable T. An example is the procyanidin polymer from the seed coat of sorghum (Fig.). Synthesis of condensed T. by biomimetic condensation reactions has been reported. The condensation sequence is initiated by flavan-3-ols acting as nucleophiles, and flavan-3,4-diols as potential 4-carbenium ions [J.J. Betha et al. *Phytochemistry* **21** (1982) 1289–1294; C. Hartisch & H. Kolodziej *Phytochemistry* **42** (1996) 191–198].

5α-Taraxastane: see Taraxasterol.

Taraxasterol, *α-lactucerol, α-anthesterol:* a simple, unsaturated, pentacyclic triterpene alcohol, M_r 426.73, m.p. 227 °C $[\alpha]_D + 50°$, structurally a derivative of the parent hydrocarbon, 5α-taraxastane. It occurs as the acetate in the latex of the dandelion *(Taraxacum officinale)* and other members of the *Compositae.*

5α-Taraxerane: see Taraxerol.

Taraxerol, *alnulin, skimmiol:* a simple, unsaturated, pentacyclic triterpene alcohol, M_r 426.73, m.p. 285 °C, $[\alpha]_D + 3°$, structurally a derivative of the parent hydrocarbon, 5α-taraxerane. It occurs in many members of the *Compositae,* e.g. dandelion *(Taraxaum officinale),* and in alder *(Alnus)* bark.

Ellagic acid (derived from hexahydroxydiphenic acid during hydrolysis of ellagitannins)

hexahydroxydiphenic acid residue glucose residue gallic acid residue

Corilagin (an elligatannin from *Caesalpinia coriaria*)

Chebulic acid (component of some hydrolysable tannins)

Condensed tannin (or procyanidin polymer) from seed coat of sorghum. $n = 4–5$. M_r 1700–2000

Brevifolin carboxylic acid (component of some hydrolysable tannins)

Gallic acid (component of gallotannins)

Chinese gallotannin ("tannic acid") (a heterogeneous tannin, in which $n = 0$, 1 or 2, and the C-1 galloyl group may be absent)

Tarichatoxin: the main toxin of North American salamanders (*Taricha torosa, T. rivularis*). It is identical with Tetrodotoxin (see).

Tartronate-semialdehyde synthase, ***glyoxylate carboligase*** (EC 4.1.1.47): a plant enzyme which converts two molecules of glyoxylate to tartronate semialdehyde and CO_2. Hydroxymethyl thiamin pyrophosphate is a reactive intermediate in the reaction. The enzyme plays a role in the biosynthesis of carbohydrates from C2 compounds.

Taurine: see L-Cysteine.

Taurocholic acid: see Bile acids.

Tay-Sachs disease: see Lysosomal storage diseases.

Taxol: a unique antineoplastic agent extracted from the bark of *Taxus brevifolia* (Pacific yew), which offers great promise as a chemotherapeutic agent against cancer [W. P. McGuire et al *Ann. Inter. Med.* **111** (1989) 273–277; L. Lenaz & M. D. DeFuria *Fitotherapia* LXIV supp. 1 (1993) 27–35]. Even higher antitumor activity is displayed by the taxol analog, Taxotere® [F. Lavelle et al. *Bull. Cancer* **80** (1993) 326–338]. Both taxol and its analogs (Fig.) can be prepared semisynthetically from 10-deacetylbaccatin III. The latter compound can be extracted from the leaves of *Taxus baccata* (European yew), but since it is also synthesized by tissue cultures of *T. baccata,* it may be possible to produce this precursor by biotechnological methods, and avoid exploitation of both trees [A. Zhiri et al. *Biol Chem Hoppe Seyler* **376** (1995) 583–586; K. Cheng et al. *Phytochemistry* **42** (1996) 73–75]. T. binds to tubulin and hyperstabilizes microtubules (see Cytoskeleton).

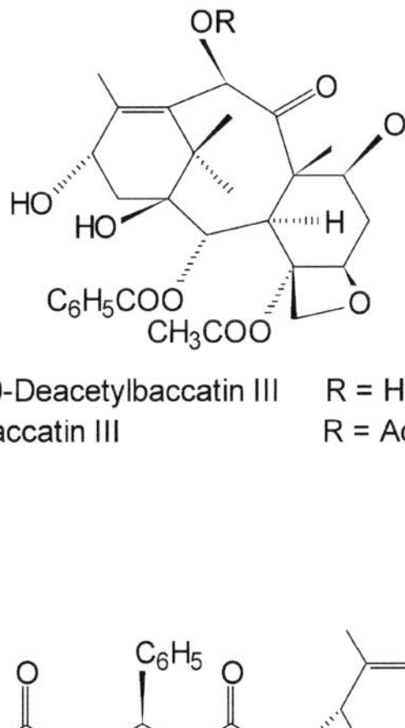

10-Deacetylbaccatin III R = H
Baccatin III R = Ac

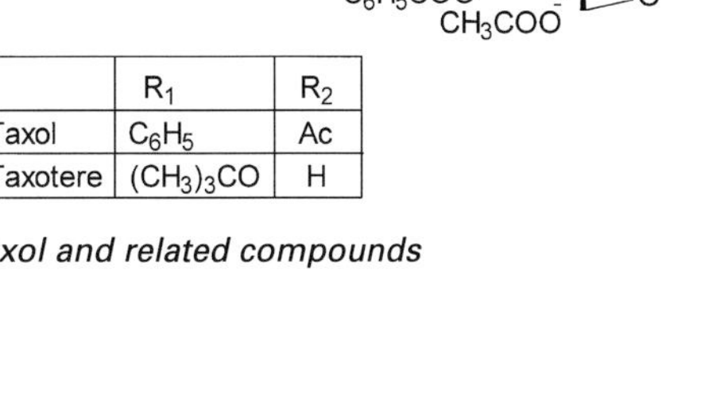

	R_1	R_2
Taxol	C_6H_5	Ac
Taxotere	$(CH_3)_3CO$	H

Taxol and related compounds

TDP: acronym of thymidine diphosphate. See Thymidine phosphates.

Tectoquinone: 2-methylanthraquinone, a yellow anthraquinone, m. p. 179 °C, present in teak wood. It is one of the few nonhydroxylated, naturally occurring anthraquinones. Owing to its content of T., teak is largely resistant to termites and fungi.

Teichmann's crystals, ***chlorhemin crystals:*** rhombic crystals formed by heating hemoglobin with sodium chloride and glacial acetic acid. T. c. are used for the microscopic detection of blood.

Teichoic acids: polymers present in the cell walls of Gram-positive bacteria. They consist of chains of glycerol or ribitol residues joined by phosphate groups; in addition sugars are linked to the glycerol or ribitol, and some of the hydroxyl groups are esterified with residues of D-alanine. For example, T. a. from *Staphylococcus aureus H* consists of eight ribitol units joined 1 → 5 by phosphodiester linkages; the sugar, *N*-acetylglucosamine, is attached to position 4 of the ribitol chiefly by β-linkages with some α-linkages. Glycerol T. a. occur more widely than ribitol T. a. In a few cases little or no sugar is present so that alkaline hydrolysis gives mainly alanine, glycerol and its phosphates; but the presence of other sugars is more usual, e. g. *N*-acetylglucosaminyl (T. a. from periplasmic space of *Staphylococcus aureus H),* glucosyl (T. a. from periplasmic space of *Lactobacillus arabinosus),* α-*N*-acetylgalactosaminyl (T. a. from wall of *Staphylococcus lactis).* The glycerol units are joined 1 → 3 by phosphodiester linkages, and the D-alanine or sugar residues are carried on C2 of the glycerol.

Biosynthesis of ribitol T. a. is by progressive transfer of D-ribitol 5-phosphate units from CDP-ribitol to position 1 of the previous unit; it is not known at which stage the sugar and alanine residues are added. Similarly, glycerol T. a. are biosynthesized from CDP-glycerol.

Telomeres: DNA sequences at the 3′ ends of linear eukaryotic chromosomes. Since the RNA primer at the 5′ end of a completed lagging strand cannot be replaced with DNA (the necessay primer would have no binding site), a special mechanism is required for the replication of T. (see Deoxyribonucleic acid; section: Replication of DNA). Telomeric DNA consists of about a 1,000 tandem repeats of a species-specific, G-rich sequence at the 3′ end of each DNA duplex. Thus, in *Tetrahymena,* the T. consist of the repeated sequence TTGGGG, whereas in the human the repeated sequence is TTAGGG. These tandem repeats are added by the action of a specific *telomerase,* a ribonucleoprotein whose RNA components contain a region complementary to the repeating telomeric sequence (i. e. the enzyme itself contains an RNA template). The telomeric DNA sequence is synthesized on the RNA template region of the enzyme (cf. reverse transcriptase) and added to the 3′ terminus; the template region then moves to the new 3′ terminus and the process is repeated up to 1,000 times or more.

If T. were not replicated, each cycle of DNA replication would shorten the chromosome by the length of an RNA primer. Essential genes would then be deleted and the descendants of the affected cells would die. There is a strong correlation between the initial T. length in cultured cells and the number of times the cells can divide before becoming senescent. Also, fibroblasts from patients with *progeria* (rapid and premature aging leading to childhood death) have abnormally short telomeres. This and other evidence strongly suggests the loss of telomerase function in somatic cells as a basis for the aging of multicellular organisms. [E.H. Blackburn *Annu. Rev. Biochem.* ***61*** (1992) 113–129]

Template RNA: see Messenger RNA.

Template strand, *codogenic strand, anticoding strand:* the strand of a double-stranded stretch of DNA that acts as the template for the generation of an RNA transcript (e.g. mRNA). It therefore has the complementary nucleotide sequence to that of the RNA transcript, save that T is in the place of U. Alternative names for it are Non-coding strand (see) and Non-sense strand (see). See Nomenclatural conventions concerned with gene transcription.

Tenascin, *cytotactin, J1, hexabrachion:* a large multisubunit glycoprotein with both adhesive and antiadhesive properties. Its structure includes EGF-like repeats (see Epithelial growth factor), an integrin-specific, cell-binding sequence (-Arg-Arg-Gly-Asp-Met-), as well as calcium-binding domains. With SPARC (see) and Thrombosondin (see), T. is classified as an extracellular protein that modulates cell-matrix interactions. All three proteins influence embryogenesis and morphogenesis, and there is a positive correlation between their synthesis and cell proliferation. Reasoning that cells that produce an extracellular matrix are also influenced by it, Sage & Bornstein coined the concept of "dynamic reciprocity", i.e. agents such as T. may regulate cell development by influencing the cell's synthesis of its own extracellular matrix. [E.H. Sage & P. Bornstein *J. Biol. Chem.* **266** (1991) 14831–14834]

Terminal: adjective for the chain end component of a biopolymer, e.g. *N*- and *C*-terminal amino acids (see Peptides).

Terminal oxidase: the terminal enzyme of the respiratory chain. In most organisms it is Cytochrome oxidase (see), but in various plant systems other T.o. are present or have been proposed. In aerobic nitrate respiration, the T.o. is nitrate reductase.

Termination: the final phase in the biosynthesis of biopolymers. See Biopolymers, Protein biosynthesis; Ribonucleic acid.

Termination codon, *stop codon, punctuation codon:* a sequence of three nucleotides in mRNA, which signals the end of polypeptide synthesis and release of the polypeptide in the process of Protein biosynthesis (see). 5′-UAA (see Ochre codon), 5′-UAG (see Amber codon) and UGA are T.c.

Termination factors: see Release factors.

Terminus: the chain of a biopolymer, e.g. *N*- or *C*-terminus of a protein, meaning the *N*- or *C*-terminal amino acid.

Terpene alkaloids, *isoprenoid alkaloids:* alkaloids containing a terpene structure, with 10–30 carbon atoms. They are conveniently classified according to the genera in which they occur (Table).

T.a. are biosynthesized from mevalonic acid, but the origin of the nitrogen is not known. They therefore differ from the iridoid, isoquinoline and indole alkaloids, in which a monoterpene is linked to an amino acid.

Classification of terpene alkaloids

Terpene type	Name
Monoterpene	Gentiana alkaloids Valeriana alkaloids
Sesquiterpene	Nuphara alkaloids Dendrobium alkaloids
Diterpene	Aconitum alkaloids Erythrophleum alkaloids
Triterpene (Steroid)	Solanum alkaloids Veratrum alkaloids Funtumia alkaloids Holarrhena alkaloids Buxus alkaloids Salamander alkaloids

Terpenes, *terpenoids, isoprenes, isoprenoids:* an extensive group of natural products whose structures are composed of isoprene units. The number of carbon atoms is usually a multiple of 5. T. are biosynthesized from the active 5-carbon unit, isopentenyl pyrophosphate. Inspection of the formulas of terpenes shows that they can be built up from a C_5 unit, which Wallach suggested might be isoprene. Ruzicka (1921) put forward the isoprene rule, in which the hydrocarbon skeleton of many open chain and cyclic terpenes is constructed from isoprene units arranged head-to-tail. The rule has proved useful in the assignment of structure, although there are exceptions, e.g. Squalene (see) contains a tail-to-tail arrangement. Free isoprene has now been shown to occur widely in plants in trace amounts. Biosynthetic studies with the biochemically equivalent isopentenylpyrophosphate ("active isoprene") have confirmed the validity of the isoprene rule.

Originally only compounds with 10 carbon atoms (monoterpenes) were considered as T., and the oxygen-containing T. were classified as camphors. According to the mechanism of biosynthesis, however, all compounds derived from "active isoprene" are now classified as T. or isoprenes, including steroids, carotenoids, etc.

Structure. T. are subdivided according to the number of C_5 units in their structure, i.e. Hemiterpenes (see), Monoterpenes (see), Sesquiterpenes (see), Diterpenes (see), Sesterterpenes (see), Triterpenes

Classification of terpenes

Group	No. of C_5-units	Examples
Hemiterpenes	1	"Active" isoprene
Monoterpenes	2	Citral, iridoids, camphor
Sesquiterpenes	3	Abscisic acid, proazulenes
Diterpenes	4	Gibberellins, resin acids
Sesterterpenes	5	Cochliobolin
Triterpenes	6	Steroids, sterols, ecdysone
Tetraterpenes	8	Carotenoids
Polyterpenes	up to 10,000	Caoutchouc, gutta percha, polyprenols

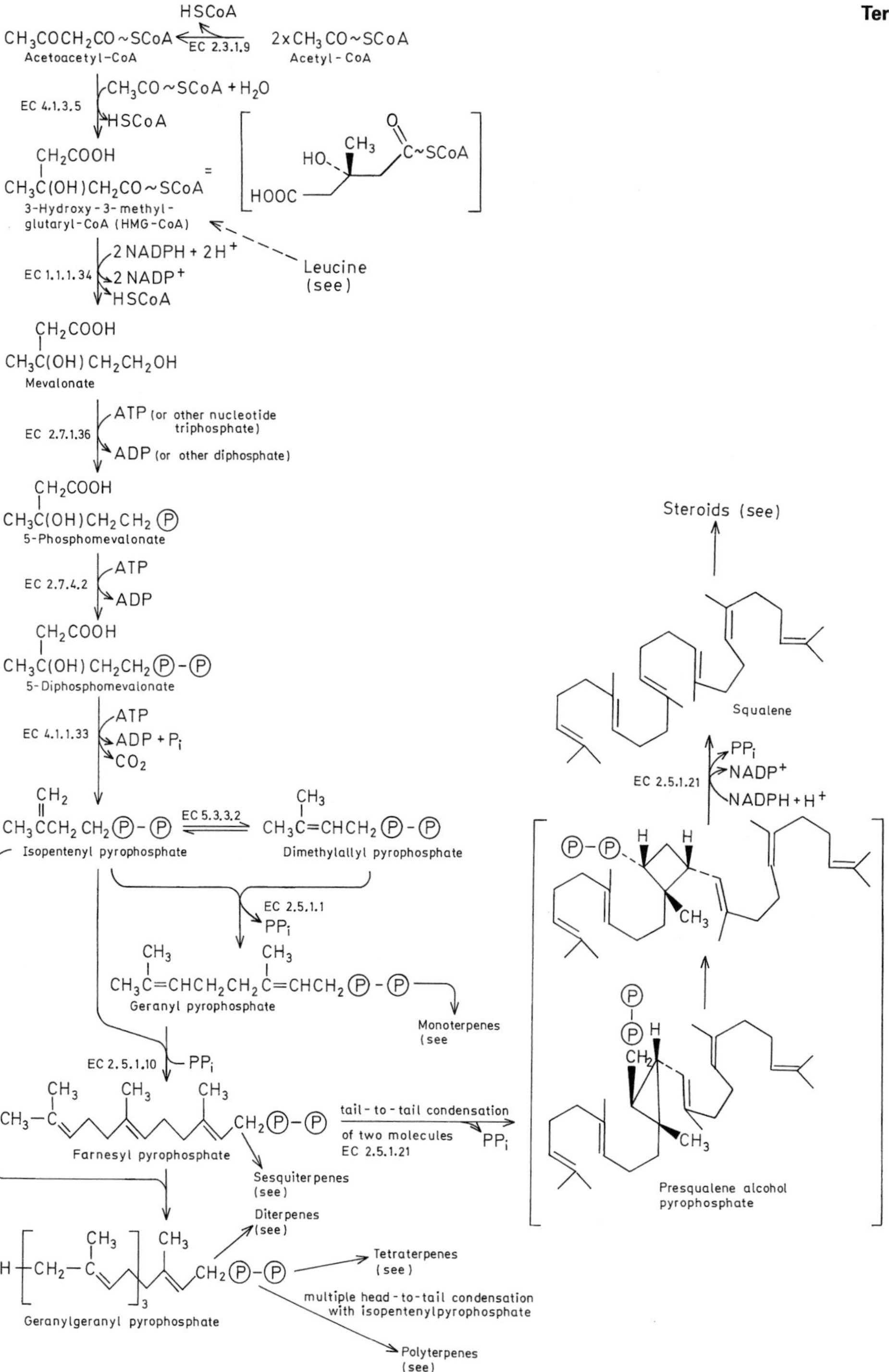

Biosynthesis of terpenes from acetyl-CoA. EC 1.1.1.34, hydroxymethglutaryl-CoA reductase (NADPH); EC 2.3.1.9, acetyl-CoA acetyltransferase; EC 2.5.1.1, dimethylallyltransferase; EC 2.5.1.10, geranyltransferase; EC 2.5.1.21, farnesyltransferase (also catalyses reduction of presqualene alcohol pyrophosphate to squalene); EC 2.7.1.36, mevalonate kinase; EC 2.7.4.2, phosphomevalonate kinase; EC 4.1.1.33, pyrophosphomevalonate decarboxylase; EC 4.1.3.5, hydroxymethylglutaryl-CoA synthase; EC 5.3.3.2, isopentenyldiphosphate Δ- isomerase.

(see) and Polyterpenes (see) (Table). Each of these groups contains different structural types, resulting from unusual linkages of C_5 precursors (e.g. head-to-head), cyclizations, various convolutions of large open chain precursors prior to cyclization, introduction of functional groups (alcohols, aldehydes, ketones, carboxylic acids, lactones), formation of epoxides, introduction of heteroatoms (e.g. terpene alkaloids contain nitrogen), cleavage of cyclic compounds, rearrangements, etc. Some T., e.g. steroids, lack certain carbon atoms, whereas others, e.g. juvenile hormones, contain additional carbon atoms. Isoprene components or small T. can also become linked to other structures, e.g. in lysergic acid, mycelianamide, humulone, indole alkaloids, chlorophyll, vitamins E and K and ubiquinone.

For conventions and rules used in the representation of structure, see Steroids.

Occurrence and Function. Structures of over 5,000 naturally occurring T. are known. They occur in all living forms, but biological functions are known for only a few. Carotenoids are important as accessory photosynthetic pigments, and several groups of T. act as hormones in plants, insects and higher animals. Many pheromones are T. Some T. are also important in medicine and as raw materials in the industrial preparation of foods, perfumes, varnishes and rubber.

Biosynthesis. Precursors of T. are biosynthesized from acetyl-CoA according to the reactions shown in the Fig. Key intermediates are mevalonate and isopentenylpyrophosphate ("active isoprene"). Isopentenylpyrophosphate is in equilibrium with dimethylallylpyrophosphate, which serves as a starter for polycondensation reactions. Thus dimethylallylpyrophosphate and isopentenylpyrophosphate form geranylpyrophosphate, the precursor of the monoterpenes. Condensation with two more molecules of isopentenylpyrophosphate produces first farnesylpyrophosphate (precursor of the sesquiterpenes) then geranylgeranylpyrophosphate (precursor of the diterpenes). Two molecules of farnesylpyrophosphate undergo tail-to-tail condensation to form the Triterpenes (see), while two molecules of geranylgeranylpyrophosphate react tail-to-tail to form the Tetraterpenes (see). Polyterpenes (see) are formed by multiple head-to-tail condensation of isopentenylpyrophosphate units.

Terpenoids: see Terpenes.

Terpinene: see *p*-Menthadienes.

Tertiary structure: see Proteins.

Testane: earlier name for 5α-androstane. See Steroids.

Testosterone: 17β-hydroxy-androst-4-en-3-one, an important androgen synthesized in the interstitial cells of the testicle. It stimulates growth of the prostate and seminal vesicles, and promotes sperm maturation and development of male secondary sexual characteristics. Apart from mammalian testes, T. also occurs in blood and urine. It was first isolated in 1935 from bovine testes. Esters of T., e.g. T. propionate, are used in the treatment of male sex hormone deficiency, endocrine disorders in gynecology and in geriatrics. For structure and biosynthesis, see Androgens.

19-nor-Testosterone: a synthetic anabolic steroid. 19-nor-T. differs from Testosterone (see) by the absence of the C19 methyl group. It has a higher anabolic but lower androgenic activity than testosterone.

Tetanus toxin: see Toxic proteins.

Tetracyclines: a group of antibiotics from various *Streptomyces* spp. T. contain four linearly fused six-membered rings; individual T. differ according to the nature of substituents (Fig. and Table). T. inhibit protein biosynthesis by preventing the binding of aminoacyl-tRNA to ribosomes. Next to the penicillins, T. were one of the most widely used antibiotics, particularly in the treatment of bronchitis, pneumonia, bile duct and urinary infections, plague and cholera. They are also widely employed as additives in animal feedstuffs. On account of side reactions and increasing resistance of bacteria to T. their use is declining.

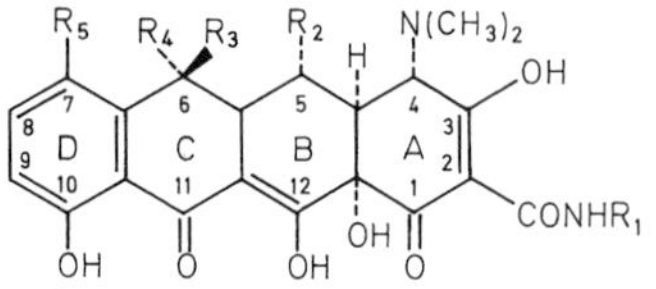

The structures of some tetracyclines

Name	R_1	R_2	R_3	R_4	R_5
Chlortetracycline (aureomycin)	H	H	OH	CH_3	Cl
Oxytetracycline (terramycin)	H	OH	OH	CH_3	H
Tetracycline	H	H	OH	CH_3	H
Methacycline (rondomycin)	H	OH	CH_2=		H
Doxycycline (vibramycin)	H	OH	H	CH_3	H

Tetrahydrobiopterin, *BH_4*: a hydrogen transfer cofactor of a number of aromatic amino acid hydroxylases, including phenylalanine, tyrosine and tryptophan hydroxylases. These are necessary for synthesis of the neurotransmitters, dopa, 5-hydroxytryptophan, dopamine, adrenalin, noradrenalin and serotonin; thus BH_4 is necessary for neurological function. Brain tissues from adults with Down's syndrome, senile dementia of the Alzheimer type and severe endogenous depression have a very low synthetic capacity for BH_4. [J. A. Blair et al. in *Biochemical and Clinical Aspects of the Pteridines* vol. 3 (de Gruyter, Berlin, 1984)]

Tetrahydrocannabinol, *THC*: see Hashish.

Tetrahydrofolic acid, *THF, folate-H_4, coenzyme F*: 5,6,7,8-tetrahydropteroylglutamic acid, the coenzyme responsible for the binding, activation and transfer of all active one carbon units, with the exception of carbon dioxide (the F in coenzyme F stands for formylation). M_r 445.4 (after chemical preparation involving acetic acid, it retains two molecules CH_3–COOH and has M_r 565.4), λ_{max} 298 nm in 0.01 M phosphate, pH 7.0, $\varepsilon_{298} \leq 28{,}000$. In the solid state, THF is slowly oxidized by air and must therefore be stored under vacuum or in an inert atmosphere. In solution, THF is oxidized rapidly to *dihydrofolic acid* (DHF, folate-H_2) (7,8-dihydropteroylglutamic acid). DHF is also produced as a byproduct in the enzymat-

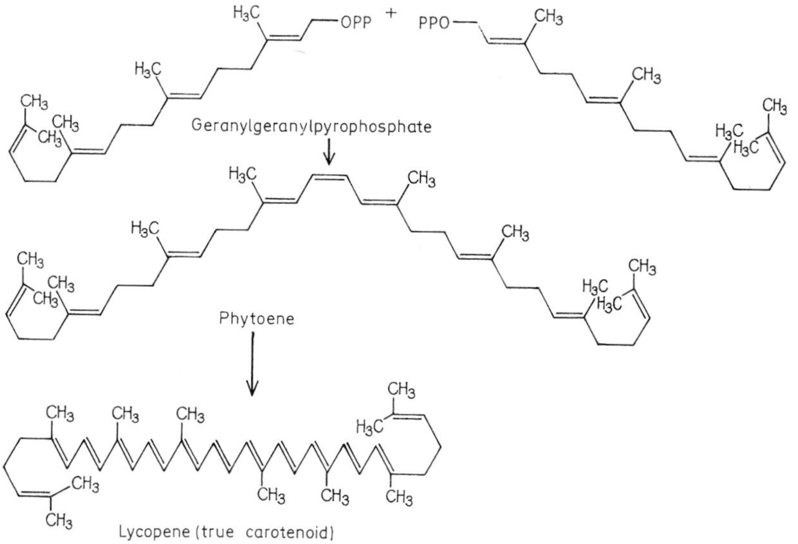

Fig. 1. *Synthesis of carotenoids from two molecules of geranylgeranylpyrophosphate.*

ic synthesis of thymidylic acid by thymidylate synthase (see Pyrimidine biosynthesis). THF/DHF forms a redox system of E_0' −0.19 V. Stability of THF is strongly influenced by pH, being highest at pH 7.4. Ascorbic acid (34 mM) and, to a lesser extent mercaptoethanol (10 mM) have a stabilizing effect on THF solutions. Autoxidation of THF or its metabolic conversion into DHF can be following spectrophotometrically by the shift of absorption from 298 nm to 282 nm. THF is formed from folic acid (for structure, see Vitamins, folic acid) by enzymatic reduction. For biological functions of THF, see Active one carbon units.

Tetrahydrofolic acid conjugates containing three to seven residues of glutamic acid function in methionine synthesis in some microorganisms.

Tetraterpenes: terpenes comprising eight isoprene units ($C_{40}H_{64}$). Naturally occurring T. are almost all carotenoids, and the group contains no polycyclic compounds.

Biosynthesis. Tail-to-tail condensation of two molecules of geranylgeranylpyrophosphate (see Terpenes) gives phytoene, which undergoes stepwise dehydrogenation to produce the all-*trans* configuration of the true carotenoids (Fig. 1). The ionone rings of cyclic carotenoids arise by addition of a proton at C_3, and formation of a bond between C2 and C7. This is followed

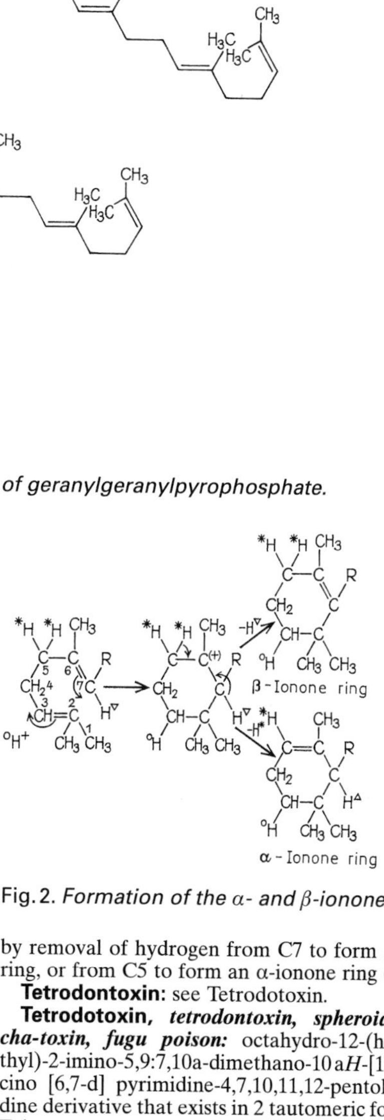

Fig. 2. *Formation of the α- and β-ionone ring.*

by removal of hydrogen from C7 to form a β-ionone ring, or from C5 to form an α-ionone ring (Fig. 2).

Tetrodontoxin: see Tetrodotoxin.

Tetrodotoxin, ***tetrodontoxin, spheroidine, taricha-toxin, fugu poison:*** octahydro-12-(hydroxymethyl)-2-imino-5,9:7,10a-dimethano-10a*H*-[1,3] dioxocino [6,7-d] pyrimidine-4,7,10,11,12-pentol, a guanidine derivative that exists in 2 tautomeric forms (Fig.). T. is an extremely potent toxin from the ovaries, liver

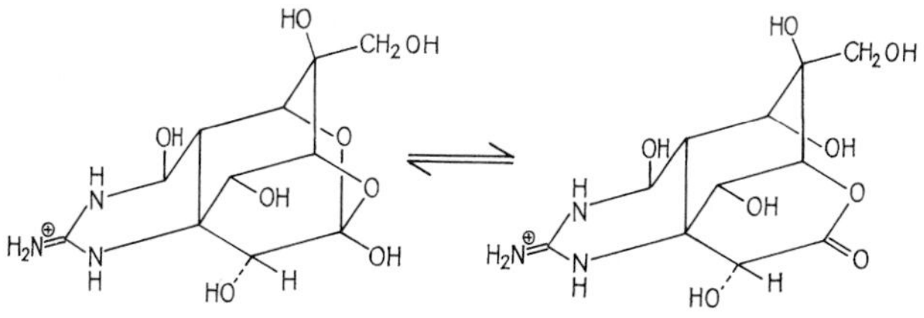

Tetrodotoxin

and skin (but not present in the blood) of many species of *Tetrodontidae,* especially the globe fish, *Spheroides rubripes.* M_r 319.28, darkens above 220 °C, $[\alpha]_D^{25}$ –8.64 ° (c = 8.55, dil. acetic acid), LD_{50} i. p. in mice 10 µg/kg. T. acts on the membranes of nerve fibers, and is an antagonist of Batrachotoxin (see).

Tetrose: a monosaccharide containing four carbon atoms, e. g. threose, erythrulose. T. occur as intermediates in carbohydrate metabolism, usually as their phosphates.

Thalassemias: a group of genetically determined disorders of hemoglobin synthesis, characterized by partial or total absence of the synthesis of one of the hemoglobin chains. The resulting imbalance in globin chain synthesis may lead to precipitation of those globins produced in excess. Such precipitated proteins appear as inclusions, which are responsible for defective maturation and decreased survival of erythroid cells. At both the genetic and molecular level, T. represent a very heterogeneous family of blood disorders. They are broadly classified according to the globin chain which is inefficiently synthesized, i. e. α, β, δβ, δ and γδβ T. Some (e. g. the δβ T.) are characterized by abnormally high levels of fetal hemoglobin ($\alpha_2\gamma_2$) and have much in common with the group of conditions known as *hereditary persistence of fetal hemoglobin* (HPFH). See Hemoglobin. [D. J. Weatherall & J. B. Clegg *The Thalassemia Syndrome* 3rd edn. (Blackwell Scientific Publications, Oxford, 1981)]

Thaumatin: a sweet tasting, strongly basic, histidine- and carbohydrate-free, single chain protein, M_r 21,000, (270 amino acid residues, 8 disulfde bridges), pI 11.5. T. is 750–1,600 times (weight basis) or 30,000100,000 times (molar basis) sweeter tasting than sucrose. It occurs in the fruits of *Thaumatococus daniellii* (a monocot of the arrowroot or Maranta family). T. shows considerable sequence homology with the B-chain of another sweet protein, Monellin (see). These two proteins are immunologically related, and it is thought that a tripeptide sequence (-Glu-Tyr-Gly-) near the surface of each molecule is a common antigenic determinant. [R. B. Iyengar et al. *Eur. J. Biochem.* **96** (1979) 193–204]

THC: acronym of Δ^1-tetrahydrocannabinol.

Thebaine: see Benzylisoquinoline alkaloids.

Theobromine: see Methylated xanthines.

Therapeutic index: see Dose.

Therapeutic range: see Dose.

Thermolysin (EC 3.4.24.4): a heat-stable, zinc- and calcium-containing neutral protease, M_r 37,500, from *Bacillus thermoproteolyticus,* with a substrate specificity similar to that of Subtilisin (see). After one hour at 80 °C, T. still has 50 % original activity. This high heat stability of T. is attributed to the large number of hydrophobic regions and the presence of four bound calcium ions, which serve in place of disulfide bridges (T. contains no disulfide bridges) to maintain the compact shape of the molecule. T. is neither a thiol nor a serine enzyme.

Thermostable enzymes, ***heat stable enzymes:*** a small number of enzymes, mostly hydrolases, which show their highest activity between 60 and 80 °C, and which are stable and catalytically active at up to ~100 °C. T. e. usually have a compact structure, stabilized by many disulfide bonds and/or extensive hydrophobic regions, and a low α-helix content. Examples are particularly common among bacterial enzymes, e. g. thermolysin and certain amylases. T. e. are used in the preparation of biological washing powders, and in the food industry.

Thetins: sulfonium compounds, e. g. dimethylthetin, which can serve as a methylating agent (see Transmethylation). T. also occur naturally, e. g. dimethyl-β-propiothetin is present in algae and higher plants. Dimethylsulfide is a decomposition product of T.

$$(H_3C)_2\overset{\oplus}{S}-CH_2-COO^{\ominus}$$

Dimethylthetin

$$(H_3C)_2\overset{\oplus}{S}-CH_2-CH_2-COO^{\ominus}$$

Dimethyl-β-propiothetin

THF: acronym of Tetrahydrofolic acid.

Thiamin: see Vitamins (vitamin B_1).

Thiamin pyrophosphate, ***TPP, aneurin pyrophosphate, APP, cocarboxylase:*** the pyrophosphoric ester of thiamin (aneurin, vitamin B_1 Fig. 1); the prosthetic group (or coenzyme) of various thiamin pyrophosphate enzymes, e. g. Pyruvate decarboxylase (see), pyruvate dehydrogenase and 2-oxoglutarate dehydrogenase (see Multienzyme complex), Transketolase (see), glyoxylate carboligase and oxalyl-CoA decarboxylase (see Oxalic acid). The free cation of TPP has M_r 425.3, while the chloride (M_r 460.8) crystallizes from ethanol with one molecule of water (M_r 478.8). M. p. of TPP 240–244 °C (d.), λ_{max} 245 and 261 nm (in phosphate buffer, pH 5.0); 231.5 and 266 nm (pH 8.0), λ_{min} 248 nm.

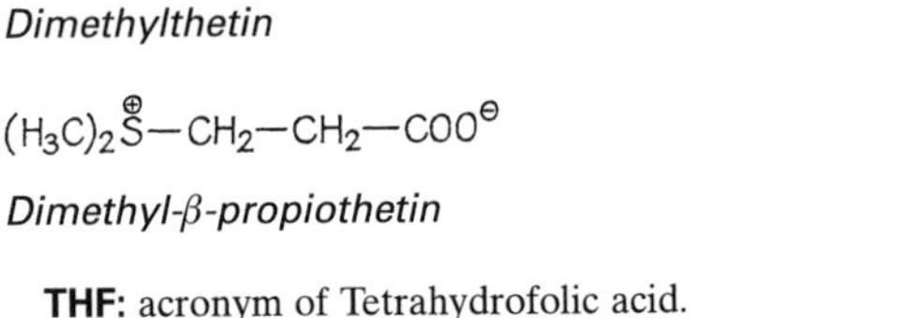

(1)
Thiamine

Fig. 1. *Thiamin pyrophosphate*

In its role as a coenzyme, TPP reacts with the substrates of TPP-enzymes to form active aldehydes: 1. Active acetaldehyde (hydroxyethylthiamin pyrophosphate, HETPP); 2. Active glycolaldehyde [2-(1,2-dihydroxyethyl)-thiamin pyrophosphate, DETPP]; 3. Active formaldehyde (2-hydroxymethylthiamin pyrophosphate, HMTPP). Active pyruvate (pyruvylthiamin pyrophosphate) and active glyoxylate [2-(hydroxycarboxymethyl)-thiamin pyrophosphate] are postulated as intermediates in the formation of active aldehydes. The hydrogen atom at position 2 of the thiazolium ring of TPP (between the sulfur and the nitrogen) (Fig. 2) has a high pK of about 12.6. It is thought that the thiazolium dipolar ion (or ylid) (i. e. C2 forms a carbanion stabilized by the positive charge on the nitrogen) is the key intermediate in the coenzyme function of TPP. Addition of the (δ +) C-atom of a substrate carbonyl group to the C2 carbanion produces an active intermediate. Electrons can then flow from the attached substrate into the ring

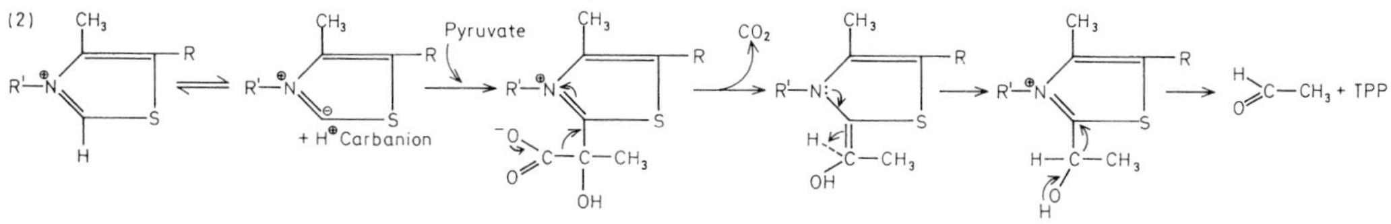

Fig. 2. *Mechanism of thiamin pyrophosphate (TPP) catalysis in the decarboxylation of pyruvate.*

system of the TPP, and bond cleavage occurs between the attached substrate carbon atom (the original carbonyl carbon) and a neighboring carbon atom. All reactions catalysed by TPP-enzymes conform to this mechanism.

Thin-layer chromatography, *TLC:* a form of chromatography (see) in which the solid carrier material is spread in a thin layer on a glass or plastic plate. The advantages of the method are the short distances required for good separation and the correspondingly short development times, high sensitivity, separation of very small amounts of substances, and, if inorganic carrier materials are used, the possibility of using caustic detection reagents. Pre-spread thin-layer plates are available commercially, as is a spreading device for spreading any desired thickness of carrier from 0 to 2 mm. With appropriate equipment, the plates can be used for ascending, descending, horizontal or multiple chromatography.

TLC can be used preparatively as an "open column" (see Column chromatography). Depending on the size of the plate, up to 100 g of material can be separated by preparative TLC while analytical TLC can be used for amounts between 10 ng and 10 mg.

TLC was originally developed for separation of lipophilic substances on inorganic carrier materials. However, if cellulose powder is used as carrier, hydrophilic substances, including amino acids, nucleotides and carbohydrates, can also be separated. Lipophilic substances can be separated on aluminum oxide, silica gel, acetylated cellulose and polyamide; hydrophilic substances on cellulose, cellulose ion exchangers, diatomaceous earth and polyamides. The cellulose ion exchangers used in TLC have shorter cellulose fibers than those used for column chromatography. Polyamide TLC makes use of hydrophilic or hydrophobic polyamides for separation of a wide range of substances. It depends on the reversible formation of hydrogen bonds between the substances and the amide groups of the carrier. The eluant (water, methanol, formamide, etc.) displaces these hydrogen bonds, forming its own with the carrier. Separation thus depends on differences in the strengths of hydrogen bonds formed by the substances to be separated.

Thiobinupharidine: see Nuphara alkaloids.

Thioctic acid: see Lipoic acid.

Thioester, *acylmercaptan:* a compound of the general formula RS ~ CO-R_1. The thioester (acylmercaptan) bond is energy-rich. All fatty acyl coenzyme A derivatives (activated fatty acids, e.g. acetyl-CoA) are T. During substrate phosphorylation on glyceraldehyde 3-phosphate dehydrogenase a thiol group of the enzyme forms an energy-rich intermediate T.

Thioethers: see Sulfur compounds.

Thioglucoside glucohydrolase, *β-thioglucosidase, myrosin, myrosinase, sinigrinase, sinigrase* (EC 3.2.3.1): a plant enzyme responsible for the conversion of glucosinolates into isothiocyanates (see Glucosinolate).

Thiol enzyme, *SH-enzyme:* an enzyme whose activity depends on the presence of a certain number of free thiol groups. T. e. are found among the hydrolases, oxidoreductases and transferases. Known T. e. are bromelain, papain, urease, various flavoenzymes, pyridine nucleotide enzymes, pyridoxal phosphate enzymes and thiolproteinases. T. e. are typically inhibited by Sulfhydryl reagents (see).

Thiolesterases: see Esterases.

Thiol group, *sulfhydryl group, mercapto group:* -SH, the functional group of thiols (mercaptans), i.e. the functional group of RSH, where R is the remainder of the molecule. T. g. may be structurally important as in Thiol enzymes (see), or functionally important as in Coenzyme A (see), Pantetheine-4′-phosphate (see), Lipoic acid (see), Thioredoxin (see), etc. The functional form of lipoic and thioredoxin is a dithiol.

Thiols: see Sulfur compounds.

2-Thiomethyl-N^6-isopentenyladenosine: 2-methylmercapto-6-isopentenyladenosine, an adenosine derivative and one of the rare nucleic acid components found in tRNA from wheat. It is an active cytokinin. The hydroxylated derivative, *2-methyl-mercapto-6-(4-hydroxy-3-methyl-*cis*-2-enylamino)-purine* has also been found in some species of tRNA.

Thioredoxin: a heat-stable, acidic, metal-free redoxin, M_r 12,000. T. is a component of deoxyribose synthase, in which T. and thioredoxin reductase form a hydrogen transfer system linked to the reduction of ribose or ribonucleoside phosphates by NADPH + H^+. T. is a single polypeptide chain of 109 amino acid residues with *N*-terminal serine. The functional group

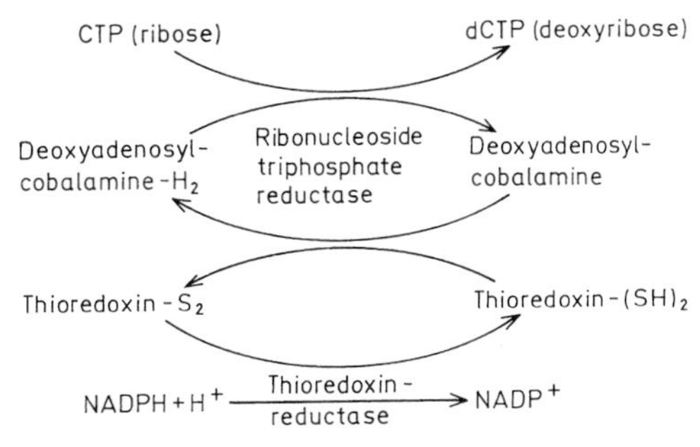

Fig. 1. *Ribonucleoside triphosphate reductase and thioredoxin reductase in* Lactobacillus leishmanii.

of the oxidized form consists of a disulfide bridge between two cysteine residues, which are separated by 10 amino acid residues. Ribonucleoside triphosphate reductase from *Lactobacillus leishmanii* (Fig. 1) also requires a cobalamine coenzyme (cobamide) as a hydrogen carrier.

The cobalamine coenzyme (DBC = oxidized, $DBCH_2$ = reduced form) mediates an intramolecular hydrogen transfer: hydrogen from T. is transferred to the DBC-coenzyme in a ternary complex of T./reductase/DBC; hydrogen is then transferred from the complex to the ribonucleoside triphosphate, e. g. cytidine triphosphate (CTP) (Fig. 2). In its action mechanism, T. resembles Lipoic acid (see).

$$\left[\begin{matrix}SH\\SH\end{matrix}\right. + Enz + DBC \longrightarrow \left[\begin{matrix}S\\ |\\ S\end{matrix}\right]\!\!-\!Enz\text{-}DBCH_2$$

Thioredoxin
(reduced)

$$\left[\begin{matrix}S\\ |\\ S\end{matrix}\right. - Enz\text{-}DBH_2 + CTP \longrightarrow \left[\begin{matrix}S\\ |\\ S\end{matrix}\right. + Enz + DBC + dCTP + H_2O$$

$$\left[\begin{matrix}S\\ |\\ S\end{matrix}\right. + NADPH + H^+ \xrightarrow[\text{reductase}]{\text{Thioredoxin}} \left[\begin{matrix}SH\\SH\end{matrix}\right. + NADP^+$$

Fig. 2. *Hypothetical role of cobalamine coenzyme (DBC) in the action of ribonucleoside triphosphate reductase.*

Thr: abb. for L-threonine.

L-Threonine, *Thr:* L-threo-α-amino-β-hydroxybutyric acid, $H_3C–CH(OH)–CH(NH_2)–COOH$, a proteogenic, essential amino acid with two asymmetric C-atoms. M_r 119.1, m. p. 253 °C (d.), $[\alpha]_D^{25}$ –28.5 (c = 1–2, water). A useful reaction for determination of L-T. is oxidation with periodate to acetaldehyde, glyoxylate and ammonia. Enzymatic hydrolysis of peptide bonds involving L-T. appears to be particularly difficult, which may be relevant to the nutritional physiology of this amino acid. The principal degradative reaction of L-T. in most organisms is conversion to 2-oxobutyrate and ammonia by the pyridoxal phosphate-dependent enzyme, L-T. dehydratase (EC 4.2.1.16). This degradative enzyme (also called L-T. deaminase) is distinct from biosynthetic L-T. dehydratase needed for the production of 2-oxobutyrate in the biosynthesis of isoleucine in *E. coli;* the latter enzyme is allosterically inhibited by isoleucine. L-T. acetaldehyde-lyase (L-T. aldolase, EC 4.1.2.5) is a pyridoxal phosphate enzyme which converts L-T. into glycine and acetaldehyde. It is present in various organisms, including mammals, and appears to be a purely degradative enzyme. Oxidation of L-T. to 2-amino-3-oxobutyrate, followed by decarboxylation, produces aminoacetone, a urinary constituent. In microorganisms, aminoacetone is converted to *R*-1-amino-2-propanol, an intermediate in the biosynthesis of vitamin B_{12}. Aminoacetone may also be oxidatively deaminated to methylglyoxal, which can be attacked by glyoxalase to form D-lactate.

In plants and microorganisms, L-T. is biosynthesized from phosphohomoserine by a γ-elimination of phosphate followed by β-replacement with an OH-group. This total reaction is catalysed by the pyridoxal phosphate enzyme, L-T. synthase (EC 4.2.99.2). The phosphohomoserine is derived from aspartate via aspartyl phosphate, aspartate semialdehyde and homoserine.

Thrombin: a blood coagulation enzyme, responsible for the conversion of fibrinogen to fibrin. T. is a glycoprotein (5 % carbohydrate), M_r 39,000, produced by activation of Prothrombin (see). It is a typical Serine protease (see) with catalytically important residues His_{58}, Asp_{102} and Ser_{195} in the B-chain, and considerable sequence homology with trypsin, chymotrypsin and elastase. Autolysis of T., leading to a decrease of M_r from 39,000 to 26,000 at the expense of the B-chain (M_r 33,000 to 19,500) causes no loss of activity.

Thrombospondin, *TS:* a large (M_r 450,000) trimeric glycoprotein, consisting of 3 disulfide-linked chains. Under the electron microscope, purified TS appears as filaments, 7 × 70 nm. It contains 1.9 % neutral sugars, 1.4 % amino sugars, 0.7 % sialic acid, and no hexuronic acid. Normal plasma contains only very low concentrations of TS. It is a major constituent of platelet α-granules, and is named from the observation that it is released from platelets in response to thrombin treatment. After release, some TS associates with platelet surfaces, where it is at least partly responsible for the lectin-like activity of platelets. It has subsequently been shown to be synthesized and secreted by a variety of cells with a broad tissue distribution. It functions in the regulation of cell growth and migration, thereby playing a role in physiological and pathological processes such as wound healing, hemostatis, morphogenesis and angiogenesis; it actually inhibits angiogenesis in vivo, by inhibiting the migration and proliferation of endothelial cells after they have been stimulated by mitogens.

TS was earlier classified as an ‘adhesive’ protein, sharing several features with other adhesive proteins, e. g. Fibrinogen (see), Fibronectin (see), von Willebrand factor (see). In addition to promoting attachment and spreading, however, TS often also displays antiadhesive properties, leading to cell rounding and detachment from a substratum. It is therefore more appropriate to regard TS as one of a family of extracellular proteins that modulate cell-matrix interactions, a family that includes Tenscin (see) and SPARC (see). All three proteins influence embryogenesis and morphogenesis, and there is a positive correlation between their synthesis and cell proliferation. Reasoning that cells that produce an extracellular matrix are also influenced by it, Sage & Bornstein coined the concept of “dynamic reciprocity”, i. e. agents such as TS may regulate cell development by influencing the cell’s synthesis of its own extracellular matrix

TS possesses multiple cell attachment sites, which enable it to associate with a variety of cells (endothelial cells, smooth muscle cells, some fibroblasts, keratinocytes, neurons, squamous carcinoma cells, malignant fibrosarcoma cells, melanoma cells), and with fibrinogen, fibronectin, laminin, plasminogen, plasminogen activator, collagen, calcium ions and glycosaminoglycans. Four sites have been implicated in cell attachment. 1. The *N*-terminal heparin-binding domain binds heparan sulfate proteoglycans and sulfatides. 2. Type I repeats of -Cys-Ser-Val-Thr-Cys-Gly- are thought to associate with CD36 (a human leukocyte differentiation antigen on the surface of platelets, monocytes and umbilical endothelium). 3. The -Arg-Gly-Asp-Ala- sequence in the last type III repeat of the calcium-binding domain interacts with β3-integrin. 4. By using a monoclonal antibody that inhibits attachment of human melanoma cells, a major cell attachment site has been identified in or near the globular *C*-terminal region (the epitope lies within the last 122 residues of the *C*-terminal domain). [J. W. Lawler et al. *J. Biol. Chem.* **253** (1978) 8609–8616; R. Wolff et al. *J. Biol. Chem.* **261**(1986) 6840–6846; E. H. Sage & P. Bornstein *J. Biol. Chem.* **266** (1991) 14831–14834; M. D. Kosfeld et al. *J. Biol. Chem.* **266** (1991) 24257–24259]

Thrombosthenin: see Muscle proteins.

Thromboxanes, *TX:* derivatives of the Prostaglandins (see). T. induce aggregation of platelets, formation of clots and smooth muscle contraction. TXA_2 is more active than TXB_2, but it decays so rapidly (Fig.) that experimental analysis is difficult. TXB_2 does not cause an increase in platelet cAMP levels (in contrast to the prostaglandins), while TXA_2 inhibits the increase of platelet cAMP caused by prostaglandins.

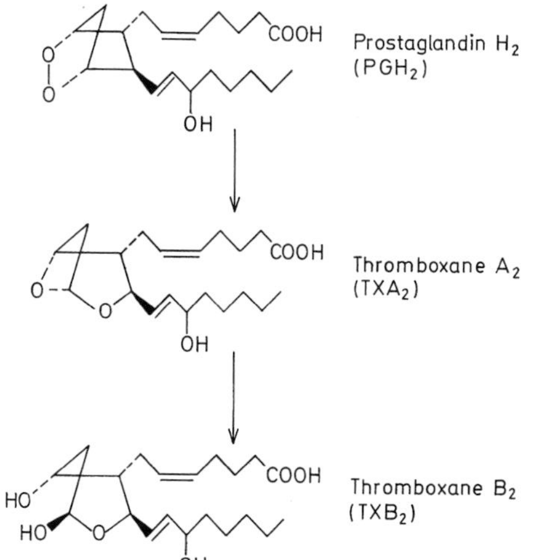

Prostaglandin H_2 (PGH_2)

Thromboxane A_2 (TXA_2)

Thromboxane B_2 (TXB_2)

The immediate precursor of the TX is prostaglandin H_2, which is derived from Arachidonic acid (see). TXA_2 is then rapidly converted to TXB_2, although this conversion is somewhat retarded by the presence of albumin. Formation of TXA_2 is inhibited by low doses of aspirin, which accounts for the anticoagulant properties of aspirin. [S. Moncada & J. R. Vane *Pharmacol. Rev.* **30** (1979) 293–331]

Thujane: see Monoterpenes (Figure).

Thy: abb. for thymine.

Thylakoids: internal membrane structures of the chloroplast. Under the electron microscope, T. appear as disk-shaped, flattened vesicles, about 600 nm diam. These are arranged in stacks, which are the grana observable under the light microscope. In addition to these granal T., there are also stromal T. which pass singly through the stroma of the chloroplast, joining together various stacks of granal T. The functional unit of T. is thought to be the Quantasome (see). The T. membrane is about 9 nm thick, enclosing a thin internal space or *loculus;* it contains approximately equal quantities of protein and lipid, and is notable for its high content of galactosyl diacylglycerol, digalactosyl diacylglycerol and sulfolipid. There is probably a greater proportion of lipid molecules on the inside of the membrane and more protein on the outside, but distinct layering into protein and lipid seems to be absent. The inside lipid layer contains the chlorophylls and carotenoids, and the chlorophyll is present largely, if not entirely, in the form of protein complexes. The protein subunits of the outer layer have a diameter of 4 nm.

Thymidine, *thymine deoxyriboside:* the nucleoside of thymine and D-2-deoxyribose. M_r 242.33, m. p. 185–186 °C, $[\alpha]_D^{16}$ +32.8 ($c = 1.04$, 1 M NaOH). Strictly speaking, it is deoxythymidine, but it is conventionally referred to as thymidine. Its phosphates are designated dTMP, dTDP and dTTP. 2′-Hydroxythymidine (not present in RNA, and of only rare occurrence) is known as ribosylthymine. For metabolic importance see Thymidine phosphates.

Thymidine phosphates: nucleotides of thymine; phosphate esters of deoxythymidine. Although T. p. contain deoxyribose, the prefix deoxy is usually omitted, because the corresponding ribose derivatives hardly ever occur naturally. ***Thymidine 5′-monophosphate (TMP, thymidylic acid, deoxythymidine 5′-monophosphate, dTMP, deoxythymidylic acid):*** a component of DNA, and an intermediate in the synthesis of TPP (see Pyrimidine biosynthesis). M_r 322.2, m. p. 225–230 °C. Stepwise phosphorylation of TMP leads to ***thymidine 5′-diphosphate (TDP, deoxythymidine 5′-diphosphate, dTDP),*** M_r 402.2, which serves as the activating group in certain Nucleoside diphosphate sugars (see); and to ***thymidine 5′-triphosphate (TTP, deoxythymidine 5′-triphosphate, dTTP),*** M_r 482.18, a substrate of DNA synthesis.

Thymidylic acid: see Ribothymidine, Thymidine phosphates.

Thymin: see Thymopoetin.

Thymine, *T* or ***Thy, 2,6-dihydroxy-5-methylpyrimidine, 5-methyluracil:*** a pyrimidine base present in DNA. M_r 126.1, m. p. 321–326 °C (d.). T. was first isolated in 1893 from thymonucleic acid. For the biosynthesis of T., see Pyrimidine biosynthesis.

UV irradiation of DNA solutions causes dimerization of adjacent T. bases in the DNA chain. Formation of T. dimers depends on the wavelength; 265 nm promotes dimerization, but at 235 nm previously formed dimers revert to the monomers. T-T can also form in living cells, where it can be excised and the damage repaired. See DNA repair.

Thymine deoxyriboside: see Thymidine.

Thymine dimer: see Dimers.

Thymonucleic acid, *thymus nucleic acid:* nucleic acid from the thymus gland, effectively an obsolete term for DNA

Thymopoietin, *thymin, thymosin:* a family of largely identical monomeric polypeptide hormones from the thymus, all consisting of 49 amino acid resi-

10
NH$_2$-Residue 1-Residue 2-Phe-Leu-Glu-Asp-Pro-Ser-Val-Leu-Thr-Lys-Glu

30 20
Glu-Gly-Ala-Pro-Leu-Thr-Val-Asn-Asn-Ala-Val-Leu-Glu-Ser-Lys-Leu-Lys

40
Gln-Arg-Lys-Residue 34-Val-Tyr-Val-Glu-Leu-Tyr-Leu-Gln-Residue 43-Leu

49
HOOC-Arg-Lys-Leu-Ala-Thr

Protein	Residue			
	1	2	34	43
TPI (thymus)	Gly	Gln	Asp	His
TPII (thymus)	Pro	Glu	Asp	Ser
TPIII (spleen)	Pro	Glu	Asp	His
Splenin (spleen)	Pro	Glu	Glu	His

Structures of thymopoietins and splenin.

dues (Fig.). T. induces differentiation of prothymocytes to thymocytes. It impairs neuromuscular transmission, and binds with high affinity to the acetylcholine binding region of the nicotinic acetylcholine receptor of *Torpedo californica.* It has been implicated in the pathogenesis of myasthenia gravis. The synthetic pentapeptide thymopentin, Arg-Lys-Asp-Val-Tyr, corresponds to residues 32–36 of T.; it possesses all the biological properties of T. and represents the active site of the latter. Splenin, a variant of T.III (Fig.), has Glu in place of Asp at position 34 and has no effect on neuromuscular transmission. [G. Goldstein et al. *Science* **204** (1979) 1309–1310; T. Audhya et al. *Biochemistry* **20** (1981) 6195 6200; K. Venkatasubramanian et al. *Proc. Natl. Acad. Sci. USA* **83** (1986) 3171–3174]

Thymosin: see Thymopoetin.

Thyrocalcitonin: see Calcitonin.

Thyroid gland, *Glandula thyreoidea:* a well vasculated gland at the front of the neck. It is paired in amphibians and birds, and unpaired in elasmobranch fish and mammals, weighing 20–60 g in the human. The T. g. synthesizes, stores (in the thyroid follicles) and secretes Thyroxin (see) and triiodothyronin, under the influence of the anterior pituitary hormone thyrotropin. It also synthesizes Calcitonin (see) in the parafollicular C-cells.

Thyroid stimulating hormone: see Thyrotropin.

Thyrotropin, *thyroid stimulating hormone, TSH:* a glycoprotein hormone, M_r 25,000 (bovine), containing 23 % carbohydrate. Primary structure of some TSH molecules is known. It consists of an α-and a β-chain, and the α-chain is identical to that of Luteinizing hormone (see). Synthesis occurs in the basophilic cells of the anterior pituitary. Both synthesis and secretion are stimulated by thyrotropin releasing hormone (see Releasing hormones) from the hypothalamus, and inhibited by thyroxin. TSH generally stimulates the thyroid gland: blood circulation of the thyroid gland is increased, uptake of iodine is promoted, the rates of synthesis of thyroglobin, triiodothyronine and thyroxin are increased, and the secretion of thyroid hormones is stimulated. Inactivation occurs in the kidney. Blood concentrations of TSH are in the order of ng/ml, and can be measured radioimmunologically.

Thyrotropin releasing hormone: see Releasing hormones.

Thyroxin, *3,5,3′,5′-tetraiodothyronine, T_4:* a hormone (M_r 776.9) produced by the thyroid gland and absolutely essential for growth and development. T_4 and the second thyroid hormone, 3,5,3′-triiodothyronine (T_3, M_r 651.0) are synthesized from L-tyrosine residues in thyroglobulin, a dimeric glycoprotein (M_r 670,000) that constitutes the bulk of the thyroid follicle. Tyrosine residues in thyroglobulin become iodinated, so that the protein contains several mono- and diiodotyrosine residues. The nature of the subsequent coupling reaction is uncertain, but it is equivalent to the transfer of iodinated rings from some iodotyrosine residues to form ether linkages by reaction with the hydroxyl functions of other iodinated tyrosine residues. T_4 and T_3 are released by proteolysis of thyroglobulin. Synthesis and release of T_4 and T_3 from thyroid epithelial cells, together with a parallel increase in the uptake of iodine by the thyroid gland, are stimulated by the hormone thyrotropin from the anterior pituitary. Both hormones are carried in the blood to all body cells, partly in the free form and partly bound to prealbumin and glycoprotein, T_4 being more tightly bound than T_3.

Metabolic action of T_3 and T_4: increased oxygen uptake by mitochondria, and increased heat production (calorigenic effect); physiological concentrations of both hormones increase synthesis of RNA and protein; in higher doses they act catabolically, causing negative nitrogen balance and mobilization of fat depots. Independently of their calorigenic effect, they increase the rate of cell differentiation and metamorphosis, e. g. development of tadpoles into frogs. Biological half life of T_4 is 7–12 days (based on the sustained activity of a single T_4 dose). Degradation consists of removal of iodine (reused by the thyroid gland), deamination and coupling with glucuronic acid or sulfate in the liver, followed by urinary excretion.

Hyperthyroidism is caused by overactivity of the thyroid gland leading to an excess of T_4 and T_3. Hypothyroidism results from decreased hormone production; this may be caused by iodine deficiency, intake of goitrogens, defective enzymes in hormone synthesis, autoimmune thyroiditis (antibodies are formed against the body's own thyroid tissue), etc.

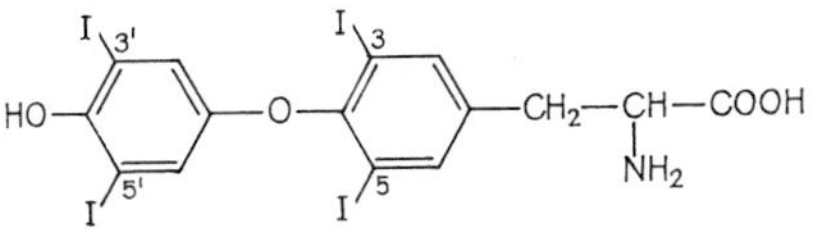

Thyroxin

Prolonged hypothyroidism may result in dwarfism, mental deficiency, goiter and myxedema.

Tin, *Sn:* a metal occurring in many tissues and dietary components. The redox potential of $Sn^{2+} \rightleftharpoons Sn^{4+}$ is 0.13 volt, near to the redox potential of the flavin enzymes, suggesting a possible biological role. It is still not certain that Sn is biologically essential, and its presence in tissue may be environmental contamination. It has been reported that Sn is essential for the growth of rats.

Tingitanin: see Guanidine derivatives.

Tissue hormones: hormones produced in specialized, single cells scattered through a tissue rather than clumped in a gland (see Hormones). They fall into three groups: 1. Secretin (see), Gastrin (see) and Cholecystokinin (see) from the gastrointestinal tract; 2. Angiotensin (see) and Bradykinin (see), which occur as inactive precursors in the blood; and 3. Biogenic amines (see), such as Histamine (see), Serotonin (see), Tyramine (see) and Melatonin (see). This last group is an exception to the rule that hormones act at sites removed from the cells which produce them, since they affect the immediately surrounding tissue.

Tlatlancuayin: 5,2′-dimethoxy-6,7-methylenedioxyisoflavone. See Isoflavone.

TMP: acronym of thymidine monophosphate. See Thymidine phosphates.

TN: abb. for troponin. See Muscle proteins.

TNF: acronym of Tumor necrosis factor.

Toad toxins: poisons found in the secretions of the skin glands of toads (*Bufonidae*) (Fig.). These are classified as 1. bufadienolides (bufogenins) with digitalis-like effects on the heart (see Cardiac glycosides), e.g. bufotoxin. They strengthen and slow the heartbeat. The bufadienolides are present in toad blood at a dilution of 1:5,000 to 1:20,000, and they are necessary for normal heart activity. 2. Alkaline toxins which are alkaloids derived from tryptamine or indole, e.g. bufotenine, dehydrobufotenine and *O*-methylbufotenine. The alkaline toxins of some toad species also contain adrenalin and similar substances. Bufotenines increase the blood pressure and have a paralysing effect on the motor centers of the brain and spinal column. As anesthetics, T.t. are several times more potent than cocaine.

Tobacco alkaloids: see Nicotiana alkaloids.

Tobacco mosaic virus, *TMV:* a worldwide distributed helical virus with a rod-shaped virion, 300 nm long and 18 nm diameter. As a rule, an infected host cell contains between 1 million and 10 million virus particles. It is readily transmitted mechanically, attacking primarily cultivated members of the *Solanaceae,* in particular, tobacco, tomato and paprika. It has also been found in low concentrations in fruit trees and grape vine. There are numerous strains of the virus, which cause a mosaic of light and dark green, a yellow mosaic, or marked leaf deformations. Infection with TMV also often leads to stunted growth. Normally (e.g. in *Nicotiana tabacum,* which includes most of the commercially grown varieties of tobacco), the virus spreads throughout the entire plant, only the meristems remaining virus-free. The infection is then said to be systemic, since the virus is transported in the vascular tissue, usually in the phloem. In other species, e.g. *Nicotiana glutinosa,* the plant reacts strongly against the infection, and the virus is restricted to (encapsulated in) certain cells. Necrotic lesions are formed at the foci of infection; these are easily counted, and under experimental conditions, the number of necrotic lesions is an index of the concentration or virulence of the virus particles in an inoculum.

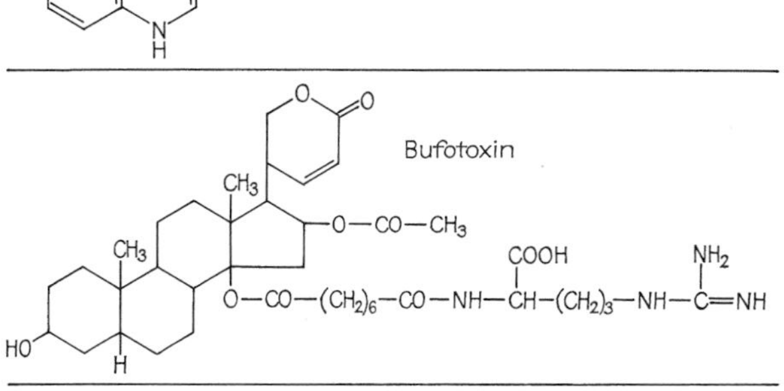

Bufotenine

Dehydrobufotenine

Bufotoxin

Toad toxins

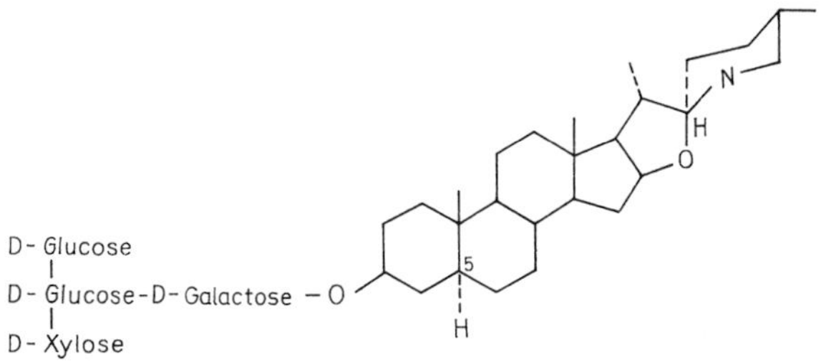

D-Glucose
|
D-Glucose-D-Galactose —O
|
D-Xylose

Tomatine

TMV is not only extremely infectious, but also very resistant to desiccation, heating and numerous chemical agents. This is one of the main reasons that it became a model for plant virus research, leading to the elucidation of many fundamental problems of molecular biology. The replication system of TMV was studied in detail, using virus-infected cultures of protoplasts, as well as synchronous virus replication. It also serves as a model RNA virus. After virus particles enter a cell (usually through a wound in the plant tissue), part of the protein of the nucleocapsid (see Viruses) is stripped off, exposing a cistron that encodes early proteins. One of these early proteins induces or accelerates the synthesis of RNA-dependent RNA polymerase. This enzyme is encoded in the genome of the host plant (earlier, it was erroneously thought to be encoded by the viral RNA). Replication of the TMV-RNA then commences with the aid of the RNA-dependent RNA polymerase. At first, mainly double stranded RNA is synthesized (known as the RF form or replicative form). At the same time, heterogeneous (with respect to size), replicative intermediates are detectable (RI forms). RI forms consist of infective template strands with attached, complementary, growing daughter stands, the two held together by replicases. Very soon, increasing amounts of replicated complete viral RNA become detectable. The ensuing exponential phase of viral RNA replication continues for about 10 hours in the host protoplasm. Translation of the coat protein starts early in the exponential phase of RNA synthesis. Very soon there are more coat protein subunits than replicase molecules. This is probably made possible by translation of the coat protein on a separate monocistronic mRNA, which is produced either by the processing of newly synthesized viral RNA or by transcription of part of the viral RNA. Production of such large amounts of coat protein leads to the incorporation of ever increasing quantities of newly synthesized TMV-RNA into nucleocapsids (each RNA strand is coated with 2,130 coat protein subunits). About 20 hours after infection of the protoplasts, practically all synthesized viral RNA strands are coated with protein. It is not clear why newly synthesized virus particles do not become uncoated like infecting particles; they are possibly protected from the relevant enzymes by inclusion in membrane vesicles. Like the particles of other viruses, TMV particles often accumulate as relatively large, light microscopically visible inclusion bodies, which in plants are also known as X-bodies. For structure, see Viruses.

Tocopherol: see Vitamins (vitamin E).

Tocoquinone: see Vitamins (vitamin E).

Tolerogens: see Immunotolerance.

Tomatidenol: see β-Solamarine.

Tomatidine: see α-Tomatine.

α-Tomatine, *tomatine:* a Solanum alkaloid, and the chief alkaloid of the tomato (*Lycopersicon esculentum*), also occurring in other *Lycopersicon* and *Solanum* spp. T. is a glycoalkaloid of the aglycon *tomatidine* [(22*S*: 25*S*)-5α-spirosolane-3β-ol, M_r 415.7, m.p. 210 °C, $[\alpha]_D$ +6.5 ° ($CHCl_3$)] and the tetrasaccharide β-lycotetraose. T. imparts a bitter taste and protects the tomato plant from attack by the Colorado potato beetle. It also has antibiotic activity against the causative agents of tomato wilt and other pathogenic fungi.

Tonoplast: see Vacuole.

Topoisomerases: enzymes which interconvert topological isomers of circular duplex DNA by altering the degree of supercoiling (superhelicity; see Superhelix). In prokaryotes, an appropriate degree of supercoiling is thought to be important in the control of replication, transcription, recombination, integration, transposition and renaturation of single-stranded DNA circles. In addition, topoisomerization may also be important in the chemomechanical activity of DNA in such processes as the filling of phage heads, transfer of DNA in bacterial conjugation, and mitosis. T. also catalyse the formation and resolution of knotted and catenated circular duplex DNA.

Type I T. transiently break one strand of the DNA double helix, so that the Linking number (see) changes in steps of one. ATP is not required. Reactions catalysed by type I T. can be explained by the rotation of a transiently broken strand around its unbroken, complementary neighbor (Fig. 1); alternatively the reaction can be represented by the passing of the unbroken strand through the transient break. Prokaryotic type I T. become covalently bound to the broken 5′-end by a phosphotyrosine bond, thus conserving the energy of the cleaved phosphodiester bond, and permitting the two ends of the broken strand to be resealed after topoisomerization. Negatively, but not positively supercoiled DNA is relaxed by prokaryotic type I T., and Mg^{2+} is required for activity.

Eukaryotic type I T. relax both positively and negatively supercoiled DNA; during catalysis, the enzyme attaches at the 3′-end of the break via a phosphotyrosine linkage. Rat liver type I T. catalyses the formation in vivo of chromatin-like material from relaxed circular DNA and core histones. This suggests a role in vivo for eukaryotic type 1 T. in the formation of chromatin, which correlates well with the fact that the eukaryotic enzymes are found almost entirely in

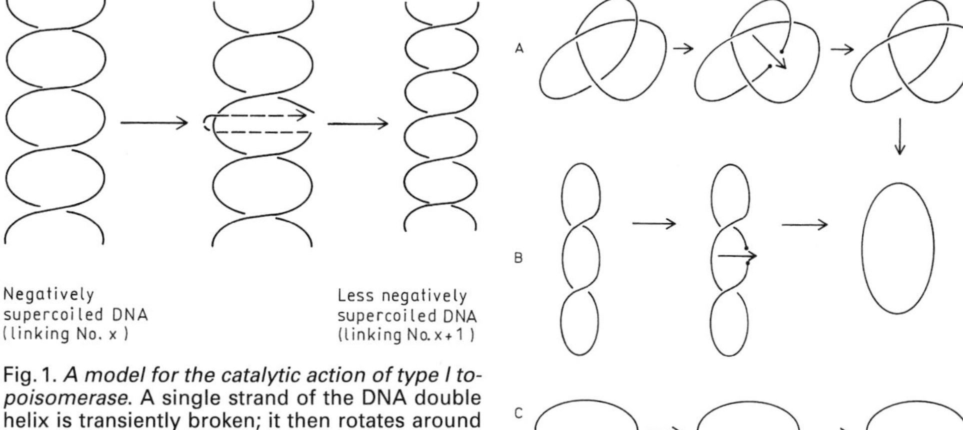

Fig. 1. *A model for the catalytic action of type I topoisomerase.* A single strand of the DNA double helix is transiently broken; it then rotates around the unbroken strand before resealing.

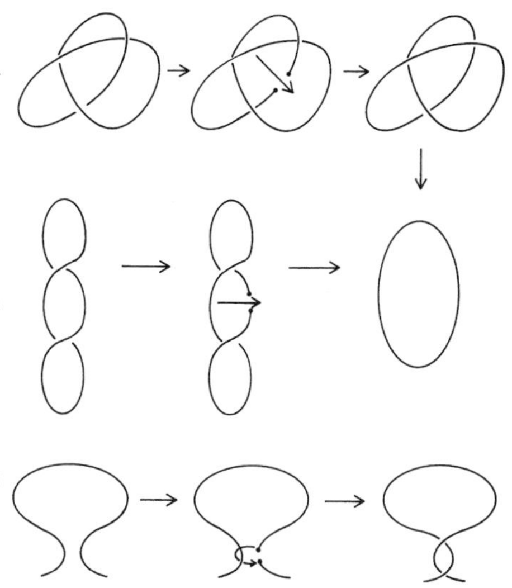

Fig. 2. *Reactions catalysed by type II topoisomerases.* In each case the single line represents duplex DNA. Transient breakage occurs in both strands of the double helix. A: resolution of a knotted circular duplex. B: relaxation of supercoiling of a circular duplex. C: one loop of a circular duplex, showing generation of supercoiling.

the chromatin fraction. It is proposed that supercoiling or unwinding of DNA within the nucleosome causes increased torsional strain on the rest of the DNA molecule, which in turn is relaxed by the action of type I T. This theory must still be reconciled with the fact that chromatin may be assembled in vivo from newly replicated, discontinuous DNA (see Chromatin).

The following enzymes described in the literature are now known to be type I T.: *E. coli* ω-protein (identical with Eco DNA T., *E. coli* swivelase, and *E. coli* type I T.), untwisting enzymes, nicking-closing enzymes, Relaxation protein (see), DNA-Relaxing enzyme (see).

Type II T. promote a double-strand breakage through which another section of unbroken double helix passes before the break is resealed (Fig. 2). The linking number is therefore changed in steps of two. Type II T. can also catenate and decatenate closed circles of DNA, as well as relaxing supercoils. DNA gyrase is a type II T., which differs from all other T. in being able to increase the torsional strain on the DNA circular duplex, i.e. it can convert relaxed circular DNA into a superhelix. This involves an increase in free energy, which is supplied by ATP. In the absence of ATP, supercoiled DNA is relaxed by DNA gyrase. Supercoiling by gyrase is always negative, i.e. the same as intracellular DNA. Other reactions catalysed by gyrase are the DNA-dependent hydrolysis of ATP, and the knotting, unknotting, catenation and decatenation of circular duplex DNA.

Gyrases are inhibited by a group of antibiotics comprising novobiocin, coumermycin and clorobiocin, and by a different group represented by nalidixic acid and oxolinic acid. Mutants of *E. coli* resistant to one or the other of these groups of antibiotics have been isolated, and the corresponding sites of mutation have been mapped. Since resistant mutants contain antibiotic-resistant DNA gyrase and the drug resistance of both the bacterium and its gyrase are cotransduced by phage P1, the site of attack of these antibiotics appears to be the gyrase enzyme itself. For this reason, the designation of the locus for nalidixic and oxolinic acid resistance (48 min on the *E. coli* map) has been changed from *nal A* to *gyr A*. Similarly, the locus for resistance to coumermycin, novobiocin and clorobiocin (82 min) has been renamed *gyr B* (formerly *cou*). Significantly, all these antibiotics are inhibitors of DNA replication, which is supporting evidence for the participation of DNA gyrase in this process. Gyrase is an equimolar complex of two proteins (A and B) and probably exists in solution as the A_2B_2 tetramer. Each protein is the site of attack of one family of antibiotics. DNA gyrase activity has not been found in any eukaryotic organism.

Other type II T. catalyse an ATP-dependent relaxation of supercoiled DNA, and are described in the literature as ATP-dependent DNA-relaxing enzymes. The first representative of this group to be described was isolated from *E. coli* infected with T4 phage [L.F. Liu et al. *Nature* **281** (1979) 456–461]. The purified enzyme has protein components of M_r 63,000 (product of phage gene 39) and 52,000 (phage gene 52). In vivo, it appears to be responsible for fork initiation (but not fork movement) in DNA replication. In vitro, it relaxes both negatively and positively supercoiled DNA. When large amounts of T4 DNA T. are incubated with circular duplex DNA in the absence of ATP, knotted DNA molecules are produced. These knotted circles are restored to a simple circular form by incubation with catalytic amounts of enzyme in the presence of ATP. Similar type II T. have now been isolated from many different eukaryotic sources (e.g. *Drosophila* embryo, *Xenopus laevis* germinal vesicles, rat liver mitochondria, calf thymus, HeLa cell nuclei, yeast).

An earlier nomenclature distinguished between type I and type II T. according to whether they relaxed positive and negative, or only negative supercoiling. This is not widely used and should be discontinued. Since all topological alterations of DNA involve transient breakage and rejoining of DNA strands, any enzyme that causes breakage of DNA may also show T. activity. For example, T. activity has been demonstrated for ΦX174 cistron A protein (which breaks the replicative form of DNA and becomes attached to it at a specific site) and for the *int* protein of phage λ (which catalyses an intermolecular strand transfer during integrative recombination). Conversely, enzymes first recognized for their T. activity may later be found to have other physiological roles involving transient DNA cleavage. [C. W. Wang & L. F. Liu, in *Molecular Genetics* part III *Chromosome structure,* J. H. Taylor, ed. (Academic Press, 1979) pp. 65–88; M. Gellert *Annu. Rev. Biochem.* **50** (1981) 879–910; K. Geider & H. Hoffmann-Berling *Annu. Rev. Biochem.* **50** (1981) 233–260; H. Peng & K. J. Marians *J. Biol. Chem.* **270** (1995) 25286–25290; J. R. Spitzner et al. *J. Biol. Chem.* **270** (1995) 5932–5943; D. B. Wigley 'Structure and Mechanism of DNA Topoisomerases' *Annu. Rev. Biophys. Biomol. Struct.* **24** (1995) 185–208; L. Stewart et al. 'The Domain Organization of Human Topoisomerase I' *J. Biol. Chem.* **271** (1996) 7602–7608; N. Tuteja & R. Tuteja 'DNA helicases: the long unwinding road' *Nature Genetics* **13** (1996) 11–12]

Topological winding number: see Linking number.

Toxalbumins: see Toxic proteins.

Toxic proteins: mostly low M_r, single chain, nonenzymatic proteins, produced especially by snakes and invertebrate animals, but also by some plants (phytotoxins) and virulent strains of bacteria. With the exception of bacterial enterotoxins, and *Botulinus* toxins, T. p. show practically no oral activity, and are only toxic when injected, i. e. when the digestive tract is bypassed.

Known ***plant toxins*** are: 1. the homologous viscotoxins (M_r 4,840; 46 amino acid residues of known sequence; 3 disulfide bridges) from leaves and branches of the European mistletoe, which have hypotensive activity and cause a slowing of the heart beat; 2. the toxalbumins Ricin (see) and abrin, which inhibit protein biosynthesis.

The best studied of the ***bacterial toxins*** are the thermolabile exotoxins of Gram-positive bacteria, which are secreted into the surrounding medium. 1. Five enterotoxins are secreted by *Staphylococcus aureus* in the gastrointestinal tract, causing diarrhea and vomiting. Enterotoxin B is of known primary structure (M_r 28,370; 293 amino acid residues; one disulfide bridge). 2. The Diphtheria toxin from *Corynebacterium diphtheriae* is an acidic, single chain protein (M_r 62,000) of high toxicity (1 µg/kg body weight is fatal). It inactivates peptidyl transferase II in eukaryotic cells, by promoting the attachment of ADP-ribose to the enzyme. 3. Tetanus toxin (M_r 150,000) from *Clostridium tetani* exists in two forms, filtrate and cell toxin. The former has two subunits (M_r 95,000 and 55,000), and the latter has one. In the mouse, 0.01 ng/kg is fatal. Tetanus toxin partially inhibits the binding of botulinum Type A neurotoxin to nerve endings. 4. The five highly toxic *Botulinus toxins* from *Clostridium botulinum* are SH proteins, and they require the presence of one free SH-group for neurotoxic activity. They are resistant to proteolytic digestive enzymes, but are destroyed by boiling; 0.03 ng/kg is fatal in the mouse. Type A (M_r 140,000) binds specifically to the membranes of peripheral nerve terminals, where it irreversibly inhibits the release of acetylcholine. [J. O. Dolly et al. *Nature* **307** (1984) 457–460]

The relatively heat-stable endotoxins are released by autolysis of the bacteria. *Cholera toxins* (M_r 84,000–102,000) are endotoxins released from the Gram-negative *Vibrio cholerae* in the intestine. They are composed of two functionally different subunits, L and H. The L subunit has a high affinity for gangliosides of the membranes of nerve cells, adipocytes, erythrocytes, etc., while the H subunit is responsible for toxicity. The colicins (M_r 60,000) are endotoxins produced by intestinal bacteria. Their toxicity is due to inhibition of cell division and inhibition of DNA and RNA degradation (colicin E_2), or to inhibition of protein biosynthesis by inactivation of the 30S ribosomal subunit (colicin E_3).

Toxoflavin: 3,8-dimethyl-2,4-dihydroxypyrimido-(5,4-e)-*as*-triazine, an antibiotic from *Pseudomonas coccovenans,* with high antibacterial activity, but no activity against fungi. In the biosynthesis of T., C8 of a purine precursor is removed, and the *as*-triazine ring is formed by introduction of the aminomethyl group of glycine (Fig.). Both methyl groups are introduced by transmethylatlon. Xanthotricin from *Streptomyces albus* is identical with T. It interferes in the transport of electrons in the cytochrome system.

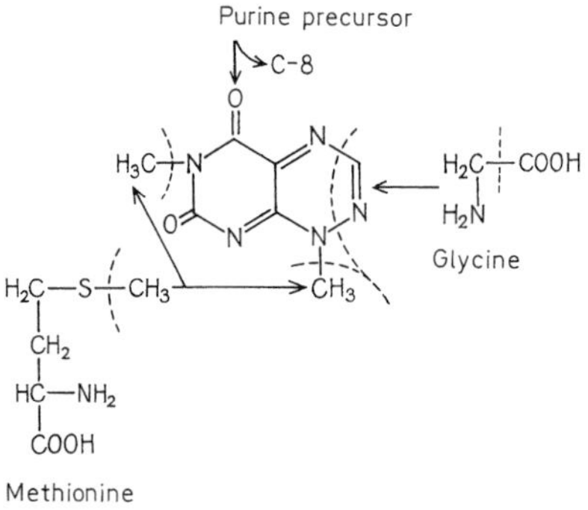

Purine precursor

C-8

H_3C H_2C—COOH H_2N Glycine

H_2C—S—CH_3 CH_3

CH_2 HC—NH_2 COOH

Methionine

Biosynthesis of toxoflavin by Pseudomonas coccovenans.

Toyocamycin: 4-amino-5-cyano-7-(D-ribofuranosyl)-pyrrolo-(2,3-d)-pyrimidine; 6-amino-7-cyano-9-β-D-ribofuranosyl-7-deazapurine, a 7-deazaadenine antibiotic from *Streptomyces toyocaensis* and *S. rimosus.* M. p. 243 °C. Biosynthesis is analogous to that of Tubericidin (see), i. e. the carbon atoms of the pyrrole ring are derived from 5-phosphoribosyl 1-pyrophosphate. T. is particularly active against *Candida albicans, Saccharomyces cerevisiae* and *Mycobacterium tuberculosis.*

TPN: acronym of triphosphopyridine nucleotide. See Nicotinamide-adenine-dinucleotide phosphate.

TPP: acronym of Thiamin pyrophosphate (see).

Trace elements, ***microelements:*** elements required in very small quantities by living organisms. They act catalytically, or are components of catalytic systems. A clear distinction between T.e. and other mineral nutrients is not always possible, e.g. in the case of iron. A further classification into T.e. and ultratrace elements is sometimes used.

Deficiency of T.e. can lead to characteristic deficiency symptoms or diseases, thus indicating the essential nature of these nutritional factors, e.g. iodine is a component of the thyroid hormones and essential for thyroid function. Iodine deficiency is responsible for endemic goiter, and certain types of cretinism; it can be avoided by addition of iodides to drinking water. Other T.e. are chromium, copper, fluoride, magnesium, manganese, nickel, vanadium, silicon, tin, selenium, zinc (see individual entries).

Trace element solution: see Nutrient medium (Table 3).

Trace nutrients, ***micronutrients:*** a general term for any essential dietary component required in small quantities, like Trace elements (see) and Vitamins (see). Deficiency of T.n. leads to deficiency symptoms, e.g. vitamin deficiency diseases. T.n. act catalytically or are precursors of catalytically active substances in the organism. Essential amino acids therefore have an equivocal status in this classification. Flavoring principles are definitely not T.n.

Tracer technique: see Isotope technique.

Transacylases: see Transacylation.

Transacylation: reversible transfer of acyl groups (R–CO–) from a donor to an acceptor, e.g. transfer of the acyl residue CH_3–CO– from acetyl-CoA to an acceptor Y: CH_3–CO ~ S–CoA + Y → CH_3–CO–Y + CoA. T. is catalysed by transacylases, which are important in the synthesis and degradation of fatty acids, synthesis of conjugated bile acids via cholic acid-CoA compounds, and other reactions such as acetylation of amino acids and amines.

Transaldolase (EC 2.2.1.2): see Transaldolation.

Transaldolation: a reaction of carbohydrate metabolism, in which a C3-unit (equivalent to a dihydroxyacetone unit) is transferred from a ketose to an aldose. T. is catalysed by transaldolase (EC 2.2.1.2). The C3-unit does not exist in the free state, but remains bound to the ε-group of a lysine residue in the enzyme (Fig.). Only fructose 6-phosphate and sedoheptulose 7-phosphate are cleaved by transaldolase. Acceptors for the C3-unit are the aldose phosphates, D-glyceraldehyde 3-phosphate, D-erythrose 4-phosphate and more rarely ribose 5-phosphate. There is no coenzyme and the mechanism of reaction is similar to that of aldolase (EC 4.1.2.13).

Transamidase: see Transamidation.

Transamidation: transfer of the amide nitrogen of Glutamine (see) as an NH_2-group. T. is catalysed by transamidases. All glutamine transamidases so far investigated have a catalytically important thiol group in their active centers and are inhibited by the glutamine analogs, azaserine, 6-diazo-5-oxonorleucine (DON) and L-2-amino-4-oxo-2-chloropentanoic acid ("chloroketone"), e.g. anthranilate synthase (EC 4.1.3.27), carbamoyl phosphate synthetase (EC 6.3.5.5), transglutaminase (EC 2.3.2.13), 5′-phosphoribosyl-*N*-formylglycinamidine synthetase (EC 6.3.5.3), glutamate synthase (EC 1.4.1.13).

Transamidinases, ***amidinotransferases:*** enzymes catalysing Transamidination (see). T. catalyse transfer of the amidine group of arginine in the synthesis of creatine and other Phosphagens (see). T. from *Streptomyces griseus* and *S. baikiniensis* catalyses amidine transfer in the biosynthesis of streptidine. T. are also involved in the synthesis of certain Guanidine derivatives (see). Transfer of the intact amidine group from L-arginine has been proved by double labeling with ^{14}C and ^{15}N. The arginine T. also has hydrolytic activity and is therefore a potential Arginase (see).

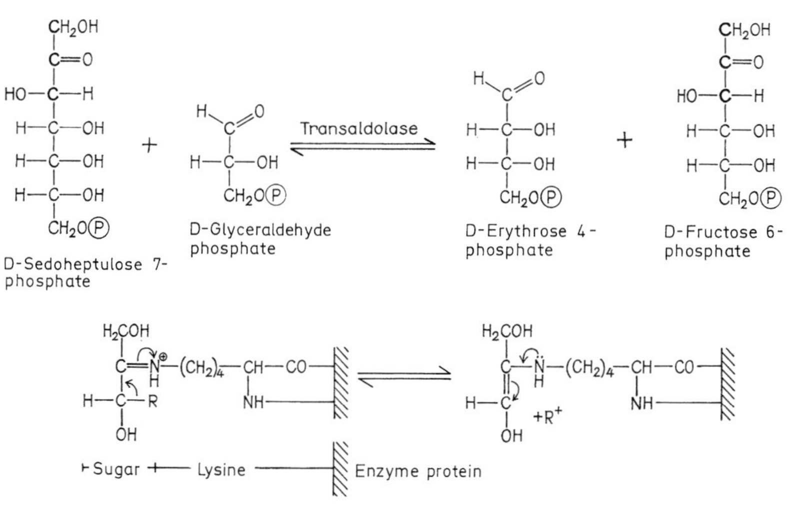

D-Sedoheptulose 7-phosphate + D-Glyceraldehyde phosphate ⇌ (Transaldolase) D-Erythrose 4-phosphate + D-Fructose 6-phosphate

Sugar — Lysine — Enzyme protein

Transaldolase reaction (above), and binding of the ketose to the ε-amino group of a lysine residue of the enzyme (below).

Transamidination: reversible enzymatic transfer of the amidine group,

$$-\overset{\displaystyle NH}{\overset{\|}{C}}-NH_2$$

between guanidines. T. is a group transfer reaction of nitrogen metabolism, which occurs in two stages and involves an intermediate enzyme-amidine complex:

$$R^1-NH-\overset{NH}{\overset{\|}{C}}-NH_2 + \text{Enzyme}-SH \rightleftharpoons R^1-NH_2 + \text{Enzyme}-S-\overset{NH}{\overset{\|}{C}}-NH_2$$

$$\text{Enzyme}-S-\overset{NH}{\overset{\|}{C}}-NH_2 + R^2-NH_2 \rightleftharpoons R^2-NH-\overset{NH}{\overset{\|}{C}}-NH_2 + \text{Enzyme}-SH$$

In the absence of a suitable acceptor, the enzyme-amidine complex is stable; on standing in aqueous solution or on heating it releases urea. T. is catalysed by Transamidinases (see). Formamidine disulfide, a SH-blocking agent, is a powerful inhibitor of T. The most important amidine donor in T. is L-arginine; biosynthesis of L-arginine is equivalent to the de novo synthesis of the amidine group. T. is important in the biosynthesis of Phosphagens (see).

Transaminases, ***aminotransferases*** (EC sub-sub-group 2.6.1): enzymes catalysing transamination, i.e. the reversible transfer of the amino group of a specific amino acid to a specific oxoacid, forming a new amino acid and a new oxoacid. Coenzyme of T. is pyridoxal 5′-phosphate, which becomes bound to the apoenzyme by condensation of its carboxyl group with the ε-amino group of a lysine residue, forming a Schiff's base or internal aldimine. During transamination, however, the coenzyme reacts with the incoming amino acid, which displaces the lysine and forms an external aldimine or *primary Schiff's base* (Figure). Formation of a chelate ring by a bridge proton between the amino nitrogen and the phenolic oxygen of the coenzyme helps to maintain the conjugated system of the Schiff's base in a planar conformation. Rearrangement, with loss of the α-hydrogen as a proton, produces a quinonoid ketimine (*transitional Schiff's base*), containing a conjugated system extending from the carboxyl group to the ring nitrogen. At this stage, the catalytic lysine residue, or another basic group is thought to act as an electron sink. Further rearrangement produces a nonquinonoid ketimine (*secondary Schiff's base*), which is hydrolysed to the new oxoacid and pyridoxamine 5′-phosphate. This represents one half of the transamination process. Another oxoacid condenses with the pyridoxamine 5′-phosphate, and the sequence of reactions is reversed to form a new amino acid, thus completing the amino group transfer.

Interconversion of the tautomeric Schiff's bases is the rate limiting step of transamination.

Practically all amino acids and oxoacids can take part in transamination. The specificity of most transaminases, however, demands that one of the reaction partners should be an acidic amino acid (i.e. glutamate or aspartate) or its corresponding oxoacid. It should be noted that transamination is a freely rever-

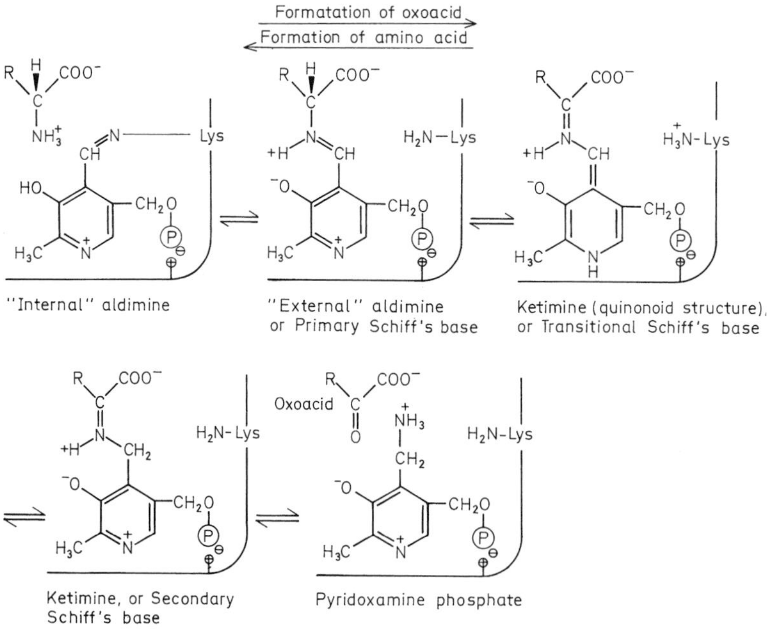

Formatation of oxoacid →
← Formation of amino acid

"Internal" aldimine

"External" aldimine or Primary Schiff's base

Ketimine (quinonoid structure), or Transitional Schiff's base

Ketimine, or Secondary Schiff's base

Pyridoxamine phosphate

Mechanism of transamination. The solid curved line represents part of the surface of the apoenzyme containing a catalytically important lysine residue.

sible, anergonic process, i.e. no high energy compound (e.g. ATP) is produced, or required, and the direction of transamination depends entirely on a mass action effect of its substrates. Thus, in the liver, when amino acids are in excess, transamination converts them to oxoacids (which enter carbohydrate metabolism) and the amino groups appear in glutamate and aspartate (and are subsequently incorporated into urea, see Urea cycle). In plants and bacteria, most pathways of amino acid synthesis involve elaboration of the oxoacid, which is finally transaminated (usually with glutamate) to the required amino acid. At no stage in transamination is free ammonia produced; the amino nitrogen is always covalently bound in an amino acid or in the pyridoxamine phosphate coenzyme.

Animal tissues, especially liver and heart muscle, contain very high activities of glutamate-oxaloacetate T. (GOT) (preferred name, aspartate aminotransferase, EC 2.6.1.1) and glutamate pyruvate T. (GPT) (preferred name, alanine aminotransferase, EC 2.6.1.2). GPT occurs in the liver as a cytosolic enzyme and shows only very low activity in heart muscle, whereas GOT is higher in heart than liver. GOT is about equally distributed between cytosol and mitochondria in both organs, M_r 90,000 (2 identical subunits, M_r 45,000). The primary sequence of pig heart cytoplasmic GOT is known (each subunit contains 412 amino acid residues). The serum activity of T. is very low, but increases markedly in certain illnesses associated with tissue damage. The value and ratio of GOT and GPT activities are used in the early diagnosis of, and for following the progress of treatment of, different liver diseases (greatly increased in acute liver inflammation, moderately increased in chronic cases, and hardly increased in obstructive jaundice) and heart muscle infarction. Both T. are determined by coupled optical tests. ***GPT:*** a serum sample is added to a buffered mixture of L-alanine, 2-oxoglutarate, lactate dehydrogenase (excess) and NADH. As pyruvate is formed it is reduced to lactate by the action of lactate dehydrogenase and NADH. The rate of formation of NAD^+, measured from the rate of decrease of absorption at 340 nm (or other suitable wavelength, e.g. 366 nm), is a measure of GPT activity. ***GOT:*** the reaction mixture contains L-aspartate, 2-oxoglutarate, malate dehydrogenase (excess) and NADH; the resulting oxaloacetate acts as the substrate of malate dehydrogenase, and the procedure is analogous to that described for GPT.

Transamination: reversible transfer of amino groups between two amino acids and their respective keto acids, catalysed by transaminases. T. is fundamentally important in the amino acid metabolism of all living organisms. For mechanism, see Transaminases.

Transcarbamylation: transfer of the carbamyl group of Carbamoyl phosphate (see).

Transcarboxylation: see Biotin enzymes.

Transcortin: see Cortisol.

Transcriptase: see RNA-polymerase.

Transcription: the DNA-dependent synthesis of RNA. See Ribonucleic acid.

Transdeamination: conversion of the amino group of an amino acid to ammonia by the combined action of a transaminase (Amino acid + 2-oxoglutarate $\rightleftharpoons$ 2-oxoacid + Glutamate) and L-glutamate dehydrogenase (Glutamate + NAD^+ + H_2O $\rightarrow$ 2-oxoglutarate + NADH + H^+ + NH_4^+). T. is an important process in ureotelic organisms, where it accounts for most of the ammonium entering the Urea cycle (see) via carbamoyl phosphate (the other nitrogen atom incorporated into urea is derived directly from aspartate, formed by transamination of amino acids with oxaloacetate). Alternatively, the ammonia may be assimilated as the amido nitrogen of L-glutamine, then used in a variety of other processes (see Ammonia assimilation, Transamidination), depending on the biosynthetic capabilities of the organism in question.

Transduction: transfer of DNA from one bacterial cell to another by bacteriophage. There are two types. In *generalized T.* the phage infects the bacterial cell (the donor) and enters a nonlysogenic cycle leading to lysis of the cell and release of phage progeny (see Phage development). Most of the phage progeny are normal, but during the process of phage assembly within the infected bacterial cell, pieces of degraded bacterial DNA occasionally become falsely packaged into phage heads. In the case of *E. coli* and phage P_1, this falsely packaged DNA cannot be larger than 3% of host genome, and it represents an entirely random sample of fragmented host DNA. The new phage population is then used to infect a second bacterial culture (recipient). The majority of phage particles, being normal, kill the bacterial cells that they infect. Under correctly chosen conditions (one phage particle per bacterial cell), most of the cells in the recipient culture are killed; the remaining viable cells are those that received falsely packed bacterial DNA instead of phage DNA during the infection process. This DNA is integrated into the recipient genome by genetic recombination, and is expressed by the recipient cell. In *specialized T.* the phage (e.g. λ) becomes integrated at a specific site in the DNA of the recipient, under lysogenic conditions. The lytic cycle is then initiated (by temperature change, UV-light, etc.), and the phage DNA of some phage progeny carries small fragments of bacterial DNA from the specific integration site. Following infection of the recipient under lysogenic conditions, phage DNA and any attached bacterial DNA become integrated into the recipient DNA. The transferred fragment of bacterial DNA does not represent a random sample, and can only be derived from donor DNA in the region of the specific integration site. Thus a λ-phage is known that integrates in the region of the histidine utilization genes (*hut*) of *Salmonella.* Similarly, studies on the organization of the tryptophan synthase operon were aided by transduction with a λ phage that specifically integrates into *E. coli* DNA in the region of the genes for tryptophan synthesis.

Transferases: see Enzymes, Table 1.

Transfer factors: see Elongation factors.

Transferrin: see Siderophilins.

Transfer-RNA, *tRNA,* (earlier names: ***soluble RNA, sRNA, acceptor RNA, transport RNA):*** the smallest known functional RNA, present in all living cells and essential for Protein biosynthesis (see). Different tRNAs contain between 70 and 85 nucleotide residues, average M_r 25,000. There is at least one specific tRNA per cell for each of the 20 protein amino acids, but there may be between 50 and 70 tRNA species within one cell. This multiplicity is the result of organelle specificity, and the fact that there may be two or more different but specific tRNAs for one amino acid. The source and specificity of a tRNA species is indicated by a code, e.g. $tRNA^{Val}{}_{yeast}$ is the valine-specific tRNA from yeast. $[^{14}C\text{-Val}]tRNA^{Val}{}_{yeast}$ represents the named tRNA esterified with ^{14}C-labeled valine.

Function. tRNA is esterified with its specific amino acid by the action of Aminoacyl-tRNA synthetase (see). The resulting aminoacyl-tRNA becomes bound to the acceptor site of the 50S-subunit of a ribosome, where antiparallel base pairing occurs between the anticodon of the tRNA and the complementary codon of the associated mRNA. The specificity of this base pairing insures that the amino acid is incorporated into the correct position in the growing polypeptide chain. During translation the deacylated tRNA is released from the ribosome and becomes available for recharging with its amino acid.

Structure. The primary structures of more than 50 different tRNAs from various organisms are known. The first tRNA structure was determined by Holley in 1965 for $tRNA^{Ala}{}_{yeast}$; indeed, this was the first reported primary sequence of any nucleic acid. Sequence determination is facilitated by the occurrence of unusual nucleic acid components, which act as markers in the identification of oligonucleotide fragments produced by nuclease degradation of the tRNA. Maximal base pairing of the primary structure gives rise to a secondary structure, known as the cloverleaf model (Fig. 1), which contains three main loops and four stems: 1. The anticodon loop contains the Anticodon (see), a sequence of 3 nucleotides specific for the relevant amino acid. Uracil is always found next to the 5′-end of the anticodon, and a purine or purine derivative is always next to the 3′-end. In tRNA from plants, this purine derivative is often N^6-(γ,γ-dimethylallyl)-adenine, also known as triacanthine. 2. The dihydrouracil (di-HU) loop always contains dihydrouracil. 3. The thymine-pseudouracil (TΨC) loop, characterized by the sequence 5′-GTΨC-3′ appears to be involved in the binding of aminoacyl-tRNA to the 50S ribosome. A fourth loop may be present, and it may even have an associated stem with base pairing, but it is not an invariable feature of every tRNA.

The 3′-terminus of all tRNAs is 3′-ACC. The 3′-terminal adenine is always separated from the first nucleoside in the TΨC loop (ribothymidine or uridine) by 21 nucleosides. The 5′-terminus is always phosphorylated. The aminoacyl-tRNA species that

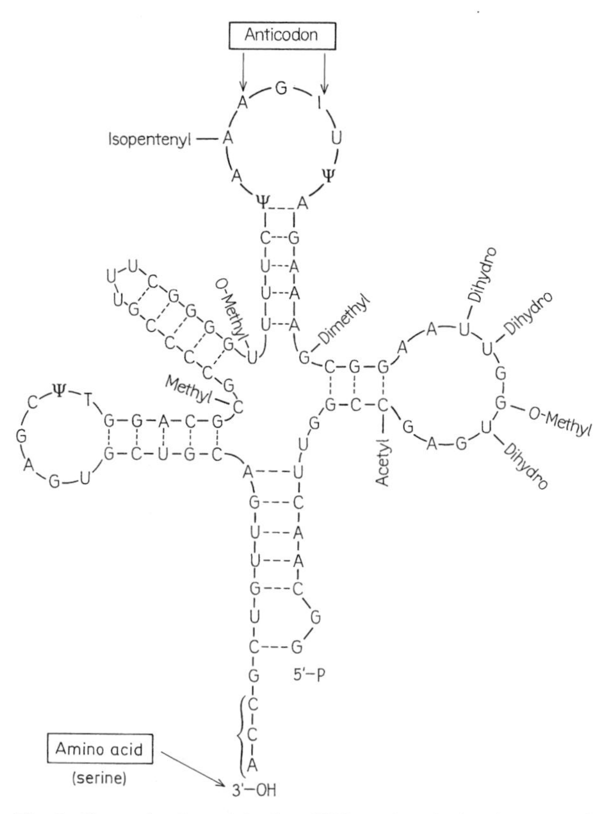

Fig. 1. *Clover leaf model of a tRNA molecule (serine-specific tRNA from yeast).* A = adenine. C = cytosine. G = guanine. I = inosine. U = uracil. T = thymine. ψ = pseudouracil.

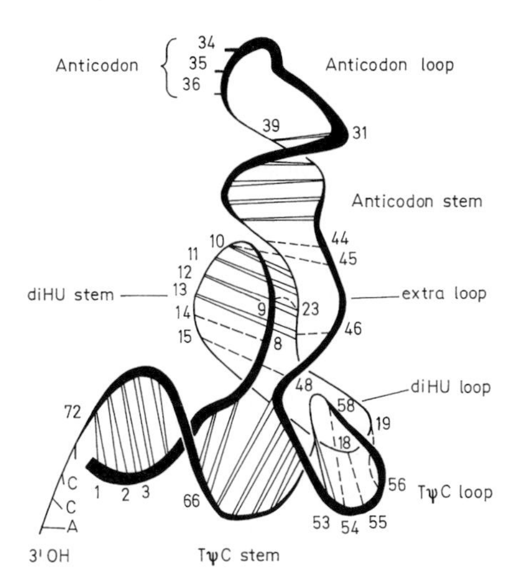

Fig. 2. *Three dimensional structure of yeast tRNA^{Phe} as determined by X-ray crystallography.* Double solid lines represent hydrogen bonds between bases in double helical stems. Dotted lines represent hydrogen bonding between bases outside the helices, i.e. 8.....14, 9.....23, 10.....45, 15.....48, 18.....55, 19.....56, 22.....46, 26.....44 and 54.....58.

binds to the ribosome, and is active in protein biosynthesis, carries the amino acid esterified at position 3′ of the 3′-terminal adenosine. The free aminoacyl-tRNA, however, represents a tautomeric mixture of 2′- and 3′-aminoacyl-tRNA. Moreover, the initial product of aminoacyl-tRNA biosynthesis is either the 2′- or the 3′-aminoacyl derivative, depending on the specificity of the Aminoacyl-tRNA synthetase (see). The highly specific recognition of a tRNA by the corresponding aminoacyl-tRNA synthetase (amino acid activating enzyme) apparently depends on the three-dimensional structure of each tRNA. X-ray analysis of crystalline tRNA confirms the existence of four regions of base pairing and three loops, and shows how these are arranged in a three-dimensional structure (Fig. 2). The TΨC and di-HU loops and stems are folded back, so that the molecule has a compact shape, 9.0 nm long and 22.5 nm wide. The anticodon loop and the 3′-ACC end are still widely separated as in the cloverleaf model. In addition to hydrogen bonding between bases in the helical stems, the structure is stabilized by an extensive network of hydrogen bonds involving specific interactions between bases and the ribose phosphate backbone, and by some hydrogen bonding between base pairs outside the helices.

Synthesis and processing. A precursor molecule of tRNA is synthesized by transcription from DNA then processed to the functional tRNA. For example, the precursor of $tRNA^{Tyr}{}_{E.\ coli}$ consists of 126 nucleotides and is 41 nucleotides longer than the mature, active molecule. The extra sequence is removed by a specific endonuclease. Certain nucleotide residues also undergo further post-transcriptional modifications, resulting in several Rare nucleic acid components (see), which are partly responsible for the specific spatial structure of each tRNA.

Transformation: conceptually the simplest form of genetic transfer. "Naked" DNA from a donor cell enters a recipient cell and is incorporated into the recipient DNA by genetic recombination. No other carrier substance or structure is involved; small fragments of the donor DNA simply penetrate the membrane (and wall if it is present) of the recipient cell. T. was first described in 1944 by Avery (USA) for the transformation of R (rough)-type nonpathogenic Pneumococci into S (smooth)-type pathogenic Pneumococci by treatment with killed S-type cells. The "transforming principle" was eventually shown to be DNA; this work gave the first proof that the genetic material of the cell is DNA. Frequency of T may be low (< 1 %) owing to rapid degradation of donor DNA before genetic recombination occurs.

Transforming growth factor-β, *TFG-β:* a cytokine originally identified in transformed fibroblasts. This 25 kd disulfide-linked dimer is now regarded as the prototype of a family of related polypeptide factors that regulate cell growth and differentiation [J. Massagué *Cell* **49** (1987) 437–438]. The family includes:

(i) Other forms of TGF-β.

(ii) Activins and Inhibins (see).

(iii) Mammalian Müllerian inhibiting substance (see).

(iv) The decapentaplegic and 60A proteins involved in *Drosophila* pattern formation, which are homologs of bone morphogenic proteins (K. A. Wharton et al. *Proc. Natl. Acad. Sci. USA* **88** (1991) 9214–9218).

(v) Cartilage-derived morphogenetic factors (CDMP; E. E. Storm *Nature* **368** (1994) 639–643).

(vi) Osteogenic bone morphogenetic proteins (OP or BMP; E. Özkaynak et al. *J. Biol. Chem.* **267** (1992) 25220–25227; A. J. Celeste et al. *Proc. Natl. Acad. Sci. USA* **87** (1990) 9843–9847).

(vii) Growth/differentiation factors (GDF; AC. McPherron *J. Biol. Chem.* **268** (1993) 2333–3449).

(viii) Mammalian protein related to *Xenopus* Vg-1 (also known as BMP-6; K. Lyons et al. *Proc. Natl. Acad. Sci., USA* **86** (1989) 4554–4558).

According to cDNA analysis, all these factors are synthesized as part of larger secretory precursor proteins.

The three identified forms of TGF-β arise from the homodimeric and heterodimeric association of subunits β1 and β2: *TGF-β1* [the original prototype TGF-β; $(\beta1)_2$], *TGF-β2* [$(\beta2)_2$] and *TGF-β1.2* [(β1, β2)]. The β1 and β2 subunits display about 70 % sequence identity in their *N*-terminal halves, and are more closely related to each other than to other known members of this gene family. In view of the high degree of conservation (e.g. 99 % sequence identity between human and mouse TGF-β1), the TGβs appear to play a crucial biological role common to all species.

High levels of TGF-βs are present in actively differentiating tissue, e.g. cartilage canals, osteocytes, Hassall's corpuscle in the thymus, bone marrow, hematopoietic stem cells, the highest levels being found in blood platelets. It is noteworthy that subcutaneous administration of TGF-β promotes the deposition of collagen, suggesting that TGK-βs can influence differentiation and morphogenesis by regulating the quantity and structure of the extracellular matrix. TGF-βs may also regulate receptors for other growth factors.

Transglycosidation: transfer of a glycosidically bound sugar residue to another molecule with a suitable recipient OH-group. T. is catalysed by transglycosidases, e. g. galactosidase catalyses transfer of a galactose residue from lactose to the C6-hydroxyl group of glucose, or to a further lactose molecule to form the trisaccharide, 6-galactosidolactose. T. is sometimes involved in the synthesis of oligo- and polysaccharides.

Transhydrogenase and Transhydrogenation see Hydrogen metabolism.

Transit peptides: amino-terminal extensions of precursors to chloroplast proteins which are encoded by nuclear DNA and synthesized in the cytoplasm. The T. p. are removed by Post-translational modification (see) before the protein takes on its mature configuration inside the chloroplast. The overall amino acid sequences of the T. p. are not conserved between species, but the positions of proline and the charged amino acids are highly conserved. The T. p. appear to mediate transport of the chloroplast protein precursors into the organelles. [G. Van den Broeck et al. *Nature* **313** (1985) 358–363.]

Transition: replacement of a purine by a different purine, or a pyrimidine by a different pyrimidine in the polynucleotide chain of DNA. T. results in a gene mutation. It may occur spontaneously, or it may be promoted experimentally with mutagens.

Transition temperature, T_c: see Membrane lipids.

Transketolase (EC 2.2.1.1): an enzyme that catalyses transketolation, an important process of carbohydrate metabolism, especially in the Pentose phosphate cycle (see) and Calvin cycle (see). T. has been found in a wide variety of cells and tissues, including mammalian liver, green plants and many bacterial species. The enzyme contains divalent metal cations and the coenzyme, thiamin pyrophosphate. Transketolation involves transfer of a C2-unit (often called active glycolaldehyde or a ketol moiety) from a ketose to C1 of an aldose. Only ketoses with L-configuration at C3 and preferably *trans* configuration on the next carbon (i. e. C1, 2, 3 and preferably 4 as in fructose) can serve as donors of the C2-unit. The acceptor is always an aldose. Transketolation is reversible. Details of the reaction in which xylulose 5-phosphate serves as the donor of

Mechanism of action of transketolase (only the thiazole group of thiamin pyrophosphate is shown).

the C2-unit, and ribose 5-phosphate as the acceptor are shown (Fig.). The C2-unit becomes bound to the thiamin pyrophosphate as 2-(α,β-dihydroxyethyl)-thiamin pyrophosphate, and the remainder of the molecule is released as glyceraldehyde 3-phosphate; transfer of the C2-unit to ribose 5-phosphate produces sedoheptulose 7-phosphate. If the ribose 5-phosphate were replaced by erythrose 4-phosphate, the products would be glyceraldehyde 3-phosphate and fructose 6-phosphate.

Transketolation: see Transketolase.

Translation: in the wider sense equivalent to Protein biosynthesis (see). In the narrower sense, T. is the decoding process whereby each codon (see Genetic code) in mRNA is translated into one of 20 amino acids during protein synthesis on polysomes.

Translocation: see Transport.

Transmethylases: see Transmethylation.

Transmethylation: transfer of a methyl group ($-CH_3$) from a physiological methyl donor to C-, O- and N-atoms of biomolecules. T. to oxygen produces the methoxy group ($-OCH_3$). The most important methyl donor is *S*-Adenosyl-L-methionine (see). Thus the methyl groups in a wide variety of methylated natural products originate from the methyl group of methionine. The thioether group of methionine itself does not participate directly in T.; it must first be activated to a sulfonium group by *S*-adenosylation, i.e. by synthesis of *S*-adenosyl-L-methionine. Betaines (see) and Thetins (see) have limited physiological significance as methyl donors. Methylated nucleic acid components are formed by methylation of the polynucleotide chains of nucleic acids by *S*-adenosyl-L-methionine, catalysed by specific transmethylases (methyltransferases). Exceptions are thymine and 5-hydroxymethylcytosine, where the methyl (or hydroxymethyl) group is derived from $N^{5,10}$-methylenetetrahydrofolic acid (see Active one-carbon units).

Transphosphatases: see Kinases.

Transplantation antigens, *histocompatibility antigens:* antigens on the surface of nucleated cells, particularly leukocytes and to a lesser extent thrombocytes. A lack of correspondence between T.a. of donor and recipient leads to transplant rejection. To insure optimal compatibility between donor and recipient, the T.a. of both are typed by the histocompatibility test.

Transport: passage of ions and certain molecules through biological membranes. Most polar molecules do not pass freely across biomembranes. Exchange of essential metabolites between a cell and its surroundings, or between cytoplasm and organelles, therefore depends on T. mechanisms within the membranes. All T. mechanisms catalysed by biomembranes have three characteristic properties: saturation, substrate specificity and specific inhibition. T. may be active or passive.

Passive T. can only operate in the presence of an appropriate concentration gradient, which acts as the driving force in transporting the material through pores in the membrane. The commonest form of passive T. involves a carrier mechanism. According to this model, the process is capable of on-off regulation, and the carrier is a specific protein which attaches to the substance to be transported on one side of the membrane, takes it across the membrane, releases it on the other side, then returns to the starting position. Depending on the concentration gradient, passive T. is reversible and can occur in either direction. Important passive T. systems in animal tissues are, e.g. the glucose carrier in human erythrocyte membranes, and the ATP-ADP carrier of the mitochondrial membrane, which normally transports one molecule of ADP into the mitochondrial matrix, and one molecule of ATP (formed by oxidative phosphorylation) from the matrix to the cytoplasm.

In ***active T.*** material is transported against a concentration gradient, and the process is linked to the cleavage of ATP. In recognition of the involvement of ATP, active T. systems ("pumps") are also called ATPases. In the model for active T., the substance in question becomes attached to a complementary binding site on the protein carrier, then transported to the other side of the membrane by diffusion, rotation, or a change of conformation. The free energy of ATP cleavage is required for release of the substrate from the carrier; this probably involves a conformational change in the carrier protein, which alters the binding affinity for the substrate; examples in animal tissues are the Na^+ and K^+ pumps, and the active transport mechanisms for glucose and other sugars, and for amino acids. Na^+K^+-ATPase also appears to be important for the T. of glucose and amino acids. Furthermore, ATPases play an important role in nerve transmission, muscle control and sensory perception. The Na^+ and K^+ transporting ATPase system (Na^+K^+ pump) transports Na^+ ions out of the cell against the electrochemical gradient (intracellular Na^+ concentration < 10 mM; extracellular conc. about 150 mM) and K^+ ions from the surrounding milieu into the cell (extracellular K^+ concentration < 4 mM, intracellular conc. 120–160 mM). The high internal K^+ concentration is biochemically important, being essential for protein biosynthesis and for maximal activity of various enzymes, etc. The Na^+/K^+ gradient across the cell membrane is necessary for the excitation response of muscle cells and for signal transmission by nerve cells. A two-stage process is postulated for the transport mechanism:

$$\mathrm{ATP} + \mathrm{N^+internal} + \text{Ⓔ} \xrightleftharpoons{Mg^{2+}} \mathrm{Na}\text{-Ⓔ} \sim \mathrm{P} + \mathrm{ADP}$$

$$\mathrm{Na}\text{-Ⓔ} \sim \mathrm{P} \rightleftharpoons \mathrm{Na}\text{-}\triangle_E \sim \mathrm{P}$$

$$\mathrm{K^+external} + \mathrm{H_2O} + \mathrm{Na}\text{-}\triangle_E \sim \mathrm{P} \rightleftharpoons \text{Ⓔ} + \mathrm{P_i} + \mathrm{Na^+external} + \mathrm{K^+internal}$$

Two different conformations (○ and △) are postulated for the ATPase. The intermediate phosphorylated form of the enzyme contains an acyl phosphate group bound to an aspartic acid residue at the active center of the ATPase. Na^+K^+-ATPase has M_r 250,000–300,000 and consists of two different subunits. Kidney and brain cells use about 70 % of their synthesized ATP for the exchange transport of Na^+ and K^+ catalysed by Na^+K^+-ATPase.

A mechanism in which glucose transport is coupled with a Na^+ gradient is postulated for the active transport of glucose, against a concentration gradient, from the small intestine into the blood stream, and from the glomerular filtrate through the epithelial cell layer of the kidney tubuli into the blood. Na^+

ions are pumped out of the cell by Na^+K^+-ATPase, thus forming a higher Na^+ concentration outside than inside cell. Glucose and Na^+ are then transported into the cell by a passive carrier that has binding sites for both glucose and Na^+ (cotransport or symport).

Active amino acid T. has certain similarities with the active T. of glucose, especially from the intestine into the blood. More than five different T. systems are known. In some cells amino acid T. also appears to be coupled with a Na^+ gradient.

The ATP-dependent, intracellular T. of Ca^{2+} from the sarcoplasm into the sarcoplasmic reticulum is an essential process in the initiation of muscle relaxation. This process depends on the activity of a Ca^{2+}-ATPase system (Ca^{2+} pump). Mitochondria of animal cells are also able to accumulate Ca^{2+} against a steep gradient.

Specific proteins in the periplasmic space of bacteria can bind sugars, amino acids and inorganic ions, and therefore play an important part as carriers associated with T. systems. Aerobic bacteria in particular possess true active T. systems.

Group translocation is a special type of membrane T. A membrane-bound enzyme catalyses a reaction between substrates on opposite sides of the membrane, and the product accumulates on one side of the membrane, e.g. various bacteria take up glucose by phosphorylating it to glucose 6-phosphate. Group translocation is also involved in the T. of amino acids. In contrast to active T., group translocation involves modification of the transported material.

Transport antibiotics: see Facilitated diffusion.

Transport RNA: see Transfer RNA.

Transposon, ***transposable genetic element:*** a mobile genetic element that can move its position independently within a chromosome, and which may carry genes ("jumping genes").

Genetic transposition was first recognized in the early 1950s by Barbara McClintock as the process responsible for the variegated pigmentation pattern of maize. At the time this report was largely ignored because it contradicted the favored Mendelian concept of genes with fixed locations. It was then found that a number of apparently random mutations in *E. coli* could arise by the insertion of relatively large sections of genetic material, known as insertion sequences. With the aid of recombinant DNA technology, it was then found that *E. coli* insertion sequences encode enzymes responsible for the insertion of an identical copy of themselves at a new site in the DNA. Thus, transposition involves both recombination and replication; two daughter copies of the insertional sequence are produced, one remaining at the parental site, the other becoming inserted at the target site. An insertion sequence always causes a deletion, rearrangement or inversion of the target gene (i.e. it causes a mutation). It has been suggested that coordinately regulated operons may have arisen by the transposon-mediated rearrangement of originally widely separated genes. It is also conceivable that transposition could lead to the formation of new proteins by bringing together formerly independent gene segments. In addition, since insertion sequences carry signals for initiation of RNA synthesis, they may also activate previously dormant genes.

It has also been shown that transposition is responsible for phase variations of bacterial phenotypic expression. One particularly well studied case is the expression of two antigenically different flagellin proteins (H1 and H2) in certain strains of *Salmonella typhimurium.* Each cell expresses only one flagellin, but after about 1000 cell divisions, *phase variation* occurs and the other type of flagellin is synthesized, possibly representing a means of evading the immunological defense of the host. The two flagellin genes are relatively widely separated on the bacterial chromosome. *H2* is linked to *rh1* (the repressor gene of H1), so that transcription of the *H2-rh1* unit leads to synthesis of H2 and repression of HI synthesis. Expression of the *H2-rh1* unit is regulated by a 995-bp DNA segment upstream of *H2.* The following three components have been identified in the 995-bp segment. 1. A promoter for *H2-rh1.* 2. The *hin* gene, encoding the 190-residue Hin DNA invertase; this enzyme (which displays 33 % sequence homology with TnpR) mediates inversion of the entire 995-bp segment. 3. Two 26-bp *hix* sites, representing the boundaries of the segment and containing its cleavage sites; each consists of two imperfect 12-bp inverted repeats separated by 2 bp. If the 995-bp segment is oriented so that the promoter is just upstream of *H2* (this is known as phase 2 orientation), then *H2* and *rh1* are coordinately expressed, and HI synthesis is repressed. If the 995-bp segment is in the opposite orientation (phase 1), neither *H2* nor *rh1* is expressed because they are separated from their promoter, and HI is synthesized.

Hin DNA invertase has been sequenced and X-ray studies have been performed on the protein on its complexes with DNA. It consists of a 52-residue C-terminal DNA-binding domain and a 138-bp catalytic domain. $\gamma\delta$-Resolvase (a dimer of identical 120-residue subunits encoded by the $\gamma\delta$ transposon) displays 40 % sequence identity with the Hin catalytic domain, and catalyses a similar reaction, i.e. a 2-bp staggered cleavage, exchange, and re-ligation of duplex DNA. Whereas Hin catalyses an inversion, $\gamma\delta$ resolvase catalyses a site-specific deletion.

Three classes of transposable elements are recognized in bacteria.

A *simple transposon* or *insertion sequence (IS)* is generally less than 2,000 bp in length. The only genetic information that it carries is that required for transposition; including specific short inverted terminal DNA repeats, and the gene for transposase which recognizes the termini and inserts the IS into its target site. An inserted IS is always flanked by a repeated segment of host DNA, implying that the IS is inserted at a staggered cut that is later filled in. Target sequences are usually 5 to 9 bp in length, and the length of this target sequence (rather than its actual nucleotide sequence) is specific to, or characteristic of, the IS.

Complex transposons (designated Tn) are larger, and they carry genetic material (e.g. functional bacterial genes such as antibiotic resistance genes) that is not required for the process of transposition. For example Tn3 contains 4957 bp with inverted terminal repeats, each of 38 bp. It also encodes three proteins: a 1015-residue transposase (TnpA); a 185-residue protein (TnpR) which functions as a repressor of *tnpA* and *tnpR,* and mediates site-specific recombination during transposition; and a β-lactamase.

Composite transposons contain a central region flanked by two identical (or nearly identical) IS-like sequences. These IS-like sequences have either the same or an inverted orientation, and are themselves flanked by inverted repeats. It therefore appears that composite transposons arose from the combination of a stretch of gene-containing DNA (the central region) with two independent insertion sequences.

Base sequences of many eukaryotic transposons bear very little similarity to those of bacterial transposons but are similar to those of retroviral genomes, suggesting that they may be degenerate retrovirosus. Eukaryotic transposons are therefore known as *retrotransposons* or *retroposons.* Transposition of a retroposon occurs in three stages. It is first transcribed to RNA; this RNA is in turn transcribed by reverse transcriptase, and the resulting DNA is inserted at random at a new site. [B. McClintock "Chromosome Organization and Gene Expression" *Cold Spring Harbor Symp. Quant. Biol* **16** (1951) 13–57; B. McClintock *Science* **266** (1984) 792–800; J.-A. Feng et al. "Proteins that promote DNA inversion and deletion" *Curr. Opin. Struct. Biol.* **4** (1994) 60–66]

Transsulfuration: exchange of sulfur between L-homocysteine and L-cysteine, with L-cystathionine as an intermediate (Fig.). Strictly speaking, T. is not a group transfer reaction, because the sulfur bond formed in the synthesis of cystathionine is different from that broken in the formation of L-cysteine. T. operates in the biosynthesis of L-cysteine from L-methionine, and in the biosynthesis of L-methionine. The methionine precursor, L-homocysteine, is formed by T. as follows:

L-Homoserine + Succinyl-CoA → CoA + *O*-Succinyl-L-homoserine

O-Succinyl-L-homoserine + L-Cysteine → L-Cystathionine + Succinate

L-Cystathionine + H_2O → L-Homocysteine + Pyruvate + NH_3

In some organisms L-homoserine is initially acylated to *O*-acetyl-L-homoserine (by acetyl-CoA), and in plants it is converted to oxalyl-L-homoserine by reaction with oxalyl-CoA.

Transvaalin: see Scillaren A.

Transversion: replacement of a pyrimidine by a purine, or a purine by a pyrimidine in the polynucleotide chain of DNA. T. results in a gene mutation. It may occur spontaneously, or it may be promoted experimentally with mutagens.

Trehalose: a nonreducing disaccharide, M_r 342.30, consisting of two glucopyranoside residues. There are three forms of T., depending on the nature of the glycosidic linkage: α,α-T. (m. p. 204 °C, $[\alpha]_D^{20}$ +197 °); α,β-T. or neotrehalose ($[\alpha]_D^{20}$ +95 °); β,β-T. or isotrehalose (m. p. 135 °C, $[\alpha]_D^{20}$ −42 °). T. is present in algae, bacteria, numerous lower and higher fungi, and occurs sporadically in nonphotosynthetic tissues of higher plants. It is the "blood sugar" of insects. T. is cleaved by many fungal enzymes, and it is fermented by certain yeasts.

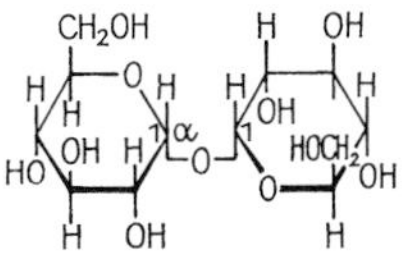

α,α-Trehalose

TRF: acronym of thyrotropin releasing factor. See Releasing hormones.

TRH: acronym of thyrotropin releasing hormone. See Releasing hormones.

Triacylglycerol: see Acylglycerols, Fats.

Triamcinolone: 16α-hydroxy-9α-fluoroprednisolone; 9α-fluoro-11β,16α,17α,21-tetrahydroxypregna-1,4-diene-3,20-dione, a synthetic steroid prepared from cortisol. The antiinflammatory activity of T. is 50 times greater than that of cortisone acetate (see Cortisone), and it does not cause undesirable salt retention. It is used in the treatment of arthritis, allergies etc.

Tricarboxylic acid cycle, *TCA-cycle, citric acid cycle, Krebs cycle:* a fundamentally important cycle of reactions in the terminal oxidation of proteins, fats and carbohydrates (Fig. 1). CO_2 is formed by decarboxylation of oxo-acids (oxalosuccinate, not shown in Fig. 1, is an intermediate between isocitrate and 2-oxoglutarate; see Isocitrate dehydrogenase). Together with the Respiratory chain (see), operation of the TCA-cycle leads to the synthesis of the energy-rich compound ATP. In addition to energy production, the TCA-cycle provides intermediates for biosynthesis. Several important groups of substances are derived from intermediates of the TCA-cycle, and various other metabolic cycles are linked with the TCA-cycle. In eukaryotes, the TCA-cycle operates in the mitochondria, where it is structurally and functionally integrated with the respiratory chain, and with the degradation of fatty acids. In prokaryotes, the enzymes of the TCA-cycle are localized in the cell cytoplasm.

The TCA-cycle is responsible for the oxidation and cleavage of the acetyl group of acetyl-CoA, with the formation of two molecules of CO_2. During one operation of the TCA-cycle, there are four dehydrogenations (each equivalent to 2 hydrogen atoms) in which the hydrogen is transferred to NAD^+ or FAD. The reduced coenzymes are then reoxidized by the respiratory chain, which catalyses the oxidation of the hydrogen to water.

Oxidation in the TCA-cycle is achieved by the addition of water and subsequent removal of hydrogen; there is no direct involvement of oxygen: $CH_3CO \sim S\text{-}CoA + 3H_2O \rightarrow 2CO_2 + 8[H] + HSCoA$. The initiating reaction of the TCA-cycle is the condensation of acetyl-CoA with oxaloacetate, catalysed by citrate synthase. One molecule of water is consumed, and the products are citrate and coenzyme A. Citrate is

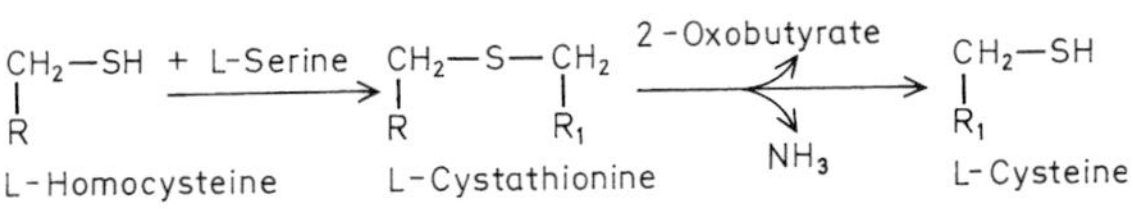

$CH_2{-}SH$ (R) L-Homocysteine + L-Serine → $CH_2{-}S{-}CH_2$ (R, R_1) L-Cystathionine → (2-Oxobutyrate, NH_3) → $CH_2{-}SH$ (R_1) L-Cysteine

Formation of L-cysteine from L-homocysteine (derived from L-methionine).

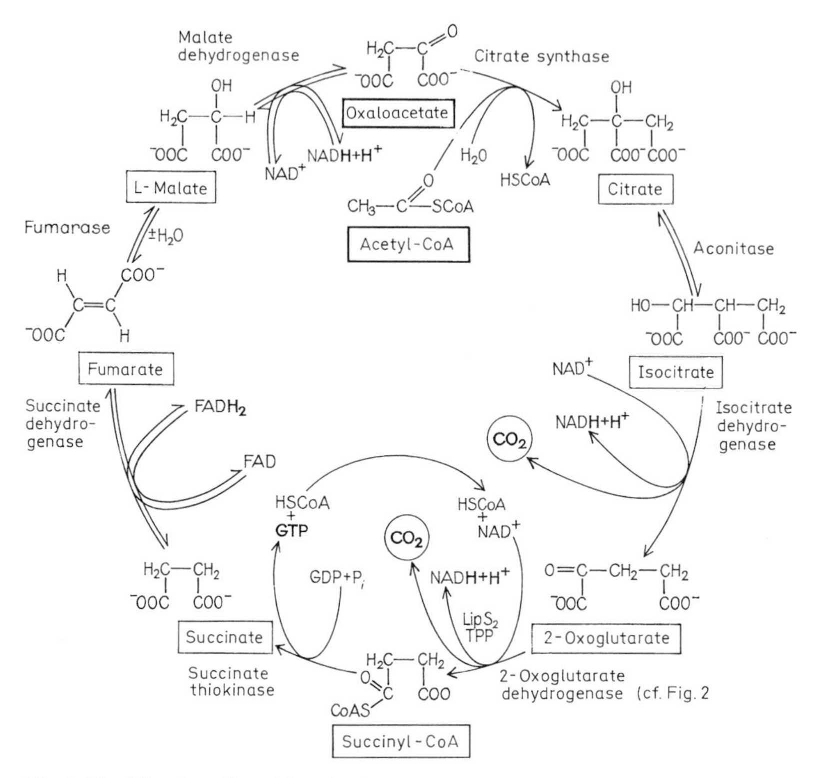

Fig. 1. *The tricarboxylic acid cycle.*

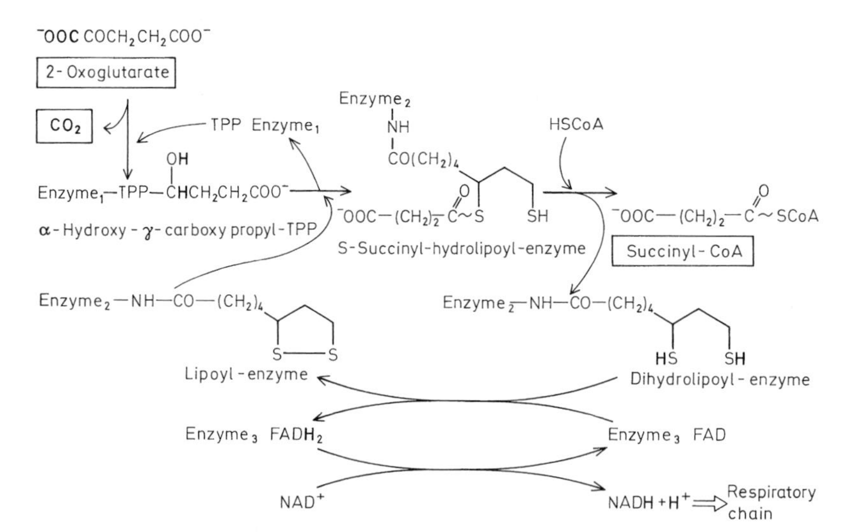

Fig. 2. *Mechanism of the oxidative decarboxylation of 2-oxoglutarate by the 2- oxoglutarate dehydrogenase complex* (EC 1.2.4.2). $Enzyme_1$ = 2-oxoglutarate decarboxylase. $Enzyme_2$ = lipoyl-reductase-transacetylase (lipoyl reductase + transsuccinylase). $Enzyme_3$ = dihydrolipoyl dehydrogenase. TPP = thiamin pyrophosphate. HSCoA = coenzyme A.

Table 1. *The reactions of the tricarboxylic acid cycle*

Reaction number	Equation	Name of enzyme (see also under separate entries)	Inhibitors	$\Delta G^{\circ\prime}$ in kJ/mol (kcal/mol)
1	Acetyl-CoA + oxaloacetate + $H_2O \rightarrow$ citrate + HSCoA + H^+	Citrate (*si*)-synthase (EC 4.1.3.7)	none	−38.04 (−9.08)
2a	Citrate $\xrightarrow{Fe^{2+},\ GSH}$ isocitrate	Aconitate hydratase (EC 4.2.1.3)	Fluorocitrate*, trans-Aconitate*	+6.66 (+1.59)
2b	Citrate $\xrightarrow{Fe^{2+},\ GSH}$ cis-aconitate	Aconitase		+8.45 (+2.04)
2c	cis-Aconitate $\xrightarrow{Fe^{2+},\ GSH}$ isocitrate	Aconitase		−1.89 (−0.45)
3	Isocitrate + NAD^+ $\xrightarrow{Mg^{2+}\ (Mn^{2+}),\ ADP}$ 2-oxoglutarate + NADH + H^+ + CO_2	Isocitrate dehydrogenase (EC 1.1.1.41)	ATP	−7.12 (−1.70)
4	2-Oxoglutarate + HSCoA + NAD^+ $\xrightarrow{Mg^{2+},\ TPP,\ LipS_2}$ succinyl-CoA + CO_2 + NADH + H^+	2-Oxoglutarate dehydrogenase complex (EC 1.2.4.2)	Arsenite, Parapyruvate*	−36.95 (−8.82)
5	Succinyl-CoA + GDP + $P_i \xrightarrow{Mg^{2+}}$ succinate + GTP + HSCoA	Succinyl-CoA synthetase (EC 6.2.1.4)	Hydroxylamine	−8.85 (−2.12)
6	Succinate + FAD $\xrightarrow{Fe^{2+}}$ fumarate + $FADH_2$	Succinate dehydrogenase (EC 1.3.99.1)	Malonate*, Oxaloactate*	~0
7	Fumarate + $H_2O \rightarrow$ L-malate	Fumarate dehydratase (EC 4.2.1.2)	meso-Tartrate*	−3.68 (−0.88)
8	L-Malate + $NAD^+ \rightarrow$ oxaloacetate + NADH + H^+	Malate dehydrogenase (EC 1.1.1.37)	Oxaloacetate*, Fluoromalate*	+28.02 (+6.69)
	Sum of equations 1–8, i.e. balance of the TCA-cycle without the respiratory chain: Acetyl-CoA + $3NAD^+$ + FAD + GDP + P_i + $2H_2O \rightarrow$ $2CO_2$ + HSCoA + 3NADH + $3H^+$ + $FADH_2$ + GTP			−60.00 (−14.32)

Abb.: HSCoA = Coenzyme A; GSH = Glutathione; AM(D)(T)P = Adenosine mono (di)(tri)phosphate; TPP = Thiamin pyrophosphate; $LipS_2$ = Lipoic acid amide; GD(T)P = Guanosine di(tri)phosphate; Pi = Inorganic phosphate; $FAD(H_2)$ = Enzyme-bound oxidized (reduced) Flavin-adenine-dinucleotide; NAD^+ (H) = Oxidized (reduced) Nicotinamide-adenine-dinucleotide. Compounds with * are competitive inhibitors.

then converted into oxaloacetate by 7 consecutive, enzyme-catalysed steps (Table 1). Reactions 3 and 4 involve decarboxylation. Fig. 2 shows the mechanism of action of the 2-oxoglutarate dehydrogenase complex.

Energy balance of the TCA-cycle. A total of about 900 kJ (215 kcal) free chemical energy is available from the oxidation of acetyl-CoA by the TCA-cycle, with the involvement of the respiratory chain. Operation of the respiratory chain alone accounts for −810 kJ (−193.4 kcal) of this free energy, i.e. three operations of the chain from NADH (NADH + $^1/_2O_2$ + H^+ $\rightarrow$ NAD^+ + H_2O, $\Delta G^{\circ\prime}$ = −219.4 kJ [−52.4 kcal]), and one operation of the chain from $FADH_2$ ($FADH_2 + {}^1/_2O_2 \rightarrow FAD + H_2O$, $\Delta G^{\circ\prime}$ = −151.6 kJ [−36.2 kcal]). A portion of this energy is used in the synthesis of 12 molecules of ATP, representing about 40 % of the total free energy: reactions 3, 4 and 8 give 3 × 3 ATP by the oxidation of 3 × NADH in the respiratory chain; reaction 6 gives 2 ATP by $FADH_2$ oxidation in the respiratory chain; and the GTP produced in reaction 5 (see Substrate level phosphorylation) is energetically equivalent to ATP, i.e. GTP + ADP $\rightleftharpoons$ GDP + ATP has an equilibrium constant of 1.0. The overall equation, including the respiratory chain and oxidative phosphorylation is:

$$CH_3\text{–}CO \sim SCoA + GDP + 11ADP + 12P_i + 2O_2 \rightarrow 2CO_2 + 13H_2O + GTP + 11ATP + HSCoA.$$

The TCA-cycle is linked to Gluconeogenesis (see) by the conversion of oxaloacetate to phospho*enol*pyr-

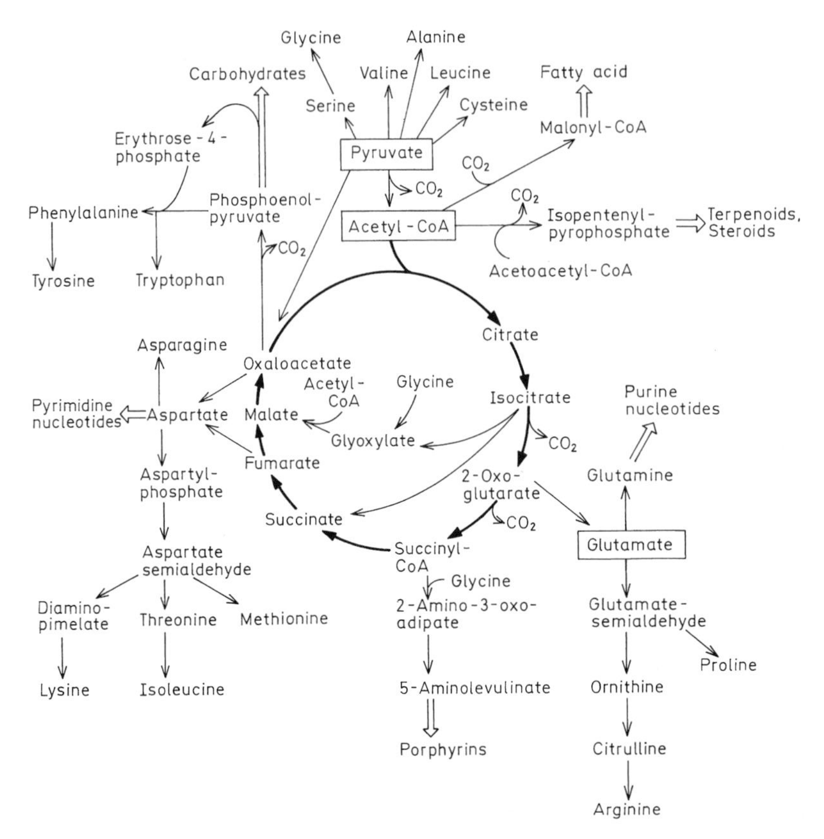

Fig. 3. *The biosynthetic functions of the tricarboxylic acid cycle.*

uvate. It is also the source of intermediates in the synthesis of many amino acids, especially aspartic and glutamic acid and other amino acids derived from them. Succinyl-CoA is a precursor of the porphyrins, e.g. heme, chlorophyll, vitamin B_{12}. The function of the TCA-cycle may be modified by integration with other pathways, e.g. γ-Aminobutyrate pathway (see), Glyoxylate cycle (see), and the Succinate-glycine cycle (see). Carboxylation of pyruvate (see Pyruvate carboxylase, Carboxylation) serves as one stage in gluconeogenesis from pyruvate; it is also an anaplerotic reaction (see Metabolic cycle) of the TCA-cycle, i.e. it maintains the level of oxaloacetate, which would otherwise be depleted by removal of intermediates of the TCA-cycle for biosynthesis. In animals the net synthesis of carbohydrate from acetyl-CoA (and therefore from fatty acids) is not possible. In plants and bacteria, however, the presence of the glyoxylate cycle permits incorporation of a second acetyl group from acetyl-CoA, resulting in the net synthesis of TCA-cycle intermediates (and therefore of carbohydrate) from two-carbon units. This is important in the utilization of oil reserves in seeds for the synthesis of carbohydrate (e.g. cellulose of cell walls) during germination, and in the growth of bacteria at the expense of simple carbon sources, such as acetate.

Reactions analogous to those of the TCA-cycle are found in the biosynthesis of leucine and lysine (Fig. 4). Similarly, the sequence of reactions represented by the dehydrogenation of succinate by a flavoenzyme, followed by hydration of fumarate to malate, then dehydrogenation of malate by a NAD-linked dehydrogenase, finds a counterpart in the initial stages of fatty acid degradation (see).

Regulation of the TCA-cycle. ADP/ATP ratios and NAD/NADH + H^+ ratios have a regulatory influence on the TCA-cycle, particularly via the regulation of the allosteric enzyme isocitrate dehydrogenase. The enzyme is activated by ADP, and inhibited by ATP and NADH (Table 2). Other regulatory sites are the syntheses of acetyl-CoA, oxaloacetate and citrate. Oxaloacetate is the catalytic intermediate of the TCA-cycle for the oxidation of acetyl-CoA to $CO_2 + H_2O$; it also inhibits succinate dehydrogenase and malate dehydrogenase. Since the TCA-cycle only works in conjunction with the respiratory chain, its activity is also regulated by the supply of oxygen. When an aerobically respiring cell is made anaerobic, the TCA-cycle comes to a halt with the accumulation of the reduced coenzymes NADH and $FADH_2$.

Asymmetrical metabolism of citrate. Although the citrate molecule has perfect bilateral symmetry, it is degraded asymmetrically (see Prochirality). According

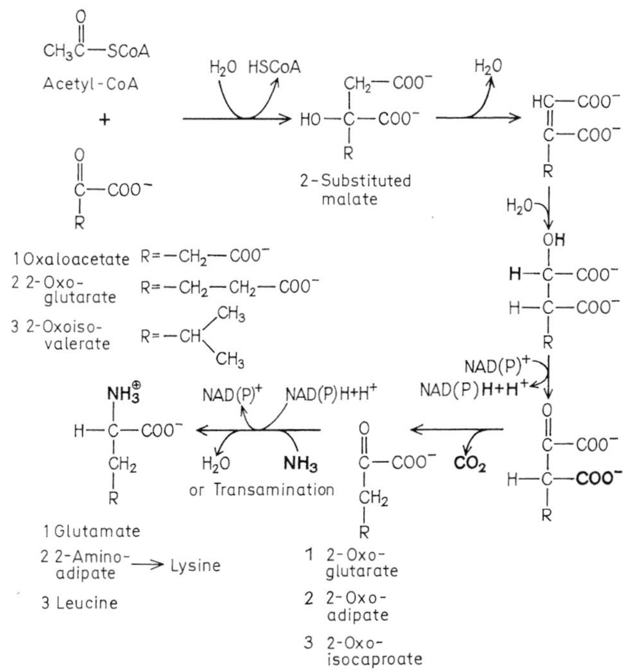

Fig. 4. *Some metabolic reaction sequences that are analogous to part of the TCA cycle.*

to the stereochemical numbering proposed by Hirschmann, C1 is at the end of the chain occupying the *pro-S* position. Citrate synthesized in the TCA-cycle from oxaloacetate and [1-^{14}C]acetyl-CoA contains ^{14}C in position 1, and is referred to as *sn*-[1-^{14}C]citrate. Aconitase catalyses the removal of the OH-group from C3 and the H_R proton from C4. After rehydration of the *cis*-aconitate (see Aconitate hydratase, Aconitic acid), the resulting isocitrate carries its OH-group on the carbon originating from C4 of citrate. The CO_2 from the conversion of isocitrate into 2-oxoglutarate is therefore derived from the original oxaloacetate and not from the acetyl group of the acetyl-CoA. Subsequent decarboxylation of succinyl-CoA removes yet another carbon of the original oxaloacetate. Thus, the new "catalytic" molecule of oxaloacetate, formed after one round of the TCA-cycle, contains only two carbons of the original oxaloacetate plus both carbon atoms of the original acetyl group. Since fumarate and succinate are metabolized symmetrically, the original acetyl C1 atom becomes equally distributed between C1 and C4, and the original acetyl C2 between C2 and C3 of the new oxaloacetate. In the next round of the cycle, all the original acetyl C1 is removed as CO_2. The C2 of the original acetyl group becomes distributed amongst all four carbons of the new oxaloacetate, so that it theoretically can never be entirely removed by decarboxylation in the TCA-cycle: it does not contribute to CO_2 until the third round of the cycle. Patterns of labeling from the incorporation of ^{14}C-labeled acetyl-CoA are therefore complex, but the fundamental experimental observation is that some ^{14}C from labeled acetyl-CoA is retained by intermediates of the TCA-cycle.

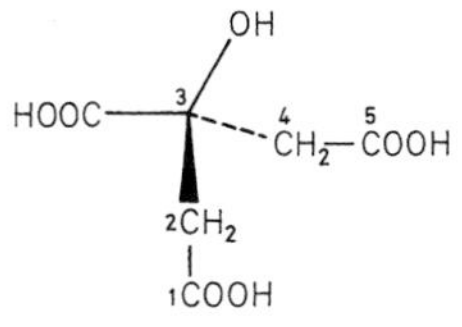

Fig. 5. *Formula of citrate with prochiral numbering of the carbon atoms.*

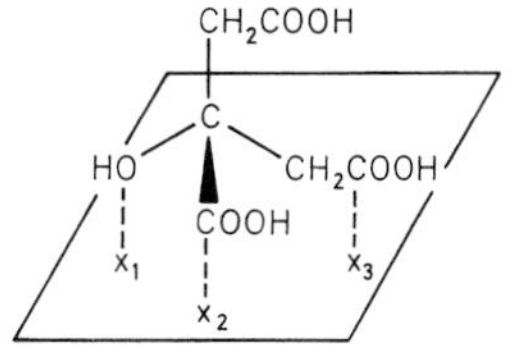

Diagrammatic representation of the three point attachment theory of Ogston. X_1, X_2 and X_3 represent sites of attachment of citrate to the active center of aconitase.

Historical. The TCA-cycle was discovered almost simultaneously in 1937 by Krebs, and by Martius and Knoop. Green coined the term "cyclophorase" for the total multienzyme complex of the TCA-cycle and the associated respiratory chain. Early studies with tissue suspensions showed that ^{14}C from C1 of acetate was found only in the 4-carboxyl carbon of 2-oxoglutarate, whereas the symmetrical metabolism of citrate

Table 2. *Sites of regulation of the TCA-cycle*

Reaction No.	Name of enzyme	Location	Requirement for	Release of	Activated by	Inhibited by	Remarks
1	Citrate synthase	Mitochondria	Acetyl-CoA, Oxaloacetate	Citrate, HSCoA		Long-chain acyl-CoA	Control point for utilization of acetyl-CoA
3a	NAD-dependent isocitrate dehydrogenase	Mitochondria	NAD^+	NADH, CO_2	ADP	ATP, NADH	For high rates of TCA-cycle see reaction 3b
3b	NADP-dependent isocitrate dehydrogenase	Cytoplasm and Mitochondria	$NADP^+$	NADPH, CO_2	Oxaloacetate?		The extramitochondrial enzyme is important in the production of NADPH
9	Glutamate dehydrogenase	Mitochondria	NADPH or NADH, NH_3	$NADP^+$ or NAD^+	ADP	GDP + NADH	
10	Pyruvate carboxylase	Cytoplasm	ATP, CO_2	ADP	Acetyl-CoA		Control of carbohydrate metabolism
11	Acetyl-CoA carboxylase	Cytoplasm	ATP, CO_2	ADP	Citrate	Long-chain acyl-CoA	Control of fat synthesis
12	Citrate lyase	Cytoplasm	Citrate	Acetyl-CoA, Oxaloacetate			Important for extramitochondrial synthesis of acetyl-CoA
13	Isocitrate lyase	Cytoplasm	Isocitrate	Glyoxylate, Dicarboxylic acids		Phospho-*enol*-pyruvate	Found only in bacteria and plants

would demand the presence of label in both the 2- and 4-carboxyl groups. It was therefore suggested that citrate could not be an intermediate in the TCA-cycle. In 1948, however, Ogston proposed his three-point attachment theory, in which the active sites of citrate synthase and aconitase are asymmetrical: three functional groups of citrate must become attached to three complementary asymmetrically arranged binding sites. With the recognition of prochirality, it is now no longer necessary to propose that the site of asymmetry lies within the enzyme. Ogston's theory is, however, not disproved, and the precise nature of citrate binding to aconitase will not be known until the active center of the enzyme has been mapped.

Tricetinidin: 3-deoxydelphinidin. See Anthocyanins.

Trichochromes: yellow-orange and violet natural pigments containing a substituted $\Delta^{2,2'}$-*bis*(1,4-benzothiazine) ring system. Color is due to the conjugated chromophore system, –S–C=C–C=N–. Biosynthetically, T. are related to the Melanins (see). Together with the phaeomelanins, T. are responsible for the red and auburn colors of human hair and bird feathers. T.B and T.C are yellow-orange; T.E and T.F are violet.

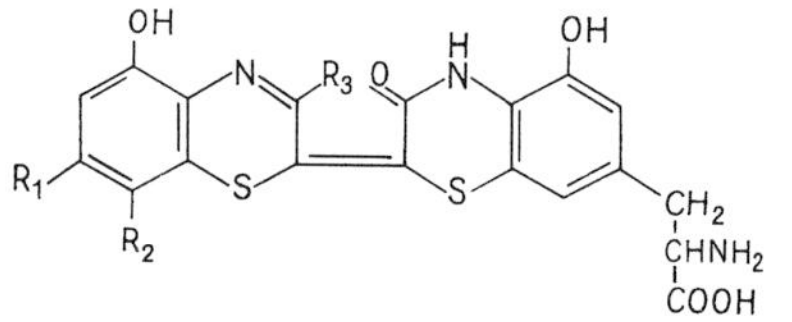

Trichochrome B: R_1=H; R_2=CH_2-$CHNH_2$-COOH; R_3=COOH

Trichochrome C: R_1=CH_2-$CHNH_2$-COOH; R_2=H; R_3=H

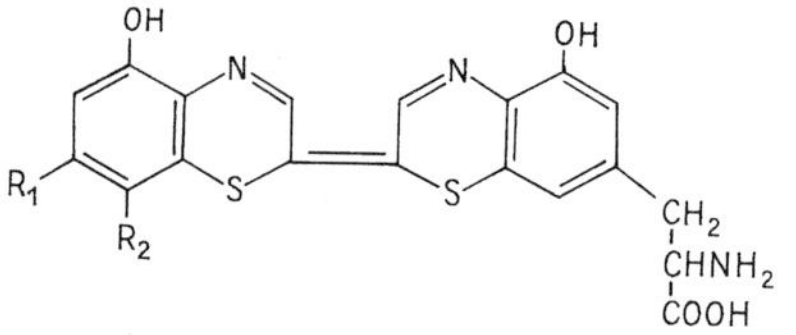

Trichochrome E: R_1=H; R_2=CH_2-$CHNH_2$-COOH

Trichochrome F: R_1=CH_2-$CHNH_2$-COOH; R_2=H

Trichochromes

Tricholomic acid: *erythro*-dihydroibotenic acid, a compound isolated from the Basidiomycete *Tricholoma muscarium*. It can also be prepared by reduction of ibotenic acid. It possesses flavor promoting activity similar to, but much more active than, that of sodium glutamate. It also has a synergistic effect on the flavor improving property of inosinic acid and guanosine 5′-phosphate.

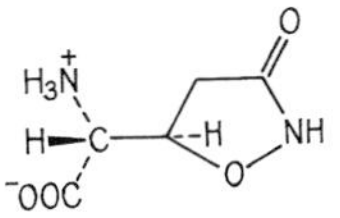

Tricholomic acid

Trifluoromethanesulfonic acid, ***triflic acid, TFMS:*** CF_3SO_3H, M_r 150.7, b.p. 162 °C. TFMS is one of the strongest acids known, and has proved effective for the simultaneous deprotection and resin cleavage at the conclusion of the solid-phase synthesis of peptides, e.g. tuftsin, enkephalin, bovine pancreatic RNase and chicken neurotensin. [H. Yajima et al. *Chem. Pharm. Bull.* **29** (1981) 2587; H. Yajima & N. Fujii *Biopolymers* **20** (1981) 1958; Y. Kiso et al. *Chem. Pharm. Bull.* **27** (1979) 1472; M. K. Chaudhuri & V. A. Najjar *Anal. Biochem.* **95** (1979) 305]

Triglyceride: see Acylglycerols, Fats.

Trigonellin: 1-methylnicotinic acid, a metabolite of nicotinic acid or nicotinamide found in many plants. It is both a hormone and a storage form of nicotinic acid. It is apparently not a niacin metabolite in animals, although it is found in the urine of coffee drinkers. Green coffee beans contain relatively large (> 500 mg/kg) amounts of T.; roasting the beans converts T. to nicotinic acid. Coffee is a significant dietary source of niacin (see Vitamins) in South and Central America.

Tri(hydroxymethyl)methylamine, ***TRIS:*** H_2N-$C(CH_2OH)_3$, M_r 121, a widely used buffer substance, suitable for the pH range 7–9. The required pH is usually obtained by adding HCl to TRIS dissolved in water. TRIS buffers have a high temperature/pH gradient, e.g. 0.05 M TRIS, pH 7.05 (adjusted with HCl) at 37 °C has pH 7.20 at 23 °C. The pH of a TRIS buffer should therefore be established at the intended working temperature.

3,5,3′-Triiodothryonin: see Thyroxin.

Trimethylglycine: see Betaines.

Triose phosphates: D-glyceraldehyde 3-phosphate (PO_3H_2–OCH_2–CHOH–CHO) and dihydroxyacetone phosphate (PO_3H_2–OCH_2–CO–CH_2OH), important intermediates in Glycolysis (see) and Alcoholic fermentation (see). The two T.p. are interconvertible via the ene-diol form, by the action of T.p. isomerase; the equilibrium mixture contains 96 % ketotriose phosphate and 4 % aldotriose phosphate. The T.p. hold a key position in carbohydrate metabolism, being intermediates of gluconeogenesis and photosynthetic CO_2 fixation.

Trioses: glyceraldehyde and dihydroxyacetone. They contain 3 C-atoms and are the simplest monosaccharides. Their phosphates are important metabolic intermediates (see Triose phosphates).

Triphosphomonoesterases: see Esterases.

Triphosphopyridine nucleotide: see Nicotinamide adenine dinucleotide phosphate.

Triple helix: see Collagen.

Triplet code: see Genetic code.

TRIS: see Tri(hydroxymethyl)methylamine.

Trisaccharides: see Carbohydrates.

Trisporic acids, ***fungal sex hormones:*** structurally similar C_{18}-terpene carboxylic acids from heterothallic fungi of the *Mucorales* type, e.g. *Blakeslea trispora* or *Mucor mucedo*. T.a. are only produced when (+)-strains are mixed with (–)-strains; a prohormone from the (–)-strain is transformed into T.a. by the (+)-strain. The T.a. then induce formation of zygospores in the (–)-cells. ***Trisporic acid C*** is the most important member of the group, M_r 306; 20 µg are sufficient to induce zygospore formation. Unlike other diterpenes, T.a. are biosynthesized by the cleavage of β-carotene.

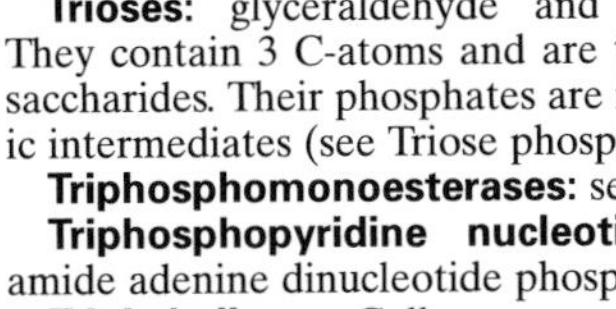

Trisporic acid C

Triterpenes: an extensive group of terpenes biosynthesized from six isoprene units. Apart from squalene, which is acyclic, most of this group are tetra- or pentacyclic hydroaromatic compounds based on the parent hydrocarbon, sterane, i.e. they are steroids. Included with T. are those terpenoid natural products with fewer than 30 C-atoms, which are biosynthesized via a C30 intermediate, but with subsequent loss of one or more C atoms. Addition of extra C atoms and incorporation of heteroatoms are also possible, e.g.

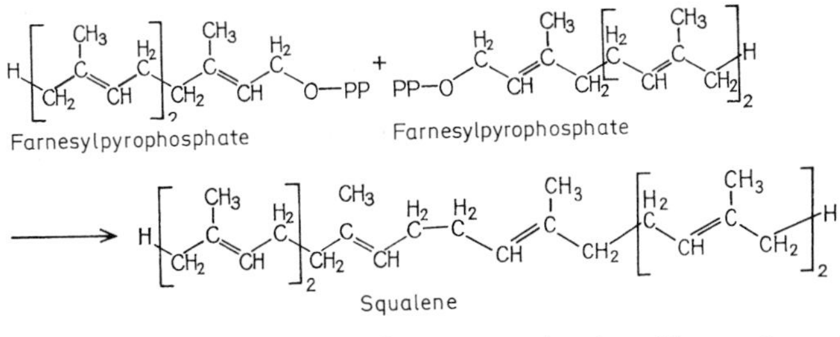

Fig. 1. *Formation of squalene from two molecules of farnesylpyrophosphate.*

Fig. 2. *Formation of pentacyclic triterpenes.*

the steroid alkaloids. Many T. have high biological activity, in particular the steroid hormones.

Biosynthesis. Tail-to-tail condensation of two molecules of farnesylpyrophosphate produces squalene. This, and subsequent reactions leading to steroids, are discussed under Terpenes (see) and Steroids (see). In the biosynthesis of pentacyclic T., the 3-hydroxysqualene cation cyclizes to a prosterol cation, which has not yet been identified in the free state. A Wagner-Meerwein rearrangement and ring closure leads to the expansion of ring D (compounds I and II, Fig.). A further Wagner-Meerwein rearrangement results in expansion of ring E in III, to form IV, which is the precursor of a large number of pentacyclic T. (germanicol, friedelin, multiflorenol, taraxerol, etc.) T. isolated between 1977 and 1981 have been reviewed by M. C. Das & S. B. Mahato *Phytochemistry* **22** (1983) 1071–1095.

Triterpenes

Type of compound	No. of carbon atoms	Examples
Sex hormones		
Estrogens	18	Estradiol
Androgens	19	Androsterone
Progestins	21	Progesterone
Adrenal corticosteroids	21	Corticosterone
Cardiac glycosides	21, 23, 24	Digitoxigenin
Steroid alkaloids	21, 27	Tomatidine
Bile acids	24, 27	Cholic acid
Sapogenins	27	Digitogenin
Vitamin D	27, 28	Vitamin D_2
Molting hormones	27	β-Ecdysone
Sterols		
Mycosterols	27, 28	Ergosterol
Zoosterols	27, 28, 30	Cholesterol
Phytosterols	28, 29, 30	β-Sitosterol
Cucurbitanes	30	Cucurbitacin C

Triterpene saponins: see Saponins.

Trityrosine: see Resilin.

tRNA: abb. for transfer-RNA.

tRNA methylases: see Polynucleotide methyltransferases.

Tropane alkaloids: esters of various amino alcohols based on a substituted tropane (*N*-methyl-8-azabicyclo-3,2,1-octane) ring system. This system is not optically active, because the two rings can only be linked in a *cis* configuration, thus resulting in a *meso* form. Introduction of a hydroxyl group at position 3 produces the geometric isomers, tropine (tropane-3α-ol; the OH-function is *trans* to the CH_3N-group, Fig. 1) and pseudotropine (Ψ-tropine; tropane-3β-ol; Fig. 1). The wide variety of T. a. is the result of substitution of the base moiety, esterification with different acids and the occurrence of isomers and demethylated derivatives. Tropane-3β-ol-2-carboxylic acid (ecgonine, Fig. 1) is important as the base component of the coca alkaloids. The acid component is usually an aromatic carboxylic acid (Table).

According to their plant of origin T. a. can be classified as Belladonna (Datura) and Coca alkaloids. The ***Belladonna*** or ***Datura*** type occur in *Atropa belladonna,* and other members of the *Solanaceae,* e. g. thorn apple (*Datura* spp.) and henbane (*Hyoscyamus* spp.); their amino alcohol component is tropine. The ***Coca alkaloids*** (e. g. cocaine and tropacocaine) are esters of ecgonine and pseudotropine, which, together with other T. a., are found in the coca shrub, *Erythroxylum coca,* cultivated in Peru and Bolivia, also in Java, and to a limited extent in Sri-Lanka. In all cases the T. a. are accompanied by pyrrolidine bases, e. g. hygrine.

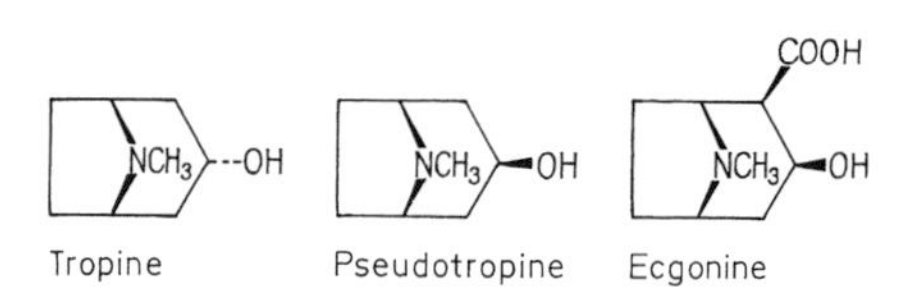

Fig. 1. *Some tropane alkaloids.*

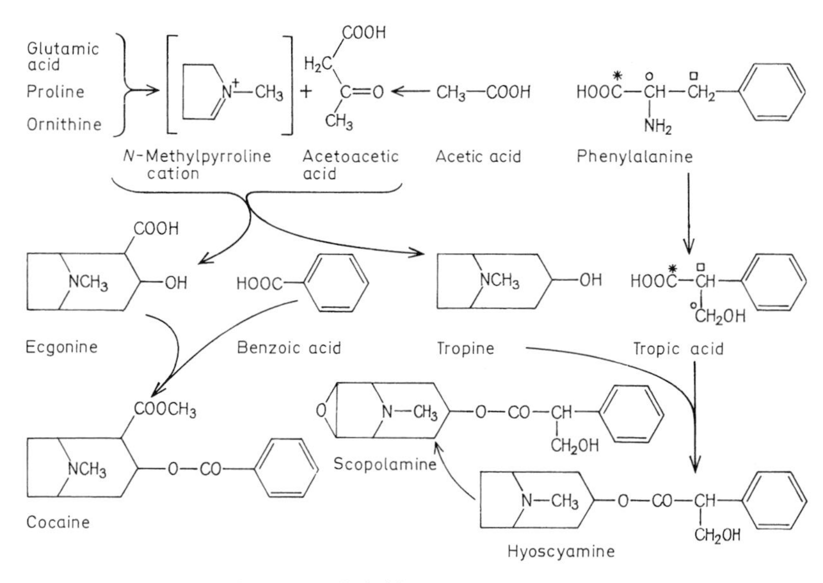

Fig. 2. *Biosynthesis of tropane alkaloids.*

Structure of tropane alkaloids

Alkaloid	Amino alcohol	Acid
Atropine	Tropine	DL-Tropic acid
L-Hyoscyamine	Tropine	L-Tropic acid
Littorine	Tropine	Phenyllactic acid
Scopolamine	Scopine	Tropic acid
Noratropine	Nortropine	Tropic acid
Tropacocaine	Pseudotropine	Benzoic acid
Cocaine	Ecgonine methyl ester	Benzoic acid

T. a. for therapeutic purposes are derived almost entirely from plant sources. The tropane ring is derived biosynthetically from compounds of the glutamic acid/proline/ornithine group, via an *N*-methylpyrroline cation which condenses with acetoacetic acid to form ketones (of the tropinone type). Following reduction of the ketone function, the resulting amino alcohol becomes esterified with an appropriate aromatic acid (ecgonine is esterified with benzoic acid, with additional methylation of the 2-carboxyl function to form cocaine; tropine is esterified with tropic acid to form hyoscyamine, which can be further epoxidized to scopolamine). See Fig. 2.

Plants containing T. a. have been used in folk medicine since ancient times. The T. a. are powerful poisons. Atropine poisoning is characterized by distension of the pupil of the eye, dryness of the mouth, delirium and double vision. Higher doses result in general paralysis (owing to the action of the drug on involuntary muscle) and finally death. Atropine is used widely in ophthalmology for dilating of the pupil of the eye (mydriatic activity). The belladonna alkaloids were therefore previously called ***mydriatic alkaloids.*** Cocaine was used widely as a local anesthetic, but has now been largely replaced by other synthetic drugs.

Tropan-3-one: see Tropinone.

Tropic acid: see Tropane alkaloids.

Tropine: see Tropane alkaloids.

Tropinone, ***tropan-3-one:*** a possible biosynthetic precursor of the Tropane alkaloids (see), present in members of the *Solanaceae*. M_r 139.19, m.p. 42 °C, b.p. 224–225 °C. Robinson's laboratory synthesis of T. (Fig.) from succindialdehyde, methylamine and the calcium salt of acetone-dicarboxylic acid in aqueous solution at ordinary temperatures (yield 42 %, 2 molecules of CO_2 are readily lost on subsequent treatment with acid), i.e. under mild or apparently physiological conditions, gave impetus to modern studies on alkaloid biosynthesis.

Succindialdehyde + Methylamine + Acetone-dicarboxylic acid → Tropinone

Robinson's laboratory synthesis of tropinone.

Tropocollagen: see Collagen.

Tropomyosin: a protein associated with Actin (see), both in muscle (see Muscle proteins) and in the cytoskeleton of other cell types. There are two very similar forms in striated muscle, α-T. and β-T. Both have 284 amino acid residues (M_r 33,000) per subunit. The molecule is a two-chain coiled-coil α-helix with the two subunits twined around one another. α-α, α-β and β-β dimers have been observed; different proportions of the two types appear in different types of muscle and may reflect specialization. T. dimers polymerize head-to-tail to form a fiber which lies in the groove of an F-actin helix. Each dimer spans 7 (muscle) or 6 actin subunits. In vertebrate striated muscle, T. binds tightly to troponins, proteins which regulate contraction. In other tissues, T. is associated with some but not all microfilaments (see Cytoskeleton), where it probably stabilizes F-actin. Non-muscle cells synthesize a number of different T. from a family of genes; the differences between the isoforms are not great but probably have regulatory function, e.g. horse platelet T. has only 247 amino acid residues, binds only slightly to troponins, and probably spans only 6 actin units instead of 7. [M.R. Payne & S.E. Rudnick in *Cell and Muscle Motility* **6** (J.W. Shay, ed.), Plenum Press, New York & London, 1985, pp. 141–184]

Troponin: see Muscle proteins.

Trp: abb. for L-tryptophan.

Trypsin (EC 3.4.21.4): a serine protease present as a zymogen (trypsinogen) in the pancreas of all vertebrates. Trypsinogen is released via the pancreatic duct into the duodenum. Conversion of trypsinogen into T. is initiated in the small intestine by enterokinase, and accelerated autocatalytically by traces of T. Activation consists of the removal of the *N*-terminal acidic hexapeptide, Val(Asp)$_4$Lys (pI 9.3), from trypsinogen. Calcium ions accelerate this process and make it more specific for cleavage of the Lys$_6$-Ile$_7$ bond, thus avoiding formation of inactive byproducts through other less selective cleavages. The resulting strongly basic β-T. (pI 10.8) is cleaved by limited autolysis of the Lys$_{131}$-Ser$_{132}$ bond to give the 2-chain structure of α-T.; further cleavage of Lys$_{176}$-Asp$_{177}$ produces a 3-chain active enzyme called pseudo T.; the individual chains are held together by 6 disulfide bridges. The amino group of the *N*-terminal Ile of T. (originally Ile$_7$ of trypsinogen) forms an ion pair with Asp$_{182}$ which is important in the conformation of T. The ion pair, Lys$_{95}$ and Asn$_{223}$, protects the *C*-terminal residue from attack by carboxypeptidase. Bovine T. has M_r 23,300 and 223 amino acid residues; 101 of these residues (41 %), including 4 of the 6 disulfide bridges, are identical with corresponding sequence positions of Chymotrypsin (see) (Fig.).

The conservative regions include the sequences around the two catalytically important residues, His$_{46}$ and Ser$_{138}$. There is also considerable correspondence between the chain conformation (secondary structure) of T. and chymotrypsin and elastase (low α-helix content, extended β-structures, similar design of the substrate cleft). T. is stable at pH 2–4 and + 4 °C for several weeks, but at pH 9 and 30 °C it undergoes progressive autolysis of its intact Lys- and Arg-bonds, and becomes totally inactive within 24 h. Calcium ions delay this process but do not prevent it. T. activity is not inhibited by 6 M urea. Of all the digestive endopeptidases, T. has the most pronounced substrate specificity, catalysing hydrolysis of only Lys- and Arg-bonds; this is due to the presence of an ionic binding site (Asp$_{189}$) in the active center. In addition to endopeptidase activity, T. displays high activity toward *N*-acylated arginyl or lysyl peptide esters and *N*-acylated arginyl- or lysylarylamides. These synthetic substrates are used for the assay of T. and in the study of T. kinetics. T. is irreversibly inhibited by many naturally occurring T. inhibitors (e.g. trasylol and soybean T. inhibitor), and by synthetic inhibitors like diisopropylfluorophosphate (forms an ester with the hydroxyl group of the reactive Ser$_{183}$) or tosyl-L-lysylchloromethylketone (alkylates N3 of the reactive His$_{46}$).

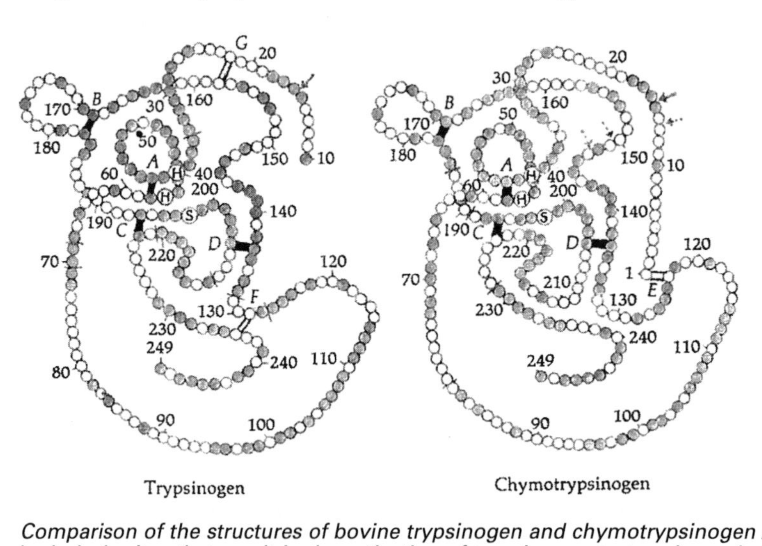

Comparison of the structures of bovine trypsinogen and chymotrypsinogen A. Solid arrows show the hydrolysis sites that result in the activation of trypsinogen to trypsin, and chymotrypsinogen to chymotrypsin. Broken arrows are additional hydrolysis sites, leading to formation of α-chymotrypsin. Shaded circles represent amino acids that are identical or similar in the two proteins. Disulfide bridges are lettered A to G. H represents the histidine residues, and S the serine residue of the active site; this serine residue reacts with diisopropylfluorophosphate. Deletions are shown by lines between the circles. To aid comparison, the residue numbers of both structures are based on those of chymotrypsinogen. [B. S. Hartley et al. *Nature 207* (1965) 1157–1159]

Enzymes similar to T. have been found in many invertebrates, like crabs and insects, e. g. cocoonase produced by silk moths to attack their proteinaceous cocoon and aid escape.

N. B. All sequence positions quoted above are for trypsinogen.

Tryptamine: β-indolyl-(3)-ethylamine, a biogenic amine, M_r 160.2, produced by decarboxylation of tryptophan. T. stimulates contraction of the smooth muscle of blood vessels, uterus and central nervous system. It is found in both plants and animals, and as a bacterial degradation product of tryptophan.

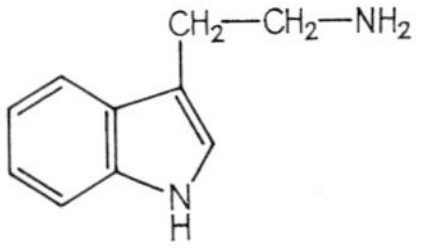

Tryptamine

L-Tryptophan, ***Trp, α-amino-β-indolepropionic acid:*** an aromatic, essential amino acid, M_r 204.2, m.p. 281–282 °C. $[\alpha]_D^{25}$ −33.7 (c = 1–2, water), or + 2.8 (c = 1–2, 1 M HCl). Trp is nutritionally very important, although it is present in relatively small amounts in proteins. It is completely destroyed by acid protein hydrolysis. *p*-Dimethylaminobenzaldehyde or xanthydrol gives a violet coloration with Trp, which is used in its determination. Trp is the precursor of several physiologically important metabolites. Cleavage of the pyrrole ring of Trp by oxidation (see L-Tryptophan 2,3-dioxygenase) to produce *N*-formylkynurenine is the first reaction in a major metabolic pathway of Trp in animal liver (Fig. 1). The open chain intermediate, 2-amino-3-carboxymuconic acid-semialdehyde (IX, Fig. 1) serves as the starting point of two different pathways. Spontaneous cyclization of IX produces quinolinic acid, which is the precursor of the nicotinamide moiety of NAD. Enzymatic conversion of IX to 2-aminomuconic acid (X, Fig. 1) initiates the pathway for the complete degradation to CO_2 and H_2O via acetyl-CoA and the TCA-cycle. The extent to which the spontaneous cyclization of IX can be exploited for the synthesis of NAD depends on its relative rates of enzymatic formation from VI and conversion

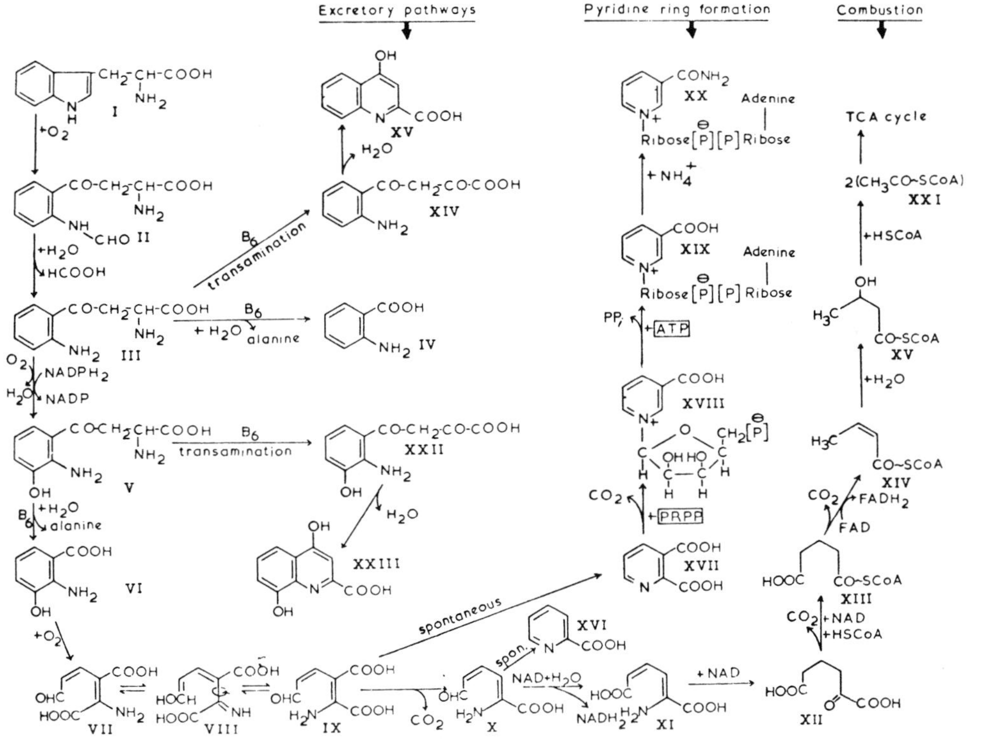

Fig. 1. *Metabolic pathways of L-tryptophan initiated by the formation of N- formylkynurenine in animal liver.*

PRPP = 5-phosphoribosyl 1-pyrophosphate. *I* L-tryptophan, *II N*-formylkynurenine, *III* kynurenine, *IV* anthranilic acid, *V* 3-hydroxykynurenine, *VI* 3-hydroxyanthranilic acid, *VII, VIII, IX* 2-amino-3-carboxymuconic acid-semialdehyde, *X* 2-aminomuconic acid-semialdehyde, *XI* 2-aminomuconic acid, *XII* 2-oxoadipic acid, *XIII* glutaryl-CoA, *XIV* crotonyl-CoA, *XV* 3-hydroxybutyryl-CoA, *XVI* picolinic acid, *XVII* quinolinic acid, *XVIII* nicotinic acid mononucleotide, *XIX* deamido-NAD, *XX* NAD, *XXI* acetyl-CoA, *XXII o*-amino-*m*-hydroxybenzoylpyruvic acid, *XXIII* xanthurenic acid, *XXIV o*-aminobenzoylpyruvic acid, *XXV* kynurenic acid.

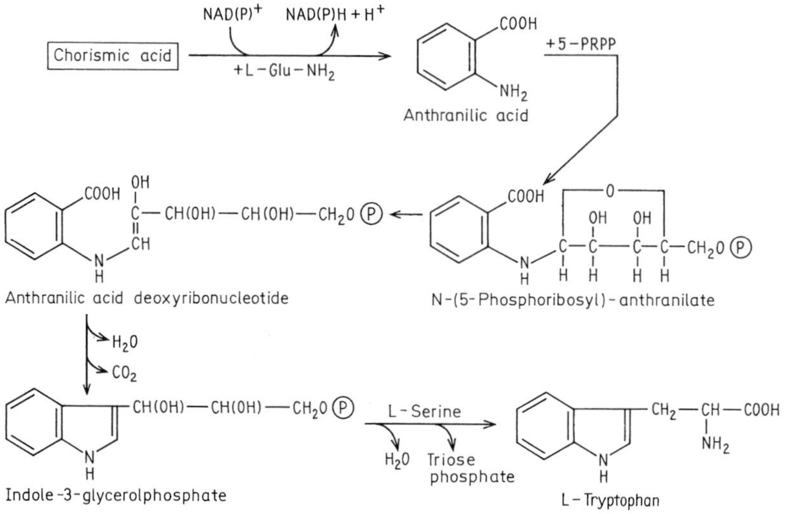

Fig. 2. *Biosynthesis of L-tryptophan from chorismic acid.*

into X. In rats, for example, nicotinic acid is not essential in the diet (i.e. it is not a vitamin) because the total requirement for NAD and NADP can be satisfied by synthesis from Trp. On the other hand, in the cat, practically no synthesis of NAD occurs from Trp, and there is a total dependence on a dietary source of nicotinic acid. In the human, part of the NAD is synthesized from Trp but the rate of cyclization of IX is insufficient for synthesis of the total NAD requirement, and the remainder must be synthesized from dietary nicotinic acid or nicotinamide (see Vitamins).

Trp degradation is markedly affected by vitamin B_6 deficiency. Kynureninase, which catalyses the cleavage of 3-hydroxykynurenine into 3-hydroxyanthranilic acid and L-alanine, is a Pyridoxal phosphate (see) enzyme. The activity of liver kynureninase (EC 3.7.1.3) decreases rapidly in vitamin B_6 deficiency, compared with the decrease shown by other pyridoxal phosphate enzymes. For example, after 14 weeks of experimental B_6 deficiency in rats, the liver kynureninase is reduced to about 16% of its original value, whereas kynurenine transaminase (EC 2.6.1.7) still shows 60% activity. Consequently, compounds derived from the Trp degradation pathway prior to hydrolysis of 3-hydroxykynurenine are excreted in greater quantity (e.g. xanthurenic acid and kynurenic acid), whereas the urinary level of nicotinic acid derivatives is decreased. Exaggeration of this picture by feeding extra Trp forms the basis of the tryptophan load test for vitamin B_6 deficiency: 2 g of Trp are given in a single dose suspended in a beverage before breakfast. The extra Trp induces synthesis of L-Tryptophan 2,3-dioxygenase (EC 1.13.11.11), so that there is a greatly increased flow of Trp metabolites into the degradation pathway. A 24-hour urine sample is analysed for Trp metabolites. Xanthurenic acid is normally measured for clinical purposes, because it is easily assayed photometrically as a green ferric-xanthurenic acid complex, or by an improved fluorimetric technique. In patients receiving deoxypyridoxine (see Vitamin B_6), the tryptophan load test reveals a 20-fold increase in the excretion of kynurenine, hydroxykynurenine and xanthurenic acid, and a 3-fold increase in the excretion of kynurenic acid, compared with normal controls. At the same time, excretion of *N*-methyl-2-pyridone 5-carboxamide (a metabolite of nicotinic acid) is decreased.

The following metabolites of Trp are listed separately: Actinomycins, Indican, Indigo, Indole alkaloids, Kynurenine, Melatonin, Ommochromes, Phallotoxins, Serotonin, Toad toxins, Tryptamine, Violacein. The status of Auxin (see) as a Trp metabolite is now in question.

Trp is synthesized in bacteria and plants via shikimate and chorismate (Fig. 2) (see Aromatic biosynthesis, L-Tryptophan synthase).

Tryptophan 2.3-oxygenase, ***tryptophan oxygenase, tryptophan pyrrolase, L-tryptophan: oxygen 2,3-oxidoreductase (decyclizing)*** (EC 1.13.11.11): an enzyme catalysing the oxidation of L-tryptophan to *N*-formylkynurenine, the first stage in the total degradation of L-Tryptophan (see) in animals and microorganisms; this is also the first stage in the conversion of L-tryptophan into the nicotinamide moiety of NAD and NADP in molds and in some mammals, and in the conversion of L-tryptophan into Ommochromes (see) in insects.

Studies with $^{18}O_2$ and $H_2{}^{18}O$ show that both atoms of oxygen appearing in the product originate from molecular oxygen and not from water. Inhibition of the reaction by superoxide dismutase suggests that the substrate oxygen is activated to the superoxide ion prior to participation in the oxygenase reaction. In *Pseudomonas,* synthesis of tryptophan oxygenase is apparently induced by L-tryptophan, but the true inducing agent is kynurenine, formed from L-tryptophan by low constituent levels of the enzyme. Normal adult mammalian liver always contains fairly high activity of the enzyme, but higher levels of synthesis (due to increased synthesis of mRNA) are induced by glucocorticoids. Administration of L-tryptophan also leads to an increase in the level of the enzyme, but this appears to be due to a decreased rate of breakdown of the enzyme protein. Tryptophan oxygenase is absent from the liver of normal rats up to the tenth postnatal day, but it can be induced at any time during this ten day period by administration of L-tryptophan or glucocorticoids.

The rat liver enzyme has M_r 167,000 (4 subunits of M_r 43,000, comprising 2 types: $\alpha_2\beta_2$). The enzyme from *Pseudomonas* has M_r 122,000 (4 subunits of M_r 31,000). Both the bacterial and mammalian enzymes contain protoporphyrin IX, which is essential for catalytic activity. Two moles of heme are present per mole of enzyme; in vitro, the enzyme is inactive unless this heme is reduced by a reducing agent such as H_2O_2, ascorbic acid, or superoxide anion. The presence of 2 g atoms of Cu per mole of enzyme has been claimed, but after some dispute it has now been proved that Cu is not essential in the enzymatic mechanism [Y. Ishimura et al. *J. Biol. Chem.* **255** (1980) 3835–3837]. The enzymatic mechanism is ordered bi-uni; L-tryptophan binds to the active (reduced) enzyme, then with molecular oxygen, forming a ternary complex of enzyme-substrate-oxygen. This ternary complex can be detected during the reaction by its spectral properties (maxima at 418, 545 and 580 nm) which are similar to those of the oxygenated ferroheme forms of hemoglobin, myoglobin or peroxidase.

Hartnup's disease, a hereditary defect associated with mental retardation, is due to a deficiency of tryptophan oxygenase. The $v^+ \rightarrow v$ mutation in *Drosophila melanogaster,* manifested as defective ommochrome synthesis, is due to a deficiency of tryptophan oxygenase; it can be reversed by the injection of kynurenine.

L-Tryptophan synthase, ***tryptophan desmolase, L-serine hydro-lyase (adding indoleglycerol-phosphate)*** (EC 4.2.1.20): the enzyme catalysing the synthesis of L-tryptophan from L-serine and indole 3-glycerol phosphate. T.s. from *E. coli* (M_r 149,000) and other prokaryotes has $\alpha_2\beta_2$ subunit composition. The enzyme separates easily into monomeric subunit α (also called protein B, M_r 29,000) and dimeric subunit β_2 (also called protein B; M_r of dimer 90,000) when eluted from DEAE cellulose with a sodium chloride gradient. The separated subunits catalyse partial reactions of L-tryptophan synthesis:

Indole 3-glycerol phosphate → indole + 3-phosphoglyceraldehyde (α-subunit)

Indole + L-serine → L-tryptophan + H_2O (β-subunit)

In the presence of the reconstituted $\alpha_2\beta_2$ complex, the rates of these partial reactions are 30 to 100 times greater than with the individual subunits. The sum of the two partial reactions: Indole 3-glycerol phosphate + L-serine → L-tryptophan + 3-phosphoglyceraldehyde + H_2O, is catalysed by the $\alpha_2\beta_2$ complex, and free indole is not detectable as an intermediate in this overall reaction. T.s. from *E. coli* and other prokaryotes therefore serves as a simple and very effective model of a multienzyme complex. Pyridoxal phosphate is essential for enzymatic activity; each β subunit binds one molecule of the coenzyme. The reactions proceed via the Schiff's base between the enzyme-bound pyridoxal phosphate and amino acrylate [structure **c** in the scheme shown in the entry: Pyridoxal phosphate (see)], which adds a nucleophile (i.e. indole). In the absence of the α-subunit, the β-subunit catalyses hydrolysis of the amino acrylate Schiff's base to pyruvic acid and ammonia, but this hydrolysis is almost totally suppressed in the $\alpha_2\beta_2$ multienzyme complex.

The primary sequence of the α-subunit (268 amino acid residues) is known [J. R. Guest et al. *J. Biol. Chem.* **242** (1967) 5442–5446]. Studies on the genetic analysis of this sequence are now classical. Outstanding among these studies was the comparison of the linear positions of mutational sites in the cistron for the α-subunit (determined by phage-mediated three point genetic crosses between mutant strains) with the linear position of the sites of amino acid replacement in the corresponding CRiM proteins. This provided the first proof of the colinearity of gene and polypeptide structure [C. Yanofsky et al. *Proc. Natl. Acad. Sci. USA* **51** (1964) 266–2272].

The primary sequence of the β-subunit (397 residues) is also known, and extensive studies have been reported on the 3-dimensional structure of the $\alpha_2\beta_2$ multienzyme complex, in particular for the enzyme from *Salmonella typhimurium.* The enzymes from *E.coli* and *S. typhimurium* are very similar; the respective α-subunits both consist 268 amino acid residues and differ at 40 positions (15 %), while the β-monomers display only 3.5 % difference in primary sequence (both consist of 397 residues, and only 14 of these are different). Crystallographic studies show that the active sites of the α- and β-subunits are 25 Å apart and connected by a tunnel, which presumably serves to carry the one metabolic intermediate (indole) from the active site of the α-subunit to the active site of the β-subunit . The kinetics of this substrate "channeling" have been studied by chemical quench-flow and stopped-flow methods. [I. P. Crawford & J. Ito *Proc. Natl. Acad. Sci. USA.* **51** (1964) 390–397; B. P. Nichols & C. Yanofsky *Proc. Natl. Acad. Sci. USA.* **76** (1979) 5244–5248; S. A. Ahmed et al. *J. Biol. Chem.* **260** (1985) 3716–3718; C. C. Hyde et al. *J. Biol. Chem.* **263** (1988) 17857–17871; K. S. Anderson et al. *J. Biol. Chem.* **266** (1991) 8020–8033]

In *Neurospora crassa,* only one T.s. protein is produced (M_r 150,000, consisting of 2 identical monomers of M_r 75,000), but genetic and biochemical analysis show that the T.s. gene in *Neurospora* is subdivided into two regions homologous with the A and B regions of *E. coli* [W. H. Matchett & J. A. DeMoss *J. Biol. Chem.* **250** (1975) 2941–2946]. Studies on T.s. from other microorganisms indicate that T.s. from *E. coli* is typical of prokaryotes, while the *Neurospora* enzyme serves as a model of the eukaryotic type. Among the prokaryotic T.s. there is some degree of cross reactivity between respective α and β subunits from different organisms.

The literature on T.s. up to 1972 has been reviewed by C. Yanofsky & I. P. Crawford in *The Enzymes,* pp. 1–31, vol VII, 3rd. edtn. (1972), edit. P. D. Boyer, Academic Press. The review by E. W. Miles (pp. 127–186 in *Advances in Enzymology* (1979) vol 49, edit. Alton Meister) deals with studies on the catalytic mechanism and structure of the enzyme up to 1979, but not with the genetic analysis.

TSH: acronym of thyroid stimulating hormone.

TTP: acronym of thymidine triphosphate. See Thymidine phosphates.

Biosynthesis of tubercidin by Streptomyces tubercidicus.

Tubercidin: 6-amino-9-β-D-ribofuranosyl-7-deazapurine, M_r 266.25, m.p. 247–248 °C (d.), $[\alpha]_D^{17}$ −67° (c = 1, 50 % acetic acid), a purine antibiotic (see Nucleoside antibiotics) from *Streptomyces tubercidicus*, and one of the group of 7-deaza-adenine-nucleoside analogs. The N7 of adenine is replaced by a methylene group. T. is biosynthesized from adenosine (Fig.): the C-atoms of the pyrrole ring are derived from a ribose moiety, which is introduced from 5-phosphoribosyl 1-pyrophosphate. As an antimetabolite of adenosine, T. interferes with purine metabolism. T. can also be converted into nicotinamide-deaza-adenine dinucleotide, which inhibits glycolysis. T. is particularly active against *Mycobacterium tuberculosis* and *Candida albicans*.

Tubulins: see Cytoskeleton.

Tuftsin: a naturally occurring phagocyte activating tetrapeptide: Thr-Lys-Pro-Arg. T. stimulates phagocytosis and pinocytosis, and promotes motility and migration of phagocytes. It causes a chemotactic response in phagocytes and generates chemiluminescence in the absence of a target particle, presumably due to formation of H_2O_2, superoxide and hydroxyl radicals. It increases the killing ability of phagocytes, and enhances the rate of clearance of bacteria from the blood by phagocytosis. T. also stimulates the immunogenic function of phagocytes, presumably by enhanced antigenic processing and signal generation to antibody-forming lymphocytes. It also stimulates macrophages and granulocytes, so that they become highly tumoricidal in vitro and vivo. All these properties are probably due to membrane activation, which is sensitive to nanomolar concentrations of T.

T. is excised from the H-chain of a precursor γ-globulin molecule, known as leukokinin, which binds to the phagocyte membrane. Release of T. occurs in two stages: the *C*-terminus is released by endocarboxypeptidase activity in the spleen, followed by release of the *N*-terminus by the action of leukokinase, an enzyme present on the outer surface of the phagocyte membrane.

Patients genetically deficient in T. (human tuftsin deficiency syndrome) have a high frequency of very severe infections. These patients produce a mutant peptide, which is a strong inhibitor of natural T. After splenectomy, leukokinase is still produced and coats the leukocytes, but it is inactive in T. formation. Activity reappears and attains normal values several months later, indicating that the necessary enzyme activity arises elsewhere in the body. [V. A. Najjar *Adv. Enzymology* **41** (1974) 129–178].

Tulipanin: see Delphinidin.

Tumor antigens: carcinoembryonic antigens which serve as an aid to the early recognition of liver carcinoma and teratoblastomas (tumors of reproductive cells, especially in the testes and ovaries). T. a. are embryonal plasma proteins produced in the placenta during pregnancy and in some organs of the embryo, but no longer detectable shortly after birth. They can be formed again later in life in response to malignant tumors. Three T. a. are important in clinical tumor diagnosis: fetoprotein (M_r 65,000), embryonogenic colon antigen (ECA), and the Regan isoenzyme of the placenta. Appearance of α-fetoprotein in the serum is a definite indication of liver carcinoma or teratoblastoma. Benign liver disorders cause no increase in tumor antigen. Determination of these T. a. permits detection in the early phase of 60- 80 % of liver carcinoma cases, and 20–25 % (80–90 % in young people) of teratoblastoma cases; 2 ng α-fetoprotein per ml serum can be detected by a radioimmunological test. The Regan isoenzyme of placental alkaline phosphatase increases particularly in cases of malignant tumors of the female genital tract.

Tumor necrosis factor, ***TNF***

TNF-α, cachectin: a protein (human, M_r 17,000, two Cys in disulfide linkage, no carbohydrate) originally found in the serum of animals sensitized with *Bacillus* Calmette-Guerin or with *Corynebacterium parvum,* and challenged with bacterial endotoxin. TNF-α produces hemorrhagic necrosis of some tumors in experimental animals, and cytolytic or cytostatic effects on tumor cells in culture. The major cellular source of TNF-α is macrophages, and it is thought to be a mediator of the cytotoxic activity of macrophages against tumor cells in culture. It has been implicated in the induction of shock and cachexia resulting from invasive stimuli, and it is a potent pyrogen, causing fever by direct action on the hypothalamic thermoregulatory centers and by induction of the synthesis of interleukin-1.

TNF-β, lymphotoxin: a glycoprotein (human, one aspartate-linked glycosylation site, M_r of the unglycosylated protein 18,664 containing 171 amino acid residues) produced by mitogen-stimulated lymphocytes, which causes cytostasis of some tumor cell lines and cytolysis of other transformed cells.

Both TNF-α and TNF-β have been purified, sequenced and cloned by recombinant DNA technology in *E. coli.* According to studies with human cervical carcinoma (cell line ME-180), TNF-α and TNF-β bind to the same receptor (2,000 receptor sites per cell). Preincubation of cells with γ-interferon increases the number of receptor sites two- to three-fold, which accords with the observation that interferons act synergistically with TNF in tumor necrosis in vitro. Studies with TNF-α show that binding is followed by receptor-mediated internalization of TNF by endocytosis. The amino acid sequences of TNF-α and TNF-β show about 50 % homology, but no significant homology with other lymphokines, e.g. γ-interferon, interleukin-2 or interleukin-3. The TNF-α gene resides on human chromosome 6, and is closely linked to the gene specifying TNF-α; presumably both genes arose from a common ancestor by tandem duplication. [B. Bharat et al. *Nature* **318** (1985) 665–667; M. Tsujimoto et al. *Proc. Natl. Acad. Sci. USA* **82** (1985) 7626–7630; B. Beutlet & A. Cerami *Nature* **320** (1986) 584–588; K. J. Tracey & A. Cerami 'Tumor Necrosis Factor, Other Cytokines and Disease' *Annu. Rev. Cell Biol.* **9** (1993) 317–343]

Tumor viruses, ***cancer viruses, oncogenic viruses:*** viruses which cause malignant tumors in animals. Only a few T. v. are known that cause malignant tumors in humans. Alternatively, the virus infection may represent a "risk factor", which together with other influences (chemical, genetic, hormonal, physical) is involved in the development of a malignant tumor, i.e. the genesis of the tumor may be multifactorial. T. v. are found among retroviruses, as well as nearly all the main groups of animal DNA viruses, in particular papova-, adeno- and herpesviruses. As a rule, T. v. also cause *malignant transformation* of cells cultured in vitro, a phenomenon associated with altered growth and multiplication. After malignant transformation in vitro, some cells develop into malignant growths when transplanted to appropriate host animals, thus providing an in vitro model of carcinogenesis. There are, however, numerous viruses (e.g. human adenoviruses), which transform human cells in vitro, but are rarely carcinogenic in the human organism. On the other hand, some oncogenic viruses (e.g. chronic oncornaviruses, or hepatitis B virus) do not transform human cells in vitro.

The genome of nearly all T. v. can be incorporated into the genome of the host cell by nonhomologous recombination (see Molecular genetics). This is also possible for oncogenic RNA viruses (oncornaviruses), which, like other retroviruses, possess an RNA-dependent DNA polymerase (revertase, reverse transcriptase) for the synthesis of virus-specific double-stranded DNA. Incorporation of viral DNA leads to the regular replication and transmission of the viral genome to all subsequent cell generations, i.e. the malignant dedifferentiation becomes genetically based. At least in the case of chronic oncornaviruses, provirus integration also appears to be necessary for the subsequent malignant transformation (see below). The pro-oncoviruses share many similarities with jumping genes. Although the proviruses of all T. v. appear to behave like the prophages of temperate bacteriophages, no analogous repressor activity has yet been demonstrated in cells infected with T. v.

Each tumor virus is oncogenic only in certain animal species, and usually only in specific organs or tissues. As a rule, when a virus is associated with malignant growth, it does not multiply in the affected cells, i.e. the cells are "nonpermissive" for the virus in question. Thus, most T. v. cause, or contribute to the development of, specific cancers, e.g. papilloma virus causes skin tumors, while hepatitis B virus is a factor in the development of liver cancer, etc. The multifactorial nature of carcinogenesis is particularly obvious in the action of the herpesvirus, *Epstein-Barr virus (EBV).* In nonimmune individuals in Europe and North America, this virus causes a benign febrile illness called infectious mononucleosis (Pfeiffer's glandular fever). In Africa and Papua New Guinea, a combination of infection by the malaria parasite, *Plasmodium falciparum,* and a chromosome mutation caused by EBV is responsible for the development of a malignant tumor, known as *Burkitt's lymphoma* (a cancer of B lymphocytes). In southern China and Greenland (Inuit tribes), EBV, in combination with other unknown factors (histocompatibility genotype? Consumption of salted fish in infancy?), causes malignant tumors of the nasopharyngeal cavity. EBV can therefore be classified as a human tumor virus, or as a contributory factor for the development of human tumors. Other human T. v. are: 1. *Human papilloma virus type 5 (HPV5)* (family *Papovaviruses),* which (in dependence on other factors) causes a very rare, wart-like malignant skin tumor. Certain papillomaviruses are thought to cause cancer of the uterine cervix. Infection by papillomavirus is usually benign, the viral DNA being maintained as plasmids in basal epithelial cells. Very occasionally, however, a fragment of a viral plasmid becomes incorporated into the host genome. The resulting unregulated production of viral replication protein drives the host cell into S-phase, with the eventual development of a cancer. 2. *Hepatitis B virus (HBV),* the causative agent of serum hepatitis. In tropical Africa and Southeast Asia, and in combination with other factors (aflatoxins in food, alcoholism, smoking, other viruses), HBV causes primary liver cancer (hepatocellular carcinoma), which is one of the commonest human cancers.

One retrovirus, *T-cell lymphoma virus (TCLV),* has been identified as a human cancer risk factor in Japan and the West Indies. Other viruses, especially other types of papilloma viruses, as well as herpes simplex virus type II, are suspected of involvement in the development of human cancers, e.g. cancers of the genital system. To the relatively small list of human T.v. must now be added the human immunodeficiency virus (HIV-1; the AIDS virus), a retrovirus which is thought to be a contributory factor in cancer of the endothelial cells of blood vessels (Kaposi's sarcoma) in Central Africa. Although relatively few human cancer-causing T.v. are known, many viruses have been identified (in particular retroviruses) which cause malignant tumors in practically all other groups of vertebrates. Some of these are therefore economically important.

There is no single uniform mechanism of malignant transformation or carcinogenesis by T.v. The process may even differ between members of the same virus genus. For example, a gene product is responsible for the oncogenic action of the polyomavirus, murine papova virus (plate 19), whereas this gene product is totally absent from simian virus SV40, which belongs to the same genus.

There are at least 3 principal mechanisms of virus-induced carcinogenesis:

1. *Virus-specific genes* and their products are directly responsible for the oncogenic action, e.g. in the case of polyoma- and adenoviruses.

2. *A cellular oncogene (c-onc)* or *proto-oncogene* is activated by the T.v. This is the mode of action of the retroviruses known as chronic oncornoviruses, which do not possess a viral oncogene. Activation occurs when the provirus is incorporated into the host genome close to or even within the proto-oncogene. This activation process of cellular oncogenes is known as *insertional mutagenesis.* Oncogenesis can also be activated by virus-induced chromosome modification. Thus, Epstein-Barr virus (see above) causes chromosome modification, and this is a contributory factor in the development of Burkitt's lymphoma.

3. The virus carries a *viral oncogene (v-onc)* into the cell, e.g. in the case of oncornaviruses. The viral oncogene is homologous with one of 15 to 20 cellular *onc* genes, and it was originally incorporated into the viral genome from a host genome during the evolution of the virus. The prototype of such an acute oncornavirus is the *Rous sarcoma virus,* the first tumor virus to be discovered (Peyton Rous, 1911; Rous received the Nobel prize for this discovery in 1966). The oncogene of the Rous sarcoma virus is the src oncogene, which codes for a protein kinase. Conversion of a proto-oncogene into an oncogene by incorporation into a retrovirus may occur by alteration or truncation of the gene sequence so that it codes for a protein with abnormal activity. Alternatively, within the viral genome, the gene may be placed under the control of powerful promoters and enhancers, resulting in excess production of gene product, or formation of gene product when it is not required. Both effects (gene modification and overproduction of gene product) are often involved.

Tunicamycin: a mixture of homologous, nucleoside antibiotics, produced by *Streptomyces lysosuperificus,* and active against viruses, Gram-positive bacteria, yeast and fungi. The structure of T. consists of one residue each of uracil, a C_{11}-aminodeoxy-dialdose (tunicamine), *N*-acetylglucosamine and a fatty acid (Fig.). T. differ from one another in the chain length of the fatty acid component. The major fatty acid components are *trans* α,β-unsaturated *iso*-acids. The names tunicamycin A, B, C, D have been proposed for those T. whose fatty acid component contains 15, 16, 17 and 18 C-atoms, respectively (i.e. n = 9, 10, 11, 12 in the formula shown) [Takatsuki et al. *Agric. Biol. Chem.* **41** (1977) 2307–2309]. High pressure liquid chromatography of T. shows two major and eight minor components, all active in the inhibition of protein glycosylation [W.C. Mahoney & D.J. Duksin *J. Biol. Chem.* **254** (1979) 6573–6576]. Four of these components presumably correspond to T.A, B, C and D of the Japanese authors. Commercially available T. is always a mixture of all possible homologous components. It is used as the unresolved mixture for biochemical studies on the inhibition of protein glycosylation. T. is reported to be a selective inhibitor of glycoprotein synthesis, and it has been shown that the transfer of *N*-acetylglucosamine to dolichylphosphate is sensitive to T. in microsomal fractions of calf liver and chicken embryo, and in yeast. The initial reaction: UDP-Glc*N*Ac + Dol-P → Dol-PP-Glc*N*Ac + UMP, is sensitive to T., but subsequent extension of the carbohydrate chain by addition of residues to Dol-PP-Glc*N*Ac is not sensitive to T. (Glc*N*Ac = *N*-acetylglucosamine, Dol = dolichyl, P = phosphate). ["Uses of Tunicamycin", a colloquium held at the 586th. meeting of the British Biochemical Society, pub. in *Biochemical Society Transactions* **8** (1980) 163–171]

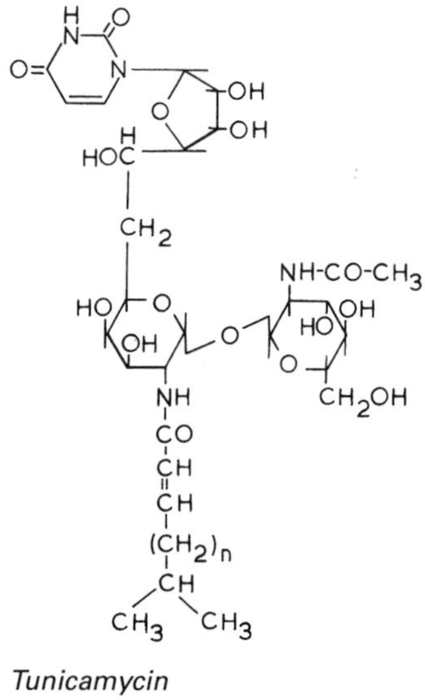

Tunicamycin

Tunichrome: a green chromogen in the blood cells of tunicates, e.g. *Ascidia nigra, Ciona intestinalis, Molgula manhattensis.*

Turbidostat: see Fermentation techniques.

Table. *Structures and sources of turgorins.* (P)LMF = (periodic) leaf movement factor

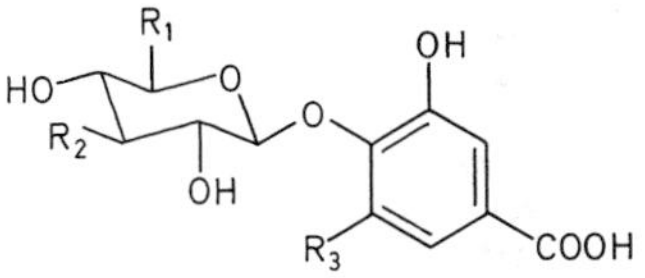

Turgorin	R_1	R_2	R_3	Plant source
PLMF 1	CH_2OSO_3H	OH	OH	*Mimosa pudica, Acacia karoo, Oxalis stricta, Robinia pseudacacia*
PLMF 2	CH_2OSO_3H	OSO_3H	OH	*Acacia karoo*
S-PLMF 2	CH_2OSO_3H	OH	H	*Oxalis stricta*
M-LMF 5	COOH	OH	OH	*Aimosa pudica*

Turgorins, ***leaf movement factors:*** plant hormones which bring about movement (nastic responses to light, temperature, physical shock, etc.) by causing changes of turgor. T. are active at concentrations of 10^{-5}–10^{-7} M or lower. The archetypal plant showing movement is the "sensitive" plant (*Mimosa pudica* L.), which serves as a test system for T. If a cut mimosa shoot is placed in a solution of T., the hormones are transported to the pulvini, which contain T. receptors. The pulvini then undergo marked turgor changes, each pair of pinnules collapses, and the entire leaf assumes the typical irritated posture. Using this system, T. have been isolated from several plants which show nastic responses (Table). Most known T. are glycosides of 3,4-di-, or 3,4,5-trihydroxybenzoic acid with an acidic monosaccharide (glucose 6′-sulfate, glucose 3′,6′-bisulfate or glucuronic acid) (Table). The leaf movement factor from *Glycine max* (Fig.) is an exception, being an ester of 4-coumaric acid and galactaric acid. [H. Schildknecht *Angew. Chem. Int. Ed. Engl.* **22** (1983) 695–710; *Endeavour* **8** (1984) 113–117.]

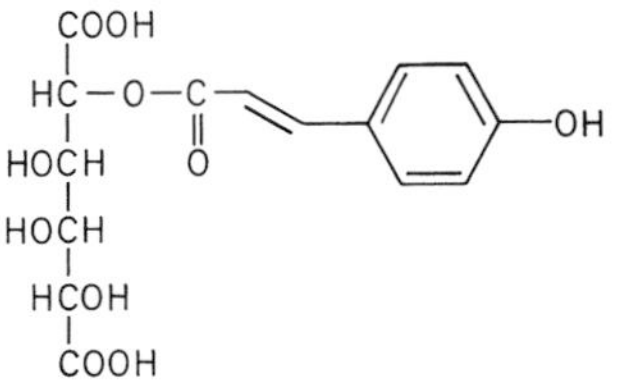

Turgorin from Glycine max.

Turnover: the balance of synthesis and degradation of biomolecules in living organisms. All cell components are subject to continual degradation and resynthesis, i. e. they are subject to T. See Steady state.

Turpentine: see Balsams.

Turpentine oil: a volatile oil obtained by steam distillation of turpentine from various *Pinus* spp. T. o. is a colorless liquid, most of which boils between 155 and 162 °C. ρ 0.865–0.870, n_D^{25} 1.465–1.480. On exposure to air, T. o. alters rapidly and finally resinifies. Its composition varies with origin, chief components being bicyclic monoterpenes of the carane and pinene type. Lower grades are obtained by steam distillation of the wood, roots, stumps, sawdust, etc. (wood turpentine), and also as a byproduct in the manufacture of sulfite cellulose (sulfite turpentine). Pinene is isolated on a large scale from T. o. for the synthesis of camphor. T. o. is used widely as a cleaning agent and solvent in the manufacture of shoe polish, varnish and paints. It is also used occasionally in pharmacy under the name Oleum terebinthinae.

Tyr: abb. for L-Tyrosine.

Tyramine: β-hydroxyphenylethylamine, a biogenic amine, M_r 137.2, found in plants (ergot, broom and other legumes) and animals (blood, urine, bile, liver), and as a bacterial degradation product of L-tyrosine. T. is produced by the decarboxylation of L-tyrosine.

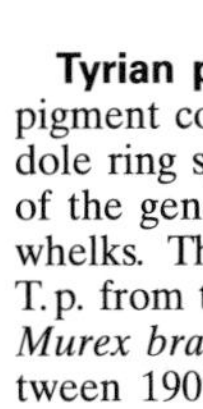

Tyramine

Tyrian purple: 6,6′-dibromoindigotin, a red-violet pigment containing two brominated and oxidized indole ring systems. T. p. is present in marine mollusks of the genera *Murex* and *Nucella,* and a few related whelks. The isolation and structural elucidation of T. p. from the hypobranchial body of the purple snail *Murex brandaris* were performed by Friedlander between 1909 and 1911. In antiquity and later in the Middle Ages, T. p. was one of the most expensive dyes.

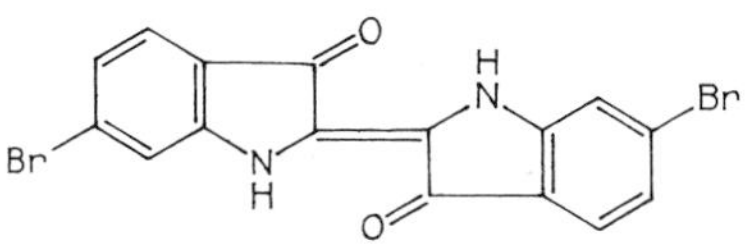

Tyrian purple

Tyrocidins: homodetic, homomeric, cyclic peptide antibiotics active against Gram-positive bacteria. Like the Gramicidins (see), T. are produced by *Bacillus brevis.* The structure of tyrocidin A, confirmed by total synthesis, is cyclo-(-Val-Orn-Leu-D-Phe-Pro-Phe-D-Phe-Asn-Gln-Tyr-). In tyrocidin B, Phe_6 is replaced by Trp. In tyrocidin C, Phe_6 is replaced by Trp, D-Phe_7 by D-Trp, and Asn_8 by Asp. Tyrocidin E differs from tyrocidin A by replacement of Asn_8 by Asp, and Tyr_{10} by Phe.

L-Tyrosine, ***Tyr:*** L-α-amino-β-(*p*-hydroxyphenyl)-propionic acid, an aromatic, ketogenic, proteogenic amino acid, M_r 181.2, m.p. 342–344°C (d.), $[\alpha]_D^{25}$ –10.0 ($c = 2$ in 5 M HCl). L-T. is nonessential in humans, since it can be synthesized by hydroxylation of L-phenylalanine. Millon's color reaction for proteins is specific for L-T. residues. The widely used Lowry protein assay also depends largely on the specific reaction of the phenolic residues of L-T. For the biosynthesis of L-T. see Aromatic biosynthesis. For all other metabolism of L-T. see L-Phenylalanine. L-T. is used therapeutically in disturbances of thyroid function.

Tyrosine kinase: see Protein-tyrosine kinase, Receptor tyrosine kinase.

Tyrosinemia: see Inborn errors of metabolism.

Tyrosinosis: see Inborn errors of metabolism.

U

U: 1. a nucleotide residue (in a nucleic acid) in which the the base is uracil. 2. Abb. for uridine (e.g. UDP is acronym of uridine diphosphate). 3. Abb. for uracil.

Ubiquinone, ***coenzyme Q:*** a low M_r electron transport component of the respiratory chain. Structurally, it is a 2,3-dimethoxy-5-methylbenzoquinone with an isoprenoid side chain. Various types are designated according to the number of carbon atoms or the number of isoprene units in the side chain: U-30 (= U-6), U-35 (= U-7), U-40 (= U-8), U-45 (= U-9) and U-50 (= U-10), where the first figure refers to the number of carbon atoms and the figure in parentheses to the number of isoprene units. Other abbreviations are, e.g. coenzyme Q_{10}, UQ-50, UQ_{10}, Q-10 and CoQ_{10} (Fig.1). U. are slowly destroyed by oxygen, UV light and sunlight, and they are rapidly oxidized in alkaline solution, except in the presence of pyrogallol to remove oxygen.

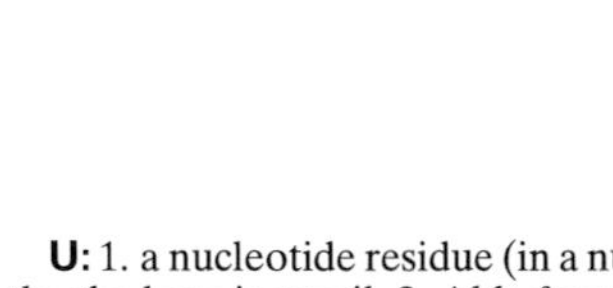

H_3CO, H_3CO, CH_3, $(CH_2-CH=C(CH_3)-CH_2)_{10}-H$

Fig. 1. *Coenzyme Q_{10}*

Hydroquinone ($\underset{-2H^+}{\overset{+2H^+}{\rightleftharpoons}}$) Hydroquinone anion (phenolate) ($\underset{+e}{\overset{-e}{\rightleftharpoons}}$) Hydroquinone radical (semiquinone) ($\underset{+e}{\overset{-e}{\rightleftharpoons}}$) Benzoquinone

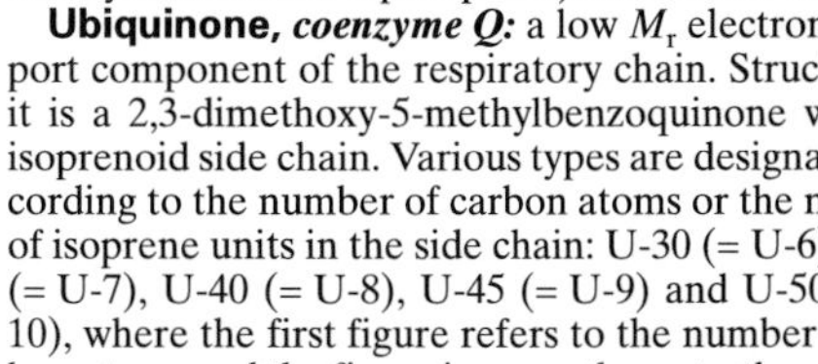

Fig. 2. *Oxidation of hydroquinone*

U./dihydroubiquinone is a redox system in the respiratory chain. Reversible reduction of the benzoquinone to hydroquinone is a stepwise reaction: the semiquinone (hydroquinone radical) is first formed by electron transfer; further transfer of a single electron forms the hydroquinone anion or phenolate, which takes up two protons to form the hydroquinone (Fig.2). The reverse reaction, dehydrogenation of the hydroquinone to the quinone, is initiated by dissociation of the hydroquinone to the hydroquinone anion by release of two protons; oxidation then proceeds in two stages by removal of electrons. Other dehydrogenations are analogous, e.g. dehydrogenation of ethanol to acetaldehyde via an intermediate alcoholate. Thus in the enzymatic dehydrogenation of ethanol by alcohol dehydrogenase (see Alcoholic fermentation) one proton (H^+) and an electron pair (2e) are transferred together as a hydride ion, and the second proton equilibrates with the protons of the aqueous medium (see Nicotinamide adenine dinucleotide).

U. is so named from its ubiquitous occurrence.

Ubiquitin, ***ATP-dependent proteolysis factor 1, APF-I:*** a small polypeptide, M_r 8,500, first isolated during the purification of polypeptide hormones from thymus. U. was subsequently found (by radioimmunoassay) in vertebrates, invertebrates, plants and yeasts. Earlier reports that it induces differentiation of thymocytes and stimulates adenylate cyclase have not been confirmed; the name, ubiquitous immunopoietic polypeptide (UBIP) is therefore inappropriate. The primary structure of U. is almost identical in insects, trout, cattle and humans.

Ubiquitination specifically targets proteins for degradation by Proteasomes (see). *N*-terminal ubiquitination appears to be necessary and sufficient for this U.-dependent degradation of proteolytic substrates. Acetylation of *N*-termini in vivo blocks U.-dependent proteolysis (measured in vitro). *N*-terminal acetylation in vivo may therefore protect proteins from degradation by preventing ubiquitination. U.-dependent proteolysis has an absolute requirement for ATP. The ATP is required for synthesis of the U.-protein complex and for subsequent degradation of the proteolytic substrate (Fig.1). As shown in Fig.1, the U. pathway for protein degradation proceeds in several discrete stages: 1. In an ATP-dependent reaction, U. is activated to a high energy intermediate by U.-activating enzyme (E_1–SH). 2. U. is transferred to U. carrier protein (E_2–SH), a reaction catalysed by U.-protein ligase (E_3). 3. The conjugated protein is then degraded by a specific high M_r protease. In the in vitro system from *Saccharomyces cerevisiae,* U. is transferred directly from its E_2-complex to histones in the absence of E_3. In the presence of E_3, U. is transferred to a much broader spectrum of target proteins. Cloning has been reported of E_2 proteins [M.L.Sullivan & R.D.Vierstra *J. Biol. Chem.* **266** (1991) 23878–23885] and of specific proteinases [J.W.Tobias & A.Varshavsky *J. Biol. Chem* **266** (1991) 12021–12028].

Chromosomal protein A24, now called uH2A (ubiquitin-H2A semihistone) is the most prominent of a family of branched proteins in which the *C*-terminal glycine (76) of U. is linked by an isopeptide bond to the ε-NH_2 of lysine 119 of histone 2A. Ubiquitination of many intracellular proteins at the ε-amino groups of their Lys residues to yield branched U.-protein conjugates may represent a physiological role for U., distinct from its established function in protein degradation. The U. of uH2A and uH2B is in rapid equilibrium with free U. in interphase cells. As mitotic chromosome condensation reaches completion, the levels of uH2A and uH2B in chromosomes show a marked decrease, probably due to enzymatic deubi-

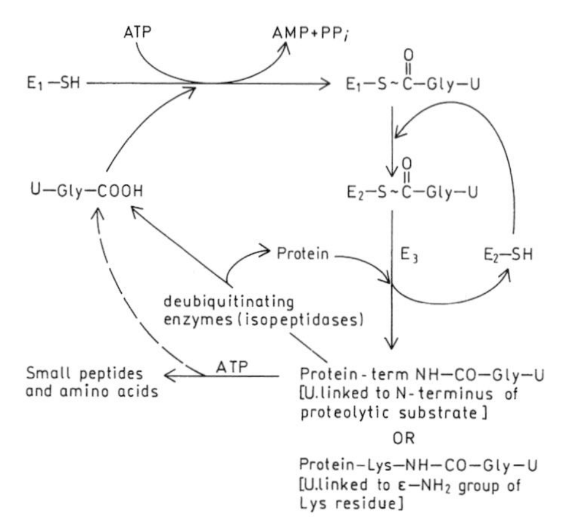

Fig. 1. *Role of ubiquitin in the degradation of proteins and branched proteins.*
E_1-SH = ubiquitin activating enzyme. E_2-SH = a transferase which transfers ubiquitin to the conjugation site. E_3 = a ligase which catalyses amide bond formation. All three enzymes have been purified [A. Hershko *J. Biol. Chem. 258* (1983) 8206–8214]. U-Gly-COOH represents ubiquitin with its *C*-terminal glycine. Deubiquitinating enzymes have also been characterized [C. M. Pickart & I. A. Rose *J. Biol. Chem. 261* (1986) 10210–10217; S.-I Matsui et al. *Proc. Natl. Acad. Sci. USA 79* (1982) 1535–1539].

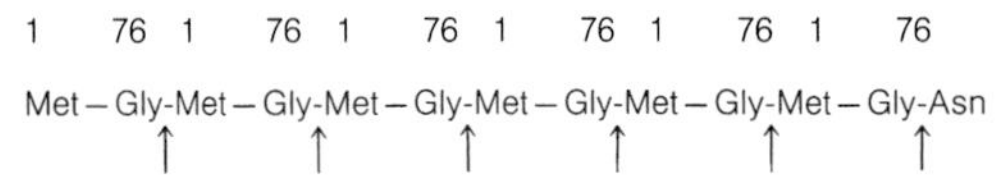

Fig. 2. *Structural organization of poly-ubiquitin precursor protein in* Saccharomyces cerevisiae, *deduced from the nucleotide sequence of the ubiquitin gene.* Arrows indicate sites of proteolytic cleavage in the maturation process. The *C*-terminus (here Asn) varies according to the organism, e. g. Val in human, Tyr in chicken.

quitination; then normal levels are rapidly restored during postmitotic chromosome decondensation.

DNA sequences for U. have been cloned from a variety of eukaryotes. U. is generated by processing of poly-U precursor protein. U-coding elements are typically organized in spacerless head-to-tail arrays, the number of coding repeats varying according to the organism, e. g. 6 in yeast, 9 in human (Fig. 2).

In addition to its well established role in targeting proteins for degradation, ubiquitination has also been shown to play a role in cellular signalling. Thus, ubiquitination of the yeast plasma membrane receptor, Ste2p (a G-protein-coupled receptor that binds a mating pheromone), acts as a signal for its ligand-stimulated endocytosis.

[J. G. Gavilanes et al. *J. Biol. Chem.* **257** (1982) 10267–10270; A. Hershko *Cell* **34** (1983) 11–12; D. Finley & A. Varshavsky *Trends Biochem. Sci.* **10** (1985) 343–347; R. Hough & M. Rechsteiner *J. Biol. Chem.* **261** (1986) 2391–2399; A. Herschko *J. Biol. Chem.* **263** (1988) 15237–15240; M. Rechsteiner (ed.) *Ubiquitin*, Plenum Publishing Corporation 1988; A. Ciechanover & A. L. Schwartz *Trends Biochem Sci.* **14** (1989) 483–488; G. Sharon et al. *J. Biol. Chem.* **266** (1991) 15890–15894; L. Hicke & H. Riezman *Cell* **84** (1996) 277–287]

UDP: acronym of uridine 5′-diphosphate. See Uridine phosphates.

UDPG: acronym of uridine diphosphate glucose. See Nucleoside diphosphate sugars.

UDP-glucose: uridine diphosphate glucose. See Nucleoside diphosphate sugars.

Ultracentrifugation: see Methods of biochemistry, Proteins.

Ultrafiltration: see Proteins.

UMP: acronym of uridine 5′-monophosphate. See Uridine phosphates.

cUMP: acronym of cyclic uridine monophosphate. See Uridine phosphates.

Uncompetitive inhibition: see Effectors.

Uncoupler: a chemical compound which prevents oxidative phosphorylation of ADP to ATP in the respiratory chain without affecting electron transport. It thus uncouples respiration and ATP formation; respiration continues or is even increased, but no energy is trapped in a utilizable form. Some important U. are 2,4-dinitrophenol (DNP), dicumarol, carbonyl cyanophenylhydrazone, salicyl-anilide, etc. These compounds affect only oxidative phosphorylation, and not substrate phosphorylation. An U. probably destroys or discharges an energy-rich state or intermediate generated by the electron transport. Interpreted in terms of the Chemiosmotic hypothesis (see), an U. acts by promoting passage of protons across the inner mitochondrial membrane, i. e. by collapsing the proton gradient that otherwise drives ATP synthesis. It is thought that a physiological process of uncoupling of oxidative phosphorylation may be involved in heat production in brown fat tissue. See Chemiosmotic hypothesis, Oxidative phosphorylation, Ionophore.

Unit membrane: see Biomembranes.

Untwisting enzymes: eukaryotic type I Topoisomerases (see).

Unusual nucleic acid components: see Rare nucleic acid components.

Upstream: a term strictly referring to any portion of the Coding strand (see) (otherwise known as the 'sense' or 'non-template' strand) of a stretch of double-stranded DNA that is on the 5′-side of the start site of transcription. The nucleotide at the start site is designated + 1 and the nucleotides upstream (i. e. on the 5′-side) of it are sequentially designated –1, –2, –3 etc, no nucleotide being designated 0. The promoter of a transcribed operon, which includes the Pribnow or TATA box (–10 region) and the –35 region, is, for instance, upstream of the initiation site of transcription. The term is sometimes used more loosely to refer to any nucleotide or region of the coding strand of DNA that is on the 5′-side of some other nucleotide or region within the strand. See Nomenclatural conventions concerned with gene transcription.

Uracil: 2,4-dihydroxypyrimidine, a 1,3-diazine which occurs as a pyrimidine base in all ribonucleic acids. M_r 112.09, m.p. 335 °C (d.). U. is formed by de-

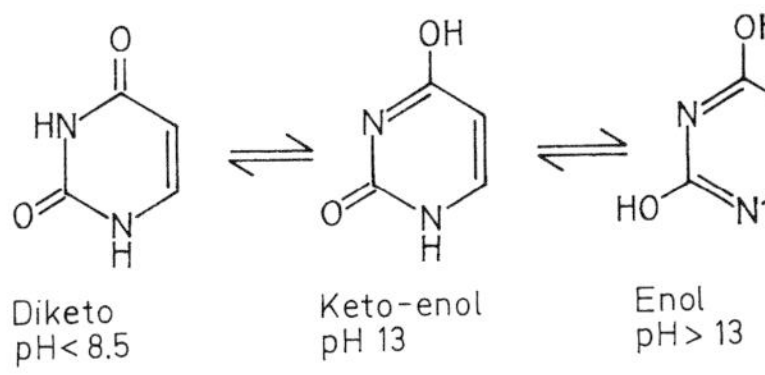

Diketo pH < 8.5 ⇌ Keto-enol pH 13 ⇌ Enol pH > 13

pH-dependent tautomeric forms of uracil.

gradation of U. nucleotides and nucleosides (see Pyrimidine biosynthesis), and it is the starting point for reductive and oxidative Pyrimidine degradation (see).

Urate oxidase: see Uricase.

Urates: salts of Uric acid (see).

Urea, ***carbamide:*** $H_2N–CO–NH_2$, the diamide of carbonic acid. M_r 60.01, m.p. 132.7 °C. Solutions of U. in water develop high concentrations of reactive cyanate ions on standing, which can be removed by acidification. U. is the product of ammonia detoxification in ureotelic animals. It is produced by several metabolic pathways: 1. via the urea cycle, 2. via oxidative purine degradation and enzymatic hydrolysis of allantoic acid and glyoxylurea (ureidoglycollate), 3. by hydrolysis of L-arginine and other guanidine derivatives by arginase and other enzymes of the EC sub-sub class 3.5.3 (see L-Arginine, Guanidine derivatives), 4. via various other metabolic pathways of limited distribution or importance, e.g. by the rare oxidative pyrimidine degradation. U. is hydrolysed by urease and urea amidolyase. Other mechanisms of urea hydrolysis are of doubtful significance and have not been enzymologically proven.

U. is accumulated by many higher fungi, particularly *Agaricus* spp., puff-balls (*Lycoperdon*) and bovists (*Bovista*). In the cultivated champignon, for example, U. is a true nitrogen excretion product, and in bovists and puff-balls it serves for the storage and translocation of nitrogen for the formation of protein and chitin in the spores (see Ammonia detoxification). U. is used for osmoregulation by marine cartilagenous fish, in which it accumulates in the tissue fluids and blood in relatively high concentrations. It is also stored in the body fluids of lung-fish (*Dipnoi*) during estivation when the animals enclose themselves in a cocoon-like structure to survive the dry period.

U. is used in high concentrations as a denaturing agent for proteins. It has a toxic effect on the skin that is not understood. It has long been used as a nitrogen fertilizer, but a urease-catalysed, sudden release of ammonia can occur in the soil, leading to nitrogen loss and poisoning. U. is therefore applied in the form of U.-aldehyde condensation compounds which are "slow-release" sources of nitrogen for plants. The difficulty can also be overcome, in principle, by a suitable application form (granulates) and by the use of urease inhibitors (see Urease). There are similar difficulties in the use of U. as a nonprotein source of nitrogen for cattle feed, and for this reason, suitable compounds, from which the urea is released slowly, must be used. In the ruminant stomach, U. is degraded by the symbiotic ruminant microorganisms (see Symbiosis).

Urea amidolyase, ***UALase, ATP: urea amidolyase, urea carboxylase (hydrolysing)*** (EC 6.3.4.6): a urea splitting enzyme present in some yeasts (*Saccharomyces, Candida,* etc.) and green algae (e.g. *Chlorella*), where it replaces urease. It is a biotin enzyme and is inhibited by avidin. The catalysed reaction is an ATP-dependent cleavage of urea to carbon dioxide and ammonia (bicarbonate and ammonium): $NH_2–CO–NH_2 + HCO_3^- + ATP \rightarrow 2HCO_3^- + 2NH_4^+ + ADP + P_i$. UALase is a Multienzyme complex (see) consisting of at least two enzyme proteins: 1. *urea carboxylase,* which catalyses carboxylation of urea to *N*-carboxyurea or allophanic acid (NH_2-CO-NH COOH), a reaction requiring biotin and ATP; 2. *allophanate hydrolase,* an amidase which cleaves the allophanate into two hydrogen carbonate ions and two ammonium ions. The regenerated hydrogen carbonate acts catalytically, being continually reused as more urea is converted. The carboxylase component appears to be constitutive, whereas the allophanate hydrolase is inducible. Allophanate is an inducer of the enzymes of Purine degradation.

Urea carboxylase: see Urea amidolyase.

Urea cycle, ***arginine-urea cycle, ornithine cycle, Krebs-Henseleit cycle:*** a metabolic cycle present in mammals and other ureotelic animals (e.g. adult amphibians), which results in the synthesis of urea from carbon dioxide, ammonia and the α-amino nitrogen of L-aspartic acid (Fig.). The process is energy-dependent; synthesis of one molecule of urea or L-arginine requires 3 molecules of ATP, and involves the expenditure of 4 high energy bonds (2 molecules of ATP are cleaved to ADP and inorganic phosphate, and one is cleaved to AMP and pyrophosphate, the latter being further hydrolysed to inorganic phosphate). The U.c. is catalytic, and depends on the recycling of the catalytic molecule, L-ornithine. The primary function of the U.c. is to convert waste nitrogen into nontoxic, soluble urea, which can be excreted. The cycle may also be completed, not by hydrolysis of L-arginine to urea, but by transfer of the amidine group to glycine, to form L-ornithine and guanidinoacetic acid (the precursor of creatine; see L-Arginine, Phosphagens). A further function is the synthesis of the proteogenic amino acid, L-arginine. The U.c. has in fact, evolved from an original pathway for L-arginine synthesis. In animals, the U.c. is more or less primed by synthesis of L-ornithine from L-glutamate, and to some extent by synthesis of L-ornithine from the products of degradation of L-proline. Dietary L-arginine may serve to supplement the cycle, or the synthesis of L-ornithine and its conversion to L-arginine in the U.c. may supplement the dietary requirement for L-arginine; the balance of these two processes depends on the animal species in question, its physiological state and its diet, e.g. many young, growing animals have a dietary requirement for L-arginine, whereas the adult appears able to synthesize its total requirement.

L-Arginine synthesis represents the primary synthesis of the amidine group; other naturally occurring guanidino compounds, chiefly Phosphagens (see), are synthesized by transfer of the amidine group from L-arginine to the appropriate amino receptor.

The chief site of the U.c. is the liver. Conversion of L-ornithine to L-citrulline and the synthesis of carba-

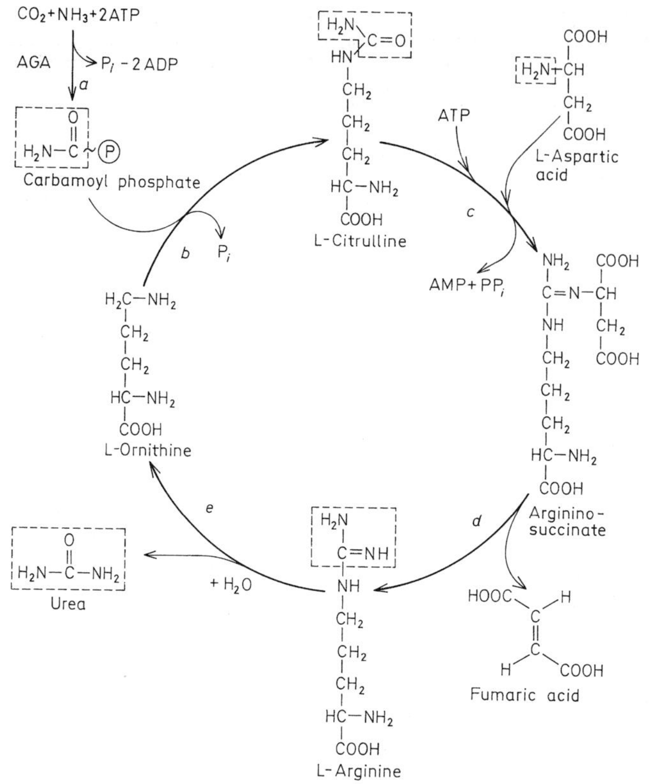

Urea cycle
a Carbamoyl-phosphate synthetase (ammonia), EC 6.3.4.16.
b Ornithine carbamoyltransferase, EC 2.1.3.3.
c Argininosuccinate synthetase, EC 6.3.4.5.
d Argininosuccinate lyase, EC 4.3.2.1.
e Arginase, EC 3.5.3.1.
AGA = *N*-Acetylglutamic acid, a stimulatory allosteric effector of carbamoyl phosphate synthetase.

moyl phosphate occur in the mitochondrial matrix, and all the other reactions of the U.c. occur in the cytoplasm. Kidney cytoplasm contains the enzymes for the conversion of L-citrulline to L-ornithine, but kidney mitochondria lack the necessary enzymes for converting L-ornithine to L-citrulline, and for synthesizing carbamoyl phosphate. Some L-citrulline is transported from the liver to the kidneys where it is converted to L-ornithine and urea. Fumarate produced by the action of argininosuccinate lyase can enter the mitochondria and be converted in the TCA cycle to oxaloacetate; transamination of the latter to aspartate represents the channeling of waste nitrogen into the amino group of aspartate, which then transfers this nitrogen to the U.c. by the action of argininosuccinate synthetase. This series of reactions links the U.c. with the TCA cycle. One reaction of this extra cycle (malate + NAD^+ → oxaloacetate + NADH + H^+) represents a source of 3 molecules of ATP by Oxidative phosphorylation (see). The chief source of ammonia consumed in the synthesis of carbamoyl phosphate is the oxidative deamination of L-glutamate by L-glutamate dehydrogenase: L-glutamate + NAD^+ + H_2O → 2-oxoglutarate + NADH + H^+ + NH_3; here again, the oxidation of NADH provides 3 molecules of ATP. The energy requirement of the U.c. is therefore amply covered by the energy production of associated processes. See also Ammonia assimilation.

Urease (EC 3.5.1.5): an enzyme (pH-optimum 7.0, pI 5.0) of high catalytic activity that catalyses the hydrolysis of urea to CO_2 and NH_3:

$$O{=}C\begin{matrix}/NH_2\\ \backslash NH_2\end{matrix} + 2H_2O \rightarrow H_2CO_3 + 2NH_3.$$

U. is found especially in plant seeds and microorganisms, as well as invertebrates (crabs, mussels), and shows a high degree of substrate specificity; apart from urea, it only attacks urea derivatives like hydroxy- and dihydroxyurea, which also act as noncompetitive inhibitors of U. The U. of soybean (M_r 489,000) was the first enzyme to be crystallized (Sumner, 1926). It consists of two enzymatically active half molecules (M_r 240,000), which separate at pH 3.5. At pH 9.0 in the presence of 0.1 % sodium dodecyl sulfate and 45 % thioglycerol, these half molecules dissociate further into quarter molecules (M_r 120,000). At neutral pH in 0.1 % sodium dodecyl sulfate, U. dissociates into its 8 subunits (M_r 60,000), which comprize two covalently bound chains (each chain of M_r 30,000). U. is remarkably resistant to denaturation by its own substrate, urea, which at 8 M denatures numerous other proteins. In 8–9 M urea, U. dissociates into M_r 60,000-subunits, which still possess urease activity. Furthermore, U.-antiurease complexes are still catalytically active. Bacterial U. is a smaller molecule than soybean U.

Ureide plants: plant families that accumulate allantoin and/or allantoic acid, and use these compounds as nitrogen reserves. U. p. are members of the *Aceraceae, Boraginaceae, Hippocastanaceae* and *Platanaceae.*

Ureotelic organisms: see Ammonia detoxification.

Uric acid, ***2,6,8-trihydroxypurine:*** an excretory product of purine metabolism in most animals. In some animals, known as uricotelic organisms (birds, reptiles, many insects), it is the main nitrogenous excretory product. U. a. (M_r 168.1, m. p. 400 °C) was discovered in urine in 1776 by Scheele, and it can be isolated in quantity from bird excrements (guano); its salts are called urates. Humans and the great apes usually excrete U. a. unchanged. In the adult human, 1–3 % of urinary nitrogen is represented by U. a.

U.a. is generated from xanthine by the enzyme xanthine oxidase (see Purine degradation). The amino nitrogen from the degradation of amino acids can also be transferred to U. a. The enzyme uricase converts U. a. to allantoin (uricolysis; see Purine degradation).

Lactim form   Lactam form

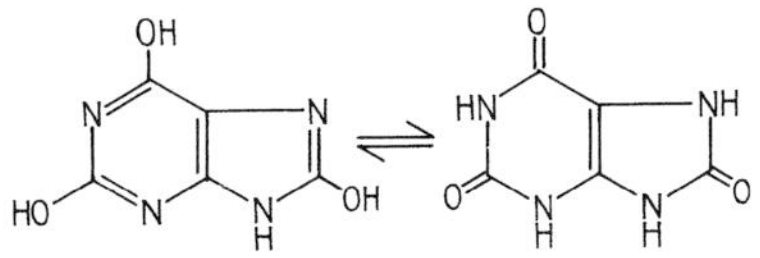

Tautomeric forms of uric acid

Uricase, ***urate oxidase*** (EC 1.7.3.3): a copper-containing aerobic oxidase, which, in the presence of oxygen, catalyses the oxidation of poorly soluble uric acid or urates to soluble allantoin, with the formation of hydrogen peroxide. U. occurs in all vertebrates and in all invertebrates with the exception of insects (only flies and related groups possess U.). The chief site of uric acid oxidation is the liver, where U. is stored in special uricase-rich microbodies called uricosomes. U. is used clinically in the enzymatic assay of uric acid (e. g. diagnosis of increased uric acid levels in gout). Porcine U. has M_r 125,000 (4 subunits, M_r 32,000), pI 6.3, pH-optimum 9–9.5. Analogs of uric acid are powerful inhibitors of U.

Uricolysis: oxidation and decarboxylation of uric acid to allantoin, catalysed by uricase as part of aerobic Purine degradation (see).

Uricotelic organisms: see Ammonia detoxification.

Uridine: 3-β-D-ribofuranosyluracil, M_r 244.20, m. p. 165–167 °C, $[\alpha]_D^{16}$ + 9.6 ° (c = 2.0, water), a β-glycosidic Nucleoside (see) of D-ribose and the pyrimidine base, uracil. See Uridine phosphates.

Uridine diphosphate glucose: see Nucleoside diphosphate sugars.

Uridine phosphates: nucleotides of uracil; phosphate esters of uridine. ***Uridine 5′-monophosphate, UMP, uridylic acid,*** M_r 324.2, m. p. 198.5 °C is produced de novo in Pyrimidine biosynthesis (see), or by degradation of nucleic acids. UMP is the starting point for the synthesis of other pyrimidine nucleotides. ***Uridine 5′-diphosphate, UDP,*** M_r 404.2, serves as the activating group of many Nucleoside diphosphate sugars (see) involved in transglycosidation. ***Uridine 5′-triphosphate, UTP,*** M_r 482.2, is a structural analog of ATP, and is required for the synthesis of uridine diphosphate sugars (see Nucleoside diphosphate sugars).

Cyclic uridine 3,5′-monophosphate, cyclo-UMP, cUMP, M_r 306.2, is a cyclic nucleotide, and like cyclic adenosine 3,5′-monophosphate (see Adenosine phosphates), it is involved in metabolic regulation. cUMP inhibits growth of some transplantable tumors. A specific cUMP-degrading enzyme is present in heart muscle.

Uridylic acid: see Uridine phosphates.

Urobilin: see Bile pigments.

Urobilinogen: see Bile pigments.

Urocanic acid: see L-Histidine.

Urokinase: see Plasmin.

Uronic acids: aldehyde carboxylic acids formed by oxidation of the terminal primary alcohol group of aldoses. U. a. are named by adding the ending "-uronic acid" to the stem of the parent monosaccharide, e. g. D-glucuronic acid, D-galacturonic acid, D-mannuronic acid. U. a. tend to form lactones, usually γ-lactones. They give the usual reactions for sugars (see Carbohydrates), and are widely distributed as components of glycosides, polyuronides, polysaccharides and mucopolysaccharides.

5α-Ursane: see Amyrin.

Ursodeoxycholic acid: 3α,7β-dihydroxy-5β-cholan-24-oic acid, a dihydroxylated steroid carboxylic acid, one of the bile acids. M_r 392.58, m. p. 203 °C, $[\alpha]_D$ + 57 ° (ethanol). U. is a characteristic component of bear bile, and is also present in human bile.

Ursolic acid: a simple unsaturated pentacylic triterpene carboxylic acid, M_r 456.71, m. p. 292 °C, $[\alpha]_D$ + 72 ° ($CHCl_3$). U. a. is a structural derivative of α-Amyrin in which the 28-methyl group is replaced by a carboxyl group (see Amyrin). It occurs widely in plants, as the free acid, esterified, or as the aglycon of triterpene saponins (see Saponins), e. g. in the wax layer of apples, pears and cherries, in the skin of bilberries and cranberries, and in the leaves of many members of the *Rosaceae Oleaceae* and *Labiatae.*

U-snRNP, ***snurp:*** one of a group of small nuclear ribonuclear proteins (snRNP). Rapid progress of elucidation began with the discovery, by Lerner and Steitz [*Proc. Natl. Acad. Sci. USA* **76** (1979) 5495–5499] that Ul, U2, U5, U4/U6 are precipitated by human systemic lupus erythematosus (autoimmune) sera of the Sm type. The RNA in U-snRNP is rich in uracil; hence the designation U. By 1988, ten types of RNA (snRNA) associated with U-snRNP had been identified and designated U1 through U10 snRNA. All these species except U6 are characterized by a trimethylguanosine "cap" at the 5′ end, and all (except U3) are precipitated by Sm antibodies. The classical U1, U2, U4/U6-snRNP are found in the nucleoplasm; U3-snRNP are found in the nucleolus. U4 and U6 are found in a single snRNP. Seven or eight proteins are associated with the particles; it seems likely that the RNA is catalytically active, while the protein provides structural stability. The U7-RNA from sea urchins has 56 or 57 nucleotides and the corresponding U7-snRNP has M_r 200,000–250,000. The *Drosophila* U7-snRNP has M_r 140,000.

Those U-snRNPs for which a biological function is known participate in the processing of RNA. U1-snRNP is required for cleavage of pre-mRNA at the 5′-splice site [A.Kramer et al. *Cell* **38** (1984) 299–307]. Antiserum specific for U2-snRNP precipitates the RNA in the complex that carries out splicing in a cell-free system; hence U2-snRNP is probably also involved in removal of introns from pre-mRNA [P.J. Grabowski et al. *Cell* **42** (1985) 345–353]. U4 and U6 RNA are associated in a single RNP. Sm antisera prevent polyadenylation of pre-mRNA in vitro, but specific anti-U1 and anti-U2 antibodies do not; hence it is likely but not proven that U4/U6 snRNPs are involved in polyadenylation of mRNA. Sea urchin U7 snRNA is required for 3′ processing of histone pre-mRNA; it contains a sequence complementary to CAAGAAAGA, which is conserved in histone genes [M.L. Birnstiel et al. *Cell* **41** (1985) 349–359]. In contrast to the other U-snRNPs, which are found in the nucleoplasm, U3-snRNP is found in the nucleolus, where it is associated with 5.8S, 18S and 28S RNA. It may therefore have a role in the processing of ribosomal RNA. [I.W. Mattaj, *Trends Biochem. Sci.* **9** (1984) 435–437; O. Georgiev & M.L. Birnstiel *EMBO J.* **4** (1985) 481–489]

Uteroferrins: a class of purple acid phosphatases from mammalian sources, including porcine uterus, bovine and rat spleen, and hairy cell leukemia cells of human spleen. They are glycoproteins, M_r 35,000–40,000, containing 2 Fe atoms per mol of enzyme. In the fully oxidized ($2Fe^{3+}$) form, they are enzymatically inactive, whereas the one-electron reduced form (Fe^{2+}, Fe^{3+}) catalyses hydrolysis of phosphate esters. Earlier controversy concerning the nature of phosphate binding is now resolved: The binuclear iron cluster binds inorganic phosphate strongly in the oxidized state; reduced U. binds phosphate weakly, accompanied by a red shift in the tyrosinate-Fe^{3+} charge transfer band and loss of characteristic EPR signals of U. The physiological significance of U. is unknown. [J.W. Pyrz et al. *J. Biol. Chem.* **261** (1986) 11015–11020]

UTP: acronym of uridine 5′-triphosphate. See Uridine phosphates.

Utter reaction: see Gluconeogenesis.

V

Vaccenic acid: Δ^{11}-octadecanoic acid, CH_3–$(CH_2)_5$–CH = CH–$(CH_2)_9$–COOH, an unsaturated fatty acid, M_r 282.5, m. p. 44 °C *(trans),* 14.5 °C *(cis).* The *trans* form occurs in the acylglycerols of animal fats, e. g. beef, mutton and butter fat, and in vegetable oils. The *cis* form is hemolytic, occurring in plasma and various animal tissues, and in *Lactobacillus.* V. a. is the principal unsaturated acid in *E. coli.*

Vacuoles: structures within plant cells composed of a three-layered membrane (tonoplast) enclosing the cell sap. In early stages of development, several small V. may be present, and they occupy a relatively small proportion of the cell volume. Mature, differentiated cells usually contain one large central V., and the cytoplasm is a relatively thin layer pressed firmly between the plant cell wall and the V. The cell sap within the V. is an aqueous solution of numerous substances in true or colloidal solution; in addition to sugars and salts these include inner secretions, so that the V. is considered as an excretory organ. Alkaloids accumulate in V. and are neutralized by salt formation with inorganic and organic anions. V. of acid and ammonium plants contain accumulated ammonium salts of organic acids. The V. is important in the maintenance of turgor (inner pressure) of the plant cell, by acting as an osmotic system (see Osmosis). The tonoplast is semipermeable; due to the high osmotic pressure of the cell sap, the V. expands and compresses the cytoplasm as a thin layer tightly against the inner surface of the cell wall; further expansion is prevented physically by the cell wall. The resulting turgor is important for the mechanical strength of herbaceous plants.

Val: abb. for L-valine.

Valepotriates: iridoids from *Valeriana* and *Kentranthus* spp. *Valeriana officinalis* contains up to 5 % V., which are responsible for the sedative properties of this drug. Hydroxyl groups on the iridoid structure are esterified with isovaleric acid. Hydrolysis of the esters with HCl causes decomposition of the unstable alcohol moiety with production of a blue color. The most important representative is Valtratum (see).

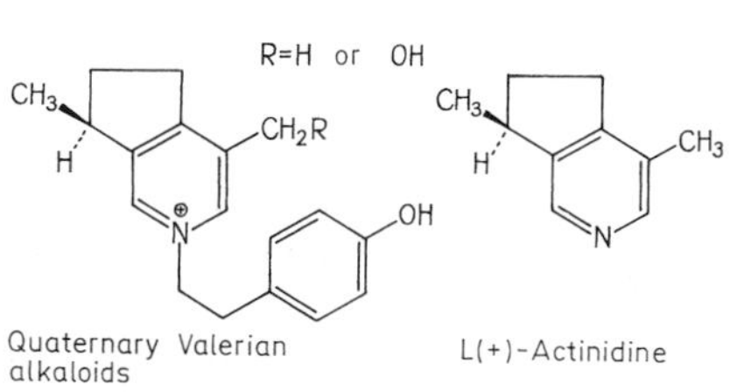

R=H or OH

Quaternary Valerian alkaloids   L(+)-Actinidine

Valepotriates

Valerian alkaloids: terpene alkaloids containing a pyridine ring (therefore also considered as pyridine alkaloids) from valerian (*Valeriana officinalis*). The quaternary V. a. (Fig.) are responsible for the excitory action of valerian on cats.

L-Valine, ***Val:*** L-α-aminoisovaleric acid, $(CH_3)_2CH$–$CH(NH_2)$–COOH, an aliphatic, neutral, essential, glucogenic, proteogenic amino acid. For biosynthesis, see L-Isoleucine. For degradation, see Leucine. The intact molecule of Val is incorporated in the biosynthesis of Penicillin.

Valinomycin: cyclo-(-D-Val-Lac-Val-D-Hyv-$)_3$, an antibiotic, cyclic depsipeptide, especially active against *Mycobacterium tuberculosis.* In addition to valine, it contains the heterocomponents, L-lactic acid (Lac) and D-α-hydroxyisovaleric acid (D-Hyv). It is an Ionophore (see), which selectively transports potassium ions across membranes.

Valtratum, ***valepotnatum, valtrate, valepotnate:*** the main member of the Valepotriates (see) from valerian root, M_r 422; an oil unstable to acid, alkali and heat, n_D^{20} 1.4906, $[\alpha]_D^{20}$ + 172.2 ° (methanol). Acid hydrolysis produces isovaleric acid and 4-acetoxymethyl-7-formylcyclopenta[c]pyran (baldrianal), the latter being formed by dehydration and rearrangement of the unstable constituent iridoid alcohol (Fig.). Baldrinal has m. p. 112–113 °C, M_r 218. Structure of V. was revised in 1968 [P. W. Thies *Tetrahedron* **24** (1968) 313–347]. Valtratum is accepted as the nonproprietary name by the World Health Organization.

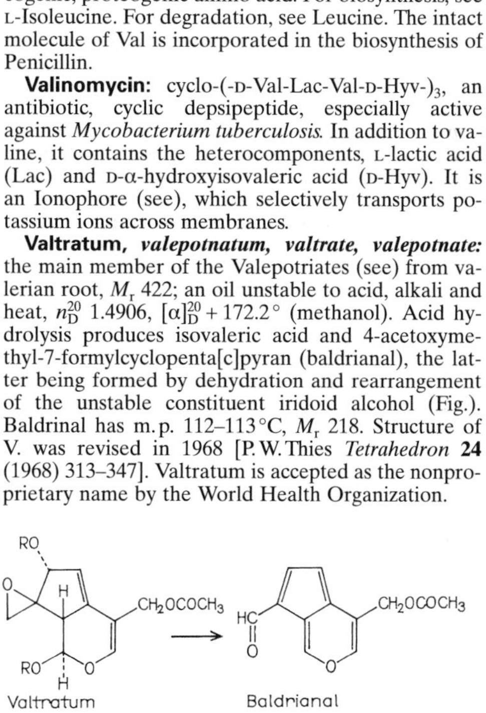

Valtratum   Baldrianal

R = —CO—CH$_2$—CH(CH$_3$)$_2$

Formation of baldrinal from valtratum (by dehydration and rearrangement).

Vanadium, ***V:*** a trace element required for normal growth by animals; most studies have been performed on rats and chicks. A reliable estimate of human V requirement is lacking, most dietary items contain less than 100 ng V/g. V is taken up as V^{5+} and reduced to V^{3+} in the cell. V-deficiency causes an increase in plasma cholesterol and triacylglycerols. V stimulates the oxidation of phospholipids and decreases cholesterol synthesis by inhibiting squalene synthase (a liver microsomal enzyme system); it also stimulates acetoacetyl-CoA deacylase in liver mitochondria. It has also been implicated in bone metabolism or formation. V-deficiency leads to abnormal bone growth, and injected radioactive V shows a high incorporation into areas of active mineralization in dentine and bone. V is required for optimal growth of some green algae, and it inhibits growth of *Mycobacterium tuberculosis.* Nitrogen fixation (see) by *Azotobacter* is increased by V. Certain ascidians show a remarkable ability to concentrate V from the surrounding sea water (see Heavy metals).

Variabilin: 3,9-dimethoxy-6a-hydroxypterocarpan. See Pterocarpans.

Vasoactive intestinal peptide, ***VIP:*** an octacosapeptide from porcine small intestine, which causes vasodilation, lowers arterial blood pressure, increases cardiac output, enhances myocardial activity, increases glycogenolysis and relaxes the smooth muscle of trachea, stomach and gall bladder. For structure, see Secretin.

Vasopressin, ***antidiuretic hormone, antidiuretin, pitressin:*** a neurohypophysial peptide hormone. V. has a direct antidiuretic action on the kidneys. It also causes vasoconstriction of peripheral vessels, with slowing of the heart beat and increase of blood pressure. The amino acids in positions 3, 4 and 8 of the nonapeptide are variable. The phylogenetic precursor of V. and of Oxytocin (see) is [8-arginine]vasotocin. V. occurs as [8-arginine]V., M_r 1,084 and [8-lysine]V., M_r 1,056. [8-Arginine]V. is physiologically and pharmacologically one of the most active substances; 2 ng are sufficient to cause pronounced antidiuresis in the human. V. is synthesized in the supraoptic nucleus of the hypothalamus then transported in granules, in combination with neurophysin II, down nerve fibers (supraoptico-hypophysis) to the posterior pituitary (neurohypophysis), where it is stored. Release of V. depends on the degree of hydration of the organism, and is stimulated by thirst and lack of water. It activates the adenyl cyclase system in the distal tubule of the kidney, resulting in increased resorption of water and increased excretion of Na^+. Determination of V. is performed biologically by measurement of blood pressure and/or diuresis, following injection into test animals. A radioimmunological assay is also used [E. G. Beardwell, *J. Clin. Endocrin. Metab.* **33** (1971) 254–260]. [8-Arginine]V. also infuences learning and memory, brain development, cardiovascular control, thermoregulation, development of tolerance to, and dependence on, opiates and ethanol, and drug-seeking behavior. Proteolysis of [8-arginine]V. to the heptapeptide pGlu-Asn-Cys-Cys-Pro-Arg-Gly-NH_2 or to the corresponding hexapeptide lacking the terminal Gly-NH_2 removes the pressor activity of V., but the derivative peptides are even more active than the parent in facilitating memory consolidation in rats [J. P. H. Burbach et al. *Science* **221** (1983) 1310–1312].

Veracevine: see Germine.

Veratramine: a Veratrum alkaloid (see) of the jerveratrum type, with C-nor-D-homo structure, M_r (anhydrous) 409, m. p. (monohydrate) 209.5–210.5 °C $[\alpha]_D^{19}$ (anhydrous) −70° (methanol), found in hellebores (*Veratrum album, V. eschscholtzii, V. viride*). *V. viride* also contains ***veratrosine,*** a glycoalkaoid in which the 3-β-hydroxyl group of veratramine is linked glycosidically to D-glucose.

Veratrosine: see Veratramine.

Veratrum alkaloids: a group of steroid alkaloids found in the *Solanaceae,* and in the genera *Veratrum* (hellebores) and *Fritillaria* (fritillary). V. a. are structural derivatives of the parent hydrocarbon cholestane (see Steroids); in some members ring C is contracted and ring D is expanded (C-nor-D-homo type). V. a. are subdivided into *jerveratrum alkaloids,* e. g. Jervine (see), Rubijervine (see), Isorubijervine (see), Veratramine (see), and *ceveratrum alkaloids,* e. g. Germine (see). Members of the first group contain 2–3 oxygen atoms and occur free or linked glycosidically to a molecule of D-glucose (glycoalkaloids). Ceveratrum alkaloids contain 7–8 oxygen atoms, and occur chiefly as esters; the commonest ester acids are acetic, angelic and veratric. V. a. have a positive ionotropic action on the heart, and cause a decrease of blood pressure by reflex inhibition of the vasomotor centers. They were used for treatment of hypertension, but have been replaced by Rauwolfia alkaloids and other drugs. V. a. have also been used as insecticides. They are biosynthesized from cholesterol.

Verbenaline: see Iridoids.

Vernaline: see Flowering hormone.

Vernine: obsolete name for guanosine.

Versene: see Ethylenediaminetetraacetic acid.

Vertebrate hormones: hormones of vertebrate animals. On the basis of studies on the phylogenetic relationships of certain proteins, proteohormones and peptide hormones, the separate classification of V.h and Invertebrate hormones (see) appears to be justified. Chemically, V. h. are a heterogeneous group, which can be subdivided into Steroid hormones (see), hormones derived from amino acids (see Thyroxin, Adrenalin, Melatonin), Peptide hormones (see), Proteohormones (see), and hormones derived from fatty acids (see Prostaglandins). There is, however, no fundamental difference between V. h. and invertebrate hormones, with respect to types of chemical structure, or biochemical mode of action.

Vestitol: 7,2′-dihydroxy-4′-methoxyisoflavan. See Isoflavan.

VHDL: see Lipoproteins.

Vimentin: a fibrous protein which makes up intermediate filaments in cells of mesenchymal origin. Subunit M_r 53,000. The structure is similar to that of the α-keratins. See Keratins, Cytoskeleton (intermediate filaments).

Vinblastine, ***vincaleucoblastine:*** a dimeric indole-indoline alkaloid, m. p. 211–216 °C (d.), $[\alpha]_D + 42°$ ($CHCl_3$). Structurally, V. is equivalent to a combination of the alkaloids vindoline and catharidine (see Vinca alkaloids). Very low concentrations of V., accompanied by vindoline and catharidine are present in *Vinca rosea.* V. is one of the most effective naturally occurring antitumor agents, and is used primarily in the treatment of Hodgkin's disease.

Vinca alkaloids, ***Catharanthus alkaloids:*** a group of about 60 iridoid indole alkaloids from *Vinca (Catharanthus)* spp. Structurally, they are tetra- or pentacyclic indole derivatives with an iridoid component, e. g. vindoline, $[\alpha]_D$ −18° ($CHCl_3$), m. p. 174–176 °C, and vincamine, $[\alpha]_D + 41°$ (pyridine), m. p. 232–233 °C. These are accompanied in the leaves by small quantities (about 0.005 %) of two dimeric V. a., i. e. Vinblastine (see) and Vincristine (see) (Fig.). Tryptophan and mevalonic acid are biosynthetic precursors of V.a. Vincamine has hypertensive activity and is used pharmaceutically, particularly in Hungary. The dimeric V.a. show good oncolytic properties and are used in the treatment of carcinomas.

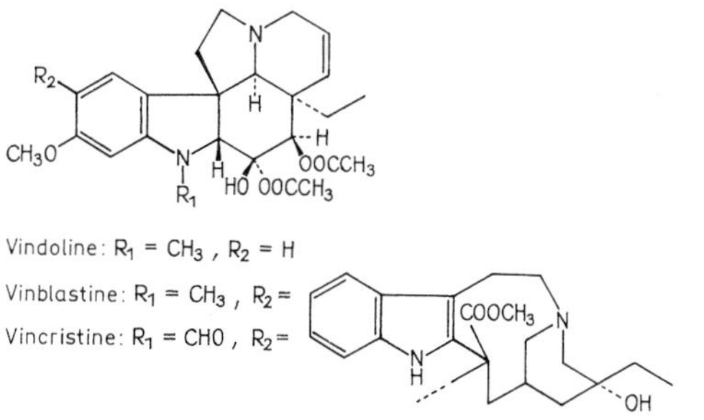

Vindoline: $R_1 = CH_3$, $R_2 = H$

Vinblastine: $R_1 = CH_3$, $R_2 =$

Vincristine: $R_1 = CHO$, $R_2 =$

Vinca alkaloids

Vincaleucoblastin: see Vinblastin.

Vincamine: see Vinca alkaloids.

Vincristine, *leurocristine:* a dimeric indole alkaloid closely related to vinblastine, from *Vinca rosea* (see Vinca alkaloids). It is used mainly for the treatment of acute leukemia in children, and against various other neoplasmic growths.

Vindoline: see Vinca alkaloids.

Violacein: the major purple pigment of *Chromobacterium violaceum,* accompanied by smaller amounts of deoxyviolacein. Every C-atom of V. is derived biosynthetically from tryptophan. DL-tryptophan labeled with ^{14}C in the ring or side chain is incorporated with very little dilution into V. by nonproliferating cells of *C. violaceum.* [R.D.DeMoss & N.R.Evans *J. Bact.* **79** (1960) 729]

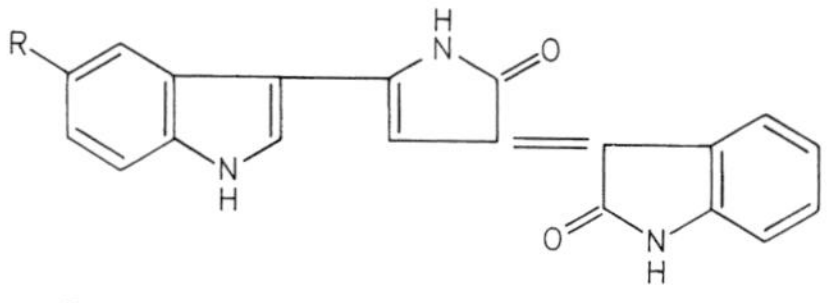

R = OH, Violacein

R = H, Deoxyviolacein

Violacein (R = OH) *and Deoxyviolacein* (R = H).

Violanin: see Delphinidin.

Violaxanthin, *3(S),3′(S)-dihydroxy-β-carotene-(5R,6S,5′R,6′R)-5,6,5′,6′-diepoxide:* a xanthophyll, M_r 600.85, m.p. 208 °C, $[\alpha]_{Cd}^{20} + 35°$ ($c = 0.08$, $CHCl_3$). V. is present as an orange or brown-yellow pigment in all green leaves, and is particularly abundant in flowers and fruits of *Viola tricolor, Taraxacum, Tagetes, Tulipa, Citrus, Cytisus,* etc. It is metabolically important as a precursor of Abscisic acid. For formula, see Abscisic acid.

Viomycin, *celiomycin, florimycin, tuberactinomycin B:* a polypeptide antibiotic [M_r 685.71, m.p. 280 °C (d.) (sulfate), $[\alpha]_D^{25} - 32°$ ($c = 1$, water).] from various *Streptomyces* spp., including *S. floridae, S. puniceus* and *S. vinaceus.* V. contains a 7-deazaadenine ring. Degradation products of V. include the guanidino compound, viomycidin. V. inhibits both nucleic acid and protein biosynthesis. It is particularly active against Gram-negative bacteria, and is used therapeutically against *Mycobacterium tuberculosis.* Structure: Noda et al. *J. antibiot.* **25** (1972) 427.

Viridicatine: a quinoline alkaloid from molds of the genus *Penicillium.* V. is biosynthesized from anthranilic acid, phenylalanine and the methyl group of methionine (Fig.). Biosynthesis is catalysed by the enzyme cyclopenase.

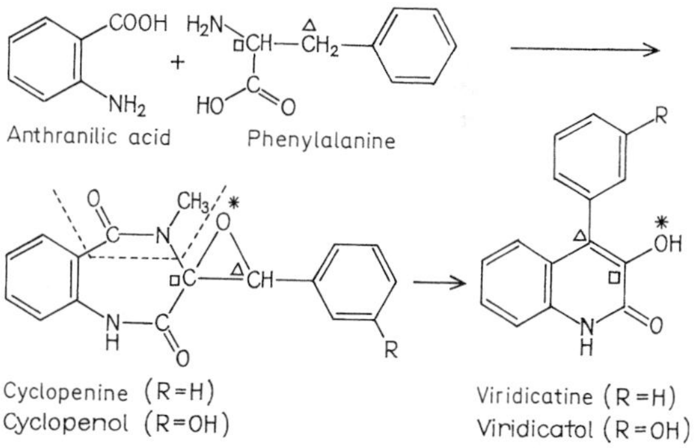

Anthranilic acid Phenylalanine

Cyclopenine (R=H)
Cyclopenol (R=OH)

Viridicatine (R=H)
Viridicatol (R=OH)

Biosynthesis of the quinoline alkaloids, viridicatine and viridicatol.

Viridicatol: see Viridicatine.

Viridogrisein: see Etamycin.

Viroids: virus-like group of plant disease agents, consisting of small, naked RNA circles, only 150–400 nucleotides in length, with no protein coat. Extensive hydrogen bonding between complementary bases causes the molecule to compress and form a rod with laterally projecting hydrogen-bonded loops. They replicate in suitable live host cells, and no protein coat is formed at any stage of replication. First of these infective agents to be called a viroid was the potato spindle tuber viroid (PSTV), which causes potatoes to become spindle-shaped, but also attacks 128 other plant species in 11 families. Other V. are listed in the Table. For their replication, V. are almost totally dependent on the replication systems of the host cell. It has been shown that the DNA of noninfected PSTV

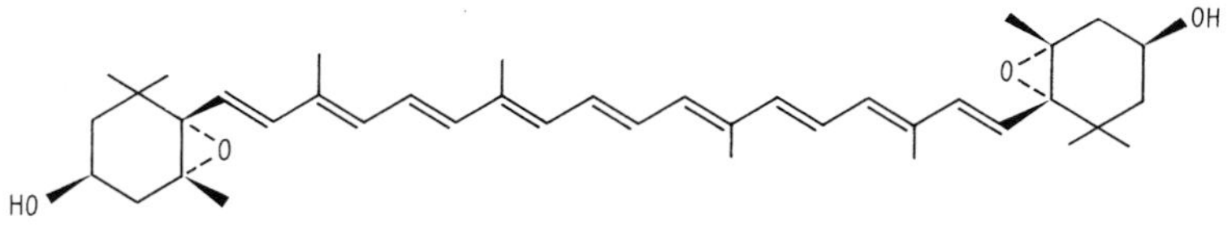

Violaxanthin

Viroids (after Gross, 1980)

Viroid	Acronym	Disease
Potato spindle tuber viroid	PSTV	Spindle tuber diseases of potatoes
Citrus exocortis viroid	CEV	Exocortis disese of citrus plants
Cucumber pale fruit viroid	CPFV	Pale fruit disease of cucumber
Chrysanthemum stunt viroid	CSV	Stunting of chrysanthemums
Chrysanthemum chlorotic mottle viroid	CCMV	Chlorotic leaf mottle of chrysanthemums
Coconut cadang-cadang viroid	CCCV	Cadang-cadang disease of coconut palms
Hop stunt viroid	HSV	Stunting of hops
Columnea erythrophae viroid	CV	
Avacado sunblotch viroid	ASV	

hosts contains nucleotide sequences complementary to those of the viroid RNA, a situation analogous to that in RNA tumor viruses. The M_r of PSTV is only about 120,000, and it is therefore much smaller than any known viral RNA. Also, it is much more infectious than any plant virus, 10 molecules being sufficient to infect a potato plant.

Virus coat proteins, ***capsids:*** proteins with the largest known M_r values (up to 40×10^6), lying on the exterior of the virus particle, enclosing the DNA or RNA. They comprise many, usually identical, subunits, called *capsomeres* (M_r 13,000–60,000). V.c.p. of tobacco mosaic virus consists of 2130 capsomers of M_r 17,500. M_r values of capsomeres from several strains of tobacco mosaic virus lie in the range 17,400–17,600 (157–158 amino acids); capsomeres of turnip yellow mosaic virus have M_r 20,000 (188 amino acids). M_r of bacteriophage capsids lie in the range M_r 5,168 (49 amino acids) to M_r 14,034 (131 amino acids).

Viruses: very small, infective particles, which are not retained by ultrafine filters that hold back the smallest bacteria. They consist of RNA or DNA enclosed in a protein coat, and they can replicate (multiply) only within a suitable live host cell. They suppress the genetic information of the host cell, and exploit the ribosomes, energy-producing mechanisms and various enzymes of the host in support of their own replication. There is a wide variety of different virus types, differing in the type of nucleic acid (RNA or DNA, single-stranded or double-stranded, linear or circular), structure of the protein coat, mode of infection, and mechanism of replication. Virus infections often cause functional disturbances of the host organism, known as *virus diseases* or *viroses.*

Occurrence and distribution. V. are found in practically all groups of organisms. The numerous V. of the Schizophyta (bacteria and cyanobacteria) are called Phages (see Phage, Phage development). Many V. are known which attack fungi (Mycoviruses), and a smaller number are known to infect green algae. No virus disease of mosses or liverworts has been identified with certainty, but several pteridophyte V. are known. Spermatophytes are attacked by hundreds of different V. Virus-like infective agents have also been identified as disease organisms of protozoa. Among invertebrates, V. have been isolated in particular from nematodes and insects, some insect V. being responsible for severe epidemics (see Insect viruses), while others hardly affect their hosts. In view of the growing list of newly discovered invertebrate V., it can be assumed that V. are abundant and widely distributed among most groups of invertebrates. Virus diseases of fishes and reptiles are also well known. Of particular importance, however, are the V. of homeothermic animals, especially those of humans, which cause hundreds of different, sometimes serious, diseases.

Structure. Virus particles display a wide range of different forms, but the particle size and shape of each species are usually constant and characteristic (Fig. 1). The diameter of isometric species ranges between 15 and 300 nm, while rod-shaped and helical particles vary in length between 180 and 2,000 nm.

Most virus particles contain a single nucleic acid molecule, but in some species (e.g. wound tumor virus) the genome is distributed among several separate nucleic acid molecules. Such V. are known as *divided genome V.* In *Multipartite viruses* (see), the genome is divided among different particles. In most plant V. (e.g. tobacco mosaic virus), many animal V. and some bacterial V., the genome consists of single-stranded RNA. In many RNA-V., however, the genome is linear, helical and double-stranded, e.g. rice dwarf virus, the reoviruses and some insect V. Other animal and bacterial V., as well as a few plant V., contain circular single-stranded DNA (e.g. bacteriophage ΦX 174), while some very small V. (e.g. parvoviruses) contain single-stranded DNA. Double-stranded linear DNA is found, e.g. in bacteriophage T4 and herpesvirus, while double-stranded circular DNA forms the genomes of, e.g. SV40 and polyoma virus. Adenoviruses contain double-stranded linear DNA with covalently linked terminal protein, while the linear double-stranded DNA of poxviruses is covalently sealed at both ends.

In most V., the nucleic acid is protected (e.g. from the action of nucleases) by a protein coat. The only exception to this structural feature is provided by the Viroids (see), which lack any proteins of their own. The *coat protein* consists of many identical subunits, e.g. the coat protein of Tobacco mosaic virus (see) contains 158 amino acid residues of known primary sequence. In the mature virus, between 2,100 and 2,700 of these subunits are arranged like the steps of a spiral staircase, and the spirally wound nucleic acid lies in a groove in each subunit. A cavity remains in the interior of the particle, which therefore has the appearance of a tube, is rod-shaped and displays helical symmetry (Figs. 1 & 2). In other V., 2 or 4 subunits combine to form *capsomers,* which in turn associate

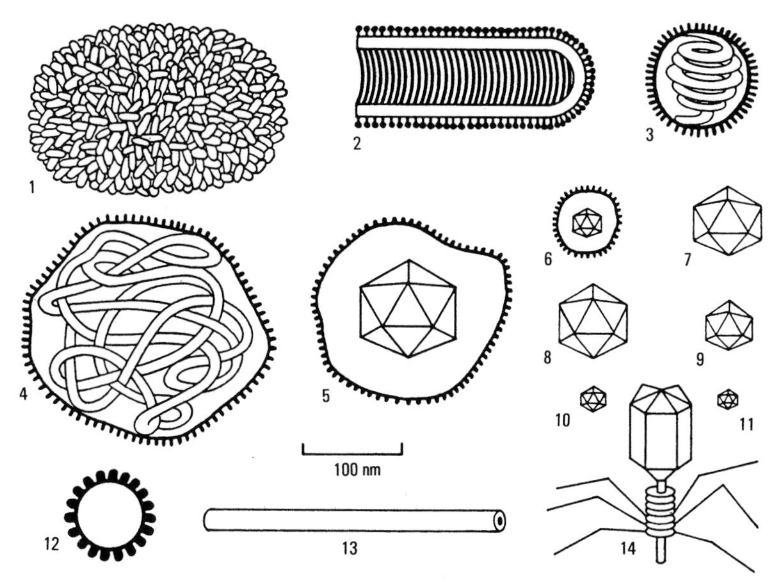

Fig. 1. *Viruses.* Schematic representation of virus particles, all drawn to scale. The type of genome (RNA or DNA) is shown in brackets. *Enveloped viruses:* 1 Pox virus (DNA). 2 Rabies virus (RNA). 3 Influenza virus (RNA). 4 Measles virus (RNA). 5 Chickenpox virus (DNA). *Naked or unenveloped viruses:* 6 Yellow fever virus (RNA). 7 Adenovirus (DNA). 8 Reovirus (RNA). 9 Wart-papilloma virus (DNA). 10 Poliomyelitis virus (RNA). 11 Parvovirus (RNA). 12 Corona virus (RNA). 13 Tobacco mosaic virus (RNA). 14 Bacteriophage T2 (DNA).

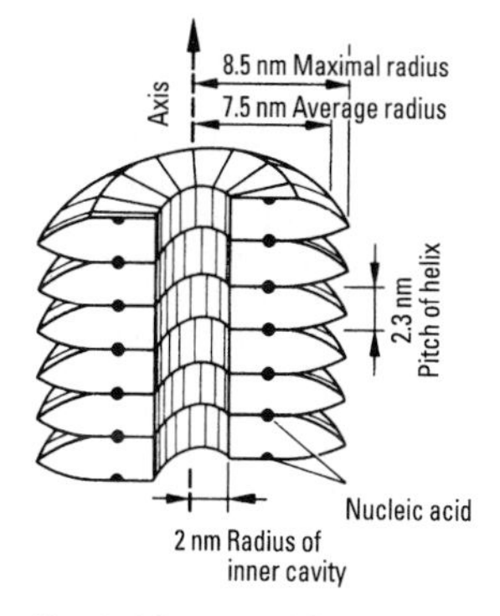

Fig. 2. *Viruses.* Model of the Tobacco mosaic virus.

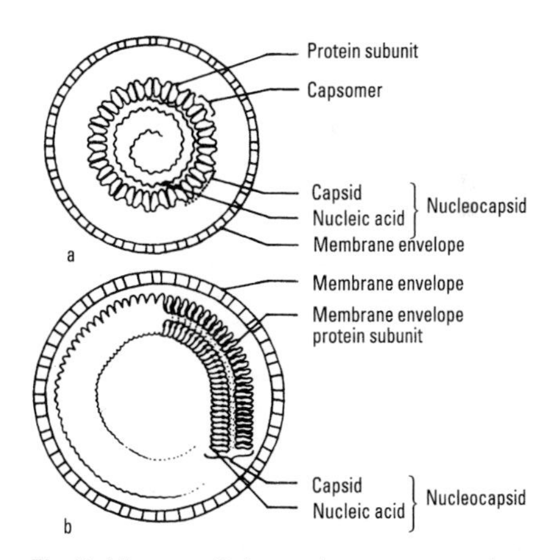

Fig. 3. *Viruses.* Schematic representation of enveloped isodiametric (a) and helical (b) viruses.

with one another to form a hollow body with cubic symmetry, known as a *capsid.* The capsid encloses and protects the nucleic acid, which is spirally wound and/or folded in various ways, in order to fit into the capsid interior. A capsid with its enclosed nucleic acid is called a *nucleocapsid.* Most V. with this type of structure are polyhedra, e.g. icosahedra, and are therefore referred to as isodiametric or isometric, and sometimes incorrectly as spherical (Fig. 3). In some cases, e.g. the adenoviruses, the capsomers forming the corners of the polyhedron possess a fibrillar appendage (a glycoprotein) which aids attachment of the virus to the host cell.

A number of V. have a secondary envelope (the *membrane envelope)* surrounding the nucleocapsid. In addition to protein, this secondary envelope usually contains carbohydrates and lipids, and many of its components are derived from the host cell membrane. The nucleocapsid of helical V. is frequently spirally wound within the membrane envelope (Fig. 3). V. with a membrane envelope are called *enveloped* V., whereas those with only a nucleocapsid are referred to as *naked* or *unenveloped.* The complete virus particle of a species is also called a *virion* (plural *viria* or *virions).* For bacteriophage structure, see Phage.

Infection. For the process of bacteriophage infection, see Phage development.

Animal V. The first stage of infection is adsorption of the virus to the exterior surface of the animal cell membrane. Adsorption occurs by interaction between a virus-coded protein on the surface of the virion and a receptor molecule on the cell membrane. Most cell receptors are glycoproteins. Moreover, they are normal membrane glycoproteins with specific functions unrelated to virus infection. The interaction is specific. Thus, the binding protein of influenza virus interacts with the $\alpha 2 \rightarrow 3$ linked terminal sialic acid residue of the host cell membrane glycoprotein; treatment of a cell with sialidase (neuraminidase) renders it resistant to infection, and glycoproteins with $\alpha 1 \rightarrow 6$ linked sialic acid do not serve as receptors. Under natural conditions, the presence of appropriate receptor molecules on the cell membrane is a precondition for virus infection, i.e. in the absence of receptors, the cell is not *permissive* for virus infection

Animal V. pass through the cell membrane, either by fusion (e.g. enveloped paramyxoviruses) or by endocytosis (e.g. Semliki Forest virus). *Fusion:* the virus binds to the cell receptor, the viral envelope and the cell membrane become perforated, the two membranes seal together, and the capsid enters the cell. *Endocytosis:* following adsorption of the virus, a depression forms on the cell surface in the receptor region (adsorption site). Continued invagination forms a coated pit (lined on the cytoplasmic side with clathrin), eventually enclosing the virus in a vesicle, which is finally released into the interior of the host cell by abstriction, still surrounded by the vesicle membrane; the latter loses its clathrin coat and fuses with other cellular vesicles to form vesicles called endosomes; viral and endosomal membranes then fuse and the nucleocapsid is released into the host cytoplasm.

Plant V. enter cells through wounds, or they are injected by biting insects, e.g. Homoptera and Heteroptera. Passage from cell to cell within the plant is then possible via plasmodesmata.

Within the host cell, the virus nucleic acid is released from both the nucleocapsid and the membrane envelope (if present). This process, known as *uncoating,* involves, inter alia, the action of specific proteases. The duration of uncoating increases with the complexity of the virus and with the tightness of the packing of the nucleic acid. Uncoating of small V., e.g. picornaviruses is complete in about 2 h., whereas the uncoating of poxviruses lasts 10–12 h.

Replication. For replication of bacteriophages, see Phage development.

Part of the uncoated nucleic acid is transcribed and translated (DNA V.) or simply translated (RNA V.), with the aid of host enzymes, to form *early proteins,* which are generally enzymes or regulatory proteins. The early proteins of many RNA V. include RNA-dependent RNA-polymerase (replicase or transcriptase), which is necessary for replication of the genome of RNA V., and which is absent from most hosts. In DNA V., other early proteins promote transcription of viral DNA rather than host DNA to form early viral mRNA, while in RNA V. early proteins determine the balance of messenger and template function. Early proteins also slow down or stop the biosynthesis of host nucleic acids and proteins. The appearance of early proteins is followed by replication of the viral genome.

Double-stranded viral DNA genomes are replicated and transcribed by mechanisms already well documented for prokaryotic and eukaryotic DNA, using the enzymatic apparatus of the host cell. Single stranded DNA genomes consist of the noncodogenic DNA strand (– strand); alternatively, the codogenic strand (+ strand) and – strand may be packaged in separate virions. Noncodogenic strands quickly become double-stranded by synthesis of a complementary strand, using host enzymes. In fact, the entire process of transcription and replication of a single-stranded DNA viral genome depends on the enzymatic apparatus of the host, the only proteins encoded by the viral genome being capsid proteins.

RNA genomes depend on different mechanisms for their replication. The single-stranded RNA of certain animal V. functions directly as mRNA, and is therefore called a *positive* or *plus* (+) *strand* genome; such (+)RNA always specifies enzymes for the replication of RNA via its complementary (–) strand. Other single-stranded RNA V. contain the nontranslatable strand of RNA, known as a *negative* or *minus* (–) *strand* genome. In this case, a plus strand of mRNA must be made from the minus strand, using virus-coded transcriptase, which is packaged in the virion, and released together with the (–)RNA during infection. Double-stranded RNA V. also carry a packaged transcriptase for the synthesis of mRNA from this type of template. Replication occurs in the host nucleus (primarily the nucleolus), in the cytoplasm, or in other organelles.

As viral genome replication slows down, *late proteins* are synthesized in increasing quantities; these include certain soluble proteins (eventually packaged with the nucleic acid in the mature phage), maturation proteins and structural proteins (e.g. capsid and envelope proteins). The phase of late protein synthesis is followed by the maturation phase, in which new virus particles are assembled from the newly synthesized nucleic acid and proteins.

Finally, in infections of bacterial and animal cells, virus progeny are released. Especially in animal enveloped V., this can occur continually by cell budding or secretion of the virus, in which the host cell remains intact, and the virus acquires its membrane envelope from the host plasma membrane. Bacteriophages are released by lysis, and therefore death, of the host cell. Newly formed plant V. often remain for months or years in their host cells, until they are transferred to a new host by wound contact or by insects.

The replication cycle varies considerably, depending on the virus, host and type of viral nucleic acid (see Phage, Tobacco mosaic virus). Occasionally, the replication cycle of certain V. cannot be completed. Thus, if the genes for early phage proteins are repressed, the viral nucleic acid becomes incorporated into the host bacterial genome (see Temperate phages). Alternatively, certain enzymes required for replication or for maturation may be absent, due to mutation (see Defective viruses, Deficient viruses). Under certain circumstances, the genome of a deficient virus may be integrated into the host genome, where it promotes tumor formation (see Oncogenes). [*Classification and Nomenclature of Viruses* – Fifth report of the International Committee on Taxonomy of Viruses, edit. R.I.B.Francki et al., Archives of Virology Supplementum 2. Springer-Verlag Wien, New York 1991; D.H.Bamford & R.B.Wickner *Semin. Virol.* **5** (1994) 61–69; S.Schlesinger et al. *Semin. Virol.* **5** (1994) 39–50]

α-Viscol: see Amyrin.

Viscotoxins: see Toxic proteins.

Visual purple: see Visual process.

Visual process: the process by which light induces a nerve impulse in a photoreceptive cell. Light is absorbed by a visual pigment, a chromoprotein consisting of an apoprotein (opsin) and the chromophore, 11-*cis*-retinal (neoretinal b). The aldehyde group of 11-*cis*-retinal forms a Schiff's base (see) with the ε-amino group of a specific lysine residue of the opsin. In vertebrate retinas, there are two classes of photoreceptors, rods and cones. The rods, which are responsible for black-and-white vision at low light intensities, have an outer segment in which there is a stack of flattened membrane discs. These and the plasma membrane contain rhodopsin (visual purple). Color vision in the cones is mediated by closely related opsins; three in the case of normal human eyes. These have absorption maxima at 430 nm, 540 nm or 575 nm, respectively. All three pigments (iodopsins) contain 11-*cis*-retinal, but their opsins (cone-type opsins) are different. Other vertebrates have different numbers of opsins with somewhat different sensitivities. Each rod or cone contains only one type of opsin. The amino acid sequences of the human color pigments have been deduced by molecular genetic methods. [J.Nathans et al. *Science* **232** (1986) 193–202; ibid. 203–210]

In the primary event of visual excitation, light isomerizes the 11-*cis*-retinal of the opsin to all-*trans*-retinal. Since all-*trans*-retinal does not fit the binding site for 11-*cis*-retinal, the opsin molecule becomes unstable and undergoes a series of conformational changes, followed by hydrolysis of the Schiff's base linkage between all-*trans*-retinal and opsin (Fig.). Further events in the process are known with some confidence only for rod cells, in which the rhodopsin is bleached to metarhodopsin I. Metarhodopsin I appears about 10^{-5} sec after illumination of rhodopsin, and metarhodopsin II at 10^{-3} sec. Red rhodopsin is thus bleached, the final mixture of opsin and all-*trans*-retinal being yellow in color (formerly called "visual yellow"). Bleached rhodopsin diffuses freely within the membrane, where it interacts with a "G-protein" (transducin). Transducin consists of three subunits, one of which (α) carries a GDP molecule. (The subunits of transducin have the following M_r: α, 39,000; β, 36,000; γ about 10,000 [B.K.-K. Fung et al. *Proc. Natl. Acad. Sci. USA.* **78** (1981) 152–156].) Activated rhodopsin binds to the transducin, causing the α-subunit to exchange the GDP for a GTP. The entire rhodopsin-transducin-GTP complex now dissociates. The rhodopsin is still catalytically active and interacts with more molecules of transducin (about 100 in 0.5 sec). The α-transducin subunit and the GTP remain associated and bind to a molecule of a cGMP-phosphodiesterase (PDE), forming the "G(α)-GTP-PDE complex". The resulting hydrolysis of cyclic GMP is observed within milliseconds of the onset of illumination.

Membrane channels in the rod cells allow a constant inflow of Na^+ ions. Light absorption by rhodopsin interrupts this flow, thus inducing a voltage impulse in the cell membrane. Both Ca^{2+} and cGMP are involved in this process. There are two hypotheses regarding the mechanism: 1. Ca^{2+} stored within the disks is released by light to block the Na^+ channels; and 2. cGMP keeps the channels open. The light-induced hydrolysis of cGMP causes the channels to close. It is known experimentally that cGMP decreases the ability of Ca^{2+} or Co^{2+} to block the Na^+ channels, and increases their ability to permeate the membrane. It is also known that several outer rod proteins are phosphorylated in the dark by a cyclic-nucleotide-dependent kinase, and that this phosphorylation is reversed in the light. It has also been found that 1 photon causes 100–300 Na^+ channels to close.

According to a combined hypothesis by E.A. Schwartz: 1. The concentration of cGMP in the cell is controlled by feedback mechanisms. 2. The Na^+ channels, which also admit a small amount of Ca^{2+}, have binding sites for Ca^{2+} on the cytoplasmic side of the membrane; when bound to these sites, Ca^{2+} causes

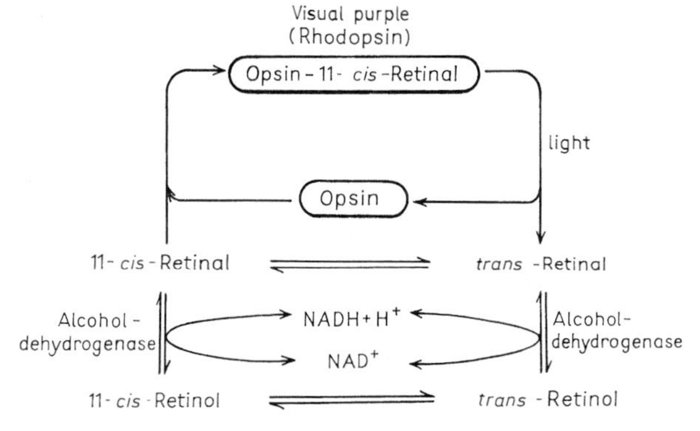

Synthesis of visual purple during the visual process.

the channels to close. 3. The affinity of the Ca^{2+} sites for Ca^{2+} is modulated by cGMP; a drop in the cGMP concentration causes the affinity of the sites for Ca^{2+} to rise. Thus hydrolysis of cGMP would indirectly cause the Na^+ channels to close. 4. The increase in extracellular Ca^{2+} observed during photoexcitation would be explained if the Ca^{2+} influx is prevented by closure of Na^+ channels, while Ca^{2+} efflux through other channels is not decreased.

In the recovery of the visual apparatus, the G(α)-GTP-PDE complex is inactivated by spontaneous cleavage of the GTP. Activated rhodopsin is the substrate for a kinase which may deactivate it. The kinase is inhibited by cGMP, so that the amount of activated rhodopsin may be subject to feedback regulation. The *trans*-retinal may be isomerized directly to 11-*cis*-retinal, which recombines with opsin to form rhodopsin; or it may first be reduced to retinol by NADH-dependent alcohol dehydrogenase, followed by isomerization of the retinol and reoxidation to 11-*cis*-retinal (Fig.). Some of the retinal is continually lost from the cells of the retina, so that the continuation of the visual cycle depends on continual replacement from the blood.

Land and marine animals possess rhodopsin. Certain freshwater fish and some amphibians have the pigment porphyropsin in place of rhodopsin. The function of porphyropsin is analogous to that of rhodopsin, and it contains 3-dehydroretinal in place of retinal. Cyanopsin (dehydroretinal + cone-type opsin) is found in the retinas of most freshwater fish.

Vitamins (Latin *vita* + amine): substances present in the animal diet in only small amounts, and indispensable for growth and maintenance of the organism. A dietary requirement is implicit in the definition of a V. Most of the substances that are V. in animals are essential for the metabolism of all living organisms, but plants and microorganisms can synthesize them (some fat-soluble V., however, may have metabolic roles unique to animals). The dietary requirement in the animal results from the evolutionary loss of this biosynthetic ability. Animals differ in their ability to synthesize certain V., and they therefore display different dietary requirements for V. For example, ascorbic acid (V.C) is a V. only for primates and a few other animals (e.g. guinea pig); most animals can synthesize it, and for them it is therefore not a V. Some V. can be synthesized from provitamins in the diet. In addition some of the V. requirement of humans and higher animals is supplied by the intestinal flora, e.g. most of the V.K required by humans is supplied in this way.

The role of V. is largely catalytic, most V. serving as coenzymes and prosthetic groups of enzymes. V.D, however, acts as a regulator of bone metabolism, and therefore resembles a hormone. As a component of the visual pigments, V.A acts as a prosthetic group, but it is not known whether it is associated with catalytic proteins in its other functions. Nicotinamide and riboflavin are constituents of hydrogen-transferring coenzymes. Biotin, folic acid, pantothenic acid, pyridoxine, cobolamin and thiamin are (or are precursors of) coenzymes of group transfer reactions. The low daily requirement for V. reflects their catalytic and/or regulatory roles. Thus, V. are nutritionally quite different from fat, carbohydrate or protein, which are required in the diet in considerable quantities as substrates of tissue synthesis and energy metabolism.

The biological activity of a pure V. can be expressed in International Units (IU). Thus, 0.3 mg V.A (retinol), 8 mg thiamin hydrochloride, 0.18 mg biotin, 50 mg L-ascorbic acid, 0.025 mg ergocalciferol or 1 mg DL-α-tocopherol acetate each corresponds to 1 IU. The system of IU is retained, even though the structures of all V. are known, because in most cases a V. is a family of closely related compounds all with the same action, but with different activities.

Lack or deficiency of a V., as a result of unbalanced nutrition, leads to characteristic metabolic disturbances. Complete absence of a V. leads to avitaminosis, with typical clinical symptoms. Relative deficiency of a V. causes hypovitaminosis. Such conditions are reversible by administration of the appropriate V. Excessive intake of certain V., e.g. V.A or V.D, can lead to hypervitaminosis. Formerly, V. were named after the diseases they cured, e.g. antiscorbutic V., antirachitic V., antiberiberi factor. Not all V., however, have such a pronounced specificity, and the clinical pictures of many avitaminoses and hypovitaminoses are complex and variable. A nomenclature based on letters of the alphabet was developed simultaneously; the designations A, B, C, D and E were applied in the historical order of discovery. Subscripts were applied as more refined chemical analysis revealed that the originally isolated substances were in fact complex mixtures. This was especially true for the B vitamins. Partly because of confusion over the "B complex", trivial names which give an indication of the chemical structure of the V. (e.g. pyridoxine or pyridoxol for V.B_6) are now preferred.

The V. represent a heterogeneous group of substances, which are classified into 2 main groups: fat-soluble and water-soluble, depending on whether they can be extracted from foodstuffs with organic solvents or water.

Retinol (V.A; axerophthol; xerophthol) (obsolete names: epithelial protection V.; growth V.) is a fat-soluble V. with polyisoprenoid structure. The alcohol, retinol, is also known as V.A_1. 3-Dehydroretinol (V.A_2) has an additional double bond between C3 and C4 in the ring. V.A is essential to the visual process, as well as for growth, skeletal development, normal reproductive function, and the maintenance and differentiation of tissues. It occurs predominantly in animal products, such as milk, butter, egg yolk, cod liver oil and the body fat of many animals. All the carotenes, which are abundant in green plants and fruits, have provitamin A activity. Conversion of carotenes to V.A occurs in the small intestine, but other organs, such as muscle, lungs and serum also have a limited ability to perform the same conversion. β-Carotene is oxidatively cleaved by the intestinal mucosa into 2 molecules of retinal (α- and γ-carotenes yield only 1 molecule of V.A), which are then reduced to all-*trans*-retinol, and esterified with a fatty acid, mainly palmitate. This V.A palmitate is transported in the lymph to the liver for storage. Free retinol is released by hydrolysis and transported from the liver by a retinol-binding plasma protein. Retinol is removed from the plasma by the cells of the retina, where it is oxidized to all-*trans*-retinal (retinaldehyde). This is iso-

Fat-soluble vitamins	First described	Recommended daily intake (mg)	Biochemical action	Clinical activity
Calciferol (V. D)	1922	0.01–0.025	Calcium and phosphate metabolism	antirachitic V.
Phylloquinone (V. K) (menaquinone)	1935	1	Cofactor for γ-carboxylation of Glu residues in coagulation proteins	antihemorrhagic V.
Retinol (V. A)	1913	2.7	Visual process	epithelial protection V., antixerophthalmic V.
Tocopherol (V. E)	1922	5	Antioxidant; otherwise unknown	antisterility V.
Water-soluble vitamins				
Ascorbic acid (V. C)	1925	75	Reducing agent for some oxygenases; cofactor for all 2-oxoacid dioxygenases, in particular that catalysing hydroxylation of proline and lysine residues in collagen	antiscorbutic V.
Biotin (V. H)	1935	0.25	Coenzyme of various carboxylation reactions	the "skin" V.
Cobalamin (V. B_{12})	1948	0.003	Coenzyme of certain methyl migrations and isomerization reactions	antianemic V., extrinsic factor
Folic acid	1941	1–2	One-carbon unit transfer	therapy of certain anemias
Niacin	1937	18	Respiration, hydrogen transfer	pellagra preventative
Pantothenic acid	1933	3–5	Transfer of acyl groups	chick antidermatitis factor; anti-gray-hair factor
Pyridoxine (V. B_6)	1936	2	Many reactions of amino acid metabolism, in particular transamination	weakness, nervous disorders, depression
Riboflavin (V. B_2)	1932	1.7	Respiration, hydrogen transfer	antidermatitis
Thiamin (V. B_1)	1926	1.2	Carbohydrate metabolism, transfer of active aldehyde	antineuritic V.

merized to 11-*cis*-retinal, which is a component of visual purple (rhodopsin).

An early symptom of V.A deficiency in humans is night blindness, caused by deficient regeneration of rhodopsin. Later, the deficiency leads to hyperkeratosis of the epithelia of the eye (xerophthalmia), skin follicles, respiratory tract and digestive tract. In children, V.A deficiency also leads to growth arrest, and in adults to resorption of fetuses, stillbirths and birth defects. The daily requirement for adults is 1.5–2.0 mg.

Thiamin (V.B_1; aneurin; antiberiberi factor; antineuritic V.) is a water-soluble V. of widespread occurrence in natural materials, especially in yeasts and germinating cereal grains. Chemically, it consists of a pyrimidine and a thiazole ring. Both ring systems are synthesized separately as phosphorylated derivatives, which then become linked via a quaternary nitrogen. The pyrophosphorylated form of V.B_1, thiamin pyrophosphate, is a coenzyme of decarboxylases, transketolases and 2-oxoacid dehydrogenases. Deficiency of V.B_1 results in disturbances of carbohydrate metabolism, accompanied by an increase in the concentration of blood oxoacids (mostly pyruvate), which reflects the role of thiamin pyrophosphate as a coenzyme of pyruvate dehydrogenase. The typical deficiency disease, beriberi, results from a diet exclusively of polished rice. It is characterized by disturbances of the central and peripheral nervous system (polyneuritis) and of cardiac function. The daily requirement for thiamin is about 1 mg.

Riboflavin (V.B_2; lactoflavin; 6,7-dimethyl-9-(D-1′-ribityl)-isoalloxazine) is a water-soluble yellow flavin derivative, occurring chiefly in a bound form in flavin nucleotides or flavoproteins in yeasts, animal products and legume seeds. Milk contains free riboflavin. It is required as a precursor of flavin mononucleotide and flavin-adenine-dinucleotide, which are coenzymes of the flavin enzymes. In rats, experimental riboflavin deficiency causes growth failure and dermatitis around the nostrils and eyes. In humans, riboflavin deficiency (ariboflavinosis) is characterized by lip

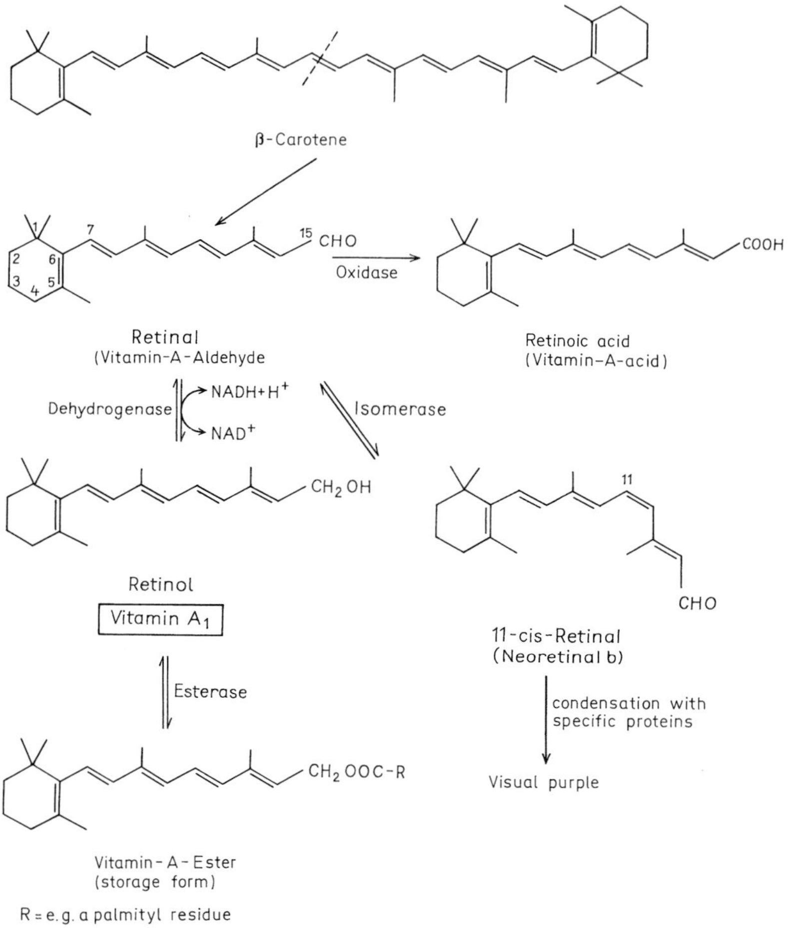

β-Carotene

Retinal
(Vitamin-A-Aldehyde

Oxidase

Retinoic acid
(Vitamin-A-acid)

Dehydrogenase
NADH+H$^+$
NAD$^+$

Isomerase

Retinol
Vitamin A_1

11-cis-Retinal
(Neoretinal b)

condensation with specific proteins

Visual purple

Esterase

Vitamin-A-Ester
(storage form)

R = e.g. a palmityl residue

Biosynthesis of vitamin A-active compounds from β-carotene.

lesions, a seborrheic dermatitis about the nose, ears and eyelids, and loss of hair. Angular stomatitis, glossitis, cheilosis, and ocular changes such as photobia, indistinct vision and corneal vascularization have also been reported as typical of human ariboflavinosis.

The biosynthetic precursors of riboflavin are a purine (probably guanine), ribitol and diacetyl. Most human diets contain adequate riboflavin. The daily requirement is about 1 mg.

Folic acid (pteroylglutamic acid) is a pteridine derivative, especially plentiful in liver, yeast and green plants. Chemically, it consists of 3 moieties: 2-amino-4-hydroxypteridine, *p*-aminobenzoic acid and one or more glutamic acid residues, linked by peptide bonds via their γ-carboxyl groups. Folic acid is a growth factor for some bacteria.

The biochemically active form of folic acid is 5,6,7,8-tetrahydrofolic acid (FH4), a coenzyme in the metabolism of one carbon units. Hydroxymethyl and formyl groups are carried on atoms N5 and N10 of FH4. For example, when the amino acid, L-serine, is converted metabolically to glycine, its hydroxymethyl group is transferred to FH4 to form $N^{5,10}$-methylene-FH4. Once attached to FH4, one carbon units can be oxidized or reduced. Thus $N^{5,10}$-methylene-FH4 can be reduced to N^5-methyl-FH4, which serves as the methyl donor in methylation reactions.

In humans, folic acid avitaminosis is more often caused by faulty uptake and/or utilization, than by dietary deficiency. It usually results in blood abnormalities, e.g. megablastic anemia and thrombocytopenia. Antimetabolites of folic acid are aminopterin and methopterin, which are used therapeutically in the treatment of leukemia. The sulfonamides are antimetabolites of *p*-aminobenzoic acid, and therefore act as inhibitors of bacterial folic acid synthesis.

Nicotinic acid (niacin) and ***Nicotinamide*** (niacinamide) (pellagra preventative factor; V.PP) are metabolically interconvertible, water-soluble, simple pyridine derivatives with equal V. activity. They are widely distributed in nature, and are especially plen-

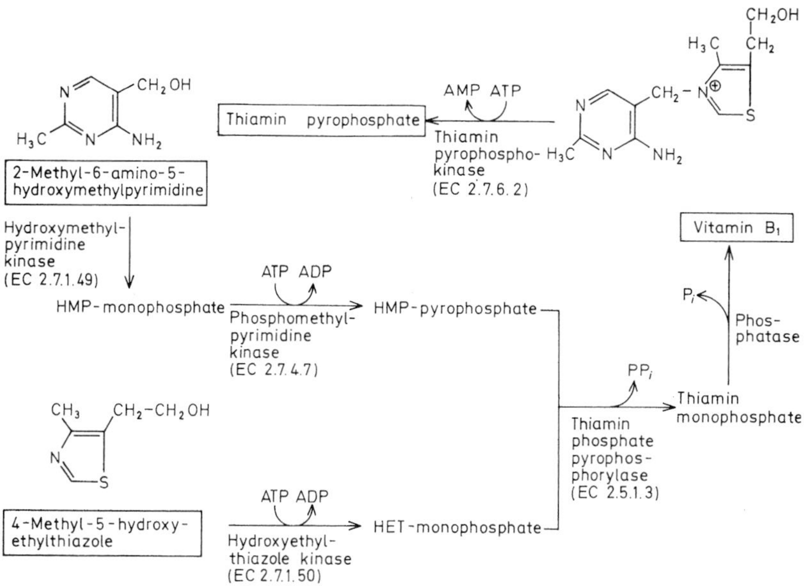

Biosynthesis of vitamin B_1 and its coenzyme form, thiamin pyrophosphate. HMP = 2-methyl-6-amino-5-hydroxymethylpyrimidine. HET = 4-methyl-5-hydroxyethylthiazole.

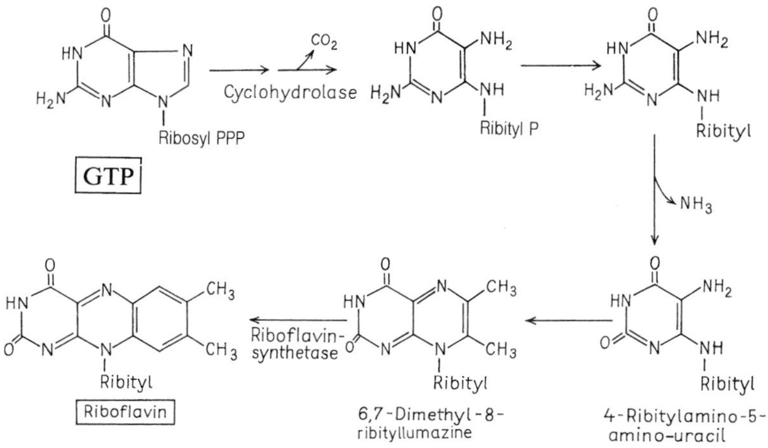

Biosynthesis of riboflavin from guanine, ribitol and diacetyl.

tiful in liver, fish, yeast and germinating cereal grains. Nicotinic acid and nicotinamide are nutritionally equivalent, since both can be assimilated for the purposes of NAD(P) synthesis. For therapeutic purposes, however, nicotinamide is preferred, because large doses of nicotinic acid may have undesirable side effects. In many mammals and in fungi, the nicotinamide moiety of NAD(P) can be derived by the oxidative degradation of L-tryptophan. The extent to which the dietary nicotinamide requirement of animals can be spared by dietary tryptophan varies according to the species. Thus, if the definition of a V. implies a dietary requirement, nicotinamide is not a V. for the rat, which can satisfy its total requirement by the degradation of tryptophan. Synthesis of the nicotinamide moiety of NAD(P) in bacteria and plants occurs by a different pathway, in which aspartic acid and dihydroxyacetone phosphate act as precursors.

Under certain nutritional conditions, e.g. when maize forms the bulk of the human diet, the deficiency of niacin leads to pellagra. Treatment of maize with lime, as practised by Central Americans, releases niacin precursors. Pellagra is therefore not common among them, in spite of their monotonous diet, but was common among poor Europeans who did

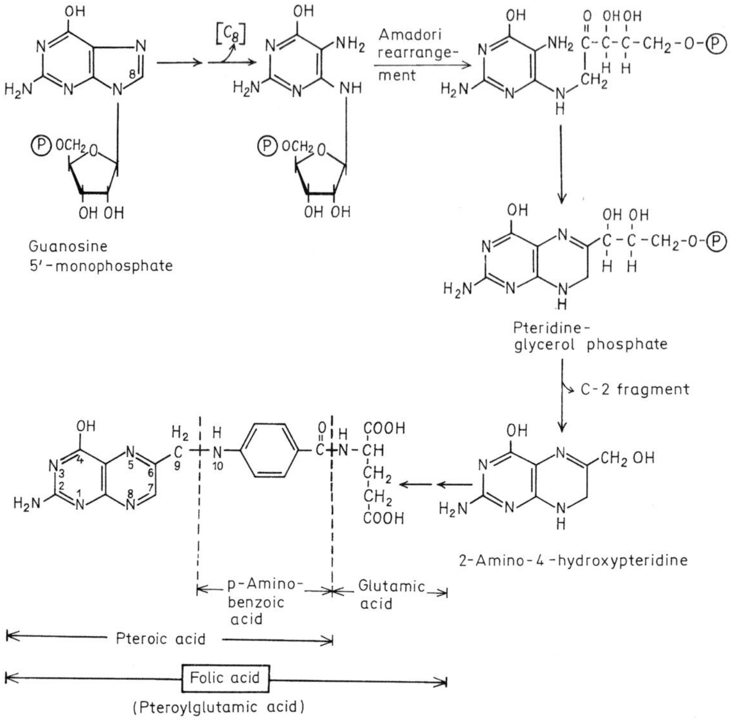

Guanosine 5′-monophosphate

Amadori rearrangement

Pteridine-glycerol phosphate

C-2 fragment

2-Amino-4-hydroxypteridine

p-Aminobenzoic acid

Glutamic acid

Pteroic acid

Folic acid

(Pteroylglutamic acid)

Biosynthesis of folic acid from guanosine monophosphate.

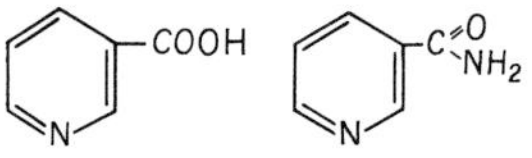

Nicotinic acid (Pyridine 3-carboxylic acid)

Nicotinamide (Pyridine 3-carboxamide)

not treat their maize in this way. This deficiency state affects the skin (brown coloration), the digestive system (diarrhea) and the nervous system (dementia). Pellagra can be cured by feeding tryptophan; or nicotinamide may be administered therapeutically. The daily requirement is 1–2 mg nicotinic acid or nicotinamide.

Pantothenic acid is a widely distributed compound in animals and plants, consisting of 2,4-dihydroxy-3,3-dimethylbutyric acid (pantoic acid) linked to β-alanine by an amide bond. Most organisms have the ability to synthesize pantoic acid from valine, and β-alanine from asparate, but humans lack the enzyme, pantothenate synthetase, which catalyses the condensation of β-alanine and pantoic acid to form pantothenic acid. Only the D(+)-form of pantothenic acid is biologically active. It is required for the synthesis of Coenzyme A (see). Non-experimental human deficiency states have not been observed, so pantothenic acid is presumably present in sufficient quantity in all diets.

Pyridoxine, Pyridoxal and ***Pyridoxamine*** (V.B_6 complex; adermine) are nutritionally equivalent, widely distributed, water-soluble V., found, e.g. in yeast, wheat, maize, liver, potatoes, vegetables, etc. All forms are metabolically interconvertible. Pyridoxal phosphate is one of the most important coenzymes of amino acid metabolism, taking part in transamination, decarboxylation and elimination reactions. Suffi-

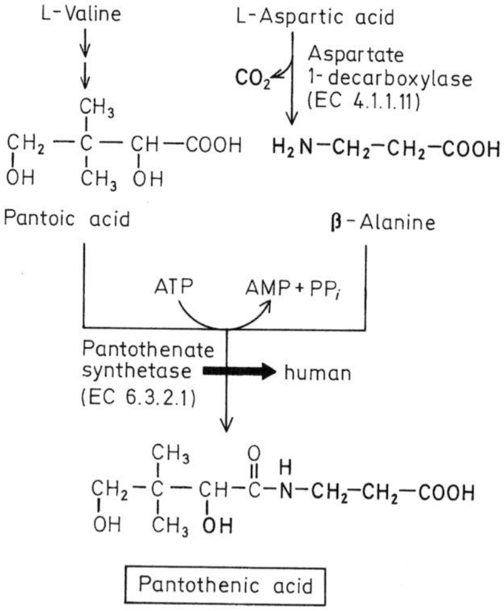

L-Valine

L-Aspartic acid

CO_2

Aspartate 1-decarboxylase (EC 4.1.1.11)

Pantoic acid

β-Alanine

$H_2N-CH_2-CH_2-COOH$

ATP

AMP + PP_i

Pantothenate synthetase (EC 6.3.2.1)

human

Pantothenic acid

Biosynthesis of pantothenic acid

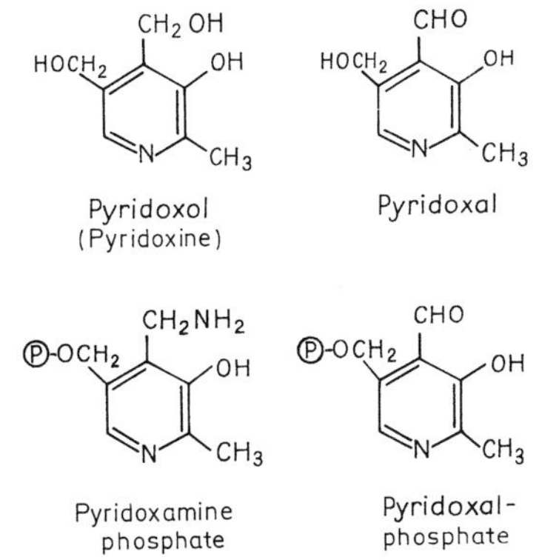

Pyridoxol (Pyridoxine) Pyridoxal

Pyridoxamine phosphate Pyridoxal-phosphate

Vitamin B_6-active compounds

cient V.B_6 is present in all basic foods, and no typical deficiency state is known in humans. The excretion of xanthurenic acid is used as an index of V.B_6 deficiency (see L-Tryptophan for more details of the tryptophan load test for V.B_6 deficiency). In the rat, V.B_6 deficiency causes a pellagra-like condition, with loss of hair, edema and red scaly skin. The daily human requirement is 1.5–2 mg.

Cobalamin (V.B_{12}; extrinsic factor; animal protein factor) represents a group of water-soluble corrinoids required in very small amounts (daily human requirement 0.005 mg V.B_{12}).

The corrinoid structure consists of a complex corrin ring system with a centrally bound trivalent cobalt atom, a base-nucleotide moiety, and a monovalent group (called the cobalt ligand) bound to the cobalt (X in the Fig.). In biological systems, the cobalt ligand is -OH, H_2O, -CH_3 or deoxyadenosine, the last two being found in cobalamin coenzymes. Extraction usually yields V.B_{12} in the form of cyanocobalamin. 5′-Deoxyadenosyl- and methyl-cobalamin are important coenzymes of rearrangement reactions.

V.B_{12} occurs predominantly in animal tissues and animal products. It is synthesized principally by bacteria. Green plants contain little or no V.B_{12}. Deficiency symptoms are sometimes observed in strict vegetarians, most often in breast-fed infants whose mothers consume no animal products. The body reserves of cobalamin are usually so large that an adult can survive for many years on them in the absence of a dietary intake.

V.B_{12} is also known as antipernicious anemia factor. Pernicious anemia is characterized by a severely reduced production of red blood cells, deficient gastric secretion and disturbances of the nervous system. It is not usually caused by dietary deficiency of V.B_{12}, but by the absence of intrinsic factor, which is required for V.B_{12} absorption. Intrinsic factor is a neuraminic acid-containing glycoprotein, normally present in the gastric mucosa, which forms a pepsin-resistant complex with V.B_{12}, and enables V.B_{12} absorption in the lower part of the intestinal tract.

Ascorbic acid (V.C; antiscorbutic V.) is a heat-sensitive, water-soluble V. with a wide natural distribution, especially in fresh vegetables and fruit. V.C is

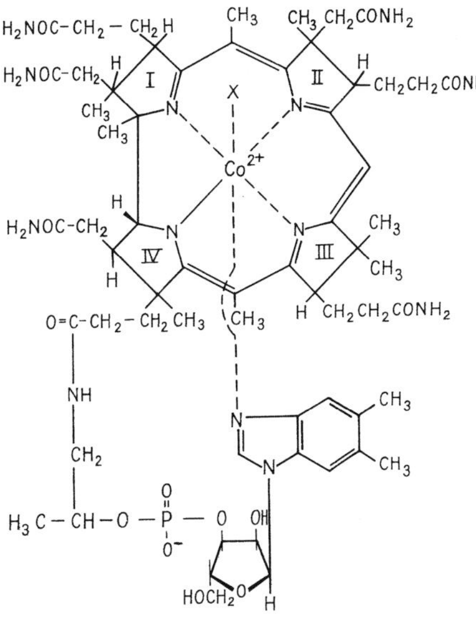

Vitamin B_{12}. X = –CN, –OH, –Cl, NO_2 or CNS. In coenzyme forms of B_{12}, X = –CH_3 or 5′-deoxyadenosyl (see 5′-Deoxyadenosylcobalamin).

the γ-lactone of 2-oxo-L-gulonic acid, derived from carbohydrate metabolism. In most mammals it is synthesized from D-glucuronate. Higher primates (including humans) and guinea pigs cannot synthesize V.C, because they lack the enzyme, L-gulonolactone oxidase. For these animals, it is therefore a true V. and must be supplied in the diet. V.C is a powerful reducing agent, and by its reversible conversion to dehydroascorbic acid it serves as a biochemical redox system. It is involved in several metabolic hydroxylation reactions, e. g. hydroxylation of the proline in collagen. Deficiency results in scurvy, a long-known avitaminosis, characterized by rupture of blood capillaries, hemorrhage of the skin and mucosas, inflammation of the gums, loosening of the teeth and painful swellings of the joints. Resistance to infectious diseases is also reduced. The recommended daily intake is 75 mg, which is considerably higher than for other V.

Calciferol (V.D; antirachitic V.) is a group of fat-soluble V. chemically related to the steroids. They are produced in the skin from provitamins ($\Delta^{5,7}$-unsaturated sterols) by UV-irradiation. If an individual receives adequate exposure to sunlight, dietary V.D is unnecessary. Calciferol is thus a V. only for people who, due to confinement indoors or highly pigmented skin at higher latitudes, are unable to synthesize a sufficient amount. In the conversion of sterols to calciferol, ring B of the steroid is opened between C9 and C10, forming precalciferol, which acts as a precursor of the V.D group. Tachysterol and lumisterol are also formed from precalciferol.

V.D_2 (ergocalciferol) is derived from ergosterol, while V.D_3 (cholecalciferol) is derived from 7-dehydrocholesterol, both conversions occurring in the skin by the action of sunlight. V.D_3 is converted enzymatically in the liver and kidneys to 25-hydroxy-

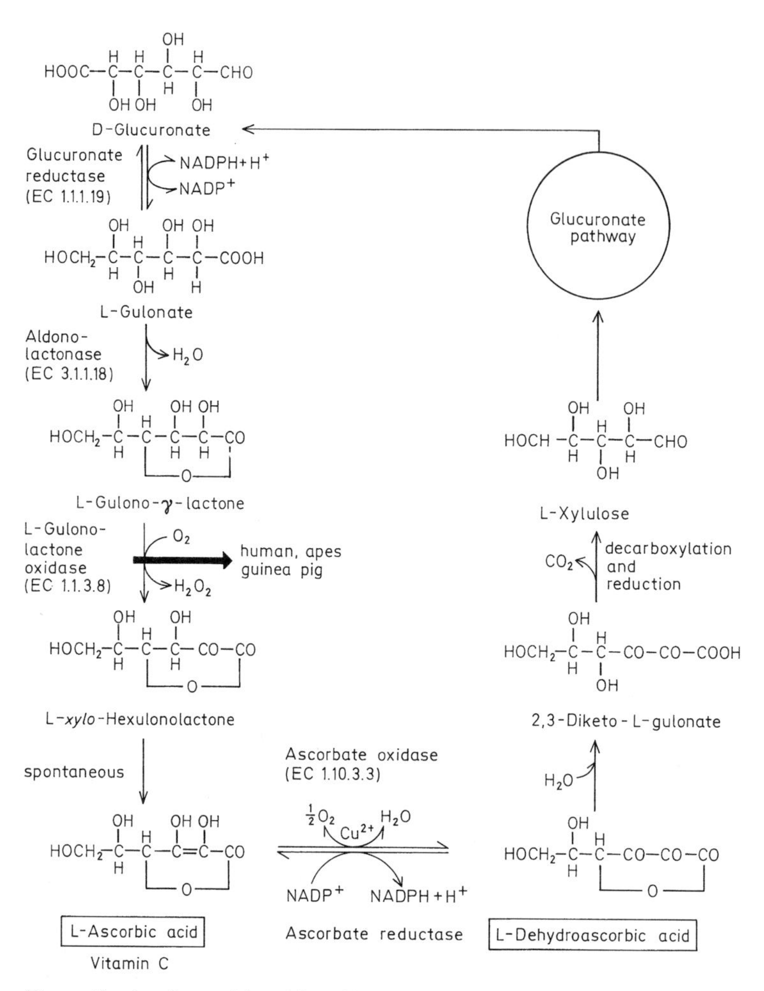

Biosynthesis of ascorbic acid and its conversion into dehydroascorbic acid. See Oxalic acid (Fig. 2.).

cholecalciferol, followed by hydroxylation to the highly active 1α,25-dihydroxycholecalciferol (1α,25-dihydroxy-V.D_3); this compound represents the active form of the V. in humans, and its behavior resembles that of a hormone, rather than a biocatalyst. V.D_3 is present in cod liver oil in particularly large quantities. The V.D complex is also present in e.g. herrings, egg yolk, butter, cheese, milk, pig liver and edible fungi.

V.D_1 is a molecular compound of lumisterol and ergocalciferol. V.D_4 is 22-dihydroergocalciferol, produced in the skin from 22-dihydroergosterol by UV-irradiation.

V.D is required for calcium absorption and the mineralization of bone. The V.D deficiency disease of children, known as rickets, is characterized by a softening and malformation of the bones. It results from poor absorption of calcium, coupled with deficient incorporation of calcium into bone tissue. Rickets can be cured by exposure to sunlight and by administration of synthetic V.D. Because V.D is stored in the liver, overdosage can lead to hypervitaminosis, with disturbances of calcium and phosphate metabolism and withdrawal of calcium from the bones. Exposure to sunlight never leads to hypervitaminosis D. The daily human requirement for V.D is 0.1 mg during growth, and 0.02 mg for adults.

Tocopherol (V.E; antisterility factor) is a group of fat-soluble V. containing a chromane ring with a polyisoprenoid side chain. Eight compounds with V.E activity are known: α-, β-, γ-tocopherol, etc., which differ in the number and positions of the methyl groups in the aromatic ring. Biologically, the most important member is α-tocopherol. Since tocopherol is easily oxidized to a quinone, V.E acts as a naturally occurring antioxidant. It prevents the spontaneous oxidation of highly unsaturated material, e.g. unsaturated fatty acids. It has other biological functions which have not been elucidated in detail. In animal experiments, V.E deficiency results in the death of the embryo in pregnant females. In the male there is atrophy of the gonads and muscle dystrophy. Neither deficien-

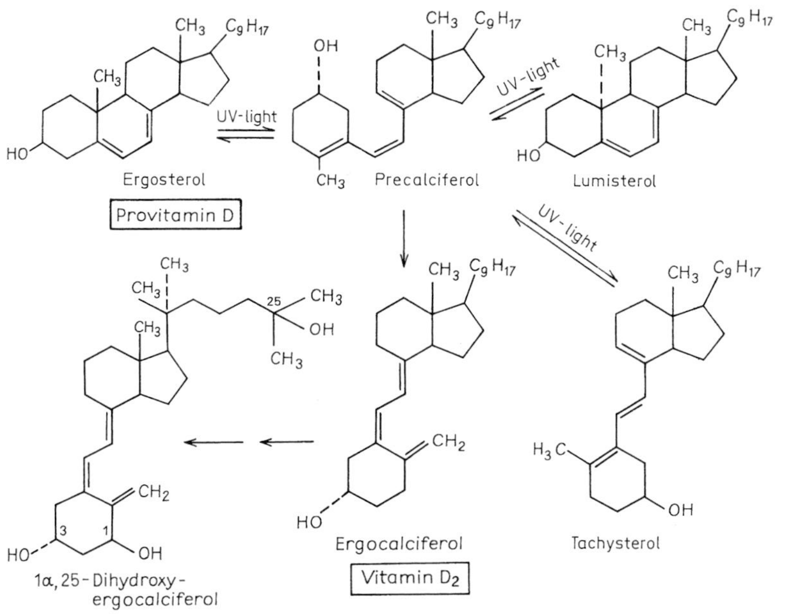

Ergosterol

Provitamin D

Precalciferol

Lumisterol

1α,25-Dihydroxy-ergocalciferol

Ergocalciferol

Vitamin D_2

Tachysterol

Biosynthesis of vitamin D_2

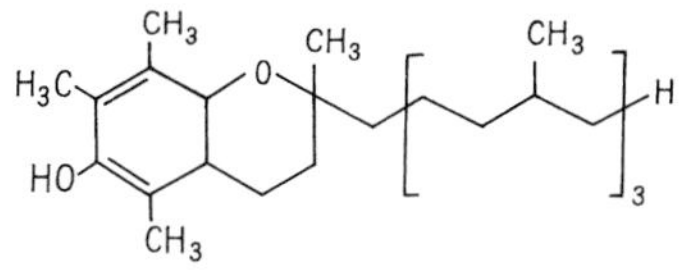

α-Tocopherol

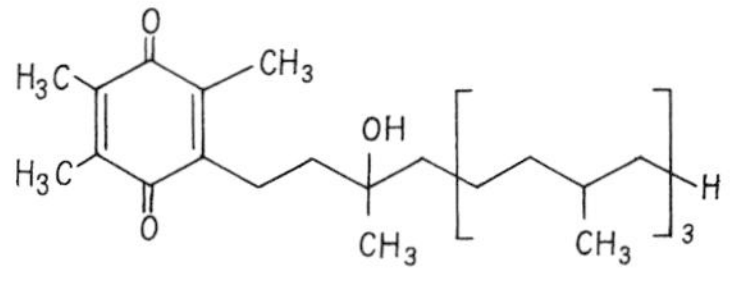

Tocoquinone

Vitamin E-active compounds

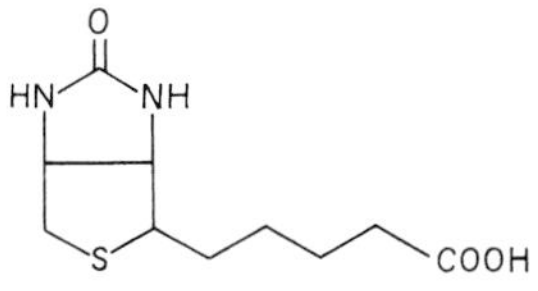

Biotin (vitamin H)

cy states nor V.E-hypervitaminosis have been described in humans. V.E occurs in wheat seedlings, and has been isolated from wheat germ oil. It is also present in lettuce, celery, cabbage, maize, palm oil, ground nuts, castor oil and butter. The daily requirement is estimated at about 5 mg.

Essential fatty acids (V.F). These are unsaturated fatty acids (in particular, linoleic, linolenic and arachidonic acid) which cannot be synthesized in the body. They serve as components of membrane lipids and as precursors of prostaglandins.

Biotin (V.H; bios II; coenzyme R) is a sulfur-containing, water-soluble V. Chemically, it contains 2 condensed 5-membered rings, and it is a cyclic urea derivative: 2′-oxo-3,4-imidazoline-2-tetrahydrothiophene-*n*-valeric acid. It was discovered as a yeast growth factor, and has been isolated from liver extracts and egg yolks. Biotin is biosynthesized from cysteine, pimelic acid and carbamoyl phosphate. There are 8 stereoisomers, the most important biological isomer being D-biotin. In animal tissues, biotin functions as a coenzyme of many important carboxylation reactions, during which it becomes covalently bound (via an amide bond) to a lysyl residue in the carboxylation enzyme. The amino acid conjugate, ε-N-biotinyllysine, is also present in animal tissues, and is known as biocytin. In animals, biotin deficiency causes skin disorders [hence V.H, where the H stands for *Haut,* the German for skin] and loss of hair. Excessive intake of raw eggs causes avitaminosis in humans; egg whites contain the glycoprotein, avidin, which specifically and tightly binds biotin and prevents its absorption. Biotin is present in practically all foods, and the daily requirement (0.25 mg) is in any case generally provided by gut bacteria. Biotin deficiency is therefore practically unknown.

Phylloquinone (V.K; antihemorrhagic V.; coagulation V.) is a group of fat-soluble naphthoquinone compounds, with varying sizes of isoprenoid side chain. Mammals can synthesize the side chain but not the naphthoquinone moiety. $V.K_1$ is plentiful in green plants. $V.K_2$ (farnoquinone; menaquinone-6; 2-methyl-1,4-naphthoquinone) is found chiefly in

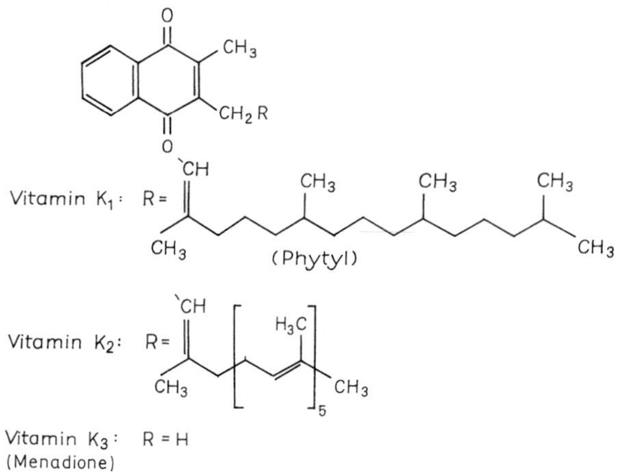

Vitamin K-active compounds

bacteria. V.K_3 (menadione; 2-methyl-1,4-naphthoquinone) is actually a provitamin.

In many bacteria, V.K is a component of the respiratory chain, in place of ubiquinone. In animals, V.K deficiency leads to deficient production of blood coagulation factors, in particular prothrombin, leading to abnormally long clotting times and a marked tendency to hemorrhage. V.K serves as a cofactor in the carboxylation of glutamic acid residues during post-translational modification blood clotting factors II (prothrombin), VII, IX and X. Avitaminosis is rare in human children and adults, because sufficient V.K is provided by the gut flora. However, it does not cross the placenta well, so that neonates are at risk of avitaminosis. Fatal hemorrhage sometimes occurs in breast-fed infants whose mother's milk does not contain adequate amounts of the vitamin. V.K_3 is used in the treatment of hemorrhages and liver diseases.

Warfarin is a V.K antagonist used as a rodenticide; it causes death by hemorrhage after the animals have fed on it repeatedly. It is also used clinically as an anticoagulant. Dicoumarol is an important antagonist of V.K; it is present in moldy clover hay, and is responsible for hemorrhage in cattle.

Vitamin solution: see Nutrient medium (Table 2).

Vitellin: a lipophosphoprotein present in egg yolk together with Phosphovitin (see). It is present in a higher concentration than phosphovitin, but in contrast to this phosphorus-rich protein, V. contains only 1 % phosphate. In yolk, and in neutral salt solutions, V. exists as a dimer (M_r 380,000; 16–22 % lipid). The monomer (M_r 190,000) consists of two dissimilar chains (L_1, M_r 31,000; L_2, M_r 130,000); only L_1 contains phosphate.

Volutin: see Blue-green bacteria.

Von Gierke's disease: see Glycogen storage disease.

von Willebrand factor: a series of self-aggregated structures, all derived from a common glycoprotein subunit, M_r 225,000, synthesized in endothelial cells and megakaryocytes. In plasma the smallest aggregates are dimers (M_r 450,000), while the largest aggregates may have M_r 20 million. Aggregates are held together by disulfide bonds. v. W. f. is also present in platelet α-granules, and it is secreted from stimulated platelets.

The primary human and bovine v. W. f. translation product in endothelial cells is an intracellular precursor, M_r 240,000–260,000, which is cleaved to the M_r 225,000 subunit immediately before secretion. v. W. f. binds strongly (but noncovalently) to circulating factor VIIIc (antihemophilic factor; see Blood coagulation), thereby stabilizing the latter. It mediates interaction of platelets with damaged epithelial surfaces, and interaction of v. W. f. with platelets is essential for normal primary hemostasis, as shown by prolonged bleeding times in patients with von Willebrand disease or Bernard-Soulier platelet defect. There is evidence for the participation of v. W. f. in platelet aggregation, but a preaggregation step promoted by ADP, thrombin or collagen may be necessary. v. W. f. has many features in common with other "adhesive" proteins: Fibrinogen (see), Fibronectin (see), Thrombospondin (see). [D. C. Lynch *J. Biol. Chem.* **258** (1983) 12757–12760; S. E. Senogles & G. L. Nelsestuen, *J. Biol. Chem.* **258** (1983) 12327–12333; J. L. Miller et al. *J. Clin. Invest.* **72** (1983) 1532–1542]

V-type ATPases: one of three classes of Ion-motive ATPases (see), and more difficult to define than the other two classes, the P-type (see) and F-type (see). However, in general, it may be said that they: (i) are H^+-transporting ATPases found in the membranes of the vacuoles (hence the designation 'V') of yeast, fungi (e.g. *Neurospora*) and higher plants (i.e. the tonoplast) and in the lysosomes, endosomes, secretory granules, hormone storage granules and clathrin-coated vesicles of higher animal cells, (ii) do not form a covalent phosphorylated intermediate (cf. the P-type) and (iii) are oligomeric, with the H^+-translocating function and the ATP hydrolytic function residing on different subunits, e.g. the yeast, *Neurospora* and tonoplast enzymes have 60–65 and 70–89 kDa components along with several other smaller peptides, one of which is a 15–19 kDa peptide that covalently binds dicyclohexyldiimide (DCCD).

They are more strongly inhibited by KNO_3, KSCN and *N*-ethylmaleimide than are the V-type and F-type ATPases, and are also inhibited by DCCD, diethylstilbestrol and tributyltin; they are not inhibited by VO_4^{3-}, a characteristic P-type inhibitor, or by oligomycin, a characteristic inhibitor of mitochondrial F-type ATPases.

The mechanism by which they couple the free energy released by ATP hydrolysis to the transmembrane transport of H^+ is not known. They catalyse the uptake of H^+ into the various vacuolar structures of plants and animals, thereby maintaining a pH that is below that of the surrounding cytoplasm and which is near the optimum pH of many of the vacuolar enzymes. They appear to be necessary for the uptake and storage of catecholamines (see) in chromaffin granules and serotonin (see) in synaptic vesicles. [P.L. Pederson & E. Carafoli *Trends Biochem. Sci.* **12** (1987) 146–150, 186–189]

Vulgaxanthin: a yellow betaxanthin from sugar beet (*Beta vulgaris*). In ***V.I,*** a betalamic acid moiety is linked to glutamic acid, in ***V.II,*** the betalamic acid is linked to glutamine.

W

Warburg-Dickens-Horecker pathway: see Pentose phosphate cycle.

Warburg's Atmungsferment: cytochrome oxidase.

Warburg's respiratory enzyme: cytochrome oxidase.

Water: H_2O, quantitatively the most important inorganic constituent of living cells. M.p. 0°C, b.p. 100°C. A normal living cell contains about 80% W. Plants can contain up to 95%, jelly fish 98%, higher animals 60–75% (Tables 1 and 2). All life processes depend on W. There are several theories regarding the structure of W. The dipole properties of W. (Fig. 1) result in an interaction of W. molecules; hydrogen bonding gives rise to molecular aggregates, such as the tetrahydrol structure (Fig. 2). Most cluster hypotheses postulate networks of four-fold linked W. molecules separated by areas of monomeric molecules (Fig. 3). The average half life of such a cluster is only 10^{-11}s. The dipole character of the water molecule determines the physical and chemical properties of W., and is essential to its biological function. Dipolar W. molecules interact with macromolecules, especially proteins and nucleic acids, to form a hydration layer. The activity of enzymes in cytoplasm is very dependent on the degree of hydration. W. dissolves organic and inorganic substances, and is responsible for their extra and intracellular transport. Organic compounds are classified as hydrophilic (dissolve in W., e.g. amino acids, proteins, nucleic acids, carbohydrates) and hydrophobic (insoluble or poorly soluble in W., e.g. fats, lipids).

W. is a reactant in enzyme (hydrolase)-catalysed hydrolytic cleavage of macromolecules (proteins, carbohydrates, fats), representing the first stage in the biological degradation of these substances. W. is formed metabolically by the operation of the respiratory chain (Table 3) (respiratory W.), and is the substrate of photosynthesis.

In animals, W. is important in the regulation of body temperature. Heat is removed by evaporation of W. from body and respiratory tract surfaces. Heavy water (D_2O) contains deuterium (^2_1H or D) in place of ^1_1H. D_2O causes a decrease in metabolic activity and leads to cytological and morphological changes, sometimes resulting in death of the organism. This isotope effect can be used to study the role of W. in biological systems. W. is formed metabolically from the reaction of oxygen with the hydrogen of substrates (see Respiratory chain), and can be produced from gaseous hydrogen and oxygen by *Hydrogenomonas* (see Hydrogen metabolism). It is also formed by dehydrases from substrates, e.g. dehydration of malic acid by the action of malate dehydrase, forming fumaric acid.

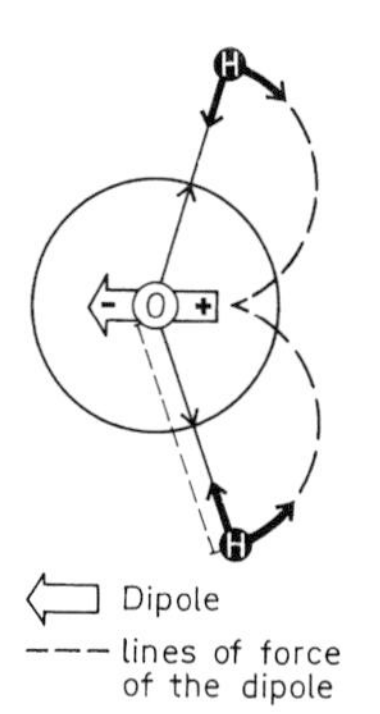

Fig. 1. *Dipole of water*

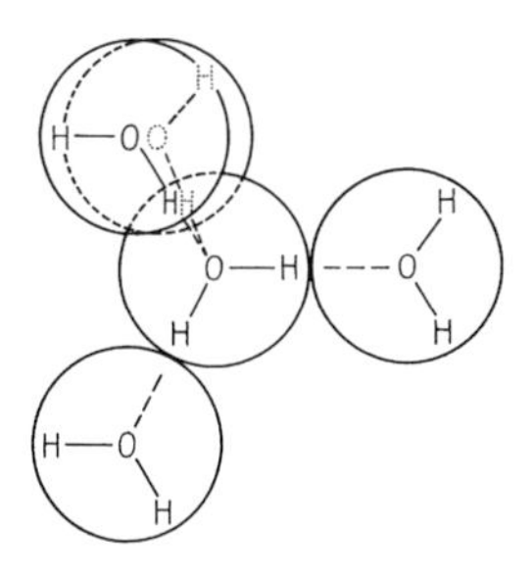

Fig. 2. *Tetrahydrol structure of water*

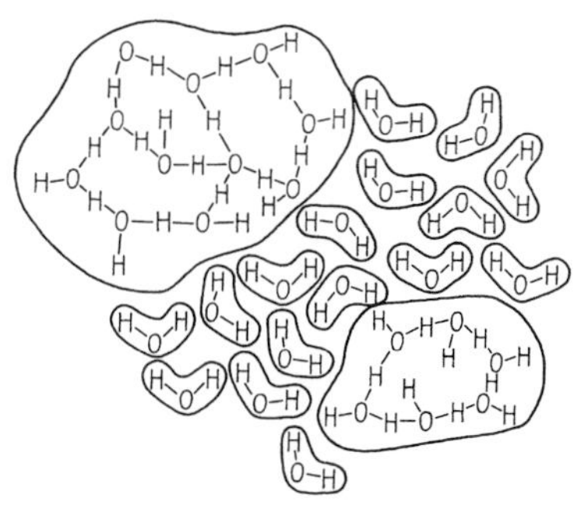

Fig. 3. *Cluster structure of liquid water* (after Némethy and Scheraga).

Watson-Crick model: see Deoxyribonucleic acid.

Wax acids: CH_3–$(CH_2)_n$–COOH, long chain even-numbered monocarboxylic acids, which occur esterified in waxes, e.g. lauric acid (C_{12}), myristic acid

Table 1. *Water content (%) of some human organs and tissues*

Total	60 (representing 42 kg)
Skin	58
Skeleton	28
Muscle	70
Adipose tissue	23
Liver	71
Brain	75

Table 2. *Distribution of water in human organs and tissues (total = 100 %)*

Muscle	50.8
Skeleton	12.5
Skin	6.6
Blood	4.7
Stomach, intestines	3.2
Liver	2.8
Brain	2.7
Lungs	2.4
Adipose tissue	2.3
Kidneys	0.6
Spleen	0.4
Remainder	11.0

Table 3. *Water formed (g) in the degradation of foodstuffs (per 100 g)*

Protein	41.3
Carbohydrate	55.9
Fats	107.0
(Alcohol	117.4)

(C_{14}), palmitic acid (C_{16}), carnaubic acid (C_{24}), cerotic acid (C_{26}), montanic acid (C_{28}), melissic acid (C_{30}) and other higher fatty acids.

Wax alcohols: long chain, monohydroxy, aliphatic alcohols of the general formula CH_3–$(CH_2)_n$–CH_2OH, which occur esterified in waxes, e. g. cetyl alcohol (C_{16}), carnaubyl alcohol (C_{24}), ceryl alcohol (C_{26}) and myricyl alcohol (C_{30}).

Waxes: esters of long chain, even-numbered fatty acids (see Wax acids) and monohydric, straight chain, aliphatic alcohols (see Wax alcohols), or sterols. W. are secreted by animals and plants, usually as ester mixtures that are difficult to resolve, often accompanied by free fatty acids, or high M_r unbranched hydrocarbons. The wax alcohols and wax acids usually have a similar number of carbon atoms, e. g. cetyl palmitate, ceryl cerotate. W. are very hydrophobic. They form a waterproof layer on the aerial parts of plants, e. g. leaves and fruits, which prevents water loss, excessive wetting of surfaces and attack by microorganisms. In animals, W. are found on skin and feathers, and are used by bees to build the honeycomb. Animal waxes are spermaceti, shellac, beeswax and wool wax (lanolin). Carnauba wax is obtained from the leaves of the palm tree, *Copernica cerifera.* W. are solid at room temperature, becoming softer at higher temperatures and eventually liquid. On the basis of their closely similar physical properties, certain mineral substances are also classified as W., e. g. solid hydrocarbons (paraffin wax), montan wax (extracted from lignite), ozocerite (a fossil resin), etc. These products and natural W. are used widely in industry.

Western blot: see Southern blot.

WGA: acronym of wheat germ agglutinin. See Lectins.

Whey proteins: see Milk proteins.

Wieland-Gumlich aldehyde: see Curare alkaloids.

Withaferol A: see Withanolides.

Withanolides: a group of C_{28} plant steroids based on the parent hydrocarbon, ergostane (see Steroids), and containing a characteristic withanolide ring system with a δ-lactone side chain. The approximately 50 known representatives have all been isolated from members of the *Solanaceae,* belonging to the genera *Withania, Dunalia. Datura* and *Nicandra.* The earliest and most important W. is ***withaferol A*** (22*R*)-4β-27-dihydroxy-1-oxo-5β,6β-epoxy-witha-2,24-dienolide (Fig.) from *Withania somnifera* and *Acnistus arborescens;* this compound is bacteriostatic and inhibits the growth of certain tumors. Other representatives are ***27-deoxywithaferol A, 27-deoxy-14α-hydroxywithaferol A*** and ***withanolide D*** (OH-group on C20 instead of C27). [M. Manickam et al. 'Withanolides from *Datura tatula*' *Phytochemistry* **41** (1996) 981–983]

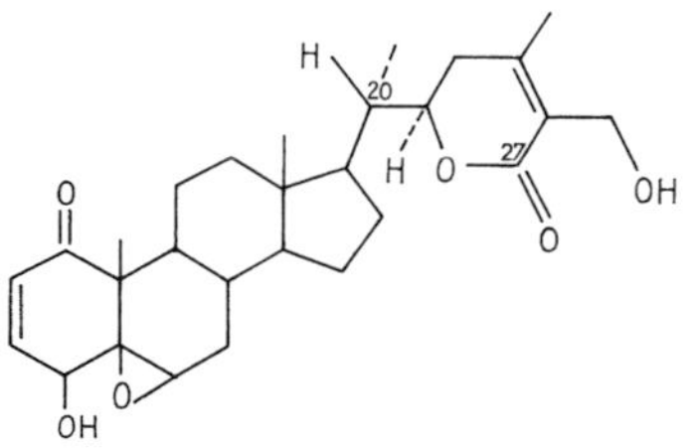

Withaferol A

Wobble base: see Wobble hypothesis.

Wobble hypothesis: a hypothesis proposed by Crick (1966) to explain the degeneracy of the Genetic code (see) with respect to codon-anticodon base pairing. The codon for practically every amino acid is specified by the first two bases, which show strict complementary base pairing with the third and second bases of the anticodon in the corre-

Comparison of the tRNA-anticodon with the corresponding codon for some amino acids

Amino acid	Anticodon (3′ → 5′)	Codon – wobble base (5′ → 3′)
Ala	CGI	GC – U, C, A
Ser	AGI	UC – U, C, A
Phe	AAG^{Me}	UU – U, C
Val	CAI	GU – U, C, A
Tyr	AΨG	UA – U, C
Met	UAC	AU – G

sponding tRNA. The 3′-codon base is less strictly specified and is called the *wobble base.* Hydrogen bonding between this base and the 5′-anticodon base shows a relative lack of base pairing specificity; thus the purine, inosine, is often found in position 5′ of the anticodon (Table). The wobble base is often a pyrimidine, which like inosine forms only two hydrogen bonds.

Wolman's disease: see Inborn errors of metabolism.

Wood: a mixed polymer with lignin as a structural component. Deposition of lignin in the cellulose matrix of the cell wall is called lignification. It is in principle comparable to an irreversible swelling. Cellulose and lignin are physically and chemically bound together in W.

Wood sugar: see D-Xylose.

Wood-Werkmann reaction: see Carboxylation.

Wool fat: see Lanolin.

Wool wax: see Lanolin.

Wyerone acid: a Phytoaloexin (see). W. a. and its methyl ester wyerone are produced by the broad bean (*Vicia faba*) in response to *Botrylis infection.* ED_{50} against various *Botrytis* spp. 9–45 μg/ml. *Vicia* also produces wyerone epoxide. [Hargreaves et al *Phytochemistry* **15** (1976) 1119–1121]

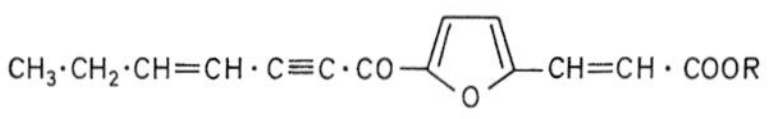

$CH_3 \cdot CH_2 \cdot CH{=}CH \cdot C{\equiv}C \cdot CO$–(furan)–$CH{=}CH \cdot COOR$

R=H : Wyerone acid
R=CH_3 : Wyerone R

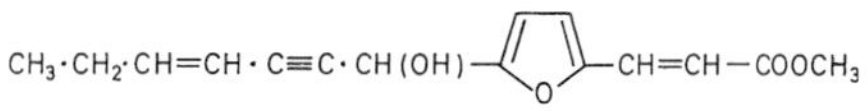

$CH_3 \cdot CH_2 \cdot CH{=}CH \cdot C{\equiv}C \cdot CH(OH)$–(furan)–$CH{=}CH{-}COOCH_3$

Wyerol

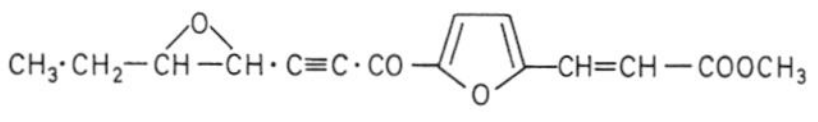

$CH_3 \cdot CH_2{-}CH{-}CH \cdot C{\equiv}C \cdot CO$–(furan)–$CH{=}CH{-}COOCH_3$ (epoxide O bridging CH–CH)

Wyerone epoxide

$CH_3 \cdot CH_2 \cdot CH_2 \cdot CH_2 \cdot C{\equiv}C \cdot CO$–(furan)–$CH{=}CH{-}COOR$

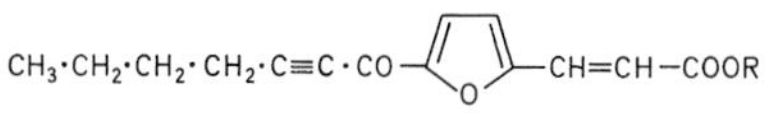

R=H : Dihydrowyerone acid
R=CH_3 : Dihydrowyerone

Wyerone acid and related acetylenic compounds produced by Vicia faba *in response to infection.* The proportions of the various compounds depends on the infected tissue, species of infecting fungus, and time after initiation of synthesis.

X

Xan: abb. for xanthine.

Xanthine, *Xan:* 2,6-dihydroxypurine, a purine and the starting point for Purine degradation (see). M_r 152.1, m.p. > 400 °C (d.). Xan was discovered in 1817 in renal stones. It occurs in the free form, accompanied by other purines. Some derivatives are physiologically important, in particular xanthosine phosphates and Methylated xanthines (see).

Xanthine dehydrogenase: see Xanthine oxidase.

Xanthine oxidase, *xanthine dehydrogenase, Schardinger enzyme:* an enzyme of aerobic purine degradation, which catalyses the oxidation of hypoxanthine to xanthine, and xanthine to uric acid: Hypoxanthine + H_2O + O_2 $\rightarrow$ Xanthine + H_2O_2; Xanthine + H_2O + O_2 $\rightarrow$ Uric acid + H_2O_2. It is a dimeric enzyme, M_r 275,000, pH-optimum 4.7, pI 5.35, containing 2 FAD, 2 Mo and 8 Fe (data for the enzyme from milk). The substrate specificity is low; it catalyses the oxidation of other purines (e.g. adenine), aliphatic and aromatic aldehydes, pyrimidines, pteridines and other heterocyclic compounds.

In animal tissues (e.g. calf liver) X.o. is in the Golgi apparatus; it is also a secretory enzyme present in milk, where its activity can be used to differentiate between fresh and heated or pasteurized milk.

X.o. and xanthine dehydrogenase are sometimes considered as separate enzymes, X.o. being the original Schardinger enzyme from milk (EC 1.2.3.2), and xanthine dehydrogenase an enzyme from chicken liver (EC 1.2.1.37). Together with aldehyde oxidase (EC 1.2.3.1), these three enzymes have very similar composition and presumably mechanism of action, although there are differences in substrate specificity. The molybdenum appears to be involved in the initial hydroxylating attack on the substrate. Electrons are then transferred to the FAD, and from the FAD to the iron which is present as a nonheme Fe-S center. Molecular oxygen is reduced by the nonheme iron center. Superoxide is produced in addition to hydrogen peroxide; these are decomposed by superoxide dismutase and catalase, respectively.

Xanthinuria: see Inborn errors of metabolism, Purine degradation.

Xanthochymusside: see Biflavonoids.

Xanthocillin: 1,4-di-(4-hydroxyphenyl)-2,3-diisonitrilobutadiene(1,3), a bacteriostatic antibiotic used against local infections by Gram-positive and Gram-negative organisms.

Xanthophylls: a group of Carotenoids (see). See also Lutein.

Xanthoprotein reaction: a qualitative test for protein, using concentrated nitric acid. The resulting yellow color is due to nitration of aromatic amino acid residues.

Xanthopterin: 2-amino-4,6-dioxotetrahydropteridine, first isolated as the yellow pteridine wing pigment of the brimstone butterfly and related species. M.p. > 400 °C. X. is also the yellow pigment of bees, wasps and hornets. It is biosynthesized from guanine and two carbon atoms of a pentose.

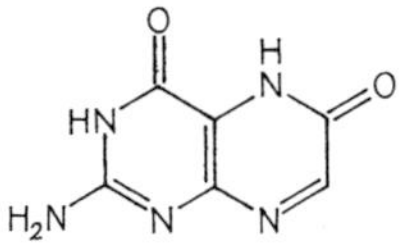

Xanthopterin

Xanthorrhone: see Flavan

Xanthosine: 9-β-D-ribofuranosylxanthine, a β-glycosidic nucleotide of D-ribose and the purine base xanthine. M_r 284.23, carbonizes at > 300 °C, $[\alpha]_D^{30}$ − 51.2° (c = 1 in 0.1 M NaOH). It is formed by the deamination of guanosine. Xanthosine 5′-monophosphate is metabolically important (see Xanthosine phosphates).

Xanthosine phosphates: phosphate esters of xanthosine, or xanthine nucleotides. ***Xanthosine 5′-monophosphate, XMP, xanthidylic acid, xanthylic acid,*** M_r 364.22 is an intermediate of Purine biosynthesis (see).

Xanthothricin: see Toxoflavin.

Xanthoxin: an endogenous growth regulator, occurring widely in higher plants and possessing inhibitory properties similar to those of abscisic acid. It occurs in both the *cis,trans,* and in the biologically less active *trans,trans* form. Plant tissues can probably convert *cis,trans*-X. into (*R*)-(+)-abscisic acid.

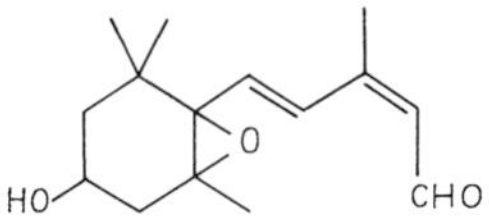

cis,trans-Xanthoxin

Xanthurenic aciduria: see Inborn errors of metabolism.

Xanthylic acid: see Xanthosine phosphates.

Xenopsin, *XP:* an octapeptide from the skin of the African clawed toad, *Xenopus laevis.* The structure of XP: Pyr-Gly-Lys-Arg-Pro-Trp-Ile-Leu-OH, is very similar to that of Neurotensin (see). Antiserum raised to the *C*-terminal region of neurotensin cross-reacts with XP. Neurotensin and XP both characteristically increase the hematocrit and induce cyanosis in anesthetized rats. XP therefore appears to be an amphibian counterpart of neurotensin. Other peptides immunologically similar to neurotensin and XP have been demonstrated in skin, brain and intestine of *Xenopus laevis, Rana catesbeiana* and *Bufo marinus.* Isolation and structure: K. Araki et al. *Chem. Pharm. Bull. (Tokyo)* **21** (1973) 2801; physiology: R. Carraway et al. *Endocrinology* **110** (1982) 1094–1101.

Xerophthol: vitamin A. See Vitamins.

XMP: acronym of xanthosine 5′-monophosphate.

X-ray crystallography: a technique for the determination of the three dimensional (3D) structure (i.e. conformation) of a given molecular species. It utilizes the diffraction of X-rays by the electrons of the atoms constituting the molecule. However, because the X-ray diffraction caused by an individual molecule is immeasurably weak, a very large number of molecules ($\geq 10^{15}$) are required. Moreover, these molecules must be arranged in a regular 3D pattern, as they are in a crystal. Thus the first requirement for X-ray crystallography is a well-ordered crystal, in which the individual molecules, each with the same conformation, are arranged in specific positions and orientations on a 3D lattice, so that their diffraction patterns add up. From this summation the conformation of the individual molecule can be deduced. X-ray crystallography was introduced about 70 years ago and has been used with ever increasing success to elucidate the conformation of a range of biologically important compounds, which includes proteins, nucleic acids and vitamins (e.g. B_{12}).

The basic unit of a crystal is the unit cell, which is defined as the smallest parallelepiped (i.e. a solid body of which each face is a parallelogram) in the crystal which, when repeated by translations that are parallel to its edges in three directions without rotation, creates the crystal lattice. The three directions define the three axes of the crystal. The unit cell may contain one or more molecules. Although the number of molecules per unit cell is always the same for all the cells of the same crystal, it may vary between different crystals of the same molecule; this is particularly true of protein crystals. The molecules constituting a multi-molecular unit cell may have quite different spacial orientations relative to each other and the crystal axes. Again this is true of globular proteins, which often have an irregular shape. This irregularity also makes it difficult for them to pack without leaving large holes or channels between individual molecules. The latter are filled with disordered solvent molecules, which often occupy more than half the crystal volume. Thus protein molecules in a crystal are in contact with one another in only a few small regions and even in these their interaction may be indirect, via one or more layers of solvent molecules. It is because of this that the conformations of proteins determined by X-ray crystallography are believed to be identical, or nearly identical, to their conformations in aqueous solution. Confirmation of this belief has come from the great similarity of the conformations of the, as yet, small number of proteins that have been deduced by both X-ray crystallography and Nuclear magnetic resonance spectroscopy (see), a procedure in which the protein is assayed in aqueous solution.

The X-rays used in X-ray crystallography are often the K_α X-rays ($\lambda = 0.1542$ nm) which are emitted when copper is bombarded with electrons. There is, however, an increasing use of X-rays produced by synchrotrons which have a much higher intensity and allow wavelengths other than that of the copper K_α line to be selected. X-rays are necessary for the elucidation of molecular conformation because their wavelength is of the same order as interatomic distances (e.g. the distance between two carbon atoms joined by a single covalent bond is 0.154 nm).

A narrow beam of X-rays is directed at the crystal. Part of the beam goes straight through the crystal, while the rest is scattered by the electrons of the atoms of the crystal. The scattering caused by each atom is directly proportional to the number of electrons that it possesses (i.e. its atomic number); the latter is low for most of the atoms of organic molecules, hydrogen, carbon, nitrogen and oxygen having 1, 6, 7 and 8 electrons each respectively. When a molecule occurs many times at equivalent locations in a 3D structure, as it does in a crystal lattice, the structure behaves like a diffraction grating and the X-ray waves are scattered in various directions. The scattered waves then recombine; if they are in phase (in step) recombination causes them to be reinforced, while if they are out of phase (out of step) it causes them to be canceled out. The way in which they recombine depends solely on the arrangement of the atoms in the molecules constituting the crystal. The recombined waves are detected either by allowing them to impinge on X-ray film, when blackened spots are seen after development, or by means of a solid state electronic detector. The diffracted beam, corresponding to each spot in the diffraction pattern, is defined by three properties, the amplitude, the wavelength and the phase, all of which are required before the position of the atoms giving rise to it can be deduced. The amplitude is given by the intensity of the spot, which can be obtained by measurement, the wavelength is known because it is that which was chosen for the incident X-ray beam, but the phase is lost. This gives rise to the 'phase problem' of X-ray crystallography. Although this has been solved for small molecules by so-called 'direct methods' based on statistical relationships between intensities, for large molecules like proteins the indirect method known as 'multiple isomorphous replacement', pioneered by Perutz and Kendrew, is used. This involves the introduction of a few atoms of a heavy metal into the crystal in such a way that that they do not change the conformation of the molecule or the unit cell. These atoms have many more electrons than the atoms of C, H, O, N and S from which a protein molecule is constructed and as a consequence scatter X-rays more strongly, thereby causing some spots in the diffraction pattern to increase in intensity. Fourier summation of these increases give maps, termed Patterson maps, of the vectors between the heavy metal atoms, from which the positions of the heavy metal atoms in the unit cell can be determined. This in turn allows the calculation of the phases and amplitudes of their contributions to the diffracted beams in which they participate. With this knowledge, along with that of the amplitudes of the protein alone and the protein-heavy metal complex (i.e. one phase and three amplitudes), it is possible to determine whether the interference of the X-ray waves scattered by the heavy metal atoms and those of the protein are additive or subtractive, and to calculate the magnitude of these two effects. This knowledge and that of the phase of the heavy metal atom enable two equally good solutions for the phase of the protein to be calculated. In order to decide which of these two is correct another, different heavy metal protein complex must be subjected to the same analysis, thereby producing a second pair of equally good solutions. However one of these has the same value as one of the first pair of solutions, and this is the correct phase. In practice several different heavy metal complexes are used so as to give good phase determination for all diffraction spots.

The amplitudes and phases of the X-ray diffraction data are then used to calculate the 3D electron density distribution of the unit cell of the crystal, from which the conformation of the protein molecule can be deduced. Although sophisticated computer graphics are now used for this purpose, the original method utilized a parallel stack of horizontal transparent plastic sheets with contours of electron density drawn on them. The contors on each sheet represented the distribution of electron density in a particular plane of the unit cell and were analogous to the contors at a particular altitude in a geological survey map. When the sheets were stacked in sequence it was possible to follow contor connections in the two vertical planes and thereby deduce the conformation of the structure responsible for them.

Although the wavelength of the X-rays used (e.g. 0.1542 nm) defines the maximum resolution that can be obtained by such X-ray analysis, the actual resolution obtained from a given diffraction pattern depends upon the number of spots analysed: the more spots the greater the resolution. At a resolution of 0.6 nm an α-helix looks like a solid cylinder, at 0.4 nm it is possible to discern the path taken by its polypeptide chain, at 0.2 nm the path of the polypeptide chain is clear, as are the positions of the amino acid side chains, but only at 0.1 nm are individual atoms resolved. In practice maximum resolution is rarely obtained because of the huge amount of data that would have to be analysed. This is due to the 3D nature of the X-ray diffraction pattern which means that there is a cubic relationship between resolution and the number of spots analysed (e.g a threefold increase in resolution requires a 3^3 increase in the number of spots analysed). Most X-ray structural work on proteins is carried out at a resolution of 0.17 – 0.3 nm at which it is not possible to differentiate such chemically different groups as OH, NH_2 and CH_3 or to deduce an unknown amino acid sequence, because several amino acid pairs are completely or almost indistinguishable (e.g. Thr & Val, Asp & Asn, Glu & Gln, His & Phe). In fact knowledge of the amino acid sequence of a protein is almost indispensable for the correct elucidation of its conformation by X-ray crystallography. Fortunately this information is readily obtained nowadays, either directly by use of a protein sequenator or indirectly from the nucleotide sequence of its gene. [K.E. van Holde (1985) *Physical Biochemistry*, 2nd Edition, pp. 253–273, Prentice Hall Inc.; J.R. Heliwell (1992) *Macromolecular Crystallography*, Cambridge University Press]

Xylans: high M_r polysaccharides of xylose. The xylose residues are in the pyranose form, and the linkages are usually β-1,4-glycosidic. Next to cellulose, X. are the commonest of all plant substances, occurring as the main constituent of hemicelluloses.

Xylitol: $CH_2OH–(CHOH)_3–CH_2OH$, an optically inactive C_5-sugar alcohol, related to xylose. M.p. 61.5 °C. X. is a byproduct of wood saccharification, and can also be prepared by catalytic hydrogenation of xylose. It is fully utilized by the human organism, and can therefore be used as a sugar substitute in diabetic diets.

D-Xylose, *wood sugar:* a monosaccharide pentose, M_r 150.13, m.p. 153 °C, $[\alpha]_D^{20}$ +94 ° → +19 °, not fermented by yeast. Reduction of X. gives xylitol, mild oxidation gives xylonic acid. X. is an important dietary component for herbivores, especially ruminants. It can be prepared by the acid hydrolysis of Xylans (see).

Xylosyl nucleosides: Nucleosides (see) in which the sugar component is xylose. They may act as analogs of purine or pyrimidine ribosides (see Pyrimidine analogs). Adenine xyloside is phosphorylated in the cell, and inhibits the growth of various transplantable animal tumors.

Xylulose: a monosaccharide pentulose, occurring naturally in both the D- and L-form. The 5-phosphate of D-X. is an important intermediate in the Pentose phosphate cycle (see) and serves as a C2-donor in Transketolation (see). L-X. is an intermediate of the Glucuronate pathway (see), and it is normally metabolized via xylitol and D-X. In pentosuria, an inherited metabolic disease, L-X. is excreted in the urine.

Y

Yeast adenylic acid: see Adenosine phosphates.

Yeast artificial chromosome, *YAC:* a vector constructed from the telomere, centromere and replication origin of yeast *(Saccharomyces cerevisiae)* chromosomes, together with selectable yeast genes and enough DNA from any source to make a total of more than 50 kb. The replication origin (see Autonomously replicating sequence) enables the vector to replicate independently of the host cell chromosomes. In mitosis, YACs replicate and segregate almost perfectly (only one daughter yeast cell in 1000 to 10,000 does not receive a YAC). In meiotic spore formation the two sister chromatids of a YAC separate correctly and produce haploid spores.

Very long lengths of DNA (up to 1000 kb) can be cloned and maintained in a YAC. In comparison, other vector types carry much shorter lengths of DNA. Thus, bacterial plasmids carry DNA up to 20 kb in length, bacteriophage λ (25 kb), cosmids (45 kb), P1 vector (100 kb). YACs can therefore be used to clone large pieces of chromosome from other species. For example, they have been used to isolate overlapping clones of human DNA that cover the entire length of individual human chromosomes (see Human genome project). By 1994, the Human Genome Database (GDB) contained about 48,371 YAC clones, of which 19,286 had been mapped: 6,374 (33 %) on the X chromosome and 3,046 (16 %) on chromosome 21. Further clones are being continually added and mapped. [D. Cohen et al. *Nature* **366** (1993) 698–701; A. J. Cuticchia (ed.) *Human Genome Mapping – A Compendium,* John Hopkins University press, Baltimore & London, 1995]

Yellow enzymes: see Flavin enzymes.

Yield coefficient: see Cultivation of microorganisms.

Yohimbine: a Rauwolfia alkaloid (see), M_r 354.45, m.p. 235–236 °C, $[\alpha]_D + 106°$ (pyridine). Y. has five chiral centers and therefore many stereoisomers; seven of these occur naturally, the most important being ***Corynanthine,*** m.p. 225–226 °C (d.),

$[\alpha]_D$ –82° (pyridine), obtained chiefly from the bark of the tropical tree, *Corynanthe yohimbe.* Y. is vasodilatory and has been used in the treatment of arteriosclerosis, in veterinary medicine and by African natives as an aphrodisiac.

Z

Zaffaroni system: see Paper chromatography.

Zeatin: 6-(4-hydroxy-3-methyl-but-*trans*-2-enyl)-aminopurine, a naturally occurring Cytokinin (see for formula). Z. occurs free in many plants, especially in immature maize kernels, and is identical with the previously described maize factor (MF). Its derivatives, Dihydrozeatin (see), Z. riboside and Z. ribotide are also cytokinins. The *cis* compound (see N^6-*cis*-γ-Methyl-γ-hydroxymethylallyladenosine) is a rare nucleic component in certain species of RNA.

Zeaxanthin: (3*R*,3′*R*)-β,β-carotene-3,3′-diol; 3,3′-dihydroxy-β-carotene, a xanthophyll, M_r 568.85, m. p. 206 °C (d.), isomeric with lutein; one of the commonest plant pigments (yellow-orange), especially plentiful in maize and fruits of the sea buckthorn; also found in algae and bacteria. It occurs free and esterified as the dipalmitate, and shows no vitamin A activity. The 5,6-monoepoxide of Z., antheroxanthin, is also a common plant pigment.

Zein: see Prolamines

Zinc, *Zn*: an essential bioelement for growth and development of plants, animals and microorganisms. Zn has a high affinity for nitrogen and sulfur ligands, and occurs in the cell in association with many different compounds, e. g. proteins (insulin), amino acids, nucleic acids. Zn is a tightly bound component of zinc-metalloenzymes, and it stimulates in vitro the activity of zinc-metal-enzyme complexes. About 20 zinc-metalloenzymes have been described, e. g. dehydrogenases, phosphatases, carboxypeptidases, carbonic anhydrase. Zn also activates the enzymatic synthesis of tryptophan. The human body contains 2–4 g Zn, the majority being intracellular. Blood contains only 7–8 μg Zn/ml, of which 85 % is in the erythrocytes, where it is required for the activity of carbonic anhydrase.

Zinc finger: a DNA binding motif first recognized in *Xenopus* transcription factor IIIA (which binds to the internal control sequence of the 5S rRNA gene). This protein contains nine 30-residue sequences, all rather similar and arranged in tandem. Each contains two invariant Cys residues and two invariant His residues, as well as several conserved hydrophobic residues. Each of these structural domains folds about a single Zn^{2+} ion, and X-ray analysis indicates that the zinc ion is tetrahedrally liganded by the invariant Cys and His residues. Such domains have been found in a variety of eukaryotic transcription factors. Zinc fingers have also been found in which the two His residues are replaced by two further Cys residues (Cys_2-Cys_2 zinc fingers, e. g. in the DNA-binding domains of the glucocorticoid receptor and the estrogen receptor), while others have been found in which six Cys residues bind two Zn^{2+} ions (binuclear Cys_6 zinc fingers, e. g. in the DNA-binding domain of the yeast protein GAL4, a transcriptional activator of several genes that encode galactose-metabolizing enzymes). [J. M. Berg 'Zinc fingers and other metal-binding domains. Elements for interaction between macromolecules' *J. Biol. Chem.* **265** (1990) 6513–6516; J. W. R. Schwabe & A. Klug *Nature Struct. Biol.* **1** (1994) 345–349; M. Schmiedeskamp & R. E. Klevit *Curr. Opin. Struct. Biol.* **4** (1994) 28–35; F. Radtke et al. *Biol. Chem. Hoppe Seyler* **377** (1996) 47–56]

Zizanin B: see Sesterterpenes.

Zoochromes: see Natural pigments.

Zoosterols: see Sterols.

Zwischenferment: Glucose 6-phosphate dehydrogenase. See Pentose phosphate cycle.

Zymase: an old name for a mixture of 11 enzymes of glycolysis isolated from yeast after mechanical disruption of the cell wall. It catalyses alcoholic fermentation.

Zymogens: inactive precursors of enzymes usually proteolytic enzymes. Z. are converted into active enzymes by limited proteolysis. Best known examples include the Z. of digestive enzymes (pepsinogen, trypsinogen, chymotrypsinogens A, B and C, proelastase, and procarboxypeptidases A and B), and of blood coagulation enzymes (prothrombin and plasminogen). Z. can be stored at their sites of synthesis without danger of causing self digestion of cells or tissues. Certain proteohormones are also produced as inactive precursors, e. g. proinsulin. Local autolysis at the synthesis sites of proteolytic hormones is also prevented by the presence of specific enzyme inhibitors, or by the storage of enzymes in particles. For further details of Z. activation, see individual entries for each Z. named above.

Zymosterol: 5α-cholesta-8(9),24-dien-3β-ol, a mycosterol (see Sterols) present in yeast. Z. is an intermediate in the biosynthesis of cholesterol from lanosterol. See Steroids.

OH

HO

Zeaxanthin

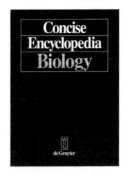

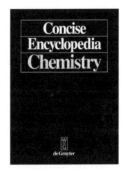

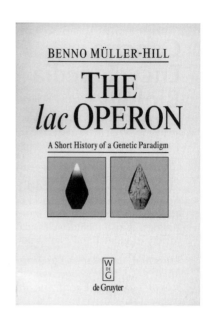